LEE COLLEGE LIBRARY
BAYTOWN, TEXAS

The World of the Cell

Related Titles of Interest from the Benjamin/Cummings Series in the Life Sciences

R. F. Boyer
Modern Experimental Biochemistry (1986)

N. A. Campbell
Biology, Second Edition (1990)

C. L. Case and T. R. Johnson
Laboratory Experiments in Microbiology (1984)

R. E. Dickerson and I. Geis
Hemoglobin (1983)

M. Dworkin
Developmental Biology of the Bacteria (1985)

P. B. Hackett, J. A. Fuchs, and J. W. Messing
An Introduction to Recombinant DNA Techniques: Basic Experiments in Gene Manipulation, Second Edition (1988)

L. E. Hood, I. L. Weissman, W. B. Wood, and J. H. Wilson
Immunology, Second Edition (1984)

C. K. Mathews and K. E. van Holde
Biochemistry (1990)

G. J. Tortora, B. R. Funke, and C. L. Case
Microbiology: An Introduction, Fourth Edition (1992)

D. E. Vance and J. E. Vance
Biochemistry of Lipids and Membranes (1985)

J. D. Watson, N. H. Hopkins, J. W. Roberts, J. A. Steitz, and A. M. Weiner
Molecular Biology of the Gene, Fourth Edition, Volumes I and II (1987)

W. B. Wood, J. H. Wilson, R. M. Benbow, and L. E. Hood
Biochemistry: A Problems Approach, Second Edition (1981)

The World of the Cell

SECOND EDITION

Wayne M. Becker
University of Wisconsin-Madison

David W. Deamer
University of California, Davis

CONTRIBUTORS

Peter B. Armstrong
University of California, Davis
Chapter 23

Joel W. Goodman
University of California, San Francisco
Chapter 24

David N. Gunn
University of Wisconsin-Madison
Chapter 22

Jeanette E. Natzle
University of California, Davis
Chapter 17

The Benjamin/Cummings Publishing Company, Inc.

Redwood City, California • Menlo Park, California
Reading, Massachusetts • New York • Don Mills, Ontario • Wokingham, U.K.
Amsterdam • Bonn • Sydney • Singapore • Tokyo • Madrid • San Juan

Sponsoring Editors: **Jane Reece and David Rogelberg**
Developmental Editor: **Jamie Northway**
Production Supervisor: **Brian Jones**
Manufacturing Supervisor: **Casimira Kostecki**
Senior Marketing Manager: **Anne Emerson**
Photo Editor: **Cecilia Mills**
Production Assistant: **Andrew Marinkovich**
Cover Designer and Artist: **Mark Stuart Ong**
Text Designer and Artist: **Mark Stuart Ong**
Copy Editor: **Mary Prescott**
Proofreader: **Melissa Andrews**
Indexer: **Katherine Pitcoff**
Second Edition Artists: **Georg Klatt, Linda McVay, Irene Imfeld**
First Edition Artists: **Greg Boren, Cyndie Clark-Huegel, Michael Fornalski, Janet Hayes, Darwen and Vally Hennings, Christina Jordan, Fran Milner, Kathy Monahan, Judy Morley, Audre Newman, Carol Verbeeck**
Composition and Film: **Syntax International**
Cover Printer: **New England Book Components**
Text Printer and Binder: **Courier/Westford**

Credits for photographs and illustrations appear after the Glossary, beginning on page 853.

Library of Congress Cataloging-in-Publication Data

Becker, Wayne M.
The world of the cell / Wayne M. Becker, David W. Deamer; contributors, Peter Armstrong . . . [et al.]. — 2nd ed.
p. cm.
Includes bibliographical references and index.
ISBN 0-8053-0870-9
1. Cytology. 2. Molecular biology. I. Deamer, D. W. II. Title.
QH581.2.B43 1991
574.87—dc20 90-28773
CIP

678910—MU—95 94

The Benjamin/Cummings Publishing Company, Inc.
390 Bridge Parkway
Redwood City, California 94065

Preface

The World of the Cell is intended as a comprehensive introduction to cellular and molecular biology for students preparing for careers in biology, medicine, and related fields. Portions of this book began as lecture notes, problem sets, and exams in Biocore 303, a cell biology course at the University of Wisconsin-Madison. These materials were expanded into the first edition of this textbook, published in 1986. Heartened by the large number of users and by the responses of instructors and students alike, we have prepared this edition jointly, with each of us bringing to the venture about twenty years of teaching experience in cell biology, including both introductory and advanced courses.

Something Old and Something New

The book is neither exhaustive nor encyclopedic. Our goal has been to sketch in as lucid a manner as possible the essential principles and processes of cell biology. Recognizing the exceptionally rapid pace of discovery in cell biology during the past several years, we have sought to weave new knowledge and insight into the fabric of the text. In doing so, we have retained the features of the first edition that readers have described as "user-friendly," while reorganizing and updating the material in ways that we hope will make the text even more useful for both students and instructors.

Features we have retained from the first edition include an organization of subject matter that is readily adaptable to a great variety of course syllabi, an illustration program that users of the first edition found very helpful, and a problem set at the end of each chapter. We have also paid careful attention to accuracy, consistency, and vocabulary, hoping thereby to maximize understanding and minimize confusion for the reader.

Additions and changes that we believe will further enhance the usefulness of the text include the following:

- Introductory unit restructured such that cell chemistry is now covered before cell structure.
- Units reorganized to cover membrane structure and function before energy flow, thereby ensuring that the reader understands membrane potential and transport before encountering mitochondrial and chloroplast function.
- Coverage of molecular biology updated and expanded, including a more extended consideration of eukaryotic gene regulation.
- Discussion of recombinant DNA technology expanded significantly and placed in an appendix for ready access from anywhere in the text.
- Coverage of the cytoskeleton expanded to an entire chapter, immediately preceding a chapter on motility and contractility that has also been updated significantly.
- Chapter on membrane receptors added to provide enhanced coverage of mechanisms of cell communication.
- Updated coverage of developmental biology, cancer biology, and immunology provided by contributors chosen for their expertise in these areas.
- Each micrograph identified in the figure legend as a light micrograph (LM), scanning electron micrograph (SEM), or transmission electron micrograph (TEM).
- Size bars used to indicate magnification of micrographs.

Techniques and Methods

Throughout the text, we have tried to explain not only *what* we know about cells but also *how* we know what we know. Toward that end, we have included descriptions of experimental techniques and findings throughout, almost always in the context of the questions they address and in anticipation of the answers they provide. For example, equilibrium density centrifugation is introduced not in a chapter concerned with how cells are studied, but in Chapter 9, where it becomes important to understand how lysosomes were originally distinguished from mitochondria and subsequently from peroxisomes as well. To help readers locate techniques out of context, an alphabetic **Guide to Techniques and Methods** follows the Acknowledgments, with page references to particular techniques.

The only exceptions to the principle of introducing techniques in context are the methodologies of microscopy and of recombinant DNA technology. Both of these topics are relevant to much of contemporary cell biology and both involve a variety of related techniques that can be logically considered as a self-contained unit. Accordingly, this edition has two appendices, **Appendix A: Principles and Techniques of Microscopy** and **Appendix B: Recombinant DNA Technology**, each fully illustrated. Both appendices are cross-referenced at numerous points in the text, to make their existence known to readers who, despite the fond hopes of the authors, are not likely to be careful readers of this Preface.

In-Text Learning Aids

To enhance the book's effectiveness as a learning tool, each chapter includes the following features:

- One or two **Boxed Essays** to help students better understand particularly important or intriguing aspects of cell biology. While some of the essays give interesting historical perspectives on how science is done (the discovery of the double helix as described in the boxed essay of Chapter 3, for example), others are intended to help students understand potentially difficult principles (the analogy of monkeys shelling peanuts to help explain enzyme kinetics, in Chapter 6). Still others provide further insights into contemporary techniques used by cell biologists (the scanning tunneling microscope, in Chapter 1) or into facets of cell biology with special medical relevance (the use of intermediate filament typing as a diagnostic tool, in Chapter 18).
- A list of **Key Terms** that includes the page number where each term first appears in boldface. Most of the key terms are included in the **Glossary** at the end of the book, which provides a definition of each term along with a reference to the chapter(s) in which that term appears most prominently.
- A **Suggested Reading** list, with an emphasis on review articles that motivated users are likely to find understandable. We have tried to avoid overwhelming readers with lengthy bibliographies of the original literature, but have referenced selected articles that are especially relevant to the topics of the chapter.
- A **Problem Set**, reflecting our conviction that we learn science not just by reading or hearing about it, but by working with it. The problems are designed to emphasize understanding and applying the principles taught in the chapter, not just rote recall. Many of the problems have been carefully selected from class-tested exams. To give instructors flexibility in their use of the problem sets, answers for all of them appear in the **Solutions Manual** described below. At the discretion of the instructor, this manual can either be made available to students through the bookstore or used by the instructor as a resource for homework and exam questions.

Supplemental Learning Aids

Supplemental learning aids available with this text include:

- A **Solutions Manual** that contains detailed answers to all of the problems in the text. As a special feature new to this edition, this **Solutions Manual** also contains black-only versions of over 150 drawings from the text. Instructors can use these illustrations as masters in creating acetate transparencies for classroom use; students can then review and annotate the illustrations while they are being discussed in class. (ISBN 0-8053-0871-7)
- A separate set of 52 two-color **acetate transparencies** corresponding to selected figures from the text but with enlarged labels to enhance their usefulness in classroom lecture. (ISBN 0-8053-0872-5)
- The **Benjamin/Cummings Micrograph Transparencies for the Life Sciences**, a set of 50 micrographs that complement (but are not usually identical to) the micrographs reproduced in the text. This set includes both electron micrographs and full-color light micrographs, many with explanatory drawings. (ISBN 0-8053-1815-1)

Comments Welcome

The real test of any textbook is how effectively it helps instructors teach and students learn. We welcome feedback from readers; please send your comments and suggestions to either of us.

Wayne M. Becker
Department of Botany
University of Wisconsin-Madison
Madison, Wisconsin 53706

David W. Deamer
Department of Zoology
University of California, Davis
Davis, California 95616

Acknowledgments

We want to acknowledge the contributions of the numerous people who have made this book possible. We are indebted to the many students in the cell biology courses at our respective institutions whose words of encouragement catalyzed the writing of these chapters and whose thoughtful comments and willing criticisms have contributed much to whatever level of lucidity the text may be judged to have.

Each of us owes a special debt of gratitude to our colleagues from whose insights and understandings we have benefitted greatly and borrowed heavily. These include Ann Burgess, Philippa Claude, Mike Hoffmann, Millard Susman, and the late Walter Plaut at the University of Wisconsin-Madison; and Roger Leslie and Jonathan Scholey at the University of California, Davis.

We are especially grateful to Jeanette Natzle (University of California, Davis), David Gunn (University of Wisconsin-Madison), Peter Armstrong (University of California, Davis), and Joel Goodman (University of California, San Francisco) for their authorship of Chapters 17, 22, 23, and 24, respectively. We also acknowledge David Spiegel, Akif Uzman, Karen Valentino, Ed Clark, and John Carson for their contributions to several chapters in the first edition. In addition, we want to express our appreciation to the many colleagues who responded so generously to our requests for micrographs and other visual aids.

The many reviewers listed below provided helpful criticisms and suggestions for revision at various stages of manuscript development. Their words of appraisal and counsel were gratefully received and greatly appreciated. Indeed, the extensive review process to which both this and the prior edition of this text have been exposed ought itself to be counted as a significant feature of its content. Nonetheless, the final responsibility for what you read here remains ours, and you may confidently attribute to us any errors of omission or commission you may encounter in these pages.

We are deeply indebted to the many people at The Benjamin/Cummings Publishing Company who made this venture a reality. Special recognition goes to Diane Bowen, Jamie Northway, Jane Reece, Brian Jones, and Cecilia Mills, whose consistent encouragement and careful attention to detail contributed much to the clarity of both the text and the art.

Finally, we are grateful beyond measure to our wives and families, without whose patience, understanding, and forbearance this book would never have been written.

Reviewers of the First and Second Editions

L. Rao Ayyagari, Lindenwood College
Margaret Beard, Columbia University
Paul Benko, Sonoma State University
Robert Blystone, Trinity University
Alan H. Brush, University of Connecticut, Storrs
Brower R. Burchill, University of Kansas
Ann B. Burgess, University of Wisconsin-Madison
Thomas J. Byers, Ohio State University
Edward A. Clark, University of Washington
Philippa Claude, University of Wisconsin-Madison
John M. Coffin, Tufts University School of Medicine
J. John Cohen, University of Colorado Medical School
Larry Cohen, Pomona College
David DeGroote, St. Cloud State
Aris J. Domnas, University of North Carolina, Chapel Hill
David R. Fromson, California State University, Fullerton
Stephen A. George, Amherst College
T. T. Gleeson, University of Colorado
Ursula W. Goodenough, Washington University
Thomas A. Gorell, Colorado State University
Marion Greaser, University of Wisconsin-Madison

Mark T. Groudine, Fred Hutchinson Cancer Research Center, University of Washington School of Medicine
Gary Gussin, University of Iowa
Laszlo Hanzely, Northern Illinois University
Bettina Harrison, University of Massachusetts, Boston
Lawrence Hightower, University of Connecticut, Storrs
Johns Hopkins III, Washington University
William R. Jeffery, University of Texas, Austin
Kwang W. Jeon, University of Tennessee
Patricia P. Jones, Stanford University
Robert Koch, California State University, Fullerton
Hal Krider, University of Connecticut, Storrs
William B. Kristan, Jr., University of California, San Diego
Frederic Kundig, Towson State University
Elias Lazarides, California Institute of Technology
John T. Lis, Cornell University
Robert Macey, University of California, Berkeley
Gary G. Matthews, State University of New York, Stony Brook
Richard Nuccitelli, University of California, Davis
Joanna Olmsted, University of Rochester
Alan Orr, University of Northern Iowa
Curtis L. Parker, Morehouse School of Medicine
Lee D. Peachey, University of Colorado
Howard Petty, Wayne State University
Susan Pierce, Northwestern University
Ralph Quatrano, Oregon State University
Gary Reiness, Pomona College
Edmund Samuel, Southern Connecticut State University
Robert D. Simoni, Stanford University
William R. Sistrom, University of Oregon
Barbara Y. Stewart, Swarthmore College
Antony O. Stretton, University of Wisconsin-Madison
Stephen Subtelny, Rice University
Millard Susman, University of Wisconsin-Madison
John J. Tyson, Virginia Polytechnic University
Akif Uzman, University of Texas, Austin
Fred D. Warner, Syracuse University
James Watrous, St. Joseph's University
Fred H. Wilt, University of California, Berkeley

Guide to Techniques and Methods

The following techniques are important to cell biologists. Each technique is described in the text at the indicated location, in the context of its actual use by researchers.

Cancer

Cell Cycle

Cytoskeleton

Embryological Development

Enzymes

Genetics

Immunology

Membranes

Microscopy

Nucleic Acids

Proteins

Recombinant DNA

Separation

Brief Contents

Detailed Contents

PART FIVE
Specific Cell Functions 553

The World of the Cell

PART ONE

INTRODUCTION

1

The World of the Cell: A Preview

The **cell** is the basic unit of biology. Every organism either consists of cells or is itself a single cell. Therefore, it is only as we understand the structure and function of cells that we can appreciate both the capabilities and the limitations of living organisms, whether animal, plant, or microorganism.

We are in the midst of a revolution in biology that has brought with it tremendous advances in our understanding of how cells are constructed and how they carry out the intricate functions necessary for life. Particularly significant is the dynamic nature of the cell, as evidenced by its capacity to grow, reproduce, and become specialized, and by its ability to respond to stimuli and to adapt to changes in its environment.

Cell biology itself is changing, as scientists from a variety of related disciplines focus their efforts on the common objective of understanding more adequately how cells work. The convergence of cytology, genetics, and biochemistry has made modern cell biology one of the most exciting and dynamic disciplines in contemporary biology.

In this chapter, we will look briefly at the beginnings of cell biology and then consider the three main historical strands that have given rise to our present-day understanding of what cells are and how they work.

The Cell Theory: A Brief History

The story of cell biology started to unfold more than 300 years ago, as European scientists began to focus their crude microscopes on a variety of biological material ranging from tree bark to human sperm. One such scientist was **Robert Hooke,** Curator of Instruments for the Royal Society of London. Hooke examined a thin slice of cork cut with a penknife and saw a network of tiny boxlike compartments that reminded him of a honeycomb. He called these little compartments *cellulae,* a Latin term meaning "little rooms." It is from this name that we get our present-day term, *cell.*

Actually, what Hooke observed were not cells at all but the empty cell walls of dead plant tissue, which is what tree bark really is. However, Hooke would not have thought of his *cellulae* as dead because he did not understand that they could be alive! Although he noticed that cells in other plant tissues were filled with what he called "juices," he preferred to concentrate on the more prominent cell walls that he had first encountered.

Meanwhile, **Antonie van Leeuwenhoek** was making an amazing series of microscopic observations that would do much to lay the foundation for our appreciation of the cellular basis of life. Van Leeuwenhoek was a Dutch shopkeeper who devoted much of his spare time to the design of simple microscopes. He then used his microscopes to examine almost anything he could get his hands on. He reported his observations to the Royal Society in a series of papers during the last quarter of the seventeenth century. His detailed reports attest to both the high quality of his lenses and his keen powers of observation.

Two factors restricted further understanding of the nature of cells. One was the limited resolution of the microscopes of the day, which even van Leeuwenhoek's superior instruments could push just so far. The second and probably more fundamental factor was the essentially descriptive nature of seventeenth-century biology. It was basically an age of observation, with little thought given to explaining the intriguing architectural details of biological materials that were beginning to yield to the probing lens of the microscope.

More than a century passed before the combination of improved microscopes and more investigatively minded microscopists resulted in a series of developments that culminated in an understanding of the importance of cells in biological organization. By the 1830s, improved lenses led to higher magnification and better resolution, such that structures only 1 micrometer (μm) apart could be resolved. (A *micrometer* is 10^{-6} m, or one-millionth of a meter; see the box on pp. 4–5 for a discussion of the units of measurement appropriate to cell biology.)

Aided by such improved lenses, the English botanist **Robert Brown** found that every plant cell he looked at contained a rounded structure, which he called a *nucleus*. In 1838, his German colleague **Matthias Schleiden** came to the important conclusion that all plant tissues were composed of cells and that an embryonic plant always arose from a single cell. Similar conclusions concerning animal tissue were reported only a year later by **Theodor Schwann,** thereby laying to rest earlier speculations that plants and animals might be structurally quite different. It is easy to understand how such speculations could have arisen. After all, plant cell walls provide conspicuous boundaries between cells that are readily visible even with a crude microscope, whereas individual animal cells, which lack cell walls, are much harder to distinguish in a tissue sample. It was only when Schwann examined animal cartilage cells that he became convinced of the fundamental similarity of plant and animal tissue, since cartilage cells, unlike most other animal cells, have boundaries that are well defined by thick deposits of collagen fibers.

It is to Schwann's credit that he was able to draw all these observations together into a single unified theory of cellular organization, illustrating the more interpretive atmosphere that was beginning to emerge. Moreover, Schwann's formulation has stood the test of time and continues to provide the basis for our own understanding of the importance of cells and cell biology.

As originally postulated by Schwann, the **cell theory** had two basic tenets:

1. All organisms consist of one or more cells.
2. The cell is the basic unit of structure for all organisms.

Less than 20 years later, a third tenet was added. This grew out of Brown's original description of nuclei, extended by Karl Nägeli to include observations on the nature of cell division. By 1855, **Rudolf Virchow,** a German physiologist, was able to conclude that cells arose in only one manner—by the division of other, preexisting cells. Virchow encapsulated this conclusion in the now-famous Latin phrase *omnis cellula e cellula,* which in translation becomes the third tenet of modern-day cell theory:

3. All cells arise only from preexisting cells.

Thus, the cell is not only the basic unit of structure for all organisms but also the basic unit of reproduction. In other words, all of life has a cellular basis. No wonder, then, that an understanding of cells and their properties is so fundamental to a proper appreciation of all other aspects of biology.

Modern Cell Biology Emerges

Modern cell biology involves the weaving together of three distinctly different strands into a single fabric. As Figure 1-1 illustrates (see p. 6), each of the strands had its own historical origins, and most of the intertwining has occurred only within the last 50 years. Each strand should be appreciated in its own right, for each contributes to the overall tapestry. The contemporary cell biologist must be adequately informed about all three strands, regardless of what his or her own immediate interests happen to be.

The first of these historical strands is **cytology,** which is concerned primarily with cellular structure. As we have already seen, cytology had its origins more than three centuries ago and depended heavily on the light microscope for its initial impetus. More recently, the advent of electron microscopy and several related optical techniques has led to considerable additional cytological activity and understanding.

The second strand represents the contributions of **biochemistry** to our understanding of cellular function. Most of the developments in this field have occurred within the last 50 years, though again the roots go back much further. Especially important has been the development of techniques such as centrifugation and chromatography for the separation of cellular components and molecules.

The third strand is **genetics.** Here, the historical continuum stretches back more than a century to Gregor Mendel. Again, however, much of our present understanding has come within the last several decades. An especially important landmark on the genetic strand came with the realization that DNA (deoxyribonucleic acid) is the bearer of genetic information in most (though not all) life forms.

To understand present-day cell biology therefore means to appreciate its diverse roots and the important contributions that each of its component strands has made to our current understanding of what a cell is and what it can do. Each of the three historical strands of cell biology will be discussed briefly here, but a fuller appreciation of each is likely to come only as various aspects of cell structure, function, and genetics are explored in later chapters. In fact, much of the rest of the text can be thought of as a further development and integration of the several historical strands that are woven into modern cell biology.

Units of Measurement in Cell Biology

The challenge of understanding cellular structure and organization is complicated by the problem of size. Cells and their organelles are so small that the units used to measure them are unfamiliar to many students and therefore often difficult to appreciate. The problem can be approached in two ways: by realizing that there are really only two units necessary to express the dimensions of most structures of interest to us and by illustrating a variety of structures that can be appropriately measured with each of these units.

The **micrometer** (μm) is the most useful unit for expressing the size of cells and larger organelles. A micrometer (sometimes also called a **micron**) corresponds to one-millionth of a meter (10^{-6} m). In general, bacterial cells are a few micrometers in diameter, and the cells of plants and animals are 10- to 20-fold larger in any single dimension. Larger organelles such as mitochondria and chloroplasts tend to have diameters or lengths of a few micrometers and are therefore comparable in size to whole bacterial cells. Smaller organelles are usually in the range 0.2–1.0 μm. As a rule of thumb, if you can see it with a light microscope, you can probably express its dimensions conveniently in micrometers, since the resolution limit of the light microscope is about 0.25 μm. Figure 1-A illustrates a variety of structures that are usually measured in micrometers.

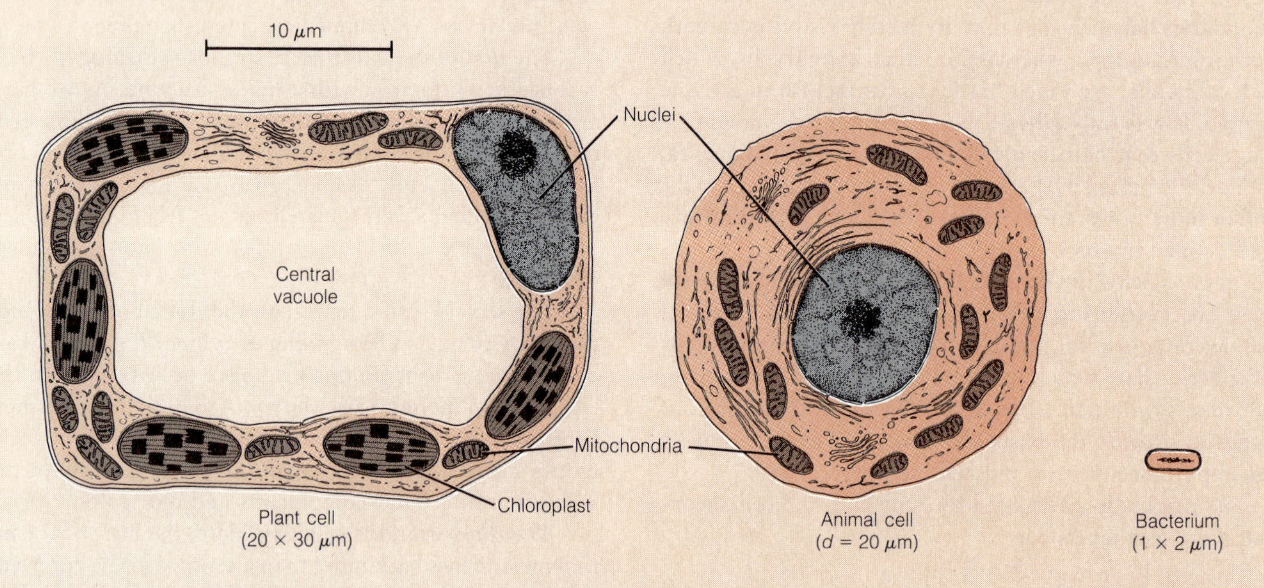

Figure 1-A The World of the Micrometer. Structures with dimensions that can be measured conveniently in micrometers include almost all cells and some of the larger organelles. (Scale: 1 in. = 10 μm; *d* = diameter.)

The **nanometer** (nm), on the other hand, is the unit of choice for molecules and subcellular structures that are too small or too thin to be seen with the light microscope. A nanometer is one-billionth of a meter (10^{-9} m). It therefore takes 1000 nanometers to equal 1 micrometer. As a benchmark on the nanometer scale, a ribosome has a diameter of about 20–25 nm. Other structures that can be measured conveniently in nanometers are microtubules, microfilaments, membranes, and DNA molecules. The dimensions of these structures are indicated in Figure 1-B.

Another unit frequently used in cell biology is the **angstrom** (Å), which corresponds to 10^{-10} m or 0.1 nm. Molecular dimensions, in particular, are often expressed in angstroms. However, since the angstrom differs from the nanometer by only a factor of ten, it adds little flexibility to the expression of dimensions at the cellular level and will therefore not be used in this text.

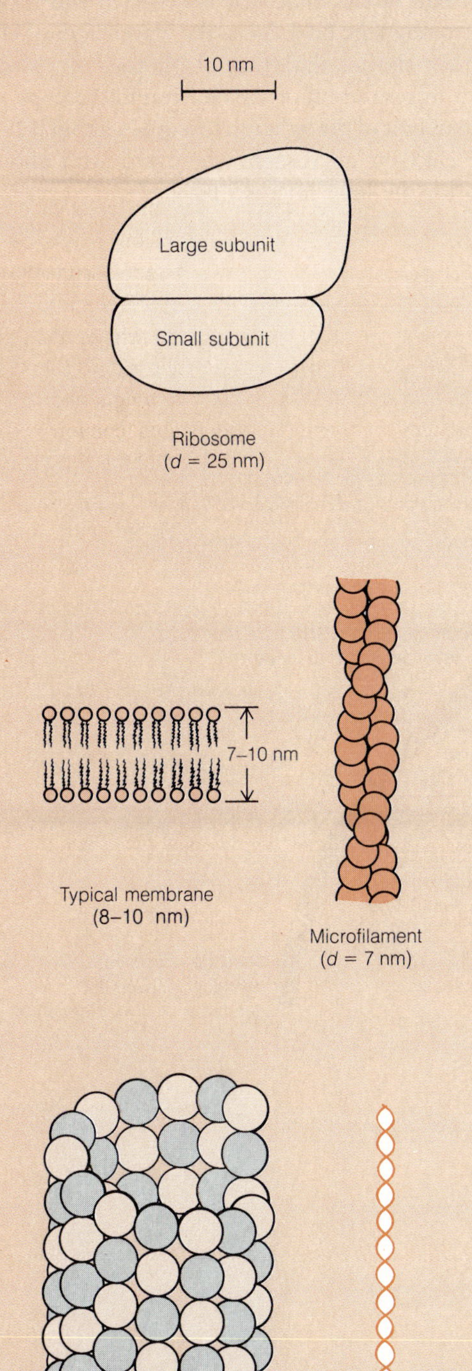

Figure 1-B The World of the Nanometer. Structures with dimensions that can be measured conveniently in nanometers include ribosomes, membranes, microfilaments, microtubules, and the DNA double helix. (Scale: 1 in. = 20 nm; d = diameter.)

The Cytological Strand

Strictly speaking, cytology is the study of cells. (Actually, the literal meaning of the Greek word *cytos* is "hollow vessel," which fits well with Hooke's initial impression of cells.) Historically, however, cytology has dealt primarily with cellular structure, mainly through the use of optical techniques.

The Light Microscope. The **light microscope** was the earliest tool of the cytologists and continues to play an important role in our elucidation of cellular structure. Light microscopy allowed cytologists to identify membrane-bounded structures such as *nuclei, mitochondria,* and *chloroplasts* within a variety of cell types. Such structures are called **organelles** ("little organs") and are prominent features of most plant and animal (but not bacterial) cells.

Other significant developments include the invention of the microtome in 1870 and the availability of various dyes and stains at about the same time. A **microtome** is an instrument for slicing thin sections of biological samples, usually after they have been dehydrated and embedded in paraffin or plastic. The technique enables rapid and efficient preparation of thin tissue slices of uniform thickness. The dyes that came to play so important a role in staining and identifying subcellular structures were primarily developed in the latter half of the nineteenth century by German industrial chemists working with coal tar derivatives.

Together with improved optics and more sophisticated lenses, these and related developments pushed light microscopy as far as it could go—to the physical limits of resolution imposed by the wavelengths of visible light. The theoretical limit of resolution is $\lambda/2$, where λ is the wavelength of the light used to illuminate the sample. For visible light in the wavelength range of 400–700 nanometers (nm), the limit of resolution is about 200–350 nm. (A *nanometer* is 10^{-9} or one-billionth of a meter; 1 nm = 0.001 μm.) Figure 1-2 illustrates the useful range of the light microscope and compares its resolving power with that of the human eye. For a more detailed discussion of microscopy, see Appendix A.

Special optical techniques, such as phase-contrast and interference microscopy, have been developed to visualize living cells and their organelles by amplifying and enhancing slight changes in the phase of transmitted light as it passes through a sample. But ultimately, these techniques, like all others involving light microscopy, are subject to the same limit of resolution imposed by the wavelength of visible light. Even the use of ultraviolet radiation with shorter wavelengths increases the resolution only by a factor of about two.

The Electron Microscope. A major breakthrough in resolving power came with the development of the **electron microscope.** The electron microscope was invented in Germany in 1932 and came into widespread biological use in the early 1950s, with G. E. Palade, F. S. Sjøstrand, and K. R. Porter among its most notable early users. In place of visible light and optical lenses, the electron microscope uses a beam of electrons that is deflected and focused by an electromagnetic field. Since the wavelength of electrons is so much shorter than that of photons of visible light, the limit of resolution of the electron microscope is much lower than that of the light microscope—about 0.1–0.2 nm for the electron microscope compared with about 200–350 nm for the light microscope.

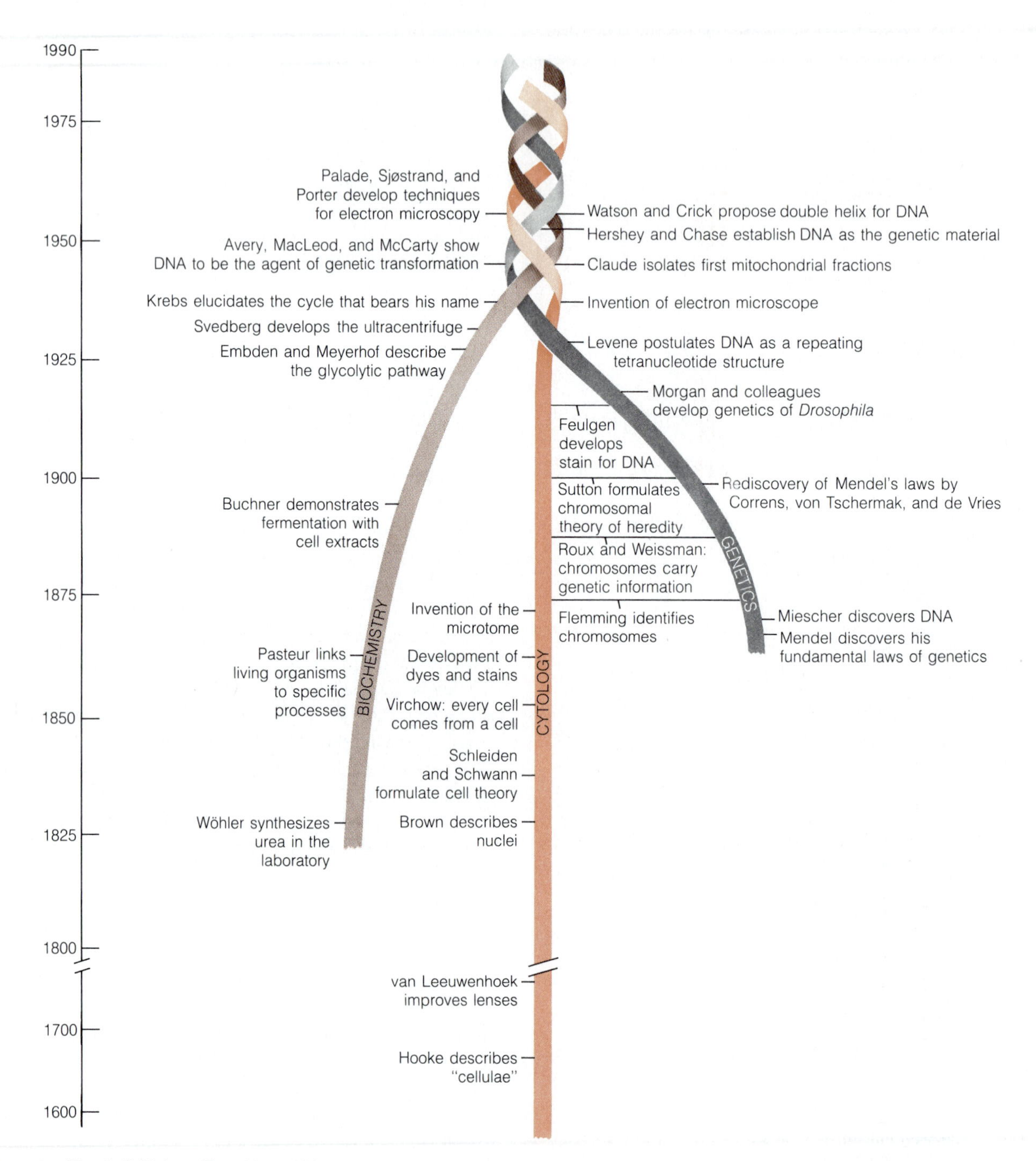

Figure 1-1 The Cell Biology Time Line. Although biochemistry, cytology, and genetics began as separate disciplines, they have increasingly merged since about the second quarter of the twentieth century.

However, for biological samples the practical limit of resolution is usually 2 nm or more. The difference is due to problems with specimen preparation and contrast. Nevertheless, the electron microscope has about 100 times more resolving power than the light microscope (Figure 1-2). The useful magnification is also greater—up to 100,000-fold for the electron microscope, compared with about 1000- to 1500-fold for the light microscope. Figure 1-3 illustrates how much more structural detail can be seen when a cell is examined with an electron microscope than with a light microscope.

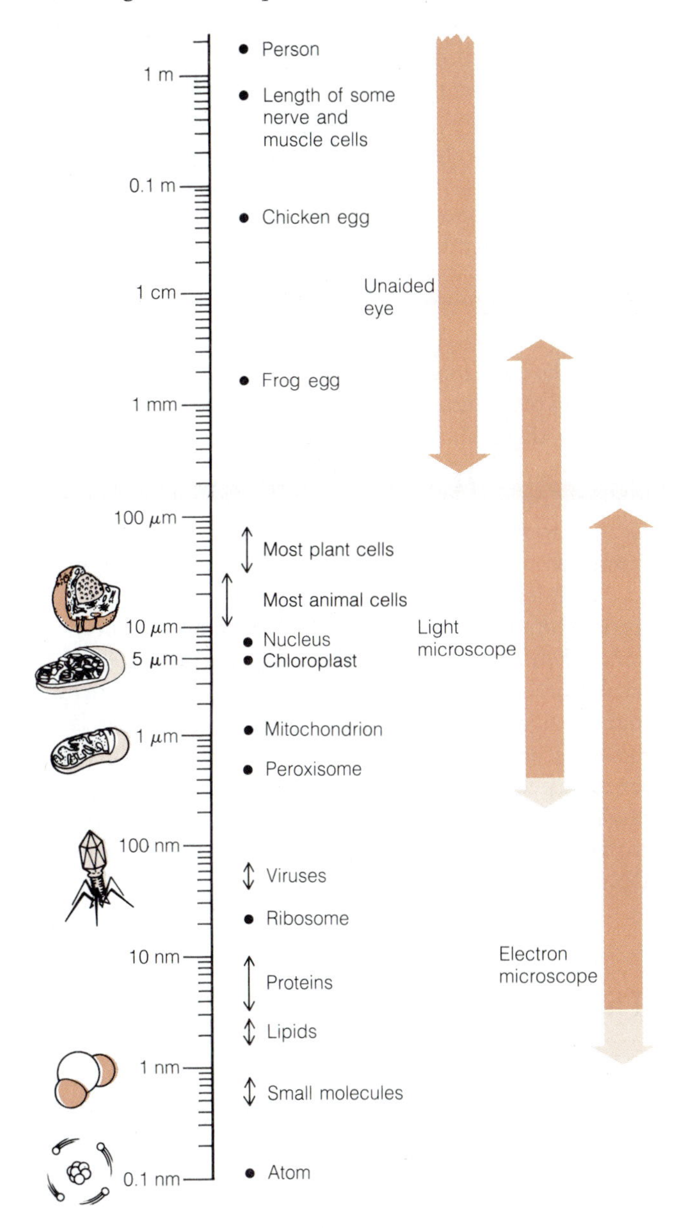

Figure 1-2 Resolving Power of the Human Eye, the Light Microscope, and the Electron Microscope. Note that the vertical axis is on a logarithmic scale to accommodate the range of sizes shown. Light-colored portions of arrows indicate the limits of resolution under special conditions.

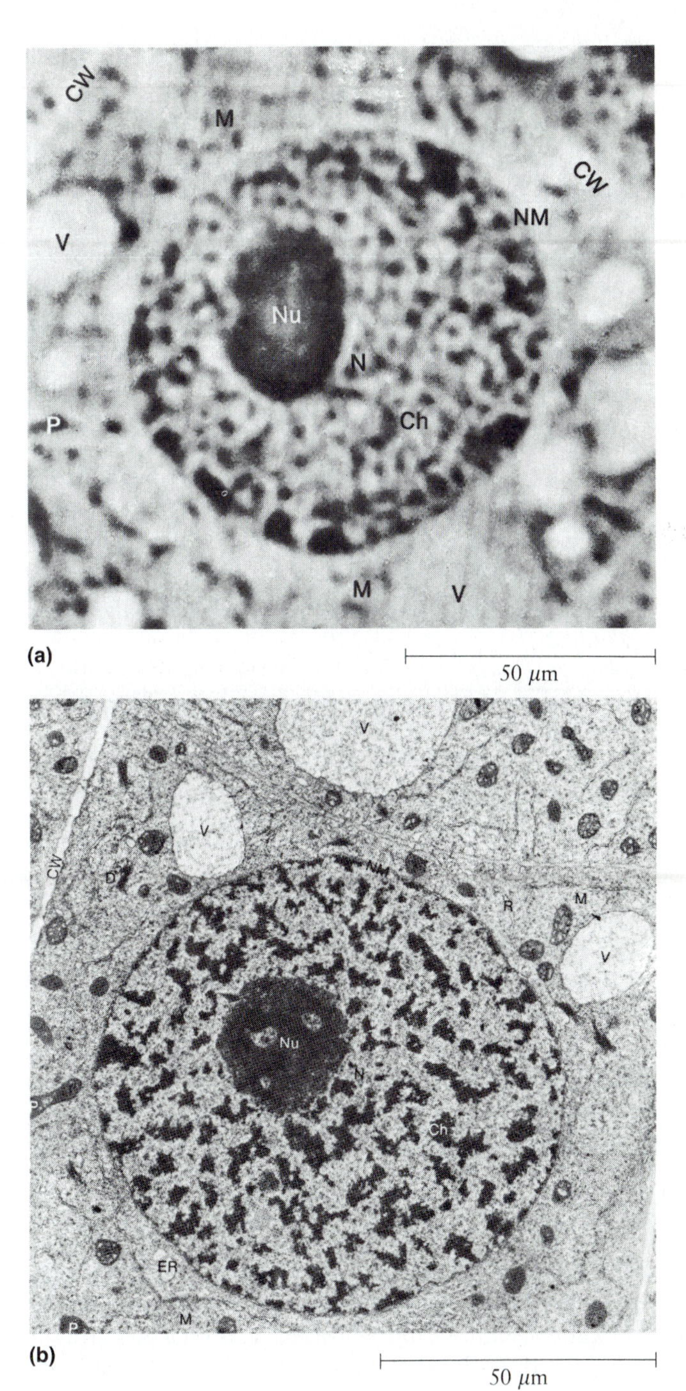

Figure 1-3 Comparison of Resolving Power. Cells can be visualized by either (a) light microscopy or (b) electron microscopy, but much more detail is seen in the latter case because of the greater resolving power of the electron microscope. Both sections were cut from the same piece of onion root tissue. For light microscopy, a 1.5-μm section was cut and photographed with phase-contrast optics. For electron microscopy, a section of 0.03 μm was used. Labeled structures include the nucleus (N), nucleolus (Nu), chromatin (Ch), nuclear membranes (NM), plastids (P), mitochondria (M), endoplasmic reticulum (ER), vacuoles (V), ribosomes (R), and cell wall (CW). (Part a, 740×; part b, 770×. A magnification of 1000× is the upper limit for the light microscope but is a very low magnification for the electron microscope.)

Because of the low penetration power of electrons, samples prepared for electron microscopy must be exceedingly thin. The instrument used for this purpose is called an **ultramicrotome.** It is equipped with a diamond knife and can cut sections as thin as 20 nm. Substantially thicker samples can also be examined by electron microscopy, but a much higher accelerating voltage is then required to increase the penetration power of the electrons adequately. Such a **high-voltage electron microscope** uses accelerating voltages up to several thousand kilovolts (kV), compared with the range of 50–100 kV common to most conventional instruments. Sections up to 1 μm thick can be studied with such a high-voltage instrument. This thickness allows organelles and other cellular structures to be examined in more depth.

Electron microscopy is discussed in detail in Appendix A. Specialized techniques described there include *freeze-fracturing, freeze-etching, negative staining,* and *scanning electron microscopy.* Scanning electron microscopy is an especially spectacular technique because of the sense of depth it gives to biological structures (Figure 1-4).

Electron microscopy has revolutionized our understanding of cellular architecture by making detailed ultrastructural investigations possible. Some organelles (such as nuclei or mitochondria) are large enough to be seen with a light microscope but can be studied in much greater detail with an electron microscope (Figure 1-3). In addition, electron microscopy has revealed cellular structures that are too small to be seen at all with a light microscope. An example is the ribosome, the site of protein synthesis in all cells. Ribosomes escape detection entirely with the light microscope because they are only 20–25 nm in diameter, depending on the type of cell in which they are found. They are readily visualized with the electron microscope, however, because of its greater resolving power.

The Scanning Tunneling Microscope. Our ability to visualize biological structures took another quantum leap forward recently when scientists in California obtained the first-ever three-dimensional images of an unaltered DNA molecule, with the structural features of the molecule shown in unprecedented detail. As described in the box on p. 11, the images were obtained using a **scanning tunneling microscope,** a novel kind of microscope invented only six years earlier. This instrument has a magnifying power that may enable scientists to visualize directly the intricate atomic structure of complex biological molecules, particularly nucleic acids and proteins.

The first direct photograph of a DNA molecule shows an isolated length of double-stranded DNA making a loop and crossing over on itself (see figure in box). Each of the two strands is only about 2 nm in diameter, so thin that it would take 50,000 of them to equal the diameter of a human hair. They are magnified about a millionfold in the figure—and with a resolution that enabled scientists to measure distances between the coils of the molecule.

The team of scientists who made the images included chemists, a biologist, a physicist, an electronics engineer, and several surface scientists. (Undergraduate readers should be encouraged to learn that the honor of actually photographing the DNA molecule was given to a 20-year-old junior in biophysics, Troy Wilson, who worked for the research team.) The instrument they used was a variation of the scanning tunneling microscope invented in 1981 by Gerd Binnig and Heinrich Rohrer, who have received a Nobel Prize for their work.

The scanning tunneling microscope features a sharp-tipped needle made of a platinum-rhodium alloy. The needle is attached to a tube-shaped piezoelectric ceramic, a material that expands and contracts in response to an electrical charge. When the tip of the needle is positioned very close to the surface of the sample, a current of electrons crosses the gap from needle to sample by means of a phenomenon of quantum mechanics known as *tunneling.* The *scanning* then follows: The tip scans across the sample, moving in and out to maintain the current at a constant level and thus tracing out the surface contours of the sample. The topographic data from the tip movements are then translated into three-dimensional images by a computer.

With its capability to probe biological molecules, the scanning tunneling microscope may well revolutionize our understanding of the structure of these molecules. The resolution seems good enough, for example, to identify individual nucleotides, the repeating units of the DNA molecule, and thus to determine the sequence in which they are arranged along the molecule. Similarly, the instrument might also prove useful for studying the three-dimensional structures of proteins as well as the interactions between DNA and proteins.

The Biochemical Strand

At about the time when cytologists were starting to explore cellular structure with their microscopes, other scientists were making observations that began to explain and clarify cellular function. Much of what is now called biochemistry dates from a discovery reported by the German chemist **Friedrich Wöhler** in 1828. Wöhler was a contemporary (as well as fellow countryman) of Schleiden and Schwann. It is doubtful, however, that he would have thought he had

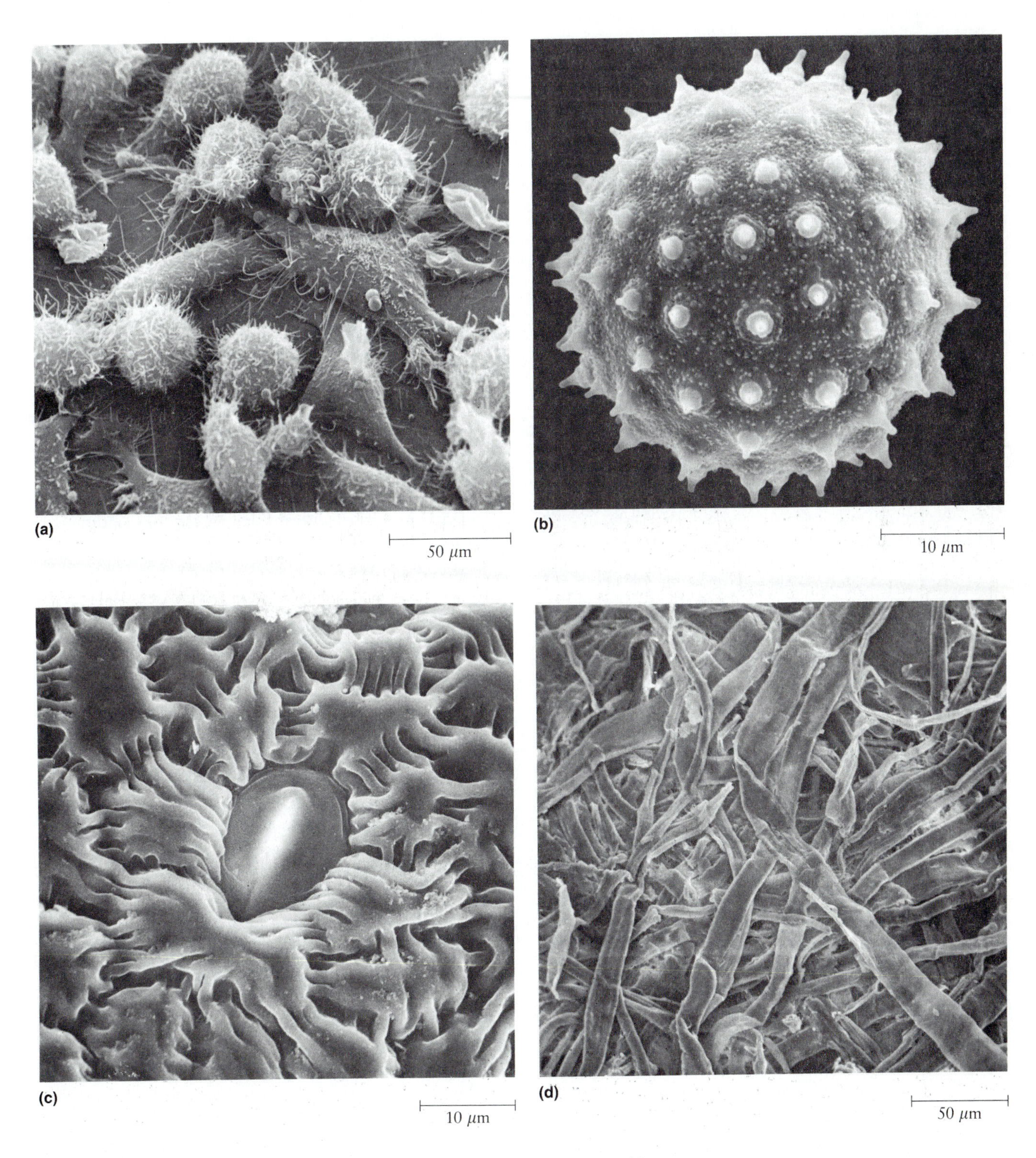

Figure 1-4 Scanning Electron Microscopy. A scanning electron microscope was used to photograph (a) cultured human neuroblastoma cells, (B) a pollen grain, (c) a stomate on the underside of a geranium leaf, and (d) the surface of a paper towel.

anything in common with a botanist and a zoologist peering at plant and animal tissues through their crude microscopes. Nor is Wöhler likely to have imagined that he would one day be placed alongside these men in a history of cellular biology. Nevertheless, Wöhler revolutionized our thinking about biology and chemistry by demonstrating that urea, an organic compound of biological origin, could be synthesized in the laboratory from an inorganic starting material, ammonium cyanate (Figure 1-5).

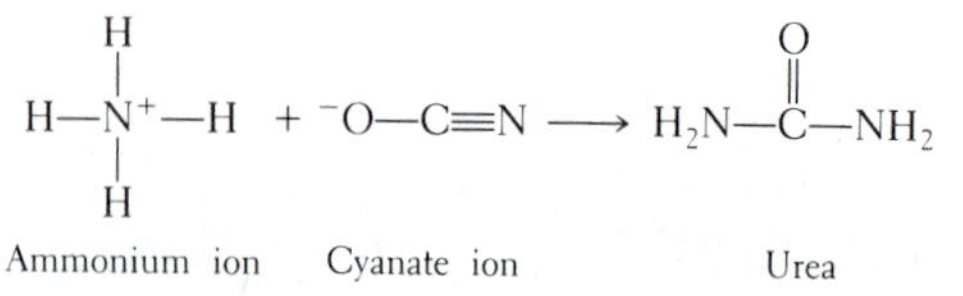

Previously, it had been widely held that living organisms were a world unto themselves, not governed by the laws of chemistry and physics that apply to the nonliving world. By showing that a compound made by living organisms—a "bio-chemical"—could be synthesized in a laboratory just like any other chemical, Wöhler helped to break down the conceptual distinction between the living and nonliving worlds and to dispel the notion that biochemical processes were somehow exempt from the laws of chemistry and physics.

Another major advance came about 40 years later, when **Louis Pasteur** linked the activity of living organisms to specific processes by showing that living yeast cells were needed to carry out the fermentation of sugar into alcohol. This observation was followed in 1897 by the finding of **Eduard Buchner** that fermentation could also take place with extracts from yeast cells—that is, the intact cells themselves were not required. Initially, such extracts were called "ferments," but gradually it became clear that the active agents in the extracts were specific biological catalysts that have since come to be called **enzymes.**

Significant progress in our understanding of cellular function came in the 1920s and 1930s as the biochemical pathways for fermentation and related cellular processes were elucidated. This was a period dominated by German biochemists such as Gustav Embden, Otto Meyerhof, Otto Warburg, and Hans Krebs. Several of these men have long since been immortalized by the pathways that have come to bear their names. For example, the *Embden-Meyerhof pathway* for glycolysis was a major research triumph of the early 1930s. It was followed shortly by the *Krebs cycle.* Both of these pathways are important because of their role in the process by which cells extract energy from foodstuffs. At about the same time, the high-energy compound *adenosine triphosphate* (*ATP*) came to be recognized as the principal energy storage compound in most cells.

Biochemistry took a major step forward with the development of the **ultracentrifuge** as a means of separating subcellular structures and macromolecules on the basis of size, shape, and density. In many ways, the ultracentrifuge was as significant for biochemistry as the electron microscope was for cytology. In fact, both instruments were

$$\mathrm{H{-}\overset{\displaystyle H}{\underset{\displaystyle H}{N^{+}}}{-}H} \;+\; {}^{-}\mathrm{O{-}C{\equiv}N} \longrightarrow \mathrm{H_2N{-}\overset{\displaystyle O}{\overset{\|}{C}}{-}NH_2}$$

Ammonium ion   Cyanate ion   Urea

Figure 1-5 The Synthesis of Urea. Urea was the first organic compound to be synthesized from inorganic reactants.

developed at about the same time, so the ability to see organelles and other subcellular structures with much greater resolution came almost simultaneously with the capability to isolate and purify them. The ultracentrifuge was developed in Sweden by The Svedberg during the period 1925–1930 and was used initially for determining the sedimentation rates of proteins. Adaptation of centrifugation techniques for the isolation of subcellular fractions came in the early 1940s, largely through the pioneering work of Albert Claude.

With an enhanced capability both to see subcellular structures and to isolate them, cytologists and biochemists began to realize the extent to which their respective observations on cellular structure and function could complement each other, thereby laying the foundations for modern cell biology.

The Genetic Strand

The third strand in the fabric of cell biology is genetics. Like the other two, this strand also has important roots in the nineteenth century. In this case, the strand begins with **Gregor Mendel,** whose contributions to science are well known. His studies with the pea plants he grew in the monastery garden must surely rank among the most famous experiments in all of biology. His major findings were published in 1866, laying out the principles of segregation and independent assortment of the "hereditary factors" that we know today as **genes.** These were singularly important principles, destined to provide the foundation for what would eventually be known as Mendelian genetics. But Mendel was clearly a man ahead of his time. His work went almost unnoticed when it was first published and was not fully appreciated until its rediscovery nearly 35 years later.

As a prelude to that rediscovery, the role of the nucleus in the genetic continuity of cells came to be appreciated in the decade following Mendel's work. Shortly thereafter, **chromosomes** were identified by Walther Flemming as threadlike bodies seen in dividing cells. Flemming called

*First Photograph Ever to Show Structure of DNA Molecule**

For more than 35 years, scientists have wondered what the molecules of human genes really look like—and now for the first time, they know.

Researchers at Berkeley and Livermore, using a revolutionary microscope invented eight years ago, reported yesterday that they have finally succeeded in photographing a tiny segment of pure DNA—the famed "double helix" that carries the genetic code for every aspect of heredity in everything that lives.

The three-dimensional images show the structural features of the DNA molecule in unprecedented detail (Figures 1-C and 1-D) and the instrument that took the pictures may soon probe even more deeply, said the team that took the pictures.

The photographs were taken with a scanning tunneling microscope, and the scientists at Berkeley and Livermore believe its power to magnify may be enough to determine the intricate atomic structure of the basic chemicals that make up the genetic material.

Miquel Salmeron, the physicist who headed the Lawrence Berkeley Laboratory's team, said the two groups of researchers normally use the microscope to probe the atomic structure of defects in the surface of metallic materials such as molybdenum and carbon films on platinum. The defects speed up important chemical reactions, but no one knows just how. "We were skeptical that the microscope could do anything good looking at biological molecules," Salmeron said, "but we decided to try anyway. Now the results have changed everything."

The change, according to Wigbert Siekhaus of the Lawrence Livermore National Laboratory's team, could significantly accelerate the international effort to map the entire sequence of 100,000 human genes.

Now that the microscope's ability to probe biological samples has been proven, the scientists said, they plan to train it directly on protein molecules. Proteins are folded into such complex three-dimensional patterns that only this new type of microscope can photograph them directly, they said.

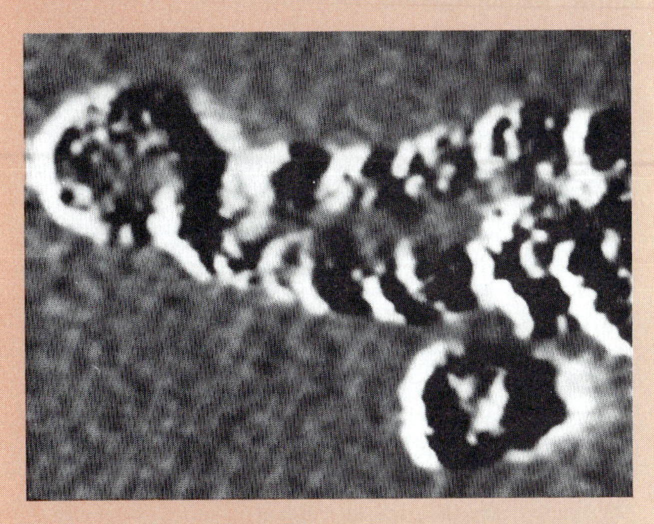

Figure 1-C First Photo of DNA. This is the first photograph ever made showing the structure of pure DNA, the double helix of linked molecules that form all the 100,000 genes governing all human heredity. The photograph shows the two strands of coiled material magnified a million times and was made through a revolutionary microscope built by scientists at the University of California's Lawrence laboratories at Livermore and Berkeley. (Scanning tunneling electron micrograph)

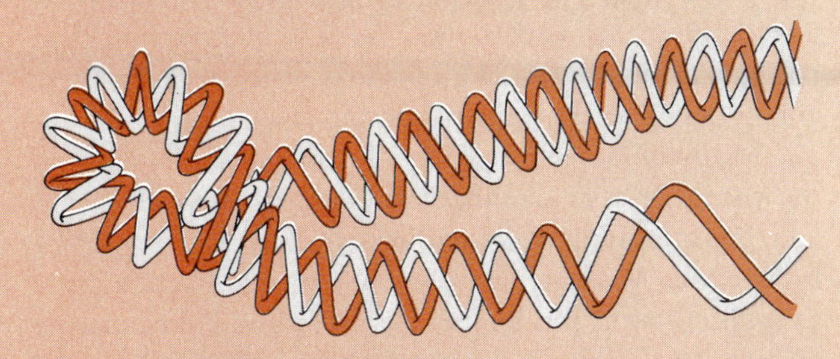

Figure 1-D The Helical Structure of DNA. The diagram shows the helical structure of the DNA; each of the two strands is so slender it would take 50,000 of them to reach the diameter of a human hair.

The system they used—and the microscope itself—was devised and built in their own laboratories by the Livermore and Berkeley teams, although the original scanning tunneling microscope was invented in 1981 by two IBM scientists in Geneva, Gerd Binnig and Heinrich Röhrer, who won a 1986 Nobel Prize for their work.

The honor of taking the prize picture of the molecule of DNA, or deoxyribonucleic acid, went to 20-year-old Berkeley junior biophysics major Troy E. Wilson who works for the team of scientists.

Wilson was excited yesterday. "We've taken a crucial first step," he said, "in understanding the real structure of the building blocks of living molecules. Its implications are unlimited, because if we can determine the structure of proteins in three dimensions too, we can have a far better understanding of their function in health and disease."

* Adapted from David Perlman, *San Francisco Chronicle*, January 20, 1989. Reprinted with permission of the author and publisher. Scientists at the University of Utah have recently reported (Science 251:640, 8 Feb. 1991) that images similar to those claimed by others to be DNA molecules can be generated from the pyrolytic graphite used as the conducting substrate in scanning tunneling microscope studies. They conclude it may be difficult to distinguish features of such artifacts from those of real biological molecules.

the division process *mitosis*, from the Greek word for thread. Chromosome number soon came to be recognized as a distinctive characteristic of a species and was shown to remain constant from generation to generation. That the chromosomes themselves might be the actual bearers of genetic information was suggested by Wilhelm Roux as early as 1883 and was expressed more formally by August Weissman shortly thereafter.

With the role of the nucleus and chromosomes established and appreciated, the stage was set for the rediscovery of Mendel's initial observations. This came in 1900, when his studies were cited almost simultaneously by three plant geneticists working independently—Carl Correns in Germany, Ernst von Tschermak in Austria, and Hugo de Vries in Holland. Within three years, the **chromosome theory of heredity** was formulated by Walter Sutton, who was the first to link the chromosomal "threads" of Flemming with the "hereditary factors" of Mendel.

Sutton's theory proposed that the hereditary factors responsible for Mendelian inheritance are located on the chromosomes within the nucleus. This hypothesis received its strongest confirmation from the work of Thomas Hunt Morgan and his students at Columbia University during the first two decades of this century. They chose as their experimental species *Drosophila melanogaster*, the common fruit fly. By identifying a variety of morphological mutants of *Drosophila*, Morgan and his co-workers were able to link specific traits to specific chromosomes.

Meanwhile, the foundation for our understanding of the chemical basis of inheritance was also slowly being laid. An important milestone was the discovery of DNA by **Friedrich Miescher** in 1869. Using such unlikely sources as salmon sperm and human pus from surgical bandages, Miescher isolated and described what he called "nuclein." But, like Mendel, Miescher was ahead of his time. It was about 75 years before the role of his nuclein as the genetic information of the cell came to be fully appreciated.

As early as 1914, DNA was implicated as an important component of chromosomes by the staining technique of Robert Feulgen, a method that is still in use today. But little consideration was given to the possibility that DNA could be the bearer of genetic information. In fact, that would have been considered quite unlikely in light of the apparently uninteresting structure of the mononucleotide constituents of DNA that were known by 1930. Until the middle of the century, it was widely held that genes were made up of proteins, since these were the only nuclear components that seemed to be able to account for the obvious diversity of genes.

A landmark experiment that clearly pointed to DNA as the genetic material was reported in 1944 by Oswald Avery, Colin MacLeod, and Maclyn McCarty. Their work focused on the phenomenon of genetic transformation in bacteria, to be discussed in Chapter 13. Their evidence was compelling, but the scientific community remained largely unconvinced of the conclusion. Just eight years later, however, a considerably more favorable reception was accorded the report of Alfred Hershey and Martha Chase that DNA, and not protein, enters a bacterial cell when it is infected by a bacterial virus.

Shortly thereafter, in 1953, **James Watson** and **Francis Crick** proposed their now-famous double helix model for DNA structure, with features that immediately suggested how replication and even mutation could occur. The *Watson-Crick model*, as it came to be known, catapulted DNA into prominence and launched an era of molecular genetics that has since revolutionized biology. In the process, the historical strand of genetics became intimately entwined with those of cytology and biochemistry, and the discipline of cell biology as we know it today came into being.

Perspective

The biological world is a world of cells. All living organisms are made up of one or more cells, each of which came from a preexisting cell. Although the importance of cells in biological organization has been appreciated for about 150 years, the discipline of cell biology as we know it today is of much more recent origin. Modern cell biology has come about by the interweaving of three historically distinct strands—cytology, biochemistry, and genetics—which in their early development probably did not seem at all related. But the contemporary cell biologist must understand all three strands because they complement one another in the quest to learn what cells are and how they function.

Key Terms for Self-Testing

cell (p. 2)
cell biology (p. 2)

The Cell Theory: A Brief History
Robert Hooke (p. 2)
Antonie van Leeuwenhoek (p. 2)
Robert Brown (p. 3)
Matthias Schleiden (p. 3)
Theodor Schwann (p. 3)
cell theory (p. 3)
Rudolf Virchow (p. 3)

Modern Cell Biology Emerges
cytology (p. 3)
biochemistry (p. 3)
genetics (p. 3)
light microscope (p. 5)
organelles (p. 5)
microtome (p. 5)
electron microscope (p. 6)
ultramicrotome (p. 8)
high-voltage electron microscope (p. 8)
scanning tunneling microscope (p. 8)
Friedrich Wöhler (p. 8)
Louis Pasteur (p. 10)
Eduard Buchner (p. 10)
enzymes (p. 10)
ultracentrifuge (p. 10)
Gregor Mendel (p. 10)
genes (p. 10)
chromosomes (p. 10)
chromosome theory of heredity (p. 12)
Friedrich Miescher (p. 12)
James Watson (p. 12)
Francis Crick (p. 12)

Units of Measurement in Cell Biology
micrometer (micron) (p. 4)
nanometer (p. 4)
angstrom (p. 4)

Suggested Reading

Modern Cell Biology Emerges

Bodenheimer, F. S. *The History of Biology*. London: Dawson, 1958.

Bradbury, S. *The Evolution of the Microscope*. New York: Pergamon Press, 1967.

Claude, A. The coming of age of the cell. *Science* 189 (1975): 433.

de Duve, C., and H. Beaufay. A short history of tissue fractionation. *J. Cell Biol.* 91 (1981): 293s.

Fruton, J. S. The emergence of biochemistry. *Science* 192 (1976): 327.

Gall, J. G., K. R. Porter, and P. Siekevitz, eds. Discovery in cell biology. *J. Cell Biol.* 91, part 3 (1981).

Hughes, A. *A History of Cytology*. New York: Abelard-Schumann, 1959.

Jacob, F. *The Logic of Life: A History of Heredity*. New York: Pantheon Press, 1973.

Judson, H. F. *The Eighth Day of Creation: Makers of the Revolution in Biology*. New York: Simon & Schuster, 1979.

Mirsky, A. E. The discovery of DNA. *Sci. Amer.* 218 (June 1968): 78.

Palade, G. E. Albert Claude and the beginning of biological electron microscopy. *J. Cell Biol.* 50 (1971): 5D.

Peters, J. A. *Classic Papers in Genetics*. Englewood Cliffs, N.J.: Prentice-Hall, 1959. (Includes classic papers by G. Mendel and W. S. Sutton.)

Stent, G. S. That was the molecular biology that was. *Science* 160 (1968): 390.

Methods in Modern Cell Biology

Bradbury, S. *An Introduction to the Optical Microscope*. Oxford, England: Oxford University Press, 1984.

Cooper, T. G. *The Tools of Biochemisty*. New York: Wiley, 1977.

Costerton, J. W. The role of electron microscopy in the elucidation of bacterial structure and function. *Annu. Rev. Microbiol.* 33 (1979): 459.

de Duve, C. Exploring cells with a centrifuge. *Science* 189 (1975): 186.

Everhart, T. E., and T. L. Hayes. The scanning electron microscope. *Sci. Amer.* 226 (January 1972): 54.

Glauert, A. M. The high-voltage electron microscope in biology. *J. Cell Biol.* 63 (1974): 717.

Meek, G. A. *Practical Electron Microscopy for Biologists*. New York: Wiley, 1976.

Pease, D. C., and K. R. Porter. Electron microscopy and ultramicrotomy. *J. Cell Biol.* 91 (1981): 287s.

Scheeler, P. *Centrifugation in Biology and Medical Science*. New York: Wiley, 1981.

Sommerville, J., and U. Scheer, eds. *Electron Microscopy in Molecular Biology: A Practical Approach*. Washington, D.C.: IRL Press, 1987.

Tanaka, K. Scanning electron microscopy of intracellular structure. *Int. Rev. Cytol.* 68 (1980): 97.

Wischnitzer, S. *Introduction to Electron Microscopy*, 3rd ed. Elmsford, N.Y.: Pergamon Press, 1981.

Problem Set

1. **The Historical Strands of Cell Biology.** For each of the following events, indicate whether it belongs to the cytological (C), biochemical (B), or genetic (G) strand in the historical development of cell biology.
   (a) Köllicker describes "sarcosomes" (now called mitochondria) in muscle cells (1857).
   (b) Hoppe-Seyler isolates the protein hemoglobin in crystalline form (1864).
   (c) Haeckel postulates that the nucleus is responsible for heredity (1868).
   (d) Ostwald proves that enzymes are catalysts (1893).
   (e) Muller discovers that X-rays induce mutations (1927).
   (f) Davson and Danielli postulate a model for the structure of cell membranes (1935).
   (g) Beadle and Tatum formulate the one gene–one enzyme relationship (1940).
   (h) Claude isolates the first mitochondrial fractions from rat liver (1940).
   (i) Lipmann postulates the central importance of ATP in cellular energy transactions (1940).
   (j) Avery, MacLeod, and McCarty demonstrate that bacterial transformation is attributable to DNA, not protein (1944).
   (k) Palade, Porter, and Sjøstrand each develop techniques for fixing and sectioning biological tissue for electron microscopy (1952–1953).
   (l) Lehninger demonstrates that oxidative phosphorylation depends for its immediate energy source on transport of electrons in the mitochondrion (1957).

2. **More on Historical Strands.** What did the development of the ultracentrifuge have in common with the invention of the electron microscope in terms of effect on their respective strands of present-day cell biology? Can you identify a comparable event (or series of events) on the genetic side of cell biology?

3. **Cell Sizes.** To appreciate the differences in cell size illustrated in Figure 1-A on p. 4, consider the following specific examples. *Escherichia coli,* a typical bacterial cell, is cylindrical in shape, with a diameter of about 1 μm and a length of about 2 μm. As a typical animal cell, consider a human liver cell, which is roughly spherical in shape and has a diameter of about 20 μm. And for a typical plant cell, consider the columnar *palisade cells* located just beneath the upper surface of many plant leaves. These cells are cylindrical in shape, with a diameter of about 20 μm and a length of about 35 μm.
   (a) Calculate the approximate volume of each of these three cell types in cubic micrometers. (Recall that $V = \pi r^2 h$ for a cylinder and that $V = 4\pi r^3/3$ for a sphere.)
   (b) Approximately how many bacterial cells would fit in the internal volume of a human liver cell?
   (c) Approximately how many liver cells would fit inside a palisade cell?

4. **Sizing Things Up.** To appreciate the sizes of the subcellular structures shown in Figure 1-B on p. 5, consider the following calculations.
   (a) All cells and many subcellular structures are surrounded by a membrane. Assuming a typical membrane to be about 8 nm wide, how many such membranes would have to be aligned side by side before the structure could be seen with the light microscope? How many with the electron microscope?
   (b) Ribosomes are the structures in cells on which the process of protein synthesis takes place. A human ribosome is a roughly spherical structure with a diameter of about 25 nm. How many ribosomes would fit in the internal volume of the human liver cell described in Problem 3 if the entire volume of the cell were filled with ribosomes?
   (c) The genetic material of the *Escherichia coli* cell described in Problem 3 consists of a DNA molecule with a diameter of 2 nm and a total length of 1.36 mm. (The molecule is actually circular, with 1.36 mm as its circumference.) To be accommodated in a cell that is only a few micrometers long, this large DNA molecule is tightly coiled and folded into a *nucleoid* that occupies a small proportion of the internal volume of the cell. Calculate the smallest possible volume into which the DNA molecule could fit, and express that as a fraction of the internal volume of the bacterial cell that you calculated in Problem 3(a).

5. **The "Facts" of Life.** Each of the following statements was once regarded as a biological fact but is now understood to be untrue. In each case, indicate why the statement was once thought to be true and why it is no longer considered a fact.
   (a) Plant and animal tissues are constructed quite differently, since animal tissues do not have conspicuous boundaries that divide them into cells.
   (b) Living organisms are not governed by the laws of chemistry and physics as is nonliving matter, but are subject to a "vital force" that is responsible for the formation of organic compounds.

(c) Genes most likely consist of proteins, since the only other likely candidate, DNA, is a relatively uninteresting molecule consisting of only four kinds of monomers (nucleotides) arranged in a relatively invariant repeating tetranucleotide sequence.

(d) The fermentation of sugar to alcohol can take place only if living yeast cells are present.

6. **More "Facts" of Life.** Each of the following statements was regarded as a biological fact until quite recently but is now either rejected or qualified to at least some extent. In each case, speculate on why the statement was once thought to be true and then try to figure out what evidence might have made it necessary to reject or at least to qualify the statement. (Note: This question requires a more intrepid sleuth than does Problem 5, but chapter references are provided to aid in your sleuthing.)

(a) A biological membrane can be thought of as a protein-lipid "sandwich" consisting of an exclusively phospholipid interior coated on both sides with thin layers of protein. (Chapter 7)

(b) The mechanism by which the oxidation of organic molecules such as sugars leads to the generation of ATP involves a high-energy phosphorylated molecule as an intermediate. (Chapter 11)

(c) When carbon dioxide from the air is "fixed" (covalently linked) into organic form by photosynthetic organisms such as green plants, the first form in which the carbon atom of the CO_2 molecule appears is always the three-carbon compound 3-phosphoglycerate. (Chapter 12)

(d) The enzymes required to catalyze the conversion of sugar into a compound called pyruvate invariably occur in the cytoplasm of the cell, rather than being compartmentalized in membrane-enclosed structures. (Chapter 10)

(e) DNA always exists as a duplex of two strands wound together into a right-handed helix. (Chapter 13)

(f) The genetic code that specifies how the information present in the DNA molecule is used to make proteins is universal in the sense that all organisms use the same code. (Chapter 16)

2

The Chemistry of the Cell

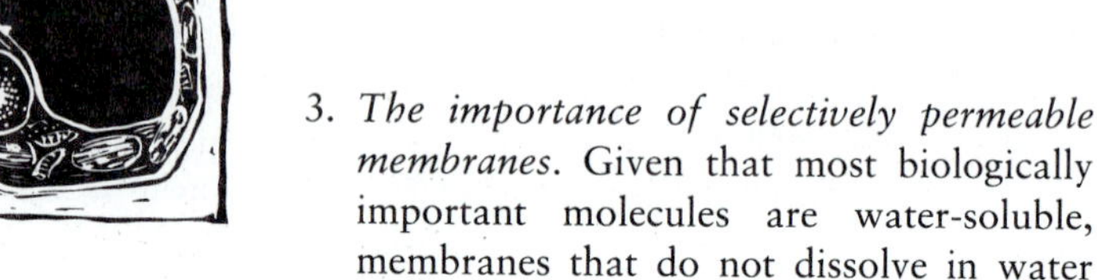

Students just beginning in cell biology are sometimes surprised—occasionally even dismayed—to find that almost all courses and textbooks dealing with cell biology involve a substantial amount of chemistry. Yet biology in general and cell biology in particular depend heavily on both chemistry and physics. After all, cells and organisms follow all the laws of the physical universe, and biology is really just the study of chemistry in systems that happen to be alive. In fact, everything cells are and do has a molecular and chemical basis. Therefore, we can truly understand and appreciate cellular structure and function only when we can describe that structure in molecular terms and express that function in terms of chemical reactions and events.

Trying to appreciate cellular biology without a knowledge of chemistry would be like trying to appreciate a translation of Goethe without a knowledge of German. Most of the meaning would probably get through, but much of the beauty and depth of appreciation would be lost in the translation. For this reason, we will concentrate on the chemical background so necessary for the cell biologist. Specifically, this chapter will focus on several principles that underlie much of cellular biology, preparing, in turn, for the next chapter, which focuses on the major classes of chemical constituents in cells.

The main points of this chapter can conveniently be structured around five themes:

1. *The importance of carbon.* The chemistry of cells is essentially the chemistry of carbon-containing compounds, since the carbon atom has several unique properties that make it especially suitable as the backbone of biologically important molecules.
2. *The importance of water.* The chemistry of cells is also the chemistry of water-soluble compounds, since the water molecule has several unique properties that make it especially suitable as the universal solvent of living systems.
3. *The importance of selectively permeable membranes.* Given that most biologically important molecules are water-soluble, membranes that do not dissolve in water and are differentially permeable are very important both in defining cellular spaces and compartments and in controlling the movements of molecules and ions into and out of such spaces and compartments.
4. *The importance of synthesis by polymerization of small molecules.* Most biologically important molecules are either small, water-soluble organic molecules that can be transported across membranes or large macromolecules that cannot in general be so transported. Biological macromolecules are polymers formed by the linking together of similar or identical small molecules. Synthesis of macromolecules by polymerization of monomeric subunits is an important principle of cellular chemistry.
5. *The importance of self-assembly.* Biological macromolecules made of repeating monomeric subunits are often capable of self-assembly into higher levels of structural organization because the information needed to specify such assembly is inherent in the chemistry of the linear array of repeating monomers formed on polymerization.

Given these five principles, we can appreciate the main topics in cellular chemistry that we need to be familiar with before venturing further into our exploration of what it means to be a cell.

The Importance of Carbon

To study cellular molecules really means to study carbon-containing compounds, since almost without exception, molecules of importance to the cell biologist have a backbone, or skeleton, of carbon atoms linked together covalently. Actually, the study of carbon-containing compounds is the domain of **organic chemistry.** In its early days, organic chemistry was almost synonymous with biological chemistry because most of the carbon-containing compounds that were first investigated were obtained from biological sources (hence the word *organic,* acknowledging the organismal origins of the compounds). The terms have long since gone their separate ways, however, because organic chemists have now synthesized a bewildering variety of carbon-containing compounds that do not occur naturally (that is, in the biological world). Organic chemistry therefore includes all classes of carbon-containing compounds, whereas **biological chemistry** (**biochemistry** for short) deals specifically with the chemistry of living systems and is, as we have already seen, one of the several historical strands that form an integral part of modern cell biology.

The **carbon atom** (C) is the most important atom in biological molecules. The diversity and stability of carbon-containing compounds are due to specific properties of the carbon atom and especially to the nature of the interactions of carbon atoms with one another as well as with a limited number of other elements found in molecules of biological importance.

The single most fundamental property of the carbon atom is its **valence** of four, which means that the outermost electron orbital of the atom lacks four of the eight electrons needed to fill it completely (Figure 2-1a). Since a complete outer orbital is required for the most stable chemical state of an atom, carbon atoms tend to associate with one another or with other electron-deficient atoms, allowing adjacent atoms to share a pair of electrons. For each such pair, one electron comes from each of the atoms. Atoms that share each other's electrons in this way are said to be

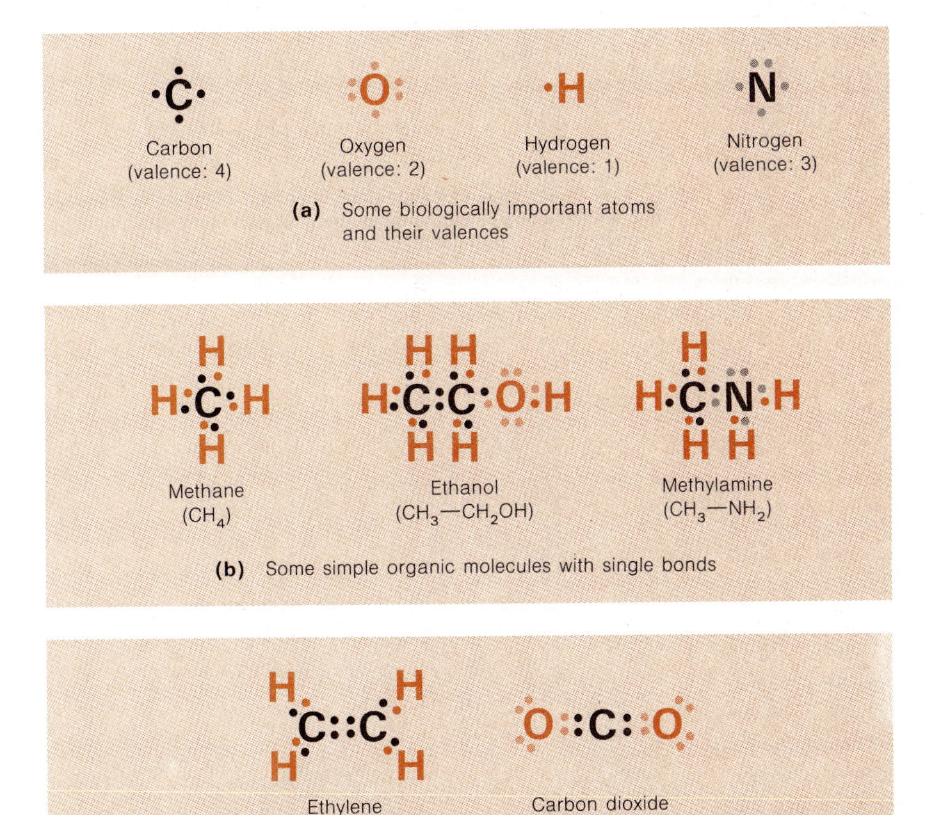

Figure 2-1 The Electron Configurations of Some Biologically Important Atoms and Molecules. Electronic configurations are shown for (a) atoms of carbon, oxygen, hydrogen, and nitrogen and for simple organic molecules with (b) single bonds, (c) double bonds, and (d) triple bonds. Only electrons in the outermost electron orbital are shown. In each case, the two electrons positioned between adjacent atoms represent a shared electron pair, with one electron provided by each of the two atoms. Electrons are color-coded; those from carbon, oxygen, hydrogen, and nitrogen are black, lightly colored, darkly colored, and gray, respectively. (All electrons are equivalent, of course; the color coding is simply to illustrate which electrons are contributed by each atom.)

joined together by a **covalent bond.** Carbon atoms are most likely to form covalent bonds with one another and with atoms of oxygen (O), hydrogen (H), nitrogen (N), and sulfur (S).

The electronic configurations of several of these atoms are shown in Figure 2-1a. Notice that in each case one or more electrons are required to complete the outer orbital. The number of "missing" electrons corresponds in each case to the valence of the atom, which indicates, in turn, the number of covalent bonds the atom can form. Carbon, hydrogen, oxygen, and nitrogen are the lightest elements that form covalent bonds by sharing electron pairs. This lightness makes the resulting compounds especially stable, since the strength of a covalent bond is inversely proportional to the atomic weights of the elements involved in the bond.

Because four electrons are required to fill the outer orbital of carbon, stable organic compounds have four covalent bonds for every carbon atom. Methane, ethanol, and methylamine (Figure 2-1b) are simple examples of such compounds, containing only **single bonds** between atoms. Sometimes, two or even three pairs of electrons can be shared by two atoms, giving rise to **double bonds** or even **triple bonds.** Ethylene and carbon dioxide (Figure 2-1c) are examples of double-bonded compounds. Triple bonds are found in molecular nitrogen (N_2) and hydrogen cyanide (Figure 2-1d).

Carbon-Containing Molecules Are Stable

As already implied, the stability of organic molecules is a property of the favorable electronic configuration of each carbon atom in the molecule. This stability is expressed in terms of **bond energy**—the amount of energy required to break 1 mole (6×10^{23}) of such bonds. Bond energies are usually expressed in *calories per mole (cal/mol)*, where a **calorie** is the amount of energy needed to raise the temperature of one gram of water one degree centigrade.

It takes a great deal of energy to break a covalent bond. For example, the carbon-carbon (C—C) bond has a bond energy of 83 kilocalories per mole (kcal/mol). The bond energies for carbon-nitrogen (C—N), carbon-oxygen (C—O), and carbon-hydrogen (C—H) bonds are all in the same range—70, 84, and 99 kcal/mol, respectively. Even more energy is required to break a carbon-carbon double bond (C=C; 146 kcal/mol) or a carbon-carbon triple bond (C≡C; 212 kcal/mol), so these compounds are even more stable.

We can appreciate the significance of these bond energies only by comparing them with other relevant energy values, as in Figure 2-2. Most noncovalent bonds in biologically important molecules have energies of only a few kilocalories per mole, and the energy of thermal vibration is even lower—about 0.6 kcal/mol. Even the ATP molecule, mentioned in Chapter 1 because of its importance in

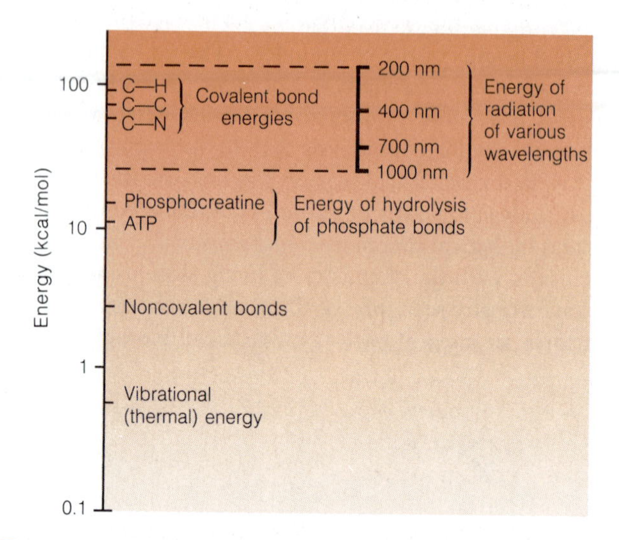

Figure 2-2 Energies of Biologically Important Transitions, Bonds, and Wavelengths of Electromagnetic Radiation. Note that energy is plotted on a logarithmic scale to accommodate the range of values shown.

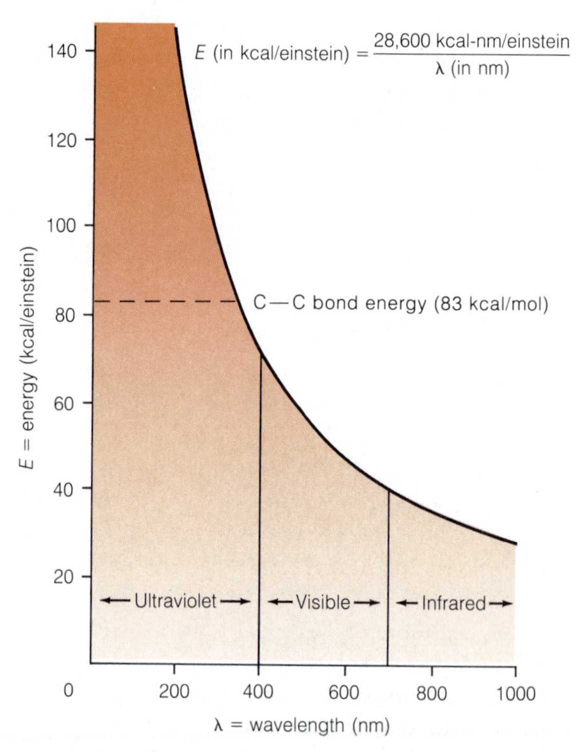

Figure 2-3 The Relationship Between Energy (*E*) and Wavelength (*λ*) for Electromagnetic Radiation. The horizontal line marks the energy content of the C—C single bond (83 kcal/mol).

cellular energy transactions, has a usable energy content of only about 10–14 kcal/mol, depending on actual conditions. Phosphocreatine, the compound used to store reserve energy in muscle cells, is in the same range. Covalent bonds are much higher in energy and therefore very stable. In fact, specific enzyme-catalyzed reactions capable of quite high energy input are usually required to break such bonds during chemical reactions in cells.

The "fitness" of the carbon-carbon bond for biological chemistry on Earth is especially clear when its energy is compared with that of solar radiation. As shown in Figure 2-3, there is an inverse relationship between the wavelength of electromagnetic radiation and its energy content. Across the visible spectrum (wavelengths of 400–700 nm), sunlight is lower in energy than the carbon-carbon bond. For example, green light with a wavelength of 500 nm has an energy content of about 57.2 kcal/einstein. (An **einstein** is equal to 1 mole of photons.) The energy of green light is therefore well below the energies of covalent bonds (Figure 2-2). If this were not the case, visible light would break covalent bonds spontaneously and life as we know it would not exist.

Figure 2-3 suggests another important point: the hazard that ultraviolet radiation poses to biological systems. At a wavelength of 300 nm, for example, ultraviolet light has an energy content of about 95.3 kcal/einstein, clearly enough to break carbon-carbon bonds spontaneously. It is this threat that has led to the current concern about pollutants that destroy the ozone layer in the upper atmosphere, since the ozone layer filters out much of the ultraviolet radiation that would otherwise reach the earth's surface.

Carbon-Containing Molecules Are Diverse

In addition to their stability, carbon-containing compounds are characterized by the great diversity of molecules that can be generated from relatively few different kinds of atoms. Again, this diversity is due to the tetravalent nature of the carbon atom and the resulting propensity of each carbon atom to form covalent bonds to four other atoms. Because one or more of these bonds can be to other carbon atoms, molecules consisting of long chains of carbon atoms can be built up. Ring compounds are also common. Further variety is possible by the introduction of branching and of double and single bonds into the carbon-carbon chains. When only hydrogen atoms are used to complete the valence requirements of such linear or circular molecules, the resulting compounds are called **hydrocarbons** (Figure 2-4). Hydrocarbons are very important economically because gasoline and other petroleum products are mixtures of short-chain hydrocarbons such as *octane,* an eight-carbon compound (C_8H_{18}).

In biology, on the other hand, hydrocarbons play only a limited role because they are essentially insoluble in water, the universal solvent in biological systems. There is an important exception to this general rule, however: The interior of every biological membrane is a nonaqueous environment from which water and water-soluble compounds are excluded by the long hydrocarbon "tails" of phospholipid molecules that project into the interior of the membrane from either surface. This feature of membranes has important implications for their role as permeability barriers, as we will see shortly.

Most biological compounds contain, in addition to carbon and hydrogen, one or more atoms of oxygen and often nitrogen, phosphorus, or sulfur as well. These atoms are usually part of various **functional groups** that confer both water solubility and chemical reactivity on the molecules of which they are a part. (Even the phospholipid molecules whose hydrocarbon tails contribute so importantly to the nonaqueous nature of the membrane interior contain atoms other than hydrogen and carbon.)

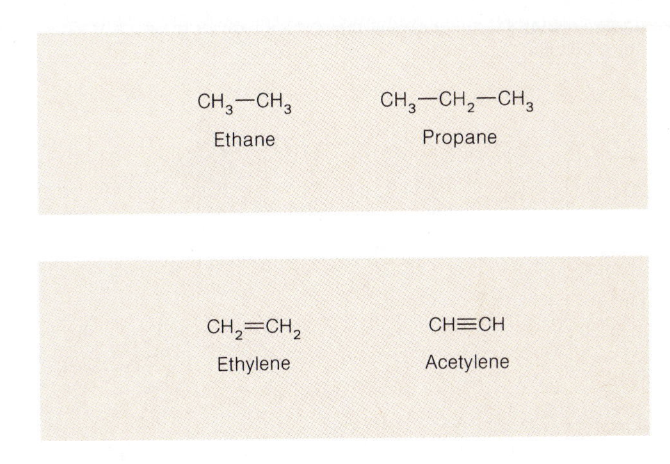

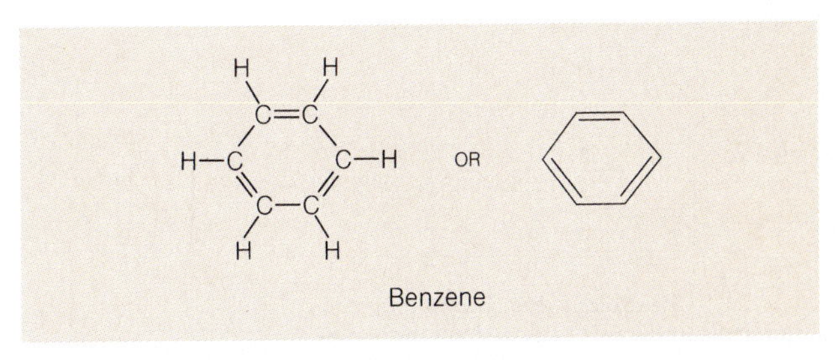

Figure 2-4 Some Simple Hydrocarbon Compounds. Compounds in the top row have single bonds only, whereas those in row two have double or triple bonds. The condensed structure shown for benzene is an example of the simplified structures that chemists frequently use for such compounds.

Some of the more common functional groups present in biological molecules are shown in Figure 2-5. Several of these groups are ionized or protonated at the near-neutral pH of most cells, including the negatively charged *carboxyl* and *phosphate* groups and the positively charged *amino* group. Other groups, such as the *hydroxyl, sulfhydryl,* and *aldehyde* groups, are uncharged at pH values near neutrality but nonetheless cause a significant redistribution of electrons within the attached molecules, thereby conferring on these molecules greater water solubility and chemical reactivity. Other less common atoms are also found in specialized molecules, such as the iron of hemoglobin, the magnesium of chlorophyll, and the several different cations present in various metalloenzymes.

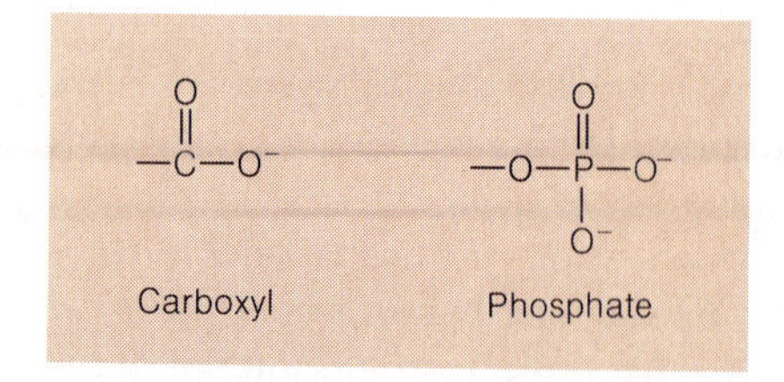

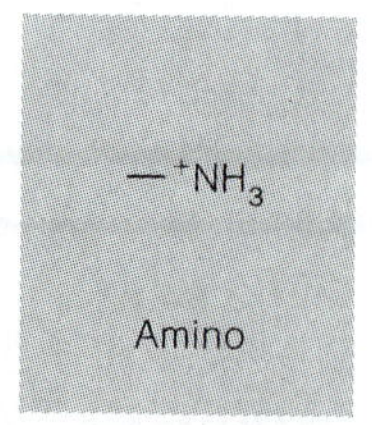

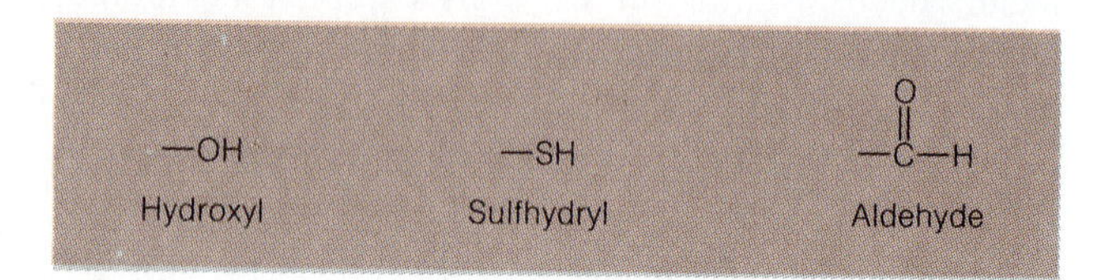

Figure 2-5 Some Common Functional Groups Found in Biological Molecules. Each functional group is shown in the form that predominates at the near-neutral pH of most cells. The carboxyl and phosphate groups are ionized and therefore have a negative charge, whereas the amino group is protonated and is therefore positively charged. Hydroxyl, sulfhydryl, and aldehyde groups are uncharged at pH values near neutrality but are much more polar than hydrocarbons, thereby conferring greater polarity and hence greater water solubility on the organic molecules to which they are attached.

Carbon-Containing Molecules Can Form Stereoisomers

Carbon-containing molecules are capable of still greater diversity because the carbon atom is a **tetrahedral** structure with *geometric symmetry* (Figure 2-6). When four different atoms or groups of atoms are bonded to the four corners of such a tetrahedral structure, two different spatial configurations are possible. Although both forms have the same structural formula, they are not superimposable but are, in fact, mirror images of each other (Figure 2-6). Such mirror-image forms of the same compound are called **stereoisomers.**

Carbon atoms that have four different substituents are called **asymmetric carbon atoms.** Since two stereoisomers are possible for each asymmetric carbon atom, a compound

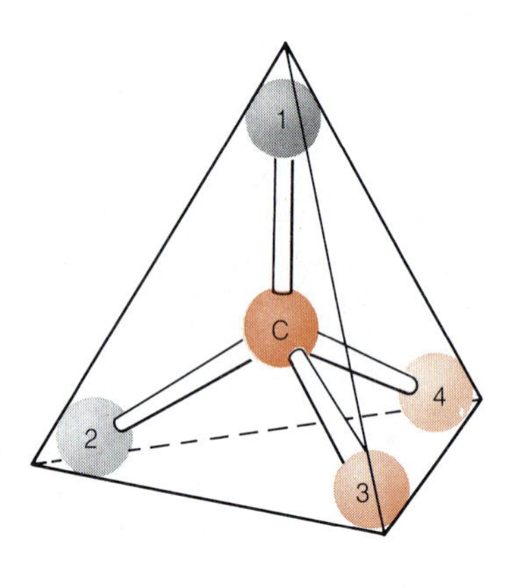

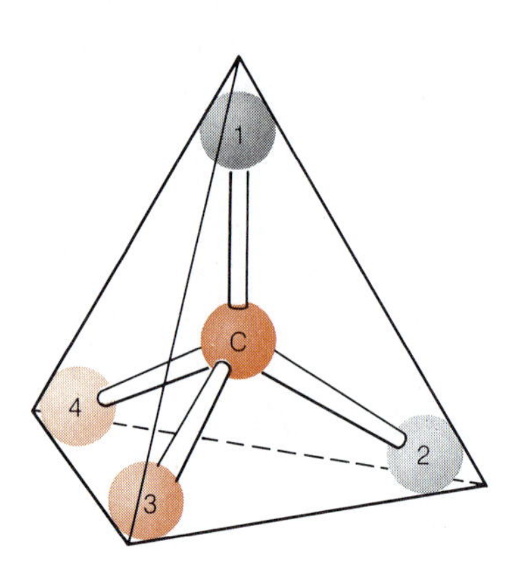

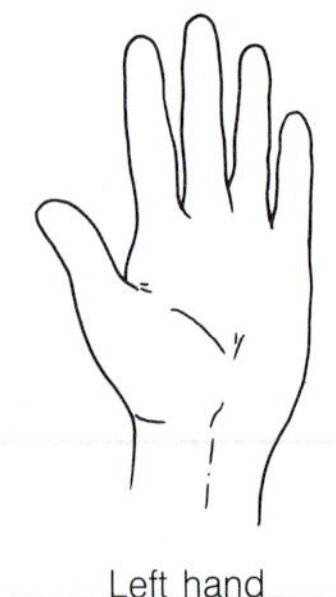

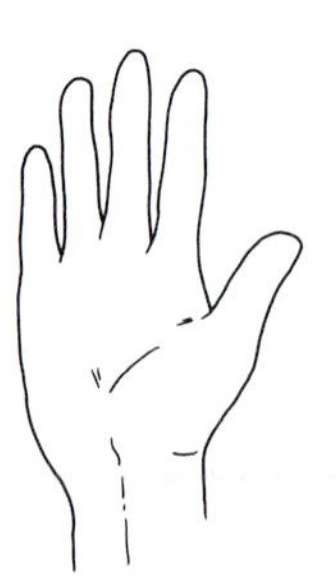

Figure 2-6 Stereoisomers. Stereoisomers of organic compounds occur when four different groups are attached to a tetrahedral carbon atom. Stereoisomers, like left and right hands, are mirror images of each other and cannot be superimposed on one another. (The dashed line down the center of the figure is the plane of the mirror.)

with n asymmetric carbon atoms will have 2^n possible stereoisomers. As shown in Figure 2-7a, the three-carbon amino acid *alanine* has a single asymmetric carbon atom (in the center) and therefore has two stereoisomers, called L-alanine and D-alanine. (It should be clear that neither of the other two carbon atoms of alanine is an asymmetric carbon atom, since one has three identical substituents and the other has two bonds to a single oxygen atom.) Both stereoisomers of alanine occur in nature, but only L-alanine is present as a component of proteins.

As an example of a compound with multiple asymmetric carbon atoms, consider the six-carbon sugar *glucose* shown in Figure 2-7b. Of the six carbon atoms of glucose, the four shown in color are asymmetric. (Again, you should be able to figure out why the other two carbon atoms are not asymmetric.) With four asymmetric carbon atoms, the structure shown, D-glucose, is only one of 2^4, or 16, possible stereoisomers of the $C_6H_{12}O_6$ molecule. In this case, however, not all of the other possible stereoisomers exist in nature, mainly because some are energetically much less favorable than others.

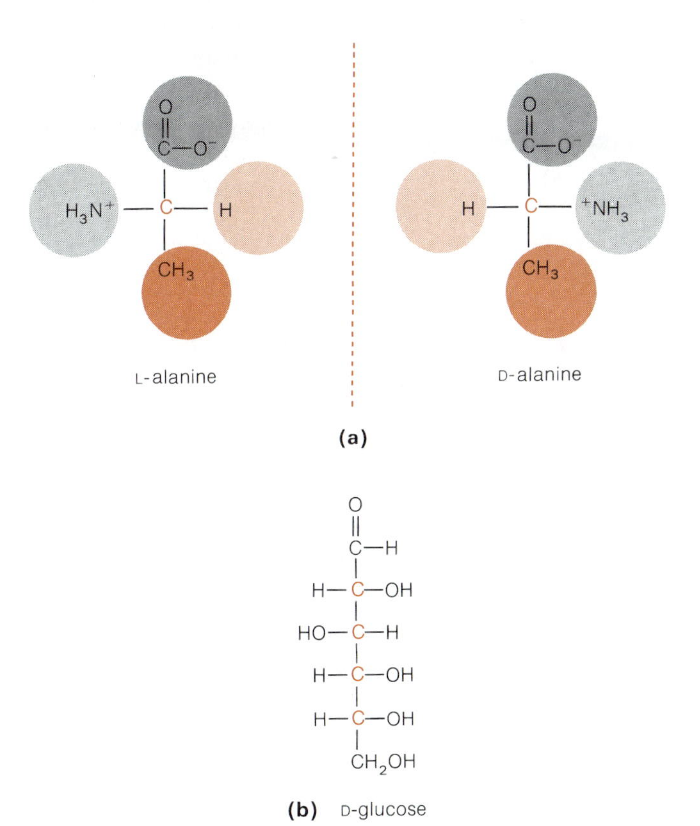

Figure 2-7 Stereoisomers of Biological Molecules. (a) The amino acid alanine has a single asymmetric carbon atom (in the center) and can therefore exist in two spatially different forms designated as L- and D-alanine. (The dashed line down the center of the figure is the plane of the mirror.) (b) The six-carbon sugar glucose has four asymmetric carbon atoms (color), so D-glucose is just one of 16 (2^4) possible stereoisomers of the $C_6H_{12}O_6$ molecule.

The Importance of Water

Just as the carbon atom is uniquely important because of its role as the universal backbone of biologically important molecules, so the water molecule commands special attention because of its indispensable role as the universal solvent in biological systems. Water is, in fact, the single most abundant component of cells and organisms. Typically, about 75–85% of a cell by weight is water, and many cells depend on an extracellular environment that is essentially aqueous as well. In some cases this is the body of water—whether an ocean, lake, or river—in which the cell or organism lives, while in other cases it may be body fluids in which the cell is suspended or with which the cell is bathed.

Water is indispensable for life as we know it. True, there are life forms that can go into a "holding action" and survive periods of severe water scarcity. Seeds of plants and spores of bacteria and fungi are clearly in this category; the moisture content of a dry seed is frequently as low as 10–20%. Some lower plants and animals, notably some of the mosses, lichens, nematodes, and rotifers, can also undergo physiological adaptations that allow them to dry out and survive in a highly desiccated form, sometimes for surprisingly long periods of time. Such adaptations are clearly an advantage in environments characterized by periods of drought. Yet all of these are at best holding actions, and resumption of normal activity always requires rehydration. Thus, an adequate water supply is needed to serve as the solvent system for the variety of activities that we associate with what it means to be alive.

To understand why water is so uniquely suitable for its role, we need to look at its chemical properties. The most critical attribute is clearly its polarity, because this property in turn accounts for its temperature-stabilizing capacity and its cohesiveness, both of which have important consequences for biological chemistry.

Water Molecules Are Polar

To understand the polar nature of water, we need to consider the shape of the molecule. As shown in Figure 2-8a, the water molecule is triangular in shape, with an oxygen "head" and two hydrogen "tails." The oxygen atom at the head of the molecule is **electronegative**; that is, it tends to draw electrons toward it, giving that end of the molecule a partial negative charge and leaving the other end of the molecule with a partial positive charge around the hydrogen atoms. This charge separation gives the water molecule its **polarity**, a property it shares with all other molecules with an asymmetric internal distribution of charge. In the case of water, the polarity of the molecule has enormous consequences.

Water Molecules Are Cohesive

Because of their polarity, water molecules have an affinity for each other and tend to orient themselves spontaneously so that the electronegative oxygen atom of one molecule is associated with the electropositive hydrogen atoms of adjacent molecules. Each such association is called a **hydrogen bond** and is frequently represented by a dotted line, as in Figure 2-8b. Since each oxygen atom can bond to two hydrogens, and both of the hydrogen atoms can associate in this way with the oxygen atoms of adjacent molecules, liquid water is characterized by an extensive three-dimensional network of hydrogen-bonded molecules (Figure 2-9). The hydrogen bonds between adjacent molecules are constantly being broken and reformed, but, on the average, each molecule of water in the liquid state is hydrogen-bonded to about $3\frac{1}{2}$ neighbor molecules at any given time. In ice, the hydrogen bonding is still more extensive, giving rise to a rigid, highly regular crystalline lattice with every oxygen hydrogen-bonded to hydrogens of two adjacent molecules.

It is this tendency to form hydrogen bonds between adjacent molecules that makes water so highly *cohesive.* This cohesiveness accounts for the high *surface tension* of water, as well as for its high *boiling point,* high *specific heat,* and high *heat of vaporization.* The high surface tension of water causes the capillary action that enables water to move up the conducting tissues of plants and allows insects such as the water strider to move across the surface of a pond without sinking.

Water Has a High Temperature-Stabilizing Capacity

An important property of water that derives directly from the hydrogen bonding between adjacent molecules is the high specific heat that gives water its **temperature-stabilizing capacity. Specific heat** is the amount of heat that a substance absorbs per gram to increase its temperature 1°C. The specific heat of water is 1.0 calorie per gram (cal/g). (This is, in fact, the way the *calorie* is defined.)

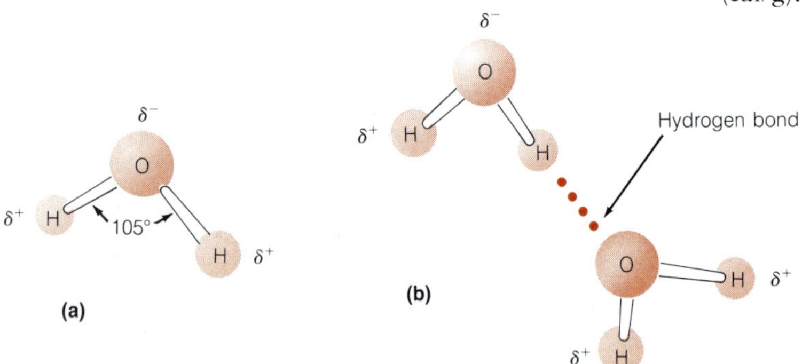

Figure 2-8 Polarity of Water Molecules. (a) The water molecule is polar because it has an asymmetric charge distribution. The two hydrogen atoms are bonded to the oxygen at an angle of 105°. The oxygen atom bears a partial negative charge (δ^-; the Greek letter delta stands for "partial") and is thus the electronegative portion of the molecule. The two hydrogen atoms are electropositive—their end of the molecule has a partial positive charge (δ^+). (b) Since the electropositive hydrogen atom of one molecule is attracted to the electronegative oxygen atom of a neighboring molecule, adjacent water molecules tend to orient themselves so that a weak hydrogen bond forms between them.

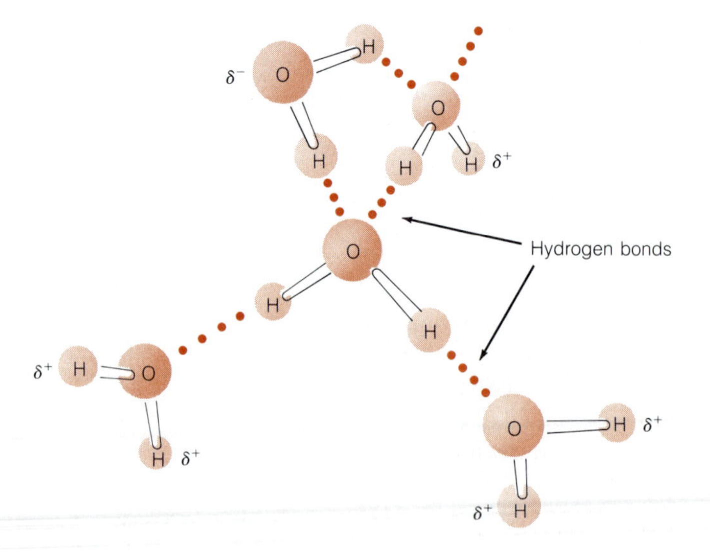

Figure 2-9 Hydrogen Bonding Between Water Molecules. The extensive association of water molecules with one another in either the liquid or solid state is due to hydrogen bonds (dotted lines) between the electronegative oxygen atom of one water molecule and the electropositive hydrogen atoms of adjacent molecules. In ice, the resulting crystal lattice is regular and complete; every oxygen is hydrogen-bonded to hydrogens of two adjacent molecules. In water, some of the structure is disrupted, but much is retained.

The heat capacity of water is much higher than that of most other liquids because of its extensive hydrogen bonding. Much of the energy that in other liquids would contribute directly to an increase in the motion of solvent molecules and therefore to an elevation in temperature is used instead to break hydrogen bonds between neighboring water molecules. In effect, by absorbing heat that would otherwise increase the temperature of the water more rapidly, hydrogen bonds buffer aqueous solutions against large changes in temperature. This capability is an important consideration for the cell biologist, since cells release large amounts of energy during metabolic reactions. This release of energy would pose a serious overheating problem for cells if it were not for the extensive hydrogen bonding and the resulting high specific heat of water molecules.

Water Is an Excellent Solvent

Probably the single most important property of water from a biological perspective is its excellence as a general solvent. A **solvent** is a fluid in which other substances, called **solutes,** can be dissolved. Water is an especially good solvent for biological purposes because of its remarkable capacity to dissolve a great variety of solutes and also because it is generally inert (does not react chemically with most solutes).

It is the polarity of water that makes it so useful as a solvent. Most of the molecules in cells are also polar and therefore interact electrically with water molecules, as do charged ions. Solutes that are polar and therefore dissolve readily in water are called **hydrophilic** ("water-loving"). Most small organic molecules found in cells are hydrophilic; examples are sugars, organic acids, and some of the amino acids. Nonpolar molecules are not very soluble in water and are accordingly termed **hydrophobic** ("water-fearing"). Among the more important hydrophobic compounds found in cells are the lipids and proteins of which membranes are made. Some molecules, as we shall see, have both hydrophobic and hydrophilic regions, so parts of the molecule may have an affinity for an aqueous environment while other parts of the molecule do not.

To understand why polar substances dissolve so readily in water, first consider a salt such as sodium chloride (NaCl) (Figure 2-10). Because it is a salt, NaCl exists in crystalline form as a lattice of positively charged sodium ions (Na^+) and negatively charged chloride ions (Cl^-). For NaCl to dissolve in a liquid, solvent molecules must overcome the attraction of the oppositely charged Na^+ cations and Cl^- anions for each other and involve them instead in interactions with the solvent molecules themselves. Because of their polarity, water molecules can form **spheres of hydration** around both Na^+ and Cl^-, thereby neutralizing their attraction for each other and lessening their likelihood of reassociation. As Figure 2-10a shows, the sphere of hydration around a cation such as Na^+ involves water molecules clustered around the ion with their negative (oxygen) ends pointing toward it. For an anion such as Cl^-, the orientation of the water molecules is reversed, with the positive (hydrogen) ends of the solvent molecules pointing in toward the ion (Figure 2-10b).

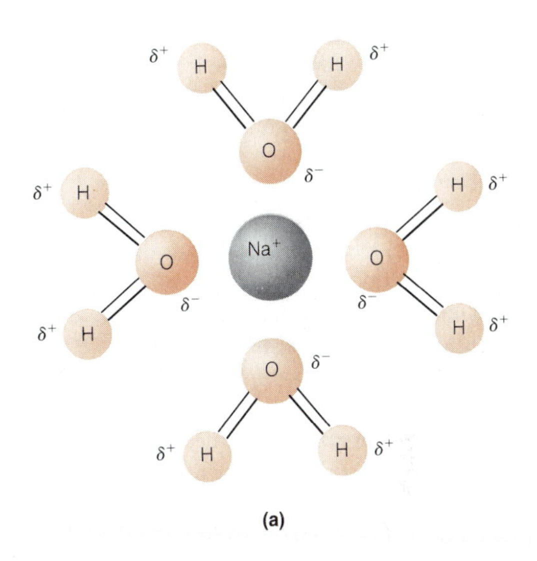

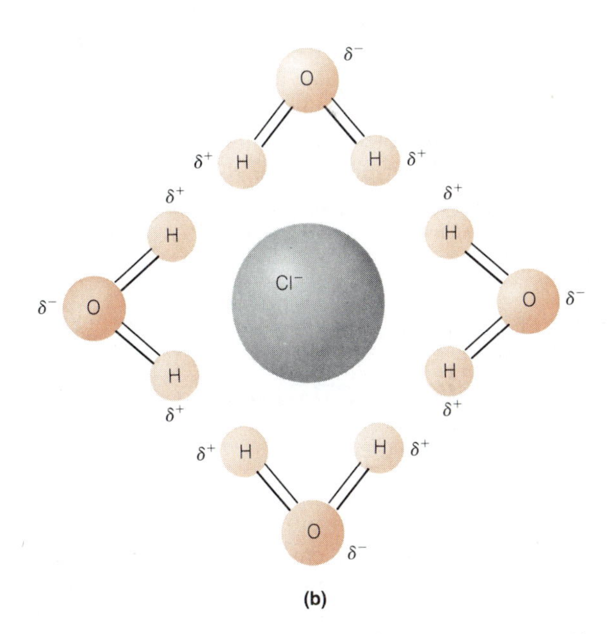

Figure 2-10 The Solubilization of Sodium Chloride. Sodium chloride (NaCl) dissolves in water because of the formation of spheres of hydration around (a) the sodium ions and (b) the chloride ions. The oxygen, sodium, and chloride ions are drawn to scale.

Some biological compounds are soluble in water because they exist as ions at the near-neutral pH of the cell and are therefore solubilized and hydrated like the ions of Figure 2-10. Compounds containing carboxyl, phosphate, or amino groups are in this category (recall Figure 2-5). Most organic acids, for example, are almost completely ionized at a pH near 7 and therefore exist as anions that are kept in solution by spheres of hydration, just as the chloride ion of Figure 2-10b is. Amines, on the other hand, are usually protonated at cellular pH and so exist as hydrated cations, like the sodium ion of Figure 2-10a.

More frequently, organic molecules have no net charge but are nonetheless hydrophilic because they have some regions that are positively charged and other regions that are negatively charged. Water molecules tend to cluster around such regions, and the resulting electrostatic interactions between solute and water molecules keep the solute molecules from associating with one another. Compounds containing the sulfhydryl, hydroxyl, or aldehyde groups shown in Figure 2-5 are usually in this category.

Hydrophobic molecules, on the other hand, have no such polar regions and therefore show no tendency to interact electrostatically with water molecules. In fact, they actually disrupt the hydrogen-bonded structure of water and, for this reason, tend to be excluded by the water molecules. Hydrophobic molecules therefore tend to coalesce in an aqueous medium, associating with one another rather than with the water. This association is driven not so much by any specific affinity of the hydrophobic molecules for one another but by the strong tendency of water molecules to form hydrogen bonds and to exclude molecules that disrupt hydrogen bonding. Such associations of hydrophobic molecules in an aqueous environment are called **hydrophobic interactions.** Hydrophobic interactions are a major driving force in the folding of molecules, the assembly of cellular structures, the organization of membranes, and the binding of substrates to enzymes.

The Importance of Selectively Permeable Membranes

Every cell and organelle needs some sort of physical barrier to keep its contents in and the external environment out, as well as some means of controlling exchange between its inside and the external environment. Ideally, such a barrier should be impermeable to most of the molecules and ions found in cells and their surroundings. Otherwise, substances could diffuse freely in and out, and the cell would not really have a defined content at all. On the other hand, the barrier cannot be completely impermeable, or else desired exchanges between the cell and its environment could not take place. Moreover, such a barrier must be insoluble in water so that it will not be dissolved by the aqueous medium of the cell. At the same time, it must be readily permeable to water, because water is the basic solvent system of the cell and must be able to flow into and out of the cell as needed.

As you might expect, the membranes that surround cells and organelles satisfy these criteria admirably. A **membrane** is essentially a hydrophobic permeability barrier consisting of hydrophobic *phospholipids*, hydrophobic *proteins*, and (in the case of human and animal cells) *cholesterol*. Actually, the phospholipids and many of the proteins in the membrane are not simply hydrophobic; they have both a hydrophilic and a hydrophobic region and are therefore referred to as **amphipathic molecules** (the Greek prefix *amphi* means "of both kinds"). The amphipathic nature of membrane phospholipids is illustrated in Figure 2-11. Each molecule consists of a polar "head" and two nonpolar hydrocarbon "tails." The polarity of the hydrophilic head is due to the presence of a negatively charged phosphate group linked to a positively charged amine.

The Membrane Bilayer

When exposed to an aqueous environment, amphipathic molecules undergo hydrophobic interactions. They spontaneously arrange themselves so that their polar heads are facing outward toward the aqueous phase but their hydrophobic tails are hidden from the water by interacting with the tails of other molecules oriented in the opposite direction. The resulting structure is the **phospholipid bilayer** shown in Figure 2-12. The heads of both layers face outward and the hydrocarbon tails extend inward, forming the continuous hydrophobic phase of the membrane. Every biological membrane has such a lipid bilayer as its basic structure. Each of the phospholipid layers is about 4–5 nm thick, so the bilayer has a width of about 8–10 nm. It is the phospholipid bilayer that gives membranes their characteristic "railroad track" appearance when seen with the electron microscope (Figure 2-13). Apparently, the osmium used in preparing the tissue for electron microscopy reacts with the hydrophilic heads of the phospholipid molecules at both surfaces of the membrane but not with the hydrophobic tails in the interior of the membrane, giving the membrane its *trilaminar* (three-layered) appearance.

Embedded within or associated with the lipid bilayer are various membrane proteins (Figure 2-14). These proteins are almost always amphipathic, and they orient themselves in the lipid bilayer accordingly. Hydrophobic regions of the protein associate with the interior of the membrane,

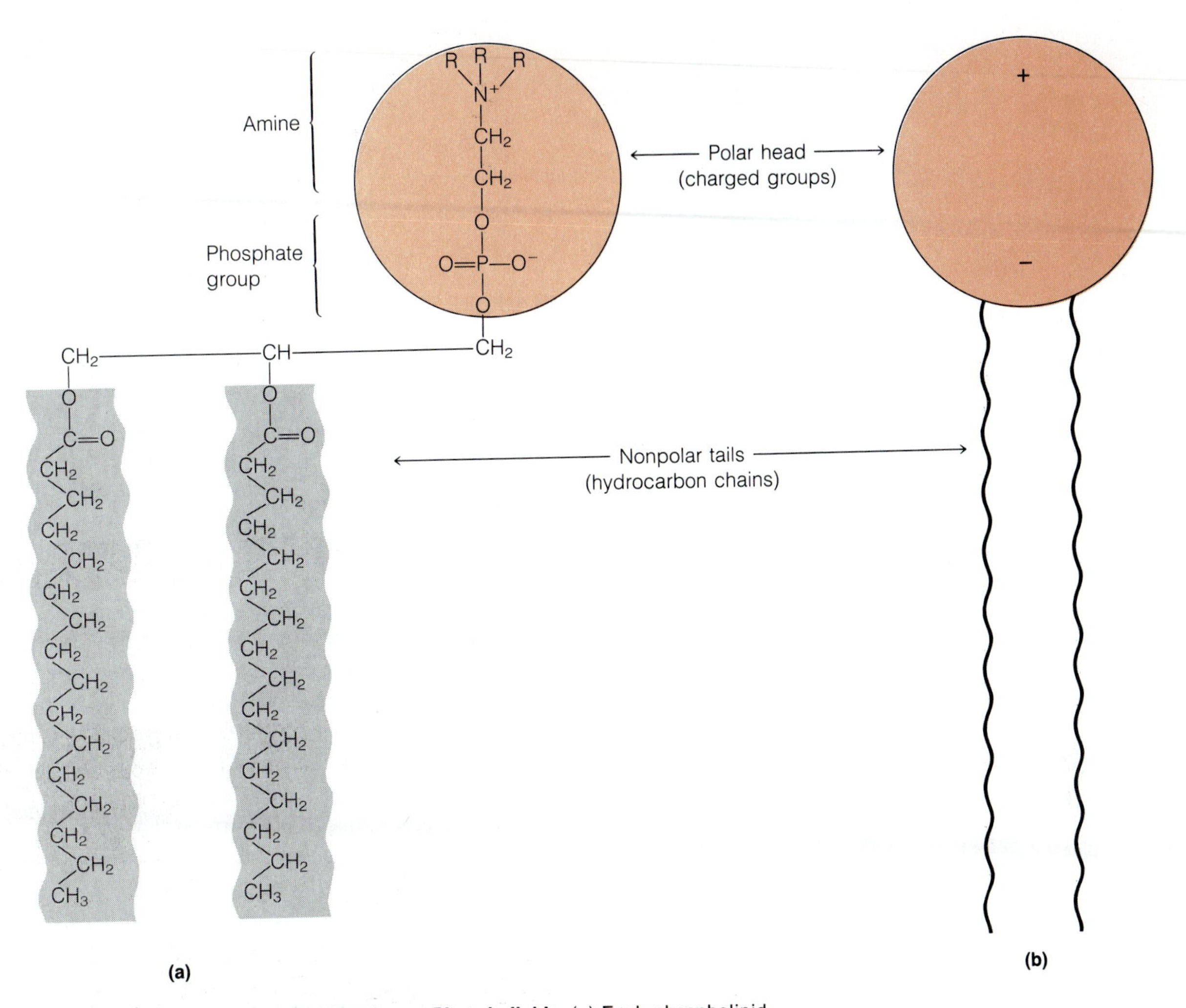

Figure 2-11 The Amphipathic Nature of Membrane Phospholipids. (a) Each phospholipid molecule consists of two long nonpolar tails (gray) and a polar head (color). The polarity of the head results primarily from a phosphate group linked to an amine. The most common R groups on the amine are hydrogen atoms or methyl groups. (b) A schematic representation of a phospholipid molecule.

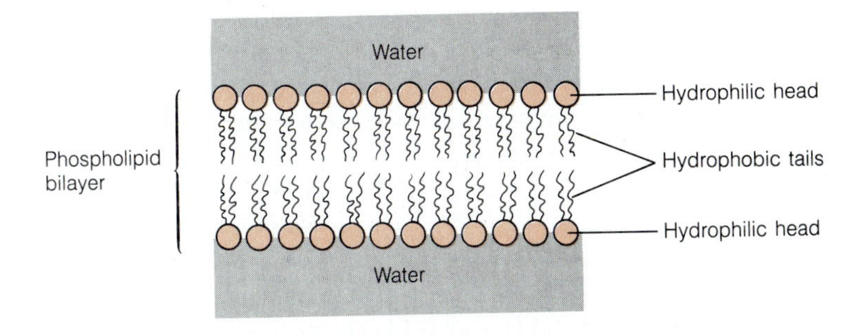

Figure 2-12 The Phospholipid Bilayer as the Basis of Membrane Structure. Due to their amphipathic nature, phospholipids in an aqueous environment orient themselves in a double layer, with the hydrophobic tails buried on the inside and the hydrophilic heads (color) pointing toward the aqueous milieu on either side of the membrane.

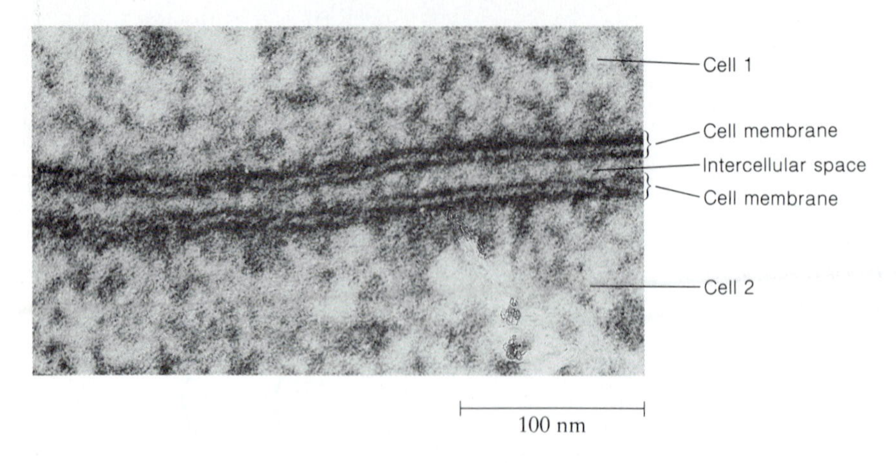

Figure 2-13 Membranes as Seen by Electron Microscopy. The plasma membranes surrounding two adjacent cells each appear as a pair of dark bands. This characteristic trilaminar, or "railroad track," appearance is thought to be due to association of osmium with the hydrophilic heads of the phospholipid molecules but not with their hydrophobic tails (TEM).

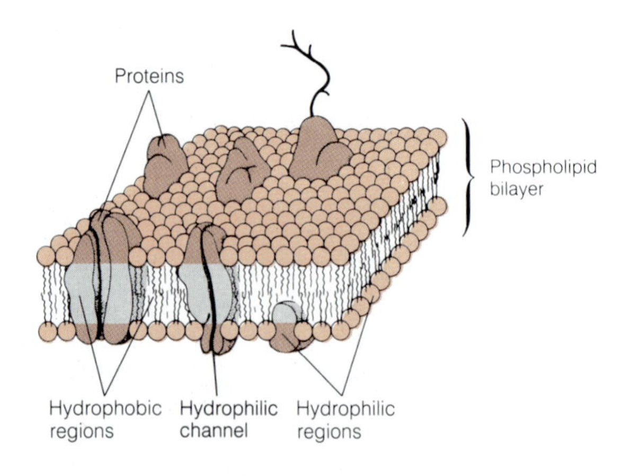

Figure 2-14 Membrane Structure. Biological membranes consist of amphipathic proteins embedded within a phospholipid bilayer. Proteins are positioned in the membrane so that their hydrophobic regions (gray) are located in the hydrophobic interior of the phospholipid bilayer and their hydrophilic regions (color) are exposed to the aqueous milieu on either side of the membrane.

whereas hydrophilic regions protrude into the aqueous environment at the surface of the membrane.

Depending on the particular membrane, the membrane proteins may play any of a variety of roles. Some are *transport proteins,* responsible for moving specific substances across an otherwise impermeable membrane. Others are *enzymes* that catalyze reactions associated with the specific membrane. Still others are the *receptors* on the outer surface of the cell membrane, the *electron transport intermediates* of the mitochondrial membrane, or the *chlorophyll-binding proteins* of the chloroplast.

Movement Across the Membrane

Since most cellular constituents are relatively hydrophilic, they have little or no affinity for the hydrophobic interior of the membrane. The membrane is relatively impermeable to such compounds and serves as a very effective barrier to any movement of these hydrophilic substances into or out of the cell or organelle. (Water, on the other hand, passes more readily across the membrane barrier, despite its highly polar nature. This may be due to channel-like imperfections in the membrane, large enough to allow the passage of water molecules but too small for anything else. Such channels are easier to postulate than demonstrate, however, and remain a matter of conjecture.)

Of course, it is essential that there be some movement of molecules and ions across this otherwise impermeable barrier. As already noted, membranes are equipped with **transport proteins** to serve this function. Transport proteins are transmembrane proteins that serve as hydrophilic channels or gates through an otherwise hydrophobic membrane (Figure 2-14). Transport proteins are highly specific for the molecules or ions that they allow to pass in and out, and their activities can be carefully regulated to meet cellular needs. As a result, biological membranes can best be described as **selectively permeable,** since only selected substances can enter or leave the cell or organelle.

The Importance of Synthesis by Polymerization

For the most part, cellular structures such as ribosomes, chromosomes, membranes, flagella, and cell walls are made up of ordered arrays of linear polymers called **macromolecules.** (Some branching occurs, notably in the polysaccharides starch and glycogen, but the linearity of biological polymers is a good first approximation.) Examples of important macromolecules in cells include *proteins, nucleic acids* (both DNA and RNA), and *polysaccharides* such as starch, glycogen, and cellulose. Macromolecules are very important in both the function and the structure of cells. To understand the biochemical basis of cell biology therefore really means to understand macromolecules—how they are made, how they are assembled, and how they function.

The Importance of Macromolecules

The importance of macromolecules in cell biology is emphasized by the cellular hierarchy shown in Figure 2-15. The compounds of which most cellular structures are made are water-soluble *organic molecules* (level 1) that cells either obtain from other cells or synthesize from simple nonbiological molecules such as carbon dioxide, ammonia, or phosphate ions available from the environment. These organic molecules polymerize to form biological *macromolecules* (level 2) such as nucleic acids, proteins, or carbohydrates. Macromolecules are then assembled into a variety of **supramolecular structures** (level 3), which in turn are components of cells and their organelles (level 4).

One of the examples in Figure 2-15 is that of cell wall biogenesis (panels on right). In plants, a major component of the cell wall (level 3) is the polysaccharide cellulose (level 2). Cellulose is in turn a repeating polymer of the simple sugar glucose (level 1), formed by the plant cell from carbon dioxide and water in the process of photosynthesis.

From such examples, a general principle emerges: The macromolecules that are responsible for most of the form and order so characteristic of living systems are generated by the polymerization of small organic molecules. This strategy of forming large molecules by joining smaller units in a repetitive manner is illustrated in Figure 2-16. The importance of this strategy can hardly be overemphasized because it is a fundamental principle of cellular chemistry. The enzymes responsible for catalysis of cellular reactions, the nucleic acids involved in the storage and expression of genetic information, the glycogen stored by your liver, and the cellulose that provides rigidity to a plant cell wall are all variations on the same design theme. Each is a macromolecule made by the linking together of small repeating units.

Examples of such repeating units, or **monomers,** are the *glucose* present in cellulose or glycogen, the *amino acids* needed to make proteins, and the *nucleotides* of which nucleic acids are made. In general, these are small, water-soluble organic molecules with molecular weights less than 350. They can be transported across most biological membranes, provided that an appropriate transport protein is present. By contrast, most macromolecules that are synthesized from these monomers cannot traverse membranes and therefore must be made in the cell or compartment in which they are needed. (Several exceptions to this rule will be seen when we get to messenger RNA and organellar proteins later in the text, but the generalization is nonetheless a good one.)

Kinds of Macromolecules

Table 2-1 indicates the major kinds of macromolecules found in the cell, along with the number and kind of monomeric subunits required for polymer formation. The distinction between *storage* or *structural macromolecules* and *informational macromolecules* is an important one, since the function of these biological polymers affects the number and order of different monomers we can expect to find in them.

Nucleic acids (both DNA and RNA) and proteins are called **informational macromolecules** because the order of the several kinds of nonidentical subunits that they contain is nonrandom and highly significant to their function. Figure 2-17 illustrates schematically the structures of these two major classes of informational macromolecules. The

TABLE 2-1 Biologically Important Macromolecules and Their Repeating Units

	Biological Polymer			
	Proteins	Nucleic Acids	Polysaccharides	
Kind of macromolecule	Informational	Informational	Storage	Structural
Examples	Enzymes, hormones, antibodies	DNA, RNA	Starch, glycogen	Cellulose
Repeating monomers	Amino acids	Nucleotides	Monosaccharides	Monosaccharides
Number of kinds of repeating units	20	4	One or a few	One or a few

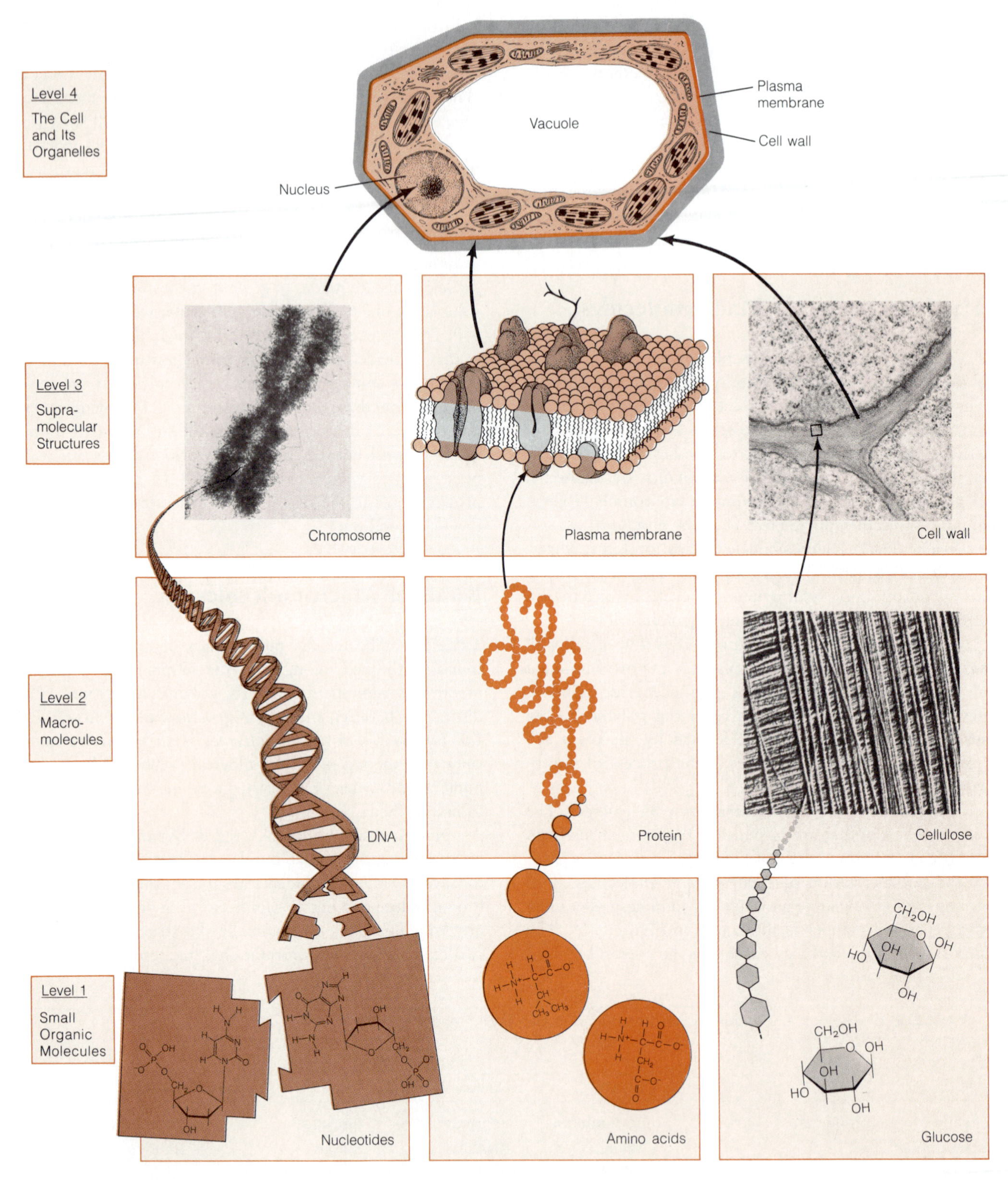

Figure 2-15 The Hierarchical Nature of Cellular Structures and Their Assembly. Small organic molecules (Level 1) are synthesized from simple inorganic substances and are polymerized to form macromolecules (Level 2). The macromolecules then assemble into the supramolecular structures (Level 3) of which the cell and its organelles are composed (Level 4).

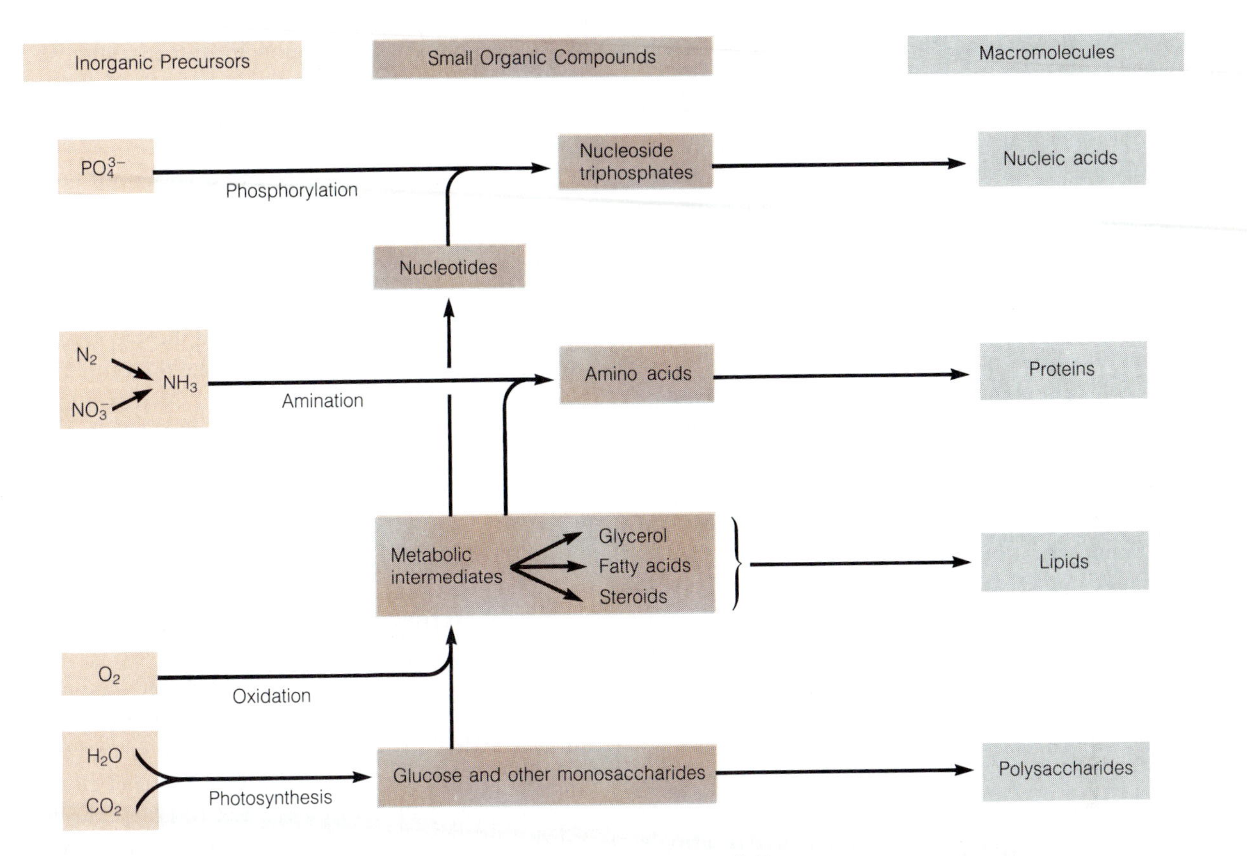

Figure 2-16 Synthesis of Biological Macromolecules. Simple inorganic precursors (left) react to form small organic molecules (center), which are then used in the synthesis of the macromolecules (right) of which most cellular structures consist.

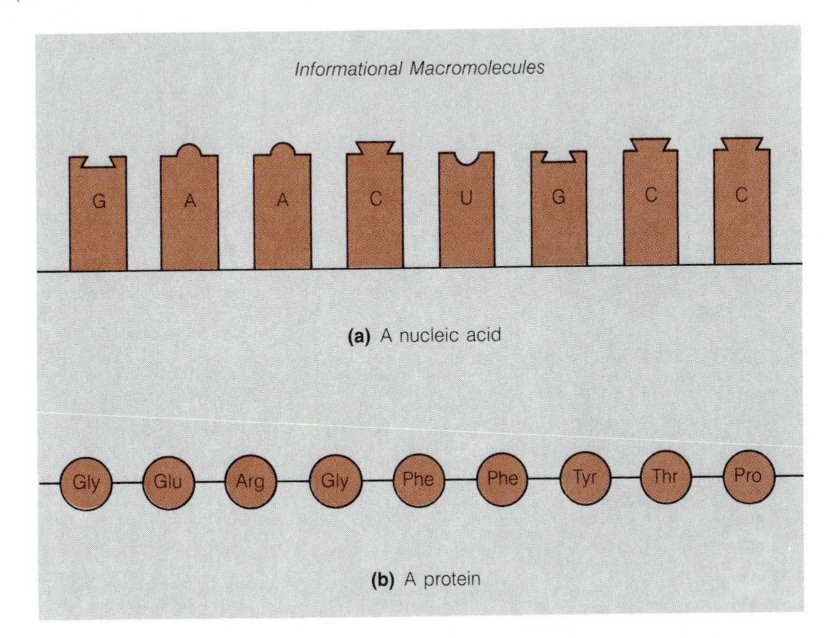

Figure 2-17 Informational Macromolecules. Informational macromolecules contain a variety of kinds of repeating units in nonrandom, genetically determined order. The two main classes of informational macromolecules are the nucleic acids DNA and RNA and the proteins. (a) A portion of the nucleotide sequence of 5S ribosomal RNA from the bacterium *Escherichia coli*. (b) A portion of the amino acid sequence of the protein insulin.

order of the monomers (nucleotides in nucleic acids and amino acids in proteins) is genetically determined and carries important information that directs the function or utilization of these macromolecules. For example, insulin (Figure 2-17b) is a protein because it is a string of amino acids linked together by peptide bonds, but it is the specific protein *insulin* because it is a particular string of amino acids in a unique, specified order. In fact, the sequence of amino acids is so important that any variation in that sequence is likely to have a deleterious effect on the ability of the protein to carry out its expected function.

For informational macromolecules, polymerization of smaller subunits is not just an economical way of building large molecules; it is an essential part of the role these macromolecules play in the cell. For example, the synthesis of nucleic acids by linkage of small repeating units (nucleotides) is important because it is in the actual order of the monomeric units that these molecules store and transmit the genetic information of the cell.

Polysaccharides, on the other hand, are not informational macromolecules. For them, the order of the monomeric units is enzymatically determined, carries no information, and is not essential to the function or utilization of the molecule in the sense that it is for proteins or nucleic acids. Instead, polysaccharides are either **storage** or **structural macromolecules** and usually consist of a single kind of repeating monomer or a strictly repetitious sequence of a few different kinds of monomers. The most familiar storage polysaccharides are the *starch* of plant cells and the *glycogen* found in animal cells. As Figure 2-18a illustrates schematically, both of these storage polysaccharides consist of a single repeating monomer, the simple sugar glucose.

The best-known example of a structural polysaccharide is the *cellulose* present in plant cell walls. Like starch and glycogen, cellulose also consists of glucose units, but the units are linked together by a somewhat different bond, as we will see in Chapter 3. The cell walls of some bacterial cells contain a somewhat more complicated kind of structural polysaccharide. In this case, the molecule consists of two different kinds of monomers, *N*-acetylglucosamine (NAG) and *N*-acetylmuramic acid (NAM), as Figure 2-18b illustrates. The two monomers occur in strictly alternating sequence, however, and carry no information.

The Synthesis of Macromolecules

In Chapter 3 we will look at each of the major kinds of biologically important polymers, called **biopolymers.** First, however, it will be useful to consider several important principles that underlie the polymerization processes by which these macromolecules arise. Although the chemistry of the monomeric units—and hence of the resulting polymers—differs markedly between such macromolecules as proteins, nucleic acids, and polysaccharides, the following basic principles apply in each case:

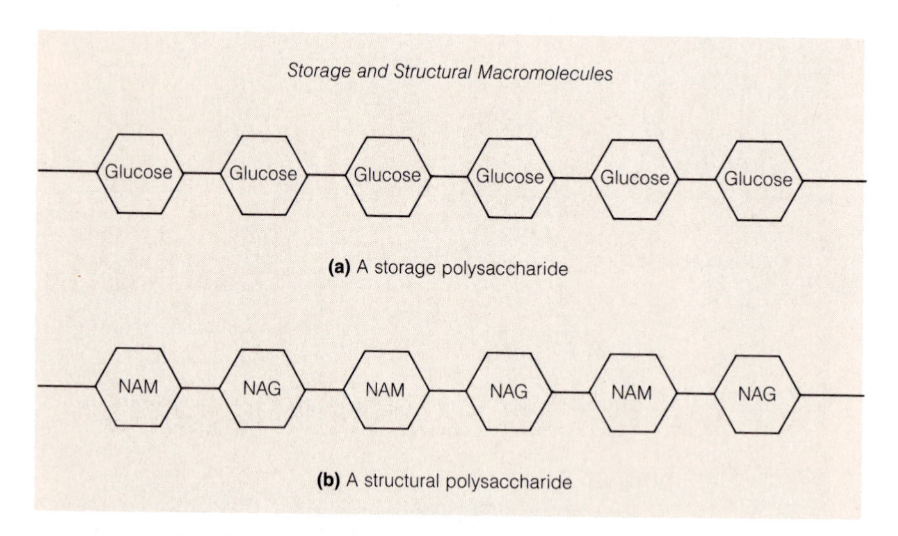

Figure 2-18 Storage and Structural Macromolecules. Storage and structural macromolecules contain one or a few kinds of repeating units in a strictly repetitious sequence. (a) A portion of the sequence of the storage polysaccharide starch or glycogen. (b) A portion of the sequence of a bacterial cell wall polysaccharide consisting of the sugars *N*-acetylglucosamine (NAG) and *N*-acetylmuramic acid (NAM) in strict alternation.

1. Biopolymers are always synthesized by the stepwise polymerization of similar or identical small molecules called monomers.
2. The addition of each monomeric unit occurs with the removal of a water molecule and is therefore termed a **condensation reaction.**
3. The monomeric units that are to be joined together must be present in an *activated form* before condensation can occur.
4. Energy to activate the monomers is provided by ATP or a closely related high-energy compound.
5. Because of the way in which they are synthesized, biopolymers have an inherent **directionality;** that is, the two ends of the polymer chain are chemically different from each other.

Since the elimination of water is essential in all biological polymerization reactions, a feature common to each of the monomeric units, regardless of their other chemical properties, is the presence of a reactive hydrogen (—H) on at least one functional group and a reactive hydroxyl group (—OH) elsewhere on the molecule. This structure is depicted schematically in Figure 2-19, which represents the monomeric units as boxes but indicates the active hydrogen and hydroxyl groups. Also clear from Figure 2-19 are the sequential nature of the addition and the directionality of the resulting polymer. All such polymerization processes take place stepwise, with a single monomeric unit added at a time.

Figure 2-19 is adequate as a representation of the overall process of polymerization by stepwise condensation of successive monomers but not as a representation of the actual mechanism of condensation. It turns out that the addition of each monomer to a growing chain is always energetically unfavorable unless the incoming molecule has been energized in a prior activation step. Inevitably, this prior step involves coupling of the monomer to some sort of carrier molecule to form a high-energy **activated monomer.** The energy to drive this activation process usually comes from the high-energy compound ATP. The chemical nature of the energized monomer, the carrier, and the actual activation process differ for each biological polymer, but the general principle always applies: Polymer synthesis always involves activated monomers, and ATP or a similar source of energy is always required to form the activated monomers.

The Importance of Self-Assembly

So far, we have seen that the macromolecules so characteristic of biological organization and function are polymers of small, hydrophilic organic molecules. The only requirements for polymerization are an adequate supply

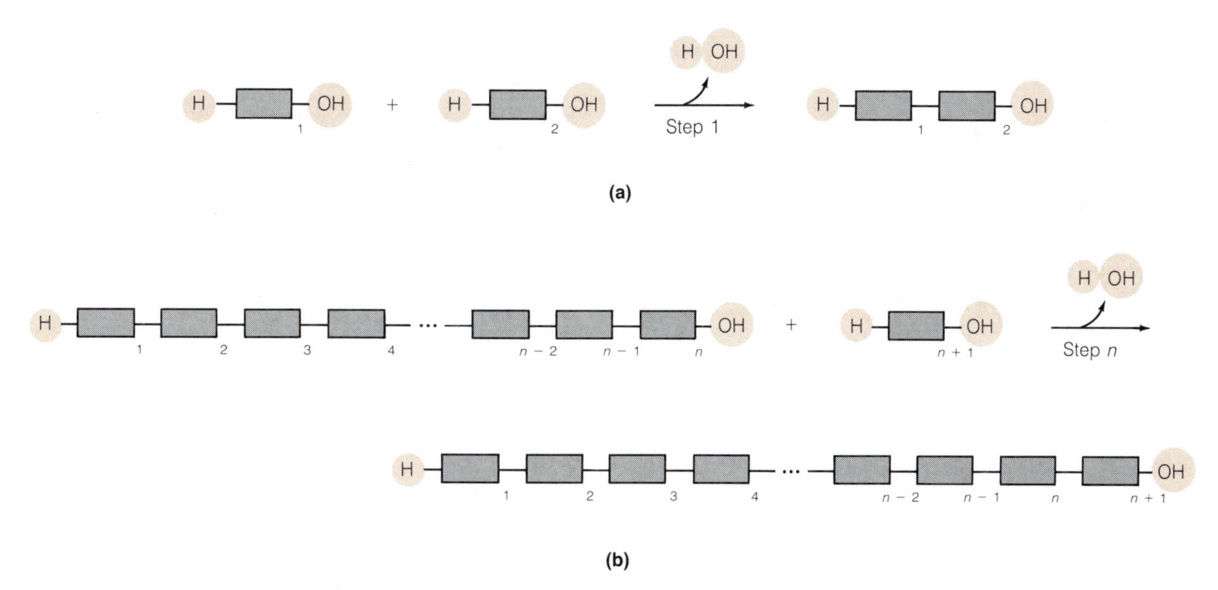

Figure 2-19 The Stepwise Addition of Monomeric Units in the Synthesis of Biopolymers. (a) The first step of polymer formation. (b) The *n*th step of polymer formation. Regardless of their chemical nature, all macromolecules in the cell are synthesized by sequential, stepwise condensation of monomeric units. However, the process is actually somewhat more complicated than suggested here because each incoming monomeric unit must first be activated, usually by coupling to a carrier molecule with input of energy from ATP or a related high-energy compound.

of the monomeric subunits, a source of energy, and, in the case of proteins and nucleic acids, sufficient information to specify the order in which the subunits (amino acids or nucleotides) are added. Still to be considered are the steps that lie beyond—the processes through which these biopolymers are organized into the supramolecular assemblies and organelles that are readily recognizable as cellular structures. Or, in terms of Figure 2-15, we need to ask how the macromolecules of level 2 are assembled into the higher-level structures of levels 3 and 4.

Crucial to our understanding of these higher-level structures is the principle of **self-assembly,** which asserts that the information required to guide and direct the ordering of macromolecules into more complicated structures is already present in the polymers themselves. Once the requisite macromolecules have been synthesized and are on hand, their assembly into more complex structures occurs spontaneously, without further input of energy or information. This means that the synthesis of a cell is in large part specified when all of its macromolecular components have been synthesized, since most subsequent assembly processes appear to be spontaneous.

The Self-Assembly of Proteins

A useful prototype for understanding such self-assembly processes is the folding necessary to make a functional, three-dimensional protein out of a linear chain of amino acids. Although the distinction is not always properly made, the immediate product of amino acid polymerization is a **polypeptide,** not a protein. To become a functional protein, one or more of such linear polypeptide chains must coil and fold in a precise, predetermined manner to assume the unique three-dimensional configuration necessary for biological activity. The striking feature of such coiling and folding is that it occurs spontaneously and progressively as the polypeptide chain reels off the ribosome during synthesis (Figure 2-20a).

By the time the fully elongated chain is released from the ribosome, it has already assumed a stable, predictable, three-dimensional structure without any input of energy or information beyond the basic polymerization process. Furthermore, the folding is unique: each polypeptide with the same amino acid sequence will fold in an identical, reproducible manner under the same conditions. And if a protein is **denatured** (treated with a chemical agent such as urea or a high salt concentration, which disrupts its three-dimensional structure and therefore its physiological function) and the resulting polypeptide is then returned to favorable conditions, it will refold spontaneously to the original (**native**) structure (Figure 2-20b). Apparently, all the information necessary to specify the three-dimensional structure of a protein is inherent in its amino acid sequence.

The key to understanding the driving force behind this spontaneous folding process lies in the recognition that proteins contain 20 different kinds of amino acids that vary greatly in affinity for water. Some amino acids are hydrophilic. Like the polar heads of phospholipid molecules, such amino acids have a strong affinity for water and tend to seek out positions near the surface of the protein, where they can interact maximally with the water molecules of the medium. Other amino acids are hydrophobic. They are essentially nonpolar, have little or no affinity for water, and tend therefore to gravitate toward the center of the protein molecule, where they can interact with one another in an essentially nonaqueous milieu.

Protein structure is the result of a balance between the tendency of the hydrophilic groups to seek an aqueous environment near the surface and that of the hydrophobic groups to minimize contact with water by associating with one another in the center of the molecule. Clearly, if all or even most of the amino acids in a protein were hydrophobic, the protein would be virtually insoluble in water and would seek out a nonpolar environment. Membrane proteins, you may recall, are localized in membranes for this very reason. Similarly, if all or most of the amino acids were hydrophilic, the polypeptide would most likely be quite content to remain in a fairly distended, random shape, allowing maximum access of each amino acid to an aqueous environment. But precisely because most polypeptide chains contain both hydrophilic and hydrophobic amino acids (and in a specified order), parts of the molecule are drawn toward the surface while other parts are driven toward the interior, and the final configuration of the protein is the inevitable outcome of these tendencies.

The same properties and interactions of amino acids that are responsible for the folding of an individual polypeptide are also important for structures that involve associations of polypeptides. These structures include the many proteins that consist of multiple polypeptide subunits, multienzyme complexes, and the various filament- and tubule-containing structures responsible for contractility and motility. In each case, the affinity of one polypeptide for another appears to be dictated by the presence of hydrophobic amino acids at or near the surface of the individual polypeptides, giving rise to hydrophobic "patches" on the surface that prefer to associate with similar regions on the surface of another polypeptide rather than with an aqueous environment.

The assembly of such structures will almost certainly turn out to be dependent on the same kinds of hydrophobic and hydrophilic interactions as are involved in the folding and association of individual polypeptides. Ultimately,

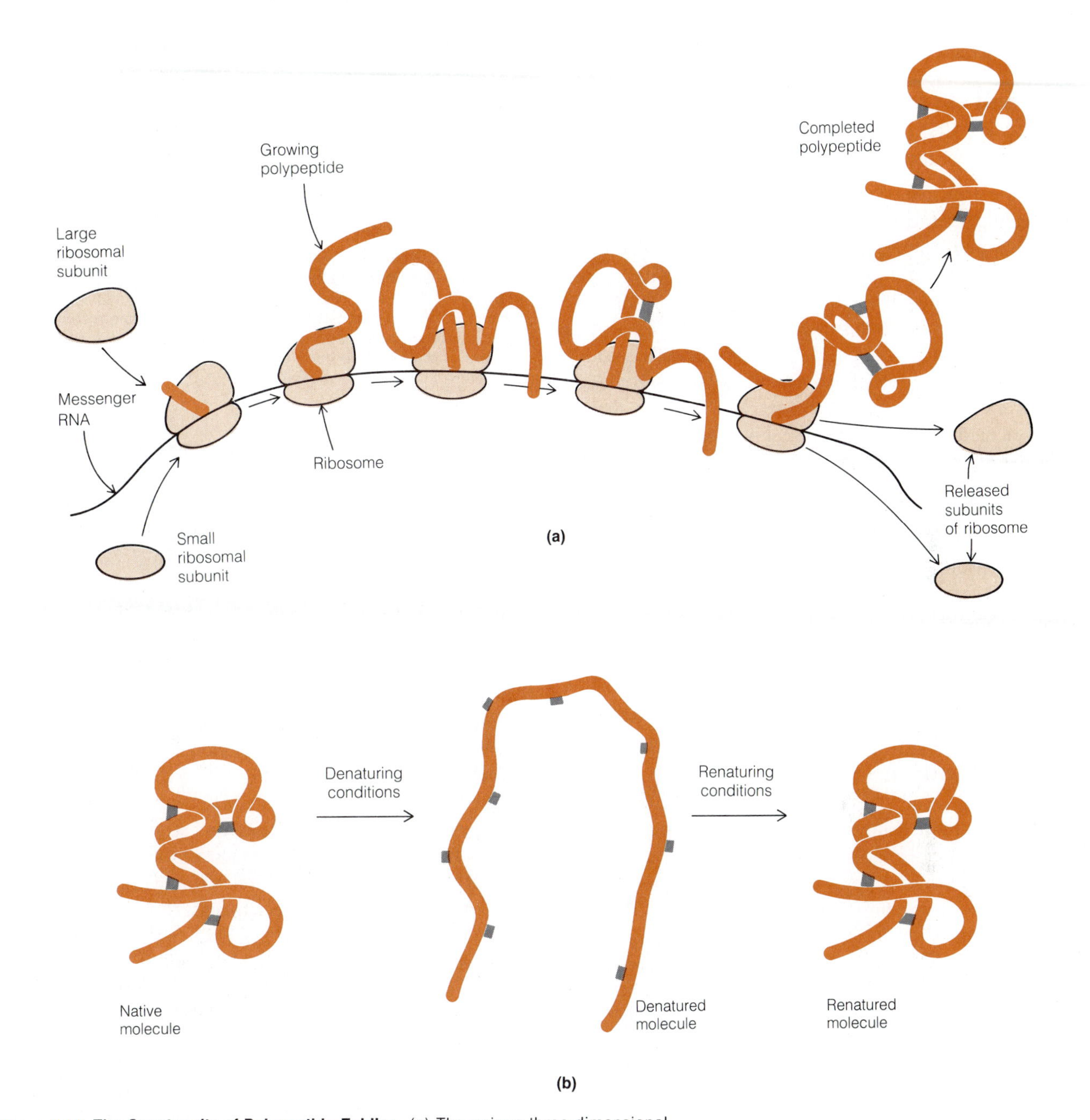

Figure 2-20 The Spontaneity of Polypeptide Folding. (a) The unique three-dimensional configuration of a protein is achieved spontaneously by progressive coiling and folding of the polypeptide chain as it is being synthesized. Synthesis is initiated on the left, and each successive ribosome from left to right has a larger portion of the total polypeptide chain completed. (b) To demonstrate that the information for the three-dimensional configuration of a protein is resident in the amino acid sequence, the intact polypeptide can be treated with denaturing reagents. This treatment results in a denatured molecule with no fixed shape. Upon gradual removal of the denaturing agents, the molecule refolds spontaneously to its original native configuration (renaturation) without the input of any additional information. The particular protein depicted here is the enzyme ribonuclease.

then, it appears that most, perhaps even all, of the interactions that a polypeptide is capable of are intrinsic in its amino acid sequence and therefore specified by the gene that codes for that sequence.

The Self-Assembly of Other Cellular Structures

The same principle of self-assembly that accounts for the folding and interactions of polypeptides may also apply to more complex cellular structures. Many of the characteristic structures of the cell are complexes of two or more different kinds of polymers and therefore clearly involve interactions chemically distinct from those of polypeptide folding and association, but the principle of self-assembly may nonetheless apply. For example, *ribosomes* contain both RNA and proteins; *membranes*, as we have already seen, are made up of both phospholipids and proteins; and even the *plant cell wall*, though composed mainly of cellulose fibrils, contains a small but apparently crucial protein component. And despite the chemical differences between such polymers as proteins, nucleic acids, and polysaccharides, the interactions that drive these supramolecular assembly processes seem to be essentially the same as those that dictate the folding of individual protein molecules.

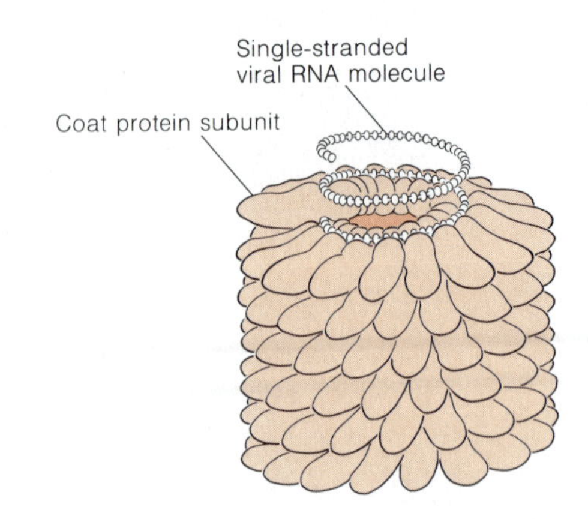

Figure 2-21 A Model of the Structure of Tobacco Mosaic Virus (TMV). A single-stranded RNA molecule is coiled into a helix, surrounded by a coat consisting of 2130 identical protein subunits. Only a portion of the entire TMV virion is shown here, and several layers of protein have been omitted from the upper end of the structure to reveal the helical RNA molecule inside.

Tobacco Mosaic Virus: A Case Study in Self-Assembly

Some of the most definitive findings concerning the self-assembly of complex biological structures have come from studies with *viruses*. As we will learn in Chapter 4, a virus is a complex of proteins and nucleic acid, either DNA or RNA. A virus is not itself alive but is able to invade and infect a living cell and subvert the synthetic machinery of the cell for the production of more viruses. Inherent in the production of more viruses are the synthesis of the viral nucleic acid and viral proteins and their subsequent assembly into the mature **virion**, or viral particle. Studies of this assembly process have provided a wealth of information on structure and assembly that exceeds what we know about any other self-assembly system.

An especially good example is **tobacco mosaic virus (TMV)**, a plant virus that has long been popular with molecular biologists. TMV is a rodlike particle about 18 nm in diameter and 300 nm in length. It consists of a single strand of RNA with about 6000 nucleotides and about 2130 copies of a single kind of polypeptide, the **coat protein**, each with 158 amino acids. The RNA molecule forms a helical core, with a cylinder of protein subunits clustered around it (Figure 2-21).

Heinz Fraenkel-Conrat and his colleagues carried out several important experiments that contributed to our understanding of self-assembly. They separated TMV into its RNA and protein components and then allowed them to reassemble in vitro. Viral particles were regenerated that were capable of infecting plant cells. This result was one of the first and most convincing demonstrations that the components of a complex biological structure can reassemble spontaneously into functional entities without external information. Especially interesting was the finding that the RNA from one strain of virus could be mixed with the protein component from another strain to form a hybrid virus that was also infective. As expected, the source of the RNA and not the protein determined the type of virus that was made by the infected cells.

The assembly process has since been studied in detail and is known to be surprisingly complex. The basic unit of assembly is a two-layered disk of coat protein, with each layer consisting of 17 identical subunits arranged in a ring (Figure 2-22a). Each disk is initially a cylindrical structure but undergoes a conformational change that tightens it into a helical shape as it interacts with a short segment (about 102 nucleotides) of the RNA molecule (Figure 2-22b). This transition allows another disk to bind (step c), and each successive disk undergoes a conformational change from a cylinder to a helix and binds to another 102 bases of the RNA. This disk-by-disk elongation process continues (step d), with the successively stacked disks creating a helical path for the RNA strand. The process eventually gives rise to the mature virion, with its RNA completely covered with coat protein.

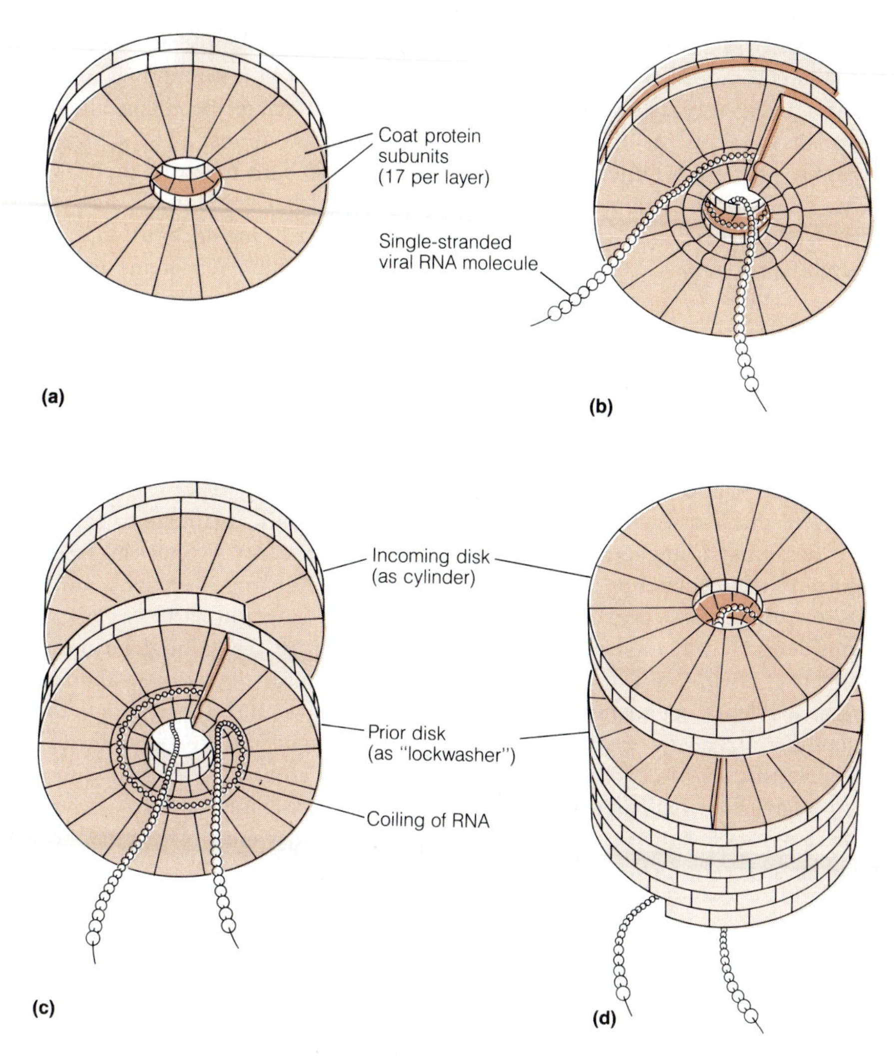

Figure 2-22 Self-Assembly of the Tobacco Mosaic Virus (TMV) Virion. (a) The unit of assembly is a double-layered disk with 17 protein subunits per layer. (b) Assembly begins as the viral RNA molecule associates with a disk, causing the RNA molecule to form a loop (involving 102 nucleotides) and the disk to change from a cylindrical to a helical conformation. (c) Another disk is then added, and another length of RNA becomes associated with it. (d) Each incoming disk adds two more layers of protein subunits and causes another length of RNA to become associated with the structure in a spiral loop. The process continues until the entire RNA molecule is involved and the virion is complete.

The Limits of Self-Assembly

In many cases, the information required to specify the exact configuration of a cellular structure seems to lie entirely within the polymers that contribute to the structure. Such self-assembling systems achieve stable three-dimensional configurations without additional information input because the information content of the component polymers is adequate to specify the complete assembly process. However, there also appear to be some assembly systems that depend, in addition, on information supplied by a preexisting structure. In such cases, the ultimate structure arises not by newly assembling the components but rather by ordering the components into the matrix of an existing structure. Examples of cellular structures that are routinely built up by addition of new material to existing structures include membranes, cell walls, and chromosomes.

On the other hand, such structures are not yet sufficiently characterized to determine whether the presence of a preexisting structure is obligatory or whether, under the right conditions, the components might be capable of self-assembly. Evidence from studies with artificial membranes and with chromatin (isolated chromosomal components), for example, suggests that a preexisting structure, though routinely present in vivo, may not be an indispensable requirement for the assembly process. Additional insight will be necessary before we can say with certainty whether, and to what extent, external information is required or exploited in cellular assembly processes.

The Advantages of Hierarchical Assembly

Each of the assembly processes we have looked at exemplifies the basic cellular strategy of **hierarchical assembly** illustrated in Figure 2-15. Biological structures are almost always constructed in a hierarchical manner, with subassemblies acting as important intermediates en route from simple starting molecules to the end products of organelles, cells, and organisms. Consider how cellular structures are built up. First, large numbers of similar, or even identical, monomeric subunits are assembled by condensation into polymers. These polymers then aggregate spontaneously but specifically into characteristic multimeric units. The multimeric units can, in turn, give rise to still more complex structures and eventually to assemblies that are recognizable as distinctive subcellular structures. This hierarchical process has the double advantage of chemical simplicity and efficiency of assembly.

To appreciate the chemical simplicity, we need only recognize that almost all structures found in cells and organisms are built up from about 30 small precursor molecules, which George Wald has called the "alphabet of biochemistry." Included in this "alphabet" are the 20 amino acids found in proteins, the 5 aromatic bases present in nucleic acids, 2 sugars, and 3 lipid molecules. We will encounter each of these in the next chapter. Given these building blocks and the polymers that can be derived from them through just a few different kinds of condensation reactions, most of the structural complexity of life can be readily elaborated by hierarchical assembly into successively more complex structures.

The second advantage of hierarchical assembly lies in the "quality control" that can be exerted at each level of assembly, allowing defective components to be discarded at an early stage rather than being built into a more complex structure that would be more costly to reject and replace. Thus, if the wrong subunit has been inserted into a polymer at some critical point in the chain, that particular molecule may have to be discarded, but the cell will be spared the cost of building up a more complicated supramolecular assembly or even a whole organelle before the defect is discovered. The story presented in the box on p. 37 is intended to illustrate this basic principle.

Perspective

In summary, the basic molecular building blocks of the cell are small organic molecules that can be strung together by stepwise polymerization to form the macromolecules that are so important to cellular structure and function. About 30 different kinds of monomeric units account for most cellular structure. These monomeric units consist primarily of carbon, hydrogen, oxygen, nitrogen, and phosphorus, all readily available to the living world in the form of inorganic compounds such as carbon dioxide, water, ammonia, nitrate, and phosphate. Because of the presence of polar functional groups, most of these small organic molecules are quite water-soluble. The cell is able to retain these molecules within defined spaces because it has selectively permeable membranes with a hydrophobic interior that polar molecules or ions cannot penetrate unless specific transport proteins are present in the membranes.

Polymerization into macromolecular form requires that the monomeric molecules be appropriately activated, usually at the expense of ATP. The resulting polymers are able to fold and coil spontaneously and then interact with one another in a unique, predictable manner to generate successively higher-order structures. Much of the information needed for these assembly processes is inherent in the chemical nature of the monomeric units and the order in which chemically different monomers are strung together. This overall strategy of hierarchical assembly from subunits has the dual advantages of chemical simplicity and efficiency of assembly.

With these principles in hand, we are now ready to proceed to Chapter 3, where we will look at the major kinds of biological macromolecules and the chemical nature of the subunits from which they are synthesized.

Tempus Fugit and the Fine Art of Watchmaking

"Drat! Another defective watch!" With a look of disgust, Tempus Fugit the watchmaker tossed the faulty timepiece into the wastebasket and grumbled to himself, "That's two out of the last three watches I've had to throw away. What kind of a watchmaker am I, anyway?"

"A good question, Fugit," came a voice from the doorway. Tempus looked up to see Caveat Emptor entering the shop. "Maybe I can help you with it if you'll tell me a bit about how you make your watches."

"Nobody asked for your help, Emptor," growled Tempus testily, wishing fervently he hadn't been caught thinking out loud.

"Ah well, you'll get it just the same," continued Caveat, quite unperturbed. "Now tell me just how you make a watch and how long it takes you. I'd be especially interested in comparing your procedure with the way Pluribus Unum does it in his new shop down the street."

At the very mention of his competitor's name, Tempus groaned again. As much as possible, he avoided even thinking about E. Pluribus Unum and that disgustingly prosperous shop he had opened recently in the next block. "Unum!" he blustered. "What does he know about the fine art of watchmaking?"

"A good deal, apparently; probably more than you do," replied Caveat. "But tell me exactly how you go about it. How many steps does it take you per watch, and how often do you make a mistake?"

"It takes exactly 100 operations to make a watch. Every step has to be done exactly right, or the watch won't work. It's tricky business, but I've got my error rate down to 1%," said Tempus with at least a trace of pride in his voice. "I can make a watch from start to finish in exactly one hour, but I only make 36 watches each week because I always take Tuesday afternoons off to play darts."

"Well now, let's see," said Caveat, as he pulled out a pocket calculator. He mumbled to himself as he punched in a few numbers, then he looked up and announced brightly, "That means, my good Mr. Fugit, that you only make about 13 watches each week that actually work. All the rest you have to throw away just like the one you pitched as I entered your shop."

"How did you know that, you busybody?" Tempus asked defensively, wondering how this know-it-all with his calculator had guessed his carefully kept secret.

"Elementary, my dear Watson," returned Caveat gleefully. "Simple probability is all it takes. You told me that each watch requires 100 operations and that there's a 99% chance that you'll get a given step right. That's 0.99 times itself 100 times, which comes out to 0.366. So only about 37% of your watches will be put together right, and you can't even tell anything about a given watch until it's finished and you test it. Want to know how that compares with Unum's shop?"

Tempus was about to protest that his name wasn't Watson and that he wasn't at all sure he wanted to know anything about that wretched Unum, but his tormentor hurried on with scarcely a pause for breath. "He can manage 100 operations per hour too, and his error rate is exactly the same as yours. He's also off every Tuesday afternoon, but he gets about 27 watches made every week. That's twice your output, Fugit! No wonder he's got so many watches in his window and so many customers at his door. I've heard that he's even thinking of expanding his shop. Want to know how he does it?"

Tempus was too depressed to even attempt a protest. And besides, although he wouldn't like to admit it, he really was dying to know how Unum managed it.

"Subunit assembly, Fugit, that's the answer! Subunit assembly! Instead of making each watch from scratch in 100 separate steps, Unum assembles the components into 10 pieces, each requiring 10 steps." Caveat began pushing calculator buttons again. "Let's see, he performs 10 operations with a 99% success rate for each, so I take 0.99 to the tenth power instead of the hundredth. That's 0.904, which means that about 90% of his subunits have no errors in them. So he spends about 33 hours each week making 330 subunits, throws the defective ones away, and still has about 300 to assemble into 30 watches during the last three hours on Friday afternoon. That takes 10 steps per watch, with the same error rate as before, so again he comes up with about 90% success. He has to throw away about 3 of his finished watches and ends up with about 27 watches to show for his efforts. Meanwhile, you've been working just as hard and just as accurately, yet you only make half as many watches. What do you have to say to that, my dear Fugit?"

Tempus buried his head in his hands as visions of E. Pluribus Unum and his subassembled watches danced through his mind. Finally, he managed a weary response. "Just one question, Emptor, and I'll probably hate myself for asking. But do you happen to know what Unum does on Tuesday afternoons?"

"I'm glad you asked," replied Caveat as he slipped his calculator back into his pocket and headed for the door. "But I don't think you'll be able to interest him in darts—he spends every Tuesday afternoon giving watchmaking lessons."

Key Terms for Self-Testing

The Importance of Carbon
organic chemistry (p. 17)
biological chemistry (p. 17)
biochemistry (p. 17)
carbon atom (p. 17)
valence (p. 17)
covalent bond (p. 18)
single bonds (p. 18)
double bonds (p. 18)
triple bonds (p. 18)
bond energy (p. 18)
calorie (p. 18)
einstein (p. 19)
hydrocarbons (p. 19)
functional groups (p. 19)
tetrahedral (p. 20)
stereoisomers (p. 20)
asymmetric carbon atoms (p. 20)

The Importance of Water
electronegative (p. 21)
polarity (p. 21)
hydrogen bond (p. 22)
temperature-stabilizing capacity (p. 22)
specific heat (p. 22)
solvent (p. 23)
solutes (p. 23)
hydrophilic (p. 23)
hydrophobic (p. 23)
spheres of hydration (p. 23)
hydrophobic interactions (p. 24)

The Importance of Selectively Permeable Membranes
membrane (p. 24)
amphipathic molecules (p. 24)
phospholipid bilayer (p. 24)
transport proteins (p. 26)
selectively permeable (p. 26)

The Importance of Synthesis by Polymerization
macromolecules (p. 26)
supramolecular structures (p. 27)
monomers (p. 27)
informational macromolecules (p. 27)
storage macromolecules (p. 30)
structural macromolecules (p. 30)
biopolymers (p. 30)
condensation reaction (p. 31)
directionality (p. 31)
activated monomer (p. 31)

The Importance of Self-Assembly
self-assembly (p. 32)
polypeptide (p. 32)
denatured (p. 32)
native (p. 32)
virion (p. 34)
tobacco mosaic virus (TMV) (p. 34)
coat protein (p. 34)
hierarchical assembly (p. 36)

Suggested Reading

General References and Reviews

Henderson, L. J. *The Fitness of the Environment.* Boston: Beacon Press, 1927, reprinted 1958.

Herriott, J., G. Jacobson, J. Marmur, and W. Parsom. *Papers in Biochemistry.* Reading, Mass.: Addison-Wesley, 1984.

Lambert, J. B. The shapes of organic molecules. *Sci. Amer.* 222 (January 1970): 58.

Lehninger, A. L. *Principles of Biochemistry.* New York: Worth, 1982.

Mathews, C. K., and K. E. van Holde. *Biochemistry.* Redwood City, Calif.: Benjamin/Cummings, 1990.

Stryer, L. *Biochemistry,* 3d ed. New York: W. H. Freeman, 1988.

Wald, G. The origins of life. *Proc. Natl. Acad. Sci. USA* 52 (1964): 595.

The Importance of Water

Franks, F., ed. *Water—A Comprehensive Treatise,* vol. 4. New York: Plenum Press, 1975.

Scholander, P. F. Tensile water. *Amer. Sci.* 60 (1972): 584.

The Importance of Membranes

Bretscher, M. S. Membrane structure: Some general principles. *Science* 181 (1973): 622.

Finean, J. B. R., R. Coleman, and R. H. Michell. *Membranes and Their Cellular Functions,* 3d ed. Oxford: Blackwell, 1984.

Lodish, H. F., and J. E. Rothman. The assembly of cell membranes. *Sci. Amer.* 240 (January 1979): 48.

Robertson, R. N. *The Lively Membranes.* Cambridge, England: Cambridge University Press, 1983.

Tanford, C. The hydrophobic effect and the organization of living matter. *Science* 200 (1978): 1012.

Yeagle, P. *The Membranes of Cells.* Orlando, Fla.: Academic Press, 1987.

The Importance of Self-Assembly

Butler, P. J. G., and A. Klug. The assembly of a virus. *Sci. Amer.* 239 (November 1978): 62.

Fraenkel-Conrat, H., and R. C. Williams. Reconstitution of active tobacco mosaic virus from its inactive protein and nucleic acid components. *Proc. Natl. Acad. Sci. USA* 41 (1955): 690.

Klug, A. The assembly of tobacco mosaic virus: Structure and specificity. *Harvey Lect.* 74 (1979): 141.

Klug, A. From macromolecules to biological assemblies. *Biosci. Rep.* 3 (1983): 395.

Lake, J., and C. F. Fox, eds. *Biological Recognition and Assembly.* New York: A. R. Liss, 1980.

Nomura, M. Assembly of bacterial ribosomes. *Science* 179 (1973): 864.

Properties of Proteins

Anfinsen, C. B. Principles that govern the folding of protein chains. *Science* 181 (1973): 223.

Creighton, T. E. *Proteins: Structure and Molecular Properties.* New York: W. H. Freeman, 1984.

Dickerson, R. E., and I. Geis. *The Structure and Action of Proteins.* Menlo Park, Calif.: Benjamin/Cummings, 1969.

Doolittle, R. F. Proteins. *Sci. Amer.* 253 (October 1985): 88.

Kendrew, J. C. The three-dimensional structure of a protein molecule. *Sci. Amer.* 205 (December 1961): 96.

Moore, S., and W. H. Stein. Chemical structures of pancreatic ribonuclease and deoxyribonuclease. *Science* 180 (1973): 458.

Richardson, J. S. The anatomy and taxonomy of protein structure. *Adv. Protein Chem.* 34 (1981): 166.

Rupler, J. A., E. Gratton, and G. Careri. Water and globular proteins. *Trends Biochem. Sci.* 8 (1983): 18.

Schulz, G. E., and R. H. Schirmer. *Principles of Protein Structure.* New York: Springer-Verlag, 1979.

Problem Set

1. **The Fitness of Carbon.** Carbon has a number of properties that make it especially fit to play a key role in biological molecules. One way to appreciate these properties is to compare and contrast them with those of silicon, the element immediately beneath carbon in the periodic table. Contrast each of the following properties of silicon with the comparable property of carbon and indicate, if appropriate, why carbon is a fitter element than silicon for biological purposes in that particular regard.
   (a) Silicon is a larger atom, with an atomic weight of 28.
   (b) Silicon combines with oxygen to form insoluble silicates or network polymers of silicon dioxide (quartz).
   (c) Silicon does not readily form double or triple bonds.
   (d) Silicon has four valence electrons and is quite abundant in the earth's crust.
   (e) Polymers of silicon are not stable in water.

2. **The Fitness of Water.** For each of the following statements about water, decide whether the statement is true and describes a property that makes water a desirable component of cells (T); is true but describes a property that has no bearing on water as a cellular constituent (X); or is false (F).
   (a) Water is a polar molecule and hence an excellent solvent for polar compounds.
   (b) Water freezes at 0°C and boils at 100°C.
   (c) The density of water is less than the density of ice.
   (d) The molecules of liquid water are extensively hydrogen-bonded to one another.
   (e) Water is a clear, colorless liquid.
   (f) Water is odorless and tasteless.

3. **Bond Energies.** Of considerable importance in assessing the fitness of the bonds in organic molecules is the relationship between the amount of energy required to break these bonds and the amount of energy available in the solar radiation to which these bonds are exposed.
   (a) Given that the carbon-carbon single bond is safe in green light (500 nm) but can be broken by the energy of ultraviolet light (300 nm, for example), where does the cutoff come? That is, what wavelength of light has just enough energy to break a carbon-carbon single bond?
   (b) Would a carbon-nitrogen bond be more or less stable than a carbon-carbon single bond at this wavelength?
   (c) What about a carbon-carbon double bond?

4. **Stereoisomers.** For each compound below, indicate how many stereoisomers exist and draw the structure of each.

(a)

```
        O
        ‖
        C—OH
        |
   H2N—C—H
        |
     H—C—OH
        |
        CH3
```

L-threonine
(an amino acid)

(b)

```
     O
     ‖
     C—H
     |
  H—C—OH
     |
  H—C—OH
     |
  H—C—OH
     |
     CH2—OH
```

D-ribose
(a monosaccharide)

(c)

```
       O
       ‖
       C—OH
       |
   (H—C—H)14
       |
       CH3
```

Palmitic acid
(a fatty acid)

(d)

```
       H  O
       |  ‖
    H—C—C—OH
       |
       |  O
       |  ‖
    H—C—C—OH
       |
       |  O
       |  ‖
   HO—C—C—OH
       |
       H
```

Isocitric acid
(a tricarboxylic acid)

5. **Solubility Properties of Biological Molecules.**
(a) Why is the organic compound hexadecane not a common chemical constituent of cells, whereas palmitic acid is?

$$CH_3{-}(CH_2)_{14}{-}CH_3$$
Hexadecane

$$CH_3{-}(CH_2)_{14}{-}\overset{\displaystyle O}{\overset{\|}{C}}{-}OH$$
Palmitic acid

(b) Which of the following compounds is more likely to occur in cells? Explain your choice.

(i) $(CH_3)_2CH{-}CH_2{-}CH_3$

(ii) $(CH_3)_2CH{-}\underset{\displaystyle H}{\overset{\displaystyle NH_2}{C}}{-}\overset{\displaystyle O}{\overset{\|}{C}}{-}OH$

6. **The Polarity of Water.** Defend the assertion that all of life as we know it depends critically on the fact that the angle between the two hydrogen atoms in the water molecule is 105° and not 180°.

7. **The Principle of Polymers.** Polymers clearly play an important role in the molecular economy of the cell. What advantage does each of the following features of biopolymers have for the cell?
(a) A particular kind of polymer is formed using the same kind of condensation reaction to add each successive monomeric unit.
(b) The bonds between monomers are formed by removal of water and are broken or cleaved by addition of water.

8. **TMV Assembly.** Each of the following statements is an experimental observation concerning the reassembly of tobacco mosaic virus (TMV) virions from TMV RNA and coat protein subunits. In each case, state as carefully as possible a reasonable conclusion that can be drawn from the experimental finding.
(a) When RNA from a specific strain of TMV is mixed with coat protein from the same strain, infectious virions are formed.
(b) When RNA from strain A of TMV is mixed with coat protein from strain B, the reassembled virions are infectious, giving rise to strain A virus particles in the infected tobacco cells.
(c) Isolated coat protein monomers can polymerize into a virus-like helix in the absence of RNA.
(d) In infected plant cells, the TMV virions that form contain only TMV RNA and never any of the various kinds of cellular RNAs present in the host cell.
(e) Regardless of the ratio of RNA to coat protein in the starting mixture, the reassembled virions always contain RNA and coat protein in the ratio of three nucleotides of RNA per coat protein monomer.

9. **Doing It Backward.** The following is the *answer;* can you figure out the *question?* "It minimizes the amount of genetic information needed and allows imperfect components to be discarded at several stages."

3

The Macromolecules of the Cell

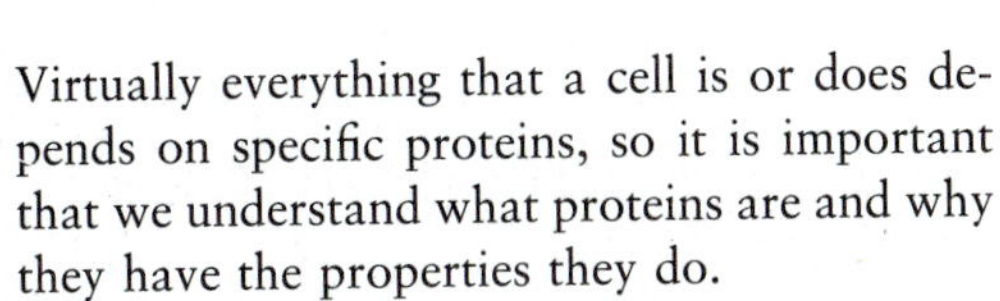

In the preceding chapter, we looked at some of the basic chemical principles of cellular organization. We saw that each of the major kinds of biopolymers—the proteins, the nucleic acids, and the polysaccharides—consists of a relatively small number (from 1 to 20) of repeating monomeric units. These polymers are synthesized by condensation reactions in which activated monomers are linked together and water is removed. Once synthesized, the individual polymer molecules fold and coil spontaneously into stable, three-dimensional shapes. These folded molecules then associate with one another in a hierarchical manner to generate higher levels of structural complexity, usually without further input of energy or information.

With these principles in hand, we are ready to look briefly at the major kinds of biological macromolecules, focusing first on the chemical nature of the monomeric components and then on the synthesis and properties of the polymer itself. We begin our survey with the proteins, because they play such an important and widespread role in cellular structure and function. We then move on to nucleic acids and polysaccharides. The tour concludes with the lipids, which do not quite fit the definition of a polymer but are important cellular components in their own right.

Proteins

Proteins are without a doubt the most important and ubiquitous macromolecules in the cell. All enzymes and antibodies are proteins, as are many hormones. Connective tissue, muscle fibrils, cilia, flagella—all are made primarily or exclusively of proteins. Whether we are talking about carbon dioxide fixation in photosynthesis, oxygen transport in the blood, or the motility of a flagellated bacterium, we are dealing with processes that depend crucially on particular proteins with specific properties and functions. Virtually everything that a cell is or does depends on specific proteins, so it is important that we understand what proteins are and why they have the properties they do.

Actually, we have picked up a few clues about proteins already. We know, for example, that proteins are linear polymers of amino acids joined in a genetically determined sequence. We also know that the properties of a particular protein depend on the proportions and sequence of the various amino acids it contains. In addition, we are aware that proteins are synthesized as polypeptide chains on "workbenches" called ribosomes under the genetic direction of a molecule of messenger RNA that is in turn derived by transcription of the genetic information stored in the DNA. The details of protein synthesis will not concern us until a later chapter. At this point, it will suffice to look briefly at some general aspects of amino acid and protein chemistry.

The Monomers Are Amino Acids

There are 20 different kinds of **amino acids** present in proteins. Most proteins contain all or most of the 20 kinds of amino acids, though the proportions vary greatly between proteins and no two proteins have the same order of amino acids. Every amino acid has the basic structure illustrated in Figure 3-1, with a carboxyl group, an amino group, a hydrogen atom, and a so-called R group all attached to a single carbon atom. Except for glycine, for which the R group is just a hydrogen atom, all amino acids have at least one asymmetric carbon atom and therefore exist in two stereoisomeric forms, called D- and L-amino acids (recall Figure 2-7a). Both kinds exist in nature, but only L-amino acids occur in proteins.

Since the carboxyl and amino groups of Figure 3-1 are common features of all amino acids, the specific properties of a particular amino acid obviously depend on the chem-

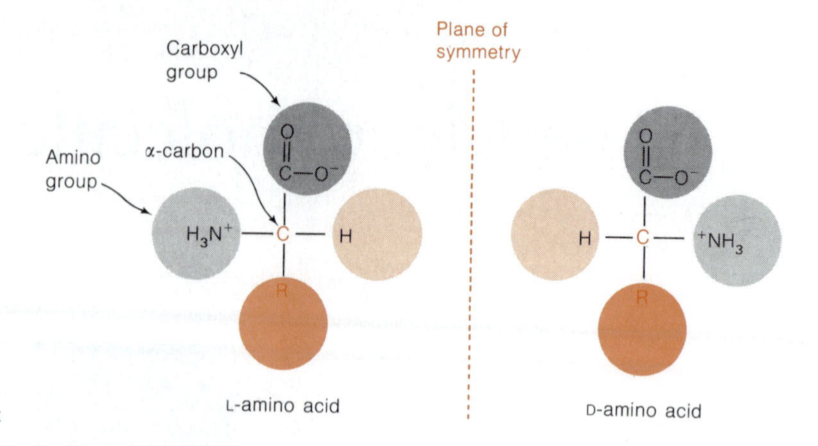

Figure 3-1 The Structure and Stereochemistry of an Amino Acid. Because the α carbon atom is asymmetric in all amino acids except glycine, most amino acids can exist in two isomeric forms, designated as L and D. Note that the L and D forms are stereoisomers, with the vertical dashed line as the plane of symmetry. Of the two forms, only L-amino acids are present in proteins.

ical nature of the R groups, which range from a single hydrogen atom to relatively complex aromatic groups. Shown in Figure 3-2 are the structures of the 20 L-amino acids found in proteins.

Eight of these amino acids have nonpolar R groups and are therefore hydrophobic (group A). As you glance at their structures, you will notice the hydrocarbon nature of the R groups, with oxygen and nitrogen conspicuous by their absence. These are the amino acids that tend to seek out the interior of the molecule as a polypeptide folds into its three-dimensional shape. If a protein (or a region of the molecule) has a preponderance of hydrophobic amino acids, the whole protein will tend to avoid aqueous environments and will instead seek out hydrophobic locations, such as the interior of a membrane.

The remaining 12 amino acids are hydrophilic, with an R group that is either distinctly polar (group B; notice the prominence of oxygen) or actually charged at the pH values characteristic of cells (group C; notice the net negative or positive charge in each case). Hydrophilic amino acids tend to cluster on the surface of proteins, thereby maximizing their interaction with the polar water molecules in the surrounding environment.

The Polymers Are Polypeptides and Proteins

The process of stringing individual amino acids together into a linear polymer involves stepwise addition of each new amino acid to the growing chain by what is essentially a condensation reaction. This process is illustrated schematically in Figure 3-3. The —H and —OH groups that are removed during any condensation reaction come in this case from the amino group of one amino acid and the carboxyl group of the other, respectively. The general term for the resulting linkage is an *amide bond,* but in the special case in which both reactants are amino acids, it is usually called a **peptide bond.** Notice that the chain of amino acids formed in this way has an intrinsic directionality, since it terminates in an amino group at one end and a carboxyl group at the other end. The end with the amino group is called the **N-** (or **amino**) **terminus,** and the end with the carboxyl group is called the **C-** (or **carboxyl**) **terminus.**

Peptide bond formation is actually more complicated than is suggested by Figure 3-3 because both energy and information are needed for the addition of each amino acid. Energy is needed to activate the incoming amino acid and link it to a special kind of RNA molecule called a *transfer RNA.* Information is needed because the order of amino acids in the growing polypeptide chain is not random but is specified genetically. Ensuring that the correct amino acid is chosen for each successive addition to the growing chain depends on a recognition process between the transfer RNA that is linked to the amino acid and the *messenger RNA* that is bound to the ribosome. We will get to all of these details in due course. Here, we need only note that as each new peptide bond is formed the growing chain of amino acids is lengthened by one and that the process requires both energy and information.

Although this process of elongating a chain of amino acids is often called *protein synthesis,* the term is not entirely accurate because the immediate product of amino acid polymerization is not a protein but a **polypeptide.** A protein is a polypeptide chain (or several such chains) that has attained a unique, stable, three-dimensional shape and is biologically active as a result. Some proteins consist of

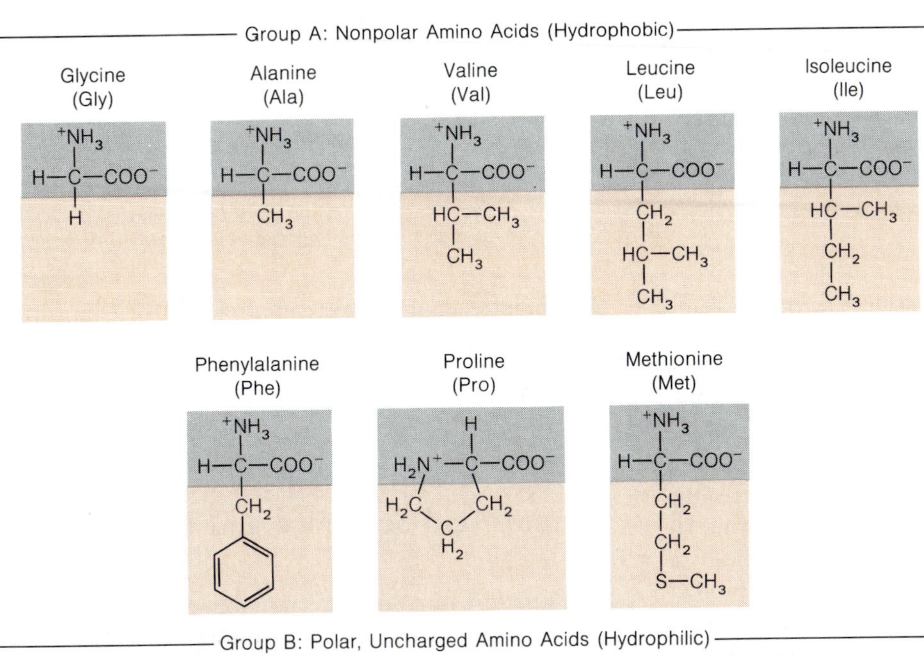

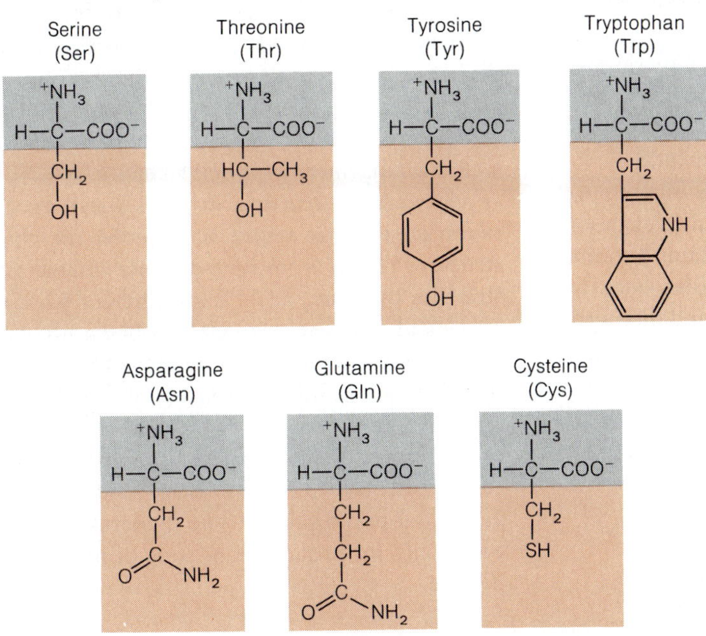

Group C: Polar, Charged Amino Acids (Hydrophilic)

Acidic

Aspartate (Asp): $^{+}NH_3$—C(H)—COO^-; CH_2—COO^-

Glutamate (Glu): $^{+}NH_3$—C(H)—COO^-; CH_2—CH_2—COO^-

Basic

Lysine (Lys): $^{+}NH_3$—C(H)—COO^-; CH_2—CH_2—CH_2—CH_2—$^{+}NH_3$

Arginine (Arg): $^{+}NH_3$—C(H)—COO^-; CH_2—CH_2—CH_2—NH—C(=NH)—$^{+}NH_3$

Histidine (His): $^{+}NH_3$—C(H)—COO^-; CH_2—imidazole (NH, HN^+)

Figure 3-2 The Structures of the Twenty Amino Acids Found in Proteins. All amino acids have a carboxyl group and an amino group on the adjacent (α) carbon (gray), but each has its own distinctive R group (color). Those in group A have nonpolar R groups and are therefore hydrophobic; note the hydrocarbon nature of their R groups. The others are hydrophilic, either because the R group is polar in nature (group B) or because the R group is protonated or ionized at cellular pH and therefore carries a formal electrical charge (group C). The three-letter abbreviations indicated in parentheses are widely used by biochemists and molecular biologists.

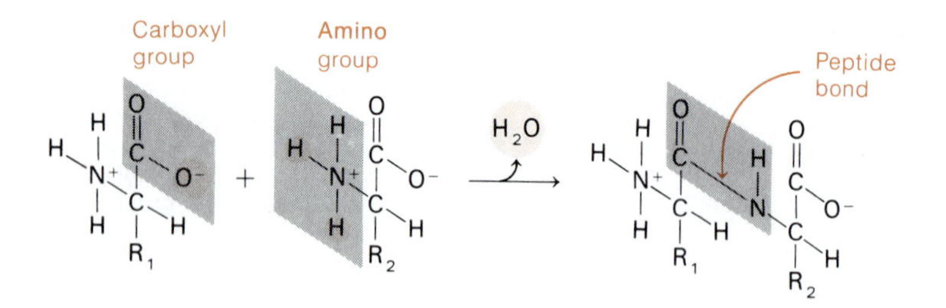

Figure 3-3 Peptide Bond Formation. Successive amino acids in a polypeptide are linked to one another by peptide bonds between the carboxyl group of one amino acid and the amino group of the next.

a single polypeptide and therefore achieve their final shape as a consequence of the folding and coiling that occur spontaneously as the chain is being formed. Such proteins are called **monomeric proteins.** (You will have to be careful with the terminology here. On the one hand, a polypeptide is a polymer, with amino acids as its monomeric repeating units; on the other hand, such a polypeptide becomes the monomer of which proteins are made.) The enzyme ribonuclease, depicted in Figure 2-20, is an example of a monomeric protein. For such proteins, the term protein synthesis is appropriate for the polymerization because the functional protein forms spontaneously as the polypeptide is elongated.

Many other proteins, however, are **multimeric proteins,** consisting of two or more polypeptides. The *hemoglobin* that carries oxygen in your bloodstream is a multimeric protein, since it contains four polypeptides, two called α chains and two called β chains (see Figure 3-4). In such cases, protein synthesis involves not only elongation and folding of the individual polypeptide chains but also their subsequent interaction. As discussed earlier, both the initial folding of the individual polypeptides and their subsequent interaction to form a multimeric protein are driven by the tendency for hydrophobic amino acids to avoid an aqueous environment.

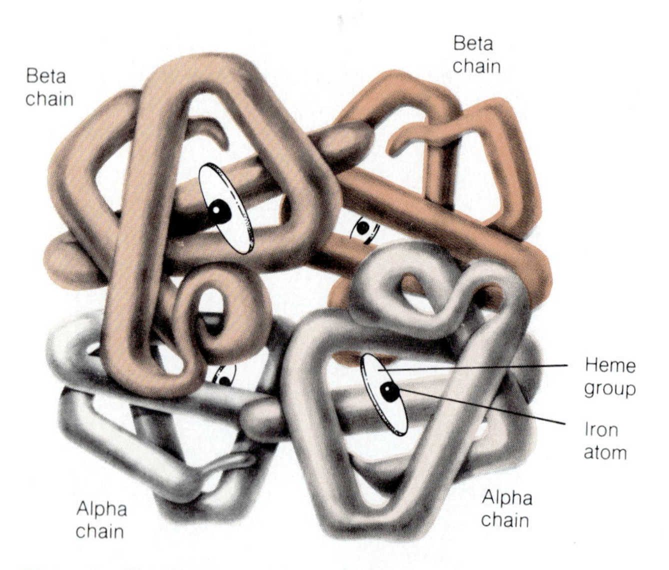

Figure 3-4 The Structure of Hemoglobin. Hemoglobin is a multimeric protein with four subunits (two α polypeptides and two β polypeptides). Each subunit contains a heme group with an iron atom. Each heme iron can bind a single oxygen molecule.

Protein Structure Depends on Amino Acid Sequence and Interactions

The structure of a protein is usually described in terms of four hierarchical levels of organization: the *primary, secondary, tertiary,* and *quaternary* structures (Figure 3-5). Primary structure refers to the amino acid sequence, while the three higher levels of organization concern the interactions between the amino acid R groups that give the protein its characteristic **conformation,** or three-dimensional arrangement of atoms in space.

Secondary structure involves local interactions between contiguous amino acids along the chain; tertiary structure results from long-distance interactions between stretches of amino acids from different parts of the molecule; and quaternary structure concerns the interaction of two or more individual polypeptides to form a single multimeric protein. All are dictated by the primary structure, but each is important in its own right in the overall structure of the protein. Secondary and tertiary structures are obviously involved in determining the conformation of the individual polypeptide, while quaternary structure is relevant only for proteins consisting of more than one polypeptide.

Primary Structure. As already noted, the **primary structure** of a protein (Figure 3-5a) is just a formal designation for the amino acid sequence of the constituent polypeptide(s). When we describe the primary structure of a protein, we are simply specifying the order in which its amino acids appear from one end of the molecule to the other. By convention, amino acid sequences are always written from the N-terminus to the C-terminus of the polypeptide.

The first complete determination of the amino acid sequence of a protein was for the hormone *insulin,* reported in 1953 by Frederick Sanger, who eventually received a Nobel Prize for this important technical advance. The next protein to be sequenced was the enzyme *ribonuclease.* Since

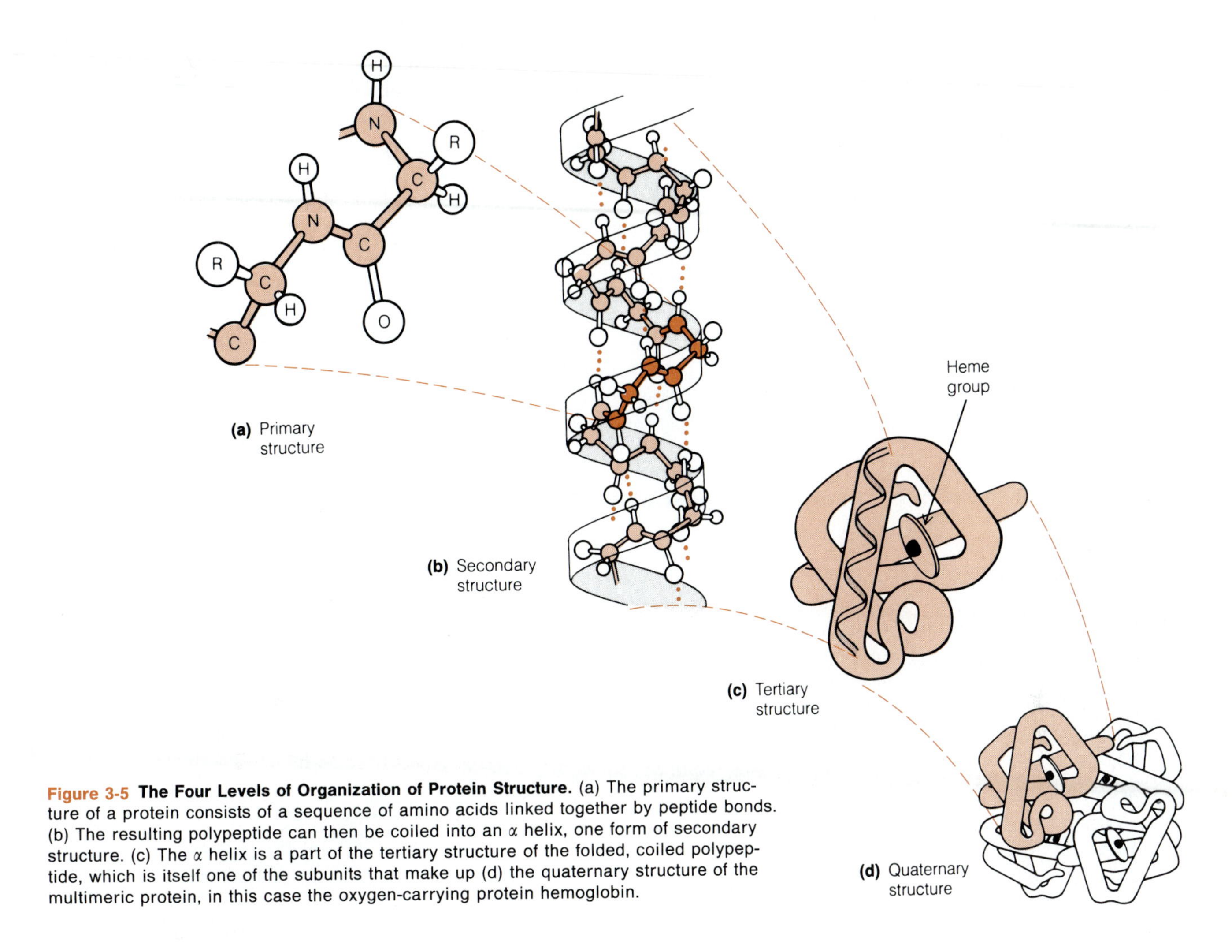

Figure 3-5 The Four Levels of Organization of Protein Structure. (a) The primary structure of a protein consists of a sequence of amino acids linked together by peptide bonds. (b) The resulting polypeptide can then be coiled into an α helix, one form of secondary structure. (c) The α helix is a part of the tertiary structure of the folded, coiled polypeptide, which is itself one of the subunits that make up (d) the quaternary structure of the multimeric protein, in this case the oxygen-carrying protein hemoglobin.

then, the sequencing of polypeptides has become an automated, routine procedure and has been successfully applied to hundreds of proteins. In fact, computerized data banks of polypeptide sequences are now available, making it easy to compare sequences and look for regions of similarity.

The primary structure of a protein is important both genetically and structurally. Genetically, it is significant because the amino acid sequence of the polypeptide derives directly from the order of nucleotides in the corresponding messenger RNA. The messenger RNA is in turn encoded by the DNA that represents the gene for this protein, so the primary structure of a protein is an inevitable result of the order of nucleotides in the DNA of the gene.

Of more immediate significance are the implications of the primary structure for higher levels of protein structure. This ground has already been covered to some degree because polypeptide folding and interaction were used as examples in our discussion of self-assembly in Chapter 2. In essence, all three higher levels of protein organization—secondary, tertiary, and quaternary structures—are direct consequences of the primary structure. This means that all the information necessary to specify completely how polypeptide chains will coil and fold and, if appropriate, how they will interact with one another is inherent in the amino acid sequence (and hence in the genetic information that dictates that sequence). Put another way, anytime a polypeptide chain is made that has the right amino acid sequence to be insulin, it will spontaneously assume the shape and function of insulin without further input of information. Similarly, if polypeptides are made that correspond in sequence to the α and β polypeptides of hemoglobin, they will assume the three-dimensional conformations appropriate to these respective subunits and then interact spontaneously in just the right way to form the $\alpha_2\beta_2$ tetramer that we recognize as hemoglobin.

Secondary Structure. The **secondary structure** of a protein (Figure 3-5b) involves local interactions between adjacent amino acids of a polypeptide chain. These interactions result in two characteristic but quite different patterns, referred to respectively as the **α helix** and **β pleated sheet** conformations.

The α helix structure was elucidated in 1951 by Linus Pauling and Robert Corey. As shown in Figure 3-6, an α helix is spiral in shape, consisting of a backbone of peptide bonds with the specific R groups of the individual amino acids jutting out from it. A helical shape is common to repeating polymers, as we will see when we get to the nucleic acids and the polysaccharides. For the α helix, there are 3.6 amino acids per turn of the helix, bringing the peptide bonds of every fourth amino acid in close proximity. The distance between these juxtaposed peptide bonds is, in fact, just right for the formation of a hydrogen bond between the imino ($-\overset{\text{H}}{\overset{|}{\text{N}}}-$) group of one peptide bond and the carbonyl ($-\overset{\text{O}}{\overset{\|}{\text{C}}}-$) group of the other (see Figure 3-6).

As a result, every peptide bond in the helix is hydrogen-bonded through its carbonyl group to the peptide bond immediately "below" it in the spiral and through its imino group to the peptide bond just "above" it. In each case, however, the hydrogen-bonded peptide bonds are separated in linear sequence by the three amino acids required to advance the helix far enough to allow the two bonds to be juxtaposed. These hydrogen bonds are all nearly parallel to the main axis of the helix and therefore tend to stabilize the spiral structure by holding successive turns of the helix together.

A major alternative to the α helix structure is the β pleated sheet shown in Figure 3-7. Instead of a spiral structure, this is an extended sheetlike conformation with successive atoms in the chain located at the "peaks" and "troughs" of the pleats. The R groups of successive amino acids jut out on alternating sides of the pleated sheet.

Like the α helix, the β pleated sheet is characterized by a maximum of hydrogen bonding. In both cases, every imino group and every carbonyl group of the peptide bonds are involved. However, hydrogen bonding in the α helix is intramolecular—between peptide bonds within the same polypeptide—whereas in the β pleated sheet, hydrogen bonding is perpendicular to the plane of the sheet, linking peptide bonds in adjacent polypeptides. Both the α helix and the β pleated sheet contribute significantly to the secondary structure of various proteins. Silk *fibroin* is a good example of a common protein with extensive pleated sheet conformation, and *α-keratin,* the protein of hair, is an example of a well-known protein that consists almost exclusively of α helix.

Tertiary Structure. The **tertiary structure** of a protein (Figure 3-5c) can probably be best understood if it is compared with secondary structure. Secondary structure is a predictable, repeating pattern that derives from the repetitive nature of the polypeptide, since it involves hydrogen bonding between peptide bonds, the common structural elements along the chain. A protein such as fibroin or keratin, which consists predominantly of a single kind of secondary structure, will look very much the same regardless of where along the amino acid chain you view it. In fact, if proteins contained only one or two kinds of very similar amino acids, virtually all aspects of protein conformation could probably be understood in terms of secondary structure, with only the slightest variation between proteins.

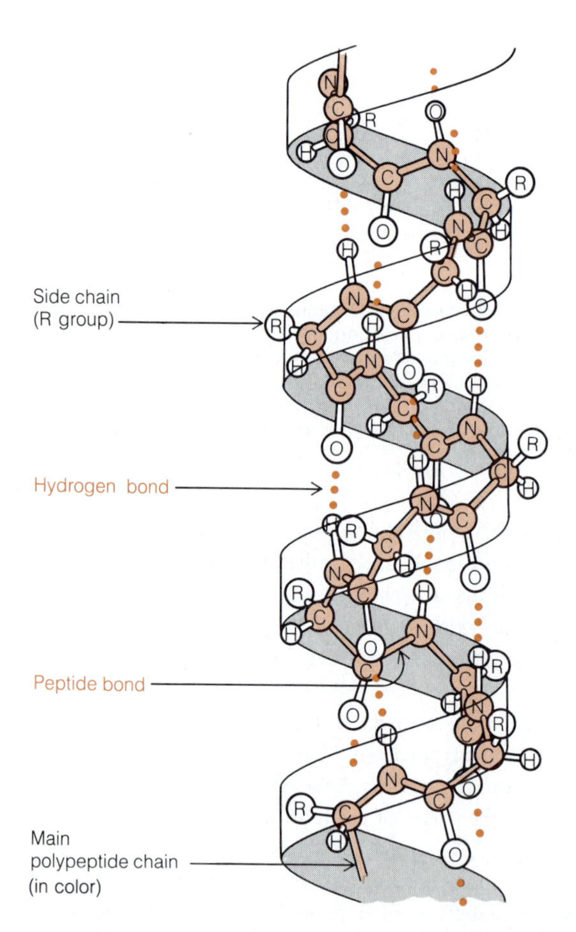

Figure 3-6 The α Helix. The α helix is an important example of secondary structure in proteins. The α helix resembles a coiled telephone cord, but with each turn of the coil stabilized by hydrogen bonds (colored dots) between each peptide bond and the peptide bonds just "beneath" and "above" it in the helix. The hydrogen bonds of an α helix are therefore within a single polypeptide chain.

Tertiary structure comes about precisely because of the variety of amino acids present in proteins and the very different chemical properties of their R groups. In fact, tertiary structure depends almost entirely on interactions between the various R groups, regardless of where along the primary sequence they happen to come. Tertiary structure therefore reflects the nonrepetitive aspect of a polypeptide, since it depends not on the carboxyl and amino groups common to all of the amino acids in the chain but on the very feature that makes each amino acid distinctive—its R group.

Tertiary structure is neither repetitive nor readily predictable, involving as it does competing interactions between side groups with different properties. Hydrophobic R groups, for example, will associate with one another and spontaneously seek out a nonaqueous environment in the interior of the molecule, whereas polar amino acids will be drawn to the surface, both by their affinity for one another and by their attraction to the polar water molecules in the surrounding milieu. As a result, the polypeptide chain will be folded, coiled, bent, and twisted into the **native conformation** that represents the most stable state for that particular sequence of amino acids.

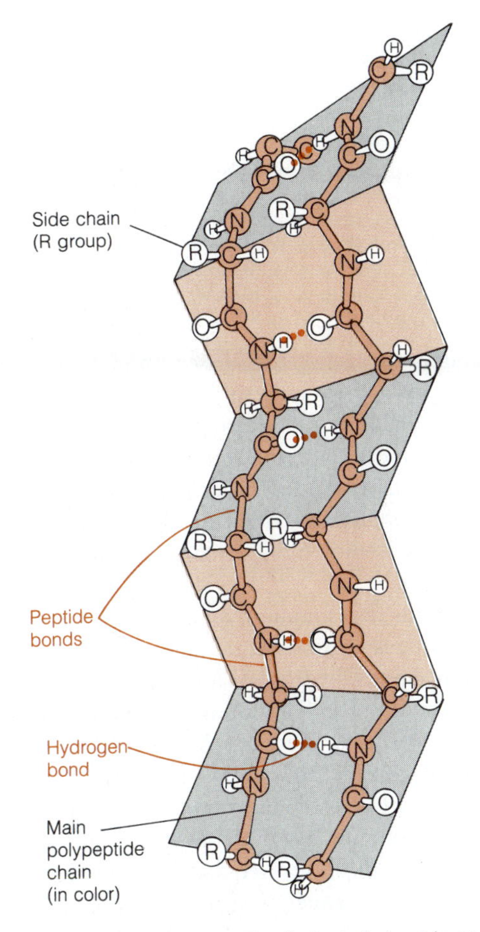

Figure 3-7 The β Pleated Sheet. The β pleated sheet is an important example of secondary structure in proteins. This structure resembles a pleated skirt, with successive atoms of each polypeptide chain located at folds of the pleats and with R groups of the amino acids jutting out of the two sides in alternation. Adjacent polypeptide chains in the sheet are stabilized by hydrogen bonds (colored dots) between each peptide bond and the peptide bonds in the immediately adjacent chains. The hydrogen bonds of a β pleated sheet are therefore between adjacent polypeptide chains.

The native conformation is highly reproducible. If the protein is **denatured** and then returned to a suitable environment, it will often spontaneously **renature** to the original shape (Figure 2-20b). We can label such a structure "unpredictable" in the sense that we cannot simply look at the amino acid sequence and deduce the most stable three-dimensional structure; but in another sense it is totally predictable, since the polypeptide always returns spontaneously to its native shape.

Once the tertiary structure of a polypeptide has been achieved, it is stabilized and maintained by several types of bonds, both covalent and noncovalent. Noncovalent bonds that are important in this regard include *hydrogen bonds* between appropriate R groups, *ionic bonds* between charged R groups, and *hydrophobic interactions* between nonpolar R groups.

The most common covalent bond that contributes to the stabilization of tertiary structure is the **disulfide bond** that is formed between two cysteines upon oxidation, as shown in Figure 3-8. The cysteines involved in a particular disulfide bond are usually quite distant from each other along the primary sequence but are juxtaposed by the folding process. Once formed, a disulfide bond confers considerable stability because of its covalent nature. A disulfide bond can be broken only by reducing it again and regenerating the two sulfhydryl groups of the cysteines. (It is specifically to reduce the four disulfide bonds of native ribonuclease that a reducing agent must be included in the denaturation step shown in Figure 2-20b.)

The relative contributions of secondary and tertiary structures to the overall shape of a polypeptide vary from protein to protein and depend critically on the relative proportions and sequence of amino acids in the chain. Broadly speaking, proteins can be divided into two categories: fibrous proteins and globular proteins. **Fibrous proteins** have extensive secondary structure (either helix or pleated sheet) throughout the molecule, giving them a highly ordered, repetitive structure. In general, secondary structure is much more important than tertiary interactions in determining the shape of such proteins. Silk fibroin and keratin are typical fibrous proteins. **Globular proteins,** on the other hand, depend much more on the kinds of inter-

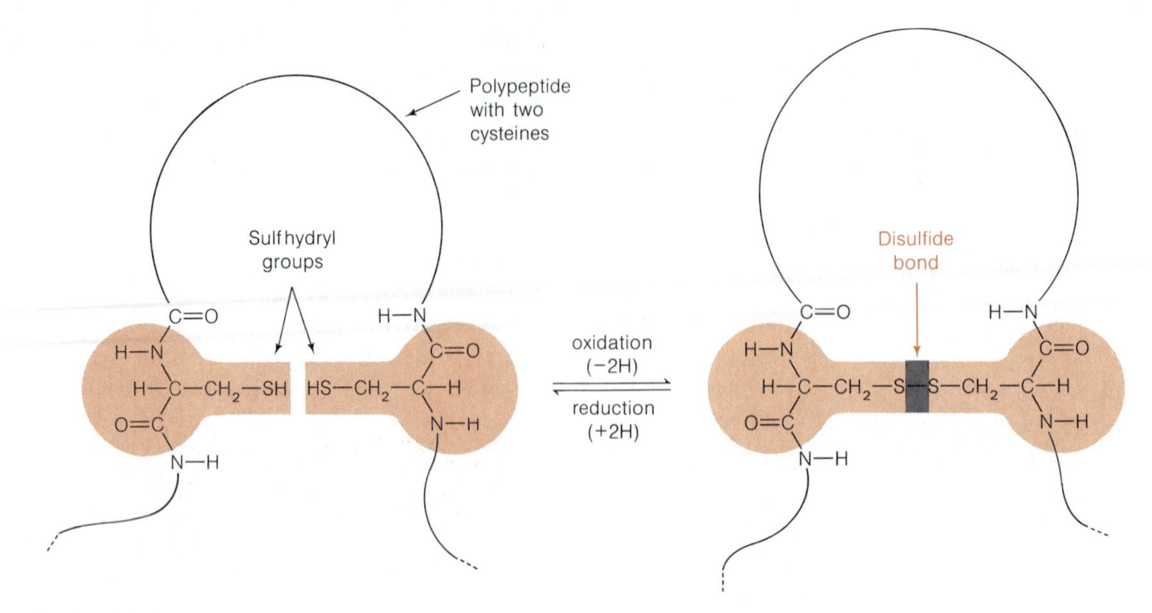

Figure 3-8 Disulfide Bond Formation. A disulfide bond (gray) is formed by oxidation of sulfhydryl (—SH) groups of two cysteines (color) located at different positions in the polypeptide chain (represented by the continuous line). A disulfide bond can be broken by reducing it to the two sulfhydryl groups from which it was originally formed. Disulfide bonds can also be formed between sulfhydryl groups in two separate polypeptide chains, thereby linking the two polypeptides covalently.

actions characteristic of tertiary structure than on the local ordering of secondary structure for their overall shape. Enzymes and antibodies are examples of globular proteins. Such proteins characteristically have local regions of secondary structure, but they also have less ordered regions called **random coils.**

Before leaving the topic of tertiary structure, we should emphasize again the dependence of these higher levels of organization on the primary structure of the polypeptide. The significance of primary structure is exemplified especially well by the inherited condition **sickle-cell anemia.** People with this trait have hemoglobin molecules with normal α chains, but their β chains have a single amino acid that is abnormal. At a specific position in the chain, the glutamate normally present at that point is replaced by valine. The other 145 amino acids in the chain are correct, but this single substitution causes enough of a difference in the tertiary structure of the β chain that the oxygen-carrying capacity of the hemoglobin molecule is impaired significantly. Not all amino acid substitutions cause such dramatic changes in structure and function, but the example serves to underscore the crucial relationship between the amino acid sequence of a polypeptide and the final shape (and often the biological activity) of the molecule.

Quaternary Structure. The **quaternary structure** of a protein (Figure 3-5d) is the level of organization concerned with subunit interactions and assembly. Quaternary structure therefore applies only to multimeric proteins. Many proteins are included in this category, however, particularly those with molecular weights above 50,000. Hemoglobin, for example, is a multimeric protein (Figure 3-4). Some multimeric proteins contain identical polypeptide subunits; others, such as hemoglobin, contain several different kinds of polypeptides.

The bonds that maintain quaternary structure are the same as those responsible for tertiary structure—hydrogen bonds, ionic bonds, hydrophobic interactions, and covalent disulfide bonds. Disulfide bonds may be either within a polypeptide chain or between chains. When they occur within a polypeptide, they stabilize tertiary structure, but when they occur between polypeptides, they help maintain quaternary structure. Again, the process of assembly is spontaneous, with all the requisite information provided by the amino acid sequence of the individual polypeptide.

Nucleic Acids

Next we come to the **nucleic acids,** macromolecules of paramount importance to the cell because of their role in the storage, transmission, and expression of genetic information. Nucleic acids are linear polymers of nucleotides, strung together in a genetically determined order that is critical to their role as informational macromolecules. The two major types of nucleic acids are **DNA** (**deoxyribonucleic acid**) and **RNA** (**ribonucleic acid**). DNA and RNA differ in terms of both chemistry and their role in the cell. As the names suggest, RNA contains the five-carbon sugar **ribose** in each of its nucleotides, while DNA contains the closely related sugar **deoxyribose.** Functionally, DNA serves primarily as the repository of genetic information, whereas RNA molecules play several different roles in the expression of that information.

RNA is transcribed from DNA in several different forms, each of which plays a specific role in protein synthesis (Figure 3-9). **Messenger RNA** (**mRNA**) provides the information that dictates amino acid sequence during polypeptide synthesis. **Transfer RNA** (**tRNA**) directs the correct amino acid to the next site in a growing polypeptide chain. **Ribosomal RNA** (**rRNA**) is the third major type of RNA. It is an important constituent of the ribosomes that serve as the site of protein synthesis.

These roles of DNA and RNA as the bearers or mediators of genetic information will be considered in detail in Part Four. For now, we will focus on the chemistry of nucleic acids and the nucleotides of which they are composed.

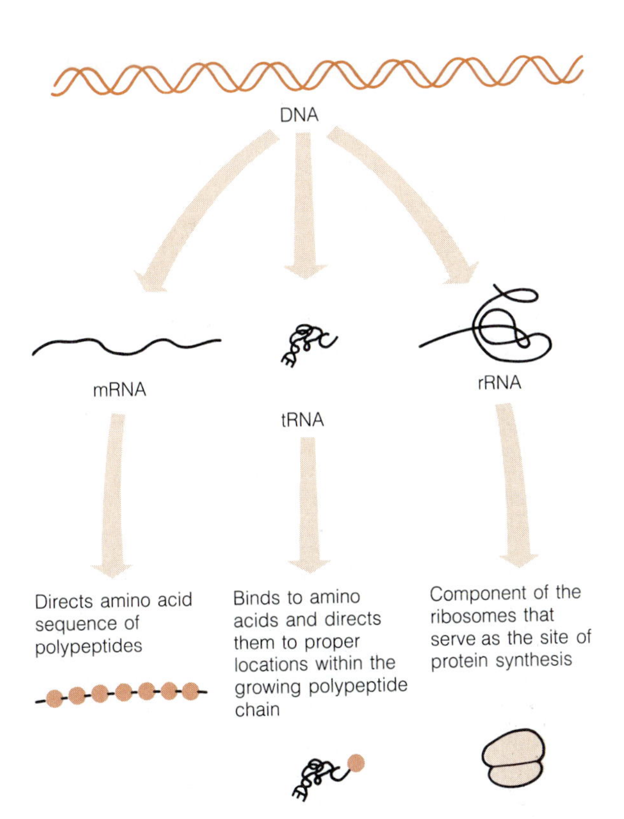

Figure 3-9 The Roles of DNA and RNA in Protein Synthesis. The genetic information in DNA is transcribed into molecules of messenger RNA (mRNA), transfer RNA (tRNA), and ribosomal RNA (rRNA). Each type of RNA plays a role in protein synthesis, as indicated in the figure.

The Monomers Are Nucleotides

Like proteins, nucleic acids are informational macromolecules and therefore contain nonidentical monomeric units in a specified sequence. The monomeric units of nucleic acids are called **nucleotides.** Nucleotides exhibit less variety than amino acids; DNA and RNA each contain only four different kinds. (Actually, there is more variety than this suggests, especially in some RNA molecules, but virtually all variant structures found in nucleic acids represent one of the four basic nucleotides that has been chemically modified after insertion into the chain.)

As shown in Figure 3-10, each nucleotide consists of a five-carbon sugar, a phosphate group, and a nitrogen-containing aromatic base. The sugar is either D-ribose (for RNA) or D-deoxyribose (for DNA). The phosphate is joined by a **phosphoester bond** to the 5′ carbon of the sugar, and the base is attached at the 1′ carbon. The base may be either a **purine** or a **pyrimidine.** DNA contains the purines **adenine** (A) and **guanine** (G) and the pyrimidines **cytosine** (C) and **thymine** (T). RNA also has adenine, guanine, and cytosine but contains the pyrimidine **uracil** (U) in place of thymine.

If the phosphate is removed from a nucleotide, the remaining base-sugar unit is called a **nucleoside.** Each pyrimidine and purine may therefore occur as the free base, the nucleoside, or the nucleotide. The appropriate names for these compounds are given in Table 3-1. Notice that nucleotides and nucleosides containing deoxyribose are specified by a lowercase *d* preceding the letter that identifies the base.

As the nomenclature indicates, a nucleotide can be thought of as a **nucleoside monophosphate,** since it is a nucleoside with a single phosphate group attached to it. This terminology can be readily extended to molecules with two or three phosphate groups attached to the 5′ carbon. For example, the nucleoside adenosine (adenine plus ribose) can have one, two, or three phosphates attached and is designated accordingly as **adenosine monophosphate** (**AMP**), **adenosine diphosphate** (**ADP**), or **adenosine triphosphate** (**ATP**). The relationship between these compounds is illustrated in Figure 3-11.

You should recognize ATP as the energy-rich compound used to drive a variety of reactions in the cell, including the activation of monomers for polymer formation. As this example suggests, nucleotides actually play two roles in cells: They are the monomeric units of nucleic acids, and they serve as intermediates in various energy-transferring reactions.

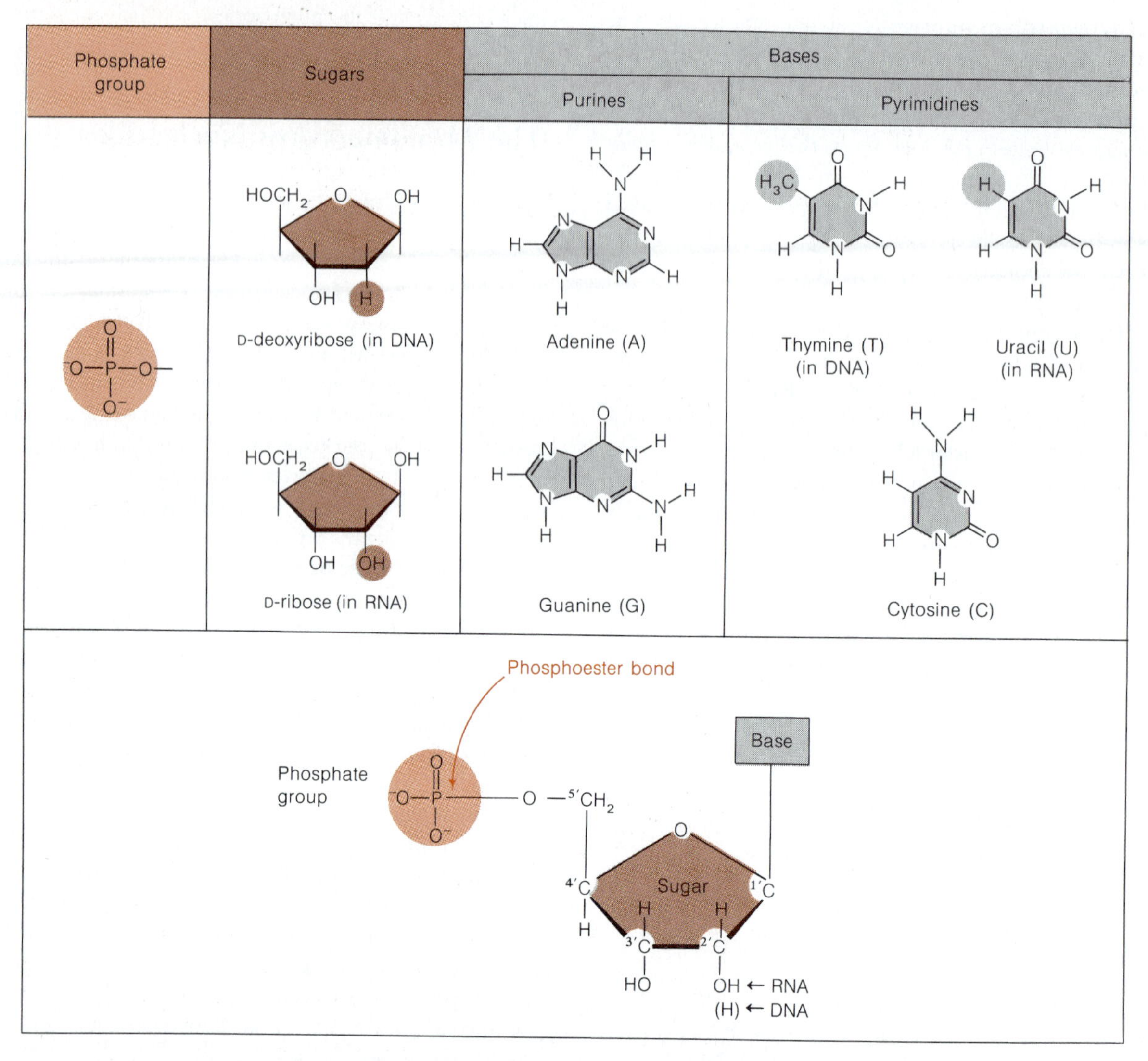

Figure 3-10 The Structure of a Nucleotide. In RNA a nucleotide consists of the five-carbon sugar D-ribose with an aromatic nitrogen-containing base attached to the 1′ carbon and a phosphate group linked to the 5′ carbon by a phosphoester bond. (Carbon atoms in the sugar of a nucleotide are numbered from 1′ to 5′ to distinguish them from those in the base, which are numbered without the prime.) In DNA, the hydroxyl group on the 2′ carbon is replaced by a hydrogen atom, so the sugar is D-deoxyribose. The bases in DNA are the purines adenine (A) and guanine (G) and the pyrimidines thymine (T) and cytosine (C). In RNA, thymine is replaced by the pyrimidine uracil (U).

The Polymers Are DNA and RNA

Nucleic acids are linear polymers formed by linking each nucleotide to the next through a phosphate group, as shown in Figure 3-12. Specifically, the phosphate group already attached by a phosphoester bond to the 5′ carbon of one nucleotide becomes linked by a second phosphoester bond to the 3′ carbon of the next nucleotide. In essence, this is a condensation reaction, with the —H and —OH groups coming from the sugar and the phosphate groups, respectively. The resulting linkage is called a 3′,5′ **phosphodiester bond.** The **polynucleotide** formed by this process has an intrinsic directionality, since it has a 5′ hydroxyl group on one end and a 3′ hydroxyl group on the other end. By convention, nucleotide sequences are always written from the 5′ end to the 3′ end of the polynucleotide.

Nucleic acid synthesis requires both energy and information, just as protein synthesis does. The energy needed for formation of the phosphodiester bond is pro-

Table 3-1 The Bases, Nucleosides, and Nucleotides of RNA and DNA

	RNA		DNA	
Bases	Nucleoside	Nucleotide	Deoxynucleoside	Deoxynucleotide
Purines				
Adenine (A)	Adenosine	Adenosine monophosphate (AMP)	Deoxyadenosine	Deoxyadenosine monophosphate (dAMP)
Guanine (G)	Guanosine	Guanosine monophosphate (GMP)	Deoxyguanosine	Deoxyguanosine monophosphate (dGMP)
Pyrimidines				
Cytosine (C)	Cytidine	Cytidine monophosphate (CMP)	Deoxycytidine	Deoxycytidine monophosphate (dCMP)
Uracil (U)	Uridine	Uridine monophosphate (UMP)	—	—
Thymine (T)	—	—	Thymidine	Thymidine monophosphate (dTMP)

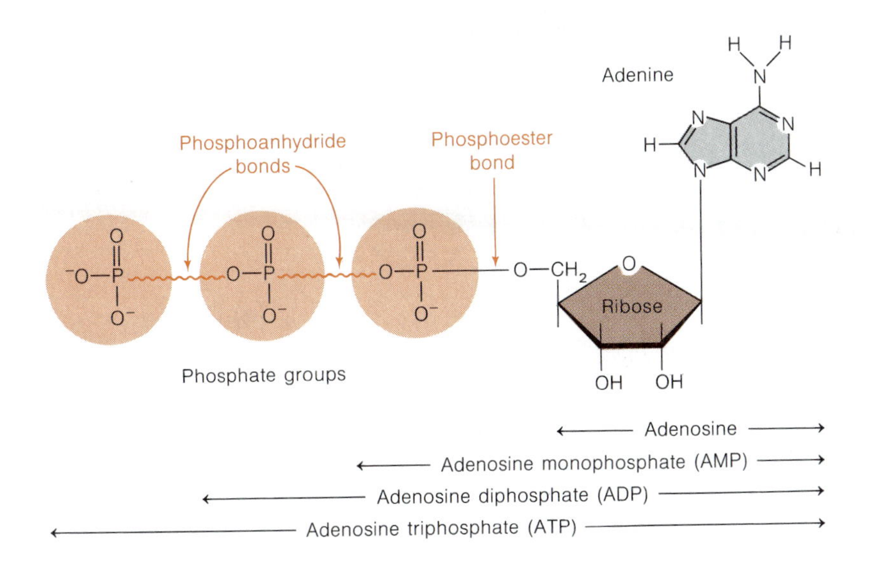

Figure 3-11 The Phosphorylated Forms of Adenosine. Adenosine occurs as the free nucleoside, the monophosphate (AMP), the diphosphate (ADP), and the triphosphate (ATP). The bond that links the first phosphate to the ribose of adenosine is a low-energy phosphoester bond, whereas the bonds that link the second and third phosphate groups to the molecule are higher-energy phosphoanhydride bonds.

vided by a nucleoside triphosphate rather than a nucleoside monophosphate (nucleotide). The precursors for DNA synthesis are therefore dATP, dCTP, dGTP, and dTTP. For RNA synthesis, ATP, CTP, GTP, and UTP are needed.

Information is required for nucleic acid synthesis because successive incoming nucleotides must be added in a specific, genetically determined sequence. For this purpose, a preexisting molecule is used as a **template** to specify nucleotide order. For both DNA and RNA synthesis, the template is usually DNA. The essence of template-directed nucleic acid synthesis is that each incoming nucleotide is selected because its base can be recognized by (will interact with) the base of the nucleotide already present at that position in the template.

This recognition process depends on an important chemical feature of the purine and pyrimidine bases shown in Figure 3-13. These bases have carbonyl groups and nitrogen atoms capable of hydrogen bond formation under appropriate conditions. Furthermore, there exist complementary relationships between purines and pyrimidines that allow A to form two hydrogen bonds with T (or U) and G to form three hydrogen bonds with C, as shown in Figure 3-13. This pairing of A with T (or U) and G with C is a fundamental property of nucleic acids. Genetically, this **base pairing** provides a mechanism for nucleic acids to recognize one another, as we will see in Part Four. For now, however, let us concentrate on the structural implications.

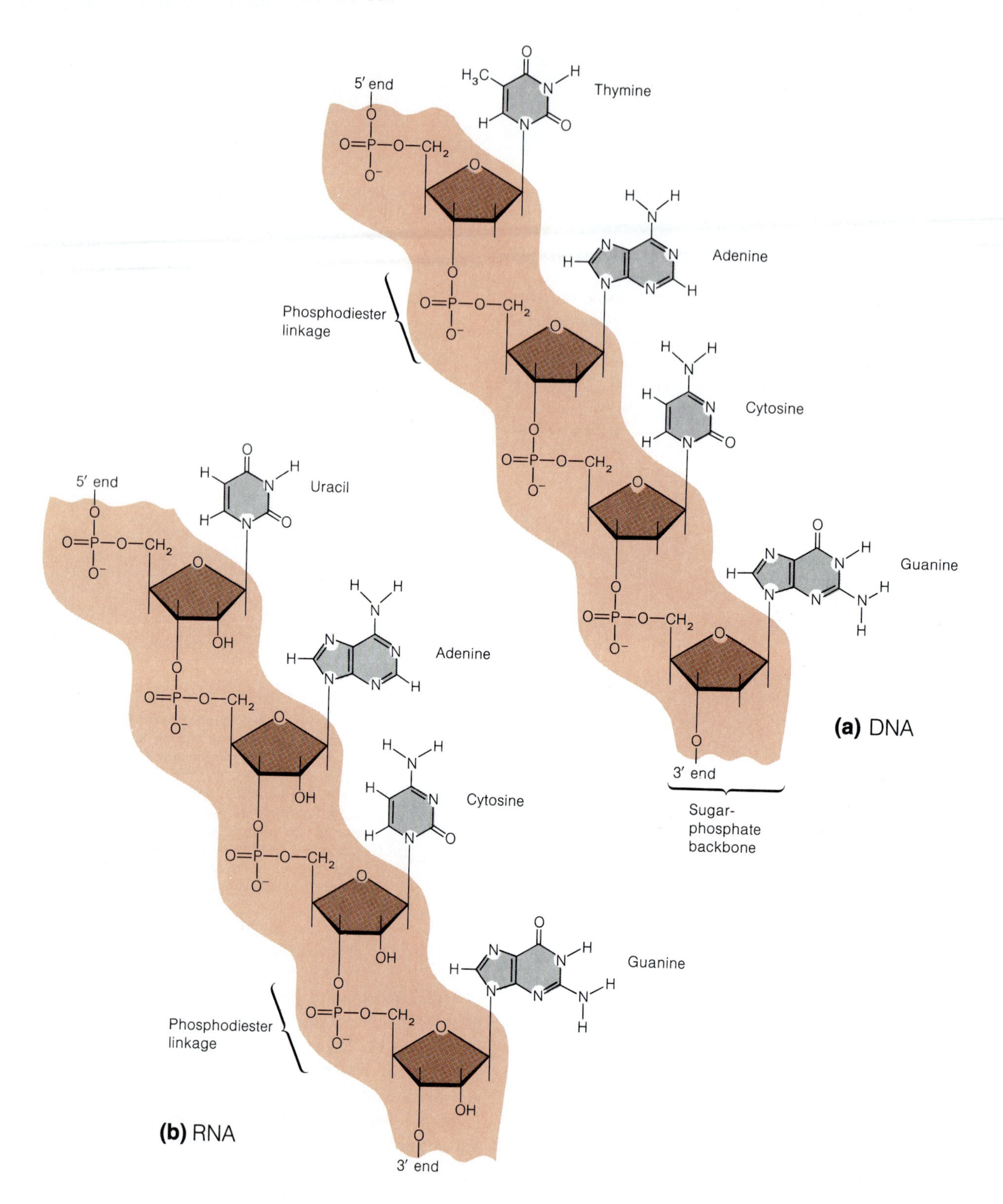

Figure 3-12 The Structure of Nucleic Acids. Nucleic acids consist of chains of nucleotides, each containing a sugar, a phosphate, and a base. The sugar is (a) deoxyribose in DNA and (b) ribose in RNA. Successive nucleotides in the chain are joined together by 3′,5′ phosphodiester linkages. The backbone of the chain (color) is an alternating sugar-phosphate sequence, from which the bases (gray) jut out.

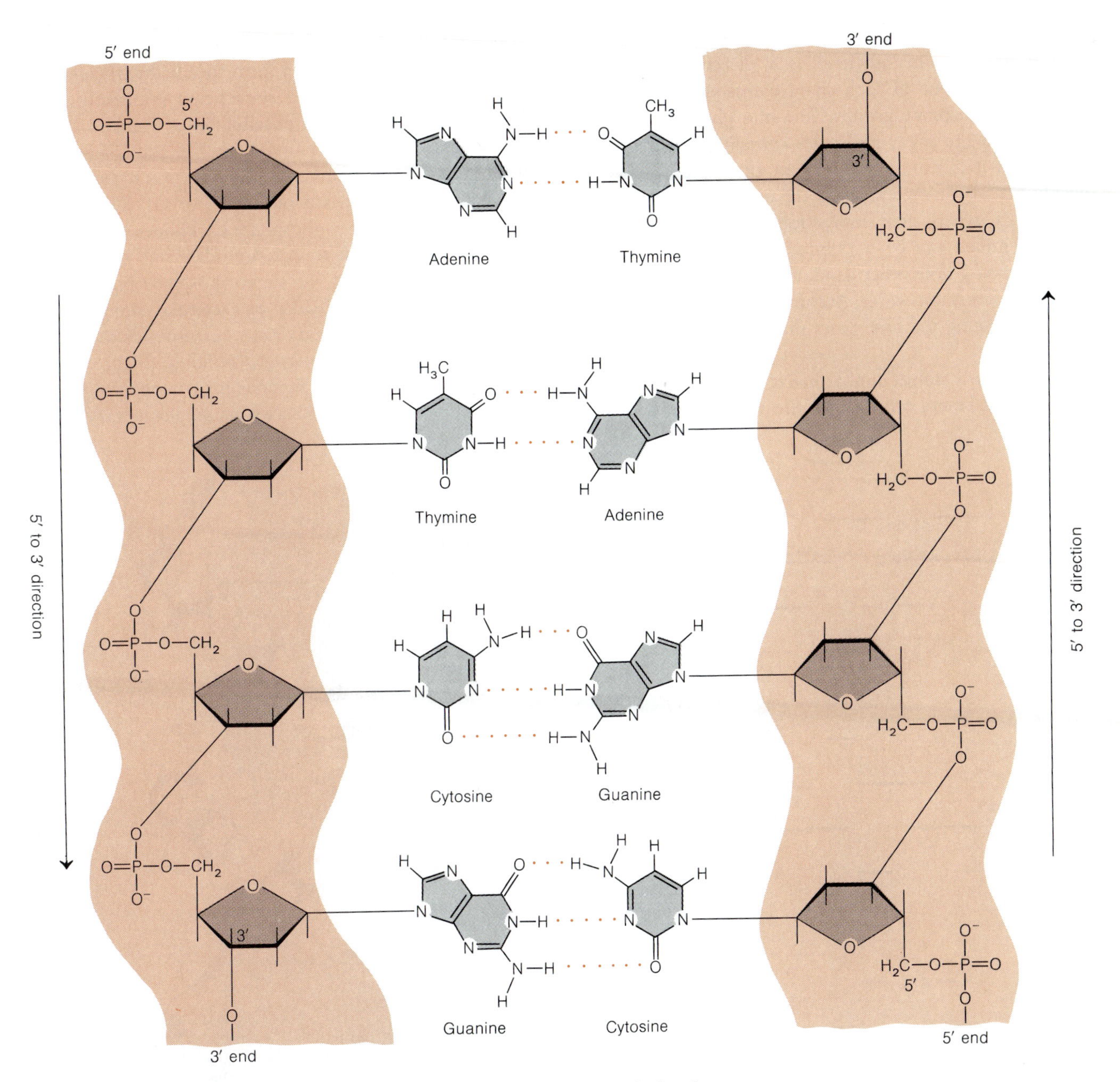

Figure 3-13 Hydrogen Bonding in Nucleic Acid Structure. The hydrogen bonds (colored dots) between adenine and thymine and between cytosine and guanine account for the A-T and C-G base pairing of DNA. If one or both strands were RNA instead, the pairing partner for adenine would be uracil (U).

A DNA Molecule Is a Double-Stranded Helix

One of the most significant biological advances of the twentieth century came in 1953 in a two-page note in the scientific journal *Nature*. In the note, Francis Crick and James Watson postulated a double-stranded helical structure for DNA—the now-famous **double helix**—that not only accounted for the known physical and chemical properties of DNA but also suggested a mechanism for replication of the structure. Some highlights of this exciting chapter in the history of contemporary biology are related in the box on pp. 56–57.

Essentially, the double helix consists of two complementary chains of DNA twisted together around a common axis to form a structure that resembles a circular staircase (Figure 3-14). The two chains are oriented in opposite directions along the helix, one running in the $5' \rightarrow 3'$ direction and the other in the $3' \rightarrow 5'$ direction (Figure 3-13). The backbone of each chain consists of alternating sugar and phosphate groups. The phosphate groups are charged and the sugar molecules contain polar hydroxyl groups, so it is not surprising that the sugar-phosphate backbones of the two strands are on the outside of the DNA helix, where their interaction with the surrounding aqueous milieu can be maximized. The pyrimidine and purine bases, on the other hand, are aromatic compounds with less affinity for water. Accordingly, they are oriented inward, forming the base pairs that hold the two chains together.

To form a stable double helix, the two component strands must be not only *antiparallel* (running in opposite directions) but also *complementary*. By this we mean that each base in one strand can form specific hydrogen bonds with the base in the other strand that is directly across from it. From the pairing possibilities of Figure 3-13, this means that A's must be paired with T's, and G's with C's. In both cases one member of the pair is a pyrimidine (T or C) and the other is a purine (A or G). The distance between the two sugar-phosphate backbones in the double helix is just sufficient to accommodate one of each kind of base. If we envision the sugar-phosphate backbones of the two strands as the sides of a circular staircase, then each step or rung of the stairway corresponds to a pair of bases held in place by hydrogen bonding (Figure 3-14).

RNA structure also depends in part on base pairing, but this pairing is usually between complementary regions

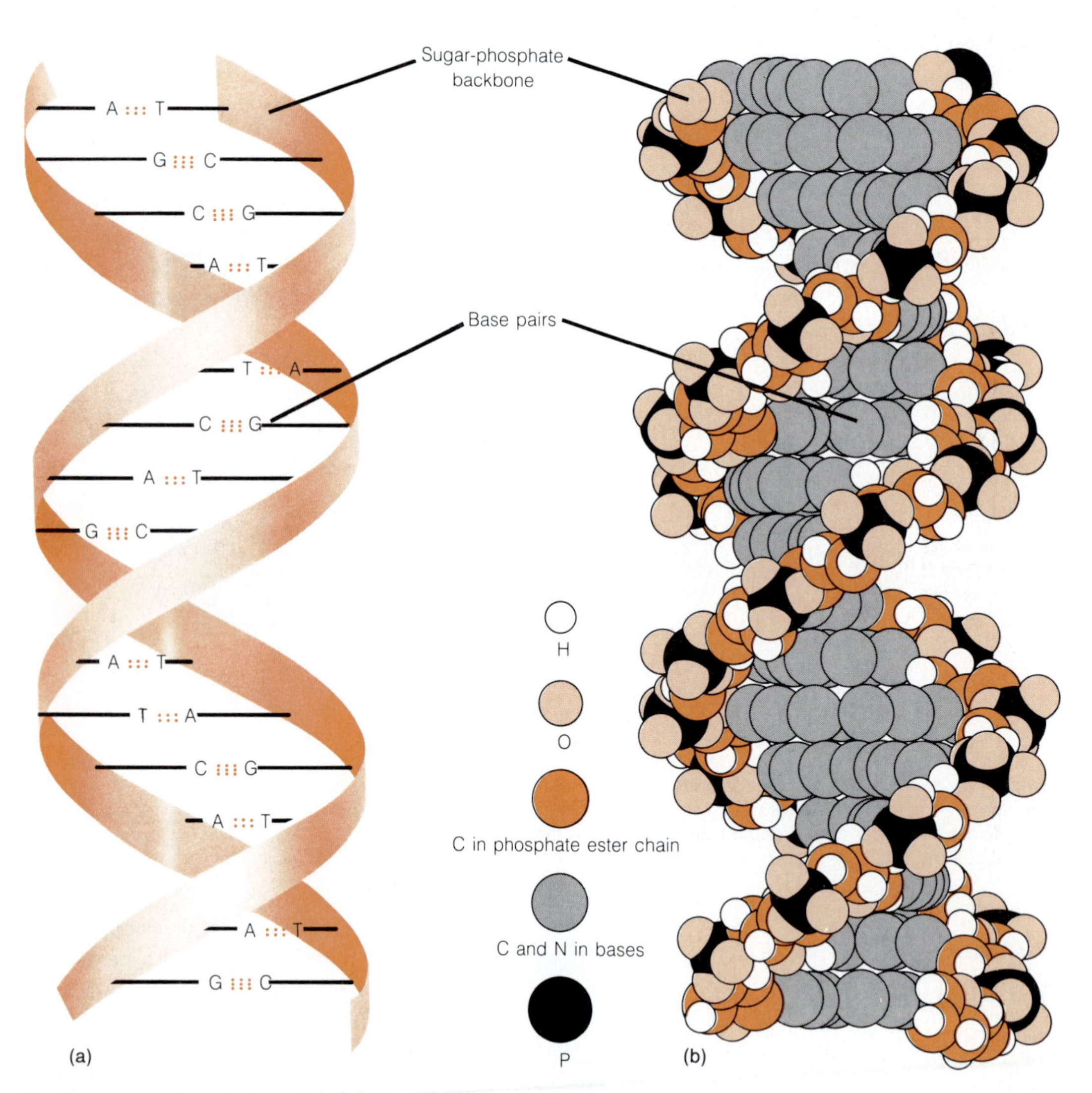

Figure 3-14 The Structure of Double-Stranded DNA. (a) A schematic representation of the double-helical structure of DNA. The continuously turning strips represent the sugar-phosphate backbones of the molecule, while the horizontal bars represent paired bases of the two strands. (b) Space-filling model of the DNA double helix.

within the same strand and is much less extensive than the interstrand pairing of the DNA duplex. Of the various RNA species, secondary and tertiary structures are well understood only for the tRNA molecules, as we will see when we get to Chapter 16.

Polysaccharides

The next class of macromolecules to be considered is the **polysaccharides.** Unlike proteins and nucleic acids, polysaccharides play no informational role in the cell. In fact, they usually consist of a single kind of repeating unit, or sometimes a strictly alternating pattern of two kinds. As noted earlier, the major polysaccharides, at least in higher organisms, are the storage polysaccharides starch and glycogen and the structural polysaccharide cellulose. Each of these polymers contains the six-carbon sugar glucose as its single repeating unit, but they differ both in the nature of the bond between successive glucose units and in the presence and extent of side branches on the chains.

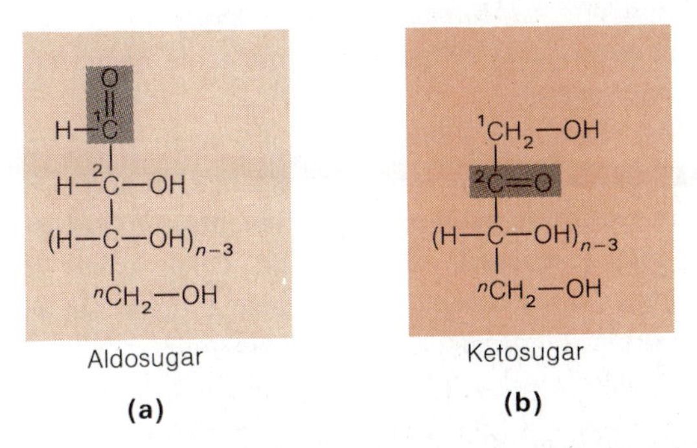

Figure 3-15 Structures of Monosaccharides. (a) Aldosugars have a carbonyl group (gray) on carbon atom 1. (b) Ketosugars have a carbonyl group (gray) on carbon atom 2. The number of carbon atoms varies from three to seven.

The Monomers Are Monosaccharides

The repeating units of polysaccharides are simple sugars called **monosaccharides** (from the Greek *mono* meaning "single" and *saccharide* meaning "sugar"). A sugar can be defined as an aldehyde or ketone that has two or more hydroxyl groups. Thus there are two categories of sugars, the **aldosugars,** with a terminal carbonyl group (Figure 3-15a), and the **ketosugars,** with an internal carbonyl group (Figure 3-15b). Within these categories, sugars are named generically according to the number of carbon atoms they contain. Most sugars have between three and seven carbon atoms and are therefore classified as a *triose* (three carbons), a *tetrose* (four carbons), a *pentose* (five carbons), a *hexose* (six carbons), or a *heptose* (seven carbons). We have already encountered two pentoses—the ribose of RNA and the deoxyribose of DNA.

The single most common sugar in the biological world is the aldohexose D-glucose, which is represented by the formula $C_6H_{12}O_6$ and has the structure shown in Figure 3-16. (The formula $C_nH_{2n}O_n$ is characteristic of sugars and gave rise to the general term **carbohydrate,** since compounds of this sort were originally thought of as "hydrates of carbon"—$C_n(H_2O)_n$. The term persists, although in no sense can a carbohydrate be thought of as hydrated carbon.) In keeping with the general rule for numbering carbon atoms in organic molecules, the carbons of glucose are numbered beginning with the most oxidized end of the molecule, the aldehyde group. Notice that glucose has four asymmetric carbon atoms (carbon atoms 2, 3, 4, and 5). There are therefore 2^4 different possible stereoisomers of the aldosugar $C_6H_{12}O_6$, but we will concern ourselves only

(a) Straight-chain form

(b) Ring form

Fischer projection

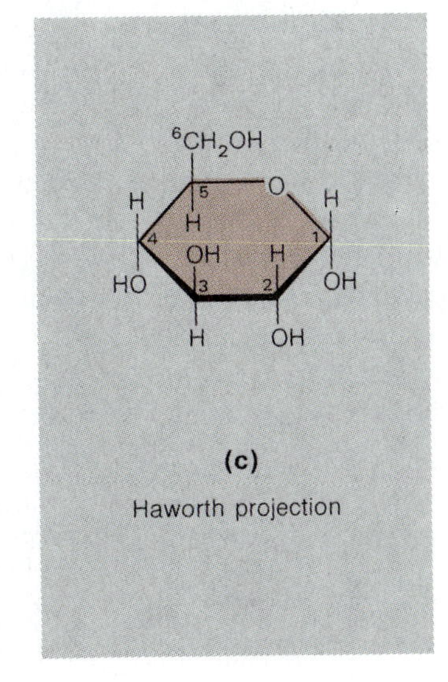

Figure 3-16 The Structure of Glucose. The glucose molecule can be represented by Fischer projections of (a) the straight-chain form or (b) the ring form of the molecule, as well as by (c) the Haworth projection of the ring form. In the Fischer projections, the —H and —OH groups are intended to be projecting slightly out of the plane of the paper. In the Haworth projection, carbon atoms 2 and 3 are intended to be jutting out of the plane of the paper, and carbon atoms 5 and 6 are behind the plane of the paper. The —H and —OH groups then project upward or downward, as indicated.

On the Trail of the Double Helix

"I have never seen Francis Crick in a modest mood. Perhaps in other company he is that way, but I have never had reason so to judge him." With this typically irreverent observation as an introduction, James Watson goes on to describe in a very personal and highly entertaining way the events that eventually led to the discovery of the structure of DNA. The account, published in 1968 under the title *The Double Helix*, is still fascinating reading for the personal, unvarnished insights it provides into how an immense scientific discovery came about. Commenting on his reasons for writing the book, Watson observes in the preface that "there remains general ignorance about how science is 'done.' That is not to say that all science is done in the manner described here. This is far from the case, for styles of scientific research vary almost as much as human personalities. On the other hand, I do not believe that the way DNA came out constitutes an odd exception to a scientific world complicated by the contradictory pulls of ambition and the sense of fair play."

As portrayed in Watson's account, Crick and Watson are about as different from each other in nature and background as they could be. But there was one thing they shared, and that was an unconventional but highly productive way of "doing" science. They did little actual experimentation on DNA, choosing instead to draw heavily upon the research findings of others and to bring their own considerable ingenuities to bear building models and exercising astute insights and hunches. And out of it all emerged, in a relatively short time, an understanding of the double-helical structure of DNA that has come to rank as one of the major scientific events of this century.

To appreciate their findings and their brilliance, we must first understand the setting in which Watson and Crick worked. The early 1950s was an exciting time in biology. It had been only a few years since Avery, MacLeod, and McCarty had published evidence on the genetic transformation of bacteria, but the work of Hershey and Chase that confirmed DNA as the genetic material had not yet appeared in print. Meanwhile, at Columbia University, Erwin Chargaff's careful chemical analyses had revealed that although the relative proportions of the four bases—A, T, C, and G—varied greatly from one species to the next, it was always the same for all members of a single species. Even more puzzling and portentous was Chargaff's second finding: For a given species, A and T always occurred in the same proportions, and so did G and C (that is, A = T and C = G).

The most important clues came from the work of Maurice Wilkins and Rosalind Franklin at King's College in London. Wilkins and Franklin were using the technique of X-ray diffraction to study DNA structure and took a rather dim view of Watson and Crick's strategy of model-building. X-ray diffraction is a useful tool for detecting regularly occurring structural elements in a crystalline substance, since any structural feature that repeats at some fixed interval in the crystal contributes in a characteristic way to the diffraction pattern that is obtained. From Franklin's painstaking analysis of the diffraction patterns of DNA, it became clear that the molecule was long and thin, with some structural element being repeated every 0.34 nm and another being repeated every 3.4 nm. Even more intriguing, the molecule appeared to be some sort of helix.

This stirred the imaginations of Watson and Crick, because they had heard only recently of Linus Pauling's α helical structure for proteins. Working with models of the bases cut from stiff cardboard, Watson and Crick came to the momentous insight that DNA was also a helix, but with an all-important difference: It was a *double* helix, with hydrogen-bonded pairing of purines and pyrimidines. The actual discovery is best recounted in Watson's own words:

> When I got to our still empty office the following morning, I quickly cleared away the papers from my desk top so that I would have a large, flat surface on which to form pairs of bases held together by hydrogen bonds. Though I initially went back to my like-with-like prejudices, I saw all too well that they led nowhere. When Jerry [Donohue, an American crystallographer working in the same laboratory] came in I looked up, saw that it was not Francis, and began shifting the bases in and out of various other pairing possibilities. Suddenly I became aware that an adenine-thymine pair held together by two hydrogen bonds was identical in shape to a guanine-cytosine pair held together by at least two hydrogen bonds. All the hydrogen bonds seemed to form naturally; no fudging was required to make the two types of base pairs identical in shape. Quickly I called Jerry over to ask him whether this time he had any objections to my new base pairs.
>
> When he said no, my morale skyrocketed, for I suspected that we now had the answer to the riddle of why the number of purine residues exactly equaled the number of pyrimidine residues. Two irregular sequences of bases could be regularly packed in the center of a helix if a purine always hydrogen-bonded to a pyrimidine. Furthermore, the hydrogen bonding requirement meant that adenine would always pair with thymine, while guanine could pair only with cytosine. Chargaff's rules then suddenly stood out as a consequence of a double-helical structure for DNA.

Even more exciting, this type of double helix suggested a replication scheme much more satisfactory than my briefly considered like-with-like pairing. Always pairing adenine with thymine and guanine with cytosine meant that the base sequences of the two intertwined chains were complementary to each other. Given the base sequence of one chain, that of its partner was automatically determined. Conceptually, it was thus very easy to visualize how a single chain could be the template for the synthesis of a chain with the complementary sequence.

Upon his arrival Francis did not get more than halfway through the door before I let loose that the answer to everything was in our hands. Though as a matter of principle he maintained skepticism for a few moments, the similarly shaped A-T and G-C pairs had their expected impact. His quickly pushing the bases together in a number of different ways did not reveal any other way to satisfy Chargaff's rules. A few minutes later he spotted the fact that the two glycosidic bonds (joining base and sugar) of each base pair were systematically related by a diad axis perpendicular to the helical axis. Thus, both pairs could be flip-flopped over and still have their glycosidic bonds facing in the same direction. This had the important consequence that a given chain could contain both purines and pyrimidines. At the same time, it strongly suggested that the backbones of the two chains must run in opposite directions.

The question then became whether the A-T and G-C base pairs would easily fit the backbone configuration devised during the previous two weeks. At first glance this looked like a good bet, since I had left free in the center a large vacant area for the bases. However, we both knew that we would not be home until a complete model was built in which all the stereochemical contacts were satisfactory. There was also the obvious fact that the implications of its existence were far too important to risk crying wolf. Thus I felt slightly queasy when at lunch Francis winged into the Eagle to tell everyone within hearing distance that we had found the secret of life.*

The rest is history. Shortly thereafter, the prestigious journal *Nature* carried an unpretentious two-page article entitled simply "Molecular Structure of Nucleic Acids: A Structure for Deoxyribose Nucleic Acid," by James Watson and Francis Crick. Though modest in length, that paper has had far-reaching implications, for the double-stranded model that Watson and Crick worked out in 1953 has proved to be correct in all its essential details, unleashing a revolution in the field of biology.

* Excerpted from *The Double Helix*, pp. 194–197. Copyright © 1968 James D. Watson. Reprinted with the permission of the author and Atheneum Publishers, Inc.

with D-glucose, which is the most stable of the 16 isomers.

Figure 3-16a shows D-glucose as it appears in what chemists call a **Fischer projection,** with the —H and —OH groups intended to be projecting slightly out of the plane of the paper. This structure indicates that glucose is a linear molecule, and it is often a useful representation of glucose for pedagogic purposes. In reality, however, glucose exists in the cell in a dynamic equilibrium between the linear (or open-chain) configuration of Figure 3-16a and the ring form shown in Figure 3-16b. The ring form is the predominant structure because it is energetically more stable and therefore favored. The ring form results from addition of the hydroxyl group on carbon atom 5 across the carbonyl group of carbon atom 1. Although the juxtaposition of carbon atoms 1 and 5 required for ring formation seems unlikely from the Fischer projection, it is actually favored by the tetrahedral nature of each carbon atom in the chain.

A more satisfactory representation of glucose is that shown in Figure 3-16c. The advantage of this **Haworth projection** is that it at least suggests the spatial relationship of different parts of the molecule and makes the spontaneous formation of a bond between carbon atoms 1 and 5 appear more likely. Any of the three representations of glucose shown in Figure 3-16 is valid, but the Haworth projection is preferred because it indicates both the ring form and the spatial relationship of the carbon atoms.

Notice that formation of the ring structure results in the generation of one of two alternative forms of the molecule, depending on the spatial orientation of the hydroxyl group on carbon atom 1. These alternative forms of glucose are designated as α and β. As shown in Figure 3-17, α-D-glucose has the hydroxyl group on carbon atom 1 pointing downward in the Haworth projection, and β-D-glucose has the hydroxyl group on carbon atom 1 pointing upward. Starch and glycogen both have α-D-glucose as their repeating unit, whereas cellulose consists of strings of β-D-glucose.

In addition to the free monosaccharide and the long-chain polysaccharides, glucose also occurs in **disaccharides,** which consist of two monosaccharide units linked cova-

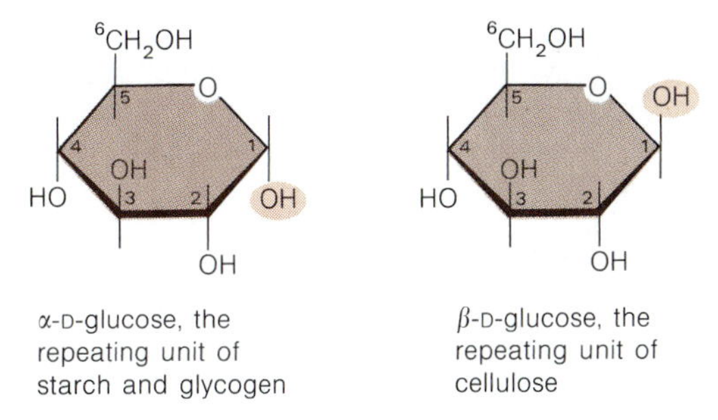

Figure 3-17 The Ring Forms of D-Glucose. The hydroxyl group on carbon atom 1 points downward in the α form and upward in the β form.

lently. Three common disaccharides are shown in Figure 3-18. *Maltose* consists of two glucose units linked together, whereas *lactose* (milk sugar) contains a glucose linked to a galactose and *sucrose* (common table sugar) has a glucose linked to a fructose. Both galactose and fructose will be discussed in more detail in Chapter 10, where the chemistry and metabolism of several sugars are considered.

Each of these disaccharides is formed by a condensation reaction in which two monosaccharides are linked together by the elimination of water. The resulting bond is called a **glycosidic bond** and is characteristic of linkages between sugars. In the case of maltose, both of the constituent glucose molecules are in the α form, and the glycosidic bond forms between carbon atom 1 of one glucose and carbon atom 4 of the other. This is called an *α glycosidic bond* because it involves a carbon atom 1 with its hydroxyl group in the α configuration. Lactose, on the other hand, is characterized by a *β glycosidic bond* because the hydroxyl group on carbon atom 1 of the galactose is in the β configuration. The distinction between α and β glycosidic bonds becomes especially critical when we get to the polysaccharides because both the three-dimensional configuration and the biological role of the polymer are dependent on the nature of the bond between the repeating monosaccharide units.

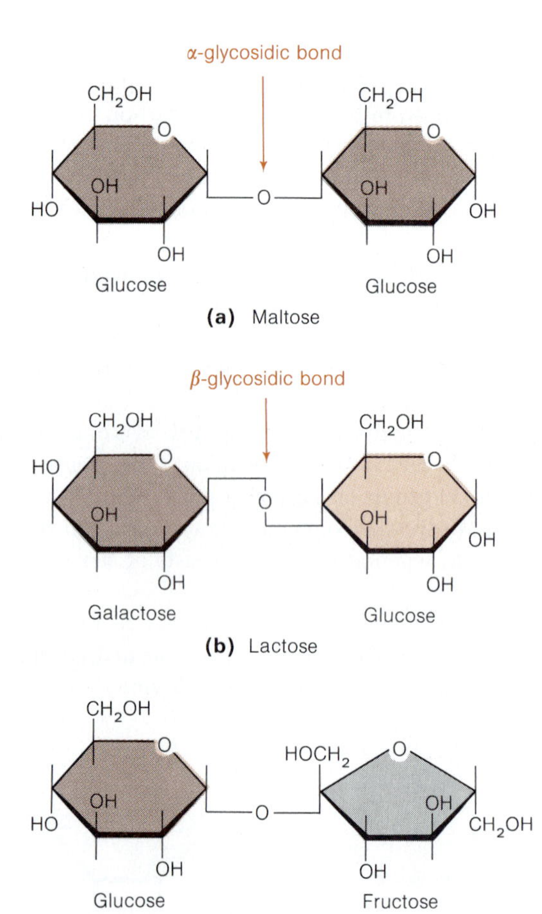

Figure 3-18 Some Common Disaccharides. (a) Maltose (malt sugar) consists of a molecule of α-D-glucose linked to a second glucose. (b) Lactose (milk sugar) consists of a molecule of β-D-galactose linked to a molecule of glucose. (c) Sucrose (common table sugar) consists of a molecule of α-D-glucose linked to a molecule of fructose, another six-carbon sugar.

The Polymers Are Storage and Structural Polysaccharides

Polysaccharides serve both storage and structural functions in cells. The most familiar *storage polysaccharides* are the **starch** of plant cells (Figure 3-19a) and the **glycogen** of animal cells (Figure 3-19b). Both of these polymers consist of α-D-glucose units linked together by α glycosidic bonds. In addition to α(1 → 4) bonds that link carbon atoms 1 and 4 of adjacent glucose units, these polysaccharides contain occasional α(1 → 6) linkages along the backbone, giving rise to side chains (Figure 3-19c). Storage polysaccharides can therefore be branched-chain polymers, depending on the presence of α(1 → 6) linkages.

Glycogen is highly branched, with α(1 → 6) linkages occurring every 8 to 10 glucose units along the backbone and giving rise to short side chains of about 8 to 12 glucose units (Figure 3-19b). Glycogen is stored mainly in the liver and in muscle tissue. In the liver it is used as a source of glucose to maintain blood sugar levels, whereas in muscle it serves as a fuel source to generate the ATP needed for muscle contraction.

Starch occurs both as unbranched **amylose** and as branched **amylopectin.** Like glycogen, amylopectin has α(1 → 6) branches, but these occur less frequently along the backbone (once every 12 to 25 glucose units) and give rise to longer side chains (lengths of 20 to 25 glucose units are common). Starch deposits are usually about 10–30% amylose and 70–90% amylopectin. Starch is stored in plant cells as *starch grains* within the plastids—either the *chloroplasts* that are the sites of carbon fixation and sugar synthesis in photosynthetic tissue (Figure 3-19a) or the *amyloplasts* that are specialized plastids for starch storage. The potato tuber, for example, is filled with starch-laden amyloplasts.

The best-known example of a *structural polysaccharide* is the **cellulose** found in plant cell walls (Figure 3-20). Cellulose is an important polymer quantitatively—more than half of the carbon in higher plants is present in cellulose! Like starch and glycogen, cellulose is also a polymer of glucose, but the repeating monomer is β-D-glucose and the linkage is therefore β(1 → 4). This linkage has

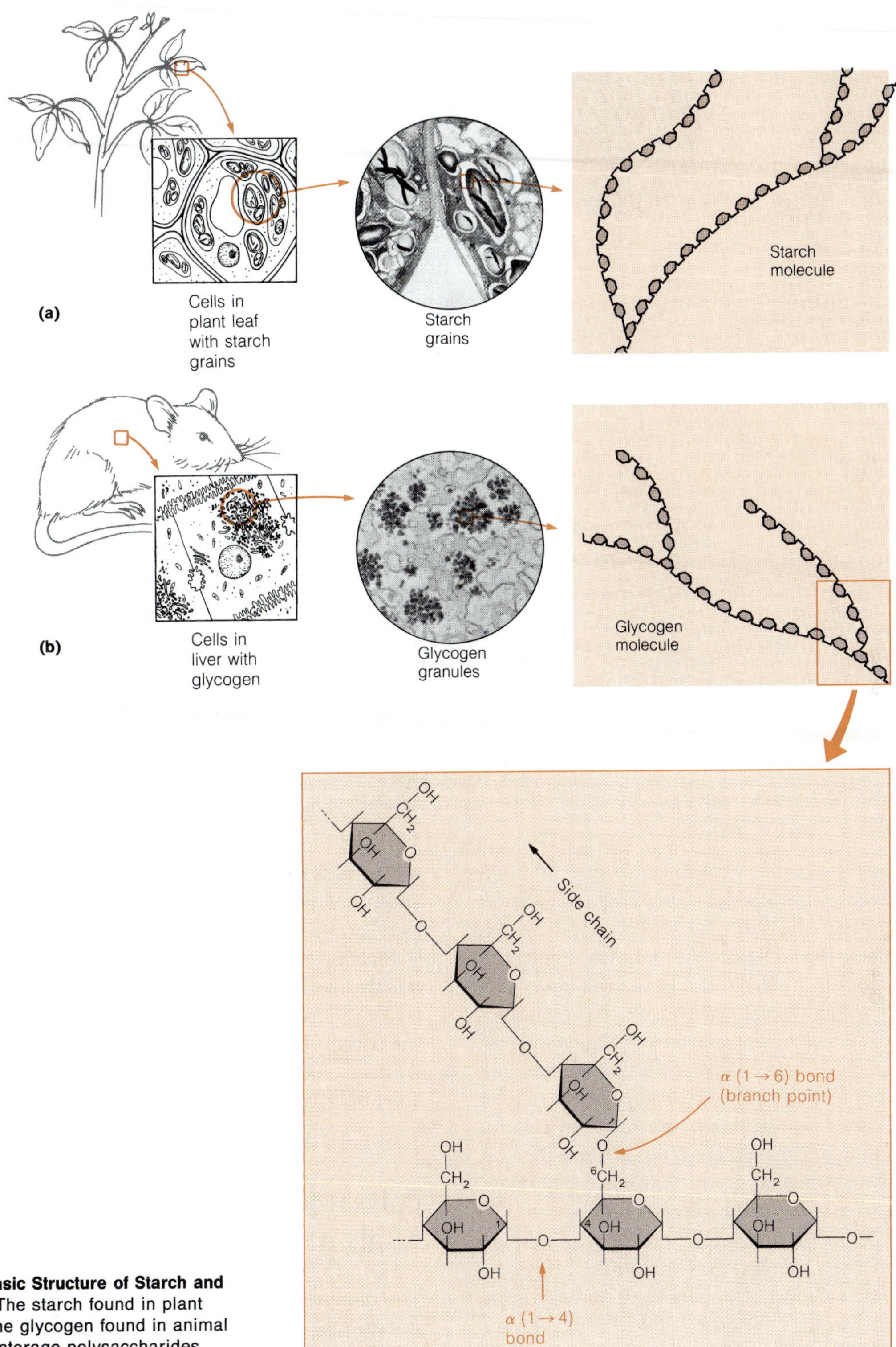

Figure 3-19 Basic Structure of Starch and Glycogen. (a) The starch found in plant cells and (b) the glycogen found in animal cells are both storage polysaccharides composed of linear chains of α-D-glucose units with occasional branch points. (c) The straight-chain portion of the starch or glycogen molecule consists of glucose units linked together by α(1 → 4) glycosidic bonds. Branch chains originate at α(1 → 6) glycosidic bonds.

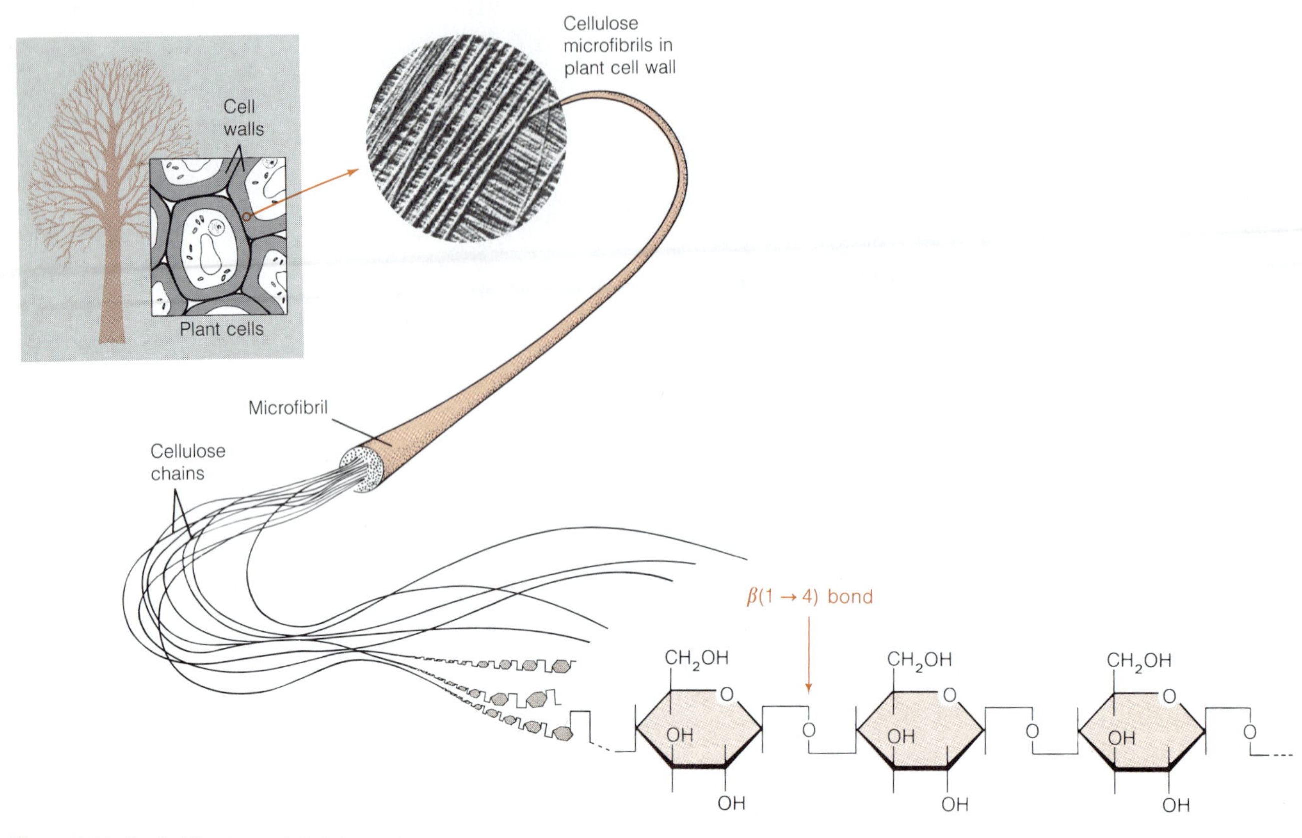

Figure 3-20 Basic Structure of Cellulose. Cellulose consists of long, unbranched chains of β-D-glucose units linked together by β(1 → 4) glycosidic bonds. Many such chains associate laterally to form microfibrils. Individual microfibrils can be seen in the electron micrograph of a plant cell wall that is shown here.

structural consequences that we will get to shortly, but it also has nutritional implications. Mammals do not possess an enzyme that can hydrolyze a β(1 → 4) bond; therefore, mammals cannot utilize cellulose as food. As a result, you can digest potatoes (starch) but not grass (cellulose). Animals such as cows and sheep might seem to be exceptions, since they clearly do eat grass and similar plant products. However, they cannot cleave β glycosidic bonds either, but depend on the population of bacteria and protozoa in their rumen (part of their compound stomach) to do this for them. The microorganisms eat the cellulose, and the host animal then obtains the end products of microbial digestion, now in a form the animal can use.

Although β(1 → 4)-linked cellulose is quantitatively the most significant structural polysaccharide, others are also known. The celluloses of fungal cell walls, for example, contain either β(1 → 4) or β(1 → 3) linkages, depending on the species. The cell wall of many bacteria is somewhat more complex, since it contains two kinds of sugars, ***N*-acetylglucosamine** (NAG) and ***N*-acetylmuramic acid** (NAM). As shown in Figure 3-21a, NAG and NAM are derivatives of *glucosamine,* a glucose molecule with the hydroxyl group on carbon atom 2 replaced by an amino group. NAG is formed by acetylation of the amino group, and NAM requires the further addition of a three-carbon lactyl group to carbon atom 3. The cell wall polysaccharide is then formed by the linking of NAM and NAG in a strictly alternating sequence with β(1 → 4) bonds (Figure 3-21b). Shown in Figure 3-21c is the structure of yet another structural polysaccharide, the **chitin** found in insect exoskeletons and crustacean shells. Chitin consists of NAG units only, joined by β(1 → 4) bonds.

Polysaccharide Structure Depends on the Kinds of Glycosidic Bonds Involved

The distinction between the α and β glycosidic bonds of storage and structural polysaccharides has more than just nutritional significance. Because of the difference in linkages and therefore in the spatial relationship of successive glucose units to each other, the two classes of polysaccharides differ markedly in secondary structure. The helical shape already seen to be a characteristic of both proteins and nucleic acids is also found in polysaccharides. Both

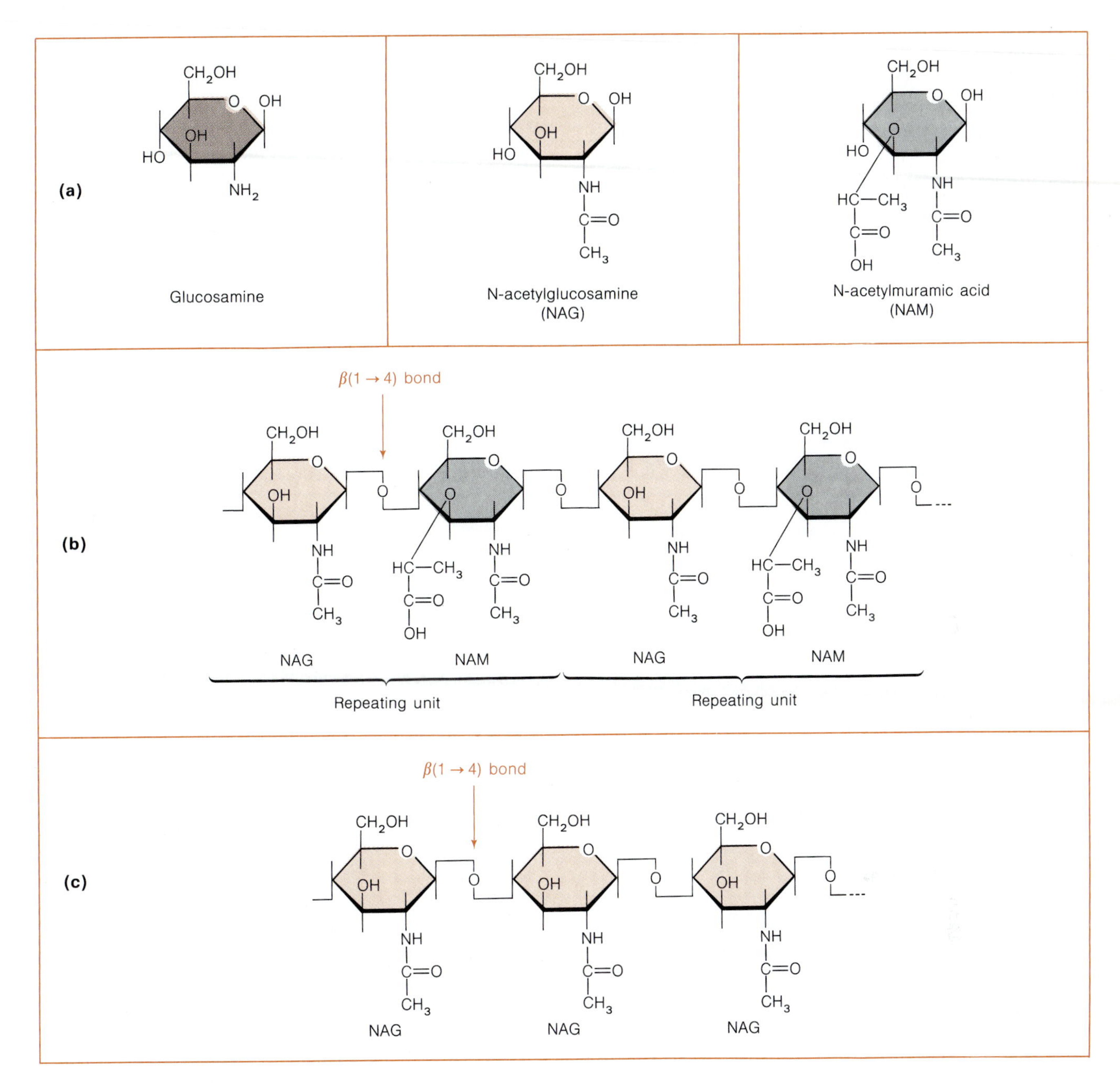

Figure 3-21 Polysaccharides of Bacterial Cell Walls and Insect Exoskeletons. (a) The subunits glucosamine, *N*-acetylglucosamine (NAG), and *N*-acetylmuramic acid (NAM). (b) A bacterial cell wall polysaccharide, consisting of alternating NAG and NAM units. (c) The polysaccharide chitin found in insect exoskeletons, with NAG as its single repeating unit.

starch and glycogen coil spontaneously into loose helices, but often the structure is not highly ordered because of the numerous side chains of amylopectin and glycogen.

Cellulose, by contrast, forms rigid, linear rods. These, in turn, aggregate laterally into **microfibrils** (Figure 3-20). Microfibrils are about 25 nm in diameter and are composed of about 2000 cellulose chains. Plant and fungal cell walls consist of these rigid microfibrils of cellulose embedded in a **noncellulosic matrix** containing a rather variable mixture of several other polymers (*hemicellulose, pectin,* and *lignin*) and a protein called *extensin* that occurs only in the cell wall. Cell walls have been rather aptly compared to reinforced concrete, in which steel rods are embedded in the cement before it hardens to confer added strength. For the cell wall, the cellulose microfibrils are the rods and the noncellulosic matrix is the "cement."

Lipids

Strictly speaking, **lipids** do not qualify for inclusion in this chapter, since they are not macromolecules and they are not formed by the kind of stepwise polymerization that gives rise to proteins, nucleic acids, and polysaccharides. Yet any discussion of cellular structure and chemical components would be incomplete without reference to this important group of molecules. Their inclusion here also seems reasonable in light of their frequent association with the macromolecules we have already discussed, especially proteins.

Lipids constitute a rather heterogeneous category of cellular components that resemble one another more in their solubility properties than in their chemical structures. The distinguishing feature of lipids is their hydrophobic nature. They have little, if any, affinity for water but are readily soluble in nonpolar solvents such as chloroform or ether. Accordingly, we can expect to find that they are rich in nonpolar hydrocarbon regions and have relatively few polar groups. Some lipids, however, are amphipathic, having both a polar and a nonpolar region. As we have already seen, this characteristic has profound implications for membrane structure. Because of their chemical heterogeneity, the lipids can be conveniently placed in four separate categories: triglycerides, phospholipids, sphingolipids, and steroids.

Triglycerides Are Storage Lipids

The **triglycerides,** or true fats, consist of a glycerol molecule with three fatty acids linked to it by ester bonds. Their primary purpose in cells is to store energy, which makes

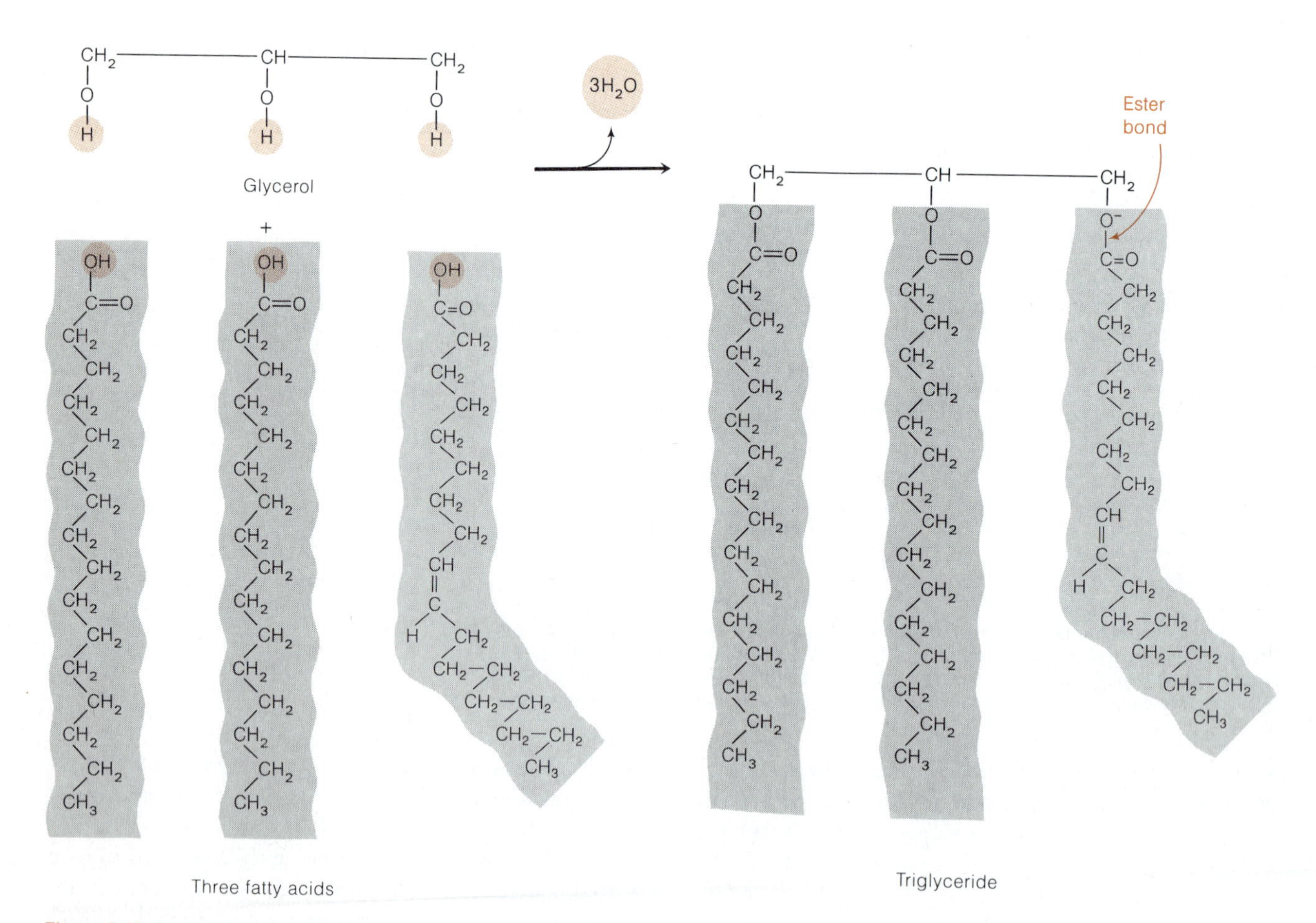

Figure 3-22 Formation of a Triglyceride. Triglycerides are synthesized by stepwise formation of ester bonds between three fatty acids and the three hydroxyl groups of glycerol.

them of special interest in our upcoming discussion of energy metabolism in Part Three. As shown in Figure 3-22, **glycerol** is a three-carbon alcohol with a hydroxyl group on each carbon. The **fatty acids** linked to these hydroxyl groups are generally long, unbranched hydrocarbon chains with a carboxyl group at one end. The fatty acid molecule is therefore amphipathic; the carboxyl group renders one end (often called the "head") polar, whereas the hydrocarbon "tail" is nonpolar. Fatty acids contain a variable, but usually even, number of carbon atoms. The usual range is from 12 to 24 carbon atoms per chain, with 16- and 18-carbon fatty acids especially common.

Table 3-2 summarizes the nomenclature of fatty acid chain length. Even numbers of carbon atoms are greatly favored because of the mode of fatty acid synthesis. Each molecule is generated by the stepwise addition of two-carbon units as acetyl-coenzyme A, a carrier of acyl groups that we will consider in more detail in Chapter 11. Once added, each new two-carbon increment is then reduced to the hydrocarbon level.

Because they are highly reduced, fats yield a great deal of energy on oxidation and are therefore compact and efficient forms of energy storage. We will see how to quantify the efficiency of storing energy as fat rather than as carbohydrate when we get to Part Three. For the present, simply note that a gram of fat contains more than twice as much usable energy as a gram of sugar or polysaccharide.

Table 3-2 also shows the variability in fatty acid structure due to the presence of double bonds between carbons. Fatty acids without double bonds are referred to as **saturated fatty acids** because every carbon atom in the chain has the maximum number of hydrogen atoms attached to it. The general formula for a saturated fatty acid with n carbon atoms is $C_nH_{2n}O_2$. By contrast, **unsaturated fatty acids** contain one or a few double bonds. The presence of such sites of unsaturation affects the shape of the molecule and therefore the kinds of structures of which it can be a part. Saturated fatty acids have long straight tails that pack together well, whereas each double bond results in a bend or kink in the molecule that prevents tight packing (Figure 3-23).

Fatty acids are linked to glycerol by **ester bonds** (Figure 3-22; note again the principle of bond formation by removal of water). *Monoglycerides* contain a single fatty acid, *diglycerides* have two, and *triglycerides* have each of the three hydroxyl groups of glycerol esterified to a fatty acid. The three fatty acids of a given triglyceride need not be identical; they can vary in both chain length and degree of saturation.

Triglycerides containing a preponderance of saturated fatty acids are usually solid or semisolid at room temperature and are called **fats.** Fats are prominent in the bodies of animals, as evidenced by the fat that you buy with most cuts of meat, by the large quantity of lard that is obtained as a by-product of the meat-packing industry, and by the widespread concern people have that they are "getting fat." In plants, most triglycerides are liquid, as the term **vegetable oil** suggests. The fatty acids of oils are predominantly unsaturated and are presumably liquid because the kinks introduced at double bonds prevent orderly packing of the molecules. Soybean oil and corn oil are two familiar examples of vegetable oil. Vegetable oils can be converted into solid products such as margarine and shortening by partial hydrogenation (saturation) of the double bonds, a process explored further in Problem 3-14 at the end of the chapter.

Table 3-2 Nomenclature of the Fatty Acids

Number of Carbons	Number of Double Bonds	Common Name	Systematic Name	Formula
12	0	Laurate	*n*-Dodecanoate	$CH_3(CH_2)_{10}COO^-$
14	0	Myristate	*n*-Tetradecanoate	$CH_3(CH_2)_{12}COO^-$
16	0	Palmitate	*n*-Hexadecanoate	$CH_3(CH_2)_{14}COO^-$
18	0	Stearate	*n*-Octadecanoate	$CH_3(CH_2)_{16}COO^-$
20	0	Arachidate	*n*-Eicosanoate	$CH_3(CH_2)_{18}COO^-$
22	0	Behenate	*n*-Docosanoate	$CH_3(CH_2)_{20}COO^-$
24	0	Lignocerate	*n*-Tetracosanoate	$CH_3(CH_2)_{22}COO^-$
16	1	Palmitoleate	*cis*-Δ^9-Hexadecenoate	$CH_3(CH_2)_5CH{=}CH(CH_2)_7COO^-$
18	1	Oleate	*cis*-Δ^9-Octadecenoate	$CH_3(CH_2)_7CH{=}CH(CH_2)_7COO^-$
18	2	Linoleate	*cis, cis*-Δ^9,Δ^{12}-Octadecadienoate	$CH_3(CH_2)_4(CH{=}CHCH_2)_2(CH_2)_6COO^-$
18	3	Linolenate	all *cis*-Δ^9, Δ^{12}, Δ^{15}-Octadecatrienoate	$CH_3CH_2(CH{=}CHCH_2)_3(CH_2)_6COO^-$
20	4	Arachidonate	all *cis*-Δ^5, Δ^8, Δ^{11}, Δ^{14}-Eicosatetraenoate	$CH_3(CH_2)_4(CH{=}CHCH_2)_4(CH_2)_2COO^-$

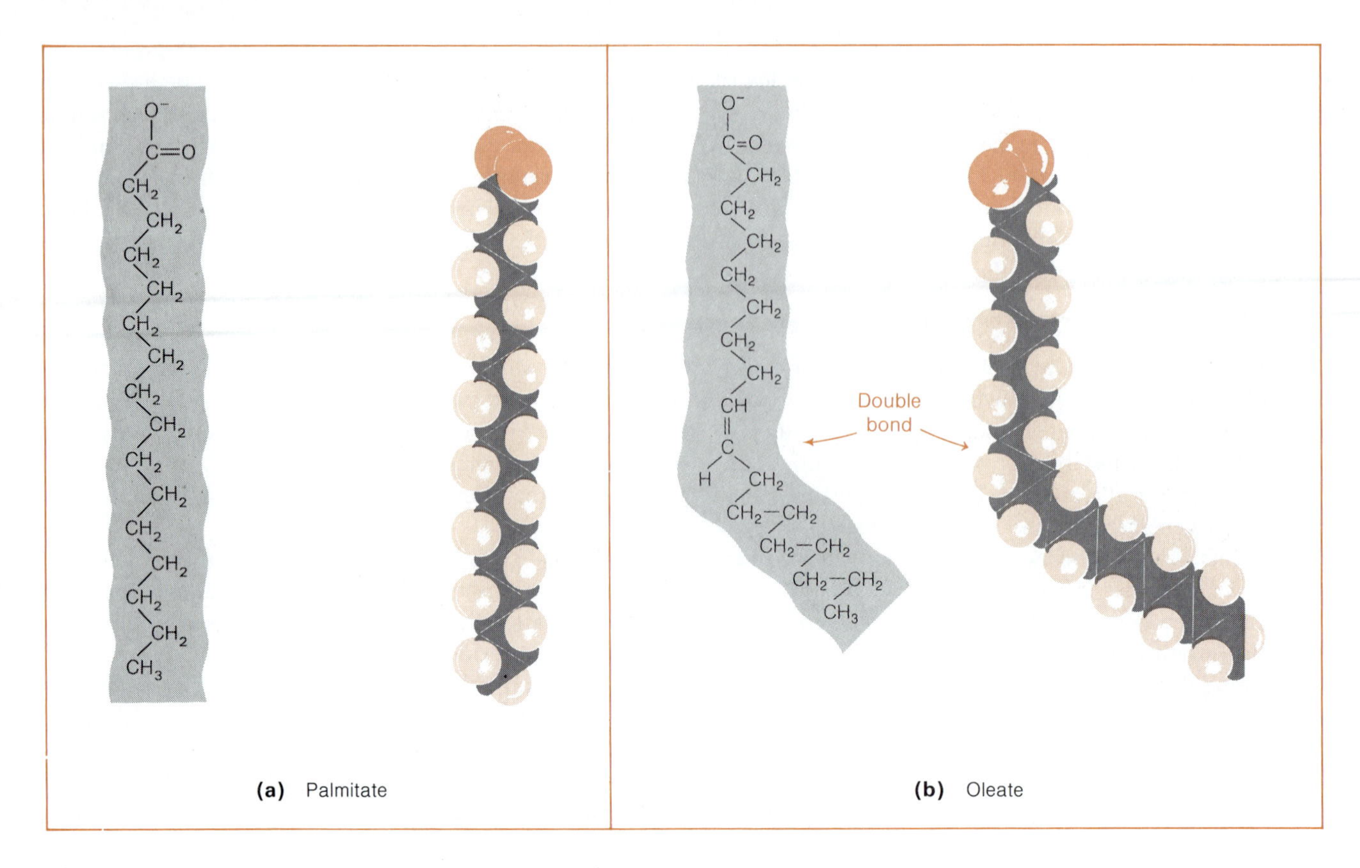

Figure 3-23 Structures of Saturated and Unsaturated Fatty Acids. (a) The saturated 16-carbon fatty acid palmitate. (b) The unsaturated 18-carbon fatty acid oleate. In each case, the structural formula is shown on the left and a space-filling model is shown on the right. The space-filling model is intended to emphasize the overall shape of the molecule. Note the kink that the double bond creates in the oleate molecule.

Phospholipids Are Important in Membrane Structure

Phospholipids are similar to triglycerides in some chemical details but differ strikingly in their properties and their role in the cell. Phospholipids were mentioned in our discussion of membrane structure in Chapter 2 because they are critical to the bilayer structure found in all membranes (recall Figure 2-12).

Membrane phospholipids resemble the fats and oils chemically, since most of them contain a glycerol molecule esterified to two fatty acids. The difference comes at the third hydroxyl group, which in a phospholipid is occupied not by a third fatty acid but by a phosphate group. This basic structure is called a **phosphoglyceride.** The structures of some common phosphoglycerides are shown in Figure 3-24. Most membrane phospholipids are phosphoglycerides. (Exceptions include the sphingomyelins, which will be discussed shortly, and the glycolipids found in plants and some bacteria.)

The basic component of the phosphoglycerides is **phosphatidic acid,** which has just two fatty acids and a phosphate group (Figure 3-24a). Phosphatidic acid is a key intermediate in the synthesis of other phosphoglycerides but is itself not at all prominent in membranes. Instead, membrane phosphoglycerides invariably have, in addition, a small hydrophilic alcohol linked to the phosphate by an ester bond. The alcohol is usually *serine, ethanolamine, choline,* or *inositol* (Figure 3-24b through e). Except for inositol, these alcohols contain an amino group that is protonated and therefore charged at cellular pH. The presence of a negatively charged phosphate and a positively charged amine in juxtaposition makes these phosphoglycerides electrically neutral but highly polar in the head region.

The presence of a highly polar head and two long

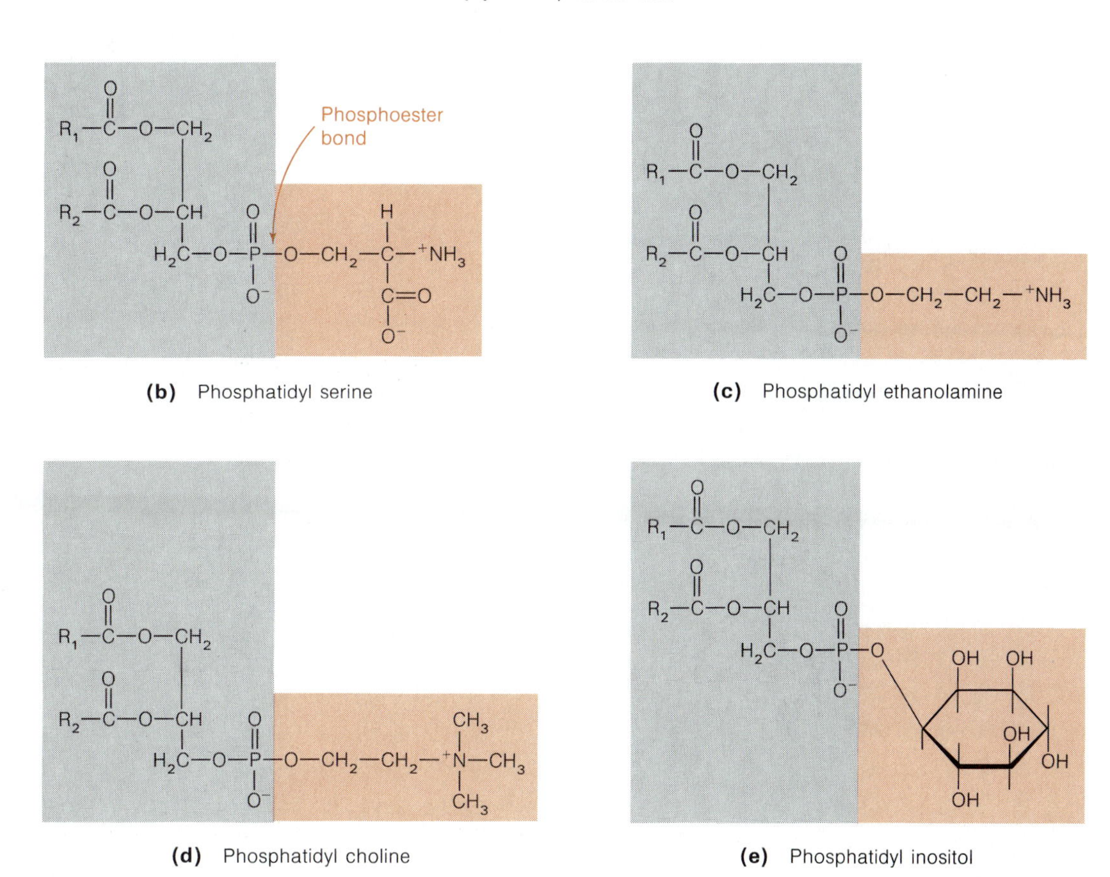

Figure 3-24 Structures of Some Common Phosphoglycerides. The starting point for phosphoglyceride synthesis is (a) phosphatidic acid, to which several different polar alcohols can be added, including (b) serine, (c) ethanolamine, (d) choline, and (e) inositol. In each case, the phosphatidic acid backbone (gray) and the polar alcohol (color) are linked by a phosphoester bond.

nonpolar chains gives the phosphoglycerides the characteristic amphipathic nature that is so critical to membrane structure. As we saw earlier, the fatty acids can vary considerably in both length and the presence and position of sites of unsaturation. In membranes, 16- and 18-carbon fatty acids are most common, and a typical phosphoglyceride molecule is likely to have one saturated and one unsaturated fatty acid. The length and the degree of unsaturation of fatty acid chains in membrane phospholipids profoundly affect membrane fluidity and can, in fact, be regulated by the cell to control this crucial membrane property.

Sphingolipids Are Also Found in Membranes

In addition to the phosphoglycerides, membranes of animal cells contain a class of lipids based not on glycerol but on the amine alcohol **sphingosine.** As shown in Figure 3-25a, sphingosine has a long hydrocarbon chain with a single site of unsaturation near the polar end. Through its amine group, sphingosine can form an amide bond to a long-chain fatty acid. The resulting molecule is called a **ceramide** and consists of a polar region flanked by two long nonpolar tails (Figure 3-25b). Because of their common nonpolar nature, the two tails tend to bend around and associate with each other, giving the molecule a hairpin bend and a shape that approximates that of the phospholipids.

The hydroxyl group on carbon atom 1 of the sphingosine then juts out from what is effectively the head of this hairpin molecule, free to accept a variety of polar groups. Actually, a whole family of **sphingolipids** exists, differing only in the chemical nature of the polar group attached to the hydroxyl group of the ceramide. The **sphingomyelins,** for example, contain phosphorylethanolamine or phosphorylcholine and therefore closely resemble phosphoglycerides in both overall shape and chemical nature of the polar head (Figure 3-25c).

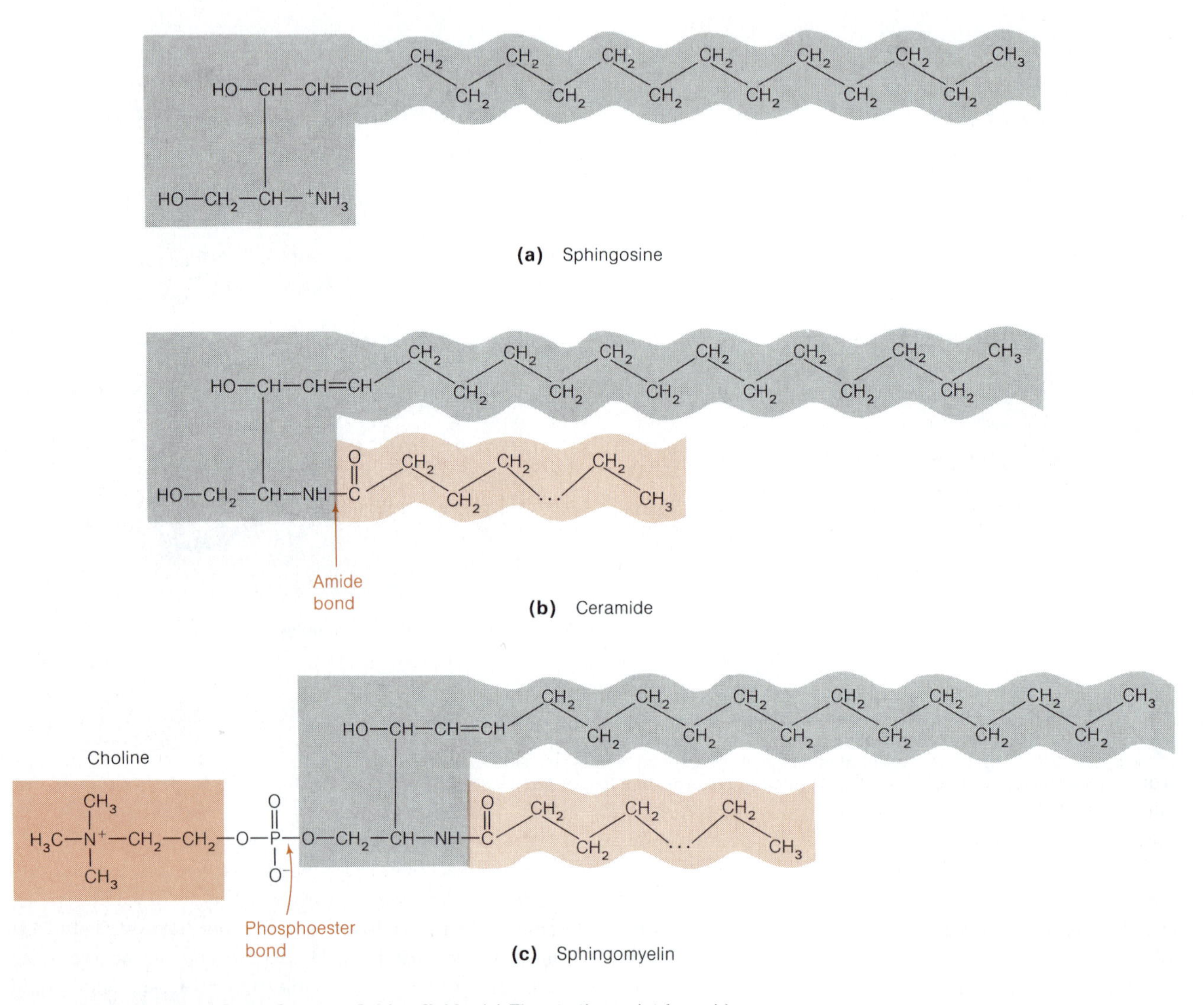

Figure 3-25 Structures of Some Common Sphingolipids. (a) The starting point for sphingolipid synthesis is sphingosine. (b) Attachment of a fatty acid to the amino group generates a ceramide. (c) Linking choline to the terminal hydroxyl group of the sphingosine by a phosphoester bond converts a ceramide to a sphingomyelin.

Steroids Are Lipids with a Variety of Functions

The final category of lipids is the **steroids.** As you can see from the structures shown in Figure 3-26, steroids have little in common chemically with the other three categories of lipids. However, they share the common property of being nonpolar and therefore hydrophobic. Most steroids contain four joined rings but differ in the number and positions of double bonds and functional groups. Steroids play a variety of roles in the cells of higher organisms but are not present in bacteria.

Several classes of mammalian hormones are steroids. These include the *adrenocortical hormones* and the *sex hormones*, including *progesterone* and *testosterone* (Figure 3-26a and b). Other familiar examples of steroids include the *bile acids* that are involved in lipid digestion in the intestine. The most common steroid in animals is **cholesterol.** Cholesterol is largely hydrophobic in nature, with only a hydroxyl group to confer a slightly hydrophilic nature on one end of the molecule (Figure 3-26d). Cholesterol is the only steroid widely found in membranes; it occurs in the plasma membrane and most of the membranes of organelles, except the inner membranes of mitochondria and chloroplasts.

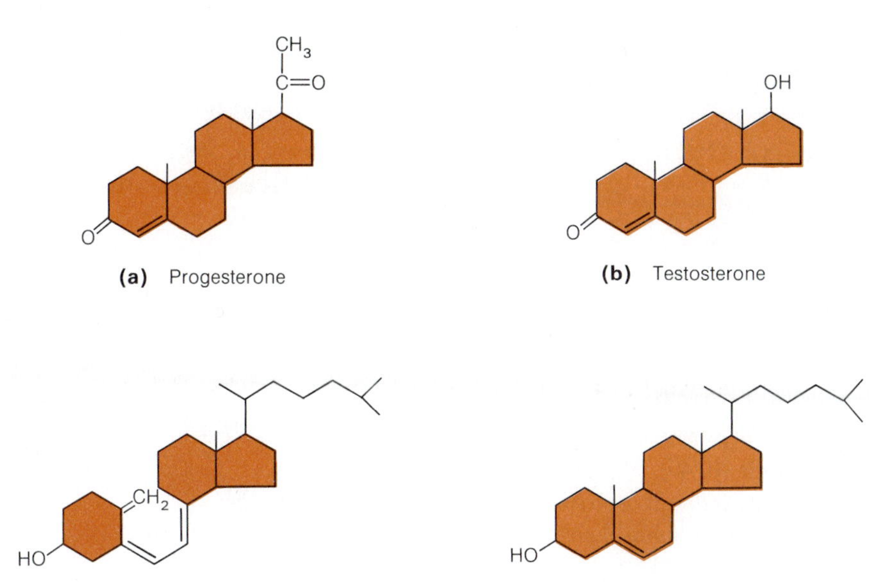

Figure 3-26 Structures of Some Common Steroids. Steroids include (a) the hormone progesterone, (b) the hormone testosterone, (c) vitamin D_3, and (d) cholesterol, a component of eukaryotic membranes. (Vitamin D_3 is technically not a steriod but a steriod derivative.)

Perspective

Three kinds of polymers characterize most of cell structure: nucleic acids, proteins, and polysaccharides. Proteins and nucleic acids are informational macromolecules that depend directly or indirectly on genetic information to determine the order of subunits that is so critical to their role in the cell. Polysaccharides, on the other hand, need no such information, since they usually contain only a single kind of repeating unit and play storage or structural roles instead.

Proteins consist of linear chains of amino acids that differ markedly in the chemical properties of their R groups. The amino acid sequence, or primary structure, of a polypeptide is all-important, since it inherently contains sufficient information to determine the local folding or orientation of the amino acid chain (secondary structure), the overall shape of the polypeptide (tertiary structure), and, in the case of multimeric proteins, the further association with other polypeptides (quaternary structure). The major force behind protein folding and polypeptide interaction is the tendency of hydrophobic amino acids to avoid an aqueous environment. Ionic bonds, hydrogen bonds, and disulfide bridges are all important in stabilizing protein structure.

Nucleic acids likewise attain a shape that is dictated by the chemical nature of the subunits. This is seen most strikingly in the DNA double helix, in which complementary base pairing (A with T, C with G) is stabilized by hydrogen bonding. The elucidation of the double-helical structure of DNA was one of the outstanding biological advances of the century.

Polysaccharides are chains of monosaccharides linked together by either α or β glycosidic bonds. The difference is critical because α linkages are readily digested by animals and are therefore suitable for storage polysaccharides such as starch or glycogen, in which glucose is stockpiled within the cell. By contrast, the β glycosidic bonds of structural polysaccharides such as cellulose or chitin are less readily digestible and give the molecule a rigid shape suitable to its function.

Lipids are not macromolecules but are included in this chapter because of their general importance as constituents of cells (especially membranes) and their frequent association with macromolecules, particularly proteins. Lipids differ substantially in chemical structure, but they all share the common property of solubility in organic solvents but not in water. The major classes of lipids include the triglycerides that make up fats and oils and the phospholipids, sphingolipids, and steroids found in membranes.

Key Terms for Self-Testing

Proteins

amino acids (p. 41)
peptide bond (p. 42)
N- (amino) terminus (p. 42)
C- (carboxyl) terminus (p. 42)
polypeptide (p. 42)
monomeric proteins (p. 44)
multimeric proteins (p. 44)
conformation (p. 44)
primary structure (p. 44)
secondary structure (p. 45)
α helix (p. 46)
β pleated sheet (p. 46)
tertiary structure (p. 46)
native conformation (p. 47)
denatured (p. 47)
renature (p. 47)
disulfide bond (p. 47)
fibrous proteins (p. 47)
globular proteins (p. 47)
random coils (p. 48)
sickle-cell anemia (p. 48)
quaternary structure (p. 48)

Nucleic Acids

nucleic acids (p. 48)
DNA (deoxyribonucleic acid) (p. 49)
RNA (ribonucleic acid) (p. 49)
ribose (p. 49)
deoxyribose (p. 49)
messenger RNA (mRNA) (p. 49)
transfer RNA (tRNA) (p. 49)
ribosomal RNA (rRNA) (p. 49)
nucleotides (p. 49)
phosphoester bond (p. 49)
purine (p. 49)
pyrimidine (p. 49)
adenine (A) (p. 49)
guanine (G) (p. 49)
cytosine (C) (p. 49)
thymine (T) (p. 49)
uracil (U) (p. 49)
nucleoside (p. 49)
nucleoside monophosphate (p. 49)
adenosine monophosphate (AMP) (p. 49)
adenosine diphosphate (ADP) (p. 49)
adenosine triphosphate (ATP) (p. 49)
phosphodiester bond (p. 50)
polynucleotide (p. 50)
template (p. 51)
base pairing (p. 51)
double helix (p. 53)

Polysaccharides

polysaccharides (p. 55)
monosaccharides (p. 55)
aldosugars (p. 55)
ketosugars (p. 55)
carbohydrate (p. 55)
Fischer projection (p. 57)
Haworth projection (p. 57)
disaccharides (p. 57)
glycosidic bond (p. 58)
starch (p. 58)
glycogen (p. 58)
amylose (p. 58)
amylopectin (p. 58)
cellulose (p. 58)
N-acetylglucosamine (NAG) (p. 60)
N-acetylmuramic acid (NAM) (p. 60)
chitin (p. 60)
microfibrils (p. 61)
noncellulosic matrix (p. 61)

Lipids

lipids (p. 62)
triglycerides (p. 62)
glycerol (p. 63)
fatty acids (p. 63)
saturated fatty acids (p. 63)
unsaturated fatty acids (p. 63)
ester bonds (p. 63)
fats (p. 63)
vegetable oil (p. 63)
phospholipids (p. 64)
phosphoglyceride (p. 64)
phosphatidic acid (p. 64)
sphingosine (p. 66)
ceramide (p. 66)
sphingolipids (p. 66)
sphingomyelins (p. 66)
steroids (p. 67)
cholesterol (p. 67)

Suggested Reading

General References and Reviews

Lehninger, A. L. *Principles of Biochemistry.* New York: Worth, 1981.

Mathews, C. K., and K. E. van Holde. *Biochemistry.* Redwood City, Calif.: Benjamin/Cummings, 1990.

Stryer, L. *Biochemistry,* 3d ed. New York: W. H. Freeman, 1988.

Tanford, C. The hydrophobic effect and the organization of living matter. *Science* 200 (1978): 1012.

Watson, J. D., N. H. Hopkins, J. W. Roberts, J. A. Steitz, and A. M. Weiner. *Molecular Biology of the Gene,* 4th ed. Menlo Park, Calif.: Benjamin/Cummings, 1987.

The Proteins

Anfinsen, C. B. Principles that govern the folding of protein chains. *Science* 181 (1973): 223.

Chothia, C. Principles that determine the structure of proteins. *Annu. Rev. Biochem.* 53 (1984): 537.

Creighton, T. E. *Proteins: Structure and Molecular Properties.* New York: W. H. Freeman, 1984.

Dickerson, R. E., and I. Geis. *The Structure and Action of Proteins.* Menlo Park, Calif.: Benjamin/Cummings, 1969.

Doolittle, R. F. Proteins. *Sci. Amer.* 253 (October 1985): 88.

Karplus, M., and J. A. McCammon. The dynamics of proteins. *Sci. Amer.* 254 (April 1986): 42.

Richardson, J. S. The anatomy and taxonomy of protein structure. *Adv. Protein Chem.* 34 (1981): 167.

Sanger, F. Sequences, sequences, and sequences. *Annu. Rev. Biochem.* 57 (1988): 1.

Schulz, G. E., and R. H. Schirmer. *Principles of Protein Structure.* New York: Springer-Verlag, 1979.

The Nucleic Acids

Cohen, J. S. DNA: Is the backbone boring? *Trends Biochem. Sci.* 5 (1980): 58.

Crick, F. H. C. The structure of the hereditary material. *Sci. Amer.* 194 (October 1954): 54.

Darnell, J. E., Jr. RNA. *Sci. Amer.* 253 (October 1985): 68.

Dickerson, R. E. The DNA helix and how it is read. *Sci. Amer.* 249 (December 1983): 94.

Felsenfeld, G. DNA. *Sci. Amer.* 253 (October 1985): 58.

Olby, R. *The Path to the Double Helix.* Seattle: University of Washington Press, 1974.

Portugal, F. H., and J. S. Cohen. *A Century of DNA: A History of the Discovery of the Structure and Function of the Genetic Substance.* Cambridge, Mass.: MIT Press, 1977.

Saenger, W. *Principles of Nucleic Acid Structure.* Berlin: Springer-Verlag, 1984.

Watson, J. D. *The Double Helix.* New York: Atheneum Press, 1968.

Watson, J. D., and F. H. C. Crick. Molecular structure of nucleic acids. A structure for deoxyribose nucleic acid. *Nature* 171 (1953): 964.

Carbohydrates and Lipids

Aspinwall, G. O., ed. *Polysaccharides,* vol. 1. New York: Academic Press, 1982.

Davison, E. A. *Carbohydrate Chemistry.* New York: Holt, Rinehart, and Winston, 1967.

Gurr, M. I., and A. T. James. *Lipid Biochemistry: An Introduction.* New York: Cornell University Press, 1971.

Hakomori, S. Glycosphingolipids. *Sci. Amer.* 254 (May 1986): 44.

Roehrig, K. L. *Carbohydrate Biochemistry and Metabolism.* Westport, Conn.: AUI Publishing, 1984.

Sharon, N. Carbohydrates. *Sci. Amer.* 243 (November 1980): 90.

Storch, J., and A. M. Kleinfeld. The lipid structure of biological membranes. *Trends Biochem. Sci.* 10 (1982): 418.

Problem Set

1. **Polymers and Their Properties.** For each of the biological polymers, (a)–(f), listed below, indicate which of the properties from the list 1–12 apply. Each polymer has multiple properties, and a given property may be used more than once.

Polymers

(a) Cellulose
(b) Messenger RNA
(c) Globular protein
(d) Amylopectin
(e) DNA
(f) Fibrous protein

Properties

1. Branched-chain polymer
2. Extracellular location
3. Aminoacyl tRNAs
4. Glycosidic bonds
5. Informational macromolecule
6. Peptide bond
7. *N*-Acetylglucosamine
8. β linkage
9. Phosphodiester bond
10. Nucleoside triphosphates
11. Helical structure possible
12. Synthesis requires a template

2. **Stability of Protein Structure.** Several different kinds of bonds are involved in generating and maintaining the structure of proteins. List four or five such bonds, give examples of amino acids that might be involved in such

bonds, and indicate which level(s) of protein structure might be generated or stabilized by those particular kinds of bonds.

3. **Amino Acid Localization in Proteins.** Amino acids tend to be localized either in the interior or on the exterior of a globular protein molecule, depending on their relative affinities for water.
   (a) Classify each of the following amino acids as likely to be found in the interior, on the exterior, or at either location, and explain.

valine	phenylalanine
glycine	alanine
aspartate	lysine

   (b) For each of the following pairs of amino acids, choose the one that is more likely to be found in the interior of a protein molecule, and explain why.

alanine; glycine	glutamate; aspartate
tyrosine; phenylalanine	methionine; cysteine

   (c) Explain why cysteines with free sulfhydryl groups tend to be localized on the exterior of a protein molecule, whereas those involved in disulfide bonds are more likely to be buried in the interior of the molecule.

4. **Myoglobin Versus Hemoglobin.** Myoglobin and hemoglobin are both oxygen-binding proteins. Myoglobin is a monomeric protein found in muscle cells, and hemoglobin is tetrameric and is found in red blood cells. The tertiary structure of myoglobin is strikingly similar to that of both the α and β subunits of hemoglobin; yet when the primary structures are compared, myoglobin can be shown to have hydrophilic amino acids at a number of positions that in the hemoglobin chains are occupied by hydrophobic amino acids. Given the extent to which tertiary structure is thought to depend on primary structure, how can a relatively hydrophobic polypeptide such as the α or β subunit of hemoglobin have a tertiary structure very much like that of the relatively hydrophilic myoglobin?

5. **Sickle-Cell Anemia.** Sickle-cell anemia is a striking example of the drastic effect that a single amino acid substitution can have on the structure and function of a protein.
   (a) Given the chemical nature of glutamate and valine, can you suggest why the substitution of the latter for the former at position 6 of the β chain would be especially deleterious?
   (b) Suggest several other amino acids that would be much less likely to cause impairment of hemoglobin function if substituted for the glutamate at position 6 of the β chain.
   (c) Can you see why in some cases two proteins could differ at a number of points in their amino acid sequence and still be very similar in structure and function? Explain.

6. **Hair Versus Silk.** The α-keratin of human hair is a good example of a fibrous protein with extensive α helical structure. Silk fibroin is also a fibrous protein, but it consists primarily of β pleated sheet structure. Fibroin is essentially a polymer of alternating glycines and alanines, whereas α-keratin contains most of the common amino acids and has many disulfide bonds.
   (a) If you were able to grab onto both ends of an α-keratin polypeptide and pull, you would find it to be both extensible (it can be stretched to about twice its length in moist heat) and elastic (when you let go, it will return to its normal length). In contrast, a fibroin polypeptide has essentially no extensibility, but it has great tensile strength. Explain these differences.
   (b) Can you suggest why fibroin assumes a pleated sheet structure, whereas α-keratin exists as an α helix and even reverts spontaneously to a helical shape when it has been stretched artificially?

7. **The "Permanent" Wave That Isn't.** The "permanent" wave that your local beauty parlor offers depends critically on rearrangements in the extensive disulfide bonds of keratin that give your hair its characteristic shape. To change the shape of your hair (to give it a wave or curl), the beautician first treats your hair with a sulfhydryl reducing agent, then uses curlers or rollers to impose the desired artificial shape, and follows this by treatment with an oxidizing agent.
   (a) What is the chemical basis of a "permanent"? Be sure to include the use of a reducing agent and an oxidizing agent in your explanation.
   (b) Why do you suppose a "permanent" isn't permanent? (Explain why the wave or curl is gradually lost during the weeks following your visit to the beautician.)
   (c) Can you suggest an explanation for naturally curly hair?

8. **The Importance of Hydrogen Bonds.** Hydrogen bonds play an important role in stabilizing the secondary structure of both proteins and nucleic acids. When a solution of either a protein or a nucleic acid is heated, one of the main effects of the thermal energy is to break hydrogen bonds; this is called *thermal denaturation.* For each statement below, decide for which, if any, of the following polymers the statement is true: the fibrous protein α-keratin (k); the silk protein fibroin (f); the DNA double helix (d).
   (a) All of the hydrogen bonds involve one nitrogen atom.
   (b) The hydrogen bonds are between units of the same polymer strand.
   (c) The hydrogen bonds are perpendicular to the main axis of the polymer.
   (d) The hydrogen bonds make the polymer more polar than it would otherwise be.

(e) The number of hydrogen bonds in a given length of the polymer does not depend on the particular amino acids or nucleotides present in that specific segment of the polymer.

(f) One of the effects of heating will be to separate polymer strands that are otherwise bonded to each other.

(g) The two ends of an individual heat-denatured polymer strand are likely to be much farther away from each other than they are in the fully hydrogen-bonded structure.

(h) Upon cooling under the appropriate conditions, the strands will return spontaneously to the original three-dimensional conformation.

(i) The renaturation process described in (h) will proceed much more slowly if the solution of denatured strands is diluted before being cooled.

9. **The Size of DNA Molecules.** The DNA double helix contains exactly ten nucleotide pairs per turn, and each turn adds 3.4 nm to the overall length of the duplex. The circular DNA molecule of *E. coli* has about four million nucleotide pairs and a molecular weight of about 2.8×10^9.

(a) What is the circumference of the *E. coli* DNA molecule? What problems might this pose for a cell that is only about 2 μm long?

(b) Human mitochondrial DNA is also a circular duplex, with a diameter of about 1.8 μm. About how many nucleotide pairs does it contain? If it takes about 1500 nucleotides of DNA to encode (specify the amino acid sequence of) a typical protein, how many proteins could mitochondrial DNA encode?

(c) What is the weight (in grams) of one nucleotide pair? How many nucleotide pairs are present in the 6 picograms (6×10^{-12} g) of double-stranded DNA present in the nucleus of most cells in your body?

(d) What is the total length of the 6 picograms of DNA present in a single nucleus?

10. **Storage Polysaccharides.** The only common examples of branched-chain polymers in cells are the storage polysaccharides glycogen and amylopectin. Both are degraded exolytically, which means by stepwise removal of terminal glucose units.

(a) Why might it be advantageous for a storage polysaccharide to have a branched-chain structure instead of a linear structure?

(b) Can you foresee any metabolic complications in the process of glycogen degradation? How do you think the cell handles this?

(c) Can you see why cells that must degrade amylose instead of amylopectin have enzymes capable of endolytic (internal) as well as exolytic cleavage of glycosidic bonds?

(d) Why do you suppose the structural polysaccharide cellulose does not contain branches?

11. **Inulin Structure.** Inulin is a rather uncommon storage polysaccharide found in artichoke tubers. It is a $\beta(1 \rightarrow 2)$ polymer of β-D-fructose units. Can you draw the structure of inulin?

12. **The Polarity of Lipids.** Arrange the following lipids in order of decreasing polarity, and explain your reasoning: triglyceride; cholesterol; phosphatidyl choline; fatty acid.

13. **Phospholipids.** Which would you expect to resemble a sphingomyelin molecule more closely, a molecule of phosphatidyl choline containing two palmitates as its fatty acid chains or a phosphatidyl choline molecule with one palmitate and one oleate as its fatty acid chains? Explain.

14. **Shortening.** A popular brand of shortening has a label on the can that identifies the product as "partially hydrogenated soybean oil, palm oil, and cottonseed oil."

(a) What does the process of "partial hydrogenation" accomplish chemically?

(b) What did the contents of the can look like before it was partially hydrogenated?

(c) What is the physical effect of partial hydrogenation?

PART TWO

CELL STRUCTURE AND FUNCTION

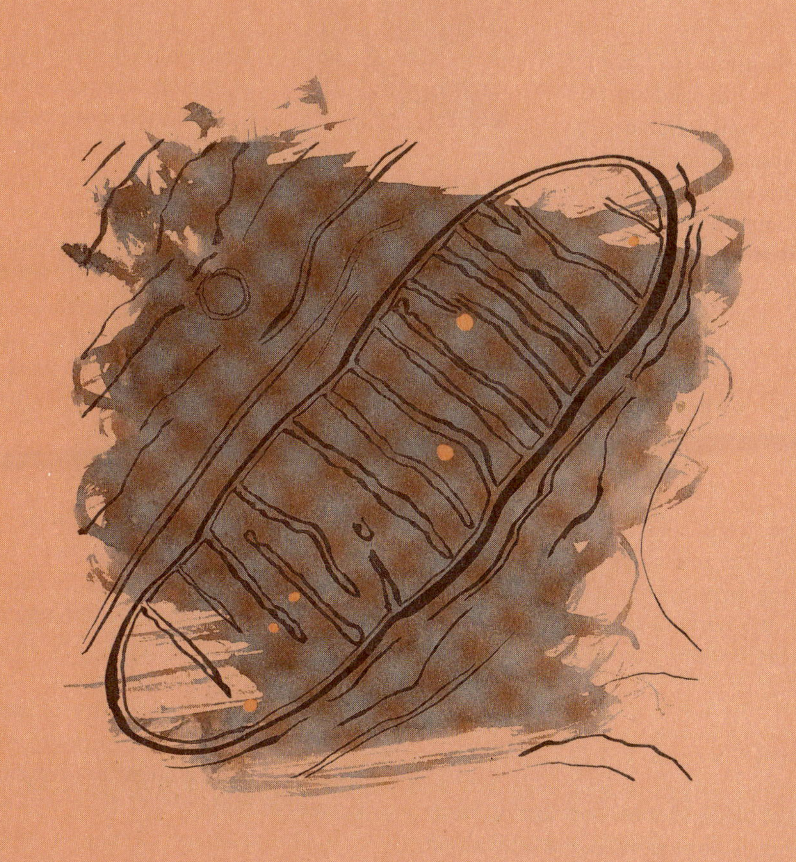

4

Cells and Organelles

In the previous two chapters, we have encountered most of the major kinds of molecules found in cells, as well as some of the principles governing the assembly of these molecules into the supramolecular structures of which cells and their organelles are composed (recall Figure 2-15). Now we are ready to focus our attention on cells and organelles directly.

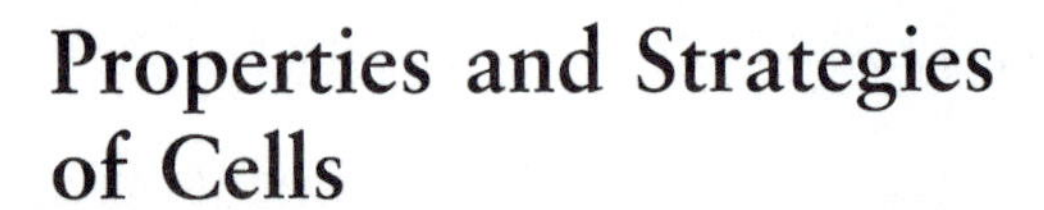

Properties and Strategies of Cells

As we begin to consider what cells are and how they function, several general characteristics of cells quickly emerge. These include the sizes and shapes of cells, the classification of cells on the basis of their organizational complexity, and the specializations that cells undergo.

Cell Sizes and Shapes

Cells come in a variety of sizes, shapes, and forms. Some of the smallest bacterial cells, for example, are only about 0.2–0.3 μm in diameter—so small that about 30,000 such cells could fit side by side across the head of a thumbtack! At the other extreme are highly elongated nerve cells, which may extend one or more meters. Those in the neck and legs of a giraffe are dramatic examples. On the other hand, the oft-cited examples of bird eggs—especially ostrich eggs—are rather misleading, since they are indeed single cells, but with the vast majority of their internal volume occupied not by living protoplasm but by yolk—large deposits of stored food intended as nourishment for the developing embryo.

Most cells, however, fall into a rather narrow and predictable range of sizes, and for several good reasons. Bacterial cells are in general about 1–5 μm in diameter, whereas most cells of higher plants and animals have dimensions in the range of 10–50 μm. Plant and animal cells, in other words, are at least an order of magnitude larger than bacterial cells in a single dimension and therefore at least three orders of magnitude (a thousandfold) bigger in three dimensions. A typical human liver cell, for example, has a diameter of about 20 μm, while a tobacco leaf cell might be about 30–40 μm in diameter. (Recall that the box on pp. 4–5 describes the units used to express cellular dimensions and illustrates the relative sizes of some cells and related structures.)

The Surface Area/Volume Ratio. The main constraint on cell size is that set by the need to maintain an adequate **surface area/volume ratio.** Surface area is important because it is here that the exchange between the cell and its environment takes place. The internal volume of the cell determines the amount of nutrients that will have to be imported and the quantity of waste products that must be excreted, but the surface area effectively measures the amount of membrane available for such uptake and excretion.

The problem of maintaining adequate surface area arises because the volume of a cell increases with the cube of the cell's length or diameter, whereas its surface area only increases with the square. Consider, for example, the cube-shaped cells shown in Figure 4-1. The cell on the left is 20 μm on a side and has a volume of 8000 μm^3 ($V = s^3$, where $s = 20$ μm) and a surface area of 2400 μm^2 ($A = 6s^2$). The surface area/volume ratio is therefore 2400 μm^2/8000 μm^3, or 0.3 μm^{-1}. When this single large cell is divided into smaller cells, the total volume remains the same but the surface area increases. Thus, the surface area/volume ratio increases as the linear dimension of the cell decreases. The 1000 cells on the right, for example, still have a total volume of 8000 μm^3 (1000×2^3) but the total

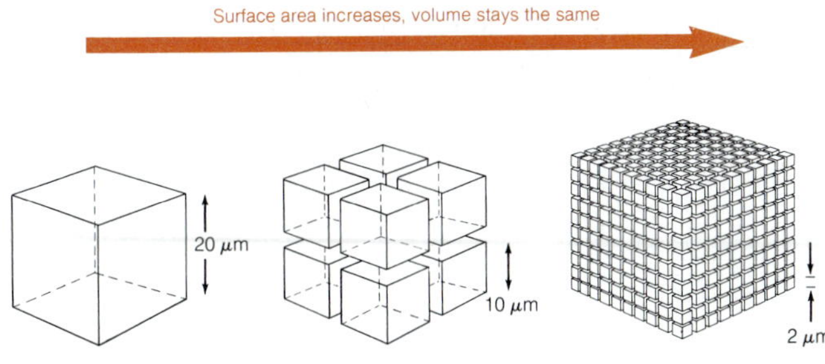

Length of one side	20 μm	10 μm	2 μm
Total surface area (height × width × number of sides × number of cubes)	2400 μm^2	4800 μm^2	24,000 μm^2
Total volume (length × width × height × number of cubes)	8000 μm^3	8000 μm^3	8000 μm^3
Surface area to volume ratio (surface area ÷ volume)	0.3	0.6	3.0

Figure 4-1 The Effect of Cell Size on the Surface Area/Volume Ratio. The single large cell on the left, the eight smaller cells in the center, and the 1000 tiny cells on the right have the same total volume (8000 μm^3), but total surface area increases as the cell size decreases. The surface area/volume ratio increases from left to right as the linear dimension of the cell decreases. Thus, 1000 prokaryotic cells with a linear dimension of 2 μm have a total surface area ten times that of a single eukaryotic cell with a linear dimension of 20 μm.

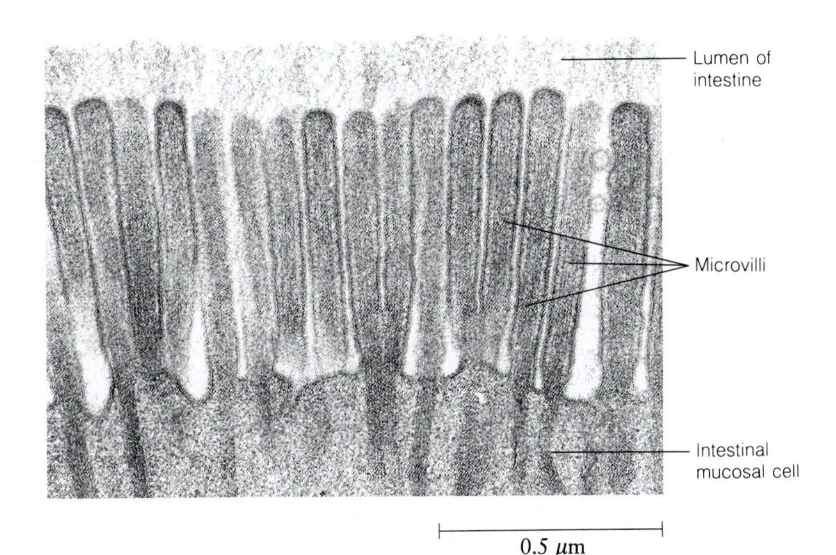

Figure 4-2 The Microvilli of Intestinal Mucosa Cells. Microvilli are fingerlike projections of the cell membrane that greatly increase the absorptive surface area of these cells. (TEM)

surface area is 24,000 μm^2 ($1000 \times 6 \times 2^2$), so the surface area/volume ratio is 24,000 μm^2/8000 μm^3, or 3.0 μm^{-1}.

This comparison illustrates a major constraint on cell size. As a cell increases in size, its surface area does not keep pace with its volume, and the necessary exchange of substances between the cell and its surroundings becomes more and more problematic. Cell size can therefore increase only over the range of values for which the membrane surface area is still adequate to meet the needs for passage of materials into and out of the cell. Once the limiting surface area/volume ratio is reached, further increases in cell size would generate more cytoplasmic volume and therefore greater exchange needs than could be met by the more modest increases in membrane surface area.

Some cells, particularly those that play a role in absorption, have ingenious ways of maximizing their surface area. Effective surface area is most commonly increased by inward folding or outward protrusion of the cell membrane. The cells that line your intestine, for example, contain many fingerlike projections called *microvilli* that greatly increase the effective membrane surface area and therefore the absorbing capacity of these cells (Figure 4-2).

Compartmentalization. Another limit on cell size is that imposed by the need to maintain adequate concentrations of the specific compounds and catalysts (enzymes) needed for the various processes cells must carry out. For a chemical reaction to occur in a cell, the appropriate reactants must collide with and bind to the surface of a particular enzyme. The frequency of such collisions will be greatly influenced by the concentrations of the reactants and of the enzyme itself. To maintain appropriate levels of reactants and enzymes as the size of a cell increases, the number of all such molecules must increase eightfold every time the length (or diameter) of the cell doubles. This increase taxes the synthetic capabilities of the cell.

The **compartmentalization** of activities within the cell is one solution to the problem of concentration. If all the enzymes and compounds involved in a specific process are localized within a specific region of the cell, a locally high concentration of those enzymes and compounds can be maintained in that region, rather than throughout the whole cell. This is what happens in plant and animal cells and is presumably the main reason they can be so much larger than bacterial cells and still function efficiently.

Plant and animal cells have a variety of *organelles*—internal compartments that are delineated by membranes and are highly specialized for specific functions. For example, the cells in a plant leaf have most of the enzymes, compounds, and pigments needed for photosynthesis compartmentalized together into structures called *chloroplasts*. Such cells can therefore maintain appropriately high concentrations of everything that is essential for photosynthesis within the chloroplasts without having to have similarly high levels of these substances elsewhere in the cell. In a similar way, other functions are localized within other compartments. This internal compartmentalization of specific functions makes it possible for the large cells of plants and animals to maintain locally high concentrations of the specific enzymes and compounds involved in particular cellular processes. Such processes can therefore proceed efficiently, even though as a whole such cells are orders of magnitude larger than bacterial cells.

Prokaryotes and Eukaryotes: An Organizational Dichotomy

Size and compartmentalization of function are just two of many features that distinguish the cells of plants and animals from those of bacteria. On the basis of these and a number of other organizational criteria, biologists have come to recognize a fundamental distinction between two broad groups of organisms, the **prokaryotes** and the **eukaryotes.** *Bacteria* and *cyanobacteria* (also called *blue-green algae*) are prokaryotes, whereas virtually all other organisms are eukaryotes, including all *plants, animals, fungi,* and *protozoa.* These two basic categories of organisms differ at the cellular level in terms of important structural, biochemical, and genetic features. Some of these differences are summarized in Table 4-1.

Table 4-1 A Comparison of Some Properties of Prokaryotic and Eukaryotic Cells

Property	Prokaryotic Cells	Eukaryotic Cells
Size	Small (a few micrometers in length or diameter)	Large (10 to 50 times the length or diameter of a prokaryotic cell)
Membrane-bounded nucleus	No	Yes
Internal membranes	No	Yes
Microtubules	No	Yes
Microfilaments	No	Yes
Intermediate filaments	No	Yes
Exocytosis and endocytosis	No	Yes
Genetic information	"Naked" DNA molecule	DNA complexed with proteins to form chromosomes
Mitosis and meiosis	No	Yes
Processing of RNA transcripts	Little	Much
Ribosomes*	Small (70S); 3 RNA molecules and 55 proteins	Large (80S); 4 RNA molecules and 78 proteins

* Ribosomes are characterized in terms of their *sedimentation coefficients* or *S values,* a measure of their sedimentation rate based on size and shape. Sedimentation coefficients are normally expressed in Svedberg units (S), where $1S = 1 \times 10^{-13}$ sec.

Eukaryotic Cells Have True Nuclei. The most fundamental difference between the cells of eukaryotes and prokaryotes is reflected in the nomenclature itself. Eukaryotic cells have a true, membrane-bounded *nucleus* (*eu* is Greek for "true" or "genuine"; *karyon* means "nucleus"), whereas prokaryotic cells lack a true nucleus and are regarded as more primitive (*pro* means "before"). The genetic information of eukaryotic cells is therefore sequestered within a membrane-bounded nucleus; in prokaryotic cells it is simply localized in a region of the cytoplasm called the *nucleoid,* without a membrane around it. Figure 4-3 compares these structural features of a typical prokaryotic cell and the nucleus of a representative eukaryotic cell.

Eukaryotic Cells Segregate Functions with Internal Membranes. Internal membranes that compartmentalize specific functions are a general feature of eukaryotic cells. The presence of a nuclear membrane is just one example. As we already know, a variety of other internal membranes allow the eukaryotic cell to compartmentalize functions that in prokaryotic cells take place within the cytoplasm or on the plasma (cell) membrane. By contrast, prokaryotic cells contain few, if any, internal membranes and do not compartmentalize specific functions.

Examples of internal membrane systems in eukaryotic cells include the *endoplasmic reticulum,* the *Golgi complex,* and the membranes that surround and delimit organelles such as *mitochondria, chloroplasts, lysosomes,* and *peroxisomes,* as well as various kinds of *vacuoles* and *vesicles.* Each of these organelles has its own characteristic membrane (or pair of membranes, in the case of mitochondria and chloroplasts), similar to other membranes in basic structure but often with its own specific chemical composition and enzymes. Localized within each such organelle are the particular cellular functions for which the structure

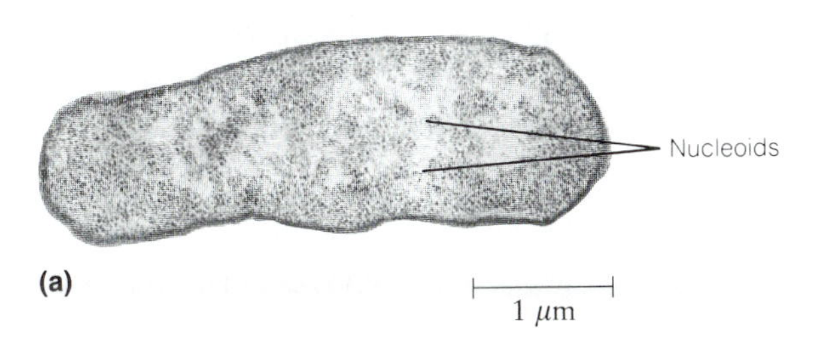

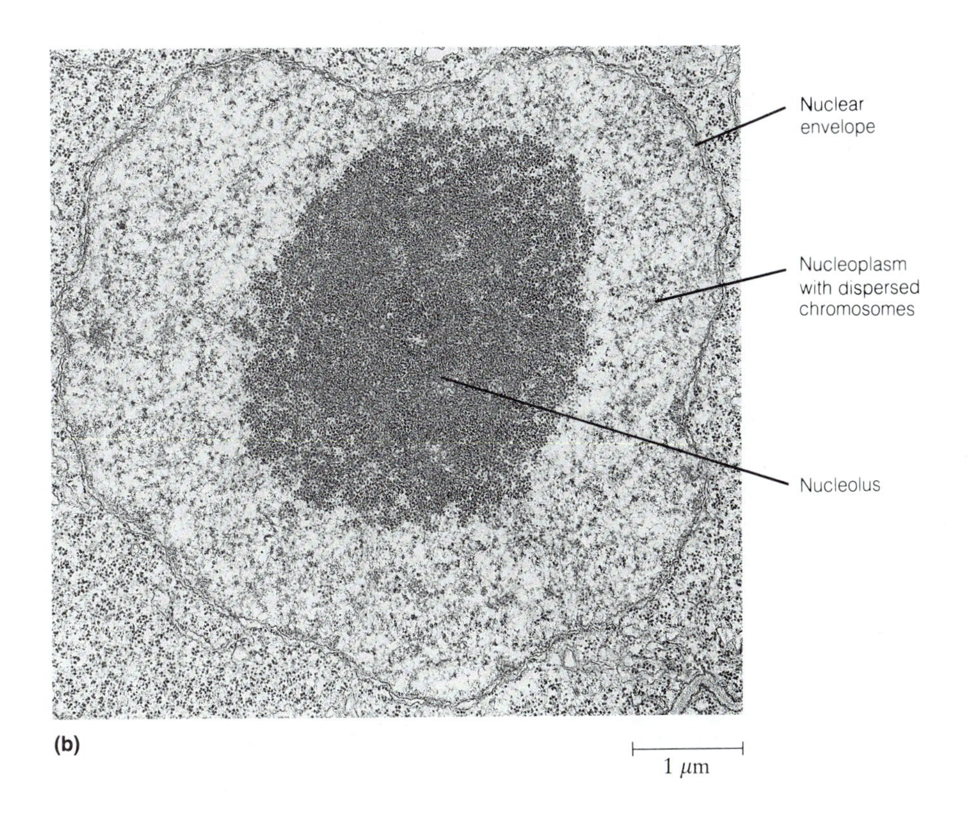

Figure 4-3 Storage of Genetic Information by Prokaryotic and Eukaryotic Cells. (a) A bacterial cell with several nucleoids, each representing a single circular molecule of DNA (TEM). (b) The nucleus of a eukaryotic cell shown at the same magnification as the bacterial cell (TEM). The pair of membranes that encloses the nucleus is called the nuclear envelope. The nucleolus is a characteristic structure within the eukaryotic nucleus that is involved in synthesis of ribosomes.

is specialized. Figure 4-4 illustrates several of these structural features in a cross-sectional view of a liver cell. We will meet each of these organelles later in this chapter and then return to each in its appropriate context in succeeding chapters.

Eukaryotic Cells Have Tubules and Filaments. Also found in the cytoplasm of eukaryotic cells, but not in their prokaryotic counterparts, are several nonmembranous structures that are involved in contraction, motility, and the establishment and support of cellular architecture. These include the *microtubules* found in the cilia and flagella of many cell types, the *microfilaments* of actin and myosin found in muscle fibrils and other structures involved in motility, and the *intermediate filaments* that are especially prominent in parts of the cell that are subject to stress. Microtubules, microfilaments, and intermediate filaments are also involved in the *cytoskeletal framework* that imparts structure and elasticity to eukaryotic cells, as we will learn shortly and explore in more detail in Chapter 18.

Eukaryotic Cells Carry Out Exocytosis and Endocytosis. A further feature of eukaryotic cells is their ability to exchange materials between the membrane-bounded compartments within the cell and the exterior of the cell. This exchange is possible because of *exocytosis* and *endocytosis,* processes that are unique to eukaryotic cells. In endocytosis, portions of the plasma membrane invaginate and are pinched off to form membrane-bounded cytoplasmic vesicles containing substances that were previously on the outside of the cell. Exocytosis is essentially the reverse of this process; membrane-bounded vesicles inside the cell fuse with the plasma membrane and release their contents to the outside of the cell.

Eukaryotic Cells Organize Their DNA into Chromosomes. Another distinction between prokaryotes and eukaryotes becomes apparent when we consider the amount and organization of the genetic material. Prokaryotes characteristically contain amounts of DNA that might be described as "reasonable"; that is, we can account for much of the DNA in terms of known proteins for which the DNA serves as a genetic blueprint. Prokaryotic DNA is usually present as a circular molecule with which few, if any, proteins are routinely associated.

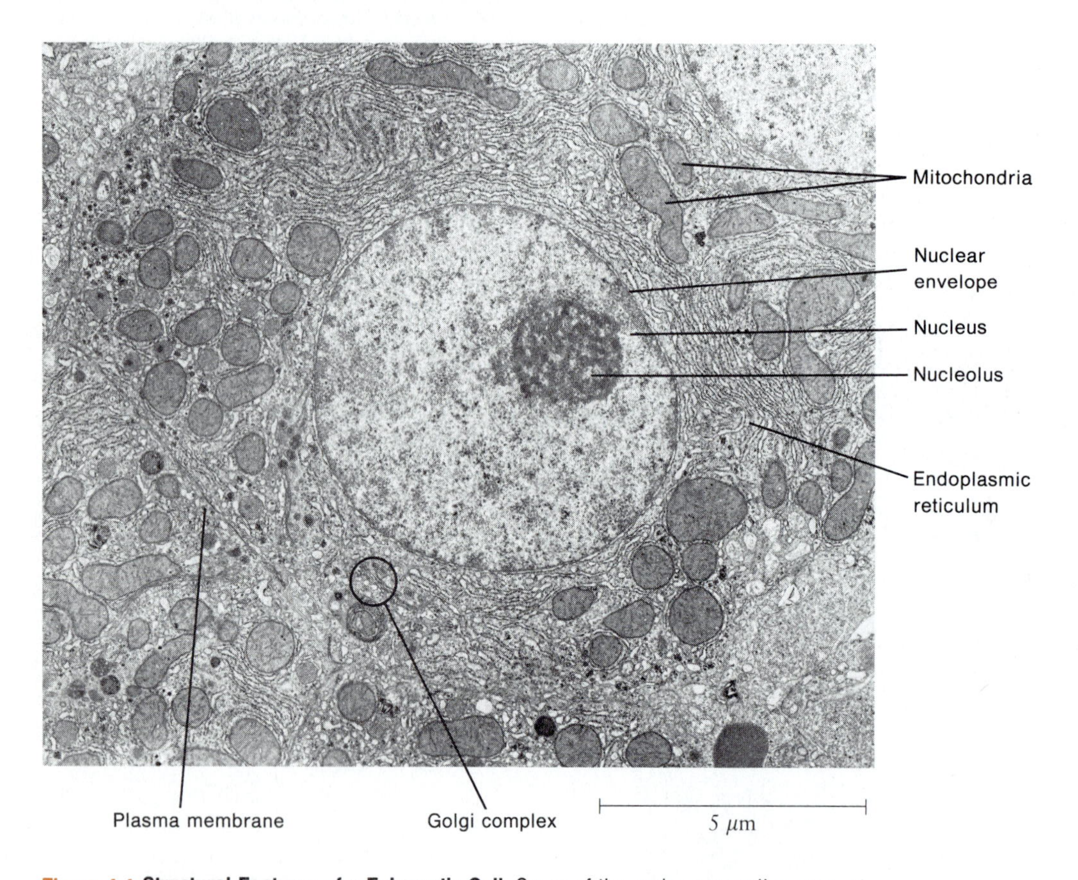

Figure 4-4 Structural Features of a Eukaryotic Cell. Some of the major organelles present in eukaryotic cells are identified in this electron micrograph of a liver cell seen in cross section. Other common features visible at this magnification but not identified in the figure include lysosomes and peroxisomes. (TEM)

Though of "reasonable" size in a genetic sense, the circular DNA molecule of a prokaryotic cell is usually much longer than the cell itself. It therefore has to be folded and packed together tightly to fit into the nucleoid of the cell. For example, the common intestinal bacterium *Escherichia coli* is only about a micrometer or two long, yet it has a circular DNA molecule that is about 1300 μm in circumference. Clearly, a great deal of folding and packing is necessary to fit that much DNA into a small region of such a small cell. By way of analogy, it is roughly equivalent to packing about 60 feet (18 m) of very thin thread into a typical thimble.

But if DNA appears to pose a packaging problem for prokaryotic cells, consider the case of the poor eukaryotic cell! Although some of the simpler eukaryotes (such as yeast and fruit flies) contain only 10 to 50 times as much DNA as bacteria, most eukaryotic cells have at least a thousand times as much DNA as *E. coli*. It is tempting to label such amounts of DNA as "unreasonable," since we cannot at present assign any known function to much of it. But that is probably a more telling commentary on cell biologists than on cells.

Whatever the genetic function of such large amounts of DNA, the packaging problem is clearly acute. It is solved universally among eukaryotes by the organization of DNA into complex structures called **chromosomes** that contain at least as much protein as DNA. (The circular molecule of DNA in prokaryotic cells is sometimes also called a chromosome, but the convention in this text will be to use that term only for the structure in the eukaryotic nucleus.) It is as chromosomes that the DNA of eukaryotic cells is packaged, segregated during cell division, transmitted to daughter cells, and transcribed as needed into the molecules of RNA that are involved in protein synthesis. Figure 4-5 shows a chromosome from an animal cell as visualized by high-voltage electron microscopy.

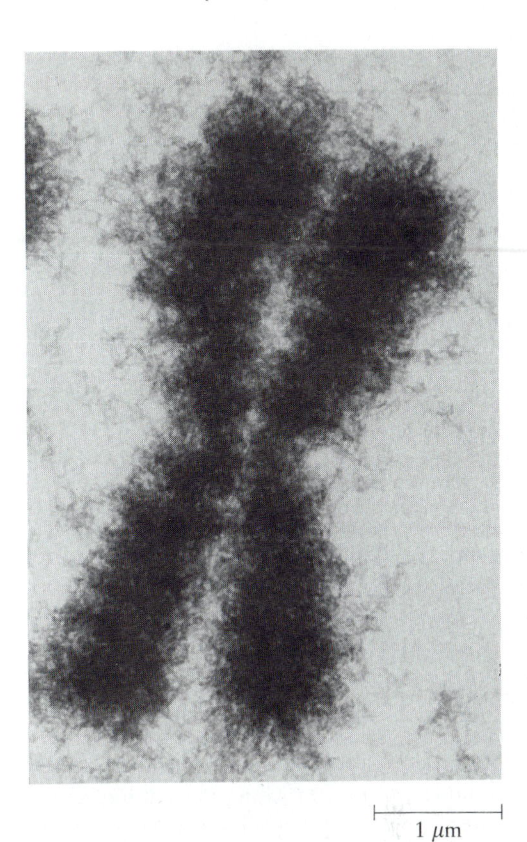

Figure 4-5 A Eukaryotic Chromosome. This chromosome was obtained from a cultured Chinese hamster cell and visualized by high-voltage electron microscopy. The cell was undergoing mitosis, and the chromosome is therefore highly coiled and condensed. If mitosis had been allowed to proceed, the chromosome would have divided into two daughter chromosomes, one destined for each of the two daughter cells that result from subsequent cell division.

Eukaryotic Cells Segregate Genetic Information by Mitosis and Meiosis. A further contrast between prokaryotes and eukaryotes is seen in the way they allocate genetic information to daughter cells upon division. Prokaryotic cells simply replicate their DNA and divide, with one daughter molecule of DNA going to each daughter cell. Eukaryotic cells also replicate their DNA, but they then use the more complex processes of *mitosis* and *meiosis* to distribute chromosomes equitably to daughter cells. The chromosome shown in Figure 4-5 was prepared from a cell that was undergoing mitosis. If the process had been allowed to proceed, the chromosome would have divided into two daughter chromosomes, each destined for one of the two daughter cells.

Eukaryotic Cells Express Their DNA Differently. The differences between prokaryotic and eukaryotic cells extend to the expression of genetic information. Eukaryotic cells tend to transcribe genetic information in the nucleus into much larger RNA molecules than they eventually use to direct protein synthesis in the cytoplasm. They depend on later processing and transport processes to deliver the desired messages selectively to the cytoplasm.

By contrast, prokaryotes seem to transcribe very specific segments of genetic information into RNA messages, and little or no processing or selection appears to be either necessary or possible. In fact, the absence of a nuclear membrane makes it possible for new messenger RNA molecules to become involved in the process of protein synthesis even before they are themselves completely synthesized (Figure 4-6). Prokaryotes and eukaryotes also differ in the size and composition of the ribosomes used to synthesize proteins (Table 4-1). We will return to this distinction in more detail later in this chapter.

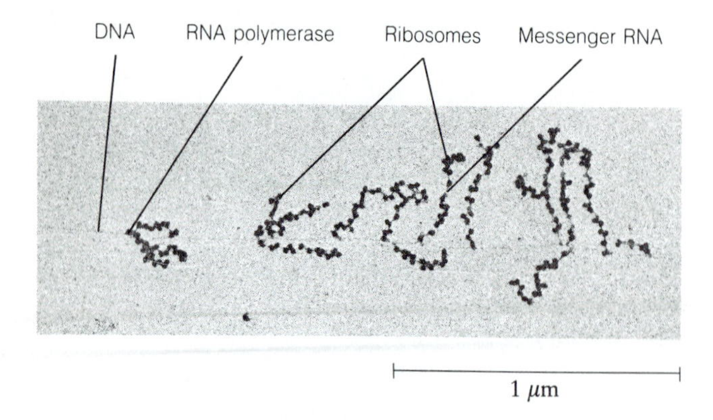

Figure 4-6 Expression of Genetic Information in a Prokaryotic Cell. This electron micrograph shows a small segment of a DNA molecule from a bacterial cell being transcribed into messenger RNA molecules by an enzyme called RNA polymerase. As soon as the messenger RNA becomes available, it associates with ribosomes, allowing protein synthesis to occur.

Cell Specialization: The Unity and Diversity of Biology

In terms of function and structure, cells are characterized by both unity and diversity, and the cell biologist needs to be aware of both. By unity and diversity, we simply mean that all cells resemble one another in some ways, yet differ from one another in other ways. We will spend most of our time in the coming chapters concentrating on aspects of structure and function common to most cell types, since these are the features of cells that are of the greatest general interest. We will find, for example, that virtually all cells oxidize sugar molecules for energy, transport ions across membranes, transcribe DNA into RNA, and undergo division to generate daughter cells. These, then, become topics of legitimate concern for us.

Much the same is true in terms of structural features, since all cells are surrounded by a selectively permeable membrane, all have ribosomes for the purpose of protein synthesis, and all contain double-stranded DNA as their genetic information. Clearly, we can be confident that we are dealing with fundamental aspects of cellular organization and function when we consider processes and structures common to most, if not all, cells.

But sometimes our understanding of cellular biology is enhanced by considering not just the unity but also the diversity of cells—not just features common to most cells, but also features that are especially prominent in a particular cell type. For example, to understand how the process of protein secretion works, it would be an advantage to consider a cell that is highly specialized for that particular function. Cells from your pancreas would be a good choice, for example, because they secrete large amounts of digestive enzymes, such as amylase and trypsin.

Similarly, to study functions known to occur in mitochondria, it would clearly be an advantage to select a cell type that is highly specialized in the energy-releasing processes that occur in the mitochondrion, since such a cell would probably have a lot of well-developed, highly active mitochondria. It was for this very reason, in fact, that Hans Krebs chose the flight muscle of the pigeon as the tissue with which to carry out the now-classic studies on the cyclic pathway of oxidative reactions that has come to bear his name.

Whenever we exploit the specialized functions of specific cell types to study a particular function, we are acknowledging the diversity of cell structure and function that arises primarily because of cellular specialization. Although we may not often realize it, we are also taking advantage of the multicellularity of many organisms, because it is usually only as part of a multicellular organism that a cell can afford to commit itself to a specialized function.

In general, unicellular organisms such as bacteria, protozoa, and some algae must be capable of carrying out any and all of the functions necessary for survival, growth, and reproduction and cannot afford to overemphasize any single function at the expense of others. Multicellular organisms, on the other hand, are characterized by a division of labor among tissues and organs that not only allows for, but actually depends on, specialization of structure and function. Whole groups of cells become highly specialized for a particular task, which then becomes their specific role in the overall economy of the organism.

The Eukaryotic Cell in Overview: Pictures at an Exhibition

From the foregoing discussion, it should be clear that all cells carry out many of the same basic functions and have some of the same basic structural features. However, cells of eukaryotic organisms are far more complicated structurally than prokaryotic cells, primarily because of the organelles and other intracellular structures eukaryotes have to compartmentalize various functions. The structural complexity of eukaryotic cells is illustrated by the typical animal and plant cells shown in Figures 4-7 and 4-8, respectively.

In reality, of course, there is no such thing as a truly "typical" cell; all eukaryotic cells have features that distinguish them from the particular cells shown in Figures 4-7 and 4-8. But most eukaryotic cells are sufficiently similar to warrant a general overview of their structural features.

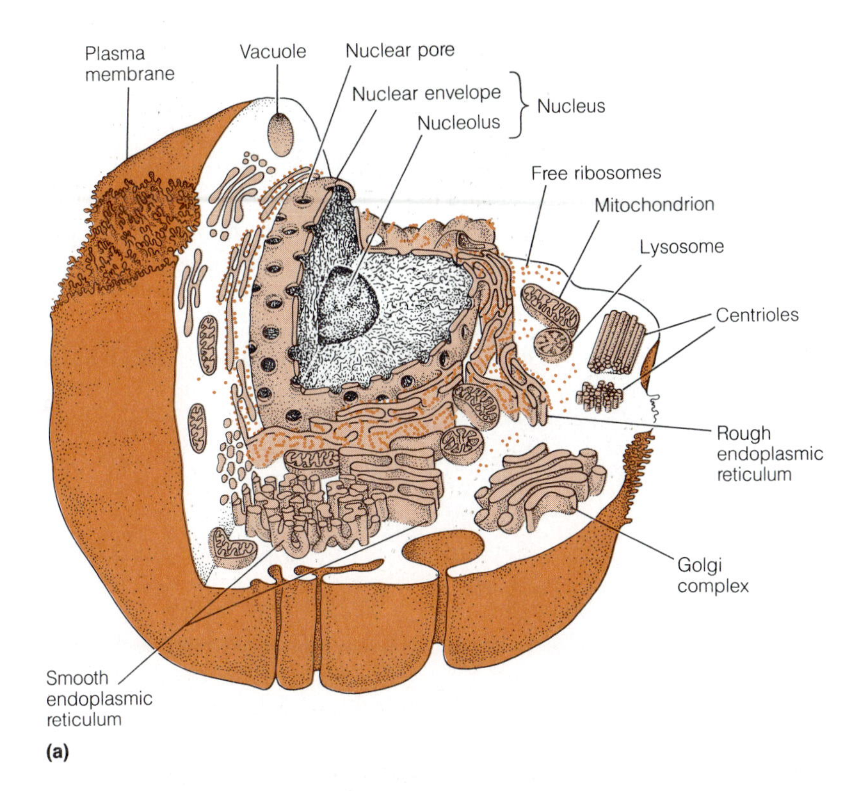

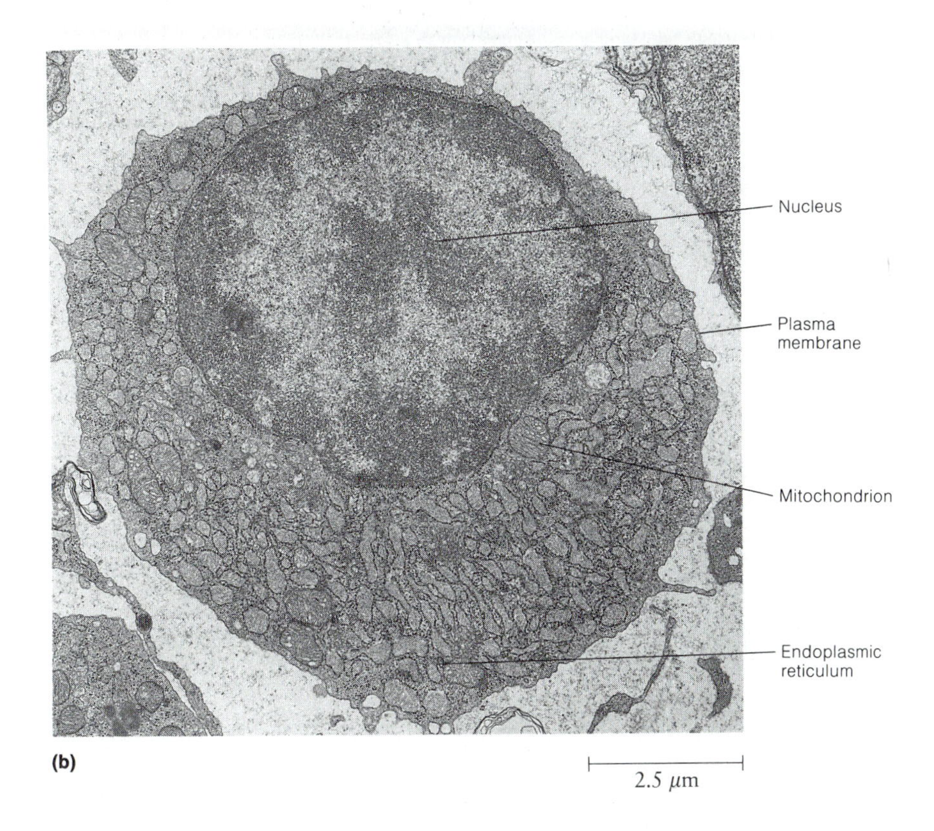

Figure 4-7 A Typical Animal Cell. (a) A schematic diagram of an animal cell to provide perspective on the relative sizes and shapes of organelles and other subcellular structures. (b) A plasma cell (white blood cell), with several subcellular structures identified (TEM).

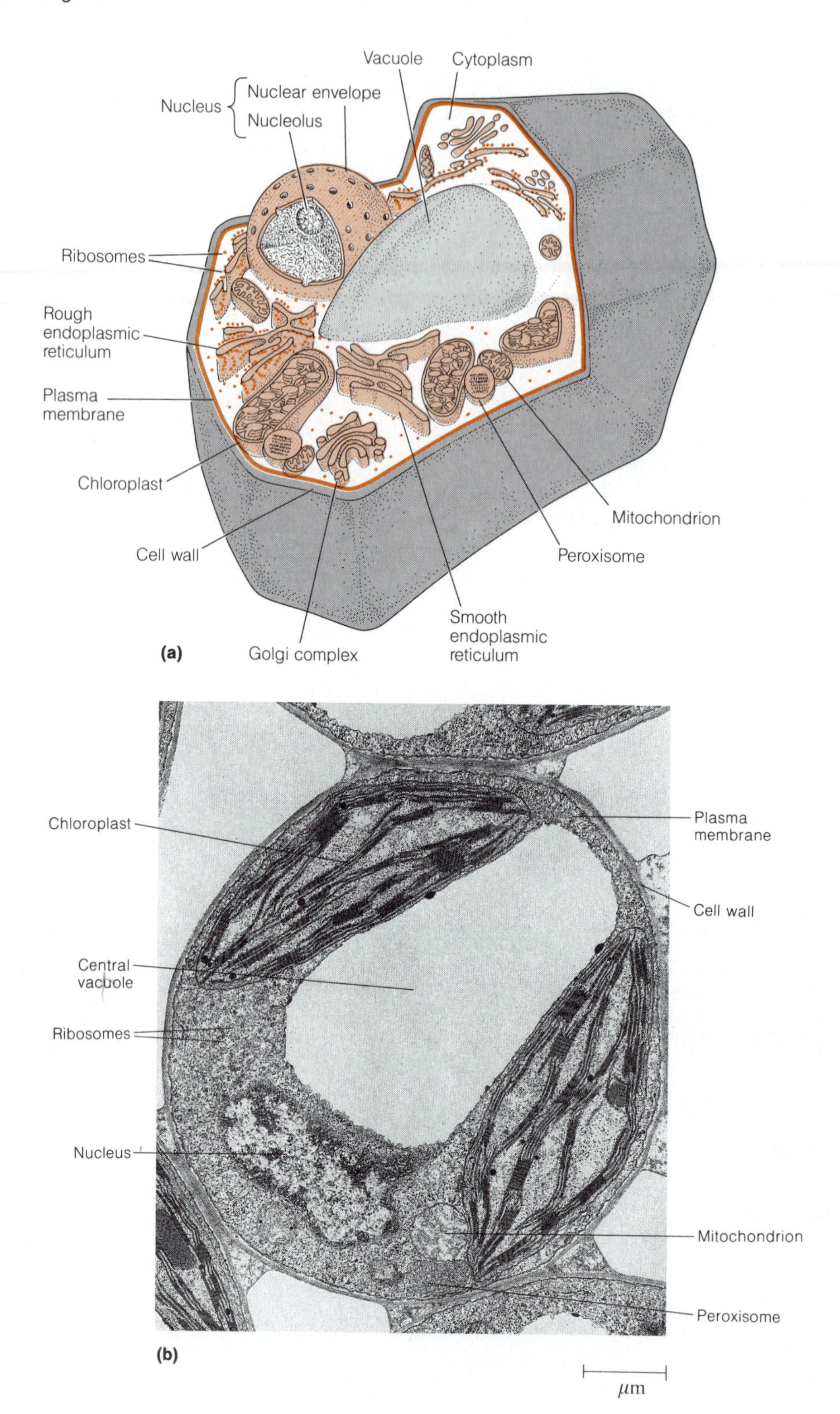

Figure 4-8 A Typical Plant Cell. (a) A schematic diagram of a plant cell. Compare this diagram with the animal cell sketched in Figure 4-7a and note that plant cells are characterized by the absence of lysosomes and the presence of chloroplasts, a cell wall, and a large central vacuole. (b) A cell from a *Coleus* leaf, with several subcellular structures identified (TEM).

In essence, a typical eukaryotic cell has at least four major structural features: a *plasma* (or *cell*) *membrane* to define its boundary and retain its contents, a *nucleus* to house the DNA that directs cellular activities, *membrane-bounded organelles* in which various cellular functions are localized, and the *cytoplasm* with its *cytoskeleton* of tubules and filaments. In addition, plant cells have a rigid *cell wall* external to the plasma membrane. Animal cells do not have a wall but are surrounded by a *cell coat* (or *glycocalyx*), which usually consists of carbohydrate chains attached to proteins and lipids in the plasma membrane.

Our intention here is to look at each of these structural features in overview, as an introduction to cellular architecture. We are not yet ready to consider any of these features in detail; that will be tackled in later chapters as we encounter the cellular processes in which the various organelles and other structures are involved. For the present, we will simply look at each structure as one might look at pictures at an exhibition. We will move through the gallery rather quickly, just to get a feel for the overall display, but with the intention of returning to each structure for a careful and more detailed examination. (If you are interested in more information on a specific topic at this point, see the reference box on p. 84 for a list of the chapters in which particular structures are considered in detail.)

The Plasma Membrane

Our tour begins with the **plasma membrane** that surrounds every cell. The plasma membrane defines the boundaries of the cell and ensures that its contents are retained. Like other biological membranes, the plasma membrane consists of both lipids and proteins organized as shown in Figure 4-9. The lipids are arranged in two layers, each about 4–5 nm thick. This **lipid bilayer** appears as a pair of dark bands when viewed with the electron microscope (Figure 4-9c). As we will see in Chapter 7, the lipid bilayer is the basic structure of all membranes and serves as a permeability barrier to most water-soluble substances.

Most of the lipids within each of the two layers are actually *phospholipids*, with a phosphate group attached to one end. The lipid portion of the molecule is *hydrophobic*, but the end with the phosphate group is *hydrophilic*. In each layer, the phospholipids are oriented such that the hydrophobic portion of the molecule faces inward, toward the other layer. The hydrophilic end of each lipid molecule faces away from the membrane, either toward the inside or toward the outside of the cell, depending on the layer in which the lipid molecule is located.

The proteins within the membrane have both hydrophobic and hydrophilic regions on their surface and orient

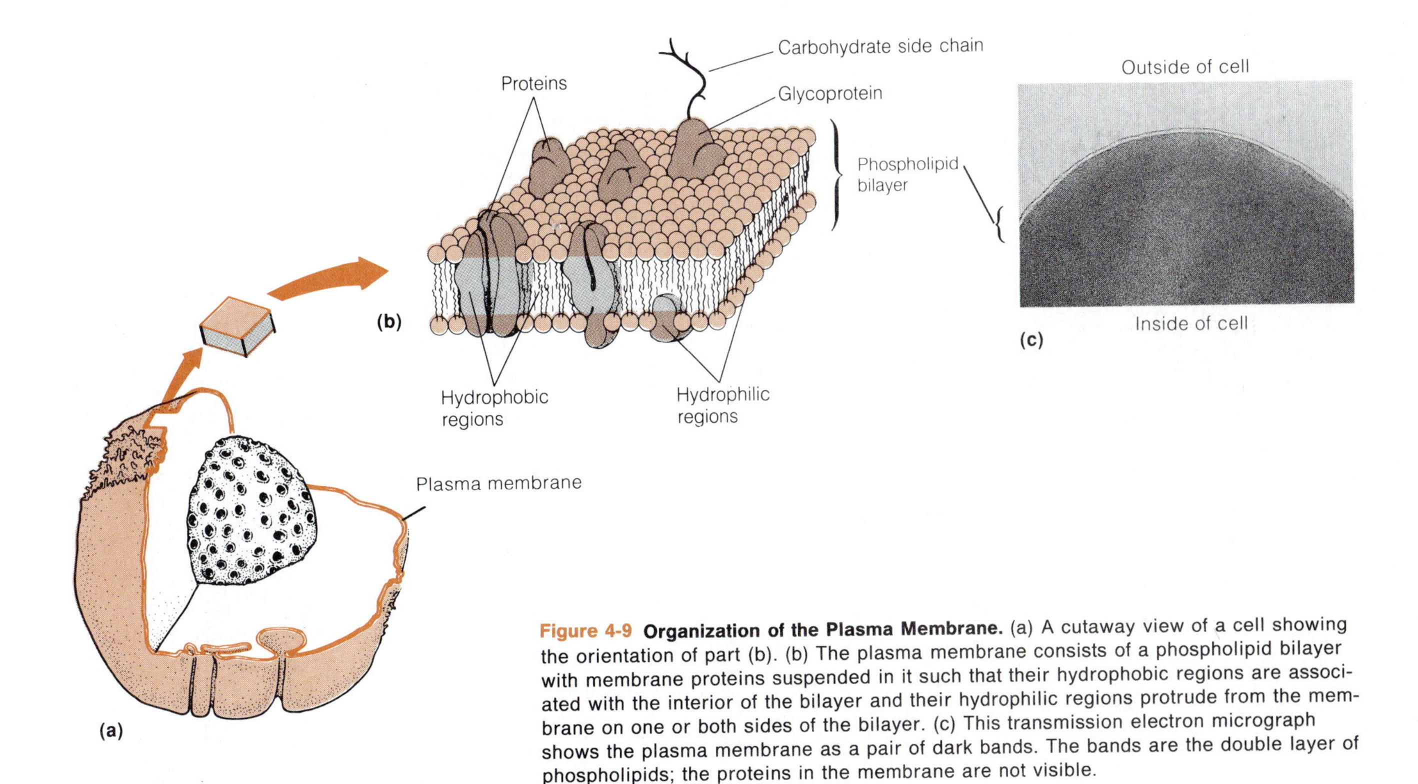

Figure 4-9 Organization of the Plasma Membrane. (a) A cutaway view of a cell showing the orientation of part (b). (b) The plasma membrane consists of a phospholipid bilayer with membrane proteins suspended in it such that their hydrophobic regions are associated with the interior of the bilayer and their hydrophilic regions protrude from the membrane on one or both sides of the bilayer. (c) This transmission electron micrograph shows the plasma membrane as a pair of dark bands. The bands are the double layer of phospholipids; the proteins in the membrane are not visible.

Reference Box

For more detailed information about cellular structures, see the following chapters:

Plasma membrane	Chapter 7
Nucleus	Chapter 13
Mitochondrion	Chapter 11
Chloroplast	Chapter 12
Endoplasmic reticulum	Chapter 9
Golgi complex	Chapter 9
Lysosome	Chapter 9
Peroxisome	Chapter 9
Ribosome	Chapter 16
Cytoskeleton	Chapter 18
Cell wall and cell coat	Chapter 7

themselves in the lipid layer accordingly. Hydrophobic regions of the protein associate with the interior of the membrane, whereas hydrophilic regions protrude into the aqueous environment at the surface of the membrane. Proteins exposed on the external side of the plasma membrane frequently have carbohydrate side chains attached to them and are therefore called *glycoproteins.*

The proteins of the plasma membrane play a variety of roles. Some are *transport proteins,* responsible for moving specific substances across an otherwise impermeable membrane. Others are *enzymes,* which catalyze reactions known to be associated with the membrane. Still others function as *receptors* for specific chemical signals that impinge on the cell from its environment. Membrane proteins are also important as "anchors" for structural elements of the cytoskeleton that we will encounter later in the chapter.

The Nucleus

If we now enter the cell, one of the most prominent structures we encounter is the **nucleus** (Figure 4-10). The nucleus serves as the control center for the entire cell. Here, separated from the rest of the cell by a membrane, are the DNA-bearing *chromosomes* of the cell. Actually, the membrane boundary around the nucleus consists of two membranes and is more properly called the **nuclear envelope.** Unique to the membranes of the nuclear envelope are numerous small openings called **pores** (Figure 4-10c). Each pore is a channel through which water-soluble molecules can move between the nucleus and cytoplasm. Ribosomes, messenger RNA molecules, chromosomal proteins, and enzymes needed for nuclear activities are also presumed to be transported across the nuclear envelope through its pores.

The number of chromosomes within the nucleus is characteristic of the species. It can be as low as two (in the sperm and egg cells of some grasshoppers, for example), or it can run into the hundreds. Chromosomes are most readily visualized during mitosis, since chromosomes in dividing cells are highly condensed and can easily be stained (Figure 4-5). During the *interphase* between divisions, on the other hand, chromosomes are dispersed as **chromatin** and are not easy to visualize (Figure 4-10b).

Also present in the nucleus are **nucleoli** (singular: **nucleolus**), structures responsible for the synthesis and assembly of the subunits that make up ribosomes. Nucleoli are usually associated with specific regions of particular chromosomes. The semifluid matrix that fills the remainder of the nucleus is called the **nucleoplasm.**

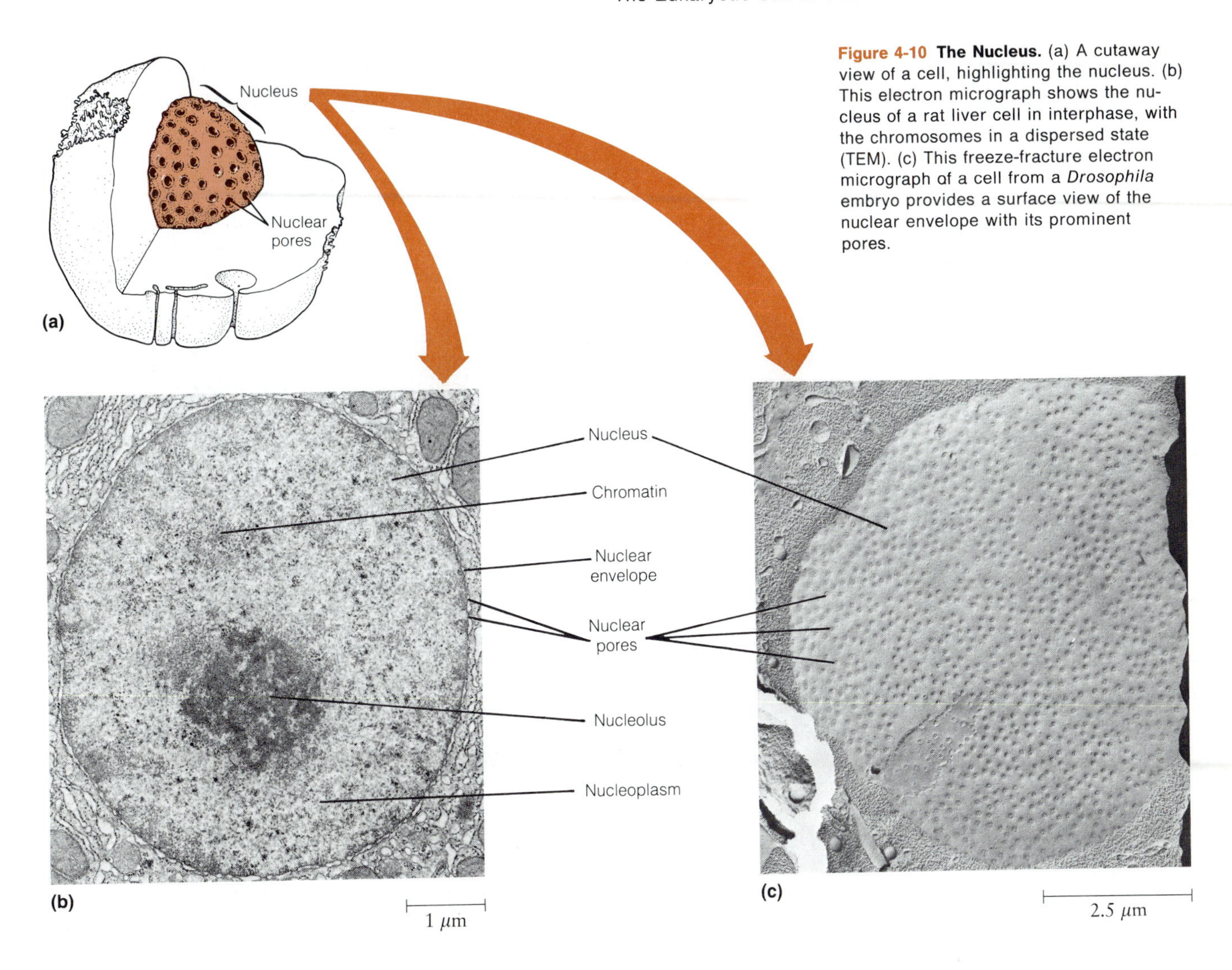

Figure 4-10 The Nucleus. (a) A cutaway view of a cell, highlighting the nucleus. (b) This electron micrograph shows the nucleus of a rat liver cell in interphase, with the chromosomes in a dispersed state (TEM). (c) This freeze-fracture electron micrograph of a cell from a *Drosophila* embryo provides a surface view of the nuclear envelope with its prominent pores.

Intracellular Membranes and Organelles

The internal volume of the cell exclusive of the nucleus is occupied by membrane-bounded *organelles* and by the *cytoplasm* in which they are suspended. In this section, we will look at each of the major eukaryotic organelles. In a typical animal cell, these compartments make up almost half of the total internal volume of the cell.

The Mitochondrion. Our tour of the eukaryotic organelles begins with a very prominent organelle, the **mitochondrion.** The structure of the mitochondrion is shown in Figure 4-11. Mitochondria are large by cellular standards—up to a micrometer across and usually a few micrometers long. A mitochondrion is therefore comparable in size to a whole bacterial cell. The fact that most eukaryotic cells contain hundreds of mitochondria, each approximately the size of an entire bacterial cell, emphasizes again the great difference in size between prokaryotic and eukaryotic cells.

The mitochondrion is surrounded by two membranes, designated the **inner** and **outer mitochondrial membranes.** Both mitochondria and the chloroplasts, which we will meet next, contain their own DNA and ribosomes and can therefore code for and synthesize some (though by no means all) of their own proteins.

The mitochondrion is often spoken of as the "powerhouse" of the eukaryotic cell, since most of the chemical reactions involved in the oxidation of sugars and other cellular "fuel" molecules occur within this organelle. The purpose of these oxidative events is to extract energy from foodstuffs and conserve as much of it as possible in the form of the high-energy compound *adenosine triphosphate* (ATP).

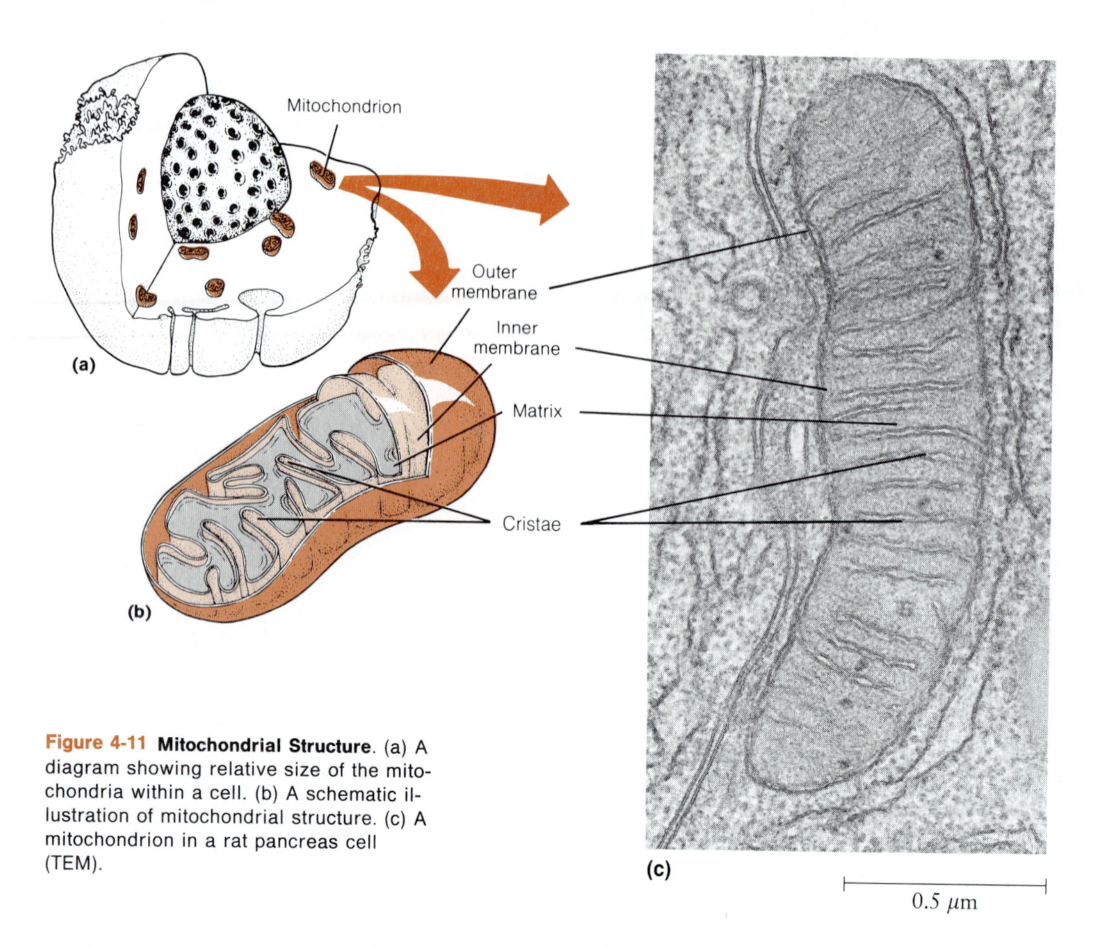

Figure 4-11 Mitochondrial Structure. (a) A diagram showing relative size of the mitochondria within a cell. (b) A schematic illustration of mitochondrial structure. (c) A mitochondrion in a rat pancreas cell (TEM).

It is within the mitochondrion that the cell localizes most of the enzymes and intermediates involved in such important cellular processes as the Krebs cycle, fat oxidation, and ATP generation. Most of the intermediates involved in the transport of electrons from oxidizable food molecules to oxygen are located in or on the **cristae,** infoldings of the inner mitochondrial membrane. Other reaction sequences, particularly those of the Krebs cycle and those involved in fat oxidation, occur in the semifluid **matrix** that fills the inside of the mitochondrion.

The number and location of mitochondria within a cell can often be related directly to their role in that cell. Tissues with an especially heavy demand for ATP as an energy source can be expected to have cells that are well endowed with mitochondria, and the organelles are usually located within the cell just where the energy need is greatest. This localization is illustrated by the sperm cell shown in Figure 4-12. As the diagram indicates, a sperm cell often has a single spiral mitochondrion wrapped around the central shaft, or *axoneme,* of the cell. The numerous mitochondrial profiles seen along the length of the sperm are therefore multiple cross sections of the same mitochondrion. Notice how tightly the mitochondrion coils around the axoneme, just where the ATP is actually needed to propel the sperm cell. Muscle cells and cells that specialize in the transport of ions also have numerous mitochondria located strategically to meet the special energy needs of such cells.

The Chloroplast. Next, our gallery tour takes in the **chloroplast,** in many ways a close relative of the mitochondrion. A typical chloroplast is shown in Figure 4-13. Chloroplasts are large organelles, typically a few micrometers in diameter and 5–10 μm long. Chloroplasts are therefore substantially bigger than mitochondria and larger than any other structure in the cell except the nucleus. Like mitochondria, chloroplasts are surrounded by both an inner and an outer membrane. In addition, they have a third membrane system consisting of flattened sacs called **thylakoid disks** and the membranes (**stroma lamellae**) that interconnect them. Thylakoid disks are stacked together to form the **grana** that are so characteristic of most chloroplasts (Figure 4-13c and d).

Chloroplasts, of course, are the site of *photosynthesis,* the light-driven process whereby carbon dioxide and water are used to manufacture the sugars and other organic compounds from which all life is ultimately fabricated. Chloroplasts are found in leaves and other photosynthetic tissues of higher plants, as well as in all of the eukaryotic algae. Located within the organelle are most of the en-

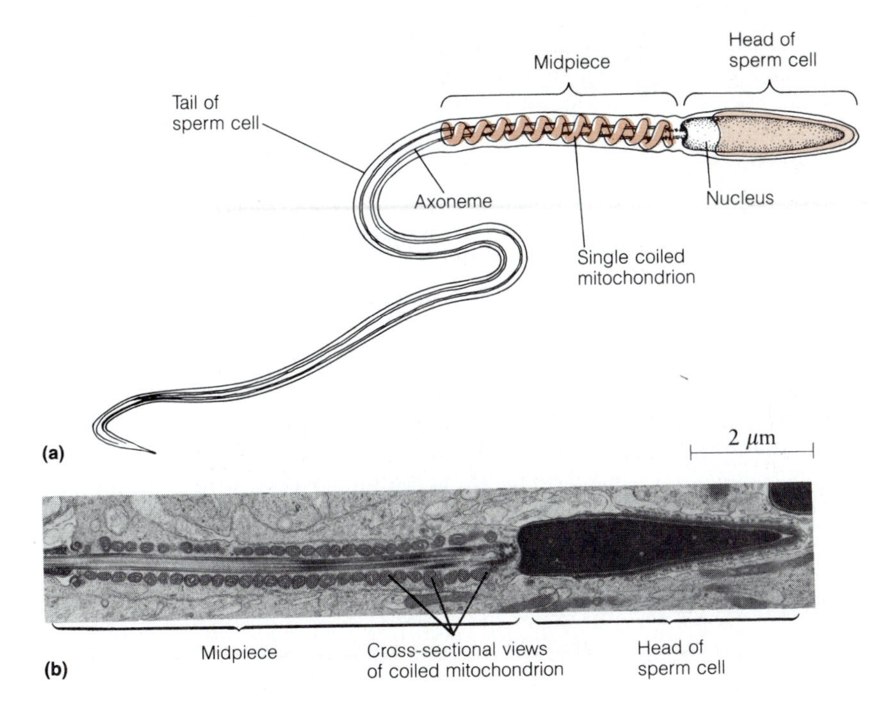

Figure 4-12 Localization of the Mitochondrion Within a Sperm Cell. The single mitochondrion present in a sperm cell is coiled tightly around the axoneme of the tail, reflecting the localized need of the sperm tail for energy. (a) A schematic diagram of a sperm. (b) An electron micrograph of a marmoset monkey sperm cell (TEM).

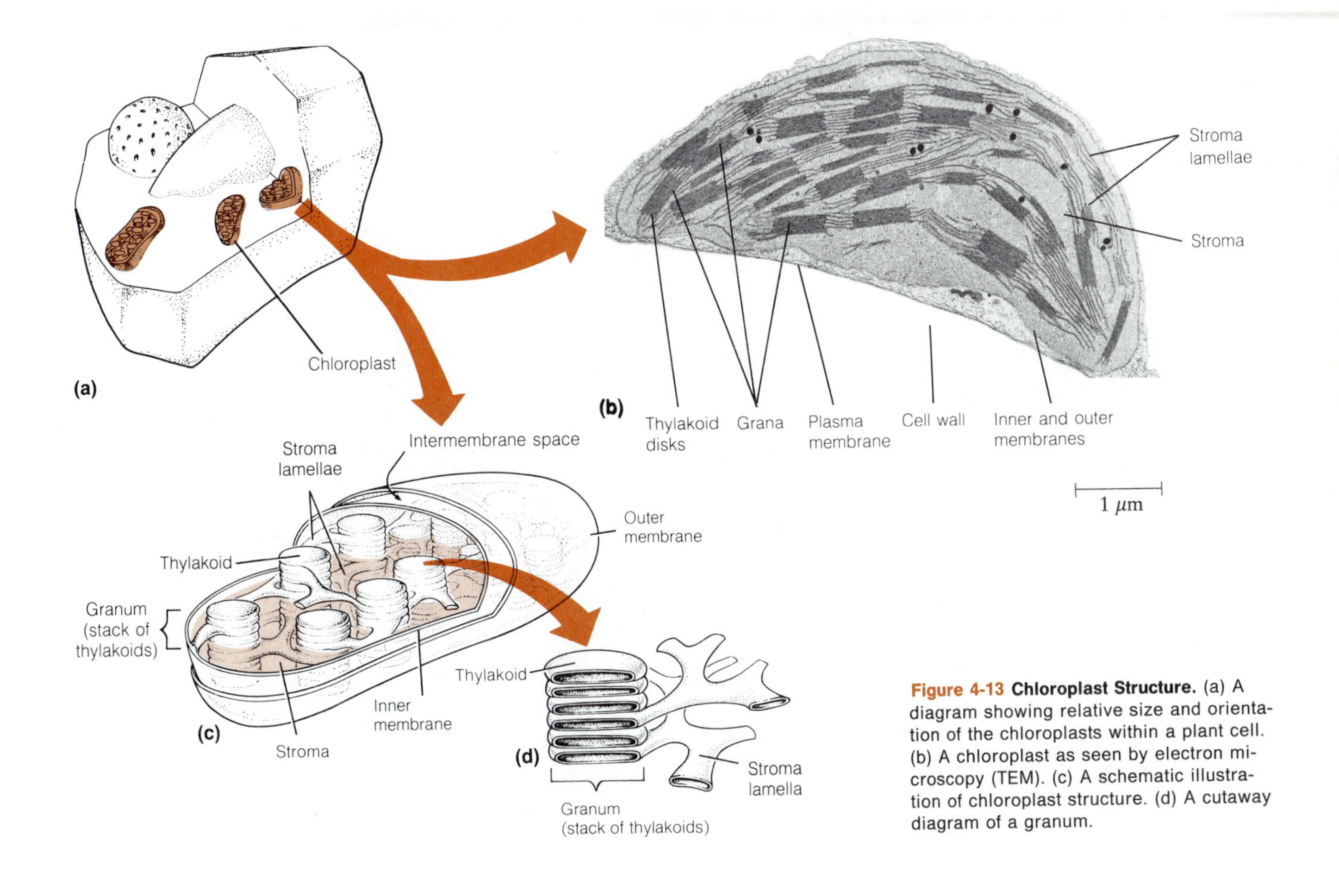

Figure 4-13 Chloroplast Structure. (a) A diagram showing relative size and orientation of the chloroplasts within a plant cell. (b) A chloroplast as seen by electron microscopy (TEM). (c) A schematic illustration of chloroplast structure. (d) A cutaway diagram of a granum.

zymes, intermediates, and pigments (light-absorbing molecules) needed to "fix" carbon from carbon dioxide into organic form and convert it reductively into sugars. The reactions that depend directly on solar energy are localized in or on the thylakoid membrane system. Reactions involved in the initial trapping of carbon dioxide into organic form and its subsequent reduction and rearrangement into sugar molecules occur within the semifluid **stroma** that fills the interior of the chloroplast.

Although known primarily for their role in photosynthesis, chloroplasts are involved in a variety of other processes as well. An important example involves the reduction of nitrogen from the oxidation level of the nitrate (NO_3^-) that plants obtain from the soil to the oxidation level of ammonia (NH_3), the form of nitrogen required for protein synthesis. Furthermore, the chloroplast is only the most prominent example of a broader class of plant organelles, the **plastids.** Plastids serve a variety of functions in plant cells. *Chromoplasts,* for example, are pigment-containing plastids that are responsible for the characteristic coloration of flowers, fruits, and other plant parts. *Amyloplasts* are plastids that are specialized for starch storage.

The Endoplasmic Reticulum. Extending throughout the cytoplasm of almost every eukaryotic cell is a network of membranes called the **endoplasmic reticulum,** or **ER** (Figure 4-14). The name sounds complicated, but *endoplasmic* just means "within the plasm" (of the cell) and *reticulum* is simply a fancy word for "network." The endoplasmic reticulum consists of tubular membranes and flattened sacs, or **cisternae,** that appear to be interconnected. The internal space enclosed by the ER membranes is called the **lumen.** The ER is continuous with the outer membrane of the nuclear envelope (Figure 4-14a and b). The space between the two nuclear membranes is therefore a part of the same compartment as the lumen of the ER.

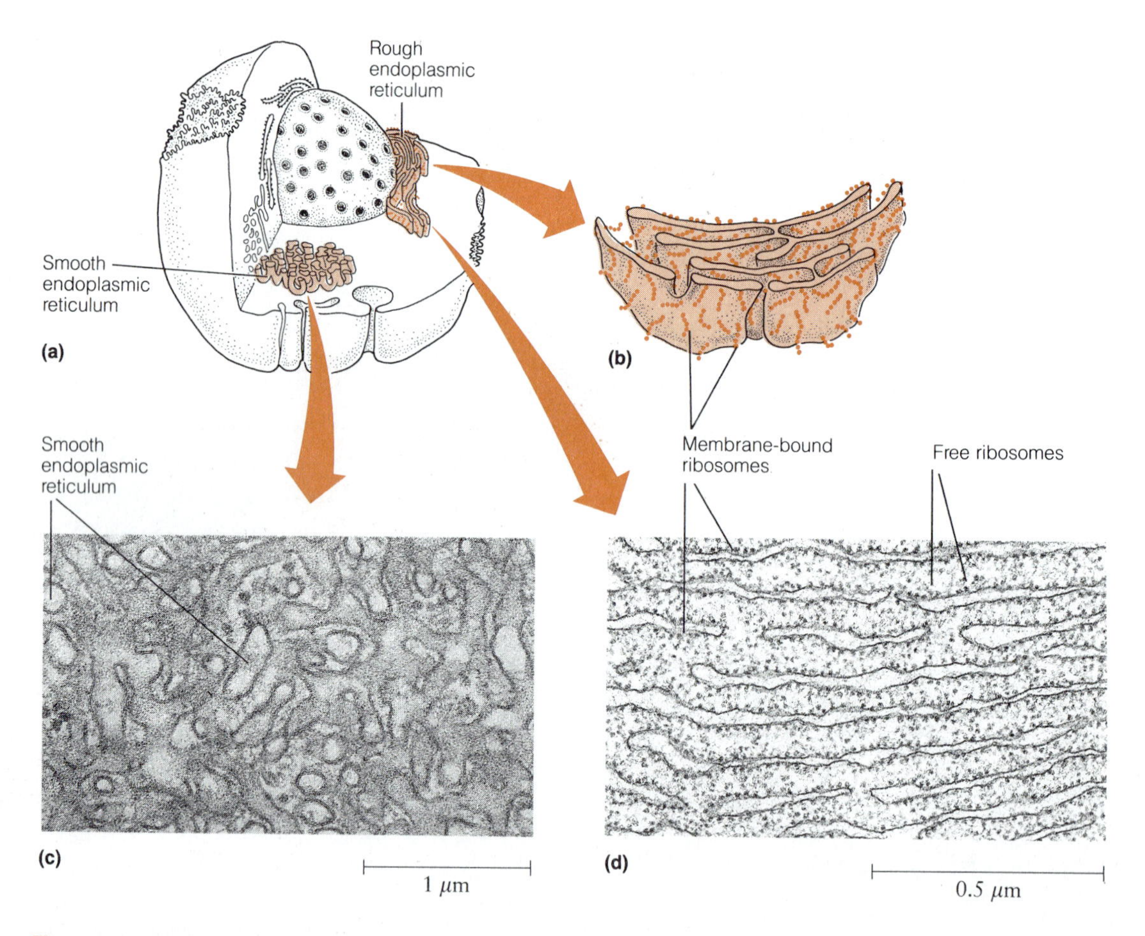

Figure 4-14 The Endoplasmic Reticulum. (a) A schematic diagram showing the location and relative size of the endoplasmic reticulum (ER) within a typical animal cell. (b) A schematic illustration depicting organization of the rough ER as layers of flattened membranes studded on the outer surface with ribosomes. (c) An electron micrograph of smooth ER in a cell from guinea pig testis (TEM). (d) An electron micrograph of rough ER in a rat pancreas cell (TEM); note that ribosomes are either attached to the ER or free in the cytoplasm.

The ER can be either *rough* or *smooth*. **Rough endoplasmic reticulum** appears "rough" in the electron microscope because it is studded with ribosomes on the side of the membrane that faces the cytoplasm (Figure 4-14d). These ribosomes are actively synthesizing proteins. Most of these proteins are transported into or across the membrane as they are synthesized, accumulating in completed form within either the membrane or the lumen of the ER. *Secretory proteins* (proteins destined to be exported from the cell) are synthesized in this way. They then make their way to the cell surface by a complex process that involves not only the rough ER but also the Golgi complex and secretory vesicles, to be described shortly.

Not all proteins are synthesized on the rough ER, however. Much protein synthesis occurs on ribosomes that are not attached to the ER but are found in the cytoplasm instead. In general, secretory proteins and membrane proteins are made by ribosomes on the rough ER, whereas proteins intended for use within the cytoplasm are made on free ribosomes.

The **smooth endoplasmic reticulum** has no role in protein synthesis and hence no ribosomes. It therefore has a characteristic "smooth" appearance when viewed by electron microscopy (Figure 4-14c). The smooth ER is involved in the packaging of secretory proteins, as well as in the synthesis of lipids. In addition, the smooth ER is responsible for the inactivation and detoxification of drugs and other compounds that might otherwise be toxic or harmful to the cell.

The Golgi Complex. Closely related to the smooth ER in both proximity and function is the **Golgi complex** (or *Golgi apparatus*), named after its Italian discoverer, Camillo Golgi. The Golgi complex consists of a stack of flattened vesicles, illustrated in Figure 4-15. The Golgi complex plays an important role in the processing and packaging of secretory proteins and in the synthesis of complex polysaccharides. Vesicles that arise by budding off the ER are accepted by the Golgi complex. Here, the contents of the vesicles (proteins, for the most part) and sometimes the vesicle membranes are further modified and processed. The processed contents are then passed on to other components of the cell by means of vesicles that arise by budding off the Golgi complex (Figure 4-15c).

Many membrane proteins and most secretory proteins are glycoproteins, meaning that they have sugar groups attached to them. The initial steps in *glycosylation* (sugar addition) take place within the lumen of the ER, but the process is usually completed within the Golgi complex. The Golgi complex should therefore be understood primarily as a processing station, with vesicles both fusing with it and arising from it. Almost everything that goes into it comes back out, but in a modified, packaged form, often ready for export from the cell.

Secretory Vesicles. Once processed by the Golgi complex, secretory proteins and other substances intended for export from the cell are packaged into **secretory vesicles.** The cells of your pancreas, for example, are likely to contain many

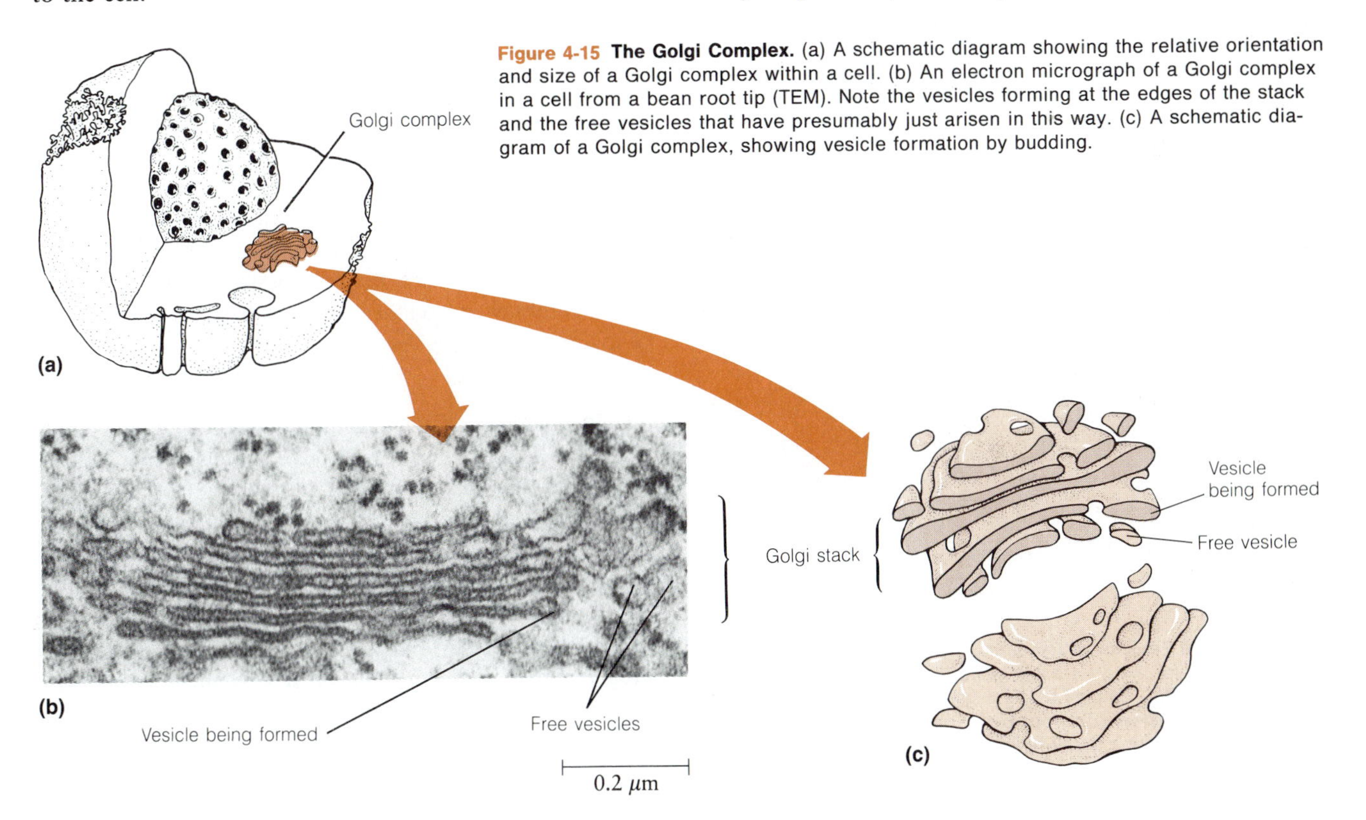

Figure 4-15 The Golgi Complex. (a) A schematic diagram showing the relative orientation and size of a Golgi complex within a cell. (b) An electron micrograph of a Golgi complex in a cell from a bean root tip (TEM). Note the vesicles forming at the edges of the stack and the free vesicles that have presumably just arisen in this way. (c) A schematic diagram of a Golgi complex, showing vesicle formation by budding.

such vesicles, since the pancreas is responsible for the synthesis of several important digestive enzymes. These enzymes are synthesized on the rough ER, packaged by the Golgi complex, and then released from the cell via secretory vesicles, as shown in Figure 4-16. These vesicles move from the Golgi region to the plasma membrane that surrounds the cell. The vesicles then fuse with the plasma membrane and discharge their contents to the exterior of the cell by the process of exocytosis. The whole process of protein synthesis, processing, and export via the ER, the Golgi complex, and secretory vesicles will be considered in more detail when we get to Chapter 16.

The Lysosome. The next picture at our cellular exhibition is that of the **lysosome,** an organelle that is about 0.5–1.0 μm in diameter and is surrounded by a single membrane (Figure 4-17). Lysosomes were discovered in the early 1950s by Christian de Duve and his colleagues. The story of that discovery is recounted in the box on pp. 94–95, both to underscore the significance of chance observations when they are made by the right people and to illustrate the importance of new techniques to the progress of science. In this case, the new technique was that of *differential centrifugation,* which allows cellular contents to be fractionated according to size and density. Use of this technique led de Duve and his colleagues to the realization that an acid phosphatase initially thought to be located in the mitochondrion was in fact associated with a class of particles that had never been reported before. Along with acid phosphatase, these organelles contained several other hydrolytic enzymes. Because of its apparent role in cellular lysis, de Duve gave the new organelle the name *lysosome.*

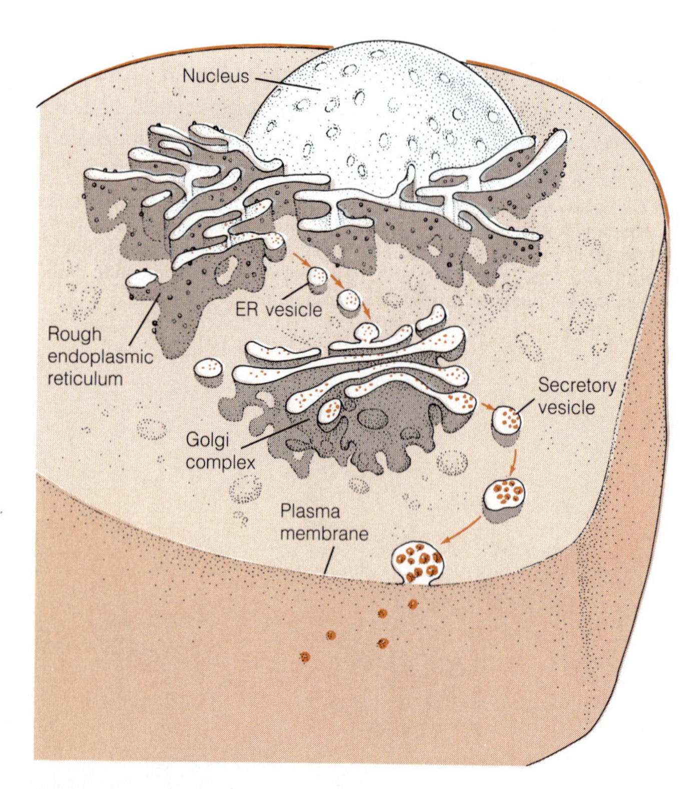

Figure 4-16 The Process of Secretion in Eukaryotic Cells. Proteins to be packaged for export are synthesized on the rough ER, passed to the Golgi complex for processing, and eventually compartmentalized into secretory vesicles. These vesicles then make their way to the plasma membrane and fuse with it to release their contents to the exterior of the cell.

Lysosomes are used by the cell as a means of storing *hydrolases,* enzymes capable of digesting specific biological molecules such as proteins, carbohydrates, or fats. It is important for cells to possess such enzymes, both to digest food molecules that the cell may acquire from its environment and to break down cellular constituents that are no longer needed. But it is also essential that such enzymes be carefully sequestered until actually needed, lest they digest cellular components that were not scheduled for destruction.

We want to consider the lysosome at this point in our overview because of its relationship to the ER and the Golgi complex. Lysosomal enzymes are somewhat similar to secretory proteins in their synthesis and packaging. They are thought to be synthesized on the rough ER and transported to the Golgi, probably in an inactive form to prevent unwanted digestion of the structures through which they pass. Lysosomes then develop by budding off the ends of the Golgi cisternae. The resulting organelle is a *primary lysosome*—a lysosome containing hydrolytic enzymes but not yet engaged in digestive activity.

To initiate the digestion process, a lysosome must encounter and fuse with a membrane-bounded vacuole containing food particles. Such fused vacuoles are called *secondary lysosomes.* The hydrolytic enzymes of the secondary lysosome are then free to digest the contents of the vacuole, breaking them down to smaller and smaller components. Eventually, the digestion products are small enough to pass through the membrane out into the cytoplasm of the cell, where they can be utilized for the synthesis of macromolecules—recycling at the cellular level!

The Peroxisome. The next organelle at our exhibition is the **peroxisome.** Peroxisomes resemble lysosomes in size, mode of origin, and general lack of obvious internal structure. Like lysosomes, they are surrounded by a single, rather than a double, membrane. Peroxisomes are found in both plant and animal cells. They carry out several distinctive functions that differ with cell type but have the common property of both generating and degrading hydrogen peroxide (H_2O_2). Hydrogen peroxide is highly toxic but can be decomposed into water and oxygen by the enzyme *catalase.* Eukaryotic cells protect themselves from the detrimental effects of hydrogen peroxide by packaging

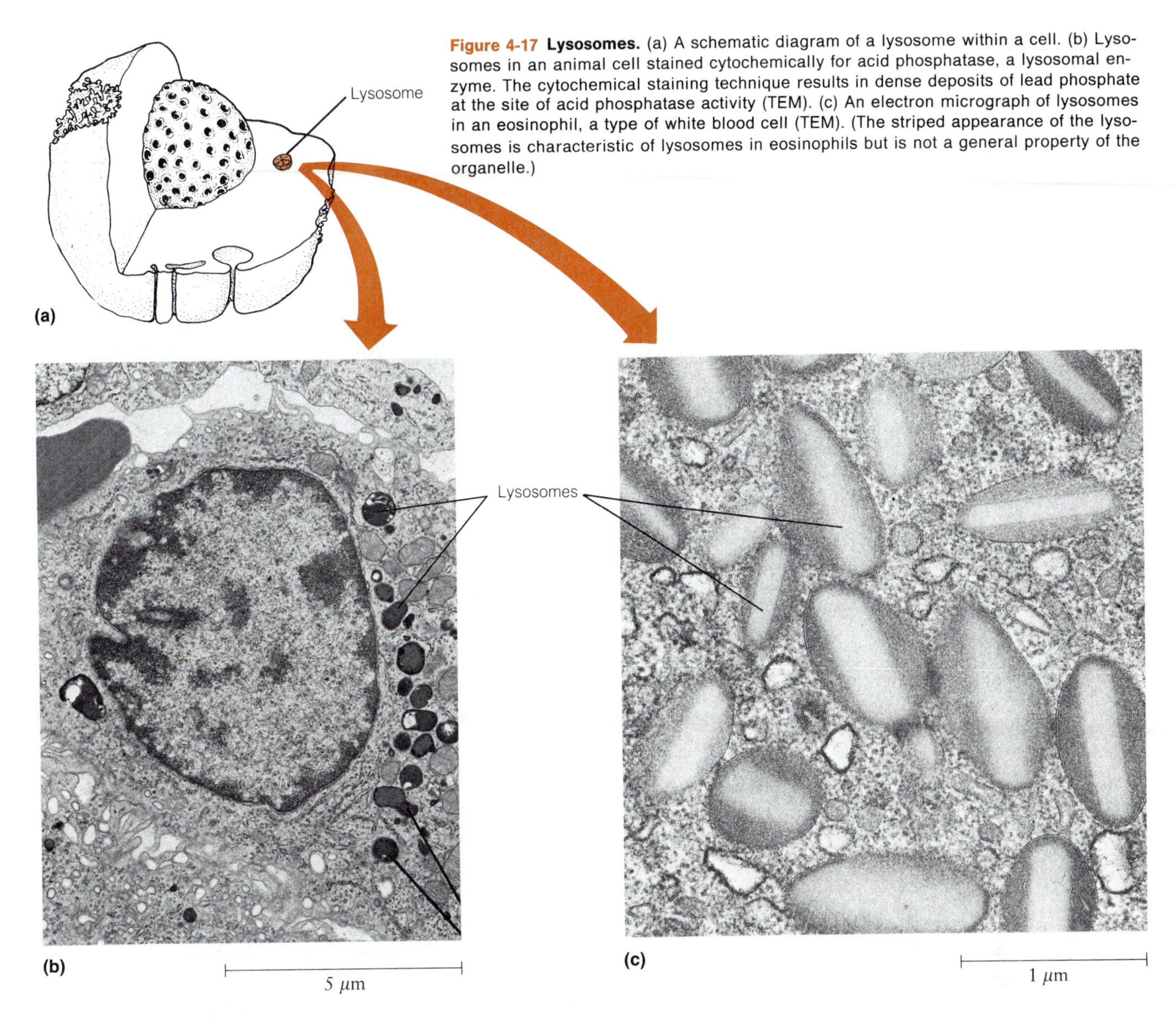

Figure 4-17 Lysosomes. (a) A schematic diagram of a lysosome within a cell. (b) Lysosomes in an animal cell stained cytochemically for acid phosphatase, a lysosomal enzyme. The cytochemical staining technique results in dense deposits of lead phosphate at the site of acid phosphatase activity (TEM). (c) An electron micrograph of lysosomes in an eosinophil, a type of white blood cell (TEM). (The striped appearance of the lysosomes is characteristic of lysosomes in eosinophils but is not a general property of the organelle.)

peroxide-generating reactions together with catalase in a single compartment, the peroxisome.

The best-understood metabolic roles of peroxisomes occur in plant cells. During the germination of fat-storing seeds, specialized peroxisomes called **glyoxysomes** play a key role in the conversion of stored fat into carbohydrate. In photosynthetic tissue, **leaf peroxisomes** are prominent because of their role in *photorespiration,* a light-dependent pathway that detracts from the efficiency of photosynthesis in many plants by "unfixing" some of the carbon that is fixed by the chloroplasts (to be discussed further in Chapter 12). The photorespiratory pathway is an example of a cellular process that involves several organelles: some of the enzymes that catalyze the reactions in this sequence occur in the peroxisome, whereas others are located in the chloroplast or the mitochondrion. This mutual involvement in a common cellular process is suggested by the intimate association of peroxisomes with mitochondria and chloroplasts in many leaf cells, as Figure 4-18 illustrates.

Vacuoles. Cells also contain a variety of other membrane-bounded organelles called **vacuoles.** In animal cells, vacuoles are frequently used for temporary storage or transport. Some protozoa, for example, take up food particles or other materials from their environment by a process called *phagocytosis* ("cell eating"). Phagocytosis is a form of endocytosis that involves an in-pocketing of the plasma membrane around the desired substance, followed by a pinching-off process that internalizes the membrane-bounded particle as a vacuole.

The term *vacuole* is also used with plant cells, but usually in reference to the large **central vacuole** that is so characteristic of most mature plant cells (Figure 4-19). Al-

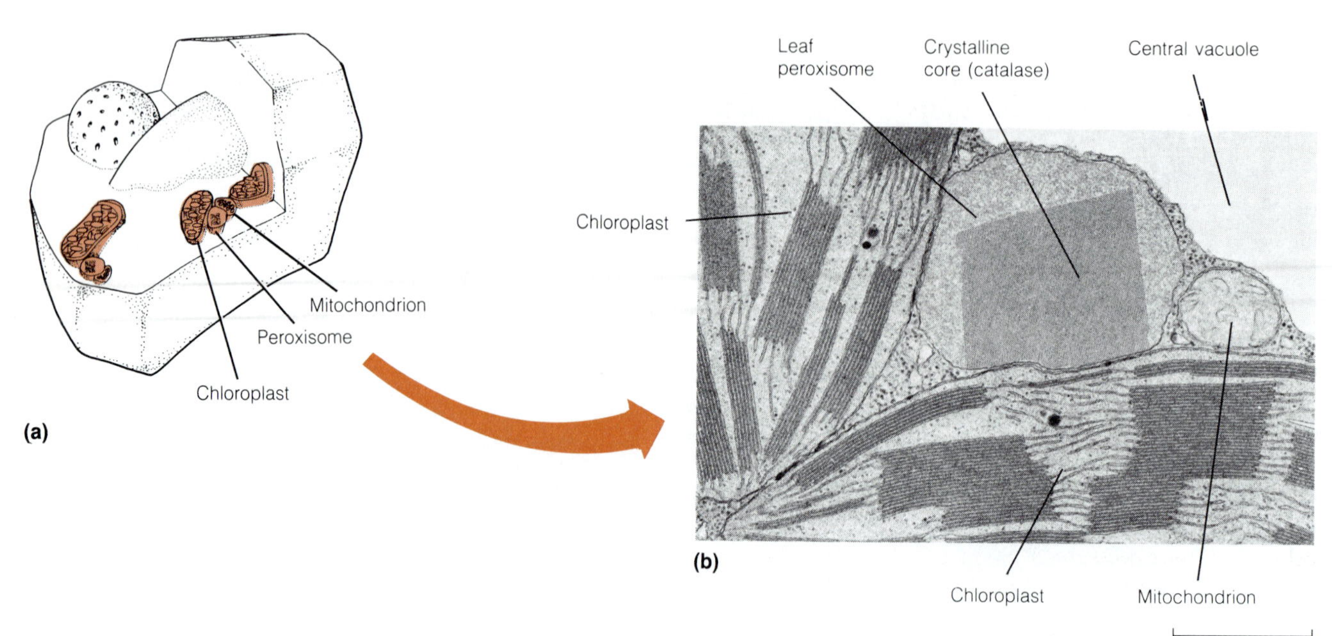

Figure 4-18 A Leaf Peroxisome and Its Relationship to Other Organelles. (a) A schematic diagram of a peroxisome, a mitochondrion, and a chloroplast within a cell. (b) A peroxisome in close proximity to both chloroplasts and a mitochondrion within a tobacco leaf cell (TEM). This is probably a functional relationship, since all three organelles participate in the process of photorespiration. The crystalline core frequently observed in leaf peroxisomes is thought to be the enzyme catalase, which catalyzes the decomposition of hydrogen peroxide into water and oxygen.

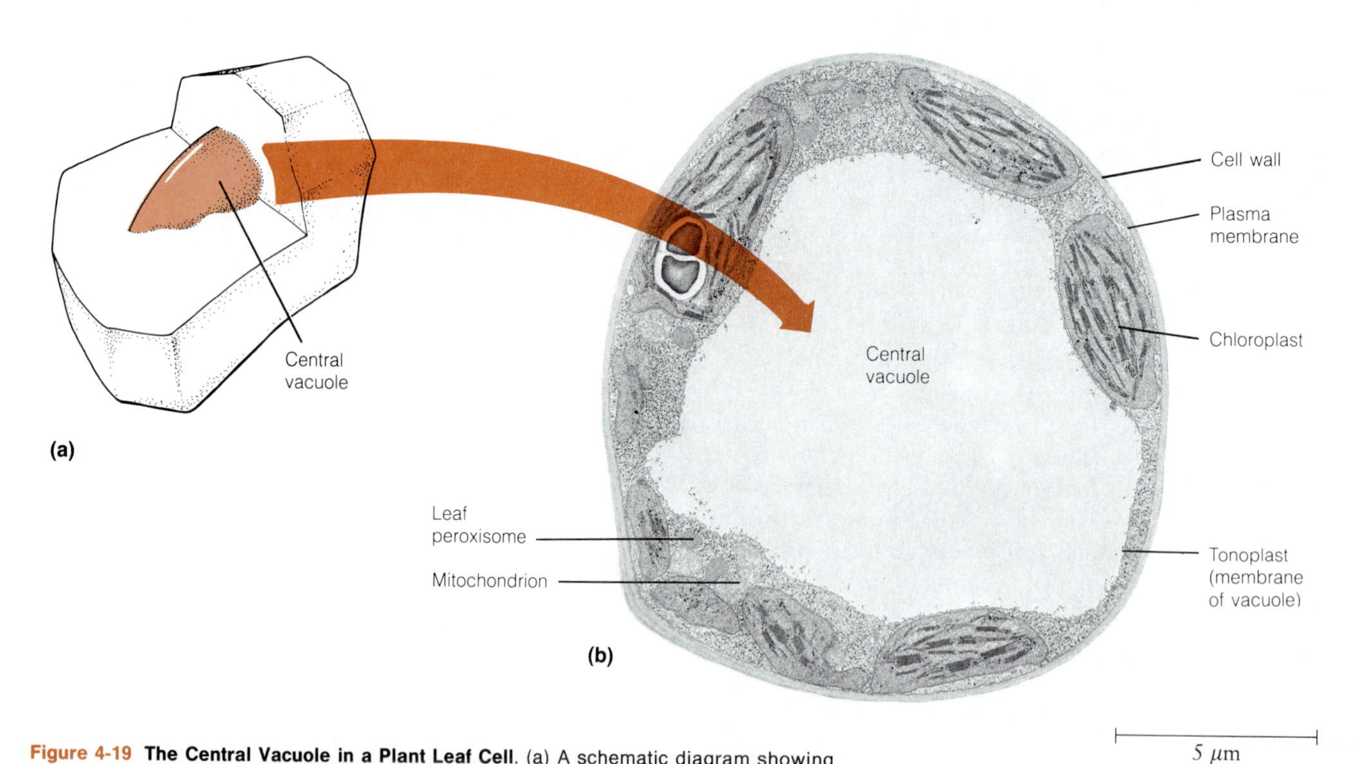

Figure 4-19 The Central Vacuole in a Plant Leaf Cell. (a) A schematic diagram showing the central vacuole in a plant cell. (b) An electron micrograph of a bean leaf cell with a large central vacuole (TEM). The central vacuole occupies much of the internal volume of the cell, with the cytoplasm sandwiched into a thin sphere between the vacuole and the plasma membrane. The membrane of the vacuole is called the tonoplast.

though the central vacuole may play a limited role in storage and appears also to be capable of a lysosomelike function in intracellular digestion, its real importance lies in the maintenance of the *turgor pressure* of the plant cell.

As you already know, a plant cell is surrounded by a rigid, nonliving cell wall. The central vacuole is like an inflatable sphere in the center of the cell that is "pumped up" with liquid, pressing the rest of the cellular constituents out against the cell wall and thereby maintaining the turgor pressure characteristic of nonwilted plant tissue. The limp, flaccid appearance associated with wilting comes about when the central vacuole does not provide adequate pressure, allowing the cell (and hence the tissue) to go limp. We can easily demonstrate this by placing a piece of crisp celery in salt water. The high concentration of salt on the outside of the cells will cause water to move out of the cells; the turgor pressure will then decrease, and the tissue will quickly become flaccid and lose its crispness.

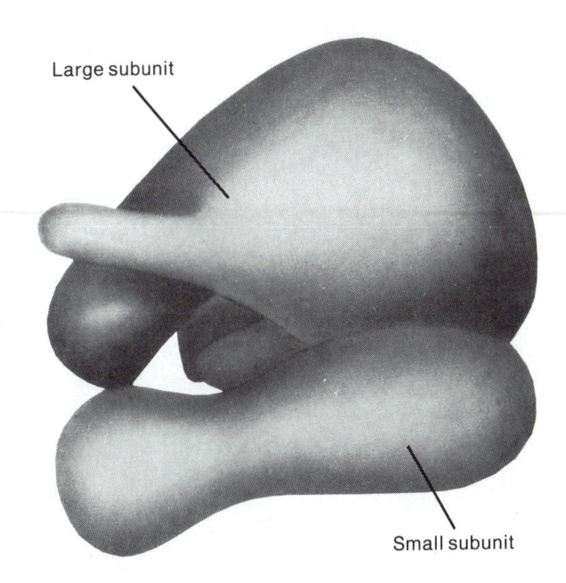

Figure 4-20 Structure of a Eukaryotic Ribosome. Each ribosome is made up of a large subunit and a small subunit that join together when they attach to messenger RNA and begin to make a protein. The large subunit of a eukaryotic ribosome consists of 45 different protein molecules and 3 molecules of RNA. The small subunit has 33 proteins and 1 molecule of RNA. The fully assembled ribosome of a eukaryotic cell measures about 20 nm by 30 nm. The ribosomes and ribosomal subunits of prokaryotic cells are slightly smaller than those of eukaryotic cells and consist of their own distinctive protein and RNA molecules. Unlike other organelles, a ribosome is not surrounded by a membrane.

Ribosomes. The last portrait in our gallery of organelles is the **ribosome,** which serves as the "workbench" for protein synthesis. Strictly speaking, a ribosome should not be considered an organelle, since it is not bounded by a membrane. But it is convenient to consider ribosomes at this time because ribosomes, like organelles, are the focal point for a specific cellular activity—in this case, protein synthesis. Unlike true organelles, ribosomes are found in both eukaryotic and prokaryotic cells. Even here, however, the dichotomy between the two basic cell types manifests itself, since prokaryotic and eukaryotic ribosomes differ characteristically in size and in the number and kinds of protein and RNA molecules that they contain. Figure 4-20 is a schematic diagram of the structure of a eukaryotic ribosome.

Compared to membrane-bounded organelles, ribosomes are tiny structures. The ribosomes of eukaryotic cells measure about 20 nm by 30 nm, and those of prokaryotic cells are slightly smaller. An electron microscope is therefore required to visualize ribosomes (recall Figures 4-6 and 4-14). To appreciate how small ribosomes are, consider that more than 350,000 ribosomes could fit inside a typical bacterial cell, with room to spare!

Another way to express the size of such a small particle is to refer to its **sedimentation coefficient.** The sedimentation coefficient of a particle or macromolecule is a measure of how rapidly the particle sediments in an ultracentrifuge and is expressed in *Svedberg units* (*S*). Sedimentation coefficients are widely used to indicate relative size, especially for large macromolecules such as proteins and nucleic acids and small particles such as ribosomes. Ribosomes from eukaryotic cells have sedimentation coefficients of about 80S; those from prokaryotic cells are about 70S (Table 4-1).

A ribosome consists of two subunits differing in size, shape, and composition (Figure 4-20). In eukaryotic cells, the **large** and **small ribosomal subunits** have sedimentation coefficients of about 60S and 40S, respectively. For prokaryotic ribosomes, the corresponding values are about 50S and 30S. (Note that the sedimentation coefficients of the subunits do not add up to that of the intact ribosome; this is because sedimentation coefficients depend critically on both size and shape and are therefore not linearly related to molecular weight.) In both eukaryotic and prokaryotic cells, ribosomal subunits are synthesized and assembled separately in the cell but come together for the purpose of making proteins.

Ribosomes are far more numerous than most other cellular structures. Prokaryotic cells usually contain thousands of ribosomes, and eukaryotic cells may have hundreds of thousands or even millions of them. Ribosomes are also found in both chloroplasts and mitochondria, where they function in organelle-specific protein synthesis. It is worth noting that the ribosomes of these eukaryotic organelles differ in size and composition from the ribosomes found in the cytoplasm of the same cell, but are strikingly similar to those found in bacteria and blue-green algae. This similarity is often cited as support for the theory that mitochondria and chloroplasts arose from

Discovering Organelles: The Importance of Centrifuges and Chance Observations

Have you ever wondered how the various organelles and other subcellular structures found within eukaryotic cells were discovered? There are almost as many answers to that question as there are kinds of organelles. In general, such structures were described by microscopists before their role in the cell was understood. As a result, the names of organelles usually reflect structural features rather than physiological roles. Thus, *chloroplast* simply means "green particle" and *endoplasmic reticulum* just means "network within the plasm (of the cell)," as we noted earlier.

Such is not the case for the *lysosome*, however. This organelle was the first to have its biochemical properties described before it had ever been reported by microscopists. Only after fractionation data had predicted the existence and properties of such an organelle were lysosomes actually observed in cells. A suggestion of its function is even inherent in the name given to the organelle, because the Greek root *lys-* means "to digest." (The literal meaning is "to loosen," but that's essentially what digestion does to chemical bonds!)

The lysosome is something of a newcomer on the cellular biology scene, since it was not discovered until the early 1950s. The story of that discovery is fascinating because it illustrates how important chance observations can be, especially when made by the right persons at the right time. The account also illustrates how significant new techniques can be, since the discovery depended on subcellular fractionation, a technique that at the time was still in its infancy.

The story begins in 1949 in the laboratory of Christian de Duve, who has since received a Nobel Prize for this work. Like so many scientific advances, the discovery of lysosomes depended on a chance observation made by an astute investigator. Because of an interest in the effect of insulin on carbohydrate metabolism, de Duve was attempting at the time to pinpoint the cellular location of *glucose-6-phosphatase*, the enzyme responsible for the release of free glucose in liver cells. As a control enzyme (that is, one not involved in carbohydrate metabolism), de Duve happened to choose *acid phosphatase*.

De Duve first homogenized liver tissue and resolved it into several fractions by the new technique of *differential centrifugation*, which separates cellular components on the basis of differences in size and density. In this way, he was able to show that the glucose-6-phosphatase activity could be recovered with the microsomal fraction. (*Microsomes* are little vesicles that form from fragments of the endoplasmic reticulum when tissue is homogenized.) This in itself was an important observation because it helped to establish the identity of microsomes, which at the time tended to be dismissed as fragments of mitochondria.

But the acid phosphatase results turned out to be even more interesting, even though they were at first quite puzzling. When de Duve and his colleagues assayed their liver homogenates for this enzyme, they found only a fraction of the expected activity. When assayed again for the same enzyme a few days later, however, the same homogenates had about ten times as much activity. Many investigators might have been tempted to dismiss the discrepancy as some sort of unfortunate error, but not de Duve. Instead, he pursued it, even though it had little or nothing to do with insulin or carbohydrate metabolism.

Speculating that he was dealing with some sort of activation phenomenon, de Duve subjected the homogenates to differential centrifugation to see with what subcellular fraction the phenomenon was associated. He and his colleagues were able to demonstrate that much of the acid phosphatase activity could be recovered in the mitochondrial fraction and that this fraction showed an even greater increase in activity after standing a few days than did the original homogenates.

To their surprise, they then discovered that upon recentrifugation, this elevated activity no longer sedimented with the mitochondria but stayed in the supernatant. They went on to show that the activity could be increased and the enzyme solubilized by a variety of treatments, including harsh grinding, freezing and thawing, or exposure to detergents or hypotonic conditions. From these results, de Duve concluded that the enzyme must be present in some sort of membrane-bounded particle that could easily be ruptured to release the enzyme. Apparently, the enzyme could not be detected within the particle, probably because the membrane was not permeable to the substrates used in the enzyme assay.

Assuming that particle to be the mitochondrion, they continued to isolate and study this fraction of their liver homogenates. At this point, another chance observation occurred, this time because of a broken centrifuge. The unexpected breakdown forced one of de Duve's students to use an older, slower centrifuge, and the result was a mitochondrial fraction that had little or no acid phosphatase in it. Others might simply have thrown out the sample and waited until the newer centrifuge was repaired, but not de Duve. He used the finding to speculate that the mitochondrial fraction as they usually prepared it might in fact contain two kinds of organelles—the actual mitochondria, which could be sedimented with either centrifuge, and some sort of more slowly sedimenting particle that came down only in the faster centrifuge.

This led them to devise a fractionation scheme that allowed the original mitochondrial fraction to be subdivided into a rapidly sedimenting component and a slowly sedimenting component. As you might guess, the rapidly sedimenting component contained the mitochondria, as evidenced by the presence of enzymes known to be mitochondrial markers. The acid phosphatase, on the other hand, was in the slowly sedimenting component, along with several other hydrolytic enzymes, including ribonuclease, deoxyribonuclease, β-glucuronidase, and a protease. Each of these enzymes showed the same characteristic of increased activity upon membrane rupture, a property that de Duve termed *latency*.

By 1955, de Duve was convinced that these hydrolytic enzymes were packaged together in a previously undescribed organelle. In keeping with his speculation that this organelle was involved in intracellular lysis, he called it a *lysosome*.

Thus, the lysosome became the first organelle to be identified entirely on biochemical criteria. At the time, no such particles had been described by microscopy. But when de Duve's lysosome-containing fractions were examined with the electron microscope, they were found to contain membrane-bounded vesicles that were clearly not mitochondria and were in fact absent from the mitochondrial fraction. Knowing what the isolated particles looked like, microscopists were then able to search for them in fixed tissue. As a result, lysosomes were soon identified and reported in a variety of animal tissues. Within six years, then, the organelle that began as a puzzling observation in an insulin experiment became established as a bona fide feature of most animal cells. And what did it take? a powerful new technique, a couple of chance observations, and an astute investigator—which is always a winning combination!

specific prokaryotic ancestors—bacteria and blue-green algae, respectively.

The Cytoplasm and the Cytoskeleton

The **cytoplasm** of a eukaryotic cell consists of that portion of the interior of the cell not occupied by the nucleus or other membrane-bounded organelles. In a typical animal cell, the cytoplasm occupies more than half of the total internal volume of the cell. Many cellular activities take place in the cytoplasm, including the synthesis of proteins, the synthesis of fats, and the initial steps in the release of energy from sugars.

Until recently, the cytoplasm was regarded as a rather amorphous, gel-like substance, and its proteins were thought to be soluble and freely diffusible. However, several new techniques have done much to change this view greatly. We now know that the cytoplasm of eukaryotic cells, far from being a structureless fluid, is permeated by an intricate three-dimensional array of interconnected filaments and tubules called the **cytoskeleton,** or the **cytoskeletal network** (Figure 4-21). Specifically, the cytoskeleton consists of a network of microtubules, microfilaments, and intermediate filaments, all of which are unique to eukaryotes.

As the name suggests, the cytoskeleton is an internal framework that gives a eukaryotic cell its distinctive shape and high level of internal organization. This elaborate array of filaments and tubules forms a highly structured yet very dynamic matrix that not only helps establish and maintain shape but also plays important roles in cell movement and cell division. As we will see in later chapters, the filaments and tubules that make up the cytoskeleton play essential roles in various kinds of cell movement, including the contraction of muscle cells, the beating of cilia and flagella, the movement of chromosomes during cell division, and in some cases the locomotion of the cell itself.

In addition, the cytoskeleton serves as a framework for positioning and actively moving organelles within the cytoplasm. The same may be true of ribosomes and enzymes. Some researchers estimate that up to 80% of the proteins of the cytoplasm are not freely diffusible but are instead associated with the cytoskeleton. Even water, which accounts for about 70% of the cell volume, may be influenced by the cytoskeleton. It has been estimated that as much as 20–40% of the water in the cytoplasm may be bound to the filaments and tubules of the cytoskeleton.

The three major structural elements of the cytoskeleton are *microtubules, microfilaments,* and *intermediate filaments*. These structures occur only in eukaryotic cells. They can be visualized by phase-contrast and immunofluores-

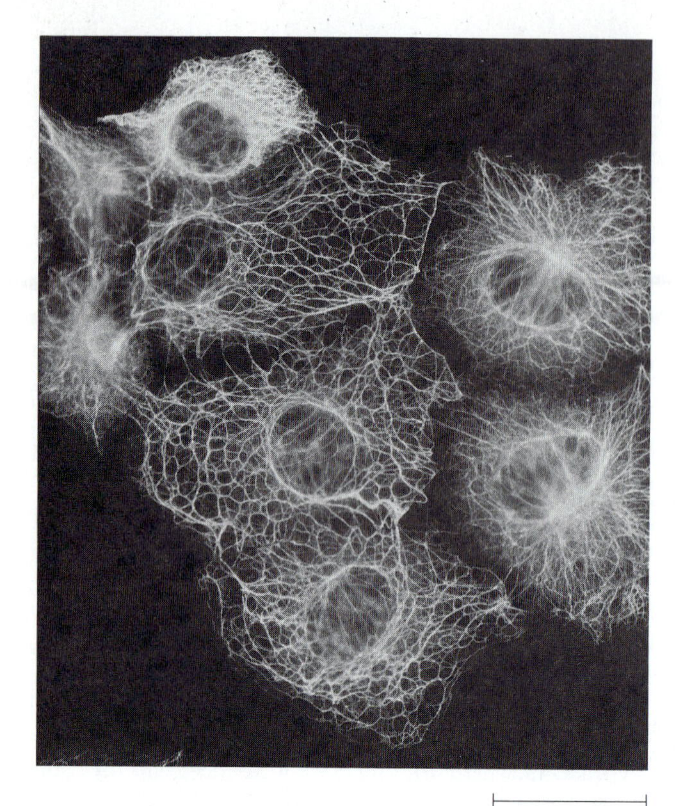

Figure 4-21 The Cytoskeleton. The cytoskeleton consists of a network of tubules and filaments that gives the cell shape, anchors organelles and directs their movement, and enables the whole cell to move or to change its shape. This micrograph shows the cytoskeleton of human amnion epithelial cells, as revealed by the technique of immunofluorescence microscopy using antibodies against keratin proteins found in the intermediate filaments of the cytoskeleton. Note that these filaments surround, but do not extend into, the space occupied by the nucleus.

cence microscopy and by electron microscopy. Some of the structures in which they are found (such as cilia, flagella, or muscle fibrils) can even be seen by ordinary light microscopy. Microfilaments and microtubules are best known for their roles in contraction and motility. In fact, these roles were appreciated well before it became clear that the same structural elements are also integral parts of the pervasive network of filaments and tubules that gives cells their characteristic shape and structure.

Chapter 18 provides a detailed description of the cytoskeleton, followed in Chapter 19 by a discussion of microtubule- and microfilament-mediated contraction and motility. Microtubules and microfilaments will also be encountered in Chapter 14 because of their roles in chromosome separation and cell division, respectively. Here, we will focus on the structural features of the three major components of the cytoskeleton—microtubules, microfilaments, and intermediate filaments.

Microtubules. Of the structural elements found in the cytoskeleton, **microtubules (MTs)** are the largest. A well-known microtubule-based cellular structure is the *axoneme* of cilia and flagella, the appendages responsible for motility of eukaryotic cells. We have already encountered an example of such a structure, since the axoneme of the sperm tail shown in Figure 4-12 consists of microtubules. Microtubules also form the *spindle fibers* needed to separate chromosomes prior to cell division, as we will see in Chapter 14.

In addition to their involvement in motility and chromosome movement, microtubules also play an important role in the organization of the cytoplasm. They contribute to the polarity and overall shape of the cell, the spatial disposition of its organelles, and the distribution of microfilaments and intermediate filaments. Examples of the diverse phenomena that are governed by microtubules include the asymmetric shapes of animal cells, the plane of cell division in plant cells, the ordering of filaments during muscle development, and the positioning of mitochondria around the axoneme of motile appendages.

As shown in Figure 4-22a, microtubules are straight, hollow cylinders with an outer diameter of about 25 nm and an inner diameter of about 15 nm. The wall of the microtubule consists of longitudinal arrays of *protofilaments,* usually 13 arranged side by side around the hollow center, or *lumen.* Each protofilament is a linear polymer of *tubulin* molecules. Tubulin is a dimeric protein, consisting of two similar but distinct polypeptide subunits, *α-tubulin* and *β-tubulin.* All of the tubulin dimers in each of the protofilaments are oriented in the same direction, such that all of the subunits face the same end of the microtubule. This uniform orientation gives the microtubule an inherent *polarity.* The polarity of microtubules has important implications for their assembly, as we will see in Chapter 18.

Microfilaments. **Microfilaments (MFs)** are much thinner than microtubules. They have a diameter of about 7 nm, which makes them the smallest of the major cytoskeletal components. Microfilaments are best known for their role in the contractile fibrils of muscle cells and will therefore be encountered again in Chapter 19. However, microfilaments are involved in a variety of other cellular phenomena as well. They can form connections with the plasma membrane and thereby influence *locomotion, amoeboid movement,* and *cytoplasmic streaming,* a cyclic or back-and-forth flow of cytoplasm seen in a variety of algal, plant, and animal cells. Microfilaments also produce the *cleavage furrows* that divide the cytoplasm of animal cells after chromosomes have been separated by the spindle fibers. In addition, microfilaments contribute importantly to the development and maintenance of cell shape.

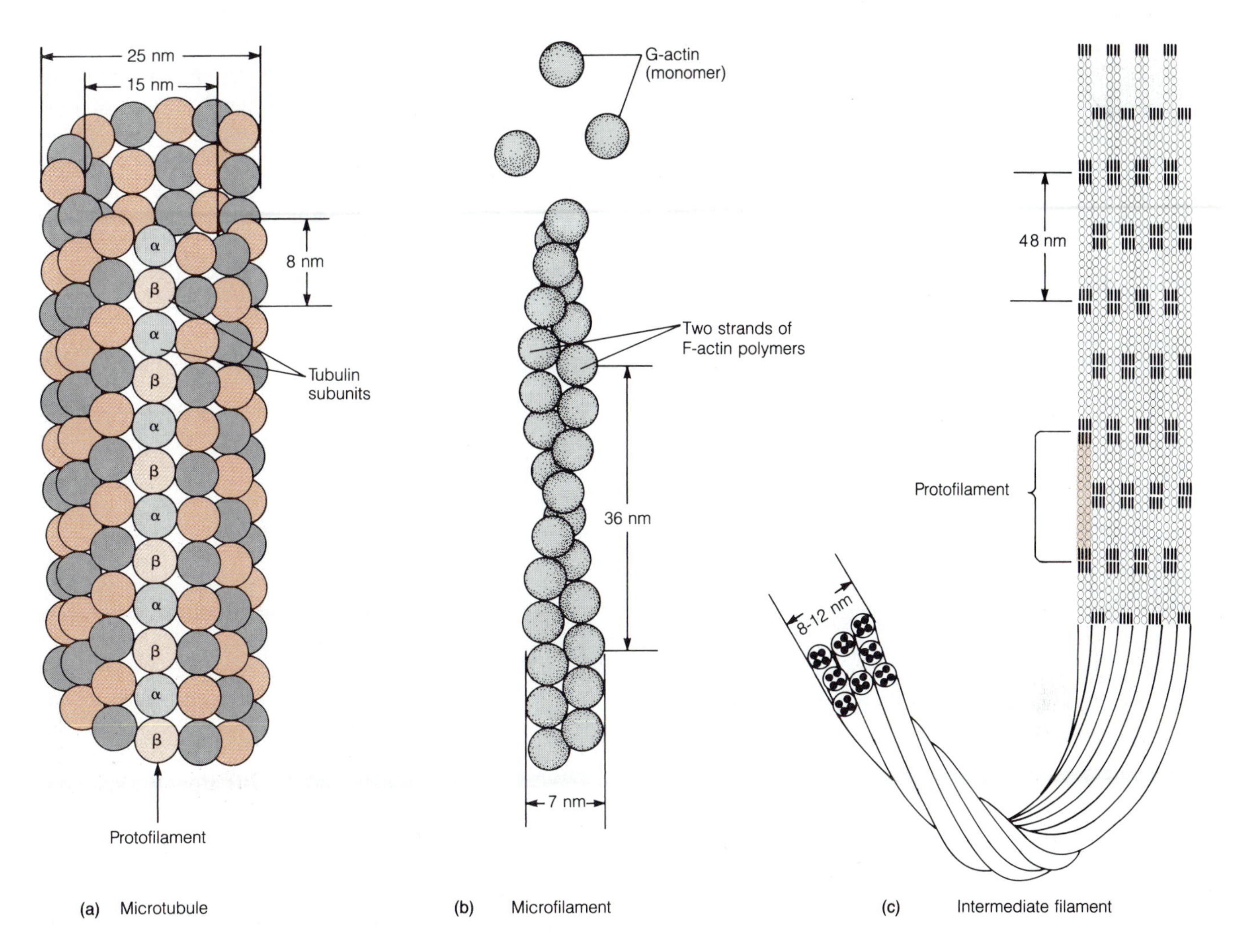

Figure 4-22 Structures of Microtubules, Microfilaments, and Intermediate Filaments. (a) A schematic diagram of a microtubule, showing 13 protofilaments arrayed longitudinally to form a hollow cylinder with an outer diameter of about 25 nm and an inner diameter of about 15 nm. Each protofilament is a polymer of tubulin dimers, each about 8 nm long. All dimers are oriented in the same direction, thereby accounting for the polarity of the protofilament and hence of the whole microtubule. (b) A schematic diagram of a microfilament, showing two strands of F-actin twisted into a right-handed double helix with a diameter of about 7 nm. A half-turn of the helix occurs every 36 nm. The F-actin polymer consists of monomers of G-actin, all oriented in the same direction to give the microfilament its inherent polarity. (c) A schematic diagram of an intermediate filament. The structural unit is the tetrameric protofilament, consisting of two pairs of coiled polypeptides, with a length of about 48 nm. Protofilaments assemble by end-to-end and side-to-side alignment to form an intermediate filament that is thought to be eight protofilaments thick at any point and has a diameter of 8–12 nm.

Microfilaments are polymers of the protein *actin* (Figure 4-22b). Actin is synthesized as a monomer called *G-actin* (G for globular). G-actin monomers polymerize reversibly into long, double-helical strands of *F-actin* (F for fibrous), each strand about 4 nm wide. A microfilament therefore consists of two strands of F-actin wrapped around each other to form a right-handed double helix with a diameter of about 7 nm (Figure 4-22b). Like microtubules, microfilaments are polar structures, since all of the subunits are oriented in the same direction. This polarity influences the direction of microfilament elongation, since assembly usually proceeds more readily at one end of the growing microfilament, whereas disassembly is favored at the other end.

Intermediate Filaments. Intermediate filaments (IFs) comprise the third structural element of the cytoskeleton. As their name suggests, IFs have a diameter of about 8–12 nm, larger than the diameter of microfilaments but smaller than that of microtubules. Intermediate filaments are the most stable and the least soluble constituents of the cytoskeleton. Because of this stability, some researchers regard IFs as a scaffold that supports the entire cytoskeletal framework. Intermediate filaments are also thought to have a tension-bearing role in some cells, since they often occur in areas that are subject to mechanical stress.

In contrast to microtubules and microfilaments, intermediate filaments differ in their composition from tissue to tissue. Based on biochemical and immunological criteria, IFs from animal cells can be grouped into five classes. A specific cell type usually contains only one or sometimes two classes of IF proteins. Because of this tissue specificity, animal cells from different tissues can be distinguished on the basis of the IF proteins present. This *intermediate filament typing* serves as a diagnostic tool in medicine.

Despite their heterogeneity of size and chemical properties, all of the IF proteins share common structural features. They all have a central rodlike segment that is remarkably similar from one IF protein to the other. Flanking the central region of the protein are N-terminal and C-terminal segments that differ greatly in size and sequence, presumably accounting for the functional diversity of these proteins.

Shown in Figure 4-22c is a possible model for IF structure. The basic structural unit is a dimer of two intertwined IF polypeptides. Two such dimers align laterally to form a tetrameric *protofilament*. Protofilaments then interact with each other to form an intermediate filament that is thought to be eight protofilaments thick at any point, with protofilaments probably joined end to end in an overlapping manner.

Outside the Cell: Walls and Coats

So far, we have considered the plasma membrane that surrounds every eukaryotic cell, the nucleus and cytoplasm within, and the variety of organelles, membrane systems, and filaments found in the cytoplasm. With that, it may seem that our tour of the cell is complete. Not so, however, for most eukaryotic cells are also characterized by one or more additional, though nonliving, structures exterior to the plasma membrane. For plant cells, this structure is the rigid, cellulose-containing **cell wall** that encases the cell (Figure 4-23); for animal cells, it is the **cell coat** (also called the **glycocalyx** because of the presence of glycoproteins and other substances) found on the cell surface (Figure 4-24).

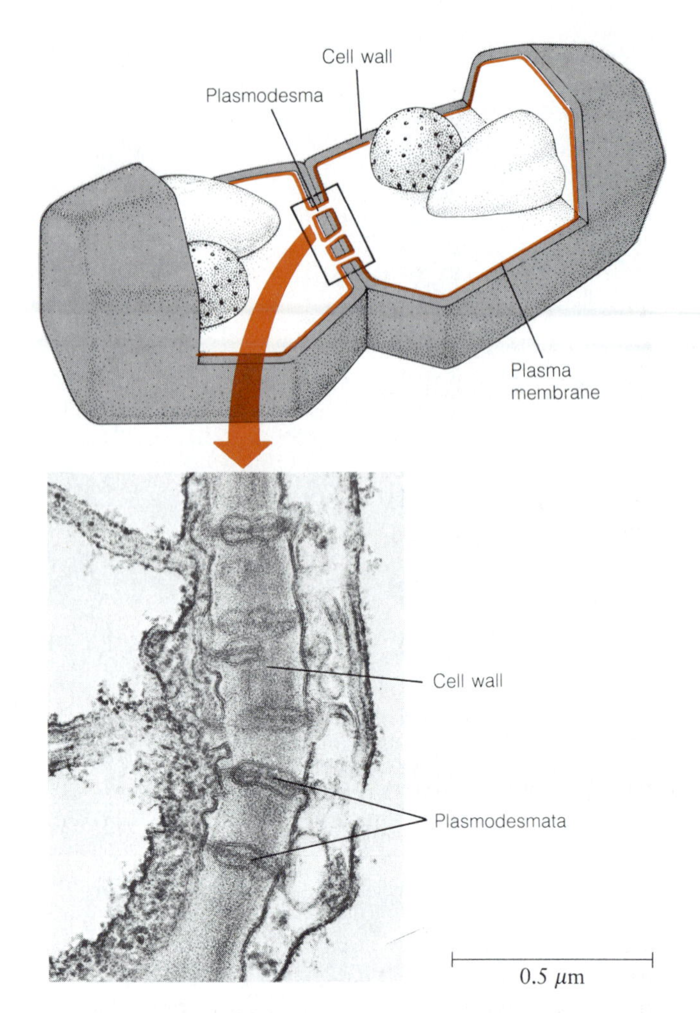

Figure 4-23 The Plant Cell Wall. The wall surrounding a plant cell consists of rigid microfibrils of cellulose embedded in a noncellulosic matrix of proteins and sugar polymers. Note how the neighboring cells are connected by the plasmodesmata. (TEM)

This difference between plant and animal cells is in keeping with the life-styles of these two broad groups of eukaryotes. Plants are generally nonmotile, a life-style that is entirely compatible with the rigidity that cell walls confer on an organism. Animal cells, on the other hand, are not encased in walls. Clearly, rigidity would be a disadvantage to an organism that depends on mobility both to find and gather food and to escape becoming food for other organisms.

The plant cell wall is not a part of the living protoplasm of the cell. Instead, the wall should be thought of as a rigid, nonliving "shell" that surrounds and confines the cell. The wall varies somewhat in chemical composition from species to species, but always consists of microfibrils of *cellulose* embedded in a matrix of other polysaccharides and small amounts of protein.

Figure 4-23 illustrates the prominence of the wall as

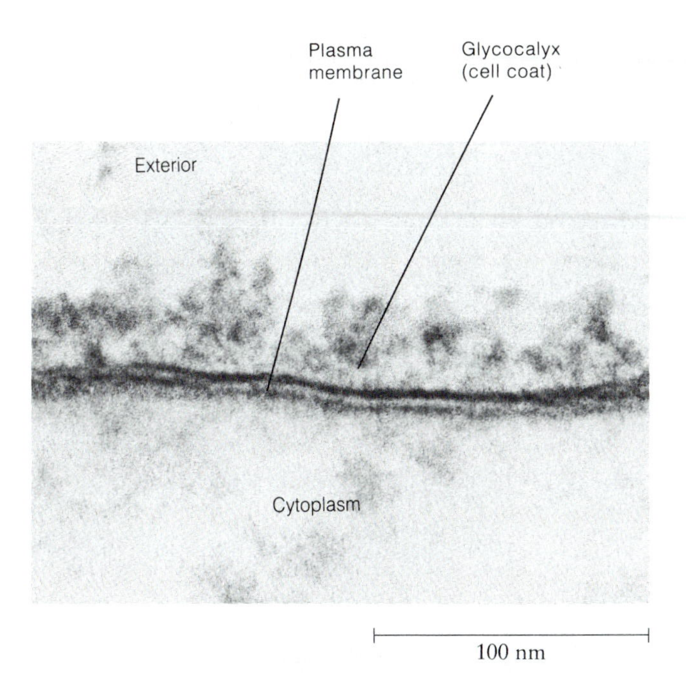

Figure 4-24 The Glycocalyx of Animal Cells. Many animal cells have a thin coat of glycoproteins and other substances on the outer surface of their plasma membrane, as shown in this electron micrograph. This glycocalyx increases the adhesiveness of adjacent cells and is particularly important in cell-cell recognition. (TEM)

a structural feature of a typical plant cell. Notice that neighboring cells, though separated by the wall, are actually connected by numerous cytoplasmic bridges, called **plasmodesmata** (singular, **plasmodesma**), which pass through the wall. The plasma membranes of adjacent cells are continuous through each plasmodesma, such that the channel is membrane-lined. The actual diameter of the plasmodesma is uncertain, but water and small solutes can pass freely from cell to cell. Most of the cells of the plant are interconnected in this way.

The glycocalyx on the surface of animal cells does not confine and restrain the cell as does the wall of plant cells. Instead, the glycocalyx serves to strengthen the cell surface and to help hold cells together. It is particularly important in cell-cell recognition, mainly because of the unique molecular structure of the carbohydrates attached to both the proteins and the lipids of the plasma membrane.

Like plant cells, animal cells can communicate with each other. But instead of plasmodesmata they have intercellular connections called **gap junctions** that are specialized for the transfer of material between the cytoplasms of adjacent cells. Two other types of intercellular junctions are characteristic of animal cells. **Tight junctions** hold cells together so tightly that transport of substances through the space between cells is effectively blocked. **Desmosomes** also link adjacent cells, but for the purpose of connecting them tightly into sturdy yet flexible sheets. Each of these types of junctions is discussed in detail in Chapter 7.

Living or Not? The Enigma of the Viruses

Before concluding this preview of cellular biology, we need to look briefly at what is often an unwelcome intruder in the cell—the **virus.** As we do, we are challenged by the question of whether or not we ought to think of viruses as living. Clearly, viruses do not qualify as cells. For example, most viruses are not membrane bounded as all cells are. Even viruses that are within a membrane do not make their own membrane but derive it from the cell in which the viral particles are made and assembled. Furthermore, viruses cannot carry on all the functions required for independent existence and must therefore depend for most of their needs on the cells they invade. Viruses are probably best thought of as subcellular parasites, incapable of a free-living existence, but able to invade and infect a living cell and subvert its synthetic machinery for the production of more viruses.

Viruses range in size from about 25 to 300 nm. The smallest viruses are therefore about the size of a ribosome, whereas the largest ones are about one-quarter the diameter of a bacterial cell. Each virus has its own characteristic shape, as shown in Figure 4-25. Despite the morphological complexity, however, viruses are chemically quite simple. Most consist only or primarily of specific proteins complexed as a "coat" around the nucleic acid (DNA or RNA) that serves as the genetic information for that particular virus.

Viruses are sometimes named for the diseases they cause. Poliovirus, influenza virus, and tobacco mosaic virus are several common examples. Other viruses have more cryptic laboratory names such as T4, Qβ, λ, Epstein-Barr, or herpes simplex virus. Viruses that infect bacterial cells are called **bacteriophages,** or often just **phages** for short. Bacteriophages and other viruses will figure prominently in our discussion of molecular genetics in Part Four, because of the striking genetic similarities between viruses and cells. In terms of the storage, expression, and transmission of genetic information, viruses seem to follow most of the same rules and use many of the same mechanisms that cells do, yet they are simpler than cells and much easier to manipulate. Viruses are therefore very useful in the study of genetics at the molecular level.

The question of whether viruses are really living de-

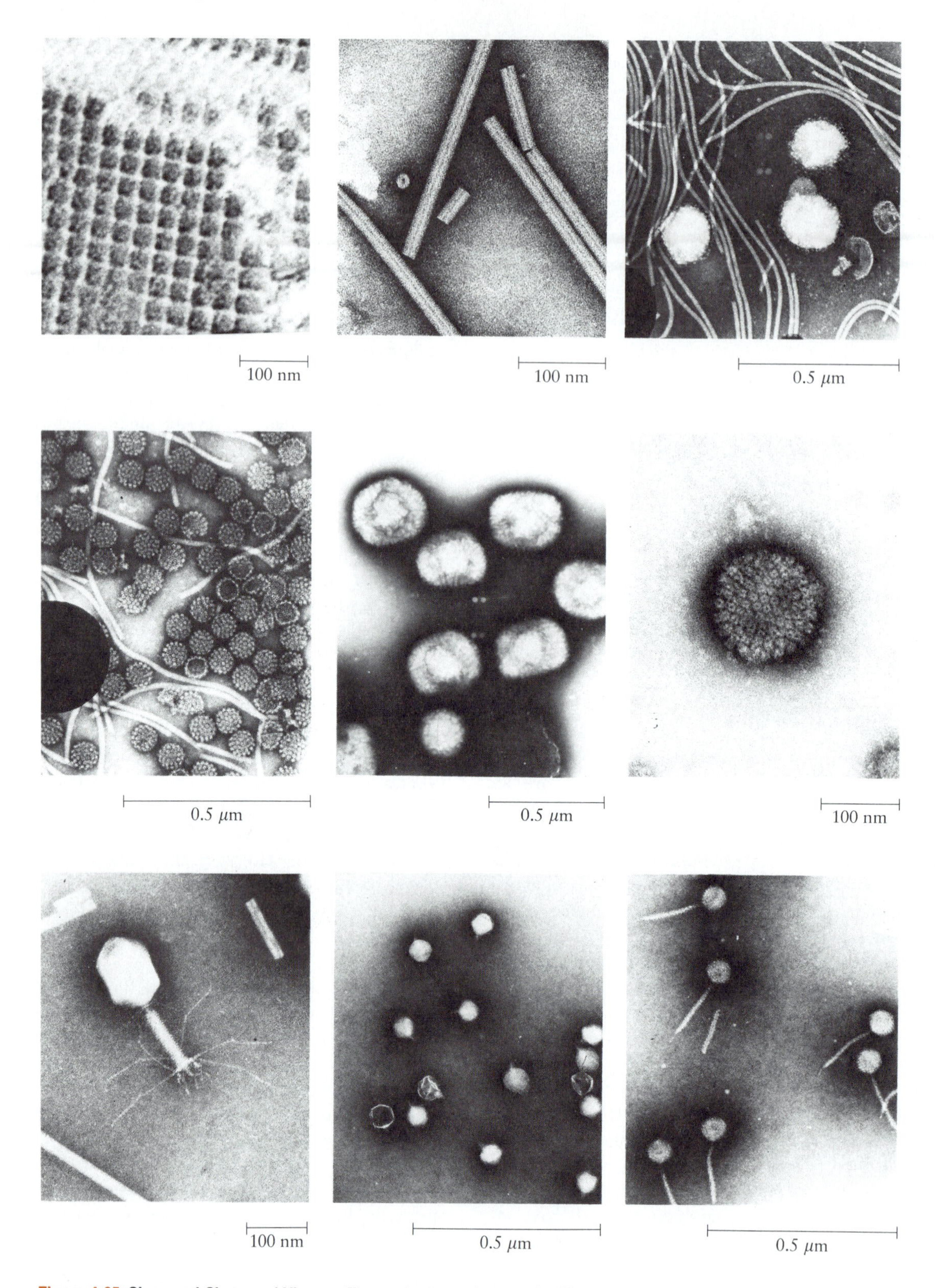

Figure 4-25 Sizes and Shapes of Viruses. These electron micrographs illustrate the morphological diversity of viruses. Top row, left to right: RNA-containing viruses of polio, tobacco mosaic, and Rous sarcoma. Middle row, left to right: DNA-containing viruses of papilloma, vaccinia, and herpes simplex. Bottom row, left to right: DNA bacteriophages T4, T7, and lambda. (all TEMs)

pends crucially on what we mean by "living" and is probably worth pondering only to the extent that it helps us to understand more fully what viruses are—and what they are not. The most fundamental properties of living things are motility, irritability, and the ability to reproduce. Viruses clearly do not satisfy the first two criteria. Outside their host cells, viruses are inert and inactive. They can, in fact, be isolated and crystallized almost like a chemical compound. It is only in the correct host cell that a virus becomes functional, and it is only in the host cell that a virus can give rise to more viruses.

Even the ability of viruses to reproduce has to be qualified carefully. A basic tenet of the cell theory is that cells arise only from preexisting cells, but this is not true of viruses. No virus can give rise to another virus by any sort of self-duplication process. Rather, the virus must subvert the metabolic and genetic machinery of the host cell, reprogramming it for synthesis of the proteins necessary to package the DNA or RNA that arises by copying the genetic information of the parent virus.

It is only in a genetic sense that one can think of viruses as living at all. One of the most fundamental properties of life is the ability to specify and direct the genetic composition of progeny, and that is an ability that viruses clearly possess. It is probably most helpful to think of viruses as "quasi-living"—satisfying part but not all of the basic definition of life.

Perspective

A fundamental distinction in biology is that between the prokaryotes (bacteria and blue-green algae) and the eukaryotes (all other living organisms). Prokaryotic cells are relatively small and structurally quite simple, lacking most of the internal membrane systems and organelles of eukaryotic cells. Ribosomes are, in fact, the only discrete intracellular particles common to both cell types. All other organelles are found only in eukaryotic cells, where they play an indispensable role in compartmentalization of function.

Eukaryotic cells have at least four major structural features: a plasma membrane to define the boundaries of the cell and retain its contents, a nucleus to house its DNA, membrane-bounded organelles, and the cytoplasm with its cytoskeleton of tubules and filaments. In addition, plant cells almost always have a rigid cell wall, and animal cells frequently have a cell coat that consists of glycoproteins.

The nucleus is surrounded by a double membrane called the nuclear envelope. The chromosomes within the nucleus contain most of the DNA of the cell. Mitochondria are often called the "powerhouses" of the cell because of their role in the oxidation of foodstuffs to release energy. Chloroplasts trap solar energy and use it to "fix" carbon from carbon dioxide into organic form and convert it into sugar.

The endoplasmic reticulum is an extensive network of membranes that are either rough (studded with ribosomes) or smooth. The rough ER is responsible for synthesis of secretory and membrane proteins, whereas the smooth ER is involved in lipid synthesis and drug detoxification. Proteins synthesized on the rough ER are further processed and packaged in the Golgi complex and then transported to the surface of the cell by secretory vesicles. Lysosomes contain hydrolytic enzymes and are involved in cellular digestion. Peroxisomes are often about the same size as lysosomes but function in the generation and degradation of hydrogen peroxide. In plants, specialized peroxisomes are involved in the conversion of stored fat into carbohydrate during seed germination and in the process of photorespiration.

Although not membrane bounded, ribosomes are included as organelles. Present in both eukaryotic and prokaryotic cells, they serve as sites of protein synthesis.

The cytoskeleton is an extensive network of filaments and tubules that gives eukaryotic cells their distinctive shapes. The filaments and tubules are also important in cellular motility and contractility, topics of later chapters. Also to be encountered in later chapters are the viruses, which satisfy some but not all of the basic criteria of living things.

Key Terms for Self-Testing

Properties and Strategies of Cells
surface area/volume ratio (p. 74)
compartmentalization (p. 76)
prokaryotes (p. 76)
eukaryotes (p. 76)
chromosomes (p. 79)

The Eukaryotic Cell in Overview: Pictures at an Exhibition
plasma membrane (p. 83)
lipid bilayer (p. 83)
nucleus (p. 84)
nuclear envelope (p. 84)
pores (p. 84)
chromatin (p. 84)
nucleolus (p. 84)
nucleoplasm (p. 84)
mitochondrion (p. 85)
inner mitochondrial membrane (p. 85)
outer mitochondrial membrane (p. 85)
cristae (p. 86)
matrix (p. 86)
chloroplast (p. 86)
thylakoid disks (p. 86)
stroma lamellae (p. 86)
grana (p. 86)
stroma (p. 88)
plastids (p. 88)
endoplasmic reticulum (ER) (p. 88)
cisternae (p. 88)
lumen (p. 88)
rough endoplasmic reticulum (p. 89)
smooth endoplasmic reticulum (p. 89)
Golgi complex (p. 89)
secretory vesicles (p. 89)
lysosome (p. 90)
peroxisome (p. 90)
glyoxysomes (p. 91)
leaf peroxisomes (p. 91)
vacuoles (p. 91)
central vacuole (p. 91)
ribosome (p. 93)
sedimentation coefficient (p. 93)
large ribosomal subunit (p. 93)
small ribosomal subunit (p. 93)
cytoplasm (p. 95)
cytoskeleton (cytoskeletal network) (p. 95)
microtubules (MTs) (p. 96)
microfilaments (MFs) (p. 96)
intermediate filaments (IFs) (p. 98)
cell wall (p. 98)
cell coat (glycocalyx) (p. 98)
plasmodesmata (p. 99)
gap junctions (p. 99)
tight junctions (p. 99)
desmosomes (p. 99)

Living or Not? The Enigma of the Viruses
virus (p. 99)
bacteriophage (phage) (p. 99)

Suggested Reading

Properties and Strategies of Cells

Bonner, J. T. *Cells and Societies.* Princeton, N.J.: Princeton University Press, 1966.

Carlile, M. Prokaryotes and eukaryotes: Strategies and successes. *Trends Biochem. Sci.* 7 (1982): 128.

Fawcett, D. W. *The Cell: Its Organelles and Inclusions,* 2d ed. Philadelphia: Saunders, 1979.

Margulis, L., and K. V. Schwartz. *Five Kingdoms.* New York: W. H. Freeman, 1982.

Roddyn, D. B., ed. *Subcellular Biochemistry.* New York: Plenum Press, 1978.

Stanier, R., E. Adelberg, and J. Ingraham. *The Microbial World,* 4th ed. Englewood Cliffs, N.J.: Prentice-Hall, 1976.

Thomas, L. *Lives of a Cell: Notes of a Biology Watcher.* New York: Viking Press, 1974.

Watson, J. D., N. H. Hopkins, J. W. Roberts, J. A. Steitz, and A. M. Weiner. *Molecular Biology of the Gene,* 4th ed. Menlo Park, Calif.: Benjamin/Cummings, 1987.

The Plasma Membrane

Bretscher, M. S. The molecules of the cell membrane. *Sci. Amer.* 253 (October 1985): 100.

Robertson, R. N. *The Lively Membranes.* New York: Cambridge University Press, 1983.

Yeagle, P. *The Membranes of Cells.* Orlando, Fla.: Academic Press, 1987.

The Nucleus

Miller, O. L. The nucleolus, chromosomes, and visualization of genetic activity. *J. Cell Biol.* 91 (1981): 15.

Newport, J. W., and D. J. Forbes. The nucleus: Structure, function, and dynamics. *Annu. Rev. Biochem.* 56 (1987): 535.

Intracellular Membranes and Organelles

Bainton, D. The discovery of lysosomes. *J. Cell Biol.* 91 (1981): 66s.

Bielka, H., ed. *The Eukaryotic Ribosome*. New York: Springer-Verlag, 1982.

Bogorad, L. Chloroplasts. *J. Cell Biol.* 91 (1981): 256s.

Bretscher, M. S., and M. C. Raff. Mammalian plasma membranes. *Nature* 258 (1975): 43.

de Duve, C. The lysosome. *Sci. Amer.* 208 (May 1973): 64.

de Duve, C. Microbodies in the living cell. *Sci. Amer.* 248 (May 1983): 74.

Dustin, P. Microtubules. *Sci. Amer.* 243 (August 1980): 66.

Ernster, L., and G. Schatz. Mitochondria: A historical review. *J. Cell Biol.* 91 (1981): 227s.

Farquhar, M., and G. Palade. The Golgi apparatus (complex)—(1945–1981) from artifact to center stage. *J. Cell Biol.* 91 (1981): 77s.

Fawcett, D. W. *The Cell: Its Organelles and Inclusions,* 2d ed. Philadelphia: Saunders, 1981.

Hoober, J. K. *Chloroplasts*. New York: Plenum Press, 1984.

Lake, J. A. The ribosome. *Sci. Amer.* 245 (August 1981): 84.

Novikoff, A. The endoplasmic reticulum: A cytochemist's view (a review). *Proc. Natl. Acad. Sci. USA* 73 (1976): 2781.

Palade, G. Intracellular aspects of the process of protein synthesis. *Science* 189 (1975): 347.

Roland, J. C., A. Szølløsi, and D. Szølløsi. *Atlas of Cell Biology*. Boston: Little, Brown, 1977.

Tolbert, N. E., and E. Essner. Microbodies: Peroxisomes and glyoxysomes. *J. Cell Biol.* 91 (1981): 271s.

Tzagoloff, A. *Mitochondria*. New York: Plenum Press, 1982.

Wolfe, S. L. *Cell Ultrastructure*. Belmont, Calif.: Wadsworth, 1985.

Wool, I. G. The structure and function of eukaryotic ribosomes. *Annu. Rev. Biochem.* 48 (1979): 719.

Cytoplasm and Cytoskeleton

Fulton, A. B. How crowded is the cytoplasm? *Cell* 30 (1980): 345.

Luby-Phelps, K., D. L. Taylor, and F. Lanni. Probing the structure of the cytoplasm. *J. Cell Biol.* 102 (1986): 2015.

Schliwa, M. *The Cytoskeleton*. New York: Springer-Verlag, 1986.

Outside the Cell: Walls and Coats

Albersheim, P. The wall of growing plant cells. *Sci. Amer.* 232 (April 1975): 80.

Hay, E. D., ed. *Cell Biology of Extracellular Matrix*. New York: Plenum Press, 1982.

Hay, E. D. Extracellular matrix. *J. Cell Biol.* 91 (1981): 205s.

Luft, J. H. The structure and properties of the cell surface coat. *Int. Rev. Cytol.* 45 (1976): 291.

Problem Set

1. **Prokaryotes and Eukaryotes.** Indicate whether each of the following statements is true (T) or false (F). If false, reword the statement to make it true.
   (a) Eukaryotic cells are in all cases larger than prokaryotic cells.
   (b) Some cells are large enough to be seen with the naked eye.
   (c) Prokaryotic cells possess none of the following features: mitochondria, membrane-bounded nucleus, plasma membrane, microtubules.
   (d) The surface area/volume ratio is generally greater for a prokaryotic cell than for a eukaryotic cell.
   (e) The ribosomes found in the mitochondria of your muscle cells are more like those of the bacteria in your intestine than they are like the ribosomes in the cytoplasm of your muscle cells.
   (f) Since prokaryotic cells have neither mitochondria nor chloroplasts, they cannot carry out either ATP synthesis or photosynthesis.

2. **That's About the Size of It.** To get some feeling for the differences in size of various cellular structures, it might be useful to compare the structures on a macroscopic scale. Listed below are various structures with their approximate dimensions. To compare their dimensions on a macroscopic scale, assume that each structure has been magnified a millionfold, using a scale such that one nanometer is represented by one millimeter. On this scale, a ribosome has a diameter of 25 mm (about 1 in.) and is therefore the size of a large marble. Convert each of the other dimensions to this macroscopic scale and suggest a physical object that has approximately the same dimensions.
   (a) Ribosome: 25 nm in diameter
   (b) Microtubule: 25 nm × 1 μm
   (c) Microfilament: 7 × 200 nm
   (d) Peroxisome: 0.5 μm in diameter
   (e) Mitochondrion: 1 × 2 μm
   (f) Chloroplast: 2 × 8 μm
   (g) Nucleus: 6 μm in diameter
   (h) Liver cell: 20 μm in diameter
   (i) Chicken egg: 4 × 6 cm
   (j) Human being: 1.8 m tall

3. **Cellular Specialization.** Each of the cell types listed here is a good example of a cell that is specialized for a specific function. Match each cell type in list A with the appropriate function from list B and explain why you matched each as you did.

List A	List B
(a) Pancreatic cell	Cell division
(b) Cell from flight muscle	Absorption
(c) Palisade cell from leaf	Motility
(d) Cell of intestinal lining	Photosynthesis
(e) Nerve cell	Secretion
(f) Bacterial cell	Transmission of impulses

4. **Cellular Structure.** Indicate whether each of the following cellular structures or components is found in animal cells (A), bacterial cells (B), and/or plant cells (P).

(a) Chloroplasts	(g) Golgi complex
(b) Cell wall	(h) Central vacuole
(c) Microtubules	(i) Thylakoids
(d) DNA	(j) Ribosomes
(e) Nuclear envelope	(k) Lipid bilayers
(f) Nucleoli	(l) Actin

5. **Matching.** For each of the structures in list A, choose the single term from list B that matches best, and explain your choice.

List A

(a) Bacterial cell wall	(e) Nucleolus
(b) Blue-green algae	(f) Rough ER
(c) Lysosome	(g) Smooth ER
(d) Nuclear envelope	(h) Virus

List B

Bacteriophage	Pores
Hydrolases	Ribosome synthesis
Cellulose	Eukaryote
Photorespiration	Svedberg
Lipid synthesis	Glyoxysomes
Muscle cells	Granum
Peptidoglycan	Interphase
Prokaryote	Secretory proteins

6. **Complete the Sentence.** Complete each of the following statements about cellular structure in ten words or less.
   (a) If you were shown an electron micrograph of a section of a cell and were asked to identify the cell as plant or animal, one thing you might do is . . .
   (b) A slice of raw apple placed in a concentrated sugar solution will . . .
   (c) A cellular structure that can be seen with the electron microscope but not with the light microscope is . . .
   (d) Ribosomes are not true organelles because . . .
   (e) One reason why it might be difficult to separate lysosomes from peroxisomes by centrifugation techniques is that . . .

7. **Structural Relationship.** For each pair of structural elements, indicate with an A if the first element is a constituent part of the second, with a B if the second element is a constituent part of the first, and with an N if they are separate structures with no particular relationship to each other.
   (a) Mitochondrion; crista
   (b) Golgi complex; nucleus
   (c) Cytoplasm; cytoskeleton
   (d) Cell wall; cell coat
   (e) Nucleolus; nucleus
   (f) Smooth ER; ribosome
   (g) Lipid bilayer; plasma membrane
   (h) Peroxisome; thylakoid
   (i) Chloroplast; granum

8. **Organellar Cooperation.** For each pair of organelles or other structures, indicate a single cellular function in which both organelles play a role, and indicate (if possible) the role of each organelle in the function.
   (a) Chloroplast; leaf peroxisome
   (b) Rough ER; Golgi complex
   (c) Nucleus; ribosomes
   (d) Mitochondrion; cytoplasm
   (e) Secretory vesicle; plasma membrane

9. **The Palisade Cell: A Look Inside.** Just under the upper surface of many plant leaves is a layer of columnar *palisade cells,* the site of much photosynthesis. A typical palisade cell is cylindrical in shape, with a diameter of 20 μm and a length of 35 μm. In round numbers, such a cell might contain 200 mitochondria, 40 chloroplasts, 100 peroxisomes, two million ribosomes, and one nucleus. The dimensions of each of these organelles are given in Problem 2. Except for the region occupied by the nucleus, the cytoplasm of the palisade cell is restricted to a 2.5-μm layer just beneath the plasma membrane because of the central vacuole, which, like the cell itself, is roughly cylindrical in shape.
   (a) Calculate the proportion of the total internal volume of the cell that is occupied by each of these populations of organelles. Consider the cell, the vacuole, the chloroplasts, and the mitochondria to be cylindrical in shape ($V = \pi r^2 h$) and the nucleus, peroxisomes, and ribosomes to be approximately spherical ($V = 4\pi r^3/3$).
   (b) What proportion of the total internal volume of the cell is not accounted for by the named organelles? What other major structural features must be accommodated in this remaining cytoplasmic volume?

10. **Protein Synthesis and Secretion.** Although we will not encounter protein synthesis and secretion in detail until later chapters, you already have enough information about these

processes to place in order the seven events that are now listed randomly. Order events (1)–(7) so that they represent the correct sequence of events corresponding to steps (a)–(g) to trace a typical secretory protein from the initial transcription (readout) of the relevant genetic information in the nucleus to the eventual secretion of the protein from the cell by exocytosis.

Transcription → (a) → (b) → (c) →
(d) → (e) → (f) → (g) → Secretion

(1) The protein is partially glycosylated within the lumen of the ER.

(2) The secretory vesicle arrives at and fuses with the plasma membrane.

(3) The RNA transcript is transported from the nucleus to the cytoplasm.

(4) The final sugar groups are added to the protein in the Golgi complex.

(5) As the protein is synthesized, it passes across the ER membrane into the lumen of a cisterna.

(6) The enzyme is packaged into a secretory vesicle and released from the Golgi complex.

(7) The RNA message associates with a ribosome and begins synthesis of the desired protein on the surface of the rough ER.

11. **The Great Debate.** A formal debate has been arranged to address the question, "Resolved, that a virus is a living entity."

(a) Would you choose to argue on the pro side or the con side of this debate?

(b) What major points would you make?

(c) What arguments would you expect your opponents to counter with?

(d) What might you offer by way of rebuttal?

5

Bioenergetics: The Flow of Energy in the Cell

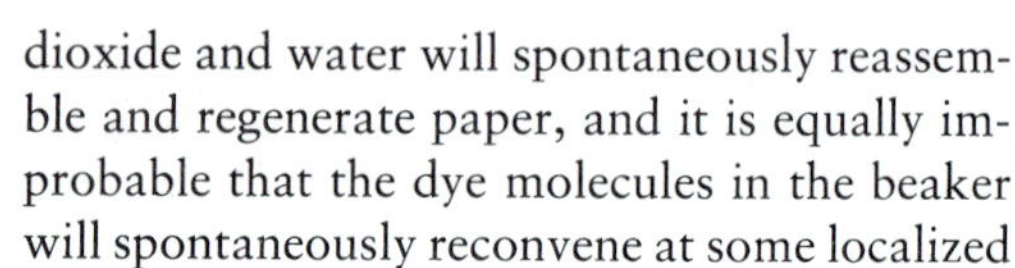

Have you ever considered how downright improbable you are? Your body contains approximately 100 trillion cells, all integrated and functioning together as a harmonious whole. Each of those cells contains trillions of molecules, every one in turn made up of dozens or hundreds or thousands of atoms, all carefully organized in a manner that is strikingly nonrandom. As you already know, your cellular machinery is composed mainly of carbon, hydrogen, oxygen, nitrogen, and phosphorus atoms, bonded together and arranged in space in such a way as to account for all the form and function, all the pattern and design that are part of you. Even as you contemplate the grandeur of all that organization, consider how utterly improbable it all is—how very unlikely that so many atoms of carbon, hydrogen, oxygen, and other elements would ever come together and be ordered into the molecules, structures, cells, tissues, and organs that make up your body.

What you are contemplating is the thermodynamic improbability that the order of the human body (or any other biological entity) could come into being spontaneously or, for that matter, could be maintained in such a highly ordered state once it had come into being. On the contrary, things in nature usually proceed from an ordered state to a less ordered one, not the other way around. For example, the carbon, hydrogen, and oxygen atoms of which this page is made tend to exist as simple inorganic molecules such as carbon dioxide and water instead of the complex cellulose microfibrils that it takes to make paper. If you doubt that tendency, just touch a match to a piece of paper and you will quickly be reminded of how strong that tendency really is (Figure 5-1a)! Similarly, you know that a drop of dye solution placed in a beaker of water will spontaneously diffuse until the dye molecules are randomized in the beaker (Figure 5-1b).

In each of our examples, it is obvious that the reverse process is highly improbable. No matter how long you wait, it is extremely unlikely that molecules of carbon dioxide and water will spontaneously reassemble and regenerate paper, and it is equally improbable that the dye molecules in the beaker will spontaneously reconvene at some localized point once they have been dispersed throughout the solution. So, too, with your body and with every cell in it: Your body is a highly ordered structure composed of large numbers of atoms that could not reasonably be expected to assemble spontaneously.

What, then, does it take to bring about the highly improbable order of biological systems? What is it that allows such complex organization to happen? Or, to put it another way, why are you possible if you are so improbable?

The Importance of Energy

The answer to our question has two components: information and energy. You are already familiar with these two important concepts, since we encountered a dual requirement for information and energy earlier, in our discussion of informational macromolecules in Chapter 3. Recall that assembly of amino acids into polypeptides or nucleotides into nucleic acids requires both a source of energy (to activate the incoming monomeric unit) and a source of information (to specify which of several alternative subunits should be added at a specific point in the chain). The same is true for biological systems in general and for cells in particular. However improbable a structure may be because of its order, it can always be generated if sufficient energy and information are available.

Energy and information are, in other words, two indispensable prerequisites for the existence of life. Order can be brought about, maintained, and even extended in biological systems provided that adequate information is on hand to specify what form that order should take and provided that adequate energy is available to drive the

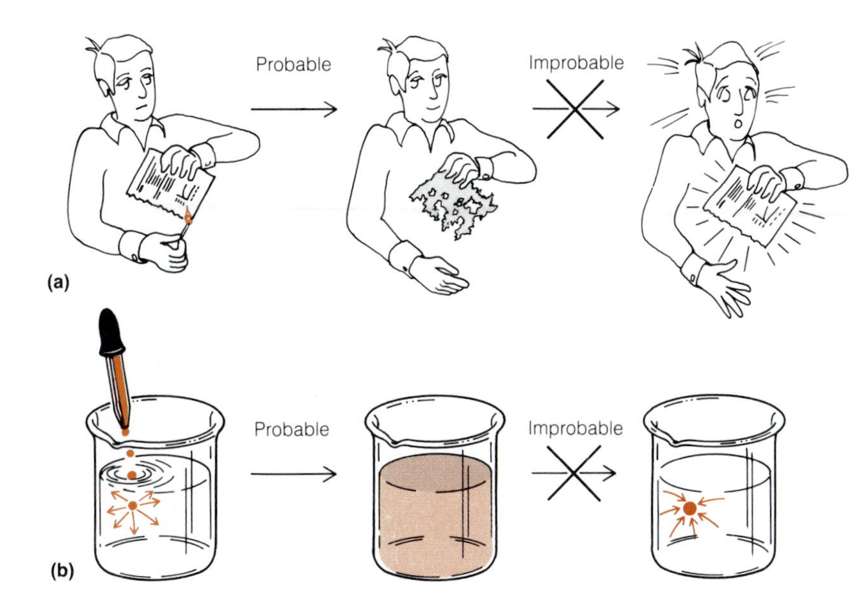

Figure 5-1 Probable and Improbable Processes. Prior experience with familiar processes such as (a) the burning of paper or (b) the diffusion of dye molecules allows us to predict that such events are probable and take place spontaneously once initiated, while their reversal is highly improbable and never occurs spontaneously. Probable, spontaneous events are associated with the evolution of heat and/or with a greater randomness of the components of the system.

reactions and processes whereby order is elaborated. We can summarize the discussion so far by saying that you are improbable because you are so highly ordered, but you are nonetheless possible because of the information available to you (in the DNA of your cells) and the copious quantities of energy at your disposal (in the bond energies of the food you eat).

Given their importance to cells, it is not surprising that the topics of *information* and *energy* are given considerable attention by cell biologists. Each is treated as a major theme in this text. Part Three deals with energy flow in detail, while Part Four is devoted to the topic of information flow.

The Need for Energy

All living systems require an ongoing supply of energy. Before discussing why cells need energy, however, it might be useful to consider what we mean by energy. Usually, energy is defined as the capacity to do work. But that turns out to be a somewhat circular definition, because work is frequently defined in terms of energy changes. A more useful definition would be that **energy** *is the ability to cause specific changes.* Since life is characterized first and foremost by change, this definition underscores the total dependence of all forms of life on the continuous availability of energy.

Now that we have defined energy in this way, we recognize that asking about cellular needs for energy really means inquiring into the kinds of changes that cells must effect, that is, the activities that cells engage in that give rise to change. Six categories of change come to mind (Figure 5-2).

Synthetic Work (Changes in Chemical Bonds). An important activity of virtually every cell at all times is the work of **biosynthesis,** resulting in the formation of new bonds and the generation of new molecules. This activity is especially obvious in a population of growing cells, where it can be shown that additional molecules are being synthesized if the cells are increasing in size or number or both. But **synthetic work** is required to maintain structures just as surely as it is needed to generate them originally. Most existing structural components of the cell are in a state of constant **turnover.** The molecules that make up the structure are continuously being degraded and replaced.

Why might cells expend energy to synthesize large molecules, only to break them up again in the turnover process? The answer has to do with molecular damage. As discussed earlier, larger molecules tend to be inherently unstable, and both physical and chemical processes are continuously damaging proteins and nucleic acids. For example, exposure to the ultraviolet component of sunlight can damage the DNA of skin cells. If the damage is too great, the cells cannot recover, and sunburn results. The skin renews itself by "turning over" the dead cells, sloughing them off and replacing them with new tissues. Something similar happens at the molecular level, except that the damaged molecules are broken down by specialized enzymes, then replaced by the ongoing synthetic processes.

In terms of the hierarchy of cellular structure shown in Figure 2-15, almost all of the energy that cells require for biosynthetic work is used to make energy-rich organic molecules from simpler starting materials and to activate such organic molecules for incorporation into macromolecules. Higher levels of structural complexity usually occur by spontaneous self-assembly, without further energy input. Of the two energy-requiring levels, synthesis of the

Figure 5-2 Several Kinds of Biological Work. The six major categories of biological work are shown here. (a) Synthetic work is illustrated by the process of photosynthesis, (b) mechanical work by the contraction of a weight lifter's muscles, and (c) concentration work by the uptake of molecules into a cell against a concentration gradient. (d) Electrical work is represented by the membrane potential of mitochondria (shown being generated by active proton transport), (e) heat production is illustrated in skunk cabbage, a plant that melts its way through spring snow (see Figure 5-6), and (f) bioluminescence is depicted by the courtship of fireflies.

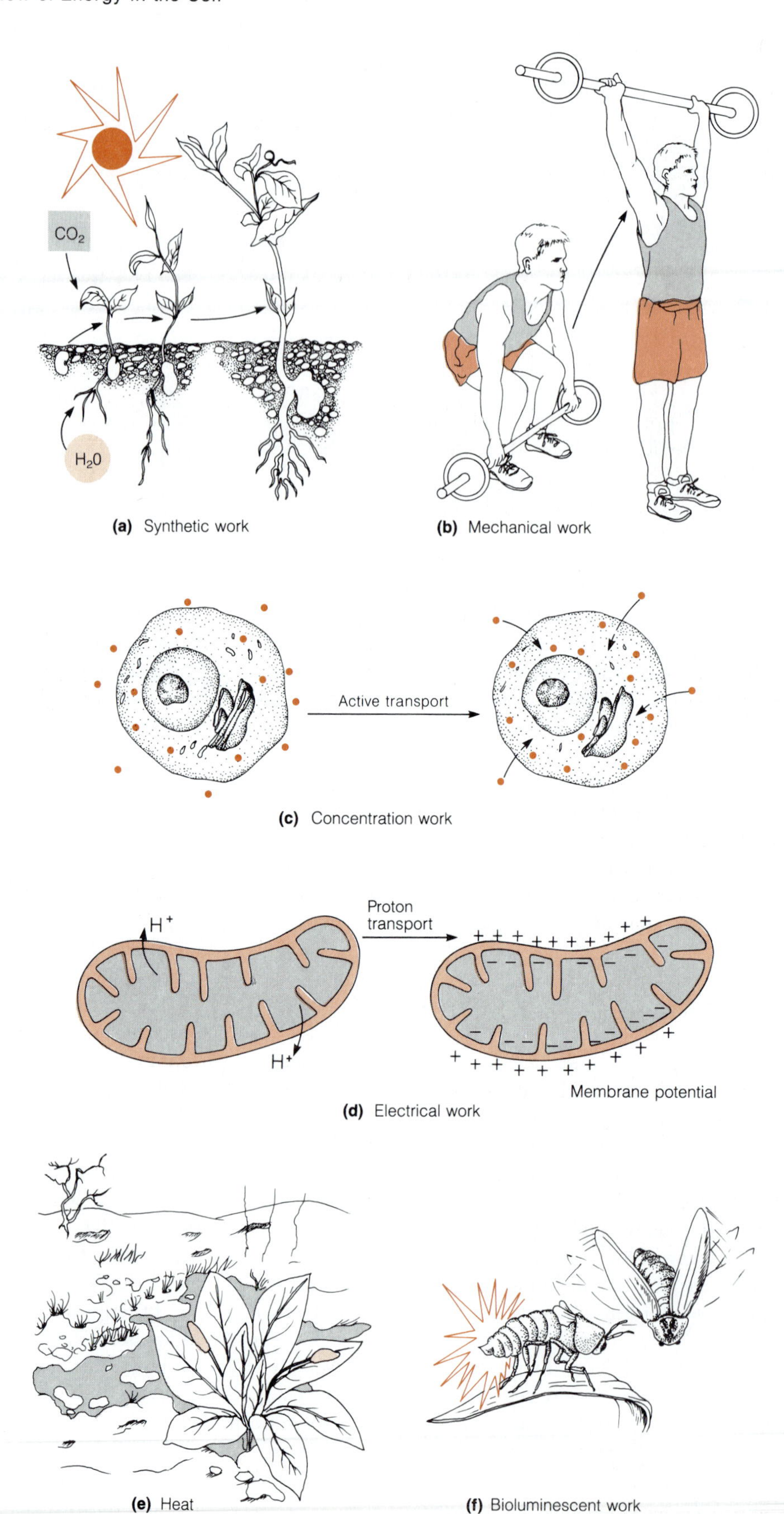

small organic molecules actually requires much more energy than their subsequent polymerization into macromolecules. As we will see in Chapter 12, it takes 18 molecules of ATP (and another 12 molecules of a reduced coenzyme called nicotinamide adenine dinucleotide phosphate) to make one molecule of glucose out of carbon dioxide and water in a photosynthetic cell. Thereafter, only two additional high-energy bonds are required to link the glucose onto the growing chain of a starch molecule. Therefore, the work of biosynthesis is essentially the work of fabricating sugars, amino acids, nucleotides, and lipid molecules from simpler starting compounds.

Mechanical Work (Changes in Location or Orientation). **Mechanical work** involves a physical change in the position or orientation of a cell or some part of it. An especially good example is the movement of a cell with respect to its environment. This movement requires the presence of some sort of motile appendage such as a flagellum or a cilium. Many prokaryotic cells propel themselves through the environment, as in the case of the flagellated bacterium in Figure 5-3. Sometimes, however, it is the environment that moves past the cell, as when the ciliated cells that line your trachea beat upward to sweep inhaled particles back to the mouth or nose, thus protecting the lungs. Muscle contraction is another good example of mechanical work, involving not just a single cell but a large number of muscle cells (Figure 5-4). Still other examples of mechanical work occur within the cell. These include the movement of chromosomes along the spindle fibers during mitosis, the streaming of cytoplasm, and the movement of a ribosome along a strand of messenger RNA.

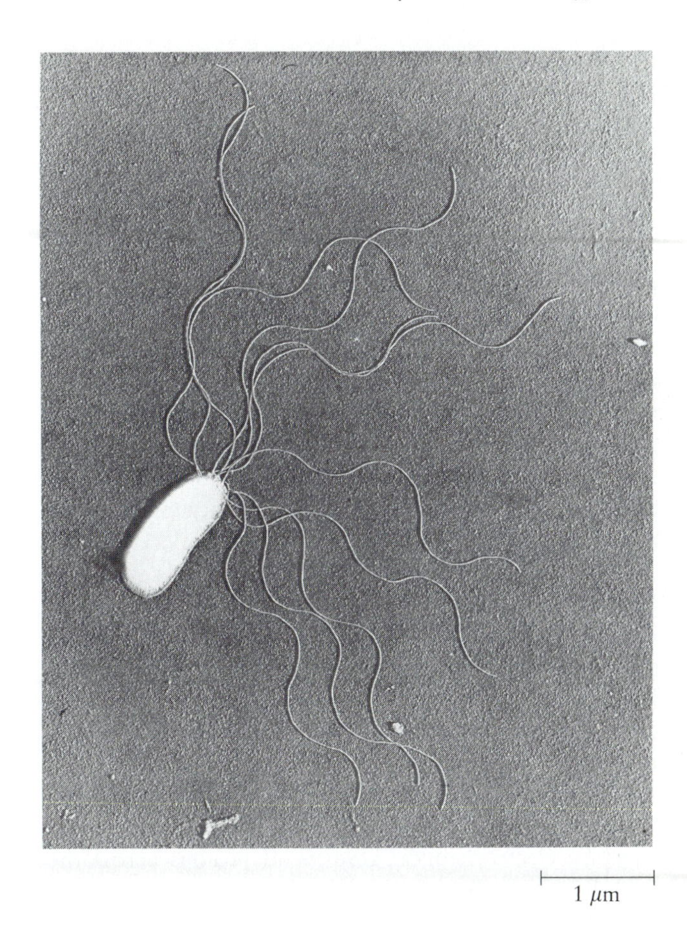

Figure 5-3 A Flagellated Bacterium. The whipping motion of bacterial flagella is driven by energy-dependent proton transport, providing motility for certain bacterial species. (TEM)

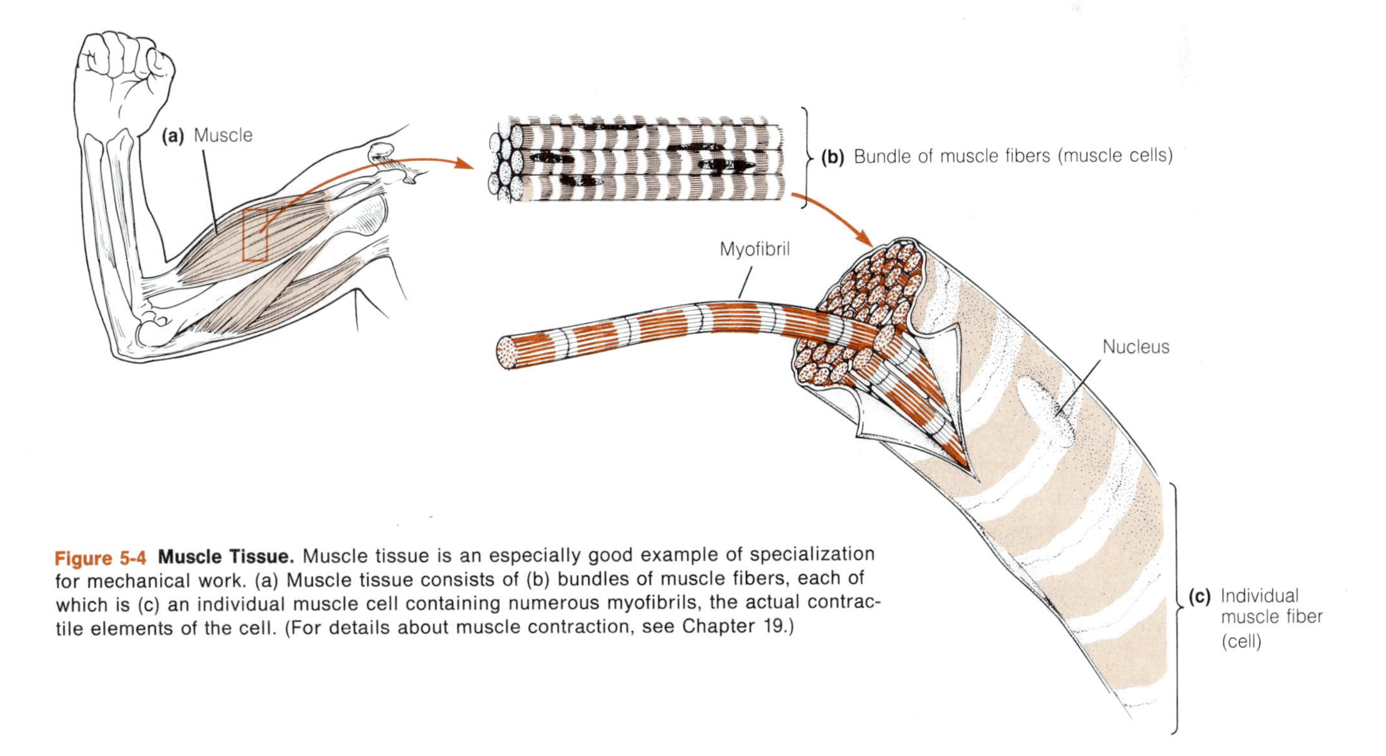

Figure 5-4 Muscle Tissue. Muscle tissue is an especially good example of specialization for mechanical work. (a) Muscle tissue consists of (b) bundles of muscle fibers, each of which is (c) an individual muscle cell containing numerous myofibrils, the actual contractile elements of the cell. (For details about muscle contraction, see Chapter 19.)

Concentration Work (Changes in Concentration Across Membranes). Less conspicuous than either of the previous two categories but every bit as important to the cell is the work of moving molecules or ions against a concentration gradient. In fact, in a resting state, about two-thirds of your energy consumption is used for the ongoing work of concentrating ions across membranes. The purpose of **concentration work** is either to accumulate substances within a cell or subcellular compartment or to remove by-products of cellular activity that cannot be further utilized by the cell and indeed might be harmful to the cell or compartment if they were to remain and build up in concentration. Examples of concentration work include the pumping of sodium and potassium ions across plasma membranes and the light-driven accumulation of protons within the chloroplasts of a plant cell.

Electrical Work (Movement of Ions Across Membranes). **Electrical work** is often considered a specialized case of concentration work, since it also involves movement across membranes. In this case, however, the species that is translocated is a charged ion, and the result is not just a change in concentration but also the establishment of a potential gradient across the membrane. Every membrane has some characteristic potential that is generated in this way. An electrochemical gradient of protons across the mitochondrial or chloroplast membrane is essential to the production of ATP in both respiration (Chapter 11) and photosynthesis (Chapter 12). Electrical work is also important in the mechanism whereby impulses are conducted in nerve and muscle cells. An especially dramatic example of electrical work is found in *Electrophorus electricus,* the electric eel. The electric organ of *Electrophorus* consists of layers of cells called *electroplaxes,* each of which can generate a membrane potential of about 150 millivolts (mV). Because the electric organ contains thousands of such cells arranged in series, the eel can develop potentials of several hundred volts.

Heat Energy and Work. It is easy to forget about **heat,** the "lowest form of energy," but in homeotherms (birds and mammals) heat production is a major use of food energy. In fact, as you read these lines, about two-thirds of your metabolic energy is being used simply to stay warm. How can two-thirds be used for heat, when we just noted above that two-thirds is used for ion transport? The answer is straightforward: the energy of ATP hydrolysis is used to pump ions, and when ATP is hydrolyzed to ADP and phosphate, heat is released as a by-product. This is the heat that warm-blooded animals use to maintain body temperature near 37°C, where their metabolism is most efficient. The relationship between work and heat energy is demonstrated when you get hot from exercise or shiver from the cold. You should now be able to understand the relationship between these seemingly unrelated processes and ATP hydrolysis, which powers muscle contraction.

Bioluminescent Work (the Production of Light). To be complete, we must also add production of light (**bioluminescence**) as yet another way in which energy is used by cells. The light produced by bioluminescent organisms is generated by the reaction of ATP with specific luminescent compounds and is usually pale blue. This is a much more specialized kind of energy use than the other five categories, and for present purposes we can leave it to the fireflies, luminous toadstools, dinoflagellates, deep-sea fish, and other creatures that live in its strange, cold light.

Using Energy: Chemotrophs and Phototrophs

The main forms of energy in our environment are listed in Table 5-1. It is immediately clear from the table that sunlight represents a major source of energy at the earth's surface. All organisms (and therefore all cells) can be classified into one of two groups, which differ in the way they take advantage of energy sources. The first group consists of organisms capable of capturing light energy by means of photosynthetic pigment systems, then storing the energy in the form of chemical bonds of organic molecules like glucose. Such organisms are called **phototrophs** (literally, "light-feeders") and include all green plants, all algae, and certain groups of bacteria that are capable of photosynthesis.

These chemical bonds then provide an energy source that can be used by a second group of organisms called **chemotrophs** (literally, "chemical-feeders") because they require the intake of chemical compounds such as carbohydrates, fats, and proteins. All animals, protists, and fungi and most bacteria are chemotrophs. The chemical energy is released by two methods. In the first, the chemical bonds are simply broken, releasing some of the stored energy. The general term for this process is **fermentation.** The more specific term **glycolysis** is used to describe the breakdown of the glucose molecule during the fermentation process. The second method involves oxidation, in which electrons are transferred from the chemical bonds to molecular oxygen. This process is often called **respiration.** During respiration, essentially all of the energy stored in chemical bonds is released.

A point that is often not appreciated about the phototrophs is that although they can utilize solar energy when it is available, they are also capable of functioning as chemotrophs and, in fact, do so whenever they are not illumi-

Table 5-1 Sources of Energy on the Earth's Surface

Energy Source	Calories/cm^2 per Year	Comments
Solar radiation	260,000	Includes infrared and ultraviolet light
Electric discharge	4	Lightning
Radioactivity	0.8	To 1 km depth in the earth's crust
Heat (volcanic)	0.13	
Chemical energy (photosynthesis)	100	
Chemical energy (stored)	—	A total of about 10^{26} calories is stored in coal, oil, and other geological deposits.

nated. Most higher plants are really a mixture of phototrophic and chemotrophic cells. A plant root cell, for example, though part of an obviously phototrophic organism, is in most cases incapable of carrying out photosynthesis and is every bit as chemotrophic as an animal cell.

The Flow of Energy in the Biosphere

So far we have seen that both chemotrophs and phototrophs depend on their environment for the energy they need, but differ in the forms of energy they can use. Chemotrophs require organic molecules, while phototrophs are uniquely equipped to trap solar radiation and transduce it into chemical bond energies.

The flow of energy through the biosphere is depicted in Figure 5-5. Solar energy is trapped by the phototrophs and used to convert carbon dioxide and water into more complex (and more reduced) cellular materials in the process of photosynthesis. As we will see in Chapter 12, the immediate products of photosynthetic carbon fixation are sugars, but in a sense we can consider the entire phototrophic organism to be the "product" of photosynthesis, since every carbon atom in every molecule of that organism is derived from carbon dioxide that is fixed into organic form by the photosynthetic process.

The chemotrophs, on the other hand, are unable to use solar energy directly, but depend on chemical energy of oxidizable molecules. The energy needs of the chemotrophs can be met either **anaerobically** (in the absence of oxygen) by a variety of fermentation processes or **aerobically** (in the presence of oxygen) by the complete oxidation of chemical compounds in the process of respiration. Chemotrophs therefore depend completely on energy that has been packaged into the bonds of fermentable or oxidizable food molecules by the phototrophs. A world composed only of chemotrophs would last only so long as food supplies held out, for even though we live on a planet that is flooded each day with solar energy, it is in a form that we cannot use to meet our energy needs.

Both the phototrophs and the chemotrophs use energy to carry out work—that is, to effect the various kinds of changes we have already catalogued. In the process, two kinds of losses occur (see Figure 5-5). It is one of the principles of energy conversion that no chemical or physical process occurs with 100% efficiency; some energy is lost as heat. In fact, most processes that involve the conversion of energy from one form to another actually dissipate more energy as heat than they succeed in converting into the desired form. An electric light bulb, for example, generates more heat than light.

As we will see in Part Three, biological processes are remarkably efficient in energy conversion. *Heat losses* are nonetheless inevitable in every biological energy transaction. Sometimes, the heat that is liberated during cellular processes is put to good use. As discussed earlier, warm-blooded animals use heat to maintain the body temperature at some constant level, usually well above ambient. Some plants use metabolically generated heat to melt the overlying snow and hasten their emergence in spring (Figure 5-6). But in general, the heat is simply dissipated into the environment and lost.

Even more fundamental is the *increase in entropy* that accompanies cellular activities. We will get to that in more detail shortly; here, we can simply note that every process or reaction that occurs anywhere in the universe always does so in such a way that the total entropy or disorder in the universe is increased. This change in entropy occurs at the expense of energy that might otherwise have been available to do useful work and is therefore an inevitable "sink" into which energy is lost. Just as the ultimate source of all energy in the biosphere is the sun, the ultimate fate of all energy in the biosphere is to become randomized in the universe as increased entropy or randomness.

Viewed on a cosmic scale, there is a continuous, massive, and unidirectional flow of energy from its source in the nuclear fusion reactions of the sun to its eventual sink,

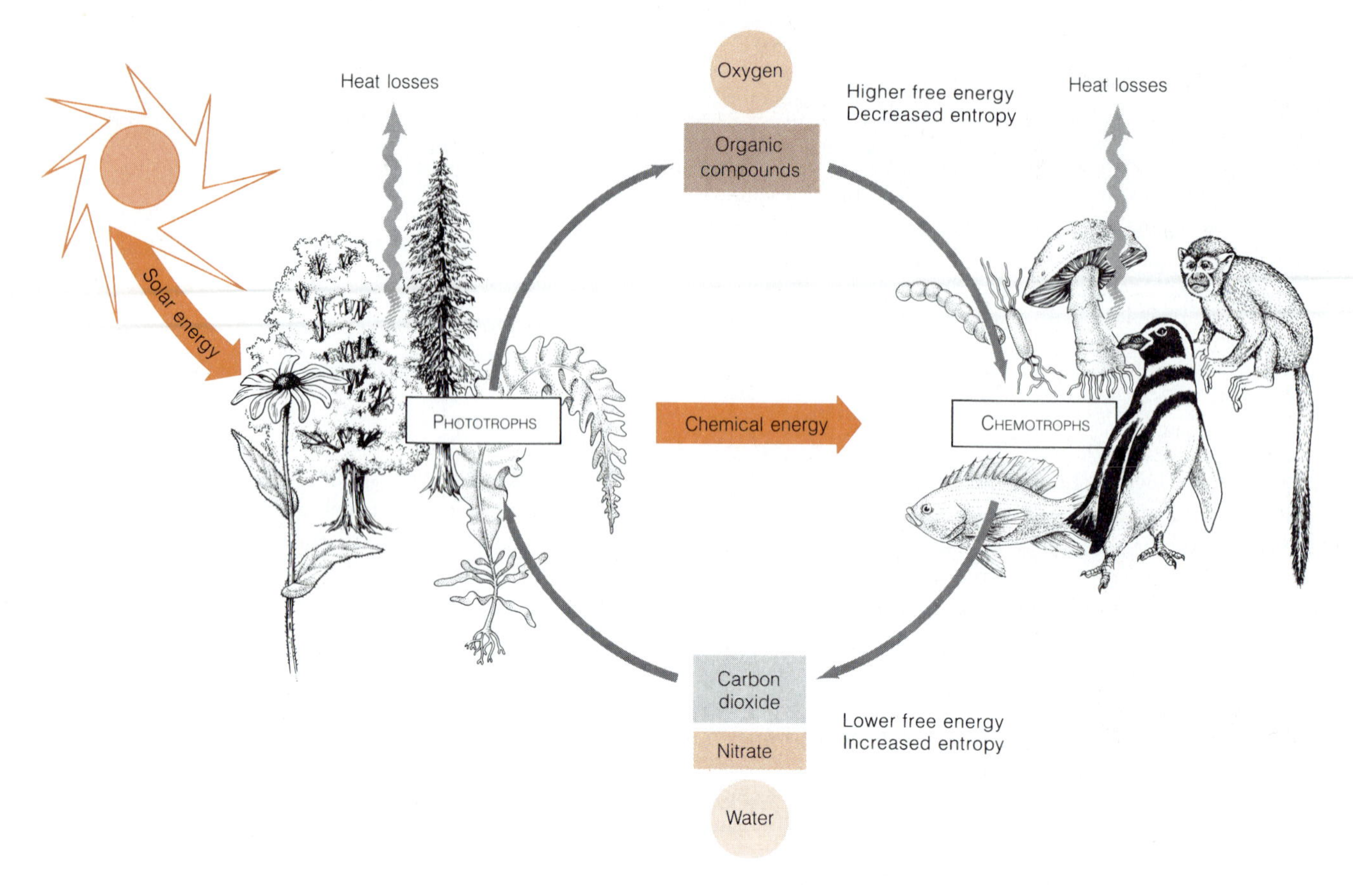

Figure 5-5 Flow of Energy Through the Biosphere. The energy in the biosphere originates in the sun and contributes eventually to the ever-increasing entropy of the universe. Accompanying the unidirectional flow of energy from phototrophs to chemotrophs is a cyclic flow of matter between the two groups of organisms.

the entropy of the universe. We here in the biosphere are the transient custodians of an almost infinitesimally small portion of that energy, but it is precisely that small but critical fraction of energy and its flow through living systems that is of concern to us. The flow begins with green plants, which use light energy to drive electrons energetically "uphill" into new chemical bonds. This energy is then released by both plants and animals in "downhill" fermentative reactions or oxidation. This flux of energy through living matter—from the sun, to phototrophs, to chemotrophs, to heat—drives the molecular machinery of all life processes.

The Flow of Matter in the Biosphere

Accompanying the flow of energy in the biosphere is a corresponding flow of matter or mass, since energy in cells and organisms is almost inevitably stored and transferred as chemical bond energies of organic molecules. Although energy enters the biosphere unaccompanied by matter (that is, as photons of light) and leaves the biosphere similarly unaccompanied (as heat losses and increases in entropy), while it is passing through the biosphere, energy exists primarily in the form of chemical bond energy. As a result, the flow of energy in the biosphere is coupled to a correspondingly immense flow of matter.

Whereas energy flows unidirectionally from sun through phototrophs to chemotrophs, matter flows in cyclic fashion between the two groups of organisms (Figure 5-5). During respiration, aerobic chemotrophs take in organic nutrients from their surroundings, usually by ingesting phototrophs or other chemotrophs that have in turn eaten phototrophs. These nutrients are oxidized to carbon dioxide and water, low-energy molecules that are returned to the environment. Those molecules then become the raw materials that the phototrophic organisms use to make new organic molecules photosynthetically, returning oxygen to the environment in the process.

In addition, there is an accompanying cycle of nitrogen. The phototrophs obtain nitrogen from the environment in inorganic form (often as nitrate from the soil, in some cases as N_2 from the atmosphere), convert it into ammonia, and use it in the synthesis of amino acids, proteins, nucleotides, and nucleic acids. Eventually, these molecules, like other components of phototrophic cells, are consumed by chemo-

Figure 5-6 Skunk Cabbage, a Plant That Depends on Metabolically Generated Heat as a Means of Emergence. The spadix, or floral spike, of species such as the western skunk cabbage, *Lysichitum americanum,* can maintain a temperature of 10 to 25°C above ambient, thereby literally melting its way through overlying snow and hastening emergence of the plant in early spring.

trophs. The nitrogen is then converted back into ammonia and eventually to nitrate, under the influence of soil microorganisms.

Carbon dioxide, oxygen, nitrogen, and water thus cycle continuously between the phototrophic and chemotrophic worlds, always entering the chemotrophic sphere as energy-rich compounds and leaving again in an energy-poor form. The two great groups of organisms can therefore be thought of as living in symbiotic relationship with each other, with a cyclic flow of matter and a unidirectional flow of energy as components of that symbiosis.

On to Cellular Energetics

When we deal with the overall macroscopic flux of energy and matter through living organisms, we find cellular biology interfacing with ecology, since the ecologist is very much concerned with cycles of energy and nutrients, with the roles of various species in these cycles, and with environmental factors that affect the flow. At the cellular level, our ultimate concern is how the flux of energy and matter that we have been considering on a macroscopic scale can be expressed and explained on a molecular scale in terms of energy transactions and the chemical processes that occur within cells. We therefore leave the macroscopic cycles to the ecologist and turn our attention to the reactions that occur within individual cells of bacteria, plants, and animals to account for those cycles. First, however, we must acquaint ourselves with the physical principles underlying energy transactions, and for that we turn to the topic of bioenergetics.

Bioenergetics

The principles that govern energy flow are incorporated in an area of science that the physical chemist calls **thermodynamics.** Although the prefix *thermo* suggests that the term is limited to heat (and that is indeed its historical origin), thermodynamics also takes into account other forms of energy and processes that convert energy from one form into another. Specifically, thermodynamics concerns the laws governing the energy transactions that inevitably accompany most physical processes and all chemical reactions. **Bioenergetics,** in turn, can be thought of as applied thermodynamics—that is, it concerns the application of thermodynamic principles to reactions and processes in the biological world.

Energy, Systems, Heat, and Work

As we have already seen, it is helpful to define energy not simply as the ability to do work but specifically as the ability to cause change. Without energy, all processes would be at a standstill, including those that we associate with living cells.

Energy exists in a variety of forms, many of them of interest to biologists. Think, for example, of the energy represented by a ray of sunlight, a teaspoon of sugar, a moving flagellum, an excited electron, or the concentration of atoms, molecules, or ions within a cell or an organelle. These phenomena are diverse, but they are all governed by certain basic principles of energetics.

Energy is distributed throughout the universe, and for some purposes it is necessary to consider the total energy of the universe, at least in a theoretical way. Usually, however, we are interested not in the whole universe but only in a small portion of it. We might, for example, be concerned with a reaction or process occurring in a beaker of chemicals, in a cell, or in a block of metal. By convention, the restricted portion of the universe that one wishes to consider at the moment is called the **system,** and all the rest of the universe is referred to as the **surroundings** (Figure 5-7). Sometimes, the system has a natural boundary, such as a glass beaker or a cell membrane. In other cases, the

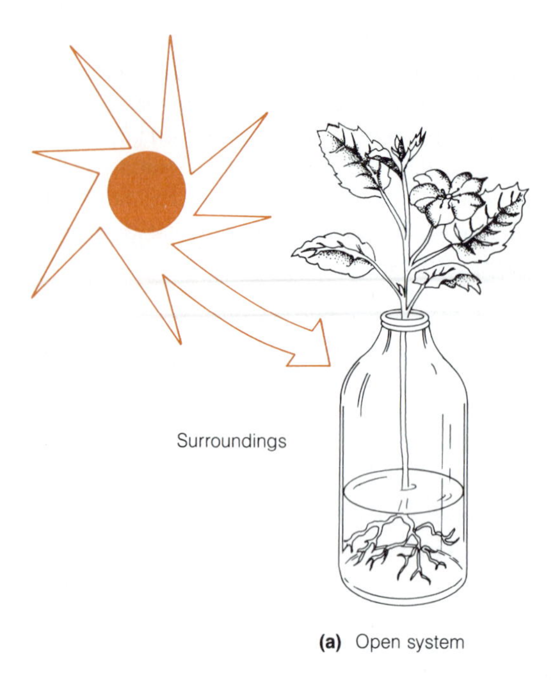

(a) Open system

(b) Closed system

Figure 5-7 System, Surroundings, and Universe. A system is that portion of the universe under consideration. The rest of the universe is called the surroundings of the system. (a) An open system can exchange energy with its surroundings, whereas (b) a closed system cannot. The open system can use incoming energy to increase its orderliness (thereby decreasing its entropy), whereas the closed system tends toward equilibrium and increases its entropy.

boundary between the system and its surroundings is a hypothetical one used only for convenience of discussion, such as the imaginary boundary around one mole of glucose molecules in a solution.

Systems can be either open or closed, depending on whether or not they can exchange energy with their surroundings (Figure 5-7). A **closed system** is sealed from its environment and can neither take in nor release energy in any form. An **open system,** on the other hand, can have energy added to it or removed from it. As we will see later, the improbable levels of organization that biological systems routinely display are possible only because cells and organisms are open systems, capable of both uptake and release of energy. Specifically, biological systems require a constant, large-scale influx of energy from their surroundings both to attain and to maintain the levels of complexity characteristic of them. That is why plants need sunlight and you need food.

Whenever we talk about a system, we have to be careful to specify the **state** of the system. A system is said to be in a specific state if each of its variable properties (such as temperature, pressure, and volume) is held constant at a specified value. In such a situation, the total energy content of the system, while not directly measurable, has some unique value. If such a system then changes from one state to another as a result of some interaction between the system and its surroundings, the change in its total energy is determined uniquely by the initial and final states of the system and is not affected at all by the mechanism by which the change occurs or the intermediate states through which the system may pass. This is a very useful property, since it allows energy changes to be determined from a knowledge of the initial and final states only.

The problem of keeping track of system variables and their effect on energy changes can be simplified if one or more of the variables are held constant. Fortunately, this turns out to be the case with most biological reactions, since they usually occur in dilute solutions within cells that are at approximately the same temperature and pressure during the entire course of the reaction. These environmental conditions, as well as the cell volume, are generally slow to change compared to the speed of biological reactions. This means that three of the most important system variables that physical chemists usually concern themselves with—temperature, pressure, and volume—are essentially constant for biological reactions.

The exchange of energy between a system and its surroundings occurs in two ways: as heat and as work. Heat is energy transfer from one place to another as a result of a temperature difference between the two places. Spontaneous transfer always occurs from the hotter place to the colder place. Heat is an exceedingly useful form of energy for many machines and other devices designed to accomplish mechanical work. However, it has only limited biological utility, because most biological systems operate under conditions of either fixed or only minimally variable temperature. Such **isothermal** systems lack the temperature gradients required to convert heat into other forms of energy. As a result, heat is not a useful source of energy for cells.

In biological systems, **work** is the application of energy from one place or form to another place or form to drive any process other than heat flow. For example, work is performed when the muscles in your arm expend chemical energy to lift this book, when a corn leaf uses light energy to synthesize sugar, or when an electric eel draws on the ion concentration gradients of its electroplax tissue to deliver a shock. It is the amount of useful energy available to do cellular work that we will be primarily interested in when we begin calculating energy changes associated with specific reactions that cells carry out.

To quantify energy changes during chemical reactions or physical processes, we need units in which energy can be expressed. The most common units are the calorie and the joule. In biological chemistry, energy changes are usually expressed in terms of the **calorie,** which we have already defined as the amount of energy required to warm 1 gram of water 1 degree centigrade (specifically, from 14.5° to 15.5°C) at a pressure of 1 atmosphere. (Again, notice that the unit of energy measurement, like the very term *thermodynamics,* is based on heat but is applied generally to all forms of energy.) Frequently, energy changes will be measured on a per-mole basis, so the most common form in which we will encounter energy units in biological chemistry will be as calories (or sometimes kilocalories) per mole (cal/mol or kcal/mol). A **joule** is equivalent to 0.239 cal. It is the preferred unit for energy among physicists, but biochemists most often use calories. (You will want to be careful, by the way, to distinguish between the calorie that the biochemist uses and the Calorie that is often used to express the energy content of foods. The "big" Calorie is really a kilocalorie and equals 1000 "small" calories.)

Conservation of Energy: The First Law of Thermodynamics

Much of what we understand about the principles governing energy flow can be summarized in the three laws of thermodynamics. Of these, only the first and second laws are of particular relevance to the cell biologist. The **first law** is called the *law of conservation of energy.* Simply put, it says that *energy can be converted from one form to another but can never be created or destroyed.* (If you are familiar with the conversion of mass to energy that occurs in nuclear reactions, you will recognize that a more accurate statement would take both mass and energy into account. For purposes of biological chemistry, however, the law is adequate as stated.)

Applied to the universe as a whole or to a closed system, the first law means that the total amount of energy present in all forms must be the same before and after any process or reaction occurs. Applied to an open system, such as a cell, the first law says that during the course of any reaction or process, the total amount of energy that leaves the system must be exactly equal to the energy that enters the system minus any energy that remains behind and is therefore stored within the system:

$$\text{energy out} = \text{energy in} - \text{energy stored} \tag{5-1}$$

Or, by simple rearrangement,

$$\text{energy stored} = \text{energy in} - \text{energy out} \tag{5-2}$$

The total energy stored within a system is called the **internal energy** of the system, represented by the symbol E. We are not usually concerned with the actual value of E for a system, because that value cannot be measured directly. However, it is possible to measure the *change in internal energy,* ΔE, that occurs during a given process. ΔE is the difference in internal energy of the system before and after the process and can be measured as the difference between the amounts of energy entering and leaving the system:

$$\Delta E = E_2 - E_1 = \text{energy in} - \text{energy out} \tag{5-3}$$

For a biological example of such an energy transaction, consider the oxidation of glucose to carbon dioxide and water:

$$C_6H_{12}O_6 + 6O_2 \longrightarrow 6CO_2 + 6H_2O + \text{energy} \tag{5-4}$$

You may recognize this as the summary equation for the process of respiration, whereby aerobic chemotrophs obtain energy from the sugar glucose. The reaction is therefore highly relevant to cellular energy flow, in addition to illustrating several points about ΔE.

By burning glucose in the laboratory, we can show that 673,000 cal (673 kcal) of energy are liberated for every mole of glucose that is oxidized. This means that 1 mole of glucose has 673 more kilocalories of internal energy than do 6 moles of CO_2 and 6 moles of H_2O. Reaction 5-4 therefore has a ΔE of -673 kcal/mol, where the negative sign indicates a *decrease* in internal energy as glucose is converted into carbon dioxide and water (Figure 5-8).

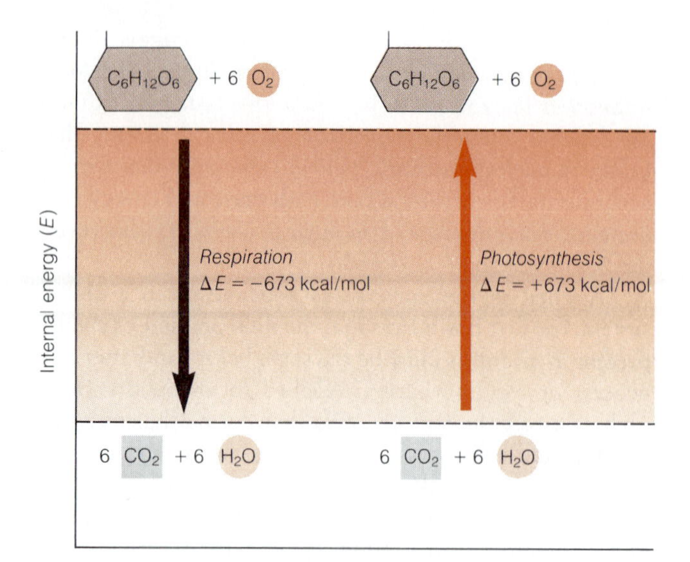

Figure 5-8 Internal Energy Changes. The internal energy E of the system decreases by 673 kcal/mol when sugar is oxidized to carbon dioxide and water (respiration) but increases by the same amount when carbon dioxide and water are converted into sugar (photosynthesis).

Now consider the reverse reaction, which turns out to be a summary equation for photosynthesis:

$$6CO_2 + 6H_2O + \text{energy} \longrightarrow C_6H_{12}O_6 + 6O_2 \quad (5\text{-}5)$$

If this reaction is carried out under the same conditions used earlier to measure ΔE for glucose oxidation, it can be shown that the ΔE value for glucose synthesis is +673 kcal/mol, where the positive number indicates an *increase* in internal energy as carbon dioxide and water are converted to glucose (Figure 5-8).

The important point is that ΔE values for reactions 5-4 and 5-5 are equal in magnitude but opposite in sign. The increase in energy as glucose is synthesized is matched exactly by the decrease in energy as glucose is oxidized. Energy is neither created nor destroyed, just as the first law predicts. The first law of thermodynamics, in other words, allows us to calculate energy changes with confidence because it assures us that energy can always be accounted for.

How to Know Which Way It Will Go: The Second Law of Thermodynamics

So far, all that thermodynamics has been able to tell us is that energy is conserved whenever a process or reaction occurs—that all of the energy going into a system must either be stored within the system or released again to the surroundings. We have seen the usefulness of ΔE as a measure of how much the total internal energy of a system would change if a given process were to occur, but we have no way as yet of predicting whether and to what extent the process will in fact occur under the prevailing conditions.

We have, at least in some cases, an intuitive feeling that some reactions or processes are possible, whereas others are not. Returning to the examples of Figure 5-1, we are somehow very sure that if we were to set a match to paper, it would burn. The oxidation of cellulose to carbon dioxide and water is, in other words, a possible reaction. Or, to use more precise terminology, it is a *thermodynamically spontaneous* reaction. Furthermore, there is clearly a directionality to the process, for we are equally convinced that the reverse process will not occur—that if we were to stand around clutching the charred remains, the paper would not spontaneously reassemble in our hands. We have, in other words, a feeling for both the possibility and the directionality of cellulose oxidation.

You can probably think of other processes for which you can make such thermodynamic predictions with equal confidence. We know, for example, that drops of dye diffuse in water, that ice cubes melt at room temperature, and that sugar dissolves in water, and we can therefore label these as thermodynamically spontaneous events. But if we ask why we recognize them as such, the answer has to do with repeated prior experience. We have seen paper burn, ice cubes melt, and sugar dissolve often enough to know intuitively that these are processes that really occur, and with such predictability that we can label them as spontaneous, provided only that we know the conditions.

However, when we move from the world of familiar physical processes to the realm of chemical reactions in cells, we quickly find that we cannot depend on prior experience to guide us in our predictions. Consider, for example, the conversion of glucose-6-phosphate into fructose-6-phosphate:

$$\text{glucose-6-phosphate} \rightleftharpoons \text{fructose-6-phosphate} \quad (5\text{-}6)$$

Glucose is a six-carbon aldosugar and fructose is its keto equivalent (recall structures of Figure 3-15). Both can form a phosphoester bond between a phosphoric acid (phosphate) molecule and the hydroxyl group on carbon 6 of the sugar, giving rise to the phosphorylated compounds. Reaction 5-6 therefore involves the interconversion of a phosphorylated aldosugar to the corresponding phosphorylated ketosugar, as shown in Figure 5-9.

This particular interconversion is a significant reaction in all cells. It is, in fact, the second step in an important and universal reaction sequence called the glycolytic pathway. In addition to illustrating an important thermodynamic principle, therefore, reaction 5-6 introduces a bit of

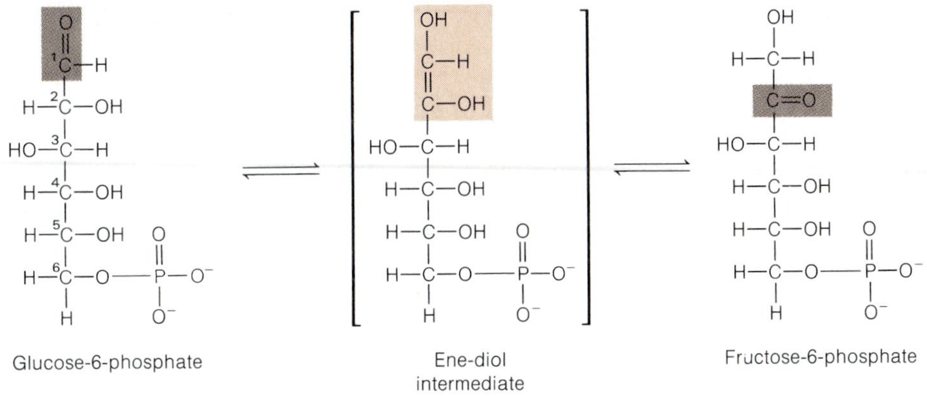

Glucose-6-phosphate

Ene-diol intermediate

Fructose-6-phosphate

Figure 5-9 Conversion of Glucose-6-Phosphate into Fructose-6-Phosphate. This reaction involves the change of an aldosugar to a ketosugar and proceeds via an intermediate that is called an ene-diol because it has a carbon-carbon double bond ("ene") with two alcohol groups ("di-ol") attached.

cellular chemistry that will come in handy later on. For present purposes, however, just focus on the reaction from a thermodynamic point of view, and ask yourself what predictions you can make about the likelihood of glucose-6-phosphate being converted into fructose-6-phosphate. You will probably be at a loss to make any predictions at all. We know what will happen with burning paper and melting ice, but we lack the familiarity and prior experience with phosphorylated sugars even to make an intelligent guess. Clearly, what we need is a reliable means of determining whether a given physical or chemical change can occur under specific conditions without having to rely on prior experience, familiarity, or intuition.

Thermodynamics provides us with exactly such a measure of spontaneity in the **second law**, or the *law of thermodynamic spontaneity*. Moreover, the second law allows us to predict not only in what direction a reaction will proceed under specified conditions but also how far from equilibrium it lies, how much energy the reaction will release or consume as it proceeds, and how the energetics of the reaction will be affected by specific changes in the conditions.

The two specific thermodynamic parameters that measure thermodynamic spontaneity are entropy and free energy. These concepts are sometimes difficult to understand because they are somewhat abstract. We will therefore limit our discussion here to their use in determining what kinds of changes can occur in biological systems. For further help, see the box on pp. 118–120 for an essay that uses "jumping beans" to introduce the concepts of internal energy, entropy, and free energy.

Entropy. Although we cannot experience **entropy** directly, we can get some feel for it by considering it to be a measure of randomness or disorder. This means that for any system, the *change in entropy*, ΔS, represents a change in the degree of randomness or disorder of the components of the system. As an example, consider again the drop of dye in Figure 5-1. As the dye molecules diffuse through the water, they become ever more random in their location in the beaker, and the entropy of the system therefore increases. Similarly, the combustion of paper involves an increase in entropy because the carbon, oxygen, and hydrogen atoms of cellulose are much more randomly distributed in space once they are converted to carbon dioxide and water. Entropy also increases as ice melts or as a volatile solvent such as gasoline is allowed to evaporate.

Entropy Change as a Measure of Thermodynamic Spontaneity. How can any of this help predict what changes will occur in a cell? As it turns out, there is a very important link between spontaneous events and entropy changes, because whenever a process occurs in nature, the randomness or disorder of the universe (that is, the entropy of the universe) invariably increases. This is one of two alternative ways to state the second law of thermodynamics (see box on p. 122). According to this formulation, *the value of* $\Delta S_{universe}$ *is positive for every real process or reaction.* Processes that would lead to a decrease in the entropy of the universe simply do not occur. Remembering that entropy is a measure of randomness, we can rephrase the second law to state that the universe becomes more random with every reaction that occurs and that reactions that would make it less random are never observed.

We have to keep in mind, however, that this formulation of the second law pertains to the universe as a whole and may not apply to the specific system under consideration. Every real process, without exception, must be

Jumping Beans and Free Energy

If you are finding the concepts of free energy, entropy, and equilibrium constants difficult to grasp, perhaps a simple analogy might help.* For this we will need an imaginary supply of jumping beans, which are really seeds of certain Mexican shrubs, with larvae of the moth *Laspeyresia saltitans* inside. Whenever the larvae inside the seed wiggle about, the seeds wiggle too. They don't really jump in the usual sense of the word, although certain wasp larvae can in fact cause the galls they inhabit (called "fleaseeds") to leap several centimeters from a standing start! The jumping action probably serves to get the larvae out of direct sunlight, which could heat them to lethal temperatures.

The Jumping Reaction

For purposes of illustration, imagine that we have some high-powered jumping beans in two chambers separated by a low partition, as shown below. Notice that the chambers have the same floor area and are at the same level, although we will want to vary both of these properties shortly. As soon as we place a handful of jumping beans in chamber 1, they begin jumping about randomly. Although most of the beans jump only to a modest height most of the time, occasionally one of them, in a burst of ambition, gives a more energetic leap, surmounting the barrier and falling into chamber 2. We can write this as the *jumping reaction:*

$$\text{Beans in chamber 1} \rightleftharpoons \text{Beans in chamber 2}$$

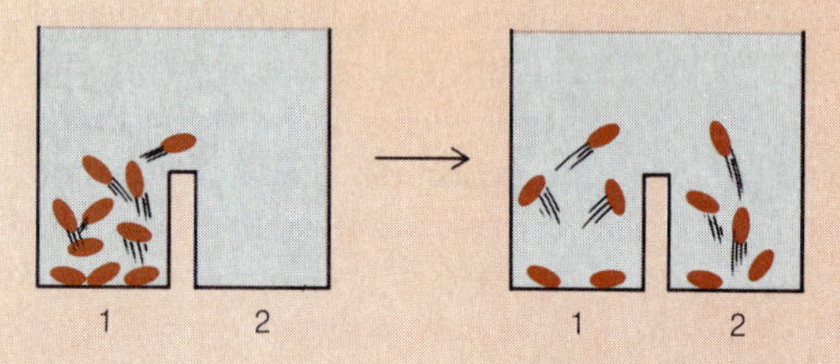

We will imagine this to be a completely random event, happening at irregular, infrequent intervals. Occasionally, one of the beans that has reached chamber 2 will happen to jump back into chamber 1, which is the *back reaction.* At first, of course, there will be more beans jumping from chamber 1 to chamber 2 because there are more beans in chamber 1, but things will eventually even out so that, on the average, there will be the same number of beans in both compartments. The system will then be at *equilibrium*. Beans will still continue to jump between the two chambers, but the numbers jumping in both directions will be equal.

The Equilibrium Constant

Once our system is at equilibrium, we can count up the number of beans in each chamber and express the results as the ratio of the number of beans in chamber 2 to the number in chamber 1. This is simply the *equilibrium constant* K_{eq} for the jumping reaction:

$$K_{eq} = \frac{\text{number of beans in chamber 2 at equilibrium}}{\text{number of beans in chamber 1 at equilibrium}}$$

For the specific case shown above, the numbers of beans in the two chambers are equal at equilibrium, so the equilibrium constant for the jumping reaction under these conditions is 1.0.

Internal Energy Change (ΔE)

Now suppose that the level of chamber 1 is somewhat higher than that of chamber 2, as shown below. Jumping beans placed in chamber 1 will again tend to distribute themselves between chambers 1 and 2, but this time a higher jump is required to get from 2 to 1 than from 1 to 2, so the latter will occur more frequently. As a result, there will be more beans in chamber 2 than in chamber 1 at equilibrium, and the equilibrium constant will therefore be greater than 1.

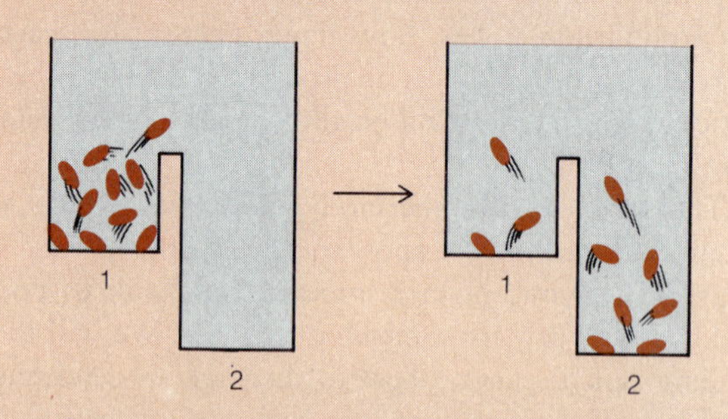

The relative positions of the two chambers can be thought of as measures of the total *internal energy* of the system, E, such that chamber 1 has a higher E value than chamber 2, and the *difference* between them is represented by ΔE. Since it is a "downhill" jump from chamber 1 to chamber 2, it makes sense that ΔE has a negative value for the jumping reaction from chamber 1 to chamber 2. By the same token, it seems

* We are indebted to Princeton University Press for permission to use this analogy, first developed by Harold F. Blum in the book *Time's Arrow and Evolution* (3d ed., 1968), pp. 17–26.

reasonable that ΔE for the reverse reaction should have a positive value, since that jump is uphill.

Entropy Change (ΔS)

So far, it might seem as if the only thing that can affect the equilibrium distribution of beans between the two chambers is the difference in internal energy, ΔE. But that is only because we have kept the floor area of the two chambers constant. Imagine instead the situation shown below, where the two chambers are again at the same height, but chamber 2 now has a greater floor area than chamber 1. The probability of a bean finding itself in chamber 2 is therefore correspondingly greater, so there will be more beans in chamber 2 than in chamber 1 at equilibrium, and the equilibrium constant will be greater than 1 in this case also. This means that the equilibrium position of the jumping reaction has been shifted to the right, even though there is no change in internal energy at all.

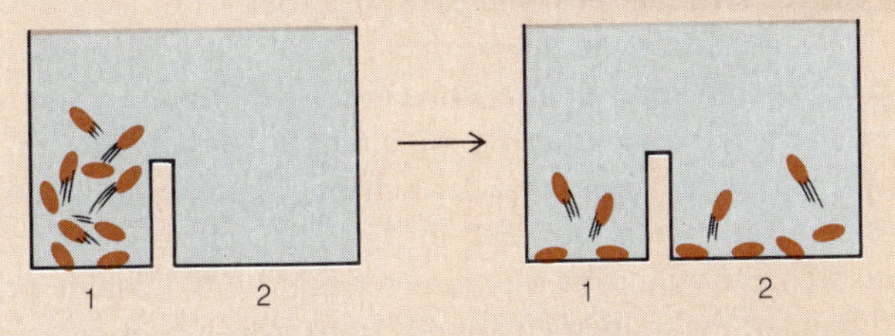

The floor area of the chambers can be thought of as a measure of the *entropy* or randomness of the system, S, and the *difference* between the two chambers can be represented by ΔS. Since chamber 2 has a greater floor area than chamber 1, the entropy change is positive for the jumping reaction as it proceeds from left to right under these conditions. Note that for ΔE, negative values are associated with favorable reactions, while for ΔS, favorable reactions are indicated by positive values.

Free Energy Change (ΔG)

So far, we have encountered two different factors that affect the distribution of beans—the difference in levels of the two chambers (ΔE) and the difference in floor area (ΔS). Moreover, it should be clear that neither of these factors by itself is an adequate indicator of how the beans will be distributed at equilibrium, since a favorable (negative) ΔE could be more than offset by an unfavorable (negative) ΔS, and a favorable (positive) ΔS could be more than offset by an unfavorable (positive) ΔE. You should, in fact, be able to design chamber conditions that illustrate both of these situations, as well as situations in which ΔE and ΔS tend to reinforce rather than counteract each other.

Clearly, what we need is some way of summing these two effects algebraically to see what the net tendency will be. The new measure we come up with is called the *free energy change*, ΔG, and is defined so that *negative* values correspond to favorable (thermodynamically spontaneous) reactions and positive values represent unfavorable reactions. Thus, ΔG should have the *same* sign as ΔE (since a negative ΔE is also favorable) but the *opposite* sign from ΔS (since for ΔS a positive value is favorable). In terms of real-life thermodynamics, the actual expression for ΔG is

$$\Delta G = \Delta E - T\,\Delta S$$

(Notice that the temperature dependence of ΔS is the only feature of this relationship between ΔG, ΔE, and ΔS that cannot be readily accounted for by our simple model—unless we are to assume that the effect of changes in room size is somehow greater at higher temperatures.)

ΔG and the Capacity to Do Work

You should be able to appreciate the difficulty of suggesting a physical equivalent for ΔG, since it represents an algebraic sum of entropy and energy changes, which may either reinforce or partially offset each other. But so long as ΔG is negative, beans will continue to jump from chamber 1 to chamber 2, whether driven primarily by changes in entropy, internal energy, or both. This means that if some sort of bean-powered "bean wheel" is placed between the two chambers as shown below, the movement of beans from one chamber

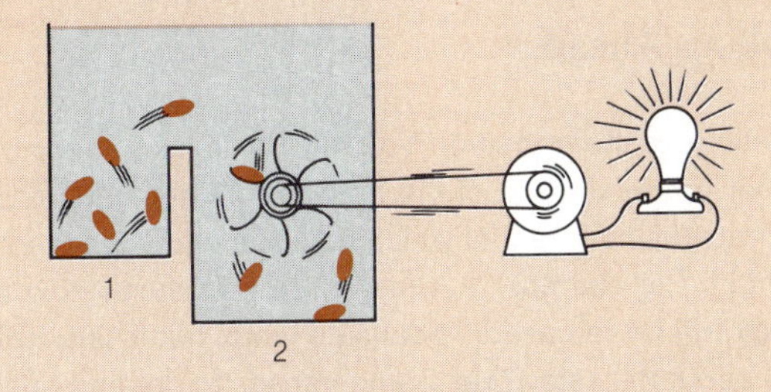

to the other can be harnessed to do work until equilibrium is reached, at which point no further work is possible. Furthermore, the greater the difference in free energy between the two chambers (the more highly negative ΔG is), the more work the system can do (see next page).

(Box continues.)

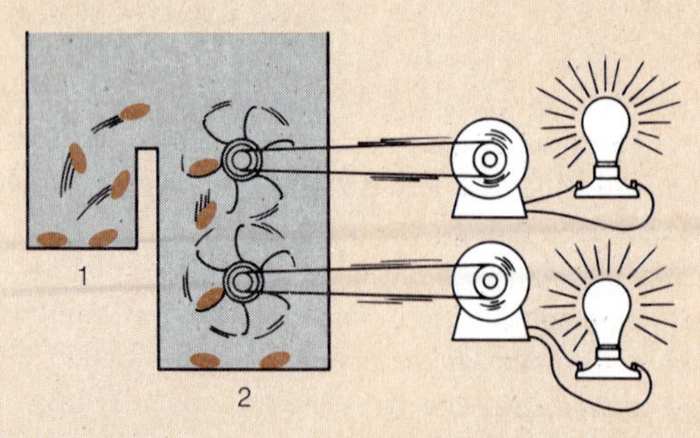

Thus, ΔG is first and foremost a measure of the capacity of a system to do work under specified conditions. You might, in fact, want to think of ΔG as free energy in the sense of energy that is free or available to do useful work. Moreover, if we contrive to keep ΔG negative by continuously adding beans to chamber 1 and removing them from chamber 2, we have a dynamic *steady state*, a condition that effectively harnesses the inexorable drive to achieve equilibrium (see below). Work can then be performed continuously by beans that are forever jumping toward equilibrium but that never actually reach it.

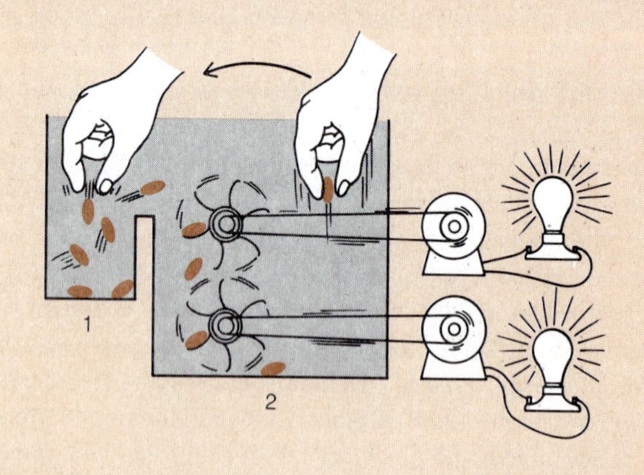

Looking Ahead

To anticipate the transition from the thermodynamics of this chapter to the kinetics of the next, you might want to begin thinking about the *rate* at which beans actually proceed from chamber 1 to chamber 2. Clearly, ΔG measures how much energy will be released if beans do jump, but it says nothing at all about the rate. That would appear to depend critically on how high the barrier between the two chambers is. Label this the *activation energy barrier,* and then contemplate means by which you might get the beans to move over the barrier more rapidly. You will probably be helped by the knowledge that the larvae inside the seeds wiggle more vigorously if they are warmed.

accompanied by an increase in the entropy of the universe, but for a given system the entropy may increase, decrease, or stay the same as the result of a specific process. For example, the oxidation of glucose to carbon dioxide and water (reaction 5-4) is spontaneous under standard conditions (25°C, 1 atmosphere pressure, and pH 7.0) and is accompanied by an *increase* in the system entropy (ΔS_{system} = +43.6 cal/mol-degree). On the other hand, the freezing of water at −0.1°C is also a spontaneous event, yet it involves a *decrease* in the system entropy (ΔS_{system} = −0.5 cal/mol-degree). This makes sense when you consider the greater ordering of water molecules in ice crystals. Thus, while the change in entropy of the universe is a valid measure of the spontaneity of a process, the change in entropy of the system is not.

To understand how the entropy of the universe can increase during a process while the entropy of the system decreases, we need only realize that the decrease in entropy of the system can be accompanied by an equal or even greater increase in the entropy of the surroundings. On the basis of the second law, such local increases in order (decreases in entropy) must be offset by an even greater decrease in the order (increase in entropy) of the surroundings. This is, in fact, the situation that normally prevails within living cells, which by their growth and reproduction produce striking local increases in order.

This means, however, that the second law, stated in terms of entropy of the universe, is of limited value in predicting the spontaneity of biological processes, for it would require keeping track of changes that occur not only within the system but also in its surroundings. Far more convenient would be a parameter that would enable prediction of the spontaneity of reactions from a consideration of the system alone.

Free Energy. As you might well guess, a measure of spontaneity for the system alone does in fact exist. It is called **free energy** and is represented by the symbol G in honor of Willard Gibbs, who first developed the concept. Because of its predictive value and its ease of calculation, the free energy function is one of the most useful thermodynamic concepts in biology. One could even make the case that all of our discussion of thermodynamics so far has really been a way of getting us to free energy, because it is here that the usefulness of thermodynamics for cell biologists becomes apparent.

Like most other thermodynamic functions, free energy is defined only in terms of a mathematical relationship. But for biological systems at constant pressure, volume, and temperature, the **free energy change,** ΔG, is related to

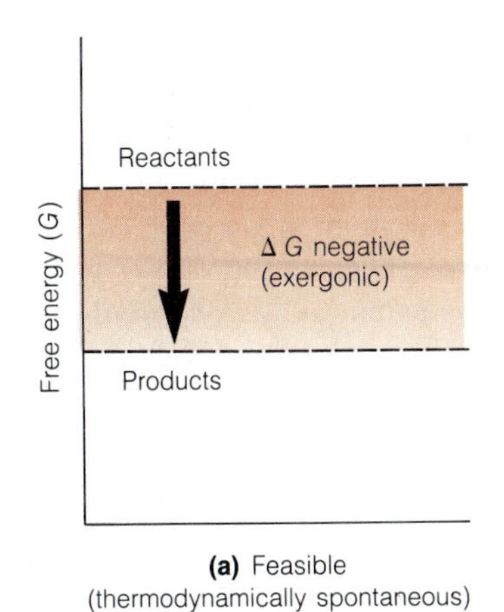

(a) Feasible
(thermodynamically spontaneous)

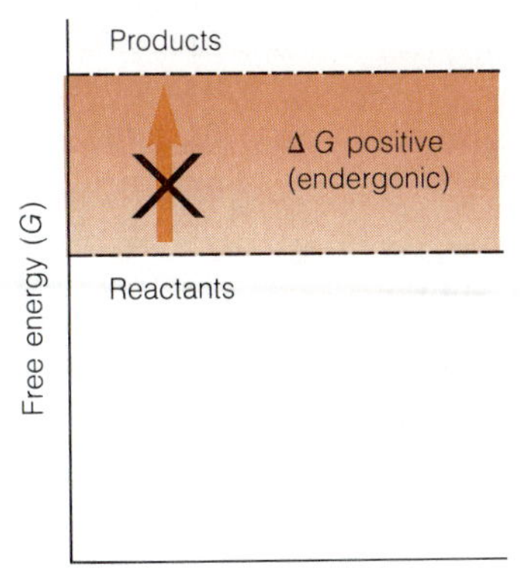

(b) Not feasible
(thermodynamically nonspontaneous)

Figure 5-10 Changes in Free Energy (ΔG) as a Measure of Thermodynamic Spontaneity. (a) Every reaction or process for which ΔG is negative is thermodynamically spontaneous and therefore feasible. (b) Any reaction or process for which ΔG is positive would yield products with a greater free energy content than the reactants and is therefore thermodynamically impossible. However, if the reaction is coupled to an exergonic reaction with a ΔG sufficiently negative to ensure that the overall ΔG for the coupled reaction is also negative, the first reaction can be "driven" by the second reaction. In the cell, the exergonic reaction of ATP hydrolysis is often used in this manner.

the changes in internal energy and entropy as follows:*

$$\Delta E = \Delta G + T\Delta S \tag{5-7}$$

or

$$\Delta G = \Delta E - T\Delta S \tag{5-8}$$

where ΔE is the change in internal energy, ΔG the change in free energy, ΔS the change in entropy, and T the temperature of the system in degrees Kelvin ($=°C + 273$).

Free Energy Change as a Measure of Thermodynamic Spontaneity. Free energy turns out to be an exceptionally useful concept as a readily measurable indicator of spontaneity. As we shall see shortly, ΔG for a reaction can easily be calculated from the equilibrium constant for the reaction and from easily measurable system variables, such as the concentrations of reactants and products. Once determined, ΔG provides exactly what we have been looking for: a measure of the spontaneity of a reaction that is based solely on the properties of the system in which the reaction is occurring.

Specifically, every spontaneous reaction is characterized by a *decrease* in the free energy content of the system ($\Delta G_{system} < 0$) just as surely as it is characterized by an *increase* in the entropy of the universe ($\Delta S_{universe} > 0$). This is true because with the temperature and pressure held constant, ΔG for the system is related to ΔS for the universe in a simple but inverse way. This gives us a second, equally valid way of expressing the second law: *All reactions that occur spontaneously result in a decrease in the free energy content of the system* (Figure 5-10a).

Such reactions are called **exergonic,** which means energy-yielding. Note carefully that the reference is specifically to the change in free energy and not to the change in total internal energy, which may be negative, positive, or zero for a spontaneous reaction and is therefore *not* a valid measure of thermodynamic spontaneity. Conversely, any process that would result in an increase in the free energy of the system is called **endergonic** (energy-requiring) and cannot proceed under the conditions used to calculate ΔG (Figure 5-10b). If, however, conditions are altered (as, for example, by increasing the concentrations of the reactants or decreasing the concentrations of the products), it is sometimes possible to convert an otherwise endergonic reaction into an exergonic reaction and thus render it spontaneous. As you will see later, this fact is fundamental to our understanding of biosynthetic reactions.

The Meaning of Spontaneity. Before looking at how we can actually calculate and use ΔG as a measure of thermodynamic spontaneity, we need to make sure we understand what is—and what is not—meant by the term

---

* Strictly speaking, the relationship between the free energy G and the internal energy E is given by the equation

$$E = G - PV + TS$$

where P is pressure and V is volume. The change in internal energy, ΔE, must therefore be expressed as

$$\Delta E = \Delta G - \Delta(PV) + \Delta(TS)$$

This relationship holds for any system and is the form most likely to be encountered in introductory thermodynamics texts. For the special case of biological systems, where P, V, and T are all constant, this equation reduces to

$$\Delta E = \Delta G + T\Delta S$$

ΔE is usually referred to as enthalpy (ΔH) under these conditions. A common form of equation 5-8 is $\Delta G = \Delta H - T\Delta S$, where ΔH is simply a measure of the heat produced or taken up by the reaction at constant pressure and volume.

Energy and Entropy: The Greek Connection

Sometimes it requires genius to see the obvious. Rudolph Clausius was just 28 when he published his first scientific paper in Berlin, in 1850. It concerned heat as an energy source, and his conclusion was that "Heat of itself cannot pass from a colder to a hotter body." This seems so simple to us now, with the understanding that heat is a form of random kinetic energy that is transferred between atoms and molecules through collision and radiation. Yet, in 1850, heat was still a mysterious quantity that scientists and engineers were struggling to understand.

In later work, Clausius went on to a second, equally powerful insight, which is not so obvious even today. The word "energy" is from the Greek, with the sense of "containing work." That is, something has "energy" if it can cause something else to move or change in some way. (One definition of work is to move a mass over a distance: moving a piano is work!) Clausius noticed that when energy was used to do work, he could never get an exact balance between the energy used and the work done. For instance, when heat was used to make steam and the steam energy was transformed into mechanical work by a steam engine, measurements showed that the work done was always less than the energy put into the system. This strange effect needed a descriptive term. Clausius reasoned that if *energy* meant "work content," he would coin the word *entropy* from the Greek words for "transformation content": the greater the number of transformations energy went through, the greater the entropy. As these ideas developed further, it became clear that this concept represented a universal law, now known as the second law of thermodynamics, in which entropy is related to the relative disorder of a system. As Clausius summed it up: "The energy of the universe is constant; the entropy of the universe tends towards a maximum." One of Murphy's laws also sums it up nicely: "If something can go wrong, it will."

spontaneous. As used in thermodynamics and bioenergetics, *spontaneous* refers to reactions that have a negative ΔG and therefore are capable of proceeding in the direction indicated. However, the term tells us only that the reaction *can* go; it says nothing at all about whether it *will* go. A reaction can have a negative ΔG value and yet not actually proceed to any measurable extent at all. The combustion of paper is a good example. The cellulose of paper obviously burns spontaneously once ignited, consistent with a highly negative ΔG value of -686 kcal per mole of glucose units. Yet in the absence of a match, paper is reasonably stable and would require thousands of years to oxidize. Thus, ΔG can really tell us only whether a reaction or process is thermodynamically possible—whether it has the *potential* for occurring. Whether an exergonic reaction will in fact proceed depends not only on its favorable (negative) ΔG but also on the availability of a mechanism or pathway to get from the initial state to the final state. Usually, an initial input of activation energy is required as well, such as the heat energy from the match that was used to ignite the piece of paper.

Thermodynamic spontaneity is therefore a necessary but insufficient criterion for determining whether a reaction will actually occur. When we get to Chapter 6, we will have an opportunity to explore the question of reaction rates in the context of enzyme-catalyzed reactions. For the moment, we need only note that when we designate a reaction as thermodynamically spontaneous, we simply mean that it is an energetically feasible event that will liberate free energy if and when it actually takes place.

Understanding ΔG

Our final task in this chapter will be to understand how ΔG is calculated and how it can then be used to assess the thermodynamic feasibility of reactions under specified conditions. For that we come back to the reaction that converts glucose-6-phosphate into fructose-6-phosphate (reaction 5-6) and ask what we can ascertain about the spontaneity of the conversion in the direction written (from left to right). Prior experience and familiarity provide no clues here, nor is it obvious how the entropy of the universe would be affected if the reaction were to proceed. Clearly, we need to be able to calculate ΔG and to determine whether it is positive or negative under the particular conditions we specify for the reaction.

The Equilibrium Constant as a Measure of Directionality

One means of assessing whether a reaction can proceed in a given direction under specified conditions involves the

equilibrium constant K_{eq}, which is the ratio of product concentrations to reactant concentrations at equilibrium. Given the equilibrium constant for a reaction, you can easily tell whether a specific mixture of products and reactants is at equilibrium and, if not, how far the reaction is away from equilibrium and in which direction it must proceed to reach equilibrium.

For example, the equilibrium constant for reaction 5-6 at 25°C is 0.5. This means that at equilibrium there should be one-half as much fructose-6-phosphate as glucose-6-phosphate, regardless of the actual magnitudes of the concentrations:

$$K_{eq} = \frac{[\text{fructose-6-phosphate}]_{eq}}{[\text{glucose-6-phosphate}]_{eq}} = 0.5 \quad (5\text{-}9)$$

If the two compounds are present in any other concentration ratio, the reaction will not be at equilibrium and will tend toward (be thermodynamically spontaneous in the direction of) equilibrium. Thus, a ratio of concentrations less than K_{eq} means that there is too little fructose-6-phosphate present, and the reaction will tend to the right to generate more fructose-6-phosphate at the expense of glucose-6-phosphate. Conversely, a ratio greater than K_{eq} indicates that the relative concentration of fructose-6-phosphate is too high, and the reaction will tend to the left.

A more general statement of this concept is given in Figure 5-11, which shows the relationship between the free energy of a chemical system and the degree of displacement from equilibrium. The free energy is lowest at equilibrium (where the concentration ratios are given as 1.0 for purposes of illustration) and increases if the system is away from equilibrium *in either direction*. Therefore, *knowing how the ratio of prevailing concentrations compares with the equilibrium concentration ratio makes it possible to predict in which direction a reaction will tend to proceed.* The tendency toward equilibrium provides the driving force for every chemical reaction, and a comparison of prevailing and equilibrium concentration ratios provides one measure of that tendency.

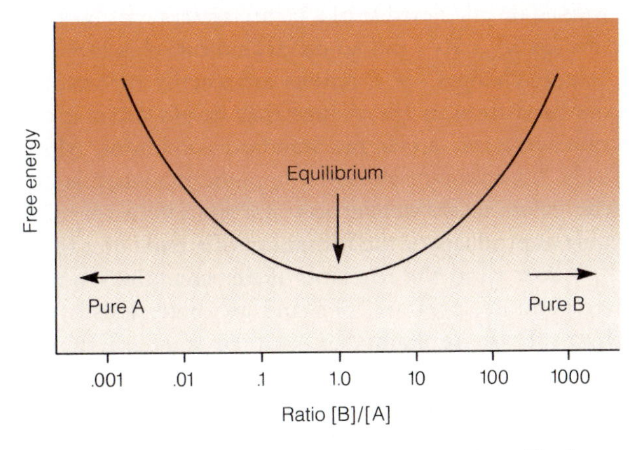

Figure 5-11 Free Energy and Chemical Equilibrium. The free energy available from a chemical reaction is a function of how far the components are from equilibrium. This is illustrated in the diagram, which shows components A and B interconverting by the reaction A ⇌ B. The total concentration [A + B] is held constant, and at equilibrium the ratio of [A] to [B] is assumed to be 1.0. Note how the free energy changes as the ratio of [A] to [B] changes on either side of equilibrium.

Calculation of ΔG

It should come as no surprise that ΔG is really just a means of calculating how far from equilibrium a reaction lies under specified conditions and how much energy will be released as the reaction proceeds toward equilibrium. Nor should it be surprising that both the equilibrium constant and the prevailing concentrations of reactants and products are needed to calculate ΔG. The equation relating these variables is as follows:

$$\Delta G = RT \ln \frac{[\text{products}]_{pr}}{[\text{reactants}]_{pr}} - RT \ln \frac{[\text{products}]_{eq}}{[\text{reactants}]_{eq}}$$

$$= RT \ln \frac{[\text{products}]_{pr}}{[\text{reactants}]_{pr}} - RT \ln K_{eq} \quad (5\text{-}10)$$

where ΔG is the free energy change, in cal/mol, under the specified conditions; R is the gas constant (1.987 cal/mol-K); T is the temperature in degrees Kelvin (use standard temperature of 298 K unless otherwise specified); ln is the natural log (to the base e), or 2.303 $\log_{10}$; $[X]_{pr}$ is the prevailing concentration of X in moles per liter; $[X]_{eq}$ is the equilibrium concentration of X in moles per liter; and K_{eq} is the equilibrium constant at the standard temperature of 298 K (25°C).

More accurately, for the general reaction in which a molecules of reactant A combine with b molecules of reactant B to form c and d molecules, respectively, of products C and D, we have

$$a\text{A} + b\text{B} \rightleftharpoons c\text{C} + d\text{D} \quad (5\text{-}11)$$

and ΔG is calculated as

$$\Delta G = RT \ln \frac{[\text{C}]^c_{pr}[\text{D}]^d_{pr}}{[\text{A}]^a_{pr}[\text{B}]^b_{pr}} - RT \ln K_{eq} \quad (5\text{-}12)$$

where all the constants and variables are as previously defined.

Returning to reaction 5-6, assume that the prevailing concentrations of glucose-6-phosphate and fructose-6-phosphate in a cell are 10 μM ($10 \times 10^{-6} M$) and 1 μM ($1 \times 10^{-6} M$), respectively, at 25°C. Since the ratio of prevailing product concentrations to reactant concentrations is 0.1 and the equilibrium constant is 0.5, there is clearly too little fructose-6-phosphate present relative to glucose-6-phosphate for the reaction to be at equilibrium.

The reaction should therefore tend toward the right (in the direction of fructose-6-phosphate generation) and we should expect ΔG to have a negative value.

The actual value for ΔG is calculated as follows:

$$\begin{aligned}\Delta G &= (1.987\ \text{cal/mol-K})(298\ \text{K})\ \ln\frac{[1\times10^{-6}]}{[10\times10^{-6}]}\\ &\quad-(1.987\ \text{cal/mol-K})(298\ \text{K})\ \ln(0.5)\\ &=(592\ \text{cal/mol})\ \ln(0.1)-(592\ \text{cal/mol})\ \ln(0.5)\\ &=(592\ \text{cal/mol})(-2.303)\\ &\quad-(592\ \text{cal/mol})(-0.693)\\ &=-1364\ \text{cal/mol}+410\ \text{cal/mol}\\ &=-954\ \text{cal/mol}\end{aligned} \qquad (5\text{-}13)$$

Note that our expectation of a negative ΔG is confirmed, and we now know exactly how much free energy is liberated upon the spontaneous conversion of 1 mole of glucose-6-phosphate into 1 mole of fructose-6-phosphate under the specified conditions. (As we will see later, the free energy liberated in an exergonic reaction can be either "harnessed" to do work or lost as heat.)

It is important to understand exactly what this calculated value for ΔG means and under what conditions it is valid. Since it is a thermodynamic parameter, ΔG can tell us whether a reaction will actually proceed but says nothing about the rate or mechanism. It simply says that if the reaction does occur, it will be to the right and will liberate 954 cal of free energy for every mole of glucose-6-phosphate that is converted to fructose-6-phosphate, *provided* that the concentrations of both the reactant and the product are maintained at the initial values (10 μM and 1 μM, respectively) throughout the course of the reaction.

More generally, ΔG is a measure of thermodynamic spontaneity for a reaction in the direction in which it is written (from left to right), at the specified concentrations of reactants and products. In a beaker or test tube, this requirement for constant reactant and product concentrations means that reactants must be added continuously and products must be removed continuously. In the cell, each reaction is part of some metabolic pathway, and its reactants and products are maintained at fairly constant, nonequilibrium concentrations by the reactions that precede and follow it in the sequence.

The Standard State and the Standard Free Energy Change

Because it is a thermodynamic parameter, ΔG is independent of the actual mechanism or pathway for a reaction, but it depends crucially on the conditions under which the reaction occurs. A reaction characterized by a large decrease in free energy under one set of conditions may have a much smaller (but still negative) ΔG or may even go in the opposite direction under a different set of conditions. The melting of ice, for example, depends on temperature, since it proceeds spontaneously above 0°C but goes in the opposite direction (freezing) below that temperature. It is therefore important to identify the conditions under which a given measurement of ΔG is made.

By convention, biochemists have agreed on certain arbitrary conditions to define the **standard state** of a system for convenience in reporting, comparing, and tabulating free energy changes in chemical reactions. For systems consisting of dilute aqueous solutions, these are usually a standard temperature of 25°C (298 K), a pressure of 1 atmosphere, and all products and reactants present in their most stable forms at a standard concentration of 1 mole per liter (or 1 atmosphere pressure for gases). The only common exception to this standard concentration rule is water. The concentration of water in a dilute aqueous solution is approximately 55.5 M and does not change significantly during the course of reactions, even when water is itself a reactant or product. By convention, biochemists do not include the concentration of water in calculations of free energy changes, even though the reaction may indicate a net consumption or production of water.

In addition to standard conditions of temperature, pressure, and concentration, biochemists also frequently specify a standard pH of 7.0, because most biological reactions occur at or near neutrality. The concentration of hydrogen ions (and of hydroxyl ions as well) is therefore 10^{-7} M, so the standard concentration of 1.0 M does not apply to H^+ or OH^- ions when a pH of 7.0 is specified. Values of K_{eq}, ΔG, or other thermodynamic parameters determined or calculated at pH 7.0 are always written with a prime (as K'_{eq}, $\Delta G'$, and so on) to indicate this fact.

Energy changes for reactions are usually reported in standardized form as the change that would occur if the reaction were run under the standard conditions. More precisely, the standard change in any thermodynamic parameter refers to the conversion of a mole of a specified reactant to products or the formation of a mole of a specified product from the reactants under conditions where the temperature, pressure, pH, and concentrations of all relevant species are maintained at their standard values. (As for the more general case, the maintenance of concentrations at 1.0 M implies that reactants are added as they are used up and that products are removed as they are formed.)

The free energy change calculated under these conditions is called the **standard free energy change,** designated as $\Delta G^{\circ\prime}$, where the superscript (°) refers to standard conditions of temperature, pressure, and concentration and the prime (′) emphasizes that the standard hydrogen ion concentration for biochemists is 10^{-7} *M*, not 1.0 *M*.

$\Delta G^{\circ\prime}$ turns out to bear a simple relationship to the equilibrium constant K'_{eq} (Figure 5-12). This relationship can readily be seen by rewriting equation 5-10 with primes and then assuming standard concentrations for all reactants and products. All concentration terms are now 1.0 and the logarithm of 1.0 is zero, so the first term in the general expression for $\Delta G^{\circ\prime}$ is eliminated, and what remains is an equation for $\Delta G^{\circ\prime}$, the free energy change under standard conditions:

$$\Delta G^{\circ\prime} = -RT \ln K'_{eq} \qquad (5\text{-}14)$$

In other words, $\Delta G^{\circ\prime}$ can be calculated directly from the equilibrium constant, provided that the latter has also been determined under the same standard conditions of temperature, pressure, and pH. This, in turn, allows equation 5-10 to be expressed in somewhat simpler form as

$$\Delta G' = RT \ln \frac{[C]^c_{pr}[D]^d_{pr}}{[A]^a_{pr}[B]^b_{pr}} + \Delta G^{\circ\prime} \qquad (5\text{-}15)$$

At the standard temperature of 25°C (298 K), the term RT becomes (1.987)(298) = 592 cal/mol, so equations 5-14 and 5-15 can be rewritten as follows, in what are the most useful formulas for our purposes:

$$\Delta G^{\circ\prime} = -592 \ln K'_{eq} \qquad (5\text{-}16)$$

$$\Delta G' = \Delta G^{\circ\prime} + 592 \ln \frac{[C]^c_{pr}[D]^d_{pr}}{[A]^a_{pr}[B]^b_{pr}} \qquad (5\text{-}17)$$

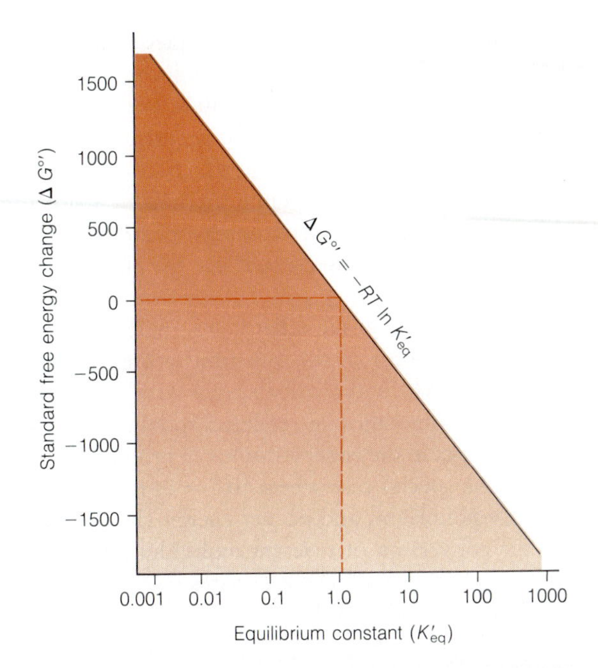

Figure 5-12 Relationship Between $\Delta G^{\circ\prime}$ and the Equilibrium Constant K'_{eq}. The standard free energy change and the equilibrium constant are related by the equation $\Delta G^{\circ\prime} = -RT \ln K'_{eq}$. Note that the equilibrium constant is plotted on an exponential scale. Note also that when the equilibrium constant is 1.0, the standard free energy change is zero, as we would expect (dashed lines).

Summing Up: The Meaning of $\Delta G'$ and $\Delta G^{\circ\prime}$

Equations 5-16 and 5-17 summarize the most important contribution of thermodynamics to biochemistry and cell biology—a means of assessing the feasibility of a chemical reaction based on the prevailing concentrations of products and reactants and a knowledge of the equilibrium constant. Equation 5-16 expresses the relationship between the standard free energy change $\Delta G^{\circ\prime}$ and the equilibrium constant K'_{eq} and allows us to calculate the free energy change that would be associated with any reaction of interest if all reactants and products were maintained at a standard concentration of 1.0 *M*.

If K'_{eq} is greater than 1.0, $\ln K'_{eq}$ will be positive and $\Delta G^{\circ\prime}$ will be negative, and the reaction can proceed to the right under standard conditions. This makes sense, since if K'_{eq} is greater than 1.0, products will predominate over reactants at equilibrium. A predominance of products can only be achieved from the standard state by conversion of reactants to products, so the reaction will tend spontaneously to the right. Conversely, if K'_{eq} is less than 1.0, $\Delta G^{\circ\prime}$ will be positive and the reaction cannot proceed to the right. Instead, it will tend toward the left, since the $\Delta G^{\circ\prime}$ for the reverse reaction will have the same numerical value but will be opposite in sign. This is in keeping with the small value for K'_{eq}, which specifies that reactants are favored over products (that is, the equilibrium lies to the left).

The $\Delta G^{\circ\prime}$ values are convenient both because of the ease with which they can be determined from the equilibrium constant and because they provide a uniform convention for reporting free energy changes. But bear in mind that a $\Delta G^{\circ\prime}$ value is an arbitrary standard in that it refers to an arbitrary state specifying conditions of concentration that cannot be achieved with most biologically important compounds. It is therefore useful for standardized reporting, but it is not a valid measure of the thermodynamic spontaneity of reactions as they occur under real conditions.

For that purpose we need $\Delta G'$, since this value provides a direct measure of how far from equilibrium a reaction is at the concentrations of reactants and products that actually prevail in the cell or other system of interest. Therefore, $\Delta G'$ is the most useful measure of thermodynamic spontaneity. If it is negative, the reaction in question is thermodynamically spontaneous and can proceed as written under the conditions for which the calculations were made. Its magnitude serves as a measure of how much free energy will be liberated as the reaction occurs. This, in turn, determines the maximum amount of work that can be performed on the surroundings, provided a mechanism is available to conserve and use the energy as it is liberated.

A positive $\Delta G'$, on the other hand, indicates that the reaction cannot occur in the direction written under the conditions for which the calculations were made. Such reactions can sometimes be rendered spontaneous, however, by changes in the concentrations of products or reactants. For the special case where $\Delta G' = 0$, the reaction is clearly at equilibrium, and no net energy change accompanies the conversion of reactant molecules into product molecules or vice versa. These features of K'_{eq}, $\Delta G^{\circ\prime}$, and $\Delta G'$ are summarized in Table 5-2.

Free Energy Change: Sample Calculations

To illustrate the calculation and utility of $\Delta G'$ and $\Delta G^{\circ\prime}$, we return once more to the interconversion of glucose-6-phosphate and fructose-6-phosphate (reaction 5-6). We already know that the equilibrium constant for this reaction under standard conditions of temperature, pH, and pressure is 0.5 (equation 5-9). This means that if the enzyme that catalyzes this reaction in cells is added to a solution of glucose-6-phosphate at 25°C, 1 atmosphere, and pH 7.0 and the solution is incubated until no further reaction occurs, both fructose-6-phosphate and glucose-6-phosphate will be present in an equilibrium ratio of 0.5. (Note that this ratio is independent of the actual starting concentration of glucose-6-phosphate and could have been achieved equally well by starting with any concentration of fructose-6-phosphate or any mixture of both, in any starting concentrations.)

The standard free energy change $\Delta G^{\circ\prime}$ can be calculated from K'_{eq} as follows:

$$\begin{aligned}\Delta G^{\circ\prime} &= -RT \ln K'_{eq} = -592 \ln K'_{eq}\\ &= -592 \ln (0.5) = -592\,(-0.693)\\ &= +410 \text{ cal/mol}\end{aligned} \tag{5-18}$$

The positive value for $\Delta G^{\circ\prime}$ is therefore another way of expressing the fact that the reactant (glucose-6-phosphate) is the predominant species at equilibrium. A positive $\Delta G^{\circ\prime}$ value also means that under standard conditions of concentration, the reaction is nonspontaneous (thermodynamically impossible) in the direction written. In other words, if we begin with both glucose-6-phosphate and fructose-6-phosphate present at concentrations of 1.0 *M*, no net conversion of glucose-6-phosphate to fructose-6-phosphate can occur.

As a matter of fact, given an appropriate catalyst, the reaction would proceed to the *left* under standard con-

Table 5-2 The Meaning of $\Delta G^{\circ\prime}$ and $\Delta G'$

	The Meaning of $\Delta G^{\circ\prime}$	
$\Delta G^{\circ\prime}$ Negative ($K'_{eq} > 1.0$)	$\Delta G^{\circ\prime}$ Positive ($K'_{eq} < 1.0$)	$\Delta G^{\circ\prime} = 0$ ($K'_{eq} = 1.0$)
Products predominate over reactants at equilibrium at standard temperature, pressure, and pH.	Reactants predominate over products at equilibrium at standard temperature, pressure, and pH.	Products and reactants are present equally at equilibrium at standard temperature, pressure, and pH.
Reaction goes spontaneously to right under standard conditions.	Reaction goes spontaneously to left under standard conditions.	Reaction is at equilibrium under standard conditions.
	The Meaning of $\Delta G'$	
$\Delta G'$ Negative	$\Delta G'$ Positive	$\Delta G' = 0$
Reaction is thermodynamically feasible as written under conditions for which $\Delta G'$ was calculated.	Reaction is not feasible as written under conditions for which $\Delta G'$ was calculated.	Reaction is at equilibrium under conditions for which $\Delta G'$ was calculated.
Work can be done by the reaction under conditions for which $\Delta G'$ was calculated.	Energy must be supplied to drive reaction under conditions for which $\Delta G'$ was calculated.	No work can be done nor is energy required by the reaction under conditions for which $\Delta G'$ was calculated.

ditions, since the $\Delta G^{\circ\prime}$ for the reaction in that direction is -410 cal/mol. Fructose-6-phosphate would therefore be converted into glucose-6-phosphate until the equilibrium ratio of 0.5 was reached. Alternatively, if both species were added or removed continuously as necessary to maintain the concentrations of both at 1.0 M, the reaction would proceed continuously and spontaneously to the left (assuming the presence of a catalyst), with the liberation of 410 calories of free energy per mole of fructose-6-phosphate converted to glucose-6-phosphate. In the absence of any provision for conserving this energy, it would be dissipated as heat.

In a real cell, it is extremely unlikely that either of these phosphorylated sugars would ever be present at a concentration even approaching 1.0 M. In fact, experimental values for the actual concentrations of these substances in human red blood cells are as follows:

[glucose-6-phosphate]: 83 μM (83×10^{-6} M)

[fructose-6-phosphate]: 14 μM (14×10^{-6} M)

Using these values, we can calculate the actual $\Delta G'$ for the interconversion of these sugars in red blood cells as follows:

$$\Delta G' = \Delta G^{\circ\prime} + 592 \ln \frac{[\text{fructose-6-phosphate}]_{pr}}{[\text{glucose-6-phosphate}]_{pr}}$$

$$= +410 + 592 \ln \frac{(14 \times 10^{-6})}{(83 \times 10^{-6})}$$

$$= +410 + 592 \ln (0.169)$$

$$= +410 + 592\,(-1.78) = +410 - 1054$$

$$= -644 \text{ cal/mol} \tag{5-19}$$

The negative value for $\Delta G^{\circ\prime}$ means that the conversion of glucose-6-phosphate into fructose-6-phosphate is thermodynamically possible under the conditions of concentration actually prevailing in red blood cells and that the reaction will yield 644 cal of free energy per mole of reactant converted to product. Thus, the conversion of reactant to product is thermodynamically impossible under standard conditions, but the red blood cell maintains these two phosphorylated sugars at concentrations adequate to offset the positive $\Delta G^{\circ\prime}$, thereby rendering the reaction possible. This adaptation, of course, is essential if the red blood cell is to be successful in carrying out the glucose-degrading process of glycolysis of which this reaction is a part.

Life and the Steady State

As this chapter has emphasized, the driving force in all reactions is their tendency to move toward equilibrium. Indeed, $\Delta G^{\circ\prime}$ and $\Delta G'$ are really nothing more than convenient means of quantifying how far and in what direction from equilibrium a reaction lies under the specific conditions dictated by standard or prevailing concentrations of products and reactants. But to understand how cells really function, we must appreciate the importance of reactions that move toward equilibrium without ever achieving it. At equilibrium, the forward and backward rates become the same for a reaction, there is no net flow of matter in either direction, and—most important—no further energy can be extracted from the reaction, since $\Delta G'$ is zero for a reaction at equilibrium.

For all practical purposes, then, a reaction at equilibrium is a reaction that has stopped. But a living cell is characterized by reactions that are continuous, not stopped. A cell at equilibrium would be a dead cell. We might, in fact, define life as a continual struggle to maintain a myriad of cellular reactions in positions far from equilibrium, since at equilibrium no net reactions are possible, no energy can be released, no work can be done, and the thermodynamically improbable order of the living state cannot be maintained.

Thus, life is possible only because living cells maintain themselves in a **steady state** far from thermodynamic equilibrium. The levels of glucose-6-phosphate and fructose-6-phosphate found in red blood cells illustrate this point. As we have seen, these compounds are maintained in the cell at steady-state concentrations far from the equilibrium condition predicted by the K'_{eq} value of 0.5. In fact, the levels are so far from equilibrium concentrations that the conversion of glucose-6-phosphate to fructose-6-phosphate can occur continuously in the cell, even though the equilibrium state has a positive $\Delta G^{\circ\prime}$ and actually favors glucose-6-phosphate. The same is true of most reactions and pathways in the cell. They can proceed and can be harnessed to perform various kinds of cellular work because reactants, products, and intermediates are maintained at steady-state concentrations far from the thermodynamic equilibrium.

This state, in turn, is possible only because a cell is an open system and receives large amounts of energy from its environment. If the cell were a closed system, all its reactions would gradually run to equilibrium and the cell would come inexorably to a state of minimum free energy, after which no further changes could occur, no work could be accomplished, and life would cease. The steady state so vital to life is possible only because the cell is able to take up energy continuously from its environment, whether in the form of light or preformed organic food molecules. This continuous uptake of energy and the flux of matter that accompanies it enable the maintenance of a steady state in which all the reactants and products of cellular chemistry are kept far enough from equilibrium to ensure that the thermodynamic drive toward equilibrium can be harnessed by the cell to perform useful work and thereby maintain and extend its activities and structural complexity.

How this is accomplished will occupy our attention in the coming chapters. In the next chapter, we will look at principles of enzyme catalysis that determine the rates of cellular reactions (that convert the "can go" of thermodynamics into the "will go" of kinetics). Then we will be ready to move on to Part Three and the functional metabolic pathways that result from series of such reactions acting in concert.

Perspective

The thermodynamically improbable level of order seen in cells is possible only because of the availability of energy from the environment. Cells require energy to carry out various kinds of work, including synthesis, movement, concentration, charge separation, and bioluminescence. The energy needed for these processes comes either from the sun or from the bonds of oxidizable organic molecules such as carbohydrates, fats, and proteins. Since chemotrophs feed directly or indirectly on phototrophs, there is a unidirectional flow of energy through the biosphere, with the sun as the ultimate source and entropy and heat losses as the eventual fate of all the energy that moves through living systems.

The flow of energy through cells is governed by the laws of thermodynamics. The first law specifies that energy can change form but must always be conserved. The second law provides a measure of thermodynamic spontaneity, although this means only that a reaction can go and says nothing about whether it actually will go, or at what rate. Spontaneous processes are always accompanied by an *increase* in the entropy of the universe and by a *decrease* in the free energy of the system. The latter is a far more practical indicator of spontaneity because it can be calculated readily from the equilibrium constant and the prevailing concentrations of reactants and products.

Cells obtain the energy they need to carry out their activities by maintaining the many reactants and products of the various reaction sequences at steady-state concentrations far from equilibrium, thereby allowing the reactions to move exergonically toward equilibrium without ever actually reaching it. A negative $\Delta G'$ is a necessary prerequisite for a reaction to proceed, but it does not guarantee that the reaction will actually occur at a reasonable rate. To assess that, we must know more about the reaction than just its thermodynamic status. We need to know whether an appropriate catalyst is on hand and at what rate the reaction can occur in the presence of the catalyst. For that we need the enzymes that await discussion in Chapter 6.

Key Terms for Self-Testing

The Importance of Energy
energy (p. 107)
biosynthesis (p. 107)
synthetic work (p. 107)
turnover (p. 107)
mechanical work (p. 109)
concentration work (p. 110)
electrical work (p. 110)
heat (p. 110)
bioluminescence (p. 110)
phototrophs (p. 110)
chemotrophs (p. 110)
fermentation (p. 110)
glycolysis (p. 110)
respiration (p. 110)
anaerobic (p. 111)
aerobic (p. 111)

Bioenergetics
thermodynamics (p. 113)
bioenergetics (p. 113)
system (p. 113)
surroundings (p. 113)
closed system (p. 114)
open system (p. 114)
state (p. 114)
isothermal (p. 115)
work (p. 115)
calorie (p. 115)
joule (p. 115)
first law of thermodynamics (p. 115)
internal energy (p. 115)
second law of thermodynamics (p. 117)
entropy (p. 117)
free energy (p. 120)
free energy change (p. 120)
exergonic (p. 121)
endergonic (p. 121)
spontaneous (p. 122)

Understanding ΔG
equilibrium constant (K_{eq}) (p. 123)
standard state (p. 124)
standard free energy change (p. 124)

Life and the Steady State
steady state (p. 127)

Suggested Reading

General References and Reviews

Barrow, G. M. *Physical Chemistry for the Life Sciences,* 2d ed. New York: McGraw-Hill, 1981.

Eisenberg, D., and D. Crothers. *Physical Chemistry with Applications to the Life Sciences.* Menlo Park, Calif.: Benjamin/Cummings, 1979.

Mathews, C. K., and K. E. van Holde, *Biochemistry.* Redwood City, Calif.: Benjamin/Cummings, 1990.

Wood, W. B., J. H. Wilson, R. M. Benbow, and L. E. Hood. *Biochemistry: A Problems Approach,* 2d ed. Menlo Park, Calif.: Benjamin/Cummings, 1981.

Zubay, G. *Biochemistry,* 2nd ed. New York: Macmillan, 1988.

Bioenergetics

Baker, J. J. W., and G. E. Allen. *Matter, Energy, and Life: An Introduction to Chemical Concepts,* 4th ed. Reading, Mass.: Addison-Wesley, 1981.

Christensen, H. N., and R. A. Cellarius. *Introduction to Bioenergetics.* Philadelphia: Saunders, 1972.

Harold, F. M. *The Vital Force: A Study of Bioenergetics.* New York: W. H. Freeman, 1986.

Klotz, I. M. *Energy Changes in Biochemical Reactions.* New York: Academic Press, 1967.

Krebs, H. A., and H. L. Kornberg. *Energy Transformation in Living Matter.* New York: Springer-Verlag, 1957.

Energy Flow

Gates, D. M. The flow of energy in the biosphere. *Sci. Amer.* 224 (September 1971): 88.

Miller, G. T., Jr. *Energetics, Kinetics, and Life—An Ecological Approach.* Belmont, Calif.: Wadsworth, 1971.

Free Energy and Entropy

Blum, H. F. *Time's Arrow and Evolution,* 3d ed. Princeton, N. J.: Princeton University Press, 1968.

Hess, B., and M. Markus. Order and Chaos in Biochemistry. *Trends Biochem. Sci.* 12 (1987): 45.

Hill, T. L. *Free Energy Transduction in Biology.* New York: Academic Press, 1977.

Morowitz, H. J. *Entropy for Biologists: An Introduction to Thermodynamics.* New York: Academic Press, 1970.

Morowitz, H. J. Entropy anyone? *Hosp. Pract.* 16 (1981): 114.

Problem Set

1. **Solar Energy.** Although we are concerned at present with a global energy crisis, we actually live on a planet that is flooded continuously with an extravagant amount of energy in the form of solar radiation. Every day, year in and year out, solar energy arrives at the upper surface of the earth's atmosphere at the rate of 1.94 cal/min per square centimeter of cross-sectional area (the *solar energy constant*).
   (a) Assuming the cross-sectional area of the earth to be about 1.28×10^{18} cm^2, what is the total annual amount of incoming energy?
   (b) A substantial portion of that energy, particularly in the wavelength ranges below 300 nm and above 800 nm, never reaches the earth's surface. Can you suggest what happens to it?
   (c) Of the radiation that reaches the earth's surface, only a small proportion is actually trapped photosynthetically by the phototrophs (you can calculate the actual value in Problem 2). Why do you think the efficiency of utilization is so low?

2. **Photosynthetic Energy Transduction.** The amount of energy trapped and the volume of carbon converted into organic form by photosynthetic energy transducers are mind-boggling—about 5×10^{16} g of carbon per year over the entire earth's surface.
   (a) Assuming that the average organic molecule in a cell has about the same proportion of carbon as glucose does, how many grams of organic matter are produced annually by the carbon-fixing phototrophs?
   (b) Assuming that all the organic matter in part (a) is glucose (or any molecule with an energy content equivalent to that of glucose), how much energy is represented by that quantity of organic matter? Assume that glucose has an energy content (free energy of combustion) of 3.8 kcal/g.
   (c) Refer to the answer for Problem 1a. What is the average efficiency with which the radiant energy incident on the upper atmosphere is trapped photosynthetically on the earth's surface?
   (d) What proportion of the net annual phototrophic production of organic matter calculated in part (a) do you think is consumed by the chemotrophs each year?

3. **Energy Flow.** A corn plant growing in a field in Wisconsin uses sunlight to convert carbon dioxide and water into sugars. The sugars are transported to the ear and deposited in the kernels as starch molecules. Sometime in summer, a horse in the adjacent pasture leans across the fence to eat one of the ears of corn. Later in the day, the horse is seen galloping across the pasture.

(a) Starting with the solar radiation incident on the corn leaf, trace the flow of energy through this biological system.

(b) Assuming that the horse continues galloping until all of the energy obtained from the ear of corn is used up, what has become of the energy?

(c) One of the things the corn plant furnishes to the horse is the chemical bond energy of starch. What are two or three other things that it provides?

(d) What, if anything, does the corn plant get, in turn, from the horse?

4. **Energy Conversion.** Most cellular activities involve the conversion of energy from one form to another. For each of the following cases, give a biological example and explain the significance of the conversion.

(a) Chemical energy into mechanical energy

(b) Chemical energy into radiant energy

(c) Radiant energy into chemical energy

(d) Chemical energy into electrical energy

(e) Chemical energy into the potential energy of a concentration gradient

5. **Equilibrium and the Steady State.** Why do reactions tend to proceed in the direction required to reach equilibrium? What is meant by "steady state" in the context of bioenergetics? What is the biological significance of this?

6. **Calculating $\Delta G^{\circ\prime}$ and $\Delta G'$.** The conversion of 3-phosphoglycerate to 2-phosphoglycerate is an important cellular reaction because it is one of the steps in the glycolytic pathway (see Chapter 10) and is therefore common to all cells:

$$\text{3-phosphoglycerate} \rightleftharpoons \text{2-phosphoglycerate} \qquad (5\text{-}20)$$

If the enzyme that catalyzes this reaction is added to a solution of 3-phosphoglycerate at 25°C and pH 7.0, the equilibrium ratio between the two species will be 0.165:

$$K'_{eq} = \frac{[\text{2-phosphoglycerate}]}{[\text{3-phosphoglycerate}]} = 0.165 \qquad (5\text{-}21)$$

Experimental values for the actual steady-state concentrations of these compounds in human red blood cells are 61 μM for 3-phosphoglycerate and 4.3 μM for 2-phosphoglycerate.

(a) Calculate $\Delta G^{\circ\prime}$. Explain in your own words what this value means.

(b) Calculate $\Delta G'$. Explain in your own words what this value means.

(c) If conditions in the cell change such that the concentration of 3-phosphoglycerate remains fixed at 61 μM but the concentration of 2-phosphoglycerate begins to rise, how high can the 2-phosphoglycerate concentration get before reaction 5-20 will cease because it is no longer thermodynamically feasible?

7. **Hydrolysis of Glucose-6-phosphate.** The hydrolysis of glucose-6-phosphate to glucose and phosphate is an important step in glycogen catabolism (Chapter 9). The free glucose formed by this reaction is released into the blood for transport to cells in need of energy.

$$\text{glucose-6-phosphate} + H_2O \longrightarrow \text{glucose} + \text{phosphate} \qquad (5\text{-}22)$$

$\Delta G^{\circ\prime}$ for this reaction is -3.3 kcal/mol at 25°C and pH 7.0.

(a) Calculate $\Delta G'$ at 25°C when the concentration of glucose-6-phosphate is 20 μM and the glucose and phosphate concentrations are 5 mM each. Assume the appropriate enzyme is present to catalyze this reaction.

(b) In which direction would you expect the reaction to proceed under the conditions defined in part (a)?

8. **Conversion of Glucose-1-phosphate to Glucose-6-phosphate.** The following reaction is another step in glycogen catabolism that precedes the hydrolysis of glucose-6-phosphate discussed in Problem 7.

$$\text{glucose-1-phosphate} \rightleftharpoons \text{glucose-6-phosphate}$$

K'_{eq} for this reaction is 19.0. Assume the appropriate enzyme is present to catalyze the reaction.

(a) In which direction is the reaction thermodynamically feasible under standard conditions?

(b) Confirm your answer to part (a) by calculating $\Delta G^{\circ\prime}$ for this reaction.

(c) If the cellular concentrations of glucose-1-phosphate and glucose-6-phosphate are equal, in which direction will the reaction proceed?

(d) Confirm your answer to part (c) by calculating $\Delta G'$.

9. **Succinate Oxidation.** The oxidation of succinate to fumarate occurs as one of the reactions in the tricarboxylic acid (TCA) cycle, an important component of chemotrophic energy metabolism (see Chapter 11). The two hydrogen atoms that are removed from succinate are accepted by a coenzyme molecule called flavin adenine dinucleotide (FAD), which is thereby reduced to $FADH_2$:

$$\text{Succinate} + \text{FAD} \rightleftharpoons \text{Fumarate} + FADH_2 \qquad (5\text{-}23)$$

The $\Delta G^{\circ\prime}$ for this reaction is 0 cal/mol.

(a) If you start with a solution containing 0.01 M each of succinate and FAD and add an appropriate amount of the enzyme that catalyzes this reaction, will any fumarate be formed? If so, calculate the resulting equilibrium concentrations of all four species. If not, explain why not.

(b) Answer part (a) assuming that 0.01 M $FADH_2$ is also present initially.

(c) Assuming that the steady-state conditions in a cell are such that the $FADH_2$/FAD ratio is 5 and the fumarate concentration is 2.5 μM, what steady-state concentration of succinate would be necessary to maintain the $\Delta G'$ for succinate oxidation at −1.5 kcal/mol?

10. **Proof of Additivity.** A useful property of thermodynamic parameters such as $\Delta G'$ or $\Delta G^{\circ\prime}$ is that they are additive for sequential reactions. Assume that K'_{AB}, K'_{BC}, and K'_{CD} are the respective equilibrium constants for reactions 1, 2, and 3 of the following sequence:

$$A \underset{\text{Reaction 1}}{\rightleftharpoons} B \underset{\text{Reaction 2}}{\rightleftharpoons} C \underset{\text{Reaction 3}}{\rightleftharpoons} D \qquad (5\text{-}24)$$

(a) Prove that the equilibrium constant K'_{AD} for the overall conversion of A to D is the *product* of the three component equilibrium constants:

$$K'_{AD} = K'_{AB} \cdot K'_{BC} \cdot K'_{CD} \qquad (5\text{-}25)$$

(b) Prove that the $\Delta G^{\circ\prime}$ for the overall conversion of A to D is the *sum* of the three component $\Delta G^{\circ\prime}$ values:

$$\Delta G^{\circ\prime}_{AD} = \Delta G^{\circ\prime}_{AB} + \Delta G^{\circ\prime}_{BC} + \Delta G^{\circ\prime}_{CD} \qquad (5\text{-}26)$$

(c) Prove that the $\Delta G'$ values are similarly additive.

11. **Thermodynamic Logic.** For a system at constant temperature, pressure, and volume, the change in internal energy ΔE is given by the expression

$$\Delta E = \Delta G + T\,\Delta S \qquad (5\text{-}27)$$

Is it also true that the total internal energy can be expressed as $E = G + TS$? Explain why or why not.

6

Enzymes: The Catalysts of Life

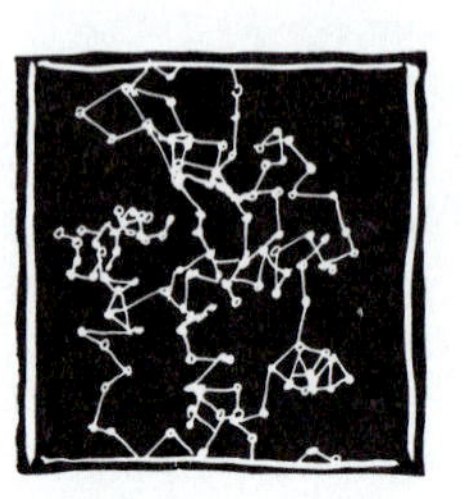

In Chapter 5, we encountered ΔG, the change in free energy, and saw its importance as an indicator of thermodynamic spontaneity. Specifically, the *sign* of ΔG tells us whether a reaction is possible in the indicated direction, and the *magnitude* of ΔG indicates how much energy can be obtained (or must be invested) as the reaction proceeds in that direction under the conditions for which ΔG was calculated. At the same time, we were careful to note that because it is a thermodynamic parameter, ΔG can provide us with no clue as to whether a feasible reaction will take place, or at what rate. In other words, ΔG tells us only whether a reaction *can* go but says nothing at all about whether it actually *will* go. For that distinction we need to know not just the direction and energetics of the reaction, but something about the mechanism and rate as well.

This brings us to the topic of **enzyme catalysis**, because almost all the reactions and processes that take place in cells are mediated by protein catalysts called **enzymes.** The only reactions that occur to any appreciable rate in a cell are those for which the appropriate enzymes are present and active. Thus, enzymes almost always spell the difference between "can go" and "will go" for cellular reactions. It is only as we explore the nature of enzymes and their catalytic properties that we begin to understand how reactions that are energetically feasible actually take place in cells and how the rates of such reactions are controlled.

In this chapter, we will look first at why thermodynamically spontaneous reactions do not usually occur at appreciable rates without a catalyst, and then we will look at the role of enzymes as specific biological catalysts. We will also see how the rate of an enzyme-catalyzed reaction is affected by the amount of substrate available to it, as well as some of the ways in which reaction rates are regulated to meet the needs of the cell.

Activation Energy and the Metastable State

If you stop to think about it, you are already familiar with many reactions that are thermodynamically feasible yet do not occur to any appreciable extent. An obvious example from Chapter 5 is the oxidation of glucose, which is highly exergonic ($\Delta G^{\circ\prime} = -686$ kcal/mol) and yet does not take place on its own. In fact, glucose crystals or a glucose solution can be exposed to the oxygen in the air indefinitely, and little or no oxidation occurs. The cellulose in the paper on which these words are printed is another example—and so, for that matter, are you, an improbable and complex collection of thermodynamically unstable molecules.

Not nearly as familiar but equally important to cellular chemistry are the myriads of thermodynamically feasible reactions in cells that could go, but seem not to on their own. As an example, consider the high-energy molecule adenosine triphosphate (ATP), which has a highly favorable $\Delta G^{\circ\prime}$ (-7.3 kcal/mol) for the hydrolysis of its terminal phosphate group to form the corresponding diphosphate (ADP) and inorganic phosphate (P_i):

$$ATP + H_2O \rightleftharpoons ADP + P_i \tag{6-1}$$

The reaction is very exergonic under standard conditions and is almost always even more so under the conditions that prevail in cells. Yet despite the highly favorable free energy change, this reaction occurs only slowly on its own, so that ATP remains stable for several days when dissolved in pure water. This property turns out to be shared by many biologically important molecules and reactions, and it is important to understand why.

Activation Energy

Molecules that should react with one another often do not because they lack sufficient energy. For every reaction, there is a specific **activation energy** (E_A), which is the minimum amount of energy that two molecules must have in order to react. Figure 6-1a shows the activation energy required for the reaction between ATP and H_2O.

In addition to $\Delta G^{\circ\prime}$, which measures the difference in free energy between reactants and products (-7.3 kcal/mol, in this case), there is also a characteristic activation energy that a given pair of ATP and H_2O molecules must have before a collision between them will be successful, leading to a reaction. The actual rate of a reaction is always proportional to the fraction of molecules that have an energy content equal to or greater than E_A. When in solution at room temperature, molecules of ATP and water move about readily, each possessing a certain amount of energy at any instant. As Figure 6-1b illustrates, the energy distribution among molecules will be bell shaped—some molecules will have very little energy, some will have a lot, and most will be somewhere near the average. The important point is that only those with an energy that exceeds the activation energy level are capable of reacting at a given instant.

The Metastable State

For most biologically important reactions at normal cellular temperatures, the activation energy is sufficiently high that the proportion of molecules possessing that much energy at any instant is extremely small. Accordingly, the rates of uncatalyzed reactions in cells are very low, and most molecules appear to be stable even though they are potential reactants in thermodynamically favored reactions. They are, in other words, thermodynamically unstable but kinetically stable, lacking adequate kinetic energy to exceed the activation energy threshold.

Such seemingly stable molecules are said to be **metastable.** For cells and cell biologists, high activation energies and the resulting metastable state of cellular constituents are crucial, since life by its very nature is a system maintained in a steady state a long way from equilibrium. If it were not for the metastable state, all reactions would proceed quickly to equilibrium, and life as we know it would be impossible. Life, then, depends critically on the high activation energies that prevent most cellular reactions from occurring at appreciable rates in the absence of a suitable catalyst.

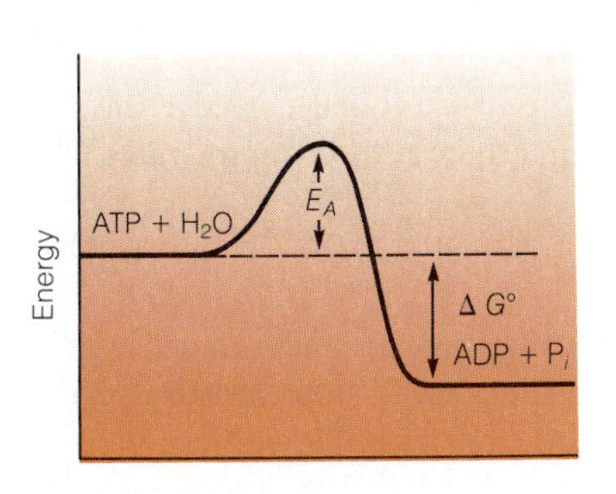

(a) Reaction sequence

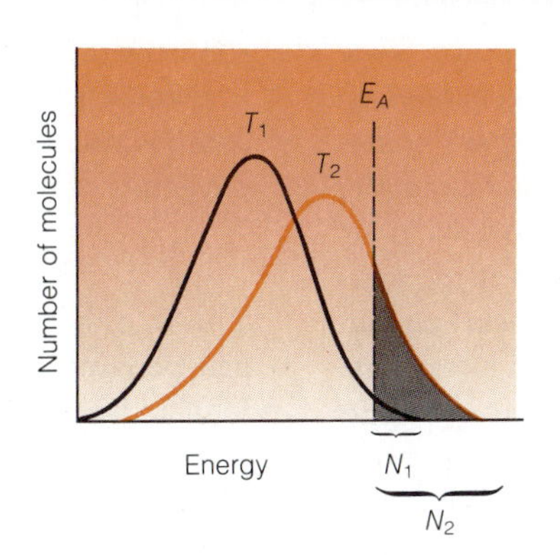

(b) Thermal activation

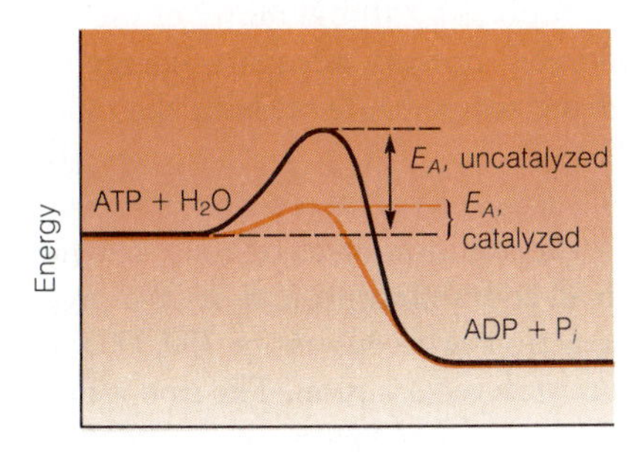

(c) Reaction sequence

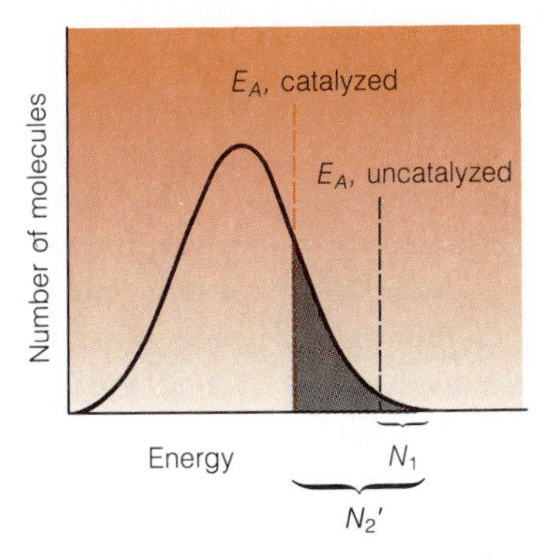

(d) Catalytic activation

Figure 6-1 Effect of Catalysis on Activation Energy and Number of Molecules Capable of Reaction. (a) The activation energy E_A is the minimum amount of kinetic energy reactant molecules must possess to permit collisions leading to product formation. After reactants overcome the activation energy barrier and enter into a reaction, the products have less energy by the amount ΔG°. (b) The number of molecules N_1 that have sufficient energy to exceed the activation energy barrier (E_A) and collide successfully can be increased to N_2 by raising the temperature from T_1 to T_2. Alternatively, the activation energy can be lowered by a catalyst (c), thereby increasing the number of molecules from N_1 to N_2', as shown in (d).

Overcoming the Activation Energy Barrier

Important as it is to the maintenance of the metastable state, the activation energy requirement is nonetheless a barrier that must be overcome if desirable reactions are to proceed at reasonable rates. Since the energy content of a given molecule must exceed E_A before that molecule is capable of undergoing reaction, the only way a reaction involving metastable reactants can be made to proceed at an appreciable rate is to increase the proportion of molecules with sufficient energy. This can be achieved either by increasing the average energy content of all molecules or by lowering the activation energy requirement.

One way to increase the energy content of the system is by the input of heat. As Figure 6-1b illustrates, simply increasing the temperature of the system from T_1 to T_2 will increase the kinetic energy of the average molecule, thereby ensuring a greater number of reactive molecules (N_2 instead of N_1). Thus, the hydrolysis of ATP could be encouraged by heating the solution, giving each ATP and water molecule more energy.

Sometimes, only a brief initial increase in temperature is enough to trigger a thermodynamically feasible reaction, because the reaction, once under way, liberates enough energy to activate additional molecules, thereby becoming self-sustaining. Lighting a match illustrates this point. Once initiated, the combustion of the match head generates so much heat that it sustains itself. An electric spark sometimes provides the same sort of initial energy. For example, the reaction of gasoline (a hydrocarbon) and oxygen in air is potentially a violently exergonic reaction, yet the activation energy is sufficiently high that the hydrocarbon and oxygen molecules can exist side by side in the metastable state. An electric spark, however, provides the energy to overcome the barrier, and the reaction becomes self-sustaining as the two gases combine explosively. When appropriately confined in a cylinder with a movable piston, this reaction provides the energy source for every gasoline engine in the world.

The problem with using a match or spark or a continuously elevated temperature is that such approaches are incompatible with life, since biological systems require a relatively constant temperature. Cells would find it extremely difficult to function if activation energy requirements for specific reactions required touching a match to the molecules in question! They are basically *isothermal* (constant-temperature) systems and require isothermal methods to solve the activation problem. Moreover, we can anticipate a later topic if we note that this approach would also suffer from lack of specificity, since the input of heat would result in many activation energy barriers being indiscriminately overcome. Yet the essence of successful regulation of cellular activities lies in the ability to facilitate specific reactions under specific conditions while leaving other metastable molecules undisturbed.

The alternative to thermal activation, of course, is to lower the activation energy requirement, thereby making it statistically more likely that a greater proportion of molecules will have sufficient energy to collide successfully and undergo reaction. This is possible because, unlike the free energy change for a given reaction, the activation energy depends not only on the initial and final states but also on the mechanism. If a reaction depends on random collision between molecules, a relatively high energy is almost always required to ensure reactivity. But if the reactants can be carefully ordered on some sort of surface in a way that brings potentially reactive portions of adjacent molecules into close juxtaposition, their interaction will be greatly favored and the activation energy will be effectively reduced.

Providing such a reactive surface is the task of a **catalyst**, an agent that enhances the rate of a reaction by lowering the energy of activation (Figure 6-1c), thereby ensuring that a higher proportion of the molecules are energetic enough to undergo reaction without the input of heat (Figure 6-1d). A primary feature of a catalyst is that it is not permanently changed or consumed as the reaction proceeds. It simply provides a suitable surface and environment to facilitate the reaction.

For a specific example of catalysis, consider the decomposition of hydrogen peroxide (H_2O_2) into water and oxygen:

$$2H_2O_2 \rightleftharpoons 2H_2O + O_2 \tag{6-2}$$

This is a thermodynamically favored reaction, yet hydrogen peroxide exists in a metastable state because of the high activation energy of the reaction. However, if we add a small number of ferric ions (Fe^{3+}) to a hydrogen peroxide solution, the decomposition reaction proceeds about 30,000 times faster than without the ferric ions. Clearly, Fe^{3+} is a catalyst for this reaction, lowering the activation energy as shown in Figure 6-1c and thereby ensuring that a significantly greater proportion (30,000-fold more) of the hydrogen peroxide molecules possess adequate energy to decompose at the existing temperature without the input of added energy.

The biological solution to hydrogen peroxide breakdown, however, is not the addition of ferric ions, but the enzyme *catalase*, an iron-containing protein. In the presence of catalase, the reaction proceeds about 100,000,000 times faster than the uncatalyzed reaction. The iron atoms in catalase are bound to chemical structures called *porphyrins*, and the combination of the iron-porphyrin complex with its surrounding protein is a much more effective catalyst for hydrogen peroxide decomposition than ferric ions by themselves. This underscores the efficacy of enzymes as catalysts and brings us to the main theme of this chapter.

Enzymes as Biological Catalysts

Regardless of their chemical nature, all catalysts share the following basic three properties:

1. A catalyst increases the rate of a reaction by lowering the activation energy requirement, thereby allowing thermodynamically feasible reactions to occur at a reasonable rate without thermal activation.
2. A catalyst acts by forming transient complexes with substrate molecules, ordering them in a manner that facilitates their interaction.
3. A catalyst changes only the *rate* at which equilibrium is achieved; it has no effect on the *position* of the equilibrium. This means that a catalyst can enhance the rate of exergonic reactions but cannot somehow drive an endergonic reaction. Catalysts, in other words, are not thermodynamic genies.

These properties are common to all catalysts, organic and inorganic alike. In terms of our example, they would apply equally to ferric ions and catalase molecules. However, biological systems rarely use inorganic catalysts. Instead, essentially all catalysis in cells is carried out by proteins called *enzymes*. Because they are proteins, enzymes are characteristically much more specific than inorganic catalysts, and their activities can be regulated much more carefully.

Enzymes as Proteins

The capacity of cellular extracts to catalyze chemical reactions has been known since the fermentation studies of Eduard Buchner. In fact, one of the first terms for what we today call enzymes was *ferments*. However, it was not until 1926 that the enzyme *urease* was crystallized and shown to be a protein. Even then, it took a while before biochemists and enzymologists appreciated that essentially all enzymes are proteins and that to understand enzymes as catalysts really means to understand them in terms of their structure and function as proteins. The one exception to this rule is discussed in the box on p. 140, which describes certain RNA molecules having catalytic activity.

The Active Site. One of the most important concepts to emerge from our understanding of enzymes as proteins is that of the **active site.** Every enzyme, regardless of the reaction it catalyzes or the details of its structure, contains somewhere within its tertiary configuration a characteristic cluster of amino acids forming the active site where the actual catalytic event, for which that enzyme is responsible, occurs. Usually, the active site is an actual groove or pocket with chemical and structural properties that accommodate the intended substrate with high specificity. Figure 6-2 shows computer graphic models of the enzymes *lysozyme* and *carboxypeptidase A*, respectively. The three-dimensional structure of the active site can be appreciated by the precise fit of the substrate molecule into pockets produced by the characteristic folding of polypeptide chains.

The amino acids that make up the active site of an enzyme are usually not contiguous to one another along the primary sequence of the protein. Instead, they are brought together in just the right configuration by the specific three-dimensional folding of the polypeptide chain (which you may recall to be a spontaneous process, dictated by the primary sequence). For the carboxypeptidase molecule, for example, the active site consists of a tightly bound zinc ion bonded to the side chains of the histidine at position 69, the glutamate at position 72, and another histidine at position 196 of the polypeptide chain. This configuration is shown in Figure 6-3a. When the substrate for the enzyme binds to the active site, the structure of the enzyme changes, bringing the arginine at position 145, the glutamate at position 270, and the tyrosine at position 248 into close proximity with the substrate, and each of these amino acids plays a critical role in the actual catalytic process (see Figure 6-3b).

For carboxypeptidase, then, 6 amino acids out of a total chain length of 307 are actually involved at the active site. This is typical of most enzymes, with the active site usually involving about 5% of the surface area of the enzyme. Also typical is the involvement of amino acids from disparate locations along the primary chain. This involvement underscores the importance of the overall tertiary structure of the protein, since it is only as an enzyme molecule attains its stable three-dimensional configuration that the specific amino acids are brought together to constitute the active site.

Of the 20 different amino acids of which proteins consist, only a few are actually involved in the active sites of the many proteins that have been studied. In most cases, these are the amino acids cysteine, histidine, serine, aspartate, glutamate, and lysine. All of these can participate in binding or bonding of the substrate to the active site during the catalytic process, and several (histidine, aspartate, and glutamate) serve as donors or acceptors of protons.

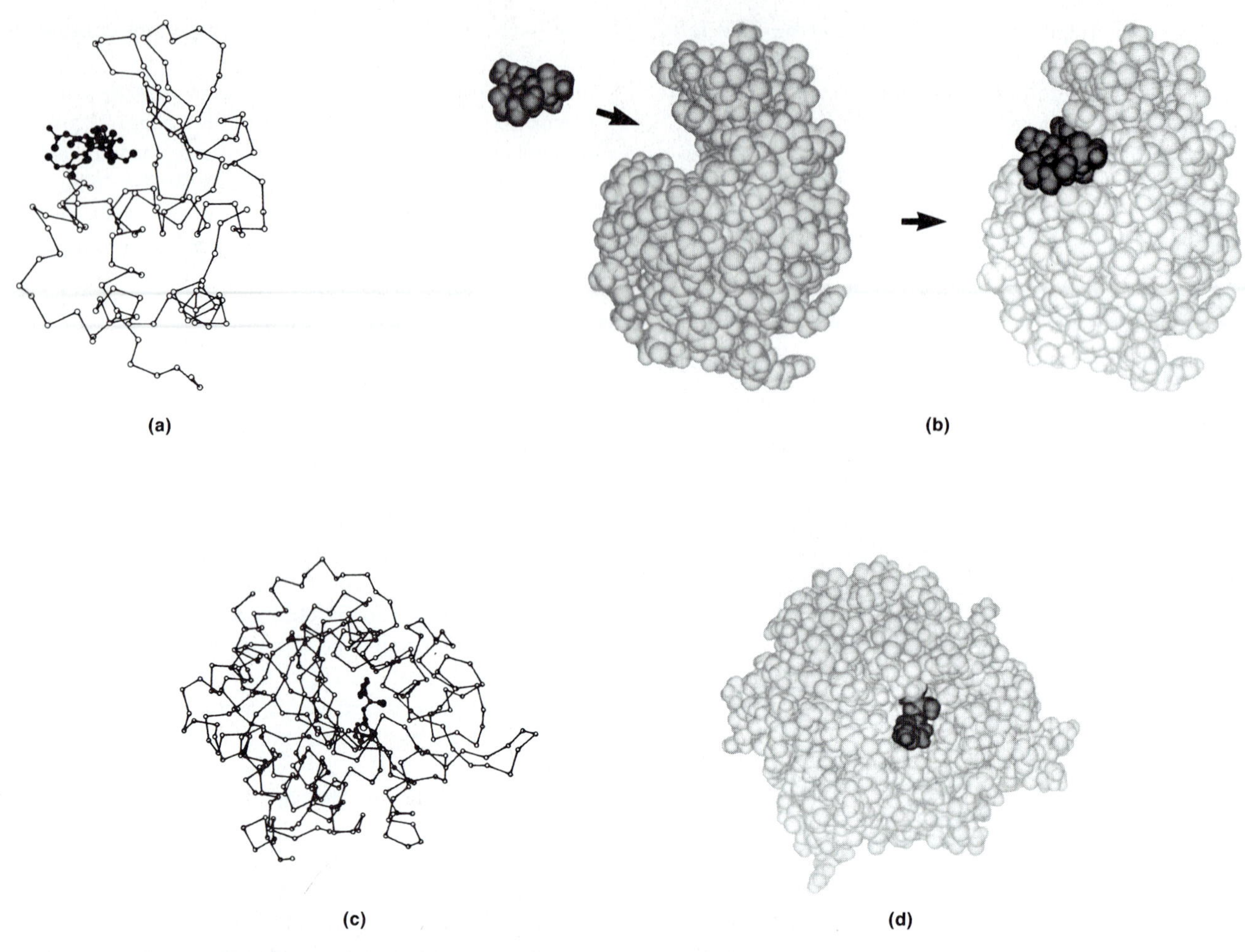

Figure 6-2 Molecular Structure of Lysozyme and Carboxypeptidase A. (a) Skeletal model, showing only the backbone of chicken lysozyme as deduced by X-ray diffraction studies of protein crystals. (b) Space-filling models produced by computer graphic techniques show the three-dimensional structure of the enzyme. A substrate molecule is shown colliding with the enzyme and binding to the active site, which appears as a cleft in the side of the enzyme molecule. For comparison, bovine carboxypeptidase A is shown below, both as a skeletal model (c) and as a space-filling model (d) containing a substrate molecule in the active site of the enzyme.

Some enzymes consist not only of one or more polypeptide chains, but of specific nonprotein components as well. These components are called **prosthetic groups** and are usually either small organic molecules or metal ions, such as the iron in catalase. Frequently, they function as electron acceptors, because none of the amino acid side chains is a good electron acceptor. Where present, prosthetic groups are located at the active site and are indispensable for the catalytic activity of the enzyme. Carboxypeptidase A is an example of an enzyme with a prosthetic group, containing a zinc atom at its active site (Figure 6-3a).

Enzyme Specificity. A consequence of the structure of the active substrate is that enzymes display a high degree of **substrate specificity,** evidenced by an ability to discriminate between very similar molecules. Specificity is probably one of the most characteristic properties of living systems, and enzymes are especially dramatic examples of biological specificity.

We can illustrate their specificity by comparing enzymes with inorganic catalysts. Most inorganic catalysts are quite nonspecific in that they will act on a variety of compounds that share some general chemical feature. Consider, for example, the *hydrogenation* of (addition of hydrogen to) an unsaturated C=C bond:

$$\mathrm{R{-}\overset{\displaystyle H}{\overset{|}{C}}{=}\overset{\displaystyle H}{\overset{|}{C}}{-}R'} + \mathrm{H_2} \xrightarrow[\text{Ni or Pt}]{} \mathrm{R{-}\underset{\displaystyle H}{\underset{|}{\overset{\displaystyle H}{\overset{|}{C}}}}{-}\underset{\displaystyle H}{\underset{|}{\overset{\displaystyle H}{\overset{|}{C}}}}{-}R'} \tag{6-3}$$

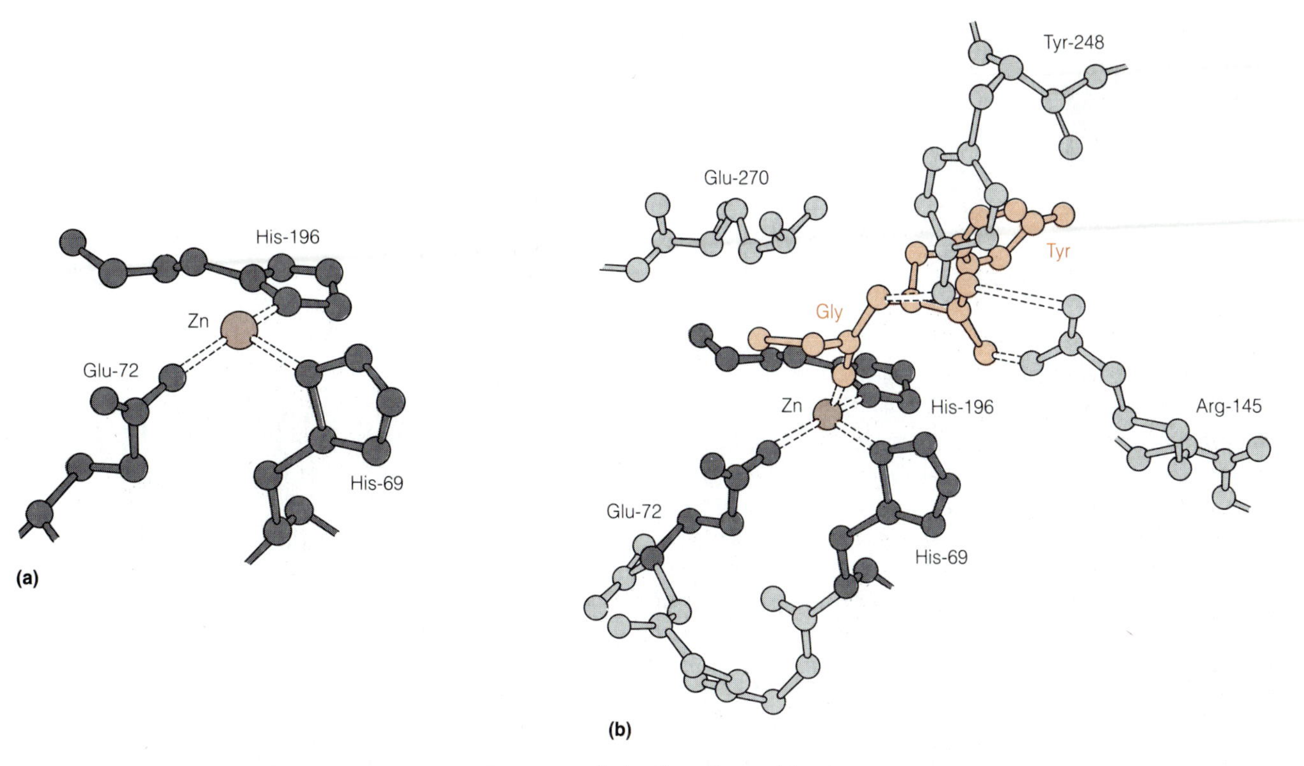

Figure 6-3 Active Site of Carboxypeptidase A. (a) The unoccupied active site consists of a zinc ion coordinated to two histidines and a glutamate. (b) Upon substrate binding (using an artificial peptide, shown in color), the active site "closes," bringing an arginine, a glutamate, and another histidine into close proximity with the substrate. The role of these amino acids in peptide bond hydrolysis is illustrated in Figure 6-10.

This reaction can be carried out in the laboratory using a platinum (Pt) or nickel (Ni) catalyst, as indicated. These inorganic catalysts are very nonspecific, however, in that they can be used to hydrogenate a wide variety of unsaturated compounds. In fact, nickel or platinum is used commercially to hydrogenate polyunsaturated vegetable oils in the manufacture of solid cooking fats or shortenings. Regardless of the exact structure of the unsaturated compound, it can be effectively hydrogenated in the presence of nickel or platinum.

By way of contrast, consider the biological example of hydrogenation involved in the conversion of fumarate into succinate, a reaction we will encounter again in Chapter 11:

$$\underset{\text{Fumarate}}{{}^{-}O{-}\overset{O}{\overset{\|}{C}}{-}CH{=}CH{-}\overset{O}{\overset{\|}{C}}{-}O^{-}} + 2H^{+} + 2e^{-} \rightleftharpoons \underset{\text{Succinate}}{{}^{-}O{-}\overset{O}{\overset{\|}{C}}{-}CH_2{-}CH_2{-}\overset{O}{\overset{\|}{C}}{-}O^{-}} \quad (6\text{-}4)$$

This particular reaction is catalyzed in cells by the enzyme *succinate dehydrogenase* (so called because it normally functions in the opposite direction during energy metabolism). This dehydrogenase, like most enzymes, is highly specific. It will not add or subtract hydrogens from any compounds except those shown in reaction 6-4. In fact, this particular enzyme is so specific that it will not even recognize maleate, a geometric stereoisomer of fumarate (Figure 6-4).

Not all enzymes are quite this specific; some accept a number of closely related substrates, and others accept any of a whole group of substrates as long as they possess some common structural feature. Such **group specificity** is seen most often with enzymes involved in the synthesis or degradation of polymers. The purpose of carboxypeptidase is to degrade polypeptide chains from the carboxyl end; thus, it makes sense for the enzyme to accept any of a wide variety of polypeptides as substrate, since it would be needlessly extravagant of the cell to require a separate enzyme for every different peptide bond that has to be hydrolyzed in polypeptide degradation.

In general, however, enzymes are highly specific with respect to substrate, such that a cell must possess almost

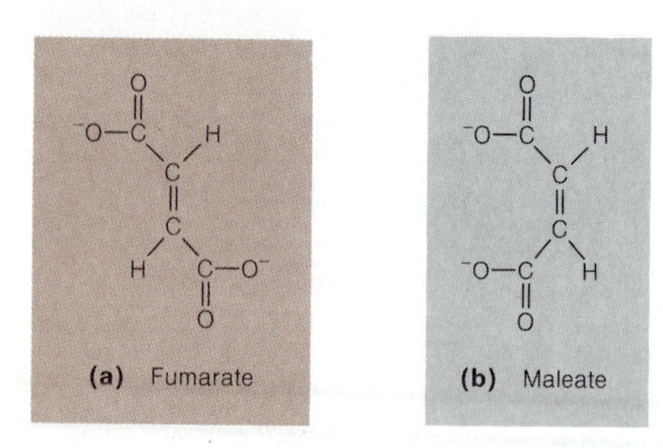

Figure 6-4 Stereoisomers: Fumarate and Maleate.

as many different kinds of enzymes as it has reactions to catalyze. For a typical cell, this means that several thousand different kinds of enzymes are necessary to carry out its full metabolic program. At first, that may seem wasteful in terms of proteins to be synthesized, genetic information to be stored and read out, and enzyme molecules to have on hand in the cell. But you should also be able to see the tremendous regulatory possibilities this suggests, a point we will return to later.

Sensitivity to Environment. Another property of enzymes that relates directly to protein chemistry is their sensitivity to their environment. Like other proteins, most enzymes are sensitive to temperature. Within limits, the rate of an enzyme-catalyzed reaction increases with temperature, primarily because the greater kinetic energy of both substrate and enzyme molecules ensures more frequent collisions, thereby increasing the chance of correct substrate binding. At some point, however, further increases in temperature are counterproductive, and the reaction rate begins to decrease rapidly because the enzyme molecule is beginning to denature. Hydrogen bonds break, hydrophobic interactions change, and the structural integrity of the active site is disrupted, causing loss of activity.

Figure 6-5 illustrates these competing effects of temperature on the activity of a typical protein. Most enzymes require temperatures of at least 55°–60°C for inactivation, but a few are remarkably sensitive to heat and are denatured and inactivated at lower temperatures.

Enzymes are also very sensitive to pH and are often active only within a fairly restricted pH range, as shown in Figure 6-6. This pH dependence is usually due to the involvement of one or more charged amino acids at the active site and the requirement that such groups be present in a specific form, either charged or uncharged. If the glutamates involved in the active site of carboxypeptidase A (Figure 6-3) must be present in the charged (ionized) form, for example, enzyme activity is likely to be adversely affected if the pH is decreased to the point where a substantial portion of the glutamate groups are protonated and therefore uncharged.

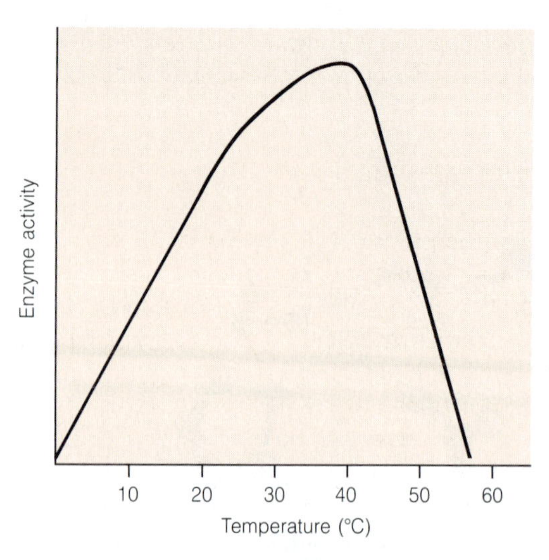

Figure 6-5 Effect of Temperature on the Activity of an Enzyme. The temperature optimum for activity results from two counteracting effects. As the temperature is increased, the chemical reaction is thermally activated as expected; but eventually an inactivating effect is seen at high temperatures as a result of thermal denaturation of protein structure.

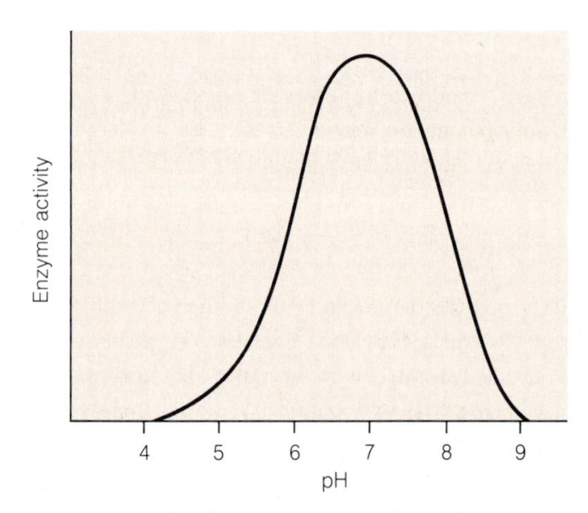

Figure 6-6 Effect of pH on the Activity of an Enzyme. The pH optimum for enzyme activity corresponds to the H^+ concentration at which ionizable groups on both the enzyme and the substrate molecules are in the most favorable form for maximum reactivity. Changes in pH away from the optimum usually reflect titration of charged groups on the enzyme, the substrate, or both.

The loss of activity as the pH is shifted away from the optimum can, in other words, usually be explained by the titration of one or more charged groups at the active site. If the substrates or products also have charged groups, activity may in addition depend on the form in which they are present at a particular pH. Most enzymes have pH optima in the range 6.5 to 7.5, but some show special adaptation to the environment in which they function naturally. *Pepsin*, for example, has a pH optimum around 2.0, consistent with its role in the digestion of proteins in the stomach. On the other hand, *trypsin* functions best at about pH 8.5, in keeping with its involvement in protein digestion in the alkaline environment of the small intestine.

Enzymes as Catalysts

Because of the precise "fit" between the active site of an enzyme and its substrates, enzymes are highly effective as catalysts. This effectiveness can be seen in the much greater extent to which reaction rates are catalyzed by enzymes as compared with inorganic catalysts. Typically, enzyme-catalyzed reactions proceed 10^8 to 10^{10} times faster than the uncatalyzed reactions, compared with a stimulation of 10^3 to 10^4 times for inorganic catalysts. As you might guess, most of the interest in enzymes focuses on the active site, where binding, activation, and chemical transformation of the substrate occur.

Substrate Binding. Initial contact between the active site of an enzyme and a potential substrate molecule depends on random collision. Once in the groove, or pocket, of the active site, however, the substrate molecules are bound temporarily to the enzyme surface in just the right orientation to one another and to specific catalytic groups on the enzyme to facilitate reaction. Substrate binding usually involves hydrogen bonds or ionic bonds (or both) to charged amino acids. These are generally weak bonds, but several bonds may hold a single molecule in place. The strength of the bonds between an enzyme and a substrate molecule is often in the range of 3 to 12 kcal/mol, less than one-tenth the strength of a single covalent bond. Substrate binding is therefore readily reversible.

For a long time, enzymologists regarded the active site as a rigid structure. They likened the fit of a substrate into the active site to that of a key into a lock, an analogy first suggested in 1890 by the German enzymologist Emil Fischer. The **lock-and-key model** explains enzyme specificity in terms of a very tight fit between the substrate and the active site. More recently, however, this model has given way to a view of enzymes that assumes greater flexibility of the active site. According to the **induced-fit model** of Daniel Koshland, the active site in its unoccupied form is already relatively specific for the right substrate, but the specificity is greatly enhanced upon binding of the proper substrate. The substrate binding induces a specific conformational change in the enzyme molecule and hence in the shape of the active site, as shown in Figure 6-7. This change results in the proper reactive groups of the enzyme being positioned maximally with respect to the substrate molecule for catalysis. Sometimes, the change involves bringing into the region of the active site specific amino acid side chains that are critical to the catalytic process but are not in the immediate area of the active site in the uninduced configuration.

In the case of carboxypeptidase A (Figure 6-3), substrate binding induces a conformational change in the enzyme molecule, bringing three critical amino acids (an arginine, a glutamate, and a tyrosine) into the active site. One of the effects of this conformational change on the substrate molecule is to twist or distort one or more of its bonds, thereby weakening the bond and making it more susceptible to catalytic attack, a point we will return to shortly.

Substrate Activation. The role of the active site is not just to recognize and bind the appropriate substrate but also to activate it by subjecting it to the right chemical environment for catalysis. For a given enzyme-catalyzed reaction, **substrate activation** may involve one or more means of activation. Three of the most common strategies are as follows:

1. As already mentioned, the change in enzyme conformation induced by initial substrate binding to the active site not only causes better complementarity and a tighter enzyme-substrate fit but also may tend to distort and deform specific bonds of the substrate. This physical stress weakens the bond, with the result that less energy is needed to break the bond.

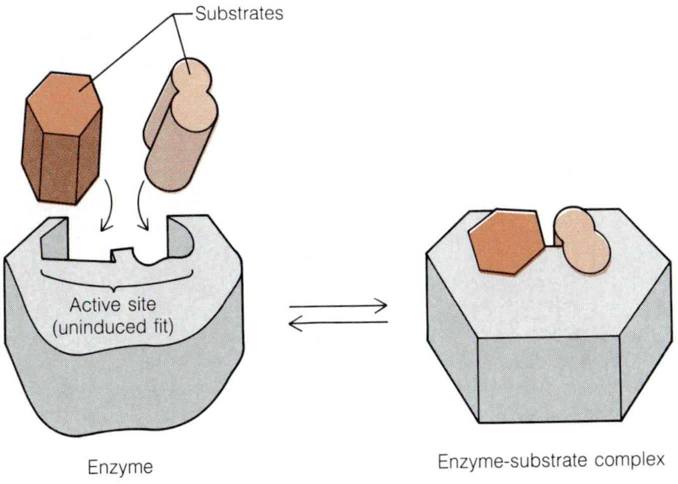

Figure 6-7 Induced-Fit Model for Enzymes and Substrates. The shape of the enzyme and its active site changes upon binding of the substrate. The active site is only generally complementary in shape to the substrate before binding, but acquires a very specific complementary fit once the substrate is in place.

Not All Enzymes Are Proteins

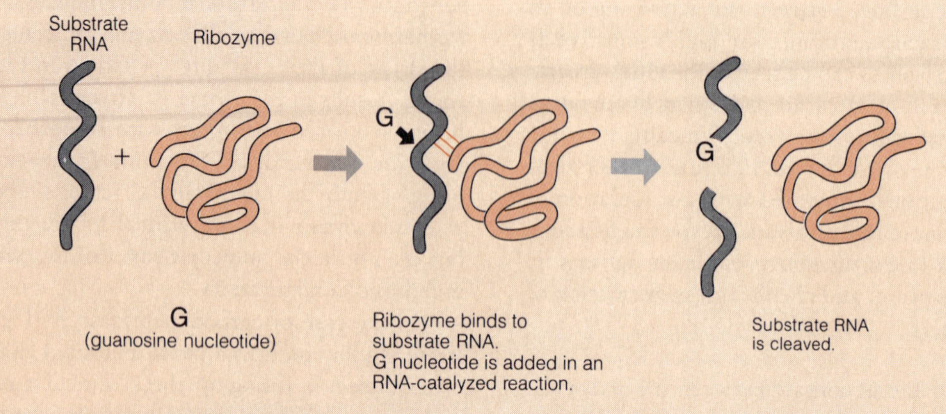

Figure 6-A Cleavage of RNA by a Ribozyme.

For many years, an enzyme was defined as a protein that had catalytic activity. But as you have seen, many different substances can act as catalysts, even those as simple as the iron ions in solution that speed up hydrogen peroxide breakdown by 30,000 times. It seems reasonable that other complex molecules might also have catalytic roles in the cell, and in 1981 Tom Cech and his colleagues discovered an important example, for which he shared a Nobel Prize in 1989. They found that certain molecules of RNA in a protozoan called *Tetrahymena* fit every definition of an enzyme, except that they were not proteins. Like enzymes, these strands of RNA could act as catalysts in certain reactions and demonstrated substrate specificity. One such reaction is shown in Figure 6-A, in which a catalytic RNA molecule catalyzes the addition of a guanosine nucleotide to another RNA molecule. As a result, the substrate RNA breaks into two fragments.

This reaction is only one step of a considerably more complex series of reactions catalyzed by a specific RNA sequence, in which a segment of RNA called an *intron* catalyzes its own removal from a longer strand and then splices the ends together. Other RNA-catalyzed reactions have recently been discovered. Most significantly, ribosomes, the molecular "machines" that string together amino acids to form proteins, appear to have active sites involving ribosomal RNA, rather than the ribosomal protein that has long been accepted as the most likely structure catalyzing the polymerization reaction.

The discovery of RNA enzymes (sometimes referred to as *ribozymes*) has markedly changed the way we think about the origin of life on earth. For many years, scientists have speculated that the first catalytic macromolecules must have been amino acid polymers resembling proteins. But this concept immediately ran into difficulties because there was no obvious way for a primeval protein to carry information or replicate itself, two primary attributes of the living state. However, if the first catalysts were RNA, it becomes conceptually simpler to imagine a system of RNA molecules acting both as catalysts and as replicating systems able to transfer information between generations.

2. The enzyme may also accept or donate protons, thereby increasing the chemical reactivity of the substrate. This is similar to the chemist's approach of changing the pH of the reaction mixture to enhance the rate of organic or inorganic reactions in the laboratory. For enzymes, the "pH change" is effected by acidic or basic groups at or near the active site. This accounts for the importance of charged amino acids in active-site chemistry, which in turn explains why enzyme activity is so often dependent on pH (Figure 6-6).
3. As a further means of substrate activation, enzymes may also accept or donate electrons, thereby forming temporary covalent bonds between the enzyme and its substrate. This mechanism requires that the substrate have a region that is either electropositive (electron-deficient) or electronegative (electron-rich) and that the active site of the enzyme have one or more groups of opposite polarity. In one case, the electronegative side group of an appropriate amino acid at the active site donates electrons to the electropositive region of the substrate, in what is called a **nucleophilic substitution** reaction (Figure 6-8a). The hydroxyl group of serine, the sulfhydryl group of cysteine, and the indole group of histidine are all active nucleophilic attacking groups. Alternatively, the electronegative group may be on the substrate and the electropositive group on the enzyme, and the reaction is then an **electrophilic substitution** (Figure 6-8b). Enzymes capable of electrophilic attack invariably require a prosthetic group at the active site, since none of the amino acid side groups is sufficiently electrophilic. Metal ions are strongly electropositive and therefore serve well in this role.

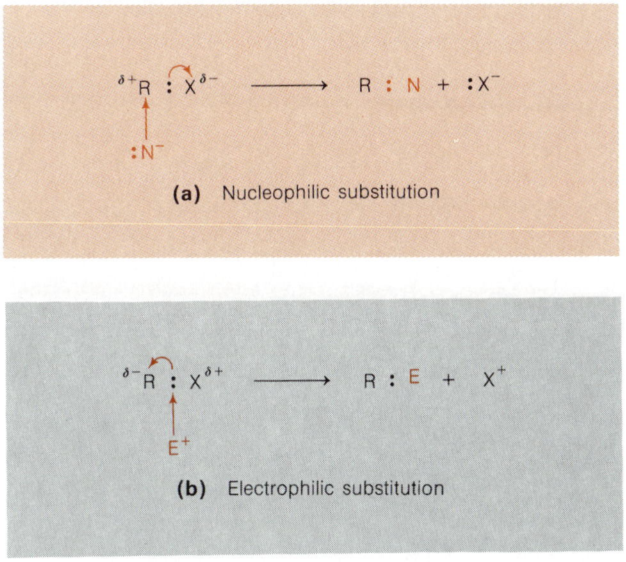

$$^{\delta+}R : X^{\delta-} \;(\uparrow :N^-) \longrightarrow R : N + :X^-$$

(a) Nucleophilic substitution

$$^{\delta-}R : X^{\delta+} \;(\uparrow E^+) \longrightarrow R : E + X^+$$

(b) Electrophilic substitution

Figure 6-8 Substitution Reactions in the Mechanism of Action of Some Enzymes. (a) Nucleophilic substitution, with N^- as the nucleophilic attacking group on the enzyme and X^- as the leaving group on the substrate. (b) Electrophilic substitution, with E^+ as the electrophilic attacking group on the enzyme and X^+ as the leaving group on the substrate. For an example of nucleophilic substitution, see the reaction mechanism illustrated for carboxypeptidase A in Figure 6-10.

The Mechanism of Enzyme Catalysis: An Example

Once the proper substrate has been bound to the active site of an enzyme and has been activated by one or more mechanisms, the catalytic reaction takes place and the products are released from the enzyme surface, all in a sufficiently short time to allow hundreds or even thousands of such reactions to occur per second. As an example of an enzyme-catalyzed reaction sequence, we will consider our earlier model, **carboxypeptidase A** (Figures 6-2 and 6-3).

Carboxypeptidase A is a digestive enzyme that degrades polypeptides by removing amino acids one at a time from the carboxyl end of the chain. It is therefore specific for carboxyl-terminal peptide bonds, and its catalytic activity can be represented as shown in Figure 6-9.

As we have already seen, carboxypeptidase A is a single polypeptide chain of 307 amino acids (Figure 6-2) with a zinc atom, a glutamate (Glu-72), and two histidines (His-

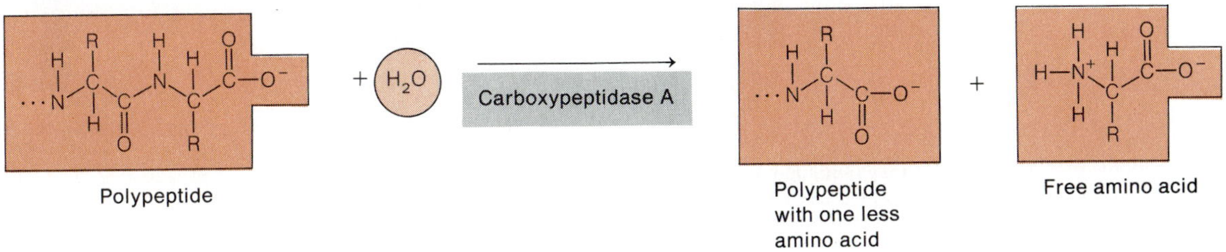

Figure 6-9 Catalytic Activity of Carboxypeptidase A. Carboxypeptidase A removes amino acids one at a time from the carboxyl end of polypeptides.

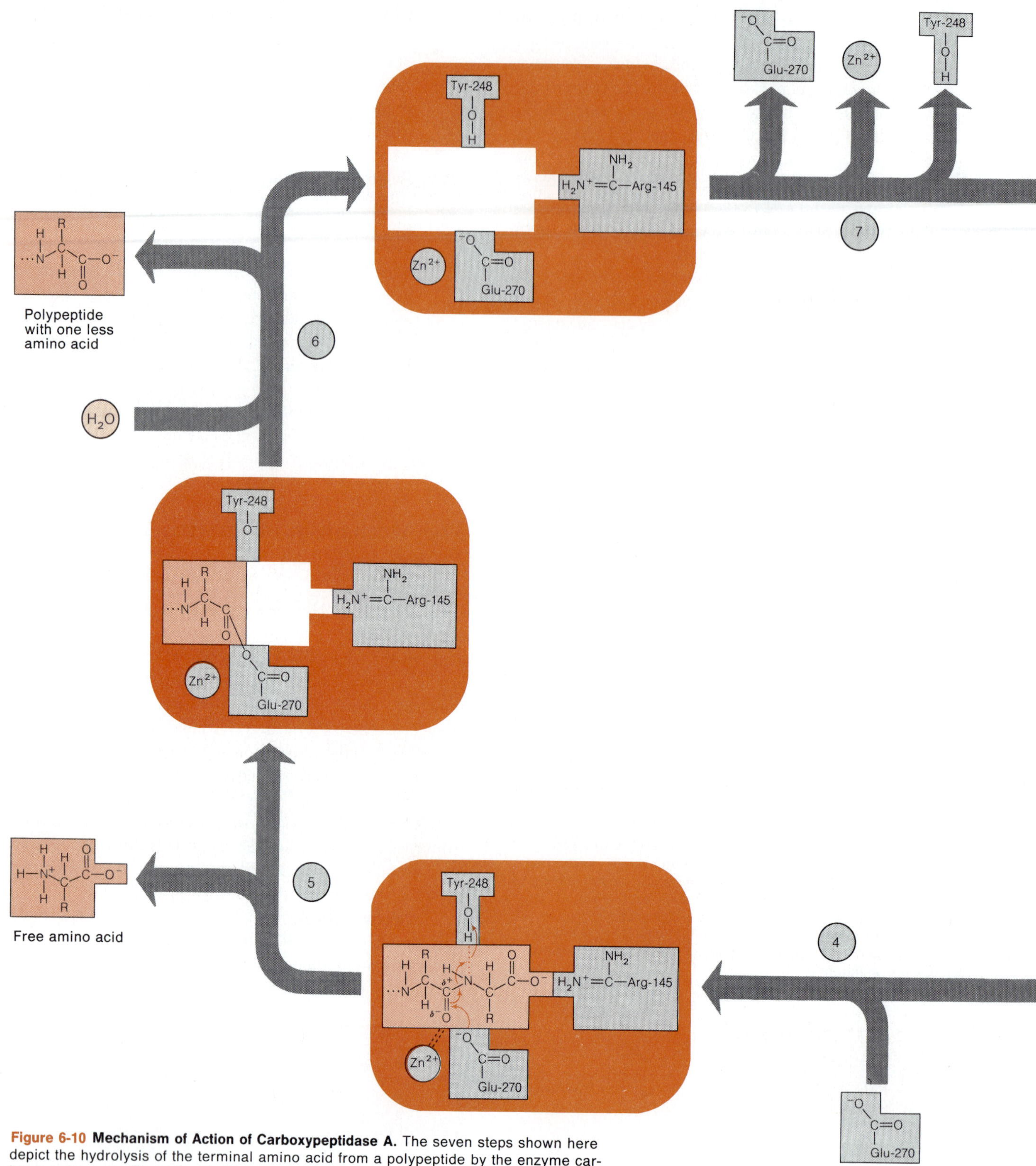

Figure 6-10 Mechanism of Action of Carboxypeptidase A. The seven steps shown here depict the hydrolysis of the terminal amino acid from a polypeptide by the enzyme carboxypeptidase A. The colored ellipse represents the active site of the enzyme. The active site contains additional amino acid groups, but the only three that are involved in the catalytic process are Arg-145, Tyr-248, and Glu-270, which are brought to the active site by the conformational change that occurs when the polypeptide substrate binds in step 1. The steps in the catalytic process are described in detail in the text.

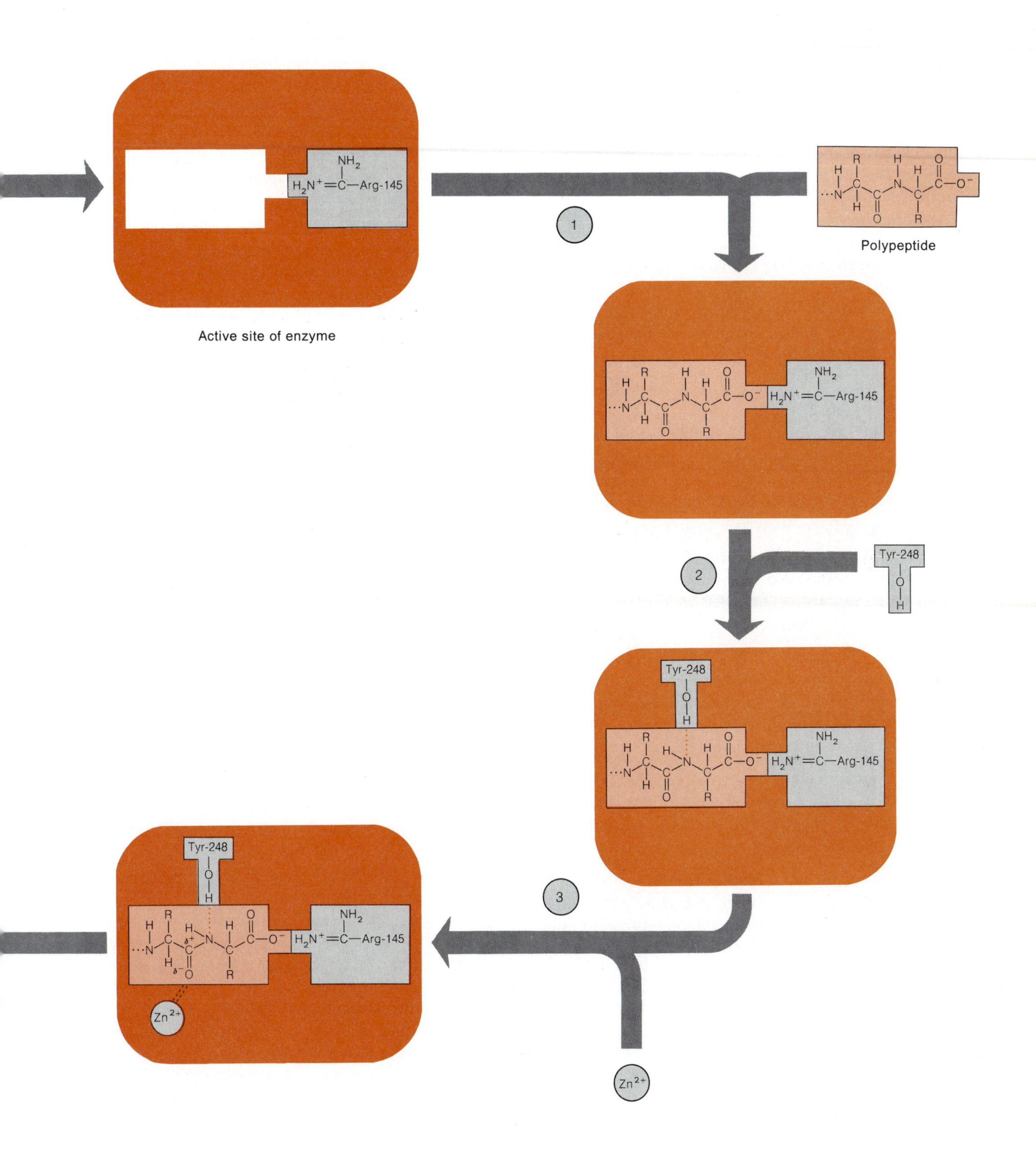

Active site of enzyme
Polypeptide
1
2
3
Tyr-248
Arg-145
Zn2+

69 and His-196) at its active site (Figure 6-3). Upon substrate binding, the enzyme undergoes a conformational change that brings an additional glutamate (Glu-270), an arginine (Arg-145), and a tyrosine (Tyr-248) into the active site. The arginine and glutamate move only about 0.2 nm each, but the tyrosine moves 1.2 nm, a distance equal to about a quarter of the entire diameter of the enzyme molecule. This enzyme therefore provides striking confirmation of the induced-fit model.

According to one proposed mechanism for carboxypeptidase catalysis, substrate binding, activation, and hydrolysis are thought to occur as follows (Figure 6-10):

Step 1: The negatively charged carboxylate group of the substrate polypeptide (color) interacts with the positively charged side chain of Arg-145 (gray) to initiate the conformational change responsible for "closing" the active site around the substrate.

Step 2: The resulting 1.2-nm shift in the position of Tyr-248 brings it into close contact with the peptide bond to be cleaved, allowing formation of a hydrogen bond between the hydroxyl group on the aromatic side chain of Tyr-248 and the amide nitrogen of the peptide bond.

Step 3: Meanwhile, the carbonyl oxygen of the peptide bond has become coordinated to the zinc atom at the active site. This polarizes the C=O bond more than usual and renders the carbonyl carbon atom especially vulnerable to nucleophilic attack.

Step 4: The attack comes from the electronegative carboxylate group of Glu-270. The resulting electron shift causes the tyrosine at position 248 to donate a proton from its hydroxyl group to the amide nitrogen.

Step 5: The peptide bond is now split, and the terminal amino acid is free to leave the enzyme surface. The remaining polypeptide chain is still bonded to Glu-270 by an anhydride linkage, however.

Step 6: The anhydride bond is subsequently hydrolyzed by a molecule of water, allowing the polypeptide to leave the enzyme surface, now one amino acid shorter than when it arrived.

Step 7: The enzyme then reverts to its original conformation and is ready for another catalytic cycle.

This is but a single example of an enzyme-catalyzed reaction for which the structure of the enzyme molecule and the catalytic events are known well enough to postulate a plausible mechanism, but it illustrates the diversity and intricacies of enzyme chemistry. As more and more enzymes are subjected to detailed mechanistic studies, we can expect to understand enzyme catalysis with greater and greater precision.

Enzyme Kinetics

So far, our discussion of enzymes has been basically descriptive. We have dealt with the activation energy requirement that prevents thermodynamically feasible reactions from occurring and with catalysts as a means of reducing the activation energy and thereby facilitating such reactions. We have also encountered enzymes as biological catalysts and have examined their structure and function in some detail. Certainly, the only reactions likely to occur in cells at reasonable rates are those for which specific enzymes are on hand, and the metabolic capabilities of cells are therefore effectively specified by the enzymes that are present.

Still lacking, however, is a means of assessing the actual rates at which enzyme-catalyzed reactions will proceed, as well as an appreciation for the factors that influence reaction rates. The mere presence of the appropriate enzyme in a cell may not necessarily ensure that a given reaction will occur at an adequate rate unless we can also be assured that cellular conditions are favorable for enzyme activity. We have already seen how factors such as temperature and pH can affect enzyme activity. Now we are ready to appreciate how critically enzyme activity also depends on the concentrations of substrates and inhibitors that prevail in the cell. In addition, we will see how at least some of these effects can be quantified.

It is as we turn our attention to these quantitative aspects of enzyme catalysis that we encounter the field of **enzyme kinetics.** The word *kinetics* is of Greek origin (*kinetikos,* meaning "moving") and, as applied to chemical reactions, concerns reaction rates and the manner in which those rates are influenced by a variety of factors, but especially by concentrations of substrates, products, and inhibitors. Most of our attention will focus on the effects of substrate concentration on the kinetics of enzyme-catalyzed reactions. Our considerations will be restricted to initial reaction rates, measured over a period of time during which the substrate concentration has not yet decreased enough to affect the rate and the accumulation of product is still too small to cause any measurable back reaction. Although this is oversimplified compared to the real-life situation, it nonetheless allows us to understand some important principles of enzyme kinetics.

Since the intricacies of enzyme kinetics can seem complex when first encountered, the analogy provided in the box on pp. 146–147 may prove to be helpful. The model used is a roomful of monkeys ("enzymes") shelling peanuts ("substrates").

Michaelis-Menten Kinetics

It has long been known that the rate of an enzyme-catalyzed reaction increases with increasing substrate concentration, but in a manner such that each additional increment of substrate results in a smaller increase in reaction rate. More precisely, the relationship between the initial reaction rate (or velocity v) and the substrate concentration [S] can be shown experimentally to be that of a hyperbola, as illustrated in Figure 6-11. An important property of this hyperbolic relationship is that as [S] tends toward infinity, v tends toward an upper limiting value that depends on the number of enzyme molecules and can be increased only by adding more enzyme.

The inability of higher and higher substrate concentrations to increase the reaction velocity beyond a finite upper value is called **saturation.** This property is a fundamental characteristic of enzyme-catalyzed reactions. Catalyzed reactions always become saturated at high substrate concentrations, whereas uncatalyzed reactions do not.

Much of our understanding of the hyperbolic relationship between [S] and v is owed to the pioneering work of two German enzymologists, Leonor Michaelis and Maud Menten. In 1913, they postulated a general theory of enzyme action that has turned out to be basic to the quantitative analysis of almost all aspects of enzyme kinetics. To understand their approach, consider one of the simplest possible enzyme-catalyzed reactions, in which a substrate S is converted into a product P:

$$S \xrightarrow[\text{Enzyme (E)}]{} P \tag{6-5}$$

According to the Michaelis-Menten hypothesis, the enzyme E that catalyzes this reaction first reacts with the substrate S to form the transient enzyme-substrate complex ES, which then undergoes the actual catalytic reaction to form free enzyme and product P, as shown in the following sequence:

$$E_f + S \underset{k_2}{\overset{k_1}{\rightleftharpoons}} ES \underset{k_4}{\overset{k_3}{\rightleftharpoons}} E_f + P \tag{6-6}$$

where E_f is the free form of the enzyme, S is the substrate, ES is the enzyme-substrate complex, P is the product, and k_1, k_2, k_3, k_4 are the rate constants for the indicated reactions.

Beginning with the definition of velocity as the disappearance of substrate or the appearance of product per unit time, as shown by

$$v = \frac{-d[S]}{dt} = \frac{+d[P]}{dt} \tag{6-7}$$

it is possible, in a series of straightforward algebraic operations, to arrive as Michaelis and Menten did at the following relationship between velocity and substrate concentration:

$$v = \frac{V_{max}[S]}{K_m + [S]} \tag{6-8}$$

Here [S] is the initial substrate concentration, v is the initial reaction velocity at that substrate concentration, and V_{max} and K_m are kinetic parameters. This is the **Michaelis-Menten equation,** the central relationship of enzyme kinetics.

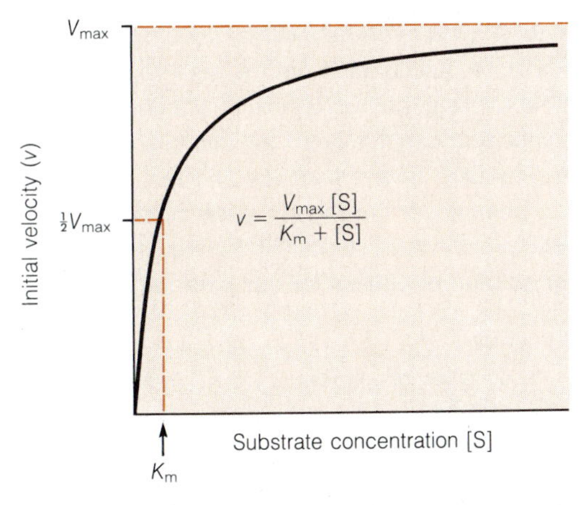

Figure 6-11 Relationship Between Reaction Velocity and Substrate Concentration. For an enzyme-catalyzed reaction that follows Michaelis-Menten kinetics, the initial velocity tends toward an upper limiting velocity V_{max} as the substrate concentration [S] tends toward infinity. The Michaelis constant K_m corresponds to that substrate concentration at which the reaction is proceeding at one-half of the maximum velocity.

The Meaning of V_{max} and K_m

To appreciate the implications of this relationship between v and [S] and to examine the meaning of the parameters V_{max} and K_m, we can consider three special cases of substrate concentration: very low substrate concentration, very high substrate concentration, and the special case of $[S] = K_m$.

Case 1: Very Low Substrate Concentration ($[S] \ll K_m$). At very low substrate concentration, [S] becomes negligibly small compared to the constant K_m in the denominator of the Michaelis-Menten equation, so we can write

$$v = \frac{V_{max}[S]}{K_m + [S]} \cong \frac{V_{max}[S]}{K_m} \tag{6-9}$$

Monkeys and Peanuts

If you found the Mexican jumping beans helpful in understanding free energy in Chapter 5, you might appreciate an approach to enzyme kinetics based on the analogy of a roomful of monkeys ("enzymes") shelling peanuts ("substrates"), with the peanuts present in varying abundance. Try to understand each step first in terms of monkeys shelling peanuts, then in terms of an actual enzyme-catalyzed reaction.

The Peanut Gallery

For our model, we need a troop of ten monkeys, all equally adept at finding and shelling peanuts. We shall assume that the monkeys are too full to eat any of the peanuts they shell, but nonetheless have an irresistible compulsion to go on shelling. To make the model a bit more rigorous, we should insist that the peanuts are a new hybrid variety that can be readily stuck back together again and that the monkeys are just as likely to put the peanuts back in the shells as they are to take them out. But these qualifications need not concern us here, since we are interested only in the initial conditions in which all the peanuts start out in their shells.

Next, we need the Peanut Gallery, a room of fixed floor space with peanuts scattered equally about on the floor. The amount of peanuts will be varied as we proceed, but in all cases there will be vastly more peanuts than monkeys in the room. Moreover, since we know the number of peanuts and the total floor space, we can always calculate the "concentration" (more accurately, the density) of peanuts in the room. In each case, the monkeys start out in an adjacent room. To start an assay, we simply open the door and allow the eager monkeys to enter the Peanut Gallery.

The Shelling Begins

Now we are ready for our first assay. We start with an initial peanut concentration of one peanut per square meter, and we assume that at this concentration of peanuts, the average monkey spends 9 seconds looking for a peanut to shell and 1 second shelling it. This means that each monkey requires 10 seconds per peanut and can consequently shell peanuts at the rate of 0.1 peanut per second. And since there are ten monkeys in the Gallery, the rate (let's call it the velocity v) of peanut-shelling for all the monkeys is 1 peanut per second at this particular concentration of peanuts (which we will call [S] to remind ourselves that the peanuts are really the substrate of the shelling reaction). All of this can be tabulated as follows:

[S] = Concentration of peanuts (peanuts/m^2)	1
Time required per peanut:	
To find (sec/peanut)	9
To shell (sec/peanut)	1
Total (sec/peanut)	10
Rate of shelling:	
Per monkey (peanut/sec)	0.10
Total (v) (peanut/sec)	1.0

The Peanuts Become More Abundant

For our second assay, we herd all the monkeys back into the waiting room, sweep up the debris, and arrange peanuts about the Peanut Gallery at a concentration of 3 peanuts per square meter. Since peanuts are now three times more abundant than previously, the average monkey should find a peanut three times more quickly than before, such that the time spent finding the average peanut is now only 3 seconds. But each peanut, once found, still takes 1 second to shell, so the total time per peanut is now 4 seconds and the velocity of shelling is 0.25 peanut per second for each monkey, or 2.5 peanuts per second for the roomful of monkeys. This generates another column of entries for our data table:

[S] = Concentration of peanuts (peanuts/m^2)	1	3
Time required per peanut:		
To find (sec/peanut)	9	3
To shell (sec/peanut)	1	1
Total (sec/peanut)	10	4
Rate of shelling:		
Per monkey (peanut/sec)	0.10	0.25
Total (v) (peanut/sec)	1.0	2.5

What Happens to v as [S] Continues to Increase?

To find out what eventually happens to the velocity of peanut-shelling as the peanut concentration in the room gets higher and higher, all you need do is extend the data table by assuming ever-increasing values for [S] and calculating the corresponding v. For example, you should be able to convince yourself that a further tripling of the peanut concentration (from 3 to 9 peanuts/m^2) will bring the time required per peanut down to 2 seconds, which will result in a shelling rate of 0.5 peanut per second for each monkey, or 5.0 peanuts per second overall.

Already you should begin to see a trend: the first tripling of peanut concentration increased the rate $2\frac{1}{2}$-fold, but the next tripling only resulted in a further doubling of the rate. There seems, in other words, to be a diminishing return on additional peanuts. You can see this clearly if you choose a few more peanut concentrations and then plot v on the y-axis (suggested scale: 0–10 peanuts/sec) versus [S] on the x-axis (suggested scale: 0–100 peanuts/m^2). What you should find is that the data generate a hyperbolic curve that looks strikingly like Figure 6-11. And if you look at your data carefully, you should see that the reason your curve continues to "bend over" as [S] gets higher (that is, the reason you get less and less additional velocity for each further increment of peanuts) is that the shelling time is fixed and therefore becomes a more and more prominent component of the total processing time per peanut as the finding time gets smaller and smaller. You should also appreciate that it is this fixed shelling time that ultimately sets the upper limit on the overall rate of peanut processing, since even when [S] is infinite (that is, in a world flooded with peanuts), there will still be a finite time of 1 second required to process each peanut.

Finally, you should realize that there is something special about the peanut concentration at which the finding time is exactly equal to the shelling time (it turns out to be 9 peanuts/m^2): this is the point along the curve at which the rate of peanut processing is exactly one-half of the maximum rate. In fact, it is such an important benchmark along the concentration scale that you might even be tempted to give it a special name, particularly if your name were Michaelis and you were monkeying around with enzymes instead of peanuts!

Thus, at very low substrate concentration, the initial reaction velocity is roughly proportional to the substrate concentration. This is therefore the **first-order region** of the Michaelis-Menten plot. As long as the substrate concentration is much lower than the K_m value, the velocity of an enzyme-catalyzed reaction increases almost linearly with substrate concentration.

Case 2: Very High Substrate Concentration ([S] $\gg K_m$). At very high substrate concentration, K_m becomes negligibly small compared to [S] in the denominator of the Michaelis-Menten equation, so we can write

$$v = \frac{V_{max}[S]}{K_m + [S]} \cong \frac{V_{max}[S]}{[S]} = V_{max} \tag{6-10}$$

This relationship means that at very high substrate concentrations, the velocity of an enzyme-catalyzed reaction is independent of variation in [S] and is therefore approximately constant. This is therefore the **zero-order region** of the Michaelis-Menten plot. As long as the substrate concentration is much higher than the K_m value, the velocity is essentially unaffected by changes in substrate concentration, remaining constant at V_{max}.

This, then, provides us with a definition of V_{max}, one of the two kinetic parameters in the Michaelis-Menten equation: V_{max} is the **maximum velocity**, or the upper limiting value, to which the initial reaction velocity v tends as the substrate concentration [S] approaches infinity. V_{max}, in other words, is the velocity at saturating substrate concentrations. Under these conditions, every enzyme molecule is occupied in the actual process of catalysis, since the substrate concentration is so high that another substrate molecule arrives at the active site almost as soon as a product molecule is released.

V_{max} is therefore an upper limit determined by (1) the time required for the actual catalytic reaction plus subsequent release of product from the surface of each enzyme molecule and (2) how many such enzyme molecules are present. Since the actual reaction rate is fixed, the only way that V_{max} can be increased is to increase enzyme concentration. In fact, V_{max} is linearly proportional to the amount of enzyme present, as shown in Figure 6-12.

Case 3: [S] = K_m. So far, we have seen the reason for first-order reaction kinetics at low substrate concentrations and for zero-order kinetics at high concentrations. We have also formulated a definition for V_{max} but have yet to discover the meaning of the second kinetic parameter, K_m. Note, however, that whatever its meaning, K_m appears to have something to do with determining how low the substrate concentration must be to ensure first-order kinetics or, alternatively, how high a concentration is required to ensure zero-order kinetics. Thus, K_m seems to be some sort of benchmark on the concentration scale that determines how high is high and how low is low. To explore its meaning more precisely, consider the special case where [S] is exactly equal to K_m. Under these conditions, the Michaelis-Menten equation can be written as follows:

$$v = \frac{V_{max}[S]}{K_m + [S]} = \frac{V_{max}[S]}{2[S]} = \frac{V_{max}}{2} \tag{6-11}$$

This equation provides us with the definition we have been looking for: K_m is that specific substrate concentration at which the reaction proceeds at one-half of its maximum (upper limiting) velocity. This specific concentration is a fixed value for a given enzyme catalyzing a specific reaction under specified conditions and is called the **Michaelis constant** (hence the designation K_m) in honor of the enzymologist who first elucidated its meaning.

In physical terms, K_m can be thought of as a measure of the affinity of an enzyme for its substrate. The *lower* the K_m value, the *greater* the affinity. Figure 6-11 illustrates the meaning of both V_{max} and K_m.

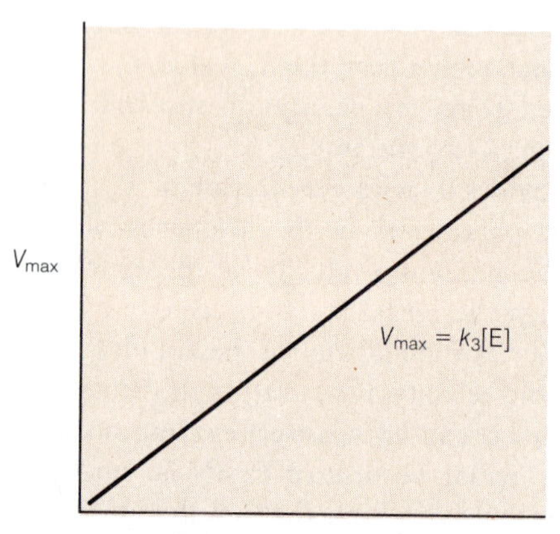

Figure 6-12 Linear Relationship of V_{max} to Enzyme Concentration. The linear increase in reaction velocity with enzyme concentration provides the basis for determining enzyme concentrations experimentally.

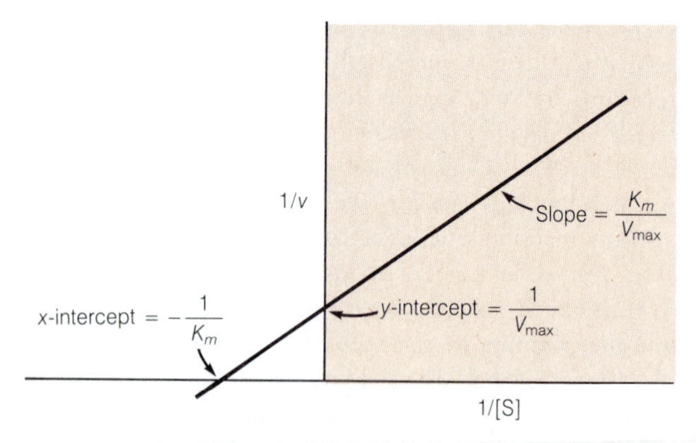

Figure 6-13 Lineweaver-Burk Double-Reciprocal Plot. The reciprocal of the initial velocity, 1/*v*, is plotted as a function of the reciprocal of the substrate concentration, 1/[S]. K_m can be calculated from the *x*-intercept and V_{max} from the *y*-intercept.

The Double-Reciprocal Plot

A classic Michaelis-Menten plot of *v* versus [S] is a faithful representation of the dependence of velocity on substrate concentration, but it is not an especially useful tool for the quantitative determination of the key kinetic parameters K_m and V_{max}. Its hyperbolic shape makes it difficult to extrapolate accurately to infinite substrate concentrations, as would be required to determine the critical parameter V_{max}, and if V_{max} is not known accurately, K_m cannot be determined. This problem is readily apparent from Figure 6-11, since it would be difficult to estimate V_{max} accurately if it were not already sketched in, and without V_{max}, K_m cannot easily be estimated either.

To circumvent this problem and provide a more useful graphic approach, H. Lineweaver and D. Burk converted the hyperbolic relationship of the Michaelis-Menten equation into a linear function by inverting both sides of equation 6-8 and simplifying the resulting expression into the form of an equation for a straight line:

$$\frac{1}{v} = \frac{K_m + [S]}{V_{max}[S]} = \frac{K_m}{V_{max}[S]} + \frac{[S]}{V_{max}[S]}$$
$$= \frac{K_m}{V_{max}}\left[\frac{1}{[S]}\right] + \frac{1}{V_{max}} \tag{6-12}$$

Equation 6-12 is the **Lineweaver-Burk equation.** If it is plotted as $1/v$ versus $1/[S]$, as in Figure 6-13, the resulting **double-reciprocal plot** is linear, with a *y*-intercept of $1/V_{max}$, an *x*-intercept of $-1/K_m$, and a slope of K_m/V_{max}. (You should be able to convince yourself of these intercept values by setting first $1/[S]$ and then $1/v$ equal to zero in equation 6-12 and solving for the other value.) Thus, once the double-reciprocal plot has been constructed, V_{max} can be determined directly from the reciprocal of the *y*-intercept and K_m from the negative reciprocal of the *x*-intercept. Furthermore, the slope can be used to check both values. In addition, the double-reciprocal plot is a useful diagnostic in analyzing for enzyme inhibition, since the several different kinds of inhibitors affect the shape of the plot in characteristic ways, as we will see shortly.

The utility of the Lineweaver-Burk plot is therefore threefold: it confirms by its linearity that the reaction in question is following Michaelis-Menten kinetics; it allows determination of the two parameters V_{max} and K_m without the complication of a hyperbolic shape; and it serves as a useful test of enzyme inhibition. Several alternatives to the Lineweaver-Burk equation for linearizing kinetic data have since come into use (such as the *Eadie-Hofstee* and *Hanes-*

Woolf plots mentioned in Problem 11), but the double-reciprocal plot of Figure 6-13 is still widely used by enzyme kineticists for determination of K_m and V_{max} values from experimental data.

Determining K_m and V_{max}: An Example

To illustrate the value of the double-reciprocal plot in determining V_{max} and K_m, consider a specific example involving the enzyme hexokinase, as illustrated in Figures 6-14 and 6-15. Hexokinase is an important enzyme in cellular energy metabolism, since it catalyzes the first reaction in the glycolytic pathway. Using ATP as a source of both the phosphate group and the energy needed for the reaction, hexokinase catalyzes the phosphorylation of glucose on carbon atom 6:

$$\text{glucose} + \text{ATP} \underset{\text{hexokinase}}{\rightleftharpoons} \text{glucose-6-phosphate} + \text{ADP} \tag{6-13}$$

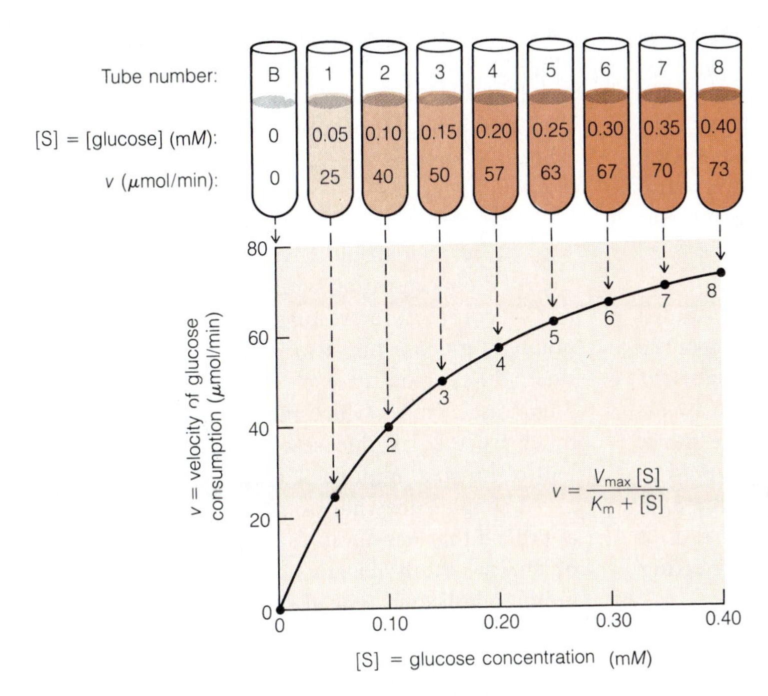

Figure 6-14 Experimental Procedure for Studying the Kinetics of the Hexokinase Reaction. Test tubes containing graded concentrations of glucose and saturating levels of ATP are incubated with a standard amount of hexokinase, and the initial rate of product appearance, *v*, is plotted as a function of the substrate concentration [S]. The curve is hyperbolic, approaching V_{max} as the substrate concentration gets higher and higher. For the double-reciprocal plot derived from these data, see Figure 6-15.

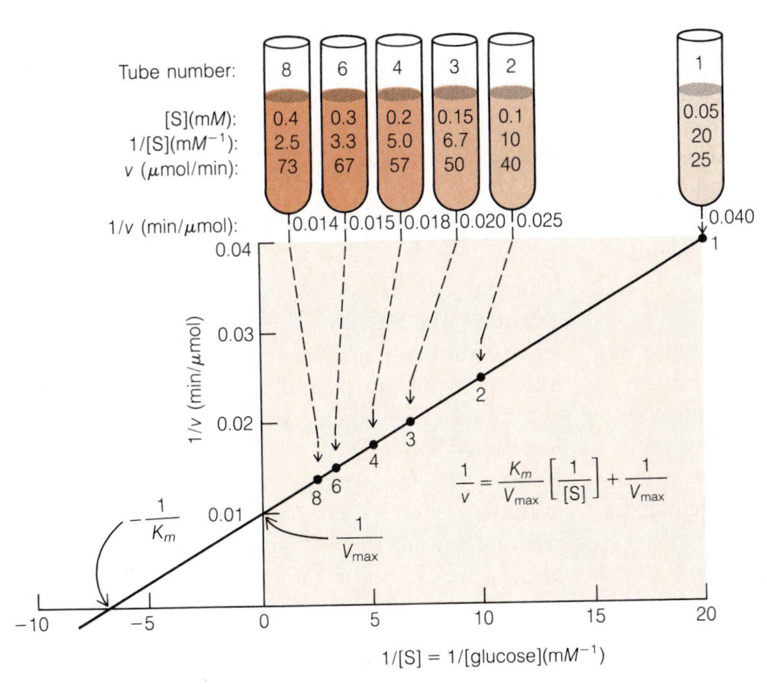

Figure 6-15 Double-Reciprocal Plot for the Hexokinase Data of Figure 6-14. For each test tube, 1/*v* and 1/[S] were calculated from the data of Figure 6-14, and 1/*v* was then plotted as a function of 1/[S]. The *y*-intercept of 0.01 corresponds to $1/V_{max}$, so V_{max} is 100 μmol/min. The *x*-intercept of −6.7 corresponds to $-1/K_m$, so K_m is 0.15 m*M*.

To analyze this reaction kinetically, the initial velocity must be determined at each of several substrate concentrations. Since the enzyme has two substrates, the usual approach is to vary the concentration of one substrate at a time, holding that of the other substrate constant at a sufficiently high (saturating) level to ensure that it does not become rate-limiting. Care must also be taken to ensure that the velocity determination is made before product accumulates to the point that the back reaction becomes significant.

In the experimental approach shown in Figure 6-14, glucose is the variable substrate, since ATP is present at a saturating concentration in each tube. Of the nine reaction mixtures set up for this experiment, one is designated the reagent blank (B), since it contains no glucose. The other eight tubes contain graded levels of glucose ranging from 0.05 to 0.40 m*M*. With all tubes prepared and maintained at some favorable temperature (25°C is often used), the reaction in each is initiated by the addition of a fixed amount of hexokinase.

The rate of product formation can then be determined either by continuous spectrophotometric monitoring of the reaction mixture (provided that one of the reactants or products absorbs light of a specific wavelength) or by allowing the reaction mixture to incubate for some short, fixed period of time, followed by chemical assay for either substrate depletion or product accumulation. In the case of the hexokinase reaction, the latter procedure is used because there is no direct photometric means of detecting any of the products or reactants.

As Figure 6-14 indicates, the initial velocity in tubes 1 through 8 ranged from 25 to 73 μmol of glucose consumed per minute, with no detectable reaction in the blank. (If any glucose consumption were noted in the blank, the values of tubes 1 through 8 would have to be corrected for that amount of noncatalytic reaction.) When these reaction velocities are plotted as a function of glucose concentration, the eight data points generate a hyperbolic curve like that of Figure 6-14. Although the data of Figure 6-14 are idealized for illustrative purposes, most kinetic data generated by this approach do, in fact, fit a hyperbolic curve unless the enzyme has some special properties that cause departure from Michaelis-Menten kinetics.

The hyperbolic curve of Figure 6-14 illustrates the need for some means of linearizing the analysis, since neither V_{max} nor K_m can be determined from the values as plotted, even though the data are known to be idealized. This need is met by the linear double-reciprocal plot of Figure 6-15. To obtain the data plotted here, reciprocals were calculated for each value of [S] and v from Figure 6-14. Thus, the [S] values of 0.05–0.40 m*M* generate reciprocals of 20–2.5 mM^{-1}, and the v values of 25–73 μmol/min give rise to reciprocals of 0.040–0.014 min/μmol. Because these are reciprocals, the data point for tube 1 is farthest from the origin, and each successive tube is represented by a point closer to the origin.

When these data points are connected by a straight line, the y-intercept is found to be 0.01 min/μmol and the x-intercept is $-6.7\ mM^{-1}$. From these intercepts, we can calculate that $V_{max} = 1/0.01 = 100$ μmol/min and $K_m = -(1/-6.7) = 0.15$ m*M*. If we now go back to the Michaelis-Menten plot of Figure 6-14, we can see that both of these values are eminently reasonable, since we can readily imagine that the plot is rising hyperbolically to a maximum of 100 μmol/min. Moreover, the graph reaches one-half of this value at a substrate concentration of 0.15 m*M*, which turns out to be the data point for tube 3. This, of course, is the K_m of hexokinase for glucose, often written $K_{m,\text{glucose}}$.

The enzyme also has a K_m for the other substrate, $K_{m,\text{ATP}}$, but that would have to be determined by varying the ATP concentration while holding the glucose concentration constant. Interestingly, hexokinase phosphorylates not only glucose but also other hexoses, and has a distinctive K_m value for each. The K_m for fructose, for example, is 1.5 m*M*, which means that it takes ten times as much fructose as glucose to sustain the reaction at one-half of its maximum velocity.

Though somewhat simplified and idealized, this is the approach that enzymologists take in studying the kinetics of enzyme-catalyzed reactions. Their analyses are often more complicated than this, and they almost always use a computer to calculate and plot double-reciprocal data and to determine K_m and V_{max} values, but the basic approach is the same.

Enzyme Inhibition and Regulation

Thus far, we have assumed that the only substances in cells that affect the activation of enzymes are their substrates. However, enzymes are also influenced by products, alternative substrates, substrate analogues, drugs, toxins, and an especially important class of regulators called allosteric effectors. With the exception of some of the allosteric effectors, most of these substances have an inhibitory effect on enzyme activity, reducing the rate of reaction toward the desired substrate. We will look first at the general phenomenon of enzyme inhibition and then turn to the specific topic of allosteric regulation.

Enzyme Inhibition

The **inhibition** of enzyme activity is important to the cell biologist for several reasons. First and foremost, it plays a vital role as a control mechanism in cells, since many enzymes are subject to regulation by specific small molecules and ions. It is also essential in understanding the mode of action of drugs and poisons, which frequently exert their effects by inhibiting enzymes. Inhibitors are also useful to enzymologists as tools in their studies of reaction mechanisms. Especially important in this latter case are **substrate analogues,** compounds that resemble the true substrate closely enough to bind to the active site but are then chemically unable to undergo reaction.

Inhibitors may be either irreversible or reversible. **Irreversible inhibitors** become permanently associated with or bound to the enzyme, causing an irrevocable loss of catalytic activity. Ions of heavy metals are often irreversible inhibitors, as are alkylating agents and nerve gas poisons. This is, in fact, the mode of action of nerve gases and the reason they are so toxic. They bind irreversibly to acetylcholinesterase, an enzyme that is vital to the process of nerve impulse transmission (see Chapter 20). One such nerve gas is diisopropylphosphofluoridate, which binds covalently to the hydroxyl group of a critical serine at the active site of the enzyme, thereby rendering it permanently inactive, as shown in Figure 6-16.

In contrast, **reversible inhibitors** undergo dissociation from their binding site on the enzyme, such that the bound and free forms of the inhibitor exist in equilibrium with each other. The two most common forms of reversible inhibitors are **competitive** and **noncompetitive** inhibitors. Competitive inhibitors bind to the active site of the enzyme and therefore compete with substrate molecules for the same site on the enzyme; noncompetitive inhibitors do not. The two forms can be distinguished from each other on the basis of their characteristically different effects on the kinetics of the reaction and hence on the shape of the double-reciprocal plot (Figure 6-17).

Allosteric Regulation

Our view of enzyme inhibition would be distorted if we were to conclude that the presence of an inhibitor and the resulting reduction in enzyme activity are invariably detrimental to the best interests of the cell. In fact, inhibition of enzyme activity is one of the most efficient ways cells have of regulating their chemical activities. Far from simply running at indiscriminately high rates, enzyme-catalyzed reactions and the biochemical sequences they are part of must be continuously regulated and adjusted to keep them finely tuned to the needs of the cell. An important aspect of that regulation and adjustment lies in the cell's ability to inhibit enzyme activities with specificity and precision. Although the types of inhibition already considered can

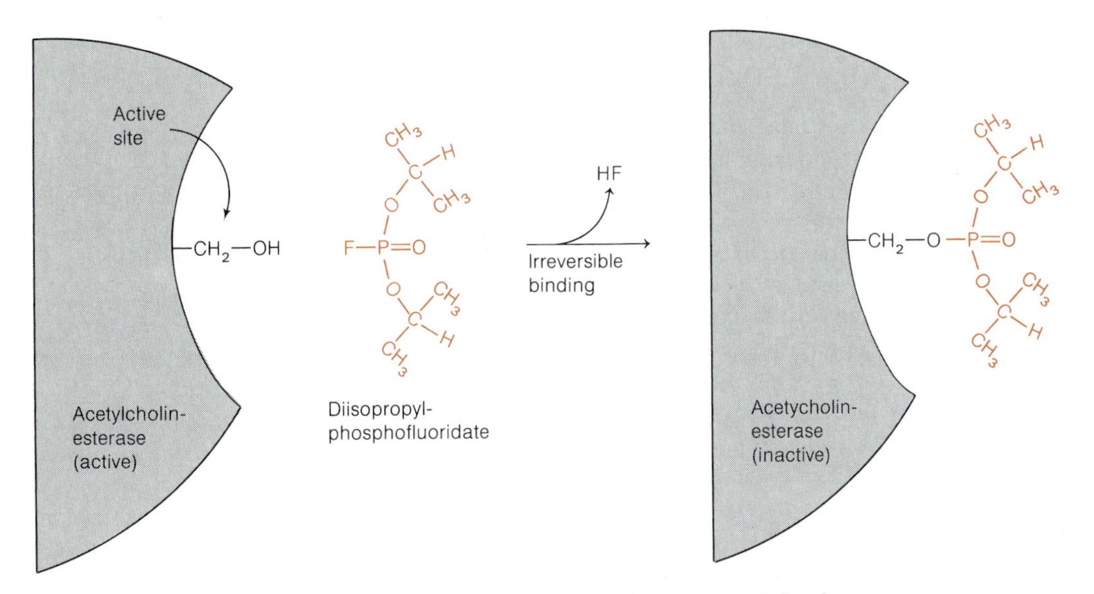

Figure 6-16 Irreversible Inhibition Caused by the Nerve Gas Diisopropylphosphofluoridate. Diisopropylphosphofluoridate (color) is an irreversible inhibitor of the enzyme acetylcholinesterase. It binds covalently to the hydroxyl group of a serine at the active site of the enzyme, blocking the active site irreversibly and resulting in permanent loss of enzyme activity.

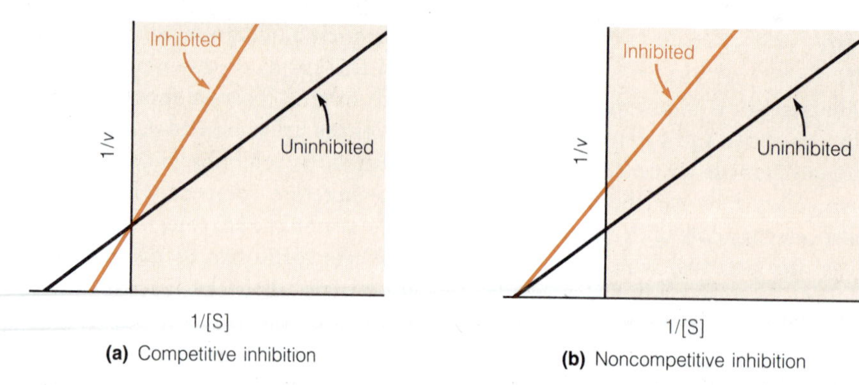

(a) Competitive inhibition

(b) Noncompetitive inhibition

Figure 6-17 Effect of Inhibition on the Double-Reciprocal Plot. (a) A competitive inhibitor increases the apparent K_m but leaves the V_{max} unchanged. The double-reciprocal plot for the inhibited reaction (colored line) therefore has a different *x*-intercept and slope than the plot for the uninhibited reaction (black line), but the *y*-intercept is unchanged. (b) A noncompetitive inhibitor reduces the V_{max} but does not affect K_m. The *x*-intercept of the double-reciprocal plot therefore remains unchanged, but the *y*-intercept and the slope change (compare colored line to black line). These characteristic changes are useful in diagnosing the nature of an inhibitory effect.

themselves be useful ways to regulate enzyme activity by effectively "throttling" the velocity of the reaction, the most important kind of regulation is that involving specific small molecules called **allosteric effectors.**

To understand this mode of regulation, consider the pathway whereby a cell converts some precursor A into some final product P via a series of intermediates B, C, and D, in reactions mediated by enzymes E_1, E_2, E_3, and E_4:

$$A \xrightarrow[E_1]{} B \xrightarrow[E_2]{} C \xrightarrow[E_3]{} D \xrightarrow[E_4]{} P \qquad (6\text{-}14)$$

Product P could, for example, be an amino acid needed for protein synthesis, and A could be some common cellular component that serves as the starting point for the specific reaction sequence leading to P.

If allowed to proceed at an unrestricted rate, this pathway clearly has the capacity to convert large amounts of A to P, with possible deleterious consequences for the cell due to an excessive accumulation of P or depletion of A. Clearly, the best interests of the cell are served when the pathway is functioning not at its maximum rate, but at a rate that is carefully tuned to the cellular need for P. Somehow, the enzymes of this pathway must be sensitive to the cellular level of the product P in somewhat the same way that a furnace is sensitive to the temperature of the rooms it is intended to heat. The desired regulation is possible in such cases because the product P is a specific inhibitor of one (or more) of the enzymes in the pathway, usually the first one. This phenomenon is called **feedback inhibition** and is illustrated schematically in Figure 6-18. Feedback inhibition is one of the most common mechanisms used by cells for adjusting pathway rates to cellular needs.

Notice from Figure 6-18 a distinctive characteristic of feedback inhibition: The enzyme E_1, which recognizes A as substrate and converts it into B, is regulated by P, which is neither its substrate nor its product. This is possible because such key regulatory enzymes always have at least *two* sites, one to recognize and bind the substrate and the other to recognize and respond to the feedback inhibitor or other regulatory substance. The site at which substrate binds and the actual chemical reaction occurs is, of course, the active site, sometimes also called the **catalytic site.** The regulatory site is always physically distinct from the active site and is called the **effector** or **modulator site.**

The distinctive feature of such enzymes is that the effector site acts as a switch to control catalytic activity at the active site. In our example (Figure 6-18b), enzyme E_1 is most active when the cellular concentration of product P is low and the effector sites of most E_1 molecules are free. At high concentrations of P, the effector sites of most E_1 molecules are occupied by P, which decreases the affinity of the active site for its substrate and therefore reduces or even eliminates enzyme activity.

Generally, we can say that such regulatory enzymes have two alternative configurations that differ in their affinity for the substrate and therefore in their activity. Binding of the regulatory substance at the effector site serves as the means of switching the enzyme reversibly from one form to the other. Such enzymes are called **allosteric enzymes** (from the Greek *allo* meaning "other" or "different" and *steric* referring to shape or form). **Allosteric regulation,** then, involves control of reaction pathways by the reversible interconversion of the two forms of one or more specific enzymes in the pathway in response to a particular effector molecule.

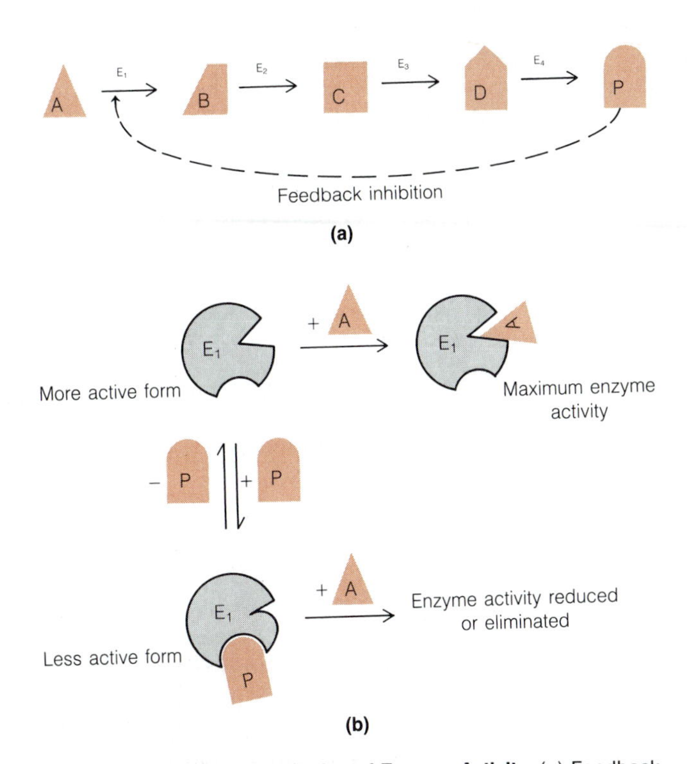

Figure 6-18 Allosteric Regulation of Enzyme Activity. (a) Feedback inhibition of a metabolic pathway usually involves allosteric inhibition of the first enzyme in the sequence (E_1) by the product of the pathway. (b) As an allosteric enzyme, E_1 has both an active site at which substrate A binds and is converted to B and an effector site that recognizes P. P binds reversibly to the effector site, stabilizing the enzyme in the form that has a reduced affinity for A and therefore a lesser activity. P therefore modulates activity of E_1. As the concentration of P increases, more and more molecules of E_1 will be allosterically inhibited, and enzyme activity will be progressively reduced or even eliminated. As the concentration of P decreases again, molecules of P will dissociate from the allosteric site of the enzyme, and E_1 activity will again increase.

Allosteric regulation may be of an "all-or-nothing" type, with one form of the enzyme active and the other form inactive. This is usually the case for end-product inhibition, which is characterized by a cessation of enzyme activity when the product concentration is high. Alternatively, allosteric enzymes can be active in both forms, but with one form clearly more active than the other. In such cases, allosteric regulation is a means of fine-tuning the level of activity by controlling the fractions of the enzyme molecules that are in the more active and the less active forms.

Although introduced here in the context of inhibition, allosteric enzymes may be subject either to **negative regulation** (*inhibition,* usually by the end product of a pathway) or to **positive regulation** (*activation,* usually by the substrate). Negative regulation is characteristic of synthetic pathways, rendering them sensitive to the level of their own end products. Degradative pathways, on the other hand, are often subject to positive regulation, allowing them to be activated as needed to carry out the breakdown of specific substances. In fact, some regulatory enzymes possess one or more specific positive effector sites and one or more specific negative effector sites simultaneously, rendering the enzyme sensitive to control in both directions by a number of cellular substances of relevance to the particular pathway. We will encounter examples of both types of allosteric regulation in coming chapters.

Perspective

We come full circle and return to the theme raised in the introduction to this chapter—namely, that thermodynamics allows us to assess the feasibility of a reaction but says nothing about the likelihood of the reaction actually occurring at a reasonable rate in the cell. To ensure that the activation energy requirement is met, a catalyst is required, which in biological systems is always an enzyme. As proteins, enzymes are chains of amino acids in a genetically programmed sequence that are sensitive to temperature and pH. They are also exquisitely specific, either for a single specific substrate or for a class of compounds. The actual catalytic process takes place at the active site, a critical cluster of amino acids responsible for substrate binding and activation and for the actual chemical reaction. Binding of the appropriate substrate at the active site often induces a more stringent fit between enzyme and substrate that facilitates substrate activation.

An enzyme-catalyzed reaction proceeds via an enzyme-substrate intermediate and follows Michaelis-Menten kinetics, characterized by a hyperbolic relationship between initial reaction velocity and substrate concentration. The upper limit on velocity is called V_{max}, and the substrate concentration needed to reach one-half of the maximum velocity is termed the Michaelis constant, K_m. The hyperbolic relationship between v and [S] can be linearized by a double-reciprocal equation and plot, from which V_{max} and K_m can be determined graphically. The double-reciprocal plot is also useful as a means of analyzing for inhibitor effects. A special kind of inhibition involves sensitivity to the level of the product of a reaction sequence. Such regulation involves allosteric enzymes, with both an active site and an effector, or modulator, site. Allosteric inhibition applies most often to synthetic pathways. Allosteric activation also occurs, usually with degradative reaction sequences.

Enzyme kinetics therefore allows us to determine the rates of cellular reactions, just as thermodynamics allows us to predict direction and to calculate energy yields and needs. Still remaining, however, is the question of route. By way of analogy, it cannot be determined when (or even if) you will arrive in New York if you simply specify that the gasoline in your tank is capable of being combusted exergonically in the cylinders of your car (thermodynamics) and that you intend to drive at a speed of 55 miles per hour (kinetics). A map and a specified route are still absent. In the living cell, routes are referred to as *metabolism*, and maps are *metabolic pathways*, as we will see in Chapter 10.

Key Terms for Self-Testing

enzyme catalysis (p. 132)
enzymes (p. 132)

Activation Energy and the Metastable State
activation energy (p. 133)
metastable (p. 133)
catalyst (p. 134)

Enzymes as Biological Catalysts
active site (p. 135)
prosthetic groups (p. 136)
substrate specificity (p. 136)
group specificity (p. 137)
lock-and-key model (p. 139)
induced-fit model (p. 139)
substrate activation (p. 139)
nucleophilic substitution (p. 141)
electrophilic substitution (p. 141)
carboxypeptidase A (p. 141)

Enzyme Kinetics
enzyme kinetics (p. 144)
saturation (p. 145)
Michaelis-Menten equation (p. 145)
first-order region (p. 147)
zero-order region (p. 147)
maximum velocity (V_{max}) (p. 147)
Michaelis constant (K_m) (p. 148)
Lineweaver-Burk equation (p. 148)
double-reciprocal plot (p. 148)

Enzyme Inhibition and Regulation
inhibition (p. 151)
substrate analogues (p. 151)
irreversible inhibitor (p. 151)
reversible inhibitor (p. 151)
competitive inhibitor (p. 151)
noncompetitive inhibitor (p. 151)
allosteric effector (p. 152)
feedback inhibition (p. 152)
catalytic site (p. 152)
effector (modulator) site (p. 152)
allosteric enzyme (p. 152)
allosteric regulation (p. 152)
negative regulation (p. 153)
positive regulation (p. 153)

Suggested Reading

General References and Reviews

Colowick, S. P., and N. O. Kaplan, eds. *Methods in Enzymology*. New York: Academic Press, 1970 (ongoing series).

Mathews, C. K., and K. E. van Holde, *Biochemistry*. Redwood City, Calif.: Benjamin/Cummings, 1990.

Stryer, L. *Biochemistry*, 3d ed. New York: W. H. Freeman, 1988.

Wood, W. B., J. H. Wilson, R. M. Benbow, and L. E. Hood. *Biochemistry: A Problems Approach*, 2d ed. Menlo Park, Calif.: Benjamin/Cummings, 1981.

Zubay, G. *Biochemistry*. 2d ed. New York: Macmillan, 1988.

Structure and Function of Enzymes

Bernard, S. A. *The Structure and Function of Enzymes*. Menlo Park, Calif.: Benjamin/Cummings, 1968.

Cori, C. F. James B. Sumner and the chemical nature of enzymes. *Trends Biochem. Sci.* 6 (1981): 194.

Dickerson, R. E., and I. Geis. *The Structure and Action of Proteins*. Menlo Park, Calif.: Benjamin/Cummings, 1969.

Ferdinand, W. *The Enzyme Molecule*. New York: Wiley, 1976.

Phillips, D. C. The three-dimensional structure of an enzyme molecule. *Sci. Amer.* 215 (November 1966): 78.

Mechanisms of Enzyme Catalysis

Kirsch, J. F. Mechanism of enzyme action. *Annu. Rev. Biochem.* 42 (1973): 205.

Mildvan, A. S. Mechanism of enzyme action. *Annu. Rev. Biochem.* 43 (1974): 357.

Neurath, H. Proteolytic enzymes, past and present. *Fed. Proc.* 44 (1985): 2907.

Stroud, R. M. A family of protein-cutting proteins. *Sci. Amer.* 231 (July 1974): 74.

Walsh, C. *Enzymatic Reaction Mechanisms*. New York: W. H. Freeman, 1979.

Enzyme Regulation

Changeux, J. P. The control of biochemical reactions. *Sci. Amer.* 212 (April 1965): 36.

Cohen, P. *Control of Enzyme Activity*. New York: Chapman and Hall, 1976.

Hammes, G. C., and C. W. Wu. Regulation of enzyme activity. *Science* 172 (1971): 105.

Koshland, D. E., Jr. Control of enzyme activity and metabolic pathways. *Trends Biochem. Sci.* 9 (1984): 155.

Pardee, A. B. Control of metabolic reactions by feedback inhibition. *Harvey Lect.* 64 (1971): 59.

Problem Set

1. **The Need for Enzymes.** You should now be in a position to appreciate the difference between the thermodynamic feasibility of a reaction and the likelihood that it will actually proceed.
   (a) Many reactions that are thermodynamically possible do not occur at an appreciable rate because of an activation energy requirement. In molecular terms, what does this mean?
   (b) One way to meet this requirement is by an input of heat, which in some cases need only be an initial, transient input. Give an example and explain what this accomplishes in molecular terms.
   (c) An alternative solution is to lower the activation energy. What does it mean in molecular terms to say that a catalyst lowers the activation energy of a reaction?
   (d) Organic chemists often use inorganic catalysts such as nickel, platinum, or cations in their reactions, whereas cells use proteins called enzymes. What advantages can you see to the use of enzymes? Can you think of any disadvantages?

2. **Properties of Enzymes.** For each of the following, indicate with a C if it is a property of enzymes as catalysts, with a P if it is a property of enzymes as proteins, and with an N if it is not a valid property of an enzyme.
   (a) Contains an active site that usually represents less than 5% of the surface area of the molecule.
   (b) Increases the rate of a chemical reaction by reducing the standard free energy change for the reaction.
   (c) Can speed up the rate of exergonic but not endergonic reactions.
   (d) Complexes transiently with reactant molecules.
   (e) Is sensitive to variations in pH of the milieu.
   (f) Displays a high degree of substrate specificity.
   (g) Distorts substrate molecules, thereby weakening bonds and increasing the likelihood of breaking bonds.
   (h) Is usually very heat-stable.

3. **Energy of Activation.** Figure 6-19 is the free energy diagram for two reaction pathways: A → C or A → B → C.
   (a) If a reaction mixture initially contains only a limited amount of pure A, which compound, A, B, or C, will eventually predominate and why? Will there be a significant quantity of either of the other compounds? Why or why not?
   (b) Which compound, A, B, or C, will eventually predominate if the reaction mixture initially contains a limited amount of compound A but this time the appropriate catalyst is present?

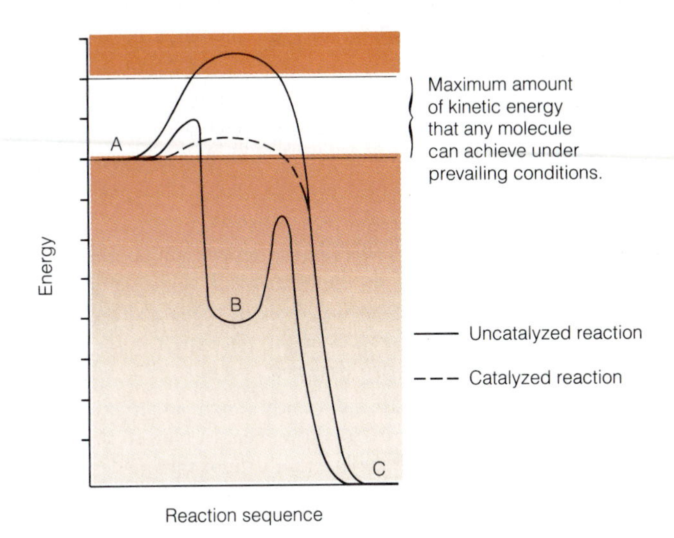

Figure 6-19 Free energy diagram for catalyzed and uncatalyzed reactions.

4. **Temperature and pH Effects.** Figures 6-5 and 6-6 illustrate enzyme activities as functions of temperature and pH, respectively. These curves are typical for most enzymes. In general, enzyme activity is highest at temperatures between 35° and 40°C and at a pH near 7.
   (a) Suggest the biological significance of these characteristic temperature and pH enzyme activity curves.
   (b) Explain the shape of each curve in terms of the major chemical or physical factors that affect enzyme activity.
   (c) Thermophilic bacteria can thrive in hot springs where temperatures can reach 85°C (185°F). How might the enzyme activity versus temperature curves for enzymes from these bacteria differ from Figure 6-5?
   (d) Figure 6-20 is a graph of enzyme activity versus pH for three enzymes: pepsin, cholinesterase, and papain. These are examples of enzymes whose activity curves deviate from Figure 6-6. For each example, suggest the adaptive advantage, to the enzyme or organism, of having the enzyme activity curve shown and explain the activity in terms of enzyme structure.

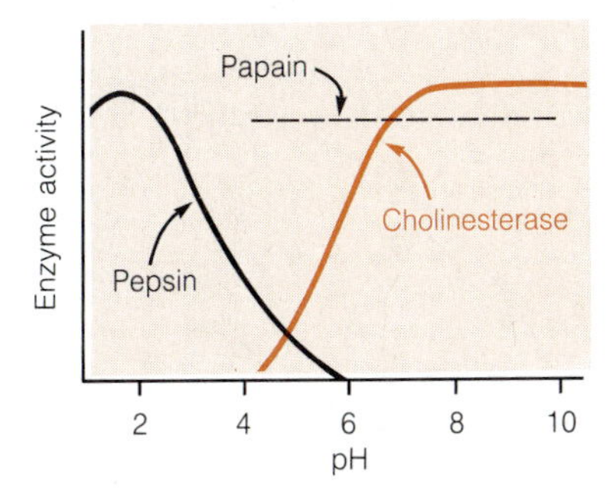

Figure 6-20 pH Dependence of Several Enzymes. Pepsin is a proteolytic enzyme found in the stomach, papain is a protease obtained from papaya fruit, and cholinesterase is an enzyme important in the transmission of nerve signals.

5. **Enzyme Specificity.** All enzymes are highly specific both in the reaction they catalyze and in their choice of substrate. Substrate specificity can be for a specific molecule or a class of molecules.
   (a) A proteolytic enzyme such as trypsin can usually degrade a variety of polypeptide chains, whereas a dehydrogenase is usually absolutely specific for a particular substrate. Explain.
   (b) Subtilisin is a bacterial protease that can cleave any peptide bond, regardless of the specific amino acids involved. Trypsin, on the other hand, splits peptide bonds only on the carboxyl side of lysine and arginine groups. What differences might you expect in the active sites of these two enzymes?
   (c) Compounds that are sufficiently similar in structure to a bona fide substrate to allow them to bind to the active site of an enzyme but that cannot then undergo the reaction catalyzed by that enzyme are usually highly effective competitive inhibitors of enzyme activity. Explain.

6. **Substrate Activation.** For each of the following, explain its role in substrate activation and indicate whether or not it is involved in the carboxypeptidase mechanism (detailed on pp. 141–144 and in Figure 6-10). Cite the example that demonstrates the specific role in the carboxypeptidase mechanism.
   (a) Induced fit
   (b) Electrophilic substitution
   (c) Nucleophilic substitution
   (d) Proton donation

7. **Enzyme Kinetics 1.** Figure 6-21 represents a Michaelis-Menten plot for a typical enzyme. Three regions of the curve are identified by the letters A, B, and C. For each of the statements that follow, indicate with a single letter which one of the three regions of the curve fits the statement best. A given letter can be used more than once.
   (a) The active site of an enzyme molecule is occupied by substrate most of the time.
   (b) The active site of an enzyme molecule is free most of the time.

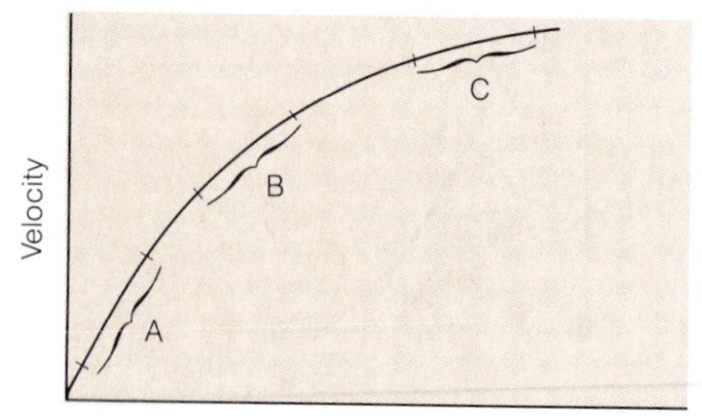

Figure 6-21 A typical Michaelis-Menten plot of reaction velocity vs substrate concentration.

   (c) The range of substrate concentration in which most enzymes usually function in normal cells.
   (d) Includes the point (K_m, $V_{max}/2$).
   (e) Reaction velocity is limited mainly by the number of enzyme molecules present.
   (f) Reaction velocity is limited mainly by the number of substrate molecules present.

8. **Enzyme Kinetics 2.** The enzyme β-galactosidase catalyzes the hydrolysis of the disaccharide lactose into its component monosaccharides:

$$\text{lactose} \xrightarrow[\beta\text{-galactosidase}]{} \text{glucose} + \text{galactose}$$

To determine V_{max} and K_m of β-galactosidase for lactose, the same amount of enzyme (1 microgram per tube) was incubated with a series of lactose concentrations under conditions where product concentrations remained negligible. At each lactose concentration, the initial reaction velocity was determined by assaying for the amount of lactose remaining at the end of the assay. The following data were obtained:

Lactose concentration (mM)	Rate of lactose consumption (μmol/min)
1	10.0
2	16.7
4	25.0
8	33.3
16	40.0
32	44.4

   (a) Why is it necessary to specify that "product concentrations remained negligible" during the course of the reaction?
   (b) Plot v (rate of lactose consumption) versus [S] (lactose concentration). Why is it that when the lactose concentration is doubled, the increase in velocity is always less than twofold?
   (c) On the same graph as part (b), plot the results you would expect if each tube contained only 0.5 μg of enzyme. Explain.
   (d) Calculate $1/v$ and $1/[S]$ for each entry on the data table and plot $1/v$ versus $1/[S]$.
   (e) Determine K_m and V_{max} from your double-reciprocal plot.
   (f) On the same graph as part (d), plot the results you would expect if the assay were conducted in the presence of a competitive inhibitor that increased the apparent K_m of the enzyme by a factor of 2.

9. **Glucose Phosphorylation.** The ATP-driven phosphorylation of glucose (reaction 6-13) can be catalyzed by either hexokinase or glucokinase. Glucokinase is quite specific for glucose, whereas hexokinase also catalyzes the phosphorylation of other sugars. Glucokinase has a much

higher K_m for glucose ($K_m = 10$ mM) than does hexokinase ($K_m = 0.15$ mM; see Figure 6-15). The V_{max} values for the two enzymes in human liver on a per-gram basis are 1.5 μmol/min for glucokinase and 0.1 μmol/min for hexokinase.

(a) Plot both a Michaelis-Menten curve and a Lineweaver-Burk double-reciprocal graph for these data. In each case, plot the data for both enzymes on the same graph.

(b) Calculate the activities of both enzymes at the blood glucose levels (i) after an overnight fast (about 80 milligrams per 100 milliliters) and (ii) after a meal (about 120 mg/100 ml), and express the activity due to each enzyme as a percent of the total activity due to both enzymes.

(c) Which enzyme accounts for most of the glucose phosphorylation in the liver under each condition? Explain why this is true.

(d) Which enzyme responds the most to the increase in blood glucose that occurs after eating a meal? Explain why this is true.

(e) Explain why the deficiency of glucokinase activity that occurs in diabetic persons is so serious.

(f) Can you think of any reason for the presence of hexokinase in the liver, given its modest role in glucose phosphorylation?

10. **Enzyme Kinetics 3.** Figure 6-22 contains the Lineweaver-Burk plots for a normal enzyme and two different mutant forms of that enzyme (mutant 1 and mutant 2).

(a) Indicate whether the K_m and V_{max} values for mutant 1 and mutant 2 are higher than, lower than, or the same as the corresponding values for the normal enzyme.

(b) Which of the two mutants is likely to increase the time required for the catalytic reaction? Explain your reasoning.

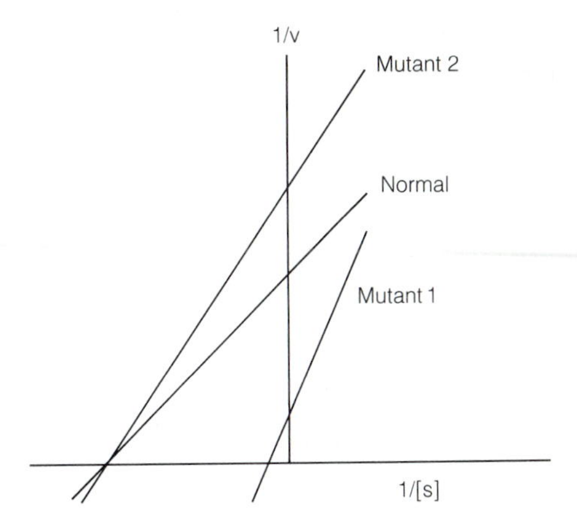

Figure 6-22 The Lineweaver-Burk plots for a normal enzyme and two mutant forms of the enzyme (Mutant 1 and Mutant 2).

(c) With the added information that V_{max}/K_m is a measure of the efficiency of the reaction (that is, V_{max}/K_m is high for an efficient enzyme and low for an inefficient enzyme), determine whether mutant 1 is more or less efficient than the normal form. Explain your reasoning.

11. **Linearizing Michaelis and Menten.** In addition to the Lineweaver-Burk plot, two other straight-line forms of the Michaelis-Menten equation are sometimes used. The *Eadie-Hofstee plot* is a graph of v versus v/[S], and the *Hanes-Woolf plot* graphs [S]/v versus [S].

(a) In both cases, show that the equation being graphed can be derived from the Michaelis-Menten equation by simple arithmetic manipulation.

(b) In both cases, indicate how K_m and V_{max} can be determined from the resulting graph.

(c) Can you suggest why the Hanes-Woolf plot is the most statistically satisfactory of the three?

7

Membranes: Their Structure and Chemistry

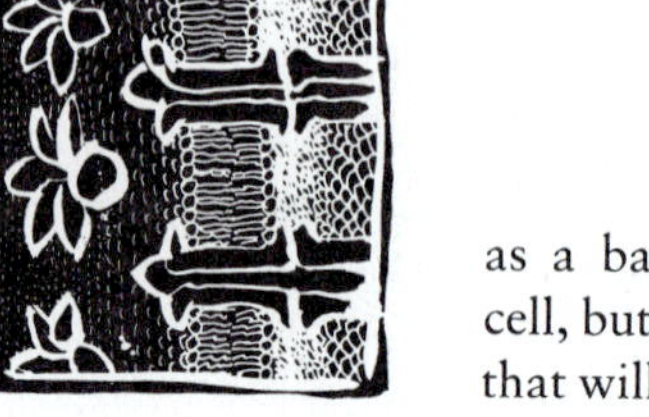

An essential feature of all cells is the presence of membranes that define and contain the cell and its various internal compartments. Even the most casual observer of electron micrographs is likely to be struck by the prominence of membranes around and within cells, especially those of eukaryotic organisms (Figure 7-1). In this chapter we will describe the molecular and supramolecular structure of membranes and outline the multiple roles they play in the life of the cell.

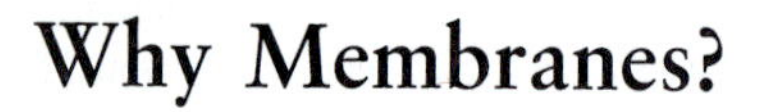

Why Membranes?

We can begin by asking "Why membranes?" The answer to this question turns out to have several parts, because membranes serve several related yet distinct functions. As Figure 7-2 illustrates, the membranes of a cell (1) define and compartmentalize the cell, (2) serve as the locus of specific functions, (3) control movement of substances into and out of the cell and its compartments, and (4) play a role in cell-to-cell communication and detection of external signals.

We will look briefly at each of these roles and then consider in detail the structural and chemical features that make membranes so well suited for them.

Definition and Compartmentation

One of the most obvious roles of membranes is to define the cell and its compartments. *Plasma* was a word originally used to describe the contents of a cell (now called *cytoplasm*), and the outer limiting membrane of cells is still called the **plasma membrane.** This membrane serves as a barrier, forming the boundary of every cell, but also has important transport functions that will be described later. In addition, various **intracellular membranes** serve to compartmentalize function in eukaryotic cells. For example, membrane-bounded compartments of mitochondria and chloroplasts contain the molecules needed for ATP synthesis and maintain the electrochemical proton gradients needed to drive this process. In the absence of such compartments, life as we know it could not exist: Plants could not carry out photosynthesis and animals could not extract energy from the foods they eat.

Locus of Function

Membranes and membranous compartments have specific functions associated with them because the molecules or structures responsible for those functions are either embedded in or localized on membranes. One of the best ways to characterize a specific membrane, in fact, is to describe the specific enzymes, transport proteins, pigments, and other molecules associated with it.

The localization of specific functions to membranes is exemplified by the roles of the inner membrane of the mitochondrion and the thylakoid membranes of the chloroplast in energy transduction, which were mentioned earlier. Membrane associations are also evident for other cellular components such as ribosomes, and a variety of enzymes are known to be localized in or on the membranes of endoplasmic reticulum, lysosomes, and peroxisomes. Such enzymes are often useful during the isolation of these organelles from suspensions of disrupted cells. For example, only lysosomes contain the enzyme *acid phosphatase.* When lysosomes are isolated from other cell components by successive purification steps, acid phosphatase activity is used as a **marker enzyme** for the lysosome distribution among various fractions and to determine the relative purity of the final preparation.

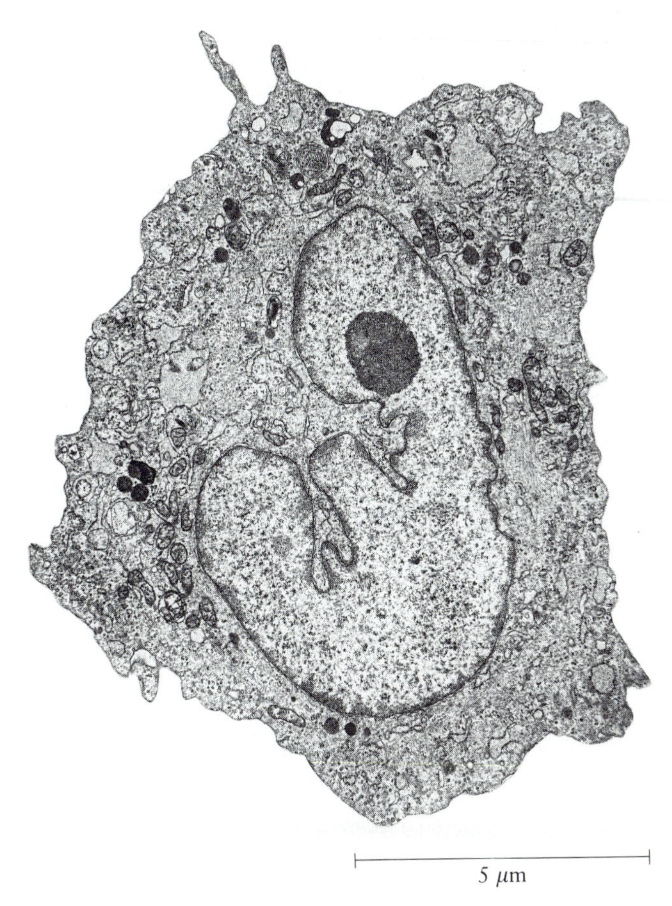

Figure 7-1 Membrane Compartments of the Cell. This micrograph shows the intracellular membrane-bounded compartments of a human white blood cell. The cell is bounded by the plasma membrane, and the large nucleus (with prominent nucleolus) is also enclosed by a membranous structure called the nuclear envelope. Endoplasmic reticulum can be seen throughout the cell, together with intracellular compartments such as mitochondria and lysosomes. (TEM)

Regulation of Transport Functions

In addition to delineating cells and their internal components, membranes also regulate the movement of substances into and out of the cell. Nutrients, ions, gases, water, and other substances are taken up by the cell, and various products and wastes must be removed. These functions are possible because membranes are *differentially permeable* to various substances, allowing some in while excluding others. As we will see in Chapter 8, the modes of entry for various substances differ. Some substances move down concentration gradients, a process called **passive transport.** For instance, molecules such as water, oxygen, and carbon dioxide readily pass through membranes because of their small size, whereas larger molecules such as glucose require special carrier proteins to facilitate their passage across the membrane. In **active transport,** substances are transported against their concentration gradients (or sometimes, against a charge gradient). This transport is carried out by enzyme systems, which typically use ATP as an energy source. Electrically charged ions such as sodium, potassium, and hydrogen are transported by this mechanism. Even molecules as large as proteins can gain entry to cells by being engulfed and incorporated into vesicles, a process called **endocytosis.**

Detection and Recognition of Signals

Cells receive information from their environment, usually in the form of electrical or chemical signals. Such signals are often crucial in changing either the nature or the rate of cellular activities. Frequently, in fact, the information that impinges on a cell initiates whole patterns or cycles of activities in which the cell was not previously engaged, such as cell division or differentiation.

The plasma membrane plays a key role in **signal recognition** and in generating a response. Sometimes, the signal is transmitted into the cell and causes an internal effect. In most cases, however, the molecules that represent the primary signal remain outside the cell, but by binding to specific receptors on the plasma membrane, they generate secondary signals on the inside. **Receptor sites** on the plasma membrane therefore allow cells to recognize and respond to a variety of specific chemical signals. We will learn more about membrane receptors in Chapter 21.

All of the above functions—compartmentation, localization of function, transport, and signal detection—depend on the chemical composition and structural features of membranes. It is to these topics that we now turn.

Membrane Structure: A Historical Perspective

In Chapter 2, the membrane was described as a semifluid "sea" of phospholipids with proteins "floating" in it. This modern concept can best be appreciated if we look first at how our present understanding of membrane structure developed. As we do so, you may also gain some insight into how such developments come about and perhaps a greater respect for the diversity of approaches and techniques that are often important in advancing our understanding of biological phenomena.

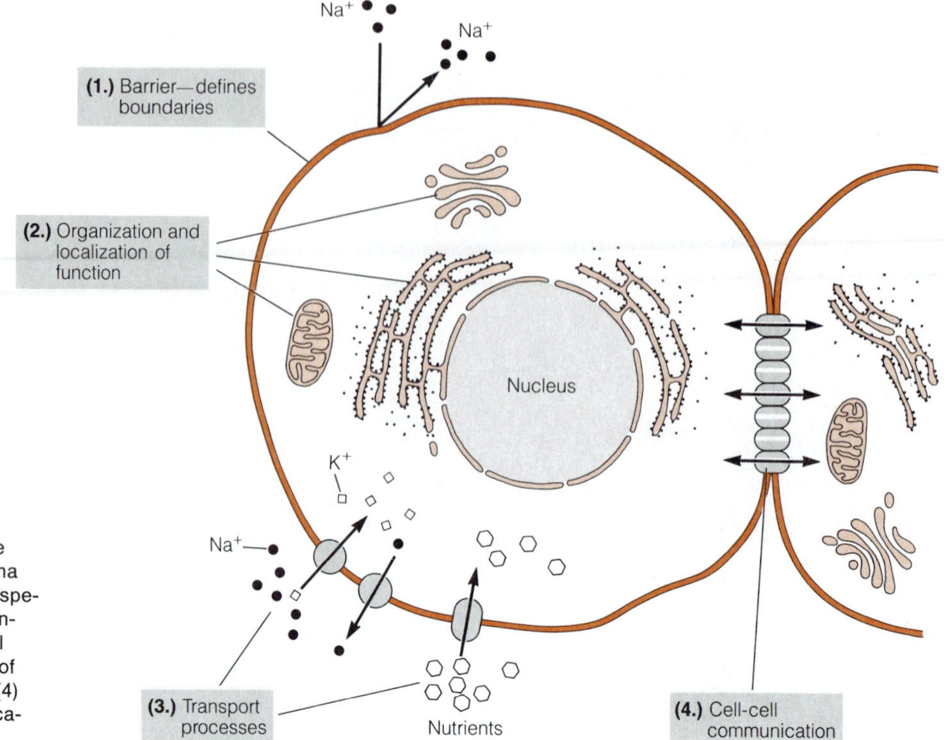

Figure 7-2 The Significance of Membranes. Membranes (1) define the boundaries of the cell (example: plasma membrane), (2) organize and localize specific cell functions (examples: mitochondrial membrane, rough ER), (3) control movement of substances into and out of the cell (example: ion transport), and (4) provide a means of cell-cell communication (example: gap junction).

Overton and Langmuir: The Importance of Lipids

Membranes are very thin structures that normally can be visualized only with an electron microscope. It is therefore not surprising that most of the early studies of membranes were indirect and did not depend on microscopy. Figure 7-3 illustrates the chronology of membrane studies, which began nearly a century ago with the understanding that lipid layers were present and continues today with investigations of membrane proteins, some of which we will encounter in this chapter.

A good starting point for a historical overview is the pioneering work of Charles E. Overton in the 1890s. Overton was aware that cells seemed to be enveloped by some sort of selectively permeable layer that allowed the passage of some substances but not of others. He reasoned that the ability of a substance to cross the membrane might be related in some way to its chemical affinity for the membrane. This, in turn, suggested that he might be able to learn something about the membrane by determining the nature of compounds that could move across it easily.

Working with cells of plant root hairs, Overton found that lipid-soluble substances penetrated readily into cells, whereas water-soluble substances in general did not. In fact, he found a good correlation between the *lipophilic* (lipid-loving) nature of a substance and the ease with which it could penetrate into the cell. From his studies, Overton concluded that the solubility properties of the cell membrane approximated those of a fatty oil, and he even suggested that such membranes were probably mixtures of *cholesterol* and *lecithins,* a suggestion that later proved to be remarkably foresighted.

The next important advance came about a decade later, through the work of Irving Langmuir, who studied the behavior of lipids such as long-chain fatty acids by spreading them out as a thin layer on a water surface in what has come to be called a **Langmuir trough** (Figure 7-4). The lipids were first dissolved in benzene, and samples of the benzene-lipid solution were then placed on the surface of the water. As the benzene evaporated, the molecules were left as a **lipid monolayer** on the water surface, with their hydrophilic heads in the water and their hydrophobic tails in the air. This lipid monolayer became the basis for further thought about membrane structure in the early years of the twentieth century.

Gorter and Grendel: The Lipid Bilayer

The next major step came in 1925 when two Dutch physiologists, E. Gorter and F. Grendel, read Langmuir's papers and thought that monolayers might help answer a question

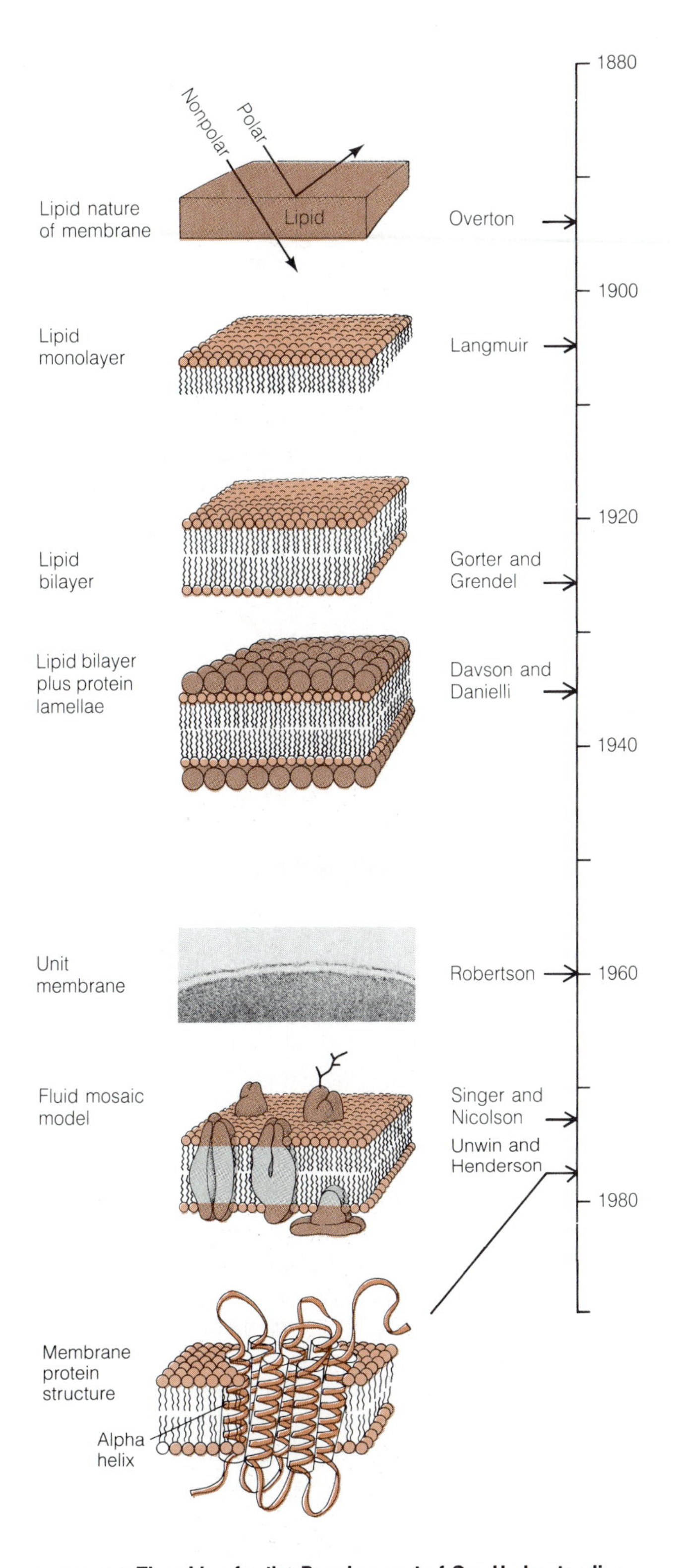

Figure 7-3 Time Line for the Development of Our Understanding of Membrane Chemistry and Structure. The final diagram in the series is a schematic illustration of a plasma membrane with a molecule of the protein bacteriorhodopsin embedded in it.

regarding the surface coating of red blood cells. Overton's earlier research had shown that lipid was present, most likely as part of the surface. How many lipid layers were in the coat? Gorter and Grendel could see in their microscope that the red cell resembled a flattened sphere about 7 μm in diameter. Therefore the surface area of a known volume of red cells could be estimated by counting the cells and multiplying by 100 μm^2, which was their best guess at the area of a human red cell calculated from its apparent diameter. They decided to extract the lipid, spread it as a monolayer by using Langmuir's methods, and compare the area with the combined area of the red cells. If lipid is present as a monolayer on the red cell surface, it would produce a monolayer of approximately the same total area on the water surface. A bilayer would produce twice the area, and so on.

When the experiment was done, they found that the area of the monolayer of lipid on the water was about twice the estimated total surface area of the red cells. Gorter and Grendel concluded, therefore, that the surface coating of red cells contained a bimolecular layer of lipid, a **lipid bilayer.**

As it turned out, Gorter and Grendel's conclusion was correct, but their experiment had two errors. To extract the lipids they used acetone, which was later found to extract only two-thirds of the lipid present. However, their measurement of erythrocyte area was taken from preparations of cells that were dried on microscope slides. Red cells shrink during drying, and Gorter and Grendel's estimate of surface area was less than the correct value (145 μm^2 for human erythrocytes), so the two errors canceled out. Despite these shortcomings, their experiment was momentous: it represented the first attempt to understand cell membranes at the molecular level and became the basic underlying assumption for each successive refinement in our understanding of membrane structure.

Davson and Danielli: The Importance of Proteins

It soon became clear that the lipid bilayer, though an essential feature of membrane structure, could not explain all the properties of membranes, particularly those related to permeability and electrical resistance. For instance, the **differential permeability** of membranes to solute molecules can depend on quite subtle chemical differences: glucose gets into cells readily, but galactose, a sugar molecule resembling glucose in size and shape, does not. It is not easy to imagine how a lipid bilayer could make such a distinction. However, it is known that enzymes demonstrate this kind of specificity, which suggested that membrane components with enzyme-like binding sites are

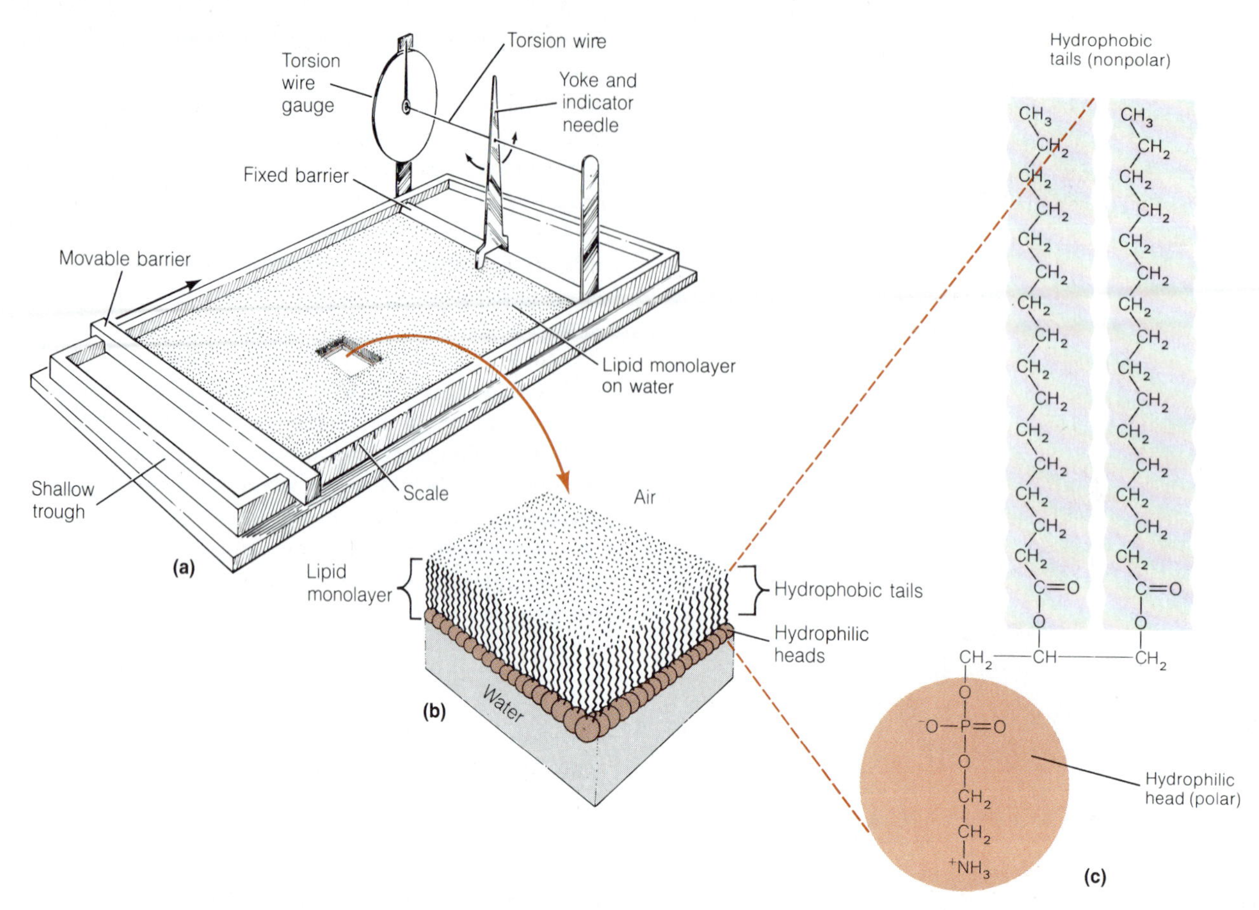

Figure 7-4 The Langmuir Trough. (a) The trough developed by Langmuir consists of a tray equipped with a movable barrier that can be used to push a phospholipid monolayer up against a fixed barrier, the latter equipped with a torsion wire gauge to allow the monolayer to be compressed to a uniform extent each time. (b) The orientation of the phospholipids in the monolayer is such that the hydrophilic heads protrude downward into the water and the hydrophobic tails stick upward into the air. (c) A structural representation of the individual phospholipid molecule, indicating the nonpolar tails (gray) and the polar head (color).

present in the membrane to facilitate glucose transport. Similarly, the permeability of biological membranes to ions is much greater than that of lipid bilayers: the leakage rate of potassium ions across bilayer membranes is measured in days, whereas the same amount of leakage across red cell membranes occurs in an hour or so.

To explain these additional features, Hugh Davson and James Danielli invoked the presence of proteins in membranes and proposed in 1935 that cell membranes consist of lipid bilayers coated on both sides with proteins, as shown in Figure 7-5a. As you can see, the original **Davson-Danielli model** was in essence a protein-lipid "sandwich." The model was modified somewhat in the years that followed. Particularly notable was the suggestion made in 1954 (and shown as Figure 7-5b) that hydrophilic proteins might penetrate the membrane in places to provide a polar pore through what was otherwise a very hydrophobic bilayer. The protein could then account for the permeability and resistivity characteristics that were not easily explained in terms of the lipid bilayer alone. Thus, the lipid interior accounted for hydrophobic properties of membranes, and the protein components explained hydrophilic properties. The real significance of the model, however, was its recognition of the importance of proteins in membrane structure. This feature more than any other made the Davson-Danielli "sandwich" the basis for much subsequent research on membrane structure.

Robertson: The Unit Membrane

All of the membrane structures discussed so far were developed as models of the plasma membrane of cells, which was the only structure large enough to be investigated by

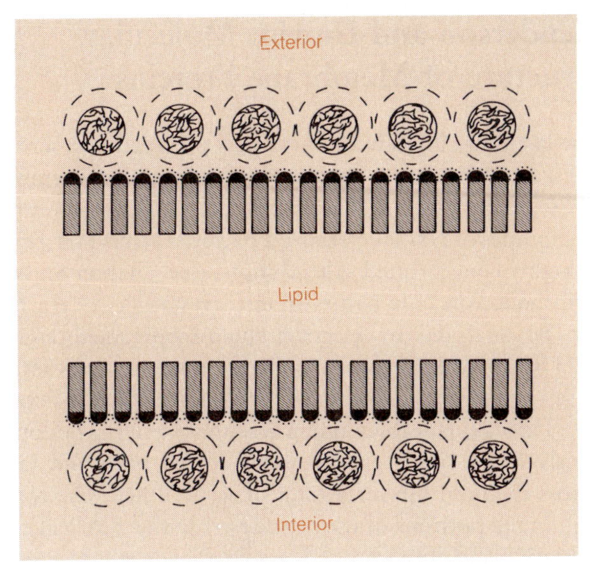

(a) The 1935 model

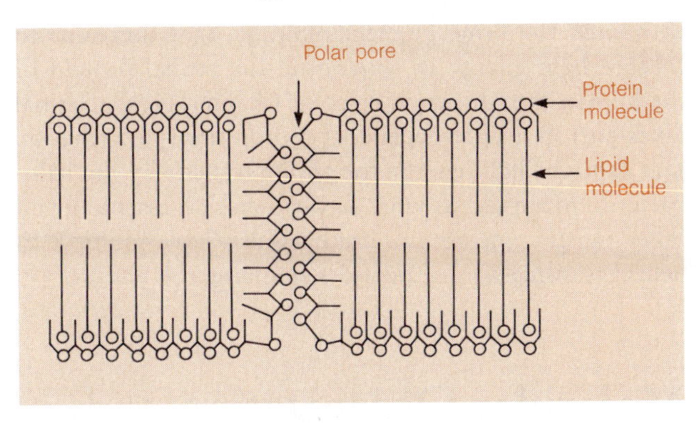

(b) The 1954 model

Figure 7-5 The Davson-Danielli Model of Membrane Structure. (a) The original model, as proposed in 1935, with a continuous layer of proteins lining both sides of the phospholipid bilayer. (b) A later version, proposed in 1954, to accommodate growing evidence of the need for some sort of hydrophilic "passageway," or pore, through the otherwise hydrophobic bilayer.

light microscopy. As the resolving power of electron microscopy improved in the 1950s, the first clear electron micrographs of sectioned cells were obtained, and it was found that many subcellular organelles had membranes as well (Figure 7-6). Furthermore, close examination at high magnification showed extensive regions of trilaminar "railroad track" structure appearing as two dark lines, each about 2 nm thick, separated by a space of 3.5 nm, for an overall thickness of about 7.5 nm (Figure 7-6, inset).

These considerations led to the first attempt to suggest a general structure for all membranes, the **unit membrane** hypothesis of J. David Robertson. In this model, membranous components of the cell were considered to be composed of a continuous lipid bilayer with functional proteins adhering to the surface. The unit membrane structure agreed remarkably well with the Davson-Danielli model. The two dark lines were equated with the outer protein layers, which appeared electron-dense because of their affinity for heavy metal stains. The space in between was thought to correspond to the hydrophobic part of the lipid molecules, which presumably did not take up stain. This latter assumption was supported by experiments that showed essentially the same staining pattern for membranes with and without prior extraction of the lipid. However, it was later found that artificial lipid bilayers also showed the characteristic "railroad track" pattern when stained with osmium, even though such bilayers contain no protein. Osmium appears to react at several sites in membranes—with unsaturated lipids in the bilayer as well as with proteins on the membrane surface. This finding will take on special significance because continuous outer layers of protein are not part of the contemporary model of membrane structure, as we are about to see.

Singer and Nicolson: The Fluid Mosaic Model

An attractive feature of the unit membrane model was its apparent universality: membranes from widely different types of cells and organelles showed the same characteristic structure. Yet, we must also recognize the distinctiveness of different kinds of membranes. Depending on their source, membranes vary considerably in thickness, in lipid composition, and especially in the ratio of protein to lipid. The protein/lipid ratio can be as low as 0.25 for the myelin sheath that serves as a membranous electrical insulation around nerve axons (see Chapter 20) or as high as 4 or more in some bacterial cells. Even the two membranes of the mitochondrion differ significantly: the protein/lipid ratio is about 1.2 for the outer membrane but about 3.6 for the inner membrane, which contains all the enzymes and proteins related to ATP synthesis and electron transport functions.

It became increasingly difficult to rationalize such variations in protein content of membranes with the unit membrane model, because the width of the "rails" simply did not vary correspondingly. Furthermore, as membrane proteins were isolated and studied, it became apparent that most of them were globular proteins with sizes and shapes that did not seem to fit with the concept of thin layers or sheets on the two surfaces of the membrane. These findings, along with others on membrane fluidity, gave rise to considerable rethinking of membrane structure and eventually led to the **fluid mosaic model** that now enjoys widespread support and acceptance.

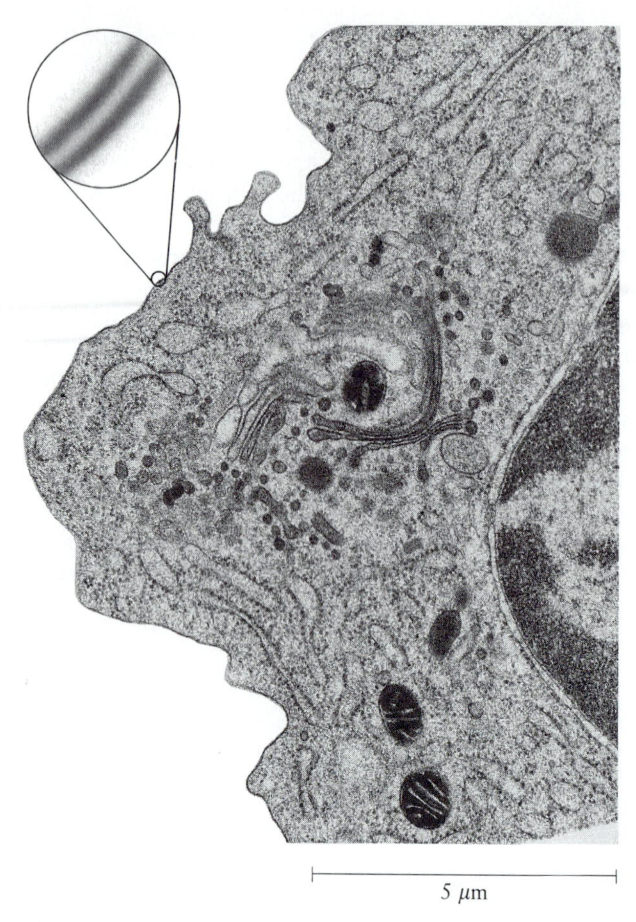

Figure 7-6 Ultrastructure of Cell Membranes. An electron micrograph of fixed and stained cells reveals ultrastructural features composed of membranes, cytoskeleton, and nucleoprotein complexes (TEM). At higher magnifications, most of the membranes have a trilaminar appearance (inset).

The fluid mosaic model was first proposed in 1972 by S. Jonathan Singer and Garth Nicolson. The model has two key features, both implied in its name. The basic lipid bilayer structure of earlier models is retained, but membrane proteins are not thought of as continuous sheets on the membrane surface. Rather, they are considered discrete globular entities that penetrate into the lipid bilayer to an extent dictated by the affinity of a particular protein for a hydrophobic environment, as shown in Figure 7-7. The membrane is therefore envisioned as a *mosaic* of proteins discontinuously embedded in a phospholipid bilayer. Also important is the *fluid* nature that the model attributes to the membrane. Rather than being rigidly anchored or embedded in the lipid bilayer, membrane proteins are considered to float in a sea of fluid lipid. Some proteins are therefore able to move around within the membrane, facilitating their lateral migration within the membrane, whereas others are anchored to cytoskeletal structures underlying the membrane, thereby limiting their mobility.

Henderson and Unwin: Molecular Structure of Membrane Proteins

The last illustration in the historical perspective described in Figure 7-3 shows a *bacteriorhodopsin* molecule embedded in a lipid bilayer. This protein is present in the plasma membranes of *Halobacterium,* a bacterial genus that grows in highly concentrated salt solutions (see Chapter 8). Bacteriorhodopsin falls naturally into crystalline arrays that can be analyzed by electron microscopic methods. In 1975 Richard Henderson and Nigel Unwin took advantage of this property to establish its three-dimensional structure.

Their remarkable finding was that bacteriorhodopsin consists of a single peptide chain folded back and forth across the lipid bilayer a total of seven times. The transmembrane portions of the peptide are in the α-helical configuration and are composed largely of hydrophobic amino acids. Buried in the protein structure is a single molecule of *retinal,* the same pigment molecule that functions to capture light energy in *rhodopsin,* the retinal protein of the human eye. As we will see in Chapter 11, the retinal takes part in a light-dependent proton transport reaction that captures light energy for photosynthesis by the halobacteria. More recent work has shown that essentially all other integral membrane proteins also have multiple α-helical transmembrane peptides spanning the bilayer.

Molecular Organization and Membrane Function

Since its formulation in 1972, the fluid mosaic model has received substantial experimental confirmation and is now the most widely accepted view of membrane structure. The essential features of membrane structure and function include the relationship of the proteins to the phospholipid bilayer, the mobility of proteins within the membrane, the asymmetric distribution of membrane components, and the fluidity of the membrane lipids. We will look at each of these features in turn, focusing both on the supporting evidence and on the implications for membrane function.

Structure of the Red Blood Cell Membrane

Most of our knowledge of membrane proteins comes from indirect methods of analysis. The red blood cell membrane has been one of the most widely studied because red blood cells (also referred to simply as red cells) are readily available, either as outdated blood from blood banks or freshly

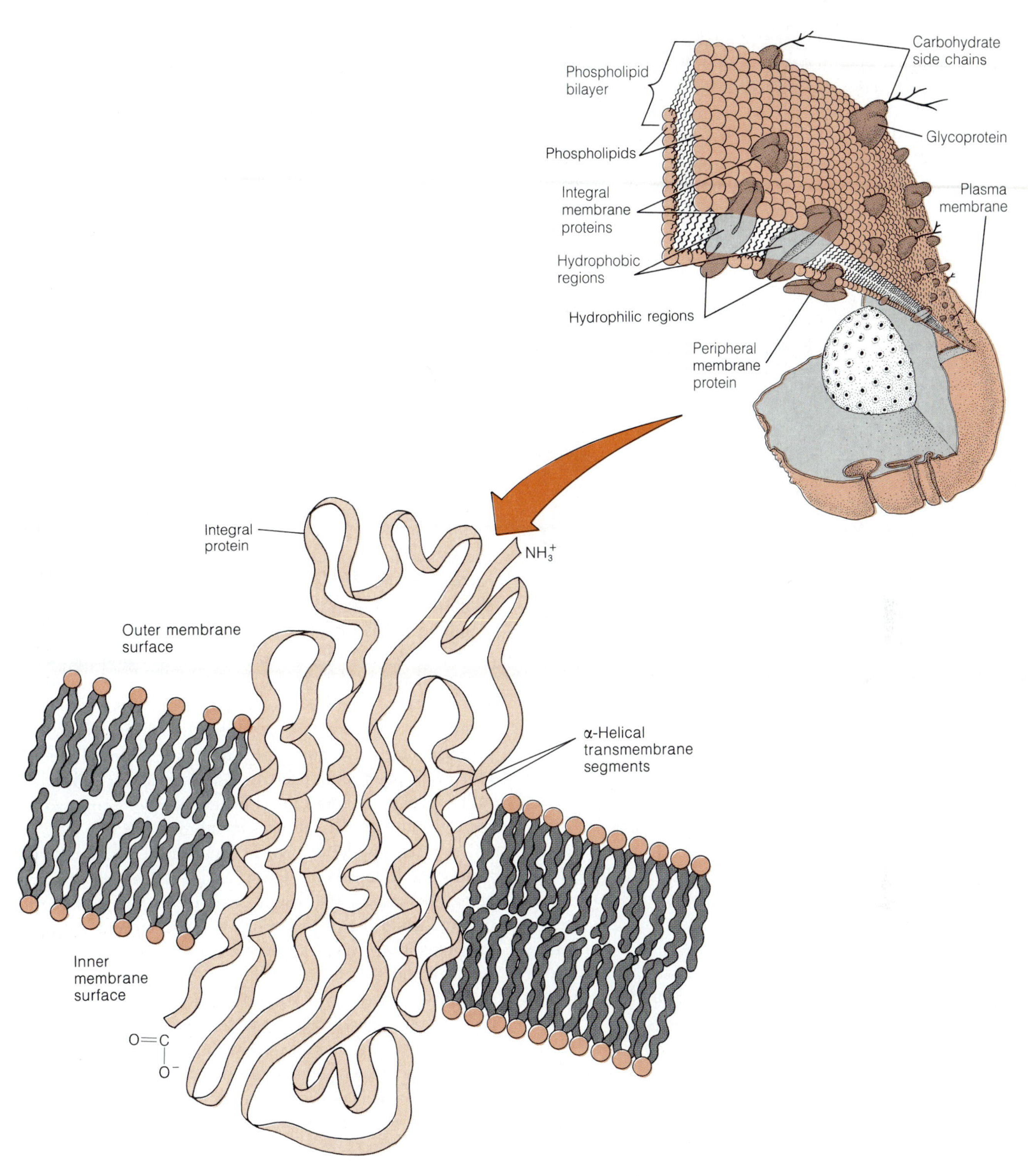

Figure 7-7 Fluid Mosaic Model of Membrane Structure. All integral proteins penetrate into (or even through) the membrane because of their affinity for the hydrophobic interior of the lipid bilayer. However, such proteins always have at least one hydrophilic region (shown in color), which protrudes into the aqueous phase at the membrane surface. Peripheral proteins are more hydrophilic and are bound to the surface.

taken from human volunteers. Equally important, a red cell is bounded by a plasma membrane, with no interior structures, so very pure membrane preparations can be obtained. For these reasons, the following discussion will focus primarily on red cell membrane structure, but the information can be generalized to most other membranes.

The fluid mosaic model recognizes two categories of membrane proteins that differ in their affinity for the hydrophobic interior of the lipid bilayer and hence in the degree to which they interact with the lipid bilayer. **Integral** (or **intrinsic**) **membrane proteins** are, as the name suggests, localized within the interior of the membrane. They typically have long sequences of hydrophobic amino acids that traverse the lipid bilayer, anchoring them to the membrane. One of the simplest examples is **glycophorin,** an integral protein of the red cell membrane, which has a single transmembrane hydrophobic sequence (Figure 7-8). However, as noted earlier, most integral proteins contain multiple transmembrane sequences, as illustrated by the examples shown in Figure 7-9.

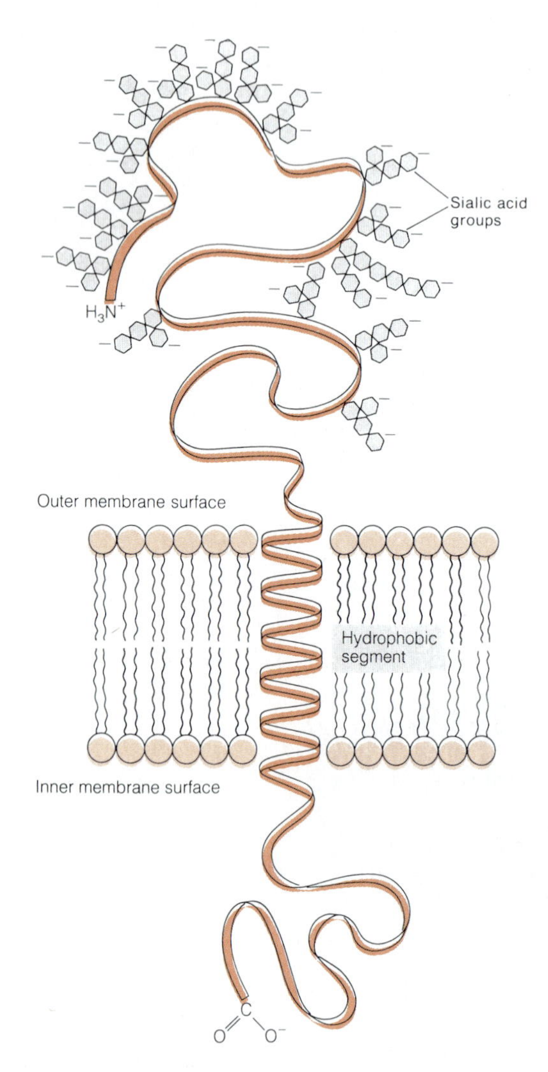

Figure 7-8 Structure of an Integral Protein. Glycophorin is a protein in red cell membranes that contains large numbers of negatively charged sialic acid groups. These anionic groups are exposed at the surface of the cell and apparently have the function of repelling other cells, thereby reducing blood viscosity. (This outer segment of the molecule also contains neutral sugar groups.) The middle segment of the glycophorin peptide chain is composed entirely of hydrophobic amino acids. This segment spans the membrane and anchors the glycophorin. Most other integral proteins have similar transmembrane segments—often six or seven in larger polypeptide chains.

Because of their hydrophobic nature, integral proteins are often difficult to isolate and study by standard protein purification techniques designed for relatively water-soluble proteins. Usually, treatment with detergents is required to solubilize integral membrane proteins. Regardless of how hydrophobic they are, however, all integral membrane proteins have one or more hydrophilic regions that protrude from the membrane into the aqueous phase.

Unlike the integral proteins, **peripheral** (or **extrinsic**) **membrane proteins** lack transmembrane segments and are located on the membrane surface (Figure 7-7). They associate with the membrane through weak electrostatic forces, binding either to the hydrophilic portions of integral proteins that protrude from the membrane or to the hydrophilic heads of membrane lipids. Peripheral proteins are much more readily removed from the membrane than integral proteins and can usually be solubilized by aqueous salt solutions, without resorting to detergents. Peripheral membrane proteins most resemble the sheets of surface proteins originally postulated by Davson and Danielli. However, peripheral proteins are individual molecules, rather than the continuum of protein strands proposed by the Davson-Danielli and unit membrane models.

A recent model of protein architecture in the red cell membrane is shown in Figure 7-10. The integral proteins (glycophorin and the anion channel) extend across the bilayer, while peripheral proteins (spectrin and ankyrin) interact with the integral proteins on the inner surface and stabilize the membrane. You can understand why this structure is necessary if you think for a moment about red cell function. A red cell exists for about four months in the bloodstream, making several hundred thousand trips between the heart, lungs, and body tissues and delivering a trillion oxygen molecules every minute across a membrane only two lipid molecules thick. It should now be clear why some sort of stabilizing architecture must be present.

Analysis of Membrane Components

One of the most important general tools in cell biology is a technique called **chromatography.** This technique takes

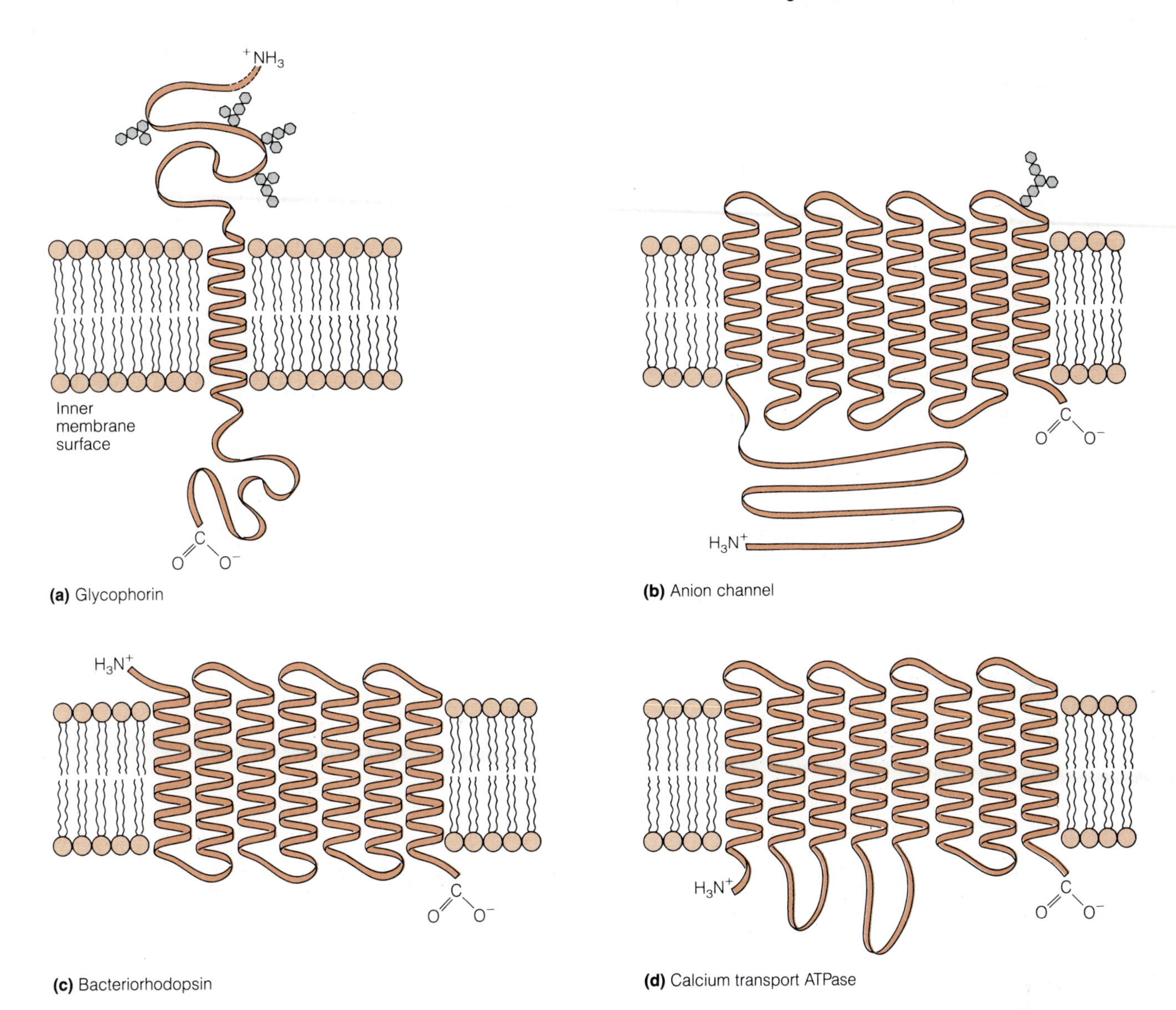

Figure 7-9 Multiple Transmembrane Sequences in Integral Proteins. Most integral proteins have multiple transmembrane sequences, often in the form of α helices. Integral proteins shown here in simplified form include (a) glycophorin, (b) the anion channel of the red blood cell, (c) bacteriorhodopsin, and (d) the calcium transport ATPase of muscle cell membranes. The cytoplasmic side of the membrane is the lower side in each case, and the amino (NH_3^+) and carboxyl (COO^-) ends of the polypeptide chains are indicated.

its name from Greek words for "color" and "writing" and was originally developed for separating plant pigments. To make a simple chromatogram, a mixture to be separated is placed at one end of a stationary phase, such as a sheet of filter paper or, more commonly, a column or thin layer of some powdered mineral like silicic acid (Figure 7-11). A fluid mobile phase is then allowed to move through the stationary phase. Because components of the mixture have different affinities for the stationary phase, they tend to separate into relatively pure fractions.

Analysis of Membrane Lipids. **Thin-layer chromatography (TLC)** can be used to separate lipids extracted from red cell membranes (Figure 7-11). Notice that cholesterol, the least polar lipid, travels near the front of the mobile phase, while phospholipids move more slowly. This is because the more polar head groups of the phospholipids interact strongly with the silicic acid, which slows their movement up the plate. When the separated lipids are eluted (dissolved in chloroform) and analyzed, it is found that approximately 60% of the lipid (molar ratio) in a

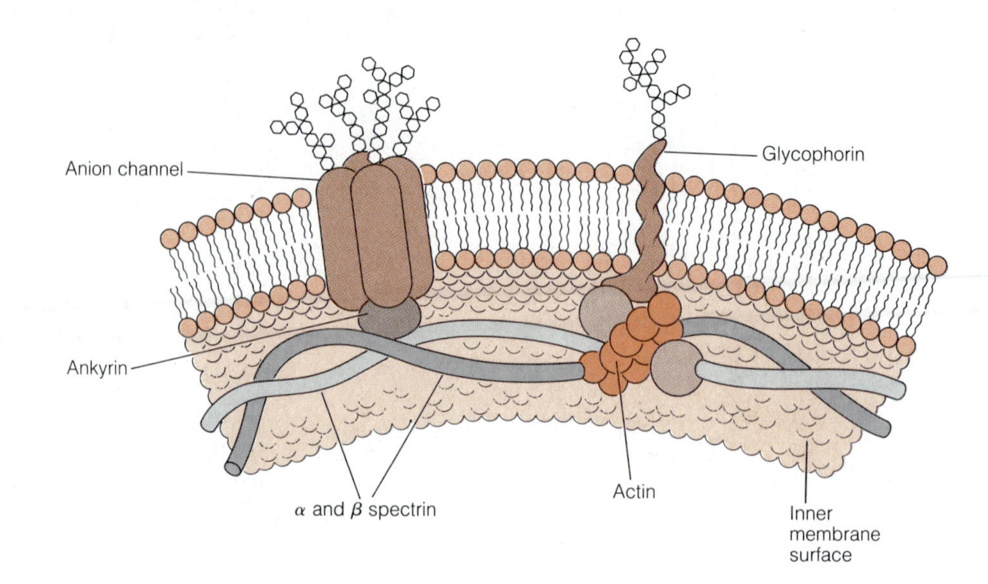

Figure 7-10 Structure of the Red Cell Membrane. The fluid mosaic structure of the red cell membrane is known in considerable detail. Note that some of the integral proteins are held in place by peripheral proteins called ankyrin (from the Greek word for anchor) and spectrin. (The latter took its name from the fact that red cell membranes are often called "ghosts" because they are nearly invisible in the light microscope.) The major integral proteins are glycophorin (see Figure 7-8) and the anion channel, which allows the exchange of chloride and bicarbonate anions across the membrane.

Figure 7-11 Chromatographic Analysis of Lipids. Thin-layer chromatography (TLC) can be used to analyze membrane lipids. (a) The extracted lipid mixture is first evaporated in a small area (the origin) of a glass or metal plate coated with silicic acid, which is a white powder similar to glass (SiO_2). The plate is then placed in a solvent system, in this case a mixture of chloroform, methanol, and water. (b) As the solvent moves up the plate by capillary action, the lipids are separated according to their polarity: nonpolar lipids like cholesterol do not adhere strongly to the silicic acid and move farther up the plate, while more polar lipids remain closer to the origin. Red cell lipids are shown in the diagram. Cholesterol represents nearly half the lipid (molar ratio). The rest is phospholipid, composed of 55% phosphatidylcholine (PC), 30% phosphatidylethanolamine (PE), and 10% phosphatidylserine (PS).

typical plasma membrane is phospholipid, the rest being cholesterol. Several kinds of phospholipid are present, but the precise function of the different head groups is still unclear.

Analysis of Membrane Proteins. Because they are bound so tightly to membranes, integral proteins are considerably more difficult to analyze than lipids. A useful technique for studying membrane proteins is **SDS-polyacrylamide gel electrophoresis,** shown in Figure 7-12. Membrane fragments (Figure 7-12a) are first solubilized with the anionic detergent *sodium dodecyl sulfate* (*SDS*), which disrupts most protein-protein and protein-lipid associations (Figure 7-12b). The solubilized proteins are then applied to the top of a gel of cross-linked polymers of acrylamide (Figure 7-12c), and a potential difference is applied across the gel. Because the proteins are surrounded by a "shell" of negatively charged SDS molecules, they migrate toward the anode at the far end of the gel, usually at a rate that is inversely related to their size (Figure 7-12d, e, f). After a few hours, the process is terminated, and the gel is stained with a dye (usually Coomassie Blue) that binds to proteins and makes them visible (Figure 7-12g).

Freeze-Fracture Analysis. An important means of visualizing and investigating the integral membrane proteins is the **freeze-fracture** technique discussed in detail in Appendix A. The preferred fracture plane in a frozen cell is down the middle of the membrane, so the membrane actually splits between the two layers of the lipid bilayer. Whenever the frozen membrane is fractured in this way, the integral membrane proteins tend to stay with one or the other of the two lipid layers, generating a hollowed-out cavity in the other layer. Usually, integral membrane proteins stay with the inner monolayer. Figure 7-13 illustrates such a freeze-fracture analysis; the particles protruding from one or the other of the two membrane faces are integral proteins that penetrate into, or even through, the membrane.

Classes and Functions of Membrane Lipids

Table 7-1 summarizes some important examples of membrane lipids. Phospholipids are primary components of all

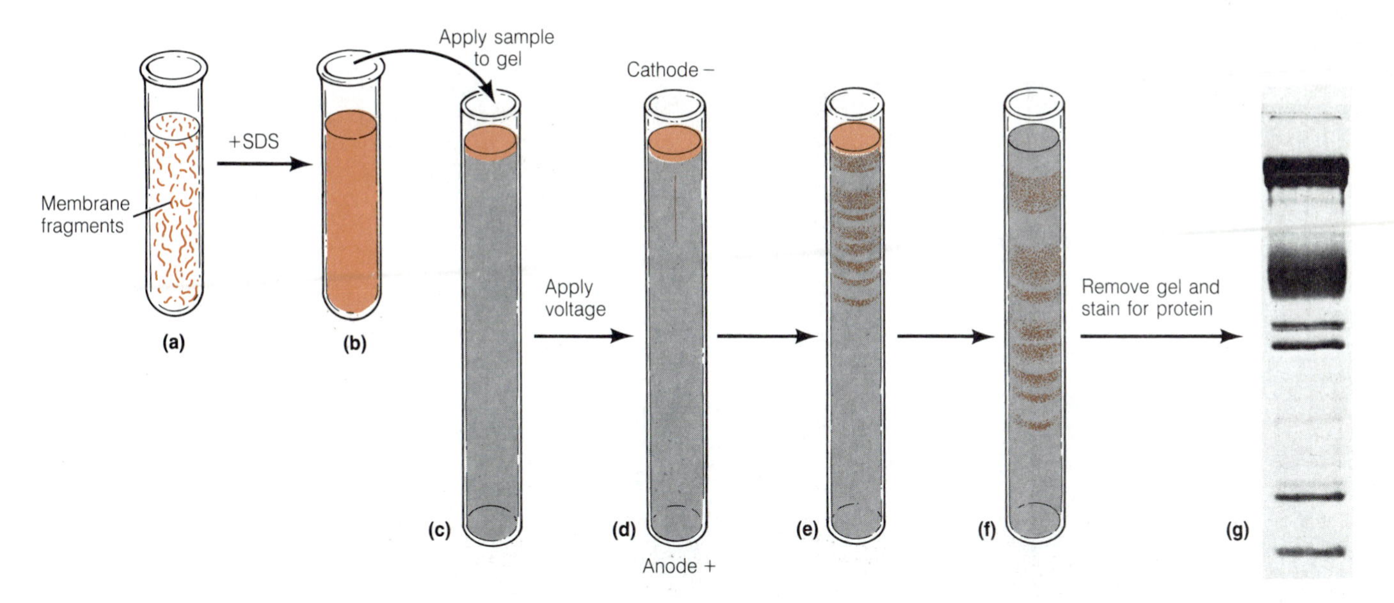

Figure 7-12 SDS-Polyacrylamide Gel Electrophoresis of Membrane Proteins. (a, b) To a suspension of membrane fragments is added a small amount of sodium dodecyl sulfate (SDS), a detergent that dissolves membranes and solubilizes their proteins. (c) A small sample of the proteins is applied to the top of a gel of polymerized and cross-linked acrylamide in a glass tube. (d) A potential is then applied across the gel, causing the detergent-coated (and hence negatively charged) protein molecules to migrate toward the anode at the far end of the gel. (e, f) The rate of migration is inversely related to size, so the smaller proteins migrate more rapidly down the gel than do the larger ones. (g) After a predetermined amount of time, the gel is removed from the glass tube and stained with a dye (usually Coomassie Blue) that binds to proteins and makes them visible. The particular profile shown in (g) is for the membrane proteins of a red blood cell. Each band corresponds to a specific protein, with the larger proteins near the top of the gel and the smaller proteins near the bottom.

Table 7-1 Summary of Typical Membrane Lipids

Lipid Class	Source	Function
Major phospholipids		
Phosphatidylcholine	Present in most membranes	Phospholipids form bilayers that provide barriers to diffusion of polar solutes
Phosphatidylethanolamine	Present in most membranes	
Phosphatidylserine	Present in most membranes	
Minor phospholipids		
Cardiolipin	Mitochondrial inner membrane	Activates cytochromes
Phosphatidylinositol	Present in most membranes	Source of inositol trisphosphate
Sphingolipids	Most mammalian cell membranes, particularly those of nervous tissue	Barrier function; activates certain enzymes
Glycolipids	Major lipid in thylakoid membranes of chloroplasts	Barrier function
Cholesterol	Most membranes, except those of prokaryotes	Reduces bilayer permeability; modulates membrane fluidity

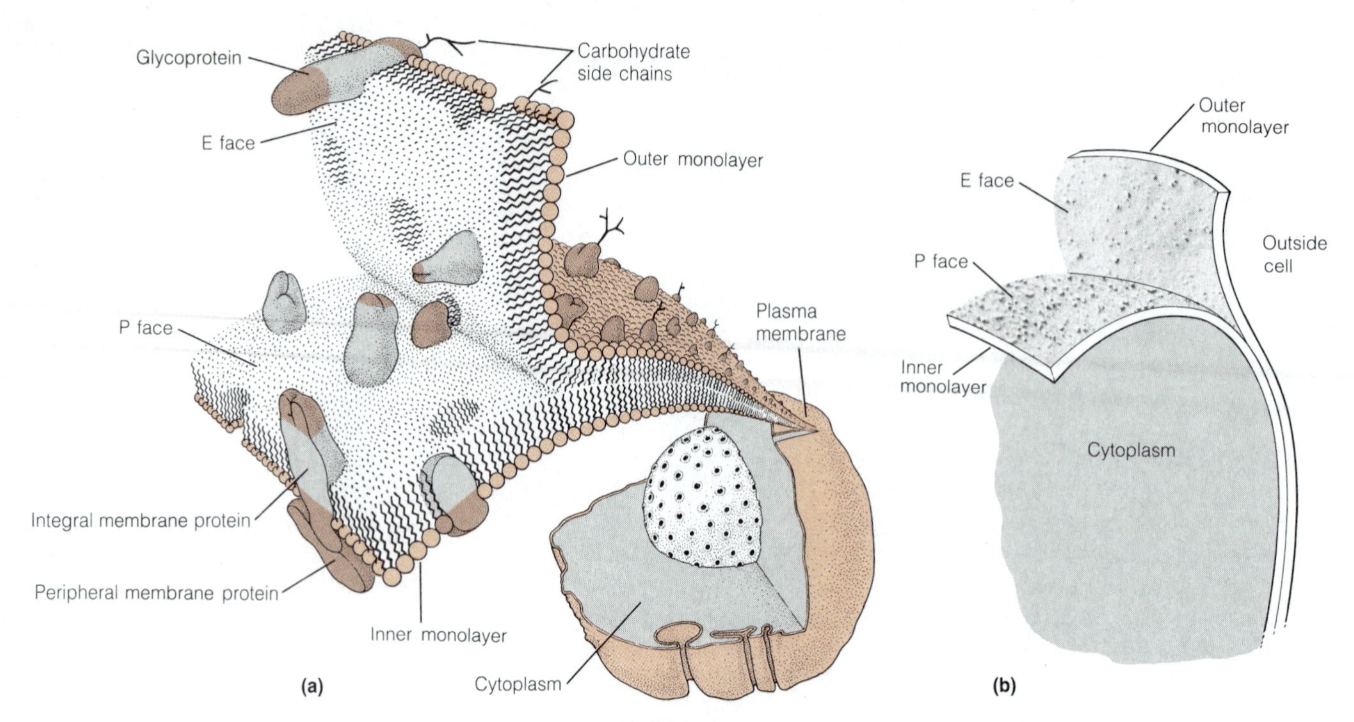

Figure 7-13 Freeze-Fracturing a Membrane. (a) Sketch of the faces of a freeze-fractured membrane, with hydrophobic regions shown in gray and hydrophilic regions in color. Proteins associated with the outer layer are seen on fracture face E (for exterior), whereas those associated with the inner layer will appear on fracture face P (for protoplasmic). (b) Sketch of a freeze-fractured membrane, with electron micrographs of the E- and P-fracture faces from the plasma membrane of a mouse kidney tubule cell superimposed on the art (TEM). The P-face of the membrane is studded with intramembranous particles, whereas the E-face is relatively smooth.

membranes and are classified according to their alcohol groups: choline, ethanolamine, serine, and inositol are common examples. (Recall Figure 3-24.) The hydrophobic portion of the molecule consists of two fatty acids linked through ester bonds to glycerol phosphate, but occasionally an ether linkage is used as well. Sphingolipids are based on the sphingosine molecule, rather than glycerol esters of fatty acids. Chloroplast membranes contain another class of lipids called glycolipids, which have galactose and glucose as head groups, rather than phosphate. Finally, cholesterol is present in most membranes, the only major exceptions being the inner mitochondrial membrane and most bacterial membranes.

A few generalizations can now be drawn about specific functions of this remarkable variety of lipids. Lipid head groups are least well understood in this regard. Inositol is known to be involved in intracellular signaling (Chapter 21), while choline activates at least one enzyme in the mitochondrial membrane.

We know more about the functions of the hydrophobic portions of lipid molecules. As discussed earlier, the primary function of the fatty acid chains of phospholipids is to form a bilayer that acts as a barrier to the diffusion of polar solutes. For this purpose, the chains must be sufficiently long, in the range of 16 to 18 carbons. Chains with fewer than 12 carbons are unable to form a stable bilayer.

A second function of the chains is to regulate membrane fluidity. This is done by introducing unsaturated bonds: Most phospholipids have a saturated fatty acid and an unsaturated fatty acid attached to the glycerol. The unsaturated bond puts a "kink" into otherwise straight chains, making close packing of the chains impossible and thereby inhibiting their solidification. As an example of this effect, consider lard and vegetable oil. Both are triglycerides containing three fatty acids attached to a glycerol, but lard contains mostly saturated fatty acids, while vegetable oil has unsaturated fatty acids. The resulting difference in their physical properties is obvious.

Cholesterol has two main functions in membranes. First, the lipid bilayer must be relatively impermeable to polar solutes, and cholesterol has the ability to act as a kind of mortar, filling in defects that otherwise are present in pure phospholipid membranes. In general, a lipid bilayer containing cholesterol is about ten times less permeable than a bilayer lacking cholesterol. A second effect of cholesterol is related to membrane fluidity. As will be seen later, membranes must maintain a certain optimal fluidity, and cholesterol has been found to act as a fluidity "buffer" for lipid bilayers.

Classes of Membrane Proteins

All of the major groups of cell membrane proteins have now been characterized (Table 7-2), and in some cases protein components seen as membrane particles by freeze-fracture analysis have been linked to specific functions. Some of the proteins that can be visualized in this way are enzymes. Others serve as transport proteins, mediating the passage of specific hydrophilic molecules or ions across the membrane. Most cells, for example, use **transport ATPases** to pump ions across membranes, as we will see in more detail in Chapter 8. Still other membrane proteins are intermediates in energy transduction. Examples include the various protein intermediates in mitochondrial electron transport, such as cytochromes, dehydrogenases and iron-sulfur proteins.

Membrane proteins also function in the recognition and mediation of specific chemical signals that impinge on the surface of the cell. Such proteins are usually located on the extracellular side of the membrane and are called **receptors.** Hormones, neurotransmitters, and growth-promoting substances are examples of chemical signals that interact with specific receptors in or on the membrane of the target cells. Usually, the binding of a hormone (or other signal) molecule to the appropriate receptor on the membrane surface triggers some sort of intracellular response that is then directly responsible for eliciting the intended effect of the signal.

A final group of membrane-associated proteins are those with structural roles in stabilizing and shaping the cell membrane. Examples include spectrin and actin in the red cell membrane, which we have already discussed, and a protein called **clathrin,** which is present in cells such as hepatocytes of liver tissue. Clathrin has the ability to coat membrane surfaces and thereby mediate transport processes such as *endocytosis*, which will be discussed in detail in Chapter 9.

Membrane Carbohydrates

In addition to lipids and proteins, most membranes contain small but significant amounts of carbohydrates. The plasma membrane of the human red blood cell, for example, contains about 52% protein, 40% lipid, and 8% carbohydrate on a weight basis. A small proportion of membrane carbohydrate is present as **glycolipids,** but most of it is in the form of **glycoproteins.** The sugar units present

Table 7-2 Classificiation of Membrane Proteins by Function

Function	Example
Structural	*Spectrin, actin, clathrin*—provide organization and flexible support to membrane components.
Transport	*Carrier proteins*—facilitate transport of nutrients such as glucose, amino acids, and other metabolites across membranes.
	Channels—provide open and gated channels across lipid bilayers. *Ion channels* are specific for sodium, potassium, and calcium ions, which are required for excitable membrane function; *connexin* produces nonspecific channels between adjoining cells.
	Active transport enzymes—uše energy, often from ATP, to drive transport of ions and metabolites across membranes. Gradients of hydrogen, sodium, potassium, and calcium ions are produced in this way in most cells.
Light transduction	*Rhodopsin*—accepts light energy in the retina and transduces it into an electrochemical signal that is recognized by the nervous system. *Bacteriorhodopsin*—occurs in certain photosynthetic bacteria and uses light energy to transport protons, which then provide an energy source. Represents the simplest example of a light-dependent ion transport process.
	Reaction center proteins—interact with pigments such as chlorophyll in membranes that trap light energy for photosynthesis.
Electron transport	Function in the transport of electrons by coupling membranes of plant and animal cells. Examples include the *cytochromes, dehydrogenases,* and *iron-sulfur proteins* of mitochondria and chloroplasts.
Receptors	Bind molecules such as hormones and neurotransmitters and elicit a cellular response. Examples include the *acetylcholine receptor,* which opens sodium channels in the neuromuscular junction if acetylcholine is present, and the *insulin receptor,* which is present in most mammalian cells that transport glucose.

in these proteins are usually short oligosaccharide chains attached to serine, threonine, or asparagine side chains.

Glycoproteins are always positioned in the plasma membrane such that the sugar chains are found only on the external surface of the cell membrane. This arrangement has been shown experimentally by using **lectins,** plant proteins that bind specific sugar groups very avidly. For example, *wheat germ agglutinin,* a lectin found in wheat embryos, binds very specifically to oligosaccharides that terminate in *N*-acetylglucosamine, whereas *concanavalin A,* a lectin from beans, recognizes mannose groups in internal positions. We can visualize these lectins in the electron microscope by linking them to *ferritin,* an iron-containing protein that shows up as a very electron-dense spot. When such ferritin-linked lectins are used as probes to localize the oligosaccharide chains of membrane glycoproteins, binding is always very specifically to the outer surface of the membrane. Surface glycoproteins are important in cell-cell recognition, such as that presumably involved in intercellular adhesion to form tissues and in antibody-antigen reactions. Further techniques for studying membrane glycoproteins are discussed in the box on pp. 174–175.

Membrane Asymmetry

Most membrane lipids are highly *asymmetric* in their distribution between the inner and outer monolayers of the bilayer. Membrane asymmetry is established during membrane biogenesis and is then maintained for hours or days, since movement of hydrophilic head groups of lipid through the hydrophobic interior of the membrane is thermodynamically unfavorable. Such "flip-flopping" or **transverse diffusion,** as it is more technically called, can actually be shown to occur in membrane lipids but is relatively slow. For instance, a typical phospholipid molecule undergoes transverse diffusion once every several hours in a lipid bilayer. By contrast, **lateral diffusion** of phospholipids is so rapid that a lipid molecule can move 10 μm (about the diameter of a typical human red cell) in a few seconds!

Examination of the plasma membranes from a variety of cell types has revealed asymmetry in the kinds of lipid present and the degree of unsaturation of the fatty acids in the phospholipids. For instance, in the red cell membrane most of the phosphatidylcholine is present in the outer monolayer, and most of the phosphatidylethanolamine and phosphatidylserine is in the inner monolayer. Cholesterol

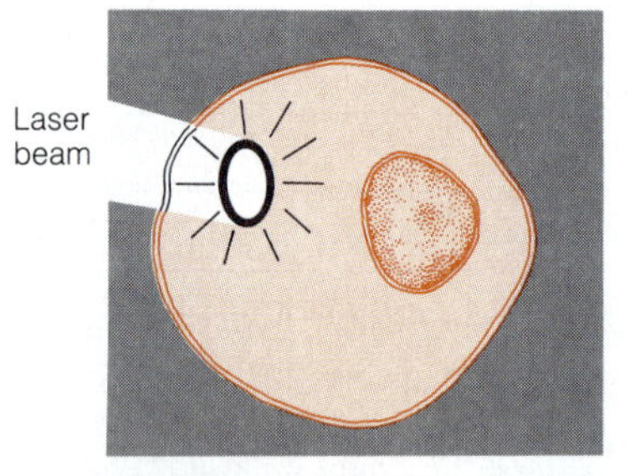

(a) Cell with fluorescent lipid in plasma membrane

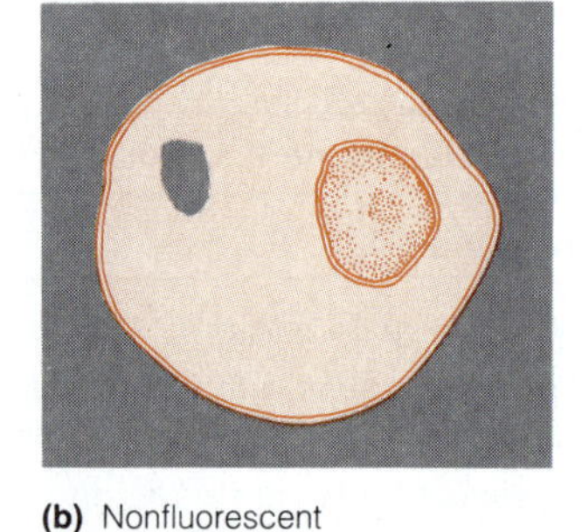

(b) Nonfluorescent bleached area

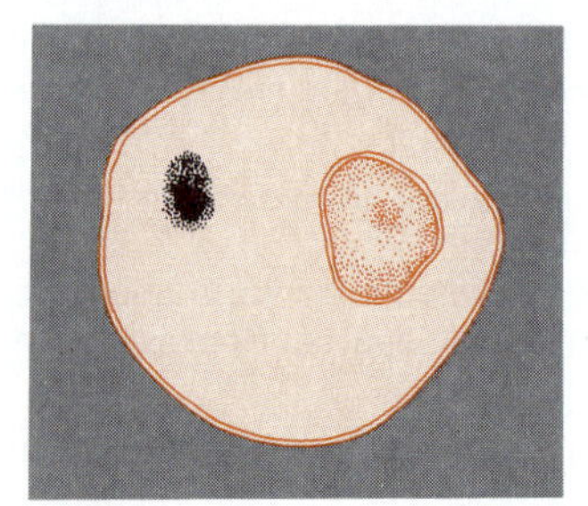

(c) Fluorescent lipid diffuses into bleached area

Figure 7-14 Measurement of Lipid Diffusion. (a) A fluorescent lipid can be added to the membrane lipids of a living cell. (b) The fluorescence can then be "bleached" by strong light from a laser. The bleached area appears dark because it does not fluoresce. (c) When bleaching is carried out, fluorescent lipid molecules from adjoining regions of the membrane diffuse into the bleached area within seconds, clearly demonstrating the fluid nature of the lipid bilayer.

is distributed approximately equally between the two layers. Especially significant is the localization of all glycolipids (sugar-containing lipids) in the outer monolayer. As a result, the carbohydrate portion of these hybrid molecules protrudes from the outer membrane surface of most cells. Such surface carbohydrates are thought to play an important role in cell-cell recognition and interaction.

Proteins are also maintained in asymmetric distributions across membranes. Peripheral proteins by definition are associated with membrane surfaces and, once in place, cannot readily move across the membrane from one surface to the other. Integral membrane proteins are also asymmetrically oriented in the membrane, often held in place by their association with peripheral proteins. The highly ordered protein complexes that make up the respiratory assemblies of the inner mitochondrial membrane and the photosystems of the chloroplast thylakoids are dramatic examples of structures that contribute to membrane asymmetry. The asymmetric distribution of membrane proteins arises at the time of protein insertion into the membrane and is then preserved by the thermodynamic restrictions on transverse diffusion. The box on pp. 174–175 provides some insights into the ingenious ways in which membrane asymmetry has been demonstrated and studied.

Membrane Fluidity

An important aspect of membrane function is the **fluidity** of lipid bilayers and the resulting freedom of motion of membrane components. This property can be observed experimentally by introducing a fluorescent lipid into the membrane and then bleaching the fluorescence with a bright flash of light focused on a small spot, much as the dye pigment in a color photograph is bleached if exposed to sunlight (Figure 7-14). When the edge of the darkened, nonfluorescent area is observed microscopically, fluorescent lipid begins to migrate into the bleached region within seconds, suggesting a fluid rather than solid state of the membrane lipid.

The mobility of membrane proteins is the second major piece of evidence in support of the fluid mosaic model. This mobility is illustrated in cell fusion experiments such as that shown in Figure 7-15, from the work of David Frye and Michael Edidin. Upon treatment with a virus known to cause cell fusion, cells from two different species are induced to form a single hybrid cell with a single fused plasma membrane (Figure 7-15a). **Fluorescent antibodies** can then be used to follow the fate of membrane proteins from the two parent cells. The antibodies needed for this purpose are prepared by injecting small quantities of purified membrane proteins into an animal such as a rabbit or goat. The animal responds immunologically to the foreign protein by producing antibodies (blood proteins called *immunoglobulins;* see Chapter 24) that react specifically with the membrane proteins used in the injections. After the antibodies have been isolated from the blood of the animal, fluorescent dyes can be covalently linked to them so that they can be seen with a fluorescence microscope.

By using two different dyes, it is possible to follow the fate of proteins from both parent cells, as shown in Figure 7-15 for the fusion of a mouse cell and a human cell. When the cells are first fused, the proteins from the mouse cell are localized on one-half of the fused membrane, whereas those from the human cell (antibodies in color) are restricted to the other half (Figure 7-15b). In a few minutes the proteins begin to intermix (Figure 7-15c), and within an hour the proteins from the two parent membranes are randomly distributed in the fused membrane of the hybrid cell (Figure 7-15d). This result would not have been observed if proteins were immobilized in the plane of the membrane.

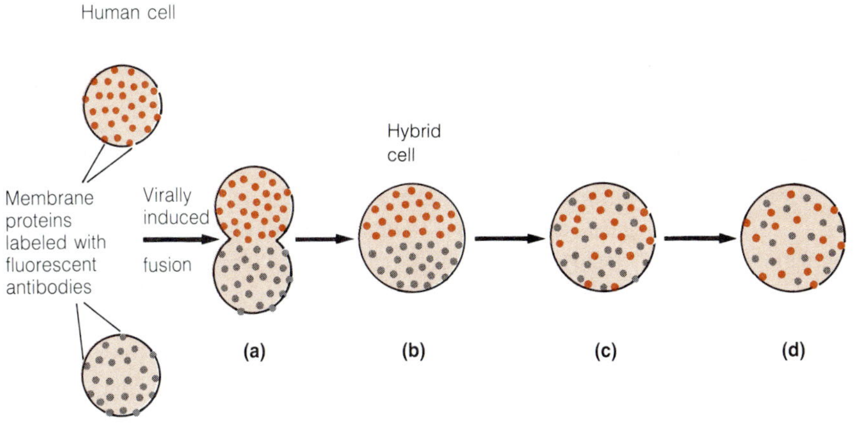

Figure 7-15 Fusion of Human Cell with Mouse Cell. Cells of different species can be induced to fuse by exposure to Sendai virus. Membrane proteins of the two cell types are marked with antibodies having different fluorescent properties, so the positions of proteins originating with one or the other of the parent cells can be followed on the surface of the hybrid cell as a function of time.

Red Blood Cells, Membranes, and Ingenuity

Do you ever stop to wonder how we know the things we know about cells? Consider, for example, the simple assertion that membranes are usually asymmetric, with some proteins associated with the external lipid layer, others with the internal layer, and still others extending into or even through both layers. How do we know these things? And how good is the evidence? In answering these questions, we will encounter some ingenious techniques and gain some fascinating insights into the way cell biology is done.

To explore membrane asymmetry, we will need some red blood cells, the enzymes lactoperoxidase (LP) and galactose oxidase (GO), some radioactive iodine (^{125}I), and the radioactively labeled reducing agent ^3H-BH_4 (tritiated borohydride). In addition, we will use the technique of SDS-polyacrylamide gel electrophoresis described in Figure 7-12. Finally, we will need some means of finding out which polypeptide bands on the polyacrylamide gels are radioactively labeled.

LP is useful because it is an enzyme capable of linking radioactive iodine atoms to tyrosine groups on a protein, provided that the reaction is carried out in the presence of hydrogen peroxide, H_2O_2. This means that a membrane protein can be labeled with ^{125}I, provided only that at least part of the molecule is exposed to the LP enzyme and that one or more tyrosine groups are present in that part of the molecule.

GO and tritiated borohydride are useful for labeling carbohydrate side chains on membrane proteins because GO can oxidize galactose (and galactosamine) groups, which then become labeled with ^3H when they are subsequently reduced with tritiated borohydride. Because LP and GO are too large to penetrate into or cross a membrane, they label only those carbohydrate chains or portions of protein molecules that are accessible on the outside of the membrane.

For example, when intact red cells are incubated with LP in the presence of ^{125}I and peroxide and the membrane proteins are then extracted and analyzed by SDS-polyacrylamide gel electrophoresis, only a few of the polypeptide bands on the gel can be shown to be radioactive. This is because the other bands represent polypeptides that do not protrude from the external surface of the cell and therefore are not accessible to the LP enzyme.

Similarly, any carbohydrate chains on the external surface can be labeled by treating intact red cells with GO (to oxidize galactose groups), followed by exposure to labeled borohydride (to reduce the galactose groups again, introducing labeled hydrogen atoms in the process). As you might guess from what you already know about the location of glycoproteins in membranes, all of the carbohydrate side chains of the membrane proteins are labeled by this procedure.

For comparison, if you first rupture the red cell membrane by subjecting the cell to hypotonic shock, the LP or GO enzymes will have access to both sides of the membrane (because the cell is no longer intact and the enzyme molecules are therefore not excluded from the interior of the cell). Now what you find on electrophoresis is that virtually all of the proteins from the LP-treated membranes are labeled with ^{125}I. This indicates that even the most hydrophobic of membrane proteins have some portion that protrudes from the membrane enough to be accessible to the LP enzyme. On the other hand, GO-borohydride treatment of disrupted membranes labels no more polypeptides than were labeled with intact cells, further strengthening the conclusion that all the carbohydrate side chains are on the external membrane surface.

The cleverest experiments, though, are those designed to label only the proteins (or carbohydrates, if there were any) that are exposed on the inner, but not the outer, surface of the red cell membrane. To understand how this might be done, you must know that once the cell is disrupted, the membrane can be fragmented into smaller pieces, and these tend to seal again spontaneously to form empty vesicles. The resealing process turns out to be random with respect to membrane orientation, so some of the vesicles are right-side-out, but others have the membrane inside-out. The two kinds of vesicles are sufficiently different in properties to allow them to be separated, so both can then be treated with either LP or GO.

As you might expect, the right-side-out vesicles show the same labeling response as intact cells. More revealing are the inside-out vesicles, because these now allow specific labeling of inside proteins only. It is interesting that almost all of the polypeptides on the gels that did not become labeled when intact cells were exposed to LP and radioactive iodine are labeled when inside-out vesicles are treated. Apparently, almost every membrane protein protrudes from the membrane on one surface or the other. In fact, a few of the red cell membrane proteins are labeled with ^{125}I whether the inside-out or the right-side-out vesicles are used. What conclusion do you think can be drawn from that observation?

From these and similar experiments, a clear picture emerges of the distribution of proteins within the red cell membrane. Most of the proteins are associated with the inner lipid layer and contain no carbohydrate chains. All glycoproteins are located in the outer lipid layer, with the carbohydrate side chain invariably protruding into the external environment. Several proteins extend all the way through the membrane and can be labeled with LP on both inner and outer surfaces, as well as with GO and borohydride on the outer surface.

These results were confirmed by another approach that used the freeze-fracture technique. Imagine that you could somehow shrink to molecular size and watch what happens when a cell is frozen in liquid nitrogen and then broken open to be examined by the freeze-fracture method of electron microscopy. You are near a plasma membrane, and all the molecules have been cooled to temperatures near 200 degrees below 0°C. Instead of the chaotic motion of the fluid state, with molecules rushing around at huge velocities, colliding, vibrating, and rotating billions of times per second, every molecule is now fixed in place, shivering a little, and only occasionally changing places with a neighbor. The lipids and proteins of the membrane are held together by hydrogen bonds and electrostatic forces between ions, and look just like the static image of the fluid mosaic model. In the interior of the lipid bilayer only weak van der Waals forces hold lipid tails together.

Suddenly you hear in the distance a loud thundering sound, and a huge metallic blade can be seen approaching, crashing through the ice and frozen cytoplasm. As it passes overhead, an immense crack descends into the cell and breaks open the membrane, easily parting the lipid bilayer along the weakly bonded tails. A brief flash of heat is felt as some of the fracture energy is turned into vibrational molecular energy, and the lipid chains melt and oscillate for a few microseconds like miniature tuning forks. What happens to the membrane proteins: Are they ripped apart by the fracture process, leaving broken peptide bonds in their amino acid chains, or do they simply pop out of the lipid like little corks?

Although we cannot shrink to molecular size to answer this question, it can be approached through biochemical analysis of membrane composition following freezing and fracturing. Knute Fischer, at the University of California, San Francisco, found that red cell membranes adhered tightly to a glass surface if it was coated with polylysine, a strongly cationic polymer of the amino acid lysine. The membranes could then be frozen together with a thin layer of water, and the glass split away from the ice. This left behind on the glass a layer of half-membranes that resulted from splitting of the frozen preparation, just as occurs in the freeze-fracture method. When the membranes were analyzed by microanalytical techniques for lipids, it was found that cholesterol was approximately evenly divided between the inner and outer leaflets of the lipid bilayer. Phosphatidylcholine was mostly in the outer leaflet, while phosphatidylethanolamine and phosphatidylserine composed the inner leaflet. The protein composition was then analyzed by gel electrophoresis. Every protein could be accounted for, even those known to be transmembrane in orientation, and no broken fragments were seen. The surprising conclusion is that proteins in frozen membranes easily break away from the lipid and ice during fracture. Therefore the particles observed in freeze-fracture images represent complete protein molecules, probably with some associated lipids, and are not fragments of proteins pulled apart like pieces of taffy.

Once again, it is a judicious combination of clever techniques and ingenious thinking that makes such experiments—and hence such findings—possible.

Another dramatic example of the mobility of membrane-associated proteins is the phenomenon of **patching and capping** in certain white blood cells (lymphocytes) associated with the immune response. Lymphocytes are blood cells that respond to foreign proteins by producing antibodies, which they carry on their plasma membrane (Figure 7-16a). One kind of foreign protein that will cause antibody production in an animal is an antibody from another species. If, for example, you inject mouse antibodies into a rabbit, the rabbit lymphocytes will produce anti-mouse antibodies. Once these antibodies have been isolated from the blood of the rabbit, they can be tagged with a fluorescent dye that can be visualized by fluorescence microscopy.

When rabbit anti-mouse antibodies are mixed with mouse lymphocytes, the labeled antibodies will be fairly evenly spread out all over the surface of the plasma membrane (Figure 7-16b). Soon, however, clusters and then patches of antibody molecules form (Figure 7-16c and d), because antibodies are *multivalent;* that is, each molecule of rabbit antibody can react with multiple mouse antibodies on the plasma membrane, and each membrane antibody can react with more than one rabbit antibody molecule. This clustering results in the buildup of a network of rabbit anti-mouse antibody molecules and membrane-bound mouse antibody molecules called "patches," which eventually aggregate as a "cap" on one side of the cell (Figure 7-16e). This patching and capping phenomenon mediated by antibodies serves as further evidence of protein mobility in membranes.

Although such experiments demonstrated that membrane proteins are capable of lateral diffusion, recall that lipids are so mobile that they can move across a cell in a few seconds, in contrast to the hour or more required for proteins to diffuse. Clearly, although proteins can move about in membranes, something limits their free diffusion. Why might proteins be restricted in this way, whereas lipids are not? Recall Figure 7-10, which showed the structure of the red cell membrane. Ankyrin links the anion channel to spectrin, thereby keeping the anion channel from diffusing in the fluid bilayer. If the spectrin is removed, the anion channel becomes freely mobile, even forming "islands" of aggregated protein that can be visualized by freeze-fracture analysis. This result suggests that a significant factor limiting protein mobility in membranes is attachment to underlying cytoskeletal structures.

Regulation of Membrane Fluidity

We have seen that membranes are best understood as existing in a fluid rather than a gel state. Furthermore, although both the lipid and protein components are mobile, the protein components are typically less so because of interactions with underlying cytoskeletal structures. The last point to be made about membranes is that they are not only in a fluid state, but are actively maintained in that state by changes in membrane composition.

Why must membranes be maintained in a fluid state? There are no precise answers to this question, but one

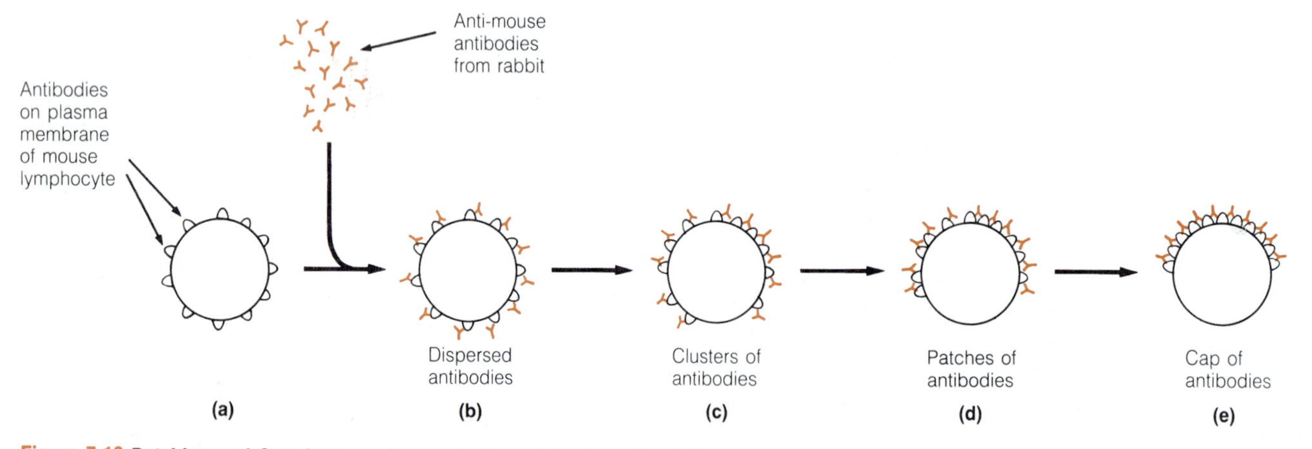

Figure 7-16 Patching and Capping as a Demonstration of the Mobility of Membrane Proteins. (a) Antibodies are normally dispersed randomly on the external surface of the plasma membrane of a mouse lymphocyte. (b) When rabbit anti-mouse antibodies are initially mixed with the mouse lymphocytes, the rabbit antibodies bind to the mouse antibodies and appear dispersed on the membrane surface of the lymphocyte. (c) Because the rabbit antibodies are capable of binding to more than one mouse antibody, antibody complexes begin to cluster on the membrane surface. (d) These clusters gradually aggregate into large patches and (e) eventually accumulate as a cap on one side of the membrane. The rapid clustering, patching, and capping of membrane proteins is strong experimental evidence for their mobility within the membrane.

reason may be to optimize the movement of substances within the membrane. Enzymes in membranes must be able to interact with one another to catalyze certain sequential reactions. For instance, the respiratory enzymes of mitochondrial membranes must literally collide with one another to transfer electrons. If the membrane lipid were a viscous gel, they would be unable to do so.

Temperature is probably the single most important environmental factor affecting membrane fluidity. Bacteria, yeasts, plants, cold-blooded animals, and other organisms that cannot regulate their temperature are called *poikilotherms*. Changes in external temperature cause difficulties for such organisms. For instance, cooler temperatures reduce the fluidity of lipids, as noted earlier, and could "freeze" membranes, thereby inhibiting their function. You may have experienced this effect, even though you are a homeotherm. On chilly days your fingers and toes can get so cold that the membranes of sensory nerve endings cease to function, and numbness results. At the other extreme, higher temperatures can make a bilayer membrane so fluid that it no longer serves as a permeability barrier. Most poikilotherms are paralyzed by temperatures much above 43°C, probably because certain nerve cell membranes become so leaky to ions that overall nervous function is disabled.

To guard against such temperature-dependent effects on function, cells have evolved mechanisms for regulating membrane fluidity by metabolically altering membrane lipids so that chains of appropriate length and degree of unsaturation are present. This mechanism follows from our understanding of the physical properties of the fatty acids composing membrane lipids. Fatty acids with shorter chains have lower melting points than those with longer chains, as might be expected. Furthermore, the melting points of saturated and unsaturated fatty acid chains are considerably different: membranes containing many saturated chains "melt" (undergo a gel-to-fluid phase transition) between 40° and 50°C, whereas membranes consisting primarily of unsaturated chains melt near 0°C or lower. The reason for this difference is that the cis double bonds of unsaturated fatty acids prevent the hydrocarbon chains from fitting together snugly, such that van der Waals forces are unable to develop fully. Saturated chains fit together precisely and can optimize their van der Waals interactions (Figure 7-17).

An interesting investigation has been carried out that clearly demonstrates the relationship between temperature and the effect of fatty acid composition on membrane fluidity. *Acholeplasma* is a genus of small bacteria that cannot manufacture their own fatty acids, but instead must use whatever is available in the environment. Their membranes then take on the physical characteristics of the nutrient fatty acids. If the bacteria take up an unsaturated fatty acid, their membranes remain fluid and the bacteria grow readily. However, if only saturated fatty acids are available, the bacteria grow only until their membrane lipids contain so many saturated chains that they undergo a phase transition to the gel state. If the temperature is then raised to induce a phase transition back to the fluid state, the bacteria are again able to grow.

Bacteria such as *Escherichia coli* and *Bacillus megaterium* can modify their membrane lipid content to adapt to different temperatures. For instance, raising the temperature of a bacterial culture triggers the synthesis and incorporation into the membrane of more saturated phospholipids with longer chains. Conversely, yeasts and plants have the ability to increase the proportion of unsaturated fatty acids in their membranes at lower temperatures. This alteration may be related to the increased solubility of oxygen at lower temperatures. Oxygen is a substrate for the *desaturase enzyme system* involved in the generation

(a) Saturated fatty acids

(b) Unsaturated fatty acids

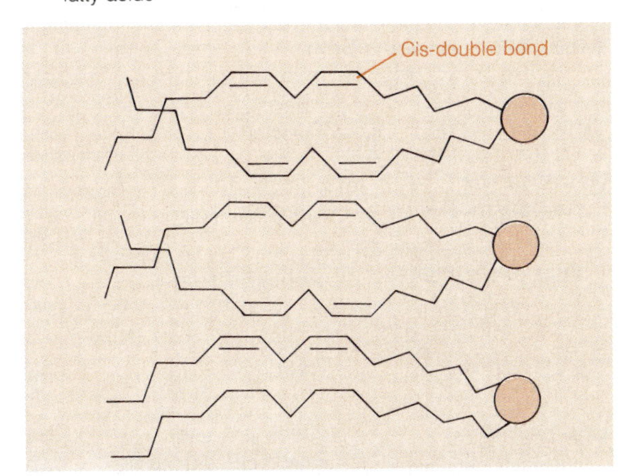

Figure 7-17 Effect of Unsaturation on Membrane Fluidity. The physical state of fatty acids in phospholipids is responsible for membrane fluidity. (a) Saturated fatty acids are able to fit together tightly and have higher melting points than (b) unsaturated fatty acids in which *cis* double bonds prevent similar close fits. Most membrane lipids contain one or more cis double bonds to ensure that they are in the fluid state at physiological temperatures.

of unsaturated fatty acids. With more oxygen available at lower temperatures, unsaturated fatty acids can be synthesized at a greater rate and membrane fluidity will increase, thereby balancing the temperature effect.

For eukaryotic cells, cholesterol also has an important influence on membrane fluidity. The cholesterol content of the plasma membrane of most eukaryotic cells is relatively high, nearly half of the total membrane lipid on a mole-to-mole basis. At such concentrations, cholesterol tends to decrease membrane fluidity of lipid bilayers at higher temperature ranges. Cholesterol also effectively prevents the hydrocarbon chains of phospholipids from aggregating as temperature is decreased, thereby reducing the tendency of membrane lipid to "freeze." Cholesterol thus acts as a kind of buffer, minimizing dramatic changes in membrane fluidity that would otherwise occur with temperature extremes.

Cell Junctions

Unicellular organisms, by definition, have no need for any permanent associations between cells, since each cell is an entity unto itself. Multicellular organisms, however, require specific means of joining cells together into the permanent associations required to constitute tissues and organs. Usually, these means involve specialized modifications of the plasma membrane at the point where two cells come together. Such specializations are called **cell junctions.**

Several general forms of junctions are recognized, differing mainly in how tightly the cells are appressed at the point of contact and what purpose the junction serves. In animal cells, the three most common kinds of junctions are *desmosomes, tight junctions,* and *gap junctions* (Table 7-3)

Desmosomes serve primarily an adhering or connecting role and are most prominent in epithelial tissue. Tight junctions function mainly as a seal between two cell-lined compartments, keeping contents such as gastric juices, urine, or other body fluids in their proper location. Gap junctions, on the other hand, provide for direct chemical and electrical communication between adajcent cells. Smooth muscle and heart muscle are examples of tissues characterized by gap junctions.

Desmosomes

Desmosomes might well be called "adhering junctions" because they are regions of tight adhesion between adjacent cells. This adherence gives the tissue structural integrity and allows the cells to function together as a single unit. Desmosomes are widely found in epithelial tissue such as skin, which must withstand considerable mechanical stress. Desmosomes typically have an extracellular space of about 25–35 nm, about that normally found between contiguous animal cells. The stucture of a typical desmosome is illustrated in Figure 7-18. The extracellular space is filled with a fine filamentous material that is probably glycoprotein in nature and that may play a cementing role.

Two types of desmosomes have been described. **Belt desmosomes** are continuous zones of attachment that en-

Table 7-3 Junctions Between Animal Cells

Type of Junction	Function	Features	Intermembrane Space	Associated Structures
Desmosomes				
Belt desmosomes	Cell adhesion	Continuous zone of attachment	25–35 nm	7-nm actin microfilaments
Spot desmosomes	Cell adhesion	Localized points of contact	25–35 nm	10-nm intermediate filaments (tonofilaments)
Tight junctions	Sealing	Membranes joined along ridges	None	None
Gap junctions	Electrical and chemical exchange	Hexagonal proteins with pores spanning membrane to provide connections	2–4 nm	7-nm-diameter particles with 1.5-nm-diameter pores; particles span membrane and make contact with similar particles in adjacent membrane

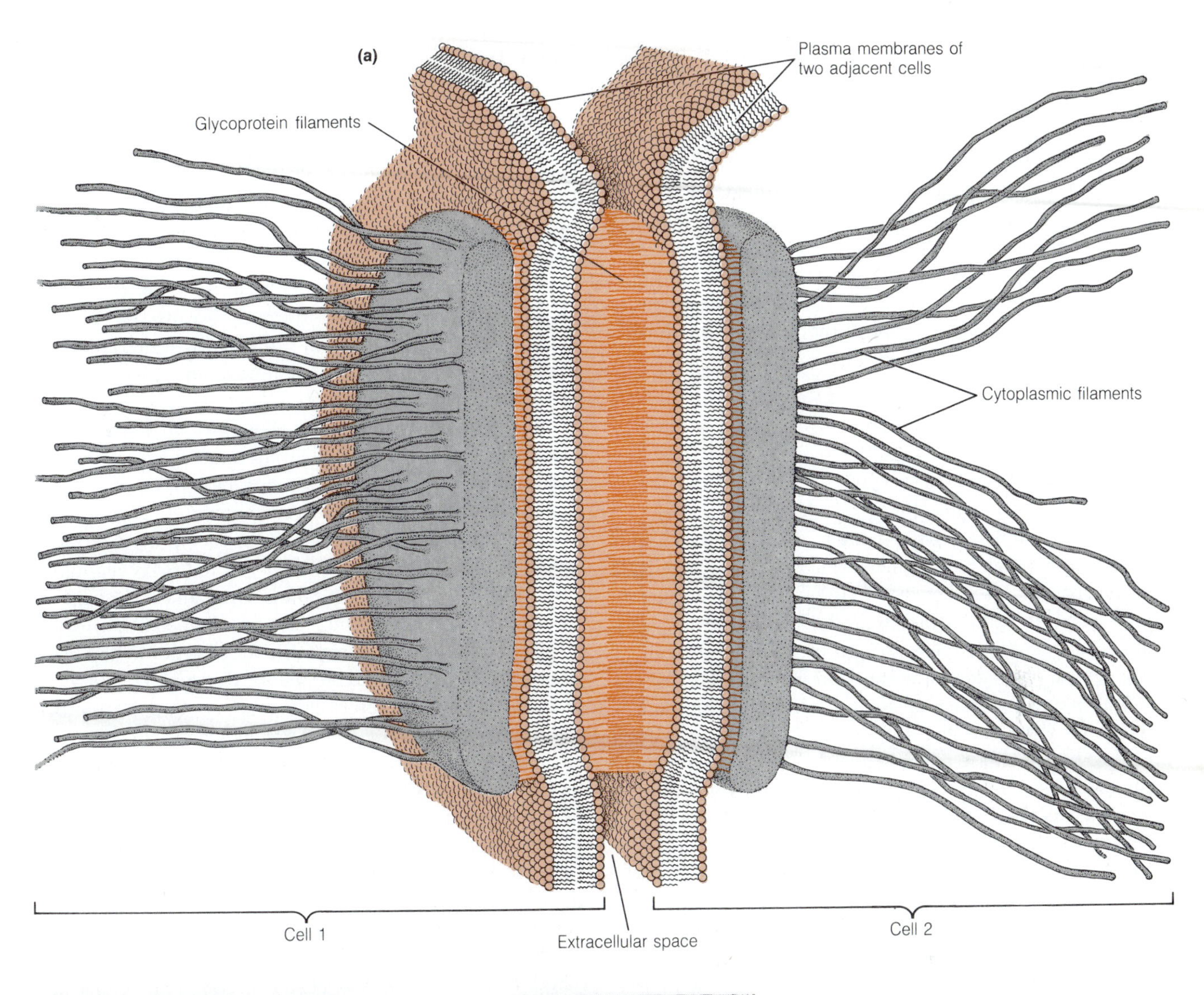

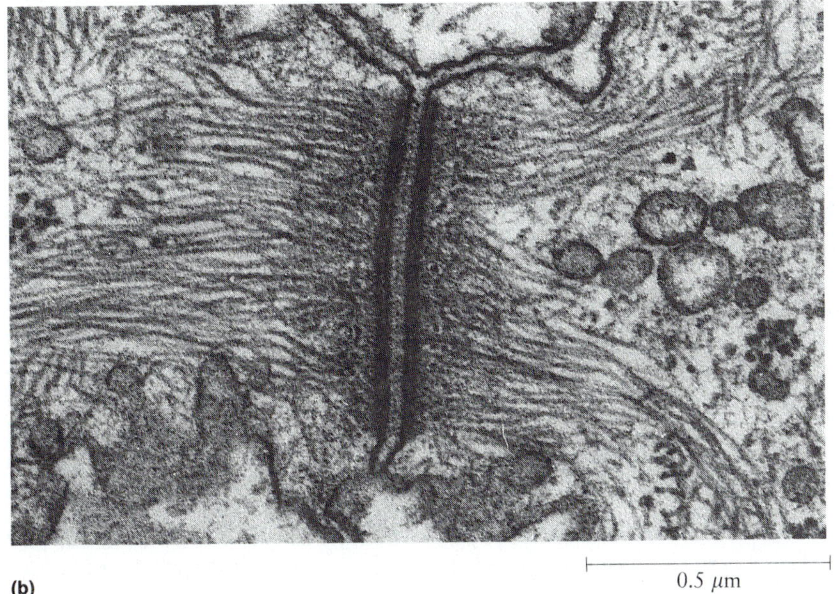

Figure 7-18 The Desmosome. (a) Schematic representation. (b) Electron micrograph of newt skin cells (TEM). The distance between cells in the desmosome region is 25–30 nm, about that for a nonjunction region. The extracellular space between membranes is filled with glycoproteins. Cytoplasmic filaments associate with the membrane in the desmosome region.

circle the cell, as the name suggests. In epithelial tissue, belt desmosomes usually occur close to the tight junction. **Spot desmosomes** are more localized points of membrane contact between adjacent cells and might be thought of as "spot welds." Both types of desmosomes have characteristic cytoplasmic filaments associated with them, the functions of which are not yet very clear. Belt desmosomes have 7-nm microfilaments of actin, suggesting a contractile role. Spot desmosomes, on the other hand, have 10-nm intermediate filaments, sometimes called **tonofilaments,** that are thought to be involved in the tensile strength of the junction.

Tight Junctions

As the name implies, **tight junctions** involve very close contact between the plasma membranes of adjacent cells, with no space at all in between (Figure 7-19). Tight junctions extend all around the circumference of the cell so that the spaces between cells are tightly sealed. As a result, the space on one side of the junction plane is effectively sealed off from the space on the other side, so that fluid cannot pass from one side to the other. The role of tight junctions in preventing unwanted exchange between body compartments is illustrated by the electron micrograph of a rat pancreas shown in Figure 7-19b. Exocrine (secretory) cells surround the lumen into which secretory proteins stored in the zymogen granules are secreted. Adjacent exocrine cells are linked together by tight junctions, thereby ensuring that the lumen is isolated from the intercellular space and hence from the space on the other side of the cells. Intestinal cells are similarly linked together by tight junctions so that liquid from the intestine cannot pass across the epithelial layer. Tight junctions are also found in the bladder, where they serve the obvious function of ensuring that the urine stored in the bladder does not seep out between cells.

Although tight junctions seal the membranes of adjacent cells together very effectively and prevent unwanted exchange of fluid between two body compartments, the membranes are not actually fused together over broad areas but rather along specific ridges of membranes, as shown in Figure 7-19a. The result is rather like placing two pieces of corrugated metal together so that each ridge of one piece is lined up with the corresponding ridge of the other and then welding the two pieces together lengthwise along each ridge of contact. Not surprisingly, there appears to be a good correlation between the number of such "welded ridges" across the junction and the tightness of the seal that the junction makes. Each such ridge is called a *tight junction element*.

Gap Junctions

The **gap junction** is the most common type of cell junction. Its basic function is to provide a point of communication between two adjacent cells through which chemicals can be exchanged and electrical signals can pass. Instead of a tight seal between two membranes, a gap junction is a region at which the membranes are brought close together, but with a gap of 2–4 nm in between. This is still an intimate region of contact, however, because the plasma membranes of adjacent cells are usually separated by a space of about 25–35 nm in nonjunctional regions of most vertebrate tissue.

Gap junction structure is illustrated in Figure 7-20. Notice from the schematic diagram (Figure 7-20a) that the two plasma membranes at a gap junction are joined by structures called **connexons.** A connexon is an assembly of six protein subunits that span the membrane and protrude into the space (or gap) between the two cells. Each connexon has a diameter of about 7 nm and is thought to have a hollow center about 1.5 nm in diameter that forms a very thin channel through the membrane. When connexons in the plasma membranes of two adjacent cells are aligned, they form a direct channel of communication between the two cells. Cells that are coupled in this way can exchange chemical substances, particularly those with molecular weights below about 1000. A single gap junction may consist of several hundred clustered connexons.

Gap junctions are sites of lesser electrical resistance than is normally found across a lipid bilayer. Gap junctions are important in heart muscle and smooth muscle in animals because they provide for the intercellular flow of potassium ions that is responsible for muscle activation. Thus, both electrical and chemical communications between adjacent cells seem to be specifically facilitated by the direct cytoplasmic connections that exist through the channels of the gap junction proteins. Gap junction permeability can be changed rapidly by experimental manipulation of the pH or the calcium concentration of the coupled cells. This raises the possibility that cells may be able to open and close their gap junctions and thus control intercellular communication.

The Cell Surface: Coats and Walls

So far, we have been treating cells as if they "end" at the outer edge of the plasma membrane. Rarely is this true, however, because most cells have some sort of surface

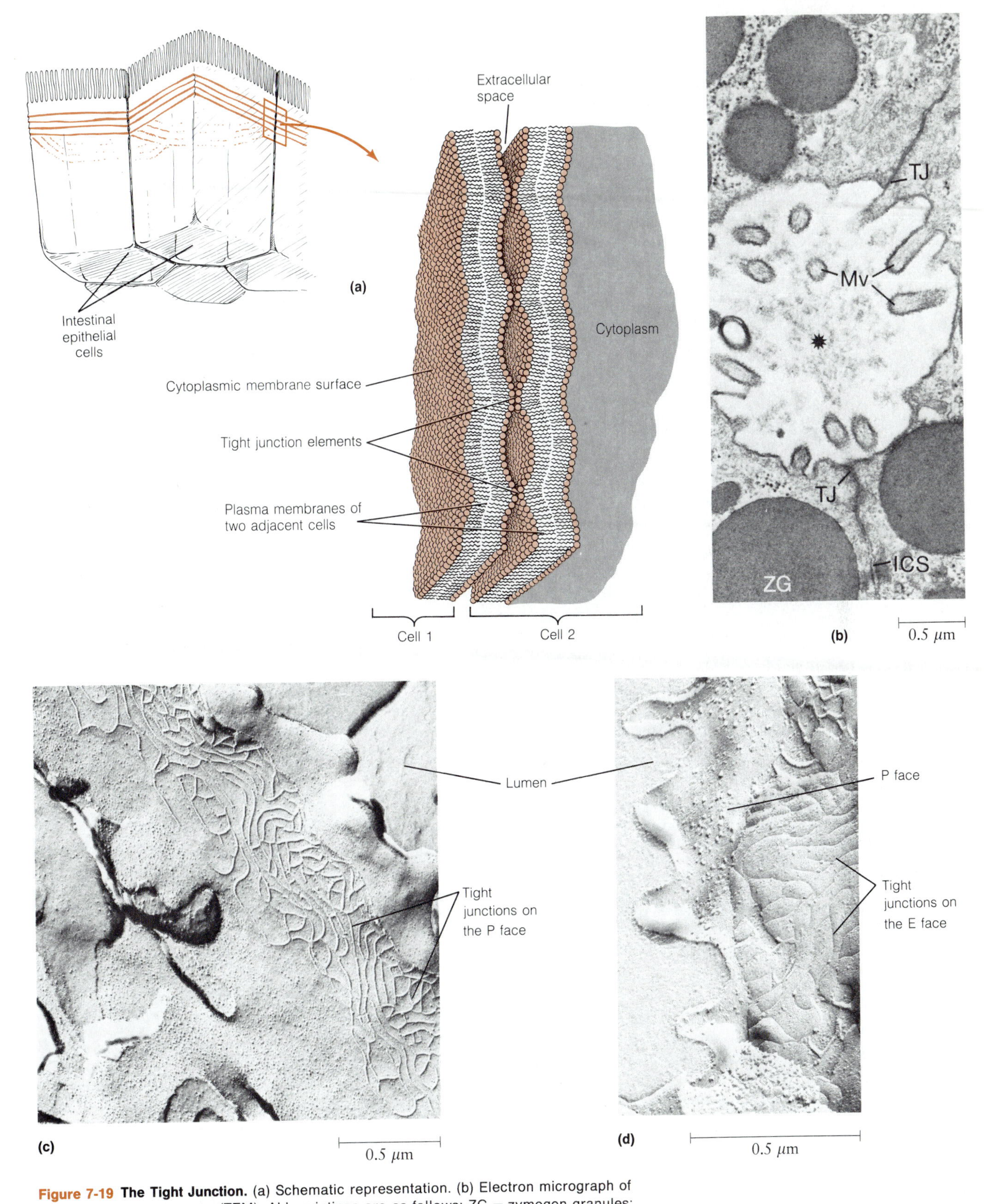

Figure 7-19 The Tight Junction. (a) Schematic representation. (b) Electron micrograph of a lumen in a rat pancreas (TEM). Abbreviations are as follows: ZG = zymogen granules; TJ = tight junction; ICS = intercellular space; Mv = short microvilli; * = lumen. (c, d) Tight junctions between cells in a frog bladder, as revealed by the freeze-fracture technique (TEM). The tight junctions appear as raised strands on the protoplasmic (P) face of the membrane and as shallow grooves on the exterior (E) face. The lumen is the cavity of the bladder.

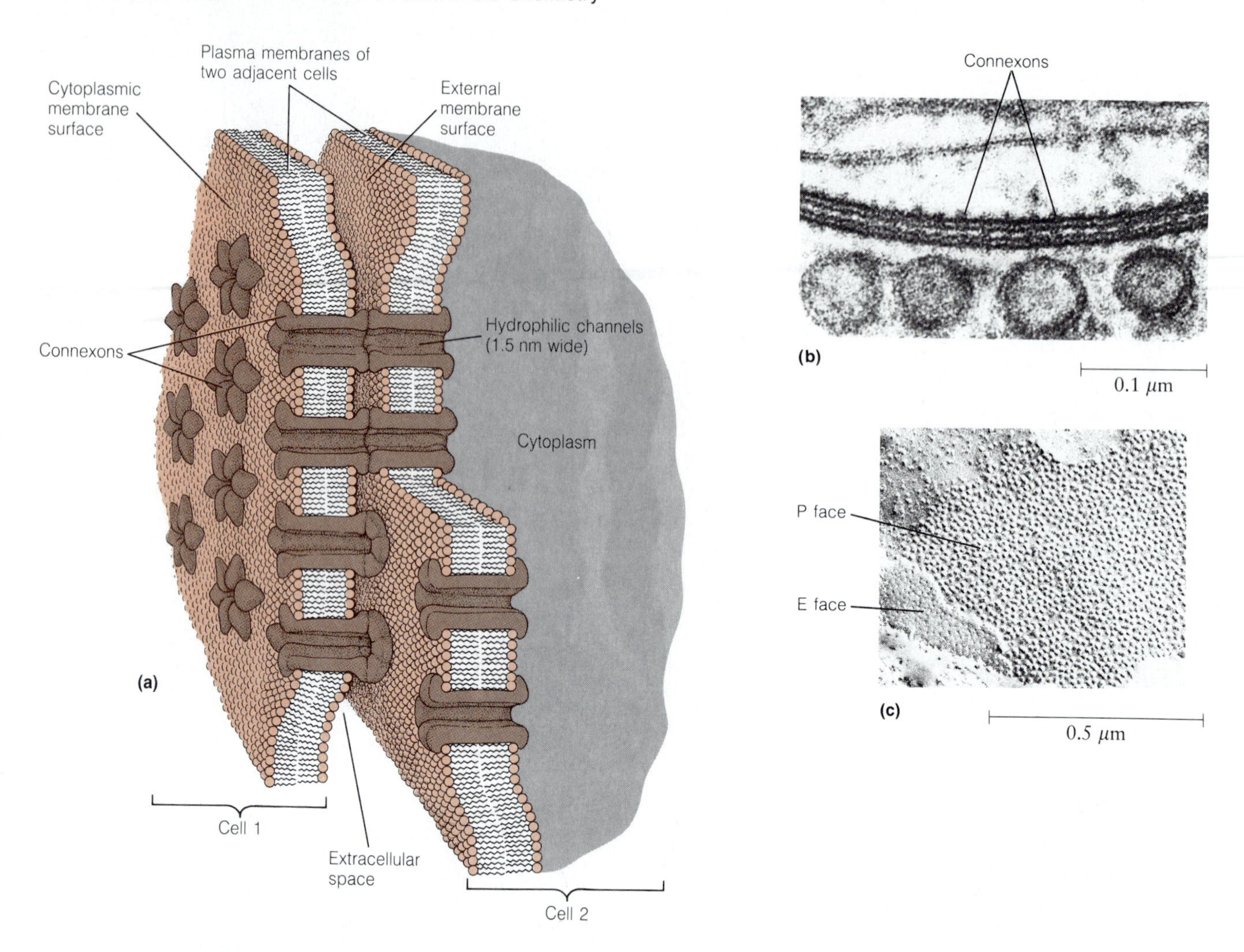

Figure 7-20 The Gap Junction. (a) Schematic representation of a gap junction. A gap junction consists of a large number of hydrophilic channels formed by the alignment of connexons in the plasma membranes of two adjoining cells. (b) Electron micrograph of a gap junction between adjacent nerve cells (TEM). The connexons that extend through the membranes are seen here as beadlike projections spaced about 17 nm apart on either side of the membrane-membrane junction. (c) A gap junction as revealed by the freeze-fracture technique (TEM). The junction appears as an aggregation of intramembranous particles on the protoplasmic fracture face (P face) and as a series of pits on the exterior (E face) of the membrane.

material that is external to the membrane and is nonetheless part of the cell in both a structural and a functional sense. Animal cells have an **extracellular matrix** of materials that usually consists of a **cell coat** and a separate **fuzzy layer** external to the coat. In plant and bacterial cells the most prominent extracellular structure is the **cell wall**, which confers both protection and rigidity on the cell it encases. We will look briefly at these features of the cell surface, noting especially the contributions that they make to the structural and functional properties of cells.

Coats and Fuzzy Layers

We are already acquainted with one structural feature of the surface of animal cells, since we encountered the carbohydrate chains of membrane glycoproteins earlier. These proteins are organized in the plasma membrane such that their oligosaccharide side chains invariably protrude out into the extracellular space. When cells are stained with heavy metal stains that complex with carbohydrates, this carbohydrate layer is visualized as an electron-dense coat on the surface of the cell. The cell coat, in other words, is nothing more than the layer of carbohydrates jutting out from the membrane. Although these are really just membrane components, the *cell coat* that they create is usually considered a distinctive feature of the cell surface.

Exterior to the cell coat is the *fuzzy layer* of material that is genuinely extracellular in location. The best-known examples of cells with a prominent fuzzy layer are epithelial cells from mammalian intestine and certain amoebas. Often, the same stains used to visualize the fuzzy layer also react with the cell coat, making it difficult to determine where one ends and the other begins. The term **glycocalyx** ("sugar coat") is sometimes used to refer to the complex of cell coat and fuzzy layer. Figure 7-21 depicts the prominent glycocalyx of an intestinal epithelial cell.

An important chemical component of the fuzzy layer and extracellular matrix is a protein called **collagen,** which composes nearly a quarter of the protein mass in mammalian tissues and is therefore the most abundant protein in the human body. Collagen is secreted by cells called *fibroblasts,* which occur throughout most tissues. In the absence of collagen, cells in tissues would not have sufficient adhesive strength to maintain a given form. Therefore tissue that contains fibroblasts is often referred to as *connective tissue.* Some specialized examples of connective tissue include bone, cartilage, tendons, and ligaments.

Collagen is actually a mixture of perhaps ten different types of closely related proteins, all taking the form of fibers with high tensile strength. In electron micrographs

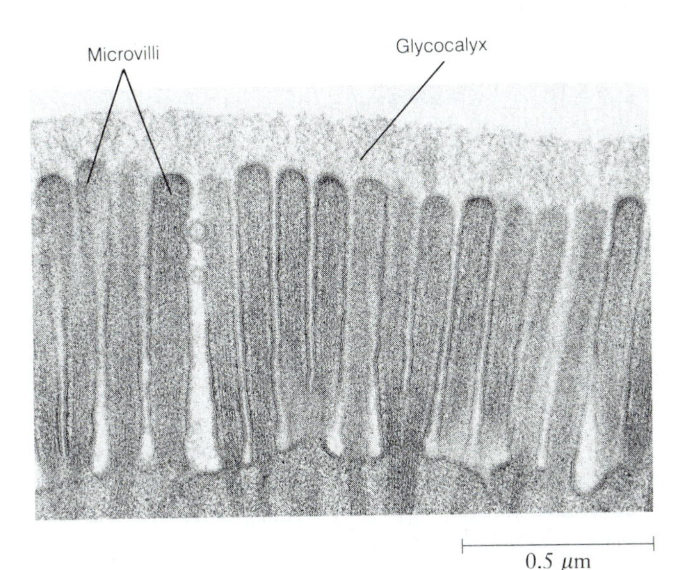

Figure 7-21 **The Glycocalyx.** This electron micrograph of a cat intestinal epithelial cell shows the microvilli that are involved in absorption and the glycocalyx on the cell surface. The glycocalyx on this cell is about 150 nm thick and consists primarily of oligosaccharide chains about 1.2–2.5 nm in diameter.

of most animal tissues, collagen fibers can be seen in bundles throughout the extracellular matrix. The structure of one such fiber is illustrated in Figure 7-22. When analyzed, the fibers are found to consist of triple helices of individual collagen molecules, which in turn are cross-linked into microfibrils of *tropocollagen.* The fibrils are grouped into individual banded fibers, the bands arising from the organization of the tropocollagen microfibrils.

Although collagen provides support to cells in tissues, its fibrous, rodlike structure is not particularly well suited to producing the elastic qualities required by tissues such as skin, lungs, and the intestine, which undergo continuous shape changes throughout life. Elasticity is provided by a second extracellular matrix protein, appropriately called **elastin,** which forms a relatively loose, flexible framework of cross-linked chains.

The important roles of collagen and elastin are clearly demonstrated during aging. Over time, collagen becomes increasingly cross-linked and inflexible, and elastin is lost from tissues like skin. As a result, in elderly individuals the skin becomes wrinkled, bones and joints are less flexible, and the skin only slowly returns to its original form when it is deformed (by gentle pinching, for instance).

Glycosaminoglycans (also called **mucopolysaccharides**) represent a third component of the fuzzy layer and extracellular matrix. Glycosaminoglycans are carbohydrates with a repeating disaccharide unit. One of the two disaccharides in the repeating unit is always an amino sugar, with *N-acetylglucosamine* and *N-acetylgalactosamine* as the two most common examples. The other repeating unit is usually a sugar acid, commonly *glucuronic acid. Hyaluronic acid, chondroitin,* and *chondroitin sulfate* are several of the most common extracellular glycosaminoglycans. The disaccharide repeating unit of each is illustrated in Figure 7-23. The glycosaminoglycans are usually complexed with proteins to form **mucoproteins.** Mucoproteins differ from the glycoproteins found in plasma membranes in that the carbohydrate chains of mucoproteins are usually very long and are highly acidic because of the glucuronate (and in some cases the sulfate groups) present in the sugar chains.

Cell Walls

Animals cells are surrounded only by the cell coats and fuzzy layers that we have just described, which allow the flexibility and motility so necessary to animals. Plant and bacterial cells, on the other hand, are almost always surrounded by a rigid, nonliving *cell wall* that serves both to protect and to contain the cell. The presence of a cell wall allows the cell to withstand osmotic and mechanical

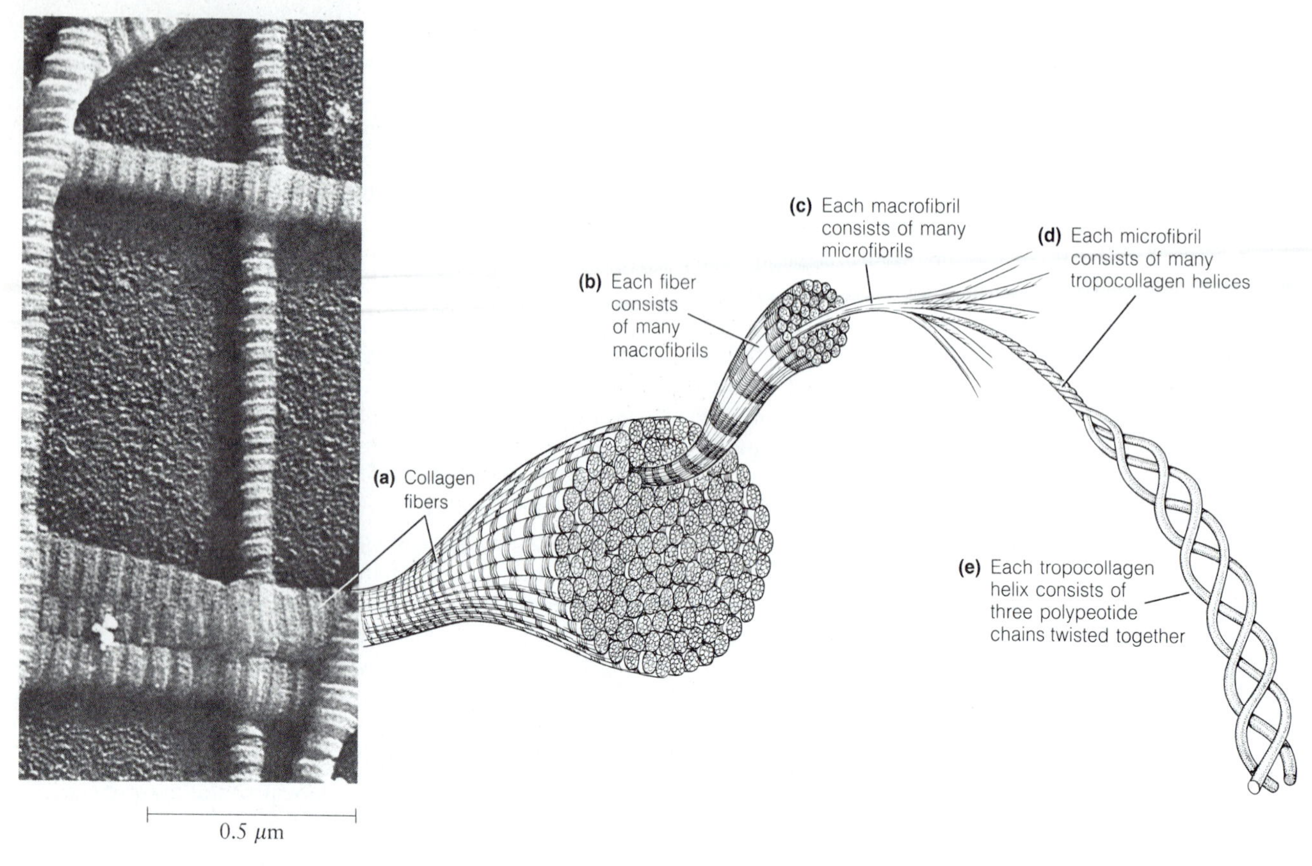

Figure 7-22 Structure of Collagen. (a) Collagen fibers as seen in the electron microscope (SEM). (b) Collagen fibers consist of many macrofibrils, (c) each in turn composed of many microfibrils. (d) Microfibrils are tiny bundles of tropocollagen helices, (e) which consist, in turn, of three polypeptide strands twisted together into a right-handed helix. The repeating bands visible on the fibers reflect the regular but offset manner in which tropocollagen helices associate laterally to form microfibrils.

stresses that would almost certainly rupture animal cells. We will look briefly at the chemical nature of the walls that surround bacterial and plant cells. In the case of plants, we will also consider specific structural adaptations called *plasmodesmata,* which allow intercellular communication despite the presence of the wall.

Bacterial Cell Walls

Bacteria are traditionally classified into two groups based on their response to the *Gram stain,* which involves the reaction of the cell wall with a colored complex of crystal violet and iodine. *Gram-positive* bacteria take up the stain, while *gram-negative* cells do not. Since the Gram stain reacts with the cell wall, it is not surprising that the two groups of bacteria have cell walls that differ significantly in their chemistry.

Gram-positive bacteria have a 50-nm wall that consists of **teichoic acid** and **peptidoglycans.** The peptidoglycan component consists of a disaccharide repeating unit containing *N-acetylglucosamine* (*NAG*) and *N-acetylmuramic acid* (*NAM*). The alternating NAG-NAM sequence is clearly the "glycan" portion of the peptidoglycan. The "peptido" component consists of short chains of D- and L-amino acids that are linked to one glycan chain but then form covalent cross-links to other peptide chains. In this way, adjacent glycan chains are linked together through their peptide side chains, as shown in Figure 7-24. This cross-linking is so pervasive that the whole cell wall can really be thought of as one giant, cross-linked molecule.

Gram-negative bacteria also contain a peptidoglycan layer, but it is thinner than that of the gram-positive bacteria. Exterior to the peptidoglycan layer is a lipid bilayer that somewhat resembles a membrane.

Plant Cell Walls

The walls that surround plant cells resemble bacterial cell walls only in that they define shape and confer rigidity. Chemically, there is little similarity at all. Plant cell walls

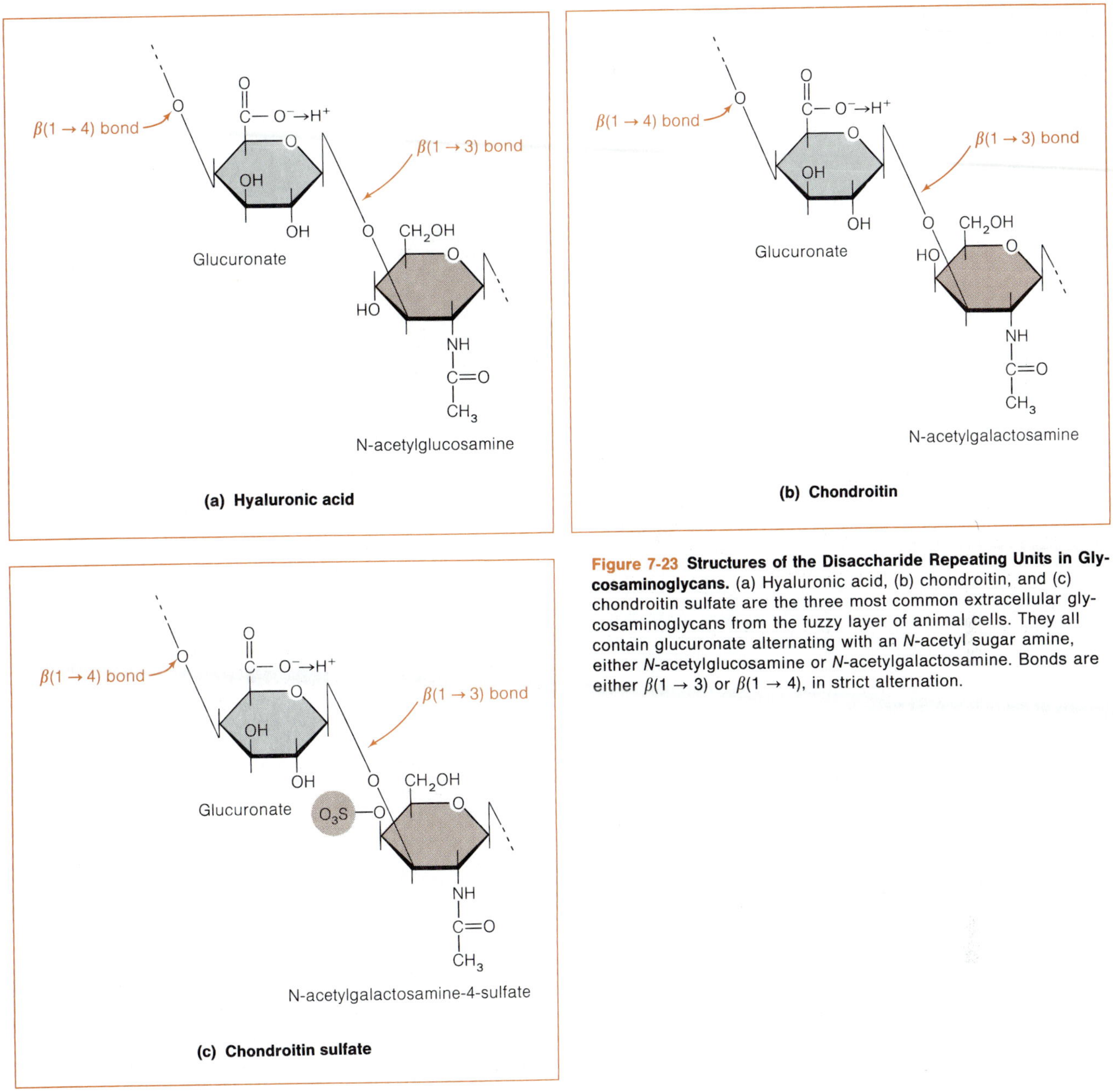

$\beta(1 \rightarrow 4)$ bond
O
C— $O^- \rightarrow H^+$
O
OH
OH
Glucuronate
$\beta(1 \rightarrow 3)$ bond
O
CH_2OH
O
HO
NH
C=O
CH_3
N-acetylglucosamine

(a) Hyaluronic acid

$\beta(1 \rightarrow 4)$ bond
O
C— $O^- \rightarrow H^+$
O
OH
OH
Glucuronate
$\beta(1 \rightarrow 3)$ bond
O
CH_2OH
HO
O
NH
C=O
CH_3
N-acetylgalactosamine

(b) Chondroitin

$\beta(1 \rightarrow 4)$ bond
O
C— $O^- \rightarrow H^+$
O
OH
OH
Glucuronate
$\beta(1 \rightarrow 3)$ bond
O
CH_2OH
O_3S—O
O
NH
C=O
CH_3
N-acetylgalactosamine-4-sulfate

(c) Chondroitin sulfate

Figure 7-23 Structures of the Disaccharide Repeating Units in Glycosaminoglycans. (a) Hyaluronic acid, (b) chondroitin, and (c) chondroitin sulfate are the three most common extracellular glycosaminoglycans from the fuzzy layer of animal cells. They all contain glucuronate alternating with an *N*-acetyl sugar amine, either *N*-acetylglucosamine or *N*-acetylgalactosamine. Bonds are either $\beta(1 \rightarrow 3)$ or $\beta(1 \rightarrow 4)$, in strict alternation.

consist of rigid microfibrils of cellulose molecules embedded in a noncellulosic, gel-like matrix in a manner that has been likened to the embedding of metal rods in concrete to reinforce it and give it added strength.

Cellulose, as you may recall from Chapter 3, is an unbranched polymer of glucose, with $\beta(1 \rightarrow 4)$ bonds. The fibrous matrix in which the cellulose microfibrils are embedded contains *hemicellulose, pectin, lignin,* and protein. Despite the name, hemicellulose bears no chemical resemblance to cellulose at all, but is in fact a pentose polymer. Pectin is a polymer of hexuronic acid, and lignin is a complex organic polymer and a major component of wood. Figure 7-25 illustrates a plant cell wall as visualized with the electron microscope.

The walls that surround growing plant cells are called **primary cell walls.** They are characterized by an extensibility that allows them to expand in response to cell growth, under the influence of the plant hormone *auxin.* Once a plant cell has achieved its final size and shape, it may deposit a **secondary cell wall** on the inner surface of the primary wall. The secondary wall usually contains more cellulose than the primary wall and often has a high lignin content as well, rendering it inextensible and specifying the ultimate size and shape of the cell definitively.

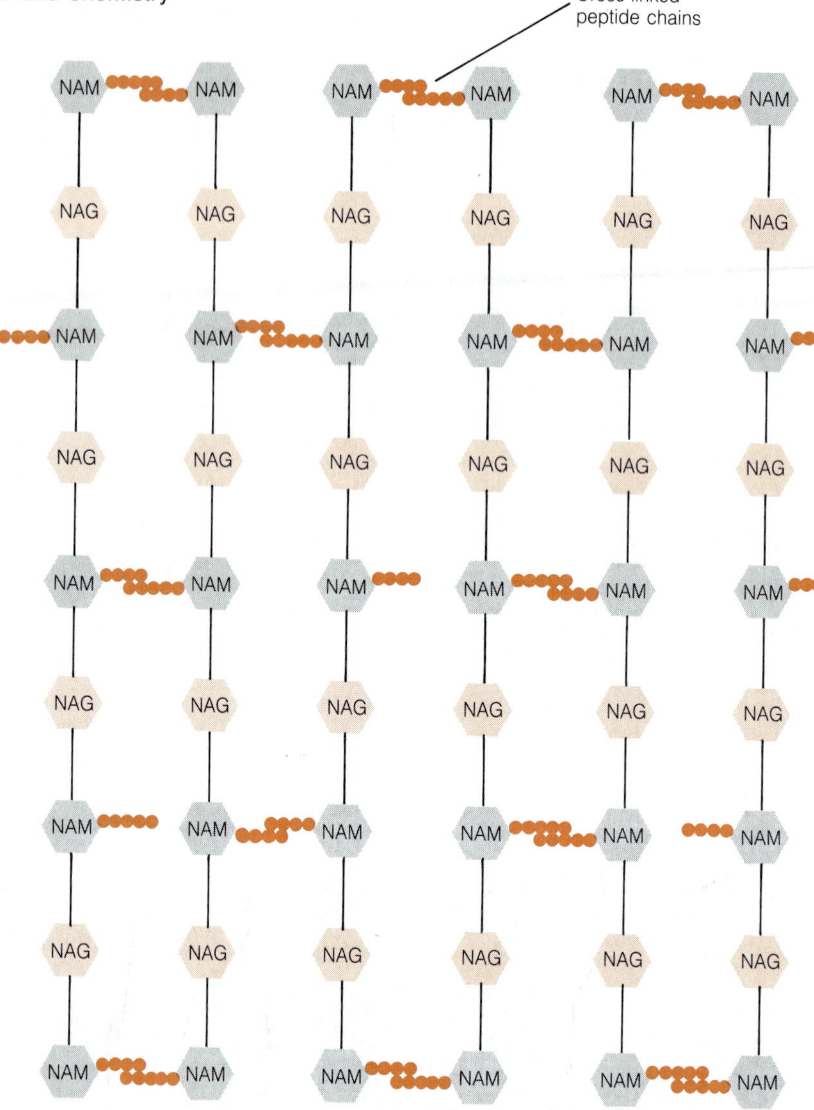

Figure 7-24 Cross-Linked Peptidoglycan Chains of the Cell Wall of Gram-Positive Bacteria. The glycan chains consist of alternating *N*-acetylglucosamine (NAG) and *N*-acetylmuramic acid (NAM) units. Short peptide chains are attached to the NAM units, and these form covalent cross-links with each other, joining the whole peptidoglycan structure into one giant, interconnected network. (See figure 3-21 for structures of NAG and NAM.)

Plasmodesmata: Bridging the Wall

Recognizing that every plant cell is surrounded by both a plasma membrane and a nonliving cellulose wall, you may think it hopeless to expect any sort of intercellular communication such as that afforded by the gap junctions of animal cells. But the **plasmodesmata** (singular: plasmodesma) accomplish this very purpose. As shown in Figure 7-26, plasmodesmata are cytoplasmic channels through pores in the cell wall, allowing fusion of the plasma membranes from two adjacent cells. Each plasmodesma is therefore lined with plasma membrane common to the two connected cells. The channel diameter varies from about 20 to about 200 nm. A single tubular structure, the **desmotubule,** usually lies in the central channel of the plasmodesma. Endoplasmic reticulum (ER) cisternae often approach the plasmodesmata on either side of the cell wall and may actually make connections between cells, possibly by means of the desmotubule.

The ring of cytoplasm between the desmotubule and the membrane that lines the plasmodesma is called the **annulus.** The annulus is thought to provide cytoplasmic continuity between adjacent cells that allows for the passage of molecules from one cell to the next, although the evidence for such passage is largely circumstantial. Even after cell division and deposition of new cell walls between the two daughter cells, cytoplasmic continuities are maintained between the daughter cells by plasmodesmata that pass through the newly formed walls. In fact, most plasmodesmata are formed at the time of cell division, when the new cell wall is being formed. Minor changes may occur later, but the number and location of plasmodesmata are largely fixed at the time of division.

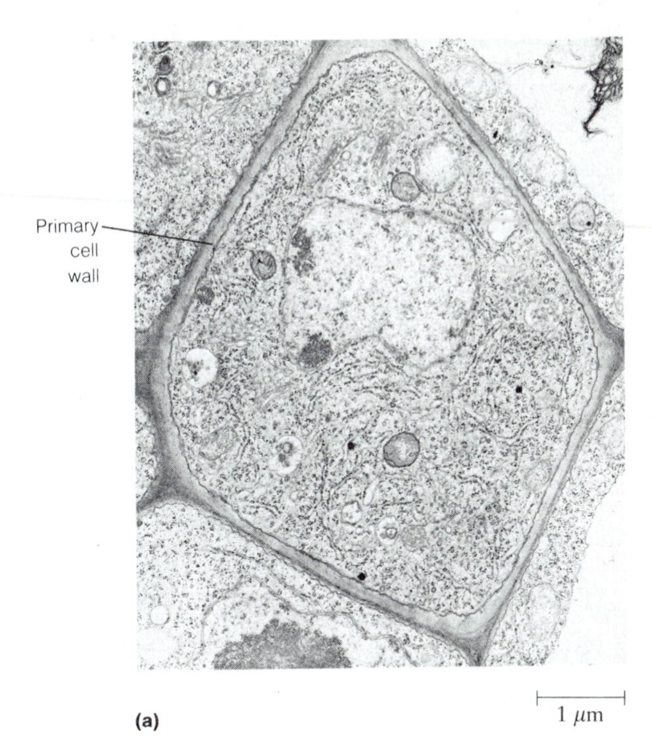

Figure 7-25 **The Plant Cell Wall.** (a) A differentiating cell from the root tip of a bean plant, surrounded by a primary cell wall (TEM). In the freeze-fracture image (b), individual cellulose fibers of the cell wall can be seen, as well as the cell membrane beneath (TEM).

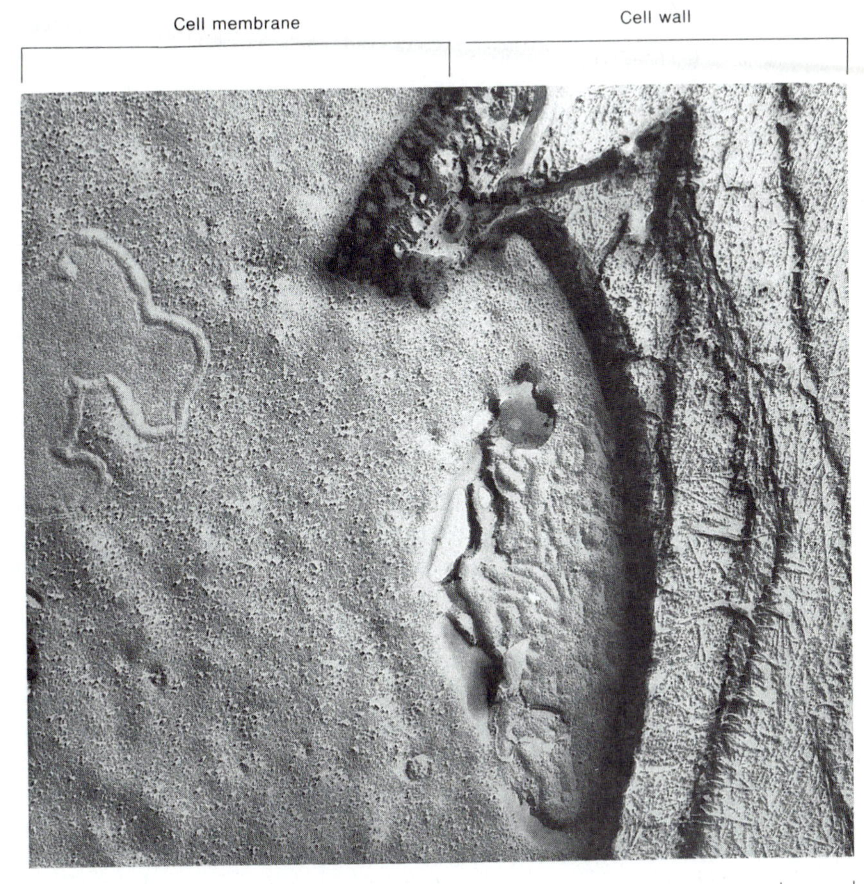

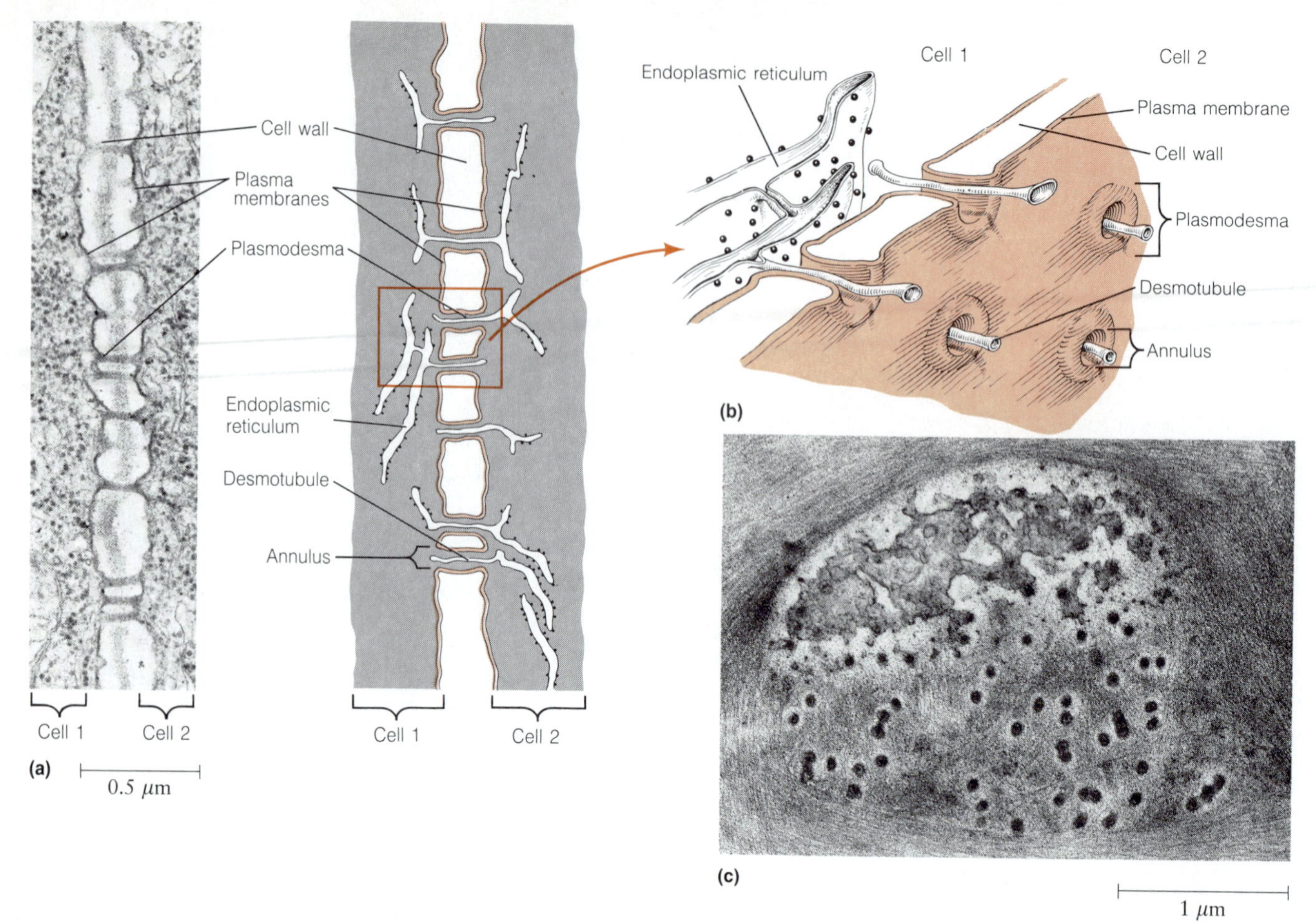

Figure 7-26 Plasmodesmata. (a) Plasmodesmata are channels or pores through the cell wall between two adjacent plant cells, allowing cytoplasmic exchange between the cells. The plasma membrane of one cell is continuous with that of the other cell at each plasmodesma. Most plasmodesmata have a narrow cylindrical desmotubule at the center, which is derived from the ER and appears to be continuous with the ER of both cells. Between the desmotubule and the plasma membrane that lines the plasmodesma is a narrow ring of cytoplasm called the annulus. This electron micrograph shows the cell wall between two adjacent root cells of timothy grass, with numerous plasmodesmata (TEM). (b) A diagrammatic view of an upright cell wall with numerous plasmodesmata. (c) This electron micrograph provides a view of a cell wall showing a number of plasmodesmata in cross section (TEM).

Perspective

Cells need membranes to define and compartmentalize space, regulate the flow of materials and information, and mediate recognition and interaction between cells. Our current understanding of membrane structure represents the culmination of almost a century of studies that began with the recognition that lipids are an important membrane component and then moved progressively from a lipid monolayer to a bilayer. Once proteins were acknowledged as important components, the Davson-Danielli "sandwich" was proposed and universalized as the basis of the unit membrane. Only recently has the fluid mosaic model emerged as the most accepted representation of membrane structure. According to the model, proteins with varying affinities for the hydrophobic membrane interior float in a sea of phospholipid. These proteins function as enzymes, transport molecules, electron carriers, and receptor sites for chemical signals such as neurotransmitters and hormones. Some are glycoproteins, with carbohydrate chains that invariably protrude out of the membrane on the external side.

In animals, membranes are joined together by desmosomes, tight junctions, and gap junctions. These serve adhering, sealing, and communicative functions, respectively. In plant cells, such membrane-mediated contact is not generally possible because of the cellulosic wall that surrounds and encases each cell. Intercellular communication still occurs, however, through the plasmodesmata that connect cells via pores in the cell walls.

Key Terms for Self-Testing

Why Membranes?
plasma membrane (p. 158)
intracellular membranes (p. 158)
marker enzyme (p. 158)
passive transport (p. 159)
active transport (p. 159)
endocytosis (p. 159)
signal recognition (p. 159)
receptor sites (p. 159)

Membrane Structure: A Historical Perspective
Langmuir trough (p. 160)
lipid monolayer (p. 160)
lipid bilayer (p. 161)
differential permeability (p. 161)
Davson-Danielli model (p. 162)
unit membrane (p. 163)
fluid mosaic model (p. 163)

Molecular Organization and Membrane Function
integral (intrinsic) membrane proteins (p. 166)
glycophorin (p. 166)
peripheral (extrinsic) membrane proteins (p. 166)
chromatography (p. 166)
thin-layer chromatography (TLC) (p. 167)
SDS-polyacrylamide gel electrophoresis (p. 168)
freeze-fracture (p. 168)
transport ATPases (p. 171)
receptors (p. 171)
clathrin (p. 171)
glycolipids (p. 171)
glycoproteins (p. 171)
lectins (p. 172)
transverse diffusion (p. 172)
lateral diffusion (p. 172)
fluidity (p. 173)
fluorescent antibodies (p. 173)
patching (p. 176)
capping (p. 176)

Cell Junctions
cell junctions (p. 178)
desmosomes (p. 178)
belt desmosomes (p. 178)
spot desmosomes (p. 180)
tonofilaments (p. 180)
tight junctions (p. 180)
gap junction (p. 180)
connexons (p. 180)

The Cell Surface: Coats and Walls
extracellular matrix (p. 182)
cell coat (p. 182)
fuzzy layer (p. 182)
cell wall (p. 182)
glycocalyx (p. 183)
collagen (p. 183)
elastin (p. 183)
glycosaminoglycans (p. 183)
mucopolysaccharides (p. 183)
mucoproteins (p. 183)
teichoic acid (p. 184)
peptidoglycans (p. 184)
primary cell wall (p. 185)
secondary cell wall (p. 185)
plasmodesmata (p. 186)
desmotubule (p. 186)
annulus (p. 186)

Suggested Reading

General Membrane Structure and Function

Branton, D., and R. Park. *Papers on Biological Membrane Structure.* Boston: Little, Brown, 1968.

Bretscher, M. The molecules of the cell membrane. *Sci. Amer.* 253 (April 1985): 100.

Finean, J. B. R., R. Coleman, and R. H. Michell, eds. *Membranes and Their Cellular Functions,* 3d ed. Oxford: Blackwell, 1984.

Gennis, R. B. *Biomembranes.* New York: Springer-Verlag, 1989.

Houslay, M. D., and K. K. Stanley. *Dynamics of Biological Membranes.* New York: Wiley, 1982.

Kleinfeld, A. Current views of membrane structure. *Current Topics Membr. Transp.* 29 (1987): 1.

Vance, D. E., and J. E. Vance. *Biochemistry of Lipids and Membranes.* Menlo Park, Calif: Benjamin/Cummings, 1985.

The Fluid Mosaic Model

de Mendoza, D., and J. E. Cronan, Jr. Thermal regulation of membrane lipid fluidity in bacteria. *Trends Biochem. Sci.* 8 (1983): 49.

Frye, C. L. D., and M. Edidin. The rapid mixing of cell surface antigens after formation of mouse-human hybrids. *J. Cell. Sci.* 7 (1970): 319.

Singer, S. J. The molecular organization of membranes. *Annu. Rev. Biochem.* 43 (1974): 805.

Singer, S. J., and G. L. Nicolson. The fluid mosaic model of the structure of cell membranes. *Science* 175 (1972): 720.

Membrane Proteins and Assembly

Byers, T. J., and D. Branton. Visualization of the protein associations in the erythrocyte membrane skeleton. *Proc. Natl. Acad. Sci. USA* 82 (1985): 6153.

Lodish, H. F., and J. E. Rothman. The assembly of cell membranes. *Sci. Amer.* 240 (January 1979): 48.

Rothman, J. E., and J. Lenard. Membrane asymmetry. *Science* 195 (1977): 743.

Unwin, N., and R. Henderson. The structure of proteins in biological membranes. *Sci. Amer.* 250 (February 1984): 78.

Junctions Between Cells

Evans, W. H. Communication between cells. *Nature* 283 (1980): 521.

Hertzberg, E. L., T. S. Lawrence, and N. B. Gilula. Gap junctional communication. *Annu. Rev. Physiol.* 43 (1981): 479.

Peracchia, C. Gap junction structure and function. *Trends Biochem. Sci.* 2 (1977): 26.

Pinto da Silva, P., and B. Kachar. On tight-junction structure. *Cell* 28 (1982): 441.

Staehelin, L. A., and B. E. Hull. Junctions between living cells. *Sci. Amer.* 238 (May 1978): 140.

Unwin, P. N. T., and G. Zampighi. Structure of the junction between communicating cells. *Nature* 283 (1980): 545.

The Cell Surface: Walls and Coats

Albersheim, P. The walls of growing plant cells. *Sci. Amer.* 232 (April 1975): 80.

Hay, E. D. Extracellular matrix. *J. Cell. Biol.* 91 (1981): 205s.

Hay, E. D., ed. *Cell Biology of Extracellular Matrix.* New York: Plenum, 1981.

Robards, A. W. Plasmodesmata. *Annu. Rev. Plant. Physiol.* 26 (1975): 13.

Problem Set

1. **Why Membranes?** For each of the following statements about membranes, indicate which of the four general functions of membranes (compartmentalization, localization of function, control of movement, or detection and recognition) the statement seems to illustrate.

   (a) The plasma membrane of a muscle cell is excitable and capable of conducting an action potential.

   (b) The membrane of a plant root cell has an ion pump that exchanges phosphate inward for bicarbonate outward.

   (c) When cells are disrupted and fractionated into subcellular components, the enzyme cytochrome *c* reductase is recovered with an endoplasmic reticulum fraction.

   (d) Cells carry tissue-specific glycoproteins on their outer surface that are responsible for adhesion.

   (e) Membranes are composed primarily of phospholipids and hydrophobic proteins.

   (f) Photosystems I and II are embedded in the thylakoid membrane of the chloroplast.

   (g) All of the acid phosphatase in a mammalian cell is found within the lysosomes.

   (h) The mitochondrial membrane is impermeable to ATP but contains an ATP-ADP carrier that couples outward ATP movement with inward ADP movement.

   (i) Insulin does not actually enter a target cell, but binds to a specific membrane receptor on the external surface of the membrane, thereby activating a specific enzyme, adenylate cyclase, on the inner membrane surface.

2. **Elucidation of Membrane Structure.** Each of the following observations played an important role in enhancing our understanding of membrane structure. Explain the significance of each, and indicate in what decade in the time line of Figure 7-3 the observation was made.

   (a) When a membrane is observed in the electron microscope, both of the thin, electron-dense lines are about 2 nm thick, but they are often distinctively different from each other in appearance.

   (b) Ethylurea penetrates much more readily into a membrane than does urea, and diethylurea penetrates still more readily.

   (c) Addition of phospholipase to living cells causes rapid digestion of the lipid bilayers of the membranes, which suggests that the enzyme has ready access to the membrane phospholipids.

   (d) When artificial lipid bilayers are subjected to freeze-fracture, no particles are seen on either face.

   (e) When the resistivity of artificial lipid bilayers is measured, it turns out to be three or four orders of magnitude greater than that of real membranes.

   (f) When artificial lipid bilayers are fixed, stained with osmium, and viewed in the electron microscope, they display the same "railroad track" appearance as real membranes.

   (g) Some membrane proteins can be readily extracted with 1 *M* NaCl, while others require the use of an organic solvent or a detergent.

   (h) When halobacteria are grown in the absence of oxygen, they produce a purple pigment that is embedded in their plasma membranes and has the ability to pump protons outward when illuminated. If the purple bacterial membranes are isolated and viewed by freeze-fracture electron microscopy, they are found to contain patches of crystallized particles.

3. **Lipid, Protein, and Bilayers in Red Cells.** A single human red cell membrane is about 8 nm thick and contains 5.2×10^{-13} g lipid and 6.0×10^{-13} g protein. The lipid is primarily phospholipid (MW 800) and cholesterol (MW 386) in a 1:1 molar ratio. In a monolayer, each phospholipid occupies a surface area of 0.55 nm^2 per molecule, and cholesterol occupies 0.38 nm^2 per molecule. A human red cell has a surface area of 145 μm^2. Using this information, you should be able to answer the following questions.

   (a) How many molecules of protein are in a single red cell membrane? (Assume an average protein has a molecular weight of 50,000.)

(b) What is the ratio of lipid molecules to protein molecules in the red cell membrane?

(c) What proportion of the total surface area of the membrane is occupied by a lipid bilayer?

4. **Lateral and Transverse Diffusion.** Lateral diffusion occurs readily and rapidly for both the proteins and lipids of the membrane, but neither can undergo transverse diffusion very easily.

   (a) Explain this difference between lateral and transverse diffusion.

   (b) What is the consequence for the cell of free lateral diffusion of membrane components?

   (c) What would be the consequence for the cell if transverse diffusion were also to occur readily?

5. **Temperature and Membrane Composition.** Which of the following responses are likely to be seen when a bacterial culture growing at 37°C is transferred to a culture room maintained at 25°C?

   (a) Initial decrease in membrane fluidity.

   (b) Gradual replacement of shorter fatty acids by longer fatty acids in the membrane phospholipids.

   (c) Gradual replacement of stearate (18 carbon atoms, saturated) by oleate (18 carbon atoms, one double bond) in the membrane phospholipids.

   (d) Enhanced rate of synthesis of unsaturated fatty acids.

   (e) Loss of integral proteins from the membrane.

6. **Cellular Junctions and Plasmodesmata.** Indicate whether each of the following statements is true of desmosomes (D), tight junctions (T), gap junctions (G), and/or plasmodesmata (P). A given statement may be true of any, all, or none of these structures.

   (a) Found in animal cells, but not plant cells.

   (b) Allow exchange of metabolites between the cytoplasms of two adjacent cells.

   (c) Membranes of two adjacent cells sealed tightly together.

   (d) Contain hexagonal particles with a central opening or core.

   (e) Involve peptidoglycan chains with *N*-acetylglucosamine repeating units.

   (f) Sites of true membrane fusion are restricted to abutting ridges of adjacent membranes.

   (g) Associated with filaments that confer either contractile or tensile properties.

7. **Membrane Densities.** The density of a membrane depends on its relative protein and lipid content, since protein has an average density of 1.33 g/cm^3 and phospholipid has an average density of about 0.92 g/cm^3. Protein/lipid ratios (on a weight basis) for several membranes are as follows: myelin, 0.25; mitochondrial inner membrane, 3.6; chloroplast lamellae, 0.8; and the plasma membrane of a liver cell, 1.2.

   (a) Calculate the density of each of the above membranes.

   (b) What protein/lipid ratio would allow a membrane to band in a sucrose density gradient at 1.18 g/cm^3?

   (c) Can you see why it is possible to separate eukaryotic organelles from one another on a sucrose density gradient? Would you expect mitochondria to band at a higher or lower density than chloroplasts on such a gradient? If you were told that lysosomes band at a density of about 1.25 g/cm^3 on such a gradient, what conclusions might you draw about these organelles?

8. **Inside or Outside?** From the box on pp. 174–175, we know that exposed regions of membrane proteins can be labeled with ^{125}I by the lactoperoxidase (LP) reaction, whereas carbohydrate side chains of membrane glycoproteins can be labeled with ^{3}H by oxidation of galactose groups with galactose oxidase (GO), followed by reduction with tritiated borohydride (^{3}H-BH$_4$). Noting that both LP and GO are too large to penetrate into the interior of an intact cell, explain each of the following observations made with intact red blood cells.

   (a) When intact cells are incubated with LP in the presence of ^{125}I and peroxide and the membrane proteins are then extracted and analyzed on SDS-polyacrylamide gels, several of the bands on the gel can be shown to be radioactive.

   (b) When intact cells are incubated with GO and then reduced with ^{3}H-BH$_4$, several of the bands on the gel can be shown to be radioactive.

   (c) All the proteins of the red blood cell membrane that are known to contain carbohydrates are labeled by the GO-^{3}H-BH$_4$ method.

   (d) None of the proteins of the red blood cell membrane that are known to be devoid of carbohydrate is labeled by the LP-^{125}I method.

   (e) If the red blood cells are ruptured before the labeling procedure, the LP procedure labels virtually all the major membrane proteins.

9. **Inside-Out Membranes.** It is technically possible to prepare sealed vesicles from red cell membranes in which the original orientation of the membrane is inverted. Such vesicles have what was originally the cytoplasmic face of the membrane facing outward.

   (a) What results would you expect if such inside-out vesicles were subjected to the GO-borohydride procedure described in Problem 8?

   (b) What results would you expect if such inside-out vesicles were subjected to the LP-iodine procedure of Problem 8?

(c) What conclusion would you draw if some of the proteins that become labeled by the LP-iodine method of part (b) were among those that had been labeled when intact cells were treated in the same way in part (a) of Problem 8?

10. **An Inside Job.** Given that it is possible to prepare inside-out vesicles from red blood cell membranes as described in Problem 9, can you think of any way to label a transmembrane protein with ^{3}H on one side of the membrane and with ^{125}I on the other side?

8

Transport Across Membranes: Overcoming the Permeability Barrier

In Chapter 7, we focused on membrane structure and chemistry. Membranes were described as forming barriers to free diffusion of molecules or ions, thereby allowing specific cellular functions to be localized within particular compartments in which the appropriate molecules, ions, enzymes, and other factors are confined. However, membranes are not simply permeability barriers. An equally important function of membranes is the ability to control the movement of selected substances across the permeability barrier and into and out of specific compartments. In other words, membranes cannot be completely impermeable; instead, they must permit certain molecules and ions to pass through and exclude others. We will now consider the means whereby substances are moved selectively across cell membranes and the significance of such transport processes.

Cells and Transport Processes

An essential feature of virtually every cell and almost every subcellular compartment is the ability to accumulate a variety of organic metabolites and inorganic ions at concentrations that are often strikingly different from those in the surrounding milieu. For most such substances, the concentration inside the cell is markedly higher than that outside. In fact, life could be defined as a state not only of improbable structural order but also of improbable solute concentrations. Very few cellular reactions or processes could occur at reasonable rates if they had to depend on the concentrations at which essential substrates were present in the cell's surroundings. A central aspect of cell function, then, is the ability to move molecules and ions across membranes, often against a concentration gradient.

Categories of Transport

When we consider the transport of small, organic metabolites and inorganic ions across membranes, three categories of transport come readily to mind. The three use similar mechanisms but accomplish different end results.

Cellular transport concerns the exchange of materials between the cell and its environment (Figure 8-1), including the uptake of nutrients and other raw materials across the plasma membrane and the removal of wastes and secretory products. The uptake of glucose from the bloodstream into the cells of your body is an example of cellular transport, as is the inward pumping of potassium ions accompanied by the outward pumping of sodium ions that is characteristic of most animal cells.

Intracellular transport, on the other hand, involves movement of substances across membranes of organelles inside the cell. Intracellular transport is a eukaryotic phenomenon and includes the molecular and ionic traffic into and out of such organelles as the nucleus, mitochondrion, chloroplast, lysosome, peroxisome, Golgi body, and endoplasmic reticulum. Several examples of intracellular transport are illustrated in Figure 8-1.

Transcellular transport, as the name implies, does not simply move a substance into the cell from the outside, but moves it in one side and out the other, thereby accomplishing net transport across the cell. In multicellular organisms, transcellular transport occurs through cell layers that act as semipermeable barriers. Examples include the epithelial cells lining the gastrointestinal tract of animals, which take up glucose, amino acids, and other nutrients during digestion and deliver them to the blood, and the cells of plant roots responsible for absorption of water and mineral salts.

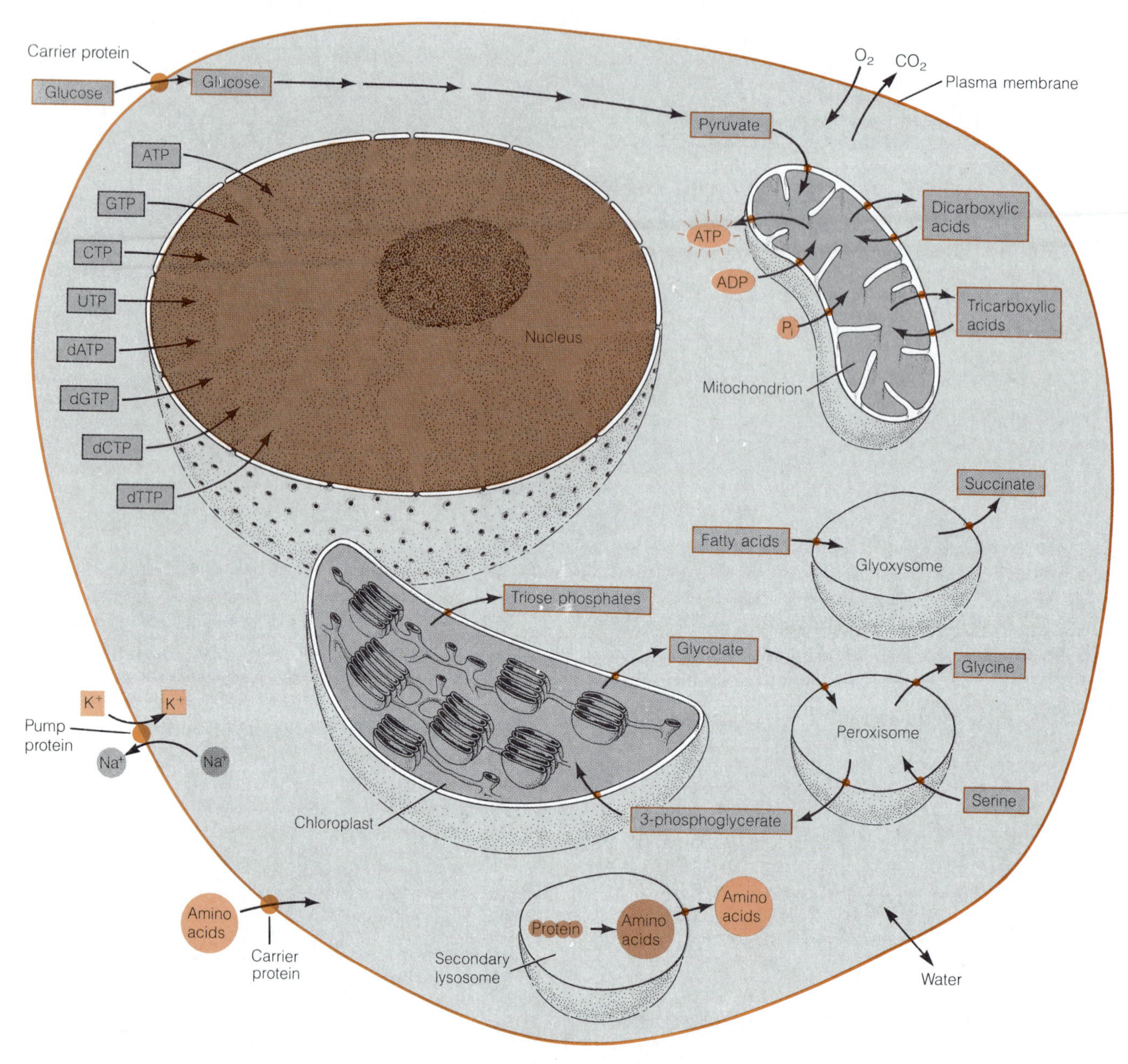

Figure 8-1 Transport Processes Operative Within a Composite Eukaryotic Cell. Common solutes that are transported across cellular or organellar membranes include water, gases, sugars, amino acids, nucleotides, ions, protons, ATP, ADP, inorganic phosphate, and various intermediates in metabolic pathways.

Mechanisms of Membrane Transport

Membranes represent significant barriers to the free movement of solutes into and out of cellular compartments. Sometimes the barrier is necessary for a cell function, such as maintenance of the ion gradients required for nerve cell function. In other cases, the barrier must be overcome so that a solute can move into or out of a cell or intracellular compartment. The latter process is called **membrane transport.**

In general, membrane transport can be either *passive* or *active*, depending on its energy requirement. **Passive transport** does not require energy. Instead, it occurs because of the tendency for dissolved molecules to move from higher to lower concentrations, as shown in Figure 8-2a. Passive transport processes usually require a **carrier protein** to help larger solutes cross the membrane barrier, although smaller molecules such as water and dissolved gases can diffuse through the barrier unaided. In contrast, **active transport** always requires the input of energy, and is catalyzed directly or indirectly by specialized membrane-associated enzymes often referred to as "pumps." Because of the energy input, active transport can result in the production of a concentration gradient (Figure 8-2b). Ions such as sodium (Na^+), potassium (K^+), calcium (Ca^{2+}), and protons (H^+) are pumped across membranes by virtually all cells. Because ions have an electrical charge, the transport process and resulting concentration gradients can produce an electrical voltage, or *membrane potential*, across the membrane. The combined concentration gradient and associated membrane potential is referred to as an **electrochemical gradient.** The stored energy of such gradients is coupled to functions such as ATP synthesis by mitochondria and chloroplasts (Chapters 11 and 12) and the action potential of nerve cell membranes (Chapter 20).

With these concepts in mind, we can go on to a more detailed discussion of passive and active transport processes.

Passive Transport

Passive transport is a common membrane transport process in the body. For instance, glucose circulates in the blood and is transported across the plasma membrane of all cells by specialized carrier proteins. This process is called **facilitated transport,** because of the involvement of carrier proteins. Inside the cell, glucose is immediately used as a nutrient, so that a concentration gradient is maintained, with the higher concentration outside the cell. It is this concentration gradient that drives the transport process. Direct input of metabolic energy is not required, so the transport of glucose and most other nutrients into cells is classified as passive transport.

Not all passive transport requires carrier proteins, however. Oxygen, for example, is small enough to cross membranes unaided. Red blood cells returning from the tissues of the body to the lungs are depleted in oxygen. As a result, oxygen continuously moves passively from the air in the lungs (higher concentration) across the red cell plasma membrane and into the intracellular volume (lower concentration) where it binds to hemoglobin. Membranes are also sufficiently permeable to water and carbon dioxide so that specialized transport proteins are not required for these small molecules either. We will refer to this transport process as **simple diffusion,** to distinguish it from facilitated transport, as described above for glucose.

Simple Diffusion and Membrane Permeability

We will return to facilitated transport shortly, but will first consider the membrane as a barrier to simple diffusion. We can begin by asking what component of a membrane is most significant in establishing the membrane as a barrier. To answer this question, scientists frequently use membrane models. An important experimental advance in understanding the barrier properties of membranes came in 1965, when Alec Bangham and his co-workers reported that membrane lipid could be extracted and dispersed as **liposomes,** small vesicles about 0.1 μm in diameter, consisting of closed lipid bilayers. Bangham found that it was possible to trap solutes such as potassium in the liposomes and then measure the rate at which they escape by passive transport across the membranes. The results were remarkable: Ions such as potassium and sodium were trapped in the vesicles for days, whereas small molecules such as water exchanged so rapidly that the rates could hardly be measured. The inescapable conclusion was that lipid bilayers represented the primary permeability barrier of all membranes. Small molecules such as water could pass through the barrier by simple diffusion, but sodium and potassium ions could hardly cross at all.

Over the years, this type of experiment was repeated for a variety of lipid bilayer systems, both synthetic and natural, and with thousands of different solutes. Now, some general rules have been established that let us predict with some confidence how permeable a membrane will be to different solutes. These rules are summarized in Table 8-1, together with examples of solutes to which bilayers are relatively permeable and relatively impermeable. The rules suggest that three primary properties of solutes must be taken into consideration: relative size, relative polarity, and ionization.

Relative Size. Generally speaking, lipid bilayers are more permeable to smaller molecules than to larger molecules. The smallest molecules relevant to cell function are water (H_2O), oxygen (O_2), and carbon dioxide (CO_2). Membranes are sufficiently permeable to such molecules so that no specialized transport processes are required to get them into and out of cells. Small molecules do not permeate freely, however. For instance, water is passively transported across a bilayer *ten thousand times slower* than it would be transported by diffusion alone in the absence of a membrane. Therefore, the best way to think about the passive transport of small molecules is to consider that their diffusion is strongly hindered by the presence of a lipid bilayer but that they can occasionally "dissolve" in the bilayer, followed by random diffusion to the other side, where they can make their exit.

The size rule holds until molecules approach glucose in molecular weight. Glucose itself is too large, and its transport too slow, for sufficient amounts to cross the lipid bilayer barrier by simple diffusion. As will be seen later on, most cells therefore include specialized carrier proteins in their plasma membranes to facilitate the entry of glucose and other large nutrient molecules.

Polarity. In general, lipid bilayers are relatively permeable to nonpolar molecules and less permeable to polar

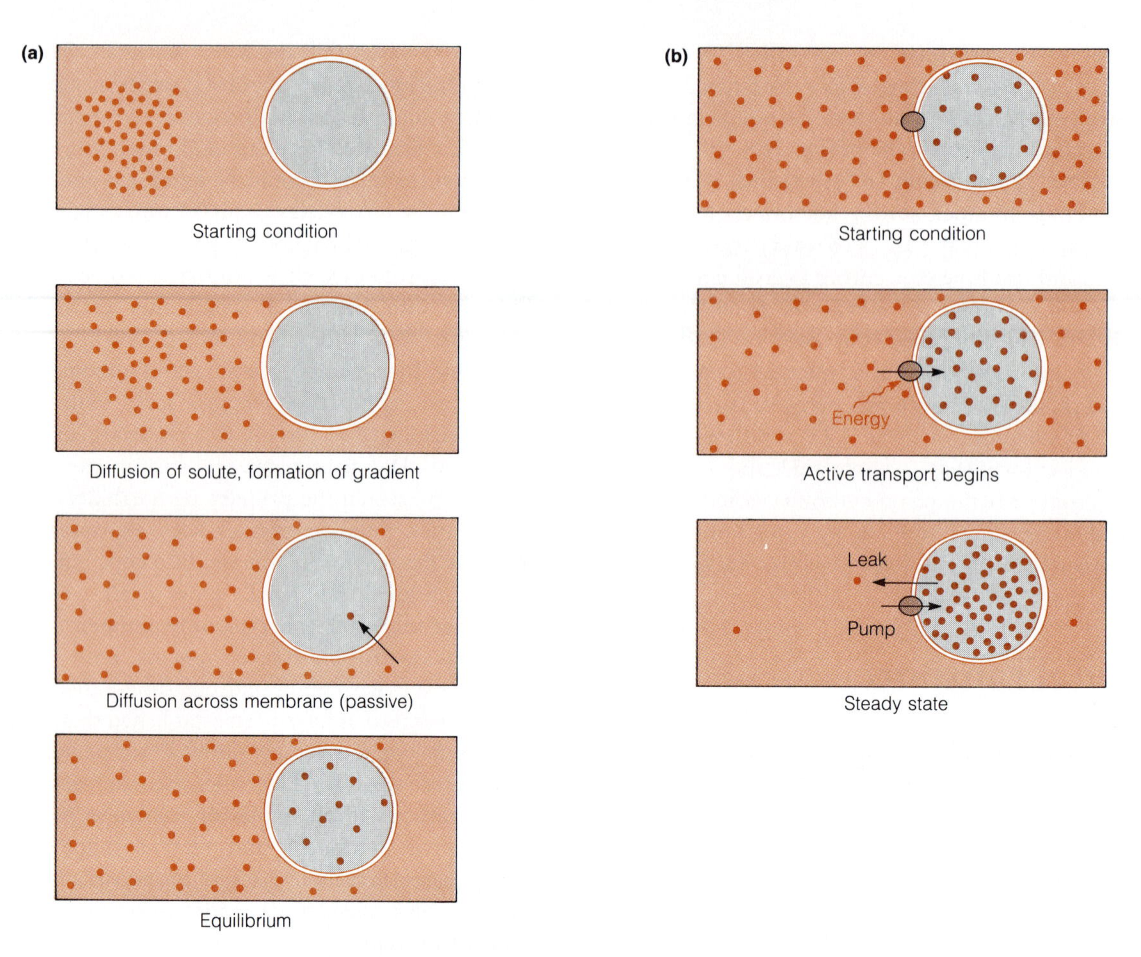

Figure 8-2 Passive and Active Transport. In passive transport processes (a) a solute is added to a volume containing a membrane-bounded space. The solute diffuses throughout the external space, but the membrane barrier keeps it from immediately entering the internal space, resulting in a concentration gradient across the membrane. Over a period of time, the solute permeates the membrane by simple diffusion or facilitated transport down its concentration gradient, and the system comes to equilibrium. Active transport is shown in (b), in which a pump is present in the membrane. The system begins at equilibrium, but when energy is provided to the pump a vectorial transport begins that produces a concentration gradient. When a sufficiently large gradient has developed, the outward leak balances the inward transport process and the system reaches a condition called steady state.

Table 8-1 Factors Governing Diffusion Across Lipid Bilayers

Rule	Examples: More Permeable	Examples: Less Permeable	Permeability Ratio
1. Size rule: bilayer more permeable to smaller molecules	H_2O (Water)	$H_2N—CO—NH_2$ (Urea)	$10^2:1$
2. Polarity rule: bilayer more permeable to nonpolar molecules	$CH_3—CH_2—CH_2—OH$ (Propanol)	$HO—CH_2—CHOH—CH_2—OH$ (Glycerol)	$10^3:1$
3. Ionic rule: bilayer highly impermeable to ions	O_2 (Oxygen)	OH^- (Hydroxide ion)	$10^9:1$

molecules. This is because nonpolar molecules more readily dissolve in the nonpolar phase of the lipid bilayer, which increases their probability of crossing the membrane barrier. As discussed in Chapter 7, the relationship between the lipid solubility of various solutes and their membrane permeability was one of the first indications that a nonpolar hydrophobic phase might be present in membranes. Figure 8-3 illustrates the relationship between the relative solubility of a compound in a nonpolar solvent (olive oil) and its rate of diffusion across a membrane. Note that a polar compound like malonamide has a relatively low membrane permeability. However, if it is made less polar by adding $—CH_2—CH_3$ (ethyl) groups to form diethylmalonamide, both its lipid solubility and membrane permeability increase. Similar relationships can be seen with the urea series: urea → ethylurea → diethylurea.

Ionization. As a general rule, lipid bilayers are not very permeable to ions. This is because a great deal of energy (about 40 kcal/mol) is required to move ions from an aqueous environment into a nonpolar environment. You would not expect to be able to dissolve salt in oil or gasoline, and by the same token, you should not expect sodium or chloride ions to dissolve in and pass through the nonpolar, "oily" portion of a lipid bilayer. The high relative impermeability of membranes to ions is very important to cell activity, because nerve cells, mitochondria, and chloroplasts must maintain large ion gradients in order to function. On the other hand, membranes must also allow ions to cross the barrier under certain circumstances, as in conduction of nerve action potentials. As we shall see, this task is carried out by specialized carrier proteins embedded in the lipid bilayer.

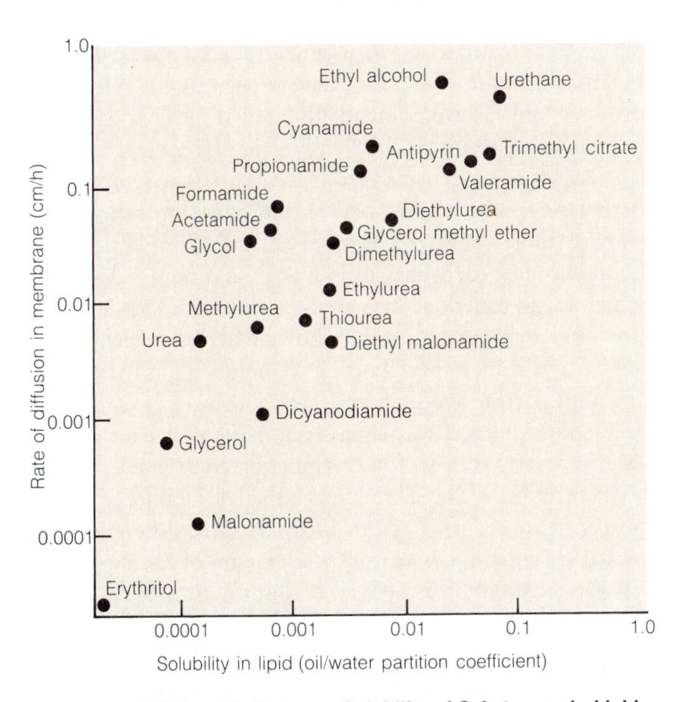

Figure 8-3 Relationship Between Solubility of Substances in Lipid and Rate of Diffusion Across a Membrane. The solubility of various solutes in a nonpolar solvent is expressed as the partition coefficient for the solute between olive oil and water increasing logarithmically from left to right. The rate of diffusion, or membrane permeability, is also plotted on a logarithmic scale. These data were determined for the plasma membrane of the alga *Chara* and are influenced by the presence of membrane proteins.

Diffusion and Passive Transport

So far, we have provided only a qualitative understanding of simple diffusion and permeability. That is, permeant solutes diffuse to a membrane from either side, dissolve in the bilayer, diffuse across the bilayer, and then emerge from the other side.

We can be more quantitative by considering the process of diffusion in further detail. Thermodynamically, simple diffusion operates in the direction dictated by the concentration gradient and is always an exergonic process, driven by the free energy inherent in the concentration gradient. Because it is exergonic, this transport requires no input of metabolic energy. Its direction is determined only by the relative concentrations of the solute on the two sides of the membrane, so there is no inherent directionality with respect to the inside of the cell or organelle.

Kinetically, a key feature of diffusion is that the net rate of transport for a specific substance is directly proportional to the concentration difference for that substance across the membrane, as indicated below:

$$v = k_D([X]_{outside} - [X]_{inside}) \tag{8-1}$$

where v is the rate of diffusion of the substance from the outside to the inside of the cell or organelle (or from inside to outside if v is negative), $[X]_{outside}$ is the concentration of substance X outside the cell, $[X]_{inside}$ is the concentration of substance X inside the cell, and k_D is the **diffusion constant.** The factors that affect the value of the diffusion constant k_D for a given solute include its relative size, polarity and ionization, and the temperature.

Passive transport by diffusion of solutes to and across lipid bilayers is characterized by a linear relationship between the concentration gradient and the rate of movement across the membrane, with no evidence of saturation at high concentrations. Simple diffusion differs in this respect from facilitated transport, which is subject to saturation and generally follows Michaelis-Menten kinetics. It is therefore possible to distinguish between the two processes kinetically, as indicated in Table 8-2.

Table 8-2 Properties of Passive and Active Transport

		Passive Transport		Active Transport
		Diffusion	Facilitated	
Solutes transported	Example			
Small nonpolar	Oxygen	Yes	No	No
Large nonpolar	Fatty acids	Yes	No	No
Small polar	Water	Yes	No	No
Large polar	Glucose	No	Yes	No
Ions	N^+, K^+, Ca^{2+}	No	Yes	Yes
Thermodynamic properties				
Direction relative to electrochemical gradient		Down	Down	Up
Effect on entropy		Increased	Increased	Decreased
Metabolic energy required		No	No	Yes
Intrinsic directionality		No	No	Yes
Kinetic properties				
Carrier mediated		No	Yes	(Pump)
Michaelis-Menten kinetics		No	Yes	Yes
Competitive inhibition		No	Yes	Yes

We can summarize our discussion of simple diffusion by saying that it applies most generally to small molecules such as water, does not require a carrier, moves a substance exergonically in the direction dictated by the concentration gradient for that substance, and results in a linear, nonsaturating relationship between rate and concentration.

Facilitated Transport

Although simple diffusion lets us understand how water and other small molecules cross the bilayer portion of membranes, most substances required by cells are polar or ionic and move across membranes at appreciable rates only if cells and organelles have specific means of facilitating that movement. Facilitated transport therefore differs from simple diffusion in that carrier proteins are required to effect the passage of substances across relatively impermeable membranes.

Facilitated transport uses the same driving force as that for simple diffusion: the movement of solutes from a region of higher concentration to a region of lower concentration. The task of the carrier proteins involved in passive transport is to facilitate the diffusion of polar solutes across an otherwise impermeable membrane. Carrier proteins are invariably integral proteins embedded in the membranes of cells and organelles and are highly specific, often for a single compound or small groups of closely related compounds.

Such carriers are sometimes called **transport proteins** or **permeases.** The latter term is especially apt because the suffix *ase* suggests a similarity between transport proteins and enzymes that turns out to be valid. Like an enzyme-catalyzed reaction, facilitated transport always involves an initial binding of carrier and substrate or solute to be transported, a process mediated on a protein surface (an enzyme-catalyzed reaction in one instance, a physical translocation of solute in the other), a subsequent release of product, and a reduction in the activation energy of the reaction as a result of the involvement of the carrier.

As might be expected from the analogy with enzymes, permeases become saturated as the concentration of the transportable solute is raised, because the number of sites is limited and each has some finite maximum velocity at which it can function. As a result, carrier-facilitated transport, like enzyme catalysis, follows Michaelis-Menten kinetics, with an upper limiting velocity and a Michaelis constant corresponding to the concentration of transportable solute needed to achieve one-half of the maximum rate of transport. A plot of transport rate versus solute concentration is therefore hyperbolic for facilitated transport instead of linear as for diffusion. This difference, depicted in Figure 8-4, is an important means of distinguishing between simple diffusion and facilitated transport.

Mechanism of Transport Protein Function. Transport proteins function by binding to the desired solute molecules in such a way as to shield the polar groups of the solute from the nonpolar interior of the membrane. As you might expect, carrier proteins are an important research topic. In no case, however, is the actual mechanism of transport known. At one time it was thought that carriers functioned by binding the solute on one side of the membrane and then either diffusing across the membrane or flipping within it to bring the solute to the other side of the membrane, where release could occur. Most membrane biochemists agree that this mechanism is no longer tenable in light of what we now know about membrane chemistry. Your understanding of membrane asymmetry from Chapter 7 should enable you to understand how unlikely such transverse or flip-flop movements would be.

Instead, transport proteins are now thought to form hydrophilic channels through the membrane that allow specific solutes to move from one side to the other. The involvement of such a hydrophilic channel is shown schematically in Figure 8-5 for the passage of a polar sugar molecule from the outside of a cell to the inside. Such models are easiest to envision if the transport proteins are transmembrane proteins, and this appears indeed to be the pattern that is emerging from contemporary studies of transport proteins.

Additional insights into the nature of membrane transport have come from studies with **ionophores,** antibiotics that greatly increase the permeability of membranes for specific ions. Ionophores function either by surrounding the ion in question with a hydrophobic "coat," thereby making it lipid-soluble, or by providing a hydrophilic channel through an otherwise hydrophobic membrane. The relevance of these intriguing compounds for our understanding of membrane transport proteins is explored further in the box on pp. 200–201.

An important property of transport proteins that is not apparent from Figure 8-5 is their specificity. Like enzymes, transport proteins catalyze the diffusion of only one or a few structurally related solutes. This specificity is a result of the precise stereochemical fit between the solute and its binding site on the carrier protein. A further similarity with enzymes is that transport proteins are often

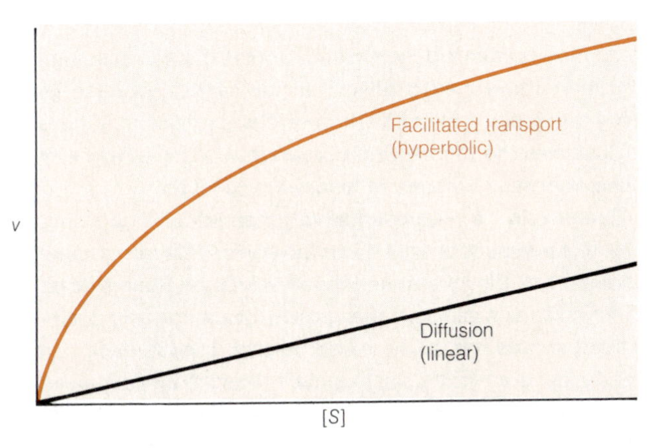

Figure 8-4 Comparison of the Kinetics of Diffusion and Carrier-Facilitated Transport. Solute transport by diffusion shows a linear relationship between transport velocity v and solute concentration [S] over a broad concentration range, whereas carrier-facilitated transport is characterized by the hyperbolic relationship typical of Michaelis-Menten kinetics. For simplicity, the initial solute concentration is assumed to be [S] on one side of the membrane and zero on the other side.

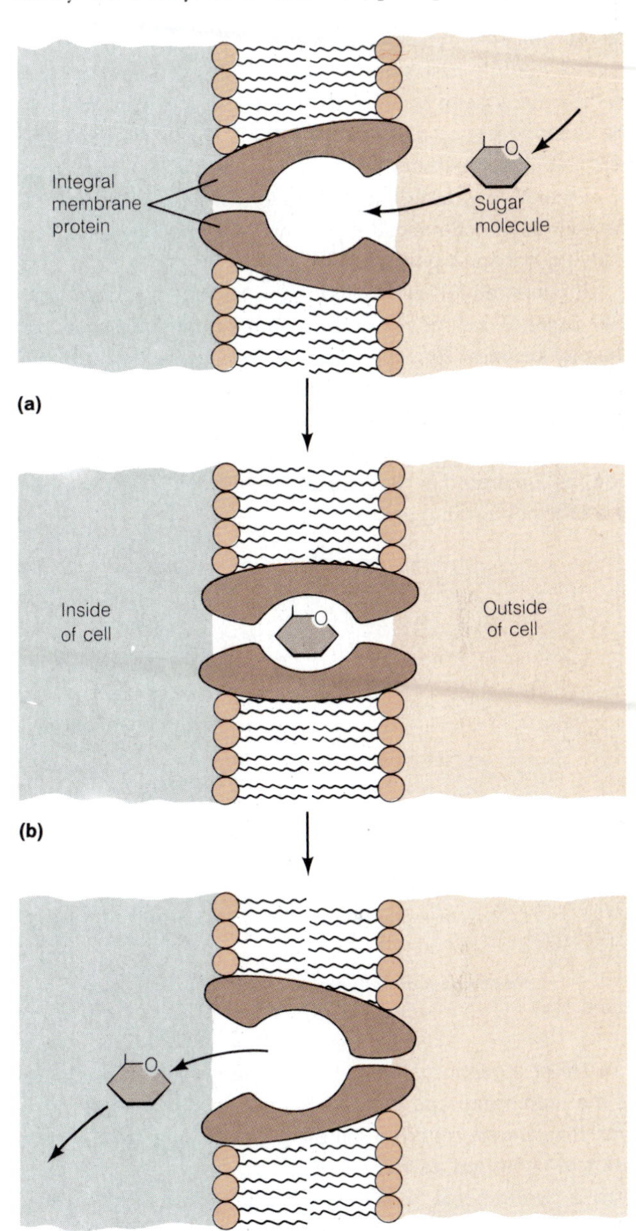

Figure 8-5 Possible Mechanism for Facilitated Transport. An integral membrane protein provides a stereospecific hydrophilic channel for a polar solute through an otherwise hydrophobic membrane. (a) Solute binds to the outer surface of the transport protein. (b) Change in protein conformation moves solute into the hydrophilic channel. (c) Further change in conformation releases solute on the other side of the membrane. Because the carrier functions equally well in both directions, the process is readily reversible. The direction of solute movement through the membrane will depend on the relative concentrations of solute on the two sides of the membrane.

Ionophores and the Study of Membranes

Microorganisms synthesize various antibiotics that are as much a boon to cell biologists and membrane biochemists as they are a bane to cells and membranes. These include the *ionophores,* small molecules that greatly increase the permeability of membranes to cations such as K^+, Na^+, and H^+. Though obviously detrimental to membrane function in the cell, such antibiotics have proved exceedingly useful in studying transport phenomena.

Ionophores fall into two categories, depending on how they facilitate ion movement across membranes. Some are **channel formers.** They organize themselves within the membrane to form a hydrophilic channel for the ion in much the way that transport proteins function. Ions enter the channel on one side of the membrane and pass through it to the other side, as illustrated in Figure 8-A. *Gramicidin* is an example of a channel-forming antibiotic.

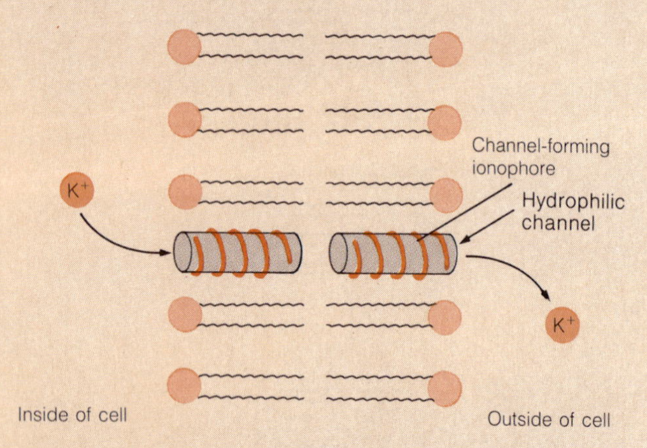

Figure 8-A Mechanism of Action of a Channel-Forming Ionophore.

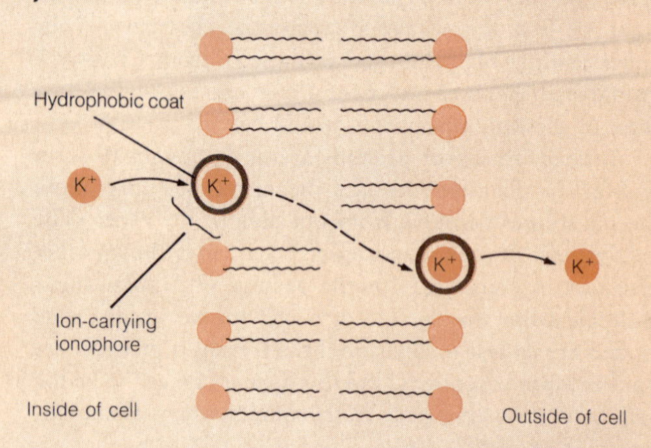

Figure 8-B Mechanism of Action of an Ion-Carrying Ionophore.

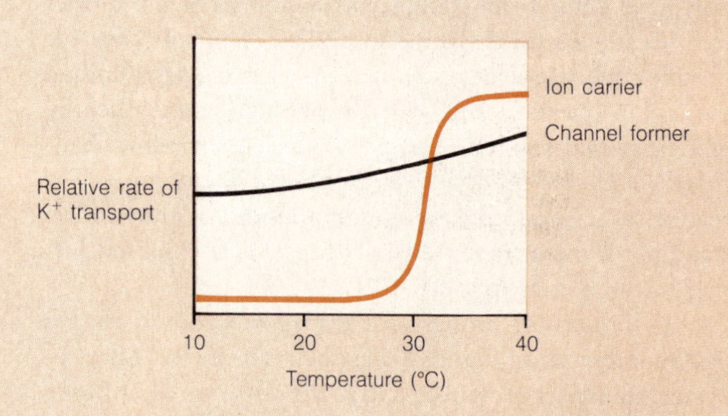

Figure 8-C Temperature Dependence of Ion-Carrying and Channel-Forming Ionophores.

Other ionophores function by binding ions on one side of the membrane and then diffusing across the membrane, with ion release occurring on the opposite side. Such membrane-soluble **ion carriers** function by surrounding an ion with a hydrophobic "coat," making the ion soluble in the phospholipids of the membrane, as shown in Figure 8-B. *Valinomycin* is an example of an ion carrier that has proved highly useful in the study of membrane transport.

It is easy to understand how ion carriers can be distinguished from channel formers, once you realize that a carrier moves across the membrane with the ion but a channel former does not. Recall that as the temperature of a membrane is lowered, the membrane phospholipids undergo a thermal transition from a highly fluid state to a much more rigidly ordered state. This has little effect on the function of a channel-forming antibiotic, since the ionophore does not have to move within the membrane. But the change in fluidity of the membrane greatly affects an ion carrier because its activity depends on diffusion through the membrane. Figure 8-C illustrates the difference in temperature dependence of the two alternative modes of ionophore function.

Gramicidin, a representative channel-forming ionophore, is a polypeptide with 15 amino acids. This remarkable substance was discovered in the late 1930s by Rene Dubos, who found that it blocked the growth of pathogenic gram-positive bacteria, hence its name. In fact, a good case can be made that gramicidin, not penicillin, was the first antibiotic to be discovered and used to treat bacterial infections. Unlike the polypeptides of proteins, gramicidin consists of L- and D-amino acids in strict alternation (Figure 8-D). When inserted into a membrane, the gramicidin polypeptide assumes a helical shape, with hydrophobic groups on the outside in contact with membrane phospholipids. Polar groups are on the inside of the helix, forming a hydrophilic "lining" for the 0.4-nm channel down the center of the helix.

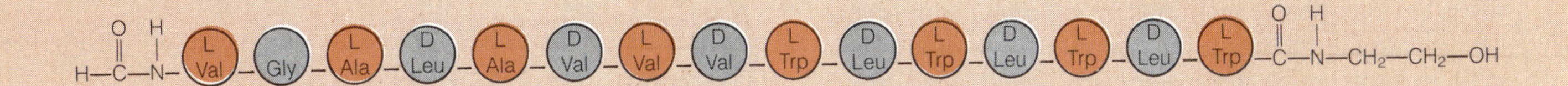

Figure 8-D The Structure of Gramicidin.

A single gramicidin helix spans only one-half of the membrane and essentially "floats" in one of the two monolayers that form the phospholipid bilayers. When the helix lines up with another such helix in the other monolayer, the two polypeptides associate end to end to form a continuous channel that spans the membrane, as shown in Figure 8-E. This channel exists only as long as the two monomers remain aligned, but during that time, potassium ions can pass through the channel at a rate exceeding ten million ions per second! Membrane biochemists are especially fascinated with channel-forming antibiotics such as gramicidin because of the attractive model they provide for transport proteins, all of which are thought to function by providing hydrophilic channels through what is otherwise a hydrophobic membrane.

Other ionophores function by providing a hydrophilic environment for an ion as it traverses a membrane, but in this case the antibiotic literally surrounds the ion and moves across the membrane with it. As shown in Figure 8-F, such ionophores are doughnut shaped, with hydrophobic hydrocarbon groups on the outside and oxygen atoms (six to eight, usually) projecting into the central cavity. These oxygen atoms bind the desired ion coordinately, stabilizing it within the cavity as the ionophore diffuses across the membrane. In the case of valinomycin (Figure 8-G), the oxygen atoms come from the carbonyl groups of the six valines in the molecule, whereas the various methyl and isopropyl side chains make up the hydrocarbon periphery. Because of its specificity for potassium ions, valinomycin is a powerful nervous system poison. Can you understand why?

Gramicidin monomer

Gramicidin monomer

K^+

Dimeric channel

K^+

Inside of cell

Outside of cell

Figure 8-E Mode of Action of Gramicidin.

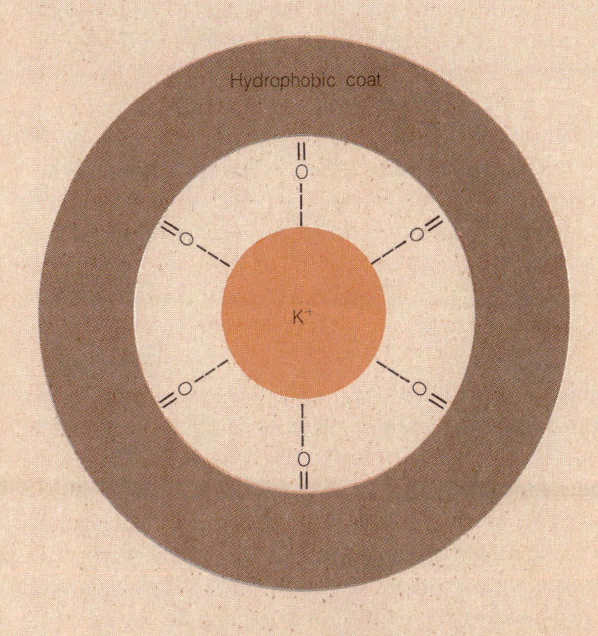

Figure 8-F The Structure of a Typical Ion-Carrying Ionophore.

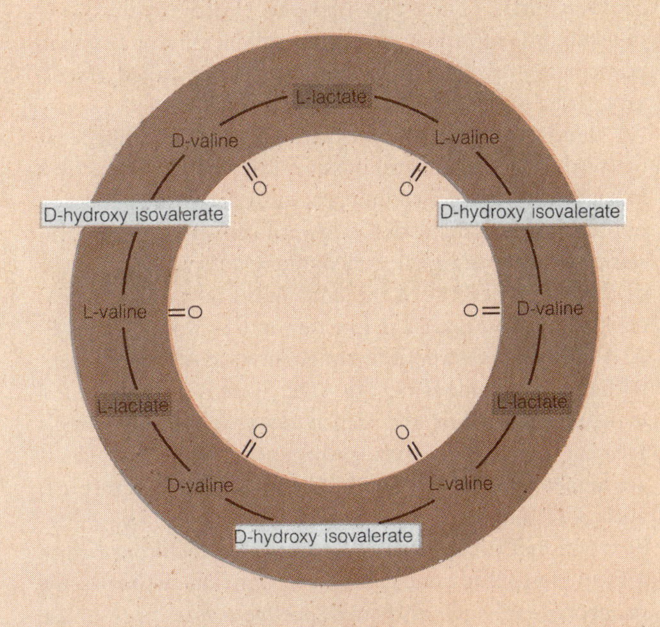

Figure 8-G The Structure of Valinomycin.

subject to competitive inhibition by molecules or ions that are structurally related to the intended "substrate." Such inhibition can be detected by the use of a double-reciprocal plot, just as is done for enzymes (recall Figure 6-17a). If the inhibition is in fact competitive, the apparent K_m value will be higher in the presence of the inhibitor and the double-reciprocal plot will be correspondingly steeper. (For an example of competitive inhibition of transport, see Problem 9 and Figure 8-16 at the end of the chapter.)

Although Figure 8-5 shows a solute molecule being transported inward by a carrier protein in the membrane, the same carrier can transport the solute outward with equal facility. A transport protein is really just a gate in an otherwise impenetrable wall, and like most gates, it facilitates traffic in either direction. The net direction of passive transport depends on the relative concentrations of solute inside and outside the cell (or organelle). If the concentration is higher outside, net flow will be inward; if the higher concentration exists inside, net flow will be outward.

Examples of Facilitated Transport. As an example of facilitated transport, consider in more detail the movement of glucose into your red blood cells. The level of glucose in blood is usually in the range 65–90 mg/100 ml, or about 3.6–5.0 m*M*. The red blood cell (or any other cell in contact with the blood, for that matter) can therefore meet its need for glucose uptake just by keeping its intracellular glucose concentration below 3.6 m*M* and equipping its plasma membrane with a passive glucose carrier. Cells, in fact, do precisely that. Glucose is phosphorylated to glucose-6-phosphate as soon as it crosses the cell membrane and then immediately enters the metabolic pathway called *glycolysis,* to be discussed in Chapter 10.

Transport of metabolites into and out of organelles is often (though not always) passive. Frequently, the carriers in such cases move two different but related substances in opposite directions, with the inward transport of one substance coupled with the concomitant outward movement of the other. Several of the mitochondrial transport proteins shown in Figure 8-1 function in this way. The ATP-ADP carrier, for example, couples outward ATP movement with the inward movement of ADP. Since ATP is generated in the mitochondrion and used in the cytoplasm, ATP levels are generally higher inside the organelle, while ADP levels are higher outside. The result is coupled passive transport, with ATP moving out and ADP moving in. The dicarboxylate and tricarboxylate carriers also shown in Figure 8-1 function in the same manner, accomplishing the passive reciprocal exchange of their respective solutes in response to the prevailing internal and external concentrations of each.

Active Transport: Energy and Gradients

Before continuing to a discussion of active transport, we will summarize the properties of passive transport and compare it to active transport (Table 8-2). Passive transport refers to movement of a solute through a membrane in the direction dictated by the existing gradient of concentration or charge—that is, *down* the electrochemical gradient. The ΔG is therefore always negative and the process always exergonic.

Active transport, on the other hand, involves the movement of substances into or out of cells or organelles *against* a concentration or electrochemical gradient. The ΔG is therefore always positive and the process always endergonic. As a result, the "pumps" involved in active transport must provide not only for translocation of the desired solute across the membrane but also for coupling of that translocation to an energy-yielding reaction, often (but not always) the hydrolysis of a high-energy phosphate bond. For example, the inward pumping of potassium and concomitant outward pumping of sodium characteristic of most animal cells is driven by ATP. In other cases, the immediate driving force for active transport is an ion gradient, usually of protons or sodium ions. Pyruvate transport into the mitochondrion, for example, is coupled to the inward movement of protons; as a result, it depends on both the existence and the magnitude of an electrochemical proton gradient across the mitochondrial inner membrane.

Active transport serves three major functions in cells and organelles:

1. It makes possible the uptake of fuel molecules and other essential nutrients from the environment or surrounding fluid, even when their concentrations in the environment are very low.
2. It allows various substances, such as secretory products, waste materials, and, as it turns out, sodium ions, to be removed from the cell or organelle, even when the concentration outside is greater than that inside.
3. It enables the cell to maintain constant, optimal internal concentrations of inorganic ions, particularly potassium, calcium, and hydrogen ions.

The ability to maintain a constant internal environment is an important aspect of active transport. Passive transport results in a greater and greater similarity between solute concentrations on both sides of the membrane. Active transport, on the other hand, represents a means of creating deliberate concentration differences across membranes and leads to a nonequilibrium steady state without which life as we know it would be impossible.

Directionality of Active Transport

Another important distinction between active and passive transport concerns the direction of transport with respect to the inside and outside of the membrane. Passive transport is inherently nondirectional with respect to the membrane; solute can move either inward or outward, depending entirely on the prevailing concentration or electrochemical gradient. Active transport, on the other hand, has an intrinsic **directionality.** An active transport system that transports a solute across a membrane in one direction will not be able to transport that solute actively in the other direction. Active transport is therefore said to be a *unidirectional,* or *vectorial,* process. However, in the absence of an energy supply, some carriers that normally function in active transport can also be used to facilitate passive transport of the same solutes. In such cases, passive transport can occur in either direction, as dictated by the gradient. Clearly, carriers that show this property are not inherently vectorial; directionality is imposed on them by the energy-yielding system to which they are coupled when involved in active transport.

A final point is that under certain conditions active transport can actually be driven in reverse. In muscle cells, for example, calcium ions are transported across the *sarcoplasmic reticulum* membrane by a specific transport ATPase. When these membranes are isolated as membranous vesicles, they maintain their ability to transport calcium, using one molecule of ATP to drive two calcium ions inward. After a high calcium concentration gradient is produced, the vesicles can be quickly placed in a reaction medium containing ADP and P_i. Surprisingly, under these conditions ATP is *synthesized* in the ratio in which it was used to pump Ca^{2+}: one ATP for every two calcium ions that leak out through the ATPase. The importance of this concept will become clear later when we discuss mitochondrial and chloroplast membranes, which use the energy of hydrogen ion gradients to synthesize ATP in a similar manner.

Energetics of Active Transport

We can now go on to a more detailed discussion of transport energetics. Every transport event is an energy transaction; energy is either released as transport occurs or is required to drive transport. To understand the energetics of transport, we need to recognize that two different factors may be involved. For uncharged species, the only variable is the concentration gradient across the membrane, and transport is either "downhill" (passive) or "uphill" (active). For charged species, however, there may be both a concentration gradient and a potential gradient across the membrane (Figure 8-6), and the two may either reinforce each other or oppose each other, depending on the charge on the ion and the direction of transport. We will first look at the transport of uncharged substances and then consider the additional complication that arises when charged species are moved across membranes.

Transport of Uncharged Species. For solutes with no net charge, we are concerned only with the concentration gradient across the membrane. We can therefore treat the transport process as a simple chemical reaction and calculate ΔG as we would for any other reaction.

The general "reaction" for transport of a solute X from the outside of a membrane-bounded compartment to the inside can be represented as

$$X_{outside} \longrightarrow X_{inside} \tag{8-2}$$

From Chapter 5, we know that the free energy change for this reaction can be written as

$$\Delta G = \Delta G^\circ + RT \ln[X]_{inside}/[X]_{outside} \tag{8-3}$$

where ΔG is the free energy change, ΔG° is the standard free energy change, R is the gas constant (1.987 cal/mol-K), T is the absolute temperature, and $[X]_{inside}$ and $[X]_{outside}$ are the prevailing concentrations of X on the inside and outside of the membrane. However, the equilibrium constant K_{eq} for the transport of an uncharged species is always 1, because at equilibrium, the solute concentrations on the two sides of the membrane will be the same:

$$K_{eq} = [X]_{inside}/[X]_{outside} = 1.0 \tag{8-4}$$

This means that ΔG° is always zero:

$$\Delta G^\circ = -RT \ln K_{eq} = -RT \ln(1) = 0 \tag{8-5}$$

So the expression for ΔG of inward transport of an uncharged species simplifies to

$$\Delta G = +RT \ln[X]_{inside}/[X]_{outside} \tag{8-6}$$

Notice that if $[X]_{inside}$ is less than $[X]_{outside}$, then ΔG will be negative, indicating that inward transport of substance X is exergonic and may occur spontaneously, as would be expected for passive transport down a concentration gradient. But if $[X]_{inside}$ is greater than $[X]_{outside}$, inward transport of X will be against the concentration gradient and the amount of energy required to effect the transport is indicated by the positive value of ΔG.

As an example, suppose that the concentration of lactose within a bacterial cell is to be maintained at 10 m*M*, while the external lactose concentration is only 0.20 m*M*. The energy requirement for inward transport of lactose at

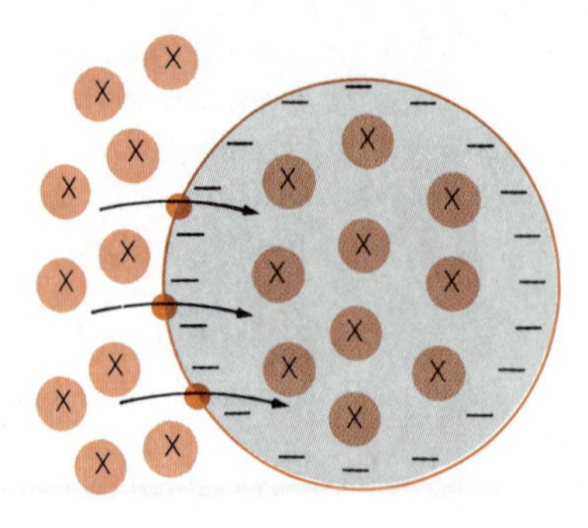

(a) Membrane potential has no effect on inward transport of uncharged solutes (zFV_m is zero)

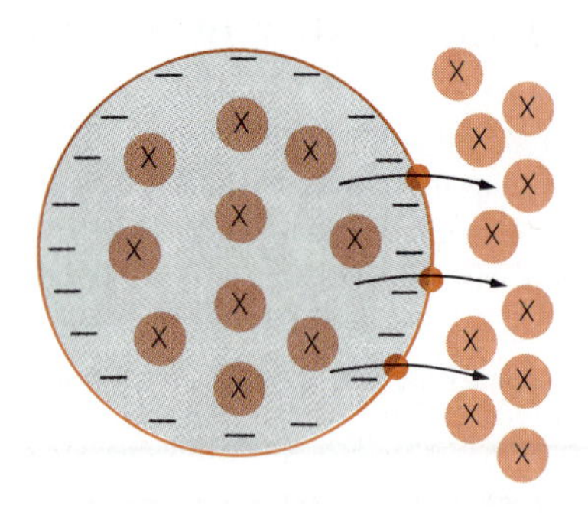

(b) Membrane potential has no effect on outward transport of uncharged solutes (zFV_m is zero)

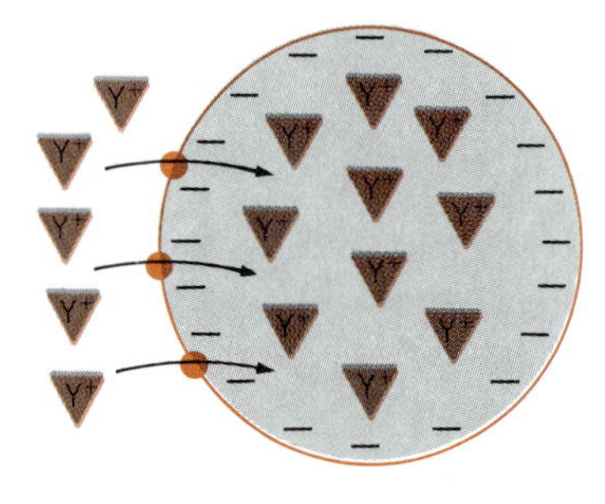

(c) Membrane potential favors inward transport of cations because of charge attraction (zFV_m is negative)

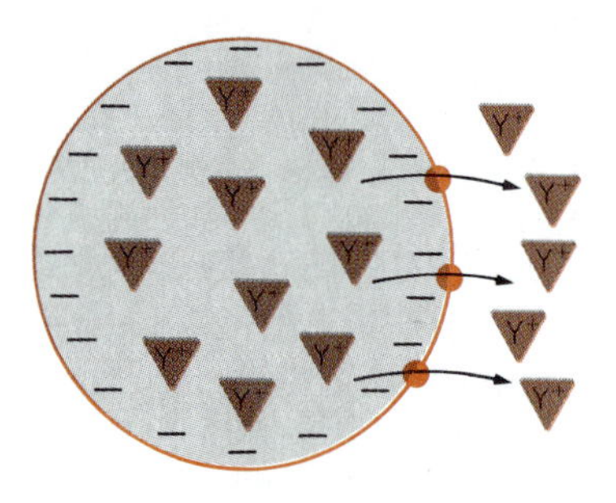

(d) Membrane potential opposes outward transport of cations because of charge attraction (zFV_m is negative)

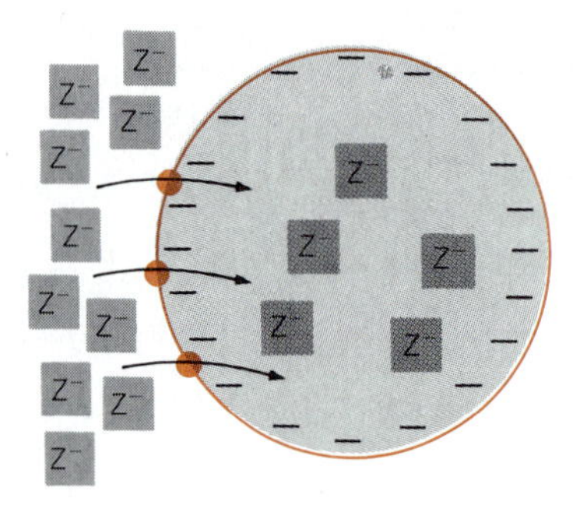

(e) Membrane potential opposes inward transport of anions because of charge repulsion (zFV_m is positive)

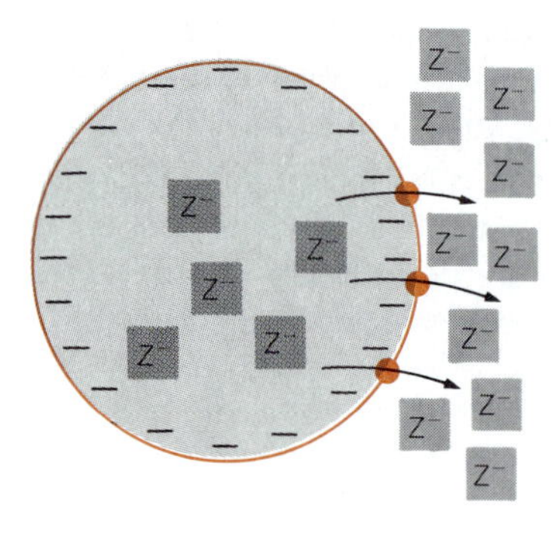

(f) Membrane potential favors outward transport of anions because of charge repulsion (zFV_m is positive)

Figure 8-6 **Effect of Negative Membrane Potential on Inward and Outward Movement of Ions and Uncharged Molecules.**

25°C can be calculated from equation 8-6 as

$$\begin{aligned}\Delta G &= +RT \ln[\text{lactose}]_{\text{inside}}/[\text{lactose}]_{\text{outside}} \\ &= +(1.987)(273 + 25) \ln(0.010)/(0.0002) \\ &= +592 \ln(50) = +2316 \text{ cal/mol} \\ &= +2.32 \text{ kcal/mol} \qquad (8\text{-}7)\end{aligned}$$

Clearly, this is an energy requirement that can be met comfortably by coupling the transport to ATP hydrolysis ($\Delta G^{\circ\prime} = -7.3$ kcal/mol), which is one of the several sources of energy that bacterial cells can use for lactose uptake.

As written, equation 8-6 applies to inward transport. For outward transport, the positions of X_{inside} and X_{outside} are simply interchanged within the logarithm. As a result, the numeric value of ΔG remains the same, but the sign is changed. As for any other process, a transport reaction that is exergonic in one direction will be endergonic to the same degree in the opposite direction. To keep signs straight, the equations for calculating ΔG of inward and outward transport of uncharged species are summarized in Table 8-3.

Transport of Charged Species. For charged species, we need to take both the concentration gradient and the potential (electrical charge) gradient into account. The latter is often important because many biological membranes have a significant **membrane potential** (V_m) that makes them negative on one side and positive on the other. The membrane potential is indicated in Figure 8-6 by the negative charges associated with the inner surface of the plasma membrane. V_m is expressed either in volts (V) or millivolts (mV). For animal cells, the plasma membrane potential is usually in the range -60 to -90 mV. In bacterial and plant cells, the plasma membrane potential is significantly more negative, often about -150 mV in bacteria and between -200 and -300 mV in plants.

The membrane potential obviously has no effect on uncharged species (Figure 8-6a and b), but it affects the energetics of ion transport significantly. Because it is almost always negative, the membrane potential *favors* the inward movement of cations (Figure 8-6c) and *opposes* their outward movement (Figure 8-6d). Conversely, anions move *against* the membrane potential when they are transported inward (Figure 8-6e) but *with* it when they are transported outward (Figure 8-6f). The net effect of both the concentration gradient and the potential gradient for an ion is usually called the electrochemical gradient for that species.

Both components of the electrochemical gradient must be considered when we are determining the energetics of ion transport. To calculate ΔG for the transport of ions therefore requires an equation with two terms, one to express the effect of the concentration gradient across the membrane and the other to take the membrane potential into account.

If we let X^z represent a solute with a charge z, then we can calculate ΔG for the inward transport of X^z as

$$\Delta G_{\text{inward}} = +RT \ln[X]_{\text{inside}}/[X]_{\text{outside}} + zFV_m \qquad (8\text{-}8)$$

where R, T, and [X] are as before and z is the charge on X (such as +1, +2, −1, −2), F is the Faraday constant (23,062 cal/mol-V), and V_m is the membrane potential, in volts. For outward transport of X, ΔG has the same value as for inward transport but is opposite in sign, so we can write

$$\begin{aligned}\Delta G_{\text{outward}} &= -\Delta G_{\text{inward}} \\ &= -RT \ln[X]_{\text{inside}}/[X]_{\text{outside}} - zFV_m\end{aligned}$$

Or, by interchanging terms within the logarithm,

$$\Delta G_{\text{outward}} = +RT \ln[X]_{\text{outside}}/[X]_{\text{inside}} - zFV_m \qquad (8\text{-}9)$$

To help you in such calculations, we include these equations for inward and outward transport of charged species along with those for the transport of uncharged species in Table 8-3, which summarizes the thermodynamic and kinetic properties of each of these processes.

Table 8-3 Calculation of ΔG for Transport of Charged and Uncharged Species

Reaction:

Membrane
Inward
X
X
Outward
Outside
Inside

ΔG for Transport of Uncharged Species:

$$\Delta G_{\text{inward}} = +RT \ln \frac{[X]_{\text{inside}}}{[X]_{\text{outside}}} \qquad R = 1.987 \text{ cal/mol-K}$$

$$\Delta G_{\text{outward}} = +RT \ln \frac{[X]_{\text{outside}}}{[X]_{\text{inside}}} \qquad T = {}^{\circ}\text{K} = {}^{\circ}\text{C} + 273$$

ΔG for Transport of Charged Species:

$$\Delta G_{\text{inward}} = +RT \ln \frac{[X]_{\text{inside}}}{[X]_{\text{outside}}} + zFV_m$$

$$\Delta G_{\text{outward}} = +RT \ln \frac{[X]_{\text{outside}}}{[X]_{\text{inside}}} - zFV_m$$

z = charge on ion
F = 23,062 cal/mol-V
V_m = membrane potential (in volts)

Mechanisms of Active Transport

Most membranes of living cells carry out one or more active transport processes. These are summarized in Table 8-4, and here we will consider several of the basic underlying mechanisms. This is an area of biology that has elicited a great deal of interest and excitement but has only recently begun to yield definitive answers. Much of the initial difficulty lay in the hydrophobic nature of most transport proteins and the problems involved in extracting these proteins from membranes in an active form. A related difficulty has been the lack of adequate assay systems once such proteins were isolated. But for at least several systems, these problems have been solved, either through the use of closed membrane vesicles that retain transport activity or by incorporation of membrane proteins into artificial phospholipid vesicles (the liposomes described earlier) so that transport activity is reconstituted. Because of these and other technical advances, substantial progress has been made toward understanding the molecular basis of active transport.

Properties of Active Transport Mechanisms

An important principle that has emerged from various studies is that all active transport does not occur by the same mechanism. Instead, several mechanisms are involved, differing primarily in terms of whether or not another solute is transported concomitantly (**cotransport**), either in the same or the opposite direction; the source of energy used to drive the process; and whether or not the solute is chemically modified (phosphorylated) in the process. As a further source of complexity, many substrates are known to be transported into the same cell by several different mechanisms. This multiplicity of uptake mechanisms is observed in all organisms, but is most striking in bacteria. *Escherichia coli,* for example, has at least five different systems for transporting galactose into the cell.

We will consider two general properties of active transport systems first and then look at several specific examples. Throughout this discussion, the term *pump* will be used to refer to the various mechanisms responsible for active transport of molecules and ions across membranes. This terminology is in keeping with general literature and textbook usage, but no functional analogy with mechanical *pumps* is intended. On the contrary, mechanical pumps invariably effect a mass flow from one location to another, whereas membrane pumps selectively transport specific components from one mass to another.

Simple Versus Coupled Transport. Some solutes are moved actively across membranes by **simple active transport**—a unidirectional pumping of the species of interest, unaccompanied by the flow of any other substance in either direction. An example of simple active transport is the calcium pump found in muscle cells. As we will see when we get to Chapter 19, calcium ions play an important role in the regulation of muscle contraction. Within the muscle cell, calcium is sequestered in the sarcoplasmic reticulum, as described earlier. When the cell is stimulated by a nerve impulse, calcium is released into the cytoplasm, triggering muscle contraction. As the cell returns to its resting state, the calcium is quickly returned to the sarcoplasmic reticulum by an ATP-driven calcium pump. This pump requires only calcium ions and ATP (plus catalytic amounts of Mg^{2+} to activate the ATPase). It therefore represents a simple active transport system, accomplishing the unidirectional transport of a single solute.

Table 8-4 Examples of Active Transport Processes

Transported Species	Example	Transport Protein	Comments
Hydrogen ion	Halobacterium	Bacteriorhodopsin	Light-dependent pump
	Chloroplast	Reaction center	Light-dependent pump
	Mitochondrion	Several sites	Uses electron transport
	Plasma membrane	E_1-E_2ATPase	Regulates pH of cell
	Lysosome	E_1-E_2ATPase	Maintains acid interior of organelle
Sodium-potassium	Plasma membrane	E_1-E_2 ATPase	Most cells, maintains gradient across plasma membrane
Calcium	Sarcoplasmic reticulum	E_1-E_2 ATPase	Regulates muscle contraction
	Endoplasmic reticulum	E_1-E_2 ATPase	Regulates cytoplasmic Ca^{2+}
	Plasma membrane	E_1-E_2 ATPase	Regulates cytoplasmic Ca^{2+}
Nutrients, metabolites	Plasma membrane	Coupled transport	Uses energy of Na^+ or H^+ gradients

More frequently, however, cotransport mechanisms are involved, such that two different solutes are moved simultaneously across a membrane in a tightly coupled manner. This process is called **coupled transport,** a term which implies an obligatory concomitant passage of two different solutes across the membrane. The two cotransported solutes may move either in the same direction (**symport**) or in opposite directions (**antiport**), as shown in Figure 8-7.

In some cases, both solutes are moved against their concentration gradient (or, for ions, their electrochemical gradient), and the energy source must therefore be adequate to pump both solutes "uphill." Such is the case with the sodium-potassium pump, which we will consider shortly. In other cases, the transmembrane gradients of the two solutes are such that exergonic transport of one solute down its concentration (or electrochemical) gradient can be used to transport the other species actively against its concentration gradient. Clearly, the gradient of the passively transported solute must exceed that of the actively transported solute; otherwise the coupled process will not be exergonic. In animal cells, active uptake of sugars, amino acids, and other metabolites is coupled to the passive inward movement of sodium ions in a process called **sodium cotransport.** Because of the large outside-to-inside sodium gradient that most cells maintain across the plasma membrane, inward sodium transport is highly exergonic and can be used to drive the active symport of organic solutes.

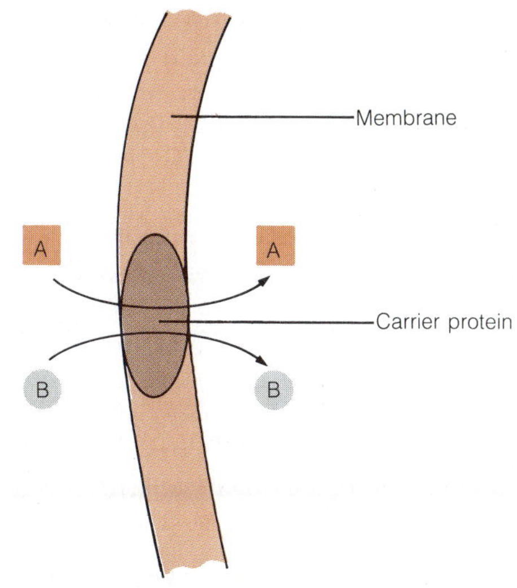

(a) Symport: Both solutes move in the same direction

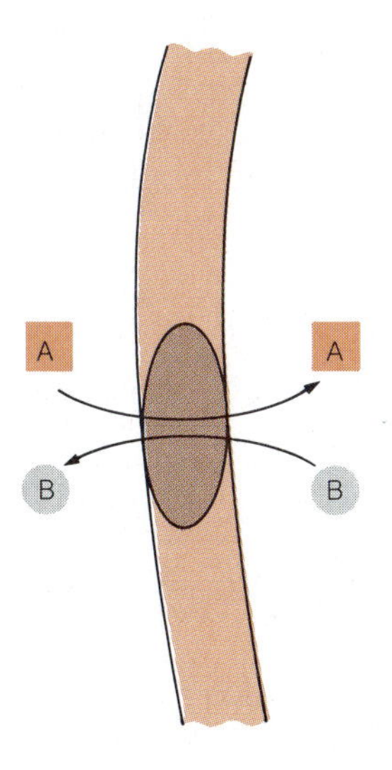

(b) Antiport: Solutes move in opposite directions

Figure 8-7 Cotransport. Cotransport involves an obligatory, coupled transport of two different solutes, A and B, across a membrane. The two solutes may move either (a) in the same direction (symport), or (b) in opposite directions (antiport).

Energy Source. The energy required for active transport is provided either by hydrolysis of a high-energy phosphate bond or by cotransport of a second species. Not surprisingly, ATP is the compound of choice for most reaction mechanisms that couple to phosphate bond hydrolysis. However, some transport systems use other compounds instead. For example, one of the mechanisms used for sugar transport by bacterial cells depends on phosphoenolpyruvate as its energy source.

Cotransport almost always depends on an electrochemical gradient of either sodium ions or protons to drive the active symport of the desired solute, which is usually a small organic molecule. For animal cells, sodium cotransport is a common option. Bacteria, fungi, and plants, on the other hand, usually depend instead on an electrochemical proton gradient as the driving force for coupled transport. In some cases, the proton gradient is harnessed directly by means of a proton symport mechanism. In other instances, coupling is indirect, involving symport or antiport with some other ion, usually sodium, the electrochemical gradient for which is in turn maintained by a proton-driven pump.

A Look at Several Active Transport Mechanisms

Having considered several features of active transport systems, we are now ready to look at three specific mechanisms. In each case, we will want to know what kinds of solutes are transported, what the driving force is, and how the energy source is actually coupled to the transport mechanism. We will look first at the *sodium-potassium pump* present in all animal cells, because this is currently the best-understood transport system. Then we will consider two alternative modes of uptake for organic molecules: *cotransport* (using sodium cotransport in animal cells as our example) and *phosphorylating transport* (in bacterial cells). We will finally introduce the concept of *proton transport* in the bacteriorhodopsin system, since this will be a central topic of Chapter 11.

The Sodium-Potassium Pump. One of the characteristic features of most cells is a high intracellular level of potassium ions and a low intracellular level of sodium ions. This feature is essential for the well-being of the cell, because potassium is required for a number of vital cellular processes (such as ribosome function and activation of a variety of enzymes) and sodium often inhibits these functions. In addition, the resulting gradients of potassium and sodium ions are essential for the transmission of nerve impulses (Chapter 20) and as the driving force for cotransport.

Potassium levels are usually maintained at about 100–150 m*M* inside the cell, whereas external levels of potassium are generally much lower than that and may fluctuate widely. Conversely, the intracellular concentration of sodium is usually considerably less than that in the surrounding medium. Both the inward pumping of potassium and the outward pumping of sodium are therefore energy-requiring processes. The significance of this coupled pumping of sodium and potassium is best appreciated by realizing that about a third of the energy expended by a resting animal is used just to maintain these gradients of sodium and potassium.

The pump responsible for this process was discovered in the mid-1950s and represented the first documented case of active transport. The pump was shown to have ATPase activity and to couple ATP hydrolysis to the inward transport of potassium ions and the outward transport of sodium ions. In fact, the ATPase requires both potassium and sodium ions for activity, differing in this respect from virtually all other potassium-activated enzymes in the cell, which are almost all inhibited by sodium.

The **sodium-potassium pump,** as this enzyme is called, occurs in the plasma membrane of virtually all animal cells but has been studied in the greatest detail in red blood cells. Like other active transport systems, the sodium-potassium pump is a vectorial system: it has inherent directionality in that potassium is pumped only inward and sodium is pumped only outward. In fact, it has been demonstrated that sodium and potassium ions activate the ATPase only on the side of the membrane from which they are transported—sodium from the inside, potassium from the outside.

Because the gradients against which sodium and potassium are pumped by animal cells seldom exceed 50:1, the energy requirement per ion moved is relatively low, usually less than 2 kcal/mol ($\Delta G = +RT \ln(50) = 2.32$ kcal/mol at 25°C). The possibility that this suggests of coupling the transport of several ions to the hydrolysis of a single ATP molecule is actually realized by the cell. The stoichiometry apparently varies somewhat with cell type, but for red blood cells, three sodium ions are moved out and two potassium ions are moved in per molecule of ATP hydrolyzed.

Figure 8-8 is a schematic illustration of the sodium-potassium pump consistent with available evidence. The pump is a tetrameric transmembrane protein, with two α and two β subunits. The α subunits are catalytically active. They have binding sites for ATP and sodium ions on the cytoplasmic (inner) side and for potassium ions on the external side of the membrane. The β subunits are glycosylated and are therefore located on the extracellular side of the plasma membrane. The function of the β subunits is not yet clear.

The sodium-potassium pump is an allosteric protein, with two alternative conformational states, referred to as E_1 and E_2. E_1 is thought to be "open" to the inside of the cell and to have a high affinity for sodium ions, whereas E_2 is regarded as "open" to the outside, with a high affinity for potassium ions. Phosphorylation of the enzyme, a sodium-triggered event, stabilizes it in the E_2 form. Dephosphorylation, on the other hand, is triggered by potassium

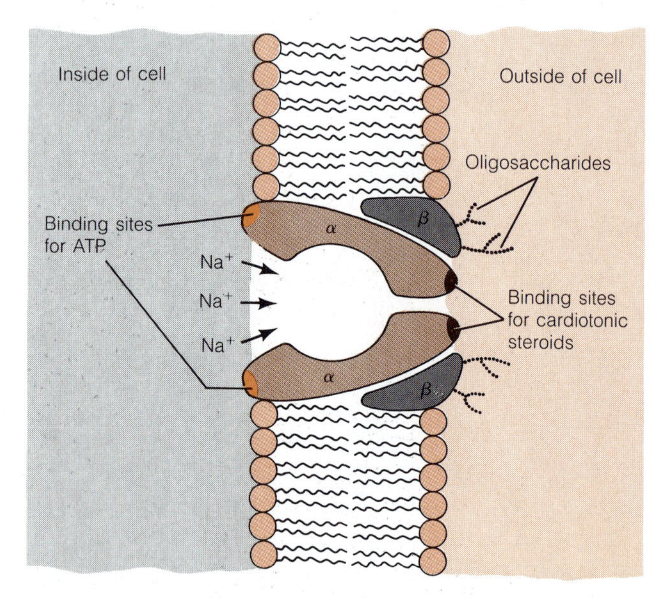

Figure 8-8 The Sodium-Potassium Pump. The sodium-potassium pump found in most animal cells consists of α and β subunits. The α subunits are transmembrane proteins, with ATP binding sites on the cytoplasmic side and binding sites for cardiotonic steroids on the outer side. The β subunits are glycosylated and are located on the outer side of the membrane. The pump is shown in the E_1 configuration, with the ion binding sites facing the inside of the cell. Binding of sodium ions will cause a conformational change to the E_2 form, which opens to the outside.

and stabilizes the enzyme in the E_1 form. Other ions transported by E_1-E_2 ATPases include Ca^{2+} in sarcoplasmic reticulum of muscle tissue and H^+ in membranes such as lysosomes, secretory granules, and the plasma membrane.

As illustrated in Figure 8-9, the actual transport mechanism probably involves an initial binding of three sodium ions to E_1 on the inner side of the membrane (step 1). The binding of sodium ions triggers phosphorylation of the enzyme by ATP (step 2), leading to a conformational change to E_2. As a result, the bound sodium ions are translocated through the membrane to the external surface, where they are released to the outside (step 3). Then potassium ions from the outside bind to the α subunits (step 4), triggering dephosphorylation and a return to the original conformation (step 5). In the process, the potassium ions are translocated to the inner surface, where they dissociate, leaving the carrier ready to accept more sodium ions (step 6).

The sodium-potassium pump is sensitive to inhibition by a class of poisonous plant steroids that includes compounds such as *ouabain* and *digitalis*. These inhibitors are quite toxic; ouabain, for example, is used by South American Indians to poison their arrowheads because of its effectiveness in killing prey. In carefully controlled doses, however, these compounds are of great medical importance because of their profound effects on the function of the heart. For this reason, they are called **cardiotonic steroids.** Digitalis, for example, is often the drug of choice in the treatment of congestive heart failure because it causes an increase in intracellular calcium concentration, thereby stimulating muscle contraction in heart cells.

The binding sites for the cardiotonic steroids are located on the external side of the α subunits, as shown in Figure 8-8. These steroids are thought to inhibit the sodium-potassium pump by blocking the dephosphorylation reaction (step 5 of Figure 8-9). The demonstration that ouabain blocks the transport of both sodium and potassium as well as ATP hydrolysis was crucial confirmation that all three of these processes are, in fact, catalyzed by a single protein.

The sodium-potassium pump is not only the best-understood transport system but also one of the most important. In addition to maintaining the appropriate intracellular concentrations of both potassium and sodium, it is largely responsible for the maintenance of the membrane potential that characteristically exists across most biological membranes. As we will see in Chapter 20, the membrane potential plays an important part in the transmission of nerve signals. The sodium-potassium pump assumes still further significance when we take into account the vital role that sodium plays in the inward transport of organic substrates, a topic to which we now come as we consider sodium cotransport.

Sodium Cotransport of Organic Molecules in Animal Cells. A unifying feature that has emerged from studies of the active uptake of sugars, amino acids, and other organic molecules by animal cells is that the inward transport of such molecules is often coupled obligatorily to the concomitant inward cotransport of sodium ions. It is, in fact, only as we consider such *sodium cotransport* that we appreciate the full significance of the continuous outward movement of sodium accomplished by the sodium-potassium pump. It is the steep electrochemical gradient of sodium ions maintained by this pump that serves as the driving force for the inward transport of a variety of sugars and amino acids. Uptake of such compounds by animal cells is therefore not coupled directly to ATP hydrolysis, but is powered instead by the sodium gradient. Ultimately, of course, such uptake is still dependent on ATP, since the sodium-potassium pump is itself an ATP-driven system.

The carriers involved in sodium-driven transport have not yet been isolated in purified form, but several of their properties can be predicted from what is already known about the process in which they are involved. Clearly, the carrier must have two kinds of sites, one specific for sodium ions and the other specific for the solute to be cotransported. Furthermore, the binding of sodium to its site almost certainly affects the affinity of the other site for the desired solute, since the solute should bind when sodium binds and dissociate when sodium dissociates if its transport is to follow the sodium gradient faithfully.

To visualize the mechanism whereby sodium-driven uptake of a compound like glucose might occur, we need an allosteric membrane protein with sites for sodium and the appropriate solute. Figure 8-10 shows such a transport protein, with glucose as the transported solute. When sodium binds to its site on the outside of the membrane, the conformation of the protein changes, giving the glucose binding site a high affinity for glucose (step 1). Binding of glucose (step 2) presumably causes a further conformational change in the protein that results in the translocation of both sodium and glucose to the inner surface of the membrane (step 3). There, the sodium dissociates in response to the low intracellular sodium concentration (step 4). This will cause glucose to be ejected from its site regardless of the cellular glucose level (step 5), provided only that the affinity of that site for glucose is greatly reduced on dissociation of the sodium ion. The carrier protein is then free to revert to its original conformation, returning the empty binding sites to the outer surface of the membrane (step 6).

Sodium cotransport is a common mechanism for uptake of organic substrates by animal cells. Other organisms also use cotransport as a means of uptake for sugars, amino acids, and other solutes, but they usually rely on a proton gradient rather than a sodium gradient to drive the uptake.

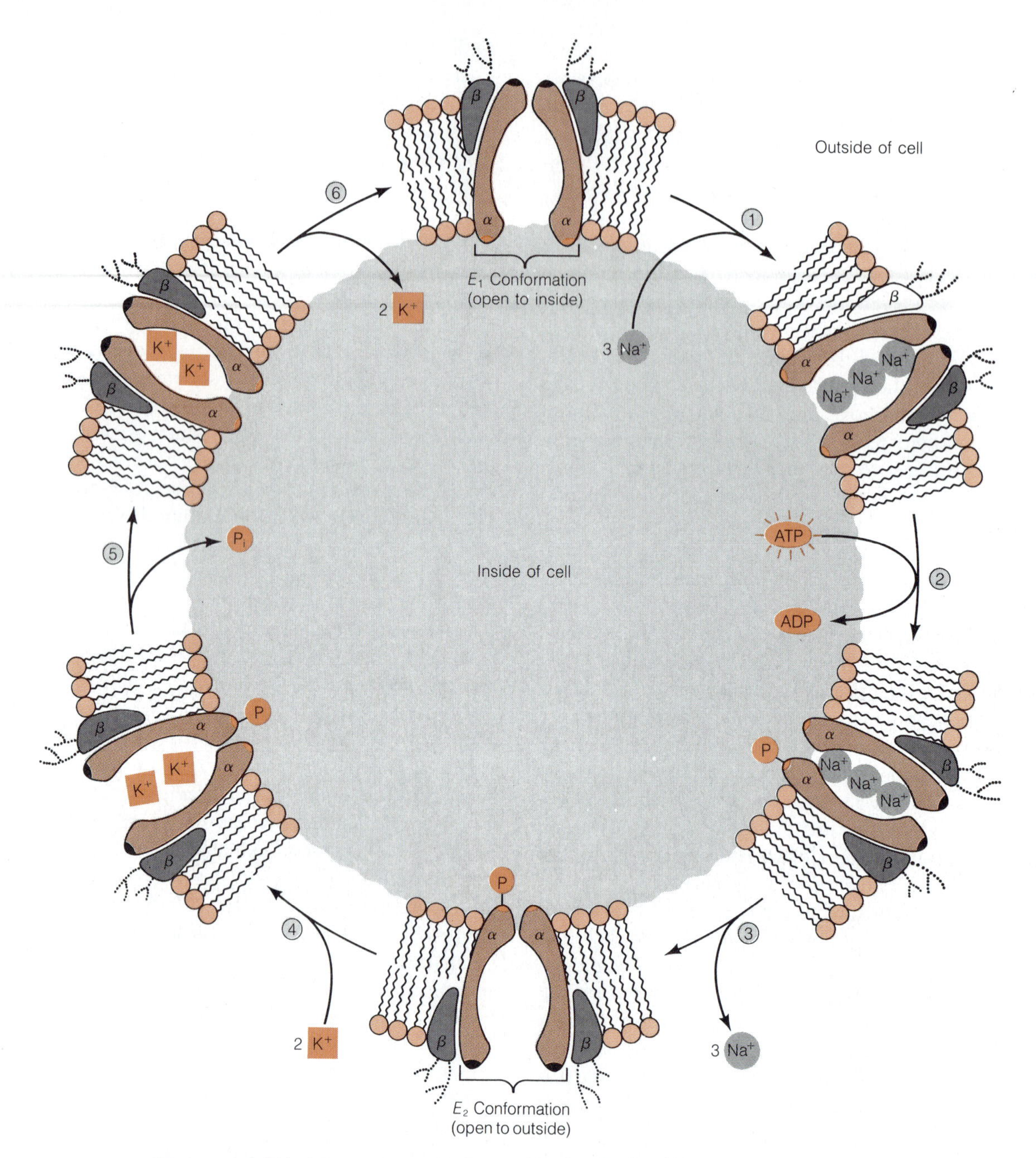

Figure 8-9 A Model Mechanism for the Sodium-Potassium Pump. The transport process is shown here in six steps arranged around the periphery of a "cell." The outward transport of sodium ions is coupled to the inward transport of potassium ions, both against their respective electrochemical gradients. The driving force is provided by ATP hydrolysis, which is required for phosphorylation of the α subunit of the permease. Selective transport of sodium ions and potassium ions in opposite directions is possible because sodium ions activate the ATPase of the carrier protein only on the inner membrane surface, whereas potassium ion activation of the dephosphorylation reaction occurs only on the outer surface. E_1 is the conformational state of the enzyme with the channel open to the inside; E_2 is the conformational state with the channel open to the outside.

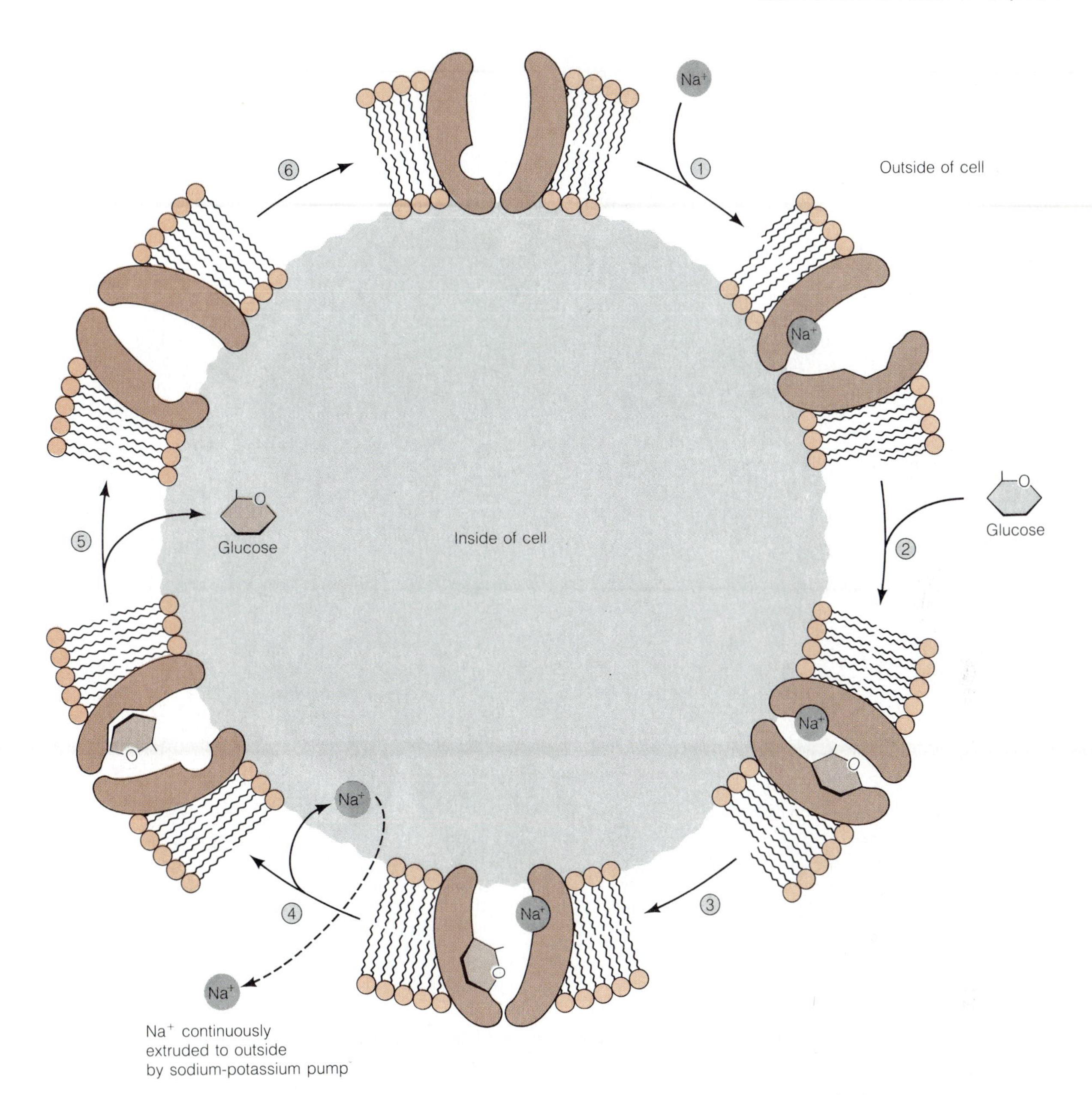

Figure 8-10 A Model Mechanism for Sodium Cotransport. The transport process is shown here in six steps arranged around the periphery of a "cell." The inward transport of an organic solute such as glucose against its concentration gradient is driven by the inward cotransport of sodium down its electrochemical gradient. The process may occur in either direction, but preferential inward movement is ensured by the high external concentration of sodium ions. The sodium gradient is in turn maintained by the continuous outward extrusion of sodium ions (dashed line) by the sodium-potassium pump of Figure 8-9.

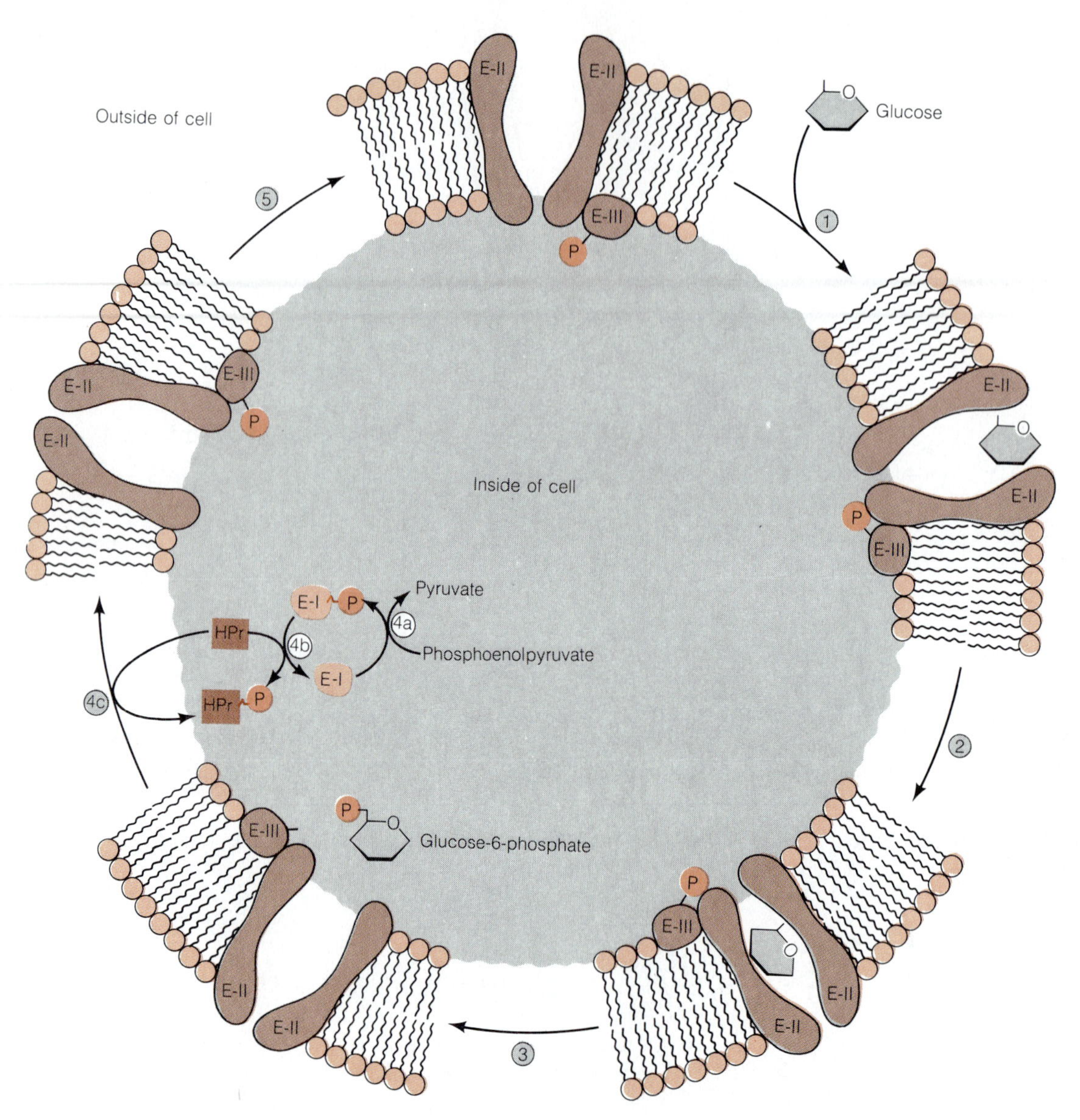

Figure 8-11 A Model Mechanism for Phosphorylating Transport of Sugars in Bacterial Cells. The transport process is shown here in five steps arranged around the periphery of a "cell." The inward transport of a solute such as glucose is driven by phosphorylation of the solute on the inner surface of the membrane, with phosphoenolpyruvate (PEP) as both the phosphate donor and the energy source. Enzymes II and III (E-II and E-III) are solute-specific membrane proteins. Enzyme I (E-I) and HPr are nonspecific soluble proteins that transfer the high-energy phosphate group of PEP to specific E-III proteins. From E-III, the phosphate group is transferred to the specific sugar that binds to the associated E-II protein.

Bacterial cells, for example, make extensive use of proton cotransport to drive the uptake of organic solutes. Fungi and plants also utilize proton symport for uptake of organic solutes, but these organisms depend on an ATP-driven proton pump for the generation and maintenance of the electrochemical proton gradient.

Phosphorylating Transport. The several types of transport we have examined so far differ in mechanism and in the immediate driving force on which they depend, but they share the property that a given molecule or ion is transported across a membrane against its concentration (or electrochemical) gradient without being chemically changed in any way. Such is not universally the case, however. In bacteria, uptake is often accomplished in a way that couples transport of a solute to its chemical modification.

The best-understood example of such a mechanism is the **phosphotransferase system** responsible for sugar uptake in *E. coli* and other bacteria. This system uses the high-energy phosphate group of a compound called phosphoenolpyruvate (PEP) to phosphorylate the sugar molecule. The process is accordingly termed **phosphorylating transport,** since the sugar is phosphorylated as an inherent part of the uptake mechanism. In glucose, for example, the sugar is taken up at the outer membrane surface as free glucose but is released at the inner membrane surface as glucose-6-phosphate:

$$\text{glucose} + \text{PEP} \longrightarrow \text{glucose-6-P} + \text{pyruvate} \qquad (8\text{-}10)$$

Since the plasma membrane is impermeable to sugar phosphates, they accumulate inside the cell.

As Figure 8-11 shows, the phosphotransferase system responsible for this process is complex. It involves four components, designated as *enzyme I (E-I)*, *enzyme II (E-II)*, *enzyme III (E-III)*, and *HPr*, a small heat-stable protein. E-II is the actual transport protein. It is an integral membrane protein that forms a transmembrane channel and catalyzes phosphorylation of the solute. There are probably as many kinds of E-II proteins in the membrane as there are different sugars to be transported by this mechanism, since the specificity of sugar recognition appears to be a property of E-II.

The molecule to be transported inward (glucose, in the example shown) binds to the appropriate E-II carrier at the outer membrane surface (step 1). As the sugar is translocated inward, presumably through the hydrophilic transmembrane channel of E-II (step 2), it receives a phosphate group from E-III, a peripheral membrane protein that interacts with E-II (step 3). The phosphate group on E-III is actually supplied by phosphoenolpyruvate, but reaches E-III via E-I and HPr (steps 4a, 4b, and 4c). As the figure suggests, E-I and HPr are both soluble proteins, having no association with the membrane. Finally, E-II reverts to its original configuration (step 5), ready for the entry of another molecule of solute.

Like E-II, E-III is specific for a particular sugar. Thus, there are different sets of E-II and E-III proteins for each sugar to be transported. E-I and HPr, however, are nonspecific proteins that serve as the common phosphorylating system for every E-III.

The Bacteriorhodopsin Proton Pump. The last active transport system we will discuss is the simplest, in that nothing more is involved than a small integral protein called **bacteriorhodopsin.** This transport process is one of the most significant recent discoveries in membrane biology, in part because bacteriorhodopsin was the first membrane protein to be analyzed as a crystal, but also because it provided convincing evidence that strongly supported the chemiosmotic theory of ATP synthesis in mitochondria and chloroplasts (see Chapter 11). Despite years of work, it is still not clear exactly how protons are pumped by bacteriorhodopsin, which shows how difficult it is to sort out chemical events at the molecular level in compounds as complex as proteins.

Bacteriorhodopsin (BR) is produced by certain primitive prokaryotes called halobacteria. These organisms require concentrated salt solutions to reproduce and are therefore classed as halophiles ("salt-loving"). They produce bacteriorhodopsin under anaerobic conditions, presumably in order to use light as an alternative energy source when they are no longer able to make ATP by aerobic metabolism. Bacteriorhodopsin is bright purple, appearing as patches in the plasma membrane. The patches are sometimes referred to as purple membranes.

The pigment of BR is **retinal,** the same carotenoid derivative that is found in human retinas in the visual pigment called rhodopsin. It is a remarkable coincidence that both human visual processes and bacterial energy transduction use the same molecule as a primary pigment! Figure 8-12 illustrates the structure of BR and its orientation in the membrane. The retinal in BR is covalently linked to an amino group of lysine, forming a positively charged *Schiff base* (Figure 8-13a). A proposed mechanism for the proton transport process is shown in Figure 8-13b. Upon illumination, one of the double bonds of retinal isomerizes from the trans to the cis form configuration (step 1), reducing the ability of the nitrogen atom to bind protons. A proton therefore leaves (step 2), permitting the double bond to relax to the trans configuration (step 3). The nitrogen again binds a proton, but from the other side (i.e., the inside) of the membrane (step 4).

The phenomenon of energy-dependent proton pumping is one of the most basic concepts in cellular bioenergetics: it occurs in all bacteria, mitochondria, and chloroplasts and represents the driving energy of life on Earth because it is an absolute requirement for the efficient synthesis of ATP. The underlying mechanisms of the proton pump and ATP formation will be discussed in Chapter 11.

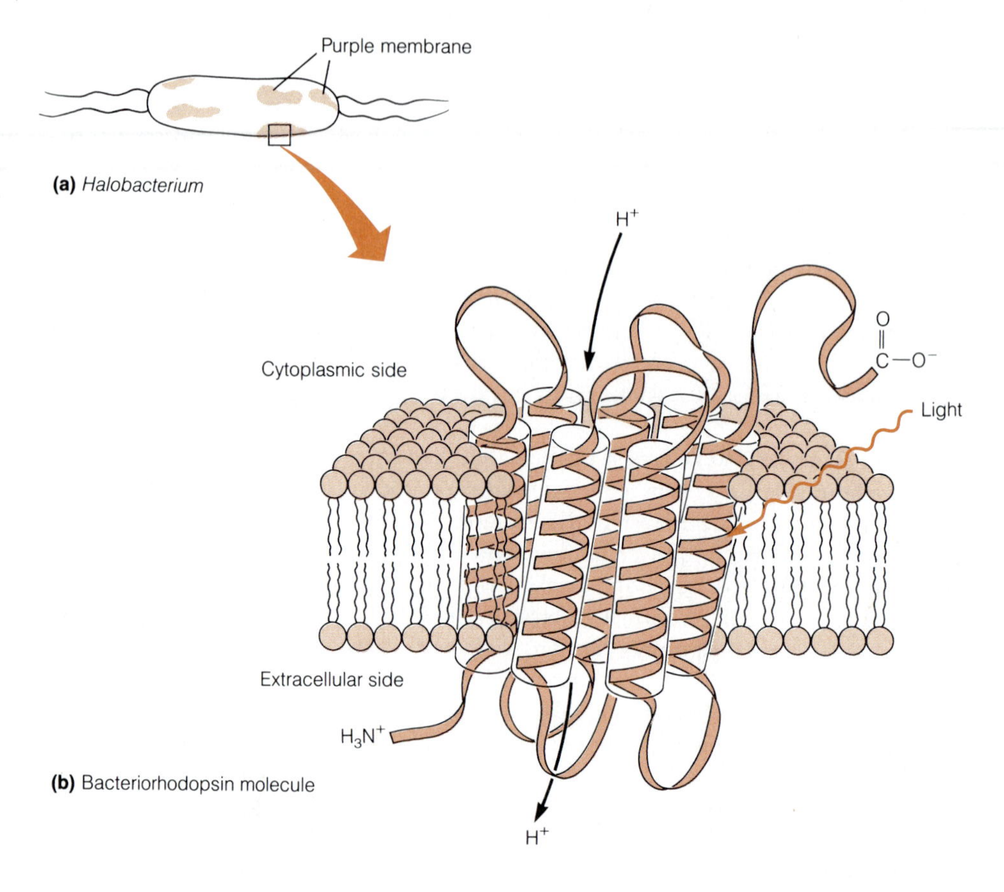

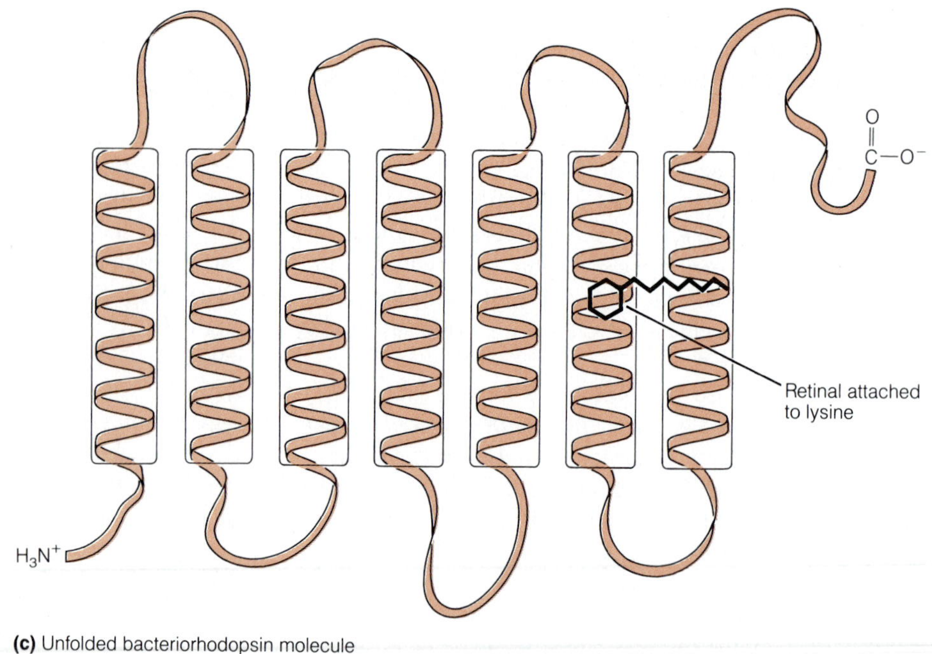

Figure 8-12 The Bacteriorhodopsin Molecule. (a) Bacteriorhodopsin (BR) is synthesized by certain halophilic bacteria under anaerobic conditions. In the early 1970s, it was found that the BR was present as bright purple patches that could be isolated in the form of "purple membrane." (b) Later work showed that during illumination, light energy is used by the BR to pump protons across the membrane, thereby producing an electrochemical proton gradient. (c) The purple pigment turned out to be a retinal molecule attached to the BR through a Schiff base, a type of covalent bond formed between an aldehyde (retinal) and an amine (a lysine in the BR polypeptide chain). (Cylinders are drawn around the helical portions of the polypeptide simply to make the diagram less confusing.)

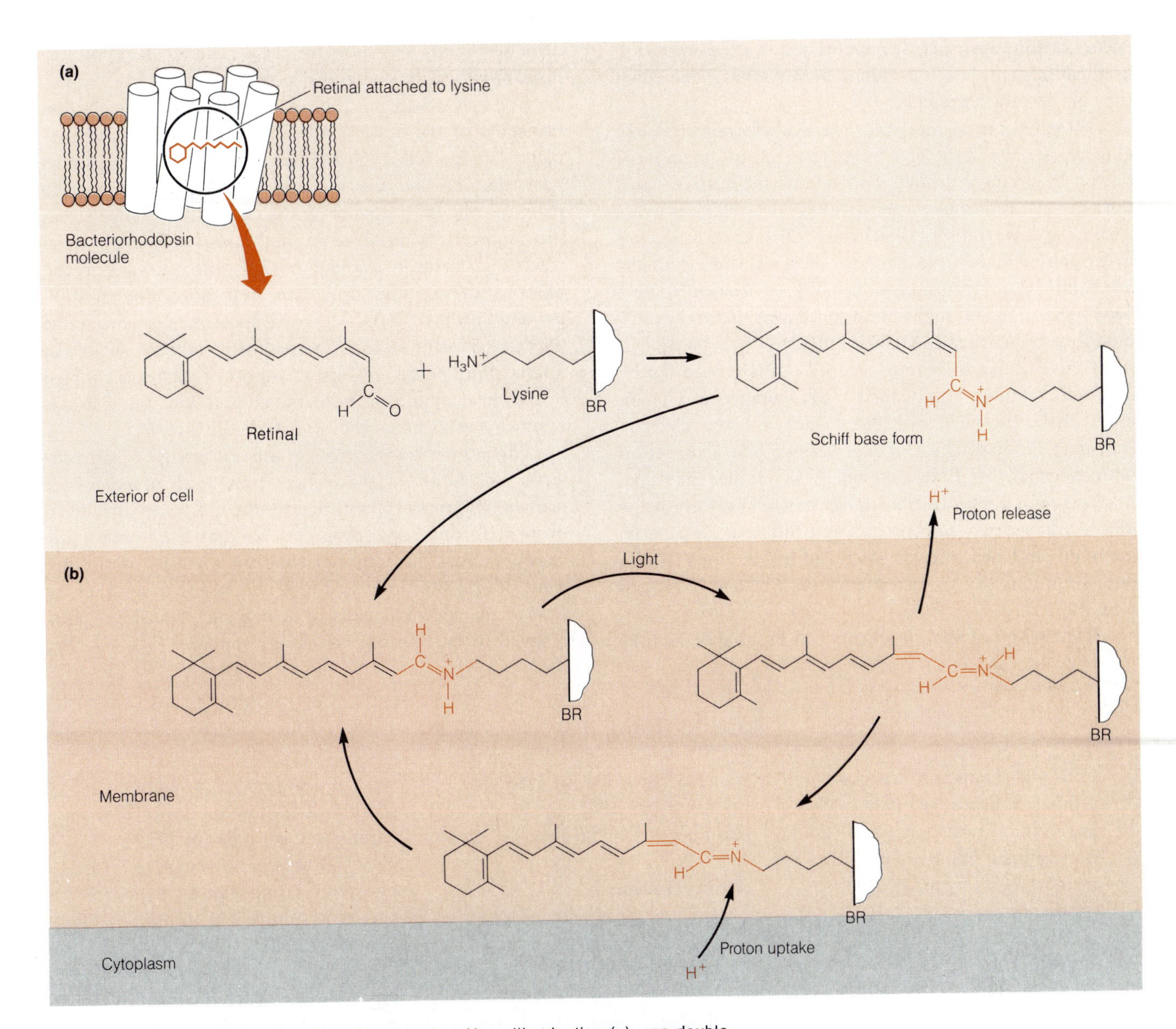

Figure 8-13 Bacteriorhodopsin and Proton Pumping. Upon illumination (a), one double bond in the retinal of the BR molecule changes from a trans to cis configuration. This change causes a proton to leave the Schiff base, which appears on the outside of the cell. (b) When the cis bond relaxes to the trans form, a second proton is taken up from the inside of the cell. This release and uptake of protons represents a proton pump, and the resulting gradient can be used as an energy source for the synthesis of ATP.

Perspective

Membrane transport by cells concerns specific mechanisms that cells have for ensuring that the right substances are moved into and out of the right compartments at the right times, in the right amounts, and at the right rates. Membrane transport is either passive or active, depending on whether the transport mechanism requires metabolic energy. Passive transport does not require a direct input of energy, and it underlies the movement of small, neutral molecules such as water, oxygen, and carbon dioxide, to which the lipid bilayer barrier of membranes is relatively permeable. Passive transport of such molecules is best understood as a diffusional process, in which flux occurs continuously across membranes. If a concentration gradient of the solute is present, net flux takes place down the concentration gradient.

Most important molecules, particularly nutrients like glucose, are too large and polar to cross the membrane barrier by simple diffusion. To facilitate the passive transport of such molecules, specific integral proteins provide a solute-specific transport mechanism through an otherwise hydrophobic membrane. Because such transport proteins are finite in number and in velocity of transport, they become saturated at high solute concentrations and can be described by Michaelis-Menten kinetics.

Active transport requires energy and is carried out by enzymatic "pumps" embedded in the membrane. The energy can be provided by sources such as light, electron transport, or hydrolysis of high-energy phosphate compounds, usually ATP. Because energy is put into the transport process, concentration or electrochemical gradients of the transported solutes can be produced. Most cells maintain electrochemical gradients of sodium ions (Na^+), potassium ions (K^+), calcium ions (Ca^{2+}), and hydrogen ions (H^+).

The energy of such gradients can be used in turn to drive secondary transport processes, often called cotransport, in which the exergonic movement of an Na^+ or H^+ down its electrochemical gradient drives the concomitant transport of the desired solute up its concentration gradient. Animal cells typically use sodium ions for this purpose and depend on the sodium-potassium pump to maintain the high outside-to-inside sodium gradient that this requires. Bacterial, plant, and fungal cells use an electrochemical proton gradient instead, with outward proton pumping driven either by electron transport or by an ATP-powered proton pump. Most transport mechanisms move the desired ion or molecule across the membrane unaltered, but in some bacteria inward transport of solutes is coupled to chemical modification of the solute, with phosphorylation of sugars being the best-studied example.

Membrane transport is an area of intense current interest. Recent technical advances in the study of membrane proteins and the reconstitution of transport activity in vitro promise further rapid progress. We may soon have some answers about the actual mechanism whereby solutes are translocated from one side of a membrane to another, as well as the specific means by which such translocation is coupled to the appropriate energy source.

Key Terms for Self-Testing

Cells and Transport Processes
cellular transport (p. 193)
intracellular transport (p. 193)
transcellular transport (p. 193)
membrane transport (p. 194)
passive transport (p. 194)
carrier protein (p. 194)
active transport (p. 194)
electrochemical gradient (p. 195)

Passive Transport
facilitated transport (p. 195)
simple diffusion (p. 195)
liposomes (p. 195)
diffusion constant (p. 197)
transport proteins (p. 198)
permeases (p. 198)
ionophores (p. 199)

Active Transport: Energy and Gradients
directionality (p. 203)
membrane potential (V_m) (p. 205)

Mechanisms of Active Transport
cotransport (p. 206)
simple active transport (p. 206)
coupled transport (p. 207)
symport (p. 207)
antiport (p. 207)
sodium cotransport (p. 207)
sodium-potassium pump (p. 208)
cardiotonic steroids (p. 209)
phosphotransferase system (p. 213)
phosphorylating transport (p. 213)
bacteriorhodopsin (p. 213)
retinal (p. 213)

Ionophores and the Study of Membranes
channel formers (p. 200)
Ion carriers (p. 200)

Suggested Reading

General References and Reviews

Bonting, S. L., and J. J. dePont, eds. *Membrane Transport.* Amsterdam: Elsevier/North Holland, 1981.

Bronner, F., and A. Kleinzeller, eds. *Carriers and Membrane Transport Proteins.* New York: Academic Press, 1980. (Volume 14 in a series entitled "Current Topics in Membranes and Transport"; see also Volume 12.)

Martinosi, A. N., ed. *Membranes and Transport,* vol. 1. New York: Plenum, 1982.

Stein, W. D. *Transport and Diffusion Across Cell Membranes.* Orlando, Fla.: Academic Press, 1986.

Stein, W.D., ed. *Ion Pumps: Structure, Function and Regulation.* New York: A. R. Liss, 1988.

Active Transport

Hobbs, A. S., and R. W. Albers. The structure of proteins involved in active membrane transport. *Annu. Rev. Biophys. Bioeng.* 9 (1980): 259.

Poole, R. J. Energy coupling for membrane transport. *Annu. Rev. Plant Physiol.* 29 (1978): 437.

Stekhoven, F. S., and S. L. Bonting. Transport adenosine triphosphatases: Properties and function. *Physiol. Rev.* 61 (1981): 1.

Specific Transport Mechanisms

Cantley, L. C. Structure and mechanism of the (Na,K)-ATPase. *Current Topics Bioenerg.* 11 (1981): 201.

Dills, S. S., A. Apperson, M. R. Schmidt, and M. H. Saier, Jr. Carbohydrate transport in bacteria. *Microbiol. Rev.* 44 (1980): 385.

Glynn, I. M. The Na^+-K^+ transporting adenosine triphosphatase. In *The Enzymes of Biological Membranes*, 2d ed. (A. Martonosi, ed.). New York: Plenum, 1985.

Jones, M. N., and J. K. Nickson. Monosaccharide transport proteins of the human erythrocyte membrane. *Biochim. Biophys. Acta* 650 (1981): 1.

Trachtenberg, M. C., D. J. Packey, and T. Sweeney. In vivo functioning of the Na,K-activated ATPase. *Current Topics Cell. Regul.* 19 (1981): 217.

Problem Set

1. **Mechanism of Transport.** For each of the following statements, answer with a D if the statement is true of passive diffusional transport, with an F if it is true of passive facilitated transport, and with an A if it is true of active transport. Any, all, or none (N) of the choices may be appropriate for a given statement.
   (a) Requires a carrier molecule localized within the membrane.
   (b) Depends primarily on solubility properties of the substance.
   (c) Doubling the concentration gradient of the molecule to be transported will double the rate of transport over a broad range of concentrations.
   (d) Applies only to transcellular transport processes.
   (e) Work is done during the transport process.
   (f) Applies only to nonpolar solutes.
   (g) Applies only to ions.
   (h) Transport can occur in either direction across the membrane, depending only on the concentration gradient prevailing at the moment.
   (i) $\Delta G^\circ = 0$.
   (j) Michaelis constant can be calculated.

2. **Transport Mechanisms.** Some microorganisms can use either ethanol (CH_3CH_2OH) or acetate (CH_3COO^-) as the sole carbon and energy source. Assume that the following data were obtained for the movement of ethanol and acetate across uniformly sized segments of the cell membrane for such an organism:

	Rate of Membrane Passage per Membrane Segment	
Concentration in Culture Medium (m*M*)	Ethanol (μmol/min)	Acetate (μmol/min)
0.1	2.5	9
1.0	25	50
10.0	250	91
100.0	2500	99

   (a) Plot the rate of uptake v as a function of carbon source concentration C for both ethanol and acetate. What can you conclude about the mode of movement of each of these compounds across the cell membrane?
   (b) In what way are your conclusions from part (a) consistent with known properties of ethanol and acetate?
   (c) Can you estimate the K_m value for acetate uptake? What is it? What about the K_m value for ethanol uptake?
   (d) Can you exclude active transport as a mechanism for ethanol uptake under the conditions used to obtain these data? What about acetate uptake?

3. **The Case of the Acid Stomach.** The gastric juice in your stomach has a pH of 2.0. This acidity is due to the secretion of hydrogen ions into the stomach by the epithelial cells of the gastric mucosa. Epithelial cells have an internal pH of 7.0 and a membrane potential of −70 mV (inside negative) and function at body temperature (37°C).
   (a) What is the concentration gradient of hydrogen ions across the epithelial membrane?
   (b) Calculate the free energy change associated with the secretion of 1 mole of protons into gastric juice at 37°C.
   (c) Do you think that hydrogen transport can be driven by ATP hydrolysis at the ratio of one molecule of ATP per hydrogen ion transported?
   (d) If hydrogen ions were free to move back into the cell, calculate the membrane potential that would be required to prevent them from doing so.

4. **Inverted Vesicles.** An important advance in transport research was development of methods for making closed membrane vesicles that retain the activity of certain transport systems. One such system uses resealed vesicles from red blood cell membranes, in which the orientation of membrane proteins may be either the same as in the intact erythrocyte (right-side-out) or inverted (inside-out). Such vesicles have demonstrable ATP-driven sodium-potassium pump activity. By resealing the vesicles in one medium and

then placing them in another, it is possible to have ATP, sodium, and potassium present inside the vesicle, present outside the vesicle, or not present at all.

(a) Suggest one or two advantages that such vesicles might have compared to intact red blood cells for studying the sodium-potassium pump. Can you think of any possible disadvantages?

(b) For the inverted vesicles, indicate whether each of the following should be present inside the vesicle (I), outside the vesicle (O), or not present at all (N) in order to demonstrate ATP hydrolysis: Na^+, K^+, ATP.

(c) If you plot the rate of ATP hydrolysis as a function of time after initiating transport in such inverted vesicles, what sort of a curve would you expect to obtain?

5. **Calcium Pump of Sarcoplasmic Reticulum.** Muscle cells use calcium ion to regulate the contractile process. Calcium is both released and taken up by the sarcoplasmic reticulum (SR). The release of calcium from the SR activates muscle contraction, a calcium-dependent process, and uptake causes the muscle cell to relax afterward. When muscle tissue is disrupted by homogenization, the SR forms small vesicles called *microsomes* that maintain their ability to take up calcium. In one experiment, 1 ml of reaction medium was made up containing 5 m*M* Mg^{2+}-ATP and 0.1 *M* KCl buffered at pH 7.5. An aliquot of SR microsomes was then added containing 1.0 mg protein, followed by 0.4 μmole of calcium. Two minutes later a calcium ionophore was added. ATPase activity was monitored during the additions, with the results shown in Figure 8-14.

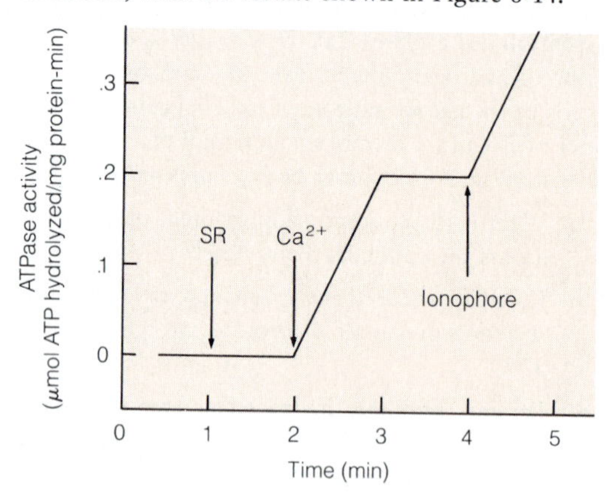

Figure 8-14 Calcium Uptake by Sarcoplasmic Reticulum. (See Problem 5 for details.)

(a) What is the ATPase activity, calculated as micromoles of ATP hydrolyzed per milligram of protein per minute?

(b) The ATPase is calcium-activated, as shown by the increase in ATP hydrolysis when the calcium was added and the decrease in hydrolysis when all the added calcium was taken up into the vesicles 1 minute after it was added. How many calcium ions are taken up for each ATP hydrolyzed?

(c) The final addition is a calcium ionophore known to carry calcium ions across membranes. Why does ATP hydrolysis begin again?

6. **Sodium Cotransport as a Mechanism of Amino Acid Uptake.** The epithelial cells that line your small intestine are thought to take up amino acids from the gut by a process of cotransport, in which the electrochemical gradient of sodium ions across the plasma membrane drives the uptake of amino acids against their concentration gradients. The following values were obtained at 37°C:

pH of epithelial cells = 7.0

V_m = membrane potential = -65 mV (inside negative)

$[Na^+]_{in} = 7.5$ m*M*	$[Na^+]_{out} = 105$ m*M*
$[\text{glycine}]_{in} = 15.0$ m*M*	$[\text{glycine}]_{out} = 0.10$ m*M*
$[\text{aspartate}]_{in} = 22.5$ m*M*	$[\text{aspartate}]_{out} = 0.15$ m*M*

The structures of glycine and aspartate at pH 7.0 are as follows:

Glycine: $H_3\overset{+}{N}-CH_2-COO^-$

Aspartate: $H_3\overset{+}{N}-CH(COO^-)-CH_2-COO^-$

(a) Calculate ΔG for the movement of 1 mole-equivalent of sodium ions into the epithelial cells.

(b) Calculate ΔG for the movement of 1 mole of glycine into the epithelial cells. What is the minimum number of sodium ions that must be cotransported with one glycine molecule to drive the uptake?

(c) Calculate ΔG for the movement of 1 mole of aspartate into the epithelial cells. What is the minimum number of sodium ions that must be cotransported with one aspartate molecule to drive the uptake?

(d) Explain why the number of sodium ions required for amino acid cotransport is different for glycine (part b) and aspartate (part c).

(e) Do you think the diffusion of aspartate back out through the epithelial cell membrane would be a problem? Why or why not? What about that of glycine?

7. **Sodium Transport.** A marine protozoan is known to pump sodium ions outward by a simple sodium pump that is driven by ATP hydrolysis and operates independently of potassium. The intracellular concentrations of ATP, ADP, and P_i are 20 m*M*, 2 m*M*, and 1 m*M*, respectively, and the membrane potential is -75 mV.

(a) Assuming that the pump transports three sodium ions outward per molecule of ATP hydrolyzed, what is the lowest internal sodium ion concentration that can be maintained at 25°C when the external sodium ion concentration is 150 m*M*?

(b) If you were dealing with a neutral molecule rather than an ion, would your answer for part (a) be higher or lower, assuming all conditions remained the same? Explain.

8. **Ouabain Inhibition.** Ouabain is a very specific inhibitor of the active transport of sodium ions out of the cell and is therefore a valuable tool in studies of membrane transport mechanisms. Which of the following processes in your own body would you expect to be sensitive to inhibition by ouabain? Explain in each case.

(a) Passive transport of glucose into a muscle cell.

(b) Active transport of dietary phenylalanine across the intestinal mucosa.

(c) Uptake of potassium ions by red blood cells.

(d) Active uptake of lactose by the bacteria living in your intestine.

9. **Dual Mechanism of Ion Uptake.** Although most of our discussion of transport has assumed a single mechanism for a given solute, this is not always the case. Figure 8-15 illustrates the rate of absorption of $^{42}K^+$, a radioactive isotope of potassium, by barley roots as a function of the potassium chloride (KCl) concentration in the solution. Both components of this biphasic uptake curve can be shown to follow Michaelis-Menten kinetics, and the overall velocity v_T can be closely approximated by summing the Michaelis-Menten equations for the two components. Shown in Figure 8-16 are separate double-reciprocal plots for the two components of the uptake curve. Also indicated on the double-reciprocal plots are the effects on the kinetics of potassium uptake of having sodium present at a fixed concentration.

(a) Describe how you think these experiments were done.

(b) Determine the V_{max} and K_m values for components 1 and 2 of the uptake curve, using the $-Na^+$ data of Figure 8-16.

(c) Calculate v_T for a series of K_m concentrations bracketing the concentration range of Figure 8-15 (0.0005–50 mM), and superimpose a plot of v_T versus potassium ion concentration on Figure 8-15. How well does your calculated curve fit the experimental data?

(d) What can you conclude about the relative effects of sodium on the potassium transport mechanism operative at low potassium concentrations (component 1)? On the high-potassium mechanism (component 2)?

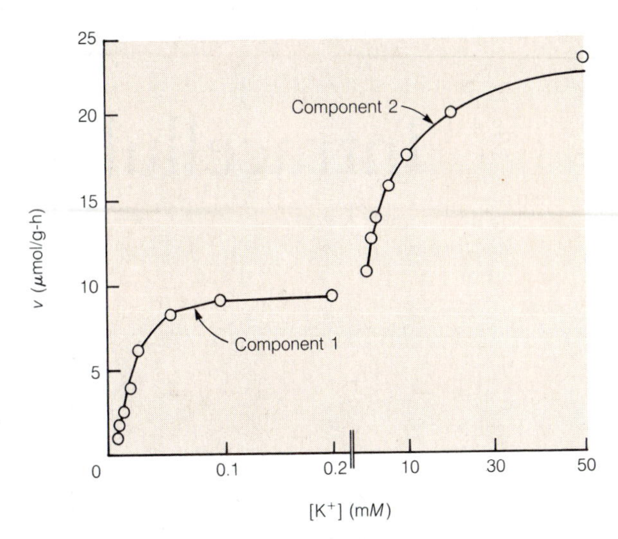

Figure 8-15 The Biphasic Nature of Potassium Uptake by Root Cells. Roots were detached from young barley seedlings and suspended in ^{42}KCl solutions at various potassium concentrations. The roots were rinsed after exposure and then assayed for radioactivity. Data were expressed as micromoles of potassium taken up per gram of root per hour.

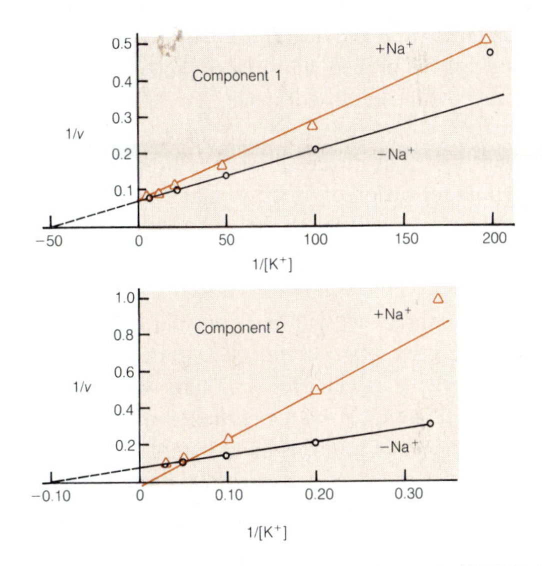

Figure 8-16 Double-Reciprocal Plots for the Biphasic ^{42}KCl Uptake Data of Figure 8-15. Potassium uptake was determined in the presence and absence of sodium at a fixed potassium concentration (0.5 m*M* for component 1, 2.0 m*M* for component 2).

9

Intracellular Compartments

The study of eukaryotic cells is really the study of an elaborate series of intracellular membranes and of the many discrete membrane-bounded compartments in which so much of the drama of cellular activity takes place. Whether we talk about storage and readout of genetic information, secretion of proteins, digestion of food particles, or generation and degradation of hydrogen peroxide, the locale for the process within a eukaryotic cell is in each case a membrane-enclosed space. To appreciate the eukaryotic cell fully, therefore, means to understand the prominent role that intracellular membranes play and the compartmentalization of function that they make possible. (Prokaryotes, you will recall, have no internal membranes or organelles, so this discussion does not apply to such cells.)

In this chapter, we will first consider the *endoplasmic reticulum* and the *Golgi complex*, organelles we have already met briefly in Chapter 4. Then, we will focus on *lysosomes* and *peroxisomes*, organelles related to the endoplasmic reticulum and the Golgi complex, but with specific properties and functions of their own. In each case, we will want to ask how the organelle was discovered and how it is studied, what we know about its structure, and what sorts of functions take place in or on it. In addition, we intend to explore the intimate relationships of origin and function between these compartments.

Much of our understanding of eukaryotic membrane systems was established by the complementary techniques of electron microscopy and subcellular fractionation. For their pioneering work in this area, Albert Claude, George Palade, and Christian de Duve shared a Nobel Prize in 1974. Claude was instrumental in developing **differential centrifugation** as a way to isolate organelles. This is a powerful technique for separating organelles that differ in size, shape, or density (Figure 9-1). Palade was quick to use differential centrifugation in studies of the endoplasmic reticulum and the Golgi complex. This enabled him to establish the role of these organelles in the synthesis, processing, and secretion of proteins. De Duve is the discoverer of both lysosomes and peroxisomes. In each case, subcellular fractionation and electron microscopy were vital techniques in the detection and characterization of the new class of organelles.

The Endoplasmic Reticulum

The **endoplasmic reticulum** (**ER**) is one of the most pervasive features of the cytoplasm of eukaryotic cells. Unlike more prominent organelles such as the mitochondrion and chloroplast, however, the ER cannot usually be seen with the light microscope. In the late nineteenth century it was noted that some eukaryotic cells, particularly those involved in secretion, contained regions that stained intensely with basic dyes. These regions were called *ergastoplasm*, but their significance remained in doubt until the advent of electron microscopy in the 1950s. With the tremendous increase in resolving power that this technique allowed, it became possible for the first time to visualize the elaborate network of intracellular membranes that came to be called the endoplasmic reticulum.

Although the name sounds formidable, it is actually quite descriptive. *Endoplasmic* simply means "within the (cyto)plasm," and *reticulum* is a Latin word meaning "network." (Actually, the term was originally coined because of the extent to which the ER resembled a reticule, a netted handbag with a drawstring that was popular in the early part of the century.)

The ER, then, is an interconnecting membranous network of vesicles, tubules, and flattened sacs. These sacs are called **cisternae,** and the internal volume that they define is called the **cisternal space.** Since the cisternal space is separated by a membrane from the surrounding **cytoplasmic space,** the ER in a eukaryotic cell effectively divides the cytoplasm into two compartments. The cisternal space

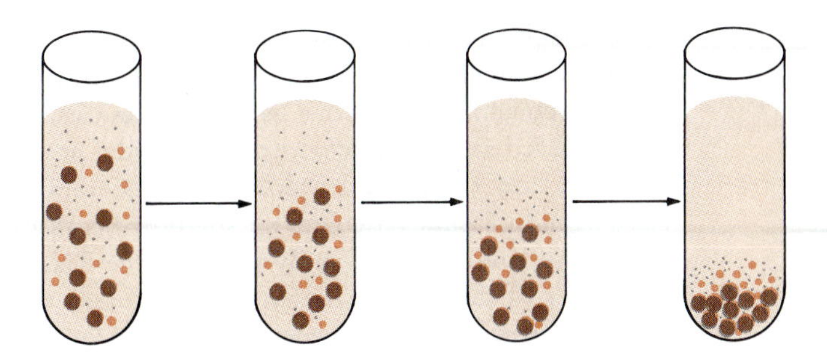

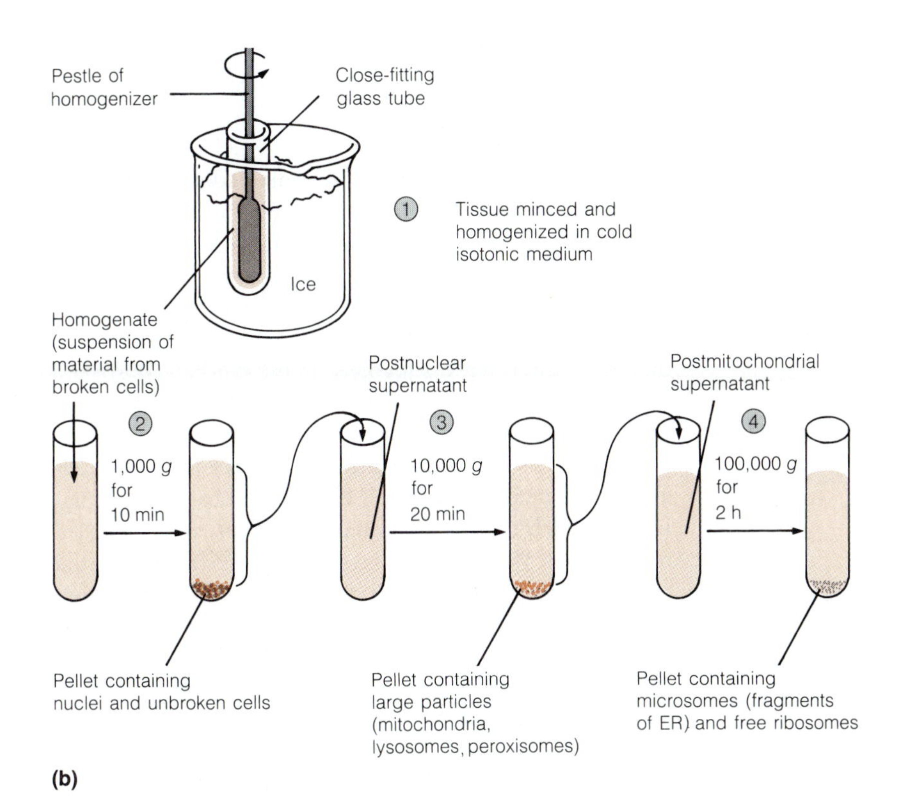

Figure 9-1 Differential Centrifugation as a Means of Isolating Organelles. (a) Differential centrifugation of organelles depends on differences in size, shape, and density that result in different rates of sedimentation in response to a centrifugal force. Particles with a high sedimentation rate reach the bottom of the tube most rapidly, whereas those with a low sedimentation rate require higher speeds or longer periods of centrifugation to do so. (b) In subcellular fractionation, the tissue of interest is first minced (often with razor blades) and homogenized (step 1). Subcellular fractions are then isolated by subjecting first the homogenate and then subsequent supernatant fractions to successively higher centrifugation speeds and/or longer centrifugation times. In the three-stage procedure shown, nuclei and unbroken cells are first sedimented at low speed (step 2), followed by large organelles such as mitochondria, lysosomes, and peroxisomes at higher speed (step 3), and finally microsomes (fragments of ER) and ribosomes at still higher speeds (step 4). In each case, the acceleration due to centrifugation is expressed as a multiple of g, the acceleration due to gravity (980 cm/sec^2).

is part of an intracellular system that also includes the internal spaces of the Golgi complex and lysosomes and the perinuclear space between the two membranes of the nuclear envelope. These relationships are illustrated in Figure 9-2.

The Membrane of the Endoplasmic Reticulum

Our knowledge of the endoplasmic reticulum comes largely from studies of hepatocytes (liver cells) because these cells are relatively accessible and can be obtained in substantial quantities. The properties of hepatocyte ER can usually be generalized to other cells, but these cells also have a few specialized functions that will be pointed out as we go along.

ER membranes are only about 5–6 nm in thickness and are therefore noticeably thinner than the plasma membrane. Another distinguishing feature of the ER membrane is a protein/lipid ratio about twice that of other membranes. The lipid content is almost entirely phospholipid, in contrast to the plasma membrane, which has significant amounts of cholesterol.

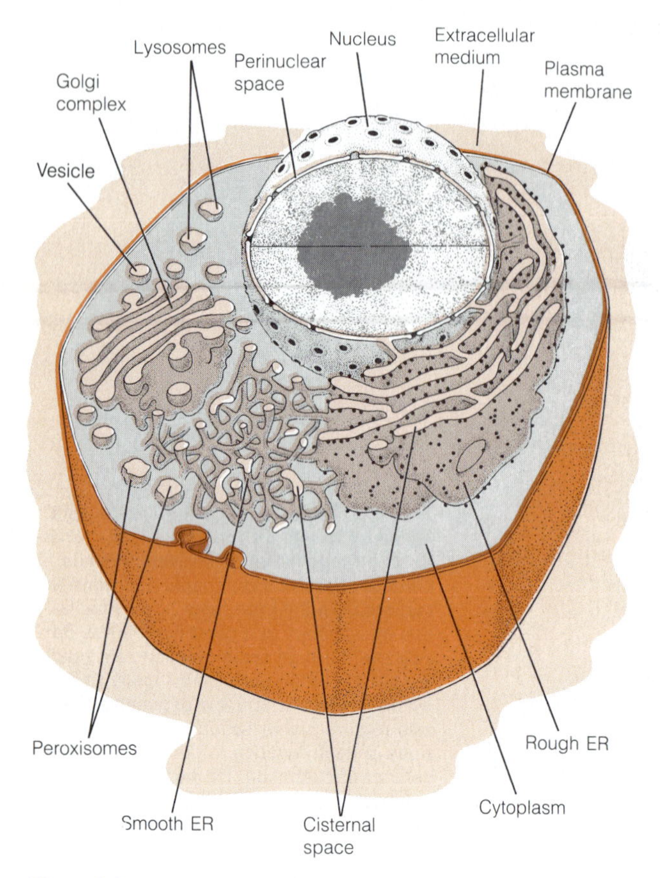

Figure 9-2 The Cisternal Space of the ER. The cisternal space of the endoplasmic reticulum (ER) is continuous with the perinuclear space between the two nuclear membranes and can also communicate with the interior of the Golgi complex, and lysosomes, by means of vesicles.

Much of the protein content can be accounted for in terms of enzymes known to be associated with the ER. Some of the best-known enzymes of the ER include cytochrome b_5 reductase, cytochrome *c* reductase, and glucose-6-phosphatase. Because these enzymes are unique to the endoplasmic reticulum, they are often used as markers to identify and keep track of the ER in subcellular fractionation experiments. Other prominent proteins include cytochrome b_5 and cytochrome P-450, both of which are components of an electron transport chain present in the ER of hepatocytes. The purpose of this transport chain is to transfer electrons stepwise from a reduced coenzyme called nicotinamide dinucleotide phosphate (NADPH) to oxygen, thereby activating the oxygen for hydroxylation steps involved in detoxification reactions.

Two Types of Endoplasmic Reticulum

The ER actually consists of two different kinds of membranous network, which can be readily distinguished from each other by the presence or absence of membrane-bound ribosomes, as illustrated in Figure 9-3. The **rough endoplasmic reticulum** consists of a network of vesicles, tubules, and flattened sacs and is characterized by the presence of ribosomes on the outer surface of the membrane. The **smooth endoplasmic reticulum** is "smooth" because of the absence of ribosomes. (It is, by the way, the RNA in the ribosomes of the rough ER that reacts strongly with the basic dyes originally used to identify the ergastoplasm seen through the light microscope. Ergastoplasm, in other words, turned out to be rough ER.)

Rough and smooth ER can also be distinguished morphologically. Membranes of the rough ER usually consist of large, flattened sheets, whereas smooth ER membranes are generally more tubular in shape. In most cells, the rough and smooth ER components appear to be continuous.

Both types of ER are present in most eukaryotic cells, but considerable variation is seen in the relative amounts, depending on the activities of the cell. Cells characterized by the synthesis of secretory proteins tend to have a very prominent rough ER, as shown in Figure 9-4 for a pancreatic cell. This is due to the involvement of the rough ER in the process of protein synthesis (which, as you have probably guessed, has something to do with the presence of all those ribosomes on the surface of the rough ER). Similarly, smooth ER is involved in the synthesis of steroid hormones and is therefore a prominent feature of steroid hormone-producing cells (Figure 9-5).

When tissue is homogenized for subcellular fractionation, the membranes of the ER break into smaller fragments, which then seal spontaneously into vesicles called **microsomes.** When microsomes are prepared by differential centrifugation, fractions can be isolated with and without attached ribosomes, depending on whether the membrane originated from the rough or smooth ER. Such preparations have proved tremendously useful in exploring both types of ER, but you need to keep in mind that microsomes do not actually exist as such in the cell; they are merely a phenomenon of the fractionation process.

Smooth Endoplasmic Reticulum

The smooth ER is involved in several different cellular processes, including drug detoxification, carbohydrate metabolism, and synthesis of neutral fats, phospholipids, and steroids. Especially important in this latter category are the male and female sex hormones and the steroid hormones of the adrenal cortex.

Hydroxylation Reactions. The smooth ER of hepatocytes plays an important role in a variety of **hydroxylation reactions,** in which molecular oxygen (O_2) is used to generate

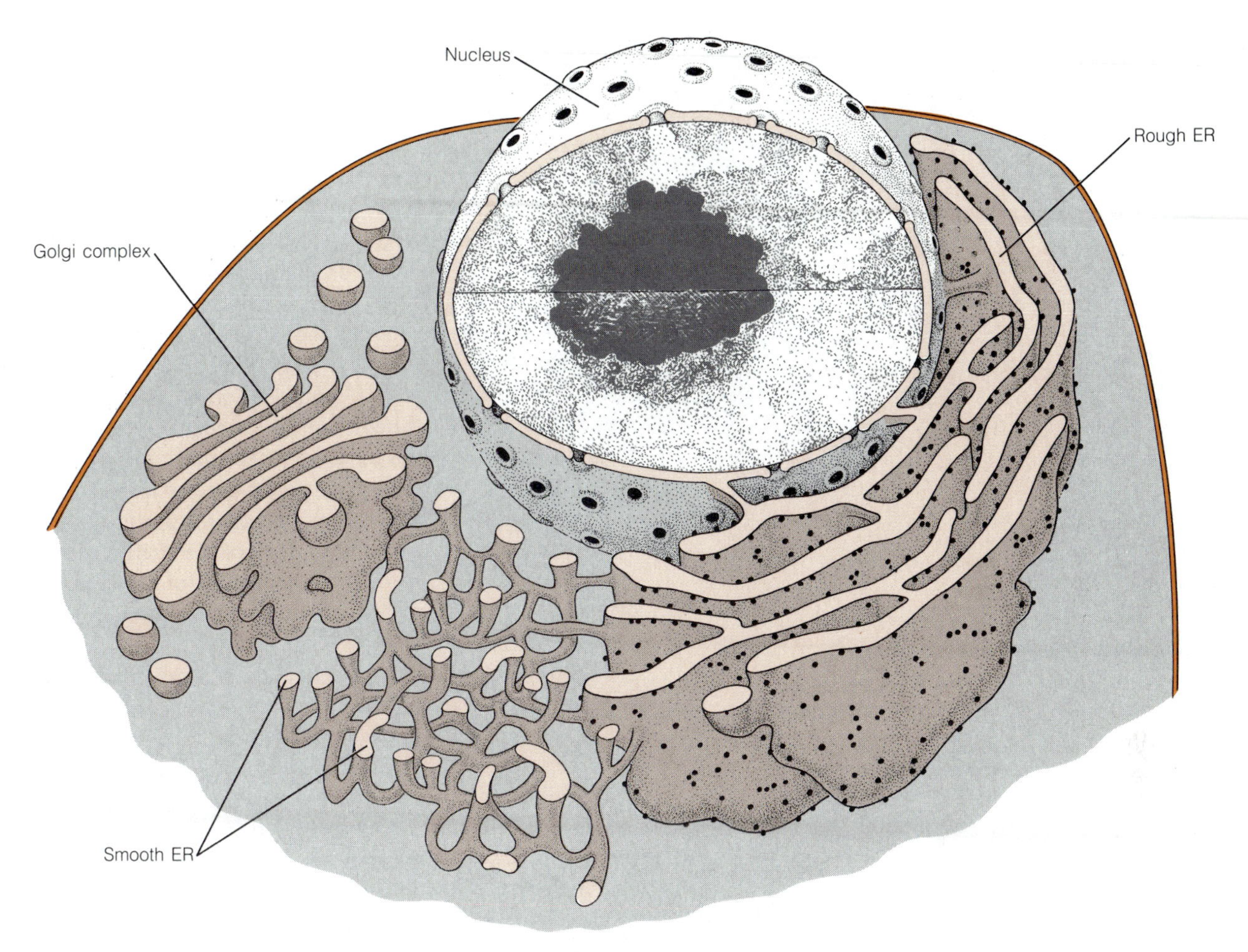

Figure 9-3 Smooth and Rough ER and Their Relation to the Nuclear Envelope and the Golgi Complex. The smooth ER is a network of branched tubules without ribosomes, whereas the rough ER is characterized by the presence of ribosomes on the outer surface. The rough ER is often physically continuous with the outer membrane of the nuclear envelope. The Golgi complex is involved in processing and packaging proteins synthesized on the rough ER.

hydroxyl groups, with electrons supplied by NADPH, the reduced coenzyme mentioned earlier. One such hydroxylation reaction converts the amino acid phenylalanine into tyrosine, the hydroxylated equivalent:

$$\underset{\text{Phenylalanine}}{C_6H_5{-}CH_2{-}CH(NH_3^{\oplus}){-}COO^{\ominus}} + O_2 + NADPH + H^+ \longrightarrow \underset{\text{Tyrosine}}{HO{-}C_6H_4{-}CH_2{-}CH(NH_3^{\oplus}){-}COO^{\ominus}} + H_2O + NADP^+ \quad (9\text{-}1)$$

All such hydroxylation reactions require both NADPH and oxygen. One atom of the oxygen molecule goes into the substrate, and the other is reduced to water. Enzymes that carry out such hydroxylation reactions are called **monooxygenases** or **mixed-function oxidases.** Although reaction 9-1 suggests a direct transfer of electrons from NADPH to oxygen, the transfer actually involves an electron transport chain found in the smooth ER, with cytochrome P-450 as the terminal component. It is the reduced form of P-450 that actually activates oxygen for hydroxylation.

Drug Detoxification and Carcinogenesis. The cytochrome P-450 found in the smooth ER is important in **drug detoxification,** especially in the liver. The mode of detoxification usually involves hydroxylation, which increases the solubility of the drug in water. This alteration is critical, because hydrophobic compounds tend to be membrane-soluble and are therefore retained in the body, whereas

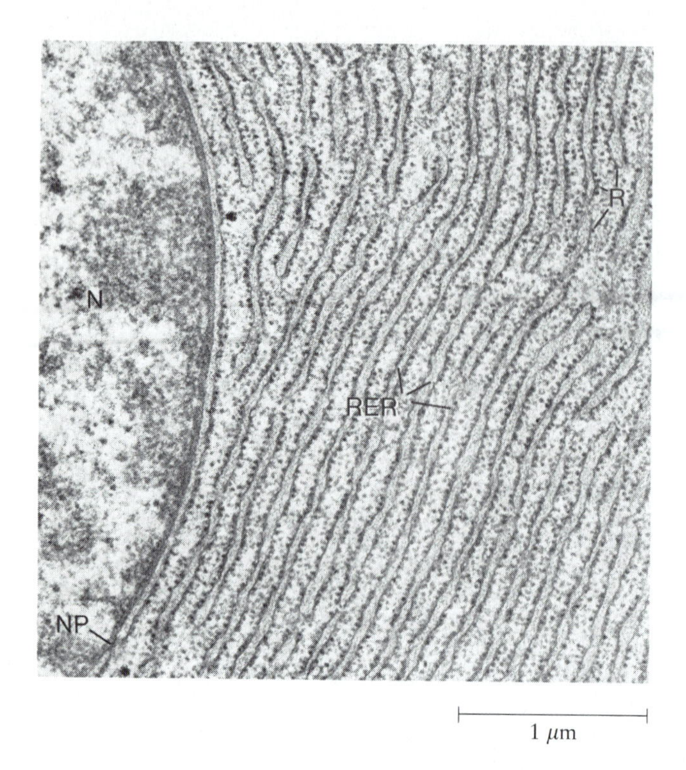

Figure 9-4 The Rough ER. This electron micrograph shows rough endoplasmic reticulum (RER) in an exocrine (secretory) cell from the pancreas of a dog. The rough ER is studded with ribosomes (R). Also visible in the micrograph are a portion of the nucleus (N) and a nuclear pore (NP). (TEM)

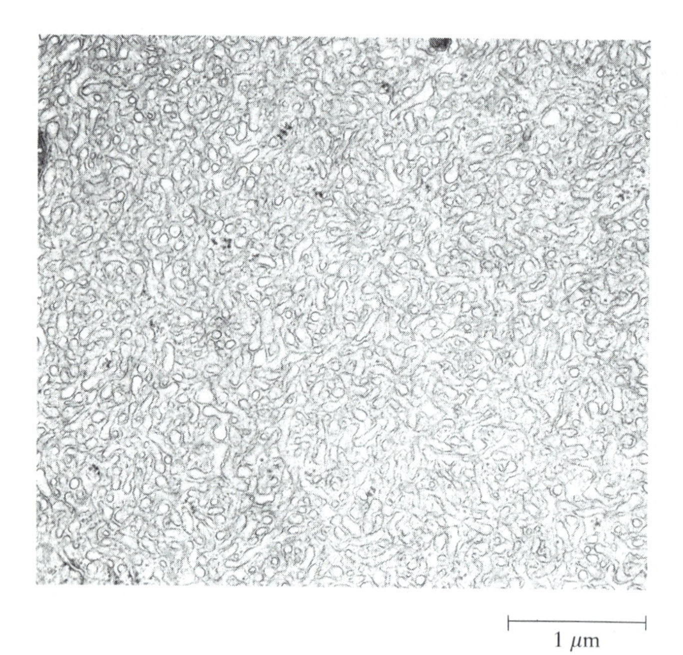

Figure 9-5 The Smooth ER. This electron micrograph shows smooth ER in a Leydig cell from the testis of a rat. The prominence of smooth ER reflects the role of Leydig cells in the biosynthesis of steroids. Steroid synthesis is important in the testes because the sex hormones are steroids. (TEM)

water-soluble compounds are more easily flushed away by the blood and subsequently excreted from the body. Barbiturate drugs, for example, can be detoxified by hydroxylation enzymes associated with the smooth ER. This can be demonstrated by injecting the sedative phenobarbital into a rat. One of the most striking effects is a rapid increase in the level of the barbiturate-detoxifying enzymes in the liver, accompanied by a dramatic proliferation of the smooth ER.

This process enhances removal of the drug from the body, but at the same time means that higher and higher doses of the drug must be administered to achieve the same sedative effect in habitual users of phenobarbital. Furthermore, the monooxygenase that is induced by phenobarbital is of such broad specificity that it can hydroxylate and therefore solubilize a variety of other drugs, including such useful agents as antibiotics, narcotics, steroids, and anticoagulants. As a result, the chronic use of barbiturates decreases the effectiveness of a host of other clinically useful drugs.

Another hydroxylase present in the smooth ER appears to have a **carcinogenic** (cancer-causing) effect. The enzyme *aryl hydrocarbon hydroxylase* is thought to be involved in converting potentially carcinogenic compounds into their chemically active form. This enzyme complex contains cytochrome P-448, a molecule closely related to cytochrome P-450. Significantly, cigarette smoke is a potent inducer of aryl hydrocarbon hydroxylase.

Glycogen Catabolism. Yet another function of the smooth ER of hepatocytes depends on the presence of the enzyme glucose-6-phosphatase in the smooth ER membrane. This enzyme catalyzes the removal of the phosphate group from glucose-6-phosphate to form free glucose:

$$\text{glucose-6-P} + H_2O \longrightarrow \text{glucose} + P_i \qquad (9\text{-}2)$$

To understand the importance of this phosphatase, we need to appreciate both the way in which liver cells store glycogen and the reason and mechanism for its subsequent breakdown.

A major role of the liver is to keep the level of glucose in the blood relatively constant. The liver stores glucose as glycogen and releases it as needed by the body, especially between meals and in response to muscular activity. Liver glycogen is stored as granules associated with the smooth ER (Figure 9-6a). The mobilization of liver glycogen is hormonally mediated. This process involves the stimulation of membrane-bound adenylate cyclase activity (Chapter 21, pp. 649–656), resulting in a transient elevation in the intracellular level of cyclic AMP. The cyclic AMP elevation triggers a complicated cascade of events, leading eventually to the activation of glycogen phosphorylase,

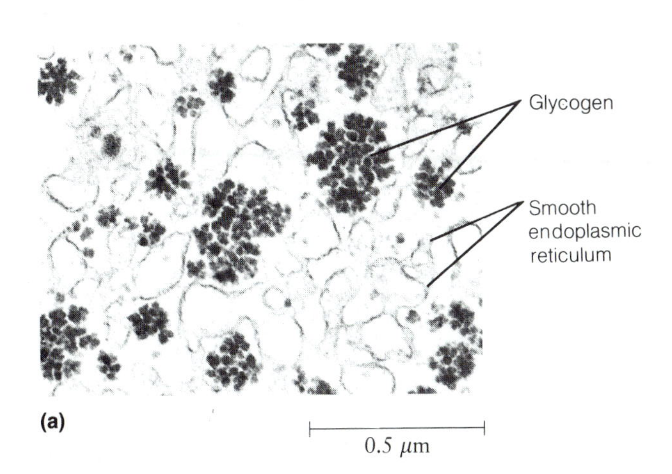

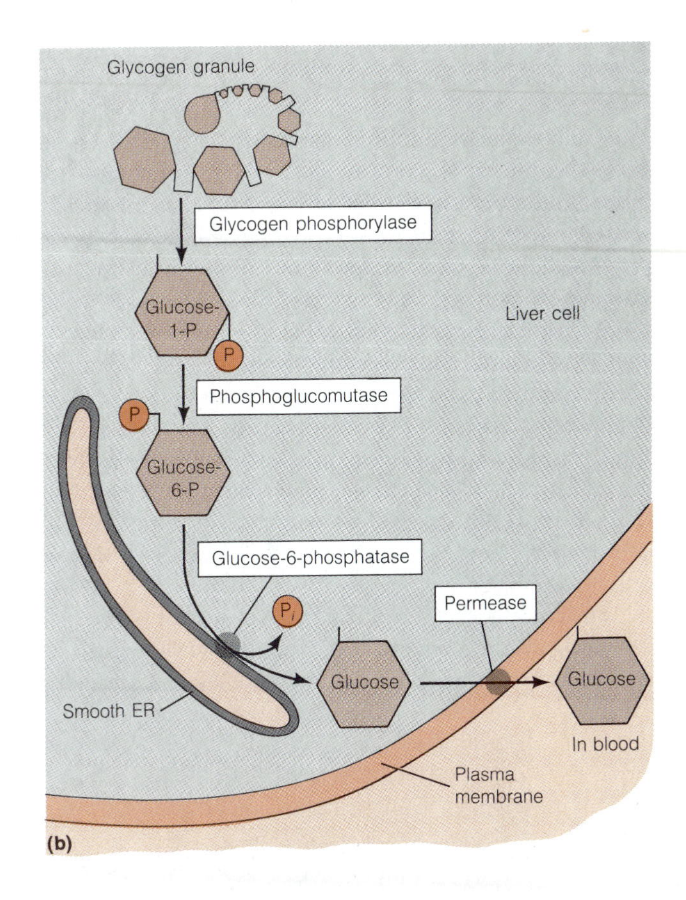

Figure 9-6 Role of Smooth ER in the Breakdown of Liver Glycogen. (a) This electron micrograph of a rat liver cell shows numerous glycogen granules in close association with the smooth ER (TEM). (b) The breakdown of glycogen involves the stepwise removal of glucose units as glucose-1-phosphate by glycogen phosphorylase, a cytoplasmic enzyme. Glucose-1-phosphate is then converted to glucose-6-phosphate by phosphoglucomutase, also in the cytoplasm. Removal of the phosphate group is accomplished by the glucose-6-phosphatase associated with the membrane of the smooth ER. The free glucose generated in this way is then transported out of the liver cell by a permease in the plasma membrane.

which cleaves glycogen into glucose-1-phosphate units (Figure 9-6b). Glucose-1-phosphate can readily be converted into glucose-6-phosphate in the cytoplasm by an enzyme called phosphoglucomutase.

To leave the liver cell and enter the bloodstream, however, the glucose-6-phosphate must be converted to free glucose, because membranes are generally impermeable to phosphorylated sugars. The purpose of the phosphatase in the smooth ER membrane is therefore to remove the phosphate group from glucose-6-phosphate to allow the glucose to move out of the liver cell into the blood for transport to cells in the body that are in need of energy (Figure 9-6b). Significantly, glucose-6-phosphatase activity is present in liver, kidney, and intestinal cells but not in muscle or brain cells. Muscle and brain cells therefore retain glucose-6-phosphate and use it to meet their high energy needs.

Rough Endoplasmic Reticulum

As already noted, rough ER is characterized by the presence of ribosomes (Figure 9-3). The ribosomes of the rough ER are always located on the side of the membrane that faces the cytoplasm and are responsible for the protein synthesis associated with the rough ER. The proteins made on the rough ER are usually destined either for export from the cell as secretory products or for incorporation into one of the several membrane systems derived directly or indirectly from the ER. In addition to protein synthesis, the rough ER is also the site of the initial steps in the addition of sugar groups to glycoproteins.

Investigators are not in full agreement concerning the means by which the ribosomes are attached to the membrane of the rough ER, nor is it entirely clear that there is a single mechanism. However, for integral membrane proteins and for secretory proteins that are inserted into the cisternal space of the rough ER as they are synthesized, it has been convincingly established that the ribosomes are bound to the rough ER membrane in part by the growing polypeptide chain. It is likely that one or more specialized ER proteins are also involved, since ribosomes remain loosely attached to the ER even when protein synthesis is not taking place. The mechanism of protein synthesis and the role of the ER in protein processing will be discussed in detail in Chapter 16.

The Golgi Complex

The **Golgi complex** (or **Golgi apparatus**) derives its name from Camillo Golgi, the Italian biologist who first described this structure in 1898. He reported that nerve cells soaked in osmium tetroxide showed a heavy metal deposit in a threadlike network surrounding the nucleus. The same staining reaction was demonstrated with other heavy metals and with a variety of cells, but it was difficult to get consistent results, and no cellular structure could be identified that might explain the staining. As a result, the nature (and very existence) of the Golgi complex remained highly controversial until the 1950s, when its existence was confirmed by electron microscopy. Since then, we have come to understand a good deal about the Golgi complex and its role in the preparation of proteins for export from the cell.

Structurally, the Golgi complex consists of a series of flattened, membranous **saccules,** disk-shaped cisternae that are stacked together as shown in Figure 9-7. Such a series of saccules is called a **Golgi stack** in animal cells and a **dictyosome** in plant cells. Usually, there are about five to eight saccules, though the Golgi stacks of some lower organisms can have several dozen saccules. The size and number of Golgi complexes vary from one type of cell to the next and also with the metabolic activity of the cell. Some cells have one, whereas others (particularly those that are especially active in secretion) have thousands.

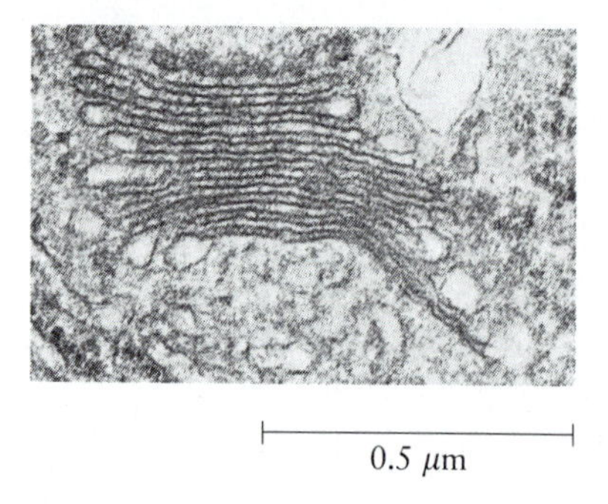

Figure 9-7 Structure of the Golgi Complex. This electron micrograph shows a Golgi stack (or dictyosome) from an algal cell. (TEM)

The saccules of the Golgi are thought to form by the fusion of **vesicles** that bud off the ER, as shown in Figure 9-8. As these vesicles fuse, they give rise to new saccules on what is termed the ***cis,*** or **forming, face** of the organelle. Meanwhile, on the ***trans,*** or **maturing, face** of the complex, vesicles appear to bud off the tips of the saccules continuously. In secretory cells at least, these vesicles give rise to the **secretory granules** that eventually carry their contents to the cell surface for release to the outside.

The two faces of a Golgi stack are biochemically distinct. Cytochemical and immunological staining techniques can be used to show that specific enzymes and receptor proteins are concentrated within the cisterna on the *cis* face of the stack, whereas other proteins are localized primarily in the cisterna on the *trans* face. This biochemical polarity of the Golgi complex is illustrated in Figure 9-9, which shows several Golgi stacks in an epithelial cell of the rat epididymis. The cell section has been stained to detect the receptor protein for the carbohydrate mannose-6-phosphate, using an antibody against the receptor protein as a probe. Clearly, the receptor protein for this carbohydrate is concentrated in the cisterna on the *cis* face. (As we will see later in this chapter, mannose-6-phosphate is a distinctive component of glycoproteins that are destined to become lysosomal enzymes. The mannose-6-phosphate is added to these proteins in the ER, and the presence of

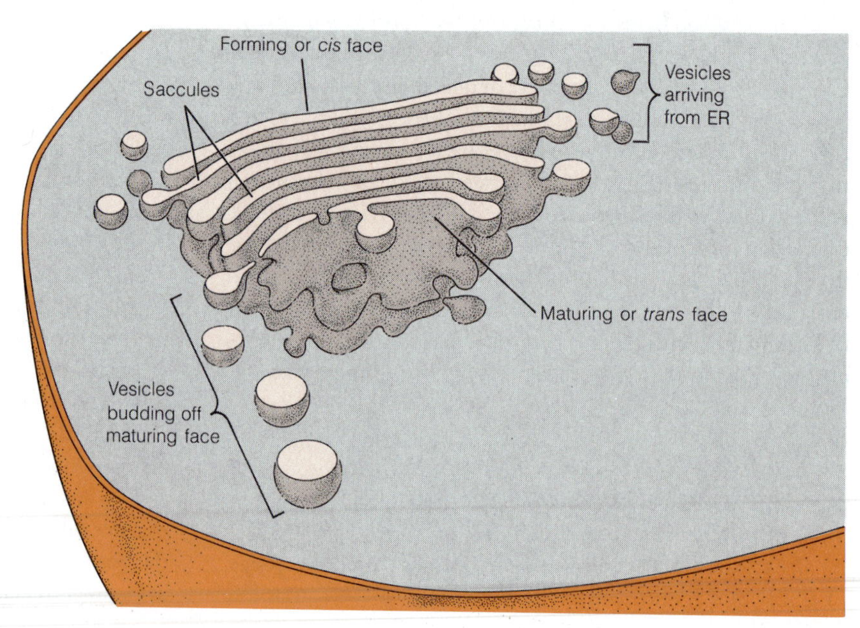

Figure 9-8 Illustration of Golgi Structure. Each Golgi stack consists of a small number of flattened saccules stacked together. On the *cis,* or forming, face, vesicles arriving from the ER fuse to form new saccules. On the *trans,* or maturing, face, vesicles arise by budding off the tips of the saccules.

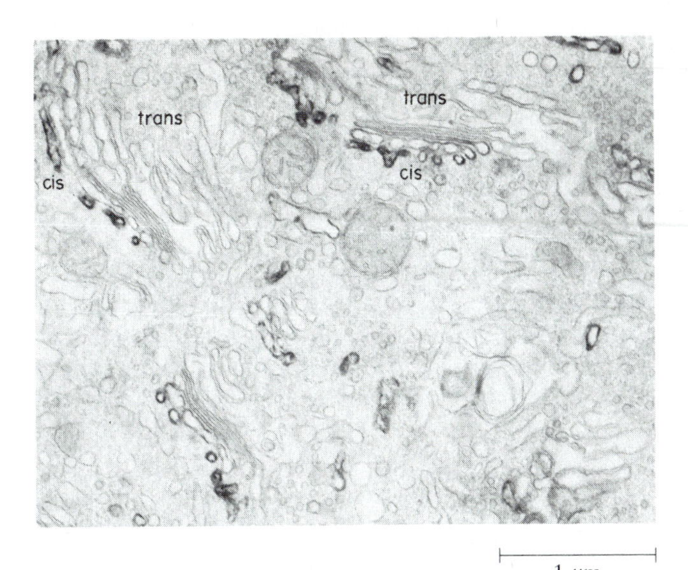

Figure 9-9 Immunochemical Staining of Golgi Stacks. This electron micrograph shows several Golgi stacks in a cell from the absorptive epithelium of the rat epididymis. The cell was stained by an immunochemical technique to detect receptor proteins specific for the sugar mannose-6-phosphate. The receptors for this sugar are concentrated in the cisterna of the *cis* face of the Golgi stack. Thus, glycoproteins with mannose-6-phosphate groups that arrive at the Golgi complex will be recognized and processed on the *cis* face of the stack. (TEM)

the mannose-6-phosphate receptor in the *cis* cisterna of the Golgi stack allows the Golgi complex to recognize such proteins and to target them to the lysosomes.)

Role of the Golgi Complex in Protein Processing

The Golgi complex serves a role in mediating the flow of secretory proteins from the ER to the exterior of the cell. Consistent with this role, the Golgi complex usually has a protein and lipid composition that is intermediate between that of the ER, from which the Golgi membrane is thought to arise, and that of the plasma membrane, with which Golgi-derived secretory vesicles eventually fuse in discharging their contents to the outside of the cell. Similarly, the thickness of the Golgi membrane is usually intermediate between the thin membrane of the ER and the thicker plasma membrane. In fact, the saccule membranes at the forming face resemble the ER in both morphology and staining properties, whereas those at or near the maturing face resemble the plasma membrane.

This characteristic suggests a flow of membrane components from the ER through the Golgi complex to the plasma membrane, a concept that is supported by studies conducted with the protozoan *Trichonympha*. When *Trichonympha* cells are deprived of nutrients, the Golgi complex and most of the ER disappear. When feeding is resumed, the membranes of the ER are the first to reappear, followed by vesicles that bud off the ER and form Golgi saccules. These saccules become ordered into stacks, and eventually there is a movement of the outermost saccules toward the plasma membrane, with which they fuse.

Within the Golgi, membrane flow through the stacks can be quite rapid. In certain mucus-secreting cells of the intestinal mucosa, it is thought to take less than 40 minutes for membrane components to move from one face of a Golgi stack to the other face.

Two enzymes that are frequently used as cytochemical markers for Golgi membranes are thiamine pyrophosphatase and nucleoside diphosphatase. The former is an especially useful marker, even though its physiological significance is not known, because it occurs only in Golgi membranes. The most important enzymes of the Golgi complex, however, are glycosyl transferases and glucan synthetases. The transferases play a role in attaching sugar groups to proteins, and the synthetases catalyze the formation of polysaccharides from sugar nucleotides. The presence of these enzymes in the Golgi membranes suggests the importance of the Golgi complex in carbohydrate metabolism and glycoprotein synthesis, topics that will be discussed in detail in Chapter 16.

Transport of proteins to the Golgi complex is mediated by **coated vesicles** that bud off the ER and then fuse with the forming face of the Golgi complex. The coat that surrounds these vesicles and gives them their name is a bristlelike, polyhedral lattice of **clathrin** subunits. Clathrin is a large (180,000 daltons) protein that is a distinctive feature of vesicles involved in a variety of intracellular transport processes. In addition to their role in bringing secretory proteins from the ER to the Golgi complex, clathrin-coated vesicles also transport membrane proteins from the Golgi to lysosomes, to the plasma membrane, and to other cellular destinations. In each case, the clathrin coat seems to form a "basket" or "cage" around the vesicle, as illustrated in Figure 9-10. Coated vesicles are explored in more detail in the box on pp. 230–231.

Two Cellular Transport Processes: Exocytosis and Endocytosis

As we have already seen, cellular products, such as secretory proteins, flow from the ER to the Golgi complex to the plasma membrane, where they are released to the outside of the cell. The process by which these materials are released is called **exocytosis.** There also exists in some cells

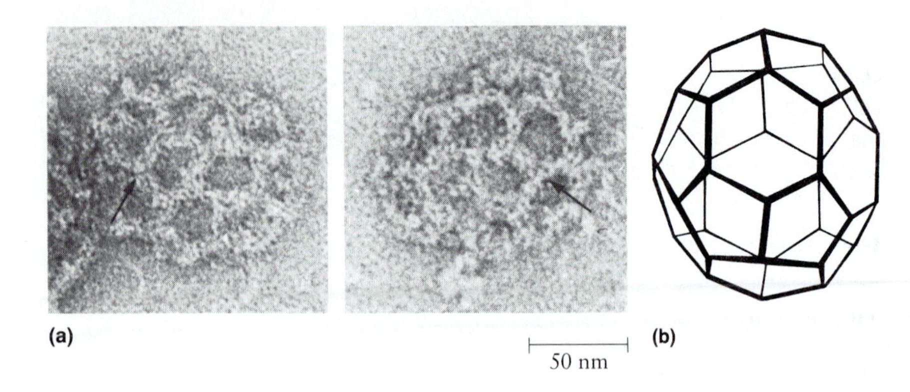

Figure 9-10 Coated Vesicles. Coated vesicles are involved in the transport of proteins from the ER to the Golgi complex. Each such vesicle is surrounded by a "basket" or "cage" of clathrin subunits. (a) Electron micrographs of clathrin cages that have been negatively stained with uranyl acetate (TEM). Arrows point to edges that exhibit a clear double- or triple-lined nature due to the overlap of the legs of adjacent clathrin molecules (see Figure 9-C, p. 231). (b) Interpretive drawing of a clathrin cage.

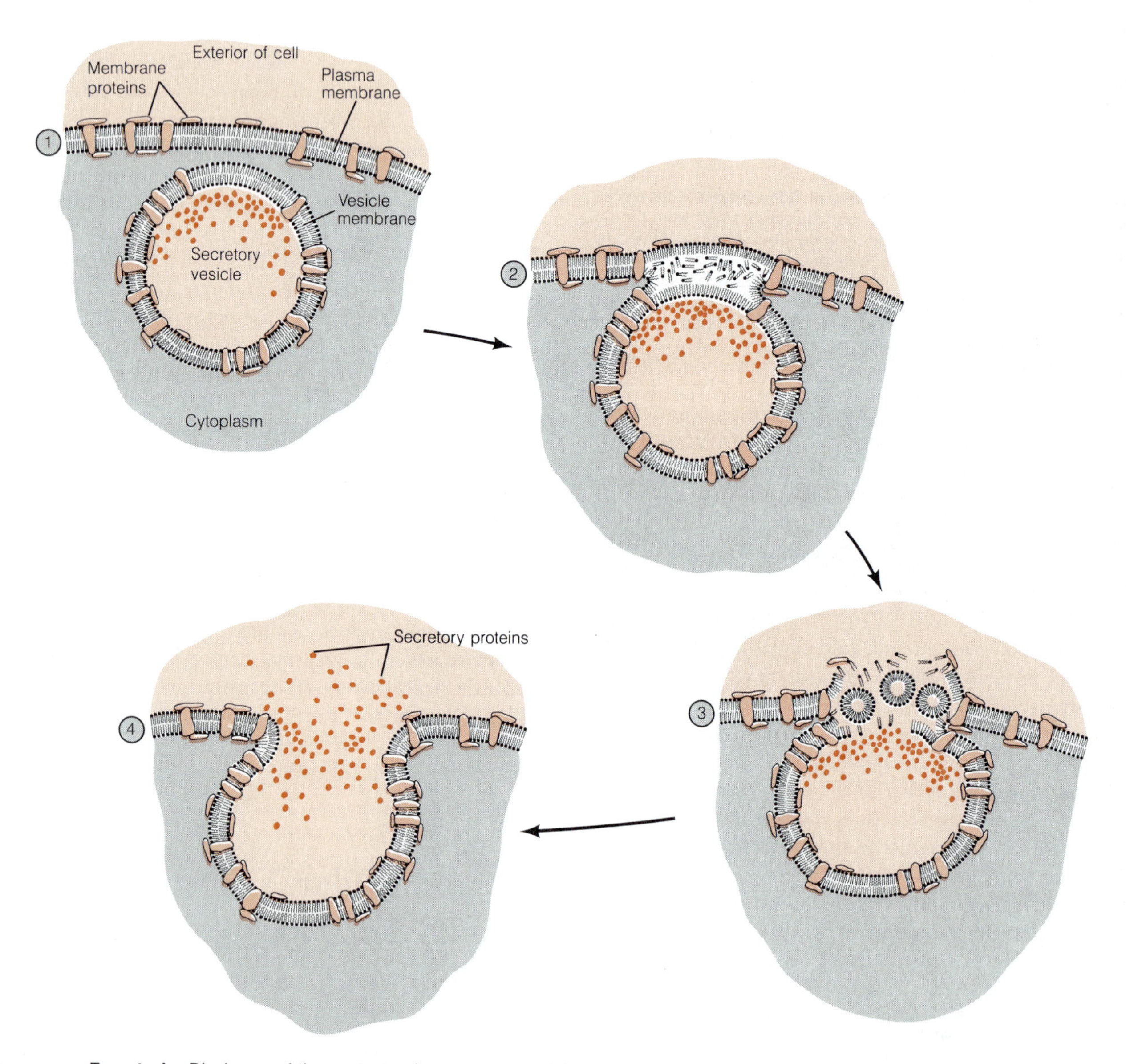

Figure 9-11 Exocytosis. Discharge of the contents of a secretory vesicle or granule to the exterior of the cell involves approach of the vesicle to the plasma membrane (step 1), lipid movement as the membranes touch (step 2), fusion of the vesicle membrane with the plasma membrane (step 3), and discharge of the contents of the vesicle to the outside (step 4).

a complementary process, called **endocytosis,** in which the cell internalizes materials that were previously outside the cell. Both these processes serve as a means of transporting macromolecules across the plasma membrane, as we will see shortly.

Exocytosis

The process of exocytosis is illustrated in Figure 9-11. Before cellular products can be discharged to the outside of the cell, they must first be concentrated and packaged. The concentration process occurs in structures called **condensing vacuoles** located on the periphery of the Golgi complex. As the vacuoles fill with concentrated protein, they fuse with one another to form large secretory granules, sometimes called **zymogen granules** if they contain enzymes. For example, the zymogen granules of pancreatic cells contain precursor forms of enzymes such as *trypsin* that are used in digestion. When triggered to discharge their contents to the exterior of the cell, secretory granules move to the cell surface, where the membrane of the granule fuses with the plasma membrane, thereby allowing expulsion of the contents of the granule to the outside of the cell. Calcium ions appear to be involved in triggering this discharge, since substances such as hormones and neurotransmitters that activate secretion all share the property of elevating the intracellular calcium level. In the case of pancreatic cells, mature secretory granules can be induced to discharge their contents by microinjection of calcium into the cells.

The actual mechanism of exocytosis is not yet clearly established, but the fusion process appears to be directed by a complementarity in structure between the two membranes. From studies with the protozoan *Tetrahymena,* the plasma membrane appears to contain recognition sites for membrane fusion. These involve rosettes of protein particles that can be seen by freeze-fracture analysis of the plasma membrane.

Membrane asymmetry is maintained throughout the secretion process, as shown in Figure 9-12. The inside of the ER membrane is preserved as the inside of the coated vesicles, the Golgi vesicles, the condensing vacuoles, and the secretory granules. When a secretory granule (or, in other processes, a coated vesicle) fuses with the plasma membrane, the inside surface of its membrane becomes the extracellular surface of the plasma membrane.

Endocytosis

In some cells, the complementary process of endocytosis is also important. In endocytosis, the cell internalizes ma-

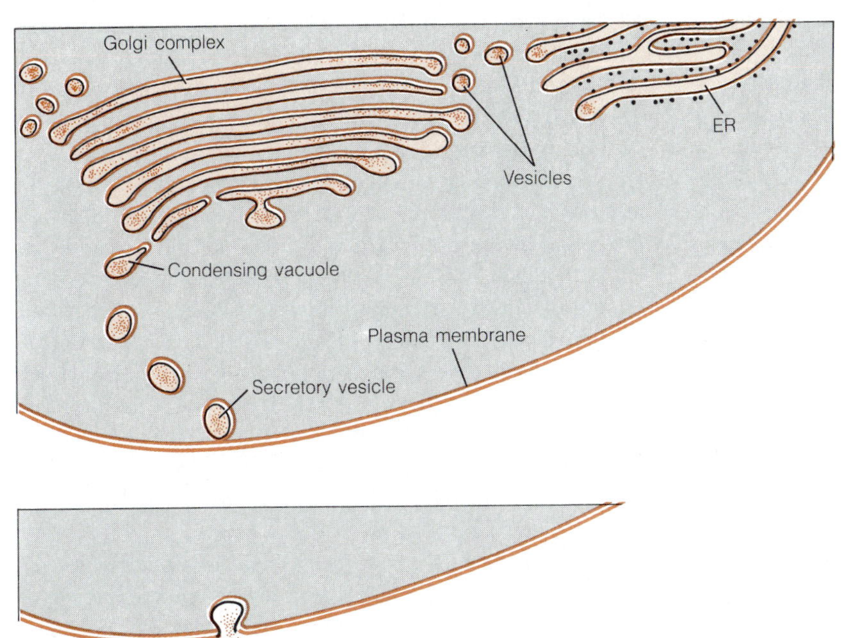

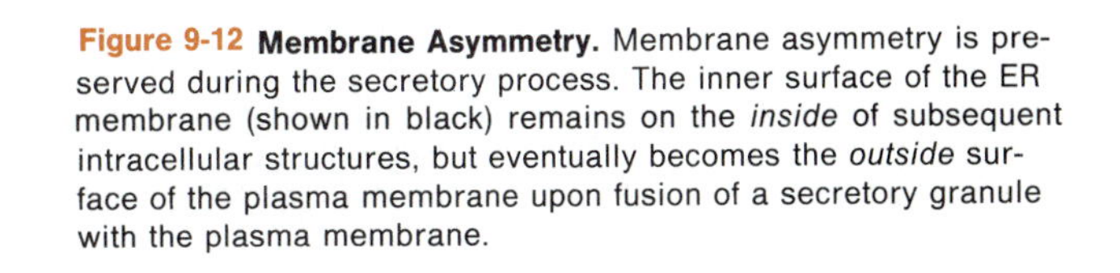

Figure 9-12 Membrane Asymmetry. Membrane asymmetry is preserved during the secretory process. The inner surface of the ER membrane (shown in black) remains on the *inside* of subsequent intracellular structures, but eventually becomes the *outside* surface of the plasma membrane upon fusion of a secretory granule with the plasma membrane.

The Intriguing World of the Coated Vesicle

Eukaryotic cells contain a variety of related membrane-bounded compartments. These include the rough and smooth endoplasmic reticulum, the Golgi complex, lysosomes, peroxisomes, various kinds of vacuoles and vesicles, and the plasma membrane itself. Until recently, most research on these structures has focused on an adequate understanding of their functions. Now, attention is turning increasingly to the interactions and transfers between these compartments that must occur when several such compartments are involved in a specific cellular activity. Even a brief consideration of the process of secretion or of lysosomal digestion emphasizes the importance to the eukaryotic cell of membrane interactions and transfers.

In recent years, it has become clear that most interactions between membranous components involve vesicles that bud off one membrane and fuse with another. And wherever such budding and fusion has been investigated in detail, *coated vesicles* seem to be a common feature. Coated vesicles, in other words, appear to play a key role in the selective transfer and exchange of substances between specific components in the cell. Because of this, many investigators believe that we will really understand how such selective interactions occur only when we understand coated vesicles better.

Coated vesicles have been observed in many different kinds of eukaryotic cells. Their characteristic feature is a "bristle" coat on the outer (cytoplasmic) surface of the membrane. They were reported in 1964 by Thomas Roth and Keith Porter, who described the involvement of these vesicles in the selective uptake of yolk protein by developing mosquito oocytes (immature egg cells). Since then, coated vesicles have been shown to play a role in such diverse cellular processes as the transport of immunoglobulins from maternal blood to fetal blood (transcytosis), the recycling of membranes at neuromuscular junctions, the uptake of substances destined for lysosomal digestion, and the delivery of newly synthesized molecules from the ER or Golgi complex to appropriate compartments within the cell. These vesicles have even been implicated in the uptake of viruses into cells and in the transport of viral proteins from the ER to the plasma membrane. In fact, it appears that coated vesicles may well be involved in all exchanges between membranes, accounting not only for the mechanism of the interaction but perhaps also for the selectivity.

Coated vesicles have been isolated from many tissues and have diameters of about 50–250 nm. They can be thought of as membranous vesicles, each packaged inside a lattice-like basket that consists mainly of pentagonal and hexagonal units of *clathrin,* a polypeptide with a molecular weight of 180,000. The clathrin lattice is flexible, accommodating itself readily to changes in vesicle size during budding, probably by varying the number of hexagonal units.

The vesicles dissociate readily into membranous components and soluble clathrin complexes, and these, in turn, can reassemble spontaneously under appropriate conditions. Clathrin complexes can even be induced to reassemble in the absence of vesicle membranes, resulting in empty shells that have been called clathrin "cages." Such reassembly occurs remarkably fast—under the right conditions, within seconds. It appears, then, that the clathrin lattice can be readily assembled and diassembled. This is probably an important feature of the clathrin coat, since fusion of the underlying membrane with another organelle or membranous structure seems to require partial or complete uncoating of the vesicle.

In 1981, E. Ungewickell and Daniel Branton succeeded in visualizing the structural units of the clathrin lattice and showed that they exist as trimers called **triskelions** (Figure 9-A). Each triskelion consists of three clathrin polypeptides radiating from a central vertex (Figure 9-B). The triskelion is thought to be the unit of assembly for clathrin coats.

By examining negatively stained preparations of partly reassembled clathrin structures, R. A. Crowther and B. M. F. Pearse were able to suggest a model for the assembly of clathrin triskelions into the characteristic hexagons and pentagons of coated vesicles, as shown in Figure 9-C. According to their model, there is one clathrin triskelion at each polyhedral vertex, with each polypeptide leg extending along one of the legs of two neighboring trimers. This allows a maximum amount of longitudinal contact between clathrin polypeptides. Such contacts may confer the mechanical strength and flexibility needed when a vesicle is pinched off from a membrane.

One of the real challenges to you as a student of cell biology should be to realize not only how many questions remain unanswered, but also how many answers seem tantalizingly close. And few areas illustrate that more readily than the intriguing world of coated vesicles.

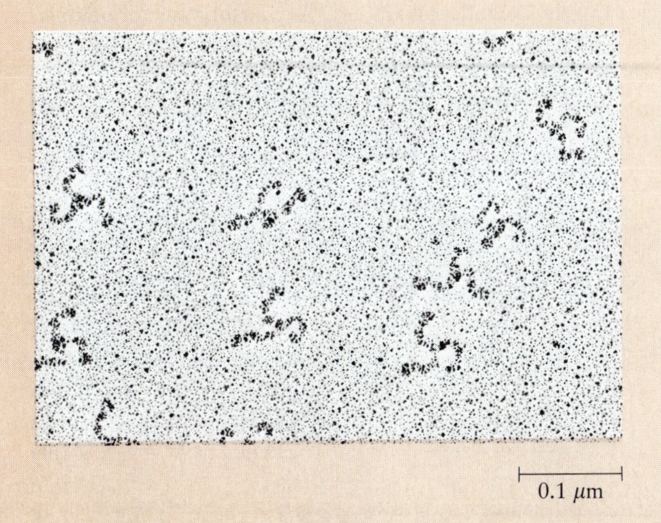

Figure 9-A Clathrin Triskelions. This micrograph shows individual triskelions of clathrin as isolated from calf brain. Each triskelion is a three-legged trimer, consisting of three clathrin polypeptides plus associated proteins. (TEM)

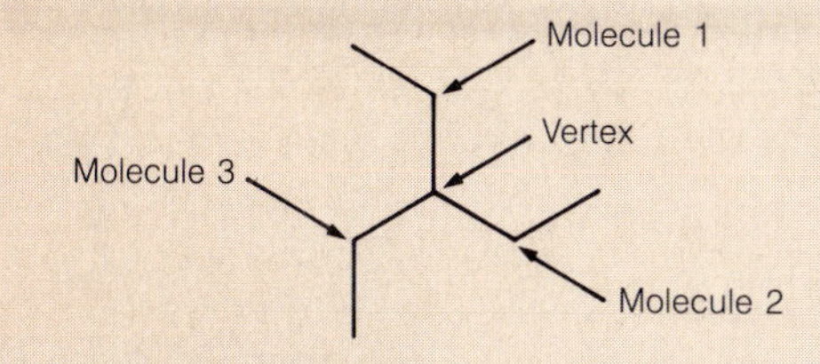

Figure 9-B The Clathrin Triskelion.

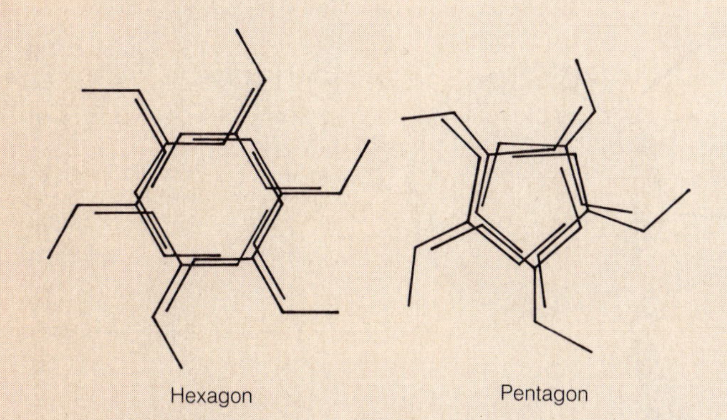

Figure 9-C Models for the Hexagons and Pentagons of Clathrin Found in Coated Vesicles.

terials by the invagination and budding of vesicles to the interior of the cell. Historically, endocytosis has received far more attention than exocytosis and is therefore better understood.

In general, endocytosis involves the formation of cytoplasmic vesicles by an infolding of the plasma membrane. Substances that were previously outside the cell are entrapped within these vesicles and brought into the cell. In terms of membrane flow, endocytosis and exocytosis have opposite effects. Endocytosis internalizes small portions of the plasma membrane as vesicles, whereas exocytosis adds to the plasma membrane by the fusion of vesicles with it. The resulting membrane exchange can be very impressive, especially in cells that carry out both processes actively. In cultured **macrophages** (large white blood cells), for example, an amount of membrane equivalent to the whole surface area of the cell is replaced every few hours.

The term *endocytosis* encompasses several different processes that vary in the nature and quantity of material taken up and the mechanism employed. Usually, a distinction is made between *phagocytosis* ("cell eating"), *pinocytosis* ("cell drinking"), and *receptor-mediated endocytosis*. In all three cases, however, the substances that are internalized remain separated from the cytoplasmic space by the membrane of the vesicle. Pinocytosis and receptor-mediated endocytosis are discussed here; phagocytosis will be considered later in the chapter in the context of lysosomal function.

Pinocytosis. **Pinocytosis** occurs in many cell types, including macrophages, leukocytes (another kind of white blood cell), kidney cells, intestinal epithelial cells, and plant root cells. Figure 9-13 depicts the steps in the pinocytotic process, which is usually triggered by the presence of specific proteins, amino acids, or ions in the medium around the cell. The process begins with the binding of the inducing substance to specific **receptors** on the plasma membrane (step 1 in Figure 9-13). The membrane then invaginates to form **pinocytotic vesicles** (steps 2 and 3). Energy is required for vesicle formation, though not for the initial binding of inducer to receptor.

The pinocytotic vesicles eventually pinch off to form free vesicles (step 4), which usually contain water, salts, and other extracellular substances, in addition to the specific molecules or ions that induced the invagination. Pinocytotic vesicles usually have diameters that are either about 0.1 μm (**micropinocytosis**) or in the range of 1–2 μm (**macropinocytosis**). Once internalized, vesicles may either fragment into smaller vesicles (step 5a) or coalesce into larger vesicles (step 5b).

Receptor-Mediated Endocytosis. Recently, interest has focused on a form of endocytosis that involves specialized membrane regions called **coated pits.** The protein coat that gives these sites their name occurs on the cytoplasmic surface of the membrane and consists of clathrin, the same protein that forms the "basket" around coated vesicles (recall Figure 9-10). In fact, the coated vesicles that form when a coated pit invaginates and detaches are strikingly similar to those involved in shuttling secretory proteins from the ER to the Golgi and from the Golgi to the plasma membrane. The electron micrographs in Figure 9-14 illustrate the sequence of events that is thought to occur in the formation of a coated pit during the uptake of yolk particles by maturing chicken oocytes.

This process of endocytosis initiated at coated pits and leading to coated vesicles is called **adsorptive pinocytosis** or **receptor-mediated endocytosis** (RME). RME is now thought to be the major mechanism for selective uptake of most macromolecules and peptide hormones that bind to the cell surface, and it is being actively investigated in

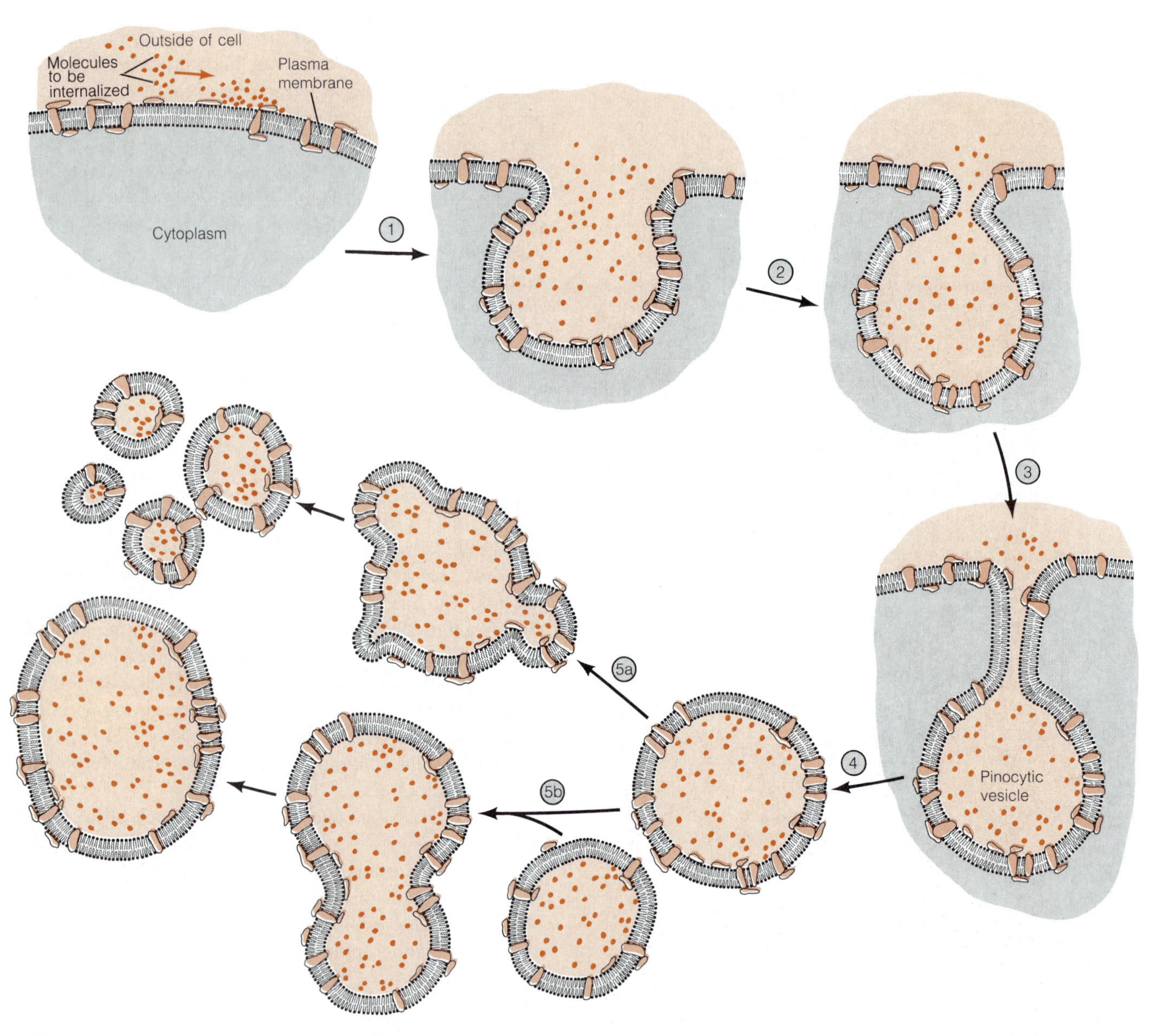

Figure 9-13 Pinocytosis. Molecules or particles to be internalized bind to receptors on the plasma membrane (step 1). The membrane then invaginates (steps 2–4) to form pinocytotic vesicles, which may either fragment into smaller vesicles (step 5a) or coalesce into large vesicles (step 5b). Steps 1–4 of this process are illustrated in the electron micrographs of Figure 9-14.

a variety of systems. At present, RME is far better understood and characterized than are the exocytotic processes in which coated vesicles are also known to be involved. In fact, most of what we know about coated vesicles to date has not come from studies of exocytosis but from studies of receptor-mediated endocytosis.

For every different kind of substance that is internalized by RME, the cell membrane contains a specific kind of receptor. Thus, there are as many different kinds of receptors in the membrane as there are substances that can be internalized by this route. Substances that bind to specific sites in this manner are called **ligands.**

RME is illustrated in Figure 9-15. The process begins with the binding of ligand molecules to their respective receptors on the outer surface of the plasma membrane (Figure 9-15, step 1). Ligand-receptor complexes diffuse laterally along the membrane to coated pits (step 2), which serve as points of collection and internalization. The actual internalization process (step 3) seems somehow to be facilitated by the presence of the clathrin coat on the cytoplasmic surface. The result of such an endocytotic event is a coated vesicle within the cytoplasm (step 4).

Once the coated pit has invaginated and broken away from the plasma membrane as a coated vesicle, the clathrin coat is released and the vesicle appears uncoated (Figure 9-15, step 5). The clathrin coat protein is thought to be recycled back to the plasma membrane (step 6), as are many of the receptors and some of the lipid (step 7). Recycling appears to depend on acidification of the vesicle, mediated by an ATP-dependent proton pump. The acid pH is apparently required to decrease the affinity of the receptor-ligand complexes, thereby freeing receptors to be recycled to the plasma membrane.

Further processing of the vesicle and its contents (Figure 9-15, step 8) depends on the nature of the ligands it contains. The uncoated vesicle may fuse with other vesicles to form an **endosome** (or **receptosome**). In other cases, the vesicle membrane may bud inward, forming a **multivesicular body.** Often, but not always, such vesicles transfer their contents to lysosomes for digestion, as will be de-

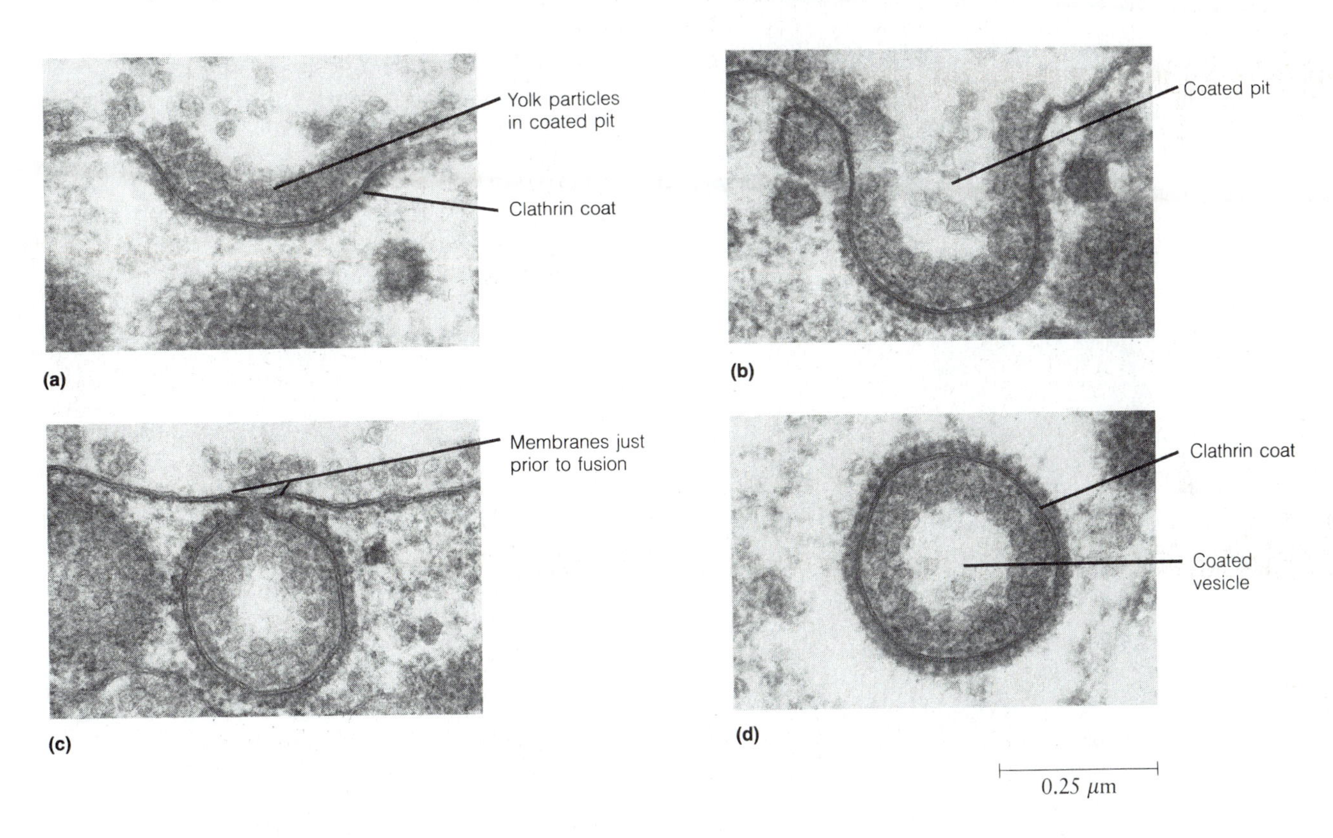

Figure 9-14 Endocytosis of Yolk Protein by a Chicken Oocyte. This series of electron micrographs illustrates the formation of a coated vesicle from a coated pit during endocytosis of yolk particles into a chicken oocyte (all TEMs). (a) Yolk particles accumulate in a coated pit, which appears initially as a shallow invagination of the plasma membrane with a clathrin coat on its inner surface. (b) A deeper coated pit containing several free particles in addition to those adhering to the membrane. (c) The final stage in the formation of a coated vesicle, just prior to the fusion of membranes at the neck of the invagination. (d) A coated vesicle that has just formed at the plasma membrane and therefore still has its clathrin coat intact.

scribed later in this chapter. In other cases, the ligand may be delivered to specific intracellular compartments by fusion of the vesicle with other membranous structures within the cell.

As a result of the secretory and endocytotic processes described here and the lysosomal digestive processes to be discussed later, there is a great deal of shuttling of membranes between various compartments of a eukaryotic cell. One of the central challenges of contemporary cell biology is to understand how these dynamic processes continue without the specific components of the various compartments becoming mixed. Only recently have we begun to appreciate the important role that coated pits and coated vesicles appear to play in this sorting. Refer to the box on pp. 230–231 for some of the fascinating aspects of coated vesicles.

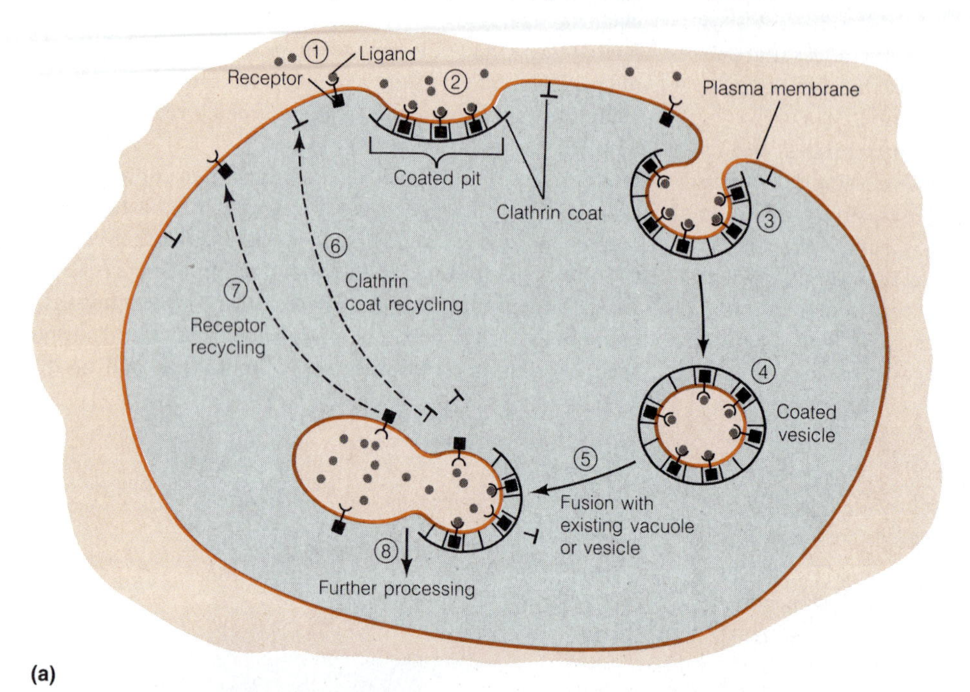

(a)

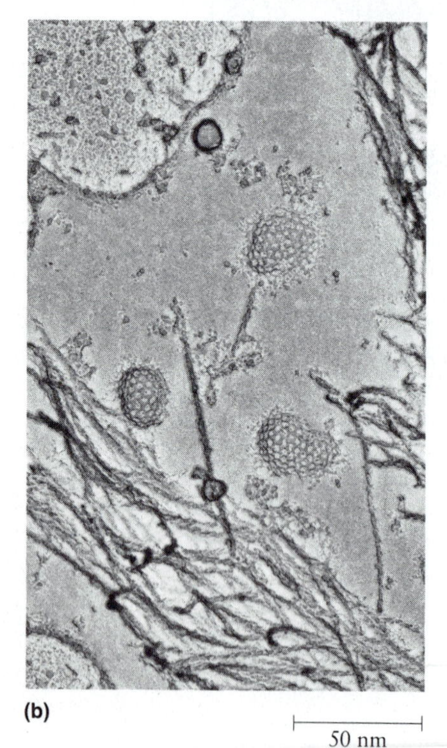

(b)

Figure 9-15 Receptor-Mediated Endocytosis. (a) Molecules to be internalized bind to specific receptors on the cell surface (step 1). Receptor-ligand complexes migrate to coated pits (step 2), where invagination is apparently facilitated by the clathrin coat on the cytoplasmic surface of the membrane (step 3). The resulting endosome is a coated vesicle (step 4), which must lose all or part of its clathrin coat before it can fuse with the membrane of an internal vesicle, vacuole, or other structure (step 5). The cell is believed to recycle the clathrin coat (step 6) and the receptors (step 7) to the plasma membrane, leaving the internalized molecules to their appropriate intracellular fate (step 8). (b) This electron micrograph of a freeze-etched cell shows a clathrin-coated area of the plasma membrane engaged in receptor-mediated endocytosis, as seen from the inside of the cell (TEM).

Membrane Biosynthesis and Turnover

Inherent in the secretory process we discussed earlier is a unidirectional flow of membrane from the ER through the Golgi complex to the plasma membrane. Every time a secretory granule fuses with the plasma membrane and discharges its contents by exocytosis, another bit of membrane that originated with the rough ER becomes a part of the plasma membrane. This membrane flow suggests that the rough ER is the precursor to both the Golgi membranes and the plasma membrane of eukaryotic cells. Studies of lipid biosynthesis and fate within the cell indicate that the rough ER may be the source of other cellular membranes as well, including the membranes of lysosomes and both the inner and outer membranes of the mitochondrion.

Since the rough ER is the source of most (perhaps even all) membranes in the eukaryotic cell, it clearly must be the site of synthesis not just of secretory proteins but of membrane components as well. This appears to be true for membrane proteins and phospholipids alike.

Biosynthesis and Processing of Membrane Proteins

Earlier in the chapter we emphasized secretory proteins, but only because these are easiest to approach experimentally. In fact, most proteins are not destined for secretion, but instead are targeted for intracellular sites such as membranes. Nonetheless, the evidence suggests that processing of membrane proteins is similar to that already described for secretory proteins. The ribosomal complex becomes anchored to the rough ER, just as it does for a secretory protein. At this point, however, an important difference is noted. Instead of passing through the ER membrane into the lumen, as secretory proteins do during synthesis, membrane proteins remain anchored in the membrane as they are synthesized.

Glycosylation of membrane proteins is also similar to the processing of secretory proteins. Glycosylation is begun in the rough ER and completed in the Golgi complex. Experiments with radioactively labeled sugars show a progressive movement of label from the Golgi to the cell surface. Glycosylation of proteins in the ER and Golgi occurs on the luminal side (i.e., the inner surface) of the membrane only. Since this side is topologically equivalent to the exterior surface of the plasma membrane (see Figure 9-12), it is easy to see why all plasma membrane glycoproteins are found on the extracellular side of the membrane.

Biosynthesis of Membrane Lipids

The ER, both rough and smooth, is the site of synthesis of most membrane lipids, including phospholipids and cholesterol. In fact, most of the enzymes needed for the synthesis of the various membrane phospholipids are found nowhere else in the cell. Labeling experiments with radioactive choline confirm the role of the rough ER in phospholipid biosynthesis.

Lipids synthesized in the ER become incorporated into all of the membranes of the cell, not just the plasma membrane. For most membranes, such as those of the Golgi complex and lysosomes, this transfer of lipids (and proteins as well) probably occurs by means of small **transport vesicles** that pinch off from the rough ER and fuse with the target membranes. Transport vesicles are probably coated vesicles similar to those involved in the transport of secretory proteins from the rough ER to the Golgi complex.

Transfer of phospholipids from the ER to the mitochondrion poses a unique problem, since mitochondria do not grow by fusion with vesicles synthesized elsewhere in the cell. Instead, special **phospholipid transfer proteins** mediate the transfer of phospholipid molecules into the mitochondrial membrane. Each such protein recognizes a specific kind of phospholipid, removes it from one membrane, and adds it to the other. These transfer proteins may be involved in the movement of phospholipids from the ER to other membrane systems of the cell as well, possibly including the plasma membrane.

Membrane Turnover

A striking feature of all membranes is the relatively high rate of turnover or replacement of membrane components. Membranes are in no sense static structures: Both the lipid and protein components are continuously removed and replaced. **Membrane turnover** is most convincingly demonstrated when cells are grown in a radioactive precursor (an amino acid for proteins, choline for phospholipids) for a period of time such that the desired molecules are adequately labeled. The cells are then washed free of radioactivity and transferred to an unlabeled medium. At various intervals thereafter, cells are fractionated and the amount of radioactivity in various fractions is measured. From the results of such experiments, we know that all membrane components—proteins and lipids alike—continuously turn over. Moreover, the rate of turnover varies for different proteins and lipids, suggesting that the replacement process is selective. For instance, the half-life of some phospholipids in membranes is measured in hours, whereas protein turnover is measured in days.

Such turnover is an essential process by which the cell can continuously remove damaged components. For instance, it is known that unsaturated fatty acid chains often become oxidized and lose their ability to function as membrane bilayer components. The turnover process removes such damaged molecules and replaces them with newly synthesized fatty acids. Protein turnover is required for the same reason and is carried out by enzymes called *proteases*. Some proteases are found in the cytoplasm, whereas others are located in lysosomes.

Lysosomes and Cellular Digestion

Lysosomes are derived from the ER and the Golgi complex but are distinctively different in their function in the cell. Basically, a **lysosome** is an organelle that contains digestive enzymes capable of degrading all major classes of biological macromolecules. These enzymes are needed both to degrade materials that are brought into the cell from the outside and to digest cellular molecules and structures that are damaged or no longer needed. We will first look at the organelles themselves and then consider the digestive processes in which they are involved, as well as some of the disease conditions that result from lysosomal malfunction.

Discovery of Lysosomes

Lysosomes were discovered in the early 1950s by Christian de Duve and his colleagues. The story of that discovery was recounted in the box on pp. 94–95 in Chapter 4. The technique that allowed de Duve to make this discovery was differential centrifugation, which allows cellular contents to be fractionated according to size and density, as already illustrated in Figure 9-1.

Use of this technique led de Duve and his colleagues to realize that an acid phosphatase initially thought to be located in the mitochondrion was in fact associated with a class of particles that had never been reported before. Along with acid phosphatase, these organelles contained several other hydrolytic enzymes, including ribonuclease, deoxyribonuclease, β-glucuronidase, and a protease. Because of its apparent role in cellular lysis, de Duve called the new organelle a *lysosome*.

Only after its existence had been predicted, its properties described, and its enzyme content specified was the lysosome actually observed in the electron microscope and recognized as a regular constituent of most animal cells. Final confirmation came from cytochemical staining reactions capable of localizing acid phosphatase and other lysosomal enzymes to specific structures that can be seen in the electron microscope, as illustrated in Figure 9-16. Lysosomes vary considerably in size and shape but are in general about 0.5 μm in diameter. They are surrounded by a single membrane and have an electron-dense, granular matrix when visualized in the electron microscope.

The list of lysosomal enzymes has expanded considerably since de Duve's original work, but all known lysosomal enzymes have the common property of being *acid hydrolases*—hydrolytic enzymes with a pH optimum around 5.0. The list includes at least 5 phosphatases, 4 proteases, 2 nucleases, 6 lipases, 12 glycosidases, and an arylsulfatase. Taken together, these lysosomal enzymes are capable of digesting all the major classes of biological molecules. No wonder, then, that they are sequestered together in a single kind of organelle, away from the rest of the cell, with which they would quickly wreak havoc.

Biogenesis of Lysosomes

The enzymes of the lysosome are synthesized on ribosomes bound to the rough ER and are then packaged together within a membrane in a manner that is probably somewhat similar to the synthesis and packaging of secretory proteins. Most investigators regard the Golgi complex as the site of lysosome formation. Lysosomal enzymes are thought to be synthesized on the rough ER and transferred to its lumen before being transported to the Golgi. Lysosomes then develop from coated vesicles that bud off the *trans*-most cisternae of the Golgi complex.

Targeting Hydrolases to the Lysosomes

Enzymes destined for the lysosomes are glycoproteins, containing oligosaccharides that are added in the ER. However, lysosomal enzymes are distinguishable from other glycoproteins by the presence of an unusual oligosaccharide containing mannose-6-phosphate. This unique feature of lysosomal enzymes appears to serve as a recognition marker or "address" that targets all such proteins to the lysosomes. The Golgi membrane, in turn, contains receptors that recognize this special oligosaccharide and ensure that all enzymes with this "address" are packaged into lysosomes. As you may recall from Figure 9-9, mannose-6-phosphate receptors are concentrated in the cisternae on the *cis* face of the Golgi stack. This localization is consistent with the role of these receptors in the detection of glycoproteins with this "address," as they arrive at the Golgi

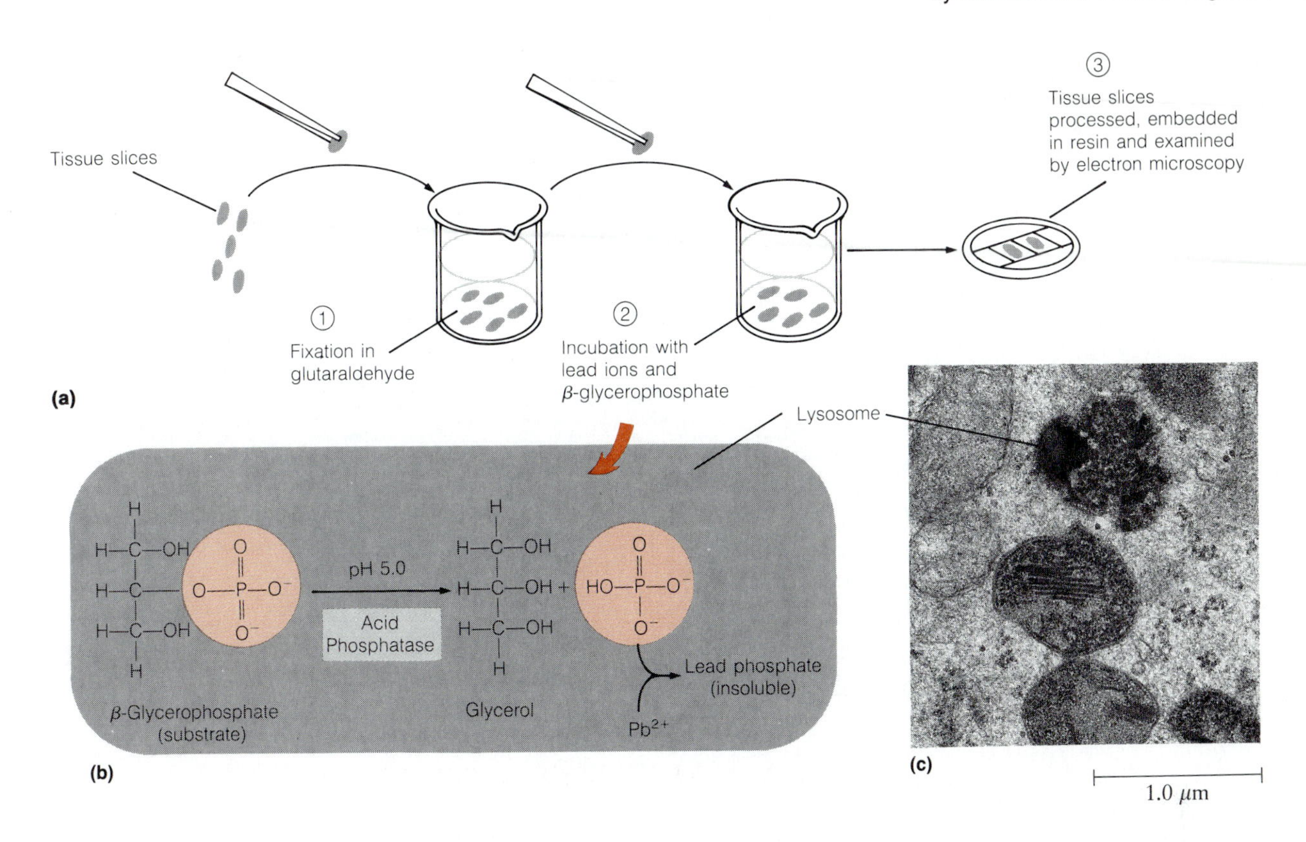

Figure 9-16 Cytochemical Localization of Acid Phosphatase. (a) Thin sections of tissue are fixed in glutaraldehyde and incubated at pH 5.0 in a medium containing β-glycerophosphate (a substrate for the enzyme acid phosphatase) and a soluble lead salt. The tissue slices are then embedded in resin and examined in the electron microscope. (b) During the incubation, the acid phosphatase within the lysosome cleaves the phosphate from the substrate. The free phosphate reacts with lead ions to form lead phosphate, which is insoluble and therefore precipitates at the site of enzyme activity. Since the lead phosphate is electron-dense, microscopy reveals the location of the enzyme acid phosphatase in the cell. (c) Cells treated as in part (a) contain lysosomes that appear darkly stained in the electron microscope, indicating the presence of lead phosphate. Thus, acid phosphatase must be located in lysosomes. (TEM)

complex from their site of synthesis on the rough ER.

Strong support for this concept of hydrolase targeting by an oligosaccharide "address" came from studies of a human genetic disorder called **I-cell disease.** Fibroblast cells from patients with this disease synthesize all the expected hydrolases in culture, but they release the enzymes into the medium instead of incorporating them into their lysosomes. The distinguishing feature of these hydrolases is that they lack the oligosaccharide that contains mannose-6-phosphate. Apparently, the genetic defect in I-cell disease involves some step in the process of adding the correct oligosaccharide "address" to all the lysosomal enzymes.

Cellular Digestion

The organelle formed as we have just described is called a **primary lysosome**—a lysosome with its full complement of lytic enzymes, but not yet engaged in digestive activity. A primary lysosome, in other words, is a collection of as-yet-unused digestive enzymes packaged together in a way that protects the cell from their activities until they are required for specific hydrolytic functions. These functions include the digestion of foreign materials, often brought into the cell for that express purpose, the breakdown of cellular components that are damaged in some way, and the removal of compounds that are no longer needed.

The several digestive processes in which lysosomal enzymes are involved can be distinguished in terms of both the site of their activity and the origin of the materials that are digested, as indicated by the several alternatives depicted in Figure 9-17. Usually, the site of activity is intracellular, though in some cases lysosomes may release their enzymes to the outside of the cell by exocytosis. The materials to be digested are often of extracellular origin, but important processes are also known involving lysosomal

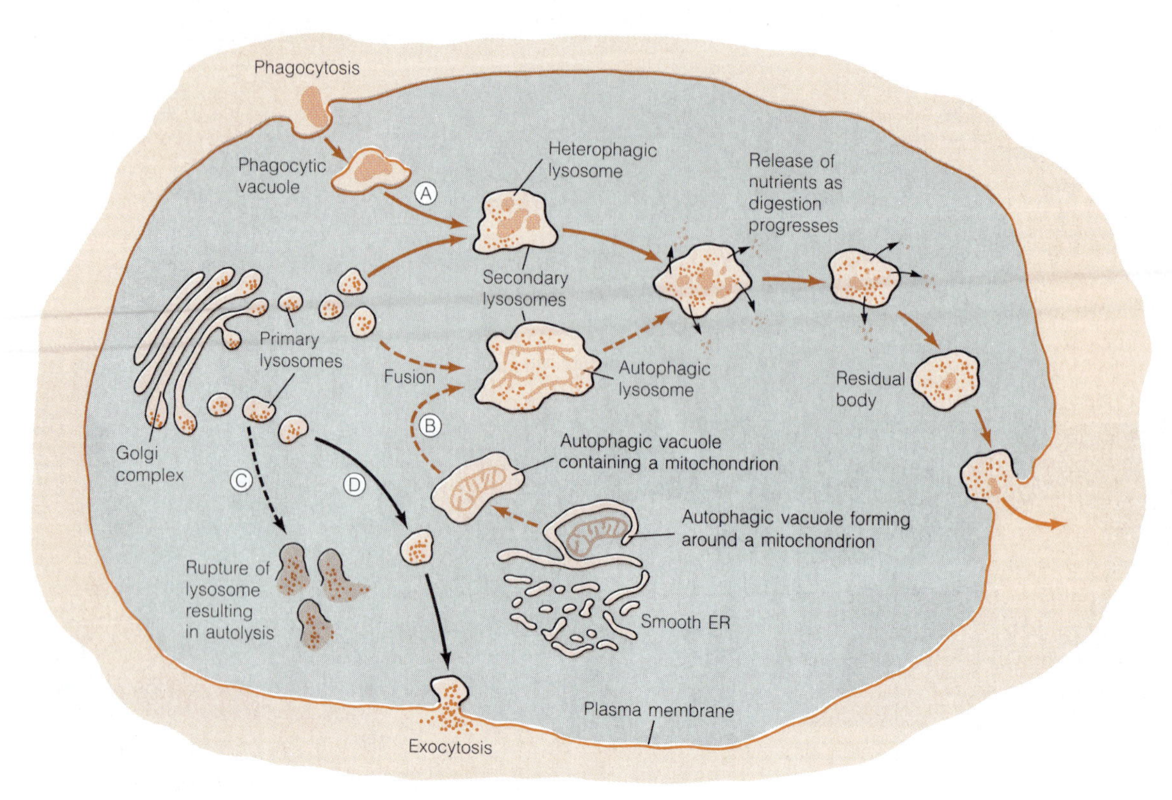

Figure 9-17 Formation of Primary and Secondary Lysosomes and Their Role in Digestive Processes. The major processes in which lysosomes are involved are illustrated in this composite cell. Pathway A depicts phagocytosis, pathway B represents autophagy, pathway C illustrates autolysis, and pathway D traces the route of lysosomes involved in extracellular digestion.

digestion of cellular components or even whole cells or tissues. As a result, lysosomes play a role in such diverse cellular activities as nutrition, defense, recycling of cellular components, differentiation, and even cell death. The specific processes in which lysosomes are involved are called *phagocytosis, autophagy, autolysis,* and *extracellular digestion.* These are illustrated in Figure 9-17 as pathways A, B, C, and D, respectively. We will now examine each of these processes in detail.

Phagocytosis: Lysosomes in Nutrition and Defense. One of the most important functions of lysosomal enzymes is the degradation of foreign materials brought into the cell by the process of endocytosis. Among the endocytotic processes contributing to the uptake of foreign substances destined for lysosomal digestion are pinocytosis and receptor-mediated endocytosis (recall Figures 9-13 and 9-15). In most cells, the majority of endocytotic vesicles formed by these processes ultimately fuse with lysosomes to initiate digestion of the endocytosed macromolecules.

In addition, some cells can take up much larger particles, including whole cells, by phagocytosis, a process seen most prominently among single-celled heterotrophs such as amoebas and ciliated protozoa. These organisms routinely engulf and internalize food particles and smaller organisms for intracellular digestion. Phagocytosis is also used by some primitive animals, notably flatworms, coelenterates, and sponges, as a means of obtaining nutrients. In higher organisms, phagocytosis is usually restricted to specialized cells called **phagocytes.** Your body, for example, contains two classes of white blood cells that function as phagocytes—*macrophages* and *polymorphonuclear leukocytes* (also called *neutrophils*). These types of cells engulf and digest invading microorganisms or foreign materials in the bloodstream. Phagocytosis therefore serves a nutritional purpose among lower organisms but becomes a means of defense in higher animals.

Phagocytosis has been studied most extensively in the amoeba. The cell surface of most amoebas is covered either with fine hairs or with a coat of glycosaminoglycans. The function is the same in either case—a means of adsorbing and trapping food particles or smaller organisms. Contact with such an object appears to trigger the onset of phagocytosis. Folds of membrane gradually surround and engulf the particle, and a **food cup** forms as the membrane invaginates and folds around the particle. Eventually, this pinches off to form a free **phagocytic vacuole** within the cell, as pathway A of Figure 9-17 shows.

Once inside the cell, phagocytic vacuoles move toward the Golgi region, where they encounter and fuse with the primary lysosomes that form in that region of the cell. (Obviously, such fusions must be highly selective, since the fusion of lysosomes with any membrane-bounded structures other than phagocytic vacuoles could have disastrous consequences for the cell.) Such fused vacuoles are called **secondary lysosomes.** Since they contain both digestive enzymes and digestible foodstuffs, secondary lysosomes are the site of digestive activity. Soluble products of digestion, such as sugars, amino acids, and nucleotides, cross the membrane and are used as a source of nutrients by the cell. Secondary lysosomes are capable of fusing with further phagocytic vacuoles, thereby acquiring further material to digest. As a result, they vary considerably in size, appearance, content of digestible material, and stage of digestion.

Eventually, however, only indigestible material remains in the vacuole, which becomes a **residual body** as digestion ceases. In protozoa, residual bodies routinely fuse with the plasma membrane and expel their contents to the outside by exocytosis, as illustrated in Figure 9-17. In vertebrates, there appears to be no such mechanism, so residual bodies accumulate in the cytoplasm. This accumulation is thought to contribute to cellular aging, particularly in long-lived cells such as those of the nervous system.

Autophagy: The Original Recycling System. A second major role of lysosomes is the breakdown of cellular components and structures that have become damaged or are no longer needed. Most cellular organelles are in a state of dynamic flux, with new organelles being synthesized and old organelles being destroyed continuously. The digestion of old or unwanted organelles or other cell structures is called **autophagy,** which literally means "self-eating." Autophagy is illustrated in Figure 9-17 as pathway B.

Autophagy begins when an organelle or other structure becomes wrapped in membranes that are thought to arise from the ER. The resulting vesicle is called an **autophagic vacuole.** It is often possible to see identifiable remains of cellular structures in these vacuoles, as shown in Figure 9-18. The fate of an autophagic vacuole closely parallels that of a phagocytic vacuole. In both cases, fusion with a primary lysosome gives rise to a secondary lysosome, in which the digestive action of the lytic enzymes is unleashed. To distinguish between secondary lysosomes, we refer to those containing substances of extracellular origin as **heterophagic lysosomes,** while those with materials of intracellular origin are called **autophagic lysosomes.**

Autophagy occurs at varying rates in most cells most of the time, but it is especially prominent in certain developmental situations. During the maturation of a red blood cell, for example, virtually all of the intracellular contents must be destroyed, including all of the mitochondria. This is accomplished by autophagic digestion. A marked increase in autophagy is also noted in cells stressed by starvation. Presumably, the process represents a rather desperate attempt on the part of the cell to continue to provide for its energy needs, even if it has to consume its own structures to do so.

Autolysis: Cellular Self-Destruction. Normally, the hydrolytic enzymes of the lysosome are kept safely packaged within the membranes of the organelle to prevent unwanted degradation of the cell itself. Materials destined for degradation, whether originating from inside the cell or from outside, are first carefully enclosed in vacuoles and then brought in contact with the digestive enzymes by fusion of the vacuoles with primary lysosomes. At no point in the process are the lytic enzymes allowed to escape from behind the membranes that confine them. If they do, as sometimes occurs in injured red blood cells, the released enzymes rapidly digest and destroy the cell. Under most conditions, this would be disastrous.

Such is not always the case, however. During the development of many multicellular organisms, there are specific processes in which such self-destruction, or **autolysis,** plays an important role. Wherever the removal of specific cells or tissues is necessary to the development of a particular structure or organ system, autolysis is the means by which the unwanted cells are destroyed. The creation of individual digits (fingers and toes) from the initially webbed hands and feet of a human embryo by selective removal of the interdigital cells is an example of such "programmed cell death" during development. The progressive loss of its tail as a tadpole undergoes metamorphosis is another case in which lysosomally mediated autolysis of cells is part of a normal developmental program. The mechanism in each instance appears to involve the deliberate release of digestive enzymes from the lysosomes, giving the organelle its reputation as a "suicide bag" in such cases. Autolysis appears in Figure 9-17 as pathway C.

Extracellular Digestion. Most of the digestive processes involving lysosomal enzymes occur intracellularly, either within vacuoles (phagocytosis and autophagy) or within the cell as a whole (autolysis). In rare cases, however, lysosomes may discharge their enzymes to the outside of the cell by exocytosis, resulting in **extracellular digestion** (Figure 9-17, pathway D). One such example occurs during

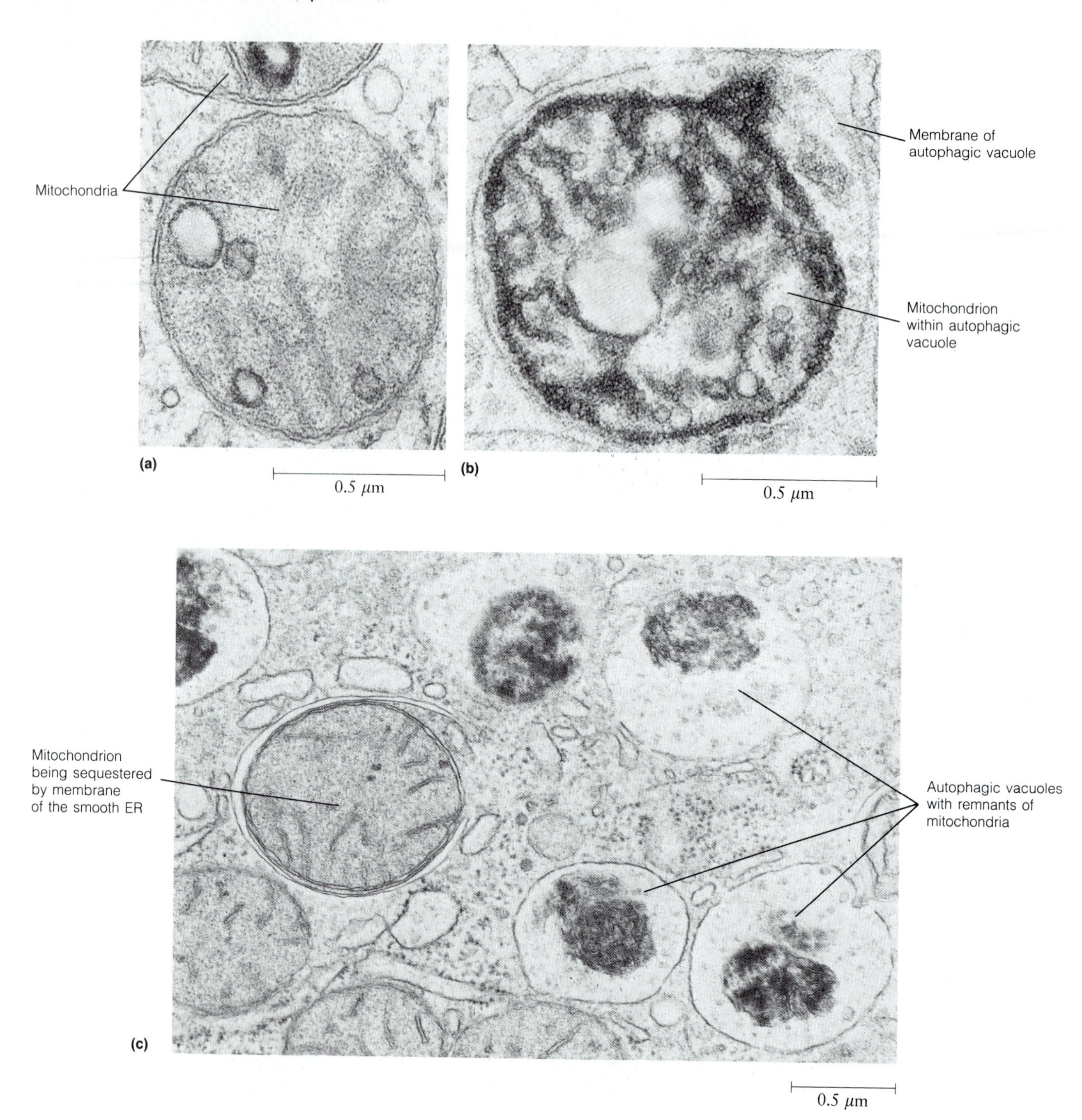

Figure 9-18 Autophagic Digestion. The electron micrographs on top illustrate (a) functional mitochondria in the cytoplasm of a rat kidney cell and (b) an autophagic vacuole in the same cell, with a mitochondrion in it that is being digested but is still readily recognizable. (c) Earlier and later stages of autophagic digestion are seen in the micrograph of a rat liver cell on the bottom, which shows an autophagic vacuole in the process of formation, as a mitochondrion is enveloped by membranes of the smooth ER. Also seen are autophagic vacuoles with remnants of mitochondria in them. (all TEMs)

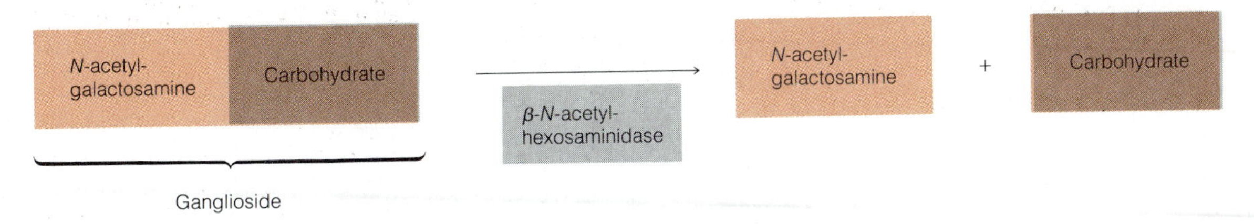

Figure 9-19 The β-N-Acetylhexosaminidase Reaction. The enzyme β-N-acetylhexosaminidase is required to remove the terminal N-acetylgalactosamine from the carbohydrate portion of gangliosides in the nervous system. This enzyme is missing or defective in individuals with Tay-Sachs disease. Infants afflicted with Tay-Sachs disease cannot carry out this reaction and therefore accumulate greatly elevated ganglioside levels in the brain.

fertilization of animal eggs. The head of the sperm releases lysosomal enzymes capable of degrading chemical barriers that would otherwise keep the sperm from penetrating the egg surface. Moreover, certain inflammatory diseases, such as rheumatoid arthritis, appear to result from the release of lysosomal enzymes into the joints. The steroid hormones cortisone and hydrocortisone are thought to be effective anti-inflammatory agents because of their role in stabilizing lysosomal membranes and thereby inhibiting enzyme release.

Lysosomal Storage Diseases

The important role of lysosomes in cellular turnover is clearly seen in diseases that result from the deficiency of specific lysosomal enzymes. There are, in fact, about 40 such **lysosomal storage diseases,** each characterized by the undesirable accumulation and storage of excessive amounts of specific substances, usually polysaccharides or lipids. The cells in which this accumulation occurs are greatly impaired in function, if not destroyed. Muscle weakness, skeletal deformities, and mental retardation commonly result, often with a fatal outcome.

In each case, a specific lysosomal enzyme is absent or deficient. The first of these storage diseases to be explained was **type II glycogenosis,** in which young children accumulate excessive amounts of glycogen in the liver, heart, and muscles and eventually die. The problem turned out to be a deficiency of the lysosomal enzyme α-glucosidase, which catalyzes glycogen hydrolysis in normal cells.

Two of the best-known lysosomal storage diseases are **Hurler syndrome** and **Hunter syndrome.** Both involve defects in the degradation of acid glycosaminoglycans. When sweat gland cells from a patient with Hurler disease were observed in the electron microscope, large numbers of atypical vacuoles were seen that stained both for acid phosphatase and for undegraded acid glycosaminoglycans. These are thought to be aberrant secondary lysosomes with contents that cannot be degraded.

Mental retardation is so common a feature of the lysosomal storage diseases because of the impaired metabolism of glycolipids, which are important in brain tissue and in the sheaths of nerve axons. One particularly well-known example is **Tay-Sachs disease,** which occurs mainly among Jews, especially those of eastern European origin. The condition is inherited as a recessive trait. Children afflicted with the condition show rapid mental deterioration after about six months of age and usually die within the first three years. The disease results from the accumulation in the brain of a particular kind of glycolipid called a *ganglioside.* The missing lysosomal enzyme in this case is β-N-acetylhexosaminidase, which is responsible for cleaving the terminal N-acetylgalactosamine from the carbohydrate portion of the ganglioside (Figure 9-19). Lysosomes from children suffering from Tay-Sachs disease are filled with membrane fragments containing undegraded gangliosides.

Peroxisomes

Peroxisomes occur in both animal and plant cells, but their physiological significance is more clearly understood in plants. We will look at plant peroxisomes specifically when we discuss photosynthesis in Chapter 12. The primary role of peroxisomes in animal cells is to detoxify certain potentially damaging products of oxidative metabolism, particularly peroxides, such as hydrogen peroxide. Before going on to their functional role, we will first look at how peroxisomes were discovered.

Discovery of Peroxisomes

During the course of de Duve's early studies on lysosomes, he became aware of at least one enzyme, urate oxidase, that seemed to be associated with his lysosomal fractions yet was not an acid hydrolase. Small differences in the

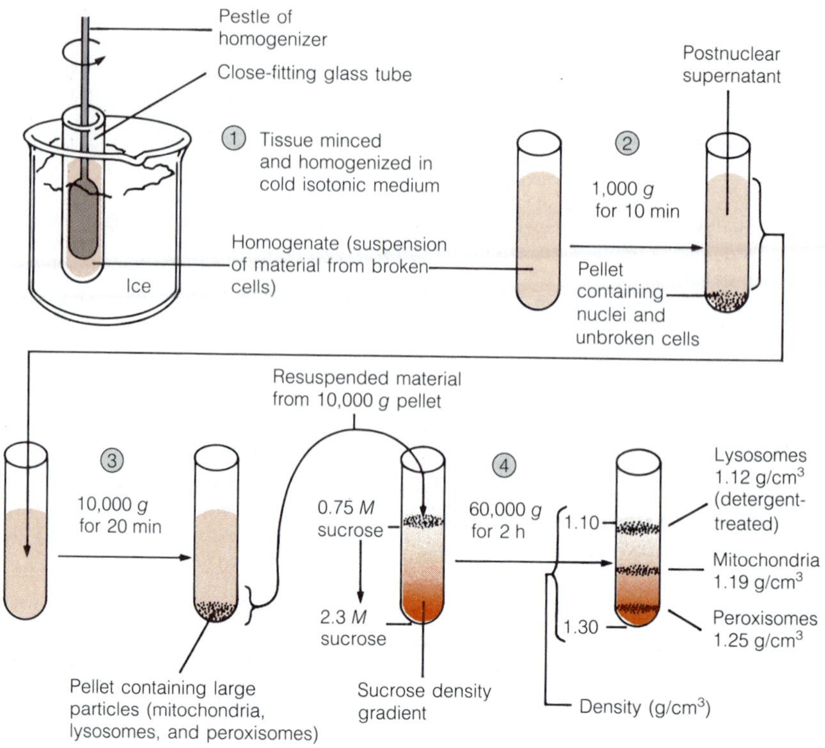

Figure 9-20 Equilibrium Density Centrifugation. To separate organelles based on differences in buoyant density, tissue is minced and homogenized (step 1), and the homogenate is then subjected to differential centrifugation (steps 2 and 3) as described in Figure 9-1. The pellet from the 10,000 *g* centrifugation step is resuspended in an isotonic solution (0.25 *M* sucrose) and applied to the top of a tube containing a preformed gradient of sucrose, from 0.75 *M* (1.10 g/cm^3) to 2.3 *M* (1.30 g/cm^3). Upon centrifugation at high speed (two hours in an ultracentrifuge; step 4), the organelles in the applied sample sediment into the gradient until each reaches a density of sucrose equal to its own density.

behavior of urate oxidase compared to known lysosomal enzymes during differential centrifugation also kept de Duve from identifying it as a lysosomal enzyme. The clue that eventually allowed de Duve and his colleagues to assign urate oxidase to a new class of previously undescribed organelles came when they employed the technique of *equilibrium density centrifugation.*

Equilibrium Density Centrifugation. The technique of **equilibrium density centrifugation** is illustrated in Figure 9-20. Organelles are centrifuged into a solution that increases in density from the top to the bottom of the centrifuge tube. This **density gradient** is often achieved using increasing concentrations of sucrose, though other solutes are also possible. The important point is that the density gradient must span the range of densities represented by the organelles that are to be separated on it. In most cases, this is the density range from about 1.10 to about 1.30 g/cm^3 corresponding to a sucrose concentration range from about 0.75 to 2.3 *M*. The sample applied to such a gradient is usually either a cellular homogenate of some sort or a subcellular fraction obtained by a prior differential centrifugation step. Such a sample typically contains a mixture of organelles, each of which has its own characteristic **buoyant density** in sucrose.

When centrifugation begins, centrifugal force acts on the organelles, driving them into the gradient. A given organelle will continue to move into the gradient until it reaches the point in the gradient at which the density of the solution surrounding it is exactly equal to the density of the organelle. At that point, there is no net force acting on the organelle, so it moves no farther. Given enough time, all organelles will reach their characteristic buoyant density positions in the gradient and will then simply remain at those positions. Because each class of eukaryotic organelle has its own distinctive buoyant density, this is a very powerful technique for resolving organelles from one another, as Figure 9-20 indicates.

Resolution of Peroxisomes from Lysosomes. By the use of equilibrium density centrifugation, de Duve and his colleagues found that the enzyme *urate oxidase* was associated with a particle that had a slightly different density from that of other organelles such as lysosomes and mitochondria. Because the density difference was so small, this new class of organelles could not be resolved from lysosomes under normal conditions, but by employing an experimental trick de Duve was able to achieve good separation. The trick was based on the chance observation that animals injected with the detergent Triton WR-1339 accumulate the detergent preferentially in their lysosomes and that these organelles have a much lower buoyant den-

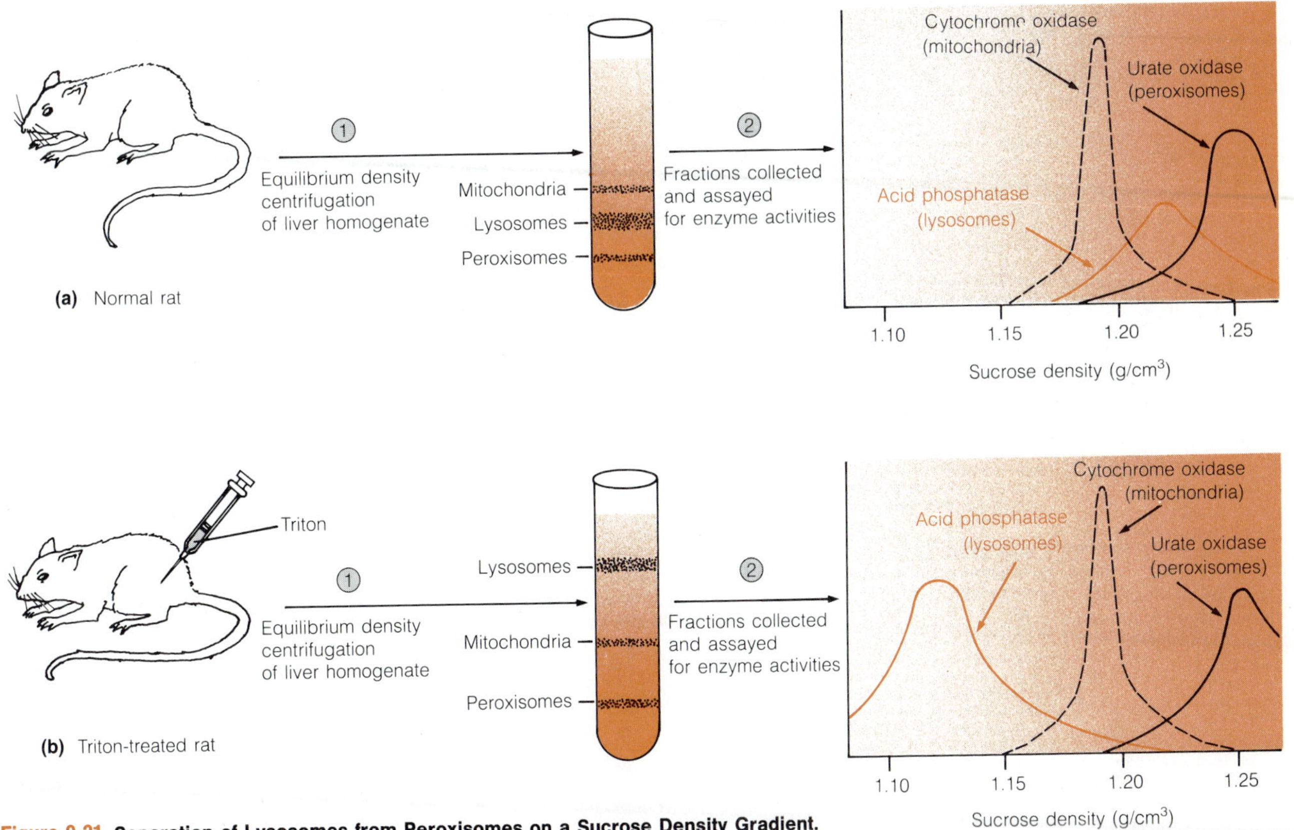

Figure 9-21 Separation of Lysosomes from Peroxisomes on a Sucrose Density Gradient. (a) For organelles obtained from the liver of a normal rat, lysosomes (marker enzyme: acid phosphatase) have a range of densities intermediate between those of mitochondria (cytochrome oxidase) and peroxisomes (urate oxidase), making resolution of the three organelles difficult. (b) For organelles obtained from the liver of a rat treated with the detergent Triton WR-1339, the lysosomes band at a much lower density (about 1.12 g/cm^3), allowing them to be resolved from mitochondria (1.19 g/cm^3) and peroxisomes (1.25 g/cm^3).

sity as a result. By using liver homogenates from detergent-treated rats, it was possible to achieve almost complete separation of urate oxidase, which continued to band at a density of about 1.25 g/cm^3, from mitochondria and lysosomes (Figure 9-21).

Once this separation was achieved, additional enzymes were shown to be present in the urate oxidase-containing particles. Two such enzymes were D-amino acid oxidase, which generates hydrogen peroxide, and catalase, which degrades hydrogen peroxide. Because of the apparent involvement of this new organelle in the metabolism of hydrogen peroxide, the organelle soon came to be known as a *peroxisome*. Other enzymes have since come to be recognized as peroxisomal in their location, and it is now clear that the enzyme complement of the organelle varies significantly from species to species and sometimes from organ to organ. However, the presence of catalase and one or more hydrogen peroxide-generating oxidases still remains a defining characteristic of all peroxisomes.

Occurrence and Properties of Animal Peroxisomes

Once peroxisomes had been identified and isolated biochemically, electron microscopy was employed to verify the existence of organelles with the expected properties, first in isolated peroxisomal fractions from density gradients and then in intact cells. Peroxisomes turned out to be the biochemical equivalent of organelles that had been seen earlier in electron micrographs of both animal and plant cells. Because their function was unknown at the time, these organelles were simply called **microbodies.** In animal tissues, microbodies are usually about 0.2–2.0 μm in diameter, have a single membrane, and generally have a finely granular matrix. Figure 9-22 shows the appearance of microbodies in a rat liver cell.

As can be seen in Figure 9-22b, animal microbodies often contain a distinctive crystalline core, which is thought

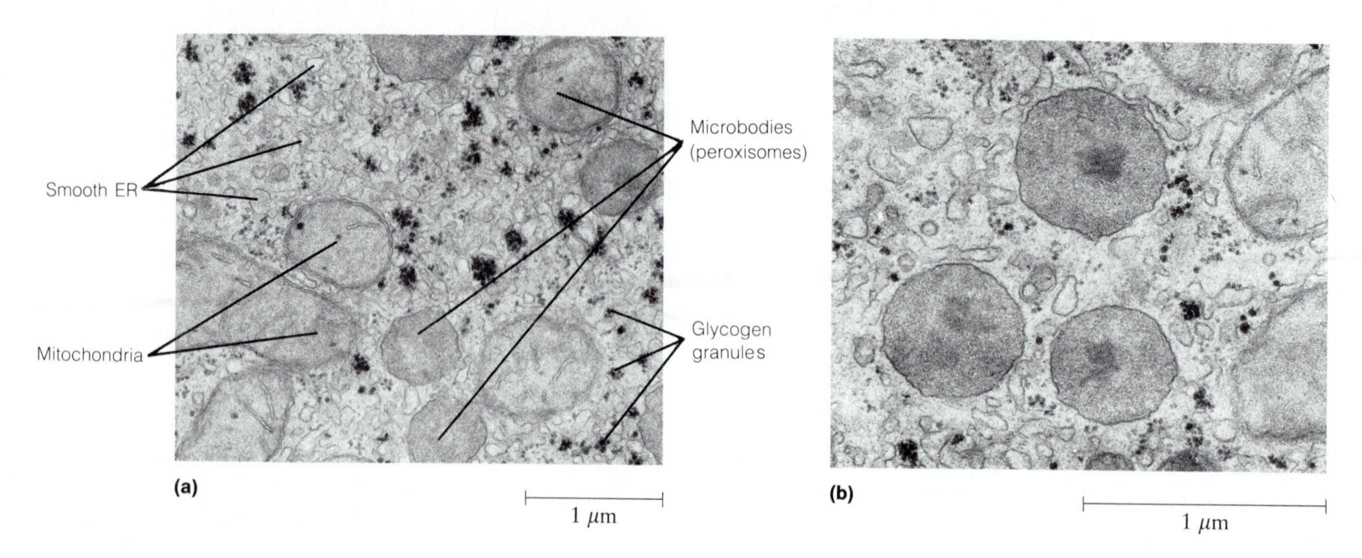

Figure 9-22 Microbodies (Peroxisomes). (a) This electron micrograph shows several microbodies (peroxisomes) in the cytoplasm of a rat liver cell. Other structural features visible in the micrograph include mitochondria, glycogen granules, and smooth ER. (b) At higher magnification, a crystalline core of urate oxidase is readily visible in each microbody. (both TEMs)

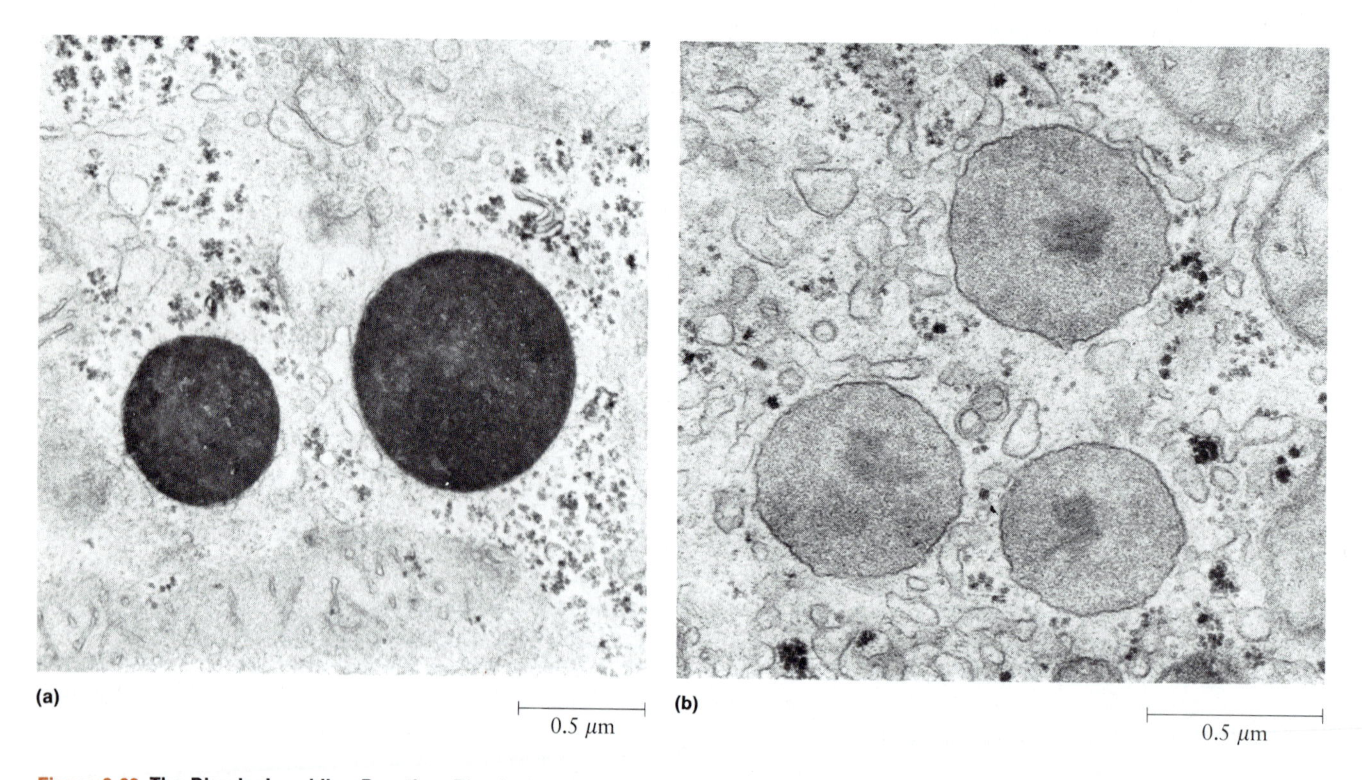

Figure 9-23 The Diaminobenzidine Reaction. Diaminobenzidine (DAB) is oxidized by catalase to a polymeric form that causes electron-dense osmium atoms to be deposited in peroxisomes of OsO_4-treated tissue. (a) Rat liver peroxisomes stained with the DAB reaction. (b) Unstained peroxisomes from the same tissue. (both TEMs)

to consist of crystals of urate oxidase. When such cores are present it is easy to identify microbodies as peroxisomes, since urate oxidase is one of the enzymes by which peroxisomes are defined. In the absence of a crystalline core, however, it is not always easy to spot peroxisomes ultrastructurally. A useful technique in such cases is a cytochemical test for catalase called the **diaminobenzidine (DAB) reaction.** This assay depends on the ability of catalase to oxidize DAB to an electron-dense polymer that is very readily seen in stained cells (Figure 9-23). Since catalase is the single enzyme present in all peroxisomes and does not occur routinely in any other organelle, the DAB reaction is a highly reliable and highly specific means of identifying organelles unequivocally as peroxisomes.

In animals, peroxisomes are most prominent in liver and kidney tissue, and for a time after their discovery it was thought that they were unique to these organs. It is now recognized, however, that they occur widely in most animal tissues, though not as prominently as in liver and kidney.

Hydrogen Peroxide Metabolism. The oxidases found in peroxisomes vary considerably in their substrate requirements, but they are all flavoproteins and they all transfer electrons from their substrate to oxygen to form hydrogen peroxide (H_2O_2):

$$RH_2 + O_2 \longrightarrow R + H_2O_2 \quad (9\text{-}3)$$

The hydrogen peroxide formed in this way is broken down by catalase in one of two ways. Usually, catalase functions in what is called its *catalatic mode,* in which one molecule of hydrogen peroxide is oxidized (to oxygen) and a second is reduced (to water):

$$2H_2O_2 \longrightarrow O_2 + 2H_2O \quad (9\text{-}4)$$

Alternatively, catalase can function in its *peroxidatic mode,* in which hydrogen peroxide is reduced to water using electrons derived from an organic donor:

$$R'H_2 + H_2O_2 \longrightarrow R' + 2H_2O \quad (9\text{-}5)$$

In either case, what is accomplished is the breakdown of hydrogen peroxide. Given the toxicity of this compound, it makes good sense for the cell to ensure that the enzymes responsible for most peroxide generation are compartmentalized together with the catalase responsible for its degradation. Indeed, catalase is the most abundant protein in most peroxisomes, representing up to 15% of the total protein content of the organelle.

Physiological Role of Animal Peroxisomes. Beyond its role in the detoxification of hydrogen peroxide, the animal peroxisome has been linked to several additional metabolic functions:

1. *Detoxification of Toxic Compounds.* In its peroxidatic mode, catalase can use as its electron donor a variety of substances, including methanol, ethanol, formate, formaldehyde, nitrites, and phenols. Since these are all deleterious to the cell, it has been suggested that their oxidative detoxification by catalase may be an important peroxisomal function. The prominent peroxisomes of liver and kidney cells are thought to be important in such detoxification reactions.
2. *Oxidation of Fatty Acids.* Most peroxisomes catalyze the breakdown of fatty acids, a process that also occurs in the mitochondria (see Chapter 11). About 25–50% of fatty acid oxidation in animal tissues occurs in peroxisomes, with the remainder localized in mitochondria.

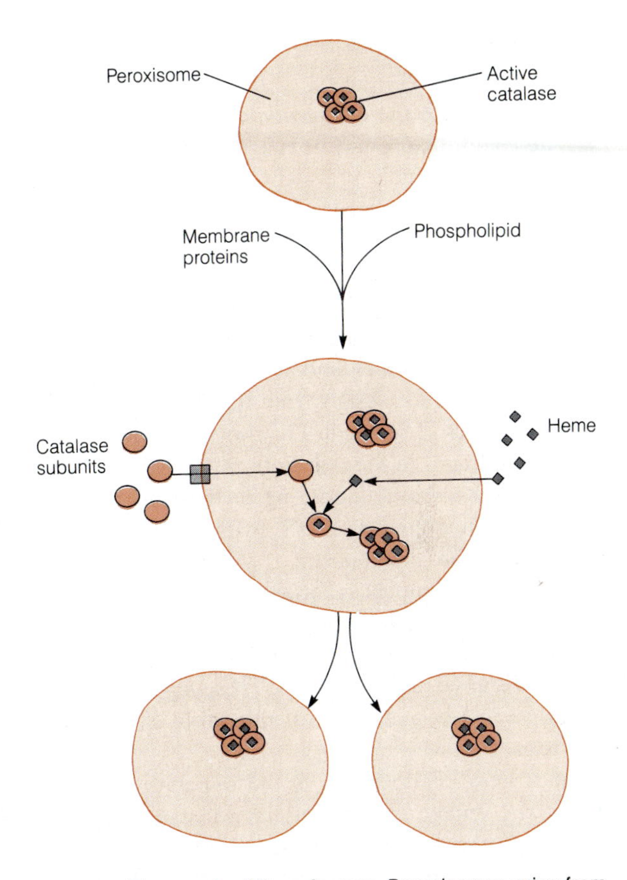

Figure 9-24 Biogenesis of Peroxisomes. Peroxisomes arise from preexisting peroxisomes rather than being products of the Golgi-ER system. Membrane proteins, lipids, and internal enzymes are added from cytoplasmic sources. For instance, catalase subunits are synthesized on cytoplasmic ribosomes and targeted for a receptor on the peroxisomal membrane, shown in the diagram as a square. Upon entering the internal volume, the catalase subunits first bind heme, then assemble into the active tetrameric structure. By continuing addition of such components, peroxisomes undergo a growth process followed by division.

3. *Catabolism of Unusual Substances.* Some of the substrates for peroxisomal oxidases are rare compounds for which the cell may have no other degradative pathway. D-Amino acids may be in this category, since they are not found in proteins and are not recognized by enzymes involved in the degradation of the more common L-amino acids.
4. *Regulation of Oxygen Tension.* The peroxisome has also been suggested as a sort of "safety valve" to protect cells from toxic levels of oxygen. Unlike mitochondrial respiration, peroxisomal oxygen consumption is directly proportional to cellular oxygen concentration. In cases of excessive oxygen levels, peroxisomal respiration will be greatly stimulated, and this may serve as a useful means of consuming excess oxygen and thereby reducing the *oxygen tension* in the cell.

Biogenesis of Peroxisomes

Peroxisomes apparently arise by growth and division of preexisting peroxisomes. However, peroxisomes have no genetic material of their own, so all of their lipid and protein components must be imported from the cytoplasm. An example is the enzyme catalase, a principal peroxisomal protein (Figure 9-24; see p. 245). Catalase is synthesized on cytoplasmic ribosomes, and the monomers are imported into the peroxisomes, presumably by some sort of receptor on the peroxisomal membrane. Once in the peroxisome, the catalase binds heme and assembles into its final tetrameric subunit structure.

Perspective

To appreciate the eukaryotic cell fully means to understand the prominent role that intracellular membranes play. Especially prevalent within most eukaryotic cells is an elaborate system of membranes and organelles derived either directly or indirectly from the endoplasmic reticulum. The ER itself is an interconnecting network of vesicles, tubules, and flattened sacs that separates the cisternal space from the cytoplasmic space. The rough ER is studded with ribosomes and is the site of synthesis of proteins destined for export from the cell or for insertion into membranes. The rough ER also synthesizes proteins for all membranes in the cell. Both rough ER and smooth ER synthesize various lipids, while the smooth ER has special roles in steroid synthesis, hydroxylation reactions, and carbohydrate metabolism.

The Golgi complex is intimately associated with the ER in the process of secretion, since secretory products, especially proteins, usually pass through the Golgi en route from the ER to secretory granules. These can then discharge their contents to the exterior of the cell by the process of exocytosis.

In a sense, endocytosis is the opposite process, involving the uptake of extracellular substances by an infolding of the plasma membrane. Receptor-mediated endocytosis occurs at clathrin-coated pits and results in the formation of coated vesicles, which recycle membrane and receptors back to the cell surface.

Lysosomes and peroxisomes share many properties, including size, density, and the presence of a single bounding membrane. They differ strikingly, however, in the enzymes they contain and therefore in the specific roles they play within the cell. As the name suggests, lysosomes contain hydrolytic enzymes and are involved in a variety of cellular digestive processes. Peroxisomes, on the other hand, are named for the hydrogen peroxide that is generated by oxidases and then destroyed by catalase.

Lysosomes are involved in several different kinds of digestive processes. In some cases, primary lysosomes containing acid hydrolases fuse with vacuoles containing digestible materials, either of extracellular (phagocytic) or intracellular (autophagic) origin. As digestion proceeds within the resulting secondary lysosome, nutrients are released to the cell until only a residual body remains. In other cases, primary lysosomes fuse with the plasma membrane and release their contents to the outside of the cell by exocytosis. Yet another mode of lysosomal function involves autolysis, or self-digestion, of the cell by intracellular rupture of primary lysosomes.

In animals, peroxisomes have several possible roles. These include detoxification of noxious compounds, and regulation of oxygen tension.

So far, we have looked at most of the major membranous structures of cells, including the ER, the Golgi complex, lysosomes, peroxisomes, and the plasma membrane itself. In each case, we have recognized that the membrane is both a permeability barrier and a selective gateway that allows specific molecules and ions in or out of cells or their compartments. Now we move on to cellular metabolic pathways and two other important intracellular structures—the mitochondrion and the chloroplast.

Key Terms for Self-Testing

differential centrifugation (p. 220)

The Endoplasmic Reticulum
endoplasmic reticulum (ER) (p. 220)
cisternae (p. 220)
cisternal space (p. 220)
cytoplasmic space (p. 220)
rough endoplasmic reticulum (p. 222)
smooth endoplasmic reticulum (p. 222)
microsomes (p. 222)
hydroxylation reactions (p. 222)
monooxygenases (mixed-function oxidases) (p. 223)
drug detoxification (p. 223)
carcinogenic (p. 224)

The Golgi Complex
Golgi complex (Golgi apparatus) (p. 226)
saccules (p. 226)
Golgi stack (p. 226)
dictyosome (p. 226)
vesicles (p. 226)
cis (forming) face (p. 226)
trans (maturing) face (p. 226)
secretory granules (p. 226)
coated vesicles (p. 227)
clathrin (p. 227)

Two Cellular Transport Processes: Exocytosis and Endocytosis
exocytosis (p. 227)
endocytosis (p. 229)
condensing vacuoles (p. 229)
zymogen granules (p. 229)
macrophages (p. 231)
pinocytosis (p. 231)
receptors (p. 231)
pinocytotic vesicles (p. 231)
micropinocytosis (p. 231)
macropinocytosis (p. 231)
coated pits (p. 232)
adsorptive pinocytosis (p. 232)
receptor-mediated endocytosis (RME) (p. 232)
ligands (p. 233)
endosome (receptosome) (p. 233)
multivesicular body (p. 233)

Membrane Biosynthesis and Turnover
transport vesicles (p. 235)
phospholipid transfer proteins (p. 235)
membrane turnover (p. 235)

Lysosomes and Cellular Digestion
lysosome (p. 236)
I-cell disease (p. 237)
primary lysosome (p. 237)
phagocytes (p. 238)
food cup (p. 238)
phagocytic vacuole (p. 239)
secondary lysosomes (p. 239)
residual body (p. 239)
autophagy (p. 239)
autophagic vacuole (p. 239)
heterophagic lysosomes (p. 239)
autophagic lysosomes (p. 239)
autolysis (p. 239)
extracellular digestion (p. 239)
lysosomal storage diseases (p. 241)
type II glycogenosis (p. 241)
Hurler syndrome (p. 241)
Hunter syndrome (p. 241)
Tay-Sachs disease (p. 241)

Peroxisomes
peroxisomes (p. 241)
equilibrium density centrifugation (p. 242)
density gradient (p. 242)
buoyant density (p. 242)
microbodies (p. 243)
diaminobenzidine (DAB) reaction (p. 245)

The Intriguing World of the Coated Vesicle
triskelions (p. 230)

Suggested Reading

General References and Reviews

Claude, A. The coming of age of the cell. *Science* 189 (1975): 433.

de Duve, C. Exploring cells with a centrifuge. *Science* 189 (1975): 186.

Watson, J. D., N. H. Hopkins, J. W. Roberts, J. A. Steitz, and A. M. Weiner. *Molecular Biology of the Gene*, 4th ed. Menlo Park, Calif.: Benjamin/Cummings, 1987.

The Endoplasmic Reticulum

Cardell, R. R., Jr. Smooth endoplasmic reticulum in rat hepatocytes during glycogen deposition and depletion. *Int. Rev. Cytol.* 48 (1977): 221.

Depierre, J., and G. Dallner. Structural aspects of the membrane of the endoplasmic reticulum. *Biochim. Biophys. Acta* 415 (1975): 411.

Kappas, A., and A. P. Alvares. How the liver metabolizes foreign substances. *Sci. Amer.* 232 (June 1975): 22.

The Golgi Complex

Farquhar, M., and G. Palade. The Golgi apparatus (complex)—(1954–1981) from artifact to center stage. *J. Cell Biol.* 91 (1981): 71s.

Northcote, D. H. The Golgi complex. In *Cell Biology in Medicine* (E. E. Bittar, ed.). New York: Wiley, 1973.

Rothman, J. E. The compartmental organization of the Golgi apparatus. *Sci. Amer.* 253 (March 1985) 74.

Tartakoff, A. M. Simplifying the complex Golgi. *Trends Biochem. Sci.* 7 (1982): 174.

Endocytosis

Dautry-Varsat, A., and H. F. Lodish. How receptors bring proteins and particles into cells. *Sci. Amer.* 250 (May 1984): 52.

Goldstein, J. L., R. G. W. Anderson, and M. S. Brown. Coated pits, coated vesicles, and receptor-mediated endocytosis. *Nature* 279 (1979): 679.

Clathrin and Coated Vesicles

Crowther, R. A., and B. M. F. Pearse. Assembly and packing of clathrin into coats. *J. Cell Biol.* 91 (1981): 790.

Pearse, B. M. F. Clathrin: A unique protein associated with intracellular transfer of membrane by coated vesicles. *Proc. Natl. Acad. Sci. USA* 73 (1976): 1255.

Pearse, P. A., and R. A. Crowther. Structure and assembly of coated vesicles. *Annu. Rev. Biophys. Chem.* 16 (1987): 49.

Ungewickell, E., and D. Branton. Triskelions: The building blocks of clathrin coats. *Trend Biochem. Sci.* 7 (1982): 358.

Woodward, M. P., and T. F. Roth. Coated vesicles: Characterization, selective dissociation and reassembly. *Proc. Natl. Acad. Sci. USA* 75 (1978): 4394.

Lysosomes

Allison, A. C. *Lysosomes.* Oxford Biology Readers, vol. 58. New York: Oxford University Press, 1974.

Bainton, D. The discovery of lysosomes. *J. Cell Biol.* 91 (1981): 66s.

Dean, R. T. *Lysosomes.* Southampton, England: Camelot Press, 1977.

de Duve, C. The lysosome. *Sci. Amer.* 208 (May 1963): 64.

de Duve, C. Exploring cells with a centrifuge. *Science* 189 (1975): 186.

Hers, H. G., and F. Van Hoof. *Lysosomes and Storage Diseases.* New York: Academic Press, 1973.

Kolodny, E. H. Lysosomal storage diseases. *N. Engl. J. Med.* 294 (1976): 1217.

Kornfeld, S. Trafficking of lysosomal enzymes. FASEB *J.* 1 (1987): 462.

Pitt, D. *Lysosomes and Cell Function.* New York: Longman, 1975.

Animal Peroxisomes

de Duve, C. The peroxisome: A new cytoplasmic organelle. *Proc. R. Soc. London. Ser. B* 173 (1969): 71.

de Duve, C. Microbodies in the living cell. *Sci. Amer.* 248 (May 1983): 74.

Fahimi, H. D., and H. Sies, eds. *Peroxisomes in Biology and Medicine.* Heidelberg: Springer-Verlag, 1987.

Lazarow, P. B. Functions and biogenesis of peroxisomes. In *International Cell Biology, 1980–1981* (H. G. Schweiger, ed.). New York: Springer-Verlag, 1981.

Problem Set

1. **Compartmentalization of Function.** Each of the following processes is associated with a specific eukaryotic organelle. In each case, identify the organelle and indicate the more general cellular function of which the process is a part. In some cases, more than one organelle may be appropriate.
   (a) Hydroxylation of phenobarbital
   (b) Synthesis of insulin
   (c) Synthesis of testosterone (male sex hormone)
   (d) Glycosylation of proteins
   (e) Detoxification of toxic compounds
   (f) Digestion of old or unwanted organelles

2. **Endoplasmic Reticulum.** Label each of the statements below with an S if it is true of the smooth ER only, with an R if it is true of the rough ER only, with an RS if it is true of both, and with an N if it is true of neither.
   (a) Consists of about 70% protein and 30% lipid by weight.
   (b) Is studded with ribosomes on the outer surface.
   (c) Is involved in the breakdown of glycogen.
   (d) Is involved in detoxification of drugs.
   (e) Is site of synthesis of secretory proteins.
   (f) Tends to be tubular in shape.
   (g) Can be seen only with the electron microscope.
   (h) Usually consists of flattened sacs.

3. **Sulfation of Secretory Products.** In addition to containing carbohydrate, a number of secretory products also contain sulfate esterified to certain sugars. Chondroitin sulfate, for example, is a complex sugar secreted by bone marrow cells that carries a sulfate group on the *N*-acetylglucosamine unit of its disaccharide repeating unit (recall Figure 7-23). The sulfotransferase enzyme required to link inorganic sulfate (SO_4^{2-}) to this sugar is thought to be located in the Golgi complex. Describe an experiment designed to test the location of this sulfotransferase activity, and indicate the results you would expect to obtain.

4. **Synthesis of Integral Membrane Proteins.** In addition to their role in cellular secretion, the rough ER and the Golgi complex are also responsible for the synthesis of integral membrane proteins. Specifically, the glycoproteins found so often in the outer phospholipid layer of many plasma membranes are synthesized by this route. In a series of diagrams, depict the synthesis and glycosylation of glycoproteins of the plasma membrane, and explain why the sugar groups

are always located on the outer surface of the plasma membrane. What assumptions about membrane asymmetry must you make?

5. **Lysosomal Enzymes.** The following are true observations about lysosomes and their enzymes. In each case, explain the observation in terms of known properties of the organelle and its enzymes.
   (a) In most classical procedures for differential centrifugation, lysosomes are found mainly in the "mitochondrial" fraction.
   (b) Substantially higher acid hydrolase activities are detected in homogenates of animal tissues if the homogenization is performed in distilled water instead of isotonic (0.25 *M*) sucrose.
   (c) Lysosomal nucleases, phospholipases, and proteases are synthesized ribosomes bound to the membranes of the rough ER, but do not digest either the ribosomes or the ER membrane.
   (d) Lysosomes can be stained cytochemically by incubating a tissue section with a phosphorylated substrate in the presence of lead ions.
   (e) Most of the lysosomal enzymes are latent in that they are inactive against their substrates as long as lysosomal membrane remains intact.

6. **Cellular Digestion.** For each of the following statements, indicate with the appropriate letter(s) the specific digestive process(es) of which a given statement is true: phagocytosis (A), autophagy (B), autolysis (C), or extracellular digestion (D).
   (a) Occurs within vacuoles called secondary lysosomes.
   (b) Involves acid hydrolases.
   (c) Important in certain developmental processes.
   (d) Essential for sperm penetration during fertilization.
   (e) Serves as a source of nutrients within the cell.
   (f) May involve exocytosis.
   (g) Material digested is of intracellular origin.
   (h) Involves fusion of primary lysosomes with the plasma membrane.
   (i) Involves fusion of primary lysosomes with a vacuole.
   (j) Involves no fusion of primary lysosomes.

7. **Silicosis and Asbestosis.** *Silicosis* is a miner's disease that results from the uptake of silica particles (such as sand or glass) by macrophages in the lungs. *Asbestosis* is a similarly debilitating condition caused by inhalation of asbestos fibers. In both cases, the particles or fibers are found in secondary lysosomes in the cells, and the eventual result is cell death. Ultimately, fibroblasts (collagen-secreting cells) in the lungs are stimulated to deposit nodules of collagen fibers, leading to reduced lung elasticity and impaired breathing.
   (a) How do you think the fibers or particles get into the secondary lysosomes?
   (b) What effect do you think fiber or particle accumulation has on the secondary lysosomes?
   (c) How might you explain the death of silica-containing or asbestos-containing cells?
   (d) What do you think happens to the silica particles or asbestos fibers when such a cell dies? Can you see how cell death could continue almost indefinitely, even after further exposure to silica dust or asbestos fibers has ceased?
   (e) Cultured fibroblast cells will secrete collagen and produce connective tissue fibers after addition of material from a culture of lung macrophages that have been exposed to silica particles. What does this tell you about the deposition of collagen nodules in the lungs of silicosis patients?

8. **Lysosomal Storage Diseases.** Despite a bewildering variety of symptoms, lysosomal storage diseases have a number of properties in common. For each of the following statements, respond with a C if you would expect the property to be common to all (or most) lysosomal storage diseases, with an S if you would expect it to be true of a specific lysosomal storage disease, or with an N if you do not consider it to be true of lysosomal storage diseases at all.
   (a) Results in accumulation of excessive quantities of glycogen.
   (b) Is a rare heritable disorder.
   (c) Symptoms include mental retardation and enlargment of the liver and spleen.
   (d) Results in proliferation of catalase-containing organelles.
   (e) Results from the genetic deficiency of an acid hydrolase.
   (f) Leads to death in early infancy, usually before the age of six months.
   (g) Results from inability to regulate synthesis of acid glycosaminoglycans.
   (h) Results in accumulation of secondary lysosomes in the cell.

9. **Enzymes and Organelles.** Most organelles are identified and described in terms of their enzyme content. Both the lysosome and the peroxisome were discovered because specific enzymes displayed properties that could not be easily reconciled with their presence in already-known structures.
   (a) What criteria would you invoke to identify a new organelle unequivocally, based on enzymes contained within the organelle?
   (b) Would your criteria have been adequate to identify lysosomes in a mitochondrial preparation?
   (c) Would your criteria have been adequate to identify peroxisomes in a lysosomal preparation?

PART THREE

ENERGY FLOW IN CELLS

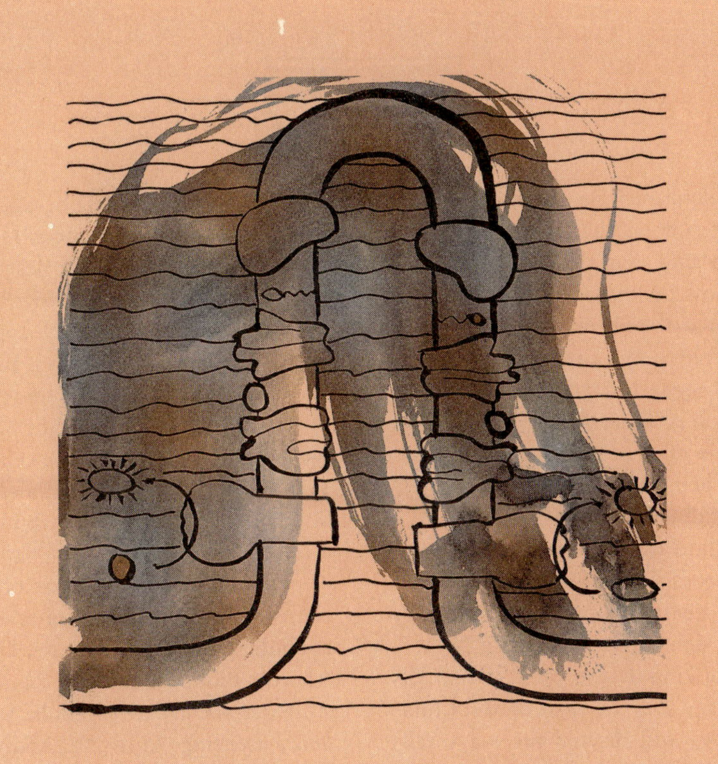

10

Energy from Chemical Bonds: The Anaerobic Mode

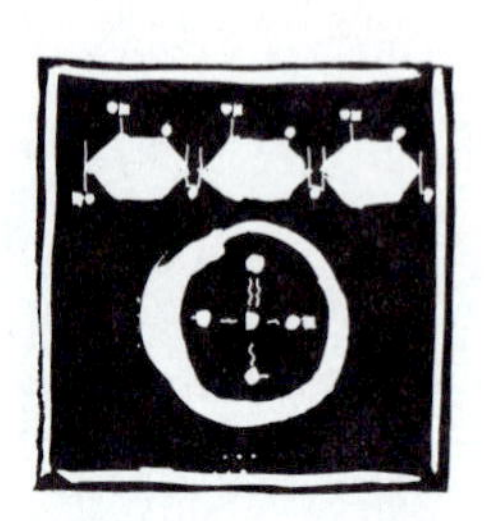

In earlier chapters, we described the basic chemical reactions relevant to life processes. Furthermore, we saw that thermodynamic considerations allow us to predict in which direction a reaction will proceed under specified conditions and how much free energy can be derived from (or must be put into) the system during the reaction. We also learned how application of kinetic theory enables us to measure and describe the rates at which enzyme-catalyzed reactions proceed. Then, in Chapters 7 and 8, we encountered the membranes that define cell boundaries and compartments and showed how energy is used to transport various solute molecules across those boundaries. Still unanswered, however, is the question of where the energy comes from and how it is delivered to the cell. We will answer this question by looking at the three ways cells have for capturing the energy they need to drive cellular reactions. In this chapter we will discuss the ways cells meet their energy needs in the absence of oxygen (*anaerobic* processes). Oxygen-dependent, or *aerobic* processes are the topic of Chapter 11 and in Chapter 12 we will discuss *photosynthesis,* the process by which green plants provide an energy source for all life on Earth.

Metabolic Pathways

So far, we have considered only individual chemical reactions and individual enzymes working in isolation. But that is not the way cells function. To accomplish any major task requires many such reactions occurring in an organized sequence. This, in turn, requires many enzymes working in concert, since most enzymes catalyze only a single reaction, and many reactions may be needed to accomplish a major biochemical operation.

When we consider all chemical reactions occurring within a cell, we are talking about **metabolism.** (The word is an apt one because it comes from two Greek words that mean literally "to throw around," which is a fair description of what happens to molecules during cellular metabolism.) The overall metabolism of a cell consists of a large number of specific **metabolic pathways.** Each pathway accomplishes a specific task, and each therefore answers the question "by what route" for a specific process. From a biochemist's perspective, then, life at the cellular level might be defined as a network of integrated and carefully regulated metabolic pathways, each contributing to the sum of activities a cell must carry out.

Metabolic pathways are of two types. Pathways concerned with the synthesis of cellular components are called **anabolic pathways** (from the Greek *ana,* meaning "up"), whereas those involved in the breakdown of cellular constituents are called **catabolic pathways** (from the Greek *kata,* meaning "down"). Anabolic pathways usually involve a substantial increase in molecular order (and therefore a local decrease in entropy) and are almost always energy-requiring (endergonic). Catabolic pathways, by contrast, are usually energy-liberating (exergonic) because they involve a release of chemical bond energies and a decrease in atomic order (increase in entropy). Catabolic metabolism thus has two products that are important to overall cell function: energy and metabolites. The energy is used to drive anabolic metabolism, by which many of the metabolites are built up into various cell structures.

ATP: The Universal Energy Coupler

The anabolic reactions of cells represent the growth and repair processes characteristic of all living systems, while catabolic reactions furnish the energy needed to drive the anabolic reactions and other forms of cellular work. The efficient linking of energy-yielding processes to energy-requiring processes is therefore crucial to the thermodynamic success of a cell. This coupling is made possible by specific intermediate molecules that store the energy derived from exergonic reactions and release it later to drive what would otherwise be an endergonic reaction. In the biological world, the common intermediate is usually a "high-energy" phosphorylated compound called **adenosine triphosphate (ATP)**, a molecule of paramount importance as the universal energy currency of all cells. Since ATP is involved in almost every cellular energy transaction, it is essential that we understand its structure and function.

ATP Structure and Function

As you may remember from Chapter 3, ATP is a complex molecule containing an aromatic base called *adenine,* a five-carbon sugar called *ribose,* and a chain of three phosphate groups linked to each other by **phosphoanhydride bonds** and to the ribose by a **phosphoester bond** (recall Figure 3-11). The compound formed by the linking of adenine and ribose is called *adenosine.* Adenosine may occur in the cell in the unphosphorylated form or with one, two, or three phosphates attached to carbon atom 5 of the ribose, forming *adenosine monophosphate* (AMP), *diphosphate* (ADP), and *triphosphate* (ATP), respectively. The ATP molecule is well suited for its role as an intermediate in cellular energy metabolism because of the unstable, energy-rich nature of the phosphoanhydride bond that links the third (outermost) phosphate to the second. (The bond between the first and second phosphates has the same properties and is critical in some reactions, but we are interested in the terminal phosphoanhydride bond at present.) The phosphoanhydride bonds that link the phosphate groups of ATP together are high in energy because of **charge repulsion** between the adjacent negatively charged phosphate groups and also because of **resonance stabilization** of both products of hydrolysis (ADP and inorganic phosphate).

Charge repulsion is easy to understand. By way of analogy, imagine holding two magnets together with like poles touching. The like poles repel each other, and you need to make an effort (put energy into the system) to force them together. If you let go, the magnets spring apart, releasing the energy. Now consider the three phosphate groups of ATP. Each group bears at least one negative charge at the near-neutral pH of the cell, and these negative charges tend to repel one another, thereby straining the covalent bond linking them together. If the link is broken (for instance, in a hydrolysis reaction catalyzed by an ATPase enzyme), the energy stored in the covalent bond is released.

A second, even more important contribution to ATP bond energy is resonance stabilization. To understand this phenomenon, we need to realize that groups like the carboxylate or phosphate groups, though formally written with one double bond and the rest as single bonds to oxygen, in reality have an unshared electron pair spread over all the bonds to oxygen (Figure 10-1). The true struc-

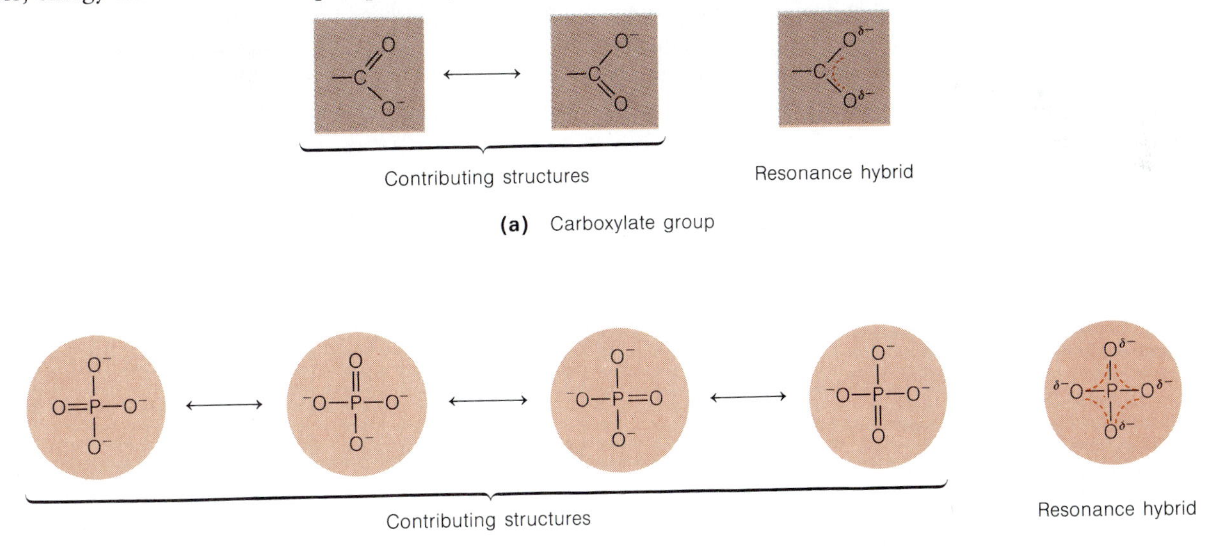

Figure 10-1 Resonance Stabilization. Both (a) the carboxylate group and (b) the phosphate ion are resonance-stabilized. In both cases, the extra electron pair that is formally localized to one of the oxygen bonds of the contributing structures is in fact delocalized over all C—O or P—O bonds, as shown in the resonance hybrid structures on the right. Delocalized electrons are shown by colored dashed lines.

ture of the carboxylate and phosphate groups is actually an average of the **contributing structures** called a **resonance hybrid,** in which the extra electrons are delocalized over all possible bonds (two carbon-oxygen bonds for the carboxylate group and four phosphorus-oxygen bonds for the phosphate group; see Figure 10-1).

When the electrons of such a compound are maximally delocalized in this way, the molecule is in its most stable (lowest-energy) configuration and is said to be *resonance-stabilized.* Consider now what happens if one or more of the oxygen atoms of such an anion become involved in a covalent linkage to an organic compound, such that the oxygen atom is no longer available for electron delocalization (Figure 10-2). In such a case, the extra electrons are delocalized over fewer than the maximum number of oxygen atoms, and the molecule is consequently locked into a higher-energy (less delocalized) configuration. This means that any chemical linkage to a carboxylate or phosphate group will result in a compound that is higher in energy than the free carboxylate or phosphate itself. Specifically, it means that esters, phosphoesters, anhydrides, and phosphoanhydrides are higher-energy compounds than the alcohol, carboxylate, and phosphate compounds that they yield on hydrolysis and will therefore liberate energy if they are hydrolyzed to these products.

For esters, the amount of energy liberated on hydrolysis is moderate, because only one of the two products can undergo resonance stabilization. For anhydrides, however, both products can be resonance-stabilized, so the hydrolysis reaction is highly exergonic. In general, esters and phosphoesters have free energies of hydrolysis around −3 to −4 kcal/mol, whereas anhydrides and phosphoanhydrides have roughly twice that free energy change. ATP is in this respect a representative phosphoanhydride, since its hydrolysis proceeds with a standard free energy change of −7.3 kcal/mol:

$$\text{ATP} + \text{H}_2\text{O} \longrightarrow \text{ADP} + \text{P}_i \qquad (10\text{-}1)$$

$$\Delta G^{\circ\prime} = -7.3 \text{ kcal/mol}$$

In fact, the free energy change under cellular conditions, $\Delta G'$, is almost always greater, often in the range of −10 to −14 kcal/mol, because of the high ATP/ADP ratio that prevails in most cells.

Although the phosphoanhydride bond itself is often referred to as "high energy" or "energy-rich," it is actually the hydrolysis of that bond that is energy-rich. To say that ATP is a high-energy compound should therefore always be understood as a shorthand way of saying that it possesses one or more bonds that are strongly exergonic when hydrolyzed.

ATP as an Intermediate in Energy Transactions

Although we often refer to ATP as a high-energy compound, we really should think of it as an intermediate-energy compound because that is where it ranks in the overall spectrum of bond energies in the cell. Figure 10-3 makes this clear by ranking some of the more common phosphorylated intermediates involved in cellular energy metabolism. Recall from Chapter 5 that the $\Delta G^{\circ\prime}$ values on the left fall on a logarithmic scale calculated from the relationship $\Delta G^{\circ\prime} = -RT \ln K_{eq}$. A $\Delta G^{\circ\prime}$ value of −7.3 kcal/mol for ATP is only about 2.2 times that of glucose-6-phosphate (−3.3 kcal/mol) but this in fact reflects a 10,000-fold difference in the equilibrium constants for the two hydrolysis reactions.

To generalize from this example, compounds with the most negative $\Delta G^{\circ\prime}$ values for hydrolysis of the phosphate group are closest to the top of the figure, so that any

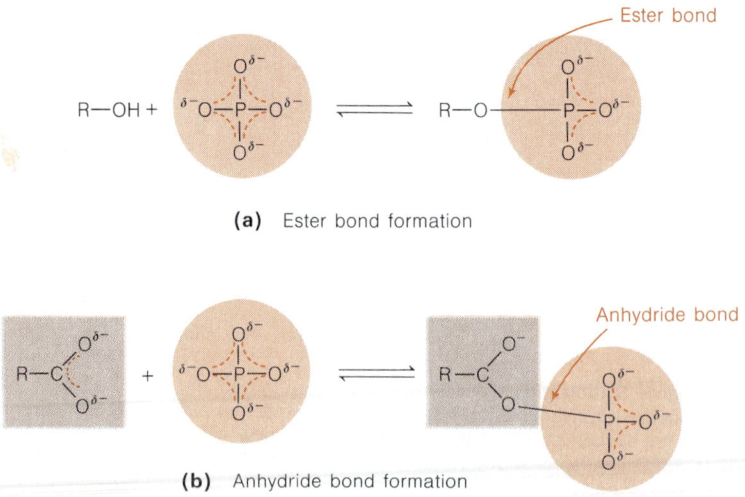

(a) Ester bond formation

(b) Anhydride bond formation

Figure 10-2 Decreased Resonance Stabilization Upon Bond Formation. Formation of (a) an ester bond or (b) an anhydride bond results in decreased opportunity for electron delocalization (shown by colored dashed lines). This makes the ester or anhydride product a higher-energy compound than the phosphate group and the alcohol or carboxylic acid involved in bond formation. An ester bond is a low-energy bond because its hydrolysis (reverse of reaction a) results in increased electron delocalization for only one of the two products. An anhydride bond is a high-energy bond because its hydrolysis (reverse of reaction b) leads to increased electron delocalization for both products.

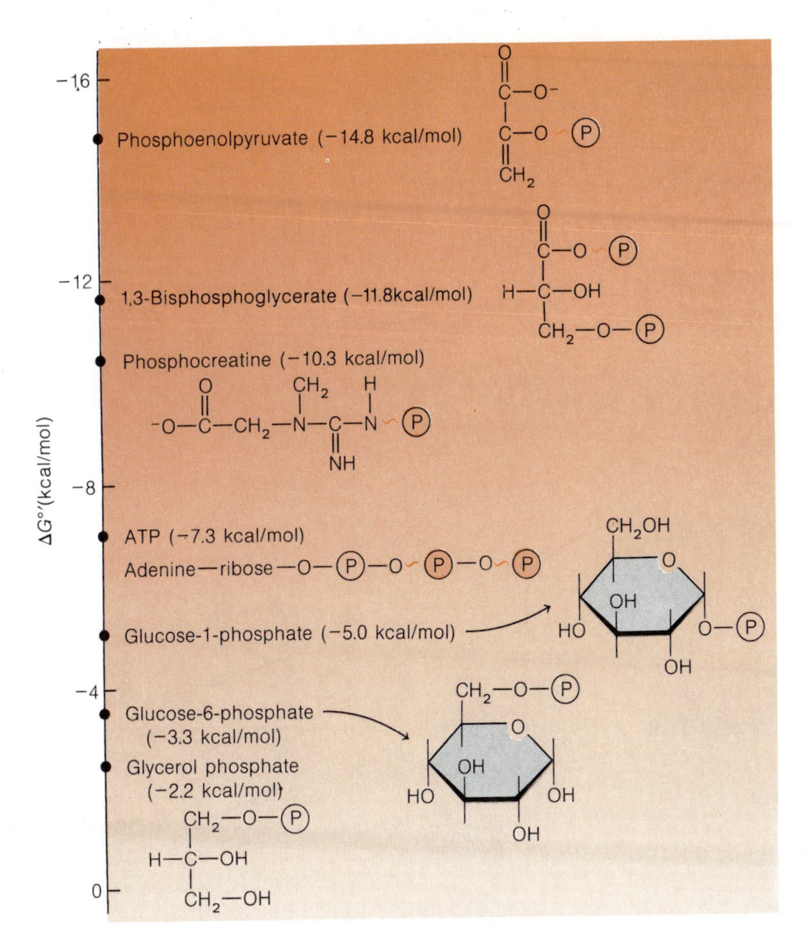

Figure 10-3 Standard Free Energies of Hydrolysis for Some Common Phosphorylated Compounds Found in Cells.

compound is capable of exergonically phosphorylating compounds lying below it (that is, they have a less negative $\Delta G^{\circ\prime}$ value), but not compounds above it. Thus, ATP can phosphorylate glucose and glycerol, but not pyruvate. Similarly, ATP can be formed from ADP by transfer of a phosphate group from 1,3-bisphosphoglycerate, but not from glucose-6-phosphate. Such reactions are called **group transfer reactions,** and they represent one of the most common processes occurring in intermediary metabolism.

The most important point to understand from Figure 10-3 is that ATP occupies a crucial intermediate position in terms of bond energies. This allows the ATP/ADP system to serve as a facile acceptor and donor of energy-rich phosphate groups, since there are compounds both above and below it in energy. We will honor biochemical convention and refer to ATP as a high-energy compound, but we will also try to emphasize that its critical role in energy metabolism really depends on it being intermediate in energy level.

In summary, the ATP/ADP system represents a reversible means of conserving, transferring, and releasing energy within the cell (Figure 10-4). ATP is the "charged," or higher-energy, form, whereas ADP is the "discharged," or lower-energy, form. When a cell carries out certain catabolic pathways, it can couple the reaction sequence to the ATP/ADP system and use the available free energy of the exergonic sequence to drive reaction 10-1 to the left, generating the "high-energy" bond of ATP. Conversely, when the cell needs energy to drive what would otherwise be an endergonic process (such as an anabolic pathway, muscle contraction, or active transport), it uses the energy of that phosphoanhydride bond to provide the driving force by allowing reaction 10-1 to move to the right.

We are dealing, in other words, with an energy-storing system that is charged during the metabolism of nutrients (or, as we will see in Chapter 12, during the photosynthetic trapping of solar energy) and is discharged during the performance of cellular work. The ratio of ATP to ADP and AMP is called the **energy charge** of the cell and is formally described by the equation given below:

$$\text{Energy charge} = \frac{[\text{ATP}] + 0.5[\text{ADP}]}{[\text{ATP}] + [\text{ADP}] + [\text{AMP}]} \qquad (10\text{-}2)$$

If permitted to go to equilibrium, a mixture of ATP, ADP, and AMP would be almost entirely hydrolyzed to AMP, and the energy charge would approach 0. If a cell could drive phosphorylation to the other extreme, so that only

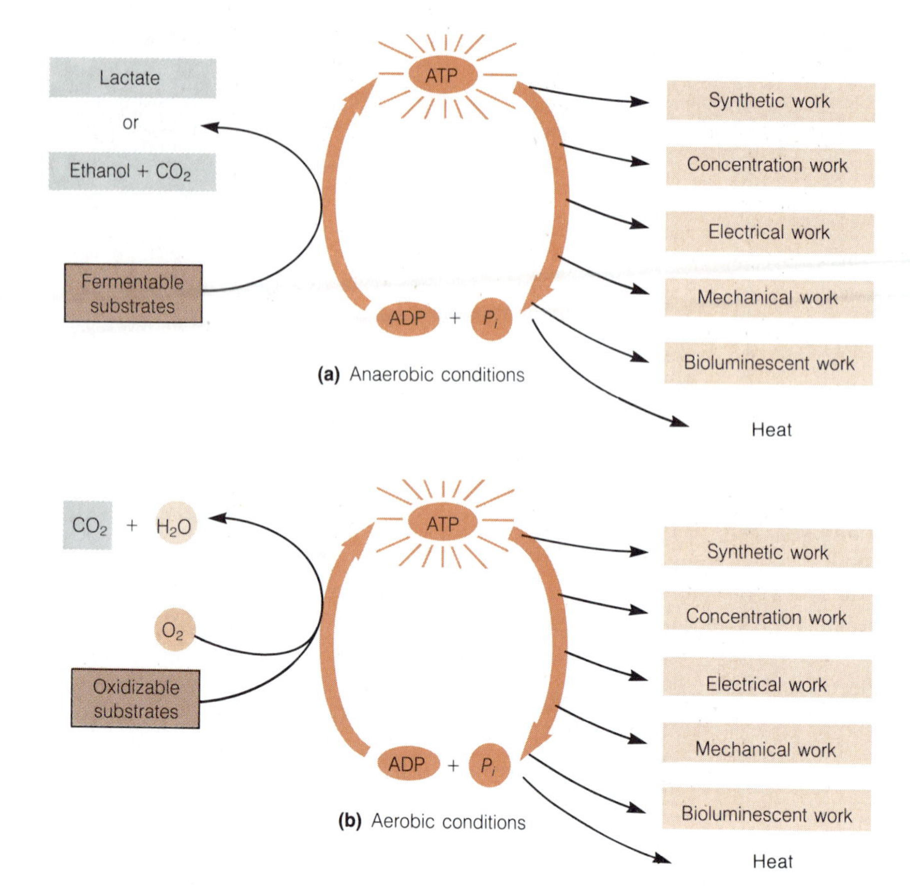

Figure 10-4 The ATP/ADP System as a Means of Conserving and Releasing Energy Within the Cell. ATP is generated during the oxidative catabolism of foodstuffs (left-hand side) and is used to do cellular work (right-hand side). (a) Under anaerobic conditions, ATP is generated by fermentation, with lactate or ethanol plus carbon dioxide as the most common end products. (b) Under aerobic conditions, ATP is generated by respiration, as oxidizable foodstuffs are catabolized completely to carbon dioxide and water.

ATP were present, the energy charge would be 1.0. Most cells maintain an energy charge in the range of 0.9, which shows how far from equilibrium cells can be. An important point to remember is that the total concentration of adenine nucleotides (AMP, ADP, and ATP) is a fixed value. It is the *ratio* of ATP to the other nucleotides that is important, because the ratio gives a quantitative measure of the force driving the cell's biochemical machinery.

Chemotrophic Energy Metabolism

Now we have the essential concepts needed to take up the main theme of this section—the means whereby the chemical energy of nutrients (or of sunlight) is conserved during cellular energy metabolism for subsequent use in driving energy-requiring processes wherever they occur in the cell. This means that we are returning to the phototrophs and chemotrophs we first met in Chapter 5 and to the overall task of understanding how solar energy is trapped as chemical bond energies by the phototrophs and how this chemical bond energy is then released to meet the cellular energy needs of both types of cells.

We will take the case of **chemotrophic energy metabolism** first. How do chemotrophic cells, which are dependent on the chemical bond energies of food molecules, actually make use of that energy? Or, to put it more personally, what are the specific metabolic processes by which the cells of your own body make use of the food you eat to meet your energy needs?

Biological Oxidation

To say that food molecules such as carbohydrates, fats, or proteins have energy stored in their chemical bonds is another way of saying that these are oxidizable organic compounds and that their oxidation is a highly exergonic process. **Oxidation** means the removal of electrons. Thus, hydrogen (H_2) and ferrous ion (Fe^{2+}) are both oxidizable species because they readily give up electrons:

$$H{:}H \longrightarrow 2H^+ + 2e^-$$

$$Fe^{2+} \longrightarrow Fe^{3+} + e^-$$

In organic and biological chemistry, the definition is exactly the same: oxidation is the removal of electrons. The only difference is that the oxidation of organic molecules frequently involves the removal of both electrons and hydrogen ions (protons), so that the process is also one of **dehydrogenation.**

Consider, for example, the oxidation of an alcohol to an aldehyde:

$$R—CH_2—OH \xrightarrow[\text{Oxidation}]{} R—\overset{\displaystyle H}{\overset{|}{C}}=O + 2e^- + 2H^+ \qquad (10\text{-}3)$$

This is clearly an oxidation, since electrons are removed. But protons are liberated as well, and since an electron plus a proton is the equivalent of a hydrogen atom, what is in effect occurring is the removal of the equivalent of two hydrogen atoms:

$$R—CH_2—OH \xrightarrow[\text{(Dehydrogenation)}]{\text{Oxidation}} R—\overset{\displaystyle H}{\overset{|}{C}}=O + 2[H] \qquad (10\text{-}4)$$

Thus, for cellular reactions involving organic molecules, oxidation is almost always manifested as a dehydrogenation reaction. Many of the enzymes that catalyze oxidative reactions in cells are in fact called **dehydrogenases.**

Conversely, **reduction** is defined as the addition of electrons, but in biological reactions this is frequently accompanied by the addition of protons. The overall effect is therefore a **hydrogenation** reaction:

$$R—\overset{\displaystyle H}{\overset{|}{C}}=O + 2[H] \xrightarrow[\text{(Hydrogenation)}]{\text{Reduction}} R—CH_2—OH \qquad (10\text{-}5)$$

Both this and the preceding reaction illustrate the further general feature that biological oxidation-reduction reactions almost always involve two-electron (and therefore two-proton) transfers.

Reactions 10-4 and 10-5 are only half-reactions, representing an oxidation and a reduction event. In real reactions, oxidation and reduction always take place simultaneously. Any time an oxidation occurs, a reduction occurs as well, since the electrons (and protons) removed from one molecule must be added to another molecule. Despite the way reactions 10-4 and 10-5 are written, hydrogen atoms are usually not simply released but instead are transferred to another molecule.

In chemotrophic energy metabolism, the ultimate acceptor of electrons is usually (though not always) oxygen. Rarely, however, are electrons ever passed directly from an oxidizable substrate to oxygen. Instead, the immediate electron acceptor in most biological oxidations (and the immediate electron donor in most biological reductions) is any of several **coenzymes.** In general, coenzymes are small molecules that function along with enzymes (hence the name), usually by serving as a carrier of electrons or functional groups.

The most common coenzymes involved in energy metabolism are **nicotinamide adenine dinucleotide** (**NAD^+**) and **flavin adenine dinucleotide** (**FAD**). Despite their formidable names, the function of these coenzymes is very straightforward. NAD^+ serves as an electron acceptor by adding two electrons and one proton to the reducible part of the molecule, thereby generating the reduced form, NADH, with concomitant release of a proton into the medium:

$$NAD^+ + 2[H] \rightleftharpoons NADH + H^+ \qquad (10\text{-}6)$$

FAD, on the other hand, accepts two electrons and both of the protons to generate $FADH_2$:

$$FAD + 2[H] \rightleftharpoons FADH_2 \qquad (10\text{-}7)$$

NAD^+ accepts electrons onto a carbon atom in the nicotinamide ring of the molecule (Figure 10-5), whereas for FAD the electrons are added to nitrogen atoms (Figure 10-6). This distinction will turn out to be important in understanding why NAD^+ is used as the electron acceptor in some reactions, whereas FAD is used in others.

Also worth mentioning is that both the nicotinamide of NAD^+ and the flavin of FAD are derivatives of compounds (*nicotinic acid* and *riboflavin,* respectively) that we recognize as *B vitamins,* essential in the diet of humans and other vertebrates. It is precisely because these compounds are involved in energy metabolism as parts of indispensable coenzymes that they are essential in the diet of any organism unable to manufacture its own supply of them.

Glucose as a Substrate

We are interested in oxidation because this is how chemotrophs meet most of their energy needs. Many different kinds of substances can be used as substrates for biological oxidation. Some microorganisms can use reduced inorganic compounds (such as reduced forms of iron, sulfur, or nitrogen) as their energy sources. These organisms play important roles in the inorganic economy of the biosphere and utilize rather specialized oxidative reactions. But most chemotrophs depend on organic food molecules, with carbohydrates, fats, and proteins as the three major categories.

To simplify our discussion initially and provide a unifying metabolic theme, we will concentrate on the biological oxidation of the six-carbon sugar **glucose** ($C_6H_{12}O_6$). This compound turns out to be a good choice for several reasons. In many vertebrates, including humans, glucose is the single most important sugar in the blood; and in plants, glucose makes up one-half of the transport sugar, sucrose. Moreover, the pathway for glucose catabolism in cells is a mainstream of energy metabolism into which other compounds are also channeled when the cell uses them for energy. Rather than looking at the fate of a single obscure compound, then, we are considering a meta-

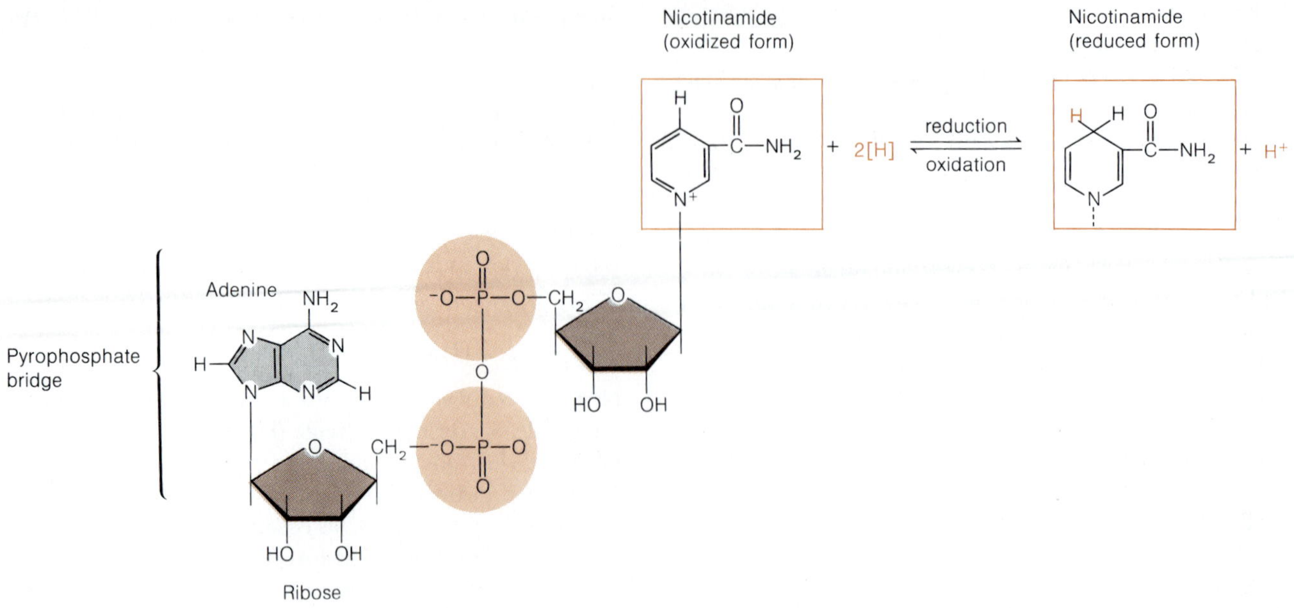

Figure 10-5 **Structure of NAD^+ and Its Oxidation and Reduction.** The portion of the coenzyme outlined in color is the vitamin nicotinamide. The hydrogens derived from an oxidizable substrate are shown in color.

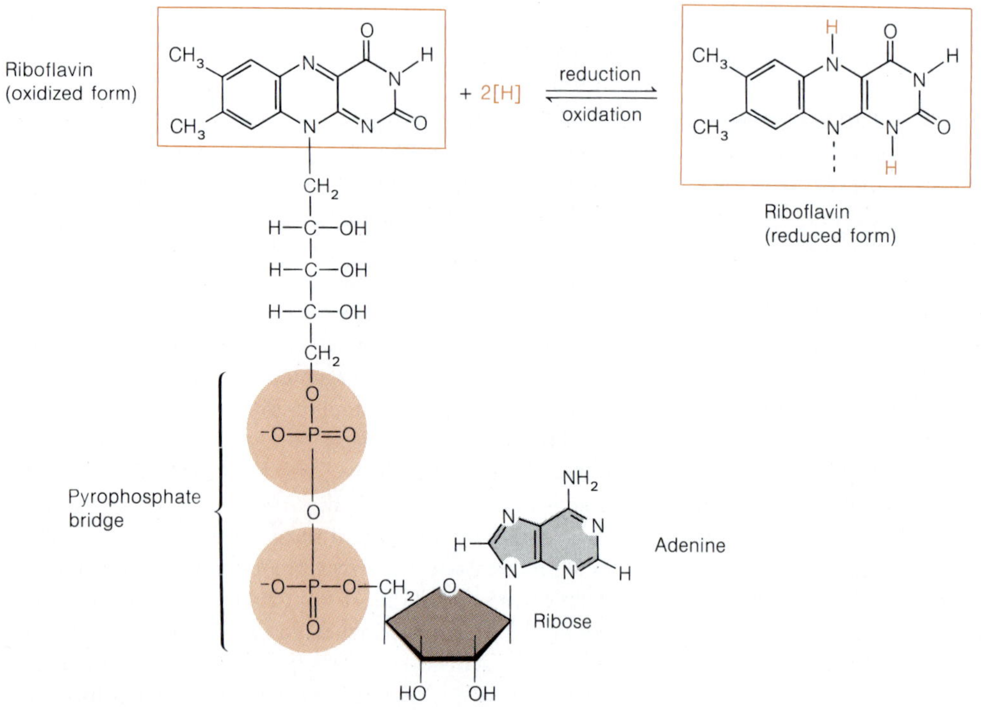

Figure 10-6 **Structure of FAD and Its Oxidation and Reduction.** The portion of the coenzyme outlined in color is the vitamin riboflavin. The hydrogens derived from an oxidizable substrate are shown in color.

bolic sequence that is at the very heart of chemotrophic energy metabolism.

Glucose is a good potential source of energy, since its oxidation is an exergonic process, with a $\Delta G^{\circ\prime}$ of -686 kcal/mol for its complete combustion to carbon dioxide and water:

$$C_6H_{12}O_6 + 6O_2 \longrightarrow 6CO_2 + 6H_2O \qquad (10\text{-}8)$$

As a thermodynamic parameter, $\Delta G^{\circ\prime}$ is unaffected by route and will therefore have the same value whether the oxidation is by direct combustion or biological oxidation, with maximum trapping of energy as ATP. For cells, however, route is critical: uncontrolled combustion would be incompatible with life, whereas controlled, stepwise oxidation mediated by a series of enzyme-catalyzed reactions represents an isothermal process that can be coupled to ATP generation and subjected to careful regulation.

In this and the following chapter, therefore, we want to look at the biological processes whereby the energy of the oxidizable bonds of glucose can be released in ways that ensure the conservation of as much of that energy as possible in the form of ATP, and under conditions compatible with life. We will also want to focus on the means by which cells continuously adjust the rate of glucose oxidation to meet actual cellular needs for ATP.

Respiration with Oxygen, Fermentation Without

To speak of glucose (or any other compound) as an oxidizable substrate assumes some sort of electron acceptor, without which there can be no oxidation. Notice, for example, that reaction 10-8 assumes the availability of oxygen. Access to the full 686 kcal/mol of free energy is possible only with oxygen as the electron acceptor and hence requires aerobic conditions. Aerobic catabolism of glucose is called **respiration.** Respiration is a complex, multistep process, as we will see when we get to Chapter 11.

Under anaerobic conditions, the cell must use an electron acceptor other than oxygen. Usually, the electrons are passed to pyruvate, a three-carbon compound generated by the catabolic process itself. The pyruvate is thereby reduced to lactate or to ethanol and carbon dioxide. Because most, if not all, of the carbon remains in organic form during anaerobic catabolism, the net energy yields are usually much lower than those of aerobic respiration. All such anaerobic processes are called **fermentations** and are usually identified in terms of the principal end product. In animal cells, the end product is usually *lactate,* so the process of anaerobic glucose catabolism is called **lactate fermentation.** In yeast, the process is termed **alcoholic fermentation** because the end product is the alcohol *ethanol.*

Aerobic and Anaerobic Organisms

Organisms can be classified in terms of their need for and use of oxygen as an electron acceptor. Many organisms have an absolute requirement for oxygen and are called **strict** or **obligate aerobes.** You look at such an organism every morning in the mirror. **Strict** or **obligate anaerobes,** on the other hand, cannot tolerate the presence of oxygen; it is toxic to them. Not surprisingly, such organisms occupy environments from which oxygen is generally excluded, such as soil, deep water, or puncture wounds. Most are bacteria, including those responsible for gangrene, food poisoning, and methane production.

Facultative organisms are those that can function in either mode. Given the availability of oxygen, most can carry out the full respiratory process. But if forced to exist anaerobically, they switch to a fermentative mode. Many bacteria and fungi are facultative organisms. Some cells or tissues of otherwise aerobic organisms can function anaerobically if required to do so. Your muscle cells are an example, because they can operate under temporarily anaerobic conditions when oxygen demand exceeds supply during periods of strenuous exercise.

The remainder of this chapter will be devoted to the anaerobic generation of ATP, with lactate and alcoholic fermentations as the main processes of interest. This will be followed in Chapter 11 by a discussion of aerobic energy metabolism. It seems appropriate to consider the anaerobic options first, since even under aerobic conditions the basic fermentation pathways are still operative. In that case, however, they serve as just the first phase of respiration, rather than representing the full catabolic capability of the cell. Thus, by beginning here with fermentation, we examine not only anaerobic energy metabolism but also the first steps that lead to aerobic respiration.

Fermentation: The Anaerobic Mode for ATP Generation

Whether obligately or facultatively anaerobic, cells that are obliged to meet their energy needs by fermentation face the special challenge of carrying out energy-yielding oxidative reactions without being able to use oxygen as an electron acceptor. To do so, they split the glucose molecule into two pieces and then carry out what is in effect an intramolecular oxidation-reduction reaction. The overall process is sufficiently exergonic to allow two ATP molecules to be formed per molecule of glucose fermented. This is the maximum possible energy yield under anaerobic conditions.

Glycolysis

The process of **glycolysis** is a ten-step reaction sequence that is without a doubt the single most ubiquitous pathway in all of energy metabolism. It is common to both aerobic and anaerobic metabolism of glucose, since the pyruvate to which it leads can be either fermented to lactate or ethanol under anaerobic conditions or oxidized further if oxygen is available. As such, glycolysis is a universal process, occurring in almost every living cell.

Historically, the **glycolytic pathway** was the first major metabolic sequence to be elucidated. Most of the decisive work was done in the 1930s by the German biochemists Gustav Embden, Otto Meyerhof, and Otto Warburg. An alternative name for the glycolytic sequence is, in fact, the *Embden-Meyerhof pathway.*

The glycolytic pathway is shown in the context of fermentation in Figure 10-7 and as a component of aerobic respiration in Figure 11-3. The essence of the process is suggested by its very name, since glycolysis comes from two Greek roots, *glykos* meaning "sweet" and *lysis* for "loosening." Literally, glycolysis is the loosening (or splitting) of something sweet, which is, of course, the starting sugar. The splitting occurs at step 4 (labeled Gly-4 in Figure 10-7), since it is at that point that the six-carbon sugar is split into two three-carbon fragments, one of which turns out to be the only oxidizable molecule in the whole pathway.

The oxidation of that molecule (called glyceraldehyde-3-phosphate) in step 6 involves NAD^+ as the electron acceptor. The reduced coenzyme generated in this step, NADH, is later used as the source of electrons for the reductive fermentation step at the end of the pathway, thereby regenerating the oxidized form of the coenzyme. At two points in the glycolytic sequence (Gly-7 and Gly-10), specific reactions are sufficiently exergonic to be coupled to the generation of ATP. These represent the energy payoff of the process, since they are the only ATP-yielding reactions in the whole pathway as it functions under anaerobic conditions.

The important features of the glycolytic pathway are therefore the sugar-splitting reaction for which the sequence is named, the oxidative event that powers the whole pathway, and the two specific steps at which the reaction sequence is coupled to ATP generation. These features will be emphasized as we consider the overall pathway in three phases: preparatory and cleavage steps (Gly-1 through Gly-5); the oxidative sequence, which includes the first ATP-generating event (Gly-6 and Gly-7); and the second ATP-generating sequence (Gly-8 through Gly-10).

Phase 1: Preparation and Cleavage. Reaction Gly-4 is the step at which the cleavage of a six-carbon compound to two three-carbon molecules occurs. It is useful to focus on the substrate for that reaction, *fructose-1,6-bisphosphate,* and ask how it can be formed from glucose, because that will undoubtedly explain the first three preparatory steps in the sequence. The crucial feature of fructose-1,6-bisphosphate is clearly the presence of two phosphate groups, one on each terminal carbon. Looking at glucose, it is easy to see how phosphorylation can take place on carbon atom 6, because the hydroxyl group there can be readily linked to a phosphate group to form a *phosphoester.* That is, in fact, what happens in reaction Gly-1, with ATP providing both the phosphate group and the driving force to render the phosphorylation reaction exergonic ($\Delta G^{\circ\prime} = -4.0$ kcal/mol). Notice, by the way, that the bond being formed is an *ester* bond, while the bond by which the terminal phosphate is linked to ATP is a *phosphoanhydride* bond, so we should have expected the transfer to be exergonic, as it is (recall Figure 10-3). In fact, the equilibrium for this reaction lies so far to the right that virtually all the glucose that enters a cell is almost immediately converted into the phosphorylated form.

In addition to activating the glucose for subsequent cleavage, phosphorylation ensures that the sugar molecule is effectively trapped within the cell once it enters, since the phosphate group is highly polar and prevents the sugar molecule from passing across the plasma membrane again. A quick look at the glycolytic pathway (Figure 10-7) will confirm that all the intermediates between glucose (which usually enters the cell initially by crossing the plasma membrane) and pyruvate (which in aerobic eukaryotic cells must cross the mitochondrial membrane for further catabolism) are present in phosphorylated form.

Looking next at the other end of the glucose molecule, we see that the carbonyl group is not as readily phosphorylated as a hydroxyl group, so the next reaction involves a conversion of the aldosugar into the corresponding ketosugar, *fructose-6-phosphate* (reaction Gly-2). This leaves a hydroxyl group on carbon atom 1 that can now be phosphorylated just as before, yielding the doubly phosphorylated sugar, fructose-1,6-bisphosphate (reaction Gly-3). The energy difference between the anhydride of ATP and the phosphoester bond on carbon atom 1 renders the reaction highly exergonic and essentially irreversible in the glycolytic direction ($\Delta G^{\circ\prime} = -3.4$ kcal/mol).

Next, the actual cleavage of sugar occurs, from which glycolysis derives its name. Fructose-1,6-bisphosphate is split reversibly to yield two three-carbon sugars, called

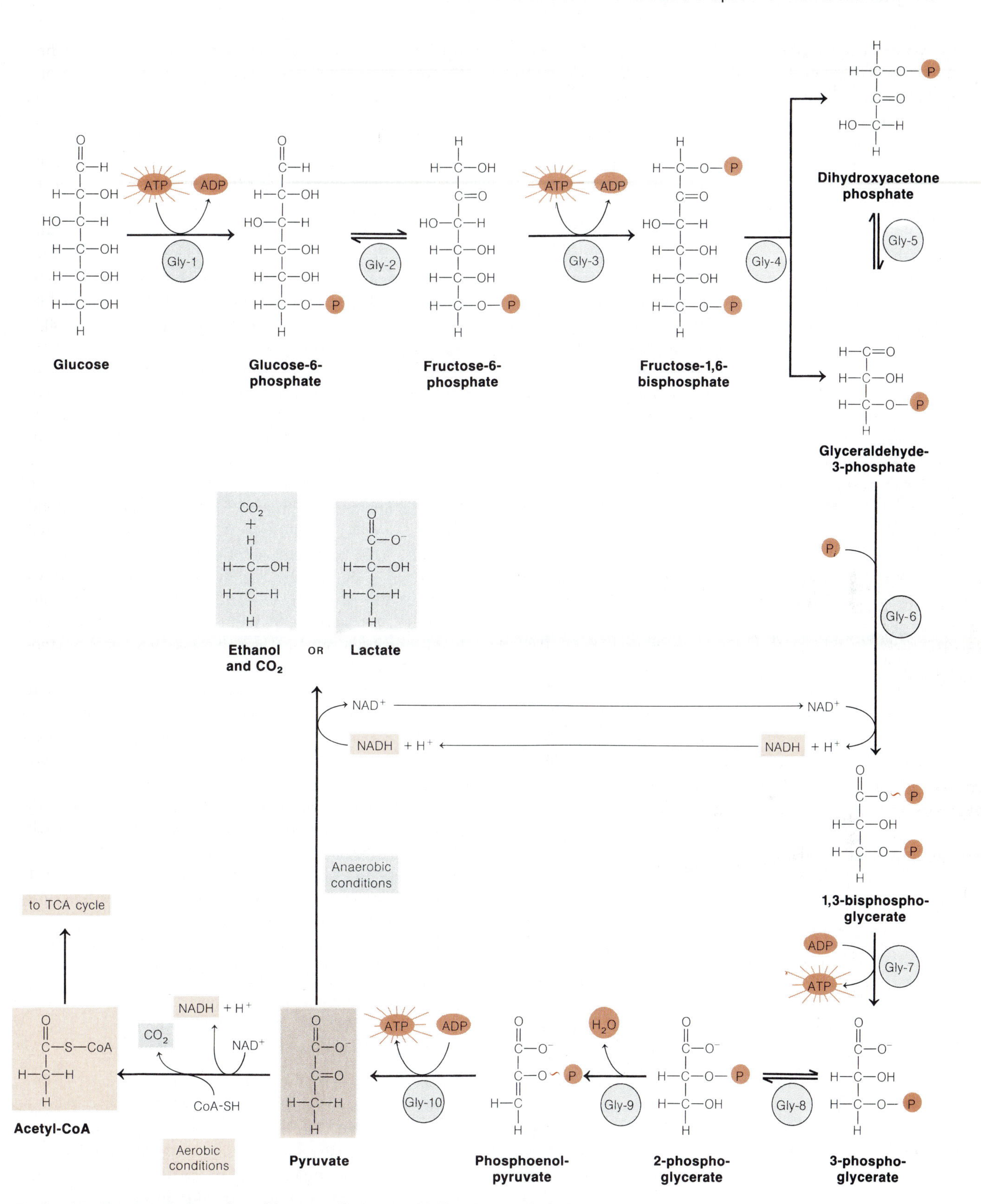

Figure 10-7 Glycolytic Pathway from Glucose to Pyruvate, with Two Anaerobic Options. ATP, ADP, NAD^+, and NADH are shown in color. Under anaerobic conditions, the NADH generated by reaction Gly-6 is reoxidized by transferring its electrons to pyruvate. The most common products are lactate or ethanol plus carbon dioxide.

dihydroxyacetone phosphate and glyceraldehyde-3-phosphate (reaction Gly-4). This cleavage, catalyzed by an enzyme called *aldolase,* is shown in more detail in Figure 10-8. Note that carbon atoms 1, 2, and 3 of the hexose give rise, respectively, to carbon atoms 3, 2, and 1 of dihydroxyacetone phosphate, whereas carbon atoms 4, 5, and 6 of the hexose become carbon atoms 1, 2, and 3 of glyceraldehyde-3-phosphate.

The two trioses formed in Gly-4 bear the same relationship to each other as do glucose-6-phosphate and fructose-6-phosphate. It is therefore not surprising that dihydroxyacetone phosphate and glyceraldehyde-3-phosphate are readily interconvertible (reaction Gly-5). Since it is only the latter of these compounds that is directly oxidizable in the next phase of glycolysis, interconversion of the two trioses enables dihydroxyacetone phosphate to be catabolized simply by conversion to glyceraldehyde-3-phosphate.

We can summarize this first phase of the glycolytic pathway as follows:

$$\text{glucose} + 2\text{ATP} \longrightarrow 2 \text{ glyceraldehyde-3-phosphate} + 2\text{ADP} \quad (10\text{-}9)$$

Phase 2: Oxidation and ATP Generation. So far, five of the ten steps of glycolysis have been accounted for, and the original glucose molecule has been doubly phosphorylated and cleaved into two interconvertible trioses, each of which can therefore be subjected to the same further catabolic fate. Note also that the net ATP yield so far is −2, since two molecules of ATP have been consumed per molecule of glucose up to this point. Now, however, the ATP debt is about to be turned into a profit as we encounter the two energy-yielding phases of glycolysis. In the first instance (reactions Gly-6 and Gly-7), ATP production is linked directly to an oxidative event, while in the second case (reactions Gly-8 through Gly-10) it is the highly unstable nature of the enol form of pyruvate that serves as the driving force behind ATP generation.

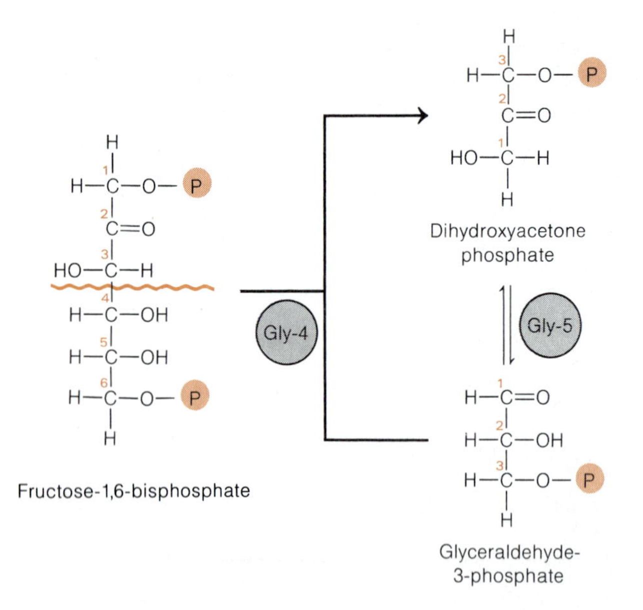

Figure 10-8 Cleavage of Fructose-1,6-Bisphosphate.

The oxidation of glyceraldehyde-3-phosphate to the corresponding acid, 3-phosphoglycerate, is highly exergonic—sufficiently so, in fact, to drive both the reduction of the coenzyme NAD^+ (Gly-6) and the phosphorylation of ADP with inorganic phosphate, P_i (Gly-7). Historically, this was the first example of a reaction sequence in which the coupling of ATP generation to an oxidative event was understood. Because of its usefulness as a prototype for understanding such coupling, this two-reaction sequence is considered in detail in the box on pp. 264–265 with special attention to the mechanism of action of *glyceraldehyde-3-phosphate dehydrogenase,* the enzyme responsible for reaction Gly-6.

The important features of the actual oxidative event (Gly-6) are the involvement of NAD^+ as the electron acceptor and the coupling of the oxidation to the formation of a high-energy doubly phosphorylated intermediate, 1,3-bisphosphoglycerate. The phosphoanhydride bond on carbon atom 1 of this intermediate has such a highly negative $\Delta G^{\circ\prime}$ of hydrolysis (−11.8 kcal/mol; see Figure 10-3) that the transfer of the phosphate to ADP is a highly exergonic, essentially irreversible reaction ($\Delta G^{\circ\prime} = -11.8 - (-7.3) = -4.5$ kcal/mol). ATP generation by direct transfer of a high-energy phosphate group from a phosphorylated compound such as 1,3-bisphosphoglycerate is called **substrate-level phosphorylation.** In the next chapter, we will encounter **oxidative phosphorylation**—ATP generation coupled to the transfer of electrons from reduced coenzymes to oxygen.

To summarize the substrate-level phosphorylation of reactions Gly-6 and Gly-7, we can write an overall reaction with a stoichiometry that accounts for *both* of the glyceraldehyde-3-phosphate molecules generated from each glucose molecule in the first phase of glycolysis:

$$2 \text{ glyceraldehyde-3-phosphate} + 2\text{NAD}^+ + 2\text{ADP} + 2\text{P}_i \longrightarrow 2 \text{ 3-phosphoglycerate} + 2\text{NADH} + 2\text{H}^+ + 2\text{ATP} \quad (10\text{-}10)$$

As the stoichiometry already makes clear, each reaction in the glycolytic pathway beyond glyceraldehyde-3-phosphate needs to occur twice per starting molecule of glucose to account for both molecules of triose phosphate generated in the cleavage reaction (Gly-4). This means that on a per-glucose basis, we will have two molecules of NADH in need of eventual reoxidation. It also means that both of the ATP molecules that were required for initial activation of the glucose molecule in phase 1 are recovered here in phase 2, so the net ATP yield is now zero.

Thus far, then, seven of the ten reactions of glycolysis have been used to convert one molecule of glucose into two molecules of 3-phosphoglycerate, but we have as yet nothing to show for it in terms of net ATP generation. That comes only as we consider the final phase of the pathway.

Phase 3: Pyruvate Formation and ATP Generation. To generate another molecule of ATP at the expense of 3-phosphoglycerate depends on the phosphate group on carbon atom 3, which at this stage is linked by a phosphoester bond with an unpromisingly low free energy ($\Delta G^{\circ\prime} = -3.3$ kcal/mol). In this last phase of the glycolytic pathway, this ester bond is converted to a high-energy *phosphoenol bond* by what is essentially a rearrangement of internal energy within the molecule. To accomplish this, the phosphate group on carbon atom 3 is first moved to carbon atom 2 (reaction Gly-8), followed by the removal of water from 2-phosphoglycerate, thereby generating phosphoenolpyruvate (PEP). If you look carefully at PEP, you will notice that, unlike the phosphoester bonds of either 3- or 2-phosphoglycerate, the phosphoenol bond has what we might define as a distinguishing characteristic of a high-energy phosphate bond: a phosphate group adjacent to a double bond. In fact, PEP has one of the highest-energy phosphate bonds known in biological systems, with a $\Delta G^{\circ\prime}$ for hydrolysis of -14.8 kcal/mol (see Figure 10-3).

To understand why PEP is such a high-energy compound, we need to recognize that the pyruvate molecule can exist in either the enol or the keto form but that the equilibrium of the interconversion greatly favors the latter (Figure 10-9). This means that the keto form of pyruvate is much more stable, whereas the enol form is a highly unstable, thermodynamically unlikely configuration. Thus, when pyruvate is essentially "trapped" in the enol form by chemical constraints that do not allow transition to the keto form, it will be highly unstable.

Such is the case with the PEP generated in reaction Gly-9. When water is removed from 2-phosphoglycerate, the product is pyruvate "locked" in the enol form by the presence of a phosphate group on carbon atom 2 that prevents transition (the proper chemical term is *tautomerization*) to the more stable keto form (Figure 10-9b). PEP is therefore especially high in energy, because in addition to the usual free energy released when the extra electron pair becomes maximally delocalized over the phosphate P—O bonds, there is also free energy released upon reversion of the pyruvate from the enol form to the keto form.

Enol form of pyruvate (unstable) — (a) — Keto form of pyruvate (stable)

Enol form of phosphopyruvate ("locked" in unstable form) — (b)

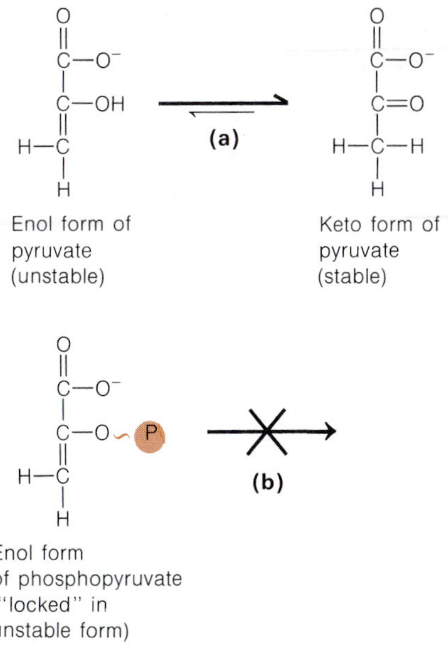

Figure 10-9 The Instability of Phosphoenolpyruvate. (a) Pyruvate can exist in either the enol or the keto form, but the equilibrium greatly favors the keto form. Conversion of the enol form to the keto form is therefore thermodynamically very favorable. (b) Phosphoenolpyruvate is in the thermodynamically unfavorable enol configuration, but cannot undergo conversion to the stable keto form because of the phosphate group covalently linked to carbon atom 2. Release of this phosphate group is highly exergonic because the pyruvate molecule is then free to assume the stable keto form instead.

To say that the phosphate bond of PEP is high in free energy of hydrolysis is to make the last step in the sequence, reaction Gly-10, entirely reasonable, since it involves transfer of that phosphate to ADP, generating another molecule (or, on a per-glucose basis, two more molecules) of ATP. That transfer is strongly exergonic ($\Delta G^{\circ\prime} = -14.8 - (-7.3) = -7.5$ kcal/mol) and therefore virtually irreversible.

To summarize this third phase of glycolysis, we can write an overall reaction, again using a stoichiometry that accounts for both of the three-carbon compounds derived from glucose:

$$2\ \text{3-phosphoglycerate} + 2\text{ADP} \longrightarrow 2\ \text{pyruvate} + 2H_2O + 2\text{ATP} \quad (10\text{-}11)$$

Summary of Glycolysis. Since the ATP initially invested in reactions Gly-1 and Gly-3 was recouped in the first phosphorylation event (Gly-7), the two molecules of ATP formed per molecule of glucose by the second phosphorylation event (Gly-10) represent the net yield of the glycolytic pathway. This becomes clear when we add up the three reactions that summarize the three phases of the pathway (reactions 10-9, 10-10, and 10-11) to obtain the

Glyceraldehyde-3-Phosphate Oxidation: A Prototype Par Excellence

The biological oxidation of glyceraldehyde-3-phosphate to 3-phosphoglycerate as it occurs in reactions Gly-6 and Gly-7 of the glycolytic pathway is of special importance from several points of view. In terms of the energetics of glycolysis, this reaction sequence is crucial because it represents the only site of coenzyme reduction and accounts for 50% of the ATP yield of the pathway. Historically, the elucidation of this sequence by Otto Warburg and his colleagues in 1938–1939 can rightfully be regarded as a major landmark in biochemistry, since it was the first example of a reaction mechanism by which the energy liberated upon oxidation of an organic molecule is conserved as ATP. Moreover, the reaction serves as a prototype par excellence for understanding ATP-coupled oxidations, because to examine the mechanism of glyceraldehyde-3-phosphate oxidation is really to come to grips with an essential aspect of cellular energy metabolism. Fortunately, the reaction sequence is very well understood, and each step can be explained in terms of straightforward principles of chemistry and energetics.

Crucial to the conservation of energy in this sequence is the coupling in reaction Gly-6 of the actual oxidative event to the formation of a high-energy phosphoanhydride bond on carbon atom 1, yielding 1,3-bisphosphoglycerate. This then allows generation of ATP in reaction Gly-7 by transfer of the phosphate to ADP.

The reaction mechanism of the enzyme glyceraldehyde-3-phosphate dehydrogenase, which catalyzes reaction Gly-6, is shown in the upper portion of Figure 10-A. The enzyme is a tetrameric protein, with an active catalytic site on each of its four identical subunits. The active site contains a critical cysteine, with its nucleophilic sulfhydryl group, and a binding site for NAD^+. The reaction sequence begins with the binding of the substrate to the active site by nucleophilic attack on the carbonyl group of glyceraldehyde-3-phosphate by the ionized form of the sulfhydryl group (Figure 10-A, step 1). This links the substrate covalently to the enzyme surface as a *hemithioacetal*. Carbon atom 1 is much less positively charged in the hemithioacetal than it was in the free carbonyl group, thereby facilitating the removal of a hydride ion, $:H^-$, in step 2. The acceptor of the hydride ion is the molecule of NAD^+ that is already tightly bound to the enzyme at the active site. (It is, in fact, the binding of NAD^+ that renders the enzyme susceptible to substrate binding.) Some of the free energy of this oxidative event is preserved as the high-energy *thioester bond* by which the oxidized molecule is attached to the enzyme.

The NADH now dissociates from the active site and is subsequently replaced on the enzyme by another molecule of NAD^+ (step 3). Finally, inorganic phosphate attacks the thioester to form the high-energy phosphoanhydride 1,3-bisphosphoglycerate, which leaves the enzyme surface as the product of the reaction (step 4).

Energetically, the essential feature of the whole sequence is that a thermodynamically unfavorable reaction, the formation of an anhydride between a carboxylic acid and inorganic phosphate, is driven by a thermodynamically favorable reaction, the oxidation of an aldehyde. The two reactions are coupled by an enzyme-bound thioester intermediate, which preserves much of the free energy that would otherwise have been released as heat in the oxidation reaction.

The high energy of the anhydride bond is then used to generate ATP in reaction Gly-7, catalyzed by the enzyme phosphoglycerokinase (Figure 10-A, lower portion). The overall result is the oxidation of glyceraldehyde-3-phosphate to 3-phosphoglycerate, the reduction of the coenzyme NAD^+, and the generation of a molecule of ATP from ADP and inorganic phosphate. The free energy changes for these events are as follows:

Partial Reaction	$\Delta G^{\circ\prime}$ (kcal/mol)
glyceraldehyde-3-phosphate $\longrightarrow$ 3-phosphoglycerate	−63.0
$NAD^+ \longrightarrow NADH$	+52.7
$ADP + P_i \longrightarrow ATP$	+7.3

The phosphoglycerokinase reaction and its free energy change can be summarized in this way:

$$\text{glyceraldehyde-3-phosphate} + NAD^+ + ADP + P_i \longrightarrow \text{3-phosphoglycerate} + NADH + H^+ + ATP$$

$$\Delta G^{\circ\prime} = -63.0 + 52.7 + 7.3 = -3.0 \text{ kcal/mol}$$

Of the total amount of free energy that could be released if glyceraldehyde-3-phosphate were oxidized directly by molecular oxygen ($\Delta G^{\circ\prime} = -63$ kcal/mol), about 95% (60 kcal/mol) is conserved by this reaction sequence, either as the energy-rich phosphoanhydride bond of ATP (7.3 kcal/mol) or in the form of the reduced coenzyme NADH, itself a high-energy compound, capable of releasing 52.7 kcal/mol of free energy upon subsequent oxidation during the process of electron transport (see Chapter 11).

The process of glyceraldehyde-3-phosphate oxidation is therefore an excellent example of energy conservation during biological oxidation and should be studied carefully as an especially good prototype for all such reactions as they occur in cells.

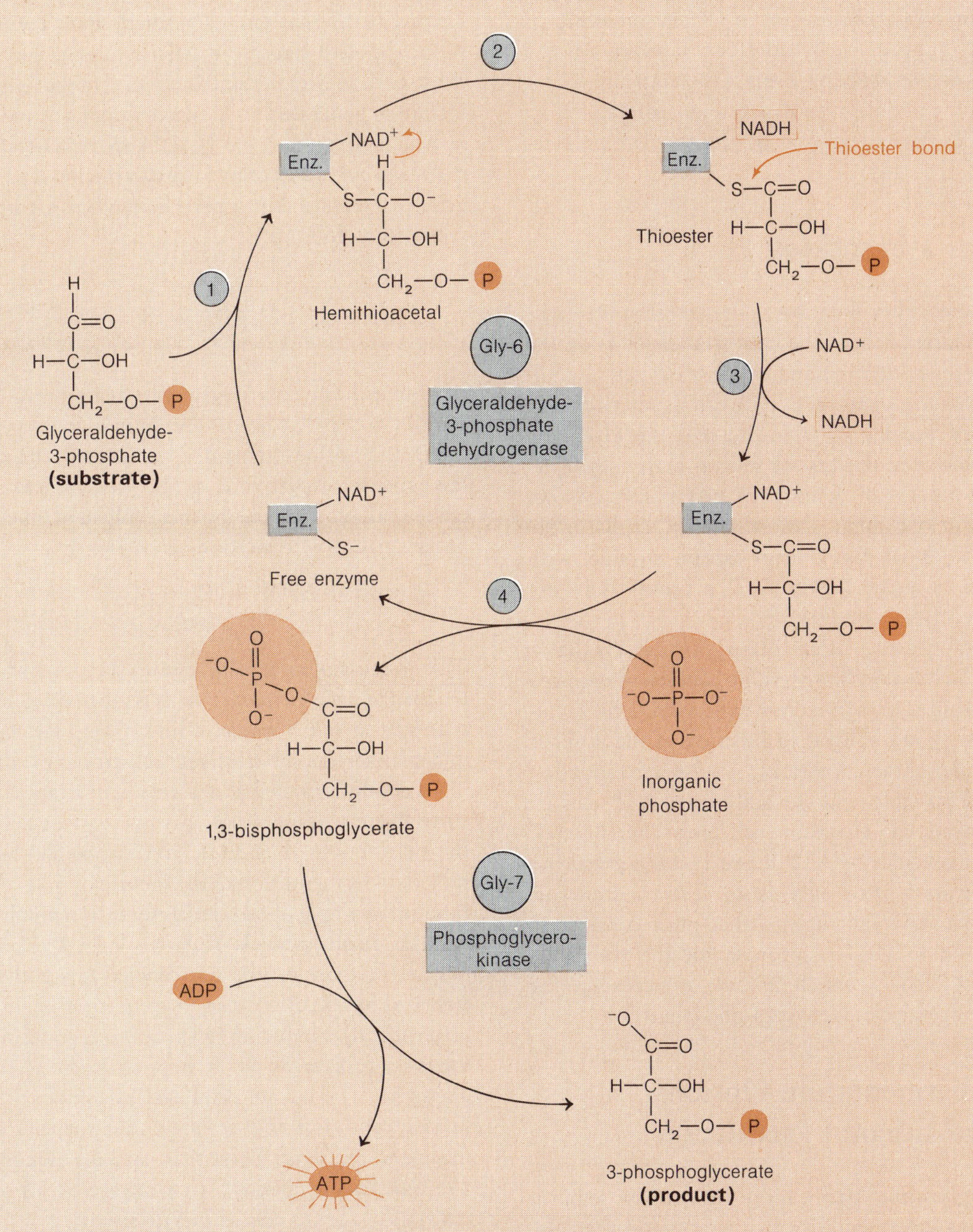

Figure 10-A Detailed Mechanisms for Reactions Gly-6 and Gly-7 of the Glycolytic Pathway. Glyceraldehyde-3-phosphate is oxidized to 3-phosphoglycerate by a two-enzyme sequence, with 1,3-bisphosphoglycerate as an intermediate. ATP is generated by substrate-level phosphorylation, for which the sequence is an excellent prototype.

following overall expression for the glycolytic sequence from glucose to pyruvate:

$$\text{glucose} + 2\text{NAD}^+ + 2\text{ADP} + 2\text{P}_i \xrightarrow{\text{Gly-1 through Gly-10}} 2\ \text{pyruvate} + 2\text{NADH} + 2\text{H}^+ + 2\text{ATP} + 2\text{H}_2\text{O} \quad (10\text{-}12)$$

The glycolytic pathway is one of the most universal metabolic pathways known, since virtually all cells possess the ability to convert glucose to pyruvate, with some of the energy of the oxidative event in the pathway trapped in the form of two molecules of ATP per molecule of glucose. What happens next, however, depends on the availability of oxygen, since pyruvate is the branching point at which aerobic catabolism takes a different direction than is possible under the anaerobic conditions we are presuming at present.

Pyruvate as a Branching Point

Pyruvate occupies a key position at the crossroads of several metabolic alternatives. The next step depends on the particular organism involved and whether oxygen is available. In the presence of oxygen, pyruvate is channeled in the direction of aerobic metabolism, and the glycolytic pathway becomes just the first of several major segments of the overall process of respiratory metabolism. As we will see in the next chapter, this eventually results in the complete oxidation of pyruvate to carbon dioxide, with the generation of much higher ATP amounts than are possible by glycolysis alone.

An important feature of glycolysis, however, is that it can also take place under the anaerobic conditions required by strict anaerobes, or sometimes experienced by facultative cells if they are obliged to function in the absence of oxygen. Under these conditions, no further oxidation of pyruvate is possible and no additional ATP can be generated. Instead, the anaerobic cell must content itself with the two molecules of ATP formed in the glycolytic pathway. Pyruvate is used only as an acceptor molecule for the electrons (hydrogen atoms) that must continually be removed from NADH to provide for the continued regeneration of NAD^+, which in turn is necessary for oxidation of more glyceraldehyde-3-phosphate.

Common Fermentation Options: Lactate and Ethanol Production

Although the glycolytic pathway as we usually define it ends with pyruvate, fermentative processes cannot, because of the net consumption of NAD^+ that occurs as glucose is converted to pyruvate (reaction 10-12). Regardless of the particular end products, then, fermentation must always terminate in some sort of reductive reaction sequence capable of regenerating NAD^+ from NADH by transfer of electrons to an organic molecule. Because of its carbonyl group, pyruvate is itself a potential electron acceptor. It is therefore not surprising that the two most common fermentation options both use pyruvate for this purpose, reducing it either to lactate or to ethanol, from which the fermentation processes derive their names (see Figure 10-7).

Lactate Fermentation. The single most common fermentation process has as its characteristic end product the three-carbon organic acid lactate. As Figure 10-7 indicates, lactate is generated in a simple, one-step reduction event in which electrons from NADH are transferred directly to the carbonyl group of pyruvate, reducing it to the hydroxyl group of lactate. On a per-glucose basis, this can be represented as follows:

$$2\ \text{pyruvate} + 2\text{NADH} + 2\text{H}^+ \xrightarrow{\text{Lactate dehydrogenase}} 2\ \text{lactate} + 2\text{NAD}^+ \quad (10\text{-}13)$$

Since the glycolytic sequence (reaction 10-12) generates NADH and pyruvate on an equimolar basis, both of the NADH molecules generated per glucose are reoxidized at the expense of pyruvate, as shown in reaction 10-13, and the overall reaction for the metabolism of glucose to lactate under anaerobic conditions is as follows:

$$\text{glucose} + 2\text{ADP} + 2\text{P}_i \longrightarrow 2\ \text{lactate} + 2\text{ATP} + 2\text{H}_2\text{O} \quad (10\text{-}14)$$

This process of anaerobic glycolysis terminating in lactate is called *lactate fermentation*. It is the major energy-yielding pathway in many bacteria and in animal cells operating under anaerobic or relatively anaerobic conditions.

Lactate fermentation is very important commercially, since bacteria capable of this process are responsible for the production of cheeses, yogurts, and other foods obtained by fermentation of the lactose of milk. An especially familiar example of lactate fermentation involves your own muscles during periods of particularly strenuous exertion. Whenever muscle cells use oxygen faster than it can be supplied by the circulatory system, they begin functioning anaerobically, reducing pyruvate to lactate instead of oxidizing it further, as would happen if oxygen supplies were adequate. The lactate produced by temporarily anaerobic muscle cells is carried by the circulatory system from the muscle to the liver. There it is used for **gluconeogenesis**—the synthesis of glucose from lactate by a process that is essentially the reverse of glycolysis.

Alcoholic Fermentation. An alternative fermentation process that also involves both NADH and pyruvate, but with different end products, is *alcoholic fermentation*, which is carried out by yeast. In this case, the pyruvate is first decarboxylated to a two-carbon compound, acetaldehyde, which then serves as the electron acceptor. Acetaldehyde reduction gives rise to ethanol, the alcohol from which the process derives its name. This reductive sequence actually involves two separate events (pyruvate decarboxylation and subsequent acetaldehyde reduction) catalyzed by two separate enzymes. The overall reaction can be summarized as follows:

$$\text{2 pyruvate} + \text{2NADH} + \text{2H}^+ \xrightarrow[\text{acetaldehyde}]{\text{Via}} \text{2 ethanol} + 2CO_2 + \text{2NAD}^+ \quad (10\text{-}15)$$

By adding this reductive step to the overall equation for glycolysis (reaction 10-12), we arrive at the following summary equation for alcoholic fermentation:

$$\text{glucose} + \text{2ADP} + 2P_i \longrightarrow \text{2 ethanol} + 2CO_2 + \text{2ATP} + 2H_2O \quad (10\text{-}16)$$

Alcoholic fermentation also has considerable economic significance, since the fermentation carried out by yeast is a key process in baking, brewing, and wine-making industries. For the baker, the carbon dioxide is the important end product. The yeast cells that are added to bread dough function anaerobically, generating both carbon dioxide and ethanol. The carbon dioxide becomes entrapped within the mass of dough, causing it to rise, and the alcohol is then driven off harmlessly during the subsequent baking process. (It is the ethanol that gives baking bread its characteristic pleasant odor.)

For the brewer, both carbon dioxide and ethanol are essential, since the ethanol is what makes beer an alcoholic beverage and the carbon dioxide is what makes it carbonated. (In modern brewing, carbon dioxide is added artificially.) In the making of wine, on the other hand, interest focuses specifically on the ethanol, and the carbon dioxide that is also produced simply helps to produce an anaerobic environment favorable to fermentation by the yeast. (Amateur wine-makers should note the importance of maintaining anaerobic conditions, since certain contaminating bacteria can make acetic acid if oxygen becomes available, and the result will be a keg of vinegar instead of wine!)

Other Fermentation Options. Although lactate and ethanol are the fermentation products of greatest economic significance, they by no means exhaust the microbial options with respect to fermentation. *Propionic acid bacteria*, for example, convert pyruvate reductively to propionate (CH_3—CH_2—COO^-), an important reaction in the production of Swiss cheese. Many bacteria that cause food spoilage do so by *butylene glycol fermentation*. Other fermentation processes yield butyrate (the cause of rancid butter), acetone, and isopropanol. All these reactions, however, are just metabolic variations on the common theme of reoxidizing NADH by transfer of electrons to an organic acceptor, which is in essence the definition of fermentation.

The Energetics of Fermentation

An essential feature of every fermentative process is that no external electron acceptor is involved and no net oxidation occurs. This is seen most clearly in the summary equations for lactate and alcoholic fermentation (reactions 10-14 and 10-16). Although NAD^+ serves as an electron acceptor within the pathway, it is regenerated stoichiometrically and therefore does not appear in the overall reaction. Because fermentation involves no net oxidation, it is a relatively low-energy process. In the case of lactate fermentation, for example, the two lactate molecules produced from every glucose molecule account for most of the 686 kcal of free energy present per mole of glucose, since the complete aerobic oxidation of lactate has a $\Delta G^{\circ\prime}$ of -319.5 kcal/mol. This means that about 93% ($2 \times 319.5/686 \times 100\%$) of the original free energy content of glucose is still present in the product molecules and that anaerobic fermentation is therefore able to tap only about 7% (47 kcal/mol) of the free energy potentially available from glucose.

Given that fermentation involves no net oxidation, you may wonder where even this relatively modest amount of free energy comes from. The oxidation of an aldehyde to an acid characteristically releases more free energy than is required to reduce a ketone to an alcohol. In terms of lactate fermentation, this means that the amount of free energy liberated upon oxidation of glyceraldehyde-3-phosphate (reaction Gly-6) is greater than the amount required to reduce pyruvate to lactate (reaction 10-13), and this difference in free energy is the driving force for ATP production during fermentation.

Put another way, the NAD^+/NADH system occupies a critical intermediate position in terms of the energetics of oxidation-reduction reactions, just as the ATP/ADP system does for phosphate transfer reactions. This allows NAD^+ to accept electrons exergonically from glyceraldehyde-3-phosphate in reaction Gly-6 while still ensuring that the subsequent transfer of electrons from NADH to pyruvate in reaction 10-13 will also be exergonic. The $\Delta G^{\circ\prime}$ values for the two reactions are, in fact, -10.3 kcal/mol for glyceraldehyde-3-phosphate oxidation by NAD^+ and -6.0 kcal/mol for pyruvate reduction by NADH. The $\Delta G^{\circ\prime}$ value for the internal oxidation-reduction process as it occurs in lactate fermentation is therefore -16.3 kcal/mol:

$$\text{glyceraldehyde-3-phosphate} + \text{pyruvate} \longrightarrow \text{3-phosphoglycerate} + \text{lactate} \quad (10\text{-}17)$$

Since the above reaction would have to be multiplied

by 2 to put it on a per-glucose basis, this coupled process of internal oxidation-reduction clearly accounts for 32.6 (2 × 16.3) kcal/mol of glucose, or about 70% of the total free energy change (47 kcal/mol) that takes place in lactate fermentation.

Especially worthy of note is the efficiency with which the anaerobic cell manages to conserve much of that 47 kcal/mol as ATP. The ATP yield, of course, is two molecules of ATP per molecule of glucose. Based on standard free energy changes (admittedly somewhat arbitrary), these two molecules of ATP represent 2 × 7.3 = 14.6 kcal/mol, which represents an efficiency of energy conservation of about 31% (14.6/47 × 100%). This compares favorably with human-made machines, which seldom exceed 25% efficiency in energy conservation. If anything, this underestimates the efficiency of fermentation, because actual ΔG values for ATP hydrolysis under cellular conditions are usually substantially more negative than −7.3 kcal/mol. If we assume a representative ΔG value of about −12.5 kcal/mol for ATP hydrolysis, for example, the efficiency of energy conservation can easily exceed 50%.

Alternative Substrates for Glycolysis

Thus far, we have assumed glucose to be the starting point for glycolysis and therefore, by implication, for all of cellular energy metabolism, both aerobic and anaerobic. While it is true that glucose represents a significant substrate for both fermentation and respiration in a variety of organisms and tissues, it is by no means the only such substrate and is not very important at all in some tissues and under some circumstances. So it is useful to ask what some of the major alternatives to glucose are and how they are handled by cells.

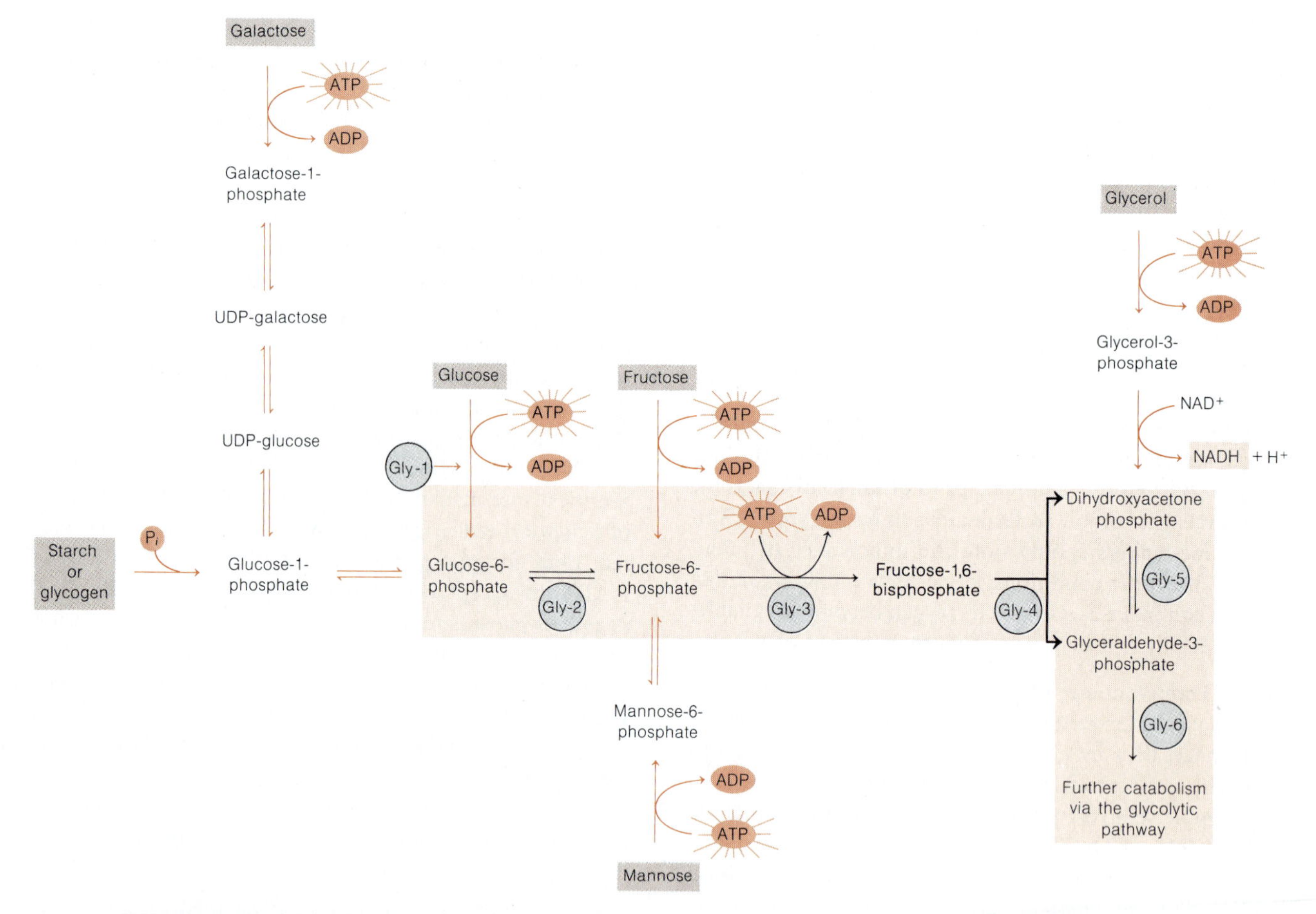

Figure 10-10 Carbohydrate Catabolism by the Glycolytic Pathway. A variety of carbohydrate substrates can be metabolized by the common mode of conversion into an intermediate in the glycolytic pathway. (The conversion reactions are shown by colored arrows, and the glycolytic pathway is highlighted by light color.)

One principle quickly emerges: Regardless of the chemical nature of the alternative substrates, they are converted as quickly as possible into a form that can be channeled into the mainstream glycolytic pathway already familiar to us (Figure 10-10). To emphasize this point, we will briefly consider two classes of alternative substrates: other simple sugars and storage carbohydrates.

Catabolism of Other Sugars

A variety of sugars other than glucose are available to cells, either by ingestion or upon degradation of storage carbohydrates. Most of these are either hexoses or pentoses, with the former predominant. Ordinary table sugar (sucrose), for example, consists of the hexoses glucose and fructose, and milk sugar (lactose) contains glucose and galactose. Mannose is also a hexose with which cells must frequently contend. In general, each of these has a specific reaction sequence that brings it as quickly as possible into the glycolytic sequence. Glucose and fructose enter most directly, since they require only phosphorylation on carbon 6, catalyzed by the hexokinase enzyme of Gly-1. Hexokinase can also phosphorylate mannose, and the resulting mannose-6-phosphate undergoes conversion (by phosphomannoisomerase) into fructose-6-phosphate, a glycolytic intermediate. Entry of galactose requires a somewhat more complex reaction sequence, the details of which are explored in Problem 7.

Phosphorylated pentoses can also be channeled into the glycolytic pathway, but only after being converted into hexose phosphates by a reaction sequence (the *phosphogluconate pathway*) that is beyond the scope of our present discussion. Just remember that the typical cell has the metabolic capabilities to convert almost any naturally occurring sugar (and a variety of other compounds as well) into one or another of the glycolytic intermediates for further catabolism under anaerobic or aerobic conditions.

Catabolism of Storage Polysaccharides

Although glucose is the immediate substrate for both fermentation and respiration in many cells and tissues, it is not present in the cell to any large extent as the free monosaccharide, but occurs instead in the form of storage polysaccharides, most commonly starch in plants and glycogen in animals. As indicated in Figure 10-10, these storage polysaccharides are mobilized by stepwise cleavage of successive glucose units in a reaction that requires inorganic phosphate to break the $\alpha(1 \rightarrow 4)$ bond between successive glucose units, liberating the glucose monomers as glucose-1-phosphate. The process is called **phosphorolytic cleavage,** and it is depicted schematically in Figure 10-11. Both glycogen and starch are cleaved in this manner, by the enzymes glycogen phosphorylase and starch phosphorylase, respectively.

The glucose-1-phosphate liberated by phosphorylase activity can be converted into glucose-6-phosphate by the enzyme phosphoglucomutase, and the glucose-6-phosphate

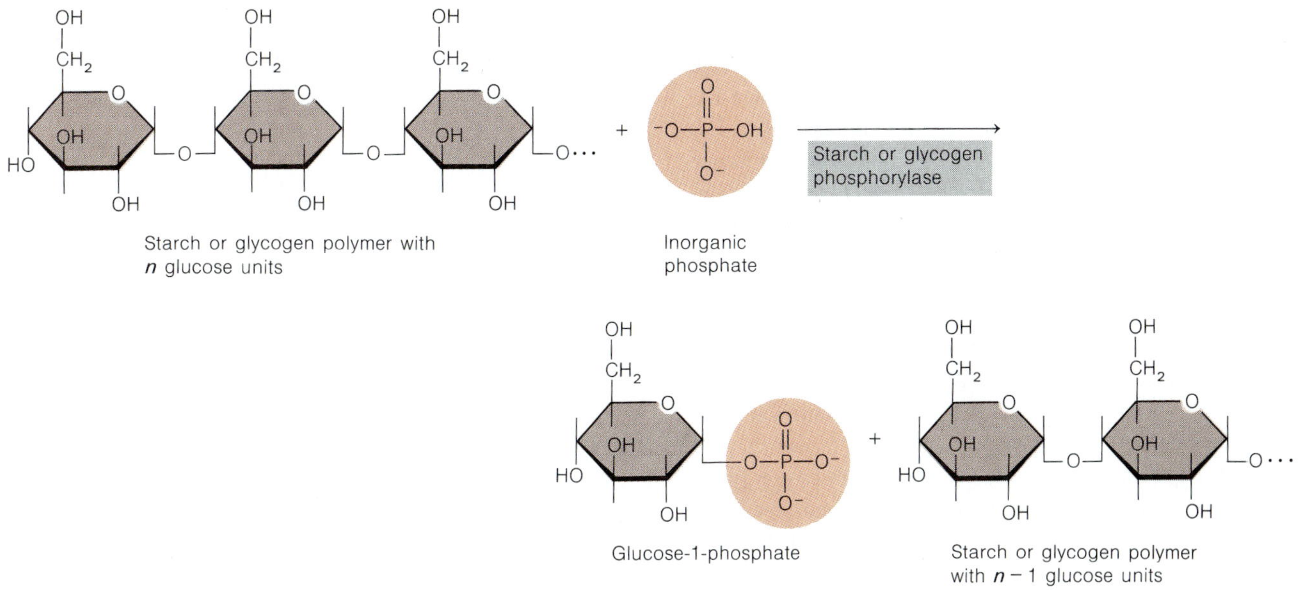

Starch or glycogen polymer with n glucose units + Inorganic phosphate → (Starch or glycogen phosphorylase) → Glucose-1-phosphate + Starch or glycogen polymer with $n-1$ glucose units

Figure 10-11 Phosphorolytic Cleavage of Storage Polysaccharides. Terminal glucose units of storage polysaccharides such as starch or glycogen are liberated by phosphorolytic cleavage.

then enters the glycolytic pathway for further catabolism. Notice, by the way, that glucose stored in polymerized form enters the glycolytic pathway as glucose-6-phosphate without the input of ATP that would be required for initial phosphorylation of the free sugar. Consequently, the overall energy yield for glucose is greater by one molecule of ATP when it is catabolized from the polysaccharide level than when it is catabolized with the free sugar as the starting substrate. Storage polysaccharides therefore represent a means of storing glucose units in a higher-energy form than is possible with the free monomer.

Regulation of Glycolysis

Since the purpose of the glycolytic pathway under either anaerobic or aerobic conditions is to generate ATP, it is crucial that the functioning of the pathway be carefully and continuously adjusted to meet actual cellular needs for ATP. This is accomplished in two major ways: (1) control of the rate at which glucose is converted into pyruvate and (2) regulation of the extent to which glycogen or starch is mobilized to release glucose.

Allosteric Regulation of Phosphofructokinase

Like every other metabolic pathway, the glycolytic sequence best serves the cell not by operating at its maximum rate, but rather by functioning at a rate that is carefully regulated to keep pace with the energy needs of the cell. This control ensures that sugar molecules will enter the pathway at a rapid rate when the ATP supply is low, but at a much reduced rate when ATP is plentiful. The major control point in the glycolytic pathway is at reaction Gly-3, catalyzed by the enzyme phosphofructokinase. This enzyme is allosterically inhibited by ATP and activated by ADP and AMP, rendering it exquisitely sensitive to the energy charge of the cell.

The choice of the phosphofructokinase reaction as the site of feedback inhibition seems at first glance to violate the general principle that such **allosteric regulation** is characteristic of the first enzyme in a reaction sequence. The principle holds, however, because glucose, glucose-6-phosphate, and fructose-6-phosphate are intermediates common to many reaction sequences in the cell, and the steps labeled Gly-1 and Gly-2 are in reality general reactions involved in a variety of hexose interconversions. The **phosphofructokinase** reaction, by contrast, is the specific step in glycolysis at which sugars become irreversibly committed to degradation; therefore, it fits the criterion for feedback inhibition and allosteric regulation.

It may also seem surprising that the main inhibitor of phosphofructokinase is ATP rather than pyruvate, lactate, or some other compound that might be identified as the end product of the pathway, in keeping with the principle of end-product inhibition shown in Figure 6-18. But in a vital sense, the real end product of glycolysis is not the organic molecule with which the sequence happens to terminate but rather the ATP that the sequence generates. It is not so strange, therefore, that ATP inhibits phosphofructokinase allosterically.

Another surprising feature of this regulatory mechanism is that ATP is apparently an allosteric inhibitor of an enzyme for which ATP is itself a required substrate (for phosphorylation of fructose-6-phosphate). This seems to create a contradiction of effects, since increasing levels of substrate should increase the rate of an enzyme-catalyzed reaction, yet increasing levels of an allosteric inhibitor should render the enzyme less active. This contradiction is resolved for phosphofructokinase because the active (or catalytic) site and the effector (or allosteric) site of the enzyme differ markedly in their affinities for ATP. The active site has a high affinity (low K_m) for ATP, while the affinity of the effector site for ATP is lower. Thus, at low ATP levels, binding occurs at the catalytic site but not at the allosteric site, so the enzyme remains in its active form and is functional. At high ATP levels, however, binding is promoted at the effector site, converting the enzyme to its inactive form and thereby serving as a throttle for the whole glycolytic sequence.

The effect of ATP level on the kinetics of the enzyme is shown in Figure 10-12, which depicts the effect of fructose-6-phosphate concentration on initial reaction velocity. At low ATP concentrations, the dependence of velocity on

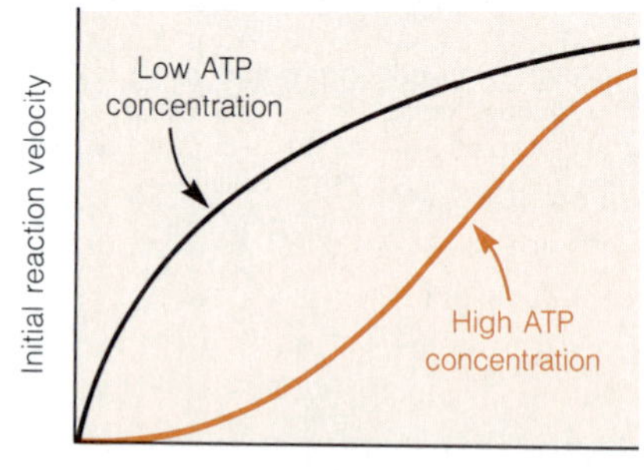

Figure 10-12 Kinetics of Allosteric Inhibition of Phosphofructokinase by ATP. At low ATP concentrations, phosphofructokinase follows classical Michaelis-Menten kinetics in the dependence of its reaction velocity on the concentration of the substrate fructose-6-phosphate (black line). At high ATP concentrations, the curve is sigmoidal (colored line), a diagnostic of allosteric regulation.

substrate concentration follows classical Michaelis-Menten kinetics; but at high ATP concentrations, the curve is sigmoidal, which enzymologists recognize as a distinguishing characteristic of an allosterically regulated enzyme.

In addition to its sensitivity to ATP, ADP, and AMP, phosphofructokinase is also allosterically inhibited by citrate and by fatty acids. The sensitivity to citrate provides a crucial regulatory link between glycolysis and the tricarboxylic acid cycle of which citrate is an intermediate, as we will see in Chapter 11. This ensures that the two stages of respiration are carefully regulated with respect to each other. When the level of citrate rises due to excessive glycolytic activity, the citrate binds to its effector site on phosphofructokinase, thereby converting the enzyme reversibly to an inactive form and ensuring that the glycolytic pathway is adjusted downward to actual cellular needs. Similarly, when fatty acids are available for use in ATP generation by the β-oxidation pathway (see Chapter 11), their inhibitory effect on phosphofructokinase activity causes the glycolytic pathway to be shut down, thereby facilitating the use of fatty acids rather than carbohydrates as an energy source.

Perspective

Metabolic pathways in cells are either anabolic (synthetic) or catabolic (degradative). The cell carries out the latter reactions to derive the ATP necessary to drive the former. ATP is a useful intermediate for this purpose because its terminal anhydride bond has a free energy of hydrolysis that allows it to serve as both a donor and an acceptor of phosphate groups. Chemotrophs derive the energy needed for ATP generation from the bond energies of organic molecules. They do this either anaerobically (by fermentative processes) or aerobically (by respiratory metabolism). Using glucose as a prototype substrate, catabolism under both anaerobic and aerobic conditions begins with the glycolytic pathway, a ten-step sequence that converts glucose to pyruvate with the net production of two molecules of ATP per molecule of glucose.

In the absence of oxygen, the reduced coenzyme NADH generated during glycolysis must be reoxidized at the expense of pyruvate, leading to end products such as lactate or ethanol and carbon dioxide. This severely limits the extent to which the free energy content of the glucose molecule can be tapped, but the 7% or so that is available is actually conserved as ATP with great efficiency. Although usually written with glucose as the starting substrate, the glycolytic sequence is also the mainstream pathway for the catabolism of a variety of related sugars, as well as for the utilization of the glucose-1-phosphate derived from storage polysaccharides such as starch or glycogen by phosphorolytic cleavage.

Like other metabolic pathways, glycolysis is carefully regulated to ensure that the rate of ATP generation for which it is responsible is carefully tuned to actual cellular needs. A key regulatory step is the phosphofructokinase reaction, which is subject to allosteric inhibition by ATP. As we will see in Chapter 21, extracellular signals also regulate glycolysis through hormonal control pathways involving phosphorylase activity and cyclic AMP.

As complex as glycolysis may seem upon first encounter, it represents the simplest possible way in which glucose can be degraded in dilute solution at temperatures compatible with life and with a large portion of the free energy yield conserved as ATP. Coupled to an appropriate reductive sequence to regenerate the coenzyme NAD^+, glycolysis serves the cell well under anaerobic conditions, meeting energy needs despite the absence of oxygen.

All we have seen so far, however, pales in comparison with the potential for energy release and conservation in the presence of oxygen, for aerobic respiration is the capstone of bioenergetics and the mainspring of cellular energy metabolism for most chemotrophic forms of life. And for that, we take the NADH and the pyruvate that glycolysis gives us and proceed not to fermentative options, but to the aerobic world of Chapter 11.

Key Terms for Self-Testing

Metabolic Pathways
metabolism (p. 252)
metabolic pathway (p. 252)
anabolic pathway (p. 252)
catabolic pathway (p. 252)

ATP: The Universal Energy Coupler
adenosine triphosphate (ATP) (p. 253)
phosphoanhydride bond (p. 253)
phosphoester bond (p. 253)
charge repulsion (p. 253)
resonance stabilization (p. 253)
contributing structures (p. 254)
resonance hybrid (p. 254)
group transfer reaction (p. 255)
energy charge (p. 255)

Chemotrophic Energy Metabolism
chemotrophic energy metabolism (p. 256)
oxidation (p. 256)
dehydrogenation (p. 257)
dehydrogenase (p. 257)
reduction (p. 257)
hydrogenation (p. 257)
coenzyme (p. 257)
nicotinamide adenine dinucleotide (NAD^+) (p. 257)
flavin adenine dinucleotide (FAD) (p. 257)
glucose (p. 257)
respiration (p. 259)
fermentation (p. 259)
lactate fermentation (p. 259)
alcoholic fermentation (p. 259)
strict (obligate) aerobe (p. 259)
strict (obligate) anaerobe (p. 259)
facultative organism (p. 259)

Fermentation: The Anaerobic Option for ATP Generation
glycolysis (p. 260)
glycolytic pathway (p. 260)
substrate-level phosphorylation (p. 262)
oxidative phosphorylation (p. 262)
gluconeogenesis (p. 266)

Alternative Substrates for Glycolysis
phosphorolytic cleavage (p. 269)

Regulation of Glycolysis
allosteric regulation (p. 270)
phosphofructokinase (p. 270)

Suggested Reading

General References and Reviews

Mathews, C. K., and K. E. van Holde. *Biochemistry.* Redwood City, Calif.: Benjamin/Cummings, 1990.

McGilvery, R. W. *Biochemistry: A Functional Approach,* 3d ed. Philadelphia: Saunders, 1983.

Stryer, L. *Biochemistry,* 3d ed. San Francisco: W. H. Freeman, 1988.

Wood, W. B., J. H. Wilson, R. M. Benbow, and L. E. Hood. *Biochemistry: A Problems Approach,* 2d ed. Menlo Park, Calif.: Benjamin/Cummings, 1981.

ATP Generation

Bridger, W. A., and J. F. Henderson. *Cell ATP.* New York: Wiley, 1983.

Kalcker, H. M., ed. *Biological Phosphorylations.* Englewood Cliffs, N.J.: Prentice-Hall, 1969.

Glycolysis and Fermentation

Bagley, S., and D. E. Nicolson. *Metabolic Pathways.* New York: Wiley, 1970.

Fothergill-Gilmore, L. A. The evolution of the glycolytic pathway. *Trends Biochem. Sci.* 11 (1986): 47.

Regulation of Glycolysis

Atkinson, D. E. *Cellular Energy Metabolism and Its Regulation.* New York: Academic Press, 1977.

Erecinska, M., and D. F. Wilson. Regulation of cellular energy metabolism. *J. Membr. Biol.* 70 (1982): 1.

Herman, R. H., R. M. Cohn, and P. D. McNamara, eds. *Principles of Metabolic Control in Mammalian Systems.* New York: Plenum Press, 1980.

Hofmann, E. The significance of phosphofructokinase to the regulation of carbohydrate metabolism. *Rev. Physiol. Biochem. Pharmacol.* 75 (1976): 1.

Ottaway, J. H., and S. Mowbray. The role of compartmentation in the control of glycolysis. *Current Topics Cell Regul.* 12 (1977): 108.

Problem Set

1. **History of Glycolysis.** Following are several observations that led to the elucidation of the glycolytic pathway. In each case, suggest a metabolic basis for the observed effect and explain the significance of the observation for the elucidation of the pathway.
   (a) Alcoholic fermentation in yeast extracts requires a heat-labile fraction originally called *zymase* and a heat-stable fraction (*cozymase*) that is necessary for activity of zymase.
   (b) Alcoholic fermentation does not take place in the absence of inorganic phosphate.
   (c) In the presence of iodoacetate, fermenting yeast extracts accumulate a doubly phosphorylated hexose.
   (d) In the presence of fluoride ion, fermenting yeast extracts accumulate two phosphorylated three-carbon acids.

2. **Alcoholic Fermentation.** Identify as true (T) or false (F) each of the following statements concerning a yeast culture carrying out alcoholic fermentation of glucose labeled in carbon atom 6 with ^{14}C.
   (a) Little if any of the CO_2 that is produced will be radioactive.
   (b) The ethanol that is produced will be radioactively labeled primarily in the carbon atom that carries the hydroxyl group.
   (c) The distribution of radioactivity would be the same if the label were in carbon atom 1 of the glucose.
   (d) None of the dihydroxyacetone phosphate molecules in the cell will be labeled with ^{14}C.

3. **Energetics of Carbohydrate Utilization.** The anaerobic fermentation of free glucose has an ATP yield of 2 molecules of ATP per molecule of glucose. For glucose units in a glycogen molecule, the yield is 3 molecules of ATP per molecule of glucose. The corresponding value for the disaccharide sucrose is $2\frac{1}{2}$ molecules of ATP per molecule of monosaccharide.
   (a) Explain why the glucose units present in glycogen have a higher ATP yield.
   (b) Based on what you know about the process of glycogen breakdown, suggest a mechanism for sucrose metabolism consistent with an energy yield of $2\frac{1}{2}$ molecules of ATP per molecule of monosaccharide.
   (c) What energy yield (in molecules of ATP per molecule of monosaccharide) would you predict for raffinose, a trisaccharide?

4. **Glucose Phosphorylation.** The direct phosphorylation of glucose by inorganic phosphate is a thermodynamically unfavorable reaction:

$$\text{glucose} + P_i \longrightarrow \text{glucose-6-phosphate} + H_2O \qquad (10\text{-}18)$$
$$\Delta G^{\circ\prime} = +3.3 \text{ kcal/mol}$$

In the cell, this is accomplished by coupling the reaction to the hydrolysis of ATP, a highly exergonic reaction:

$$ATP + H_2O \longrightarrow ADP + P_i \qquad (10\text{-}19)$$
$$\Delta G^{\circ\prime} = -7.3 \text{ kcal/mol}$$

Typical concentrations of these intermediates in human red blood cells are as follows: [glucose-6-phosphate]: 0.08 m*M*; [ATP]: 1.8 m*M*; [ADP]: 0.15 m*M*; [P_i]: 1.0 m*M*.
   (a) What minimum concentration of glucose would have to be maintained in the red blood cell for direct phosphorylation (reaction 10-18) to be thermodynamically spontaneous? Is this physiologically reasonable?
   (b) What is the overall equation for the coupled (ATP-driven) phosphorylation reaction? What is its $\Delta G^{\circ\prime}$ value?
   (c) What minimum concentration of glucose would have to be maintained in the red blood cell for the coupled reaction to be thermodynamically spontaneous? Is this physiologically reasonable?
   (d) By about how many orders of magnitude is the minimum required glucose concentration changed by coupling the phosphorylation of glucose to the hydrolysis of ATP?
   (e) Assuming the red blood cell to have a glucose concentration of 5.0 m*M*, what is $\Delta G'$ for the coupled phosphorylation reaction?

5. **Balanced Fermentation.** We speak of a fermentation process as "balanced" if there is no net oxidation or reduction of the substrate (that is, if there is no net involvement of a coenzyme in the overall reaction). Fermentation of glucose to lactate is clearly a balanced process, since no net oxidation or reduction of coenzyme occurs, even though NAD^+ is involved in the process.
   (a) How can lactate fermentation be a balanced process when one of the reactions (Gly-6) involves the oxidation of an aldehyde to an acid, with the net reduction of NAD^+ to NADH?
   (b) Although no net oxidation or reduction of a coenzyme occurs in the pathway, the lactate molecule is clear evidence that an intramolecular transfer of electrons has occurred. Explain.
   (c) Write a reaction that summarizes the glycolytic pathway as a balanced oxidation-reduction process involving two three-carbon substrates, one serving as the electron donor and the other as the electron acceptor.

6. **Propionate Fermentation.** Although lactate and ethanol are the best-known products of fermentation, other options are known, some with important commercial application. Swiss cheese production, for example, depends on the bacterium *Propionibacterium freudenreichii,* which converts pyruvate to propionate ($CH_3—CH_2—COO^-$). Fermentation of glucose to propionate always generates at least one other product as well.
   (a) Why is it not possible to devise a scheme for the balanced fermentation of glucose with propionate as the sole end product?
   (b) Suggest an overall scheme for propionate production that generates only one additional product, and indicate what that prodnct might be.
   (c) If you know that Swiss cheese production actually requires both propionate and carbon dioxide and that both are produced by *Propionibacterium* fermentation, what else can you now say about the fermentation process that this bacterium carries out?

7. **Galactose Metabolism.** The glycolytic pathway is usually written with glucose as the starting substrate, since this is the single most important sugar for most organisms. But cells can utilize a variety of sugars. The diet of young mammals, for example, consists almost entirely of milk, which contains as its principal carbohydrate the disaccharide lactose. When lactose is hydrolyzed in the intestine, it yields one molecule each of the hexoses glucose and galactose, so the cells of these animals (or of human babies) must metabolize just as much galactose as glucose. Galactose is metabolized by phosphorylation and conversion to glucose. The reaction sequence is a bit complicated, though, because the conversion to glucose (an *epimerization* reaction on carbon atom 4) occurs while the sugar is attached to the carrier *uridine diphosphate* (*UDP*, a close relative of ADP). The reactions are as follows:

$$\text{galactose} + \text{ATP} \longrightarrow \text{galactose-1-phosphate} + \text{ADP} \quad (10\text{-}20)$$

$$\text{galactose-1-phosphate} + \text{UDP-glucose} \longrightarrow \text{glucose-1-phosphate} + \text{UDP-galactose} \quad (10\text{-}21)$$

$$\text{UDP-galactose} \longrightarrow \text{UDP-glucose} \quad (10\text{-}22)$$

$$\text{glucose-1-phosphate} \longrightarrow \text{glucose-6-phosphate} \quad (10\text{-}23)$$

   (a) Write a reaction for the overall conversion of galactose to glucose-6-phosphate.
   (b) How do you think the $\Delta G^{\circ\prime}$ for the overall reaction of part (a) compares with that for the hexokinase reaction (reaction Gly-1)?
   (c) If you know that the epimerase reaction (reaction 10-22) has an absolute requirement for the coenzyme NAD^+ and involves 4-ketoglucose as an enzyme-bound intermediate, can you suggest a logical reaction sequence to explain the conversion of galactose to glucose?
   (d) The congenital disease *galactosemia* is caused by a genetic absence of the enzyme that catalyzes reaction 10-21. The symptoms of galactosemia, which include mental disorders and cataracts of the eye, are thought to result from high levels of galactose in the blood. Why does this seem a reasonable hypothesis?

8. **Arsenate Poisoning.** Arsenate ($HAsO_4^{2-}$) is a potent poison in almost all living systems. Among other effects, arsenate is known to uncouple the phosphorylation event from the oxidation of glyceraldehyde-3-phosphate. This uncoupling occurs because the enzyme involved, glyceraldehyde-3-phosphate dehydrogenase, can utilize arsenate instead of inorganic phosphate, forming glycerate-1-arseno-3-phosphate. This product is a highly unstable compound, which immediately undergoes nonenzymatic hydrolysis into glycerate-3-phosphate and free arsenate.
   (a) In what sense might arsenate be called an uncoupler of substrate-level phosphorylation?
   (b) Why is arsenate such a toxic substance for an organism that depends critically on glycolysis to meet its energy needs?
   (c) Can you think of other reactions that are likely to be uncoupled by arsenate in the same way as the glyceraldehyde-3-phosphate dehydrogenase reaction?

9. **Enzyme Tests in Clinical Medicine.** Lactate dehydrogenase (LDH) catalyzes the reduction of pyruvate to lactate in anaerobic metabolism. LDH activity in the blood is elevated following a heart attack, and the degree of elevation can be used to diagnose the severity of tissue damage. Other soluble enzymes of cells such as creatine phosphokinase and the transaminases show similar patterns. Why might there be a relationship between a heart attack and metabolic enzymes circulating in the blood?

11

Energy from Chemical Bonds: The Aerobic Mode

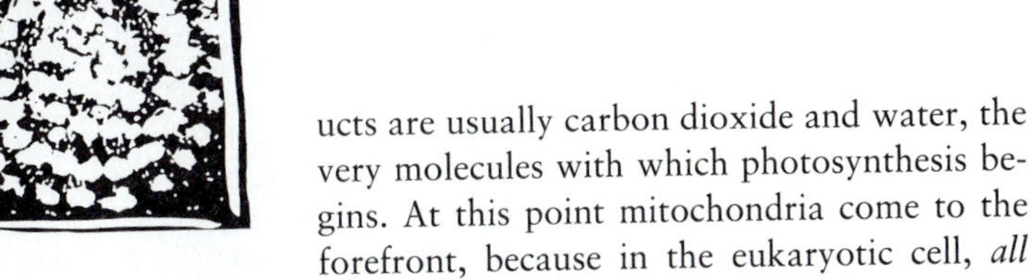

When you look at a eukaryotic cell through the microscope, what catches your eye? The nucleus is most prominent, to be sure—and we will be getting to that in later chapters. But nearly as prominent are the numerous **mitochondria,** shown in the electron micrograph in Figure 11-1. Mitochondria function as the powerhouses of the eukaryotic cell, serving the cell's energy needs under aerobic conditions.

As we saw in the last chapter, some cells meet all of their energy needs by anaerobic fermentation processes, either because they are strict anaerobes and cannot tolerate oxygen or because they are facultative cells functioning temporarily in the absence of oxygen. We also noted, however, how limited the energy yields of fermentation are. In the absence of oxygen, the electrons that are removed from one organic compound (glyceraldehyde-3-phosphate) must eventually be returned to another organic compound (pyruvate), and the difference in energy is such that only 2 moles of ATP can be generated per mole of glucose.

In short, anaerobic metabolism is a feasible way of meeting energy needs, but the ATP yield is low, and the process is energetically wasteful in that the cell has access to only a limited portion of the total free energy stored in the oxidizable molecules it uses as substrate. Moreover, fermentation always results in the accumulation of waste products such as ethanol or lactate that cannot be used as starting materials by the phototrophs, thereby failing to complete the chemotrophic portion of the cyclic flow of energy shown in Figure 5-5.

The Aerobic Mode

All of this changes dramatically when we come to the aerobic process of **respiratory metabolism,** or **respiration** for short. With oxygen available as the final electron acceptor, complete substrate oxidation becomes possible, maximum energy yields can be realized, and the end products are usually carbon dioxide and water, the very molecules with which photosynthesis begins. At this point mitochondria come to the forefront, because in the eukaryotic cell, *all aerobic energy metabolism beyond pyruvate occurs within the inner mitochondrial membrane* (Figure 11-2). As a result, most aerobic ATP production in eukaryotic cells takes place within the mitochondrion. For example, the complete aerobic oxidation of glucose by a eukaryotic cell yields 36 moles of ATP per mole of glucose, and 34 of these (all but the 2 of glycolysis) are generated within mitochondria. For fats, the proportion of ATP produced per mole of fatty acid in the mitochondria is even higher.

Our goal for this chapter is to understand the fate of pyruvate under aerobic conditions and how alternative substrates, such as fats and amino acids, can be catabolized to generate ATP. We will also see how the large amount of free energy that is not available under anaerobic conditions gets released during respiration and how it comes to be conserved as ATP. In addition, we will establish the cellular context for aerobic energy metabolism by looking closely at mitochondrial structure and function and by considering the prokaryotic alternative to mitochondrial compartmentalization.

Respiratory Metabolism: An Overview

The crucial feature that allows a chemotrophic cell to move beyond glycolysis and its fermentative options is the availability of oxygen and the respiratory metabolism it allows. With oxygen as the electron acceptor, the carbon atoms of glucose (or any other oxidizable substrate) can be oxidized fully to carbon dioxide, with all of the electrons that are removed transferred ultimately to oxygen.

The oxygen that makes all of this possible serves only as a terminal electron acceptor, providing a means for the

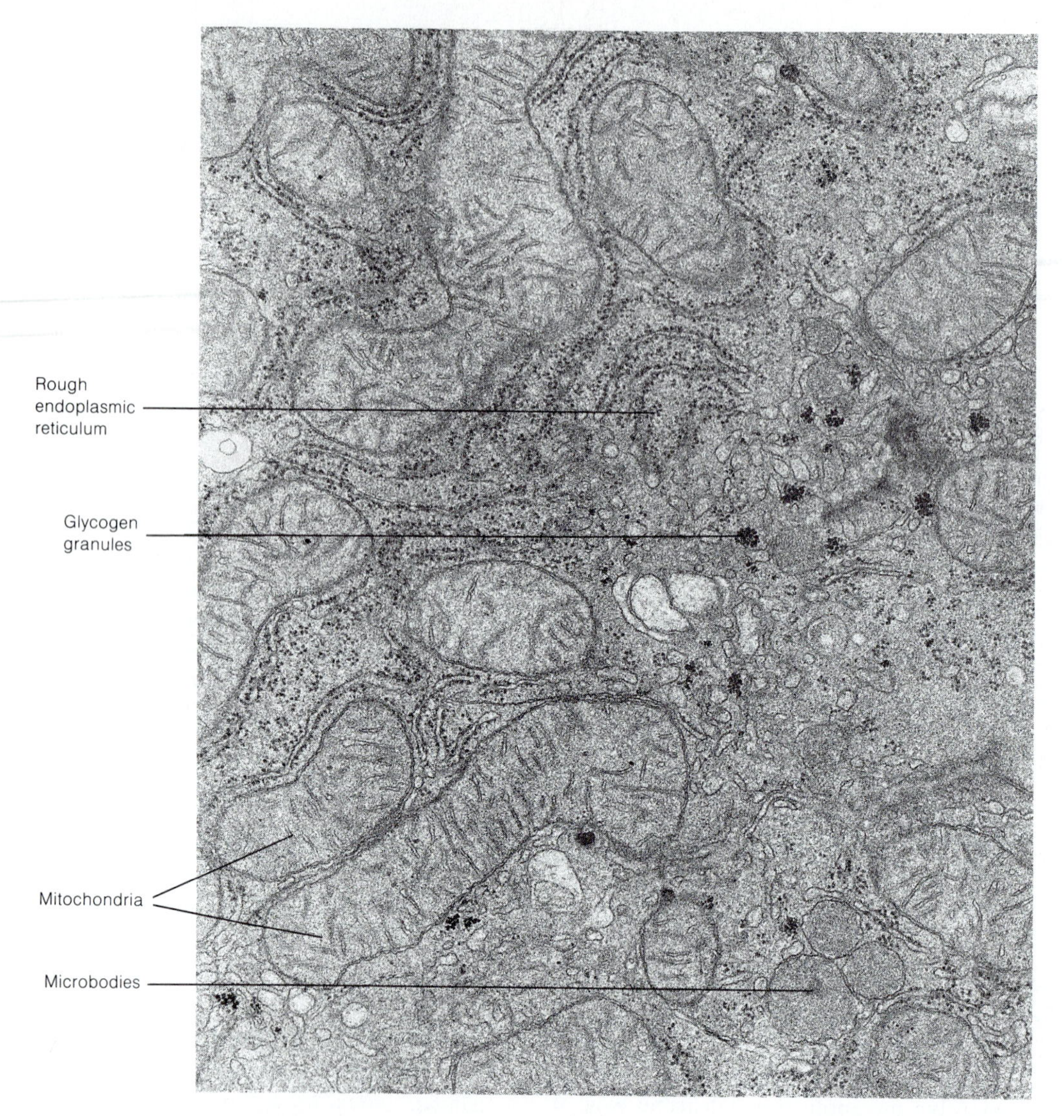

Figure 11-1 The Prominence of Mitochondria. Mitochondria are prominent features in this electron micrograph of a rat liver cell. Other cellular components shown include rough endoplasmic reticulum, glycogen granules, and microbodies. (TEM)

continuous reoxidation of the reduced coenzyme molecules—NADH and $FADH_2$. It is these coenzyme molecules, in their oxidized forms, that accomplish the stepwise oxidation of organic intermediates derived from pyruvate. Respiratory metabolism involves both the oxidative steps in which electrons are removed from organic substrates and transferred to coenzyme carriers and the concomitant processes whereby reduced coenzymes are reoxidized by transfer of electrons to oxygen, accompanied indirectly by the generation of ATP.

Respiratory metabolism can be considered in five stages, two concerned with coenzyme-mediated oxidative events and three involved in coenzyme reoxidation, proton pumping, and ATP production. These stages are shown in Figure 11-3. The initial stage is the glycolytic pathway that we encountered in Chapter 10. Under aerobic conditions, the function of glycolysis is the same—conversion of glucose to pyruvate—but the fate of the pyruvate is different. Instead of being used as an electron acceptor, pyruvate is further oxidized in the *tricarboxylic acid (TCA) cycle,* which becomes the second stage of respiration (Figure 11-3).

The third stage is the *electron transport chain* required to reoxidize coenzyme molecules at the expense of oxygen. Electron transport is coupled to the pumping of protons in the fourth stage, generating an electrochemical proton gradient across the mitochondrial membrane. In the fifth stage, this electrochemical proton gradient is used to drive the *phosphorylation* of ADP to ATP.

We will take a closer look at each of these latter three

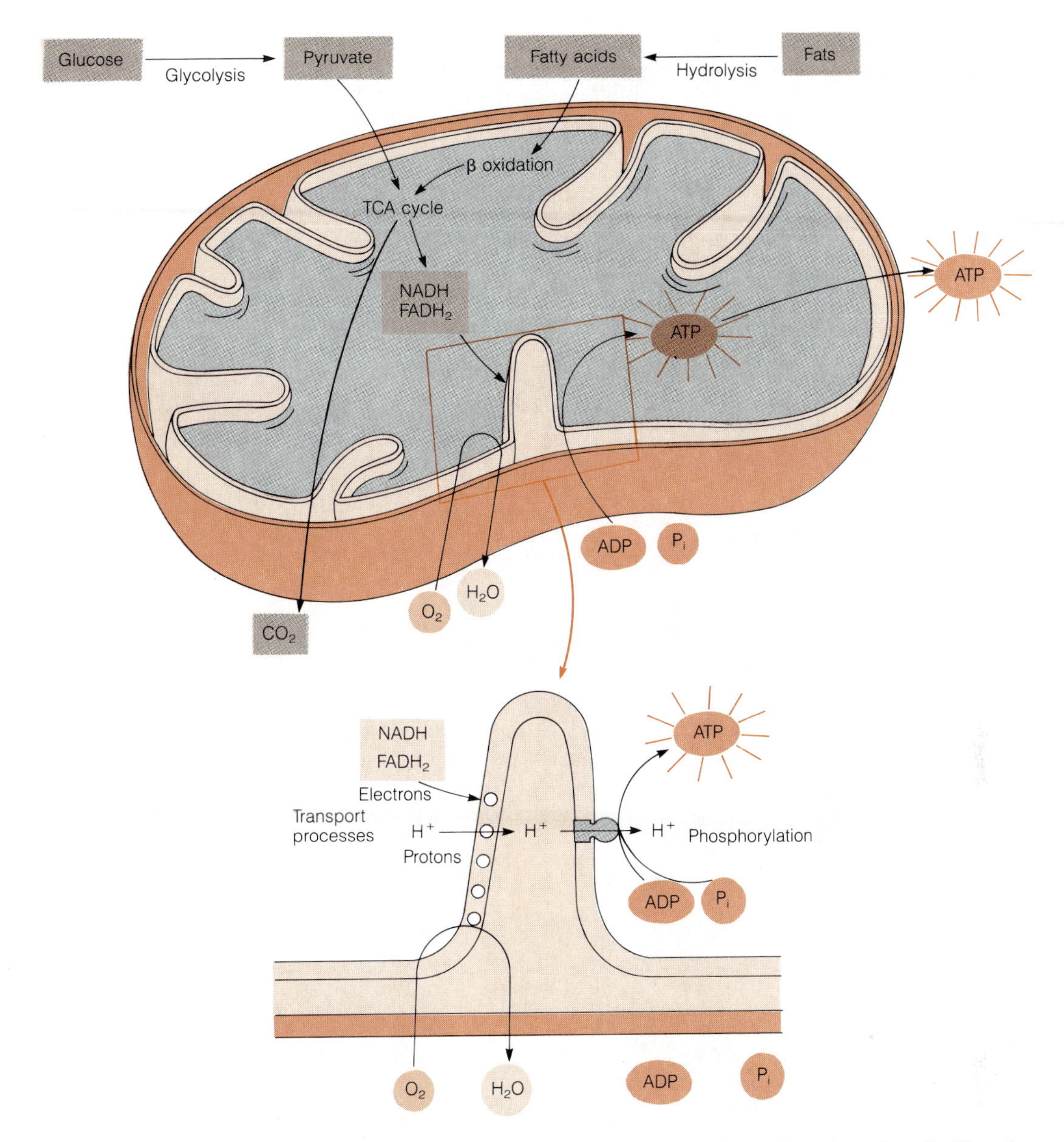

Figure 11-2 Central Role of the Mitochondrion in Aerobic Respiratory Metabolism. Pyruvate from glycolysis and fatty acids from hydrolysis of triglycerides enter the mitochondrion, where all further metabolism is localized in eukaryotic cells. The other major substrates are oxygen, ADP, and inorganic phosphate, and the products are water, carbon dioxide, and ATP.

stages in due course. First, however, we will consider the mitochondrial context in which these processes occur in eukaryotic cells.

Mitochondrial Structure and Function

To understand respiratory metabolism, we must begin with an appreciation of mitochondrial structure, since much of aerobic energy metabolism is localized within this organelle in eukaryotic cells. In fact, except for the ten reactions of the glycolytic pathway, all of the ATP-generating capacity of eukaryotes takes place within the mitochondria. It is no wonder that the mitochondrion is called the "energy powerhouse" of the cell.

Mitochondria have been known and studied for more than a century. As early as 1850, Rudolph Kölliker described the presence of what he called "ordered arrays of particles" in muscle cells. When these particles were isolated, they were found to swell in water, suggesting the presence of a limiting membrane with osmotic activity. A

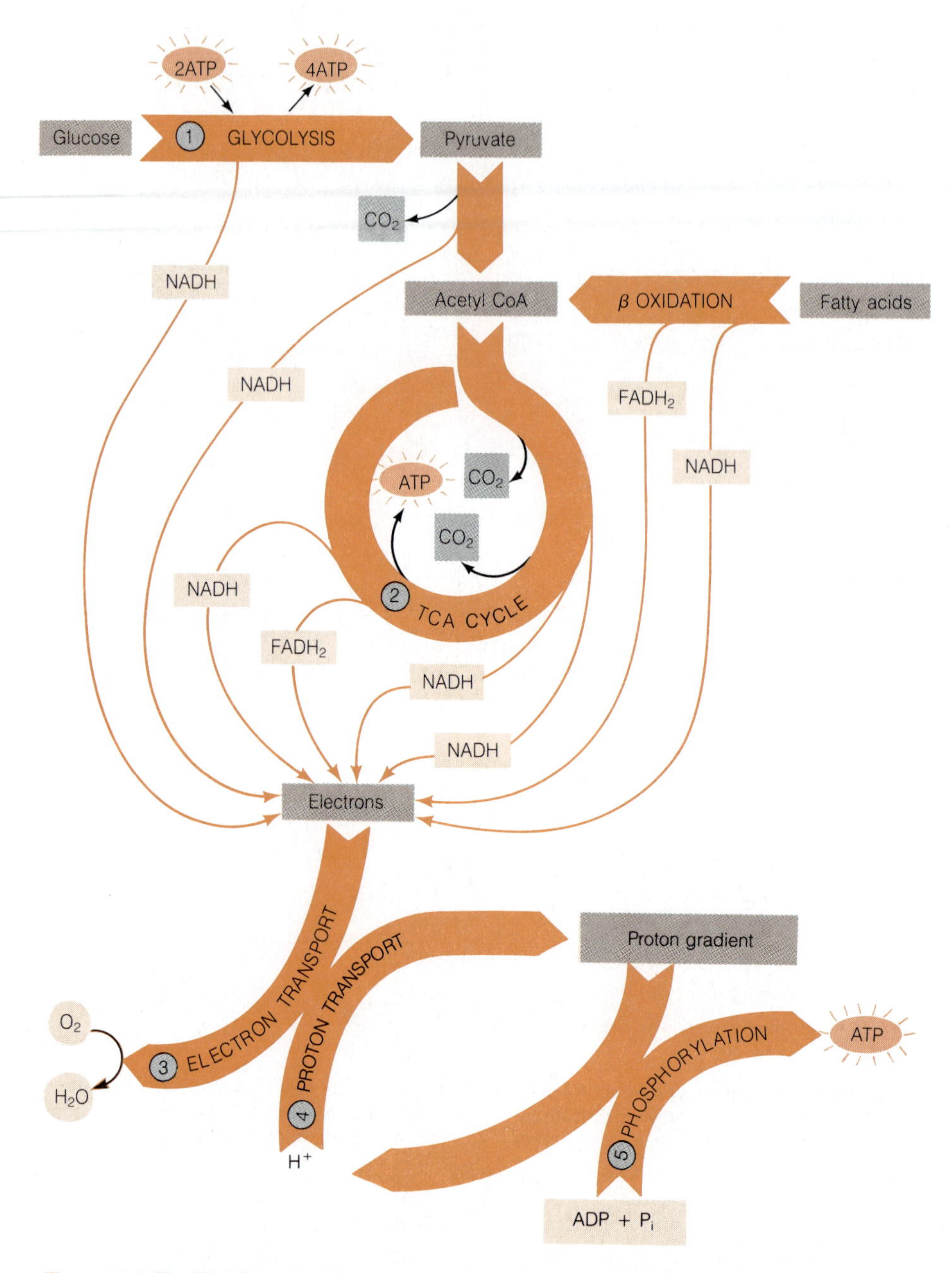

Figure 11-3 The Five Stages of Respiratory Metabolism. Oxidation of organic substrates occurs during glycolysis (stage 1), which produces pyruvate. Pyruvate gives rise to acetyl CoA, the primary substrate of the TCA cycle (stage 2). An alternative source of acetyl CoA is β-oxidation of fatty acids. Electron transport (stage 3) is coupled to proton transport (stage 4), and the energy is conserved as an electrochemical proton gradient across the mitochondrial membrane. Stage 5 is ATP synthesis, driven by flow of proton equivalents through ATP synthase enzymes that are components of the membrane.

variety of names were given to such particles in early work, but the term *mitochondrion* (meaning "threadlike granule"), first introduced in 1898, gradually replaced other names and is now the universally recognized term.

Evidence suggesting a role for the organelle in oxidative events began to accumulate early in the present century. In 1913, for example, Otto Warburg showed that cellular oxygen consumption could be associated with particles obtained by the filtering of homogenates. Not surprisingly, however, most of our understanding of the role of mitochondria in energy metabolism dates from the development of differential centrifugation, pioneered by Albert Claude. Intact mitochondria were first isolated in 1948 and were subsequently shown by Eugene Kennedy, Albert Lehninger, and others to be capable of carrying out all the reactions of the TCA cycle, electron transport, and oxidative phosphorylation.

Structural Features of Mitochondria

Mitochondrial structure is illustrated in Figure 11-4. A distinctive feature is the presence of two membranes. The **outer membrane** is smooth, has no folds, and is readily permeable to molecules with molecular weights of 5000 or less. This size range includes all metabolites pertinent to mitochondrial function. The outer membrane is therefore not regarded as a significant permeability barrier, and it

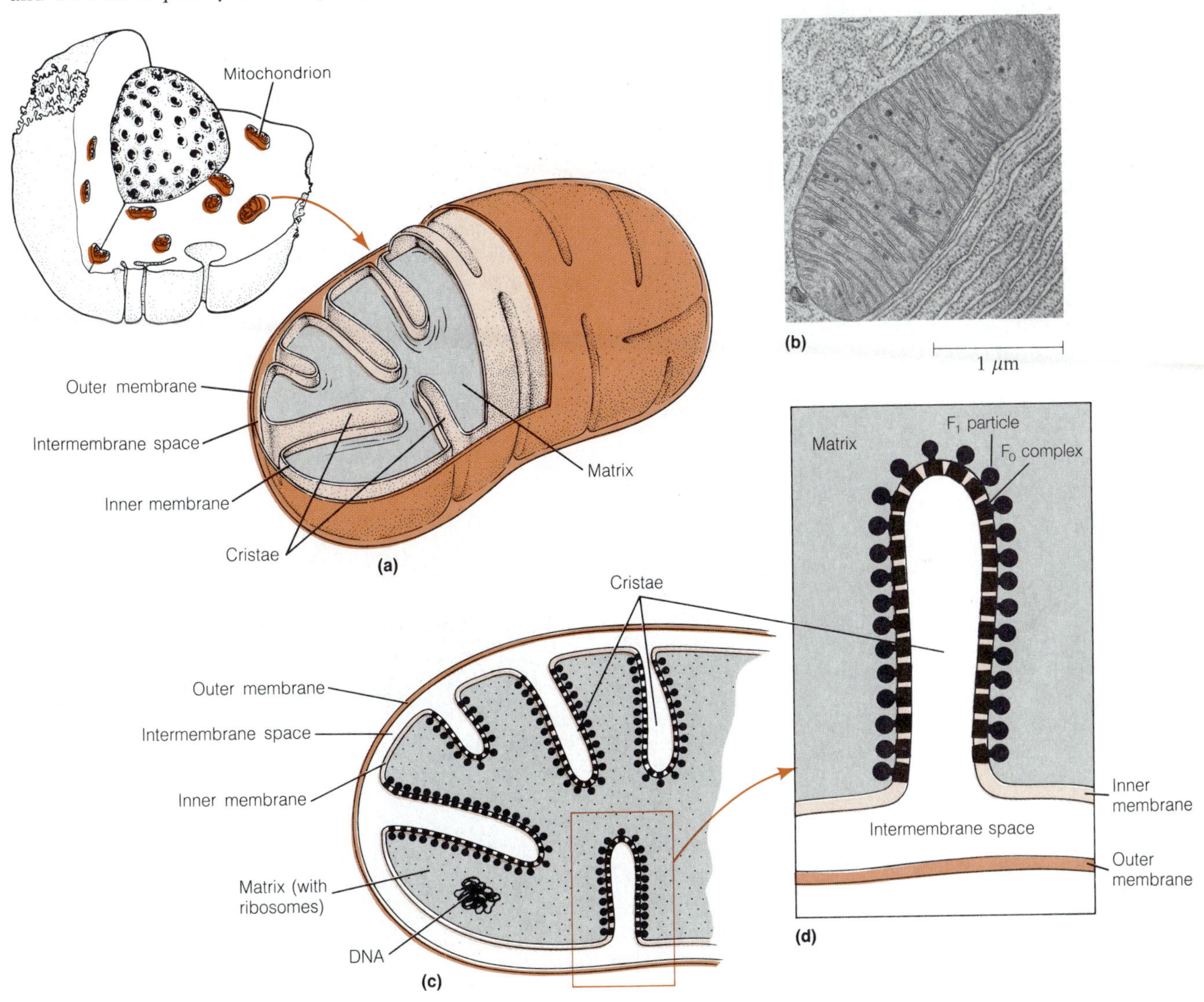

Figure 11-4 Mitochondrial Structure. (a) Mitochondrial structure is illustrated schematically in this cutaway view. (b) A mitochondrion of a bat pancreas cell as seen by electron microscopy (TEM). (c) Cross section of a mitochondrion. (d) An enlargement of a single crista, showing the F_1 (ATP synthase) particles that line the inner membrane on the matrix side.

has few enzyme activities associated with it. Because of the permeability of the outer membrane, the **intermembrane space** between the inner and outer membranes is essentially continuous with the cytoplasm of the cell.

In contrast to the outer membrane, the **inner membrane** of the mitochondrion is highly folded and presents a permeability barrier to most solutes. The inner membrane is also the locale of electron transport and ATP synthesis, as we will see later. The distinctive infoldings of the inner membrane are called **cristae** (singular: *crista*). Cristae greatly increase the surface area of the inner membrane, thereby enhancing the capacity of the mitochondrion for ATP generation. Figure 11-5 illustrates structural details of the inner and outer mitochondrial membranes as revealed by the freeze-fracture technique described in Appendix A. Note especially the high density of protein particles associated with both fracture faces of the inner membrane. These represent the transmembrane portions of enzymes associated with mitochondrial function.

The interior of the mitochondrion is filled with a semifluid **matrix** that is actually more like a gel than a solution. Within the matrix are the DNA molecules and ribosomes that give the organelle the capacity to make some of its own proteins. In most mammals, the mitochondrial genome consists of a circular DNA molecule with a molecular weight of about 11 million, coding for about a dozen polypeptides. (See Chapter 13.)

Occurrence and Size of Mitochondria

Mitochondria are found in virtually all aerobic cells of eukaryotes. They are common to both chemotrophic and phototrophic cells. Their presence in phototrophic cells serves as a reminder that photosynthetic organisms are capable of respiration, an option they depend on to meet energy needs during periods of darkness. The number of mitochondria per cell is highly variable, ranging from one or a few per cell in some protists, fungi, and algae up to several hundred or even a few thousand per cell in higher plants and animals. Human liver cells, for example, contain about 500 to 1000 mitochondria each.

After the nucleus, the mitochondrion is the largest organelle in most animal cells. Typically, a mitochondrion has a diameter of 0.5–1.0 μm and a length of several micrometers (more rarely, up to 10 μm). A mitochondrion is therefore similar in size to an entire bacterial cell.

Localization of Mitochondria

The crucial role of the mitochondria in meeting cellular ATP needs is reflected both in the localization of mitochondria within the cell and in the extent to which the inner membrane is folded into cristae. Frequently, mitochondria are clustered within the cell in regions of most intense metabolic activity, where the ATP need is greatest. An especially good example of mitochondrial localization

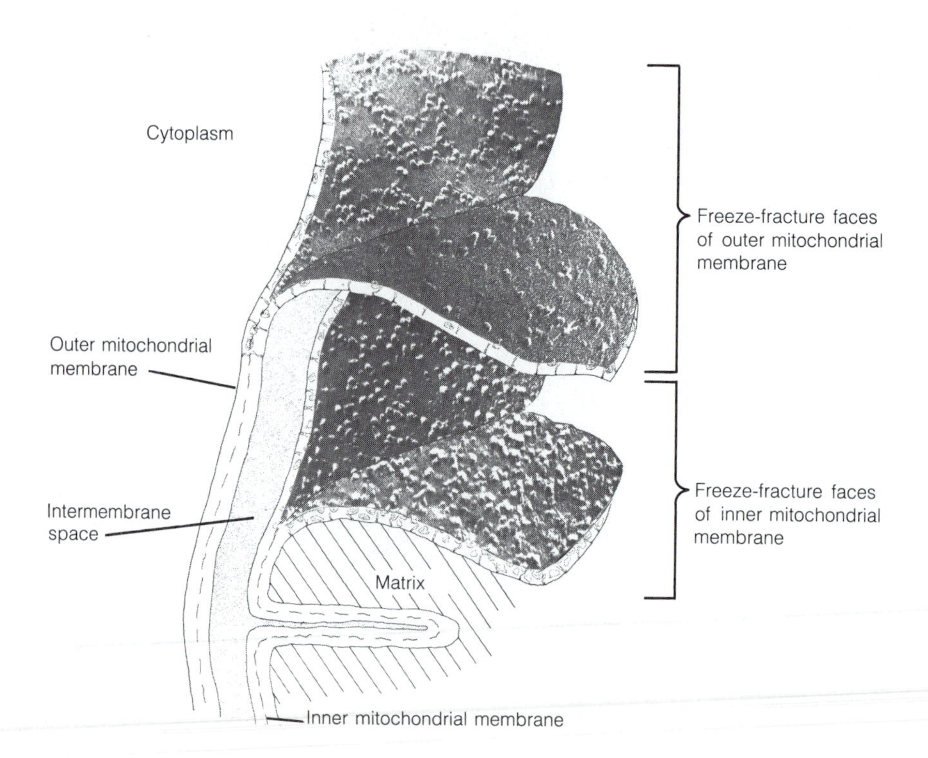

Figure 11-5 Structure of the Inner and Outer Mitochondrial Membranes. When the inner and outer mitochondrial membranes are subjected to freeze-fracturing, the membranes are split along their hydrophobic interior, separating each membrane into two fracture faces. Segments of electron micrographs of the two fracture faces for both inner and outer membranes are superimposed on a schematic diagram of the freeze-fractured membranes to illustrate the density of protein particles in each membrane.

is seen in muscle cells (Figure 11-6). Just as Kölliker originally reported, the mitochondria in such cells are organized in rows along the fibrils responsible for contraction. Presumably, this association minimizes the distance that ATP molecules must diffuse to get from their site of generation in the mitochondrion to the site of ATP utilization in the contracting fibrils. A similar localization of mitochondria is seen in sperm tails (recall Figure 4-12), in flagella and cilia, and at the base of kidney tubule cells, where exchange with the blood is most rapid.

The prominence of cristae within the mitochondrion frequently reflects the relative metabolic activity of the cell or tissue in which the organelle is located. Heart, kidney, and muscle cells have high respiratory activities, and their mitochondria have correspondingly large numbers of prominent cristae. The flight muscles of birds are especially high in respiratory activity and have mitochondria that are exceptionally well endowed with cristae. Plant cells, on the other hand, are generally characterized by lower rates of respiratory activity and by relatively fewer mitochondrial cristae.

Localization of Function Within the Mitochondrion

Specific functions and pathways have been localized within the mitochondrion by disruption of the organelle and fractionation of the various components. Enzymes and other proteins that are readily solubilized upon disruption are assumed to be either in the matrix or only loosely associated with the membrane. Most of the mitochondrial enzymes involved in pyruvate oxidation, in the TCA cycle, and in catabolism of fatty acids and amino acids are soluble matrix enzymes.

On the other hand, most of the intermediates in the electron transport chain are integral components of the inner membrane, including large protein complexes such as the dehydrogenases, the cytochromes, and cytochrome oxidase. A single mitochondrion has 10,000 or so complexes in its inner membrane. Unlike the situation in the plasma membrane, the integral protein complexes of the mitochondrial inner membrane are mobile and free to diffuse. The diffusion can be visualized by freeze-fracture analysis. In one study, Charles Hackenbrock and his coworkers placed inner mitochondrial membranes in an electrical field and found that the field caused the freeze-fracture particles (representing the protein complexes) to accumulate at one end of the membrane by a kind of electrophoresis. When the field was turned off, the particles returned to a random distribution within seconds, clearly demonstrating their freedom to diffuse in a fluid lipid bilayer. Further studies compared the diffusional mobility with electron transport rate and demonstrated that electron transport was limited by collision between diffusing integral proteins. This work represents some of the strongest evidence that the electron transport complexes are *not* lined up in the orderly fashion we tend to show in diagrams, but instead resemble the random array of particles shown in Figure 11-5.

Protruding from the membranes of the cristae are knoblike spheres called F_1 (or **coupling factor**) **particles** (Figure 11-4d). Each F_1 particle is part of the **ATP synthase** complex. The other part of the complex is the F_0 structure embedded in the inner membrane. The F_1 particles are therefore the actual sites of ATP generation within the

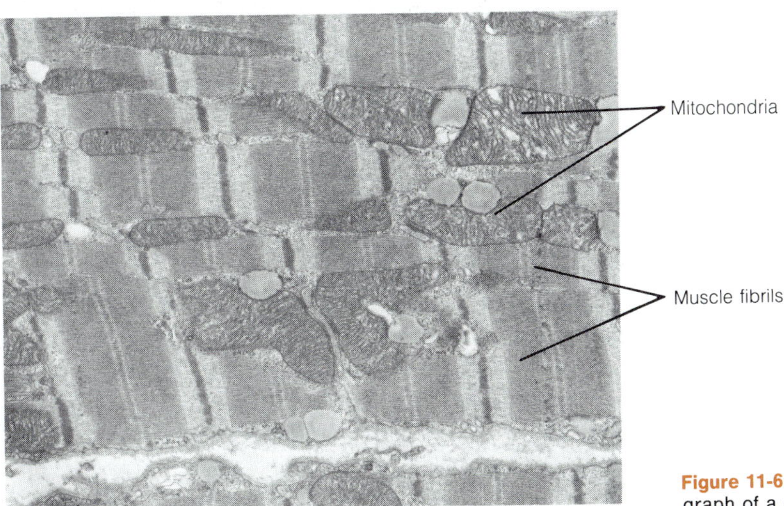

Figure 11-6 Mitochondria in a Muscle Cell. This electron micrograph of a cat heart muscle cell shows the intimate association of mitochondria with contractile fibrils (TEM). For details of muscle structure, see Chapter 19.

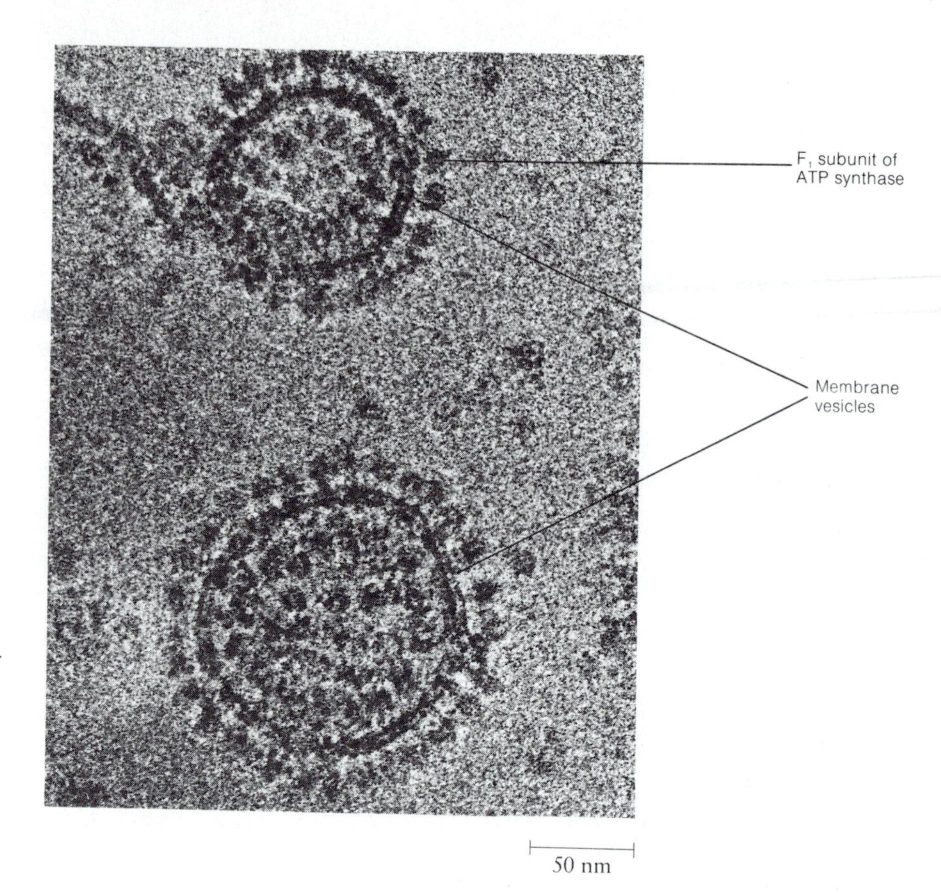

Figure 11-7 F_1 Particles. Because they are relatively simple to isolate, bacterial membranes are commonly used to investigate the properties of the ATP synthase. Isolated bacterial membranes are shown in this electron micrograph, and the F_1 subunits of the ATP synthase can be clearly seen lining the outer surface.

mitochondrion. The energy required for ATP generation by the F_1 particles is provided by an electrochemical gradient of protons. The energy to generate and maintain this gradient is in turn provided by the exergonic transport of electrons from reduced coenzymes to oxygen within the inner membrane. F_1 particles have been visualized on membranes isolated from bacteria (Figure 11-7) and closely resemble those of the mitochondrial inner membrane.

The Tricarboxylic Acid Cycle

Having looked in some detail at mitochondrial structure, we are now ready to consider the several stages of respiratory metabolism that occur within this organelle. For that, let us follow a molecule of pyruvate across the inner membrane of the mitochondrion and see what fate awaits it inside.

In the presence of oxygen, pyruvate is oxidized fully to carbon dioxide in a cyclic process that is at the very heart of energy metabolism in almost all aerobic chemotrophs. An important intermediate in this cyclic series of reactions is citrate, which has three carboxylic acid groups. For this reason, the cyclic pathway is usually called the **tricarboxylic acid** or **TCA cycle,** though sometimes (originally, in fact!) it is referred to as the *citric acid* (or *citrate*) *cycle*. Yet another name for it is the *Krebs cycle,* in honor of Sir Hans Krebs, who received a Nobel Prize in 1953 for his efforts in working out this metabolic sequence.

The TCA cycle is shown in Figure 11-8. In outline form, the cycle always begins with **acetyl coenzyme A (acetyl CoA)**, its only real substrate. As we will see, acetyl CoA arises either by oxidative decarboxylation of pyruvate or by oxidative cleavage of fatty acids. Regardless of the source, acetyl CoA transfers its two-carbon acetate group to a four-carbon acceptor (oxaloacetate), thereby generating the citrate for which the cycle was originally named. In a cyclic series of reactions, citrate is subjected to two successive decarboxylations and several oxidative events, leaving a four-carbon compound from which the starting oxaloacetate is eventually regenerated.

Each turn of the cycle involves the entry of two carbons as the acetate from acetyl CoA and the release of two carbons as carbon dioxide, with provision for regeneration of the oxaloacetate with which the cycle was initiated. Oxidation occurs at four steps in the cycle and in the preliminary step by which acetyl CoA is generated from pyruvate, with coenzymes serving as electron acceptors (FAD at one point, NAD^+ at all others). The substrate for the TCA cycle is therefore always acetyl CoA, and the products are carbon dioxide and the reduced coenzyme molecules, NADH and $FADH_2$.

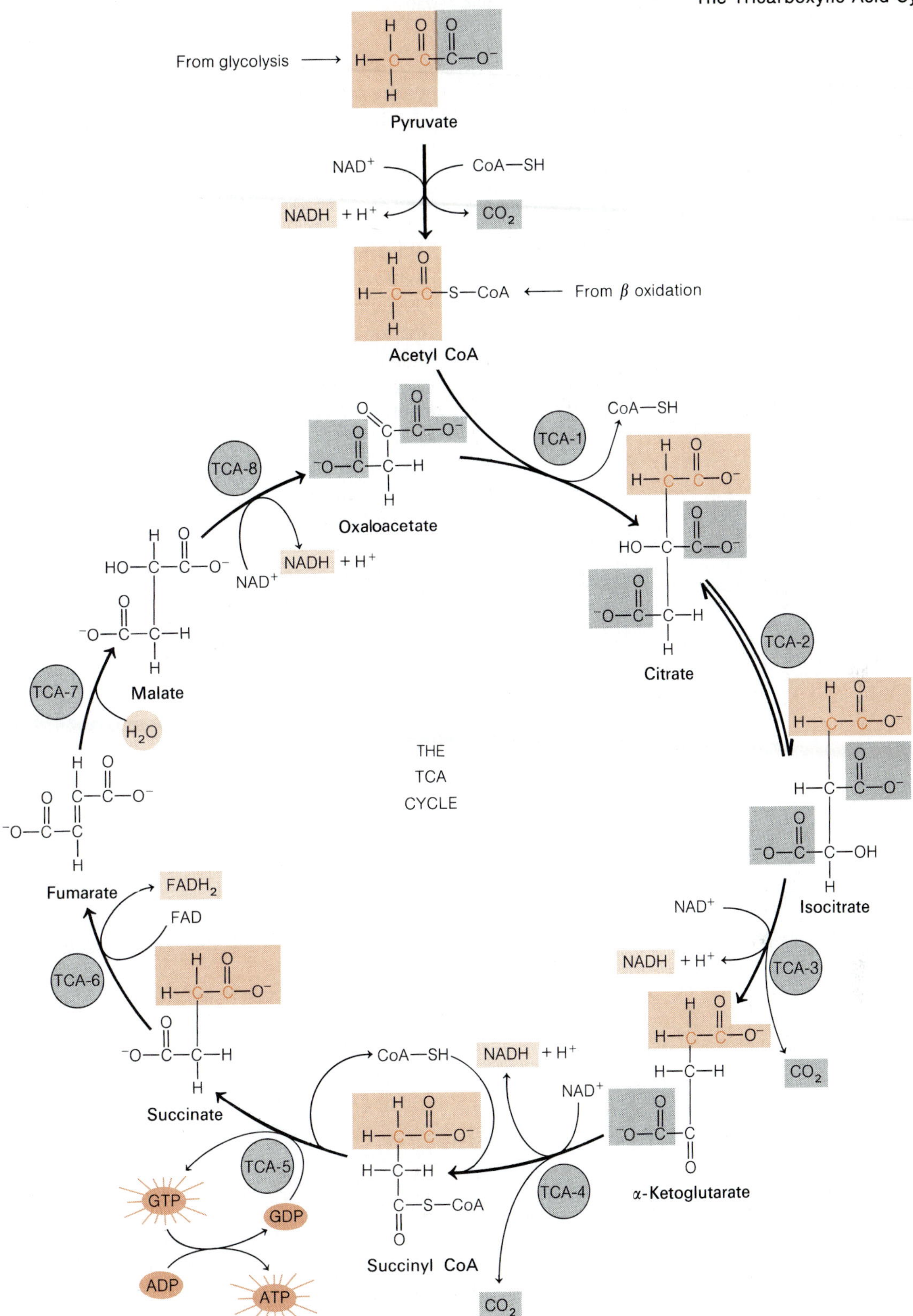

Figure 11-8 Tricarboxylic Acid (TCA) Cycle. The two carbon atoms of pyruvate that enter the cycle via acetyl CoA are shown in color in citrate and subsequent molecules until they are randomized by the symmetry of the fumarate molecule. The carbon atom of pyruvate that is lost as carbon dioxide is shown in gray, as are the two carbon atoms of oxaloacetate that give rise to carbon dioxide in steps TCA-3 and TCA-4.

Oxidative Conversion of Pyruvate to Acetyl Coenzyme A

Acetyl CoA is the necessary starting point for the TCA cycle, but from Chapter 10 it is clear that the glycolytic pathway ends with pyruvate, not acetyl CoA. To get from the pyruvate with which glycolysis ends to the acetyl CoA with which the TCA cycle begins requires the activity of pyruvate dehydrogenase, the enzyme complex that both oxidizes and decarboxylates pyruvate, as follows:

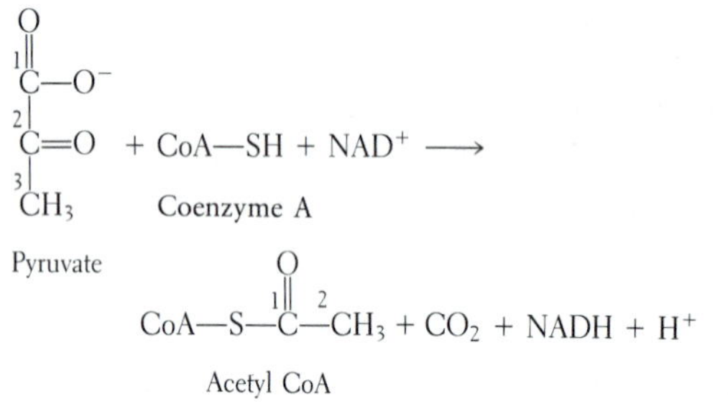

$$\underset{\text{Pyruvate}}{{}^{-}O-\overset{O}{\overset{\|}{C}}_{1}-\underset{2}{C}(=O)-\underset{3}{CH_3}} + \underset{\text{Coenzyme A}}{CoA-SH} + NAD^{+} \longrightarrow \underset{\text{Acetyl CoA}}{CoA-S-\overset{O}{\overset{\|}{C}}_{1}-\overset{2}{CH_3}} + CO_2 + NADH + H^{+} \quad (11\text{-}1)$$

This is called a *decarboxylation reaction* because one of the carbons of pyruvate (carbon atom 1) is liberated as carbon dioxide. As a result, carbon atoms 2 and 3 of pyruvate become, respectively, carbon atoms 1 and 2 of acetate. In addition, reaction 11-1 is also an *oxidative reaction,* since two electrons are transferred from the substrate to the coenzyme acceptor, NAD^+. The actual oxidation occurs on carbon atom 2 of pyruvate, which is oxidized from a keto to a carboxylic acid group. This oxidation is possible only because of the concomitant elimination of carbon atom 1 as carbon dioxide. The oxidation is highly exergonic ($\Delta G^{\circ\prime} = -7.5$ kcal/mol) and could be used to generate ATP if that were desired. Instead, the energy of oxidation is used to energize or activate the acetate molecule for further metabolism by linking it to a carrier called **coenzyme A** (CoA, or sometimes CoA—SH to emphasize the importance of its sulfhydryl group).

As shown in Figure 11-9, coenzyme A is a complicated molecule containing the vitamin *pantothenic acid.* (Like nicotinamide and riboflavin, pantothenic acid is classified as a vitamin because humans and other vertebrates need

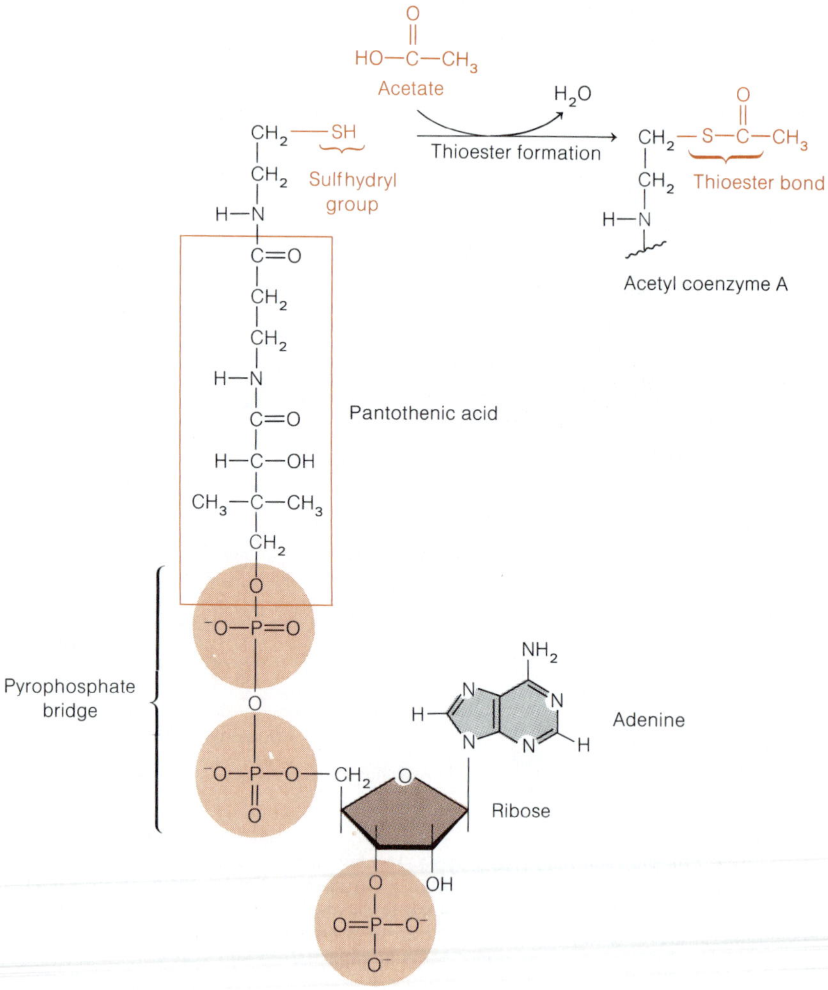

Figure 11-9 Structure of Coenzyme A. Note the chemistry of thioester formation to generate acetyl coenzyme A. The portion of the coenzyme outlined in a colored box is pantothenic acid, a B vitamin.

it as an essential coenzyme component but cannot synthesize it themselves.) Fortunately, we can ignore the detailed molecular structure of coenzyme A and concentrate on the free sulfhydryl group on the end of the molecule. It is this sulfhydryl group that can form a high-energy *thioester bond* with organic acids such as acetate. Just as NAD^+ and FAD are coenzymes adapted for transfer of electrons, so coenzyme A is well suited as a carrier of acyl groups such as acetate. The advantage of having acetate in this activated, high-energy form is about to become apparent.

The Entry of Acetate into the TCA Cycle

As Figure 11-8 illustrates, the TCA cycle is essentially a disassembly-line-in-the-round. With each turn of the cycle, two carbon atoms enter in organic form (as acetate) and two carbon atoms leave in inorganic form (as carbon dioxide). Acetate enters the TCA cycle bound to coenzyme A because the energy of the thioester bond is needed to drive the condensation of the acetate onto oxaloacetate (reaction TCA-1 of Figure 11-8). Since acetate has two carbon atoms and oxaloacetate has four, the product, *citrate,* is a six-carbon tricarboxylic acid. The colored carbon atoms in Figure 11-8 will help you keep track of the incoming atoms during subsequent steps in the cycle.

The Oxidative Decarboxylation Steps of the Cycle

Since the purpose of the TCA cycle is to accomplish the complete oxidation of organic substrates, it should not be too surprising to find that four out of the eight steps in the cycle are oxidative events. This can be seen in Figure 11-8, since four steps (TCA-3, TCA-4, TCA-6, and TCA-8) involve coenzymes that enter in the oxidized form and leave in the reduced form. The first two of these reactions, TCA-3 and TCA-4, are also decarboxylation steps: one molecule of carbon dioxide is eliminated in each step, reducing the number of carbons first from six to five and then from five to four.

Prior to these events, however, reaction TCA-2 converts citrate to the related compound isocitrate. If you look closely at the structures of these two compounds, you will see why this conversion is essential. Citrate is a *tertiary alcohol* and is therefore not oxidizable. Isocitrate, on the other hand, is a *secondary alcohol,* with a hydroxyl group that can be oxidized.

The enzyme that converts citrate to *isocitrate* is called aconitase. It actually carries out successive dehydration and rehydration reactions, in which the elements of water are removed from citrate to generate an unsaturated intermediate called aconitate, to which the —H and —OH of water are added back again, but in opposite positions.

The hydroxyl group of isocitrate is now the target of the first oxidative event of the cycle. Isocitrate is oxidized to the corresponding six-carbon keto compound, with NAD^+ serving as the electron acceptor. The six-carbon product is unstable, however, and undergoes spontaneous loss of carbon dioxide to form the five-carbon compound *a-ketoglutarate.* This reaction, TCA-3, is therefore one of the two decarboxylation steps of the cycle.

The second decarboxylation event occurs in the next step, reaction TCA-4. Again, it is an oxidative reaction, with NAD^+ as the electron acceptor. To understand this reaction, compare the structure of α-ketoglutarate with that of pyruvate. Both compounds are α keto acids, so it should not be surprising that the mechanism of oxidation is the same for both, complete with decarboxylation (from pyruvate to acetate in one case, and from α-ketoglutarate to succinate in the other) and linkage of the oxidized product to coenzyme A as a thioester. Thus, α-ketoglutarate is oxidized to *succinyl CoA,* a four-carbon compound with a high-energy thioester bond.

The ATP-Generating Step of the Cycle

Since we are already halfway around the cycle, let us pause a moment to take stock. Note that the carbon balance of the cycle is already satisfied: two carbon atoms entered as acetyl CoA and two carbon atoms have now been lost as carbon dioxide. (But notice carefully from Figure 11-8 that the two carbon atoms that leave in a given cycle are *not* the same two that entered in reaction TCA-1 of that cycle. Instead, they arise from what were initially the two carboxyl carbons of oxaloacetate.) We have also encountered two of the four oxidative steps of the TCA cycle and therefore have two molecules of NADH ready for eventual reoxidation via the electron transport chain. In addition, we recognize succinyl CoA as a compound that, like acetyl CoA, has a high-energy thioester bond.

Unlike acetyl CoA, however, succinyl CoA is not destined to be condensed onto another molecule, so the energy of the thioester bond is not needed for that purpose. Instead, the energy is used to generate a molecule of ATP. In fact, it is because succinyl CoA is formed rather than free *succinate* in reaction TCA-4 that the energy of α-ketoglutarate oxidation is conserved in a form suitable for ATP production. The high-energy intermediate generated upon hydrolysis of the succinyl CoA in reaction TCA-5 is not ATP but guanosine triphosphate (GTP), a closely related compound (recall Figure 3-10). However, the terminal phosphate of GTP can be readily transferred to ADP, so the net result of succinyl CoA hydrolysis is the generation of one molecule of ATP.

The Final Oxidative Sequence of the Cycle

Of the remaining three steps in the TCA cycle, two are oxidations. In reaction TCA-6, the free succinate formed in the previous step, is oxidized to *fumarate*. This reaction is unique in that both electrons come from carbon atoms, generating a carbon-carbon double bond. The reaction is not sufficiently energetic to allow transfer of electrons to NAD^+, so the acceptor for this dehydrogenation is FAD. The lower energy of $FADH_2$ compared to NADH is related to the fact that FAD accepts electrons on nitrogen atoms, whereas NAD^+ accepts electrons on carbon atoms (compare Figures 10-5 and 10-6).

The double bond of fumarate is then hydrated to produce *malate* (reaction TCA-7). Since fumarate is a symmetric molecule, the hydroxyl group of water has an equal chance of adding to either of the internal carbon atoms. As a result, the colored carbon atoms of Figure 11-8, which we are using to keep track of the most recent acetate group to enter the cycle, are randomized at this step between the "upper" and "lower" two carbon atoms of malate and are therefore omitted from the figure from this point on. (The discerning chemist will note that the citrate of reaction TCA-2 is also a symmetric molecule, and its hydroxyl group should, by the same argument, be moved either "up" or "down" the molecule, thereby randomizing carbon atoms at an early stage in the cycle. However, the enzyme that catalyzes that reaction is capable of distinguishing between the two ends of the molecule and moves the hydroxyl group in one direction only. By contrast, the enzyme responsible for reaction TCA-7 cannot distinguish between the two inner carbon atoms.)

In reaction TCA-8, the hydroxyl group of malate becomes the target of the final oxidative event in the cycle. Again, NAD^+ serves as the electron acceptor, and the product is the corresponding keto compound, *oxaloacetate*.

The TCA Cycle in Summary

With the regeneration of oxaloacetate, one full cycle is complete. We can summarize what has been accomplished by noting the following properties of the TCA cycle:

1. Acetate enters the cycle as acetyl CoA by condensation onto a four-carbon acceptor molecule, forming citrate, a six-carbon compound.
2. Decarboxylation occurs at two steps in the cycle so that the input of two carbons as acetate is balanced by the loss of two carbons as carbon dioxide.
3. Oxidation occurs at four steps, with NAD^+ serving as the electron acceptor in three cases and FAD serving as the electron acceptor in one case.
4. ATP is generated at one point, with GTP as an intermediate.
5. The cycle is completed upon regeneration of the original acceptor.

By summing the eight component reactions of the TCA cycle as shown in Figure 11-8, we arrive at the following overall summary reaction:

$$\text{acetyl CoA} + 3H_2O + 3NAD^+ + FAD + ADP + P_i \longrightarrow 2CO_2 + 3NADH + 3H^+ + FADH_2 + CoA{-}SH + ATP + H_2O \quad (11\text{-}2)$$

Since the cycle must, in effect, occur twice to metabolize both of the acetyl CoA molecules derived from a single molecule of glucose, the overall reaction on a per-glucose basis can be obtained by multiplying all the coefficients in reaction 11-2 by 2. If we add to this reaction the overall expression for glycolysis through pyruvate (reaction 10-12 from the previous chapter) and then add the overall expression for the oxidative decarboxylation of pyruvate to acetyl CoA (reaction 11-1 multiplied by 2), we arrive at the following overall equation for the entire sequence from glucose through the TCA cycle:

$$\text{glucose} + 6H_2O + 10NAD^+ + 2FAD + 4ADP + 4P_i \longrightarrow 6CO_2 + 10NADH + 10H^+ + 2FADH_2 + 4ATP + 4H_2O \quad (11\text{-}3)$$

Looking at this summary reaction, you may be struck by how little evidence it provides for the substantially greater ATP yield that is supposed to be characteristic of respiratory metabolism. As it stands, the reaction suggests only a modest enhancement in ATP yield over glycolysis alone (four instead of two molecules of ATP per molecule of glucose). This hardly seems to justify the metabolic jungle we have just been through. Where, one might ask, is all the energy?

The answer, of course, is that it is right there in the reaction—stored in the reduced coenzyme molecules NADH and $FADH_2$, which are high-energy compounds in their own right. As we will see later, the energy released upon reoxidation of these reduced coenzymes by molecular oxygen is substantial indeed. For the release of that energy, we must look to the remaining stages of respiratory metabolism—electron transport and oxidative phosphorylation. It is only as electrons are transferred stepwise from these coenzymes to oxygen that the coupled generation of ATP occurs. This final train of oxidations accounts for the vast majority of ATP molecules produced during the complete oxidation of glucose.

Before moving on to these final stages of respiration and the ATP yields that they promise, we will look briefly at several additional features of the TCA cycle. These include its centrality in energy metabolism, its role in other metabolic pathways, and its regulation.

The Centrality of the TCA Cycle

It is important to understand the central role of the TCA cycle in all of aerobic energy metabolism. Thus far, we have regarded glucose as the main substrate for respiration. True, a variety of alternative carbohydrate substrates were considered in Chapter 10 (see Figure 10-10), but most summary reactions written for chemotrophic energy metabolism assume glucose as the starting compound, and reaction 11-3 is no exception. In a sense, the assumption is very reasonable, since glucose is the single most important source of energy for the chemotrophic world. However, it is also important to note the role of other substrates in cellular energy metabolism and the centrality of the TCA cycle in the catabolism of a variety of alternative fuel molecules, especially fats and proteins. Far from being a minor pathway for the catabolism of a single sugar, the TCA cycle represents the very heart of aerobic energy metabolism.

Fat as a Source of Energy. When we first encountered fats in Chapter 3, we noted their role in energy storage and observed that because they are highly reduced, they liberate more energy per gram upon oxidation than do carbohydrates. For this reason, fats are the most important long-term energy storage form of most organisms. Storage polysaccharides such as starch and glycogen are important as localized, mobilizable energy reserves, but for long-term energy stores, fat is almost always used. Fat reserves are especially important in hibernating animals and migrating birds and are also a common form in which energy and carbon are stored by plants in their seeds. Fats are well suited for this storage function because they allow the packaging of a maximum of calories in a minimum of volume and weight, a feature of obvious significance for both animal motility and seed dispersal.

Most fat is stored as deposits of **triglycerides,** which are neutral triesters of **glycerol** and long-chain **fatty acids** (recall Figure 3-22). Triglycerides may be obtained either directly from the diet or by mobilization of stored fat. In either case, triglyceride catabolism begins with hydrolysis of the triglyceride to yield glycerol and free fatty acids. The glycerol is channeled into the glycolytic pathway by oxidative conversion to dihydroxyacetone phosphate, as shown in Figure 10-10.

To understand the catabolism of fatty acids derived from triglycerides, it is important to recall that the primary substrate entering the TCA cycle is acetyl CoA. With that in mind, it makes sense that fatty acids are degraded by a stepwise process involving successive removal of two-carbon units as acetyl CoA. Similarly, it should now be clear why the entire process is localized in the mitochondrion, where the TCA cycle is present to accept acetyl CoA (Figure 11-2).

The sequential process of fatty acid catabolism to acetyl CoA is called **β oxidation** because the initial oxidative event in each successive cycle occurs on the carbon atom in the β position of the fatty acid (that is, the second carbon from the carboxylic acid group). The process involves successive cycles of oxidative attack on the fatty acid, as shown in Figure 11-10. Each cycle begins with oxidation of the β carbon, results in the release of two carbon atoms as acetyl CoA, and leaves the fatty acid shortened by two carbon atoms but ready for another round of oxidation.

In brief, β oxidation of a fatty acid starts with an activation step (reaction FA-1 of Figure 11-10) in which the energy of ATP is used to form the fatty acyl CoA derivative. This activated form of the fatty acid is then oxidized by a series of four reactions, three of which (FA-2, FA-3, and FA-4) parallel exactly the sequence whereby succinate is converted into oxaloacetate in the TCA cycle (recall reactions TCA-6, TCA-7, and TCA-8 of Figure 11-8). Just as that series of reactions begins with the FAD-mediated dehydrogenation of succinate, here fatty acyl CoA is dehydrogenated to the corresponding α,β unsaturated acyl CoA (reaction FA-2). Water is added across the double bond, generating the β-hydroxyl fatty acyl CoA (reaction FA-3), similar to the formation of malate from fumarate.

The hydroxyl group is then oxidized in an NAD^+-dependent manner to form the corresponding β keto acid (reaction FA-4). This compound is not very stable; it is split in the final step of the sequence with the uptake of another molecule of coenzyme A into acetyl CoA and an acyl CoA compound with two fewer carbon atoms than the original molecule (reaction FA-5).

The newly formed acyl CoA compound is identical to the activated starting compound that served as the substrate for reaction FA-2, except that it is two carbons shorter. Clearly, this new acyl CoA molecule can undergo the same series of reactions represented by reactions FA-2 through FA-5, thereby removing another two-carbon unit

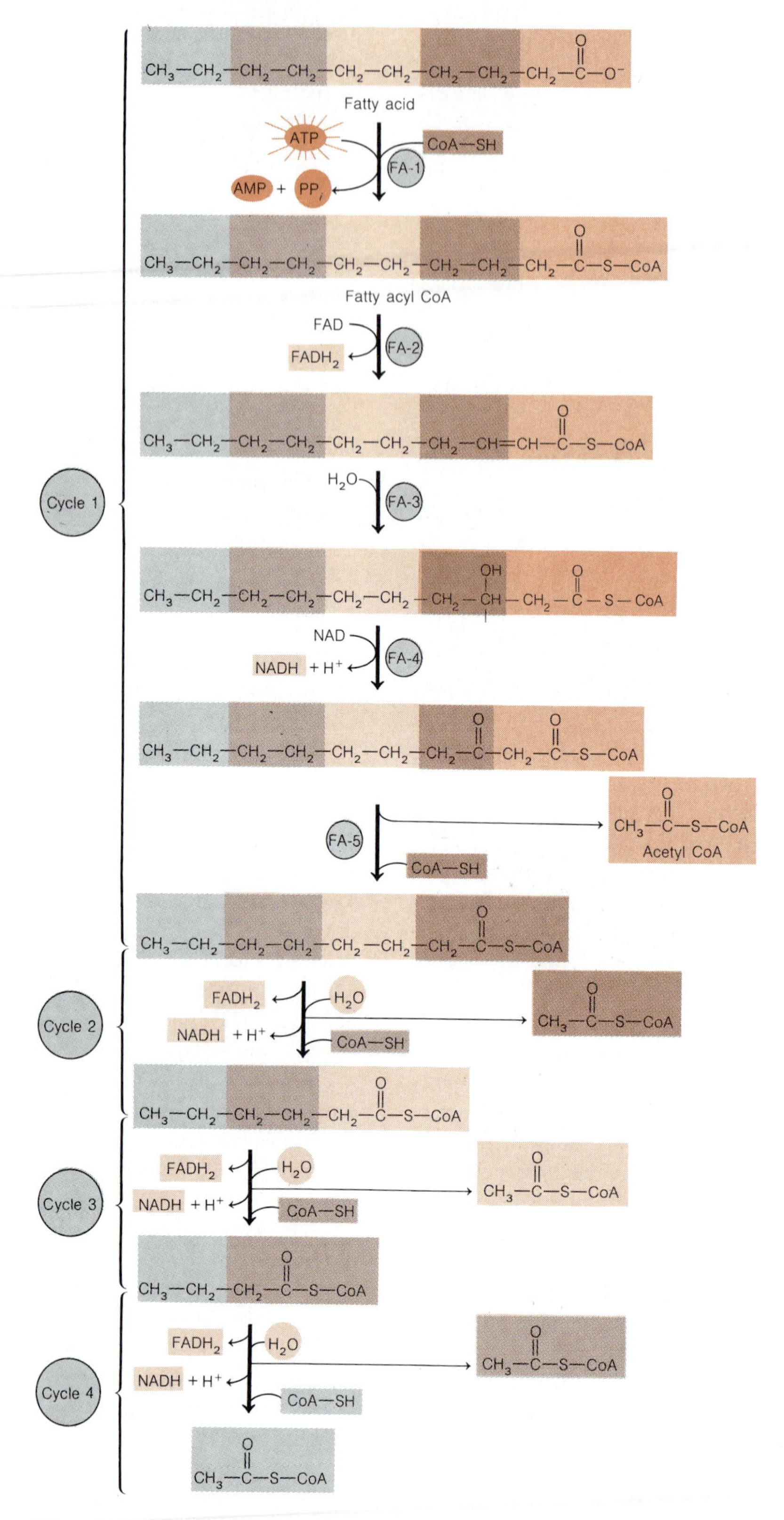

Figure 11-10 The Process of β Oxidation. Reaction numbers and chemical details are shown for the first cycle only. The fatty acid shown here is decanoate, a saturated fatty acid with ten carbon atoms, so four cycles of β oxidation are required, yielding four molecules of NADH, four molecules of $FADH_2$, and generating five molecules of acetyl CoA, which are subject to further catabolism by the TCA cycle.

as acetyl CoA. This establishes the cyclic, or repetitive, nature of the process: oxidation continues down the fatty acid backbone, removing two carbon atoms at a time. The complete β oxidation of a fatty acid with $2n$ carbon atoms requires $n - 1$ such cycles of oxidation, yields $n - 1$ molecules each of NADH and $FADH_2$, and gives rise eventually to n molecules of acetyl CoA, each of which can be catabolized by the TCA cycle to carbon dioxide and water. The repetitive nature of β oxidation is illustrated in Figure 11-10 for a fatty acid with ten carbon atoms.

Protein as a Source of Energy. Proteins are not considered primarily an energy source because they have more fundamental roles in the cell. But they, too, can be catabolized to generate ATP if necessary. In animals, protein catabolism is prominent under conditions of starvation or when dietary intake of proteins exceeds cellular needs for amino acids. In plants, catabolism of proteins is especially important during germination of protein-storing seeds. In addition, all cells undergo metabolic turnover of most proteins and protein-containing structures, and the amino acids to which the proteins are degraded can either be recycled into proteins or degraded oxidatively to yield energy.

As shown in Figure 11-11, protein catabolism begins with hydrolysis of the peptide bonds that link amino acids together in the polypeptide chain. The process is called **proteolysis,** and the enzymes responsible for it are called **proteases.** The products of proteolytic digestion are small peptides and free amino acids. Further digestion of peptides depends on **peptidases,** which either hydrolyze internal peptide bonds (*endopeptidases*) or remove successive amino acids from the end of the peptide (*exopeptidases*). Exopeptidases are in turn either *aminopeptidases* or *carboxypeptidases,* depending on the end of the peptide from which digestion proceeds (Figure 11-11).

Free amino acids, whether ingested as such or obtained by digestion of proteins, can be catabolized for energy. The pathways by which they are degraded illustrate the general principle that alternative substrates are converted to intermediates of mainstream catabolism in as few steps as possible. In spite of their number and chemical diversity, all of these pathways eventually lead to a few key intermediates in the TCA cycle, notably acetyl CoA, α-ketoglutarate, oxaloacetate, fumarate, and succinyl CoA. Most pathways for amino acid catabolism begin with removal of the amino group, either by the process of transamination (Figure 11-12a) or by direct oxidative deamination (Figure

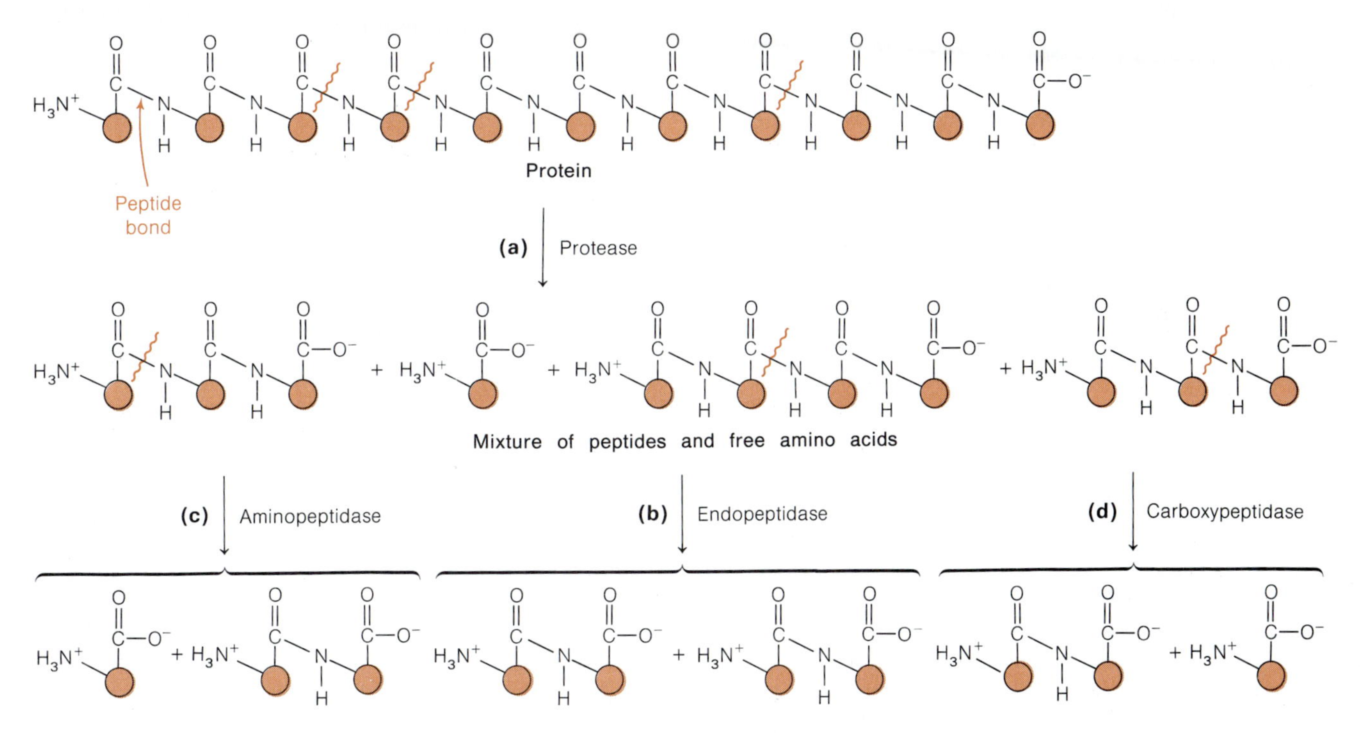

Figure 11-11 Hydrolysis of Proteins and Peptides. Amino acids are represented by circles; only peptide bonds are shown in detail. (a) The protein shown at the top is digested by protease activity to a mixture of peptides and free amino acids by hydrolysis of the peptide bonds at the sites indicated by the wavy lines. Peptides are further digested either by internal cleavage (b, endopeptidase activity) or by successive removal of amino acids from either the amino end (c, aminopeptidase activity) or the carboxyl end (d, carboxypeptidase activity) of the chain. Eventually, proteins and polypeptides are degraded in this way to free amino acids, which can then be further catabolized by the cell.

Figure 11-12 Transamination and Oxidative Deamination. The amino group of an amino acid can be removed by (a) transamination or (b) oxidative deamination. Transamination involves transfer of the amino group to an α keto acid acceptor. In oxidative deamination the amino group is liberated as ammonia or ammonium ion. All amino acids can undergo transamination, but only a few can be oxidatively deaminated.

$$R_1{-}CH(NH_3^+){-}C(=O){-}O^- + R_2{-}C(=O){-}C(=O){-}O^- \rightleftharpoons R_1{-}C(=O){-}C(=O){-}O^- + R_2{-}CH(NH_3^+){-}C(=O){-}O^-$$

(a) Transamination

$$R{-}CH(NH_3^+){-}C(=O){-}O^- + NAD^+ + H_2O \longrightarrow R{-}C(=O){-}C(=O){-}O^- + NH_4^+ + NADH + H^+$$

(b) Oxidative deamination

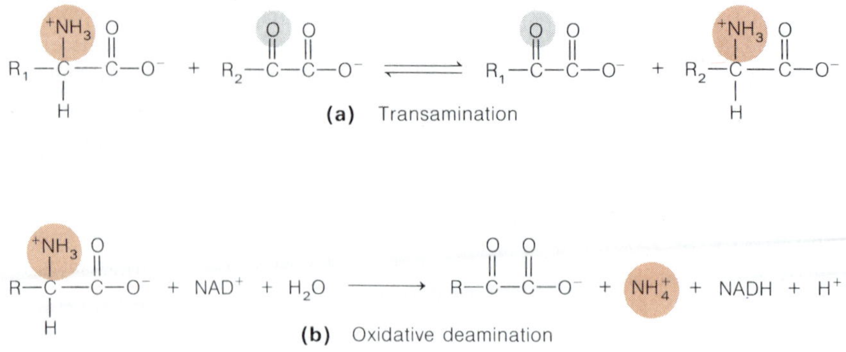

11-12b). **Transamination** involves the transfer of an amino group from an amino acid to an α keto acid acceptor. The *deamination* of the amino acid is therefore accompanied by the *amination* of the α keto acid. Transamination is a common means of shifting amino groups between carbon skeletons and is used by the cell not only in degradative processes but also in synthetic pathways.

Transamination does not effect the net removal of nitrogen from the pool of organic molecules, as must happen if net catabolism of amino acids is to occur. It does, however, allow the transfer of amino groups to several common carbon skeletons, especially glutamate. The amino group can then be liberated as free ammonia by a process called **oxidative deamination.** All amino acids can undergo transamination, but only a few can be oxidatively deaminated.

Of the 20 amino acids, 3 give rise to TCA cycle intermediates or precursors directly upon transamination or deamination. All the others require more complicated pathways, often with many intermediates. You will probably encounter these pathways at some point in a biochemistry course. When you do, be sure to note how many of them have end products that are TCA cycle intermediates, because that will help emphasize the centrality of the TCA cycle for all of cellular energy metabolism, regardless of the starting substrate.

The Amphibolic Role of the TCA Cycle

The only purpose of the TCA cycle when it is functioning as such is to oxidize acetyl CoA to carbon dioxide, with concomitant conservation of energy as reduced coenzymes. Strictly speaking, then, acetyl CoA is the only substrate the TCA cycle can accept; it is not possible for the cycle to accomplish the net intake and catabolism of any other substance. Yet in most cells, there is a considerable flow of four-, five-, and six-carbon intermediates into and out of the cycle. This flux takes place in addition to the primary catabolic function of the cycle. These "side reactions" replenish and augment the supply of intermediates in the cycle as needed, as well as provide for the synthesis of compounds derived from any of several intermediates of the cycle. Because the TCA cycle can function both in a catabolic mode and as a source of precursors for anabolic pathways, it is often called an **amphibolic pathway** (from the Greek *amphi*, meaning "both").

A variety of reactions and pathways could be cited to illustrate the amphibolic role of the TCA cycle. For example, two of the intermediates of the cycle, α-ketoglutarate and oxaloacetate, are the α keto equivalents of the amino acids glutamate and aspartate, respectively. They can be converted to these amino acids by transamination, as depicted in Figure 11-12. Glutamate and aspartate are both found in proteins, so in this way the TCA cycle becomes involved in protein synthesis. Other amphibolic precursors in the cycle include succinyl CoA, which is a precursor for the biosynthesis of heme (shown in Figure 11-16), and citrate, which can be transported out of the mitochondrion and used as a source of acetyl CoA for synthesis of fatty acids in the cytoplasm.

In each of these cases, intermediates of the cycle are drawn off for biosynthetic purposes. This, in turn, dictates the need for mechanisms whereby the cycle can be continuously replenished. The most important replenishment reactions are those in which oxaloacetate is formed by carboxylation of either pyruvate or phosphoenolpyruvate (PEP). Several such possibilities are shown in Figure 11-13. Transamination reactions can also serve as a source of TCA intermediates by functioning in the direction of α-ketoglutarate and oxaloacetate synthesis at the expense of the analogous amino acids.

By a variety of such side reactions, the TCA cycle acquires a metabolic versatility beyond its primary catabolic function. Even anaerobic organisms use portions of the TCA pathway for synthetic purposes, despite the lack of TCA cycle activity as such in the absence of oxygen. The common feature of these amphibolic roles is that no cyclic function is involved. One intermediate is removed for synthetic purposes, another is replenished by a side reaction, and only that portion of the cycle that links the two intermediates actually operates. In a way, it is mere

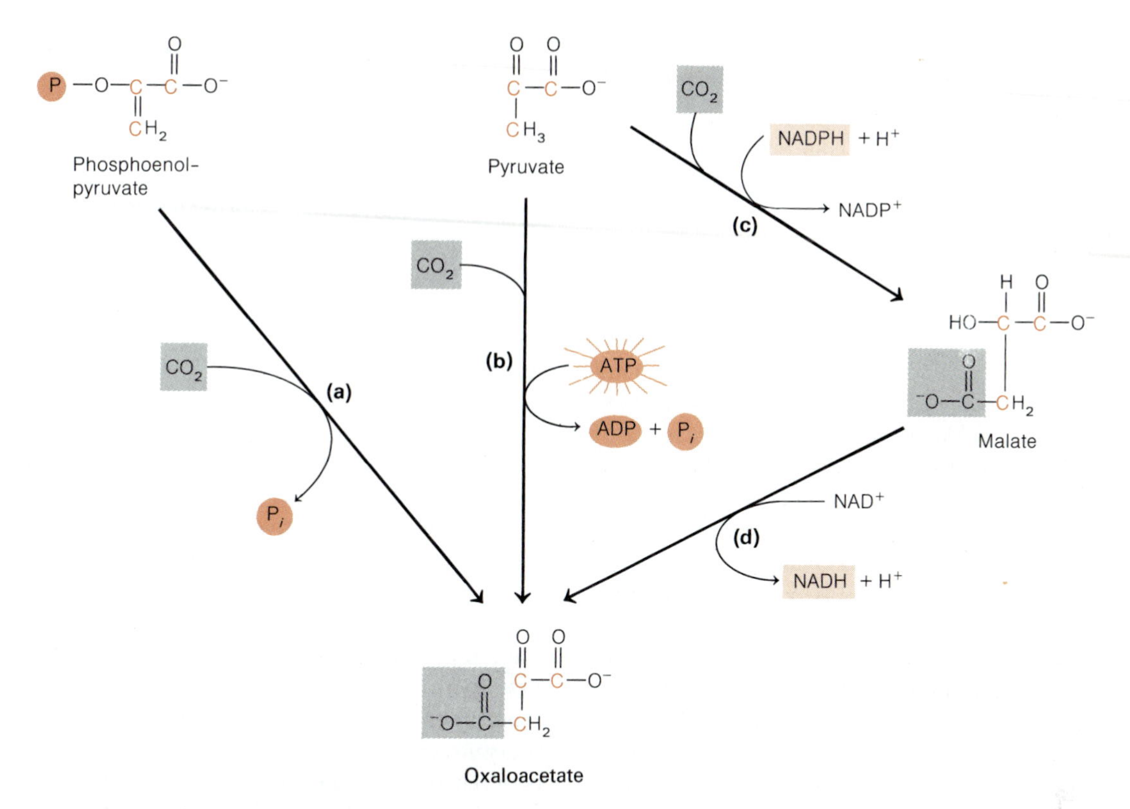

Figure 11-13 Replenishment of Oxaloacetate in the TCA Cycle. The level of oxaloacetate may be augmented (a) by carboxylation of phosphoenolpyruvate or (b) by ATP-driven carboxylation of pyruvate. Alternatively, pyruvate may be (c) reductively carboxylated to yield malate, which can then give rise to oxaloacetate upon (d) NAD-dependent oxidation.

coincidence that several of the enzymes in the biosynthetic pathway also happen to have a role in the TCA cycle.

Regulation of TCA Cycle Activity

Like all metabolic pathways, the TCA cycle must be carefully regulated to ensure that its level of activity corresponds closely to cellular needs. Moreover, since the cycle serves both catabolic and amphibolic roles, we might expect both of these functions to be reflected in its regulation. The key energy-linked regulatory factor is the NAD^+/NADH ratio, both because NAD^+ is a required substrate for several dehydrogenase reactions in the cycle and because several enzymes are subject to allosteric regulation by NAD^+ and NADH.

Citrate synthetase, the enzyme responsible for citrate formation from oxaloacetate and incoming acetyl CoA (TCA-1), is allosterically inhibited by physiological concentrations of NADH and succinyl CoA. It is also controlled by the levels of both of its substrates, since acetyl CoA concentrations vary greatly with the metabolic status of the cell, and the concentration of oxaloacetate is frequently in the regulatory range. Both isocitrate dehydrogenase (TCA-3) and α-ketoglutarate dehydrogenase (TCA-4) are subject to product inhibition by NADH. In addition, isocitrate dehydrogenase is allosterically activated by ADP, rendering the cycle sensitive to the ATP/ADP ratio of the cell.

The overall availability of acetyl CoA is regulated primarily by pyruvate dehydrogenase (reaction 11-1). This enzyme complex is deactivated by an ATP-dependent phosphorylation when the ATP level rises and is reactivated by dephosphorylation as the ATP level falls again. The reaction is also inhibited by high ratios of acetyl CoA to free CoA and of NADH to NAD^+. Feedback control from the TCA cycle to the glycolytic pathway is provided by the inhibitory effect of citrate on phosphofructokinase, as we saw in Chapter 10.

Electron Transport

In all that we have seen so far, chemotrophic energy metabolism has been characterized by the stepwise oxidation of organic molecules, with each oxidation accompanied by the reduction of a coenzyme, either NAD^+ or FAD. The catabolism of glucose, for example, involves six different oxidative events—one during glycolysis, one in the conversion of pyruvate to acetyl CoA, and the other four as part of the TCA cycle. In the process, all six carbon atoms of the glucose molecule are oxidized completely to carbon dioxide, and 12 pairs of electrons (hydrogen atoms, actually) are transferred to NAD^+ (in 10 of the 12 cases) or to FAD (in the remaining 2 cases). Similarly, oxidation of fatty acids involves the reduction of one molecule each of NAD^+ and FAD per cycle of oxidation and of additional coenzyme molecules as the resulting acetyl CoA molecules are oxidized further in the TCA cycle. Amino acids are also catabolized by oxidative pathways that use these same coenzymes as electron acceptors.

Clearly, these oxidative events can be sustained only if oxidized coenzyme molecules are available as electron acceptors, and that, in turn, depends on the continuous reoxidation of reduced coenzymes. It is in this process of coenzyme reoxidation that the oxygen requirement for aerobic energy metabolism finally becomes apparent, because oxygen is the terminal electron acceptor. Moreover, the transfer of electrons from NADH or $FADH_2$ to oxygen is highly exergonic, so the process of coenzyme oxidation is also responsible for most of the ATP yield of respiratory metabolism.

The process of coenzyme reoxidation by transfer of electrons to oxygen is called **electron transport.** As Figure 11-3 indicates, electron transport is the third stage of respiratory metabolism. The accompanying process of ATP generation is called **oxidative phosphorylation.** For discussion purposes, we will consider the processes separately. However, remember that they occur in the cell not as isolated events, but as integral parts of respiratory metabolism.

The concept of electron transport can at first seem fairly abstract, but in fact there are many common examples of such reactions. For instance, when iron rusts, electrons are transferred from metallic iron to oxygen. Another example of electron transport occurs in a car battery, where metallic lead is a donor of electrons. In this case the electrical voltage of the reaction is converted into an actual electrical current, the energy of which starts the car's engine. The fact that every electron transport reaction can be turned into an electrical voltage is an important concept, because the voltage is one way to measure the free energy available in the reaction. As we will see later, mitochondrial electron transport generates a voltage of approximately 200 mV, and this is the primary driving force of ATP synthesis.

Since electron transport involves the reoxidation of coenzymes at the expense of molecular oxygen, we can write summary reactions as follows:

$$NADH + H^+ + \tfrac{1}{2}O_2 \longrightarrow NAD^+ + H_2O \qquad (11\text{-}4)$$

$$FADH_2 + \tfrac{1}{2}O_2 \longrightarrow FAD + H_2O \qquad (11\text{-}5)$$

Electron transport therefore accounts not only for the reoxidation of coenzymes and the consumption of oxygen but also for the formation of water, which we recognize as one of the two end products of aerobic energy metabolism—the other being carbon dioxide. The carbon dioxide is generated in the TCA cycle, but the water is formed upon reduction of oxygen here in the process of electron transport.

The most important aspect of reactions 11-4 and 11-5 is the large amount of free energy released upon oxidation of NADH and $FADH_2$ by oxygen. The $\Delta G^{\circ\prime}$ values for reactions 11-4 and 11-5 are -52.7 kcal/mol and -43.4 kcal/mol, respectively. These $\Delta G^{\circ\prime}$ values make it clear that the oxidation of a coenzyme is a highly exergonic process. It should therefore not surprise you that electrons are not passed directly from reduced coenzymes to oxygen. Rather, the transfer is accomplished as a multistep process that involves a series of reversibly oxidizable electron acceptors functioning together in what is called an **electron transport chain.** In this way, the total free energy difference between reduced coenzymes and oxygen is parceled out among a series of electron transfers and is released in increments, maximizing the opportunity for ATP generation.

Reduction Potentials

To understand the order of intermediates in the electron transport chain requires an acquaintance with the **standard reduction potential,** E_0, a convention used to quantify the electron transfer potential of **oxidation-reduction (redox) couples.** The standard reduction potentials (at 25°C, 1 *M* concentration, 1 atmosphere pressure, and pH 7.0) for a number of biologically important redox couples are given in Table 11-1.

We can think of a reduction potential as a measure of how easily a compound can be reduced—or, in other words, how good an electron acceptor it is. A good electron acceptor has a *positive* reduction potential, whereas a good electron donor has a *negative* reduction potential. The H^+/H_2 redox couple is used as a standard and is assigned the value 0.0 V (Table 11-1). For a redox couple to have a positive reduction potential therefore means that the oxidized form of the couple is a better electron acceptor—and therefore a better oxidizing agent—than is H^+ (or, alternatively, that the reduced form of the couple is a less willing electron donor—and hence a poorer reducing agent—than is H_2). Conversely, a negative reduction potential means that the oxidized form of the couple has less affinity for electrons than does H^+ or that the reduced form has a greater potential for donating electrons than does H_2.

The redox couples of Table 11-1 are arranged with the most negative reduction potentials (that is, the best electron donors and hence the strongest reducing agents) first. For any two couples on the table, the direction of the reaction under standard conditions can be predicted by inspection of the table, since the reduced form of any couple will spontaneously reduce the oxidized form of any couple below it on the table. Thus, NADH will spontaneously reduce oxaloacetate or pyruvate, but not α-ketoglutarate or acetate.

For any oxidation-reduction reaction, the standard free energy change, $\Delta G^{\circ\prime}$, is linearly related to ΔE_0, the difference between the reduction potentials of the two half-reactions:

$$\Delta G^{\circ\prime} = -nF\,\Delta E_0 = -nF(E_{0,\text{acceptor}} - E_{0,\text{donor}}) \qquad (11\text{-}6)$$

where n is the number of electrons transferred (see Table 11-1), F is the Faraday constant (23,062 cal/mol-V), and ΔE_0 is the difference in reduction potentials between the two redox couples (in volts).

For example, consider the NADH-mediated reduction of pyruvate to lactate as it occurs during anaerobic fermentation (reaction 10-13):

$$\text{Pyruvate} + \text{NADH} + \text{H}^+ \longrightarrow \text{lactate} + \text{NAD}^+ \qquad (11\text{-}7)$$

Since the standard reduction potential of the NAD^+/NADH couple is more negative than that of the pyruvate/lactate couple (−0.32 versus −0.185 V), the reaction is thermodynamically spontaneous. (That is, NADH is a better reducing agent than lactate, so NADH can reduce pyruvate to lactate under standard conditions.) Because NADH is the donor and pyruvate is the acceptor, ΔE_0 is calculated as follows:

$$\Delta E_0 = E_{0,\text{acceptor}} - E_{0,\text{donor}}$$
$$= -0.185 - (-0.32) = +0.135 \text{ V} \qquad (11\text{-}8)$$

Table 11-1 Standard Reduction Potentials for Redox Couples of Biological Relevance

Half-Cell Equation	E_0*	n†
	Volts	
acetate + CO_2 + $2H^+$ + $2e^-$ ⟶ pyruvate + H_2O	−0.70	2
succinate + CO_2 + $2H^+$ + $2e^-$ ⟶ α-ketoglutarate + H_2O	−0.67	2
acetate + $2H^+$ + $2e^-$ ⟶ acetaldehyde + H_2O	−0.58	2
3-phosphoglycerate + $2H^+$ + $2e^-$ ⟶ glyceraldehyde-3-P + H_2O	−0.55	2
α-ketoglutarate + CO_2 + $2H^+$ + $2e^-$ ⟶ isocitrate	−0.38	2
NAD^+ + $2H^+$ + $2e^-$ ⟶ NADH + H^+	−0.32	2
1,3-bisphosphoglycerate + $2H^+$ + $2e^-$ ⟶ glyceraldehyde-3-P + P_i	−0.29	2
S + $2H^+$ + $2e^-$ ⟶ H_2S	−0.23	2
acetaldehyde + $2H^+$ + $2e^-$ ⟶ ethanol	−0.197	2
pyruvate + $2H^+$ + $2e^-$ ⟶ lactate	−0.185	2
FAD + $2H^+$ + $2e^-$ ⟶ $FADH_2$	−0.18‡	2
oxaloacetate + $2H^+$ + $2e^-$ ⟶ malate	−0.166	2
fumarate + $2H^+$ + $2e^-$ ⟶ succinate	−0.031	2
$2H^+$ + $2e^-$ ⟶ H_2	0.000§	2
cytochrome *b* (Fe^{3+}) + e^- ⟶ cytochrome *b* (Fe^{2+})	+0.06	1
ubiquinone + $2H^+$ + $2e^-$ ⟶ dihydroubiquinone	+0.10	2
cytochrome *c* (Fe^{3+}) + e^- ⟶ cytochrome *c* (Fe^{2+})		
cytochrome *a* (Fe^{3+}) + e^- ⟶ cytochrome *a* (Fe^{2+})	+0.29	1
Fe^{3+} + e^- ⟶ Fe^{2+}	+0.77	1
$\frac{1}{2}O_2$ + $2H^+$ + $2e^-$ ⟶ H_2O	+0.816	2

* E_0 values are determined at pH 7.0 and 25°C, relative to the standard hydrogen half-cell. For a two-electron reaction, a difference in E_0 values of 0.10 V corresponds to a $\Delta G^{\circ\prime}$ of −4.6 kcal/mol.

† n represents the number of electrons involved in the half-cell reaction.

‡ The value for the FAD/$FADH_2$ couple presumes the free coenzyme; when bound to a flavoprotein, the coenzyme has an E_0 value in the range of 0.0 to +0.3 V, depending on the specific protein.

§ This is the standard hydrogen half-cell; it requires that $[H^+] = 1.0\ M$ and therefore specifies pH 0.0. At pH 7.0, the value for the $2H^+/H_2$ couple is −0.421 V.

The standard free energy change is then calculated according to equation 11-6:

$$\Delta G^{\circ\prime} = -nF\,\Delta E_0 = -2(23{,}062)(+0.135)$$
$$= -6227 \text{ cal/mol} \qquad (11\text{-}9)$$

Note that $\Delta G^{\circ\prime}$ and ΔE_0 have opposite signs. A reaction that is spontaneous under standard conditions has a negative $\Delta G^{\circ\prime}$ but a positive ΔE_0.

The Electron Transport Chain

Figure 11-14 shows most of the major components of the electron transport chain from NADH ($E_0 = -0.32$ V) to oxygen ($E_0 = +0.816$ V), ordered according to their standard reduction potentials. We will consider the chemical nature of these electron carriers shortly. For the moment, simply note that the position of each in the sequence is determined by its E_0 value. The transport chain, in other words, can be thought of as a series of electron carriers, with the reaction sequence determined by their relative oxidation-reduction potentials. All of the carriers are membrane components, localized within the inner mitochondrial membrane of eukaryotic cells and in the plasma membrane of prokaryotes.

Figure 11-14 also illustrates another important feature of the electron transport chain. At several points, the difference in free energy between successive carriers is large enough to drive the phosphorylation of ADP to ATP. Three such sites can be readily identified. This agrees with the observation that three ATP molecules are formed as each pair of electrons flows exergonically from NADH to oxygen. However, the linkage between electron transport and ATP synthesis is indirect, as we will see.

The Electron Carriers of the Transport Chain

The carriers that make up the electron transport chain include **flavoproteins, cytochromes,** and a quinone called **coenzyme Q (CoQ,** also known as ubiquinone because of its ubiquitous occurrence), as well as several **iron-sulfur proteins.** Except for coenzyme Q, all of these carriers are proteins with specific prosthetic groups capable of being reversibly oxidized and reduced. Most of the events of electron transport occur within membranes, so it is not surprising that, except for cytochrome *c*, these carriers are all hydrophobic molecules. We will look briefly at their chemistry and then see how they are ordered into a sequence.

Flavoproteins. Several membrane-bound enzymes participate in energy metabolism, using either **flavin mononucleotide (FMN)** or **flavin adenine dinucleotide (FAD)** as the prosthetic group. One such enzyme is NADH dehydrogenase, which uses FMN to transfer electrons from NADH to coenzyme Q. Another example, already familiar

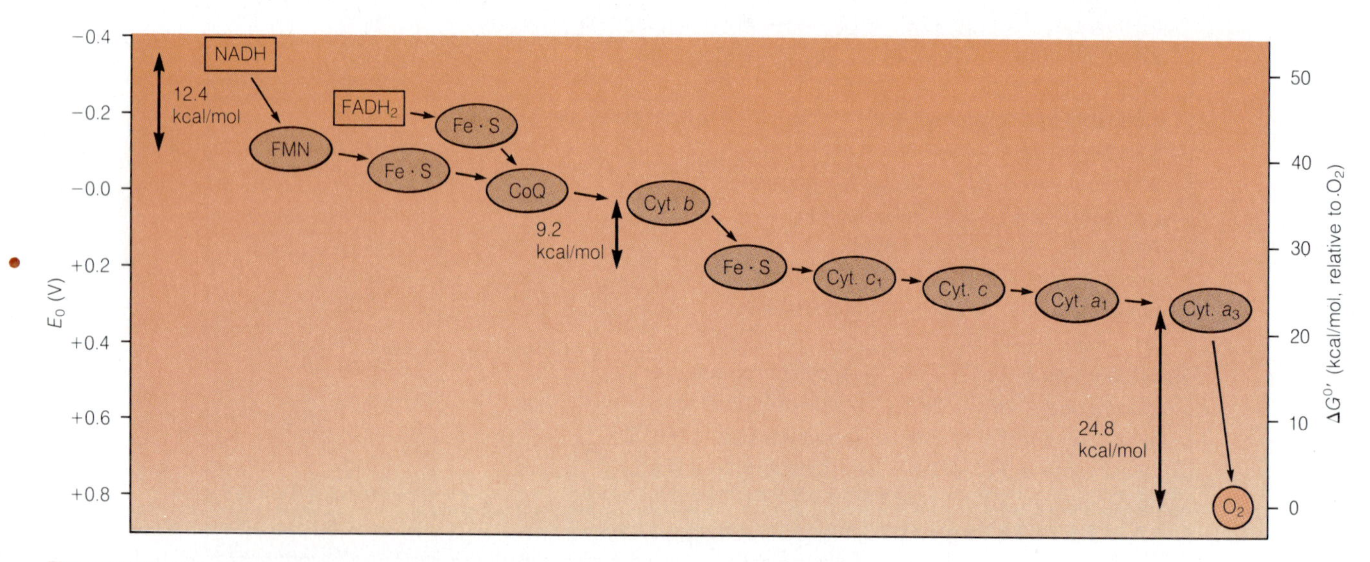

Figure 11-14 Energetics of Electron Transport. The major intermediates in the transport of electrons from NADH (−0.32 V) to oxygen (+0.816 V) are shown at positions appropriate to their energy level, as measured by the standard reduction potential. On the right-hand axis are shown the free energy levels relative to oxygen. The vertical arrows indicate three points along the chain at which the electron transfer between successive carriers is sufficiently exergonic to drive the transport of protons upon which the generation of ATP depends.

to us from the TCA cycle, is succinate dehydrogenase, which has FAD as its prosthetic group. Succinate dehydrogenase is the only enzyme of the TCA cycle that is found in the inner membrane rather than in the matrix of the mitochondrion.

Coenzyme Q. Coenzyme Q is the only nonprotein component of the electron transport chain. As Figure 11-14 indicates, coenzyme Q occupies a central position in the electron transport chain, since it is the collection point for electrons from both NADH and the $FADH_2$-linked dehydrogenases of the cell.

Figure 11-15 illustrates both the structure of coenzyme Q and its reversible reduction from the ubiquinone form to the dihydro- form. Note that coenzyme Q must take two protons into its chemical structure when it accepts a pair of electrons during electron transport, then release those protons when it is oxidized. Now imagine a closed membrane vesicle in which coenzyme Q could accept electrons from a donor on the inner surface, then diffuse to the outer surface to be oxidized by an electron acceptor. Under these conditions, protons would be picked up by the coenzyme Q on the inside and delivered to the outside, thereby providing a proton pump coupled to electron transport. As we will see, this kind of pump is used by mitochondria, chloroplasts, and bacteria to store the energy of electron transport as an electrochemical proton gradient, which in turn is used to synthesize ATP.

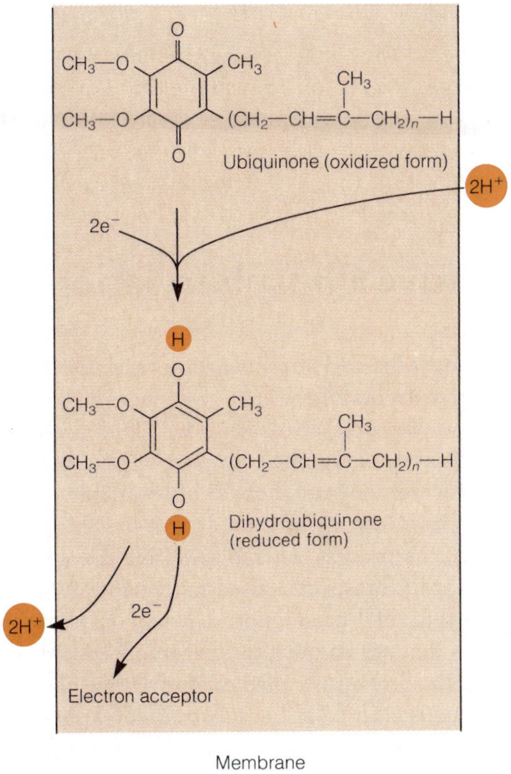

Figure 11-15 Oxidized and Reduced Forms of Coenzyme Q. Ubiquinone (the oxidized form of coenzyme Q) can be reversibly reduced to dihydroubiquinone. When ubiquinone accepts electrons from a donor in the electron transport chain, it necessarily accepts protons as well. If it gives the electrons to an acceptor on the other side of the membrane, the protons are released, thereby providing a proton pump coupled to the electron transport reaction.

Iron-Sulfur Proteins. Iron-sulfur proteins are a family of proteins containing iron and sulfur atoms complexed with four cysteine groups of the protein. The iron and sulfur atoms are usually present in equimolar amounts, with Fe_2S_2 and Fe_4S_4 the most common forms. The iron atoms are reversibly oxidizable and are therefore the actual electron acceptors and donors. NADH dehydrogenase has already been identified as a flavoprotein, but it is also an iron-sulfur protein. Electrons are initially accepted by the enzyme on its FMN group and then transferred via its iron-sulfur centers to coenzyme Q. Iron-sulfur proteins are also involved in the transfer of electrons from succinate to coenzyme Q and from coenzyme Q to cytochrome *c*.

Cytochromes. Like the iron-sulfur proteins, cytochromes also contain iron, but always as part of a porphyrin prosthetic group called **heme.** The basic structure of heme is shown in Figure 11-16. Cytochromes play an important role in the transfer of electrons from coenzyme Q to oxygen. There are at least five different cytochromes in the chain, designated as cytochromes b, c, c_1, a_1, and a_3. Cytochromes b, c, and c_1 all contain *iron-protoporphyrin IX*, the form of heme found in hemoglobin (Figure 11-16a). Cytochromes a_1 and a_3, on the other hand, contain a modified prosthetic group called *heme A* (Figure 11-16b).

Cytochromes b, c_1, a_1, and a_3 are integral membrane proteins, the latter two occurring together at the end of the transport chain as a complex called *cytochrome c oxidase*. Cytochrome c, on the other hand, is hydrophilic and is only loosely associated with the membrane. Like the iron of the iron-sulfur complex, the iron of the heme prosthetic group is reversibly oxidizable and serves as the electron acceptor for the cytochromes. Both cytochromes and iron-sulfur proteins are therefore one-electron carriers.

Organization and Function of the Electron Transport Chain

The organization of the electron transport chain is shown in detail in Figure 11-17. The transfer of electrons from NADH to molecular oxygen is thought to occur via three successive membrane-bound complexes called **NADH dehydrogenase, coenzyme Q-cytochrome *c* reductase,** and **cytochrome *c* oxidase.** As the figure indicates, NADH,

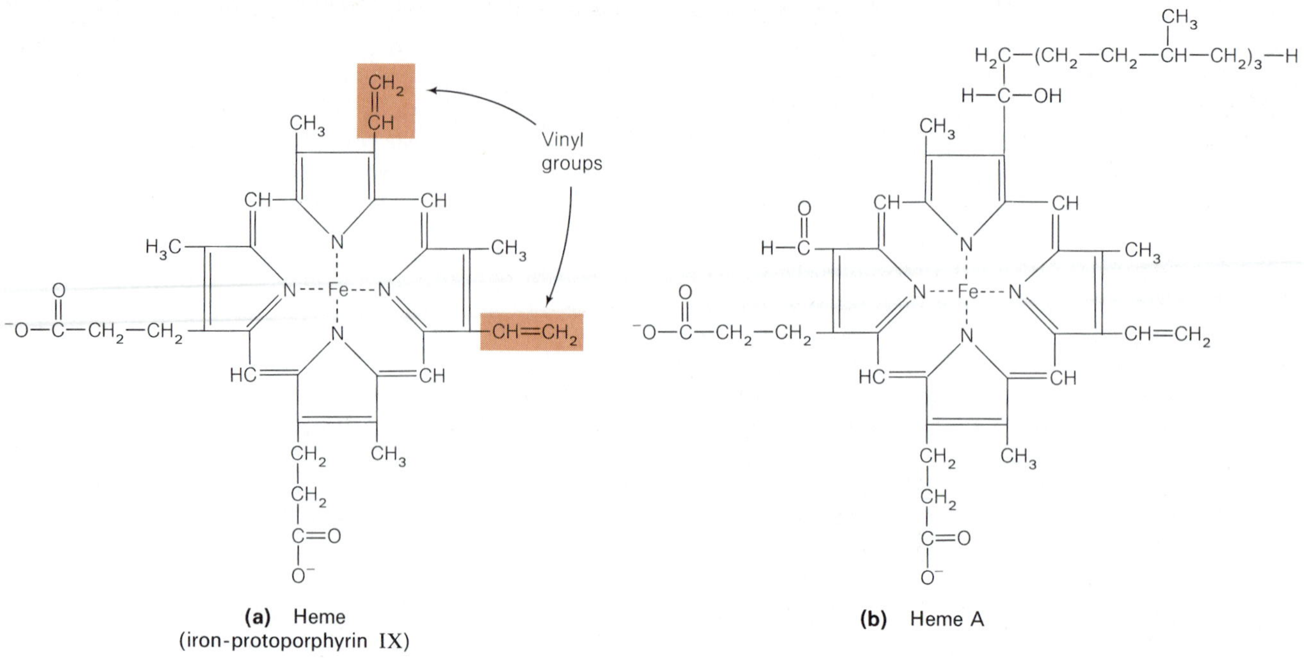

(a) Heme (iron-protoporphyrin IX)

(b) Heme A

Figure 11-16 Basic Structure of Heme. (a) Heme (iron-protoporphyrin IX), as found in cytochromes *b*, *c*, and c_1. (b) Heme A, as found in cytochromes a_1 and a_3. The heme of cytochromes *c* and c_1 is covalently attached to the protein by thioether bonds between the sulfhydryl groups of two cysteines in the protein and the vinyl ($—CH{=}CH_2$) groups of the heme (shown in color).

coenzyme Q, and cytochrome *c* are key connecting links in the electron transfer process. NADH links the chain to the TCA cycle (and other oxidative processes), while coenzyme Q and cytochrome *c* serve as electron transfer links between the complexes of the chain.

In addition to complexes I, III, and IV that transport electrons from NADH to oxygen, another enzyme complex, called complex II, transfers to coenzyme Q the electrons derived from succinate oxidation in reaction TCA-6. This complex is called the **succinate-coenzyme Q reductase** (Figure 11-17). Similar but separate complexes are required to transfer electrons to coenzyme Q from other FAD-linked dehydrogenases, such as that involved in the oxidation of fatty acids (reaction FA-2, Figure 11-10).

Of all the electron-transferring intermediates involved in respiratory metabolism, only cytochrome a_3 at the end of the transport chain is a *terminal oxidase,* capable of direct transfer of electrons to oxygen. Therefore, almost every electron extracted from any oxidizable organic molecule anywhere in an aerobic cell must eventually pass through cytochrome a_3, for this is the only link between respiratory metabolism and the oxygen that makes it all possible.

Oxidative Phosphorylation

So far, we have learned how coenzymes are reduced during the oxidative events of the first two stages of respiratory metabolism. We have also seen how reduced coenzymes are reoxidized by exergonic transport of electrons to oxygen via a membrane-bound chain of reversibly oxidizable intermediates. Now we are ready for the final stage of respiration, the process whereby the free energy released during electron transport is used to generate ATP.

Since this ATP production depends on phosphorylation events that are coupled to oxygen-dependent electron transport, the process is called **oxidative phosphorylation.** This distinguishes it from the **substrate-level phosphorylation** that occurs as an integral part of a specific reaction in a metabolic pathway. (The generation of ATP at steps Gly-7 and Gly-10 of the glycolytic pathway and at reaction TCA-5 during the TCA cycle is an example of substrate-level phosphorylation.)

Oxidative phosphorylation has been a highly controversial topic over the last 40 years. Much of the controversy has focused on the mechanism responsible for the actual

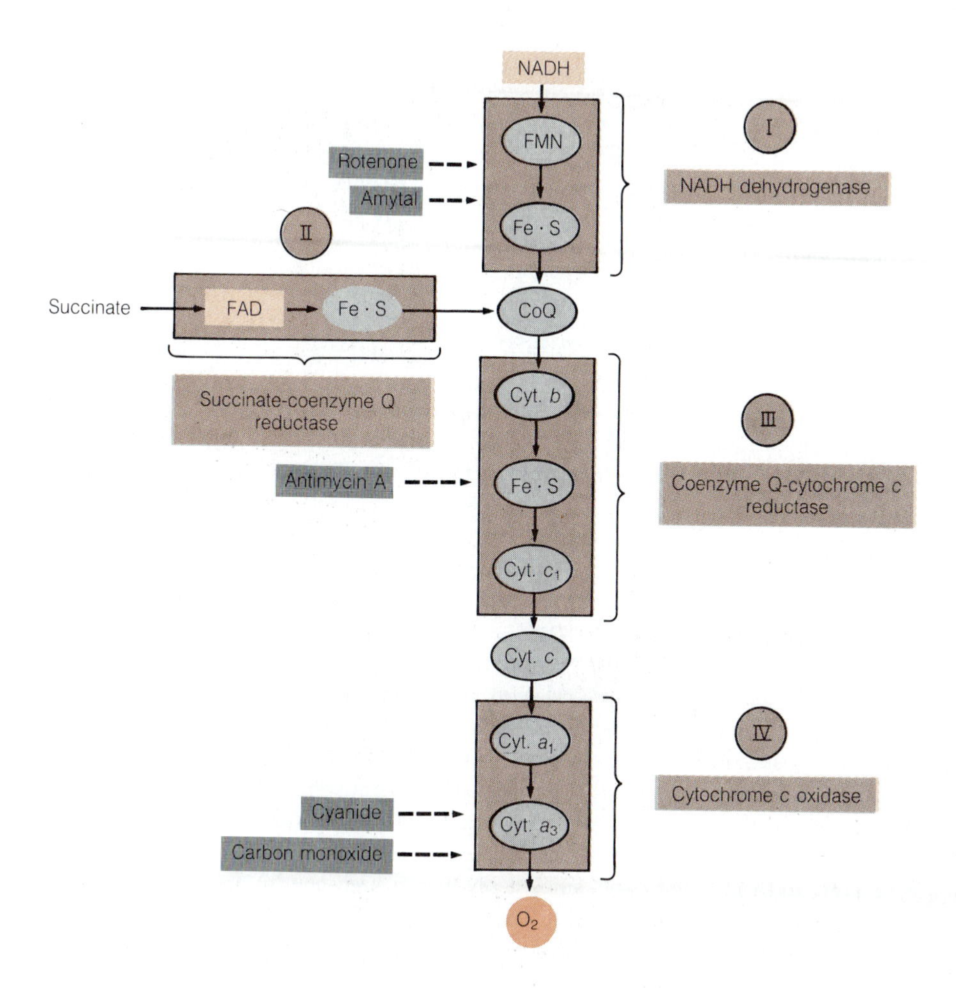

Figure 11-17 Order of Intermediates in the Electron Transport Chain. The four major enzyme complexes are indicated in boxes, with coenzyme Q and cytochrome *c* as strategic intermediates. Dashed arrows indicate sites of action of frequently used inhibitors of electron transport. Roman numerals identify the four enzyme complexes, three of which include sites of coupling to ATP synthesis.

coupling of electron transport to ATP generation, but there is now general agreement that the crucial link between electron transport and ATP production is an **electrochemical proton gradient** established by the directional pumping of protons across the membrane in which electron transport is occurring.

We will come to the details of this model shortly, but first we need some appreciation for the historical background out of which the model arose. Specifically, we want to consider the evidence for the coupling of ATP synthesis to electron transport and then look at what is known about the specific sites along the chain at which such coupling is thought to occur.

Coupling of ATP Synthesis to Electron Transport

Normally, electron transport is tightly coupled to phosphorylation. By **coupling,** we mean not only that ATP generation depends critically on electron flow but also that electron flow is possible only when ATP can be synthesized. Coupling is an important regulatory mechanism, since it ensures that the ATP/ADP ratio is a significant factor in regulating activity of the electron transport chain (and therefore of all of respiratory metabolism).

If one process can be coupled to another, one might imagine that they could also be uncoupled. In fact, chemical compounds called *uncouplers* exist and have proved to be useful probes of oxidative phosphorylation mechanisms. A classic uncoupler is 2,4-dinitrophenol (DNP). If DNP is added to respiring mitochondria while they are synthesizing ATP, the respiration rate suddenly increases and ATP synthesis drops to zero. The DNP has uncoupled electron transport from ATP synthesis. In a sense, this is like putting your car's automatic transmission in neutral: the engine speeds up somewhat, and forward motion ceases. Any mechanism put forward to account for the coupling between electron transport and ATP synthesis must also explain how uncouplers exert their effect.

Respiratory Control

Because electron transport is coupled to ATP synthesis, the availability of ADP regulates the rate of oxidative phosphorylation and therefore of electron transport. This is called **respiratory control.** Its physiological significance is easy to appreciate, since electron transport and ATP generation will be favored when the ADP level is high (that is, when the ATP level is low) and inhibited when the ADP level is low (when the ATP level is high). Oxidative phosphorylation is therefore regulated by cellular ATP needs, such that electron flow from organic fuel molecules to oxygen is adjusted to the energy needs of the cell. This regulatory mechanism becomes apparent during exercise, when ADP accumulating in muscle tissue increases electron transport rates, followed by a dramatic rise in the need for oxygen.

Sites of Synthesis

We now want to know where along the electron transport chain the actual events occur that account for the coupling of ATP generation to electron flow. Severo Ochoa in the early 1940s showed that a fixed relationship usually exists between the number of moles of ATP generated and the number of moles of oxygen atoms consumed in respiration. This relationship is expressed as the **P/O ratio**—the number of molecules of ATP generated as a pair of electrons passes along the chain from reduced coenzyme (or other electron donor) to oxygen. For NADH oxidation, the P/O ratio is 3, whereas for the oxidation of $FADH_2$ or $FMNH_2$, the ratio is 2.

These numbers suggest the existence of three regions of the electron transport chain that are related to ATP generation. These regions have been localized in the transport chain by a variety of ingenious experiments that use artificial electron donors and acceptors and specific inhibitors of electron transport to ensure that electrons flow only through a selected portion of the chain. From such studies, three sites of ATP generation have been identified. One site occurs between NADH and coenzyme Q, the second between coenzyme Q and cytochrome *c*, and the third between cytochrome *c* and oxygen. There is, in other words, one ATP-generating event associated with each of the complexes (I, III, and IV) that transfer electrons from NADH to oxygen. However, ATP is not synthesized specifically at each site. Instead, a site is simply a portion of the chain in which an electron transport complex generates sufficient energy to synthesize ATP.

Confirmation of these ATP-generating sites was provided by Ephraim Racker and his colleagues, who succeeded in incorporating each of the three complexes into synthetic phospholipid vesicles along with the mitochondrial ATP-synthesizing system. Upon addition of the appropriate oxidizable substrate, such vesicles are capable of generating one molecule of ATP per pair of electrons that passes through the complex.

Mechanism of Coupling

How can ATP generation, a dehydration reaction, be tightly coupled to electron transport, which involves the sequential oxidation and reduction of various protein complexes in a lipid bilayer? This is a fascinating question, and it is instructive to follow the evolution of ideas as the puzzle was solved by hypothesis and experimentation. An early concept was the **chemical coupling model,** first proposed in 1953. According to this scheme, oxidative phosphorylation occurs in a manner similar to substrate-level phosphorylation, involving the formation of a high-energy phosphorylated intermediate followed by transfer of the activated phosphate to ADP. The oxidation of glyceraldehyde-3-phosphate to 3-phosphoglycerate in the glycolytic pathway (reactions Gly-6 and Gly-7; recall Figure 10-7 and the box in Chapter 10) was an attractive prototype, since it involves a high-energy phosphorylated intermediate, 1,3-bisphosphoglycerate.

Despite years of attempts, no such high-energy intermediates were demonstrated for oxidative phosphorylation, and the hypothesis therefore failed an important test. Nonetheless, it played an important role in stimulating research, and an alternative hypothesis, the **chemiosmotic coupling model,** was proposed in 1961 by Peter Mitchell. According to this hypothesis, each of the three sites of coupling along the transport chain involves an electron transfer event that is accompanied by the unidirectional pumping of protons across the membrane in which the transport chain is localized. For mitochondria and bacterial membranes, protons are pumped outward. The electrochemical proton gradient that is generated and maintained in this way provides the driving force for ATP synthesis, as Figure 11-18 illustrates.

The chemiosmotic model therefore postulates an electrochemical potential across a membrane as the link between electron transport and ATP formation, rather than the high-energy phosphorylated intermediate required by the chemical hypothesis. Mitchell's model has turned out to be exceptionally useful not only because of the very plausible explanation it provides for coupled ATP generation but also because of its pervasive influence on the way we think about energy conservation in biological systems. To understand the chemiosmotic model, we will look

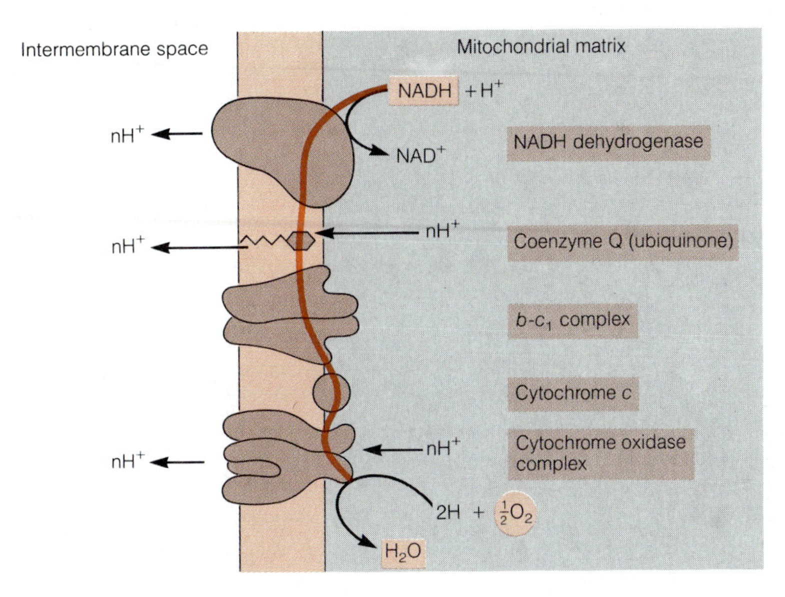

Figure 11-18 Vectorial Pumping of Protons Across the Inner Mitochondrial Membrane. Intermediates of the electron transport chain are arranged in the membrane to provide for outward proton pumping as electrons flow from reduced substrates (NADH is shown here) to oxygen. The path of electrons is shown in color. The number of protons (nH^+) pumped varies from site to site and in some cases is not known with certainty. One estimate is that 10 to 12 protons are pumped for each pair of electrons traveling down the chain.

briefly at each of its major features: the existence of a transmembrane electrochemical proton gradient, the vectorial pumping of protons that causes the gradient, and the ability of the gradient to drive ATP synthesis.

The Electrochemical Proton Gradient

According to the chemiosmotic hypothesis, the energy liberated during electron transport is conserved as an electrochemical proton gradient across the membrane in which the electron transport intermediates are embedded. This electrochemical gradient exerts what is frequently called a **proton motive force,** or **pmf.** This is analogous to *electromotive force,* or *emf,* which is used to describe the electrical potential produced by electrons. In fact, both pmf and emf are measured in volts. If you think about the terms, a motive force is simply the capacity to "move" something, in one case protons (or their equivalent) and in the other electrons. *The fundamental concept of chemiosmosis is that a pmf is produced by a proton pump coupled to electron transport. The energy available in the pmf is then used to drive an enzymatic synthesis of ATP.* It does this by reversing an ATPase activity, so that instead of ATP being hydrolyzed (by water addition) to ADP and P_i, the components of water are removed from ADP and P_i to produce ATP.

The idea that ATP can be synthesized by "reversing" an ATPase enzyme is best understood by remembering that, in thermodynamics, any reaction has an equilibrium point and is therefore reversible in principle. The main concept of the chemiosmotic theory is shown in Figure 11-19. In part (a), an imaginary lipid vesicle is shown with one F_0F_1 ATPase embedded in the membrane. ATP is presented to the vesicle, and the ATPase begins to hydrolyze it to ADP and phosphate. Peter Mitchell proposed that such ATPases have the ability to pump protons, using ATP as an energy source. For convenience, the drawing shows protons being pumped outward, so that a gradient of protons is produced. The gradient, or pmf, includes the membrane potential produced when the positively charged protons are pumped out, and the concentration gradient (pH gradient) if other ions move across the membrane to balance the charge.

The energy of the ATP hydrolysis reaction in such a system is conserved by the formation of the membrane potential and pH gradient, and it follows that at some point such a large proton gradient will be produced that no further net hydrolysis can occur. This is analogous to a water turbine pump that has driven water so far up a standpipe that no more can be pumped against the back pressure. To continue the analogy, imagine that water is poured into the top of the standpipe at this point, further increasing the back pressure: the pump will actually begin to run backward. If there were some way to put a ther-

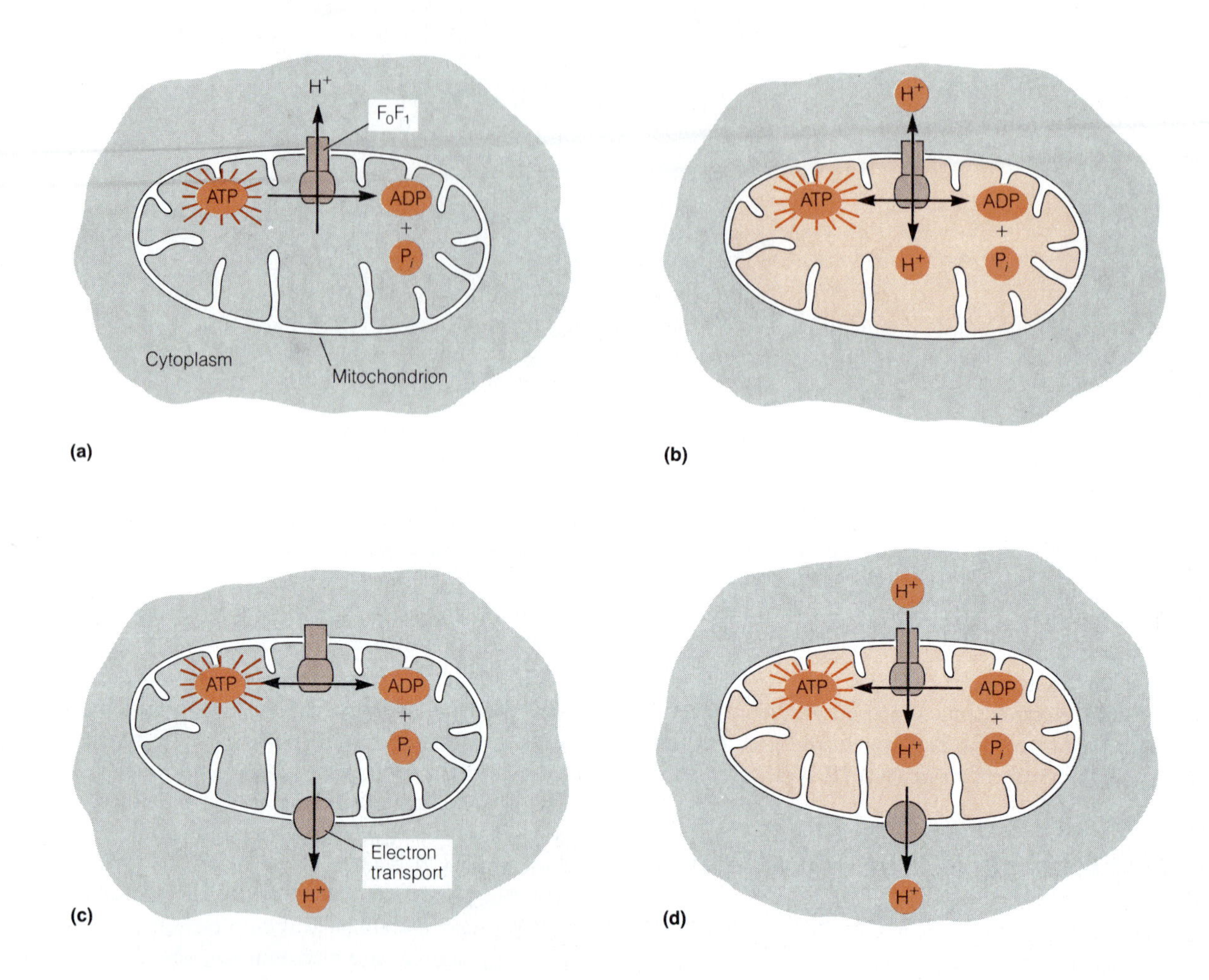

Figure 11-19 The Chemiosmotic Mechanism of ATP Synthesis. According to the chemiosmotic theory, ATP synthesis involves the "reversal" of ATP hydrolysis, so that ATP is produced from ADP and P_i. The process is catalyzed by an F_0F_1 ATPase component of the membrane. Such an ATPase is shown in part (a) for a mitochondrial membrane. When ATP is hydrolyzed by the enzyme, the energy is used to pump protons outward and is stored as a proton gradient, indicated by a lighter shade of color. (b) After a few seconds, the magnitude of the proton gradient becomes so large that the system comes to steady state, indicated by arrows in both directions. If the proton gradient were further increased at this point, it is clear that the ATPase reaction would be driven in the opposite direction, so that ATP is synthesized from ADP and P_i. This is shown in part (c), in which electron transport produces a proton gradient large enough to drive the reverse ATPase reaction, thereby synthesizing ATP (d). The proton gradient is measured as pmf (proton motive force). It can be expressed as either a concentration gradient of protons (pH gradient) or an electrical potential (membrane potential) produced when positively charged protons are pumped across the membrane, or some mix of membrane potential and pH gradient.

modynamic "back pressure" on the ATPase, it should run backward, thus synthesizing ATP from ADP and phosphate.

How can such a back pressure be exerted? The answer, shown in Figure 11-19c, is to have a second system pump protons out of the lipid vesicle. The second system, as you have probably guessed, is the proton pump coupled to electron transport. As protons are pumped outward, the proton gradient becomes so great that the ATPase can run in reverse, synthesizing ATP if ADP and phosphate are present (Figure 11-19c).

Calculating the Proton Motive Force

As noted above, the pmf is the sum of the pH gradient across the membrane and the membrane potential. The proton motive force is calculated as

$$\begin{aligned} \text{pmf} &= (RT/F)\ln([\text{H}^+]_{\text{outside}}/[\text{H}^+]_{\text{inside}}) + V_m \\ &= 2.303(RT/F)(\text{pH}_{\text{inside}} - \text{pH}_{\text{outside}}) + V_m \end{aligned} \quad (11\text{-}10)$$

where pmf is the proton motive force (in volts), R is the gas constant (1.987 cal/mol-degree), T is the temperature in degrees Kelvin (equal to degrees Celsius + 273°), F is the Faraday constant (23,062 cal/mol-V), $[H^+]_{inside}$ is the H^+ concentration inside the membrane, $[H^+]_{outside}$ is the H^+ concentration outside the membrane, pH_{inside} is the pH inside the membrane, $pH_{outside}$ is the pH outside the membrane, and V_m is the membrane potential (in volts).

A typical mitochondrion actively involved in ATP generation at 37°C might have a membrane potential of 0.16 V and a matrix pH that is 1.0 units higher than that of the surrounding cytoplasm. The pmf can therefore be calculated as

$$\begin{aligned} \text{pmf} &= (2.303)(1.987)(273 + 37)(1.0)/23{,}062 + 0.16 \\ &= 0.06 + 0.16 = 0.22 \text{ V} \end{aligned} \quad (11\text{-}11)$$

From equation 11-6, this corresponds to a free energy change of about 5.1 kcal per mole of protons ($0.22 \times 23{,}062 = 5074$ cal, or about 5.1 kcal/mol). When measured, the stoichiometry of oxidative phosphorylation is approximately one molecule of ATP generated per pair of protons. These conditions of pH and membrane potential therefore provide a driving force of about 10 kcal per mole of ATP, which is certainly in the right range for ATP synthesis.

Testing Chemiosmosis: The Unidirectional Pumping of Protons

A basic concept of the chemiosmotic model is that the establishment and maintenance of an electrochemical proton gradient occurs through **unidirectional pumping** of protons across the membrane, in which electron carriers pick up protons at one surface of a membrane and release them on the other side.

The first breakthrough relative to this prediction occurred in 1963, when André Jagendorf and Geoffrey Hind discovered that isolated spinach chloroplasts pumped protons inward upon illumination. Similar observations were soon made in mitochondrial and bacterial membranes (Figure 11-19), except that the protons were pumped in the opposite direction, from inside to out, and the energy source was respiration rather than light. The salt-loving bacterium *Halobacterium halobium* provides an especially interesting and instructive example of proton pumping, since the proton pumps of this organism can be driven either by respiratory electron transport or by light. The details of this fascinating system are discussed in the box on pp. 305–306.

ATP Synthase and the Proton Translocator

The second prediction of the chemiosmotic hypothesis was that a reversible, proton-translocating ATPase must be present in coupling membranes. We can now return to the F_1 particles of Figure 11-7. These knoblike spheres can be seen along the cristae when mitochondrial membranes are examined with the electron microscope. The particles have a diameter of about 8.5 nm, and because of their appearance they are sometimes called "lollipops."

Efraim Racker and his co-workers found that the particles could be dislodged from the membrane by mechanical agitation. Without the particles, the membrane vesicles could still carry out electron transport but could no longer synthesize ATP: they became "uncoupled" (Figure 11-20). The capacity for ATP generation was restored by adding the particles back to the membranes, suggesting that the spherical projections are an important part of the ATP-generating complex of the membrane. These particles were therefore referred to as coupling factors, and are now known to be the mitochondrial *ATP synthase*.

As shown in Figure 11-21, F_1 is only part of the ATP-synthesizing complex. F_1 is attached by a "neck" to F_0, which is embedded in the inner mitochondrial membrane, at the base of the F_1 head. F_0 serves as the **proton translocator,** the channel through which protons flow when the electrochemical gradient across the membrane is being used to drive ATP synthesis by F_1. However, it is not yet clear how the exergonic flow of protons through F_0 drives the otherwise endergonic phosphorylation of ADP to ATP by F_1.

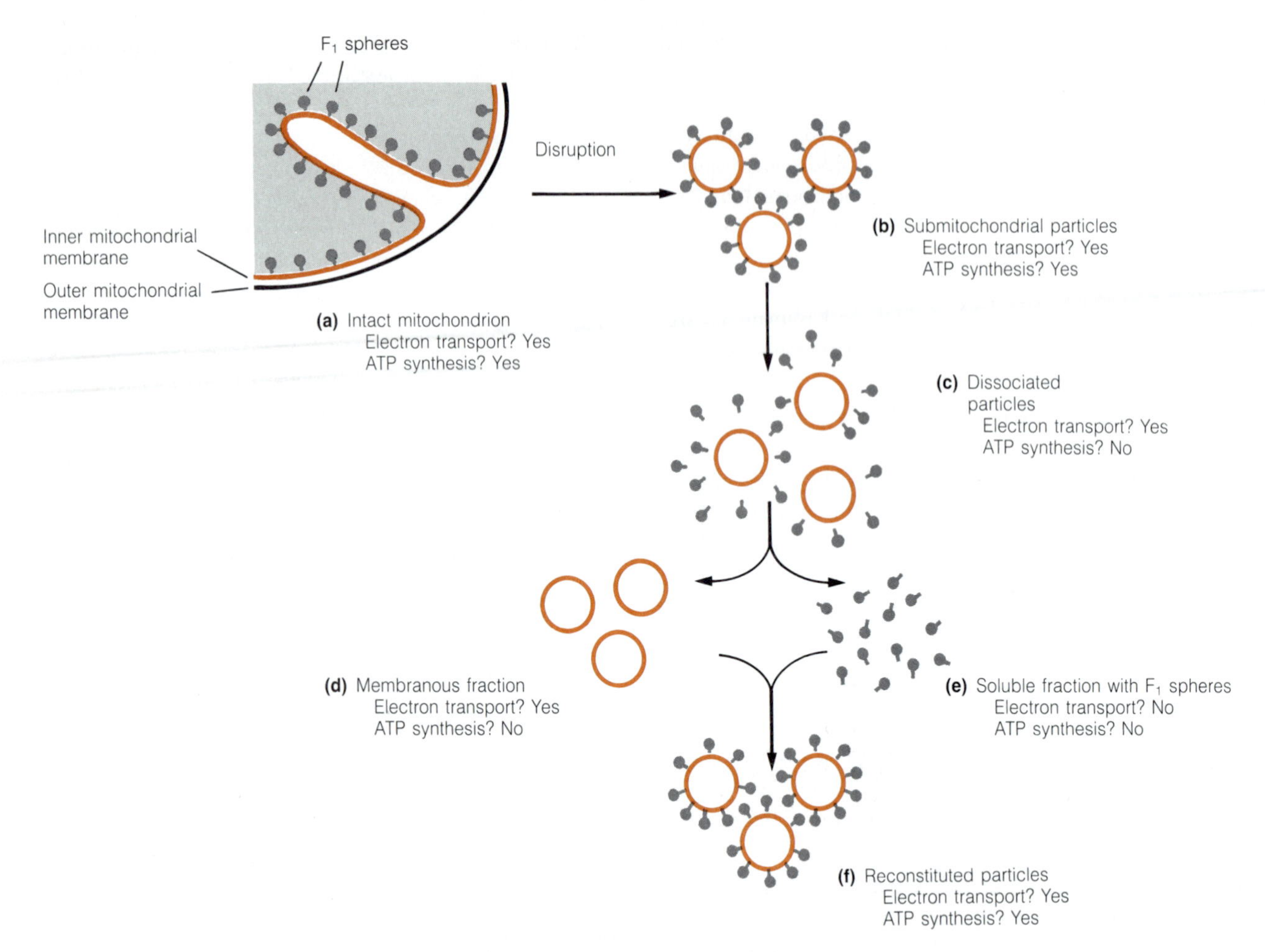

Figure 11-20 Dissociation and Reconstitution of the Mitochondrial ATP-Synthesizing System. (a) Intact mitochondria can be disrupted so that fragments of the inner membrane form (b) submitochondrial particles, capable of both electron transport and ATP synthesis. When these particles are dissociated by mechanical agitation or enzyme treatment, (c) the components can be separated into (d) a membranous fraction devoid of ATP-synthesizing capacity and (e) a soluble fraction with the F_1 spheres. (f) Mixing the two fractions reconstitutes the particles and restores ATP synthase activity.

The Role of the Electrochemical Gradient: ATP Synthesis

If the electrochemical proton gradient is the primary link between electron transport and phosphorylation of ADP, it should be possible to show not only that the gradient is generated during electron transport but also that the gradient is both necessary and sufficient to account for ATP generation. A crucial test of this came in 1966, when Jagendorf went on to demonstrate that a pH gradient imposed on isolated chloroplasts kept in the dark was adequate to cause ATP synthesis. In these experiments, chloroplasts were first incubated in an acidic buffer for several hours so that internal membranous compartments contained substantial amounts of buffer at pH 4.0. When the equilibrated chloroplasts were quickly transferred to a solution at pH 8 with ADP and inorganic phosphate present, a pH gradient of 4 units was produced, and a burst of ATP synthesis was observed as the pH gradient decayed. In other experiments, it was shown that the energy of membrane potentials artificially imposed across mitochondrial membranes could also be used to synthesize ATP. These results strongly supported the chemiosmotic concept that electrochemical proton gradients in coupling membranes drive ATP generation.

The dynamics of the proton motive force are illustrated in Figure 11-22. The electron transport chain pumps protons from the inside to the outside of the mitochondrion, and the resulting electrochemical gradient then drives ATP generation by means of the F_1F_0 systems associated with the same membrane. Presuming only that the number of

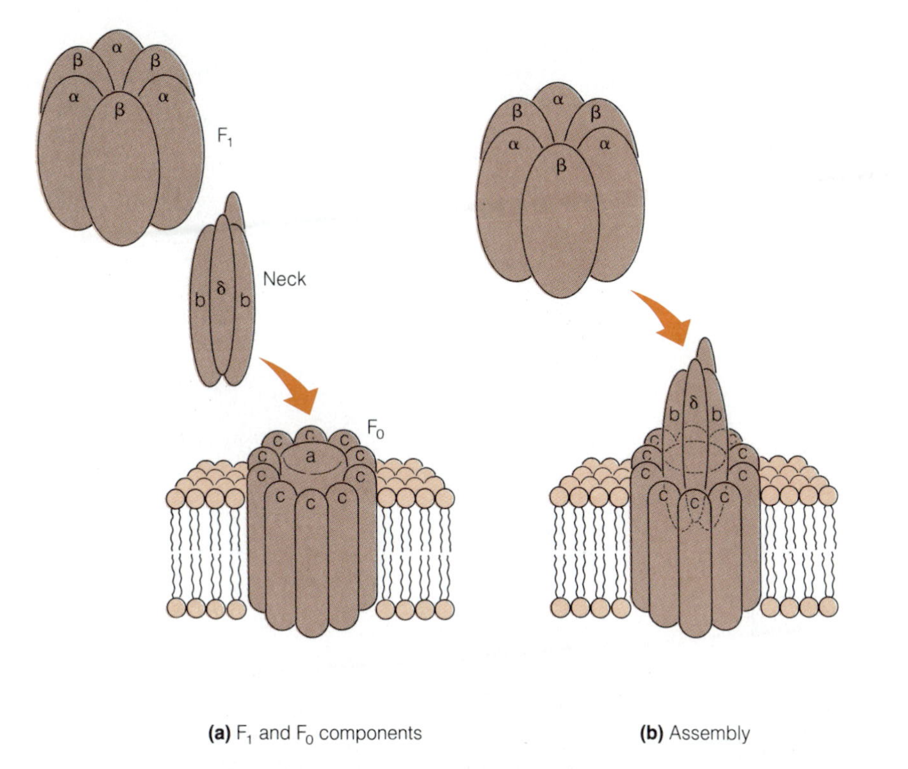

(a) F_1 and F_0 components

(b) Assembly

(c) Assembled ATP synthase

Figure 11-21 F_1 and F_0 Components of the Mitochondrial ATP-Synthesizing Enzyme. F_1 is a knoblike projection on the inner mitochondrial membrane that is responsible for ATP generation. F_0 is embedded in the lipid bilayer of the inner membrane and provides a channel for proton translocation from the cytoplasmic to the matrix side of the membrane. (a) The F_1 particle is a complex of three α and three β subunits and is shown on the upper left. The F_0 complex is composed of smaller c subunits surrounding a larger a subunit. F_1 and F_0 are attached through a neck consisting of b and δ subunits (b). The assembled ATP synthase is shown in (c). The basic reaction of ATP synthesis is the opposite of ATP hydrolysis. That is, ADP and P_i bind to the F_1 complex and form ATP by loss of water. By a still-unknown mechanism, the energy of proton flux through the F_0 complex is used to dislodge the ATP from the F_1 complex so that the active sites can bind the next round of ADP and P_i.

protons extruded by the passage of a pair of electrons through a coupling site of the transport chain is the same as the number of protons that must pass through the F_0 channel to cause one phosphorylation event, then the stoichiometry of coupling represented by the P/O ratio is clear.

It now becomes clear how uncouplers might work, according to the chemiosmotic theory. Compounds like DNP are actually weak acids that are soluble in membranes. If DNP is present when a proton gradient begins to build up from pump activity, DNP molecules can continually associate with protons on one side of the membrane, carry them across, and release the protons on the other side. If enough DNP is present, proton pumps associated with electron transport may not be able to keep up with this constant "leak" and therefore cannot produce a proton gradient of sufficient magnitude to drive ATP synthesis. The result is that the membranes are uncoupled: electron transport continues, but ATP cannot be synthesized because the protons are carried by the DNP as fast as they are pumped.

Summary of Respiratory Metabolism

To summarize respiratory metabolism, we need only place each of the components in its proper perspective. As the glycolytic pathway and the TCA cycle (or other catabolic pathways, such as β oxidation) proceed, coenzymes are continuously reduced. These reduced coenzymes represent a storage form of much of the energy of substrate oxidation—energy that can be tapped and released as the

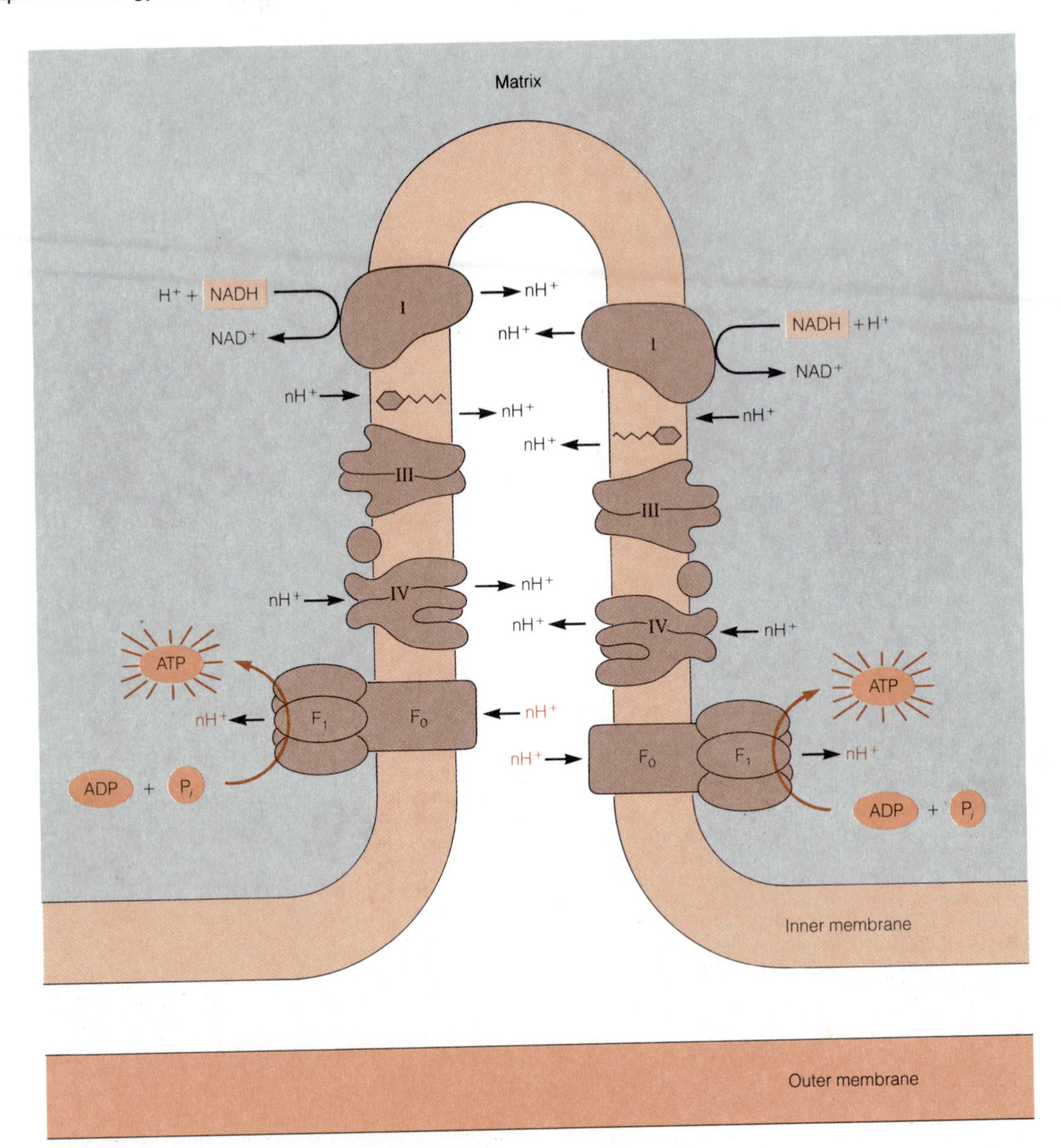

Figure 11-22 Dynamics of the Electrochemical Proton Gradient. Electron transport assemblies of complexes I, III, and IV are integrated into the inner mitochondrial membrane to provide for directional pumping of protons from the matrix to the cytoplasmic side of the membrane. The resulting electrochemical proton gradient then drives ATP synthesis by F_1 as protons are translocated back across the membrane by the F_0 complex also embedded in the inner membrane.

coenzymes are themselves reoxidized by molecular oxygen by means of the electron transport system.

As electrons are transported from NADH or $FADH_2$ to oxygen, they pass through several complexes of carriers. Each complex represents a site at which the energy of electron transport is conserved by virtue of its being coupled to the directional pumping of protons across the membrane. The resulting electrochemical gradient exerts a pmf that serves as the driving force for ATP synthesis. Under most conditions, a steady-state pmf will be maintained across the membrane, with the transfer of electrons from coenzymes to oxygen carefully and continuously adjusted so that the outward pumping of protons balances the inward flux of protons necessary to synthesize ATP at the desired rate.

The ATP Yield of Respiratory Metabolism

Now we can return to the question of the total ATP yield per molecule of glucose under aerobic conditions. Recall

Salt-Loving Bacteria, Purple Membranes, and the Chemiosmotic Model

If you have ever been swimming in the Great Salt Lake or the Dead Sea, you will probably agree that such bodies of water are aptly named, because they are very salty, and they certainly appear to be "dead." Yet as devoid of life as they may seem, such saltwater bodies are actually the only possible habitat for ***Halobacterium halobium,*** a fascinating bacterium that has shed important light on the chemiosmotic model of ATP generation. *Halobacterium* is a halophilic (salt-loving) organism that not only thrives in a high-salt environment but actually requires it. In fact, the bacterium grows best in about 4.3 *M* NaCl and cannot survive in an environment below about 3.0 *M* NaCl. (Ordinary seawater is only about 0.6 *M* NaCl.)

Since our primary interest in *Halobacterium* is bioenergetic, we want to focus not on its fondness for salt, but on its ability to meet its energy needs in either of two ways. Provided with an oxidizable substrate and an aerobic environment, *Halobacterium* cells will carry out respiratory metabolism as expected, using oxidative phosphorylation to make ATP. However, Walther Stoeckenius and his colleagues have shown that, under anaerobic conditions, *Halobacterium* synthesize **bacteriorhodopsin,** the purple membrane pigment that is capable of pumping protons across the bacterial cell membrane to create a light-dependent electrochemical proton gradient. (Recall Figure 8-12.) This led them to postulate that the energy of the gradient might be used to drive ATP synthesis directly,

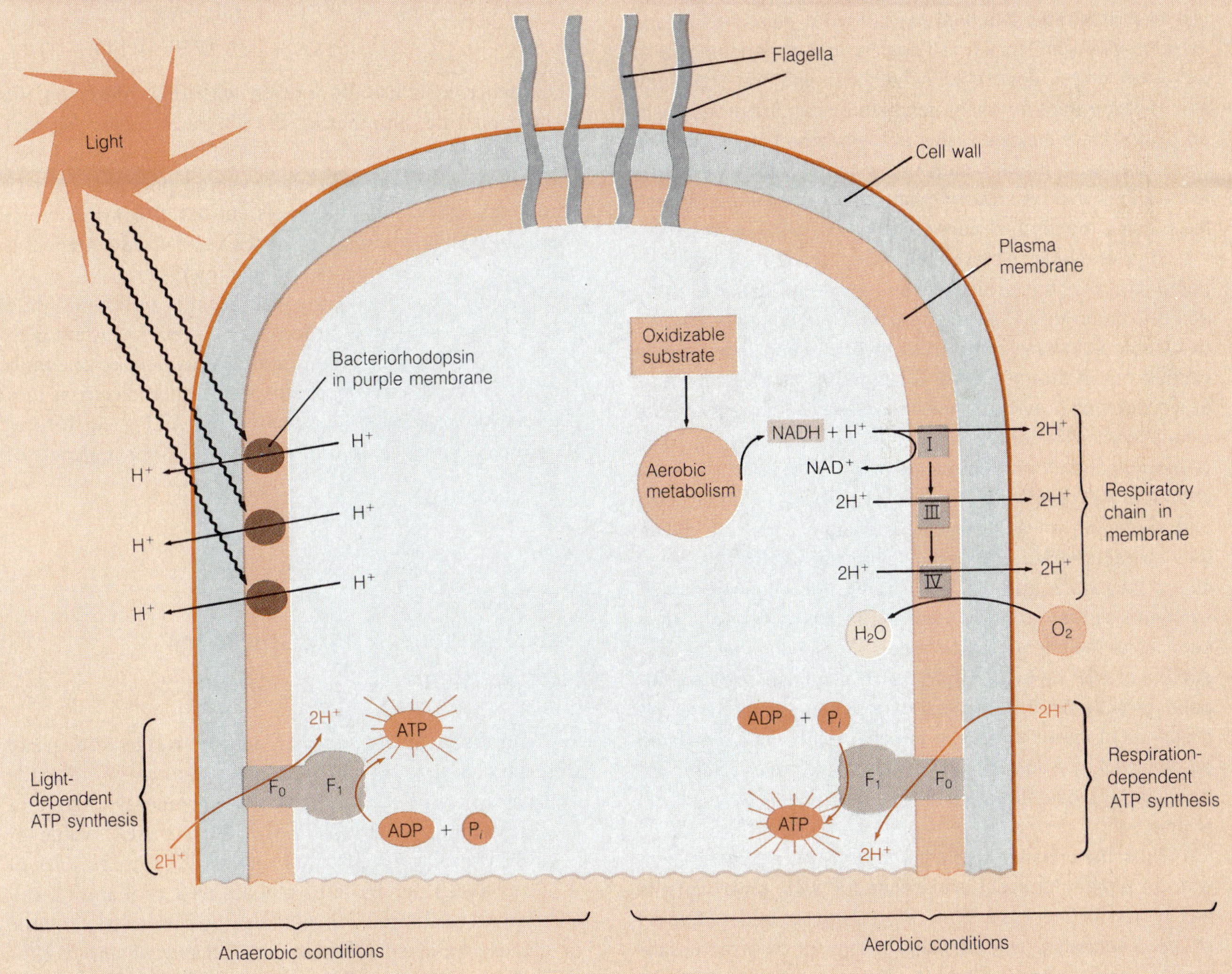

Figure 11-A The Purple Membrane of Halobacterium.

an interpretation prompted by Mitchell's chemiosmotic model of energy transduction.

They tested this idea by using known inhibitors of membrane ATP synthase activity as well as uncouplers of oxidative phosphorylation, classes of compounds well known from mitochondrial studies. Both results came out as expected: The inhibitors prevented ATP synthesis, and the uncouplers not only inhibited ATP accumulation but also prevented the establishment of a proton gradient. Furthermore, the inhibitors and the uncouplers were equally effective in preventing ATP generation by either light or respiration, strongly suggesting that an electrochemical proton gradient is essential for ATP generation in both cases. Significantly, inhibitors that interfered specifically with the electron transport chain blocked both acidification of the medium and ATP generation in respiring cells, but were without effect on either process when light was the only source of energy.

These findings are consistent with the concept of the purple membrane as a light-driven proton pump, capable of extruding protons into the medium and thereby generating an electrochemical gradient indistinguishable from that generated by electron transport during aerobic respiration. Figure 11-A illustrates this dual capability for pumping protons in *Halobacterium*. The left half depicts light-dependent proton extrusion under anaerobic conditions, and the right half shows respiration-driven proton extrusion in the presence of an oxidizable substrate under aerobic conditions. Although these are two different means of pumping protons, both lead to the same result: an electrochemical proton gradient across the plasma membrane, capable of driving ATP synthesis.

Further evidence that bacteriorhodopsin acts as a proton pump and that bacteriorhodopsin plus light has the same ability to generate ATP as electron transport components plus oxidizable substrate came from reconstitution experiments carried out in Efraim Racker's laboratory at Cornell, Racker's group had previously shown that they could reconstitute a functional ATP-generating system by incorporating parts of the mitochondrial electron transport chain and the mitochondrial ATP synthase complex into artificial lipid vesicles and supplying an appropriate oxidizable substrate. When they substituted bacteriorhodopsin for the electron transport chain, they were able to demonstrate light-dependent ATP synthesis by their artificial vesicles. This provided important support for the chemiosmotic model by showing that the production of ATP by mitochondrial ATP synthase does not depend directly on the process of electron transport or the presence of intermediates in the chain, but can be effected equally well by a light-sensitive membrane protein from bacterial cells, provided only that light is available as a source of energy.

Thus, the unusual purple membrane of a microorganism with an equally unusual preference for salty environments has played an important role in enhancing our understanding of the mechanism whereby the energy of oxidizable substrates or the energy of light is used to synthesize ATP.

from reaction 11-3 that the complete oxidation of glucose to carbon dioxide results in the generation of 4 molecules of ATP by substrate-level phosphorylation, with most of the remaining free energy of glucose oxidation stored in the 12 coenzyme molecules—10 of NADH and 2 of $FADH_2$. The electrons from NADH pass through all three complexes of the electron transport chain and therefore give rise to three ATP-generating events, while for $FADH_2$ the ATP yield is only 2 molecules. The reoxidation of the 12 coenzyme molecules can be represented as follows:

$$10NADH + 10H^+ + 5O_2 + 30ADP + 30P_i \longrightarrow 10NAD^+ + 10H_2O + 30ATP + 30H_2O \quad (11\text{-}12)$$

$$2FADH_2 + O_2 + 4ADP + 4P_i \longrightarrow 2FAD + 2H_2O + 4ATP + 4H_2O \quad (11\text{-}13)$$

Addition of reaction 11-12 and reaction 11-13 to the summary reaction for glycolysis and the TCA cycle (reaction 11-3) leads to the following overall expression for the complete aerobic respiration of glucose in a prokaryotic cell:

$$C_6H_{12}O_6 + 6O_2 + 38ADP + 38P_i \longrightarrow 6CO_2 + 38ATP + 44H_2O \quad (11\text{-}14)$$

(This reaction is usually written with only $6H_2O$ on the right-hand side, eliminating the 38 water molecules that result from phosphorylation of ADP. Keep in mind that for each glucose molecule, electron transport generates 12 water molecules, the TCA cycle consumes 6 water molecules, and ATP synthesis produces another 38 water molecules for a net balance of 44, as shown here.)

For eukaryotic cells, the ATP yield is actually slightly lower, because the two molecules of NADH that are generated in the cytoplasm require the equivalent of one molecule of ATP each to effect transport of their electrons into the mitochondria. As a result, the net ATP yield for eukaryotic cells is only 36 molecules of ATP per molecule of glucose.

The Efficiency of Respiratory Metabolism

To determine the efficiency of aerobic respiration, recall that the complete oxidation of glucose to carbon dioxide and water occurs with a standard free energy change of −686 kcal/mol (reaction 10-8). Given a $\Delta G^{\circ\prime}$ value of −7.3 kcal/mol for the hydrolysis of ATP, the 38 moles of ATP produced per mole of glucose in a prokaryotic cell correspond to about 277 kcal of energy conserved per mole of glucose oxidized. This is an efficiency of about 40% (277/686 × 100%), well above that obtainable with the most efficient machines. And when we recognize that actual $\Delta G^{\circ\prime}$ values for ATP hydrolysis under cellular conditions are often substantially more negative than -7.3 kcal/mol, the efficiency of the process becomes even more impressive.

Transport Across the Mitochondrial Membrane

To function properly, a mitochondrion must be able to move a number of substances across its membranes selectively. The outer membrane poses no problem—most molecules can pass through it freely. The inner mitochondrial membrane, on the other hand, is a significant permeability barrier. In general, the only molecules that readily cross the inner membrane are those for which specific transport proteins are present in the membrane. The most important of these transport mechanisms are summarized in Figure 11-23.

Pyruvate and fatty acids must be transported into the mitochondrion because these substances are required for respiratory metabolism. For amphipathic functions, dicarboxylic and tricarboxylic acids must be able to move into or out of the organelle. In addition, ATP that is generated within the mitochondrion must get out into the cytoplasm, and ADP and P_i must get back into the organelle. Moreover, since at least some NADH is formed in the cytoplasm, we also need to ask how the electrons from cytoplasmic coenzymes are passed into the mitochondrion and why the ATP yield from cytoplasmically generated NADH molecules is two molecules per molecule of coenzyme instead of three.

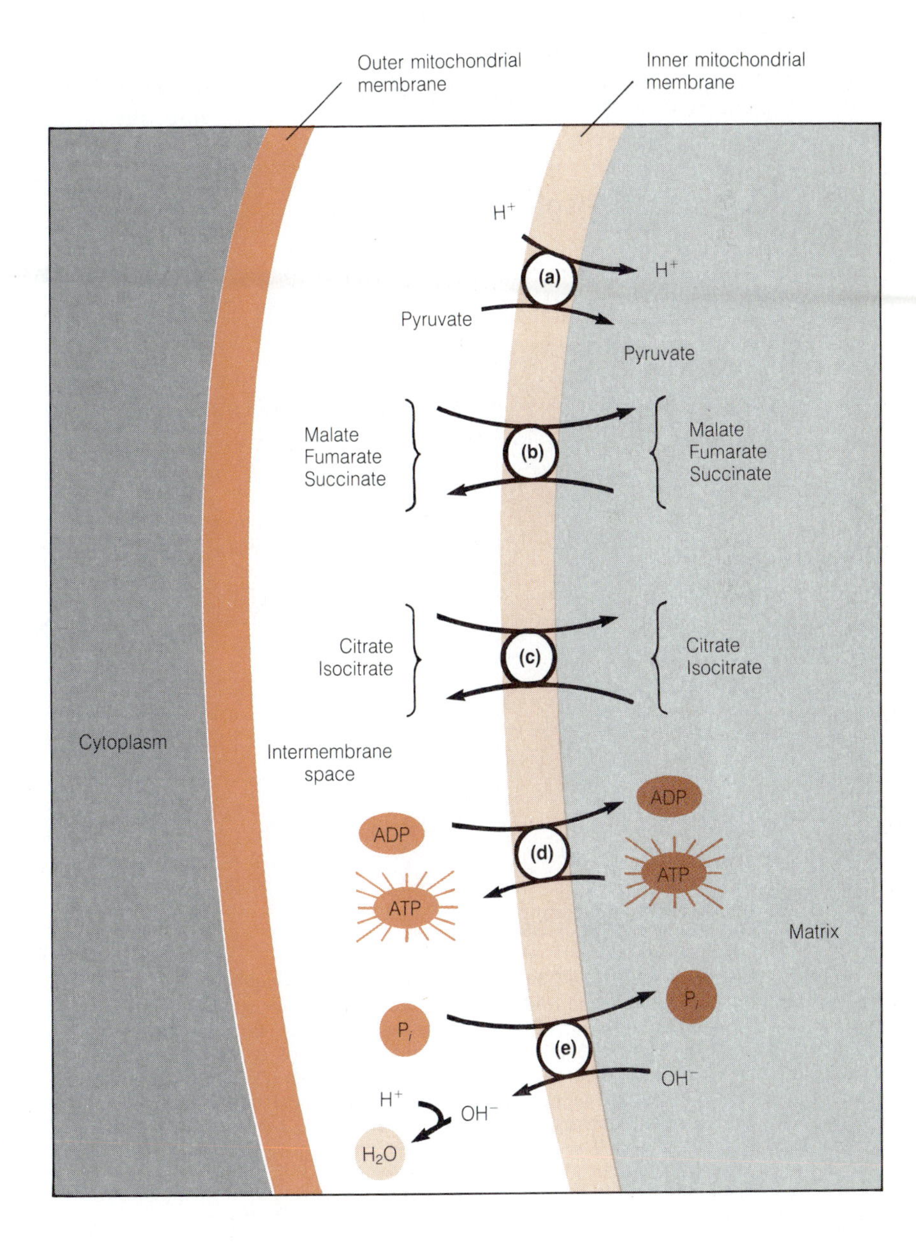

Figure 11-23 Major Transport Systems of the Inner Mitochondrial Membrane. (a) The pyruvate carrier cotransports pyruvate and protons inward, driven by the proton gradient. (b) The dicarboxylate and (c) tricarboxylate carriers exchange organic acids across the membrane. (d) The ATP-ADP carrier exchanges ATP outward for ADP inward, and (e) the phosphate carrier couples inward movement of phosphate with the outward movement of hydroxyl ions, which are neutralized by protons in the intermembrane space. For the mechanism of inward electron transport, see Figure 11-24.

Metabolites

Pyruvate is the major metabolite that must cross the membrane when carbohydrates are being metabolized aerobically. The transport protein that facilitates movement of pyruvate across the inner membrane requires the simultaneous inward transport of protons (Figure 11-23a). Pyruvate transport therefore depends on both the existence and size of the electrochemical proton gradient across the membrane. In other words, not all of the pmf generated by electron transport is used for ATP generation; some of it is required to drive the inward transport of pyruvate.

Other organic molecules move across the membrane in such a way that the inward transport of one solute is coupled with the concomitant outward movement of another. Thus, the *dicarboxylate carrier* shown in Figure 11-23b allows for the reciprocal exchange of malate, succinate, and fumarate with one another or for the exchange of any of these acids with inorganic phosphate. Similarly, the *tricarboxylate carrier* provides for the exchange of citrate and isocitrate with one another or with a dicarboxylic acid (Figure 11-23c).

ATP, ADP, and Phosphate

Since most of the ATP in an aerobic eukaryotic cell is made in the mitochondria but is used elsewhere in the cell, we should expect an outward flux of ATP across the inner mitochondrial membrane. This flux is mediated by the *ATP-ADP carrier*, which couples outward ATP movement with the inward movement of ADP (Figure 11-23d). The total concentration of ADP + ATP within the organelle therefore remains constant. Transport of phosphate is mediated by the *phosphate carrier*, which exchanges phosphate inward for hydroxyl ion outward, as Figure 11-23 illustrates. Since each hydroxyl ion neutralizes a hydrogen ion when it reaches the outside of the membrane, this carrier is in effect also driven by the proton gradient.

Electrons from Cytoplasmically Generated NADH

Finally, we come to the NADH that is generated in the cytoplasm but must be reoxidized within the mitochondrion. In contrast to the other molecules, neither NADH nor NAD^+ can cross the inner mitochondrial membrane, since the membrane has no carrier for these coenzymes. Instead, the electrons and H^+ ions are passed inward to the electron transport chain without the physical movement of coenzyme molecules across the membrane. This is accomplished by a mechanism called the **glycerol-3-phosphate–dihydroxyacetone phosphate shuttle.** This shuttle involves cytoplasmic and mitochondrial forms of the en-

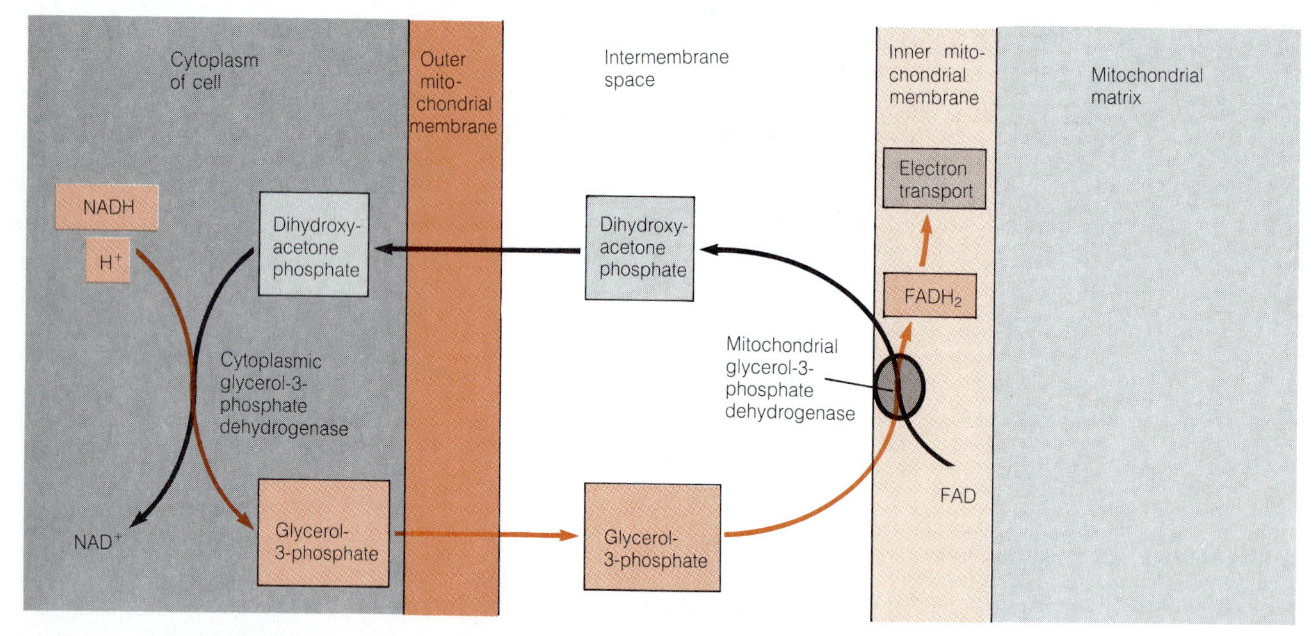

Figure 11-24 Glycerol-3-Phosphate/Dihydroxyacetone Phosphate Shuttle. Electrons (and protons) from cytoplasmic reduced NADH are used to reduce dihydroxyacetone phosphate to glycerol-3-phosphate, which then becomes the vehicle by which the electrons are moved inward. The glycerol-3-phosphate can be oxidized back to dihydroxyacetone phosphate by an FAD-linked glycerol-3-phosphate dehydrogenase (GPDH) in the inner membrane. Because NADH is a more energy-rich compound than $FADH_2$, the inward transport of electrons is exergonic, driven by the difference in E_0 between the two coenzymes.

zyme glycerol phosphate dehydrogenase (GPDH), which reversibly oxidizes glycerol-3-phosphate to dihydroxyacetone phosphate (Figure 11-24).

The cytoplasmic form of the enzyme transfers electrons from NADH to dihydroxyacetone phosphate, reducing it to glycerol-3-phosphate. The glycerol-3-phosphate then passes into the mitochondrion, where it is reoxidized to dihydroxyacetone phosphate by the mitochondrial form of the enzyme. Mitochondrial GPDH is localized within the inner membrane. It uses FAD instead of NAD^+ as its electron acceptor and therefore transfers electrons directly to coenzyme Q. As a result, these electrons bypass the first energy-conserving site of the transport chain, generating only two molecules of ATP instead of three. The ATP yield of NAD^+-linked oxidations is therefore three molecules for mitochondrial dehydrogenases but only two molecules for cytoplasmic dehydrogenases. The difference in energy between the NADH in the cytoplasm and the $FADH_2$ in the mitochondrion provides the driving force for the shuttle that brings the electrons into the organelle.

Perspective

Compared to fermentative processes, aerobic respiration gives the cell access to much more of the energy of oxidizable organic substrates. The complete catabolism of carbohydrates begins with the glycolytic pathway, but the pyruvate that is formed is passed into the mitochondrion, where it is oxidized fully by means of the TCA cycle. Fatty acids are alternative substrates for energy metabolism in many cells. Their catabolism begins with β oxidation to acetyl CoA, which then enters the TCA cycle.

Reduced coenzymes are reoxidized by the electron transport chain embedded in the inner mitochondrial membrane (or, in the case of prokaryotes, in the plasma membrane). Oxygen is the ultimate electron acceptor, and water is the final product. As electrons move exergonically down the transport chain, they pass through three coupling sites. These are transfer events at which protons are either released or taken up. The carriers at each such junction are ordered in the membrane so that protons are drawn from the inside of the organelle as needed and extruded to the exterior when released. This establishes an electrochemical proton gradient that is the driving force for ATP generation. The ATP-synthesizing system consists of a proton translocator, F_0, embedded in the membrane, and an ATP synthase, F_1, a knoblike structure that projects from the inner membrane on the matrix side.

Mitochondria are universally the site of respiratory metabolism in eukaryotic cells and are prominent organelles in terms of both size and numbers. They are usually several micrometers long and range in abundance from a few to a few thousand per cell. Specific carriers are required to effect the inward transport of pyruvate, fatty acids, and other organic molecules across the inner membrane of the organelle. ATP transport outward is coupled to the inward movement of ADP, and phosphate exchanges with hydroxyl ion, driven by the proton gradient. The electrons of coenzyme molecules that undergo reduction in the cytoplasm are passed inward to the electron transport chain by a shuttle system, since the membrane is not permeable to the coenzymes themselves.

This, then, is aerobic energy metabolism. No transistors, no mechanical parts, no noise, no pollution—and all done in units of organization that require an electron microscope to visualize. Yet the process goes on continuously in almost every living cell with a degree of integration, efficiency, fidelity, and control that we can scarcely understand well enough to appreciate fully, let alone aspire to reproduce in our test tubes.

Key Terms for Self-Testing

mitochondria (p. 275)

The Aerobic Mode
respiratory metabolism (p. 275)
respiration (p. 275)

Mitochondrial Structure and Function
outer membrane (p. 279)
intermembrane space (p. 280)
inner membrane (p. 280)
cristae (p. 280)
matrix (p. 280)
F_1 (coupling factor) particles (p. 281)
ATP synthase (p. 281)
F_0 complex (p. 281)

The Tricarboxylic Acid Cycle
tricarboxylic acid (TCA) cycle (p. 282)
acetyl coenzyme A (acetyl CoA) (p. 282)
coenzyme A (CoA) (p. 284)
triglycerides (p. 287)
glycerol (p. 287)
fatty acids (p. 287)
β oxidation (p. 287)
proteolysis (p. 289)
proteases (p. 289)
peptidases (p. 289)
transamination (p. 290)
oxidative deamination (p. 290)
amphibolic pathway (p. 290)

Electron Transport
electron transport (p. 292)
oxidative phosphorylation (p. 292)
electron transport chain (p. 292)
standard reduction potential (p. 292)
oxidation-reduction (redox) couples (p. 292)
flavoproteins (p. 294)
cytochromes (p. 294)
coenzyme Q (CoQ) (p. 294)
iron-sulfur proteins (p. 294)
flavin mononucleotide (FMN) (p. 294)
flavin adenine dinucleotide (FAD) (p. 294)
heme (p. 295)
NADH dehydrogenase (p. 295)
coenzyme Q-cytochrome *c* reductase (p. 295)
cytochrome *c* oxidase (p. 295)
succinate-coenzyme Q reductase (p. 296)

Oxidative Phosphorylation
oxidative phosphorylation (p. 296)
substrate-level phosphorylation (p. 296)
electrochemical proton gradient (p. 297)
coupling (p. 297)
respiratory control (p. 298)
P/O ratio (p. 298)
chemical coupling model (p. 298)
chemiosmotic coupling model (p. 298)
proton motive force (pmf) (p. 299)
unidirectional pumping (p. 301)
proton translocator (p. 301)

Transport Across the Mitochondrial Membrane
glycerol-3-phosphate–dihydroxyacetone phosphate shuttle (p. 308)

Salt-Loving Bacteria, Purple Membranes, and the Chemiosmotic Model
Halobacterium halobium (p. 305)
bacteriorhodopsin (p. 305)

Suggested Reading

General References and Reviews

Harold, F. M. *The Vital Force: A Study of Bioenergetics.* New York: W. H. Freeman, 1986.

Mathews, C. K., and K. E. van Holde. *Biochemistry,* Redwood City, Calif.: Benjamin/Cummings, 1990.

Racker, E. From Pasteur to Mitchell: A hundred years of bioenergetics. *Federation Proc.* 39 (1980): 210.

Stryer, L. *Biochemistry,* 3d ed. New York, W. H. Freeman, 1988.

Wood, W. B., J. H. Wilson, R. M. Benbow, and L. E. Hood. *Biochemistry: A Problems Approach,* 2d ed. Menlo Park, Calif.: Benjamin/Cummings, 1981.

Mitochondrial Structure

Attardi, G., and G. Schatz. Biogenesis of mitochondria. *Annu. Rev. Cell Biol.* 4 (1988): 289.

Ernster, L., and G. Schatz. Mitochondria: An historical overview. *J. Cell Biol.* 91 (1981): 227s.

Srere, P. A. The structure of the mitochondrial inner membrane-matrix compartment. *Trends Biochem. Sci.* 7 (1982): 375.

Tzagoloff, A. *Mitochondria.* New York: Plenum Press, 1982.

Whittaker, P. A., and S. M. Danks. *Mitochondria: Structure, Function, and Assembly.* New York: Longman, 1979.

The Tricarboxylic Acid Cycle

Krebs, H. A. The history of the tricarboxylic acid cycle. *Perspect. Biol. Med.* 14 (1970): 154.

Electron Transport and Oxidative Phosphorylation

Boyer, P. O., B. Chance, L. Ernster, P. Mitchell, E. Racker, and E. C. Slater. Oxidative phosphorylation and photophosphorylation. *Annu. Rev. Biochem.* 46 (1977): 935.

Cross, R. L. The mechanism and regulation of ATP synthesis by F_1-ATPases. *Annu. Rev. Biochem.* 50 (1981): 681.

Fillingame, R. The proton-translocating pumps of oxidative phosphorylation. *Annu. Rev. Biochem.* 49 (1980): 1079.

Hinkle, P. C., and R. E. McCarty. How cells make ATP. *Sci. Amer.* 238 (March 1978): 104.

Mitchell, P. Keilin's respiratory chain concept and its chemiosmotic consequences. *Science* 206 (1979): 1148.

Schneider, E., and Altendorf, K. The proton-translocating portion (F_0) of the E. coli ATP synthase. *Trends Biochem. Sci.* 9 (1984): 51.

Regulation of Respiratory Metabolism

Atkinson, D. E. *Cellular Energy Metabolism and Its Regulation.* New York: Academic Press, 1977.

Erecinska, M., and D. F. Wilson. Regulation of cellular energy metabolism. *J. Membr. Biol.* 70 (1982): 1.

Herman, R. H., R. M. Cohn, and P. D. McNamara, eds. *Principles of Metabolic Control in Mammalian Systems.* New York: Plenum Press, 1980.

The Purple Membrane of Halophilic Bacteria

Racker, E., and W. Stoeckenius. Reconstitution of purple membrane vesicles catalyzing light-driven proton transport and adenosine triphosphate formation. *J. Biol. Chem.* 249 (1974): 662.

Stoeckenius, W. The purple membrane of salt-loving bacteria. *Sci. Amer.* 234 (June 1976): 38.

Stoeckenius, W., and D. Oesterhelt. Light energy transduction by the purple membrane of halophilic bacteria. *Federation Proc.* 36 (1977): 1797.

Problem Set

1. **True or False.** Decide whether each of the following statements is true (T) or false (F).
   (a) The orderly flow of carbon through the TCA cycle is possible because each of the enzymes of the cycle is embedded in the inner mitochondrial membrane in a manner such that their order in the membrane is the same as their sequence in the cycle.
   (b) Thermodynamically, acetyl CoA should be capable of driving the phosphorylation of ADP (or GDP), just as succinyl CoA does.
   (c) It is impossible for the carboxyl carbon of pyruvate to enter the TCA cycle because it is evolved as carbon dioxide when pyruvate is oxidized to acetyl CoA.
   (d) Because of the cyclic nature of the pathway, the TCA cycle can continue to operate even though the $\Delta G'$ for one or more of its reactions is positive.
   (e) Unlike NAD^+, the coenzyme FAD tends to be tightly associated with dehydrogenase enzymes that require it as an electron acceptor.
   (f) Nine cycles of β oxidation are required to degrade the 18-carbon fatty acid oleate completely to acetyl CoA.

2. **Carbon Flow Through the TCA Cycle.** In each of the following cases, assume that the radioactively labeled compound is fed to an aerobic bacterial culture from which malate is isolated a short time later. Which carbon atom(s) of malate would you expect to find labeled *first* with ^{14}C in each case?
   (a) Acetyl CoA with ^{14}C in the methyl carbon of the acetate
   (b) Palmitate-1-^{14}C (palmitate is a 16-carbon saturated fatty acid)
   (c) Fructose-1-^{14}C
   (d) Pyruvate-1-^{14}C

3. **More Carbon Flow.** In which carbon atom(s) of glucose fed to an aerobic bacterial culture would ^{14}C have to be located to ensure that labeled CO_2 was *first* released at the step catalyzed by
   (a) Pyruvate dehydrogenase
   (b) Isocitrate dehydrogenase
   (c) α-Ketoglutarate dehydrogenase

4. **Glutamate Synthesis.** Synthesis of the amino acid glutamate can be effected from pyruvate and alanine by a metabolic sequence that illustrates the amphibolic role of the TCA cycle. Devise such a pathway, assuming the availability of whatever additional enzymes may be needed.

$$^-O{-}\overset{O}{\overset{\|}{C}}{-}CH_2{-}CH_2{-}\overset{^+NH_3}{\overset{|}{CH}}{-}\overset{O}{\overset{\|}{C}}{-}O^- \qquad CH_3{-}\overset{^+NH_3}{\overset{|}{CH}}{-}\overset{O}{\overset{\|}{C}}{-}O^-$$

Glutamate Alanine

5. **Lethal Synthesis.** The leaves of *Dichapetalum cymosum*, a South African plant, are very poisonous. Animals that eat the leaves have convulsions and usually die shortly thereafter. One of the most pronounced effects of poisoning is a marked elevation in citrate concentration and a blockage of the TCA cycle in many organs of the affected animal. The toxic agent in the leaves of the plant is fluoroacetate, but the actual poison in the tissues of the animal is fluorocitrate. If fluoroacetate is incubated with purified enzymes of the TCA cycle, it has no inhibitory effect.
   (a) Why might you expect fluorocitrate to have an inhibitory effect on one or more of the TCA cycle enzymes

if incubated with the purified enzymes in vitro, even though fluoroacetate has no such effect?

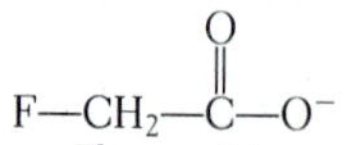

$$\mathrm{F{-}CH_2{-}\overset{\overset{\large O}{\|}}{C}{-}O^-}$$

Fluoroacetate

$$\mathrm{F{-}CH{-}\overset{\overset{\large O}{\|}}{C}{-}O^-}$$
$$\mathrm{^-O{-}\overset{\overset{\large O}{\|}}{C}{-}C{-}OH}$$
$$\mathrm{CH_2{-}\overset{\overset{\large O}{\|}}{C}{-}O^-}$$

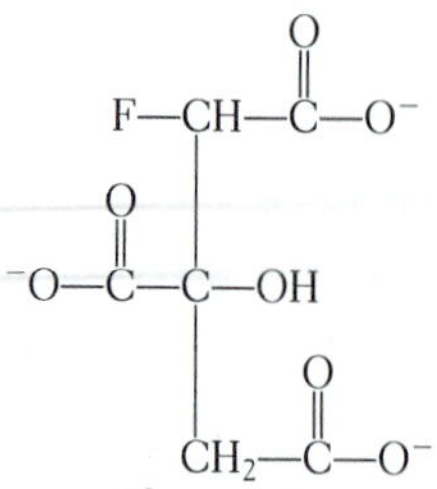

Fluorocitrate

(b) Which enzyme in the TCA cycle do you suspect is affected by fluorocitrate? Give two reasons for your answer.

(c) How do you suppose fluoroacetate gets converted to fluorocitrate?

(d) Why is this phenomenon referred to as *lethal synthesis*?

6. **Regulation of Catabolism.** Explain the advantage to the cell of each of the following regulatory mechanisms.

   (a) Isocitrate dehydrogenase (reaction TCA-3) is allosterically activated by ADP.

   (b) Citrate synthetase (reaction TCA-1) is inhibited by NADH.

   (c) Pyruvate dehydrogenase (reaction 11-1) is inactivated by ATP.

   (d) Phosphofructokinase (reaction Gly-3) is allosterically inhibited by citrate.

   (e) Pyruvate carboxylase (Figure 11-13, reaction b) is allosterically activated by acetyl CoA.

7. **Palmitate Oxidation.** Palmitate is a 16-carbon saturated fatty acid ($C_{16}H_{32}O_2$) that can be oxidized completely to carbon dioxide and water by a combination of β oxidation and the TCA cycle.

   (a) Calculate the net number of ATP molecules generated by the complete catabolism of palmitate to the two-carbon (acetyl CoA) level. (Be sure to remember the ATP that is used in the initial activation of the fatty acid in reaction FA-1!)

   (b) Next, calculate the number of ATP molecules generated by the further oxidation of the eight resulting acetyl CoA molecules to carbon dioxide and water.

   (c) Calculate the total number of ATP molecules generated by the complete oxidation of one molecule of palmitate to carbon dioxide and water.

   (d) Write a balanced overall reaction for the complete oxidation of one molecule of palmitate to carbon dioxide and water.

8. **Comparative Energy Content of Carbohydrates and Fats.** Higher plants and animals store energy reserves preferentially as fat rather than as carbohydrate because fat has a higher energy content per unit weight. The calculations specified here are designed to quantify this difference.

   (a) Consider first the utilization of glucose ($C_6H_{12}O_6$; molecular weight, 180) as an energy source and calculate the moles of ATP generated during the complete oxidative metabolism of 1 g of glucose. Assuming that the $\Delta G'$ value for the hydrolysis of ATP under physiological conditions is -12 kcal/mol, how much free energy is conserved as ATP upon aerobic catabolism of 1 g of glucose?

   (b) Now repeat the calculations of part (a), but for the complete oxidation of 1 g of palmitate ($C_{16}H_{32}O_2$; molecular weight, 256). [See part (c) of Problem 7.]

   (c) On a per-gram basis, how much more efficient is fat than carbohydrate as a form of energy storage, assuming the values for glucose and palmitate are representative of carbohydrates and fats in general? Why do organisms as diverse as castor beans and humans prefer fat as a means of storing energy reserves?

   (d) Bearing in mind that respiratory metabolism is essentially an oxidative process, how might you explain the difference in energy content of carbohydrate and fat on a per-gram basis?

9. **The Energetics of Malate Oxidation.** Malate dehydrogenase is the enzyme that catalyzes the oxidation of malate to oxaloacetate at step 8 of the TCA cycle, with NAD^+ as the electron acceptor.

   (a) Without doing any calculations, predict from Table 11-1 in which direction reaction TCA-8 should be thermodynamically spontaneous under standard conditions.

   (b) Calculate ΔE_0 and $\Delta G^{\circ\prime}$ for reaction TCA-8 in the direction required for TCA cycle activity. What do the values of these parameters tell you about the reaction? What is it that makes the reaction possible under cellular conditions when the $\Delta G^{\circ\prime}$ is so unfavorable?

   (c) Again, without doing any calculations, predict whether the reactions TCA-3 and TCA-6 will be thermodynamically feasible in the direction required for TCA cycle activity under standard conditions. Why is it not possible to make a similar prediction for reaction TCA-4 without any calculations?

10. **Inhibitors of Electron Transport.** Although not discussed explicitly in the text, several specific inhibitors of electron transport are shown in Figure 11-17, with the known site of action indicated in each case by an arrow. These inhibitors have proved especially useful in studies designed to elucidate the order of carriers in the chain and the sites of coupling to oxidative phosphorylation.

(a) Antimycin A is known to block electron transport between cytochromes *b* and c_1. Explain how this inhibitor could be used to localize one coupling site to the segment of the chain between NADH and coenzyme Q, assuming the availability of ferricyanide as an artificial electron acceptor.

(b) What independent evidence also supports this localization for one of the three coupling sites?

(c) Describe a means by which a second site might be localized to the segment of the chain between coenzyme Q and cytochrome *c*.

(d) How could the position of a third coupling site be shown, given the availability of ascorbate, an oxidizable substrate that has been shown to pass electrons directly to cytochrome *c*?

(e) Why is cyanide ion such a potent poison for virtually all aerobic forms of life? Would you expect it to be toxic to an anaerobic organism also?

11. **Rotenone Poisoning.** *Rotenone* is an extremely potent insecticide and fish poison. As shown in Figure 11-7, its mode of action is to block electron transport from the FMN of NADH dehydrogenase to coenzyme Q.

(a) Why do fish and insects die after digesting rotenone?

(b) Would you expect the use of rotenone as an insecticide to be a potential hazard to other forms of animal life (people, for example)? Explain.

(c) Would you expect the use of rotenone as a fish poison to be a potential hazard to aquatic plants that might be exposed to the compound? Explain.

12. **The Importance of Iron.** The electron transport chain consists of some intermediates (such as $FADH_2$-containing proteins and coenzyme Q) that transfer both protons and electrons and other intermediates that transfer only electrons. Iron-sulfur proteins and cytochromes are in this latter category.

(a) Why is it essential to have both types of intermediates present in the chain?

(b) Why do you suppose iron is the atom of choice for carriers that transfer only electrons?

13. **Dinitrophenol as an Uncoupler.** Normally, oxidative phosphorylation is tightly coupled to electron transport in that no electrons will flow if ATP synthesis cannot occur concomitantly (a result, for example, of low levels of either ADP or P_i). Uncoupling agents dissociate ATP synthesis from electron transport, allowing the latter to occur in the absence of the former. One such uncoupling agent is *2,4-dinitrophenol* (DNP), which is highly toxic to humans, causing a marked increase in metabolism and temperature, profuse sweating, collapse, and death. For a brief period in the 1940s, however, sublethal doses of DNP were actually prescribed as a means of weight reduction in humans.

(a) Why would an uncoupling agent like DNP be expected to cause an increase in metabolism, as evidenced by consumption of oxygen or catabolism of foodstuffs?

(b) Based on what you know about allosteric regulation of the glycolytic pathway and the TCA cycle, why would an uncoupling agent like DNP be likely to have greater, more far-reaching effects on respiratory metabolism than might be predicted by the simple lack of control of the rate of electron transport?

(c) Why would consumption of DNP lead to an increase in temperature and to profuse sweating?

(d) DNP has been shown to carry protons across biological membranes. How might this observation be used to explain its uncoupling effect?

(e) Why would DNP have been considered a drug for weight reduction? Can you guess why it was abandoned as a reducing aid?

14. **Localization of Functions Within Mitochondria.** Indicate whether you would expect to find each of the following mitochondrial constituents in the matrix (M), the inner membrane (IM), the outer membrane (OM), or the intermembrane space (IS).

(a) Coenzyme A

(b) Coenzyme Q

(c) Malate dehydrogenase

(d) Succinate dehydrogenase

(e) Fatty acyl CoA dehydrogenase

(f) High proton concentration

(g) Dicarboxylate carrier

(h) ATP-ADP carrier

(i) ATP synthetase

(j) Site of cyanide binding

15. **Mitochondrial Transport.** For aerobic energy metabolism to proceed, a variety of substances must be in a state of flux across the inner mitochondrial membrane. Assuming a cell in which glucose is the sole energy source, indicate for each of the following substances whether you would expect a net flux across the membrane and, if so, in which direction and with what stoichiometry (on a per-glucose basis).

(a) Pyruvate

(b) Oxygen

(c) ATP

(d) ADP

(e) Acetyl CoA

(f) Glycerol-3-phosphate

(g) NADH

(h) $FADH_2$

(i) Oxaloacetate

(j) Water

(k) Electrons

(l) Protons

12

Energy from the Sun: Photosynthesis

The land masses of our planet are dominated by the color green, a pleasant distinction arising from the chlorophyll-containing organisms that cover the Earth's surface in temperate and tropical regions. These organisms use sunlight as a source of energy and are therefore our only link with the solar energy that makes life on this planet possible.

to trap carbon dioxide in the form of reduced organic compounds. We therefore have two general questions to deal with: how solar energy is trapped and converted to chemical energy that can be used to drive biosynthetic processes, and how carbon dioxide is used as the sole carbon source in that biosynthesis.

The Photosynthetic Mode

In the preceding two chapters, we described the first of two major solutions to the universal problem of meeting energy needs of living cells. The chemotrophic solution depends on the energy contained in chemical compounds, often organic molecules. Thus, the starting materials for the energy metabolism discussed in Chapters 10 and 11 were organic molecules such as sugars, fatty acids, or amino acids.

We now come to the second major solution, that of the **phototrophs.** As the name suggests, phototrophs are organisms that are capable of using light energy of solar radiation for ATP synthesis. There are two types of phototrophs, depending on the way they meet their carbon needs. **Photoheterotrophs** get their energy from the sun but depend on organic compounds for their carbon. **Photoautotrophs** also depend on solar radiation for their energy, but they are able to use carbon dioxide to meet their carbon needs. Of these two categories, photoautotrophs are the more familiar. All higher plants, all algae, and all photosynthetic bacteria are photoautotrophs. The nonsulfur purple bacteria, on the other hand, are examples of photoheterotrophs.

In this chapter, we will focus on the photoautotrophs, organisms in which light-driven reactions provide energy

Photosynthesis Defined

In broad terms, the process of respiration that was described in the previous chapter is really the *oxidative decarboxylation* of organic substrates carried out by chemotrophic cells to obtain the energy they need. Similarly, photosynthesis can be defined as the *reductive carboxylation* of organic substrates carried out by chlorophyll-containing cells capable of using light as their energy source. Fully oxidized carbon atoms in the form of carbon dioxide are "fixed" (covalently linked) to organic acceptor molecules and are subsequently reduced and rearranged into sugars and other organic molecules, with light energy used to drive the fixation and provide the reducing power.

This means that there are both a "photo" and a "synthesis" component to our definition of photosynthesis. In terms of energy transactions (the "photo" component), we are dealing with the photochemical events whereby the energy of visible light is converted into chemical energy in a biological system. In terms of carbon metabolism (the "synthesis" component), photosynthesis can be thought of as nonphotochemical, ATP-driven events whereby photoautotrophs fix carbon dioxide into organic form and then reduce the carbon to the oxidation level of sugar molecules.

The electrons necessary for this reduction come from light-driven oxidation of a suitable donor. Photosynthetic bacteria use a variety of donors, including hydrogen gas (H_2), hydrogen sulfide (H_2S), and ammonia (NH_3). In plants and algae, however, the universal donor is water, which is oxidized to oxygen in the process. Since these organisms produce oxygen as a product of photosynthesis, they are called **oxygenic photoautotrophs** (or *oxygenic phototrophs* for a less accurate but more manageable name).

In general, photosynthesis can be viewed as an oxidation-reduction process: Carbon dioxide is reduced to the oxidation level of the carbon atoms in a sugar molecule, and the appropriate electron donor is concomitantly oxidized. The overall process can be written as

$$CO_2 + 2H_2A \longrightarrow [CH_2O] + 2A + H_2O \qquad (12\text{-}1)$$

where H_2A is the electron donor, A is the oxidized form of the donor, and $[CH_2O]$ represents organic molecules with carbon at the oxidation level of carbohydrate.

By expressing photosynthesis in this way, we avoid perpetuating the incorrect notion that water is the only electron donor for photosynthetic organisms. Although the remainder of this chapter is devoted to oxygenic phototrophs with water as the electron donor, the proper context for the discussion should always be the broader biological perspective, which recognizes a variety of alternative electron donors.

Photosynthesis in Oxygenic Phototrophs

When we focus on the oxygenic phototrophs with water as the electron donor, reaction 12-1 can be rewritten in the following more specific (and more familiar) form:

$$6CO_2 + 12H_2O \longrightarrow C_6H_{12}O_6 + 6O_2 + 6H_2O \qquad (12\text{-}2)$$

Notice that water appears on both sides of the reaction, since it is both the source of electrons for the reduction of fixed carbon and a product of the reductive process.

As you might guess, the actual chemistry and photochemistry are far more complex than suggested by this summary reaction. It will be our goal in this chapter to consider in detail both the "photo" and the "synthesis" components of the process. First, however, we will examine the structure and function of the chloroplast, since this is the site of photosynthetic action in every eukaryotic phototroph.

Chloroplast Structure and Function

Just as most of aerobic energy metabolism is localized within the mitochondria of eukaryotic cells, most of the events of photosynthesis occur within the **chloroplasts.** And just as the mitochondrion is thought of as the "energy powerhouse" of the cell, the chloroplast qualifies as the "energy transducer" because of its role in converting light energy into chemical energy.

Since chloroplasts are larger than all other organelles except the nucleus, it is hardly surprising that they were described and studied early in the history of cell biology. In fact, the earliest descriptions go back to the work of Antonie van Leeuwenhoek and Nehemiah Grew in the seventeenth century, at the very dawn of microscopic observation. The first clear association of chloroplasts with oxygen evolution came in 1880, when George Engelmann observed that small oxygen-seeking bacteria migrate toward the surface of the alga *Spirogyra* near the location of its single spiral chloroplast, but only when the algal cell is illuminated. The bacteria in question require oxygen for motility and seek it out. Their affinity for the chloroplast-containing region of the algal cell suggested to Engelmann that the chloroplast is the site of oxygen production.

Further advances in our understanding of chloroplast function came as centrifugation techniques were developed, allowing chloroplasts to be isolated for study. Robert Hill, for example, was able to demonstrate oxygen evolution directly by isolating chloroplasts and then adding a suitable electron acceptor to the preparation. This finding confirmed a proposal made in 1931 by Cornelius van Niel that the oxygen evolved during photosynthesis comes from water rather than carbon dioxide.

The advent of electron microscopy was an important landmark in our understanding of chloroplast structure. Progress in this area has been especially rapid since the 1960s, when the techniques of freeze-fracturing and negative staining began to be applied to the internal membrane organization of the organelle.

Occurrence and Size

Chloroplasts occur in all eukaryotic phototrophs, though with considerable variation in number and shape. Plant leaves usually have about 20 to 50 chloroplasts per cell. The prominence of chloroplasts in the cells of higher plants is seen clearly in the cross-sectional view of a *Coleus* leaf cell shown in Figure 12-1. Algae frequently have fewer chloroplasts per cell—sometimes only one. Figure 12-2, for example, shows the internal structure of the flagellated green alga *Chlamydomonas reinhardtii. Chlamydomonas* has a single chloroplast that occupies much of the cell, as indicated by the color in Figure 12-2. In higher plants, chloroplasts are usually discoidal, or lens shaped. Among the algae, considerable variation in shape is observed. Chloroplasts are often 2–4 μm in diameter and 5–10 μm in

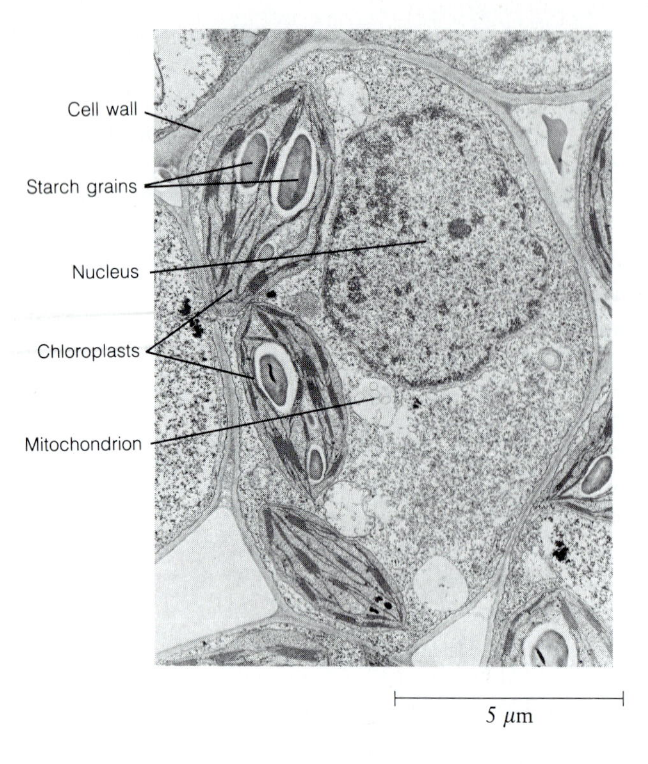

Figure12-1 Chloroplasts in a Leaf Cell. The prominence of chloroplasts in leaf cells of higher plants is illustrated by this electron micrograph of a parenchyma cell from a *Coleus* leaf. The cell contains many chloroplasts, three of which are seen in this particular cross section. The presence of large starch grains in the chloroplasts indicates that the cell was photosynthetically active just prior to fixation for electron microscopy. (TEM)

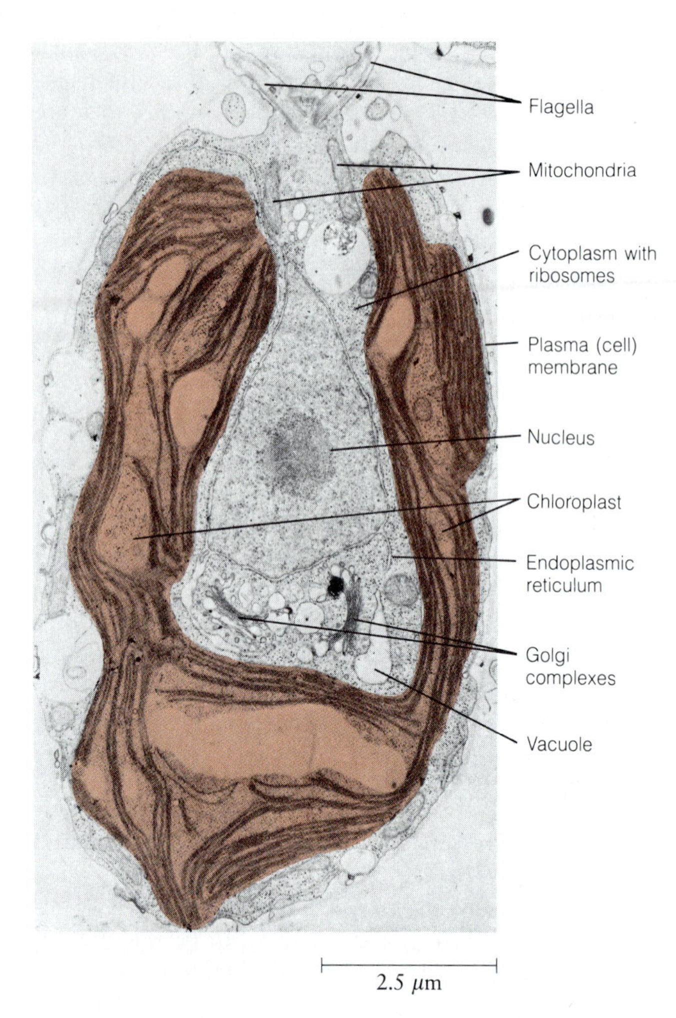

Figure 12-2 The Chloroplast of an Algal Cell. This electron micrograph illustrates the internal structure of the flagellated eukaryotic alga *Chlamydomonas reinhardtii.* Note the single chloroplast that occupies much of the cell. (TEM)

length, making them about 50 times larger than a typical mitochondrion and almost as large as a whole red blood cell.

Structural Features

A typical chloroplast from a plant cell is shown in the electron micrograph of Figure 12-3a. Like the mitochondrion, the chloroplast has both an **outer** and an **inner membrane,** often separated by a narrow **intermembrane space** (Figure 12-3b). The granular, unstructured matrix within the organelle is called the **stroma.** The outer membrane is freely permeable to most small organic molecules, whereas the inner membrane forms the primary permeability barrier of the organelle. Transport proteins in the inner membrane accomplish the movement of metabolites such as ATP and dicarboxylic acids into and out of the organelle.

In addition to an inner and an outer membrane, the chloroplast contains an extensive internal network of membranes. These membranes consist primarily of flattened sacs called **thylakoids,** often stacked together to form the **grana** (singular: *granum*) that resemble stacks of coins and are so distinctive a feature of the internal organization of the chloroplast. The thylakoids are interconnected by a network of membranes called **stroma lamellae** that extend from one granum to the next (Figures 12-3c and 12-3d).

The membranes of the stroma lamellae and the thylakoids are thought to arise from invaginations of the inner membrane during chloroplast development and may in that sense be analogous to the cristae of the mitochondrion. In the chloroplast, however, this internal membrane system is not physically contiguous with the inner membrane and is best thought of as an entity in its own right. Certainly, its lipid and protein content is different from that of either the inner or outer membrane.

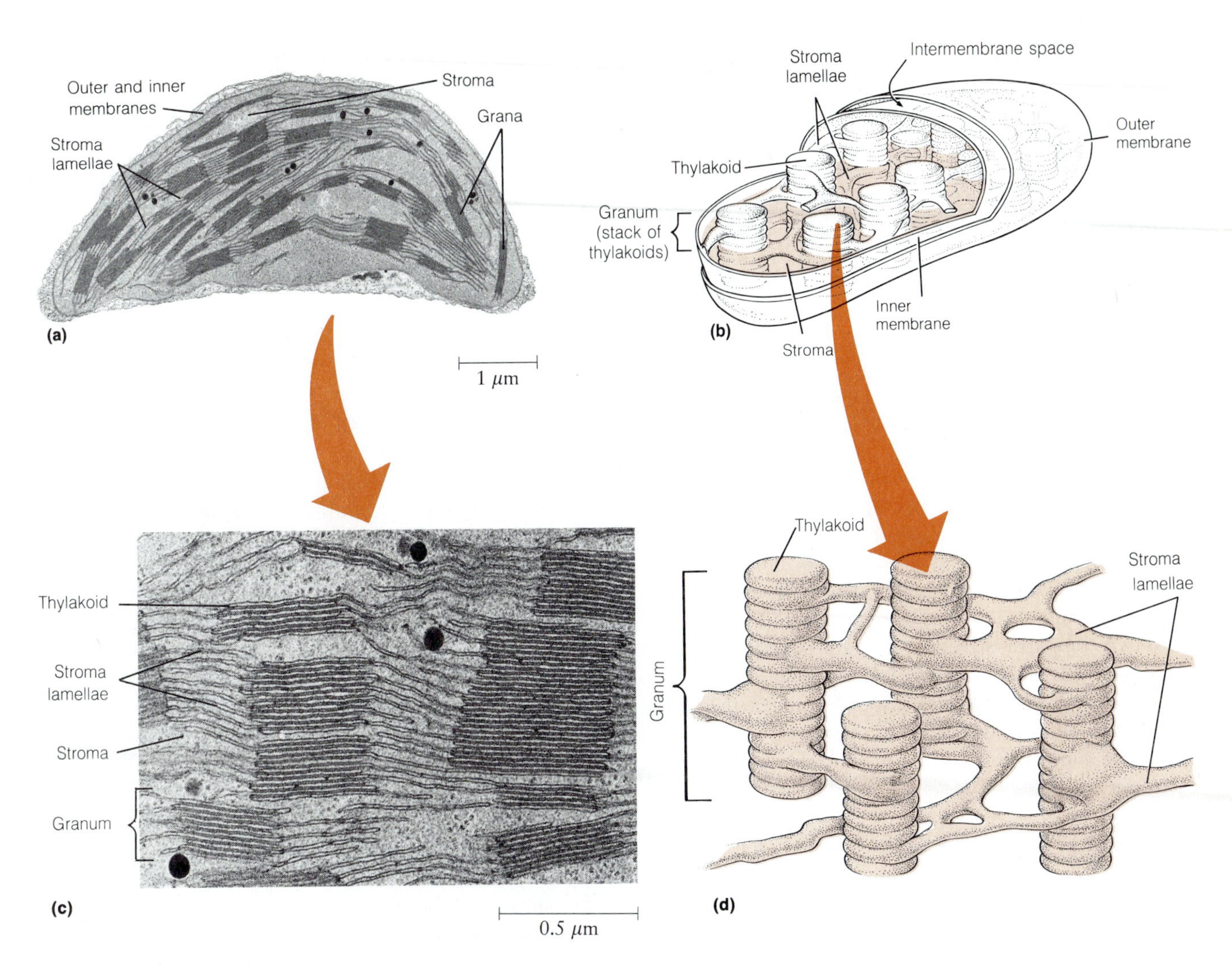

Figure 12-3 Structural Features of the Chloroplast. (a) A transmission electron micrograph of a chloroplast from a leaf of timothy grass (*Phleum pratense*). (b) A sketch showing the three-dimensional structure of a typical chloroplast. (c) A more highly magnified electron micrograph of the chloroplast in (a), showing the arrangement of thylakoid membranes and the stroma lamellae. (d) A sketch depicting the interconnections between the thylakoids and the stroma lamellae. Note that the thylakoid membranes are actually stacked disks called grana.

Localization of Function Within the Chloroplast

Similarities between mitochondria and chloroplasts are seen in the localization of the component parts of the respiratory and photosynthetic processes, as indicated by Table 12-1. In both cases, the reactions involved in carbon metabolism (the TCA cycle and β oxidation in the case of the mitochondrion, the pathway for carbon fixation and sugar synthesis in the case of the chloroplast) occur in the soluble matrix or stroma of the organelle.

Components of electron transport and phosphorylation, on the other hand, are localized in or on internal membranes. For the mitochondrion, the inner membrane is the site of these activities, but in the chloroplast, the thylakoid membranes play this role. Essentially all of the photosynthetic pigments, the enzymes required for the photochemical events, the carriers involved in electron transport, and the factors that couple transport to phosphorylation are located in or on the thylakoid membranes. In photosynthetic prokaryotes, these components are associated with the plasma membrane, since prokaryotes have no chloroplasts.

Table 12-1 Localization of Processes Within Mitochondria and Chloroplasts

		Carbon Metabolism		Electron Transport	
Process	Organelle	Pathway	Localization	Process	Localization
Respiratory metabolism	Mitochondrion	TCA cycle, β oxidation	Matrix	Oxidative phosphorylation	Cristae
Photosynthesis	Chloroplast	Calvin cycle	Stroma	Photoreduction and photophosphorylation	Grana

Thylakoid Structure

When serial sections of chloroplasts are examined in the electron microscope, the thylakoids appear to be linked by stroma lamellae in such a way that the internal space (or lumen) of every thylakoid is part of an interconnecting network called the **intrathylakoid space.** The inner volume of the chloroplast, in other words, appears to be divided into two compartments—the intrathylakoid space within the thylakoids and the stroma lamellae, and the stroma in which these membranes are suspended and bathed. This compartmentalization plays an important role in proton pumping and ATP generation in chloroplasts, as we will see later.

The Reactions of Photosynthesis

The easiest way to think about energy transduction in photosynthesis is to understand that light energy is first absorbed by the electronic structure of *chlorophyll,* driving the molecule from a low-energy ground state to a high-energy activated form. The energy is captured when an electron is transferred from the activated chlorophyll to an electron transport chain. As the electron "falls" back downhill, some of its energy is stored as ATP and NADPH. (NADPH, like NADH, is a source of reducing power, and will be described in detail later in the chapter.) The stored energy is then used in a series of metabolic steps that capture carbon dioxide in the form of reduced organic compounds such as carbohydrates. Since they require light energy, the reactions in which ATP and NADPH are synthesized are referred to as the **light-dependent reactions.** The reactions underlying carbon dioxide fixation do not directly require light energy and are therefore called the **light-independent reactions.** These CO_2-fixation reactions are also sometimes referred to as *dark reactions.* In the discussion to follow, we will discuss the light-dependent and light-independent reactions in that order.

The Importance of Chlorophyll

An appropriate place to begin our discussion of the light-dependent reactions is with **chlorophyll,** the primary link between the sunlight that floods our planet and the life that inhabits it. In all photosynthetic organisms, from bacteria to higher plants, chlorophyll is the only **pigment** (light-absorbing compound) that can donate photoenergized electrons to organic molecules, thereby initiating the photochemical events that lead to NADPH and ATP generation.

Absorption of light is not in itself a unique property of chlorophyll. Cells contain a variety of pigments capable of absorbing light energy to *photoexcite* electrons. But only chlorophyll can pass the photoexcited electron (via several membrane-bound intermediates) to a molecule of the coenzyme $NADP^+$ to reduce it to NADPH. Therein lies the significance of the chlorophyll molecule: no other pigment is capable of mediating light-dependent $NADP^+$ reduction.

Shown in Figure 12-4 is the structure of chlorophyll *a*, the form common to all oxygenic phototrophs. In addition, all such organisms contain a second kind of chlorophyll. This may be chlorophyll *b* (in green plants), *c* (in brown algae, diatoms, and dinoflagellates), or *d* (in red algae). Each of these forms is a chemical variant of the basic structure shown in Figure 12-4; in chlorophyll *b,* for example, the CH_3 group shown in color is replaced by a CHO group.

Photosynthetic bacteria are nonoxygenic; they cannot extract electrons from water and consequently do not evolve oxygen. They contain only a single type of chlorophyll, called **bacteriochlorophyll.** This, too, is a variation of the chlorophyll shown in Figure 12-4. One form of bacteriochlorophyll, for example, differs from chlorophyll *a* in having all the carbon-carbon bonds of one ring saturated (fully hydrogenated).

Photoexcitation of an electron from one energy level to another always requires the absorption of a photon of light with just the right energy content, since electron orbitals are characterized by discrete energy levels. The en-

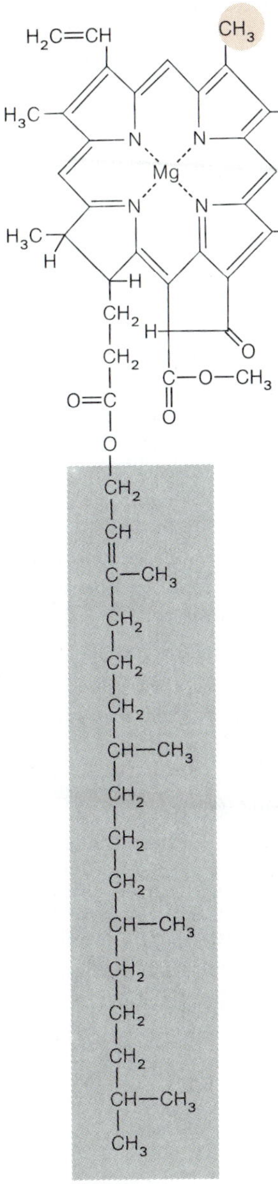

Figure 12-4 **Structure of Chlorophyll *a*.** The structure of chlorophyll *b* is similar; the methyl group shown in color is replaced by a formyl (—CHO) group.

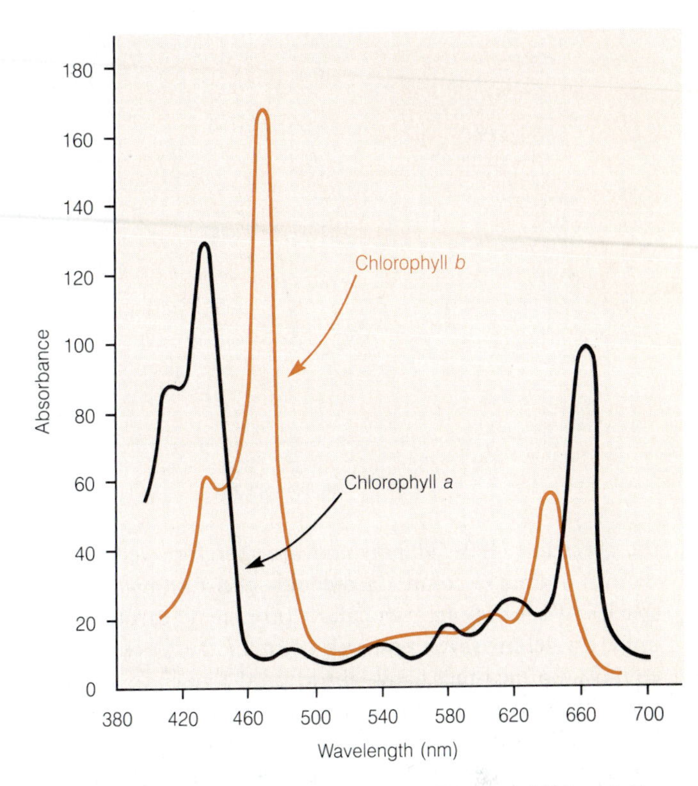

Figure 12-5 **Absorption Spectra for Ether Extracts of Chlorophylls *a* and *b*.**

ergy content of photons of light is in turn related to the wavelength of the light in the inverse manner depicted earlier in Figure 2-3. As a result, a pigment molecule will absorb only light with wavelengths corresponding in energy to the orbital transitions that its electrons can undergo. For chlorophyll, these wavelengths lie in the blue-green (420–480 nm) and the red (620–680 nm) regions of the spectrum, as shown by the absorption spectra for chlorophylls *a* and *b* in Figure 12-5.

It is worth noting that absorption spectra such as those shown in Figure 12-5 are frequently determined for pure chlorophyll dissolved in an organic solvent such as ether. In the cell, however, the chlorophyll is mixed with other pigments and embedded in the thylakoid membranes, so its absorption characteristics may vary slightly from that of the purified compound.

Red light is strongly absorbed by chlorophyll and is responsible for most of photosynthesis in nature. Using Figure 2-3, we can readily calculate the energy content of red light. At 670 nm, the approximate absorption maximum for chlorophyll *a* in most cells, light has an energy content of about 43 kcal/einstein (recall that an einstein is just a "mole" of photons). Keep this number in mind, because it will turn out to be important in understanding why all oxygenic phototrophs need two separate light-absorbing events operating in series to accomplish the reduction of $NADP^+$.

Accessory Pigments

In addition to chlorophyll, many photosynthetic organisms also contain **accessory pigments** capable of absorbing light in wavelength ranges at which chlorophyll is not effective. These pigments include the **carotenoids** and the **phycobilins.** Carotenoids absorb in the violet to green region of

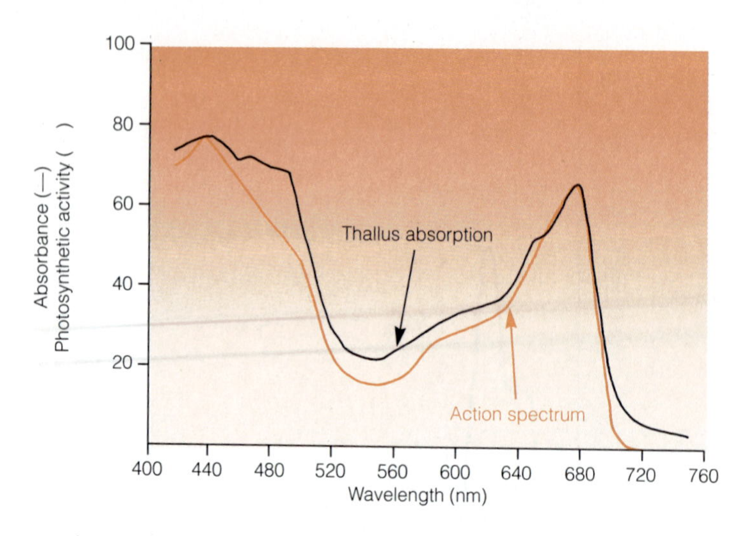

Figure 12-6 Absorption Spectrum and Action Spectrum for the Photosynthetic Thallus of *Ulva taeniata*. Note the much broader absorption range compared to chlorophyll (Figure 12-5) and the good agreement between absorption of light and its physiological effectiveness.

the spectrum (400–500 nm) and are therefore yellow or reddish orange in color. Carotenoids occur in most plant species. When they are particularly prominent, carotenoids impart a yellowish or orangish color to the leaves. Phycobilins, on the other hand, absorb in the green to orange range (550–630 nm) and are found only in red algae and the cyanobacteria (formerly called blue-green algae). It is the phycobilins that give these microorganisms the characteristic colors for which they are named.

Accessory pigments enable photosynthetic organisms to absorb light over a much broader range of wavelengths than would be possible with chlorophyll alone. This is illustrated in Figure 12-6 for the photosynthetic thallus of the green alga *Ulva taeniata*. The **absorption spectrum** indicates the relative extent to which light of different wavelengths is absorbed by the thallus. The **action spectrum** depicts the wavelength dependence of the photosynthetic process. Both spectra show clear peaks corresponding to the absorption maxima of chlorophyll; but in both its absorption and its physiological activity, the alga is able to use a broader spectrum of light than could be accounted for by chlorophyll alone, especially in the range 500–620 nm. Clearly, the accessory pigments confer enhanced light-gathering properties upon photosynthetic tissue. Much of the adaptation of phototrophs to specific conditions of illumination is in fact the result of differences in the numbers, kinds, and properties of accessory pigments they contain.

Photoreduction (NADPH Generation)

We will see that the photochemical events of photosynthesis can be considered as two discrete though closely related processes, photoreduction and photophosphorylation. **Photoreduction** is the light-dependent reduction of **nicotinamide adenine dinucleotide phosphate** ($NADP^+$) to NADPH using electrons derived from either water (for oxygenic phototrophs) or alternative donors (for photosynthetic bacteria). $NADP^+$ is the coenzyme of choice in a large number of synthetic pathways in biological systems. It is a derivative of NAD^+, and the two coenzymes differ only by the presence in $NADP^+$ of a third phosphate group attached to the hydroxyl group on carbon atom 2′ of the ribose group (see Figure 12-7). **Photophosphorylation** refers to the light-dependent generation of ATP driven by an electrochemical proton gradient. The photochemical events that account for both of these processes are depicted in Figure 12-8. We will deal with photoreduction first and then consider photophosphorylation.

Photosystem I and the Generation of NADPH

When a pigment such as chlorophyll absorbs a photon of light, the energy excites an electron from the *ground state* to a higher-energy *excited state*. Excited-state electrons are unstable; either the electron is passed to another molecule to initiate a *photochemical reduction,* or the energy is quickly reemitted as heat or light (usually the latter, as a *fluorescent emission*). Photochemical reactions are remarkably fast. In fact, excited electrons usually return to the ground state in less than a nanosecond (10^{-9} sec)!

To produce NADPH and thus supply the reduced coenzymes required by the light-independent reactions, the energized electrons of photoexcited chlorophyll molecules must be passed to $NADP^+$. This transfer involves several membrane-bound intermediates and results in reduction of the coenzyme and oxidation of the chlorophyll. The chlorophyll molecules involved in $NADP^+$ reduction are des-

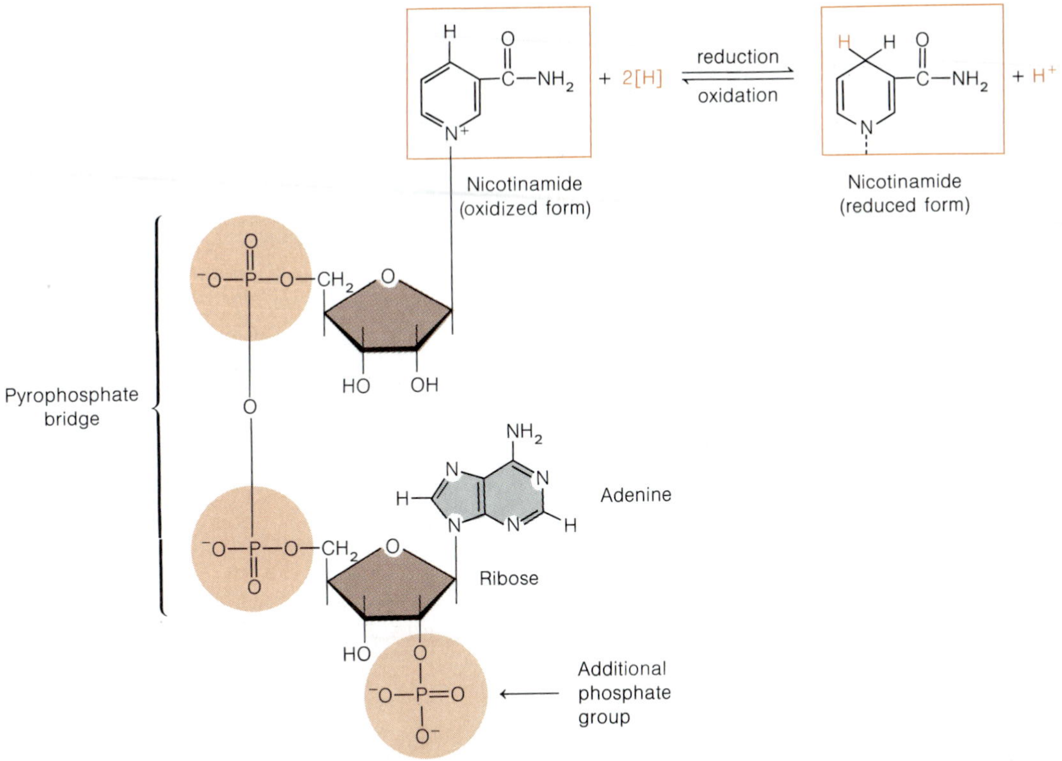

Nicotinamide (oxidized form)

Nicotinamide (reduced form)

reduction

oxidation

Pyrophosphate bridge

Adenine

Ribose

Additional phosphate group

Figure 12-7 Structure of $NADP^+$ and Chemistry of Its Oxidation and Reduction. The extra phosphate that distinguishes $NADP^+$ from NAD^+ is labeled and the hydrogens derived from an oxidizable substrate are depicted in color.

ignated as **photosystem I** in Figure 12-8. (A **photosystem** is a cluster or assembly of chlorophyll and other pigment molecules that function as a unit, as well as the proteins that anchor them tightly to the thylakoid membrane. Photosystem I represents the cluster responsible for coenzyme reduction.)

The vertical arrows in Figure 12-8 indicate the difference in energy between the unexcited, or ground-state, chlorophyll molecule and the photoexcited state. Upon absorption of light, the chlorophyll molecule has sufficient energy to donate an excited electron to an intermediate called **bound ferredoxin.** Ferredoxin, in turn, is sufficiently electronegative to ensure that the subsequent electron transfer to $NADP^+$ will be thermodynamically spontaneous. As indicated in Figure 12-8, the transfer involves bound and free forms of ferredoxin, as well as a flavoprotein.

This chlorophyll-mediated, light-driven reduction of $NADP^+$ is called *photoreduction.* Like most biological reductions, photoreduction requires two electrons per molecule of coenzyme reduced, so the process can be summarized as follows, using chlorophyll* to represent the photoexcited state of the pigment:

$$2 \text{ chlorophyll} + 2 \text{ photons} \longrightarrow 2 \text{ chlorophyll}^* \quad (12\text{-}3)$$

$$2 \text{ chlorophyll}^* + \text{NADP}^+ + \text{H}^+ \longrightarrow 2 \text{ chlorophyll}^+ + \text{NADPH} \quad (12\text{-}4)$$

Photosystem II and the Oxidation of Water

For continuous net reduction of coenzyme to take place, oxidation must occur concomitantly, since the chlorophyll molecules that lose electrons to $NADP^+$ must regain them if the process is to continue. In the case of photosynthetic bacteria, the source of electrons is quite straightforward. The single photosystem of these organisms contains bacteriochlorophyll that is energetically the equivalent of photosystem I in eukaryotes, with an E_0 value of about $+0.4$ V. The electron donors used by photosynthetic bacteria are more electronegative than this (that is, they have E_0 values that are more negative than $+0.4$ V) and can therefore transfer electrons spontaneously to oxidized bacteriochlorophyll molecules.

For oxygenic phototrophs, however, photoreduction is more complex because the electron donor is water, with a highly positive E_0 value ($+0.816$ V). Water is a very unwilling electron donor in photosynthesis. In fact, the amount of energy required to extract an electron from

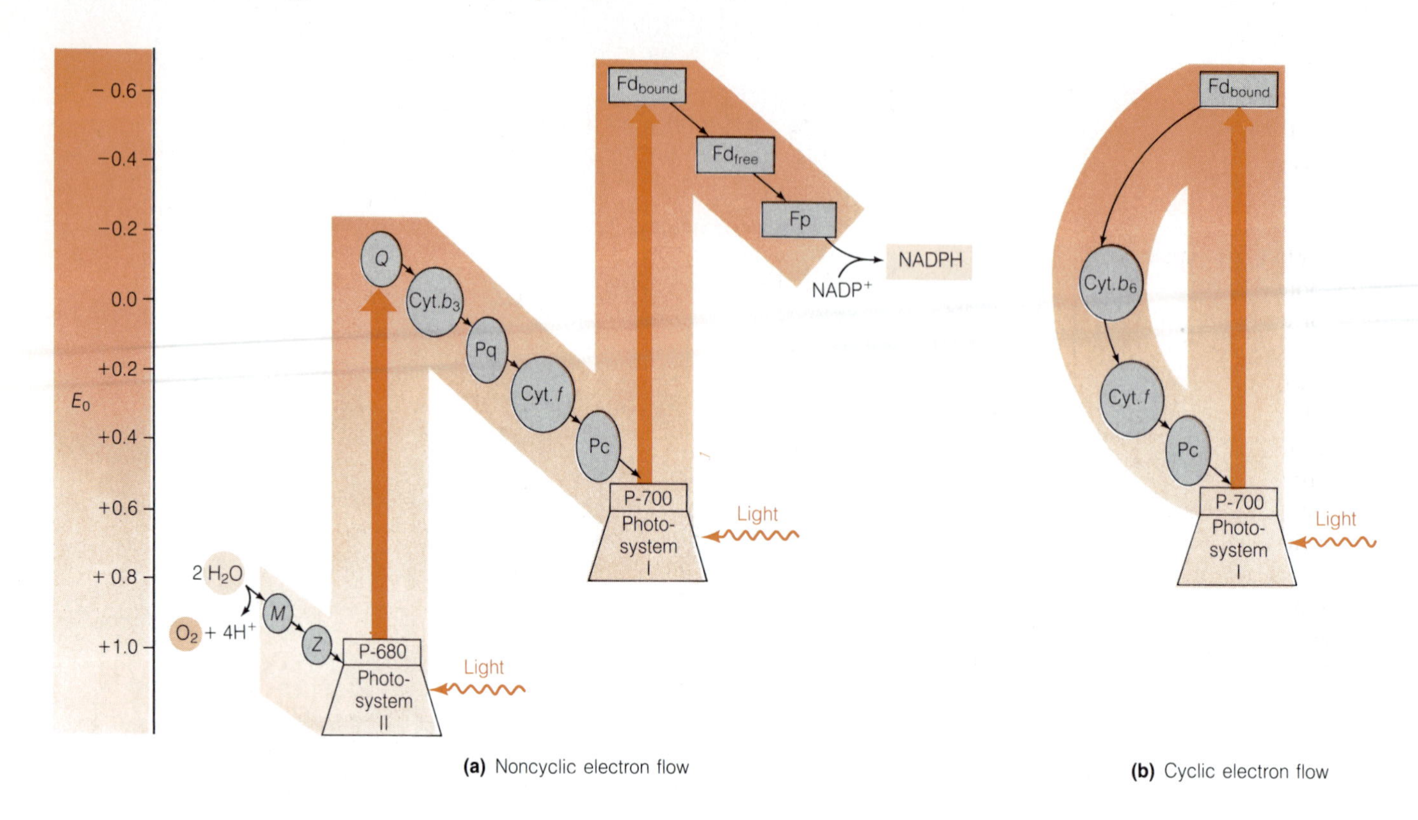

(a) Noncyclic electron flow

(b) Cyclic electron flow

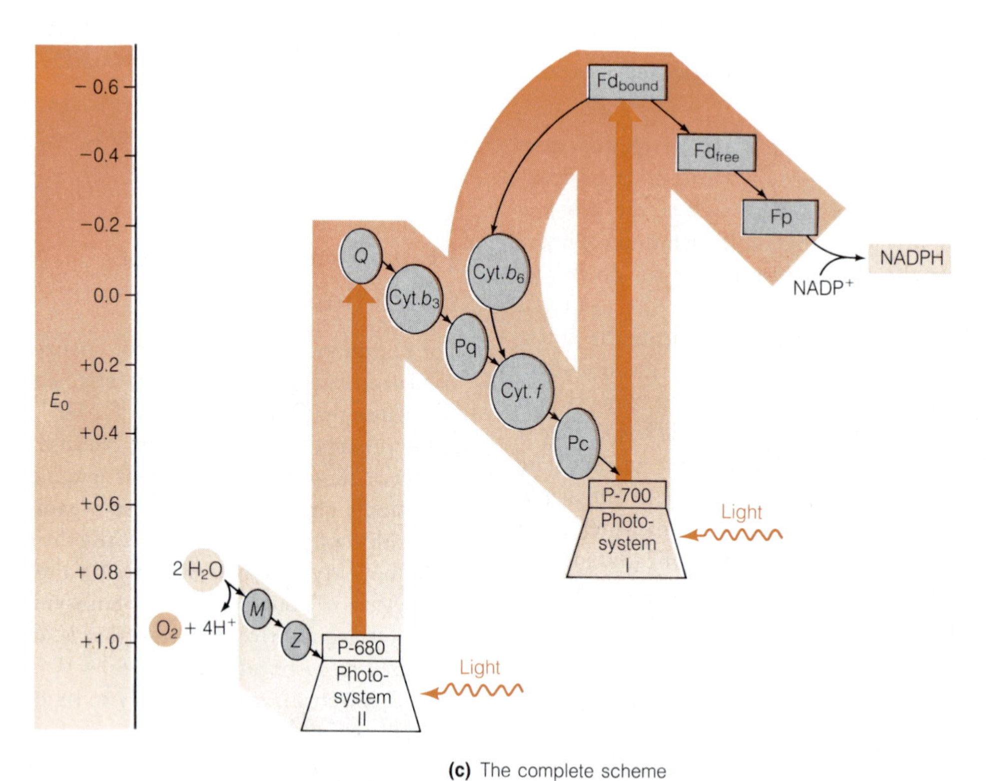

(c) The complete scheme

Figure 12-8 Photochemical Events and Electron Flow of Photosynthesis. (a) The major reactions involved in the light-driven, noncyclic transfer of electrons from water to $NADP^+$ (b) Cyclic electron flow. (c) The complete system, with cyclic and noncyclic electron flow. Cytochromes b_3, b_6, and f are intermediates, as are plastoquinone (Pq), plastocyanin (Pc), ferredoxin (Fd), and flavoprotein (Fp). *Q* is the unknown electron acceptor for photosystem II. *M* and *Z* are unknown intermediates in the transfer of electrons from water to photosystem II.

water and donate it to ferredoxin exceeds the light-driven excitation energy of a single chlorophyll molecule. Specifically, the energy difference between water and ferredoxin is at least 1.4 V (from +0.816 to about −0.6 V; see Figure 12-8a). By contrast, a chlorophyll molecule can excite an electron only through a potential difference of about 1.0 V. A single photosystem is therefore incapable of spanning the potential difference between so unwilling an electron donor as water and so unwilling an electron acceptor as ferredoxin.

For this reason, all oxygenic phototrophs have not one, but two, photosystems. Photosystem I is electronegative enough to "reach" ferredoxin, and **photosystem II** is electropositive enough to "reach" water. In fact, photosystem II is sufficiently electropositive (an E_0 value for the unexcited state of at least +0.9 V) to ensure that the transfer of electrons from water will be thermodynamically spontaneous. The transfer of electrons from water to oxidized chlorophyll molecules of photosystem II can therefore be written as

$$H_2O + 2\ \text{chlorophyll}^+ \longrightarrow 2\ \text{chlorophyll} + \tfrac{1}{2}O_2 + 2H^+ \quad (12\text{-}5)$$

As shown in Figure 12-8a, this transfer of electrons from water to the chlorophyll of photosystem II involves at least two intermediates, designated *M* and *Z*. *M* is a manganese-containing protein capable of accumulating four positive charges as four electrons are transferred successively via *Z* to chlorophyll. Once *M* has accumulated its four charges, it can catalyze the reaction whereby two water molecules are oxidized to form a molecule of oxygen, with the release of $4H^+$. This process of light-dependent oxidative splitting of water is sometimes called **photolysis.**

Noncyclic Electron Flow

So far, we have seen how electrons from photosystem I are passed by means of ferredoxin to $NADP^+$ and how electrons enter photosystem II from water via *M* and *Z*. All that remains is to put the pieces of Figure 12-8a together and order them physically in the thylakoid membrane. Figure 12-9 is a schematic diagram of how the intermediates in photosynthetic electron transport might be arranged in the membrane.

Since electrons are being acquired from water by one assembly of chlorophyll molecules but are being donated to $NADP^+$ by another assembly, some sort of transport chain must be responsible for passing electrons from one system to the other. The transport chain involves a variety of intermediates that are capable of being reversibly oxidized and reduced. These include several **cytochromes** (iron-containing compounds very similar to those involved in electron transport in the mitochondrion), **plastoqui-**

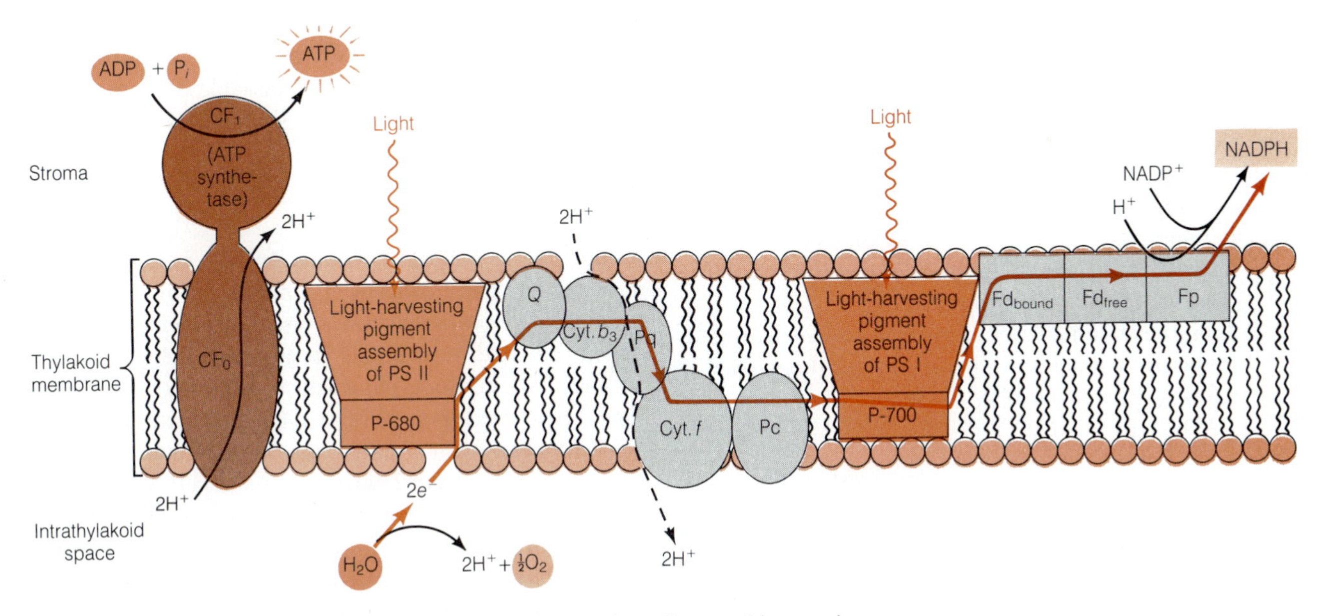

Figure 12-9 Model of the Thylakoid Membrane. The diagram shows the possible organization of the pigments and transport intermediates involved in photoreduction and photophosphorylation. The light-harvesting pigments and reaction centers of photosystems I and II are shown in color. The solid colored line depicts a possible route for the noncyclic flow of electrons from water to $NADP^+$. The dashed line represents the pumping of protons across the thylakoid membrane upon reduction and oxidation of plastoquinone, as well as the release or uptake of protons on either side of the membrane. CF_1 is the ATP synthase of the chloroplast, and CF_0 is the proton translocator.

nones, copper-containing **plastocyanins,** and an elusive electron acceptor for photosystem II identified only as Q.

As with the mitochondrial electron transport chain, these intermediates differ in reduction potentials and can be arranged in order of descending electronegativity (Figure 12-8a). Most likely, the physical arrangement of these intermediates in the membrane follows the same order, as shown in Figure 12-9. Some of the intermediates transport both electrons and protons, whereas others pass electrons only. The result is a mechanism for the directional pumping of protons across the membrane from the stroma into the intrathylakoid space, as Figure 12-9 also illustrates. The electrochemical proton gradient that is generated is used to drive ATP synthesis, as we will see shortly.

The essential features of the complete system for transfer of electrons from water to $NADP^+$ can therefore be summarized in terms of the following component parts:

1. A light-harvesting assembly of chlorophyll molecules, photosystem II, with a sufficiently electropositive reduction potential to allow it to receive electrons exergonically from water. Electrons are then photoexcited from +0.9 V to −0.1 V, using the energy of one photon of red light for every electron so activated.
2. A second assembly of chlorophyll molecules, photosystem I, with a sufficiently electronegative reduction potential to enable it to accept electrons exergonically from plastocyanin (about +0.4 V) and is, when photoexcited, sufficiently electronegative to pass them to bound ferredoxin (about −0.6 V), again requiring one photon of light per electron so activated.
3. An exergonic series of electron carriers linking the electron acceptor for photosystem II (Q; $E_0 = -0.1$ V) with the electron donor for photosystem I (plastocyanin), with provision for transport-driven pumping of protons across the thylakoid membrane to establish and maintain an electrochemical gradient capable of generating ATP.
4. A short exergonic series of electron carriers linking the electron acceptor for photosystem I (bound ferredoxin; $E_0 = -0.6$ V) with the ultimate acceptor, $NADP^+$ ($E_0 = -0.32$ V).

Functioning together within the thylakoid membrane, these components provide for a continuous, unidirectional flow of electrons from water to $NADP^+$, as indicated by the arrows of Figure 12-8a and the solid line of Figure 12-9. This is referred to as **noncyclic electron flow,** primarily to distinguish it from the cyclic flow of electrons shown in Figure 12-8b. Regardless of the organism or the electron donor, photoreduction always involves a noncyclic, or unidirectional, flow of electrons from a relatively electropositive donor to a relatively electronegative acceptor, with light providing the energy to drive the transfer.

Because coenzyme reduction is a two-electron event and each electron coming from water must be activated twice, the overall equation for photoreduction in oxygenic phototrophs is

$$H_2O + NADP^+ + 4 \text{ photons} \longrightarrow \tfrac{1}{2}O_2 + NADPH + H^+ \quad (12\text{-}6)$$

The Emerson Enhancement Effect

Experimental evidence for the existence of two separate light-dependent events during photoreduction in oxygenic phototrophs came from work published by Robert Emerson and his co-workers in 1957. Their basic observation was that a greater photosynthetic activity could be achieved with red light of two slightly different wavelengths than is possible by summing the activities obtained with either wavelength separately. This finding, called the **Emerson enhancement effect,** can be explained in terms of the sensitivity of photosystem I to red light of slightly longer wavelengths (690–720 nm) than those to which photosystem II is maximally sensitive (below 690 nm). This wavelength differential is indicated in Figure 12-8a, where the P-700 designation of photosystem I represents a chlorophyll molecule with an absorption maximum of 700 nm, and the P-680 of photosystem II stands for a chlorophyll molecule that absorbs maximally at 680 nm. Since the light-harvesting pigment assemblies of the two photosystems respond to slightly different wavelengths, maximum activity of both photosystems will be possible only upon illumination with both wavelengths simultaneously.

Photosynthetic Units

Along with his colleague William Arnold, Emerson is also responsible for the concept of a photosystem as a cluster of chlorophyll molecules and accessory pigments organized structurally and functionally for photosynthetic light harvesting. By saturating chloroplasts with flashes of light, these researchers were able to show that with photosynthesis functioning at a maximum, one molecule of oxygen was released for every 2500 molecules of chlorophyll present. If we know (as we will by the end of the chapter) that eight to ten photons of light are required for the evolution of one molecule of oxygen, it appears that only one or two out of every 250 to 300 chlorophyll molecules are capable of actually donating electrons to the appropriate acceptors and thereby initiating the photochemical events illustrated in Figure 12-8. This finding gave rise to the concept of the **photosynthetic unit**—a cluster of about 250 to 300 chlorophyll molecules, each capable of absorbing light, but with only one or a few molecules that actually participate in photochemical reactions.

Most of the chlorophyll molecules and accessory pigments in a photosynthetic unit serve only as light-gathering "antennae," absorbing photons of light and passing the energy on to an adjacent chlorophyll molecule in the unit. Eventually (within 10^{-9} second, in fact!), the energy reaches the special chlorophyll molecules at the **reaction center** of the unit, where it can be used to initiate the photochemical events that lead to coenzyme reduction and ATP generation. Each kind of chlorophyll molecule in the unit has its own characteristic absorption maximum, depending on the kind of chlorophyll and its immediate environment. Passage of energy is always from the molecules that absorb light of the shortest wavelengths (greatest energy) to those that absorb light of the longest wavelengths (least energy), since transfer must always be thermodynamically "downhill."

The chlorophyll molecules at the reaction center absorb light of the longest wavelength of all, thereby acting as a "sink" or "trap" into which all the energy gathered by the photosynthetic unit is eventually funneled. For photosystem I, the absorption maximum at the reaction center is about 700 nm, whereas for photosystem II, the absorption maximum is 680 nm. Photosystem I can therefore be thought of as an assembly of several hundred chlorophyll molecules (plus accessory pigments) functioning as a photosynthetic unit, with a molecule of chlorophyll P-700 at the reaction center (Figure 12-8). Photosystem II is a similar unit of membrane-bound pigment molecules, but with chlorophyll P-680 at its center. In both cases, the pigment molecules of the photosystem are anchored to the thylakoid membrane by special **chlorophyll-binding proteins.**

Visualization of Photosystems

Particles that correspond to photosystems I and II can be visualized by subjecting the thylakoid membrane to the technique of freeze-fracturing, as described in Appendix A and illustrated in Figure 12-10 for thylakoid membranes. The fracture plane is thought to run along the interior of the membrane but may jump from one thylakoid to the next in regions of stacking (Figure 12-10a). As a result, the interior and exterior faces of both stacked and unstacked regions can be observed (Figure 12-10b). The thylakoid membrane contains at least two different kinds of particles that are separated from each other by the fracture plane, as shown especially clearly in Figure 12-10c. The smaller particles (diameter of about 11 nm) on the internal face are thought to be the structural units of photosystem I, whereas those of a larger diameter (about 14 nm) on the external face may correspond to photosystem II. It is likely that some of the particles also represent cytochromes and the ATP synthase.

An exciting recent discovery by Hartmut Michel, Johann Deisenberger, and Robert Huber, is that the photosystems of certain photosynthetic bacteria can be crystallized and analyzed by X-ray diffraction methods. This work has given us the first close look at how pigment molecules are arranged to capture light energy. (See Appendix A for discussion of X-ray diffraction methods.) A diagrammatic representation of a bacterial reaction center is shown in Figure 12-11, together with a magnified view of the pigment molecules that capture light energy. The reaction center consists of four protein subunits. The first is a cytochrome, which is bound to the outer membrane surface. The second and third subunits (L and M) span the membrane and contain a total of four bacteriochlorophyll molecules, two bacteriopheophytin molecules, and two plastoquinones. A fourth subunit (H) is bound to the cytoplasmic surface of the membrane. The bacteriochlorophyll molecules in the L and M subunits are arranged so that one pair is able to absorb light energy and produce an excited electron. The electron is rapidly transferred through intermediate bacteriochlorophylls and bacteriopheophytin molecules to a bound plastoquinone, leaving a positive charge (an electron "hole") in the original pigment molecule. The hole is filled by an electron from the cytochrome, and the electron on the bound quinone is transferred to a mobile plastoquinone, thereby producing a source of chemical reducing power. The reduced plastoquinone can then do one of two things. It can donate its electron through a cytochrome *b-c* complex to the original bound cytochrome, a cyclic process that pumps protons. Alternatively, it can reduce NAD^+ to NADH, which in turn transfers electrons to $NADP^+$. The resulting NADPH is a source of reducing power, as it is in eukaryotic plant cells.

In summary, the bacterial reaction center represents a somewhat simplified version of photosystem II in higher plants. The main product of the reaction center is reduced quinone, which in turn is used to drive a proton pump or synthesize NADH. The proton pump generates a proton gradient that is used in chemiosmotic ATP synthesis. We will now return to the thylakoid membrane system of chloroplasts to consider in more detail how ATP synthesis occurs.

ATP Synthesis

Noncyclic Photophosphorylation

As we saw in Chapter 11, ATP generation in mitochondria depends on an electrochemical proton gradient maintained

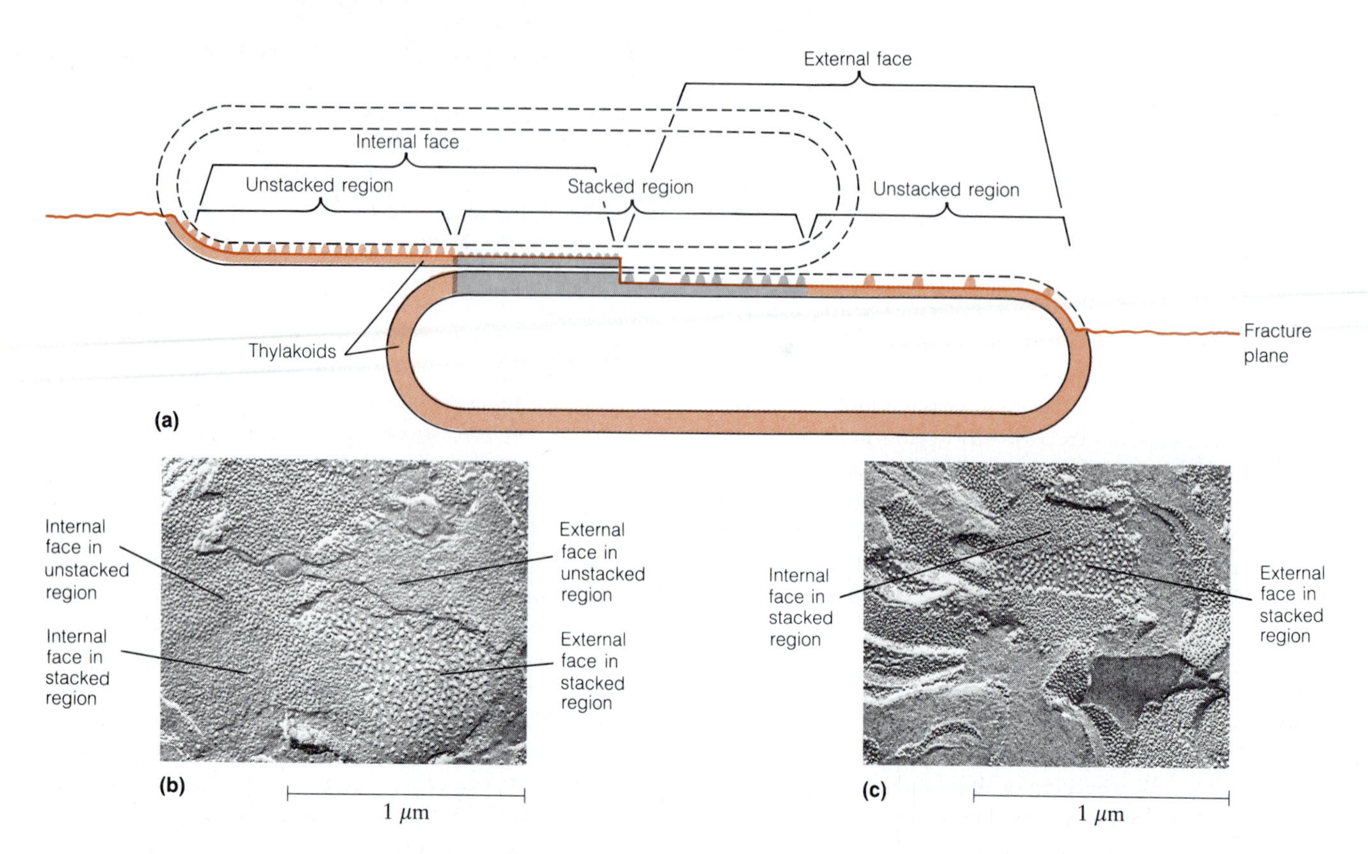

Figure 12-10 Freeze-Fracturing of Thylakoid Membranes. (a) Artist's sketch illustrating the four faces of thylakoid membranes that are revealed by freeze-fracturing of stacked and unstacked thylakoids. As the fracture plane (color line) proceeds down the hydrophobic interior of the thylakoid membranes, it exposes the external face and the internal face in both stacked regions and unstacked regions of the granum. These faces are typical of photosynthetic membranes from higher plants. The dashed lines indicate portions of the granum that are removed in the freeze-fracturing process. (b) Electron micrograph illustrating all four faces of freeze-fractured thylakoid membranes from a barley leaf chloroplast (TEM). (c) Electron micrograph illustrating only the internal and external faces of thylakoid membranes freeze-fractured in the stacked region of the granum of a barley leaf chloroplast (TEM). Note the presence of small particles on the internal face and large particles on the external face.

by the pumping of protons across the inner mitochondrial membrane. Proton pumping depends, in turn, on the transport of electrons down a series of carriers with different reduction potentials. Such a transport chain clearly exists in the scheme depicted in Figure 12-8a, since electrons flow downhill between the electron acceptor for photosystem II (Q) and the immediate electron donor for photosystem I (plastocyanin).

As photosynthetic electron transport occurs within the thylakoid membranes, protons are pumped across the membrane from the stroma into the intrathylakoid space. An **electrochemical proton gradient** is therefore generated in chloroplasts as it is in mitochondria, but across the thylakoid membranes instead of the inner membrane. You will recall that mitochondria pump protons from the matrix *outward* into the intermembrane space; however, chloroplasts pump protons *inward* from the stroma into the intrathylakoid space.

The details of the electron transport chain are not as well understood for chloroplasts as for mitochondria. However, it is clear that sites for coupling of electron transport to proton pumping exist along the chain. Plastoquinone is similar in structure to coenzyme Q and transfers both electrons and protons, but it is flanked in the transport series by intermediates that require only electrons (Figure 12-9). This arrangement drives unidirectional proton pumping, assuming only that these intermediates are oriented asymmetrically in the thylakoid membrane, as is the case in mitochondria. Such asymmetry has in fact been demonstrated, such that protons are pumped out of the stroma and into the intrathylakoid space. The dashed line of Figure 12-9 represents the pumping of protons across

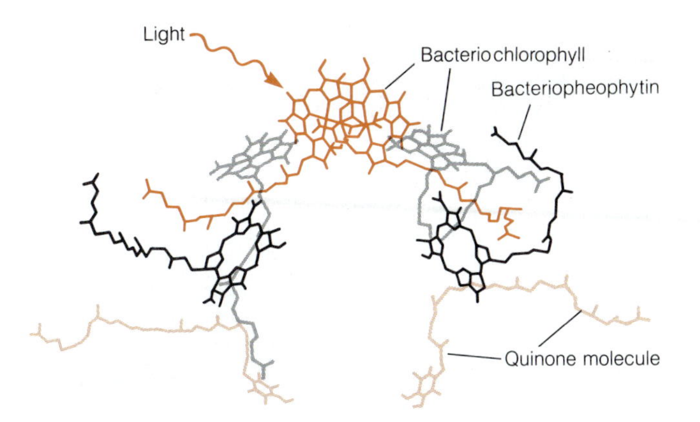

(a) Reaction center pigments

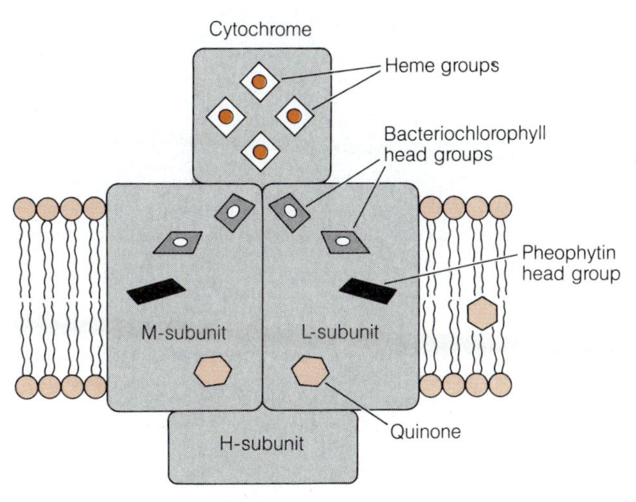

(b) Reaction center protein subunits

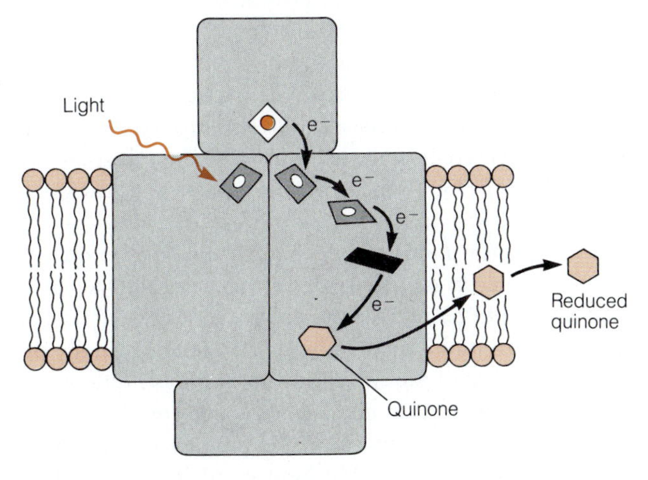

(c) Reaction center electron transport

Figure 12-11 Schematic Diagram of the Bacterial Photosynthetic Reaction Center. Michel and colleagues were able to crystalize the reaction center and analyze it by X-ray crystallography. They found that it contained four subunits and that the placement of pigment molecules (bacteriochlorophyll and bacteriopheophytin) made sense with respect to the reaction center's function. The specific arrangement of pigment molecules is shown in (a), and their functional placement in the protein subunits is shown in (b). For simplicity, only the head groups of the bacteriochlorophyll and bacteriopheophytin molecules are indicated. When light is absorbed by the upper pair of bacteriochlorophyll molecules, an electron is released that is rapidly passed down to quinone (c). Loss of the electron from the bacteriochlorophyll leaves a positive "hole," which is immediately filled by an electron from the cytochrome subunit. The reduced quinone is then used either to pump protons or as source of reducing power.

the thylakoid membrane as electrons pass through the cytochrome *b-f* complex from plastoquinone to plastocyanine.

Investigators do not yet agree on the number of sites along the chain at which electron transport is coupled to proton pumping and hence to ATP generation. Most early experiments suggested a single ATP synthesized per electron pair. More recently, results have indicated fractional values between one and two, leading to the hypothesis that the real number may be two. For our purposes, we will assume a single site of proton pumping between photosystems I and II and therefore presumably a single ATP generated per electron pair.

Each pair of electrons that flows through the noncyclic pathway of Figure 12-8a therefore generates one molecule of NADPH and leads indirectly to the formation of at least one molecule of ATP. This process of ATP formation is called **noncyclic photophosphorylation,** because the driving force comes from the unidirectional flow of electrons from water to NADP. The use of two separate photosystems by oxygenic phototrophs therefore accomplishes two things: It successfully spans the potential difference from water to bound ferredoxin to allow electron flow and coenzyme reduction, and it provides for the downhill flow of electrons between the two photosystems, which is the driving force for ATP synthesis.

To summarize noncyclic electron flow, we can rewrite reaction 12-6, taking ATP generation into account:

$$H_2O + NADP^+ + ADP + P_i + 4\text{ photons} \longrightarrow \tfrac{1}{2}O_2 + NADPH + H^+ + ATP + H_2O \qquad (12\text{-}7)$$

Cyclic Photophosphorylation

Having seen how both ATP and NADPH are generated in photon-driven electron flow, we are near the end of our discussion of the light-dependent reactions, since these are the only two requirements for the carbon dioxide fixation reactions to be discussed next. The stoichiometry of reaction 12-7 presents a problem, however. ATP and NADPH are required in a 3:2 ratio for carbon dioxide fixation, as we will see later, yet appear to be generated in equimolar amounts by noncyclic photophosphorylation. Where do the extra molecules of ATP come from?

The answer is provided by a photochemical option that makes it possible for photosystem I to generate ATP *without* concomitant $NADP^+$ reduction. As shown in Figure 12-8b, there is an alternative fate for the electrons of bound ferredoxin. Instead of being passed from ferredoxin to $NADP^+$, electrons can flow down the same electron transport chain as for noncyclic flow and return to an oxidized chlorophyll molecule in photosystem I. As before, this flow of electrons is coupled to the generation of ATP at one point at least.

In this way, ATP is produced but no water is oxidized, no oxygen is evolved, and no NADPH is formed. Light is simply used to excite electrons within photosystem I, and ATP is formed as these electrons pass along a transport chain that returns them to chlorophyll molecules in the photosystem from which they came. This is the **cyclic electron flow** mentioned earlier, and the ATP synthesis that it supports is called **cyclic photophosphorylation.** Cyclic photophosphorylation can be summarized as follows, assuming two electrons (and hence two photons) per molecule of ATP:

$$ADP + P_i + 2 \text{ photons} \longrightarrow ATP + H_2O \qquad (12\text{-}8)$$

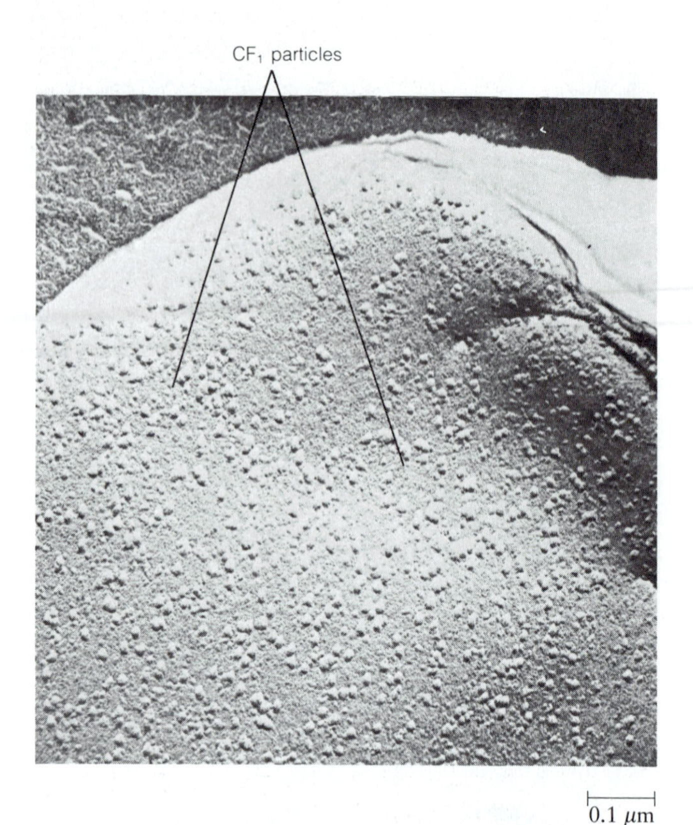

Figure 12-12 CF_1(ATP Synthase) Particles on the Thylakoid Membrane. Particles believed to represent CF_1, the chloroplast coupling factor, are seen in this electron micrograph of the outer surface of a thylakoid from a barley leaf chloroplast. (SEM)

CF_1 and ATP Synthesis

The synthesis of ATP is driven in the chloroplast as it is in the mitochondrion—by being coupled to the "downhill" flow of protons across the membrane in which electron transport occurs. ATP is synthesized by **chloroplast coupling factor,** or **CF_1,** a particle that protrudes from the thylakoid membrane in complete analogy with the F_1 particle of the mitochondrion. Particles thought to represent CF1 can be seen lining the freeze-etched thylakoid membrane of Figure 12-12. (In freeze-etching, the frozen specimen is exposed to vacuum for a few minutes after fracturing. Some of the ice evaporates, thereby exposing true membrane surfaces, rather than the membrane interior structures visualized by freeze-fracture.) The CF_1 particles face outward toward the compartment from which protons are pumped during electron transport, as do the F_1 particles of the mitochondrion.

Like its mitochondrial counterpart, CF_1 is only one part of the ATP-generating system of the organelle. The other part of the system, CF_0, is embedded in the thylakoid membrane and serves as the **proton translocator** (Figure 12-9). It is the flow of protons through CF_0 from the inside of the thylakoid to the stromal side that is believed to drive ATP synthesis by CF_1. Since the CF_1 particles protrude into the stroma, ATP is synthesized on the stromal side of the membrane, where it is needed to drive carbon fixation and reduction in the light-independent reactions to be discussed shortly.

The **proton motive force (pmf)** generated across the thylakoid membrane consists of a membrane potential term and a pH term, as expressed by equation 11-10. For the mitochondrion, the membrane potential is usually the more significant of the two terms, since the pH differential across the membrane is only about 1.0 unit (recall equation 11-11). For the chloroplast, on the other hand, nearly all of the pmf is due to the light-induced pH gradient, which can be as steep as 3.5 units. It is not unusual, for example, for the intrathylakoidal space of a chloroplast suspension to have a pH of 4.5, while the pH of the stroma is at 8.0.

Summary of the Light-Dependent Reactions

In the functional chloroplast, both cyclic and noncyclic pathways of electron flow are operative, as depicted by the complete scheme of Figure 12-8c. A great deal of flexibility is therefore possible in the relative amounts of NADPH and ATP that are generated. ATP can be produced on an equimolar basis with respect to NADPH if the noncyclic pathway is operating alone, but can be generated in any desired excess simply by shunting more and more of the electrons from bound ferredoxin to the cyclic rather than the noncyclic "side" of photosystem I. This flexibility is important, because chloroplasts use ATP and NADPH for a variety of functions other than carbon fixation and must therefore be able to generate them in a variety of ratios. For example, NADPH is also required for the reduction of both nitrate and sulfate, and ATP is used in the transport of solutes across the inner membrane of the chloroplast.

Photosynthetic Carbon Metabolism: The Calvin Cycle

With the light-dependent reactions and the structural context of the chloroplast in mind, we are now ready to look at the light-independent reactions in detail. From the standpoint of carbon metabolism, the purpose of photosynthesis is to take carbon dioxide, one of the two end products of respiration, and recycle it back into organic form by fixing (linking) it covalently to acceptor molecules. This completes the cyclic flow of carbon between the chemotrophic and phototrophic worlds shown earlier in Figure 5-5. Photosynthesis, in other words, fixes what respiration unfixes. Another way to think about it is to consider that photosynthesis drives a set of reactions "uphill," thereby storing energy, while respiration lets a second set of reactions run "downhill," releasing the stored energy for use in chemotrophic metabolism.

The process of photosynthetic carbon fixation and metabolism can be considered in four steps:

1. The initial fixation of carbon dioxide into organic form by carboxylation of an appropriate acceptor molecule.
2. The reduction of the fixed carbon from the oxidation level of an acid to that of an aldehyde, as found in sugars.
3. The subsequent synthesis of carbohydrate or other cellular components from the fixed and reduced carbon.
4. The eventual regeneration of the original acceptor molecule, thereby allowing continuation of the fixation process.

Keep in mind that these steps are not really separate processes that occur in sequence, but instead occur as integral parts of an overall pathway, as made clear in Figure 12-13. The overall pathway is called the **Calvin cycle,** named after Melvin Calvin, who received the Nobel Prize for the work that he and his colleagues Andrew Benson and James Bassham did to elucidate this process.

Carbon Fixation

The Calvin cycle of photosynthetic carbon metabolism begins with the covalent fixation of carbon dioxide into organic form. For many photosynthetic organisms, this process involves the carboxylation of the five-carbon phosphorylated sugar ribulose-1,5-bisphosphate. The exciting series of events that led up to this discovery are described in the box on pp. 334–335 to illustrate the way progress in science often depends on the ability of investigators to bring new techniques together to bear on old problems in ingenious ways. As Calvin and his colleagues were able to show, the first detectable product of carbon fixation by this pathway is 3-phosphoglycerate (PGA). The six-carbon intermediate that is postulated to be the immediate product of the carboxylation reaction has never been isolated, presumably because it exists only as an enzyme-bound intermediate, which breaks down almost immediately into two molecules of 3-phosphoglycerate. This initial carboxylation step is shown in Figure 12-13 as reaction Cal-1.

As in the case of the glycolytic pathway (recall reaction Gly-4 from Chapter 10), the need to split a larger molecule (a hexose in the case of glycolysis, a carboxylated pentose in the Calvin cycle) into two smaller molecules is anticipated by the use of a doubly phosphorylated species, such that each of the cleavage products has a phosphate group. As indicated by the carbon atoms appearing in color in Figure 12-13, the newly fixed carbon atom appears as the carboxyl group of one of the two molecules of 3-phosphoglycerate. Although reaction Cal-1 is the means by which carbon enters the Calvin cycle in all photosynthetic organisms, some phototrophs have an alternative primary carbon-fixing process that gives them greater photosynthetic efficiency, as we will see later.

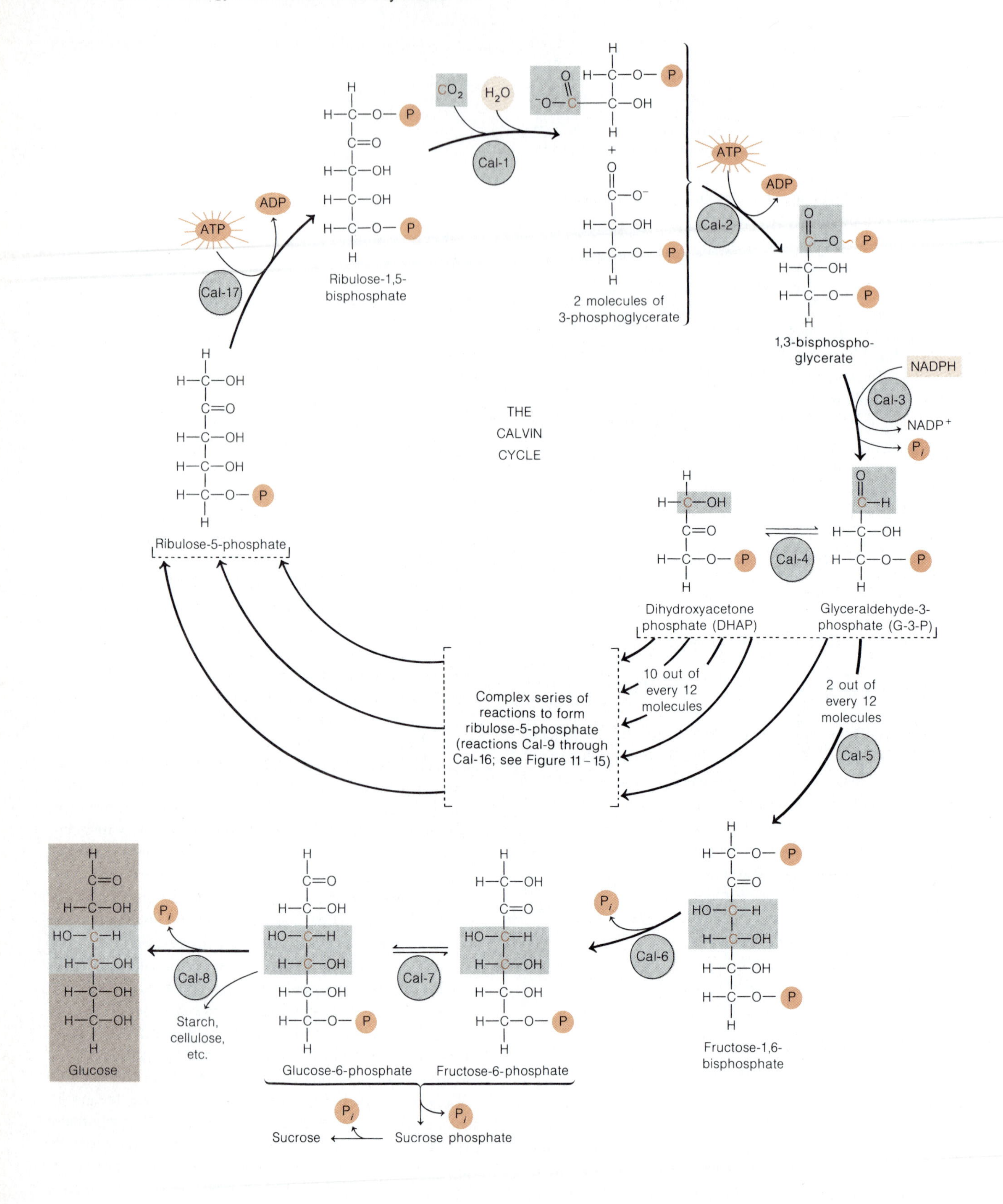

THE CALVIN CYCLE
Ribulose-1,5-bisphosphate
CO_2
H_2O
Cal-1
2 molecules of 3-phosphoglycerate
ATP
ADP
Cal-2
1,3-bisphosphoglycerate
NADPH
Cal-3
$NADP^+$
P_i
Dihydroxyacetone phosphate (DHAP)
Cal-4
Glyceraldehyde-3-phosphate (G-3-P)
10 out of every 12 molecules
2 out of every 12 molecules
Cal-5
Complex series of reactions to form ribulose-5-phosphate (reactions Cal-9 through Cal-16; see Figure 11 – 15)
Ribulose-5-phosphate
Cal-17
Fructose-1,6-bisphosphate
Cal-6
Fructose-6-phosphate
Cal-7
Glucose-6-phosphate
Cal-8
Starch, cellulose, etc.
Glucose
Sucrose
Sucrose phosphate

Figure 12-13 (opposite) Calvin Cycle for Photosynthetic Carbon Metabolism. Reactions Cal-1, Cal-2, Cal-3, and Cal-4 provide for the initial fixation and reduction of carbon, generating the interconvertible trioses, glyceraldehyde-3-phosphate (G-3-P) and dihydroxyacetone phosphate (DHAP). On the average, 2 out of every 12 triose molecules are used in the synthesis of glucose or glucose-derived compounds by reaction sequence Cal-5 through Cal-8. The most common products of photosynthesis are the disaccharide sucrose and the polysaccharide starch. Ten out of every 12 triose molecules are used to generate ribulose-5-phosphate by the complex series of reactions (Cal-9 through Cal-16) shown in Figure 12-15. The ribulose-5-phosphate is then phosphorylated at the expense of ATP in Cal-17 to form the acceptor molecule with which the sequence began.

Reduction of 3-Phosphoglycerate

The 3-phosphoglycerate formed upon fixation of carbon dioxide is reduced to glyceraldehyde-3-phosphate (G-3-P) by a reaction sequence that is the reverse of the oxidative sequence in glycolysis (reactions Gly-6 and Gly-7), except that the coenzyme involved is not NAD^+ but $NADP^+$. The NADPH-mediated reductive sequence is shown in Figure 12-13 as reactions Cal-2 and Cal-3. The sequence requires ATP (to activate 3-phosphoglycerate for reduction) and NADPH (as the source of reducing power). It is therefore at this point (as well as in the eventual regeneration of ribulose-1,5-bisphosphate in reaction Cal-17) that the Calvin cycle couples to the light-dependent steps of photosynthesis. Take careful note of the stoichiometry also at this point: For every molecule of carbon dioxide that is fixed, 2 molecules of 3-phosphoglycerate are generated (reaction Cal-1), necessitating the input of 2 molecules each of ATP (Cal-2) and NADPH (Cal-3). Since the net synthesis of 1 molecule of glucose will require the fixation of 6 molecules of carbon dioxide to maintain carbon balance, the requirement per glucose molecule thus far is 12 molecules of ATP and 12 molecules of NADPH.

Carbohydrate Synthesis

Formation of carbohydrate from the glyceraldehyde-3-phosphate of reaction Cal-3 can be understood in terms of glycolysis, since the pathway involved (Cal-4 through Cal-9) is essentially the reverse of the initial steps of glycolysis (reactions Gly-1 through Gly-5) with one small but important difference. In the glycolytic sequence, the conversion of glucose to glyceraldehyde-3-phosphate is rendered exergonic by the input of two molecules of ATP, one each in steps Gly-1 and Gly-3. To reverse the sequence as it occurs in glycolysis would result in a highly endergonic pathway, because it would require the generation of two molecules of ATP. Instead, the conversion of glyceraldehyde-3-phosphate to glucose in the Calvin cycle involves the simple hydrolysis of phosphate groups. Thus, instead of the conversion of fructose-1,6-bisphosphate to fructose-6-phosphate (reaction Cal-6) being coupled to the phosphorylation of ADP, the phosphate group on carbon atom 1 is removed by hydrolysis. The same is true for reaction Cal-8, which generates free glucose.

Glucose Formation. Summing the sequence from Cal-4 through Cal-8 makes it clear that two molecules of glyceraldehyde-3-phosphate can give rise to one glucose molecule. Glucose is often represented in this way as the end product of photosynthetic carbon metabolism. This, however, is more a definition of convenience for writing summary equations than a metabolic fact, since very little free glucose is actually generated in photosynthetic cells. In reality, the designation of an end product for the Calvin cycle becomes rather arbitrary once you get past fructose-1,6-bisphosphate, the form in which carbon exits from the actual cycle (Figure 12-13). In a sense, we can regard the whole phototrophic organism—plant, algal cell, or bacterium—as the "end product" of photosynthesis, since the Calvin cycle is the sole source of all fixed carbon from which every molecule and structure in the organism or cell is created. Even in a much more immediate sense, glucose is at best a formal end product of photosynthesis, since most of the carbon is converted either into transport forms such as *sucrose* or into storage carbohydrates such as *starch*.

Sucrose Formation. As you may recall from Chapter 3, sucrose is a disaccharide consisting of one molecule each of glucose and fructose linked together (see Figure 3-18). For sucrose formation (Figure 12-14), glucose-6-phosphate generated photosynthetically must be converted to glucose-1-phosphate. The glucose is then transferred to a carrier molecule called uridine triphosphate (UTP), generating the activated carrier form **UDP-glucose.** Finally, the glucose is linked to fructose-6-phosphate to form the disaccharide sucrose-6'-phosphate. Hydrolysis of the phosphate group leaves free sucrose, the most common transport carbohydrate in most plant species. An alternative route for sucrose formation in some species of plants involves the direct transfer of glucose from UDP-glucose to free fructose, yielding sucrose directly.

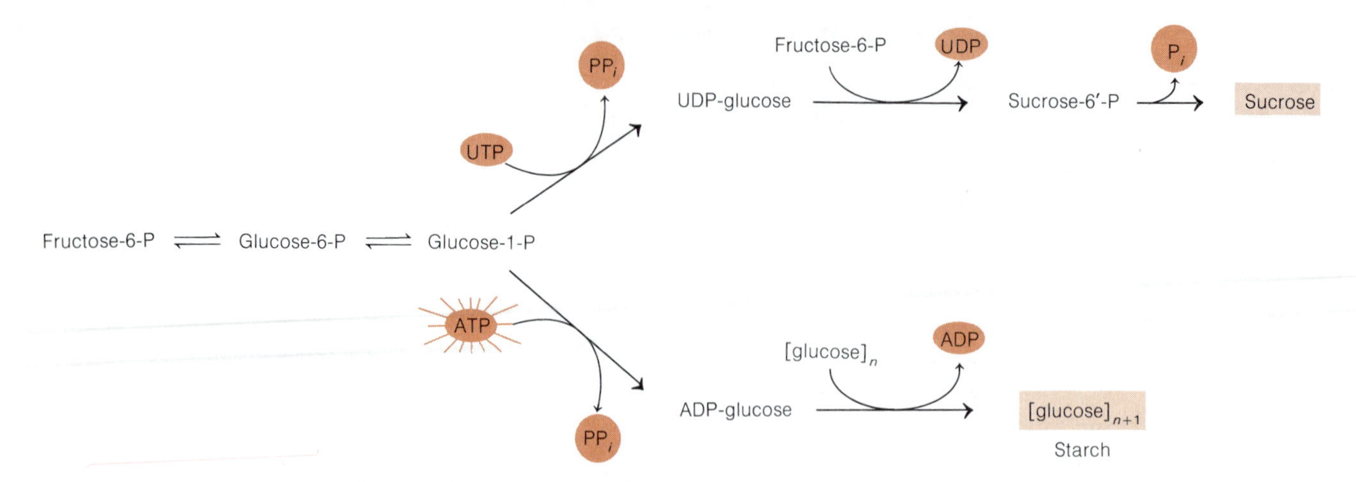

Figure 12-14 Synthesis of Sucrose and Starch from Intermediates of the Calvin Cycle.

Starch Formation. Starch is also formed by a mechanism that involves an activated form of glucose, but in this case ADP-glucose is the form of choice (see Figure 12-14). **ADP-glucose** is generated from glucose-1-phosphate in a mechanism analogous to UDP-glucose formation. The activated glucose is then added directly to the growing starch chain by the enzyme starch synthetase, resulting in stepwise elongation of the polysaccharide.

Thus, glucose may be regarded formally as the end product of photosynthesis for ease of balancing equations and to emphasize the complementary relationship between photosynthesis and respiration; but it is, in fact, the phosphorylated sugars and compounds derived from them that represent the real fate and flux of fixed carbon in photosynthetic cells.

Regeneration of Ribulose-1,5-Bisphosphate

It may seem at first glance that the metabolic sequence represented by reactions Cal-1 through Cal-8 should be all of the carbon chemistry necessary to understand photosynthesis. We have, after all, observed both the reductive fixation with which the process begins and the formation of the several carbohydrate products to which it leads. Notice the stoichiometry, however: So far, we have simply added one carbon atom to a five-carbon sugar and converted the products to a six-carbon sugar. The sequence from Cal-1 through Cal-8, in other words, accounts for the synthesis of a six-carbon sugar but involves the fixation of only a single carbon atom.

Missing, of course, is the regeneration of the starting compound, without which the sequence could not continue beyond the exhaustion of the initial supply of that compound. It is the regeneration of the ribulose-1,5-bisphosphate (RuBP) that is accomplished in the remainder of the Calvin cycle (and this, in fact, is what makes it a cycle). On the average, only 2 out of every 12 molecules of glyceraldehyde-3-phosphate are used to synthesize glucose by the sequence Cal-5 through Cal-8; the other 10 trioses are required within the cycle for the regeneration of the starting compound.

To see the stoichiometry clearly, consider reaction 12-9, which summarizes reactions Cal-1 through Cal-3 with the coefficients necessary to fix 6 molecules of carbon dioxide and generate 12 molecules of glyceraldehyde-3-phosphate:

$$\begin{aligned} &6\text{RuBP} + 6\text{CO}_2 + 12\text{NADPH} + 12\text{H}^+ \\ &\quad + 12\text{ATP} + 6\text{H}_2\text{O} \longrightarrow 12 \text{ glyceraldehyde-3-P} \\ &\quad + 12\text{NADP}^+ + 12\text{ADP} + 12\text{P}_\text{i} \end{aligned} \qquad (12\text{-}9)$$

On average, 2 of these 12 glyceraldehyde-3-phosphate molecules will be used for the synthesis of a glucose molecule by the sequence of reactions Cal-4 through Cal-8, summarized as follows:

$$2 \text{ glyceraldehyde-3-P} + 2\text{H}_2\text{O} \longrightarrow \text{glucose} + 2\ \text{P}_\text{i} \qquad (12\text{-}10)$$

The consequence of reactions 12-9 and 12-10 is that one glucose molecule is formed for every six carbon dioxide molecules fixed, which is consistent with the carbon balance necessary for the process to be self-sustaining.

The remaining ten molecules of glyceraldehyde-3-phosphate from reaction 12-9 serve as the starting point for regeneration of the six molecules of ribulose-1,5-bisphosphate with which the cycle commenced. The problem, then, is to regenerate 6 five-carbon molecules from 10 three-carbon molecules. The sequence of reactions required to do this is shown as reactions Cal-9 through Cal-16 in Figure 12-15 (and in less detail in Figure 12-13).

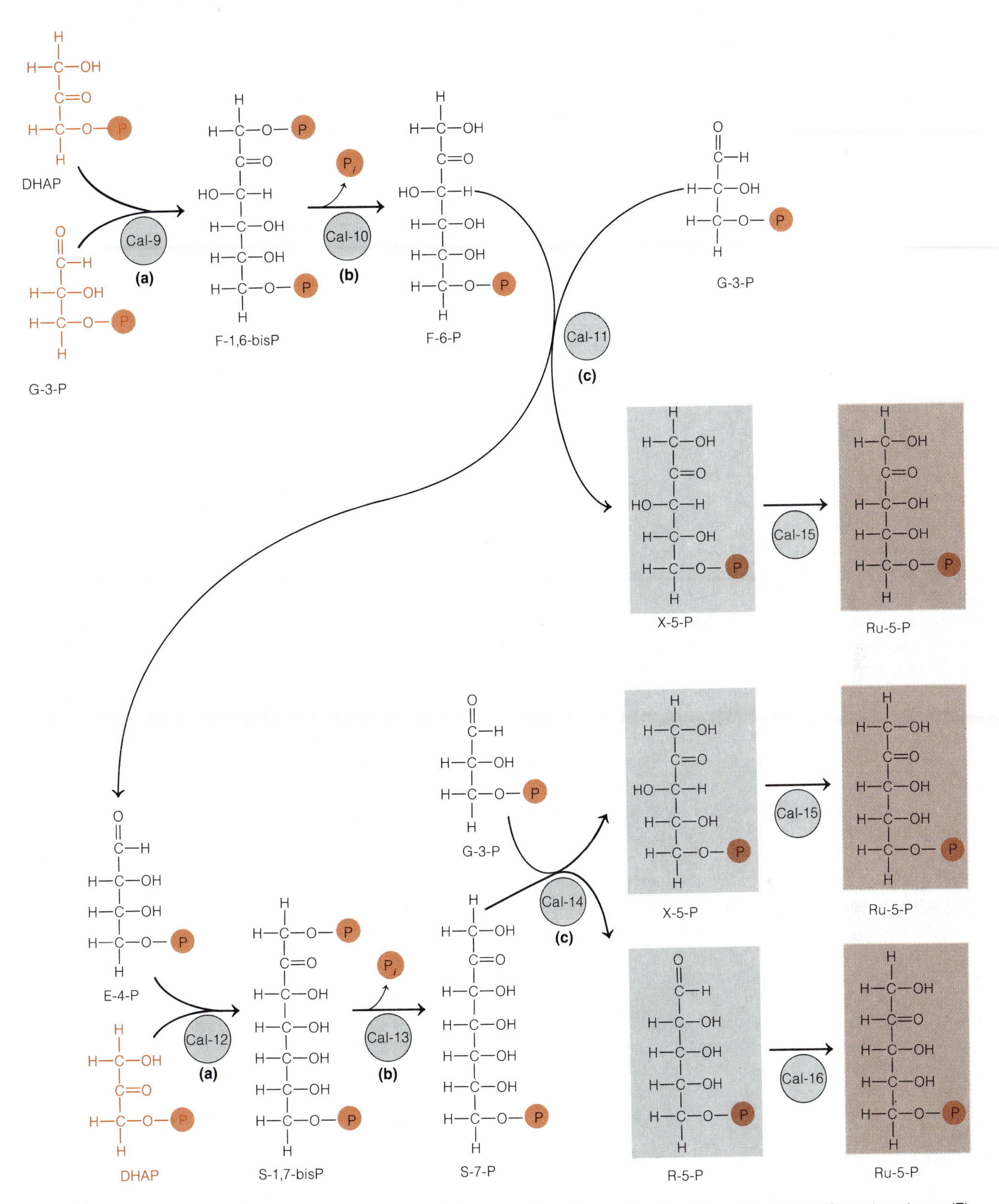

Figure 12-15 Regeneration Sequence of the Calvin Cycle. The starting trioses (in colored type) are glyceraldehyde-3-phosphate (G-3-P) and dihydroxyacetone phosphate (DHAP). The pentose products (gray) are xylulose-5-phosphate (X-5-P) and ribose-5-phosphate (R-5-P), both of which can be converted to ribulose-5-phosphate (Ru-5-P) (shaded). Other intermediates are phosphorylated forms of fructose (F), sedoheptulose (S), and erythrose (E). The sequence involves two series of reactions (Cal-9 through Cal-11 and Cal-12 through Cal-14), each consisting of (a) a condensation step (b) a phosphate-cleaving step, and (c) a transketolation step that shifts a two-carbon fragment from one sugar to another.

Carbon-14, Paper Chromatography, and the Calvin Cycle

One of the most reliable marks of scientific genius is an ability to bring new techniques to bear on old problems in a way that pushes the frontiers of science ahead in quantum leaps. An especially good example of this can be seen in the pioneering work of Melvin Calvin and his colleagues at Berkeley, work that eventually led to the elucidation of the cycle that now bears Calvin's name.

It had long been known that atmospheric carbon dioxide could be somehow incorporated into organic form by green plants, but the mechanism behind that "somehow" was hidden in a maze of enzyme-catalyzed biochemical reactions. This was the problem that Calvin and his co-workers Andrew Benson and James Bassham set out to solve. One measure of their eventual success can be seen in the Nobel Prize that was later awarded to Calvin for this work. Aspiring scientists would do well to take note of their approach, for much of their success depended on the ingenuity with which they were able to tackle the old problem of carbon dioxide fixation with several powerful new techniques.

Their story also illustrates the impact that political and social developments have on science, because the success of their approach depended on the use of ^{14}C, a radioactive isotope of carbon that had just become available in the 1940s as a result of research undertaken in connection with the Second World War. The basic approach in Calvin's laboratory was to expose illuminated cells of the green alga *Chlorella* to $^{14}CO_2$ for a short period of time and then extract and analyze the various molecules in the algal cells to see in which ones the labeled carbon atoms appeared. For that analysis, they successfully adapted the newly developed technique of two-dimensional *paper chromatography*, which essentially allows compounds of different solubilities in one or both of two solvent systems to be resolved from each other based on their differential mobility as the solvent is allowed to move up a sheet of filter paper. To locate the resolved spots in which the ^{14}C was localized, they used yet another new technique called autoradiography. This technique depends on the ability of decaying radioactive atoms such as ^{14}C to expose an X-ray film if the film is pressed tightly against the paper chromatogram for a sufficient period of time.

The *Chlorella* cells were grown in illuminated chambers into which radioactive carbon dioxide was then injected. After the desired incubation period (measured usually in seconds or minutes), a valve at the bottom of the chamber was opened and the algae were collected in a beaker of hot alcohol. The alcohol served both to kill the algae and to extract soluble organic compounds. Extracts of the cells were then spotted onto a sheet of filter paper and subjected to chromatography, first in one dimension with one solvent system, then in a second dimension (at 90° to the first) with a second solvent system. Exposure of the dried chromatogram to X-ray film allowed radioactive compounds to be located, and these were then identified by chemical analysis.

When the algae were allowed to incubate with $^{14}CO_2$ for more than a few minutes, many of the molecules in the cell were found to be radioactively labeled upon extraction. This is hardly surprising, since every organic compound in the cell is synthesized from photosynthetically fixed carbon. Far more instructive, of course, would be to know which molecules are labeled *first* upon exposure to the $^{14}CO_2$. For this, very short incubation times were essential. Even with an incubation time of only 30 seconds, for example, the chromatogram turned out to have a large number of radioactive spots, as shown in Figure 12-Aa.

With shorter and shorter incubation times, fewer and fewer radioactive compounds were detected (Figure 12-Ab). By 5 seconds, only about a half-dozen spots were found (Figure 12-Ac), and with incubation times of 2 seconds or less, a single labeled compound predominated. This turned out to be 3-phosphoglycerate, a compound that had earlier been identified as one of the intermediates in glycolysis. By painstakingly identifying each of the additional compounds that became labeled with slightly longer incubations and by determining where within each such molecule the labeled carbon was located, Calvin and his colleagues were eventually able to piece together the complex pathway of photosynthetic carbon metabolism.

Something of the perspiration that inevitably accompanies inspiration in any definition of genius can be sensed in Calvin's own words when he noted in his Nobel laureate lecture that the data on which their conclusions were based came mainly from "the number, position, and intensity—that is, radioactivity—of the blackened areas. The paper ordinarily does not print out the names of these compounds, unfortunately, and our principal chore for the succeeding ten years was to properly label those blackened areas on the film" (Calvin 1962, 881).

Perspiration, then, to be sure—ten years of it, by Calvin's report. But inspiration, as well, for without the foresight to bring together radioactive carbon, paper chromatography, and X-ray autoradiography, there would have been no blackened areas to label, no pathway to name, and no prize to claim.

(a)

(b)

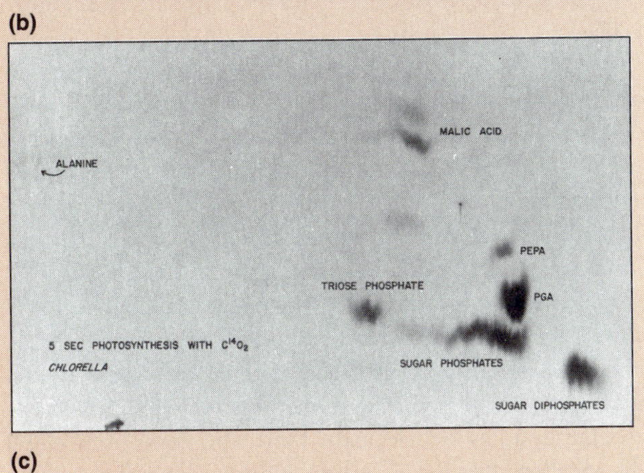

(c)

Figure 12-A Autoradiographic Identification of Radioactive Compounds in *Chlorella* Cells After Exposure to $^{14}CO_2$. Algal cells were incubated with $^{14}CO_2$ (as the bicarbonate ion, $H^{14}CO_3^-$) for (a) 30 sec, (b) 10 sec, and (c) 5 sec and were then killed by being plunged into boiling alcohol. Extracts were then chromatographed, and radioactive compounds were located by exposure of the paper chromatogram to X-ray film. Radioactive compounds were identified by chemical analysis and co-chromatography of known compounds.

The key to understanding the sequence lies in three kinds of reactions, each of which occurs twice. The first reaction involves a *condensation* of one three-carbon unit onto a second three- (or four-) carbon unit to yield a single six- (or seven-) carbon compound. The second is a *phosphate-splitting* reaction to eliminate the second phosphate group that results when two monophosphorylated sugars are so condensed. The third reaction, called *transketolation,* involves the transfer of a two-carbon unit from a ketosugar to an aldosugar. This sequence occurs twice, once as the series of reactions Cal-9 to Cal-11 and again as Cal-12 to Cal-14.

In the process, five trioses—three molecules of glyceraldehyde-3-phosphate and two molecules of dihydroxyacetone phosphate (DHAP)—are consumed and three pentoses are generated. These include two molecules of xylulose-5-phosphate (X-5-P) and one molecule of ribose-5-phosphate (R-5-P), both of which are readily converted into ribulose-5-phosphate (Ru-5-P) by reactions Cal-15 and Cal-16, respectively. Recognizing the need for a stoichiometry that accounts for the conversion of ten triose molecules into six pentose molecules, we can write the following summary reaction, which represents the reaction sequence of Figure 12-15 multiplied by 2:

$$\text{10 glyceraldehyde-3-phosphate} + 4H_2O \longrightarrow \text{6 ribulose-5-phosphate} + 4P_i \quad (12\text{-}11)$$

Finally, the resulting six molecules of ribulose-5-phosphate must be converted to the doubly phosphorylated form that serves as the carbon dioxide acceptor in the cycle. This process is accomplished by a phosphorylation reaction, with ATP as the phosphate donor and energy source:

$$\text{6 ribulose-5-phosphate} + 6ATP \longrightarrow \text{6 ribulose-1,5-bisphosphate} + 6ADP \quad (12\text{-}12)$$

Summary of the Calvin Cycle

To summarize the carbon metabolism of the Calvin cycle, we can simply add reactions 12-9 through 12-12, since these, in turn, summarize all the chemistry of the cycle from carbon fixation and reduction through glucose synthesis and regeneration of the acceptor molecule. The resulting reaction is

$$6CO_2 + 12NADPH + 12H^+ + 18ATP + 12H_2O \longrightarrow C_6H_{12}O_6 + 12NADP^+ + 18ADP + 18P_i \quad (12\text{-}13)$$

Note the actual requirements for NADPH and ATP: The Calvin cycle requires 12 molecules of NADPH and 18 molecules of ATP for every molecule of glucose synthesized. For each CO_2, this becomes 2 NADPH and 3 ATP.

This is the ratio we accounted for earlier by including cyclic photophosphorylation as an essential light reaction.

Some Do It Differently: The C_4 Plants

The Calvin cycle occurs universally in all photosynthetic organisms for the reduction and processing of fixed carbon. However, the reaction that initiates the cycle (Cal-1) is only one of two major carbon-fixing mechanisms in plants. It is certainly the more common mechanism and is in fact the only means of photosynthetic carbon fixation in the preponderance of plant species. However, some species preface the Calvin cycle with a short carbon-fixing sequence usually referred to as the **Hatch-Slack pathway,** in honor of M. D. Hatch and C. R. Slack, two Australian plant physiologists who played key roles in its elucidation. The Hatch-Slack pathway is common among tropical grasses, including such economically important species as corn, sorghum, and sugarcane. Figure 12-16 depicts the Hatch-Slack pathway and its relationship to the Calvin cycle.

The Hatch-Slack Pathway

The fixation step in the Hatch-Slack (HS) cycle involves the formation of oxaloacetate by carboxylation of phosphoenolpyruvate (PEP) (reaction HS-1). Both phosphoenolpyruvate and oxaloacetate should be familiar to you already; in fact, the carboxylation reaction of the Hatch-Slack pathway has also been encountered earlier as one of the means of replenishment of oxaloacetate for the TCA cycle (see Figure 11-13a).

Oxaloacetate is the immediate product of carboxylation in such species, but it is rapidly converted into either malate (by NADPH-mediated reduction; reaction HS-2) or aspartate (by transamination; reaction HS-3), depending on the species. Because the immediate products of carboxylation in plants with this pathway are four-carbon compounds, such plants are called **C_4 plants.** This distinguishes them from C_3 **plants,** in which the immediate product of fixation is the three-carbon compound 3-phosphoglycerate, as Calvin and his colleagues originally demonstrated for the C_3 alga with which they did their initial pioneering work on carbon fixation.

Although the newly fixed carbon of C_4 plants appears initially in oxaloacetate and then in malate or aspartate, it eventually finds its way into all intermediates of the Calvin cycle. This is because the initial fixation sequence is followed by a decarboxylation reaction (HS-4), and the liberated carbon dioxide (containing the same carbon atom fixed initially) is then refixed by the carboxylating enzyme of the Calvin cycle. The pyruvate that remains after malate decarboxylation is phosphorylated at the expense of ATP to regenerate the initial acceptor molecule, phosphoenolpyruvate (reaction HS-5).

The overall process (HS-1 through HS-5) is therefore cyclic, and the net result is a "feeder system" that provides for the initial entrapment of carbon dioxide, which is then passed to the Calvin cycle. Thus, the Hatch-Slack pathway is not a substitute for the Calvin cycle, but simply an initial carbon-fixing sequence that precedes the Calvin cycle in certain species. Moreover, the ATP required in step HS-5 is above and beyond that needed for Calvin cycle activity, so the overall energy requirement for photosynthesis is inherently greater in C_4 plants than in C_3 plants.

The Advantage of Being a C_4 Plant

Since the Hatch-Slack pathway makes photosynthesis more "expensive," we need to ask what advantage it confers on the C_4 plants in which it is found. To appreciate the advantage of being a C_4 plant, we need to consider the spatial arrangement of the Hatch-Slack and Calvin cycles within the leaves of C_4 plants. Unlike C_3 plants, C_4 species have two distinctive types of photosynthetic cells in their leaves, and the two component cycles of carbon metabolism in these plants are actually located in two different types of cells. These features of C_4 leaf anatomy are shown in Figure 12-17.

Initial fixation of carbon dioxide occurs via the oxaloacetate-generating reaction (HS-1) of the Hatch-Slack pathway, which is localized in **mesophyll cells** that have ready access to atmospheric carbon dioxide. (The term *mesophyll* comes from Greek roots meaning "middle of the leaf." The first layer of cells in a leaf consists of epidermal cells that do not carry out photosynthesis, but the mesophyll cells are usually located directly beneath, as Figure 12-17 shows.) Subsequent carbon metabolism by the Calvin cycle, on the other hand, takes place in internal **bundle sheath cells** found, as the name suggests, in close proximity to the vascular bundles of the leaf. Thus, the outer mesophyll cells in effect "collect" carbon dioxide from the air and move it (as four-carbon compounds) to the inner bundle sheath cells, where it can be passed to the Calvin cycle with minimal danger of inadvertent escape and return to the environment.

To appreciate this arrangement fully, we need to know that phosphoenolpyruvate carboxylase, the carboxylating enzyme of the Hatch-Slack cycle, is an excellent "scavenger'' of carbon dioxide. In other words, the enzyme has a high affinity (low K_m) for carbon dioxide and can operate efficiently at the low concentration of carbon dioxide found in air, 380 ppm. Ribulose-1,5-bisphosphate carboxylase (RuBPCase), on the other hand, is less suited for carbon dioxide fixation at low carbon dioxide concentrations. In addition to functioning as a *carboxylase*, which it does under conditions of high carbon dioxide and low oxygen, this enzyme has the unusual property of functioning as an *oxygenase* at low carbon dioxide and high oxygen. The oxygenase function leads to oxygen consumption and carbon dioxide evolution, a counterproductive, light-dependent carbon dioxide loss called *photorespiration*, which will be described later in this chapter. The critical function of

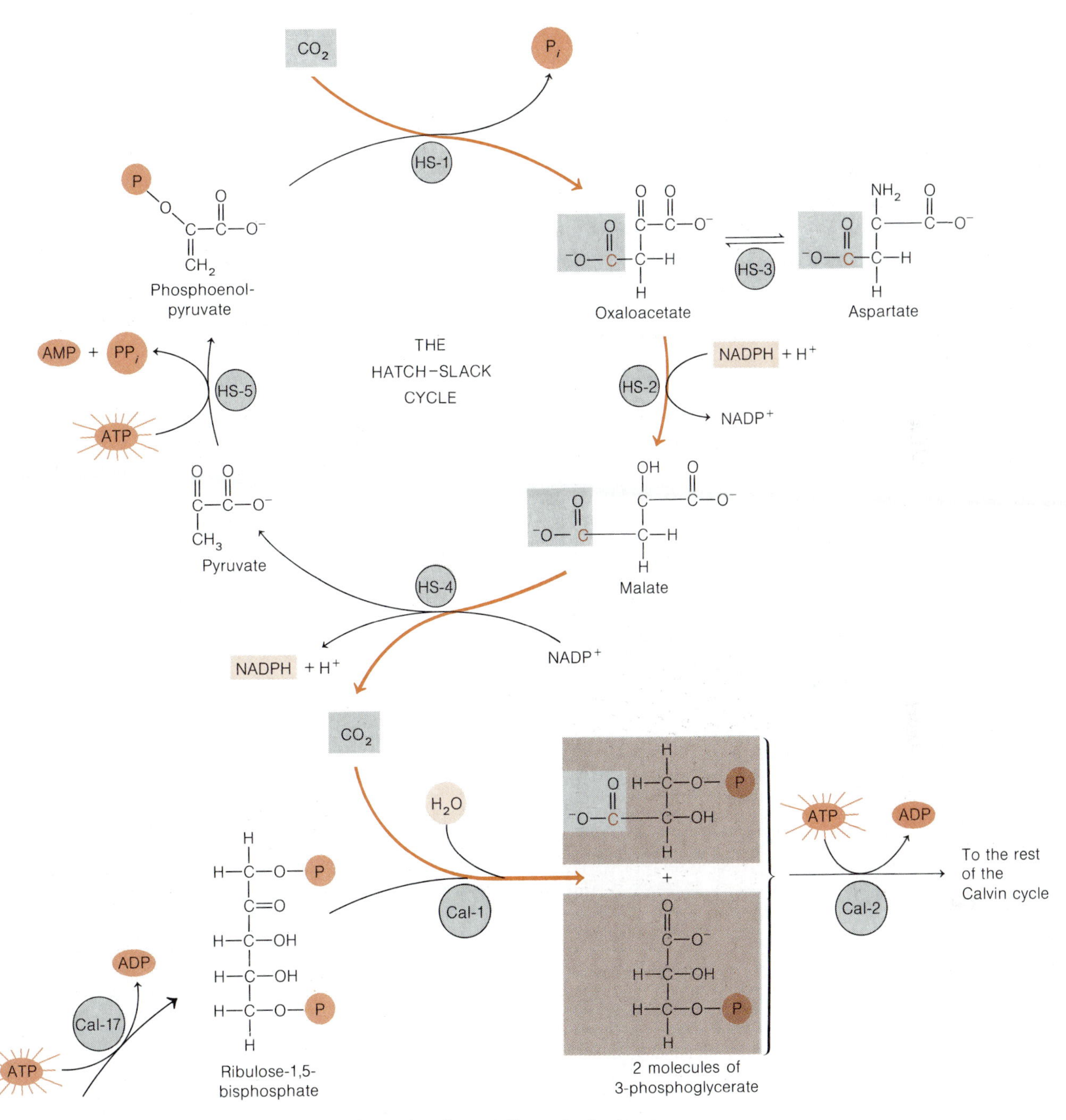

Figure 12-16 Hatch-Slack Pathway. To aid in tracing carbon through the cycle, the incoming carbon (from the newly fixed carbon dioxide molecule) is shown in colored type. Note that the carbon atom released to the Calvin cycle in step HS-4 is the same as that fixed as carbon dioxide in step HS-1.

the Hatch-Slack pathway, therefore, is to maintain the carbon dioxide concentration in the bundle sheath cells at a sufficiently high level to maximize the carboxylation function of ribulose-1,5-bisphosphate carboxylase and minimize its oxygenase function. As we will see when we discuss photorespiration, C_4 plants are virtually devoid of the light-dependent release of carbon dioxide that detracts so significantly from the photosynthetic efficiency of C_3 plants.

In a sense, the Hatch-Slack pathway can be thought of as a means of concentrating carbon dioxide, fixing it efficiently in the mesophyll cells and transporting the fixed carbon inward to the bundle sheath cells. In this way, the carbon dioxide concentration in the bundle sheath cells is maintained at a high enough level to favor the carboxylase

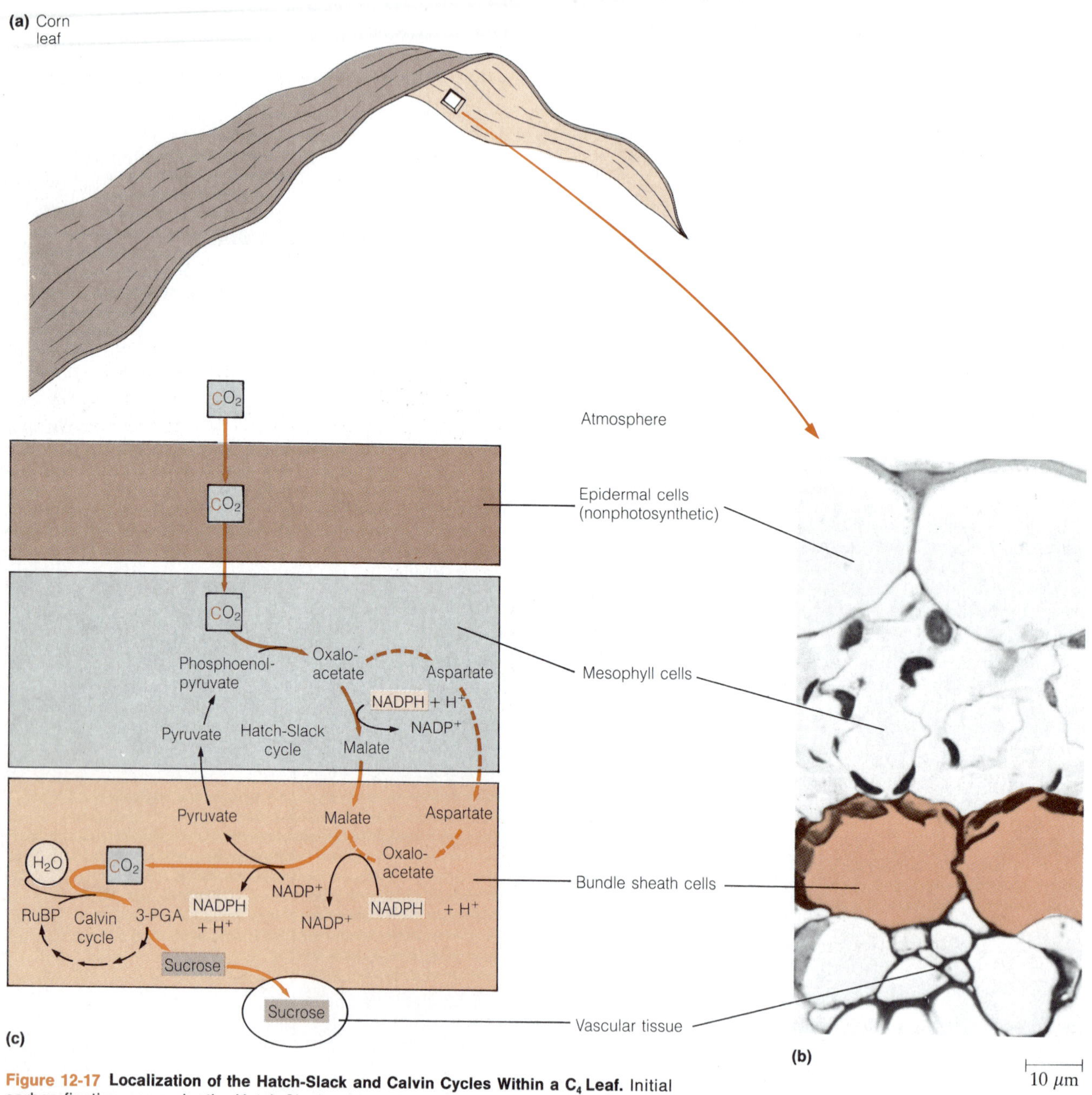

Figure 12-17 Localization of the Hatch-Slack and Calvin Cycles Within a C_4 Leaf. Initial carbon fixation occurs by the Hatch-Slack cycle within the mesophyll cells. Depending on the species, carbon is passed inward to the bundle sheath cells as either malate or aspartate, which is then decarboxylated. The carbon dioxide is refixed by the Calvin cycle, yielding sucrose, which passes into the adjacent vascular tissue for transport to other parts of the plant. The particular C_4 leaf shown is that of corn, *Zea mays*. (LM)

rather than the oxygenase function of ribulose-1,5-bisphosphate carboxylase. As a result, C_4 plants such as corn and sugarcane are characterized by net photosynthetic rates that are often two to three times those of C_3 plants such as the cereal grains.

This higher productivity comes at a price, however. Because of the additional ATP needed to drive the Hatch-Slack pathway, the amount of energy required to fix carbon is greater for a C_4 plant than for a C_3 plant. Still, the difference in productivity is significant, and the agricultural importance of such a difference can hardly be overestimated. Little wonder, then, that crop physiologists and plant breeders have devoted so much attention to these species and to the question of whether it is possible to improve on the inherently inefficient carbon fixation of the C_3 plants. Some scenarios of genetic engineering even depict the genetic "conversion" of C_3 plants into C_4 plants, but for now, such procedures exist only in the realm of the hypothetical.

Summary of Photosynthesis

Now we are ready to write an overall reaction for photosynthesis that takes both the light-dependent and light-independent reactions into consideration. Considering just the needs of the Calvin cycle, the requirements are clear: The synthesis of 1 molecule of glucose requires 12 molecules of NADPH and 18 molecules of ATP (reaction 12-13). To meet this requirement, we need only specify the flow of 12 pairs of electrons through the noncyclic pathway (12 × reaction 12-7) and 6 pairs of electrons through the cyclic option (6 × reaction 12-8). This stoichiometry can be summarized as follows:

$$12H_2O + 12NADP^+ + 18ADP + 18P_i + 60 \text{ photons} \longrightarrow 6O_2 + 12NADPH + 12H^+ + 18ATP + 18H_2O \quad (12\text{-}14)$$

By adding the summary equation for the Calvin cycle (reaction 12-13) to that for the light-dependent reactions (reaction 12-14), the overall expression becomes

$$6CO_2 + 12H_2O + 60 \text{ photons} \longrightarrow C_6H_{12}O_6 + 6O_2 + 6H_2O \quad (12\text{-}15)$$

This summary reaction is just like the one with which we began our discussion of photosynthesis (reaction 12-2). Now, however, we are in a much better position to understand some of the metabolic and photochemical complexity behind what might otherwise appear to be a simple reaction.

According to our summary expression, the synthesis of one molecule of glucose requires 60 photons, one for each of 60 individual excitation events (24 in photosystem I and another 24 in photosystem II to account for the noncyclic component of the overall reaction, and 12 more in photosystem I only for the cyclic portion). This is equivalent to 10 photons absorbed per molecule of carbon dioxide fixed, a figure consistent with the experimental estimates of 8 to 10 photons per molecule of carbon dioxide for the **photosynthetic quantum requirement.**

For red light (wavelength 670 nm), 60 photons represent about 60 × 43 kcal/einstein, or 2580 kcal of energy. Since glucose differs in free energy from carbon dioxide and water by 686 kcal/mol, the overall efficiency of photosynthetic energy transduction is about 27% (686/2580 × 100%). Although efficiencies of this order can be observed in the laboratory with algal cells or isolated chloroplasts, photosynthesis under field conditions occurs with a much lower overall efficiency of energy utilization. Only a few percent of the solar energy incident on a green plant during its growing season is actually used for photosynthesis, and only about 0.2%–0.4% of the total radiant energy reaching the Earth's surface is stored by phototrophs annually. Problem 11 at the end of the chapter will provide you with an opportunity to quantify the practical efficiency of photosynthesis for an agricultural crop and to consider some of the reasons for the discrepancy between theoretical laboratory values and practical agricultural experience.

Leaf Peroxisomes: Glycolate Oxidation and Photorespiration

In the 1960s, small, densely staining organelles were discovered in both plant and animal cells. Later investigations showed that these organelles contained oxidase enzymes that generate hydrogen peroxides. They also contain catalase, which decomposes hydrogen peroxide to water and oxygen. Therefore such organelles were called **peroxisomes.** In animal cells, peroxisomes serve a protective function, as described in Chapter 9. In plants, peroxisomes play an important role in the metabolism of a specific by-product of photosynthetic carbon fixation, the two-carbon compound glycolate.

Leaf peroxisomes are found not only in leaf cells but in all green, photosynthetic tissue. A typical leaf peroxisome is shown in the electron micrograph of Figure 12-18. Wherever leaf peroxisomes occur, they are always in close proximity to the chloroplasts, reflecting the metabolic association between the two organelles. The crystalline core

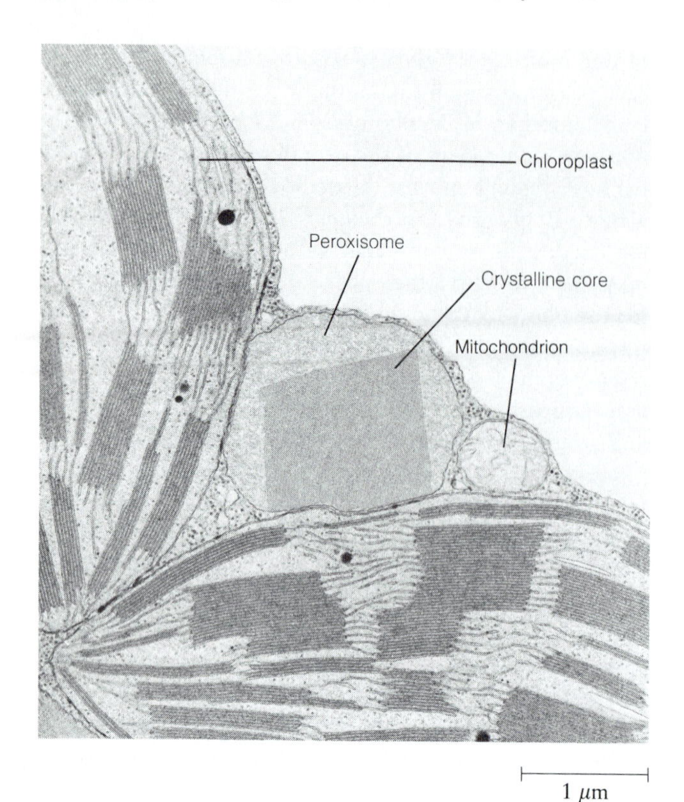

Figure 12-18 Leaf Peroxisomes. This electron micrograph of a leaf mesophyll cell shows a peroxisome with a crystalline core, a mitochondrion, and portions of two chloroplasts. The close association of these organelles probably reflects their mutual involvement in the glycolate pathway of photorespiration. The crystalline core in the peroxisome is thought to be catalase. (TEM)

seen within the matrix of the organelle is a common feature of leaf peroxisomes and represents crystallized catalase.

The Source of Glycolate

Before continuing our discussion, we will briefly review the chemical structure and source of glycolate. As shown in Figure 12-19, glycolate is a two-carbon organic acid. It arises during photosynthetic carbon fixation as a result of an unusual property of the enzyme ribulose-1,5-bisphosphate carboxylase, the CO_2-fixing enzyme of the Calvin cycle (recall reaction Cal-1 of Figure 12-13). The main reaction catalyzed by RuBPCase involves the fixation of carbon dioxide onto the five-carbon sugar ribulose-1,5-bisphosphate to form two molecules of 3-phosphoglycerate. This reaction, called the **carboxylase function** of the enzyme, is favored by high carbon dioxide concentrations and low oxygen concentrations (that is, by a high CO_2 ratio).

As mentioned earlier, RuBPCase is unusual in that it is capable of an alternative reaction in which ribulose-1,5-bisphosphate is split without prior carboxylation, but with the uptake of oxygen. This reaction, called the **oxygenase function** of the enzyme, is favored by a low CO_2 ratio. Instead of two three-carbon molecules, the result is one three-carbon product (3-phosphoglycerate) and one two-carbon product (phosphoglycolate), as shown in Figure 12-19.

Although particularly favored under conditions of high oxygen concentration, the oxygenase function is in fact active to some extent under almost all conditions of photosynthesis. This means that whenever the Calvin cycle is operative and carbon is being fixed, there is always some oxygenase activity and hence some generation of phosphoglycolate.

As we consider the metabolic fate of phosphoglycolate in what is usually called the **glycolate pathway,** we can understand the function of peroxisomes in glycolate metabolism and the close physical association between chloroplasts and peroxisomes that is frequently seen in photosynthetic cells.

The Glycolate Pathway

The phosphoglycolate generated in the chloroplast during Calvin cycle activity diffuses out of the chloroplast and in the process loses its phosphate group as a result of a phosphoglycolate phosphatase enzyme in the chloroplast membrane (Figure 12-19). The free glycolate that results passes into the peroxisome, where it is further metabolized by the enzymes of the glycolate pathway that are localized in this organelle. Presumably, it is the need to pass glycolate from the chloroplast to the peroxisome that dictates the close juxtaposition of these organelles in photosynthetic cells.

The glycolate pathway (GP) is shown in Figure 12-20. Notice that the pathway begins with the activity of an oxidase in reaction GP-1, resulting in the generation of hydrogen peroxide expected of a peroxisomal process. As you might also expect, the peroxide is immediately degraded by the catalase that is also present in abundance in leaf peroxisomes. Meanwhile, the glyoxylate that results from glycolate oxidation is transaminated in reaction GP-2 to glycine.

In the next reaction (GP-3), two glycine molecules are condensed to form a single serine. (Somewhat surprisingly, this reaction occurs in mitochondria, so the complete glycolate pathway actually requires the metabolic cooperation and presumably the physical juxtaposition of not two, but three, organelles.) The carbon atom that must be eliminated to form one three-carbon compound from two two-carbon compounds is liberated as carbon dioxide. (Notice

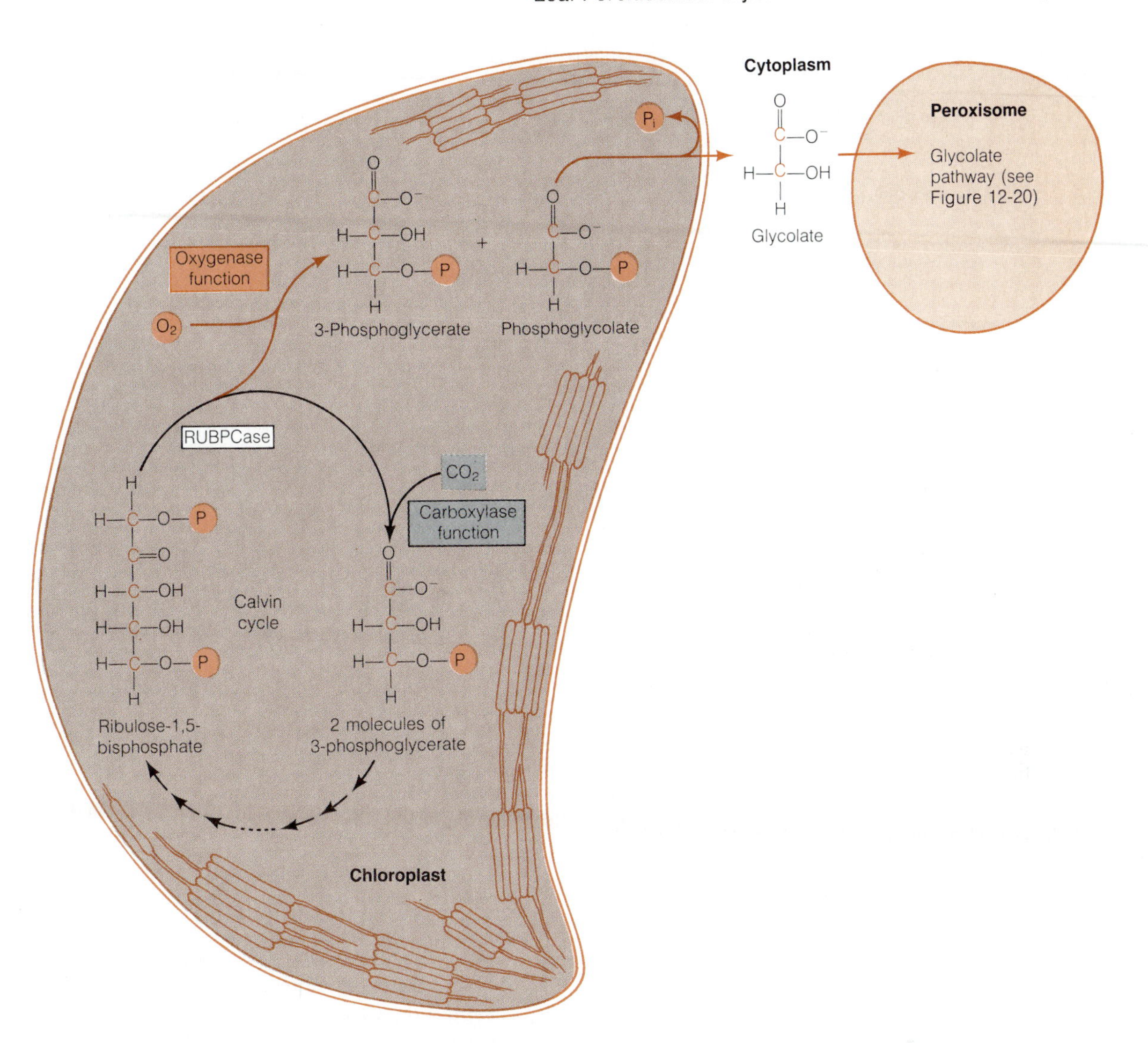

Figure 12-19 Formation and Transport of Glycolate. Glycolate arises in the chloroplast as a result of the oxygenase activity of ribulose-1,5-bisphosphate carboxylase (RuBPCase). The immediate product is phosphoglycolate, which is converted to free glycolate by a phosphoglycolate phosphatase enzyme located in the chloroplast membrane. Free glycolate diffuses through the cytoplasm to the peroxisome, where it is metabolized by the glycolate pathway shown in Figure 12-20. The oxygenase activity of RuBPCase is favored by a low carbon dioxide-to-oxygen ratio. A high carbon dioxide-to-oxygen ratio favors the carboxylase activity needed for the Calvin cycle.

that this pathway is characterized by both the uptake of oxygen (reaction GP-1) and the liberation of carbon dioxide (reaction GP-3); we will return to this feature of glycolate metabolism shortly.)

Back in the peroxisome, the serine is transaminated to a compound called hydroxypyruvate (reaction GP-4), and this is reduced by NADH to glycerate (reaction GP-5). The glycerate is returned to the chloroplast, where it can be phosphorylated to generate 3-phosphoglycerate, which will then be treated just like the 3-phosphoglycerate generated by the Calvin cycle.

What, then, does the glycolate pathway accomplish? Apparently, it exists in effect to "salvage" as much carbon as possible from the glycolate that results from the oxygenase activity of the RuBPCase enzyme. It is tempting to speculate that if this enzyme were a "straight" carboxylase without the oxygenase activity, there would probably be no need for a glycolate pathway and perhaps no need for leaf peroxisomes.

But since RuBPCase does have oxygenase activity, and since glycolate is in fact produced, the peroxisome is needed to salvage the carbon that would otherwise be lost. If you

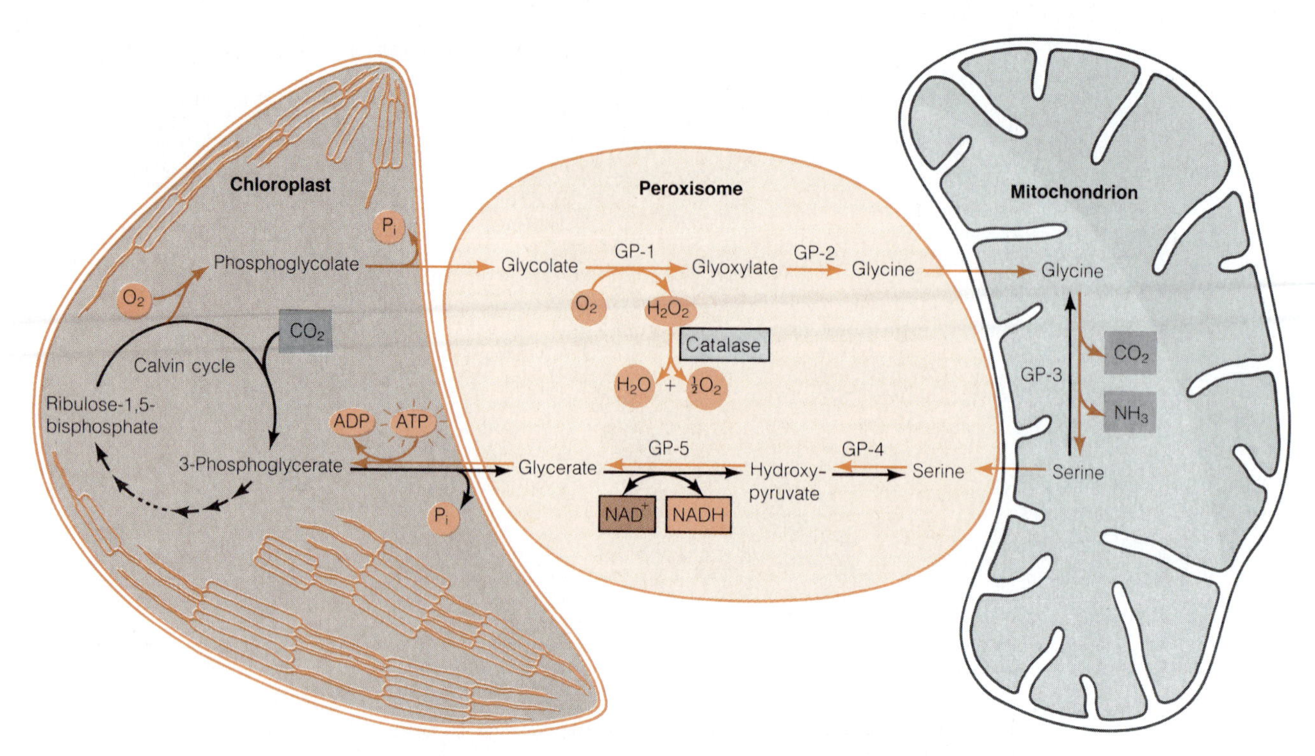

Figure 12-20 The Glycolate Pathway. Glycolate arises as phosphoglycolate during Calvin cycle activity in the chloroplast and is then metabolized by a five-step pathway (GP-1 through GP-5) that occurs partially in the peroxisome and partially in the mitochondrion. The end product is 3-phosphoglycerate, which can be further metabolized within the chloroplast. The oxygen uptake and carbon dioxide evolution characteristic of photorespiration occur in the peroxisome (reaction GP-1) and mitochondrion (reaction GP-3), respectively.

consider Figure 12-20 carefully, you will notice that the salvage operation is quite effective, since three of the four carbons present in every two glycolate molecules are conserved as glycerate, and only one is lost as carbon dioxide in reaction GP-3.

Photorespiration

We can now summarize the role of glycolate in plant cell metabolism. The glycolate pathway is characterized by an uptake of oxygen (reaction GP-1) and an evolution of carbon dioxide (reaction GP-3) that is light-dependent because it depends on the photosynthetic enzyme RuBPCase for the generation of its substrate. This exchange of gases is called **photorespiration,** since it resembles mitochondrial respiration but is dependent on light. Unlike mitochondrial respiration, however, photorespiration is not coupled at any point to ATP generation and is therefore a wasteful process in terms of energy. Moreover, photorespiration is directly counterproductive to the photosynthetic activity of the plant, because photorespiration results in the release of carbon dioxide instead of its uptake. Photorespiration, in other words, tends to "unfix" the carbon dioxide that photosynthesis fixes. Although the glycolate pathway manages to conserve 75% of the carbon that gets diverted to glycolate by the oxygenase activity of RuBPCase, the other 25% of the carbon is unavoidably lost, so the overall process detracts significantly from the photosynthetic productivity of the plant.

For this reason, photorespiration is generally considered an undesirable characteristic of plants—an unproductive loss of carbon that would otherwise remain in organic form.

As discussed earlier, C_4 plants such as corn, sorghum, and sugarcane have a leaf anatomy such that the initial fixation of carbon dioxide by the Hatch-Slack pathway takes place in the outer mesophyll cells, whereas Calvin cycle activity occurs in the inner bundle sheath cells. Since photorespiration accompanies the metabolism of glycolate, and since glycolate is produced only in cells with Calvin cycle activity, photorespiratory carbon dioxide release occurs only in the bundle sheath cells. As this carbon dioxide diffuses outward, it is very likely to be trapped and refixed by phosphoenolpyruvate carboxylase, the carboxylating enzyme of the Hatch-Slack pathway, which is an excellent scavenger of carbon dioxide because of its low K_m.

The result is that little, if any, of the carbon dioxide released by glycolate oxidation in peroxisomes of the bundle sheath cells actually escapes from the plant. C_4 plants are therefore photosynthetically more efficient than C_3 plants. The differences are often significant, since C_4 plants such as corn and sugarcane are capable of outyielding some of the C_3 cereal grains by a factor of two or three. And they do it by improving still further on what the peroxisome seems to be intended to do in its own right: minimize the carbon losses to which every photosynthetic organism is prone.

Glyoxysomes: The Glyoxylate Cycle and Gluconeogenesis

At about the same time that peroxisomes were discovered in plant leaves, another group of investigators reported that seedlings of fat-storing plant species have organelles that look like peroxisomes but serve a quite different metabolic purpose. These organelles are called **glyoxysomes** because they contain the enzymes of the **glyoxylate cycle,** a metabolic pathway required for the conversion of fat into carbohydrate. This pathway is not directly involved in photosynthesis but does play an important role in energy metabolism of plants.

Many plant species use fat (mainly triglycerides) as an

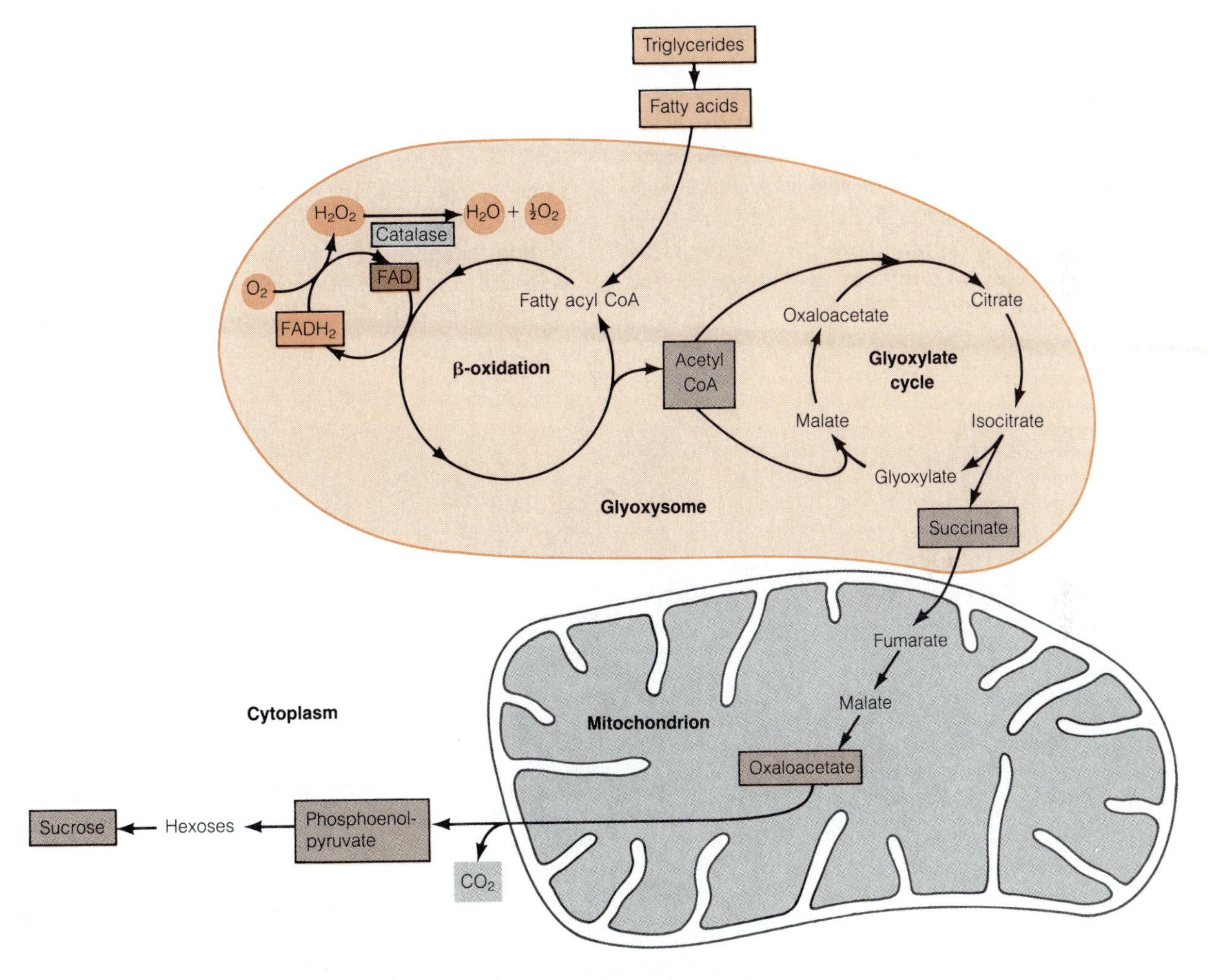

Figure 12-21 Gluconeogenesis in Fat-Storing Seedlings. Seedlings of fat-storing species are capable of synthesizing sugar from stored fat. Fatty acids derived from storage triglycerides are oxidized to acetyl CoA by the process of β oxidation. Acetyl CoA is then converted into succinate by the glyoxylate cycle. All the enzymes of β oxidation and the glyoxylate cycle are located in the glyoxysome. Conversion of succinate to oxaloacetate occurs within the mitochondrion, whereas the further metabolism of oxaloacetate via phosphoenolpyruvate (PEP) to hexoses and hence to sucrose takes place in the cytoplasm.

important means of storing carbon and energy reserves in the seed. You are probably familiar with some of the better-known fat-storing species, since they are commercially important as sources of vegetable oil (vegetable oil is just fat that is liquid at room temperature). Common examples are corn, peanuts, soybean, sunflower, and safflower seeds. The advantage of using fat for this purpose is clear when you consider that fat contains about 2.25 times more energy per gram than does carbohydrate. This is especially important for seeds, because they are the major means of propagation for most plant species and therefore need to be light enough to be readily dispersed by wind and animals.

Conversion of fat into carbohydrate is not possible for most organisms. Higher animals (including humans), for example, readily convert carbohydrate into fat, but most eukaryotes cannot carry out the reverse process. Yet for the seedlings of fat-storing plants, generation of carbohydrate from the storage triglycerides is essential, since it is mainly as sugars that carbon and energy reserves are translocated to the growing shoot and root tips of the plant.

Synthesis of sugars from noncarbohydrate starting compounds is called **gluconeogenesis,** though in the case of plants it might more appropriately be thought of as "sucroneogenesis," since the desired end product is the disaccharide sucrose.

Gluconeogenesis from triglycerides proceeds in several stages, as shown in Figure 12-21 (p. 343). First, the triglycerides must be hydrolyzed to fatty acids. These are then transported into the glyoxysomes, where they are degraded to acetyl CoA by β oxidation, as already discussed in Chapter 11. The glyoxylate cycle becomes relevant at this point, because it is the pathway by which the acetyl CoA derived from fatty acids is metabolized further within the glyoxysomes. The glyoxylate cycle is a modified version of the TCA cycle that bypasses the steps at which carbon dioxide would be evolved and takes in not one but two acetyl CoA units per cycle, generating the four-carbon compound succinate in the process. The succinate is converted to oxaloacetate in the mitochondria by means of reactions already familiar to us from the TCA cycle. Oxaloacetate is then used to form phosphoenolpyruvate, from which hexoses and eventually sucrose can be generated by a sequence that is essentially the reverse of the glycolytic pathway.

As Figure 12-21 indicates, all the reactions of β oxidation and of the glyoxylate cycle occur within the glyoxysomes in the cells of fat-storing seedlings. Glyoxysomes closely resemble leaf peroxisomes in size, shape, morphological appearance, and buoyant density, but the two classes of organelles have different metabolic roles. Peroxisomes are characteristic of green, photosynthetic tissue

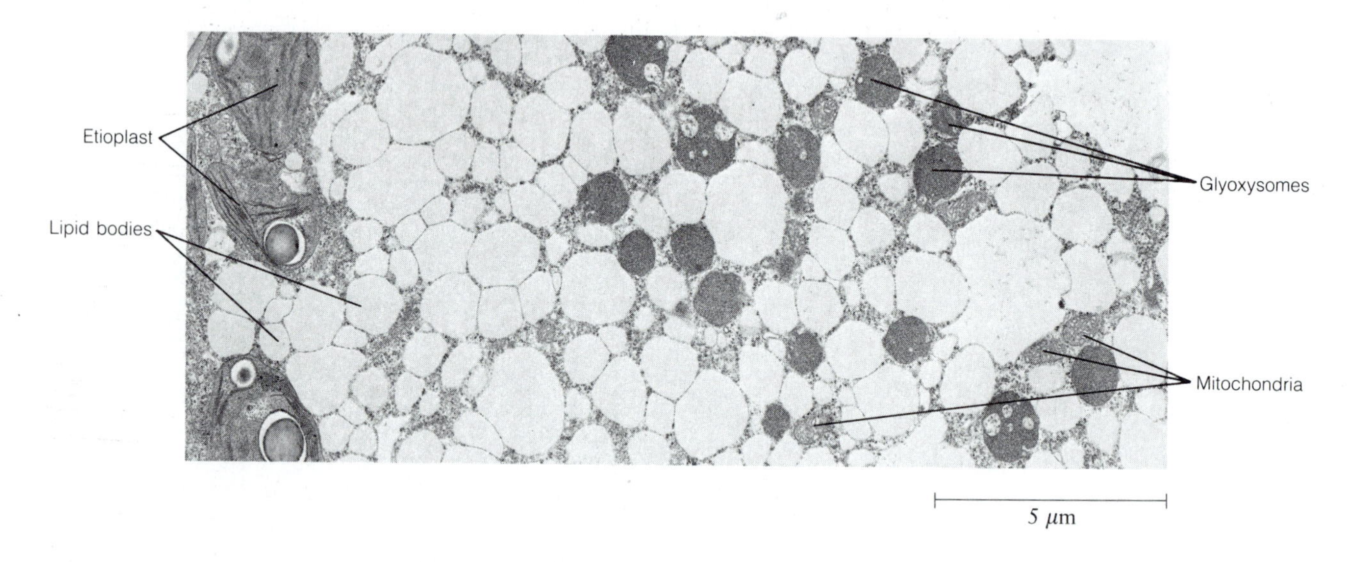

Figure 12-22 Association of Glyoxysomes and Lipid Bodies in Fat-Storing Seedlings. Shown in the electron micrograph is a cell from the cotyledon of a cucumber seedling during early postgerminative development. The glyoxysomes involved in fat mobilization and gluconeogenesis are intimately associated with the lipid bodies in which the fat is stored. Evidence that the cotyledon is not yet photosynthetically active and is therefore still heterotrophic in its nutritional mode can be seen in the presence of etioplasts instead of mature chloroplasts. (TEM)

(usually leaves) and are involved in glycolate metabolism and photorespiration. Glyoxysomes, on the other hand, are found in heterotrophic, fat-storing tissue (usually endosperm or cotyledons) and carry out the portion of the gluconeogenic pathway that converts fatty acids into succinate. The involvement of glyoxysomes in fat catabolism is reflected in their intimate association with *lipid bodies,* the form in which fat is stored in plant cells. This association, shown in the electron micrograph of Figure 12-22, presumably facilitates the transfer of fatty acids from lipid bodies to glyoxysomes.

Perspective

In a very real sense, photosynthesis is the single most vital metabolic process for all forms of life on this planet, because all of us, whatever our immediate sources of energy, are ultimately dependent on the radiant energy of the sun. Photosynthesis involves both light-dependent reactions and light-independent reactions. In the light-dependent reactions, photons of light are absorbed by chlorophyll or accessory pigment molecules within the thylakoid membranes, and the energy is passed rapidly to the special long-wavelength chlorophyll molecules at the reaction center. There, the energy is used to excite and eject an electron, which, in the case of photosystem I, is passed via ferredoxin to $NADP^+$, generating the reduced coenzyme required for carbon fixation and reduction.

The source of electrons in oxygenic phototrophs is water. Electron transfer from water to $NADP^+$ requires two photosystems acting in series, with photosystem II responsible for the oxidation of water and photosystem I responsible for the reduction of $NADP^+$. Electron flow between the two photosystems (or in cyclic fashion around photosystem I) drives the pumping of protons into the lumen of the thylakoid disks. The resulting proton motive force across the thylakoid membrane is due largely to the pH differential and is used to drive ATP synthesis by the CF_1 particles that protrude outward from the thylakoid membranes into the stroma of the chloroplast.

In the stroma, ATP and NADPH are used in the reductive fixation of carbon dioxide into organic form. In C_3 plants, fixation is directly onto ribulose-1,5-bisphosphate, generating phosphoglycerate. In C_4 plants, a preliminary CO_2-fixing sequence called the Hatch-Slack pathway is used to trap carbon dioxide in organic form. The eventual product of carbon dioxide fixation in both cases is phosphoglycerate, which is reduced to glyceraldehyde-3-phosphate. Some of these triose molecules are used to synthesize carbohydrate, and the remainder are used to regenerate the acceptor molecule with which the cycle began. The synthesis of 1 glucose molecule requires the fixation of 6 carbon dioxide molecules and uses 12 molecules of NADPH and 18 molecules of ATP. Generation of NADPH and ATP in this ratio requires both cyclic and noncyclic electron flow.

This transduction of solar energy into chemical energy is crucial to the continued existence of the biological world. All the energy of oxidizable foodstuffs on which chemotrophs depend represents the energy of sunlight, originally entrapped within the molecules of organic compounds during photosynthesis. We have not yet discovered anything particularly unique about photoautotrophs, since carbon dioxide fixation (carboxylation) occurs in animals and other nonphotosynthetic organisms. What is remarkable about photosynthetic organisms is their ability to carry out sustained net fixation and reduction of carbon using the energy of solar radiation to drive what would otherwise be a highly endergonic process. Only phototrophs can utilize solar energy to extract electrons from such poor (electropositive) donors as water and use them to reduce organic compounds. And they can do so because of the photochemical events that are initiated whenever light of the appropriate wavelength is absorbed by chlorophyll, a remarkable molecule that has transformed the biosphere of an entire planet, Earth.

Key Terms for Self-Testing

The Photosynthetic Mode
phototrophs (p. 314)
photoheterotrophs (p. 314)
photoautotrophs (p. 314)
oxygenic photoautotrophs (p. 315)

Chloroplast Structure and Function
chloroplasts (p. 315)
outer membrane (p. 316)
inner membrane (p. 316)
intermembrane space (p. 316)
stroma (p. 316)
thylakoids (p. 316)
grana (p. 316)
stroma lamellae (p. 316)
intrathylakoid space (p. 318)

The Reactions of Photosynthesis
light-dependent reactions (p. 318)
light-independent reactions (p. 318)
chlorophyll (p. 318)
pigment (p. 318)
bacteriochlorophyll (p. 318)
accessory pigments (p. 319)
carotenoids (p. 319)
phycobilins (p. 319)
absorption spectrum (p. 320)
action spectrum (p. 320)

Photoreduction (NADPH Generation)
photoreduction (p. 320)
nicotinamide adenine dinucleotide phosphate ($NADP^+$) (p. 320)
photophosphorylation (p. 320)
photosystem I (p. 321)
photosystem (p. 321)
bound ferredoxin (p. 321)
photosystem II (p. 323)
photolysis (p. 323)
cytochromes (p. 323)
plastoquinones (p. 323)
plastocyanins (p. 324)
noncyclic electron flow (p. 324)
Emerson enhancement effect (p. 324)
photosynthetic unit (p. 324)
reaction center (p. 325)
chlorophyll-binding proteins (p. 325)

ATP Synthesis
electrochemical proton gradient (p. 326)
noncyclic photophosphorylation (p. 327)
cyclic electron flow (p. 328)
cyclic photophosphorylation (p. 328)
chloroplast coupling factor (CF_1) (p. 328)
proton translocator (p. 328)
proton motive force (pmf) (p. 328)

Photosynthetic Carbon Metabolism: The Calvin Cycle
Calvin cycle (p. 329)
UDP-glucose (p. 331)
ADP-glucose (p. 332)

Some Do It Differently: The C_4 Plants
Hatch-Slack pathway (p. 336)
C_4 plants (p. 336)
C_3 plants (p. 336)
mesophyll cells (p. 336)
bundle sheath cells (p. 336)

Summary of Photosynthesis
photosynthetic quantum requirement (p. 339)

Leaf Peroxisomes: Glycolate Oxidation and Photorespiration
peroxisomes (p. 339)
carboxylase function (p. 340)
oxygenase function (p. 340)
glycolate pathway (p. 340)
photorespiration (p. 342)

Glyoxysomes: The Glyoxylate Cycle and Gluconeogenesis
glyoxysomes (p. 343)
glyoxylate cycle (p. 343)
gluconeogenesis (p. 344)

Suggested Reading

General References and Reviews

Clayton, R. K., and W. R. Sistrom, eds. *The Photosynthetic Bacteria.* New York: Plenum Press, 1978.

Harold, F. M. *The Vital Force: A Study in Bioenergetics.* New York: W. H. Freeman, 1986.

Mathews, C. K., and K. E. van Holde. *Biochemistry.* Redwood City, Calif.: Benjamin/Cummings, 1990.

Stryer, L. *Biochemistry,* 3d ed. New York: W. H. Freeman, 1988.

Youvan, D. C., and B. L. Marrs. Molecular mechanisms of photosynthesis. *Sci. Amer.* 249 (June 1987): 42.

Chloroplast Structure and Function

Bogorad, L. Chloroplasts. *J. Cell Biol.* 91 (1981): 256s.

Cramer, W. A., W. R. Widger, R. G. Herrmann, and A. Trebst. Topography and function of thylakoid membrane proteins. *Trends Biochem. Sci.* 10 (1985): 125.

Haliwell, B. *Chloroplast Metabolism—the Structure and Function of Chloroplasts in Green Leaf Cells.* Oxford: Clarendon Press, 1982.

Heber, U., and D. A. Walker. The chloroplast envelope—barrier or bridge? *Trends Biochem. Sci.* 4 (1979): 252.

Kirk, J. T. O., and R. A. E. Tilney-Bassett. *The Plastids: Their Chemistry, Structure, Growth, and Inheritance,* 2d ed. Amsterdam: Elsevier, 1979.

The Light-Dependent Reactions

Bennett, J. The protein that harvests sunlight. *Trends Biochem. Sci.* 4 (1979): 268.

Blankenship, R. E., and R. C. Prince. Excited state redox potentials and the Z scheme of photosynthesis. *Trends Biochem. Sci.* 10 (1985): 382.

Deisenhofer, J., and H. Michel. The photosynthetic reaction center from the purple bacterium *Rhodopseudomonas viridis. Science* 245 (1989): 1463.

Govindjee, ed. *Bioenergetics of Photosynthesis.* New York: Academic Press, 1975.

Miller, K. R. The photosynthetic membrane. *Sci. Amer.* 241 (April 1979): 102.

Skulachev, V. P. Membrane bioenergetics—should we build the bridge across the river or alongside of it? *Trends Biochem. Sci.* 9 (1984): 182.

The Calvin Cycle

Bassham, J. A. The control of photosynthetic carbon metabolism. *Science* 172 (1971): 526.

Calvin, M. The path of carbon in photosynthesis. *Science* 135 (1962): 879.

Clayton, R. K. *Photosynthesis: Physical Mechanisms and Chemical Patterns.* Cambridge, U.K.: Cambridge University Press, 1980.

Ellis, R. J. The most abundant protein in the world. *Trends Biochem. Sci.* 4 (1979): 241.

Jensen, R. G., and J. T. Bahr. Ribulose-1,5-bisphosphate carboxylase-oxygenase. *Annu. Rev. Plant Physiol.* 28 (1977): 379.

C_4 Plants and Photorespiration

Bjorkman, O., and J. Berry. High-efficiency photosynthesis. *Sci. Amer.* 229 (October 1973): 80.

Hatch, M. D., and C. R. Slack. Photosynthetic CO_2-fixation pathways. *Annu. Rev. Plant Physiol.* 21 (1970): 141.

Heber, U., and G. H. Krause. What is the physiological role of photorespiration? *Trends Biochem. Sci.* 4 (1979): 32.

Plant Peroxisomes and Glyoxysomes

Beevers, H. Microbodies in higher plants. *Annu. Rev. Plant Physiol.* 30 (1979): 159.

Lazarow, P. B., and H. Kindl, eds. *Peroxisomes and Glyoxysomes* (Symposium Proceedings). Ann. N.Y. Acad. Sci., vol. 286, 1982.

Tolbert, N. E. Metabolic pathways in peroxisomes and glyoxysomes. *Annu. Rev. Biochem.* 50 (1981): 133.

Vigil, E. L. Structure and function of plant microbodies. *Subcell. Biochem.* 2 (1973): 137.

Problem Set

1. **True, False, or ?.** Identify each of the following statements as true (T), false (F), or (?) if you cannot tell whether the statement is true or false on the basis of the information given.
   (a) Although sometimes called the dark reactions of photosynthesis, the Calvin cycle actually depends indirectly on light and will not continue to function very long in the dark.
   (b) The pathway from 3-phosphoglycerate to glucose in the Calvin cycle is identical to the route from glucose to 3-phosphoglycerate in glycolysis, except for its localization in eukaryotic cells.
   (c) If the time interval between $^{14}CO_2$ exposure and extraction of intermediates is short enough, the first compound to be radioactively labeled in a plant leaf will be 3-phosphoglycerate.
   (d) The energy requirement expressed as ATP consumed per molecule of carbon dioxide fixed is higher for a C_3 plant than for a C_4 plant.
   (e) The ultimate electron donor in photosynthetic NADPH generation is water.
   (f) The enzyme ribulose-1,5-bisphosphate carboxylase is atypical in that it exhibits two different enzymatic activities, depending on conditions.

2. **Photosynthetic Carbon Metabolism.** If the leaf of a C_3 plant is exposed to radioactive carbon dioxide ($^{14}CO_2$) in the light for a few seconds and then extracted to isolate various compounds, in which carbon atoms of the following compounds would you expect most of the ^{14}C to be localized?
   (a) 3-Phosphoglycerate
   (b) Glucose-6-phosphate
   (c) Sedoheptulose-7-phosphate
   (d) Erythrose-4-phosphate

3. **The Advantage of Being C_4.** A C_4 plant is more efficient at fixing carbon, an advantage that becomes more evident as the carbon dioxide concentration decreases.
   (a) Explain in your own words why a C_4 plant is inherently more efficient at carbon fixation than a C_3 plant.
   (b) Why would this advantage be more apparent at an atmospheric carbon dioxide concentration of 50 ppm than at the normal level of 380 ppm?
   (c) If a C_4 plant and a C_3 plant are grown under constant illumination in a sealed container with an initial carbon dioxide concentration of 380 ppm, the C_4 plant will eventually kill the C_3 plant. Explain!

4. **The Role of Sugar.** A plant can be viewed as a system that uses solar radiation to make ATP, which then drives the synthesis of sugar in the leaves so that the sugar can be translocated to other, nonphotosynthetic parts of the plant (roots, stems, flowers, fruits) for use as an energy source. Thus, ATP is used to make sugar, and the sugar is then used to make ATP. It would seem simpler for the plant just to make ATP and translocate the ATP itself directly to other parts of the plant, thereby completely eliminating the need for a Calvin cycle, a glycolytic pathway, and a TCA cycle (and making life a lot easier for cell biology students in the process). Why do you suppose this is not the way a plant operates its energy economy? (You should be able to think of at least two major reasons.)

5. **Effects on Photosynthesis.** Assume that you have an illuminated suspension of *Chlorella* cells carrying out photosynthesis in the presence of 1000 ppm (0.1%) carbon dioxide and 20% oxygen. What effect would the following changes in conditions have on the levels of 3-phosphoglycerate and ribulose-1,5-bisphosphate?
   (a) Carbon dioxide concentration is suddenly reduced 1000-fold.
   (b) Light is suddenly switched off.
   (c) Inhibitor of photosystem II is added.
   (d) Oxygen concentration is reduced from 20% to 1%.

6. **Overall Reactions.** Eukaryotic phototrophs use water as their electron donor, but prokaryotic phototrophs depend instead on a variety of organic and inorganic electron donors. Reaction 12-1 is intended as a general reaction to cover all cases. Some bacteria use H_2S as their source of electrons, generating elemental sulfur.
   (a) Write a balanced overall reaction for photosynthesis based on H_2S as the electron donor.
   (b) Why are photosynthetic bacteria able to use H_2S as an electron donor, but not H_2O?
   (c) Do you think that photosynthetic bacteria are able to generate reduced coenzyme by oxidizing ferrous ions to ferric ions? (E_0 for the Fe^{3+}/Fe^{2+} redox couple: +0.77 V.)

7. **Energy Flow in Photosynthesis.** The energy that arrives at the surface of a plant leaf in the form of sunlight eventually appears as chemical bond energies in the sugars that are generally regarded as the end products of photosynthesis. In between the photons and the sugar molecules, however, the energy exists in a variety of forms. Trace the flow of energy from photon to sugar molecule, assuming the wavelength of light to be in the absorption range of one of the accessory pigments rather than that of chlorophyll.

8. **The Mint and the Mouse.** A prominent name in the early history of research in photosynthesis is that of Joseph Priestley, a British clergyman. In 1771, Priestley wrote these words:

   > One might have imagined that since common air is necessary to vegetable as well as to animal life, both plants and animals had affected it in the same manner; and I own that I had that expectation when I first put a sprig of mint into a glass jar standing inverted in a vessel of water; but when it had continued growing there for some months, I found that the air would neither extinguish a candle, nor was it at all inconvenient to a mouse which I put into it.

   Explain the basis of Priestley's observations, and indicate their relevance to the early understanding of the nature of photosynthesis.

9. **The Hill Reaction.** A highly significant advance in our understanding of photosynthesis came in 1937 when Robert Hill showed that isolated chloroplasts, though not capable of fixing carbon dioxide, were able to produce molecular oxygen, but only if they were illuminated and provided with an artificial electron acceptor (ferricyanide was the acceptor of choice). Which of the following statements are valid conclusions from Hill's experiment? (Some, all, or none may be correct.)
   (a) The oxygen evolved during photosynthesis apparently does not come from carbon dioxide, as was believed earlier.
   (b) These findings are in accord with a proposal made six years earlier by C. B. van Niel that the oxygen evolved during photosynthesis comes from water.
   (c) Ferricyanide is probably one of the reversibly oxidizable electron carriers in the transport chain that links photosystems I and II.
   (d) Reaction 12-2 would be less indicative of what happens during photosynthesis in eukaryotic phototrophs if six molecules of water were subtracted from each side of the reaction.

10. **Photophosphorylation.** ATP generation in higher plants can occur as the result of either cyclic or noncyclic electron flow. To sustain the ATP/NADPH ratio of 3:2 required by the Calvin cycle, the ratio of noncyclic to cyclic electron flow is thought to be 2:1 (see reaction 12–14). For each of the following conditions, indicate whether you would expect the noncyclic/cyclic ratio to be higher than 2:1, lower than 2:1, or unchanged.
    (a) Chloroplasts carrying out not only the Calvin cycle but also the reduction of nitrite (NO_2^-) to ammonia (NH_3), a process that requires NADPH but not ATP.
    (b) Chloroplasts carrying out not only the Calvin cycle but also extensive active transport across the inner membrane, a process requiring ATP but not NADPH.
    (c) Chloroplasts carrying out not only the Calvin cycle but also the Hatch-Slack pathway.

(d) Plants treated with DCMU, a herbicide that prevents the oxidation of water to oxygen.

(e) Plants treated with ferricyanide, which accepts electrons from the "bottom" of the electron transport chain, becoming reduced to ferrocyanide (E_0 for the ferricyanide/ferrocyanide redox couple: +0.4 V).

11. **The Practical Efficiency of Photosynthesis.** Since we of the chemotrophic world depend on phototrophs as our energy link to the sun, it becomes important to know how efficient the phototrophs are at trapping and storing solar energy. Consider, for example, a field of sugar beets growing in Wisconsin, where the growing season for sugar beets is from about May 15 to September 1. During this period, the average length of daylight is about 15 h/day. A respectable yield is 35 metric tons per hectare (about 15.6 tons per acre), of which about 20% is sucrose. Assume that of the total solar radiation impinging on the upper atmosphere (1.94 cal/cm^2-sec, the solar energy constant), the amount actually reaching the earth's surface is 1.4 cal/cm^2-sec. Assume also that the quantum requirement of photosynthesis is about 10 photons of light absorbed per molecule of carbon dioxide fixed and reduced.

(a) What is the minimum amount of light energy in einsteins that would have to be absorbed during the growing season by a hectare of sugar beets to produce the 7 metric tons of sucrose actually obtained?

(b) What is the maximum theoretical amount of light energy in einsteins available to the sugar beet crop during the growing season? (Assume an average energy content of 55 kcal/einstein for visible light.)

(c) What is the efficiency of conversion of available light energy to sugar?

(d) Can you suggest three or four major reasons why the efficiency calculated in part (c) is so low?

12. **Chloroplast Structure.** Where in the chloroplast is each of the following substances or processes located? (Be as specific as possible.)

(a) Phosphoglycerate reduction

(b) Cyclic electron flow

(c) Carotenoid molecules

(d) Photophosphorylation

(e) Proton pumping

(f) Molecules of chlorophyll P-700

(g) Transketolase

(h) Plastoquinone

13. **Metabolite Transport Across Membranes.** For each of the following metabolites, indicate whether you would expect it to be in steady-state flux across one or more membranes in a photosynthetically active chloroplast, and, if so, indicate which membrane(s) the metabolite must cross.

(a) Carbon dioxide

(b) Protons

(c) Electrons

(d) Starch

(e) Ribose-5-phosphate

(f) $NADP^+$

(g) ATP used for the Calvin cycle

14. **Glyoxysomal Function.** The glyoxysomes of fat-storing seedlings contain all the enzymes necessary to degrade fatty acids completely to acetyl CoA and to synthesize succinate from the resulting acetyl CoA. Fatty acid oxidation (Figure 11-10) begins with ATP-dependent formation of the fatty acyl CoA thioester (reaction FA-1). Each cycle of β oxidation that follows generates one molecule each of $FADH_2$ (FA-2) and NADH (FA-4), culminating with the release of two carbons as acetyl CoA (FA-5). In the glyoxysome, $FADH_2$ (but not NADH) is reoxidized by direct, oxidase-mediated transfer of electrons to oxygen, generating hydrogen peroxide, without ATP synthesis. Synthesis of succinate from the acetyl CoA occurs by means of the glyoxylate cycle. The further steps required to convert succinate to hexoses (and then to sucrose) occur elsewhere in the cell.

(a) What are the main products of β oxidation of a fatty acid in the glyoxysome? What happens to each?

(b) How many molecules of succinate are produced from a single molecule of palmitate ($C_{16}H_{32}O_2$)? How many molecules of $FADH_2$ and NADH are generated in the process?

(c) How many hexose molecules can be formed from the succinate of part (b)? Where in the cell does this process occur? What other products are formed in the conversion of succinate to hexoses?

(d) How many molecules of sucrose can be produced from a single molecule of palmitate? How many of the original 16 carbon atoms of palmitate eventually appear in sucrose? What happens to the others?

(e) Assuming quantitative conversion, how much sucrose can a fat-storing seedling produce from a gram of stored palmitate?

(f) In addition to fatty acids and succinate, what other substance(s) would you predict have to be transported across the glyoxysomal membrane? In which direction would you expect these substance(s) to move?

PART FOUR

INFORMATION FLOW IN CELLS

13

The Flow of Information: DNA, Chromosomes, and the Nucleus

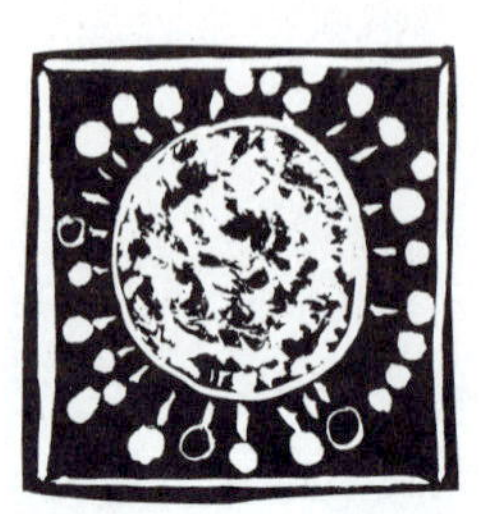

Most of the discussion in the preceding chapters has focused on cellular structure and function—what cells look like and what they do. Implicit in much of what we have considered so far has been a sense of predictability, order, and control. We have come to expect that organelles and other cellular structures will have a predictable appearance and function, that metabolic pathways and other cellular activities will proceed in an orderly fashion in specific intracellular locations, and that all of this will be carried out in a carefully controlled, highly efficient, and heritable manner.

Such expectations express our confidence that cells have "instructions" that specify their structure, dictate their functions, and regulate their activities and that these instructions can be passed on faithfully to daughter cells. These instructions are resident in the **genetic information** of the cell, which is almost universally represented by molecules of *DNA (deoxyribonucleic acid).*

In viruses and prokaryotes, the genetic material is stored as molecules of DNA (or RNA in some viruses) called **genophores.** By contrast, all eukaryotic organisms organize and store their DNA in more complex structures called **chromosomes,** which contain at least as much protein as DNA. Chromosomes are always localized in the **nucleus** of the eukaryotic cell, so much of our attention in this and the next several chapters focuses on nuclear activities.

DNA, chromosomes, and the nucleus are the topics of discussion as we introduce *information flow in cells* (Figure 13-1). We will begin by considering the chemical nature of the gene. And for that, a good starting point is with Friedrich Miescher, the Swiss biologist who discovered DNA.

The Chemical Nature of the Gene

The history of *genetics* begins with Gregor Mendel, the Augustinian monk whose experiments with peas laid the foundations for our understanding of heredity. In fact, Mendel was the first of the early geneticists and cytologists to grapple with the physical basis of inheritance, and his work eventually led to an understanding of chromosomes and genes.

But while Mendel and others focused on the *physical* basis of inheritance, Friedrich Miescher and those who followed him were interested in the *chemical* nature of the gene and were therefore instrumental in weaving a second strand into the tapestry of genetics. Both strands date from about the same time, however, because Miescher published his first findings in 1868, just a few years after Mendel's initial report of his findings on the heritability of specific traits in peas.

Pus, Fish Sperm, and the Discovery of DNA

In his initial experiments, Miescher isolated nuclei from the white blood cells present in pus recovered from surgical bandages in the clinic where he worked. By treating the nuclei with alkali, he was able to prepare a nuclear extract containing a substance that he called "nuclein" but that we now know as DNA. Miescher then went on to study "nuclein" in salmon sperm. Fish sperm may seem like a somewhat unusual biological material until we realize that the nucleus accounts for more than 90% of the mass of a

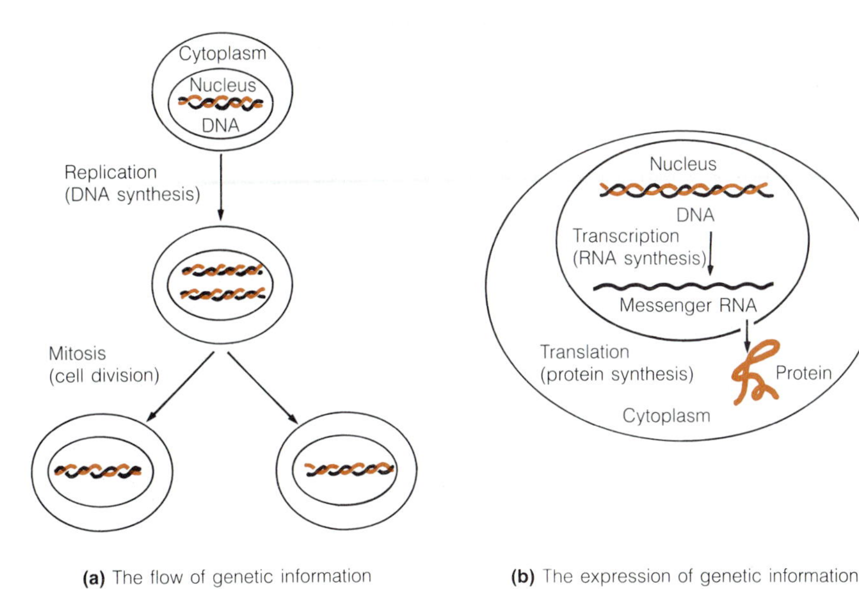

Figure 13-1 The Flow of Information in Cells. (a) Genetic information encoded in DNA molecules is passed on to successive generations of cells by the processes of *replication* and *mitosis*—the DNA is first duplicated, and then divided equally between the two daughter cells. In this way, each daughter cell is assured of having the same genetic information as the cell from which it arose. (b) Within each cell, genetic information encoded in the DNA is *expressed* through the processes of *transcription* (RNA synthesis) and *translation* (protein synthesis). Transcription involves the use of selected segments of the DNA as templates for the synthesis of messenger RNA and other RNA molecules. Translation is the process whereby amino acids are joined together in an order dictated by the order of nucleotides in the RNA.

typical sperm cell. Sperm cells are therefore rich in DNA, and fish sperm is a widely used source of DNA. From sperm nuclei, Miescher isolated not only "nuclein" but also a substance that he called **protamine.** We now recognize protamine as a protein that is unique to sperm nuclei, containing high concentrations of the basic amino acids lysine and arginine. Miescher was convinced of the importance of his "nuclein," but it was to be more than 80 years before DNA would be established as the genetic material. Ironically, much of the delay was due to the implications of Miescher's other finding—that nuclei contain not only "nuclein" but also proteins.

Most scientists believed that nuclear proteins, rather than the "nuclein," were involved in transmitting genetic information. Proteins were already appreciated as complex substances and were therefore considered the more logical candidates for genes. DNA, by contrast, was widely believed to be a simple polymer consisting of a single tetranucleotide sequence repeated over and over, almost certainly without the complexity expected of genetic material. This view prevailed until two key experiments were reported—one in 1944 and the other in 1952—that resolved the matter definitively in favor of DNA and in so doing ushered in the age of molecular genetics.

DNA as the Transforming Principle of *Pneumococcus*

The first experiment was reported by Oswald Avery, Colin MacLeod, and Maclyn McCarty. They were studying a pathogenic strain of the bacterium *Pneumococcus,* which causes a fatal pneumonia in animals. When grown in culture in the laboratory, the bacteria produce smooth, shiny colonies because of the mucous coat that each cell secretes. Such strains are designated as S (**smooth**) **strains,** to distinguish them from **R** (**rough**) **strains,** which have lost the ability to synthesize the mucous coat and which therefore produce colonies that have a rough surface.

Significantly, the R strains of *Pneumococcus* are nonpathogenic. When injected into a mouse, for example, the bacteria are killed by antibodies that the mouse produces in response to the infection, and the mouse usually recovers from the infection. Injection of S strain bacteria, on the other hand, leads almost inevitably to pneumonia and to the death of the mouse. Pathogenicity is directly related to the presence of the S strain's mucous coat, since it is the mucous coat that protects the bacterial cell against the antibodies produced by the mouse.

This famous experiment of Avery, MacLeod, and McCarty depended critically on an earlier, less-known finding. In 1928, Frederick Griffith showed that mice could be killed by infection with cells of an R strain of *Pneumococcus,* provided that heat-killed cells of an S strain were also injected (Figure 13-2). Since live S-type cells could be recovered from the dead mouse, Griffith concluded that nonpathogenic R cells could be converted into pathogenic S cells by something present in the heat-killed S cells that were co-injected. He called the phenomenon **transformation** and referred to the active (though still unknown) substance in the S cells as the **transforming principle.**

It was then found that the transformation of R cells

into S cells could be accomplished in vitro by the addition of an extract of heat-killed S cells to a culture of R cells. This finding set the stage for Avery, MacLeod, and McCarty, who pursued the investigation to its logical conclusion by asking which component of the heat-killed S cells was actually responsible for the transforming activity. They fractionated cell-free extracts of S cells and found that only the nucleic acid fraction was capable of causing transformation. Moreover, the activity was specifically eliminated by treatment with deoxyribonuclease, an enzyme that degrades DNA.

The investigators therefore concluded that DNA is the transforming principle of *Pneumococcus*. This was the first rigorously documented assertion that DNA could carry genetic information. Some of the excitement of that discovery and a glimpse into Avery's appreciation of its implications can be found in a letter that Avery wrote to his brother Roy in May 1943. Here is an excerpt from that letter:

> For the past two years, first with MacLeod and now with Dr. McCarty, I have been trying to find out what is the chemical nature of the substance in the bacterial extract which induces this specific change. The crude extract of Type III is full of capsular polysaccharide, C (somatic) carbohydrate, nucleoproteins, free nucleic acids of both the yeast and thymus type, lipids, and other cell constituents. Try to find in the complex mixtures the active principle! Try to isolate and chemically identify the particular substance that will by itself, when brought into contact with the R cell derived from Type II, cause it to elaborate Type III capsular polysaccharide and to acquire all the aristocratic distinctions of the same specific type of cells as that from which the extract was prepared! Some job, full of headaches and heartbreaks. But at last perhaps we have it.
>
> . . . If we prove to be right—and of course that is a big if—then it means that both the chemical nature of the inducing stimulus is known and the chemical structure of the substance produced is also known, the former being thymus nucleic acid, the latter Type III polysaccharide, and both are thereafter reduplicated in the daughter cells and after innumerable transfers without further addition of the inducing agent and the same active and specific transforming substance can be recovered far in excess of the amount originally used to induce the reaction. Sounds like a virus—may be a gene. But with mechanisms I am not now concerned. One step at a time and the first step is what is the chemical nature

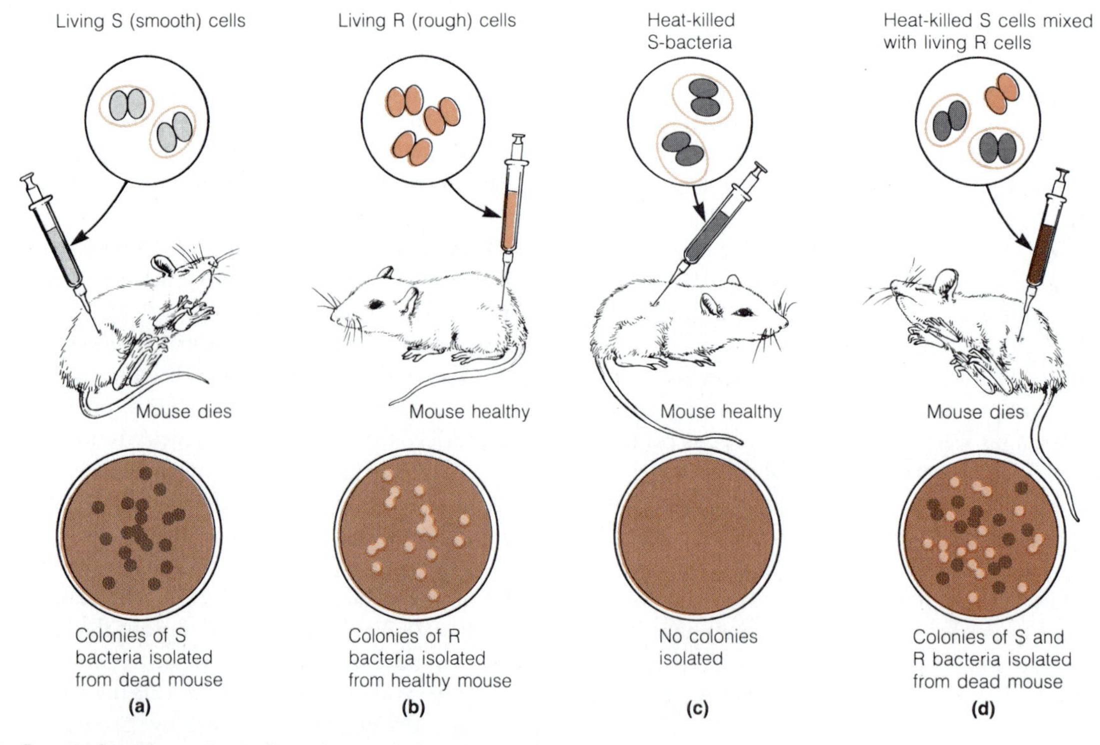

Figure 13-2 Genetic Transformation in *Pneumococcus*. S (smooth) *Pneumococcus* cells are pathogenic in mice; R (rough) cells are not. (a) Injection of living S cells into a mouse results in pneumonia and death. (b) Injection of living R cells leaves the mouse healthy. (c) Heat-killed S cells have no effect when injected alone. (d) When injected with living R cells, heat-killed S cells cause pneumonia and death of the mouse. The finding that pathogenic S-type bacteria could be recovered from the blood of the mouse in case (d) suggested that some chemical factor from the heat-killed S cells was able to cause a heritable change (transformation) of nonpathogenic R cells into pathogenic S cells. The chemical factor was later identified as DNA.

of the transforming principle? Some one else can work out the rest. Of course the problem bristles with implications. It touches the biochemistry of the thymus type of nucleic acids which are known to constitute the major part of chromosomes but have been thought to be alike regardless of origin and species. It touches genetics, enzyme chemistry, cell metabolism and carbohydrate synthesis. But today it takes a lot of well documented evidence to convince anyone that the sodium salt of deoxyribose nucleic acid, protein free, could possibly be endowed with such biologically active and specific properties and that is the evidence we are now trying to get. It is lots of fun to blow bubbles but it is wiser to prick them yourself before someone else tries to.*

Though their methods were rigorous and their findings definitive, the conclusion of Avery and his colleagues that a genetic role could be assigned to DNA did not meet with immediate acceptance. Skepticism was due in part to the widespread conviction that DNA lacked the necessary complexity for such a role. In addition, many scientists questioned whether genetic information in bacteria had anything at all to do with heredity in higher organisms. All remaining doubts were removed only eight years later, however, when DNA was conclusively shown to be the genetic material of another newcomer on the scene—the bacteriophage T2.

DNA as the Genetic Material of Viruses

Bacteriophages—or phages, for short—were studied by geneticists even before bacteria were. As a result, a great deal is known about phages, and much of our understanding of molecular genetics has come from experiments with these organisms. The box on pp. 364–366 highlights some of the advantages of bacteriophages for genetic studies.

Bacteriophage T2 is one of the most thoroughly studied of the phages that infect the intestinal bacterium *Escherichia coli.* That was, in fact, a distinction it already enjoyed in 1952, when it was used by Alfred Hershey and Martha Chase to prove that DNA, and not protein, is the genetic material of bacteriophages. Hershey and Chase took advantage of the fact that the protein of T2 contains the element sulfur (in the amino acids methionine and cysteine) but not phosphorus and that DNA contains phosphorus (in its sugar-phosphate "backbone"; recall Figure 3-13) but not sulfur. Hershey and Chase were therefore able to label the protein of some phages with the radioactive isotope ^{35}S and to label the DNA of other phages with the isotope ^{32}P.

By using radioactive isotopes in this way, Hershey and Chase were able to trace the fate of both protein and DNA during the infection process (Figure 13-3a). The experiment began with adsorption of the phages to the surface of the bacterial cells (step 1) and transfer of genetic information from the phage into the host cell (step 2). At this point, Hershey and Chase found that the empty protein coats (or *phage ghosts*) could be effectively removed from the surface of the bacterial cells by agitating the suspension vigorously in an ordinary blender and then recovering the bacterial cells by centrifugation (step 3).

The critical observation that Hershey and Chase made was that most (80%) of the ^{35}S could be dislodged from the bacterial cells but most (65%) of the ^{32}P remained with the cells (Figure 13-3b). They also showed that the ^{32}P gets transferred from parental to offspring phage but the ^{35}S does not (Figure 13-3a, steps 4–6). This finding led them to the conclusion that DNA, not protein, is the actual genetic material of phage T2.

The Hershey-Chase experiment further strengthened the case for DNA as the genetic material. This experiment came only eight years after Avery's work on transformation in *Pneumococcus,* but it received a much readier welcome. An important factor was the use of a bacteriophage rather than a bacterium, since phages were already accepted as having genetic properties similar to those found in higher organisms. In addition, however, careful chemical studies on DNA from a wide variety of organisms showed that the proportions of purine and pyrimidine bases varied greatly from species to species. This in turn did much to show that DNA was indeed capable of the chemical complexity expected of genetic material.

DNA Base Composition and Chargaff's Rules

The chemical studies on DNA composition were carried out between 1944 and 1952, largely by Erwin Chargaff and his colleagues. These investigators used quantitative chromatographic methods to separate and analyze the individual purine (adenine and guanine) and pyrimidine (cytosine and thymine) bases present in the DNA isolated from a variety of organisms. From their analyses came several important conclusions. First, they showed clearly that DNA from the same species always had the same percentage of each of the four bases and that this did not vary

---

* *DNA as the Transforming Principle of Pneumococcus.* Excerpt from a letter written by Oswald Avery to his brother Roy, capturing some of the excitement and implications of Avery's findings on the transforming principle in *Pneumococcus.* Reproduced by R. D. Hotchkiss in *Age and the Origins of Molecular Biology,* J. Cairns, G. S. Stent, and J. D. Watson, eds. Cold Spring Harbor, New York: Cold Spring Harbor Laboratory, 1966. Reprinted by permission.

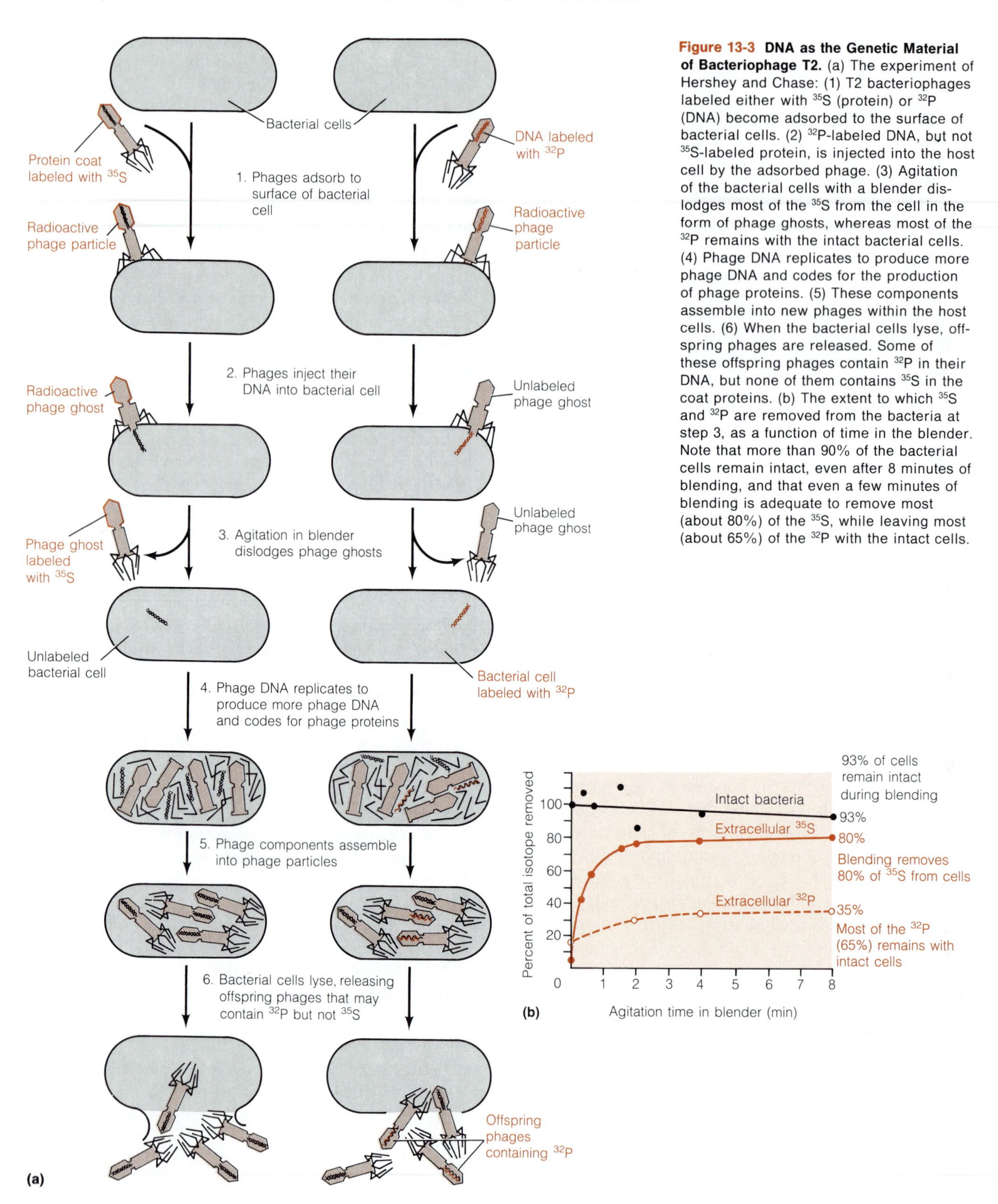

Figure 13-3 DNA as the Genetic Material of Bacteriophage T2. (a) The experiment of Hershey and Chase: (1) T2 bacteriophages labeled either with ^{35}S (protein) or ^{32}P (DNA) become adsorbed to the surface of bacterial cells. (2) ^{32}P-labeled DNA, but not ^{35}S-labeled protein, is injected into the host cell by the adsorbed phage. (3) Agitation of the bacterial cells with a blender dislodges most of the ^{35}S from the cell in the form of phage ghosts, whereas most of the ^{32}P remains with the intact bacterial cells. (4) Phage DNA replicates to produce more phage DNA and codes for the production of phage proteins. (5) These components assemble into new phages within the host cells. (6) When the bacterial cells lyse, offspring phages are released. Some of these offspring phages contain ^{32}P in their DNA, but none of them contains ^{35}S in the coat proteins. (b) The extent to which ^{35}S and ^{32}P are removed from the bacteria at step 3, as a function of time in the blender. Note that more than 90% of the bacterial cells remain intact, even after 8 minutes of blending, and that even a few minutes of blending is adequate to remove most (about 80%) of the ^{35}S, while leaving most (about 65%) of the ^{32}P with the intact cells.

with individual, tissue, age, nutritional state, or environment. However, the percentage composition of DNA varied from species to species. In general, DNAs from closely related species were shown to have similar base compositions, whereas those from very different species were likely to have quite different base compositions.

At least as significant were their observations that in all of the DNAs they examined, the number of adenines was always equal to the number of thymines (A = T), and the number of guanines was always equal to the number of cytosines (G = C). This, in turn, meant that the number of purines always equaled the number of pyrimidines (A + G = C + T). These observations, known as **Chargaff's rules,** remained an enigma until the double-helical model of DNA was deduced from X-ray diffraction studies.

The Structure of DNA

The experiments of Avery and his colleagues with *Pneumococcus* and of Hershey and Chase with bacteriophage infection provided strong biological support for DNA as the genetic material. For many, however, the evidence that clinched the argument came in 1953, when James Watson and Francis Crick published their model of DNA as a double helix. We caught a glimpse of the double helix in Chapter 3, but we now return to it for a more detailed look at DNA structure.

Watson, Crick, and the Double Helix

Watson and Crick deduced their model for the double-stranded structure of DNA from its X-ray diffraction pattern, as determined by Rosalind Franklin. They were aided in their model building by the realization that the specific forms in which A, G, C, and T exist at physiological pH permit specific hydrogen bonds to form. As we saw in Chapter 3, the Watson-Crick model envisions DNA as a duplex of two strands wound together into a right-handed *helix* (recall Figure 3-14). The backbone of each strand is a repeating sequence of deoxyribose and phosphate groups, with a purine or pyrimidine base attached to each sugar. The bases face inward toward the center of the helix, forming the "steps" of the "circular staircase" that the structure resembles.

Figure 13-4a illustrates several structural features of the DNA double helix. The helix contains about ten nucleotide pairs per turn and advances 0.34 nm per nucleotide pair. Consequently, each complete turn of the helix adds about 3.4 nm to the length of the molecule. The diameter of the helix is about 2 nm. This distance turns out to be too small for two purines and too great for two pyrimidines, but it accommodates a purine and a pyrimidine well, consistent with Chargaff's rules. Pyrimidine-purine pairing, in other words, was necessitated by steric considerations in the model building of Watson and Crick, but was then found to be in accord with the chemical findings that Chargaff had already reported.

Also shown in Figure 13-4a are the **major** and **minor grooves** that result from the way the two strands are twisted around each other. These are significant in the interactions of a variety of molecules with the DNA helix.

A further important structural feature of DNA is the *antiparallel orientation* of the two strands, as shown in Figure 13-4b. This means that as you move along one strand in a given direction, successive nucleotides are linked together by phosphodiester bonds that have a $5' \rightarrow 3'$ orientation. The complementary nucleotides of the other strand, however, are joined by bonds with a $5' \rightarrow 3'$ orientation that runs in the *opposite* direction. The two strands therefore differ in their **polarity** or directionality along the length of the double helix, a feature with important implications for both replication and transcription of double-stranded DNA, as we will see in Chapters 14 and 16.

Z-DNA

Until quite recently, it was thought that all DNA would have the structure predicted by the Watson-Crick model. Since 1979, however, it has become clear that segments of a DNA duplex may have an alternative configuration, with a left-handed instead of the usual right-handed helix. Although Alexander Rich and his colleagues originally detected this configuration in a synthetic oligonucleotide, it is now clear that such a structure exists in segments of naturally occurring DNA molecules as well.

Because the sugar-phosphate backbones follow a zigzag pattern around the axis, this left-handed structure has come to be called **Z-DNA.** The right-handed Watson-Crick helix is called **B-DNA.** (The A-DNA that this terminology implies is also known; A-DNA is a right-handed helical configuration induced by dehydration of B-DNA.) Both B-DNA and Z-DNA are shown in Figure 13-5. The Z form of DNA is a considerably slimmer helix than the B form, with 12 nucleotide pairs instead of 10 per turn and a linear distance of 4.5 nm instead of 3.4 nm per turn.

Recent observations suggest that the Z form coexists in the same DNA molecule with the B form. Some investigators speculate that segments of DNA in the Z form may be important in the regulation of gene expression, since they represent sites at which DNA is less likely to open and allow entry of RNA polymerase, the enzyme

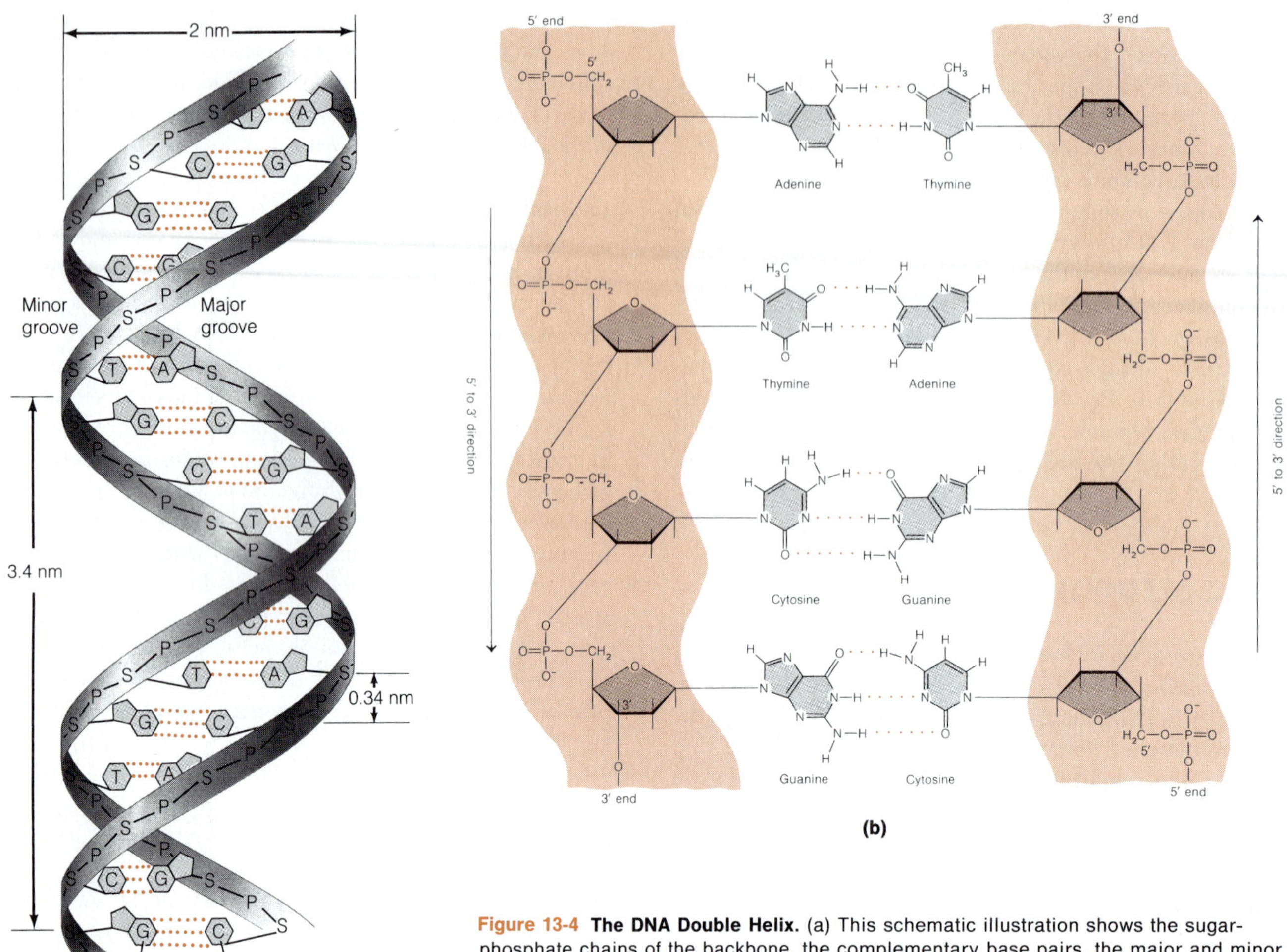

Figure 13-4 The DNA Double Helix. (a) This schematic illustration shows the sugar-phosphate chains of the backbone, the complementary base pairs, the major and minor grooves, and several important dimensions. A = adenine, G = guanine, C = cytosine, T = thymine, P = phosphate, and S = sugar (deoxyribose). (b) One of the strands of a DNA duplex is oriented 5′ → 3′ in one direction, whereas its complement has a 5′ → 3′ orientation in the opposite direction.

responsible for transcription of DNA to RNA. To appreciate such speculation, we need to understand what supercoiled DNA is, because supercoiling is thought to play an important role in opening up the DNA molecule for both replication and transcription.

Supercoiled DNA

Most of the DNA present in the chromosomes of eukaryotes is believed to be present as linear molecules. However, electron microscopic observations have established that DNA molecules from many sources are circular. In particular, many bacteria and viruses have circular DNA, as do the mitochondria and chloroplasts of eukaryotic cells. *E. coli,* for example, has as its genophore a circular DNA molecule with a contour length of about 1.36 mm. Some viruses have circular DNA, others have linear DNA. In some cases, such as bacteriophage λ, the DNA molecule exists in linear and circular forms during different parts of the life cycle.

In addition to the normal helical configuration typical of all DNA molecules, a circular DNA molecule can be twisted upon itself to form a new, higher-order helix—a **supercoiled** or **supertwisted** molecule. A circular DNA without supercoils is said to be in its **relaxed state** (Figure 13-6).

To understand supercoiling, you might perform the following exercise. Start with a length of rope consisting of two strands twisted together into a right-handed coil; this is the equivalent of a relaxed DNA molecule. Just joining the ends of the rope together changes nothing; the rope is still in a relaxed state. But if before sealing the ends you first give the rope a right-handed twist (that is, a twist in the direction in which the strands are already entwined about each other), the rope becomes *overwound* and is

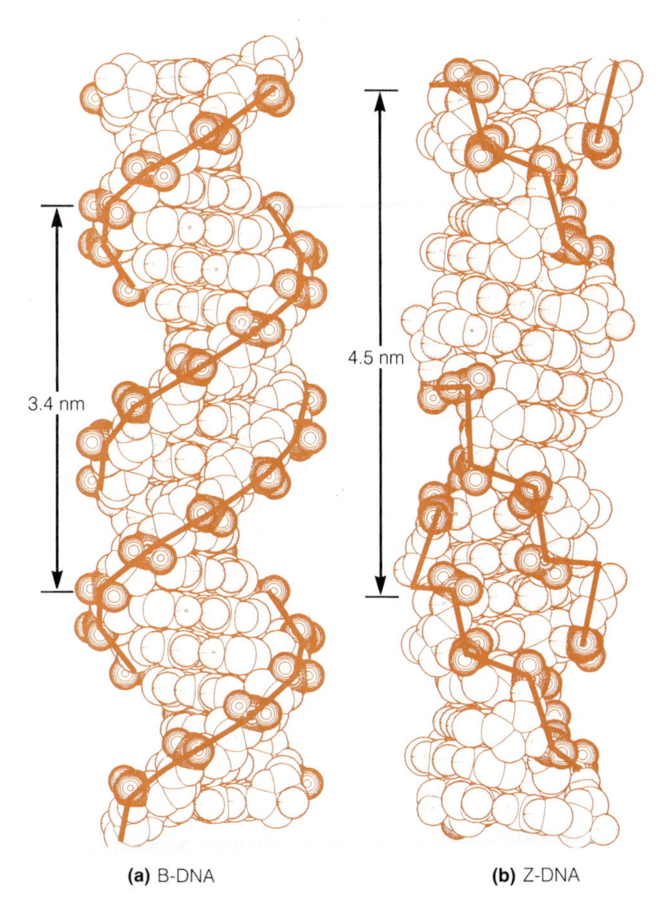

Figure 13-5 The Structures of B-DNA and Z-DNA. (a) B-DNA has a right-handed helix and a highly regular sugar-phosphate backbone, as originally predicted by Watson and Crick. (b) Z-DNA has a left-handed configuration and an irregular backbone, as traced by the heavy lines that zigzag along the strands from phosphate to phosphate.

thrown into a **positive supercoil.** Conversely, if the rope is given a left-handed twist before sealing (that is, twisted in the direction opposite to that in which it is wound), it becomes *underwound* and is thrown into a **negative supercoil.**

In the same manner, a relaxed DNA molecule can be converted into a negative supercoil by a left-handed twist and into a positive supercoil by a right-handed twist (Figure 13-6). Circular DNA molecules found in nature, including those of bacteria, viruses, and organelles, are invariably negatively supercoiled. Moreover, supercoiling is not limited to circular DNA molecules. It also occurs in linear eukaryotic DNA, provided only that the ends of the molecule are not free to rotate. Supercoiling makes a circular DNA molecule more compact. This compactness causes the DNA to sediment more rapidly upon centrifugation and may play a role in the packaging of DNA in viruses, bacteria, and organelles.

Supercoiling also may affect the ability of a circular DNA molecule to interact with other molecules. Positive supercoiling favors tighter winding of the double helix and therefore reduces opportunities for interaction. Conversely, negative supercoiling tends to unwind the double helix, thereby increasing access to its strands for enzymes involved in DNA replication or transcription. Z-DNA is considered a possible regulator of gene expression because a segment of left-handed helix in an otherwise right-handed molecule will uncoil the DNA. Thus, in negatively supercoiled DNA, a segment of Z-DNA relieves the tendency of the DNA to open and thereby reduces interaction with other molecules, which the negative supercoiling otherwise induces.

Molecules that differ only in their state of supercoiling are called **topological isomers** of one another. Correspondingly, the enzymes that carry out the interconversion of relaxed and supercoiled forms of DNA are called **topoisomerases.** Most topoisomerases are classified as *type I* or *type II*. Both types catalyze the stepwise relaxation of supercoiled DNA, but type I enzymes relax one coil at a time, whereas type II enzymes relax two coils at a time (Figure 13-7). Topoisomerases function by attaching to a supercoiled duplex and producing a transient break, or nick, in either one (type I) or both (type II) of the strands. The two strands can then rotate around each other, followed by resealing of the nick. This relaxation reaction requires no energy source, suggesting that the phosphoester bond energy that is liberated when a strand is nicked must somehow be conserved and used to form the new bond.

DNA gyrase is a type II topoisomerase that can induce as well as relax supercoiling. DNA gyrase is one of several enzymes required for DNA replication, as we will see in Chapter 14. It can relax the positive supercoiling that results from partial unwinding of the double helix, or it can actively introduce negative supercoils that then promote strand separation, allowing access of the other enzymes involved in the replication process. It is not surprising that DNA gyrase requires ATP to generate supercoiling but not to relax an already-supercoiled molecule.

DNA Denaturation and Renaturation

Because the two strands of the DNA duplex are held together only by hydrogen bonds, they can be readily separated under appropriate conditions. Strand separation is an integral part of both the replication of DNA and the synthesis of RNA, as we will see in coming chapters. Strand separation can also be induced experimentally, resulting in the **denaturation** of the DNA. One way to denature DNA is to raise the temperature. If this is done slowly, the DNA retains its double-stranded, or **native,** state until

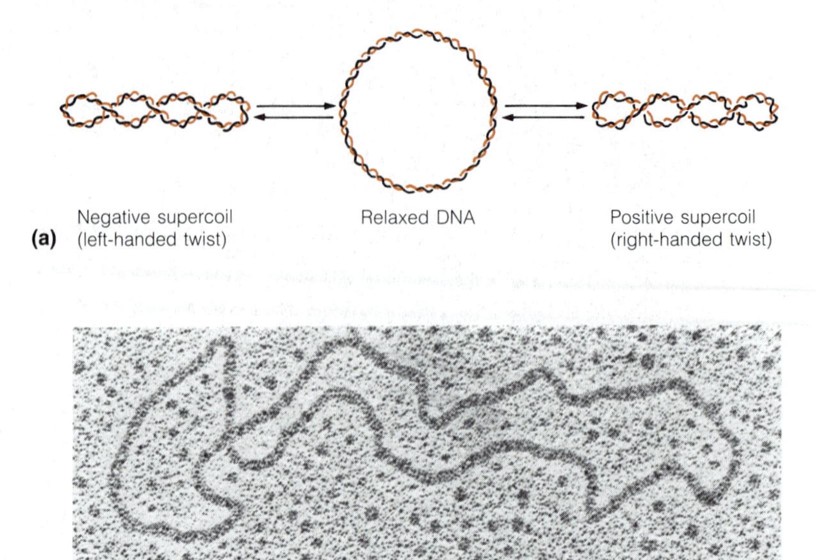

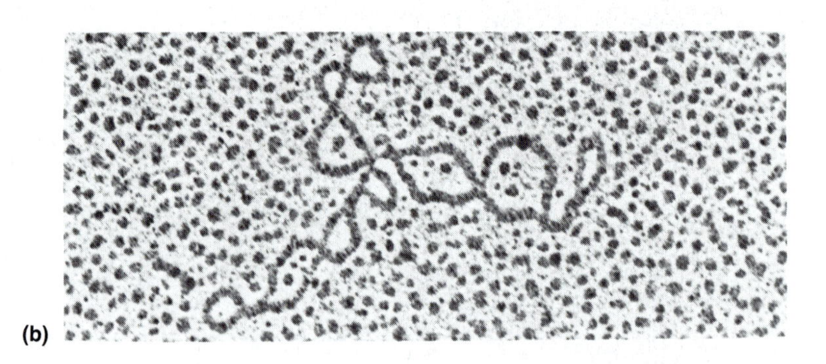

Figure 13-6 Interconversion of Relaxed and Supercoiled DNA. (a) Schematic diagram of a relaxed circular DNA molecule and its conversion into supercoiled forms with a left-handed twist (negative supercoiling) or a right-handed twist (positive supercoiling). (b) Electron micrographs of circular molecules of DNA from a bacteriophage called PM2, with a relaxed molecule on the top and a negatively supercoiled molecule on the bottom. (TEMs)

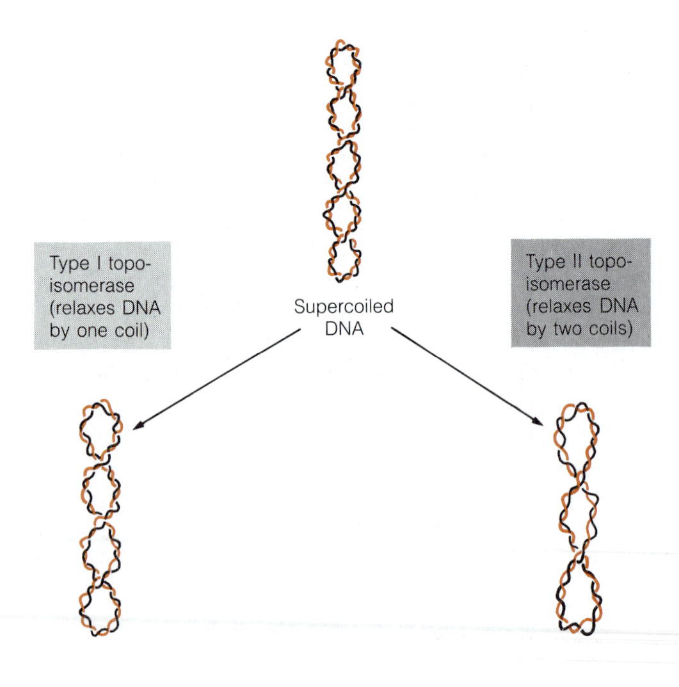

Figure 13-7 Mechanism of Topoisomerase Action. Topoisomerases relax supercoiled DNA by one coil (type I) or two coils (type II) at a time. Type I topoisomerases cause a single-strand nick, whereas type II enzymes cause transient breakage of both strands.

some critical temperature is reached, at which point the duplex thermally denatures, or "melts," into its component strands.

The melting process can be monitored by the difference in absorbance between double-stranded and single-stranded DNA. DNA absorbs ultraviolet light, with an absorption maximum around 260 nm (Figure 13-8). (RNA has a similar absorption profile; the absorbance at 260 nm is routinely used to determine concentrations of both DNA and RNA.) When the temperature of a DNA solution is raised, the absorbance remains relatively constant until the duplex structure "melts" into its component strands. At that point the absorbance of the solution increases rapidly because of the higher intrinsic absorption of single-stranded DNA (Figure 13-9).

The temperature at which one-half of the absorbance change has been achieved is called the **melting temperature** (T_m) of the DNA. The melting temperature is a sensitive measure of the base composition of DNA. G-C pairs are more resistant to denaturation because they are held together by an extra hydrogen bond (three for G-C pairs, two for A-T pairs; recall Figure 13-4). The melting temperature therefore increases linearly with the proportion of G-C base pairs in the DNA (Figure 13-10).

Renaturation of DNA. Denatured DNA will **renature,** or reestablish its double-stranded structure, provided only that conditions are favorable for the reestablishment of hydrogen bonding. The renaturation process is illustrated in Figure 13-11. DNA strands collide randomly until a small region of one strand encounters a complementary region of another strand and hydrogen bonds are formed between base pairs. This initial **nucleation event** is then followed by a **"zipping-up"** of the two strands in both directions. Since the nucleation event depends on random collision, it is the rate-limiting step in the reassociation process. The frequency of collision is in turn dependent on the number of strands in solution, so the overall process is concentration-dependent.

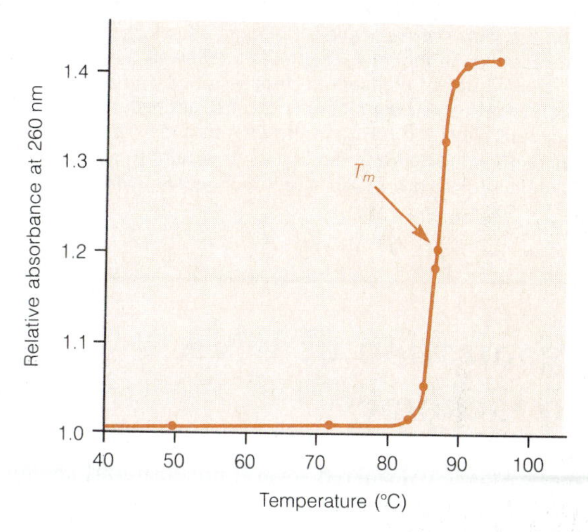

Figure 13-9 Thermal Denaturation Profile for DNA. As the temperature of a solution of double-stranded (native) DNA is raised, a point is reached at which the DNA denatures into single strands, accompanied by a characteristic increase in absorbance called the hyperchromic shift. The temperature at which the midpoint of this increase occurs is called the melting temperature, T_m. For the sample shown, the T_m value is about 86°C.

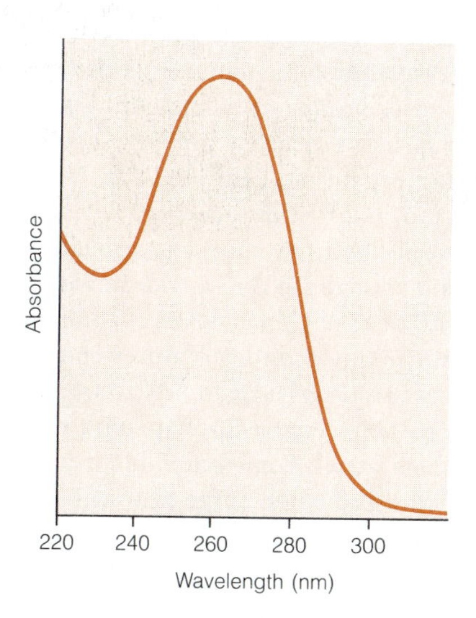

Figure 13-8 Ultraviolet Absorption Spectrum of DNA. Nucleic acids absorb ultraviolet light with an absorption maximum at about 260 nm because of the absorption by the pyrimidine and purine bases.

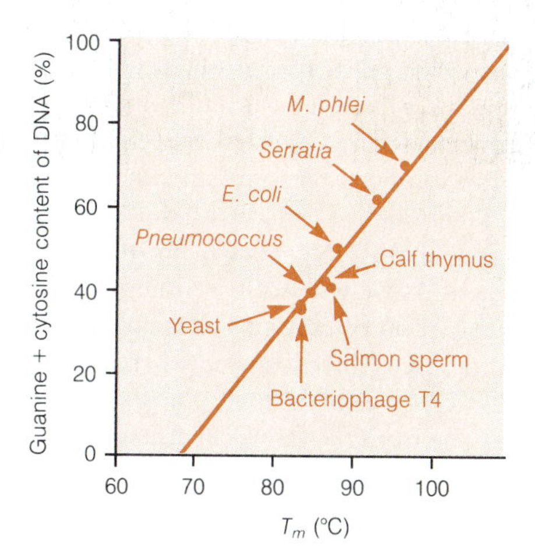

Figure 13-10 Dependence of Melting Temperature T_m on Base Composition. The melting temperature of DNA increases linearly with its G + C content, as illustrated by the relationship between T_m and G + C content for DNA samples from a variety of organisms.

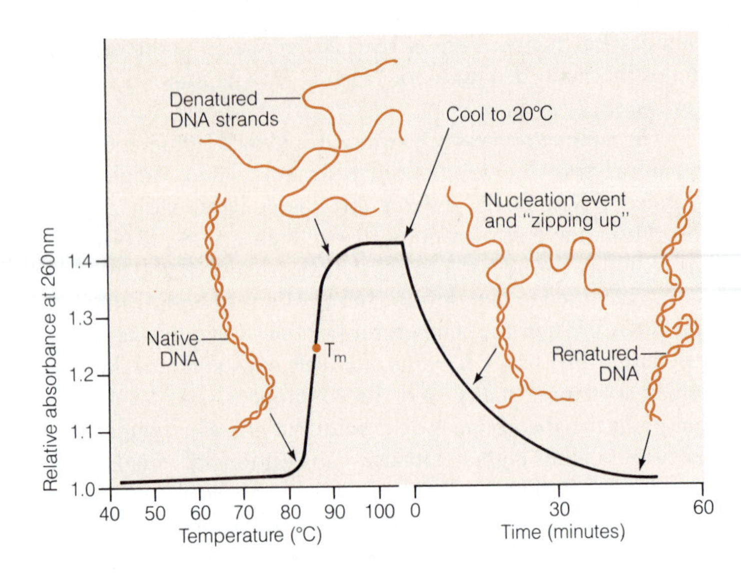

Figure 13-11 DNA Reassociation. If a solution of native (double stranded) DNA is heated slowly under carefully controlled conditions, the DNA "melts" at a fairly precise temperature, with an increase in absorbance at 260 nm as shown in Figure 13-9. When the solution is allowed to cool, the separated DNA strands recombine with kinetics that depend on the initial concentration. Complementary strands collide randomly in the nucleation event, followed by a rapid "zipping up" of adjacent nucleotide pairs. The reassociation requires varying amounts of time, depending on both concentration and length of the DNA strands. This dependence provided some of the first estimates of relative genome size in viruses, bacteria, and eukaryotic cells.

Organization of DNA into Genomes

So far, we have considered several physical and chemical properties of DNA. But as cell biologists, we are primarily interested in its importance to the cell. We therefore want to know how much DNA cells have, how and where they store it, and how they use the genetic information it contains. We begin by inquiring about the amount of DNA present, because that determines the maximum information content of the cell.

The **genome** of an organism or virus consists of the DNA (or RNA, in the case of some viruses) that is present in one complete copy of all the genetic information of that organism or virus. For many viruses and prokaryotes, the genome consists of one or a small number of linear or circular DNA molecules. Eukaryotic cells have a nuclear genome, a mitochondrial genome, and, in the case of plants and algae, a chloroplast genome as well. The nuclear genome consists of the DNA present in one haploid set of chromosomes. (As we will explore in more detail in Chapter 15, a *haploid* set of chromosomes consists of one representative of each chromosome, whereas a *diploid* set consists of two copies of each chromosome, one from the mother and one from the father. Sperm and egg cells have a haploid set of chromosomes, whereas most other eukaryotic cells have a diploid chromosome set.)

Genome Size

Genome size is usually expressed in numbers of nucleotide pairs. The circular DNA molecule present in an *E. coli* cell, for example, contains about four million nucleotide pairs. Table 13-1 presents the genome size of a variety of viruses and organisms both as nucleotide pairs per haploid genome and as fractions or multiples of the size of the *E. coli* genome. Even with the limited number of species included in the table, there is a spread of almost eight orders of magnitude in genome size, from 5000 nucleotide pairs in the case of virus SV40 to 10^{11} nucleotide pairs for the nuclear genome of *Trillium*, a common spring wildflower.

Table 13-1 also gives the actual physical length of the DNA. For the genomes shown, this ranges from less than 2 μm in the case of SV40 to about 34 m (more than 100 feet!) for *Trillium*. For viruses and bacteria, all of this DNA is usually a single molecule, but for eukaryotes it is always dispersed among a number of chromosomes and is therefore never present as a single molecule.

Broadly speaking, genome size increases with the complexity of the organism (Figure 13-12). Viruses contain enough DNA to code for only a few or a few dozen proteins, bacteria can specify a few thousand proteins, and eukaryotic cells have enough DNA to encode millions of proteins, at least theoretically. Upon closer examination of such data, however, many puzzling features emerge. Notably, there are great variations in genome size among eukaryotic species that do not seem to correlate with the complexity of the organism at all. Some amphibians and plants, for example, have inordinately large genomes—tens or even hundreds of times larger than those of mammalian species. *Trillium*, for example, is a member of the lily family, with no obvious need for exceptional amounts of genetic information. Yet its genome size is more than 20 times that of humans, and we have no idea what it does with all that DNA.

Table 13-1 DNA Content and Genomic Complexity

Species	Genome Size (Nucleotide Pairs)	Relative Size (*E. coli* = 1)	Length of DNA (mm)
Viruses			
SV40	5×10^3	0.00125	0.0017
T7	4×10^4	0.01	0.014
T2	2×10^5	0.05	0.068
Prokaryotes			
Mycoplasma	3×10^5	0.075	0.10
Bacillus	3×10^6	0.75	1.02
E. coli	4×10^6	1.00	1.36
Animals			
Fruit fly	2×10^8	50	68
Chicken	2×10^9	500	680
Human	3×10^9	750	1,020
Fungi			
Yeast	2×10^7	5	6.8
Plants			
Peas	9×10^9	2,250	3,100
Trillium	1×10^{11}	30,000	34,000

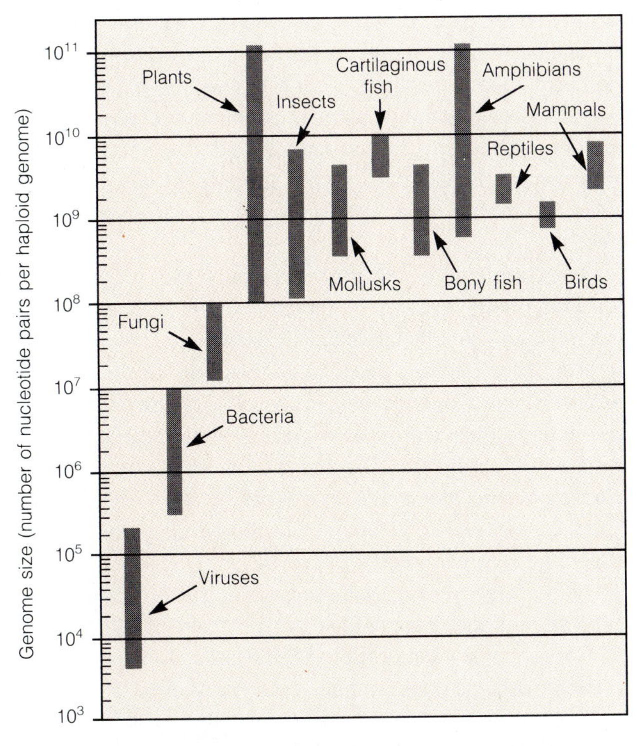

Figure 13-12 The Relationship Between Genome Size and Complexity of the Organism. For each group of organisms shown, the bar represents the approximate range in genome size, measured as the number of nucleotide pairs per haploid genome.

DNA Sequencing

Our understanding of the molecular biology of DNA was given an enormous boost in the 1970s by Frederick Sanger, Walter Gilbert, and their colleagues, who devised methods for the rapid determination of the nucleotide sequence of DNA fragments. At about the same time, Stanley Cohen, Paul Berg, Herbert Boyer, and their colleagues developed the DNA cloning techniques that are essential not only to DNA sequencing, but to the whole field of **recombinant DNA technology.** Recombinant DNA techniques are described in detail in Appendix B, which should be read as an integral part of this and the next several chapters on information flow in cells. Here, we will note briefly the techniques needed to prepare DNA fragments for sequencing and will then turn directly to the topic of sequence determination.

Preparing DNA for Sequencing. Most DNA molecules are far too large to be studied as intact molecules. Until the early 1970s, in fact, DNA was the most difficult biological molecule to analyze biochemically. Eukaryotic DNA seemed especially intimidating, given the size and complexity of the genomes of most eukaryotes. The prospect of ever being able to identify, isolate, sequence, or manipulate specific eukaryotic genes seemed unlikely at best. Yet in less than a decade, from the early to the late 1970s, DNA became one of the easiest biological molecules to work with.

Phages: The Best-Understood "Organisms" in the World

From its inception in the mid-nineteenth century, genetics has drawn upon a wide variety of organisms for its experimental materials. Initially, attention focused on plants and animals such as Mendel's peas and the fruit flies popularized by later investigators. Since the 1950s, however, bacteria and viruses have come into their own, providing geneticists with experimental systems that have literally revolutionized the science.

Bacteriophages have been especially important. Bacteriophages, or **phages,** for short, are viruses that infect bacterial cells. It is easy to obtain large numbers of phages in a short time, which greatly facilitates screening for mutants. In addition, the genomes of most phages are much smaller than those of even the simplest bacteria, making them much more amenable to physical, chemical, and genetic analysis. As a result, more is known about the storage, transmission, and expression of genetic information in bacteriophages than in any other genetic system. One might even say that from a genetic perspective, phages are the best-understood "organisms" in the world!

In terms of structure, some of the best-known phages are the T2, T4, and T6 (the so-called T-even) bacteriophages that infect the bacterium *E. coli*. The structure of T4 is shown in the top box of Figure 13-A. The *head* of the phage is filled with DNA, surrounded by a protein capsule that is shaped like an icosahedron (a 20-sided figure). The head is attached by a *neck* to a *tail* that consists of a hollow core surrounded by a contractile sheath and terminating in an *end plate* to which the *tail fibers* are attached. Figure 13-B is an electron micrograph of a T4 bacteriophage.

Figure 13-A depicts the events in the attachment of a bacteriophage such as T4 to the wall of the cell it is about to infect. As an unattached phage approaches the cell, it anchors itself to the wall by means of its tail fibers (step a). Next, the phage "squats down" until the tail pins on its end plate make contact with the wall. The tail sheath then contracts, driving the hollow core through the cell wall. This core forms the "needle" through which the DNA is "injected" into the bacterium (step b). Figure 13-C shows many viruses attached to the surface of a bacterial cell, injecting their DNA molecules into the bacterium.

Adsorption of the phage and injection of its DNA into the host cell are the first two steps in the "life cycle" of a bacteriophage. Once the bacteriophage has gained entry to the bacterial cell, the genetic information of the phage is expressed, giving rise to a few key products that succeed in subverting the metabolic machinery of the host cell for the phage's purpose, which is usually its own rapid multiplication. Since the phage consists simply of a DNA molecule surrounded by a protein capsule, much of the metabolic activity in the infected cell is channeled toward the synthesis of phage DNA and capsule proteins (Figure 13-A, step c).

These components then self-assemble into a large number of new phages. Within about half an hour, the infected cell lyses, releasing into the medium several hundred phages (steps d and e). Each of these can infect another bacterial cell, making it possible to obtain enormous populations of phage—as many as 10^{11} phages per milliliter in infected bacterial cultures.

To determine the number of phages in a culture, a measured volume of the culture is diluted appropriately and mixed with a few drops of growing bacterial culture and a small volume of melted nutrient medium. The melted agar is then poured into a Petri dish, where it hardens. Upon incubation, the bacteria multiply to produce a dense **lawn** of cells on the agar. Wherever a virus particle has adsorbed to and infected a bacterial cell, however, a hole is visible in the lawn because the bacterial cells have been killed by the multiplying phage population. Such holes are called **plaques.** The number of plaques on the bacterial lawn represents the number of phage particles added to the original bacteria-agar mix, provided only that the initial number of phages was small enough to ensure that each would give rise to a separate plaque. Figure 13-D shows plaques of varying sizes that were formed by a bacteriophage on a lawn of *E. coli* cells.

The course of events shown in Figure 13-A is caused by a **virulent phage** such as T4. Infection of bacterial cells by a virulent phage usually leads to **lytic growth,** characterized by the production of many progeny phages and resulting in the death and lysis of the host cell. In contrast, a **temperate phage** can either produce lytic growth, just as a virulent phage does, or simply integrate its DNA into the bacterial chromosome without causing any immediate harm to the host cell. An especially well-studied example of a temperate phage is bacteriophage λ, which infects *E. coli* cells.

In the integrated or **lysogenic state,** the viral DNA, now called a **prophage,** is replicated along with the bacterial DNA, often through many generations of host cells. During this time, the bacteriophage genes, though potentially lethal to the host, are inactive, or *repressed*. Under appropriate conditions, the prophage DNA is excised from the host chromosome and again enters a lytic cycle, producing progeny phages and lysing the host cell.

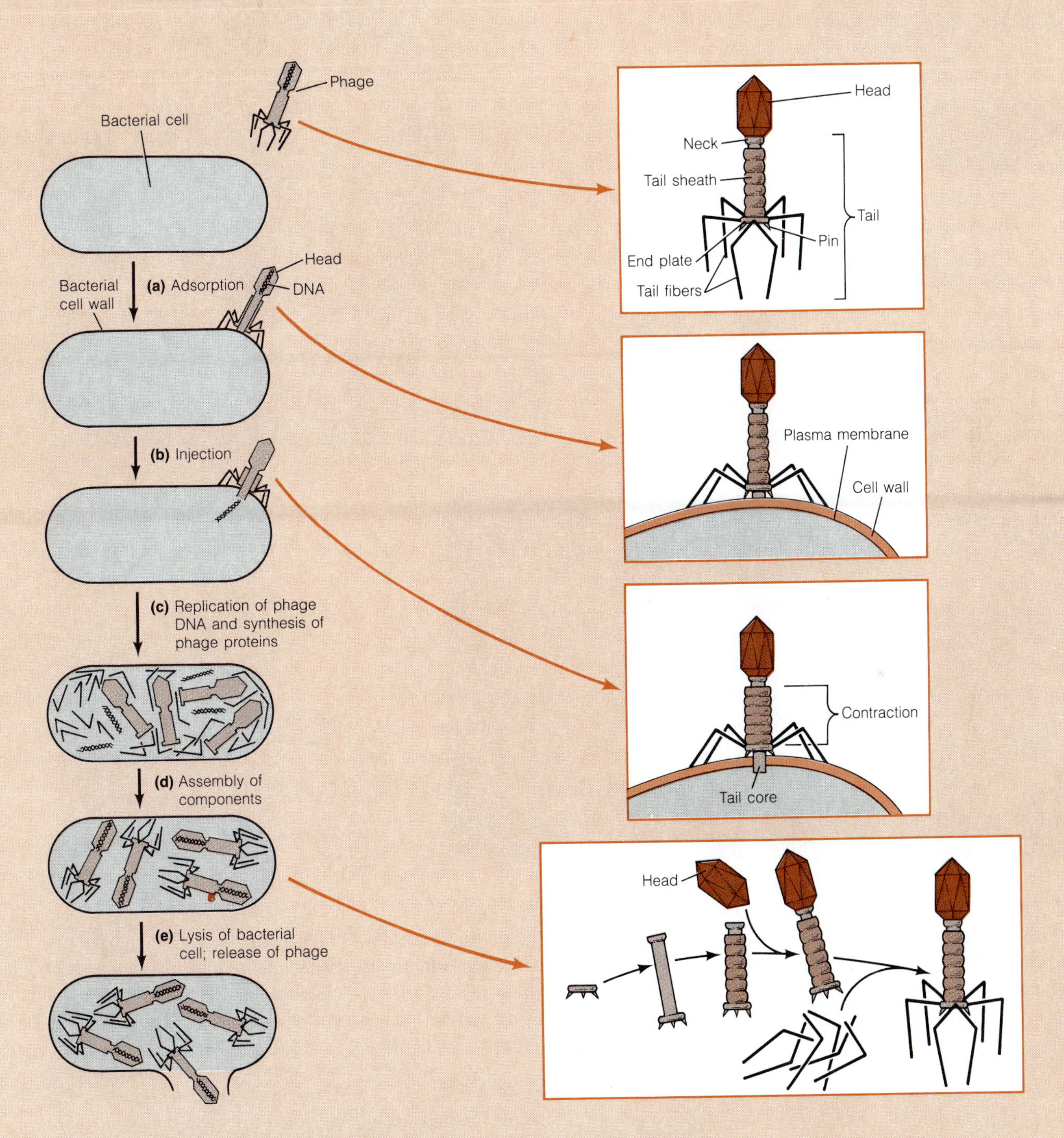

Figure 13-A The Life Cycle of a Bacteriophage. The life cycle of a bacteriophage begins when a phage (a) becomes adsorbed to the surface of a bacterial cell and (b) injects its DNA into the cell. (c) The phage DNA replicates in the host cell and codes for the production of phage proteins. (d) These components assemble into new phages. (e) Eventually, the host cell lyses, releasing offspring phages that can infect additional bacteria.

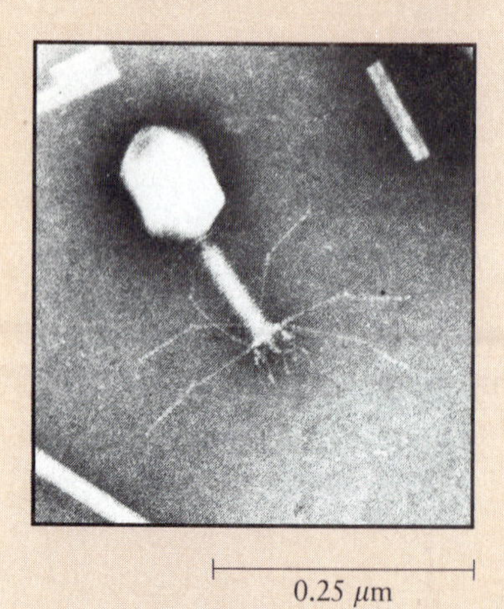

Figure 13-B Electron Micrograph of a T4 Bacteriophage. (TEM)

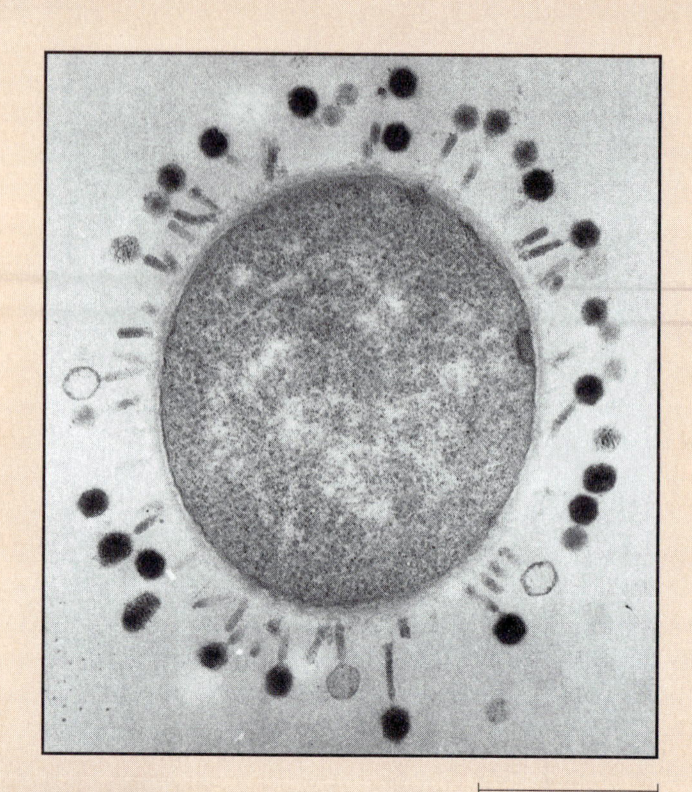

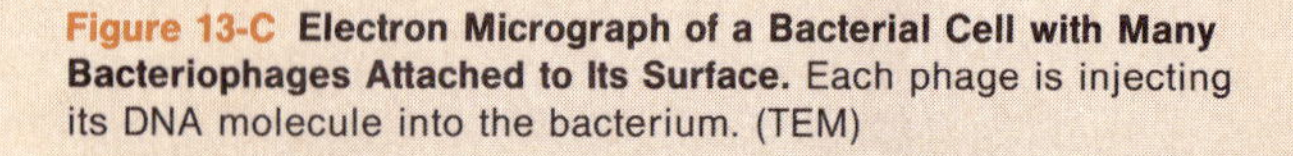

Figure 13-C Electron Micrograph of a Bacterial Cell with Many Bacteriophages Attached to Its Surface. Each phage is injecting its DNA molecule into the bacterium. (TEM)

Figure 13-D Bacteriophage Plaques on a Lawn of Bacteria. In the left-hand Petri dish, a culture of *E. coli* cells has been infected by the bacteriophage λ. Clear phage plaques have been formed on the lawn of *E. coli* cells. In the right-hand Petri dish, uninfected *E.coli* cells form an uninterrupted lawn.

One reason bacteriophages are so attractive to geneticists is the small size of their genomes. This makes it possible to identify all or most of their genes and to understand the genetic organization and regulation of the genome in great detail. Bacteriophage λ, for example, has a single DNA molecule with a molecular weight of 32 million. This DNA codes for fewer than 60 genes, compared to the several thousand genes in a bacterium such as *E. coli*. Other phages are still smaller. A phage called ϕX174, for example, has only 1.8 million daltons of DNA encoding only 9 genes.

ϕX174 is also noteworthy because its DNA is not double-stranded as in virtually all organisms and many other viruses, but is instead a single-stranded circle, containing 5375 nucleotides. Furthermore, the complete nucleotide sequence of ϕX174 DNA is known. This was, in fact, the first complete genome to be sequenced, a feat accomplished in 1977 by Frederick Sanger and his colleagues.

Largely because of their simple genomes, their rapidity of multiplication, and the enormous numbers of phages that can be produced in a small volume of culture medium, bacteriophages are among the best-understood of all "organisms." They have proved exceedingly useful as model systems in our continuing quest to understand the much more complex genomes of bacteria, plants, and animals.

This breakthrough was made possible largely by the discovery of **restriction enzymes.** As you will read in Appendix B, restriction enzymes are bacterial nucleases that cleave DNA molecules at specific sites, thereby generating a series of specific pieces called *restriction fragments.* The site at which a given restriction enzyme cleaves double-stranded DNA consists of a short segment of nucleotide pairs (four or six, usually) called the *recognition sequence.* The frequency with which such recognition sequences occur along a DNA molecule is such that restriction enzymes cleave most DNA into fragments that vary in length from a few hundred to a few thousand nucleotide pairs. Fragments of these lengths are far more amenable to further manipulation than the enormously long DNA molecules from which they were generated.

To work further with DNA fragments obtained in this way, researchers face three main challenges: generating the larger amounts of DNA needed for most biochemical studies; separating fragments of different lengths; and identifying specific fragments from among the large number generated by restriction enzymes. Amplifying a fragment of DNA and thereby meeting the first of these needs is called **DNA cloning,** and the techniques used are at the heart of recombinant DNA technology. DNA cloning is commonly done in the bacterium *E. coli,* with an appropriate *cloning vector* used to carry the DNA fragments into the bacterial cells. Alternatively, DNA can be amplified in vitro using the *polymerase chain reaction.*

To separate DNA fragments of different lengths, *gel electrophoresis* is the method of choice, just as it is for the separation of polypeptides (see Figure 7-12). Once separated, specific fragments can be identified in two alternative ways. One approach is to test for complementarity between the fragments and known DNA or RNA sequences. The other approach is to test for the production of specific polypeptide products encoded by the fragments, when introduced into a bacterial cell. All of these techniques are described in more detail in Appendix B.

DNA Sequence Determination. Recognizing that procedures are available to generate, clone, separate, and identify DNA fragments, we are now ready to consider how the nucleotide sequence of a specific DNA segment can be determined. Two different **DNA sequencing techniques** are in widespread use, a chemical method pioneered by Allan Maxam and Walter Gilbert, and an enzymatic method developed by Fred Sanger. The *chemical method* is based on the use of chemicals that cleave DNA preferentially at specific bases, whereas the *enzymatic method* (also called the *dideoxy method*) is based on the incorporation of chain-terminating dideoxyribonucleotides. We will describe the chemical method here.

The basis of the chemical sequencing technique is that a single-stranded DNA fragment can be treated with specific chemical reagents that selectively delete one of its four nucleotides, thereby cleaving the nucleotide chain at that point. When separate samples of an end-labeled DNA strand are treated with reagents specific for each of the four nucleotides and the resulting fragments of DNA are fractionated in adjacent lanes on a polyacrylamide gel, the nucleotide sequence of the DNA can be deduced from the labeled bands present in the four lanes.

Figure 13-13 illustrates how a DNA fragment is pre-

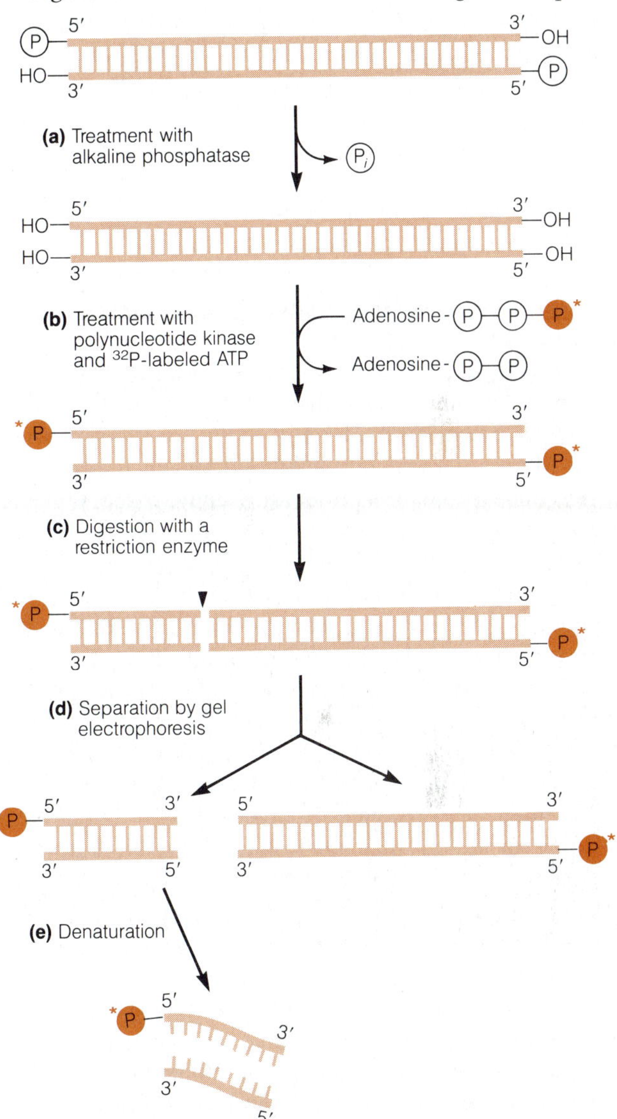

Figure 13-13 Preparation of DNA for Sequencing by the Chemical Method. DNA molecules or fragments to be sequenced by the chemical method are (a) treated with alkaline phosphatase to remove phosphate groups from the 5′ ends of each strand and (b) incubated with ^{32}P-labeled ATP and polynucleotide kinase, which phosphorylates only the 5′ ends of DNA strands (*P = ^{32}P). (c) The end-labeled DNA is then digested with a restriction enzyme that is known to cleave it into two unequal pieces, which can then be (d) separated by gel electrophoresis. (e) The piece to be sequenced is denatured into its two component strands, only one of which is radioactively labeled. The labeled strand can then be sequenced using the approach shown in Figure 13-14.

pared for sequencing by this method. The double-stranded fragment is first labeled with ^{32}P on the 5′ end of each strand. To do so, the fragment is treated with the enzyme *alkaline phosphatase* to remove any nonradioactive phosphate groups from the 5′ ends (step a) and then incubated with ^{32}P-labeled ATP and *polynucleotide kinase,* an enzyme that transfers the ^{32}P-labeled phosphate (designated as *P) from ATP to the dephosphorylated 5′ ends (step b). Next, the end-labeled fragment is cleaved with a restriction enzyme known to cut it into two (or more) different-sized segments (step c). These segments are separated by gel electrophoresis (step d), and the segment to be sequenced first is denatured into its two component strands, only one of which is end-labeled (step e).

The sequence of such an end-labeled strand is then determined as shown in Figure 13-14 for a DNA strand containing nine nucleotides (*P-AGTACATAT). Equal portions of the DNA are incubated with one of four reagents, each of which is specific for the deletion of a different nucleotide (step a). Tube 3, for example, contains a reagent that eliminates cytosine (C). Since there is a single C in the fragment, this reagent will split the DNA strand into two pieces, *P-AGTA and ATAT, only one of which (*P-AGTA) is radioactively labeled and therefore detectable upon subsequent autoradiography. Tube 4, on the other hand, contains a reagent that eliminates thymine (T). Since there are three Ts in the DNA sequence, the strand can be split into six different possible pieces. All of these pieces will be present in the incubation mixture, but only three of them (*P-AG, *P-AGTACA, and *P-AGTACATA) will be radioactive and hence detectable on the autoradiogram.

Samples from each of the four incubations are then subjected to electrophoresis in adjacent lanes of a polyacrylamide gel capable of resolving oligonucleotides consisting of up to nine nucleotides (step b), and autoradiography is used to visualize the radioactively-labeled oligonucleotides in each of the lanes (step c).

The sequence of the strand can now be deduced by considering the bands present in each of the four lanes. First, look at lane 1, which contains the pieces generated by guanine deletion. Only a single band appears in lane 1, at the position corresponding to a mononucleotide. This indicates that a guanine must have been deleted from position 2, so that the labeled nucleotide at position 1 was released. Lane 3 also contains a single band, but at a position corresponding to a tetranucleotide. Since this tube was treated with an agent that deletes cytosines, we can conclude that cytosine was present in the original fragment at position 5. By similar logic, we can conclude (from the bands in lane 2) that adenines were present at positions 4, 6, and 8; and (from lane 4) that thymines were at positions 3, 7, and 9.

By analyzing the banding patterns in the four lanes, we have deduced the sequence to be

X—G—T—A—C—A—T—A—T

where X is a base that cannot be identified without additional data. Now notice an important feature of the autoradiogram in Figure 13-14: By starting at the bottom of the autoradiogram and recording the letter (G, A, C, or T) of the lane in which each successive band appears (step d), we can read the nucleotide sequence of the DNA strand directly! This is obviously a very attractive feature of the technique, since it is so easy to deduce the sequence once the autoradiogram is in hand. (As you might guess, however, the data shown in Figure 13-14 are idealized; the chemical treatments are less specific in actual practice, and the autoradiograms are not always as unambiguous as our example. Also, a very short fragment was shown here for purposes of illustration. Usually the fragments contain hundreds of nucleotides.)

From Genes to Genomes. The significance of the techniques that we have just been exploring can scarcely be overestimated. DNA sequencing is now so routine and so automated that virtually any DNA fragment can be sequenced easily and rapidly. Complete DNA sequences are already available for hundreds—perhaps even thousands—of genes, from microorganisms, plants, animals, and humans; and computerized data banks have been established to store and analyze sequence information. Researchers have virtually abandoned the sequencing of proteins, since it is generally much easier to isolate and sequence the DNA that encodes a protein and to deduce the amino acid sequence of the protein from the nucleotide sequence of the DNA.

Increasingly, sequencing techniques are being applied not just to individual genes, but to whole genomes. Genome sequencing is already underway, with the complete sequence known for the circular chloroplast genomes of several plants, as well as for the genomes of several viruses. For example, the complete sequence of the DNA of the Epstein-Barr virus, which is 172,282 nucleotides in length, is known. Much larger genomes are now being tackled as well, including those of the bacterium *E. coli* (genome size: 4×10^6 nucleotide pairs), the plant *Arabidopsis thaliana* (7×10^7 nucleotide pairs, one of the smallest plant genomes known), and the fruit fly *Drosophila* (1.8×10^8 nucleotide pairs).

Plans are also underway to turn the tools and techniques of DNA sequencing to the ultimate challenge, the "Mount Everest" of molecular biology—determining the nucleotide sequence of the entire human genome. How awesome a challenge is that? To answer that question, we need to consider the amount of DNA involved. But how

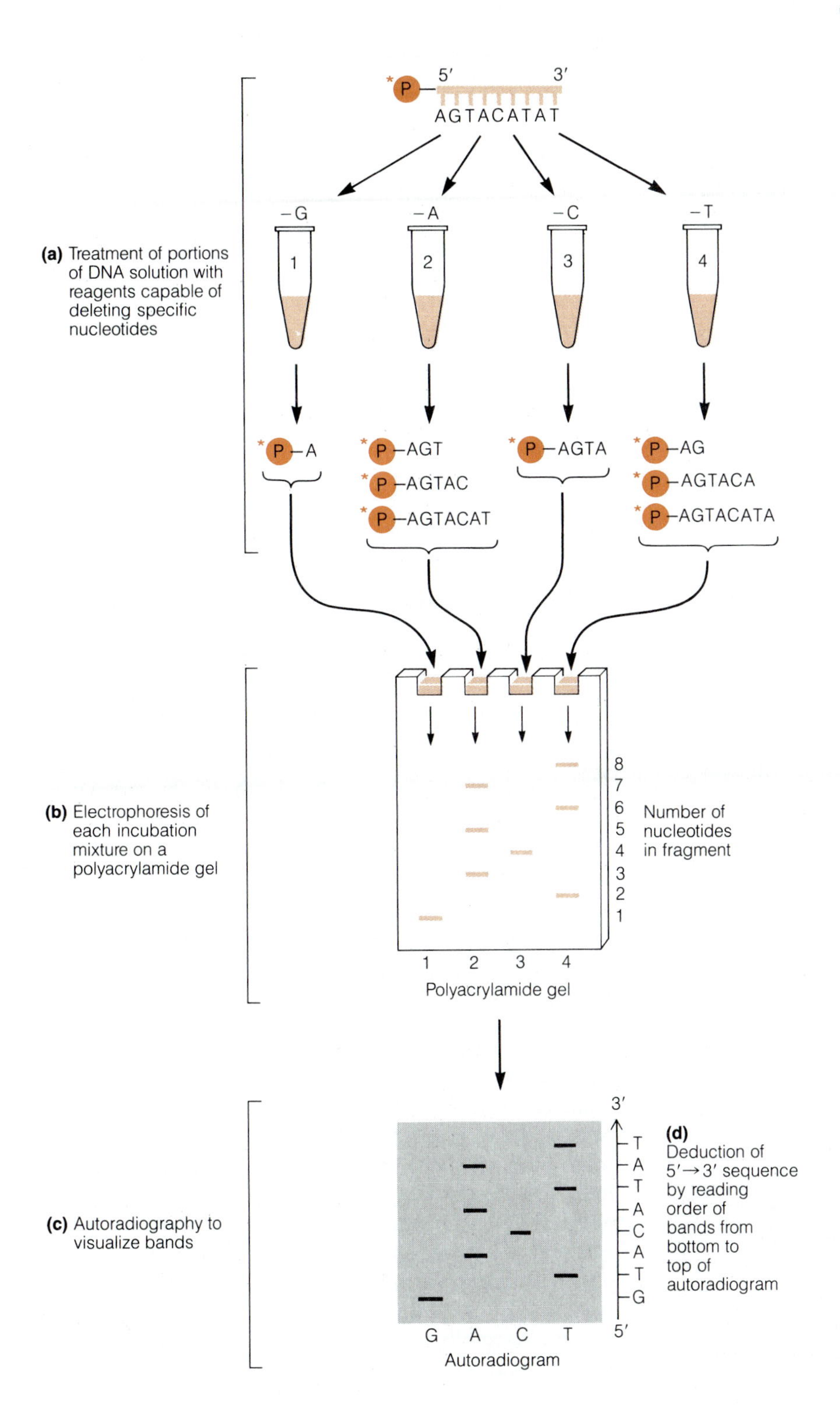

Figure 13-14 DNA Sequencing. The chemical sequencing technique is illustrated here for a short (nine-nucleotide) strand of DNA that has been end-labeled and prepared for sequencing as described in Figure 13-13. The radioactive phosphate group at the 5′ end is indicated by the symbol *P. (a) Equal portions of the fragment are treated with one of four specific chemical agents, each known to break DNA fragments by deleting a different nucleotide from the chain. Tubes 1 through 4 contain agents known to break DNA at G, A, C, and T, respectively. (b) Samples from each of the four incubations are then subjected to electrophoresis in adjacent lanes of a polyacrylamide gel that is capable of resolving oligonucleotides of up to nine nucleotides. (c) The gel is then exposed to photographic film to visualize the radioactively-labeled oligonucleotides in each of the four lanes. (d) The sequence of the starting strand can then be deduced directly from the autoradiogram by reading the banding pattern upward, recording in order the letter of the lane (G, A, C, or T) in which each successive band appears. (In practice, the initial DNA fragments are much longer than this example and have overlapping regions that would allow any nucleotides to the right or left of our sample sequence to be identified by matching the sequences in other fragments.)

do we comprehend the 3 billion nucleotide pairs in the haploid nuclear genome? We can describe it as about a thousandfold more DNA than is present in an entire *E. coli* cell, or at least a hundred thousand times more DNA than most viruses have, but even these comparisons fail to capture the magnitude of the project. Perhaps the best way to comprehend the challenge is to note that the project is likely to require ten years of intense international effort, and the cost may run well over three billion dollars—about a dollar per nucleotide pair! However, there is general agreement that the benefits will be worth the cost in time and effort. Alterations at specific sites in the human genome are involved in many diseases, including cystic fibrosis, muscular dystrophy, and several forms of leukemia. Knowledge of the sequences at such sites and of the alterations that cause these diseases will permit significant progress in prevention and treatment.

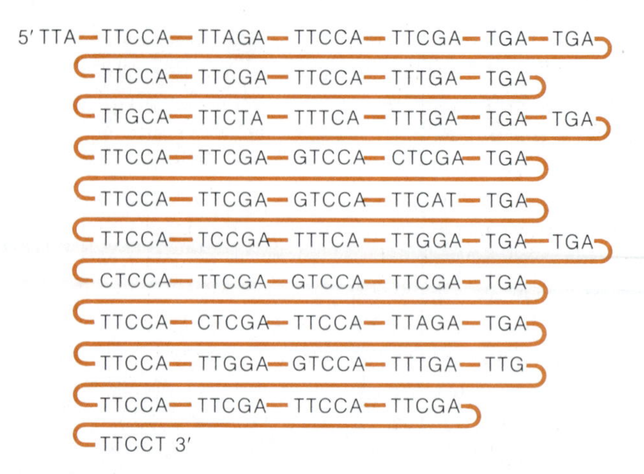

Figure 13-15 Sequence of a Human Satellite DNA. This distinctive pattern of repeating sequences, called BLUR 17, is present in the human genome. Segments of chromosomal DNA that have distinctive repeating sequences, such as the 3- and 5-base repeats of BLUR 17, often appear as a separate "satellite" band when total genomic DNA is cleaved to short lengths and centrifuged to equilibrium in a density gradient of cesium chloride.

Repeating Sequences in Eukaryotic Genomes

With the ability to cut DNA molecules at specific sites with restriction enzymes, researchers immediately applied these techniques to the analysis of DNA derived from a variety of sources. We noted earlier that restriction enzymes cleave DNA at specific sequences of 4–6 base pairs. Such sequences would be expected to occur only once in several thousand base pairs if DNA sequences were randomly distributed, and DNA fragments would be in the range of a few thousand nucleotides in length. Much of the DNA does, in fact, end up in such long, random fragments, but a surprising amount was found to occur as fragments of shorter, defined lengths. This could only mean that repeating sequences of specific lengths are present throughout human DNA.

It is now clear that all known eukaryotic genomes contain repeating DNA sequences. The proportion of the genome represented by such sequences varies greatly among species, as does the degree of sequence reiteration. Now we can explain, at least in part, the mystery of the seemingly excess amount of DNA in species such as *Trillium,* which contains a relatively high proportion of repeating DNA sequences. Using the sequencing techniques we have just described, researchers were able to determine the nucleotide structure of various repeating sequences. Figure 13-15 shows an example of one such sequence called BLUR 17, which is composed of 3- and 5-base repeats. These sequences are found specifically in material called **satellite DNA,** so named because its distinctive base composition causes it to appear in a separate "satellite" band in certain DNA preparation procedures.

What could be the function of such sequences? Certainly they are not coding for protein, because the triplet nucleotide sequences that are known to code for individual amino acids are absent. One explanation is that the highly repetitive sequences are responsible for imparting special physical properties to some portion of the DNA. It has been demonstrated that the centromere regions of chromosomes are particularly rich in satellite DNA, and it is probable that the repeating sequences provide a specialized structure for the chromatin of the centromere.

Another type of repeating sequence is called **interspersed repeats.** These are more extensive, about 300 nucleotides long. An example of this type of sequence in humans is called the *Alu family* after the restriction enzyme originally used to isolate it. These sequences are somewhat more puzzling than satellite DNAs because they do not have an obvious association with any specialized portion of the chromosome, but instead are found throughout the genome, composing as much as 5% of human DNA. One possible explanation is that the interspersed repeating sequences, at least in mammals, are a kind of molecular parasite resembling *transposable elements,* those portions of DNA that normally move between chromosomes.

DNA Organization and Packaging

The amount of DNA that a cell must accommodate is awesome, even for organisms with genomes of modest size.

Figure 13-16 illustrates the logistic problem posed by the amounts of DNA that must fit in a virus (T2) and a bacterium (*E. coli*). In each case, the total length of DNA greatly exceeds the dimensions of the space into which it must fit. The typical *E. coli* cell, for example, is about 1 μm in diameter and 2 μm in length, yet it must accommodate a circular DNA molecule with a length of about 1360 μm—enough DNA to encircle the cell more than 400 times!

Eukaryotic cells face an even greater challenge. A typical human liver cell, for example, has a diameter of about 35 μm and contains enough DNA to wrap around the cell more than 15,000 times! Drawn to scale on Figure 13-16, the cell would have a diameter of 35 cm, and the DNA would have a total length of about 20,400 m—slightly more than 12 miles! Clearly, the packaging of DNA is an important topic for all forms of life. We will look first at how viruses and prokaryotes organize their DNA, and then we will consider the eukaryotic answer to the same problem.

Organization of DNA in Viruses and Prokaryotes

As we are already aware, the organization and packaging of DNA and RNA are quite different in viruses and prokaryotes than in eukaryotic cells. Almost without exception, DNA-containing viruses and prokaryotes have one or a small number of DNA molecules, either linear or circular, depending on the species. These are typically "naked" DNA molecules; that is, they do not have large amounts of protein permanently associated with them, as in eukaryotes. Such "naked" DNA molecules are sometimes referred to as viral or prokaryotic "chromosomes," but such terminology is potentially confusing. Many geneticists therefore restrict this term to true eukaryotic chromosomes, using the term *genophore* to refer to the genomic DNA of a virus or prokaryote.

Packaging of the Bacterial Genophore. Even the relatively modest amounts of DNA present in prokaryotes pose prob-

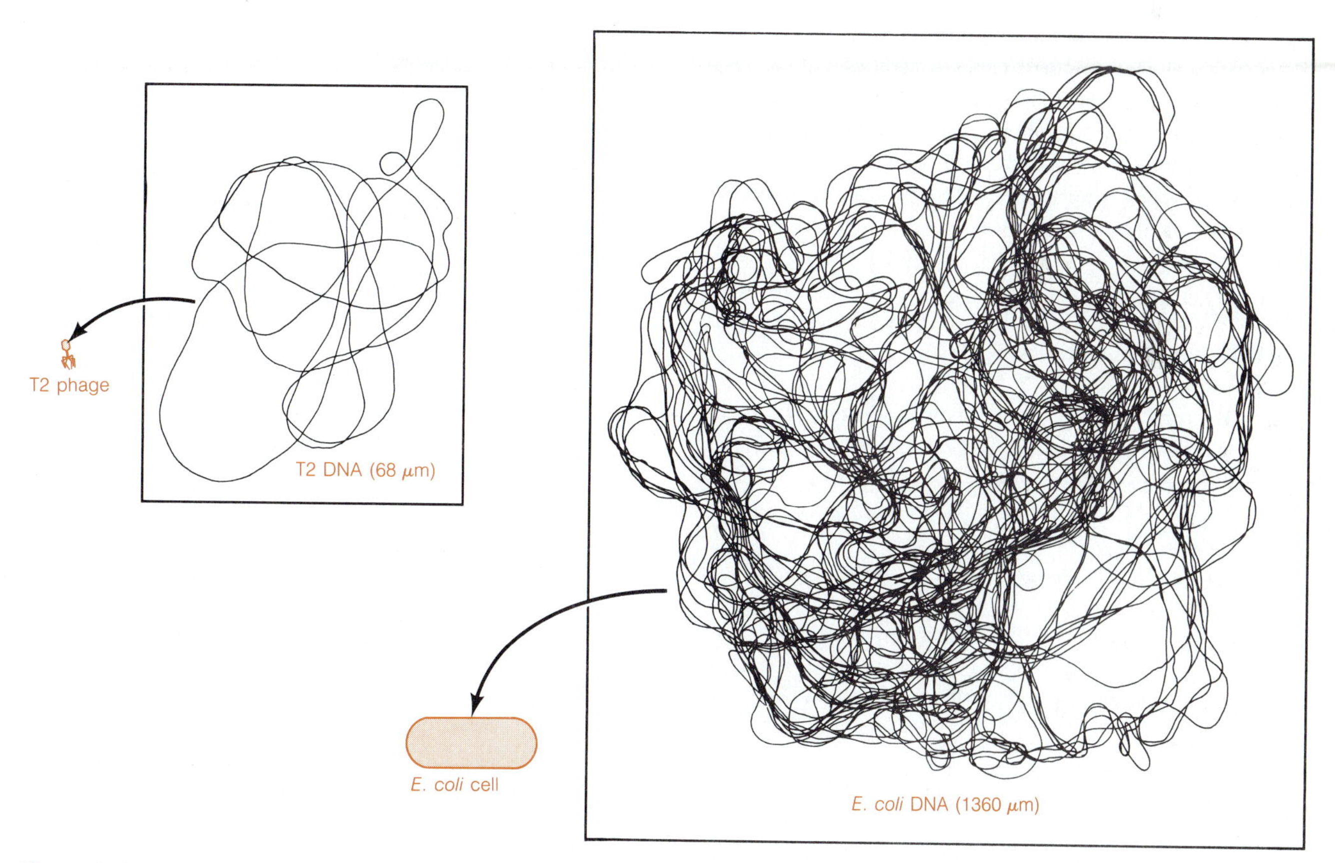

Figure 13-16 DNA Packaging in Bacteria and Viruses. The bacteriophage T2, the bacterium *E. coli*, and the DNA molecule that must fit in each are shown to scale; 1 cm in the figure represents 1 μm of actual dimensions (10,000 ×). On this scale, the T2 DNA molecule is 68 cm long, and the *E. coli* genome is 1360 cm in length.

lems of organization and packaging. It is, for example, no small feat for an *E. coli* cell to accommodate a DNA molecule that has a contour length more than a thousand times that of the entire cell. The DNA molecule is highly folded and coiled in a complex but orderly manner. Figure 13-17 illustrates how this is accomplished. The molecule is negatively supercoiled and is thought to be folded into a number of loops that are held in place by RNA and possibly also protein molecules. As Figure 13-17 shows, treatment with ribonuclease releases some of the folds but does not relax the supercoiling. Nicking the DNA with a topoisomerase, on the other hand, relaxes the supercoiling but does not disrupt the folds. The folded, supercoiled DNA molecule forms a compact structure called a **nucleoid.**

Bacterial Plasmids. In addition to its genophore, a bacterial cell may contain one or more plasmids. Most plasmids are supercoiled, giving them a condensed, compact shape. They replicate autonomously, but usually in sufficient synchrony with the genophore to ensure a roughly comparable number of plasmids from one generation to the next. In *E. coli* cells, three classes of plasmids are recognized. **F (fertility) factors** are involved in the process of sexual conjugation, which we will discuss later. **R (resistance) factors** carry genes that confer drug resistance on the bacterial cell. **Col (colicinogenic) factors** allow the bacterium to secrete **colicins,** compounds that kill other bacteria lacking the col factor. In addition, at least one strain of *E. coli,* strain 15T, contains **cryptic plasmids,** which have no known function.

Organization of DNA in Eukaryotic Chromosomes

When we turn from viruses and prokaryotic cells to eukaryotic cells, the problem of DNA organization and packaging becomes even more complicated. First, substantially greater amounts of DNA must be accommodated, and second, greater structural complexity is introduced by the association of the DNA with proteins in chromosomes. We will look first at the basic structural unit of eukaryotic chromosomes and then at the means by which higher levels of structure and DNA packing are achieved.

The Nucleosome as the Basic Structural Unit of Chromosomes. Our understanding of how such amazingly long DNA molecules are packaged into the compact structures we recognize as chromosomes is still far from complete. However, substantial progress has been made in recent years with the recognition that the basic structural unit of the chromosome is a complex called the **nucleosome.** Nucleosomes consist of DNA associated with a class of basic proteins called **histones.** Histones are present in all eukaryotic chromosomes except those of sperm cells, where they are replaced by the protamines originally isolated and described by Miescher. There are five major classes of histones, identified as histones H1, H2A, H2B, H3, and H4. Table 13-2 gives several properties of these classes of histones. Histone H1 is especially rich in lysine, H2A and H2B are slightly lysine-rich, and H3 and H4 are rich in arginine instead.

The presence of histones in chromosomes has been known since the 1940s. In fact, chromosomes have long been recognized as essentially a nucleoprotein complex consisting of approximately equal amounts by weight of DNA and histone, with a small but highly significant proportion of acidic proteins called **nonhistone** or **acidic chromosomal proteins.** Because DNA is negatively charged and histones are positively charged, their association is stabilized by ionic bonds.

When eukaryotic nuclei are washed free of all soluble substances, the nucleoprotein complex that makes up the chromosomes remains as an insoluble complex called **chro-**

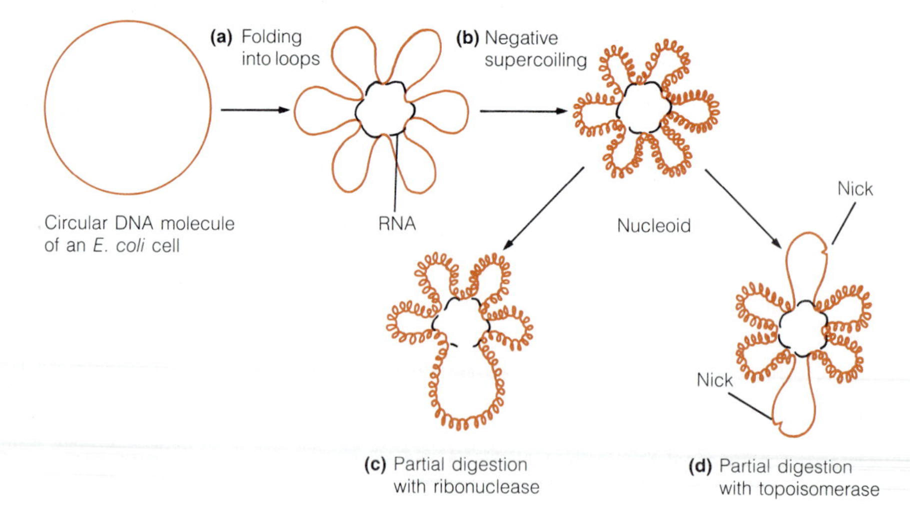

Figure 13-17 Packaging of the Bacterial Genophore. The circular DNA molecule of a prokaryote such as *E. coli* is (a) folded into loops and (b) negatively supercoiled to form the nucleoid, a compact structure that occupies less than 0.2% of the internal volume of the cell. (c) Partial digestion with ribonuclease releases some of the folds but does not relax the supercoiling. (d) Partial digestion with topoisomerase relaxes some of the supercoiling but does not disrupt the folds.

Table 13-2 Properties of Histones

Class	Chemistry	Molecular Weight	Variation Among Species
H1	Lysine-rich	19,500–21,000	Highly variable*
H2A, H2B	Slightly lysine-rich	13,000–17,000	Moderately variable
H3, H4	Arginine-rich	11,500–15,000	Highly invariant*

* To illustrate the difference in variation between histone classes, there are 40 differences in the amino acid sequence of histone H1 between calf and *Drosophila*, but only 2 differences in the amino acid sequence of histone H4 between calf and pea.

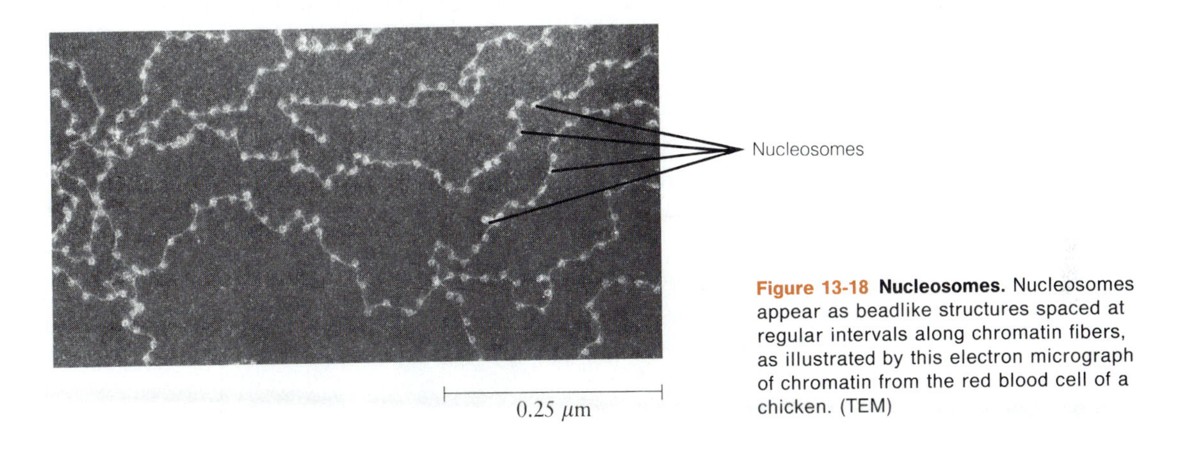

Figure 13-18 Nucleosomes. Nucleosomes appear as beadlike structures spaced at regular intervals along chromatin fibers, as illustrated by this electron micrograph of chromatin from the red blood cell of a chicken. (TEM)

matin. Electron microscopic studies have shown that isolated chromatin consists of repeating structural units, the nucleosomes, spaced regularly along the DNA rather like beads on a string. This structure is shown in Figure 13-18. Each such nucleosome consists of an **octamer** of histones associated with 146 nucleotide pairs of DNA (Figure 13-19). The octamer involves two each of histones H2A, H2B, H3, and H4. These form a cylindrical core of about 5.5 × 10 nm, around which the DNA is wrapped to form a left-handed superhelix with almost two full turns. Histones H3 and H4 are believed to be in the center of the core and therefore associated with the main loop of DNA, while dimers of H2A and H2B bind to the DNA at either end of the loop.

As Figure 13-19 also shows, nucleosomes are separated from each other by a length of **spacer** or **linker DNA,** at least in isolated chromatin. Treatment of chromatin fibers with deoxyribonuclease results in preferential digestion of the spacer DNA and release of free nucleosomes. Initially, each nucleosome released in this way has about 200 nucleotide pairs of DNA associated with it, but on further digestion the nuclease degrades the spacer DNA completely, leaving a **core particle,** which is the histone octamer with its 146 nucleotide pairs of DNA.

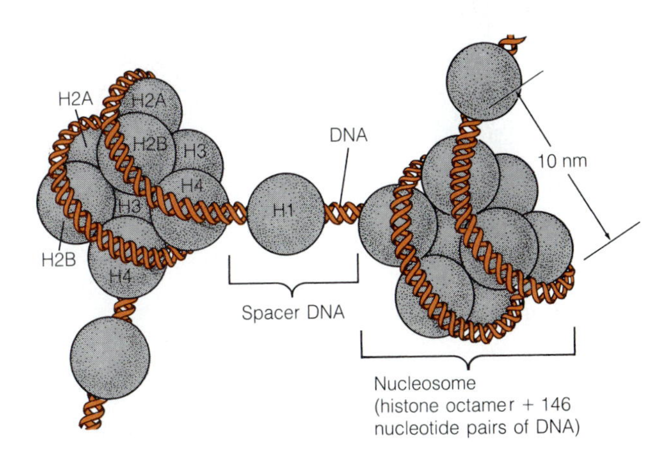

Figure 13-19 Nucleosome Structure. Each nucleosome consists of an octamer of histones (two each of histones H2A, H2B, H3, and H4) associated with 146 nucleotide pairs of DNA. Histone H1 is not a part of the nucleosome core, but is thought to be associated with the spacer DNA between successive nucleosomes. The diameter of each nucleosome is about 10 nm.

The amount of spacer DNA between successive nucleosomes averages about 50 to 60 nucleotide pairs but varies with the source of the DNA, ranging from as few as 20 nucleotide pairs to as many as 100. The significance of this variation is not known, nor is it yet clear whether the nucleosomes are spaced at regular intervals along the DNA or whether the histones associate preferentially with specific nucleotide sequences.

Chromosomal Structure. Histone H1 is not a part of the nucleosomal structure but associates instead with the spacer DNA between nucleosomes (Figure 13-19). In so doing, H1 is thought to bring adjacent nucleosomes into close juxtaposition, thereby forming a **chromatin fiber,** as shown in Figure 13-20a. Such fibers are about 10 nm thick, since that is the diameter of the individual nucleosomes of which each fiber consists. These fibers are regarded as the basic structural element of eukaryotic chromosomes.

Under the right conditions, the chromatin fiber forms a helical coil called a **solenoid,** with a thickness of about 30 nm. Each turn of the solenoid coil consists of about six nucleosomes, as shown in Figure 13-20b. Histone H1 is important in stabilizing this solenoid structure, since no such ordered structures are formed in its absence. The solenoid structure itself coils to form a hollow tube about 200 nm in diameter (Figure 13-20c). In cells that are not dividing, this is the extent of chromosome condensation.

As a cell prepares to undergo division, its chromosomes become more highly condensed. This condensation involves a still higher order of coiling in which the 200-nm tubes of interphase (between-division) chromosomes form the final 600-nm chromatids present at metaphase, one of the stages in the division process (Figure 13-20d).

The extent of DNA coiling is quantitated by the **pack-**

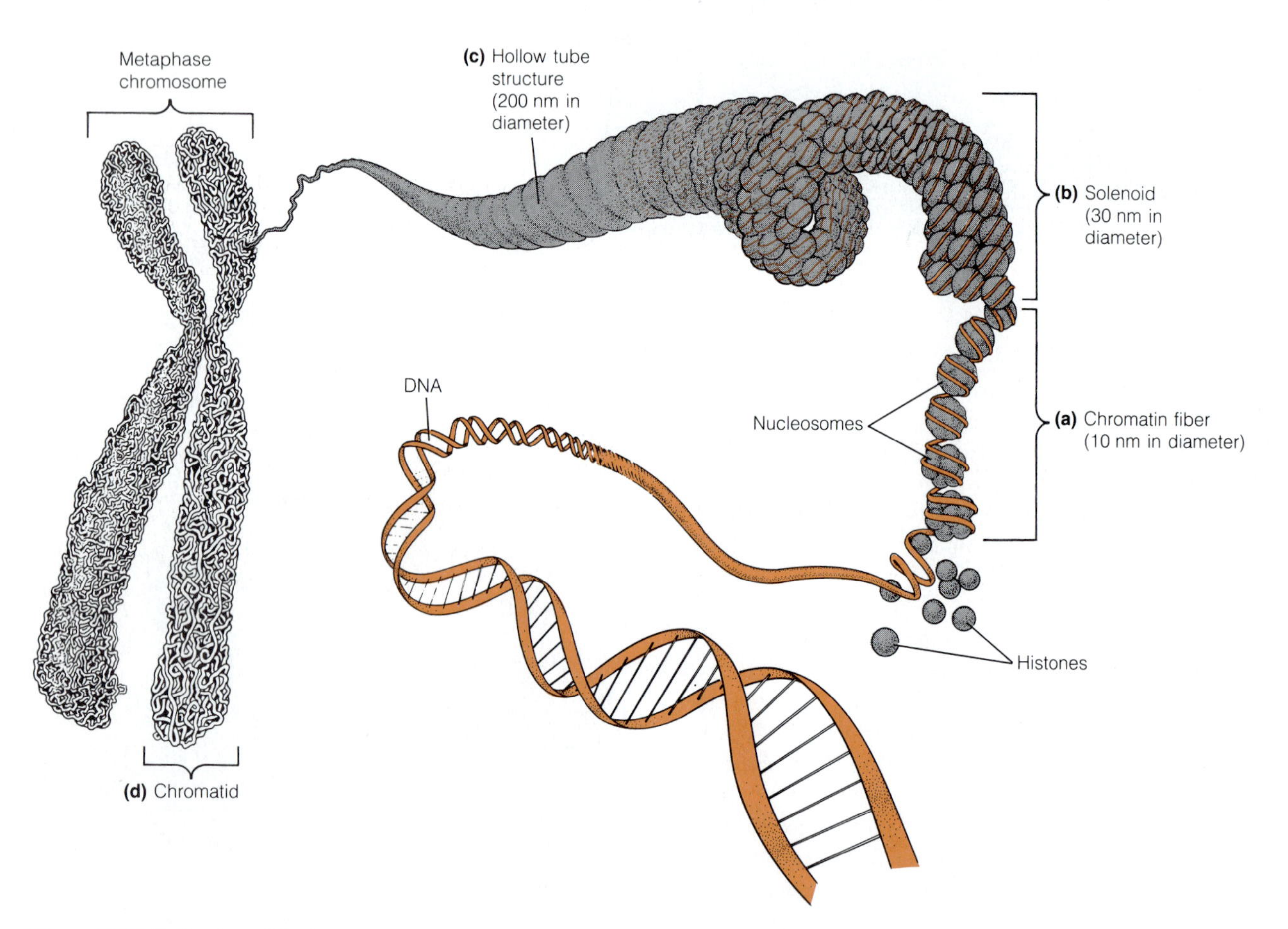

Figure 13-20 Packaging of Nucleosomes into Chromosomes. (a) In chromosomes, nucleosomes are closely juxtaposed into chromatin fibers, each with a thickness of about 10 nm—the diameter of the nucleosomes themselves. (b) Each chromatin fiber can form a helical coil called a solenoid, with about six nucleosomes per coil and a thickness of about 30 nm. (c) These solenoid structures then coil further to form the 200-nm tubes characteristic of interphase chromosomes. (d) As the cell enters mitosis, a still higher degree of coiling generates the 600-nm chromatids characteristic of metaphase chromosomes.

ing ratio, which is defined as the length of the linear DNA molecule divided by the length of the chromosome or fiber into which it is packaged. The initial coiling of the DNA around the histone cores of the nucleosomes reduces the length by a factor of seven, and solenoid formation results in a further sixfold condensation. The packing ratio of the solenoid structure (Figure 13-20b) is therefore about 42. Solenoid coiling into the 200-nm tubes of Figure 13-20c condenses the length of the DNA by an additional factor of 18, making the overall packing ratio from linear DNA molecule to the hollow tube of interphase chromosomes about 750.

For metaphase chromosomes, the packing ratio is still higher because of the further condensation that takes place as the nucleus enters division. For example, a typical human chromosome contains about 75 mm of DNA and has a length of about 4 or 5 μm at metaphase. The overall packing ratio must therefore be somewhere in the range of 15,000 to 20,000.

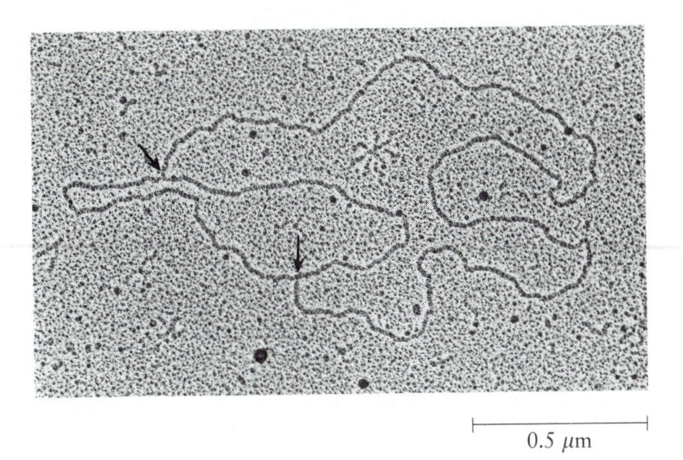

Figure 13-21 Mitochondrial DNA. Mitochondrial DNA from most organisms is circular, as seen in this electron micrograph of a mitochondrial DNA molecule from a rat liver cell. The molecule was caught in the act of replication; the arrows indicate the points at which replication was proceeding when the molecule was fixed for electron microscopy. (TEM)

The DNA of Mitochondria and Chloroplasts

Not all of the DNA of a eukaryotic cell is present in the nucleus. Nuclear DNA certainly accounts for most of the genetic information of the cell, but both mitochondria and chloroplasts contain some DNA of their own, as well as the machinery necessary to replicate, transcribe, and translate the information encoded by their DNA. For mitochondria and chloroplasts, the genome is prokaryotic in nature, consisting of "naked" DNA devoid of histones and usually circular (Figure 13-21). Moreover, the base compositions of the mitochondrial and chloroplast genomes often differ significantly from that of the nuclear genome.

In both organelles, the genome is small, comparable in size to a viral genome. Both organelles can therefore code for some, but by no means all, of their own polypeptides. They are therefore **semiautonomous organelles,** able to specify some of their polypeptides but dependent on the nuclear genome to encode most of them.

The genome of the human mitochondrion, for example, consists of a circular DNA molecule containing about 15,000 base pairs and having a contour length of about 5 μm. This amount of DNA can code for only about a dozen products, just a small fraction (about 5%) of the number of RNA molecules and proteins needed by the mitochondrion. This is nonetheless a vital genetic contribution, because those products include the RNA molecules present in mitochondrial ribosomes, all the transfer RNA molecules required for mitochondrial protein synthesis, and subunits for several mitochondrial proteins, including cytochrome *b*, cytochrome *c* oxidase, and ATPase (Figure 13-22).

The size of the mitochondrial genome varies considerably with species. Mammalian mitochondria typically have only about 15,000 nucleotide pairs of DNA, yeasts have about five times that much, and plants have large and variable amounts, including both circular and linear molecules. It is not at all clear, however, that larger mitochondrial genomes necessarily code for correspondingly more polypeptides. A comparison of yeast and human mitochondrial DNA molecules, for example, suggests that most of the additional DNA present in the yeast mitochondrion consists of long, noncoding sequences.

Chloroplasts typically contain circular DNA molecules with about 130,000 nucleotide pairs. In addition to ribosomal and transfer RNA, the chloroplast genome also codes for a number of polypeptides, including one of the two subunits present in ribulose-1,5-bisphosphate carboxylase, the carbon-fixing enzyme of the Calvin cycle (reaction Cal-1 of Figure 12-13).

Each polypeptide encoded by either the mitochondrial or the chloroplast genome is invariably part of a multimeric protein that also contains subunits coded for by the nuclear genome. Every organellar protein that contains subunits encoded within the organelle is, in other words, a "hybrid" consisting of some polypeptides encoded and synthesized within the organelle and others encoded by the nuclear genome and synthesized by cytoplasmic ribosomes. This poses intriguing questions as to how the proteins synthesized in the cytoplasm enter the organelle and how their uptake and availability are synchronized with the synthetic activities of the organelle. We will return to these questions in Chapter 16.

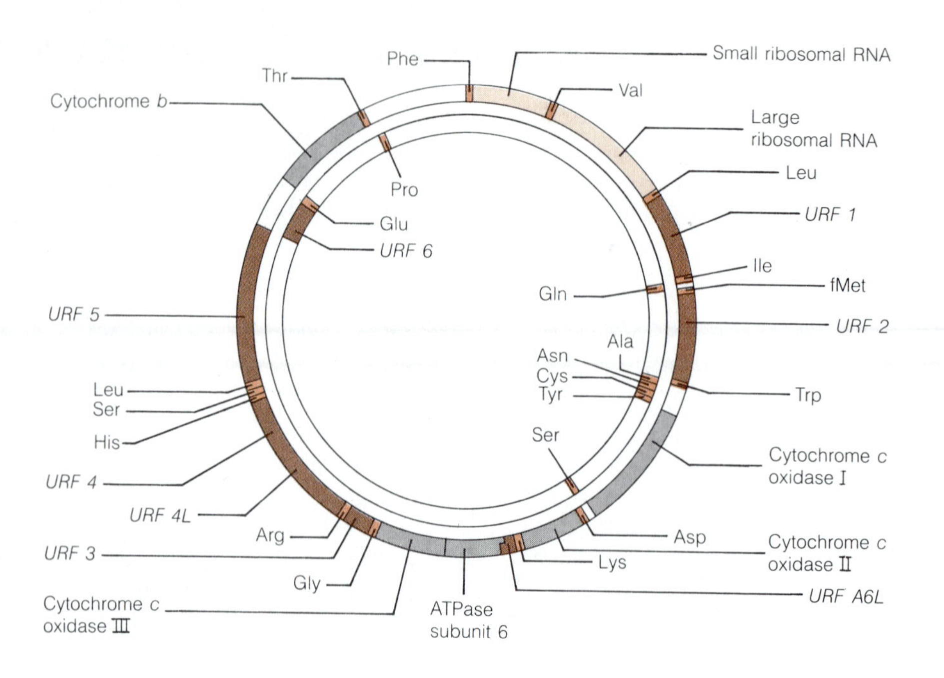

Figure 13-22 The Genophore of the Human Mitochondrion. The double-stranded DNA molecule of the human mitochondrion is circular and contains about 15,000 nucleotide pairs. Some genes are located on one of the two strands (inner circle) and others on the complementary strand (outer circle). The genophore codes for large and small ribosomal RNA molecules, transfer RNA molecules (each identified by the three-letter abbreviation for the amino acid it carries), and subunits of several mitochondrial proteins, including cytochrome *c* oxidase, cytochrome *b*, and ATPase. In addition, the genome contains six unassigned (open) reading frames (URF) that presumably code for proteins that have not yet been identified. The genome is extremely compact, with little noncoding DNA between genes.

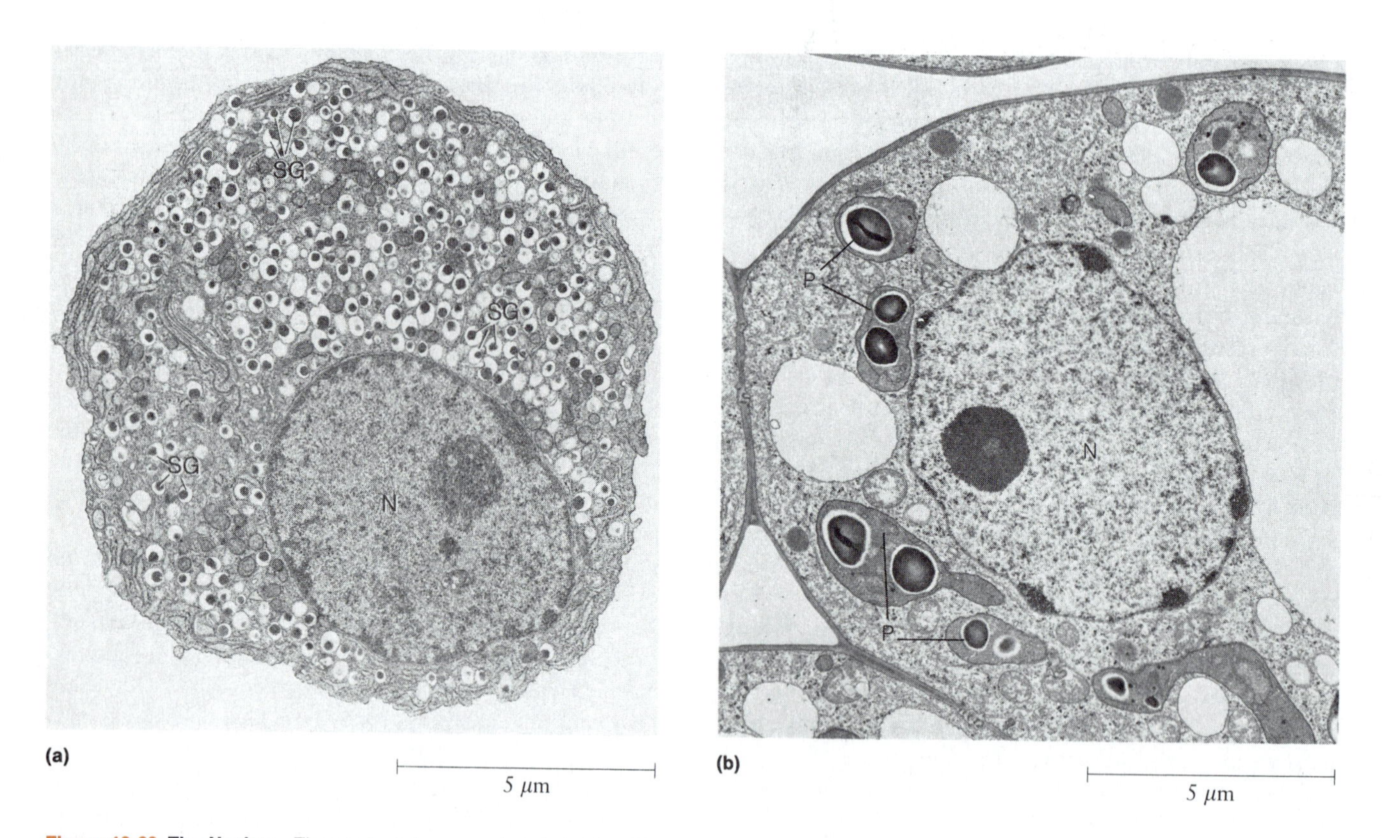

Figure 13-23 The Nucleus. The nucleus is a prominent structural feature in most eukaryotic cells. (a) The nucleus (N) of an animal cell. This is an insulin-producing cell from a rat pancreas; hence, the prominence of secretory granules (SG). (b) The nucleus (N) of a plant cell. This is a cell from a soybean root nodule. The prominence of plastids (P) reflects their role in the storage of starch granules. (TEMs)

The Nucleus

So far, we have considered DNA as the genetic material of the cell, the genome as the total amount of DNA available to the cell of a particular species, and the chromosome or genophore as the physical means of packaging the DNA within the cell. Now we come to the **nucleus,** the site within the eukaryotic cell where the chromosomes are localized and replicated and the DNA that they contain is selectively transcribed. The nucleus is therefore both the repository of most of the genetic information for the cell and the control center for the expression of that information.

The nucleus is one of the most prominent and characteristic features of eukaryotic cells (Figure 13-23). Recall that the term *eukaryon* means "true nucleus." The very essence of a eukaryote is its true membrane-bounded nucleus. In the following discussion, we will focus first on the envelope of membranes that surrounds the nucleus. Then we will turn our attention to several other aspects of the nucleus, including the pores that perforate the envelope, the nuclear cortex that lies just inside the envelope, the nucleoplasm, the chromatin, and the nucleolus (Figure 13-24).

The Nuclear Envelope

The existence of a membrane around the nucleus was suggested in the late nineteenth century, primarily on the basis of observed osmotic properties of the nucleus, and was confirmed by phase-contrast microscopy during the first half of the present century. Further investigation gradually led to the concept of a double membrane with a space between the two phospholipid bilayers. Additional insight came with the advent of electron microscopy, which verified the presence of a double membrane and added much to our understanding of its structure.

The **nuclear envelope** consists of two unit membranes, the **inner** and **outer nuclear membranes,** with a **perinuclear space** between them (Figure 13-24). Both membranes are

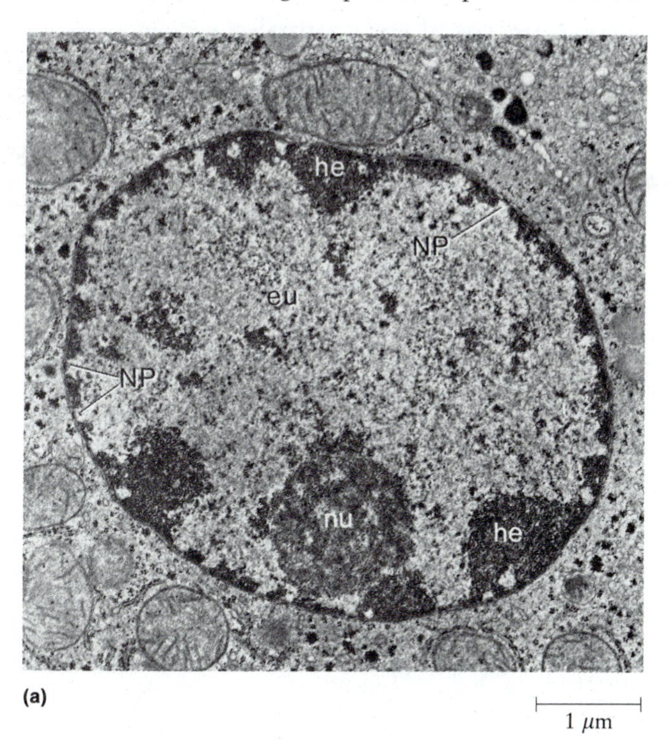

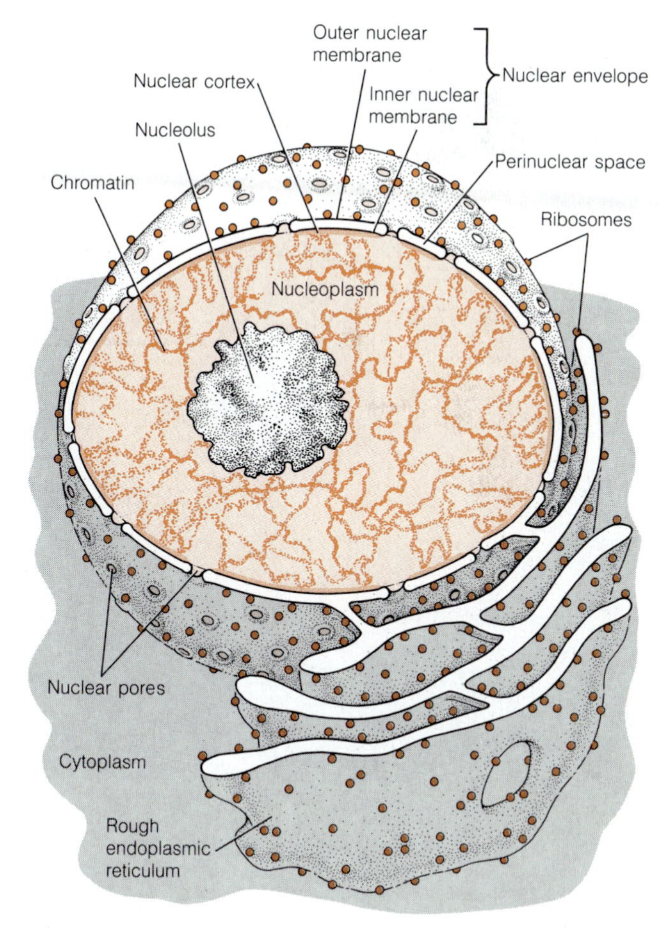

Figure 13-24 The Structure of the Nucleus. (a) An electron micrograph of the nucleus from a mouse liver cell with prominent structural features labeled (TEM). The nuclear envelope is a double membrane perforated by nuclear pores (NP). Internal structures include the nucleolus (nu), euchromatin (eu), heterochromatin (he), and the nucleoplasm. (b) A diagram of a typical nucleus. Structural features included on the diagram but not shown in the electron micrograph include the nuclear cortex, the presence of ribosomes on the outer nuclear membrane, and the continuity between the outer nuclear membrane and the rough endoplasmic reticulum.

about 7–8 nm wide and have the same trilamellar appearance as most other membranes. The outer nuclear membrane is occasionally continuous with the endoplasmic reticulum and is thought to derive from the ER. Like the membrane of the rough ER, the outer membrane is often studded on its outer surface with ribosomes involved in protein synthesis. In some cells, intermediate filaments extend outward from the outer membrane into the cytoplasm, possibly anchored on the other end to the plasma membrane or other organelles. These filaments probably position the nucleus firmly within the cell.

The perinuclear space between the two membranes ranges in width from 10 to 70 nm but is usually a gap of about 20–40 nm. This fluid-filled compartment may be continuous with the cisternae of the ER and the cisternae of the Golgi complex. Little is known of the biochemical composition of the perinuclear fluid or the significance of this compartment.

The most distinctive feature of the nuclear envelope is the presence of numerous **nuclear pores,** small cylindrical channels that extend through both membranes, providing direct contact between the cytoplasm on the outside and the nucleoplasm within. Nuclear pores are readily visible when the nuclear envelope is examined by freeze-fracture microscopy (Figure 13-25). Each pore represents a point of fusion of the inner and outer nuclear membranes. The two membranes of the nuclear envelope are therefore physically continuous, although they differ distinctively in their biochemical properties and functional roles. Because the outer nuclear membrane is continuous both with the ER and with the inner membrane, both membranes of the nuclear envelope can exchange material directly with the ER membrane, thereby allowing the nuclear envelope to expand and contract as needed.

Nuclear pores represent more than points of continuity between the two membranes. They have a structural complexity that is still being unraveled, and they pose intriguing possibilities for transport between the cytoplasm and the nucleoplasm. Because of the importance of this transport role, we will look at the nuclear pore complex in some detail.

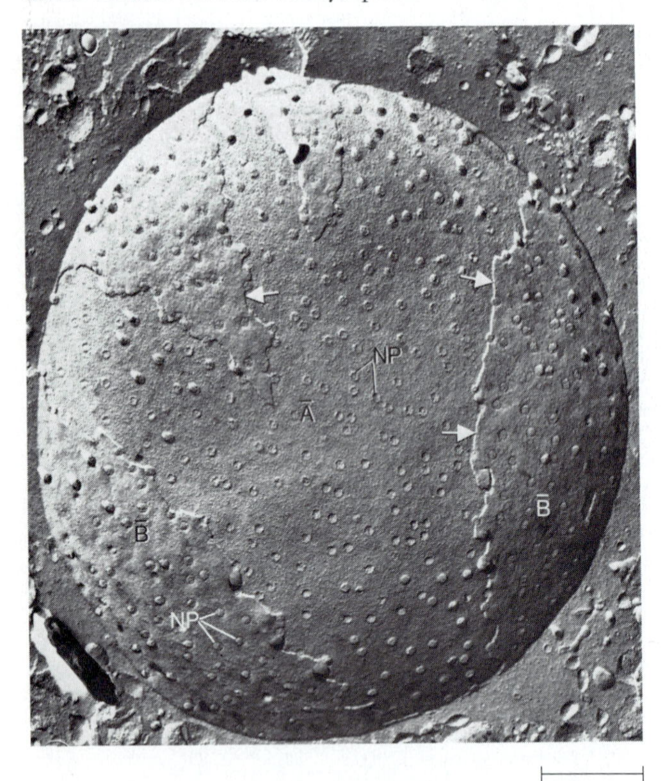

Figure 13-25 Nuclear Pores. Numerous nuclear pores (NP) are readily visible in this freeze-fracture micrograph of the nuclear envelope of an epithelial cell from a rat kidney. The fracture plane reveals faces of both the inner membrane (A̅) and the outer membrane (B̅). The ridges to which the arrows point represent the perinuclear space delimited by the two membranes. (TEM)

The Nuclear Pore Complex

Nuclear pores were first reported in 1950 and are now known to be a property of the nuclear envelope in all eukaryotic cells, both plant and animal. As Figure 13-26

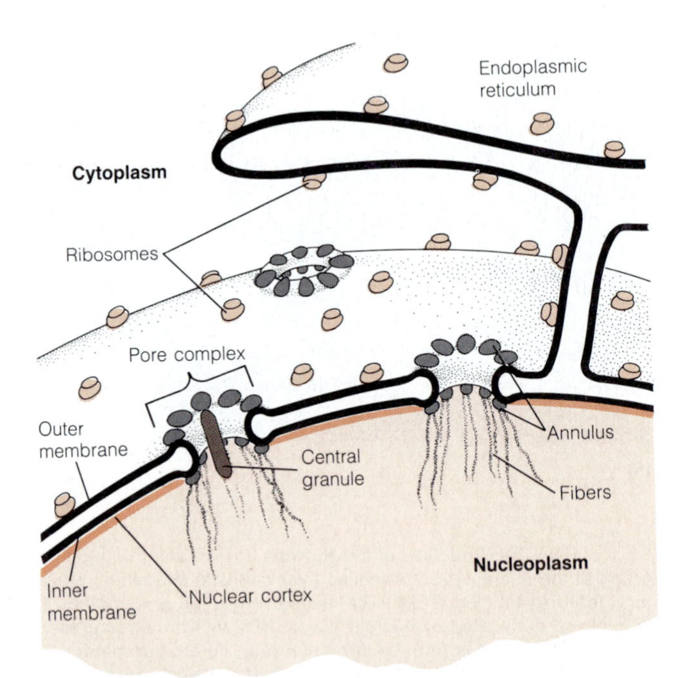

Figure 13-26 Structure of the Nuclear Pore Complex. A pore is formed by the fusion of the inner and outer nuclear membranes and is lined by a nonmembranous annulus. In some cases, a central granule is present in the pore.

illustrates, a pore is formed by the fusion of the two membranes, but has nonmembranous components associated with it that form the **nuclear pore complex.** The diameter of the entire pore complex is usually about 70–90 nm, although the channel that links cytoplasm and nucleoplasm is thought to be no larger than about 9 nm.

Pore Density. The density of pores (number per unit surface area of the nuclear envelope) varies greatly, depending mainly on cell type and the amount of RNA being transported from the nucleus to the cytoplasm. Values range from 3 to 4 pores per square micrometer in some white blood cells up to 50 or more pores per square micrometer in amphibian oocytes. This represents a range from about 2% to about 25% of the total surface area of the envelope, since each pore represents about 0.005 μm^2. For highly active, differentiated cell types such as liver and kidney cells, the nuclear envelope usually has a pore density of about 15 to 20 pores per square micrometer, whereas actively dividing cells have about half this many. A typical mammalian cell has about 3000 to 4000 pores on its nuclear surface, a density of about 10 to 12 pores per square micrometer.

The maximum density that can be achieved is about 60 pores per square micrometer. The nuclei of amphibian oocytes come close to this value. Such nuclei are also exceptionally large, so the total number of pores is incredibly high. For example, an oocyte of *Xenopus laevis* (the South African clawed toad) at an advanced stage in its development has a nuclear envelope with a pore density of about 58 pores per square micrometer. Because the oocyte nucleus is so large and the pore density is so high, the envelope from a single nucleus may contain more than ten million pores. Needless to say, such oocytes have been used to great advantage in studies of nuclear pores.

Pore Structure. The morphology of nuclear pores is not yet well understood. It appears to depend to some extent on cell type and on the way the sample is prepared for electron microscopy. The inside of the pore is ringed by an **annulus** of electron-dense nonmembranous material that appears in micrographs to protrude on both the cytoplasmic and nucleoplasmic sides of the envelope (Figure 13-27). A **central granule** (or **central plug**) sometimes lies inside the annulus. Since the central granule is seen in some pores but not others, it is not yet certain whether it is an integral part of the pore structure. Some investigators believe that the granule consists of newly synthesized ribosomal subunits or other particles that were in transit through the pore at the time of fixation. In some instances, fibers are seen extending into the nucleoplasm or the cytoplasm from the annulus or central granule.

The annulus appears to consist of a 30-nm ring of eight subunits arranged in an octagonal pattern. This pattern is especially striking in negative stains of oocyte nuclear envelopes because of the high pore density of oocyte nuclei (Figure 13-27). It is possible to dissolve the membranes of the nuclear envelope with a nonionic detergent and show that the nuclear pore complex with its annular pattern remains, apparently because the complex is anchored to the nuclear cortex just inside the inner membrane. Under the proper conditions of detergent and salt, pore complexes can be released from the envelope onto an electron microscope grid and examined directly. This procedure should greatly facilitate further study of the nuclear pore complex.

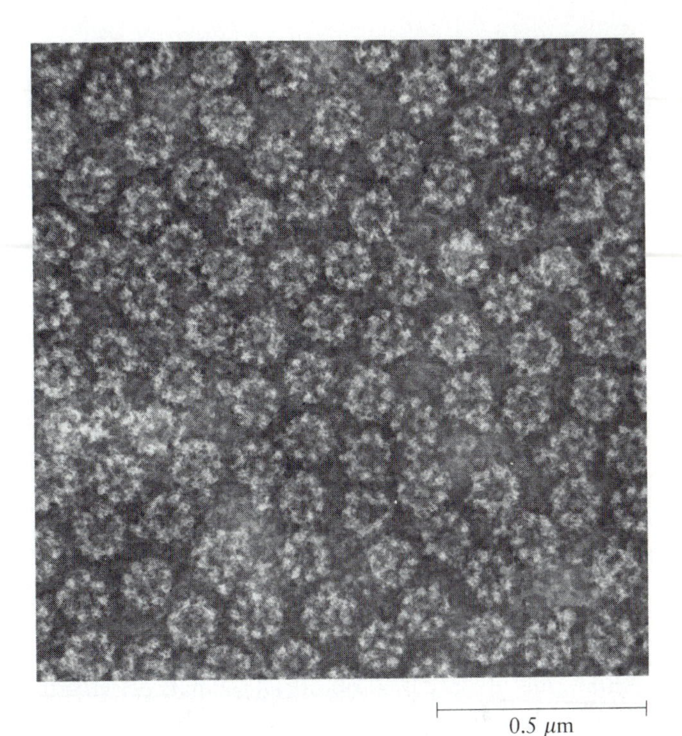

Figure 13-27 Nuclear Pore Complexes. Negative staining of an oocyte nuclear envelope reveals the octagonal pattern of the annulus that surrounds each pore. The nuclear envelope is from an oocyte of the newt *Taricha granulosa*. (TEM)

Transport Across the Nuclear Envelope

The nuclear envelope is both a solution to one problem and the source of another. As a means of restricting chromosomes to one part of the cell, it is an example of the general eukaryotic strategy of compartmentalization. Presumably, it is advantageous for a nucleus to possess a barrier that keeps things like chromosomes in and things like mitochondria, lysosomes, ribosomes, and microtubules out.

At the same time, however, the nuclear envelope creates for the eukaryotic cell several formidable transport problems that are unknown to prokaryotes. Specifically, all the enzymes and other proteins required for chromosome replication and DNA transcription in the nucleus must be imported from the cytoplasm, and all the RNA molecules and partially assembled ribosomes needed for protein synthesis in the cytoplasm must be obtained from the nucleus (Figure 13-28). It is perhaps to overcome these problems that the nuclear envelopes of all eukaryotes have pores, because all this macromolecular transport is believed to be mediated by the pores. To get some idea of how much traffic is involved, consider just the flow of ribosomes from the nucleus to the cytoplasm. As we will see shortly, ribosomes are synthesized in the nucleus as two classes of subunits, each of which is a complex of RNA and proteins. These subunits move to the cytoplasm as such and are assembled into functional ribosomes when needed for protein synthesis in the cytoplasm.

An actively growing mammalian cell can easily be synthesizing about 20,000 ribosomal subunits per minute. We already know that such a cell has about 3000 to 4000 nuclear pores, so ribosomal subunits must be transported to the cytoplasm at a rate of about 5 to 6 subunits per minute per pore. Traffic in the opposite direction is, if anything, even heavier. When chromosomes are being replicated, histones are needed at the rate of about 300,000 molecules per minute. The rate of inward movement must therefore be about 100 histone molecules per minute per pore!

Permeability Characteristics of the Nuclear Envelope. The channel in the center of each nuclear pore is considered a direct connection between the nucleoplasm and cytoplasm. The channels appear to be freely permeable to small molecules and ions, because such substances cross the nuclear envelope very quickly when they are injected into cells. For example, the nucleoside triphosphates required for DNA and RNA synthesis probably pass freely through the pores, as do substrates needed for any metabolic pathways that may function within the nucleus.

Ions are also thought to pass freely across the nuclear envelope. At one time this was questioned because of the potential difference that can be demonstrated across the nuclear envelope in electrophysiological studies. However, the nucleus has many negatively charged macromolecules in it that bind cations, and it is primarily these bound cations that account for the observed charge gradient across the envelope.

To study the ability of macromolecules to traverse the nuclear envelope, investigators have injected proteins of various sizes into cells and observed how long it takes for the proteins to appear in the nucleus. Many proteins can cross the envelope, presumably by means of its pores. But passage becomes more difficult with increasing molecular weight. A globular protein with a molecular weight of 20,000 takes only a few minutes to equilibrate between the nucleus and the cytoplasm, but proteins of 60,000 daltons or more seem barely able to penetrate at all.

The effect of particle size has been further assessed using colloidal gold particles of various sizes. Particles with diameters up to 5–6 nm can be detected in the nucleus within minutes; those with diameters of 9–10 nm take several hours; and above 15 nm, the particles do not enter at all. These and other transport measurements indicate that the nuclear pore complex has an aqueous channel about 9 nm in diameter. Apparently, a channel of this size is adequate to allow the necessary macromolecules to enter and leave the nucleus, while still ensuring that chromosomes are kept in and that organelles and other cytoplasmic structures are kept out.

Nucleocytoplasmic Transport of Proteins, Nucleic Acids, and Ribosomes. Most of the proteins involved with DNA replication and transcription are small enough to pass through a 9-nm channel. Histones, for example, have molecular weights of 21,000 or less (Table 13-3) and should therefore traverse the envelope with little problem. Some nuclear proteins are very large, however. The polymerases involved in both DNA and RNA synthesis, for example, have molecular weights in excess of 100,000 and would presumably not fit easily through a 9-nm orifice. However, such proteins consist of multiple subunits and could conceivably be assembled in the nucleus from subunits that diffuse in separately. Another possibility is that the proteins undergo changes in shape to allow passage through the pore.

Ribosomal subunits also pose a special problem. They are known to be assembled in the nucleus and exported to the cytoplasm, yet they have a diameter of about 15 nm. Some investigators believe that these particles are transported through the pores, but by an active transport process that greatly distorts their shape as they pass through. In fact, the central granules seen in some nuclear pores are thought to be ribosomal subunits in transit through the pores. Messenger RNA also poses a special challenge, because mRNA molecules are believed to leave the nucleus complexed with proteins in the form of **ribonucleoprotein particles**—complexes of RNA and protein—up to 40 or even 50 nm in diameter. Clearly, the transport of particles of this size through a 9-nm channel poses a significant challenge!

The Nuclear Cortex

The **nuclear cortex** is an electron-dense layer of fibrous material on the nucleoplasmic side of the inner nuclear

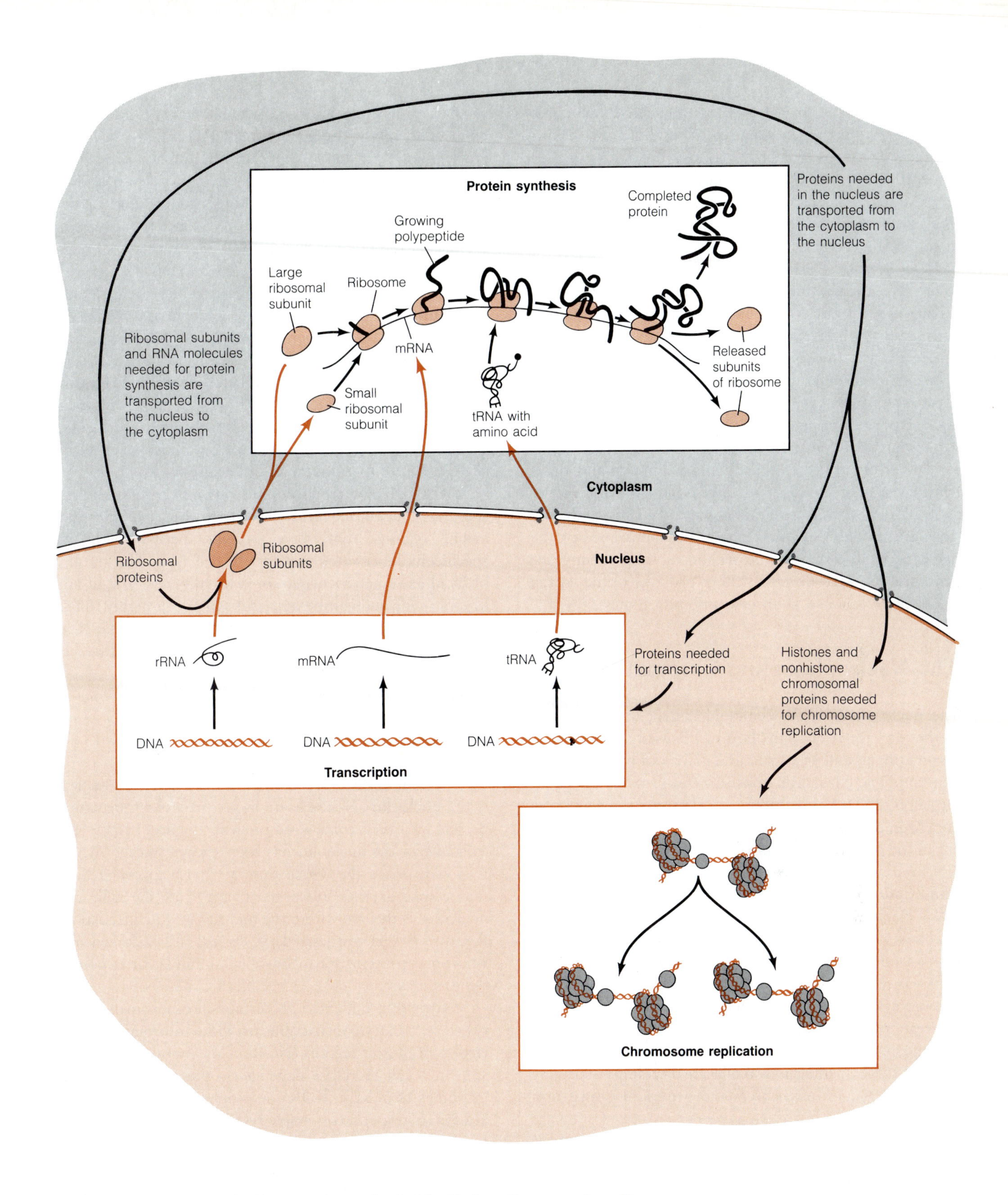

Figure 13-28 Macromolecular Transport into and out of the Nucleus. Because eukaryotic cells store their genetic information in the nucleus but synthesize proteins in the cytoplasm, all the proteins needed in the nucleus must be transported inward from the cytoplasm, and all the RNA molecules and ribosomal subunits needed for protein synthesis in the cytoplasm must be transported outward from the nucleus. The three kinds of RNA molecules required for protein synthesis are ribosomal RNA (rRNA), messenger RNA (mRNA), and transfer RNA (tRNA).

membrane. The cortex, also called the **nuclear lamina,** is up to 30–40 nm thick in some cells but difficult to detect in others. It consists of proteinaceous fibers arranged in whorls that some investigators believe may serve to funnel materials to the nuclear pores for passage to the cytoplasm. The cortex may also be involved in pore formation and in conferring shape to the nuclear envelope.

On the nucleoplasmic side, the nuclear cortex appears to organize the chromosomes by binding interphase chromatin to special sites on the inner nuclear membrane. The chromatin that binds in this way is highly condensed and can, in at least some cells, be identified with specific regions on the chromosomes. The chromatin-binding sites on the cortex seem to be excluded from the immediate vicinity of nuclear pores, presumably to ensure unobstructed access to the pores for transport purposes.

In vertebrates, the cortex contains three major polypeptides that are thought to bind to integral membrane proteins of the inner nuclear membrane. These polypeptides are probably involved in the disappearance of the nuclear envelope during mitosis and its subsequent reappearance after cell division. Other proteins in the cortex are believed to serve as chromatin-binding sites.

The Nucleoplasm and Chromatin

The **nucleoplasm** is the semifluid matrix in the interior of the nucleus. It contains both condensed and extended chromatin, as well as a structural matrix of nonchromatin material. The chromatin consists of approximately equal measures of DNA and chromosomal proteins, whereas the nonchromatin matrix is largely protein. The proteins present in the matrix include enzymes and factors involved in DNA replication, RNA synthesis (transcription), and post-transcriptional processing, packaging, and transport of RNA. Each of these processes will be considered in later chapters.

The chromatin represents chromosomes as they exist between cell divisions. As we noted earlier, chromosomes assume a highly condensed form as the cell prepares to undergo mitosis. Immediately after division, however, the chromosomes once again assume a highly extended state. Chromosomes in interphase nuclei are therefore not usually detectable by microscopy because they exist in diffuse form as chromatin.

Actually, we need to qualify our statement somewhat because it is true of only one of the two general classes of chromatin. **Euchromatin** is the term used to define diffuse, uncondensed chromatin and in fact describes most of the chromatin in most interphase nuclei. However, virtually all eukaryotic nuclei contain some **heterochromatin,** which consists of chromosomes or chromosomal regions that remain condensed during interphase (Figure 13-24a). Heterochromatin appears to retain the highly condensed structure that other chromatin assumes only during division. The DNA in heterochromatin is genetically inactive and is replicated late in the cell cycle, after replication of the rest of the DNA is complete (see Chapter 14).

Heterochromatin can be categorized as either constitutive heterochromatin or facultative heterochromatin. **Constitutive heterochromatin** represents chromosomal regions that exist in this permanently condensed form in all cells of the organism and is therefore never genetically active in any cell. In most cases, heterochromatin contains very simple DNA sequences that repeat serially. In human chromosomes, the constitutive heterochromatin is located at the **kinetochore,** the region of the chromosome to which spindle microtubules attach during division.

Facultative heterochromatin varies from tissue to tissue and appears to represent chromosomes or chromosomal regions that have become permanently but specifically inactivated in a specific cell type. The amount of facultative heterochromatin is usually very low in embryonic cells and can be quite substantial in highly differentiated cells. Facultative heterochromatin may therefore be an important means of inactivating entire blocks of genetic information during development.

The Nucleolus

The remaining structural feature of the eukaryotic nucleus is the **nucleolus.** Nucleoli are large, prominent structures present in every eukaryotic nucleus (Figure 13-29). The existence of the nucleolus has been known since 1774, but it is only since the 1960s that we have appreciated its importance as the "ribosome factory" of the cell. The nucleolus is the site within the nucleus at which ribosomal RNA (rRNA) is synthesized, processed, and packaged with ribosomal proteins into ribosomal subunits for export to the cytoplasm.

Nucleoli are located on certain chromosomes at sites called **secondary constrictions** or, more informatively, **nucleolus organizer regions (NORs).** The number of nucleoli seen in a cell therefore depends on the number of chromosomes that have NORs. A single nucleolus does not always correspond to a single NOR, however. The human genome contains five NORs per haploid chromosome set, or ten per diploid nucleus, each located near the tip of a chromosome. But instead of ten separate nucleoli, the typical human nucleus contains a single large nucleolus representing the fusion of loops of chromatin from ten separate chromosomes.

The DNA at the NOR of a chromosome contains the genes for ribosomal RNA. These genes are present in multiple copies in all eukaryotic genomes. The number of copies varies greatly, depending on the species, but animal cells often contain hundreds of copies and plant cells usually contain thousands. This DNA is present as fibrils of chromatin associated with the nucleolus. Some of these fibrils surround the nucleolus and are called the **perinucleolar chromatin;** others penetrate it to varying degrees and are referred to as the **intranucleolar chromatin.**

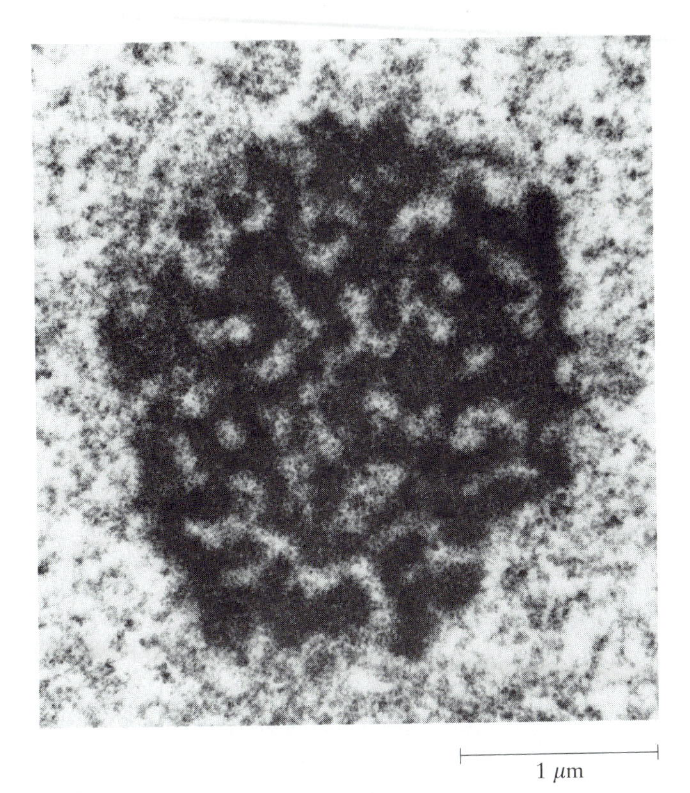

Figure 13-29 The Nucleolus. The nucleolus is a prominent intranuclear structure. Shown here is a nucleolus of a cat nerve cell. (TEM)

In addition to DNA, the nucleolus consists of a *granular component* and a *fibrillar component* embedded in an amorphous matrix. The fibrillar component consists of rRNA molecules that have already become associated with proteins to form fibrils with a thickness of about 5 nm. The granular component consists of 15-nm particles that are ribosomal subunits in the process of maturation. Not surprisingly, nucleoli stain intensely for ribonucleoprotein and become heavily labeled when the cell is exposed to radioactive precursors to RNA (Figure 13-30).

The size of the nucleolus is correlated with its level of activity. In cells characterized by a high rate of protein synthesis and hence by the need for many ribosomes, the nucleolus can account for 20–25% of the total volume of the nucleus. In less active cells, it is much less significant. The main difference is in the amount of granular component present. Cells that are producing many ribosomes transcribe, process, and package large quantities of rRNA and have higher steady-state levels of partially complete ribosomal subunits on hand in the nucleolus, thus accounting for the prominent granular component.

An interesting feature of the nucleolus is the disappearing act that it does during mitosis, at least in the cells of higher plants and animals. As the cell approaches division, the condensation of chromosomes is accompanied by a remarkable change in the nucleolus, which decreases in size and then disappears altogether. A cell undergoing

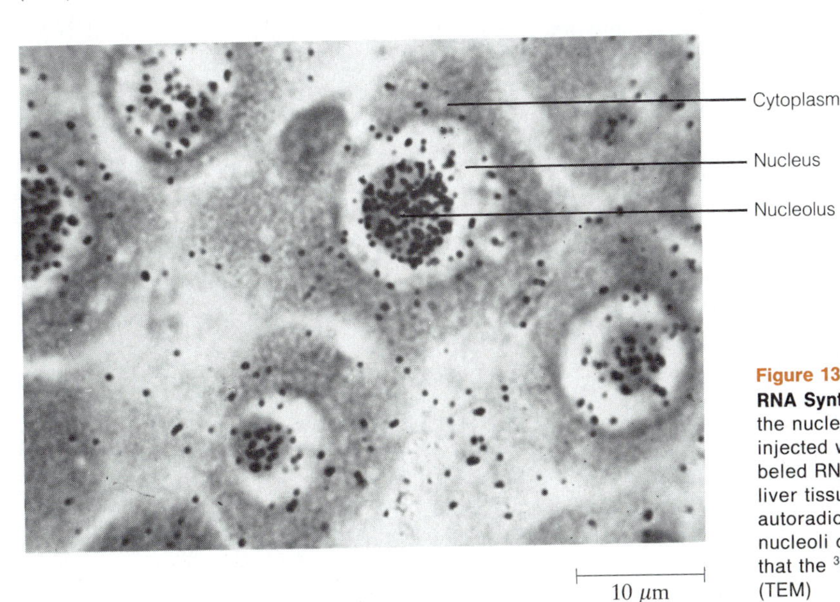

Figure 13-30 The Nucleolus as a Site of RNA Synthesis. To demonstrate the role of the nucleolus in RNA synthesis, a rat was injected with ^{3}H-cytidine, a radioactively labeled RNA precursor. Five hours later, liver tissue was removed and subjected to autoradiography. The black spots over the nucleoli on this autoradiograph indicate that the ^{3}H is concentrated in the nucleoli. (TEM)

mitosis therefore has no nucleolus (and, as a corollary, synthesizes no rRNA). When mitosis is complete, the nucleolus reappears and rRNA synthesis resumes. This is the only point in the cell cycle at which the multiple NORs of the human genome are apparent. As rRNA synthesis begins again, ten tiny nucleoli become visible, one near the tip of each of ten different chromosomes. As these nucleoli enlarge, they quickly fuse into the single large nucleolus characteristic of the interphase human nucleus.

Perspective

Our knowledge of DNA dates back to the early studies of Miescher, but it was not until the mid-twentieth century that experiments with pneumococcal bacteria and bacteriophage T2 convinced the skeptics that DNA was indeed the genetic material of most organisms. This was followed closely by Watson and Crick's elucidation of the double-helical structure of DNA, one of the truly significant landmarks in twentieth-century biology. This early discovery has matured into the discipline of molecular biology, which has some of the most powerful tools ever made available to cell biologists. We can now amplify a single molecule of DNA a millionfold, determine its nucleotide base sequence, then experimentally alter the sequence and reinsert a synthetic version into the genome of a living cell. An unprecedented expansion of biotechnology has accompanied the basic scientific advances. For instance, if the synthetic DNA is inserted into bacterial or yeast cells, the microorganisms can be grown in large quantities and the resulting gene product isolated for clinical applications. Proteins such as insulin, somatotropin, and an enzyme that dissolves blood clots have been produced in this way.

The genome consists of the DNA (or, in some cases, RNA) present in one complete set of the genetic information of an organism or virus. For most viruses and prokaryotes, the genome consists of one or a small number of DNA molecules. Most eukaryotes have a nuclear genome, a mitochondrial genome, and, in the case of plants, a chloroplast genome as well. Genome size is generally related to the complexity of the organism, but variations are known that cannot be explained in this way.

The physical length of the DNA molecules present in cells and viruses necessitates considerable packaging. In prokaryotes and viruses, DNA is organized into genophores. In eukaryotes, DNA is complexed with proteins and organized into chromosomes. The structural unit of chromosomes is the nucleosome, an octet of histones associated with a short length of DNA. Adjacent nucleosomes are brought together to form a 10-nm chromatin fiber, which can then coil in a 30-nm helical solenoid. Further coiling generates a hollow tube of about 200 nm.

In the eukaryotic cell, all the chromosomes are localized and replicated within the nucleus. The double membrane, or envelope, that surrounds the nucleus is perforated with nuclear pores that allow passage of molecules or particles up to about 9 nm in diameter. The pores are important for the inward movement of histones and other proteins and the outward movement of RNA and ribosomes. The chromosomes in the nucleus condense into visible structures during mitosis, but exist as extended, highly diffuse euchromatin during the remainder of the cell cycle. Heterochromatin consists of chromosomes or chromosomal regions that remain highly condensed at all times. The nucleolus is the structure within the nucleus responsible for rRNA synthesis and ribosome assembly.

Key Terms for Self-Testing

genetic information (p. 352)
genophores (p. 352)
chromosomes (p. 352)
nucleus (p. 352)

The Chemical Nature of the Gene
protamine (p. 353)
S (smooth) strains (p. 353)
R (rough) strains (p. 353)
transformation (p. 353)
transforming principle (p. 353)
Chargaff's rules (p. 357)

The Structure of DNA
major groove (p. 357)
minor groove (p. 357)
polarity (p. 357)
Z-DNA (p. 357)
B-DNA (p. 357)
supercoiled (supertwisted) DNA (p. 358)
relaxed state (p. 358)
positive supercoil (p. 359)
negative supercoil (p. 359)
topological isomers (p. 359)
topoisomerases (p. 359)
DNA gyrase (p. 359)
denaturation (p. 359)
native DNA (p. 359)
melting temperature (p. 361)
renature (p. 361)
nucleation event (p. 361)
zipping-up (p. 361)

Organization of DNA into Genomes
genome (p. 362)
recombinant DNA technology (p. 363)
restriction enzymes (p. 367)
DNA cloning (p. 367)
DNA sequencing techniques (p. 367)
satellite DNA (p. 370)
interspersed repeats (p. 370)

DNA Organization and Packaging
nucleoid (p. 372)
F (fertility) factors (p. 372)
R (resistance) factors (p. 372)
col (colicinogenic) factors (p. 372)
colicins (p. 372)
cryptic plasmids (p. 372)
nucleosome (p. 372)
histones (p. 372)
nonhistone (acidic) chromosomal proteins (p. 372)
chromatin (p. 372)
octamer (p. 373)
spacer (linker) DNA (p. 373)
core particle (p. 373)
chromatin fiber (p. 374)
solenoid (p. 374)
packing ratio (p. 374)
semiautonomous organelles (p. 375)

The Nucleus
nucleus (p. 377)
nuclear envelope (p. 377)
inner nuclear membrane (p. 377)
outer nuclear membrane (p. 377)
perinuclear space (p. 377)
nuclear pores (p. 378)
nuclear pore complex (p. 379)
annulus (p. 379)
central granule (central plug) (p. 379)
ribonucleoprotein particles (p. 380)
nuclear cortex (p. 380)
nuclear lamina (p. 382)
nucleoplasm (p. 382)
euchromatin (p. 382)
heterochromatin (p. 382)
constitutive heterochromatin (p. 382)
kinetochore (p. 382)
facultative heterochromatin (p. 382)
nucleolus (p. 382)
secondary constrictions (p. 382)
nucleolus organizer regions (NORs) (p. 382)
perinucleolar chromatin (p. 383)
intranucleolar chromatin (p. 383)

Phages: The Best-Understood "Organisms" in the World
bacteriophages (phages) (p. 364)
lawn (of cells) (p. 364)
plaques (p. 364)
virulent phage (p. 364)
lytic growth (p. 364)
temperate phage (p. 364)
lysogenic state (p. 364)
prophage (p. 364)

Suggested Reading

General References and Reviews

Alberts, B., D. Bray, J. Lewis, M. Raff, K. Roberts, and J. D. Watson. *Molecular Biology of the Cell*, 2d ed. New York: Garland Publishing, 1989.

Carlson, E. A. *The Gene: A Critical History*. New York: W. B. Saunders, 1966.

Hess, E. L. Origins of molecular biology. *Science* 168 (1970): 664.

Judson, H. F. *The Eighth Day of Creation: Makers of the Revolution in Biology*. New York: Simon and Schuster, 1979.

Lewin, B. *Genes*, 3d ed. New York: Wiley, 1987.

Singer, M., and P. Berg. *Genes and Genomes*. Mill Valley, Calif: University Science Books, 1990.

Stent, G. S. That was the molecular biology that was. *Science* 160 (1968): 390.

Watson, J. D., N. Hopkins, J. Roberts, J. Steitz, and A. Weiner. *Molecular Biology of the Gene,* 4th ed. Menlo Park, Calif.: Benjamin/Cummings, 1987.

The Chemical Nature of the Gene

Avery, O. T., C. M. MacLeod, and M. McCarty. Studies on the chemical nature of the substance inducing transformation of pneumococcal types. Induction of transformation by a desoxyribonucleic acid fraction isolated from *Pneumococcus* Type III. *J. Exp. Med.* 79 (1944): 137.

Chargaff, E. Preface to a grammar of biology: A hundred years of nucleic acid research. *Science* 172 (1971): 637.

Hershey, A. D., and M. Chase. Independent functions of viral protein and nucleic acid in growth of bacteriophage. *J. Gen. Physiol.* 36 (1952): 39.

Mirsky, A. E. The discovery of DNA. *Sci. Amer.* 218 (June 1968): 78.

The Structure of DNA

Bauer, W. R., F. H. C. Crick, and J. H. White. Supercoiled DNA. *Sci. Amer.* 243 (July 1980): 118.

Crick, F. H. C. The structure of the hereditary material. *Sci. Amer.* 194 (October 1954): 54.

Dickerson, R. E. The DNA helix and how it is read. *Sci. Amer.* 249 (December 1983): 94.

Dickerson, R. E., H. R. Drew, B. N. Conner, R. M. Wing, A. V. Fratini, and M. L. Kopka. The anatomy of A-, B-, and Z-DNA. *Science* 216 (1982): 475.

Maxam, A.M., and W. Gilbert. A new method of sequencing DNA. *Proc. Natl. Acad. Sci. USA* 74 (1977): 560.

Sanger, F. Determination of nucleotide sequences in DNA. *Science* 214 (1981): 1205.

Smith, G. R. DNA supercoiling: Another level for regulating gene expression. *Cell* 24 (1981): 599.

Wang, J. C. DNA topoisomerases. *Sci. Amer.* 247 (July 1982): 94.

Watson, J. D., and F. H. C. Crick. A structure for deoxyribose nucleic acid. *Nature* 171 (1953): 737.

———. Genetical implications of the structure for deoxyribonucleic acid. *Nature* 171 (1953): 964.

Watson, J. D., J. Tooze, and D. T. Kurtz. *Recombinant DNA: A Short Course.* New York: W. H. Freeman, 1983.

Organization of DNA into Genomes

Britten, R. J., and D. E. Kohne. Repeated sequences of DNA. *Science* 161 (1968): 529

Gall, J. G. Chromosome structure and the C-value paradox. *J. Cell. Biol.* 91 (1981): 3s.

Long, E. O., and I. B. Dawid. Repeated genes in eucaryotes. *Annu. Rev. Biochem.* 49 (1980): 727.

DNA Organization and Packaging

Adolph, K. W., ed. *Chromosomes and Chromatin,* vols. 1–3. Boca Raton, Fla.: CRC Press, 1988.

Borst, P., L. A. Grivell, and G. S. P. Groot. Organelle DNA. *Trends Biochem. Sci.* 9 (1984): 128.

Kornberg, R. D., and A. Klug. The nucleosome. *Sci. Amer.* 244 (February 1981): 52.

McGhee, J. D., and G. Felsenfeld. Nucleosome structure. *Annu. Rev. Biochem.* 49 (1980): 1115.

The Nucleus

De Robertis, E. M. Nucleocytoplasmic segregation of proteins and RNAs. *Cell* 32 (1983): 1021.

Franke, W. W., U. Scheer, G. Kroline, and E.-D. Jarasch. The nuclear envelope and the architecture of the nuclear periphery. *J. Cell Biol.* 91 (1981): 39s.

Jordan, E.-C., and C. A. Cullis, eds. *The Nucleolus.* Cambridge, England: Cambridge University Press, 1982.

Maul, G. G., ed. *The Nuclear Envelope and the Nuclear Matrix.* New York: A. R. Liss, 1982.

Miller, O. L., Jr. The nucleolus, chromosomes, and visualization of genetic activity. *J. Cell Biol.* 91 (1981): 15s.

Newport, J. W., and D. J. Forbes. The nucleus: Structure, function and dynamics. *Annu. Rev. Biochem.* 56 (1987): 535.

Problem Set

1. **The Genetic Material.** Label each of the following statements concerning the chemical nature of the genetic material with a B if the statement is most appropriately dated to the period *before* 1944, with an I if it belongs to the *interim* period 1944–1952, with an A if it is most appropriate to the period *after* 1952, and with an N if it was *never* a widely held concept.
   (a) DNA may not be simply a repeating tetranucleotide sequence after all.
   (b) The genetic material in higher organisms is most likely protein.
   (c) DNA is the genetic information in both bacteria and their phages.
   (d) DNA is chemically too simple to be considered the genetic information of any cell.
   (e) Nuclein is an important component of virtually every bacterial cell, even though we do not yet know what it does.
   (f) DNA may well be the genetic material of bacterial cells, but it is still an open question what that means for higher organisms.
   (g) Smooth (S) strains of *Pneumococcus* are capable of converting nonpathogenic rough (R) strains into pathogenic S strains, but we do not yet know which component of the S cells effects this transformation.
   (h) Once adsorbed onto a host cell, a bacteriophage injects its proteins into the bacterium.

2. **Prior Knowledge.** Virtually every experiment performed by biologists builds upon knowledge that has resulted from prior experiments.

   (a) Of what significance to Avery and his colleagues was the finding (made in 1932 by J. L. Alloway) that the same kind of transformation of R cells into S cells that Griffith observed to occur in mice could also be demonstrated in vitro with cultured *Pneumococcus* cells?

   (b) Of what significance to Hershey and Chase was the following suggestion (made in 1951 by R. M. Herriott)? "A virus may act like a little hypodermic needle full of transforming principles; the virus as such never enters the cell; only the tail contacts the host and perhaps enzymatically cuts a small hole through the outer membrane and then the nucleic acid of the virus head flows into the cell."

   (c) Of what significance to Watson and Crick were the data of their colleagues at Cambridge that the specific forms in which A, G, C, and T exist at physiologic pH permit the formation of specific hydrogen bonds?

   (d) How did the findings of Hershey and Chase help explain an earlier report (by T. F. Anderson and R. M. Herriott) that bacteriophage T2 loses its ability to reproduce when it is burst open osmotically by suspending the viral particles in distilled water prior to their addition to a bacterial culture?

3. **DNA Structure.** Carefully inspect the DNA molecule shown here and note that it has two-fold rotational symmetry. Label each of the following statements as T if true and F if false.

   3′—A—G—C—G—C—T—A—T—A—G—C—G—C—T—5′

   5′—T—C—G—C—G—A—T—A—T—C—G—C—G—A—3′

   (a) There is no way to distinguish the right end of the double helix from the left end.

   (b) If a solution of these molecules were denatured, every single-stranded molecule in the solution would be capable of annealing with every other molecule.

   (c) If the molecule were cut at its midpoint into two halves, it would be possible to distinguish the left half from the right half.

   (d) If the two single strands were separated from each other, it would not be possible to distinguish one strand from the other.

   (e) In a single strand from this molecule, it would be impossible to determine which was the 3′ end and which was the 5′ end.

4. **DNA Sequencing.** The same DNA molecule as that shown in Problem 3 was labeled at the 5′ end with radioactive phosphate, then treated with dimethyl sulfate-piperidine, which breaks the chain by deleting guanine. Afterwards, the radioactive fragments were separated by gel electrophoresis. Draw the gel pattern that would be observed, and indicate the nucleotide composition of each radioactive spot.

5. **Genome Size.** The bacterium *E. coli* has as its genome a circular DNA molecule containing about four million nucleotide pairs. The bacterial cell itself can be thought of as a cylinder with a diameter of 1 μm and a length of 2 μm.

   (a) What is the length of the DNA molecule?

   (b) If the molecule existed as a fully extended circle, what would its radius be? Do you think it is likely that it exists as a fully extended circle in the bacterial cell? Explain.

   (c) Assuming a molecular weight of 330 for the average nucleotide, what is the molecular weight of the molecule? What is its actual weight in grams?

   (d) Knowing that the density of DNA is about 1.7 g/cm^3, calculate the volume that the circular *E. coli* DNA molecule would occupy if tightly packed together. What proportion of the total internal volume of the *E. coli* cell does this represent?

   (e) At what rate (in nucleotides added per second) must the replication of *E. coli* DNA proceed in a culture with a generation time of 20 minutes?

   (f) At what rate (in revolutions per second) must the double helix of DNA unwind to sustain the rate of replication calculated in part (e)? How does this compare with the speed of a 33 rpm phonograph record?

6. **DNA Melting.** Shown in Figure 13-31 are the melting curves for two DNA samples that were thermally denatured under the same conditions.

   (a) What conclusion can you draw concerning the base compositions of the two samples? Explain.

   (b) How might you explain the steeper slope of the melting curve for sample A than for sample B?

   (c) Formamide and urea are agents known to form hydrogen bonds with pyrimidines and purines. What effect, if any, would the inclusion of a small amount of formamide or urea in the incubation mixture have on the melting curves?

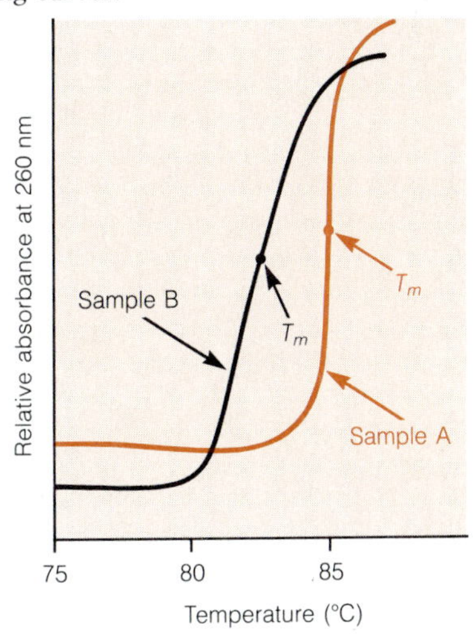

Figure 13-31 Thermal Denaturation Profiles for Two DNA Samples.

7. **Nuclear Structure and Function.** Indicate the implications for nuclear structure or function of each of the following experimental observations.

   (a) Sucrose crosses the nuclear envelope so rapidly that its rate of movement cannot be accurately measured.

   (b) Colloidal gold particles with a diameter of 5.5 nm equilibrate rapidly between the nucleus and cytoplasm when injected into an amoeba, but gold particles with a diameter of 15 nm do not.

   (c) Nuclear pore complexes stain heavily for ribonucleoprotein.

   (d) Many of the proteins of the nuclear envelope appear from electrophoretic analysis to be the same as those found in the endoplasmic reticulum.

   (e) Ribosomal proteins are synthesized in the cytoplasm but are packaged with rRNA into ribosomal subunits in the nucleus.

   (f) Mutants in maize that lack secondary chromosomal constrictions also fail to form nucleoli.

   (g) If nucleoli are irradiated with a microbeam of ultraviolet light, synthesis of ribosomal RNA is inhibited.

   (h) Treatment of nuclei with the nonionic detergent Triton X-100 solubilizes the nuclear envelope but leaves an intact nucleus devoid of inner and outer membranes.

8. **Nuclear Pores.** At one stage in its development, the oocyte of *Xenopus laevis* has a diameter of 0.3 mm and a nuclear surface area of 200,000 μm^2. The pore density is 50 pores per square micrometer, and each pore has a diameter of about 75 nm.

   (a) What is the total number of nuclear pores on the envelope of this oocyte?

   (b) Approximately how far apart are the pores, assuming an even distribution over the membrane?

   (c) What proportion of the total surface area of the nucleus do the pores occupy? Is that a high or low value, compared to most other cell types?

   (d) Can you guess why an oocyte might be so well endowed with nuclear pores?

14

The Cell Cycle, DNA Replication, and Mitosis

A fundamental property of living organisms is their capacity to grow. At the cellular level, this is accomplished by the synthesis of additional molecules of proteins, nucleic acids, carbohydrates, lipids, and other cellular constituents. As a cell grows, the plasma membrane must expand to allow the internal volume of the cell to increase. However, a cell cannot continue to enlarge indefinitely because there is a concomitant decrease in its surface area/volume ratio and hence in its capacity for effective exchange with the environment (recall Figure 4-1). For this reason, cell growth must be accompanied by **cell division,** whereby one cell gives rise to two **daughter cells.**

For single-celled organisms, cell division achieves an increase in the total number of individuals in a population. In multicellular organisms, cell division either increases the number of cells in the organism during growth or replaces cells in adults. In an adult human, for example, about two million stem cells in bone marrow divide every second to maintain a constant number of red cells in the body.

An important feature of the division process is its genetic fidelity. Both daughter cells are faithful genetic duplicates of the parent cell, containing the same (or almost the same) DNA sequences. Clearly, the generation of two identical daughter cells from a single parent cell means that all the genetic information present in the nucleus of the parent cell must be duplicated and carefully parceled out to the daughter cells. The sequence of events by which this is accomplished in eukaryotic cells is called the **cell cycle.**

Overview of the Cell Cycle

The eukaryotic cell cycle involves both nuclear and cytoplasmic events. The cytoplasmic events are usually divided into two overall phases—the **division phase** and a much longer **growth phase,** during which the cell doubles its mass and duplicates all of its contents in preparation for the next round of division (Figure 14-1). The division phase actually consists of two overlapping processes in which the nucleus divides first, followed by division of the cell itself. Nuclear division is called **mitosis** (Figure 14-1, inner circle), and the subsequent division of the cytoplasm into two daughter cells is called **cytokinesis** (outer circle).

Mitosis and cytokinesis commanded most of the attention of early investigators because they are dramatic events that can be studied easily by direct microscopic observation. However, these division events account for a relatively small proportion of the total cell cycle. The typical cell spends most of its time in the growth phase between divisions. The nucleus, in turn, spends most of its time in **interphase,** directing the synthetic activities of the cell and preparing for the next mitotic division. It is in interphase that all of the nuclear contents are duplicated, including the DNA. Interphase is in turn divided into three phases, called G-1, S, and G-2 (Figure 14-1). We will first consider the three stages of interphase and then look at the division event itself.

The S Phase

Since the purpose of mitosis is to generate two daughter nuclei that are identical to the parent nucleus in both amount and kind of genetic information, the parceling out of chromosomes in the division phase of the cell cycle must be offset by a prior replication of chromosomes. Most cellular contents are synthesized continuously during interphase, so that cell mass increases gradually as the cell approaches division. However, DNA synthesis is restricted to a limited portion of interphase, called the **S (synthetic) phase.**

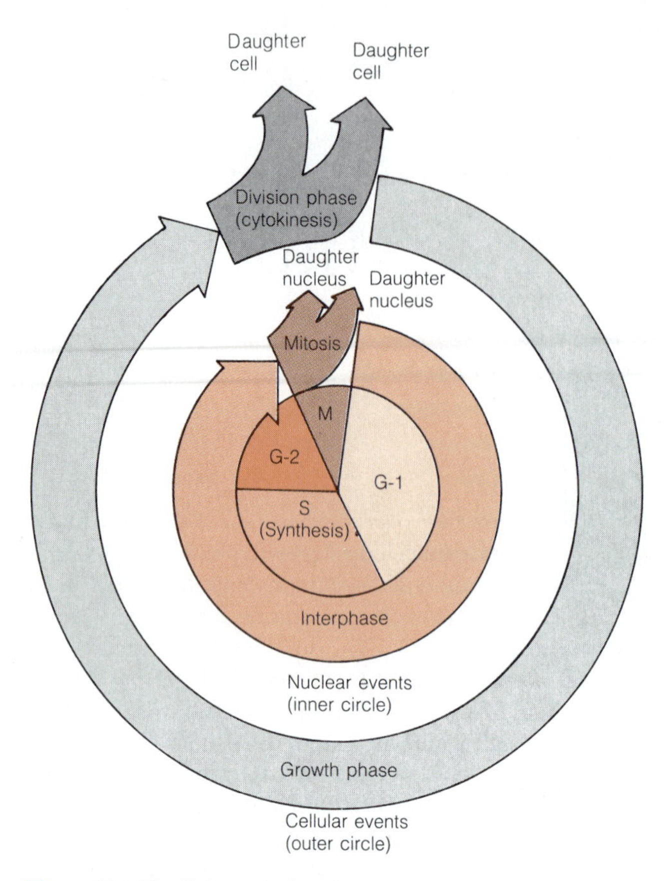

Figure 14-1 The Eukaryotic Cell Cycle. The cycle consists of nuclear events (inner circle) and cellular events (outer circle). The cellular events include a growth phase and a short division phase (cytokinesis). The nuclear events include mitosis (M) and an interphase period divided into a synthetic (S) phase and two "gaps," G-1 and G-2. At the end of each cell cycle, the cell has divided into two daughter cells, each with a nucleus containing the same genetic material as the original cell.

During the S phase, the amount of DNA in the nucleus doubles from what is called the *2C amount,* present in a cell immediately after division, to the *4C amount,* present at the end of the S phase. The doubling of the DNA content of the nucleus reflects the replication of chromosomes during the S phase, with each chromosome giving rise to two **sister chromatids.** Sister chromatids remain attached to each other until they are parceled out to the two daughter nuclei during mitosis.

We can quantify the increase in DNA that occurs during the S phase by using microspectrophotometric methods to determine the total amount of nuclear DNA. The cells are stained with a DNA-specific dye, and the amount of dye associated with a specific nucleus is quantified by measuring the amount of light it absorbs. As Figure 14-2 shows, the amount of DNA doubles from the 2C level to the 4C level during S phase, then returns abruptly to the 2C level during the separation of chromosomes in the division event that follows. This pattern is repeated with each cell cycle.

DNA replication is a highly ordered process. The DNA of each chromosome in the nucleus is organized into a series of replicons that replicate independently of one another, but with a definite sequence. Similarly, DNA synthesis does not start simultaneously on all chromosomes within the nucleus, but again there appears to be an order to its onset. Except for the replacement of damaged DNA through what is called **repair synthesis,** DNA synthesis is restricted to the S phase of the cell cycle. We will examine the process of DNA replication in detail in a later section.

The G-1 Phase

Interphase includes intervals of time both before and after the S period in which little or no DNA synthesis occurs.

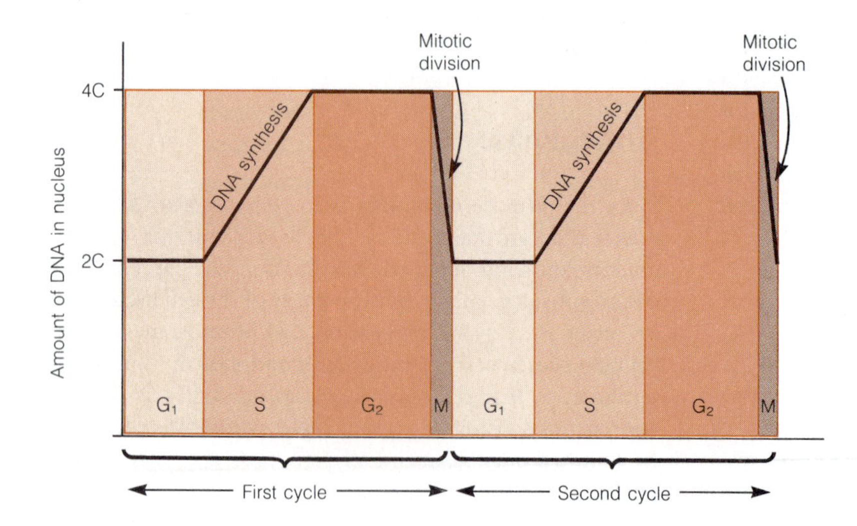

Figure 14-2 Nuclear DNA Content During the Cell Cycle. DNA content doubles from the 2C level to the 4C level during the S phase of the cell cycle. DNA content returns abruptly to the 2C level when chromosomes separate at mitosis (M).

These intervals are called the **G-1** and **G-2 phases**, respectively. The G stands for "gap," because these stages represent gaps or interruptions in DNA synthesis. G-1, the period between the end of the previous division and the onset of chromosomal DNA synthesis, appears to be the time when the "decision" is made as to whether and when the cell is to divide again.

The length of G-1 is quite variable, depending on the cell type. In the same organism, some cells may spend only minutes or hours in G-1, whereas others spend weeks, months, or years. Cells that are destined never to divide again are permanently arrested in G-1. Most of the nerve cells in your body are in this state.

The G-2 Phase

The postsynthetic gap phase begins when the 4C amount of DNA has been reached and ends with the first visible sign that mitosis is beginning. There is, as yet, little information on the cellular or nuclear processes that occur specifically in this stage, nor is it clear what determines the length of the G-2 stage. In general, G-2 is shorter than G-1 and is usually much more uniform in duration among the different cell types of an organism. However, G-2 arrest is known to occur in some cells, leaving the cell with its DNA fully replicated but without entry into mitosis.

Mitosis

Mitosis (M) is nuclear division, involving separation of duplicated chromosomes into two clusters that are genetically identical to each other and to the parent nucleus. Figure 14-3 depicts the mitotic process as it occurs in cells of higher animals. Mitosis begins with the progressive condensation (thickening and coiling) of the duplicated interphase chromosomes, accompanied by the migration of structures called centrioles to opposite ends of the cell (Figure 14-3a, b). Microtubules then become organized into the fibers of the **mitotic spindle**, which is responsible for directing chromosome movement.

The nuclear envelope fragments, and chromosomes are drawn into position at the center of the cell (c). Sister chromatids then move toward opposite ends of the cell (d). Chromosomal movement continues until the two sets of chromosomes are completely separated (e), by which point cell division is usually already under way. The nuclear envelope then re-forms, delimiting two daughter nuclei of identical genetic makeup (f). As mitosis is completed, the chromosomes decondense and revert to the extended form characteristic of interphase chromatin (g).

Cytokinesis

In most cases, nuclear division is followed quickly by cell division. In fact, division of the cytoplasm usually begins before mitosis is complete. In animal cells and protists, this occurs by the formation of a cleavage furrow in the plasma membrane, which gradually deepens until the daughter cells are completely separated (Figure 14-3f, g). In higher plants, a primary cell wall must also be laid down between the two daughter cells. Division of the cytoplasm is at best only approximately equal and in some cases is distinctly unequal.

Length of the Cell Cycle

The cells in a multicellular organism divide at varying rates, but most studies of the cell cycle have been done with cells in culture, where the length of the cycle is similar for different cell types. We can easily determine the length of the cell cycle (also called the **generation time**) for cultured cells either by counting the cells at intervals under a microscope or by monitoring the cell mass spectrophotometrically. For mammalian cells in culture, the cycle usually takes about 18–24 hours (Figure 14-4).

Once the total length of the cycle is known, the length of specific phases can also be determined. The S and mitosis (M) phases are easiest to measure. To determine the length of the S phase, we expose cells to a radioactively labeled DNA precursor (^{3}H-thymidine, usually) for a short period of time and then examine them by autoradiography. The relative number of cells with exposed silver grains over their nuclei is a measure of the fraction of cells that were somewhere in S phase when the radioactive compound was available. When this fraction is multiplied by the total length of the cell cycle, the result is an estimate of the average length of the S phase. For mammalian cells in culture, this fraction is often around 0.33, and the S phase is therefore about 6–8 hours in length.

Similarly, the length of the M phase can be estimated by multiplying the generation time by the fraction of the cells that are actually in mitosis at any point in time. This fraction is called the **mitotic index.** The mitotic index for cultured cells is often about 0.03–0.05. G-1 and G-2 must be determined by less direct methods, because there is no way to label or identify cells in these phases specifically.

The various phases of the cell cycle have been measured in many types of cultured cells. For typical mammalian cells, G-1 lasts about 8–10 hours, S is completed in about 6–8 hours, G-2 takes about 4–6 hours, and mitosis requires an hour or less—usually about 30–45 minutes (Figure 14-4).

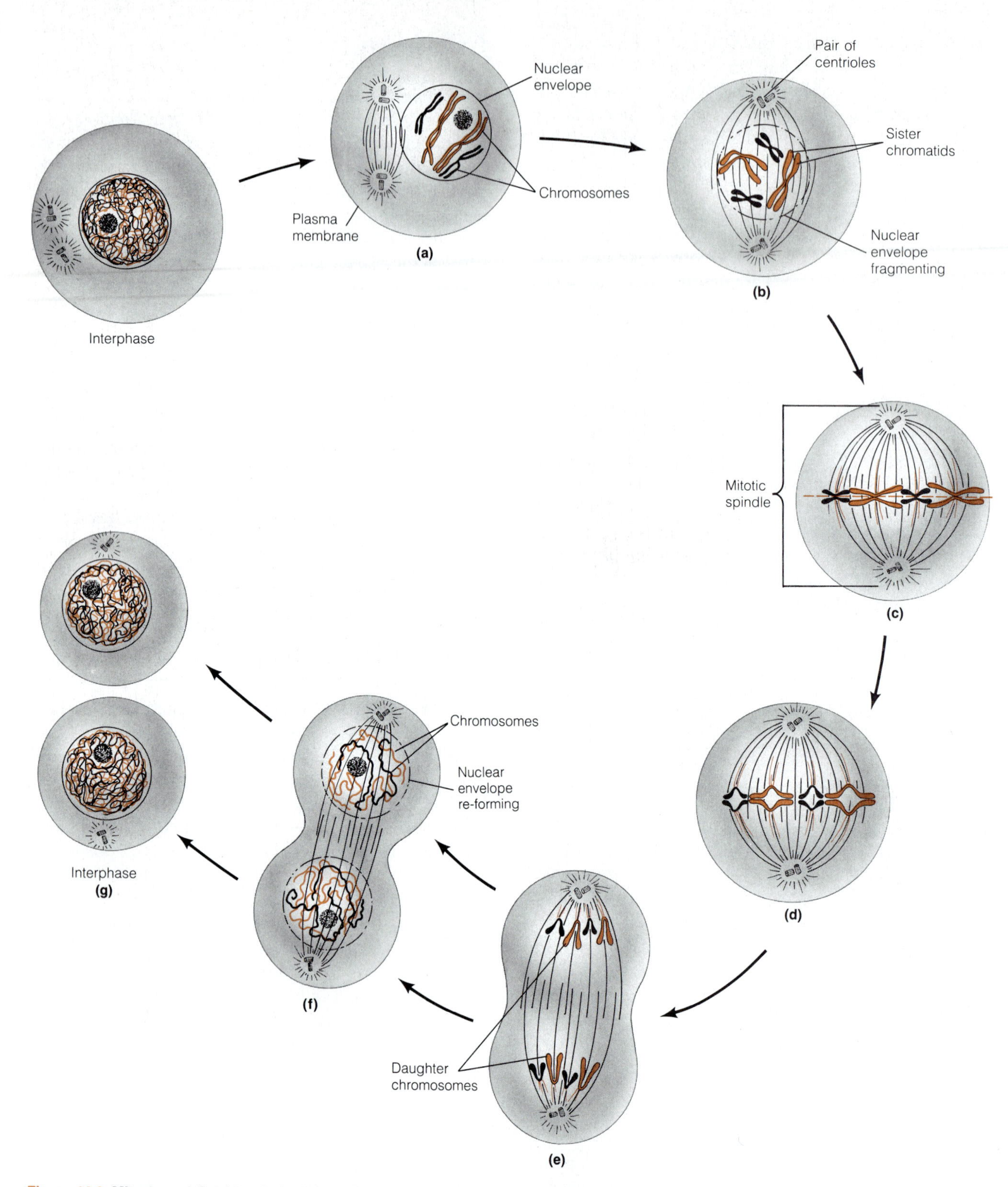

Figure 14-3 Mitosis and Cytokinesis in Animal Cells. (a) As a cell moves from interphase into division, the chromosomes become visible and gradually condense. (b) As they do, the centrioles migrate to opposite ends of the cell. (c) The nuclear envelope fragments, and condensed chromosomes are drawn into position at the center of the cell. (d) Sister chromatids then move to opposite ends of the cell. (e) As the two sets of chromosomes separate, (f) the cell divides, giving rise to (g) two daughter cells of identical genetic composition, with their chromosomes again in the dispersed interphase state.

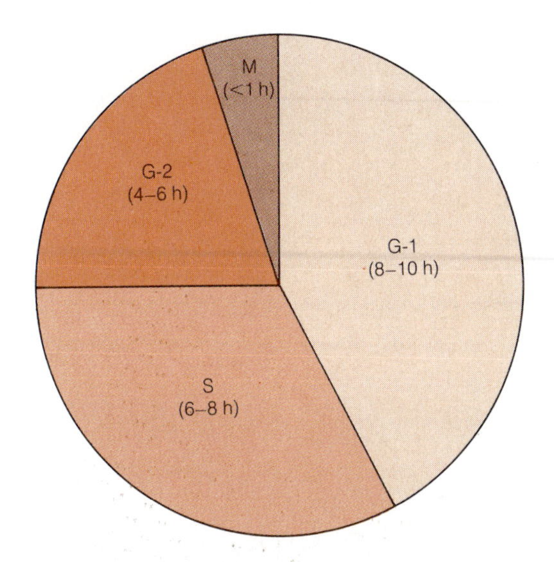

Figure 14-4 Cell Cycle Length. The length of the cell cycle varies greatly for eukaryotic cells, but cultured mammalian cells often have a generation time of 18–24 hours, with about 8–10 hours spent in G-1, 6–8 hours in S, 4–6 hours in G-2, and less than an hour in mitosis (M).

Now that we have given an overview of the entire cell cycle, we will look at two phases in more detail. First, we will examine the main feature of the S (synthetic) phase—DNA replication. Later in the chapter we will discuss mitosis in more detail.

DNA Replication

One of the most significant features of the double-helical model of DNA is that it immediately suggests a mechanism for **DNA replication.** In fact, a month after Watson and Crick published their now-classic paper postulating a duplex structure for DNA, they followed it with an equally important paper suggesting how such a base-paired structure might duplicate itself. Here, in their own words, is the basis of that suggestion (Watson and Crick 1953, p. 966):

> Now our model for deoxyribonucleic acid is, in effect, a pair of templates, each of which is complementary to the other. We imagine that prior to duplication the hydrogen bonds are broken, and the two chains unwind and separate. Each chain then acts as a template for the formation onto itself of a new companion chain, so that eventually we shall have two pairs of chains, where we only had one before. Moreover, the sequence of the pairs of bases will have been duplicated exactly.

The model they proposed for DNA replication is reproduced in Figure 14-5. The essence of their suggestion is that one of the two strands of every daughter DNA molecule is derived from the parent molecule, whereas the other strand is newly synthesized. This is called **semiconservative replication,** since half of the parent molecule is retained by each daughter molecule.

Proof of the Semiconservative Model

Within five years of its publication, the Watson-Crick model for DNA replication was tested and proved correct by Matthew Meselson and Franklin Stahl. The ingenuity of their contribution lay in the method they devised, in collaboration with Jerome Vinograd, for distinguishing semiconservative replication from other possibilities. Their approach was to grow prokaryotic cells (*Escherichia coli*) in a medium containing ^{15}N, a heavy (but nonradioactive) isotope of nitrogen. Because all the DNA synthesized by *E. coli* cells under these conditions will have the ^{15}N isotope instead of ^{14}N, it will have a higher density than ordinary DNA. After many generations of growth on the ^{15}N-containing medium, the cells were transferred to a medium containing only ^{14}N, and the density of the DNA was examined after one or several successive cycles of replication.

The density of DNA molecules was assessed by **equilibrium density centrifugation,** a technique that we encountered earlier in its application to the separation of organelles (recall Figure 9-20). Briefly, the technique allows organelles or macromolecules to be separated on the basis of differences in buoyant density by centrifugation in an appropriate solution that increases in density from the top of the tube to the bottom. In response to the centrifugal force, the organelles or molecules migrate "down" the tube (actually, they move *outward,* away from the axis of rotation) until they reach a density equal to their own. They then remain at this equilibrium density and can be recovered as a "band" at that position in the tube after centrifugation.

A popular solute for equilibrium density centrifugation of macromolecules such as DNA is cesium chloride (CsCl), because it is highly soluble, very dense, and chemically inert toward organic molecules. The DNA to be analyzed is mixed with an initially homogeneous solution of cesium chloride of an appropriate density. The solution is then centrifuged at high speed for a relatively long time (several days at 50,000 rpm, for example).

As a gradient of cesium chloride establishes itself during the centrifugation, the DNA molecules move "up" or "down" the gradient to reach their equilibrium density position. The exact position depends on the average base composition of the DNA molecules and on whether the DNA is single-stranded or double-stranded, but most DNA molecules band in the range 1.65–1.75 g/cm^3.

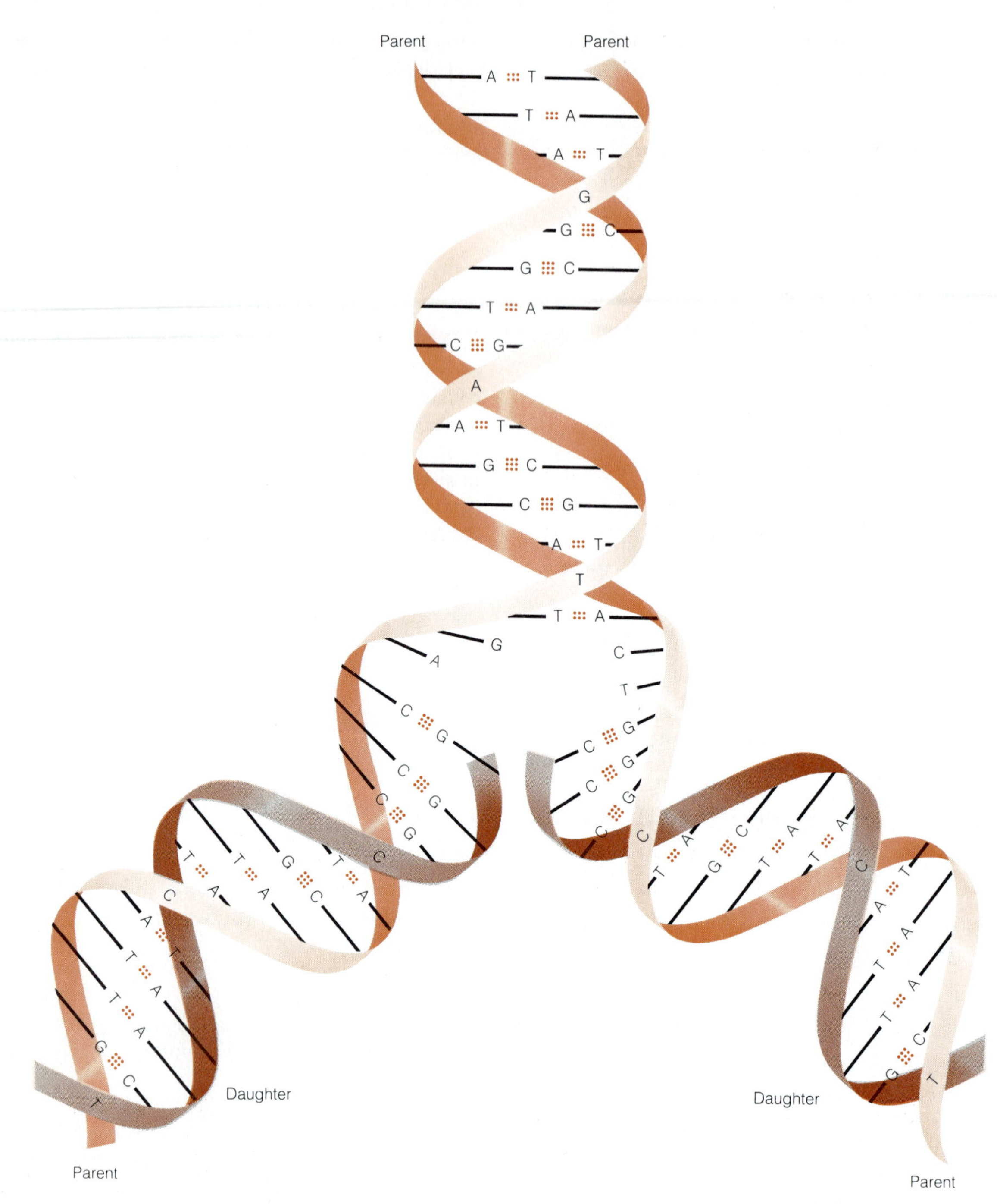

Figure 14-5 The Watson-Crick Model of DNA Replication. In 1953, Watson and Crick proposed that the DNA double helix replicates semiconservatively, using this model to illustrate the principle.

Meselson and Stahl were able to separate the DNA of ^{15}N-grown cells from that of ^{14}N-grown cells because the difference in density between "heavy" (^{15}N-containing) and "light" (^{14}N-containing) DNA is sufficiently great to allow the two kinds of DNA to be resolved as separate bands in cesium chloride (Figure 14-6). This density difference made possible the critical experiment shown in Figure 14-7, where equilibrium density centrifugation was used to determine the banding pattern of DNA from cells that were first grown in ^{15}N (Figure 14-7a) and then transferred to ^{14}N for one or more cycles of replication.

After one cell cycle in the ^{14}N medium, a single band of DNA was observed in the cesium chloride gradient, with a density exactly halfway between that of ^{15}N DNA and that of ^{14}N DNA (Figure 14-7b). Because they saw no band at the density of "heavy" DNA, Meselson and Stahl concluded that the original double-stranded parent DNA was not preserved as such in the replication process. Similarly, the absence of a band at the density of "light" DNA indicated that no daughter DNA molecules consisted exclusively of newly synthesized nucleotides. Instead, it appeared that a part of every daughter DNA molecule was newly synthesized, while another part was derived from the parent molecule. In fact, a density midway between those of ^{14}N DNA and ^{15}N DNA meant that these hybrid DNA molecules were one-half parent and one-half newly synthesized, just as predicted by the Watson-Crick model of semiconservative replication (Figure 14-5).

The data from cells allowed to grow on ^{14}N for several generations provided further confirmation. After two cycles of DNA replication, for example, Meselson and Stahl saw two equal bands, one at the hybrid density of the previous cycle and one at the density of ^{14}N DNA (Figure 14-7c). As the figure illustrates, this is also consistent with a semiconservative mode of replication.

From these findings, Meselson and Stahl (1958, p. 682) concluded "that the nitrogen of a DNA molecule is divided equally between two physically continuous subunits; that, following duplication, each daughter molecule receives one of these; and that the subunits are conserved through many duplications." Further experimentation was then required to prove that the "physically continuous subunits" into

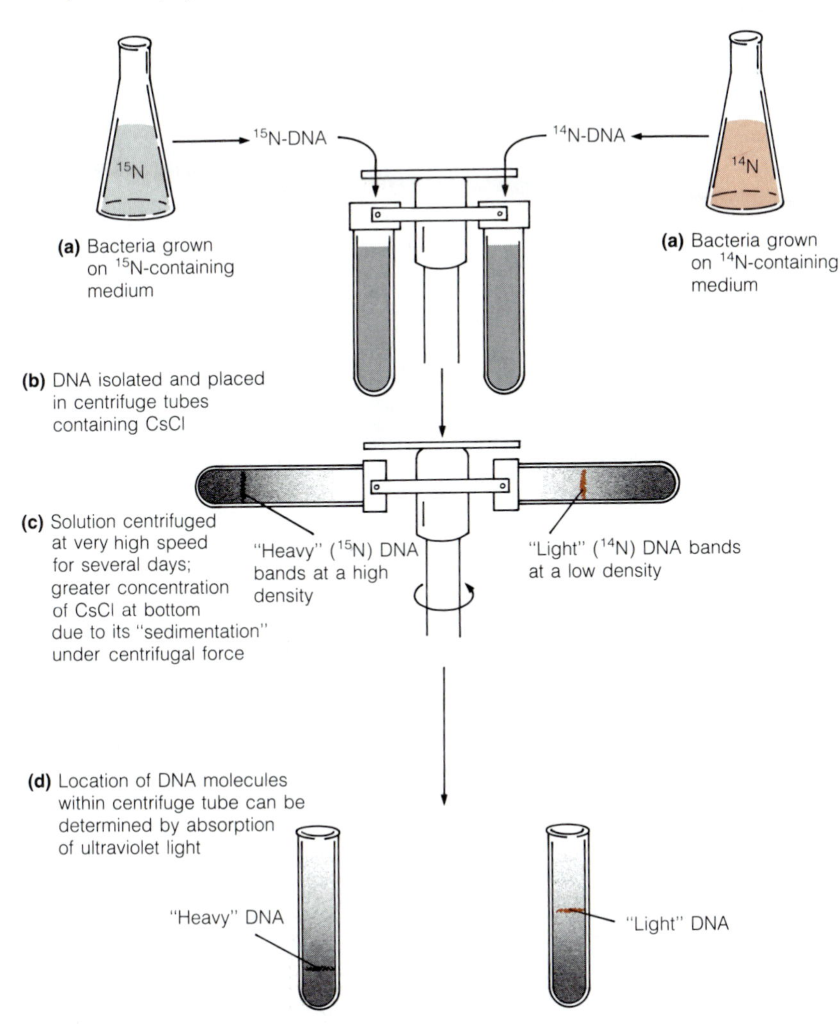

Figure 14-6 Equilibrium Density Centrifugation in DNA Analysis. Equilibrium density centrifugation can be used to distinguish between heavy (^{15}N-containing) and light (^{14}N-containing) DNA. (a) If bacterial cells are grown on either ^{15}N or ^{14}N medium for many generations, we can distinguish the DNA from the two cultures by (b) placing it in tubes containing cesium chloride at the appropriate concentration and (c) centrifuging the tubes at a very high speed until the DNA reaches its equilibrium, or buoyant, density. Heavy DNA bands at a higher density than does light DNA because of the presence of the ^{15}N atoms in its structure. (d) After centrifugation, the DNA bands can be visualized by their absorption of ultraviolet light. The density difference (about 1%) is sufficient not only to resolve the two bands, but also to detect hybrid DNA molecules of an intermediate density, as in Figure 14-7c.

which DNA is partitioned are indeed separate DNA strands. In an extension of the cesium chloride technique, $^{14}N/^{15}N$ hybrid DNA was separated into single strands, and these were shown to be either all "heavy" or all "light," demonstrating that the semiconservative partitioning of nitrogen reported by Meselson and Stahl extends over long stretches of DNA. In the end, Watson and Crick were right—DNA is replicated semiconservatively.

The Universality of Semiconservative Replication

Researchers have demonstrated semiconservative replication with eukaryotic DNA as well, but using a different experimental approach. In a classic study published a year before Meselson and Stahl's report, J. H. Taylor, P. Woods, and W. Hughes had already shown that whole chromosomes of the bean *Vicia faba* are replicated semiconservatively over their entire length. Their approach was to grow bean seedlings in a medium containing thymidine that was labeled with tritium (^3H), a radioactive isotope of hydrogen. The seedlings were exposed to the ^3H-thymidine for a long enough period to ensure that most of the cells had replicated their DNA and undergone one round of cell division. (Note that at this point only one DNA strand of each double helix should have been labeled, since this is equivalent to the first generation of bacteria in the Meselson-Stahl experiment.)

The seedlings were then transferred to a nonradioactive medium containing *colchicine,* a plant alkaloid that binds to tubulin and prevents its polymerization into microtubules. Since the spindle fibers required to draw chromosomes apart during mitosis consist of microtubules, colchicine inhibits mitosis by interfering with microtubule-mediated chromosome movements. In the presence of colchicine, chromosomes replicate but cannot separate, such that sister chromatids remain paired. When such chromosomes are subjected to autoradiography, the distribution of silver grains over paired sister chromatids can be compared directly, as shown in Figure 14-8.

Mitotic chromosomes fixed and examined immediately after DNA replication in ^3H-thymidine were uniformly labeled in both chromatids (Figure 14-8a). After one round of cell division in the unlabeled medium, however, each chromosome pair was labeled in only one of its two chromatids at any point along its length (Figure 14-8b). This finding established that each chromatid must consist of two subunits, presumably single strands of DNA, and that during the formation of daughter chromosomes, one of the strands is conserved in each of the daughter chromatids, whereas the other strand in each case is newly synthesized.

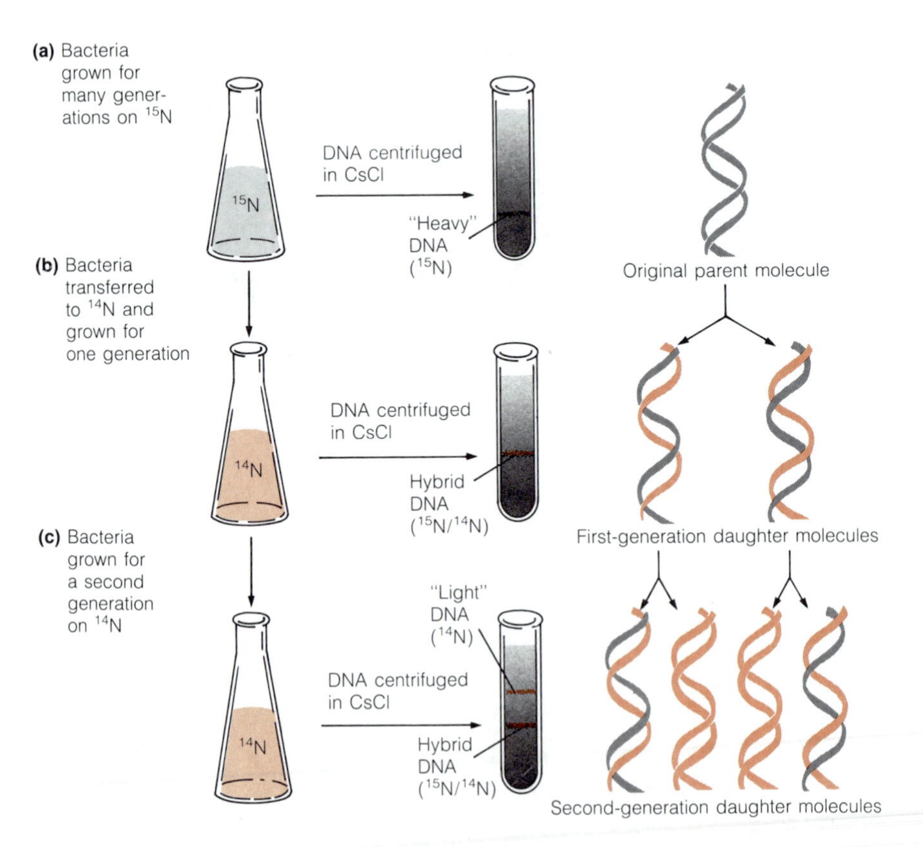

Figure 14-7 Semiconservative Replication of Density-Labeled DNA. Meselson and Stahl (a) grew bacteria for many generations on an ^{15}N-containing medium and then transferred the cells to an ^{14}N-containing medium for (b) one or (c) two further cycles of replication. In each case, DNA was extracted from the cells and centrifuged to equilibrium on cesium chloride as described in Figure 14-6. Bacterial cultures appear on the left, cesium chloride gradients in the middle, and schematic illustrations of the DNA molecules on the right. Dark gray strands contain ^{15}N, whereas color strands are synthesized with ^{14}N.

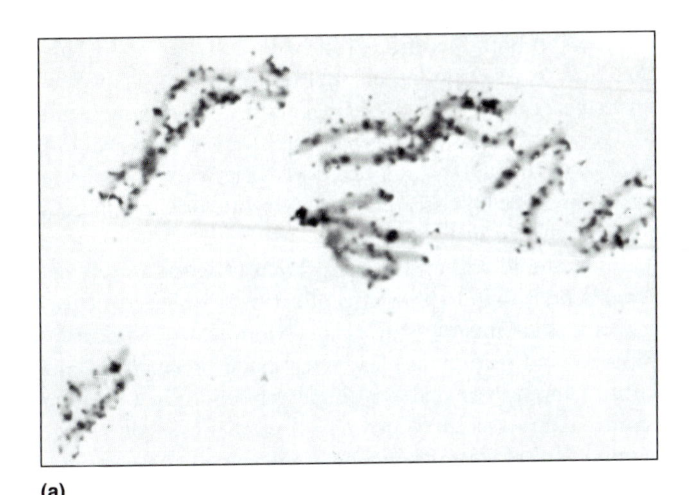

(a)

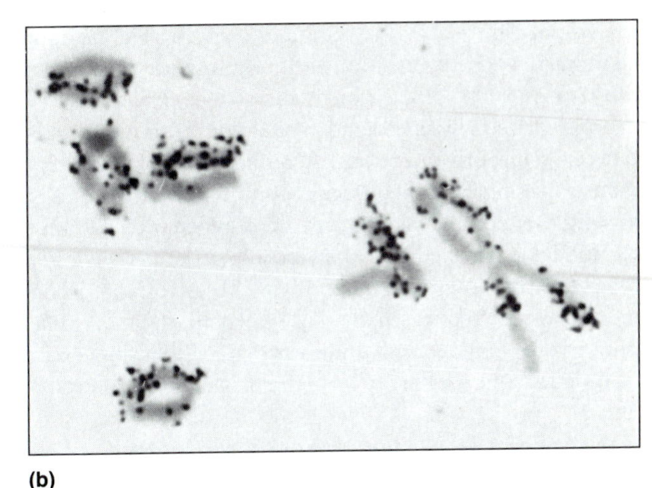

(b)

Figure 14-8 Autoradiographic Evidence for Semiconservative DNA Replication. Seedlings of the bean *Vicia faba* were incubated in ^{3}H-thymidine, and mitotic chromosomes of root-tip cells were examined autoradiographically, either after (a) one round of cell division or (b) an additional cycle of replication in a nonradioactive medium. The drug colchicine was used to prevent chromosome separation, so that the distribution of silver grains could be examined over paired sister chromatids. Notice that in (a), silver grains are frequently found over both chromatids at a given point along their length, but in (b), only one of the two sister chromatids is labeled at any point.

Since the work of Meselson and Stahl and of Taylor and his colleagues, similar techniques have been used to extend the conclusions to the replication of DNA in a variety of other organisms. These studies have shown that semiconservative DNA replication is universal. Even viruses that use single-stranded DNA for their genome synthesize a complementary second strand to generate the replicative form of the virus.

The Mechanism of DNA Replication

Conceptually, the overall process of DNA replication is as simple as the Watson-Crick model implied. The same cannot be said of the actual mechanism, however. Replication has been shown to be a complex process requiring a number of enzymes. Furthermore, RNA is used as a primer. Replication involves continuous synthesis in one direction along one of the two strands being copied but discontinuous synthesis in the opposite direction along the other strand. We will first look at some of the early evidence bearing on the replicative process and then focus in detail on the mechanism itself.

The first direct visualization of DNA replication was provided by John Cairns, who grew *E. coli* cells for varying lengths of time in a medium containing the DNA precursor ^{3}H-thymidine and then used autoradiography to look for DNA molecules caught in the act of replication. One such molecule is seen in the autoradiograph of Figure 14-9. The forklike structures represent the sites at which the DNA duplex was being replicated. These **replication forks** are now known to result from a replication process that begins at a specific point and moves along the DNA strand, unwinding the helix and copying both strands as it goes.

For a circular DNA molecule such as that of *E. coli,* replication is initiated at a discrete point on the circle called

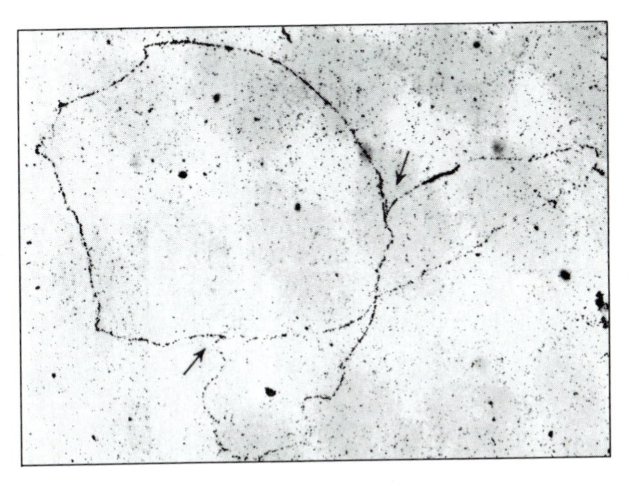

Figure 14-9 Autoradiographic Visualization of DNA Replication. This autoradiograph shows an *E. coli* DNA molecule caught in the act of replication. The bacterium from which the molecule came had been grown in a culture containing ^{3}H-thymidine, thereby ensuring that the DNA molecule could be visualized by autoradiography. The arrows point to the replication forks.

the **origin.** Although Cairns originally interpreted his data as indicating one replication fork, we now know that in most cases DNA replication is **bidirectional:** Two replication forks are formed, and replication proceeds simultaneously in both directions. A **replication bubble** forms that grows in size as replication continues bidirectionally around the circle. This is called **theta replication** because it produces intermediates that look like the Greek letter theta (θ), as shown in Figure 14-10. Most circular DNA molecules are replicated in this way, although in some prokaryotes replication is unidirectional, with a single replication fork moving away from the origin in one direction only.

The linear DNA molecules of some viruses and of all eukaryotic chromosomes are also replicated bidirectionally from initiation points. Small viral DNAs have a single such point, but the large DNA molecules present in eukaryotic chromosomes may have hundreds or even thousands of initiation points.

Figure 14-11 depicts the replication process for a small length of a eukaryotic DNA molecule. Replication bubbles form at multiple initiation sites (a) and grow as replication proceeds in both directions (b). Eventually, adjacent bubbles fuse to generate larger bubbles (c), and Y-shaped intermediates form as a bubble reaches the end of a molecule (d). Finally, the strands separate, forming two separate double-stranded DNA molecules, each with one parent strand and one new strand (e). Eukaryotic replication bubbles are shown in Figure 14-12.

The individual unit of replication is called a **replicon.** A replicon is therefore a self-replicating segment of a chromosome that includes an origin from which replication proceeds bidirectionally. For viruses and prokaryotes, the entire genome is usually a single replicon. On the other hand, eukaryotic chromosomes always consist of a large number of replicons. Replicons in eukaryotic chromosomes are usually between 15 and 100 μm (about 50,000 to 300,000 nucleotide pairs) in length. Not all replicons begin replicating at the same time during the S phase of the cell cycle. Some always replicate early and others always replicate late. Neither the mechanism nor the significance of this is yet understood, but it is clear that all the replicons must be replicated before the daughter chromatids can separate.

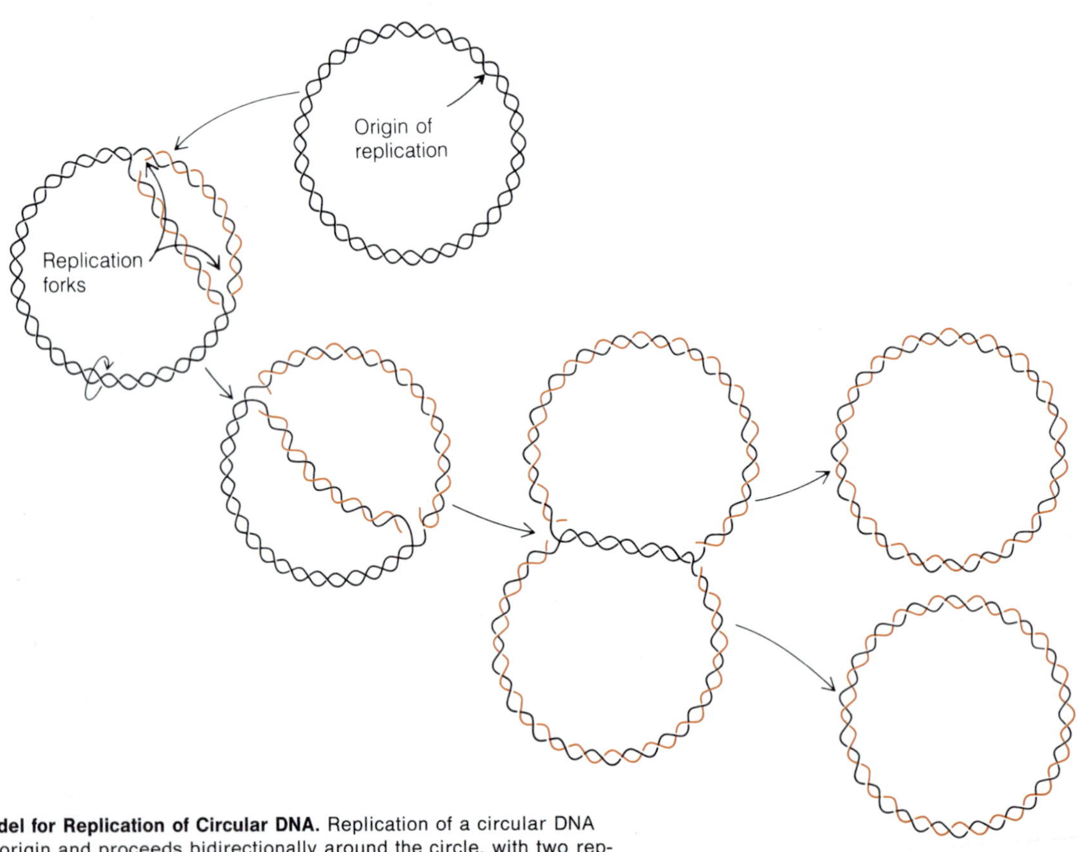

Figure 14-10 Theta Model for Replication of Circular DNA. Replication of a circular DNA molecule begins at an origin and proceeds bidirectionally around the circle, with two replication forks moving in opposite directions. The new strands are shown in color. This generates intermediates that resemble the Greek letter theta (θ), from which the model derives its name.

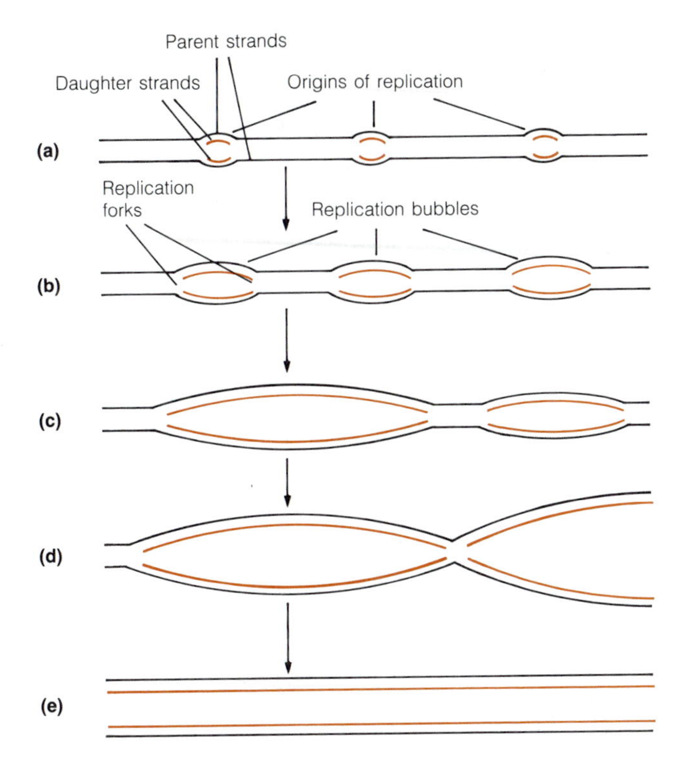

Figure 14-11 Multiple Origins of Replication for Eukaryotic DNA. Replication of the linear DNA molecules of eukaryotic chromosomes is initiated at numerous points of origin along the chromosome, with a timing that is specific for each origin. (a) Replication bubbles form at sites of initiation. (b) The bubbles grow in size as the replication forks move along the chromosome in both directions from each origin. (c) Eventually, individual bubbles meet and fuse. (d) Y-shaped structures form as a replication fork reaches the end of a chromosome. (e) When all bubbles meet, replication is complete and the two daughter molecules separate.

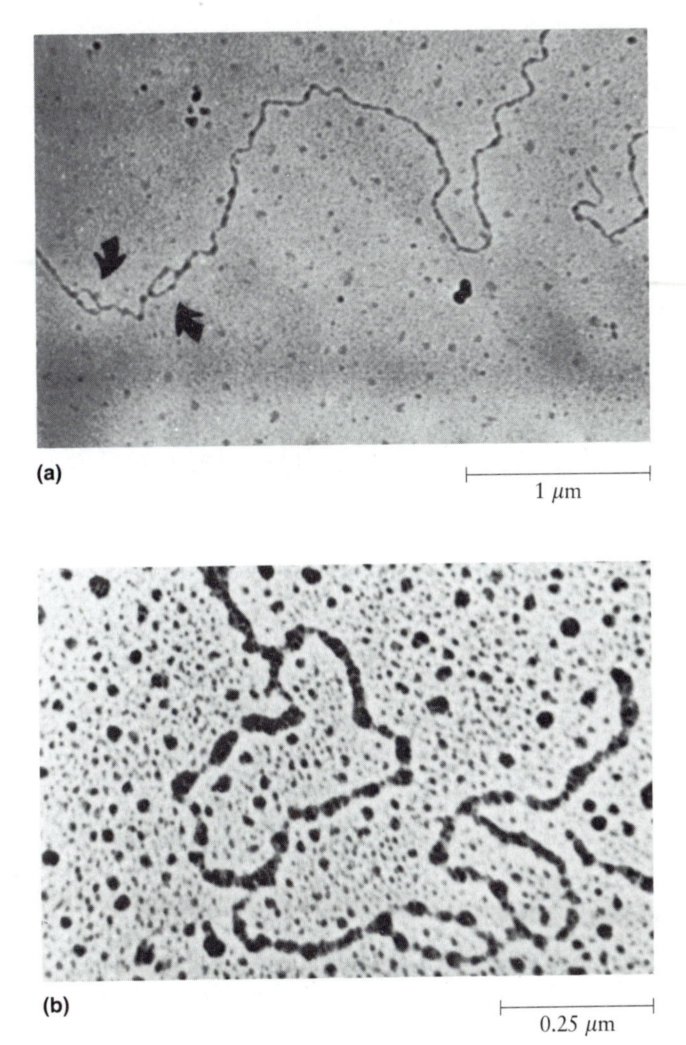

Figure 14-12 Replication Bubbles of Eukaryotic DNA. (a) Two replication bubbles (arrows) can be seen in replicating DNA from cultured Chinese hamster cells. (b) These replication bubbles are from the same cells but are shown at higher magnification. (TEMs)

DNA Polymerases

Enzymes capable of adding successive nucleotides to a growing DNA strand are called **DNA polymerases.** All known DNA polymerases require a template, all use deoxynucleoside triphosphates as their substrates, and all add nucleotides to the 3′-hydroxyl end of the growing chain. Each successive nucleotide is added to the growing chain by a phosphoester bond between the phosphate group on its 5′ carbon and the hydroxyl group on the 3′ carbon of the nucleotide added in the previous step (Figure 14-13). Chain elongation therefore always occurs at the 3′ end of a DNA strand, such that the strand grows in the 5′ → 3′ direction.

Multiple DNA polymerases have been found in both prokaryotic and eukaryotic cells. In *E. coli,* three DNA polymerases have been described, but only DNA polymerases I and III have been studied in any detail. **DNA polymerase I** was discovered first (by Arthur Kornberg and his colleagues, in 1956) and is the most abundant DNA polymerase in *E. coli.* However, the main replication enzyme in *E. coli* cells is **DNA polymerase III.** This enzyme was first discovered in mutant cells that had lost most (all but 2%) of their polymerase I activity, but with no noticeable impairment in their ability to grow. DNA polymerase III is a complicated enzyme. The complete enzyme, called the **holoenzyme,** consists of at least seven different polypeptide subunits.

DNA polymerase I, on the other hand, functions primarily as an editing and repair enzyme. For this function, DNA polymerase I is equipped with an exonuclease activity that functions in the 3′ → 5′ direction. This activity allows the enzyme to remove a nucleotide from the 3′ end of the strand in what is effectively a reversal of the polymerization

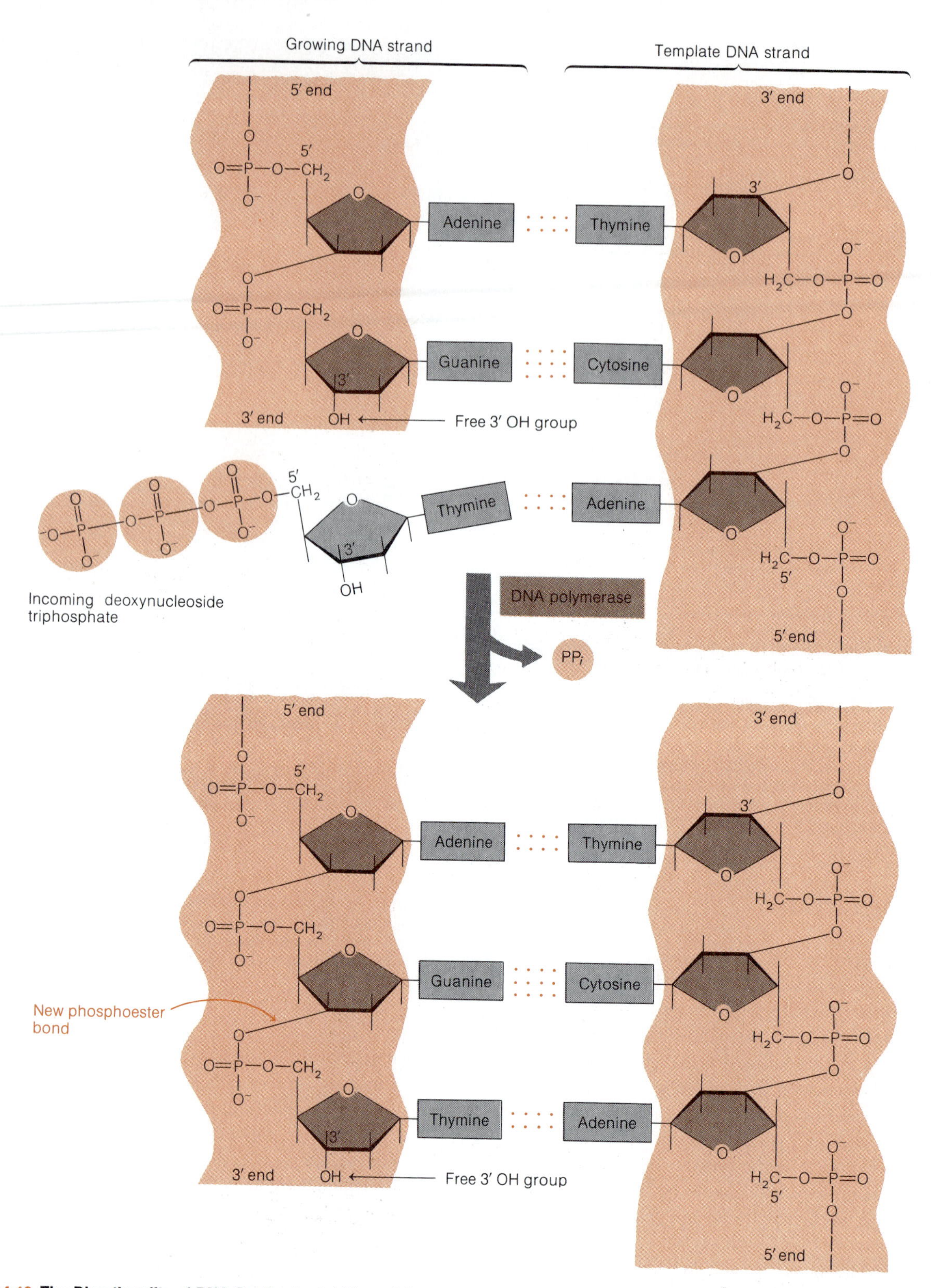

Figure 14-13 The Directionality of DNA Synthesis. Addition of the next nucleotide to a growing DNA strand is catalyzed by the enzyme DNA polymerase and always occurs at the 3′ end of the strand. A phosphoester bond is formed between the 3′ OH of the terminal nucleotide and the 5′ phosphate of the incoming deoxynucleoside triphosphate, extending the growing chain by one nucleotide, liberating pyrophosphate (PP_i), and leaving the 3′ end of the strand with a free hydroxyl group to accept the next nucleotide.

process. Coupled to the $5' \rightarrow 3'$ polymerase activity, the $3' \rightarrow 5'$ exonuclease activity gives DNA polymerase I the capacity to excise an incorrect nucleotide and insert the correct nucleotide in its place. The enzyme therefore has an editing function that enhances the fidelity of the replication process. DNA polymerase I also has a $5' \rightarrow 3'$ exonuclease activity that seems to be part of a repair process for removing and replacing damaged bases in DNA. In addition, the $5' \rightarrow 3'$ exonuclease activity is involved in removing the small pieces of RNA that are needed as primers in DNA replication, as we will see shortly.

In eukaryotic cells, three classes of DNA polymerases have also been described. Polymerase α is found in the nucleus only, polymerase β occurs in both the nucleus and the cytoplasm, and polymerase γ is localized in the mitochondria and is most likely involved only in the replication of mitochondrial DNA. DNA polymerase α is thought to be responsible for most replicative synthesis, since it fluctuates in activity during the cell cycle. Moreover, the drug *aphidocolin,* which stops DNA synthesis in growing cells, has been shown to inhibit polymerase α but not β or γ, and mutant cells that can grow in the presence of aphidocolin have an α polymerase that is resistant to the drug.

Discontinuous Synthesis

When it became clear that DNA replication always involves the addition of nucleotides to the 3′ end of each growing nucleotide chain, a conceptual problem became evident. Given the antiparallel orientation of the two strands in a DNA molecule, continuous nucleotide addition along both strands of a replication fork would require synthesis in the $5' \rightarrow 3'$ direction on one strand and in the $3' \rightarrow 5'$ direction on the other strand. But all known DNA polymerases function only in the $5' \rightarrow 3'$ direction.

The dilemma was resolved when it was realized that synthesis at each replication fork is *continuous* in the direction of fork movement for one strand, but *discontinuous* in the opposite direction for the other strand (Figure 14-14). The two parent strands can therefore be distinguished based on their mode of replication. The **leading strand** is the one to which nucleotides can be added continuously, because it is growing in the $5' \rightarrow 3'$ direction as the replication fork advances. The **lagging strand,** on the other hand, grows by synthesis of short pieces in the reverse direction, followed eventually by joining of the pieces.

The short pieces that serve as replication intermediates for the lagging strand were first described by Reiji Okazaki and his colleagues and are therefore called **Okazaki fragments.** Okazaki fragments are about 1000 to 2000 nucleotides long in viral and bacterial systems, but only about one-tenth this length in animal cells. Okazaki fragments are joined together by an enzyme called **DNA ligase,** which catalyzes the formation of phosphoester bonds between the nucleotides at the 3′ and 5′ ends of adjacent fragments.

The Role of RNA in Priming DNA Replication

A surprising feature of the Okazaki fragments came to light soon after their discovery. When first synthesized, each fragment has a short piece of RNA (four to ten nucleotides) at its 5′ end. In fact, the little piece of RNA is synthesized first and serves as a **primer** of DNA synthesis—a short stretch of oligonucleotides onto which deoxyribonucleotides are added. These RNA primers are synthesized

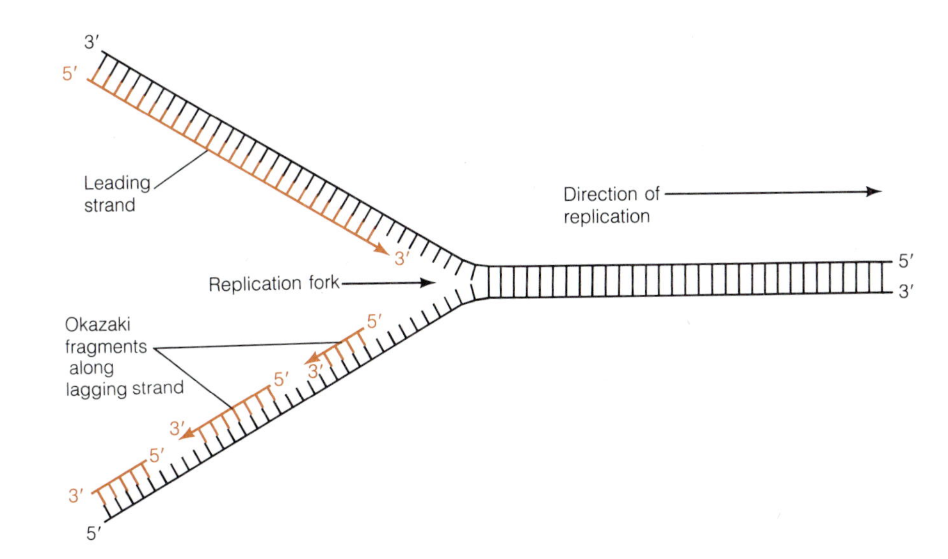

Figure 14-14 Direction of DNA Synthesis at a Replication Fork. Since DNA polymerases function only in the $5' \rightarrow 3'$ direction, synthesis at each replication fork is continuous in the direction of fork movement for the leading strand, but discontinuous in the opposite direction for the lagging strand. The intermediates in discontinuous synthesis, called Okazaki fragments, are synthesized as short segments and joined later by the enzyme DNA ligase. Parent DNA is shown as black lines, newly synthesized DNA as color lines.

by a DNA-dependent RNA polymerase called **primase,** as Figure 14-15a illustrates. Primase is a specific RNA polymerase that is involved only in the process of DNA replication.

Primases vary greatly in structure and specificity. Some primases find and transcribe a specific sequence of four or so nucleotide pairs. Others are not sequence-specific but simply copy a fixed stretch of template—usually about ten nucleotide pairs. In *E. coli,* primase is accompanied by a complex of proteins called the **primosome.** The primosome complex consists of at least six proteins, designated as i, n, n′, n″, dnaC, and dnaB. The n proteins are **primer recognition factors** that cause priming at specific sites.

Once the RNA primer is available, DNA synthesis can proceed, with DNA polymerase III adding successive deoxyribonucleotides to the 3′ end of the primer (Figure 14-15b). Unlike primase and other RNA polymerases, all known DNA polymerases require a primer when assayed in vitro. In the absence of the RNA primer, DNA polymerase III is inactive and DNA replication does not take place. When this finding was first reported, it was difficult to imagine how DNA synthesis could ever be initiated by such enzymes, until it became clear that the synthesis of short RNA primers is an integral part of DNA replication in vivo for both the leading strand and the lagging strand.

After serving its priming function, an RNA primer does not remain attached to the DNA fragment that has been added to it but is removed by the 5′ → 3′ exonuclease activity of DNA polymerase I (Figure 14-15c). This enzyme apparently replaces the ribonucleotides of the RNA primer with deoxyribonucleotides, thereby filling the gap with DNA as it goes. Adjacent fragments can then be linked together covalently by DNA ligase.

Unwinding the DNA

One aspect of DNA replication remains to be considered—the unwinding of the double helix to expose the single strands to the enzymes responsible for copying them. This is the most complicated and least understood part of the process, involving several proteins with very distinct functions. In prokaryotes, these include a **helicase,** a **gyrase,** and **single-strand binding (SSB) protein** also called **helix destabilizing protein.** These proteins are shown in Figure 14-16a and b.

The helicase, also called **unwinding protein** or "**unwindase,**" uses the energy of ATP to cause unwinding of the parent helix as the enzyme moves down the DNA duplex in the 3′ → 5′ direction. Two ATP molecules are

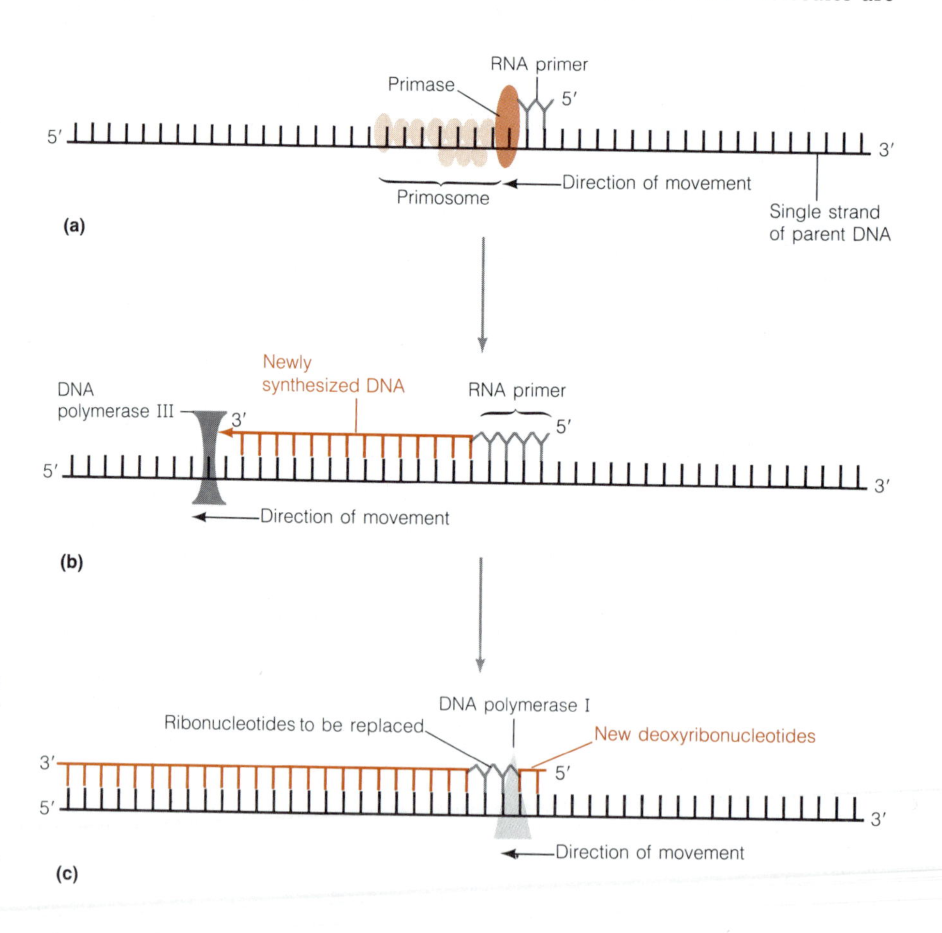

Figure 14-15 The Role of RNA Primer in DNA Replication. DNA synthesis is always initiated with a short RNA primer. (a) The primer is synthesized by primase, an RNA polymerase that uses a single strand of DNA as its template. In *E. coli,* primase is aided by the primosome complex. (b) Once the short stretch of RNA is available, DNA polymerase III uses it as a primer to initiate DNA synthesis, which proceeds in the 5′ → 3′ direction. (c) The RNA primer is eventually removed by the 5′ → 3′ exonuclease activity of DNA polymerase I, which replaces ribonucleotides with deoxyribonucleotides as it proceeds.

required for each nucleotide pair that is separated. DNA gyrase promotes unwinding of the helix by introducing negative supercoils ahead of the replicative fork. As a type II topoisomerase, gyrase cuts both strands and uses the energy of ATP to create double negative supercoils, which then favor strand separation.

Once strand separation has begun, single-strand binding protein interacts with the exposed single strands at the replication fork to prevent the separated strands from forming a duplex again.

Putting It All Together: Replication in Summary

Figure 14-16 summarizes much of what we currently understand about the process of DNA replication in *E. coli* cells. Conceptually, at least, replication involves the following steps:

1. Helicase binds to double-stranded DNA at an origin of replication, separating parent strands. Strand separation requires ATP and results in a replication fork. DNA gyrase promotes strand separation by inducing negative supercoils in advance of the replication point (Figure 14-16a).
2. Strand separation is maintained by single-strand binding protein, which stabilizes the unwound region and allows the separated strands to serve as templates (Figure 14-16b).
3. Primase molecules bind to the single strands and synthesize short RNA primers that are complementary to the DNA template. The leading strand requires a single such priming event, because DNA synthesis is continuous thereafter. For the lagging strand, however, RNA primers are needed at regular intervals along the strand, and the entire primosome complex is somehow required for their formation (Figure 14-16c).
4. DNA polymerase III uses the primers to initiate DNA synthesis, which proceeds continuously in the direction of replication for the leading strand and discontinuously in the opposite direction along the lagging strand. On the lagging strand this generates short Okazaki fragments, each of which begins at an RNA primer and continues until it meets the adjacent fragment (Figure 14-16d).
5. RNA primers are removed by the $5' \rightarrow 3'$ exonuclease activity of DNA polymerase I. This enzyme is also believed to use its polymerase activity to fill in the resulting gap with deoxyribonucleotides (Figure 14-16e).
6. Finally, adjacent fragments are linked together by covalent phosphoester bonds formed by DNA ligase (Figure 14-16f).

Much remains to be discovered concerning DNA replication, especially in eukaryotic cells. Nevertheless, this is in essence the way DNA is thought to be copied and passed on to daughter chromosomes and cells in all forms of life.

DNA Repair

An important feature of DNA is the faithfulness with which its sequences are maintained from one generation of cells to the next. This genetic fidelity requires not only that the DNA be replicated very accurately, but also that provision be made for repairing the many changes that occur spontaneously in DNA. A low rate of spontaneous changes is actually desirable, since such changes, called **mutations,** provide the genetic variability that is the raw material of evolutionary selection. The mutation rate is very low, however; according to some estimates, an average gene accumulates one mutation every 200,000 years.

The rate of spontaneous DNA damage is far greater than the mutation rate would suggest, because most damage is repaired shortly after it occurs and therefore does not affect the genetic record of the organism. As we noted earlier, DNA polymerase I plays an important role in the repair of damaged DNA in bacterial cells.

Kinds of DNA Damage

DNA can be damaged by a variety of chemical and physical agents, but most spontaneous changes result from depurination of purine nucleotides, deamination of cytosines and other bases, and formation of pyrimidine dimers (Figure 14-17, p. 406). The first two categories are due to random thermal interactions between DNA and water molecules and are therefore spontaneous hydrolytic reactions. Pyrimidine dimer formation, on the other hand, is induced by ultraviolet light.

Depurination involves the loss of a purine base (either adenine or guanine) by spontaneous hydrolysis of the glycosidic bond that links it to deoxyribose (Figure 14-17a). This glycosidic bond is intrinsically unstable and is in fact so susceptible to random thermal cleavage that the typical human cell may lose as many as 5000 to 10,000 purine bases every day.

A second common source of DNA damage is the spontaneous **deamination** of cytosine, adenine, or guanine. Of the three bases, cytosine is most susceptible to deamination,

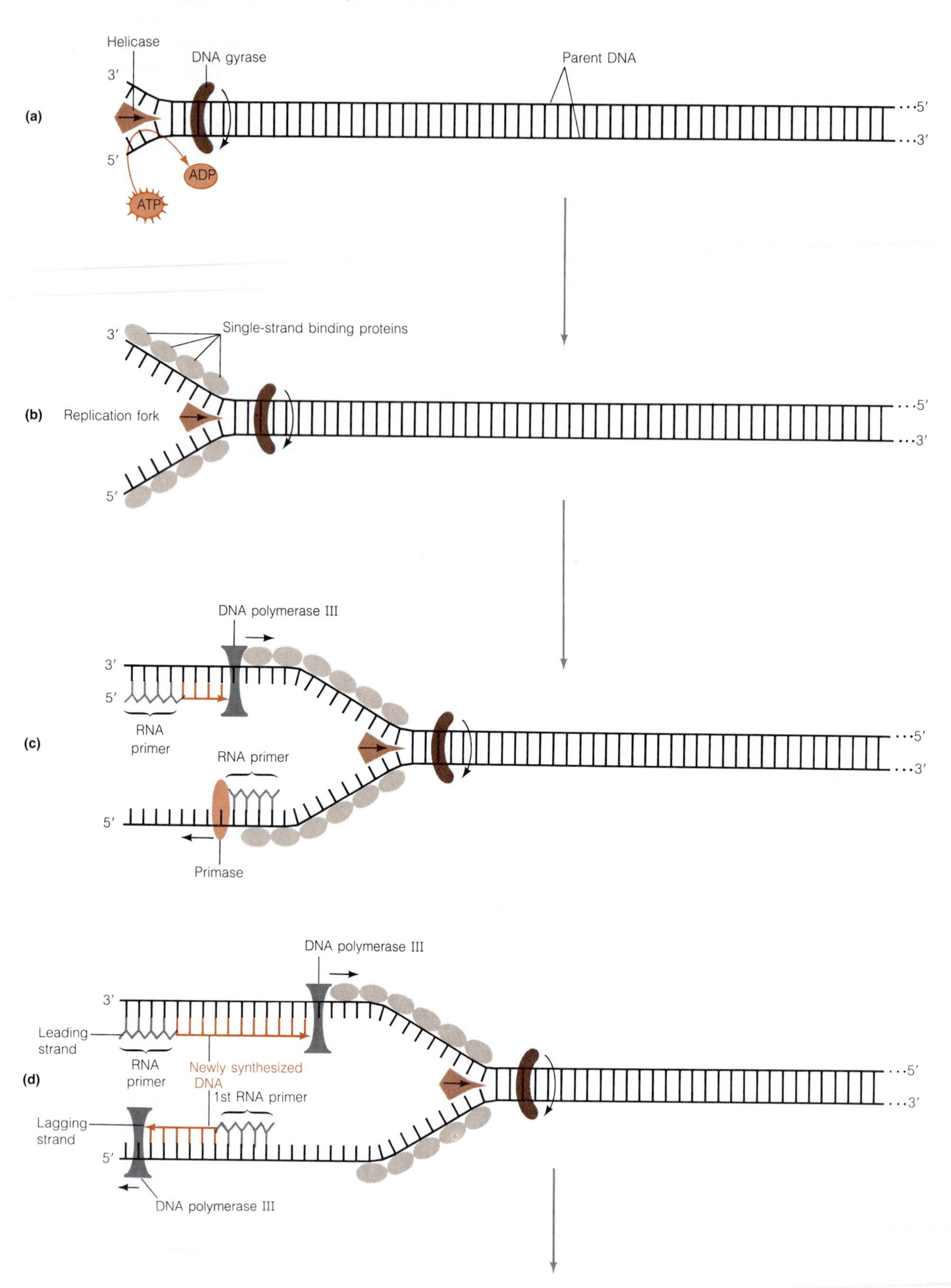
Helicase
DNA gyrase
Parent DNA
3′
5′
(a)
...5′
...3′
ADP
ATP
Single-strand binding proteins
(b)
Replication fork
DNA polymerase III
RNA primer
(c)
RNA primer
Primase
DNA polymerase III
Leading strand
RNA primer
Newly synthesized DNA
(d)
1st RNA primer
Lagging strand
DNA polymerase III

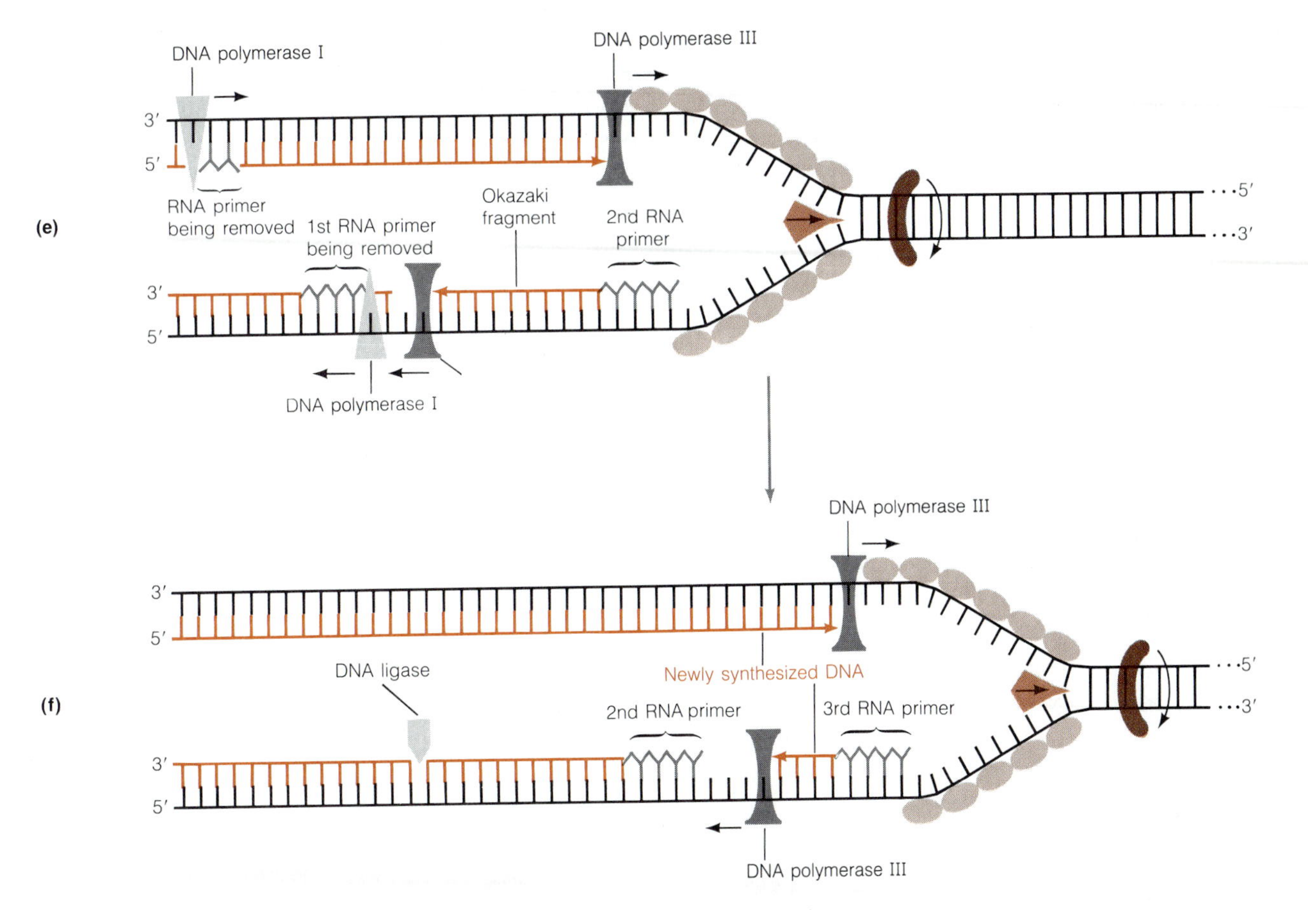

Figure 14-16 The Events of Bacterial DNA Replication. The synthesis of *E. coli* DNA at a replication fork is depicted here in six steps. (a) Strand separation is effected by helicase and gyrase and (b) is stabilized by single-strand binding proteins. (c) Once initiated with a short RNA primer, DNA synthesis on the leading strand proceeds continuously in the direction of replication. (d) On the lagging strand, DNA synthesis proceeds discontinuously in the opposite direction, giving rise to short fragments of DNA. (e) Synthesis of each such Okazaki fragment begins with a short RNA primer and continues until the previous fragment is encountered. RNA primers are removed by the 5′ → 3′ exonuclease activity of DNA polymerase I, which replaces ribonucleotides with deoxyribonucleotides as it proceeds. (f) The next Okazaki fragment on the lagging strand can then be joined by DNA ligase to the discontinuously growing DNA strand. Meanwhile a new Okazaki fragment has begun farther down the DNA molecule.

giving rise to uracil (Figure 14-17b). Like depurination, deamination is a hydrolytic reaction, caused by random thermal collision of a water molecule with the bond that links the amino group of the base to the pyrimidine or purine ring. The rate of damage to the human genome by this means is about 100 deaminations per day.

Pyrimidine dimer formation is yet another form of spontaneous damage to DNA. It is promoted by ultraviolet light from the sun and involves the formation of a covalent bond between two adjacent pyrimidine bases, usually two thymines (Figure 14-17c). Both replication and transcription are blocked by such dimers, presumably because the enzymes involved in these functions cannot cope with a dimer structure.

Repair Mechanisms

DNA repair occurs in three steps: excision of the defective nucleotide(s) from one strand of the duplex, replacement of the missing nucleotides, and ligation of the newly inserted nucleotides to the remainder of the strand. Different enzymes are involved in identifying the defect and nicking the damaged strand, depending on the nature of the damage to be repaired. The defective nucleotides are then excised and replaced with the correct ones by DNA polymerase I. The nucleotide sequence of the other strand ensures correct base insertion, just as it does in DNA replication. DNA ligase then seals the nick by forming the missing phosphoester bond.

Repair of depurination damage is shown in Figure 14-18. Excision of the depurinated sugar requires a special **repair endonuclease** that detects the absence of a base and nicks the phosphodiester backbone of the strand at that point. DNA polymerase I and DNA ligase then act to restore the strand.

Repair of deaminated DNA follows a similar course, except that deaminated bases must first be detected and removed. This is accomplished by specific **DNA glycosidases,** enzymes that recognize a specific deaminated base and remove it from the DNA molecule by cleaving the glycosidic bond between the base and the deoxyribose

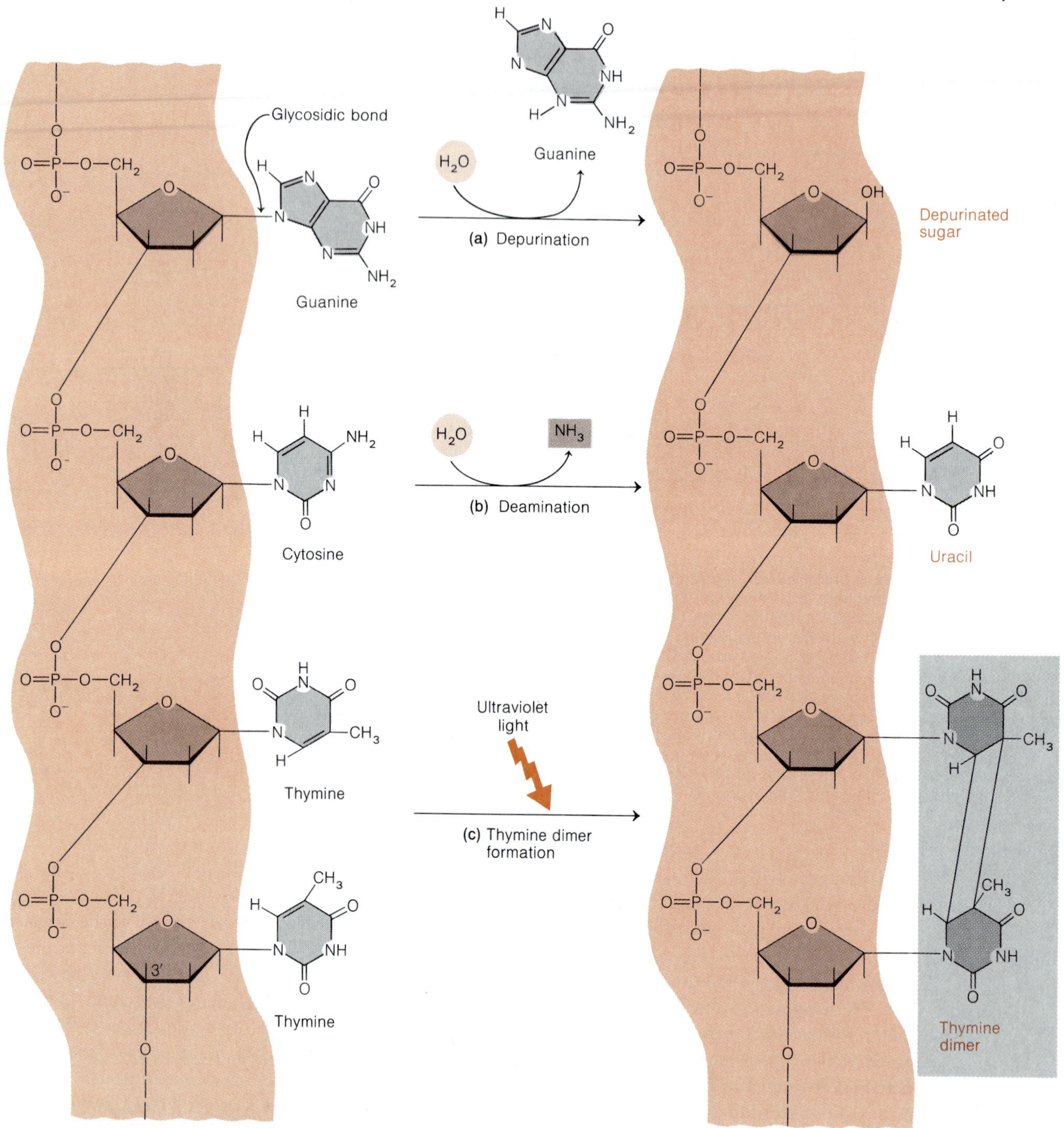

Figure 14-17 Spontaneous Damage to DNA. The most common kinds of chemical changes that can damage DNA are (a) depurination, (b) deamination, and (c) formation of thymine dimers. Depurination and deamination are spontaneous hydrolytic reactions, whereas dimers result from covalent bonds induced by ultraviolet light.

sugar to which it is attached. The sugar with the missing base is then recognized by the same repair endonuclease that detects depurination damage. The nuclease nicks the defective strand at that point, DNA polymerase I replaces the faulty nucleotide, and DNA ligase seals the strand again. Removal of pyrimidine dimers involves a similar mechanism, except that a *UV-specific endonuclease* is involved in nicking the defective region near the dimer.

Why DNA Contains Thymine Instead of Uracil

Until recently, it was not clear why DNA contains thymine instead of the uracil found in RNA. Both bases pair with adenine, although thymine contains a methyl group not present on uracil (Figure 14-19). In terms of energy, it would

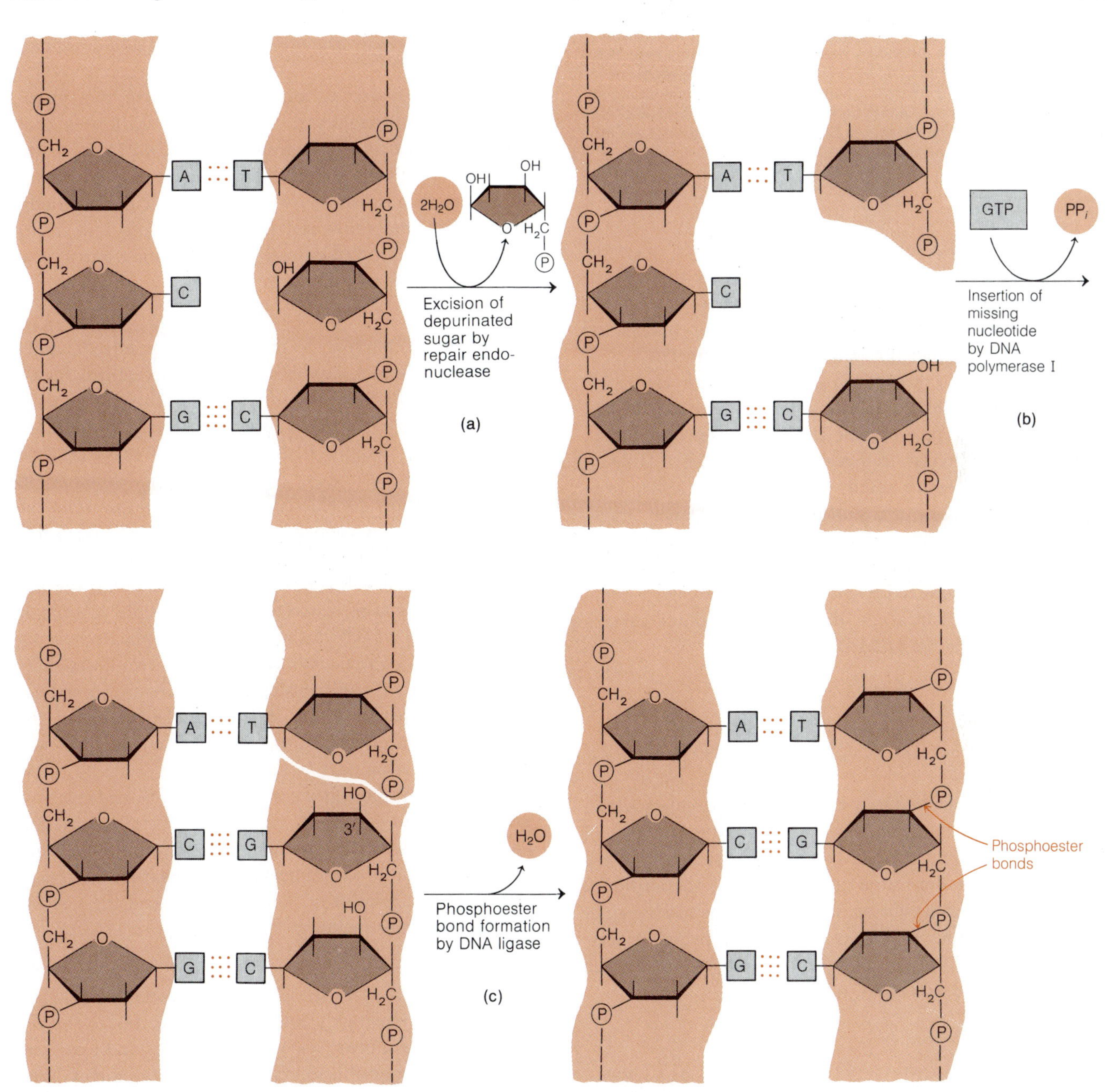

Figure 14-18 Repair of Depurinated DNA. Depurinated DNA is repaired by (a) excision of the depurinated sugar by a repair endonuclease, (b) insertion of the correct nucleotide by DNA polymerase I, and (c) ligation of the strand by DNA ligase. P = phosphate, A = adenine, G = guanine, C = cytosine, and T = thymine.

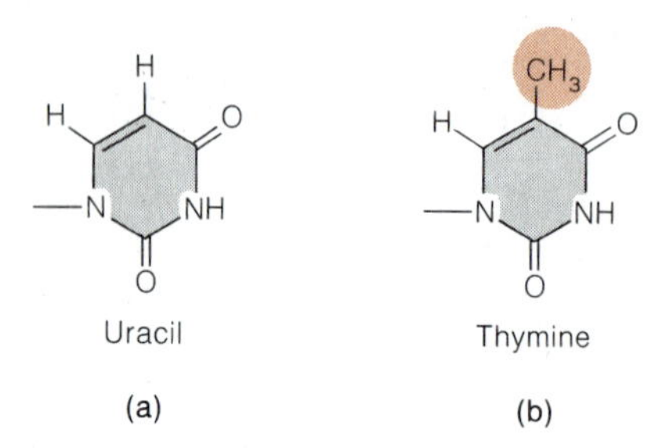

Figure 14-19 Comparison of Uracil and Thymine. (a) Any uracil present in DNA was formed by deamination of cytosine (see Figure 14-17b) and is removed by a specific glycosidase repair enzyme. (b) Thymine contains a methyl group not present on uracil. Unlike uracil, thymine cannot be generated by deamination of a base that is normally present in DNA and is therefore not the target of a specific glycosidase repair enzyme.

make more sense for DNA to contain uracil, since the methylation of deoxyuridylate to form deoxythymidylate is energetically expensive. But now that we understand how deamination damage is repaired, we also understand why thymine, rather than uracil, is present in DNA.

Deamination of cytosine generates uracil (Figure 14-17b), which is then detected and removed by **uracil-DNA glycosidase,** a repair enzyme that recognizes only uracil. If uracil were present as a normal component of DNA, it is not clear how such uracils would be distinguished from those generated by cytosine deamination. Since uracil-DNA glycosidase recognizes uracil but not thymine, the methyl group on thymine is apparently the "label" that tells the glycosidase to leave it alone, thereby ensuring that deaminated cytosines can be replaced without causing other changes in the DNA molecule.

Nuclear and Cell Division

Having seen how the DNA in the nucleus is replicated, we return now to the processes whereby the two copies of each chromosome that have been generated during the prior S phase are separated from each other and partitioned into daughter nuclei during mitosis and then into daughter cells during cytokinesis. We considered the overall process earlier (Figure 14-3), but now we are ready to look at both mitosis and cytokinesis in more detail.

Mitosis

Mitosis has been known and studied for a long time, but only recently has significant progress been made toward understanding the mitotic process at the molecular level. We will first consider the morphological events involved in chromosome separation and then turn to the molecular mechanisms that underlie mitotic chromosomal movements.

Morphologically, mitosis is usually described as a series of four phases, based primarily on the appearance and behavior of the chromosomes. It is convenient to consider mitosis in discrete steps, but as with any dynamic process, we must remember that the phases are arbitrary conventions and serve only as a means of studying and describing the process.

The four phases of mitosis are **prophase, metaphase, anaphase,** and **telophase.** Figure 14-20 illustrates each phase using schematic diagrams. Figure 14-21 depicts the sequence of mitotic stages as they appear during division of a living plant cell. As we look at each phase briefly, keep in mind that the overall purpose of mitosis is to ensure that one copy of each duplicated chromosome is partitioned into each of the two daughter nuclei.

Prophase. Preparatory to their segregation into daughter cells, chromosomes condense from the extended, highly diffuse form characteristic of interphase chromosomes to the very dense, coiled structures characteristic of the division process (Figure 14-20a). Although the transition from interphase to mitosis is not a sharply defined event, a cell is generally considered to be in prophase when the chromosomes have condensed to the point of being visible in the light microscope. Rather than appearing as an amorphous mass of dispersed chromosomal material, the nucleus now contains discrete, readily visible structures (Figure 14-20a). Each chromosome has duplicated during the preceding S phase and now consists of two sister chromatids (Figure 14-20b). Sister chromatids are joined together at a specific point along their length called the **centromere** and are usually twisted loosely around each other for the remainder of their length.

As the chromosomes condense, nucleoli become progressively more dispersed, so that cells entering division are characterized by the gradual disappearance of nucleoli. By late prophase, chromosome condensation is almost complete, nucleoli are no longer visible, and the nuclear envelope has also begun to fragment (Figure 14-20b).

Metaphase. As the cell approaches metaphase, chromosomes become highly condensed (Figure 14-22) and the *mitotic spindle* appears (Figure 14-23). The spindle consists of the fibers responsible for chromosomal movements. Each spindle fiber is a bundle of microtubules, assembled from tubulin molecules that become available as the microtubules of the cytoskeleton disassemble. The focal point on either end of the spindle is a **spindle pole,** sometimes

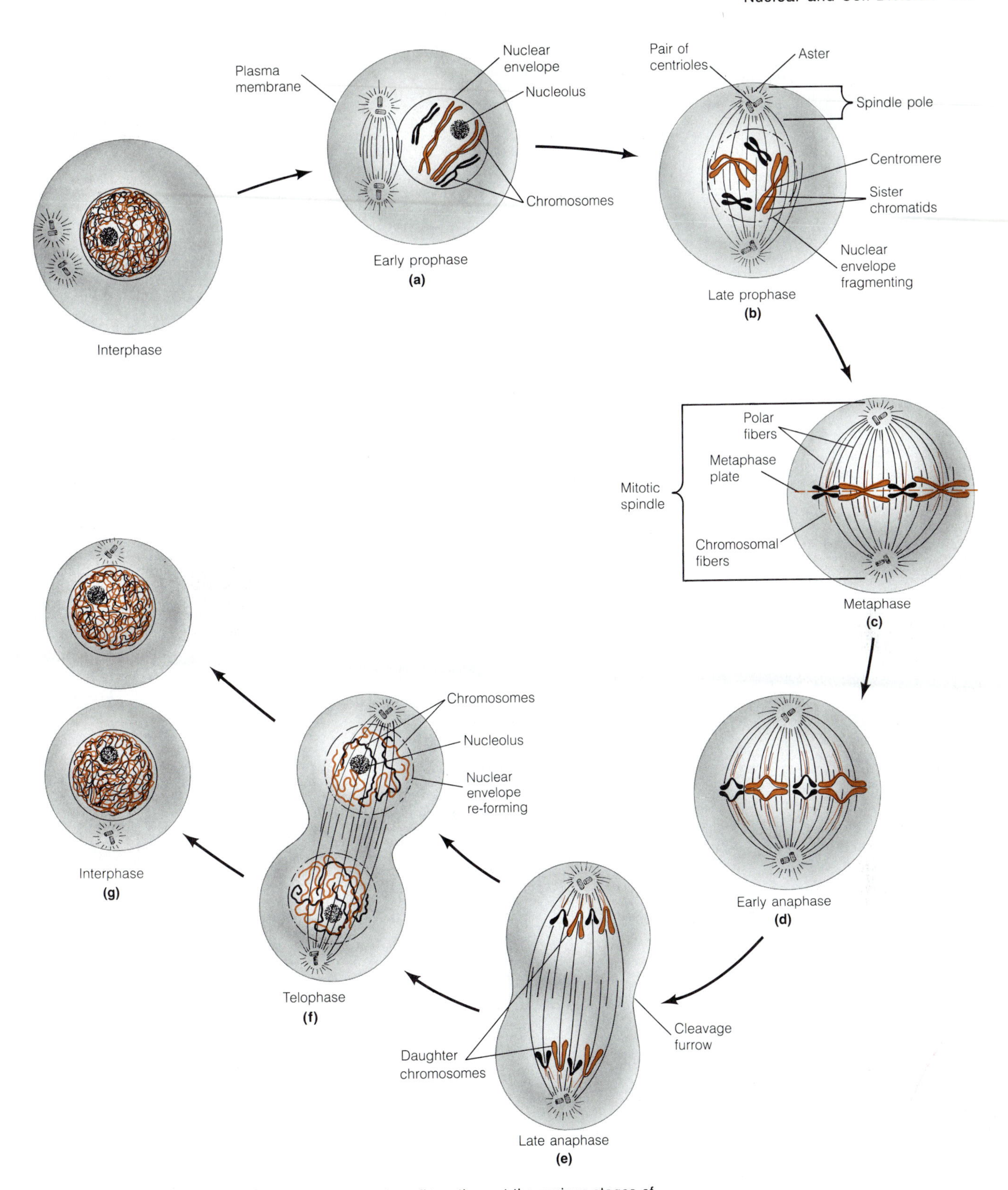

Figure 14-20 The Phases of Mitosis. Chromosomal configurations at the various stages of mitosis are shown schematically.

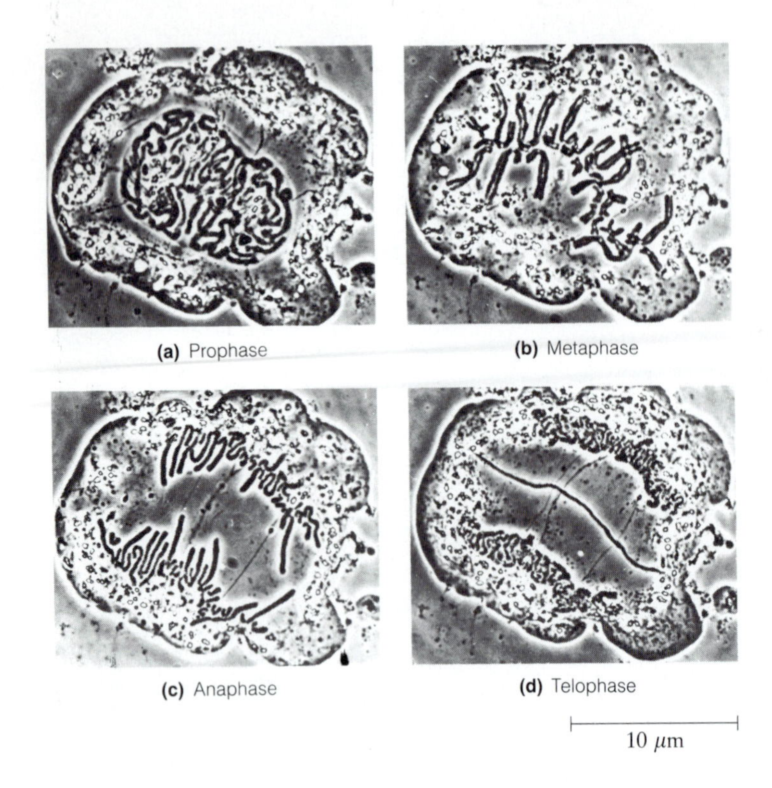

Figure 14-21 Mitosis in a Living Cell. These light micrographs show a living cell of the African blood lily, *Haemanthus katherinae,* at the four phases of mitosis.

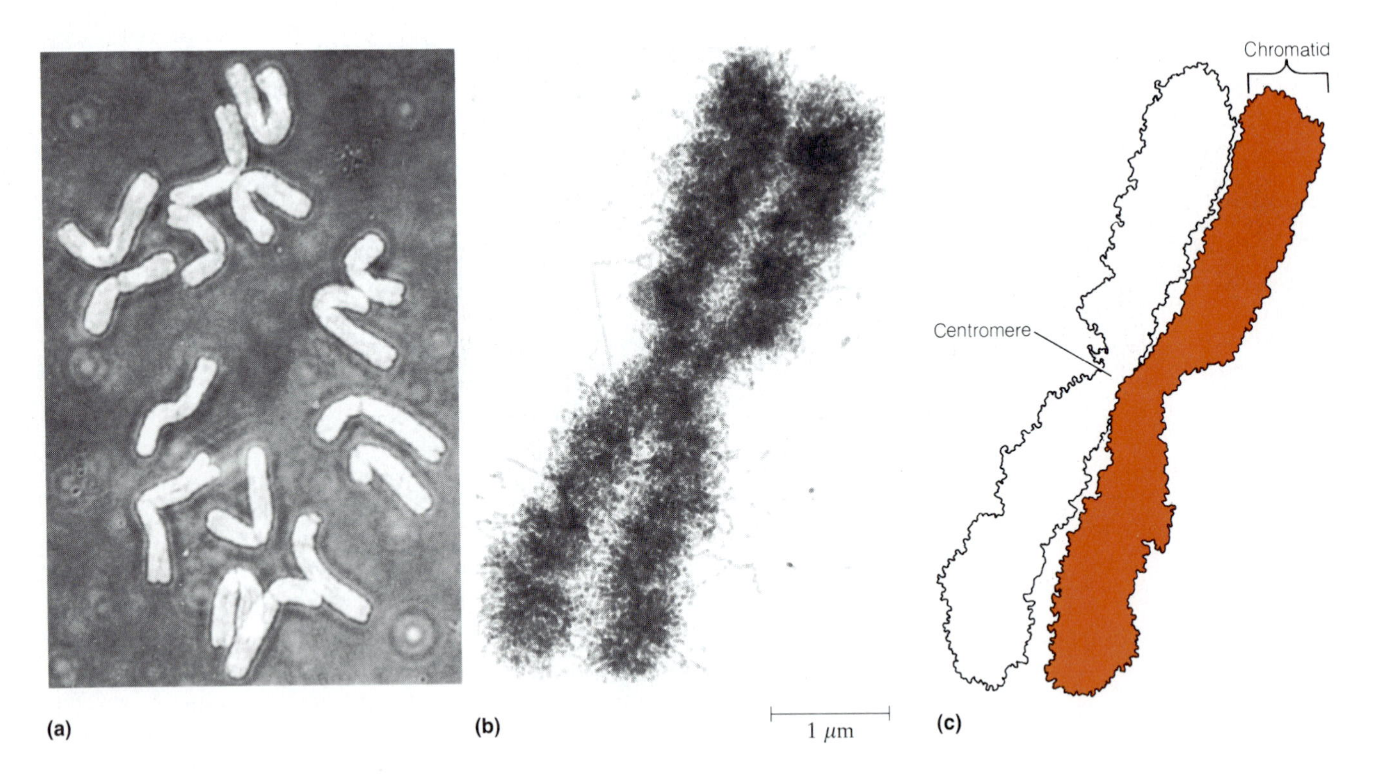

Figure 14-22 Metaphase Chromosomes. Chromosomes become highly condensed during metaphase. (a) In this light micrograph of an onion cell at metaphase, each of the 16 chromosomes is present as a pair of sister chromatids. (b) An electron micrograph of a human metaphase chromosome, showing the two sister chromatids joined at the centromere (TEM). (c) Sketch of the chromosome shown in (b).

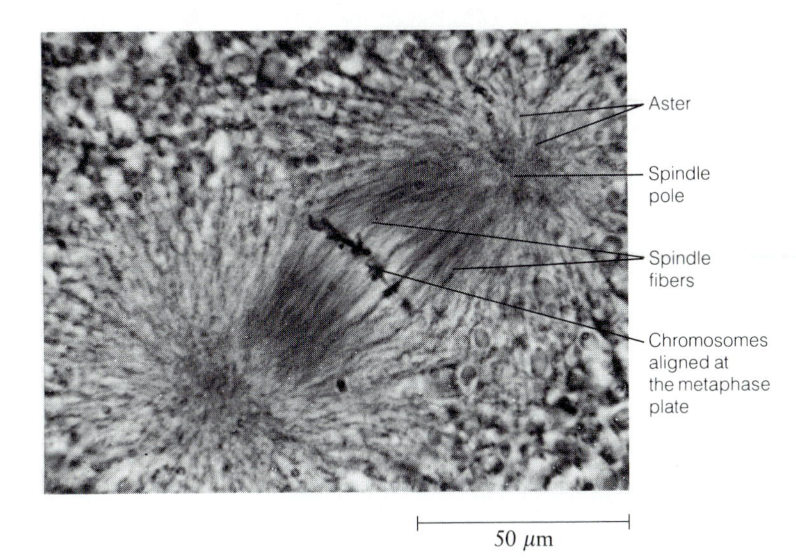

Figure 14-23 The Mitotic Spindle. This micrograph shows the mitotic spindle of a dividing cell from a *Xenopus* embryo. The chromosomes are aligned at the metaphase phage. The two poles of the spindle and the spindle fibers are clearly visible, as are the asters characteristic of the mitotic spindle of an animal cell.

also called a **mitotic center.** In animal cells, the spindle pole consists of a pair of **centrioles** surrounded by an **aster,** or radial array of microtubules. Centrioles are self-replicating structures about 0.15 μm in diameter and about 0.3–0.5 μm long.

Plant cells do not have centrioles, and their mitotic spindles are less focused at the poles, without the astral array of microtubules seen in animal cells. Consequently, a distinction is made between **astral mitosis,** which occurs in cells of higher animals, and **anastral mitosis,** characteristic of higher plant cells.

The fibers of the mitotic spindle are of two types. **Polar fibers** extend inward from the spindle pole toward the equator, whereas **chromosomal fibers** extend from individual chromosomes outward toward the poles. Chromosomal fibers are attached to the chromosomes at the **kinetochores,** specialized structures that form on opposite sides of the centromere of each chromosome. Chromosomal fibers are therefore also commonly called **kinetochore fibers.**

During metaphase, the chromosomes, which are now maximally condensed, gradually become aligned at the **metaphase plate** through the activity of the spindle apparatus. Both in their movements to the metaphase plate and in their subsequent movements toward one end of the cell or the other, chromosomes seem to be pulled by the chromosomal fibers.

Anaphase. Anaphase is usually the shortest phase of mitosis, typically lasting only a few minutes. It is characterized by the movement of the two sister chromatids of each metaphase pair toward opposite spindle poles and therefore toward opposite ends of the cell (Figure 14-20d). All chromatids begin to separate at the same time and are pulled toward the spindle poles at a rate of about 1 μm/min. Once they have separated, sister chromatids are referred to as **daughter chromosomes** (Figure 14-20e).

The position of the centromere and the relative lengths of the two chromosome **arms** are especially obvious during anaphase, since the chromosomes appear to be drawn toward the spindle poles at the centromere, with the arms trailing behind. A **metacentric chromosome** (one with the centromere at or near the middle) appears V-shaped, whereas an **acrocentric chromosome** (centromere closer to one end) has the shape of an uppercase L, and a **telocentric chromosome** (centromere at or near the end) appears to have a single arm.

Anaphase is characterized by two different kinds of movements: the chromosomes move toward the spindle poles, and the poles themselves move away from each other. This can be seen clearly in Figure 14-24, which is a series of phase-contrast micrographs showing the sequence of chromosomal movements during anaphase in a cultured epithelial cell. Each micrograph in the series is labeled with the time in minutes that has elapsed since the beginning of anaphase. At the start of the sequence ($t = 0$ min), the chromosomes are aligned at the equator of the spindle, along the metaphase plate. By the end of anaphase ($t = 18$ min), chromosome separation is complete, and the cell itself has begun to divide.

The separation of chromosomes seen in Figure 14-24 results not only from the chromosomes being drawn to the spindle poles but also from an increase in the distance between the poles. We can see the movement of the poles away from each other by comparing the vertical lines that mark the positions of the poles in the initial and final micrographs. The poles are almost twice as far apart at 18 min as they were at the beginning of anaphase.

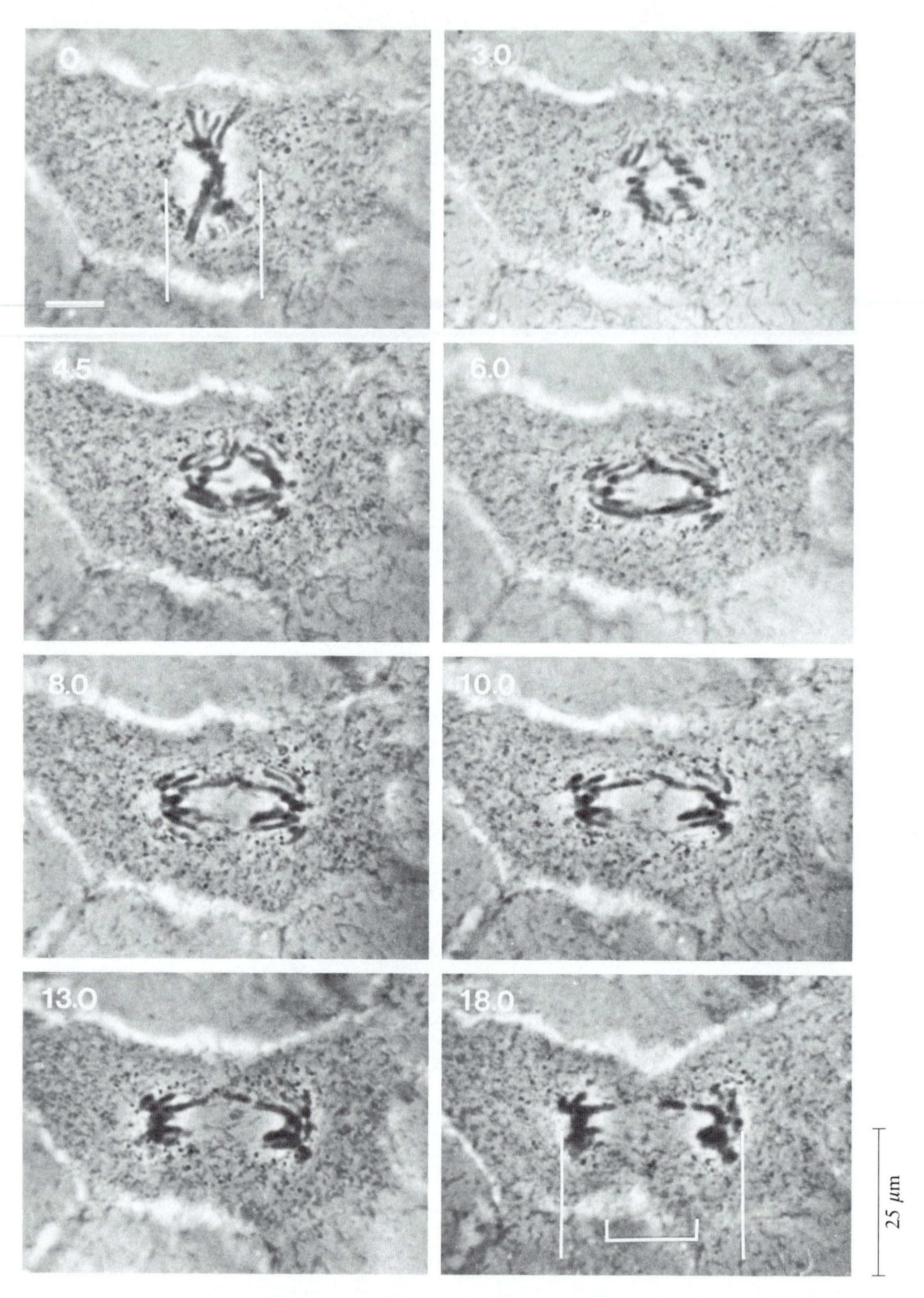

Figure 14-24 Sequence of Anaphase Chromosome Movements. This is a series of phase-contrast micrographs showing mitotic chromosome movements during anaphase in the tissue culture cell line Ptk_1. Times are indicated in minutes from the onset of anaphase. As the cell enters anaphase ($t = 0$), the chromosomes are aligned on the spindle equator, forming the metaphase plate. The separation of chromatids is accomplished by two kinds of movements. At the same time that the chromatids move toward opposite poles, the spindle poles themselves move away from each other. The bars in the first and final micrographs indicate the position of the poles at the beginning and end of anaphase. The brackets in the final micrograph show the original position of the poles to demonstrate the extent of pole movement. By the end of anaphase ($t = 18$ min), the cleavage furrow has begun to constrict the cell. Subsequently, the cell will be pinched in two and the two sets of chromosomes will be segregated into separate daughter nuclei. The Ptk_1 cell line is derived from kidney epithelial tissue of the rat kangaroo. (all LMs)

Telophase. By the beginning of telophase, the daughter chromosomes have arrived at the poles of the spindle (Figure 14-20f). During telophase, they uncoil and revert to the extended form and homogeneous appearance of interphase chromosomes. The chromosomal fibers disappear, nucleoli develop at the nucleolar organizing sites, and nuclear envelopes form around the two daughter nuclei, completing the mitotic process.

The Mechanism of Mitosis: Chromosome Alignment and Movement

Attachment of the chromosomal fibers to the kinetochores of sister chromatids occurs early in metaphase. These fibers serve two purposes: They orient the chromatids with respect to the spindle at metaphase and they are responsible for the separation of sister chromatids at anaphase. The attachment of fiber to kinetochore is random; either kinetochore can end up facing either pole.

Once aligned on the metaphase plate, chromosomes appear to be stationary. This appearance is somewhat misleading, however, because both sister chromatids are already being actively tugged toward their respective poles, even at metaphase. They appear stationary because the forces acting on the two chromatids are equal in magnitude and opposite in direction. Each chromosome on the metaphase plate is therefore in a state of active tension, held in place by opposing forces.

When sister chromatids separate at anaphase, the two chromatids are free to be pulled toward opposite poles, thereby accomplishing the chromosomal movement and separation that are characteristic of anaphase (Figure 14-24). Several of the models that have been proposed to explain chromosomal movement are discussed in the box on pp. 414–415, both to acquaint you with various theories of chromosome movement and to illustrate the importance of controversy in the development of scientific understanding.

Cytokinesis

Nuclear division and cytoplasmic division are not always linked events. Some fungi and algae undergo many rounds of nuclear division unaccompanied by cell division, resulting in large multinucleated cells. In most cases, however, cytokinesis closely follows or even accompanies nuclear division, such that each of the daughter nuclei acquires its own cytoplasm and becomes a separate cell. Cytokinesis usually starts during late anaphase or telophase, as the nuclear envelope and nucleoli are re-forming and the chromosomes are beginning to decondense again (Figure 14-20f).

Cytokinesis in Animal Cells. Cytokinesis occurs by quite different mechanisms in plant and animal cells. In animals, cell division is called **cleavage** and has been studied most extensively in species such as frogs or sea urchins, which fertilize their eggs externally. Cleavage begins as a slight indentation or puckering of the cell surface, which deepens into a **cleavage furrow** that encircles the egg, as shown in Figure 14-25 for a fertilized frog egg. The plane of the cleavage furrow is always perpendicular to the long axis of the mitotic spindle, thereby ensuring that the two sets of chromosomes will be segregated into the two daughter cells. The furrow continues to deepen until opposite surfaces make contact and the cell is split in two (Figure 14-20g).

Cleavage depends on a beltlike bundle of actin microfilaments called the **contractile ring,** which forms beneath the plasma membrane during early anaphase. Examination of the cortical cytoplasm of the cleavage furrow with the electron microscope reveals large numbers of microfilaments, oriented with their long axes parallel to the plane of the furrow. As cleavage progresses, the ring of micro-

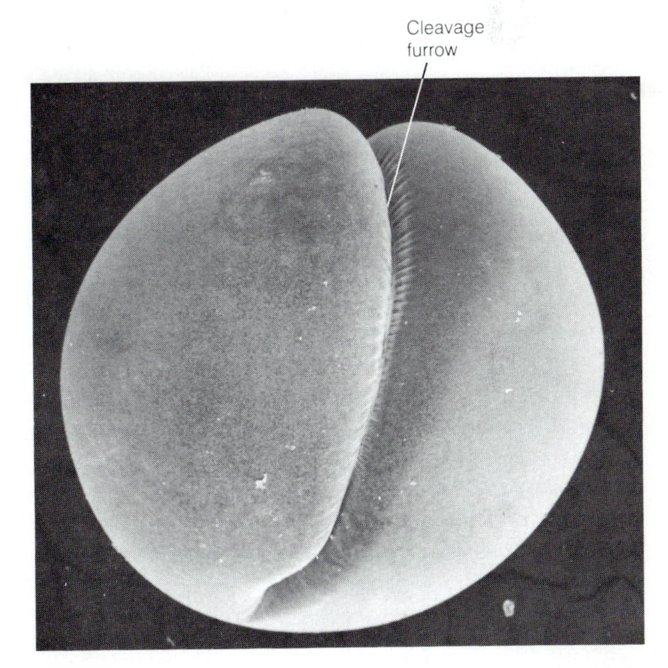

Figure 14-25 Cytokinesis in an Animal Cell. A scanning electron micrograph of a frog zygote (fertilized egg) caught in the act of dividing. The cleavage furrow is clearly visible as an inward constriction of the plasma membrane. Within the cell, mitosis is nearly complete, so the cleavage furrow will separate the two sets of chromosomes as it continues to constrict the membrane.

The Mitotic Spindle: A Case Study in Controversy

Controversy and dispute are often the raw materials out of which scientific progress is made. Whenever real progress is being made in a specific area, researchers in that area often espouse opposing views, suggest alternative models, and defend both their models and their views passionately. Some of the most interesting and best-attended sessions at scientific meetings are those in which the major researchers in a controversial area confront one another with their most recent data, their latest models, and their current views, often expressed with great conviction and usually disputed with equal ardor. And out of the controversy, models, and data there frequently emerges a better understanding of the problem.

Much of the information that you read about in a biology text such as this one went through such periods of uncertainty and controversy before emerging as "facts" with enough support to warrant inclusion in a textbook. Examples of present-day "facts" that arose out of especially memorable controversies include the role of DNA as the bearer of genetic information, the fluid mosaic model of membrane structure, and the chemiosmotic model of oxidative phosphorylation. In each of these cases, the controversy is largely over. Researchers are in general agreement on these topics, the main questions are answered, and alternative explanations and models have historical interest only.

One question that is not yet answered and is therefore still a controversial topic concerns the mechanism by which the mitotic spindle accomplishes chromosome separation at anaphase. That the mitotic spindle is involved in directing the movement of sister chromosomes to opposite poles during anaphase has long been known. Indeed, the synchronous poleward migration of chromosomes as they leave the metaphase plate has been the subject of some of the most dramatic time-lapse cinematography in cell biology. Each chromosome appears to be pulled toward its respective pole, with the kinetochore at the leading edge. At the same time, the two poles actually move farther apart, ensuring that the resulting sets of chromosomes are well separated, in anticipation of cell division. This movement involves an elongation of the spindle, which in some species is so extensive that the poles are separated by a distance that is from 10 to 15 times their original separation at metaphase.

It is well established that microtubules are involved in both chromosome movement and pole separation. Both the polar fibers and the chromosomal fibers consist of microtubules, and anything that disrupts the microtubules—whether by physical or chemical means—stops the movement of chromosomes. An influential early model was proposed by Shinya Inoué, who studied mitosis by polarization microscopy and observed that chromosomes seemed to be tugged toward the poles by the fibers attached to their kinetochores. He proposed that the fibers are shortened by a disassembly process involving the progressive removal of tubulin dimers from the fibers at the pole end.

Some of the strongest support for this model comes from studies of mutations in tubulin genes that produced abnormally stable microtubules. Such mutants are arrested in metaphase, presumably because the microtubules are unable to disassemble. If the cells are then treated with drugs that destabilize microtubules, normal anaphase ensues. Recently, other investigators have made the surprising observation that most of the microtubules in the spindle are oriented with the fast-growing ends *away* from the pole. If microtubule disassembly is responsible for shortening of fibers and movement of chromosomes, the disassembly must occur at the point of attachment of the fiber to the kinetochore, rather than at the spindle pole. This observation was confirmed by other investigations of microtubule association with chromosomes in vitro, in which it was observed that microtubules assemble and disassemble from the kinetochore end. Moreover, it has also been shown that the magnitude of the pole-directed force exerted on a chromosome by the fiber attached to its kinetochore is proportional to the length of the fiber. This finding suggests that "mitotic motors" are distributed along the length of the chromosomal fibers rather than at a site of microtubule disassembly at the end of the fiber and that the poleward force cannot be due to the disassembly process per se.

Part of the problem in explaining the spindle-mediated movements of chromosomes at anaphase is that two different sets of events are involved: the movement of chromosomes *toward* the two poles and the movement of the poles *away* from each other. The movement of poles has been reactivated in vitro by Zac Cande, using isolated spindles from diatoms, where overlapping half spindles are known to generate force by sliding microtubules. Roger Leslie and Jeremy Pickett-Heaps showed that the motor for spindle elongation is in the overlap region of the microtubules, suggesting that some as yet unidentified motor molecule is active in that region.

The controversy concerning this mechanism is far from over, however. An alternative model proposes that spindle asters are in a sense automotive, "crawling" away from each other at anaphase. For instance, when the spindle of a fungal cell is broken by a laser microbeam, the spindle poles move apart faster than normal. This suggests that microtubules extending from the spindle poles into the cytoplasm, rather than motors acting between sliding microtubules, are responsible for polar movement.

In summary, research has shown that chromosomes are *pulled* toward their respective poles by disassembly and shortening of microtubules, whereas the poles are *pushed* apart by a second mechanism, perhaps involving microtubules sliding past one another. These are only the best guesses of the moment, however. The uncertainties persist, the definitive experiments remain to be done, and the controversy goes on. The spindle and chromosome movements during mitosis provide an excellent example of the excitement associated with unanswered questions in biology. For in biology, as in every other science, the controversies of today become the accepted facts of tomorrow through definitive experiments that are as yet undone. And these experiments, in turn, are almost always spurred on by the uncertainties the investigators try to resolve, the colleagues they engage in controversy, and the scientific problems they seek thereby to settle.

filaments tightens around the cytoplasm, rather like a belt around the waist.

The force needed to tighten the contractile ring and divide the cytoplasm is thought to be generated by the interaction of the actin microfilaments with the protein myosin. Myosin is the protein with which actin interacts in muscle tissue to produce the ratchetlike sliding of protein filaments that accounts for muscle contraction (Chapter 19). Unlike actin microfilaments, myosin filaments cannot be seen directly in the contractile ring, but their presence can be deduced from antibody studies. The filaments in the contractile ring apparently constrict the cleavage furrow progressively in a manner that depends on the sliding of actin and myosin filaments to generate the needed force.

The contractile ring of dividing cells provides one of the most dramatic examples of the transitory nature of actin-myosin structures in most nonmuscle cells and the rapidity with which they can be assembled and disassembled. Apparently, polymerization of actin occurs just before the initial indentation becomes visible, and the entire structure is dismantled again shortly after cell division is complete. The actin monomers needed to assemble the microfilaments of the contractile ring are obtained by disassembly of the actin filaments of the cytoskeleton, just as the tubulin needed for spindle fibers is derived from cytoskeletal microtubules.

The intermediate filaments are the only major components of the cytoskeleton that are not disassembled during cell division. Instead, the network of intermediate filaments elongates during mitosis and is parceled out into the two daughter cells as cleavage continues. All the organelles and other components of the cell are similarly segregated into the two daughter cells during division, so that daughter cells usually contain about the same numbers of components.

Cytokinesis in Plant Cells. In cells of higher plants, cytokinesis occurs by a different mechanism, presumably because of the presence of a cell wall. Plant cells divide not by pinching the cytoplasm in half with a contractile ring, but by assembling a new cell wall between the two daughter nuclei (Figure 14-26). A cell wall in the process of formation is called the **cell plate.**

The cell plate is formed as small membranous vesicles bearing cell wall precursors align themselves along the midplane of the cell, usually during late anaphase or early telophase. These vesicles are derived from dictyosomes of the Golgi complex. They appear to be guided to the midplane by microtubules that are derived from the polar spindle fibers and that are oriented perpendicular to the developing cell plate. The parallel array of microtubules forms the **phragmoplast,** an open, cylindrical structure re-

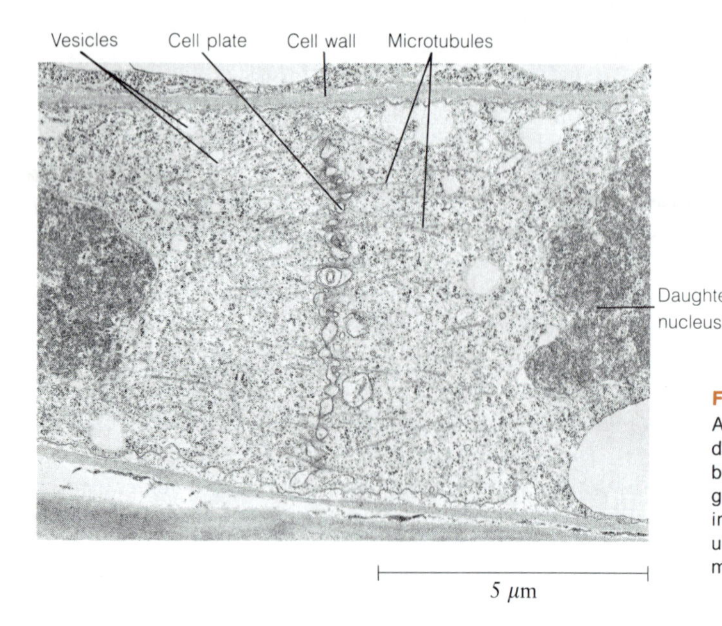

Figure 14-26 Cytokinesis and Cell Plate Formation in a Plant Cell. An electron micrograph of a plant cell at late telophase. The daughter nuclei with their sets of chromosomes are partially visible as the dark material on the right and far left of the micrograph, and the developing cell plate is seen as a line of vesicles in the midregion of the cell. Microtubules are oriented perpendicular to the cell plate. The cell is from *Acer saccharinum,* the soft maple. (TEM)

stricted initially to the central region of the cell. Fusion of the vesicles forms a large, flattened sac called the **early cell plate.** The contents of the vesicles assemble to form the noncellulosic components of the primary cell wall, which expands outward as clusters of microtubules and vesicles form at the lateral edges of the advancing cell plate.

Eventually, contact is made with the original cell wall, and the two daughter cells are completely separated from each other. The new cell wall is then completed by deposition of cellulose microfibrils. The plasmodesmata that provide channels of continuity between the cytoplasms of adjacent plant cells are also present in the cell plate and the new wall as it forms. They extend through the plate and the wall, connecting the daughter cells and presumably acting as communication channels.

Variations in the Cell Cycle

Figure 14-1 suggests that eukaryotic cells proceed continuously through predictable cycles of growth and division, with G-1, S, G-2, and M always following one another in unbroken progression and with every nuclear division accompanied by cytokinesis. Such is often the case, of course, particularly in growing organisms or cultured cells that have not run out of nutrients or space. But many variations are also possible, especially in terms of the relative length of time spent in various phases of the cycle and in the immediacy with which mitosis and cytokinesis are coupled.

Variations in Cell Cycle Length

Some of the most common variations in the cell cycle in vivo involve differences in generation time between different cell types. Within the same organism, some cells divide at approximately the same rate as cells in culture, but others differ greatly, depending on their role in the organism (Figure 14-27). Some cells divide rapidly and continuously throughout the life of the organism as a means of replacing cells that are lost or destroyed during the normal functioning of the organism. Included in this category are the cells that lead to sperm formation and the precursor cells (or **stem cells**) that give rise to blood cells, skin cells, and the epithelial tissues that line the inner surfaces of body cavities such as the lungs and intestines. Human stem cells may have generation times as short as 8 hours.

Cells of slow-growing tissues, on the other hand, may have generation times of several days or more, and some cells, such as those of nerve or muscle tissue, do not divide at all. Still other cell types do not divide under normal conditions, but can be induced to begin dividing again by an appropriate stimulus. Liver cells are in this category, since they do not normally proliferate in the mature liver but can be induced to do so if a portion of the liver is removed surgically. Lymphocytes (white blood cells) are another example; when exposed to a foreign protein, they begin dividing as part of the immune response, as we will see in Chapter 24.

Most of these variations in generation time involve differences in G-1, although S and G-2 can also vary somewhat. Cells that divide very slowly can spend days, months, or even years in G-1, whereas cells that divide very rapidly have almost no G-1 phase at all. In fact, it is even possible for cells to begin DNA synthesis before mitosis is complete, eliminating G-1 entirely.

The embryos of insects, amphibians, and other lower animals are dramatic examples of very short cell cycles, with no G-1 phase and a very short S phase. During early embryonic development in amphibians such as *Xenopus laevis,* for instance, cell division can take less than 30 minutes, even though the normal length of the cell cycle in adult tissues is about 20 hours. Under these conditions, the S phase is completed in less than 3 minutes, at least 100 times faster than in adult tissues. The incredible rate of DNA synthesis needed to sustain such a rapid cell cycle

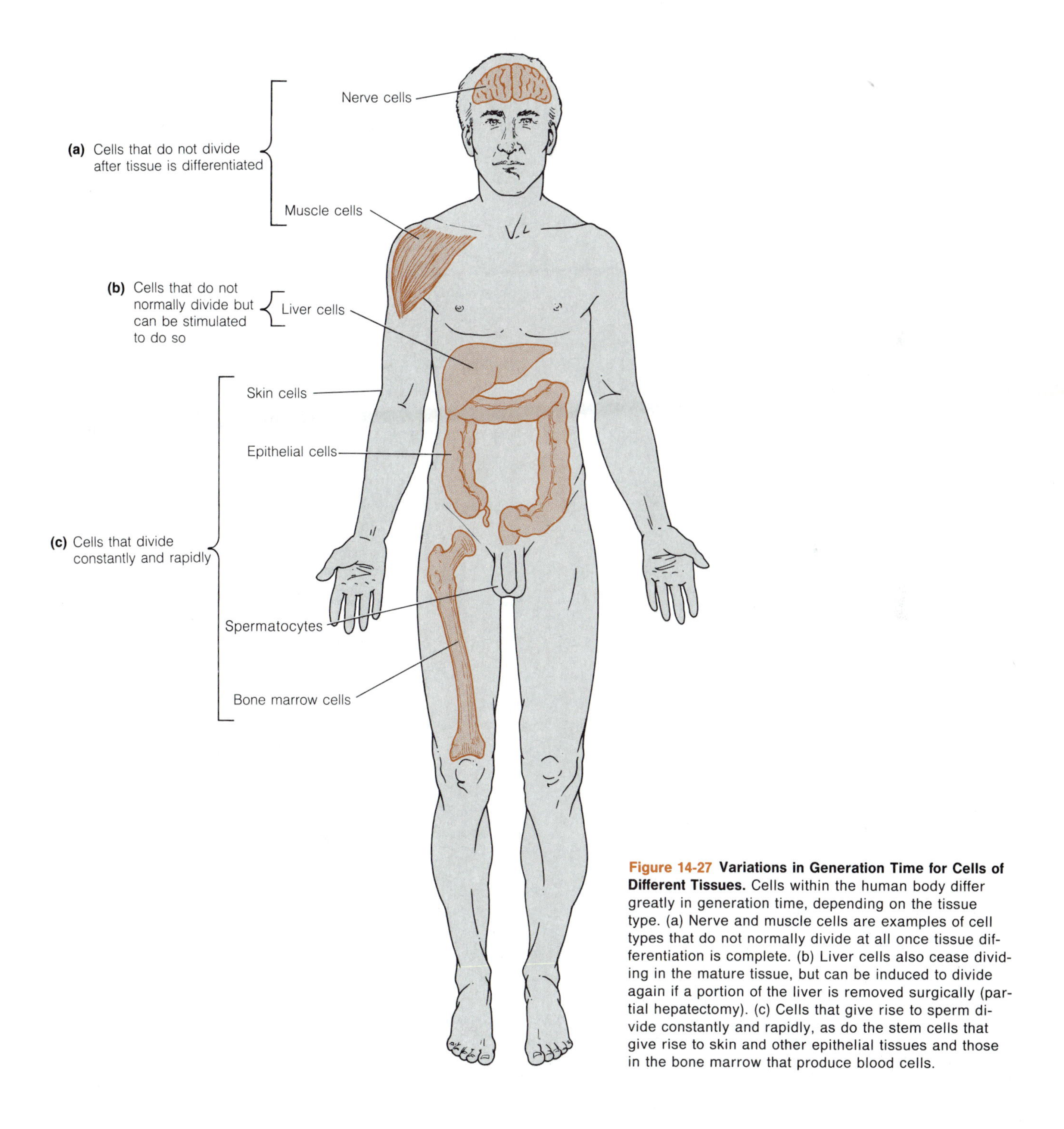

Figure 14-27 Variations in Generation Time for Cells of Different Tissues. Cells within the human body differ greatly in generation time, depending on the tissue type. (a) Nerve and muscle cells are examples of cell types that do not normally divide at all once tissue differentiation is complete. (b) Liver cells also cease dividing in the mature tissue, but can be induced to divide again if a portion of the liver is removed surgically (partial hepatectomy). (c) Cells that give rise to sperm divide constantly and rapidly, as do the stem cells that give rise to skin and other epithelial tissues and those in the bone marrow that produce blood cells.

is possible because virtually all replicons are active at the same time, in contrast to the sequential activation seen in adult tissues.

In addition, these embryonic cells of lower animals have little or no need for synthesis of other cellular components because the fertilized egg is a very large cell, with enough cytoplasm to sustain many rounds of cell division. Each round of division subdivides the initial cytoplasm into smaller cells, until the cell size characteristic of adult tissues is reached (Figure 14-28). In *Xenopus* embryos, for example, there is no G-1 and only a very short G-2 during the early cleavage division, so that cells go almost directly from DNA synthesis to mitosis and back to DNA synthesis. In fact, the S phase in such cells begins even before mitosis is complete. From such examples, we know that cell growth during the G-1 and G-2 phases is not a prerequisite for cell division, even though growth and division are usually coupled processes.

Variations in Timing of Mitosis and Cytokinesis

Two other processes that are usually closely associated, but that need not be, are mitosis and cytokinesis. In most cases, nuclear division is closely followed by cytokinesis, resulting in daughter cells that contain a single nucleus each. However, DNA replication and mitosis occasionally occur without subsequent cytokinesis, giving rise to multinucleate cells. In some cases, such as the fungal and algal cells mentioned earlier, this is a permanent condition. In other organisms, however, the multinucleate state is only a temporary phase of development.

In the development of endosperm tissue in certain species of plants, such as the cereal grains, nuclear division occurs for a time unaccompanied by cell division, generating many nuclei in a common cytoplasm. Successive rounds of cytokinesis then occur without mitosis, walling off the many nuclei into separate endosperm cells. A similar process occurs in developing insect eggs. The early embryo undergoes mitosis but not cytokinesis and soon consists of hundreds of nuclei in the same cytoplasm until later, when cytokinesis "catches up."

Regulation of the Cell Cycle

The variability in generation time for cells of the same organism makes it clear that the cell cycle must somehow be regulated, but the biochemical nature of this control is not yet well understood. A great deal of research is currently focused on this question, however, because of its fundamental importance to our understanding of how normal cells work and how cancer cells manage to escape normal control mechanisms to proliferate. To date, no single regulatory mechanism has been identified, but several intriguing possibilities have been suggested. To ap-

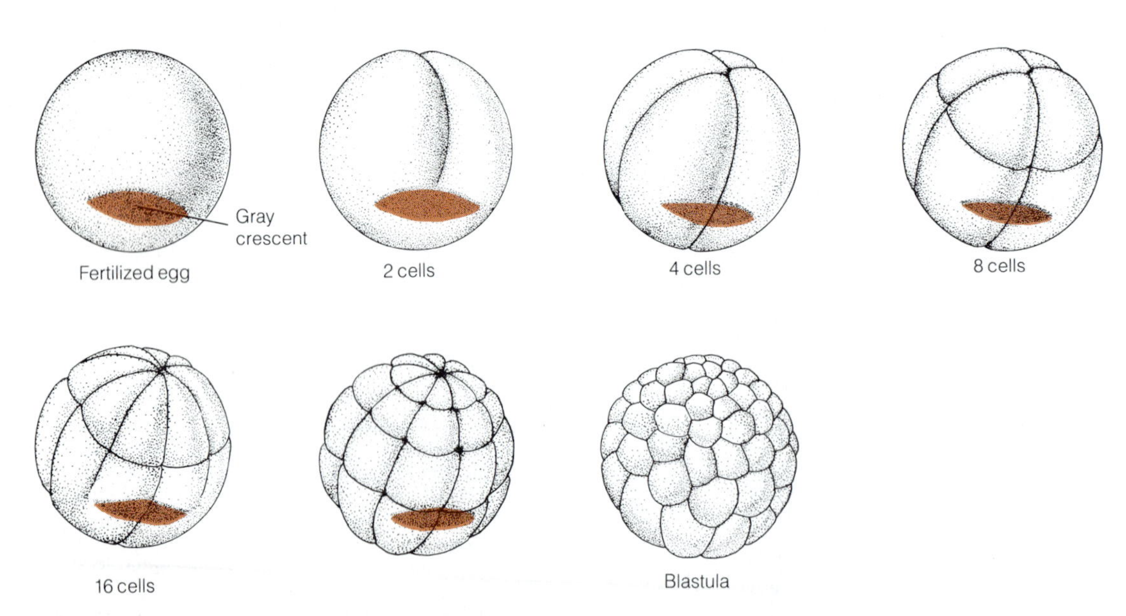

Figure 14-28 Cleavage of a Fertilized Egg into Progressively Smaller Cells. Amphibian eggs are very large, with enough cytoplasm to sustain many rounds of cell division after fertilization. Each round of division during early development parcels the cytoplasm into smaller cells.

preciate their significance, we must first recognize that the critical control point for the cell cycle appears to be in the G-1 phase.

G-1 as a Likely Control Point

Much evidence points to G-1 as the most critical phase in the cell cycle for regulatory purposes. We have already seen that this is the phase that varies most among cell types. Moreover, cells that have stopped dividing are almost always arrested in the G-1 phase. For example, we can stop or slow down the process of cell division in cultured cells by allowing the cells to run out of either nutrients or space or by adding inhibitors of vital processes such as protein synthesis. In all such cases, the cells are arrested in G-1.

These findings suggest that when a cell leaves G-1 and enters the S phase, it is committed to completing the cycle. Therefore, the release of cells from G-1 appears to be a critical control mechanism. More specifically, researchers have identified a "point of no return" in late G-1 called the **restriction point.** Cells that have passed this point are committed to division, whereas those that have not passed this point can remain in G-1 indefinitely.

Possible Regulatory Mechanisms

Many mechanisms have been suggested to explain cell cycle regulation, but it is often difficult to distinguish between causal relationships and simple correlations. Just because an event usually happens at a specific point in the cell cycle does not necessarily mean it is involved in regulation of the cycle. For example, it was long thought that the critical regulatory factor was the ratio of cytoplasmic mass to nuclear mass and that when the cytoplasm reached a certain size, cell division would be triggered. It is certainly true that cell division is generally correlated with an increase in cytoplasmic mass, but there is no evidence to suggest a causal relationship. Indeed, the cleavage of a fertilized amphibian or insect egg into many smaller cells by a series of cell divisions that are unaccompanied by cell growth directly contradicts such a suggestion.

Similarly, there appears to be no validity to the suggestion that the transition from G-1 to S is controlled by the availability of specific enzymes or factors required for DNA replication. The most conclusive evidence to dispute this model comes from experiments in which DNA is injected into unfertilized frog eggs. Such exogenous DNA is replicated efficiently, so all the necessary machinery must be on hand. Yet the frog egg remains arrested in G-1 until it is fertilized.

Recent research has provided several important leads that aid our understanding of regulatory aspects of cell division. These leads followed technical breakthroughs that make it possible for two different cells to be fused, forming a new cell with two nuclei. Such a cell is called a **heterokaryon** (Figure 14-29). Cells can be caused to fuse by several methods. For instance, certain viruses contain *fusion proteins* and, when added to the medium, will induce cells in tissue culture to fuse. High concentrations of *polyethylene glycol* are also commonly used to cause cell fusion. A relatively new method is called *electroporation.* In this technique, an electrical field is discharged through the medium. The electrical pulse causes membranes to destabilize momentarily, so that neighboring cells undergo fusion.

By fusing two cells at different stages of the cell cycle—for example, a cell in S phase and a cell in G-1—researchers can ask in which phase the regulatory mechanisms occur. In this example, for instance, is the G-1 nucleus activated by the S phase activity of the first cell, or does it continue in its original cycle? As depicted in Figure 14-29, experimental evidence showed that the G-1 nucleus immediately enters S phase, as though a signal present in the first cell initiates the process. The signal must disappear rapidly following mitosis, however, because fusion of a G-1 cell and a G-2 cell does not activate S phase.

Similar experimental approaches established that another powerful intracellular signal initiates mitosis. If a cell undergoing mitosis is fused with another cell in any stage of its cell cycle, the second nucleus is immediately driven into the preparatory steps for mitosis, including dissolution of the nuclear envelope, condensation of dispersed interphase chromatin into chromosomes, and spindle formation. This remarkable observation provides clear evidence for the existence of **maturation-promoting factors (MPF)**, which can activate the nuclear mitotic process at any point in the cell cycle. MPF activity has now been demonstrated in a broad range of eukaryotic cells. The protein responsible for MPF activity has been isolated and found to consist of two polypeptides with molecular masses of 32 and 45 kilodaltons. A similar protein with a molecular

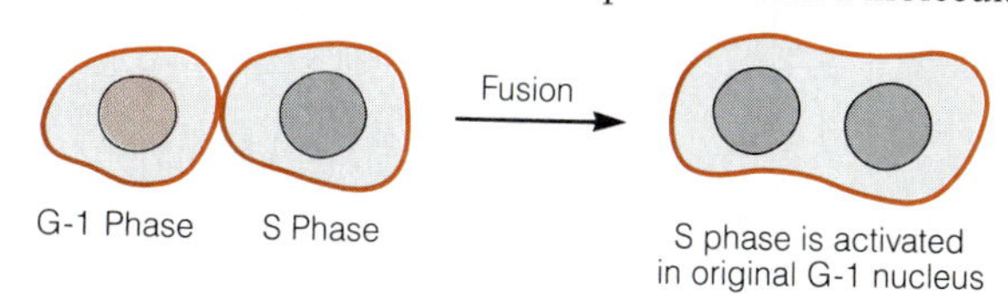

Figure 14-29 Cell Fusion Methods. Important information can be obtained from experiments in which two different cells are induced to fuse. Cell fusion can be brought about by any of several methods, including addition of fusigenic viruses or polyethylene glycol, or by a brief electrical pulse (electroporation). If cells in G-1 phase and S phase are fused, DNA synthesis begins in the original G-1 nucleus, suggesting that an S-phase activating substance is present in the S-phase cell.

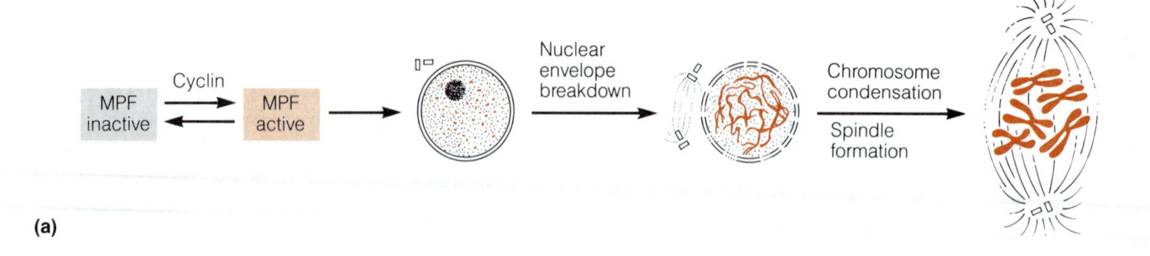

(a)

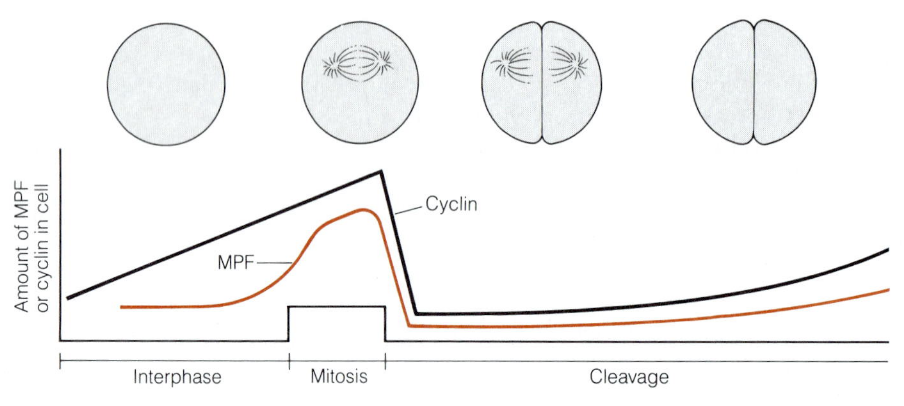

(b) Embryonic cell cleavage

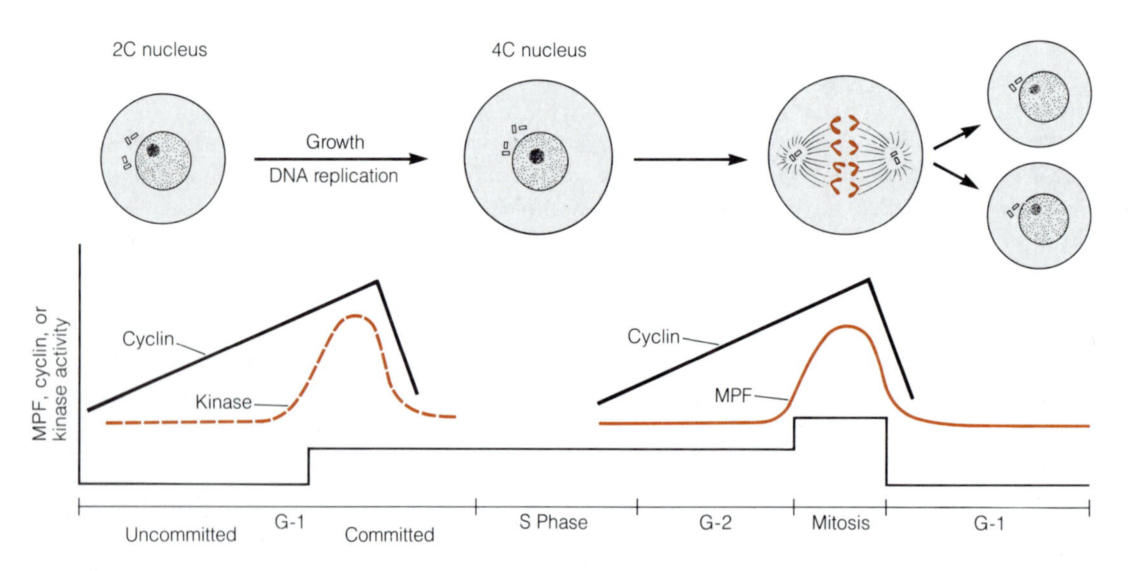

(c) Somatic cell cycle

Figure 14-30 Regulation of Cell Cycling. (a) Eukaryotic cells contain maturation-promoting factor (MPF), a protein kinase that is normally present in an inactive form. A second protein, cyclin, has the ability to activate MPF, which in turn activates a cascade of events leading to mitosis. (b) In embryonic cells of a fertilized egg, cyclin and MPF oscillate to control the simplified cell cycle of cleavage, which does not include an S phase. (c) In the somatic cell cycle, one form of cyclin appears to control a kinase that commits G-1 cells to enter S phase. After S phase is complete, a second form of cyclin activates MPF, which initiates mitosis. Following mitosis, cyclin is degraded by proteolysis and MPF returns to an inactive form.

mass of 34 kilodaltons initiates DNA synthesis and regulates mitosis in yeast. This protein is encoded by the *cdc* 2 gene (*c*ell *d*ivision *c*ontrol). Significantly, both MPF and the *cdc2* gene product are protein kinases and most likely activate a cascade of reactions that regulate mitosis.

Although such activating substances are important factors in the cell cycle, it is clear that they themselves must somehow be placed in a proper sequence and regulated. For instance, if MPF suddenly initiated mitosis in G-1, before chromosomes had duplicated their DNA in S phase, the result would be disastrous for the cell.

One attractive model to explain the control of cell division suggests that cells must accumulate a critical amount of a specific protein, which then causes the cell to pass the restriction point in G-1, enter S phase, and eventually divide. Such a **trigger protein** would have to accumulate to a sufficient extent during G-1, then rapidly be degraded following mitosis. Unlike proteins or factors actually involved in the process of DNA replication, the trigger protein would not be required for subsequent processes in the cell cycle but would function only to initiate them. One such protein was discovered recently in fertilized eggs undergoing cleavage. This protein increases and decreases in concentration markedly during the cell cycle and is therefore called **cyclin.** Significantly, when cyclin is introduced experimentally into a cell in G-2, mitosis begins almost immediately. Current evidence suggests that cyclin activates the protein kinase activity of MPF.

We now have enough information to establish a simple model for regulation of the cell cycle, as shown in Figure 14-30. This model depends on an oscillatory feedback control mechanism in which intracellular cyclin levels control the activity of MPF (Figure 14-30a). Cyclin is normally synthesized at a constant rate in a cell and accumulates slowly in the cytoplasm. At a certain cyclin concentration, the cells first become committed to S phase, followed by MPF activation and mitosis (Figure 14-30c). The mitotic process briefly activates cyclin degradation by proteolysis, so that MPF activity is turned off. Both proteins fall to minimal levels following cytokinesis, and the cell enters G-1, where the slow rise of cyclin begins again.

This is a simplified model, derived from investigations of cleaving amphibian eggs and yeast mutants. Further investigations will be required to establish more precisely the link between S phase and the oscillatory regulators, and the mechanism of the activating substances. Despite the uncertainties, we are now able to identify some of the major features controlling cell division, representing enormous progress in one of the most complex problem areas of cell biology.

Perspective

The eukaryotic cell cycle involves both nuclear and cellular events. Cellular division is called cytokinesis and is usually preceded by a growth phase. The nuclear division event is called mitosis, and the interphase period that follows is divided into an S phase and two "gap" phases, G-1 and G-2. Cultured mammalian cells often have generation times of about 18–24 hours, but in multicellular organisms, cells differ greatly in generation time, ranging from stem cells that divide rapidly and continuously to differentiated cells that do not normally divide at all.

DNA synthesis occurs during the S phase of the cell cycle, when the entire chromosome replicates. DNA replicates by a semiconservative mechanism that is catalyzed by DNA polymerase, using a short piece of RNA as a primer. DNA synthesis is continuous in the direction of replication along the leading strand but discontinuous in the opposite direction along the lagging strand. The Okazaki fragments generated on the lagging strand are joined together by DNA ligase. In *E. coli* cells, DNA polymerase III is the enzyme of replication, whereas DNA polymerase I is involved in repair of damage caused by spontaneous deamination, depurination, and dimer formation.

Mitosis consists of four phases: prophase, metaphase, anaphase, and telophase. During prophase, replicated chromosomes condense as sister chromatids, joined at the centromere. As metaphase approaches, the mitotic spindle forms by polymerization of tubulin into the microtubules of which the chromosomal fibers and spindle fibers consist. During metaphase, chromosomes are drawn into place along the equator of the spindle by the tension of being tugged in opposite directions at the same time. When the centromere divides, each pair of sister chromatids separates into daughter chromosomes, one drawn to each pole as the chromosomal fibers shorten during anaphase. The underlying mechanism is still somewhat controversial, but it probably involves disassembly of microtubules at the point of attachment of the chromosomal fiber to the kinetochore. During telophase, the separated chromosomes decondense, the nuclear membrane reappears, and nucleoli form again.

Cytokinesis usually begins before mitosis is complete. In animal cells, a cleavage furrow forms, which progressively restricts the cell at the midline and eventually separates the cytoplasm into two daughter cells. In plant cells, a cell wall forms through the middle of the parent cell.

Regulation of the cell cycle is a topic of much current research. An attractive model suggests that trigger proteins accumulate in cells, causing them first to pass the restriction point in G-1 and enter the S phase, then to undergo mitosis and eventually divide. One pair of trigger proteins—maturation-promoting factor (MPF) and cyclin—has been demonstrated to function in a variety of eukaryotic cells.

Key Terms for Self-Testing

cell division (p. 389)
daughter cells (p. 389)
cell cycle (p. 389)

Overview of the Cell Cycle
division phase (p. 389)
growth phase (p. 389)
mitosis (p. 389)
cytokinesis (p. 389)
interphase (p. 389)
S (synthetic) phase (p. 389)
sister chromatids (p. 390)
repair synthesis (p. 390)
G-1 phase (p. 391)
G-2 phase (p. 391)
mitotic spindle (p. 391)
generation time (p. 391)
mitotic index (p. 391)

DNA Replication
DNA replication (p. 393)
semiconservative replication (p. 393)
equilibrium density centrifugation (p. 393)
replication forks (p. 397)
origin (p. 398)
bidirectional replication (p. 398)
replication bubble (p. 398)
theta replication (p. 398)
replicon (p. 398)
DNA polymerases (p. 399)
DNA polymerase I (p. 399)
DNA polymerase III (p. 399)
holoenzyme (p. 399)
leading strand (p. 401)
lagging strand (p. 401)
Okazaki fragment (p. 401)
DNA ligase (p. 401)
primer (p. 401)
primase (p. 402)
primosome (p. 402)
primer recognition factors (p. 402)
helicase (p. 402)
gyrase (p. 402)
single-strand binding (SSB) protein (p. 402)
helix-destabilizing protein (p. 402)
unwinding protein (unwindase) (p. 402)

DNA Repair
mutations (p. 403)
depurination (p. 403)
deamination (p. 403)
pyrimidine dimer formation (p. 405)
repair endonuclease (p. 406)
DNA glycosidase (p. 406)
uracil-DNA glycosidase (p. 408)

Nuclear and Cell Division
prophase (p. 408)
metaphase (p. 408)
anaphase (p. 408)
telophase (p. 408)
centromere (p. 408)
spindle pole (p. 408)
mitotic center (p. 411)
centrioles (p. 411)
aster (p. 411)
astral mitosis (p. 411)
anastral mitosis (p. 411)
polar fibers (p. 411)
chromosomal fibers (p. 411)
kinetochores (p. 411)
kinetochore fibers (p. 411)
metaphase plate (p. 411)
daughter chromosomes (p. 411)
arms (of chromosome) (p. 411)
metacentric chromosome (p. 411)
acrocentric chromosome (p. 411)
telocentric chromosome (p. 411)
cleavage (p. 413)
cleavage furrow (p. 413)
contractile ring (p. 413)
cell plate (p. 415)
phragmoplast (p. 415)
early cell plate (p. 416)

Variations in the Cell Cycle
stem cells (p. 416)

Regulation of the Cell Cycle
restriction point (p. 419)
heterokaryon (p. 419)
maturation-promoting factor (MPF) (p. 419)
trigger protein (p. 421)
cyclin (p. 421)

Suggested Reading

General References and Reviews

Ayala, F. J., and J. A. Kiger, Jr. *Modern Genetics,* 2d ed. Menlo Park, Calif.: Benjamin/Cummings, 1984.

Watson, J. D., N. Hopkins, J. Roberts, J. Steitz, and A. Weiner. *Molecular Biology of the Gene,* 4th ed. Menlo Park, Calif.: Benjamin/Cummings, 1987.

The Cell Cycle

Dunphy, W. G., and J. W. Newport. Unraveling of mitotic control mechanisms. *Cell* 55 (1988): 925.

John, P. C. L., ed. *The Cell Cycle.* Cambridge, England: Cambridge University Press, 1981.

Murray, A. W., and M. W. Kirschner. Cyclin synthesis drives the early embryonic cell cycle. *Nature* 339 (1989) 275, 280.

Pardee, A. B., et al. The cell cycle. *Science* 246 (1989): 603–640. (Six articles on the cell cycle by leading investigators.)

Prescott, D. M. *Reproduction of Eukaryotic Cells.* New York: Academic Press, 1976.

Yanishevsky, R. M., and G. H. Stein. Regulation of the cell cycle in eukaryotic cells. *Int. Rev. Cytol.* 69 (1981): 223.

DNA Replication

Abdel-Momen, M., and H. Hoffman-Berling. DNA unwinding enzymes. *Trends Biochem. Sci.* 5 (1980): 128.

Alberts, B. M., and R. Sternglanz. Recent excitement in the DNA replication problem. *Nature* 269 (1977): 655.

DePamphilis, M. L., and P. M. Wassarman. Replication of eukaryotic chromosomes: A close-up of the replication fork. *Annu. Rev. Biochem.* 49 (1980): 627.

Ferscht, A. R. Enzymatic editing mechanisms in protein synthesis and DNA replication. *Trends Biochem. Sci.* 5 (1980): 262.

Gellert, M. DNA topoisomerases. *Annu. Rev. Biochem.* 50 (1981): 879.

Harland, R. Initiation of DNA replication in eucaryotic chromosomes. *Trends Biochem. Sci.* 6 (1981): 71.

Kornberg, A. *DNA Replication,* 2d ed. San Francisco: Freeman, 1982.

Kornberg, A. DNA replication. *Trends Biochem. Sci.* 9 (1984): 122.

Laskey, R. A., and R. M. Harland. Replication origins in the eukaryotic chromosome. *Cell* 24 (1981): 283.

Meselson, M., and F. W. Stahl. The replication of DNA in *E. coli. Proc. Natl. Acad. Sci. USA* 44 (1958): 671.

Ogawa, T., and R. Okazaki. Discontinuous DNA replication. *Annu. Rev. Biochem.* 49 (1980): 421.

Scovassi, A. I., P. Plevani, and U. Bertazzoni. Eukaryotic DNA polymerases. *Trends Biochem. Sci.* 5 (1980): 335.

Wang, J. C. DNA topoisomerases. *Sci. Amer.* 247 (January 1982): 94.

Watson, J. D., and F. H. C. Crick. Genetical implications of the structure of deoxyribonucleic acid. *Nature* 171 (1953): 964.

DNA Repair

Howard-Flanders, P. Inducible repair of DNA. *Sci. Amer.* 245 (November 1981): 72.

Kenyon, C. J. The bacterial response to DNA damage. *Trends Biochem. Sci.* 8 (1983): 84.

Sancar, A. Z., and G. B. Sancar. DNA repair enzymes. *Annu. Rev. Biochem.* 57 (1988): 29.

Nuclear and Cell Division

Bajer, A. S., and J. Molé-Bajer. Spindle dynamics and chromosome movements. *Int. Rev. Cytol.,* Suppl. 3 (1972).

Inoué, S. Cell division and the mitotic spindle. *J. Cell Biol.* 91 (1981): 132s.

McIntosh, J. R. Mechanisms of mitosis. *Trends Biochem. Sci.* 9 (1984): 195.

Pickett-Heaps, J. D., H. Tippit, and K. R. Porter. Rethinking mitosis. *Cell* 29 (1982): 79.

Rappaport, R. Cytokinesis in animal cells. *Int. Rev. Cytol.* 31 (1971): 169.

Yeoman, M. M., ed. *Cell Division in Higher Plants.* New York: Academic Press, 1976.

Problem Set

1. **The Cell Cycle.** Indicate whether each of the following statements is true of the G-1 phase of the cell cycle (G1), the G-2 phase (G2), the S phase (S), mitosis (M), or cytokinesis (C). A given statement may be true of any, all, or none of the phases.
   (a) The amount of nuclear DNA in the cell doubles.
   (b) The nuclear envelope is not visible.
   (c) Sister chromatids become separated from each other.
   (d) Cells that will never divide again are most likely arrested in this phase.
   (e) The primary cell wall of plant cells forms.
   (f) Chromosomes are present in their diffuse, extended form.
   (g) This phase is part of interphase.

2. **More Cell Cycle.** For each of the following pairs of phases from the cell cycle, indicate how you could tell which of the two phases a specific cell is in.
   (a) G-1 and G-2
   (b) G-1 and S
   (c) G-2 and mitosis
   (d) Mitosis and cytokinesis

3. **Meselson and Stahl Revisited.** Indicate the distribution on a cesium chloride gradient that you would expect for DNA molecules isolated from bacterial cells that had been grown for many generations on an ^{15}N medium and then switched to an ^{14}N medium for:
   (a) Three generations
   (b) Four generations
   (c) Two generations, followed by one generation on the ^{15}N medium again

4. **DNA Replication.** Sketch a replication fork of DNA in which one strand is being replicated discontinuously and the other is being replicated continuously. List six different enzyme activities associated with the replication process, identify the function of each, and indicate on your sketch where each would be located on the replication fork. Identify, in addition, the following features on your sketch: DNA template, RNA primer, Okazaki fragments, and single-strand binding protein.

5. **More DNA Replication.** Following are the results of five experiments carried out to determine the mechanism of DNA replication in the hypothetical organism *Fungus mungus*. For each experiment, indicate whether the results support (S), refute (R), or have no bearing (NB) on the hypothesis that this fungus replicates its DNA by the same mechanism as that known for *E. coli*. Explain your reasoning in each case.
   (a) Neither of the DNA polymerases of *F. mungus* appears to have an exonuclease activity.
   (b) Replicating DNA from *F. mungus* DNA shows continuous synthesis on both strands of the replication fork.
   (c) Some of the DNA sequences from *F. mungus* are present in multiple copies per genome, whereas other sequences are unique.
   (d) Short fragments of *F. mungus* DNA isolated during replication contain both ribose and deoxyribose.
   (e) When *F. mungus* cells are grown in the presence of the heavy isotopes ^{15}N and ^{13}C for several generations and then grown for one generation in normal (^{14}N, ^{12}C) medium, single-stranded DNA isolated from these cells yields a single band in a cesium chloride density gradient.

6. **DNA Damage and Repair.** Indicate whether each of the following statements is true of depurination (DP), deamination (DA), or dimer formation (DF). A given statement may be true of any, all, or none of these processes.
   (a) This process is caused by spontaneous hydrolysis of a glycosidic bond.
   (b) This process is induced by ultraviolet light.
   (c) This can happen to guanine but not cytosine.
   (d) This can happen to thymine but not adenine.
   (e) This can happen to thymine but not cytosine.
   (f) Repair involves a DNA glycosidase.
   (g) Repair involves a UV-specific nuclease.
   (h) Repair involves DNA ligase.
   (i) Repair depends on the existence of separate copies of the genetic information in the two strands of the double helix.
   (j) Repair depends on cleavage of both strands of the double helix.

7. **Deamination and Its Repair.** Shown in Figure 14-31a are three bases that are formed by deamination of DNA.
   (a) Indicate which base in DNA must be deaminated to form each of these bases.
   (b) Why are there only three bases shown, when DNA contains four bases?
   (c) Why is it important that none of the bases shown in Figure 14-31a occurs naturally in DNA?
   (d) Why is the presence of 5-methylcytosine (Figure 14-31b) in the DNA sequence likely to increase the probability of a mutation at that site?

8. **Xeroderma Pigmentosum.** *Xeroderma pigmentosum* is a rare skin disease in humans characterized by extreme sensitivity to sunlight, severe changes in the skin, and abnormally high rates of skin cancer. Fibroblasts (skin cells) from patients with xeroderma pigmentosum are almost completely incapable of excising pyrimidine dimers caused by ultraviolet light. Normal fibroblasts show a marked reduction in the molecular weight of single-stranded DNA within a few hours of irradiation with ultraviolet light, but xeroderma pigmentosum fibroblasts do not.
   (a) Why would an inability to excise pyrimidine dimers be likely to lead to severe abnormalities?
   (b) From what you know of the dimer excision process, why would you expect single-stranded DNA extracted from normal fibroblasts to have a lower molecular weight after exposure of the fibroblasts to ultraviolet light than before exposure?
   (c) On what basis might one conclude that xeroderma pigmentosum probably involves a defect in the UV-specific endonuclease involved in dimer excision?

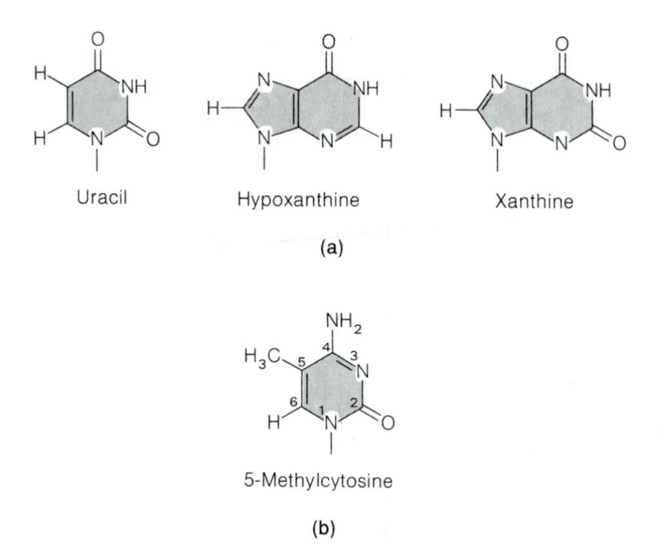

Figure 14-31 Structures of Several Purines and Pyrimidines. (a) The purines xanthine and hypoxanthine and the pyrimidine uracil are formed by deamination of naturally occurring bases in DNA. (b) 5-Methylcytosine is a naturally occurring pyrimidine in DNA.

(d) Would you expect xeroderma pigmentosum to be transmitted as a dominant or recessive genetic trait?

(e) What is the most important advice you can give to a person afflicted with xeroderma pigmentosum?

9. **The Mitotic Index and the Cell Cycle.** The mitotic index is a measure of the amount of mitotic activity in a cell. It is calculated as the percentage of cells in division (prophase, metaphase, anaphase, or telophase) at any one time. Assume that upon examination of a sample of 1000 cells, you find 50 cells in prophase, 20 in metaphase, 10 in anaphase, 20 in telophase, and 900 in interphase. Of those in interphase, 400 are found (by microspectrophotometric analysis after staining the cells with a DNA-specific stain) to have an X amount of DNA, 200 to have a 2X amount, and 300 cells to be somewhere in between. Autoradiographic analysis indicates that the G-2 period lasted 4 hours.

(a) What is the mitotic index for this population of cells?

(b) Specify the proportion of the cell cycle spent in each of the following phases: prophase, metaphase, anaphase, telophase, G-1, S, G-2.

(c) What is the total length of the cell cycle?

(d) What is the actual amount of time (in hours) spent in each of the phases of part (b)?

(e) To measure the G-2 period, radioactive thymidine (a DNA precursor) is added to the culture at some time *t*, and samples of the culture are analyzed autoradiographically for labeled nuclei at regular intervals thereafter. What specific observation would have to be made to assess the length of the G-2 period?

(f) What proportion of the interphase cells would you expect to exhibit labeled nuclei in autoradiographs prepared shortly after exposure to the labeled thymidine? (Assume a labeling period just long enough to allow the thymidine to get into the cells and begin to be incorporated into DNA.)

10. **Cell Cycle Variations.** A normal diploid cell from the insect *Imagineria diploides* contains ten chromosomes and 0.02 nanogram (ng) of DNA in its nucleus. Elevated amounts of DNA are found in certain *Imaginera* cells, as follows:

Cell type	DNA/cell	Nuclei/cell	Chromosomes/ nucleus
Embryo	20 ng	1000	10
Liver	20 ng	1	10,000
Salivary gland	20 ng	1	10

(a) Explain how the elevated DNA content of each of these cell types might have arisen.

(b) Which of these cell types would you expect to have the shortest generation time? Explain.

(c) Would you expect the salivary gland cells to be dividing actively? Explain.

(d) How might you induce the liver cells to begin dividing again?

11. **Cell Cycle Regulation.** One approach to the study of cell cycle regulation has been to fuse cultured cells that are at different stages of the cell cycle and observe the effect of the fusion on the nuclei of the fused cells. When cells in G-1 were fused with cells in S, the nuclei from the G-1 cells were observed to begin DNA replication earlier than they would have if they had not been fused. In fusions of cells in G-2 and S, however, nuclei continued their previous activities, apparently uninfluenced by the fusion. Fusions between mitotic cells and interphase cells always led to chromatin condensation in the nonmitotic nuclei. Based on these results, identify each of the following statements about cell cycle regulation as probably true (T), probably false (F), or not possible to conclude from the data (NP).

(a) The activation of DNA synthesis may result from the positive action of one or more cytoplasmic factors.

(b) The transition from S to G-2 may result from the presence of a cytoplasmic factor that inhibits DNA synthesis.

(c) The transition from G-2 to mitosis may result from the presence in the G-2 cytoplasm of one or more factors that induce chromatin condensation.

(d) G-1 is not an obligatory phase of all cell cycles.

(e) Like the transition from G-1 to S, the transition from G-2 to mitosis appears to be under positive control.

(f) The transition from mitosis to G-1 appears to be under negative control.

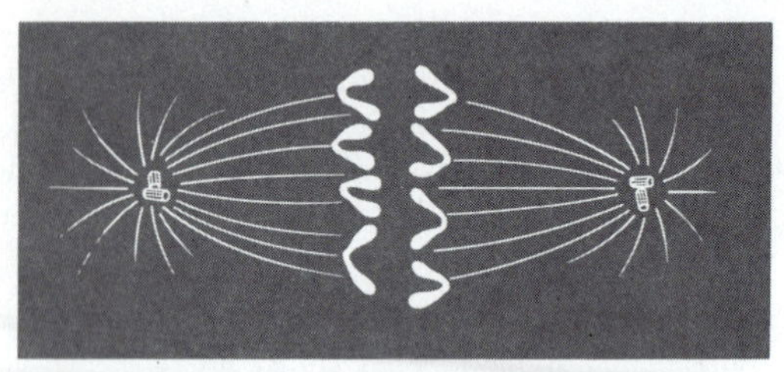

Microtubules are stained with a fluorescent antibody during anaphase.

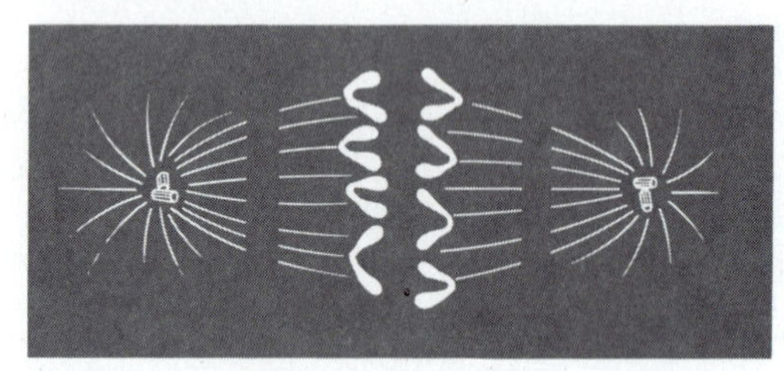

A laser microbeam is used to bleach the fluorescent marker (dark bands).

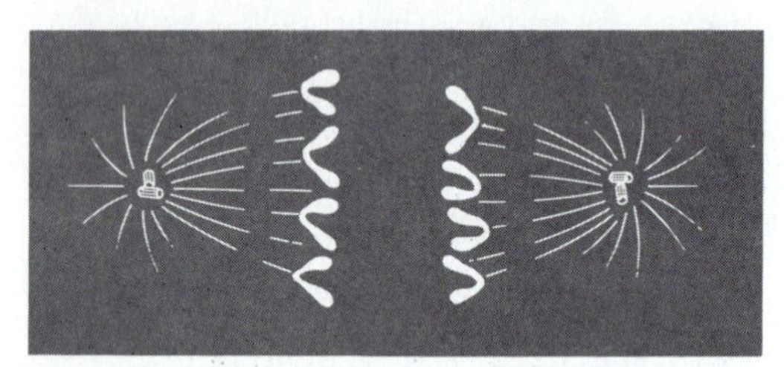

The chromosomes are observed to move toward the dark bands.

Figure 14-32 Chromosome Movement During Mitosis. Laser microbeam photobleaching can provide information about the mechanism of chromosome movement during mitosis.

12. **Chromosome Movement.** It is possible to "mark" the microtubules of a spindle by laser photobleaching (Figure 14-32). When this is carried out, it is found that chromosomes move *toward* the marked area during anaphase. Are the following statements consistent or inconsistent with this experimental result?

    (a) Microtubules move chromosomes by disassembling at the spindle poles.

    (b) Chromosomes move by disassembling microtubules with their kinetochore region.

    (c) Chromosomes are moved by a springlike elastic property of microtubules.

    (d) Chromosomes are moved along microtubules by a "motor" protein.

15

Sexual Reproduction, Meiosis, and Genetic Variability

The process of mitotic cell division that we encountered in the preceding chapter is responsible for the proliferation of eukaryotic cells, leading either to more organisms or to more cells per organism. Since mitosis involves the segregation of identical chromatids, the daughter cells of every mitotic division are genetically identical, or very nearly so. By starting with a single cell and allowing mitotic division to proceed for a number of generations, an organism can generate many cells, all of them genetically similar to the initial cell.

Mitotic division is an excellent way to perpetuate specific genetic traits faithfully, and it is the basis of all **asexual reproduction** in eukaryotes. In asexual reproduction, new individuals are generated by a single parent organism, whether unicellular or multicellular. Asexual reproduction is widespread in nature and is accomplished by different means in different groups of organisms (Figure 15-1). Examples include *mitotic division* of unicellular organisms (Figure 15-1a), *budding* of offspring as outgrowths from the parent's body (Figure 15-1b), and *regeneration* of whole organisms from pieces of a parent organism (Figure 15-1c). In plants, whole organisms can even be regenerated from single cells or clusters of cells isolated from the adult plant (Figure 15-1d).

Sexual Reproduction

The fundamental characteristic of asexual reproduction is that all progeny are genetically similar to the single parent organism from which they arise mitotically. **Sexual reproduction,** on the other hand, involves the mixing of genetic information from two parent organisms and results in offspring that are genetically dissimilar, both from each other and from the parents. Moreover, the offspring are unpredictably dissimilar; that is, we cannot anticipate exactly how the genetic information of the two parents will be rearranged in any particular offspring. Since most plants and animals—and even many microorganisms—reproduce sexually, this type of reproduction must provide some distinct advantages. We will look first at the advantages of sexual reproduction and then at the mechanisms whereby it is accomplished.

Advantages of Sexual Reproduction

Sexual reproduction has two important advantages over asexual reproduction: a competitive advantage when the organism is faced with environmental changes and an ability to combine into a single genome favorable mutations that arise in separate genomes.

The most fundamental advantage of sexual reproduction is the increased likelihood that the species will survive unpredictable changes in the environment. As long as the environment remains constant, asexual reproduction is not only adequate, it is preferable. If a species is already well adapted to its environment, then the genetic predictability of asexual reproduction fits well with the predictability of a static environment. If the environment changes, however, a genetically static organism may be ill suited to cope with the new conditions and will have no dependable mechanism for generating new combinations of genetic information that might enable it to adapt.

Sexual reproduction, on the other hand, involves a constant reshuffling of genetic information, thereby providing a range of individual variations within a given population. It follows that at least some of these variations will confer advantages on some members of a sexually reproducing population such that they will be better suited to changing and stressful conditions. Of course, other members of the population will be less suited for survival. For the species as a whole, however, it is clearly preferable that

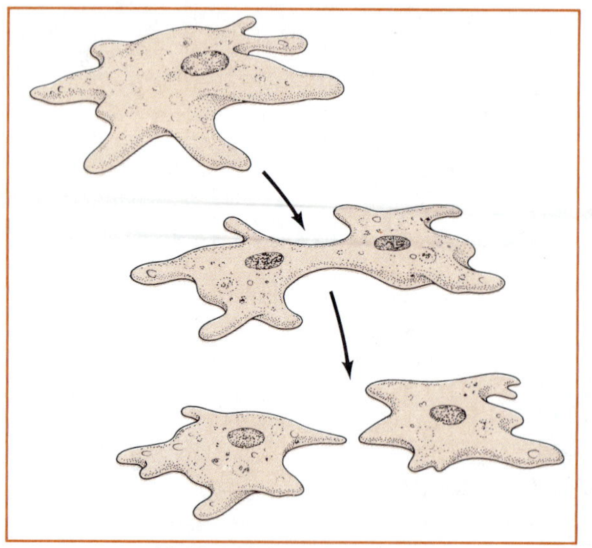

(a) Mitotic division of an amoeba

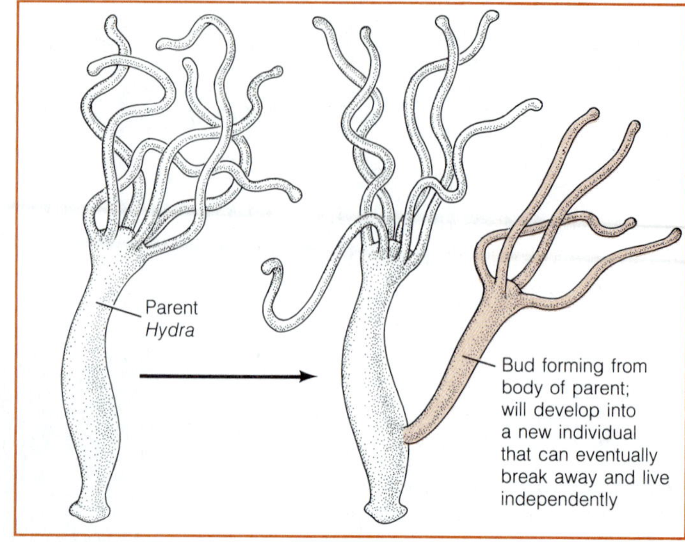

(b) Budding in the freshwater polyp *Hydra*

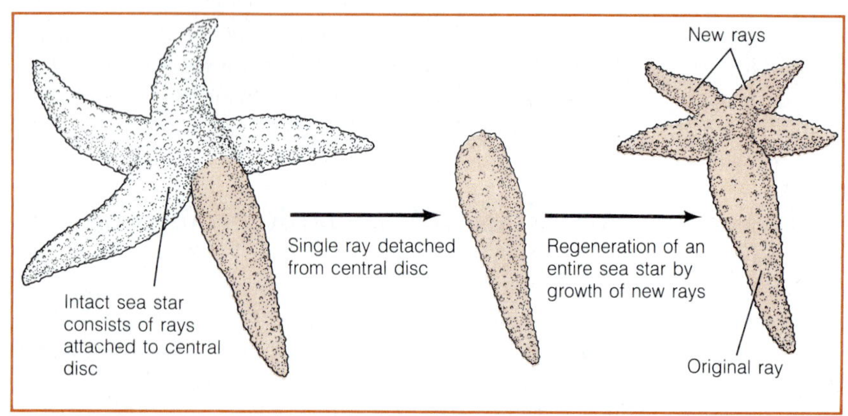

(c) Regeneration of a sea star from a detached ray of the parent organism

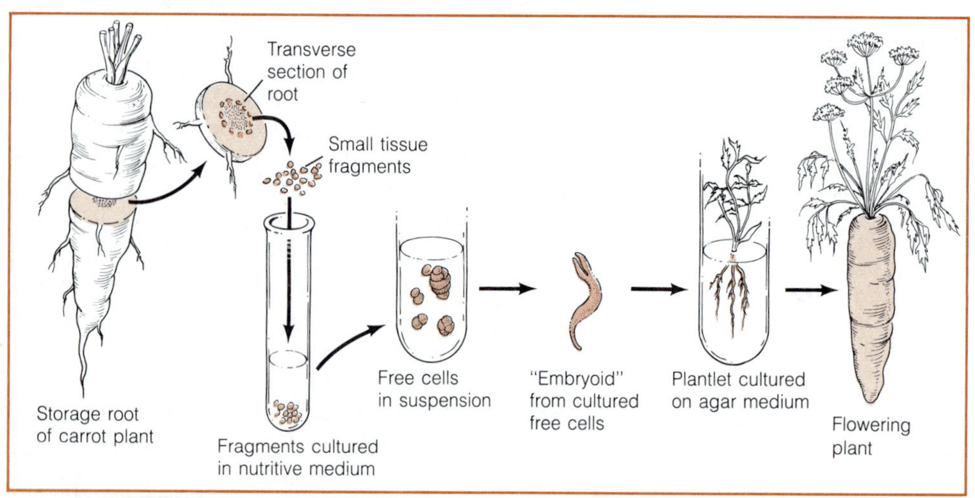

(d) Regeneration of an entire carrot plant from cultured cells obtained from a section of root tissue

Figure 15-1 Asexual Reproduction. In asexual reproduction, new individuals are generated from a single parent. Examples of asexual reproduction include (a) simple mitotic division of an amoeba cell; (b) budding of offspring as outgrowths from the body of a parent organism in *Hydra,* a freshwater polyp; (c) regeneration of an entire sea star (genus *Linckia*) from a single ray that was detached from a parent organism; and (d) regeneration of an entire carrot plant from a single cell obtained from an adult plant.

genetic variations are distributed throughout a population so that some members will be able to adapt to a changing environment.

The second major advantage of sexual reproduction is that it provides a means for combining in a single organism desirable genetic changes that arise in separate organisms. Genetic modification of organisms depends on the occurrence of **mutations,** which are spontaneous, unpredictable alterations in the genome, usually involving changes in specific bases or rearrangements in the order of nucleotides. Mutations are very rare events, and beneficial mutations are even rarer. When a beneficial mutation occurs, however, it is clearly to the advantage of the species to preserve the mutation in the genome. It is even more advantageous to combine several desirable mutations in a single genome—and therein lies the advantage of sexual reproduction. Although mutations can occur in both sexual and asexual species, only sexual reproduction can readily bring together beneficial mutations that arise in two separate organisms.

Sexual Reproduction and the Diploid Genome

By definition, sexual reproduction involves the combination in one organism of genetic information from two parents. At some point in its life cycle, every sexually reproducing species must therefore have a genome with two sets of chromosomes, one set inherited from each parent. Such a genome is said to be **diploid** (from the Greek word *diplous,* meaning "double"). A diploid genome contains two copies, or **alleles,** of the genetic information for a particular trait. A genome with a single set of chromosomes—and therefore a single gene per trait—is **haploid** (from the Greek word *haplous,* meaning "single"). By convention, the haploid chromosome number for a species is designated as $1n$ and the diploid number as $2n$.

In a diploid organism, the two alleles that govern a single trait may be either identical or different. In garden peas, for example, seeds may be either green or yellow, depending on which alleles are present in a particular genome (Figure 15-2). An organism with two identical alleles for a given trait is said to be **homozygous** for that trait. Thus, a pea plant that inherited an allele for yellow seed color from both of its parents is homozygous for yellow seed color. An organism with two *different* alleles for the same trait is said to be **heterozygous** for that trait. A pea plant with one allele that specifies yellow seed color and a second allele that specifies green seed color is therefore heterozygous for seed color.

In a heterozygous individual, one of the two alleles is often **dominant** (physically expressed in the organism) and the other allele is **recessive** (not physically expressed,

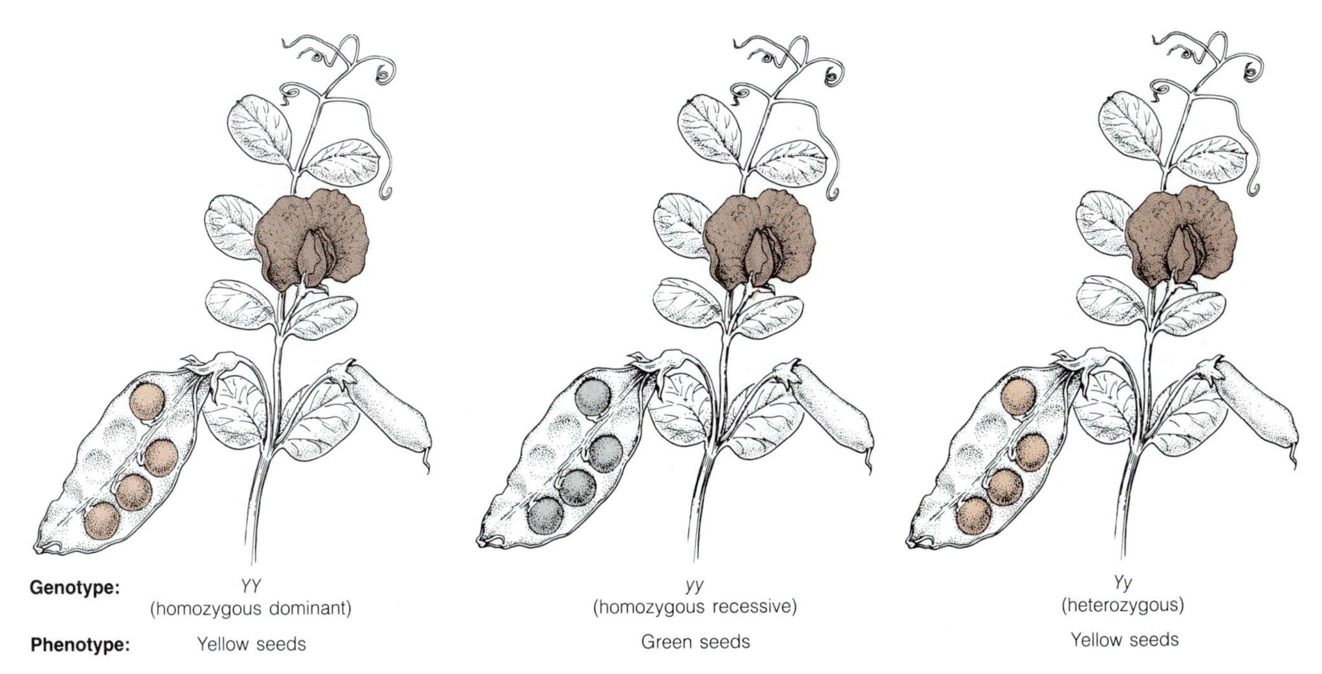

Figure 15-2 Genotype and Phenotype. In garden peas, seed color *(phenotype)* can be either yellow or green. The seed color alleles are *Y* (yellow, dominant) or *y* (green, recessive). Since the pea plant is a diploid organism, its genetic makeup *(genotype)* for seed color may be homozygous dominant (*YY*), homozygous recessive (*yy*), or heterozygous (*Yy*).

though present in the genome). In peas, for example, yellow seed color is dominant over green, so pea plants that are heterozygous for this trait will have yellow seeds. By convention, a dominant allele is designated by an uppercase letter that stands for the characteristic, and a recessive allele is represented by the same letter lowercased. Thus, the alleles for seed color in peas are represented by *Y* for yellow and *y* for green, because yellow is the dominant characteristic. As Figure 15-2 illustrates, a pea plant can have a genome that is homozygous for the dominant allele (*YY*), homozygous for the recessive allele (*yy*), or heterozygous (*Yy*).

It is important to distinguish between the **genotype,** or genetic makeup of an organism, and its **phenotype,** or physical appearance (Figure 15-2). The phenotype of an organism is always a consequence of its genotype and can usually be determined by inspection. Genotype, on the other hand, can usually be deduced only from information about the parents of the organism or from analysis of the organism's offspring. All organisms having the same phenotype do not necessarily have identical genotypes. In the example of Figure 15-2, pea plants with yellow seeds (phenotype) can be either *YY* or *Yy* (genotype).

Diploidy is a necessary feature of sexually reproducing species and is a distinct advantage to them because of the greater genetic flexibility it allows. In most cases, a diploid organism can function adequately with a single allele for a given trait, even though it has two alleles for every trait. In a sense, then, a diploid genome contains a spare set of genes that is not immediately necessary to the organism and is therefore available for mutation and genetic innovation. Changes in a spare gene, in other words, will not threaten the survival of the organism even if the mutation is deleterious to the original function of that particular gene.

Gametogenesis and Fertilization

The hallmark of sexual reproduction is that it brings together in a single organism genetic information contributed by two parents, one usually identified as the male and the other as the female. (The distinction between male and female parents is not universal, however; among some of the lower eukaryotes, parent organisms differ not in sex but in mating type.) Because sexual reproduction results in offspring that are diploid, the contribution from both the male and female parents must be haploid. Haploid cells that are specialized for sexual reproduction are called **gametes,** and the process whereby they arise is called **gametogenesis.** In humans, as in most other higher animals, gametogenesis occurs in the **gonads.** Female gametes, called **eggs** or **ova** (singular: ovum), are produced in the **ovaries;** male gametes, called **sperm** or **spermatozoa,** are produced in the **testes.** Eggs are typically large and nonmotile, whereas sperm are usually small and motile.

The union of a sperm and an egg is called **fertilization.** The fertilized egg that results is diploid, having received one chromosome set from the sperm and one from the egg. This fertilized, diploid egg is called a **zygote.** In the case of multicellular organisms, fertilization is followed by **embryonic development**—a series of mitotic divisions and progressive specialization of various groups of cells to form the multicellular embryo. Gametogenesis, fertilization, and embryonic development will be considered in more detail in Chapter 22. For a discussion of what determines the sex of offspring in mammals, see the box on p. 432.

Meiosis

Gametes are haploid cells and therefore cannot be generated from diploid precursors by mitosis, since mitotic division always results in daughter cells that are genetically identical to the parent cell. If gametes were formed mitotically, both sperm and egg would have a diploid chromosome number, and the zygote that resulted from their fusion would be tetraploid. In fact, the chromosome number would then double every generation, which is clearly not the case. Thus, another kind of division must occur to maintain the constancy of chromosome number from generation to generation. That process, first described in the 1880s, is meiosis.

Meiosis can be defined as two successive nuclear divisions with only one duplication of chromosomes. The products are four daughter nuclei (each in a separate daughter cell, usually) with only one set of chromosomes per nucleus. Figure 15-3 provides an overview of the meiotic process by showing a single period of DNA replication followed by two division events, *meiosis I* and *meiosis II,* which result in the formation of four haploid daughter cells.

Meiosis and the Life Cycle

Both meiosis and fertilization are indispensable components of the life cycle of every sexually reproducing organism, because the doubling of chromosome number that takes place at fertilization is balanced by the halving that occurs during meiosis. As a result, the life cycle of sexual organisms is divided into two phases: a diploid ($2n$) phase

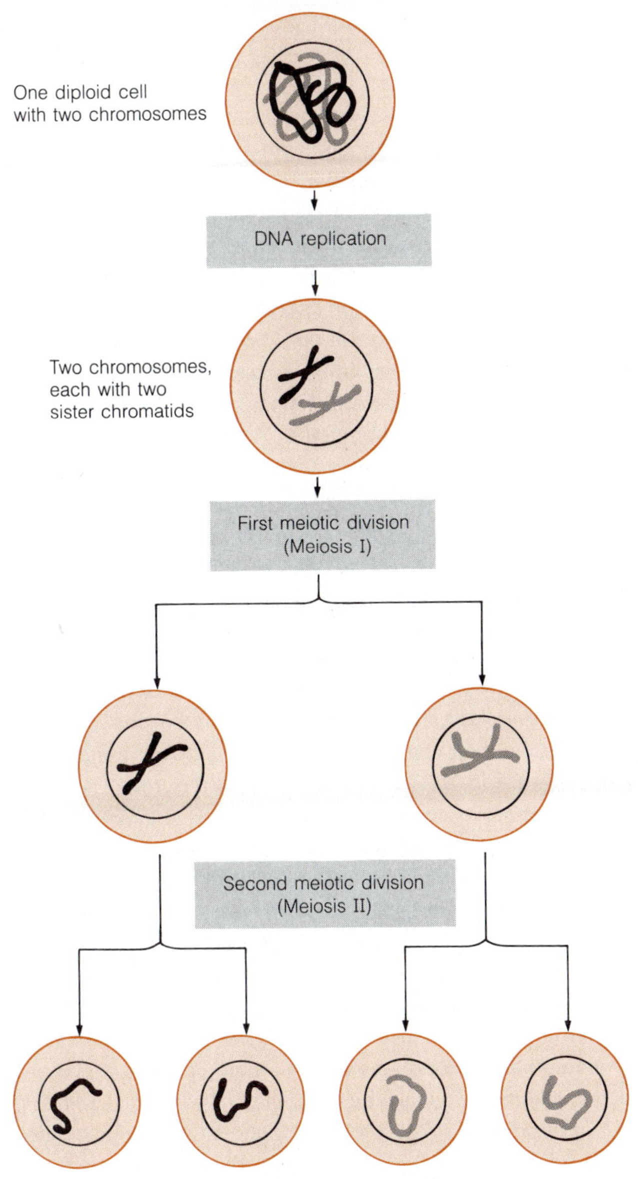

Figure 15-3 Overview of Meiosis. Meiosis involves a single period of DNA replication followed by two successive division events. In this example, the diploid cell has only two chromosomes, one shown in black and the other gray. After replication of the DNA, each chromosome consists of two sister chromatids. During the first meiotic division (meiosis I), homologous chromosomes separate, but sister chromatids remain attached. During the second meiotic division (meiosis II), sister chromatids separate, resulting in four haploid daughter cells with one chromosome each.

and a haploid (1*n*) phase. The diploid phase begins at fertilization and extends until meiosis, whereas the haploid phase is initiated at meiosis and ends in fertilization.

Organisms vary greatly in the relative prominence of the haploid and diploid phases of their life cycles (Figure 15-4). In general, bacteria reproduce asexually and therefore do not have a diploid phase in their life cycle (Figure 15-4a). (As we will learn later, some bacteria have a means of transmitting DNA from one cell to another that is usually labeled "sexual." However, such transmission of genetic information is not an inherent part of the life cycle in these bacteria, and it often involves only a portion of the genome; therefore, it does not qualify as true sexual reproduction.)

Fungi are also haploid organisms, but they can reproduce sexually; their life cycle therefore includes a diploid phase that begins with gamete fusion (the fungal equivalent of fertilization) and ends with meiosis (Figure 15-4b). However, meiosis usually occurs almost immediately after gamete fusion, so the diploid phase is very short and diploid cells are accordingly very rare.

Mosses and ferns are probably the best examples of organisms in which both the haploid and diploid phases are prominent features of the life cycle. Every species of these plants has two alternative, morphologically distinct forms, one haploid and the other diploid (Figure 15-4c). For mosses, the haploid form is the more prominent, and the diploid form is a modest, rather inconspicuous structure. For ferns, it is the other way around.

Organisms that alternate between haploid and diploid forms in this way are said to display an **alternation of generations** in their life cycles. In addition to mosses and ferns, eukaryotic algae and all higher plants also have an alternation of diploid and haploid generations in their life cycles. In all such organisms, the products of meiosis are always haploid **spores,** which, upon germination, give rise to the haploid form of the plant. The haploid form in turn produces the gametes, which, upon fertilization, give rise to the diploid form. Because the diploid form produces spores, it is called a **sporophyte** ("spore-producing plant"). The haploid form produces gametes and is therefore called a **gametophyte.** All plants, including the angiosperms (seed-bearing plants), have an alternation of generations. Among the higher plants, however, the sporophyte generation is predominant, and the gametophyte generation is an almost vestigial structure located in the appropriate floral part (the *male gametophyte* in the anther, the *female gametophyte* in the carpel).

The best examples of a life cycle dominated almost completely by the diploid form are found among the higher animals, including humans (Figure 15-4d). In such organisms, meiosis gives rise not to spores but to gametes directly, so the haploid phase of the life cycle is represented only by the gametes. Meiosis in such species is called **gametic meiosis** to distinguish it from the **sporic meiosis** that occurs in spore-producing organisms with an alternation of generations. Meiosis is therefore gametic in animals and sporic in plants.

Male or Female?

What determines the sex of an organism? At least for human beings, the distinction of being male or female depends on the "sex" of the sperm cell that fertilizes the ovum at conception. Haploid sperm contain 23 chromosomes, of which one is either an X chromosome or Y chromosome. The haploid ovum, on the other hand, always contains an X chromosome. The diploid cell that results from fertilization therefore has either an XX or XY pair of chromosomes. The Y chromosome, when present, somehow confers maleness on the infant that grows from the fertilized egg.

And yet, there is a surprising exception to this rule: a few rare individuals with the XX chromosome pair grow up as functioning males. This fact forces us to think more deeply about the genetic control of maleness, and particularly about where the male/female switch might reside in the human genome. There is no doubt that in normal men maleness is conferred by genetic information contained in the Y chromosome. What appears to happen in the XX men is that one of their X chromosomes carries a small segment that was transposed from their father's Y chromosome during meiosis. This segment is now referred to as the testis-determining factor (*TDF*). If it is present, the child is indisputably male, even though he has what appear to be normal XX chromosome pairs.

The existence of XX males in the human population turns out to be a valuable tool for furthering our understanding of sexual development. If maleness depends only on a small segment of the Y chromosome, some obvious questions follow: What is the actual size of the segment? Where on the Y chromosome does it reside? Does the segment contain one, several, or many genes?

Although the *TDF* has been known since 1966 to be located on the short arm of the Y chromosome, it is only in recent years that its location has been pinned down more precisely. In 1990, Andrew Sinclair and co-workers isolated a *TDF*-containing region about 35,000 nucleotides long and determined its nucleotide sequence. Numerous repeating sequences were found, confusing the search, but finally a single-copy gene less than 250 nucleotides long was discovered.

Could this be the long-sought-after *TDF*? It is still not certain, but supportive evidence is accumulating. For instance, one would expect such an important gene to be conserved within mammalian species, and, in fact, the sequence of the human gene has 80% homology with the equivalent region of the mouse Y chromosome. Furthermore, the gene is expressed in the mouse about 11.5 days after fertilization, just when sexual differentiation begins in the embryo. This evidence is sufficiently strong that the new gene is now called the *sex determining region of the Y chromosome*, or *SRY* for short.

Proof of *SRY*'s function will probably come from further research with mice. One approach is to use a method by which a gene can be transferred from one animal to another. This method was first demonstrated in 1982 by Richard Palmiter and Ralph Brinster, who showed that it is possible to insert single genes for rat growth hormone into fertilized mouse ova. A few of the adult mice expressed the gene, producing growth hormone at levels hundreds of times greater than their control littermates. As shown in the box in Appendix B (pp. 814–815), the extraordinary result was rat-sized mice! Such animals are now referred to as *transgenic*, and the procedure has become fairly commonplace.

We can now see how transgenic methods could be applied to the *SRY* question. If the *SRY* gene could be inserted into the genome of fertilized XX mouse ova, one would expect that some of the mice would express the gene and grow into males. A positive result in such an experiment would be highly convincing. Most remarkably, it would mean that a fundamental aspect of human life—being male or female—is determined by a cascade of regulatory genetic events that is triggered by a single DNA sequence only 250 nucleotides long.

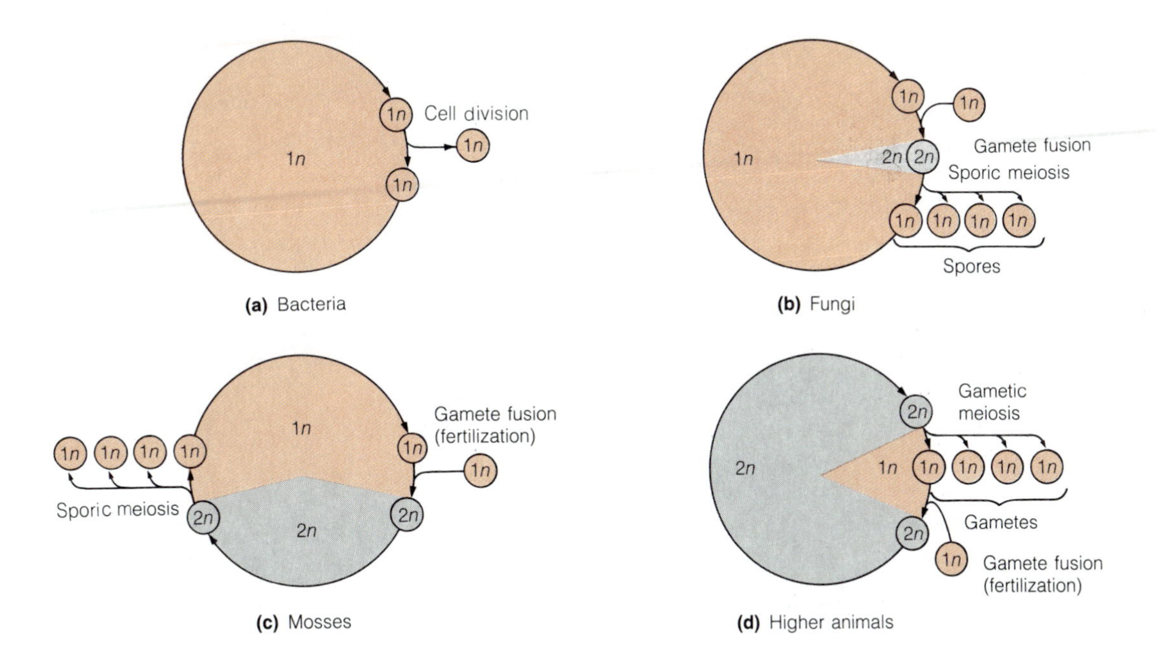

Figure 15-4 Life Cycles. The relative prominence of the haploid (1*n*) and diploid (2*n*) phases of the life cycle differ greatly, depending on the organism. (a) Bacteria exist exclusively in the haploid state. (b) Fungi exemplify a life form that is predominantly haploid but has a brief diploid phase. Because the products of meiosis are haploid spores, this type of meiosis is called sporic meiosis. (c) Mosses alternate between haploid and diploid forms, both of which are significant components in the life cycle of these organisms. (d) Higher animals are the best examples of organisms that are predominantly diploid, with the gametes representing the only haploid phase of the life cycle. Animals have a gametic meiosis, since the products of meiosis are haploid gametes.

The Process of Meiosis

Wherever it occurs in the life cycle, meiosis is always characterized by an initial chromosome duplication followed by two successive division events (Figure 15-5). Meiosis begins with a pairing of each chromosome from the maternal set with the related, or *homologous,* chromosome from the paternal set. In the first division, called **meiosis I,** homologous chromosomes **synapse** (come together) to form structures called **bivalents,** which become aligned at the metaphase plate in the middle of the cell. Each bivalent contains a pair of homologous chromosomes and is therefore a bundle of four chromatids. Homologous chromosomes then separate in such a way that each daughter nucleus from the first meiotic division contains one of the two chromosomes of each original bivalent. The formation of a second metaphase plate follows almost immediately. **Meiosis II** ensues, resulting in the separation of sister chromatids in a manner that is analogous to the mitotic process. Figure 15-6 compares meiosis with mitosis to illustrate both the similarities and the differences between the two processes.

We can also compare meiosis with mitosis by considering the amount of DNA per nucleus. As we saw in the preceding chapter, the DNA level in mitotically dividing cells rises slowly to the 4C level during S phase, then returns to the 2C level upon division (recall Figure 14-2). In meiosis, a single duplication process followed by two successive divisions results in a rise from the 2C level to the 4C level followed by a reduction in the amount of DNA per nucleus from the 4C level back to the 2C level and then to the 1C level (Figure 15-7). Upon fertilization, the amount of DNA is restored to the 2C level in the zygote.

Each meiotic division consists of the same four stages as mitosis: *prophase, metaphase, anaphase,* and *telophase.* However, prophase I is much more complicated than mitotic prophase, and prophase II, by contrast, is often very short or even absent. The interphase prior to meiosis is quite similar to mitotic interphase in that the main characteristics are the duplication of chromosomes and the consequent doubling of the DNA content.

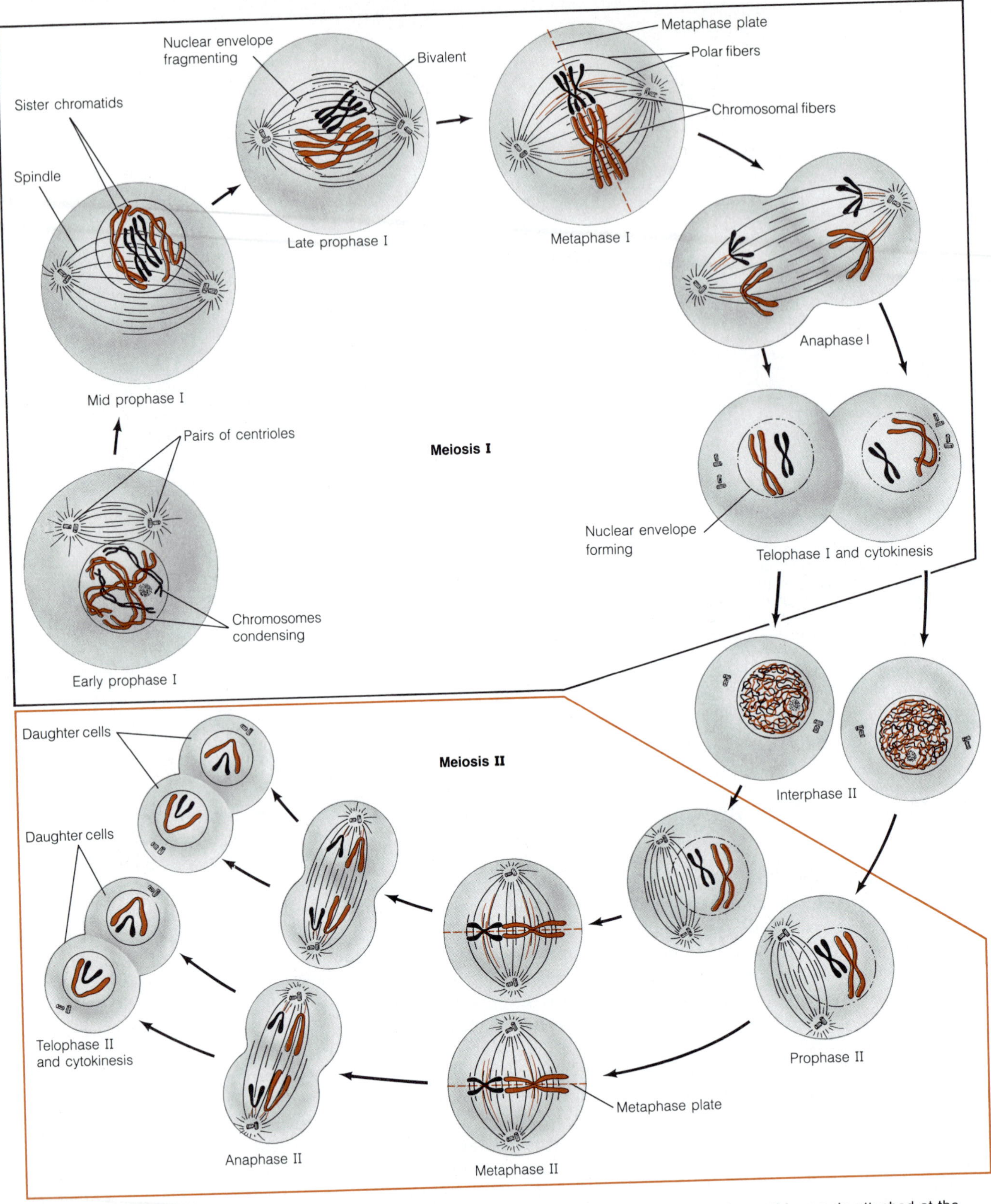

Figure 15-5 Meiosis. Meiosis consists of two successive divisions with no intervening DNA synthesis or chromosome replication. After DNA replication, chromosomes begin to condense and centrioles migrate to opposite sides of the nucleus (early prophase I). Each chromosome (four, in this example) consists of two sister chromatids (mid prophase I). Homologous chromosomes pair to form bivalents (late prophase I), which become aligned at the metaphase plate (metaphase I). Homologous chromosomes separate during meiosis I, but sister chromatids remain attached at the centromere (anaphase I). Following a short interphase (interphase II), chromosomes recondense (prophase II), a second metaphase plate forms (metaphase II), and sister chromatids now separate (anaphase II). The result is four haploid daughter cells, each containing one copy of each chromosome (telophase II). Prophase I is a complicated process shown in more detail in Figure 15-8. Prophase II is either very short or absent entirely.

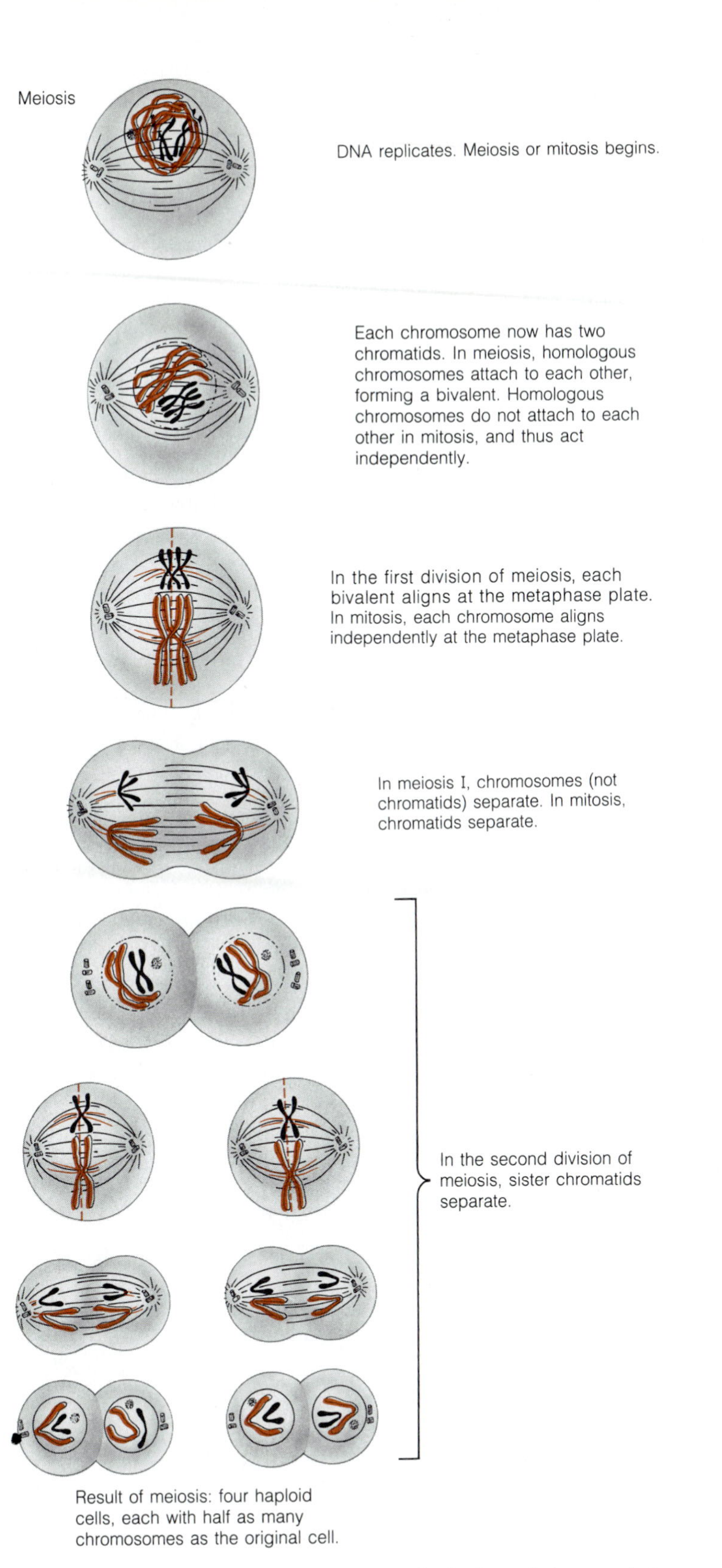

Figure 15-6 Comparison of Meiosis and Mitosis. Meiosis and mitosis are both preceded by DNA replication, resulting in two chromatids per chromosome. Meiosis involves two divisions, halving the chromosome number to the haploid level. Mitosis involves only a single division, producing two diploid cells, each with the same number of chromosomes as the original cell.

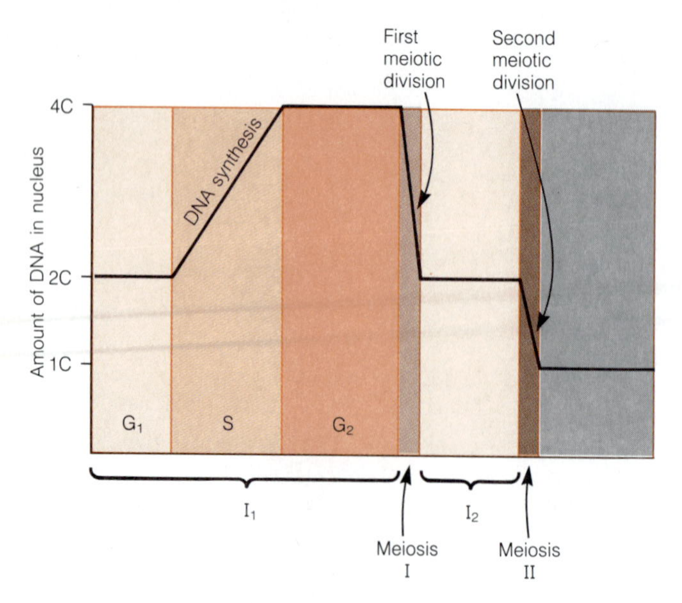

Figure 15-7 Nuclear DNA Content During Meiosis. DNA content doubles from the 2C level to the 4C level during the S phase preceding division, is then halved from 4C to 2C during the first meiotic division, and is halved again from 2C to 1C during the second meiotic division, resulting in haploid gametes. I_1 and I_2 are the interphases that precede meiosis I and meiosis II, respectively. I_1 consists of a synthetic (S) phase flanked by two gaps, G_1 and G_2.

Meiosis I

The first meiotic division results in the separation of homologous chromosomes. This is the feature of meiosis that is of greatest genetic significance, since it is at this point in the life cycle that the two alleles for each genetic locus part company. Also of considerable significance is the localization of genetic recombination to this event in the life cycle. **Recombination,** as we will learn shortly, involves a physical exchange of genetic information between homologous chromosomes as they are paired during prophase I.

Prophase I. **Prophase I** is a particularly complex stage, rendered even more so by the complicated terminology conferred on it by early cytologists. They divided prophase I into five substages, called the *leptotene, zygotene, pachytene, diplotene,* and *diakinesis* phases (Figure 15-8). The **leptotene** phase is characterized by the initial condensation and coiling of chromosomes, which appear as very long, threadlike structures. At **zygotene,** homologous chromosomes have begun to pair and synapse laterally into bivalents. Keep in mind that each such bivalent contains four chromatids and two centromeres. Bivalent formation is of great genetic significance because it is between the chromatids of bivalents that **crossing over** occurs. Crossing over is the physical exchange of genetic information between homologous chromosomes that accounts for recombination.

By the **pachytene** phase, synapsis is complete, and the bivalents have become still shorter and thicker. Homologous chromosomes are held in tight apposition by a specialized structure called the **synaptonemal complex** (Figure 15-9). The synaptonemal complex forms just before pachytene and disappears shortly thereafter. It appears to be essential for crossing over to occur. At the **diplotene** phase, the chromosomes are still more highly condensed, and the chromosomes of each bivalent begin to separate from each other, particularly near the centromere. As a result of this partial separation, **chiasmata** (singular: *chiasma*) become visible. Each chiasma represents a zone of contact along the chromosome pair at which a crossover event has occurred between two chromatids, one from each of the two homologous chromosomes. A chiasma is therefore the physical manifestation of what the geneticist recognizes as recombination.

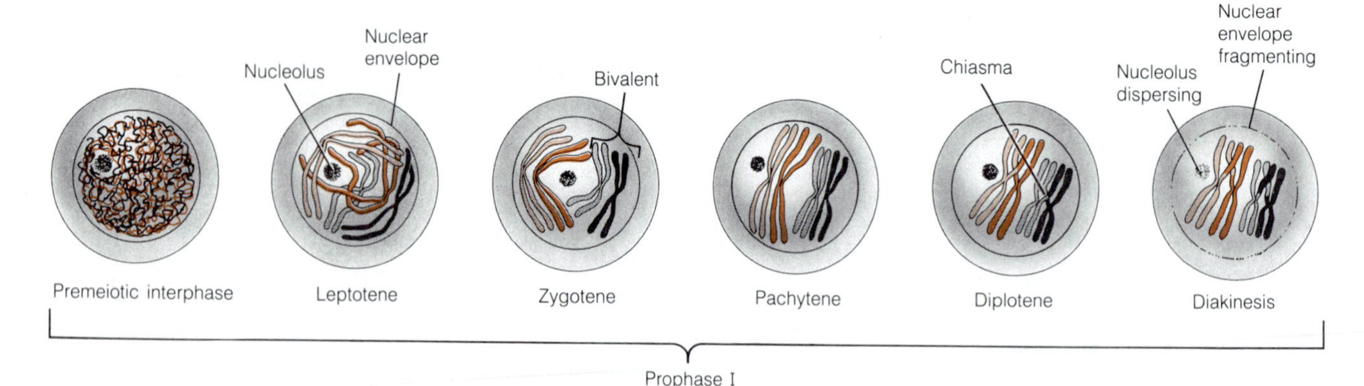

Figure 15-8 Meiotic Prophase I. Prophase I is a complicated process that is subdivided into five stages.

Diakinesis is the final phase of prophase I and is characterized by maximally condensed and coiled chromosomes. As the centromeres continue to separate, the zones of contact appear to move toward the ends of the bivalents, eventually becoming the final points of contact between the separating homologues. Diakinesis is also the stage at which the nuclear envelope fragments and the nucleoli disperse.

Metaphase I. At **metaphase I,** the bivalents become attached to the spindle fibers by their centromeres and line up on the metaphase plate, with the two centromeres of each pair on opposite sides of the plate. The number of bivalents is equal to the haploid chromosome number for a given species. The pairing of homologous chromosomes is a crucial distinction between metaphase I of meiosis and the metaphase of mitosis, where such pairing does not occur.

Anaphase I. At **anaphase I,** the centromeres of each bivalent now begin to move toward opposite poles, pulling the homologous chromosomes of each bivalent apart. At this stage, the recombined genetic information becomes physically separated. Again, note the distinction between meiosis and mitosis. In mitotic anaphase, centromeres and sister chromatids separate. In anaphase I of meiosis, homologous centromeres and chromosomes separate, but sister chromatids remain together.

Telophase I. The movement of homologous chromosomes to opposite poles is completed during **telophase I.** The cell then divides into two daughter cells, each ready for the second division.

Meiosis II

The second meiotic division usually follows almost immediately; the interphase between the two divisions is either very short or nonexistent, depending on the species. Chromosomes are already duplicated as meiosis II begins, but only one set of chromosomes is present because of the prior separation of homologues. **Prophase II** is very brief. If detectable at all, it is much like mitotic prophase. **Metaphase II** also resembles the equivalent stage in mitosis, except that only half as many chromosomes are present on the metaphase plate. **Anaphase II** follows, effecting the separation of sister centromeres. What were until this point sister chromatids now become chromosomes and move toward opposite spindle poles. This movement is completed in **telophase II,** and a nuclear envelope re-forms around each of the haploid nuclei.

The micrographs of Figure 15-10 illustrate each of the four phases of both meiotic divisions as they occur in the male grasshopper.

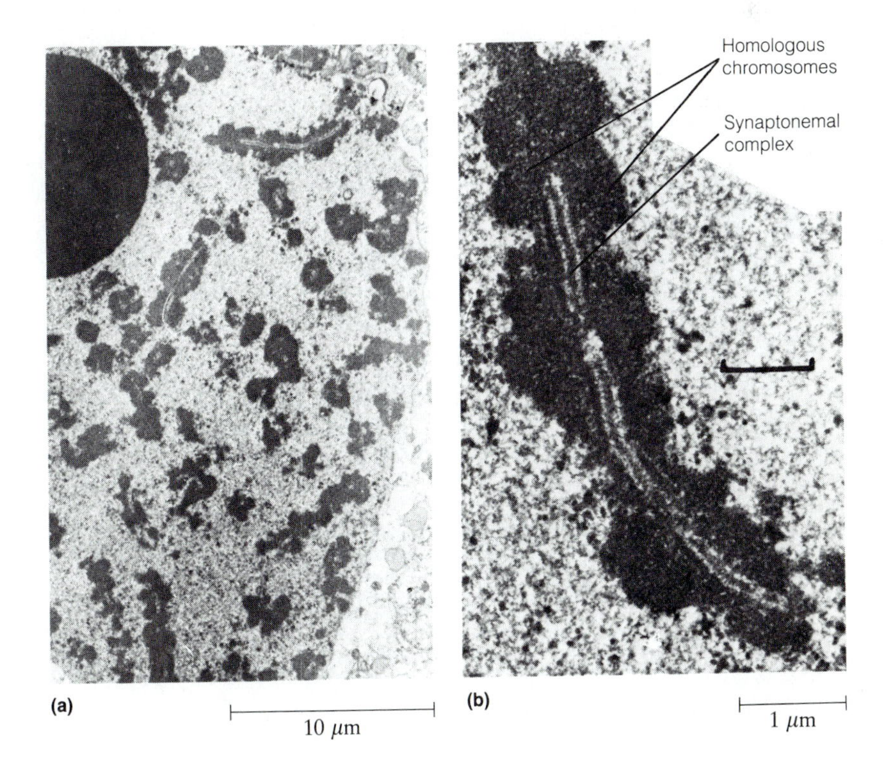

Figure 15-9 The Synaptonemal Complex. The synaptonemal complex is a structure that draws homologous chromosomal regions together at the pachytene stage of meiotic prophase I, just prior to the initiation of crossing over. These electron micrographs illustrate synaptonemal complexes in the nuclei of cells from a lily. (both TEMs)

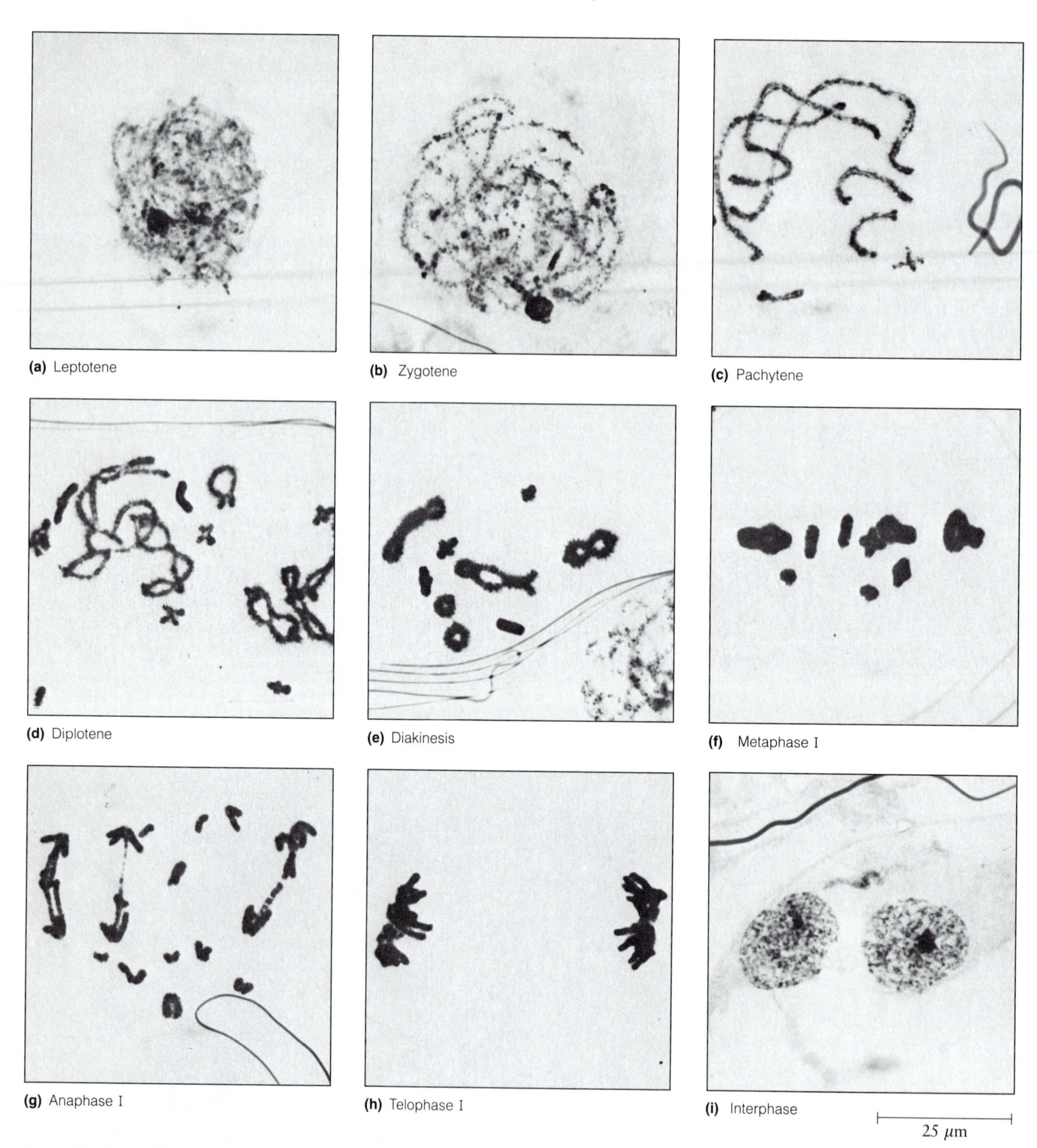

Figure 15-10 Meiosis in the Male Grasshopper. The micrographs illustrate meiosis as it occurs in the testes of a male grasshopper, *Chorthippus parallelus* (all LMs). Each of the phases of meiosis I and meiosis II is shown, including the five stages of prophase I.

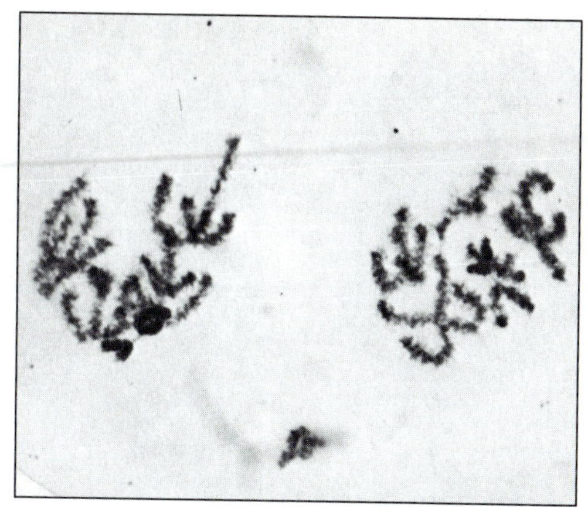

(j) Prophase II

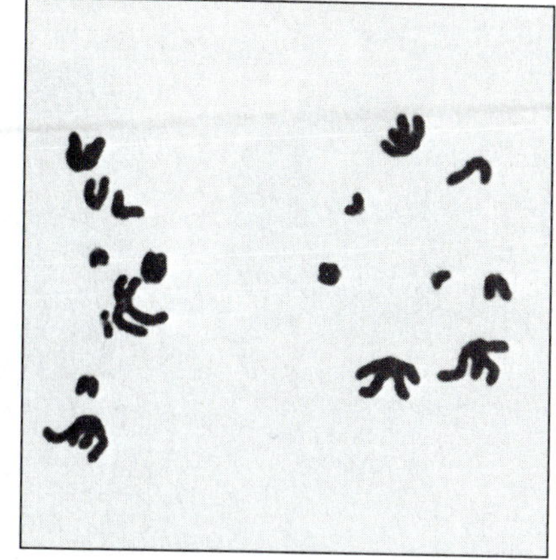

(k) Metaphase II

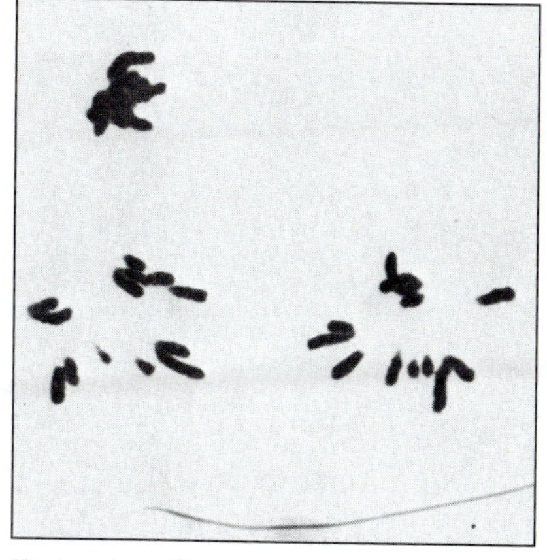

(l) Anaphase II

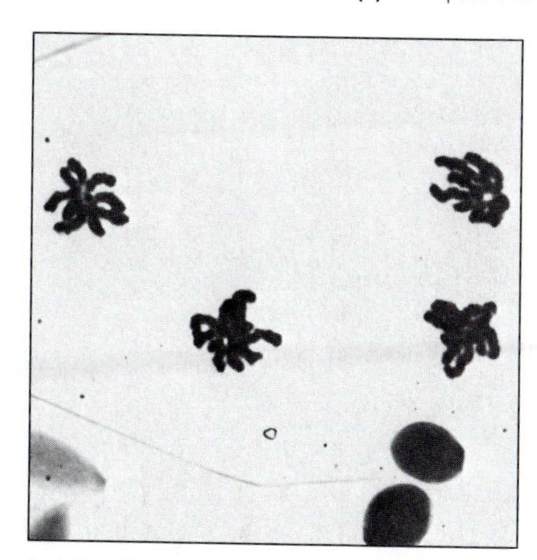

(m) Telophase II

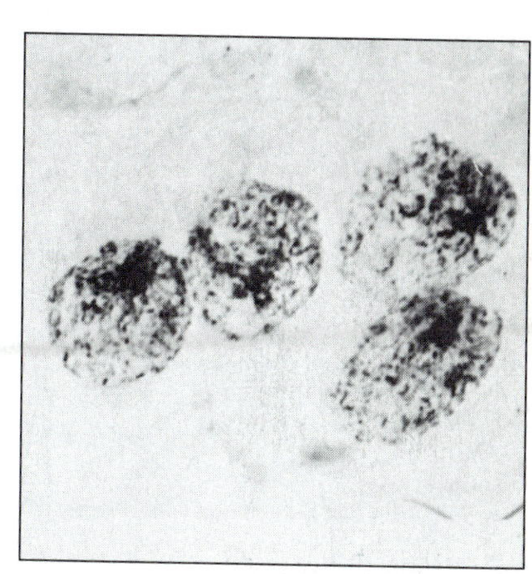

(n) Interphase

25 μm

The Products of Meiosis

Figure 15-11 summarizes meiosis and depicts its relationship to gametogenesis in animals. Meiosis always begins with a diploid cell containing duplicated chromosomes. The first meiotic division segregates homologous chromosomes and ends with two cells (or at least two nuclei), each having the haploid number of duplicated chromosomes. Sister chromatids are segregated in the second meiotic division, resulting in four cells, each having the haploid number of single chromosomes (Figure 15-11a). In male animals, all four of these haploid cells give rise to spermatozoa (Figure 15-11b). In female animals, however, only one of the four haploid nuclei survives and gives rise to a functional egg. This is because the two cell divisions are very unequal, with only one of the four daughter cells containing the bulk of the cytoplasm of the original diploid cell; the other three smaller cells, called **polar bodies,** fail to function as gametes (Figure 15-11c).

The Significance of Meiosis

Meiosis preserves the constancy of chromosome number in sexually reproducing organisms. If it were not for meiosis, gametes would have as many chromosomes as all the other cells in the body, and the chromosome number would double with every new generation.

Meiosis is also significant as the point in the flow of genetic information at which alleles separate randomly, allowing paternal and maternal chromosomes to be combined in the gametes. Whenever geneticists calculate the

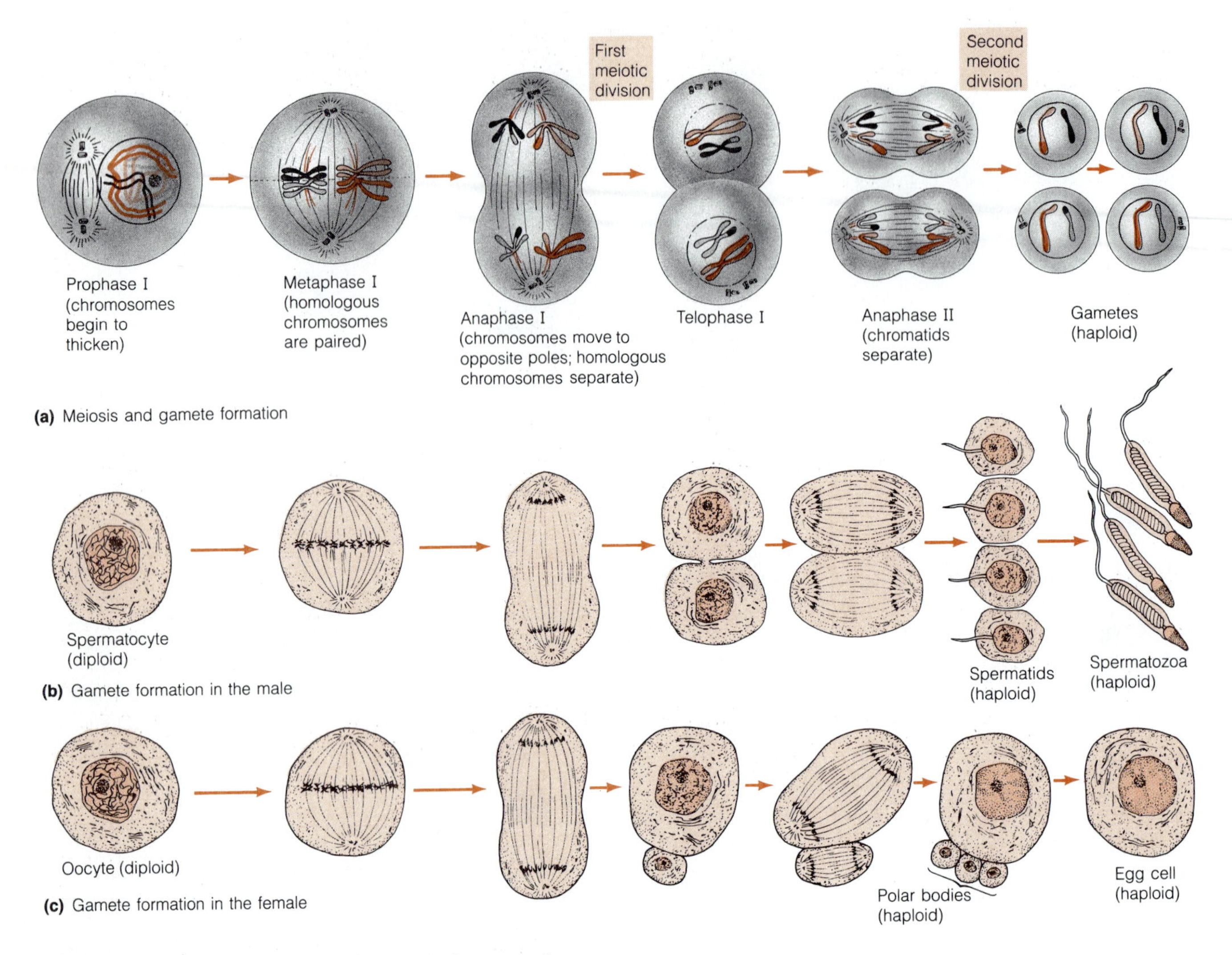

Figure 15-11 Meiosis and Gamete Formation. (a) Summary of meiosis and its relationship to gamete formation in animals. (b) In the male, all four haploid products of meiosis are retained and mature into spermatozoa. (c) In the female, both meiotic divisions are asymmetric, forming one large egg cell and three (in some cases only two) polar bodies that do not give rise to functional gametes.

expected frequency of gametes with specific combinations of genes, they are essentially expressing confidence in the randomness with which the paternal and maternal chromosomes segregate into one or the other of the daughter nuclei in meiosis I.

Finally, meiosis is of great genetic significance because of the crossing over that is possible between the synapsed homologous chromosomes during prophase I. Crossing over allows the recombination of paternal and maternal traits in the offspring, even when those traits are determined by genes on homologous chromosomes.

Genetic Variability: Segregation and Assortment of Alleles

The genetic variability that we associate with sexual reproduction is due in part to the random manner in which genes on different chromosomes segregate during meiosis and in part to the recombination between genes on homologous chromosomes during meiotic prophase I. We will consider the phenomenon of recombination shortly. First, however, we will look at the random segregation and independent assortment of genes during meiosis. For this, we turn to Gregor Mendel and the laws he deduced from his classical studies with pea plants.

Mendel's Experiments

Most students of biology have heard of Gregor Mendel and the genetics experiments he conducted in a monastery garden. Mendel's findings, first published in 1865, laid the foundations for what we now know as **Mendelian genetics.** Working with the common garden pea, Mendel chose seven readily identifiable traits and selected in each case two strains of peas that displayed alternative forms of the trait. Seed color was one such trait, because Mendel was able to identify one variety of peas that always had yellow seeds and another that always had green seeds (recall Figure 15-2).

Mendel's experimental approach was simple. He first showed that each of the seven traits was **true-breeding** upon self-fertilization. For example, plants grown from yellow seeds produced only yellow seeds, and plants grown from green seeds produced only green seeds. These results are depicted in Figure 15-12a.

Having established that the seven traits he had chosen were true-breeding (*homozygous,* in present-day terminology), Mendel cross-fertilized his plants to produce **hybrids** between the two parent strains for a given trait, hoping to deduce from his results the principles that govern the inheritance of such traits. The results must have seemed mystifying at first: In every case, the resulting offspring (which we now call the **F_1 generation**) regularly showed one or the other of the parental (or P_1) traits, but never both. In other words, one parental trait was always dominant and the other was always recessive. In the case of seed color, for example, all the progeny (offspring) had yellow seeds (Figure 15-12b), because yellow is dominant.

The next summer, Mendel allowed all the F_1 hybrids to self-fertilize. For each of the seven traits under study, he made the same surprising observation: The recessive trait that had seemingly been lost in the F_1 generation appeared among the progeny in what is now called the **F_2 generation!** Moreover, for each of the traits, the ratio of dominant to recessive offspring was always about 3:1 (Figure 15-12c). In the case of seed color, for example, plants grown from the yellow F_1 seeds produced both yellow and green seeds in a ratio of about 75% yellow seeds and 25% green.

This outcome, of course, was quite different from that obtained with the true-breeding yellow seeds of the parent strain. Clearly, there was something different about the yellow seeds of the parent stock and the yellow seeds of the F_1 generation. They looked alike, but the former bred true, whereas the latter did not.

Next, Mendel investigated the F_2 plants through self-fertilization (Figure 15-13). The F_2 plants that showed the recessive trait (green, in the case of seed color) gave the simplest results: They always bred true (Figure 15-13c),

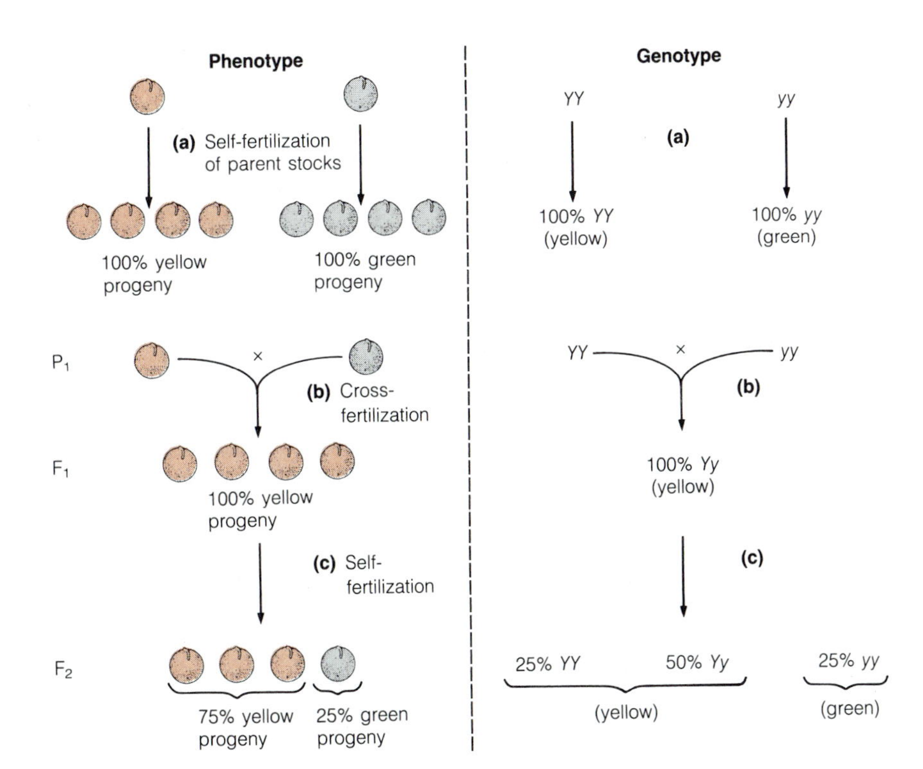

Figure 15-12 Genetic Analysis of Seed Color in Peas. Genetic crosses were made with peas having either yellow (*Y*, dominant) or green (*y*, recessive) seeds. The phenotypic results are shown on the left and the genotypes on the right. (a) The parent stocks are homozygous for either the dominant (*YY*) or recessive (*yy*) trait and breed true upon self-fertilization. (b) When crossed, the parent stocks yield an F_1 hybrid that is heterozygous and therefore shows the dominant trait. (c) Upon self-fertilization, the F_1 plants produce yellow and green seeds in the F_2 generation in a ratio of 3:1. See Figures 15-13 and 15-14 for further analyses.

suggesting their genetic identity with the recessive parent strains with which Mendel had begun. F_2 plants with the dominant trait yielded a more complex pattern, however. One-third bred true for the dominant trait (Figure 15-13a) and therefore seemed to be identical to the dominant parent plants. The other two-thirds of the dominant F_2 hybrids, however, produced both dominant and recessive progeny, in the same 3:1 ratio produced by the F_1 hybrids (Figure 15-13b).

The consistency with which Mendel obtained these results led him to conclude that the recessive trait must somehow be present in at least some of the hybrid seeds and plants, even though it does not manifest itself. The same conclusion arose from parallel experiments in which F_1 hybrids were crossed with the original parent stocks, a process called **backcrossing** (Figure 15-14). The dominant F_1 hybrids when backcrossed to the dominant parent stock always resulted in progeny with the dominant trait (Figure 15-14a), but when backcrossed to the recessive parent they resulted in a 1:1 mixture of dominant and recessive plants (Figure 15-14b). Moreover, the dominant progeny from the latter cross behaved just like the dominant F_1 hybrids: Upon self-fertilization, they gave rise to a 3:1 mixture of dominant and recessive traits (Figure 15-14c), and upon backcrossing to the recessive parent they again displayed the 1:1 ratio of dominant to recessive progeny (Figure 15-14d).

Mendel's Laws

After a decade of careful work, Mendel came to several important conclusions that have since come to be known as Mendel's **laws of inheritance.** The first principle to emerge from Mendel's work was that phenotypic traits are determined by discrete "factors" that are present in most organisms as pairs of "determinants." Today, we recognize

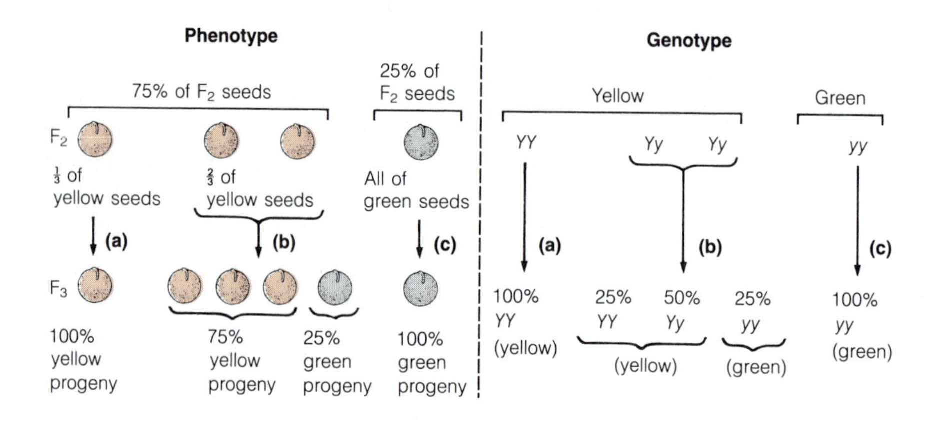

Figure 15-13 Analysis of F_2 Hybrids by Self-Fertilization. The F_2 hybrids of Figure 15-12 were analyzed through self-fertilization. The phenotypic results are shown on the left and the genotypes on the right. (a) One-third of the yellow F_2 progeny of Figure 15-12 (25% of the total F_2 progeny) breed true for yellow seed color upon self-fertilization, because they are genotypically *YY*. (b) Two-thirds of the yellow F_2 progeny (50% of the total F_2 progeny) yield yellow and green seeds upon self-fertilization, in a ratio of 3:1, just as the F_1 plants of Figure 15-12 did upon self-fertilization. (c) All the green F_2 seeds (25% of the total F_2 progeny) breed true for green seed color upon self-fertilization, because they are genotypically *yy*.

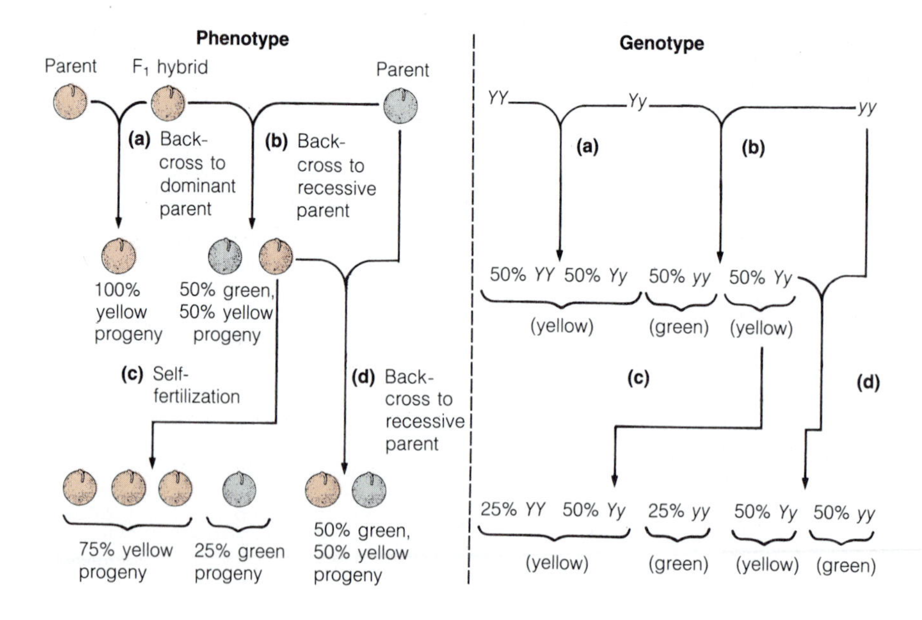

Figure 15-14 Analysis of F_1 Hybrids by Backcrossing. The F_1 hybrids of Figure 15-12 were analyzed by backcrossing to the parent strains. The phenotypic results are shown on the left and the genotypes on the right. (a) Upon backcrossing of the F_1 hybrid (*Yy*) to the dominant parent (*YY*), all the progeny are phenotypically yellow, although the genotype is either *YY* or *Yy*. (b) Backcrossing to the recessive parent (*yy*) yields yellow (*Yy*) or green (*yy*) seeds in equal proportions. (c) These yellow progeny will give rise, upon self-fertilization, to the same 3:1 mixture of yellow and green seeds seen with the F_1 hybrids (Figure 15-12). (d) Backcrossing of the yellow progeny to the homozygous recessive parent again yields a 1:1 mixture of yellow and green seeds, as in the backcross of part (b).

these "factors" as *genes* and the "determinants" as *alleles*. Mendel's conclusion seems almost self-evident to us, but it was an important assertion in his day. At that time, most scientists favored a *blending theory of inheritance*, one that viewed traits such as seed color rather like cans of paint that are poured together to yield intermediate results. Other investigators had described the nonblending nature of inheritance before Mendel, but without the accompanying data and mathematical analysis that were Mendel's special contribution.

The Law of Segregation. Of singular importance to much of what was to follow in genetics was Mendel's conclusion concerning the way in which factors are parceled out during gamete formation. According to the **law of segregation,** the two alleles governing a specific trait segregate, or separate from each other (at anaphase I of meiosis, as we now know), without having influenced each other during their joint presence in the organism. *The two alleles are, in other words, discrete entities that retain their identity when paired with a dissimilar allele and therefore emerge from the hybrid unchanged.*

The law of segregation is illustrated in Figure 15-15 for the alleles governing seed color in a heterozygous pea plant with the genotype *Yy*. (Peas have seven pairs of chromosomes, but only the pair bearing the alleles for seed color is shown.) During meiosis, the two homologous chromosomes, each with two chromatids, pair at metaphase I and then segregate into separate cells. The second division separates sister chromatids, so that each haploid cell has one allele for seed color, either *Y* or *y*.

The Law of Independent Assortment. Mendel's second law, the **law of independent assortment,** states that pairs of alleles governing different traits segregate independently of each other, generating gametes with all possible combinations of alleles. Mendel came to this conclusion by studying multifactor crosses between plants that differed in several characteristics. In addition to seed color, for example, he studied seed shape, stem length, and flower position. As in his single-factor crosses, he used parent plants that were true-breeding (homozygous) for two or more traits and generated F_1 hybrids heterozygous for each trait. He then self-fertilized these hybrids and determined the frequency with which the dominant and recessive forms of the several traits appeared among the progeny.

Since all possible combinations occurred in the F_2 progeny, Mendel concluded that all possible combinations of the different factors must be present in the F_1 gametes. Furthermore, from the proportions in which the various phenotypes were found in the F_2 generation, Mendel was also able to deduce that all possible combinations of factors occur in the gametes with the same frequency. In other words, *the two alleles that determine one genetic trait*

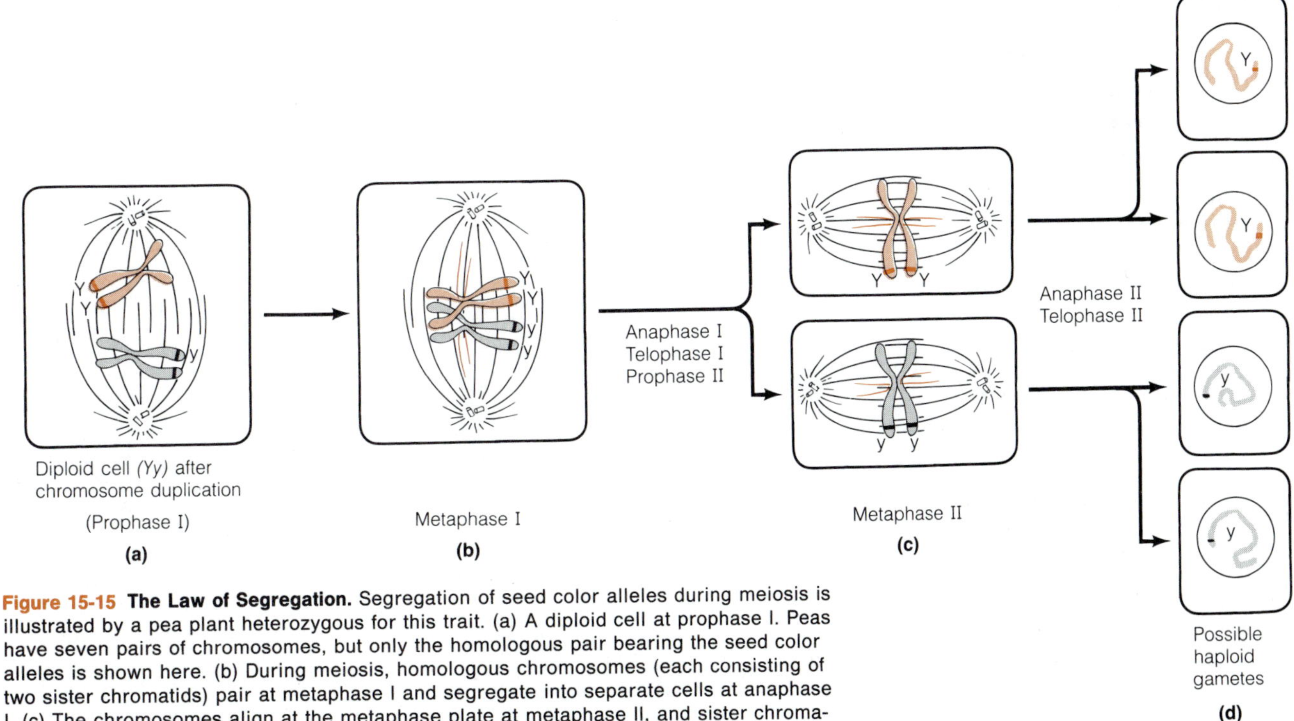

Figure 15-15 The Law of Segregation. Segregation of seed color alleles during meiosis is illustrated by a pea plant heterozygous for this trait. (a) A diploid cell at prophase I. Peas have seven pairs of chromosomes, but only the homologous pair bearing the seed color alleles is shown here. (b) During meiosis, homologous chromosomes (each consisting of two sister chromatids) pair at metaphase I and segregate into separate cells at anaphase I. (c) The chromosomes align at the metaphase plate at metaphase II, and sister chromatids segregate at anaphase II. (d) The result is four haploid daughter cells, each of which has one allele for seed color.

segregate independently of the alleles that govern another trait.

This is the law of independent assortment, another cornerstone of genetics. Expressed in terms of meiotic events, this law means that alleles on different chromosomes segregate randomly at anaphase I, as shown in Figure 15-16 for the chromosomes carrying the alleles for seed color (*Y* and *y*) and seed shape (*R* and *r*, where *R* stands for dominant round seeds and *r* for recessive wrinkled seeds). Note that independent assortment holds only for traits on *different* chromosomes. It is remarkable that Mendel happened to choose seven independently assorting traits in an organism that has only seven pairs of chromosomes.

The Cytological Basis of Inheritance

Mendel's findings lay dormant in the literature until 1900, when his paper was rediscovered almost simultaneously by three different European botanists. In the meantime, much had been learned about the cytological basis of inheritance. By 1875, for example, scientists had identified chromosomes with the help of stains produced by the developing aniline dye industry. At about the same time, fertilization was shown to involve the fusion of two parent nuclei, one contained within the egg and the other contributed by the sperm.

The first suggestion that chromosomes might be the bearers of genetic information was made in 1883. Within ten years, chromosomes had been studied in dividing cells and were known to split longitudinally into two apparently identical daughter chromosomes. This led to the realization that equal chromosome numbers were maintained in the development of an organism. With the development of better optical systems, even more detailed analysis of chromosomes became possible. In this way, cell division was shown to involve the movement of identical daughter chromosomes to opposite poles, thereby ensuring the presence of one chromosome from each pair in the resulting daughter cells.

Against this backdrop came the rediscovery of Mendel's paper, followed almost immediately by three crucial studies that established chromosomes as the carriers of Mendel's factors. The investigators were Edward Montgomery, Theodor Boveri, and Walter Sutton. Montgo-

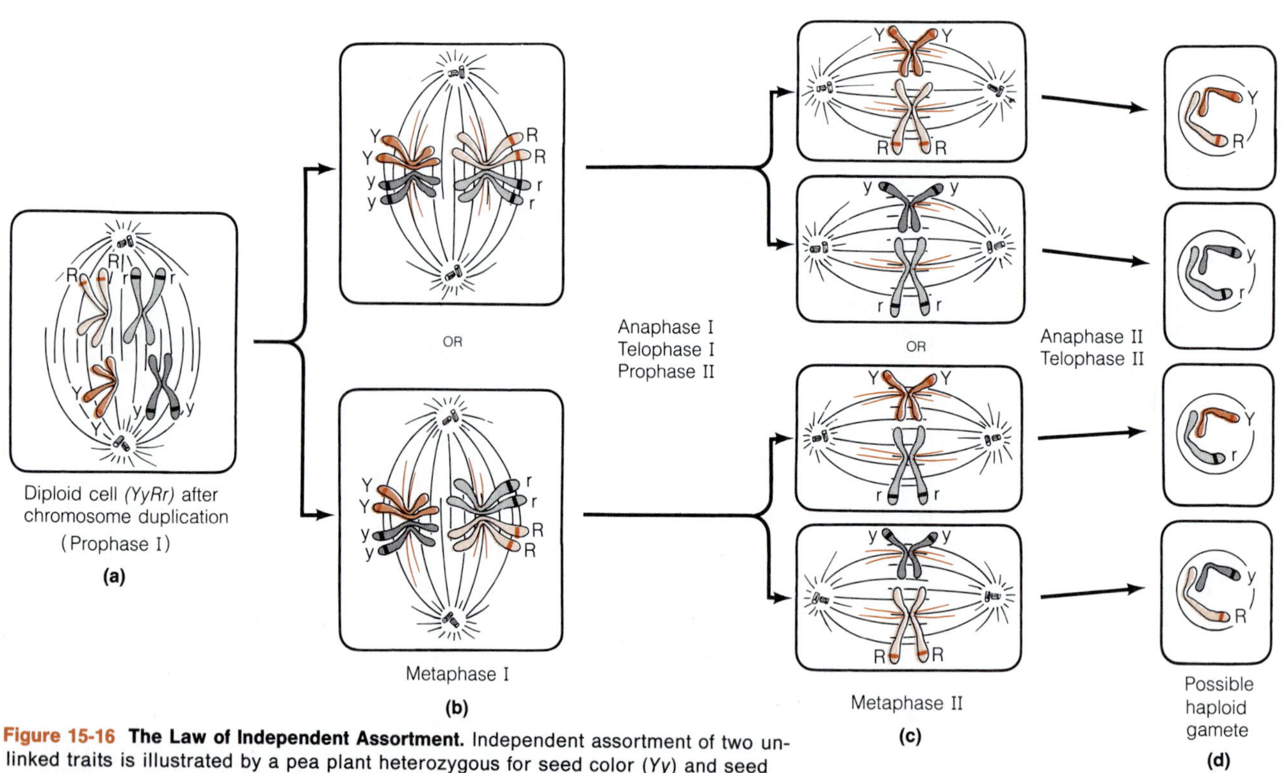

Figure 15-16 The Law of Independent Assortment. Independent assortment of two unlinked traits is illustrated by a pea plant heterozygous for seed color (*Yy*) and seed shape (*Rr*). (a) A diploid cell at prophase I. Pea plants have seven pairs of chromosomes, but only the two pairs bearing the seed color alleles and the seed shape alleles are shown. (b,c) During meiosis, the segregation of the seed color alleles occurs independently of the segregation of the seed shape alleles. (d) All possible combinations of alleles therefore occur with equal probability in the gametes.

mery's contribution was to recognize the existence of homologous chromosomes. From careful observations of insect chromosomes, he concluded that chromosomes were present in pairs, one of maternal origin and the other of paternal origin. Furthermore, the two chromosomes of each type synapsed (came together) in the "reduction division" (now called meiosis) of gamete formation, a process that had been reported only a decade or so earlier. To this, Boveri added the observation, based on his studies of fertilization in sea urchins, that chromosomes have functional individuality and that each chromosome plays a specific role in development.

Sutton, meanwhile, was studying meiosis in the grasshopper. In 1902, he made the important observation that the position of the chromosome pairs on the metaphase plate at metaphase I is purely a matter of chance. Any chromosome pair, in other words, may lie with the maternal or paternal chromosome toward either pole, regardless of the positions of other pairs. Many different combinations of maternal and paternal chromosomes are therefore possible in the mature gametes of an individual.

The Chromosomal Theory of Inheritance

In the period 1902–1903, Sutton put all these observations together into a coherent theory describing the role of chromosomes in inheritance. Sutton's theory can be summarized as six main points:

1. Nuclei of all cells except those of the germ line (sperm and eggs) contain two sets of homologous chromosomes, one of maternal origin, the other of paternal origin.
2. Chromosomes retain their individuality and are genetically continuous throughout the life cycle of an organism.
3. Maternal and paternal homologues synapse during the reduction division preceding gamete formation, pass to opposite poles of the division spindle, and are thus separated in the process.
4. Each chromosome (of a haploid set) plays a specific role in development.
5. The sets of homologues in a diploid are equivalent in a functional sense, each carrying one complement of genetic determinants.
6. The maternal and paternal members of different homologous pairs separate randomly at reduction division.

Thus, chromosomes became the physical basis for understanding how Mendel's factors could be carried, transmitted, and segregated. The presence of two sets of homologous chromosomes in the cell parallels Mendel's suggestion of two determinants for each phenotypic trait. Moreover, the separation of homologous chromosomes during the process of gamete formation provides the basis for Mendel's law of segregation, and the randomness with which homologous chromosomes align and separate accounts for his law of independent assortment.

Genetic Variability: Recombination and Crossing Over

Genes on different chromosomes become randomly assorted during sexual reproduction because of the segregation and independent assortment of homologous chromosomes during the first meiotic division. If *A* and *B* are genes on two different chromosomes, then an organism that is heterozygous for both traits (*AaBb*) will produce gametes in which allele *A* is just as likely to occur with allele *B* as it is with allele *b* (Figure 15-17a). But what happens if the two genes are on the same chromosome? In that case, alleles *A* and *B* will routinely occur together in the same gamete but alleles *A* and *b* will not (Figure 15-17b).

However, even for genes on the same chromosome, some mixing or scrambling of alleles occurs because of the phenomenon of *recombination.* Recombination involves an interchange of genetic information between homologous chromosomes and occurs while homologues are synapsed during prophase I of meiosis. Recombination therefore contributes further to the genetic variability seen in sexually reproducing species by providing a mechanism for rearranging genetic information on the same chromosome (Figure 15-17c).

Recombination was originally discovered from studies with the fruit fly, *Drosophila melanogaster,* conducted by Thomas H. Morgan and his colleagues beginning around 1910. We will therefore turn to Morgan's work to understand recombination. As we do so, we also encounter the concept of linkage groups.

Linkage Groups

Unlike Mendel's peas, the fruit flies with which Morgan and his colleagues worked did not come with any ready-made variant phenotypes. Whereas Mendel was able to purchase seed stocks of different varieties, the only type

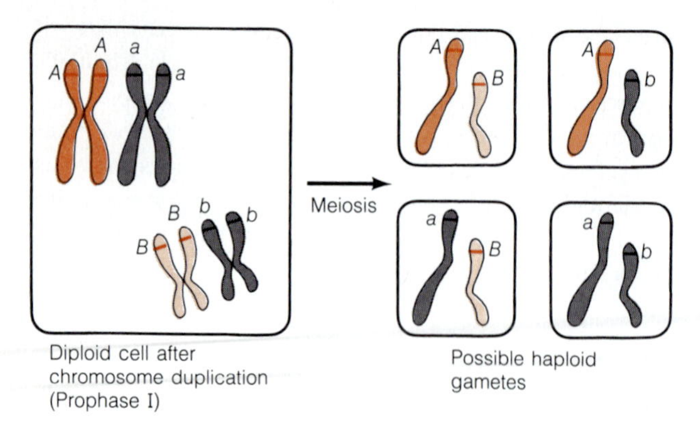

(a) Unlinked genes segregate randomly

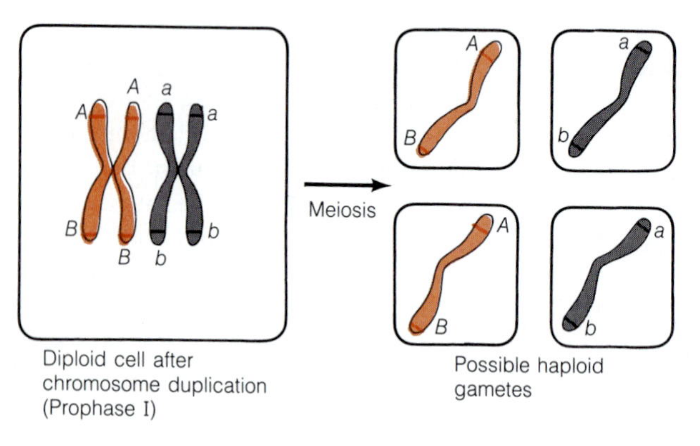

(b) Linked genes segregate together in the absence of recombination

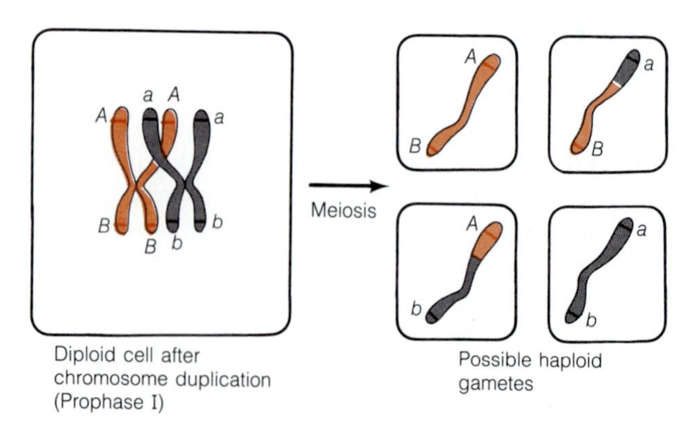

(c) Linked genes do not segregate together in the presence of recombination

Figure 15-17 Segregation of Linked and Unlinked Genes. (a) Genes on different chromosomes segregate randomly during meiosis; allele *A* is as likely to occur in the gametes with allele *B* as it is with allele *b*. (b) Genes on the same chromosome remain linked during meiosis; in the absence of recombination, allele *A* will occur routinely with allele *B*, but not with allele *b*. (c) Genes on the same chromosome become interchanged when recombination takes place, so that allele *A* can occur not only with allele *B* but also with allele *b*.

of fruit fly available initially was what has come to be known as the **wild type**, or "normal" organism. Morgan and his colleagues therefore had to generate all the variants or mutants with which they worked.

To do so, they bred large numbers of flies and selected mutant individuals—flies having detectable phenotypic modifications that were heritable. (Later, X-irradiation was used to enhance the mutation rate, but in their early work Morgan's group depended entirely on spontaneous mutations.) Within five years, they were able to identify about 85 different mutants. Each of these mutants could be propagated as a laboratory stock and could be used for matings as needed.

One of the first things Morgan and his colleagues realized as they began analyzing their mutants was that, unlike the traits Mendel chose to study in his peas, the mutations did not all assort independently. Instead, some mutations behaved as if they were linked together, and the new combinations predicted by Mendel were infrequent or even nonexistent. In fact, it was soon recognized that each of the mutations belonged to one of four **linkage groups.** Each group consisted of a collection of mutations that were transmitted, inherited, and assorted together. Morgan quickly realized that the number of linkage groups correlated exactly with the number of different chromosomes in the organism, since the haploid chromosome number for *Drosophila* is four. The conclusion was obvious: Each chromosome is the physical basis for a specific linkage group.

Recombination, Crossing Over, and Chiasmata Formation

Although all their mutations could be ordered into linkage groups, Morgan and his colleagues also found that the linkage within such groups was incomplete. Most of the time, genes that were known to belong to the same linkage group (and therefore localized to the same chromosome) assorted together, as would be expected. Sometimes, however, two or more such traits would appear in the offspring in nonparental combinations. This phenomenon of less-than-complete linkage was called *recombination,* since the genes appeared in new and unexpected associations in the offspring.

To explain such recombinant offspring, Morgan proposed that homologous chromosomes could exchange segments, presumably by some sort of breakage-and-fusion event, as illustrated in Figure 15-18a. By this process, which Morgan termed *crossing over,* a particular gene or group of genes initially present on one member of a homologous pair of chromosomes could be transferred to the other chromosome in a reciprocal manner.

In the example of Figure 15-18a, two homologous chromosomes, one with alleles *A* and *B* and the other with alleles *a* and *b*, lie side by side at synapsis. Portions of an *AB* chromatid and a nonsister *ab* chromatid then exchange positions, producing two recombinant chromatids, one with the alleles *A* and *b* and the other with the alleles *a* and *B*. Each of the four chromatids ends up in a different gamete at the end of the second meiotic division, so the products of meiosis will include two *parental* gametes and two *recombinant* gametes, assuming a single crossover event (Figure 15-18b).

We now know that crossing over occurs during the pachytene stage of meiotic prophase I, at a time when sister chromatids are packed tightly together. Later in prophase, as the chromatids begin to separate at diplotene, each of the four chromatids in a bivalent can be identified as belonging to one or the other of the two homologues. Wherever crossing over has occurred between nonsister chromatids, the two homologues remain attached to one other, forming an X-shaped figure called a *chiasma*. A chiasma, in other words, is the cytological manifestation of a crossover event that occurred earlier in prophase.

Homologous chromosomes are usually held together by at least one chiasma. Many bivalents contain multiple chiasmata. Human bivalents, for example, contain two or three chiasmata per bivalent, because multiple crossover events routinely occur between paired homologues. To be of genetic significance, crossing over must involve nonsister chromatids. In some species, sister chromatid exchange can also occur, but such exchanges are of no genetic consequence because sister chromatids are genetically identical. Figure 15-19 illustrates several types of chiasmata involving nonsister chromatids.

Recombination Frequency and Gene Mapping

Eventually, it became clear to Morgan and others that the frequency of occurrence of recombinant progeny differed

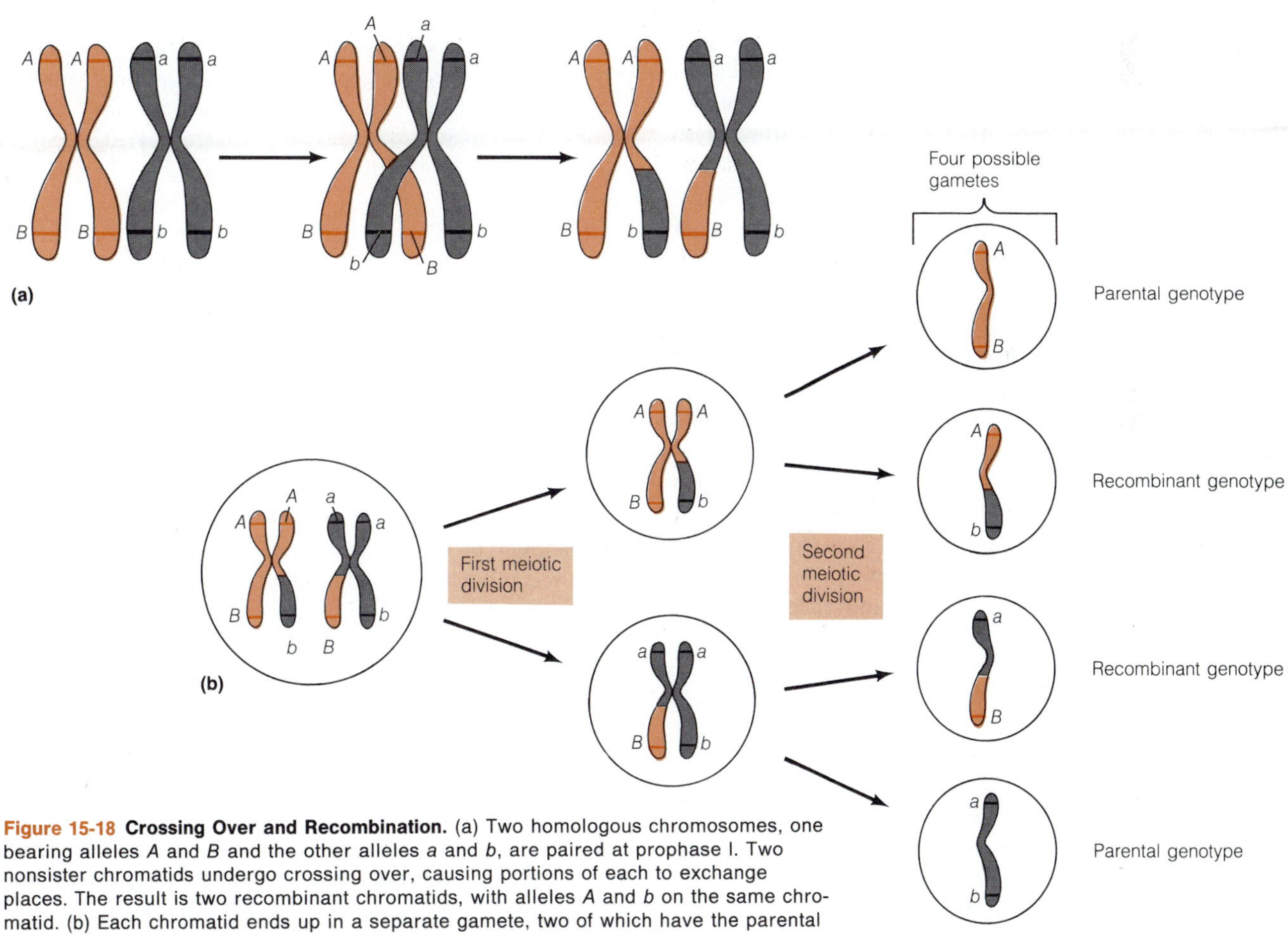

Figure 15-18 Crossing Over and Recombination. (a) Two homologous chromosomes, one bearing alleles *A* and *B* and the other alleles *a* and *b*, are paired at prophase I. Two nonsister chromatids undergo crossing over, causing portions of each to exchange places. The result is two recombinant chromatids, with alleles *A* and *b* on the same chromatid. (b) Each chromatid ends up in a separate gamete, two of which have the parental genotypes (*AB* and *ab*) and two the recombinant genotypes (*Ab* and *aB*).

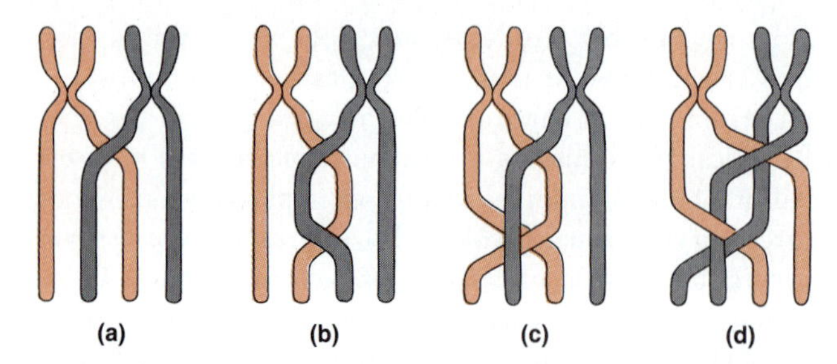

Figure 15-19 Several Types of Chiasmata. (a) A single chiasma. (b) Two chiasmata involving two chromatids. (c) Two chiasmata involving three chromatids. (d) Two chiasmata involving all four chromatids.

for different pairs of genes within the various linkage groups, but that for a specific pair of genes it was remarkably constant. This led to the suggestion that the frequency of recombination between two genes might be a measure of how far the markers are from each other along the chromosome. Genes located very close to each other would be less likely to become separated by a crossover event between them than would genes that are far apart.

It was, in fact, quickly realized that the frequency of recombination, expressed as the percentage of recombinant progeny, was an accurate means of quantifying distance between genetic markers. This led Alfred Sturtevant to suggest in 1911 that recombination data could be used to order genes along the chromosomes of *Drosophila*. Thus, a chromosome was now being viewed as a linear string of genes that could be ordered and spaced along its length on the basis of recombination data.

The ordering and spacing of genes on a chromosome on the basis of crossover frequencies is called **chromosome mapping.** In the construction of such maps, crossover frequency becomes the map distance. If, for example, alleles *A* and *B* of Figure 15-18b appear in the progeny in their parental combination 85% of the time and as recombinants 15% of the time, we conclude that the two genes are linked (are on the same chromosome) and are 15 **map units** apart. This approach has been used to map the chromosomes of many species of plants and animals, and of bacteria and viruses as well.

Recombination in Bacteria and Viruses

From what we have learned so far, we might expect recombination to be restricted to organisms that reproduce sexually, since recombination depends on crossover events between homologous chromosomes. As we have seen, sexual reproduction brings with it an opportunity for recombination once every generation, because the necessary juxtaposition of homologous chromosomes is an intrinsic part of the meiotic process. Viruses and prokaryotes, on the other hand, might seem to be poor candidates for recombination. They reproduce asexually and have haploid genomes, with no obvious mechanisms for bringing genomes from two different parents together regularly.

Yet viruses and bacteria are also capable of recombination. In fact, recombination data have been very useful in the extensive mapping of viral and bacterial genomes. To understand how recombination can occur despite a haploid genome, we need to examine the several mechanisms by which two haploid genomes (or portions of genomes) can be brought together to allow recombination. Some of these mechanisms occur naturally, whereas others are laboratory techniques.

Bacteriophages: Recombination by Co-Infection

Bacteriophages have proved to be especially important to our understanding of genes and gene mapping. Much of what we currently know about genes, as well as the vocabulary in which we express that knowledge, comes from experiments with phages, particularly with the T4 phage.

In bacteriophages, the standard method for creating recombinants and thereby allowing genetic mapping is **co-infection** of bacterial cells with two stocks of phages that differ in their genotypes, as shown in Figure 15-20. The two parent phage stocks are mixed together, and bacteria are then infected with enough phages so that virtually every bacterium will be infected with some phages of each genotype. The host cell therefore contains complete copies of the genomes of both parent phages. As the phages replicate in the bacterial cell, their genophores occasionally become juxtaposed in ways that allow recombination to occur. The recombinant progeny are generated with frequencies that depend on the physical distance between the markers under study, just as in diploid organisms.

Bacteria: Recombination by Transformation and Transduction

We have already encountered one means of getting DNA from one bacterial cell into another. Recall from Chapter 13 the experiments with rough and smooth *Pneumococcus* bacteria that led Frederick Griffith to postulate the exis-

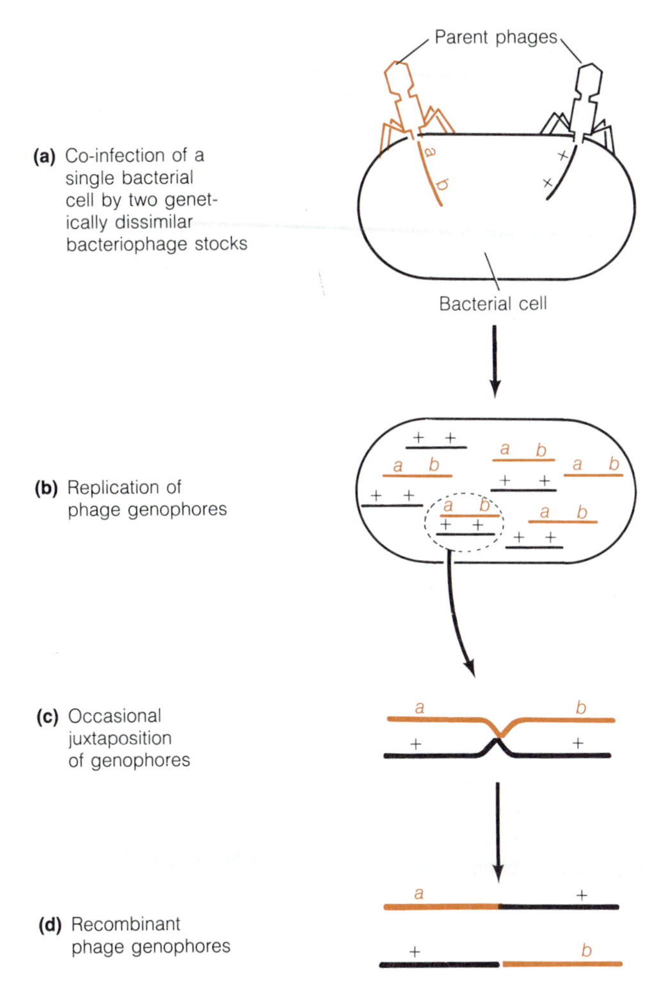

Figure 15-20 Co-Infection in Bacteriophages. (a) Two parent phage stocks with different genotypes are allowed to co-infect a bacterial cell, thereby ensuring the presence of both phage genomes in a single bacterial cell. One of the parent stocks carries the mutants *a* and *b*, while the other is wild type (+) for both genes. As the phage genophores replicate (b), they occasionally become juxtaposed in a way that allows crossing over and recombination to occur (c), giving rise to recombinant phage genophores (d). The frequency of occurrence of recombinant genotypes provides a measure of the distance between markers.

tence of some sort of *transforming principle,* subsequently identified as DNA by Oswald Avery and his colleagues. The ability of a bacterial cell to take up DNA and to incorporate segments of that DNA heritably into its own genome is called **transformation** (Figure 15-21a).

Although initially described for a single trait in a specific bacterium, transformation is now recognized as a general mechanism whereby some (though by no means all) genera of bacteria can acquire genetic information from other cells and pass it on to offspring. Transformation can occur naturally whenever bacterial cells have access to DNA from other cells, but it is much more common as a laboratory technique, as described in Chapter 13. However, it occurs more readily with some bacteria than with others. For example, *Pneumococcus* cells are easily transformed, but until quite recently it was not possible to transform *Escherichia coli* cells.

A related process is **transduction,** in which DNA is brought into a bacterial cell not as an isolated molecule or fragment, but by a bacteriophage. Most of the time, phages contain only their own DNA, but some phages occasionally incorporate a bit of host DNA into their coat. When such a phage infects another bacterium, it injects the fragment of DNA that it picked up from the previous host cell, acting in effect like a syringe carrying DNA from one bacterial cell to the next (Figure 15-21b). Phages capable of doing this are called **transducing phages.** Bacteriophage P1 is one of the most frequently used transducing phages.

The amount of DNA that will fit into a bacteriophage head is quite small compared with the size of the bacterial genome. Two markers must therefore be close together for both of them to be carried into a bacterial cell by a single phage. This is the basis of **cotransductional mapping,** in which the proximity of one marker to another is determined by quantifying the frequency with which they accompany each other in a transducing phage. In general, the closer two markers are, the more likely they are to be cotransduced into a bacterial cell. When such studies are carried out with the transducing phage P1, it is found that the markers cannot be cotransduced if they are separated on the bacterial genophore by more than about 10^5 nucleotide pairs. This finding agrees with the observation that the P1 phage carries a DNA molecule of about this size.

Sex and the Single Cell: Conjugation in Bacteria

In addition to transformation and transduction, some bacteria can also transfer genetic information from one cell to another by a process called **conjugation.** As the name suggests, conjugation resembles the sexual process in that one bacterium is clearly identifiable as the donor ("male") and another as the recipient ("female"), and the whole genophore can be transferred. The existence of such a process was postulated in 1946 by Joshua Lederberg and Edward L. Tatum, who were the first to show that recombination could occur in bacteria. They also established that physical contact was necessary for conjugation to take place. We now understand that conjugation involves the directional transfer of genetic material from the male bacterium to the female bacterium. Recombinant cells always arise from the female parent, not from the male.

The F Factor. Maleness in bacteria is due to the presence of a plasmid called the **F factor,** which can either exist as an independent replicating entity in the cell or integrate itself into the genophore. Male bacteria can therefore be designated as F^+ and females as F^-. Male cells have long, hairlike projections called **F pili** (singular: pilus) on their surface (Figure 15-22). Each pilus is a slender tube with a sticky end that adheres to the surface of a female cell. Pili are thought to facilitate formation of the **mating bridges** that are required for transfer of DNA from the male cell to the female cell.

When present as the free plasmid, the F factor of the male cell is transferred very effectively and efficiently during conjugation, converting the recipient cell from F^- to F^+,

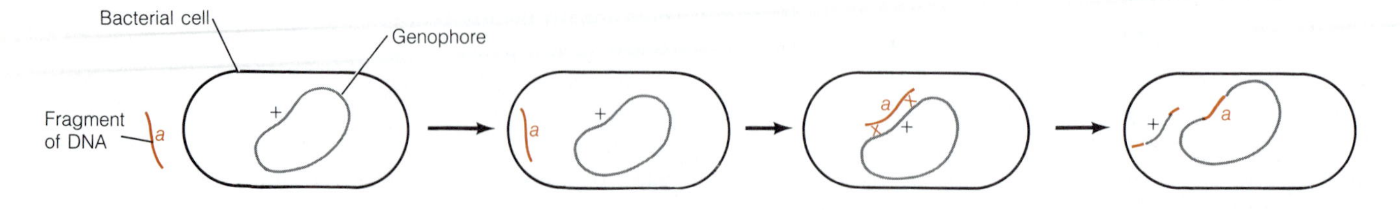

(a) Transformation of a bacterial cell by exogenous DNA

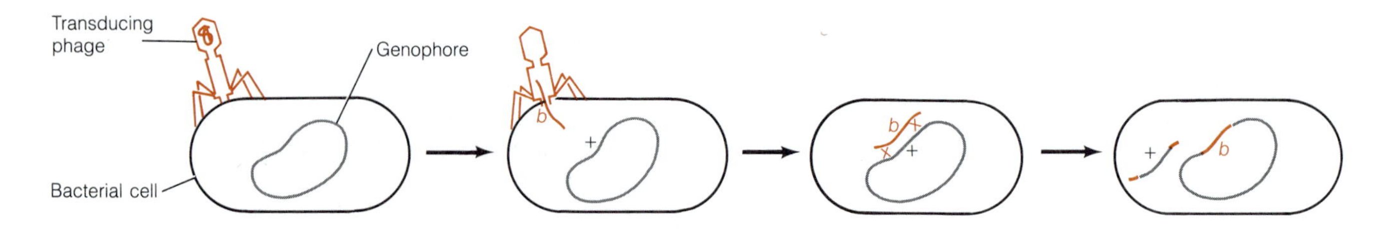

(b) Transduction of a bacterial cell by a transducing phage

Figure 15-21 Transformation and Transduction in Bacterial Cells. (a) Transformation involves the uptake by the bacterial cell of exogenous DNA, which occasionally becomes integrated into the bacterial genome by two crossover events. The exogenous DNA will be detectable in progeny cells only if integrated into the genome, because the fragment of DNA initially taken up does not normally have the capacity to replicate itself autonomously in the cell unless it is an intact plasmid. (b) Transduction involves the introduction of exogenous DNA into a bacterial cell by a phage. Once injected into the host cell, the DNA can become integrated into the genome in the same manner as in transformation.

Figure 15-22 F Pili. The male (F^+) bacterial cell (right) has numerous pili on its surface. F pili are slender appendages that are thought to help the male cell get close to and attach to a female (F^-) cell (left), thereby facilitating the unidirectional passage of DNA from the male to the female cell through the mating bridge that forms subsequently. (TEM)

as shown in Figure 15-23a. Transfer always begins at a point on the plasmid called the **origin of transfer,** shown in the figure as an arrowhead. As the arrowhead implies, the transfer process has an inherent directionality. Although conjugation involves transfer of an F factor to the female cell, the original male cell does not lose its maleness in the process; it always retains an F factor, while passing a copy of it on to the recipient. A cross between an F^+ and an F^- strain therefore results in an entire culture of male cells, such that all progeny will be F^+. Maleness, in other words, is infective, and the F factor is the infective agent.

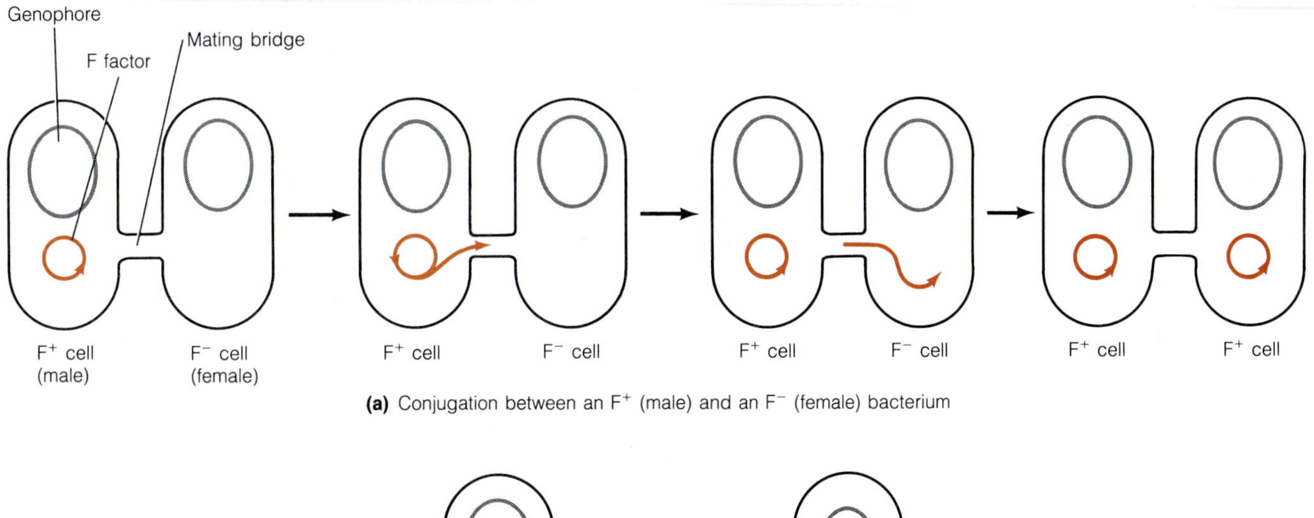

(a) Conjugation between an F^+ (male) and an F^- (female) bacterium

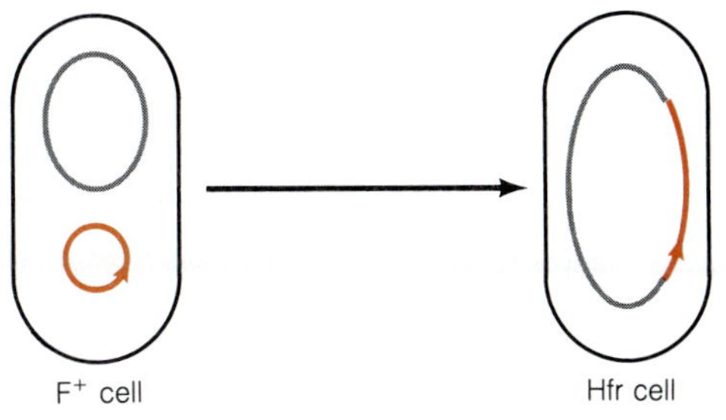

(b) Conversion of an F^+ male into an Hfr male by integration of the F factor into the genophore

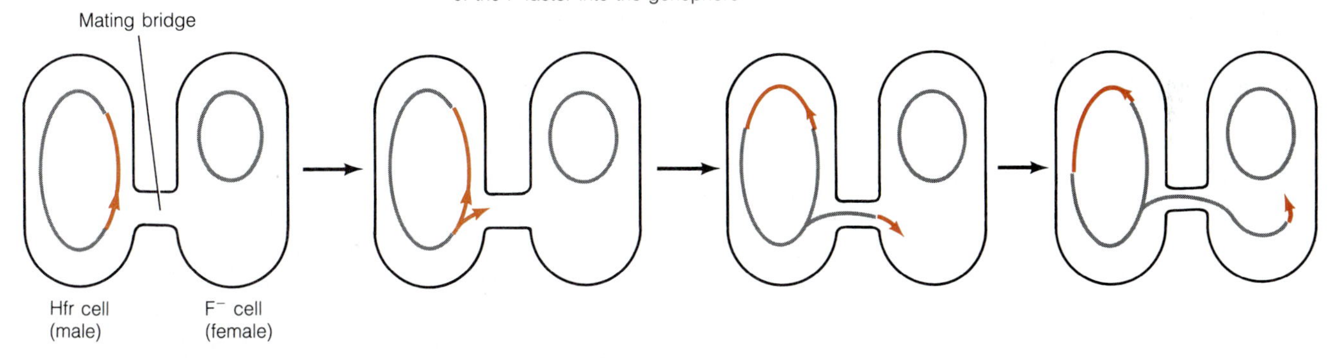

(c) Conjugation between an Hfr (male) and an F^- (female) bacterium

Figure 15-23 Bacterial Conjugation. (a) Conjugation between an F^+ (male) bacterium and an F^- (female) bacterium involves transfer of a copy of the F factor from the male to the female, thereby converting the female into a male. Transfer of the F factor occurs through a mating bridge and always begins at the origin of transfer, indicated by the arrowhead on the plasmid. (b) Conversion of an F^+ male into an Hfr male by integration of the F factor into the genophore. (c) Conjugation between an Hfr (male) bacterium and an F^- (female) bacterium involves transfer of a copy of the male genome through a mating bridge into the female cell, beginning with the origin of transfer on the integrated F factor. Transfer is usually incomplete, since cells rarely remain in mating contact long enough for the entire genophore to be transferred.

Hfr Cells and Genophore Transfer. So far, we have seen that male and female bacterial cells are defined in terms of the presence or absence of the F factor, and we have seen that the F factor is transmissible by conjugation. But how do recombinant progeny arise by this means?

Although usually present as a plasmid, the F factor can become integrated into the bacterial genophore, as shown in Figure 15-23b. This converts an F^+ cell into an **Hfr bacterium,** which is capable of producing a *h*igh *f*requency of *r*ecombination in further matings because it is now capable of transferring genomic DNA during conjugation.

When an Hfr male bacterium is mated to a female cell, mating bridges form, and DNA is transferred into the female cell. Instead of transferring just the F factor, however, the Hfr bacterium transfers at least part (and sometimes all) of its genophore (or, more accurately, a copy of its genophore, since the Hfr cell always retains the original genophore). Transfer begins at the origin of transfer within the integrated F factor and proceeds in a direction dictated by the orientation of the F factor within the genophore (Figure 15-23c). Notice that, for transfer, the genophore is converted from its normal circular form into a linear form, with part of the F factor at the leading end and the remainder at the trailing end.

Because the F factor is split in this way during the transfer process, only the female cells that receive a complete genophore from the Hfr donor actually become Hfr cells themselves. However, transfer of the whole genophore is quite rare, since it takes about 90 minutes to complete. Usually, mating contact is spontaneously lost and the cells separate before transfer is complete, leaving the female cell with only a portion of the Hfr genophore. As a result, genetic markers located close to the origin of transfer on the Hfr genophore are much more likely to be transmitted to the female cell than are markers farther away.

This correlation between position on the genophore and likelihood of transfer turns out to be a useful way of mapping markers with respect to the origin of transfer and therefore with respect to one another. Once a portion of the Hfr genophore has been introduced into a female cell by conjugation, it can recombine with homologous regions of the genophore of the female cell. The recombinant genophores that arise in this way contain some information from the donor cell and some from the recipient. These recombinant genophores appear among daughter cells of the female bacterium and become the basis of genetic analysis. Typically, a cross is made between an Hfr male strain and a female strain that differ in at least two genetic properties. After conjugation has taken place in the mixed culture, the cells are plated on a medium on which recombinants can grow but parent strains cannot, thereby allowing the number of recombinants to be determined.

Perspective

Asexual reproduction is based on mitosis and results in offspring that are genetically identical to the single parent. By contrast, sexual reproduction involves two parents and leads to an unpredictable mixing of parental traits in the offspring. Sexual reproduction allows organisms to respond adaptively to environmental changes, enables desirable mutations to be combined in a single individual, and promotes genetic flexibility by maintaining a diploid genome.

The life cycle of every sexually reproducing species includes both a haploid phase and a diploid phase. In animals, haploid gametes are generated by meiosis and fuse at fertilization to restore the diploid chromosome number. Meiosis consists of two successive divisions without an intervening duplication of chromosomes. In the first division, homologous chromosomes separate and segregate into the two daughter cells. In the second division, sister chromatids separate, forming four haploid daughter cells. Meiosis differs from mitosis both in the ploidy level of the daughter cells and in the synapsis of homologous chromosomes during prophase I. It is during synapsis that crossing over occurs between nonsister chromatids, leading to genetic recombination. The chiasmata observed during meiosis I are morphological manifestations of crossover events.

Mendel's laws describe the genetic consequences of chromosome behavior during meiosis, even though chromosomes had not yet been discovered at the time of Mendel's experiments. According to his laws, maternal and paternal alleles segregate randomly during meiosis, and alleles on separate chromosomes (or linkage groups) segregate independently of one another. The genetic variability that characterizes sexual reproduction arises in part from the random assortment of chromosomes during anaphase I and in part from the recombination that occurs as a result of crossing over during prophase I.

The frequency with which recombination is observed between genetic markers on the same chromosome or genophore is a measure of the distance between the two markers and can be used to map genes in both eukaryotes and prokaryotes. Prokaryotes do not reproduce sexually, but partially diploid organisms can be produced by transformation, transduction, and conjugation.

Key Terms for Self-Testing

asexual reproduction (p. 427)

Sexual Reproduction
sexual reproduction (p. 427)
mutations (p. 429)
diploid (p. 429)
alleles (p. 429)
haploid (p. 429)
homozygous (p. 429)
heterozygous (p. 429)
dominant (p. 429)
recessive (p. 429)
genotype (p. 430)
phenotype (p. 430)
gametes (p. 430)
gametogenesis (p. 430)
gonads (p. 430)
eggs (p. 430)
ova (p. 430)
ovaries (p. 430)
sperm (p. 430)
spermatozoa (p. 430)
testes (p. 430)
fertilization (p. 430)
zygote (p. 430)
embryonic development (p. 430)

Meiosis
meiosis (p. 430)
alternation of generations (p. 431)
spores (p. 431)
sporophyte (p. 431)
gametophyte (p. 431)
gametic meiosis (p. 431)
sporic meiosis (p. 431)
meiosis I (p. 432)
synapse (p. 432)
bivalents (p. 432)
meiosis II (p. 433)
recombination (p. 436)
prophase I (p. 436)
leptotene (p. 436)
zygotene (p. 436)
crossing over (p. 436)
pachytene (p. 436)
synaptonemal complex (p. 436)
diplotene (p. 436)
chiasmata (p. 436)
diakinesis (p. 437)
metaphase I (p. 437)
anaphase I (p. 437)
telophase I (p. 437)
prophase II (p. 437)
metaphase II (p. 437)
anaphase II (p. 437)
telophase II (p. 437)
polar bodies (p. 439)

Genetic Variability: Segregation and Assortment of Alleles
Mendelian genetics (p. 441)
true-breeding (p. 441)
hybrids (p. 441)
F_1 generation (p. 441)
F_2 generation (p. 441)
backcrossing (p. 442)
laws of inheritance (p. 442)
law of segregation (p. 443)
law of independent assortment (p. 443)

Genetic Variability: Recombination and Crossing Over
wild type (p. 446)
linkage groups (p. 446)
chromosome mapping (p. 448)
map units (p. 448)

Recombination in Bacteria and Viruses
co-infection (p. 448)
transformation (p. 449)
transduction (p. 449)
transducing phages (p. 449)
cotransductional mapping (p. 449)
conjugation (p. 449)
F factor (p. 450)
F pili (p. 450)
mating bridges (p. 450)
origin of transfer (p. 451)
Hfr bacterium (p. 452)

Suggested Reading

General References and Reviews

Ayala, F. J., and J. A. Kiger, Jr. *Modern Genetics,* 2d ed. Menlo Park, Calif.: Benjamin/Cummings, 1984.

Mendel, G., H. de Vries, C. Correns, and E. Tschermak. The birth of genetics. *Genetics* 35 (1950, Suppl.): 1. (Original papers in English translation.)

Singer, M. and P. Berg. *Genes and Genomes.* Mill Valley, Calif.: University Science Books, 1990.

Sturtevant, A. H. *A History of Genetics.* New York: Harper & Row, 1965.

The Advantages of Sexual Reproduction

Crow, J. F. The importance of recombination. In *The Evolution of Sex: An Examination of Current Ideas* (Michod, R. E., and B. R. Levin, eds.). Sunderland, Mass.: Sinauer, 1988.

Lewis, J., and L. Wolpert. Diploidy, evolution, and sex. *J. Theor. Biol.* 78 (1979): 425.

Maynard Smith, J. *Evolution of Sex.* New York: Cambridge University Press, 1978.

Meiosis

Chandley, A. C. Meiosis in man. *Trends Genet.* 4 (1988): 79.

Evans, C. W., and H. G. Dickinson, eds. Controlling Events in Meiosis. *Symp. Soc. Exp. Biol,* vol. 38. Cambridge, England: The Company of Biologists, 1984.

John, B. Myths and mechanisms of meiosis. *Chromosoma* 54 (1976): 295.

John, B., and K. R. Lewis. *The Meiotic Mechanism.* Oxford Biology Readers, edited by J. J. Head. New York: Oxford University Press, 1976.

Bacterial and Viral Genetics

Berg, P. Dissections and reconstructions of genes and chromosomes. *Science* 213 (1981): 296.

Hayes, W. *The Genetics of Bacteria and Their Viruses,* 2d ed. New York: Wiley, 1968.

Transgenic Processes

Brinster, R. L., and R. D. Palmiter. Induction of foreign genes in animals. *Trends Biochem. Sci.* 7 (1982): 438.

Palmiter, R. D., R. L. Brinster, R. E. Hammer, M. E. Trumbauer, M. G. Rosenfeld, N. C. Birnberg, and R. M. Evans. Dramatic growth of mice that develop from eggs microinjected with metallothionein-growth hormone fusion genes. *Nature* 300 (1982): 611.

Problem Set

1. **The Truth About Sex.** For each of the following statements, indicate with an S if it is true of sexual reproduction, with an A if it is true of asexual reproduction, and with an N if it is not true of either.
   (a) Desirable traits from two different parents can be combined in a single offspring.
   (b) Each generation of offspring is genetically identical to the previous generation.
   (c) Mutations do not get propagated to the next generation.
   (d) Some offspring in every generation will be less suited for survival than the parents, but others may be better suited.
   (e) Mitosis is not involved in the life cycle.

2. **Ordering the Stages of Meiosis.** Shown in Figure 15-24 are sketches of several stages of meiosis in an organism, labeled a through f.
   (a) What is the diploid chromosome number in this species?
   (b) Order the six stages chronologically and identify each.
   (c) Between which two stages do homologous centromeres separate?
   (d) Between which two stages does recombination occur?

3. **Telling Them Apart.** Describe briefly how you might distinguish between each of the following stages in the same organism:
   (a) Metaphase of mitosis and metaphase I of meiosis.
   (b) Metaphase of mitosis and metaphase II of meiosis.
   (c) Metaphase I and metaphase II of meiosis.
   (d) Telophase of mitosis and telophase II of meiosis.
   (e) Pachytene and diplotene phases of meiotic prophase I.

4. **Your Centromere Is Showing.** Suppose you have a diploid organism in which all the chromosomes contributed by the sperm have cytological markers on their centromeres that allow you to distinguish them from the chromosomes contributed by the egg.
   (a) Would you expect all the somatic cells of the organism to have equal numbers of maternal and paternal centromeres? Explain.
   (b) Would you expect equal numbers of maternal and paternal centromeres in each gamete produced by that individual? Explain.

5. **How Much DNA?** Let X be the amount of DNA present in the gamete of an organism that has a diploid chromosome number of 4. Assuming all chromosomes to be of approximately the same size, how much DNA (X, 2X, $\frac{1}{2}$X, and so on) would you expect in each of the following?
   (a) A zygote immediately after fertilization
   (b) A single sister chromatid
   (c) A daughter cell following mitosis
   (d) A single chromosome following mitosis

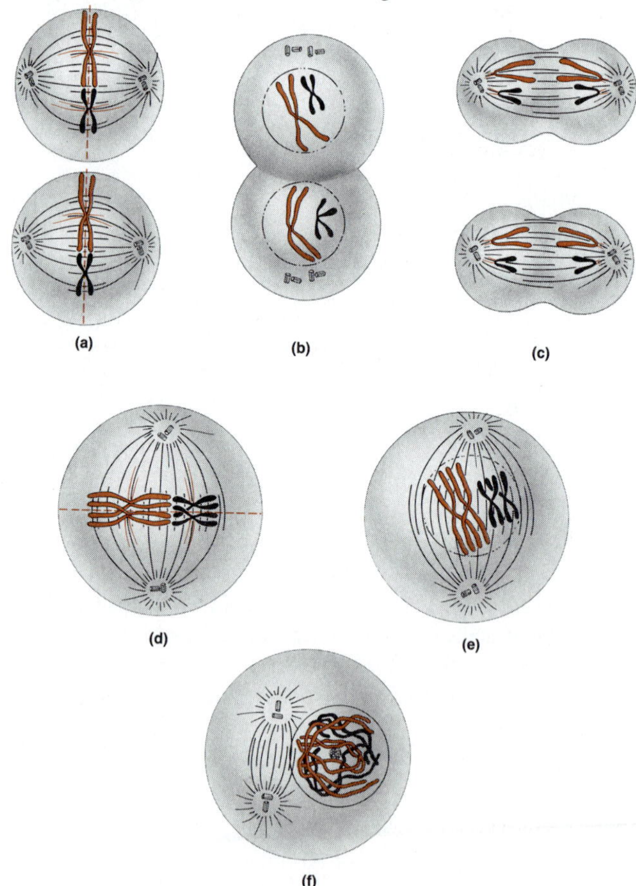

Figure 15-24 Six Stages of Meiosis to Be Ordered and Identified.

(e) A nucleus in mitotic prophase

(f) The cell during metaphase II of meiosis

(g) One bivalent

6. **Punnett Squares.** A *Punnett square* is a diagram representing all possible outcomes of a genetic cross. The genotypes of all possible gametes from the male and female parents are arranged along two sides of a square, and each box in the matrix is then used to represent the genotype that would be produced by the union of the two gametes at the head of the respective row and column. Figure 15-25 illustrates the Punnett squares for (a) a cross between two parent pea plants that are each heterozygous for seed color ($Yy \times Yy$) and (b) a cross between plants that are each heterozygous for both seed color (Yy) and seed shape (Rr).

   (a) Using the Punnett square of Figure 15-25a, explain the 3:1 phenotypic ratio that Mendel observed for the offspring of such a cross.

   (b) Explain why the Punnett square of Figure 15-25b is a 4×4 matrix with 16 genotypes. In general, what is the mathematical relationship between the number of heterozygous allelic pairs in the parents and the number of different gametes?

   (c) How does the Punnett square of Figure 15-25b reflect Mendel's law of independent assortment?

   (d) Complete the Punnett square of Figure 15-25b by indicating each of the possible progeny genotypes. How many different genotypes will be found in the progeny? In what ratios?

   (e) For the case of Figure 15-25b, how many different phenotypes will be found in the progeny? In what ratios?

7. **More Punnett Squares.** Three of the traits that Mendel chose to study in peas were *flower position, stem length,* and *seed shape.* Each of these traits is controlled by a gene that is located on a different chromosome, and each has a dominant and a recessive expression, as follows:

Trait	Dominant	Recessive
Flower position	Axial (A)	Terminal (a)
Stem length	Long (L)	Short (l)
Seed shape	Round (R)	Wrinkled (r)

   Use a Punnett square as described in Figure 15-25 to answer the following:

   (a) If you allow a plant that is heterozygous for all three traits to self-fertilize, what proportion of the offspring would you expect to be homozygous for all three traits?

   (b) If you cross a plant that is heterozygous for all three traits with one that is heterozygous for the first two but homozygous for round seeds, what proportion of the offspring would you expect to be homozygous for all three traits?

   (c) What would have to be true of the genetic makeup of the two parent plants if you found that a particular genetic cross yielded no offspring that were homozygous for all three traits?

   (d) Of what significance is it that all three of these traits are controlled by genes located on different chromosomes?

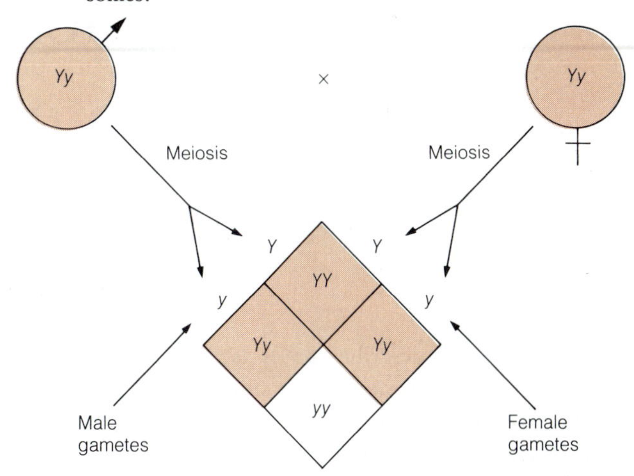

(a) One-factor cross

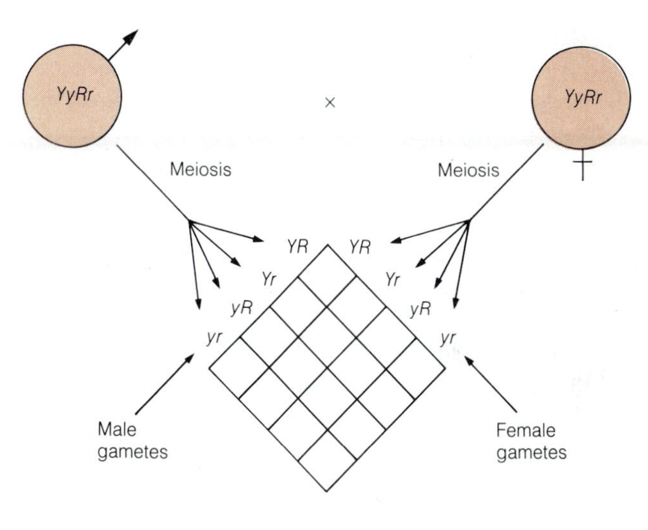

(b) Two-factor cross

Figure 15-25 The Punnett Square as a Genetic Tool. The Punnett square is used to illustrate all possible combinations of genes in a genetic cross. For every cross, all possible male gametes are represented along one side of the square, and all possible female gametes along a second side. Shown in the box at each point of intersection is the genotype that results from a specific fertilization event. By the law of independent assortment, all possible combinations are equally likely, so the frequency of occurrence of a given genotype among the boxes represents the frequency of occurrence of that genotype among the progeny of the genetic cross represented by the square. (a) A one-factor cross between two pea plants, each heterozygous for seed color. (b) A two-factor cross between parent plants heterozygous for each of two traits, seed color and seed shape. Seed shape can be either round (R, dominant) or wrinkled (r, recessive).

16

From Genes to Proteins: The Genetic Code and Protein Synthesis

So far, we have described DNA as the genetic material of most forms of life. We have come to understand its structure, chemistry, and replication, as well as the way it is packaged into chromosomes. Now we are ready to explore how that genetic information is read out, or expressed, by the cell to code for protein synthesis.

The flow of genetic information from DNA to protein is always a two-step process, with RNA as an intermediate. The nucleotide sequence of DNA is first used to direct the synthesis of an RNA molecule, and the RNA molecule then serves as a template for protein synthesis. This trio of macromolecules—*DNA, RNA,* and *protein*—constitutes the machinery that stores, expresses, and transmits genetic information in all cells.

In this chapter, we will consider the mechanism of gene expression. First we will focus on transcription (DNA → RNA), and then we will look at translation (RNA → protein) and how proteins are processed by the cell. In the following chapter, we will go on to explore the ways in which gene expression is regulated to meet cellular needs.

The Central Dogma of Molecular Biology

The flow of information through macromolecules has come to be known as the **central dogma of molecular biology,** a term coined by Francis Crick. As Figure 16-1 shows, the central dogma involves the processes of (a) *replication,* (b) *transcription,* and (c) *translation.* Having looked at DNA replication in Chapter 14, we turn now to the sequence of information flow from DNA via RNA to protein.

The RNA intermediate in this sequence is called **messenger RNA (mRNA)** because in essence it carries the genetic message from the DNA to the ribosomes, where protein synthesis actually occurs. In other words, mRNA is encoded by specific DNA sequences (genes) and used to direct the synthesis of proteins by the ribosomes. In addition to mRNA, two other classes of RNA are required for protein synthesis. **Ribosomal RNA (rRNA)** is an integral component of the ribosomes, and **transfer RNA (tRNA)** functions as an intermediate between the nucleotide sequence of the messenger RNA and the amino acid sequence of the protein. The involvement of all three major classes of RNA in the overall flow of information from DNA to protein is illustrated in Figure 16-2.

The synthesis of RNA using DNA as a template is called **transcription,** emphasizing that this phase of gene expression is simply a transfer of information from DNA to RNA. Both are nucleic acids, so the "language" is the same; all that is involved is a change in "script." By contrast, the process of protein synthesis is called **translation** because it involves a change in the language itself, from the nucleotide sequence of the mRNA to the amino acid sequence of the protein.

Also shown in Figure 16-1 is the process of **reverse transcription,** whereby genetic information can, under certain circumstances, flow "backward" from RNA to DNA. Reverse transcription is an important process for certain viruses and may occur in cells when certain DNA sequences are rearranged and modified.

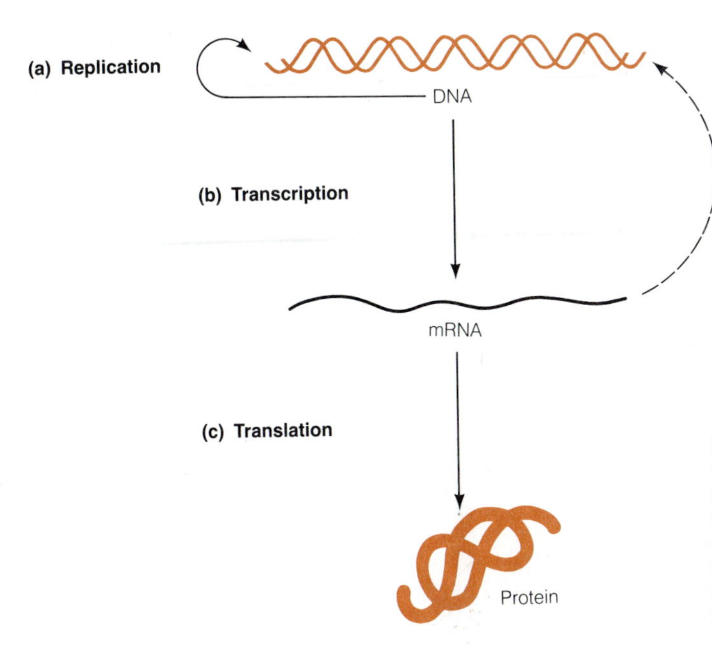

Figure 16-1 The Central Dogma of Molecular Biology. Genetic information is stored as DNA in all cells, and in many viruses as well. (a) Replication of the genetic information involves DNA-directed DNA synthesis. When genetic information is expressed in a cell, it flows unidirectionally (b) from DNA to RNA (transcription) and then (c) from RNA to protein (translation). (d) A reverse flow of information from RNA to DNA (reverse transcription) is possible under certain circumstances. Note that DNA serves as the template both for its own replication and for RNA synthesis.

The Genetic Code

We can now go on to look at the **genetic code**—the set of rules that governs the relationship between the order of nucleotides in a DNA molecule and the order of amino acids in proteins. In that relationship lies the essence of gene expression, and the "cracking" of the genetic code is one of the real landmarks of twentieth-century biology.

One of the definitions of a code is "a system of symbols (as letters, numbers, or words) used to represent assigned and often secret meanings" (*Webster's Ninth New Collegiate Dictionary,* 1987). Consider, for example, the International Morse Code (Figure 16-3), a system of dots, dashes, and spaces used primarily to send messages by telegraph and shortwave radio. To the uninitiated, the coded message in Figure 16-3a is just a series of dots and dashes, with no obvious meaning. But given access to the Morse Code (Figure 16-3b), anyone can convert the sequence of dots and dashes into a string of letters, and the message becomes recognizable as an English sentence (Figure 16-3c). Thus, a sequence of dots and dashes carries useful information but must be translated into letters of the alphabet before the message makes sense to the reader.

So, too, with the genetic code. To the uninitiated, the message shown in Figure 16-4a is just a series of nucleotides in a DNA molecule, with no obvious meaning. But given access to the genetic code (Figure 16-4b; see also Figure 16-9), anyone can convert the sequence of purines and pyrimidines into a string of amino acids, and the message becomes recognizable as a polypeptide (Figure 16-4c). Thus, a nucleotide sequence contains useful information but must be translated into amino acids before the message makes sense to the cell.

What is needed in both cases, of course, is a knowledge of the appropriate code—the set of rules that determines how many and which dots and dashes (or purines and pyrimidines) correspond to which letters (or amino acids). The encoded message can then be translated into sentences that make sense to the reader (or into proteins that make sense to the cell). In the case of the Morse Code, one has only to take the trouble to learn the code and all messages are readily understandable. Most codes, however, are not public, but are specifically intended to convey secret messages that cannot be understood by outsiders unless someone manages to crack the code.

In a sense, the genetic code was a secret code, and all of us were outsiders until the code was cracked in the early 1960s. In fact, the genetic code was a secret in a double sense: before scientists could figure out the exact relationship between the nucleotide sequence of a DNA molecule and the amino acid sequence of a protein, they first had to become aware that such a relationship existed at all. And that awareness depended, in turn, on the recognition that mutations could be linked to changes in proteins.

Mutations Cause Changes in Proteins

The link between mutations and proteins was first made by George Beadle and Edward Tatum, based on the results of experiments with the common bread mold, *Neurospora crassa.* They induced mutations in *Neurospora* by X-ray treatment and then isolated nutritional mutants that could

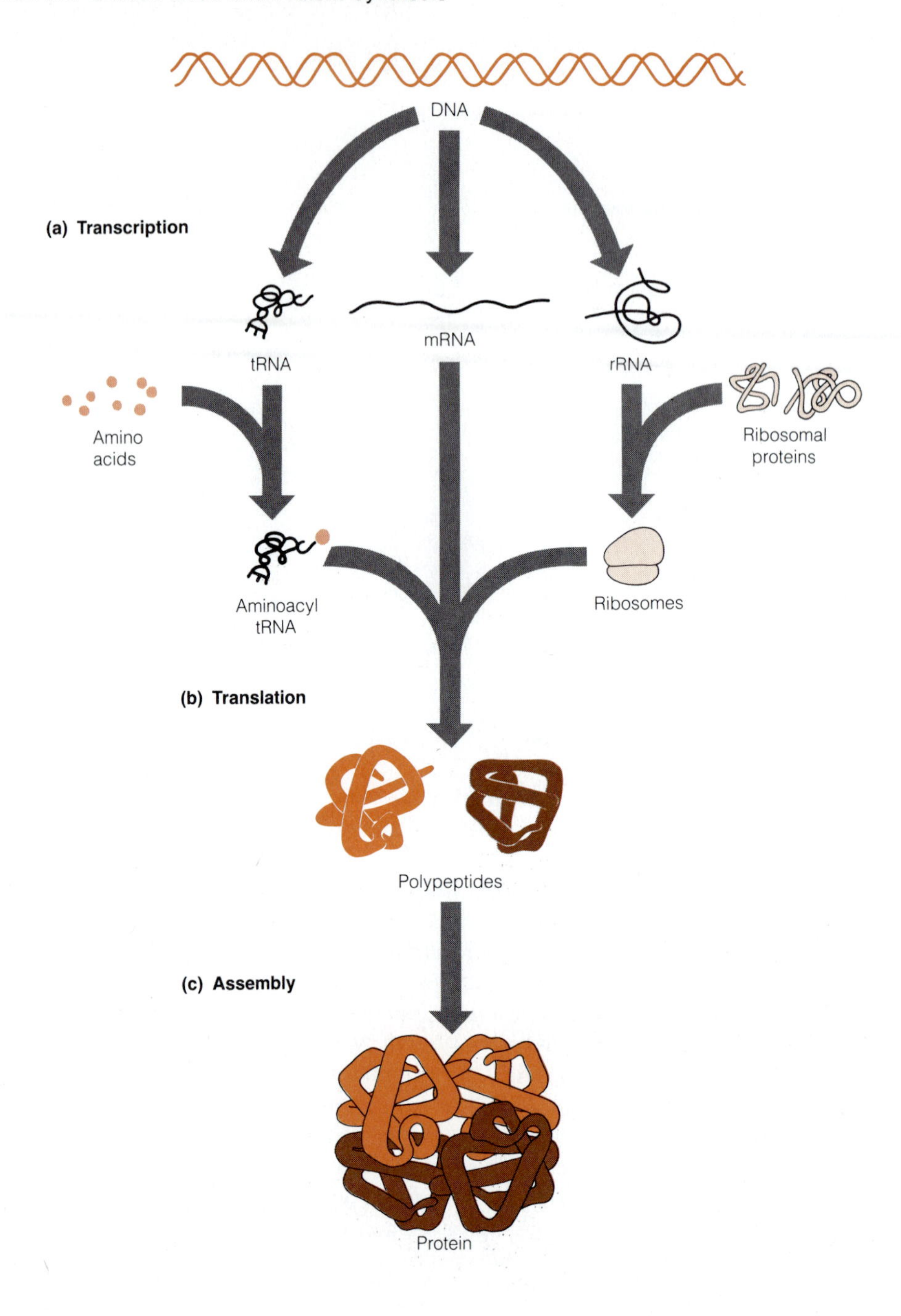

Figure 16-2 RNAs as Intermediates in Information Flow. All three major species of RNA—tRNA, mRNA, and rRNA—are (a) synthesized by transcription of the appropriate DNA sequences and (b) involved in the subsequent process of translation (polypeptide synthesis). (c) Polypeptides then fold and assemble into functional proteins. The specific polypeptides shown are the globin chains of the protein hemoglobin. (For clarity, the processing of initial transcripts into mature tRNA, mRNA, and rRNA is omitted from the diagram, as are all details of the transcriptional and translational processes.)

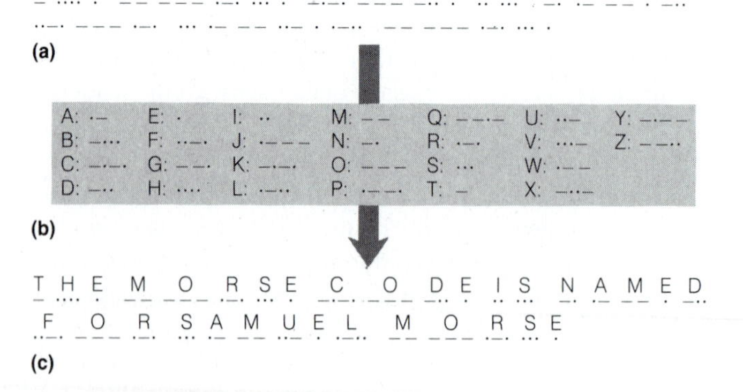

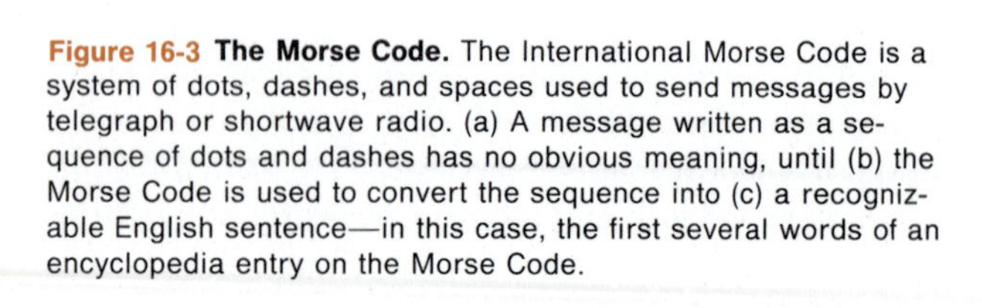
Figure 16-3 The Morse Code. The International Morse Code is a system of dots, dashes, and spaces used to send messages by telegraph or shortwave radio. (a) A message written as a sequence of dots and dashes has no obvious meaning, until (b) the Morse Code is used to convert the sequence into (c) a recognizable English sentence—in this case, the first several words of an encyclopedia entry on the Morse Code.

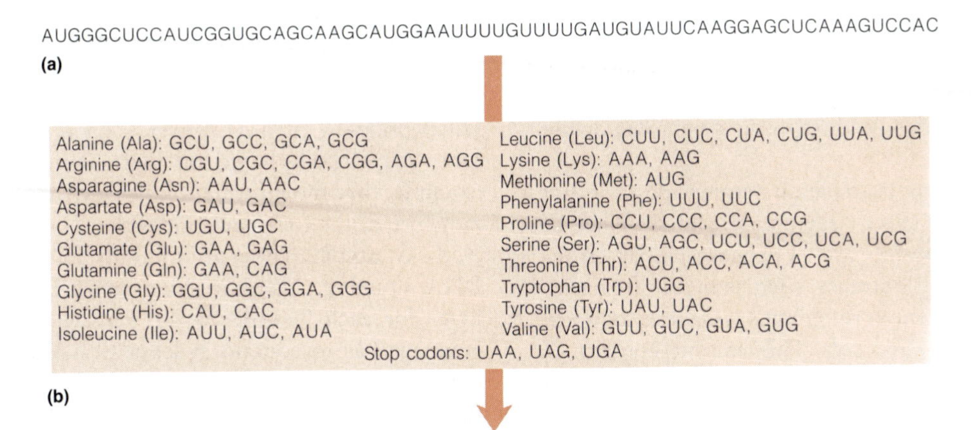

Figure 16-4 The Genetic Code. The genetic code is a system of purines and pyrimidines used to send messages from the genome to the ribosomes. (a) A message written as a sequence of nucleotides in an mRNA molecule has no obvious meaning, until (b) the genetic code is used to convert the sequence into (c) a recognizable polypeptide—in this case, the first 22 of the 385 amino acids in ovalbumin, the major protein of egg white.

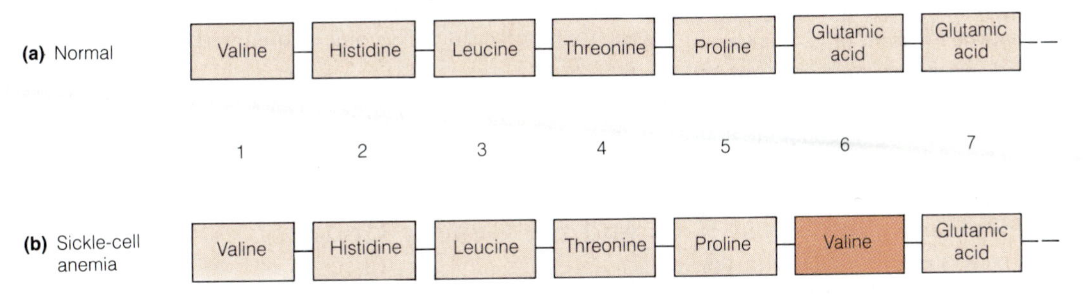

Figure 16-5 Sickle-Cell Anemia. The β chain of human hemoglobin consists of 146 amino acids. Shown here are the first seven amino acids of the β chain from (a) a normal individual and (b) an individual with sickle-cell anemia. The only difference between the two polypeptides is the substitution of a valine for glutamic acid at the sixth position. This single amino acid difference is sufficient to impair hemoglobin function and cause sickle-cell anemia.

grow only if specific supplementary compounds were added to the growth medium. They were able to show that in each such mutant strain a particular enzyme-catalyzed step leading to the synthesis of a specific compound was blocked. There was, in other words, a one-to-one correspondence between a genetic mutation and the lack of a specific enzyme required in a biochemical pathway. From these findings, Beadle and Tatum formulated the *one gene–one enzyme hypothesis*. Their hypothesis was thoroughly confirmed in later studies and now is more precisely expressed as **one gene–one polypeptide.**

A better understanding of the nature of this relationship came as it became clear that mutations almost inevitably result in changes in the amino acid sequence of specific polypeptides. The first suggestion of this arose from studies of **sickle-cell anemia,** a severe human blood disease characterized by the unusual "sickle" shape of the red blood cells of afflicted individuals. This condition was traced to a difference in the oxygen-carrying protein, *hemoglobin,* which consists of two α chains and two β chains ($\alpha_2\beta_2$).

In 1957, Vernon Ingram showed that normal hemoglobin and sickle-cell hemoglobin have identical α chains but have β chains that differ in a single amino acid. Normal hemoglobin has *glutamic acid* at the sixth amino acid position of the β chain, whereas sickle-cell hemoglobin (hemoglobin S) has *valine* at that position (Figure 16-5). That single change is enough to affect the way in which hemoglobin molecules pack into red blood cells. Normal hemoglobin is jelly-like in its consistency, but hemoglobin S tends to form a kind of crystal when it delivers oxygen to and picks up carbon dioxide from tissues. The crystalline array deforms the cell into a sickled shape, which then blocks blood flow in capillaries, leading to the symptoms of the disease.

From this and other examples, it became apparent that genes must somehow specify the amino acid sequence of proteins. This relationship was even more clearly established by further genetic studies, such as those of Charles Yanofsky on the enzyme *tryptophan synthetase,* found in the bacterium *Escherichia coli.* This enzyme consists of two polypeptides, A and B, coded for by the genes *trpA* and *trpB,* respectively. Yanofsky mapped many mutants lacking tryptophan synthetase activity and found that most of them could be ordered linearly within the *trpA* gene. He also discovered that a number of these mutants produced an altered form of polypeptide A that was enzymatically inactive but could nonetheless be isolated and sequenced.

From the sequence data for many such mutants, Yanofsky was able to show that each of the mutant polypeptides differed from the wild type by a single amino acid, but that different mutants had different amino acid substitutions. When he compared the location of each altered amino acid in the polypeptide with the map position of the mutation that caused that alteration, he found a close correlation. This correlation led Yanofsky to conclude that there is a **colinearity** between a gene and its polypeptide, further substantiating the relationship between the nucleotide sequence of DNA and the amino acid sequence of proteins, which the genetic code seeks to explain. In retrospect, it was fortunate that Yanofsky was studying bacterial genes. As we will see, most eukaryotic genes contain noncoding sequences interspersed among the coding regions of the gene, and therefore do not usually show strict colinearity between a gene and its polypeptide product.

The Nature of the Genetic Code

Given a sequence relationship between DNA and proteins, the next question concerned the number of nucleotides necessary to specify each amino acid. In the Morse Code, more than one symbol is clearly required per letter, because only three different kinds of symbols—dots, dashes, and spaces—are used to represent 26 letters, 10 numbers, and various punctuation marks. In fact, the number of symbols per letter varies, ranging from a single dot for the letter E to various combinations of four symbols for some of the less common letters (such as X, Y, and Z). Fortunately, the genetic code is more orderly, and the number of nucleotide pairs per amino acid is always three. Thus, three nucleotide pairs in the genome are required to specify each amino acid in a polypeptide.

Most of the initial evidence suggesting a *triplet code* was genetic in nature, though biochemical confirmation quickly followed once it became possible to synthesize polypeptides in vitro. However, a good case can be made for a triplet code based on logic alone. We know, for example, that the information in DNA must reside in the sequence of the four nucleotides that constitute the DNA: A, T, G, and C. These are the only four "letters" of the DNA alphabet. Therefore, there must be more than one letter for each "word" in the message, since the DNA language has to contain at least 20 words, one for each amino acid. A doublet code would not be adequate, because four nucleotides taken two at a time can generate only $4^2 = 16$ different combinations.

With three letters per word, however, the number of different words that can be produced with an alphabet of just four letters is $4^3 = 64$. This appears to be more than adequate, and it raises the obvious question of whether every one of the 64 possible combinations is actually used. The answer is yes, although several of them have a special "punctuation" function instead of coding for an amino acid, as we will see later. First, however, we will look at the genetic evidence for the triplet nature of the code. And for that, we need to become acquainted with *frameshift mutations*.

Frameshift Mutations and the Genetic Code

In 1961, Francis Crick, Sydney Brenner, and their colleagues deduced the triplet nature of the genetic code by studying the effects of the mutagen *proflavin* on the bacteriophage T4. Their work is well worth considering, not just because of the critical genetic evidence it provided concerning the nature of the code but also because of the ingenuity of the deductive reasoning that was necessary to understand the significance of their findings.

Proflavin is one of several *acridine dyes* used as **mutagens** (mutation-inducing agents) in genetic research. The acridines are interesting mutagens because they act by causing the addition or deletion of single nucleotide pairs in the DNA. Sometimes, mutants of this sort appear to "revert" to the wild type in the presence of the mutagen. Closer examination, however, shows that this apparent reversion is not a true reversal of the original event, but the acquisition of a second mutation that maps very close to the first. In fact, the two mutations are always in the same gene, and each displays the mutant phenotype when they are separated by recombination.

These mutations show an interesting kind of arithmetic. If the first is called a "plus" (+) mutation, then the second can be called a "minus" (−) mutation. By itself, each is a mutation, but when they occur together as a double mutation, they cancel each other out.

This finding can be explained schematically by the example in Figure 16-6. Line 1 is a wild-type "gene," written in a language that uses three-letter words. When we "translate" the line by starting at the beginning and reading three letters at a time, the message of the gene is readily comprehensible. A + mutation involves the addition of a single letter to the message (line 2). That change may seem minor, but since the message is always read three letters at a time, the insertion of an extra letter early in the sequence means that all the remaining letters are read "out of phase." There is, in other words, a shift in the *reading frame,* and the result is a garbled message from the point of the insertion onward. A − mutation can be explained in a similar way, since the deletion of a single letter also causes the reading frame to shift, resulting in another garbled message (line 3). Such **frameshift mutations** are characteristic effects of the acridine dyes and other mutagens that cause the insertion or deletion of individual nucleotide pairs.

By themselves, + and − mutations always change the reading frame and garble the message. When a + and a − mutation occur in the same gene, however, they can cancel each other out, particularly if they are located in close proximity. The insertion caused by the + mutation compensates for the deletion caused by the − mutation, and the message is intelligible from that point on (line 4). Notice, however, that double mutations with either two additions (+/+; line 5) or two deletions (−/−; line 6) do not compensate in this way. They simply stay out of phase for the remainder of the message.

Crick and Brenner made similar observations with their proflavin-induced T4 mutants. They were able to isolate revertants from their − mutants and show that these always contained a + mutation at a site different from, but close to, the original mutation. New + and − mutations could, in fact, be localized to a variety of positions within the gene, and double +/− mutants could be constructed in many combinations. Many of these had wild-type (or more properly, *pseudo-wild-type*) phenotypes. But when Crick and Brenner generated +/+ or −/− double mutants by recombination, no wild-type phenotypes were ever seen.

Crick and Brenner also constructed triple mutants of the same type (+/+/+ or −/−/−) and found, probably to their surprise, that many of these had wild-type phenotypes. This finding, of course, can be readily understood by consulting lines 7 and 8 of Figure 16-6: The reading frame can be restored by either adding or removing three letters. The portion of the message between the first and third mutations will be garbled, but provided these are sufficiently close to each other, enough of the sentence may remain to convey an intelligible message.

Keep in mind, however, that Crick and Brenner did not have Figure 16-6 to assist them. Their ability to deduce the correct explanation from their analysis of acridine-induced mutations is an especially inspiring example of the careful, often ingenious reasoning that almost inevitably accompanies significant advances in science.

Figure 16-6 Frameshift Mutations. The effect of frameshift mutations can be illustrated with an English sentence. The wild-type sentence (line 1) consists of three-letter words. When read in the correct frame, it is fully comprehensible. The addition (line 2) or deletion (line 3) of a single letter shifts the reading frame and garbles the message from that point onward. (Garbled words due to shifts in the reading frame are underscored.) Double mutants containing a deletion that "cancels" a prior addition have a restored reading frame from the point of the second mutation onward (line 4). However, double additions (line 5) or double deletions (line 6) do not have the same effect. Triple additions or deletions (lines 7 and 8) garble part of the message, but restore the reading frame with the net addition or deletion of a single word.

The Genetic Code Is a Triplet Code

Because the wild-type phenotype was restored by three additions (or deletions) but not by one or two, Crick and Brenner concluded that the nucleotide pairs of the DNA must be read in groups of three (or perhaps multiples thereof). They reasoned that the reading of a message must begin at a specific starting place on the DNA (to ensure the proper reading frame) and must then proceed three nucleotides at a time, with each such triplet translated into the appropriate amino acid, until the end of the message is reached. The genetic code, in other words, is a **triplet code.**

Adding or deleting one nucleotide pair shifts the reading frame and garbles the message from that point onward, as does adding or deleting a second nucleotide pair. After a third shift, however, the reading is back in phase again, and the only segment of the message that is translated incorrectly is the segment between the first and third mutations. Such errors can apparently be tolerated in many proteins, provided that the affected region is short and the change in amino acid sequence in that region does not destroy protein function. Presumably, this is why the individual mutations in a triple mutant with wild-type phenotype map so closely together (Figure 16-6, lines 7 and 8).

The Genetic Code Is Nonoverlapping

A further conclusion from Crick and Brenner's work is that the genetic code is **nonoverlapping.** An overlapping code would result if the reading frame was advanced only one or two nucleotides at a time, so that each nucleotide was used two or three times. But when the reading frame is advanced three nucleotides at a time, each nucleotide is used only once, and the code is therefore nonoverlapping.

Figure 16-7 compares a nonoverlapping code with an overlapping code in which the reading frame advances only one nucleotide at a time. With such an overlapping code, the addition or deletion of a single nucleotide pair in the gene would lead to the addition or deletion of one amino acid at one point in the protein and would change several amino acids at that point, but would not affect the reading frame of the remainder of the gene. (You might want to explore for yourself what the consequences would be of an overlapping code in which the reading frame advanced two nucleotides at a time.) Since the frameshift mutations that Crick and Brenner found would not be possible with an overlapping code, their results clearly indicated the nonoverlapping nature of the code: Each nucleotide pair is a part of one, and only one, triplet.

The Genetic Code Is Degenerate

Because so many of their triple mutants with out-of-phase segments were viable, Crick and Brenner were able to draw an additional conclusion: Most of the 64 possible nucleotide triplets must be used by the cell, not just the 20 that would be necessary to specify 20 amino acids. If only 20 of the 64 possible combinations of nucleotides made sense, the chances of a meaningless triplet appearing in the out-of-phase stretches would be high, and frameshift mutants would "revert" only rarely.

Since they detected such revertants frequently and found that any + mutation had a fairly high probability of suppressing any − mutation if the two were close together, Crick and Brenner reasoned that most of the 64 possible triplets must be meaningful. This, in turn, suggested to them that the genetic code is **degenerate**—that is, a given amino acid may be specified by more than one triplet. (The term *degenerate* has a foreign sound to many biologists; it was actually borrowed from quantum mechanics and probably reflects the influence of physics on some of the early investigators in the emerging field of molecular biology.)

From Triplets to Codons

From the publication of Crick and Brenner's historic findings in 1961, it took only five years for the genetic code to be cracked and the meaning of each of the 64 triplets to be elucidated. We are about to look at how that was done. First, however, it is important to understand that the genetic code as we usually describe it refers not to the order of nucleotide pairs in double-stranded DNA, but to the order of nucleotides in the single-stranded mRNA molecules that are actually used in protein synthesis.

As we will see in the next section, transcription of DNA into RNA proceeds by a complementary base-pairing mechanism that is analogous to DNA replication, except that only one of the two DNA strands is copied. The strand of the DNA duplex that serves as the template for RNA synthesis is called the **template strand;** its complement is usually referred to as the **nontemplate strand.** The RNA product is therefore always single-stranded and is complementary to the template strand of the DNA (Figure 16-8).

As a result of the transcription process, the genetic information represented as triplets of nucleotide pairs in DNA is present in mRNA as triplets of nucleotides. These triplets are called **codons** because they are the actual coding units read by the translational machinery during protein synthesis. The genetic code is always expressed in terms of triplet codons in mRNA molecules. The four bases present in RNA are the purines **adenine** (A) and **guanine** (G) and the pyrimidines **cytosine** (C) and **uracil** (U), so the 64 codons of the genetic code consist of all possible permutations of these four "letters" taken three at a time. And since mRNA molecules are synthesized and translated from the 5′ to the 3′ end, codons are always written in the 5′ → 3′ orientation.

Codon Assignments

The year 1961 is especially important in the history of the genetic code, for it brought not only the publication of Brenner and Crick's findings but also the development of the first of several methods for assigning specific codons to particular amino acids. Pioneered by Marshall Nirenberg and J. Heinrich Matthei, the original technique depended on the ability of an enzyme called *polynucleotide phosphorylase* to make synthetic mRNA of predictable nucleotide composition. Unlike the enzymes involved in transcription, polynucleotide phosphorylase does not require a template, but simply assembles available nucleotides into a random linear chain in the proportions in which they are present in the incubation mixture.

Homopolymers as mRNA. If polynucleotide phosphorylase were given access to all four nucleotides (ATP, GTP, CTP, and UTP), all possible codons would be made randomly, and no useful information could be obtained. But by using only one or two nucleotides, only a restricted number of codons can be generated. The simplest case

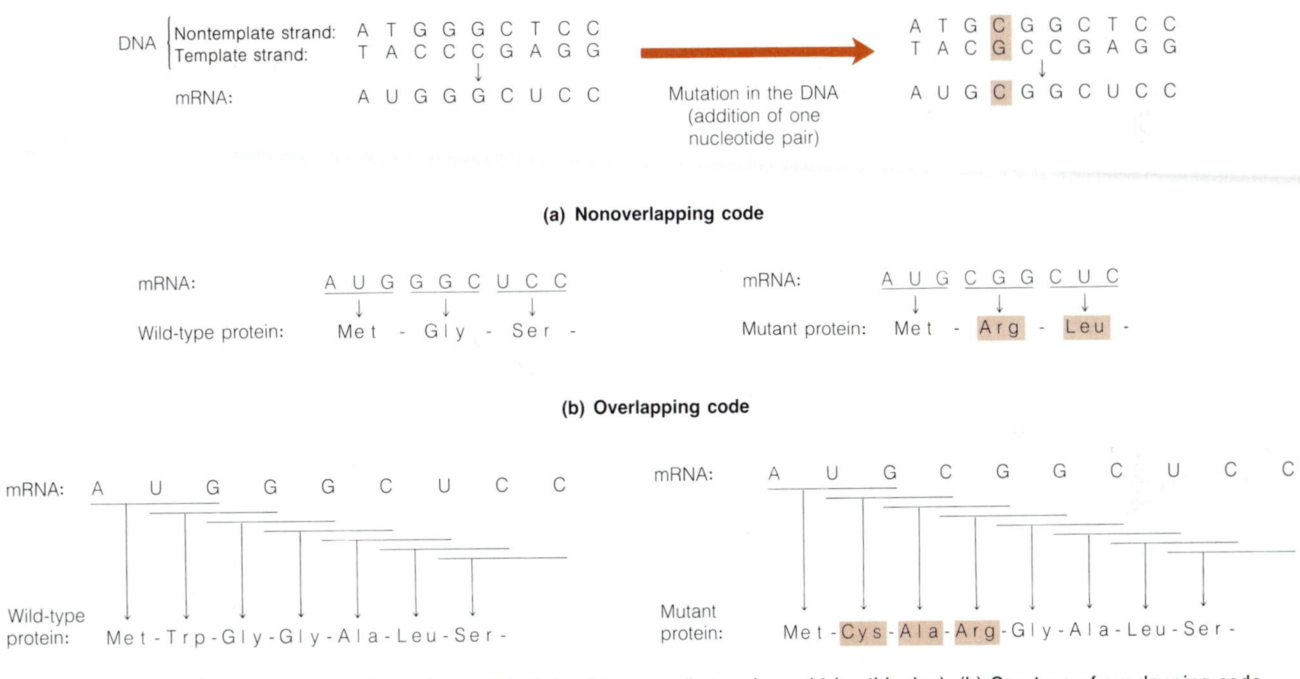

Figure 16-7 Effect of the Addition of a Single Nucleotide Pair with Overlapping and Nonoverlapping Genetic Codes. The template strand of the DNA duplex at the top is transcribed into the 9-nucleotide segment of mRNA shown. (a) With a nonoverlapping code, the reading frame advances three nucleotides at a time, and this mRNA segment is therefore read as three successive triplets, coding for the amino acids methionine, glycine, and serine. (See Figure 16-4 for amino acid coding rules.) If the DNA duplex is mutated by addition of a single nucleotide pair (the CG pair in the color box), the mRNA will have an additional nucleotide (also in color). This insertion alters the reading frame beyond that point, so that the remainder of the mRNA is read incorrectly and all amino acids are wrong. In the example shown, the insertion occurs near the beginning of the message, and the only similarity between the wild-type protein and the mutant protein is the first amino acid (methionine). (b) One type of overlapping code can be generated by assuming that the reading frame advances only one nucleotide at a time. The wild-type protein will therefore contain three times as many amino acids as would a protein generated from the same mRNA using a nonoverlapping code. Insertion of a single nucleotide pair in the DNA again results in an mRNA molecule with one extra nucleotide. However, in this case the effect of the insertion on the protein is modest; two amino acids in the wild-type protein are replaced by three different amino acids in the mutant protein, but the remainder of the protein is normal. The frameshift mutations that Crick and Brenner found in their studies with the mutagen proflavin would not occur if the genetic code were overlapping. Accordingly, their data indicated the code to be nonoverlapping.

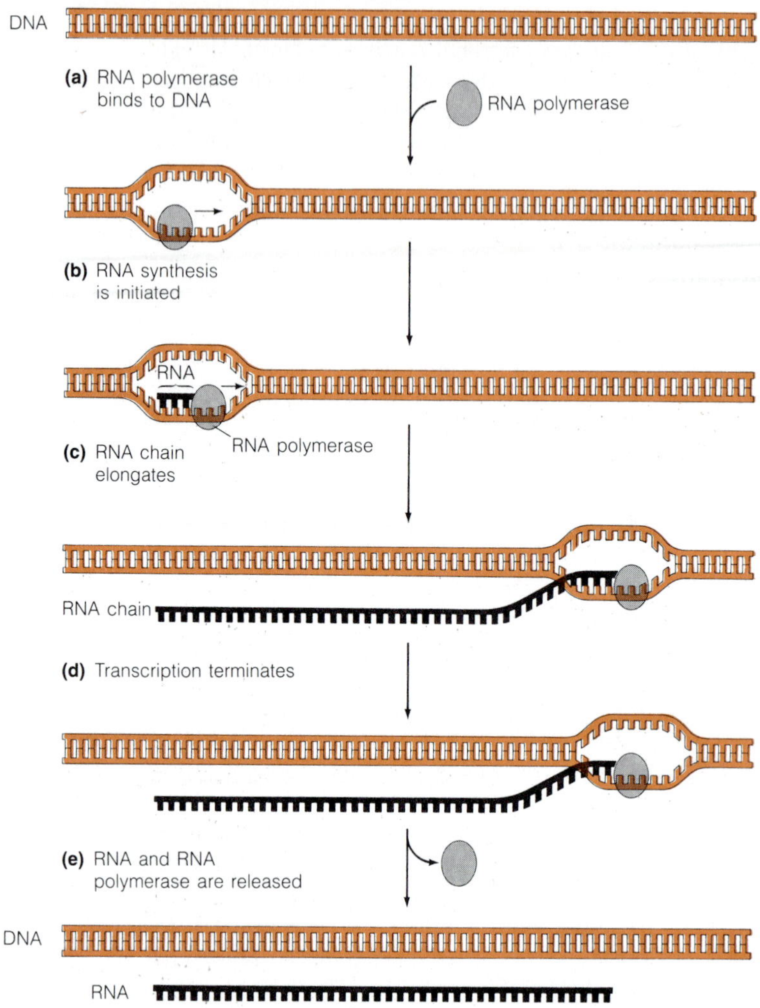

Figure 16-8 The Transcription of Single-Stranded RNA from a Double-Stranded DNA Template. Transcription resembles DNA replication (recall Figure 14-5), except that only one of the two strands is copied, and the newly synthesized strand does not remain associated with its template. (a) RNA polymerase binds to DNA and recognizes the template strand. (b) RNA synthesis is initiated as RNA polymerase begins transcribing the template strand. (c) The RNA chain elongates as transcription continues. (d) Transcription terminates when the RNA chain reaches the desired length. (e) The RNA polymerase and the completed RNA molecule are then released from the DNA.

results when a single nucleotide is used, because the only possible product is a **homopolymer** consisting of a single repeating nucleotide. Not surprisingly, the first synthetic mRNA molecules that Nirenberg and Matthei made were homopolymers, and the first codon assignments were therefore easy.

For example, when polynucleotide phosphorylase was incubated with UTP as the only precursor, the product was a homopolymer of uracil called **poly U.** Nirenberg and Matthei then added this synthetic mRNA to an in vitro protein-synthesizing system and showed that the only product was a polypeptide containing phenylalanine as its only amino acid. Poly U, in other words, codes for polyphenylalanine. From this observation, Nirenberg and Matthei deduced that the sequence UUU on the mRNA is read as a signal for phenylalanine insertion during protein synthesis. They therefore made the first codon assignment: UUU = phenylalanine. Synthesis of other homopolymers quickly revealed that AAA codes for lysine and CCC for proline. (Poly G turned out not to be a good messenger because of unexpected structural complications and therefore could not be tested.)

Codon Binding Studies. After the homopolymers were tested, the method was extended to more complex polymers, but quickly proved to be less definitive because of the random choice of nucleotides by polynucleotide phosphorylase. For example, the **copolymer** synthesized by incubating the enzyme with the precursors CTP and ATP contains C's and A's, but in no predictable order. Such a copolymer contains eight different codons: CCC, CCA, CAC, ACC, AAC, ACA, CAA, and AAA. When the copolymer was used to direct protein synthesis in vitro, the polypeptides that were produced contained 6 of the 20 possible amino acids. The codons for two of these (CCC and AAA) were known from the homopolymer studies, but the other four could not be unambiguously assigned.

Further progress depended on an alternative means of

codon assignment devised by Nirenberg's group. Instead of using long polymers, they synthesized all possible codons (that is, 64 short mRNA molecules, each only three nucleotides long) and used these in binding studies to see which amino acid binds to the ribosome in response to each codon. With this approach, they were able to determine the majority of the codon assignments.

Copolymers as mRNA. Meanwhile, a refined method of polymer synthesis had been devised in the laboratory of H. Gobind Khorana. Khorana's approach was similar to that of Nirenberg and Matthei, but with the important difference that the polymers he synthesized had defined sequences. Thus, he could produce a synthetic mRNA molecule with the strictly alternating sequence UAUA. . . . Such an RNA copolymer has only two codons, UAU and AUA, and they alternate in strict sequence. Knowing that the polypeptide product contained only tyrosine and isoleucine, Khorana was able to narrow the choices for UAU and AUA to these two amino acids. When the results obtained with various such synthetic polymers were combined with the findings of Nirenberg's binding studies, most of the codons could be assigned unambiguously.

Codon Assignments Complete. By 1966, just five years after the first codon was identified, all 64 codons had been assigned and the entire genetic code had been deciphered. Codon assignments are shown in Figure 16-9. As the code was elucidated, several properties that had been deduced earlier from indirect evidence were shown to be correct. All 64 codons are in fact used. Three of these are **stop signals,** and the other 61 translate into specific amino acids. The degenerate nature of the code is clear, since many of the amino acids are specified by more than one codon. There are, for example, two codons for histidine (His), four for threonine (Thr), and six for leucine (Leu). Also clear from Figure 16-9 is the **unambiguous** nature of the code: Every codon has one and only one meaning.

Second Position

First Position	U		C		A		G		Third Position
	UUU	Phe	UCU		UAU	Tyr	UGU	Cys	U
U	UUC		UCC	Ser	UAC		UGC		C
	UUA	Leu	UCA		UAA	Stop	UGA	Stop	A
	UUG		UCG		UAG	Stop	UGG	Trp	G
	CUU		CCU		CAU	His	CGU		U
C	CUC	Leu	CCC	Pro	CAC		CGC	Arg	C
	CUA		CCA		CAA	Gln	CGA		A
	CUG		CCG		CAG		CGG		G
	AUU		ACU		AAU	Asn	AGU	Ser	U
A	AUC	Ile	ACC	Thr	AAC		AGC		C
	AUA		ACA		AAA	Lys	AGA	Arg	A
	AUG	Met	ACG		AAG		AGG		G
	GUU		GCU		GAU	Asp	GGU		U
G	GUC	Val	GCC	Ala	GAC		GGC	Gly	C
	GUA		GCA		GAA	Glu	GGA		A
	GUG		GCG		GAG		GGG		G

Figure 16-9 The Genetic Code. The code consists of three-letter codons present in the nucleotide sequence of mRNA, as read in the 5′ → 3′ direction. Letters represent the nucleotide bases uracil (U), cytosine (C), adenine (A), and guanine (G). Each codon specifies either an amino acid or a stop signal (color boxes). To decode a codon, read down the left-hand column for the first letter, then across the grid for the second letter, and then down for the third letter. For example, the codon AUG represents methionine (gray box).

The (Near) Universality of the Genetic Code

A final property of the genetic code that is worth noting is its near-universality. All organisms studied so far, both prokaryotes and eukaryotes, use the same code for synthesis of proteins by cytoplasmic ribosomes. The only known exception occurs in mitochondria. In common with chloroplasts, mitochondria contain their own genetic information and can carry out both transcription and translation. They differ from all other known genetic systems, however, in that they deviate from the standard genetic code in several ways (see Table 16-1). For example, the codon UGA, which is otherwise a stop codon, is read in mitochondria as tryptophan. As Table 16-1 indicates, there are even differences in codon assignments between the mitochondria of yeast and those of mammals. It is not yet clear why mitochondria have a genetic code that differs from that of all other systems.

Table 16-1 The Genetic Code of the Mitochondrion

		Meaning in Mitochondrion	
Codon	Normal Meaning	Mammalian	Yeast
UGA	Stop	Tryptophan	Tryptophan
AUA	Isoleucine	Methionine	Methionine
CUA	Leucine	Leucine	Threonine
AGA	Arginine	Stop	Arginine

Transcription: From DNA to RNA

Having looked in detail at the way in which the amino acid sequence of a protein is specified genetically, we now come to the first of the two major processes in the flow of genetic information from DNA to protein: the transcription of the nucleotide sequence of DNA into that of RNA. RNA is chemically similar to DNA, except that it contains ribose instead of deoxyribose as its sugar and has the base uracil in place of thymine. RNA is almost always a single-stranded molecule, though often with some intramolecular double-strandedness. Our discussion of transcription will include mRNA, tRNA, and rRNA synthesis, since each of these classes of RNA is involved in protein synthesis (Figure 16-2) and each is transcribed as a single-stranded complement to the template strand of the appropriate DNA sequence.

RNA Polymerases: The Enzymes of Transcription

The enzyme responsible for catalyzing the synthesis of RNA from a DNA template is **RNA polymerase.** Bacterial cells have a single kind of RNA polymerase, which must therefore be capable of synthesizing all three classes of RNA. The RNA polymerase from *E. coli* has been especially well characterized. It consists of two α subunits, two β subunits that differ enough to be identified as β and β', and a dissociable subunit called the **sigma (σ) factor.** The sigma subunit ensures that RNA synthesis is initiated at the right place on the right DNA strand and is therefore important in the specificity of transcription. The **core enzyme** ($\alpha_2\beta\beta'$) lacks the sigma subunit but is nonetheless competent to carry out RNA synthesis. However, the **holoenzyme** ($\alpha_2\beta\beta'\sigma$) is required to ensure correct initiation. Moreover, bacteria contain several different sigma factors that can be utilized in special situations, as we will see in Chapter 17.

In eukaryotic cells five different kinds of RNA polymerases have been identified. Three of these are located in the nucleus and the other two are organellar, with one in the mitochondrion and the other in the chloroplast. Table 16-2 summarizes some of the properties of these enzymes. The nuclear enzymes are designated as *RNA polymerases I, II,* and *III.* As Table 16-2 indicates, these enzymes differ in their location within the nucleus and in the kinds of RNA they make. In addition, the nuclear polymerases can be distinguished from each other based on their sensitivity to *α-amanitin,* a deadly toxin produced by the poisonous mushroom *Amanita phalloides,* known commonly as the "death cap" or the "destroying angel."

RNA polymerase I is located in the nucleolus. It is responsible for the synthesis of the precursor form of rRNA and is insensitive to α-amanitin. The association of this enzyme with the nucleolus is understandable, since the nucleolus is the site of ribosomal RNA synthesis and of ribosomal assembly, as we learned in Chapter 13. RNA polymerase II is found in the nucleoplasm and synthesizes the precursors to mRNA. It is highly sensitive to α-amanitin, which explains the toxicity of this compound to humans and animals. RNA polymerase III is also a nucleoplasmic enzyme, but it is responsible for synthesis of a variety of small RNAs, including tRNA precursors and a species of ribosomal RNA called 5S rRNA. Polymerase III is sensitive to α-amanitin, but only at substantially higher levels of the toxin than those required to inhibit activity of RNA polymerase II.

The Steps of Transcription

Transcription involves the synthesis of an RNA molecule that is complementary in base sequence to the template strand of the DNA duplex. The overall process of transcription is illustrated in Figure 16-8. RNA polymerase binds to the DNA duplex, causing a localized unwinding of the helix at which point RNA synthesis is initiated. The RNA molecule is elongated by polymerization of nucleotides in an order determined by their base pairing to the template strand of the DNA template. The RNA polymerase molecule proceeds along the template molecule,

Table 16-2 Properties of Eukaryotic RNA Polymerases

RNA Polymerase	Location	Products	α-Amanitin Sensitivity
RNA polymerase I	Nucleolus	Pre-rRNA	Resistant
RNA polymerase II	Nucleoplasm	Pre-mRNA (hnRNA)	Very sensitive
RNA polymerase III	Nucleoplasm	Pre-tRNA, 5S rRNA	Moderately sensitive
Mitochondrial	Mitochondrion	Mitochondrial RNA	Resistant
Chloroplast	Chloroplast	Chloroplast RNA	Resistant

unwinding the DNA helix and elongating the RNA chain as it goes, until a termination signal is encountered. At that point, the completed RNA molecule is released and the enzyme dissociates from the template.

Transcription is a complicated process but can be thought of in four steps: *binding, initiation, elongation,* and *termination.* Figure 16-10 illustrates these steps as they occur in *E. coli.* We will now look at each in detail.

Binding. **Binding** of RNA polymerase to DNA occurs at specific asymmetric sequences called **promoters** (Figure 16-10a). Each transcription unit has a promoter located near the beginning of the sequence that is actually to be transcribed. In *E. coli,* the sigma subunit of RNA polymerase is responsible for promoter recognition and therefore for specificity of binding. Once transcription has been initiated at the proper site by the holoenzyme, however, the sigma subunit dissociates, leaving the core enzyme to continue the transcription process (Figure 16-10b).

Several of the promoter sequences recognized by the *E. coli* enzyme have been well characterized. Each consists of about 40 nucleotide pairs. A key feature of each promoter is the presence of a specific hexanucleotide sequence called the **Pribnow box,** after its discoverer. The Pribnow box is responsible for determining the precise nucleotide at which transcription will begin. It is located on the nontemplate strand about six nucleotides upstream from the point at which transcription is actually initiated.

Figure 16-11 illustrates the Pribnow box sequences for several bacterial and viral promoters. The sequences do not vary much between promoters and are therefore said to be *evolutionarily conservative.* When variants of a highly conserved sequence are studied, it is usually possible to identify a **consensus sequence** that indicates the most likely and hence presumably the preferred nucleotide at each position. For the Pribnow box, the consensus sequence is TATAAT (Figure 16-11). Changes in the Pribnow box sequence are likely to affect promoter function. Mutations toward the consensus sequence can enhance promoter activity, whereas mutations away from the consensus sequence are likely to reduce the strength of the promoter and may even eliminate promoter activity entirely.

In addition to the Pribnow box, a second critical sequence located about 35 nucleotides upstream from the point at which transcription begins is probably also involved in the initial polymerase-binding event in *E. coli.*

Each of the three RNA polymerases in eukaryotic nuclei appears to require its own specific promoter sequence, although so far these have not been as well characterized as prokaryotic promoters. For RNA polymerase II, transcription seems to be promoted by an oligonucleotide sequence about 25 nucleotides upstream from the initiation site. The key feature of this sequence is the tetranucleotide TATA, usually called the **TATA box.** The TATA box is thought to be important in positioning the RNA polymerase molecule correctly on the DNA template. A single nucleotide change in the TATA box can drastically decrease transcription in vitro.

An additional sequence about 80 nucleotides upstream from the initiation site may also be a part of the promoter site for RNA polymerase II. For the globin genes, the sequence CAAT is common to this region. When many different genes are examined, a recognizable pattern emerges, although more variation is seen in this region than in the region of the TATA box.

A functional test for the sequences needed for promoter activity involves the deletion of specific sequences from a cloned DNA molecule, which is then tested for its ability to direct normal transcription, either in vitro or in cultured cells. When the gene for β globin was tested in this way, deletion of either the TATA box or the CAAT box reduced the level of transcription at least tenfold.

Initiation. Once an RNA polymerase molecule is bound to a promoter, **initiation** of RNA synthesis is possible. Initiation requires an unwinding by the enzyme of about one turn of the DNA helix (Figure 16-10b). The short segment of single-stranded DNA that is exposed in this way serves as the template for the synthesis of RNA from incoming **ribonucleoside triphosphate molecules (rNTPs).** As soon as the first two rNTPs are in place, RNA polymerase catalyzes the formation of a phosphodiester bond between the 3′ hydroxyl group of the first rNTP and the 5′ hydroxyl group of the second, with the removal of pyrophosphate, PP_i.

Elongation. **Chain elongation** occurs as the RNA polymerase moves along the DNA molecule, untwisting the helix bit by bit and adding one complementary rNTP at a time (Figure 16-10c). The enzyme moves along the template strand from the 3′ to the 5′ end. Because nucleotide pairing is antiparallel, the RNA strand is elongated in the $5' \rightarrow 3'$ direction as each successive nucleotide is added to the 3′ end of the growing chain. A short stretch of RNA-DNA hybrid is formed as the RNA chain grows, but such a duplex is thermodynamically less stable than a DNA-DNA duplex. As a result, the DNA returns to its double-stranded form as the polymerase moves on, displacing the growing RNA chain from its template.

Termination. **Termination** of transcription occurs when the RNA polymerase encounters a special sequence in the DNA called the **termination signal.** In prokaryotes, the termination signal is a DNA sequence that gives rise in the RNA product to a hairpin helix followed by a string of U's (Figure 16-12). The actual sequence is apparently

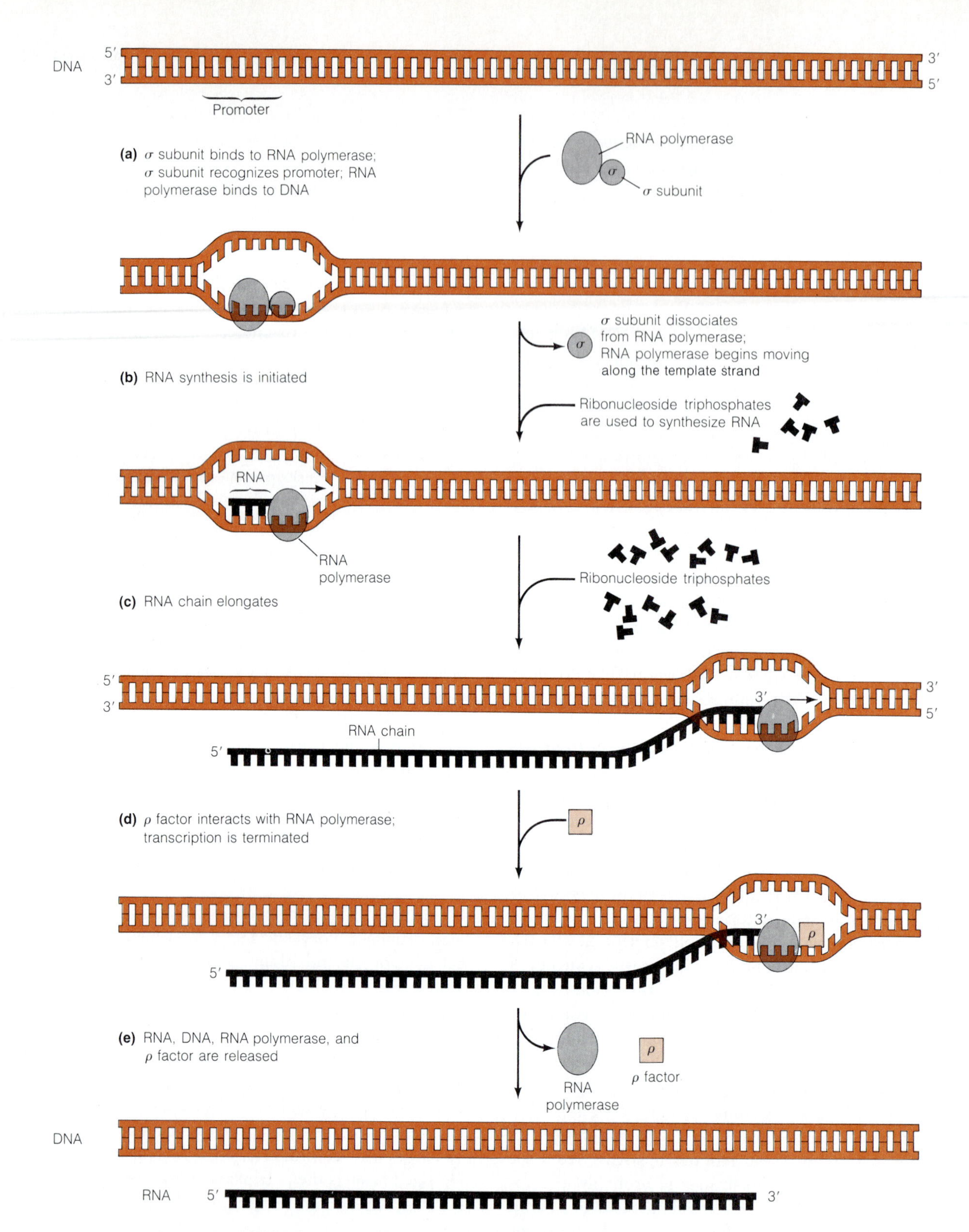

Figure 16-10 Transcription in Bacterial Cells. Transcription of genetic information occurs in four steps: (a) binding of RNA polymerase to DNA at a promoter, (b) initiation of transcription, (c) subsequent chain elongation, and (d) eventual termination of transcription, followed by (e) release of RNA polymerase and the completed RNA product from the DNA template. Sigma factor (σ) is required for promoter recognition, and rho factor (ρ) is in some cases involved in the release of the RNA transcript upon termination of synthesis. RNA polymerase moves along the template strand of the DNA in the 3′ → 5′ direction such that the RNA molecule is synthesized in the 5′ → 3′ direction.

less significant than the self-complementarity that allows the RNA to fold spontaneously into a hairpin loop. In *E. coli,* termination at other regions is brought about by a factor called **rho (ρ),** which interacts with the polymerase to prevent further transcription (Figure 16-10d). Termination results in the release of the completed RNA molecule and of the free RNA polymerase (Figure 16-10e), which is then available to bind sigma factor again and reinitiate synthesis at another promoter.

Consensus sequence: TATAAT

(a) 5′-T [TATAAT] G C C G C (G) - 3′

(b) -T [TATAAT] G G T T A C (A) -

(c) -A [TATGAT] G C G C C C C (G) -

(d) -G [TATGTT] G T G T G G (A) -

(e) -C [TATGGT] T A T T T C (A) -

(f) -T [TTTCAT] G C C T C C (A) -

(g) -T [GATACT] G A G C A C (A) -

(h) -G [TTAACT] A G T A C G (A) -

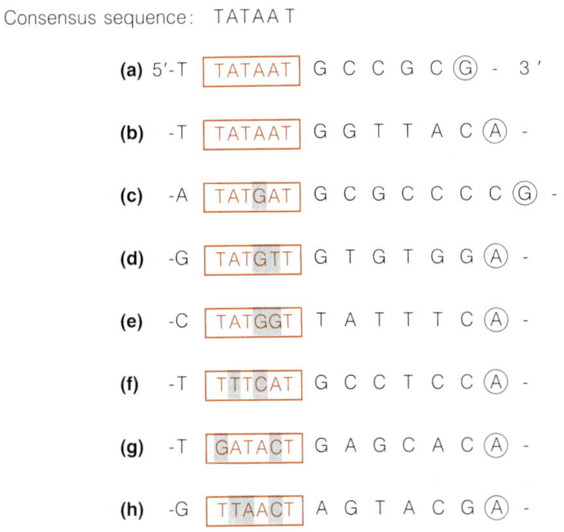

Figure 16-11 Pribnow Box Sequences. A portion of the nucleotide sequence is shown for eight different bacterial and phage promoters, with the Pribnow box in color. In each case, the circled purine (A or G) on the right indicates the start of the transcribed sequence, which extends to the right and is therefore not included in the figure. The promoter is located upstream from (i.e., on the 5′ side of) the transcribed sequence and represents the binding site for RNA polymerase. The Pribnow box is a highly conserved hexanucleotide sequence involved in promoter function. The consensus (preferred) sequence for the Pribnow box is shown at the top, and nucleotides that vary from this consensus sequence are in gray boxes. The sequences are from (a) the spc (ribosomal protein) operon of *E. coli,* (b) the virus SV40, (c) the gene for *E. coli* tyrosine tRNA, (d) the lactose operon of *E. coli,* (e) the galactose operon of *E. coli,* (f) the bacteriophage ϕX174, (g) the bacteriophage λ, and (h) the tryptophan operon of *E. coli.* Shown in each case is the nontemplate strand of the DNA duplex.

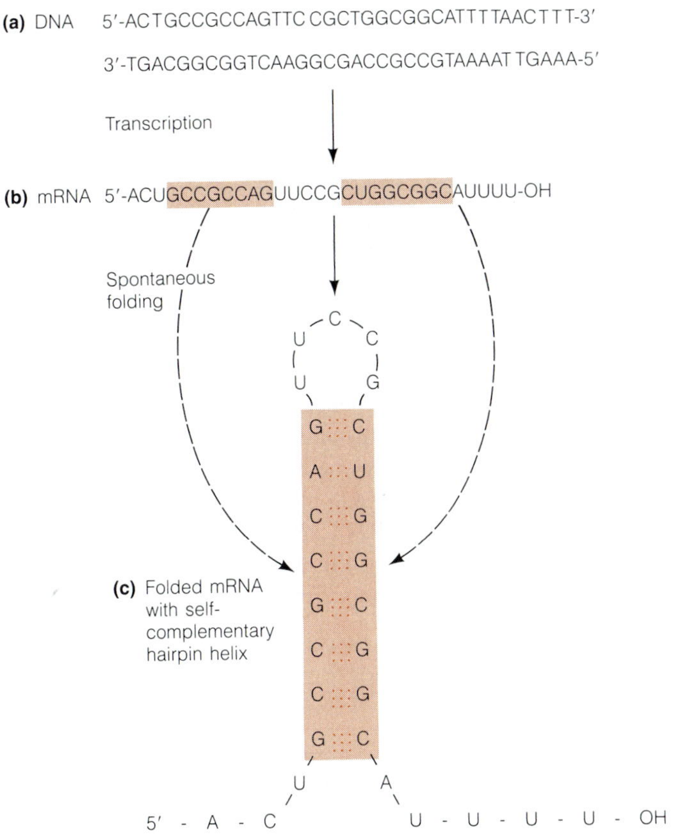

Transcription Units and Cistrons

A segment of DNA that is transcribed as a single, continuous RNA molecule is called a **transcription unit.** Every transcription unit is delimited by a promoter on one end

(a) DNA 5′-ACTGCCGCCAGTTCCGCTGGCGGCATTTTAACTTT-3′

3′-TGACGGCGGTCAAGGCGACCGCCGTAAAATTGAAA-5′

Transcription

(b) mRNA 5′-ACUGCCGCCAGUUCCGCUGGCGGCAUUUU-OH

Spontaneous folding

(c) Folded mRNA with self-complementary hairpin helix

C
U C
U G
G:::C
A:::U
C:::G
C:::G
G:::C
C:::G
C:::G
G:::C
U A
5′ - A - C U - U - U - U - OH

Figure 16-12 Termination Signals. (a) The DNA sequence encoding the termination signal for the tryptophan operon of *E. coli.* (b) The 3′ terminus of the mRNA, with the self-complementary regions of symmetry shown in the shaded blocks, followed by a short sequence of U's. (c) The RNA transcript folds rapidly and spontaneously into a hairpin helix, stabilized by hydrogen bonds between the short stretch of nucleotide pairs in the two self-complementary regions of the molecule. Termination seems to be determined by the hairpin structure itself rather than by any particular nucleotide sequence, since a variety of self-complementary sequences can serve as termination signals.

and a termination signal on the other. (In bacteria, the unit of transcription is also called an **operon,** but the operon concept relates more to regulation than to transcription.)

A **cistron** (or *gene,* if one is less fastidious about terminology), on the other hand, is a segment of DNA that carries the information for a single polypeptide, a single rRNA molecule, or a single tRNA molecule. In bacteria, a transcription unit frequently includes more than one cistron, resulting in **polycistronic mRNA** molecules that yield more than one polypeptide product upon translation. In contrast, most units of transcription in eukaryotes are **monocistronic,** giving rise to a single polypeptide chain. However, the transcription units in eukaryotic cells are often much larger than this suggests, because the initial transcripts often contain substantial regions that are later removed during processing.

Reverse Transcriptase

Most transcription in the cell goes on just as described above, with DNA serving as a template for RNA synthesis. However, RNA can also serve as a template for DNA polymerization, a process called *reverse transcription,* which is catalyzed by an enzyme known as **reverse transcriptase.** This reaction was first discovered by Howard Temin and David Baltimore in certain viruses that use RNA, rather than DNA, to store genetic information. Such viruses are commonly referred to as *retroviruses.* Examples of retroviruses include such important pathogens as the human immunodeficiency virus (HIV, the AIDS virus) and viruses that produce leukemia in animals.

In order to infect a cell, a virus that uses RNA to store genetic information must be able to change that information into DNA, which the cell then uses to synthesize new viruses (Chapter 13, pp. 364–366). Therefore, when a retrovirus infects a cell, it provides not only the RNA containing the information needed to produce new virus particles but also the reverse transcriptase enzyme required to transcribe the RNA into DNA (Figure 16-13). The DNA then integrates into the host cell DNA. When the DNA of viral origin is transcribed, the resulting viral RNA codes for three proteins—capsid, envelope, and reverse transcriptase—which assemble into new virus particles.

The ability of the viral genome to insert into the host cell DNA is an important aspect of our present knowledge of cancer. Certain RNA viruses induce tumors in animals by causing a permanent genetic change when the resulting DNA inserts into the host cell DNA. The viral DNA is called an *oncogene* (cancer-causing gene) because it turns on the synthesis of a normal growth factor, which then causes the uncontrolled cell proliferation that is characteristic of cancerous growth. Such genes will be discussed in Chapter 23.

A certain amount of reverse transcription may also take place in normal eukaryotic cells in the absence of viral infection. For example, yeast cells contain a segment of their DNA called a *retrotransposon* that has the remarkable ability to make a copy of itself in much the same manner as a retrovirus would. That is, transcription of the DNA segment results in an RNA that codes for a reverse transcriptase. When the enzyme is translated from the message, it catalyzes the synthesis of DNA from its own RNA. The DNA then inserts back into one of the yeast chromosomes to begin the cycle again.

Similar segments called *transposable elements* have been discovered throughout the DNA of all higher organisms so far examined, reproducing by normal replication rather than reverse transcription. Retrotransposons and transposable elements are apparently a kind of molecular parasite, using the cell's replicative or transcriptional machinery to reproduce, but never escaping the confines of the cell.

Processing of RNA

Before we can move on from the synthesis of mRNA, rRNA, and tRNA to their utilization in the process of translation, we need to recognize that the primary products of transcription frequently require **posttranscriptional processing** before they can be used in protein synthesis. When we refer to the processing of RNA, we mean all the chemical modifications necessary to generate the final RNA product from the primary transcript that serves as its precursor. Since most precursor molecules are larger than the final product, processing usually involves hydrolytic removal of portions of the precursor.

Processing may also include the addition or chemical modification of specific nucleotides. Methylation of bases or ribose groups is a common example, but other modifications are also known. Other posttranscriptional events that precede the involvement of RNA in translation include their association with specific proteins and, in eukaryotic cells, their passage from the nucleus to the cytoplasm.

In this section, we will look at some of the most important properties of rRNA, tRNA, and mRNA and at the mechanisms involved in the processing of each from the primary products of transcription.

Processing of Ribosomal RNA

Ribosomal RNA is by far the most abundant and most stable form of RNA in most cells. This can easily be demonstrated by extracting total cellular RNA and fractionating it according to size, either by sedimentation through

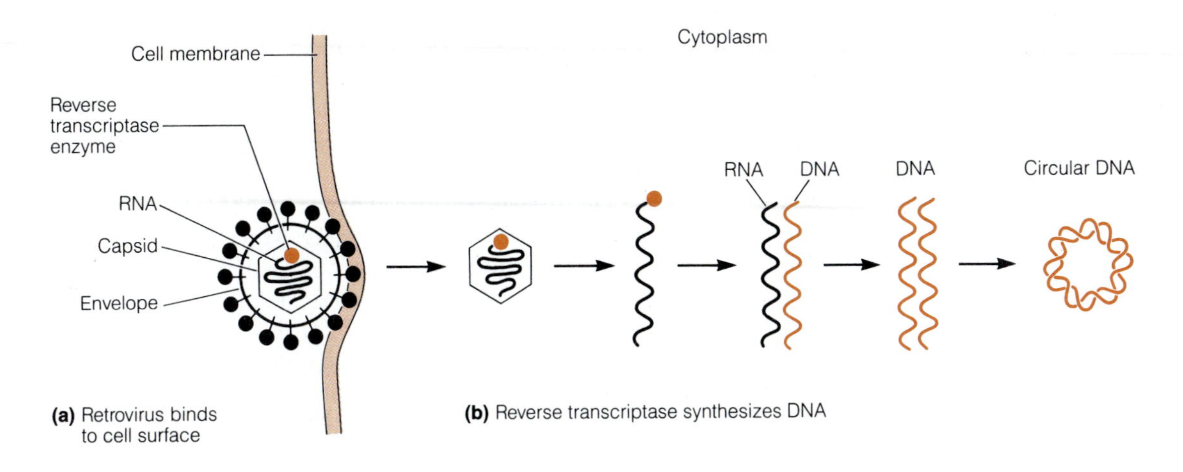

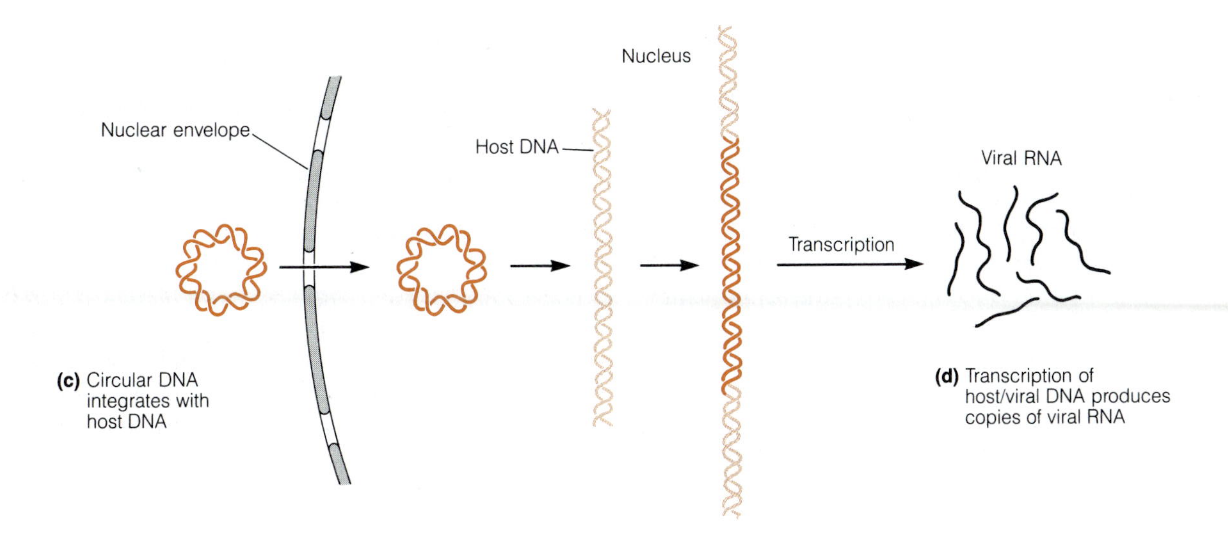

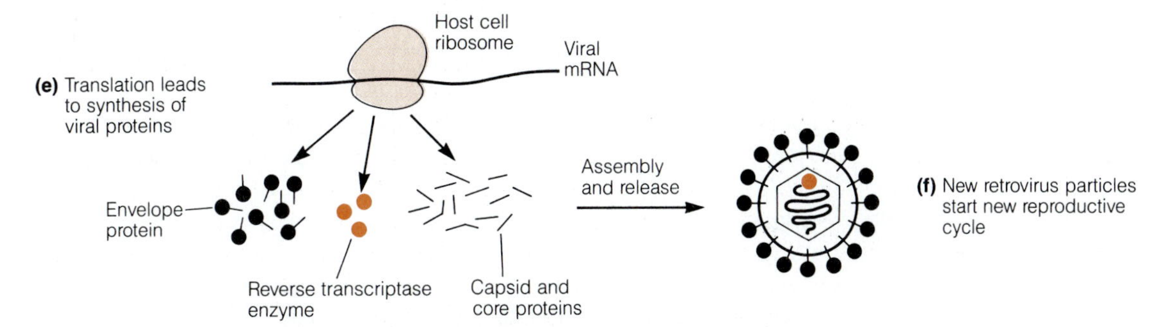

Figure 16-13 Retrovirus Infection and Reverse Transcription. (a) When a retrovirus particle infects a cell, it first binds to the cell surface, then inserts the viral genome and reverse transcriptase into the cytoplasm. (b) The reverse transcriptase uses cellular nucleotides to synthesize DNA with the RNA as a template. The DNA assumes a circular form, which (c) enters the nucleus and integrates with the host DNA. (d) When the DNA is transcribed to RNA, multiple copies of the viral RNA are made. (e) These code for the synthesis of new viral proteins, which assemble into the capsid and envelope that encapsulate the viral RNA. (f) The multiple copies of the virus then leave the cell to start a new reproductive cycle.

sucrose gradients or by electrophoresis on polyacrylamide gels. As Figure 16-14 shows, most of the RNA detectable on the gradient or gel is either rRNA or tRNA. In most cells, rRNA represents about 70–80% of the total RNA, tRNA represents about 10–20%, and mRNA accounts for about 5–10%.

Eukaryotic cells have four species of rRNA, usually identified according to their sedimentation coefficients, or **S values** (Table 16-3). The smaller of the two ribosomal subunits has a single rRNA molecule with a sedimentation coefficient of about 18S. The larger of the two subunits contains three molecules of RNA, one with a sedimentation coefficient in the range 25S–28S (depending on the species) and the other two with S values of about 5S and 5.8S. In prokaryotes, only three species of rRNA are present: a 16S molecule associated with the small subunit, and molecules with S values of 23S and 5S associated with the large subunit.

The processing of rRNA has been studied most extensively in eukaryotes. Of the four species of eukaryotic rRNA, the three larger ones (5.8S, 18S, 25S–28S) are usually coded for by contiguous genes and are synthesized as part of a single large precursor in most, if not all, eukaryotic nuclei. 5S rRNA, on the other hand, is encoded by DNA that is located elsewhere in the genome. Figure 16-15 illustrates the processing events necessary to generate mature rRNA molecules from their precursor form in human cells.

The initial transcript in most mammalian cells contains about 13,000 nucleotides and has a sedimentation coefficient of about 45S. The three molecules that are generated from it contain about 4700 nucleotides (28S rRNA), 1900 nucleotides (18S rRNA), and 160 nucleotides (5.8S rRNA). As a result, only about 52% of the primary transcript is preserved in the final products. The remaining 48% (about 6200 nucleotides) consists of transcribed **spacer sequences** that are excised during the processing steps and degraded in the nucleus.

The rRNA precursors of lower animals and plants contain smaller amounts of spacer sequences, so a correspondingly smaller proportion of the precursor molecule is degraded. In all cases, however, the precursor is larger than the aggregate size of the three rRNA molecules that are formed from it, so at least some processing is necessary. Even in bacterial cells, rRNA formation involves processing of precursor forms that are slightly larger than the end products.

Processing of Transfer RNA

Transfer RNA is also synthesized in a precursor form in both eukaryotic and prokaryotic cells. Processing of tRNA precursors involves several distinctly different events, as Figure 16-16 indicates for yeast tyrosine tRNA. At the 5′ end, a short **leader sequence** (of 16 nucleotides, in this case) is removed (Figure 16-16a). At the 3' end, the two terminal nucleotides (UU) of the precursor are removed and replaced with the trinucleotide sequence CCA, which is a distinguishing characteristic of functional tRNA molecules (Figure 16-16b). A number of bases are modified chemically (Figure 16-16c), particularly by methylation; tRNA molecules are especially noted for the presence of a higher proportion of modified bases.

Table 16-3 Properties of Prokaryotic and Eukaryotic Ribosomes

	Size of Ribosomes			Subunit Size			Subunit RNA	
Source	S Value*	Mol. Wt.	Subunit	S Value	Mol. Wt.	Subunit Proteins	S Value	Nucleotides
Prokaryotic cells	70S*	2.5×10^6	Large	50S	1.6×10^6	34	23S	2900
							5S	120
			Small	30S	0.9×10^6	21	16S	1540
Eukaryotic cells	80S*	4.2×10^6	Large	60S	2.8×10^6	45	25S–28S	≤4700
							5.8S	160
							5S	120
			Small	40S	1.4×10^6	33	18S	1900

* If you are surprised that the S values of the subunits do not add up to that of the ribosome, recall that an S value is a measure of the velocity at which a particle sediments upon centrifugation and is only indirectly related to the mass of the particle.

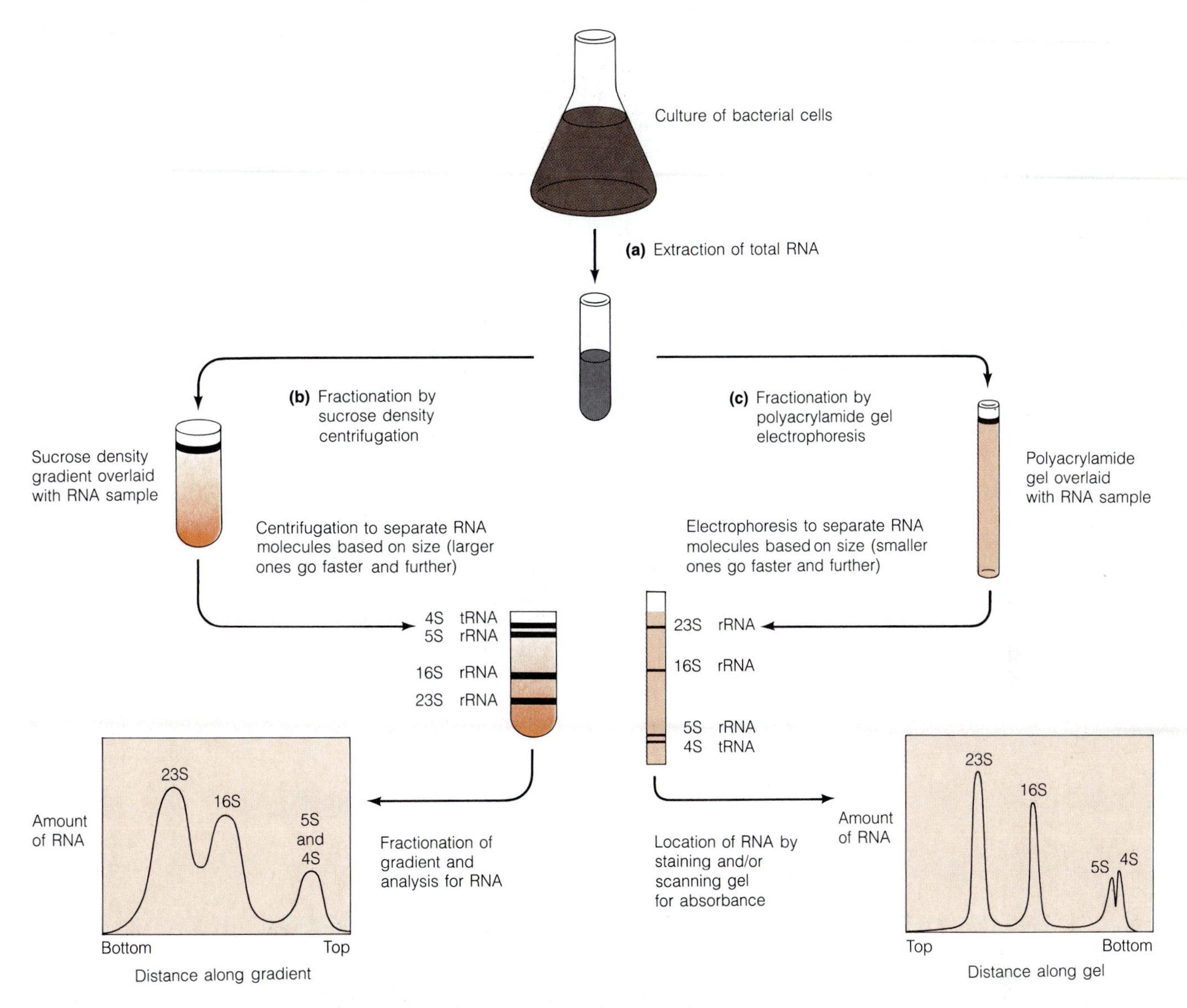

Figure 16-14 The Prominence of rRNA and tRNA. When total RNA is (a) extracted from bacterial (or other) cells, and either (b) fractionated by equilibrium density centrifugation on a sucrose gradient or (c) fractionated by electrophoresis on a polyacrylamide gel, most of the RNA seen on the gradient or gel is rRNA (5S, 16S, and 23S for bacteria) and tRNA (4S); specific mRNA molecules are usually too heterogeneous in size and too low in concentration to be seen by these techniques.

The processing of yeast tyrosine tRNA is also characterized by the removal of an internal 14-nucleotide sequence (Figure 16-16d), although that kind of excision is by no means common to all tRNA molecules. An internal segment of an RNA transcript that must be removed to form the functional RNA product is called an **intervening sequence.** We will consider intervening sequences in more detail in the discussion of mRNA processing, since they are an almost universal feature of mRNA precursors in eukaryotic cells. For the present, simply note that some tRNA precursors also contain intervening sequences and that these must be eliminated during processing by a precise mechanism that cuts and splices the precursor molecules at exactly the same location every time.

It is striking that the enzyme system responsible for this cutting and splicing of tRNA precursors appears to be very similar even among species that are phylogenetically distant from one another. For example, cloned yeast genes for the tyrosine tRNA shown in Figure 16-16 were microinjected into eggs of *Xenopus laevis*, the South African clawed toad. Despite the difference between fungi and amphibians, the yeast genes were transcribed in the toad eggs and were processed properly, including the removal of the 14-nucleotide intervening sequence.

Processing of Messenger RNA

Prokaryotic mRNA seems to be an exception to the generalization that RNA requires processing before it can be used by the cell. Most bacterial messages are synthesized in a polycistronic form that is ready for translation immediately. Moreover, the nucleoid of a bacterial cell is not separated by a membrane barrier from the ribosomes responsible for protein synthesis, as is the case with the nucleus of a eukaryotic cell. As a result, bacterial mRNA molecules that are still being elongated by RNA polymerase can often be shown to have ribosomes already associated with them, as illustrated by the electron micrograph of Figure 16-17. Transcription and translation are therefore coupled processes in prokaryotic cells.

In eukaryotic cells, on the other hand, transcription occurs in the nucleus, but translation takes place in the cytoplasm. The two processes are therefore separated in both time and space. Part of the posttranscriptional processing that eukaryotic mRNA undergoes is associated with the physical transport of the message out of the nucleus. In addition, substantial chemical processing is necessary to form mature cytoplasmic mRNA from the primary transcripts in the nucleus.

Although all transcription units for eukaryotic mRNA are monocistronic, the initial transcripts are often very large, ranging from about 2000 to 20,000 nucleotides. This size heterogeneity is reflected in the term **heterogeneous nuclear RNA (hnRNA)** that is applied to the nonribosomal, nontransfer RNA that is such a prominent product of

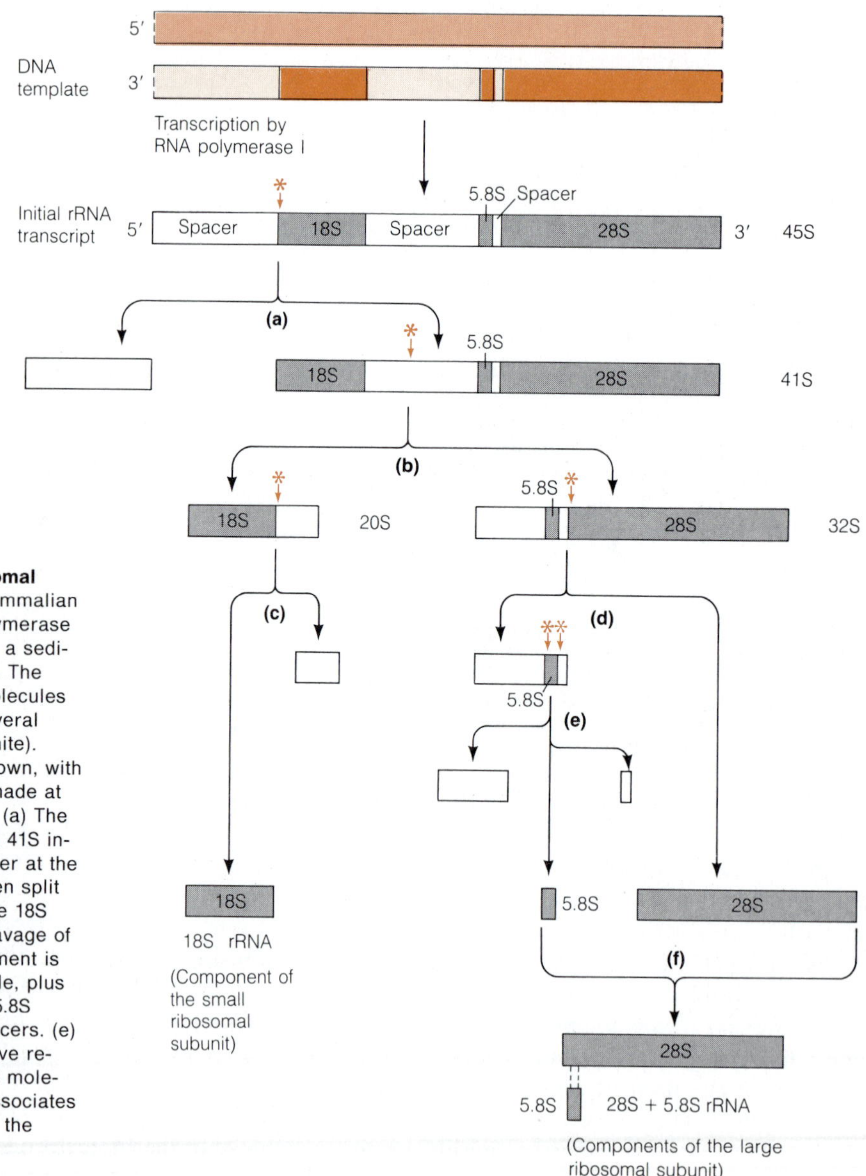

Figure 16-15 Processing of Ribosomal RNA. The rRNA genes of most mammalian cells are transcribed (by RNA polymerase I) as a single large precursor with a sedimentation coefficient of about 45S. The precursor contains three rRNA molecules (18S, 5.8S, and 28S; gray) plus several lengths of transcribed spacers (white). Processing occurs stepwise as shown, with each successive cleavage event made at the point indicated by an asterisk. (a) The 45S precursor is first trimmed to a 41S intermediate by removal of the spacer at the 5′ end. (b) The 41S molecule is then split into 20S and 32S segments. (c) The 18S rRNA molecule is obtained by cleavage of the 20S segment. (d) The 32S segment is cleaved into the 28S rRNA molecule, plus a segment of RNA containing the 5.8S rRNA molecule flanked by two spacers. (e) Subsequent trimming events remove remaining spacer RNA from the 5.8S molecule. (f) The 5.8S molecule then associates noncovalently with the 28S RNA in the large ribosomal subunit.

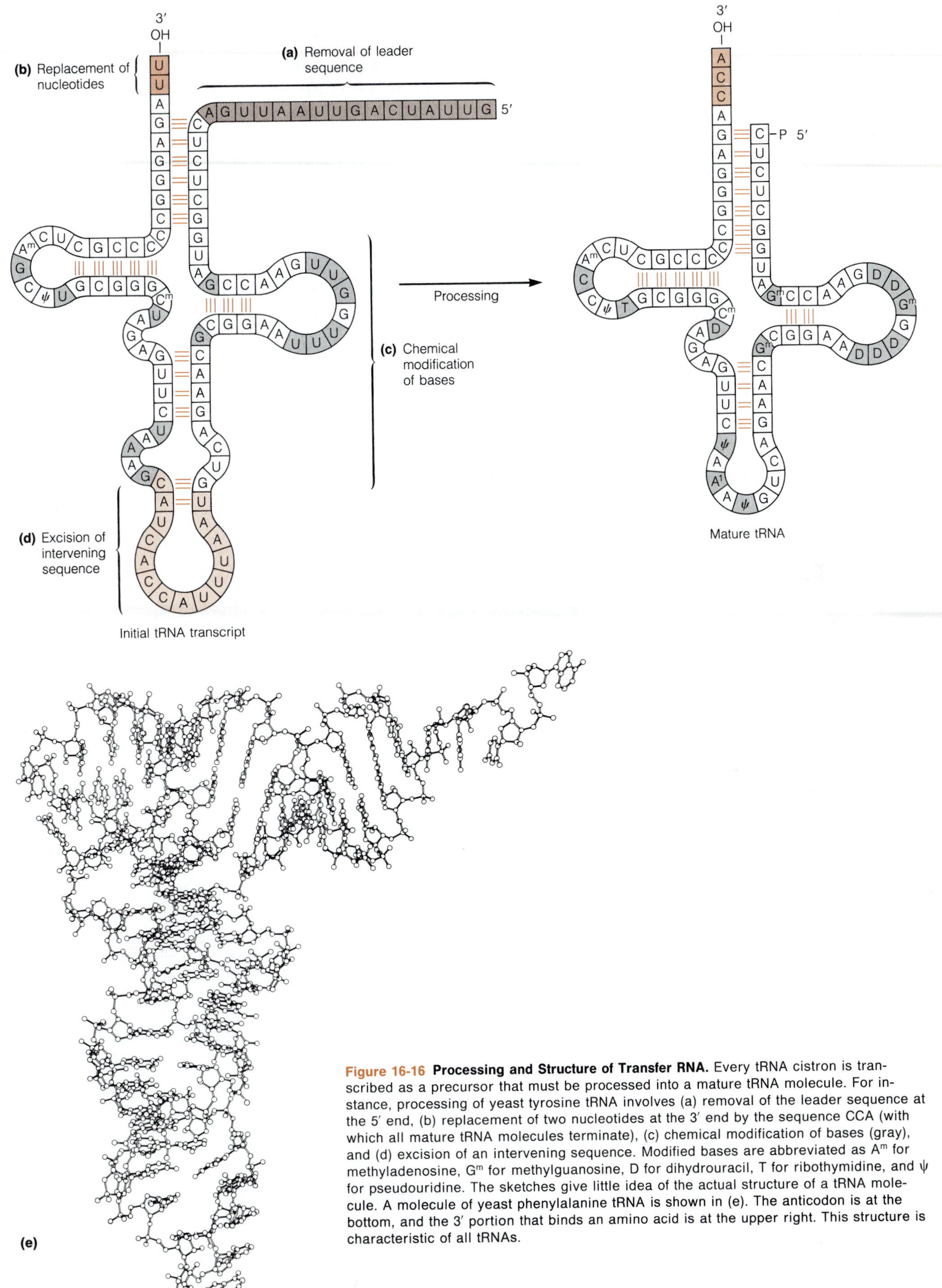

Figure 16-16 Processing and Structure of Transfer RNA. Every tRNA cistron is transcribed as a precursor that must be processed into a mature tRNA molecule. For instance, processing of yeast tyrosine tRNA involves (a) removal of the leader sequence at the 5′ end, (b) replacement of two nucleotides at the 3′ end by the sequence CCA (with which all mature tRNA molecules terminate), (c) chemical modification of bases (gray), and (d) excision of an intervening sequence. Modified bases are abbreviated as A^m for methyladenosine, G^m for methylguanosine, D for dihydrouracil, T for ribothymidine, and ψ for pseudouridine. The sketches give little idea of the actual structure of a tRNA molecule. A molecule of yeast phenylalanine tRNA is shown in (e). The anticodon is at the bottom, and the 3′ portion that binds an amino acid is at the upper right. This structure is characteristic of all tRNAs.

transcription in eukaryotic nuclei. It is not yet clear whether all hnRNA molecules are in fact precursors to cytoplasmic messengers, nor is it known whether hnRNA molecules have any function other than as precursors of mRNA.

The processing of hnRNA into mRNA molecules may represent a level of **posttranscriptional regulation** of gene expression in eukaryotic cells, a possibility to which we will return in the next chapter. Eukaryotic cells may be able to control not only which genes are transcribed but also which of the primary transcripts are degraded and discarded and which are processed into functional mRNA and exported to the cytoplasm.

Introns, Exons, and Splicing. In eukaryotic cells, the precursors for many mRNAs (and for some tRNAs and rRNAs) contain intervening sequences that are a part of the initial transcript but not of the functional RNA. Intervening sequences are also called **introns,** thereby distinguishing them from **exons,** the portions of the initial transcript that are preserved in the mature mRNA (Figure 16-18).

Introns are therefore regions of the initial transcript that must be excised during RNA processing. The excision process is called **RNA splicing** and is illustrated in Figure 16-19. Splicing enzymes bring the intron-exon junctions on

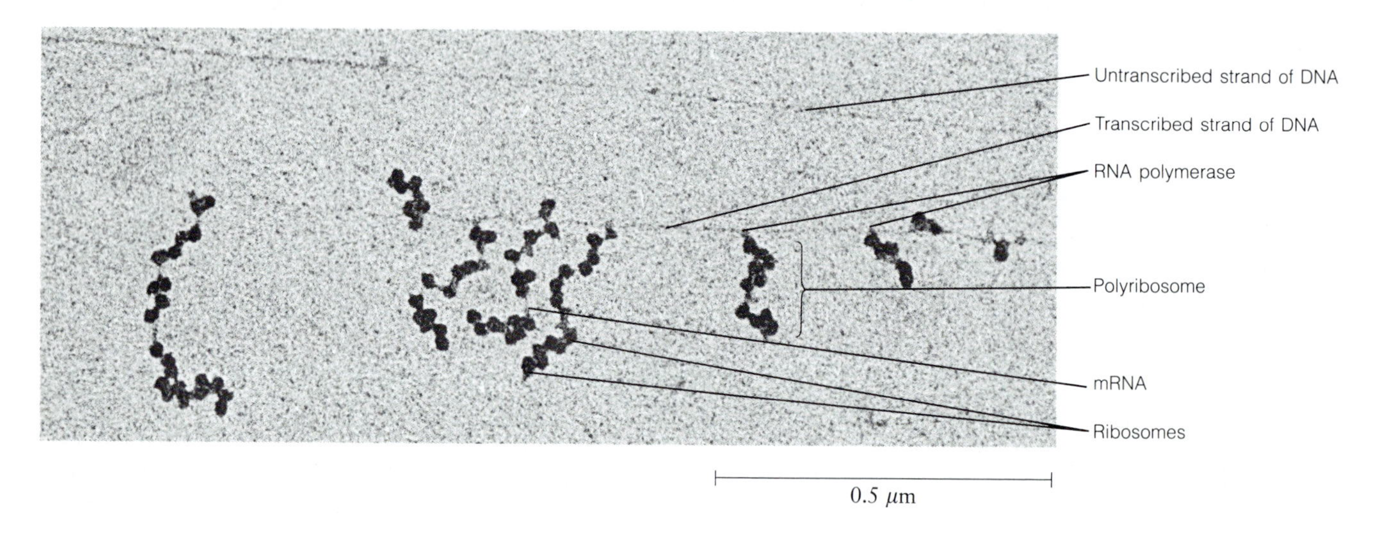

Figure 16-17 Coupling of Transcription and Translation in Bacterial Cells. This electron micrograph shows a strand of *E. coli* DNA being transcribed by RNA polymerase molecules that are moving from right to left. Attached to each polymerase molecule is a strand of mRNA still in the process of being transcribed. The dark spheres attached to each growing mRNA strand are ribosomes that are actively translating the partially complete mRNA. A cluster of ribosomes attached to a single mRNA strand is called a polyribosome. (TEM)

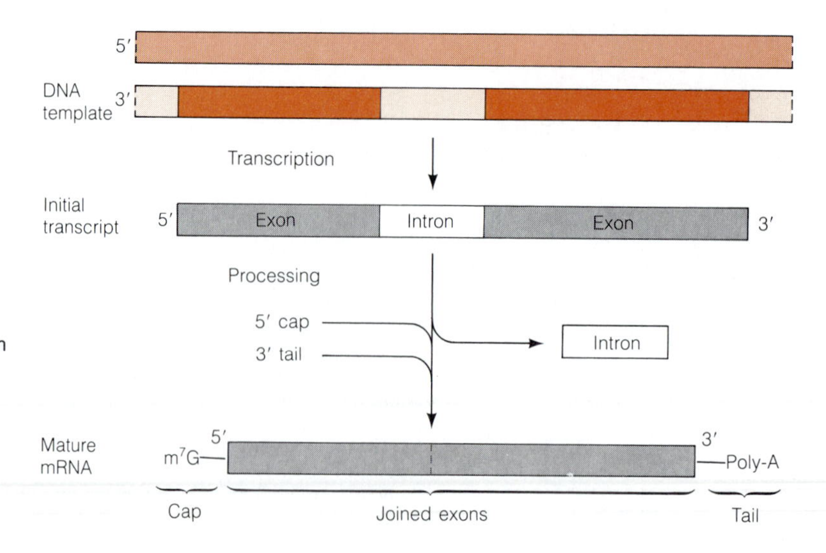

Figure 16-18 Processing of Eukaryotic mRNA. The initial transcripts that give rise to most eukaryotic mRNA molecules contain sequences to be preserved (exons) and sequences to be excised (introns). The mature mRNA is the result of several processing steps. During processing, the introns are removed by splicing, an m^7G cap is added at the 5′ end, and a poly-A tail is added at the 3′ end.

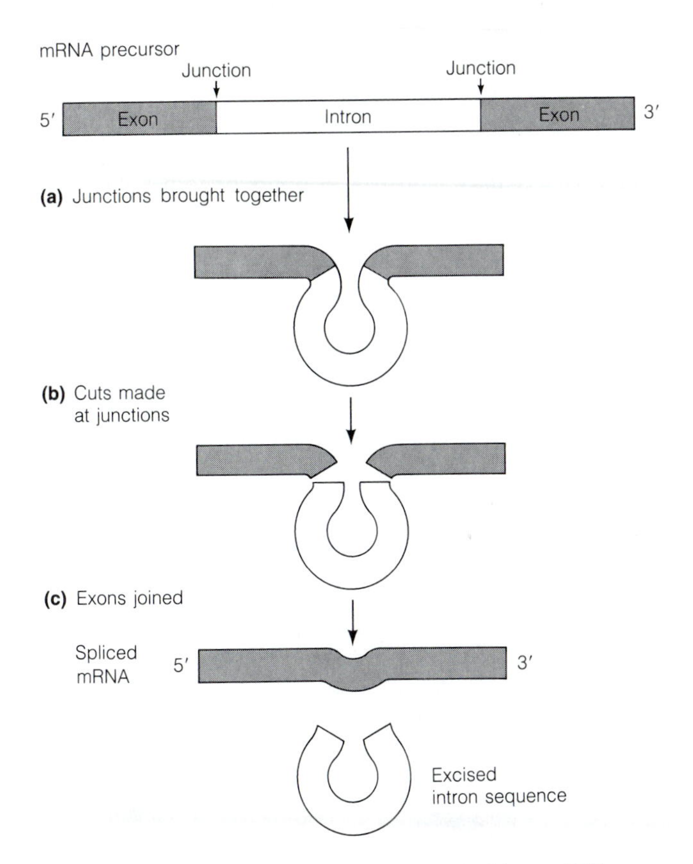

Figure 16-19 RNA Splicing. Introns are removed from mRNA precursors by RNA splicing. (a) The intron-exon junctions on either end of the intron are brought together, and (b) cuts are made at very precise points in both junctions. (c) The two exons are then joined to each other, creating a spliced mRNA with the intron removed.

either end of the intron together, causing the intron to form a loop (Figure 16-19a). Two cuts are then made in the phosphodiester backbone (b), and the two flanking exons are joined to each other, creating a contiguous molecule with the intron removed (c). Current evidence suggests the involvement of a small ribonucleoprotein particle called *U1,* which apparently functions by holding the two junctions together in the correct configuration to allow the splicing events to occur. Significantly, U1 contains a small RNA molecule, a part of which is complementary to the nucleotide sequence that is common to intron-exon junctions.

Most eukaryotic mRNA is derived from precursors that contain introns, but little else can be said about introns or their likely significance, because both the number and the size of introns vary so greatly for the mRNA that has been studied to date. The discovery, implications, and current fascination with intervening sequences are explored in more detail in the box on pp. 480–481.

Caps and Tails. In addition to being decidedly smaller than hnRNA, most eukaryotic mRNA molecules bear at least two other distinctive marks from having been processed before being transported out of the nucleus. All mRNA molecules contain a modified nucleotide **cap** at their 5′ end. At their 3′ end, most mRNA molecules have a long repeating sequence of adenylate (A) groups called a **poly-A "tail."** As Figure 16-18 indicates, neither of these structural features is present on the primary transcript.

The cap on the 5′ end of mRNA molecules is really a guanosine that is methylated at position 7 and is "on backward"; that is, the bond joining it to the mRNA is a 5′ → 5′ pyrophosphate linkage, instead of the usual 3′ → 5′ bond. This distinctive feature of eukaryotic mRNA is added posttranscriptionally to the primary transcript (Figure 16–20).

If the only methyl group is on the N-7 of the guanine, the structure is called *cap 0* (Figure 16-20b). However, adjacent nucleotides can be methylated on their ribose groups, generating *cap 1* (next nucleotide only) or *cap 2* (next two nucleotides methylated on the ribose), as shown in Figure 16-20c and d. These cap structures protect the mRNA from degradation by 5′ nucleases, thereby contributing to the stability of the mRNA. In addition, the cap is important in aligning the messenger for translation. However, many of the viruses that infect eukaryotic cells have RNA molecules that lack caps and are nonetheless able to be translated.

The poly-A tails found on most eukaryotic mRNA molecules are usually 100 to 200 A's in length. They are clearly added posttranscriptionally, since the DNA strands coding for such mRNA molecules do not contain corresponding stretches of thymidylate (T) groups. The enzyme responsible for the sequential addition of the adenylate groups is *polyadenylate polymerase* (*poly-A polymerase*). To be polyadenylated, a primary transcript must have a GC dinucleotide at its 3′ end and the sequence AAUAA about 11 to 30 nucleotides to the 5′ side of the polyadenylation site.

The presence of a poly-A tail protects the mRNA from nuclease attack but, like the 5′ cap, does not seem to be indispensable for translation. It has been suggested that the poly-A tail may be involved in the export of mRNA precursor molecules from the nucleus, and specific poly-A-binding proteins have been reported. However, mRNA molecules are known that do not have this structural feature, so polyadenylation is apparently not a universal requirement for transport.

Properties of Messenger RNA

Before moving on to translation, we should note several important properties of mRNA beyond the distinctive features of a 5′ cap and a 3′ tail. These include its rapid turnover, its nucleotide composition and sequence complementarity, its size heterogeneity, and its role in genetic amplification.

Turnover

A characteristic property of mRNA is its rapid *turnover*, or rate of synthesis and degradation. In this respect, mRNA contrasts with the other major forms of RNA in the cell, rRNA and tRNA, which are notable for their stability. Because of its more rapid turnover, mRNA accounts for most of the RNA synthetic activity in many cells, even though it represents only a small fraction of the total RNA content of the cell. Turnover is usually measured in terms of the **half-life** of the molecule in question: the length of time required for degradation of 50% of the molecules on hand at a given time. The mRNA molecules of bacterial cells generally have half-lives that are measured in minutes. The mRNA molecules of eukaryotic cells, on the other hand, usually have half-lives of hours or sometimes even days.

Nucleotide Composition

Messenger RNA has a nucleotide composition that more closely approximates that of total genomic DNA than does tRNA or rRNA. This is not necessarily true for a specific mRNA molecule, but is almost always the case when nucleotide composition is averaged over the thousands of mRNA species in a typical cell. In fact, it was this property

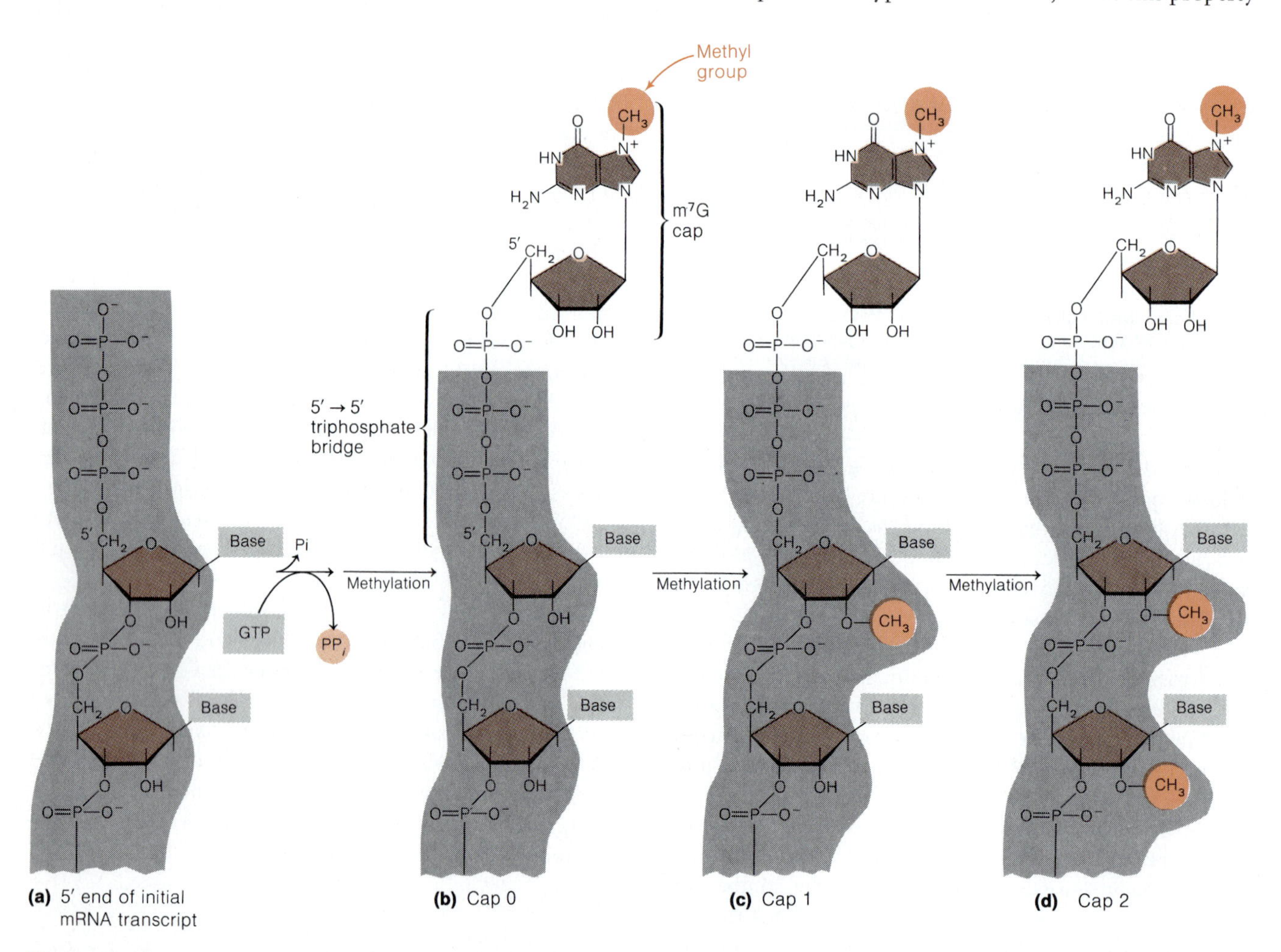

Figure 16-20 The m^7G Cap of Eukaryotic mRNA Molecules. The processing of eukaryotic mRNA molecules involves both the addition of an m^7G cap at the 5′ end of the initial transcript and one or more subsequent methylation events. (a) The initial transcript has a triphosphate group at its 5′ end but is unmodified. (b) Cap 0 is generated by the addition of a methylated guanosine (m^7G) cap with a 5′ → 5′ linkage. (c) Cap 1 is formed by the addition of a methyl group on the ribose of the adjacent nucleotide. (d) Cap 2 requires methylation of the ribose of the next nucleotide as well.

that first led to the suggestion that a "DNA-like RNA" might be involved in protein synthesis.

Sequence Complementarity

In addition to base composition, mRNA molecules can be characterized in terms of the proportion of genomic DNA to which they are complementary. Complementarity is assessed by a technique called **RNA-DNA hybridization.** Hybridization is similar in principle to DNA-DNA annealing (recall Figure 13-11), except that the association is between RNA and DNA strands instead of complementary DNA strands.

RNA is allowed to associate with denatured DNA, either in solution or after immobilization of the DNA on filter paper. Radioactively labeled RNA is used in sufficient excess to ensure that all complementary sites on the DNA are occupied; thereafter the fraction of the genome to which RNA molecules are complementary can be quantified. This fraction is invariably much higher for mRNA than for tRNA or rRNA because of the much larger number of different sequences in the mRNA population.

Two molecules of RNA can also have complementary sequences that hybridize. A potentially important application of RNA–RNA hybridization would be its use to inhibit the expression of specific genes. When DNA is transcribed into RNA, only one strand of the DNA is used. The messenger RNA that is produced "makes sense" with respect to the protein synthesis it encodes and is therefore called *sense RNA*. The DNA strand from which it is produced is called the *template strand*. But what about the other strand? If RNA could be synthesized using its sequence, the resulting **antisense RNA** would have some very interesting properties. In particular, antisense RNA would hybridize to sense RNA and thereby inhibit its ability to direct protein synthesis (Figure 16-21).

This opens up the possibility of using antisense RNA to inhibit the expression of specific gene products. To imagine how important this could be, consider that every viral infection and every cancerous transformation involves the expression of specific genes. If those genes could be repressed by antisense RNA that binds to the mRNA coding for their gene products, significant therapeutic results would follow.

The feasibility of using antisense RNA to inhibit gene expression has been demonstrated in several recent studies. For example, Harold Weintraub and his colleagues developed an antisense RNA for actin, the major protein of microfilaments. When it was present in cells, the antisense RNA effectively inhibited actin synthesis, so that actin microfilaments disappeared and the cells were unable to maintain their normal shape. In other work by Paul Nie-

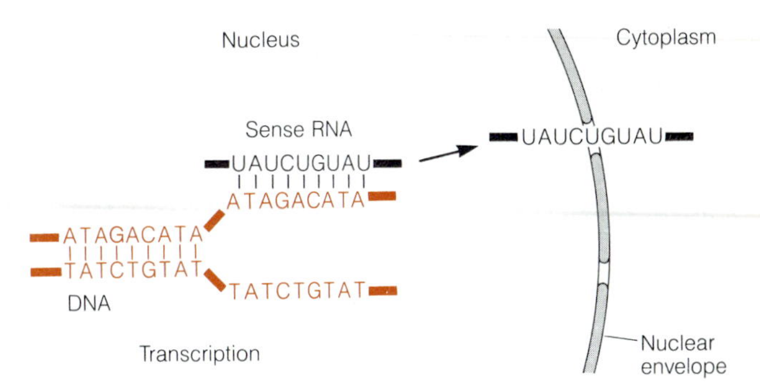

(a) Normal transcription produces sense RNA

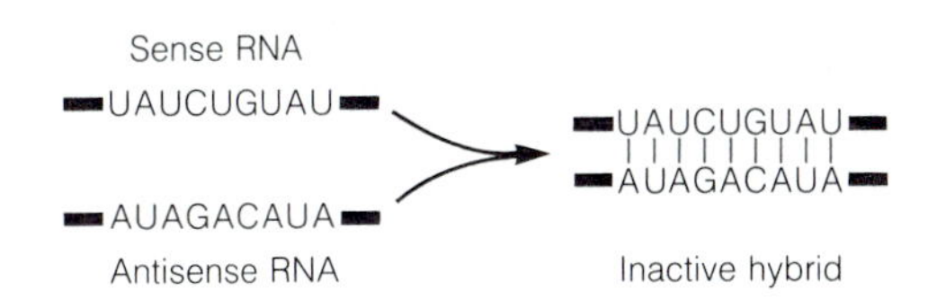

(b) Antisense RNA hybridizes with sense RNA in cytoplasm and inhibits translation

Figure 16-21 Antisense RNA. (a) The normal product of transcription is sense RNA. (b) If a synthetic antisense RNA is present, it hybridizes with the sense RNA and inhibits translation.

man and his co-workers, viruses were developed that were able to express a specific antisense RNA for a cancer-causing gene. (See discussion of oncogenes, Chapter 23.) Cells were infected with this virus, then exposed to a second virus known to cause transformation into cancer-like cells. The cells were unaffected by the cancer-inducing virus, presumably because the antisense RNA inhibited synthesis of the gene product leading to cancerous transformation.

Size Heterogeneity

As a class, mRNA is much more heterogeneous in size than rRNA or tRNA. Of course, the mRNA molecule coding for a specific polypeptide will have a very predictable size; but mRNA molecules coding for various polypeptides differ greatly in length, reflecting the broad range of polypeptide sizes. Because of their size heterogeneity and low concentration in the cell, mRNA molecules cannot normally be detected by techniques such as centrifugation or electrophoresis, which are useful for visualizing rRNA and tRNA (Figure 16-14). A particular mRNA species can usually be detected only if complementary DNA (cDNA) is available as a radioactively labeled **cDNA probe.** Specific fractions from a gradient or bands from a gel can then be analyzed for a particular mRNA by measuring the extent to which the labeled cDNA probe will hybridize to the RNA in each fraction or band.

Introns, Exons, and the Excitement of Interrupted Genes

It has been known for many years that in bacteria, the order of amino acids in a polypeptide correlates exactly with a contiguous sequence of nucleotide pairs in the DNA. This relationship was shown in the colinearity experiments of Yanofsky and others in the early 1960s and was confirmed by direct comparisons of nucleotide and amino acid sequences as rapid and accurate methods became available for sequencing DNA and proteins. It was quite naturally assumed that the same would turn out to be true for eukaryotes, once it became possible to isolate and sequence specific segments of eukaryotic DNA.

It therefore came as a surprise when several laboratories reported almost simultaneously in 1977 that at least some eukaryotic genes did not follow this pattern, but were in fact interrupted by "extra" sequences of nucleotides that were not represented in either the functional mRNA or the protein product. This startling discovery resulted from electron microscopic studies of hybrids between mRNA molecules and segments of DNA known to contain the genes coding for these mRNA molecules. Figure 16-A depicts the expected result when a prokaryotic mRNA is allowed to hybridize with double-stranded DNA corresponding to the gene from which the mRNA was transcribed. The mRNA hybridizes to the template strand of the DNA to which it is complementary, leaving the other strand as a free, displaced loop that is readily identifiable as such in the electron microscope.

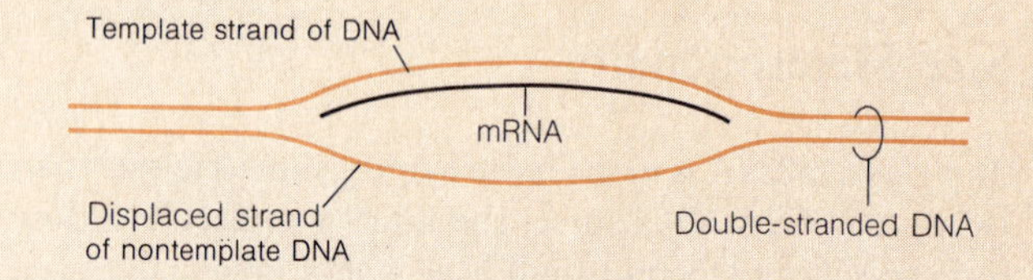

Figure 16-A Hybridization of an mRNA Molecule to Double-Stranded DNA from a Gene with No Introns.

If eukaryotic genes were contiguous sequences of nucleotides, they would be expected to show the same profile. However, when such experiments were actually done with mRNA and genes for proteins such as human β globin, the surprising result was that *multiple* loops were seen. Characteristically, these involve a loop of double-stranded DNA flanked by two loops of displaced single strands, as illustrated by Figure 16-B. This pattern is now known to result from the presence within the gene of one or more *intervening sequences* that are not present in the mRNA and therefore do not become involved in hybrid formation. The double-stranded DNA loop corresponds to the intervening sequence, and the two loops of RNA-DNA hybrid represent the surrounding sequences preserved in the mRNA.

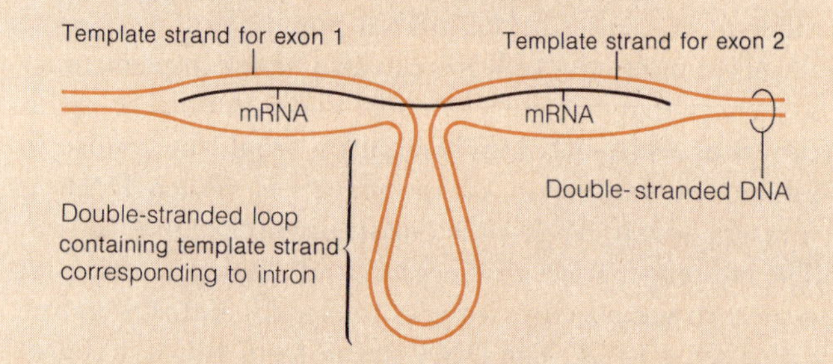

Figure 16-B Hybridization of an mRNA Molecule to Double-Stranded DNA from a Gene with a Single Intron.

Intervening sequences are also referred to as *introns,* and the coding sequences around them are called *exons.* Eukaryotic genes are therefore mosaics of exons and introns—of translated and untranslated DNA sequences. Once they had been reported for a few genes, introns began popping up everywhere, especially as restriction enzyme techniques and DNA sequencing began to be applied to a wide variety of eukaryotic genes. To date, introns have been found in every eukaryotic gene that has been sequenced except for the histone genes. Such "split" genes are therefore clearly the rule, not the exception.

The split genes that have been studied so far show a remarkable variety with respect to the number and size of introns they contain, as well as the total proportion that the introns represent of the whole gene. The β globin gene, for example, has only two introns, one of 120 nucleotide pairs and the other of 550 nucleotide pairs. Together, these account for about 40% of the total length of the gene, as Figure 16-C illustrates. Chick ovalbumin, on the other hand, contains seven introns, representing more than 60% of the total gene (Figure 16-D). Even more dramatic examples are the chick conalbumin gene with 13 introns and the gene for procollagen α chain, which has more than 50 introns!

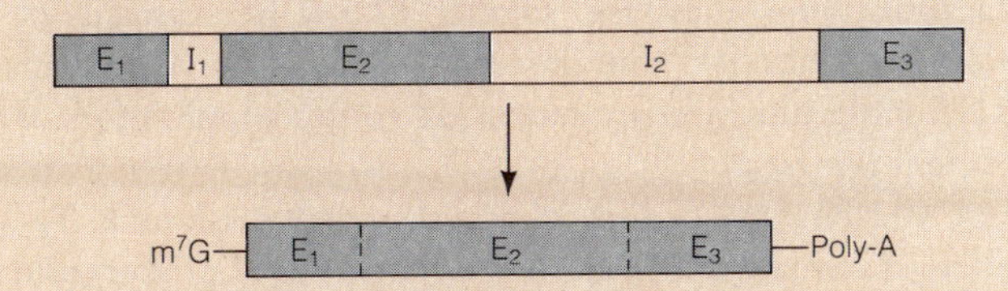

Figure 16-C Structure and Processing of a Transcript from the Human β Globin Gene. The human β globin gene has two introns. Processing involves excision of both introns, addition of the m^7G cap at the 5′ end, and polyadenylation at the 3′ end.

Although not preserved in the functional message, introns are actually transcribed along with the exons. As a result, the initial transcript of a split gene contains sequences that must be removed as the transcript is processed into the final mRNA product. The removal of these segments is called *RNA splicing*. Splicing enzymes cleave the RNA at the intron-exon junctions flanking an intron and ligate the 3′ and 5′ ends of the two adjacent exons to create a continuous coding sequence. Clearly, the alignment and splicing mechanisms must be very precise, because a shift of a single nucleotide would alter the reading frame and render the mRNA useless. The intron-exon junctions have been studied for a number of mRNA molecules, and a common pattern has emerged: An intron always begins with GU at the 5′ terminus and ends with AG at the 3′ end.

The burning question, of course, is *why*. Why do nearly all genes in eukaryotes have so much DNA that seems to serve no coding function? Why, in generation after generation of cells, is so much energy invested in making segments of DNA molecules—and of RNA transcripts—that are apparently destined only for the splicing scrap heap? Several suggestions have been made, but no compelling answers have yet emerged. One possibility is that each exon represents a separate *domain*, or unit of structure or function, and that proteins with multiple exons have been "assembled" evolutionarily from what were originally separate entities. Or perhaps the splicing step is a key regulatory mechanism in determining which nuclear transcripts are actually processed into mRNA and exported to the cytoplasm.

Easily the most exciting possibility is that the RNA splicing mechanism may confer on the cell the flexibility of generating several different proteins from the same RNA transcript. This flexibility is possible because each intron is flanked by similar or identical junction sequences, thereby allowing the exons to be assembled in different combinations by juxtaposition of different junctions. This has already been shown to occur in the processing of RNA transcripts of a large animal virus called adenovirus as well as in antibody formation. We will return to the latter example in Chapter 24 because of its importance in the immune system. In both of these cases, the splicing and ligating operations are such that different protein products can be generated from the transcripts of a single gene, and that may well be the most exciting advantage that split genes can confer on a cell.

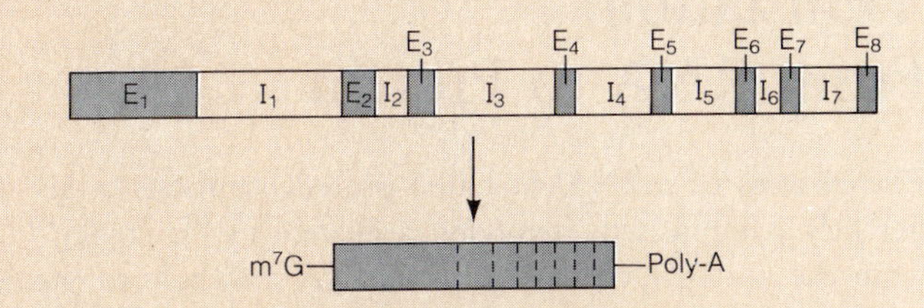

Figure 16-D Structure and Processing of a Transcript from the Chick Ovalbumin Gene. The chick ovalbumin gene has seven introns. Processing involves excision of all seven introns, addition of the m^7G cap at the 5′ end, and polyadenylation at the 3′ end.

Messenger RNA as a Means of Genetic Amplification

A singularly important feature of mRNA as an intermediate between a gene and its product is the opportunity for **message amplification** that it affords. If DNA were used directly as the template for protein synthesis, a single gene would be quite limited in terms of the ultimate number of protein molecules that could be translated from it. But with mRNA as an intermediate, multiple copies of the information can be made, and each of these can in turn be used to synthesize many copies of the protein product.

As an especially dramatic example of this amplification effect, consider the synthesis of *fibroin,* the major protein of silk. The haploid genome of the silkworm has only one copy of the fibroin gene, but about 10^4 copies of fibroin mRNA are transcribed from a single gene in the silk gland. Each of these mRNA molecules in turn directs the synthesis of about 10^5 fibroin molecules, resulting in the production of about 10^9 molecules of fibroin per cell—all in a period of four days! Without mRNA as an intermediate, the haploid genome of the silkworm would need 10^4 copies of the fibroin gene (or about 40,000 days!) to accomplish this.

Significantly, most genes that code for proteins (and thus have mRNA as an intermediate) are present in only one or a few copies per haploid genome. Genes that code for rRNA and tRNA, however, are always present in multiple copies in the genome, presumably because there is no opportunity for amplification when the RNA transcript is used directly, rather than being translated into a protein product.

Translation: From RNA to Protein

The process of mRNA-mediated protein synthesis is inherently complicated, involving a change in "language" from the nucleotide order in an mRNA molecule to the amino acid sequence of a polypeptide. In essence, the order of nucleotides, read as triplet codons, specifies the order in which incoming amino acids are added to the growing polypeptide chain. Ribosomes serve as the intracellular site of translation, and RNA molecules are the agents that ensure insertion of the correct amino acids at each position in the polypeptide. First we will look at the cast of characters, and then we will consider the translation process in some detail.

Translation: The Cast of Characters

The main elements of the translation apparatus are the *mRNA* to be translated, the *ribosomes* responsible for the process, the *tRNA* molecules that direct activated amino acids into their correct locations, the *aminoacyl-tRNA synthetases* that add amino acids to their appropriate tRNA molecules, and several *protein factors* that participate at several stages in the translation process. We have already considered the properties of mRNA, so we can now concentrate on the other elements involved in translation.

Ribosomes. **Ribosomes** are particles found in the cytoplasm of both prokaryotic and eukaryotic cells, as well as in the matrix of mitochondria and the stroma of chloroplasts. Ribosomes are often referred to as organelles, but they differ from true organelles because they are not surrounded by a membrane. Ribosomes serve as the site of protein synthesis in all cells, but those found in prokaryotic cells differ somewhat from those of eukaryotic cells. Prokaryotic ribosomes are smaller (2.5×10^6 instead of 4.2×10^6 daltons), contain fewer proteins, have smaller RNA molecules, and are sensitive to different inhibitors of protein synthesis.

All ribosomes consist of two dissociable subunits. Table 16-3 summarizes some of the properties of prokaryotic and eukaryotic ribosomes and their subunits. A prokaryotic ribosome has a sedimentation coefficient of about 70S and consists of a 30S subunit and a 50S subunit; its eukaryotic equivalent is an 80S particle made up of a 40S subunit and a 60S subunit.

Each subunit contains one or more rRNA molecules and a number of proteins, as shown in Figure 16-22. For prokaryotic ribosomes, the 30S subunit contains a 16S rRNA molecule and 21 proteins. The total molecular weight of the particle is about 0.9×10^6. The 50S subunit contains two RNA molecules (5S and 23S) and 34 proteins, for a molecular weight of 1.6×10^6. As Table 16-3 indicates, the subunits of eukaryotic ribosomes contain correspondingly larger rRNA molecules and more proteins.

Functionally, ribosomes are sometimes called the "workbenches" of protein synthesis, although their involvement in translation is not as passive as the term "workbench" might suggest. Each ribosome has two sites involved in protein synthesis: an **A (aminoacyl) site,** at which the incoming tRNA-linked amino acid binds, and a **P (peptidyl) site,** to which the growing polypeptide chain is attached. The involvement of these two sites in the process of translation will become clear shortly.

Transfer RNA. Transfer RNA molecules are small RNA molecules that serve as the actual intermediates (or **adap-**

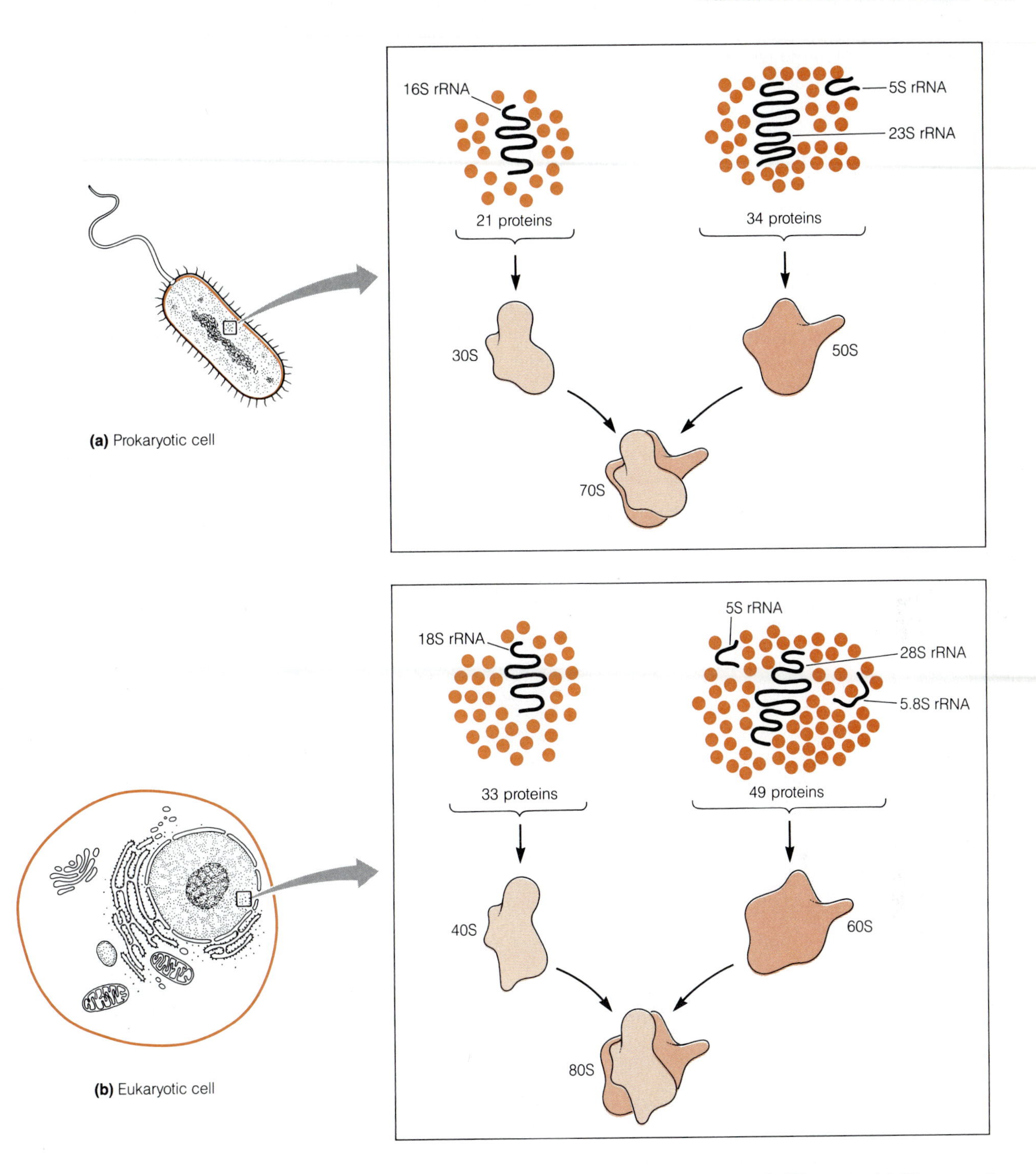

Figure 16-22 Composition of Prokaryotic and Eukaryotic Ribosomes. (a) Ribosomes from prokaryotes contain two subunits—30S and 50S—each composed of RNA molecules and proteins. The subunits in turn assemble into a 70S ribosome, which functions in translation. (b) Eukaryotic ribosomes are somewhat larger, reflecting their increased content of RNA and protein molecules.

tors, in Crick's original terminology) between the nucleotide sequence of the messenger and the amino acid sequence of the polypeptide. Appropriate to this role, tRNA molecules have two kinds of specificity. Each tRNA molecule recognizes a specific codon on the mRNA, and each carries an amino acid that is specifically chosen to match that codon according to the dictates of the genetic code.

As Figure 16-23 indicates, the amino acid is always linked by an ester bond to the 3′ OH group of the nucleotide (always an A) at one end of the molecule. Selection of the appropriate amino acid for a given tRNA is the responsibility of the enzyme that catalyzes formation of the ester bond between the two. The amino acid for which a given tRNA is specific is indicated by a superscript. The tRNA molecules specific for alanine, for example, are identified as $tRNA^{Ala}$.

The other aspect of tRNA specificity involves recognition of the appropriate codon. This is the role of the **anticodon,** a trinucleotide sequence located on one of the loops of the tRNA (Figure 16-23). The anticodon is always complementary to the codon, which it therefore recognizes by hydrogen-bonded base pairing. Take careful note of the convention used in representing codons and anticodons: Codons on mRNA molecules are usually written in the $5' \rightarrow 3'$ direction, whereas anticodons on tRNA molecules are usually represented in the $3' \rightarrow 5'$ orientation. Thus, one of the codons for alanine is 5′-GCC-3′, and the corresponding anticodon is therefore 3′-CGG-5′.

Since the genetic code has 61 different codons that specify amino acids, you might expect to find 61 different tRNA molecules involved in protein synthesis, each responsible for recognizing a separate codon. However, the

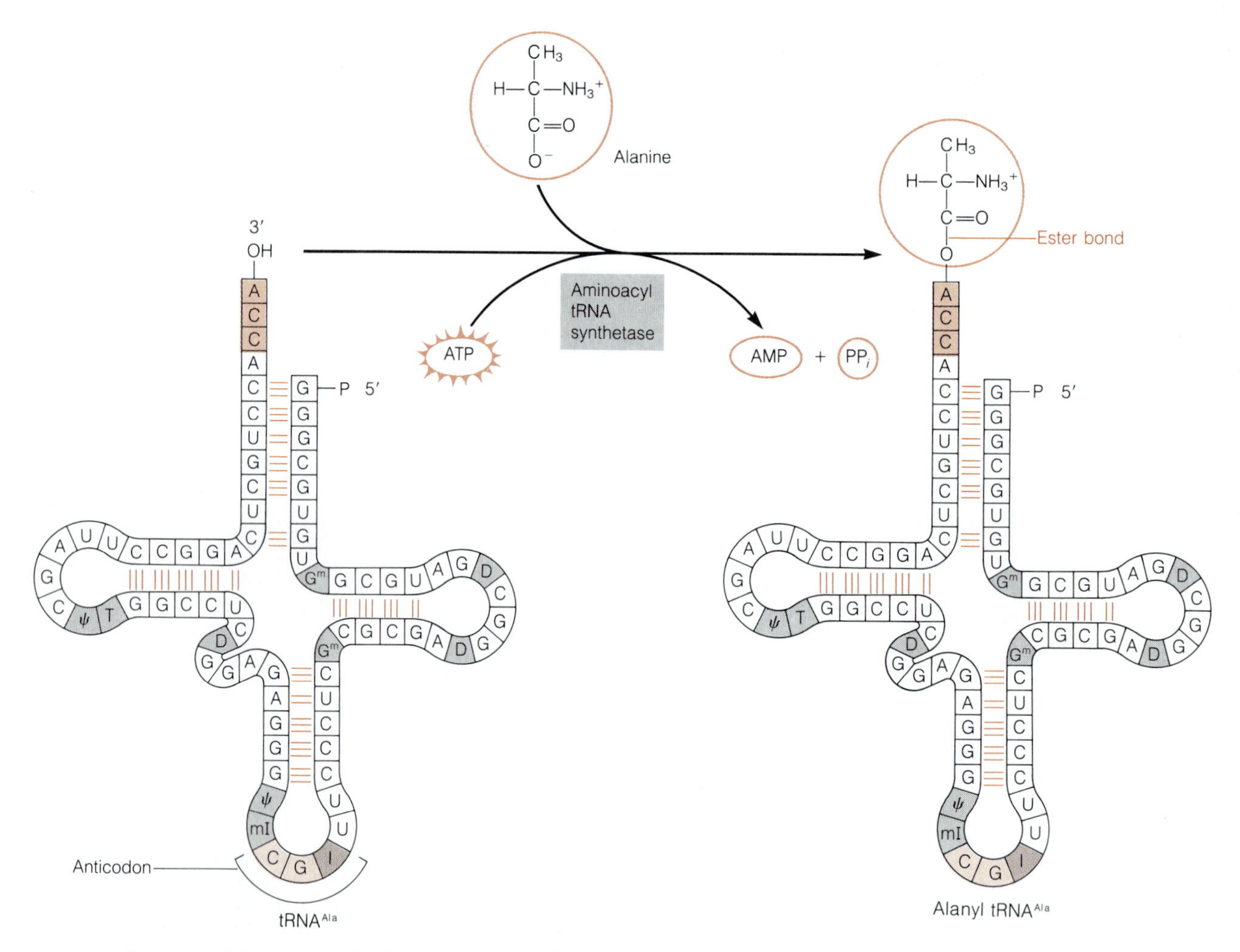

Figure 16-23 Sequence, Structure, and Aminoacylation of a tRNA. Yeast alanine tRNA, like all tRNA molecules, contains three loops, several paired regions, an anticodon triplet, and a 3′ terminal sequence of CCA, to which the appropriate amino acid can be attached by an ester bond. Modified bases are abbreviated as I for inosine, mI for methylinosine, D for dihydrouridine, T for ribothymidine, ψ for pseudouridine, and G^m for methylguanosine. For the significance of an inosine in the anticodon, see Figure 16-24 and the wobble rules of Table 16-4.

number of different tRNA molecules is significantly less than 61 because some tRNA molecules are able to recognize several different codons. This in turn is possible because the complementary binding of codon to anticodon is not as strict as base pairing is in other contexts, such as the DNA duplex.

Specifically, there is some flexibility in the binding of the base in the third position (the 5′ end) of the anticodon to that in the complementary position (the 3′ end) of the codon. Francis Crick referred to this flexibility as "**wobble**" when he first postulated it, and the term has stuck. The wobble rules are shown in Table 16-4. Notice that wobble is allowed with G or U in the third position of the anticodon, but not with C or A.

Notice also that tRNA molecules contain **inosine** (I), an unusual base not found in DNA or in other RNA molecules. The structure of inosine is shown in Figure 16-24a. Inosine is often present in the third position of the anticodon. In *E. coli,* in fact, A is never found in the third position of anticodons but is almost always modified to I. I is the "wobbliest" of all third-position bases, since it can pair with U, C, or A (Figure 16-24b). For example, a tRNA with the anticodon UAI could recognize the codons AUU, AUC, and AUA, which code for the amino acid isoleucine (Ile; see Figure 16-9).

As a result of wobble, fewer tRNA molecules are required for some amino acids than the number of codons that specify those amino acids. In the case of isoleucine, a cell can translate all three codons with a single tRNA

Table 16-4 The Wobble Rules for Codon-Anticodon Pairing

Normal Base-Pairing Rules	
Base Present in Anticodon of tRNA	Base Recognized in Codon of mRNA
C	G
A	U
U	A
G	C
Wobble Base-Pairing Rules	
Base Present in 5′ Position of tRNA Anticodon	Bases Recognized in 3′ Position of mRNA Codon
C	G
A	U
U	A, G
G	C, U
I*	U, C, A

*I stands for inosine; see Figure 16-24 for structure

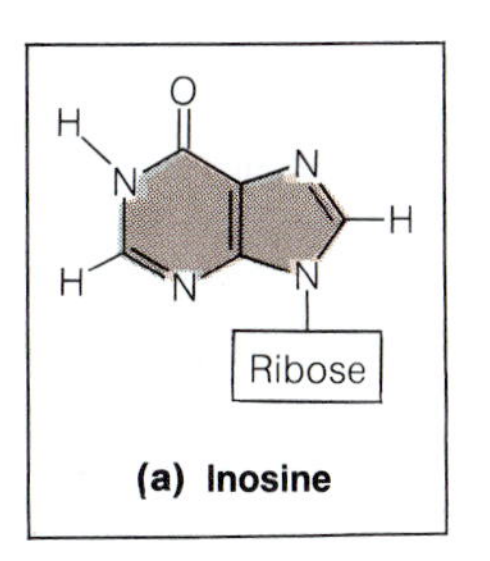

Figure 16-24 The Structure of Inosine and Its Base-Pairing Possibilities. (a) Inosine (I) is an unusual base found in tRNA molecules. (b) Its structure allows it to base-pair with U, C, or A, such that tRNA molecules with an I at the 5′ position of the anticodon can recognize codons that end with any of these three nucleotides. For example, the tRNA shown in Figure 16-23 can recognize the codons GCU, GCC, or GCA, all of which code for alanine.

molecule provided only that the tRNA has UAI as its anticodon. Similarly, the six codons for the amino acid leucine (UUA, UUG, CUU, CUC, CUA, and CUG) require only three tRNA molecules because of wobble.

Aminoacyl-tRNA Synthetases. The enzymes responsible for linking amino acids to their cognate tRNA molecules are called **aminoacyl-tRNA synthetases.** The linkage is an ester bond, and the reaction is driven by the hydrolysis of ATP to AMP and pyrophosphate:

$$\text{R—C}(\text{H})(\overset{\oplus}{\text{N}}\text{H}_3)\text{—C}(=\text{O})\text{—O}^{\ominus} + \text{HO—[tRNA]} \xrightarrow[\text{Aminoacyl-tRNA synthetase}]{\text{ATP} \quad \text{AMP, PP}_i} \text{R—C}(\text{H})(\overset{\oplus}{\text{N}}\text{H}_3)\text{—C}(=\text{O})\text{—O—[tRNA]} \qquad (16\text{-}1)$$

The driving force for the reaction is provided by the subsequent hydrolysis of pyrophosphate (to 2 P_i). The product is an **aminoacyl tRNA,** with sufficient energy in its ester bond to drive formation of the peptide bond that will eventually join the amino acid to a growing polypeptide chain. The process of aminoacylation of a tRNA molecule is therefore also called **amino acid activation,** because it not only links an amino acid to its proper tRNA but also activates it for subsequent peptide bond formation.

Protein Factors. Translation also requires the participation of a number of protein factors, some involved in initiation, others in elongation, and still others in chain termination. We will mention these factors as we encounter them during our discussion of the actual mechanism of protein synthesis, to which we are now ready to turn.

The Steps of Translation

Like transcription, translation is easier to understand if we consider it in steps. The major stages in translation are *initiation, elongation* of the polypeptide chain, and chain *termination.* These steps are similar in prokaryotes and eukaryotes.

Initiation. The initiation of translation in prokaryotes is illustrated in Figure 16-25. It begins (Figure 16-25a) with the formation of an **initiation complex** between the mRNA molecule and a 30S ribosomal subunit, mediated by *IF3,* one of several **initiation factors.** (Unfortunately, the initiation factors are numbered in order of discovery rather than order of function.) Proper alignment of the mRNA with respect to the 30S subunit is ensured by a specific nucleotide sequence (often AGGA) on the mRNA molecule that base-pairs to a short complementary sequence on the 3' end of the 16S rRNA molecule.

In the next step (Figure 16-25b), this mRNA-30S complex binds a special tRNA carrying ***N*-formylmethionine,** a modified methionine that has been formylated on its nitrogen atom, as shown in Figure 16-26. *N*-Formylmethionine (f-Met) is the amino acid with which every polypeptide is initiated in prokaryotes. (In eukaryotes, protein synthesis is also initiated with methionine, but the amino acid is not formylated in these organisms.) The $\text{tRNA}^{\text{f-Met}}$ that carries *N*-formylmethionine recognizes the codon AUG; this is therefore the universal **initiation codon.**

N-Formylmethionyl tRNA (f-Met-$\text{tRNA}^{\text{f-Met}}$) is the only aminoacyl tRNA that can bind to the P site on the ribosome. Binding of f-Met-$\text{tRNA}^{\text{f-Met}}$ requires the presence of the AUG codon at the P site of the ribosome and is mediated by a second initiation factor, *IF2.* Also required is the binding of a molecule of GTP, which serves as the energy source for the next step.

The **30S initiation complex** formed in this way now binds to a free 50S subunit (Figure 16-25c), generating the **70S initiation complex** needed for the onset of protein synthesis. Binding of the 50S subunit requires yet another initiation factor, *IF1,* and is driven by hydrolysis of the GTP that was bound in the previous step. Concomitant with the hydrolysis of GTP, all three initiation factors are released.

We can summarize the complicated process of initiation by representing the three steps of Figure 16-25 as summary equations:

$$\begin{aligned}
&\text{mRNA} + \text{30S subunit} + \text{IF3} \longrightarrow \text{mRNA-IF3-30S complex}\\
&\text{mRNA-IF3-30S complex} + \text{f-Met-tRNA}^{\text{f-Met}} + \text{IF2} + \text{GTP} \longrightarrow\\
&\qquad\qquad \text{30S initiation complex}\\
&\text{30S initiation complex} + \text{50S subunit} + \text{IF1} \longrightarrow\\
&\qquad\qquad \text{70S initiation complex} + \text{IF1} + \text{IF2} + \text{IF3} + \text{GDP} + \text{P}_i\\
\hline
&\text{mRNA} + \text{30S} + \text{50S} + \text{f-Met-tRNA}^{\text{f-Met}} + \text{GTP} \longrightarrow\\
&\qquad\qquad \text{70S initiation complex} + \text{GDP} + \text{P}_i \qquad (16\text{-}2)
\end{aligned}$$

Chain Elongation. Once initiated, a polypeptide chain is elongated by successive additions of amino acids. Each such addition involves the formation of a new peptide bond and proceeds in prokaryotes via the series of steps shown in Figure 16-27. In the first round of elongation, the incoming aminoacyl tRNA (alanyl tRNA^{Ala} in the example shown) is brought into position at the A site on the ribosome (Figure 16-27a). Specificity is ensured by the complementary match between the anticodon of the tRNA and the codon immediately adjacent to the AUG codon that directed f-Met-$\text{tRNA}^{\text{f-Met}}$ to the P site. The binding of the new aminoacyl tRNA requires the elongation factor *EF-tu* and is driven by the hydrolysis of GTP to GDP and P_i.

The binding actually proceeds in two steps. First, the

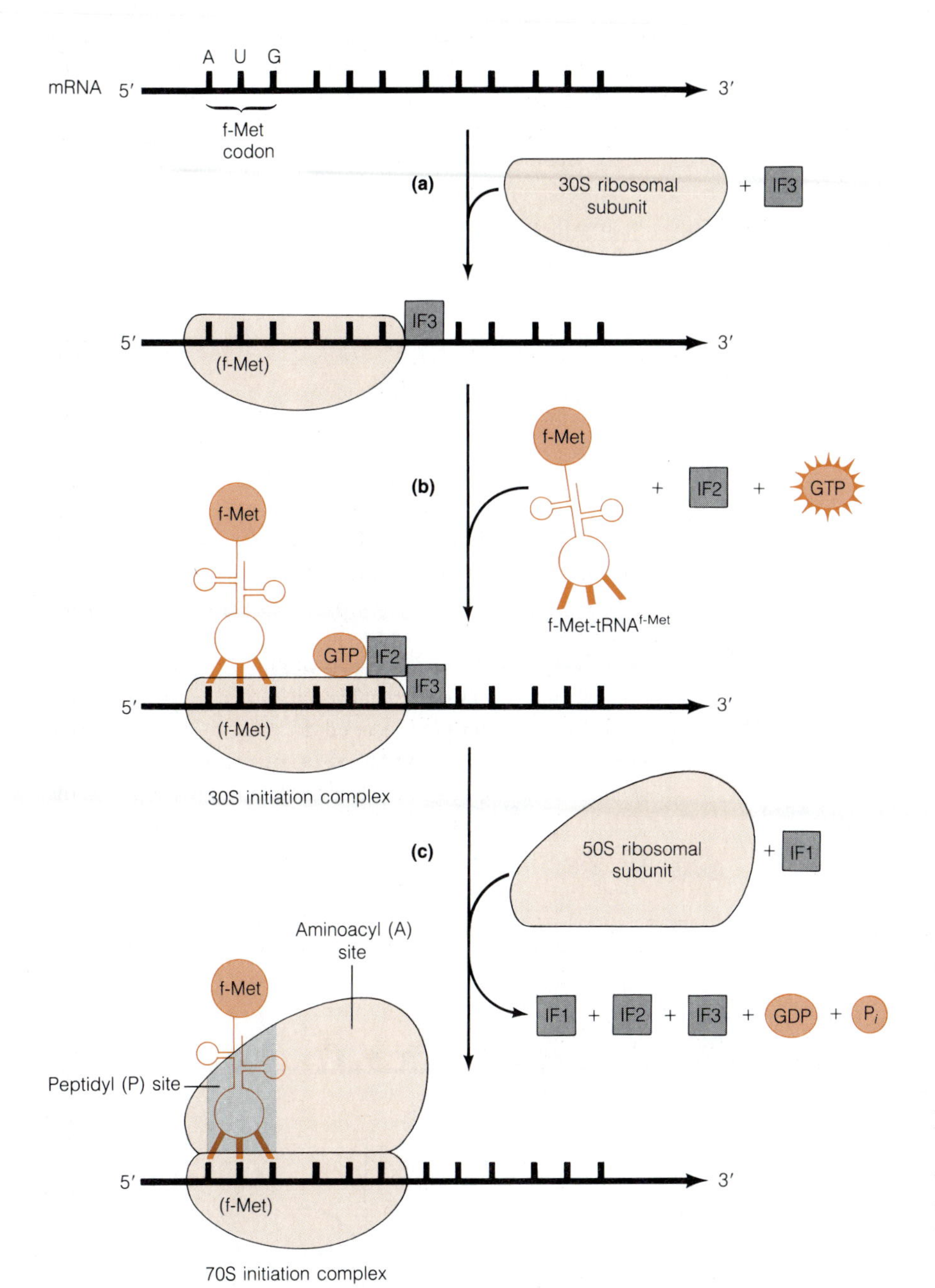

Figure 16-25 The Initiation Complex as a Prerequisite for Protein Synthesis in Prokaryotes. (a) Formation of an initiation complex between an mRNA molecule and a 30S ribosomal subunit is mediated by initiation factor IF3. (b) To form a 30S complex, a molecule of f-Met-tRNA$^{f\text{-}Met}$ is added. This requires the binding of IF2 and a molecule of GTP. (c) The binding of a 50S ribosomal subunit to the 30S complex follows, mediated by IF1 and driven by the hydrolysis of the GTP that was bound in the previous step. The result is a 70S initiation complex with f-Met-tRNA$^{f\text{-}Met}$ in place at the peptidyl (P) site, and the aminoacyl (A) site available for an incoming aminoacyl tRNA, as illustrated in Figure 16-27.

aminoacyl tRNA (aa-tRNA) forms a complex with EF-tu and GTP. The aa-tRNA is then transferred to the 70S initiation complex, with concomitant hydrolysis of GTP. To regenerate free EF-tu from the residual EF-tu-GDP complex requires a second elongation factor, EF-ts. This sequence can be summarized as follows:

$$\begin{aligned}
&\text{EF-tu} + \text{aa-tRNA} + \text{GTP} \longrightarrow \text{EF-tu-aa-tRNA-GTP}\\
&\text{EF-tu-aa-tRNA-GTP} + \text{70S complex} \longrightarrow\\
&\qquad\qquad \text{70S complex-aa-tRNA} + \text{EF-tu-GDP} + P_i\\
&\underline{\text{EF-tu-GDP} \longrightarrow \text{EF-tu} + \text{GDP}}\\
&\text{70S complex} + \text{aa-tRNA} + \text{GTP} \longrightarrow\\
&\qquad\qquad \text{70S complex-aa-tRNA} + \text{GDP} + P_i \qquad (16\text{-}3)
\end{aligned}$$

Formyl group

$$CH_3{-}S{-}CH_2{-}CH_2{-}\underset{H}{\overset{NH{-}C(H){=}O}{C}}{-}\overset{O}{\overset{\|}{C}}{-}O^-$$

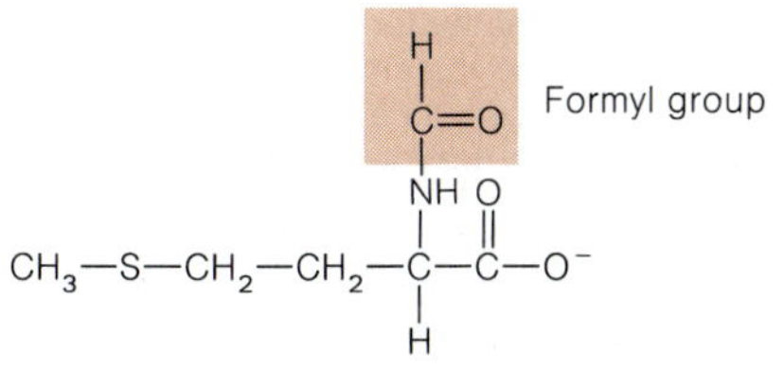

Figure 16-26 The Structure of *N*-Formylmethionine. *N*-Formylmethionine (f-Met) is the modified amino acid with which every polypeptide is initiated in prokaryotes.

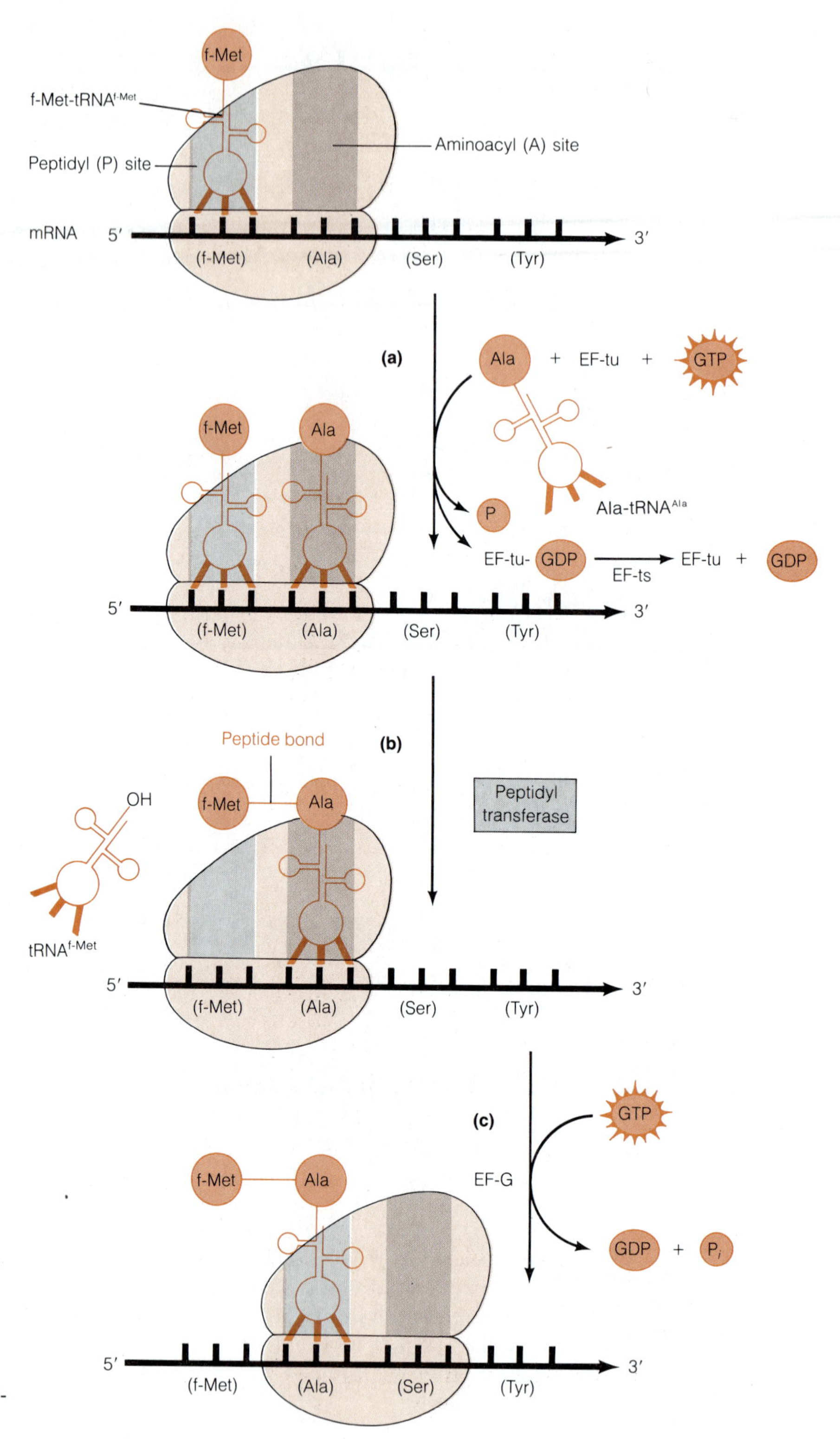

Figure 16-27 Polypeptide Chain Elongation. Chain elongation during protein synthesis requires the presence of a peptidyl tRNA (or, in the first step, an f-Met-tRNA$^{f\text{-}Met}$) molecule at the peptidyl (P) site. (a) Elongation begins with the arrival of the correct aminoacyl tRNA at the aminoacyl (A) site. This step requires the action of elongation factors EF-tu and EF-ts and is driven by the hydrolysis of a molecule of GTP. (b) A peptide bond is then formed between the carboxyl group of the terminal amino acid at the P site and the amino group of the newly arrived amino acid at the A site, catalyzed by the enzyme peptidyltransferase. (c) The elongated peptide is transferred from the A site to the P site, leaving the A site open for the next incoming amino acid. This translocation step requires the factor EF-G and is driven by the hydrolysis of a second molecule of GTP. This series of events is repeated for each amino acid that is added.

With f-Met-tRNA$^{f\text{-}Met}$ now at the P site and alanyl tRNA^{Ala} at the A site, the stage is set for peptide bond formation between the activated carboxyl group of f-Met and the amino group of alanine (Figure 16-27b). The reaction is catalyzed by the enzyme **peptidyltransferase,** a component of the large subunit. The energy to drive peptide bond formation is provided by hydrolysis of the ester bond by which f-Met was attached to its tRNA. Once the peptide bond has been formed, the tRNA$^{f\text{-}Met}$ that carried the formylmethionine is released from the P site of the ribosome, leaving behind a dipeptide attached to the tRNA in the A site.

This dipeptide is now shifted to the P site, as the ribosome moves three nucleotides along the mRNA molecule (Figure 16-27c). As a result, the next codon on the messenger is positioned under the A site to direct the binding of the next aminoacyl tRNA. This process of **translocation** requires yet another elongation factor, *EF-G*, as well as the energy supplied by hydrolysis of another molecule of GTP to GDP and P_i.

The ribosome is now set to receive the next incoming aminoacyl tRNA and repeat the three-step sequence of events shown in Figure 16-27. Each successive amino acid is added in this way, as the ribosome progresses along the mRNA in the 5′ → 3′ direction, translating one codon at a time. In each cycle, the growing polypeptide chain is always bound at the P site by the most recent codon-anticodon interaction, and the incoming amino acid is attached to the A site.

Chain Termination. The process depicted in Figure 16-27 continues in cyclic fashion, reading one codon after the other and adding successive amino acids to the polypeptide chain until **termination of translation** occurs when one of the three possible **stop codons** (UAG, UAA, or UGA; see Figure 16-9) is encountered on the mRNA. There are no tRNA molecules that recognize these codons. Instead, they are recognized by **release factors** that trigger the hydrolysis of the peptidyl tRNA, thereby releasing the completed polypeptide from the ribosome-mRNA complex.

Nonsense Mutations and Suppressor tRNA

Mutations sometimes convert a codon that codes for an amino acid into a stop codon. Figure 16-28 illustrates such a case, in which the change of a single nucleotide pair in the DNA results in the replacement of a lysine codon in the mRNA (Figure 16-28a) with the stop signal UAG (Figure 16-28b). Such mutations normally lead to the production of short, nonfunctional peptides that end at the stop codon, as Figure 16-28b shows. These mutations were first studied in the bacteriophage T4 and were called **nonsense mutations.**

Nonsense mutations are usually lethal, but phages that have such mutations can nonetheless grow in certain strains of bacteria. The bacteria seem to "suppress" the normal chain-terminating effect of the mutation. This is possible because such **suppressor strains** of bacteria have a mutant tRNA that recognizes what would otherwise be a stop codon and inserts an amino acid at that point (Figure 16-28c). In the example shown, the codon UAG is read as a codon for tyrosine by a mutant tRNA that has an altered anticodon. The inserted amino acid is almost inevitably different from the amino acid that would be present at that position in the wild-type protein, but the crucial feature of suppression is that chain termination is averted and a full-length polypeptide can be made. A tRNA molecule that functions in this way is called a **suppressor tRNA.** Suppressor tRNAs must be relatively inefficient; otherwise, the protein-synthesizing apparatus would be unable to terminate most proteins.

The Energetics of Translation

The energy to drive peptide bond formation is provided by hydrolysis of four phosphoanhydride bonds per amino acid added. Two of these bonds are supplied by the ATP that is hydrolyzed to AMP in the aminoacyl-tRNA synthetase reaction. The remaining two are accounted for by the two GTP molecules that are hydrolyzed per amino acid, one as the incoming aminoacyl tRNA is bound to the A site and the other as the ribosome is translocated along the mRNA. Assuming that each of these phosphoanhydride bonds has a $\Delta G^{\circ\prime}$ value of -7.3 kcal/mol, the four bonds represent a standard free energy input of 29.2 kcal per mole of amino acid inserted. Clearly, protein synthesis is a very expensive process energetically, accounting for a substantial fraction of the total energy budget of most cells. If we consider the energy required to synthesize the messenger RNA as well, the cost of protein synthesis becomes even greater.

Summary of Translation

The translation process can be summarized as a mechanism for converting the information of strings of RNA codons into a chain of amino acids linked by peptide bonds. As the ribosome moves along the mRNA codon by codon in the 5′ → 3′ direction, successive amino acids are brought into place by complementary base pairing between the codons on the mRNA and the anticodons on specific aminoacyl tRNA molecules. When a stop codon is encountered, the completed polypeptide is released, and the mRNA and ribosomal subunits become available for further cycles of translation.

Most messages are read by many ribosomes at once, one following closely behind the other. Such clusters of ribosomes and mRNA molecules are called **polyribosomes.** Several polyribosomes are seen in the electron micrograph of Figure 16-17. By allowing the synthesis of many polypeptides from the same message simultaneously, cells maximize the efficiency of mRNA utilization.

Posttranslational Processing

Like RNA, polypeptides may be modified after synthesis. Removal of *N*-formylmethionine from the amino end of the chain is a particularly common modification. As a result, relatively few mature polypeptides have methionine at their amino terminus, even though all of them started out that way. Sometimes, whole blocks of amino acids are removed from one end of the polypeptide or the other. Some enzymes, for example, are synthesized as inactive precursors and must be activated by removal of a specific sequence at one end or the other.

The transport of proteins across membranes also may involve the removal of a terminal signal sequence, as we will see shortly. Other common processing events include chemical modifications of individual amino acid groups (such as methylation, phosphorylation, and acetylation reactions), glycosylation (addition of oligosaccharides), binding of prosthetic groups, and association with other polypeptides into functional proteins or specific supramolecular complexes.

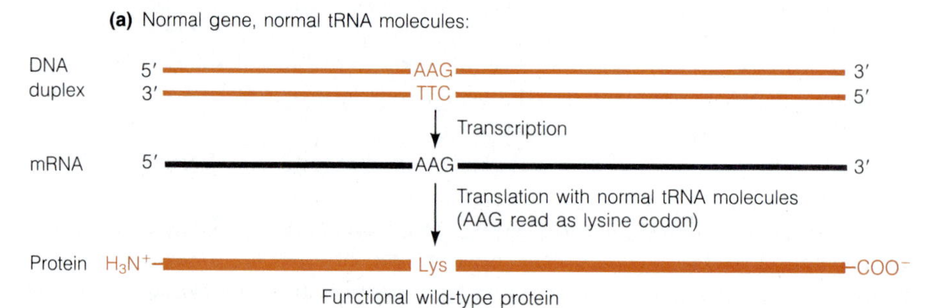

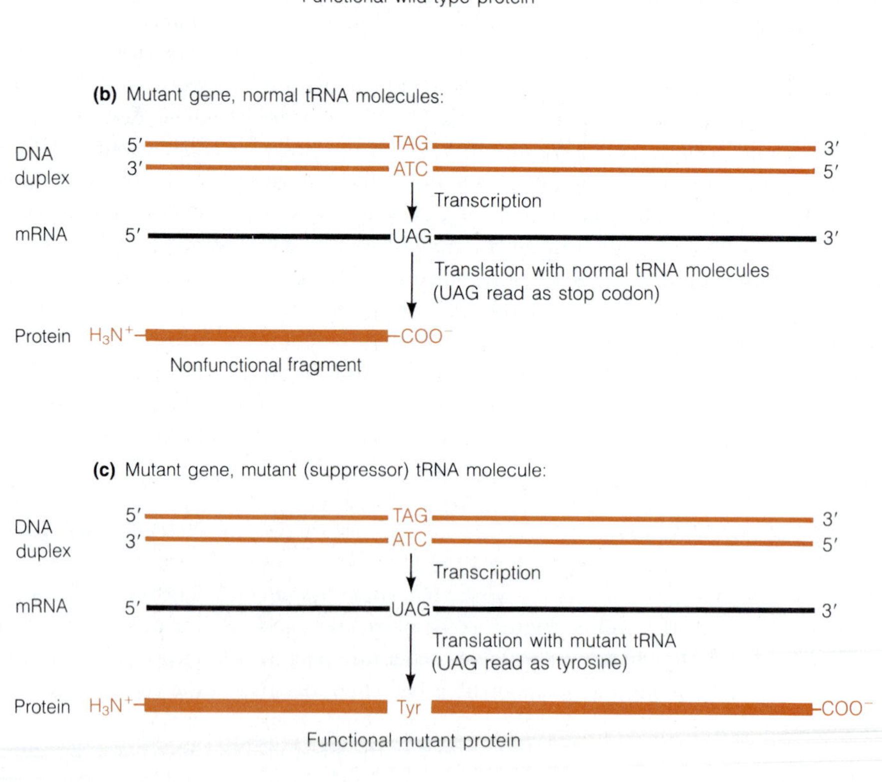

Figure 16-28 Nonsense Mutations and Suppressor tRNAs. (a) A normal gene is transcribed into an mRNA molecule that contains the codon AAG at one point. Upon translation, this codon specifies the amino acid lysine (Lys) at one point in the functional, wild-type protein. (b) If a mutation occurs in the DNA such that the AAG codon in the mRNA is changed to UAG, the UAG codon will be read as a stop signal, and the translation product will be a short, nonfunctional polypeptide. Mutations of this sort are called nonsense mutations. (c) In the presence of a mutant tRNA molecule that reads UAG as a codon instead of a stop signal, an amino acid will be inserted, and protein synthesis will continue. In the example shown, UAG is read as a codon for tyrosine because the mutant tyrosine tRNA has as its anticodon 3′-AUC-5′ instead of the expected 3′-AUG-5′, which would otherwise recognize the tyrosine codon 5′-UAC-3′. The resulting protein will be mutant, since a lysine has been replaced by a tyrosine at one point along the chain. The protein may still be functional, however, provided that its biological activity is not adversely affected by the amino acid substitution.

Protein Targeting and Sorting

So far, we have considered the flow of genetic information from genes to proteins, including the intricacies of transcription, RNA processing, translation, and polypeptide processing. Not yet explored, however, are the mechanisms needed to ensure that each protein synthesized within the cell reaches its correct intracellular location. Think for a moment about a typical eukaryotic cell with its diversity of organelles and compartments, each containing its own unique set of proteins. Such a cell is likely to have about 10 billion protein molecules, representing at least 10,000 different kinds of polypeptides, each of which must find its way to the appropriate intracellular location.

A limited number of these polypeptides are encoded by the genome of the mitochondrion (and, for plant cells, by the chloroplast genome as well), but most are encoded by nuclear genes and are synthesized by a process that begins in the cytoplasm. Each of these polypeptides must then be directed to the right cellular compartment and must therefore have some sort of molecular "address" that ensures its delivery to the proper location. As our final topic for this chapter, we will now discuss this process, which is called **protein** (or, more specifically, **polypeptide**) **targeting and sorting.**

We can begin by grouping the various compartments of eukaryotic cells into three categories: (1) the cytoplasm; (2) nucleus, mitochondria, chloroplasts, peroxisomes, and related organelles; and (3) the endoplasmic reticulum and all other organelles with which it is in contact. Included in this third category are the Golgi complex, lysosomes, and secretory vesicles discussed in Chapter 9, and the plasma membrane. Recall that the interiors of all these compartments can communicate with one another and with the outside of the cell by means of clathrin-coated vesicles.

As Figure 16-29 indicates, polypeptides encoded by nuclear genes follow one of two divergent pathways, with several alternative destinations in each case. Synthesis of all such polypeptides is directed by mRNA molecules from the nucleus and begins on ribosomes in the cytoplasm (Figure 16-29a). Shortly thereafter, however, the two pathways diverge. Ribosomes involved in the synthesis of polypeptides destined for the endoplasmic reticulum or any of the membrane systems derived from it become associated with the membrane of the ER early in the translational process. Such polypeptides are transferred across (or, in the case of integral membrane proteins, inserted into) the membrane of the ER as synthesis proceeds (Figure 16-29b).

Polypeptides destined for the cytoplasm, nucleus, mitochondria, chloroplasts, or peroxisomes follow an alternative pathway (Figure 16-29c). Ribosomes involved in the synthesis of these polypeptides remain free in the cytoplasm, unattached to any membrane. The polypeptides are released from the ribosomes after synthesis, and either remain in the cytoplasm or are taken up by the appropriate organelle.

This latter mechanism of polypeptide uptake into organelles is called **posttranslational import,** since this process involves the transfer across membranes of polypeptides that have already been released from the ribosome. Posttranslational import is characteristic of proteins targeted to the nucleus, mitochondrion, chloroplast, and peroxisome. Transfer of polypeptides into the ER is called **cotranslational import,** since uptake or insertion is coupled directly to the translational process. Cotranslational import is characteristic of proteins targeted to the ER or any of its derivatives.

We will consider both of these processes in some detail. First, we will discuss cotranslational import, using secretory proteins as our example. Then we will turn to posttranslational import, focusing on uptake of proteins into mitochondria and chloroplasts.

Cotranslational Import of Secretory Proteins

Secretory proteins provide a good model system for understanding cotranslational import. In fact, much of the research that led to our current understanding of this mode of import was carried out to explain secretory processes in eukaryotic cells. We have already considered several aspects of cellular secretion in Chapter 9, including the sequential involvement of the endoplasmic reticulum, the Golgi complex, and secretory granules. Here, we will examine the molecular mechanisms involved. To begin, we will look at the experimental evidence that established the pathway of secretory proteins from their synthesis on the rough ER to their eventual release from the cell by exocytosis.

To elucidate this pathway, George Palade and his colleagues injected animals with a radioactive amino acid and then used autoradiography to trace the fate of the radioactivity through the cells in the parotid gland. Figure 16-30 illustrates the results of such an experiment. For the first few minutes, the label is found only in the rough ER. Within 10 minutes radioactivity begins to appear in the Golgi complex, and by 30 minutes it can be detected in the vesicles (also called *condensing vacuoles*) that bud off the Golgi complex. After three hours, most of the label is found in *secretory granules,* vesicles containing materials

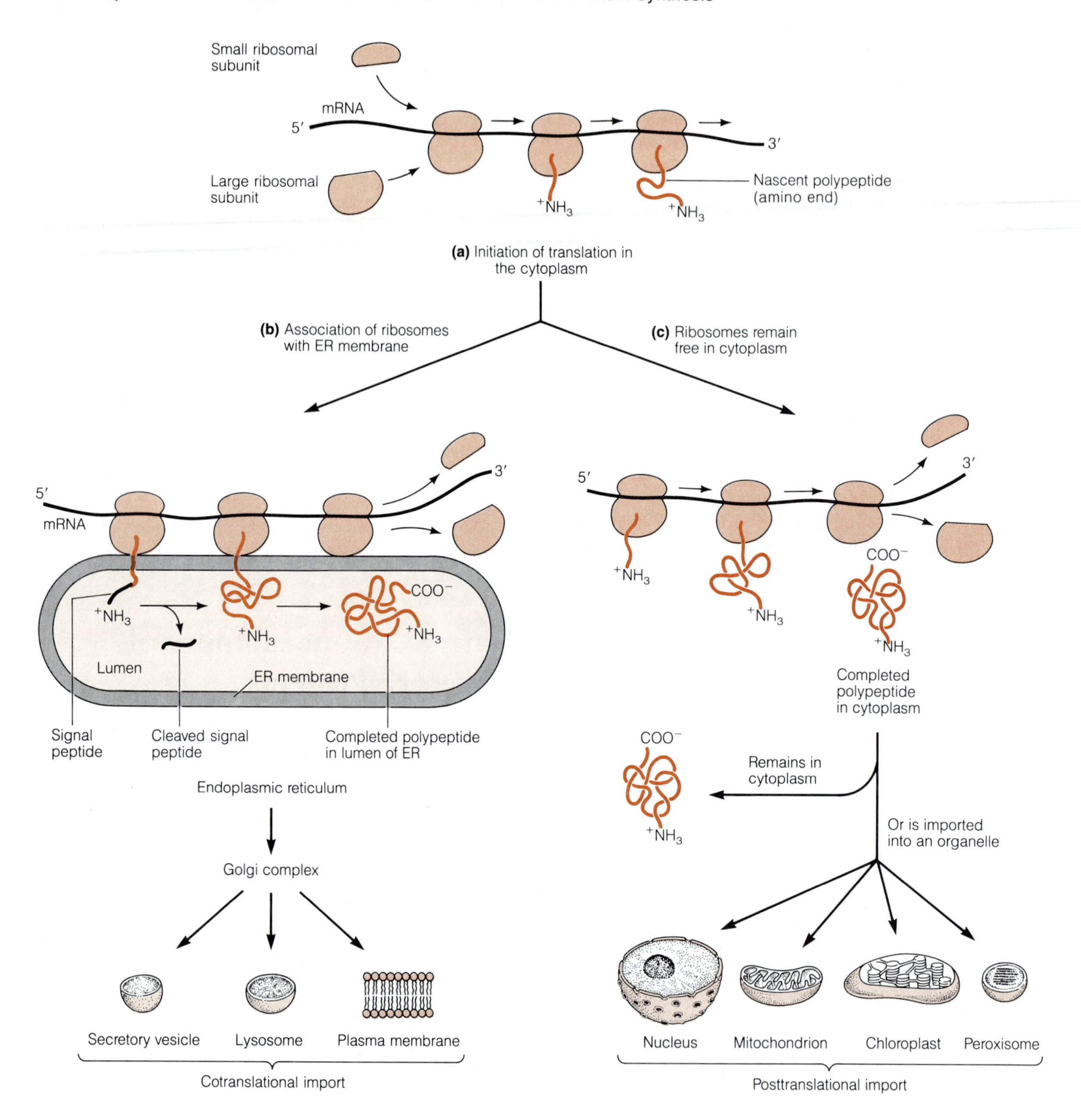

Figure 16-29 Intracellular Sorting of Proteins. Polypeptides synthesized on cytoplasmic ribosomes follow one of two alternative pathways. (a) Synthesis of all polypeptides encoded by nuclear genes begins in the cytoplasm as the large and small ribosomal subunits associate with each other and with the 5′ end of an mRNA molecule, to form a functional ribosome. (b) Polypeptides destined for any of the compartments derived from the endoplasmic reticulum become associated with the ER membrane shortly after translation is initiated, and are transferred across the membrane into the lumen of the ER as synthesis continues (cotranslational import). (c) Polypeptides destined either for the cytoplasm or for import into the nucleus, mitochondrion, chloroplast, or peroxisome are synthesized on ribosomes that remain free in the cytoplasm. Completed polypeptides are eventually released from the ribosomes and either remain in the cytoplasm or are transported into the appropriate organelle shortly after synthesis (posttranslational import).

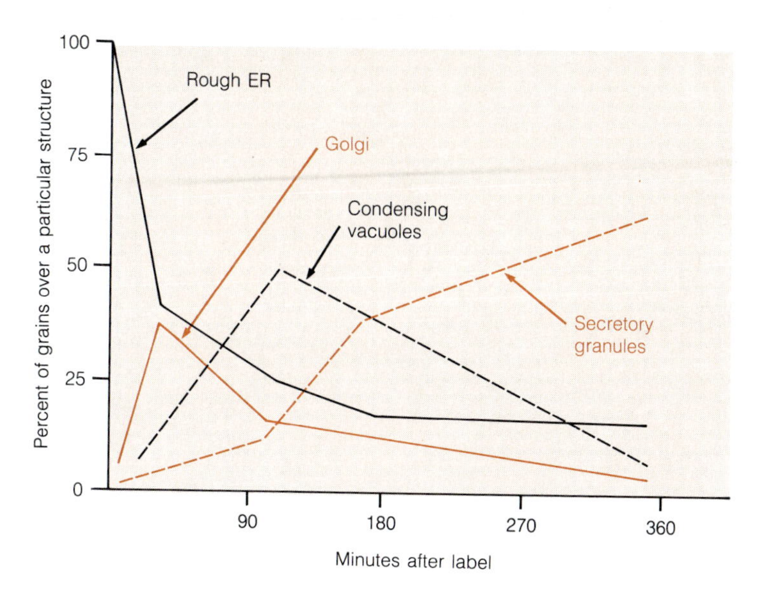

Figure 16-30 Autoradiographic Tracing of Newly Synthesized Protein Through Secretory Cells. Rabbits were injected with a radioactively labeled amino acid, and autoradiography was then used to assess the distribution of radioactivity within parotid gland cells at various times thereafter. Initially, most of the radioactivity was in the rough ER. Then it moved through the Golgi complex to vesicles called condensing vacuoles, and finally to mature secretory (or zymogen) granules.

destined for export from the cell. Shortly thereafter, the secretion of radioactive proteins from the cell begins.

Thus, secretory proteins are synthesized on the rough ER, pass through the Golgi, and are packaged into mature secretory granules. We can follow a secretory protein through this process in a series of steps, beginning with its synthesis on the membrane-bound ribosomes and ending with its discharge from the cell.

Cotranslational Transfer and the Signal Mechanism

Proteins destined to be secreted from the cell are synthesized on ribosomes attached to the membrane of the rough ER when translation is occurring. As synthesis proceeds, the proteins are transported across the membrane into the cisternal space. A model to explain this cotranslational transfer of secretory proteins into the cisternae of the rough ER was put forward in 1971 by Gunter Blobel and David Sabatini. Their model was called the **signal hypothesis** because it assumed that some sort of signal is required to single out these proteins from the many proteins destined to remain in the cytoplasm. A working model of the signal hypothesis is depicted in Figure 16-31.

According to the model, the synthesis of secretory and membrane proteins begins in the cytoplasm, as free ribosomes (or, more accurately, free ribosomal subunits) associate with the messenger RNA for a specific polypeptide (Figure 16-31, step 1). Peptide bond formation then begins, giving rise to a **nascent** (growing) **polypeptide** attached to the ribosome (step 2). Thus far, the process is the same as for the synthesis of all proteins, including those destined to remain in the cytoplasm. However, if the nascent polypeptide is part of a secretory protein, then the first 15 to 30 amino acids are a **signal peptide** (or **signal sequence**) that plays a key role in directing the ribosome-mRNA-polypeptide complex to the surface of the rough ER (step 3). It is now clearly established that only peptides with such signal sequences are capable of being inserted into or across the membrane of the rough ER as synthesis proceeds. Signal sequences therefore represent the first level of targeting during protein synthesis.

Segregation of Proteins into the ER

Once anchored to the ER membrane, the ribosomal complex continues to add amino acids to the growing chain, causing the nascent polypeptide to be translocated across the membrane as synthesis proceeds. In effect, the process of chain elongation causes secretory polypeptides to be "injected" progressively across the membrane and into the **lumen**, or inner cavity, of the ER cisterna (Figure 16-31, steps 3 to 6). After crossing the membrane, the signal peptide is removed from the polypeptide by a **signal peptidase** (step 4). Chain elongation continues (step 5), with the growing polypeptide gradually assuming its final three-dimensional configuration within the cisterna. When synthesis is complete (step 6), the ribosomal subunits dissociate from the mRNA (step 7), leaving the newly synthesized polypeptide sequestered within the lumen of the ER. As the polypeptide folds spontaneously into its three-dimensional form (and, in some cases, associates with other polypeptides), it can no longer traverse the membrane and is therefore effectively "locked" into the lumen of the ER.

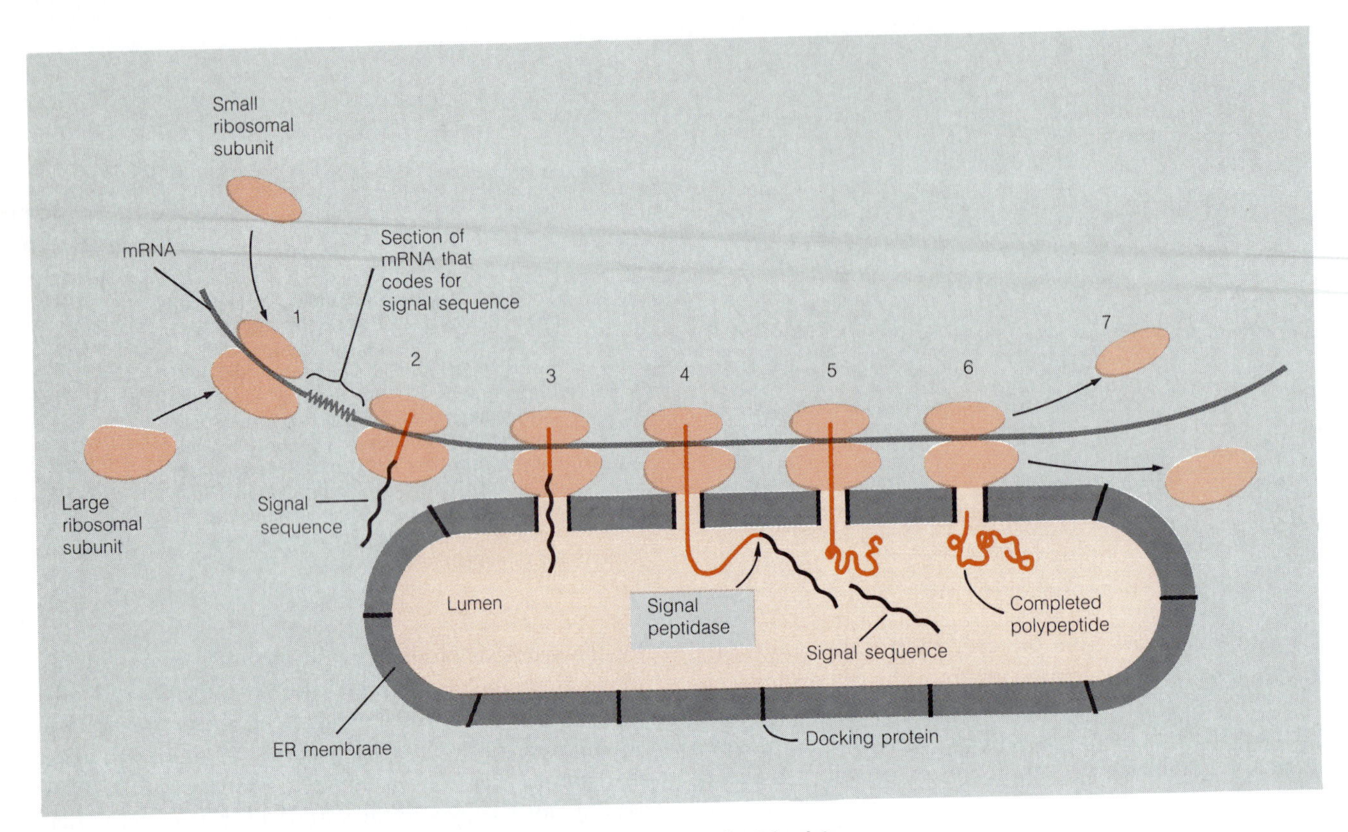

Figure 16-31 Role of the Signal Sequence in the Segregation of Secretory Proteins into the ER. According to the signal hypothesis, it is the signal sequence at the amino terminus of a nascent polypeptide that causes the ribosome-mRNA complex to bind to the ER membrane. The growing polypeptide is transported across the membrane and inserted into the lumen of the ER as synthesis continues. The signal sequence is excised by a specific peptidase, leaving the growing polypeptide to attain its final three-dimensional configuration as synthesis continues.

Recognition and Docking

Once the existence and role of signal sequences were established, it quickly became clear from further studies that nascent polypeptides had to become attached to the ER membrane just after the signal sequence emerged from the ribosome and before further chain elongation could occur. If translation were allowed to continue without attachment, the protein would presumably begin to fold, and the hydrophobic signal sequence would probably become buried within the growing peptide chain. Initially, this was a puzzling realization, since it was not at all clear how such timely attachment could be ensured. Recently, however, at least a part of the puzzle has been solved by the finding that two newly discovered factors are involved in the process—one to block translation until the ribosomal complex reaches and binds to the rough ER, the other to overcome the block and allow translation to resume once the ribosomal complex has been "docked" on the ER.

The first of these factors is called the **signal recognition particle (SRP)** because it recognizes and binds to the signal sequences of nascent secretory or membrane proteins. At first, SRP was thought to be a protein (in fact, the P in its name originally stood for protein). More recently, however, it has been shown to contain an RNA molecule as well, and current speculation suggests a role for the RNA in recognizing the translation complex. As the signal sequence of a nascent secretory or membrane polypeptide emerges from the ribosome, SRP binds to it and blocks further translation. The ribosomal complex then has whatever time it needs to contact the membrane, since the block persists until that contact is made.

Release of the translation block is triggered by a second factor, called **docking protein (DP)**, a membrane protein of the rough ER that serves as the receptor for SRP-blocked ribosomal complexes. Docking protein presumably binds to such complexes, thereby "docking" them onto the ER. In the process, the translation block is overcome and synthesis of the secretory or membrane polypeptide resumes. Because the ribosome is now "docked" on the ER, the polypeptide is transferred into or across the ER membrane as translation proceeds, in accord with Figure 16-31.

Glycosylation of Proteins in the ER

Once a protein has been sequestered into the ER, several chemical modifications can occur, including disulfide bond formation to maintain protein conformation. For proteins destined to have carbohydrate side chains added to them, the initial stages of **glycosylation,** or sugar addition, also occur in the ER. Only the initial or "core" portions of the carbohydrate side chains are added here, however; the remaining events in glycosylation occur in the Golgi complex.

Most glycoproteins have the sugar groups linked to the free amino group of the amino acid asparagine. Invariably, the sugar that is directly linked to asparagine is ***N*-acetylglucosamine** (**NAG**). Despite the variety of oligosaccharides found in glycoproteins, all of the sugar groups attached to asparagines have a common core structure consisting of three glucose units, nine mannose units, and two NAG units. This high-mannose **core oligosaccharide** is built up by sequential addition of monosaccharide units to an activated lipid carrier called **dolichol phosphate.** Dolichol phosphate is shown in Figure 16-32a, and the stepwise addition of monosaccharides to form the oligosaccharide is illustrated in Figure 16-32b. The core oligosaccharide is then transferred as a block to an asparagine side chain of the nascent polypeptide (Figure 16-32c).

Transport of Proteins to the Golgi Complex

Proteins synthesized and partially glycosylated by the rough ER are transported to the Golgi complex, where further processing and packaging take place. This transport is mediated by the clathrin-coated vesicles we learned about in Chapter 9. In addition to their role in bringing secretory proteins from the ER to the Golgi complex, clathrin-coated vesicles transport membrane proteins from the Golgi to lysosomes, to the plasma membrane, and to other cellular destinations.

Transport of secretory proteins out of the ER is blocked by inhibitors of ATP synthesis, indicating that the process requires energy at some point. Transport is not blocked by inhibitors of protein synthesis, however, so it does not appear to depend on a concentration gradient of accumulated protein in the ER.

Modification and Concentration of Proteins by the Golgi Complex

Upon arrival at the forming surface of the Golgi complex, coated vesicles fuse with the Golgi membrane, discharging their contents into the lumen of the Golgi saccules. Here, glycosylation is completed in a sequence of events that always involves the removal of some of the sugars of the oligosaccharide originally added in the ER. For some glycoproteins, nothing further is required. For others, additional sugars are added to the trimmed core, generating a more complex oligosaccharide. This final sequence of events is called **terminal glycosylation** to distinguish it from the **core glycosylation** that occurs in the ER.

Formation, Accumulation, and Discharge of Secretory Granules

After newly synthesized proteins are transported to the Golgi complex, they must be concentrated into the highly condensed form present in mature secretory granules. This concentration usually takes place in the **condensing vacuoles** located on the periphery of the Golgi complex. As the condensing vacuoles of the Golgi fill with concentrated protein, they fuse with one another to form large **secretory** (or **zymogen**) **granules.** In some secretory organs, such as the liver, secretory products are released continuously, so secretory granules tend not to accumulate. In other organs, such as the pancreas, secretion is intermittent rather than continuous, and secretory granules tend to accumulate in the cytoplasm until the actual secretory event is triggered. Zymogen granules are therefore prominent features of pancreatic cells (Figure 16-33). When triggered to discharge their contents to the exterior of the cell, secretory granules move to the cell surface and fuse with the plasma membrane, allowing the contents of the granule to be released by exocytosis.

Import and Sorting of Proteins into Other ER-Related Compartments

Our discussion of protein import into the ER has focused on secretory proteins, because much of our current understanding of the signal peptide strategy of protein import derives from studies of cellular secretion. However, we now recognize that this strategy applies not only to secretory proteins, but also to proteins intended for import into lysosomes or for insertion into the plasma membrane. Moreover, the same mechanism is used by prokaryotic cells

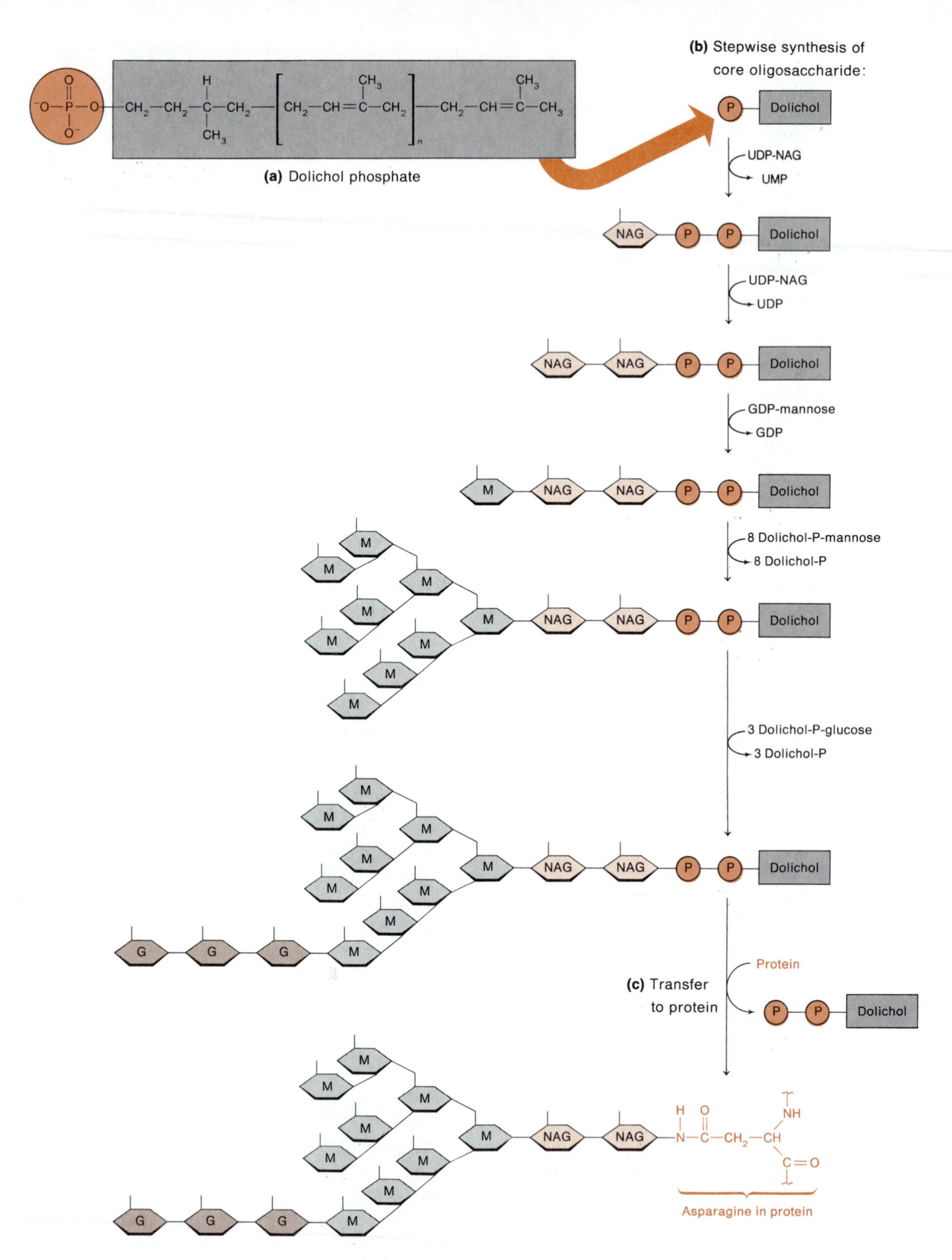

Figure 16-32 Glycosylation of Proteins in the ER. (a) Dolichol phosphate, the carrier of oligosaccharide units in glycosylation reactions. (b) Stepwise synthesis of the common high-mannose oligosaccharide by successive addition of *N*-acetylglucosamine (NAG), mannose (M), and glucose (G) units to the growing chain, with UDP-NAG, GDP-mannose, and dolichol phosphate-glucose as the sugar donors. (c) When the oligosaccharide is complete, it is transferred from dolichol phosphate to the asparagine side chain of the protein that is to be glycosylated.

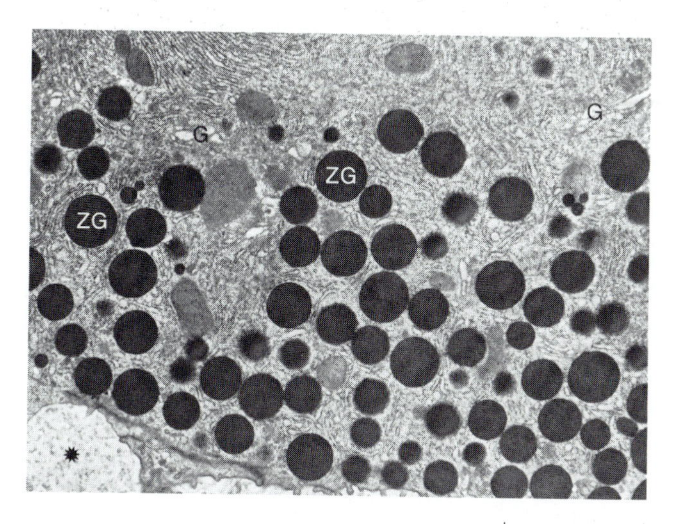

Figure 16-33 Zymogen Granules. This electron micrograph of an acinar (secretory) cell from the exocrine pancreas of a rat illustrates the prominence of zymogen (secretory) granules (ZG) in such cells. Zymogen granules are usually concentrated in the region of the cell between the Golgi complexes (G) from which they arise, and the portion of the plasma membrane bordering the acinar lumen (star) into which the contents of the granules are eventually discharged by exocytosis. (TEM)

to translocate proteins into or across the plasma membrane. Polypeptides destined for any of these locations are all targeted in the same way, by a short hydrophobic signal peptide at the start (the amino end) of the molecule.

Since the ER serves as the initial acceptor of polypeptides destined for a variety of locations, the subsequent *sorting* of these polypeptides is clearly an important concern. How, for example, are polypeptides that are intended for such diverse locations as the plasma membrane, a secretory granule, or the interior of a lysosome recognized and sorted appropriately?

At present, we understand this sorting best for the hydrolytic enzymes that are targeted to the lysosome. As we learned in Chapter 9, mannose-6-phosphate (M6P) is a distinctive component of these proteins, added during the glycosylation process. (Many of the polypeptides that pass through the ER/Golgi system are glycosylated, but only those destined for the lysosome acquire M6P groups.) Located in the Golgi membrane are **mannose-6-phosphate receptors**, transmembrane proteins with binding sites for M6P on the inner surface of the membrane. Because of their specificity for M6P, these receptors recognize and bind lysosomal hydrolases exclusively, thereby separating these polypeptides from all others within the lumen of the Golgi. The receptor-bound polypeptides are concentrated in clathrin-coated vesicles, which then deliver them to the lysosomes.

Although we understand how hydrolases destined for the lysosome reach their location, we do not yet know how the glycosylating enzymes that catalyze the addition of M6P recognize these polypeptides specifically. Since all glycoproteins entering the Golgi from the ER have identical oligosaccharide chains, the information that marks these proteins for M6P addition must reside within the amino acid sequence of each such protein.

Presumably, other proteins that are cotranslationally transported into the ER lumen also acquire specific characteristics that allow them to be recognized and sorted to the various sites shown in Figure 16-29b. However, we do not yet know the molecular basis for the sorting of proteins to locations other than the lysosome.

Posttranslational Import of Proteins

In contrast to the cotranslational import of secretory and other proteins into the ER, proteins destined for the nucleus, mitochondrion, chloroplast, or peroxisome are imported into these organelles after translation. We will focus on import into mitochondria and chloroplasts, since we know the most about these organelles at present.

Import of Polypeptides into Mitochondria and Chloroplasts

Although mitochondria and chloroplasts both contain their own DNA and protein-synthesizing machinery, they encode and synthesize relatively few of the polypeptides that they require. Most of the proteins in these organelles, like all of the proteins in the nucleus and the peroxisome, are encoded by nuclear genes and synthesized on cytoplasmic ribosomes (Figure 16-29c). (The polypeptides encoded by the organellar genome are located mainly in the mitochondrial inner membrane or in the thylakoid membrane of the chloroplast. Almost without exception, these polypeptides are subunits of multimeric proteins, with the other subunits encoded by the nuclear genome and imported from the cytoplasm.)

Polypeptides intended for the mitochondrion or chloroplast are completed and released from the ribosome before import, but are then taken up by the organelle within minutes. The targeting signal for such polypeptides is called a **transit peptide.** Like the signal peptide of ER-targeted polypeptides, the transit peptide is located at the amino end of the polypeptide. Once inside the mitochondrion or chloroplast, the transit peptide is removed by a **transit**

peptidase present in the interior of the organelle. Removal of the transit peptide usually occurs before the transport process is complete.

Unlike signal peptides, the transit peptides of mitochondrial or chloroplast polypeptides are not simply hydrophobic sequences, but usually consist of both hydrophobic and hydrophilic amino acids. The order of the amino acids suggests that the secondary structure of the transit peptide may be more important than the amino acid sequence itself. For example, some mitochondrial signal peptides have positively charged amino acids interspersed with hydrophobic amino acids. When the transit peptide is coiled into an α helix, it forms an amphipathic structure, with most of the positively charged amino acids on one side of the helix and the hydrophobic amino acids on the other. Such an amphipathic helix may be important for moving the polypeptide across the membrane, but the actual mechanism is not yet known.

Initial recognition of their respective polypeptides by the mitochondrion or chloroplast is thought to involve receptor proteins on the outer membrane that bind transit peptides, though such receptors have yet to be character-

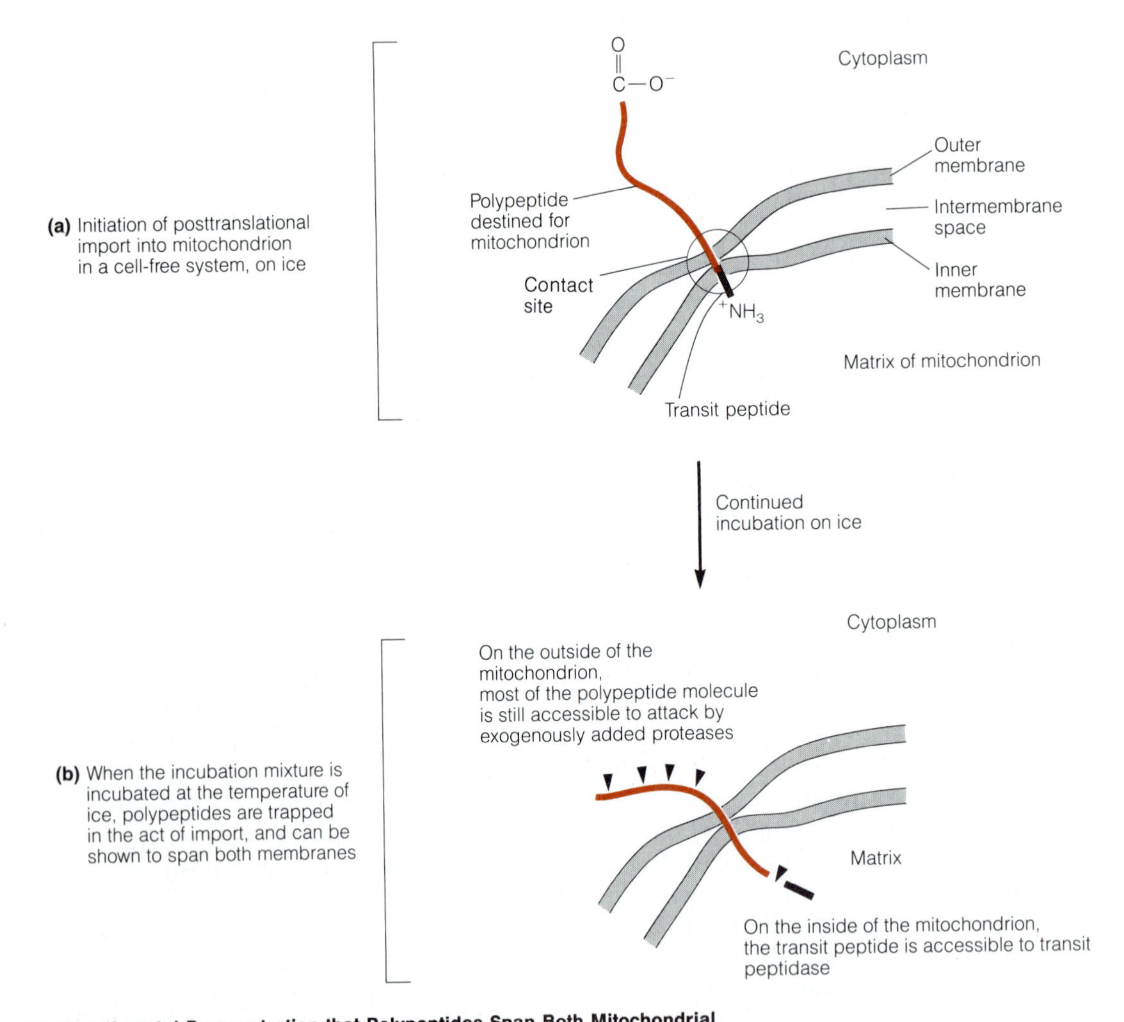

Figure 16-34 Experimental Demonstration that Polypeptides Span Both Mitochondrial Membranes During Import. (a) To demonstrate that polypeptides being imported into the mitochondrion span both membranes at the same time, a cell-free import system is incubated on ice instead of at the usual temperature of 37°C. Polypeptides can penetrate the mitochondrion, but, because of the low temperature, they then stall. (b) Under these conditions, the transit peptide can be cleaved by the transit peptidase present in the matrix, indicating that the amino end of the polypeptide is within the mitochondrion. At the same time, most of the polypeptide molecule is readily accessible to attack by exogenously added proteolytic enzymes on the outside of the mitochondrion. Therefore, we conclude that the polypeptide must span both membranes transiently during import, presumably at one of the contact sites between the two membranes.

ized. Once bound to the outer membrane, a polypeptide apparently passes through both the inner and outer membrane at once, presumably at a **contact site,** where the two membranes appear to touch.

Evidence for this model comes not only from electron microscopy, which reveals many sites of apparent contact between the two membranes, but also from biochemical experiments, in which a cell-free mitochondrial import system is incubated on ice to trap polypeptides in the act of translocation (Figure 16-34). Such polypeptides have already had their transit peptides removed by the transit peptidase in the matrix, but they can still be attacked by exogenously added proteolytic enzymes. These results indicate that polypeptides span both membranes transiently during import. The amino end is clearly within the matrix of the mitochondrion, while the rest of the molecule is still on the outside of the organelle.

Polypeptides almost certainly unfold before (or perhaps as) they are imported into the mitochondrion or chloroplast. It is unlikely that a polypeptide could maintain its three-dimensional conformation while crossing a membrane, especially when one considers the wide range of polypeptide sizes. It is equally unlikely that any sort of pore in the membrane could be large enough to allow a fully folded polypeptide through without disrupting such essential membrane functions as the maintenance of an electrochemical gradient. We know that polypeptide folding is an exergonic process, since polypeptides fold spontaneously as they reel off the ribosome. The unfolding of a polypeptide is therefore an endergonic process, suggesting an energy requirement for polypeptide import.

Both chloroplasts and mitochondria are known to require energy for import of polypeptides. In the case of mitochondria, import requires not only the hydrolysis of ATP, but also an electrochemical gradient across the inner membrane. The electrochemical gradient seems to be necessary for the initial penetration of the transit peptide into the membrane. The energy of ATP hydrolysis, on the other

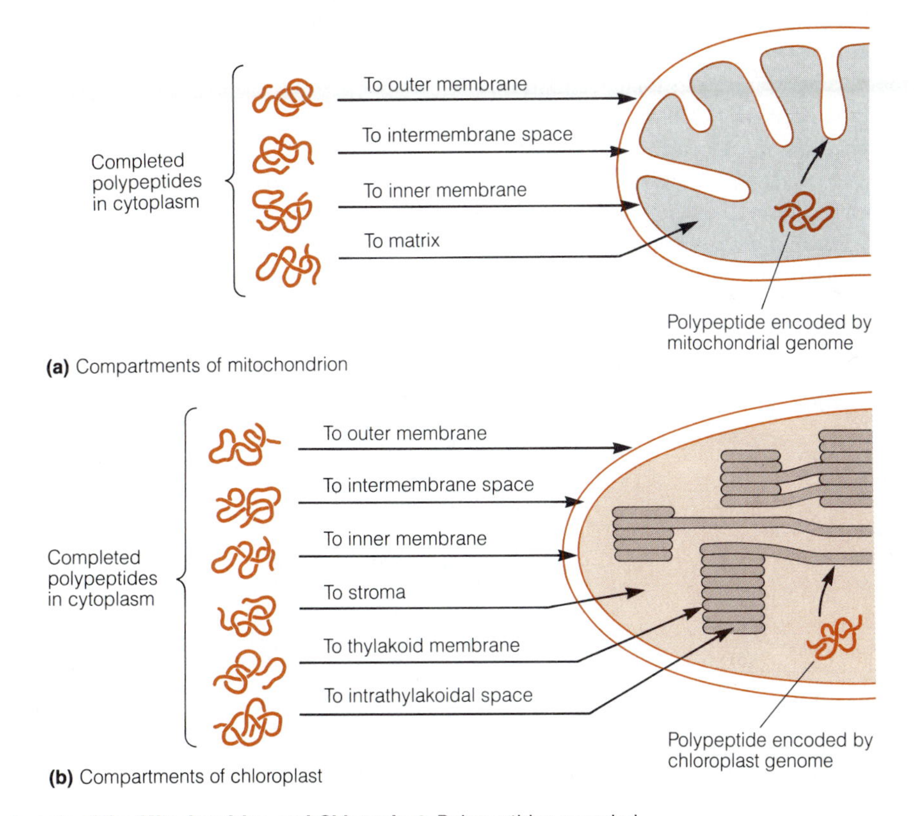

Figure 16-35 Compartments of the Mitochondrion and Chloroplast. Polypeptides encoded by nuclear genes and synthesized on cytoplasmic ribosomes are targeted to each of the compartments of the mitochondrion and chloroplast. (a) Mitochondria have four compartments: the outer membrane, the intermembrane space, the inner membrane, and the matrix. Polypeptides encoded by the mitochondrial genome and synthesized within the organelle are targeted mainly to the inner membrane. (b) Chloroplasts have six compartments: the outer membrane, the intermembrane space, the inner membrane, the stroma, the thylakoid membrane, and the intrathylakoidal space. Polypeptides encoded by the chloroplast genome and synthesized within the organelle are targeted mainly to the thylakoid membrane.

hand, is needed to unfold the polypeptide, thereby allowing it to pass through the membrane. Chloroplasts, on the other hand, maintain an electrochemical gradient across the thylakoid membrane, but not across the inner membrane. Presumably, their energy needs for import are met by ATP alone.

Targeting Polypeptides to Organellar Compartments

Because of the structural complexity of mitochondria and chloroplasts, proteins to be imported from the cytoplasm must be targeted not only to the right organelle, but also to the appropriate compartment within the organelle. Mitochondria have four compartments: the outer membrane, the intermembrane space, the inner membrane, and the matrix (Figure 16-35a). Chloroplasts have four similar compartments (but with the term stroma substituted for matrix), and two additional compartments as well: the thylakoid membrane and the intrathylakoidal space (Figure 16-35b). Note that a polypeptide may have to cross one, two, or even three membranes to reach its final destination.

Given the structural complexity of both organelles, it is probably not surprising that at least some mitochondrial and chloroplast polypeptides require more than one signal to arrive at their proper locations. For example, targeting of a polypeptide to the intermembrane space or to the inner membrane of the mitochondrion requires both a transit peptide to direct the polypeptide to the mitochondrion and a second, very hydrophobic sequence to deliver it to its final destination (Figure 16-36).

Once the polypeptide has been imported into the matrix (at a contact site, most likely; see Figure 16-36a), its transit peptide is cleaved, revealing an adjacent sequence of hydrophobic amino acids that resembles the signal peptide required for cotranslational transport (Figure 16-36b). In fact, this sequence functions as a signal peptide, inserting the polypeptide into the inner membrane in the same way that mitochondrially encoded polypeptides are known to be inserted into this membrane (Figure 16-36c). With its hydrophobic signal peptide anchored in the inner membrane, the polypeptide is then either transported across the inner membrane into the intermembrane space (Figure 16-36d) or into the inner membrane, if that is its final destination. Once in the intermembrane space, the polypeptide has its signal peptide removed by another peptidase (Figure 16-36e). The signal peptide remains behind in the inner membrane, while the mature polypeptide is released into the intermembrane space.

Multiple signals and transport steps are also required to direct some chloroplast polypeptides to their final destination. Polypeptides intended for insertion into (or transport across) the thylakoid membrane, for example, must first be targeted to the chloroplast and transported into the stroma, presumably crossing the inner and outer membranes at a contact site. In the stroma, the transit peptide used for this first step is cleaved from the polypeptide,

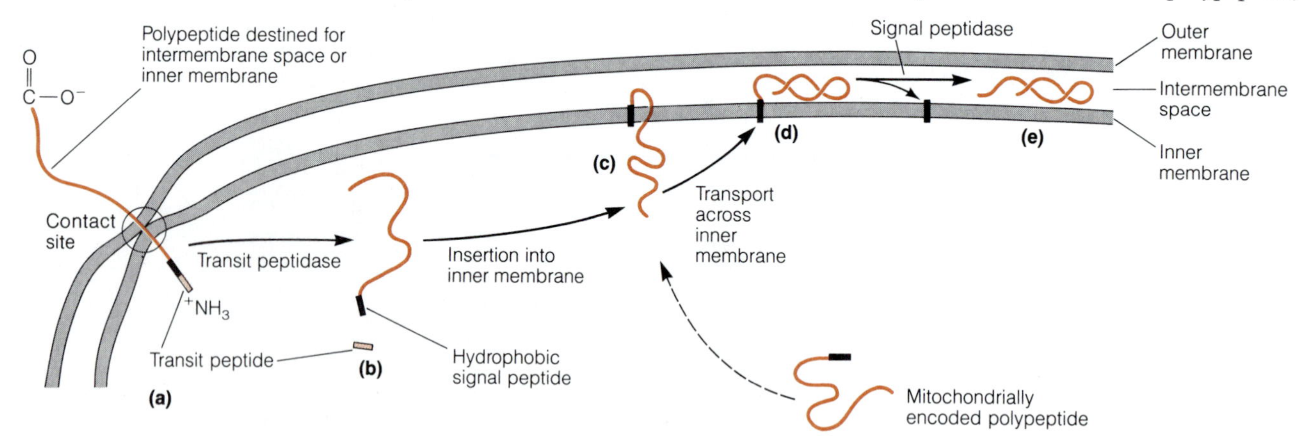

Figure 16-36 Targeting of Polypeptides to the Intermembrane Space or the Inner Membrane of the Mitochondrion Requires Multiple Signals and Multiple Translocation Events. Polypeptides synthesized on cytoplasmic ribosomes but destined either for the intermembrane space or the inner membrane of the mitochondrion require two separate targeting sequences, both located at the amino end. (a) The polypeptide is directed to the mitochondrion by an amphipathic transit peptide. (b) Cleavage of the transit peptide by a peptidase in the matrix reveals a sequence of hydrophobic amino acids. (c) This hydrophobic signal peptide causes the polypeptide to be inserted into the inner membrane in the same way that mitochondrially encoded polypeptides are targeted to this membrane. (d) The polypeptide is then moved across the membrane into the intermembrane space (or into the inner membrane, if that is the final destination of the polypeptide). (e) Cleavage by a second peptidase releases the polypeptide into the intermembrane space, leaving the signal peptide behind in the inner membrane.

unmasking a hydrophobic thylakoid signal peptide. This hydrophobic sequence is then used to insert the polypeptide into the thylakoid membrane in the same way that chloroplast-encoded polypeptides are known to be inserted into this membrane. A similar mechanism is thought to be involved in targeting polypeptides to the intrathylakoidal space, except that in this case the polypeptide moves across the thylakoid membrane instead of simply inserting into it.

Perspective

The expression of genetic information is one of the fundamental activities of all cells. Instructions stored in the DNA of the chromosomes or genophore must be first transcribed and processed into molecules of mRNA, rRNA, and tRNA. These then play specific roles in the synthesis of proteins, with the nucleotide sequence of the mRNA providing the actual information that dictates the order of amino acids in the polypeptide product. The mRNA sequence is read as triplet codons, each recognized by a tRNA molecule with the appropriate complementary anticodon. The genetic code specifies which amino acid corresponds to each of the triplet codons. The code is unambiguous, nonoverlapping, degenerate, and nearly universal.

Transcription involves the synthesis of RNA by a mechanism that depends on complementary base pairing between incoming nucleotides and the template strand of the DNA. This function is carried out by a single kind of RNA polymerase in bacterial cells but by a multiplicity of enzymes in eukaryotes. In retroviruses, the process of reverse transcription from RNA to DNA is used to insert the viral genes into the host cell DNA. Once synthesized, most gene transcripts must be processed to generate functional rRNA, tRNA, or mRNA. Processing can involve cleavage of polycistronic transcription units, removal of noncoding terminal sequences, excision of introns, and addition of special structural features at the 3′ and/or the 5′ terminus, as well as chemical modification of specific nucleotides. In addition, RNA molecules synthesized in eukaryotic nuclei must be transported to the cytoplasm for participation in protein synthesis.

Translation occurs on ribosomes, with the involvement of aminoacyl tRNA, mRNA, and protein factors that trigger specific events in the initiation, elongation, and termination phases. The specificity required to link the right amino acids to the right tRNA molecules is a property of the aminoacyl-tRNA synthetases that catalyze this reaction. Translation involves the recognition of successive codons on the mRNA by the appropriate aminoacyl tRNA molecules, with each amino acid linked covalently to the growing polypeptide chain. Chain termination occurs when one of the stop codons is encountered, and the completed polypeptide is then released from the ribosome.

Proteins destined to be secreted from the cell by exocytosis are cotranslationally imported into the ER, as are lysosomal and plasma-membrane proteins. Targeting of polypeptides to the ER depends on a signal peptide of hydrophobic amino acids on the amino end of the nascent polypeptide. Once anchored to the ER membrane, the ribosome–mRNA–polypeptide complex "injects" the nascent polypeptide across (or into) the membrane as synthesis proceeds. The signal peptide is then cleaved by a signal peptidase, leaving the polypeptide to fold into its final three-dimensional shape. Glycoproteins have a core oligosaccharide added in the ER, with further additions and modifications made in the Golgi complex. Secretory proteins are concentrated and packaged within condensing vacuoles, finally appearing in secretory granules, which then discharge their contents to the exterior of the cell by exocytosis. Lysosomal hydrolases are tagged with mannose-6-phosphate, which enables them to bind specifically to mannose-6-phosphate receptors on the inner surface of the Golgi, for eventual transport to lysosomes.

Proteins destined for the chloroplast, mitochondrion, nucleus, or peroxisome are synthesized on cytoplasmic ribosomes, and are then imported posttranslationally into the respective organelle. Polypeptides are targeted to the mitochondrion or chloroplast by an amphipathic transit peptide at the amino end. Polypeptides appear to be transported into the organelle at contact sites where the inner and outer membranes are joined. The energy needed for polypeptide unfolding and transport into the mitochondrion is provided by ATP hydrolysis and by the electrochemical gradient across the inner membrane. In the case of the chloroplast, unfolding and transport are driven by ATP hydrolysis alone. Mitochondria and chloroplasts have multiple compartments (four and six, respectively) to which polypeptides must be targeted. Import into some compartments requires multiple signals, which are located at the amino end of the polypeptide and are cleaved successively as the polypeptide moves first into the matrix (or, in the case of the chloroplast, the stroma) and then to its final location within the organelle.

Having seen how the complex process of gene expression, protein synthesis, and protein targeting functions, we now move on to Chapter 17, where we will look at some of the aspects of gene regulation in both prokaryotes and eukaryotes.

Key Terms for Self-Testing

The Central Dogma of Molecular Biology
central dogma of molecular biology (p. 456)
messenger RNA (mRNA) (p. 456)
ribosomal RNA (rRNA) (p. 456)
transfer RNA (tRNA) (p. 456)
transcription (p. 456)
translation (p. 456)
reverse transcription (p. 456)

The Genetic Code
genetic code (p. 457)
one gene–one polypeptide (p. 459)
sickle-cell anemia (p. 459)
colinearity (p. 460)
mutagens (p. 460)
frameshift mutations (p. 461)
triplet code (p. 462)
degenerate (p. 462)
nonoverlapping (p. 462)
template strand (p. 462)
nontemplate strand (p. 462)
codons (p. 463)
adenine (p. 463)
guanine (p. 463)
cytosine (p. 463)
uracil (p. 463)
homopolymer (p. 464)
poly U (p. 464)
copolymer (p. 464)
stop signals (p. 465)
unambiguous (p. 465)

Transcription: From DNA to RNA
RNA polymerase (p. 466)
sigma (σ) factor (p. 466)
core enzyme (p. 466)
holoenzyme (p. 466)
binding (p. 467)
promoters (p. 467)
Pribnow box (p. 467)
consensus sequence (p. 467)
TATA box (p. 467)
initiation (p. 467)
ribonucleoside triphosphate molecules (rNTPs) (p. 467)
chain elongation (p. 467)
termination (p. 467)
termination signal (p. 467)
rho (ρ) factor (p. 469)
transcription unit (p. 469)
operon (p. 470)
cistron (p. 470)
polycistronic mRNA (p. 470)
monocistronic mRNA (p. 470)
reverse transcriptase (p. 470)

Processing of RNA
posttranscriptional processing (p. 470)
S values (p. 472)
spacer sequences (p. 472)
leader sequence (p. 472)
intervening sequence (p. 473)
heterogeneous nuclear RNA (hnRNA) (p. 474)
posttranscriptional regulation (p. 476)
introns (p. 476)
exons (p. 476)
RNA splicing (p. 476)
cap (p. 477)
poly-A "tail" (p. 477)

Properties of Messenger RNA
half-life (p. 478)
RNA-DNA hybridization (p. 479)
antisense RNA (p. 479)
cDNA probe (p. 479)
message amplification (p. 482)

Translation: From RNA to Protein
ribosomes (p. 482)
A (aminoacyl) site (p. 482)
P (peptidyl) site (p. 482)
adaptors (p. 482)
anticodon (p. 484)
wobble (p. 485)
inosine (p. 485)
aminoacyl-tRNA synthetases (p. 486)
aminoacyl tRNA (p. 486)
amino acid activation (p. 486)
initiation complex (p. 486)
initiation factors (p. 486)
N-formylmethionine (p. 486)
initiation codon (p. 486)
30S initiation complex (p. 486)
70S initiation complex (p. 486)
peptidyltransferase (p. 489)
translocation (p. 489)
termination of translation (p. 489)
stop codons (p. 489)
release factors (p. 489)
nonsense mutations (p. 489)
suppressor strains (p. 489)
suppressor tRNA (p. 489)
polyribosomes (p. 490)

Protein Targeting and Sorting
protein (polypeptide) targeting and sorting (p. 491)
posttranslational import (p. 491)
cotranslational import (p. 491)
signal hypothesis (p. 493)
nascent polypeptide (p. 493)
signal peptide (signal sequence) (p. 493)
lumen (p. 493)
signal peptidase (p. 493)
signal recognition particle (SRP) (p. 494)
docking protein (DP) (p. 494)
glycosylation (p. 495)
N-acetylglucosamine (NAG) (p. 495)
core oligosaccharide (p. 495)
dolichol phosphate (p. 495)
terminal glycosylation (p. 495)
core glycosylation (p. 495)
condensing vacuoles (p. 495)
secretory (zymogen) granules (p. 495)
mannose-6-phosphate receptor (p. 497)
transit peptide (p. 497)
transit peptidase (p. 497)
contact site (p. 499)

Suggested Reading

General References and Reviews

Alberts, B., D. Bray, J. Lewis, M. Raff, K. Roberts, and J. D. Watson. *Molecular Biology of the Cell,* 2d ed. New York: Garland Press, 1989.

Ayala, F. J., and J. A. Kiger, Jr. *Modern Genetics,* 2d ed. Menlo Park, Calif.: Benjamin/Cummings, 1984.

Lewin, B. *Genes III.* New York: Wiley, 1987.

Pfeffer, S. R., and J. E. Rothman. Biosynthetic protein transport and sorting by the endoplasmic reticulum and Golgi. *Annu. Rev. Biochem.* 56 (1987): 829.

Rothman, J. E., and J. Lenard. Membrane traffic in animal cells. *Trends Biochem. Sci.* 9 (1984): 176.

Sharon, N. Glycoproteins. *Trends Biochem. Sci.* 9 (1984): 198.

Singer, M. and P. Berg. *Genes and Genomes.* Mill Valley, Calif.: University Science Books, 1990.

Spirin, A. S. Ribosome Structure and Protein Synthesis. Menlo Park, Calif.: Benjamin/Cummings, 1986.

Stent, G. S. *Molecular Genetics: An Introductory Narrative.* San Francisco: W. H. Freeman, 1971.

Verner, K., and G. Schatz. Protein translocation across membranes. *Science* 241 (1988): 1307.

Watson, J. D., N. H. Hopkins, J. W. Roberts, J. A. Steitz, and A. M. Weiner. *Molecular Biology of the Gene,* 4th ed. Menlo Park, Calif.: Benjamin/Cummings, 1987.

The Genetic Code

Crick, F. H. C. The genetic code. *Sci. Amer.* 207 (October 1962): 66.

Crick, F. H. C. The genetic code III. *Sci. Amer.* 215 (October 1966): 55.

Grantham, R. Workings of the genetic code. *Trends Biochem. Sci.* 5 (1980): 327.

Khorana, H. G. Nucleic acid synthesis in the study of the genetic code. In *Nobel Lectures: Physiology or Medicine (1963–1970)*, 341. New York: American Elsevier, 1973.

Nirenberg, M. W. The genetic code II. *Sci. Amer.* 208 (March 1963): 80.

Transcription

Chambon, P. Eucaryotic nuclear RNA polymerases. *Annu. Rev. Biochem.* 44 (1975): 613.

Darnell, J. E., Jr. Transcription units for mRNA production in eukaryotic cells and their DNA viruses. *Prog. Nucleic Acid Res. Mol. Biol.* 22 (1981): 326.

Losick, R., and M. Chamberlin, eds. *RNA Polymerase.* Cold Spring Harbor, N.Y.: Cold Spring Harbor Laboratory, 1976.

McClure, W. Mechanism and control of transcription initiation in prokaryotes. *Annu. Rev. Biochem.* 54 (1985): 171.

Rodriguez, R. L., and M. J. Chamberlin, eds. *Promoters: Structure and Function.* New York: Praeger, 1982.

Rosenberg, M., and D. Court. Regulatory sequences involved in the promotion and termination of RNA transcription. *Annu. Rev. Genet.* 13 (1979): 319.

Varmus, H. Reverse transcription. *Sci. Amer.* 30 (September 1987): 56.

RNA Processing

Chambon, P. Split genes. *Sci. Amer.* 224 (May 1981): 60.

Crick, F. Split genes and RNA splicing. *Science* 204 (1979): 264.

Darnell, J. E., Jr. The processing of RNA. *Sci. Amer.* 249 (October 1983): 90.

Greer, C. L., and J. Abelson. RNA splicing: Rearrangement of RNA sequences in the expression of split genes. *Trends Biochem. Sci.* 9 (1984): 139.

Perry, R. P. RNA processing comes of age. *J. Cell Biol.* 91 (1981): 28s.

Weintraub, H. M. Antisense RNA and DNA. *Sci. Amer.* 262 (January 1990): 40.

Translation

Bielka, H. *The Eukaryotic Ribosome.* New York: Springer Verlag, 1982.

Hershey, J. W. B. The translational machinery. In *Cell Biology: A Comprehensive Treatise,* vol. 4 (D. M. Prescott and L. Goldstein, eds.) New York: Academic Press, 1980.

Hunt, T. The initiation of protein synthesis. *Trends Biochem. Sci.* 5 (1980): 178.

Jimenez, A. Inhibitors of translation. *Trends Biochem. Sci.* 1 (1976): 28.

Lake, J. A. Evolving ribosome structure: Domains in archaebacteria, eubacteria, eocytes and eukaryotes. *Annu. Rev. Biochem.* 54 (1985): 507.

Nomura, M. The control of ribosome synthesis. *Sci. Amer.* 250 (January 1984): 102.

Rich, A., and S. H. Kim. The three-dimensional structure of transfer RNA. *Sci. Amer.* 238 (January 1978): 52.

Siekevitz, P., and P. C. Zamecnik. Ribosomes and protein synthesis. *J. Cell Biol.* 91 (1981): 53s.

Cotranslational Protein Import

Blobel, G., P. Walter, G. N. Chang, B. M. Goldman, A. H. Erickson, and V. R. Lingappa. Translocation of proteins across membranes: The signal hypothesis and beyond. *Symp. Soc. Exp. Biol.* 33 (1979): 9.

Burgess, T. L., and R. B. Kelly. Constitutive and regulated secretion of proteins. *Annu. Rev. Cell Biol.* 3 (1987): 243.

Castle, J. D., J. D. Jamieson, and G. E. Palade. Radioautographic analysis of the secretory process in the parotid acinar cell of the rabbit. *J. Cell Biol.* 53 (1973): 290.

Davis, B. D., and P. C. Tai. The mechanism of protein secretion across membranes. *Nature* 283 (1980): 433.

Kornfeld, S. Trafficking of lysosomal enzymes. FASEB J. 1 (1987): 462

Lodish, H. F. Transport of secretory and membrane glycoproteins from the rough endoplasmic reticulum to the Golgi. *J. Biol. Chem.* 263 (1988): 2107.

Palade, G. Intracellular aspects of protein synthesis. *Science* 189 (1975): 347.

Walter, P., and V. R. Lingappa. Mechanism of protein translocation across the endoplasmic reticulum membrane. *Annu. Rev. Cell Biol.* 2 (1986): 499.

Weidmann, M., T. V. Kurzhchalia, E. Hartmann, and T. A. Rapoport. A signal sequence receptor in the endoplasmic reticulum membrane. *Nature* 332 (1987): 830.

Posttranslational Protein Import

Attardi, G., and G. Schatz. Biogenesis of mitochondria. *Annu. Rev. Cell Biol.* 4 (1988): 289.

Hartl, F. U., J. Ostermann, B. Guiard, and W. Neupert. 1987. Successive translocation into and out of the mitochondrial matrix: targeting of proteins to the intermembrane space by a bipartite signal peptide. *Cell* 51 (1987): 1027.

Hurt, E. C., and A. P. G. M. van Loon. How proteins find mitochondria and intramitochondrial compartments. *Trends Biochem. Sci.* 11 (1986): 204.

Pain, D., Y. S. Kanwar, and G. Blobel. Identification of a receptor for protein import into chloroplasts and its localization to envelope contact zones. *Nature* 331 (1988): 232.

Roise, D., S. J. Horvath, J. M. Tomich, J. H. Richards, and G. Schatz. A chemically synthesized pre-sequence of an imported mitochondrial protein can form an amphiphilic helix and perturb natural and artificial phospholipid bilayers. *EMBO J.* 5 (1986): 1327.

Schleyer, M., and W. Neupert. Transport of proteins into mitochondria: translocation intermediates spanning contact sites between outer and inner membranes. *Cell* 43 (1986): 339.

Schmidt, G. W., and M. L. Mishkind. The transport of proteins into chloroplasts. *Annu. Rev. Biochem.* 55 (1986): 879.

Smeekens, S., C. Bauerle, J. Hageman, K. Keegstra, and P. Weisbeek. The role of the transit peptide in the routing of precursors toward different chloroplast compartments. *Cell* 46 (1986): 365.

van Loon, A. P. G. M., A. W. Brandli, and G. Schatz. The presequences of two imported mitochondrial proteins contain information for intracellular and intramitochondrial sorting. *Cell* 44 (1986): 801.

Problem Set

1. **Codes and Coding.** One way to help you think about the genetic code is by asking you to compare it with the Morse Code. Explain the meaning of each of the following terms, and indicate with an M if it is true of the Morse Code and with a G if it is true of the genetic code.

   (a) Degenerate (d) Universal

   (b) Unambiguous (e) Nonoverlapping

   (c) Triplet

2. **Frameshift Mutations.** Each of the following mutants contains one or more nucleotide additions (+) or deletions (−) of the type caused by the acridine dyes. Assume in each case that all the mutants are very near the beginning of the gene coding for protein X. In each case, indicate with an "OK" if you would expect the mutant protein to be nearly normal and with a "NO" if you would not expect it to be nearly normal.

   (a) + (d) −/+/− (g) −/−/+/−

   (b) +/− (e) −/+/−/+ (h) +/+/−/+/+

   (c) +/+ (f) −/−/− (i) +/+/+/+/+/+

3. **Life with an Overlapping Code.** Assume that the genetic code of all forms of life on the planet QB9 consists of overlapping triplets, such that the translation apparatus shifts only one base at a time. Thus, the nucleotide sequence ABCDEF would be read on earth as two codons (ABC, DEF), but on QB9 as four codons (ABC, BCD, CDE, DEF) and the start of two more. For each of the following kinds of mutations, describe briefly the effect it would have on the amino acid sequence of the coded protein (i) on earth and (ii) on QB9. (Assume that the mutation occurs near the middle of the gene.)

   (a) Single-base substitution

   (b) Single-base deletion

   (c) Deletion of three consecutive bases

4. **Translating the Code.** The following is the actual sequence of a small stretch of human DNA:

   3′ AATTATACACGATGAAGCTTGTGA-
   CAGGGTTTCCAATCATTAA 5′

   5′ TTAATATGTGCTACTTCGAACACT-
   GTCCCAAAGGTTAGTAATT 3′

   (a) What are the two possible RNA molecules that could be transcribed from this DNA?

   (b) Only one of these two RNA molecules can actually be translated. It is in fact the mRNA for the hormone *vasopressin.* What is the apparent amino acid sequence for vasopressin?

   (c) In its active form, vasopressin is a nonapeptide (that is, it has nine amino acids) with cysteine at the N terminus. How can you explain this in light of your answer to part (b)?

   (d) A related hormone, *oxytocin,* has the following sequence:

   Cys-Tyr-Ile-Glu-Asp-Cys-Pro-Leu-Gly

   Where and how would you change the DNA given at the beginning of this problem so that it would code for oxytocin instead?

5. **The Case of the Economical Phage.** The bacteriophage ϕX174 is marvelously economical with its genetic information in that it contains some *overlapping genes*. For example, cistron D is entirely contained within cistron E, but the nucleotides constituting cistron D are read in a different reading frame from those in cistron E, as shown in Figure 16-37. With reference to the figure, label each of the following statements as either true (T) or false (F).
   (a) A point mutation that causes an amino acid substitution in the D protein is likely also to cause an amino acid substitution in the E protein.
   (b) Some nonsense ("stop") mutations in gene D will also be nonsense mutations in gene E.
   (c) In virus-infected cells, some protein molecules will be made that correspond to the E protein at the amino end and to the D protein at the carboxyl end.
   (d) There should be ribosome-binding sites at positions x and y on the mRNA.
   (e) The deletion of a single nucleotide in the region between x and y is more likely to affect the E protein than the D protein.

6. **RNA Polymerases.** For each of the following statements about RNA polymerases, indicate with a B if the statement is true of the bacterial enzyme and with a I, II, or III if it is true of the respective eukaryotic RNA polymerases. A given statement may be true of any, all, or none of these enzymes.
   (a) The enzyme is insensitive to α-amanitin.
   (b) The enzyme catalyzes an exergonic reaction.
   (c) All the primary transcripts must be processed before being used in translation.
   (d) The enzyme may sometimes be found attached to an RNA molecule that also has ribosomes bound to it.
   (e) The enzyme synthesizes rRNA.
   (f) The enzyme adds a poly-A sequence to mRNA.
   (g) The enzyme moves along the DNA template strand in the $3' \rightarrow 5'$ direction.
   (h) The enzyme synthesizes a product that is likely to acquire a 7-methyl G cap.

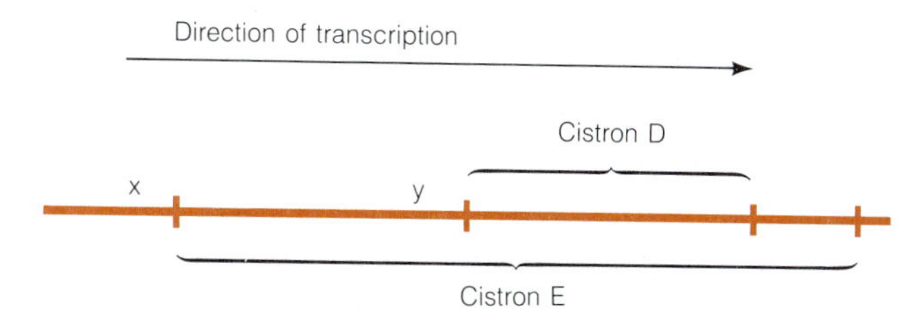

Figure 16-37 The Overlap of Genes in the Bacteriophage ϕX174. Cistron D of the ϕX174 genome is entirely contained within cistron E, but the two cistrons are read in different reading frames.

7. **RNA Processing.** The three major species of RNA found in the cytoplasm of a typical eukaryotic cell are rRNA, tRNA, and mRNA. For each, indicate:
   (a) Two or more kinds of processing to which that RNA has almost certainly been subjected.
   (b) A processing activity unique to that RNA species.
   (c) A processing activity that you would also expect to find for the same species of RNA from a bacterial cell.

8. **Antibiotic Inhibitors of Transcription.** *Rifamycin* and *actinomycin D* are two antibiotics derived from the fungus *Streptomyces*. Rifamycin binds to the β subunit of *E. coli* RNA polymerase and interferes with the formation of the first phosphodiester bond in the RNA chain. Actinomycin D binds to double-stranded DNA by *intercalation* (slipping in between neighboring base pairs).
   (a) Which of the four steps in transcription would you expect rifamycin to affect the most?
   (b) Which of the four steps in transcription would you expect actinomycin D to affect the most?
   (c) Which of the two inhibitors is more likely to affect RNA synthesis in cultured human liver cells?
   (d) Which of the two inhibitors would be more useful for an experiment in which it is necessary to block the initiation of new RNA chains without interfering with the elongation of chains that are already being synthesized?

9. **An Antibiotic Inhibitor of Translation.** *Puromycin* is a powerful inhibitor of protein synthesis. It is an analogue of the 3′ end of aminoacyl tRNA, as Figure 16-38 reveals. When puromycin is added to an in vitro system containing all the necessary machinery for protein synthesis, incomplete polypeptide chains are released from the ribosomes. Each such chain has puromycin covalently attached to one end.
   (a) Explain these results.
   (b) To which end of the polypeptide chains would you expect the puromycin to be bound? Explain.
   (c) Would you expect puromycin to bind to the A or P site on the ribosome, or to both? Explain.
   (d) Assuming that it can penetrate into the cell equally well in both cases, would you expect puromycin to be a better inhibitor of protein synthesis in a eukaryotic cell or in a prokaryotic cell? Explain.

10. **Copolymer Analysis.** In their initial attempts to determine codon assignments, Nirenberg and Matthei first used homopolymers and then used copolymers synthesized by the enzyme polynucleotide phosphorylase. This enzyme adds nucleotides randomly to the growing chain, but in propor-

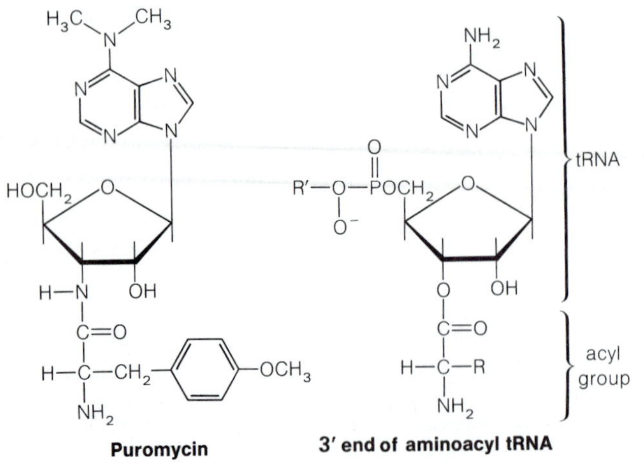

Puromycin 3′ end of aminoacyl tRNA

Figure 16-38 The Structure of Puromycin. Puromycin is an analogue of the 3′ end of aminoacyl tRNA, and a potent inhibitor of protein synthesis in both prokaryotic and eukaryotic cells. (R represents the functional group of the amino acid; R′ represents the remainder of the tRNA molecule.)

tion to their presence in the incubation mixture. By varying the ratio of precursor molecules in the synthesis of copolymers, Nirenberg and Matthei were able to deduce base compositions (but usually not actual sequences) of the codons that code for various amino acids. Suppose you carry out two polynucleotide phosphorylase incubations, with UTP and CTP present in both, but in different ratios. In incubation A the precursors are present in equimolar concentrations, but in incubation B there is three times as much UTP as CTP. The copolymers generated in both incubation mixtures are then used in an in vitro protein-synthesizing system, and the resulting polypeptides are analyzed for amino acid composition.

(a) What are the eight possible codons represented by the nucleotide sequences of the resulting copolymers in both incubation mixtures? What amino acids do these codons code for?

(b) For every 64 codons in the copolymer formed in incubation A, how many of each of the eight possible codons would you expect on the average? How many for incubation B?

(c) What can you say about the expected frequency of occurrence of the possible amino acids in the polypeptides obtained upon translation of the copolymers from incubation A? What about the polypeptides that result from translation of the incubation B copolymers?

(d) Explain what sort of information can be obtained by this technique.

(e) Would it be possible by this technique to determine that codons with 2 U's and 1 C code for phenylalanine, leucine, and serine?

(f) Would it be possible by this technique to decide which of the three codons with 2 U's and 1 C (UUC, UCU, CUU) correspond to each of the three amino acids mentioned in part (e)?

(g) Suggest a way to assign the three codons of part (f) to the appropriate amino acids of part (e).

11. **Antisense RNA.** One of the early problems in working with antisense RNA was simply making amounts sufficient to perform experiments. Assume that you could buy any of the enzymes discussed in Chapters 13 and 16. From what you know about their properties, how would you go about synthesizing sense and antisense RNA?

12. **Signal Sequences.** Although initially discovered in studies of secretory proteins, signal sequences are now recognized as a common property of almost all proteins that are destined to be segregated into a membranous compartment. What specific property of a signal sequence is suggested by each of the following observations?

(a) Signal sequences seem to exist for the purpose of facilitating the transport of polypeptides across membranes.

(b) It is often the case that several proteins with different functions and properties are segregated into the same membrane-bounded compartment.

(c) Although a number of proteins must often be segregated into the same compartment, a single peptidase appears to be adequate to remove the signal sequence from each of them.

13. **Tunicamycin Inhibition.** Tunicamycin is an antibiotic that inhibits the secretion of glycoproteins. Structurally, it is an analogue of UDP-*N*-acetylglucosamine.

(a) At what specific step in the synthesis and secretion of glycoproteins would you expect tunicamycin to exert its inhibitory effect? Explain.

(b) Would you expect tunicamycin to be more effective in blocking core glycosylation or terminal glycosylation? Explain.

(c) Briefly describe an experimental approach that might be used to confirm the predictions you made in parts (a) and (b).

14. **Insulin Secretion.** Insulin is a polypeptide hormone secreted by the pancreas and involved in the control of blood glucose levels. Shortly after a carbohydrate-rich meal, insulin levels in the blood may rise three or four times above normal levels. Assume that an experimental animal has been injected with the radioactive amino acid ^{3}H-leucine just prior to receiving a carbohydrate-rich meal. Within a few minutes after the meal, the level of insulin in the blood rises dramatically, as expected. None of this insulin is radioactively labeled, however. In fact, it takes about an hour before labeled insulin is found in the blood. Explain these results.

15. **Vesicular Stomatitis Virus.** When vesicular stomatitis virus (VSV) infects a mammalian cell, it causes the cell to begin synthesizing a new glycoprotein called *G-protein.* G-protein is synthesized on membrane-bound ribosomes and eventually can be found as an integral membrane protein spanning the plasma membrane. The orientation of the protein in the membrane is such that its amino end is exposed to the extracellular space and its carboxyl end is exposed to the cytoplasm. A *pulse-chase experiment* is performed in which VSV-infected cells are exposed to ^{14}C-labeled amino acids for 5 minutes (the "pulse"), followed by incubation with a large excess of unlabeled amino acids (the "chase"). At various time intervals thereafter, aliquots of cells are removed and homogenized. The separation of Golgi membranes, plasma membranes, and rough endoplasmic reticulum (rough ER) is accomplished by centrifugation, and the amount of labeled G-protein present in each fraction is then measured. The results are shown in Figure 16-39a. When the experiment is repeated, but with ^{3}H-galactose instead, the labeling pattern shown in Figure 16-39b is obtained.

   (a) Briefly describe the synthesis of G-protein and the route by which it gets to the plasma membrane.

   (b) To which compartments of the cell are the amino end and the carboxyl end of the G-protein exposed as it moves to the plasma membrane?

   (c) Where in the cell does the galactose become attached to the G-protein? Explain. Why does labeled galactose appear in the Golgi fraction before labeled amino acids do?

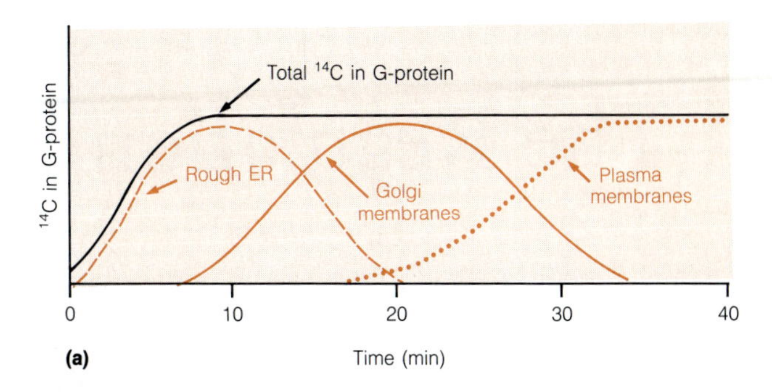

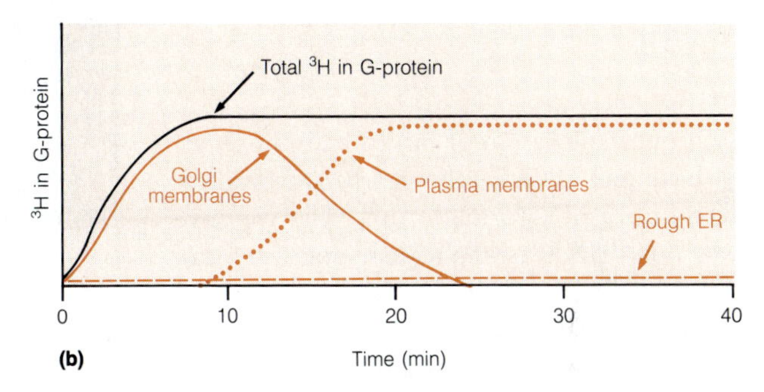

Figure 16-39 G-Protein in the Membranes of VSV-Infected Cells. These graphs depict the amount of radioactively labeled G-protein in the Golgi membranes, plasma membranes, and rough ER of VSV-infected cells after a five-minute pulse label. (a) Cells labeled with ^{14}C-amino acids. (b) Cells labeled with ^{3}H-galactose.

17

The Regulation of Gene Expression

In our exploration of information flow, we have met DNA as the nearly universal repository of information in cells, we have seen how DNA is replicated, and we have looked at the processes involved in expression of genetic information. However, we have not yet considered how that expression is regulated to meet the needs of cells. It should be no surprise, therefore, to find that this concluding chapter on information flow deals with the regulation of gene expression.

Regulation is an important part of almost any process. Rarely, if ever, is it adequate simply to describe how a process works or how a product is made. Invariably, we must also ask how to initiate the process, what is required to turn it off, at what rate it proceeds, and how that rate can be adjusted and controlled. This is certainly true of gene function. Most genes are not used all the time, and we need to understand how their expression is turned on and off, providing the regulation that is so fundamental a part of cellular function.

As you might expect, we know far more about the regulation of gene expression in prokaryotes than in eukaryotes. Bacteria are, after all, far more amenable to the kinds of genetic and biochemical manipulations necessary to study control mechanisms. But considerable progress is now being made with eukaryotes as well, aided particularly by recombinant DNA techniques and rapid methods for determining DNA and RNA nucleotide sequences. We will look first at some of the regulatory features that operate in prokaryotes. Then we will turn to the eukaryotes to explore some of the additional control mechanisms found in these organisms.

Gene Regulation in Prokaryotes

Of the several thousand genes present on the typical bacterial genophore, few, if any, are used all the time. Instead, their expression is controlled so that the amount of a given gene product in a cell is carefully tuned to the cell's need for that product. The intracellular concentrations of many of the enzymes involved in a variety of metabolic processes are regulated by the starting and stopping of enzyme synthesis in response to cellular needs. This is called **adaptive enzyme synthesis.**

Strategies of Adaptive Synthesis

Bacterial cells use two different kinds of adaptive strategies, depending on whether a given set of enzymes is involved in an anabolic (synthetic) or catabolic (degradative) pathway. Figure 17-1 illustrates a typical anabolic pathway—the synthesis of the amino acid tryptophan from the starting compound chorismate. The enzymes required to carry out this synthesis are regulated coordinately, as we will see later in the chapter.

Figure 17-2 depicts the first stages in a typical catabolic pathway—the degradation of the disaccharide lactose to obtain energy. The actual breakdown of lactose begins with its hydrolysis into the monosaccharides glucose and galactose, a reaction catalyzed by the enzyme *β-galactosidase.* Before it can be catabolized by the cell, however, lactose must be transported across the plasma membrane. The *galactoside permease* responsible for this transport is regulated coordinately with β-galactosidase. (The further catabolism of galactose and glucose is also illustrated in Figure 17-2, but enzyme names are not included because these enzymes are not part of the same regulatory unit as β-galactosidase and galactoside permease.)

Anabolic Pathways. For anabolic pathways, enzyme synthesis is usually controlled by the level of the end product in the cell. As the cellular concentration of tryptophan rises, for example, it makes sense for the cell to reduce the rate at which it is synthesizing the several enzymes involved in tryptophan production, or perhaps even to cease synthesizing them entirely. But it is equally important that the cell be able to turn the synthesis of these enzymes back on when the level of tryptophan decreases again. This kind of control is characterized by the ability of the end product of an anabolic pathway to *repress* the further synthesis of the enzymes involved in its formation and is therefore called **end-product repression** or often just **repression.** Most biosynthetic pathways in bacterial cells are regulated in this way.

It is important to note that true repression always involves an effect on enzyme synthesis, and not just on enzyme activity. As you may recall from Chapter 6, end products of biosynthetic pathways often have an inhibitory effect on enzyme activity as well. This is called **feedback**

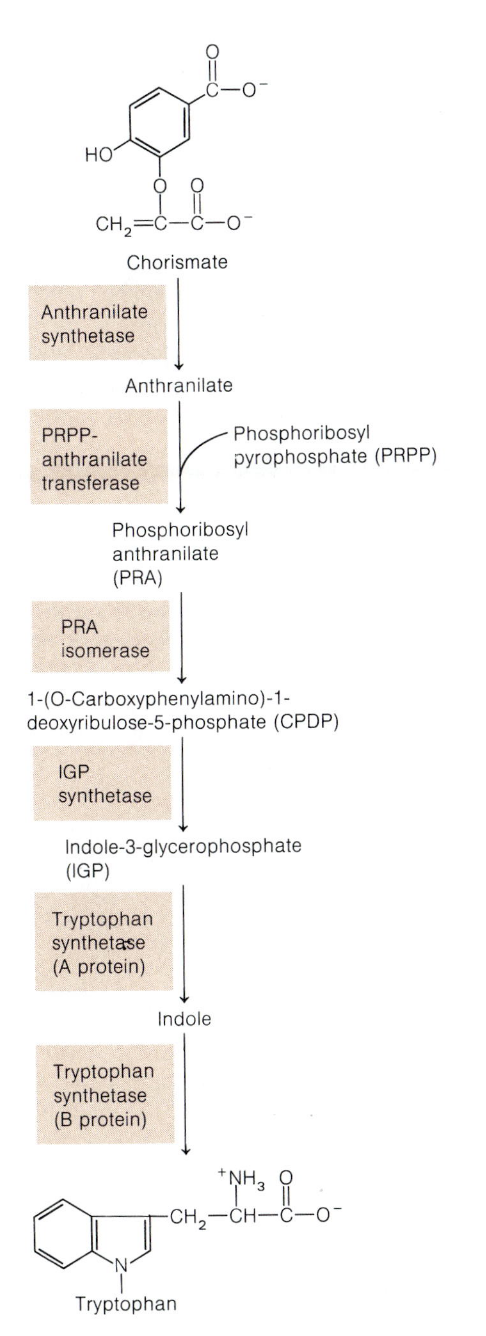

Figure 17-1 A Typical Anabolic Pathway. Synthesis of the amino acid tryptophan from the starting compound chorismate involves a set of enzymes (color) that are synthesized and regulated coordinately. See Figure 17-6 for the organization and regulation of the genes that code for these enzymes.

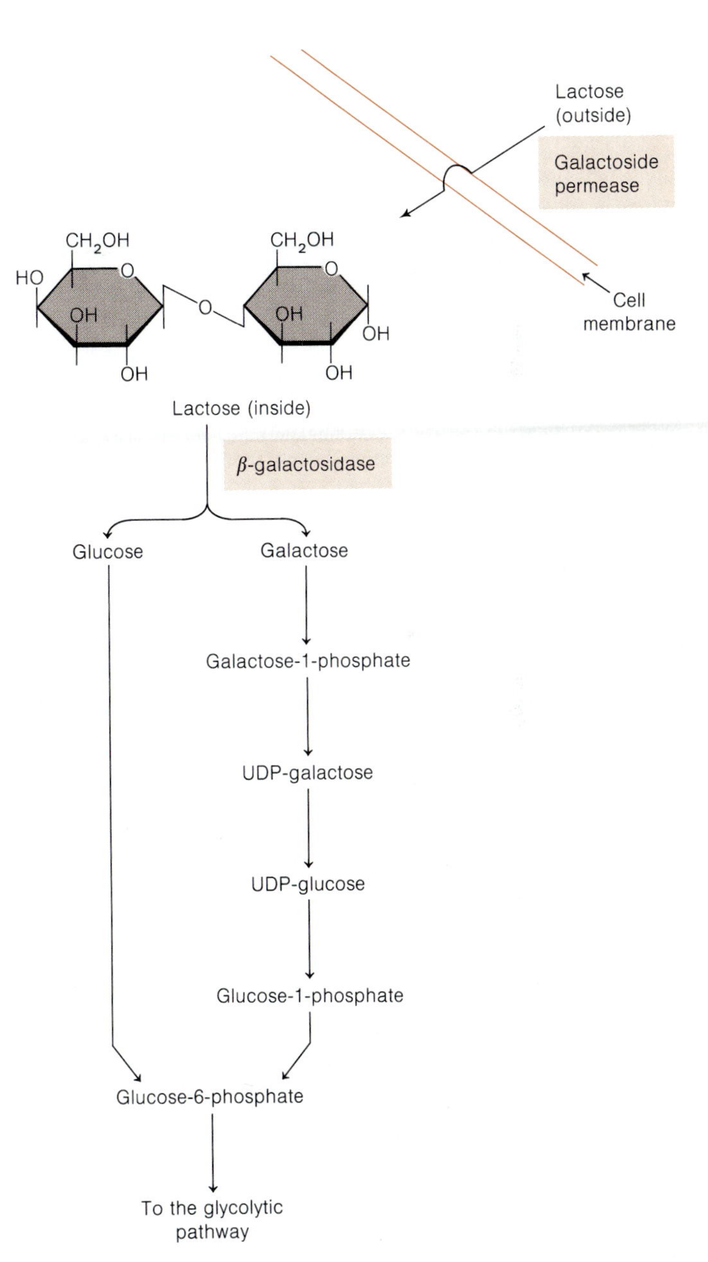

Figure 17-2 A Typical Catabolic Pathway. Degradation of the disaccharide lactose involves enzymes (color) that are synthesized and regulated coordinately. See Figure 17-3 for the organization and regulation of the genes that code for these enzymes. (The enzymes responsible for the subsequent catabolism of the monosaccharides glucose and galactose are not a part of the same regulatory unit and are therefore not identified here.)

inhibition and should be clearly distinguished from repression, since the two differ in both mechanism and result. The term *repression* should be carefully reserved for changes in enzyme levels due to decreased rates of enzyme synthesis, and should not be used indiscriminately to describe decreases in activity for which the underlying mechanism is not yet known.

Catabolic Pathways. The rationale for regulation of catabolic pathways is in a sense just the opposite of that for repressible anabolic pathways. Catabolic enzymes exist for the primary purpose of degrading specific substrates, usually as a means of obtaining energy. Such enzymes are therefore needed by the cell only when it is confronted with the relevant substrate. The enzyme β-galactosidase, for example, is useful only when the cell is actually using lactose as an energy source; in the absence of lactose, the enzyme is unessential. Accordingly, it makes sense for the synthesis of β-galactosidase to be turned on, or *induced,* in the presence of lactose, but to be turned back off in its absence. This is called **substrate induction,** and enzymes whose synthesis is regulated in this way are referred to as **inducible enzymes.** Most catabolic pathways in bacterial cells are subject to substrate induction.

Effectors: Regulatory Molecules. The feature common to repressible and inducible enzymes is that enzyme synthesis is regulated by small organic molecules available to the cell from its surroundings. Molecules that function in this way are called **effectors** because of the effect they have on the allosteric proteins that control enzyme synthesis. For catabolic pathways, effectors are almost always substrates, and they function as inducers of enzyme synthesis. For anabolic pathways, effectors are usually end products, and they usually lead to the repression of enzyme synthesis.

The Lactose System of *Escherichia coli*

The classic example of an inducible enzyme system is the set of enzymes involved in lactose catabolism in the intestinal bacterium *Escherichia coli.* Much of what we currently know about the regulation of gene expression in bacteria and most of the vocabulary used to express that knowledge have come from studies of this system by French molecular geneticists François Jacob and Jacques Monod. Their classic paper, published in 1961, has probably had more influence on our whole concept of gene regulation than any other work.

The lactose system involves three enzymes, each encoded by a separate gene (Figure 17-3). The ***z* gene** codes for the enzyme *β-galactosidase,* which hydrolyzes β-galactosides such as the sugar lactose. A *galactoside permease* that transports lactose across the bacterial cell membrane is the product of the ***y* gene.** The ***a* gene** produces a *transacetylase* enzyme that acetylates lactose as it is taken up

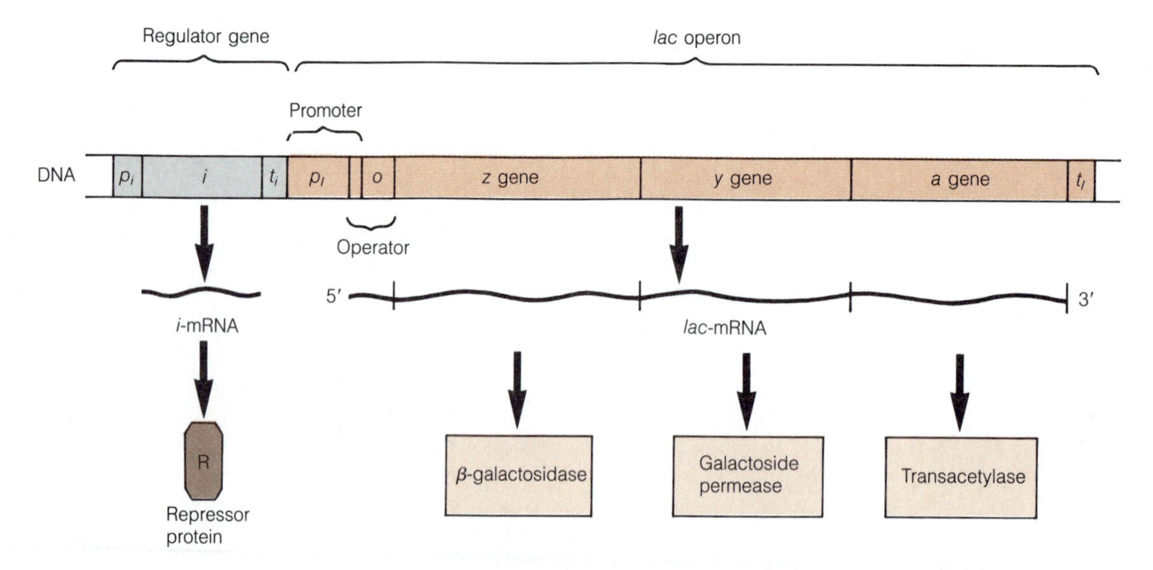

Figure 17-3 The Lactose (*lac*) Operon of *E. coli.* The *lac* operon consists of a segment of double-stranded DNA that includes three contiguous structural genes (*z*, *y*, and *a*) that are transcribed and regulated coordinately. The adjacent regulator gene *i* codes for the repressor protein R. Both the regulator gene and the *lac* operon itself contain promoters (p_i and p_l) at which RNA polymerase binds and terminators (t_i and t_l) at which transcription ceases. The *lac* promoter overlaps partially with the operator site (*o*) to which the active form of the repressor protein binds. For details of regulation, see Figure 17-4.

by the cell. Mapping studies showed that the genes *z*, *y*, and *a* lie adjacent to one another on the genetic map of *E. coli* as Figure 17-3 illustrates. This led Jacob and Monod to suggest that the three genes might all belong to a single regulatory unit, or, as they called it, an **operon**—a cluster of genes with related functions, regulated in such a way that all the genes in the cluster are turned on and off together. Genes such as *z*, *y*, and *a* that actually code for polypeptide products are called **structural genes,** to distinguish them from **regulatory genes,** examples of which we are about to encounter. It is important to note that the organization of several functionally related genes into an operon is a feature commonly observed in prokaryotes but rarely observed in eukaryotic cells, in which individual genes are typically regulated as independent transcription units. Nevertheless, the operon model proposed by Jacob and Monod established several basic principles that have shaped our understanding of transcription regulation in both prokaryotic and eukaryotic systems.

The Operon Model

Jacob and Monod went on to propose a mechanism by which such coordinated regulation might be accomplished. Their operon model, illustrated in Figure 17-3, represents one of the truly important conceptual advances in biology. According to the model, bacterial genes coding for proteins with metabolically related functions are likely to occur in clusters and may therefore be transcribed together into a single **polycistronic messenger RNA** molecule. In such a system, transcription begins at the **promoter (*p*),** the site of RNA polymerase attachment, and ends at a **termination signal.** An operon, in other words, is really just a transcriptional unit bounded by a promoter at one end and a termination signal at the other.

However, the essence of the operon is that it is not just a unit of transcription, but also a unit of regulation. To understand this, we need to consider three additional elements of the system: the operator (a nucleotide sequence located in front of the *z* gene), the repressor protein (a DNA-binding protein that can bind to the operator sequence), and the regulator gene that codes for the repressor. Figure 17-3 depicts each of these elements for the specific case of the lactose (*lac*) operon.

The **operator,** or ***o*** **site,** overlaps with the promoter and is recognized specifically by the **repressor protein,** usually designated as **R.** The repressor, in turn, is produced by a **regulator gene** located outside the operon. In the lactose system, the regulator gene is called the ***i*** **gene** (for "inducibility") and is located immediately adjacent to the *lac* operon that it regulates. Like every gene that codes for a protein and must therefore be transcribed, the *i* gene has its own promoter and terminator. These elements—structural genes, promoter, operator, regulator gene, and repressor protein—are fundamental to the operon model as we now understand it (Figure 17-3). However, the promoter site was not part of the original model of Jacob and Monod because its existence was not recognized until later.

The repressor protein plays a key role in regulating expression of the structural genes in the operon, as Figure 17-4 illustrates. It does so by binding reversibly to the operator site. When the repressor is bound to the operator (Figure 17-4a), RNA polymerase cannot bind to the promoter, and transcription of the structural genes is not possible. Binding of the repressor to the operator therefore inactivates the operon and keeps its structural genes turned off. Without repressor bound to it, however, the operator site is open (Figure 17-4b). RNA polymerase can therefore bind to the promoter and proceed down the operon, transcribing the *z*, *y*, and *a* genes into a single polycistronic message.

The Repressor Is an Allosteric Protein

It should be clear from Figure 17-4 that the crucial feature of a repressor protein is its ability to exist in two forms, only one of which binds to the operator. This is possible because the repressor is an **allosteric protein.** As we learned in Chapter 6, an allosteric protein can exist in either of two conformational states, depending on whether or not the appropriate effector molecule is present. In one state the protein is active, and in the other state the protein is inactive. The effector molecule binds reversibly to one of the conformational states of the protein, but not to the other. In the presence of its effector, therefore, an allosteric protein is effectively "locked" into the form to which the effector binds. The binding is readily reversible, however, so that removal of the effector results in a rapid return to the alternative form.

As Figure 17-5 shows, the repressor of the *lac* operon exists in two forms, depending on the presence of the effector, *allolactose,* an isomeric form of lactose produced after lactose enters the cell. The free (unbound) form of the repressor recognizes and binds to the operator, whereas the form with effector bound to it does not. This means that the repressor protein will be active (will bind to the operator) in the absence of lactose, but not in its presence. Lactose therefore induces enzymes of the *lac* operon by converting the repressor to the inactive form, which cannot recognize the operator and hence cannot prevent the transcription of the structural genes by RNA polymerase (Figure 17-4).

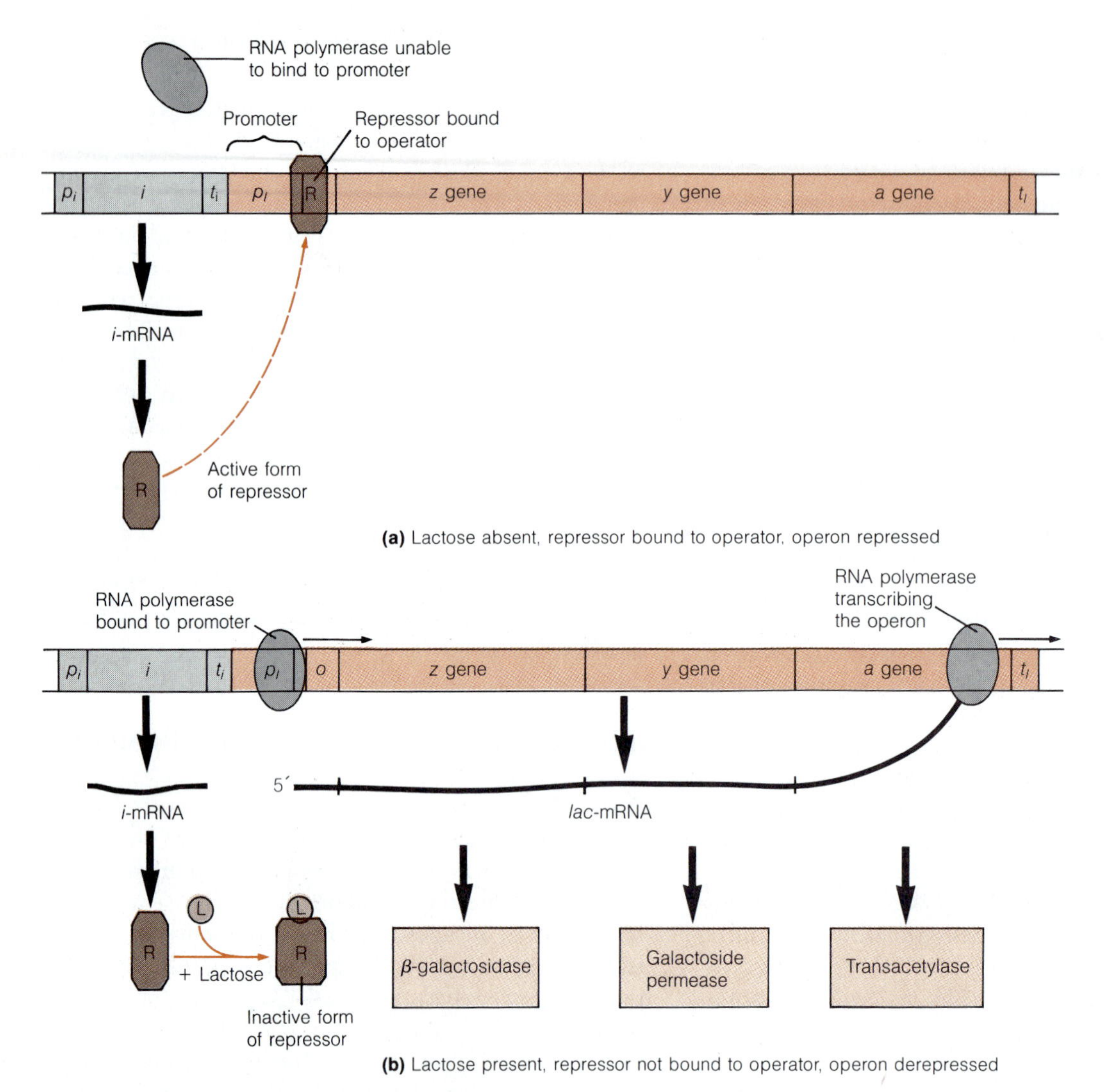

Figure 17-4 Regulation of the *lac* Operon. Transcription of the *lac* operon is regulated by binding of the repressor protein to the operator. (a) In the absence of lactose, the repressor remains bound to the operator and RNA polymerase cannot gain access to the promoter. Transcription is therefore blocked and the operon remains repressed. (b) In the presence of lactose, the repressor is converted to the inactive form that no longer binds the operator. RNA polymerase is therefore able to bind to the promoter and transcribe the structural genes *z*, *y*, and *a* into a single polycistronic message.

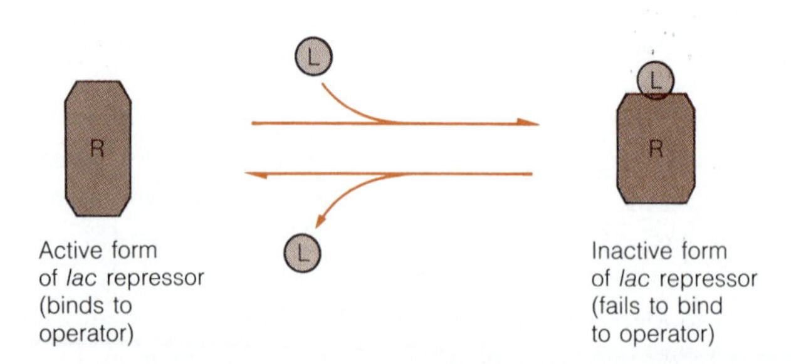

Figure 17-5 Allosteric Regulation of the Lactose Repressor. The lactose repressor (R) is an allosteric protein, capable of reversible conversion between two alternative forms. In the absence of the effector allolactose (L), the protein is stabilized in the form that is active in binding to the *lac* operator. In the presence of the effector, the protein exists preferentially in the alternative conformational state, which does not recognize the operator and is therefore inactive as a repressor of transcription.

Genetic Analysis of the Lactose Operon

Most of the initial evidence in support of the operon model was based on genetic analyses of mutant cells that either produced abnormal amounts of the enzymes involved in lactose catabolism or showed abnormal responses to the addition or removal of lactose. These mutations were found to map either to the structural genes (z, y, or a) or to the regulatory elements of the system (o, p, or i). These two classes of mutations can be readily distinguished, because mutations in a structural gene affect only a single protein, whereas mutations in regulatory regions almost always affect expression of all the structural genes coordinately.

Table 17-1 summarizes these mutations and their phenotypes, including the inducible phenotype of the wild-type cell (line 1). We will consider each of the mutations in turn in the sections below. Experiments such as those depicted in Table 17-1 are usually carried out using the synthetic β-galactoside *isopropylthiogalactoside* (IPTG) rather than lactose as the inducer. This is done because IPTG is a good inducer of the system but cannot itself be metabolized by the cell, so its use avoids possible complications due to changes in the level of inducer caused by its catabolism. Considering the types of mutations described in Table 17-1 in turn should help us understand the operon model and should also illustrate the power of genetic analysis. It is a tribute to Jacob and Monod that they were able to formulate a model for gene regulation based entirely on genetic data and that the operon model as we understand it today is essentially that which they laid out in their landmark paper of 1961.

Mutations in the Structural Genes. Mutations in the y or z genes of the *lac* operon can result in production of altered gene products with little or no biological activity, even in the presence of inducer (Table 17-1, lines 2 and 3). Such mutants are therefore unable to utilize lactose as a carbon source, either because they cannot transport lactose into the cell (y^-) or because they cannot cleave the glycosidic bond between galactose and glucose in the lactose compound (z^-).

Mutations in the Operator. Mutations in the operator lead to a *constitutive* phenotype; that is, the mutant cells produce the *lac* enzymes whether inducer is present or not (Table 17-1, line 4). These mutants result from changes in the nucleotide sequence of the operator DNA that reduce the strength of the binding between the operator and the repressor. Such **operator-constitutive mutants** are represented as o^c. As is expected for mutations in a regulatory site, o^c mutations simultaneously affect the synthesis of all three structural genes in the same way.

Mutations in the Promoter. The promoter was not a part of the original operon model, but it is now known to be the binding site for RNA polymerase and therefore an essential feature of the operon. Promoter mutations decrease the affinity of the polymerase enzyme for the promoter, which can be measured experimentally as a change in the binding constant for the interaction between the promoter and RNA polymerase. This, in turn, reduces the number of RNA polymerase molecules that bind per unit time and hence reduces the rate of mRNA synthesis. Promoter mutations (p^-) therefore decrease both the elevated level of enzyme produced in the presence of inducer *and* the already low basal level of *lac* enzyme production that the cell manages to achieve in the absence of inducer (Table 17-1, line 5).

Mutations in the Regulator Gene. Mutations in the i gene are of two types. Some mutants fail to produce any of the *lac* enzymes, regardless of whether inducer is present, and are therefore called **superrepressor mutants** (i^s in Table 17-1, line 6). The repressor molecule in such mutants either has lost its ability to recognize and bind the inducer but can still recognize the operator or has a high affinity for

Table 17-1 Genetic Analysis of Mutations in the *lac* Operon

Line Number	Genotype of Bacterium	− Inducer		+ Inducer	
		β-Galactosidase	Permease	β-Galactosidase	Permease
1	$i^+p^+o^+z^+y^+$	−	−	+	+
2	$i^+p^+o^+z^+y^-$	−	−	+	−
3	$i^+p^+o^+z^-y^+$	−	−	−	+
4	$i^+p^+o^cz^+y^+$	+	+	+	+
5	$i^+p^-o^+z^+y^+$	↓	↓	↓	↓
6	$i^sp^+o^+z^+y^+$	−	−	−	−
7	$i^-p^+o^+z^+y^+$	+	+	+	+

the operator regardless of whether inducer is bound to it. In either case, the repressor binds tightly to the operator and represses enzyme synthesis under all conditions.

The other class of *i* gene mutations is characterized by a mutant repressor that is not able to recognize the operator (or, in some cases, is not synthesized at all). The *lac* operon in such i^- mutants cannot be turned off, and the enzymes are therefore synthesized constitutively (Table 17-1, line 7).

These i^- mutants, along with the o^c and p^- mutants, illustrate the importance of specific recognition of DNA sequences by regulatory proteins during regulation of gene transcription. A small change caused by a mutation either in the DNA sequence or in the protein that binds to it can dramatically affect gene expression.

The Cis-Trans Test Using Partially Diploid Bacteria

The existence of two different kinds of constitutive mutants, o^c and i^-, raises the question of how one type might be distinguished from the other. The **cis-trans test** is used to distinguish between **trans-acting mutations,** which affect proteins (for example, i^-), and **cis-acting mutations,** which affect DNA binding sites (for example, o^c).

The basis of the test is construction of a partially diploid bacterial cell that contains two genetically different copies of the *lac* operon. Jacob and Monod accomplished this by placing a second copy of only the *lac* portion of the bacterial genome onto the F factor of male (F^+) bacteria. This second copy could be transferred by conjugation into a host bacterium of any desired *lac* genotype to create partial diploids such as the types listed in Table 17-2. If only one copy of the operon contains an i^- or o^c regulatory mutation, one can determine whether the mutation acts in *trans* (both copies of the structural genes of the operon are affected) or in *cis* (only the structural genes physically linked to the mutant locus are affected). In order to determine which of the two copies of the *lac* operon is being expressed in each combination, one copy contains a defective *z* gene while the other contains a defective *y* gene.

You can see that in the examples in Table 17-2, a diploid that has both an i^+ allele and an i^- allele in the same cell is inducible for both the β-galactosidase and permease, even though the functional gene for one enzyme is physically linked to a defective *i* gene (Table 17-2, line 3). We now know that this occurs because the one functional *i* gene present in the cell produces active repressor molecules that can diffuse through the cytoplasm and bind to *both* operator sites in the absence of lactose.

Quite different results are obtained with partial diploids containing both the o^+ and o^c alleles (Table 17-2, line 4). In this case, structural genes linked to the o^c allele are constitutively transcribed, whereas those linked to the wild-type allele are inducible. The o^c locus, in other words, acts only in *cis;* it affects the behavior of structural genes only in the operon of which it is physically a part. Such *cis* specificity is characteristic of mutations that affect DNA binding sites rather than protein products.

The Operon Model as an Example of Negative Transcriptional Control

From the results of these and similar genetic analyses of partially diploid strains, Jacob, Monod, and their colleagues formulated the operon model essentially as we know it today. Specifically, they postulated that the operator locus (and later the promoter site as well) serves only as a recognition site, whereas the regulator gene is responsible for production of a repressor substance that was eventually shown to be a protein.

Although much of the initial work that led to the operon concept was done with the *lac* system of *E. coli*, other regulatory systems in this and other species of bac-

Table 17-2 Diploid Analysis of Mutations in the *lac* Operon

Line Number	Genotype of Diploid Bacterium	− Inducer		+ Inducer	
		β-Galactosidase	Permease	β-Galactosidase	Permease
1	$i^+p^+o^+z^+y^+/i^+p^+o^+z^+y^+$	−	−	+	+
2	$i^+p^+o^+z^-y^+/i^+p^+o^+z^+y^-$	−	−	+	+
3	$i^+p^+o^+z^-y^+/i^-p^+o^+z^+y^-$	−	−	+	+
4	$i^+p^+o^+z^-y^+/i^+p^+o^c z^+y^-$	+	−	+	+
5	$i^+p^+o^+z^-y^+/i^s p^+o^+z^+y^-$	−	−	−	−

teria have since been shown to follow the same general pattern. This almost always means that the genes coding for enzymes of a particular metabolic pathway are clustered into a region of the bacterial genophore that serves as a unit of both transcription and regulation. One or more operators, promoters, and regulator genes are usually involved, although there is sufficient variation in theme from one operon to another to preclude too many generalizations.

The *lac* operon, like many others, regulates the appearance of enzymes involved in a specific degradative pathway. Recall that such operons are inducible; they are turned on by a specific allosteric effector, usually the substrate for that pathway. Now we can understand why: Such operons are regulated by a repressor protein that can bind to a specific DNA sequence in the absence of the effector and is therefore active in its free form. The repressor becomes inactive due to loss of the ability to bind to the DNA sequence when repressor is complexed with the effector (Figures 17-4 and 17-5).

In contrast, operons that regulate enzymes involved in synthetic pathways are repressible; they are turned off allosterically, usually by the end product of the pathway. The tryptophan (*trp*) operon shown in Figure 17-6 is a good example. This operon contains the structural genes for the enzymes that catalyze the reactions in the pathway for tryptophan synthesis, as well as the control elements necessary to regulate the synthesis of these enzymes.

Expression of the enzymes of the *trp* operon is repressed in the presence of tryptophan (Figure 17-6a) and derepressed in its absence (Figure 17-6b). In this case, the regulator gene for this operon (*trpR*) codes for a repressor protein that is active when complexed with the effector and inactive in its free form when effector interaction does not occur. The effector in such systems (in this case the amino acid tryptophan) is sometimes called a **corepressor** because it is required along with the repressor protein to shut off transcription from the operon. In other words, repressor proteins control the transcription of genes for enzymes of both catabolic and anabolic pathways, but the active (DNA-binding) form of the repressor is the effector-free molecule in a catabolic pathway and the repressor-effector complex in an anabolic pathway.

Regardless of which is the active form of the repressor, the effect of repressor action is always the same: It prevents transcription of the operon under appropriate conditions by blocking the movement of RNA polymerase along the DNA. Repressors, in other words, never specifically turn anything on; their effect is always to turn off gene expression (or to keep it turned off). The operon model is therefore a system of **negative control,** because the repressor protein, its key regulatory element, always acts by turning off expression of the operon.

Catabolite Repression: Positive Control of Transcription

The *lac* and *trp* operons are good examples of negative control mechanisms in bacteria, but some operons are also under **positive control.** Such operons usually code for catabolic enzymes and respond to a positive regulatory protein that is in turn controlled by the level of glucose in the cell. To understand this sensitivity to glucose, we need to recognize that glucose is the preferred energy source for almost all prokaryotic cells (and for most eukaryotic cells too). The enzymes of the glycolytic and tricarboxylic acid (TCA) pathways are present constitutively in most cells, so glucose can be catabolized at any time without the synthesis of additional enzymes. Although other molecules can also be used for energy, their catabolism always requires synthesis of one or more specific enzymes. Consequently, when presented with a choice of carbon sources, cells usually show a selective preference for glucose. For example, *E. coli* cells grown in the presence of both glucose and lactose use the glucose preferentially and have very low levels of the enzymes encoded by the *lac* operon, despite the presence of the inducer for that operon.

This effect of glucose on inducible catabolic enzymes is called **catabolite repression.** It guarantees that other carbon sources will be used only when glucose is not available. Like the other regulatory systems we have encountered, catabolite repression depends on an allosteric regulatory protein and a small organic effector. In this case, however, the effector is not glucose itself, but a secondary signal that reflects the level of glucose in the cell. This secondary signal is a modified form of AMP called **3′,5′-cyclic AMP** or **cAMP,** which we will describe in more detail in Chapter 21, and will encounter again in Chaprer 22. To understand how catabolite repression works, we must look at the roles of these regulatory elements.

Cyclic AMP as a Mediator of Catabolite Repression. To understand the glucose effect, we need to recognize that cAMP is synthesized from ATP by the enzyme *adenylate cyclase* and degraded by *phosphodiesterase* action (Figure 17-7). The intracellular concentration of cAMP is therefore a balance of these two activities. The level of cAMP is high in the absence of glucose and low in its presence. Whether this is because of an inhibitory effect of glucose on adenylate cyclase or a stimulatory effect on phosphodiesterase is not clear.

The Role of Catabolite Activator Protein. Despite its name, catabolite repression is actually a positive mode of control at the molecular level. The cellular level of cAMP is translated into an effect on gene transcription via interaction of cAMP with an allosteric regulatory protein. The regulatory protein involved is called the **catabolite activator protein (CAP)**. Like repressor proteins, CAP is an allosteric protein, as Figure 17-8 indicates. By itself, CAP is inactive, but when complexed with cAMP the protein conformation changes, and it binds to a CAP recognition site (c) adjacent to the promoter for a variety of inducible operons. The DNA sequence in the CAP recognition site differs from the repressor recognition site, and the CAP site is located in

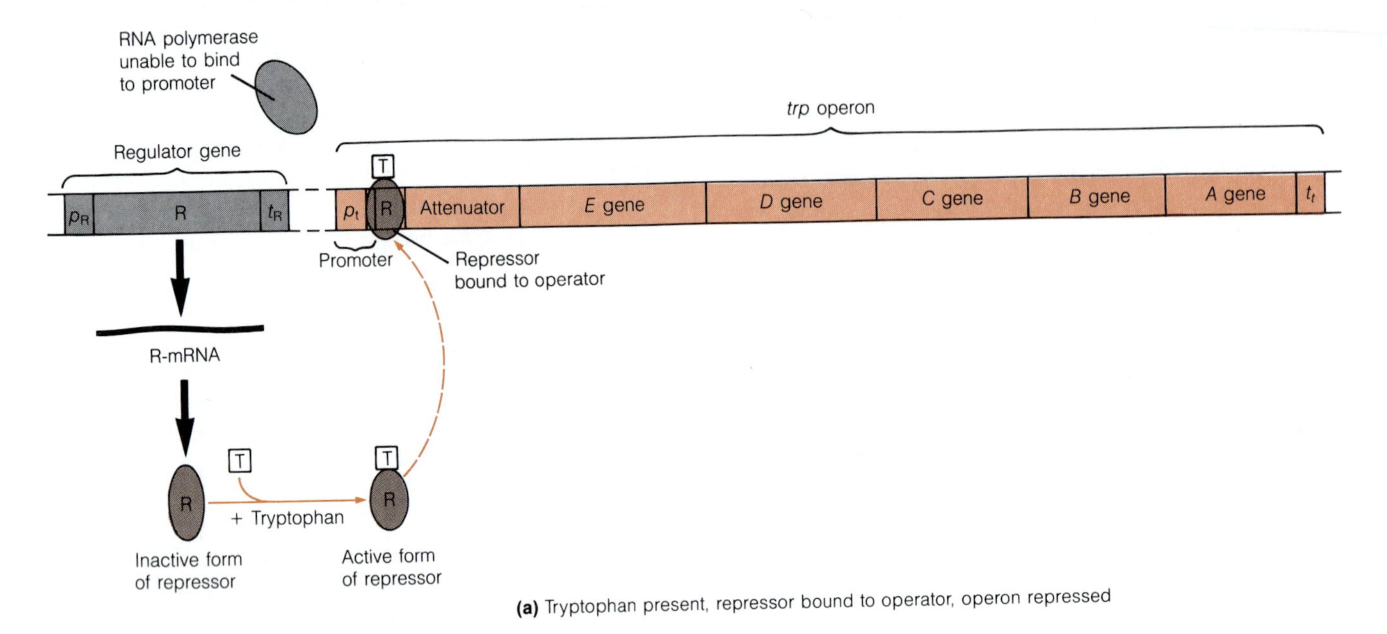

(a) Tryptophan present, repressor bound to operator, operon repressed

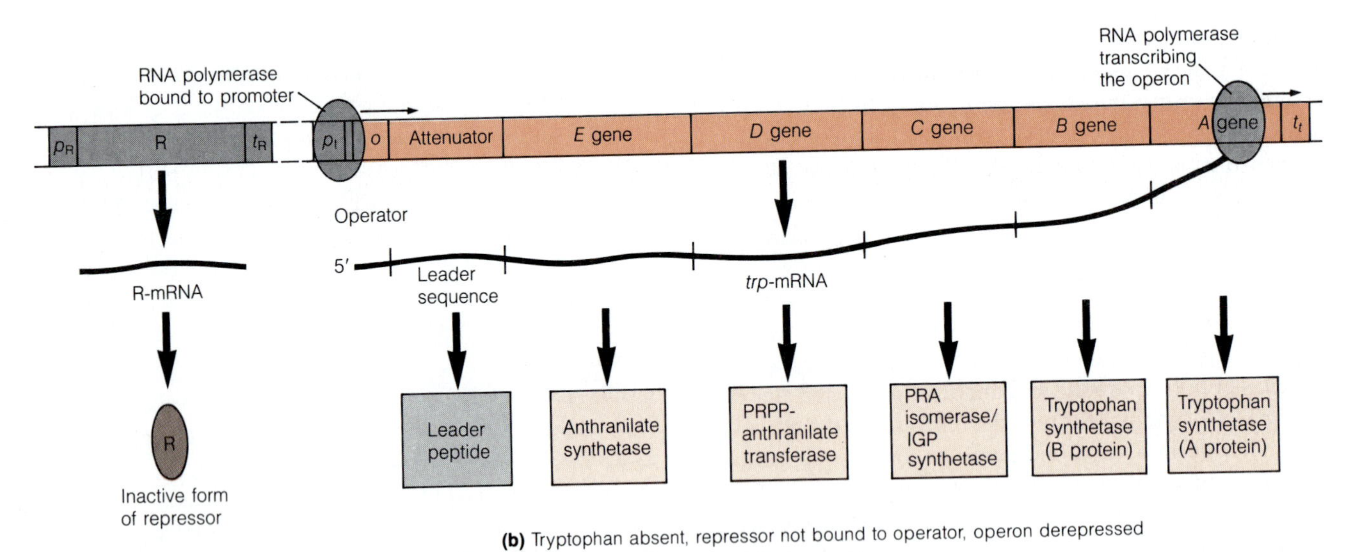

(b) Tryptophan absent, repressor not bound to operator, operon derepressed

Figure 17-6 The Tryptophan (*trp*) Operon of *E. coli*. (a) The *trp* operon consists of a segment of double-stranded DNA that includes five contiguous structural genes (*A*, *B*, *C*, *D*, and *E*) as well as promoter (p_t), terminator (t_t), and attenuator sites. The structural genes are transcribed and regulated as a unit. The resulting polycistronic message codes for the enzymes of the tryptophan pathway (see Figure 17-1). The repressor protein encoded by the *R* gene is inactive (cannot recognize the operator site) in the free form but is active when complexed with tryptophan. In the presence of tryptophan, the repressor is converted to the active form and binds tightly to the operator, thereby blocking access of RNA polymerase to the promoter and keeping the operon repressed. (b) In the absence of tryptophan, the repressor remains in the free form and does not bind to the operator site. RNA polymerase is therefore able to bind to the promoter and transcribe the structural genes, giving the cell the capability to synthesize tryptophan.

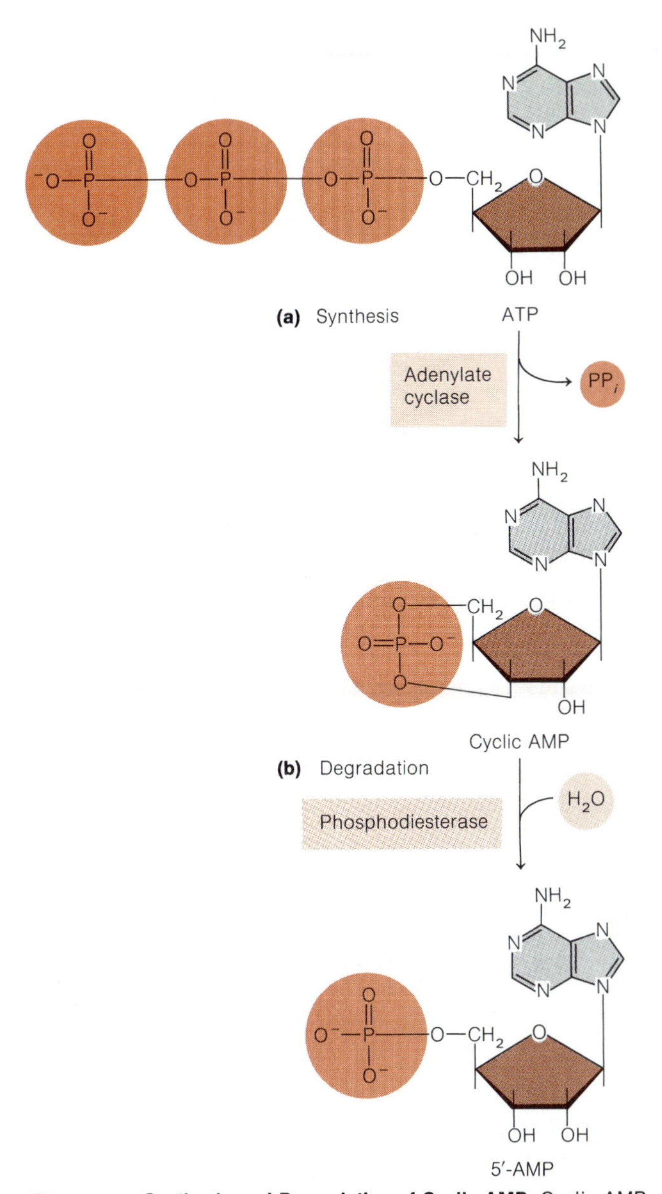

Figure 17-7 Synthesis and Degradation of Cyclic AMP. Cyclic AMP is (a) synthesized from ATP by the action of adenylate cyclase and (b) degraded to 5′-AMP by phosphodiesterase. Presumably, glucose acts to reduce the cAMP level in the cell either by inhibiting adenylate cyclase or by stimulating phosphodiesterase, or both.

front of the RNA polymerase binding site rather than between the polymerase binding site and the structural gene. Unlike a repressor, CAP has a *positive* effect on gene regulation. When CAP is bound to its recognition site, the binding of RNA polymerase to the promoter is greatly enhanced, thereby *stimulating* initiation of transcription. For the *lac* operon, transcription is increased 50-fold in this way.

As the level of glucose in the cell rises, the concentration of cAMP falls, CAP is converted primarily to its inactive form, and its stimulatory effect on the transcription of inducible operons is abolished. Conversely, when the

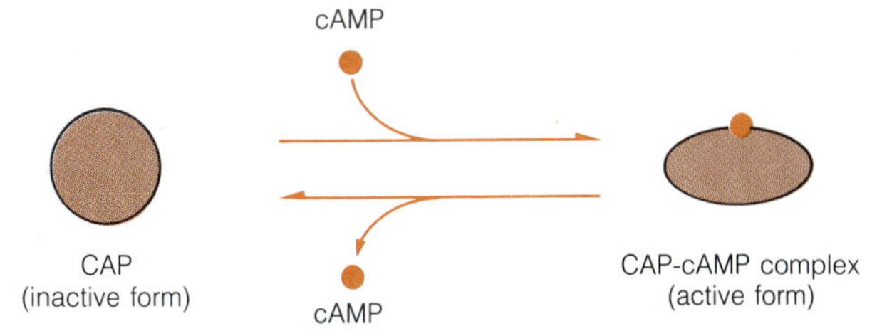

Figure 17-8 Allosteric Regulation of the Catabolite Activator Protein by Cyclic AMP. Catabolite activator protein (CAP) is an allosteric protein that is inactive in the free form but is converted to the active form by binding to cyclic AMP (cAMP). The CAP-cAMP complex binds to the promoter of a variety of inducible operons, increasing the affinity of the promoter for RNA polymerase and thereby stimulating transcription.

glucose level falls, the cAMP level rises, and CAP is activated by formation of the CAP-cAMP complex. This greatly enhances transcription of an inducible operon such as the *lac* operon, *provided*, of course, that the repressor for that operon has been inactivated by the presence of its effector. This dual mode of regulation is illustrated in Figure 17-9.

Dual Control of Inducible Operons. As we have seen, inducible operons are often under two kinds of control, rendering such operons sensitive to two kinds of signals. The repressor-operator interaction provides an "all-or-nothing" sensitivity to the presence of an alternative energy source (such as lactose) that turns on a particular operon. The effect of CAP, on the other hand, is to render the level of transcription of that operon sensitive to the glucose concentration of the cell, mediated by the cAMP level. An *E. coli* cell might therefore have very low levels of β-galactosidase either because it is growing in the absence of lactose and its *lac* operon is fully repressed or because it has access to both lactose and glucose. In the latter case, the high glucose concentration will suppress the cAMP level, and CAP will therefore be inactive and incapable of stimulating transcription of the *lac* operon even though the operon is otherwise derepressed.

It is an unfortunate feature of genetic nomenclature that the repressor-operator system of negative control has so positive-sounding a name as "induction," whereas the CAP-cAMP system of positive control has so negative-sounding a name as catabolite "repression." In each case, the name reflects the observed effect of a particular environmental signal: Lactose induces the appearance of specific enzymes, and glucose represses the same enzymes. At the molecular level, however, lactose is really just alleviating the negative effect of a repressor protein, and glucose is suppressing the positive effect of an activator protein.

To keep negative and positive controls clear in your mind, always ask about the primary effect of the DNA-binding protein. If, in binding to the DNA, the regulatory

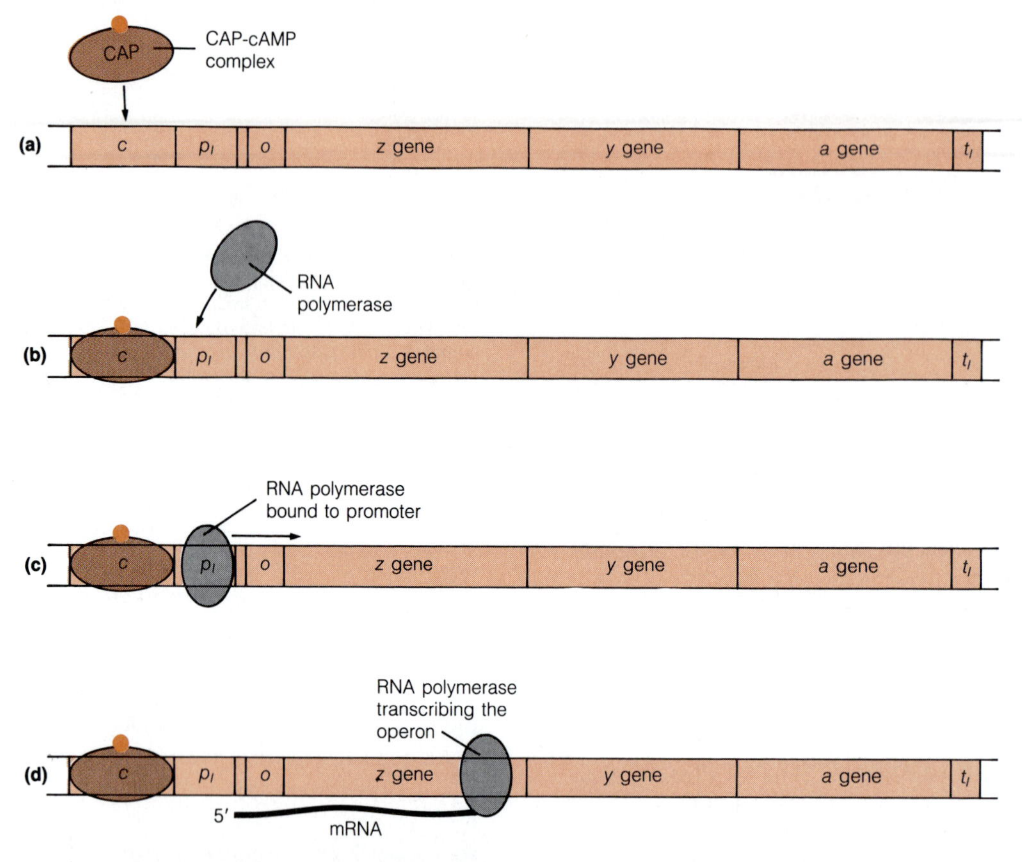

Figure 17-9 Mode of Action of Catabolite Activator Protein. CAP is inactive until cAMP binds to it (see Figure 17-8). (a) The CAP-cAMP complex binds to the CAP (*c*) site at or near the promoter region, thereby (b) modifying the adjacent RNA polymerase binding site, (c) allowing RNA polymerase to bind to the promoter, and so stimulating transcription of the operon (d).

protein prevents or turns off transcription, then it is part of a negative control mechanism. If, on the other hand, its binding to DNA results in the activation or enhancement of transcription, then the regulatory protein is part of a positive control mechanism.

Alternative Sigma Factors Modify Utilization of Promoters by RNA Polymerase

In addition to regulation by proteins that bind to DNA near the promoter and act as activators or repressors of transcription, specificity of promoter recognition can be modified as a result of the binding of the sigma (σ) factor to RNA polymerase core enzyme (see Chapter 16 for a discussion of sigma factors). Each bacterial cell contains several different types of sigma factors that can be utilized in special situations. The major form of prokaryotic RNA polymerase holoenzyme contains the σ^{70} subunit (molecular weight 70 kilodaltons) and initiates transcription at the prokaryotic consensus promoter site, present at the 5′ end of most bacterial operons. However, alterations in cellular environment, such as elevated temperature (heat shock), can favor binding of an alternative sigma factor (σ^{32}) to RNA polymerase. With this sigma factor bound, promoter recognition is slightly altered to favor a shift in initiation of transcription to a special set of genes encoding proteins that help the cell adapt to the altered environment. Yet another sigma factor (σ^{54}) enables RNA polymerase to recognize promoters that regulate genes involved in nitrogen utilization (such as the gene for glutamine synthetase). In both of these cases, the promoter site DNA sequence is slightly different from the consensus sequence, and the interaction with an alternate sigma factor results in an adjustment in the DNA sequence binding specificity of RNA polymerase.

Attenuation: Regulation After Transcription Initiation

Prokaryotes also employ regulatory mechanisms that operate following transcription initiation, such as the mechanism called **attenuation**. This phenomenon was discovered when Charles Yanofsky and his colleagues made the unexpected observation that some mutants in the *trp* operon containing deletions *within* a specific region of the polycistronic *trp* mRNA actually show increased synthesis of the *trp* mRNA. They then examined the mRNA carefully and found it to contain a string of 162 nucleotides at its 5′ end before the initiation codon for the *trpE* gene is encountered. This transcribed region between the operator and the first structural gene is called the **leader sequence** (Figure 17-6b). All the mutations that resulted in increased levels of *trp* mRNA mapped about 30 to 60 nucleotides in advance of *trpE* in the region coding for the leader sequence.

Yanofsky and his collaborators then showed that in addition to the expected production of full-length *trp* mRNA under conditions of tryptophan scarcity, a short (130-nucleotide) transcript is produced when tryptophan levels are high. This transcript was found to be the 5′ end of the leader sequence. From these findings, Yanofsky concluded that the tryptophan operon must contain an additional control element located between the operator and the first structural gene that is sensitive to tryptophan levels and somehow determines not whether but how much of the operon will be transcribed. He called this the **attenuator site** because of its apparent role in reducing the synthesis of mRNA.

Since the initial work with the *trp* operon, attenuators have also been found in several other operons that code for enzymes involved in amino acid biosynthesis, including the phenylalanine and histidine operons. In some cases, attenuation appears to be the only means of regulation for the operon. In other cases, the attenuator site complements the operator in the regulation of gene expression. When the cellular level of the relevant amino acid is high, the repressor-effector complex binds to the operator, effectively blocking initiation of transcription. As the level of effector decreases, the operon becomes derepressed and transcription begins. Fine-tuning of the system now occurs at the attenuator site, allowing greater numbers of RNA polymerase molecules to proceed past the attenuator site as the effector becomes scarcer.

Attenuation Depends on the Coupling of Transcription and Translation in Prokaryotes. Yanofsky and his colleagues noted several interesting facts about the leader sequence. The leader sequence begins with an initiation codon AUG and apparently codes for a **leader peptide** of 14 amino acids. Within this leader peptide are two important adjacent codons for the amino acid tryptophan. This region of the mRNA also contains a string of nucleotides capable of forming a distinctive secondary structure that resembles a bacterial transcription termination signal, as in Figure 17-10. Apparently, when tryptophan

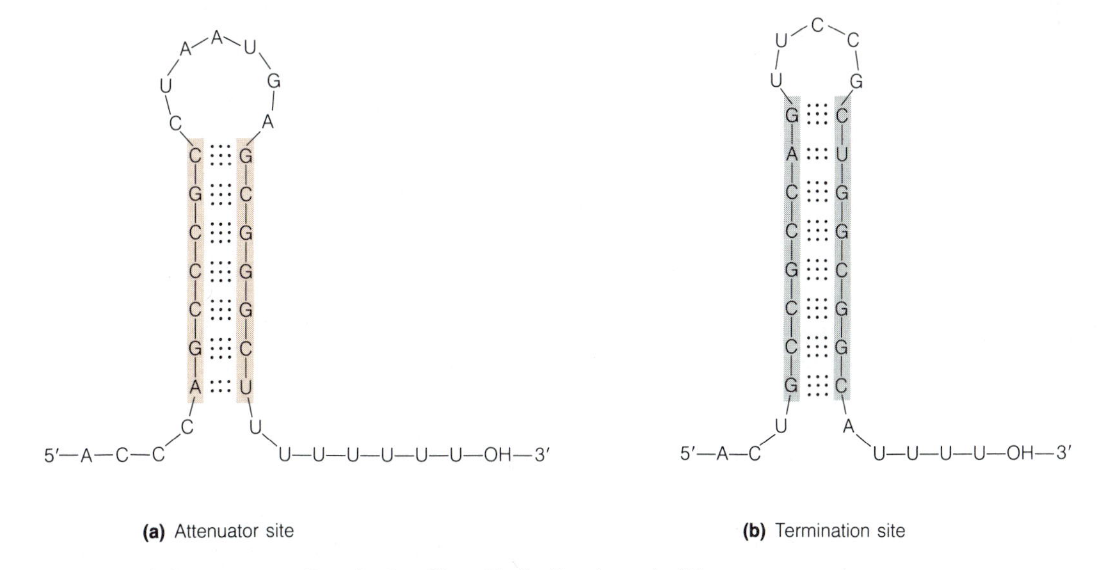

Figure 17-10 Attenuators as Termination Sites. Similarities in nucleotide sequence and possible secondary structure are seen when (a) the attenuator portion of the leader sequence at the 5′ end of the mRNA of the tryptophan operon is compared with (b) the termination site at the 3′ end of the same mRNA.

levels are low, leading to a low concentration of tryptophanyl tRNA, the bacterial ribosomes "stall" at the tryptophan codons. The presence of the stalled ribosome destabilizes the potential terminator structure and allows the polymerase to continue producing transcripts of the *trp* operon. This leads to increased production of tryptophan. On the other hand, if tryptophanyl tRNA levels are high, the ribosome does not stall and the terminator structure can form, causing the polymerase to produce only the short leader transcript rather than transcribing the entire operon. Although most of the work on attenuation has been done with the *trp* operon, the same mechanism seems to apply to other operons where attenuation has been observed, such as phenylalanine and histidine.

The proposed mechanism for attenuation clearly presumes a close coupling between transcription and translation, with a ribosome following closely behind the RNA polymerase molecule, translating the nascent mRNA. Such coupling is indeed the case for prokaryotic cells, but it cannot occur in eukaryotes, because transcription in eukaryotic cells occurs in the nucleus whereas translation is a cytoplasmic process. Apparently, then, attenuation is an exclusively prokaryotic process, at least as presently understood.

Gene Regulation in Eukaryotes

An adage that became popular during the early days of molecular biology insisted that "what is true of *E. coli* is also true of elephants." This maxim aptly expressed the initial conviction of many microbial geneticists that virtually everything learned from studies of bacteria would also be applicable to eukaryotes. Sometimes, the conviction seemed even stronger—that virtually everything we needed to know about eukaryotes could be learned by studying bacteria.

However, the adage is only partly true. In terms of basic metabolic pathways, mechanisms for transport of solutes across membranes, and such fundamental features as DNA structure, protein synthesis, and enzyme function, there are many similarities between the prokaryotic and eukaryotic worlds, and findings from studies with prokaryotes can be extrapolated to eukaryotes with considerable confidence. But when the discussion turns to the regulation of gene expression, the comparison needs to be scrutinized. There are certainly many similarities between prokaryotes and eukaryotes in gene expression and its control, but there also appear to be significant differences. Especially significant is the multiplicity of levels at which gene expression is controlled in eukaryotes. In bacteria, much (though by no means all) regulation is effected at the level of transcription, with allosteric proteins as the regulatory elements. Transcriptional regulation is clearly important in eukaryotes also, but eukaryotes appear to depend on a variety of posttranscriptional controls as well.

The greater diversity of regulatory mechanisms in eukaryotic cells probably reflects basic differences between prokaryotes and eukaryotes in the structure, organization, and compartmentalization of the genome, as well as some fundamental features of the multicellular way of life. We will first look at some of these differences and then consider possibilities for regulation of gene expression at a variety of points in the overall flow of genetic information from a gene in the nucleus to an active enzyme (or other gene product) in the cytoplasm or elsewhere in the cell.

Prokaryotes and Eukaryotes Differ in Important Ways

In previous chapters, we encountered many fundamental distinctions between the genetic systems of prokaryotes and eukaryotes, but it might be helpful to review the most important ones in a context that focuses explicitly on the regulation of gene expression.

Genome Size and Complexity. As we already know, eukaryotic genomes are almost always much larger than those of prokaryotes—often by two or three orders of magnitude or more. For most eukaryotes, the actual coding sequences seem to constitute only a small portion of the genome, leaving the genetic role of the bulk of the DNA, including most of the repeated sequences, unexplained. It has frequently been suggested that much of this "extra" DNA has a regulatory function, but that is still largely speculative. In many eukaryotic genomes, repeated sequences have been shown to be interspersed with unique sequences, suggesting a role for the former in regulating the latter.

The unexpected finding of intervening sequences in eukaryotic DNA and their presence in most genes studied to date suggest further genomic complexities that may be exploited by the cell for regulatory purposes. Especially intriguing is the possibility that specific segments of genetic information may be spliced together in different combinations to generate related proteins with different properties. We will consider a specific example of this in Chapter 24, when we look at the means by which the immune system of our body generates a diversity of antibodies against foreign proteins.

Genomic Compartmentalization. One of the fundamental differences between prokaryotes and eukaryotes is that most of the genetic information of eukaryotes is present in the nucleus, segregated from the site of synthesis of most proteins in the cytoplasm. Transcription and translation are therefore separated in both time and space in eukaryotic cells by the presence of the nuclear envelope.

This separation has profound implications for gene regulation, especially by allowing the possibility for selectivity in the processing and transport of nuclear transcripts. It is clear, for example, that primary transcripts are extensively modified, cleaved, and spliced in eukaryotic nuclei and that most nuclear RNA never actually appears in the cytoplasm as a functional message. The nuclear envelope may therefore serve as a means of "screening" transcripts for selective passage to the protein-synthesizing system of the cytoplasm. On the other hand, the separation of transcription and translation makes control by attenuation impossible in eukaryotes, since the presumed mechanism requires a close physical proximity of RNA polymerase molecules and ribosomes.

Structural Organization of the Genome. The chromosomes of a eukaryotic cell and the genophore of a prokaryote differ strikingly, both in chemistry and in structure. Indeed, recent work indicates that bacterial DNA is complexed with small basic proteins that may interact with DNA in much the same way as histones do in eukaryotic chromatin. However the packaging is not based on the structural unit observed in eukaryotic chromosomes, and the intermediate levels of compaction are not present. Because the DNA of eukaryotic chromosomes is intimately associated with histone and nonhistone proteins and is highly folded into several different levels of structural organization, mechanisms that rely on the binding of allosteric proteins to regulatory sites on the DNA may not be the only means of regulating transcription. It seems likely that the desired region of the chromosome must first be uncoiled before the DNA becomes accessible for transcription.

It is also remarkable that genes for related proteins tend not to be physically clustered into units of transcription or regulation in eukaryotes, a fact that argues against eukaryotic operons, at least in the sense of coordinated regulation of contiguous structural genes. Current data suggest that eukaryotes employ different strategies for coordinating gene transcription. Specifically, they have similar regulatory elements near physically unlinked genes that must be transcribed at the same time or in response to the same signals.

Stability of Messenger RNA. Messenger RNA molecules in prokaryotes and eukaryotes differ in their average stability, or lifetime in the cell before degradation. Although the values vary with specific mRNA molecules and with species and tissues, the mRNA molecules of eukaryotic cells have in general a much greater longevity than their prokaryotic counterparts. This can probably best be understood in light of what we might call the "life-styles" of these organisms.

Prokaryotes are unicellular organisms with little or no assurance of environmental constancy and an ongoing requirement for rapid adaptation to new conditions. Consistent with this need is a short life span for most mRNA molecules and rapid transcriptional regulation via effector molecules and allosteric regulatory proteins. Not surprisingly, the effector molecules are almost always sugars, amino acids, and other small molecules—the very environmental factors to which the cells must be responsive. This environmental responsiveness provides the ability to rapidly change the population of cellular mRNAs, and therefore the spectrum of proteins that are translated, in response to external conditions.

Eukaryotic cells, on the other hand, are often (though by no means always) part of a multicellular organism and can usually depend on a much more predictable internal environment. Moreover, the many different cell types of multicellular eukaryotes are organized in tissues and organs that are usually quite highly differentiated and specialized in function, reflecting the division of labor that is so characteristic of the multicellular way of life. Most eukaryotic cell efforts are therefore committed to the synthesis of a specific collection of gene products, usually with little likelihood of dramatic changes in that commitment. This commitment to the synthesis of a predictable set of proteins allows such cells to use longer-lived mRNA molecules than is typical for bacterial cells. The presence of stable mRNA suggests, in turn, a greater dependence on posttranscriptional regulatory mechanisms that affect the processing, persistence, and utilization of an mRNA after it is transcribed. While these generalizations are true for many eukaryotic mRNAs and cell types, there are also numerous examples (some of which we will discuss later in this chapter) of significant variations in the stabilities of different mRNAs within the same cell type or as a result of different environmental or developmental conditions. Single-celled eukaryotes, such as yeast, may not fit these generalizations because, like bacteria, they are sensitive to environmental conditions and in general have mRNAs that are relatively unstable.

Protein Turnover. Another difference that arises from the contrasting life-styles of most prokaryotes and eukaryotes concerns the relative importance of enzymatic hydrolysis as a means of eliminating proteins that are no longer needed or wanted by the cell. Prokaryotes have proteolytic enzymes and can degrade defective or unwanted proteins, but this is not their only option. Because most prokaryotes continue to grow and divide, they can get rid of unwanted proteins simply by ceasing to synthesize them and allowing them to be diluted out by successive cell divisions. As a result, prokaryotic cells are in general not dependent on proteolytic enzymes as their only means of eliminating proteins.

Eukaryotic cells, on the other hand, are much more likely to cease dividing and to persist as discrete metabolic entities for relatively long periods of time thereafter. The nerve cells in your body are an especially good example. They are as old as you are and will never divide again. Clearly, such cells cannot depend on dilution by division to get rid of unwanted or defective proteins. Instead, they have specific proteolytic enzymes to degrade proteins selectively. The **turnover,** or regulated synthesis and degradation, of proteins is therefore a much more prominent feature of eukaryotic cells than it is of most prokaryotes. This, in turn, suggests yet another level of regulation at which eukaryotes may differ from prokaryotes.

Multiple Levels of Control

All the differences described in the preceding paragraphs underscore the difficulty of explaining eukaryotic regulation of gene expression in terms of known prokaryotic mechanisms. Some of the basic principles of prokaryotic gene regulation are probably applicable to eukaryotic systems, but additional levels and mechanisms of eukaryotic control are almost certainly involved as well. What is true of *E. coli* may be true—to a degree—of elephants, but with respect to regulation of gene expression, we may need to be quite careful in our extrapolation. If we are ever to understand regulation in eukaryotic cells, we must approach the topic with a eukaryotic perspective.

An important part of that perspective is that gene expression, ultimately measured in terms of activity of gene products in eukaryotic cells, is probably controlled at several different levels. This is illustrated in Figure 17-11, which traces the flow of genetic information from the DNA of chromosomes to the functional proteins of the cell, indicating possibilities for regulation at each of several levels. There are, in theory, at least five different levels in the chain of events from DNA to functional proteins at which control can be exerted. These are usually referred to as *genomic control, transcriptional control, control of processing and translocation, translational control,* and *posttranslational control.* The latter three are sometimes considered together under the category *posttranscriptional control,* a term that therefore encompasses a wide variety of quite different phenomena. Experimental means of distinguishing among these levels of control are described in the box on pp. 532–534.

We will look at several of these levels of control, emphasizing as we do the multiplicity of regulatory mechanisms that eukaryotic cells seem to employ. Examples will be included as we go along. However, it is important to remember that much of the current excitement concerning the regulation of gene expression in eukaryotes derives not so much from what we already know about control mechanisms at the molecular level as from the rapid technical advances of the past few years and the quantum leaps in our understanding that are taking place right now.

Genomic Control

Regulatory mechanisms have been suggested for multicellular organisms at the level of the genome itself. Such **genomic control** could presumably involve either the selective loss or the selective augmentation of specific genetic information, and both possibilities have been suggested. Although examples of genomic loss or amplification exist, these mechanisms apparently do not constitute a generally applicable, major mechanism of genetic regulation in eukaryotes. One important exception, the production of active immunoglobulin genes during immune cell differentiation, will be discussed in Chapter 24.

The Totipotency of Differentiated Cells. Early in the history of animal embryology, August Weismann postulated that the differentiation and specialization of cells during embryonic development could be explained in terms of a differential and irreversible parceling out of genetic information into daughter cells. The fertilized egg, in other words, would contain all the genetic information of the organism, and that full complement would be maintained in the cells of the **germ line,** which give rise to the gonads. But in the **somatic** (normal body) **cells,** Weismann imagined a progressive segregation of genes, such that a muscle cell becomes a muscle cell because it loses the genetic competence to specify any functions other than those required of muscle cells. Apparent support for Weismann's hypothesis was seen in the phenomenon of **chromosome diminution,** in which the chromosomes of some less complex species of animals (such as the roundworm *Ascaris*) undergo a progressive decrease in size and number during development.

However, it has long since become clear that no such physical parceling out of genetic information occurs in most species of plants and animals. For animals, proof of this was provided by John Gurdon and colleagues using amphibians in which nuclei from differentiated cells have been physically transplanted into oocytes (unfertilized

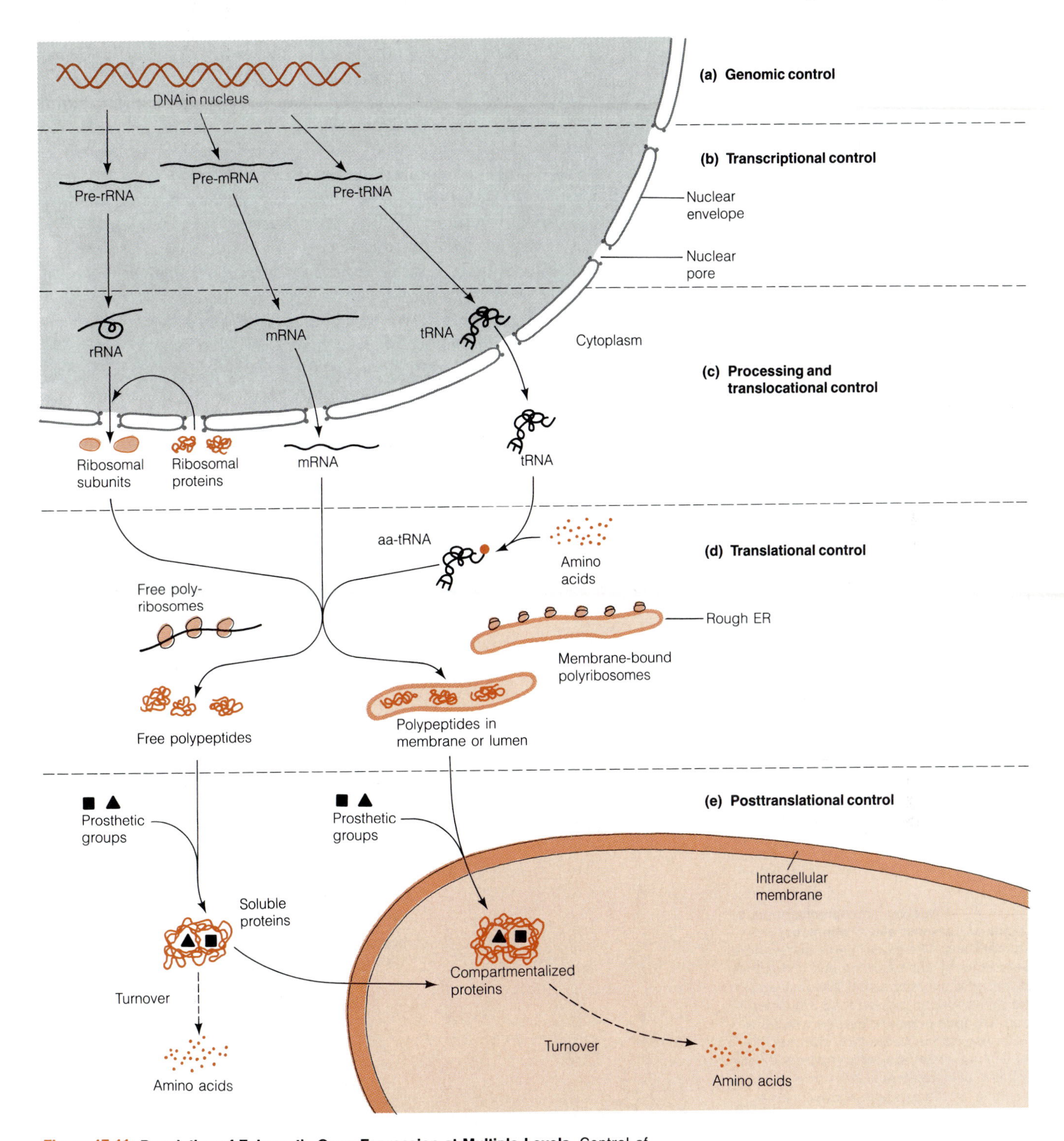

Figure 17-11 Regulation of Eukaryotic Gene Expression at Multiple Levels. Control of gene expression can be exerted at a number of levels between the genome and production of functional proteins. These include (a) genomic control (selective loss or amplification of genetic information prior to expression); (b) transcriptional control (selective transcription of genetic information into nuclear RNA); (c) control at the level of processing and translocation of nuclear transcripts (selective processing of transcripts into functional RNA and/or selective transport of nuclear RNA species to the cytoplasm); (d) translational control (selective translation of cytoplasmic mRNA molecules into polypeptide products); and (e) a variety of posttranslational controls (regulation of polypeptide assembly, biochemical protein modifications, compartmentalization of gene products, and protein turnover).

eggs) that have been deprived of their own nuclei (Figure 17-12). Although the frequency of success is low, perhaps reflecting the difficulty of reprogramming a differentiated nucleus, some eggs do give rise to viable, swimming tadpoles, indicating that nuclei derived from differentiated tissues are capable of directing the development of the whole organism. Such a nucleus is said to be **totipotent**: It has the full genetic potential expected of the species.

The totipotency of plant cells has been demonstrated even more directly in experiments pioneered by F. C. Steward at Cornell University. As Figure 17-13 shows, pieces of differentiated tissue can be obtained from a mature plant and dissociated into single cells, and individual cells can be used to regenerate an entire adult plant. Such regenerated plants are genetically identical to the original plant and are called **clones.** This means of plant propagation avoids the genetic variability inherent in sexual reproduction and is of great commercial interest because it ensures the maintenance of desirable traits.

Gene Amplification. Selective losses of genetic information do not seem to occur routinely during plant or animal development, but what about selective increases? It would, for example, be tempting to suggest that a cell characterized by the synthesis of a large amount of a single gene product might selectively replicate the appropriate DNA, a phenomenon called **gene amplification.** While several examples of gene amplification exist, it is not a routinely used mechanism for differential gene expression.

The best-studied example is amplification of ribosomal RNA genes in *Xenopus laevis,* the organism used by Gurdon in his nuclear transplantation experiments. The haploid genome of *Xenopus* normally contains about 500 copies of the genes that code for 5.8S, 18S and 28S ribosomal RNA. During **oogenesis** (development of the egg prior to fertilization), this whole set of genes is selectively amplified about 4000-fold, such that the mature oocyte contains about 2×10^6 copies of the genes for ribosomal RNA. Apparently, this level of amplification is necessary to ac-

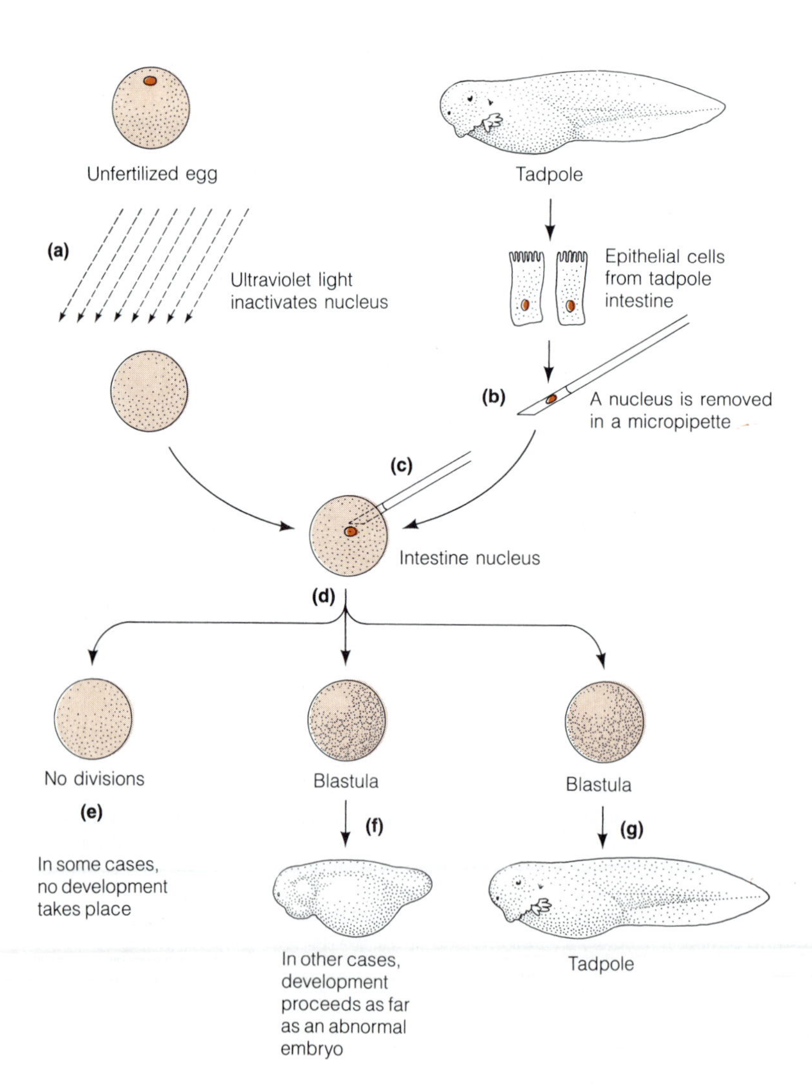

Figure 17-12 Nuclear Transplantation as a Means of Demonstrating Totipotency in Animal Cells. The experiments shown here were done by John Gurdon and colleagues at Oxford University, using *Xenopus laevis,* the South African clawed toad. (a) Unfertilized *Xenopus* eggs are treated with ultraviolet light to inactivate their own nuclei. (b) Diploid nuclei are removed from differentiated cells (epithelial cells from tadpole intestine, in the example shown) with a micropipette and are (c) injected into the enucleated eggs. (d) The eggs are then activated and allowed to develop. (e) Some of the eggs never divide at all. (f) Other eggs begin to develop, but give rise to abnormal embryos. (g) However, some eggs with transplanted nuclei develop into normal embryos or tadpoles. These results show that the nucleus of a differentiated cell possesses all the genetic information necessary to direct the development of an entire organism.

commodate the intense synthesis of ribosomes that occurs during oogenesis, which in turn is required to provide sufficient ribosomes to sustain the high rate of protein synthesis characteristic of early embryogenesis.

The several thousand extra copies of the ribosomal genes are present in amplified oocytes as extrachromosomal circles of DNA. These are located in the hundreds of nucleoli that appear in the nucleoplasm of the oocyte as amplification progresses. The prominence of these nucleoli in the oocyte nucleus is shown in the micrograph of Figure 17-14.

Another example of specific gene amplification involves the genes for the *chorion proteins* that form the hard coat of insect eggs. These proteins are synthesized and secreted by *follicle cells* that surround the eggs. In *Drosophila,* the genes for the chorion proteins are amplified about 30-fold just before the proteins are needed. In contrast to ribosomal gene amplification, this amplification is thought to involve successive lateral replications of a localized region of the chromosome. The extra copies of the chorion gene therefore remain a part of the chromosome.

Despite these interesting examples, however, most genes are not amplified during development. Even in differentiated cells devoted almost exclusively to the synthesis of one or a few proteins, there is but a single copy of the genes for those proteins per haploid genome. For example, the silk gland cells of the silk moth contain only one functional gene for silk fibroin per haploid genome. The same is true for the ovalbumin (egg white protein) gene in cells of the chick oviduct and for the globin genes in chick red blood cells. (Unlike the red blood cells of humans and other mammals, those of birds do not undergo a loss of nuclei during differentiation.)

In general, then, it seems that most DNA sequences in an organism are present at the same levels in various differentiated cells. Selective gene amplification is therefore not likely to be a critical control mechanism for most genes.

Transcriptional Control

It is at the level of transcription that most of the regulation of gene expression occurs in prokaryotes. Not surprisingly, **transcriptional control** is also important in eukaryotic cells. The best evidence for this comes from direct comparisons of the newly synthesized transcripts still present in nuclei

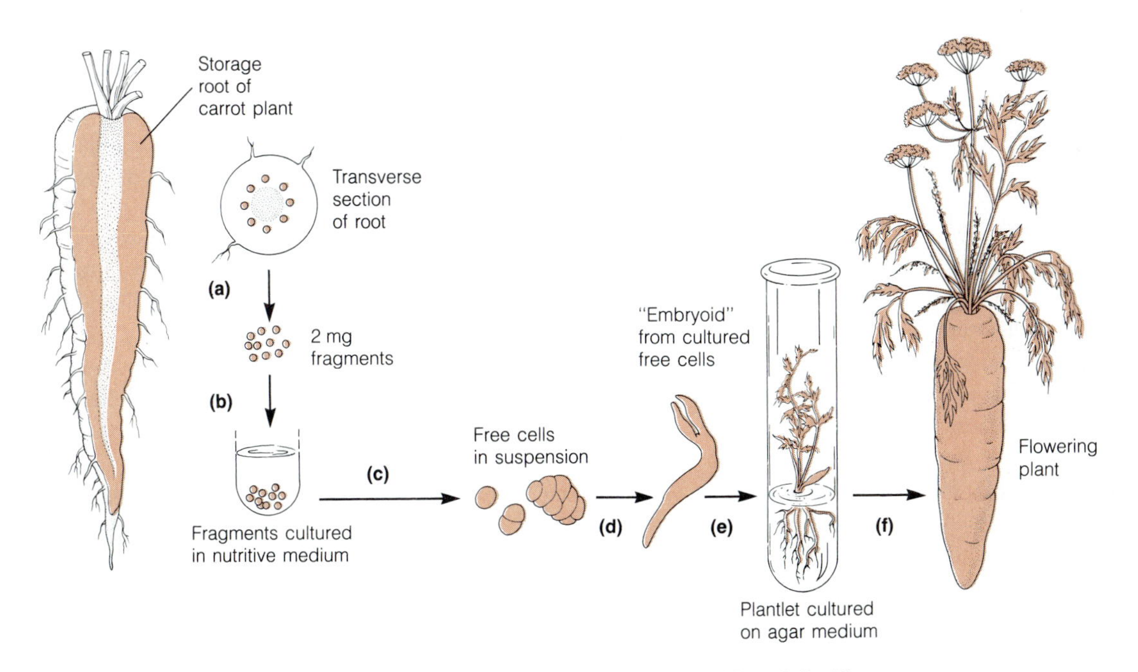

Figure 17-13 Tissue Culture as a Means of Demonstrating Totipotency in Plant Cells. The experiments shown here were done by F. C. Steward and colleagues at Cornell University, using *Daucus carota,* the wild carrot. (a) A section of differentiated tissue from a carrot root is cut into small fragments, which are (b) cultured on an agar medium containing an appropriate balance of the plant hormones cytokinin and auxin. (c) The cultured tissue is then dissociated into cells and cell clusters. (d) Cell division leads to a mass of cells that can organize and go through embryogenesis, forming embryo-like structures called embryoids. (e) Embryoids develop further when explanted onto an agar medium, giving rise eventually to (f) mature, fertile carrot plants.

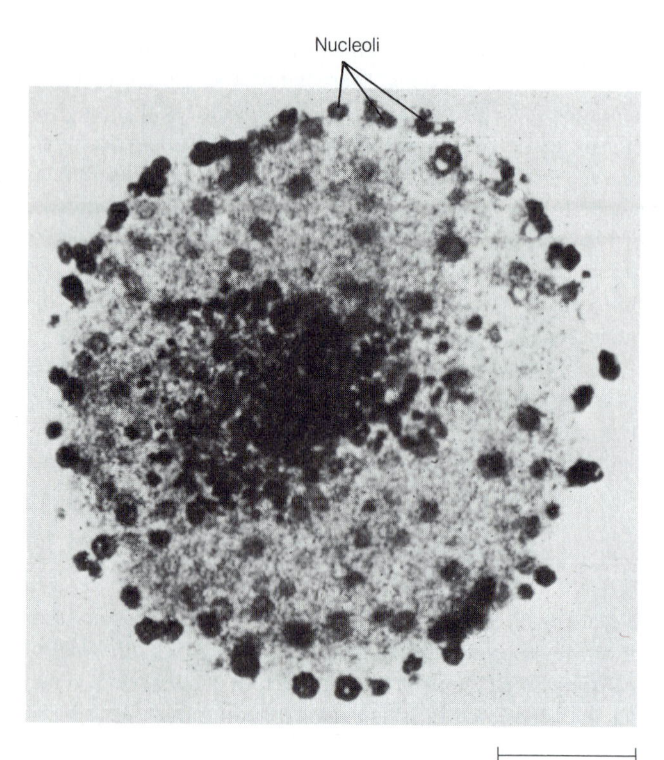

Figure 17-14 Amplification of the Genes for Ribosomal RNA in an Amphibian Ooctye. This micrograph shows an isolated nucleus from a *Xenopus* oocyte, stained to reveal the many nucleoli that are formed during oogenesis by the amplification of the genes for ribosomal RNA. Each nucleolus contains several copies of the ribosomal RNA genes, present as extrachromosomal circles of DNA.

of different tissues. For example, liver cells and brain cells synthesize an overlapping but nonidentical set of proteins, reflecting the availability of different populations of mRNA for translation in these two tissues. If these differences reflect differential gene transcription in the nucleus, we should see corresponding differences between the populations of nuclear RNA derived from brain and from liver tissue. On the other hand, if all genes are equally transcribed in liver and in brain cells, we would find no differences between the populations of nuclear RNA from the two tissues and we would conclude that the tissue-specific differences in protein synthesis were due to post-transcriptional mechanisms.

The experiment diagrammed in Figure 17-15 demonstrates that the differences in composition between the mRNA population present in the cytoplasm of liver cells and that in brain cells can be explained by **differential transcription** of genes in the nuclei of cells of these two tissues. As shown in Figure 17-15a, liver-specific mRNAs were prepared from total cytoplasmic liver mRNAs by selectively removing all mRNA species common to both liver and brain cells. The remaining liver-specific mRNAs from the cytoplasm were used as a template for synthesis of complementary DNA (cDNA) by reverse transcription. These complementary DNAs were used as **probes** to test for the presence of the liver-specific sequences in *nuclear* RNA preparations from liver cells (part b) or from brain cells (part c). The liver-specific probes were able to reanneal with complementary RNA in the liver nuclei, but no complementary mRNA sequences were present in the nuclei from brain cells. These results suggest that differential transcription may be an important mechanism for generating specific differences between liver and brain cell mRNA populations.

The same conclusion may also be reached by employing a different technique, **nuclear run-on transcription.** This procedure provides a "snapshot" of the transcriptional activity occurring in a nucleus (Figure 17-16). Transcriptionally active nuclei are gently isolated from cells and allowed to complete synthesis of nascent transcripts in vitro in the presence of radiolabeled nucleoside triphosphates. The presence of labeled nucleotides in an mRNA thus indicates that it was being transcribed in the nucleus at the time of isolation. The resulting set of newly transcribed, radiolabeled mRNAs in liver nuclei includes transcripts from the liver-specific genes, but these sequences are not detected among the radiolabeled mRNAs synthesized by isolated brain nuclei.

Regulation of Transcription Is a Two-Stage Process. The protein packaging of the eukaryotic chromosome adds an additional level of complexity not encountered in our analysis of prokaryotic gene regulation. It is likely that the eukaryotic RNA polymerase enzyme must recognize and interact with a complex matrix of DNA and specific DNA-binding proteins in the promoter region of a eukaryotic gene. This promoter region is embedded within a highly folded and ordered chromosomal superstructure. Models for regulation of transcription in eukaryotes must therefore consider the effect of chromatin structure on the process of transcription initiation.

One current model describes transcriptional regulation in eukaryotes as a two-stage process, as diagrammed in Figure 17-17. Diagrams such as Figure 17-17 should be viewed as a guide for thinking about the regulatory issues rather than as an accurate structural representation of the chromosome, because we do not yet have a clear understanding of the nature of the structural changes in chromatin that accompany transcription activation. The first stage would involve a localized change in chromatin structure, causing a decondensation of one chromosomal domain to allow access to the transcriptional machinery. During the second stage, initiation of transcription is regulated by interactions between RNA polymerase and

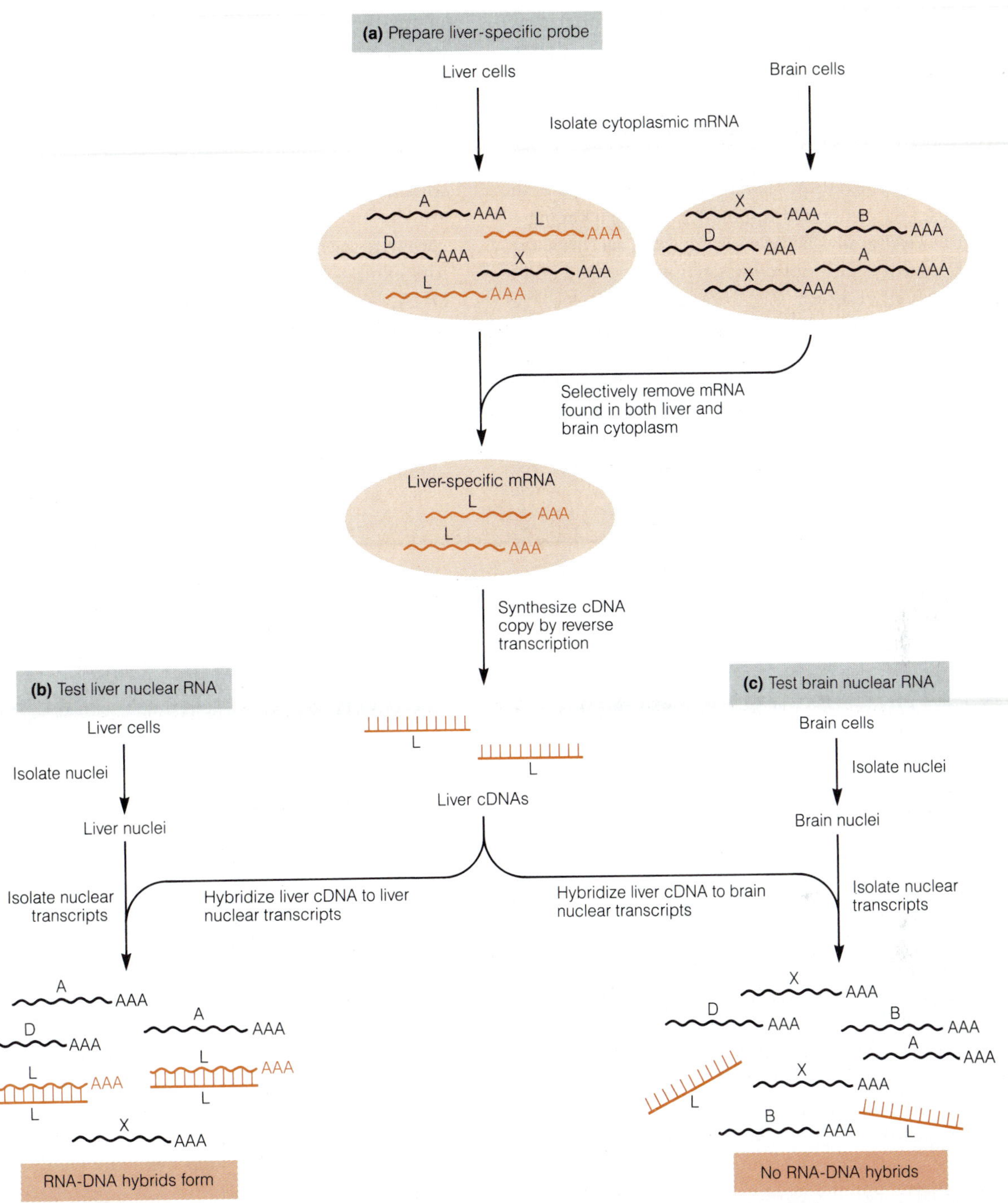

Figure 17-15 Evidence for Transcriptional Control of Gene Expression in Eukaryotes. Hybridization studies have been used to show that specific differences between the populations of cytoplasmic mRNA sequences observed in liver and in brain tissue, for instance, reflect corresponding differences observed in nuclear RNA populations. These differences in nuclear RNA populations presumably result from differential transcription of specific genes in different tissues. (a) Synthesis of complementary DNA probes representing liver-specific transcripts. Messenger RNA is isolated from the cytoplasm of liver cells, and the mRNA molecules also represented in brain cytoplasm are removed by hybridization to complementary copies of brain mRNA sequences. The remaining liver-specific mRNAs are used as templates for synthesis of complementary DNA (cDNA) copies using reverse transcriptase enzyme. (b) Liver nuclear RNA and (c) brain nuclear RNA are then tested for binding to liver-specific probes. RNA isolated from the nucleus (rather than cytoplasm) of liver or brain tissue is tested for its ability to form hybrids with the liver-specific probes. As expected, the liver nuclear RNA contains the liver-specific sequences (positive control experiment). However, the brain nuclear RNA cannot form hybrids with liver-specific probes, suggesting that selective transcription of different genes occurs in different tissues. If posttranscriptional mechanisms accounted for the observed differences in cytoplasmic RNA, you might have expected "liver-specific" genes to be equally transcribed in nuclei of the two tissues.

sequence-specific DNA binding proteins, **transcription factors,** that recognize cis-acting control regions within the DNA sequences surrounding the structural gene. We will first consider the aspects of eukaryotic transcription regulation that are most similar to regulation of prokaryotic transcription: the interactions of the regulatory proteins with signals in the DNA. Then we will investigate the additional levels of control imposed by the packaging of DNA into chromatin. The relationship between these two stages of control is still not completely understood and is an active area of research.

Binding of Transcription Factors Regulates Transcription Initiation. Earlier in the chapter we learned that the binding of proteins to DNA sequences can either reduce (*lac* repressor) or increase (CAP protein) transcription from prokaryotic genes. In eukaryotes, binding of specific regulatory proteins to DNA sequence elements surrounding eukaryotic genes can also determine *when* (temporal control), *where* (spatial or tissue-specific control), and *how often* (level of expression) the genes are transcribed.

Transcription factors belong to a more general class of proteins known as **nonhistone chromosomal proteins,** so called because they are associated with chromatin but are distinct from the well-defined histone proteins that comprise the nucleosome core structure. Transcription factors exhibit a non-sequence-specific, weak affinity for all DNA, but their role as gene regulators depends on very specific, high-affinity binding to characteristic, short nucleotide sequences near the genes that they regulate. The average nucleotide sequence recognized by a particular transcription factor is known as the **consensus binding site** for that regulatory protein. Deviations from the consensus sequence can decrease the binding interaction of the tran-

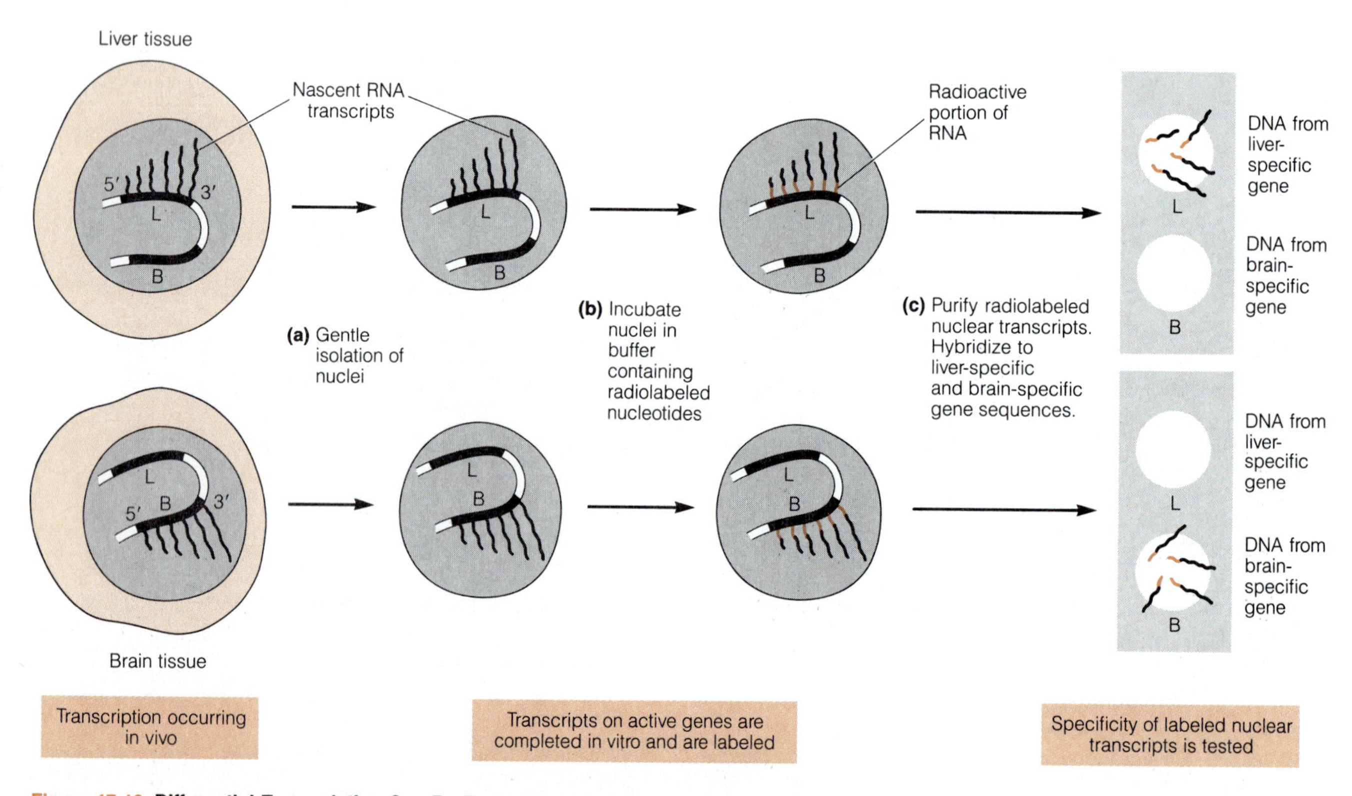

Figure 17-16 Differential Transcription Can Be Demonstrated by Nuclear Run-on Transcription Assays. Nuclei gently isolated from tissues (step a) are incubated in a transcription buffer containing radiolabeled ribonucleotides (step b). These labeled nucleotides can enter isolated nuclei through nuclear pores and become incorporated in mRNA being synthesized on active genes. If different genes are active in liver and in brain tissue, some labeled sequences in the liver nuclear transcripts will not be present in brain transcripts, and vice versa. The composition of the labeled RNA population is assayed (step c) by allowing the labeled RNA to form hybrids with DNA sequences representing different genes that have been attached to a filter paper support. Labeled transcripts from liver can form hybrids with a different set of genes than labeled transcripts from brain, indicating that the identity of active genes in the two tissues differs.

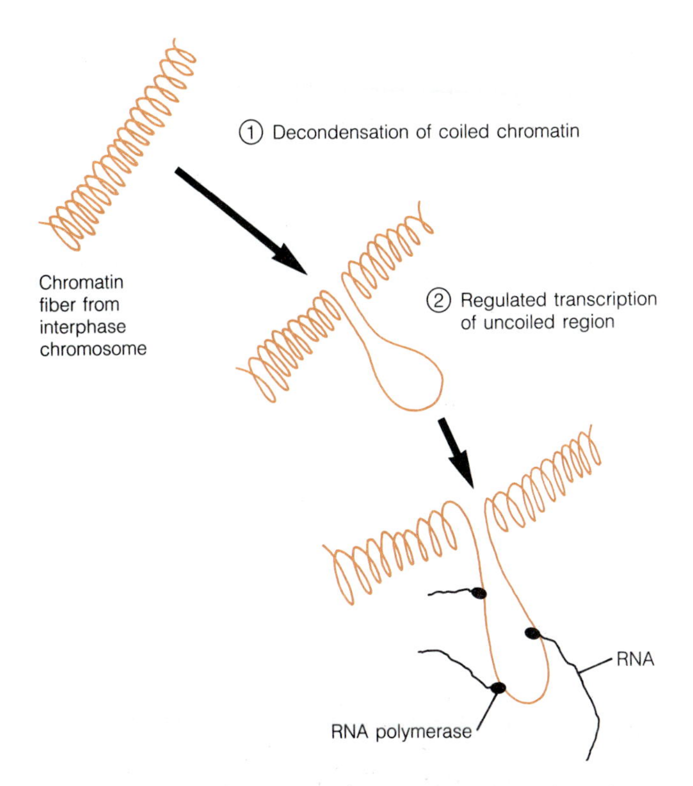

Figure 17-17 Two Stages of Regulation of Eukaryotic Transcription. Transcription in eukaryotes may be regulated in two different stages. (1) Localized changes in chromatin structure cause a decondensation of selected regions of the chromatin fibers of interphase chromosomes. (2) The DNA in the loops is then accessible for transcription by RNA polymerase, presumably under the control of regulatory proteins.

scription factor and reduce the frequency of transcription initiation from the associated gene. In patients with some forms of the genetic disorder *thalassemia,* for example, defects in the synthesis of the α or β subunit of hemoglobin can be traced to alterations in the consensus binding sites for important transcription factors required for proper expression of one of the globin genes. Just as the binding of the positive regulator, CAP protein, is necessary to activate transcription of the *lac* operon in bacteria, the binding of transcription factors seems to be an essential step in activation of eukaryotic genes.

In general, the simultaneous binding of more than one type of transcription factor is required for proper regulation of transcription in eukaryotes. This observation has led to a **combinatorial model for gene regulation.** This model proposes that by using a relatively small number of different transcription factors in alternative combinations it should be possible to establish highly specific and precisely controlled patterns of gene expression. A gene could be expressed when the particular set of transcription factors produced by a cell corresponded to the cluster of *cis*-acting regulatory DNA sequence elements near the gene. As shown schematically in Figure 17-18, this model would allow liver cells to produce proteins such as albumin but would prevent expression of these proteins in other tissues such as the brain. The model implies that some types of transcription factors are present in many cell types because some genes, such as those that encode proteins used for basic cellular metabolism and structure (components of the cytoskeleton, energy production and glycolysis, etc.), must be transcribed in most cells. On the other hand, expression of genes that encode specialized products produced by particular tissues may depend on the presence of unique transcription factors or unique *combinations* of transcription factors in those cell types. Data collected in many eukaryotic model systems support this idea. What are some of the signals and the factors that interact with them?

Cis-Acting Elements: Eukaryotic Promoters and Enhancers. Recombinant DNA techniques have been used in recent years in conjunction with genetic analysis of mutations to identify DNA sequences that are important for proper regulation of eukaryotic gene function. Results of these analyses allow us to make some generalizations about the types of *cis*-acting DNA sequences used to regulate eukaryotic genes. We will focus primarily on regulation of the transcriptional activity of the enzyme RNA polymerase II, the polymerase that transcribes protein-coding genes into mRNA. The regulation of RNA polymerase I activity (rRNA transcription) and RNA polymerase III activity (tRNA, 5S rRNA transcription) will be briefly contrasted later in the chapter.

The region within about 100 nucleotide pairs in front of the point where transcription begins may contain several types of DNA sequence elements that are found in similar relative locations in a large number of different genes. Figure 17-19 illustrates some features of this region, the **upstream promoter region,** of a generalized eukaryotic RNA polymerase II transcription unit. Some of the more important elements of the upstream promoter region are the "TATA box" at position -25, the "CAAT box" at position -80, and, in some genes, a GC-rich element. Removing these elements by deletion or by a mutation within the consensus region reduces the frequency and accuracy of transcription initiation. Transcription factors that bind specifically to each of these sequences have been identified (Table 17-3), and it is thought that these factors must bind to the promoter before the RNA polymerase–promoter interaction can occur. The fact that many different genes, each with a unique pattern of expression, utilize the same set of components in the upstream promoter region suggests that this region is important for basic promoter function. However, we must look elsewhere to discover the basis of differential gene transcription.

Another type of discrete regulatory sequence element,

the **enhancer,** may help to provide the specificity of time, place, and level of expression of a specific gene. The main function of an enhancer is to increase the level of transcription of the associated gene in cells containing the transcription factor or factors that bind to the enhancer sequence. One of the first enhancer elements studied was identified in the genome of the virus *SV40* (simian virus 40). This enhancer plays an important role in ensuring high levels of production of viral products following infection of a host cell.

Enhancer elements, varying in specific sequence but sharing common properties, have now been identified in association with many eukaryotic genes. Unlike upstream promoter elements, enhancers can function properly when

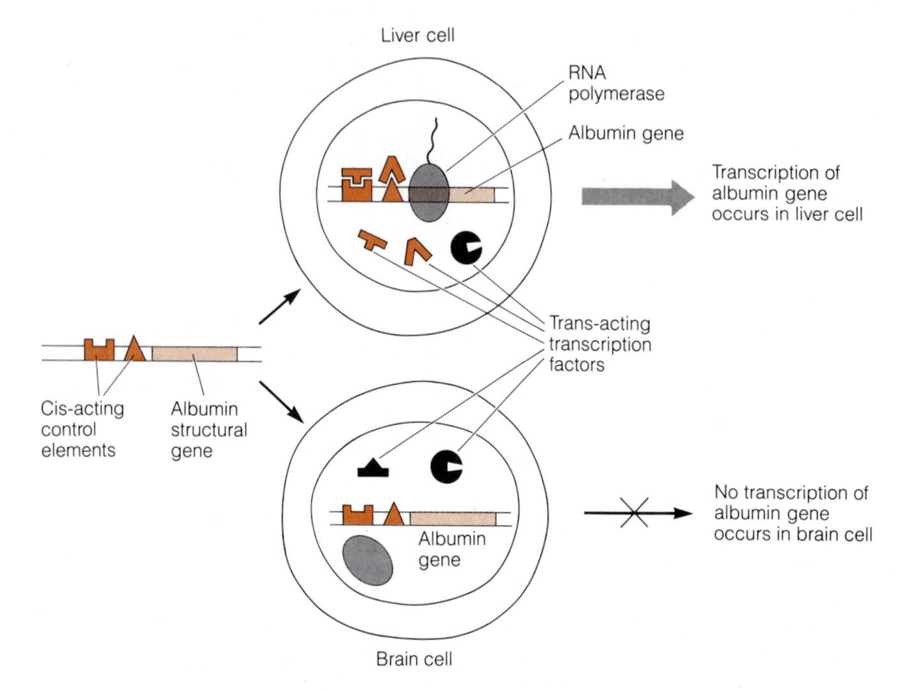

Figure 17-18 Combinatorial Model for Gene Expression. Each structural gene is associated with an array of regulatory DNA elements. In this case, for the liver protein albumin, the set of transcription factors present in liver nuclei includes those required for binding of RNA polymerase and facilitation of transcription of the albumin gene. However, because brain cell nuclei contain an overlapping but different set of transcription factors, the complete array of specific factors required to establish an active transcription complex on the albumin gene is not present, and albumin is not transcribed in brain cells.

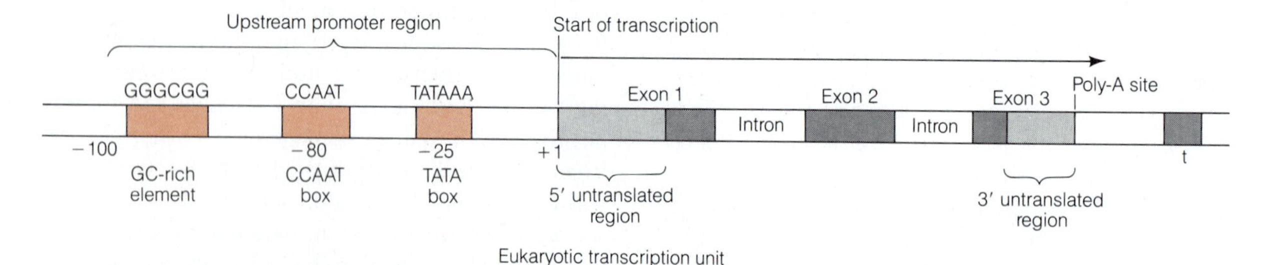

Figure 17-19 Diagram of a Generalized Eukaryotic Upstream Promoter Region and Transcription Unit. A generalized eukaryotic *transcription unit* contains a single structural gene and its associated transcriptional control regions. In addition to the coding region of the gene (codons that specify the gene product's amino acid sequence), the transcription unit includes a 5′ untranslated region and a 3′ untranslated region that are included in the mRNA transcript but do not contribute primary sequence information for the gene product. Downstream (to the 3′ side) of the structural gene is a *transcription termination region* (t), and upstream (to the 5′ side) of the site of transcription initiation is the *upstream promoter region.* Nearly all upstream promoter regions contain the TATA box at position −25 nucleotides, and many genes also contain the CAAT box farther upstream. The GC box is a common but not universal component of upstream promoter regions. These DNA sequence elements are binding sites for specific protein transcription factors required for formation of an active transcription complex with RNA polymerase.

relocated at variable distances from the start of transcription, as long as the promoter is present (Figure 17-20). Although most are found within several thousand nucleotides of the promoter they regulate, some enhancers can function from distances of 10,000 to 20,000 nucleotide pairs. A second characteristic of enhancers is that activity is retained even when the orientation of the enhancer element relative to the beginning of the gene is reversed (Figure 17-20d and f). Moreover, even though the upstream promoter elements of genes transcribed by RNA polymerase

Table 17-3 Cis- and Trans-Acting Eukaryotic Regulatory Elements

Gene	Consensus Binding Site*	Name of Factor†
Many	TATAAAA	TFIID
Many	CCAAT	CTF
Many	GGCGGG	Spl
Heat shock genes	CnnGAAnnTTCnnG	HSTF
Mouse mammary tumor virus	GGTACAnnnTGTTCT	Glucocorticoid receptor
SV40 virus	TGCTTTGCAT	Octamer binding factor
Ovalbumin	AGGGTCAnnnTGACCT	Estrogen receptor
Kappa light chain of immunoglobolin	GGGGACTTTCC	NF-KB

* This is the *cis*-acting regulatory element. Some variation can occur in the sequence of the consensus binding sites. Lowercase n signifies that any nucleotide can be located at that position.

† This is the *trans*-acting regulatory element. In some cases, other names have also been assigned to the factors listed here.

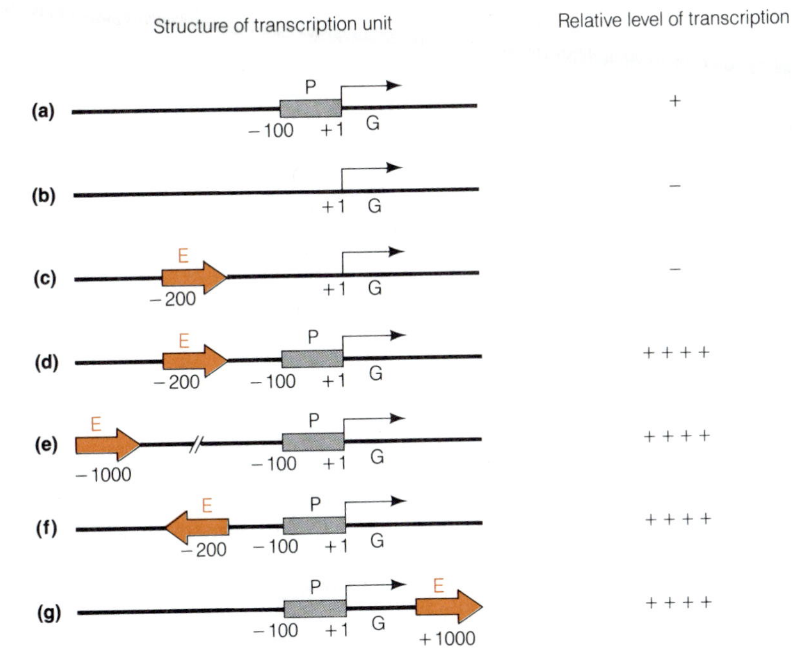

Figure 17-20 Properties of Eukaryotic Enhancer Elements. Recombinant DNA techniques can be used to alter the orientation and location of control regions and study the effect of the change on the level of transcription of the gene. (a) The upstream promoter region (P), in its typical location just upstream of the structural gene (G), allows a basal level of transcription to occur. (b) When the promoter region is removed from the structural gene, transcription levels drop. (c) An enhancer element (E) alone cannot substitute for the promoter region, but combining an enhancer element with an intact upstream promoter region (d) results in a significant increase in transcription of the structural gene. This increase in transcription is observed when the enhancer is moved farther upstream (e), when it is inverted in orientation (f), and even when it is moved to the 3′ end of the structural gene (g). The black arrows indicate the direction of transcription of gene G, with the first transcribed nucleotide labeled "+1." The other numbers give the positions of nucleotides relative to the +1 nucleotide.

Gene Expression in Eukaryotes: Multiple Levels of Control

Gene expression in eukaryotic cells is a topic of great interest and excitement, mainly because of new techniques that make it possible to dissect and analyze the complex series of events that lead from a specific gene to the functional protein for which that gene codes. In bacteria, the expression of most genes is regulated at the level of transcription, by varying the kinds and amounts of mRNA molecules that are synthesized. For eukaryotic cells, however, we already recognize several different points along the pathway from DNA to proteins at which gene expression can be regulated, and many scientists are now attempting to distinguish carefully between the various levels at which control can be exerted. Especially important in these studies are techniques that make it possible to identify and quantify specific proteins and specific mRNA molecules.

Levels of Gene Expression

The expression of a specific gene or group of genes usually results in an increase in the levels of a particular gene product, commonly an enzyme that can be detected because of its activity. Thus, the first indication of the expression of a particular gene is often an increase in the activity of a specific enzyme. Examples of such changes in enzyme activities include the induction of β-galactosidase activity in *Escherichia coli* cells (Figure 17-A), the appearance of tyrosine aminotransferase activity in hormone-treated rat hepatoma cells (Figure 17-B), and the increase in β-amylase activity that occurs in light-grown but not in dark-grown mustard seedlings (Figure 17-C).

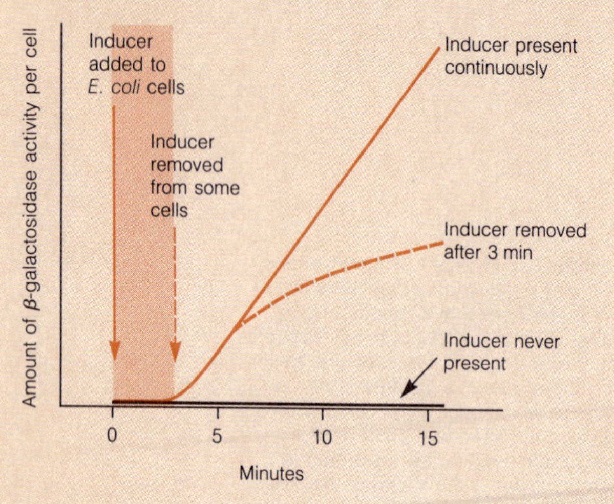

Figure 17-A

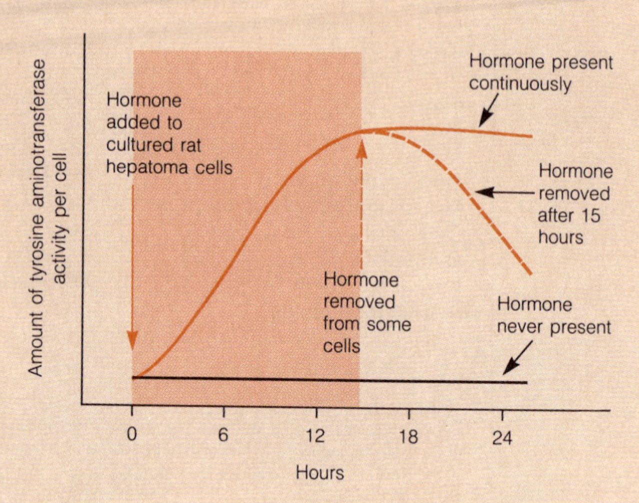

Figure 17-B

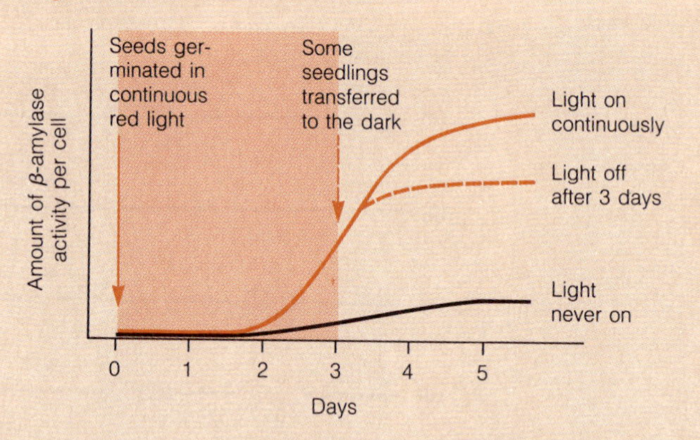

Figure 17-C

The time scales vary greatly in the three examples, but the effect is very similar in each case: An enzyme activity that would otherwise remain low (black line) increases many times in response to the appropriate stimulus (color line). Note also that if the stimulus is removed, the amount of enzyme stops increasing, or even decreases (dashed line).

An increase in the activity of a specific enzyme indicates the expression of a specific gene, but tells us little about the level at which that expression is regulated, particularly in eukaryotic cells. From what we know about the control of gene expression in bacterial cells, we might be tempted to regard an increase in enzyme activity as evidence that transcription of a specific gene has been turned on. But other explanations would be equally consistent with the data and need to be considered seriously. For example, the enzyme we are interested in could be already present in the cell in an inactive form, such that the observed increase in activity in response to some specific stimulus involves the activation of existing enzyme molecules. Alternatively, perhaps messenger RNA molecules for the desired enzyme are already on hand in the cell, but in an untranslated form. In this case, the appearance of enzyme activity would reflect the translation of existing mRNA molecules.

Distinguishing among these several possibilities requires an ability to detect specific proteins and mRNA molecules. We will consider each, in turn.

Detecting and Quantifying Protein Molecules

Detection of specific proteins almost always depends on the use of *antibodies,* because of the high specificity of the *antibody-antigen reaction.* As we will learn in Chapter 24, antibodies are blood proteins called *immunoglobulins* that are highly specific with respect to the particular *antigens* (proteins, usually) to which they will bind. To prepare antibodies against a specific protein, it is usually necessary to isolate the protein, purify it, and inject it into an experimental animal such as a rabbit. The blood that is subsequently withdrawn from the animal then has a high concentration, or *titer,* of antibody capable of reacting specifically with the protein that was injected.

The antibody can then be used to detect and quantify that specific protein by any of a variety of immunological techniques. An especially useful technique is called *immunoblotting.* A homogenate or extract of cells thought to contain the protein is subjected to SDS polyacrylamide gel electrophoresis, as illustrated in Figure 7-12. The proteins in the gel are then transferred to a sheet of nitrocellulose paper, and the paper is exposed to the antibody preparation. Because of the specificity of the antibody-antigen reaction, the antibody will bind only to regions of the paper at which the protein of interest is located. In this way, the protein can be localized to a specific band on the polyacrylamide gel, and the amount of protein in that band can be quantified.

Figure 17-D illustrates how we can use the results of an immunological assay such as immunoblotting to analyze gene expression. In addition to assaying for enzyme activity at regular time intervals (color line), we can use the immunoblotting technique to quantify the amount of enzyme protein present at the same time points (black lines). If the protein level increases concomitantly with enzyme activity (a), we

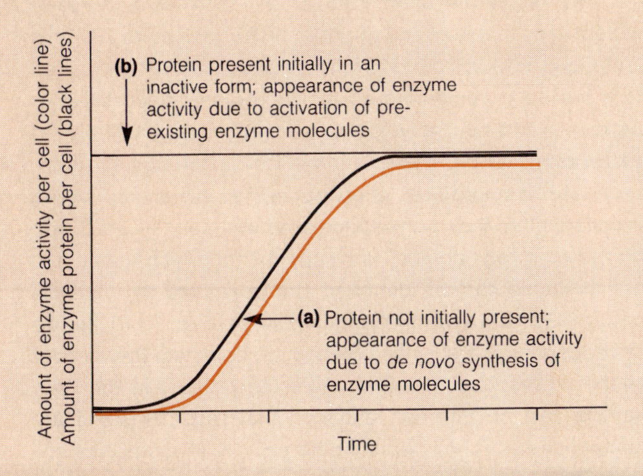

Figure 17-D

can conclude that the appearance of enzyme activity is almost certainly due to the accumulation of newly synthesized enzyme molecules.

However, if the protein is already present at early time points and the amount remains constant as enzyme activity increases (b), we can conclude instead that the appearance of enzyme activity is due to the activation of preexisting enzyme molecules. This is an example of *posttranslational control,* since the regulatory events that lead to activation of the enzyme occur at some point after synthesis of the enzyme molecule. Common mechanisms of activation include the removal of a short peptide from one end of an inactive precursor, phosphorylation of specific amino acid side chains, and transport of the protein into the appropriate organelle.

Detecting and Quantifying Messenger RNA Molecules

Many examples of posttranslational control of enzyme activity are known, but for most enzymes, activity increases concomitantly with enzyme protein. In such cases, the appearance of enzyme activity could be regulated at the level of either transcription or translation, depending on whether the messenger RNA preexists or must be synthesized as needed. Thus, to distinguish between these possibilities, we must be able to detect and quantify specific mRNA molecules. Two general techniques are available for this purpose. One depends on the ability of an mRNA molecule to direct the synthesis of the appropriate protein in vitro, and the other takes advantage of the sequence homology between the mRNA molecule and a complementary DNA probe.

The technique of *in vitro protein synthesis* (*IVPS*) is widely used to detect and quantify mRNA molecules. For this technique, RNA is isolated from the cells of interest and added to a test tube containing all the ingredients necessary for protein synthesis, including one or more radioactively labeled amino acids. The solution is then incubated at an appropriate temperature, allowing mRNA molecules to be translated into their polypeptide products, each of which will be radioactively labeled. Polypeptides for which antibodies are available can be identified and quantified by *immunoprecipitation* from the IVPS incubation mixture. From the relative amount of a specific polypeptide detected, the amount of the corresponding mRNA molecule present in the RNA sample can be inferred, although that requires some assumptions about how readily different mRNA molecules are translated by the IVPS system.

Alternatively, an mRNA molecule can be identified and quantified by its nucleotide sequence rather than by its ability to direct translation. This has the advantage of detecting possible precursor forms of mRNA that have not yet been activated for translation. For this purpose, a *complementary DNA* (*cDNA*) *probe* is necessary. Such probes are synthesized by reverse transcription of mRNA molecules. For detection of a specific mRNA, an RNA sample is size-fractionated by gel electrophoresis and transferred to special paper resembling the approach used for proteins in the immunoblotting technique. The paper is exposed to a solution of radioactively labeled cDNA, which hybridizes only to regions on the paper that contain RNA molecules with a sequence complementary to that of the cDNA probe. The amount of RNA is then quantified by determining the amount of radioactive cDNA bound to a given RNA band.

Either the IVPS technique or hybridization with a cDNA probe can be used to test for a specific mRNA, although the IVPS technique will detect only translatable mRNA molecules. Figure 17-E illustrates the results that might be obtained when the same samples used to demonstrate an increase in enzyme activity (color line) are also assayed for mRNA (black lines). If the mRNA level is low initially and increases just in advance of the enzyme activity (diagram a) we can conclude that either (1) the appearance of enzyme activity is dependent on newly synthesized mRNA molecules that are beginning to be translated as they accumulate in the cell (an example of *transcriptional control*) or (2) stabilization of mRNA molecules allows increased accumulation within the cell even though transcription rates have not changed (an example of *posttranscriptional control*). Application of the *nuclear run-on transcription* technique would allow us to distinguish between these two possibilities.

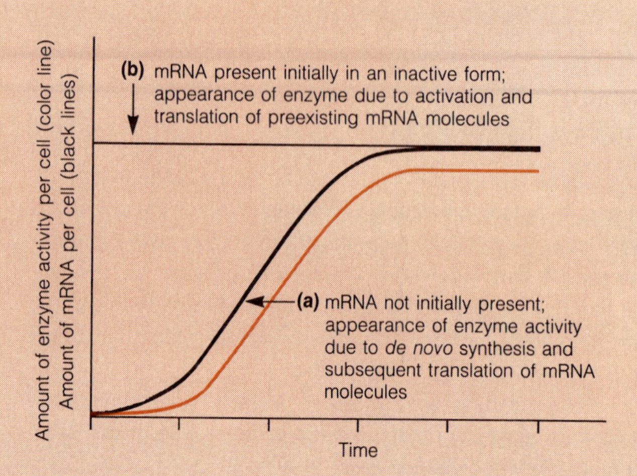

Figure 17-E

Alternatively, the mRNA may already be on hand at early time points, such that the amount remains constant as enzyme level increases (diagram b). In this case, the appearance of enzyme activity could be due to (1) activation and translation of preexisting mRNA molecules (*translational control*) or (2) stabilization of a protein that initially demonstrated a short half-life (*posttranslational control*).

Thus, the expression of a particular gene may be regulated transcriptionally (neither mRNA nor protein present beforehand), translationally (activation or stabilization of preexisting mRNA), or posttranslationally (activation or stabilization of preexisting protein). Techniques that can detect and quantify specific mRNAs and proteins at very low concentrations are currently being extended to many eukaryotic genes in attempts to determine the point at which expression is controlled in each case. If present progress is a fair indication, we can look forward to a further expansion in our understanding of the regulation of gene expression in eukaryotes.

II are located at the 5′ end of the gene, enhancer elements have been found near the promoter, as well as beyond the 3′ end of the gene, and even within an intron in the case of the immunoglobulin gene enhancer (Figure 17-20g). Although enhancers share these general characteristics, individual types of enhancer elements differ in the nucleotide consensus sequence that is important for mediating the effect on gene transcription. This difference reflects the fact that different transcription factors bind to different types of enhancers. Although the original activity associated with cis-acting upstream regulatory elements was enhancement or an increase in the level of transcription, negative regulatory elements have also been characterized. Binding of transcription factors to these negative sites reduces the level of transcription of the associated gene sequence.

Trans-Acting Factors: Regulatory Proteins That Bind to Promoters and Enhancers. From the multitude of transcription factors that have now been identified, some common patterns are emerging. First, like many other important cellular proteins, transcription factors possess several distinct functional domains. In general, the portion of the transcription factor that recognizes and interacts with the DNA consensus binding site (the *DNA binding domain*) is distinct from the portion required for activation of transcription (the *transcription activation domain*). Second, most transcription factors fall into one of several categories based on the type of protein structural motifs present in the DNA binding domain (Figure 17-21).

A common category, identified in both eukaryotic and prokaryotic transcription factors, is the *helix-turn-helix* structural motif (Figure 17-21a). The *lacI*, *trpR* and CAP proteins are prokaryotic examples of this category, and a set of transcription factors that regulate developmental decisions in *Drosophila* (the factors encoded by the homeotic genes, Chapter 22) are eukaryotic examples. These transcription factors contain two α helices separated by a turn (see Chapter 3 for a discussion of protein secondary structure), although the specific primary amino acid sequence can differ from one particular factor to the next. A second general category, initially identified in a transcription factor for the 5S rRNA genes (TFIIIA), is the *zinc finger* motif (Figure 17-21b). These proteins contain several protein loops held in place by the interaction of precisely positioned cysteine or histidine residues with a zinc atom. A third structural motif, the *leucine zipper* (Figure 17-21c) includes proteins that may use a leucine-rich surface for protein-protein interactions to produce a transcription factor dimer active in DNA binding and transcription activation. The list will undoubtedly grow as more transcription factors are identified and characterized.

Some eukaryotic transcription factors have the ability to change their binding affinity in response to cellular signals, in much the same way as in the case of the prokaryotic regulatory proteins that we considered earlier in the chapter. These eukaryotic transcription factors include the family of nuclear receptors for **steroid hormones** such as progesterone, estrogen, and glucocorticoids. They belong to the zinc finger class of transcription factors. When the small steroid hormone binds to the receptor protein, the receptor structure is altered to a form with high affinity for specific enhancer elements called **hormone response elements (HRE).** Binding of the activated receptor to these enhancers stimulates transcription of the adjacent gene. Steroid hormones are synthesized by cells in endocrine tissues and released into the circulation, and are then taken up by target tissues that contain the appropriate intracellular hormone receptor protein. In this way, a general signal can result in the specific activation of genes in these target tissues, a topic to which we will return in Chapter 21.

The Mechanism of Action of Enhancers and Transcription Factors. While the ultimate effect of an enhancer on the

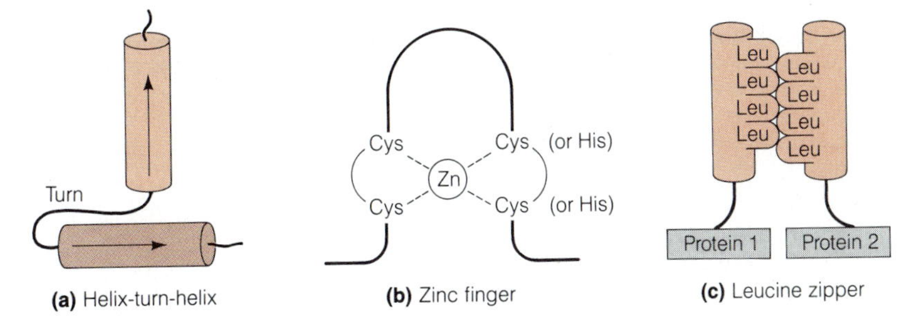

Figure 17-21 Common Structural Motifs in Transcription Factors. Several types of protein structural motifs near the DNA binding domain have been identified in regulatory DNA-binding proteins. These include the (a) *helix-turn-helix* class, in which two α-helical domains are joined by a flexible loop or turn; (b) the *zinc finger* class, in which a protein loop is formed as a result of interaction of four cysteine (Cys) or two cysteine and two histidine (His) residues with a zinc atom; and (c) the *leucine zipper* class, in which an α helix with regularly arranged leucine residues (Leu) in one protein interacts with a similar region in a second protein.

transcriptional activity of a gene cannot be disputed, there is still disagreement about the mechanism by which enhancers and their bound transcription factors act. Several models consistent with experimental evidence are currently under consideration (Figure 17-22). One model proposes that the binding of a transcription factor to an enhancer triggers an alteration in chromatin structure that can be propagated over some distance along the chromosome, resulting in activation of nearby genes (Figure 17-22a). A second possibility is that an enhancer occupied by a transcription factor can act as an attractive "gateway" into the chromatin for RNA polymerase (part b). The resulting local increase in concentration of polymerase means that more enzymes can move along the DNA helix and eventually encounter and interact specifically with the nearby promoter and its bound transcription factors. Finally, an idea that has gained increasing support in recent years suggests that a "looping" mechanism might bring two linearly distant transcription factor binding sites into close proximity (part c). In such a configuration, proteins bound at two separated sites (such as a distant enhancer and the upstream promoter region) could be close enough to allow protein-protein interactions. Together, the complex of transcription factors might produce a recognition target for RNA polymerase interaction and transcription initiation. The bending or looping could even occur as a natural consequence of coiling of the DNA backbone around nucleosomes or in higher-order chromatin structures.

Coordinate Gene Regulation in Eukaryotes. It is often necessary for a cell to activate several different genes in a coordinated fashion in response to a signal or stimulus. In prokaryotes, genes are organized in operons, and we have seen that several genes can be transcribed as a single reg-

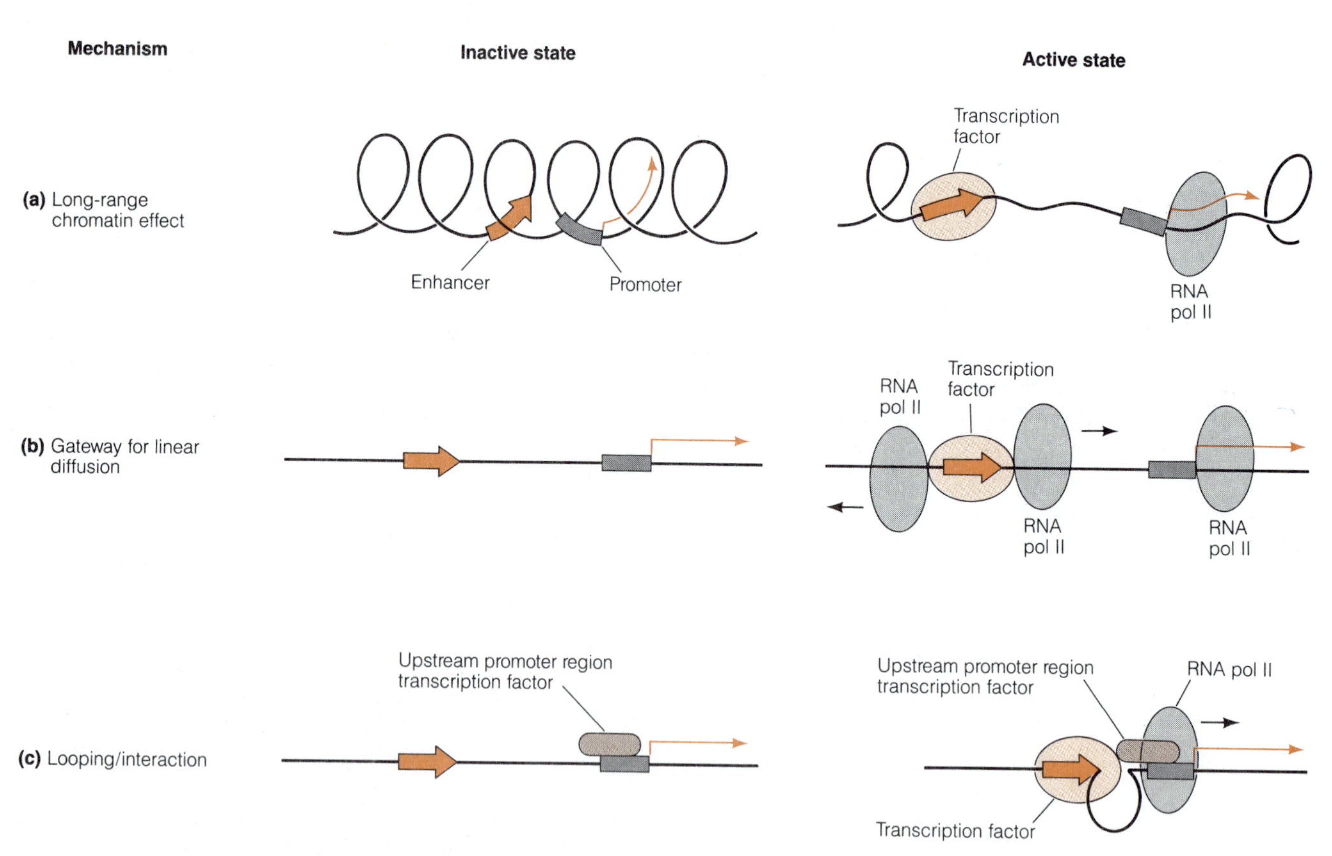

Figure 17-22 Models for Enhancer/Transcription Factor Action. Several models have been proposed to explain the increase in transcription as a result of enhancer action. (a) Binding of a transcription factor to an enhancer exerts an effect by altering some aspect(s) of chromatin structure that is propagated across a distance to facilitate RNA polymerase II (Pol II) interaction with the promoter region. (b) A transcription factor bound to an enhancer serves as a marker or gateway for entry of RNA Pol II near a region containing a transcription unit that should be activated. Linear diffusion of polymerase along the DNA occurs, leading to interaction with the promoter. (c) RNA Pol II recognizes a complex formed by protein-protein interactions between a transcription factor bound to the enhancer and factors bound to the upstream promoter region. The DNA in between the enhancer and the promoter can bend or loop to facilitate the interaction.

ulatory unit. For instance, in the case of the *lac* operon, the genes that encode a permease, an acetylase, and β-galactosidase are coordinately induced by the presence of a lactose metabolite in the absence of glucose. Because genes in eukaryotes are not organized in operons, a different strategy has been adopted to allow **coordinate regulation of genes** that must be activated at the same time and place. The requirement for coordinated gene regulation is critical for events such as differentiation of tissues during development, when cells that will become part of a specialized tissue, such as muscle, must synthesize a specific set of proteins required for the function of that tissue.

Analysis of one eukaryotic system has given us a clue about strategies for coordinate regulation of genes that are located at many different chromosomal sites. In this system an environmental stimulus simultaneously activates several different genes. These genes, the *heat shock genes,* are expressed in many eukaryotes when the external temperature is increased over the normal optimal temperature for that organism. While their function is not completely understood, it appears that at least some of the heat shock gene products minimize cellular damage that results from thermal denaturation of important proteins at various sites within the cell. Analysis of the 5′ regulatory regions of different members of the heat shock gene family revealed that each gene shares with other members of the set a specific *cis*-acting regulatory element, the **heat shock element.** Heat-induced activation of the heat shock transcription factor leads to binding of the factor to the heat shock element, thereby activating transcription simultaneously from the entire set of different heat shock genes. This system illustrates a basic principle of eukaryotic coordinate regulation: Different genes can be activated by the same signal by placing the same regulatory elements near them. As we will see below, the answer is not entirely that simple, but this basic principle appears to be utilized for coordinate transcription regulation both in simpler systems such as heat shock genes and in more complex situations such as tissue differentiation.

Regulation of Transcription by RNA Polymerases I and III. Transcriptional regulation by RNA polymerases I and III also depends on specific DNA-protein interactions between *cis*-acting sequence elements and *trans*-acting factors. However, the identity of the regulatory molecules and their arrangement relative to the start of transcription differ in these two cases from the examples we have considered up to now.

The result of RNA polymerase I activity is production of the two large 18S and 28S rRNAs and the small 5.8S rRNA. These three products are transcribed as a single precursor that is processed to yield equimolar amounts of the individual components (recall Figure 16-15). In most eukaryotes, the multiple copies of these rRNA genes are arranged in tandem arrays, separated by nontranscribed DNA spacer regions containing the promoter (nucleotide position -45 to $+6$) and a variable number of repetitions of a unit of 60 or 81 base pairs. Increasing the number of these repeats increases the level of transcription of the rRNA gene, suggesting that these units act as enhancers, to ensure high levels of transcription of these genes in many different cell types.

RNA polymerase III is responsible for transcription of small RNAs such as tRNA and 5S rRNA. This enzyme utilizes yet another type of promoter structure and a different set of transcription factors, although recent experiments suggest that some factors can interact with both RNA polymerase II and III. Unlike the RNA polymerase I and II promoter regions, which are located adjacent to the beginning of the transcript, the RNA polymerase III promoter is located internally, *within* the transcribed portion of the gene. Included in the set of transcription factors that bind to this internal promoter is TFIIIA (transcription factor, polymerase III, factor A), one of the first transcription factors purified and the first example of the zinc-finger class of transcriptional activator proteins.

Chromosome Decondensation and DNA Accessibility. Because of the way in which DNA is condensed and packaged into chromosomes, the selective decondensation of chromatin is almost certainly an important regulatory step in activation of eukaryotic genes. The best evidence for selective alteration of chromatin structure comes from experiments in which chromatin is treated with *DNase I,* a DNA-degrading enzyme from the pancreas. At low enzyme concentrations, DNase I preferentially degrades transcriptionally active DNA in chromatin. Presumably, the increased accessibility of this DNA to the DNase I enzyme, called **DNase I sensitivity,** reflects increased access for transcription factors and RNA polymerase as well.

When chromatin from different tissues of the same organism is compared, the pattern of DNase sensitivity for specific genes varies depending upon the potential for transcriptional activity of the gene in that tissue. In general, when a gene is transcriptionally active, the DNA is more sensitive to DNase I digestion. For example, the globin gene is transcribed in chicken erythrocyte nuclei. (Avian red blood cells retain nuclei in contrast to many other vertebrate species.) In these nuclei the globin gene is completely digested at DNase I concentrations that do not affect globin gene integrity in tissues such as brain, where globin is not transcribed (Figure 17-23). As you might predict, a gene that is *not* active in erythrocytes (for example, ovalbumin, an egg-white protein) is not digested by DNase I. The data suggest, therefore, that transcriptionally active eukaryotic DNA can be correlated with DNase accessibility

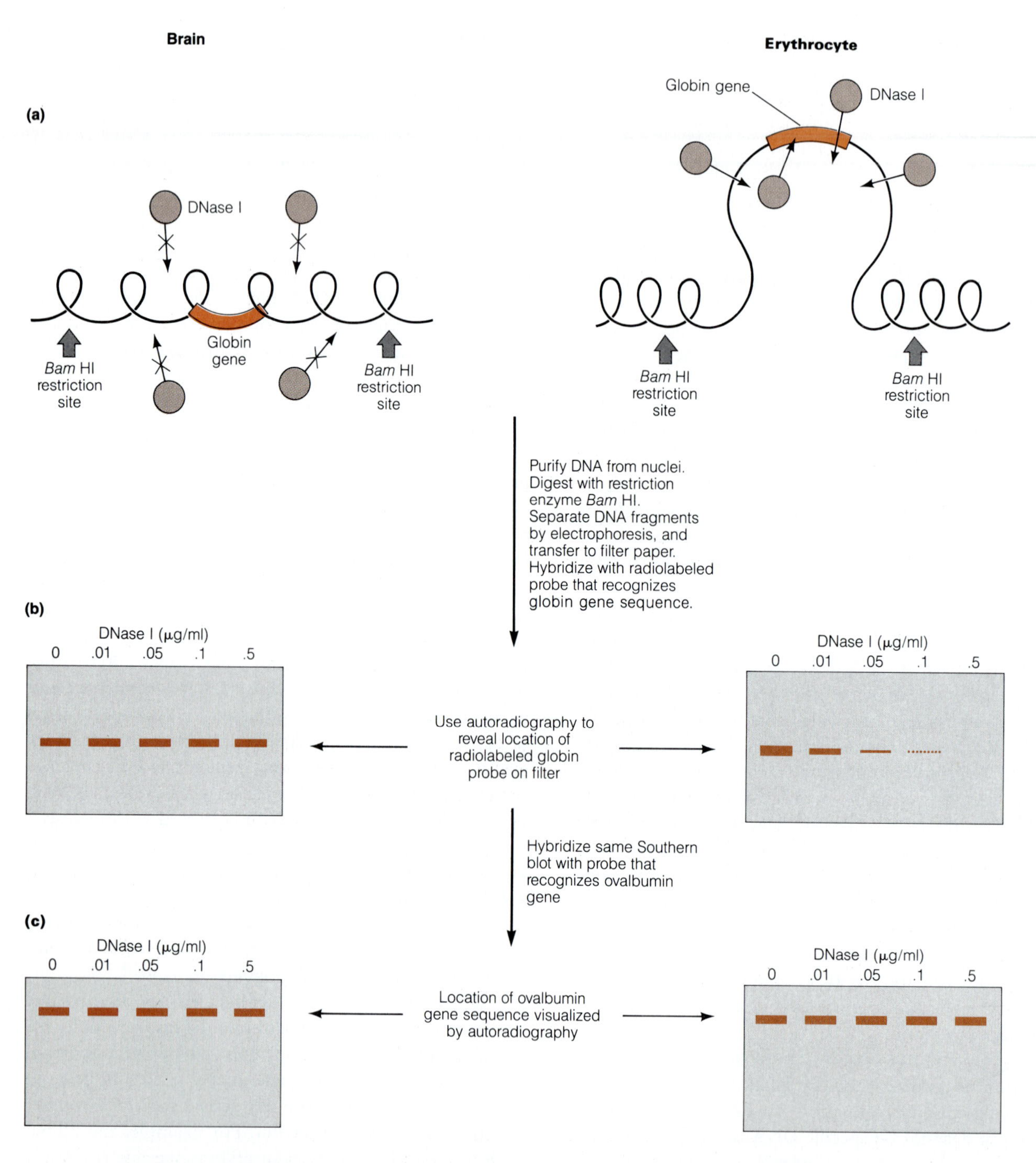

Figure 17-23 Detection of DNase I Sensitivity of Active Genes in Chromatin. The chromatin configuration of active genes can be assayed by digesting chromatin in cell nuclei with a nuclease, DNase I. (a) If the gene is inactive (for instance, a globin gene in brain tissue) the gene will still be present intact in DNA purified from the nuclease-digested nuclei. However, if the gene is active (for instance, a globin gene in erythrocytes) the digestion will nick the DNA backbone in the region of the globin gene. (b) The condition of the gene is assayed by digestion of the purified nuclear DNA with a restriction enzyme, which should release the globin gene (if it has not been digested) from the rest of the DNA as a restriction fragment of a characteristic size. The presence of that restriction fragment is detected by hybridization of a radiolabeled DNA probe for the globin gene to the nuclear DNA that is immobilized on a filter paper (Southern blot). DNA isolated from erythrocyte nuclei treated with increasing amounts of DNase I contains less of the intact restriction fragment with the globin gene. However, high concentrations of DNase I have no effect on the globin gene in brain nuclei. (c) An important control experiment shows that a gene such as ovalbumin that is inactive in both erythrocyte and brain nuclei is resistant to DNase I digestion in both cases.

and that the state of the chromatin may be critical to this accessibility. While it is not absolutely clear whether the observed chromatin changes are a cause or an effect of transcriptional activity the general DNase sensitivity of the globin gene can be observed during the period prior to and following actual transcription of the gene, suggesting that active transcription is not absolutely required for the structural change reflected by DNase sensitivity.

DNase I Hypersensitive Sites. When nuclei are treated with very low concentrations of DNase I, it is possible to detect specific locations in the chromatin that are exceedingly sensitive to digestion. Nicks in the DNA at these regions appear before the general digestion of a larger active chromatin domain. These **DNase I hypersensitive sites** can be reproducibly located in the vicinity of active genes. The positions of hypersensitive sites often correlate with specific binding sites for transcription factors in the enhancer or promoter regions of genes. Some hypersensitive sites are said to be constitutive—that is, they are present before, during, and after actual periods of gene activity, like general DNase sensitivity. Other hypersensitive sites are observed in the chromatin only when the adjacent gene is transcriptionally active. The hypersensitive site at the hormone response element (HRE) for glucocorticoid (steroid) hormone receptor binding is an excellent example of the latter class. This hypersensitive site is present in chromatin only when the steroid-bound form of the receptor is also present. It is possible that hypersensitive sites mark areas where the binding of a transcription factor(s) alters or displaces normal DNA/nucleosome structure, increasing the sensitivity of the less protected DNA helix to nuclease attack.

Visualization of Chromosome Decondensation. Visual evidence for the decondensation of chromatin comes from two sources: polytene chromosomes of insects such as the fruit fly *Drosophila melanogaster,* and so-called **lampbrush chromosomes** of amphibian oocytes. The precise relationship between these visually observable changes and the changes detected biochemically in other systems is unclear, but these examples support the idea that chromosomes are packaged in structural domains that can be physically reorganized during periods of active transcription.

Polytene (multistranded) chromosomes are formed by repeated replication of the chromosome without separation of the daughter chromatids. Such **endoreplication,** as it is called, results in a large number of parallel chromatids in register. The polytene chromosomes found in the salivary glands of *Drosophila* larvae, for example, are generated by ten rounds of endoreplication and therefore have 1024 (2^{10}) chromatids in lateral register. Figure 17-24a shows several polytene chromosomes present in the nucleus of a *Drosophila* salivary gland cell, as seen by phase-contrast microscopy. Each of these polytene chromosomes has a characteristic pattern of **bands,** several of which are identified by number in the figure. Each such band is thought to represent a relatively condensed structural unit (containing one or more genes) that can be uncoiled during transcription. Activation of a given band results in a **chromosome puff** that labels heavily with a radioactive RNA precursor such as ^{3}H-uridine (Figure 17-24c). When such puffs are examined with the electron microscope, they are found to consist of loops of DNA that are much less condensed than the DNA of other bands within the rest of the chromosome. The less condensed, lighter staining interband regions of the chromosome also contain genes, some of which are actively transcribed. Therefore, while puffs are by no means the only sites of active transcription along the polytene chromosome, the physical change in chromosome condensation correlates well with an enhancement of transcriptional activity.

As the *Drosophila* larva proceeds through development, each of the chromosomes visible in salivary gland nuclei undergoes very reproducible changes in puffing patterns, under the control of the insect steroid hormone *ecdysone.* This insect steroid hormone functions as the ligand for a nuclear transcription factor in much the same way that vertebrate steroid hormones act. It appears, in other words, that the characteristic puffing patterns seen during development of *Drosophila* larvae are direct visual manifestations of the selective activation, decondensation, and transcription of specific genes according to a genetically controlled developmental program.

An alternative way to demonstrate the transcriptional activity of specific bands is by the use of fluorescent antibodies. In Figure 17-24b, the polytene chromosomes of *Drosophila* have been exposed to fluorescent antibodies against RNA polymerase II, the enzyme responsible for synthesis of mRNA precursors. Bands that were being transcribed at the time the chromosomal preparation was made can be readily seen because the binding of the fluorescent antibodies to RNA polymerase molecules causes such bands to "light up" under the fluorescence microscope.

Another good source for the visualization of gene activation and transcription is the amphibian oocyte. Prior to the meiotic divisions with which oogenesis concludes in amphibian species, the oocyte persists for relatively long periods of time (often months and sometimes even years) at the tetraploid level, with four paired chromatids per chromosome. This is a period of intense gene activity, so RNA synthesis can easily be monitored. When active, such chromosomes have extended loops of uncoiled DNA that have been shown to be specific sites of RNA synthesis.

When first discovered in the nineteenth century, these

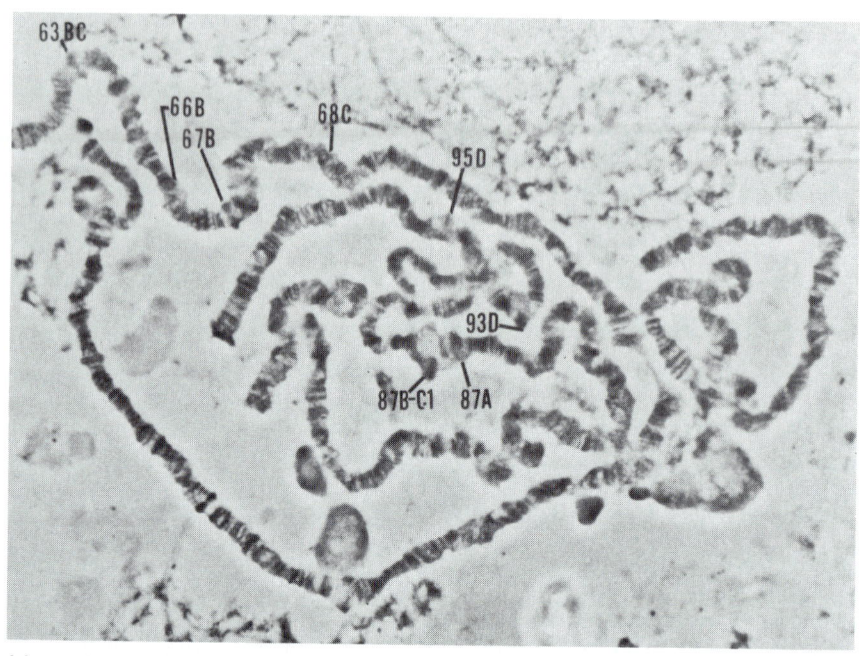

(a)

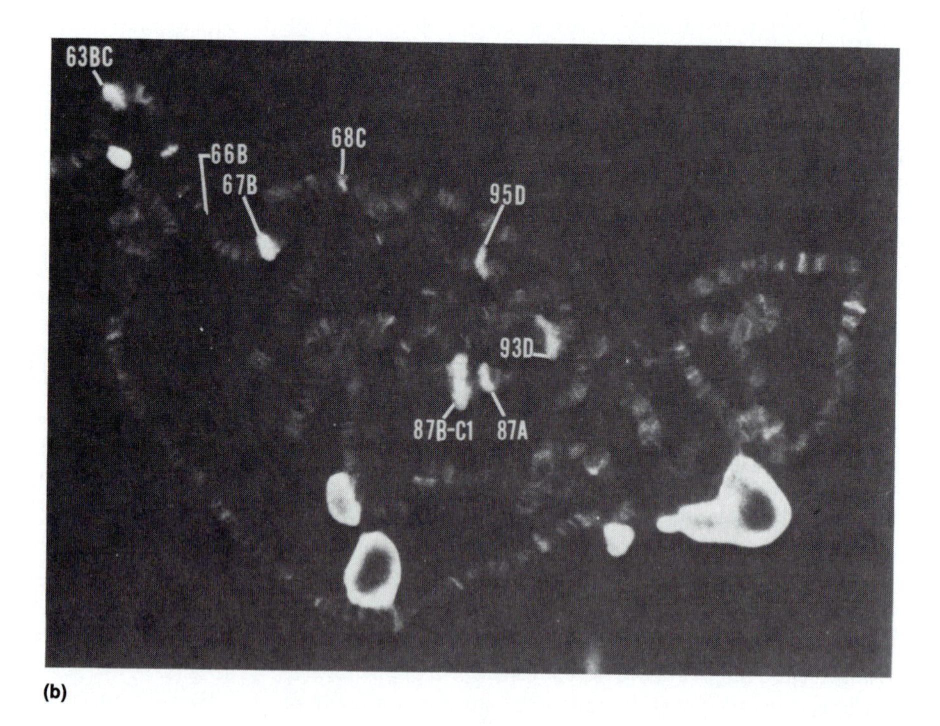

(b)

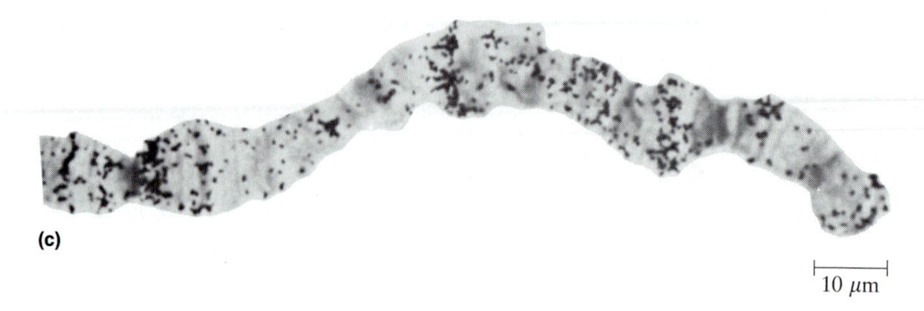

(c)

Figure 17-24 Transcriptional Activity of Polytene Chromosomes. (a) This phase-contrast micrograph shows several polytene chromosomes from the salivary gland of *Drosophila*. The banding pattern is a characteristic property of each chromosome, such that individual bands can be identified by number. (b) The same chromosomes seen under the fluorescence microscope. The fluorescent antibodies used recognize RNA polymerase II and cause the chromosomes to "light up" wherever RNA polymerase II molecules are located. The larva from which these chromosomes were obtained had been subjected to an elevated temperature for a short time to activate genes that code for specific "heat shock" proteins. (c) Differential RNA synthesis between bands can be shown by autoradiographic analysis of polytene chromosomes after exposure of the larva to ^{3}H-uridine, an RNA precursor. The high densities of silver grains over specific puffs identify these as the most active sites of RNA synthesis.

chromosomes were called lampbrush chromosomes because they resembled the brushes used to clean kerosene lamps; the name has persisted, even though the lamps have not. Figure 17-25 illustrates the highly extended lampbrush chromosomes of an amphibian oocyte. Like the puffs on a polytene chromosome, these loops seem to correspond to units of chromatin folding and DNA transcription.

Mechanism of Decondensation. Although most chromosomes do not have the lateral replication that allows such puffs or loops to be readily visualized, they may undergo an analogous process of selective decondensation to activate a unit of DNA by making it accessible for transcription.

Several features seem to be important in facilitating DNA accessibility. Nucleosomes isolated from transcriptionally active chromatin have lower levels of histone H1. Since H1 is thought to be involved in packing nucleosomes tightly together, it may be that a decrease in H1 is important in unpacking them again. In addition, the histones of the nucleosome core are highly acetylated in active chromatin, which also renders them less likely to pack tightly together. Furthermore, the nucleosomes of active chromatin are rich in two nonhistone proteins, *HMG 14* and *HMG 17* (HMG stands for *high-mobility group*). All these features seem to favor the decondensation of chromatin and therefore the accessibility of specific DNA segments. Not yet clear, however, is what renders specific regions of the chromosome susceptible to these modifications to ensure that the correct regions are transcribed at the appropriate times. Recently, a region of the human chromosome has been identified near the β-globin gene cluster that functions as a chromatin **dominant control region** to affect the degree of nuclease sensitivity and the level of transcription of the globin gene cluster. Further analysis of DNA regions that appear to regulate chromatin structure may provide insight into regulation of this critical first step of gene activation.

Possible Role of DNA Methylation in Regulating DNA Availability. One additional means of regulating DNA sequence availability is by **methylation** of selected cytosine groups in the DNA. Many eukaryotes have a small portion of their cytosine groups methylated. Moreover, the methylation pattern of DNA is heritable, since the enzyme that methylates the DNA is specific for cytosines that are linked to guanines (CpG) and is active only if the CpG sequence is base-paired to a CpG sequence in the other strand that is already methylated.

This mechanism may be useful to maintain a specific DNA sequence in either a methylated or an unmethylated state. It is not yet clear how important methylation is in the regulation of gene activation, but the DNA of inactive genes does tend to be more heavily methylated than the DNA of active genes, perhaps most significantly in control regions such as promoters. Examples are known in which the activation of a gene containing methylated DNA correlates with the loss of methyl groups, but it is not known whether the change in methylation is a cause or an effect of the activation process. In some cases, in vitro methylation of a transcription factor binding site can reduce bind-

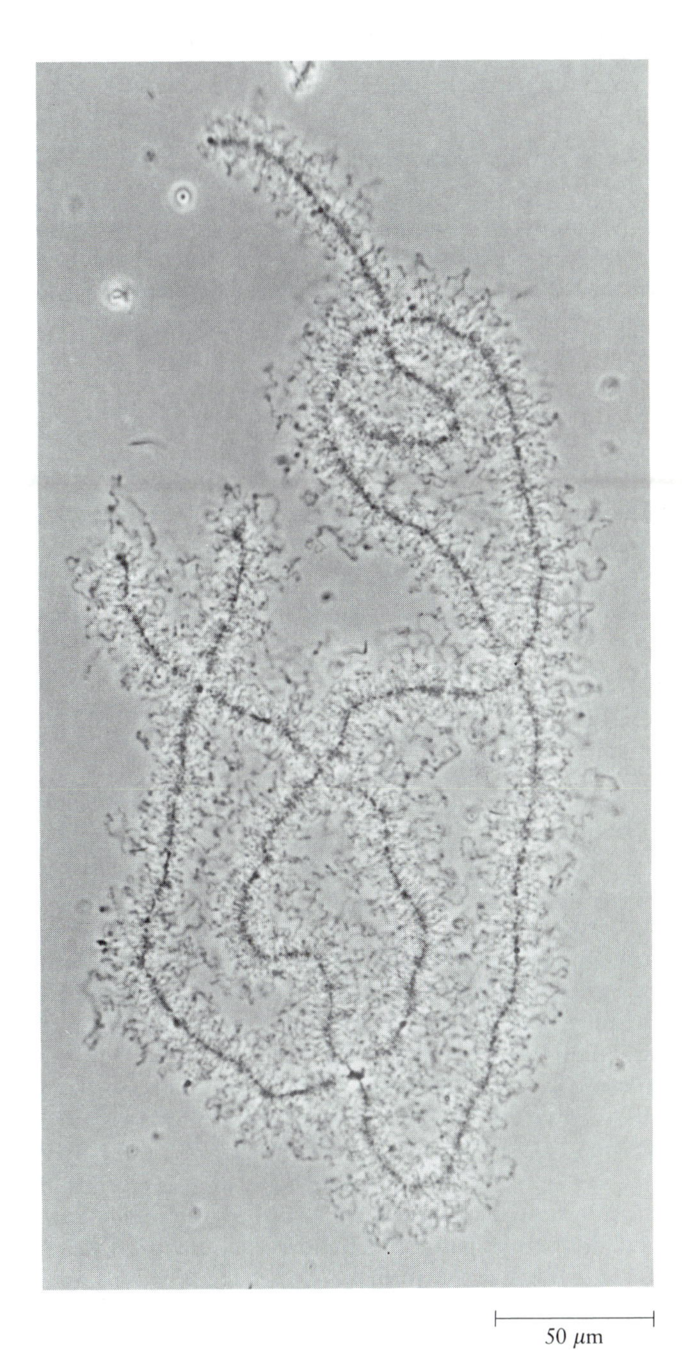

Figure 17-25 Lampbrush Chromosomes. This phase-contrast micrograph shows lampbrush chromosomes from an oocyte of the amphibian *Notophthalmus viridescens.* Hundreds of highly extended loops can be seen protruding from the central axis; these are thought to represent transcriptionally active chromatin.

ing of the factor and subsequent transcription levels of the gene, but in other examples methylation seems to have little effect. Moreover, some invertebrates such as *Drosophila* have almost no methylated DNA but employ other gene regulatory mechanisms similar to those used by all other eukaryotes. This information suggests that while methylation (when it occurs) may reinforce patterns of gene regulation, it may be less important as a primary regulatory mechanism.

Posttranscriptional Controls

The flow of genetic information in eukaryotic cells involves a series of posttranscriptional events, any or all of which may also turn out to be critical regulatory points. Several interesting examples of **posttranscriptional control** have been identified in recent years that lead us to consider posttranscriptional events in our overview of eukaryotic gene regulation. In particular, posttranscriptional events may provide a way to change the details of the gene expression pattern rapidly to respond to the intracellular or extracellular environment without changing overall cellular transcription patterns.

RNA Processing and Translocation. It is clear from the previous chapter that virtually all primary transcripts in eukaryotic nuclei undergo substantial processing, with most of it occurring before the transcripts leave the nucleus. This processing includes splicing of intervening sequences, addition of poly-A tails at the 3′ terminus, capping at the 5′ terminus, and chemical modifications such as methylation.

Control of RNA splicing patterns to create variant forms of a single gene product is a fascinating, yet poorly understood, regulatory mechanism that operates after gene transcription but prior to the point at which the transcript leaves the nucleus. In one well-documented example, **alternative splicing** of the 3′ end of the immunoglobulin gene transcript alters the exon utilization pattern to produce the secreted version rather than the plasma membrane-bound form of an antibody during the B-cell antigenic response (Figure 17-26a). The change in splicing patterns removes the membrane anchor, the hydrophobic amino acid domain that anchors the membrane-bound form of the immunoglobulin heavy chain at the cell surface. Alternative splicing also generates variant forms of important cell and substrate adhesion molecules such as neural cell adhesion molecule (N-CAM, Figure 17-26b) and fibronectin. Recently, an even more spectacular example of RNA processing as an important regulatory mechanism has come to light in the form of a cascade of alternative RNA-splicing events that regulate somatic sex determination in *Drosophila*. At present, our understanding of these events is descriptive; we can observe the alternative outcomes but have a relatively poor understanding of how they are achieved. Current attention, however, is focused on the role of specific RNA-binding proteins that may interact with portions of the nuclear mRNA to produce selectivity in the choice of 5′ and 3′ splice sites that are utilized during intron removal.

Translational Control. Various regulatory possibilities can also be envisioned at the translational level. Selective utilization of specific mRNA molecules, variations in rates of mRNA degradation, and differential availability of specific tRNA molecules or tRNA synthetases are some of the more obvious examples of **translational control.** In some cases, translation is known to depend on the availability of prosthetic groups. To establish these processes as regulatory events, it is necessary to demonstrate that specific mRNA molecules are selected for translation in a way that defines the metabolic or developmental capabilities of the cell more precisely than would otherwise be the case.

Probably the best example of such a regulatory process is the known dependence of globin synthesis on the availability of **hemin,** the required prosthetic group in hemoglobin formation by red blood cells. In the absence of hemin, the initiation of protein synthesis in such cells is inhibited and the polyribosomes disaggregate. In this case, the molecular mechanism is known. A deficiency of hemin results in the phosphorylation of *eIF2* (eukaryotic initiation factor 2), one of several proteins specifically required for initiation of protein synthesis in eukaryotes. Phosphorylated eIF2 cannot form a complex with GTP and methionyl tRNA and is therefore inactive in initiation.

As outlined in Figure 17-27, the actual mechanism by which hemin causes the phosphorylation of eIF2 involves the phosphorylation and activation of a specific eIF2 kinase. The kinase then phosphorylates and inactivates eIF2. Translational inhibitors with eIF2 kinase activity have been reported in a variety of eukaryotic tissues, suggesting that phosphorylation of a specific initiation factor may be a means of regulating translation in other tissues as well.

Another well-documented case of translational control of gene expression is seen during early embryonic development of the sea urchin (and other species, presumably). Protein synthesis occurs at a very low rate in the unfertilized egg, but is activated very dramatically upon fertilization. This increase is seen in both the presence and absence of *actinomycin D,* an inhibitor of RNA synthesis, so new RNA molecules appear not to be needed. Instead, the fertilized egg uses preformed mRNA molecules that were synthesized during oogenesis but kept in an inactive form until fertilization.

The reason for the translational inactivity of these mRNA molecules during oogenesis is not clear; however,

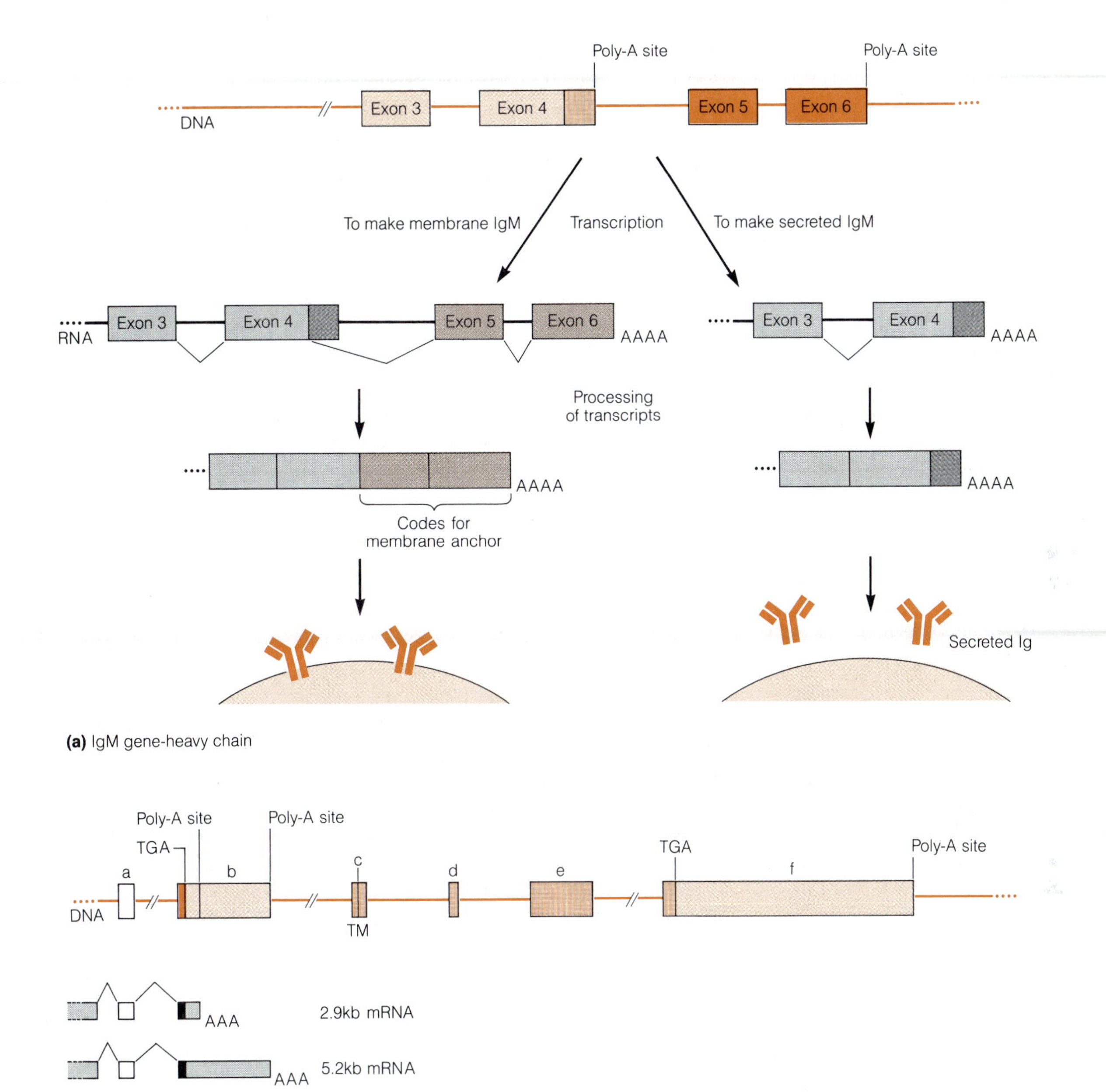

Figure 17-26 Alternative Splicing to Produce Variant Gene Products. (a) Alternative arrangements of exons in the carboxyl terminal region of the immunoglobulin M (IgM) gene. A splicing pattern that uses a splice junction within exon 4 and retains exons 5 and 6 results in synthesis of IgM molecules that are held in the plasma membrane of the B cell by a membrane anchor. The alternative product is secreted because a splice within exon 4 is not made and the transcript is terminated after exon 4. (b) The mouse gene encoding the cellular adhesion protein N-CAM (neural cell adhesion molecule) can produce at least four different cytoplasmic transcripts that produce three different protein products. One of these forms lacks the transmembrane domain encoded by exon c.

it seems to involve their inability to associate with ribosomes, since polyribosomes are virtually nonexistent in the unfertilized egg but appear rapidly after fertilization. This inability to form polyribosomes seems to be a property of the mRNA molecules and not the ribosomes, because ribosomes from unfertilized eggs translate other mRNA molecules readily. Most likely, the preformed mRNA molecules in the oocyte are complexed with proteins in a way that keeps them unavailable for translation until the oocyte is activated by fertilization. A similar phenomenon occurs in plant embryos upon seed germination. In this case, the rapid resumption of cellular activity is accompanied by the formation of polyribosomes and the resumption of protein synthesis, which does not seem to depend on RNA synthesis.

An example of specific, rather than general, translational control in response to cellular environment is the increase in synthesis of the intracellular iron-storage protein, *ferritin,* when excess iron is present in the extracellular medium. Under these conditions, ferritin synthesis increases dramatically although levels of ferritin mRNA in the cell remain constant, suggesting that translational regulation is occurring (Figure 17-28a). In the absence of iron, an allosteric translation repressor protein binds to a stem-loop structure formed by intrachain base pairing in the 5′ untranslated region of the ferritin mRNA. Binding of the protein to this **iron response element** (**IRE**) blocks formation of an active polyribosome complex. When iron levels increase, the repressor protein binds iron and undergoes a conformational change that prevents binding to the

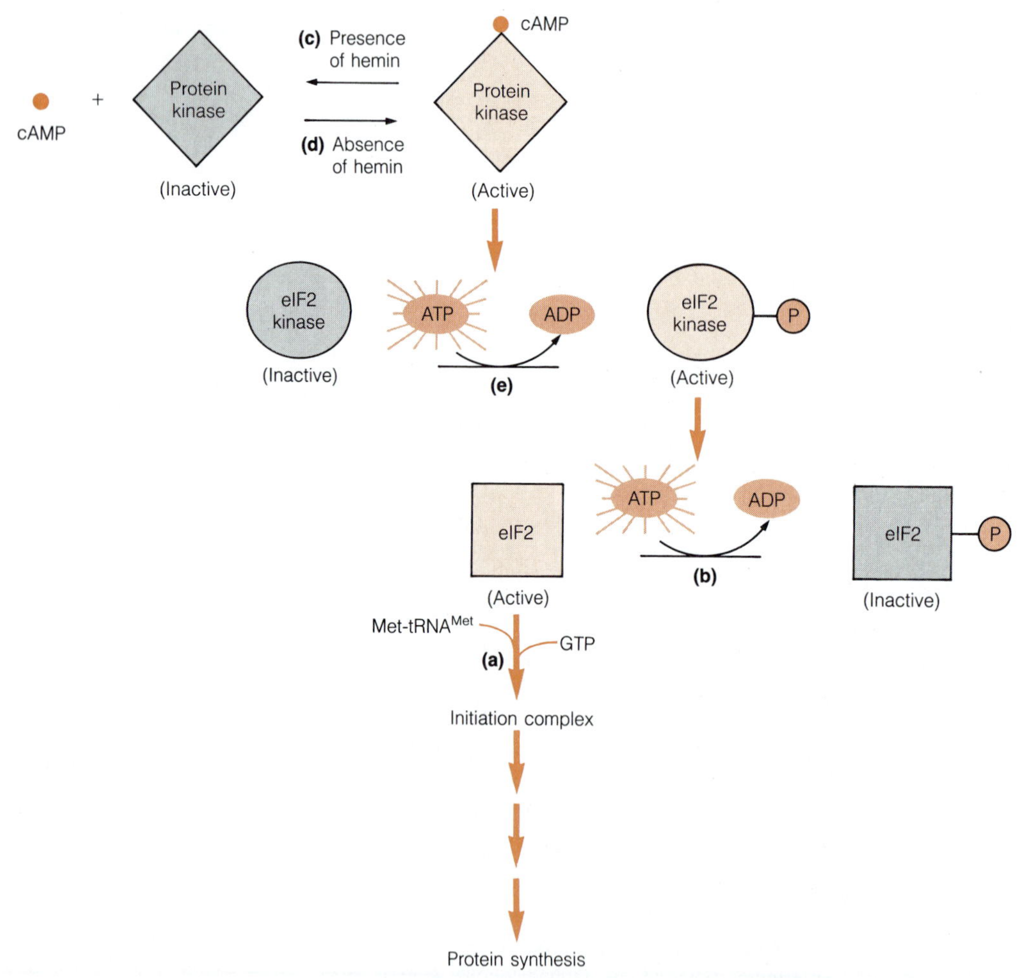

Figure 17-27 Regulation of Translation by Hemin in Red Blood Cells. (a) Protein synthesis in eukaryotic cells is initiated by formation of a ternary complex between Met-tRNA^{Met}, GTP, and the eukaryotic initiation factor eIF2. (b) Phosphorylation of eIF2 (by eIF2 kinase) causes its inactivation and leads to dissociation of the initiation complex. (c) In the presence of hemin, eIF2 remains in the active (unphosphorylated) form because eIF2 kinase is kept in its inactive form. (d) In the absence of hemin, an otherwise inactive protein kinase is activated by cyclic AMP (cAMP). (e) This, in turn, triggers the phosphorylation of eIF2 kinase, converting it to the active form. Active eIF2 kinase then causes the phosphorylation of eIF2, inactivating it and preventing further translation.

IRE, allowing translation of the ferritin message to occur. This type of rapid response to cellular conditions can be accomplished much faster at the level of translational control than by transcriptional control.

The availability of an mRNA for translation in the cytoplasm can also be regulated by altering mRNA stability. The *half-life,* or time required for 50% of the initial amount of specific transcript to be degraded, varies widely among different mRNA transcripts in eukaryotic cells, ranging from 30 minutes or less for some growth factor transcripts to over 10 hours for the mRNA encoding β-globin. The degree of polyadenylation of a cytoplasmic message has been postulated to play a general role in controlling mRNA stability, in that messenger RNAs with very short poly-A tails are postulated to be less stable. The issue is still controversial and has not yet been adequately resolved.

In some cases, the characteristic stability of a messenger RNA has been associated with specific features of the sequence in the 3′ untranslated region of the mRNA. Messages for several growth factors with very short half-lives of 30 minutes or less share an AU-rich sequence in the 3′ untranslated region of the transcript. When that region of the transcript is transferred via recombinant DNA techniques to the 3′ end of a normally stable globin message, the hybrid mRNA acquires the short half-life of the growth factor mRNA donor.

Another example of mRNA stability changes involves an interesting turn of events in which low levels of iron stabilize an mRNA to protect it from degradation and allow enhanced translation (Figure 17-28b). In this case, the same IRE (iron response element) that we discussed above is located in the 3′ untranslated region of the mRNA for *transferrin receptor,* a plasma membrane protein

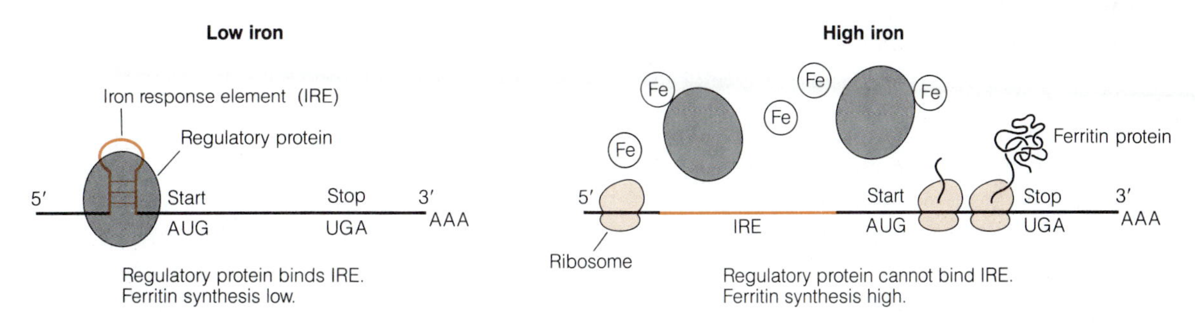

(a) Iron increases translation initiation from ferritin mRNA

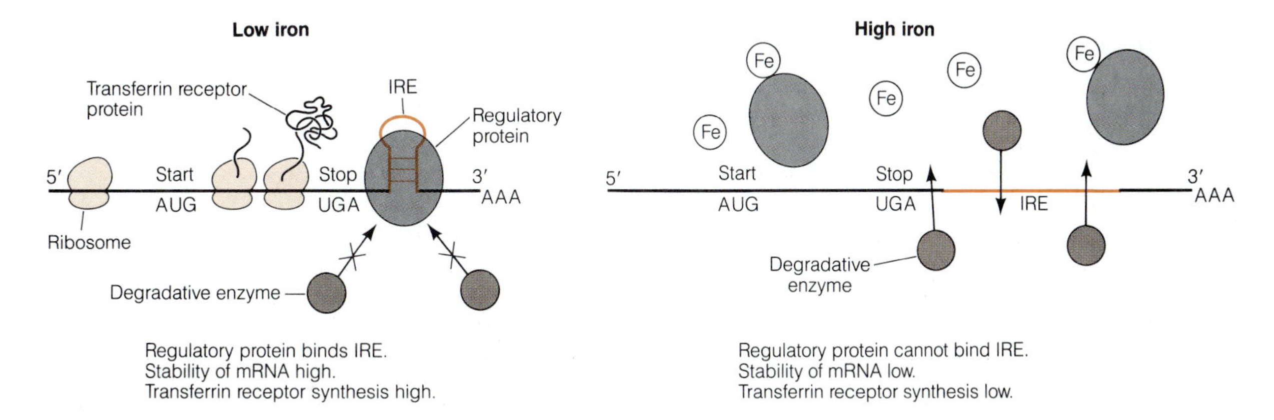

(b) Iron decreases stability of transferrin receptor mRNA

Figure 17-28 Two Levels of Translational Control in Response to Iron. (a) Control of initiation of translation of the ferritin mRNA is influenced by binding of a regulatory protein to an iron response element (IRE) in the 5′ untranslated leader sequence of the mRNA. In the presence of iron, an allosteric regulatory protein will not bind to the IRE, allowing ribosomes to assemble and move to the coding region to translate more ferritin. (b) Control of degradation of the transferrin receptor mRNA by regulatory protein binding to an IRE in the 3′ untranslated region. When intracellular iron concentration is low, the allosteric regulatory protein remains bound to the 3′ region of the mRNA, protecting it from degradation and allowing more transferrin receptor to be produced.

important in uptake of iron from the extracellular fluid. When intracellular iron levels are low and increased uptake of iron is necessary, the regulatory protein is able to bind to the 3′ IRE and therefore protect the transcript from attack by degradative machinery. Thus, more transferrin receptor mRNA is available for translation. When iron levels in the cell are high and additional uptake is not necessary, dissociation of the regulatory protein bound to iron destabilizes the mRNA and thus ultimately decreases transport of iron.

Posttranslational Control. Once a polypeptide is synthesized, further possibilities for regulation can be suggested. These **posttranslational control** mechanisms may involve many different events that modulate protein function via relatively permanent (e.g., proteolysis, glycosylation) or reversible (e.g., phosphorylation) structural modification. Other control pathways include interaction with regulatory molecules such as cAMP or Ca^{2+}, guiding protein assembly, targeting proteins to the appropriate intracellular or extracellular site of action, and regulating the turnover of proteins within the cell. Many of these events have been discussed already, in the context of other aspects of cellular regulation, but they deserve brief mention again here since the flow of information in the cell is not complete until a functional gene product has been produced.

An especially interesting example that points out the importance of considering regulation at this level concerns ribulose-1,5-bisphosphate carboxylase (RuBPCase), the carbon-fixing enzyme of the Calvin cycle. The intact enzyme is very large, consisting of eight identical **large subunits** (**LSU**, molecular weight about 52,000 each) and eight identical **small subunits** (**SSU**, molecular weight about 13,000 each). The enzyme is located in the chloroplast, and its large subunit is encoded by the chloroplast genome and synthesized within the organelle on chloroplast ribosomes. However, the small subunit is encoded by the nuclear genome, synthesized on cytoplasmic ribosomes, and imported into the chloroplast posttranslationally. Possible mechanisms for protein targeting and organelle import are discussed in Chapter 16.

Within the organelle, large and small subunits associate with each other to generate the intact enzyme. Significantly, neither subunit accumulates in excess of the other within the chloroplast. Somehow, the levels of the two subunits within the organelle are continuously adjusted to meet the needs for enzyme assembly, without an imbalance in either direction. The underlying regulatory mechanism is not yet known, but it is clearly of great importance, since RuBPCase is the single most abundant protein in nature and all of the RuBPCase in eukaryotic phototrophs must be assembled in this way from subunits synthesized in two different cellular compartments.

Finally, regulation of protein turnover within the cell can be an important mechanism to enable adaptation to changing cellular environments or interactions. For example, many proteins involved in conveying signals within or between cells (such as transcription factors, proteins involved in regulating cell division or critical metabolic processes) are relatively short-lived. One mechanism for designating proteins for intracellular degradation involves attachment of a small protein called *ubiquitin* to one or more lysine residues of the targeted protein. This signal is recognized by an intracellular ATP-dependent protease system that is also responsible for destruction of misfolded or denatured cellular proteins. The critical step in this process, the mechanism of recognition of proteins that need to be modified by ubiquitin addition, is not yet well understood.

Perspective

In addition to knowing how genetic information is expressed in cells, it is also important to understand how that expression is regulated. In prokaryotes, most regulation is effected at the level of transcription and is mediated by allosteric proteins. Some genes are expressed constitutively, but many are turned on and off in response to cellular needs. In general, enzymes of anabolic pathways are subject to repression. Their synthesis is specifically turned off in the presence of the end product.

Catabolic enzymes, on the other hand, are inducible. Their synthesis is specifically activated by the presence of substrate. Both types of regulation are effected by allosteric repressor proteins that function by binding to the operator locus and preventing transcription of the clustered structural genes in the operon. Operons for amino acid biosynthesis may also have an attenuator site that renders expression of the operon sensitive to the cellular level of the aminoacyl tRNA.

In addition to the repressor-operator interaction that renders operons sensitive to specific substrates, operons coding for some catabolic enzymes have adjacent or nearby sites that respond to the catabolite activator protein. This response allows the operon to be shut down in the presence of glucose, ensuring preferential utilization of this sugar.

The effect of glucose is mediated by cyclic AMP, the allosteric effector of the catabolite activator protein.

Regulation is more complicated in eukaryotes because of the greater structural and organizational complexity of the eukaryotic genome and because of the progressive, long-term nature of the developmental processes that take place in multicellular organisms. Transcriptional regulation is important in eukaryotes but is probably more complex than in prokaryotes. Activation of genes most likely involves a selective decondensation of chromatin. This structural change can be seen most readily in polytene chromosomes of insect salivary glands and in the lampbrush chromosomes of amphibian oocytes, but it is probably a general phenomenon. Once uncoiled, the DNA in such loops or puffs becomes accessible to RNA polymerases and may be subject to regulation by binding of specific transcription factors, some of which are allosteric proteins.

In addition to transcriptional regulation, eukaryotic cells depend on a variety of posttranscriptional controls as well, including generation of variant forms of a gene product by alternative splicing, differences in mRNA stability, general, as well as transcript-specific, mechanisms to regulate translational activity, and posttranslational modulation of protein activity. Eukaryotic gene regulation is currently an area of intense research and rapid progress, and we are likely to see some exciting developments within the next several decades.

Key Terms for Self-Testing

regulation (p. 508)

Gene Regulation in Prokaryotes

adaptive enzyme synthesis (p. 508)
end-product repression (p. 509)
repression (p. 509)
feedback inhibition (p. 509)
substrate induction (p. 510)
inducible enzymes (p. 510)
effectors (p. 510)
z gene (p. 510)
y gene (p. 510)
a gene (p. 510)
operon (p. 511)
structural genes (p. 511)
regulatory genes (p. 511)
polycistronic messenger RNA (p. 511)
promoter (*p*) (p. 511)
termination signal (p. 511)
operator (*o* site) (p. 511)
repressor (R) protein (p. 511)
regulator gene (p. 511)
i gene (p. 511)
allosteric protein (p. 511)
operator-constitutive mutants (p. 513)
superrepressor mutants (p. 513)
cis-trans test (p. 514)
trans-acting mutation (p. 514)
cis-acting mutation (p. 514)
corepressor (p. 515)
negative control (p. 515)
positive control (p. 515)
catabolite repression (p. 515)
3′,5′-cyclic AMP (cAMP) (p. 515)
catabolite activator protein (CAP) (p. 516)
attenuation (p. 519)
leader sequence (p. 519)
attenuator site (p. 519)
leader peptide (p. 519)

Gene Regulation in Eukaryotes

turnover (p. 522)
genomic control (p. 522)
germ line (p. 522)
somatic cells (p. 522)
chromosome diminution (p. 522)
totipotent (p. 524)
clones (p. 524)
gene amplification (p. 524)
oogenesis (p. 524)
transcriptional control (p. 525)
differential transcription (p. 526)
probes (p. 526)
nuclear run-on transcription (p. 526)
transcription factors (p. 528)
nonhistone chromosomal proteins (p. 528)
consensus binding site (p. 528)
combinatorial model for gene regulation (p. 529)
upstream promoter region (p. 529)
enhancer (p. 530)
steroid hormone (p. 535)
hormone response element (HRE) (p. 535)
coordinate regulation of genes (p. 537)
heat shock element (p. 537)
DNase I sensitivity (p. 537)
DNase I hypersensitive sites (p. 539)
lampbrush chromosomes (p. 539)
endoreplication (p. 539)
bands (p. 539)
chromosome puff (p. 539)
dominant control region (p. 541)
methylation (p. 541)
posttranscriptional control (p. 542)
alternative splicing (p. 542)
translational control (p. 542)
hemin (p. 542)
iron response element (IRE) (p. 544)
posttranslational control (p. 546)
large subunits (LSU) (p. 546)
small subunits (SSU) (p. 546)

Suggested Reading

General References and Reviews

Leighton, T. J., and W. F. Loomis, eds. *The Molecular Genetics of Development: An Introduction to Recent Research on Experimental Systems.* New York: Academic Press, 1981.

Lewin, B. *Genes,* 3d ed. New York: Wiley, 1987.

Watson, J. D., et al. *Molecular Biology of the Gene,* 4th ed. Menlo Park, Calif.: Benjamin/Cummings, 1987.

Negative Control in Prokaryotes: The Operon Model

Dickson, R., J. Abelson, W. Barnes, and W. Reznikoff. Genetic regulation: The *lac* control region. *Science* 187 (1975): 27.

Jacob, F., and J. Monod. Genetic regulatory mechanisms in the synthesis of proteins, *J. Mol. Biol.* 3 (1961): 318.

Maniatis, T., and M. Ptashne. A DNA operator-repressor system. *Sci. Amer.* 234 (January 1976): 64.

Miller, J. H., and W. S. Reznikoff. *The Operon.* Cold Spring Harbor, N.Y.: Cold Spring Harbor Laboratory, 1978.

Ptashne, M. Repressors. *Trends Biochem. Sci.* 9 (1984): 142.

Reznikoff, W. S., D. A. Siegele, D. W. Cowing, and C. Gross. The regulation of transcription initiation in bacteria. *Annu. Rev. Genet.* 19 (1985): 355.

Positive Control in Prokaryotes: Catabolite Repression

Englesberg, E., and G. Wilcox. Regulation: Positive control. *Annu. Rev. Genet.* 8 (1974): 219

Pastan, I., and S. Adhya. Cyclic adenosine 3′,5′-monophosphate in *Escherichia coli. Bacteriol. Rev.* 40 (1976): 527.

Raibaud, O., and M. Schwartz. Positive control of transcription initiation in bacteria. *Annu. Rev. Genet.* 18 (1984): 173.

Attenuation

Holmes, W. M., T. R. Platt, and M. Rosenberg. Termination of transcription in *E. coli. Cell* 32 (1983): 1029.

Oxender, D. L., G. Zurawski, and C. Yanofsky. Attenuation in the *Escherichia coli* tryptophan operon: The role of RNA secondary structure involving the tryptophan codon region. *Proc. Natl. Acad. Sci. USA* 76 (1979): 5524.

Yanofsky, C. Operon specific control by transcription attenuation. *Trends Genet.* 3 (1987): 356.

Regulation of Gene Expression in Eukaryotes

Brown, D. D. Gene expression in eucaryotes. *Science* 211 (1981): 667.

Darnell, J. E., Jr. Variety in the level of gene control in eucaryotic cells. *Nature* 297 (1982): 365.

Gurdon, J. B. *The Control of Gene Expression in Animal Development.* Cambridge, Mass.: Harvard University Press, 1974.

Kozak, M. A profusion of controls. *J. Cell Biol.* 107 (1988): 1.

Singer, M., and P. Berg. *Genes and Genomes.* Mill Valley, Calif.: University Science Books, 1990.

Gene Amplification in Eukaryotes

Chisholm, R. Gene amplification during development. *Trends Biochem. Sci.* 7 (1982): 161.

Schimke, R. T., ed. *Gene Amplification.* Cold Spring Harbor, N.Y.: Cold Spring Harbor Laboratory, 1982.

Transcription in Eukaryotes

Derman, E., K. Krauter, L. Walling, C. Weinberger, M. Ray, and J. E. Darnell, Jr. Transcriptional control in the production of liver-specific mRNAs. *Cell* 23 (1981): 731.

Maniatis, T., S. Goodbourn, and J. A. Fischer. Regulation of inducible and tissue specific expression. *Science* 236 (1987): 1237.

Yamamoto, K. Steroid receptor regulated transcription of specific genes and gene networks. *Annu. Rev. Genet.* 19 (1985): 209.

Transcription Factors and Enhancers

Atchison, M. L. Enhancers: Mechanisms of action and cell specificity. *Annu. Rev. Cell. Biol.* 4 (1988): 127.

Dynan, W. S., and R. Tjian. Control of eukaryotic messenger RNA synthesis by sequence-specific DNA-binding proteins. *Nature* 316 (1985): 774.

Parker, C. S. Transcription factors. *Curr. Opinion Cell Biol.* 1 (1989): 512.

Parker, C. S., and J. Topol. A *Drosophila* RNA polymerase II transcription factor binds to the regulatory site of an hsp 70 gene. *Cell* 37 (1984): 273.

Ptashne, M. Gene regulation by proteins acting nearby and at a distance. *Nature* 322 (1986): 697.

Struhl, K. Helix-turn-helix, zinc-finger, and leucine-zipper motifs for eukaryotic transcriptional regulatory proteins. *Trends Biochem. Sci.* 14 (1989): 137.

Gene Expression and Chromatin Structure

Ashburner, M. Temporal control of puffing activity in polytene chromosomes. *Cold Spring Harbor Symp. Quant. Biol.* 38 (1973): 655.

Cedar, H. DNA methylation and gene activity. *Cell* 53 (1988): 3.

Dynan, W. S. Understanding the molecular mechanism by which methylation influences gene expression. *Trends Genet.* 5 (1989): 35.

Elgin, S. C. R. DNAse I-hypersensitive sites of chromatin. *Cell* 27 (1981): 413.

Gross, D., and W. Garrard. Nuclease hypersensitive sites in chromatin. *Annu. Rev. Biochem.* 57 (1988): 159.

Patient, R. K., and J. Allan. Active chromatin. *Curr. Opinion Cell Biol.* 1 (1989): 454.

Weisbrod, S. Active chromatin. *Nature* 297 (1982): 289.

Zaret, K., and K. Yamamoto. Reversible and persistent changes in chromatin structure accompany activation of a glucocorticoid-dependent enhancer element. *Cell* 38 (1984): 29.

Posttranscriptional Control in Eukaryotes

Bernstein, P., and J. Ross. Poly(A), poly(A) binding protein and the regulation of mRNA stability. *Trends Biochem. Sci.* 14 (1989): 373.

Boggs, R., P. Gregor, S. Idriss, J. M. Belote, and M. McKeown. Regulation of sexual differentiation in *D. melanogaster* via alternative splicing of RNA from the transformer gene. *Cell* 50 (1987): 739.

Breitbart, R., A. Andreadis, and B. Nadal-Ginard. Alternative splicing: A ubiquitous mechanism for the generation of multiple protein isoforms from single genes. *Annu. Rev. Biochem.* 56 (1987): 467.

Casey, J., M. W. Hentze, D. M. Koeller, S. W. Caughman, T. A. Rouault, R. D. Klausner, and J. B. Harford. Iron responsive elements: Regulatory RNA sequences that control mRNA levels and translation. *Science* 240 (1988): 924.

Early, P., J. Rogers, M. Davis, K. Calame, M. Bond, R. Wall, and L. Hood. Two mRNAs can be produced from a single immunoglobulin *mu* gene by alternative RNA processing. *Cell* 20 (1980): 313.

Shaw, G., and R. Kamen. A conserved AU sequence from the 3′ untranslated region of GM-CSF mRNA mediates selective mRNA degradation. *Cell* 46 (1986): 659.

Problem Set

1. **Laboring with *lac*.** Most of what we know about the *lac* operon of *E. coli* has come from genetic analysis of various mutants, using either haploid strains or partial diploids constructed by mating two mutant strains. In the following list are the genotypes of ten strains of *E. coli*. For each strain, indicate whether the *z* gene product will be expressed (i) in the presence of lactose and (ii) in the absence of lactose. Explain your reasoning in each case.

   (a) $i^+p^+o^+z^+$
   (b) $i^s\,p^+o^+z^+$
   (c) $i^+p^+o^c\,z^+$
   (d) $i^-p^+o^+z^+$
   (e) $i^s\,p^+o^c\,z^+$
   (f) $i^+p^-o^+z^+$
   (g) $i^-p^+o^+z^+/i^s\,p^+o^+z^+$
   (h) $i^s\,p^+o^+z^+/i^+p^+o^c\,z^-$
   (i) $i^-p^+o^+z^-/i^+p^+o^+z^+$
   (j) Same as (a), but with glucose present

2. **More Laboring with *lac*.** The genes *a*, *b*, and *c* in the following table are from the *lac* operon of *E. coli*. One represents the repressor (*i*); another, the operator (*o*); and the third, the structural gene for β-galactosidase (*z*). Study the data and decide which gene (*o*, *i*, or *z*) each of the three is, and explain briefly in each case how you decided.

Genotype of Bacterium	Activity of the *z* Gene	
	− Inducer	+ Inducer
$a^-b^+c^+$	+	+
$a^+b^+c^-$	+	+
$a^+b^-c^+$	−	−
$a^+b^-c^+/a^-b^+c^-$	+	+
$a^+b^+c^+/a^-b^-c^-$	−	+
$a^+b^+c^-/a^-b^-c^+$	−	+
$a^-b^+c^+/a^+b^-c^-$	+	+

3. **The Pickled Prokaryote.** *Pickelensia hypothetica* is an imaginary prokaryote that converts a wide variety of carbon sources into ethanol when cultured anaerobically in the absence of ethanol. When ethanol is added to the culture medium, however, the organism obligingly shuts off its own production of ethanol and makes lactate instead. Several mutant strains of *Pickelensia* have been isolated that differ in their ability to synthesize ethanol. Class I mutants cannot synthesize ethanol at all. Mutations of this type map at two loci, *a* and *b*. Class II mutants, on the other hand, are constitutive for ethanol synthesis: They continue to produce ethanol whether it is present in the medium or not. Mutations of this type map at loci *c* and *d*. Strains of *Pickelensia* constructed to be diploid for the ethanol operon have the following phenotypes:

   $a^+b^-c^+d^+/a^-b^+c^+d^+$: inducible
   $a^+b^+c^+d^-/a^+b^+c^+d^+$: inducible
   $a^+b^+c^+d^+/a^+b^+c^-d^+$: constitutive

   (a) Identify each of the four genes of the ethanol operon.
   (b) Indicate the expected phenotype for each of the following partially diploid strains:
   (i) $a^-b^+c^+d^+/a^-b^+c^+d^-$
   (ii) $a^+b^-c^+d^+/a^-b^+c^+d^-$
   (iii) $a^+b^+c^+d^-/a^-b^-c^-d^+$
   (iv) $a^-b^+c^-d^+/a^+b^-c^+d^+$

4. **Regulation of Bellicose Catabolism.** The enzymes bellicose kinase and bellicose phosphate dehydrogenase are coordinately regulated in the bacterium *Hokus focus*. The genes encoding these proteins, *bel-A* and *bel-B*, are contiguous segments on the genetic map of the organism. In their pioneering work on this system, Professors Susan Bright and Edgar Early established that the bacterium can grow with the monosaccharide bellicose as its only carbon and energy source, that the two enzymes involved in bellicose catabolism are synthesized by the bacterium only when bellicose is present in the medium, and that enzyme production is turned off in the presence of glucose. They identified a number of mutations that reduce or eliminate enzyme production and showed that these could be grouped into two classes on the basis of the *cis-trans* test. Class I mutants act only in *cis* in diploids, whereas those in class II all map at a distance from the structural genes and are trans-acting. Thus far, the only constitutive mutations that Bright and Early have found are deletions that connect *bel-A* and *bel-B* to new DNA at the "left" side of *bel-A*. The following is a list of conclusions that Bright and Early would like to draw from their observations. Indicate in

each case whether the conclusion is consistent with the data (C), inconsistent with the data (I), or irrelevant to the data (X).

(a) Bellicose can be metabolized by *H. focus* cells to yield ATP.

(b) Enzyme production by genes *bel-A* and *bel-B* is under positive control.

(c) Phosphorylation of bellicose makes the sugar less permeable to transport across the plasma membrane.

(d) The operator for the bellicose operon is located to the left of the promoter.

(e) Some of the mutations in class I may be in the promoter.

(f) Some of the mutations in class II may be in the operator.

(g) Class II mutations may include mutations in the gene that makes CAP.

(h) The constitutive deletion mutations connect genes *bel-A* and *bel-B* to a new promoter.

(i) The constitutive deletion mutations are *trans*-acting in diploids.

(j) Expression of the bellicose operon is subject to regulation by attenuation.

5. **Attenuation in 25 Words or Less.** Complete each of the following statements about attenuation in 25 words or less.

(a) Attenuators can also be called *conditional terminators* because . . .

(b) When the gene product of the operon is not needed, the 5′ terminus of the mRNA forms a secondary structure that . . .

(c) Implicit in our understanding of the mechanism of attenuation is the assumption that ribosomes follow RNA polymerase very closely. This "tailgating" is essential to the model because . . .

(d) Measurements of rates of synthesis indicate that translation normally proceeds faster than transcription. This supports the proposed mechanism for attenuation because . . .

(e) If the operon for amino acid X is subject to attenuation, the leader sequence probably encodes a polypeptide that . . .

(f) Attenuation is probably a property of operons that code for enzymes involved in amino acid biosynthesis because . . .

(g) Attenuation is not a likely mechanism for regulating gene expression in eukaryotes because . . .

6. **Positive and Negative Control.** Assume that you have a culture of *E. coli* cells growing on medium B, which contains both lactose and glucose. At time *t*, 33% of the cells are transferred to medium L, which contains lactose but not glucose; 33% are transferred to medium G, which contains glucose but not lactose; and the remaining cells are left in medium B. For each of the following statements, indicate with an L if it is true of the cells transferred to medium L, with a G if it is true of the cells transferred to medium G, with a B if it is true of the cells left in medium B, and with an N if it is true of none of the cells. In some cases, more than one letter may be appropriate.

(a) The rate of glucose consumption per cell is approximately the same after time *t* as before.

(b) The rate of lactose consumption is higher after time *t* than before.

(c) The intracellular cAMP level is lower after time *t* than before.

(d) Most of the *lac* operator sites have *lac* repressor proteins bound to them.

(e) Most of the catabolite activator proteins exist as a CAP-cAMP complex.

(f) Most of the *lac* repressor proteins exist as a repressor-glucose complex.

(g) The rate of transcription of the *lac* operon is greater after time *t* than before.

(h) The *lac* operon has both the catabolite activator protein and the repressor protein bound to it.

7. **Polytene Chromosomes.** The fruit fly *Drosophila melanogaster* has about 2×10^8 nucleotide pairs of DNA per haploid genome, of which about 75% is unique-sequence DNA. The DNA is distributed over four chromosomes, which have a total of about 5000 visible bands in the case of the polytene chromosomes of the salivary gland. The number of complementation groups estimated from mutational studies is also about 5000, although recent molecular studies indicate that many bands contain more than one gene.

(a) Why was it tempting to speculate that each band corresponds to a single gene? Why might the mutational and molecular studies reach different conclusions about gene number? What does this suggest about the number of different proteins that *Drosophila* can make? Does that seem like a reasonable number to you?

(b) Assuming that all the unique-sequence DNA is uniformly distributed in the chromosomes, how much unique-sequence DNA (in nucleotide pairs) is there in the average band?

(c) How much DNA would it take to code for a single protein with a molecular weight of 50,000? (Assume that amino acids have an average molecular weight of 110.) What proportion does that represent of the total unique-sequence DNA in the average band, assuming a gene number of 5000 or of 15,000?

(d) How do you account for the discrepancy in part (c)?

8. **Gene Amplification.** The best-studied example of gene amplification is that of the ribosomal genes during oogenesis in *Xenopus laevis*. The unamplified number of genes is about 500 per haploid genome. After amplification, the tetraploid (premeiotic) oocyte contains about 2,000,000 genes (that is, 500,000 genes per haploid genome). This level of amplification is apparently necessary to allow the egg cell to synthesize the 10^{12} ribosomes that accumulate during the two months of oogenesis in this species. Each ribosomal gene consists of about 13,000 nucleotide pairs, and the genome size of *Xenopus* is about 2.7×10^9 nucleotide pairs per haploid genome.
   (a) What fraction of the total haploid genome do the 500 copies of the ribosomal genes represent?
   (b) What is the total size of the amplified genome (in nucleotide pairs)? What proportion of this do the amplified ribosomal genes represent?
   (c) Assume that all the ribosomal genes in the amplified oocyte are transcribed continuously to generate the needed number of ribosomes during the two months of oogenesis. How long would oogenesis have to extend if the genes had not been amplified?
   (d) Why do you think genes have to be amplified when the gene product needed by the cell is an RNA, but not usually when the desired gene product is a protein?

9. **Levels of Control.** Assume that liver and kidney tissues from the same mouse contain about 10,000 species of cytoplasmic mRNA, but that only about 25% of these are common to the two tissues.
   (a) Suggest an experimental approach that might have been used to establish that some of the mRNA molecules were common to both tissues, but others were not.
   (b) One possible explanation for the data is that differential transcription occurs in liver and kidney nuclei. What is another possible explanation? Describe an experiment that would enable you to distinguish between these possibilities.
   (c) If all mRNA molecules had been shown to be common to both liver and kidney and yet the two tissues were known to be synthesizing different proteins, what level of control would you have to assume?

10. **Messenger RNA Complexity in Sea Urchins.** Figure 17-29 depicts the complexity of mRNA molecules found in oocytes, embryonic stages, and adult tissues of the sea urchin. The solid portion of each bar indicates the fraction of the gastrula-stage mRNA also found in mRNA from other sources. The open portion of each bar indicates the amount of mRNA not shared in common with the gastrula. Label each of the following conclusions as C if it is consistent with the data in the figure, I if it is inconsistent, and X if you cannot tell from the data.
   (a) A small fraction of RNA species is common to all tissues examined.
   (b) The mRNA molecules common to all tissues probably code for "housekeeping" proteins responsible for metabolic processes common to all cells.
   (c) The mRNA molecules present at the gastrula stage of embryogenesis represent less than 3% of the unique-sequence DNA in the genome.
   (d) The mRNA molecules present in the coelomocyte of the adult can code for about 30,000 average-sized proteins.
   (e) Differentiation of adult tissues appears to be accompanied by a decrease in the number of genetic functions being expressed in the cells.
   (f) All the genes that are being actively transcribed in the intestine are also turned on at the pluteus stage of embryogenesis.
   (g) The oocyte is expressing genetic information that is not being expressed in the intestine.
   (h) The intestine is expressing genetic information that is not being expressed in the oocyte.

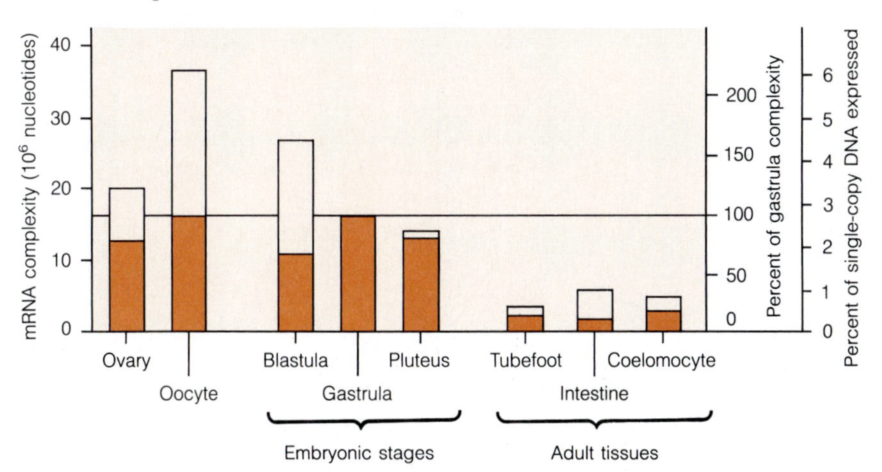

Figure 17-29 Sequence Complexity of mRNA Molecules. The mRNA molecules found in the cytoplasm in oocytes, embryos, and adult tissues of the sea urchin represent a considerable diversity of sequences. The solid portion of each bar indicates the amount of mRNA shared between gastrula mRNA and mRNA from other sources. The open portion of each bar shows the amount of mRNA present in the various sources that is not present in gastrula mRNA. Blastula, gastrula, and pluteus are successive stages in embryogenesis: tubefoot, intestine, and coelomocyte are adult tissues.

PART FIVE

SPECIFIC CELL FUNCTIONS

18

Cytoskeletal Structure and Function

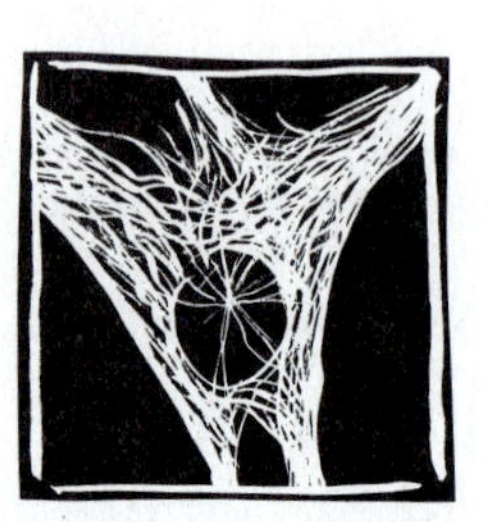

In the foregoing chapters, we have examined a variety of cellular processes and pathways, many of which occur in the organelles of eukaryotic cells. We come now to the **cytoplasm,** which is of much current interest because of its intricate network of filaments and tubules that impart shape to and facilitate various kinds of movements in eukaryotic cells.

Until relatively recently, the cytoplasm of the eukaryotic cell was regarded as a generally uninteresting, gel-like substance in which the nucleus and other organelles were suspended. Cell biologists knew that the cytoplasm was quite rich in proteins—about 20–30%, usually—but these proteins were thought to be soluble and freely diffusible. Except for those of known enzymatic activity, little was understood about the structural or functional significance of the cytoplasmic proteins. The advent of electron microscopy made us aware of structures such as microtubules and microfilaments in the cytoplasm, but initially this knowledge did little to change the prevailing view of the cytoplasm as an amorphous matrix without much fascination for cell biologists.

Within recent years, however, that view has changed dramatically. Several new microscopic techniques have revealed that the interior of a eukaryotic cell is highly structured. A complex network of interconnected filaments and tubules called the **cytoskeleton** extends throughout the cytoplasm, from the nucleus to the inner surface of the plasma membrane. This elaborate array of filaments and tubules forms a highly structured yet very dynamic matrix that helps to establish the shape of the cell and plays important roles in cell movement and cell division.

In addition, the cytoskeleton serves as a framework for positioning and actively moving organelles within the cytoplasm and may even play a similar role for ribosomes. For example, clusters of ribosomes are frequently seen in association with filaments and remain behind with the cytoskeleton after extraction of cells with nonionic detergents. The same may be true of enzymes and other soluble proteins. Some researchers think that many, perhaps even most, enzymes in the cytoplasm are not really soluble at all, but are physically clustered and attached to the cytoskeleton in close proximity to other enzymes involved in the same pathway, thereby facilitating the channeling of intermediates within each pathway.

The term "cytoskeleton" expresses well the role of this proteinaceous matrix in providing an architectural framework for eukaryotic cells. This framework, or internal scaffolding, confers a high level of internal organization on such cells and enables them to assume and maintain complex shapes that would not otherwise be possible. However, the term is less adequate at conveying the functional significance of the cytoskeleton, which is probably more important than its structural role.

In a sense, the cytoskeleton might be more aptly regarded as the "muscles" of the eukaryotic cell than its "skeleton," since the latter term implies a static framework, whereas the cytoskeleton is a dynamic, changeable matrix. The various components of the cytoskeleton are, in fact, involved in a great diversity of cellular shape changes and movements. Examples of such movements include muscle contraction, beating of cilia and flagella, amoeboid movement, cell division, chromosome movements, endo- and exocytosis, changes in cell shape, and the active movement of vesicles and other organelles within the cell.

Some of these examples obviously involve movement of whole cells, whereas others entail motion of intracellular components such as membranes, chromosomes, or organelles. All, however, are mediated by the structural elements of the cytoskeleton, which are the subject of this and the next chapter.

Structural Elements of the Cytoskeleton

The three major structural elements of the cytoskeleton are *microtubules, microfilaments,* and *intermediate filaments* (sometimes also called *intermediate-sized filaments*). The existence of three distinct systems of filaments and tubules was first revealed by electron microscopy. Biochemical and immunological studies then identified the distinctive proteins of each system. The technique of *immunofluorescence microscopy,* to be described in the next section, was especially important in localizing specific proteins within the cytoskeletal network.

Each of the structural elements of the cytoskeleton has a characteristic size, structure, and intracellular distribution, and each is formed by polymerization of a different kind of protein monomer (Table 18-1). Microtubules consist of *tubulin* and are about 25 nm in diameter. Microfilaments are polymers of the protein *actin,* with a diameter of about 7 nm. Intermediate filaments have diameters in the range 8–12 nm and contain any of several related but tissue-specific proteins. In addition to its major protein component, each structural element of the cytoskeleton also has a number of other proteins associated with it. These *accessory proteins* account for the remarkable structural and functional diversity of the cytoskeletal elements.

Microtubules, microfilaments, and intermediate filaments are all unique to eukaryotic cells, as are their respective monomers; prokaryotes apparently contain no such filaments, tubules, or proteins. The absence of such molecules and structures, and hence of the cytoskeleton itself, from prokaryotic cells suggests that the cytoskeleton may have played a crucial role in the evolution of eukaryotic cells.

Microtubules and microfilaments are best known for their roles in contraction and motility. Microfilaments are essential components of *muscle fibrils,* and microtubules are the structural elements of *cilia* and *flagella,* appendages that enable certain cells either to propel themselves through a fluid environment or to move fluids past the cell. These structures are large enough to be seen with the light microscope and were therefore known and studied long before it became clear that the same structural elements are also integral parts of the cytoskeleton.

Motility and contractility are very important topics in cell biology and will therefore be discussed in detail in the next chapter. Here, we will focus specifically on the involvement of filaments and tubules in cytoskeletal structure and on the molecular basis for the general functions in which they are involved. We will look first at several of the techniques currently in use to study cytoskeletal organization and then consider each of these structural elements in detail.

In doing so, we will be discussing microtubules, microfilaments, and intermediate filaments as though they were separate entities with their own structure and function. But to appreciate their role in giving cells shape and in mediating various types of cell movement, we need to understand that the several components of the cytoskeleton are linked together structurally and function not as independent entities, but as an integral meshwork of filaments, tubules, and their associated proteins. Moreover, the cyto-

Table 18-1 Properties of Microtubules, Microfilaments, and Intermediate Filaments

Property	Microtubules	Microfilaments	Intermediate Filaments
Structure	Hollow tube with a wall consisting of 13 protofilaments	Two intertwined chains of F-actin	Eight protofilaments joined end-to-end with staggered overlaps
Diameter	Outer: 25 nm Inner: 15 nm	7 nm	8–12 nm
Monomers	α-tubulin β-tubulin	G-actin	Several proteins; see Table 18-3 on page 572
Functions	Cell motility Chromosome movements Disposition and movement of organelles Determination of cell shape Maintenance of cell shape	Muscle contraction Amoeboid movement Cell locomotion Cytoplasmic streaming Cell division Maintenance of cell shape	Structural support Maintenance of cell shape Formation of nuclear lamina and scaffolding Strengthening of nerve cell axons (NF protein) Keeping muscle fibers in register (desmin)

skeleton of one cell can influence that of neighboring cells through intercellular junctions and by means of the extracellular matrix that many animal cells secrete.

Techniques for Studying the Cytoskeleton

The cytoskeleton is a topic of much current research interest to eukaryotic cell biologists. Most of the recent progress in our understanding of cytoskeleton structure is owed to three powerful microscopic techniques: *immunofluorescence microscopy, quick-freeze deep-etch microscopy*, and *high-voltage electron microscopy*. We will look at each technique briefly, both to appreciate its potential for revealing cell structure and to understand the ways in which the three techniques provide us with complementary insights into the cytoskeleton. In addition, we will consider the use of specific drugs and antibodies in analyses of cytoskeletal function.

Immunofluorescence Microscopy

Immunofluorescence microscopy is a very useful technique for studying the overall distribution of tubules and filaments within whole cells. It is also valuable as a means of determining which accessory proteins are associated with the different systems. This technique usually requires two sets of antibodies, designated respectively as *primary antibodies* and *secondary antibodies*. The primary antibodies are specific for a particular protein (actin, for example). They are allowed to bind to the target protein in fixed cells that have been treated with an organic solvent or a detergent to make the membranes permeable to the antibodies. The cells are then exposed to secondary antibodies that recognize the primary antibodies and are covalently linked to a fluorescent dye such as *fluorescein*, thereby enabling them to be localized by fluorescence microscopy.

Thus, the primary antibodies bind to the target protein wherever it occurs in the cell and the secondary antibodies bind to the primary antibodies, causing the appropriate structures to "light up" in the fluorescence microscope. Figure 18-1 illustrates this technique for each of the three structural elements of the cytoskeleton, using antibodies against tubulin, actin, and the protein *vimentin* to visualize microtubules (Figure 18-1a), microfilaments (Figure 18-1b), and intermediate filaments (Figure 18-1c), respectively.

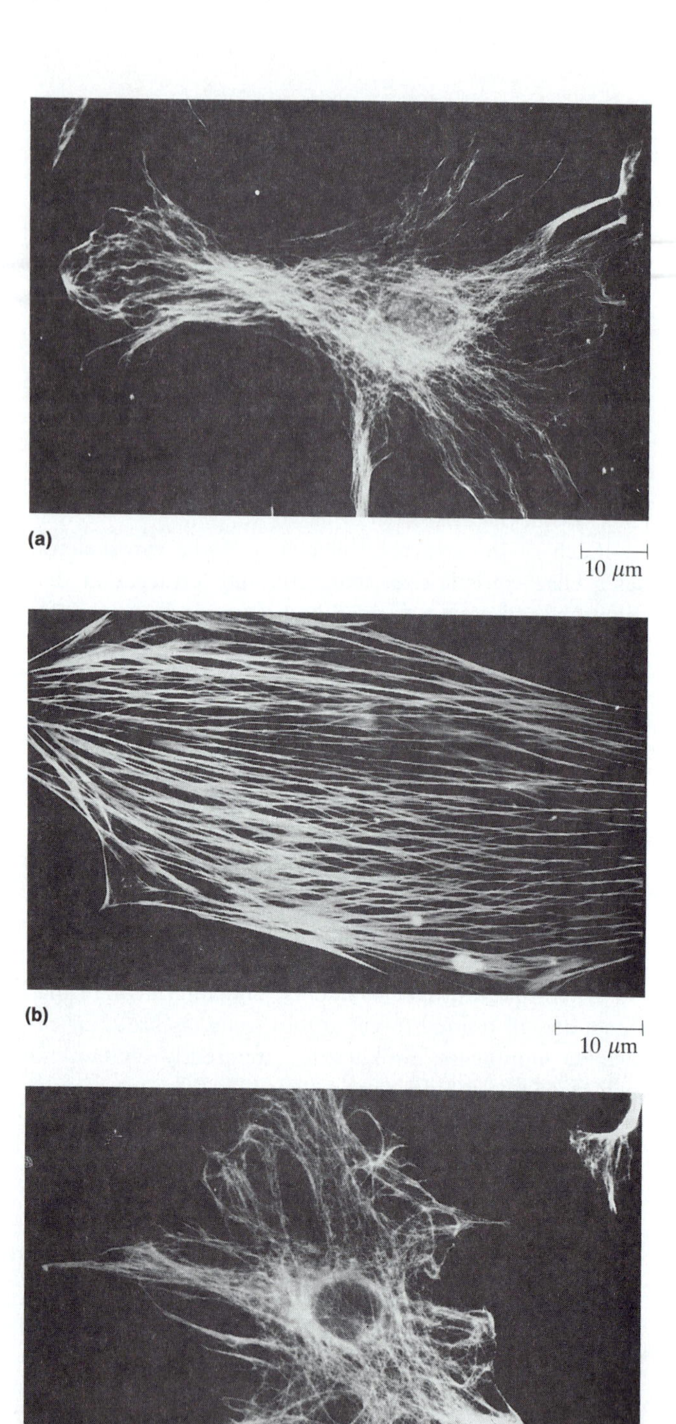

Figure 18-1 The Cytoskeleton as Revealed by Immunofluorescence Microscopy. Mouse fibroblast cells were prepared for microscopy, then subjected to the double-antibody procedure described in the text, with the fluorescent dye *fluorescein* bound to the secondary antibodies (goat anti-rabbit antibody). (a) Microtubules as revealed by the use of rabbit antibodies against tubulin. (b) Microfilaments as revealed by the use of rabbit antibodies against actin. (c) Intermediate filaments as revealed by the use of rabbit antibodies against vimentin. (all LMs)

Quick-Freeze Deep-Etch Microscopy

Quick-freeze deep-etch microscopy, or **deep-etching** for short, is a modification of the freeze-etching technique described in Appendix A. In freeze-etching, cells are frozen very quickly at the temperature of liquid nitrogen (−196°C), and the frozen tissue is cracked with a knife blade. Ice is then removed by sublimation in a vacuum, thus making it possible to see more deeply into the cell. Finally, a platinum replica of the exposed surface is prepared and examined in the electron microscope.

The biggest limitation of freeze-etching in the past has been the formation of ice crystals, which cause distortions in the cytoskeletal architecture. In the deep-etching technique, a copper block cooled with liquid helium (−269°C) is used to freeze the sample so quickly (within milliseconds) that ice crystals do not form, thereby avoiding structural distortions. The frozen sample is then subjected to fracturing and sublimation, and coated with a thin molecular layer of platinum to visualize filamentous material. Deep-etching provides fascinating glimpses into the interior of the cell. It has been very useful in exploring the cytoskeleton and especially in examining connections with other structures of the cell (Figure 18-2).

High-Voltage Electron Microscopy

Our understanding of the cytoskeleton has also been aided by the advent of **high-voltage electron microscopy.** As we learned in Chapter 1, a high-voltage electron microscope can develop an accelerating potential of several thousand kilovolts, much higher than is possible with conventional instruments. As a result, much thicker specimens can be examined (up to 10–20 μm), allowing an in-depth view of cells that was not previously possible (Figure 18-3).

This technique has also been responsible for a controversy concerning the cytoskeleton, mainly because investigators differ both in the procedures they use and in the way they interpret their findings. At issue is the significance of cytoskeletal fibrils called *microtrabeculae* that can be seen under certain conditions and appear to interconnect the various tubules and filaments of the cytoskeleton. These fibrils are highly variable in diameter and appear to branch with each other at smooth, continuous junctions rather than at single points of crossover as seen with other techniques.

Initially, these microtrabecular fibrils were interpreted by some microscopists to represent a cytoskeletal component distinct from microtubules, microfilaments, and intermediate filaments. Now, however, most investigators consider them artifacts caused by the precipitation of soluble proteins during the fixation procedure. It seems likely that the soluble proteins precipitate onto the accessory proteins that link microtubules, microfilaments, and intermediate filaments within the cytoskeleton. If so, the microtrabecular fibrils, though artifacts in a sense, may help to visualize the normal connections between the several components of the cytoskeleton.

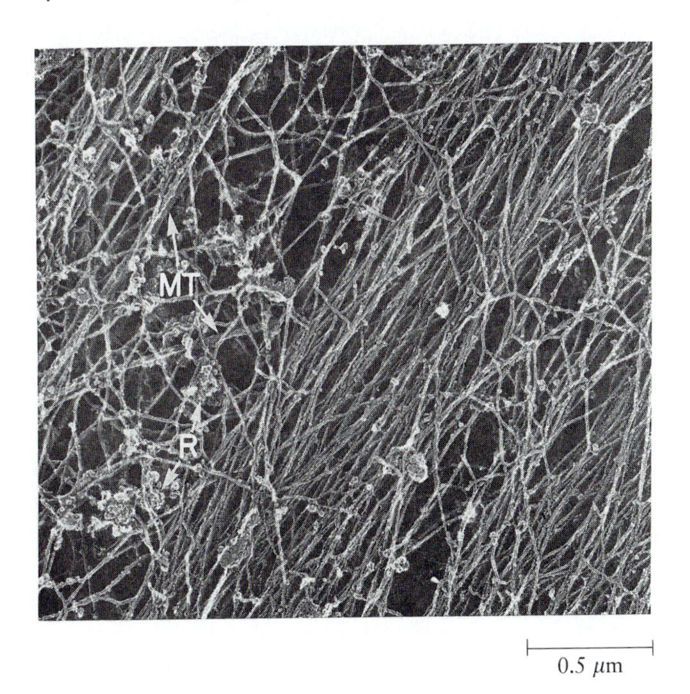

Figure 18-2 The Cytoskeleton as Revealed by Quick-Freeze Deep-Etch Microscopy. A fibroblast cell was extracted with a nonionic detergent to remove soluble cytoplasmic proteins and integral membrane proteins. A platinum replica of the cytoskeleton was then prepared by the quick-freeze deep-etch technique. The bundles of actin microfilaments in the lower right are part of the cell's stress fibers. Microtubules (MT) are visible in the upper left, and several ribosomes (R) are seen lining some of the filaments.

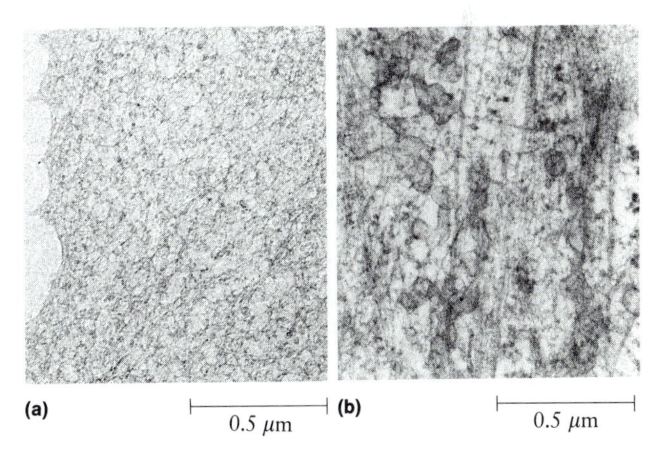

Figure 18-3 The Cytoskeleton as Revealed by High-Voltage Electron Microscopy. The cytoskeleton of a kidney cell in culture, as visualized by high-voltage electron microscopy. (a) A cell grown on an electron microscope support grid, fixed in formaldehyde-glutaraldehyde, and prepared for electron microscopy by a technique called critical point drying. (b) A thick section (0.25 μm) of the cytoplasm reveals a complex system of microtubules, intermediate filaments, and actin microfilaments, as well as connecting filaments with a diameter of about 3 nm.

Drugs and Antibodies as Tools for Studying Cytoskeletal Function

Microscopic techniques can reveal much about the structure of the cytoskeleton but do not usually allow us to deduce much about its function, since these techniques provide only a static view of the cytoskeleton—a glimpse of the cell at the moment it was fixed for observation. In some cases, microscopic techniques can be extended to living cells. For example, a cytoskeletal protein can be "tagged" with a fluorescent compound and injected into a living cell with a fine glass needle. An *image-intensifying video camera* can then be used to monitor the compound in the cell.

In general, however, alternatives to microscopic techniques are needed for functional analyses. One such approach is the use of *drugs* that are known to bind specifically to a particular cytoskeletal protein. By studying the effects of such drugs on specific cellular processes, it is often possible to identify, at least tentatively, functions in which the particular protein might be involved.

For example, the drugs *colchicine* and *taxol* disrupt microtubule function in distinctively different ways. Colchicine binds to tubulin, strongly inhibiting its further assembly into microtubules and fostering the disassembly of existing microtubules. Taxol, in contrast, binds tightly to microtubules and stabilizes them, causing much of the free tubulin in the cell to assemble into microtubules. Sensitivity of a cellular process to colchicine or taxol is therefore a good first indication that microtubules may mediate that process within the cell. In a similar manner, the drug *cytochalasin B* inhibits the polymerization of actin microfilaments, whereas *phalloidin* blocks the depolymerization of actin, thereby stabilizing microfilaments. Obviously, processes that are disrupted in cells treated with either of these drugs are likely to be dependent in some way upon microfilaments.

Antibodies can also be used to assess the involvement of cytoskeletal proteins in cellular functions, since an antibody against a specific protein will often inactivate that protein when bound to it. As with exposure to a drug, injection of a specific antibody into a living cell can provide evidence to link a particular protein to a certain cellular function.

With these techniques in mind, we are now ready to look at each of the three major components of the cytoskeleton. In each case, we will consider the chemistry of the monomer, the structure of the polymer, the polymerization process, the role of accessory proteins, and some of the structural and functional roles attributable to that component. We will look first at microtubules, because of their key role in organizing both microfilaments and intermediate filaments in the cytoskeletal array. Then we will turn to microfilaments and finally to intermediate filaments.

Microtubules

With a diameter of about 25 nm, **microtubules (MTs)** are the largest of the cytoskeletal elements (Table 18-1). The microtubules in eukaryotic cells can be classified into two general groups, which differ both in degree of organization and in structural stability. One group includes the highly organized, stable microtubules found in specific subcellular structures associated with cellular movement, including cilia, flagella, and the *basal bodies* to which these appendages are attached. The central shaft, or *axoneme,* of a cilium or flagellum consists of a highly ordered bundle of **axonemal microtubules** and associated proteins. The other group consists of the more loosely organized, dynamic network of **cytoplasmic microtubules.**

Given their order and stability, it is not surprising that the axonemal microtubules were the first of the two groups to be recognized and studied. We have already encountered an example of such a structure, since the axoneme of the sperm tail shown in Figure 4-12 consists of microtubules. As already noted, we will consider axoneme structure and microtubule-mediated motility further in Chapter 19.

The occurrence of cytoplasmic microtubules in eukaryotic cells was not recognized until the early 1960s, when the introduction of gentler fixation techniques permitted the direct visualization of the network of microtubules now known to pervade the cytoplasm of most eukaryotic cells. Since then, immunofluorescence microscopy has revealed that microtubules radiate out as lacelike threads toward the periphery of the cell from a **microtubule-organizing center (MTOC)** near the nucleus. In animal cells, the best-known MTOC is the **centrosome** (or **cell center**), which consists of granular material surrounding two *centrioles,* the MT-containing structures that we encountered in Chapter 14. Other examples of MTOCs include the *kinetochore* and the poles of the *mitotic spindle,* structures that are also familiar from Chapter 14.

In general, cellular structures consisting of cytoplasmic microtubules are much less stable and hence more dynamic than those involving axonemal microtubules. But even among cytoplasmic microtubules, some are more stable than others. For example, those that are acetylated may remain intact longer than their nonacetylated counterparts.

Cytoplasmic microtubules are responsible for a variety of functions (Table 18-1). For example, they define and maintain the overall shape and architecture of the cell, confer polarity on the cell, and determine the distribution of the microfilaments and intermediate filaments. While all components of the cytoskeleton contribute to the overall shape of the cell, microtubules appear to play a unique role in determining it. By providing a radiating system of fibers to guide the movement of vesicles and other organelles, microtubules also contribute to the spatial disposition and directional movement of these subcellular structures.

Specific examples of the diverse phenomena that are thought to be governed by cytoplasmic microtubules include the asymmetrical shape of cells, the configuration of the plasma membrane, the plane of cell division in higher plant cells, the orientation with which cellulose microfibrils are deposited during growth of plant cell walls, the ordering of filaments during muscle development, the location of the Golgi complex and the endoplasmic reticulum, and the active movement of vesicles and other organelles. Cytoplasmic microtubules are also involved in many of the cell movements and changes in cell shape that occur during embryonic development in animals.

Structure of Microtubules

Microtubules are straight, hollow cylinders with an outer diameter of about 25 nm and an inner diameter of about 15 nm (Figure 18-4a). Microtubules vary greatly in length. Some are less than 200 nm long, while others, particularly those in nerve cells, can be as long as 25μm. Figure 18-4b shows microtubules of an axoneme as revealed by electron microscopy.

The wall of the microtubule consists of longitudinal arrays of **protofilaments**, usually 13 arranged side-by-side around the hollow center, or **lumen** (Figure 18-4a). Each protofilament is a linear polymer of **tubulin** molecules. Tubulin is a dimeric protein, consisting of two similar but distinct polypeptide subunits, **α-tubulin** and **β-tubulin,** linked to each other covalently. Each of these polypeptides has a diameter of about 4–5 nm and a molecular weight of about 50,000. Bound to the $\alpha\beta$ dimer are two molecules of the high-energy nucleotide GTP.

Within a protofilament, all of the tubulin dimers are oriented in the same direction, such that all of the α subunits face the same end. This uniform orientation of tubulin dimers means that one end of the protofilament differs chemically and structurally from the other, giving the protofilament an inherent **polarity.** And since the orientation of the tubulin dimers is the same for all of the protofilaments in a microtubule, the microtubule itself is also a polar structure.

This polarity is an important feature of the microtubule because of its implications for microtubule formation in the cell. If a fragment of a ciliary axoneme is incubated in vitro with tubulin molecules, one end of the axoneme will elongate more rapidly than the other. This faster-

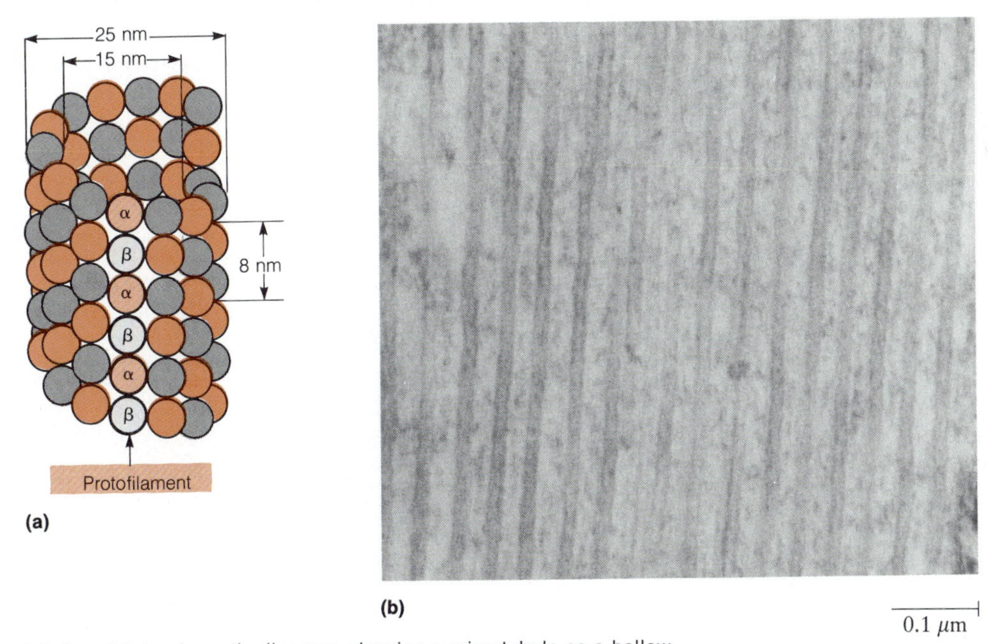

Figure 18-4 Microtubules. (a) A schematic diagram, showing a microtubule as a hollow cylinder with an outside diameter of about 25 nm and an inside diameter of about 15 nm. The wall of the cylinder consists of 13 protofilaments, one of which is noted by an arrow. A protofilament is a linear polymer of tubulin dimers, each of which consists of two polypeptides, α-tubulin and β-tubulin. All dimers in the protofilaments have the same orientation, thereby accounting for the polarity of the microtubule. (b) Microtubules as seen in a longitudinal section of an axoneme (TEM).

growing end is called the **plus end** and the other is designated as the **minus end.** In the cell, the ends of the microtubules farthest away from the center of the cell are always the plus ends, such that microtubule growth occurs in an outward direction.

As we will see shortly, microtubule elongation is a complex process. Whether assembly or disassembly occurs at a given plus end appears to depend on how rapidly the GTP on the newly added tubulin dimer is hydrolyzed to GDP. *Microtubule-associated proteins* are also important in regulating the assembly process, most likely by binding to the growing plus end of the microtubule, thereby stabilizing it against disassembly.

Genetics of Tubulin

Most organisms have several closely related genes that encode each of the tubulin subunits, thereby allowing for possible tubulin heterogeneity. For example, both the fruit fly *Drosophila* and the chicken have four genes that encode α-tubulin and another four for β-tubulin, whereas the green alga *Chlamydomonas reinhardtii* has two genes for each polypeptide. In general, it is not clear whether all of the genes are actually expressed, nor is it known whether specific genes are expressed only in specific cell types. In the chicken, each of the four genes for β-tubulin is known to be functional, but that does not necessarily mean that all four are actually necessary.

Different MT-containing structures appear to contain different kinds of α- and β-tubulin molecules, at least in some species. In *Drosophila,* for example, the tubulin subunits present in the axoneme of the sperm flagellum are different from those in axonemes elsewhere in the organism. It is not yet known, however, whether the observed heterogeneity in tubulin polypeptides is due to the presence of different gene products for each subunit or is instead the result of chemical modifications (such as acetylation) of a single gene product. Nor is it yet clear how significant this heterogeneity is for microtubule function, since some species appear not to require tubulin heterogeneity. Yeast cells, for example, have a single β-tubulin gene, and the two β-tubulin genes of *Chlamydomonas* are known to encode identical proteins. These organisms must therefore be able to assemble all of the necessary MT-containing structures using a single kind of β-tubulin.

When the tubulins of diverse species are compared, they are found to be remarkably similar in sequence and structure—though not as strikingly as in the case of actin, as we will see later. In fact, all known tubulins will polymerize into microtubules if mixed together in vitro, whether they come from different organisms or from different tissues within the same organism.

Assembly of Microtubules

Microtubules form by the reversible polymerization of tubulin dimers, a process that has been studied extensively in vitro. A schematic representation of microtubule assem-

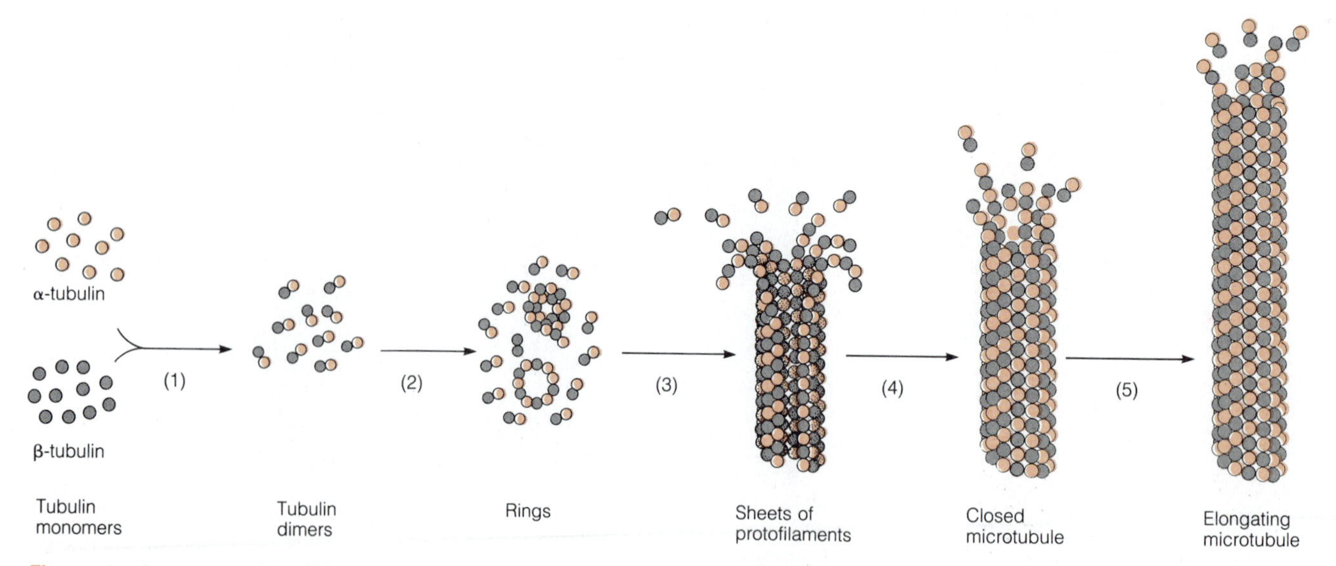

Figure 18-5 Assembly of Microtubules. (1) Microtubules are assembled from tubulin synthesized as separate α and β polypeptides, which associate spontaneously to form stable αβ dimers. (2) These dimers polymerize into rings, which (3) uncoil into linear protofilaments that align laterally into sheets. (4) The growing sheet rolls up and forms a closed microtubule with 13 protofilaments, and (5) the closed microtubule continues to elongate by further addition of dimers to the ends of the protofilaments.

bly is shown in Figure 18-5. The α and β subunits are synthesized as separate polypeptides, which associate stably to form dimeric tubulin molecules (step 1). As the dimers polymerize (step 2), they form rings, which appear to be obligatory intermediates in the assembly process, at least as it occurs in vitro. The rings then uncoil into protofilaments, which associate side-by-side into sheets (step 3).

As the sheet grows, it begins to roll up, eventually closing to form an intact microtubule when the number of laterally aligned protofilaments reaches 13 (step 4). Thereafter, the microtubule elongates by addition of dimers to the ends of the protofilaments (step 5). Figure 18-6 shows microtubules that have been assembled in vitro, as they appear in both longitudinal section (Figure 18-6a) and cross section (Figure 18-6b).

As studied in vitro, microtubule assembly is characterized by a **lag phase** that is thought to reflect a need for dimers and rings to come together in the right configuration to initiate tubule formation. This initial **nucleation step** is then followed by a rapid **elongation phase** as additional dimer units are added to the growing microtubule. As long as the concentration of free tubulin dimers is high, net assembly will occur and the microtubules will continue to elongate. As the concentration of tubulin dimers falls, the rate of polymerization declines, and the rate of net elongation declines. Eventually, a **critical concentration** of tubulin dimers is reached below which no further assembly will occur.

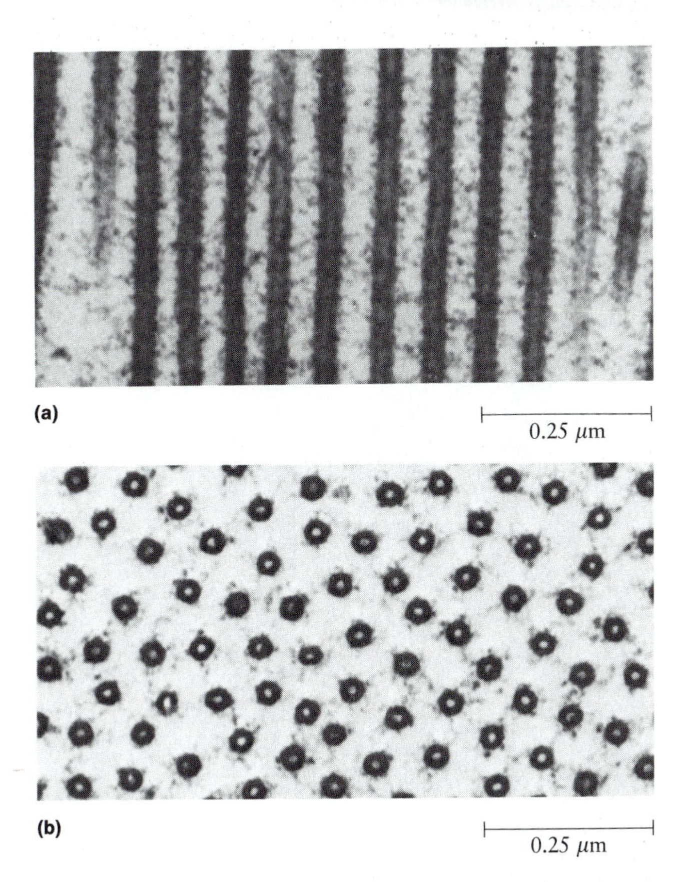

Figure 18-6 Microtubules Assembled in Vitro. These electron micrographs illustrate brain microtubules that have been assembled in vitro and examined (a) in longitudinal section and (b) in cross section. (both TEMs)

The addition of a tubulin dimer at the plus end of a growing microtubule is usually followed by the hydrolysis of one of the two molecules of GTP that are normally bound to each dimer. However, the energy of GTP hydrolysis does not appear to be necessary for the polymerization process, since assembly can occur even if a nonhydrolyzable analogue of GTP is present instead. Rather, the hydrolysis of GTP appears to change the affinity of the microtubule end for incoming tubulin dimers in a way that has important consequences for one model of microtubule assembly and disassembly.

Tubulin assembly has a directionality to it due to the uniform orientation of the dimers, which gives rise to the polarity of the protofilaments and of the microtubules themselves. Directionality is a property of microtubule assembly not only in vitro but also in the intact cell. Here, the microtubule-organizing centers mentioned earlier appear to serve as natural nucleation sites. Microtubules then grow outward from such sites by the addition of tubulin subunits to their distal ends. In other words, the end of a microtubule farthest away from the MTOC (and hence, in most cases, from the nucleus) is always the plus end. The minus ends of the microtubules are anchored in the granular material of the centrosome.

The Dynamic Instability Model of Microtubule Assembly. The polymerization of tubulin appears to be readily reversible both in vitro and in vivo. In fact, most MT-containing structures in the cytoplasm are assembled rapidly from a pool of preexisting tubulin monomers and can be disassembled with equal rapidity. The assembly process is understood quite well, but scientists are less certain about the disassembly process. As a result, two alternative models of microtubule elongation have been proposed.

According to the *treadmilling model,* individual microtubules are thought to assemble preferentially at the plus end and to disassemble preferentially at the minus end. The net result is a "treadmilling" effect, with a given tubulin molecule incorporated at the plus end, transferred progressively along the microtubule, and eventually lost by depolymerization at the opposite end. This model was proposed in the late 1970s by Robert Margolis and Leslie Wilson, based on results obtained when they exposed a cell-free microtubule polymerizing system to radioactive GTP. As we will see later in the chapter, a treadmilling model may, in fact, be relevant for the assembly of actin into microfilaments. But for microtubule assembly, this model has been called into serious question by recent evidence suggesting that microtubules can behave differently under the same conditions, depending on whether the bound nucleotide is present as GTP or GDP.

Specifically, studies on microtubule assembly using isolated centrosomes as nucleation sites showed that one set of microtubules can grow by polymerization while another set shrinks by depolymerization. As a result, one set (or perhaps one type of MT-containing structure) effectively enlarges at the expense of another, at least as studied in vitro. That such concomitant growth and shrinkage also occurs in vivo is clear; as a cell approaches mitosis, for example, the mitotic spindle assembles at the same time that cytoplasmic microtubules are disassembling, presumably to provide the tubulin dimers needed to make the spindle fibers.

To explain how both polymerization and depolymerization might occur simultaneously, Tim Mitchison and Marc Kirschner proposed a **dynamic instability model,** which presumes two populations of microtubules, one growing in length by continued polymerization at the plus end, and the other shrinking in length by depolymerization (Figure 18-7). The distinction between the two populations is that growing microtubules have GTP bound to the tubulin at the plus end, while shrinking microtubules have GDP instead. Because GTP-tubulin molecules are thought to have a greater affinity for each other than for GDP-tubulin, the presence of a GTP "cap" at the plus end is postulated to create a stable microtubule tip to which further dimers can add (Figure 18-7a). A GDP "cap," on the other hand, results in an unstable tip at which depolymerization occurs rapidly (Figure 18-7b).

Whether GTP or GDP will be present depends, in turn, on whether the GTP on the newly added tubulin dimer has time to hydrolyze before the next dimer adds. Thus, dynamic instability is thought to be the result of the delayed hydrolysis of GTP after polymerization of a tubulin dimer. When dimers are being added quickly, the GTP does not have a chance to hydrolyze, and the end remains a stable growing end. But if the rate of dimer addition slows down, hydrolysis is likely to occur, converting what was a growing end into an unstable tip, which then begins to undergo rapid depolymerization.

Crucial to this model are the concentrations of tubulin and GTP. When GTP-tubulin is readily available, polymerization will occur so rapidly at the growing tip that

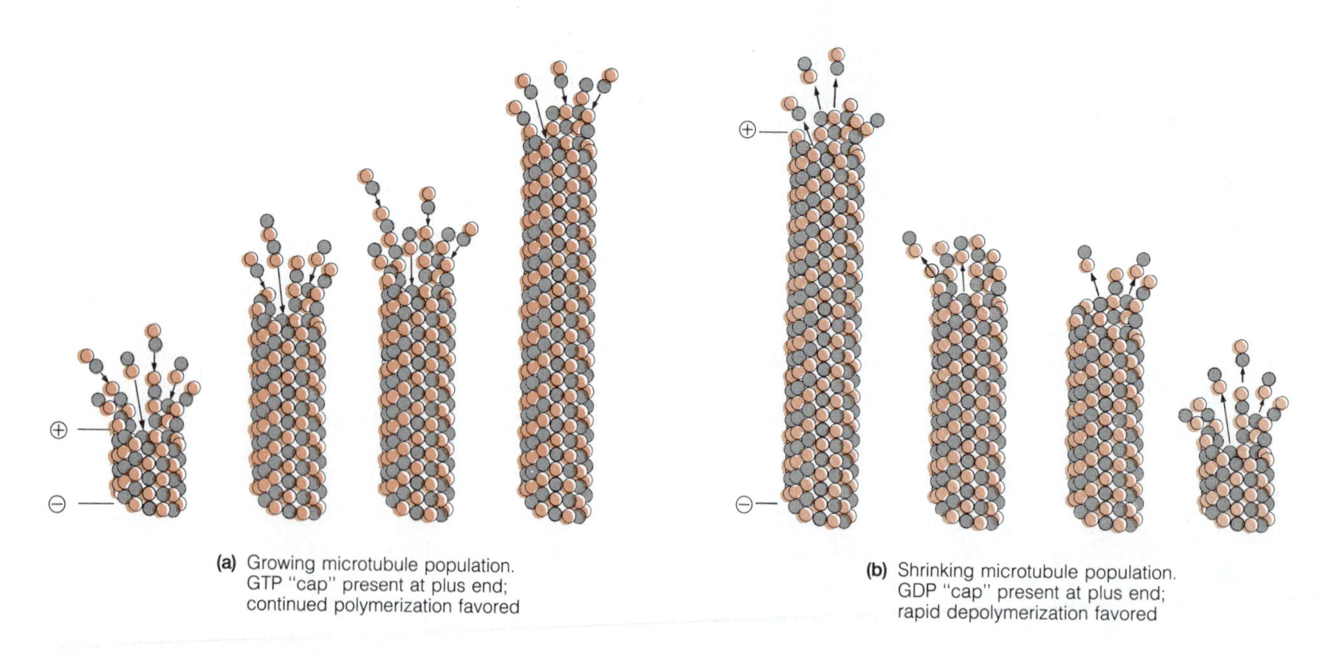

Figure 18-7 Dynamic Instability Model of Tubulin Assembly and Disassembly. This model assumes two populations of microtubules, one growing in length and the other shrinking in length. (a) The growing microtubules have GTP bound to the newly added tubulin dimers at the plus end. Because GTP-tubulin dimers have a high affinity for each other, the GTP "cap" creates a stable microtubule tip to which further tubulin dimers can add. (b) The shrinking microtubules have GDP bound to the newly added tubulin dimers instead, because the GTP originally present on each dimer has already hydrolyzed. The presence of a GDP "cap" makes the tip unstable, and rapid depolymerization is favored.

GTP hydrolysis is unlikely, and the tip continues to grow. If the concentration of GTP-tubulin falls, however, the rate of polymerization will decrease, hydrolysis of the GTP bound to already-added dimers is likely to occur, and the tip will become unstable, with depolymerization favored.

Drug Sensitivities of Microtubule Assembly. As noted earlier, several drugs are known to have specific effects on microtubule assembly. The best-known is *colchicine*, a plant alkaloid that acts by binding to tubulin dimers. The resulting tubulin-colchicine complex can still add to the growing end of a microtubule, but it then prevents any further addition of tubulin molecules and destabilizes the structure, thereby promoting microtubule disassembly. *Vinblastine* and *vincristine* are related compounds that cause tubulin to aggregate instead of assembling into microtubules.

All three of these compounds are known as *antimitotic drugs* because they cause the mitotic spindle of dividing cells to disappear, blocking the further progress of mitosis in the cells. The sensitivity of the mitotic spindle to these drugs is understandable, since the spindle fibers are microtubular structures. Vinblastine and vincristine also find application in medical practice as *anticancer drugs*. They are useful for this purpose because cancer cells divide rapidly and are therefore preferentially susceptible to drugs that interfere with the mitotic spindle.

Taxol has the opposite effect on microtubules: It binds to them and stabilizes them. Within cells, it causes free tubulin to assemble into microtubules and arrests dividing cells in mitosis. Thus, both taxol and colchicine block cells in mitosis, but they do so by opposite effects on microtubules and hence on the fibers of the mitotic spindle.

Microtubule-Associated Proteins

Although tubulin is the main protein component of microtubules, a variety of other proteins are known to be involved in microtubule structure, assembly, and function. These **microtubule-associated proteins (MAPs)** account for 10–15% of the microtubule mass. Two major classes of MAPs present in microtubules from brain cells are the *high-molecular-weight (HMW) proteins*, with molecular weights in the range 200,000 to 300,000, and the smaller *tau proteins*, which have molecular weights in the range 40,000 to 60,000. Both types of MAPs are known from immunological studies to bind along the entire length of cytoplasmic microtubules. One of the HMW proteins, MAP-2, forms regularly spaced sidearms along microtubules, with one end of the MAP-2 molecule bound to the microtubule and the other projecting away from the microtubule.

One of the functions of MAPs is to enhance the assembly of microtubules, probably by binding to several tubulin molecules simultaneously. Thus, the addition of MAPs to isolated tubulin dimers promotes their conversion to ring forms (recall Figure 18-5, step 2), which in turn stimulates the overall assembly process (Figure 18-8). Covalent modification of MAPs may regulate the microtubule assembly process. For example, phosphorylation of MAP-1 and MAP-2 has been shown to reduce the rate of microtubule formation (Figure 18-8), most likely by shifting the assembly/disassembly balance in favor of disassembly.

Other MAPs are almost certainly involved in conferring specific properties on different microtubular structures. It is, in fact, generally assumed that all microtubules have functionally equivalent tubulin molecules, but are endowed with distinctive properties because of the specific MAPs they contain. At least two MAPs are known to use the energy of ATP hydrolysis to cause movement along microtubules. We will encounter these proteins shortly, because of their role in the process of *axonal transport*.

Cytoplasmic Microtubules and Cell Shape

Cytoplasmic microtubules play a variety of structural and functional roles in cells. We will consider an example of each, focusing first on the involvement of microtubules in

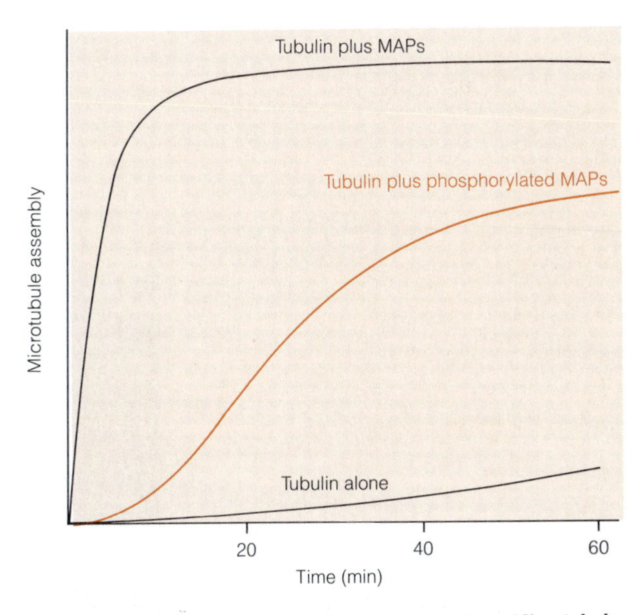

Figure 18-8 Stimulation of Microtubule Assembly by Microtubule-Associated Proteins. The assembly of microtubules occurs very slowly in vitro in the presence of purified tubulin only (lower line). The process is stimulated significantly by the presence of microtubule-associated proteins (MAPs; upper line). When present in phosphorylated form, however, MAPs are less effective at promoting assembly (middle line).

the maintenance of cell shape and then on their role in the transport of vesicles in the axons of nerve cells.

Microtubules contribute importantly to the structure and shape of animal cells, but play a less important role in plant cells, where cell shape is determined largely by the rigid cell wall. In animal cells, microtubules appear to determine shape in at least two different ways. The initial establishment of asymmetric shape seems to require the presence of a correctly ordered array of microtubules. And once established, such asymmetry depends critically on microtubules for its maintenance. The erythrocytes of most nonmammalian vertebrates, for example, have a prominent ring of microtubules just beneath the plasma membrane that is apparently vital to the maintenance of the oval shape of these cells. (Even in plant cells, microtubules have at least some influence on cell shape, since they are involved in determining the plane of cell division as well as the orientation of cellulose deposition during cell wall formation.)

The ongoing dependence of cell shape on microtubule integrity can be demonstrated by treating cultured animal cells with colchicine to disrupt microtubules. This treatment causes the cells to lose their characteristic ellipsoid or elongated appearance and to assume a more spherical shape instead. The disruption is reversible; removal of the drug leads to the reassembly of microtubules and the reestablishment of the original asymmetric shape of the cell.

Cytoplasmic Microtubules and Axonal Transport

In addition to the establishment and maintenance of cell shape, cytoplasmic microtubules also play an important part in the movement of organelles. This is most likely a general phenomenon, involving a variety of cellular organelles. For example, microtubules appear to serve as "tracks" for the outward movement of the endoplasmic reticulum in growing cells (see Problem 8 at the end of the chapter). However, the role of microtubules in organelle movement is best documented in the case of nerve cells, which transport membrane-bound vesicles in both directions along the *axons* that connect the *cell body* of the nerve cell with the *synaptic knobs*, or nerve endings (see Figure 20-2 on p. 618).

The need for such transport arises because ribosomes are present only in the cell body, so no protein synthesis occurs in the axons or synaptic knobs. Instead, proteins and membranous vesicles are synthesized in the cell body and transported along the axons to the synaptic knobs. The process is called **axonal transport** and appears to involve the movement of protein-containing vesicles and other organelles along "tracks" of microtubules.

Microtubules were initially implicated because axonal transport can be inhibited by colchicine and other drugs

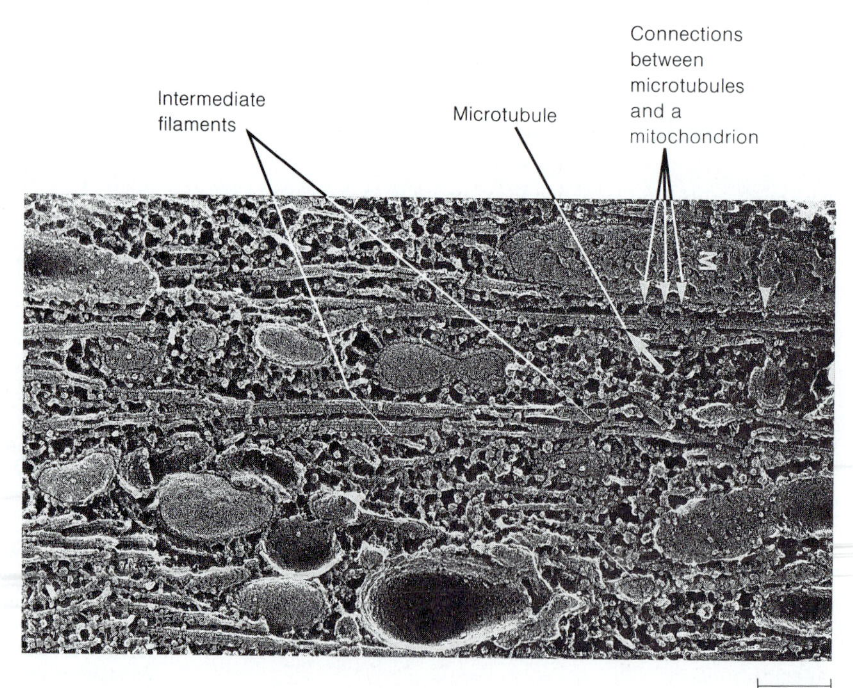

Figure 18-9 The Prominence of Microtubules in Axonal Cytoplasm. The microtubules in a frog axon have been visualized by deep-etch microscopy. Microtubules (large arrow) run longitudinally along the axon, as do intermediate filaments (small arrows). Mitochondria (M) and other membranous vesicles are embedded in the cytoskeletal matrix and appear to be connected to the microtubules (arrowheads).

that impair microtubule function, but is insensitive to drugs such as cytochalasin B that affect microfilaments. Since then, microtubules have been visualized along the axon by the deep-etching technique described earlier and have been shown to be prominent features of the axonal cytoskeleton. Moreover, axonal microtubules have small membranous vesicles and mitochondria associated with them (Figure 18-9).

Still more recently, a group of investigators found that fine filamentous structures present in exuded *axoplasm* (the cytoplasm of axons) could direct the movement of organelles in the presence of ATP. The organelle movement was visualized by *video-enhanced differential interference microscopy* and shown to occur at a rate comparable to the axonal transport rate in intact neurons, about 2 μm/sec. The researchers then used a combination of immunofluorescence and electron microscopy to demonstrate that the "tracks" along which the organelles move are single microtubules. They concluded that axonal transport depends on an interaction between microtubules and organelles.

The energy to drive axonal transport is provided by the hydrolysis of ATP, catalyzed by a protein called **kinesin.** Although vesicle transport occurs in both directions in crude axoplasmic extracts, purified kinesin drives transport in one direction only—toward the plus ends. Since the microtubules in the axon have their plus ends pointing outward toward the synaptic knob, kinesin-mediated movement accounts for the outward movement of vesicles and organelles. Kinesin is therefore an ATPase that uses the energy of ATP hydrolysis to move vesicles unidirectionally along a microtubule.

Another ATPase, originally designated as MAP-1C and now called **cytoplasmic dynein,** is responsible for transport in the other direction—that is, inward toward the cell body. As the name suggests, this protein is related to *axonemal dynein,* the ATPase responsible for the ATP-driven movement of cilia and flagella. Kinesin and dynein are two of three classes of *motor molecules*—ubiquitous ATPases that account for a wide variety of movements in eukaryotic cells. The third class of motor molecule is *myosin,* the ATPase involved in microfilament-based movement. Each of these motor molecules—kinesin, dynein, and myosin—will be discussed in detail in the next chapter.

Microfilaments

With a diameter of about 7 nm, **microfilaments** (**MFs**) are the smallest of the cytoskeletal filaments (Table 18-1). Microfilaments are best known for their role in the *contractile fibrils* of muscle cells, where they interact with thicker filaments of myosin to cause the contractions characteristic of muscle. This role will be explored in detail in Chapter 19. However, microfilaments are not confined to muscle cells; they occur in almost all eukaryotic cells and are involved in numerous other phenomena as well, including a variety of locomotory and structural functions.

Examples of cell movements in which microfilaments play a role include *amoeboid movement, locomotion* of cultured cells (fibroblasts, most commonly) over a substratum, and *cytoplasmic streaming,* a regular pattern of cytoplasmic flow that includes both the *cyclosis* of algal cells and higher plant cells and the *shuttle streaming* of animal cells. Microfilaments also produce the *cleavage furrows* that divide the cytoplasm of animal cells after chromosomes have been separated by the spindle. All of these phenomena will also be discussed in Chapter 19.

In addition to mediating a variety of cell movements, microfilaments are important structurally in the development and maintenance of cell shape. Most animal cells, for example, have a dense network of microfilaments called the **cell cortex** just beneath the plasma membrane. The cortex confers structural rigidity on the cell surface and facilitates shape changes and cell movement. In some cells, microfilaments are ordered into long parallel bundles called **stress fibers,** which often span the length of the cell (Figure 18-10). Parallel bundles of microfilaments also make up the structural core of **microvilli,** the fingerlike extensions found on the surface of many animal cells (recall Figure 4-2). Several of these structural roles will be discussed later in this chapter.

Structure of Microfilaments

Microfilaments are polymers of a protein called **actin.** Although actin is best known for its role in muscle contrac-

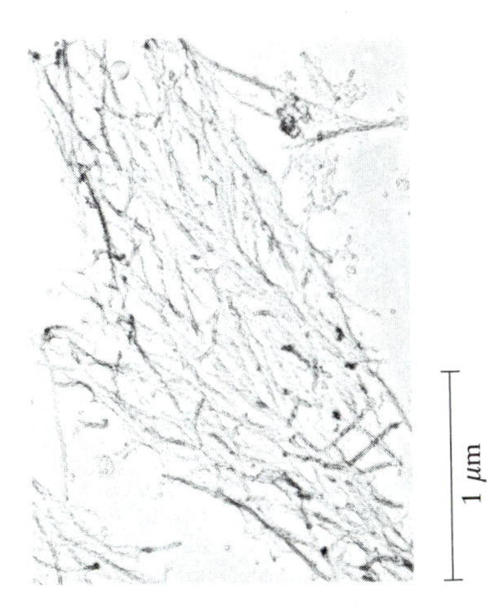

Figure 18-10 Stress Fibers. This micrograph shows stress fibers in a cultured fibroblast cell. The stress fibers consist of linear bundles of actin filaments and associated proteins and are prominent features of the cytoskeleton in many cultured cells. (Deep-etch micrograph)

tion, it is present in virtually all eukaryotic cells, including those of plants, algae, and fungi. In fact, actin is the single most abundant protein in most cells, usually comprising more than 5% of the total cellular protein.

Actin is synthesized as a monomer called **G-** (for **globular**) **actin.** The G-actin molecule is a single polypeptide consisting of 375 amino acids, with a molecular weight of about 42,000. Each molecule of G-actin has a molecule of ATP associated with it noncovalently. Several forms of actin can be distinguished, based on slight differences in amino acid sequence. Muscle cells contain α-actin, whereas nonmuscle cells have two similar but distinct forms, β-actin and γ-actin. These different forms of actin are very similar. They can, for example, polymerize into a single microfilament in vitro. Similarly, antibodies directed against α-actin bind to microfilaments of nonmuscle cells as well.

G-actin monomers polymerize reversibly into long helical strands of **F-** (for **filamentous**) **actin** (Figure 18-11). The exact structure of F-actin is somewhat uncertain. According to the model shown in Figure 18-11, G-actin molecules are considered to be roughly spherical in shape and to polymerize into double-helical strands, with each strand about 4 nm in diameter. An F-actin microfilament therefore consists of two strands of polymerized monomers wrapped around each other to form a right-handed double helix with a diameter of about 7 nm (Figure 18-11a).

Alternatively, some scientists consider the G-actin monomer to be dumbbell-shaped and the F-actin filament to consist of a single chain of monomers ordered into a tight right-handed helix, with about two actin monomers per turn. We will know which of these models is the better representation of the actin microfilament once the three-dimensional molecular structure of actin is determined by X-ray crystallography.

Within a microfilament, all of the actin monomers are oriented in the same direction, such that a microfilament, like a microtubule, has an inherent *polarity,* with one end differing chemically and structurally from the other end. This polarity can be readily demonstrated by incubating microfilaments with *heavy meromyosin,* a fragment of the myosin molecule that binds tightly to actin. The meromyosin molecules bind to, or "decorate," the microfilaments in a distinctive arrowhead pattern, with all of the meromyosin molecules pointing in the same direction (Figure 18-12).

This polarity is an important feature of the microfilament, because the polymerization of G-actin subunits usually occurs more rapidly at one end (called the *plus end;* see Figure 18-11a), while disassembly occurs mainly at the other end (the *minus end*). As a result, the microfilament elongates preferentially in one direction. Microfilament elongation is actually more complicated than Figure 18-11a suggests, however. Both assembly and disassembly can occur at either end, depending on local concentrations of actin and actin-binding regulatory proteins. Under appropriate in vitro conditions, both processes occur preferentially at the plus end, with assembly favored by a high ATP-actin/ADP-actin ratio and disassembly favored by a low ratio. Moreover, growth can occur at the minus end as well, albeit more slowly than at the plus end.

Genetics of Actin

Actin varies little in its amino acid sequence from species to species. In fact, the actin genes are the most highly

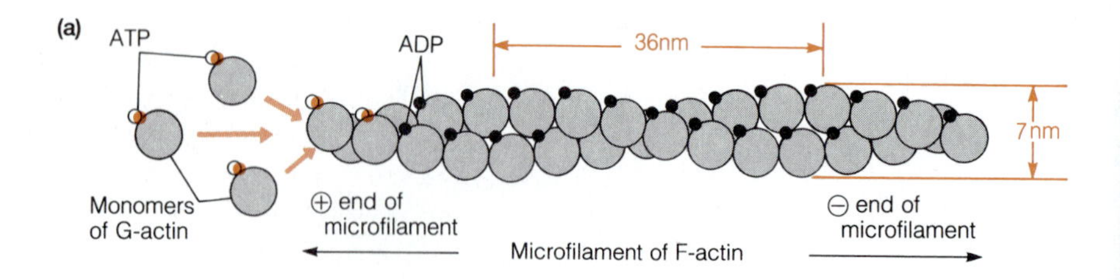

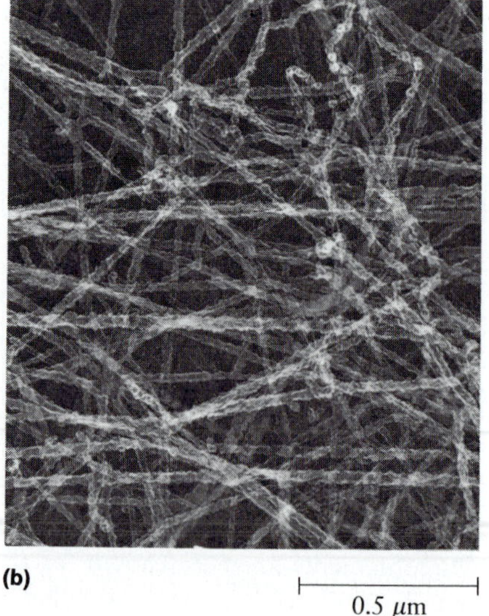

Figure 18-11 Assembly of Microfilaments. (a) Monomers of G-actin polymerize into long double-stranded filaments of F-actin with a diameter of about 7 nm. A half-turn of the helix occurs every 36 nm, with 13.5 monomers required for a full turn. The addition of each G-actin monomer is usually accompanied or followed by the hydrolysis of the ATP molecule that is tightly bound to the monomer (colored circles), although the energy of ATP hydrolysis is not required to drive the polymerization reaction. (b) A micrograph of purified F-actin (TEM).

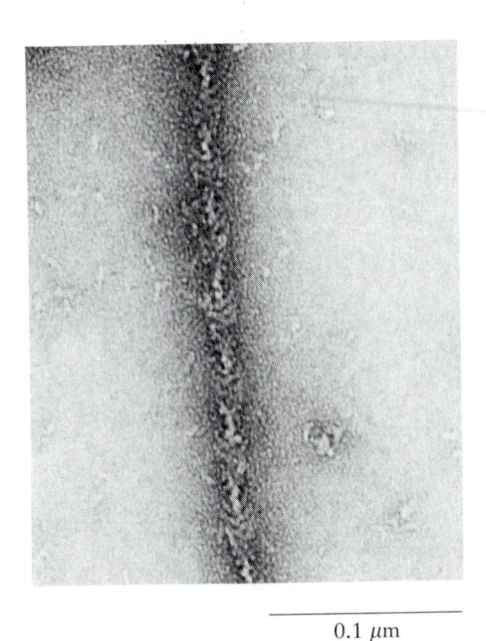

Figure 18-12 Meromyosin Binding to Actin Microfilaments. Actin microfilaments were incubated with meromyosin fragments containing the actin-binding head of the myosin molecule, and then examined in the electron microscope. The bound meromyosin fragments that "decorate" the microfilaments look like arrowheads. They are all oriented in the same direction, indicating the polarity of the microfilament. The arrowheads point toward the minus end of the microfilament and away from the plus end. (TEM)

conserved of the genes encoding cytoskeletal proteins. For example, the actins from yeast and chicken are identical at more than 90% of their amino acids, whereas tubulins from the two species show only about 70% identity. The minor differences in actin sequences between species do not appear to have much functional significance, since actin molecules from widely divergent species are indistinguishable when used for in vitro assays. The evolutionary conservatism of actin, and of tubulin as well, most likely reflects the constraints imposed on these molecules by the variety of other proteins with which they must interact in the cell.

Assembly of Microfilaments

Microfilaments form by the spontaneous assembly of G-actin monomers into F-actin. As studied in vitro, actin polymerization begins with a *lag phase* that reflects a slow *nucleation step*. Thereafter, polymerization occurs rapidly, with additional actin monomers added to both ends of the growing filament. Addition of a G-actin monomer is usually accompanied by hydrolysis of the molecule of ATP bound to it, with the resulting ADP remaining attached to the actin (Figure 18-11a). However, the polymerization process does not require energy, since G-actin monomers can polymerize spontaneously, albeit more slowly, under appropriate conditions even if the bound nucleotide on the G-actin monomers is ADP or a nonhydrolyzable analogue of ATP.

The assembly process is readily reversible; both polymerization and depolymerization can be demonstrated in vitro. The net rate of microfilament growth is determined by the concentration of G-actin monomers. As long as the monomer concentration is high, net assembly will occur and the microfilaments will continue to elongate. As assembly continues and the concentration of G-actin falls, the rate of polymerization decreases, and the rate of depolymerization increases. Eventually, a *critical concentration* is reached at which the two rates are equal, and no further elongation occurs.

Although microfilament assembly in vitro occurs at both ends of the filament, filament elongation in vivo occurs preferentially at the plus end. This directionality can be demonstrated in the test tube by incubating fragments of microfilaments with radioactively labeled actin monomers under conditions favorable for polymerization, and then using autoradiography to localize the labeled actin. When ionic conditions are similar to those in the cell, the radioactivity is found preferentially associated with the plus end of the elongating microfilaments. In fact, growth under these conditions is five to ten times faster at the plus end than at the minus end.

Under these conditions, a steady state is reached at which actin is added preferentially at the plus end and removed predominantly at the minus end (Figure 18-13a). The net result is that, at appropriate G-actin concentrations, the microfilament acts as a "treadmill," in the sense that a given actin monomer is incorporated at the plus end, transferred along the microfilament from one end to the other, and eventually lost by depolymerization at the opposite end (Figure 18-13b through e). This **treadmilling** of actin subunits may be a significant part of the mechanism by which microfilaments generate movements in cells.

Two features of microfilament assembly in cells suggest that the treadmilling model, although observed initially in vitro, may in fact apply in vivo. In general, microfilaments, like microtubules, are oriented in cells with their minus ends facing toward the center of the cell and their plus ends facing outward. This agrees well with the higher ATP-actin concentration in the peripheral cortex of the cell compared to the center. Thus, the relative rates of assembly and disassembly at the plus and minus ends of the microfilaments may be regulated by changes in the differential concentrations of ATP-actin monomers in the two regions of the cell. To date, however, treadmilling has been linked specifically to motility only in the case of fibroblast-like movement.

Actin-Binding Proteins

Many proteins are known to bind to actin, thereby conferring specific properties upon actin that regulate its assembly and function. These **actin-binding proteins** serve a variety of functions, but they can be grouped into three major classes, as shown in Table 18-2. Some are involved in regulating the length of microfilaments. They bind preferentially to one end of the microfilament, thereby "capping" it and causing a decrease in the rate of polymerization. Other actin-binding proteins favor the de-

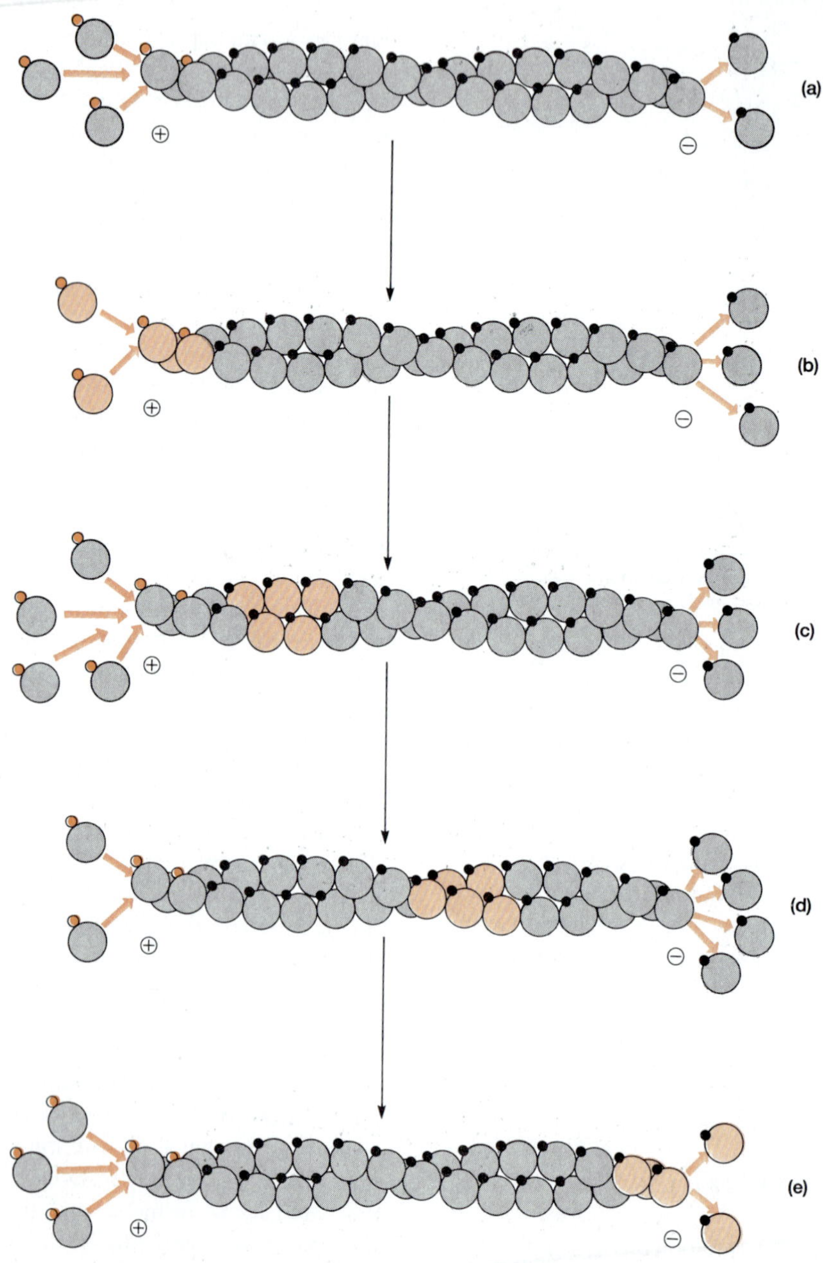

Figure 18-13 Treadmilling of Actin Microfilaments. (a) Even when undergoing no net change in length, microfilaments continue to incorporate G-actin monomers at the plus end and to lose monomers at the minus end. The result is a "treadmilling" of actin monomers along the microfilament. Actin monomers added at one point in time (b; colored spheres) start out at the plus end, but are displaced progressively along the microfilament (c and d) as assembly and disassembly continue at the respective ends, and (e) are eventually lost by depolymerization at the minus end.

Table 18-2 Major Classes of Actin-Binding Proteins

Length-Regulating Proteins
β-Actinin (from muscle)
Gelsolin (from macrophages)
Villin (from intestinal epithelium)
Depolymerizing Proteins
γ-Actinin (from muscle)
Profilin (from sea urchin sperm)
Cross-Linking and Bundling Proteins
α-Actinin (from muscle)
Fascin (from sea urchin eggs)
Filamin (from fibroblasts)
Fimbrin (from intestinal epithelium)
Fodrin (from intestinal epithelium)
Spectrin (from erythrocytes)
Villin (from intestinal epithelium)
Vinculin (from fibroblasts)

polymerization of microfilaments by shifting the equilibrium in the direction of actin monomers. One of the proteins in this class, *profilin,* sequesters monomers of G-actin in the sperm of some invertebrates, thereby preventing formation of F-actin until polymerization is needed to drive the *acrosomal reaction* during fertilization (see Problem 7 at the end of the chapter).

Still other actin-binding proteins cross-link adjacent microfilaments to one another. Depending on the protein, this cross-linkage may result in the formation of either loose networks or tight bundles of microfilaments. *Filamin* is an example of the former category, whereas *fimbrin* and *fascin* are examples of the latter group. Actin-binding proteins are also known that link microfilaments to other cellular structures, including the plasma membrane. We will encounter several different kinds of actin-binding proteins as we consider the involvement of microfilaments in cell structure and shape.

Microfilaments and Cell Shape

In addition to their involvement in the various kinds of cellular movement to be discussed in the next chapter, microfilaments play important roles in the development and maintenance of asymmetric cell shapes. This involvement can be demonstrated readily by exposing elongated cells (such as cultured fibroblasts) to cytochalasin B, which disrupts the bundles of microfilaments seen in such cells. The cells lose their characteristic shape and round up. Microfilaments stabilize not only the overall shape of the cell but also local projections of the cell surface such as the *microvilli* mentioned earlier. We will examine both roles briefly, looking first at the cell cortex that underlies and supports the plasma membrane and then at the microvillus as an example of a cell surface projection.

Microfilaments and the Cell Cortex. The *cell cortex* is a three-dimensional meshwork of actin microfilaments and associated proteins located just under the plasma membrane of most animal cells. The cortex supports the plasma membrane, confers strength on the cell surface, and facilitates shape changes and cellular movement. An important function of certain actin-binding proteins in the cortex is to link the microfilaments into a stable network with gel-like properties. One of these **microfilament cross-linking proteins** is *filamin,* a long molecule consisting of two identical polypeptides joined head to head, with an actin-binding site at each tail. Molecules of filamin act as "splices," joining two adjacent microfilaments together as shown in Figure 18-14. In this way, actin microfilaments are linked to form large three-dimensional networks.

Other cortex proteins play the opposite role of breaking up the microfilament network, thereby causing the actin gel of the cortex to soften and liquefy. These **microfilament-severing proteins** are important in mediating the gel-to-sol transition associated with the phenomenon of cytoplasmic streaming to be encountered in the next chapter. One such protein is *gelsolin,* which functions by breaking actin

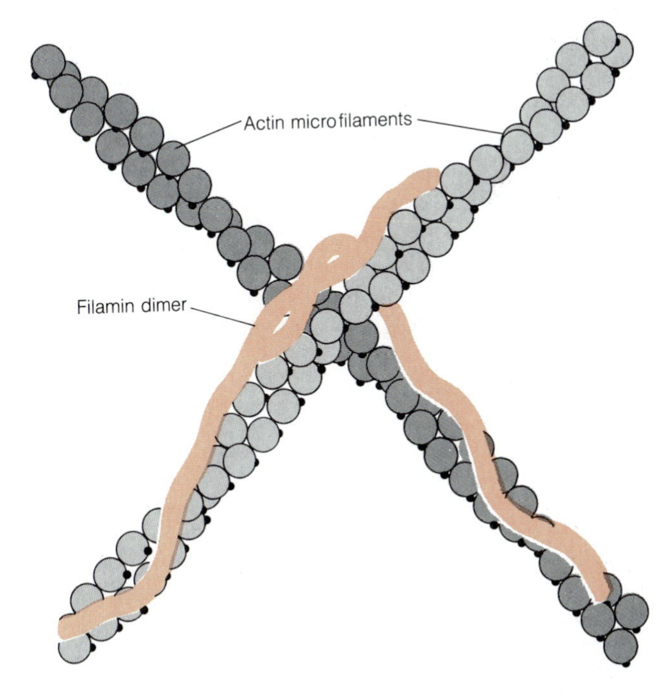

Figure 18-14 Cross-Linking of Actin Filaments by Filamin. Filamin is one of several cross-linking proteins that link actin filaments into an extensive three-dimensional network. Filamin is a dimer of two identical polypeptides joined head to head into a long flexible molecule. The tail of each polypeptide contains a binding site for actin filaments.

microfilaments, and "capping" the newly exposed plus ends of the microfilaments, thereby preventing further polymerization. Gelsolin and other microfilament-severing proteins are activated by Ca^{2+}, which renders them sensitive to extracellular signals that result in transient changes in cytoplasmic Ca^{2+} levels.

In addition to proteins such as filamin and gelsolin that determine the degree of microfilament cross-linking in the cortex, other proteins mediate the association of the cortex microfilaments with the plasma membrane. The erythrocyte proteins *spectrin* and *ankyrin* mentioned in Chapter 7 are examples of this category. As illustrated in Figure 18-15, the plasma membrane of the erythrocyte is supported by a network of spectrin filaments that are cross-linked by very short actin chains. This network is connected to the plasma membrane by molecules of ankyrin that link the spectrin filaments to a specific transmembrane protein. Although spectrin and ankyrin are specific to erythrocytes, the presence of similar proteins in the cortex of other animal cells suggests that the plasma membrane of other, perhaps even all, animal cells is also supported by a cortical network of microfilaments.

Microfilaments and Microvilli. *Microvilli* are cell surface projections that serve both to increase the surface area of the cell and to mediate changes in cell shape and behavior. Microvilli are especially prominent features of intestinal mucosal cells (Figure 18-16a). A single mucosal cell in your small intestine, for example, has several thousand microvilli, each about 1–2 μm long and about 0.1 μm in diameter, which increase the surface area of the cell about 20-fold. This increase in surface area is important to intestinal function, because the uptake of digested food depends on extensive absorptive surface.

As illustrated in Figure 18-16b, the core of the microvillus consists of a tight bundle of microfilaments. The microfilaments are all oriented with their plus ends pointing toward the tip, where they are attached to the membrane through an amorphous electron-dense plaque. The microfilaments in the bundle are also connected to the plasma membrane by lateral cross-links, which extend outward about 20 to 30 nm from the bundle and make contact with electron-dense patches on the inner membrane surface. The microfilaments in the bundle are held together at regular intervals by **actin-bundling proteins,** of which *fimbrin* and *villin* are examples. Unlike actin cross-linking proteins, which connect microfilaments into loose networks, these actin-bundling proteins bind adjacent microfilaments together tightly. For example, each molecule of fimbrin binds two or three actin subunits, linking the microfilaments into a tightly packed bundle.

At the base of the microvillus, the bundle of microfilaments extends into a network of filaments called the **terminal web** (Figure 18-17). The filaments of the terminal web are composed mainly of myosin and spectrin, which connect the microfilaments to each other, to the plasma membrane, and perhaps also to the network of intermediate filaments beneath the terminal web. The purpose of the terminal web is apparently to give rigidity to the microvilli by anchoring their microfilament bundles securely so that they project straight out from the cell surface.

Intermediate Filaments

Intermediate filaments (IFs) have a diameter of about 8–12 nm, which makes them intermediate in size between microtubules and microfilaments (Table 18-1), or between the thin (actin) and thick (myosin) filaments in muscle cells, where IFs were first discovered. To date, most studies have focused on animal cells, where IFs occur singly or in bundles and appear to play a structural or tension-bearing

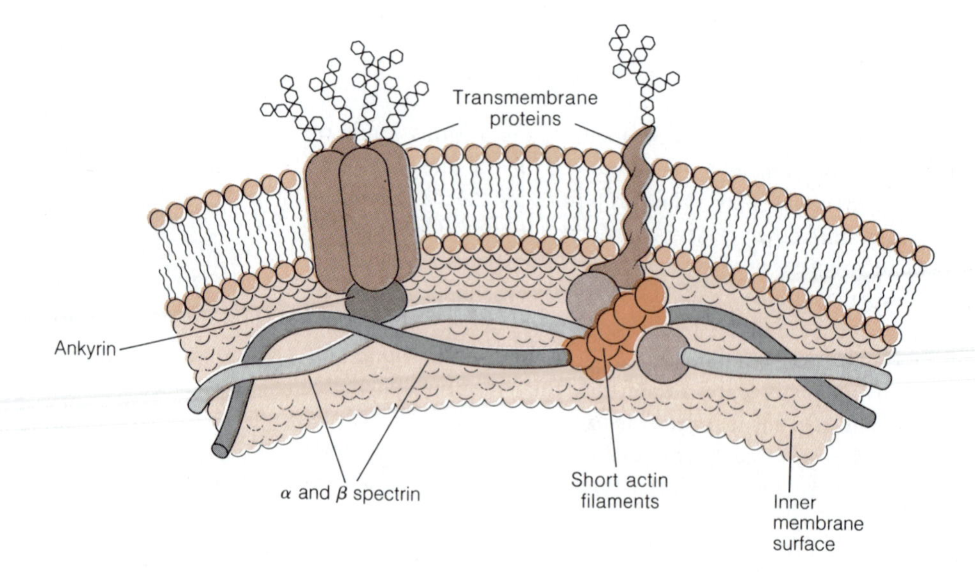

Figure 18-15 Support of the Erythrocyte Plasma Membrane by a Spectrin-Ankyrin-Actin Network. The plasma membrane of the red blood cell is supported on its inner surface by a filamentous network that gives the cell both strength and flexibility. Long filaments of spectrin are cross-linked by short actin filaments. The network is anchored to a specific transmembrane protein by molecules of the protein ankyrin.

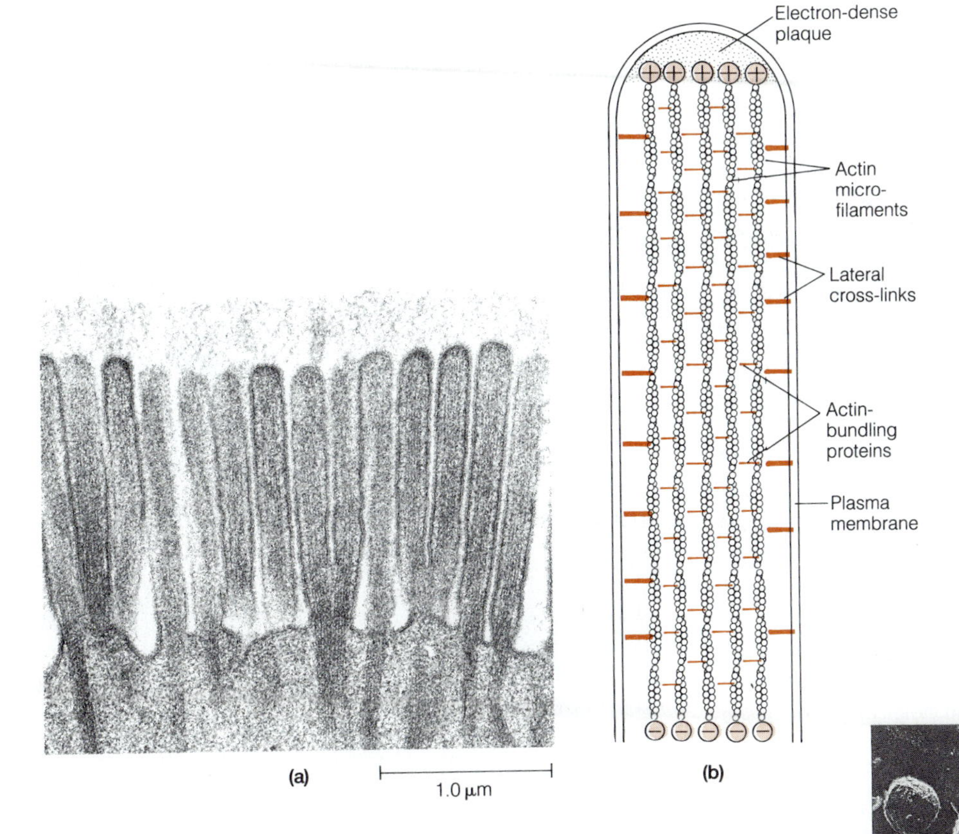

Figure 18-16 Structure of a Microvillus. (a) Electron micrograph of microvilli from intestinal mucosa cells (TEM). (b) Schematic diagram of a single microvillus, showing the core of microfilaments that gives the microvillus its characteristic stiffness. The core consists of several dozen microfilaments oriented with their plus ends facing outward toward the tip and their minus ends facing toward the cell. The microfilaments are tightly linked together by actin-bundling proteins and are connected to the inner surface of the plasma membrane by lateral cross-links composed of calmodulin and a protein thought to have ATPase activity.

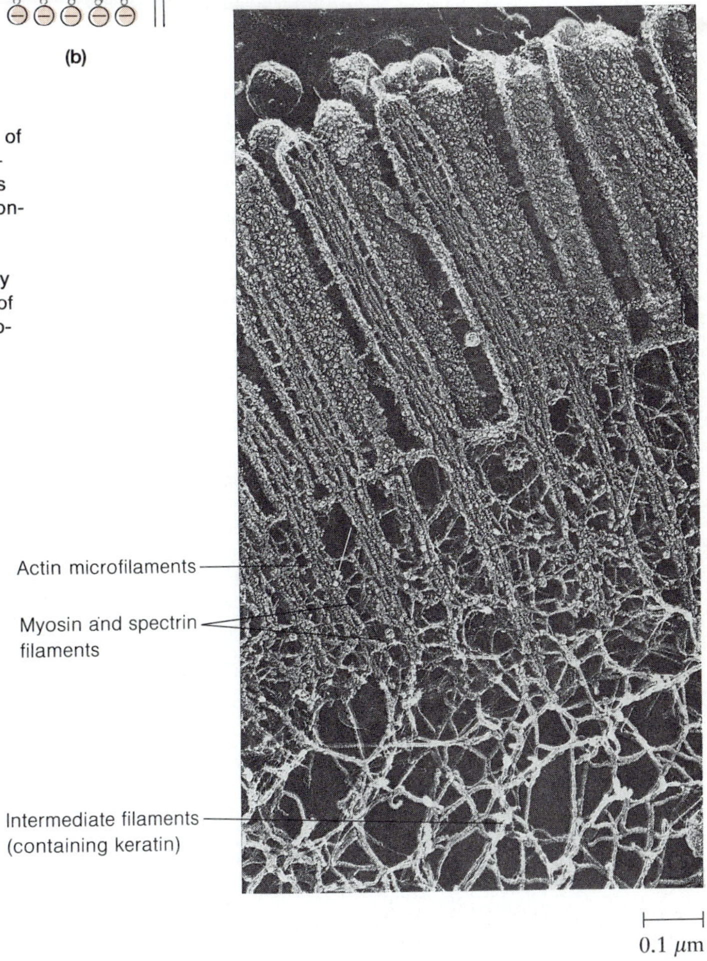

Actin microfilaments

Myosin and spectrin filaments

Intermediate filaments (containing keratin)

0.1 μm

Figure 18-17 The Terminal Web of an Intestinal Mucosa Cell. The terminal web beneath the plasma membrane is seen in this freeze-etch electron micrograph of an intestinal epithelial cell. Bundles of microfilaments that form the core of microvilli extend into the terminal web, where they are linked to each other and to the plasma membrane by filaments of myosin and spectrin. These are supported in turn by a network of intermediate filaments located beneath the terminal web. The intermediate filaments contain keratin, since this is an epithelial cell. (Freeze-etch micrograph)

role. Figure 18-18 is an electron micrograph of IFs from a cultured human fibroblast cell.

Intermediate filaments are the most stable and the least soluble constituents of the cytoskeleton. Treatment of cells with detergents or with solutions of high or low ionic strength removes most of the microtubules, microfilaments, and other proteins of the cytoplasm, but leaves networks of intermediate filaments that retain their original shape. (In fact, the original structure to which the term "cytoskeleton" was applied was really a residual network of intermediate filaments from which the microtubules and microfilaments now considered an integral part of the cytoskeleton had already been removed.) Because of the stability of the intermediate filaments, some investigators suggest that they serve as a scaffold to support the entire cytoskeletal framework.

Tissue Specificity of Intermediate Filaments

In contrast to microtubules and microfilaments, intermediate filaments differ in their composition from tissue to tissue. Based on the cell type in which they are found, IFs and their proteins can be grouped into five classes (Table 18-3). **Keratins** are the proteins of the **tonofilaments** found in the epithelial cells that cover the body surfaces and line its cavities. (The intermediate filaments seen beneath the terminal web in the intestinal mucosa cell of Figure 18-17 consist of keratin, since the intestinal lining is an epithelial tissue.) Keratins, in turn, are subdivided into **acidic keratins** and **basic** or **neutral keratins,** with at least 15 different keratins in each subclass. **Vimentin** is present in connective tissue and other cells of mesenchymal origin. Vimentin-containing filaments are often prominent features in cultured fibroblast cells, where they form a network that radiates from the center out to the periphery of the cell. **Desmin** is found in muscle cells, **neurofilament**

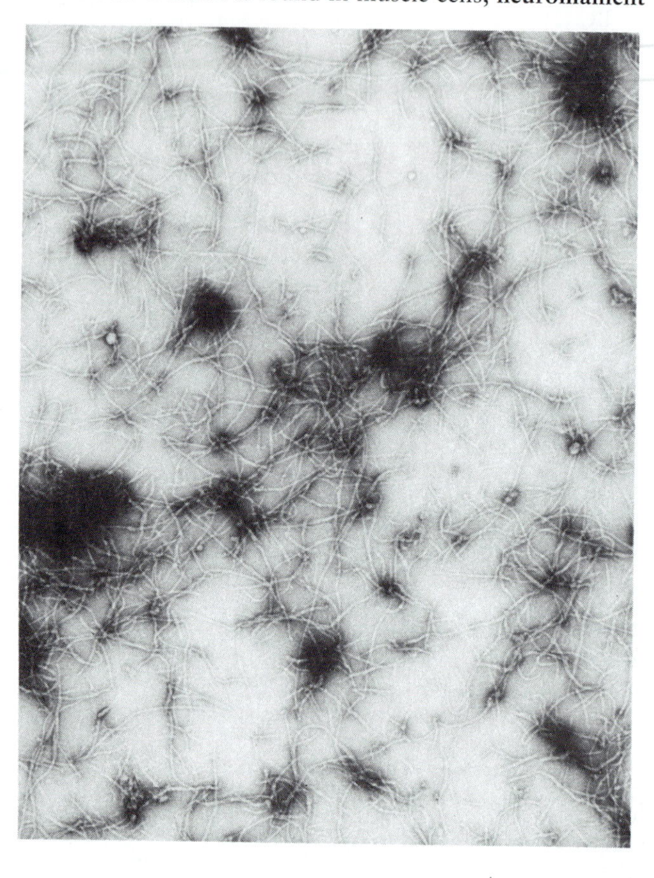

0.5 μm

Figure 18-18 Intermediate Filaments. An electron micrograph of negatively stained intermediate filaments from a cultured human fibroblast cell. (TEM)

Table 18-3 Major Types of Intermediate Filaments and Their Proteins

Tissue	IF Protein	Molecular Weight	Amino Acid Sequence Type
Epithelial	Keratins		
	Acidic keratins	44–60 K	I
	Basic and neutral keratins	50–70 K	II
Mesenchymal	Vimentin	53 K	III
Muscle	Desmin	52 K	III
Glial	GFA (glial fibrillary acidic) protein	51 K	III
Neurons	NF (neurofilament) protein	60 K, 100 K, 130 K	IV
Nuclear lamina of all cells	Nuclear lamins A, B, and C	60–75 K	V

(NF) **protein** occurs in the **neurofilaments** of nerve cells, and **glial fibrillary acidic** (GFA) **protein** is characteristic of the glial cells that surround and insulate nerve cells.

Because of the tissue specificity of intermediate filaments, animal cells from different tissues can be distinguished on the basis of the IF protein present, as determined by immunofluorescence microscopy. This **intermediate filament typing,** as it is called, serves as a diagnostic tool in medicine. It is especially useful in the diagnosis of cancer, because tumor cells are known to retain the IF proteins characteristic of the tissue of origin, regardless of where the tumor occurs in the body. Since appropriate treatment of cancer often depends on knowing the tissue of origin, IF typing is a valuable diagnostic aid, especially in cases where diagnosis using conventional microscopic techniques is difficult. This medical application of basic research on intermediate filaments is explored in more detail in the box on pp. 575–576.

In addition to five categories of tissue-specific IF proteins, Table 18-3 includes the **nuclear lamins** (**A**, **B**, and C), which are located in the nucleus of animal cells but are similar in amino acid sequence to the proteins of cytoplasmic IFs. The nuclear lamins are the protein components of a dense network of intermediate filaments called the **nuclear lamina** or "karyoskeleton," located on the inside surface of the nuclear envelope. Unlike the IF proteins of the cytoplasm, nuclear lamins are common to most animal cell types. Most animal cells therefore contain at least two different kinds of IF proteins—the lamins of the nucleus and the cytoplasmic IF protein appropriate to the specific cell type.

The classification of IFs by cell type is biologically useful, because the genes for IF proteins are expressed in accordance with the pattern of tissue differentiation during embryogenesis. As IF proteins and their genes have been sequenced, however, it has become clear that these proteins are encoded by a single (though large) family of related genes and can therefore be classified according to sequence relatedness as well. On this basis, five types of IF proteins are identified, as shown in the right-hand column of Table 18-3.

The acidic keratins are *type I* IF proteins, whereas the neutral and basic keratins are *type II*. Keratins within each of these classes have a high degree of sequence homology. Keratin filaments are always heteropolymers consisting of equal numbers of type I and type II subunits. *Type III* IF proteins include vimentin, desmin, and GFA protein. These proteins are about 70% homologous in sequence. They can form homopolymers, but will also copolymerize with other type III IF proteins (but not with IF proteins of other types). The three NF proteins of neurofilaments—the so-called neurofilament triplet—are *type IV* IF proteins, and the nuclear lamins are *type V*.

Structure of Intermediate Filaments

As products of a family of related genes, all of the IF proteins share some common features, although they differ significantly in size and chemical properties. In contrast to actin and tubulin, all of the IF proteins are fibrous, rather than globular, proteins. All IF proteins have a homologous central rodlike domain of about 310–315 amino acids that has been remarkably conserved in size, in secondary structure, and to some extent in sequence. As shown in Figure 18-19, this central domain consists of four segments of coiled helices interspersed with three short linker regions. Flanking the central helical domain are N-terminal and C-terminal domains that differ greatly in size, sequence, and function among IF proteins, presumably accounting for the functional diversity of these proteins.

A possible model for IF assembly is shown in Figure 18-20. The basic structural unit of intermediate filaments consists of two IF polypeptides (a) intertwined into a *coiled coil* (b). The two polypeptides have their central helical domains aligned in parallel, with the N- and C-terminal regions protruding as globular domains at each end. Two such dimers align laterally to form a tetrameric *protofilament* (c), the basic structural unit of intermediate filaments. Protofilaments then interact with each other, associating in an overlapping manner to build up a filamentous structure both laterally and longitudinally (d). When fully assembled, an intermediate filament is thought to be eight protofilaments thick at any point, with protofilaments probably joined end to end in staggered overlaps (e).

Functions of Intermediate Filaments

Intermediate filaments are considered to be principal structural determinants in many cells and tissues. Because they often occur in areas of the cell that are subject to mechanical stress, IFs are thought to have a tension-bearing role.

A specific function of IFs may be to maintain the position of the nucleus in the cell. Intermediate filaments form a ring around the nucleus with branches extending outward through the cytoplasm. These filaments have been observed to extend into the pores of the nuclear envelope, possibly connecting with the filaments of the nuclear lamina that may play a similar structural role within the nucleus. Intermediate filaments may interact with other elements of the cytoskeleton as well. For example, when microtubules are disassembled, the intermediate filaments that normally extend throughout the cytoplasm collapse into a ring around the nucleus.

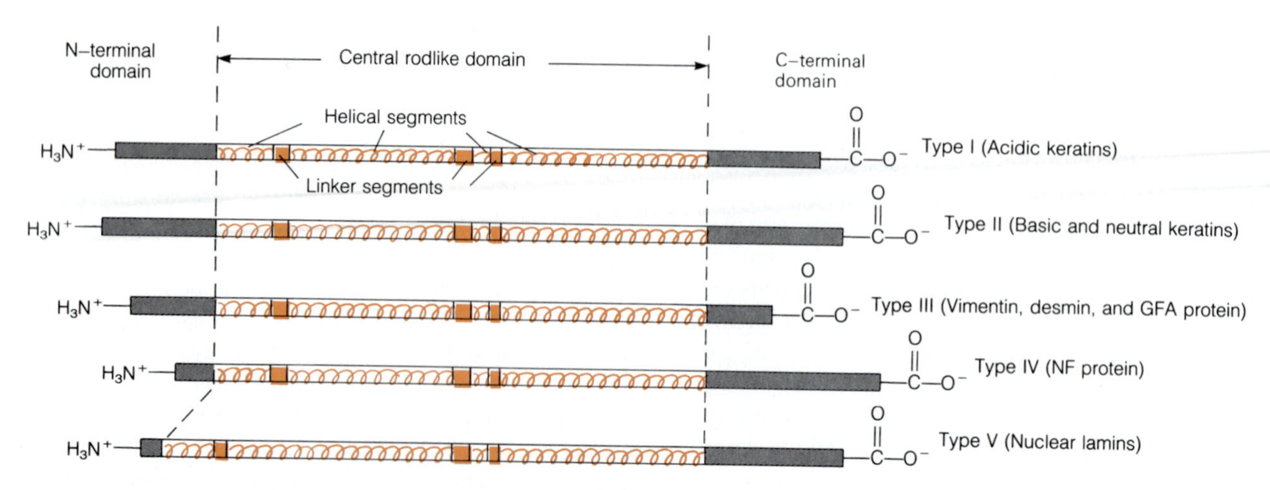

Figure 18-19 Structural Similarities of IF Proteins. All five types of IF proteins have a central rodlike domain consisting of four helical segments interrupted by three linker segments. The central domain is thought to be important for filament assembly (see Figure 18-20). This domain is highly conserved in size, secondary structure, and sequence, though sequence homologies are confined to the helical regions. In types I–IV, the helical segments contain a total of 276 amino acids and the linker segments are nonhelical. In type V (the nuclear lamins), the helical segments contain 318 amino acids and the linker segments are also helical. The N-terminal and C-terminal domains that flank the central section are nonhelical and are much more variable in size and sequence.

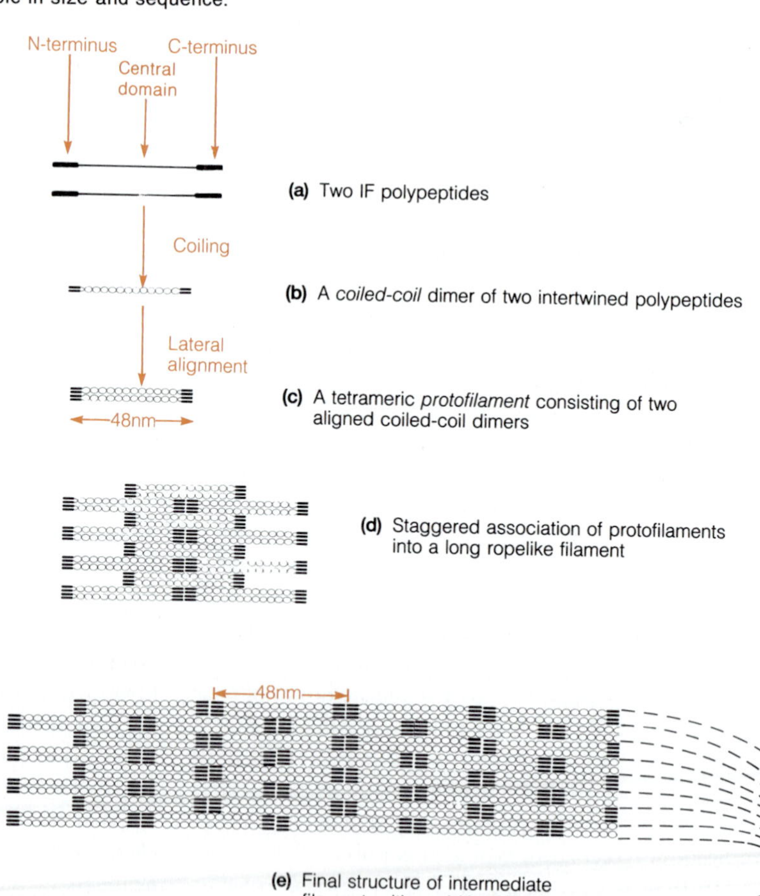

Figure 18-20 Model for Intermediate Filament Assembly. One current suggestion for the assembly of intermediate filaments from IF polypeptides is shown here. (a) The starting point for assembly is a pair of IF polypeptides. The two polypeptides are identical for all IFs except keratin filaments, which are obligate heterodimers with one each of the type I and type II polypeptides. (b) The two polypeptides twist around each other to form a two-chain coiled coil, with their conserved central domain aligned in parallel. (c) Two dimers then align laterally to form a tetrameric protofilament. (It is not yet clear whether the dimers are in parallel or antiparallel orientation in the protofilament.) (d) Protofilaments then assemble into larger filaments by end-to-end and side-to-side alignment. (e) The fully assembled intermediate filament is thought to be eight protofilaments thick at any point.

Intermediate Filaments and the Diagnosis of Tumors: A Medical Application of Basic Research

"But how useful is your research? What good will it do anyone?" These are questions that are often directed to scientists, usually by citizens whose tax dollars support the research and who therefore deserve honest answers to their questions. The problem with such questions, however, is their inherent implication that research is worthwhile only if it is directed toward the solution of some particular problem or the achievement of some specific benefit, usually defined in human terms. When we invoke such criteria, we fail to distinguish between *applied research*, which is in fact intended to solve a specific problem or to yield a particular benefit, and *basic research*, which is not. Even more important, we may fail to realize that applied research almost always depends critically on basic research, without which we would have no findings to apply.

Thus, both kinds of research are important. We need applied researchers who are motivated by the satisfaction of seeing a problem solved or a benefit realized. But we also need basic researchers whose main reason for studying a specific phenomenon is the same reason people give for climbing Mount Everest—"just because it's there!" For it is out of such basic research that intriguing applications often emerge, usually in ways that could not have been anticipated or even guessed in advance.

Research on the cytoskeleton illustrates this point well. Most of what you have been reading about in this chapter is a result of investigations conducted by basic researchers, who study the cytoskeleton and its proteins "just because they're there." But out of their work have come some important applications that could not have been easily predicted beforehand, much less deliberately planned. One such example is the recent application of basic findings on intermediate filaments to the practical problem of distinguishing tissue types and identifying tumors, thereby providing pathologists with a valuable new tool for cancer diagnosis.

Underlying this application is the recognition, from the results of basic research, that cells from different kinds of vertebrate tissues have remarkably similar microtubules and microfilaments, but distinctively different kinds of intermediate filaments (IFs). In fact, the kind of IF—and hence the kind of IF protein—present in the cells of a particular tissue is a distinguishing characteristic of that tissue. Thus, epithelial cells contain mainly *keratin*, muscle cells have *desmin*, nerve cells have *NF* (*neurofilament*) *protein*, glial cells have *GFA* (*glial fibrillary acidic*) protein, and cells of mesenchymal origin are characterized by the presence of *vimentin* (recall Table 18-3). Because of these characteristic differences in IF proteins, it is possible to differentiate between various tissues by identifying, or *typing*, the intermediate filaments using the technique of *immunofluorescence microscopy*.

Intermediate filament typing is, in turn, a useful tool for further basic research. Developmental biologists, for example, find the technique useful for tracing cell lineages during embryonic development. But the technique also has important practical implications in the diagnosis of cancer because of its capability for distinguishing between different kinds of tumors. Moreover, IF typing is also a useful diagnostic tool for certain other human diseases as well, including prenatal detection of congenital defects.

The diagnosis of cancer relies crucially on the ability of pathologists to classify tumors according to the tissue of origin. Such distinctions are important because the appropriate treatment of the cancer depends, sometimes critically, on knowing the cellular origin of the tumor. Yet tumors often appear in the body far from the tissue in which they originated, because of the tendency of cancer cells to *metastasize*, or become dislodged from the original tumor and to be carried to distant parts of the body. (As we will learn in Chapter 23, a mass of cancer cells in their original location in the body is called a *primary tumor*, whereas a mass of cancer cells that has spread to a distant part of the body and is proliferating there is called a *metastasis*.)

In most cases, tumors can be successfully diagnosed by light microscopy, using standard staining techniques. However, about 5–10% of all tumors are difficult to diagnose by conventional means. These include small cell tumors in children, tumor samples that contain only a few tumor cells, or poorly differentiated tumors. In such cases, IF typing enables pathologists to characterize the tumor rapidly and unambiguously. Moreover, the technique is incredibly sensitive and can therefore be used to detect tiny tumors containing only a few cells.

IF typing depends not only on the presence of only one (or at most two) kind of IF protein in a given cell type but also on the maintenance of that specificity in *neoplastic* (cancerous) cells. The evidence that tumors maintain the IF specificity of the original tissue is strong. Consider, for example, the *carcinomas*, which are tumors of epithelial origin and the most common malignant tumors. Virtually all carcinomas react positively with an antibody specific for keratin, the IF protein characteristic of epithelial tissues. Few, if any, carcinomas respond positively when tested with antibodies specific for vimentin, desmin, GFA, or NF protein. Thus, carcinomas not only retain the keratin characteristic of epithelial tissues but also do not acquire additional IF types.

Similarly, *sarcomas* of muscle-cell origin stain positively and specifically for desmin, *lymphomas* of mesenchymal origin stain for vimentin, *gliomas* of glial origin stain for GFA protein, and *neuroblastomas* of nervous system origin stain for NF protein. Moreover, these same specificities are generally true for primary tumors and also for metastases. In the case of carcinomas, tumors can be further subtyped using either two-dimensional gel electrophoresis or monoclonal antibodies to identify subsets of keratins, thereby allowing finer diagnostic distinctions. For example, *adenocarcinomas* contain the same subset of keratins as the epithelial cells of the small intestine and colon from which these carcinomas are derived, while *hepatocellular carcinomas* contain only those keratins found in normal liver cells from which these tumors originate.

IF typing has other medical applications as well. Combined with *amniocentesis* (sampling of the *amniotic fluid* surrounding the fetus in the womb), the technique is of potential use in prenatal diagnosis of birth defects. Normally, most of the cells in the amniotic fluid react positively with keratin antibodies, presumably because they are of epithelial origin. Most also test positive for vimentin, but not usually for desmin, GFA, or NF proteins. However, in the case of a fetus with anencephaly, a severe congenital defect in which the brain fails to develop properly, the amniotic fluid contains many cells positive for GFA or NF protein, both of which are derived from cell types related to the nervous system.

IF typing can also be used to detect and study abnormalities of IF organization in other diseases. Already it has been used to demonstrate IF anomalies in muscle cells of patients with certain muscle disorders and possibly also in the brain of people with Alzheimer's disease.

Cancer, prenatal diagnosis, muscle disorders, Alzheimer's disease—who could have guessed in advance that basic research on intermediate filaments would lead to medical applications in areas such as these? And who can predict what unexpected applications the basic research that is currently under way in other areas of cell biology may have in the future?

"But how useful is your research? What good will it do anyone?" Good questions to ask, and deserving of answers. But for the basic researcher, the appropriate response is to continue climbing Mount Everest, because you never know what you might find when you get to the top—or even along the way.

Perspective

The cytoskeleton is a structural feature of eukaryotic cells revealed by high-voltage electron microscopy, deep-etching, and immunocytochemistry. It consists of extensive three-dimensional networks of microtubules, microfilaments, and intermediate filaments that determine cell shape and facilitate a variety of cell movements.

Microtubules (MTs) are hollow tubes with walls consisting of tubulin dimers polymerized linearly into protofilaments. They are polar structures and elongate preferentially from one end. Microtubules were first identified as components of the axonemal structures of cilia and flagella and the mitotic spindle of dividing cells, but are now recognized as a general cytoplasmic constituent of most eukaryotic cells. Cytoplasmic microtubules govern the asymmetrical shape of cells, the configuration of the plasma membrane, the plane of cell division in plant cells, and the changes in cell position and shape that occur during embryonic development.

Microfilaments (MFs) are double-stranded polymers of actin that were discovered initially because of their role in the contractile fibrils of muscle cells but are now recognized as a component of virtually all eukaryotic cells. Like microtubules, microfilaments are polar structures, with actin monomers added preferentially to one end and removed from the other. Microfilaments are involved in a variety of locomotory functions to be discussed in Chapter 19. They are also important structurally in the development and maintenance of cell shape, as evidenced by their presence in the cell cortex, stress fibers, and the core of microvilli.

Intermediate filaments (IFs) are the most stable and least soluble constituents of the cytoskeleton. They appear to play a structural or tension-bearing role. Unlike microtubules and microfilaments, IFs are tissue-specific and can be used to identify cell type. Such typing is useful in the diagnosis of cancer, since tumor cells are known to retain the IF proteins of their tissue of origin. Intermediate filaments can also be classified by amino acid sequence, since they are encoded by a family of related genes. All IF proteins have a highly conserved central domain flanked by terminal regions that differ in size and sequence, presumably accounting for the functional diversity of IF proteins.

With this background, we are now ready to proceed to the next chapter, which will explore in more detail the role of microtubules and microfilaments in cellular motility and contractility.

Key Terms for Self-Testing

cytoplasm (p. 554)
cytoskeleton (p. 554)

Techniques for Studying the Cytoskeleton
immunofluorescence microscopy (p. 556)
quick-freeze deep-etch microscopy (deep-etching) (p. 556)
high-voltage electron microscopy (p. 557)

Microtubules
microtubules (MTs) (p. 558)
axonemal microtubules (p. 558)
cytoplasmic microtubules (p. 558)
microtubule-organizing center (MTOC) (p. 558)
centrosome (cell center) (p. 558)
protofilaments (p. 559)
lumen (p. 559)
tubulin (p. 559)
α-tubulin (p. 559)
β-tubulin (p. 559)
polarity (p. 559)
plus end (p. 560)
minus end (p. 560)
lag phase (p. 561)
nucleation step (p. 561)
elongation phase (p. 561)
critical concentration (p. 561)
dynamic instability model (p. 562)
microtubule-associated proteins (MAPs) (p. 563)
axonal transport (p. 564)
kinesin (p. 565)
cytoplasmic dynein (p. 565)

Microfilaments
microfilaments (MFs) (p. 565)
cell cortex (p. 565)
stress fibers (p. 565)
microvilli (p. 565)
actin (p. 565)
G- (globular) actin (p. 566)
F- (filamentous) actin (p. 566)
treadmilling (p. 567)
actin-binding proteins (p. 568)
microfilament cross-linking proteins (p. 569)
microfilament-severing proteins (p. 569)
actin-bundling proteins (p. 570)
terminal web (p. 570)

Intermediate Filaments
intermediate filaments (IFs) (p. 570)
keratins (p. 572)
tonofilaments (p. 572)
acidic keratins (p. 572)
basic or neutral keratins (p. 572)
vimentin (p. 572)
desmin (p. 572)
neurofilament (NF) protein (p. 572)
neurofilaments (p. 573)
glial fibrillary acidic (GFA) protein (p. 573)
intermediate filament typing (p. 573)
nuclear lamins (A, B, and C) (p. 573)
nuclear lamina (p. 573)

Suggested Reading

Birchmeier, W. Cytoskeleton structure and function. *Trends Biochem. Sci.* 9 (1984): 192.

Borisy, G. G., D. W. Cleveland, and D. B. Murphy. *Molecular Biology of the Cytoskeleton.* Cold Spring Harbor, N.Y.: Cold Spring Harbor Laboratory, 1984.

Cohen, C. Cell architecture and morphogenesis. I. The cytoskeletal proteins. *Trends Biochem. Sci.* 4 (1979): 73.

Fulton, A. B. How crowded is the cytoplasm? *Cell* 30 (1982): 345.

Fulton, A. B. *The Cytoskeleton. Cellular Architecture and Choreography.* London: Chapman and Hall, 1984.

Luby-Phelps, K., D. L. Taylor, and F. Lanni. Probing the structure of the cytoplasm. *J. Cell Biol.* 102 (1986): 2015.

Mooseker, M. S. Organization, chemistry, and assembly of the cytoskeletal apparatus of the intestinal brush border. *Annu. Rev. Cell Biol.* 1 (1985): 209.

Porter, K. R., and J. B. Tucker. The ground substance of the living cell. *Sci. Amer.* 244 (March 1981): 56.

Schliwa, M. *The Cytoskeleton: An Introductory Survey.* Cell Biology Monographs, Vol. 13. New York: Springer-Verlag.

Shay, J. W., ed. *Cell and Molecular Biology of the Cytoskeleton.* New York: Plenum Press, 1986.

Weber, K., and M. Osborn. The molecules of the cell matrix. *Sci. Amer.* 253 (October 1985): 110.

Techniques for Studying the Cytoskeleton

Bridgman, P. C., and T. S. Reese. The structure of cytoplasm in directly frozen cultured cells. 1. Filamentous meshworks and the cytoplasmic ground substance. *J. Cell Biol.* 99 (1980): 1655.

Heuser, J., and M. W. Kirschner. Filament organization revealed in platinum replicas of freeze-dried cytoskeletons. *J. Cell Biol.* 86 (1980): 212.

Hirokawa, N., and J. E. Heuser. Quick-freeze, deep-etch visualization of the cytoskeleton beneath surface differentiations of intestinal epithelial cells. *J. Cell Biol.* 91 (1981): 399s.

Ris, H. The cytoplasmic filament system in critical point-dried whole mounts and plastic-embedded matrix. *J. Cell Biol.* 100 (1985): 1474.

Microtubules

Allen, R. D. The microtubule as an intracellular engine. *Sci. Amer.* 256 (February 1987): 42.

Brinkley, B. R. Microtubule organizing centers. *Annu. Rev. Cell Biol.* 1 (1985): 145.

de Brabander, M. Microtubules, central elements of cellular organization. *Endeavor* 6 (1982): 124.

Dustin, P. *Microtubules*, 2d ed. New York: Springer-Verlag, 1984.

Haimo, L. T., and J. L. Rosenbaum. Cilia, flagella, and microtubules. *J. Cell Biol.* 91 (1981): 125s.

Margolis, R. L., and L. Wilson. Microtubule treadmills—possible molecular machinery. *Nature* 293 (1981): 705.

Mitchison, T., and M. Kirschner. Dynamic instability of microtubule growth. *Nature* 312 (1984): 237.

Olmstead, J. B. Microtubule-associated proteins. *Annu. Rev. Cell Biol.* 2 (1986): 421.

Soifer, D., ed. *Dynamic Aspects of Microtubule Biology. Ann. New York Acad. Sci.*, vol. 466, 1986.

Wiche, G. High-molecular weight microtubule associated proteins (MAPs): A ubiquitous family of cytoskeletal connecting links. *Trends Biochem. Sci.* 10 (1985): 67.

Microfilaments

Geiger, B. Microfilament-membrane interactions. *Trends Biochem. Sci.* 10 (1985): 456.

Pollard, T. D., and S. W. Craig. Mechanism of actin polymerization. *Trends Biochem. Sci.* 7 (1982): 55.

Weeds, A. Actin-binding proteins—regulators of cell architecture and motility. *Nature* 296 (1982): 811.

Intermediate Filaments

Geiger, B. Intermediate filaments: Looking for a function. *Nature* 329 (1987): 392.

Lazarides, E. Intermediate filaments as mechanical integrators of cellular space. *Nature* 283 (1980): 249.

Steinert, P. M., and D. R. Roop. Molecular and cellular biology of intermediate filaments. *Annu. Rev. Biochem.* 57 (1988): 593.

Traub, P. *Intermediate Filaments: A Review.* New York: Springer-Verlag, 1985.

Wang, E., D. Fischman, R. K. H. Liem, and T.-T. Sun. *Intermediate Filaments. Ann. New York Acad. Sci.*, vol. 455, 1985.

Boxed Essay

Osborn, M., and K. Weber. Tumor diagnosis by intermediate filament typing: A novel tool for surgical pathology. *Lab. Invest.* 48 (1983): 372.

Problem Set

1. **Filaments and Tubules.** Indicate whether each statement is true of microtubules (MT), microfilaments (MF), intermediate filaments (IF), or none of these (N).
   (a) Involved in muscle contraction.
   (b) Involved in the movement of cilia and flagella.
   (c) More important for chromosome movements than for cell division.
   (d) More important for cell division than for chromosome movements in animal cells.
   (e) Most likely to remain when cells are treated with solutions of nonionic detergents or high ionic strength.
   (f) Found in bacterial cells.
   (g) Differ in composition in muscle cells versus nerve cells.
   (h) Can be detected by immunofluorescence microscopy.
   (i) Play well-documented roles in cell movement.
   (j) Assembled from protofilaments.

2. **True, False, or Maybe.** Identify each of the following statements as true (T), false (F), or maybe (M), where M indicates a statement that may be either true or false, depending on the circumstances. If you answer F or M, explain why.
   (a) The most abundant cytoskeletal protein in most eukaryotic cells is actin.
   (b) The energy required for tubulin and actin polymerization is provided by hydrolysis of a nucleoside triphosphate.
   (c) Microtubules, microfilaments, and intermediate filaments all exist in the cell in dynamic equilibrium with a pool of subunit proteins.
   (d) Cytochalasin B inhibits chromosome movement but not cytokinesis (cell division).
   (e) An algal cell contains neither tubulin nor actin.
   (f) Despite their structural variability, all of the proteins of intermediate filaments are encoded by genes in the same gene family.
   (g) Most eukaryotic cells contain only one kind of intermediate filament.
   (h) As long as actin monomers continue to be added to the plus end of a microfilament, the microfilament will continue to elongate.

3. **Weighty Matters.** A microfilament exists as two helical strands of actin molecules twisted around each other. One full turn includes 13.5 monomers in each strand and extends the microfilament by 72 nm. A microtubule exists as a hollow tube with a wall of 13 protofilaments, each a linear polymer of tubulin dimers. Each dimer extends a protofilament by 8 nm. Assume that intermediate filaments are constructed of tetrameric protofilaments with a length of 48 nm, and that each intermediate filament has 8 protofilaments in cross section.
   (a) Calculate the mass (in daltons) of a 100-nm microfilament, assuming the presence of actin only.
   (b) Calculate the mass (in daltons) of a 100-nm microtubule, assuming the presence of tubulin only.
   (c) How much heavier, on a length basis, is a microtubule compared to a microfilament?
   (d) Calculate the mass (in daltons) of a 100-nm intermediate filament from a fibroblast cell, assuming the presence of vimentin only.
   (e) How heavy, on a length basis, is a vimentin filament compared to a microfilament or a microtubule?

4. **Cytoskeletal Sources.** Each of the tissues below is an especially good source of a particular kind of cytoskeletal protein. For each of the tissues, decide for which of the three possible proteins it is likely to be a good source.
   (a) Brain: actin/tubulin/desmin
   (b) Intestinal mucosa cells: tubulin/NF protein/keratin
   (c) Stomach muscle: tubulin/vimentin/desmin
   (d) Skeletal muscle: actin/tubulin/GFA protein
   (e) Spinal cord: actin/tubulin/NF protein

5. **Cytoskeletal Studies.** Described below are the results of several recent studies on the proteins of the cytoskeleton. In each case, state the conclusion(s) that can be drawn from the findings.
   (a) Filaments of the same polarity were generated by polymerization of purified actin and were attached to a carbon film. Small polystyrene beads with myosin molecules linked to them were placed on the actin. When ATP was added to the system, the beads were observed to move along the filaments.
   (b) When an animal cell is treated with colchicine, its microtubules depolymerize and virtually disappear. If the colchicine is then washed away, the microtubules appear again, beginning at the centrosome and elongating outward at about the rate (1 μm/min) at which tubulin polymerizes in vitro.
   (c) When a gene for heart-muscle actin is introduced into and expressed in a cultured fibroblast cell that normally synthesizes only skeletal-muscle actin, the actin produced by the "foreign" gene combines readily with the indigenous actin molecules without any adverse effects on the shape or function of the cell.
   (d) Homogenates from dividing clam eggs, but not from nondividing eggs, were found to contain sedimentable structures that could induce the polymerization of tubu-

lin into microtubules in vitro, regardless of whether the tubulin was prepared from dividing or nondividing eggs. When examined with the electron microscope, this structure was shown to consist of the amorphous granular material that surrounds the centrioles in the centrosome.

6. **Actin Polymerization.** Shown in Figure 18-21 is a time course for actin polymerization in vitro. Polymerization was initiated by increasing the ionic strength of a solution of G-actin at time 0, and the concentration of F-actin was determined at time intervals, as shown. Note that three points on the abscissa are labeled as times A, B, and C, respectively. Each of the statements below may apply to any, all, or none of these three time points. Indicate for each statement whether it is true of the actin solution at time A, B, and/or C. Use N if the statement is not true at any point.
   (a) The main process occurring in the solution is nucleation.
   (b) G-actin is present at its critical concentration.
   (c) Actin monomers are polymerizing.
   (d) Actin monomers are adding to the microfilament faster than they are dissociating.
   (e) Actin monomers are dissociating from the microfilament faster than they are polymerizing.
   (f) If the actin solution is suddenly diluted, the microfilament will begin to depolymerize and will continue to do so until the critical concentration is restored.

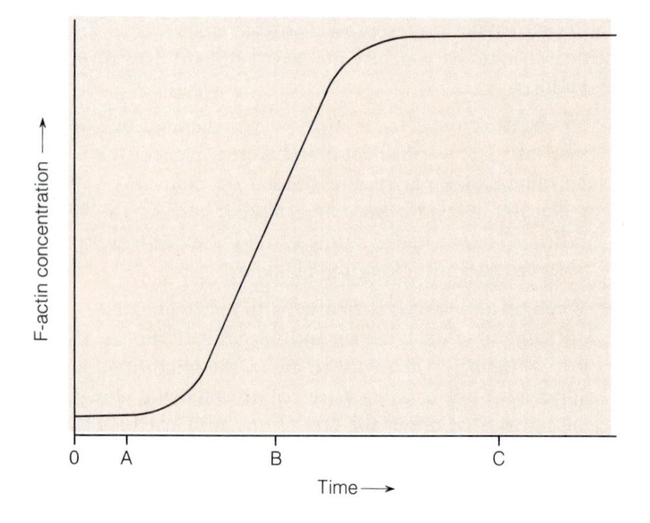

Figure 18-21 Time Course for Actin Polymerization in Vitro. Polymerization of G-actin was initiated at time 0, and the concentration of F-actin was determined at time intervals thereafter.

7. **Profilin and the Acrosomal Reaction.** Not much is known about how the polymerization of G-actin monomers is initiated except in the case of the *acrosomal reaction* in the sperm of some invertebrates. When a sperm cell makes contact with the outer coat of an egg cell, a long, thin, membrane-covered protrusion called the *acrosomal process* shoots out explosively from the sperm cell, puncturing the egg coat and allowing the sperm and egg membranes to fuse. Prior to such contact, the sperm can be shown to be crammed with unpolymerized actin molecules, but with little or no F-actin. Also present in the sperm cell is a protein called *profilin,* which is equal in abundance to actin on a molecular basis.
   (a) Assuming that the formation of F-actin in the core of the acrosomal process drives its rapid elongation, postulate a mechanism to explain the sudden onset of actin polymerization necessary to accomplish the rapid extension of the acrosomal process upon contact with an egg cell. Suggest a way to test your hypothesis.
   (b) The first measurable response of sperm upon contact with an egg cell is a very rapid rise in the intracellular pH of the sperm cell, which occurs even before the polymerization of actin begins. Postulate a possible link between the pH change and the initiation of actin polymerization. Suggest a way to test your hypothesis.

8. **Microtubules and the Endoplasmic Reticulum.** Kinesin is an ATPase thought to be involved in the transport of vesicles along axonal microtubules. It may also play a part in the location and shape of the endoplasmic reticulum (ER). Recent studies have demonstrated that kinesin can attach to the ER membrane in vitro and stretch it by pulling it along microtubules. In intact cells, the ER is known to be stretched outward from the centrosome.
   (a) Suggest at least two different ways in which you could test whether microtubules are responsible in part for the location and shape of the ER in vivo.
   (b) How might you establish more directly the involvement of kinesin in vivo?

19

Cellular Movement: Motility and Contractility

In the previous chapter, we considered cytoskeletal components and their basic functions in eukaryotic cells. We saw that an important function of the cytoskeleton is to provide an intracellular scaffolding that organizes structures within the cell and shapes the cell itself. We also noted that this scaffolding is as much "cytomuscle" as cytoskeleton, because of its role in cell motility. In the present chapter we will explore how cells use their cytoskeletal elements to perform mechanical work and thereby accomplish movement. This may involve the movement of a cell (or a whole organism) through its environment, the movement of the environment past or through the cell, the movement of components within the cell, or the shortening of the cell itself. In each case, we are dealing with some aspect of **motility.** The term **contractility** is used to describe the shortening in muscle cells and can be considered a specialized form of motility.

Systems of Motility

Motility is an especially intriguing use of energy by cells because it involves mechanisms that convert chemical energy directly into mechanical energy. In contrast, most mechanical devices that produce movement from chemicals (such as a steam engine, which depends on the combustion of coal) require an intermediate form of energy, usually heat or electricity. Motility occurs at the tissue, cellular, and subcellular levels, with the most conspicuous examples at the tissue level, particularly in the animal world. The muscle tissues common to most animals consist of cells specifically adapted for contraction, and the movements produced are often obvious, whether manifested as the bending of an arm or leg, the beating of a heart, or uterine contraction during birth. In humans, about 40% of body weight is accounted for by skeletal muscles, which consume a significant proportion of our total energy budget.

At the cellular level, the emphasis is on the movement of the cell through its environment or, in some cases, the movement of the environment past the cell. Cellular motility is a phenomenon observed most often in organisms that consist of one or only a few cells. It occurs among cell types as diverse as flagellated bacteria, ciliated protozoa, and motile sperm. In each of these cases some sort of cellular appendage adapted for propulsion is present. In eukaryotic cells, these may be cilia or flagella, which have a similar structure and mechanism of movement. Other examples of motility at the cellular level include amoeboid movement, cell migration during animal embryogenesis, and the invasive action of cancer cells in malignant tumors.

Equally important is the movement of intracellular components, which might be regarded as motility at the subcellular level. For example, highly ordered structures in the cytoplasm play a key role in the separation of chromosomes during cell division, as we saw in Chapter 14. In addition, some cells display a phenomenon called *cytoplasmic streaming,* in which the cytoplasm undergoes rhythmic patterns of flow, probably as a mixing and stirring function. Other examples of mechanical work at the subcellular level include the characteristic movements of molecular structures that take place during cell growth and differentiation. An example of such a process is the transport of cellulose myofibrils to the growing wall of a dividing or differentiating plant cell.

The Molecular Basis of Motility

The microfilaments and microtubules of the cytoskeleton provide a basic scaffolding for specialized **motor molecules,** which interact with the cytoskeleton to produce motion at the molecular level (Table 19-1) . The molecular motions are summed up to produce motion at the cellular level,

Table 19-1 Motor Molecules of Eukaryotic Cells

Molecules	Role
Microfilament-based (actin)	
Myosin-1, monomer	Motion along actin filaments
Myosin-2, filament	Thick filament of muscle cell
Microtubule-based (tubulin)	
Cytoplasmic dynein	Motion toward minus end of microtubule
Axonemal dynein	Activation of sliding in flagellar microtubule
Kinesin	Motion toward plus end of microtubule

and in some cases, as in muscle contraction, the cellular motions are summed to produce motion at the tissue level. We will discuss each of these motor molecules in detail later in this chapter.

The major motility systems can be classified into three groups. The first type of motility is filament-based, depending on interactions between microfilaments and motor molecules. The most common example of **microfilament-based movement** is muscle contraction. This occurs when filaments of *actin* slide along thicker filaments composed of *myosin*—a mechanism that is commonly referred to as the *sliding-filament model.* Actin and myosin are also active in nonmuscle cells in other filament-based movements, including amoeboid movement, cytoplasmic streaming, and cell division.

The second type of motility involves interactions between specialized motor molecules and microtubules. Microtubules are abundant in interphase cells and are responsible for a variety of intracellular movements. A specific example of **microtubule-based movement** is axonal transport, the process by which a nerve cell transports materials between the central part of the cell and the outlying regions. In this case the motor molecule is *kinesin,* which attaches to structures such as axoplasmic vesicles and "walks" them along microtubules. A second example of microtubule-based movement is the motility produced by the *cilia* and *flagella* of certain eukaryotic cells. Microtubules in cilia or flagella function by a relative sliding between adjacent tubules, and the sliding mechanism is activated by another motor molecule called *dynein.*

The third type of motility is unique to prokaryotes and involves the rotation of **bacterial flagella.** Despite the similarity in terminology, bacterial flagella are chemically and structurally very different from eukaryotic flagella. Bacterial flagellar rotation is driven by the flux of protons through a motor at the base of the flagella. We will now take a closer look at these three motility systems.

Nonmuscle Filament-Based Motility

Although actin and myosin are best known as the major components of the thin and thick filaments of muscle cells, to be discussed later in this chapter, it is now clear that these proteins occur in almost all eukaryotic cells, as both filaments and monomers. Actin and myosin are especially characteristic of cells that carry out what is usually called **nonmuscle motility,** and their role in muscle contraction is simply a prominent example of a more general role in cellular motility.

Actin and Myosin in Nonmuscle Cells

Actin occurs ubiquitously in eukaryotic cells and usually represents 5–10% of the protein in a nonmuscle cell. Typically, about half of the actin in such cells is present as filaments of polymerized F-actin (recall Figure 18-11), whereas the other half is monomeric but usually complexed with other proteins. Actins from phylogenetically diverse organisms are remarkably similar in amino acid sequence and structural properties. Such evolutionary conservation argues for all parts of the molecule being important for its function.

Myosin also occurs widely in nonmuscle cells, but usually at a lower concentration than actin. In general, myosin accounts for less than 0.5% of the protein of nonmuscle cells. The greater prominence of actin in such cells probably reflects a dual role: It is involved with myosin in filament-based motility, but it is apparently also a part of other motility systems in which myosin plays no role.

The myosins from nonmuscle sources are considerably more diverse in structure than are the actins. Nonetheless, almost all of them share three critical properties: They have a rod-shaped tail and a globular head with ATPase activity (Figure 19-1); they are capable of assembling into bipolar filaments; and they bind strongly to actin filaments. The filaments of myosin isolated from nonmuscle sources are by no means as large or as prominent as the thick filaments of muscle cells. In fact, the myosin filaments in nonmuscle cells are usually so small that they cannot be

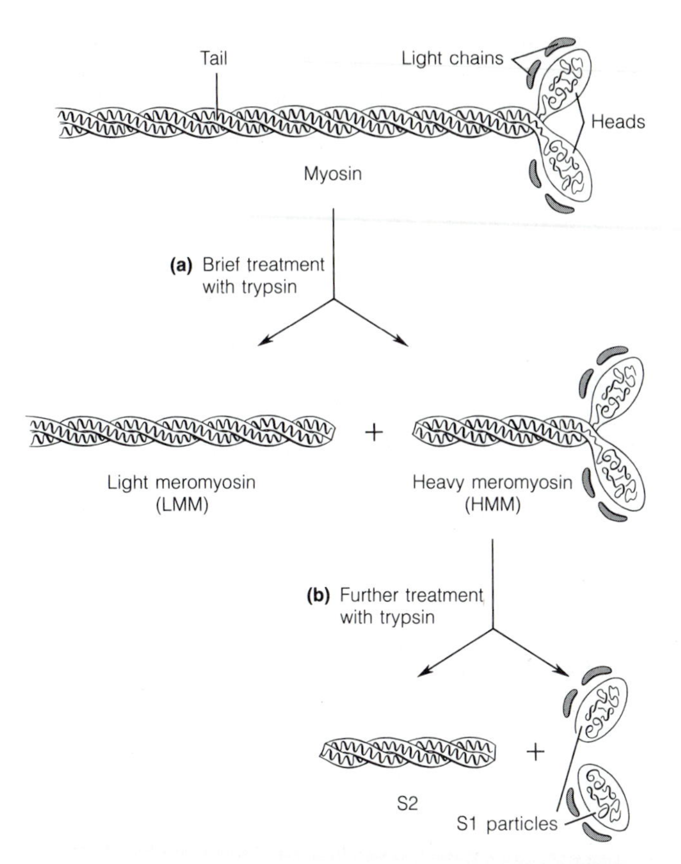

Figure 19-1 Molecular Structure of Myosin. Myosin is the motor molecule of filament-based motility and is composed of four light chains and two heavy chains of approximately 110–140 kDa (cytoplasmic monomers) or 170–240 kDa (thick filaments). The myosin molecule can be cleaved by the hydrolytic enzyme trypsin to light meromyosin (LMM) and heavy meromyosin (HMM). More extensive tryptic digestion splits the HMM fragment into two smaller particles called S1 and S2. The S1 particles contain an ATPase activity, and both HMM and S1 have a high affinity for actin.

visualized directly by microscopy but must be detected using fluorescent antibodies.

Myosin of nonmuscle cells is activated by phosphorylation. Upon phosphorylation, the myosin molecules aggregate into small bipolar filaments, as Figure 19-2 shows. The *myosin light-chain kinase* that catalyzes this phosphorylation is activated by the *calcium-calmodulin complex,* a property that nonmuscle cells share with smooth muscle cells, as we will see later.

The fact that myosin binds so avidly to actin filaments means that it can be used to mark actin in the cell. The affinity is for the head of the myosin molecule, so if myosin is treated briefly with the proteolytic enzyme trypsin, the molecule is split into two fragments, called **heavy meromyosin** and **light meromyosin** (Figure 19-1a). The heavy meromyosin (HMM) fragment contains the heads of the myosin molecule and is therefore capable of interacting with actin. More extensive digestion with trypsin (or with the enzyme papain) cleaves the HMM fragment at a point where the heads are attached, releasing the heads as **S1 (subfragment 1) particles** (Figure 19-1b). Purified S1 particles have been shown to have ATPase activity and to contain an actin binding site.

When tissue sections or preparations of cells are incubated in the absence of ATP with HMM fragments or, more commonly, with S1 particles, the fragments or particles will attach to any actin filaments that are present and "decorate" them in a highly characteristic manner that can be readily identified in the electron microscope. Figure 19-3 shows actin filaments that have been decorated with HMM fragments. This kind of localization is also possible with the light microscope if the fragments are first labeled by linking them to a fluorescent molecule. Decoration of actin with HMM fragments or S1 particles provides a very sensitive means of identifying and localizing actin filaments within the cell. Moreover, the polarity of the actin filaments can be determined from the polarity of the actin-HMM or actin-S1 complex.

Evidence for the Involvement of Actin and Myosin in Nonmuscle Motility

The mere presence of actin and myosin in nonmuscle cells does not, of course, prove that they are involved in motility. The most convincing evidence for such a role is the localization of both actin and myosin filaments at sites of known contractile activity, as demonstrated by fluorescent antibodies. In addition, actin itself can be identified at such sites by decoration with HMM fragments or S1 particles.

Further evidence for the involvement of actin in motility comes from studies with a family of related drugs called the **cytochalasins,** which were described in Chapter 18. These drugs are produced by various species of molds and have the common property of paralyzing many different kinds of cell movements in vertebrates. At the molecular level, cytochalasins act by binding specifically to one end of F-actin filaments, thereby preventing any further addition of G-actin monomers to the filament. The net result is the prevention of actin polymerization.

The disruptive effect of cytochalasins on nonmuscle motility can be explained in terms of the transient nature of the actin filaments in these systems. Unlike the myofibrils of muscle cells, most nonmuscle contractile structures are temporary: They are formed only when needed and are disassembled again thereafter. This explains why at least half of the actin in nonmuscle tissue exists in nonpolymerized form and implies a dynamic relationship between the F-actin filaments and unpolymerized subunits. Clearly, any process that is dependent on de novo assembly of filaments from subunits will be sensitive to inhibition by cytochalasins.

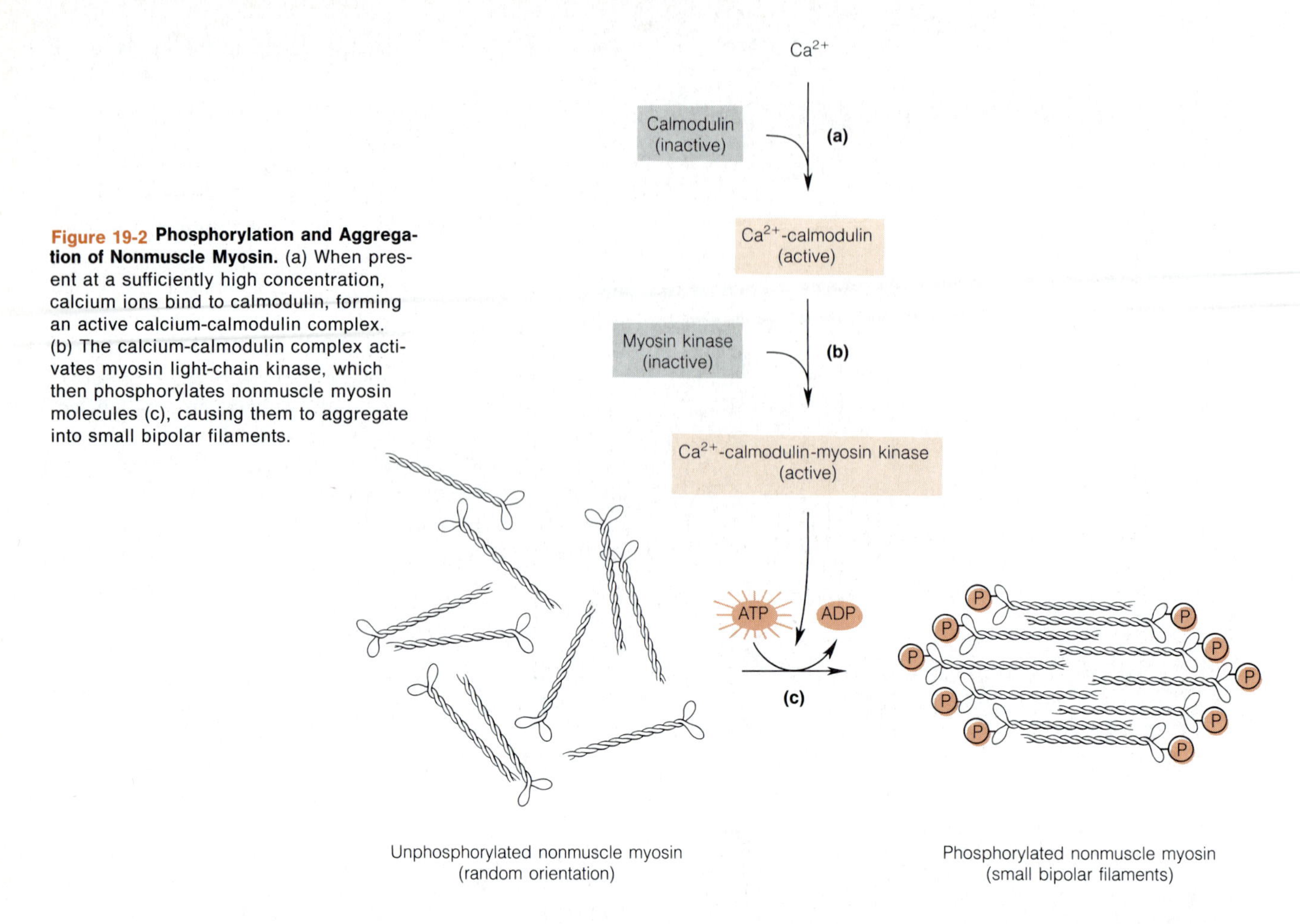

Figure 19-2 Phosphorylation and Aggregation of Nonmuscle Myosin. (a) When present at a sufficiently high concentration, calcium ions bind to calmodulin, forming an active calcium-calmodulin complex. (b) The calcium-calmodulin complex activates myosin light-chain kinase, which then phosphorylates nonmuscle myosin molecules (c), causing them to aggregate into small bipolar filaments.

Cell movements that are sensitive to the cytochalasins are in general also inhibited by phalloidin, even though the effect of phalloidin is to stabilize actin filaments and prevent their depolymerization, as described in Chapter 18. This dual sensitivity of actin-linked motility to both cytochalasins and phalloidin is further evidence of the importance of the dynamic assembly and disassembly of actin filaments in a variety of nonmuscle movements.

Intracellular Microtubule-Based Motility: Dynein and Kinesin

The interaction of motor molecules with the cytoskeleton was briefly outlined in Chapter 18. Before continuing with a detailed description of their role in cell motility, it is worth reviewing some of the basic concepts.

It has been known for some years that the motor molecule **dynein** is directly involved in the movement of flagella and cilia, as we will see later. More recently it has become clear that there is also a cytoplasmic form of dynein that uses microtubules as tracks to move structures within cells. Even more remarkable was the discovery of **kinesin,** a second motor molecule that also moves along microtubular tracks, but in the opposite direction! This means that cells have the capability to produce bidirectional movement of intracellular structures.

Such motion is particularly important for neurons. A receptor protein or neurotransmitter might be synthesized in the cell body of the neuron and then might have to be transported over distances up to a meter between the cell body and the nerve ending. Some form of energy-dependent transport must be available for this to occur, and microtubule-based motility provides the required mechanism.

The evidence for dynein- and kinesin-mediated motility is dramatic. For instance, in one experimental system the cytoplasm of a squid nerve cell is squeezed out (see Chapter 20) and examined in a video-enhanced light microscope. Under these conditions, small vesicles can be observed moving along microtubule tracks at rates up to 2 μm per second. If artificial particles such as polystyrene beads are added, these move as well, suggesting that some sort of motor molecule attaches even to these plastic surfaces and tugs the particles along the microtubule. Significantly, the motion is ATP-dependent.

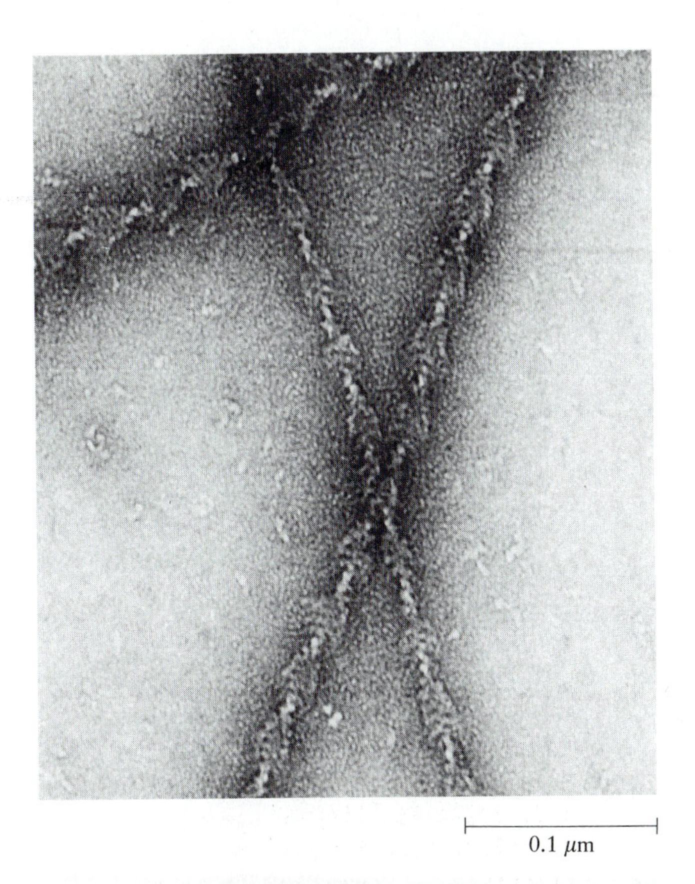

Figure 19-3 Actin Filaments Decorated with HMM Fragments. Incubation of actin-containing intracellular components with HMM results in formation of a characteristic "arrowhead" complex that is readily visible in the electron microscope. Decoration of actin with HMM provides a sensitive method for identifying and localizing actin filaments within a cell. (TEM)

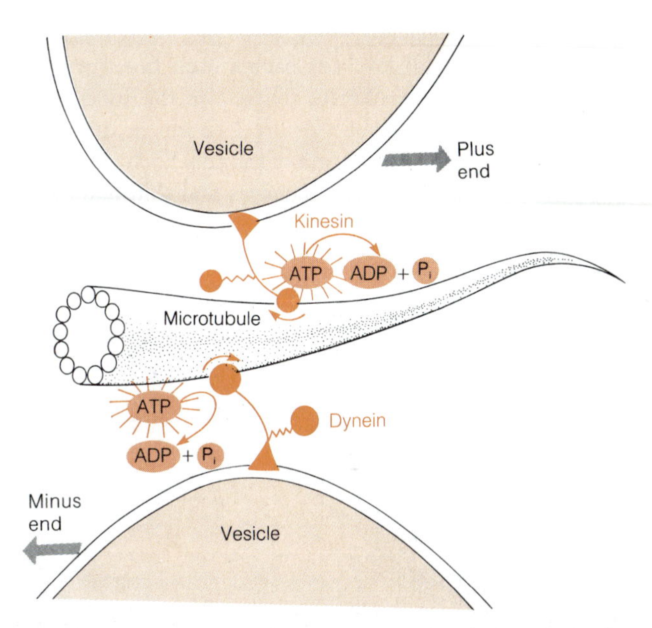

Figure 19-4 Microtubule-Based Motility. A proposed mechanism is that the "motor molecules" dynein and kinesin use the energy of ATP hydrolysis to change shapes. In the process, they move intracellular structures along microtubules. Here kinesin moves toward the plus end of microtubules, and dynein toward the minus end. The mechanism is not yet established.

In what direction does the movement occur in the cell? Recall from Chapter 18 that microtubules have plus and minus ends and that centrosomes act as organizing centers, polymerizing tubulin so that the plus end is outward. When polystyrene beads and purified kinesin are added to microtubules polymerized by centrosomes, the beads move toward the plus end. This finding means that in a nerve cell, kinesin activates transport from the cell body down the axon to the nerve ending. When similar experiments are carried out with dynein, particles are moved in the *opposite* direction, toward the minus end of the microtubules.

These results are summarized in Figure 19-4, together with a proposed mechanism for the movement generated by kinesin and dynein. The motion is thought to develop from a cycle of attachment, conformational change, and release that is driven by ATP hydrolysis. As will be described in the next section, myosin also undergoes a similar ATP-driven cycle when it interacts with actin filaments to produce muscle contraction.

Filament-Based Movement in Muscle

Muscle contraction is the most familiar example of mechanical work mediated by intracellular filaments. Mammals have several different kinds of muscles, including skeletal muscle, cardiac (heart) muscle, and smooth muscle. We will first consider skeletal muscle, because much of our knowledge of the contractile process grew out of early investigations of its molecular structure and function.

Skeletal Muscle Cells

Skeletal muscles are responsible for voluntary movement. The structural organization of skeletal muscle is shown in Figure 19-5. A muscle (a) consists of bundles of parallel **muscle fibers** (b) joined by tendons to the bones that the muscle must move. Each fiber (c) is actually a long, thin multinucleate cell, highly specialized for its contractile function. The multinucleate state arises from end-to-end fusion of embryonic cells called *myoblasts* during muscle differentiation. This cell fusion also accounts at least in part for the striking length of muscle cells, which may be measured in centimeters.

At the subcellular level (d), each muscle fiber (or cell) contains numerous **myofibrils.** These are the functional units of contraction. Myofibrils are 1–2 μm in diameter and may extend the entire length of the cell. Each myofibril (e) contains bundles of **thick filaments** and **thin filaments** (f). Thick filaments consist of myosin, whereas thin filaments consist mainly of actin (g), although several other important proteins are also present. The thin filaments are arranged around the thick filaments in a hexagonal pattern, as can be seen when the myofibril is viewed in cross section (Figure 19-6).

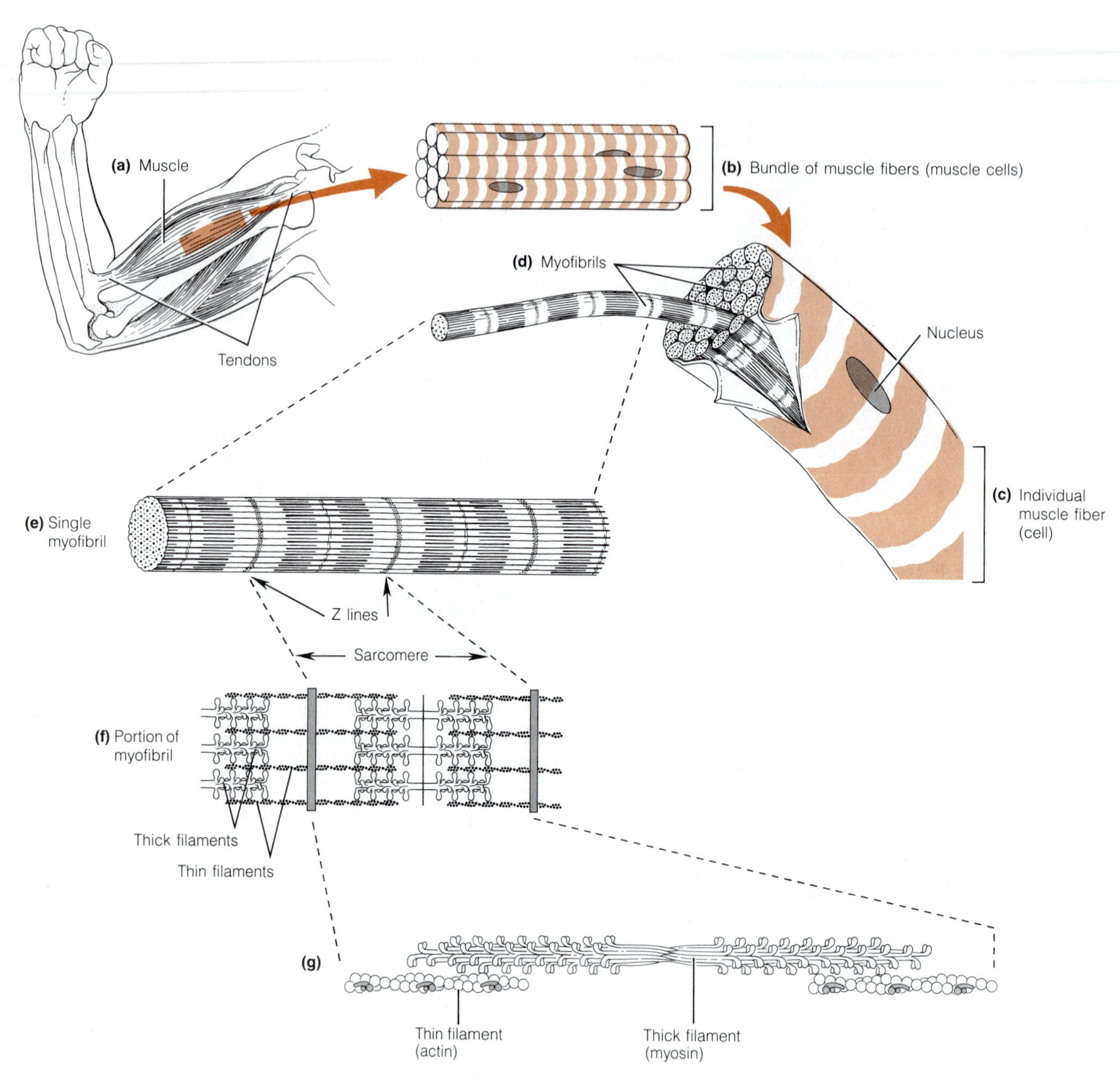

Figure 19-5 Levels of Organization of Skeletal Muscle Tissue. (a) Muscle tissue is attached by means of tendons to the specific bones it must move. (b) The tissue consists of bundles of muscle fibers, each of which is a long, thin multinucleate cell (c). (d) Within each cell are many myofibrils. (e) Each myofibril consists of bundles of filaments aligned laterally to give skeletal muscle its striated appearance. (f) Within the myofibril, thick filaments interdigitate with thin filaments that are attached to the Z line. The unit of contraction along each myofibril is the sarcomere, defined as the distance from one Z line to the next. (g) The thick and thin filaments consist primarily of myosin and actin, respectively.

The filaments in skeletal muscle are aligned in lateral register, giving the myofibrils a pattern of alternating dark and light bands (Figure 19-7). This pattern of bands, or *striations*, is characteristic of skeletal and cardiac muscle, which are therefore referred to as *striated muscle*.

The basic nomenclature for skeletal muscle is summarized in Figure 19-8. The dark bands are called **A bands** and the light bands are called **I bands.** (The terminology for the structure and appearance of muscle myofibrils was developed from observations originally made with the polarizing light microscope. *I* stands for *isotropic*, and *A* for *anisotropic*, terms related to the appearance of these bands when illuminated with plane-polarized light.)

The lighter region in the middle of each A band is called the **H zone**, from the German word *hell*, which means "light." In the middle of each I band appears a dense **Z line** (the Z is for the German word *zwischen*, meaning "between"). The distance from one Z line to the next defines a **sarcomere**, the basic repeating unit along the myofibril. A sarcomere is about 2.5–3.0 μm long in the relaxed state, but shortens progressively as the muscle contracts.

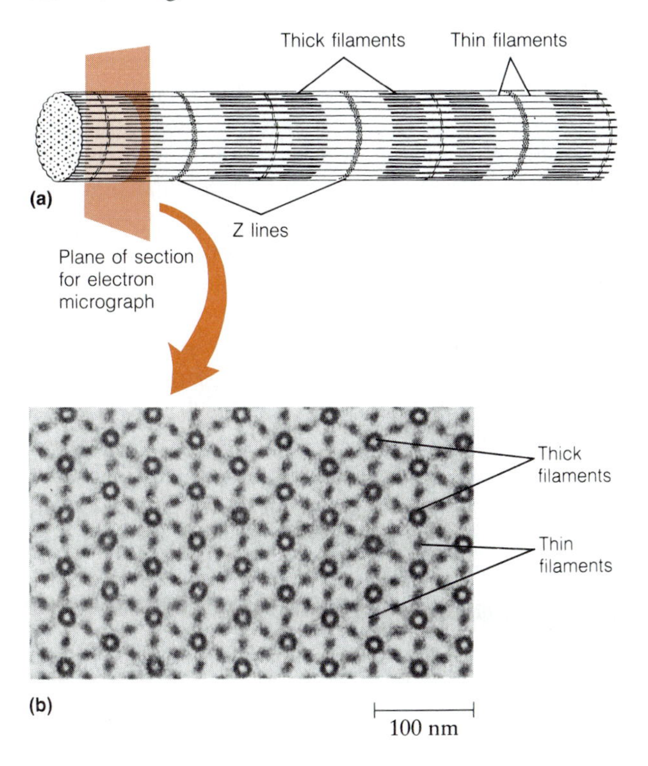

Figure 19-6 Arrangement of Thick and Thin Filaments in the Myofibril. (a) A myofibril consists of interdigitated thick and thin filaments. (b) The thin filaments are arranged around the thick filaments in a hexagonal pattern, as seen clearly in this cross section of a flight muscle from the fruit fly *Drosophila melanogaster* viewed in a high-voltage electron microscope (TEM).

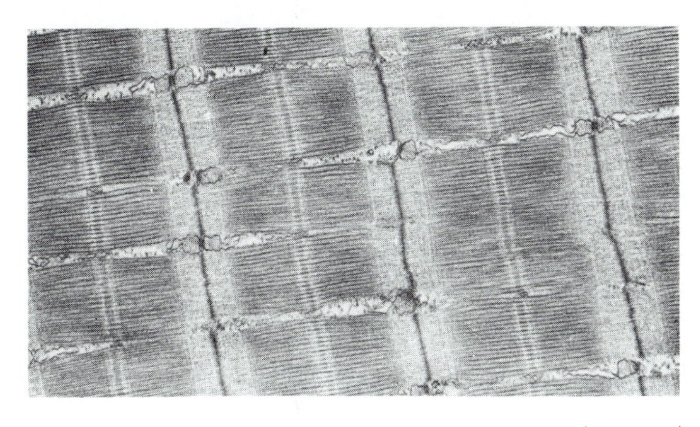

Figure 19-7 Skeletal Muscle. The striated pattern of muscle is seen in this electron micrograph of a fiber from frog skeletal muscle. (TEM)

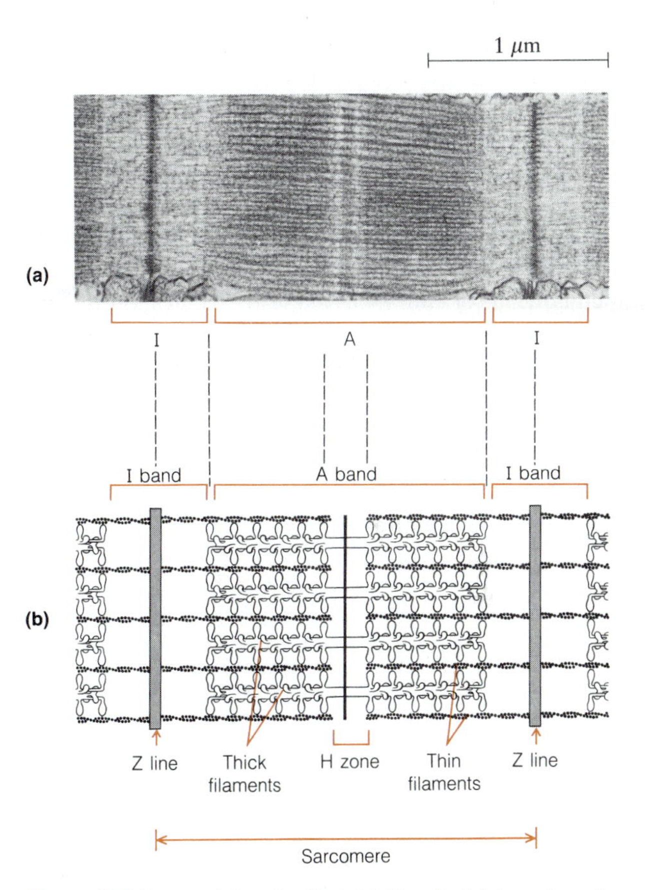

Figure 19-8 Nomenclature for Skeletal Muscle. (a) A portion of Figure 19-7 showing a single sarcomere (TEM). (b) A schematic diagram to interpret the repeating pattern of bands in striated muscle in terms of the interdigitation of thick and thin filaments. An A band corresponds to the length of the thick filaments, whereas an I band represents that portion of the thin filaments that does not overlap with thick filaments. The lighter area in the center of the A band is called the H zone. The dense zone in the center of each I band is called the Z line. A sarcomere, the basic repeating unit along the myofibril, is defined as the distance between successive Z lines.

Muscle Filaments and Their Proteins

The striated pattern of skeletal muscle and the observed shortening of the sarcomeres during contraction can be explained in terms of the thick and thin filaments that make up the myofibrils. We will therefore look in some detail at both types of filaments and then come back to the contraction process in which they play so vital a role.

The Thick Filaments. The thick filaments of myofibrils are about 15 nm in diameter and about 1.6 μm long. They lie parallel to one another in the middle of the sarcomere, thus forming the dark A bands of the myofibril (Figure 19-8). Each thick filament consists of many molecules of myosin, which are oriented in opposite directions in the two halves of the filament. Each myosin molecule is long and thin, with a molecular weight of about 520,000.

As Figure 19-9a illustrates, myosin has a complex structure, consisting of two identical **heavy chains** (223,000 daltons each) and four **light chains,** two of about 20,000 daltons each and two of about 16,000 daltons each. These six polypeptides are organized into two globular heads and a narrow tail. The tail of the molecule consists of the two heavy chains coiled around each other. At its amino terminus, each of these chains is coiled into a globular head, with which a pair of light chains is associated. Part of the globular head contains the ATPase activity responsible for the ATP hydrolysis that drives the contraction process.

Every thick filament consists of hundreds of myosin molecules organized in a staggered array such that the heads of successive myosin molecules project out of the thick filament in a repeating pattern, as shown in Figure 19-9b. The heads occur in pairs, which protrude from the thick filament facing away from the center of the filament. Projecting pairs of heads are spaced 14.3 nm apart along the thick filament, with each pair displaced one-third of the way around the filament from the previous pair. These protruding heads can make contact with adjacent thin filaments, forming the cross-bridges between thick and thin filaments that are so essential to the mechanism of muscle contraction.

The Thin Filaments. The thick filaments of the myofibril interdigitate with the thin filaments. The thin filaments are about 7 nm in diameter, have a length of about 1 μm, and are the only filaments in the I bands of the myofibril. In fact, each I band consists of *two* sets of thin filaments, one set on either side of the Z line, with each filament attached to the Z line and extending toward and into the A band in the center of the sarcomere. This accounts for the length of almost 2 μm for I bands in extended muscle.

The structure of thin filaments is shown in Figure 19-10. A thin filament consists of at least three proteins, the most prominent of which is actin. Recall from Chapter 18 that actin is synthesized as **G-actin** but polymerizes into long, linear strands of **F-actin.** A thin filament consists of two strands of F-actin assembled into a helix. In addition, thin filaments contain small amounts of the proteins **tropomyosin** and **troponin.** Tropomyosin is a long, rodlike molecule, similar to the myosin tail, and fits in the groove on either side of the actin helix (Figure 19-10). Each tropomyosin molecule stretches for about 38.5 nm along the filament and associates along its length with seven actin monomers.

Troponin is actually a complex of three polypeptide chains, called **TnT, TnC,** and **TnI.** TnT binds to tropomyosin and is thought to be responsible for positioning the complex on the tropomyosin molecule. TnC binds cal-

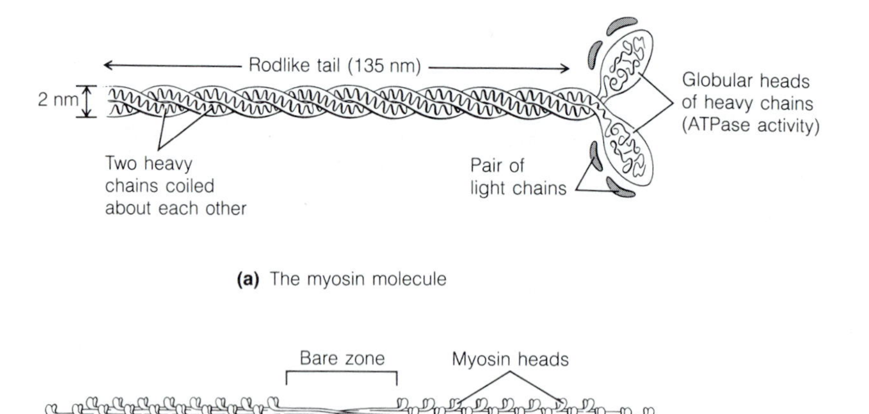

Figure 19-9 Organization of the Thick Filament of Skeletal Muscle. As described in Figure 19-1, a myosin molecule contains both heavy chains and light chains, shown here in (a). The long, rodlike myosin tail, about 135 nm in length and 2 nm in diameter, consists of the helical tails of the two heavy chains coiled around each other. At its amino end, each heavy chain terminates in a globular head associated with the light chains. (b) Organization of myosin molecules into the thick filaments of muscle cells. The thick filament of the myofibril consists of hundreds of myosin molecules organized in a repeating, staggered array. A typical thick filament is about 1.6 μm long and about 15 nm in diameter. Individual myosin molecules are integrated into the filament longitudinally, with their ATPase-containing heads oriented away from the center of the filament. The central region of the filament is therefore a bare zone containing no heads.

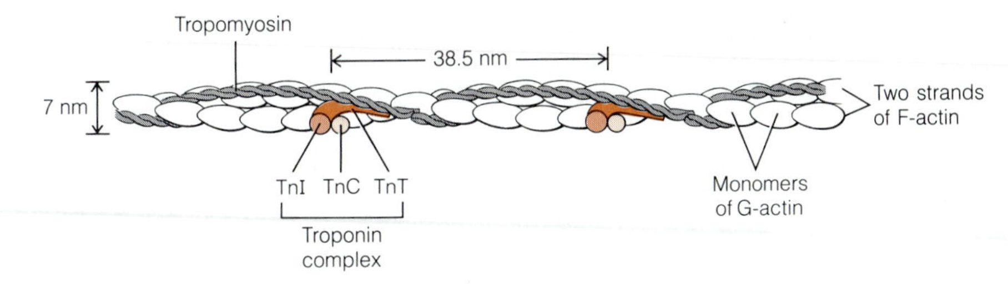

Figure 19-10 The Thin Filament of Striated Muscle. Each thin filament consists of two strands of F-actin twisted together into a helix with 13.5 G-actin units per turn. Long, rod-shaped molecules of tropomyosin lie along the F-actin grooves. Each tropomyosin molecule consists of two α helices wound about each other to form a coil about 2 nm in diameter and 38.5 nm long. Associated with each tropomyosin molecule is a troponin complex consisting of the three polypeptides TnT, TnC, and TnI.

cium ions, and TnI binds to actin. (The "I" in TnI stands for "inhibitory," since TnI inhibits an ATPase enzyme activity involved in muscle contraction, as we will see later.) One troponin complex is associated with each tropomyosin molecule, so the spacing between successive troponin complexes along the thin filament is 38.5 nm. Troponin and tropomyosin constitute a calcium-sensitive switch that turns on contraction in both skeletal and heart muscle.

How can the filamentous proteins of muscle fibers maintain such a precise organization, when in other cells microfilaments are relatively disorganized? The answer has to do with structural proteins that play a central role in maintaining the architectural relationships of muscle proteins (Figure 19-11). For instance, **α-actinin** keeps actin filaments bundled into parallel arrays and attaches them to the Z line. A myosin-binding protein called **myomesin** is present at the H zone of the thick filament arrays and performs the same bundling function for the myosin molecules composing the arrays. A third structural protein, **titin,** attaches the thick filaments to the Z lines. Titin is highly flexible, so that during contraction-relaxation cycles it can keep thick filaments in the correct position relative to the thin filaments.

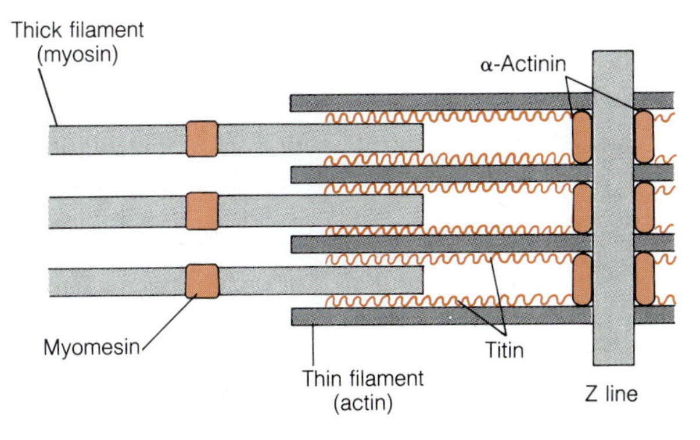

Figure 19-11 Structural Proteins of the Sarcomere. The thick and thin filaments require structural support to maintain their precise organization in the sarcomere. The support is provided by two proteins—actinin and myomesin—which bundle actin and myosin filaments, respectively. Titin attaches thick filaments to the Z line, thereby maintaining their position within the thin filament array.

The protein components involved in muscle contraction are summarized in Table 19-2. The contractile process involves the complex interaction of all of these proteins. Our understanding of the underlying molecular mechanisms is a remarkable example of the power of modern cell biology. We will now go on to describe contraction in terms of interacting molecular systems of proteins, and the equally important regulatory mechanisms that control contraction and relaxation of the muscle cell.

The Sliding-Filament Model of Muscle Contraction

With our understanding of muscle structure, we can now consider what happens during the contraction process. From electron microscopic studies, it is clear that the A bands of the myofibrils remain fixed in length during contraction, whereas the I bands shorten progressively and virtually disappear in the fully contracted state. To explain these observations, the **sliding-filament model** shown in Figure 19-12 was proposed in 1954. According to this model, muscle contraction is due to thin filaments sliding past thick filaments, with no change in the length of either type of filament. The sliding-filament model not only proved to be correct, but was instrumental in focusing attention on the molecular interactions between thick and thin filaments that underlie the sliding process.

As Figure 19-12 indicates, contraction involves a sliding of thin filaments such that they are drawn progressively into the spaces between adjacent thick filaments, overlapping more and more with the thick filaments and narrowing the H zone in the process. The result is a shortening of

Table 19-2 Protein Components of Skeletal Muscle Contractile Apparatus

Protein	Molecular Weight	Function or Role
Actin	42,000	Composes thin filaments.
Myosin	510,000	Composes thick filaments.
Tropomyosin	64,000	Bound to thin filament.
Troponin	78,000	Bound to thin filament. Mediates calcium regulation of contraction.
Titin	2,500	Links thick filaments to Z band.
α-Actinin	190,000	Bundles actin filaments in Z band.
Myomesin	—	Bundles myosin in thick filament.
Calcium pump	115,000	Major protein of sarcoplasmic reticulum (SR). Transports Ca^{2+} into SR to relax muscle contraction.
Calcium gate	—	Releases Ca^{2+} from SR to initiate contraction.

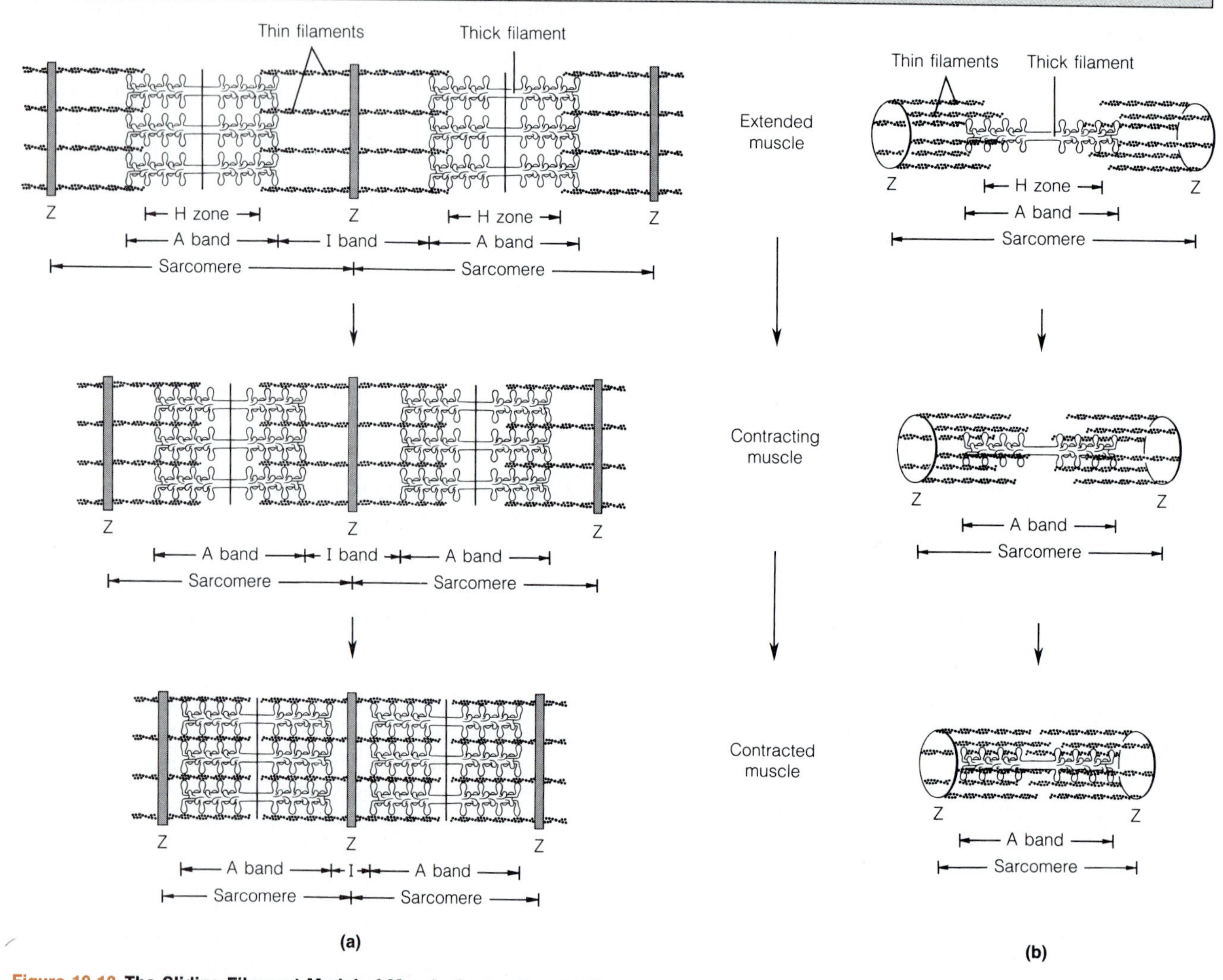

Figure 19-12 The Sliding Filament Model of Muscle Contraction. (a) Two sarcomeres of a myofibril during the contraction process. (b) A single thick filament and the six surrounding thin filaments during the contraction process. In both (a) and (b), the extended configuration is shown at the top, while the middle and lower views represent progressively more contracted myofibrils. Shortening of a myofibril is accomplished by a progressive sliding of thick and thin filaments with respect to each other. The result is a greater interdigitation of filaments without any change in length of individual filaments. The increasing overlap of thick and thin filaments leads to a progressive decrease in the length of the I band as interpenetration continues during contraction.

the individual sarcomeres and fibrils and hence also of the muscle cell and the whole tissue. This, in turn, causes the movement of the body parts attached to the muscle.

The sliding of thin filaments past thick filaments requires energy, which gives rise to three questions: By what mechanism are the thin filaments drawn or pulled progressively into the spaces between thick filaments to cause the actual contraction? How is the energy of ATP used to drive this process? And what keeps the partially interdigitated thick and thin filaments associated with each other, so that they do not simply fall apart?

Cross-Bridge Formation. We can begin by taking the third question first and asking what holds interdigitated thick and thin filaments together. This is the role of the heads of the myosin molecules that project outward at regular intervals along each thick filament. These heads represent binding sites for the F-actin of the thin filaments. Regions of overlap between thick and thin filaments, whether extensive (in the contracted muscle) or minimal (in the relaxed muscle), are always characterized by the presence of transient **cross-bridges** between the myosin heads of the thick filaments and the actin of the thin filaments. Numerous cross-bridges can be seen in Figure 19-13, which is a high-voltage electron micrograph of a single contracted sarcomere from an insect flight muscle.

Cycles of Cross-Bridge Formation. For actual contraction, it is not enough that cross-bridges be present. They must form and dissociate repeatedly and in such a manner that each cycle of cross-bridge formation causes the thin filaments to interdigitate with the thick filaments more and more, thereby shortening the individual sarcomeres and causing the muscle fiber to contract. A given myosin head on the thick filament repeatedly undergoes a cycle of events whereby it binds to a specific actin site on the thin filament, undergoes an energy-requiring change in shape that pulls the thin filament, then breaks that particular association with the thin filament and associates with another actin site farther along the thin filament toward the Z line.

In this way, the protruding heads of the myosin molecules in the thick filaments draw each thin filament along unidirectionally toward the center of the A band. Overall contraction, then, is the net result of the repeated making and breaking of many such cross-bridges, with each cycle of bridge formation causing the translocation of a small length of thin filament of a single fibril in a single cell. What about the direction of contraction? Recall from Chapter 18 that actin filaments have plus and minus ends. We know from the above description that contraction of a muscle cell involves a motor molecule—myosin—"walking" along the actin filament in an ATP-dependent process, but does contraction occur toward the plus or minus end of actin thin filaments? The observation is that myosin always walks toward the *plus* end of the thin filament, thereby establishing the direction of contraction.

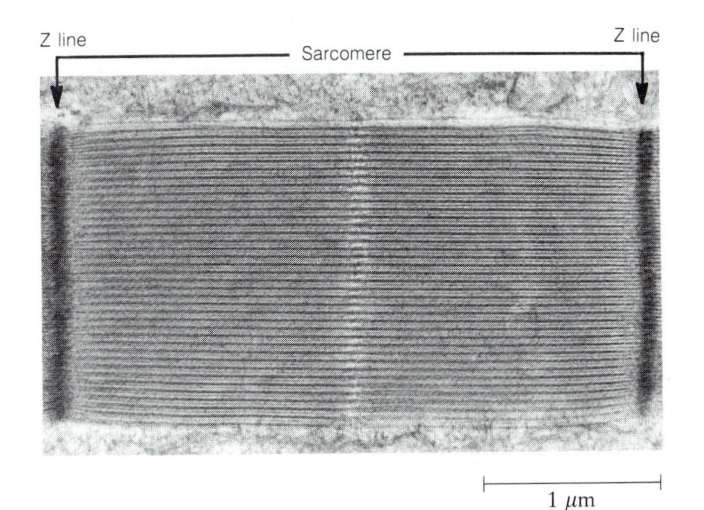

Figure 19-13 Cross-Bridges. The cross-bridges between thick and thin filaments formed by the projecting heads of myosin molecules can be readily seen in this electron micrograph of a single contracted sarcomere from a flight muscle of *Drosophila melanogaster*. A section of 0.5 μm was photographed in a high-voltage electron microscope. (TEM)

ATP and the Contraction Cycle

The driving force for this cyclic formation of cross-bridges in skeletal muscle is the hydrolysis of ATP, catalyzed by the actin-activated ATPase located strategically in the myosin head. This requirement for ATP can be demonstrated in vitro, since isolated muscle fibers can be shown to contract in response to added ATP.

The mechanism whereby this is accomplished is depicted in Figure 19-14 as a four-step cycle. In step 1, a specific myosin head in a high-energy configuration binds to a specific actin site on an adjacent thin filament. The needed flexibility is thought to be provided by a hinge region of the myosin molecule. Step 2 is then the "power stroke": As the myosin head that forms the cross-bridge reverts from a high-energy to a low-energy configuration, it undergoes a change, causing the thick filament to pull against the thin filament, which then moves with respect to the thick filament. Concomitantly, the ADP and P_i generated in the previous contraction cycle dissociate from the myosin head.

Cross-bridge dissociation follows in step 3, as ATP binds to the myosin head in preparation for the next step. (In the absence of adequate ATP, cross-bridge dissociation does not occur, and the muscle becomes locked in a stiff, rigid state called **rigor.** The *rigor mortis* associated with death results from the depletion of ATP and the progressive accumulation of cross-bridges in the configuration shown at the bottom of Figure 19-14.) Note that once detached, the thick and thin filaments would be free to slip back to their previous position *except* that they are held together at all times by the many other cross-bridges along their length at any given moment, just as at least some legs of a millipede are always in contact with the surface on which it is walking. In fact, each thick filament has about 350 myosin heads, and each of these attaches and detaches about five times per second during rapid contraction, so there are always many cross-bridges intact at any time.

Finally, in step 4, the energy of ATP hydrolysis is used to reposition the myosin head for the next cycle by returning it to the high-energy configuration necessary for the next round of cross-bridge formation and filament sliding. This brings us back to where we started, for the myosin head is now activated and ready to form a bridge to the actin again. But the new bridge will be formed with an actin site farther along the thin filament, since the first cycle resulted in a net displacement of the thin filament with respect to the thick filament. In succeeding cycles, the particular myosin head shown in Figure 19-14 will draw the thin filament in the direction of further contraction.

Of particular interest in this contraction cycle are the separation of the actual hydrolysis of ATP (step 4) from

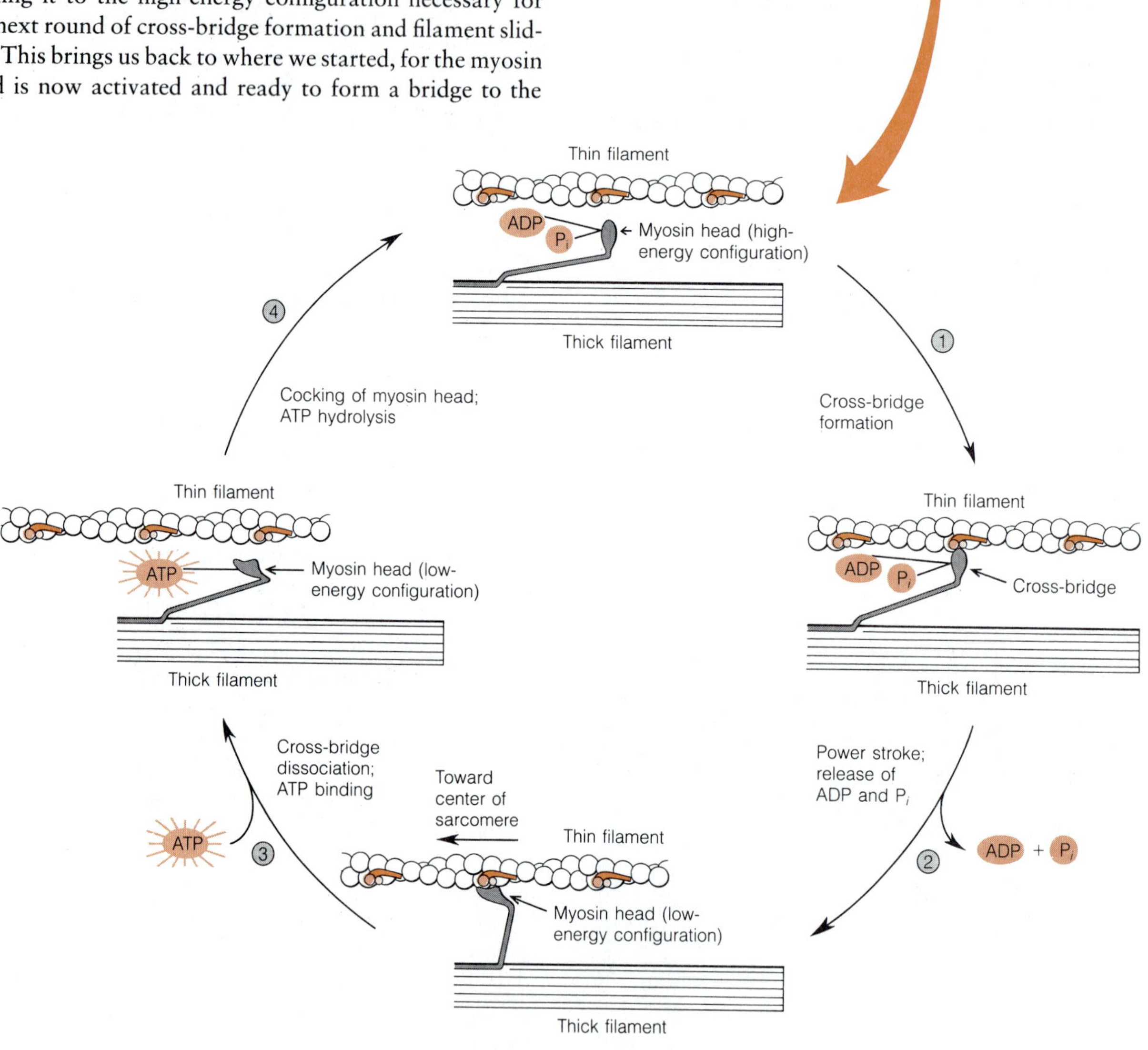

Figure 19-14 The Cyclic Process of Muscle Contraction. A small segment of adjacent thick and thin filaments (see inset) is used to illustrate the series of events whereby the cross-bridge formed by a myosin head is used to draw the thin filament toward the center of the sarcomere, thereby causing the myofibril to contract. The configuration shown at the top of the figure is that of relaxed muscle. When a muscle is in rigor, its cross-bridges have the configuration shown at the bottom of the figure.

the contraction event that it drives (step 2) and the use of an energized configuration of the myosin head as the interim "carrier" of the energy. Note also that the ADP and P_i formed by ATP hydrolysis remain bound to the myosin as long as the protein is in the energized form but are released upon reversion to the low-energy form in the power stroke of step 2.

The Activation of Muscle Contraction

The sequence of events depicted in Figure 19-14 shows how muscle contraction is accomplished, but provides no insight into how the process is regulated. Indeed, the description provided so far implies that skeletal muscle ought to contract continuously, as long as there is sufficient ATP. Yet experience tells us that most skeletal muscles spend more time in the relaxed state than they spend in contraction. Contraction and relaxation must therefore be regulated to result in the coordinated movements associated with muscle activity.

The Role of Calcium in Contraction. The key to regulation lies in the critical dependence of the contraction process on free calcium ions (Ca^{2+}) and in the ability of the muscle cell to raise and lower calcium levels rapidly in the cytoplasm around the myofibrils. (In muscle cells, the cytoplasm is also called the **sarcoplasm,** from the Greek word *sarkos,* meaning "flesh.") The thin filaments of skeletal muscle contain, in addition to actin, the regulatory proteins tropomyosin and troponin. These molecules act in concert to regulate the availability of myosin binding sites on the actin filament in a way that depends critically on the level of calcium in the sarcoplasm.

To understand this, we must recognize that the myosin binding sites on actin can be blocked by tropomyosin and that this blocking effect of tropomyosin depends on calcium, but in a way that is mediated by troponin. Figure 19-15 illustrates how this occurs. When the level of calcium in the sarcoplasm is relatively low ($<10^{-7}$ *M*), tropomyosin blocks the binding sites on the actin filament, effectively preventing their interaction with myosin (Figure 19-15a). As a result, cross-bridge formation is inhibited and the muscle becomes or remains relaxed. However, at higher calcium concentrations ($>10^{-6}$ *M*), tropomyosin molecules shift their position slightly, allowing myosin heads to make contact with the binding sites on the actin filament and thereby initiate contraction (Figure 19-15b).

This sensitivity to calcium is a property not of the tropomyosin molecule itself, but of the TnC polypeptide of the troponin molecule with which it is complexed. When a calcium ion binds to it, TnC undergoes a conformational change that is transmitted to the tropomyosin molecule, causing it to move toward the center of the helical groove of the thin filament, out of the blocking position. The binding sites on actin are then accessible to the myosin heads, so contraction can proceed.

Thus, an increase in the sarcoplasmic calcium concentration stimulates the contraction of skeletal muscle by triggering the following series of events:

1. Calcium binds to troponin and induces a conformational change in the complex.
2. This change in troponin causes a shift in the position of the tropomyosin with which it is complexed.

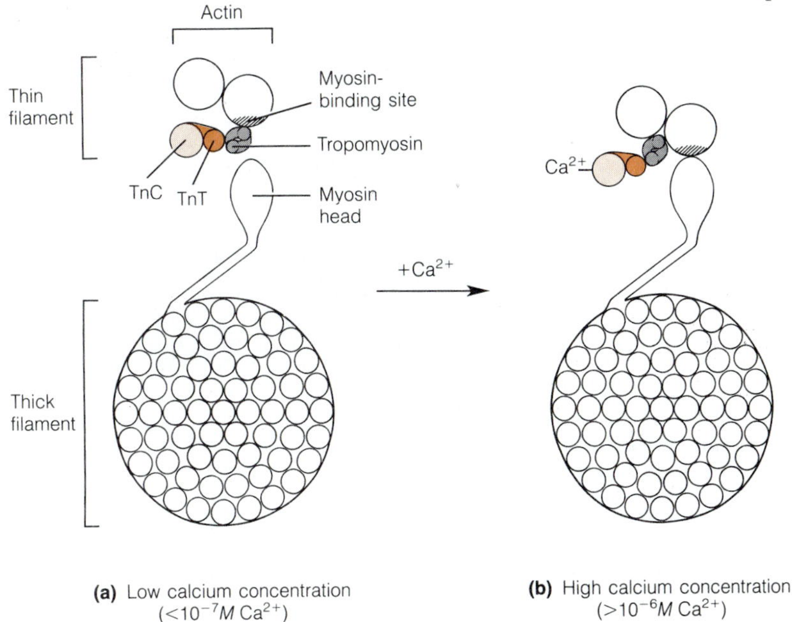

Figure 19-15 Regulation of Contraction in Striated Muscle. (a) At low concentrations ($<10^{-7}$ *M* Ca^{2+}), calcium is not bound to the TnC subunit of troponin, and tropomyosin blocks the binding sites on actin, preventing access by myosin and thereby maintaining the muscle in the relaxed state. (b) At high concentrations ($>10^{-6}$ *M* Ca^{2+}), calcium binds to the TnC subunit of troponin, inducing a conformational change that is transmitted to tropomyosin. The tropomyosin molecule moves toward the center of the groove in the thin filament, allowing myosin to gain access to the binding sites on actin, thereby triggering contraction.

3. The binding sites on actin become available for interaction with myosin.
4. Cross-bridges form, setting in motion the sequence of events depicted in Figure 19-14 and resulting in contraction.

When the calcium concentration falls again, the troponin-calcium complex dissociates and the tropomyosin moves back to the blocking position. Myosin binding is therefore inhibited, further cross-bridge formation is prevented, and the contraction cycle ends.

Regulation of Calcium Levels in Skeletal Muscle Cells. We now have some understanding of what occurs during contraction and can consider in more detail a second aspect of muscle cell function. Think for a moment what must happen when you move any part of your body, such as when you flex an index finger. A nerve impulse is generated in the brain and transmitted down the spinal column to the nerve cells (motor neurons) that control a small muscle in your forearm. The motor neurons activate the appropriate muscle cells, which contract and relax, all within about 100 milliseconds. Two main questions arise from this description: How is the muscle cell activated, and how does it relax after the contractile cycle?

From the earlier discussion, you know that calcium ions are required to activate the contractile process. Therefore, if there were some way to increase and decrease calcium ion concentration in the muscle cell, we would have a possible regulatory mechanism for contraction. Evidence for this mechanism was first obtained in the early 1960s, when it was discovered that a particulate fraction obtained from homogenized muscle tissue was able to cause relaxation of specially treated muscle fibers. It was later found that the particles were in fact membrane vesicles derived from the **sarcoplasmic reticulum** (SR) of muscle cells (Figure 19-16). As the name suggests, the SR is similar to the endoplasmic reticulum found in nonmuscle cells.

An important property of the SR is the ability to store calcium and to release it when activated by a motor neuron. The activation involves depolarization of the muscle cell plasma membrane, which is excitable in the same way that

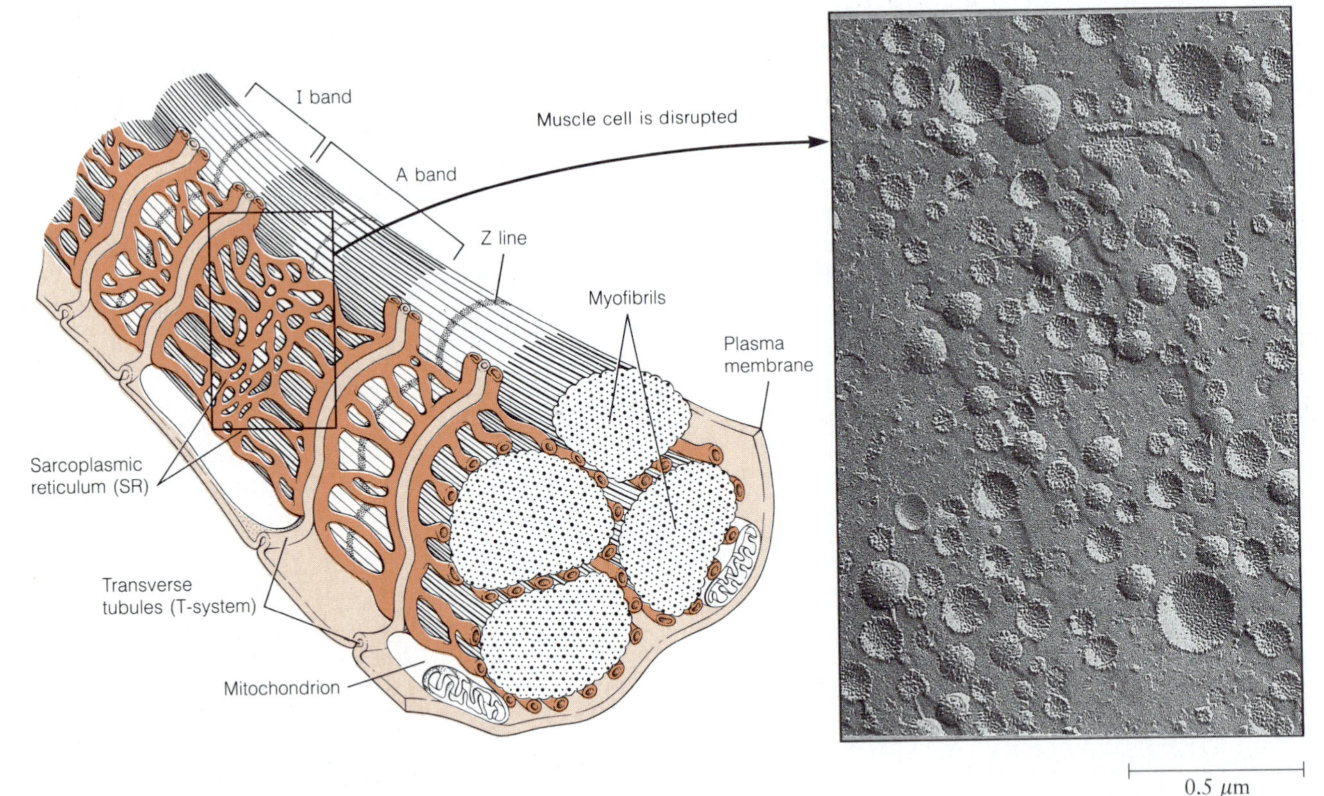

Figure 19-16 The Sarcoplasmic Reticulum and the Transverse Tubule System of Skeletal Muscle Cells. The sarcoplasmic reticulum (SR) is an extensive network of specialized ER that accumulates calcium and releases it on signal. The transverse tubules of the T-system are invaginations of the plasma membrane that relay the contraction signal from the plasma membrane to the interior of the cell. If the muscle cell is disrupted, the SR breaks up into membranous vesicles that are capable of active calcium transport. The pump enzyme, a Ca^{2+} ATPase, can be seen as small particles in the membranes of SR visualized by the freeze-fracture technique (TEM).

a nerve cell membrane is (see Chapter 20). How does a nerve signal arriving at the surface of the muscle cell then transmit its signal quickly to the SR membranes in the cell so that calcium release occurs simultaneously throughout the sarcoplasm? Rapid transmission of the signal is made possible by a series of **transverse tubules** (the **T-system**) that are actually invaginations of the cell membrane (Figure 19-16). These tubules penetrate into the cell and allow rapid conduction of electrical impulses into the interior of the cell. By a poorly understood mechanism, the depolarization of the T-tubules causes calcium channels in the SR to open, thereby increasing calcium concentrations in the sarcoplasm from 10^{-7} to 10^{-6} *M*, activating contraction.

The second question concerns the manner in which the muscle cell relaxes after the contractile cycle is completed. In the membrane of the SR, there is an active transport enzyme called **calcium ATPase,** which can pump calcium ions from the sarcoplasm into the cisternae of the SR. The calcium ATPase can be seen clearly on freeze-fractured SR membranes (Figure 19-16). In 1969 this was the first freeze-fracture particle to be linked to a specific membrane function. The calcium pump from mammalian muscle tissue is a single peptide chain with a molecular weight of approximately 115,000, and a known amino acid sequence. From the sequence and biochemical results, David Maclennan and Michael Green proposed a three-dimensional structure for the protein that suggests a possible pump mechanism. This structure and mechanism are shown in Figures 19-17 and 19-18, representing the first ATP-driven pump to be analyzed in such detail.

The idea is similar to that discussed in Chapter 8 for the sodium-potassium pump. ATP-Mg^{2+} is normally bound to the nucleotide-binding portion of the enzyme, but the ATP is not hydrolyzed in the absence of calcium. When a sufficient amount of calcium is present—for instance, following a muscle cell contraction—two calcium ions bind to high-affinity sites on the ATPase and activate the pump cycle shown in Figure 19-18. A phosphate group is transferred from the bound ATP to the enzyme, producing the E1-P form. This causes a conformational change to occur in which the calcium binding site simultaneously becomes a low-affinity site (E2-P) and exposes bound calcium to the interior of the sarcoplasmic reticulum, releasing the calcium ions. The ATPase then binds another ATP and cycles back to the E1 form.

The continued pumping of calcium from the cytoplasm back into the SR cisternae quickly lowers the cytoplasmic calcium level to the point where troponin releases calcium, tropomyosin again moves back to the blocking position on actin sites, and further cross-bridge formation is prevented. Cross-bridges therefore disappear rapidly as actin dissociates from myosin and becomes blocked by tropomyosin. This leaves the muscle relaxed and free to be reextended, since the absence of cross-bridge contacts allows the thin filaments to slide out from between the thick filaments. In some cases extension is caused by contraction of an opposing muscle, and in other cases by the tension of elastic connective tissue.

Meeting the Energy Needs of Muscle Contraction

Because muscle contraction involves the hydrolysis of an ATP molecule for every cycle of attachment and detachment, muscle cells need ways to regenerate ATP continuously. Muscle cells actually have a variety of mechanisms to ensure maximum ATP availability, even during prolonged periods of intense activity.

ATP Generation: The Aerobic Mode. The ATP needs of muscle cells are met either by glycolysis or by mitochondrial respiration. The extent to which one or the other of these pathways is favored depends on the kind of muscle involved and whether it is functioning under anaerobic or aerobic conditions. Skeletal muscles that are characterized by frequent use and high activity usually rely on complete respiratory metabolism. The flight muscles of birds are a good example. Such muscles draw both glucose and oxygen from the circulatory system, oxidizing the glucose completely to carbon dioxide and water and generating ATP by oxidative phosphorylation in the mitochondrion. These muscles are characterized by an abundance of mitochondria and by a red color. Mitochondria occur in close association with the myofibrils in almost all aerobic muscle cells (Figure 19-19). The red color of such tissue is due to the high degree of vascularization, to the cytochromes present in the mitochondria, and to the presence of **myoglobin,** a protein related in structure and function to hemoglobin, but localized in muscle cells and used to bind and store oxygen.

ATP Generation: The Anaerobic Mode. During periods of intense exercise, the demand for ATP regeneration may exceed the rate at which oxygen can be supplied to the tissue by the circulatory system. After depletion of the reserve oxygen available from myoglobin, the tissue begins to function anaerobically, converting glucose to lactate. Because the ATP yield of glucose is greatly reduced under anaerobic conditions (from 36 to 2 molecules of ATP per molecule of glucose; recall Chapter 10), much more glucose is required per unit time under these conditions. The extra glucose is supplied by degradation of *glycogen,* the storage carbohydrate of muscle cells. The lactate formed under these anaerobic conditions is usually released into the blood and eventually reaches the liver, where it is either oxidized fully to carbon dioxide and water or used for the resynthesis of glucose (**gluconeogenesis**). A cycle is therefore established, with glucose moving from the liver via the blood to the anaerobic muscle and lactate returning from the muscle to the liver. This cyclic process is called the **Cori cycle,** in honor of the husband-and-wife team credited with its discovery.

Intense muscular activity cannot be sustained long under anaerobic conditions, both because of the rapid depletion of glycogen stores and because of the buildup of lactate. However, this anaerobic option is useful for short bursts of activity when oxygen cannot be supplied fast enough.

The Role of Creatine Phosphate. Although ATP clearly serves as the immediate source of energy to drive muscle contraction and is also the form in which energy is conserved during glycolysis and respiratory metabolism, the muscle cell does not use ATP as its major storage form of energy. A surprising observation made early in muscle research was that the ATP content of working muscle remains remarkably constant until the muscle is near exhaustion. What decreases instead during prolonged exer-

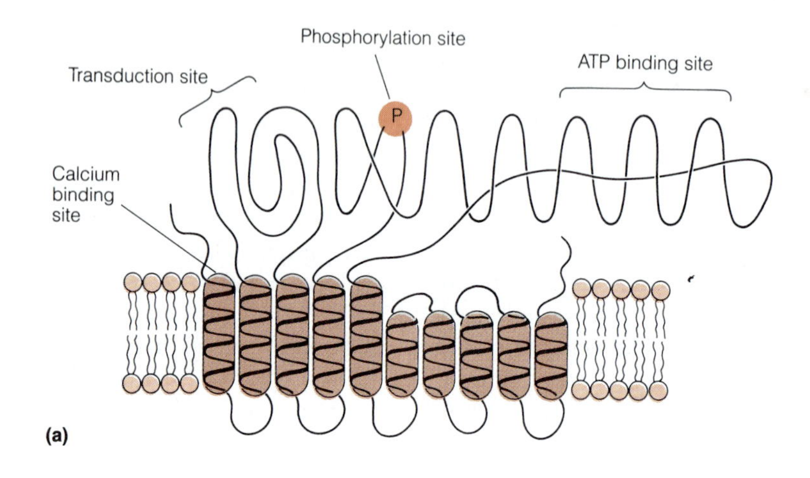

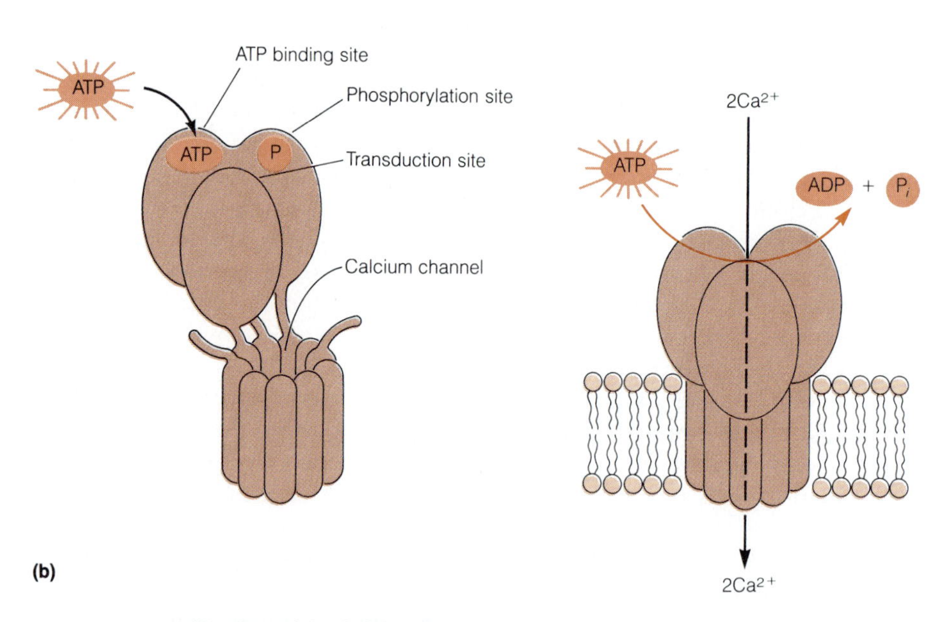

Figure 19-17 Structure of the Calcium Pump Protein. The calcium pump protein is a single polypeptide chain containing 1001 amino acids. (a) A partially unfolded chain indicates the main features. Ten α helices are embedded in the lipid bilayer, and perhaps five or six of the helices cluster to form a transmembrane calcium channel. (b) The head group has three significant features, including an ATP binding site, a phosphorylation site, and a transduction site, where the energy of ATP hydrolysis is made available for the pumping of calcium ions. When calcium is present, the ATPase activity is activated, and two calcium ions are transported inward for every ATP hydrolyzed.

tion is the cellular level of **creatine phosphate,** a high-energy compound with the structure shown in Figure 19-20.

Creatine phosphate represents a reservoir of high-energy phosphate that can be used to recharge ADP, as shown in the reaction of Figure 19-20. The enzyme that catalyzes this reaction is called *creatine kinase*. The hydrolysis of creatine phosphate is highly exergonic ($\Delta G^{\circ\prime} = -10.3$ kcal/mol), which ensures that the equilibrium for the creatine kinase reaction lies far to the right. The recharging of ADP is therefore driven effectively, maintaining a high ATP/ADP ratio in the muscle cell.

The Role of Myokinase. As a final "backup" system, muscle cells also contain an enzyme called *myokinase*, which is capable of phosphorylating one ADP molecule at the expense of another, as follows:

$$2\text{ADP} \longrightarrow \text{ATP} + \text{AMP} \qquad (19\text{-}1)$$

The myokinase reaction provides a means of extracting energy from the remaining acid anhydride bond of ADP.

Meeting Energy Needs: A Summary. The variety of mechanisms available to skeletal muscle cells to meet their energy needs for continued contraction under a wide variety of conditions can be summarized as follows:

1. The primary energy storage form in muscle cells is creatine phosphate, and the energetics of phosphate transfer from creatine phosphate to ADP are sufficiently favorable to maintain a high ATP/ADP ratio despite the considerable depletion of creatine phosphate that may occur during muscle activity.
2. With adequate oxygen and glucose available from the circulatory system, complete respiratory metabolism is possible, and the muscle functions aerobically.
3. As a safeguard, oxygen can be stored within the cell as a complex with myoglobin and can be released to prolong respiratory metabolism for a short time, even when the circulatory system is unable to supply oxygen at an adequate rate.
4. When oxygen supplies become limited, the muscle cell can still function anaerobically to meet its needs by lactate formation.
5. If the supply of blood glucose becomes limited, stored glycogen can be used as a source of glucose to continue glycolysis.
6. If all else fails and ADP begins to accumulate, the myokinase reaction provides a last-ditch means of regenerating ATP.

The overall picture that emerges of the skeletal muscle cell is one of almost incredible specialization for its role in contraction—specialization in design of the contractile elements as well as in the mechanism available to ensure

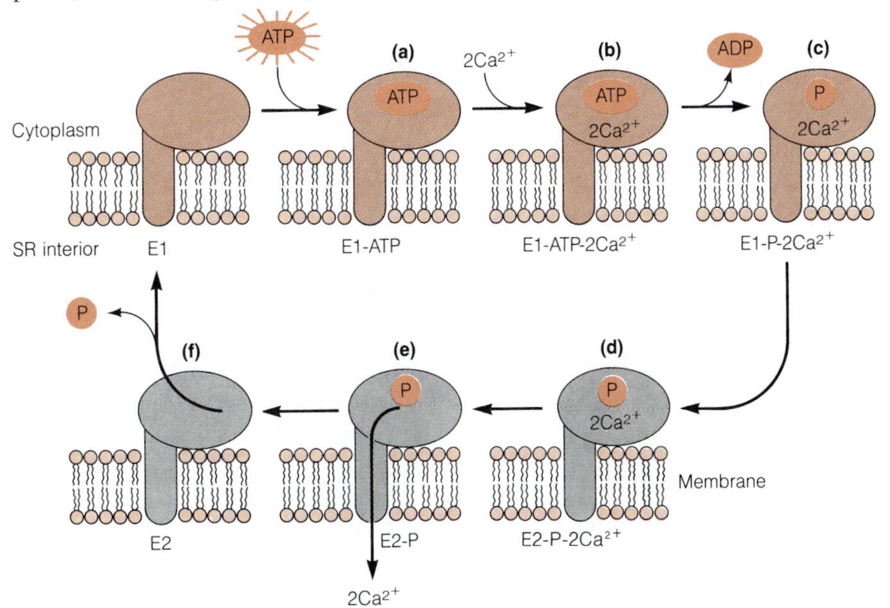

Figure 19-18 Mechanism Proposed for Active Calcium Ion Transport. The calcium pump protein, indicated by E, has two conformations, E1 (high affinity for calcium) and E2 (low affinity for calcium). Two calcium ions and one ATP bind to the protein in the cytoplasmic side of the SR membrane (a, b) followed by phosphorylation as ATP donates its phosphate to an aspartic acid in the protein (c). The protein shifts to its low-affinity form (d), exposing the bound calcium on the SR interior. The calcium ions dissociate (e), the phosphate group is hydrolyzed (f), and the protein returns to its high-affinity state for the next round of calcium transport.

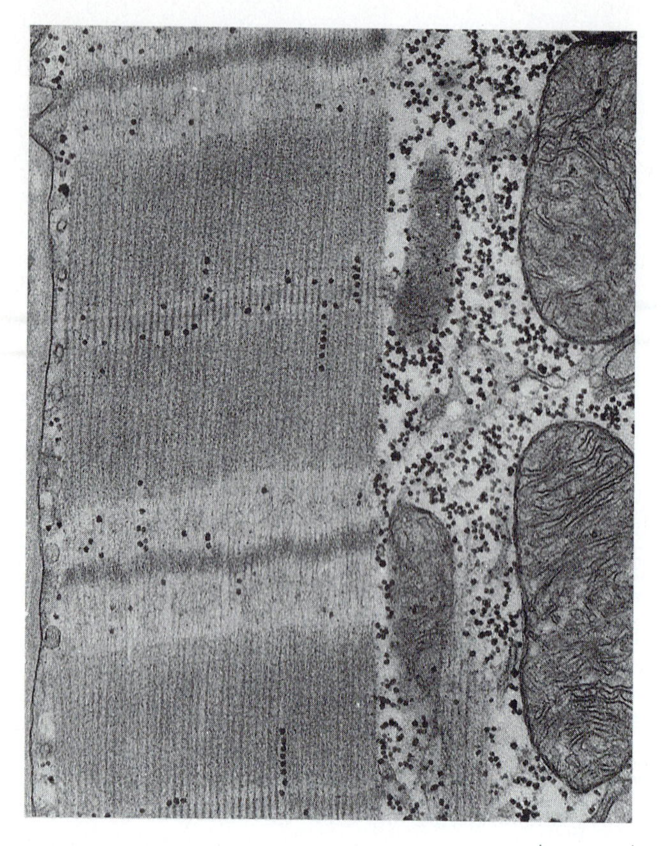

Figure 19-19 The Prominence of Mitochondria in Aerobic Muscle Cells. The cells of aerobic muscle tissue are well endowed with mitochondria (M), which associate closely with the myofibrils to which they supply ATP, as this electron micrograph of rat cardiac muscle illustrates.

that the ATP needed for contraction can be supplied under virtually any condition.

Cardiac Muscle

Cardiac (heart) muscle is responsible for the beating of the heart and therefore for the pumping of blood through the body's circulatory system. Cardiac muscle functions continuously; in one year, your heart beats about 40 million times! Cardiac muscle is very similar to skeletal muscle in organization of actin and myosin filaments and has the same striated appearance (see Figure 19-19). However, the two kinds of muscle differ significantly in their metabolism. Cardiac muscle is highly dependent on aerobic respiration; only in emergencies is glycolysis used as an energy source. Accordingly, it has much less glycogen and many more mitochondria than are usually found in skeletal muscle. In fact, mitochondria represent about 30–40% of the total volume of heart muscle.

Most of the energy required for the beating of the heart under resting conditions is provided not by blood glucose, but by free fatty acids that are transported from adipose (fat storage) tissue to the heart by serum albumin, a blood protein. These fatty acids are degraded by β oxidation, and the resulting acetyl coenzyme A is oxidized via the TCA cycle. When a heavy work load is suddenly imposed on the heart, the consumption of blood glucose and muscle glycogen increases significantly. Shortly thereafter, however, fatty acid oxidation is greatly accelerated and the dependence on glucose utilization again becomes minimal.

A second difference between cardiac and skeletal muscle is that heart muscle cells are not multinucleate. Instead, cells are joined end to end through structures called *intercalated disks*. The disks have gap junctions that electrically couple neighboring cells, thereby providing a way for depolarization waves to spread throughout the heart during its contractile cycle. The reason for this arrangement is that the heart does not require activating nerve impulses, as skeletal muscle does, but contracts spontaneously once every second or so. The heart rate is controlled by a "pacemaker" region in the upper right atrium, which contracts 70–80 times per minute, slightly faster than other heart tissues. The depolarization wave initiated by the pacemaker then spreads to the rest of the heart to produce the familiar heart beat.

Smooth Muscle

Smooth muscle is responsible for involuntary contractions such as those of the stomach, intestine, uterus, and blood vessels. In general, such contractions are slower and of greater duration. Smooth muscle is not able to contract rapidly but is well adapted to maintain tension for long periods of time, as is required in these organs and tissues.

Unlike skeletal or heart muscle, smooth muscle has no striations. Thick and thin filaments are both present, but they are not organized into myofibrils and are not in register. In this respect, smooth muscle cells resemble nonmuscle cells more closely than they do skeletal muscle. The thin filaments of smooth muscle cells contain actin and tropomyosin, but no troponin. They are aligned with the long axis of the cell, with thick myosin filaments dispersed among them.

The organization of myosin molecules in the thick filaments of smooth muscle also differs from that of skeletal muscle. Instead of the bare central zone characteristic of thick filaments from skeletal muscle, smooth muscle thick

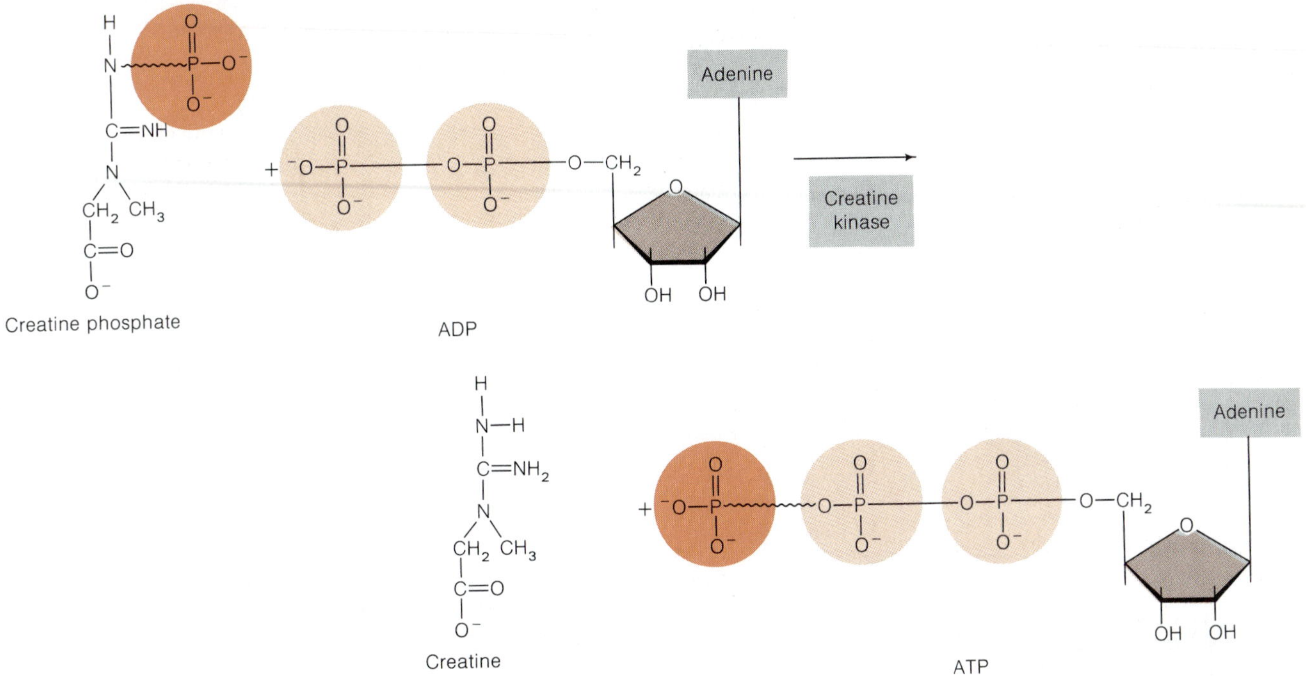

Figure 19-20 The Structure of Creatine Phosphate and the Creatine Kinase Reaction. The transfer of the high-energy phosphate group from creatine phosphate to ADP is thermodynamically favorable, since the $\Delta G^{\circ\prime}$ for the hydrolysis of creatine phosphate is substantially more negative than that for the hydrolysis of ATP (-10.3 versus -7.3 kcal/mol).

filaments have myosin heads distributed along the entire length of the filament. Cross-bridges connect thick and thin filaments in smooth muscle, but not in the regular, repeating pattern seen in skeletal muscle.

Regulation of Contraction in Smooth Muscle. Unlike the myosin of skeletal muscle, the myosin of smooth muscle (and that of nonmuscle cells as well) can interact with actin only if its light chains are phosphorylated. Smooth muscle myosin therefore exists in two forms, a phosphorylated form that has actin-binding activity and a nonphosphorylated form that is inactive. The enzyme that phosphorylates the light chains of smooth muscle myosin is called **myosin light-chain kinase.**

The activity of this kinase—and hence the actin-binding activity of the myosin—is regulated by the concentration of calcium ions in the cell. However, the effect of calcium is indirect, mediated by **calmodulin,** a calcium-binding protein that occurs universally in eukaryotic cells and is involved in many calcium-regulated processes. Calmodulin is structurally and functionally very similar to the troponin present in skeletal muscle cells. The two molecules are closely related in amino acid sequence, and both have four calcium binding sites. Like troponin, calmodulin undergoes major conformational changes when it binds calcium ions.

The cascade of events involved in activation of smooth muscle myosin is shown in Figure 19-21. In response to a nerve impulse or hormonal signal reaching the smooth muscle cell, an influx of extracellular calcium ions occurs (Figure 19-21a), increasing the intracellular calcium concentration and causing contraction. In smooth muscle, the effect of the increased calcium concentration on muscle contraction is mediated by the binding of calcium to calmodulin (Figure 19-21b). The calcium-calmodulin complex (but not free calmodulin) is able to bind to myosin light-chain kinase, thereby activating the enzyme (Figure 19-21c). As a result, myosin light chains become phosphorylated (Figure 19-21d) and are therefore able to interact with actin to cause contractions (Figure 19-21e).

Thus, both skeletal muscle and smooth muscle are activated to contract by calcium ions, but from different sources and by different mechanisms. In skeletal muscle, the calcium comes from the sarcoplasmic reticulum. Its effect on the actin-myosin interaction is mediated by troponin and is very rapid because it depends on conformational changes only. In smooth muscle, the calcium comes from outside the cell, and its effect is mediated by calmodulin. The effect is much slower in this case because it involves a covalent modification of the myosin molecule. The two kinds of muscle also differ in the signal that initiates calcium release from the sarcoplasmic reticulum.

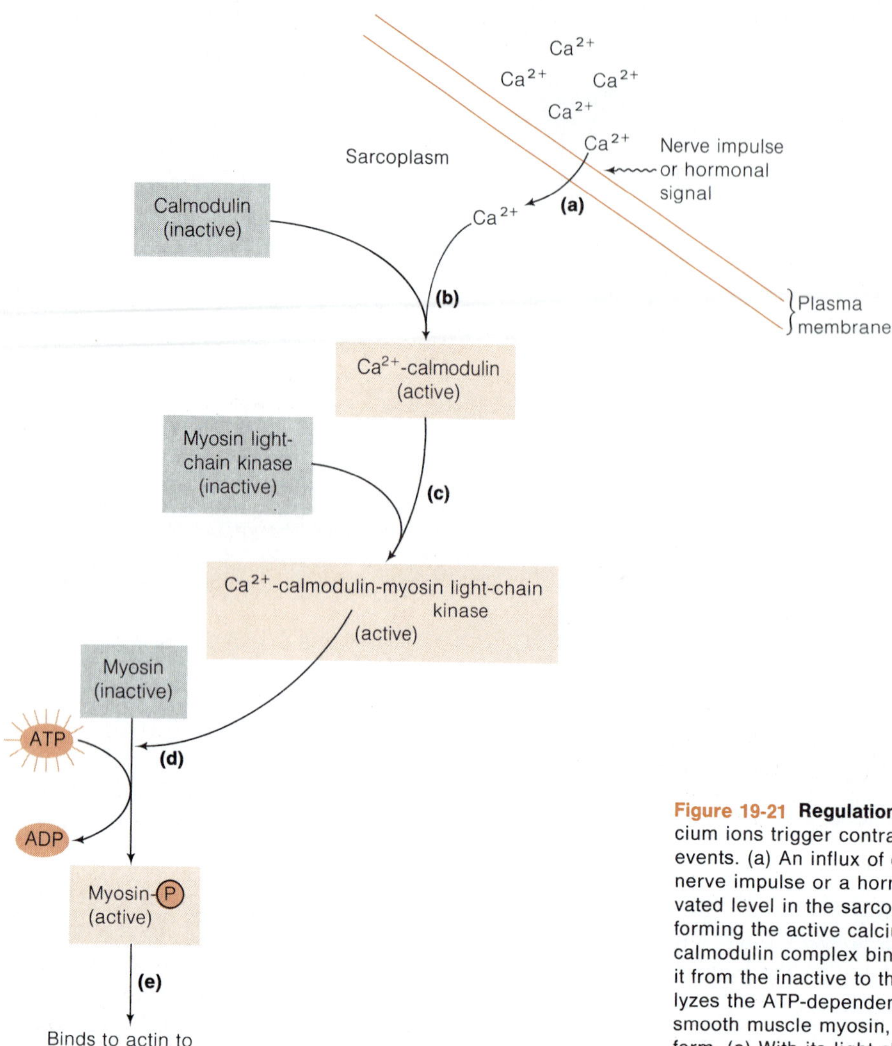

Figure 19-21 Regulation of Contraction in Smooth Muscle. Calcium ions trigger contraction of smooth muscle by a cascade of events. (a) An influx of calcium ions into the cell is triggered by a nerve impulse or a hormonal signal. (b) When present at an elevated level in the sarcoplasm, calcium ions bind to calmodulin, forming the active calcium-calmodulin complex. (c) The calcium-calmodulin complex binds to myosin light-chain kinase, converting it from the inactive to the active form. (d) The active kinase catalyzes the ATP-dependent phosphorylation of the light chains of smooth muscle myosin, which is inactive in the unphosphorylated form. (e) With its light chains phosphorylated, myosin is able to interact with actin, causing contraction.

In skeletal muscle, the signal comes from voluntary nerves, whereas smooth muscle is activated by nerves of the autonomic (involuntary) component of the nervous system or by hormonal stimuli.

Specialized Filament-Based Motility Systems

Besides their role in muscle contraction, interactions between myosin and actin microfilaments are important in other forms of cell motility. The mechanisms are not as well understood, but the lessons learned from skeletal muscle contraction have turned out to be generally applicable in approaching topics as diverse as amoeboid motion, cytoplasmic streaming, and cell division.

Amoeboid Movement

Amoebas, slime molds, leukocytes, and a number of other types of cells depend for movement (and sometimes also for food gathering) on cytoplasmic extensions called **pseudopodia** (from the Greek, meaning "false feet"). The progressive extension of a pseudopod is accomplished by a unidirectional streaming of cytoplasm. This mode of locomotion is called **amoeboid movement.** The protozoan *Amoeba proteus,* shown in Figure 19-22a, is a good example of a cell with this type of movement. These cells can be up to 500 μm in diameter, with pseudopodia of more than 100 μm. In Figure 19-22b, an *Amoeba* cell is using pseudopodia to phagocytose a ciliated cell, on which the *Amoeba* feeds.

Cells that undergo amoeboid movement have an outer, or *cortical,* layer of thick, gel-like cytoplasm called the

ectoplasm and an inner layer of more fluid, sol-like cytoplasm called the **endoplasm.** As a pseudopod is extended from the cell, fluid endoplasm streams forward in the direction of extension and appears to congeal into ectoplasm at the tip of the pseudopod. Meanwhile, at the rear of the moving cell, ectoplasm appears to be changing into more fluid endoplasm and streaming toward the pseudopod (Figure 19-22c).

These sol-gel transitions are thought to be the key to amoeboid movement, but the underlying mechanism is not yet understood. One possibility has been suggested by R. D. Allen and colleagues, who proposed that the endoplasm in the tip region of the pseudopod contracts as it is converted into ectoplasm, thereby pulling the endoplasm forward into the pseudopod. There is good evidence that microfilaments are involved in amoeboid movement, which would then represent another example of actin-based motility. For instance, both actin and myosin filaments are prominent in the ectoplasm, although not always demonstrably interacting with each other. Furthermore, amoeboid movement is inhibited by cytochalasin B and phalloidin (see Chapter 18), and crude extracts of cytoplasm from these cells undergo contractions in vitro in response to ATP and calcium. It seems probable that actin-binding proteins are also involved in such a mechanism. For example, one can imagine ways in which calcium ion concentrations are regulated in the region of the pseudopod, so that higher concentrations increase the level of interaction between the proteins and actin filaments. This in turn would produce transient networks of filamentous proteins, with contraction being produced as myosin motor molecules move along actin filaments. (See Box on p. 604)

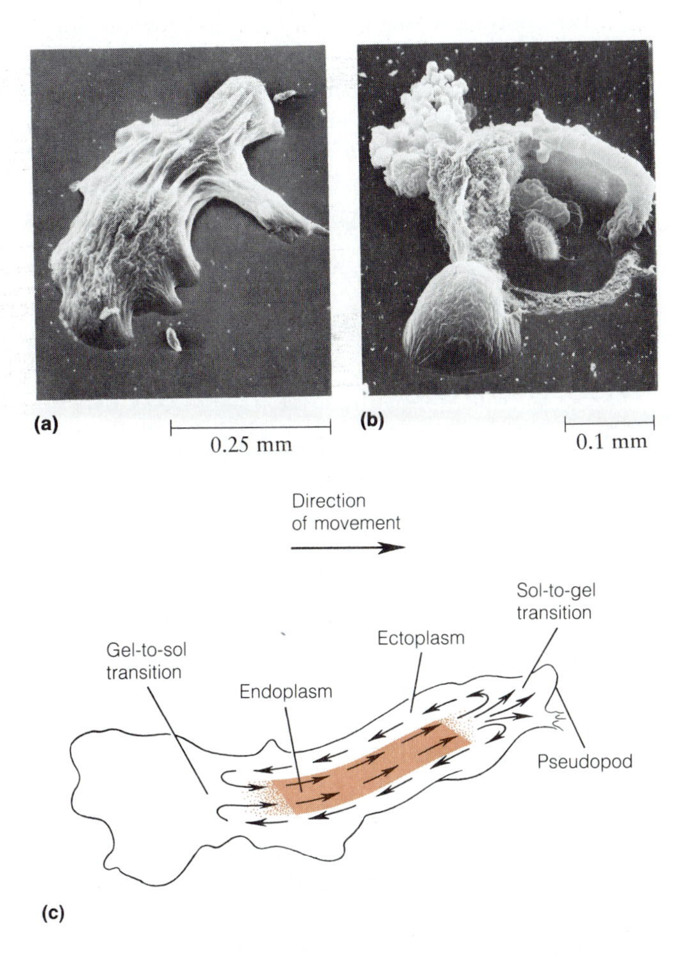

Figure 19-22 Amoeboid Movement. (a) A scanning electron micrograph of *Amoeba proteus*, a protozoan that moves by extension of pseudopodia. (b) This scanning electron micrograph shows an *Amoeba* cell using its pseudopodia to engulf a smaller ciliated cell on which *Amoeba* feeds by phagocytosis. (c) Amoeboid movement as a sol-gel transition. Amoeboid movement depends on sol-like endoplasm streaming forward in the direction of extension, driven by the pull exerted on the endoplasm by protoplasmic contraction at the tip of the pseudopod as endoplasm is converted into ectoplasm. Filaments of actin and myosin in the cortical ectoplasm are thought to be responsible for generating the driving force.

Cytoplasmic Streaming

Cytoplasmic streaming is seen in a variety of organisms that do not display amoeboid movement. In slime molds like *Physarum polycephalum*, cytoplasm streams back and forth in the branched network of protoplasm of which the *Physarum* cell mass consists. The flow of cytoplasm reverses direction with predictable periodicity and is therefore called **shuttle streaming.** The purpose seems to be both nourishment and locomotion, since the streaming process is correlated with the further extension of the fingerlike projections by which the slime mold reaches out into its environment in search of nutrients.

Many plant cells display a circular flow of cell contents around a central vacuole. This streaming process, called **cyclosis,** has been studied most extensively in the giant algal cell *Nitella*. In this case, the movement seems designed to circulate and mix cell contents.

The unifying feature in these and other cases of cytoplasmic streaming seems to be the presence of filaments of actin in the cytoplasm. These filaments are thought to be integrally involved in the streaming mechanism, and the supporting evidence, as in the case of amoeboid movement, is increasingly convincing.

Cytokinesis

Cell division involves not just the parceling out of chromosomes to the daughter nuclei, but also the physical separation of the cytoplasm into two compartments, each of which gives rise to a daughter cell. This process, called **cytokinesis,** has been studied most extensively in the cleavage of animal eggs, with the sea urchin egg often the system of choice. As Figure 19-23 shows, cytokinesis begins as a slight indentation of the cell surface, which deepens into a **cleavage furrow** that encircles the egg. The plane of the cleavage furrow is perpendicular to the long axis of the mitotic spindle, thereby ensuring that the two sets of chromosomes will be separated into the two daughter cells.

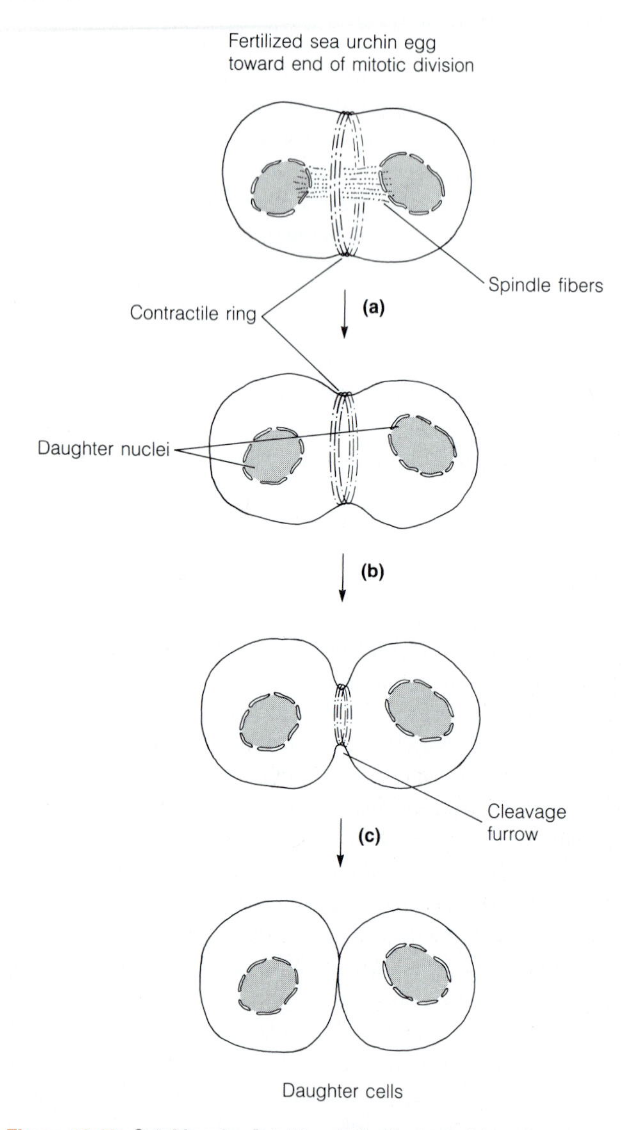

Figure 19-23 Cytokinesis. Cytokinesis is illustrated here for the first division of a fertilized sea urchin egg. (a) As the cell approaches the end of mitosis, cytokinesis begins with an indentation on the cell surface due to the contractile ring that encircles the cell in the midregion. (b) The indentation gradually deepens to form a cleavage furrow, and (c) eventually the cell divides. The contractile ring responsible for the progressive constriction of the cleavage furrow is thought to consist of actin microfilaments and myosin in the cortical cytoplasm just beneath the plasma membrane.

The furrow continues to deepen until opposite surfaces make contact and the cell is split in two.

The underlying mechanism depends on a beltlike bundle of actin filaments and myosin called the **contractile ring,** which forms during cell division in the cortical cytoplasm just beneath the plasma membrane. Examination of the cortical cytoplasm of the cleavage furrow with the electron microscope reveals large numbers of microfilaments, oriented with their long axes parallel to the plane of the furrow. Myosin filaments cannot be seen directly, but their presence has been deduced by use of antibodies. Apparently, the filaments in the cortical ring constrict the cleavage furrow progressively in a manner that depends on an actin-myosin interaction to generate the needed force.

The contractile ring of dividing cells provides one of the most dramatic examples of the transitory nature of actin-myosin structures in most nonmuscle cells and the rapidity with which they can be assembled and disassembled. Polymerization of actin apparently occurs just before the initial indentation becomes visible, and the entire structure is dismantled again shortly after cell division is complete.

Microtubule-Based Motility: The Motile Appendages of Eukaryotic Cells

We now come to a set of specialized motility mechanisms that build on the basic motor molecules and cytoskeletal structures we described earlier. We will first discuss flagella and cilia, the motile appendages of eukaryotic cells.

Cilia and Flagella

Cilia and flagella share a common structural basis and differ from one another only in relative length and number and in mode of beating. **Cilia** have a diameter of about 0.25 μm, are about 2–10 μm long, and tend to occur in large numbers on the cell surface. Each cilium is bounded by an extension of the plasma membrane and is therefore an intracellular structure. Cilia occur in both unicellular and multicellular eukaryotes. Unicellular organisms, such as the protozoa, use cilia both for locomotion and for the collection of food particles. A ciliated protozoan called *Tetrahymena* is shown in Figure 19-24.

The cilia of multicellular organisms serve primarily to move the environment past the cell rather than to propel the cell through the environment. The cells that line the air passages of your respiratory tract, for example, have several hundred cilia each. (This means, by the way, that every square centimeter of epithelial tissue lining your respiratory tract has about a billion cilia!) It is the coordinated, wavelike beating of these cilia that carries mucus, dust, dead cells, and other foreign matter out of the lungs. One of the health hazards of cigarette smoking, in fact, lies in the inhibitory effect that smoke has on normal ciliary beating. Moreover, certain respiratory ailments can be traced to defective cilia.

Flagella, although of the same diameter as cilia, are often much longer (from 1 μm to several millimeters, though usually 10–200 μm) and may be limited to one or a few per cell. Like cilia, flagella are bounded by an extension of the plasma membrane.

Flagella differ distinctively from cilia in the nature of their beat. Flagella move with a propagated bending motion that is usually symmetrical and undulatory and may even have a helical pattern. This type of beat generates a force parallel to the flagellum, such that the organism moves in approximately the same direction as the axis of the flagellum. The locomotory pattern of most flagellated cells usually involves propulsion of the cell by the trailing flagellum, but examples are also known in which the flagellum actually precedes the cell. Figure 19-25a illustrates the swimming movement of a sperm cell as visualized by photographing the cell at intervals of 5 milliseconds.

Cilia, on the other hand, display an oarlike pattern of beating with a power stroke that is perpendicular to the cilium, thereby generating a force that is parallel to the cell surface. The numerous cilia on the cell surface usually beat in a coordinated manner, ensuring a steady movement of fluid past the cell surface. The cycle of beating for epithelial cilia is shown in Figure 19-25b. Each cycle requires about 0.1–0.2 sec and involves an active forward stroke followed by a recovery phase.

Structure of Motile Appendages

Cilia and flagella have a common structure consisting of an **axoneme,** or main cylinder of tubules, about 0.25 μm

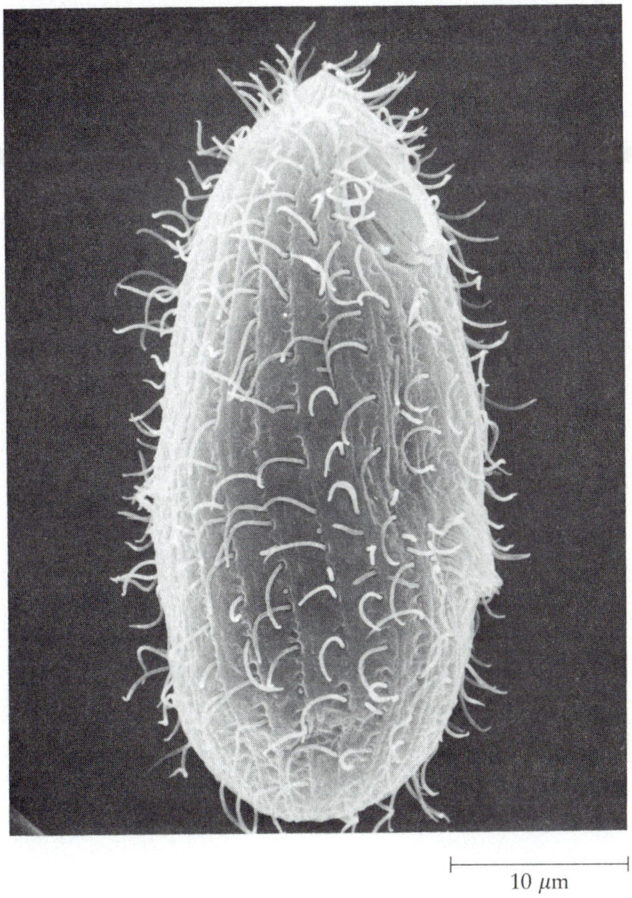

10 μm

Figure 19-24 A Ciliated Protozoan. The protozoan *Tetrahymena* has numerous cilia on its surface, as this scanning electron micrograph of a living cell reveals.

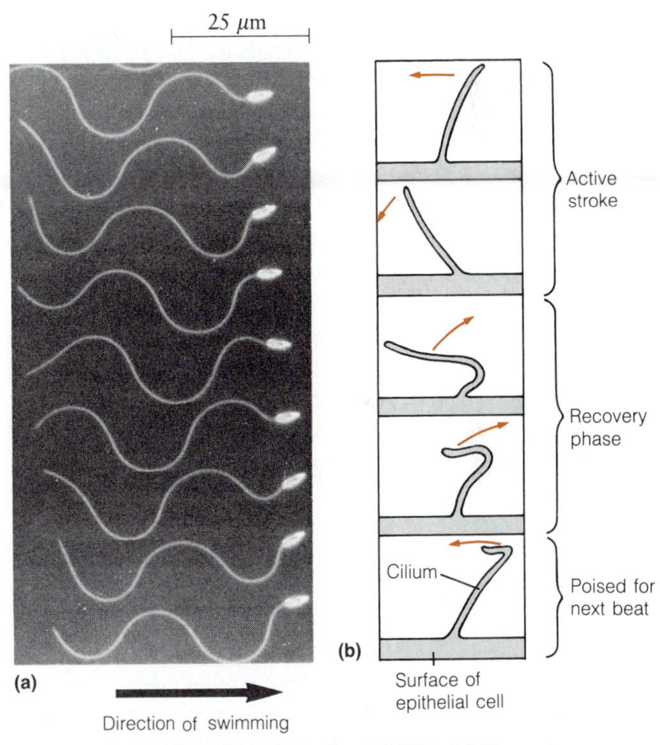

Figure 19-25 The Beating of Flagella and Cilia. (a) The swimming movement of a sperm cell of *Ciona intestinalis,* a tunicate. A single sperm cell swimming in seawater was visualized by multiple-flash photography on moving film with a flash rate of 200 flashes/sec. The sequence of images is from top to bottom and the direction of swimming is from left to right. (LM) (b) The beat of a cilium on the surface of an epithelial cell from the human respiratory tract. The beat begins with an active stroke that sweeps fluid over the cell surface. A recovery phase follows, leaving the cilium poised for the next beat. Each cycle requires about 0.1–0.2 sec.

How Is an Amoeba Like an Elephant?

The lowly slime mold *Dictyostelium discoideum* spends much of its life cycle in a unicellular amoeboid form. However, under certain conditions the amoebae undergo a startling transformation in which a million cells gather to form a multicellular slug. After a migratory phase, the slug in turn differentiates into a fruiting body that disperses spores to initiate a new generation. (For a more detailed look at this cycle, see Figure 22-4.)

Because of its relative simplicity, the eukaryotic *Dictyostelium* cell is an excellent model system for studying cell motility. Several recent investigations using *Dictyostelium* have provided significant new insights into amoeboid motion. Myosin and actin are present in all eukaryotic cells, including *Dictyostelium,* and until recently it was supposed that both proteins are somehow involved in amoeboid motion. This hypothesis led to a critical experiment in which cells were experimentally produced that lacked a functional myosin gene product, with the expectation that the cells would be nonmotile. Surprisingly, the mutation had little effect. The cells still produced pseudopodia and could move, although not as well as normal cells. What could be generating the force required to produce pseudopodia?

Further work showed that *Dictyostelium* actually has two forms of myosin, now called myosin I and myosin II. Myosin II is the form found in the muscle sarcomere, a motor molecule having two ATPase head groups with actin binding sites, and a long α-helical tail (recall Figure 19-9). Myosin I has only a single head lacking ATPase activity, and a short tail that binds to actin. The *Dictyostelium* cells described above lack myosin II, but myosin I is still present. Apparently the presence of myosin I is sufficient to produce pseudopodia and partial mobility.

The next advance came when researchers used fluorescent antibodies to localize myosin I and II in motile *Dictyostelium* (For a discussion of this technique, see Appendix A and Figure A-12.) The results were astonishing: myosin I was concentrated in the leading edges of the pseudopodia, while myosin II was confined to the trailing posterior region. In dividing cells, myosin II was localized in the ring structure that forms during cytokinesis and contracts to produce two new cells. Significantly, the mutant cells lacking myosin II are unable to undergo cytokinesis, so myosin I cannot substitute for myosin II in the contractile ring.

We can now begin to formulate testable hypotheses to explain the molecular mechanism of amoeboid motion. The location and properties of myosin I suggest that it generates the force required to push out a pseudopodium, perhaps by binding to neighboring actin microfilaments and "walking" them past each other. Myosin II, on the other hand, may function by the usual sliding-filament mechanism to cause contraction of the posterior region of a motile cell, so that cytoplasm is forced forward into the developing pseudopodia.

There is a moral to this story. Scientific research is often like the fable of the blind men who are trying to understand an elephant, each feeling a different part. If only a single leg is felt, it is difficult to imagine how an elephant might move. But as soon as we imagine (or discover) the possibility of more than one leg (or more than one form of myosin) it becomes easier to understand how elephants can walk—and amoebae creep.

in diameter, connected to a **basal body** and surrounded by an extension of the cell membrane (Figure 19-26). Cross-sectional views of the axoneme, transition zone, and basal body are shown in Figure 19-26 as parts b–d. The basal body is identical in appearance to the centriole, a structure we encountered in our discussion of the mitotic spindle (Chapter 14). Centrioles and basal bodies are usually considered to be two different functional manifestations of the same structure. A basal body consists of nine sets of tubular structures arranged around its circumference. Each set is called a *triplet* because it consists of three tubules that share common walls. Each triplet has one complete microtubule and two incomplete tubules.

The axoneme, on the other hand, has a characteristic "9 + 2" pattern, with nine **outer doublets** of tubules and two additional microtubules in the center, often called the **central pair.** Figures 19-27 and 19-28 illustrate these structural features in greater detail. The nine outer doublets of the axoneme are thought to be extensions of two of the three subfibers from each of the nine triplets of the basal body. Each outer doublet of the axoneme therefore consists of one complete microtubule, called the **A tubule,** and one incomplete microtubule, the **B tubule.** The A tubule has 13 protofilaments, whereas the B tubule has only 10 or 11. The A tubule shares 5 protofilaments with the otherwise unclosed B tubule. The tubules of the central pair, on the other hand, are both complete, with 13 protofilaments each. All of these structures contain *tubulin,* together with a second protein called **tektin.** Tektin is related to intermediate filament proteins (Chapter 18) and is required as a structural subunit of the axoneme, which cannot be produced from tubulin alone. For instance, the A and B tubules share a wall that appears to contain tektin as a major component.

In addition to microtubules, axonemes contain several other key structures (Figure 19-28). The most important of these are the sets of **sidearms** that project out from each of the A tubules of the nine outer doublets. Each sidearm reaches out clockwise toward the B tubules of the adjacent doublet. These arms consist of **axonemal dynein,** which has an ATPase activity that is critical in motility, just as in cytoplasmic dynein function. The dynein arms occur in pairs, one **inner arm** and one **outer arm,** spaced along the microtubule at regular intervals. At less frequent intervals, adjacent doublets are joined by **interdoublet links.** These links are thought to limit the extent to which doublets can move with respect to each other as the axoneme bends.

At regular intervals, **radial spokes** project inward from each of the nine microtubule doublets, terminating near a set of projections that extend outward from the central pair of microtubules. Researchers believe that these spokes are important in translating the sliding of adjacent doublets into the bending motion that characterizes the beating of these appendages. Besides the radial spoke attachment to the central pair, a protein called **nexin** links adjacent doublets to one another and probably also plays a role in translating sliding into bending motion.

The Sliding-Microtubule Model for Motile Appendages

The mechanism of motility in microtubule-based systems is not as well understood as muscle contraction, but it is clear that ATP hydrolysis is the driving force in both cases. Axonemes of isolated cilia and flagella lacking their basal bodies and membranes can be induced to beat if ATP is added to the medium, so it appears likely that the appendages are actively involved in their own motility. In fact, it can be demonstrated that energy is expended all along the length of the appendage.

The mechanism proposed to explain microtubule-based motility is similar to that of muscle contraction, but modified in a way that accomplishes bending instead of contraction. According to the **sliding-microtubule model** for cilia and flagella, the length of the microtubules remains unchanged, but adjacent outer doublets slide past each other in an ATP-dependent process. However, instead of causing contraction, this sliding movement is converted into a localized bending, because the doublets of the axoneme are connected radially to the central pair and circumferentially to one another and are therefore not able to slide past each other freely. The resultant bending takes the form of a wave that begins at the base of the organelle and proceeds toward the tip.

Dynein Arms Are Responsible for Sliding. The driving force for microtubule sliding is provided by ATP hydrolysis, catalyzed by the ATPase activity of the dynein arms. The importance of the dynein arms is indicated by two kinds of evidence. When dynein is selectively extracted from isolated axonemes, the arms disappear from the outer doublets, and the axonemes lose both their ability to hydrolyze ATP and their capacity to beat. Furthermore, the effect is reversible: If purified dynein is added to isolated outer doublets, the sidearms reappear and ATP-dependent sliding is restored.

A second kind of evidence comes from studies of nonmotile mutants in such normally motile species as *Chlamydomonas.* In some of these mutants, flagella are still present, but they are nonfunctional. Such nonmotile flagella lack either dynein arms, radial spokes, or the central pair of microtubules. These structures are therefore almost certainly essential mechanistically.

Dynein is a very large protein, with a molecular weight exceeding 10^6. It has multiple subunits, the three largest

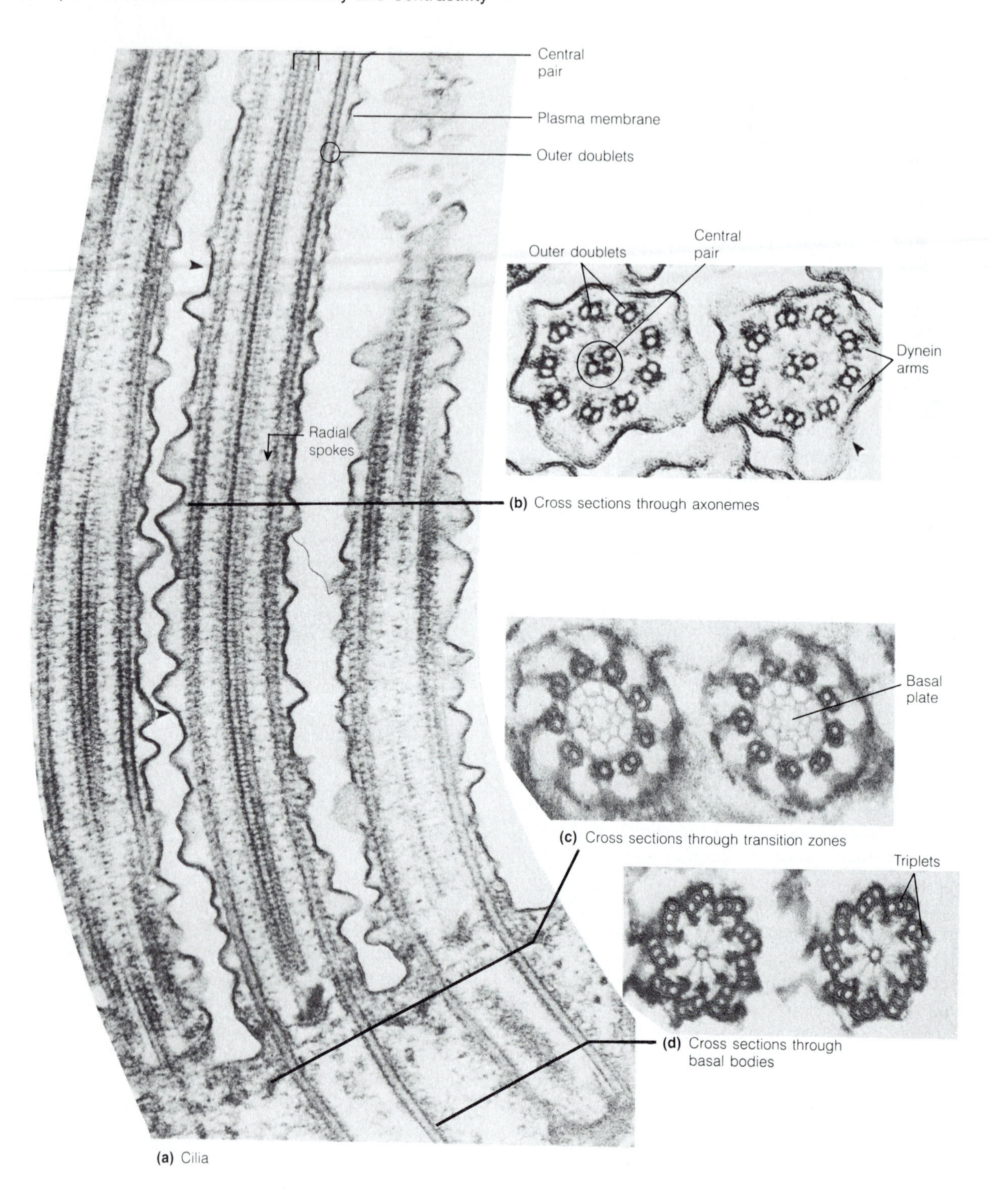

Figure 19-26 Structure of a Cilium. (a) These longitudinal sections of cilia from the protozoan *Tetrahymena thermophila* illustrate several structural features of cilia, including the central pair and outer doublet microtubules, the radial spokes, the dynein arms, and the basal body structures. Cross-sectional views are shown for (b) cilia, (c) the transition zone between cilia and basal bodies, and (d) basal bodies. Notice the triplet pattern of the basal body and the "9 + 2" pattern of tubule arrangement in the axoneme of the cilium. (all TEMs)

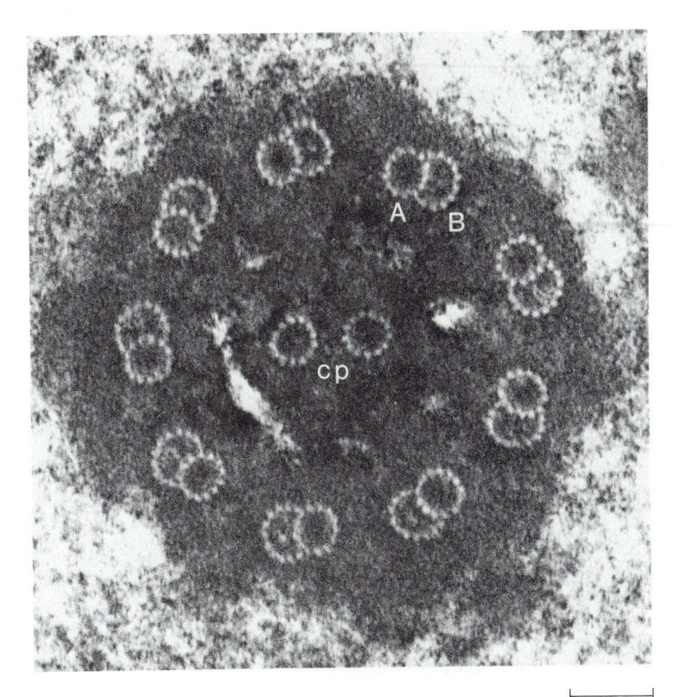

Figure 19-27 Cross Section of an Axoneme at High Magnification. The electron micrograph shows an axoneme from a flagellum of the alga *Chlamydomonas reinhardtii*. The microtubules of the central pair (cp) have 13 protofilaments each, as do the A tubules of the outer doublets. Each B tubule has 11 protofilaments of its own and shares 5 protofilaments with the A tubule. (TEM)

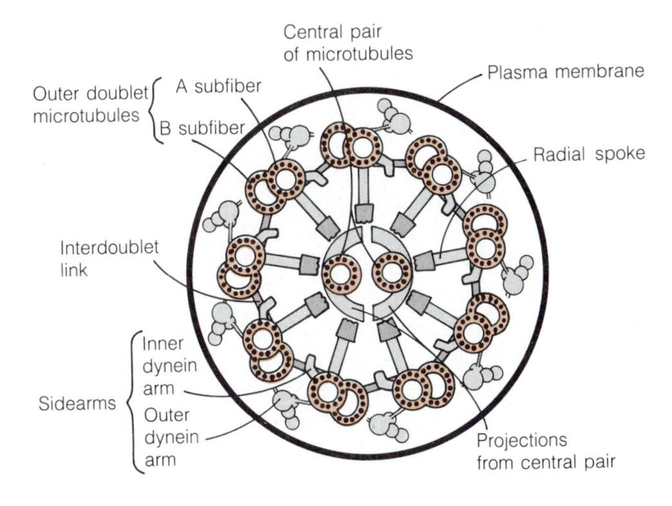

Figure 19-28 Diagram of an Axoneme in Cross Section. The dynein arms have ATPase activity and are thought to be responsible for the sliding of adjacent doublets. The interdoublet links join adjacent doublets, and the radial spokes project inward, terminating near projections that extend outward from the central pair of microtubules.

having ATPase activity and molecular weights of about 450,000. During the sliding process, the stalk of the dynein arm apparently attaches and detaches to and from the B tubule in a cyclic manner similar to the contraction cycle of muscle (Figure 19-29). Each cycle requires the hydrolysis of ATP and leaves the dynein arm of one doublet farther along the next doublet. In this way, the dynein arms of one doublet move the neighboring doublet, resulting in a relative displacement of the two.

Cross-Links and Spokes Are Responsible for Bending. To convert the dynein-mediated displacement of doublets into a bending motion, the doublets must be restrained in a way that resists sliding but allows deformation. This resistance is provided by the radial spokes that connect the doublets to the central pair of microtubules and possibly also by the nexin cross-links between doublets. If these cross-links and spokes are removed (by partial digestion with proteolytic enzymes, for example), the resistance that translates doublet sliding into a bending action is absent, and sliding is uncoupled from bending. Instead, the free doublets move with respect to each other just as actin and myosin filaments do, and the axonemes simply become longer and thinner as the microtubules slide apart (Figure 19-29a).

The importance of the radial spokes is shown by electron microscopic observations on bent and straight regions of a cilium. The radial spokes are oriented perpendicular to the doublets in straight regions, but are tilted at an angle in bent regions. The spokes, in other words, are deformed by the sliding process and provide the resistance that translates the sliding of doublets into the bending of the cilium or flagellum. Eventually, though, the sliding movement of the doublets overcomes the resistance and causes displacement of the radial spokes with respect to the central sheath. The progressive displacement of the spokes on the inside of the bend indicates that the two doublets on opposite sides of the axoneme are in fact sliding with respect to each other.

The Bacterial Flagellum

One of the most remarkable motile appendages in all of nature is the **bacterial flagellum,** an appendage strikingly dissimilar to the flagella of eukaryotic cells. A flagellated bacterium is shown in Figure 19-30. Unlike eukaryotic appendages, bacterial flagella are not membrane-bounded and are therefore extracellular structures. As Figure 19-31a illustrates, the flagellum is a spiral filament, usually about 15 nm in diameter and about 10–20 μm long. The filament

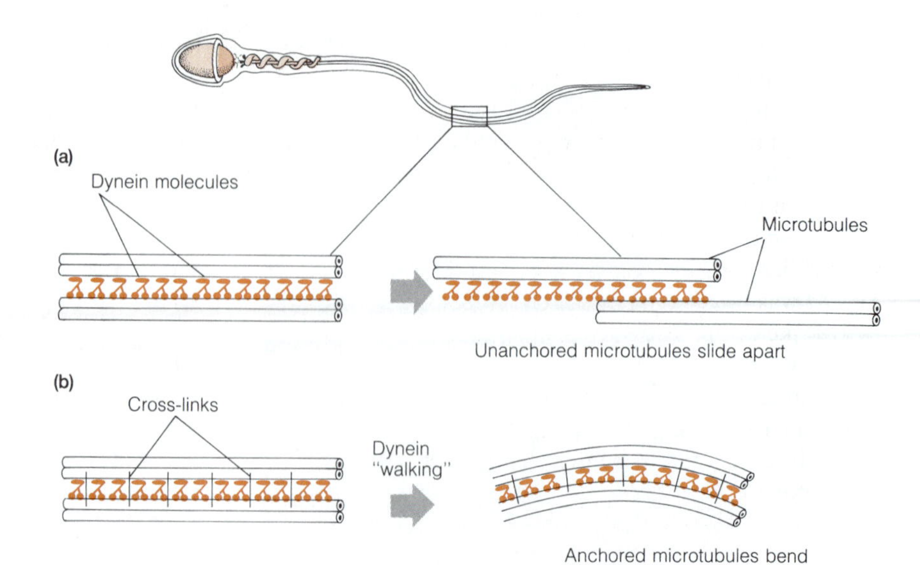

Figure 19-29 Dynein and Flagellar Motion. (a) If the microtubules on a flagellar axoneme are treated to remove cross-linking proteins, addition of ATP causes the microtubule doublets to slide apart. (b) This result suggests that cross-linking proteins like nexin serve to translate the sliding motion into a bending motion. The precise mechanism underlying dynein interaction with microtubules at the molecular level is still not understood.

is attached to a **hook,** and this, in turn, is connected by a **rod** to four ringlike structures in the base. The rod penetrates the outer membrane, the peptidoglycan wall, and the inner (plasma) membrane. Two of the rings are anchored in the outer membrane, and two in the inner (plasma) membrane. The two rings in the plasma membrane are called the **S** (for **stator**) **ring** and the **M** (for **motor**) **ring** (Figure 19-31b). Chemically, bacterial flagella consist of parallel strands of protein coiled around each other. The common subunit is **flagellin,** with a molecular weight of 40,000 in the monomeric form. Flagellin is not an ATPase; in fact, it has no known enzymatic activity.

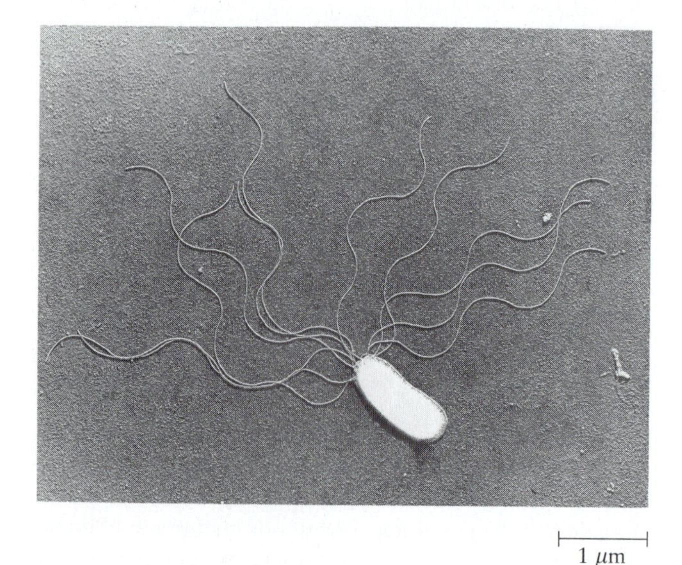

Figure 19-30 A Flagellated Bacterium. An electron micrograph of the flagellated bacterium *Pseudomonas marginalis.* (SEM)

Nature's Wheel: Locomotion by Rotation

The bacterial flagellum has been recognized as a locomotory structure for more than a hundred years, but its mode of propulsion has been elucidated only recently. Initially, the characteristic helical motion of the flagellum was thought to result from waves of bending, originating from the base and propagated along the length of the flagellum. It is now quite clear, however, that the flagellum actually *rotates* as a rigid, helical structure, driven by a rotary "motor" at the base of each filament. The S ring in the plasma membrane is anchored to the peptidoglycan layer and is regarded as the nonrotating stator, as the name suggests. The M ring in the plasma membrane rotates against the S ring and is in effect the motor that drives the flagellum (Figure 19-31b).

Requiring, as it does, the structural equivalents of a rotor, a stator, and rotary bearings, such a mechanism was originally considered highly unlikely, and certainly without precedent in the biological world. But a series of ingenious experiments provided conclusive evidence for just such a propeller-like rotary drive. Much of this evidence is owed to the availability of antibodies that react specifically with flagellar filaments. In a particularly clever experiment, Michael Silverman and Melvin Simon used antibodies to link, or "tether," flagellar filaments to a glass surface, thereby preventing normal rotation of the filament. In so doing, they made the remarkable observation that if the filament cannot rotate, the whole body of the cell rotates instead!

The same researchers were also able to visualize flagellar rotation directly. Using antibodies as "glue," they succeeded in linking microscopic beads of polystyrene-latex to individual filaments as visible markers along the filament surface. Viewed under the microscope, the beads were seen to remain fixed in location with respect to one another but to revolve together around the filament. Apparently, we must credit nature with the invention of the wheel, for one exists in the rotary motor at the base of every flagellar filament in every motile bacterial cell!

Once it became clear that the driving force for flagellar rotation lies in the basal motor rather than in the flagellum itself, it was no longer surprising that the bacterial flagellum, unlike the motile appendages or contractile filaments of eukaryotic cells, possesses no ATPase activity along its length. In fact, it is now clear that ATP is not required at all as an energy source for flagellar rotation. Instead, the driving force appears to be the energy of the electrochemical proton gradient across the plasma membrane, but it is not yet clear how the proton gradient is coupled to flagellar rotation.

Flagellar Rotation and Chemotaxis

Flagellar rotation in the bacterium *E. coli* can be either clockwise or counterclockwise (Figure 19-32). The spiral shape of the flagellum is such that clockwise rotation results in each flagellum acting independently, each *pulling* the cell in its own direction (Figure 19-32a). Coherent motion of the cell is therefore impossible, and the bacterium displays a chaotic *tumbling* behavior. Counterclockwise rotation, on the other hand, results in each flagellum *pushing* on the cell. The flagella associate into a single coherent bundle and function together in a concerted manner, causing the bacterium to swim in a straight line (Figure 19-32b).

An *E. coli* cell reverses the rotation of its flagella quickly and frequently, such that the cell alternates between tumbling and smooth swimming. Characteristically, the tumbling lasts for only a few tenths of a second; then the flagella are reversed, and the cell swims off in a randomly chosen direction. In the absence of any changes in environmental stimuli, this results in a characteristic motility pattern in which short periods of undirected tumbling are interspersed with periods (or "runs") of swimming, but in randomly chosen directions (Figure 19-33a).

When exposed to a specific chemical attractant or repellent, the bacterial cell modifies this pattern to allow net movement toward the attractant or away from the repellent. *Attractants* include oxygen and nutrients such as sugars or amino acids. *Repellents* include waste substances excreted from the cell and various noxious chemicals such as phenol. The ability to move toward a chemical attractant or away from a repellent is called **chemotaxis.**

A bacterial cell responds not only to the presence or absence of an attractant or repellent, but to a concentration gradient of the chemical. The chemotactic response involves a relative increase in the length of a run when the cell is moving in the right direction, either toward an attractant (Figure 19-33b) or away from a repellent (Figure 19-33c). In other words, when a bacterial cell is moving in the right direction with respect to a concentration gradient, it will tend to tumble little and swim much, thereby accomplishing net movement in the desired direction. Chemotaxis therefore depends on the ability of the bacterial cell to change the direction of flagellar rotation and thereby either to swim smoothly in a desired direction or tumble randomly before striking out in a new direction. And the bacterial cell accomplishes all of this with tiny helical propellers driven by rotary "motors" so small that they can barely be seen with an electron microscope!

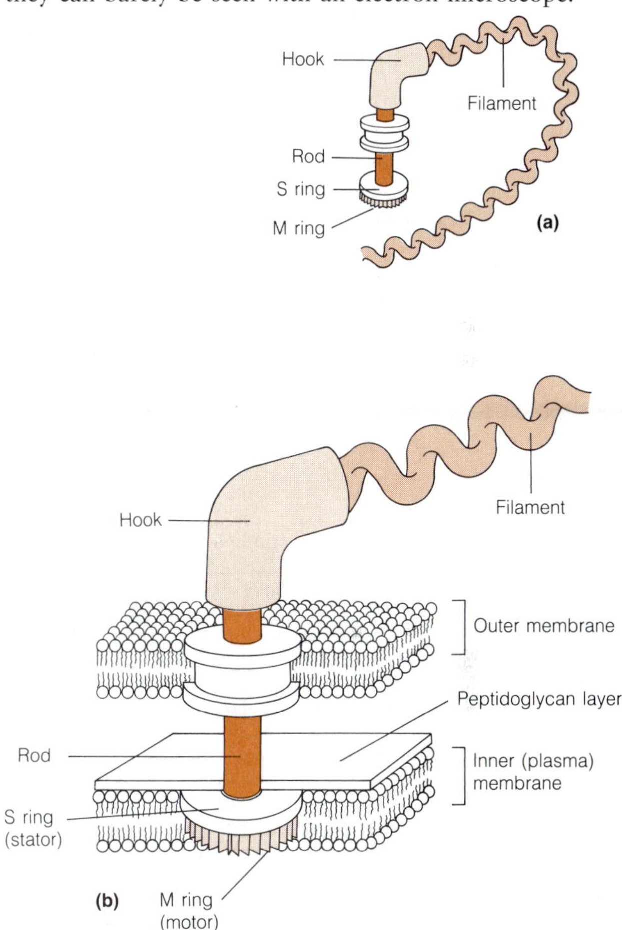

Figure 19-31 Structure of a Bacterial Flagellum and the "Motor" Responsible for Its Rotation. (a) The helical filament of each flagellum is attached to a hook structure, and this, in turn, is joined by a rod to four ringlike structures in the base. (b) Two of the rings are anchored in the outer membrane of the bacterial cell, and two in the inner (plasma) membrane. The S ring in the plasma membrane is anchored to the peptidoglycan layer and is regarded as the nonrotating stator. The M (motor) ring in the plasma membrane rotates against the S ring, with the torque provided by the proton motive force generated by the electrochemical proton gradient across the plasma membrane. Structures that rotate are shown in color.

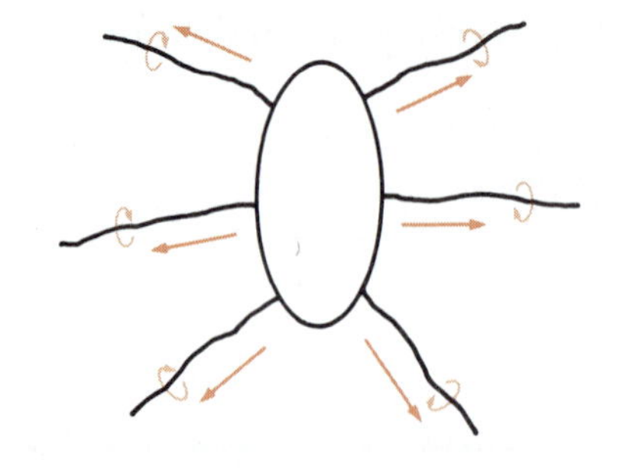

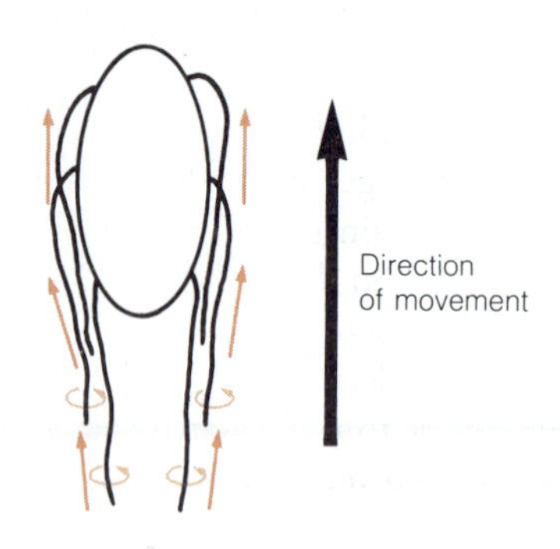

(a) Clockwise rotation of flagella (tumbling movement)

(b) Counterclockwise rotation of flagella (swimming movement)

Figure 19-32 Effect of Direction of Rotation on Cell Movement. (a) When the flagella of an *E. coli* cell rotate clockwise (small circular arrows), each flagellum pulls outward (large arrows) because of the helical shape of the flagellum. The bacterium is therefore drawn in several directions at once and displays an undirected tumbling movement. (b) When the flagella of the cell rotate counterclockwise, each flagellum pushes instead of pulls, and the flagella are drawn together into a bundle. Since each flagellum is now pushing in the same direction, the cell is propelled along with a smooth swimming movement.

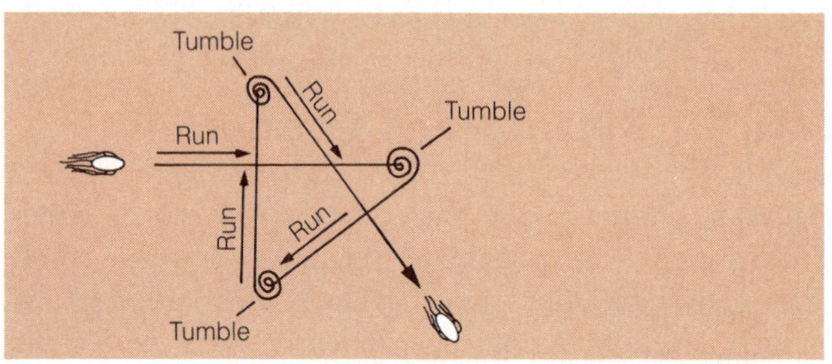

(a)

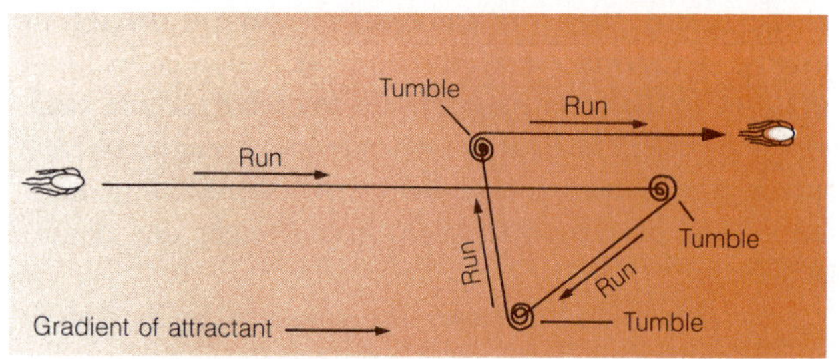

(b)

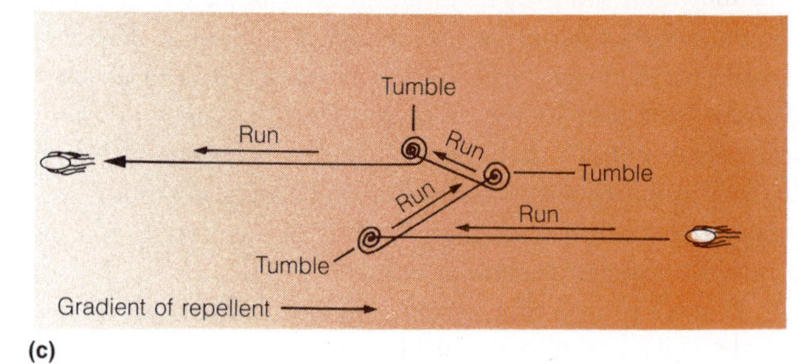

(c)

Figure 19-33 Chemotaxis in Bacteria. (a) In the absence of a chemotactic stimulus, the bacterium *E. coli* reverses the direction of flagellar rotation regularly. The flagella rotate counterclockwise for a few seconds, for a short period (or "run") of smooth swimming, then reverse to clockwise rotation for a few tenths of a second, for a short period of tumbling. This is followed by another reversal of direction, causing the bacterium to swim in a straight line again, but in a randomly chosen direction. The overall pattern is short runs in randomly chosen directions interspersed with brief periods of tumbling. (b) In the presence of a gradient of an attractant, the flagella rotate counterclockwise much longer when the bacterium senses it is moving up the attractant gradient than when it is swimming away from the attractant. Net movement is therefore toward the attractant, even though the direction of swimming chosen after each tumble is still random. (c) In the presence of a gradient of a repellent, the runs carrying the bacterium down the concentration gradient are much longer than those that move it up, so the net movement is away from the repellent.

Perspective

Motility is a major theme in cell biology. Our knowledge of mechanisms underlying motility has increased considerably over the past decade, and we now understand that it is driven at the molecular level by a set of ATP-dependent motor molecules. In eukaryotic cells, these molecules use portions of the cytoskeleton as a kind of track to pull subcellular components into position.

Two major eukaryotic motility systems are known, based on the interaction of myosin with actin microfilaments and that of dynein or kinesin with microtubules. Actin and myosin are found widely distributed in nonmuscle cells, where they are involved in a variety of motility mechanisms, including amoeboid movement, cytoplasmic streaming, and cytokinesis. Skeletal muscle contraction is a specialization of this more general motility process, and is the best-understood example. Muscle contraction involves a progressive sliding of thin actin filaments past thick myosin filaments, driven by the interaction between the ATPase head of the myosin molecules and successive myosin binding sites on the actin filaments. Contraction is triggered by the release of calcium from the sarcoplasmic reticulum and ceases again as the calcium is actively pumped back into the SR. In skeletal muscle, calcium binds to troponin and causes a conformational change in tropomyosin, which opens myosin binding sites on the thin filament. In smooth muscle, the effect of calcium is mediated by calmodulin, which activates myosin light-chain kinase, leading to the phosphorylation of myosin.

The second major motility system in eukaryotic cells is that based on microtubules. Cytoplasmic kinesin and dynein are motor molecules that move intracellular structures in opposite directions along microtubule tracks. The axoneme present in both cilia and flagella is a highly specialized example of dynein-tubulin interaction. The nine outer doublets of the axoneme are connected laterally to one another and radially to the central pair of single microtubules. Dynein arms project out from one microtubule doublet to the next. Dynein has ATPase activity and is thought to be involved in the sliding of one set of microtubules past the next. This sliding is opposed by the radial spokes between the doublets and the central pair of tubules and by the connections between adjacent doublets. As a result, the sliding is converted into a bending motion.

Bacterial cells have neither actin-myosin-based nor tubulin-based motility systems. They depend instead on flagella that bear no structural resemblance to the eukaryotic appendages of the same name. Bacterial flagella propel the cell by actually rotating like small propellers, using the energy of a proton gradient to drive the rotor. Depending on the direction of rotation, they can cause the cell to move smoothly in a fixed direction or to tumble randomly, and this process underlies the phenomenon of bacterial chemotaxis.

Key Terms for Self-Testing

motility (p. 581)
contractility (p. 581)

Systems of Motility
motor molecules (p. 581)
microfilament-based movement (p. 582)
microtubule-based movement (p. 582)
bacterial flagella (p. 582)

Nonmuscle Filament-Based Motility
nonmuscle motility (p. 582)
actin (p. 582)
myosin (p. 582)
heavy meromyosin (HMM) (p. 583)
light meromyosin (p. 583)
S1 (subfragment 1) particles (p. 583)
cytochalasins (p. 583)

Intracellular Microtubule-Based Motility: Dynein and Kinesin
dynein (p. 584)
kinesin (p. 584)

Filament-Based Movement in Muscle
skeletal muscles (p. 585)
muscle fibers (p. 585)
myofibrils (p. 586)
thick filaments (p. 586)
thin filaments (p. 586)
A bands (p. 587)
I bands (p. 587)
H zone (p. 587)
Z line (p. 587)
sarcomere (p. 587)
heavy chains (p. 588)
light chains (p. 588)
G-actin (p. 588)
F-actin (p. 588)
tropomyosin (p. 588)
troponin (p. 588)
TnT (p. 588)
TnC (p. 588)
TnI (p. 588)
α-actinin (p. 589)
myomesin (p. 589)
titin (p. 589)
sliding-filament model (p. 589)
cross-bridges (p. 591)
rigor (p. 591)
sarcoplasm (p. 593)

sarcoplasmic reticulum (SR) (p. 594)
transverse tubules (T-system) (p. 595)
calcium ATPase (p. 595)
myoglobin (p. 595)
gluconeogenesis (p. 596)
Cori cycle (p. 596)
creatine phosphate (p. 597)
cardiac (heart) muscle (p. 598)
smooth muscle (p. 598)
myosin light-chain kinase (p. 599)
calmodulin (p. 599)

Specialized Filament-Based Motility Systems
pseudopodia (p. 600)
amoeboid movement (p. 600)
ectoplasm (p. 601)
endoplasm (p. 601)
cytoplasmic streaming (p. 601)
shuttle streaming (p. 601)
cyclosis (p. 601)
cytokinesis (p. 602)
cleavage furrow (p. 602)
contractile ring (p. 602)

Microtubule-Based Motility: The Motile Appendages of Eukaryotic Cells
cilia (p. 602)
flagella (p. 603)
axoneme (p. 603)
basal body (p. 605)
outer doublets (p. 605)
central pair (p. 605)
A tubule (p. 605)
B tubule (p. 605)
tektin (p. 605)
sidearms (p. 605)
axonemal dynein (p. 605)
inner arm (p. 605)
outer arm (p. 605)
interdoublet links (p. 605)
radial spokes (p. 605)
nexin (p. 605)
sliding-microtubule model (p. 605)

The Bacterial Flagellum
bacterial flagellum (p. 607)
hook (p. 608)
rod (p. 608)
S (stator) ring (p. 608)
M (motor) ring (p. 608)
flagellin (p. 608)
chemotaxis (p. 609)

Suggested Reading

General References and Reviews

Bretscher, M. S. How animal cells move. *Sci. Amer.* 257 (December 1987): 72–90.

Huxley, A. F. *Reflections on Muscle.* Princeton, N.J.: Princeton University Press, 1980.

Lackie, J. M. *Cell Movement and Cell Behaviour.* London: Allen and Unwin, 1985.

Nonmuscle Motility

Goldman, R., T. Pollard, and J. Rosenbaum, eds. *Cell Motility.* Cold Spring Harbor, N.Y.: Cold Spring Harbor Laboratory, 1976.

Komnick, H., W. Stockem, and K. E. Wohlfarth-Bottermann. Cell motility: Mechanisms in protoplasmic streaming and ameboid movement. *Int. Rev. Cytol.* 34 (1973): 169.

Korn, E. D., M.-F. Carlier, and D. Pantaloni. Actin polymerization and ATP hydrolysis. *Science* 238 (1987): 638–644.

Lazarides, E., and J. P. Revel. The molecular basis of cell movement. *Sci. Amer.* 240 (May 1979): 110.

Scholey, J. M. Multiple microtubule motors. *Nature* 343 (1990): 118.

Taylor, D. L., and J. S. Condeelis. Cytoplasmic structure and contractility in amoeboid cells. *Int. Rev. Cytol.* 56 (1979): 57.

Weeds, A. Actin-binding proteins—regulators of cell architecture and motility. *Nature* 296 (1982): 811.

Muscle Contraction

Eisenberg, E., and L. E. Greene. The relation of muscle biochemistry to muscle physiology. *Annu. Rev. Physiol.* 42 (1980): 293.

Franzini-Armstrong, C., and L. D. Peachey. Striated muscles—contractile and control mechanisms. *J. Cell Biol.* 91 (1981): 166s.

Huxley, H. E. The mechanism of muscular contraction. *Science* 164 (1969): 1356.

Pollard, T. D. The myosin crossbridge problem. *Cell* 48 (1987): 909.

Squire, J. *The Structural Basis of Muscle Contraction.* New York: Plenum Press, 1981.

Taylor, K. A., and L. A. Amos. A new model for the geometry of the binding of myosin crossbridges to muscle thin filaments. *J. Mol. Biol.* 147 (1981): 297.

Warrick, H. M., and J. A. Spudich. Myosin structure and function in cell motility. *Annu. Rev. Cell Biol.* 3 (1987): 379–421.

Microtubule-Based Motility

Brokaw, C. J., D. J. L. Luck, and B. Huang. Analysis of the movement of *Chlamydomonas* flagella: The function of the radial-spoke system is revealed by comparison of wild-type and mutant flagella. *J. Cell Biol.* 92 (1982): 722.

Dustin, P. *Microtubules,* 2d ed. New York: Springer-Verlag, 1984.

Gibbons, I. R. Cilia and flagella of eukaryotes. *J. Cell Biol.* 91 (1981): 107s.

Goodenough, U. W., and J. E. Heuser. Substructure of the outer dynein arm. *J. Cell Biol.* 95 (1982): 795.

Satir, P. How cilia move. *Sci. Amer.* 231 (October 1974): 44.

Satir, P., J. K. Wais-Steider, S. Lebduska, A. Nasr, and J. Avolio. The mechanochemical cycle of the dynein arm. *Cell Motil.* 1 (1981): 303.

Warner, F. D., and D. R. Mitchell. Dynein, the mechano-chemical coupling adenosine triphosphatase of microtubule-based sliding filament mechanisms. *Int. Rev. Cytol.* 66 (1980): 1.

Wilson, L. Action of drugs on microtubules. *Life Sci.* 17 (1975): 303.

Calmodulin

Adelstein, R. S. Calmodulin and the regulation of the actin-myosin interaction in smooth muscle and nonmuscle cells. *Cell* 30 (1982): 349.

Cheung, W. Y. Calmodulin. *Sci. Amer.* 246 (June 1982): 48.

Bacterial Flagella

Adler, J. The sensing of chemicals by bacteria. *Sci. Amer.* 234 (April 1976): 40.

Berg, H. C. How bacteria swim. *Sci. Amer.* 233 (August 1975): 36.

Hazelbauer, G. L. Bacterial chemotaxis: Molecular biology of a sensory system. *Endeavour* 4 (1980): 67.

Macnab, R. M. The bacterial flagellar motor. *Trends Biochem Sci.* 9 (1984): 185.

How is an Amoeba Like an Elephant?

Fukui, Y., T. J. Lynch, H. Brzeska, and E. D. Korn. Myosin I is located at the leading edges of locomoting Dictyostelium amoebae. *Nature* 341 (1990): 328–330.

Problem Set

1. **Muscle Structure.** Frog skeletal muscle consists of thick filaments with a length of about 1.6 μm and thin filaments with a length of about 1 μm.
   (a) What is the length of the A band and the I band in a muscle with a sarcomere length of 3.2 μm? Describe what happens to the length of both bands as the sarcomere length decreases during contraction from 3.2 μm to 2.0 μm.
   (b) The H zone is a specific portion of the A band. (The H stands for *hell,* the German word for light.) If the H zone of each A band decreases in length from 1.2 to 0 μm as the sarcomere length contracts from 3.2 to 2.0 μm, what can you deduce about the physical meaning of the H zone?
   (c) What can you say about the distance from the Z line to the edge of the H zone during contraction?

2. **Energy Requirement of Muscle Contraction.** During contraction, mammalian skeletal muscle hydrolyzes ATP at the rate of 1 mmol/min per gram of muscle tissue. Muscle concentrations of ATP and creatine phosphate are about 5 μmol/g and 25 μmol/g, respectively.
   (a) How long could muscle contraction continue if it depended only on existing ATP supplies in the tissue? How long could it continue if it depended solely on existing supplies of ATP and creatine phosphate? Can you think of circumstances where these reserves of immediate energy would be essential despite the short times over which they can sustain contraction?
   (b) Assuming that the need for ATP is in fact being met by aerobic respiration, at what rate must the tissue be supplied with oxygen (in milliliters of oxygen per minute per gram of tissue) to sustain contraction? (Recall that 1 mol of a gas occupies 22.4 liters at standard temperature and pressure.)
   (c) Now assume that the need for ATP is met by anaerobic glycolysis, with stored glycogen as the only energy source. If muscle glycogen reserves are equal to 1% of the tissue by weight and all of this is available for catabolism, how long can anaerobic activity be sustained at the expense of glycogen?
   (d) Finally, assume that oxygen is not available, glycogen has been consumed, creatine phosphate stores are depleted, and all ATP has been converted to ADP, leaving the myokinase reaction as the cell's last resort. How long can contraction go on? Are the assumptions of the question realistic?

3. **ATP and Calcium Transport.** Another energy requirement of muscle is to drive the calcium pump. The calcium ATPase has a molecular weight of 115,000 and consumes ATP at the rate of about 10 micromoles ATP per minute per mg of enzyme. The enzyme represents about 1% of total muscle protein, and protein comprises about 30% of the muscle cell by weight.
   (a) Calculate the turnover number for the calcium ATPase. (The turnover number for an enzyme is the number of substrate molecules consumed per second per enzyme molecule.)
   (b) What proportion of the total ATP hydrolysis is used to transport calcium during muscle activity? (See Problem 2 for the total ATP hydrolysis rate of skeletal muscle.)

4. **Creatine Phosphate.** The $\Delta G^{\circ\prime}$ of hydrolysis is -10.3 kcal/mol for creatine phosphate versus -7.3 kcal/mol for ATP. Long after its involvement in muscle energetics was first realized, creatine phosphate was thought to be the immediate source of energy for muscle contraction. It was not until inhibitors of creatine kinase were discovered that the mechanical work of muscle contraction could be correlated directly with ATP consumption and the true nature of creatine phosphate as a reservoir of high-energy phosphate for recharging ADP was realized.
   (a) What is the $\Delta G^{\circ\prime}$ for the creatine kinase reaction in the direction of ATP generation? What is the value of the equilibrium constant K'_{eq} at 25°C?
   (b) Assuming conditions under which $\Delta G' = \Delta G^{\circ\prime}$, calculate the ratio ATP/(ATP + ADP) when the following

percentages of creatine are phosphorylated: 90%, 50%, 10%, and 1%.

(c) If you were assaying for creatine phosphate and ATP levels during the process of muscle contraction, why might you conclude, as the early researchers did, that the creatine phosphate rather than the ATP was the immediate source of the energy needed for contraction?

(d) If, on the other hand, you were to assay for creatine phosphate and ATP levels during contraction in the presence of an inhibitor (such as 2,4-dinitrofluorobenzene) that inactivates creatine kinase in intact muscle, how would your results differ from those of part (c)?

(e) How does this result support the "reservoir" nature of creatine phosphate?

5. **Rigor Mortis and the Contraction Cycle.** At death, the muscles of the body become very stiff and inextensible, and the corpse is said to go into *rigor*.

   (a) Explain the basis of rigor. Where in the contraction cycle is the muscle arrested? Why?

   (b) Would you be likely to go into rigor faster if you were to die while racing to class or while sitting in lecture? Explain.

   (c) What effect do you think the addition of ATP might have on muscles in rigor?

6. **AMPPCP and the Contraction Cycle.** AMPPCP is the abbreviation for a structural analogue of ATP in which the third phosphate group is linked to the second by a CH_2 group instead of an oxygen atom. AMPPCP binds to the ATP binding site of virtually all ATPases, including myosin. It differs from ATP, however, in that its terminal phosphate cannot be removed by hydrolysis. When isolated myofibrils are placed in a flask containing a solution of calcium ions and AMPPCP, contraction is quickly arrested.

   (a) Where in the contraction cycle will contraction be arrested? Draw the arrangement of the thin filament, the thick filament, and a cross-bridge in the arrested configuration.

   (b) Do you think contraction would resume if ATP were added to the flask containing the AMPPCP-arrested myofibrils? Explain.

   (c) What other processes in a muscle cell do you think are likely to be inhibited by AMPPCP?

7. **Tetanus.** Single nerve impulses to skeletal muscle normally cause separate twitches, interspersed by periods of relaxation. When a string of impulses arrives in rapid succession, however, it results in a summation of twitches and steady contraction. This leads to *tetanus,* a condition of rigidity, with the muscles temporarily "locked" in the fully contracted state. So defined, tetanus is a normal physiological phenomenon, since this kind of extreme contraction occurs naturally under conditions of intense muscular effort and is followed by relaxation as soon as the nerve stimulation stops. However, the same term is used in a pathological sense to describe a disease caused by the bacterium *Clostridium tetani,* a spore-producing anaerobe. *C. tetani* is normally a nonpathogenic inhabitant of the intestinal tract of animals. But if it gains entry into tissue (usually through a wound), it can produce a powerful exotoxin (a poison liberated from the bacterial cell) capable of inducing an abnormal state of extreme muscle contraction.

   (a) Explain the physiological phenomenon of tetanus in terms of stimulation of muscle contraction by nerve impulses.

   (b) What is a likely explanation for the ability of *C. tetani* to produce a pathological tetanus condition?

   (c) Why is vaccination with an attenuated strain of *C. tetani* (a "tetanus shot") especially recommended for wounds involving skin puncture?

8. **A Moving Experience.** For each of the following statements, indicate with the appropriate letter(s) whether it is true of the motility system that you use to lift your arm (A), to cause your heart to beat (H), to move ingested food through your intestine (I), or to sweep mucus and debris out of your respiratory tract (R).

   (a) It depends on muscles that have a striated appearance when examined with an electron microscope.

   (b) It would probably be affected by the same drugs that inhibit motility of a flagellated protozoan.

   (c) It requires ATP.

   (d) It involves calmodulin-mediated calcium stimulation.

   (e) It involves interaction between actin and myosin filaments.

   (f) It depends heavily on fatty acid oxidation for energy.

   (g) It is under the control of the voluntary nervous system.

9. **The Shear Pleasure of *Chlamydomonas*.** *Chlamydomonas reinhardtii* is a flagellated green alga. Its flagella can be readily sheared off by agitating the cells in a common kitchen blender. However, the *Chlamydomonas* quickly regenerates a new set of flagella.

   (a) If the *Chlamydomonas* cells are transferred to a colchicine-containing medium immediately after the flagella are sheared off, new flagella are not regenerated. Explain.

   (b) If the *Chlamydomonas* cells are transferred to a medium containing cycloheximide (an inhibitor of protein synthesis) immediately after the flagella are sheared off, new flagella are regenerated, but they are shorter than normal. Explain.

   (c) If only one of the two flagella is sheared off, the remaining flagellum is observed to shorten as the second flagellum grows out again. When they reach the same length, they both continue to grow out at the same rate. Explain.

10. **Microtubule Structure.** The major proteins making up the outer doublets have been isolated from the cilia of the protozoan *Tetrahymena*. These proteins were found to have the following properties:

    Protein A: mol. wt. of 50,000; no ATPase activity

    Protein B: mol. wt. of 400,000; has ATPase activity

    Protein C: mol. wt. is some small multiple of 400,000; has ATPase activity

    When a solution of protein C is mixed with a suspension of isolated *Tetrahymena* outer doublets from which the arms have previously been removed, protein C binds to the doublets, and electron microscopy shows that the arms of the outer doublets are thereby restored.

    (a) Explain the likely nature and purpose of proteins A, B, and C in outer doublet structure.

    (b) Suggest additional experiments to test your explanation.

11. **Sterility, Bronchitis, and Sinusitis.** Some forms of sterility in human males are due to nonmotile sperm. Upon cytological examination, the sperm of such individuals are found to have tails (i.e. flagella) that lack one or more of the normal structural components. Such individuals are also likely to have histories of respiratory tract disease, especially recurrent bronchitis and sinusitis, caused by an inability to clear mucus from the lungs and sinuses.

    (a) What is the likely explanation for nonmotility of sperm in such cases of sterility?

    (b) Why is respiratory tract disease linked with sterility in affected individuals?

    (c) Would you expect male offspring of an affected individual to have the same defects?

20

Electrical Signals: Nerve Cell Function

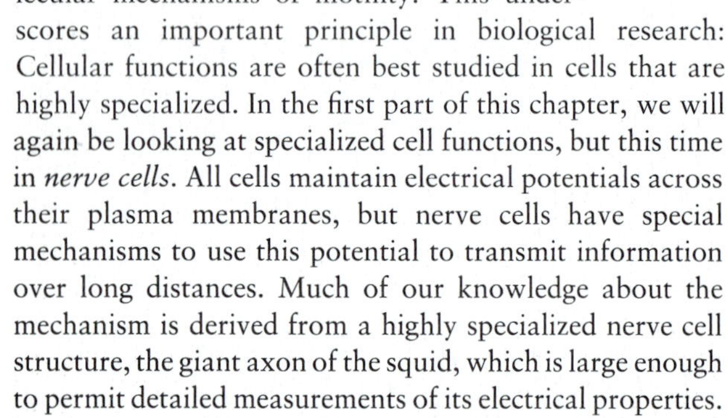

In the previous chapter we saw how studies of cytoskeletal structures in muscle cells led to more general understanding of the molecular mechanisms of motility. This underscores an important principle in biological research: Cellular functions are often best studied in cells that are highly specialized. In the first part of this chapter, we will again be looking at specialized cell functions, but this time in *nerve cells*. All cells maintain electrical potentials across their plasma membranes, but nerve cells have special mechanisms to use this potential to transmit information over long distances. Much of our knowledge about the mechanism is derived from a highly specialized nerve cell structure, the giant axon of the squid, which is large enough to permit detailed measurements of its electrical properties.

In the second part of the chapter, we will discuss the process by which information is passed between cells. Communication of information always involves at least two parties, the sender and the receiver. In the nervous system, the sender is a nerve cell or sensory cell, and the receiver is a second nerve cell or a muscle cell. We will see that nerve cells use a specialized process to deliver information across their junctions with other cells. The process involves the exocytotic release of chemical *neurotransmitters* at the junction, followed by binding of the neurotransmitters to receptors on the plasma membrane of the second cell. Our knowledge of neurotransmitters lets us understand how pharmaceutical agents interact with the brain, and how certain highly toxic compounds affect the nervous system.

The Nervous System

All animals have a **nervous system** in which electrical impulses are transmitted along specialized plasma membranes of nerve cells. The nervous system serves three functions: It *collects information* from the environment ("the light just turned green"), it *processes* that information ("green means go"), and it *responds* to that information by triggering specific effectors, usually muscle tissue or glands ("push on the accelerator").

To accomplish these functions, the nervous system has special components for sensing and processing information and for triggering the appropriate response (Figure 20-1). In vertebrates, the nervous system is divided into two components, the **central nervous system (CNS)** and the **peripheral nervous system (PNS)**. The CNS consists of the brain and the spinal cord, including both sensory and motor cells. The PNS consists of all other sensory and motor components, including the **somatic nervous system,** which controls voluntary movements of skeletal muscles, and the **autonomic nervous system,** which controls involuntary activities of the heart muscle, the smooth muscles of the gastrointestinal tract and blood vessels, and a variety of secretory glands.

Sensory nerve cells are specialized for receiving information from the environment via sensory receptors and transmitting that information inward to the central nervous system (Figure 20-1a,b). Sensory receptors are found in the eyes, ears, nose, taste buds, and all body surfaces capable of sensing touch. Sensory cells also play an important role in providing information about internal processes of the body, such as joint position, blood pressure, and temperature (Figure 20-1c). In response to sensory input, motor neurons in the central nervous system transmit regulatory signals to the appropriate muscles or glands, resulting in either the movement of a muscle or the secretion of a hormone (Figure 20-1d–g).

Intricate networks of neurons make up the complex tissues of the brain that are responsible for coordinating nervous function. Perhaps 10 billion nerve cell bodies are involved in the networks, which sometimes form layers of tissue. Each neuron has input from thousands of other neurons, so the brain's connections easily number in the trillions. By comparison, computers don't even come close to this number of connections, despite their seeming complexity.

We will not be concerned here with the overall functioning of the nervous system. Instead, we will focus on the cellular mechanisms whereby the electrical signals called *nerve impulses* are propagated. In addition to understanding how nerve cells work, we will acquire a better appreciation for several fundamental aspects of membrane function, of which nerve cells are a specialized example.

Structure of the Neuron

The nervous system consists of a variety of cell types, but the most important of these is the **neuron,** or nerve cell.

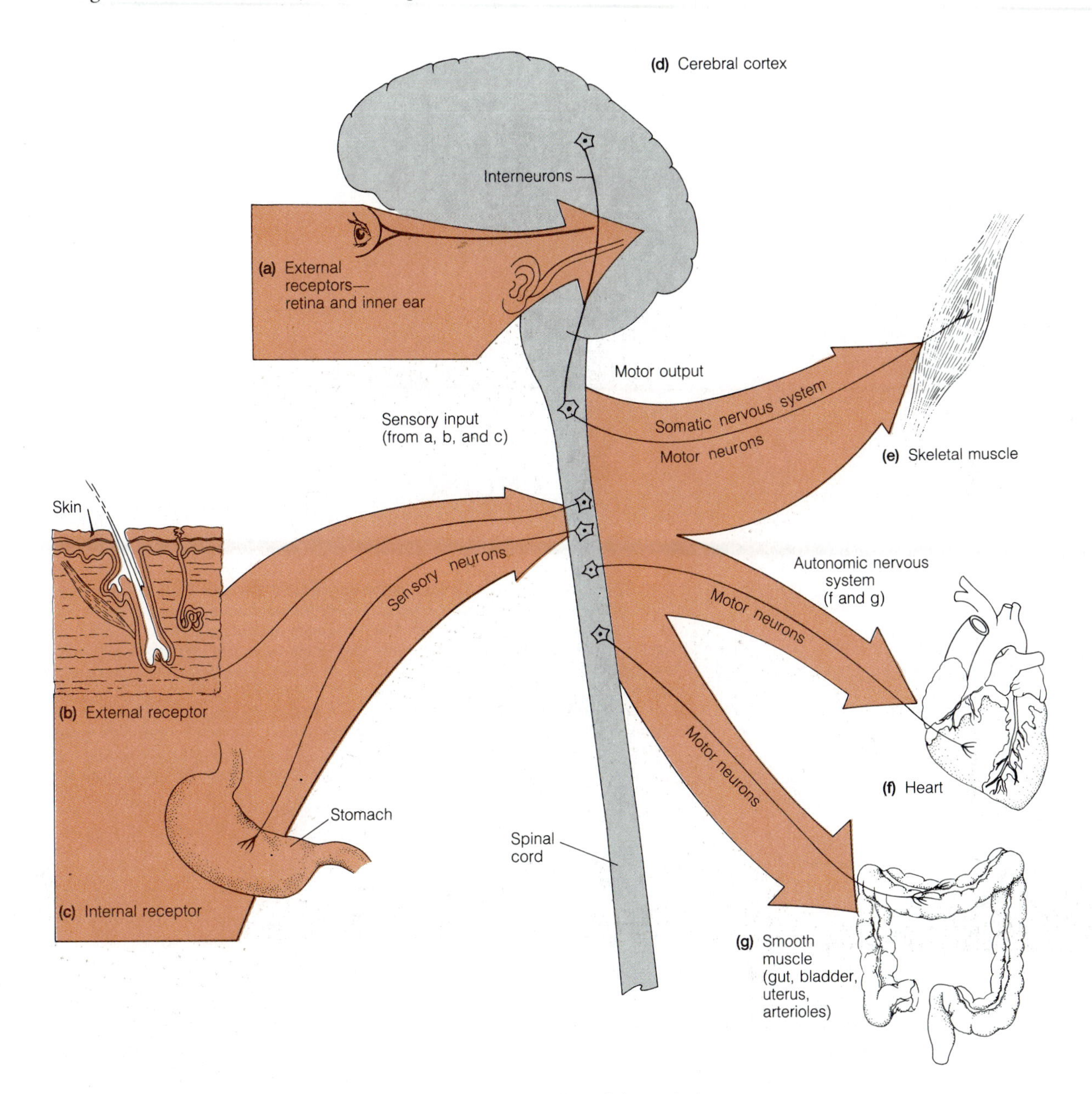

Figure 20-1 The Vertebrate Nervous System. The nervous system consists of the central nervous system (gray) and the peripheral nervous system (color). The sensory neurons of the peripheral nervous system receive information from both external and internal receptors (a, b, c) and transmit it to the central nervous system. Interneurons in the central nervous system integrate and coordinate response to the sensory input, a primary example being cerebral cortical neurons (d). Motor responses originate in the central nervous system and are transmitted to skeletal muscles by neurons of the somatic nervous system (e) and to involuntary muscles (f, g) and glands by means of the autonomic nervous system.

The neuron is the only cell type that is directly involved in the conduction and transmission of information through the nervous system. Much as in a computer, neural information consists of individual "bits." For the purposes of this discussion, we will describe these bits of neural information as "signals."

Neurons have evolved to fulfill specific functions in the nervous system. **Sensory neurons** carry information about the external or internal environment to the central nervous system, whereas **motor neurons** transmit signals from the central nervous system to muscles or glands. Most of the cells in the central nervous system are **interneurons,** which process and integrate the information received from different sensory and motor neurons. Nervous tissue also contains *glial cells* that support the intricate network of neurons, but these are not directly involved in transmission.

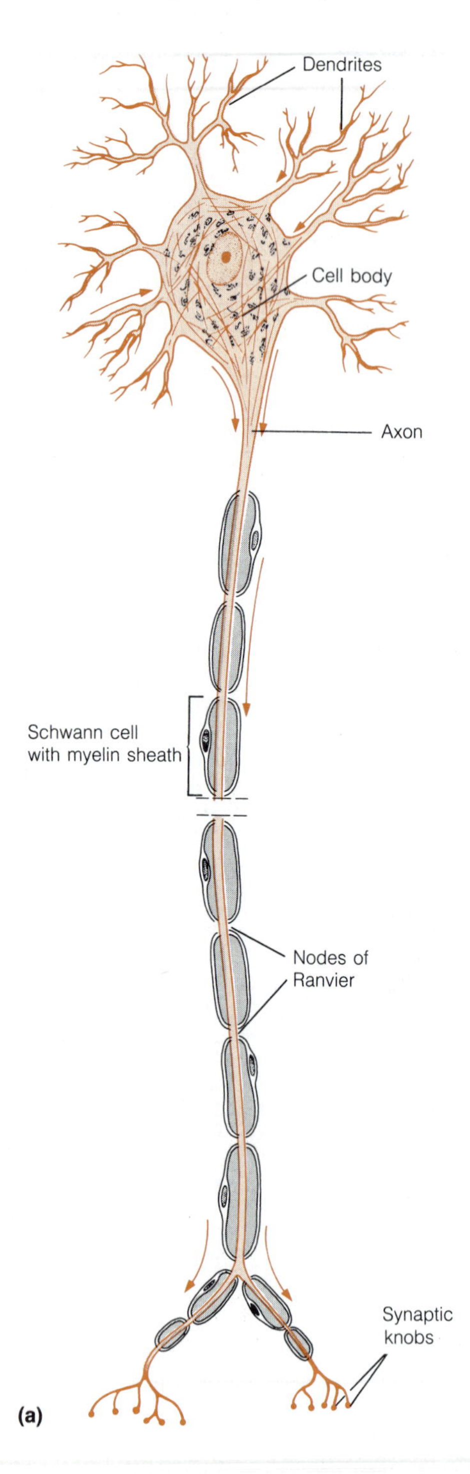

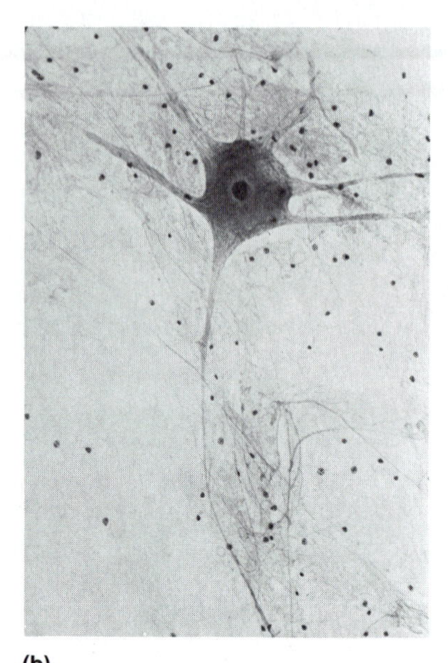

Figure 20-2 Structure of a Typical Motor Neuron. (a) The cell body contains the nucleus and most of the other organelles. Dendrites conduct signals passively inward to the cell body, whereas the axon transmits signals actively outward (direction of transmission shown by color arrows). At the end of the axon are numerous synaptic knobs (also called nerve terminals). Some, although not all, neurons have a discontinuous myelin sheath around their axons to insulate the axon electrically. Each segment of the sheath consists of a concentric layer of membranes wrapped around the axon by a Schwann cell (or an oligodendrocyte, in the case of the central nervous system). The breaks in the myelin sheath, called nodes of Ranvier, are concentrated regions of electrical activity. (b) A light micrograph of a neuron from the human spinal cord (TEM).

The structure of a typical motor neuron is shown in Figure 20-2. The **cell body** of most neurons is similar to other cells, with a nucleus and most of the expected organelles present. In addition, however, neurons contain extensions, or branches, called **processes.** It is these processes that make neurons so easy to distinguish from almost any other cell type. There are two types of processes. Those that receive signals and transmit them inward to the cell body are called **dendrites.** Processes that conduct signals away from the cell body are called **axons.** The cytoplasm within an axon is commonly referred to as **axoplasm.** A **nerve** is simply a tissue composed of bundles of axons.

Neurons display more structural variability than Figure 20-2 suggests. Some sensory neurons have only one process, which conducts signals both toward and away from the cell body. Moreover, the structure of the dendritic processes is by no means random; many different classes of neurons in the central nervous system can be identified by structure alone (Figure 20-3).

As Figure 20-2 illustrates, a motor neuron has multiple, branched dendrites and a single axon leading away from the cell body. The axon of a typical neuron is much longer than the dendrites and forms multiple branches. Each branch terminates in structures called **synaptic knobs,** also known as **nerve terminals.** The synaptic knobs are responsible for transmitting the signal to the next cell, which may be another neuron or a muscle or gland cell. In each case, the junction is called a **synapse.** For neuron-to-neuron junctions, synapses may occur between an axon and a dendrite, between an axon and a cell body, or even between an axon and an axon. Typically, neurons have synapses with many other neurons. The transmission of a signal across a synapse may be by direct electrical connection, but more often it involves a variety of chemicals called **neurotransmitters.**

Axons can be very long—up to several thousand times longer than the diameter of the cell body. A motor neuron that innervates your foot has its cell body in the spinal cord, and its axon extends approximately a meter down a nerve tract in your leg.

The Myelin Sheath

Most axons are surrounded by a discontinuous **myelin sheath** consisting of many concentric layers of membrane. The myelin sheath is a very effective electrical insulation for the segments of the axon that it envelops. In the central nervous system, the myelin sheath is formed by glial cells called **oligodendrocytes.** In the peripheral nervous system, the myelin is made by other specialized glial cells called **Schwann cells.** Figure 20-4 shows a cross section of a myelinated nerve axon in the peripheral nervous system, and Figure 20-5 illustrates the process of myelination. Schwann cells envelop the axon and wrap layer after layer of their own plasma membrane around the axon in a tight spiral.

Each Schwann cell is responsible for the myelin sheath around a short segment—about 1 mm—of a single axon. Numerous Schwann cells are therefore required to encase an axon of the peripheral nervous system with discontin-

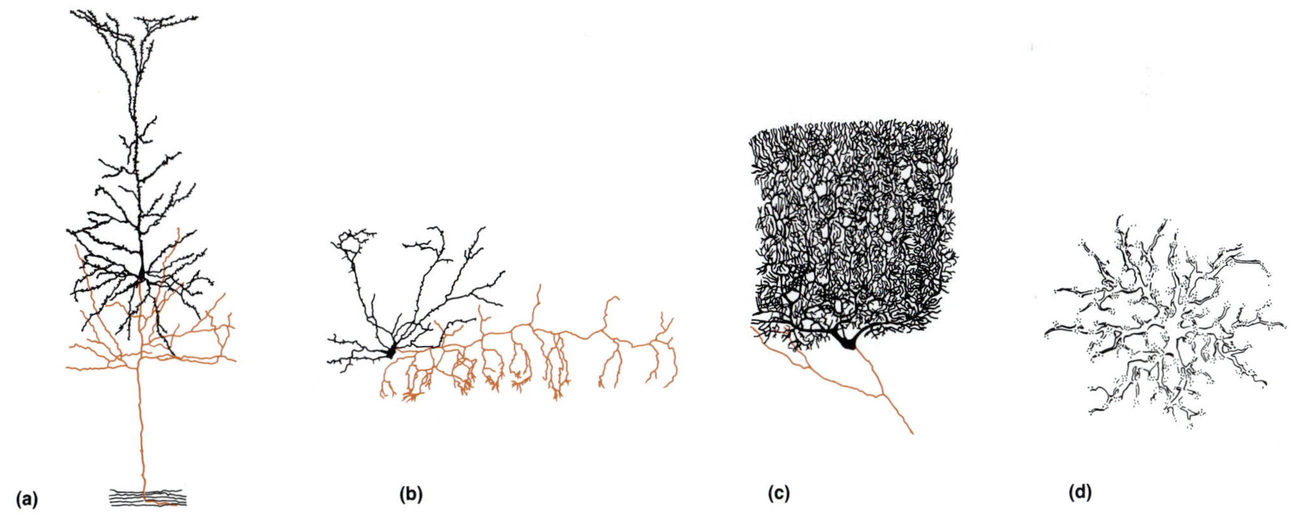

Figure 20-3 Neuron Shapes. The neurons of the central nervous system display a wide variety of characteristic shapes. (a) A pyramidal neuron from the cerebral cortex. (b) A number of short-axon cells in the cerebral cortex. (c) A Purkinje cell in the cerebellum. (d) An axonless horizontal cell in the retina of the eye. Axons are shown in color and dendrites in black.

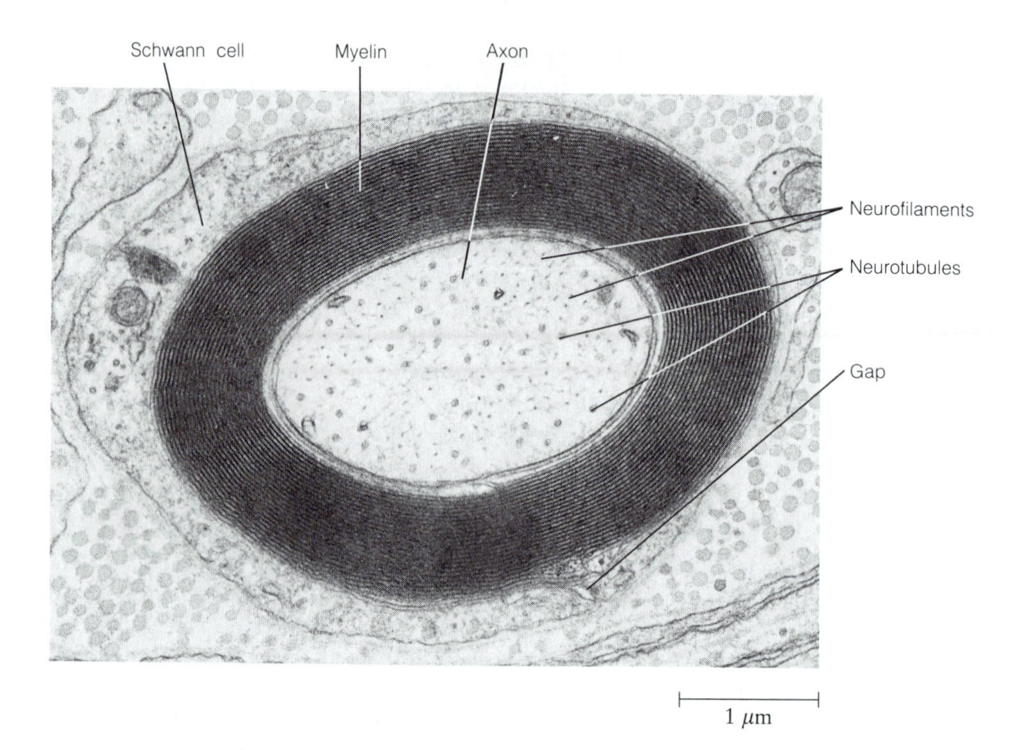

Figure 20-4 A Myelinated Axon. This cross-sectional view of a myelinated axon from the nervous system of a cat shows the concentric layers of unit membrane that have been wrapped around the axon by the Schwann cell that envelops it. The gap in the plasma membrane of the Schwann cell is the point at which the membrane initially invaginated to begin enveloping the axon. Notice also the neurotubules and neurofilaments in the axoplasm of the myelinated axon. (TEM)

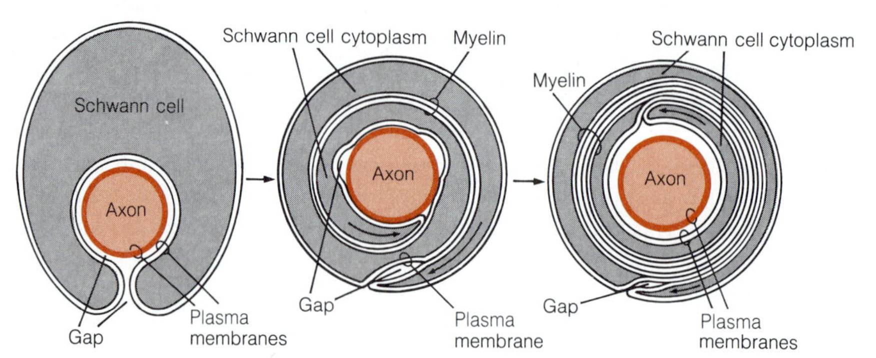

Figure 20-5 Origin of the Myelin Sheath. Myelinated axons of the peripheral nervous system are enveloped by Schwann cells at intervals along the length of the axon. Each Schwann cell gives rise to one segment of myelin sheath by wrapping its own plasma membrane concentrically around the axon. The myelin layer gets progressively thicker as more layers of unit membrane derived from the plasma membrane of the Schwann cell are added to it, accompanied by the gradual loss of the Schwann cell cytoplasm.

uous sheaths of myelin. In the central nervous system, however, a single oligodendrocyte myelinates many axons. The small regions of bare axon between successive segments of the myelin sheath are called **nodes of Ranvier** (see Figure 20-2). These nodes are only about 0.5 μm long, but they are concentrated regions of electrical activity and therefore play an important role in the transmission of signals.

The Resting Potential of Nerve Cells

Since the middle of the nineteenth century, it has been known that nervous information is conducted as changes in the electrical potential at the cell surface. A fundamental property of essentially all cells is the presence of an electrical potential (membrane potential) across the plasma membrane. In the nerve cell the membrane potential is called the **resting potential** (V_m), because the cell is "at rest" (not transmitting a nerve signal). A few cell types have the ability to change the electrical properties of their plasma membranes in response to a stimulus, a capacity called **excitability.** Examples include the islet cells of the pancreas, a water plant called *Nitella,* and, of course, nerve cells. When a membrane displays excitability in response to a stimulus, the resting potential undergoes a transient change called the **action potential.** As we will see, in nerve cells the action potential has the specific function of transmitting an electrical signal along the axon. This signal represents the unit of information by which cells communicate in the nervous system, and it is therefore essential to understand how nerve signals are generated. We will begin with a discussion of resting potentials.

Establishing the Resting Potential

In 1902, Julius Bernstein proposed that the resting potential of nerve cells might have something to do with an unequal distribution of potassium ions across the cell membrane. Charles Overton went on to suggest that an electrical potential could arise if the nerve cell membrane was *selectively permeable* to potassium ions. If the nerve cell was then stimulated, the selective permeability might temporarily "collapse." The change in ion permeability could then travel along the axon of a nerve cell, providing a mechanism for transmitting a nerve impulse.

Although the concepts of Overton and Bernstein were not completely correct, the idea that nerve cell function is somehow related to the ion permeability of a membrane was crucial. To understand how an electrical potential might arise from the flow of ions across a membrane, consider a model system consisting of two compartments separated by a lipid bilayer (Figure 20-6). This experimental device, referred to as a *planar lipid membrane,* is commonly used in membrane biology to test ideas about how membranes function. The value of this system is that electrical properties of the bilayer can be investigated simply by placing electrodes, one in each compartment, so that the voltage associated with ionic currents across the bilayer can be measured with a sensitive amplifier.

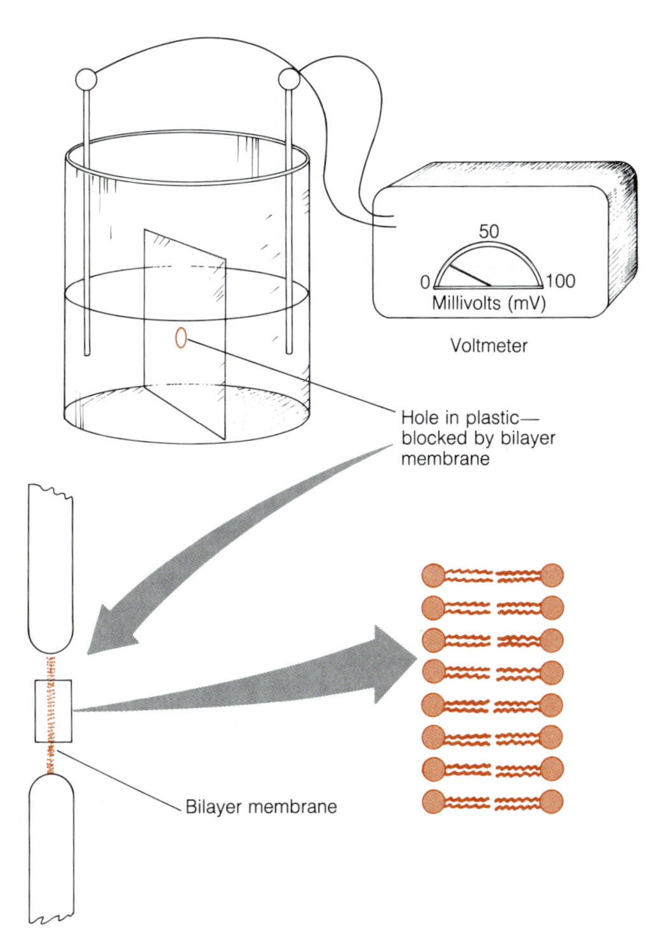

Figure 20-6 A Lipid Bilayer Model Membrane. To investigate properties of membranes, a lipid bilayer is established across a small hole in a thin plastic sheet used to separate two compartments. An electrode is inserted into each compartment, so that electrical parameters such as membrane potential and ionic current can be measured.

Next, imagine that we place 100 m*M* potassium chloride (KCl) in each compartment, and measure the voltage (Figure 20-7). As you might expect, the voltage would be 0 millivolts (mV), and no electrical current would be measurable because no concentration gradient is present. The system is at equilibrium.

We can determine the effect of an ionic gradient on the system simply by reducing the KCl concentration in one compartment to 10 m*M* so that a tenfold concentration gradient is produced. Again, no voltage or current can be measured. The reason is that the bilayer is relatively impermeable, so that no electrical current in the form of ions can cross the bilayer barrier. However, if we add an antibiotic compound called *gramicidin* to the solution with a KCl gradient present, suddenly our amplifier indicates a large increase in voltage. This is because gramicidin has the ability to dissolve in lipid bilayers and form channels that allow cations (but not anions) to cross the bilayer (recall Figure 8-E, p. 201). The measured voltage is about 60 mV, in conjunction with an ionic current of several picoamperes (1 picoampere = 10^{-12} ampere) through the open channels of the antibiotic.

Understanding Membrane Potentials in Neurons

The flow of potassium across the model membrane described above is driven by its **electrochemical gradient.** This gradient is the sum of two components, which may either oppose or reinforce each other. The *electro* portion of the gradient refers to the potential difference across the membrane. The *chemical* part of the gradient refers to the concentration gradient of the ion across the membrane. In the model system, gramicidin forms a potassium-specific channel, which allows positively charged potassium ions to undergo a net flux from higher to lower concentrations, leaving negatively charged chloride ion behind. This separation of positive and negative charges *polarizes* the membrane, and the degree of **polarization** is measured as millivolts.

It is important to understand that as a result of potassium efflux, the negative charges left on the inside of the membrane will increasingly draw potassium inward and impede its flow outward. Potassium, in other words, tends to be drawn *inward* by the negative membrane potential but *outward* by its own concentration gradient. When these two competing tendencies balance each other exactly, the electrochemical gradient is zero, and there will be no net flow of potassium across the membrane. This means that for any specific concentration gradient for potassium (such as that which exists across the membrane of an axon), there will always be a specific membrane potential at which the electrochemical gradient for potassium is zero. The particular membrane potential that will exactly offset the effect of the concentration gradient for a given ionic species is called the **equilibrium potential** for that species.

The Nernst Equation

The equilibrium potential for a particular species of ion can be calculated using the **Nernst equation:**

$$\begin{aligned} E_X &= RT/zF \ln [X]_o/[X]_i \\ &= 2.303RT/zF \log_{10} [X]_o/[X]_i \end{aligned} \quad (20\text{-}1)$$

where E_X is the equilibrium potential for ion X (in volts), R is the gas constant (1.987 cal/mol-degree), T is the absolute temperature in degrees Kelvin, z is the valence of the ion, F is the Faraday constant (23,062 cal/V-mol), $[X]_o$ is the concentration of X outside the cell (in M), and $[X]_i$ is the concentration of X inside the cell (in M).

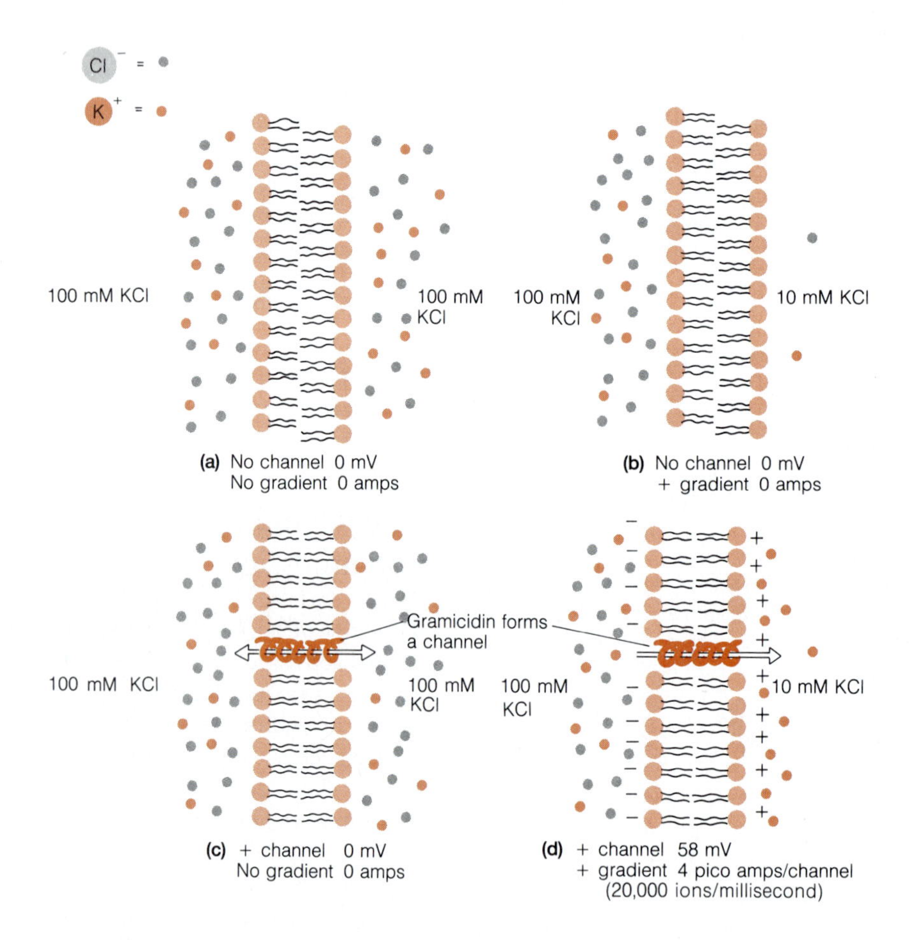

Figure 20-7 Development of Membrane Potential Across a Model Membrane. Bilayer membranes such as that shown in Figure 20-6 can be set up with a single ion-conducting channel present. In the absence of a channel, no voltage or ionic current can be measured (a, b) because the membrane is essentially impermeable to ion flux. When the antibiotic gramicidin is added, it forms a channel through the membrane. If only a channel is present (c), the system remains at equilibrium. However, if a channel and concentration gradient are both present (d), ionic current and associated voltage are measurable. Model membrane systems have significantly advanced our understanding of membrane potentials, ionic currents, and nerve cell membranes at the molecular level.

For a univalent cation at room temperature, the expression $2.303RT/zF$ is equal to about 0.058 V or 58 mV, so the Nernst expression under these conditions simplifies to

$$E_X = 58 \log_{10} [X]_o/[X]_i \qquad (20\text{-}2)$$

where E_X is in millivolts.

We can now go on to discuss how this conceptual approach helps us to understand the function of nerve cell membranes.

The Squid Giant Axon as an Experimental System

Evidence for the existence and nature of the resting potential of nerve membranes came as electrophysiologists began to devise experimental means of measuring transmembrane electrical potentials directly. Their research was greatly aided in the 1930s, when axons from the *giant neurons* of the squid were discovered. Many invertebrates have very large axons that are especially amenable to electrophysiological measurements. The giant nerve fibers of the squid innervate muscles that are used to expel water explosively from the mantle cavity of the animal, enabling the squid to propel itself backward to escape predators (Figure 20-8). The axons involved in triggering this "jet propulsion" system are very large, with diameters of about 0.5–1.0 mm. This large size allows electrodes to be inserted readily into the axon, thereby making it possible to control and measure electrical potentials and ionic currents across the axonal membrane.

An apparatus for measuring potential differences across the membrane of such an axon is shown in Figure 20-9. With a recording electrode implanted in the axon and a reference electrode in the electrolyte solution surrounding the membrane, potential differences between the electrodes can be measured directly, using either a voltmeter or an oscilloscope. Furthermore, a sophisticated electronic feedback circuit called a *voltage clamp* can be used to set the voltage at a given value. As we will see, this provides a way to measure the ionic currents related to a given ion, such as sodium or potassium, during nerve cell function.

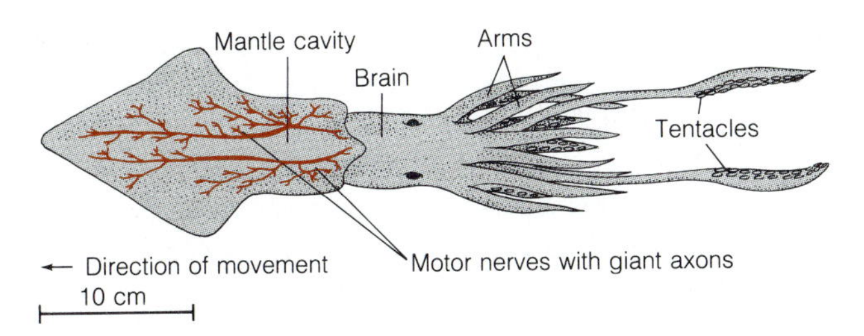

Figure 20-8 Squid Axons. The squid nervous system includes motor nerves that control swimming movements. The nerves contain axons with diameters ranging up to a millimeter, providing a convenient system for studying resting and action potentials in a biological membrane.

Steady-State Ion Concentrations and the Resting Potential

Table 20-1 shows approximate steady-state concentrations of sodium, potassium, and chloride ions across the membrane of the squid giant axon and, for comparison, in a mammalian neuron. By convention, the concentration gradient for each ion is expressed as a ratio of the concentration of that ion on the outside (numerator) to the concentration on the inside (denominator). If electrodes are placed across the membrane of an axon in an apparatus like the one shown in Figure 20-9, the voltmeter will record a steady-state potential difference of about −60 mV, expressed as the voltage inside with respect to the voltage outside. This is the resting potential of the membrane, V_m.

The resting potential of a nerve cell can be explained in terms of the model system described earlier. The plasma membranes of most cells have channels that make them selectively permeable to potassium ions, just as the gram-

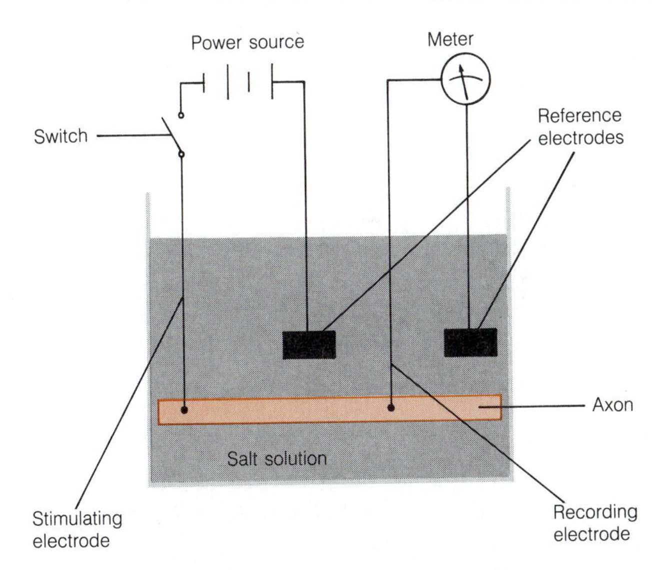

Figure 20-9 An Apparatus for Initiating and Measuring Changes in Membrane Potential. The stimulating electrode is connected to a power source, which delivers a pulse of current to the axon when the switch is closed momentarily. The nerve impulse this generates is propagated down the axon and can be detected a few milliseconds later by the recording electrode, which is connected to a meter or an oscilloscope. The impulse is detected as a transient change in transmembrane potential, measured with respect to a reference electrode that is placed in the electrolyte solution surrounding the membrane.

Table 20-1 Ionic Concentrations Inside and Outside Axons, Concentration Ratios,* and Resulting Equilibrium Potentials

	Squid Axon				Mammalian Neuron			
Ion	Inside (mM)	Outside (mM)	Ratio*	Potential (mV)	Inside (mM)	Outside (mM)	Ratio*	Potential (mV)
Na^+	50	440	8.8	55	10	145	14.4	68
K^+	400	20	0.05	−75	140	5	0.035	−85
Cl^-	50	560	0.09	−40	10	110	0.09	−60

* Concentration ratios are outside/inside for cations, inside/outside for anions.

icidin does in our model system. But where does the gradient come from? This is an essential point and can be understood in terms of our knowledge of other cells. As discussed in Chapter 8, the plasma membranes of all cells have an ATP-driven sodium-potassium pump that transports potassium ions inward and sodium ions outward, thereby maintaining steep gradients of both cations across the membrane. This gradient provides the energy necessary for the resting potential of neurons and, as we will see later, for the action potential as well.

Having described the membrane potential arising from the movement of a single ionic species—potassium—through channels in a model membrane, we can now go on to the more complex biological membrane. The membrane potential of a neuron is essentially the weighted sum of individual equilibrium potentials for the three most relevant ions—potassium, sodium, and chloride. Potassium is a good ion with which to begin, because it plays a major role in determining the resting potential of most membranes.

Using the actual concentrations for potassium on the inside and outside of the squid axon (Table 20-1), we can use the Nernst equation to calculate the equilibrium potential for this ion as follows:

$$E_K = 58 \log_{10} (20/400) = 58 \log_{10} (0.05) = -75 \text{ mV} \qquad (20\text{-}3)$$

This means that for the particular gradient of potassium concentration that is maintained across the membrane of the squid axon, a membrane potential of −75 mV is required to set the electrochemical gradient for potassium to zero and ensure that no net flow of potassium occurs across the membrane. If the actual membrane potential is more negative than −75 mV, it will more than offset the concentration gradient and there will be a net inward flow of potassium. On the other hand, if the actual membrane potential is less negative than −75 mV, it will not completely counteract the concentration gradient and there will be a net outward flow of potassium. This is what must happen in the squid axon, since the resting potential is only about −60 mV.

The Nernst equation can, of course, be used to determine the equilibrium potential for sodium and chloride ions also. In contrast to the potassium concentration, the sodium concentration of the squid axon is high outside and low inside (Table 20-1). The equilibrium potential for sodium can be calculated as

$$E_{Na} = 58 \log_{10} (440/50) = 58 \log_{10} (8.8) = +55 \text{ mV} \qquad (20\text{-}4)$$

The equilibrium potentials calculated for potassium (equation 20-3) and sodium (equation 20-4) are very close to the membrane potentials determined experimentally for a variety of cells. For a typical cell, E_{Na} is somewhere between +50 mV and +65 mV, and E_K is between −70 mV and −100 mV.

What do these values tell us about the tendencies of sodium and potassium ions to move across a membrane with a resting potential of about −60 mV? For potassium, there is a modest tendency to move outward, with the driving force proportional to a difference of about 10–30 mV between the equilibrium potential for potassium and the actual membrane potential. But for sodium the tendency is to move inward, and the driving force is much greater, since the difference between the actual membrane potential and the equilibrium potential for this ion is about 110–125 mV. This means that if the axonal membrane of a nerve cell is suddenly rendered permeable to both sodium and potassium, there will be a slight tendency for potassium to flow out of the cell but a much greater tendency for sodium to rush in. Keep that in mind, because we will return to it later.

Understanding the Resting Potential

If the sodium-potassium pump maintains a steep gradient of both sodium and potassium across the axonal membrane, why is the resting potential so close to the equilibrium potential for potassium and so different from the equilibrium potential for sodium? The answer to this question lies in the difference in permeability of the membrane for the two cations. By definition, the resting potential of a membrane is that potential at which the net flow of charged species across the membrane is zero, because a net flow would result in a change in membrane potential. In other words, sodium, potassium, chloride, and other ions may still move across the membrane at the resting potential, but the net flow of charge must be zero, or the resulting current would change the potential to a different value.

If the axonal membrane were equally conductive to both sodium and potassium ions (and to no other ions), the resting potential for the membrane would be midway between the equilibrium potentials for these two species. In point of fact, however, a resting nerve cell is only about 4% as permeable to sodium as it is to potassium. This means that the membrane potential of the resting cell is maintained nearer the equilibrium potential for potassium. Otherwise, the tendency for potassium to leave the cell would be too great to be offset by other ion movements, and it would not be possible to maintain a net ion flow of zero.

One way to test this concept is to alter ionic concentrations outside a squid axon and look for any effect on the resting potential. For instance, the potassium ion concentration outside the neuron can be increased until it equals the internal concentration. When this experiment is carried out, it is apparent that equalizing the internal and external potassium ion concentrations completely abolishes the resting potential, whereas similar alterations of other ionic concentrations have little effect.

Potassium is therefore the critical ion. Whenever the membrane potential differs much from the equilibrium potential for potassium, potassium ions will flow in or out to offset the potential difference because of the selective permeability of the membrane for this ion. This movement of potassium ions is too large to be balanced or offset by a counterflow of other ions and will therefore tend to return the membrane potential to the original resting value. The resting potential of a membrane, in other words, is determined to a large extent by the equilibrium potential for the ionic species to which the membrane is most permeable, and for most membranes that ion is potassium.

The Goldman-Hodgkin-Katz Equation

Although heavily influenced by potassium, the resting potential in fact involves all three monovalent ions—potassium, sodium, and chloride. A more complete description of cellular membrane potentials is given by the **Goldman-Hodgkin-Katz equation** (usually called simply the Goldman equation), developed by and named after several pioneering investigators in neurobiology. This is a version of the Nernst equation that adds in the effects of potassium, sodium, and chloride ions, each weighted for its relative permeability:

$$V_m = 2.303RT/F \log_{10} \frac{P_K[K^+]_o + P_{Na}[Na^+]_o + P_{Cl}[Cl^-]_i}{P_K[K^+]_i + P_{Na}[Na^+]_i + P_{Cl}[Cl^-]_o} \tag{20-5}$$

Here P_K, P_{Na}, and P_{Cl} are the relative permeabilities of the membrane for the respective ions, and all other terms are as previously defined. (Remember that chloride ion has a negative valence, so that $[Cl^-]_i$ appears in the numerator and $[Cl^-]_o$ in the denominator.) The permeability of the membrane to other ions is very small compared with that of potassium, sodium, and chloride ions. As a result, other ions rarely contribute significantly to the resting potential of the neuronal membrane and are usually ignored in calculating V_m.

We have already noted that the permeability for sodium is only about 4% of that for potassium. For chloride ion, the estimated value is 45%. Relative values of P_K, P_{Na}, and P_{Cl} are therefore 1.0, 0.04, and 0.45, respectively. Using these values and assuming a chloride concentration of 50 m*M* inside the cell, the membrane potential of the resting squid axon can be calculated as

$$\begin{aligned} V_m &= 58 \log_{10} \frac{(1.0)(20) + (0.04)(440) + (0.45)(50)}{(1.0)(400) + (0.04)(50) + (0.45)(560)} \\ &= 58 \log_{10} (60/654) = -60 \text{ mV} \end{aligned} \tag{20-6}$$

This value closely approximates measured resting potentials of the squid axonal membrane.

The Generation of Action Potentials

A resting neuron is a system poised for electrical action. Although potassium is near electrochemical equilibrium, a large electrochemical gradient exists for sodium across the plasma membrane of the axon. Both the electrical and the concentration components of the gradient strongly favor inward movement of sodium. In fact, the ΔG for the inward flux of sodium can be calculated (from equation 8-8, on p. 205) to be about -2.65 kcal/g-equivalent for the squid axon, assuming a resting potential of about -60 mV. But little inward movement of sodium ions actually occurs across the resting membrane, because the membrane is relatively impermeable to this ion in the resting state.

The Action Potential

The development of an *action potential* is possible because the axonal membrane is not impermeable to sodium under all conditions. The membrane actually possesses one set of channels that are specific for sodium ions and another set of channels with a specificity for potassium ions. Some of the potassium channels are voltage-insensitive. They are always open and account for the leakage of the ion across the membrane. Of specific interest to us at the moment, however, are the voltage-sensitive sodium and potassium channels—those for which the ionic permeability changes as the voltage changes.

In the resting state, the voltage-sensitive channels are closed. The passage of a nerve impulse along the axonal membrane results from a transient change in the permeability of the membrane, first to sodium and then to potassium. This, in turn, results in a predictable pattern of electrical changes in the membrane that is propagated along the membrane as an action potential.

Measuring the Action Potential. The development and propagation of an action potential can be readily studied in large axons such as those of the squid by using the apparatus shown earlier in Figure 20-9. To excite the axon, an electrode is connected to a power source and positioned some distance from the recording electrode. An impulse can then be initiated in the axonal membrane by a brief electrical voltage applied at the *stimulating electrode,* provided only that the voltage is of sufficient magnitude to depolarize the membrane by about 20 mV (that is, from −60 to about −40 mV). The wave of electrical changes that this induces in the membrane will travel along the membrane in both directions. As the wave passes the *recording electrode,* the voltmeter or oscilloscope will display the characteristic pattern of potential changes shown in Figure 20-10a. In less than a millisecond, the membrane potential rises dramatically from the resting potential of the membrane to about +40 mV, which means that the interior of the membrane actually becomes positive for a brief period. The potential then falls somewhat more slowly, dropping to about −75 mV before stabilizing again at the resting potential of about −60 mV. As the figure indicates, all these changes occur within a few milliseconds.

Note that at its peak (about +40 mV), the action potential approaches, though it does not actually reach, the equilibrium potential for sodium (about +55 mV). Similarly, the "undershoot" or **hyperpolarization** that occurs just prior to resumption of the normal resting potential closely approximates the equilibrium potential for potassium (about −75 mV). The hyperpolarization is caused by additional potassium channels opening during the action potential and remaining open even after the sodium channels close.

The Refractory Period. The response shown in Figure 20-10a can be elicited again and again by repeated stimulation of the membrane, as long as the time interval between successive stimuli is long enough to allow the axon membrane to return to its resting potential following the undershoot. About 2–3 msec is usually required for recovery. During this interval, the membrane is hyperpolarized, as described above, and the sodium channels are transiently inactivated. The combination of a hyperpolarized state and inactivated sodium channels causes a **refractory period** during which the axon is unable to respond to a second depolarizing stimulus.

The Action Potential as an All-or-Nothing Response. The action potential of a membrane is an "all-or-nothing" response. It usually occurs with the same characteristic amplitude and time course, regardless of the magnitude of the initial stimulus. If the stimulus is sufficiently

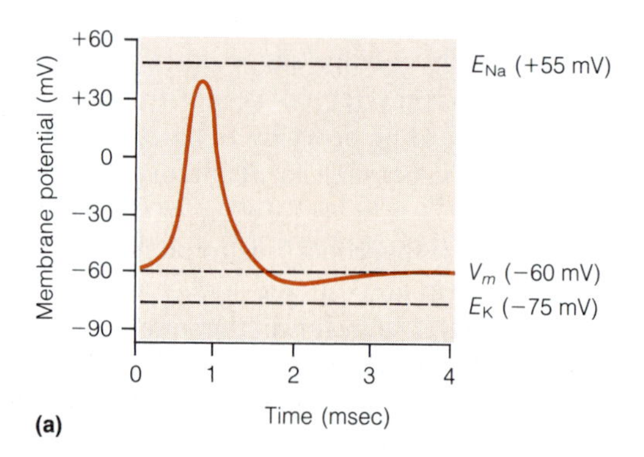

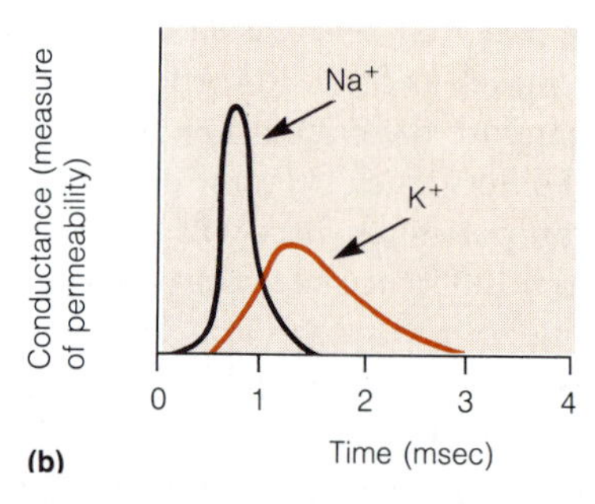

Figure 20-10 Electrical Changes in the Membrane of an Axon as a Nerve Impulse Is Transmitted Along the Membrane. (a) The change in membrane potential. Upon stimulation, the membrane becomes depolarized as its potential rises from about −60 mV (V_m, the resting potential) to about +40 mV in less than a millisecond. The membrane potential then decreases again, with the membrane becoming hyperpolarized until it stabilizes a few milliseconds later at the resting potential. (b) The change in membrane conductance (permeability of the membrane to specific ions). The depolarized membrane initially becomes very permeable to sodium ions, facilitating a large inward rush of sodium that causes the membrane potential to approach the E_{Na} value of about +55 mV. Thereafter, as permeability to sodium declines, the permeability of the membrane to potassium increases transiently, causing the membrane potential to undershoot nearly to the E_K value of about −75 mV.

strong to depolarize the membrane by at least 20 mV (the **threshold stimulus**) and thereby initiate an action potential, the resulting pattern of potential changes always has the same shape and height. It follows then, that the intensity of a sensory signal is not detected by the nervous system as a difference in the amplitude or time course of an individual action potential. Instead, the nervous system depends on the total *number* of action potentials arriving per unit time for its measure of intensity. This phenomenon, known as *summation,* will be discussed in a later section.

The Ionic Events of the Action Potential. Much of what we understand about the ionic events that take place during the generation of an action potential was worked out by Alan Hodgkin and Andrew Huxley, who received the Nobel Prize in 1963 for their efforts. In an extended series of experiments that have now become classic, Hodgkin and Huxley established that the changes in membrane potential occurring during nerve excitation are due to transient changes in the permeability of the membrane, first to sodium and then to potassium ions.

These changes in membrane permeability are shown in Figure 20-10b. The first event in the generation of an action potential is that the membrane becomes depolarized enough to render it temporarily very permeable to sodium ions. Sodium ions therefore rush inward, down the steep electrochemical gradient that normally exists for this ion across the axonal membrane. Because of this influx of sodium ions, the interior of the membrane becomes positive for an instant, as evidenced by the rapid change in membrane potential to about +40 mV. This change in membrane potential causes an increase in the permeability of the membrane to potassium ions, to an even higher level than in the resting state (Figure 20-10b).

The outward flow of potassium ions leads to a rapid reestablishment of a negative membrane potential. After a short period of depolarization, the sodium channels become *inactivated,* a condition different from the resting state. They remain in this state for a few milliseconds before they are able to respond again to depolarization. This period of sodium inactivation is known as the **absolute refractory period.** Eventually, the membrane potential becomes sufficiently negative that the permeability for potassium ions is restored to its normal value, at which point the membrane has returned to its resting potential.

To demonstrate the importance of sodium ions in the generation of an action potential, Hodgkin and Huxley removed the sodium ions from the solution in which the axon was bathed and replaced them with choline, an organic cation to which the axonal membrane is impermeable. Under these conditions, the initial inward flow of cations did not occur, and no action potential could be measured. However, the outward flow of cations was still apparent. To demonstrate that this outward flow was due to potassium ions, a radioactive isotope of potassium, $^{42}K^+$, was included in the internal solution. The $^{42}K^+$ appeared in the external medium in the expected amounts when the axon was stimulated.

Ion Fluxes During the Action Potential. The action potential clearly involves the flow of both sodium and potassium ions across the membrane. However, the actual fluxes of these cations that are necessary to account for the observed changes in membrane potential are incredibly small and make no detectable difference in the actual concentrations of these ions on either side of the membrane. For the change from −60 mV to +40 mV that characterizes a typical action potential, for example, it has been calculated that about 1 picomole (10^{-12} mol) of sodium ions must enter the cell per square centimeter of cell surface. An equal quantity of potassium ions must then leave the cell to return the membrane potential to its resting level. This turns out to involve about one-millionth of the total potassium content of a typical cell per impulse. Clearly, a nerve cell can sustain a large number of impulses without incurring significant changes in its monovalent cation content. Moreover, the continual functioning of the sodium-potassium pump ensures that such changes as do occur are countered promptly.

The Ion Channels of Nerve Cell Membranes

Once the movements of sodium and potassium ions that are responsible for the action potential had been deduced, researchers' attention focused on the means by which these ions traverse the membrane. The changes in membrane permeability that occur in response to depolarization are not superimposable in time (Figure 20-10b). For this reason, it was suggested that sodium and potassium ions pass through the membrane by means of two different sets of *channels.* The sudden inward rush of sodium ions across a depolarized axonal membrane could then be attributable to the selective opening of sodium channels, and the subsequent repolarization could be explained by the closing of those channels and the opening of channels specific for potassium.

Specificity of the Ion Channels. The movements of sodium and potassium ions that characterize an action potential must occur through relatively specific channels. What imparts such specificity? For the sodium channel, size appears to be the main criterion; molecules or ions larger than the size of a monohydrated sodium ion (about 0.3–0.5 nm^2) do not pass readily through the sodium channel. Potassium channels have a more complex specificity, since they can discriminate against both smaller ions, such as those of sodium, and larger ions, such as cesium (Cs^+) or tetraethylammonium ions. In this case, specificity appears to depend on both the diameter of the channel and specific chemical interactions of channel components with the monohydrated potassium ion.

Inhibitors of Ion Transport. Confirming evidence for the existence of separate channels for sodium and potassium ions came with the discovery of compounds that specifically block the passage of one or the other of these ions through the membrane. Two agents that inhibit the transmembrane movement of sodium ions specifically are *tetrodotoxin* and *saxitoxin*. Tetrodotoxin is an extremely toxic compound found in the liver and ovaries of the Japanese puffer fish, *Spheroides rubripes*. It is so poisonous that less than 0.01 μg is sufficient to kill a mouse. Saxitoxin is produced by marine dinoflagellates found in "red tide," seawater whose color is altered by dense populations of dinoflagellates. Saxitoxin is a hazard to human health because shellfish that feed on such dinoflagellates accumulate saxitoxin. A small mussel, for example, can accumulate enough saxitoxin to kill 50 people! Tetrodotoxin and saxitoxin are both thought to bind to the outside of the nerve cell membrane in a way that blocks the sodium channels without affecting the membrane's permeability to potassium.

The outward flow of potassium ions is specifically blocked by *tetraethylammonium ion*. The ethyl groups of this ion sterically hinder its passage through the potassium channels, causing the channels to become blocked and thereby inhibiting potassium transport. Tetraethylammonium ions bind to the membrane at different sites than either tetrodotoxin or saxitoxin, indicating the existence of separate channels for the two cations.

The high affinity and specificity of tetrodotoxin for the sodium channel made it possible to determine the densities of these channels in axonal membranes. In nonmyelinated nerve cells, the density of sodium channels is typically about 20 channels per square micrometer, so that the channels are about 200 nm apart. In striking contrast, the density of sodium channels at the nodes of Ranvier in myelinated axons is about 10,000 channels per square micrometer. These nodes are almost "wall-to-wall" sodium channels and are therefore highly specialized for the development of action potentials.

Studying a Single Channel: The Patch-Clamp Method

The study of individual membrane channels has been made possible by a technique called **patch clamping** (Figure 20-11). A glass micropipette with a tip diameter of approximately 1 μm is placed on the surface of a cell. Gentle suction is then applied, and a tight seal forms between the pipette and the cell membrane, allowing a "patch" of membrane to be removed from the cell. Current can then enter and leave the pipette only through the small number of channels (sometimes only one) in the patch of membrane sealed to the pipette tip. The device is called a "clamp" because the amplifier has a voltage clamp circuit that maintains voltage at a set value while measuring actual ionic currents through individual channels.

The ability to "visualize" the activity of a single channel in a patch of excitable membrane has provided new

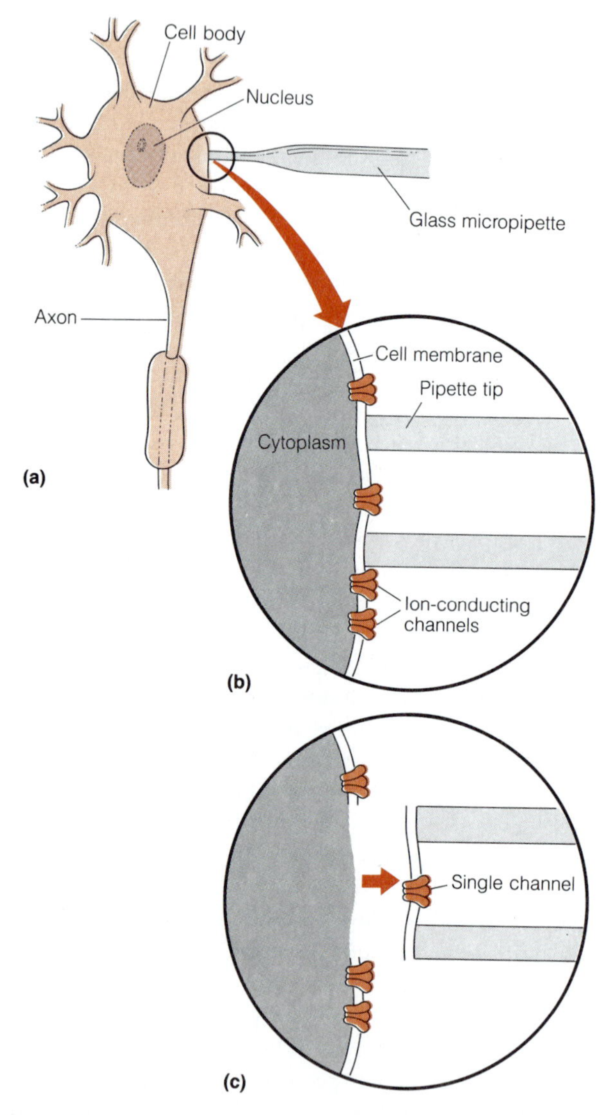

Figure 20-11 Patch Clamping. In this method, (a) a micropipette with a diameter of about 1 μm is carefully placed against a cell, such as the neuron shown here, and (b) a patch of membrane is removed by gentle suction. The patch contains any channels that happened to be in that area. In this case, a single ion-conducting channel was removed (c), and its electrical properties can now be studied.

insights into the events occurring during an action potential. For instance, we now know that the sodium channel itself is *voltage gated,* meaning that it opens more frequently for longer periods of time as the depolarizing voltage increases. When the channel is open, an electrical current of approximately 2 picoamps in the form of sodium ions flows through, equivalent to 12 million ions per second.

Mode of Action of the Sodium Channel. At present, much research with patch clamps focuses on the mechanism whereby ion channels of excitable membranes can sense changes in the membrane potential and respond to them by opening or closing at the right times to ensure propagation of an action potential. Our understanding in this area has been greatly enhanced by studies of **gating currents.** These are small currents that last only about 0.1 msec and precede the opening of the sodium channels. They reflect the rearrangement of charged groups within the sodium channel protein in response to membrane depolarization. The sodium channel can be thought of as having a built-in "voltage sensor" that undergoes a change in orientation when the membrane becomes depolarized. This change in orientation then serves as the trigger that opens the sodium channel, probably by causing a change in the conformation of the sodium channel protein.

Isolation of the sodium channel protein has been greatly aided by the availability of tetrodotoxin and saxitoxin. Because of the affinity and specificity of these compounds for the sodium channel protein, they are very useful in assaying for the presence of the protein during its purification. This has allowed the isolation of a tetrodotoxin-binding protein from the electric organ of the electric eel (*Electrophorus electricus*), a source that is rich in sodium channels. Similarly, a saxitoxin-binding protein has been isolated from rat brain. Both proteins are similar in size and complexity. The protein from rat brain, for example, consists of two polypeptide subunits with molecular weights of about 270,000 and 38,000.

The amino acid sequence of these subunits has been determined, and the order of amino acids in the larger subunit suggests the model shown in Figure 20-12. The proposed structure has four regions that readily form transmembrane α helices, with six helical segments in each (Figure 20-12a). Although it is still uncertain how these might form channels, it is not difficult to imagine that the helical regions aggregate in the membrane to produce a transmembrane channel, shown in Figure 20-12b. One configuration of the helices is relatively impermeable to sodium ions, whereas another configuration forms an open channel that allows sodium ions to pass freely. The two configurations respond to membrane potential and thereby provide a gating mechanism.

From these studies, a picture of the sodium channel protein has emerged that envisions several alternative conformational states, as shown in Figure 20-13. In the resting state, the sodium channel is *closed* (Figure 20-13a), because that is the thermodynamically favored conformation at the prevailing voltage. Upon membrane depolarization, the channel snaps to the *open* conformation (Figure 20-13b), which is the lowest-energy form at that voltage. The channel remains open only momentarily, however, before undergoing a transition to the *inactivated* state (Figure 20-13c). The inactivated state is functionally different from the closed state, since an inactivated channel is not capable of opening again immediately. Patch-clamp studies have shown that sodium channels can open and close spontaneously. Transformations from one state to another therefore occur randomly, but large changes in membrane potential can also cause conformational changes.

Potassium Channels. Several categories of potassium channels have also been identified by patch-clamp studies. These include early, delayed, and calcium-activated potassium channels. We have already described the delayed potassium channels in our discussion of the action potential. These channels respond to depolarization as the sodium channels do, but with a longer delay time. The early potassium channels open rapidly in response to depolarization and then become inactivated. The early channels

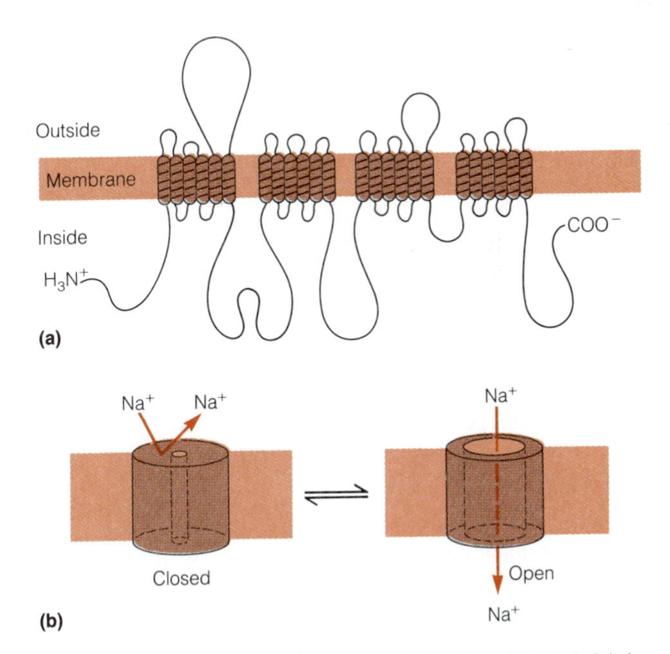

Figure 20-12 The Molecular Structure of a Sodium Channel. (a) A subunit of the sodium channel has six regions of helical structure that span the membrane. (b) It is likely that transmembrane ion-conducting channels are somehow composed of such α helices, which have the ability to assume open and closed configurations in response to membrane potential. When the channel is open, ionic currents in the picoampere range are able to cross.

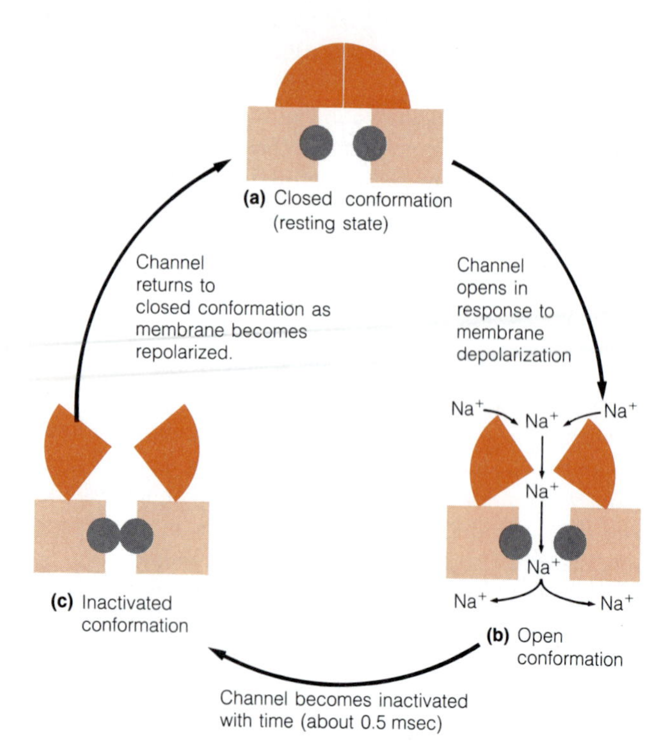

Figure 20-13 Conformational States of the Sodium Channel Protein. The sodium channel protein can exist in several alternative conformational states. (a) In the resting state, the closed conformation is favored, and sodium movement is blocked. (b) Upon depolarization of the membrane, the channel protein flips to the open conformation, allowing passage of sodium ions. (c) After about 0.5 msec, the channel undergoes a transition to the inactivated state, a conformation that cannot immediately undergo opening again. Only as the membrane becomes repolarized is the sodium channel protein restored to its closed conformation, at which point it again becomes responsive to membrane stimulation.

help make the firing rate of a cell proportional to the intensity of stimulation. Other potassium channels are activated by internal calcium and may play a role in the regulation of membrane potential and secretion.

The Propagation of Action Potentials

So far, we have looked at what happens at one point on an excitable membrane in response to stimulation. An actual impulse, however, involves a wave of depolarization-repolarization events that is propagated along the membrane of the axon. The series of events shown in Figure 20-10a occurs at one point on the membrane over a period of a few milliseconds. If a second set of recording electrodes were inserted farther along the axon, these electrodes would detect the same characteristic pattern of potential changes, but after a time delay that is proportional to the distance between the two sets of recording electrodes.

The mechanism by which a signal is propagated along the membrane of a nerve cell is illustrated in Figure 20-14. Stimulation of a resting membrane (a) at point *P* results in a depolarization at that point and a sudden inward rush of sodium ions into the axon at that location (b). Membrane polarity is temporarily reversed at that point, and this, in turn, is the trigger that initiates membrane depolarization in the region of adjacent point *Q*, allowing sodium ions to rush in there (c). By this time, the membrane at point *P* has become highly permeable to potassium ions. As potassium ions rush out of the cell at that point, the negative polarity is restored and that portion of the membrane returns to its resting status (d).

Meanwhile, the events at *Q* have stimulated the membrane in the neighboring region at *R*, initiating the same sequence of events there (e). In this way, the signal moves along the membrane as a ripple of depolarization-repolarization events, with the membrane polarity reversed in the immediate vicinity of the signal, but returned to normal again as the signal passes on down the axon. The propagation of this cycle of events along the nerve fiber is called a **nerve impulse.** The nerve impulse can move in only one direction, because the sodium channels that have just been depolarized are in the inactivated state and cannot respond immediately to further stimulation.

The Energetics of Impulse Transmission

An interesting feature of nerve impulses is that the signal is transmitted along the membrane with no decrease in its strength. This can be shown with the apparatus of Figure 20-9 by using two recording electrodes, each inserted at a different distance from the stimulating electrode. The magnitude of response detected by the two electrodes will be the same, even though the signal has had to travel farther along the membrane to reach the second electrode.

A nerve impulse can be propagated along the membrane with no reduction in amplitude because it is constantly being renewed along the way. When an electric current is transmitted along a wire, the pulse decreases in strength with distance because of resistance. But in the nerve cell, the action potential initiated at each successive point is a new event, equal in magnitude to the action potentials at preceding points. A nerve impulse, in other words, is not just conducted along the membrane passively. It is generated anew as an "all-or-nothing" event at each successive point along the membrane, using energy provided by the electrochemical ion gradients. Thus, a nerve impulse can be transmitted over any distance with no decrease in strength.

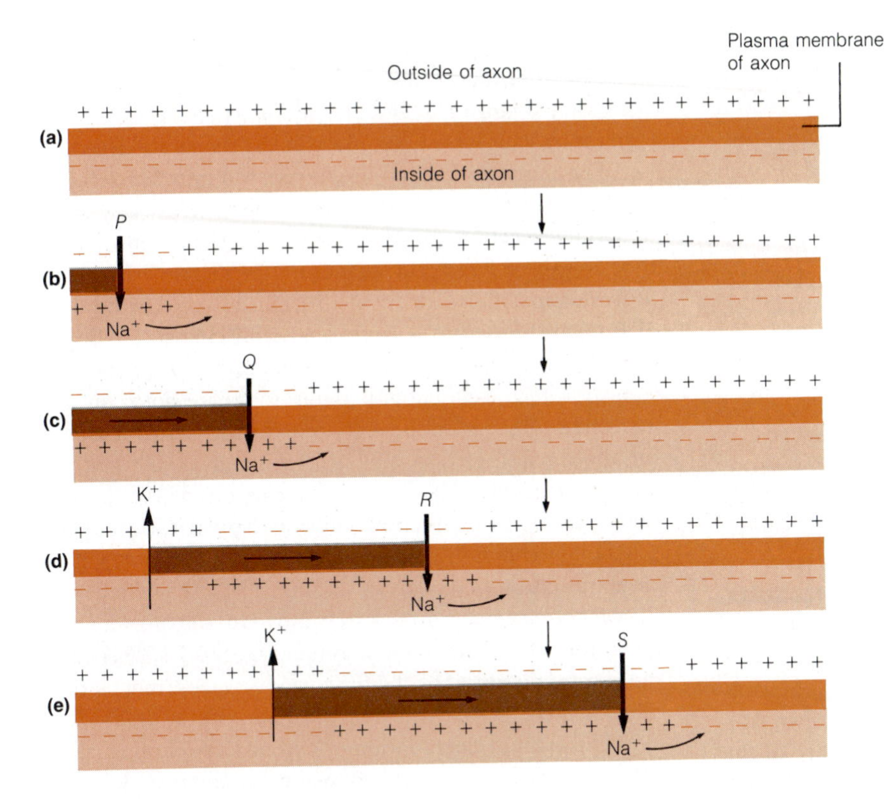

Figure 20-14 Transmission of a Nerve Impulse Along a Nonmyelinated Nerve Axon. Points *P*, *Q*, *R*, and *S* represent adjacent regions of excitability along the plasma membrane of the axon. A nerve impulse consists of the wave of depolarization-repolarization events that is propagated along the axon.

Sheaths, Nodes, and Saltatory Transmission

For an uninsulated nerve cell, transmission of an impulse along the membrane occurs from one adjacent region to the next, as shown in Figure 20-14. However, much more rapid transmission is possible along axons that are surrounded by discontinuous sheaths of myelin. The nodes of Ranvier that interrupt the myelin insulation are concentrated regions of membrane excitability, where the cytoplasmic membrane is exposed to the surrounding fluid. Because of the myelin sheath, the current can be conducted efficiently within the axoplasm between nodes without loss of signal. Action potentials therefore "jump" from node to node along such axons, rather than moving as a steady ripple along the membrane. Nerve impulses are therefore transmitted in a *saltatory* (from the Latin word for dancing) manner along myelinated axons, as illustrated in Figure 20-15. Saltatory propagation is much more rapid than the continuous propagation that occurs in nonmyelinated axons.

Synaptic Transmission

Nerve cells interact with one another and with glands and muscles at junctions called **synapses.** In some cases, the signal is passed from one cell to the next through a gap junction that allows direct cell-cell contact. Such **electrical synapses** provide for transmission with no delay and tend to occur in places in the nervous system where speed of transmission is of essence. Most synapses, however, are not electrical but chemical. **Chemical synapses** depend on specific chemicals called **neurotransmitters** to conduct the signal across the junction. The neurotransmitter binds to a receptor on the membrane of the second cell, and typically the binding produces a change in ionic conductance of the receptor, as we will see. However, neurotransmitters can also work through *second messenger* systems, which will be described in Chapter 21. In this case, the neurotransmitter binds to a receptor, but the signal is transmitted to the cytoplasm of the second cell, rather than to its membrane.

The Chemical Synapse: Structure and Function

The structure of a typical chemical synapse is shown in Figure 20-16. At the synapse, there is a gap of about 20–50 nm between the membranes of the two cells. This gap is called the **synaptic cleft.** A nerve signal arriving at the **presynaptic membrane** cannot bridge the synaptic cleft as an electrical impulse, but is transmitted across to the **postsynaptic membrane** by means of a neurotransmitter, the best-known of which is a chemical called **acetylcholine.** In vertebrates, acetylcholine is the most common neurotransmitter for synapses between neurons outside the central nervous system, as well as for neuromuscular junctions. Figure 20-17 depicts the structure of acetylcholine, as well as its synthesis from choline and acetyl coenzyme A and its hydrolysis into choline and acetate.

Transmission of a Signal Across a Chemical Synapse

The transmission of a signal across a chemical synapse is shown in Figure 20-16b and in greater detail in Figure 20-18. The particular example shown in Figure 20-18 is a **cholinergic synapse,** that is, a synapse for which acetylcholine is the neurotransmitter. Clustered near the terminus of the presynaptic axon are numerous **synaptic vesicles,** each containing several thousand molecules of acetylcholine. The release of acetylcholine depends critically on calcium ions (Ca^{2+}), which are present in higher concentrations outside the axon than inside. The arrival of a nerve impulse causes a large but transient increase in the permeability of the presynaptic membrane to calcium ions, so that calcium ions flow down their electrochemical gradient into the axoplasm (Figure 20-18a).

Once in the cell, calcium ions cause synaptic vesicles to move to the presynaptic membrane and fuse with it, in what is essentially an exocytotic process. The acetylcholine from the synaptic vesicles is thereby released into the synaptic cleft (Figure 20-18b). Incredible as it may seem, the entire release process takes place in less than a millisecond!

The acetylcholine molecules released into the cleft diffuse across to the postsynaptic membrane, where they bind to specific receptor proteins (Figure 20-18c). Neuroreceptors are always integral membrane proteins with a high degree of specificity for the neurotransmitter. The binding of transmitter causes a conformational change in the receptor that permits increased ionic conductance. For instance, when acetylcholine binds to its receptor in the neuromuscular junction, a channel in the receptor opens for about a millisecond, and approximately 30,000 sodium ions pass through. This occurs simultaneously in thousands of receptors at the junction, causing depolarization of the postsynaptic membrane. The depolarization initiates muscle contraction, as we saw in Chapter 19. An enzyme called *acetylcholinesterase* immediately begins to degrade the acetylcholine into acetate and choline (Figure 20-17b) so that the resting potential of the postsynaptic membrane is rapidly restored.

The prompt degradation of excess acetylcholine is absolutely essential. If it were not destroyed, the acetylcholine would continue its stimulatory effect until the acetylcholine diffused away, and control of nervous activity would quickly be lost. Some organic phosphate compounds are particularly potent inhibitors of acetylcholinesterase and

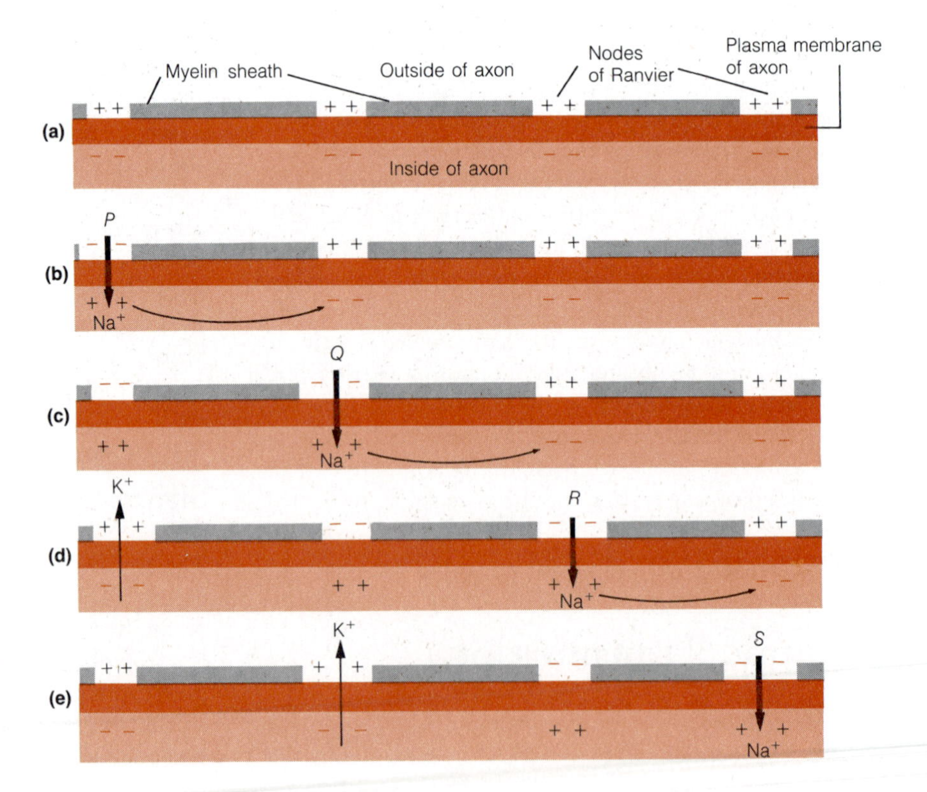

Figure 20-15 Transmission of a Nerve Impulse Along a Myelinated Nerve Axon. Points *P*, *Q*, *R*, and *S* represent successive nodes along the axon. A nerve impulse consists of a wave of depolarization-repolarization events that is propagated along the axon from node to node.

are therefore especially potent **neurotoxins.** Such compounds have been synthesized for use as insecticides in agricultural practice and as nerve gases for chemical warfare. Their mode of action is considered in the box on pp. 638–639, along with a number of other toxins that interfere with neuromuscular transmission.

Mode of Action of Acetylcholine

An important question that we have not yet considered is how the binding of a neurotransmitter such as acetylcholine to its receptor causes a depolarization of the postsynaptic membrane and therefore a propagation of the signal. Bernard Katz and his collaborators were the first to make the important observation that acetylcholine increases the permeability of the postsynaptic membrane to both sodium and potassium ions within 0.1 msec of binding to its receptor. This triggers an inward rush of sodium ions (Figure 20-18d) and a simultaneous (though lesser) outward movement of potassium ions. The inward current of sodium ions depolarizes the membrane, thereby creating an **excitatory postsynaptic potential** (**EPSP**), not unlike the experimental stimulation of an axon by the transient flow of current through the stimulating electrode of Figure 20-9. Provided only that the EPSP exceeds a threshold level, an action potential may be triggered that can be transmitted as a nerve signal along the membrane of the recipient cell toward the next synapse (Figure 20-18e). Figure 20-19 depicts the change in potential of the postsynaptic membrane upon binding of the neurotransmitter to its receptor.

How the neurotransmitter effects a change in membrane permeability is under intensive investigation. For cholinergic receptors, two molecules of acetylcholine bind to the receptor, causing a conformational change that results in the opening of a cation channel. Whether the resulting EPSP will be great enough to trigger an action potential depends entirely on how many such channels are opened, and that, in turn, depends on how many receptors bind the neurotransmitter. The number of occupied receptors, in other words, determines the magnitude of the inward flow of sodium ions and hence the magnitude of change in the membrane potential.

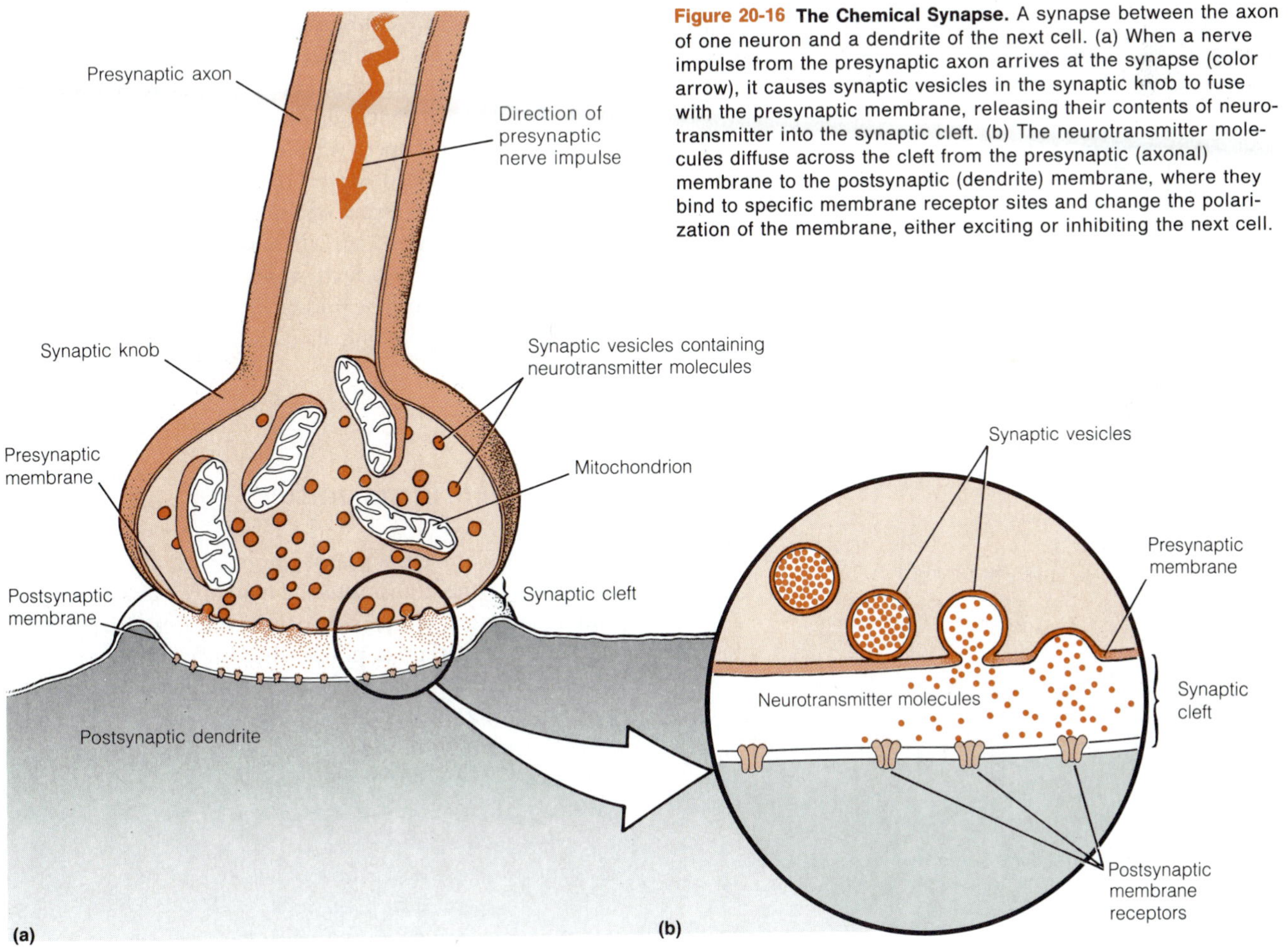

Figure 20-16 The Chemical Synapse. A synapse between the axon of one neuron and a dendrite of the next cell. (a) When a nerve impulse from the presynaptic axon arrives at the synapse (color arrow), it causes synaptic vesicles in the synaptic knob to fuse with the presynaptic membrane, releasing their contents of neurotransmitter into the synaptic cleft. (b) The neurotransmitter molecules diffuse across the cleft from the presynaptic (axonal) membrane to the postsynaptic (dendrite) membrane, where they bind to specific membrane receptor sites and change the polarization of the membrane, either exciting or inhibiting the next cell.

The Acetylcholine Receptor

Our understanding of synaptic transmission has been greatly aided by the ease with which membranes rich in **acetylcholine receptors** can be isolated from the electric organs of the electric ray (*Torpedo californica*), an organism that is also useful as a source of sodium channel protein. The electric organ consists of *electroplaxes*—stacks of cells that are innervated on one side but not on the other. The innervated side of the stack can undergo a potential change from about −90 mV to about +60 mV upon excitation, whereas the noninnervated side stays at −90 mV. A potential difference of about 150 mV can therefore be built up across a single electroplax at the peak of an action potential. Because the electric organ contains thousands of electroplaxes arranged in series, their voltages are additive, allowing the organism to deliver a jolt of several hundred volts.

When electroplax membranes are examined under the electron microscope, they are found to be rich in rosettelike particles about 8 nm in diameter (Figure 20-20). Each such particle consists of five subunits arranged around a central axis, which is assumed to be the ion channel. Their size and reaction with antibodies indicate that these structures are the acetylcholine receptors.

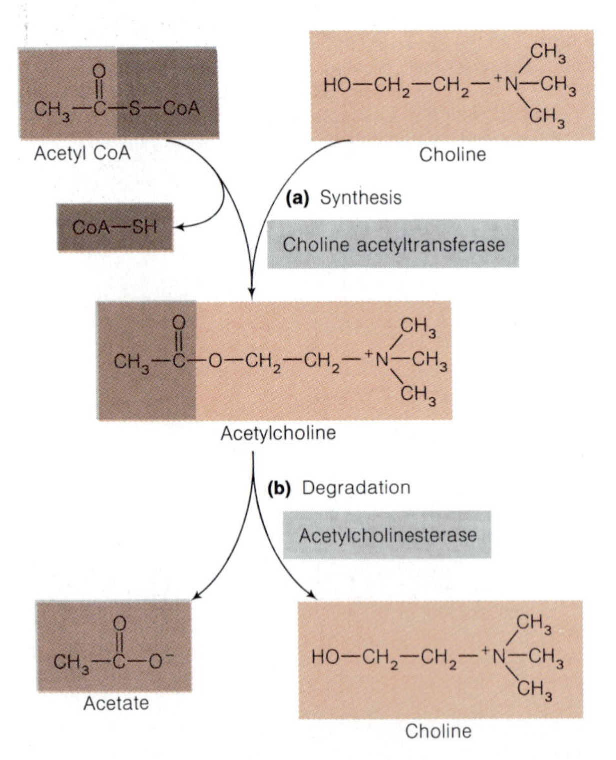

Figure 20-17 Structure, Synthesis, and Hydrolysis of Acetylcholine. Acetylcholine is (a) synthesized from acetyl CoA and choline by choline acetyltransferase, and (b) degraded into acetate and choline by acetylcholinesterase.

The acetylcholine receptor from the electric ray can be purified by solubilizing electroplax membranes with nonionic detergents, followed by several chromatographic procedures. Purification of the acetylcholine receptor was greatly aided by the availability of several neurotoxins from snake venom, including *α-bungarotoxin* and *cobratoxin.* These toxins serve as a highly specific means of locating and quantifying acetylcholine receptors, since they can be made highly radioactive and they bind to the receptor protein very tightly and specifically. The radioactive toxin can therefore be used as an assay for the acetylcholine receptor after each step in the purification procedure.

The purified receptor has a molecular weight of about 300,000 and consists of four kinds of subunits—α, β, γ, and δ—each containing about 500 amino acids. The transmembrane portion of the subunits includes sequences of relatively hydrophobic amino acids, which probably form α helices grouped together in the plane of the bilayer. The intact receptor contains the subunits in the ratio 2:1:1:1, so the simplest empirical formula for the receptor protein is $\alpha_2\beta\gamma\delta$.

Figure 20-21 shows one postulated model of the acetylcholine receptor that takes into account all of the above information. The five subunits aggregate in the membrane to produce a transmembrane channel, which is normally in a closed configuration. The α subunits have binding sites, and when acetylcholine binds to the sites the transmembrane channel opens transiently, allowing sodium ions to cross. As described earlier, the sodium influx results in a transient depolarization of the postsynaptic membrane. Within a millisecond, the channel undergoes a process called *desensitization,* which causes it to close, even though acetylcholine may still be present. The acetylcholine is then released from the binding sites and hydrolyzed by acetylcholinesterase.

Other Neurotransmitters

Compounds other than acetylcholine are also known to function as neurotransmitters. To qualify as a neurotransmitter, a compound must satisfy several criteria: It must elicit the appropriate response when microinjected into the synaptic cleft; it must be found to occur naturally in the presynaptic axon; and it must be released at the right time when the presynaptic membrane is stimulated. Compounds that satisfy these criteria include a number of **catecholamines,** certain amino acids and their derivatives, and several peptides.

Catecholamines are derivatives of the amino acid tyrosine and include *dopamine,* as well as the hormones *norepinephrine* and *epinephrine.* The structures of these

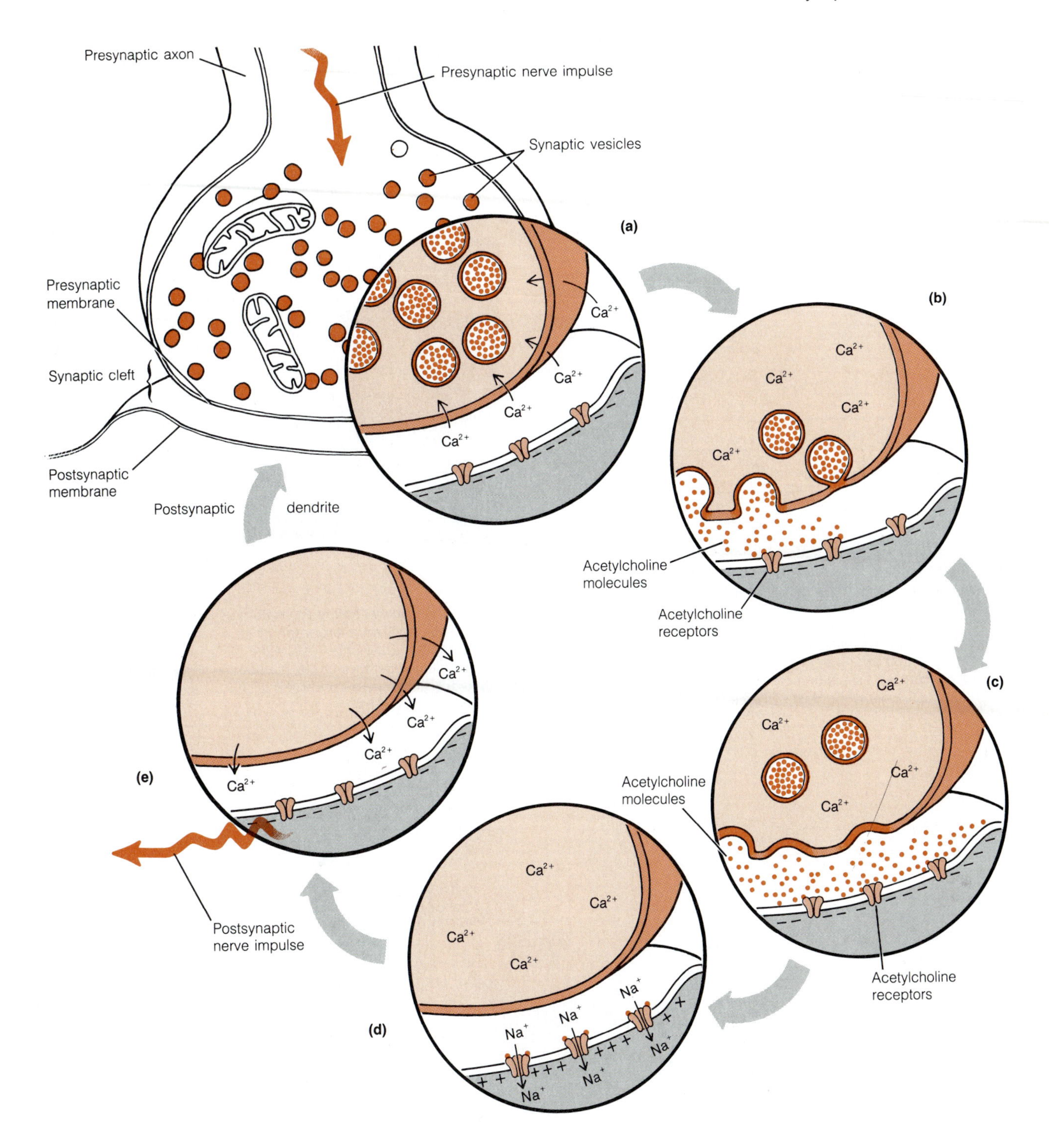

Figure 20-18 Transmission of a Signal Across a Cholinergic Synapse. (a) Arrival of a nerve impulse at the end of an axon causes a change in the permeability of the presynaptic membrane to calcium ions, such that calcium ions diffuse inward. (b) Once in the axon, calcium ions cause synaptic vesicles to move toward the presynaptic membrane, fuse with it, and empty their contents of acetylcholine into the synaptic cleft. (c) The acetylcholine molecules diffuse across to the postsynaptic membrane, where they bind to receptor sites. (d) Binding of acetylcholine causes ion channels in the postsynaptic membrane to open, allowing an inward rush of sodium. (e) The inward flow of sodium ions depolarizes the membrane, causing an excitatory postsynaptic potential that can trigger an action potential in the postsynaptic membrane. Meanwhile, the acetylcholine is degraded, the calcium is pumped back out of the presynaptic axon, and further synaptic vesicles form, leaving the synapse ready for another impulse.

catecholamines are shown in Figure 20-22a. Because these hormones are also synthesized in the adrenal gland, synapses that use them as neurotransmitters are termed **adrenergic synapses.** Adrenergic synapses are found at the junctions between nerves and smooth muscles in internal organs such as intestine, as well as at nerve-nerve junctions in the brain.

Other amino acid derivatives that function as neurotransmitters include *histamine, serotonin, glycine,* and *γ-aminobutyric acid,* or *GABA.* Their structures are shown in Figure 20-22b. Moreover, peptides such as somatostatin and the enkephalins also serve as neurotransmitters.

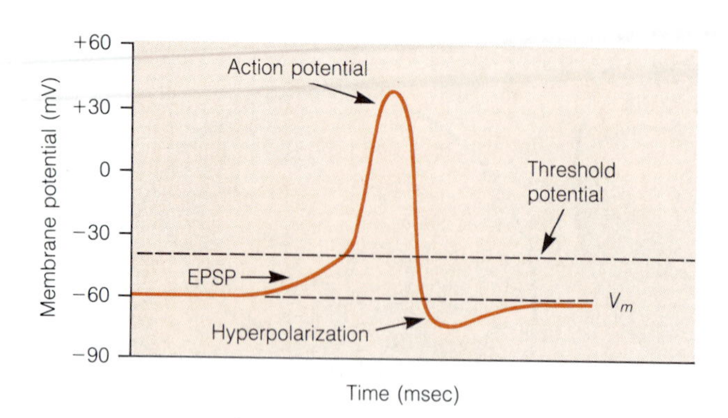

Figure 20-19 Initiation of an Action Potential in the Postsynaptic Membrane of a Chemical Synapse. An action potential is initiated in the postsynaptic membrane upon binding of an excitatory neurotransmitter to its receptor. The transmitter triggers an inward rush of sodium ions that depolarizes the membrane slightly, creating an excitatory postsynaptic potential (EPSP). If this exceeds the threshold potential of the membrane, an action potential may be triggered that is then propagated along the postsynaptic cell toward the next synapse.

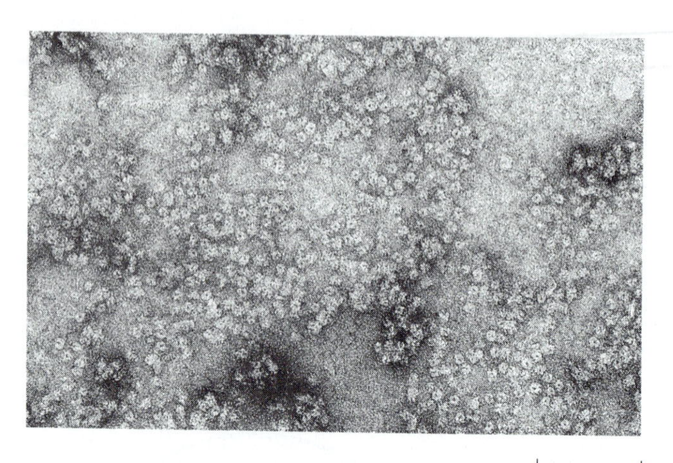

Figure 20-20 Acetylcholine Receptors of an Electroplax Membrane. This electron micrograph of an electroplax postsynaptic membrane illustrates the rosettelike particles thought to be the acetylcholine receptors of the membrane. Each particle consists of five subunits clustered around a central axis that is probably the ion channel. (TEM)

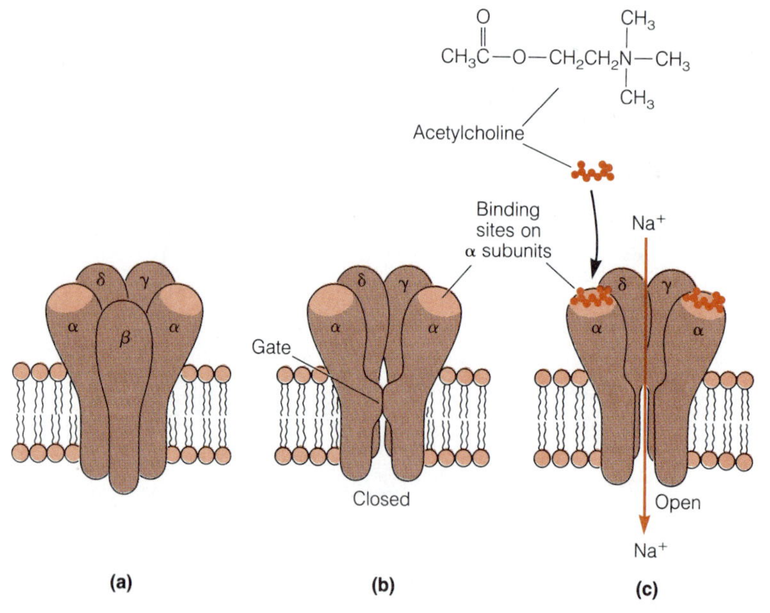

Figure 20-21 Molecular Structure of the Acetylcholine Receptor. (a) This receptor contains five subunits, including two α subunits with binding sites for acetylcholine, and one each of β, γ and δ. (b) The subunits aggregate in the lipid bilayer in such a way that the transmembrane portions form a channel, shown here with the β subunit removed. (c) The channel is normally closed, but when acetylcholine binds to the two sites on the α subunits, the subunits are altered in such a way that the channel opens to allow sodium ions across.

Most tissues in the human body are innervated by several types of nerve cells, each using a different neurotransmitter. This allows a greater variety of signals and responses. It seems likely that further neurotransmitters remain to be discovered, considering the complexity of the nervous systems of higher animals and the chemical diversity of compounds already known to play such a role.

Inhibitory Neurotransmitters

Some neurotransmitters act not to excite the postsynaptic cell, but to inhibit it. In fact, *inhibition* may play as important a role in nervous system function as excitation. The mechanism of inhibition involves a permeability change to potassium or chloride ion instead of to sodium ion as in excitation. As a result, the membrane potential is stabilized near the equilibrium potential for these ions, which is near the resting potential. The change in membrane potential caused by the synaptic permeability change (usually hyperpolarization) is called an **inhibitory postsynaptic potential** (**IPSP**). This IPSP reduces the amplitude of an EPSP, thereby preventing the firing of an action potential. GABA and glycine are two important inhibitory transmitters: about 40% of brain receptors are GABAergic, and glycine serves a similar function in the spinal cord.

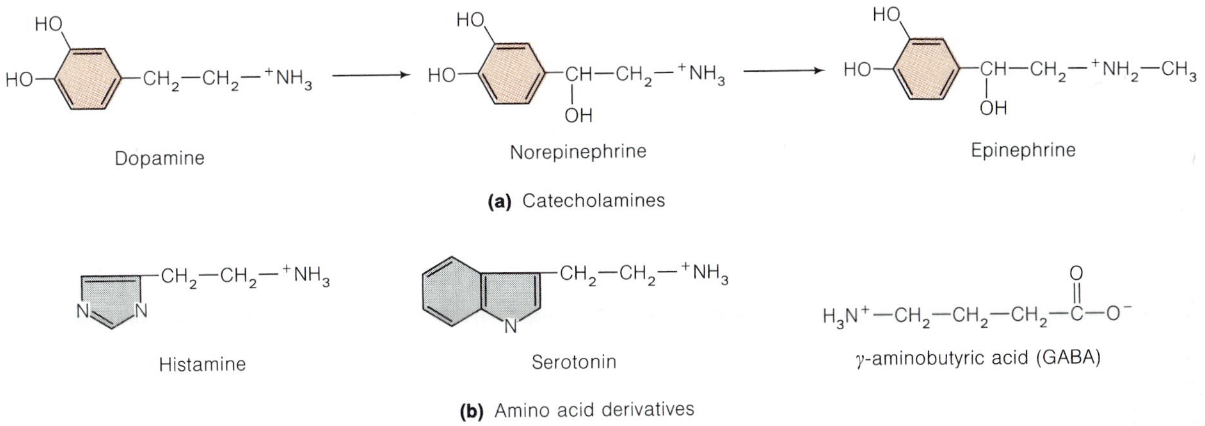

HO, HO— $CH_2—CH_2—^+NH_3$ → HO, HO— $CH—CH_2—^+NH_3$, OH → HO, HO— $CH—CH_2—^+NH_2—CH_3$, OH

Dopamine | Norepinephrine | Epinephrine

(a) Catecholamines

$CH_2—CH_2—^+NH_3$ (N, N) | $CH_2—CH_2—^+NH_3$ (N) | $H_3N^+—CH_2—CH_2—CH_2—C(=O)—O^-$

Histamine | Serotonin | γ-aminobutyric acid (GABA)

(b) Amino acid derivatives

Figure 20-22 Structures of Several Neurotransmitters. (a) Dopamine, norepinephrine, and epinephrine are catecholamines. They are synthesized from the amino acid tyrosine and are inactivated by the enzyme monoamine oxidase. Dopamine can be converted into norepinephrine, and norepinephrine into epinephrine, as indicated by the arrows. (b) Histamine, serotonin, and γ-aminobutyric acid (GABA) are other amino acid derivatives. The amino acids glycine and glutamine (not shown) are also neurotransmitters.

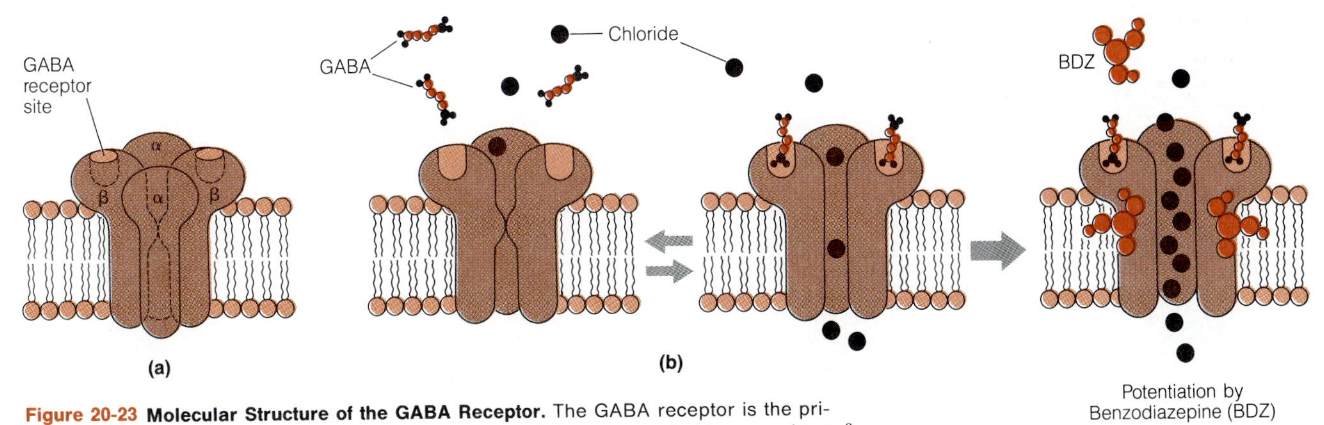

Figure 20-23 Molecular Structure of the GABA Receptor. The GABA receptor is the primary inhibitory receptor of the central nervous system. It is composed of two α and two β subunits (a) and each β subunit has binding sites for GABA. When GABA is present (b), the channel assumes a configuration that permits chloride ion to enter the cell down its electrochemical gradient. This hyperpolarizes the cell membrane, thereby reducing excitability and inhibiting nervous activity. Several pharmacologically active agents act at the GABA receptor. For instance, when benzodiazepines (BDZ) bind to the receptor (c), the effect of GABA is enhanced and the overall level of excitability is reduced. Presumably, this produces the tranquilizing effect of benzodiazepines such as Valium and Librium.

Poisoned Arrows, Snakebites, and Nerve Gases

Because the coherent functioning of the human body depends so critically on its nervous system, anything that disrupts transmission of nerve impulses is likely to be very toxic. And because of the importance of acetylcholine as a neurotransmitter, any substance that interferes with its function is almost certain to be lethal. Various substances are known that disrupt nerve and muscle function by specific effects on cholinergic synapses. We will consider several of these *neurotoxins*, not only to underscore the serious threat they pose to human health, but also to illustrate how clearly their modes of action can be explained once the physiology of synaptic transmission is understood. We will also see how useful such compounds can be as research tools in studying the very phenomenon they disrupt so effectively.

Once acetylcholine has been released into the synaptic cleft and depolarization of the postsynaptic membrane has occurred, excess acetylcholine must be rapidly hydrolyzed. If it is not, the membrane cannot be restored to its polarized state, and further transmission will not be possible. The enzyme acetylcholinesterase is therefore essential, and substances that inhibit its activity are usually very toxic.

One such family of acetylcholinesterase inhibitors consists of *carbamoyl esters*. These compounds inhibit acetylcholinesterase by covalently blocking the active site of the enzyme, effectively preventing the breakdown of acetylcholine. An example of such an inhibitor is *physostigmine* (sometimes also called *eserine*), a naturally occurring alkaloid produced by the Calabar bean. Once used as a poison in

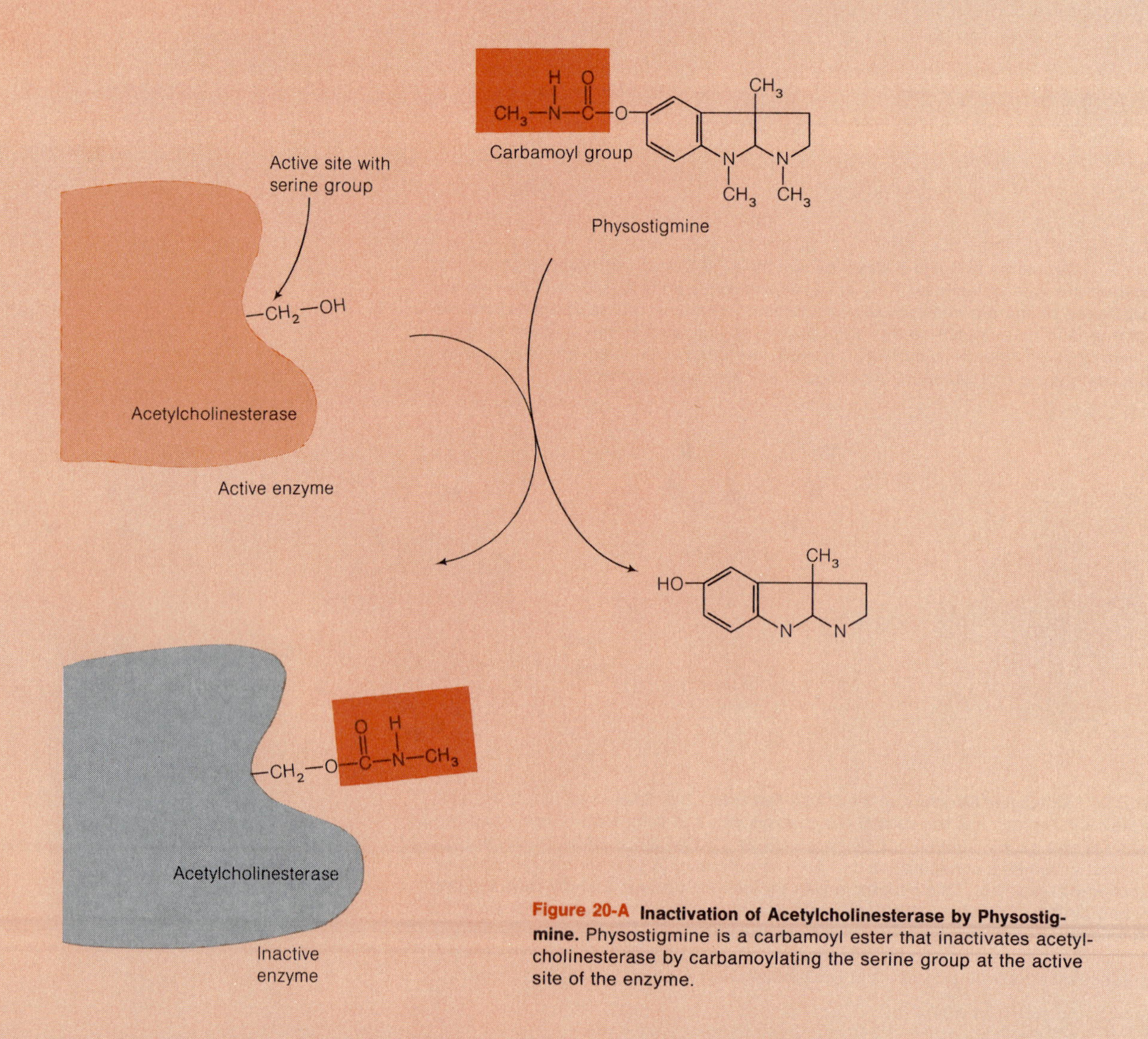

Figure 20-A Inactivation of Acetylcholinesterase by Physostigmine. Physostigmine is a carbamoyl ester that inactivates acetylcholinesterase by carbamoylating the serine group at the active site of the enzyme.

witchcraft trials, physostigmine now finds use as an acetylcholinesterase inhibitor in studies of cholinergic transmission. Figure 20-A shows the structure of physostigmine and illustrates how it inhibits the enzyme by forming a stable carbamoyl-enzyme complex at the active site.

Many synthetic organic phosphates form even more stable covalent complexes with the active site of acetylcholinesterase and are therefore still more potent inhibitors. Included in this class of compounds are the widely used insecticides *parathion* and *malathion*, as well as nerve gases such as *tabun* and *sarin*. The structures of several such poisons are shown in Figure 20-B. The primary effect of these compounds is muscle paralysis, caused by an inability of the postsynaptic membrane to regain its polarized state.

Nerve transmission at cholinergic synapses can be blocked not only by inhibitors of acetylcholinesterase, but also by substances that compete with acetylcholine for binding to its receptor on the postsynaptic membrane. A particularly notorious example of such a poison is *curare*, a plant extract used by South American Indians to poison arrows. One of the active factors in curare is *d-tubocurarine* (Figure 20-C). Snake venoms act in the same way. Both α-bungarotoxin (from snakes of the genus *Bungarus*) and *cobratoxin* (from cobra snakes) are small, basic proteins that bind noncovalently to the acetylcholine receptor, thereby blocking depolarization of the postsynaptic membrane.

Neurotoxins that function in this way are referred to as *antagonists* of cholinergic systems. Other compounds, called

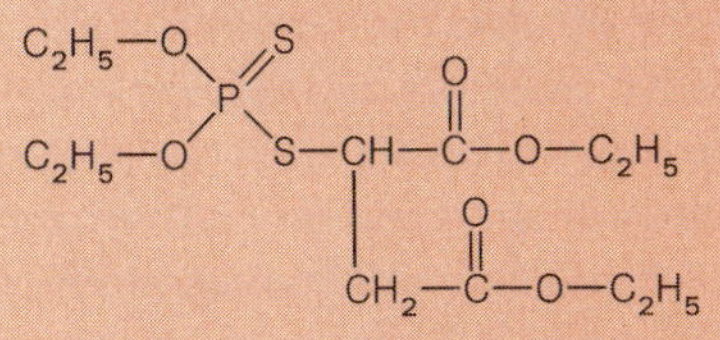

Malathion

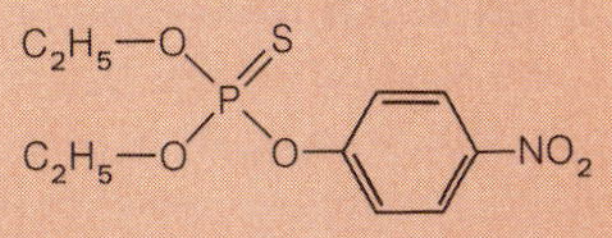

Parathion

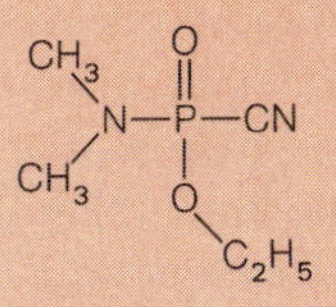

Tabun

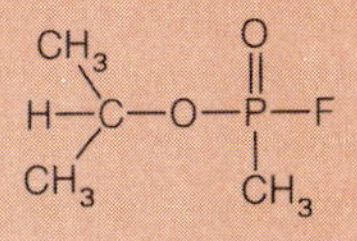

Sarin

Figure 20-B Structures of Several Organophosphate Inhibitors of Acetylcholinesterase.

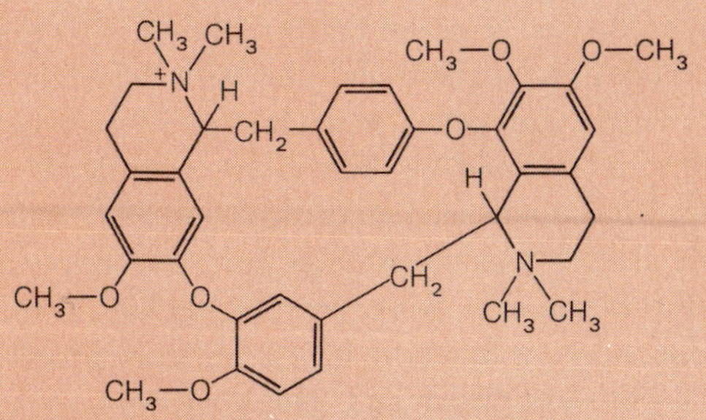

Figure 20-C Structure of d-Tubocurarine.

agonists, have just the opposite effect. Agonists also bind to the acetylcholine receptor, but in so doing they mimic acetylcholine, causing depolarization of the postsynaptic membrane. Unlike acetylcholine, however, they cannot be rapidly inactivated, so the membrane does not regain its polarized state.

As these effects imply, agonists effectively lock the acetylcholine receptor in its "open" state, whereas antagonists essentially lock it in its "closed" state. These substances have therefore proved inordinately valuable in studying the receptor and especially the effects of its open and closed states on membrane permeability. In addition, several of these toxins have been useful in purification of the receptor protein because of their great specificity for this single protein.

An analogue of acetylcholine called *succinylcholine* (Figure 20-D) is an agonist that is medically useful as a muscle relaxant. Compared with acetylcholine, succinylcholine is hydrolyzed more slowly in vivo, resulting in a persistent depolarization of the postsynaptic membrane. The muscle relaxation that this depolarization produces at neuromuscular synapses is especially useful in surgical procedures.

Though of disparate origins and uses, poisoned arrows, snake venoms, nerve gases, and surgical muscle relaxants all turn out to have some features in common. Each interferes in some way with the normal functioning of acetylcholine, and each is therefore a neurotoxin, since it disrupts the transmission of nerve impulses, usually with lethal consequences. And each has turned out to be useful as an investigative tool, illustrating once again the strange but powerful arsenal of exotic tools upon which biologists and biochemists are able to draw in their continued probings into the intricacies of cellular function.

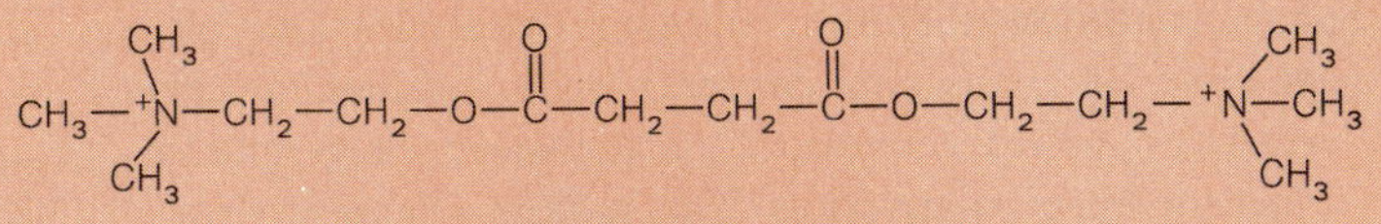

Figure 20-D Structure of Succinylcholine.

Why do you suppose that inhibition is so important for nervous function? A useful analogy is to imagine driving a car with a gas pedal only: In the absence of brakes (inhibition) it would be difficult to control the car's response. You could speed up or slow down but would have trouble making fine adjustments (not to mention stopping entirely). In the same way, most physiological functions, including nervous function, have regulatory processes that include both excitatory and inhibitory controls to "fine-tune" the response.

A clear example of the importance of inhibition is the effect of *strychnine*, a plant toxin, on motor control. Strychnine blocks the glycine receptors of the spinal cord, so that their normal inhibitory function is lost. As a result, excitatory motor neurons take over, producing uncontrolled convulsions that often lead to death. Another interesting example of an inhibitory response is the effect of *benzodiazepines* on the brain. The benzodiazepines are a family of pharmacologically active compounds that include such familiar substances as Valium and Librium. GABA receptors have highly specific binding sites for benzodiazepines, which produce an enhancement of the hyperpolarizing chloride ion flux that inhibits excitability of cells with GABA receptors (Figure 20-23; see p. 637).

Perspective

All cells maintain an electrical potential across their membranes, but neurons have specialized to use membrane potentials as a means for transmitting signals from one part of an organism to another. For this function they possess slender processes (dendrites and axons) that either receive transmitted impulses or conduct them to the next cell. The membrane of an axon may or may not be encased in a myelin sheath. The giant axons of invertebrates such as the squid have been especially useful in electrophysiological studies of nerve cell function.

The resting potential of a membrane is in effect the algebraic sum of the equilibrium potentials for sodium, potassium, and chloride ions, each weighted for the relative permeability of the unstimulated membrane for that ion. The resting potential for the plasma membrane of most animal cells is in the range −40 to −90 mV. These values are near the equilibrium potential for potassium ion (usually about −75 mV) but very far from that for sodium ion (about +55 mV), reflecting the greater permeability of the resting membrane for potassium.

The action potential of a neuron represents a transient depolarization and repolarization of its membrane, requiring only a few milliseconds. An action potential is initiated by a stimulus, from either a sensory cell in an organism or an electrical voltage in the laboratory. Upon stimulation, the membrane becomes locally depolarized. Voltage-sensitive sodium channels in the membrane open, allowing an influx of sodium ions. The current that results from the sodium ion flow triggers a depolarization in adjacent regions of the membrane, and the depolarization is propagated along the axon. Following depolarization, voltage-sensitive potassium channels open, causing an increased outward movement of potassium ions and a transient hyperpolarization. The sodium channels close, and the membrane returns to its original resting state.

Upon reaching a synapse, or junction, between a nerve cell and a second cell with which it communicates, the electrical impulse increases the permeability of the membrane to calcium. As calcium ions cross the presynaptic membrane, they cause synaptic vesicles to fuse with the membrane. The synaptic vesicles contain neurotransmitter molecules, which are released into the synaptic cleft by the fusion event. Neurotransmitter molecules migrate across the cleft to the postsynaptic membrane, where they bind to specific receptors.

The best-understood receptor is the acetylcholine receptor of the neuromuscular junction, in which binding of acetylcholine causes a sodium channel to open. The resulting sodium influx produces a local depolarization of the postsynaptic membrane, which in turn can initiate an action potential in the postsynaptic cell. Following depolarization, the enzyme acetylcholinesterase hydrolyzes the acetylcholine to return the synapse to its resting state.

Thirty or more specific receptors and neurotransmitters have been identified in the central nervous system. These produce both excitatory and inhibitory postsynaptic potentials, so that a nerve cell body must integrate thousands of incoming action potentials before becoming sufficiently depolarized to initiate a new action potential of its own. An example of an inhibitory receptor is the GABA (γ-aminobutyric acid) receptor, which represents a gated channel for chloride ion. Upon binding GABA, this receptor allows increased chloride flux, leading to hyperpolarization of the postsynaptic membrane and reduced excitability.

The action of pharmaceutically useful drugs can be understood in terms of synaptic function. Valium, for example, is a tranquilizer that binds to GABA receptors and enhances their inhibitory effect on the central nervous system. A variety of neurotoxins also act at the synapse. For instance, certain organophosphate compounds inhibit acetylcholinesterase, producing a complete disruption of normal synaptic transmission. Other toxins like curare bind to the acetylcholine receptor itself, causing paralysis by inhibiting its ability to interact with acetylcholine.

Key Terms for Self-Testing

The Nervous System
nervous system (p. 616)
central nervous system (CNS) (p. 616)
peripheral nervous system (PNS) (p. 616)
somatic nervous system (p. 616)
autonomic nervous system (p. 616)
neuron (p. 617)
sensory neurons (p. 618)
motor neurons (p. 618)
interneurons (p. 618)
cell body (p. 619)
processes (p. 619)
dendrites (p. 619)
axons (p. 619)
axoplasm (p. 619)
nerve (p. 619)
synaptic knobs (p. 619)
nerve terminals (p. 619)
synapse (p. 619)
neurotransmitters (p. 619)
myelin sheath (p. 619)
oligodendrocytes (p. 619)
Schwann cells (p. 619)
nodes of Ranvier (p. 620)

The Resting Potential of Nerve Cells
resting potential (p. 620)
excitability (p. 620)
action potential (p. 620)
electrochemical gradient (p. 621)
polarization (p. 621)
equilibrium potential (p. 622)
Nernst equation (p. 622)
Goldman-Hodgkin-Katz equation (p. 625)

The Generation of Action Potentials
hyperpolarization (p. 626)
refractory period (p. 626)
threshold stimulus (p. 627)
absolute refractory period (p. 627)
patch clamping (p. 628)
gating currents (p. 629)

The Propagation of Action Potentials
nerve impulse (p. 630)

Synaptic Transmission
synapses (p. 631)
electrical synapses (p. 631)
chemical synapses (p. 631)
neurotransmitters (p. 631)
synaptic cleft (p. 631)
presynaptic membrane (p. 631)
postsynaptic membrane (p. 631)
acetylcholine (p. 632)
cholinergic synapse (p. 632)
synaptic vesicles (p. 632)
neurotoxins (p. 633)
excitatory postsynaptic potential (EPSP) (p. 633)
acetylcholine receptors (p. 634)
catecholamines (p. 634)
adrenergic synapses (p. 636)
inhibitory postsynaptic potential (IPSP) (p. 637)

Suggested Reading

General References and Reviews

Kandel, E. R., and J. H. Schwartz. *Principles of Neural Science*, 2d ed. New York: Elsevier, 1985.

Katz, B. *Nerve, Muscle and Synapse*. New York: McGraw-Hill, 1966.

Kuffler, S. W., J. G. Nicholls, and A. R. Martin. *From Neuron to Brain*, 2d ed. New York: Sinauer Press, 1984.

Neurons

Morell, P., and W. T. Norton. Myelin. *Sci. Amer.* 242 (May 1980): 88.

Stevens, C. F. The neuron. *Sci. Amer.* 240 (March 1979): 55.

Ion Channels and Membrane Excitation

Armstrong, C. M. Sodium channels and gating currents. *Physiol. Rev.* 61 (1981): 644.

Baker, P. F., A. L. Hodgkin, and T. Shaw. The effects of changes in internal ionic concentrations on the electrical properties of perfused giant axons. *J. Physiol.* 164 (1962): 355–374.

Catterall, W. A. Structure and function of voltage-sensitive ion channels. *Science* 242 (1988): 50–61.

Hodgkin, A. L., and A. F. Huxley. A quantitative description of membrane current and its application to conduction and excitation in nerve. *J. Physiol.* 117 (1952): 500.

Hille, B. *Ionic Channels of Excitable Membranes*. Sunderland, Mass.: Sinauer, 1984.

Keynes, R. D. Ion channels in the nerve-cell membrane. *Sci. Amer.* 240 (March 1979): 126.

Synaptic Transmission

Axelrod, J. Neurotransmitters. *Sci. Amer.* 230 (June 1974): 59.

Bloom, F. E. Neuropeptides. *Sci. Amer.* 245 (October 1981): 148.

Guy, H. R., and F. Hucho. The ion channel of the nicotinic acetylcholine receptor. *Trends Neurosci.* 10 (1987): 318.

Kelly, R. B., J. W. Deutsch, S. S. Carlson, and J. A. Wagner. Biochemistry of neurotransmitter release. *Annu. Rev. Neurosci.* 2 (1979): 399.

Kistler, J., R. M. Stroud, M. W. Klymkowsky, R. A. Lalancette, and R. H. Fairclough. Structure and function of an acetylcholine receptor. *Biophys. J.* 37 (1982): 371.

Lester, H. A. The response to acetylcholine. *Sci. Amer.* 236 (February 1977): 106.

Problem Set

1. **The Truth About Nerve Cells.** For each of the following statements, indicate whether it is true of all nerve cells (A), of some nerve cells (S), or of no nerve cells (N).
   (a) The axonal endings make contact with muscle or gland cells.
   (b) An electrical potential is maintained across the axonal membrane.
   (c) The axon is surrounded by a discontinuous sheath of myelin.
   (d) The resting potential of the membrane is much closer to the equilibrium potential for potassium ion than to that for sodium ion because the sodium-potassium pump maintains a much larger transmembrane gradient for potassium than for sodium.
   (e) Excitation of the membrane results in a permanent increase in its permeability to sodium ions.
   (f) The electrical potential across the membrane of the axon can be easily measured using electrodes.
   (g) Both the sodium and potassium concentration gradients completely "collapse" every time a nerve impulse is transmitted down the axon.
   (h) Upon arrival at a synapse, a nerve impulse causes the secretion of acetylcholine into the synaptic cleft.

2. **The Equilibrium Potential.** Answer each of the following questions with respect to E_{Cl}, the equilibrium potential for chloride ions. The chloride ion concentration inside the squid axon can vary from 50 to 150 mm.
   (a) Before doing any calculations, predict whether E_{Cl} will be positive or negative. Explain.
   (b) Now calculate E_{Cl}, assuming an internal chloride concentration of 50 m*M*.
   (c) How much difference would it make in the value of E_{Cl} if the internal chloride concentration were 150 m*M* instead?
   (d) Why do you suppose the chloride concentration inside the axon is so variable?

3. **The Effect of Temperature.** In contrast to the convention we adopted in Chapter 11 of using 25°C as the standard temperature for electrochemical determinations, membrane electrophysiologists often use 18°C, reflecting the somewhat lower temperature of laboratories in England, where most of the initial work on the squid axon was done.
   (a) How much difference will it make in the E_K for the membrane of the squid axon if the temperature is changed from 18°C to 25°C? Practically speaking, is this difference significant?
   (b) How much difference will it make in E_K if the temperature is changed from 18°C to 37°C? Practically speaking, is this difference significant? For what neuron sources might 37°C be a more meaningful choice of temperature than 18°C or 25°C?

4. **The Resting Membrane Potential.** The Goldman equation is used to calculate V_m, the resting potential of a biological membrane. As presented in the chapter, it contains terms for sodium, potassium, and chloride ions only.
   (a) Why do only these three ions appear in the Goldman equation as it applies to nerve transmission?
   (b) Suggest a more general formulation for the Goldman equation that would be applicable to membranes that might be selectively permeable to other monovalent ions as well.
   (c) Would the version of the Goldman equation you suggested in part (b) be adequate for calculating the resting potential of the membrane of the sarcoplasmic reticulum in muscle cells? Explain.
   (d) How much would the resting potential of the membrane change if the relative permeability for sodium ions were 1.0 instead of 0.04?
   (e) Would you expect a plot of V_m versus the relative permeability of the membrane to sodium to be linear? Why or why not?

5. **The "All-or-Nothing" Response of Membrane Excitation.** A nerve cell membrane exhibits an "all-or-nothing" response to excitation; that is, the magnitude of the response is independent of the magnitude of the stimulus, once a threshold value is exceeded.
   (a) Explain in your own words why this is so.
   (b) Why is it necessary that the stimulus exceed a threshold value?
   (c) If every neuron shows an "all-or-nothing" response, how do you suppose the nervous system of an animal can distinguish different intensities of stimulation? How do you think your own nervous system can tell the difference between a warm iron and a hot iron, or between a chamber orchestra and a rock band?

6. **The Action Potential.** Figure 20-24 depicts the change in membrane potential of the squid giant axon when a nerve cell is stimulated. Several regions of the action potential are identified by uppercase letters. For each of the statements below, indicate with the appropriate letter(s) the specific part(s) of the action potential of which the statement is true. For a given statement, one, several, or no letters may be appropriate.
   (a) The membrane is temporarily much more permeable to sodium ions than it is in the resting state.
   (b) The membrane is temporarily much more permeable to potassium ions than it is in the resting state.

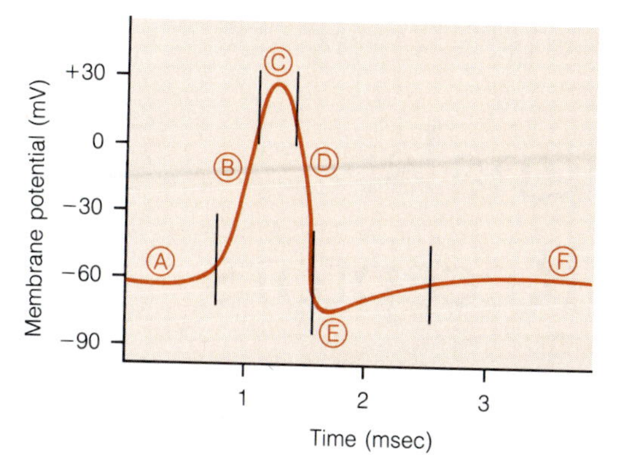

Figure 20-24 **Action Potential of the Giant Axon of the Squid.**

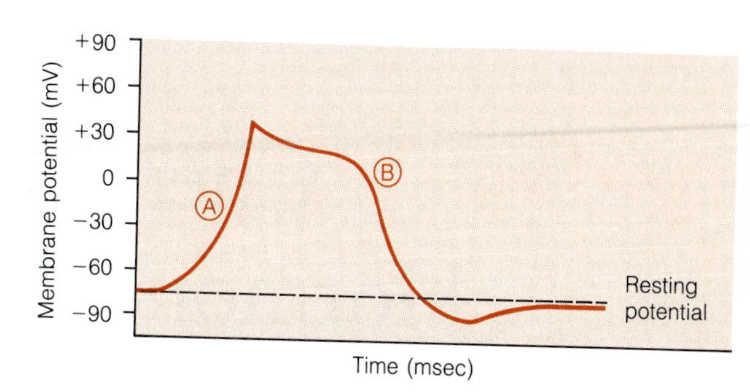

Figure 20-25 **Action Potential of a Muscle Cell from the Human Heart.**

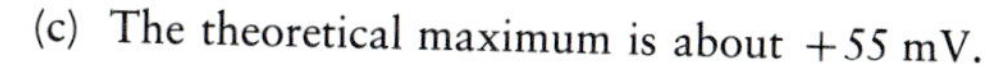

(c) The theoretical maximum is about +55 mV.

(d) The membrane is at its resting potential.

(e) The concentration of sodium inside the cell is temporarily almost as great as it is outside the cell.

(f) The sodium-potassium pump is functioning.

(g) Sodium gates are becoming inactivated and potassium gates are opening.

(h) The membrane is becoming depolarized.

(i) The membrane has become hyperpolarized.

7. **Patch Clamping.** Patch-clamp instruments permit measurements of a single channel opening and closing in a membrane. It has been determined that a typical acetylcholine receptor channel passes about 5 pA of ionic current (1 picoampere = 10^{-12} ampere) at −60 mV over a period of about 5 milliseconds.

(a) Given that an electrical current of one ampere is about 6.2×10^{18} electrical charges per second, how many ions (potassium or sodium) pass through the channel during the time that it is open?

(b) Do you think that the opening of a single receptor channel would be sufficient to depolarize a postsynaptic membrane? Why or why not?

8. **Heart Throbs.** An understanding of muscle cell stimulation involves some of the same principles as nerve cell stimulation, except that calcium ions play an important role in the former. The following ion concentrations are typical of those in human heart muscle and in the serum that bathes the muscles:

$[K^+]$: 150 m*M* in cell, 4.6 m*M* in serum

$[Na^+]$: 10 m*M* in cell, 145 m*M* in serum

$[Ca^{2+}]$: 0.001 m*M* in cell, 6 m*M* in serum

Figure 20-25 depicts the change in membrane potential with time upon stimulation of a heart muscle cell.

(a) Calculate the equilibrium potential for each of the three ions, given the concentrations listed.

(b) Why is the resting potential of the membrane significantly more negative than that of the squid axon (−75 mV versus −60 mV)?

(c) The increase in membrane potential in the region of the graph marked by the letter A could in theory be due to the movement across the membrane of one or both of two cations. Which cations are they, and in what direction would you expect each to move across the membrane?

(d) How might you distinguish between the possibilities suggested in part (c)?

(e) The rapid decrease in membrane potential that is occurring in the region marked by the letter B is caused by the outward movement of potassium ions. What are the driving forces that cause potassium to leave the cell at this point? Why aren't the same forces operative in the region of the curve marked by the letter A?

(f) People with heart disease often take drugs that can double or triple their serum potassium levels to around 10 m*M* without altering intracellular potassium levels. What effect should this have on the rate of potassium ion movement across the heart cell membrane during muscle stimulation? What effect should it have on the resting potential of the muscle?

9. **Trouble at the Synapse.** Transmission of a nerve impulse across a cholinergic synapse is subject to inhibition by a variety of neurotoxins. Indicate, as specifically as possible, what effect each of the following poisons or drugs has on synaptic transmission and what effect each has on the polarization of the postsynaptic membrane.

(a) The snake poison α-bungarotoxin

(b) The insecticide malathion

(c) Succinylcholine

(d) The carbamoyl ester neostigmine

21

Chemical Signals: Hormones and Receptors

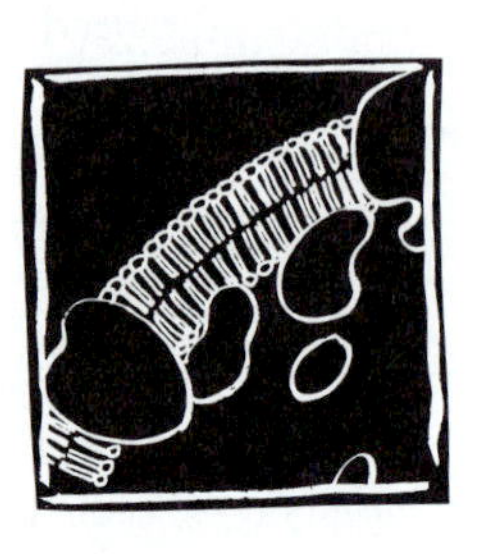

In the previous chapter, we learned how nerve cells are able to communicate with one another and with other types of cells by means of electrochemical changes in their membranes. We saw how the arrival of an action potential at a synapse causes release of *neurotransmitters*, which bind in turn to *receptors* on the postsynaptic cell membrane, thereby passing on the signal. Now we are ready to explore a second major means of intercellular communication that also involves interactions between chemicals and receptors. In this case, however, the signal is transmitted by regulatory chemicals called *hormones* and the receptors are located on cells that may be quite distant from the hormone-secreting cells. Thus, the animal body has two different but complementary systems of communication and control, and receptors play a crucial role in both systems.

Why do organisms need two different means of signal transduction? One way to answer this question is to consider how information is transmitted in human societies. The electrochemical communication between neurons discussed in Chapter 20 is like a telephone call between two people: The target is very specific, and the information transfer is precise. In contrast, the hormonal signals to be considered in this chapter are more like broadcast news: The signals are transmitted to the entire system, but they affect only those cells that are equipped to listen. Both means of communication have specific uses in human societies, and the same is true among cells. And in both cases, receptors are an important component of the system.

In this chapter, we will encounter several general mechanisms by which receptors detect chemical signals impinging upon the cell and transmit those signals inward to the cytoplasm, where they trigger specific cellular responses. Some of the most interesting receptor systems are those that respond to hormones circulating in the bloodstream. Because of the importance of such receptors for human medicine, they are the subject of much current research and will therefore play an important role in our discussion of chemical signal transduction.

Chemical Signals and Membrane Receptors

Although much of this chapter will focus on membrane-bound receptors involved in hormonal function in humans and other mammals, the detection of **chemical signals** by **membrane receptors** is a feature common to all cells. Prokaryotic cells, for instance, have membrane-bound receptor molecules on the cell surface that allow these cells to respond by chemotaxis to chemicals in their environment, as described in Chapter 19. As another example, amoeboid cells of the slime mold *Dictyostelium* secrete a compound called *cyclic AMP* at one stage in their life cycle. The cyclic AMP binds to receptors on the surfaces of neighboring cells, triggering a process whereby thousands of individual amoeboid cells aggregate and differentiate into a multicellular organism. Plant cells also use membrane receptors to detect hormones such as *auxin* and *ethylene* that regulate growth and development.

The **chemical receptors** in the human body detect a variety of different chemical signals. In most cases, we are not specifically aware of receptor activity, but there are some notable exceptions. Our senses of smell and taste, for example, involve receptors whose activities we experience directly. In both cases, chemical substances from the outside world are detected by specific *sensory cells*, either those of the *olfactory tissues* in the nose or those of the *taste buds* on the tongue. The substances bind to receptor molecules in the sensory cells, producing changes in the electrical activity of the cells and thereby causing an electrical impulse to be transmitted to the brain. Most chemical receptors work in a similar fashion, except that the responses they mediate are related not to nervous activity but to other cellular functions, such as secretion, transport, and differentiation.

Endocrine and Paracrine Hormone Systems

Since our discussion of chemical receptors will focus mainly on hormonal regulation in the human body, we can begin by asking what hormones are and where they are made. An important distinction between hormones is based on the distances over which they operate. An **endocrine hormone** is a chemical that is released by one set of cells and travels by means of the circulatory system to other sets of cells, where it regulates one or more specific functions. (The word *endocrine* comes from Greek roots meaning "to secrete into.") A **paracrine hormone,** on the other hand, is a more local signal that is taken up, destroyed, or immobilized so rapidly that it can act only on cells in the immediate environment. (The word *paracrine* means "to secrete around.")

Endocrine hormones are synthesized by the **endocrine tissues** of the body and are secreted directly into the bloodstream. This mode of secretion distinguishes endocrine tissues from **exocrine tissues,** which secrete their products into ducts that then transport the secretions to other parts of the body. Some organs of the body have both endocrine and exocrine tissues. The *pancreas* is a good example of such an organ. Cells of the pancreatic endocrine tissue secrete two hormones, *glucagon* and *insulin,* that regulate the concentration of glucose circulating in the blood. Cells of the pancreatic exocrine tissue, on the other hand, secrete *digestive enzymes* that are collected and sent to the small intestine through the pancreatic duct. Figure 21-1 depicts the major endocrine tissues of the human body.

Once secreted into the circulatory system, endocrine hormones have a limited life span, ranging from a few seconds for epinephrine (a product of the adrenal gland) to many hours for insulin. As they circulate in the bloodstream, hormone molecules come into contact with cell surfaces in tissues throughout the body. This contact is the means by which most endocrine hormones regulate cell function in a variety of tissues. A tissue that is specifically affected by a particular hormone is called a **target tissue** for that hormone (Figure 21-2). For example, the heart and the liver are target tissues for epinephrine, whereas the liver and skeletal muscle are targets for insulin action.

The Chemistry of Hormones

When first encountered, the hormones may seem confusing, given the number of them and the diversity of their effects. However, our task of understanding the hormones is made somewhat easier by recognizing that they can be classified according to their chemical properties (Table 21-1) and their functions in the body (Table 21-2).

Chemically, the endocrine hormones fall into four categories: amino acid derivatives, peptides, proteins, and steroid hormones (Figure 21-3). An example of an amino acid derivative is *epinephrine,* derived from tyrosine. *Antidiuretic hormone* (also called *vasopressin*) is an example of a peptide hormone, whereas *insulin* is a protein. *Testosterone* is an example of a steroid hormone. The steroid hormones are derivatives of cholesterol that are synthesized either in the gonads (the *sex hormones*) or in the adrenal cortex (the *corticosteroids*).

As already mentioned, paracrine hormones are chemicals that act locally, rather than throughout the body. *Histamine* is a paracrine hormone, as are the *prostaglandins* (Table 21-1). Histamine is produced by decarboxylation of the amino acid histidine and is the primary agent responsible for local inflammatory responses. The prostaglandins are produced by most cells, particularly in response to stress or injury. They are synthesized by

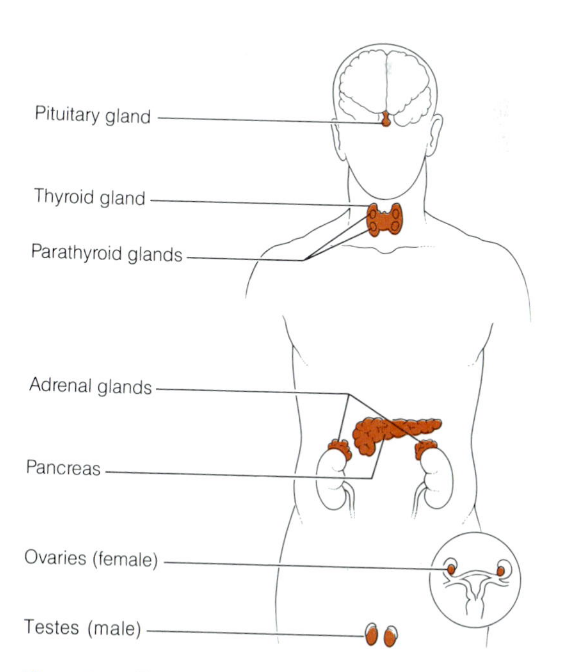

Figure 21-1 The Major Endocrine Tissues of the Human Body. The cells of endocrine tissues secrete hormones directly into the bloodstream, effectively "broadcasting" the hormone throughout the body. The hormones act as chemical signals, binding to receptors of specific target cells and thereby regulating the function of such cells.

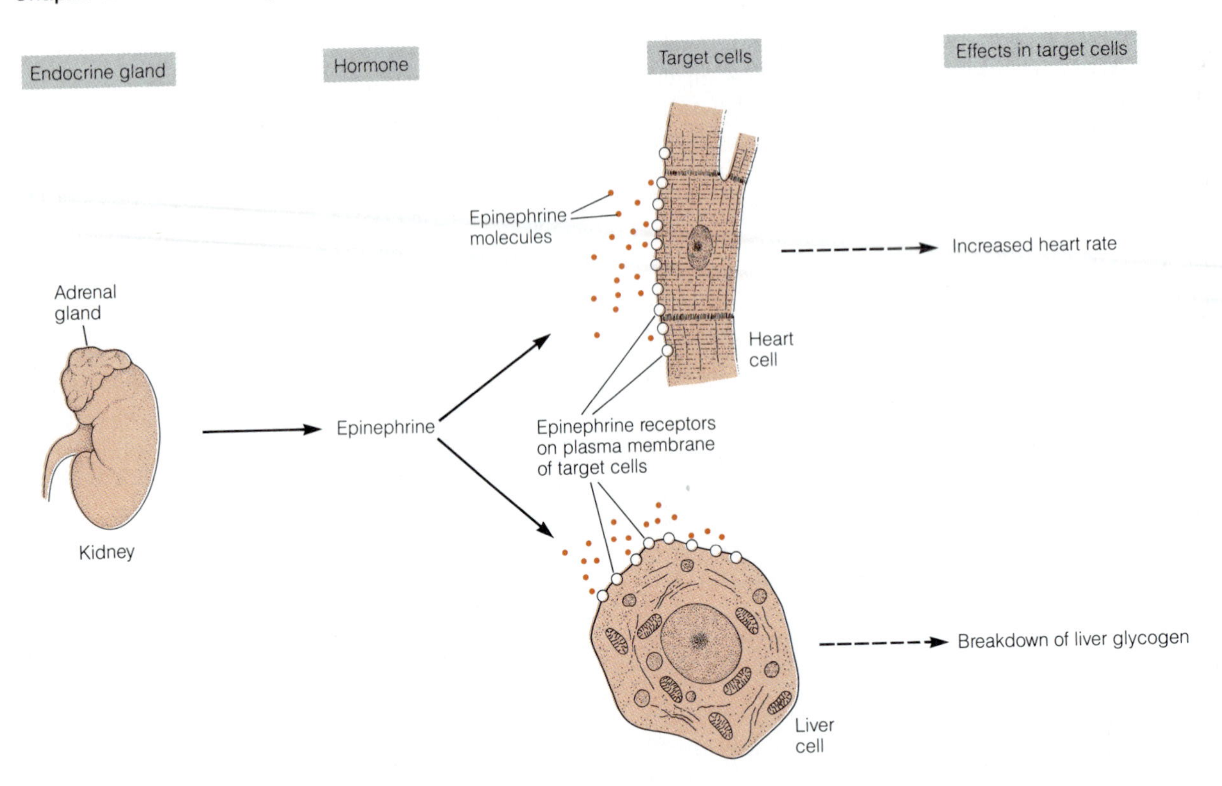

Figure 21-2 Target Tissues for Endocrine Hormones. A tissue that responds specifically to a particular hormone is called a target tissue for that hormone. Cells in a target tissue have hormone-specific receptors embedded in their plasma membranes (or, in the case of the steroid hormones, present in their cytoplasm). Heart and liver cells can respond to epinephrine synthesized by the adrenal gland because these cells have epinephrine-specific receptors on their outer surfaces. A specific hormone may elicit different responses in different target cells, depending on the enzymes present in the cells. Epinephrine causes an increase in the rate of the heart, but stimulates glycogen breakdown in the liver.

oxidation of a fatty acid called *arachidonic acid,* as will be described later.

Hormonal Control of Physiological Functions

Table 21-2 summarizes some of the major physiological functions of the human body that are known to be regulated by hormones. As with the chemical properties, hormone functions are a little less confusing when we classify them. In this case, four general categories of hormonally regulated functions can be identified: *growth and development, rates of body processes, concentrations of substances,* and *responses to stress.* For example, *somatotropin* is involved in the regulation of overall growth of the body, whereas the sex hormones (*androgens* and *estrogens*) control differentiation of tissues into secondary sex characteristics. *Thyroxine* is an example of a rate-controlling hormone, since it regulates the rate at which the body makes energy available. Examples of hormones that control concentrations within the body include *insulin* (control of blood glucose level) and *parathyroid hormone* (control of blood calcium level). The body's response to stress is regulated by *epinephrine* and by *cortisol,* and its response to local injury is regulated by release of *histamine* and production of *prostaglandins.*

Feedback Regulation of the Endocrine System

Although hormones serve as the primary agents of endocrine control, the secretory activity of the individual endocrine glands must also be regulated. This regulatory function is provided by the **anterior pituitary,** which secretes a series of *tropic hormones* that regulate other endocrine glands through **feedback control loops.**

A feedback control loop always contains a **sensor** and an **effector.** This type of control can be illustrated by the

Table 21-1 Chemical Classification and Function of Hormones

Chemical Classification	Example	Function
	Endocrine Hormones	
Amino acid derivatives	Epinephrine (adrenalin) and norepinephrine (both derived from tyrosine)	Stress responses: regulation of heart rate and blood pressure; release of glucose and fatty acids from storage sites
	Thyroxine (derived from tyrosine)	Regulation of metabolic rate
Peptides	Antidiuretic hormone (ADH)	Regulation of body water and blood pressure
	Hypothalamic hormones (releasing factors)	Regulation of tropic hormone release from pituitary gland
Proteins	Anterior pituitary hormones	Regulation of other endocrine systems
Steroids	Sex hormones (androgens and estrogens)	Development and control of reproductive capacity
	Corticosteroids	Stress responses; control of blood electrolytes
	Paracrine Hormones	
Amino acid derivative	Histamine	Local responses to stress and injury (both histamine and the prostaglandins)
Arachidonic acid derivatives	Prostaglandins	Local responses to stress and injury

relationship between a thermostat and a furnace: When the temperature falls below a set level on the thermostat (sensor), the thermostat sends an electrical signal to the furnace (effector), which turns on until the set temperature is reached and then switches off. The resulting on-off cycle produces a characteristic oscillation around the set temperature.

A classic example of a feedback loop in the endocrine system is the control of thyroxine production. When the thyroxine concentration in the bloodstream falls below a certain level, the anterior pituitary secretes *thyroid stimulating hormone (TSH),* a tropic hormone that activates the production and release of thyroxine by the thyroid gland. TSH is clearly the effector in this system, but where is the sensor that controls the anterior pituitary? The sensor is a specialized set of cells in the **hypothalamus,** a portion of the brain that is located just above the pituitary, with a microcirculation that carries blood downward to the anterior pituitary. Specialized cells in the hypothalamus respond to low levels of circulating thyroxine by secreting a **releasing factor,** which circulates to the anterior pituitary and stimulates release of TSH, thereby closing the sensor-effector loop. Similar feedback loops control production of most other tropic hormones by the anterior pituitary.

Second Messengers and Membrane Receptors

How does a signal molecule impinging on the surface of a cell regulate a specific function within that cell? The answer to this fundamental question involves interactions between chemical signals (hormones, in most cases), protein receptors (embedded in the plasma membrane, in most cases), and changes in intracellular concentrations of specific substances, such as cyclic AMP and calcium ions, that mediate the effect within the cell. Such intracellular mediators are called **second messengers** because they mediate (and often amplify) the impinging hormonal signal (the "first messenger") within the cell.

Most hormones are relatively polar molecules that do not readily cross the lipid bilayer barrier of the plasma membrane. As a result, the only signal that most hormones can provide to the cell is a change in hormone concentration at the cell surface, usually as a result of the release of hormone into the bloodstream elsewhere in the body. (The steroid hormones are an exception to this generalization, since they are lipophilic and can therefore cross the plasma membrane. As we will see later, these hormones interact

Table 21-2 Physiological Functions of Hormones

Function Under Hormonal Control	Hormone	Source (Endocrine Tissue)
	1. Growth and Development	
Body size	Somatotropin	Anterior pituitary
Sexual development	Androgens (males)	Testes
	Estrogens (females)	Ovaries
Reproductive cycle	Luteinizing hormone	Anterior pituitary
	Follicle stimulating hormone	Anterior pituitary
	Chorionic gonadotropin	Follicle
	2. Rates of Body Processes	
Hormone secretion	Tropic hormones	Anterior pituitary
Basal metabolism	Thyroxine	Thyroid
Glucose uptake	Insulin	Pancreas
Kidney filtration	Antidiuretic hormone	Posterior pituitary
Uterine contraction	Oxytocin	Posterior pituitary
	3. Concentrations of Substances	
Blood glucose	Glucagon, insulin	Pancreas
Mineral balance	Corticosteroids	Adrenal cortex
Blood calcium	Parathyroid hormone	Parathyroid
	4. Responses to Stress	
Heart rate	Epinephrine	Adrenal medulla
Blood pressure	Epinephrine	Adrenal medulla
Inflammation	Histamine	Mast cells
	Prostaglandins	All tissues
	Corticosteroids	Adrenal cortex

with soluble receptors within the cell rather than with membrane-bound receptors on the cell surface.)

Cells are able to respond to changes in the concentrations of hormones or other substances at the cell surface because of highly specific **receptor proteins** embedded in their plasma membranes (Table 21-3). Substances that bind specifically to such receptors are called **ligands,** from a Latin word meaning "to bind." When a ligand binds to its receptor protein, it fits into a specific site on the protein, just as a substrate molecule fits into the active site of an enzyme (recall Figure 6-7).

The transmission of a signal across the membrane can involve any of several underlying mechanisms, including activation of cyclic AMP-generating enzymes, a transient increase in the calcium ion concentration within the cell, or phosphorylation of specific cytoplasmic proteins. In each case, the essential feature of the response system is the generation or release of a second messenger within the cell. An important feature of such second messengers is the *amplification* of the original signal that they make possible. Binding of a few ligand molecules to their receptor sites on the membrane is adequate to stimulate the production or release of a large number of second messenger molecules or ions inside the cell. This amplification effect allows hormones or other substances to be effective at very low concentrations.

Table 21-3 Examples of Receptors

Hormones That Bind to Membrane Surface Receptors	Hormones That Bind to Intracellular Receptors
Insulin	Steroid hormones (androgens and estrogens)
Glucagon	Thyroxine
Epinephrine	
Tropic hormones	
Somatotropin	
Growth factors	
Antidiuretic hormone	
Parathyroid hormone	

We are now ready to look at several different mechanisms of chemical signal transduction, each involving a specific second messenger. In each case, we will look at the regulatory pathway that links the ligand-receptor interaction on the cell surface to the activation of specific processes within the cell.

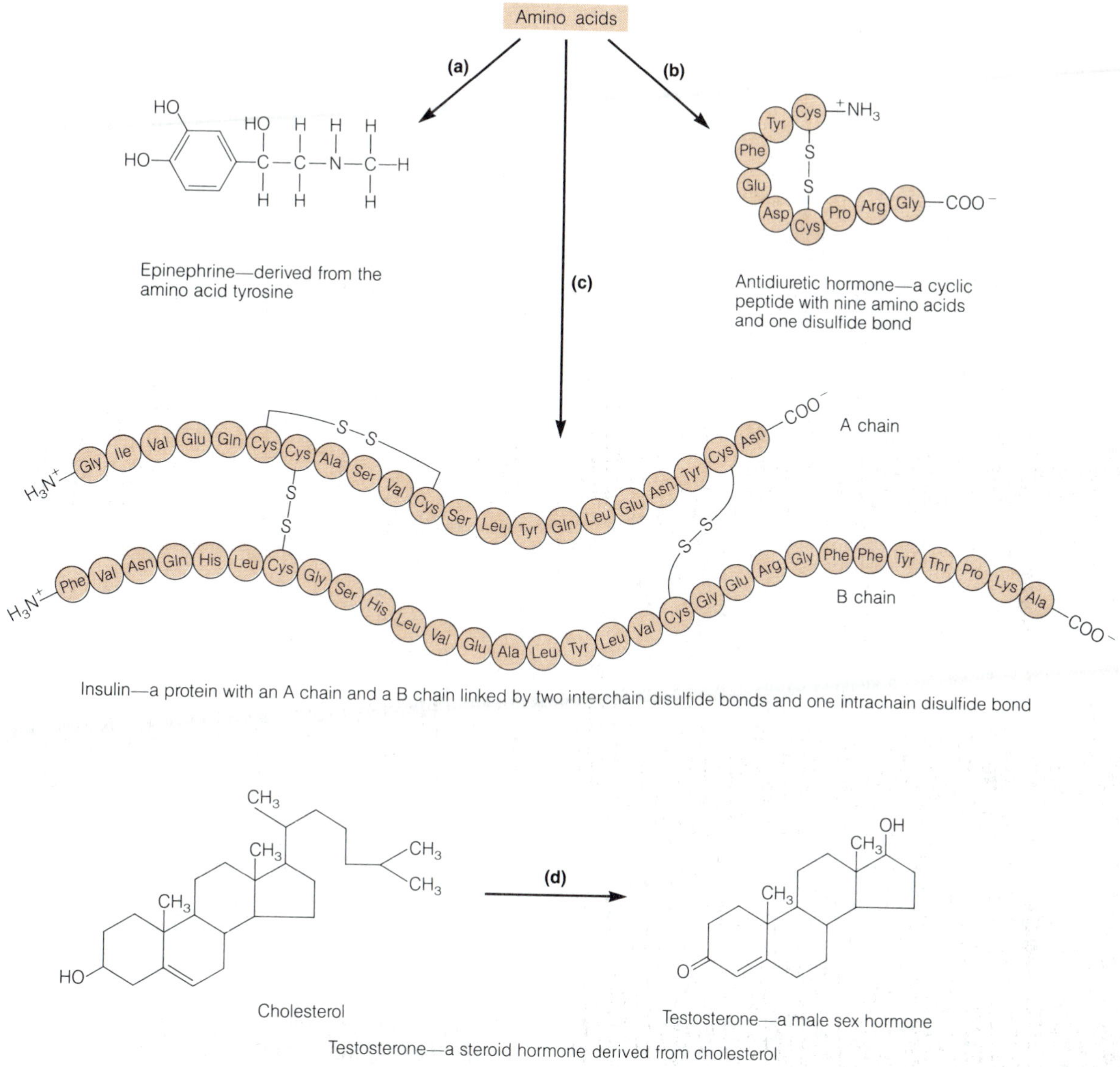

Amino acids

Epinephrine—derived from the amino acid tyrosine

Antidiuretic hormone—a cyclic peptide with nine amino acids and one disulfide bond

Insulin—a protein with an A chain and a B chain linked by two interchain disulfide bonds and one intrachain disulfide bond

Cholesterol

Testosterone—a male sex hormone

Testosterone—a steroid hormone derived from cholesterol

Figure 21-3 Chemistry of Hormones. Most endocrine hormones are (a) derivatives of amino acids, (b) peptides consisting of a small number of amino acids, (c) proteins containing many amino acids, or (d) steroids (either sex hormones or corticosteroids) derived from cholesterol. (Note that insulin is actually a globular protein and is shown here in extended form only for purposes of illustration. For full names of amino acids, see Figure 3-2 on p. 43)

Cyclic AMP as a Second Messenger

An agent that mediates many hormone-dependent cell functions is **cyclic AMP** (adenosine-3′,5′-cyclic monophosphate). Cyclic AMP is generated from ATP by **adenylate cyclase** and is hydrolyzed to AMP by **phosphodiesterase** (Figure 21-4). Adenylate cyclase is located on the inner surface of the plasma membrane and generates cyclic AMP in response to ligand-receptor binding on the outer surface of the membrane. Cyclic AMP is therefore a second messenger, since it is formed within the cell in response to a hormonal signal impinging on the surface of the cell.

The involvement of cyclic AMP as an intracellular mediator of hormonal effects was suggested by experiments in which direct injection of cyclic AMP into a cell triggered a cascade of enzyme activities, thereby activating specific

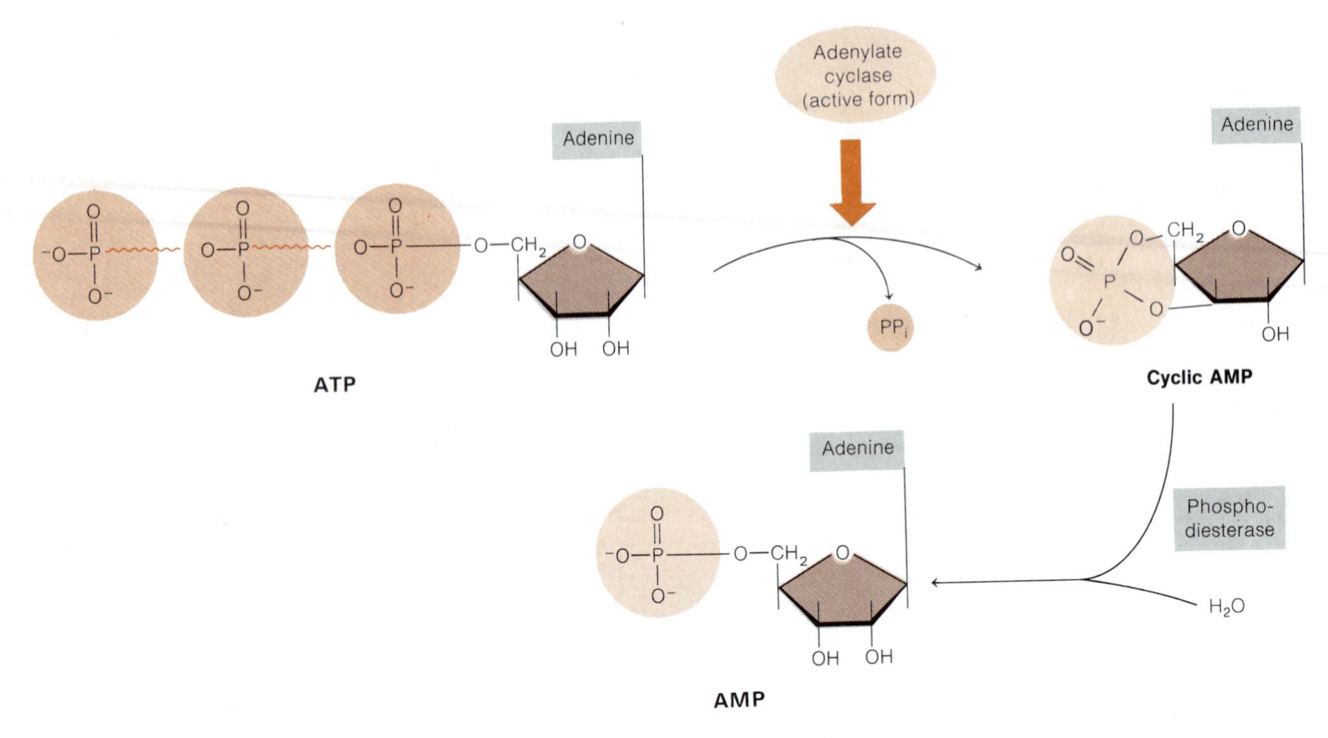

Figure 21-4 Structure and Metabolism of Cyclic AMP. Cyclic AMP (adenosine-3′,5′-cyclic monophosphate) is generated from ATP in a reaction catalyzed by the active form of the enzyme adenylate cyclase and is inactivated by hydrolysis to AMP, a reaction catalyzed by the enzyme phosphodiesterase. Adenylate cyclase is a membrane-bound enzyme, whereas phosphodiesterase is located in the cytoplasm.

cell functions. Until recently, however, it was not clear how the binding of a ligand to a membrane surface on the *outside* of a cell could lead to the formation of cyclic AMP *inside* the cell.

G Proteins and Cyclic AMP Synthesis

Our understanding of cyclic AMP-mediated responses increased dramatically with the recent discovery of a class of membrane proteins that provide the link between ligand-receptor binding and adenylate cyclase activity. These proteins bind guanosine nucleotides and are therefore called *GTP-binding regulatory proteins*, or **G proteins.** G proteins were first discovered as mediators of cyclic AMP production, but are now known to be involved in a variety of *signal transduction systems*. A G protein called *transducin* is even involved in the signal transduction pathway required for vision, as described in the box on pages 652–653.

Each G protein has a binding site that can be occupied by either GDP or GTP, and its activity depends on which nucleotide is bound. In general, G proteins are inactive when GDP is bound and are activated when the GDP is displaced by GTP. The displacement of GDP by GTP occurs only when a neighboring receptor in the membrane is activated by a chemical signal such as a change in the extracellular concentration of a hormone. When present in the active form, G proteins can activate membrane-bound adenylate cyclase, thereby triggering the production of cyclic AMP. G proteins, in other words, are the "missing link" between ligand binding on the outer surface of the plasma membrane and an increase in cyclic AMP concentration within the cell. The overall scheme therefore looks like this:

$$\text{Ligand} \longrightarrow \text{receptor} \longrightarrow \text{G protein} \longrightarrow \text{adenylate cyclase} \longrightarrow \text{cyclic AMP} \quad (21\text{-}1)$$

Keep this overall scheme in mind as we proceed now to a more detailed description of the events that underlie receptor function.

Regulation of G Proteins

To be effective, every regulatory system must be capable of both **activation (up-regulation)** and **inactivation (down-regulation)**. Figure 21-5 illustrates the process whereby G proteins are activated. Each G protein consists of three

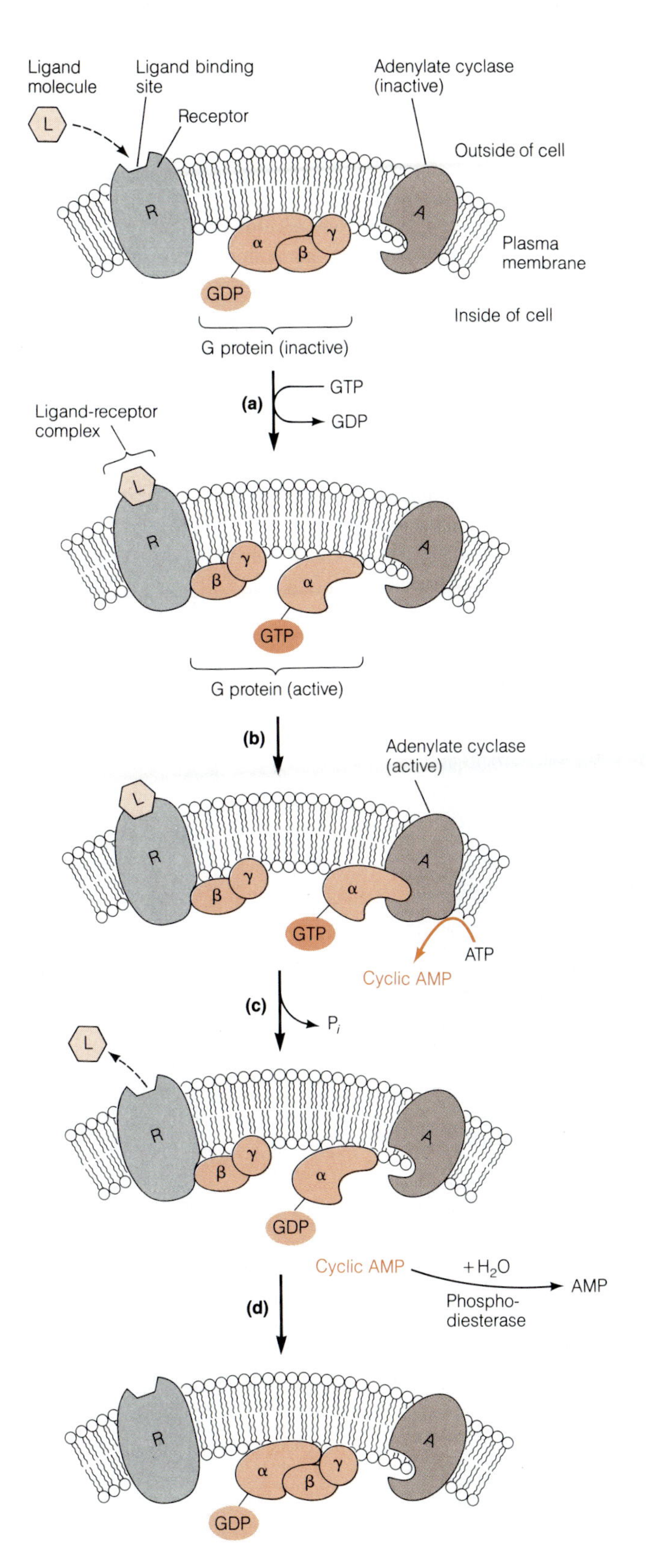

subunits, called α, β, and γ. The α subunit binds and hydrolyzes GTP and activates adenylate cyclase, whereas the β and γ subunits form a complex that anchors the G protein to the cytoplasmic side of the plasma membrane. In the inactive state, all three subunits are present as a complex in the membrane, with GDP bound to the subunit (Figure 21-5, top). When a neighboring receptor molecule is activated by binding of its specific ligand, the nucleotide binding site on the G protein is altered, causing GTP to replace GDP and the α subunit-GTP complex to dissociate (Figure 21-5, step a). The α subunit-GTP complex then binds tightly to a membrane-bound adenylate cyclase molecule, activating it for production of cyclic AMP (step b).

Inactivation, or down-regulation, of this system is essentially the reverse of the activation process. When the concentration of the hormone on the outside of the cell decreases again (by metabolism of the hormone in the liver or kidney, in most cases), the membrane receptor reverts to its inactive form and the GTP is hydrolyzed by a GTPase activity on the α subunit. The α subunit then returns to its original conformation, causing it to dissociate from the adenylate cyclase molecule (step c). Adenylate cyclase reverts to its inactive form, and the α subunit reassociates with the βγ complex (step d) until or unless it is activated again by another round of steps a and b. Meanwhile, the cyclic AMP in the cytoplasm is inactivated by the enzyme phosphodiesterase, which hydrolyzes it to AMP.

G Proteins and Diseases

What would happen if the G protein/adenylate cyclase system could not be down-regulated? We can answer this question by considering several human diseases that can now be explained at the molecular level because of what

Figure 21-5 The Role of G Proteins and Cyclic AMP in Signal Transduction. A typical G protein is associated with the inner surface of the plasma membrane, where it mediates the interaction between a specific receptor (R) and adenylate cyclase (A), both of which are integral membrane proteins. In the inactive state, the α, β, and γ subunits are present as a complex, with GDP bound to the α subunit. (a) When a neighboring receptor is activated by binding of its specific ligand (L) on the outer surface of the plasma membrane, the receptor-ligand complex associates with the G protein, causing the displacement of GDP by GTP and the dissociation from the G protein of the α subunit-GTP complex. (b) The α subunit-GTP complex then binds tightly to a molecule of membrane-bound adenylate cyclase, activating it for synthesis of cyclic AMP. (c) The system is down-regulated when the ligand leaves the receptor, the GTP is hydrolyzed to GDP by the GTPase activity of the α subunit, and the α subunit dissociates from the adenylate cyclase. (d) Adenylate cyclase then reverts to the inactive form, the α subunit reassociates with the βγ complex, and cyclic AMP molecules in the cytoplasm are hydrolyzed to AMP by the enzyme phosphodiesterase.

G Proteins and Vision

At first glance, a relationship between vision and receptors for endocrine hormones might seem highly unlikely. Why would photons of light striking a sensory cell in the retina of the eye have anything to do with hormonal receptors? Yet, it is now clear that light sensing by retinal cells depends directly on a G protein-mediated mechanism that resembles somewhat the hormonal regulatory cascades encountered in this chapter.

The retinal cell that is most sensitive to light is called the *rod cell* (Figure 21-A). The rod cell membrane, like other excitable membranes, contains gated sodium channels. Unlike the channels in the nerve cell membrane, however, the sodium channel of the rod cell membrane has a binding site for *cyclic GMP.* When bound to this site, cyclic GMP keeps the channels open, so that the cell is normally depolarized in the resting state (that is, in the dark). The rod cell has a synaptic connection to a sensory neuron and continuously releases a neurotransmitter that is *inhibitory* for the neuron. The neuron is thereby maintained in a state of relatively low activity as long as no light strikes the rod cell and the membrane remains depolarized.

What happens when light strikes the rod cell? The stacked membrane disks in the outer segment of the cell are packed with molecules of a light-sensitive protein called *rho-*

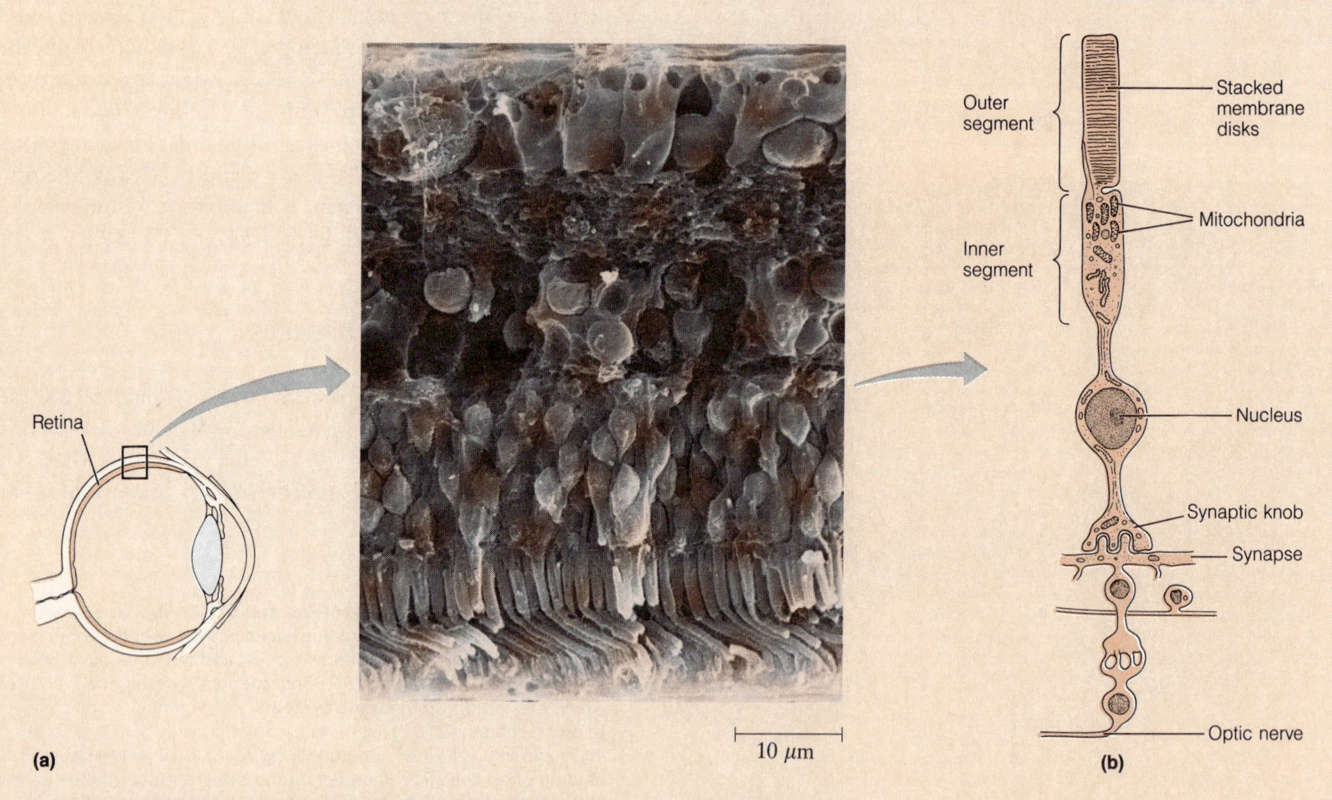

Figure 21-A Structure of Rod Cells. (a) A scanning electron micrograph of *rod cells,* light-sensitive cells in the retina of the human eye . Rod cells are especially useful for vision in dim light. (b) A sketch of an individual rod cell, showing the *outer segment,* the *inner segment,* the *nucleus,* and the *synaptic knob.* The outer segment consists of a stack of about a thousand flattened and tightly packed membrane disks. Embedded within these membranes are molecules of *rhodopsin,* a complex protein consisting of the pigment *retinaldehyde* attached covalently to a transmembrane glycoprotein called *opsin.* The inner segment of the rod cell contains a variety of cell organelles, predominantly mitochondria. The synaptic knob is filled with synaptic vesicles that contain an inhibitory neurotransmitter. Signal transmission via the optic nerve is suppressed in the dark but is activated in the light by a cyclic AMP-mediated decrease in the release of the inhibitory neurotransmitter.

dopsin. Rhodopsin has as its prosthetic group a derivative of vitamin A called *retinal* (Figure 21-B). This molecule is the primary visual pigment of the retina. When retinal absorbs light energy, it undergoes a conformational change from the 11-*cis* to the all-*trans* configuration, thereby inducing a conformational change in the rhodopsin molecule of which it is a part. This change causes the rhodopsin in the rod cell membrane to bind to a G protein called *transducin*. The transducin-G protein complex in turn activates *cyclic GMP diesterase,* a membrane-bound enzyme that hydrolyzes cyclic GMP to GMP monophosphate. The resulting decrease in the cyclic GMP concentration in the cytoplasm causes cyclic GMP to be released from its binding sites on the sodium channels. The channels then close, the cell becomes polarized, and less inhibitory neurotransmitter is released, thereby activating the postsynaptic sensory neurons. As a result, increasing numbers of action potentials reach the brain, where they are interpreted as light.

This remarkable cascade of events goes on continuously in every one of your rod cells whenever your eyes are open. And the regulatory mechanisms that underlie this process are strikingly similar to those involved in a wide variety of other G protein-mediated processes in your body, underscoring the importance of G proteins and the excitement that their discovery has generated among cell biologists.

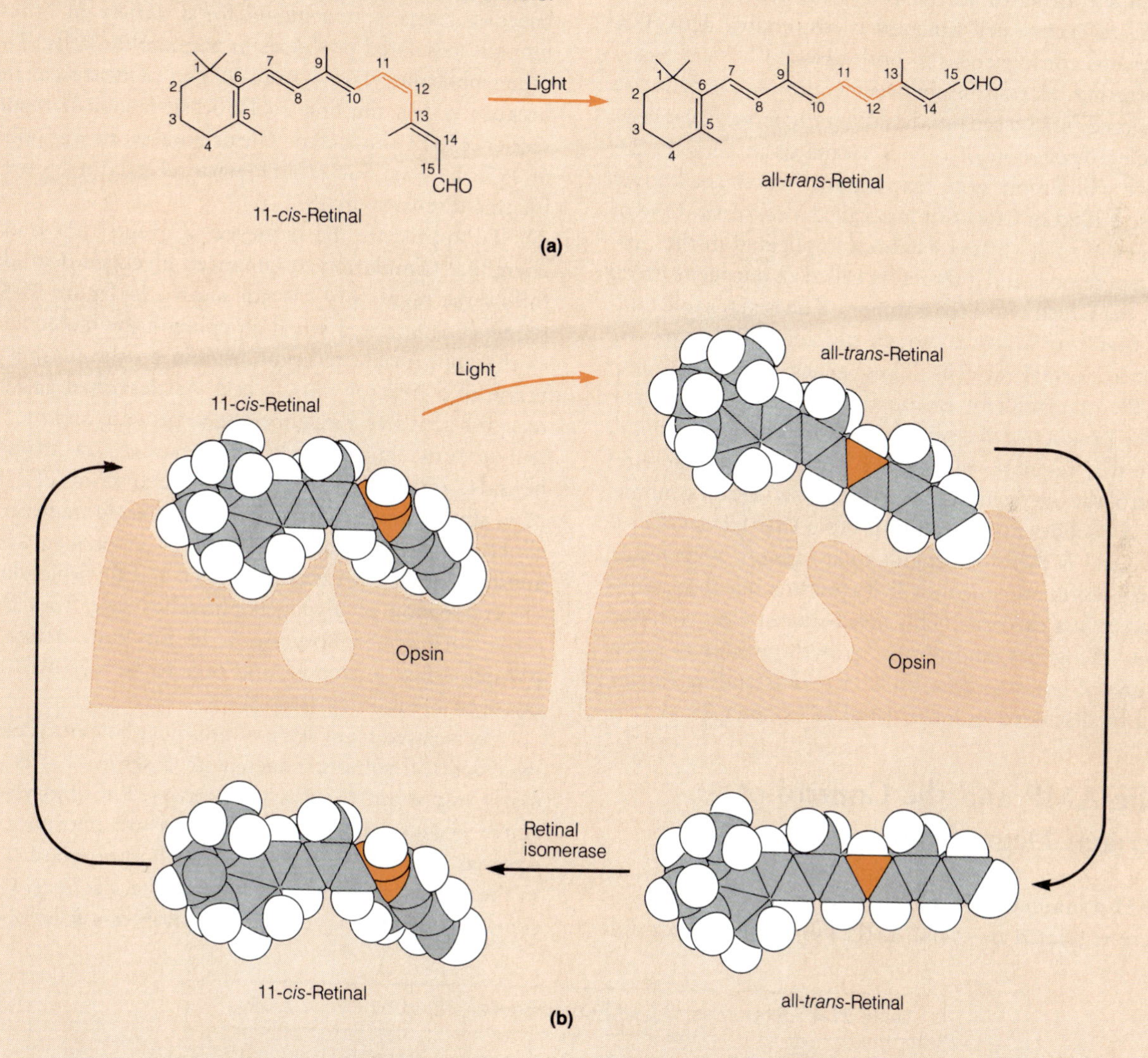

Figure 21-B Effect of Light on Retinal. (a) The primary visual pigment of the rod cell is 11-*cis*-retinal, the light-sensitive prosthetic group of the rhodopsin molecule. When retinal absorbs a photon of light, it isomerizes from the 11-*cis* form to all-*trans*-retinal. (b) As these space-filling models show, 11-*cis*-retinal is attached to the protein opsin, such that the light-induced isomerization to the all-*trans* form leads to a conformational change in the opsin protein. The shape change initiates a signal transduction pathway that includes the G protein trasducin, cyclic AMP, and the sodium channels in the plasma membrane of the rod cell. The *trans*-retinal dissociates from the opsin and is released into the cytoplasm, where it is converted by the action of retinal isomerase to the 11-*cis* form that is again capable of association with opsin to reinitiate the cycle.

we have learned about G proteins. *Cholera bacteria* and *pertussis bacteria* both cause pathological responses in epithelial secretory cells. Infection with cholera bacteria causes intestinal epithelial cells to secrete large amounts of fluid, primarily water and sodium ions, giving rise to a *diarrhea* that can lead to death by dehydration. A pertussis infection causes *whooping cough* by a similar effect on respiratory epithelia.

In both cases, a *toxin* secreted by the bacteria affects the G protein/cyclic AMP regulatory cascade. **Cholera toxin** inhibits the GTPase activity of the G proteins, whereas **pertussis toxin** inactivates the process that normally down-regulates adenylate cyclase activity. As a result, the respective cell function is irreversibly activated, with serious consequences in both cases.

A further relationship between G proteins and disease has emerged from recent studies of virally induced cancers. A detailed discussion of cancer awaits us in Chapter 23; here, we will simply note that certain viruses are known to cause a kind of cancer in rats called a *sarcoma*. One of the genes in such viruses has been implicated in the production of tumors and is therefore called an **oncogene** (from *onkos*, the Greek word for a tumor). The remarkable finding is that very similar, though not identical, genes are present in normal rat cells. These genes encode a protein called the *ras* (from *rat* sarcoma) *protein*.

The product of the *ras* gene is now known to be a G protein that regulates normal cell proliferation. The viral version of the oncogene is able to insert into the rat genome, where it encodes an altered G protein that lacks GTPase activity and is therefore unable to switch off. As a result, the cell goes into an uncontrolled growth mode that results ultimately in a sarcoma. This link between viral transformation, oncogenes, and normal growth promoters represents one of our strongest leads in understanding cancer, as will be discussed in more detail in Chapter 23.

Cyclic AMP and the Control of Glycogen Degradation

Table 21-4 summarizes some major physiological functions that are regulated by cyclic AMP. For a specific example of cyclic AMP-mediated regulation, we will consider the control of glycogen degradation in liver or muscle cells by the hormone **epinephrine.** (Epinephrine is also called **adrenalin;** the two words are of Greek and Latin derivation, respectively, and mean "above, or near, the kidney," referring to the location in the body of the *adrenal glands*, which synthesize this hormone.)

The breakdown of glycogen involves the enzyme *glycogen phosphorylase*, which cleaves glucose units from glycogen as glucose-1-phosphate by the addition of inorganic phosphate (P_i). The sequence of events from hormonal stimulation of the tissue to enhanced glycogen degradation is complex, but is well worth understanding because it serves as a model for a variety of other hormonally regulated processes in mammalian cells. The glycogen phosphorylase system is also of historical interest because it was the first cyclic AMP-mediated regulatory sequence to be elucidated. The original work was published in 1956 by Earl Sutherland, who received a Nobel Prize for this discovery in 1971.

To appreciate the sequence of events that lead from hormonal stimulation to enhanced glycogen degradation, follow the regulatory cascade shown in Figure 21-6. The sequence is initiated when an epinephrine molecule binds to the appropriate receptor protein on the plasma membrane of a liver or muscle cell. As described earlier, the receptor activates a neighboring G protein, and the G protein in turn activates adenylate cyclase, the membrane-bound enzyme that generates cyclic AMP from ATP (Figure 21-6, steps a and b). The resulting transient increase in the concentration of cyclic AMP in the cytoplasm activates another regulatory cascade (steps c and d), which leads to the conversion of glycogen phosphorylase from the less active form, phosphorylase *b*, to the more active form, phosphorylase *a* (step e), and thus to an increased rate of glycogen breakdown (step f).

As isolated from skeletal muscle, phosphorylase *a* has two identical subunits, each with a serine that is particularly important for enzyme activity. Phosphorylase *a* is active because these crucial serines are phosphorylated, whereas the relative inactivity of phosphorylase *b* is due to the lack of phosphate groups on these serines. Conversion of phosphorylase *a* to phosphorylase *b* involves hy-

Table 21-4 Examples of Cell Functions Regulated by Cyclic AMP

Regulated Function	Target Tissue	Hormone
Glycogen degradation	Muscle, liver	Epinephrine
Fatty acid production	Adipose	Epinephrine
Heart rate, blood pressure	Cardiovascular	Epinephrine
Water reabsorption	Kidney	Antidiuretic hormone
Bone resorption	Bone	Parathyroid hormone

drolysis of the phosphate group of each subunit by a *phosphatase* enzyme. Conversion of the phosphorylase *b* back to the *a* form requires phosphorylation of the serine groups by ATP, catalyzed by the enzyme *phosphorylase kinase*. Phosphorylase kinase also exists in an active and an inactive form, with the conversion from the inactive to the active form mediated by the enzyme *protein kinase*. Activation of this enzyme is effected by cyclic AMP.

Thus, the signal transduction chain for epinephrine-stimulated glycogen breakdown (Figure 21-6) involves activation, in order, of (a) a membrane-bound G protein, (b) adenylate cyclase, (c) protein kinase, (d) phosphorylase kinase, and (e) phosphorylase, which catalyzes the phosphorolytic cleavage of glycogen into glucose-1-phosphate. Since cyclic AMP also stimulates the inactivation of the enzyme system responsible for glycogen synthesis, the overall effect of cyclic AMP involves both an increase in glycogen breakdown and a decrease in its synthesis.

To summarize, cyclic AMP is a second messenger in numerous cell functions that are under hormonal control.

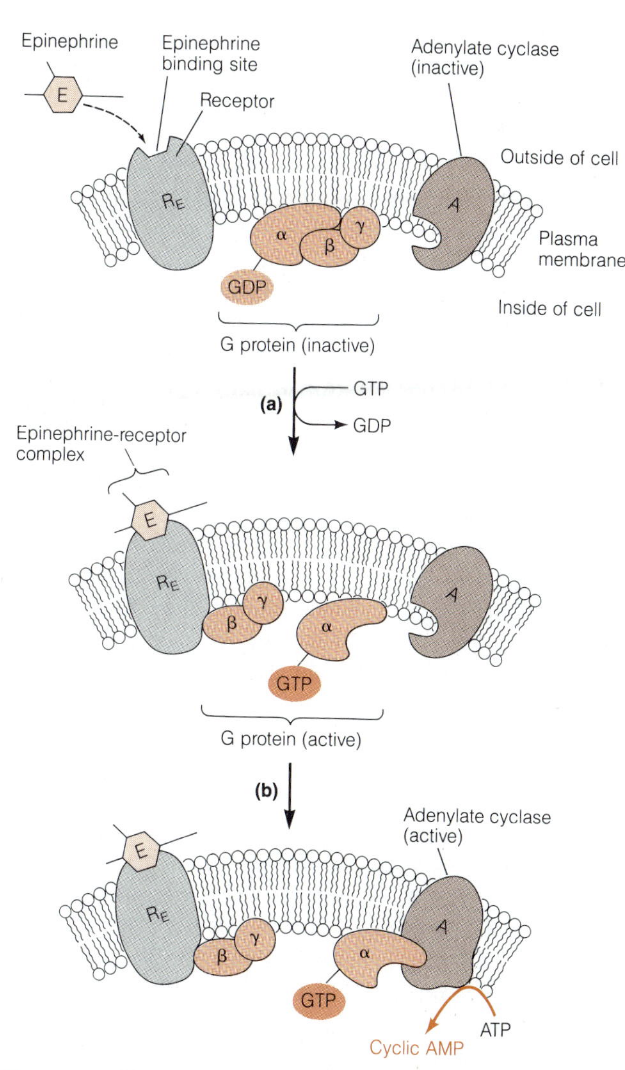

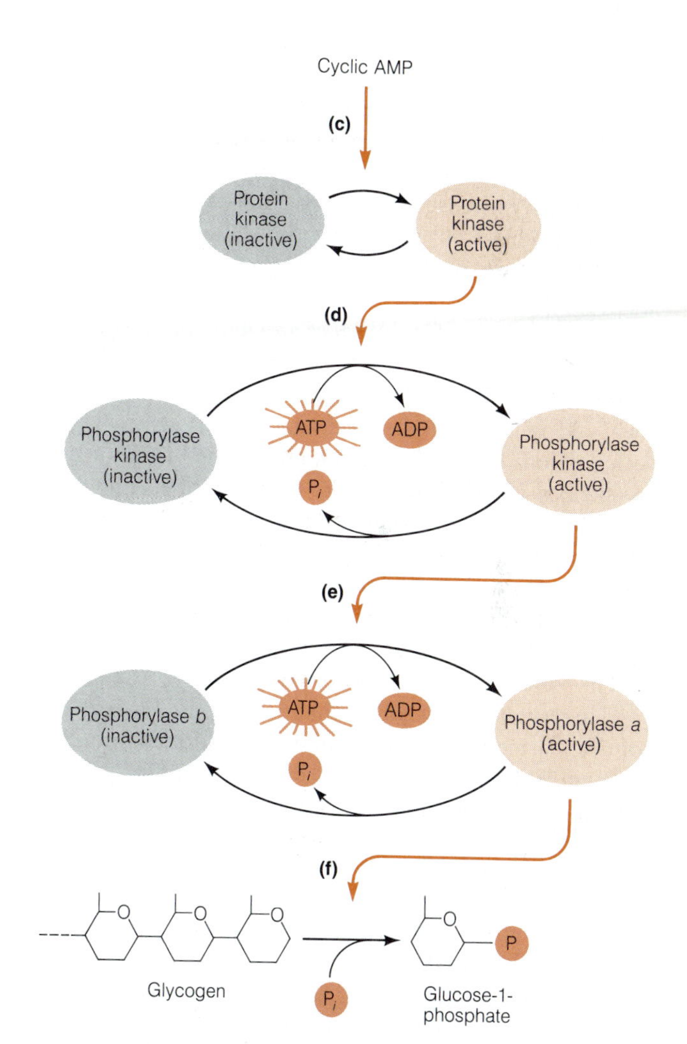

Figure 21-6 Stimulation of Glycogen Breakdown by Epinephrine. Muscle and liver cells respond to an increased concentration of epinephrine in the blood by increasing their rate of glycogen breakdown. The stimulatory effect of extracellular epinephrine on intracellular glycogen catabolism is mediated by a G protein/cyclic AMP regulatory cascade. In this case, the ligand is epinephrine (E) and the membrane protein to which it binds is an epinephrine-specific receptor (R_E). (a) Binding of epinephrine to its receptor activates the G protein, causing dissociation of the subunit-GTP complex. (b) The α subunit-GTP complex then binds to and activates adenylate cyclase (A). (c) As the intracellular level of cyclic AMP rises, cyclic AMP initiates a further cascade of regulatory events that begins with the activation of protein kinase. (d) Protein kinase then converts inactive phosphorylase kinase into the active form by ATP-dependent phosphorylation. (e) Active phosphorylase kinase in turn phosphorylates phosphorylase *b*, converting it into phosphorylase *a*, the active form of the enzyme. (f) Phosphorylase *a* then catalyzes the phosphorolytic cleavage of glycogen into molecules of glucose-1-phosphate.

Cyclic AMP is produced when a hormone binds to its membrane receptor, activating adenylate cyclase for cyclic AMP production. Cyclic AMP then activates a protein kinase, which in turn phosphorylates the enzyme system under hormonal control—glycogen breakdown, in the case of Figure 21-6, as well as the enzymes involved in a variety of other cellular functions (Table 21-4).

Calcium as a Second Messenger

Cyclic AMP is only one of several second messengers that act in the cytoplasm of cells. Although chemically quite different from a cyclic nucleotide, **calcium ion** (Ca^{2+}) also plays a very important role in regulating a variety of cellular functions in response to external signals. In this case, regulation is achieved by changes in the calcium concentration of the cytoplasm, mediated in turn by changes in membrane permeability.

Normally, most membranes are quite impermeable to calcium ions. The intracellular calcium concentration in most mammalian cells is maintained at a low level (10^{-7} *M* or less) by the continual outward pumping of calcium ions across the plasma membrane or, in the case of skeletal muscle cells, by the pumping of calcium into the cisternae of the sarcoplasmic reticulum. In both cases, the pumping is due to the activity of calcium ATPases embedded in the membrane. Because extracellular calcium concentrations are usually in the millimolar range, the gradient of calcium across the plasma membrane is very steep—often several thousandfold or more.

In response to an extracellular signal, the permeability of the plasma (or sarcoplasmic reticulum) membrane for calcium is increased transiently by the opening of calcium channels, thereby allowing calcium ions to enter the cytoplasm. Even a relatively small influx of calcium ions produces a large relative change in the calcium concentration within the cell, because the cytoplasmic calcium concentration is normally so low. Moreover, only modest increases in the intracellular calcium concentration are required for physiological effects within the target cell, since most calcium-dependent responses are triggered whenever the calcium concentration in the cytoplasm gets much above 10^{-7} *M*. Depending on the cell type, these responses include such diverse functions as muscle contraction, microtubule polymerization and disassembly, secretion by exocytosis, and regulation of intracellular levels of cyclic nucleotides.

The Role of Calmodulin

Like cyclic AMP, calcium does not exert its effect on physiological processes directly, but only when complexed with a specific protein that has a high affinity for it. The best-studied calcium-binding protein is **calmodulin.** Calmodulin was discovered in the late 1960s and is now known to play a pivotal role in the regulation of a variety of cellular processes.

Calmodulin is a small cytoplasmic protein present in eukaryotic (but not prokaryotic) cells. Calmodulin consists of a single polypeptide chain with 148 amino acids. The amino acid sequences of calmodulins from many different species of plants, animals, and eukaryotic microorganisms are strikingly similar. Moreover, the amino acid sequence of calmodulin is closely related to that of the calcium-binding subunit of the troponin complex discussed in Chapter 19. These findings suggest that calcium-binding proteins have evolved from a common molecular ancestor and serve an important function in all of these organisms.

How does calmodulin mediate calcium-activated events in the cell? The calmodulin molecule contains four binding sites for calcium ions (Figure 21-7). As the cytoplasmic calcium concentration increases, calmodulin binds calcium ions at each of its four sites successively, undergoing extensive conformational changes in the process. These changes increase the affinity of calmodulin for a variety of **calmodulin-binding proteins** that depend on the **Ca^{2+}-calmodulin complex** for activation. Most calmodulin-binding proteins are enzymes. In some cases, calmodulin is a permanent regulatory subunit of the enzyme, whereas in other cases the Ca^{2+}-calmodulin complex binds reversibly to a regulatory subunit. In both cases, the result is the same: An enzyme that is inactive at a low intracellular calcium concentration is activated as the calcium concentration increases.

The response of a target cell to an increase in calcium concentration depends critically on the particular calmodulin-binding proteins that are present in the cell. Thus the same change in calcium concentration can produce distinctly different effects in two target cells if the cells possess different calmodulin-sensitive enzyme systems.

The role of the Ca^{2+}-calmodulin complex in activating myosin light-chain kinase in smooth muscle has already been discussed (recall Figure 19-2). In some invertebrates, the contraction of striated muscle is also regulated by calmodulin, rather than by troponin. For example, the adductor muscle that closes the valve, or "shell," of a clam is under calmodulin control. Calmodulin is also thought to be important in the disassembly of spindle microtubules during cell division. When fluorescent antibodies are used

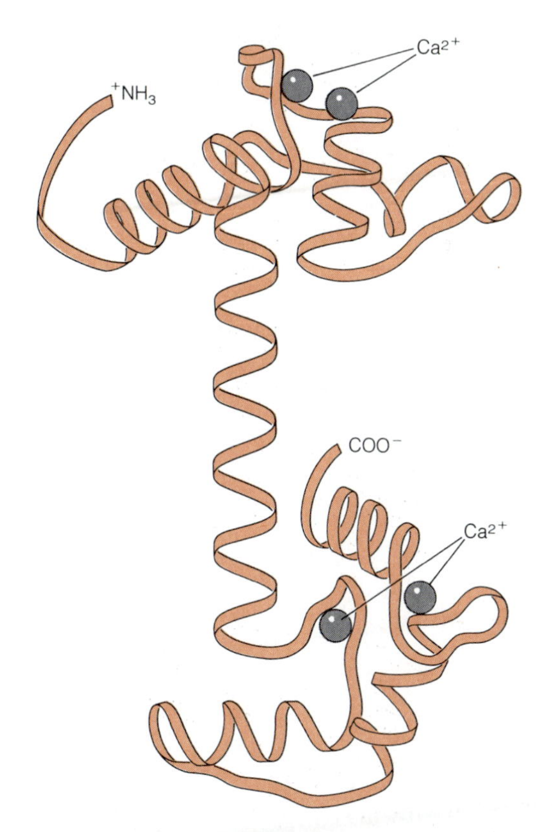

Figure 21-7 Structure of Calmodulin. Calmodulin is a cytoplasmic calcium-binding protein. This model of its molecular structure is based on data from X-ray crystallography. The molecule consists of two globular ends joined by a helical region. Each end has two calcium binding sites. It is likely that the helical region undergoes length changes as a result of calcium binding, thereby increasing the affinity of the Ca^{2+}-calmodulin complex for calmodulin-binding proteins.

to detect calmodulin in the spindle, the calmodulin is localized in the region between the kinetochores and the poles.

Calmodulin also mediates the effect of calcium on the intracellular levels of cyclic nucleotides. Recall that cyclic AMP is formed from ATP by the action of adenylate cyclase and is hydrolyzed to AMP by phosphodiesterase (Figure 21-4). Phosphodiesterase and at least some forms of adenylate cyclase are active only when the Ca^{2+}-calmodulin complex is bound to them. Since calcium affects both the formation and hydrolysis of cyclic AMP, the net effect of an increased concentration of calcium on the cyclic AMP level depends on the relative abundance and activities of the adenylate cyclase and phosphodiesterase enzymes.

Calcium also affects the process of exocytosis, apparently by facilitating the fusion of the membranes of secretory vesicles with the plasma membrane. Again, the effect is not due to calcium directly but is mediated by the Ca^{2+}-calmodulin complex. This complex also stimulates the calcium transport system of the plasma membrane, apparently by activating membrane-bound Ca^{2+}-ATPases.

Thus, calcium plays an important role in a remarkable variety of cellular activities. In each case the effect is elicited not by the calcium ion itself but by the Ca^{2+}-calmodulin complex. In these functions, calcium shares several properties with cyclic AMP: Its level is determined by events occurring within a membrane, and its effect depends on allosteric activation of an intracellular protein. Like cyclic AMP, therefore, calcium is a second messenger—a means of transmitting an impinging signal to the interior of the cell, thereby triggering specific intracellular events that are as important as they are diverse.

Inositol Trisphosphate and Diacylglycerol as Second Messengers

In the early 1980s, Michael Berridge noticed that the salivary secretion of certain insect larvae was activated by **inositol trisphosphate ($InsP_3$)**. We now know that $InsP_3$ is also a second messenger, functioning in a signal transduction pathway that involves a specific G protein, a regulatory kinase, and a flux of calcium ions. $InsP_3$ is generated from *phosphatidylinositol (PI)*, a relatively uncommon membrane phospholipid, by the reaction sequence shown in Figure 21-8. Two phosphate groups are first added to the inositol group of PI by successive kinase-catalyzed reactions to form PI-4,5-bisphosphate (PIP_2). This compound is then hydrolyzed by the membrane-bound enzyme **phospholipase C** to produce $InsP_3$, which is released into the cytoplasm. The other product of the hydrolysis, **diacylglycerol (DAG)**, remains membrane-bound but also acts as a second messenger, since it is an activator of a specific protein kinase that can phosphorylate a variety of proteins with different functions, depending on the target cell.

The $InsP_3$/DAG signal transduction pathway is shown in Figure 21-9. The sequence begins with the binding of a ligand to its membrane receptor, leading to the activation of a specific G protein called G_p (step a). G_p then activates phospholipase C, thereby generating both $InsP_3$ and DAG (step b). The $InsP_3$ is released into the cytoplasm, where it triggers the release of calcium from intracellular stores (step c). Calcium then binds to calmodulin, and the Ca^{2+}-calmodulin complex activates the desired physiological process, such that calcium serves in a sense as a "third messenger" in this system. The DAG that is also generated by phospholipase C activity remains in the membrane,

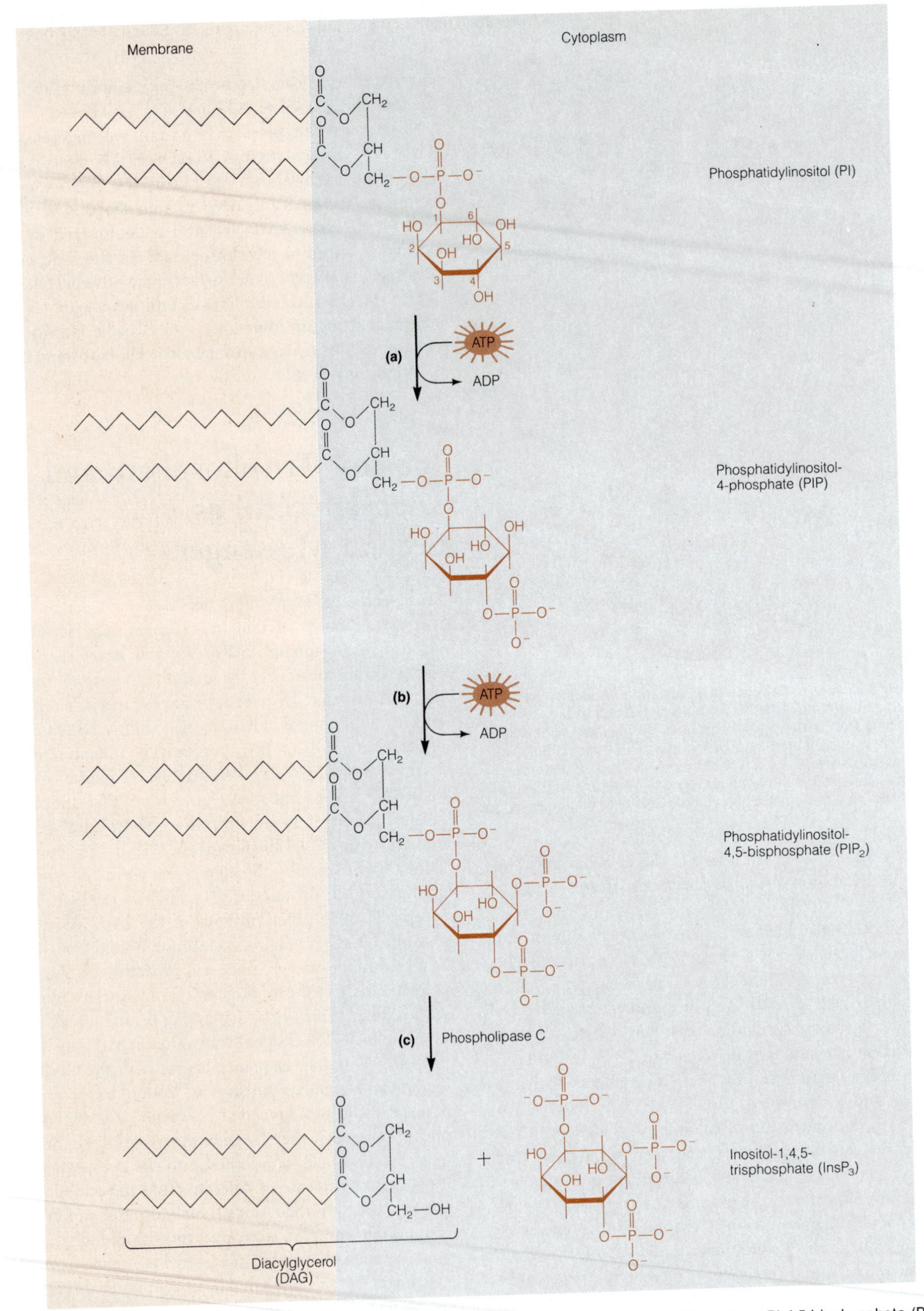

Figure 21-8 Synthesis of Inositol Trisphosphate and Diacylglycerol. Inositol trisphosphate ($InsP_3$) and diacylglycerol (DAG) are generated from phosphatidylinositol (PI), one of the phospholipids present in membranes. The reaction sequence includes (a) the phosphorylation of PI to form PI-4-phosphate (PIP), (b) the subsequent phosphorylation of PIP to form PI-4,5-bisphosphate (PIP_2), and (c) the cleavage of PIP_2 by phospholipase C to generate $InsP_3$ and DAG. $InsP_3$ is released into the cytoplasm, whereas DAG remains within the membrane. Both $InsP_3$ and DAG are second messengers in a variety of cellular processes.

where it activates **protein kinase C** (step d). This enzyme can then phosphorylate specific serine and threonine groups on a variety of target proteins, depending on the cell type. Cell functions that are activated in this way include sodium transport across excitable membranes and the sodium-hydrogen exchange mechanism in plasma membranes, which regulates intracellular pH.

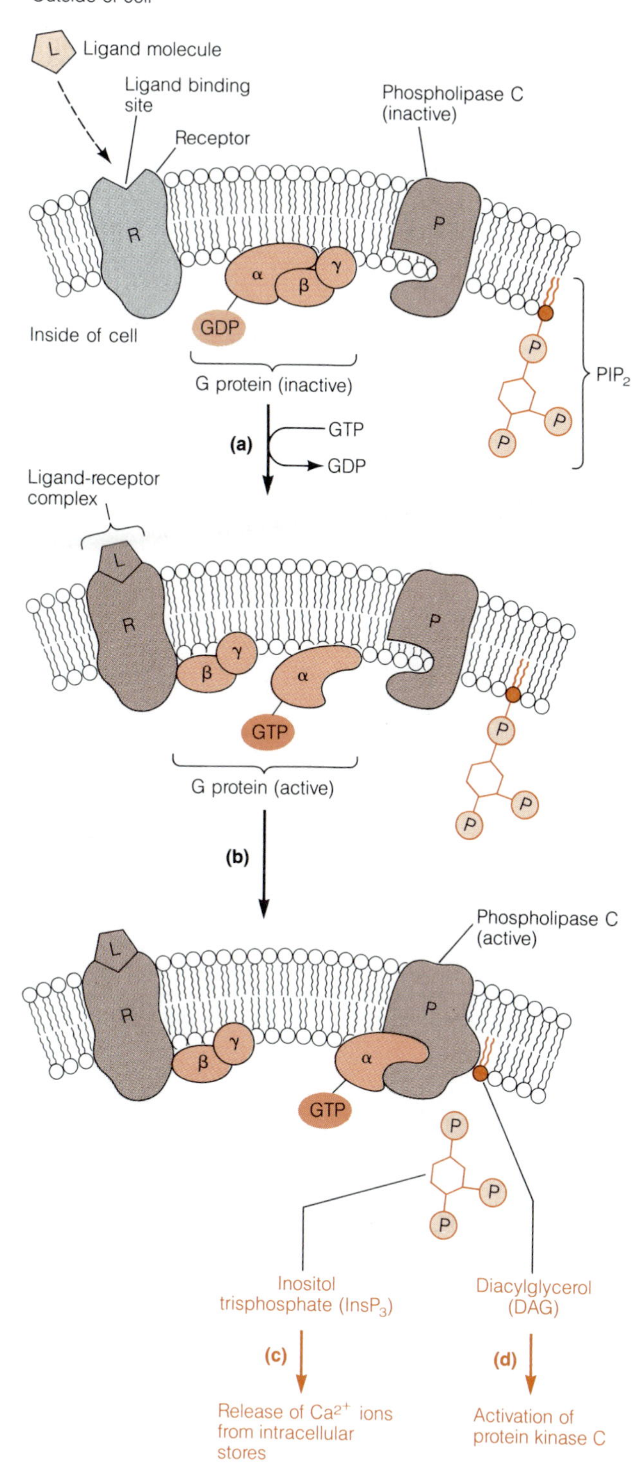

How do we know that the sequence of events shown in Figure 21-9 actually occurs in cells? The answer to this question begins with the original observation that a specific physiological phenomenon—salivary secretion, in this case—could be activated by the products of phospholipase action, $InsP_3$ and DAG. Evidence for the role of calcium was provided by a quite different experimental approach, involving the injection into a target cell of a calcium-dependent fluorescent dye. Because the fluorescence of the dye increases with the calcium concentration, the dye serves as a sensitive indicator of the intracellular calcium concentration. By measuring the increase in fluorescence in response to activation by a ligand or by $InsP_3$ directly, investigators were able to establish a link between ligand binding, $InsP_3$ generation, and calcium concentration.

To complete the sequence of events, the increase in calcium concentration had to be linked to the actual physiological response of the target cell. This link was established by treating target cells with a *calcium ionophore* (such as *ionomycin*) in the absence of extracellular calcium. The ionophore renders membranes permeable to calcium, thereby releasing intracellular stores of calcium. Treatment with the calcium ionophore mimicked the effect of $InsP_3$, thus implicating calcium as an intermediary in the $InsP_3$ signal transduction chain.

The role of DAG was established experimentally by showing that its effects can be mimicked by *phorbol esters*, plant metabolites that bind to protein kinase C, thereby activating it directly. Using ionomycin, phorbol esters, and other pharmacological agents, researchers showed that $InsP_3$-stimulated calcium release and DAG-mediated kinase activity are both required to produce a full response in target cells.

If the sequence of events shown in Figure 21-9 occurred only in insect salivary glands, Berridge's observations would have warranted no more than a modest note in the story of signal transduction. Instead, $InsP_3$ and DAG were quickly shown to be second messengers in a variety of regulated cell functions, some of which are indicated in

Figure 21-9 The Role of Inositol Trisphosphate and Diacylglycerol in Signal Transduction. Both inositol trisphosphate ($InsP_3$) and diacylglycerol (DAG) serve as second messengers, with a specific G protein (called G_p) as mediator. (a) When a neighboring receptor (R) is activated by binding of its specific ligand (L) on the outer surface of the plasma membrane, the receptor-ligand complex associates with the G_p protein, causing the displacement of GDP by GTP and the dissociation from the G protein of the α subunit-GTP complex. (b) The α subunit-GTP complex then binds tightly to a molecule of phospholipase C (P), activating it for cleavage of phosphatidylinositol bisphosphate (PIP_2) into one molecule each of $InsP_3$ and DAG. (c) The $InsP_3$ is released into the cytoplasm, where it triggers the release of calcium from intracellular stores. (d) The DAG remains in the membrane, where it activates protein kinase C, an enzyme that can phosphorylate a variety of target proteins.

Table 21-5. To date, more than 25 different cell-surface receptors are known to activate this transduction pathway, with new examples being reported each year.

Phospholipase A and the Prostaglandins

Just as phospholipase C generates products involved in intracellular regulation, a related enzyme, **phospholipase A,** is involved in the generation of a specific unsaturated fatty acid that leads in turn to another family of regulatory agents, the prostaglandins. **Prostaglandins** are paracrine hormones that activate contraction of smooth muscles. As the name implies, these substances were first thought to be produced only by the prostate gland, but we now know that prostaglandins are synthesized in most mammalian tissues. Processes under prostaglandin control include uterine contraction during labor, gastric and intestinal function, and control of blood pressure, mediated by the constriction or dilation of smooth muscles in small arteries called *arterioles*.

Figure 21-10 depicts the synthesis of prostaglandins from membrane phospholipids. Phospholipase A catalyzes the hydrolysis of ester bonds that link fatty acids to glycerol in phospholipids, thereby releasing free fatty acids. Cells usually have a low level of phospholipase A activity, but this activity increases greatly in response to stress or injury. One of the products of phospholipase A activity is **arachidonic acid,** a long-chain unsaturated fatty acid with four double bonds. Arachidonic acid can be converted into prostaglandins by an oxidative reaction sequence called the *cyclooxygenase pathway*.

Two other related groups of compounds, the *thromboxanes* and the *leukotrienes,* are also formed by oxidation of arachidonic acid. The thromboxanes and prostaglandins are both involved in the blood clotting reaction, as we will see shortly, whereas the leukotrienes are activators of smooth muscle contraction. An example of leukotriene activity is the constriction of air passages in the lungs in response to irritants such as smoke or pollen, an effect that sometimes leads to asthma.

The Role of Inositol Trisphosphate, Diacylglycerol, and Prostaglandins in the Regulation of Platelet Function

One of the most dramatic effects of prostaglandins is seen in the process of **blood clotting,** which involves the activation of *blood platelets*. Regulation of platelet function entails several different kinds of membrane receptors and second messengers and is therefore an especially good example of a cellular response coordinated by chemical signals.

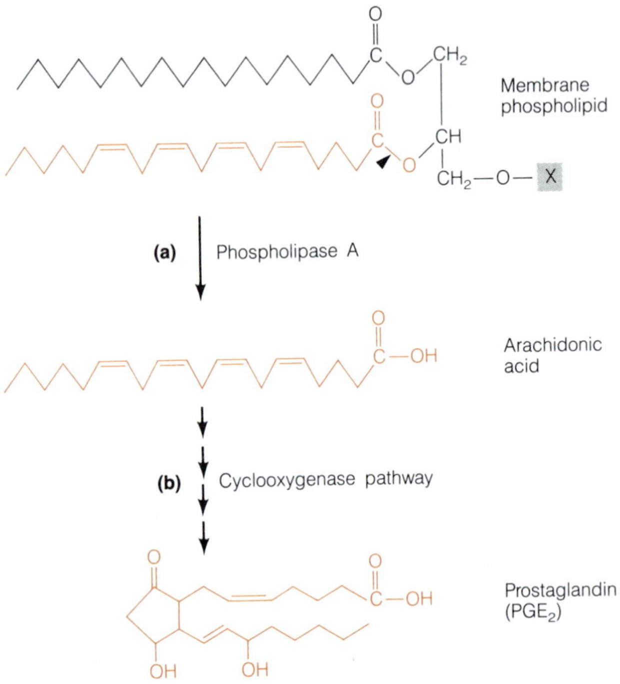

Figure 21-10 Phospholipase A and the Synthesis of Prostaglandins. (a) The synthesis of prostaglandins begins with the hydrolysis by phospholipase A of phospholipids to release fatty acids, including arachidonic acid. (b) Arachidonic acid is then oxidized by cyclooxygenase enzymes to produce the cyclic prostaglandins. Prostaglandins differ in the number and position of double bonds and the oxidation state of the oxygen-containing functional groups. The particular prostaglandin shown here is PGE_2.

Table 21-5 Examples of Cell Functions Regulated by Inositol Trisphosphate and Diacylglycerol

Regulated Function	Target Tissue	Hormone
Platelet activation	Blood platelets	Thrombin
Muscle contraction	Smooth muscle	Acetylcholine
Insulin secretion	Pancreas, endocrine	Acetylcholine
Amylase secretion	Pancreas, exocrine	Acetylcholine
Glycogen degradation	Liver	Antidiuretic hormone

Blood platelets are essential components of the blood-clotting mechanism. Platelets are not true cells, but instead are produced by fragmentation of bone-marrow cells called *megakaryocytes.* A platelet has no nucleus, mitochondria, or endoplasmic reticulum, but it does contain a cytoskeletal structure and numerous *dense granules* containing *platelet factors* that are involved in the clotting process.

Clotting involves a complex cascade of enzymatic events that is triggered by injury and leads rapidly to the formation of a clot, which is a tangled mass of red cells, platelets, and fibers of a protein called *fibrin.* The clotting mechanism must be carefully regulated, since inappropriate blood clots lead quickly to life-threatening situations. The coronary arteries of the heart are only a millimeter or so in diameter, and most heart attacks result from a small blood clot blocking the coronary circulation. Similarly, clots in the venous system can produce *phlebitis* (inflammation of the veins) in the lower limbs and *embolisms* (obstructions) in the lungs.

Clotting is initiated by tissue injury, which triggers a complex cascade of events leading to the formation of *thrombin* from its precursor, *prothrombin,* a plasma protein that is synthesized in the liver. Blood platelets play a central role in this cascade, first by aggregating in the injured region, then by secreting platelet factors that speed the formation of thrombin. Thrombin is actually a protease with the specific function of cleaving molecules of *fibrinogen* into fibrin molecules that stick together in a tangled array to form the clot.

Platelets can be activated experimentally by several substances, including epinephrine, ADP, and thrombin itself. The initial response is aggregation, followed by exocytotic release of platelet factors from the dense granules of the platelets. The generally accepted activation sequence is shown in Figure 21-11. Binding of the ligand to its receptor in the plasma membrane triggers two events: release of calcium from intracellular storage sites and activation of a sodium-hydrogen ion carrier in the plasma membrane, leading to a decrease in cytoplasmic H^+ concentration. The resulting increase in both calcium ion concentration and pH within the cell activates phospholipase A (probably by means of a G protein), resulting in the hydrolysis of membrane phospholipids and the release of arachidonic acid. A cyclooxygenase enzyme system converts the arachidonic acid into thromboxane and prostaglandins, which in turn activate phospholipase C.

Phospholipase C cleaves PIP_2 to $InsP_3$ and DAG, which then initiate the exocytotic process, probably by triggering a further release of calcium ions. The membranes of dense granules fuse with the plasma membrane, releasing platelet factors into the bloodstream. These platelet factors interact with other components of the blood-clotting process, increasing the rate at which a clot is produced. Meanwhile, the DAG is further acted upon by lipases to provide more arachidonic acid, and the resulting thromboxane is released from the platelets to engage in further platelet recruitment.

Knowledge of the regulatory pathways involved in platelet activation has enabled us to understand the mechanism of action of certain drugs. It was known, for example, that *aspirin* reduces significantly the incidence of heart attacks in older patients, apparently by decreasing the likelihood of small internal blood clots. We now understand that aspirin specifically blocks cyclooxygenase enzymes, thereby inhibiting synthesis of prostaglandins from arachidonic acid (Figure 21-11, step g). This inhibition slows platelet activation and recruitment, reducing the probability of small blood clots that might precipitate a heart attack by blocking blood flow through coronary arteries in the heart.

Growth Factors and Protein Kinases

Yet another aspect of membrane receptors concerns a remarkable relationship that links membrane receptors, regulatory proteins called *growth factors,* and the uncontrolled proliferation characteristic of tumor cells. **Growth factors** are highly specific serum proteins that stimulate cell division in particular types of cells. Examples of such proteins include *platelet-derived growth factor (PDGF), epidermal growth factor (EGF), nerve growth factor (NGF),* and *interleukin-2 (IL-2).* Most of the growth factors were discovered by cell biologists attempting to explain why vertebrate cells require serum to grow and divide when cultured in the laboratory. From such studies, we now understand that the requirement for serum is really a requirement for specific growth factors present in the serum, with different sets of these proteins required for the proliferation of different types of cells.

Each growth factor binds to a specific kind of membrane receptor on the surface of the target cell. However, these receptors differ from those we have considered so far in an important aspect: The receptors themselves transmit the signal across the membrane, without the need for G proteins. Such a receptor is a **transmembrane protein kinase,** capable of phosphorylating tyrosine, threonine, or serine groups of specific proteins on the cytoplasmic side of the membrane, but only when activated by ligand binding on the outer membrane surface.

Ligand binding causes the receptor protein to "self-activate" by catalyzing the addition of phosphate groups to its own tyrosine, threonine, or serine side groups, a

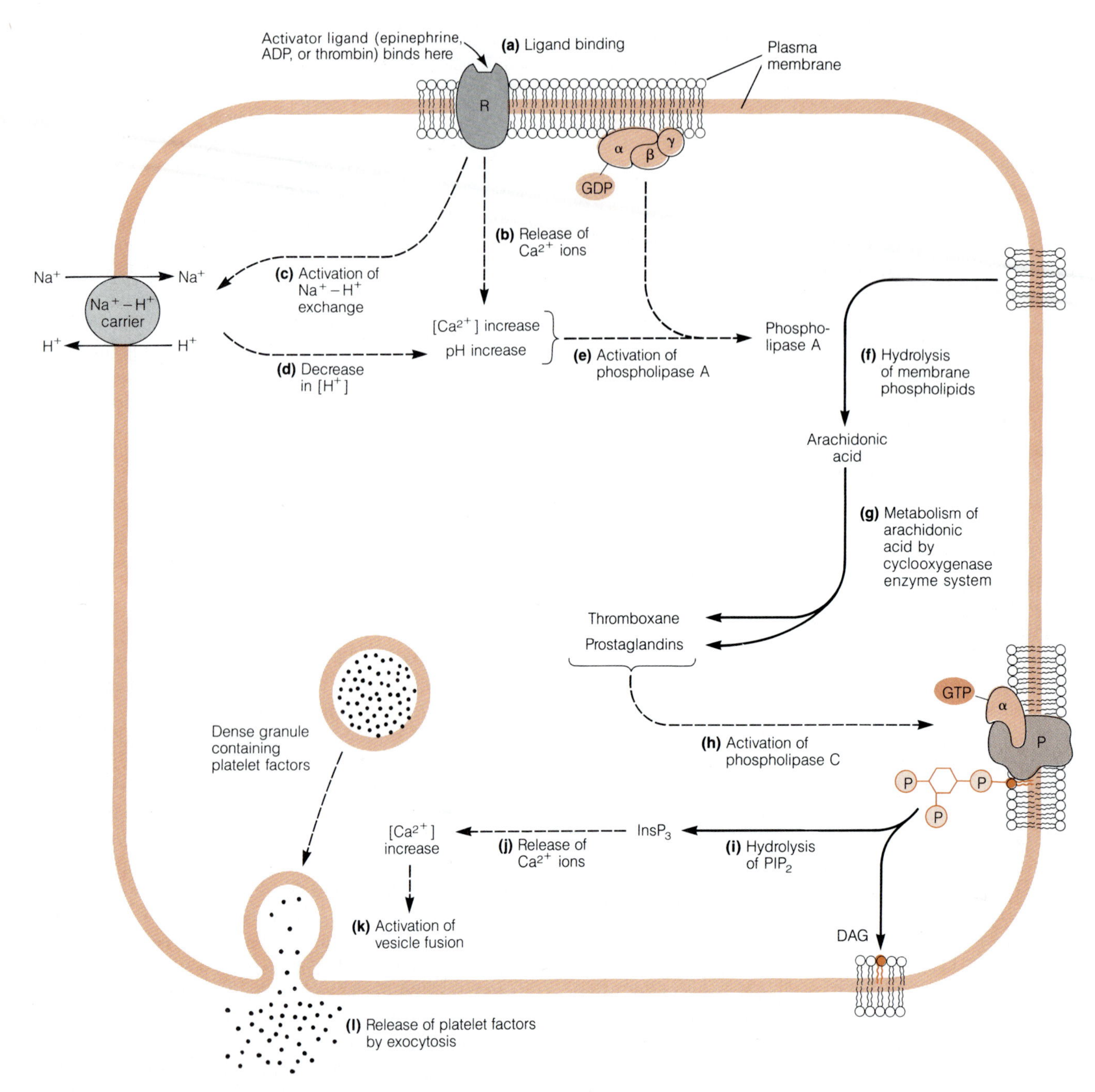

Figure 21-11 The Process of Platelet Activation. (a) Platelet activation begins with binding of an activator ligand (epinephrine, ADP, or thrombin) to an appropriate membrane receptor on the outer surface of the plasma membrane. (b) Ligand binding triggers the release of calcium from intracellular storage sites, causing an increase in the cytoplasmic Ca^{2+} concentration. (c) Ligand binding also activates an Na^+-H^+ carrier in the plasma membrane that exchanges sodium ions inward for protons outward, leading to (d) a decrease in the intracellular H^+ concentration and therefore to an increase in the pH within the platelet. (e) The increase in calcium concentration and pH within the platelet activates phospholipase A, probably mediated by a G protein. (f) Phospholipase A catalyzes the hydrolysis of membrane phospholipids, resulting in the release of arachidonic acid, which is (g) converted by the cyclooxygenase enzyme system into thromboxane and prostaglandins. (h) The latter substances activate phospholipase C, which (i) cleaves membrane-bound phosphatidylinositol-4,5-bisphosphate (PIP_2) into inositol trisphosphate ($InsP_3$) and diacylglycerol (DAG). (j) These second messengers then trigger a further release of calcium ions, such that (k) the elevated calcium concentration activates fusion of dense granules with the plasma membrane, thereby (l) releasing platelet factors from the cell by exocytosis.

process called *autophosphorylation.* (Actually, such receptors have two different types of subunits, one of which is a protein kinase capable of phosphorylating the other.) Once activated in this way, the receptor phosphorylates specific intracellular proteins, thereby activating them. The activated proteins in turn are responsible for the actual changes in cell function—in this case, those required for cell proliferation.

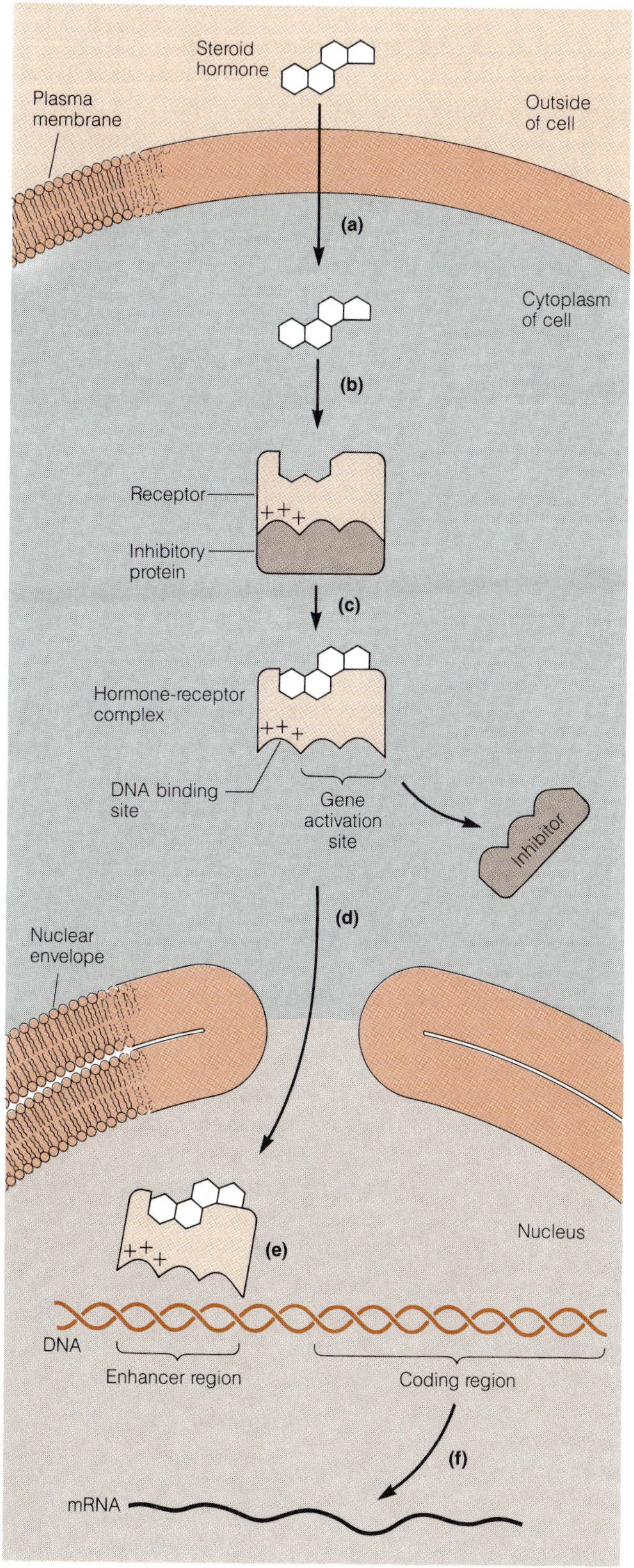

Recently, scientists have made the exciting discovery that the proteins encoded by certain *oncogenes* have amino acid sequences that are closely related to those of EGF and PDGF. In each case, the oncogene product is a protein kinase that phosphorylates tyrosine side groups on specific proteins, thereby stimulating cell proliferation. However, the oncogene-encoded kinase lacks the down-regulating control mechanisms of the normal receptor, leading to uncontrolled proliferation. Thus, oncogenes cause uncontrolled cell division in at least two ways—by encoding G proteins that lack GTPase and are therefore unable to switch off, as we saw earlier for the *ras* oncogene, and by encoding protein kinases that lack down-regulation. These observations point to an evolutionary relationship between viral reproduction and the regulatory processes controlling cell growth.

Figure 21-12 Activation of Steroid Hormone Receptors. Unlike most other hormones, steroid hormones enter the cell and act at the level of gene regulation. (a) Because of their lipophilic nature, steroid hormones can diffuse through the plasma membrane. (b) Once in the cell, a hormone molecule can bind to an intracellular receptor. (c) Binding of the hormone causes the release of an inhibitory protein, thereby exposing both a DNA binding site and a gene activation site on the receptor. (d) The activated hormone-receptor complex then enters the nucleus, where it (e) binds to an enhancer region of nuclear DNA. (f) Binding of the hormone-receptor complex to the enhancer activates transcription of the messenger RNAs that encode the proteins involved in the cell functions known to be regulated by the hormone.

Intracellular Receptors

The final group of receptors we will consider are located within the cytoplasm rather than at the cell surface and are therefore called **intracellular receptors.** As you might expect, the difference in location between surface receptors and intracellular receptors is related to the difference in the chemical properties of the hormones to which they respond. Most hormones or other ligands do not cross the plasma membrane and must therefore interact with their receptors on the cell surface, as we have already seen. However, a few hormones are able to cross membranes and can therefore interact directly with regulatory proteins inside the cell. The **steroid hormones** are in this category, presumably because of their lipidlike nature (recall Figure 3-26 on p. 67).

In contrast to other hormones that activate enzymes or membrane channels, the steroid hormones carry out their regulatory function at the level of the gene. For example, introduction of the steroid estrogen into cells in culture leads to an increase in the rate of synthesis of several proteins. This effect is highly specific, since a typical eukaryotic cell synthesizes as many as 10,000 different proteins, yet only a few are affected by the hormone.

How might a steroid affect the rate of gene expression within the nucleus? Experiments on the mechanism of action of estrogen and other steroid hormones have led to the scheme shown in Figure 21-12. In this model, a steroid hormone such as estrogen controls a specific region of DNA that is normally down-regulated—that is, the region is functional, but only at a low level, such that only a small amount of the protein product(s) is synthesized. When the appropriate steroid hormone becomes available, through either experimental manipulation or normal growth and development, the hormone diffuses across the plasma membrane and binds to specific intracellular receptor molecules in the cytoplasm of target cells. As a result of this binding, an **inhibitory protein** is released from the receptor, exposing a DNA binding site and a gene activation site on the receptor.

The activated receptor-hormone complex crosses the nuclear envelope into the nucleus, where it binds to DNA more or less indiscriminately. However, the DNA of target cells (mammary gland cells, in the case of estrogen) has specific exposed regions that serve as *enhancers* of transcription for specific genes. As we learned in Chapter 17, enhancers activate promoters by an unknown mechanism, thereby initiating transcription of the requisite genes. The result is selective expression of specific gene products in the target cells.

Perspective

Most cells are able to respond to hormones and other substances present in the extracellular fluid. Such responses are mediated by receptor proteins, either at the cell surface or in the cytoplasm. Each receptor protein has a binding site that is specific for its particular ligand. In the case of membrane receptors, ligand binding is followed by transmission of the signal to the cytoplasm, thereby regulating specific intracellular events.

Several different mechanisms for signal transduction are known. One of the most important pathways involves G proteins, which are activated when ligand binding to a neighboring receptor causes a conformational change in the G protein, resulting in displacement of GDP by GTP. The G protein then activates an enzyme system that produces intracellular chemical signals called second messengers. One of the most common second messengers is cyclic AMP, synthesized when the enzyme adenylate cyclase is activated by a G protein. In an alternative pathway, the second messengers are inositol trisphosphate and diacylglycerol, which are produced from phosphatidylinositol bisphosphate when a G protein activates the enzyme phospholipase C. Regardless of the pathway, second messengers mediate specific intracellular responses by activating specific enzymes or enzyme cascades.

Calcium ion also plays a central role in cellular activation processes and can be considered a second messenger as well. Calcium effects are mediated by calmodulin, a calcium-binding protein that is activated when calcium ions bind to it. Depending on the target cell, the Ca^{2+}-calmodulin complex can activate any of a variety of enzymes, thereby modulating enzyme activity in response to the calcium concentration.

Yet another regulatory pathway involves prostaglandins, which are classified as paracrine hormones because they are involved in local tissue responses, such as the reaction of cells to stress and injury. Prostaglandins are synthesized from arachidonic acid, which is in turn released from phospholipid by the action of phospholipase A.

Other receptors, notably those for the various growth factors, are protein kinases. Upon binding of the appropriate ligand, such receptors undergo phosphorylation by ATP. The phosphorylated receptor is then able to activate other enzymes by phosphorylation, thus controlling a specific metabolic pathway or growth process. Some viral oncogenes encode proteins that are closely related to normal growth factors. However, the oncogene product is not subject to down-regulation and therefore stimulates cells to undergo the continuous, uncontrolled growth characteristic of cancer.

Because of their lipophilic nature, steroid hormones can diffuse across the plasma membrane, interacting with cytoplasmic rather than membrane-bound receptors. The hormone-receptor complex then enters the nucleus, where it activates expression of specific genes by binding to their enhancer regions.

Key Terms for Self-Testing

Chemical Signals and Membrane Receptors
chemical signal (p. 644)
membrane receptor (p. 644)
chemical receptor (p. 644)

Endocrine and Paracrine Hormone Systems
endocrine hormone (p. 645)
paracrine hormone (p. 645)
endocrine tissue (p. 645)
exocrine tissue (p. 645)
target tissue (p. 645)
anterior pituitary (p. 646)
feedback control loop (p. 646)
sensor (p. 646)
effector (p. 646)
hypothalamus (p. 647)
releasing factor (p. 647)

Second Messengers and Membrane Receptors
second messenger (p. 647)
receptor protein (p. 648)
ligand (p. 648)

Cyclic AMP as a Second Messenger
cyclic AMP (p. 649)
adenylate cyclase (p. 649)
phosphodiesterase (p. 649)
G protein (p. 650)
activation (up-regulation) (p. 650)
inactivation (down-regulation) (p. 650)
cholera toxin (p. 654)
pertussis toxin (p. 654)
oncogene (p. 654)
epinephrine (adrenalin) (p. 654)

Calcium as a Second Messenger
calcium ion (Ca^{2+}) (p. 656)
calmodulin (p. 656)
calmodulin-binding protein (p. 656)
Ca^{2+}-calmodulin complex (p. 656)

Inositol Trisphosphate and Diacylglycerol as Second Messengers
inositol trisphosphate ($InsP_3$) (p. 657)
phospholipase C (p. 657)
diacylglycerol (DAG) (p. 657)
protein kinase C (p. 659)
phospholipase A (p. 660)
prostaglandin (p. 660)
arachidonic acid (p. 660)
blood clotting (p. 660)
blood platelet (p. 661)

Growth Factors and Protein Kinases
growth factor (p. 661)
transmembrane protein kinase (p. 661)

Intracellular Receptors
intracellular receptor (p. 663)
steroid hormone (p. 663)
inhibitory protein (p. 664)

Suggested Reading

General References and Reviews

Berridge, M. The molecular basis of communication within the cell. *Sci. Amer.* 253 (October 1985): 142.

Levitsky, A. *Receptors: A Quantitative Approach.* Menlo Park, Calif: Benjamin/Cummings, 1984.

Nishizuka, Y. Protein kinases in signal transduction. *Trends Biochem. Sci.* 9 (1984): 163.

Synder, S. H. The molecular basis of communication between cells. *Sci. Amer.* 253 (October 1985): 132.

Norman, A. W., and G. Litwack. *Hormones.* San Diego, Calif: Academic Press, 1987.

Cyclic AMP

Levitsky, A. From epinephrine to cyclic AMP. *Science* 241 (1988): 800.

Schramm, M., and Z. Selinger. Message transmission: Receptor-controlled adenylate cyclase system. *Science* 225 (1984): 1350.

Sutherland, E. W. Studies on the mechanism of hormone action. *Science* 177 (1972): 401.

G Proteins

Freissmuth, M., P. J. Casey, and A. G. Gilman. G proteins control diverse pathways of transmembrane signaling. *FASEB J.* 3 (1989): 2125.

Gilman, A. G. G proteins: Transducers of receptor-generated signals. *Annu. Rev. Biochem.* 56 (1987): 615.

Lai, C.-Y. The chemistry and biology of cholera toxin. *CRC Crit. Rev. Biochem.* 9 (1987): 171.

Stryer, L. The molecules of visual excitation. *Sci. Amer.* 257 (July 1987): 42.

Stryer, L., and H. R. Bourne. G proteins: A family of signal tranducers. *Annu. Rev. Cell Biol.* 2 (1986): 391.

Calmodulin

Carafoli, E., and J. T. Penninston. The calcium signal. *Sci. Amer.* 253 (November 1985): 70.

Cheung, W. Y. Calmodulin plays a ulation. *Science* 207 (1980): 19.

Klee, C. B., T. H. Crouch, and P. G. R *Rev. Biochem.* 49 (1980): 489.

Inositol Trisphosphate and Diacylglycerol

Berridge, M. J. Inositol lipids and calcium signalling. *Proc. Roy. Soc. London* 234 (1988): 359.

Berridge, M. J. Inositol trisphosphate and diacylglycerol: Two interacting second messengers. *Annu. Rev. Biochem.* 56 (1987): 159.

Hokin, L. E. Receptors and phosphoinositide-generated second messengers. *Annu. Rev. Biochem.* 54 (1985): 205.

Nishizuka, Y. Studies and perspectives of protein kinase C. *Science* 233 (1986): 305.

Intracellular Receptors

Gehring, U. Steroid hormone receptors: Biochemistry, genetics, and molecular biology. *Trends Biochem. Sci.* 12 (1987): 399.

Yamamoto, K. R. Steroid receptor regulated transcription of specific genes and gene networks. *Annu. Rev. Genet.* 19 (1985): 209.

Problem Set

1. **Chemical Signals and Second Messengers.** Fill in the blanks with the appropriate terms:
   (a) ______________ is an intracellular protein that binds calcium and activates enzymes.
   (b) The ______________ is a gland that regulates other endocrine tissues of the body.
   (c) A substance that fits into a specific binding site on the surface of a protein molecule is called a ______________.
   (d) Two products of phospholipase activity that serve as second messengers are ______________ and ______________.
   (e) Cyclic AMP is produced by the enzyme ______________ and degraded by the enzyme ______________.

2. **Receptors and Hormones.** Indicate whether each of the following statements is true (T) or false (F).
   (a) Receptors are found only in multicellular organisms.
   (b) Receptors are not present in plant cells.
   (c) Some hormones are released into the bloodstream through specialized ducts.
   (d) The contraction of smooth muscles is hormonally controlled.
   (e) Most hormones are synthesized from carbohydrates.
   (f) Steroid hormones are considered second messengers, since they cross the cell membrane to bind to intracellular receptors.
   (g) Calmodulin is an example of a peptide hormone.
   (h) Inositol trisphosphate is considered a lipid.

3. **Intercellular Communication.** Animal cells depend on three major communication systems: the nervous system, the endocrine system, and the paracrine system.
   (a) List a major advantage and a limitation for each of these systems.
   (b) Why do you suppose several different mechanisms for intercellular communication have evolved?

4. **Control of Insulin Release.** The release of insulin from pancreatic islet (endocrine) cells is regulated by a feedback mechanism. These cells have excitable membranes that respond to blood glucose concentrations by repetitive depolarization, the rate depending on the glucose level. From your knowledge of exocytosis, excitable membranes, and membrane receptors, sketch a feedback loop that would allow islet cells to regulate blood glucose.

5. **Neuroendocrine Response.** The nervous and endocrine systems often act in concert to regulate a specific physiological function. Imagine that you have just been frightened by something—the sudden thought of a cell biology exam that you forgot to study for, perhaps. Sketch the pathway in the nervous system by which this mental event causes the adrenal gland to release epinephrine. Include the following elements in your pathway: brain, spinal cord, peripheral nervous system, adrenal gland, bloodstream, and receptors on cardiac (heart) cells.

6. **Antidiuretic Hormone and Kidney Function.** Antidiuretic hormone (ADH) controls the process in the kidney whereby water is either reabsorbed into the blood or lost to the urine. Reabsorption of water involves the filtration of fluid from the blood into small U-shaped structures called kidney tubules. The tubules lead directly to collecting ducts, which in turn carry fluid to the ureter and bladder. Water diffuses across the membrane of the kidney tubule down an osmotic gradient and is returned to the blood. To diffuse fast enough, most of the water passes through small channels in the membranes of the kidney tubule cells. Given these facts and considering the antidiuretic property of ADH, propose a mechanism by which this hormone could regulate water loss from the blood to the urine.

7. **G Proteins and Disease.** Whooping cough, or pertussis, is caused by a bacterial infection of the respiratory tract. The pertussis toxin is known to bind to a G protein, and the secretory cells of the respiratory tract then produce copious quantities of mucus. Sketch a regulatory pathway whereby pertussis toxin elicits this response.

8. **Why Calcium?** Calcium is a universal second messenger, in the sense that it is used for regulatory purposes by all cells. Why do you think cells use calcium ion as a second messenger, rather than other common ions such as sodium, potassium, chloride, magnesium, or hydrogen ions?

9. **Membrane Receptors and Medicine.** Hypertension, or high blood pressure, is often seen in elderly people. A typical prescription to reduce a patient's blood pressure includes compounds called *beta-blockers,* which block beta-adrenergic receptors throughout the body. These receptors bind epinephrine, thereby activating a cellular response. Why do you think that beta-blockers are effective in reducing blood pressure?

10. **G Proteins in Plants?** To date, G proteins have been reported in animal cells, and bacterial proteins called *elongation factors* that are known to bind and hydrolyze GTP have been shown to have some amino acid sequence homologies with mammalian G protein, but the presence of G proteins in plant cells has not been definitively proved. Describe how you might determine whether plant cells contain G proteins.

PART SIX

SPECIAL TOPICS IN CELL BIOLOGY

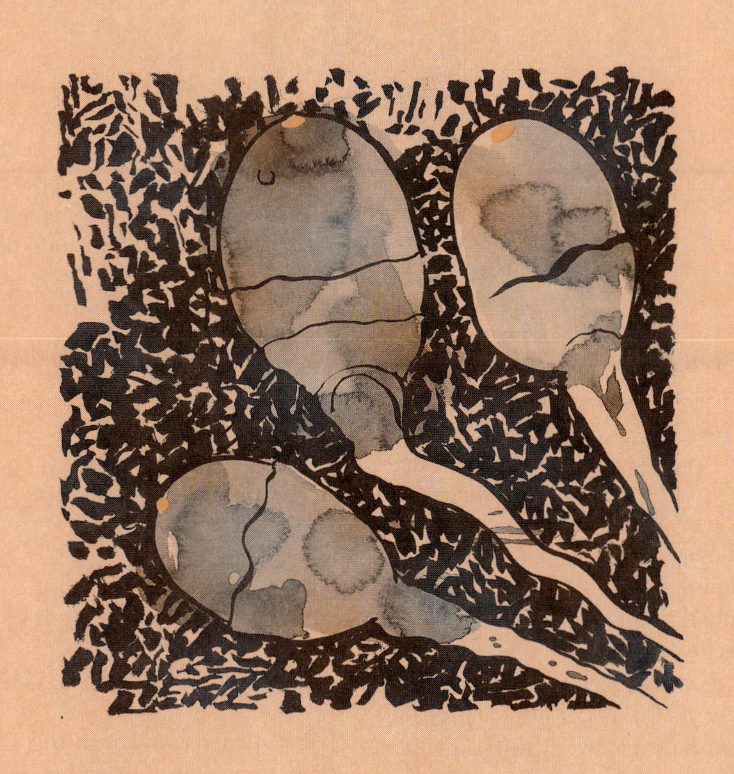

22

Cellular Aspects of Embryonic Development

In the preceding chapters we have studied the basics of cell biology, from cell structure to the flow of genetic information to cell signaling. We have seen how some cells use these basic processes to carry out specialized functions, such as muscle contraction, motility, and nervous function. No organism, however, can be satisfactorily understood as a single, static cell, or even as a group of specialized cells. Rather we must try to understand how the basic processes and specialized cell types interact to form an intact organism, and how that organism uses the same processes to build a *life cycle*.

In simple, single-celled organisms such as bacteria, the life cycle may simply be the cell cycle that we studied in Chapter 14. But the cell cycle itself is a remarkably intricate series of events. In multicellular organisms, cells must specialize, and groups of cells must interact to form tissues and organs. The life cycle of multicellular organisms thus involves temporally and spatially organized cell divisions, as well as changes in *cell fate*. For example, one daughter cell of a particular cell division may eventually give rise to muscle cells, while the other daughter cell gives rise to gut cells. Ultimately, we would like to know all the changes that each cell in the body goes through as it *differentiates* into its final form.

Yet even with such knowledge, our understanding of the life cycle would be incomplete. At some point in its life cycle every multicellular organism that reproduces sexually is represented by a single cell. That single cell—the fertilized egg produced by the fusion of sperm and egg—will develop into an adult without help from the outside world (other than nutritional support in live-bearing species). How are top and bottom, front and back, distinguished in an apparently homogeneous egg? How does a fertilized egg give rise to discrete groups of cells? Do cells of the developing embryo act independently, or does cell-cell communication play an important role in embryonic development? What forms of gene regulation mediate the changes in protein expression that occur as a cell becomes specialized?

Questions such as these make up the exciting field of **developmental biology.** To the newcomer, this field may seem as diverse as the forms of life itself; but developmental processes are often united by common principles, whether those processes are studied in bacteria or worms, fruit flies or frogs, yeast or human cells. Furthermore, you have already mastered many of the concepts and molecular mechanisms that underlie the varied phenomena studied by developmental biologists. Thus, as you work through this chapter, you will continually encounter familiar topics set in the new and broader context of developmental biology.

Studying Development in Unicellular Eukaryotes

Our primary concern in this chapter will be cellular and molecular aspects of embryonic development in multicellular eukaryotes of the animal kingdom. We begin, however, with a look at two unicellular eukaryotes, *yeast* and *slime mold*. To complete the life cycle, the cells of these simpler species execute sophisticated cell biological changes and cell-cell interactions. These events have much in common with, and thus provide important models for, the cellular basis of developmental events in higher eukaryotes.

Yeast: A Useful Model for Studying Development

Biochemists, geneticists, and cell biologists have studied *Saccharomyces cerevisiae*, more commonly known as baker's yeast, for a long time. Recently, however, developmental biologists have joined the ranks of scientists studying yeast. Much of their interest focuses on the *changes in cell type* and *cell-cell interactions* that occur in the sexual life cycle of yeast.

Yeast Has Three Different Cell Types. Yeast cells exist in either the haploid or the diploid genetic state. Haploid yeast cells can grow vegetatively or they can mate, thereby initiating the sexual life cycle (Figure 22-1). The resulting diploid cells also have two possible fates: *vegetative growth* as diploids or *sporulation*, the meiotic process that returns the organism to the haploid state (see Chapter 15). Not any two haploid cells can mate, however; rather, haploid yeast cells exist as one of two **mating types** known as **a** and **α** (alpha), and only cells of *opposite* mating type can fuse to generate the diploid, which is designated **a**/**α** (Figure 22-1). Thus, the mating types **a** and **α** are the yeast equivalents of two kinds of gametes, and the **a**/**α** diploid is the equivalent of a fertilized egg.

To survive in unfavorable conditions, vegetatively growing haploid yeast must be able to mate and sporulate. Thus cells of both mating types must be present in every yeast colony. To meet this requirement, yeast cells often switch their mating type as they divide. Mating can therefore occur in a colony that arose from a single haploid cell, a phenomenon known as *homothallism*. (Note, however, that homothallic mating involves no exchange of genetic information.)

This **mating-type switch** causes many changes in the yeast cell. Physical rearrangements of the DNA occur in the yeast's chromosomes, leading to changes in transcription patterns. As a result, cell surface receptors of one kind are replaced by a different class of receptors. The two cell types also secrete different proteins. All in all, the cell biology of **a** and **α** cells is very different. Furthermore, the

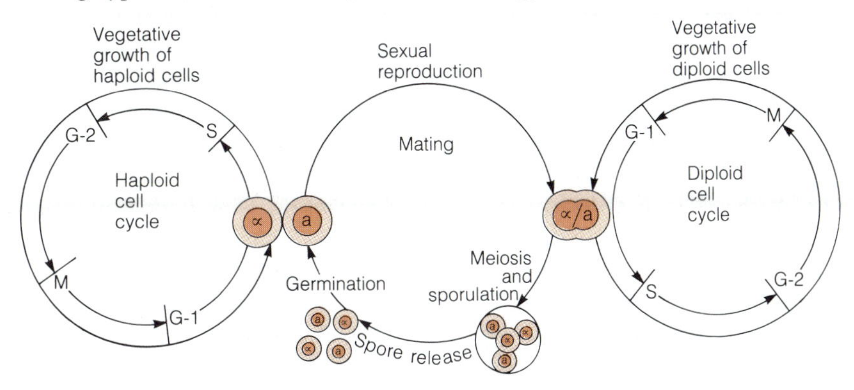

Figure 22-1 The Life Cycle of *Saccharomyces cerevisiae*. Yeast can grow vegetatively as either a haploid (left) or a diploid (right). The shift from haploid to diploid is achieved through mating between two cells of opposite mating type. The return to the haploid state occurs when conditions for vegetative growth are unfavorable, and the diploid cell undergoes meiosis and sporulation to produce four haploid spores. Upon germination, the dormant spores reenter the haploid life cycle. G-1, S, G-2, and M are the phases of the cell cycle.

Table 22-1 The Haploid and Diploid Cell Types of Yeast

	Cell Type		
Cell Trait	**a**	α	**a**/α
MAT allele	**a**	α	**a**/α
Ploidy	haploid	haploid	diploid
Ability to mate	Yes	Yes	No
Secreted mating factor	**a** factor	α factor	None
Cell surface receptor	α factor	**a** factor	None
Genes expressed	**a**-specific and haploid-specific genes	α-specific and haploid-specific genes	**a**/α-specific genes
Possible cell fates	Vegetative growth or mating with α cell	Vegetative growth or mating with **a** cell	Vegetative growth or meiosis and sporulation

cell formed by the mating of a and α cells, the a/α diploid cell, is a third distinct cell type, with its own characteristic traits (Table 22-1, p. 671).

The Molecular Mechanics of the Mating-Type Switch. Of the many changes that occur when a yeast cell switches mating type, the most important is the physical rearrangement of the DNA. A single genetic locus, designated ***MAT,*** specifies a haploid cell's mating type. The *MAT* locus exists in one of two genetic states, or *alleles:* a-mating type cells contain the DNA of the a allele at the *MAT* locus (*MAT*a), and α-mating type cells contain the DNA of the α allele (*MAT*α). Thus, to switch mating type, an a cell must replace the DNA at its *MAT* locus—the a allele—with the DNA encoding the α allele.

But how is such a switch possible? The answer lies in the fact that every yeast cell has two extra copies of the *MAT* locus: One copy contains all the α-specific DNA sequences, and the other copy contains all the a-specific DNA sequences. As shown in Figure 22-2, the *MAT* locus itself, containing either the *MAT*a or *MAT*α allele, is located between the two extra copies of *MAT* on chromosome 3. *HML*α (for *ho*mothallic *l*eft) is located 180 kilobase pairs to the left of *MAT,* and *HMR*a (for *hm*-right) is located 150 kilobase pairs to the right of *MAT.*

When a cell switches mating type, a new DNA copy of the information from either *HML*α or *HMR*a is made, and this "cassette" is inserted into the *MAT* locus. Before the new cassette can be inserted, however, a *site-specific endonuclease* must excise the old DNA cassette at *MAT.* Thus, when an a cell switches, the "a" DNA cassette is cut out of the *MAT* locus and discarded, and the "α" DNA cassette is inserted into the *MAT* locus. Unlike the DNA at the *MAT* locus, the DNA at *HML*α and *HMR*a never changes.

As we said earlier, the proteins produced by the *MAT* locus determine a cell's mating type. But the extra copies of *MAT* at *HML*α and *HMR*a present yeast cells with a problem: If the cell contains complete copies of *both* the α and the a alleles, why aren't both sets of proteins made? The answer is that a set of regulatory genes, known as the *silent information regulator* (*SIR*) genes, act together to prevent expression of the genetic information at *HML*α and *HMR*a (Figure 22-2). The proteins encoded by the *SIR* genes block transcription of *HML*a and *HML*α by binding to specific DNA sequences surrounding the a and α DNA cassettes at *HML*α and *HMR*a. The function of the SIR proteins is thus analogous to that of the *lac* repressor, which we encountered in Chapter 17.

Cell-Cell Communication Initiates Mating. Having met the requirement of separate cell types for mating, we now turn to the initiation of the mating process itself, where cell-cell communication plays a critical role. Each cell type secretes a **peptide mating factor.** The a cells make the 12-amino-acid a factor, while the α cells make the 13-amino-acid α factor. In addition, cells of each mating type express a cell surface receptor that recognizes the opposite cell type's mating factor. Since they carry out chemical signaling *between* organisms, the mating factors are ***pheromones.*** (In contrast, *hormones* carry out intercellular chemical signaling *within* multicellular organisms, as discussed in Chapter 21.)

The immediate result of the binding of a mating factor to its cell surface receptor is **cell cycle arrest:** The cell cycle stops until the cell encounters an arrested cell of the opposite mating type. In addition, the cell begins making proteins that will be needed for cell fusion. The signal for cell cycle arrest is relayed from the cell surface to the rest of the cell by *G proteins,* whose role in intracellular signaling we studied in Chapter 21. Figure 22-3 presents a model of the events that initiate mating in an α cell: Binding of a factor to the a-factor receptor on the surface of the α cell stimulates the activity of the G proteins, which then

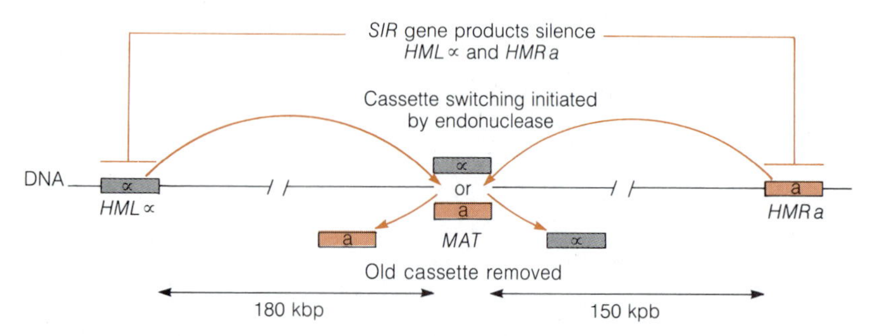

Figure 22-2 DNA Map of the Yeast *MAT* Locus. Chromosome 3 of *Saccharomyces* contains three copies of the mating type information. The central *MAT* locus contains either the **a**- or the α-specific DNA and specifies the mating type of the cell. The *HMRa* locus, 150 kilobase pairs (kbp) to the right of the *MAT* locus, contains a complete copy of the **a**-mating type DNA. The *HML*α locus, located 180 kilobase pairs to the left of *MAT,* contains a complete copy of the α-mating type DNA. *HMRa* and *HML*α are transcriptionally "silenced" by the action of the products of the *SIR* genes. When an **a** or an α cell switches mating types, the **a** or α DNA at the *MAT* locus is excised by a site-specific endonuclease, and a DNA "cassette" copy of the alternate mating type DNA is inserted in its place.

signal the cell to cease cycling (that is, to stop the cell cycle) and to prepare for fusion with an **a** cell.

When two arrested cells of opposite mating type come in contact, the cells fuse, and cell division resumes. Whereas the presence of either *MATa* or *MATα* specifies the characteristics of haploid cells, the presence of *both MATa* and *MATα* in a diploid cell specifies the **a**/**α** cell type (Table 22-1). Diploid cells will continue dividing until they are deprived of suitable carbon and nitrogen sources. At that point, **a**/**α** diploids will undergo the final developmental switch of the yeast life cycle—meiosis and the return to haploidy in the form of dormant spores (Figure 22-1).

Dictyostelium: Multicellular Development in a Unicellular Organism

Many researchers have studied development in another unicellular eukaryote—the cellular slime mold *Dictyostelium discoideum*. Although formally a unicellular organism, *Dictyostelium* spends much of its life cycle as an aggregate of individual amoeboid cells. As Figure 22-4 shows, *spore germination* gives rise to unicellular *amoebas*, which aggregate into a *mound* of up to a million cells. **Cyclic AMP** (cAMP), a molecule we have seen in a variety of biological roles, triggers this aggregation. Spiral waves of high cAMP concentrations emanate from the focus of the developing mound, serving as a *chemoattractant* for the outlying cells (Figure 22-5). As in the response of yeast cells to the mating pheromones, cell surface receptors (in this case specific for cAMP) and cytoplasmic G proteins relay the environmental stimulus to the rest of the cell. The cell responds by migrating toward the source of the cAMP (Figure 22-5). Researchers have identified several specific genes whose level of expression changes in response to the higher levels of cAMP during aggregation.

Once aggregation is complete, a tip elongates to form the leading edge of the migrating *pseudoplasmodium*, or *slug*. The pseudoplasmodium eventually becomes stationary and then develops into the mature *fruiting body*, consisting of two cell types—*stalk cells* and *spore cells*. Again, researchers have identified a key molecule in this cell-type switch. **Differentiation-inducing factor** (**DIF**) is a lipid-like compound that is synthesized by prespore cells in the posterior of the slug. Unlike cAMP, the lipid-like DIF molecule represents a novel class of bioactive molecules. When anterior cells of the slug encounter DIF from the posterior prespore cells, they respond by differentiating into stalk cells (Figure 22-4). Again, differentiation involves changes in the level of expression of several specific genes.

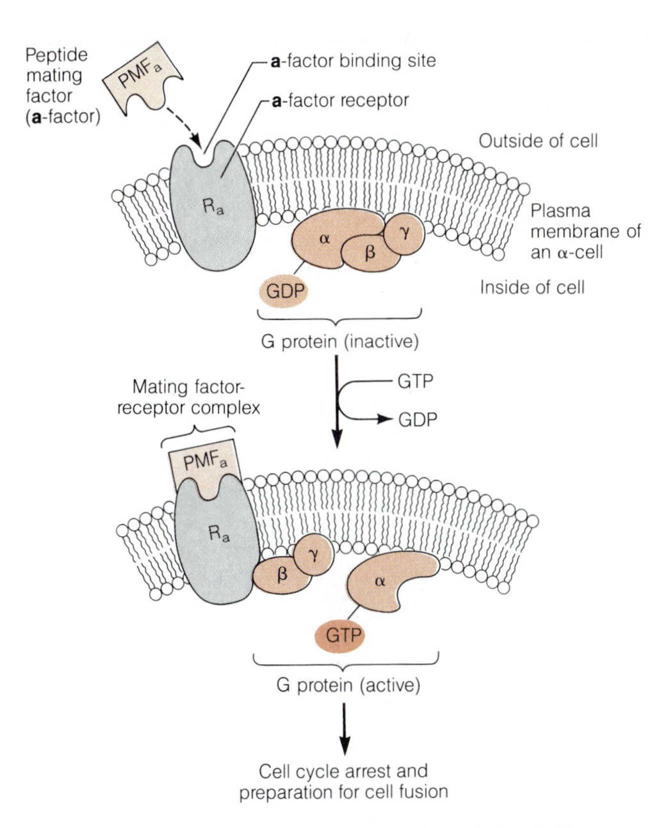

Figure 22-3 The Molecular Events That Initiate Mating in Yeast. Mating begins when the **a**-factor polypeptide binds to a specific receptor protein on the surface of an α cell. This triggers the dissociation of the subunits of the specific G protein associated with the receptor. The activated G protein subunits signal the cell to arrest the cell cycle and prepare for fusion with an **a** cell. For the arrested α cell to mate, a reciprocal set of events must occur in an **a** cell.

Animal Development: Into the Forefront of Modern Biology

The developmental events of yeast cells and slime molds foreshadow the intricate changes in cell biology and cell-cell interactions that occur during animal development. Unfortunately, our understanding of what goes on in the cells and tissues of a developing animal is seldom as complete as our comprehension of these simpler systems. But decades of intense scrutiny of animal development have provided a solid foundation of descriptive embryology and morphology, and equally intense biochemical and genetic analysis of embryogenesis in recent years has thrust the field of developmental biology into the forefront of modern biology.

We will begin our tour of animal development with the events of *gametogenesis, fertilization,* and *activation*. We will then look at events that are common to embryogenesis in all animals. These include *cleavage, gastrulation,* and *formation of the germ layers*. We will focus on

Figure 22-4 Dictyostelium discoideum Shows Multicellular Development. (a) In the beginning of the life cycle, germinated spores become single-celled amoebas, each about 8 μm long. (b) Upon nutrient starvation, about a million amoebas begin to interact, aggregating into a mound, or pseudoplasmodium, with an apical tip. (c) The apical tip then elongates, forming the leading edge of the migrating slug, which can be up to 1.5 mm long. (d) The slug may migrate for several hours before becoming a stationary pseudoplasmo dium. (e) Finally, to form the 1- to 2-mm-tall mature fruiting body, anterior cells of the pseudoplasmodium differentiate into stalk cells, while posterior cells become spores. (f) Spore dispersal then paves the way for another generation of amoebas.

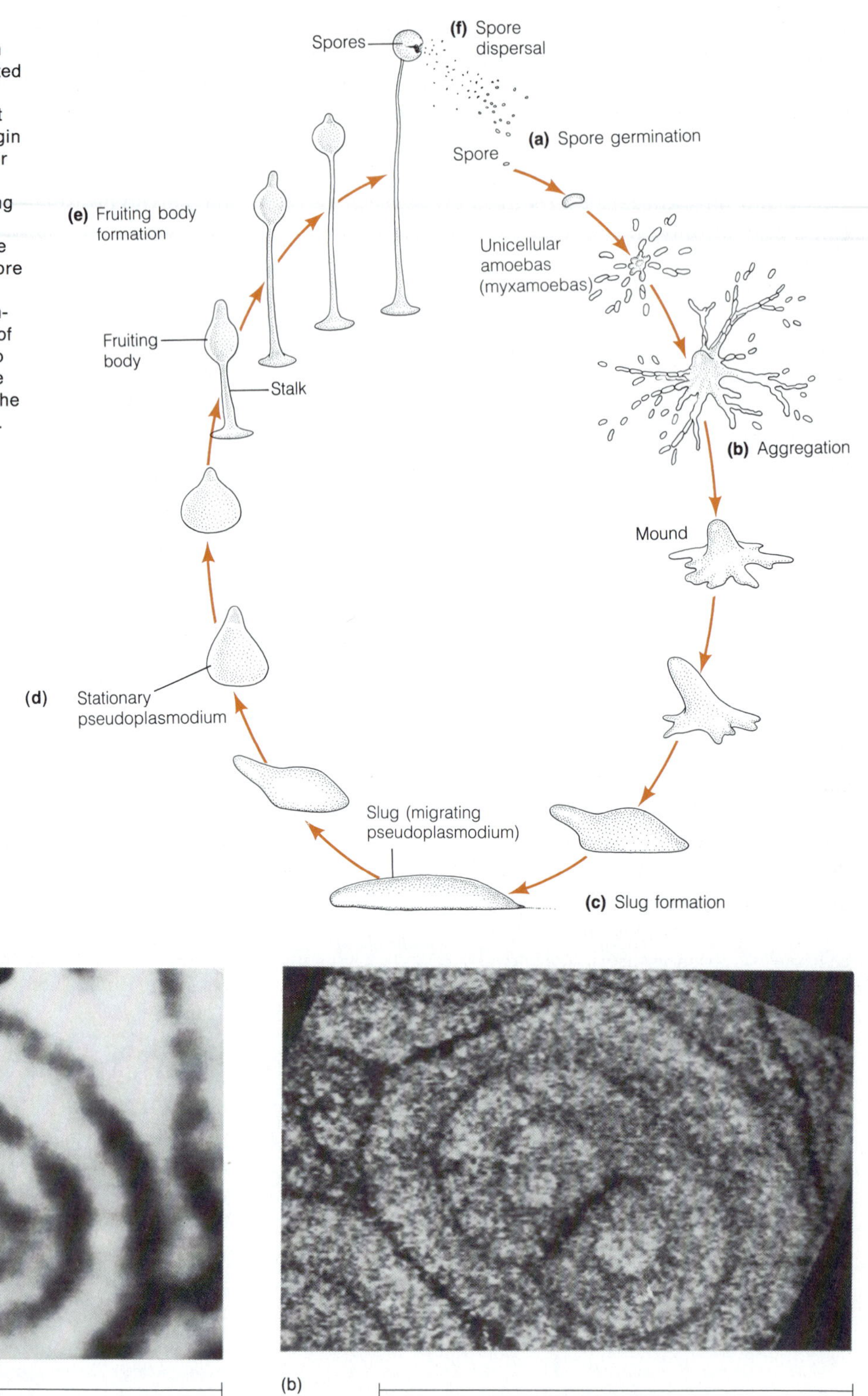

(a) 1 cm

(b) 1 cm

Figure 22-5 Cyclic AMP Waves Direct Aggregation in *Dictyostelium*. (a) A photograph of aggregating amoebas. The light areas contain moving amoebas, while the dark areas contain stationary amoebas. This pattern of behavior is organized by waves of cyclic AMP, which can be "seen" in (b), where a filter paper soaked with radioactive cAMP was placed on colonies of aggregating amoebas. The (nonradioactive) cAMP produced by the colony diluted the label, and when the filter paper was placed on X-ray film, areas of high (dark) and low (light) cAMP concentrations were captured on the film. (LMs)

vertebrates as much as possible, but will draw freely upon findings from studies with lower animals as well. Especially well-studied organisms include the sea urchin, the frog, the fruit fly *Drosophila melanogaster,* and the clawed toad *Xenopus laevis.*

Gametogenesis: Recharging the Life Cycle

Like the union of **a** and **α** cells to form the diploid yeast cell, animal development begins with the union of **gametes** to form a diploid **zygote.** In animals, the gametes are the **egg** produced by the female and the **sperm** provided by the male. The process of gamete formation is called **gametogenesis.**

The **primordial germ cells** are set aside early in development. They eventually become the **germ cells,** which will form the *gametes* that give rise to the next generation. All the other cells in the individual organism are **somatic cells.** Soon after the events of gastrulation establish the basic body plan, the primordial germ cells migrate to a mesodermal tissue called the **genital ridge.** There the primordial germ cells and the somatic cells of the genital ridge develop into an **indifferent gonad.** Unlike the primordial stage of other organs, the indifferent gonad has the potential to form two very different organs: It will become the **testes** in males and the **ovaries** in females. In humans, the indifferent gonad differentiates into the appropriate organ at about the seventh week of embryogenesis.

To produce gametes in the gonad, the diploid germ cells first proliferate mitotically and then switch to meiosis to generate the haploid state. Finally, the haploid germ cells differentiate into mature gametes. In females, this program of cell division commences shortly after the indifferent gonad begins developing into an ovary in the embryo. In males, however, the germ cells wait until the onset of sexual maturity to start dividing. The processes of egg and sperm development are called **oogenesis** and **spermatogenesis,** respectively.

Oogenesis

Egg development varies from species to species in details, but usually follows the general scheme shown in Figure 22-6. Oogenesis begins after the primordial germ cells migrate to the gonad and become **oogonia.** A finite number of mitotic divisions occur; in the human female, these divisions are complete before birth. Many more potential eggs are generated than will ever be shed; thus most oogonia regress and die before they complete oogenesis. When the final number of oogonia is obtained, the chromosomes replicate and the cells differentiate into **primary oocytes,** which enter prophase of the first meiotic division. At this stage, meiosis is usually *arrested,* and each oocyte enters a resting state until the female becomes sexually mature. This can take anywhere from a few days to many years, depending on the species.

The Stages of Ooctye Development. At sexual maturity, hormones direct primary oocytes to mature one (or a few) at a time. The nuclear membrane of the oocyte is known as the *germinal vesicle.* The steroid hormone *progesterone* triggers oocyte maturation, causing **germinal vesicle breakdown** and the completion of meiosis I. Nuclear division is equal, but the cytoplasm divides very unequally, producing one large **secondary oocyte,** from which the egg will eventually arise, and a small **polar body.** Division of the large secondary oocyte is again very unequal, as meiosis II generates another small polar body and the large **ootid,** which develops into the **ovum** (Figure 22-6; recall also Figure 15-11).

Because both meiotic divisions are very unequal, the ovum retains more than 95% of the cytoplasm of the primary oocyte. Oocytes are one of the largest cell types. This makes sense because the ovum must supply almost all of the cytoplasm and initial food supply for the embryo that is formed upon fertilization. In contrast, the only known contribution of the sperm cell is its genetic material and the minimal contents of the *male pronucleus.*

Oocyte Growth. Each primary oocyte is surrounded by **follicle cells.** The follicle cells may nourish the growing oocyte, since they are connected to the oocyte through *gap junctions.* They also play a critical role in the hormonal regulation of oogenesis. The cluster of follicle cells and the primary oocyte is called the **primary follicle** (Figure 22-6b). The growth of a primary oocyte is marked by a massive accumulation of carbohydrate and protein. The major protein present in yolk is **vitellogenin,** and the phase of oogenesis during which yolk accumulates is called **vitellogenesis.** In vertebrates, the liver is the site of vitellogenin synthesis. The follicle cells secrete the hormone *estrogen,* which stimulates the liver to synthesize and secrete masses of vitellogenin. The protein then travels through the bloodstream to the ovaries, where it is absorbed by the growing oocyte and stored in the form of **yolk granules.** A mature sea urchin egg is shown schematically in Figure 22-7.

In many insects, specialized cells known as **nurse cells** assist in the growth of the oocyte. Early in oogenesis, each oogonium undergoes extra mitotic cell divisions to form the nurse cells, which remain attached to the developing oocyte by *cytoplasmic bridges.* Like the oocyte itself, the nurse cells synthesize mRNA, ribosomes, and other proteins that will be needed for embryogenesis. These materials are transported into the oocyte through the cytoplasmic bridges. As the oocyte matures, it sheds the

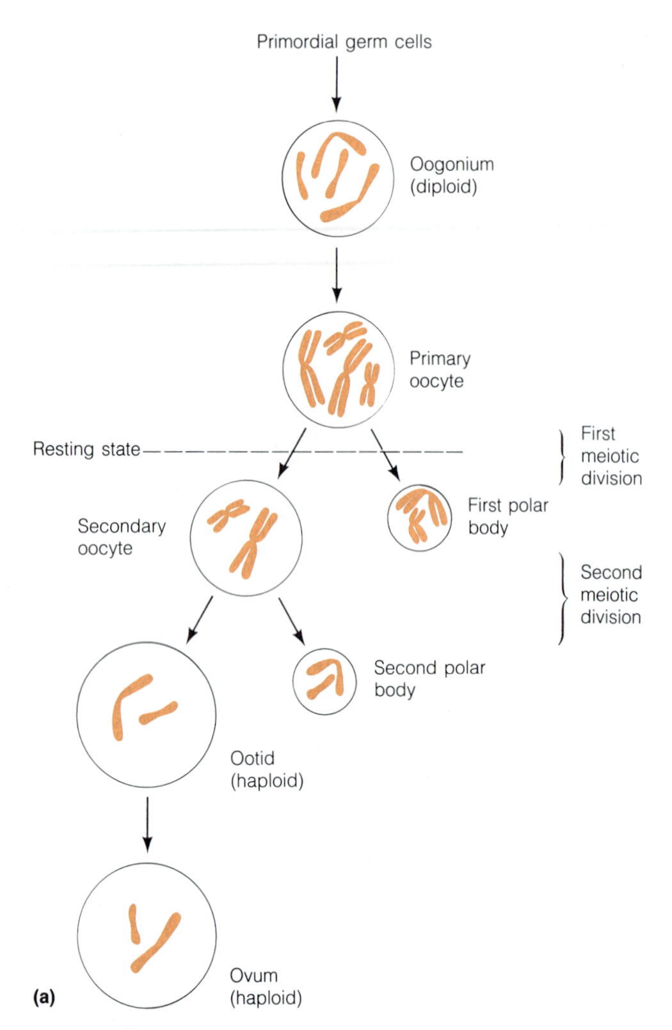

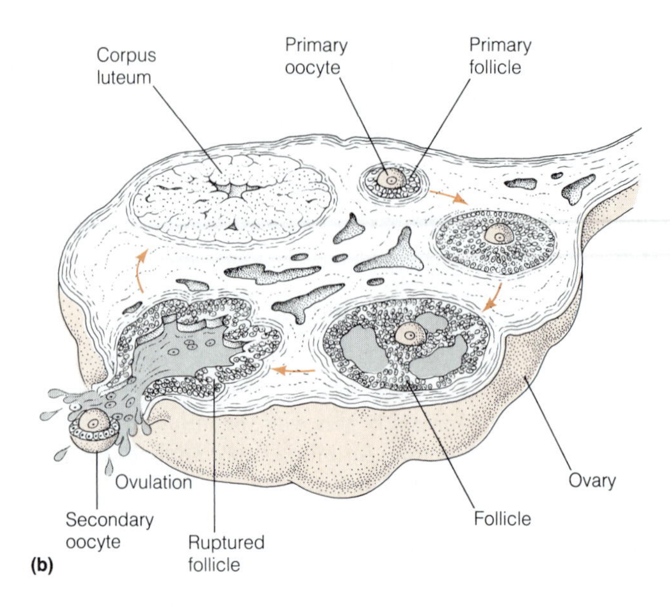

Figure 22-6 Oogenesis: Specialized Cell Divisions in the Ovary. (a) Oogonia derived from primordial germ cells differentiate into primary oocytes, which become arrested in prophase of the first meiotic division and enter an extended quiescent phase. Upon sexual maturity, hormones direct oocytes, one at a time, to complete meiosis, forming a small polar body and a large secondary oocyte. Meiosis II, which is completed after fertilization, is also unequal and produces a second nonfunctional polar body and the ootid. Finally, the ootid differentiates into the ovum. (b) The developmental stages of a single oocyte are depicted in a cross section of a human ovary. Each primary oocyte is surrounded by a spherical cluster of cells called a follicle. In response to hormonal activation, the primary oocyte completes meiosis I and becomes a secondary oocyte. The oocyte grows, accumulating the materials that will be needed to support embryogenesis. Finally, the follicle ruptures, releasing the secondary oocyte into the oviduct. The corpus luteum is the remains of the ruptured follicle.

nurse cells.

During vitellogenesis, the number of mitochondria also increases greatly, as much as 100,000-fold in the case of *Xenopus*. It is also during this stage of oogenesis that extended *lampbrush chromosomes* synthesize RNA (see Chapter 17; Figure 17-25). Although a great deal of messenger-like RNA is stored in the unfertilized egg, most of this mRNA is present in the egg cytoplasm before the lampbrush stage. We are therefore unsure of the biological significance of the RNA associated with lampbrush chromosomes. Curiously, somewhat smaller lampbrush chromosomes have been observed in the sperm of *Drosophila*.

To meet the translational demands of early embryogenesis, the growing oocyte must build a large supply of ribosomes in a short period of time. In doing so, the ooctye far exceeds its normal capacity for ribosomal RNA synthesis. In *Xenopus,* this very rapid synthesis of rRNA is possible only because of *gene amplification:* Several thousand extra copies of the 28S and 18S rRNA genes are made during oogenesis as extrachromosomal elements (recall Chapter 17). In contrast, several thousand copies of the *Xenopus* 5S rRNA gene exist permanently in the genome. Whereas the extrachromosomal 18S and 28S genes are

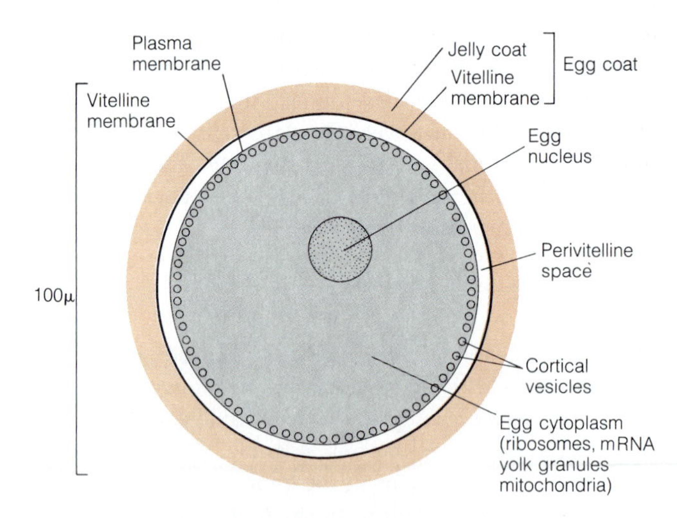

Figure 22-7 The Well-Stocked Egg Is Ready to Support Embryogenesis. At the end of oogenesis the sea urchin oocyte is a storehouse of mitochondria, ribosomes, mRNAs, and proteins and yolk granules (stored carbohydrates). Cortical vesicles line the inside of the plasma membrane, which is covered by an egg coat consisting of a tough vitelline membrane and a thick jelly coat.

simply lost during development, transcription of the excess 5S rRNA genes in the chromosomes must be specifically repressed during later development.

A final, distinctive feature of the mature egg is the presence on the plasma membrane of a specialized **egg coat** of glycoproteins that the sperm must penetrate before fertilization can occur (Figure 22-7). In invertebrates such as the sea urchin, the tough inner layer of this coat is the **vitelline membrane.** In mammals, the comparable layer is the **zona pellucida.** Examples of the outer layers of the egg coat are the whites of chicken eggs and the *jelly coat* of sea urchin eggs.

By the time an oocyte is ready to complete meiosis, it contains all the nutrients and developmental information needed for early embryonic development. In nonmammalian species, the egg must supply *all* the necessary nutrients once it is shed by the female. In mammals, however, after the embryo gets through the early stages of embryogenesis the mother provides the embryo with nutrients.

Ovulation Releases the Mature Ovum. Most species store many ova in the ovary. The release of a mature ovum from the ovary is called **ovulation.** In humans, ovulation occurs once a month when one follicle is hormonally activated to develop further in women of reproductive age. The primary oocyte completes meiosis I to form a secondary oocyte, and the surrounding follicle enlarges rapidly (Figure 22-6). Ovulation takes place when the follicle ruptures at the surface of the ovary, releasing the secondary oocyte into the oviduct. In most mammals, the secondary oocyte is arrested in the prophase of meiosis II, and completes meiosis II to become a mature ovum only after fertilization.

Oogenesis is an extravagant process in that far more eggs are produced than will ever be used. In the human female, for example, there are several thousand primordial germ cells in the developing ovaries during embryogenesis. Several million oogonia then develop, and these produce about 2 million primary oocytes in the ovaries at the time of birth. Many of these degenerate spontaneously, so that only about 300,000 remain at puberty. Of these, only about 400 (one per month for approximately 30 years) will complete meiosis and develop into ova. Thus a woman who has two children has used only two of the several million oogonia that her body originally produced! But as extravagant as oogenesis may seem, it is modest compared to sperm production. A human male releases several hundred million sperm during a single ejaculation—theoretically enough to repopulate the entire United States!

Spermatogenesis

Sperm production, or *spermatogenesis,* is depicted in Figure 22-8. In contrast to oogenesis, spermatogenesis does not begin until sexual maturity and proceeds continuously thereafter. Primordial germ cells enter the testes early in embryogenesis, but remain as inactive **spermatogonia** until puberty. After sexual maturation, spermatogonia divide mitotically in the **seminiferous tubules** of the testes (Figure 22-8b and c). The spermatogonia are associated with the basal lamina of the tubules' epithelial lining, and continue to divide mitotically throughout the reproductive life of the male. Thus the population of spermatogonia is continuously renewed.

Each mitotic division of a spermatogonium is polarized in such a way that one daughter cell remains attached to the basal lamina; the other daughter cell enters the interior of the tubule, becoming a **primary spermatocyte.** Primary spermatocytes proceed through the first meiotic division to form **secondary spermatocytes** (Figure 22-8d). The second meiotic division gives rise to haploid **spermatids,** which then undergo a dramatic morphogenetic process called **spermiogenesis** to produce mature sperm cells, or **spermatozoa.** As spermatozoa mature, they are released into the lumen of the seminiferous tubule. Since both meiotic divisions are equal and all four products are viable, every primary spermatocyte produces four functional gametes instead of the one produced by primary oocytes in females. In the human male, it takes about two months for a spermatogonium to develop into four mature sperm.

In contrast to the stockpiling of RNA, proteins, and nutrients during oogenesis, spermiogenesis is a process of stripping away many cellular structures and customizing the sperm cell for its task of delivering DNA to the egg. As the sperm matures, it modifies its Golgi complex and mitochondria and loses organelles such as the endoplasmic reticulum, peroxisomes, vacuoles, and ribosomes. The structure of a sperm cell is illustrated in Figure 22-9. The two most obvious features of the sperm cell are the *head,* which contains the haploid DNA complement packaged for delivery to the egg, and the *tail,* which propels the sperm to the egg. A single plasma membrane surrounds the entire structure.

The head of the sperm consists of a haploid nucleus, with the DNA in a highly condensed, inactive form. Instead of the histones found in the chromosomes of somatic cells, sperm chromosomes have their DNA complexed with small, positively charged proteins called **protamines** that allow for exceptionally dense packing of the genetic material. The **acrosomal vesicle,** a giant lysosome derived from the Golgi complexes of the spermatocyte, lies just beneath the plasma membrane at the tip of the sperm cell. It contains hydrolytic enzymes that enable the sperm to penetrate the outer coat of the egg during fertilization. The other organelles and most of the spermatid cytoplasm are lost during spermiogenesis.

The tail of the sperm is in essence a long flagellum (see Chapter 19). It consists of a **central axoneme** surrounded by nine **dense fibers** and a **fibrous sheath** (Figure 22-9b). The axoneme, in turn, consists of a central pair of microtubules surrounded by nine microtubule doublets, the characteristic "9 + 2" structure found in most animal fla-

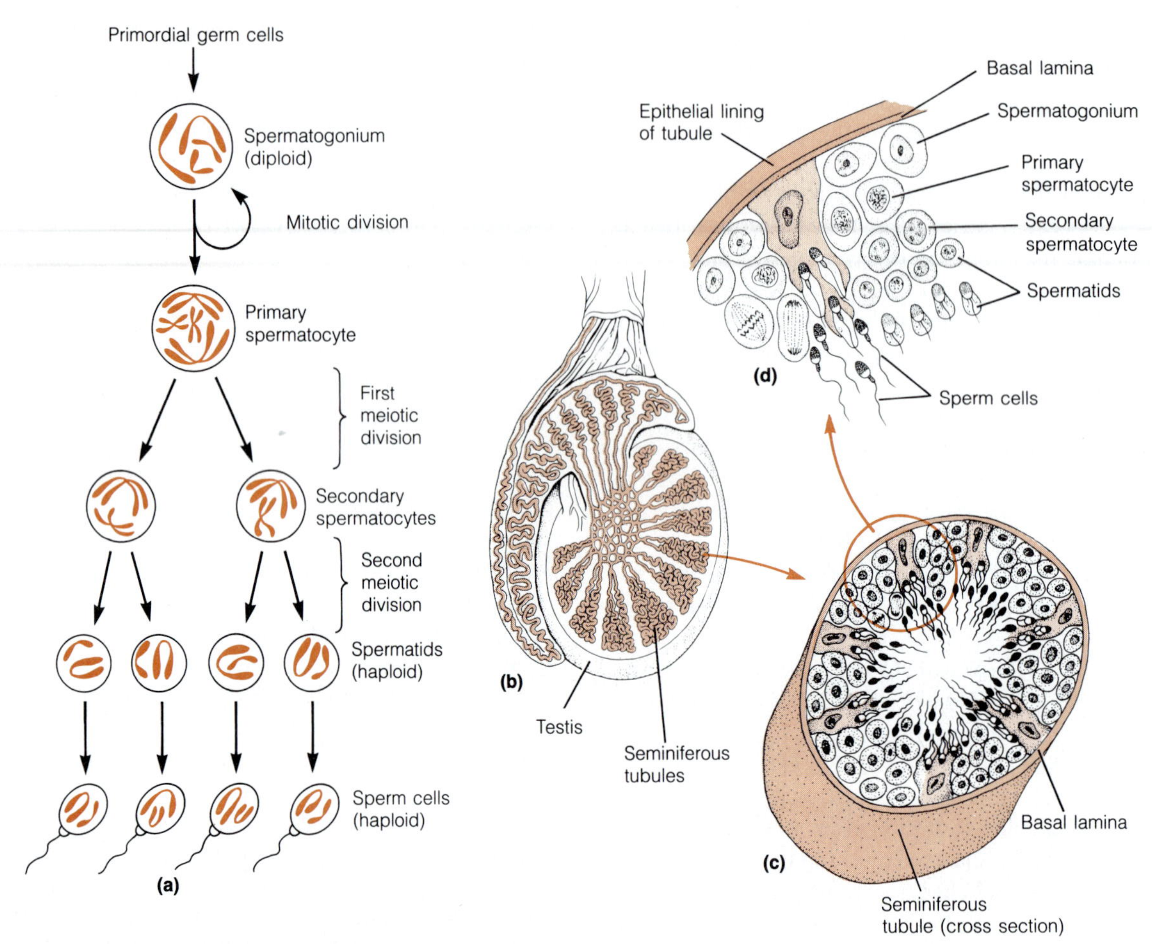

Figure 22-8 Spermatogenesis: Specialized Cell Divisions in the Testes. (a) Spermatogonia derived from primordial germ cells proliferate mitotically. After puberty, one daughter cell of each division remains a spermatogonium, while the other daughter cell becomes a primary spermatocyte. Each primary spermatocyte then undergoes meiosis, producing four spermatids, which differentiate into mature sperm cells. The testes (b) contain the seminiferous tubules (c), where spermatogenesis actually occurs. The spermatogonia remain attached to the basal lamina of the epithelial lining of the tubule (d), while each primary spermatocyte enters the interior of the tubule and continues through meiosis, forming four haploid spermatids and then four mature sperm cells.

gella. The axial filaments of the sperm tail arise from one of the two centrioles of the spermatocyte. The other centriole remains between the nucleus and the flagellum. The considerable ATP needed to power the sperm tail as it swims in search of the egg is supplied by the large mitochondrion located in the **midpiece** of the sperm cell (Figure 22-9a).

Fertilization and Activation: The Life Cycle Renewed

During **fertilization,** the male and female gametes meet and fuse, producing the diploid zygote and initiating embryogenesis. The early events triggered by the contact between sperm and egg feature several different kinds of membrane interactions. These are followed by the activation of specific biosynthetic processes, including ATP generation, protein synthesis, and DNA replication. In addition, a region of the egg cytoplasm just under the cell membrane, called the **cortical cytoplasm** or **cortex,** undergoes critical rearrangements that have profound developmental consequences. This series of events following fertilization and preceding the first cell division is known as **activation.**

Internal Versus External Fertilization

Fertilization is accomplished in one of two ways. Animals that live in aquatic environments usually depend on **ex-**

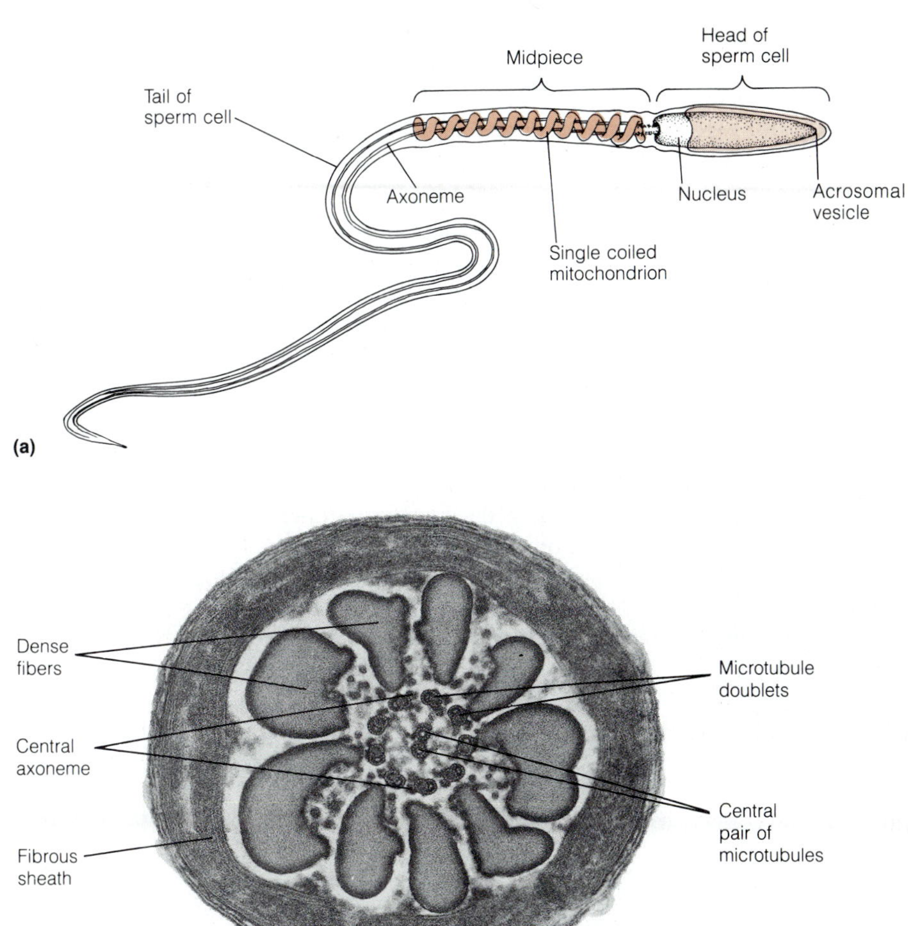

Figure 22-9 The Economical Structure of the Sperm Cell. (a) The major structural features of a sperm cell are the head, the midpiece, and the tail. The head contains the nucleus with a haploid complement of chromosomes and is capped by the acrosomal vesicle. The midpiece contains a spiral mitochondrion, which supplies the quantities of ATP needed for sperm motility during mating. (b) Sperm movement is powered by the "9 + 2" arrangement of microtubules in the central axoneme, seen in cross section in an electron micrograph (TEM). The central axoneme is surrounded by dense fibers and a fibrous sheath.

ternal fertilization to unite sperm and egg. During mating, both sperm and eggs are released into the water. The sperm then swim to the eggs or are carried there by water currents. Species that depend on external fertilization produce sperm and eggs in enormous quantities, presumably because fertilization by this means is so uncertain. For example, a male sea urchin releases billions of sperm at a time, making the output of human males seem paltry. The females are similarly productive. A typical female contains several million mature eggs at one time.

Terrestrial animals, on the other hand, usually depend on **internal fertilization.** In this case the eggs are fertilized within the reproductive tract of the female by motile sperm introduced by the male during mating. Here the sperm cells are provided with a fluid environment that is not available outside the body. In the case of birds, reptiles, and many insects, the fertilized egg is then enclosed within a protective shell and released by the female. In live-bearing species, the fertilized egg proceeds through embryogenesis within the body of the mother. The scanning electron micrographs in Figure 22-10 show the fertilization of a hamster oocyte by human sperm.

The Events of Fertilization in the Sea Urchin

We will consider the sea urchin, long a favorite organism of embryologists, as our general example of the events of fertilization. Figure 22-11 provides a timetable of the changes that occur in the sea urchin egg immediately after it comes in contact with a sperm.

The Acrosomal Reaction and the Fast Block to Polyspermy. The sea urchin egg is surrounded by a gelatinous matrix of polysaccharides and glycoproteins called the **jelly coat.** When a sperm makes contact with the egg jelly, the jelly induces a response in the sperm cell known as the **acrosomal reaction** (Figure 22-12, a–d).

The acrosomal reaction begins with the exocytotic release of hydrolytic enzymes and other proteins from the *acrosomal vesicle* at the tip of the sperm head (Figure 22-12b). These enzymes soften the jelly coat, preparing the way for the **acrosomal process,** which penetrates the jelly coat (Figure 22-12c). The acrosomal process is formed by the rapid polymerization of a pool of monomeric actin located just behind the acrosomal vesicle. One of the proteins released from the acrosomal vesicle during the acrosomal reaction is a carbohydrate-binding protein called **bindin.** Bindin coats the surface of the acrosomal process and mediates the adhesion of the sperm to the vitelline membrane of the egg, presumably through specific protein contacts with receptors on the egg surface (Figure 22-12c). In mammals, this receptor is a protein called **ZP3** (for zona pellucida protein 3). The specificity of these interactions between proteins on the sperm and egg surfaces prevents eggs of one species from being fertilized by sperm of another species.

As the tip of the acrosomal process makes contact with the egg plasma membrane, a fusion of the two membranes begins (Figure 22-12d). As the membranes fuse, specific **ion channels** in the membrane open, allowing a momentary influx of sodium ions and causing a large, very rapid depolarization of the egg membrane. This change in membrane potential renders the membrane impenetrable to

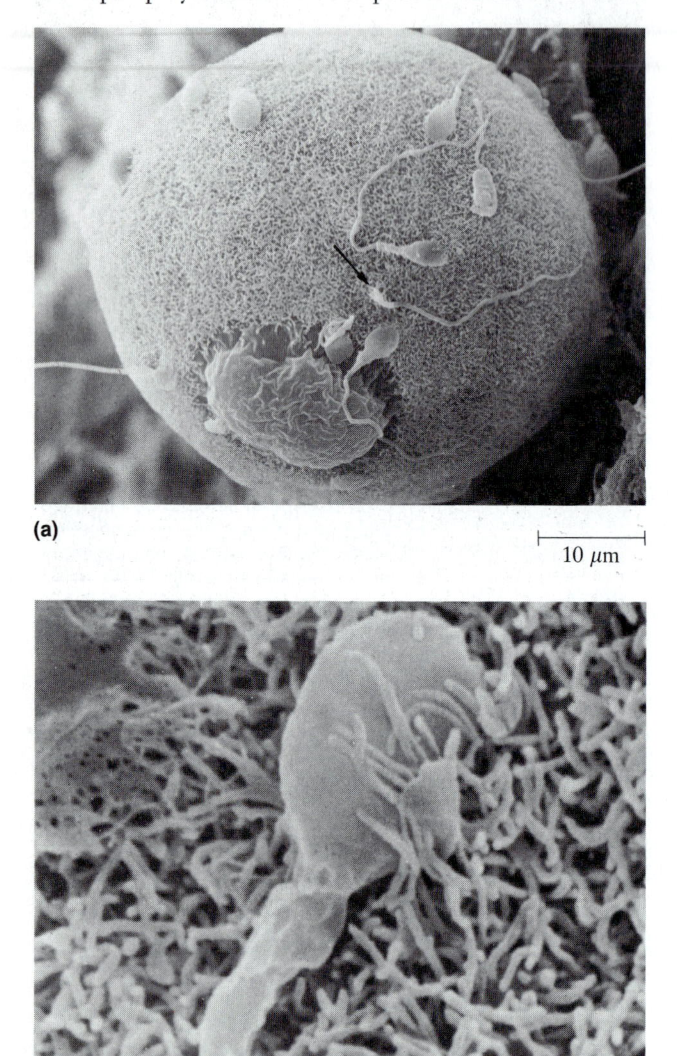

Figure 22-10 The Life Cycle Renewed: Fertilization of an Oocyte by Sperm. These scanning electron micrographs illustrate the progressive penetration of human spermatozoa into hamster oocytes from which the zona pellucida has been removed. In (a), note how numerous spermatozoa adhere to the surface of the oocyte, which is covered by microvilli. These microvilli trap and immobilize the sperm heads. (b) A single sperm cell eventually succeeds in penetrating the layer of microvilli, and the entire sperm head of that cell disappears into the ooplasm, leaving only the sperm tail to mark the site of entry (arrow in (a)). The first polar body, already released and free of microvilli, can be seen on the lower left of the egg in (a).

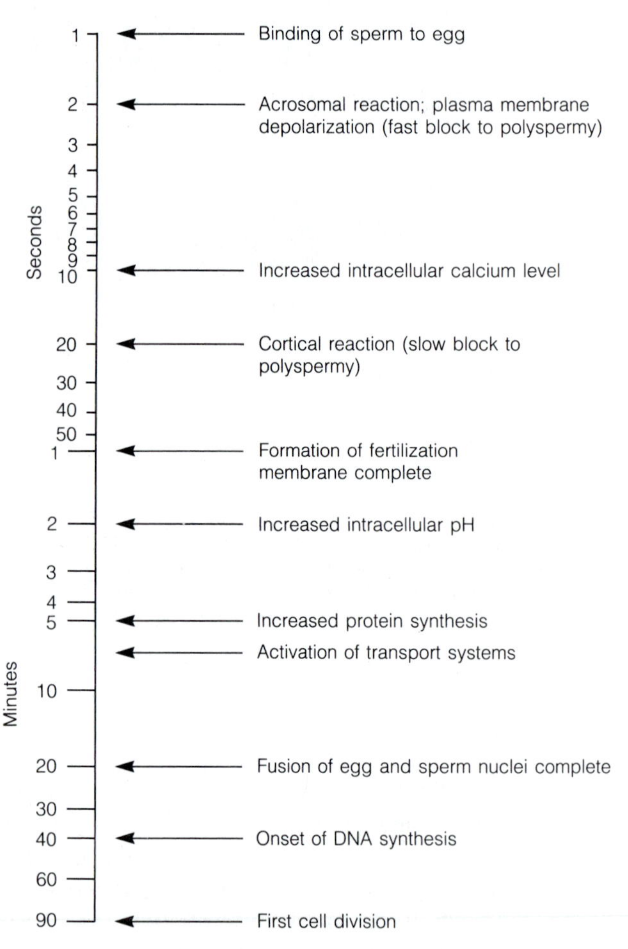

Figure 22-11 The Timetable for Fertilization in the Sea Urchin. The arrows show the approximate time at which each process begins. Note the logarithmic time scale.

sperm, thereby preventing **polyspermy**—the fertilization of an egg by multiple sperm. This **fast block to polyspermy** operates within seconds.

The Cortical Reaction and the Slow Block to Polyspermy. The change in membrane potential also causes the release of calcium ions, probably stored in the endoplasmic reticulum, into the cytoplasm of the egg cell. The release of calcium begins about 10 to 20 seconds after fertilization. From the initial site of calcium release, a remarkable wave of high calcium concentration proceeds across the egg over the next 2.5 minutes (Figure 22-13). This **calcium wave** stimulates an influx of sodium ions, which is coupled to an efflux of protons from the cell. This exchange of sodium ions and protons begins about 20 seconds after fertilization, and the net result is a permanent increase in intracellular pH from about 6.8 to 7.3 (Figure 22-14).

The calcium wave is also thought to cause the **cortical reaction**—the exocytotic release of proteases and *mucopolysaccharides* from **cortical vesicles** found in the cortex of the egg cell (Figure 22-12, d–f). The mucopolysaccharides cause a thickening of the perivitelline space that separates the plasma and vitelline membranes, and the proteases mediate the subsequent hardening of the vitelline membrane, which blocks further sperm penetration (Figure 22-12e). Because the cortical reaction is a chemical rather than an electrical event, it takes longer than the initial change in membrane potential and is therefore called the **slow block to polyspermy.**

A few minutes after the acrosomal reaction has taken place, fusion of the egg and sperm plasma membranes is complete, and the sperm nucleus is drawn into the egg (Figure 22-12f). Once in the egg, the sperm nucleus swells and its chromatin decondenses as protamines are replaced by chromatin proteins stored in the egg.

Egg Activation

In addition to triggering the cortical reaction, the increase in cytoplasmic calcium levels also directs the egg to initiate

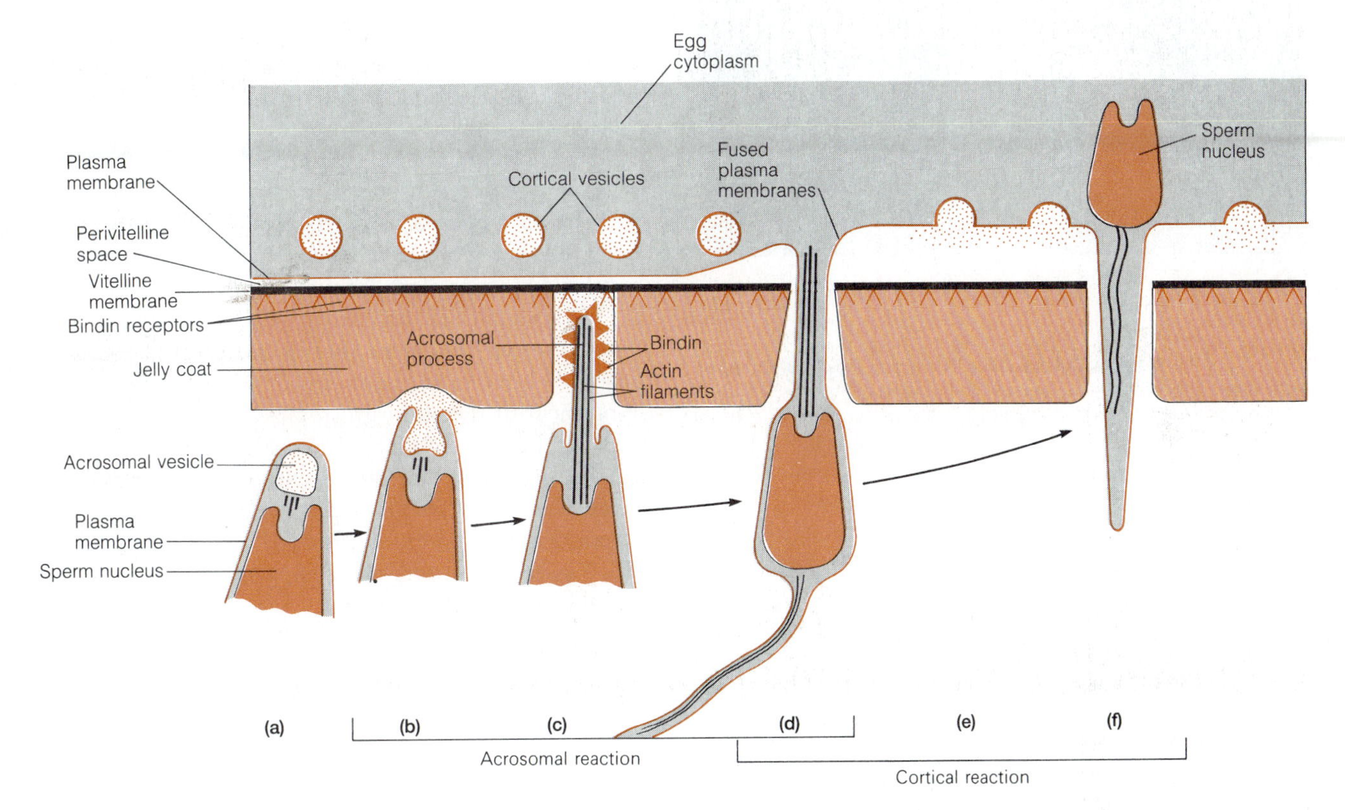

Figure 22-12 The Acrosomal and Cortical Reactions Overlap During Fertilization of Sea Urchin Eggs. Many critical events, including the acrosomal reaction and the cortical reaction, occur during the first minute of the time line shown in Figure 22-11. The acrosomal reaction (b–d) begins when the tip of the sperm cell touches the egg's jelly coat, triggering exocytosis of the acrosomal vesicle. The vesicle releases enzymes, which help clear a path for the acrosomal process, and bindin, which coats the growing acrosomal process (c). Growth of the acrosomal process through the jelly coat is driven by actin polymerization. When the acrosomal process comes in contact with the vitelline membrane (d), the bindin molecules bind to specific bindin receptors, triggering the fusion of the egg and sperm membranes and the cortical reaction (d–f). Upon membrane fusion, the cortical vesicles exocytose, releasing their enzymes into the perivitelline space, which thickens and hardens. Thus, the slow block to polyspermy is established as the sperm nucleus enters the egg cytoplasm.

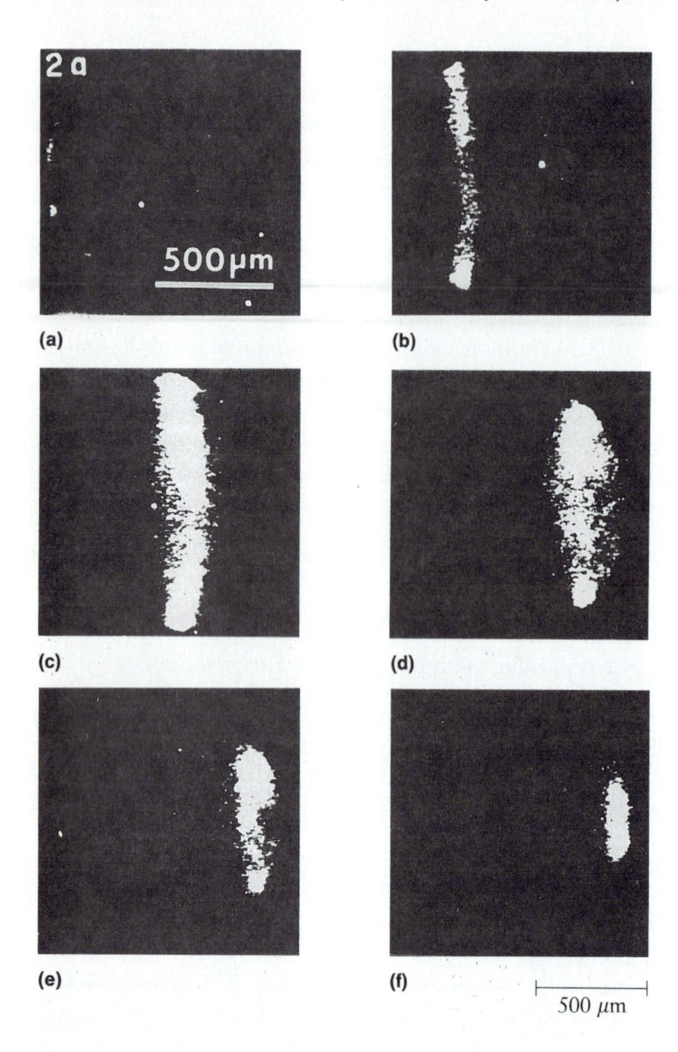

Figure 22-13 A Wave of Calcium Ions Traverses the Embryo upon Activation. This remarkable series of photographs, taken 20 seconds apart, was made by injecting fish eggs with aequorin, a protein that luminesces upon binding calcium ions. (LMs)

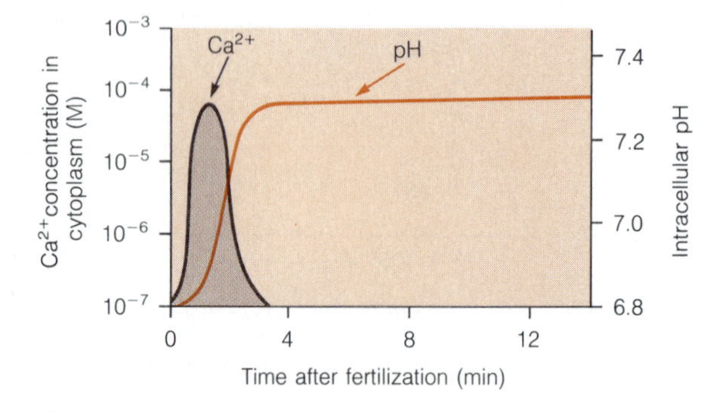

Figure 22-14 The Transient Increase in Intracellular Calcium Triggers a Permanent Rise in Intracellular pH. The stable increase in intracellular pH is due to a calcium-triggered influx of sodium ions that is coupled to an efflux of protons.

its developmental program. Egg activation is due specifically to the elevated calcium level and not to the initial change in membrane potential: In all species tested, injection of calcium ions into unfertilized eggs stimulates egg development, but artificial membrane depolarization does not. The activating effect of calcium appears to be mediated through the calcium-triggered change in intracellular pH discussed above.

Egg activation involves increases in ATP generation, protein synthesis, and DNA synthesis (Figure 22-11). In sea urchin eggs, protein synthesis can be detected within 5 to 8 minutes, and transport systems begin operating shortly thereafter. Fusion of the egg and sperm nuclei is usually complete by about 20 or 30 minutes, and DNA synthesis begins soon thereafter. The first cell division occurs at about 90 minutes and is followed by a series of rapid *cleavage* divisions. The rapid pace of cell division during cleavage places a great demand on the synthetic capabilities of the embryo, a need that is met primarily by the ribosomes and mRNA molecules that were stored during oogenesis.

In many species, dramatic rearrangements of the egg's cytoplasm occur after egg activation and before the first cell division. These are readily visible in the eggs of many amphibian species because the top half, or *animal hemisphere*, of the egg is darkly pigmented, whereas the bottom half, or *vegetal hemisphere*, is unpigmented. The animal hemisphere gives rise to the tissues of the developing embryo, whereas the vegetal hemisphere primarily provides nutrients needed to support development.

Shortly after sperm entry, the *cortical cytoplasm* just below the cell membrane rotates relative to the *inner cytoplasm*. Tubulin filaments apparently mediate this rearrangement of the cortical cytoplasm, because it does not occur in the presence of *colchicine*, an inhibitor of microtubule formation. The movement of the pigmented areas relative to each other generates a lightly pigmented region opposite the point of sperm entry. This is the **gray crescent** (see Figure 22-15), the site of important events that occur a little later in development.

Stored mRNA and Translational Regulation

The presence of **stored mRNA** in unfertilized eggs has been amply documented. In the early 1960s, Jean Brachet and his collaborators and Paul Denny and Albert Tyler centrifuged sea urchin eggs at high speed, rupturing them into nucleated and enucleated (nucleus-free) fragments. After activating the egg fragments, they demonstrated that the enucleated fragments could carry out protein synthesis. Thus mRNA must be present in some inactive form in the cytoplasm of the unfertilized egg.

Although this is the most dramatic case of *translational control* in eukaryotes, we still do not know the molecular mechanism that keeps mRNA from being translated prior to activation. One hypothesis is that ribosome binding sites on the RNA are *masked* by proteins that are bound to the 5′ end of the mRNA molecule. These proteins would then release the **masked mRNA** in response to some aspect of activation, such as the rise in intracellular calcium or pH. A second hypothesis holds that egg activation increases the **translational capacity** of stored ribosomes, possibly by causing the covalent addition or removal of a phosphate group—*phosphorylation* or *dephosphorylation,* respectively—on various initiation factors (see Chapter 16). This would be only one of many examples in development of the regulation of a protein's activity by phosphorylation (see Chapter 21).

In addition to mRNA molecules and ribosomes, many other proteins that are necessary for early development are stored in the unfertilized egg. For example, in 1983 Douglas Forbes, John Newport, and Marc Kirschner simply and elegantly showed that unfertilized eggs contain all the proteins necessary to form extra nuclei. Forbes and his colleagues injected DNA from a bacteriophage into unfertilized eggs of *Xenopus.* Remarkably, normal nuclei formed around the foreign DNA in the absence of protein synthesis! Thus, the egg has a store of the nuclear proteins required for the initial rounds of cell division.

Ribosomes, chromosomal proteins, and nuclear proteins are necessary players in development, but their importance stems from the general functions they fulfill in *all* cells. As we turn to the developmental processes that follow fertilization and egg activation, we will encounter many proteins with very specific roles in development. Like the stockpile of other molecules in the egg, many of these specialized proteins, or their mRNAs, are already in place when the egg is fertilized. For these maternally supplied molecules, egg activation is the signal to set about the task of carving up the embryo into its various parts.

Early Embryogenesis: Cleavage and Pattern Formation

Embryogenesis encompasses the events that immediately follow egg activation and give rise to the primary structural

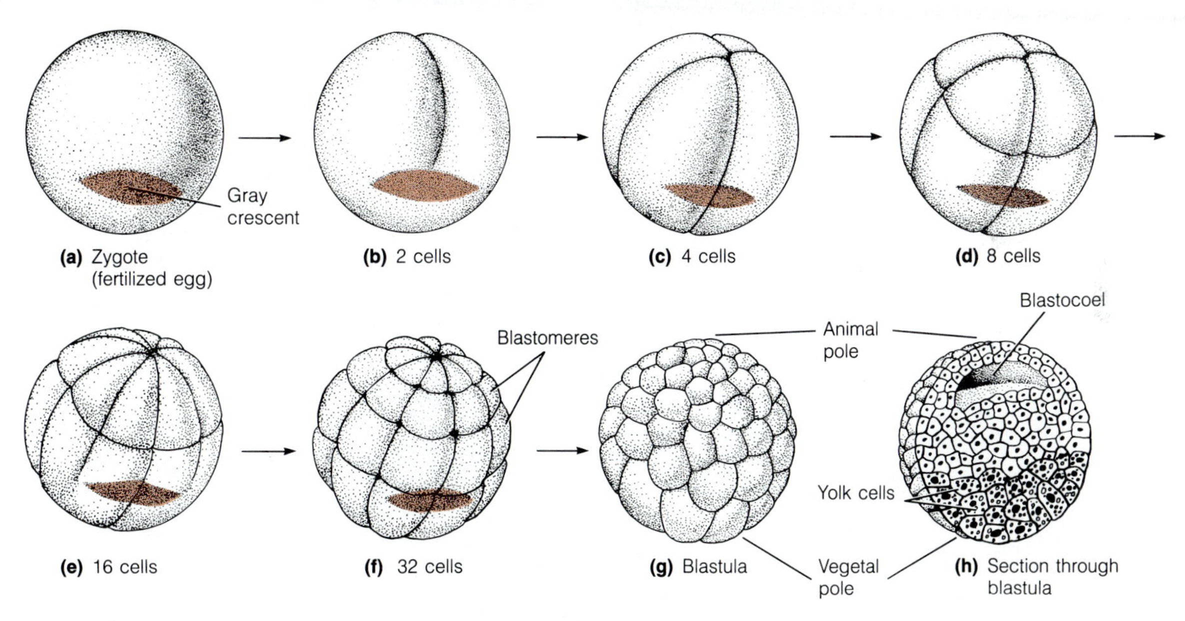

Figure 22-15 From Fertilized Egg to Blastula: The Early Development of the Frog Embryo. Rapid, synchronous, and symmetric cell division progressively divides the fertilized egg into smaller and smaller cells during cleavage (a–f). Note the precise spatial orientation of the first few cleavage divisions. At the 32-cell stage the embryo appears to be a solid ball of cells (f). By the 128-cell stage (g), however, the blastocoel cavity is developing in the animal hemisphere, and the embryo is now a blastula. The unequal division of cytoplasm during cleavage and the off-center location of the blastocoel leave the bulk of the egg yolk in the vegetal half of the embryo (h). Much of the variation in embryogenesis between species is due to differing amounts of yolk cytoplasm. The gray crescent (a–f) is formed opposite the point of sperm entry by the rearrangement of the cortical cytoplasm. The next stage of embryogenesis, gastrulation, will begin with blastopore lip formation in the gray crescent region (see Figure 22-22).

features of the organism. The events of embryogenesis will be our primary concern for the rest of this chapter.

In higher animals such as mammals and birds, much (though by no means all) of development takes place during embryogenesis. At birth or hatching the organism is essentially a miniature adult. Other than the development of the *secondary sex traits*, the primary developmental activity is the growth of an already well-defined form. In lower animals such as insects and amphibians, however, embryogenesis gives rise to a *larva*. The larva then undergoes a series of dramatic changes known as **metamorphosis** to produce the adult form.

We will examine the events of early embryogenesis—*cleavage* and *gastrulation*—as they occur in the frog. To illustrate specific concepts, we will also take side trips into the embryogenesis of several other well-characterized species, including nematodes, sea urchins, fruit flies, salamanders, chickens, and mice.

Cleavage Partitions the Embryo

Although embryonic development varies greatly between animal phyla, some events are common to all multicellular animals. After fertilization and activation, **cleavage,** a series of rapid cell divisions, progressively subdivides the diploid zygote into a mass of smaller cells. Cleavage in the frog embryo is shown in Figure 22-15. The cells created by cleavage are called **blastomeres.** Cleavage produces a superficially homogeneous mass of blastomeres known as the **blastula.** The hollow inside the blastula is known as the **blastocoel** cavity. Although cleavage patterns vary considerably between taxonomic families, Figure 22-15 provides a fairly general model for higher animals.

In *Drosophila* the blastula stage is known as the **cellular blastoderm.** The events that lead up to the cellular blastoderm are quite unusual: The first 13 rounds of *nuclear*

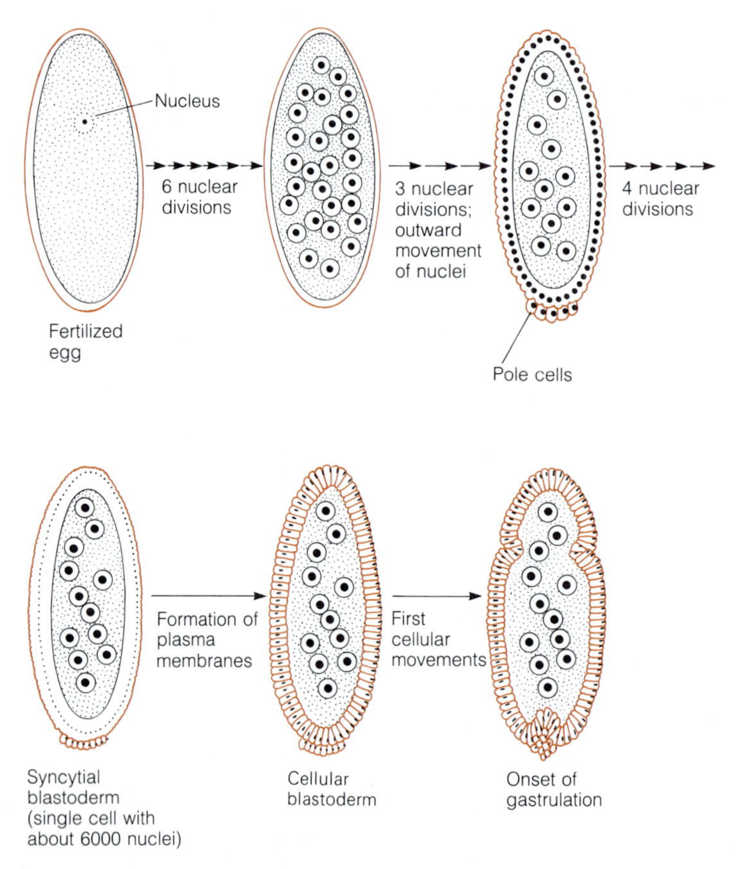

Figure 22-16 Formation of the Syncytial and Cellular Blastoderms in the *Drosophila* Embryo. Nuclear division in the *Drosophila* embryo is synchronous and very rapid (10 minutes each for rounds 1–9). During nuclear cycle 9, most of the nuclei migrate to the cortical cytoplasm immediately beneath the egg membrane. Soon thereafter the pole cells form at the posterior end of the embryo. Since there is no cellular division during cleavage in *Drosophila,* after 13 nuclear cycles the embryo is a single cell, which contains nearly 6000 nuclei and is known as the syncytial blastoderm. At this point, plasma membranes form around the individual nuclei, giving rise to the cellular blastoderm. The final embryo in the series is just beginning the cellular movements that mark the onset of gastrulation.

division occur in the absence of cellular division, producing a single *syncytial* cell containing 6000 nuclei! Only then do cell membranes form around the nuclei, and the syncytial blastoderm becomes the cellular blastoderm (Figure 22-16). As we shall see later, the cellular blastoderm is a critical stage of *Drosophila* development.

Pattern Formation

Pattern formation describes any process in which a homogeneous group of cells or a tissue develops an ordered spatial arrangement. Pattern formation can be observed throughout development and in most multicellular organisms. Aggregation and fruiting body formation in *Dictyostelium* (see p. 673) are relatively simple examples of pattern formation. Later in this chapter we will look at two more complex examples: the development of limbs in vertebrates and the creation of a segmental body plan in *Drosophila*. At this point, however, we are concerned with the blastula and how this mass of cells begins to divide into distinct parts.

Axes of Polarity. A fundamental (and very difficult) question for embryologists is that of *polarity:* How does the early embryo, a largely homogeneous mass of cells, set about distinguishing dorsal from ventral (that is, top from bottom) and anterior from posterior (front from back)? Early in embryogenesis, asymmetric cell divisions occur in well-defined patterns. Generally, two longitudinal divisions are followed by an equatorial division as depicted in Figure 22-15. For such divisions to occur reliably, developmental information must be present in the early embryo. That information, known as an **axis of polarity,** directs cells to behave in a particular way based on their position and orientation along one axis of the embryo.

Exactly how the primary axes are established is largely unknown. In *Xenopus* and other amphibians, the first axis to be formed is the animal-vegetal axis, seen in the darkly pigmented animal and lightly pigmented vegetal hemispheres of the unfertilized egg. Early asymmetric cell divisions along this axis separate the animal cells, which will go on to form the tadpole, from the vegetal cells, which store the yolk granules that fuel embryogenesis. In contrast, the dorsoventral orientation of the embryo is apparently determined by the site of sperm entry into the egg. In *Drosophila,* the egg shape is itself asymmetric and may influence axis formation. In other invertebrates and in many vertebrates, there is apparently no polarity until after the first cleavage, and in some species polarity is established only after many rounds of cleavage.

While we can deduce that developmental information must exist along each axis of polarity, we are just beginning to unravel the molecular nature of that information. Furthermore, different organisms and different axes within the same organism probably utilize different molecular mechanisms. We will return to the molecular basis of axis formation shortly when we discuss *positional information* and *morphogens*.

As the early division of *Xenopus* embryos into animal and vegetal hemispheres suggests, early cell divisions along any given axis of polarity can result in cell populations with different **developmental potentials.** This was elegantly demonstrated in sea urchin embryos by Hans Driesch in the 1890s and by Sven Hörstadius several decades later. In the sea urchin, the first two cleavages are through the animal-vegetal axis, but the third division is perpendicular to that axis. (The same order occurs in the frog; see Figure 22-15.) Driesch found that after two rounds of vertical cell division, individual blastomeres from the four-cell embryo could develop into larvae when cultured in isolation. Such a cell is **totipotent,** meaning that its developmental potential is unrestricted and it can direct complete embryonic development. But Driesch found that after the next round of cell division (the first perpendicular division), single blastomeres from an eight-cell embryo were no longer capable of developing into a normal embryo.

Hörstadius extended these findings to the 64-cell stage by splitting embryos in half along the animal-vegetal axis and showing that such half-embryos could develop into normal larvae (Figure 22-17a). But when he split the embryo *across* the animal-vegetal axis, two incomplete embryos developed (Figure 22-17b and c). Furthermore, the two partial embryos were quite different, demonstrating that the two halves contain different developmental information. Because each half lacked information that was present in the other half, the developmental potential of each was limited.

Finally, Hörstadius combined cells from the animal poles with *micromeres* from the vegetal pole, leaving out the bulk of the vegetal hemisphere. Unlike the result obtained with isolated animal poles, nearly normal larvae developed (Figure 22-17d). From these results, Hörstadius concluded that developmental potential seems to be polarized along the animal-vegetal axis at these early stages of development and that information from both ends of the axis is necessary for development. (The dramatic influence of the micromeres on the animal cells in this last experiment is an example of an *inductive interaction.* We will return to the phenomenon of induction when we discuss *gastrulation.*)

Positional Information: Morphogens and Determinants. We have seen evidence for the differing developmental potentials of cells along the axes. Clearly, molecular agents must exist that influence the developmental program of particular cells based on the cell's position within the developing embryo. Such developmental instructions are called **positional information,** and the most commonly proposed forms of positional information are *morphogens* and *cytoplasmic determinants.*

A **morphogen** is a diffusible compound that emanates from a source at one end of the egg (or one end of a particular tissue) and that instructs cells to adopt a developmental fate specified by the local concentration of morphogen. The source could either be a site of active synthesis or a sequestered supply of active morphogen that is released upon activation. As the morphogen diffuses, a concentration gradient of the morphogen is created. Along that **morphogen gradient,** different concentrations of morphogen specify different developmental fates. Morphogen gradients provide perhaps the simplest model for axis formation. Early embryonic morphogens are still largely mysterious, but the box on pp. 688–689 describes how a well-defined morphogen, retinoic acid, establishes an axis of polarity during limb development in chickens.

Cytoplasmic determinants represent a second form of positional information. In contrast to a morphogen gradient, a cytoplasmic determinant is highly localized and its presence confers a *single* developmental fate on affected cells. A well-known case of cytoplasmic determinants involves the pole plasm of *Drosophila* eggs. **Pole plasm** is the cytoplasm located at the posterior end of the egg, and the *pole cells* that develop in this portion of the egg are the fly's primordial germ cells. In the mid-1970s, Anthony Mahowald and Karl Illmensee showed that the pole plasm of the *Drosophila* egg directly influences the development of nuclei and the cells that form around them, causing them to become pole cells.

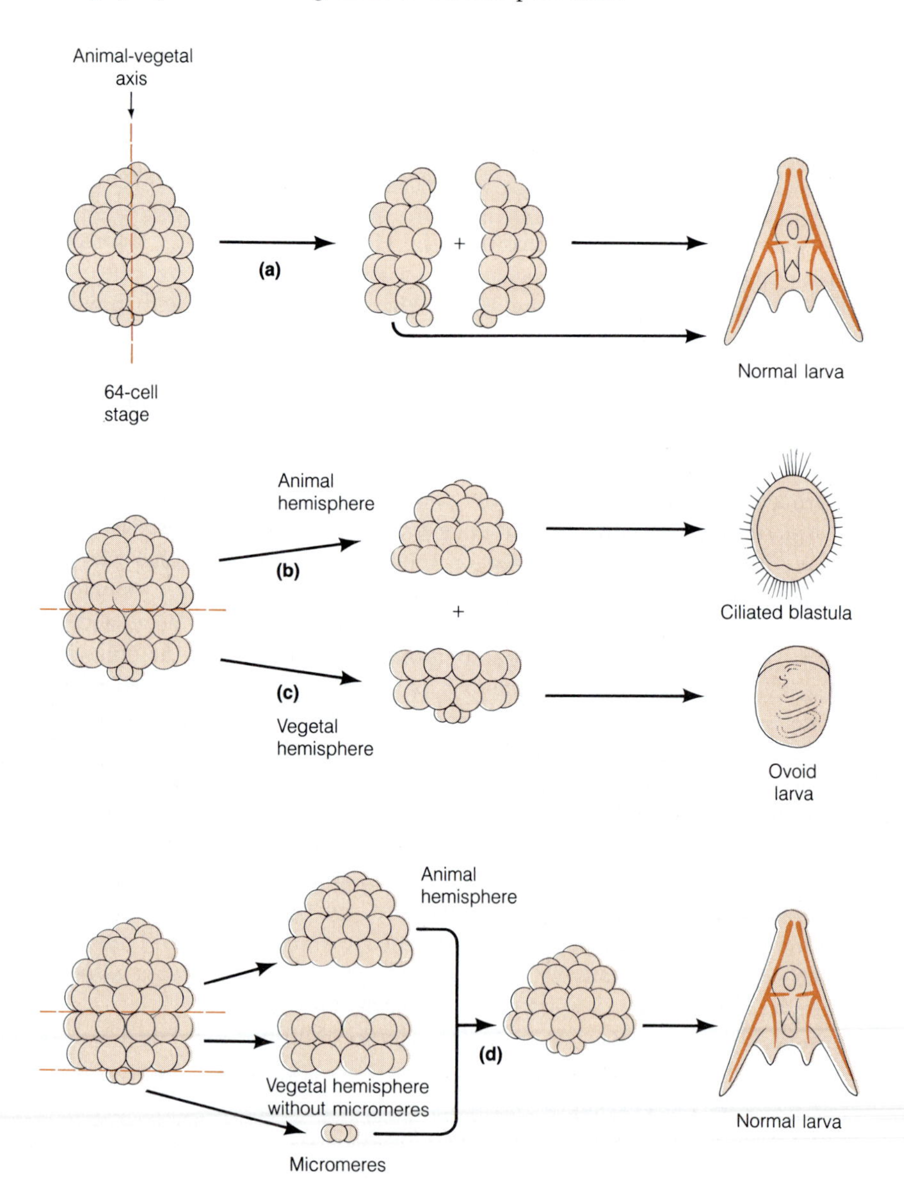

Figure 22-17 Differing Developmental Potentials and Inductive Interactions Along the Animal-Vegetal Axis of Sea Urchin Embryos. (a) When a 64-cell stage sea urchin embryo is split in half along the animal-vegetal axis, each half-embryo develops into a normal larva. But when an identical embryo is split in half *across* the animal-vegetal axis, two incomplete embryos getal axis, two incomplete embryos develop, each with some features of the normal embryo. The animal hemisphere forms a ciliated blastula (b), and the vegetal hemisphere develops into an ovoid larva with gutlike endothelium (c). However, when the animal hemisphere is combined with the micromeres from the vegetal pole (d), a nearly normal larva develops. These experiments reveal the differing developmental potentials of cells along the animal-vegetal axis. The implied interaction between micromeres and animal cells that is missing in (b) and (c) and seen in (d) is an example of induction.

Recall that until cellularization and formation of the cellular blastoderm, the early *Drosophila* embryo is a single syncytial cell with many nuclei (Figure 22-16). If the posterior end of a syncytial *Drosophila* embryo is irradiated with ultraviolet light, the pole cells fail to develop and the adult flies are sterile. Mahowald and Ilmensee believed that the UV radiation inactivated cytoplasmic determinants that specify the fate of the pole cells. To test their hypothesis, they withdrew cytoplasm from the posterior end of a wild-type egg and injected it into the anterior end of a syncytial embryo from a genetically marked strain (Figure 22-18a). If cytoplasmic determinants in the pole plasm direct the formation of pole cells, then pole cells should form in the *anterior* end of the injected eggs. Mahowald and Ilmensee did indeed see pole cells where the pole plasm was injected. To test whether these cells could develop into functional germ cells, they then transferred the induced pole cells into the posterior end of a second host embryo (22-18b). The genetically marked pole cells migrated to the gonadal primordium and gave rise to functional gametes!

Thus we can conclude that the pole plasm of insect eggs contains cytoplasmic determinants that profoundly influence the developmental fate of nuclei. The chemical nature of the determinants is still unknown. Ultraviolet sensitivity and biochemical experiments have suggested a *ribonucleoprotein,* but positive identification is still lacking.

Many embryologists believe that the rearrangement of the cortical cytoplasm that we discussed earlier in this chapter plays a crucial role in activating the various forms of positional information stored in the egg. The exact nature of that activation is as unclear as the biochemical nature of the morphogens, axes, and determinants themselves. Perhaps these movements put determinants into their correct place or cause the release of a morphogen. What *is* clear is that if the cytoplasmic rearrangements do not occur, these other mechanisms often fail.

Cellular Fate and the Loss of Developmental Potential

As cleavage progresses, more and more cells are produced and embark on particular developmental pathways. As this occurs, it becomes possible to talk about the *fate* of specific cells. At the same time, we must distinguish the *commitment* of a cell and its descendants to an eventual fate from the actual events of *differentiation* that mark the end of cellular development. The committed and differentiated states are often far separated in time, and the progress from the former to the latter is marked by the progressive loss of developmental potential.

Fate Mapping and Determination. Think about all the positional information—axes, morphogens, determinants—that is present in the early embryo and that apparently specifies the developmental fate of embryonic

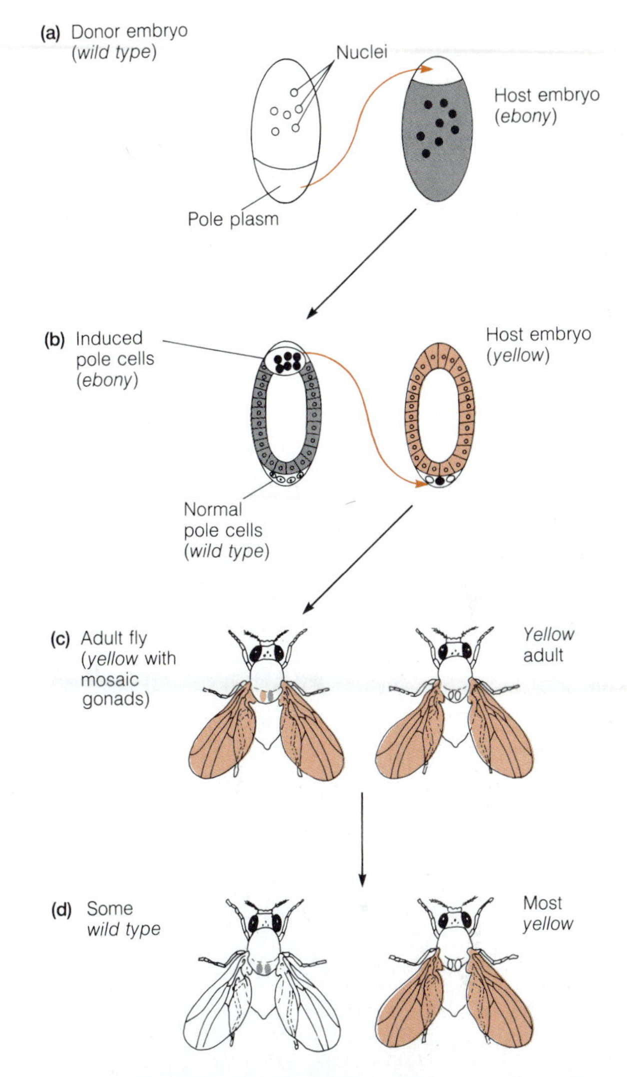

Figure 22-18 The Presence of Cytoplasmic Determinants in *Drosophila* Is Revealed Through Transplantation of the Pole Plasm. K. Illmensee and A. Mahowald transplanted posterior cytoplasm from a wild-type donor embryo into the anterior of a host embryo (a). The host embryo carried the genetic marker *ebony* body color. The transplanted pole plasm induced some host nuclei to form extra pole cells in the *anterior* end of the host embryo (b), revealing the presence of *cytoplasmic determinants* in the transplanted cytoplasm. To test whether these pole cells were functional, they transplanted them into their normal position in a second host embryo (b), which carried the *yellow* body color marker. These embryos developed into adult *yellow* flies that had mosaic gonads (c); some germ cells were of the *yellow* genotype, some were of the *ebony* genotype. To see this mosaicism, Illmensee and Mahowald mated these flies to more *yellow* flies and examined the progeny (d). Although most had *yellow* bodies, some had wild-type body color, proving that the transplanted *ebony* pole cells were functional. (The second class of flies are not *ebony* because *ebony* is a recessive mutation.)

Retinoic Acid: A Morphogen in the Chick Limb Bud

For the best-characterized example of a morphogen, we have to look at a different organism and a later stage of development—the $3\frac{1}{2}$-day-old chick embryo. At this point the rudimentary body plan has been laid out through the processes of gastrulaton, neurulation, and somite formation, and organogenesis and morphogenesis are beginning. In particular, small bulges of mesodermal and ectodermal cells arise at two pairs of symmetric locations on the side of the embryo. These bulges will develop into the limbs of the chicken and are therefore called the **limb buds.**

Figure 22-A depicts the chick limb bud and its three primary axes: *proximal-distal* from the base to tip, *dorsal-ventral* from top to bottom, and *anterior-posterior* from front to back (or thumb-to-pinky in your hand). The most dramatic growth in the bud is along the proximal-distal axis. The first distinct feature of the developing limb bud is the **apical ectodermal ridge,** or **AER.** In 1976 John Saunders, a pioneer of limb bud development, and his colleagues demonstrated that the AER is required for the outward or distal growth of the limb bud: If the AER is removed, limb bud growth ceases.

While the AER leads and is necessary for outward growth, a group of mesenchymal cells immediately beneath the AER drives the outward growth through the process of cell division. These mesenchymal cells form the **progress zone** of the limb bud. As early as 1918, Ross Harrison knew of the importance of these mesenchymal cells, but it was

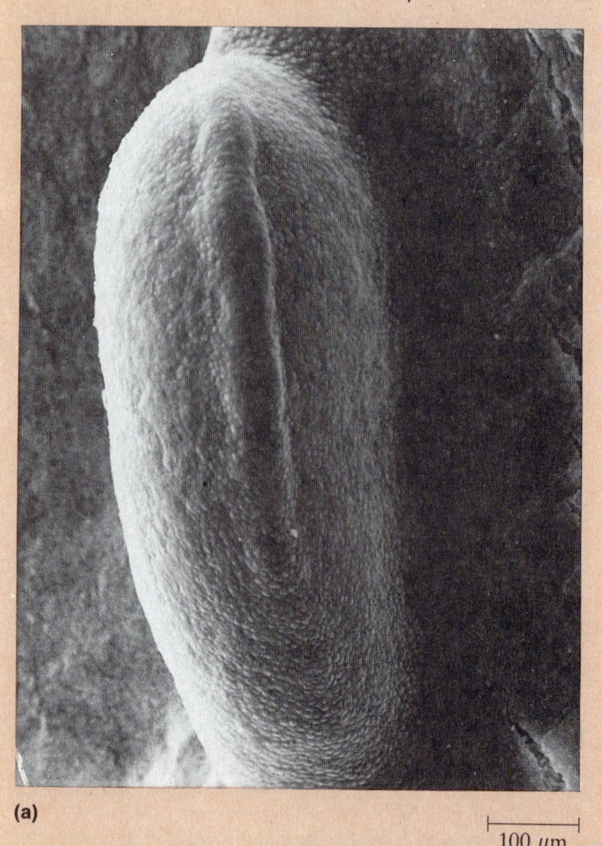

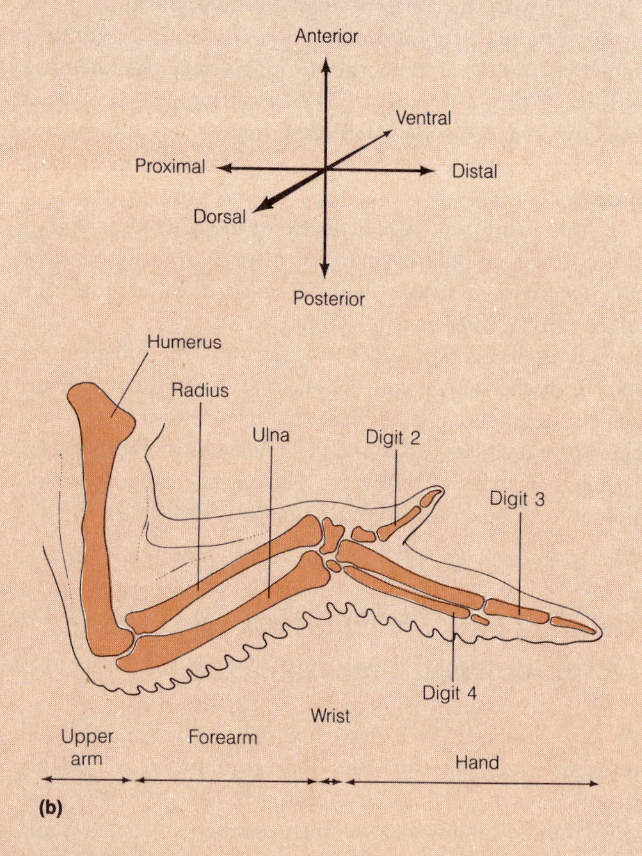

Figure 22-A The Chick Limb Bud, Its Three Primary Axes, and the Resulting Limb. (a) This scanning electron micrograph shows the chick limb bud oriented anterior up and posterior down, dorsal left and ventral right, and with the distal tip coming out of the page at the viewer. Note the apical ectodermal ridge (AER) along the distal margin of the limb bud. The mesodermal cells just below the AER form the progress zone; cell division in the progress zone drives the distal growth of the limb bud. (b) The three-dimensional system of coordinates used to describe the developing limb is shown above a mature wing. Proximal means close to the body, distal means distant from from the body; dorsal means top, ventral means bottom; anterior means front, posterior means back.

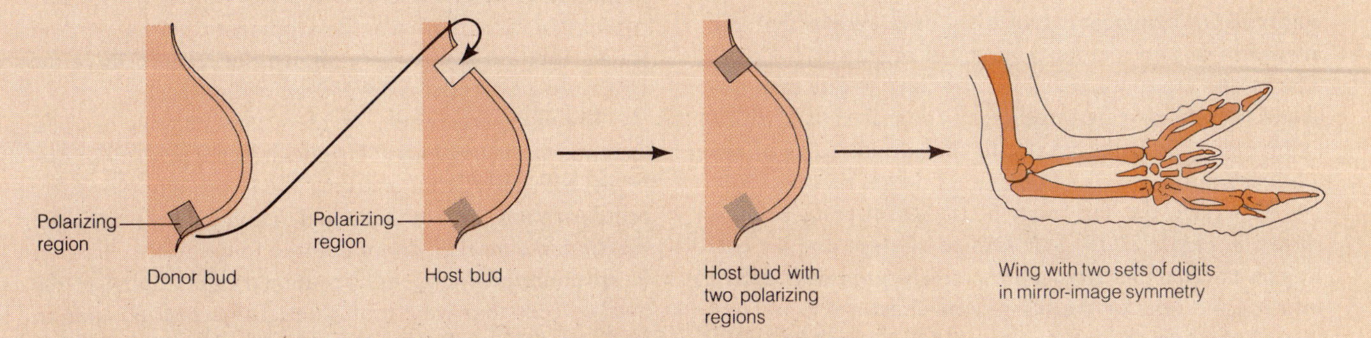

Figure 22-B Transplantation of the Zone of Polarizing Activity (ZPA) Causes a Duplication of the Digits in a Developing Chick Limb Bud. When the mesodermal cells from the zone of polarizing activity (ZPA) are transplanted from their normal posterior location in a donor limb bud to an anterior location in a host limb bud, the presence of two ZPAs causes a mirror-image duplication of digits.

Saunders, in 1948, who provided the explanation. Time provides a fourth dimension in the outward growth of the bud—cells that leave the progress zone early in development form the proximal structures at the base of the limb (the upper arm, for example), while cells that leave the progress zone later in development form the more distal structures of the limb (first the forearm, then the hand, and finally the digits). Transplantation experiments confirm this: When the progress zone from a late-stage limb bud is transplanted to the tip of an early-stage limb bud, only distal structures develop. Conversely, when an early-stage progress zone is transplanted to the tip of a late-stage limb bud, extra proximal structures are formed before any distal structures appear.

Our real interest, however, is in development along the anterior-posterior axis, for it is here that a true *morphogen* has been identified. First let's consider some more transplantation experiments that set the stage for the identification of that morphogen.

In 1968, Saunders and associates transplanted various pieces of a donor limb bud to different positions in a host limb bud. They observed that when mesodermal cells from a region at the posterior base of the limb bud are transplanted to the anterior base of the host limb bud, a mirror-image duplication occurs. Whereas the normal pattern of digits in the limb bud is 2-3-4 (anterior to posterior, Figure 22-A), the pattern of digits after transplantation is 4-3-2-2-3-4 (Figure 22-B). The researchers named the critical region of mesoderm the **zone of polarizing activity,** or **ZPA.** In 1975, Cheryll Tickle and colleagues hypothesized that the ZPA was the source of a diffusible morphogen. The concentration of the hypothesized morphogen would increase from anterior to posterior, and the digital identity of a particular group of primordial digit cells would be specified by a specific concentration of morphogen.

And so the search began for the morphogen. In 1982, Tickle, Lewis Wolpert, and colleagues showed that **retinoic acid,** or **RA,** could mimic the effects of a transplanted ZPA. Few scientists, however, believed that RA, which has many unusual effects on animal cells in culture, was the in vivo morphogen. In 1987 Christina Thaller and Gregor Eichele proved the skeptics wrong. By conducting sensitive biochemical analysis of 5536 tediously dissected limb buds, they showed that RA was indeed present in the limb bud and distributed in a gradient decreasing from posterior to anterior. Furthermore, RA's biological precursor—retinol or vitamin A—is evenly distributed at a much higher concentration throughout the limb bud. Thus, it appears that an enzymatic activity in the posterior margin of the limb bud is synthesizing RA. The cells producing this enzymatic activity, then, are the true source of the ZPA.

The embryonic chick limb bud is not an isolated case. The process of *limb regeneration* in salamanders is apparently very similar. When a salamander limb is cut off, a *regeneration blastema* forms. Left unperturbed, the blastema will regenerate a completely normal limb. However, if the regeneration blastema of a salamander limb is cut off at its base, rotated 180°, and replaced on the limb stump, the resulting limb has duplicated structures.

Is retinoic acid, the proven morphogen in the chick limb bud, also the key player in salamander limb regeneration? The answer seems likely to be "yes." Like those who study the chick limb bud, scientists studying regeneration blastemas can mimic the duplications caused by surgical manipulations through topical application of retinoic acid!

Finally, researchers interested in the various effects of RA on human cells in culture recently identified the gene for a cellular receptor for RA. The RA receptor has many similarities to the steroid receptor discussed in Chapter 21. While its precise roles in human development have yet to be fully elucidated, it seems a foregone conclusion that retinoic acid's role as a developmental regulator has been strongly conserved during evolution.

cells. Do you suppose that it is possible to assign a particular developmental fate based on a cell's position in the embryo? Developmental biologists have shown that the answer to this question in a variety of organisms is "yes." The results are summarized in *fate maps*. A **fate map** is a sketch of the blastula-stage organism that shows which embryonic cells will give rise to the various parts of the adult organism.

To create the fate map of the *Drosophila* embryo shown in Figure 22-19, Jose Campos-Ortega and his colleagues injected the enzyme horseradish peroxidase (HRP) into specific cells of the blastoderm-stage embryo. They then allowed development to proceed through embryogenesis and stained the mature larvae for enzymatic activity of HRP. By recording where the HRP was injected and what larval tissues were marked, they were able to derive the fate map. Note that a fate map is in many respects a summary of **cell lineages**—the entire program of cell divisions that a single cell and its descendants go through, and the cell types that result from those divisions.

The fate map in Figure 22-19 tells us how cells in any particular area of the blastoderm will develop, assuming those cells are not disturbed during embryogenesis. But are the cells of the blastoderm strictly limited to the fates depicted in the fate map? How would a cell develop if we moved it to a different location in the blastoderm? These

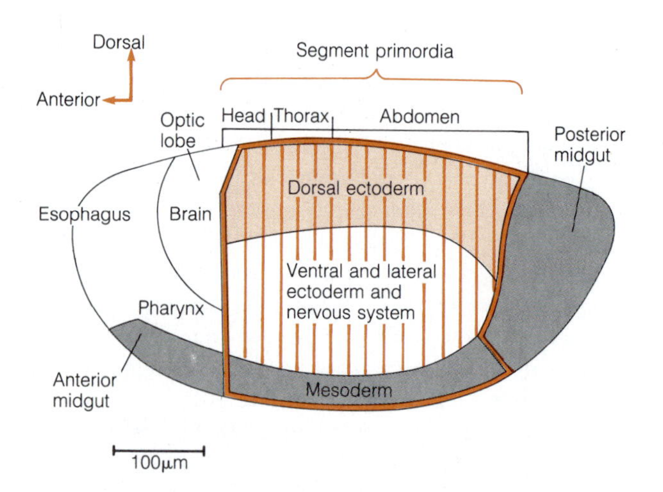

Figure 22-19 Fate Map of the *Drosophila* Embryo at the Cellular Blastoderm Stage. Fate maps of many embryonic organisms have been generated through lineage tracer and transplantation experiments, as well as more complicated genetic analysis. Here we see a plan of the larval and adult tissues that are eventually produced by the cells in the different parts of the cellular blastoderm of *Drosophila*. Cells inside the heavy line will give rise to the head, thoracic, and abdominal segments of both the larval and adult stages (see Figure 22-31). Within the segmental primordia, cells above the horizontal line will form dorsal ectoderm, whereas cells below the line will form ventral and lateral ectoderm and the nervous system. The shaded regions will invaginate during gastrulation and become the mesoderm and midgut, whereas the dorsal- and anterior-most cells will involute at the end of gastrulation to form the internal structures of the head.

questions introduce the very important concept of **determination:** A determined cell is a cell that is *irreversibly committed* to a particular developmental pathway. And transplantation studies have shown that most cells of the *Drosophila* blastoderm are indeed determined.

Knowing the results of their fate-mapping studies, Campos-Ortega and his colleagues injected *precellular* embryos with HRP, thus marking the entire embryo. After cellularization, they transplanted individual cells from one location in a marked donor embryo to a different location in an unmarked host embryo. Remarkably, many of the marked cells that were transplanted to a new position in the host embryo developed as though they were in their original position. Thus the developmental fate of these cells was already determined at the cellular blastoderm stage.

Determination has been seen in many other systems and is a crucial developmental step. Note that determination, or **commitment,** can be a progressive process. For example, in *Drosophila* a cell may become committed first to a general ventral fate, then to a general neural fate, then specifically to the central nervous system, and finally to a particular axonal cell fate. Remember, however, that we must be careful to distinguish between determination and *differentiation:* A cell can be determined without being differentiated. Such a cell shows no physical signs of its determined state and is indistinguishable from cells that have much different fates. Differentiation occurs when a cell develops the physical characteristics that are necessary for it to carry out its eventual function and that physically distinguish it from other cell types.

Selective Gene Expression and Nuclear Equivalence. Commitment involves many changes in the biology of a cell. Most obviously, the set of proteins made in a cell changes as that cell becomes committed to a particular cell fate. Therefore some genes must be inactivated as others are activated. Earlier we said commitment was irreversible, but changes in gene expression hardly seem irreversible.

There are a few cases in which commitment is obviously irreversible. When mammalian red blood cells differentiate, they break down their nuclei and so lose their entire genome. Another less dramatic case involves the somatic cells of the nematode worm *Ascaris,* which fragment the two chromosomes found in germ cells into many smaller chromosomes during development.

Such extreme cases are rare, however. In most cells the changes brought on by commitment are mediated by **selective gene expression**—changes in the subset of genes that a cell expresses at any given point in development. Because no genetic information is lost in this process, commitment by selective gene expression, and the resulting changes in cell biology, often can be reversed *experimentally.*

Hans Spemann first demonstrated this reversibility in 1938. He used a loop of fine baby hair to divide the zygote of the salamander *Triturus* into two connected lobes just prior to cleavage (Figure 22-20). Since the nucleus could go into only one of the lobes, cleavage proceeded entirely in that half of the constricted embryo. Then, at the 32-cell stage, he allowed a single nucleus to slip through to the enucleated (nucleus-free) lobe (Figure 22-20c). Spemann then separated the two lobes and observed their development. Remarkably, both lobes developed into normal tadpoles, although the lobe that was originally enucleated lagged behind the other lobe. This experiment demonstrated the phenomenon of **nuclear equivalence:** Up to the 32-cell stage, embryonic nuclei of *Triturus* are equivalent in their ability to direct subsequent development of a complete embryo.

In 1952, Robert Briggs and Thomas King showed that commitment can be experimentally reversed in much older nuclei from gastrula-stage embryos of the frog *Rana pipiens*. When transplanted with a fine syringe into an enucleated egg, such nuclei were able to support development to the tadpole stage, and in a few cases all the way through metamorphosis to the adult frog. Nuclei obtained from later-stage embryos, however, could support only partial embryonic development.

When John Gurdon repeated these experiments in *Xenopus*, he found that by subjecting older nuclei to *serial transfer* through egg cytoplasm, he could overcome some of the limitations on developmental potential observed by Briggs and King. A nucleus from an intestinal cell, for example, was injected into an enucleated egg and allowed to divide. The recipient eggs cleaved only a few times, at which point one nucleus was isolated and transplanted into a new recipient. After this second transfer, a significant proportion of the recipients developed much further. When a nucleus was transferred a *third* time, some embryos gave rise to swimming tadpoles!

These experiments show that the ability of a nucleus to promote embryonic development in enucleated eggs becomes progressively restricted over the course of devel-

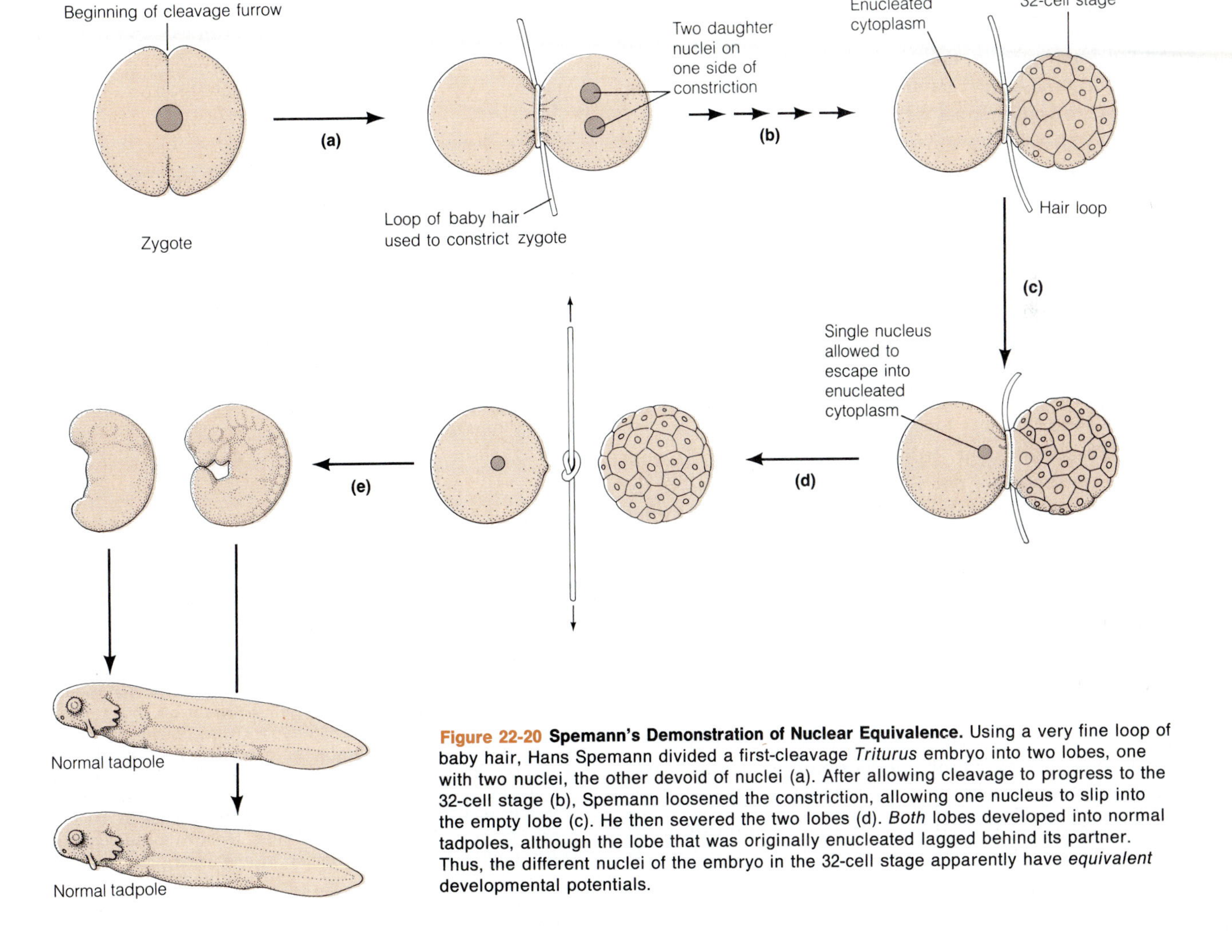

Figure 22-20 Spemann's Demonstration of Nuclear Equivalence. Using a very fine loop of baby hair, Hans Spemann divided a first-cleavage *Triturus* embryo into two lobes, one with two nuclei, the other devoid of nuclei (a). After allowing cleavage to progress to the 32-cell stage (b), Spemann loosened the constriction, allowing one nucleus to slip into the empty lobe (c). He then severed the two lobes (d). *Both* lobes developed into normal tadpoles, although the lobe that was originally enucleated lagged behind its partner. Thus, the different nuclei of the embryo in the 32-cell stage apparently have *equivalent* developmental potentials.

opment. Yet serial transfer through egg cytoplasm can restore the developmental potential of much older nuclei. How does the egg cytoplasm rejuvenate the specialized nucleus of a differentiated cell? Recall that much of development occurs through selective gene expression without the loss of any genetic material. In older nuclei the DNA is intact, but the nuclear proteins that regulate gene expression are appropriate only for the restricted fates of the older cell. Presumably the serial transfer dilutes out these regulatory proteins, restoring the chromosomes to their original, prefertilization condition.

The Cell Cycle During Cleavage

At no time in development is the cell cycle more rapid than during blastula formation. The *Drosophila* embryo offers an extreme example, as the first nine syncytial cycles take only 10 minutes apiece. Furthermore, in most organisms the cell divisions leading up to the blastoderm stage are synchronous. Obviously, the cell cycle must be very precisely controlled during embryogenesis.

In *Xenopus,* this precise cell cycle regulation is mediated by **maturation-promoting factor (MPF)**. MPF is a multisubunit protein, and one subunit, known as p34, is a *protein kinase*. In Chapter 21 we learned that protein kinases covalently attach phosphate groups to other proteins. We also saw how the hormone epinephrine activates glycolysis through a series of such protein phosphorylation events. Similarly, oscillations in MPF protein kinase activity regulate the cell cycle. When the kinase is "on" the cell goes through G-2 and then mitosis. After mitosis is completed, the MPF kinase activity is turned off, and the cell cycle stops early in G-2 until the cell is again ready for mitosis. Interestingly, *histones* and *lamins,* structural proteins of chromatin and the nuclear cytoskeleton, undergo changes in phosphorylation as the cell progresses through the cell cycle.

MPF gets its name from the critical role it plays in oocyte maturation. Prior to oocyte maturation, MPF is present in an inactive form in arrested oocytes. When the MPF protein kinase becomes active, the oocyte completes G-2, germinal vesicle breakdown (p. 675), and meiosis I. Here, too, the MPF protein kinase regulates the cell cycle.

Xenopus MPF is part of a system of cell cycle regulation that is highly conserved from yeast to humans. By looking for mutations that cause yeast cells to stop growing at defined points in the cell cycle, geneticists have identified many yeast genes that encode regulators of the cell cycle. These genes are called the **cell division cycle genes,** or ***cdc* genes.** If the *cdc2* gene is genetically inactivated, the cell cycle stops in early G-2 phase. Thus, like MPF, the *cdc2* gene product must be active for the cell to proceed through the cell cycle. DNA sequence analysis has now shown that, like the p34 subunit of MPF, the protein produced by the *cdc2* gene is a protein kinase and that the two proteins are *homologues*. Thus, the **MPF/cdc2 protein kinase** controls the cell cycle in both species.

Final and remarkable proof that cell cycle regulation is a highly conserved process came when a human gene was found that had a nucleic acid sequence very similar to that of the yeast *cdc2* gene. Researchers then transferred the human *cdc2*-like gene into yeast cells and inactivated the yeast *cdc2* gene. The human protein correctly regulated the cell cycle in the mutant yeast cells, thus "rescuing" the *cdc2* defect! Similarly, the *Drosophila* version of the *cdc2* gene can also substitute for the yeast *cdc2* gene.

The MPF/cdc2 protein kinase is phosphorylated by other protein kinases and dephosphorylated by *protein phosphatases*. When the correct combination of sites on MPF/cdc2 is phosphorylated, its protein kinase activity turns on and activates the structural and enzymatic proteins that carry out the events of G-2 and mitosis (Figure 22-21). Thus, these other kinases and phosphatases act together to regulate the MPF/cdc2 protein kinase. In yeast, these regulators of MPF/cdc2 respond to cellular and en-

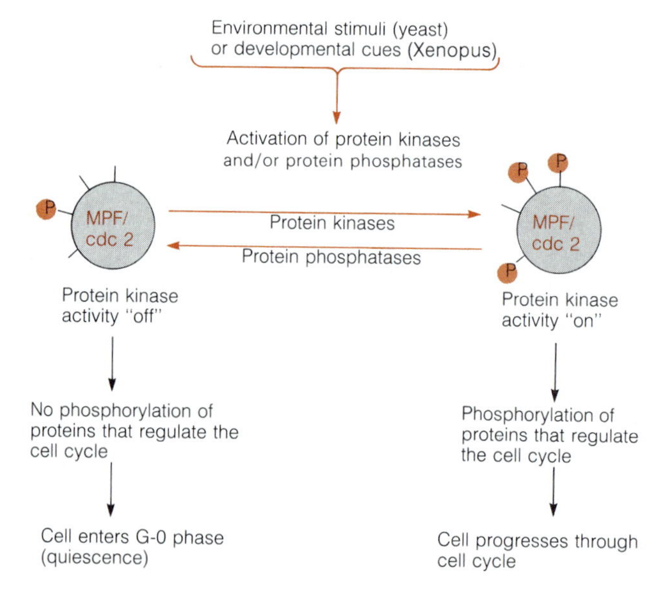

Figure 22-21 A Model for the Function and Regulation of the MPF/cdc2 Protein Kinase. At the start of G-2 in the cell cycle, the activity of cdc2 protein in yeast and MPF in *Xenopus* determines whether or not the cell will proceed through the cell cycle or enter the quiescent phase G-0. If the MPF/cdc2 protein kinase is on, it phosphorylates and thereby activates the proteins that regulate the cell cycle; if the MPF/cdc2 protein kinase is off, those proteins remain inactive. The MPF/cdc2 protein kinase is itself regulated by phosphorylation at several sites; only when the correct combination of sites is phosphorylated does the protein become active. Various other protein kinases and protein phosphatases carry out this regulation of MPF/cdc2, presumably in response to environmental stimuli and developmental cues. Many of these *cdc2* regulators in yeast have been identified by mutations.

vironmental cues that determine whether or not conditions are right for cell growth. In the *Xenopus* embryo, this phosphorylation and dephosphorylation is part of the developmental program. As long as cell division continues, MPF/cdc2 activity oscillates. But when cells undergo terminal differentiation, MPF/cdc2 is presumably turned off permanently, and the cell enters G-0—the resting state of the cell cycle.

Gastrulation: Laying Out the Body Plan

Once the blastula has formed, the embryo undergoes a series of coordinated cell movements called **gastrulation** in which a portion of the outer cell layer of the blastula invaginates into the embryo. The details of gastrulation vary widely from one organism to the next and can be extremely complex. But the essential features of gastrulation are the same: an invagination of cells on the surface of the blastula to form the primitive gastrointestinal tract, and the concomitant formation of the *primary germ layers*. The transformation from blastula to *gastrula* is a fundamental step in **morphogenesis,** the process that gives rise to the final shape of an organism. As the relatively simple mass of cells known as the blastula rearranges itself into higher levels of organization, the basic body plan of the organism is laid out. By the end of gastrulation the stage is set for the next phase of development—**organogenesis,** the process that gives rise to the different tissues and organs within that basic body plan.

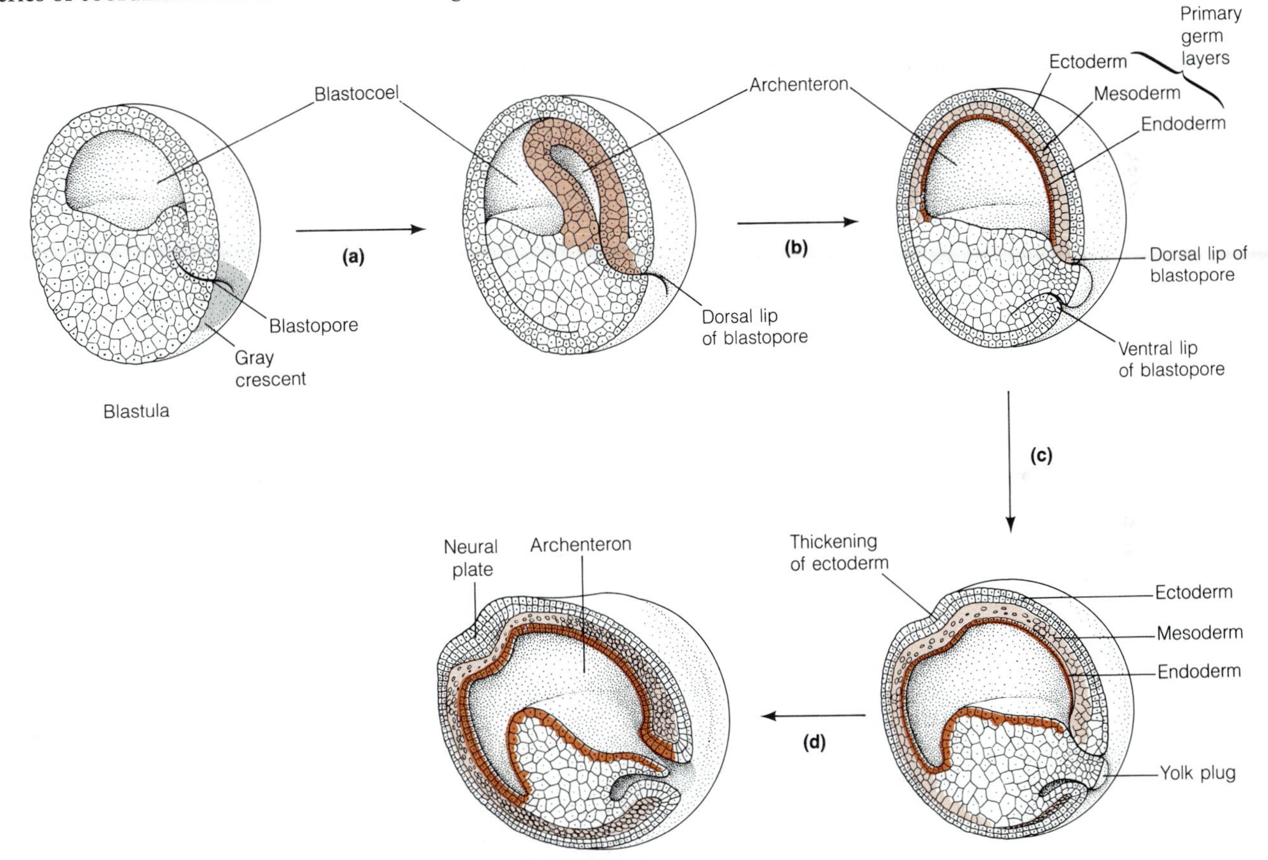

Figure 22-22 Gastrulation and Formation of the Primary Germ Layers in the Frog Embryo. Gastrulation involves a series of coordinated cell migrations in which a portion of the outer cell layer invaginates into the interior of the embryo. Early in the process, the primary germ layers become defined and move into the spatial relationships that are important for later interactions. Gastrulation begins as the blastopore lip forms in the gray crescent region. The invagination of cells through the blastopore lip produces the archenteron, or primitive gut. As more cells move inward from the surface of the embryo, the archenteron expands and displaces the blastocoel cavity (a). As the germ layers become more clearly defined, the ectodermal cells migrate over the vegetal hemisphere toward the ventral blastopore lip (b). As gastrulation continues, a yolk plug fills the blastopore and the ectoderm on top of the embryo thickens (c). Gastrulation is over when the inward cell migration has stopped and the germ layers are fully formed (d). The beginning of the next stage, neurulation, can be seen in the incipient neural plate.

The Events of Gastrulation in Amphibians

In the case of the frog, gastrulation begins with an invagination of cells in the vicinity of the *gray crescent,* as shown in Figure 22-22. Recall that the gray crescent is formed opposite the point of sperm entry by the rearrangement of the cortical cytoplasm. This involution of cells forms the *dorsal lip* of the **blastopore** on the surface of the gastrulating embryo, and the **archenteron,** or **primitive gut,** within the blastocoel cavity (Figure 22-22a). As more cells migrate to the interior of the embryo, the archenteron enlarges and the blastopore widens (Figure 22-22b). The widening blastopore exposes a patch of cells from the inner cell mass of the vegetal hemisphere. These cells, known as the **yolk plug,** fill the blastopore opening.

As the invagination of the archenteron progresses, three distinct cell layers form. These are the **primary germ layers.** This division into three general cell types is a critical step, since the distinction between cells of the different germ layers is maintained throughout development. The cells of the outermost layer, the **ectoderm,** will give rise primarily to the epidermis and the central nervous system of the organism. The innermost layer, composed of the cells that migrated inward from the surface of the embryo, is the **endoderm.** As we have already noted, the endodermal cells will give rise to the digestive tract. Finally, the **mesoderm** is sandwiched between the endoderm and the ectoderm. Mesodermal cells will give rise to skeletal elements, muscle, connective tissue, kidneys, gonads, and blood.

Cellular Changes During Gastrulation

We will return shortly to the germ layers and a discussion of some critical interactions that occur between them. First, however, we should examine how some other aspects of cell biology can play important roles in development. For example, changes in cell shape and interactions at the cell surface accompany, and in some cases may direct, gastrulation. The examples below only suggest the many ways in which cell shape, the cell surface, and the extracellular matrix participate in all of development.

The Role of Cell Shape. What directs and powers the complex cellular movements of gastrulation? The answers to this question are bound to draw on many aspects of cell biology. Obviously, differentiated cell types come in all shapes and sizes, but is this variety simply the result of differentiation, or do some changes in cell shape play a causative role in development? In the case of the very first movement of gastrulation, the invagination by cells at the blastopore lip, changes in cell shape *do* play an important developmental role.

To initiate gastrulation, the cells at the blastopore lip undergo some very dramatic changes in cell shape, becoming **bottle cells.** This reshaping is presumably mediated by the cytoskeleton. The bottle-like shape of these cells squeezes them to the interior of the embryo, where they continue to lead the inward march of the archenteron (Figure 22-23a).

The shape of the bottle cells was thought for a while to be the primary requirement for blastopore formation. But experiments have shown that once bottle cells have formed, their removal does not prevent the further inward

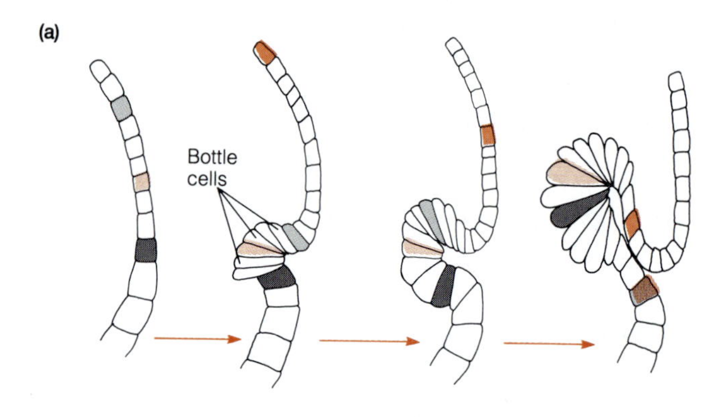

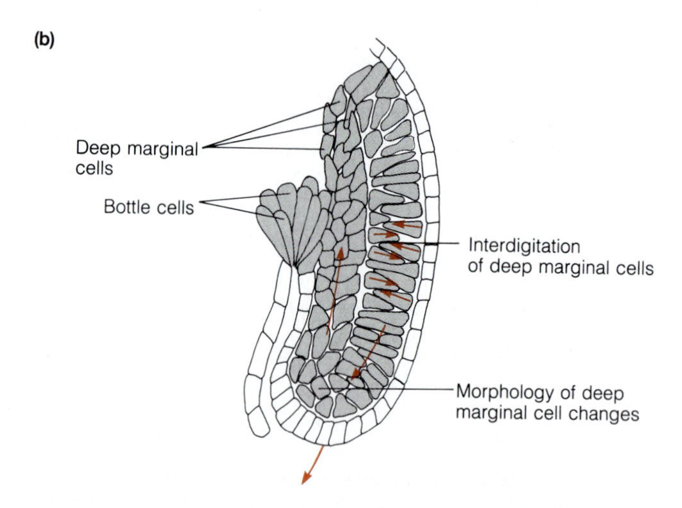

Figure 22-23 The Bottle Cells and the Deep Marginal Cells Start and Drive the Archenteron Invagination. (a) At the onset of gastrulation, the bottle cells elongate into a characteristic wedgelike shape. As they first become thinner and then migrate inward, surrounding cells are drawn toward and into the invagination. In (b), however, we see that the underlying deep marginal cells provide much of the force that drives the cellular migrations. Like the bottle cells, the deep marginal cells carry out their role through changes in cell morphology, interdigitating and thinning in such a way that the overlying cell layer has no choice but to follow the bottle cells into the archenteron.

migration of cells into the archenteron. The same cannot be said of the underlying **deep marginal cells,** which interdigitate and flatten as gastrulation proceeds. If these cells are removed, gastrulation stops. As with the bottle cells, the changes in the shape of the deep marginal cells aid in blastopore formation and power the movement of cells into the archenteron (Figure 22-23b).

Cell Surface Properties Help Define the Germ Layers. We have to this point ignored the role of the cell surface in embryonic development. The surface of different cell types varies depending on the lipid and protein composition of the plasma membrane and on the different proteins and polysaccharides on the surface of the membrane. These factors combine to regulate aspects of cell biology ranging from cell rigidity to the affinity of one cell's surface for the surface of other specific cells.

In 1955, Phillip Townes and Johannes Holtfreter published some simple but elegant experiments that indicated that cell surface affinity may indeed be important in development. They dissected frog embryos into mesodermal, ectodermal (epidermal and neural), and endodermal tissues (Figure 22-24). Next they dissociated the dissected tissues into cell suspensions and mixed the different cell populations. After first aggregating, the cells sorted themselves out into rudimentary germ layers. Furthermore, the layers formed with ectodermal cells on the outside of the cell mass, endodermal cells on the inside, and mesodermal in between. Thus the cells of the different germ layers have a natural affinity for their own kind. That affinity is presumably mediated by specific protein-protein interactions at the cell surface. Similar interactions occur throughout development, and may be particularly important in the process of *organogenesis*.

Other experiments have confirmed and extended this suggestion. For example, cell surface proteins are believed to guide many critical **cell migrations.** We have already seen one dramatic example of cell migration—that of the archenteron into the interior of the embryo. Another occurs later in development when nerve cells extend axons, often over great distances, toward the tissues that they will eventually innervate. **Neural cell adhesion molecule,** or **N-CAM,** is a protein on the surface of nerve cells. When migrating axons are treated with antibodies to N-CAM,

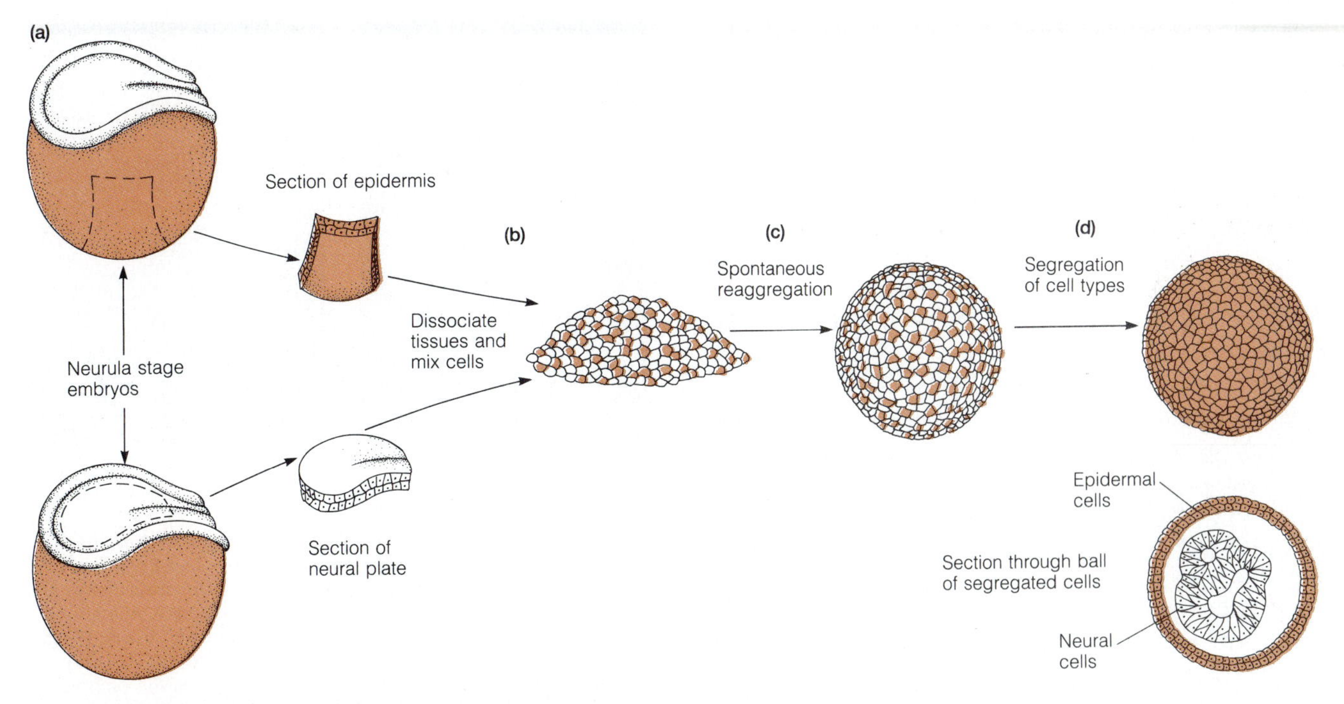

Figure 22-24 Reaggregation Experiments with Dissociated Germ Layer Cells Reveal a Role for Cell Surface Interactions in Development. (a) Pigmented ectodermal and unpigmented neural plate tissues are dissected from embryos that have completed gastrulation and formed neurulas. The two tissues are then dissociated and the cells mixed (b). The cells spontaneously reaggregate (c) and then sort themselves out by cell type (d). Similar results are seen with ectoderm and mesoderm, mesoderm and endoderm, and all three germ layers at once. In each case, like associates with like and, remarkably, the spatial relationships of the re-formed cell layers mimic the structure of the intact embryo. These results indicate that the cell surface plays a role in forming and maintaining the germ layers.

this migration is blocked. Thus N-CAM is implicated in the cellular migration of axons.

Components of the Extracellular Matrix. Like the proteins of the cell surface, we might expect that proteins of the *extracellular matrix* (Chapter 7) have a role in many aspects of development, including cell migrations. In 1987, Jean-Paul Thiery and his colleagues showed that this is indeed the case with a very simple experiment in gastrulating *Drosophila* embryos.

Many proteins of the extracellular matrix contain a conserved sequence of three amino acids: *arginine-glycine-aspartic acid,* abbreviated **RGD.** The RGD sequence is a site for specific interactions between various proteins of the extracellular matrix. To test whether the interactions mediated by RGD play an important role in embryogenesis in the fruit fly, Thiery and his colleagues injected concentrated solutions of *synthetic tripeptides* into *Drosophila* embryos. They injected some embryos with RGD, while they injected other embryos with a random sequence of three amino acids as a control experiment. Figure 22-25 shows the striking result. In embryos injected with RGD, gastrulation was blocked (b and d); in embryos injected with control tripeptides, even in 50-fold excess, development proceeded normally (a and c). A single amino acid change in the RGD sequence destroyed the tripeptide's effect. They concluded that the injected RGD tripeptide was disrupting interactions between extracellular matrix and cell surface proteins. Clearly, given the impact of injected RGD on gastrulation, these interactions must be critical for the cellular migrations of gastrulation.

Interactions Between Tissues: Induction

The formation of primary germ layers ushers in a new phase of developmental events characterized by inductive interactions between tissues. As we have seen, the different germ layers consist of distinct cell types. The cellular movements of gastrulation bring cell types not previously adjacent to one another into contact, where they begin to influence each other's development. Such interactions between cells from different germ layers and in developing tissues and organs are important during gastrulation and later developmental stages. The general term for such interactions is *induction.*

An **induction** is characterized by one cell or tissue causing an adjacent cell or tissue to develop into a particular differentiated structure. For an interaction to be truly inductive, two conditions must be met. First, the differentiation of the *induced* tissue must be wholly dependent on the presence of the underlying *inducing* tissue. In other words, in the absence of the inducing cells, the responding cells would develop differently. Second, the inducing tissue, when transplanted to a different location in the embryo, must be capable of causing induction in equivalent tissue. (Equivalent tissue means cells of the same germ layer as those that are induced in the normal situation.)

Induction: The Classical Phenomenon. Hans Spemann and Hilde Mangold first described induction in 1924, with embryos of the salamander *Triturus.* Embryogenesis in *Triturus* is very similar to that in *Xenopus.* Spemann and

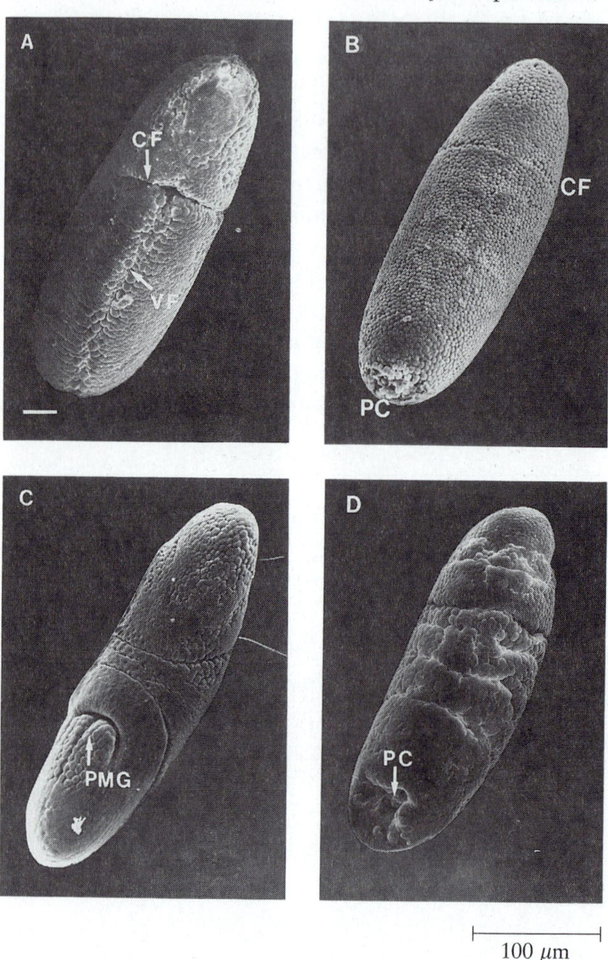

Figure 22-25 The Tripeptide RGD, an Extracellular Matrix Attachment Site, Interferes with Gastrulation in *Drosophila* Embryos. Cellular blastoderm-stage embryos were injected with either a control tripeptide (a and c) or with the RGD tripeptide, containing the amino acids arginine (R), glycine (G), and aspartic acid (D) (b and d). The ventral furrow (VF) of *Drosophila* is roughly equivalent to the blastopore lip of the *Xenopus* embryo. In the embryo injected with control tripeptide the ventral furrow is normal as the embryo begins gastrulation at 3.5 hours (a), while in the RGD-injected embryo there is no sign of the ventral furrow (b). At 4.25 hours, gastrulation continues normally in the control embryo (c). Notice the characteristic pattern of invaginations and cellular migrations: the pole cells (PC) have descended into the embryo's interior at the tip of the posterior midgut invagination (PMG). But in the 4.25-hour RGD-injected embryo (d), the normal invaginations are thoroughly disrupted and the pole cells remain at the posterior tip of the embryo. (CF designates the cephalic furrow, a second invagination that occurs early in *Drosophila* gastrulation.) (SEMs)

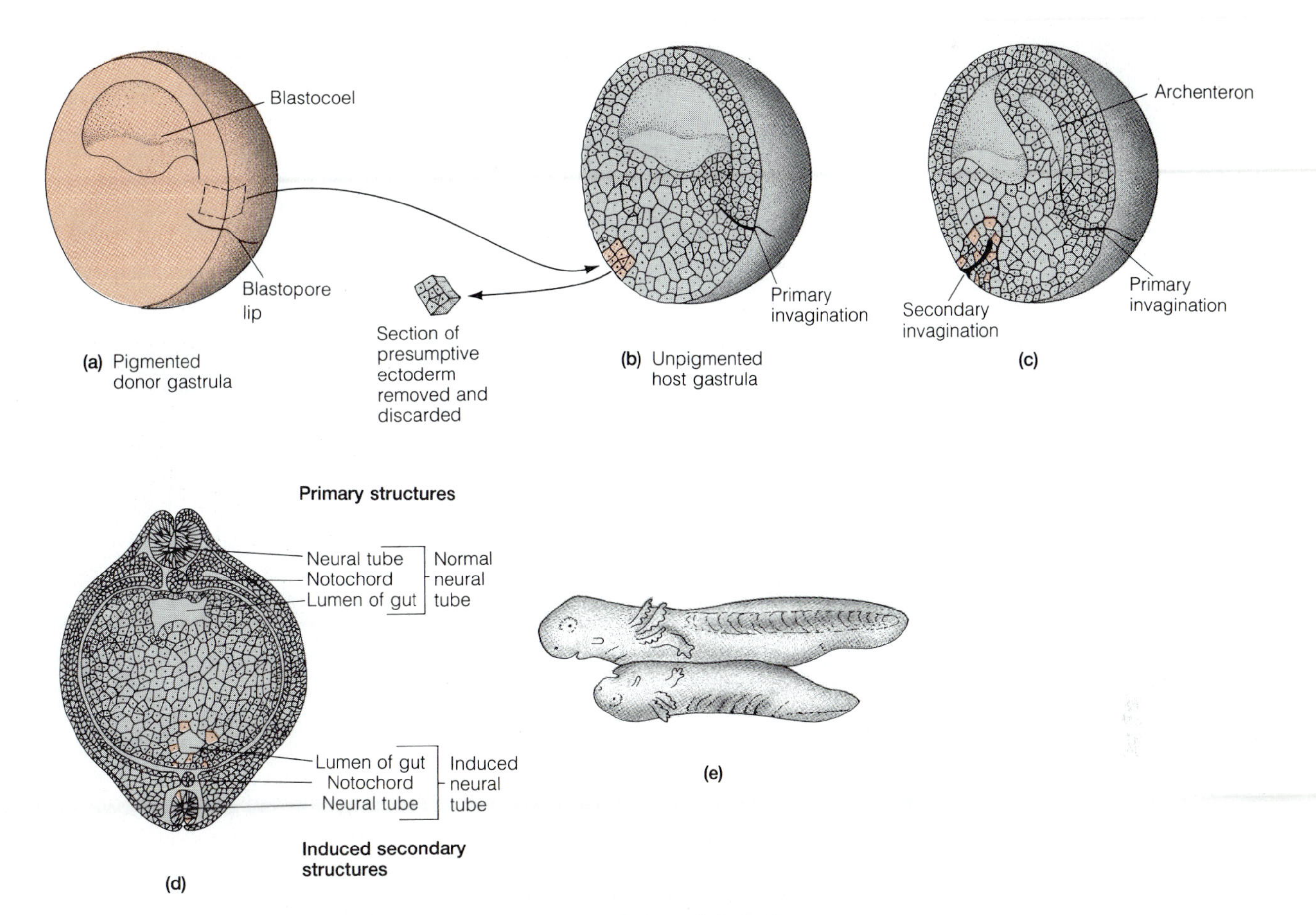

Figure 22-26 Spemann and Mangold's Demonstration of Primary Embryonic Induction. Hans Spemann and Hilde Mangold transplanted the dorsal blastopore lip region of an early gastrula of the darkly pigmented salamander *Triturus taeniatus* (a) into a ventral position on an early gastrula of its unpigmented cousin, *Triturus cristatus* (b). The transplanted tissue induced the invagination of a second archenteron (c), leading to formation of a double neurula in which the duplicated neural tube was derived from both darkly pigmented donor cells and lightly pigmented host cells (d, shown in cross section; compare to Figure 22-27c). Eventually they observed "twinning"—the development of two complete embryos, fused along their ventral surfaces (e).

Mangold transplanted the dorsally located blastopore lip from one species of *Triturus* to a ventral location in embryos of a differently colored species. Figure 22-26 shows the remarkable result: The small patch of transplanted dorsal lip cells induced cells of the host embryo to form a second embryonic axis and archenteron invagination. In some cases an entire second embryo formed. Note that the host cells that form the duplicated structures would have had an entirely different (epidermal) fate in the absence of induction by the transplanted tissue.

The critical event in the development of the second embryo was later shown to be the induction of the **neural tube,** a structure that eventually differentiates into the brain and spinal cord. To form the neural tube, mesodermal cells of the transplanted blastopore induce ectodermal cells of the host to thicken and roll up. This induction, called **neurulation,** occurs in every embryo. It is depicted as it normally occurs in Figure 22-27. Neurulation is also called the **primary embryonic induction** because it is the first of a series of dramatic inductive interactions that lead the embryo out of gastrulation and into morphogenesis and organogenesis. As in the case of the primary embryonic induction, mesodermal tissue is generally the inducing tissue in these later inductions. The formation of the *limb buds* and the *apical ectodermal ridge,* discussed in the box on pp. 688–689 is an example of an induction that occurs later in development.

Induction: Some Molecular Explanations. The mechanism of induction—how the cells of the different tissues communicate and what molecular changes occur in the induced tissue—is the subject of intense investigation. Small, diffusible molecules must mediate many inductive interactions, since these inductions can still occur when

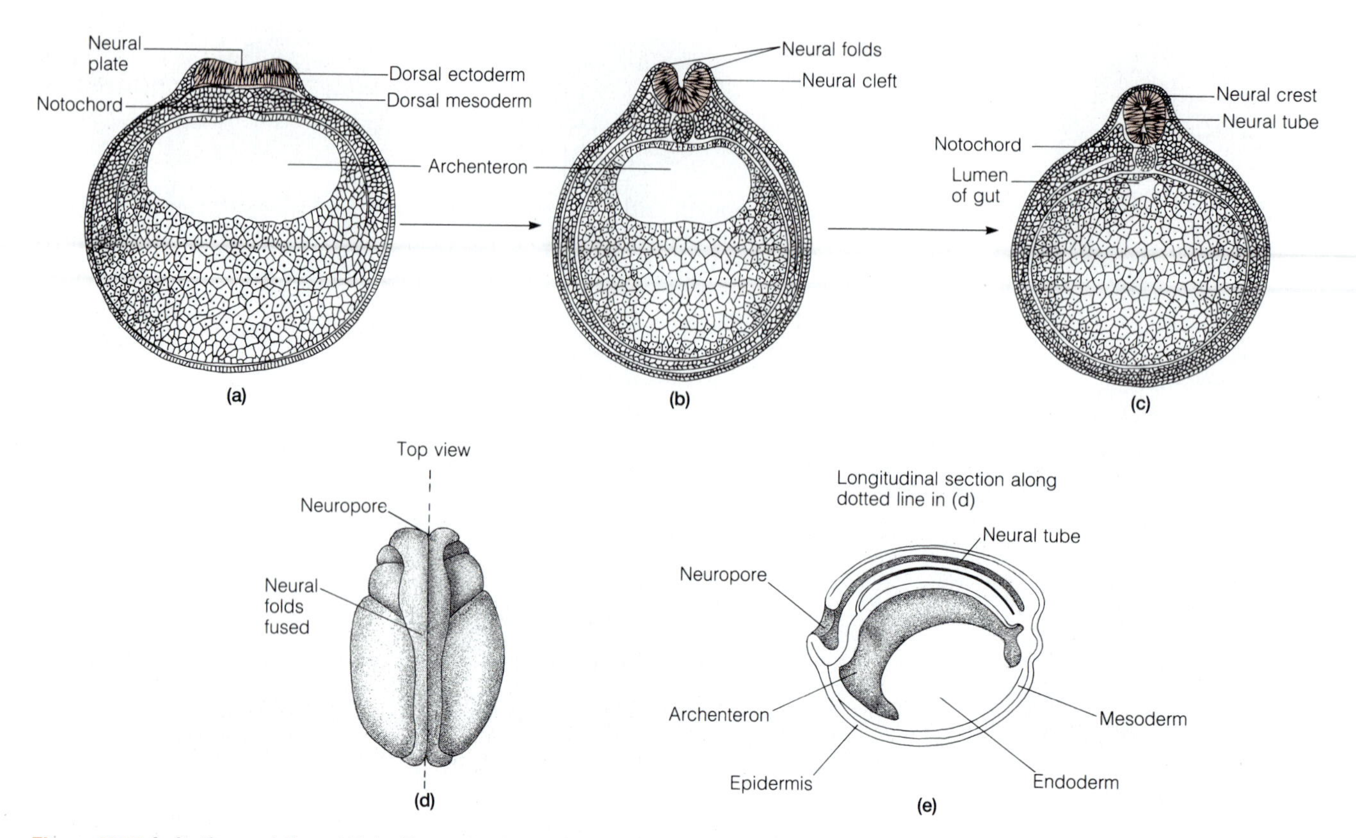

Figure 22-27 Induction and Neural Tube Formation in the Frog Embryo. The neural tube is the rudimentary central nervous system. At the onset of neurulation (a), the dorsal mesodermal cells of the archenteron roof induce the overlying ectodermal cells to thicken and roll up, forming the neural folds (b). The surrounding neural folds then seal the neural cleft, pinching off the completed neural tube from the underlying epidermis (c). The cross sections in (a)–(c) were cut at right angles to those in Figure 22-22, and the embryo in (a) above is at a slightly later stage than the last embryo in Figure 22-22. (d) and (e) are different views of the embryo seen in (c), now known as a neural tube-stage embryo, or neurula. Compare this frog neurula to the human neurula in Figure 22-30a.

the two tissues are separated by a semipermeable membrane. In other cases, cell surface proteins must interact directly for induction to occur, although it is not clear whether cell surface proteins are the inducing factors per se. It is only in the last few years that some of the molecular players in various inductions have been positively identified.

To look more closely at the molecular explanation of induction, let's turn back to the beginning of gastrulation, when the cells that will soon become the mesoderm migrate to the interior of the blastula. In the 1960s embryologist P. D. Nieuwkoop showed that these cells form mesoderm in response to an induction by the underlying cells of the vegetal hemisphere. He demonstrated this by first separating embryos into thirds and then recombining the animal and vegetal thirds, leaving out the middle third. In the recombined embryos, the animal cells that were now next to vegetal cells quickly took on mesodermal characteristics.

John Gurdon and associates later repeated these experiments and found that, in addition to simply looking like mesodermal cells, these animal cells began expressing the mRNA from muscle-specific actin genes (Figure 22-28a and b). (Recall that muscle is a mesodermal tissue.) This observation provided researchers with a molecular assay with which to test possible inducers of mesoderm differentiation: apply the candidate substance to animal hemispheres and see if the cells begin making muscle-specific actin mRNA. Two different research groups soon tried treating isolated animal hemisphere cells with *fibroblast growth factor* (FGF), a human protein that causes fibroblast cells to grow (Figure 22-28c). The cells of the animal hemisphere responded by making muscle-specific actin. The effect was enhanced in some instances by the addition of *transforming growth factor-β* (TGF-β), another human protein that regulates cell growth and differentiation. Perhaps *Xenopus* versions of these proteins are the natural mesoderm inducers.

Another set of experiments had actually set the stage

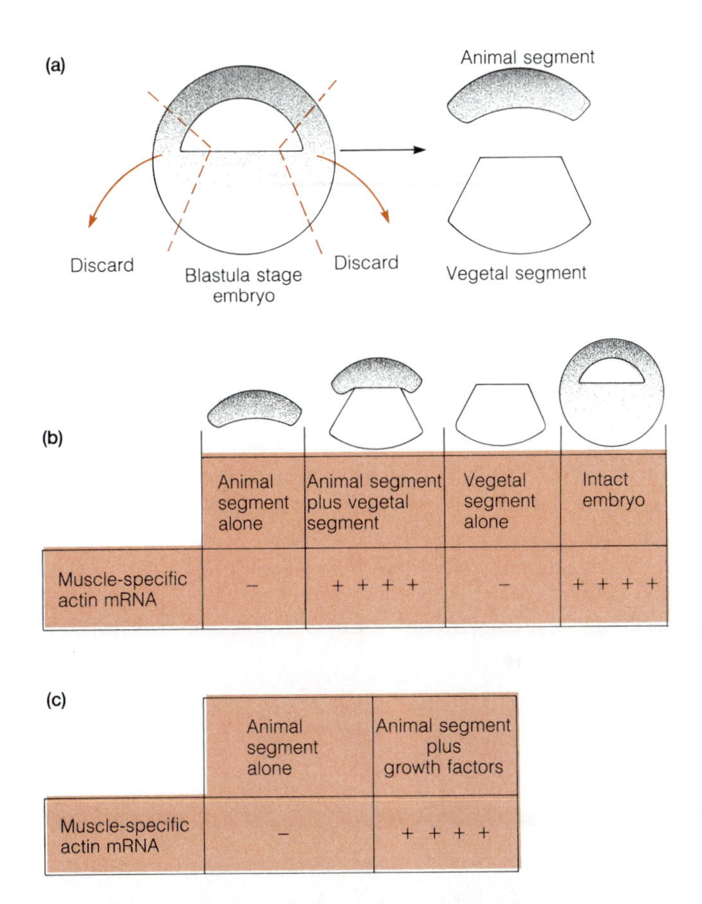

Figure 22-28 Induction of Muscle-Specific Actin mRNA in Mesoderm in Animal-Vegetal Conjugates and Growth Factor-Treated Animal Caps. (a) *Xenopus* blastulas were dissected into animal and vegetal segments, which were either combined or kept separate for RNA analysis at a later stage. The results of this RNA analysis are summarized in (b): Whereas neither animal nor vegetal segments produced the muscle-specific actin mRNA, the animal-vegetal combinations did express the muscle-specific actin mRNA, in an amount similar to that found in intact embryos. When isolated animal segments are treated with fibroblast growth factor (FGF) and transforming growth factor-β (TGF-β), a similar induction of muscle-specific actin is observed (c).

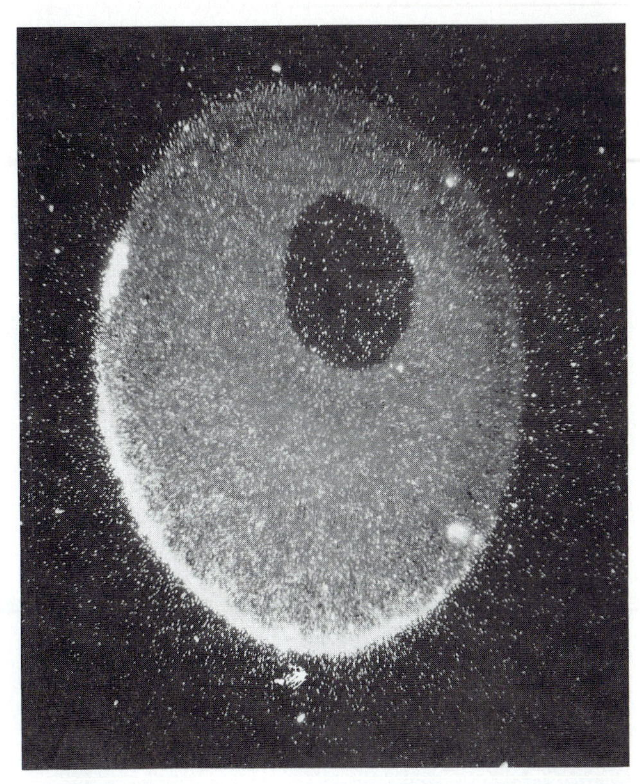

Figure 22-29 Vegl, an Inducer of the Mesodermal Cell Fate, Is Localized in the Vegetal Hemisphere of *Xenopus* Eggs. Vegl encodes a TGF-β-like protein (see Figure 22-28) and is localized in the vegetal hemisphere of the *Xenopus* oocyte. This dark-field micrograph shows a *Xenopus* embryo that had been probed with labeled RNA complementary to the Vegl mRNA. Presumably the Vegl protein from vegetal cells participates in the induction of the mesodermal cell fate in invaginating cells during gastrulation. (TEM)

for this exciting result. Specifically, researchers had been characterizing mRNAs that are expressed only in the vegetal hemisphere of the *Xenopus* embryo. They did this by dissecting the vegetal third of many embryos and then making DNA copies of the mRNAs from the vegetal material. When they determined the DNA sequence of these **cDNAs,** they found that one, designated **Vegl,** encoded a protein very similar to human TGF-β, one of the inducers of actin expression in Figure 22-28. They then used a technique called RNA in situ hybridization to confirm that the Vegl mRNA is located in the vegetal half of the embryo (Figure 22-29). Thus it seems likely that Vegl protein secreted by the vegetal cells induces the invaginating animal cells to become mesodermal cells. A *Xenopus* version of FGF has now been isolated, and although its localization in the embryo is not known, it does induce muscle-specific actin mRNA in cells of the animal hemisphere.

Later Developmental Stages in Vertebrates

At this point you should understand the sequence of events that give rise to the neurula-stage *Xenopus* embryo in Figure 22-27d. Obviously, much is yet to come, but the study of development after gastrulation relies more on descriptive morphology and less on experimental cell biology. Consequently, our understanding of the cell biology involved in later developmental stages is in many cases incomplete. Furthermore, there are fewer general rules, since similarity among species diminishes as development progresses and taxonomic differences become evident.

For better or for worse, we humans are fascinated with ourselves, and the goal for many developmental biologists is to understand human development. Essentially all of the events that we have discussed occur in the first *three weeks* of human development. For the neurula-stage human embryo shown in Figure 22-30a, 33 weeks of development in utero remain. Compare this human embryo to the *Xenopus* neurula in Figure 22-27d. The three-week-old human neurula is at a slightly later stage, as demonstrated by the formation of *somites* along the primitive nerve cord. The **somites** are mesodermal structures that will give rise to repeated structures such as the vertebrae. By five weeks (Figure 22-30b), organogenesis is obviously well under way. As organogenesis and morphogenesis proceed, the basic body plan shared by all vertebrates takes on the specific characteristics of the human form (Figure 12-30b–d). After the conclusion of organogenesis and morphogenesis at roughly nine weeks, the primary developmental task is growth.

Since many experiments cannot be performed in the human system, developmental biologists often turn to the mouse as a model for mammalian development. Even the mouse, however, has severe experimental limitations. Unlike *Xenopus* and *Drosophila,* embryogenesis takes place in utero, and so cannot be observed without terminating development. Furthermore, embryogenesis is much slower in mammals, and experiments therefore take much longer to perform. This is particularly true of genetic analysis, which we will study in the last section of this chapter.

Despite the difficulties, there are many excellent studies

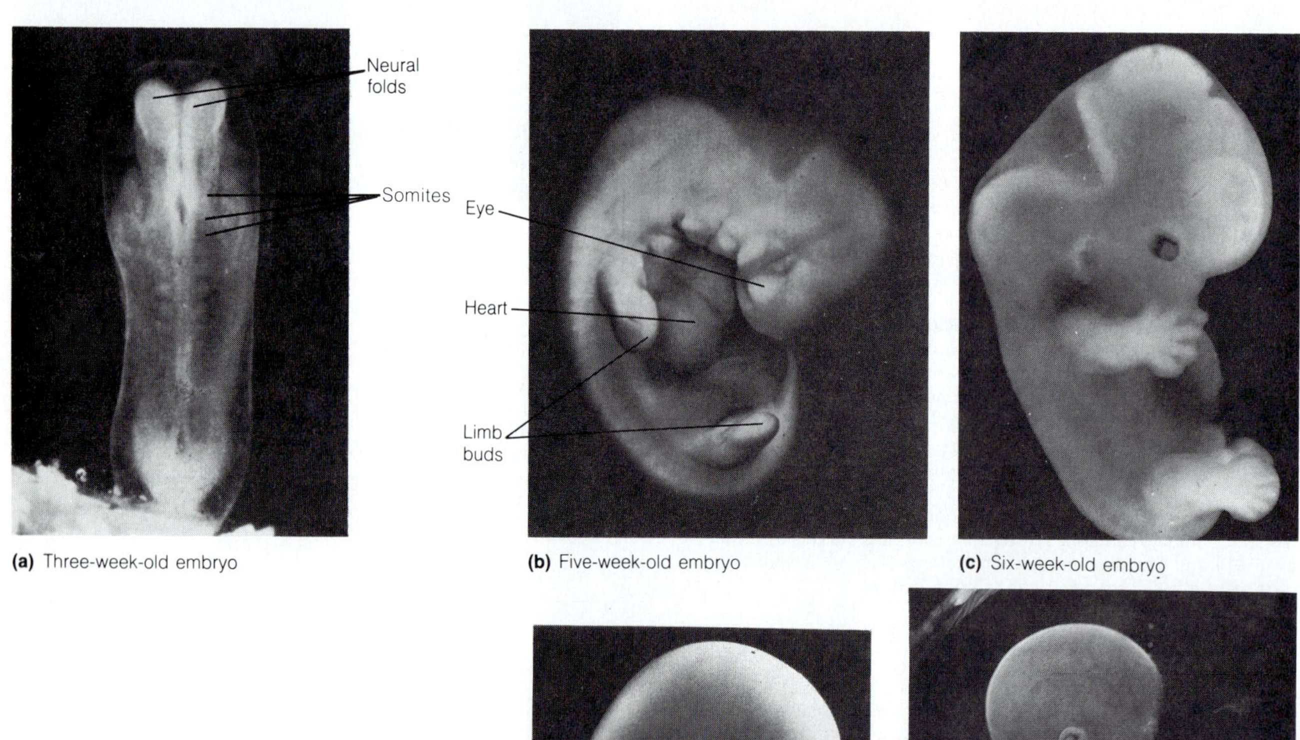

(a) Three-week-old embryo

(b) Five-week-old embryo

(c) Six-week-old embryo

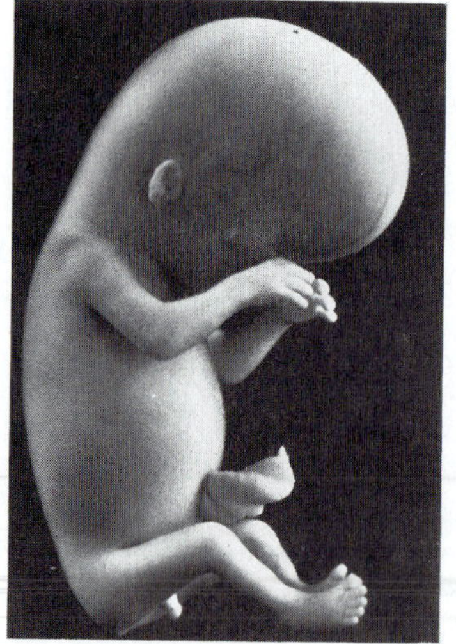

(d) Nine-week-old fetus

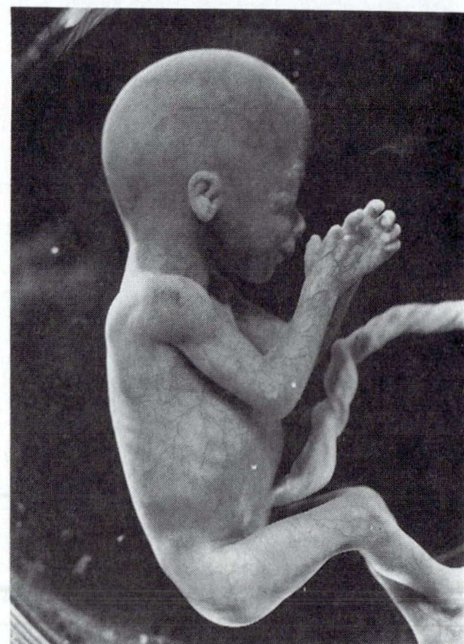

(e) 28-week-old fetus

Figure 22-30 The Later Stages of Human Fetal Development. Most of the developmental events discussed in this chapter have already occurred in the three-week-old, neurula-stage human embryo seen in (a). The neural folds will grow inward to form the neural tube, which is the forerunner of the brain, the spinal cord, and the rest of the central nervous system. Note the somites on either side of the neural tube. The remaining pictures are of a 5-week-old embryo (b), a 6-week-old embryo (c), a 9-week-old fetus (d), and a 28-week-old fetus (e). As is the case for many animals, most human development takes place before birth, so the fetus is in many respects a miniature of the adult. Postnatal development consists primarily of growth and the development of secondary sex traits.

of postgastrulation developmental events in mammals. An important event is the migration of the *neural crest cells.* After neurulation, the **neural crest cells** migrate away from their original position along the neural tube to specific spots inside the embryo and along its periphery. Once they have reached their various destinations, they form a remarkable variety of cells, including pigment cells, nerve cells, endocrine cells, and cells that produce cartilage and bone.

To explore differentiation at the cellular level, mammalian developmental biologists study different cell types in vitro in the laboratory dish. Through various techniques, many cell types can be grown indefinitely as **cell lines;** some cell types, however, can be studied only as *primary cultures* taken directly from the animal. One example of how cell lines can be used to study development involves the muscle cell precursors known as **myoblasts.** Under certain conditions, myoblasts differentiate into muscle cells. With the techniques of molecular biology, researchers have been able to identify specific genes whose expression induces myoblast differentiation. At least one of these genes, known as *MyoD,* encodes a *sequence-specific transcription factor.* A **sequence-specific transcription factor** regulates the expression of other genes by binding to a specific site in the DNA near their promoters. This binding can either activate or repress transcription, depending on the particular gene and transcription factor.

In Chapter 24, you will study some of the most thoroughly understood cells in the human body in terms of differentiation—the cells of the human immune system. Cell specialization and migration, DNA rearrangements, cell-cell signaling, and selective gene expression are just some of the themes from this chapter that recur during *lymphocyte* development.

Figure 22-30 provides a fitting conclusion to our tour of embryogenesis. Recall our understanding of the mating-type switch in yeast, of the morphogen retinoic acid in the chick limb bud, of induction in the *Xenopus* embryo. How have our studies of these simpler organisms enhanced our understanding of human development? Can our understanding of human development ever approach the sophistication we have seen in these simpler systems?

Genetic Analysis of Embryonic Development

Having completed our tour of embryogenesis, we are now ready to look at development, and particularly at embryogenesis, using a different approach. Over the past 20 years, the use of genetics has led to some of the most important advances in developmental biology. Today, no discussion of the field would be complete without some examples of how genetics can be used as a tool for dissecting biological processes. The general method consists of looking for mutations that disrupt the process of interest and then characterizing the genes affected by those mutations.

We have already encountered one organism in which genetics played a key role in our understanding of developmental events—yeast. To identify the genes that regulate the events of switching, mating, and the cell cycle, yeast geneticists looked for mutations that disrupted switching, mutations that prevented mating, and mutations that stopped the cell cycle. When they characterized the genes altered by these mutations they discovered the structures of *MAT, HMR***a**, and *HML*$\boldsymbol{\alpha}$, the mating factors and their cell surface receptors, and the kinase encoded by the *cdc2* gene.

We now turn to two higher eukaryotes that are ideally suited for **developmental genetics.** These are the now-familiar fruit fly *Drosophila melanogaster* and the nematode worm *Caenorhabditis elegans.* In *Drosophila* we will look at the genetic analysis of *pattern formation.* In *Caenorhabditis* we will see how genetics can be used to dissect interactions between specific cells.

Pattern Formation in *Drosophila*

We examined *pattern formation* in our earlier discussions of positional information, morphogens, and the chick limb bud. While pattern formation occurs in all multicellular eukaryotes, nowhere has it been more thoroughly analyzed than in *Drosophila.*

A fertilized *Drosophila* egg completes embryogenesis in less than 24 hours, hatching as a segmented larva. The larva is covered with a tough *cuticle,* shown in Figure 22-31b (and in Figure 22-D). Along the anterior-posterior axis, the larva is divided into three thoracic segments and eight abdominal segments. The top, or dorsal, side of each segment is essentially bare, but the bottom, or ventral, side of each segment has a band of small, teeth-like *denticles* that the larva uses to move. Each segment is uniquely identified by its denticle pattern. The segments form during gastrulation (hours 3–6), persist throughout development, and are readily visible in the adult fly (Figure 22-31c).

Anterior/posterior (or A/P) pattern formation is executed in three sequential steps during embryogenesis: the establishment of the A/P axis, the division of the embryo into segments, and the assignment of a particular segmental identity, such as first thoracic or third abdominal, to each of the individual segments.

Polarity Gene Products Establish the A/P Axis. In the 1970s, Christianne Nüsslein-Volhard and her colleagues began systematically inducing mutations throughout the fly's four chromosomes. The box on pp. 706–707 describes how such a mutagenesis experiment is performed. Nüsslein-Volhard and her colleagues looked for mutations that altered the larval cuticle without totally destroying the pattern. They screened hundreds of thousands of embryos and found some larvae with half the normal number of segments, others missing a specific part of every segment. They found some missing large blocks of segments and others with two tails (or two heads), one at either end. Nüsslein-Volhard and her co-workers grouped these mutations into four classes: *polarity* mutations, *gap* mutations, *pair-rule* mutations, and *segment-polarity* mutations. Figure 22-32 shows examples of these different classes of mutant embryos, all of which die as unhatched larvae.

Mutations that produced two-headed or two-tailed larvae identified candidates for **polarity genes,** the genes encoding the proteins that establish the A/P axis. Earlier research had suggested that the anterior and posterior ends of embryos contain *polar centers.* The polar centers are regions of terminal cytoplasm—perhaps morphogen

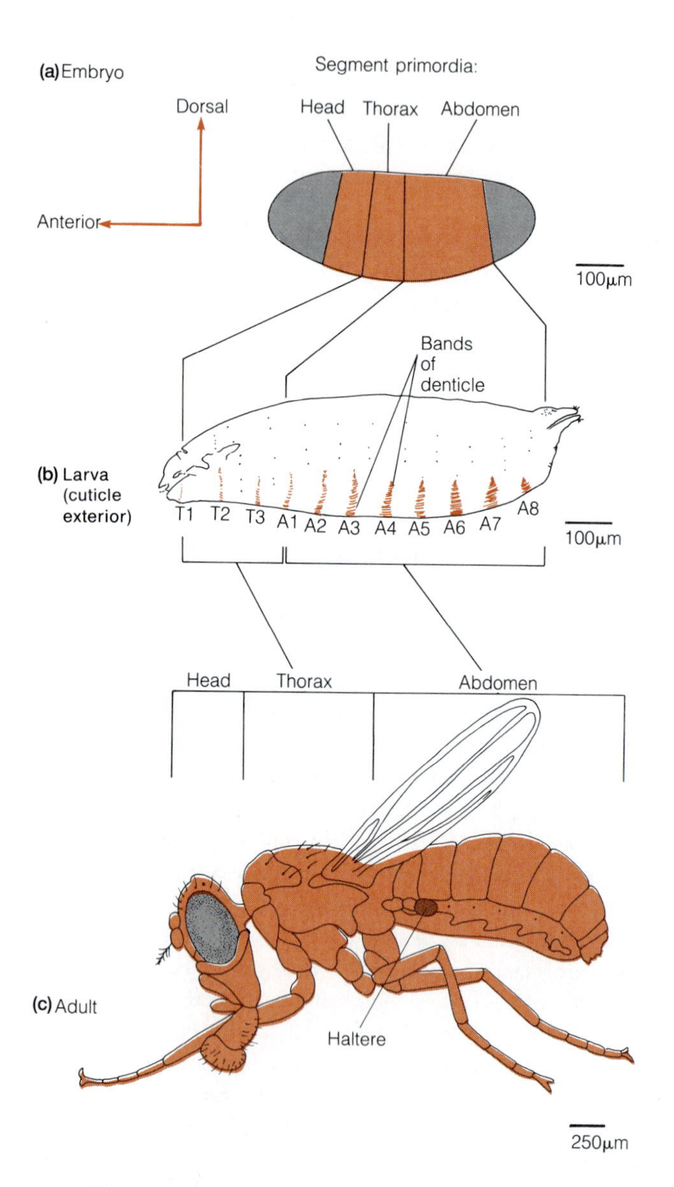

Figure 22-31 Three Stages of the *Drosophila* Life Cycle: The Embryo, the Larva, and the Adult. (a) A simplified version of the embryonic fate map of *Drosophila* (see Figure 22-19) is shown as it relates to the larval (b) and adult (c) stages. The middle portions of the embryo give rise to the thoracic (T) and abdominal (A) segments of the larva, each distinguished by a characteristic pattern of denticles on the ventral surface of the cuticle. The segmental body plan is conserved in the adult. The complete life cycle includes three progressively larger larval stages and a pupal stage between (b) and (c).

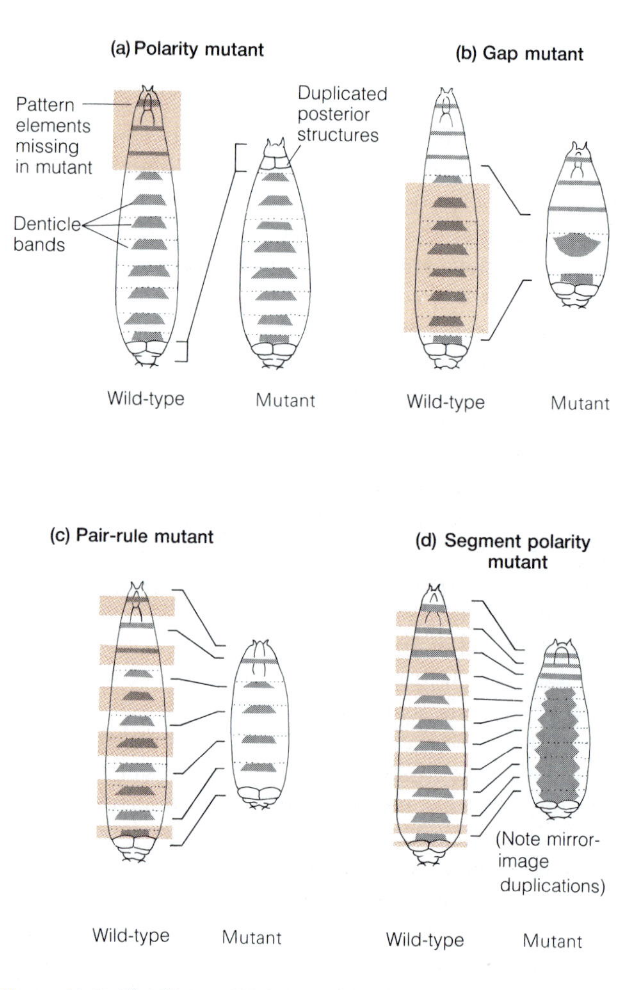

Figure 22-32 The Mutant Phenotypes of Polarity, Gap, Pair-Rule, and Segment-Polarity Genes. Mutations in the polarity genes (such as *bicoid*) cause loss of polarity. This in turn causes loss of terminal structures at one end of the embryo, and the lost structures are replaced by structures normally found at the opposite end of the embryo (a). When a gap gene is mutated, continuous blocks of segments are deleted from the larval body plan (b). When mutations disrupt a pair-rule gene, segment-wide units are lost from the pattern (c). Mutations in one of the segment-polarity genes cause the loss of pattern elements from every segment, and the missing pattern is replaced by a mirror-image duplication of part of the remaining pattern (d). All embryos are shown in ventral view with deleted structures in color.

sources—that organize the A/P axis by providing two counteracting activities, one for "anteriorness" and one for "posteriorness." Transplantation experiments had provided much of the evidence for polar centers. For example, when cytoplasm from the anterior polar center of a donor embryo is transplanted to a new site in a host embryo, the host embryo develops extra head structures at the injection sites.

A mutation in a polarity gene encoding anteriorness has the opposite effect: When the source of anterior activity in an embryo is removed by such a mutation, an extra set of *posterior* structures develops in the anterior, creating a two-tailed embryo (Figure 22-32a). But if anterior cytoplasm from a normal embryo is injected into the anterior of such a mutant embryo, the anterior polar center is partially restored and some head structures develop.

Most polarity genes are *maternal effect* genes. This means that the polarity gene products that are active in the embryo were supplied by the mother during oogenesis. Maternal effect mutations can be confusing, because their effect is not seen until the next generation. Thus a female that has a mutation in a polarity gene will be normal, but all her embryos will die because they lack the product of the mutant polarity gene. The missing mRNA or protein *should* have been put in the egg during oogenesis, but the female's gene for that molecule was mutated. Maternal effect genes are discussed further in the box on pp. 706–707.

Segmentation by Gap, Pair-Rule, and Segment-Polarity Genes. So polarity genes encode two opposing activities that set up the A/P axis. But what gene products respond to this field of positional information? The answer is that three groups of **segmentation genes** act in series to divide the embryo into segments: the *gap genes,* the *pair-rule genes,* and the *segment-polarity genes.*

Figure 22-32b shows the result of a mutation in a **gap gene.** A contiguous block of segments is lost from the embryo. In response to the maternal A/P positional information, at least seven such genes turn on in distinct but overlapping regions of the embryo, thus dividing the embryo into large blocks along the A/P axis.

The next class of segmentation genes includes at least nine **pair-rule genes.** The pair-rule gene products divide the large blocks defined by the gap gene products into the actual segments. When an embryo is mutant for a pair-rule gene, segment-wide pattern elements are deleted from the larval cuticle, and the embryo develops with half the normal number of segments (Figure 22-32c).

Finally, the nine or more **segment-polarity genes** divide each segment into an anterior and a posterior *compartment.* When an embryo is mutant for a segment-polarity gene, a pattern element is missing in every segment and is replaced by a duplication of part of the remaining pattern (Figure 22-32d).

In Situ RNA Analysis Reveals the Steps of Segmentation. Many of the polarity and segmentation genes have been cloned, enabling researchers to look at the expression of the gene products in the embryo. This can be done either by antibody staining or by in situ RNA hybridization analysis. The results of such experiments vividly support our model of segmentation.

To perform **in situ RNA hybridization analysis,** the investigator makes a single-stranded radioactive RNA or DNA *probe*—copies of a gene that are *complementary* to the mRNA of that gene. Embryos are affixed to microscope slides and then treated with the probe, which hybridizes to the native mRNA. When the slide is treated with a photographic emulsion, exposed, and developed, dark silver grains readily reveal the location of the mRNA of interest.

Figure 22-33 shows a series of RNA in situ experiments that graphically illustrate how the different classes of pattern formation genes subdivide the embryo. For example, the mRNA from an anterior polarity gene is prelocalized to the anterior of the embryo before any of the segmentation genes are expressed (Figure 22-33, column a). It disappears as the embryo begins to express the segmentation genes. The gap genes are the first of the segmentation genes to turn on, and they do so in blocks (Figure 22-33, column b). Other gap genes are expressed in different blocks. The gap genes are closely followed by the pair-rule genes, which are expressed in staggered, repeating patterns: one segment on, one segment off (Figure 22-33, column c). The pair-rule genes thus define the actual segments. Finally, the segment-polarity genes are each expressed in a portion of *every* segment (not shown). The events shown in Figure 22-33 occur in less than an hour during the late syncytial and cellular blastoderm stages.

Segment-Identity Genes Distinguish the Individual Segments. Although the embryo has been divided into segments at this point, A/P pattern formation is not yet complete. Without the action of one more class of genes, the segments would develop as identical units. The next question in A/P pattern formation is "How does each segment get its individual identity?"

To answer this question we turn to the work of Edward Lewis and colleagues, who studied the ***Bithorax* gene complex.** This is a group of closely-linked genes that are involved in A/P pattern formation. When these genes are mutated, specific segments are transformed into other segments. For example, when three mutations in the *Bithorax* complex are combined, the entire third thoracic segment, designated T3, is converted into a complete, extra second thoracic segment (T2). T3 normally does not have wings, but since T2 does have wings, the T3-to-T2 transformation produces a four-winged fly (Figure 22-34a). Other workers have shown that an equally bizarre fly results from a mutation in the *Antennapedia* (*Ant*) gene (in the *Antennapedia* gene complex). That mutation converts an antenna into a leg (Figure 22-34b).

Such mutations that transform cells or tissues in one part of the developing organism into cells or tissues normally found somewhere else in the organism are called **homeotic mutations.** Homeotic mutations disrupt the normal function of the **segment-identity genes**—that of assigning a particular developmental fate to each individual segment. In a wild-type embryo, the protein encoded by each segment-identity gene is present at high levels in a single segmental unit. For example, the Ubx protein, which is encoded by the *Ultrabithorax* gene of the *Bithorax* gene complex, is made in the T3 segment and directs that segment to make the third leg and the haltere. (The haltere is a small, club-like appendage thought to counteract wing movement during flight; see Figure 22-31c.) Similarly, cells of the T1 segment normally make the Ant protein and so produce the first leg. When both copies of a segment-identity gene are inactivated, the segment that normally expresses that gene develops like its anterior neighbor, as in the T3-to-T2 transformation of haltere to wing seen with the *bithorax* mutations (Figure 22-34a). The unique character of every segment of the developing fly is specified in a similar way by the expression of a particular segment-identity gene in that specific segment.

The remarkable *Ant* transformation of antenna into leg has a different explanation. In this case, the normal Ant protein is made in its normal location in the thorax, which is unaffected by the mutation. Instead, the mutation

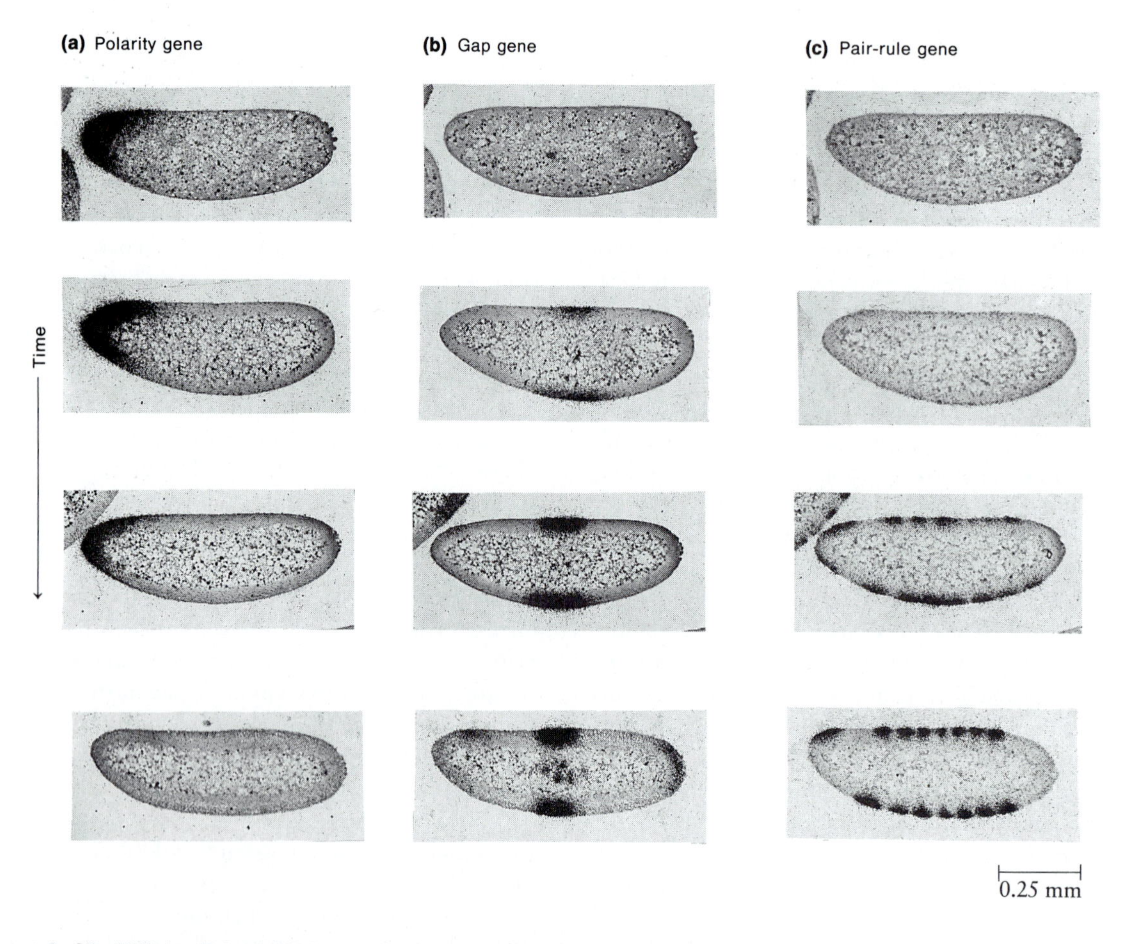

Figure 22-33 In Situ RNA Analysis of A/P Pattern Formation Genes Chronicles the Establishment of the Segmental Body Plan. Each row shows three slices of the same embryo. The embryo in each row is roughly 10 minutes older than the embryo in the previous row. The embryos in each column have been probed with RNA complementary to a polarity gene (a), and a gap gene (b), and a pair-rule gene (c). Note that the temporal order in which these mRNAs are expressed reflects their role in A/P pattern formation. Maternally supplied polarity gene products organize the A/P axis of polarity and activate the gap genes. The gap genes, in turn, regulate the pair-rule genes in segment-wide stripes that define the segments. (LMs)

causes expression of the Ant protein in a *head* segment, where it is not normally expressed. There the Ant protein directs cells to adopt a thoracic cell fate and produce a leg (Figure 22-34b).

Although we see the effects of these mutations in the adult fly, the segment-identity genes begin their work in the embryo as soon as the process of segmentation is complete. The individual segments express the mRNAs from the appropriate homeotic genes early in gastrulation. Thus, although larval differentiation is still several hours away and metamorphosis several days away, the entire process of segmentation and the assignment of segmental identity is complete early in gastrulation, a mere 3 to 4 hours after fertilization.

The Molecular Basis of A/P Pattern Formation. Many of the A/P pattern formation genes have been cloned and sequenced, and many of their proteins have been studied biochemically. This work has shed light on the molecular mechanisms that underlie A/P pattern formation in *Drosophila*.

Many A/P pattern formation proteins contain a sequence of 60 amino acids, known as the **homeo box,** which is highly conserved among these proteins. The homeo box is a **DNA-binding domain,** and proteins containing homeo boxes appear to be *sequence-specific transcription factors* (see p. 701). Although the homeo domains of different A/P pattern formation proteins are similar, critical amino acid differences confer different DNA specificities on each protein.

What is that specificity? Several of the A/P pattern formation transcription factors regulate *other* pattern formation genes. This is not surprising, given the hierarchical manner in which they act. For example, the protein made by the anterior polarity gene *bicoid* (*bcd*) appears to be a transcriptional activator that functions as a *morphogen gradient*. High levels of bcd protein presumably activate the anterior gap genes, whereas lower levels activate the posterior gap genes. Similarly, a combination of polarity and segmentation transcription factors may interact to turn on the appropriate segment-identity genes in each segment. Finally, the segment-identity transcription factors somehow activate the structural genes that carry out differentiation in each segment.

Although transcriptional regulation is clearly a critical component of A/P pattern formation, it is not the whole story. Some A/P pattern formation genes do not encode transcription factors. Furthermore, the genes that direct dorsal-ventral pattern formation encode a much different set of proteins (see the box on pp. 706–707). At least two of these appear to be regulators of transcription, and others encode growth factors, integral membrane proteins, and proteases. Our understanding of *Drosophila* pattern formation will be incomplete until we know the mechanisms by which these other proteins execute their role.

The Homeo Box Is Conserved. Using DNA from some of the *Drosophila* homeo box genes as a molecular probe, researchers have found homeo box genes in the genomes

(a)

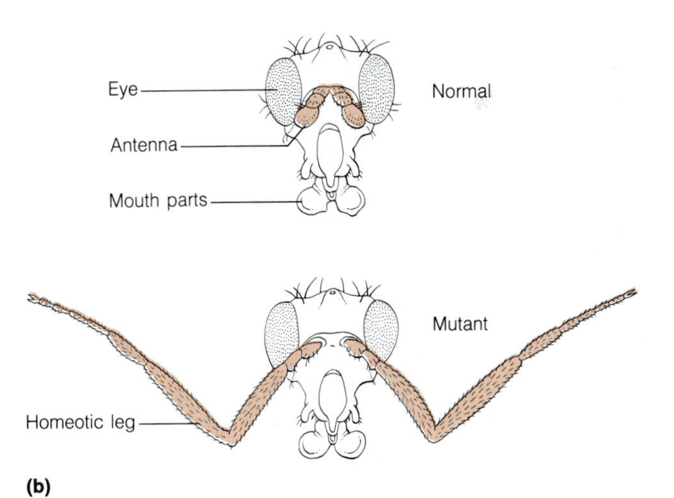

(b)

Figure 22-34 Mutations in the *Bithorax* and *Antennapedia* Gene Complexes Cause Dramatic Homeotic Transformations. (a) *Drosophila* is normally a two-winged insect, but when several mutations in the *Bithorax* gene complex are combined, the third thoracic segment (T3) is converted into a second thoracic segment (T2), and the haltere of T3 becomes the wing of T2. Note also the duplication of the T2 bristle pattern in T3. (b) The bizarre *Antennapedia* mutation converts part of a head segment into part of a thoracic segment, and the antenna therefore becomes a leg. These mutations are caused by either loss or inappropriate expression of the segment-identity genes.

Using Genetics to Probe Development

Chapter 13 introduced us to the principles of genetics. In the final section of this chapter we use genetics as a tool to dissect development. This approach, called *developmental genetics,* requires a way of thinking different from that in other biological approaches.

A biochemist typically begins by breaking the system down into its constituent parts—organelles, proteins, nucleic acids, membranes. The biochemist next characterizes the individual parts and then attempts to rebuild the system. A geneticist, however, works by removing one part at a time, while leaving the rest of the system intact. When the geneticist removes a part, he or she usually has little or no idea about the part's identity or biochemical function.

To put yourself in the right frame of mind for using genetics, think for a moment about auto mechanics. Imagine that you know nothing about how a car engine works and have only a few simple tools to help you learn. Lacking a teacher or a manual, about the only thing you can do is start taking the engine apart bit by bit. By removing one thing at a time and seeing what effect that has on the engine's function, you will eventually learn a great deal about auto mechanics.

When Christianne Nüsslein-Volhard and colleagues set out in the late 1970s to study pattern formation in *Drosophila,* they were basically doing auto mechanics. Using "genetic wrenches," they took apart the engine of the *Drosophila* embryo, one part at a time. Each "part," of course, was a gene that functions during embryogenesis. Most of the genes involved in *Drosophila* pattern formation were identified in this manner, many of which are discussed in this chapter. As each new pattern formation gene is cloned and its protein product characterized, we learn more and more about the "spark plugs," "fan belts," and "piston rings" of the *Drosophila* embryo. Eventually, we may be able to understand how the whole thing runs!

The discerning student will pause here to ask a question: How is it possible to study a mutation that causes a grossly deformed embryo? Wouldn't the mutation be lost with the dead larva? The answer, of course, is no, and the reason becomes clear when we consider how the mutagenesis is performed. Figure 22-C describes a mutagenesis to isolate new mutations on the fly's second chromosome. In this case,

P EMS-treated $\frac{cn}{cn}$ males × females $\frac{CyO}{Gla}$

f1 (pair-mating) $\frac{dpp^{*}\,cn}{CyO}$ males × females $\frac{CyO}{Gla}$

f2 (sib-mating) $\frac{dpp^{*}\,cn}{CyO}$ males × females $\frac{dpp^{*}\,cn}{CyO}$

f3 Look for recessive embryonic phenotype in $\frac{dpp^{*}\,cn}{dpp^{*}\,cn}$ flies.

Preserve mutation in stock of $\frac{dpp^{*}\,cn}{CyO}$ flies.

Figure 22-C **A Mutagenesis to Identify Lethal Mutations on the Second Chromosome of *Drosophila*.** The *CyO* (*curly* wings) and *Gla* (*glazed* eyes) mutations are lethal when homozygous. The *cn* mutation causes *cinnabar* eyes when homozygous. In the P generation, *cn/cn* males are fed the mutagen EMS (ethyl methanesulfonate) and mated to *CyO/Gla* females. Single F_1 males are then mated to *CyO/Gla* females to generate F_2 males and females carrying the same mutagenized second chromosome. These F_2 brothers and sisters are then mated. If no *cn/cn* adults are seen among the progeny, the researcher looks for unhatched larvae with an interesting cuticle. If such dead embryos are found, the mutation is preserved in a stock generated from the *cn/CyO* siblings. In this case, a new mutation (*) in the *decapentaplegic* (*dpp*) gene has been identified.

the mutagenesis has generated a *recessive, embryonic lethal* mutation in the *decapentaplegic* (*dpp*) gene, and the *dpp/dpp* animal dies. But the mutation is not lost, since the *dpp/CyO* animal is perfectly healthy. Thus, even though *dpp* is absolutely required for an essential developmental process, we can still use genetics to study the gene and the process in which it is involved.

Diploidy thus allows the geneticist to "maintain" a mutant allele with a wild-type allele on the homologous chromosome. But diploidy is not the only way to disrupt essential processes genetically, and in haploid organisms it is not an option. For example, the *cdc* mutations discussed on page 692 were isolated in haploid yeast strains. Such mutations would normally kill the cell because the only source of a protein that is essential for the cell cycle has been destroyed. But this problem can be solved by looking for **temperature-sensitive (*Ts*) mutations.** A *Ts* mutant allele makes a protein that is normal at one temperature but unstable, and hence mutant, at a higher temperature. Such mutations can be identified by growing and mutagenizing cells at the lower temperature, then shifting to the higher, **restrictive temperature** and looking for cells with defective cell cycles. Once identified, these cells can be returned to the lower, **permissive temperature,** where they grow normally. *Ts* mutations are one class of **conditional mutations,** in which the phenotype of the mutation is observed only under certain conditions.

A final nuance of developmental genetics is the distinction between *maternal effect* genes and *zygotically acting* genes. In learning genetics we generally study **zygotically acting genes,** where the genotype of the individual animal determines the phenotype of that same animal. The *dpp* gene shown in Figure 22-C is a zygotically acting gene. The *dpp* gene encodes a protein similar to the Vegl protein in *Xenopus* (see page 699) and is normally expressed in dorsal embryonic tissue. When both copies of the zygote's *dpp* gene are mutant, however, the dorsal tissue develops denticles as though it were ventral tissue. Figure 22-D shows such a "ventralized" *dpp/dpp* embryo.

As we saw in our discussion of oogenesis and cleavage, many of the most critical events of early embryonic development are mediated by molecules put in the egg during oogenesis. Examples include the stockpiling of critical mRNAs and proteins and the asymmetric placement of cytoplasmic determinants. To use genetics to study these *maternal contributions* to embryogenesis, the researcher must look for **maternal effect** mutations. A female homozygous for a maternal effect mutation is herself unaffected by the mutation, but all her eggs are defective. Thus, the mutant phenotype is a defect seen in the next generation.

Again an example may help. Any fly that is homozygous for a mutation in the *dorsal* gene (genotype *dl/dl*) develops normally. But *dl/dl* females are sterile because the crucial *dorsal* gene product is missing from all their eggs, even if those eggs are fertilized by sperm that carry a wild-type copy of the *dorsal* gene. In other words, the embryo may be +/*dl*, but the *dl/dl* genotype of the mother means the *dorsal* gene product, an important transcriptional regulator, is missing from the embryo. In embryos from a *dl/dl* female, ventral tissue develops as dorsal tissue and there are *no* denticle belts. In contrast to the "ventralized" embryo in Figure 22-D, these embryos are "dorsalized."

Most of the mutations that disrupt polarity in a global manner are maternal effect mutations. Thus many of the early events in pattern formation are carried out by gene products, whether mRNAs or proteins, that are present in the egg prior to fertilization.

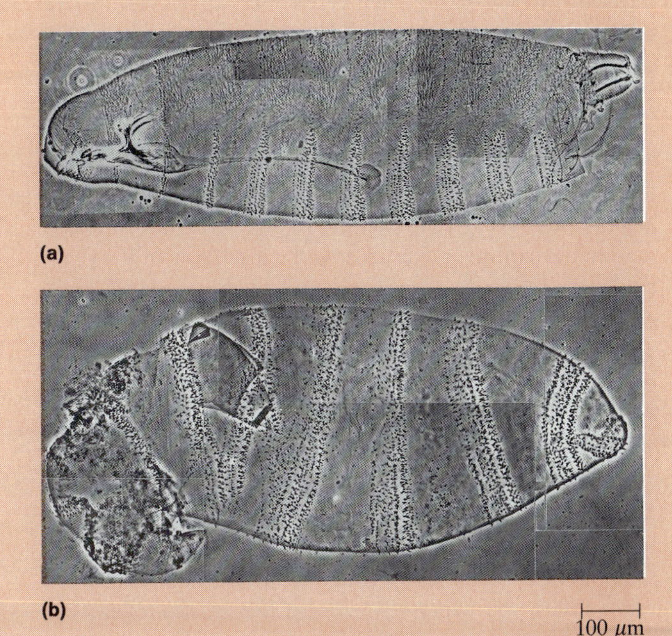

Figure 22-D A Mutation Affecting Dorsal/Ventral Pattern Formation in the *Drosophila* Embryo. (a) A photograph of the wild-type larval cuticle at the end of embryogenesis. Note the bands of denticles on the ventral surface of the larva. (b) The cuticle of an unhatched larva that is homozygous for a mutation in the *decapentaplegic* (*dpp*) gene. The denticles now wrap all the way around the larval cuticle. This embryo is therefore "ventralized." (LMs)

of many other organisms, including *Xenopus, Caenorhabditis,* human beings and mice. Although the mouse, like other mammals, does not have a segmental body plan, both the somites and the central nervous system are segmented. Interestingly, when researchers did RNA in situ analysis of the mouse homeo box genes (designated ***Hox* genes**), they found that many are expressed in the somites and central nervous system at very early stages of development (Figure 22-35).

Like the conservation of the homeo box, many things first observed in simpler systems such as *Xenopus, Drosophila,* and even yeast later prove to be relevant to mammalian development. Recall how the discovery of the yeast *cdc2* gene led to the identification of the human *cdc2* gene, and how the human gene can regulate the yeast cell cycle. This sort of extrapolation is expanding rapidly, largely due to the emerging tools of genetic engineering, discussed in Appendix B.

Cellular Interactions in *Caenorhabditis*

Compared to *Drosophila,* which has been studied since the early 1900s, the nematode worm *Caenorhabditis elegans*

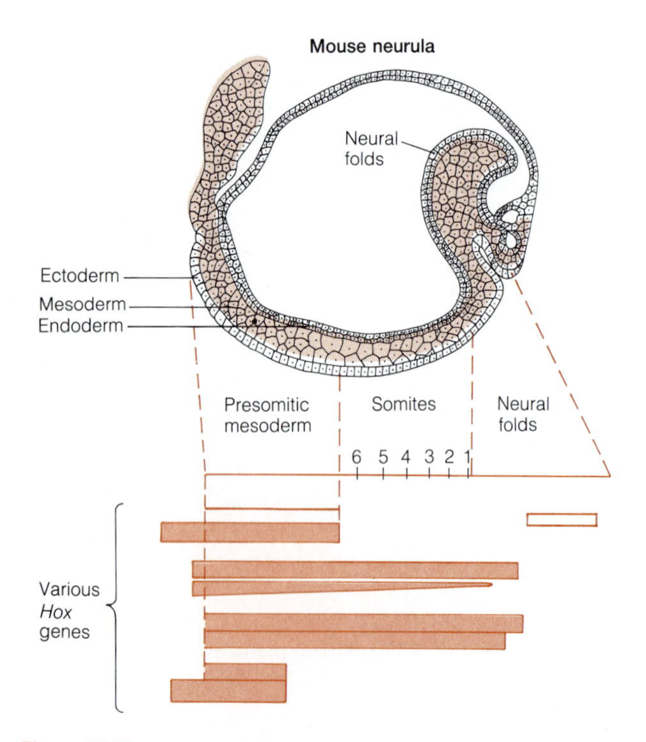

Figure 22-35 The Pattern of mRNA Expression from Five Homeo Box Genes in the Neurula-Stage Mouse Embryo. More than 20 mouse genes containing homeo boxes have been identified. Here the expression pattern of five of these *Hox* genes in the neurula-stage embryo is summarized. The boxes show *Hox* mRNA expression in the mesoderm (shaded boxes) and the ectoderm (open boxes).

(Figure 22-36) is a relative newcomer in the study of developmental biology. In a classic 1974 paper, Sydney Brenner demonstrated the power of genetic analysis in *Caenorhabditis.* In the ensuing 15 years, this tiny worm has become one of the most thoroughly characterized eukaryotic organisms. Pattern formation, the development and function of the muscular and nervous systems, and sexual differentiation are just a few of the problems that *Caenorhabditis* researchers have tackled. After a brief overview of *Caenorhabditis,* we will focus on germ cell development, where genetics has been used to probe the molecules involved in cell-cell communication.

As with *Drosophila,* genetic analysis of *Caenorhabditis* development is complemented by powerful experimental techniques in cell biology and biochemistry. One of the chief experimental advantages of *Caenorhabditis* is its transparent nature: Every cell of the organism can be seen readily under a high-resolution light microscope. Taking advantage of this fact and the worm's small size (the adult has less than 1000 somatic cells), John Sulston and his colleagues have been able to describe every cell and every cell division in the developing worm! Furthermore, they have demonstrated that the pattern of cell division is identical in every worm. In other words, *Caenorhabditis* has an **invariant cell lineage.** (Recall that we first encountered *cell lineages* in our discussion of fate mapping on page 690.) The embryo's first six cell divisions produce six **founder cells,** depicted in Figure 22-37. The pie chart beneath each founder cell summarizes the number and type of cells in each founder cell's lineage. In total, 671 somatic cells are produced during embryogenesis. Of these, 111 undergo **programmed cell death,** producing a larva with 560 somatic cells.

An important tool in *Caenorhabditis* research is **laser ablation,** in which a highly focused laser beam is used to destroy a specific cell. The researcher can then determine the consequences of the loss of that particular cell. Much as Campos-Ortega confirmed the fate map of the *Drosophila* embryo with transplantation studies, *Caenorhabditis* researchers have confirmed many individual cell lineages with laser ablation. Finally, *Caenorhabditis* researchers, like those studying other developmental systems, now have powerful tools for doing molecular biology.

A Gene Controlling Cellular Interactions: *glp-1.* *Caenorhabditis* has proved to be an excellent organism for genetic analysis of cell-cell interactions at the single-cell level. One such interaction occurs in the worm's gonad.

Caenorhabditis exists either as a male or as a **hermaphrodite.** A hermaphrodite can either self-fertilize or mate; when it mates it can do so either with males or with other hermaphrodites. In the hermaphroditic gonad, germ cells migrate from the distal portion of the gonad toward the vulva, where the hermaphrodite releases both sperm and oocytes. As they migrate along the arm of the gonad, the germ cells switch from mitotic cell division to meiotic cell division (Figure 22-38a). The **distal tip cell,** or DTC, regulates this switch from mitosis to meiosis in germ cells. The DTC is a somatic cell located, as the name suggests, at the distal tip of the gonad. Germ cells in the vicinity of the DTC normally divide mitotically, thus renewing the germ cell population, whereas cells closer to the vulva divide meiotically and thus complete gametogenesis.

In 1981 Judith Kimble showed that when the DTC is destroyed by laser ablation, all the germ cells in that arm of the gonad immediately enter into meiosis (Figure 22-38b). Furthermore, if the DTC is moved to a more proximal position in the developing gonad, germ cells near the transplanted DTC divide mitotically, while those near the position of the missing DTC divide meiotically. These results suggest that the DTC is the source of an intercellular signal that inhibits the switch from mitosis to meiosis.

Kimble and her associates then turned to genetics to identify the genes involved in that intercellular signal. Using the standard approach of the developmental geneticist, they looked for mutations that disrupted this intercellular signal. They found several mutations in a gene they called *germ-line proliferation-1,* or ***glp-1.*** Like DTC ablation, *glp-1* mutations cause the entire germ cell population to enter meiosis prematurely (Figure 22-38c). Further analysis indicated that the glp-1 protein functions in the germ cells as part of the signal-receiving mechanism that responds to the meiosis-inhibiting signal produced by the DTC.

Molecular Analysis of *glp-1.* The cloning and sequencing of the *glp-1* gene showed that, consistent with its proposed role in signal reception, the glp-1 protein is a transmembrane protein. The extracellular portion of the protein

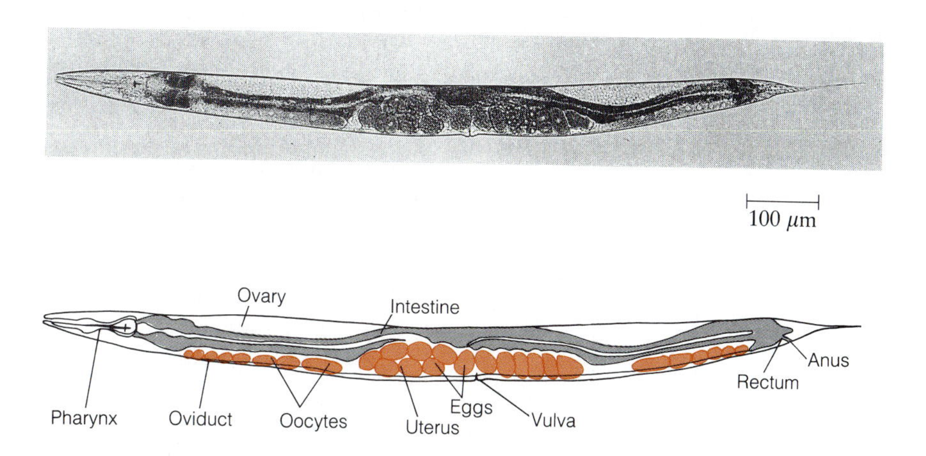

Figure 22-36 The Adult Hermaphrodite of *Caenorhabditis elegans.* Through prodigious and patient observation, every cell of this worm has been given both a name and an exact genealogy of the cell divisions that produced it. With greater magnification and a different microscope, every cell of the worm can be identified. The sperm cells made by the hermaphrodite cannot be seen in this micrograph. (LM)

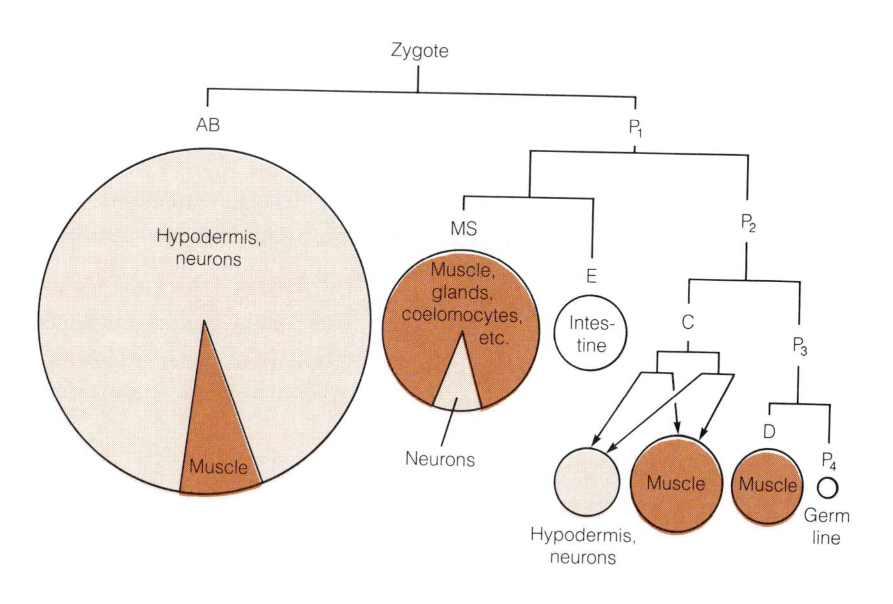

Figure 22-37 The Invariant Cell Lineage of *Caenorhabditis.* The entire development of this worm has been precisely charted. The first six cell divisions give rise to the *founder* cells of six separate lineages. The pie chart beneath each of the six founder cells summarizes the rest of the worm's developmental program, indicating the type (pie slices) and relative number (pie size) of cells that will be derived from each founder cell. Many of the cell lineages have been confirmed by *laser ablation experiments.* For example, if the P_4 cell is destroyed, an adult with normal gonads develops but the gonads are devoid of germ cells.

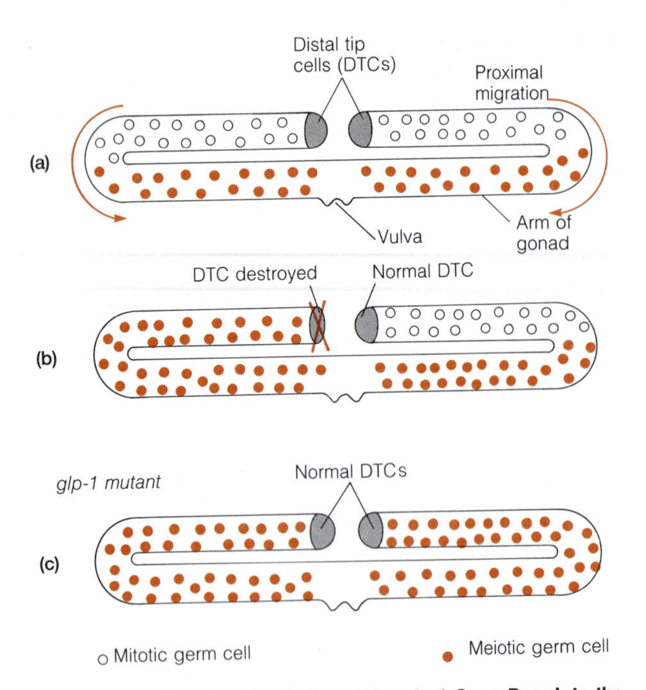

Figure 22-38 The Distal Tip Cell and the *glp-1* Gene Regulate the Germ Line Switch from Mitosis to Meiosis. (a) This schematic diagram depicts a normal gonad in a *Caenorhabditis* hermaphrodite. Germ cells divide mitotically near the distal tip of each arm of the gonad, but as germ cells migrate proximally toward the vulva, they begin to divide meiotically, thus initiating gametogenesis. In (b), the distal tip cell of one arm has been destroyed with a laser, causing the germ cell population of that arm to enter meiosis; note that the other arm is unaffected. In (c), the animal is mutant for the *glp-1 gene*, and the germ cells in *both* arms of the gonad enter meiosis prematurely. Can you tell if the *glp-1* defect occurs in the germ cells or in the distal tip cell?

contains a sequence of approximately 40 amino acids that is repeated 10 times. This repeated sequence is similar to a portion of *epidermal growth factor*, an extracellular mammalian protein that causes certain cell types to grow and that is found in other cell surface receptors. The intracellular portion of the glp-1 protein has sequences similar to one of the *cdc* genes that regulate the cell cycle in yeast. Thus cell cycle control appears to be linked to cell-cell communication in germ cells by the glp-1 protein.

The gene that *glp-1* most closely resembles, however, is another well-studied *Caenorhabditis* gene known as *lin-12*. The glp-1 and lin-12 proteins are similar along their entire lengths! Furthermore, the similarity between the two proteins is functional as well as structural.

Like *glp-1*, ***lin-12*** mediates a specific cell-cell interaction in the developing worm. In the vulval primordium of the developing worm there are two equivalent cells. Each of these cells can become *either* the ventral uterine precursor cell *or* the anchor cell. Both cells apparently express the lin-12 protein. When one cell begins to differentiate as the *anchor cell*, it sends a signal that directs the other cell to become the *ventral uterine precursor cell*. Like glp-1 in germ cells, the lin-12 protein is on the receiving end of that intercellular signal. When the lin-12 protein on the cell surface receives that signal, it directs the cell to become the ventral uterine precursor cell. But in animals with a *lin-12* mutation, the signal-receiving mechanism is destroyed and *both* cells develop into anchor cells.

Perspective

Simply stated, development consists of the entire set of changes that every organism undergoes as it proceeds through its life cycle. Developmental events can be studied in both unicellular and multicellular organisms. Simpler organisms, of course, afford a greater degree of understanding; witness how well we understand the events of the mating-type switch and the process of mating in yeast.

Although development varies a great deal among higher animals, most organisms share some common features. Beginning, somewhat arbitrarily, with the adult organism, the first step in development is the preparation for the next generation. Oogenesis and spermatogenesis create two truly remarkable haploid cell types—the oocyte and the sperm. Each is uniquely prepared for its role in the life cycle: the sperm for delivering a complete haploid genome to the oocyte, and the oocyte for supporting the ensuing development of the diploid zygote. Fertilization begins with the fusion of the sperm and egg plasma membranes and can be divided into two stages: the acrosomal reaction and the cortical reaction. Activation follows, with its characteristic change in pH and the rearrangement of the cortical cytoplasm, thereby launching the zygote into embryogenesis.

During early embryogenesis, cleavage subdivides the embryo into a blastula. Many events must occur in precise temporal and spatial relation to each other during embryogenesis. The positional information that regulates these events includes axes of polarity, morphogens, cytoplasmic determinants, and the cell cycle. The activity of the morphogen retinoic acid in the chick limb bud is an excellent case study of the importance of positional information in pattern formation. As the embryo develops, more and more cells become determined and cell lineages begin to unfold according to fate maps. As the complex invaginations and cellular migrations of gastrulation proceed, the primary germ layers—ectoderm, mesoderm, and endoderm—are defined and begin to exert inductive influences on each other. Changes in cell biology, ranging from changes in gene expression to changes in cytoskeletal and cell surface properties, play critical roles in the developing embryo.

By the end of neurulation, the basic body plan has been established. As morphogenesis and organogenesis proceed, the differences between species become obvious. When developmental biologists look at later developmental stages, they focus on individual tissues such as muscle or blood cells, examining the changes that cells of these tissues go through, alone and in concert with other cells.

In addition to the familiar tools of cell biology, biochemistry, and molecular biology, developmental biologists frequently use genetics to dissect embryogenesis one piece at a time. In the field of pattern formation, systematic genetic analysis has revealed how the segmental body plan of *Drosophila* is established through the progressive action of many genes, some of which encode transcription factors. Similarly, in *Caenorhabditis* genetic analysis has identified proteins that carry out intercellular communication at the level of single cells.

In the final two chapters you will study certain mammalian cells in great detail. As you delve into the cell biology of cancer in Chapter 23, compare the changes that cancerous cells undergo to the normal developmental events you have studied in this chapter. Finally, as you study the immune system in Chapter 24 consider the ways in which it, too, is a developmental system.

Key Terms for Self-Testing

developmental biology (p. 670)

Studying Development in Unicellular Eukaryotes
mating type (p. 671)
mating-type switch (p. 671)
MAT locus (p. 672)
peptide mating factor (p. 672)
pheromone (p. 672)
cell cycle arrest (p. 672)
cyclic AMP (p. 673)
differentiation-inducing factor (DIF) (p. 673)

Gametogenesis Recharging the Life Cycle:
gametes (p. 675)
zygote (p. 675)
egg (p. 675)
sperm (p. 675)
gametogenesis (p. 675)
primordial germ cells (p. 675)
germ cells (p. 675)
somatic cells (p. 675)
genital ridge (p. 675)
indifferent gonad (p. 675)
testes (p. 675)
ovaries (p. 675)
oogenesis (p. 675)
spermatogenesis (p. 675)
oogonia (p. 675)
primary oocyte (p. 675)
germinal vesicle breakdown (p. 675)
secondary oocyte (p. 675)
polar body (p. 675)
ootid (p. 675)
ovum (p. 675)
follicle cells (p. 675)
primary follicle (p. 675)
vitellogenin (p. 675)
vitellogenesis (p. 675)
yolk granules (p. 675)
nurse cells (p. 675)
egg coat (p. 677)
vitelline membrane (p. 677)
zona pellucida (p. 677)
ovulation (p. 677)
spermatogonia (p. 677)
seminiferous tubules (p. 677)
primary spermatocyte (p. 677)
secondary spermatocyte (p. 677)
spermatid (p. 677)
spermiogenesis (p. 677)
spermatozoa (p. 677)
protamines (p. 677)
acrosomal vesicle (p. 677)
central axoneme (p. 677)
dense fibers (p. 677)
fibrous sheath (p. 677)
midpiece (p. 678)

Fertilization and Activation: The Life Cycle Renewed
fertilization (p. 678)
cortical cytoplasm (cortex) (p. 678)
activation (p. 678)
external fertilization (p. 679)
internal fertilization (p. 679)
jelly coat (p. 679)
acrosomal reaction (p. 679)
acrosomal process (p. 680)
bindin (p. 680)
ZP3 (p. 680)
ion channels (p. 680)
polyspermy (p. 681)
fast block to polyspermy (p. 681)
calcium wave (p. 681)
cortical reaction (p. 681)
cortical vesicles (p. 681)
slow block to polyspermy (p. 681)
gray crescent (p. 682)
stored mRNA (p. 682)
masked mRNA (p. 683)
translational capacity (p. 683)

Early Embryogenesis: Cleavage and Pattern Formation
embryogenesis (p. 683)
metamorphosis (p. 684)
cleavage (p. 684)
blastomeres (p. 684)
blastula (p. 684)
blastocoel (p. 684)
cellular blastoderm (p. 684)
pattern formation (p. 685)
axis of polarity (p. 685)
developmental potential (p. 685)
totipotent (p. 685)
positional information (p. 686)

morphogen (p. 686)
morphogen gradient (p. 686)
cytoplasmic determinants (p. 686)
pole plasm (p. 686)
fate map (p. 690)
cell lineage (p. 690)
determination (p. 690)
commitment (p. 690)
selective gene expression (p. 690)
nuclear equivalence (p. 691)
maturation-promoting factor (MPF) (p. 692)
cell-division cycle (*cdc*) genes (p. 692)
MPF/cdc2 protein kinase (p. 692)

Gastrulation: Laying Out the Body Plan
gastrulation (p. 693)
morphogenesis (p. 693)
organogenesis (p. 693)
blastopore (p. 694)
archenteron (primitive gut) (p. 694)
yolk plug (p. 694)
primary germ layers (p. 694)
ectoderm (p. 694)
endoderm (p. 694)
mesoderm (p. 694)
bottle cells (p. 694)
deep marginal cells (p. 695)
cell migrations (p. 695)
neural cell adhesion molecule (N-CAM) (p. 695)
RGD sequence (p. 696)
induction (p. 696)
neural tube (p. 697)
neurulation (p. 697)
primary embryonic induction (p. 697)
cDNA (p. 699)
Vegl (p. 699)
somites (p. 700)
neural crest cells (p. 701)
cell line (p. 701)
myoblast (p. 701)
sequence-specific transcription factors (p. 701)

Genetic Analysis of Embryonic Development
developmental genetics (p. 701)
polarity gene (p. 702)
segmentation gene (p. 703)
gap gene (p. 703)
pair-rule gene (p. 703)
segment-polarity gene (p. 703)
in situ RNA hybridization analysis (p. 703)
Bithorax gene complex (p. 703)
homeotic mutation (p. 704)
segment-identity gene (p. 704)
homeo box (p. 705)
DNA-binding domain (p. 705)
Hox genes (p. 708)
invariant cell lineage (p. 708)
founder cells (p. 708)
programmed cell death (p. 708)
laser ablation (p. 708)
hermaphrodite (p. 708)
distal tip cell (DTC) (p. 709)
glp-1 gene (p. 709)
lin-12 gene (p. 710)

Retinoic Acid: A Morphogen in the Chick Limb Bud
limb bud (p. 688)
apical ectodermal ridge (AER) (p. 688)
progress zone (p. 688)
zone of polarizing activity (ZPA) (p. 689)
retinoic acid (RA) (p. 689)

Using Genetics to Probe Development
temperature-sensitive (*Ts*) mutation (p. 706)
restrictive temperature (p. 706)
permissive temperature (p. 706)
conditional mutation (p. 707)
zygotically acting gene (p. 707)
maternal effect (p. 707)

Suggested Reading

General References and Reviews

Balinsky, B. I. *An Introduction to Embryology,* 5th ed. Philadelphia: Saunders, 1981.

Bonner, J. T. *On Development: The Biology of Form.* Cambridge, Mass.: Harvard University Press, 1974.

Browder, L. W. *Developmental Biology,* 2d ed. Philadelphia: Saunders, 1984.

Fulton, C., and A. O. Klein. *Explorations in Developmental Biology.* Cambridge, Mass.: Harvard University Press, 1976.

Gilbert, S. F. *Developmental Biology,* 2d ed. Sunderland, Mass.: Sinauer Associates, 1988.

Karp, G., and N. J. Berrill. *Development,* 2d ed. New York: McGraw-Hill, 1981.

Malacinski, G. M., ed. *Pattern Formation: A Primer in Developmental Biology.* New York: Macmillan, 1984.

Trends in Genetics, vol. 5, no. 8 (1989).

Trinkaus, J. P. *Cells into Organs: The Forces That Shape the Embryo,* 2d ed. Philadelphia: Saunders, 1984.

Watson, J. D., N. Hopkins, J. Roberts, J. Steitz, and A. Weiner. *Molecular Biology of the Gene,* 4th ed. Menlo Park, Calf.: Benjamin/Cummings, 1987.

Wolpert, L. Pattern formation in biological development. *Sci. Amer.* 239 (October 1978): 154.

Historical References

Spemann, H. *Embryonic Development and Induction.* New Haven, Conn.: Yale University Press, 1938.

Wilson, E. B. *The Cell in Development and Inheritance.* New York: Macmillan, 1896.

Wolpert, L. Positional information and the spatial pattern of cellular formation. *J. Theor. Biol.* 25 (1969): 1–47.

Development in Unicellular Organisms

DeVreotes, P. N. Cell-cell interactions in *Dictyostelium* development. *Trends Genet.* 5 (1989): 242–245.

Herskowitz, I. Life cycle of the budding yeast *Saccharomyces cerevisiae. Mirobiol. Rev.* 52 (1988): 536–553.

Lee, M., and P. Nurse. Cell cycle control genes in fission yeast and mammalian cells. *Trends Genet.* 4 (1988): 287–290.

Losick, R., and L. Shapiro. *Microbial Development.* Cold Spring Harbor, N.Y.: Cold Spring Harbor Press, 1984.

Gametogenesis and Fertilization

Epel, D. *Fertilization.* Endeavour (New Series) 4 (1980): 26.

Epel, D. The program of fertilization. *Sci. Amer.* 237 (November 1977): 128.

Fawcett, D. W. The mammalian spermatozoon. *Dev. Biol.* 44 (1975): 394.

Metz, C. B. Sperm and egg receptors involved in fertilization. *Current Topics Dev. Biol.* 12 (1978): 107.

Shapiro, B. M., and E. M. Eddy. When sperm meets egg: Biochemical mechanisms of gamete interaction. *Int. Rev. Cytol.* 66 (1980): 257.

Winkler, M. M. Translational regulation in sea urchin eggs: A complex interaction of biochemical and physiological regulatory mechanisms. *BioEssays* 8 (1988): 157–161.

Topics in Embryogenesis

Eichele, G. Retinoids and vertebrate limb pattern formation. *Trends Genet.* 5(1989): 246–251.

Gerhart, J. C. The primacy of cell interactions in development. *Trends Genet.* 5(1989): 233–236.

Harland, R. Growth factors and mesoderm induction. *Trends Genet.* 4 (1988): 62–63.

Illmensee, K., and A. P. Mahowald. Transplantation of posterior pole plasm in *Drosophila.* Induction of germ cells at the anterior pole of the egg. *Proc. Natl. Acad. Sci. USA* 71 (1974): 1016–1020.

Slack, J. M. W. *From Egg to Embryo: Determinative Events in Early Development.* Cambridge, England: Cambridge University Press, 1983.

Wessells, N. K. *Tissue Interactions and Development.* Menlo Park, Calif.: Benjamin/Cummings, 1977.

Genetic Analysis of Pattern Formation

Brenner, S. The genetics of *Caenorhabditis elegans. Genetics* 77 (1974): 71–94.

Greenwald, I. Cell-cell interactions that specify certain cell fates in *C. elegans* development. *Trends Genet.* 5 (1989): 237–241.

Ingham, P. W. The molecular genetics of embryonic pattern formation in *Drosophila. Nature* 335 (1988): 25–34.

Lewis, E. B. A gene complex controlling segmentation in *Drosophila. Nature* 276 (1978): 565–570.

Nüsslein-Volhard, C. V., and E. Wieschaus. Mutations affecting segment number and polarity in *Drosophila. Nature* 287 (1980): 795–801.

Nüsslein-Volhard, C. V., H. G. Fronhofer, and R. Lehman. Determination of anteroposterior polarity in *Drosophila. Science* 238 (1987): 1675 –1681.

Problem Set

1. **Distinctions.** Distinguish between the following pairs of terms or processes. Use examples as needed.
   (a) Germ cell; somatic cell
   (b) Acrosomal reaction; cortical reaction
   (c) Blastula; blastocoel
   (d) Cytoplasmic determinant; morphogen gradient
   (e) Differentiation; determination
   (f) Segmentation gene; segment-identity gene
   (g) Homeo protein; cdc2/MPF protein kinase
   (h) Induction; positional information

2. **Cellular Mechanisms in Development.** Give examples of developmental events in which the following cellular structures or processes are involved.
   (a) DNA rearrangement
   (b) DNA amplification
   (c) Exocytosis
   (d) Translational control
   (e) Microtubules
   (f) Actin polymerization
   (g) Protein kinases and phosphatases
   (h) Cellular migration
   (i) Transcription factors
   (j) Cell-cell communication

3. **Gametogenesis.** Dick (D) and Jane (J) are 8-year-old playmates. Ken (K) and Barbie (B) are 18-year-old students, and Ma (M) and Pa (P) are both 38 years old. Indicate with the appropriate letters (D, J, K, B, M, or P) to which person(s) each of the following statements applies. Use N for any statements that apply to no one.
   (a) Spermatogonia are present in his testes.
   (b) Primary oocytes are present in her ovaries.

(c) Spermatocytes are present in his seminiferous tubules.
(d) She has fewer primary oocytes than Barbie does.
(e) He has fewer primary spermatocytes than Ken does.
(f) Secondary oocytes are present in her ovaries.
(g) No haploid cells are present in the gonads.
(h) She will never have more oogonia than are currently present in her body.
(i) He will never have more spermatogonia than are currently present in his body.
(j) Meiosis will resume at puberty.

4. **The Events of Fertilization.** Each of the following events is part of the process of fertilization in a sea urchin egg, although not in the order given. List the correct order of the steps to trace the sequence of events from initial contact of sperm and egg to the first division of the zygote.
(a) The sperm makes contact with the plasma membrane of the egg.
(b) Protein synthesis is initiated, apparently in response to an increase in cellular pH.
(c) The plasma membrane of the egg cell undergoes a rapid change in membrane potential.
(d) The sperm and egg nuclei fuse, followed by the initiation of DNA replication.
(e) The acrosomal reaction releases hydrolytic enzymes from the acrosomal vesicle.
(f) Calcium is released from intracellular storage sites into the cytoplasm of the egg cell, triggering an efflux of protons out of the cell.
(g) Contact between the sperm and the plasma membrane of the egg occurs as a result of the digestion of the vitelline layer.
(h) The first cell division occurs.

5. **Amphibian Embryology.** Eight stages in the early embryogenesis of the frog are shown in Figure 22-39, but not in chronological order.
(a) Which stage shows a blastula before the onset of gastrulation?
(b) Which two stages show gastrulation in progress?
(c) At which stage are all three primary cell layers first clearly visible?
(d) What postgastrulation event has already occurred by the final stage in the series?
(e) What is the correct chronological order for the eight stages?

6. **Fate Mapping**
(a) Describe how you might use a low-molecular-weight chemical dye to trace cell lineages in the eight-cell *Xenopus* embryo.
(b) Some of the cells of the embryo are joined by gap junctions. How will this affect your result? Why is this not a problem with horseradish peroxidase?
(c) This experiment would not work with an eight-cell mouse embryo. Why not?

7. **Gene Regulation in Development.** Comment on the role of transcriptional regulation of gene expression in each of the following. Is transcriptional regulation involved? If so, is the regulation negative or positive? Specify any uncertainties.
(a) The *HMLα* and *HMR**a*** loci in yeast
(b) The 18S and 28S rRNA genes in oocytes
(c) The 5S rRNA genes in adult somatic cells
(d) The *Vegl* gene in the *Xenopus* oocyte
(e) Pair-rule gene expression in the *Drosophila* embryo

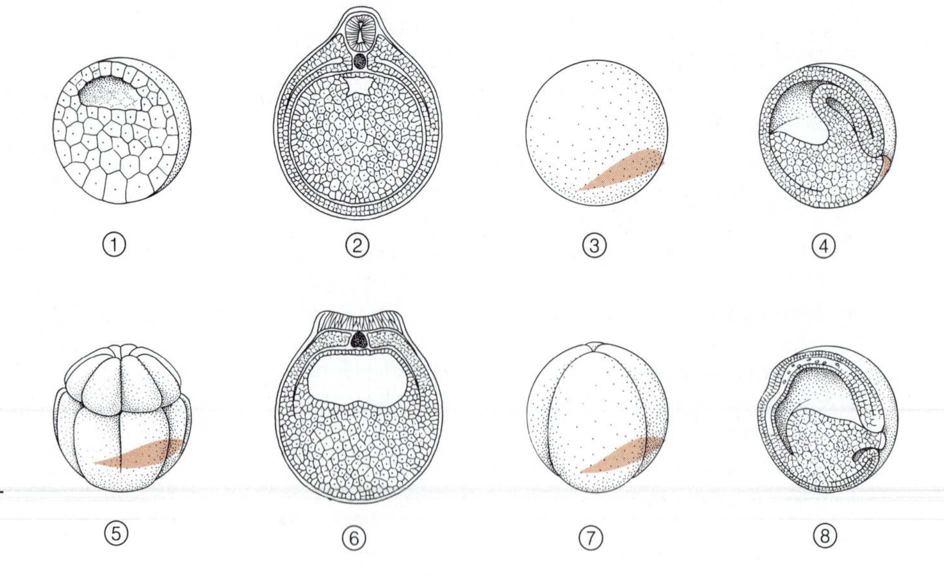

Figure 22-39 Stages of Frog Embryogenesis. Eight stages in early frog embryogenesis are shown, but not in the correct chronological order.

8. **Kinases, Phosphatases, and the Cell Cycle.** Phosphorylation by protein kinases plays an important role in cell cycle regulation. The critical cell cycle regulator MPF/cdc2 is a kinase, and as Figure 22-21 indicates, the MPF/cdc2 kinase activity is itself modulated by other kinases. But Figure 22-21 also shows a role for protein phosphatases. Assuming that the MPF/cdc2 protein is stable through many cell cycles, would you have predicted that protein phosphatases would be required for cell cycle regulation? Why or why not?

9. **Mechanisms of Localization.** When Vegl mRNA is synthesized in vitro and then is injected into developing *Xenopus* oocytes, it is soon found only in the vegetal hemisphere of the oocyte, just like native Vegl mRNA (see Figure 22-29). When globin mRNA is injected, however, it is distributed throughout the oocyte.
   (a) Why is the injection of globin mRNA important?
   (b) Suggest two mechanisms by which Vegl mRNA could be localized to the vegetal hemisphere.
   (c) A further experiment showed that the injected Vegl mRNA is very stable. What localization mechanism does this rule out?
   (d) When the experiment was repeated with Vegl mRNA from which the ribosome binding site had been removed, the Vegl mRNA was still correctly localized. What does this experiment tell us?
   (e) In preparations of the cytoskeleton from *Xenopus* oocytes, most mRNAs remain in the soluble cytoplasmic fraction. The Vegl mRNA, however, precipitates with the cytoskeletal components. What localization mechanism does this support? What experiment could you perform to test your hypothesis?

10. **Studying Molecular Gradients.** Suppose that you have identified *Xenopus* gradient protein, or XGP. Using an antibody that recognizes XGP, you have shown that the protein is made within minutes of fertilization and that the cytoplasm around the point of sperm entry is the focus of a gradient of XGP protein. Remembering that the point of sperm entry determines dorsal/ventral polarity in *Xenopus,* your colleague declares, "We have found the morphogen that determines dorsal/ventral polarity!"
   (a) Do you agree with your colleague? If so, why? If not, what words of caution do you offer?
   (b) Propose an experiment for determining whether or not XGP is involved in establishing dorsal/ventral polarity.
   (c) Recalling the various examples of polarity in this chapter, propose three models to explain how XGP might effect polarity.
   (d) Suppose you discover that XGP mRNA is distributed *throughout* the *Xenopus* egg. What kind of mechanism must regulate the distribution of XGP protein? How is this regulation different from that governing distribution of the Vegl protein (Problem 9)?

11. **Looking for Inducers.** The mRNA for N-CAM (neural cell adhesion molecule) is expressed early in primary embryonic induction.
   (a) Do you think N-CAM is involved in the response to inducer?
   (b) Suggest an experiment in which this observation could be used to test possible inducers of the neuronal cell fate.
   (c) Suppose that, using the assay you devised in (b), you identify a small polypeptide as a possible neural inducer. Would you expect this inducer to be made in mesodermal cells or in ectodermal cells of the neural plate region?
   (d) Describe two experiments that would be necessary to confirm that your candidate is indeed the neural inducer.

12. **Induction of Morphogenesis in the Culture Dish.** Salivary epithelial cells secrete a class of polysaccharides known as glycosylaminoglycans, or GAGs, which are a normal part of the extracellular matrix. Together with secreted proteins such as collagen, the GAGs form the basal lamina lining that coats the epithelium. If isolated salivary gland epithelium is enzymatically stripped of its GAGs and cultured in the absence of the mesenchymal cells that normally surround it (Figure 22-40a), the cells form a spherical cluster but fail to form the characteristic branched structure of normal salivary gland epithelium (Figure 22-40b). If, on the other hand, such cells are cultured with mesenchyme, the sphere of cells differentiates into branched epithelium (Figure 22-40c). Branching occurs, however, only after the GAG layer has been restored.
   (a) This has been described as an inductive interaction. Why? What tissue is induced, and what tissue is the inducer?
   (b) Do you think the extracellular matrix plays a role in this induction? Why? What could that role(s) be?
   (c) Salivary epithelial cells cultured in the absence of mesenchyme do not branch (Figure 22-40b). Nevertheless, they illustrate the principles of commitment and the heritability of the differentiated state. How?
   (d) What are some of the cellular structures and mechanisms that respond to induction and carry out terminal differentiation?

13. **Developmental Genetics**
   (a) To study the *cdc* genes in haploid yeast, it is necessary to work with temperature-sensitive mutations. Why?
   (b) Consider the myosin heavy chain gene. Describe three mutations that could disrupt myosin function. Two of these should affect the protein-coding regions of the gene; one should not.
   (c) Consider a mutant actin monomer that can polymerize to the growing end of a filament of normal actin, but with no more actin added on to the mutant protein. In

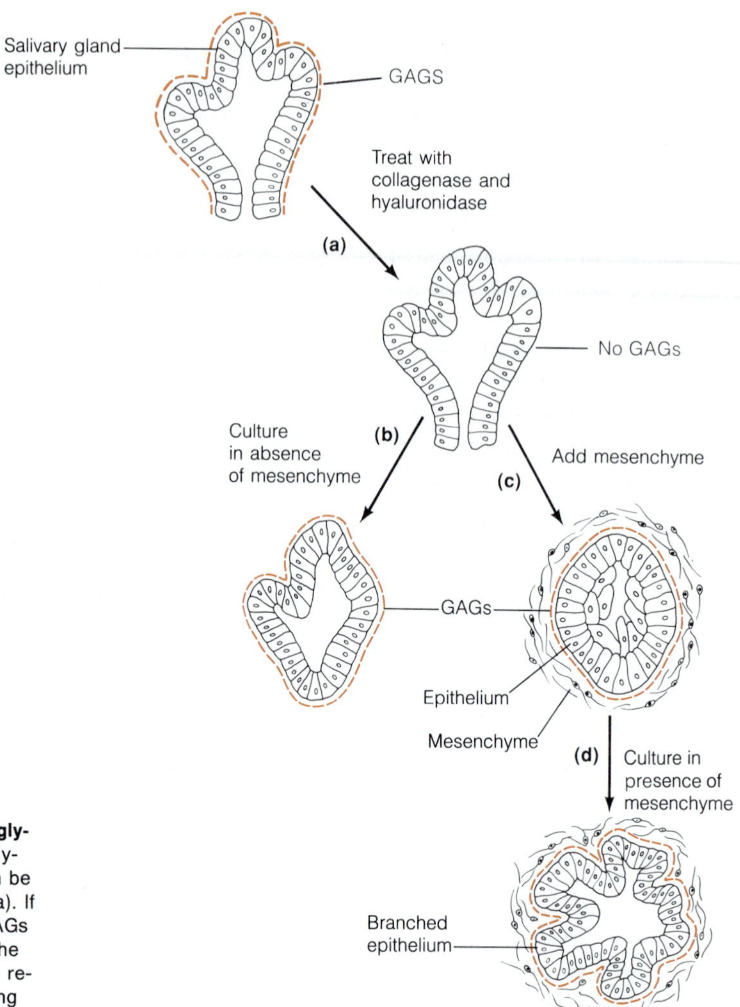

Figure 22-40 The Effects of Mesenchyme and Glycosylaminoglycans on Salivary Gland Morphogenesis. The glycosylaminoglycans, or GAGs, that coat isolated salivary epithelial cells can be removed with the enzymes collagenase and hyaluronidase (a). If the cells are cultured in the absence of mesenchyme, the GAGs are restored but no further differentiation takes place (b). If the cells are cultured with mesenchyme, however, the GAGs are restored (c), and the cells differentiate as indicated by branching (d).

a diploid cell with a normal actin gene and the mutant actin gene, do you think actin filaments would be normal? What if the mutation destroyed the ability of actin monomers to add on to growing filaments?

(d) Consider the mutagenesis described in Figure 22-C to look for autosomal mutations. It is much simpler to look for zygotically acting mutations on the X chromosome. To convince yourself that this is true, design a mutagenesis to look for zygotically acting lethal mutations on the X chromosome. Use the *white* mutation, a recessive eye color mutation on the X chromosome of *Drosophila* to follow your progeny classes.

(e) *Caenorhabditis elegans* can live as a self-fertilizing hermaphrodite. To convince yourself that this simplifies genetic analysis, design a mutagenesis to look for zygotic lethals in *Caenorhabditis*. Design your mutagenesis to look for mutations that are linked to the *unc* gene. When homozygous, this recessive mutation results in *uncoordinated* adult worms that are easily recognized in a crowd of normal worms.

14. **Interpreting Mutant Phenotypes.** The *dorsal* gene of *Drosophila* encodes a transcriptional regulator. A mutation in the maternal effect *dorsal* gene "dorsalizes" the embryo; that is, it causes the entire ectoderm to differentiate as though it were dorsal tissue. Conversely, a mutation in the zygotically acting gene *dpp* gene "ventralizes" the embryo (see Figure 22-D). The *dpp* mRNA is localized in the dorsal half of the embryo and encodes a growth factor-like protein.

(a) What do you think the role of the *dpp* "growth factor" is in dorsal/ventral pattern formation?

(b) Which of these genes is likely to play an earlier role in development and why?

(c) Propose a model for how the *dl* and *dpp* genes, or their proteins, might interact.

15. **Control of Sexual Differentiation in *Drosophila*** (Details have been simplified). Whether a cell differentiates into one cell type or another often depends on the presence or

absence of regulatory proteins. Such appears to be the case with sexual differentiation in *Drosophila:* In females the *presence* of a particular regulatory protein in a cell confers the female cell fate, while in the male the *absence* of that protein in a cell confers the male cell fate. The size of the mRNA produced by certain genes in males differs from the size of the mRNA in females. One of the genes with different-sized mRNAs in males and females is known as *Sex-lethal*. Females that lack the *Sex-lethal* gene differentiate as "males," even though they have two X chromosomes. The *Sex-lethal* gene gives rise to a 4.4-kb mRNA in males and a 3.7-kb mRNA in females. Figure 22-41 shows the mature *Sex-lethal* mRNA found in the cytoplasm of males.

(a) Recalling the lessons of Chapter 17, suggest at least two ways in which one gene could give rise to mRNAs of differing sizes.

(b) Suggest an explanation for the size difference of the *Sex-lethal* mRNA between males and females, and sketch the mRNA you expect to find in females.

(c) How might your explanation account for the difference in *Sex-lethal* gene function in males and females?

(d) Animals of one sex are totally unaffected by mutations that remove the *Sex-lethal* gene. Which sex is unaffected and why?

(e) As mentioned above, there are several other genes that give rise to different-sized transcripts in males and females. When the *Sex-lethal* gene is removed, the mRNAs of these other genes in females are the same size as they are in males. What does this suggest about how the *Sex-lethal* protein might execute its regulatory function?

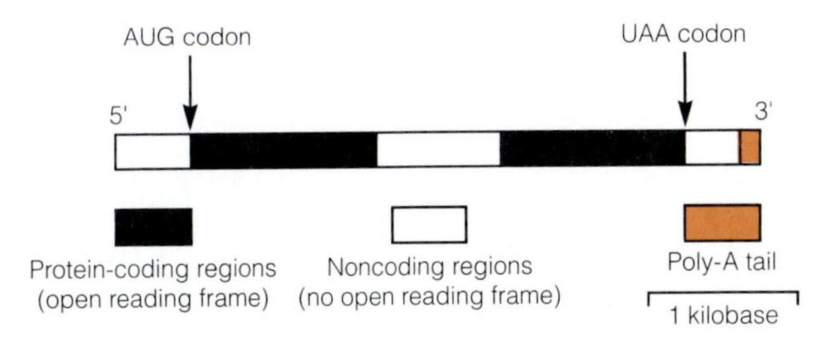

Figure 22-41 Structure of the Male-Specific mRNA of the *Drosophila Sex-lethal* Gene. The male *Sex-lethal* mRNA, shown here, is 0.7 kb longer than the female mRNA. (AUG and UAA are translational start and stop codons, respectively.)

23

Cellular Aspects of Cancer

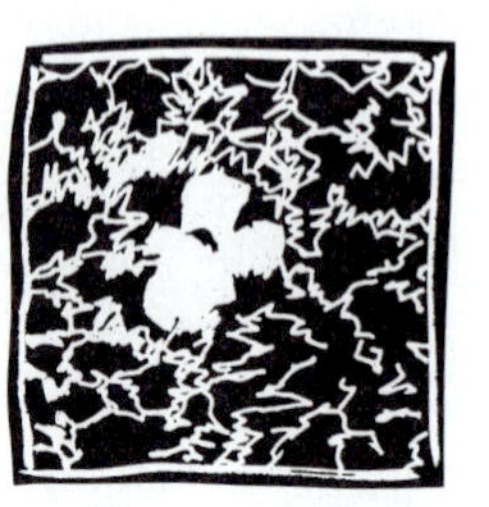

So far, our discussion of cellular structure and function has focused on normal cells. Knowing how complex cells are, we can appreciate how remarkable it is that they look and function normally so much of the time. This is particularly striking in the development of higher organisms, which requires the careful orchestration of the activities of large numbers of cells that perform different functions. A good example is the human body. In most cases, the anatomical organization of the body is stable over time. The volumes of the different tissues and organs are constant and appropriately proportioned with respect to the size of the body. Furthermore, the organization of the tissues within organs and the differentiated character of individual cells of the different tissues are also stable. This stability of tissue size, differentiation, and organization is essential for the maintenance of function.

However, in all organisms, from plants to humans, the regulatory mechanisms that control cell division sometimes go awry. Uncontrolled growth begins, resulting in loss of tissue stability, reduced tissue function, and even death of the organism. We are just beginning to understand the molecular mechanisms that cause cells to become transformed in this manner. With understanding comes the hope that we may someday be able to inhibit the transformation process, or, failing that, to control the rampant growth that leads all too often to a fatal outcome.

Cancer: A Loss of Normal Growth Regulation

If the normal stability of the organization of tissues and organs is disturbed, a variety of disease states can occur. One example is a tissue in which the control of growth becomes defective, called a **tumor** or **neoplasm** (literally "new growth"). Neoplasms can be classified as **benign** or **malignant**, and the common term for the malignant tumor is **cancer.** The word *cancer* is taken from the Latin term for "crab" because early physicians noticed that certain skin cancers had a crablike appearance.

Tumor Dissemination: Benign or Malignant?

In almost every case, malignant tumors are *monoclonal,* meaning that they develop from a single cell. The progenitor cell has undergone a permanent, heritable change that is transmitted to all its progeny, a process called the **neoplastic transformation.** Clinically, the distinction between benign and malignant is based on the effect of the neoplasm on the survival of the host organism: The host can die from the effects of the malignant tumor but will survive the presence of the benign tumor.

At the cellular level, the most important difference is the likelihood that the tumor will spread. Benign tumors typically are *encapsulated nodules* of neoplastic tissue and therefore do not spread, whereas malignant tumors often spread to neighboring tissues and even other parts of the body. This spread to neighboring tissues is called **invasion;** the spreading to distant organs is called **metastasis,** and the tumor nodules resident at sites distant from the parent tumor are referred to as *metastases.*

In summary, cancers are tissues in which normal growth and positional regulation are lost. In this chapter, we will focus on the underlying mechanisms that regulate growth and tissue organization for normal cells and tissues and the manner in which such mechanisms lose their effect in cancer. But first, we will explore the agents that can initiate the transformation of normal cells into neoplastic cells and the nature of the changes that form the genetic basis for neoplasia.

Classification of Tumors

Tumors may arise from mature, differentiated cells or from mitotically active stem cell populations. **Stem cells** are relatively undifferentiated, mitotically active cells from which some more highly differentiated cell "stems." An example is the mitotically active hematopoietic ("blood-forming") stem cell population found in bone marrow that gives rise to red and white blood cells. The stem cells of various tissues often are designated by the suffix-*blasts*. For example, **neuroblasts** are the mitotic stem cells for neurons; **myoblasts** are the stem cells for myocytes; and **fibroblasts** are the stem cells for fibrocyte connective tissue cells. Cancers of a stem cell may be designated by the suffix "*-omas*": **neuroblastomas** for tumors of neuroblastic origin; **myelomas** for tumors of myeloblasts, which are the stem cells of the granular leukocytes (white blood cells).

Another important element in tumor categorization is the tissue class of origin. **Carcinoma** refers to a tumor derived from an epithelial tissue, and **sarcoma** refers to a tumor derived from connective tissue. As indicated above, the hematopoietic system includes stem cell populations that grow and divide throughout life. All of these populations are subject to neoplasia: **lymphomas** are tumors of the lymphocytic lineage, **leukemias** and **myelomas** are tumors of the leukocytic system (the granulocytes, eosinophils, and basophils), and **erythroblastomas** are tumors of the red cell lineage.

Causes of Neoplastic Transformation

The three causative factors that have received the most study as initiators of the neoplastic transformation are genetic and chromosomal alterations, infection by oncogenic viruses, and exposure to chemical carcinogens.

Chromosomal Abnormalities and Neoplasia

Chromosomal alterations such as *deletions, translocations* (exchanges of segments between chromosomes), and *inversions* of chromosomal components are involved in both the initiation and progression of selected human tumors. One of the best-studied examples is the *Philadelphia chromosomal abnormality,* which plays a role in the initiation of the cancer *chronic myelogenous leukemia* (CML). CML is a tumor of a class of white blood cells referred to as *granulocytes* because of the granular appearance of their cytoplasm under the microscope. The mitotically active stem cells of normal granulocytes (known as myelocytes) are present in bone marrow. The Philadelphia chromosome results when a portion of the long arm of chromosome 22 is transferred to chromosome 9 (Figure 23-1). The strongest evidence that this chromosomal abnormality initiates CML is that 85% of CML patients have the Philadelphia chromosome. The 15% of CML patients lacking the Philadelphia chromosome appear to have a different form of the disease that typically develops in elderly individuals.

Since tissues other than the leukemic cells of patients with CML lack the Philadelphia chromosome, it can be concluded that CML results from a *somatic chromosomal alteration.* That is, the alteration occurs in the somatic myelogenous stem tissue of the adult and is not transferred to offspring by egg or sperm. Chromosomal abnormalities occur with low frequency in any mitotically active tissue. When they occur in the *germ cell* lineage, they have the potential to be inherited by the next generation; however, when they occur in a somatic tissue cell, they are restricted to the progeny of that cell and do not appear in the germ cells.

Tumor progression is the incremental development of increasingly malignant states by a tumor. Typically, tumors in the early stages of development are relatively benign. The tumor grows slowly and is either weakly invasive or noninvasive. With time, however, tumors can enter a phase of increasingly rapid growth, becoming highly invasive and metastatic.

Tumor progression is exemplified when a hidden metastatic nodule suddenly develops into a life-threatening tumor several years after the **primary tumor** has been surgically removed. In such cases, the metastatic nodule must have originated before surgery, but it had remained dormant for those several years before progressing to the fully malignant state. Primary tumors presumably also follow this course from a benign state to states of progressively greater malignancy. This course is suggested by the relatively high incidence of benign tumors found upon autopsy of individuals who died of other causes.

Chromosomal changes accompany, and appear to play an important role in, tumor progression. Chronic myelogenous leukemia is a good example. As the name suggests, CML is chronic and usually is not quickly fatal. The tumor progresses after an extended chronic period into the *blast cell crisis* stage, which is characterized by rapid growth of the leukemic tissue and is soon followed by death. The blast cell crisis is often accompanied by a characteristic chromosomal abnormality: the appearance of leukemic

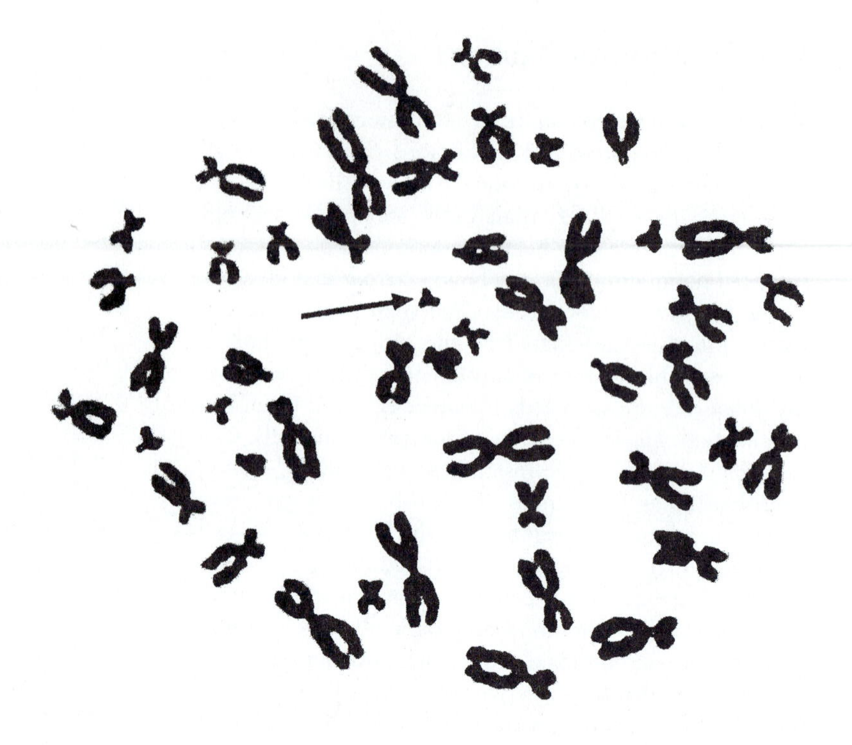

Figure 23-1 The Philadelphia Chromosome. The Philadelphia chromosome is human chromosome number 22 that has lost a portion of its long arm to chromosome number 9. The arrow points to a Philadelphia chromosome in this metaphase chromosome preparation of a bone marrow cell from a patient with chronic myelogeneous leukemia.

cells with an additional Philadelphia chromosome. This chromosomal abnormality appears to precipitate the highly malignant phenotype of the blast cell and is used by clinicians as a diagnostic marker.

Oncogenic Viruses

Certain families of viruses are able to trigger a neoplastic transformation in the cells that they infect. Members of both the DNA and RNA classes of viruses (viruses that have, respectively, DNA and RNA as the genetic material) can cause infected cells to become cancerous (Table 23-1). One piece of evidence for a viral etiology is that patients with tumors may have antibodies to antigens associated with the suspected virus (Table 23-2). Furthermore, cancerous cells often are found to contain nucleic acid sequences that are diagnostic of the virus. The experimental demonstration of the ability of particular viruses to cause tumors begins with the isolation of virus particles from tumors of mice and other experimental animals. When inoculation of this preparation of viruses into healthy, tumor-free individuals results in the development of tumors, the tumorigenic potential of the virus is confirmed. When the cycle can be repeated by isolating the same type of virus from the experimentally infected individual and using the virus to induce tumors in healthy hosts, the evidence becomes very strong that the particular virus plays a significant role in the initiation of the particular tumor.

Viruses may have either RNA or DNA as the genetic material of the viral chromosome (Table 23-1). The principal class of **RNA tumor viruses** that can cause cancer is the **retrovirus** family (Fig. 23-2). When one of these viruses infects a cell, it stimulates the production of a complementary DNA copy of the viral chromosome, called the **provirus,** which then becomes incorporated into one of the chromosomes of the host cell (Fig. 23-3). Synthesis of the DNA copy from the viral RNA template is catalyzed by **reverse transcriptase,** an enzyme that is unique to this family of viruses. The provirus replicates with the host cell's DNA, synthesizing RNA copies of the provirus by transcription. These viral RNA copies then code for the viral proteins that are needed to encapsulate new virus particles which contain the RNA as genetic material. The virus particles then bud from the surface of the infected cell (Fig. 23-4).

Two basic patterns of infectivity of oncogenic viruses have been described: **horizontal transmission,** in which virus particles are transferred between unrelated members of a population, and **vertical transmission** from parent to child. An example of *horizontal transmission* is the Marek disease virus (MDV), a herpes virus that causes a form of leukemia in chickens. The virus multiplies in the epidermis, which is a tissue that undergoes continual cellular turnover. The mitotically active basal cells produce daughter cells that differentiate into mature keratinocytes, which eventually are shed as clumps of cells known as *dander flakes*. When MDV-containing dander flakes are inhaled by healthy chickens, viral infection and leukemia may follow.

Vertical transmission occurs in the mouse mammary tumor virus. The pattern of infectivity was first established in inbred mouse strains that showed low or high incidences of mammary carcinoma. Some strains of mice showed an incidence rate of mammary tumors of less than 1% in females at 1 year of age, whereas high-incidence strains showed close to 100% incidence. It was discovered that newborn daughters of several of the strains with high tumor

Table 23-1 Oncogenic Viruses

Virus Family	Specific Virus	Tumor
I. DNA Tumor Viruses		
Pox viruses	Shope's rabbit fibroma virus	Epidermal carcinomas
Herpes viruses	Marek's disease virus	Marek's disease (a lymphoma of chickens)
	Lucké tumor virus	Lucké renal adenocarcinoma (frog)
	Epstein-Barr virus	Burkitt's lymphoma and nasopharyngeal carcinoma (human)
Papilloma viruses	Shope's rabbit papilloma virus	Sarcomas (rabbit)
	Human papilloma virus	Human cervical carcinoma
Adenoviruses	Ad	Tumors produced only under laboratory conditions
Polyoma viruses	Py, SV40	Tumors produced only under laboratory conditions
II. RNA Tumor Viruses		
B-type mammary tumor viruses	Mouse mammary tumor virus	Mammary carcinoma
C-type leukoviruses (retroviruses)	Avian (Rous) sarcoma viruses Murine sarcoma viruses Feline sarcoma viruses Human T-cell leukemia virus-I	Leukemias and sarcomas

Table 23-2 How to Demonstrate a Viral Etiology for a Specific Tumor

I. Associate a virus with the tumor by:
  1) Production of viral particles by tumor cells. Characterize the particles as viruses by:
    a) Morphology (electron microscopy)
    b) Physical properties:
      Buoyant density
      Molecular weight of nucleic acid
    c) Biochemical properties:
      Viral coat antigens
      Virus-specific enzymes (reverse transcriptase)
    d) Infectivity
  2) Virus-specific antigens within or on the surface of tumor cells
  3) Detection of viral nucleic acid in tumor cells by molecular hybridization
  4) Systemic effects on the diseased host:
    a) Excretion of viral particles by the host
    b) Presence of antiviral antibodies in the serum
II. Experimental demonstration:
  1) Isolate a particular virus from hosts with specific tumor and demonstrate tumor production by virus when inoculated into disease-free hosts.
  2) Repeat virus isolation and infection cycle with second set of disease-free hosts, showing that the same tumor is produced.

incidence could be converted into strains with low tumor incidence in the adult if, as pups, they were separated at birth from their mothers and were foster-nursed by mothers of the low-incidence strain. The milk of the high-incidence mothers contained viral particles that also were found in the mammary carcinoma cells, providing strong evidence for a vertical transfer route via the milk.

Certain other high-incidence strains of mice have the mammary tumor virus in the milk of afflicted mothers, but newborn daughters still develop the tumor when foster-nursed by mothers with low tumor incidence. This indicates that these high-incidence strains transfer the virus both through the milk and in the germ cells. In contrast to the high-incidence strains that experience virus transfer only through the milk, the strains that show germ-line transfer develop mammary tumors if sired by a high-incidence father mated with a low-incidence mother. Thus, the mammary tumor virus shows various routes of vertical transfer from parent to child.

Environmental Carcinogens

Environmental carcinogens include physical factors that can provoke the neoplastic transformation, such as ultraviolet light, which is involved in the development of skin cancer and melanoma, and ionizing radiation. Even more important are **chemical carcinogens,** which have been implicated in the development of many important human tumors. The evidence for the **oncogenic** ("tumor-causing") potential of particular chemicals includes experimental studies with animals and **epidemiologic** studies with hu-

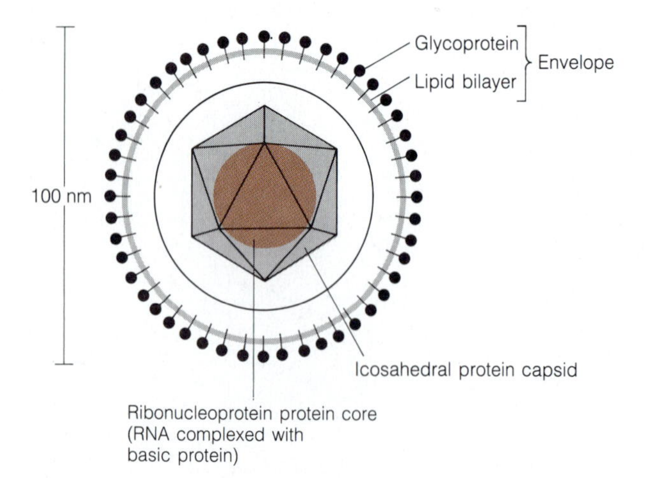

Figure 23-2 The Structure of a Retrovirus. A typical retrovirus consists of a ribonucleoprotein core, an icosahedral protein capsid, and an envelope containing viral glycoprotein molecules in a lipid bilayer that is derived from the plasma membrane of the host cell. Two copies of the viral RNA genome are packaged in the core, as are molecules of reverse transcriptase.

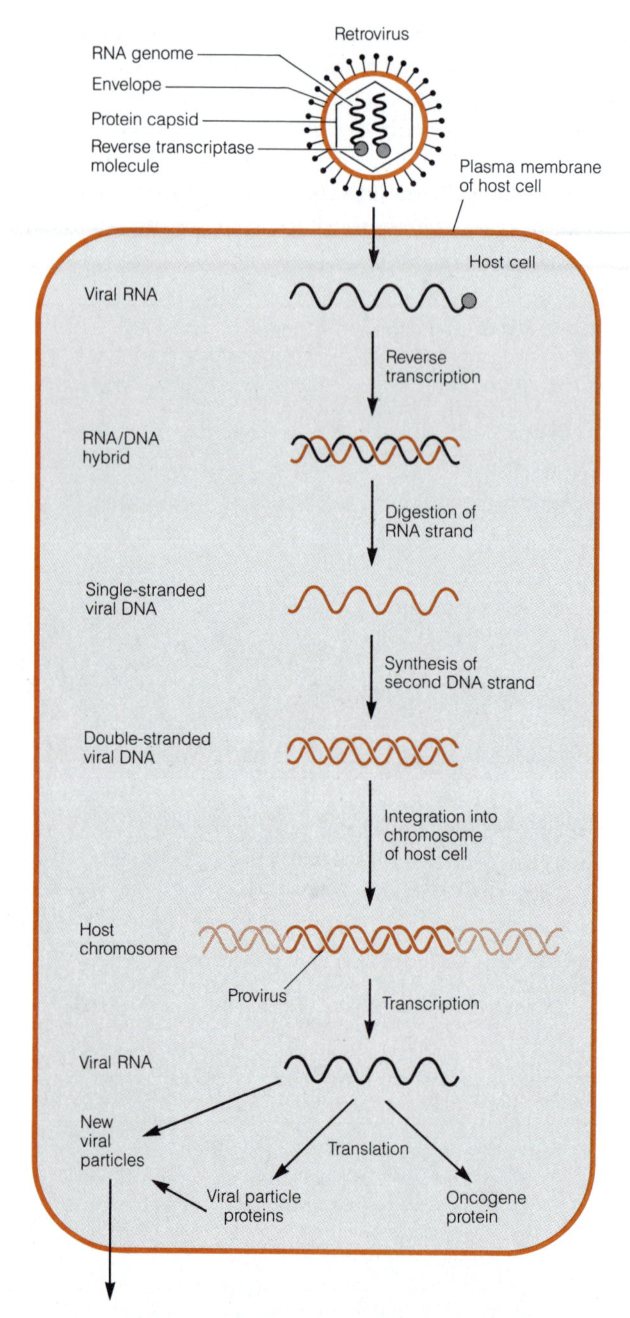

Figure 23-3 Replication Cycle of a Retrovirus. The virus introduces two copies of the viral RNA chromosome into the cell. Attached to each viral genome is a molecule of the viral enzyme reverse transcriptase, which immediately directs the synthesis of a strand of DNA that is complementary to the viral RNA strand. Still under the influence of the same enzyme, the viral RNA strand is digested, and a complementary DNA strand is synthesized to complete replication of the viral genome as a duplex strand of DNA. This then integrates into the host cell chromosome. From there it can direct the production of multiple single-stranded RNA copies, which are packaged into new infectious viral particles. When the infecting viral genome includes an oncogene, in addition to the genes necessary for viral reproduction, the oncogene protein that is synthesized from the secondary viral RNA transcripts can induce the neoplastic transformation of the infected cell.

mans. The extensive body of evidence demonstrating that noxious chemicals in cigarette smoke cause lung cancer is one of the best-publicized examples of an epidemiologic study that ties a particular cancer to exposure to a particular chemical. Experimental studies involve the determination of the incidence of cancers in groups of test animals that have been deliberately exposed to suspected carcinogens. Both experimental and epidemiological studies have indicated an oncogenic potential for a number of chemicals (Fig. 23-5).

It has been estimated that exposure to an environmental carcinogenic stimulus is implicated in the initiation of 50–90% of all human cancers. Extensive cross-cultural studies of the differences in tumor incidence between different national groups support these figures. Tumor types that are rare in one national group may be abundant in another. Colon cancer is common in the United States, for example, but rare in Japan, while stomach cancer is common in Japan and comparatively uncommon in the United States (Figure 23-6). These profound differences might be based on genetic differences between different national groups or might result from different patterns of exposure to environmental carcinogens.

Genetic differences between human populations are known to be important in development of certain types of cancer. For example, dark-skinned people are less susceptible to ultraviolet light-induced cases of skin cancer and melanoma than fair-skinned individuals. One of the selective advantages of skin pigmentation for life in the tropics may be the increased resistance to sunlight-induced damage to the skin, which can lead to skin cancer and melanoma.

However, exposure to environmental carcinogens appears to be the dominant factor in the ethnic differences in the incidence of most tumors. Emigrant populations typically develop the pattern of tumor incidence characteristic of the new host country as they adopt the diet and life style of the host country, even when immigrants show only a low frequency of intermarriage with members of the majority population of the host country (Figure 23-6). Since there are profound ethnic differences in the relative frequencies of most important human cancers, and since these differences are largely a result of local differences in exposure to environmental carcinogens, environmental carcinogens must be primary factors in the development of most cancers in humans. This implies that the incidence of cancer could be decreased significantly by reducing the exposure to major carcinogens. Certainly, elimination of the smoking of tobacco products would markedly reduce the incidence of lung cancer, a major cancer in many countries.

Most chemical carcinogens require metabolic conversion for the initial chemical (the **procarcinogen**) to be activated to the form that actually produces the neoplastic conversion of the target cell (the **ultimate carcinogen**) (Fig. 23-7). An important pathway of activation involves the introduction of reactive electrophilic groups, such as epoxide, ester, and carbonium groups, onto the procarcinogen by an enzymatic system known as the *mixed-function oxidase* or *aryl hydroxylase system*. This system of enzymes contributes to the ability of the liver to remove toxins from the body by increasing the solubility of toxic organic molecules to facilitate their removal by the kidney. It does so by chemically modifying the organic molecules, adding the electrophilic groups that have the negative side effect of

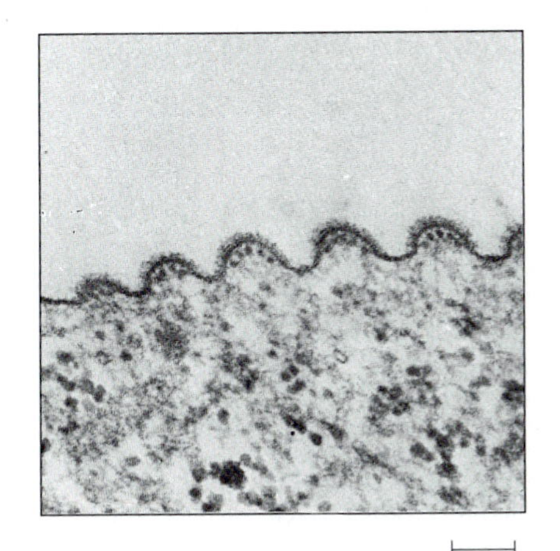

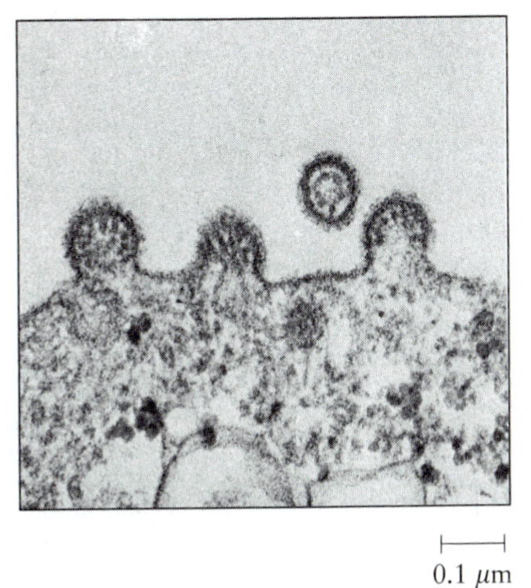

Figure 23-4 Budding of a Retrovirus from Its Host Cell. Cultured green monkey kidney cells were infected with respiratory syncytial virus, a retrovirus. The electron micrographs show viral particles at successive stages of budding from the plasma membrane of the host cell. Each viral particle is enveloped by a membrane derived from the plasma membrane of the host cell, but with viral-specific glycoproteins at the surface. (TEMs)

converting the organic procarcinogen molecule into an ultimate carcinogen. The trade-off is between acute toxic damage and chronic damage due to carcinogenesis.

1. Polycyclic aromatic hydrocarbons

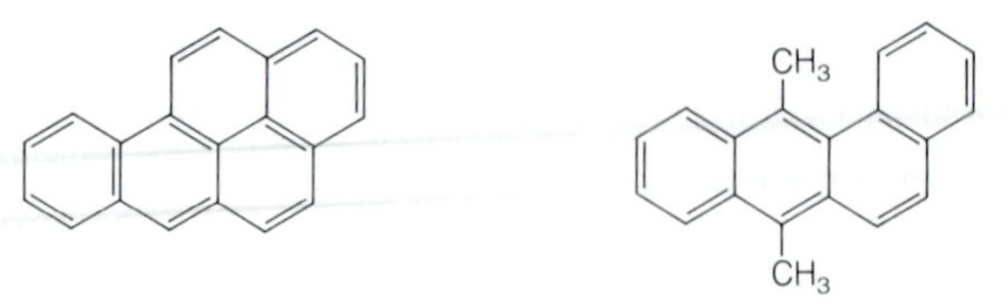

Benzo [*a*] pyrene   7,12-Dimethylbenz [*a*] anthracene

2. Aromatic amides

NH_2

2-Naphthylamine

3. Azo dyes

N=N N CH_3 CH_3

N-dimethylaminoazobenzene (DAB) = butter yellow

4. Nitrosamines

H_3C N—N—O H_3C

Dimethyl nitrosamine

5. Halogenated hydrocarbons

Cl—C—Cl (Cl, Cl)   H—C—Cl (Cl, Cl)   Cl—C—Cl, Cl, Cl—C—Cl, H

Carbon tetrachloride   Chloroform   Dichlorodiphenly-trichloroethane (DDT)

6. Alkylating agents

O HN N CH_2—CH_2—Cl CH_2—CH_2—Cl O N H

Uracil mustard

7. Metal ions:

Be^{2+}, Ca^{2+}, Co^{2+}, Ni^{2+}, Pb^{2+}

Figure 23-5 Examples of Chemical Carcinogens. A variety of chemicals have been identified as carcinogens, based on experimental studies with laboratory animals and epidemiological studies with humans. Most important are the aromatic organic compounds (groups 1, 2, and 3), the nitrosamines (group 4), and the halogenated hydrocarbons (group 5).

Chemical Carcinogens Act by Producing Genetic Mutations. The principal action of most environmental agents in the induction of cancer is the production of **genetic mutations.** Evidence includes the observation that many chemical carcinogens in the form of the ultimate carcinogen bind covalently to the nucleotide bases of DNA. This chemical modification of DNA is potentially damaging in a number of ways. Modification may disturb hydrogen bonding between the two DNA strands and may result in incorrect base pairing during replication of the affected DNA strand. Modification may also affect interactions of the DNA with regulatory DNA-binding proteins, which can then affect the level of transcription of the gene.

Evidence for the mutagenic activity of carcinogens is provided by the correlation between the carcinogenic potential of a given agent and its activity as a **mutagen.** A simple and sensitive assay to demonstrate this correlation has been developed by Bruce Ames, who compared the mutagenic activity of various chemicals on the bacterium *Salmonella* with their carcinogenicity, as demonstrated by experimental studies in mice and epidemiologic studies with humans. This mutagenesis assay (now called the **Ames test**) involves specially engineered mutant strains of *Salmonella* that require the amino acid histidine for growth. When plated on histidine-free culture medium, the only bacteria that grow are those that experience a back-mu-

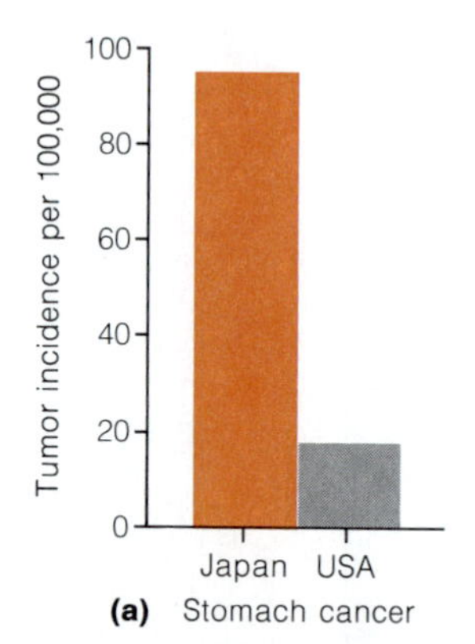

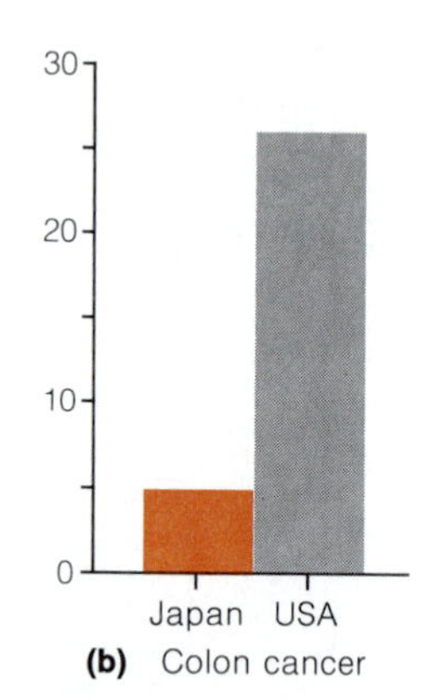

Figure 23-6 National Differences in Incidence of Different Cancers. Different national groups show marked differences in the incidence rates for different cancers. A tumor such as stomach cancer (a) may be of high incidence in one nation, such as Japan, and of low incidence in another, such as the United States. Conversely, colon cancer (b) is frequent in the United States and relatively rare in Japan. Since immigrant populations, such as Japanese immigrants to the United States, typically acquire the pattern of relative tumor frequency of the new host nation at a rate too high to be a result of intermarriage, it is concluded that the national differences in incidence are principally due to different national patterns of exposure to environmental carcinogens rather than to genetic differences between the populations.

tation of a mutant gene in the histidine biosynthetic pathway. In the absence of mutagens in the system, very few bacteria experience the requisite back-mutation to enable them to grow on agar that lacks histidine. When an appropriate mutagen is present, back-mutation occurs with higher frequency, and larger numbers of bacterial colonies are present after several days of incubation (Fig. 23-8). The potency of the mutagen can be calculated from the fraction of the bacteria that experience back-mutation and grow into colonies on the histidine-free culture medium.

When using the Ames test to investigate the mutagenic potential of procarcinogens, it is necessary to include both the procarcinogen and a source of the mixed-function oxidase enzyme system to convert the procarcinogen into the ultimate carcinogen. Ames included extracts of rat liver tissue with the bacterial culture for this purpose (Fig. 23-8). The experimental results were striking: In one test, only 13 of 108 "noncarcinogenic" chemicals had mutagenic activity in the bacterial assay (several of these chemicals were subsequently demonstrated to be weak carcinogens) and 158 of 176 of the chemicals known to be carcinogenic were mutagenic. The false-negatives (carcinogens that failed to show mutagenic activity) included a few chemicals, such as the steroid hormone diethylstilbestrol, that do not interact directly with DNA and that stimulate the development of cancer by different pathways. Thus, there is a strong correlation between two properties of these chemicals assessed by two very different assays: the mutagenic activity of a given ultimate carcinogen (assayed by its ability to produce gene mutations in bacteria) correlates with its activity as a carcinogen in mammals. The data indicate that, with only a few exceptions, a chemical's mutagenic activity in an animal underlies its ability to function as a carcinogen.

The mutagenicity tests are important as a first screen of untested chemicals that are suspected to be carcinogens. The test is rapid, relatively inexpensive, and readily quantifiable and results in low levels of false-positives and false-negatives. In contrast, animal tests of potential carcinogens are slow, inaccurate, and prohibitively expensive. Approximately 50,000 synthetic chemicals are produced and used in significant quantities by industry. Testing all of

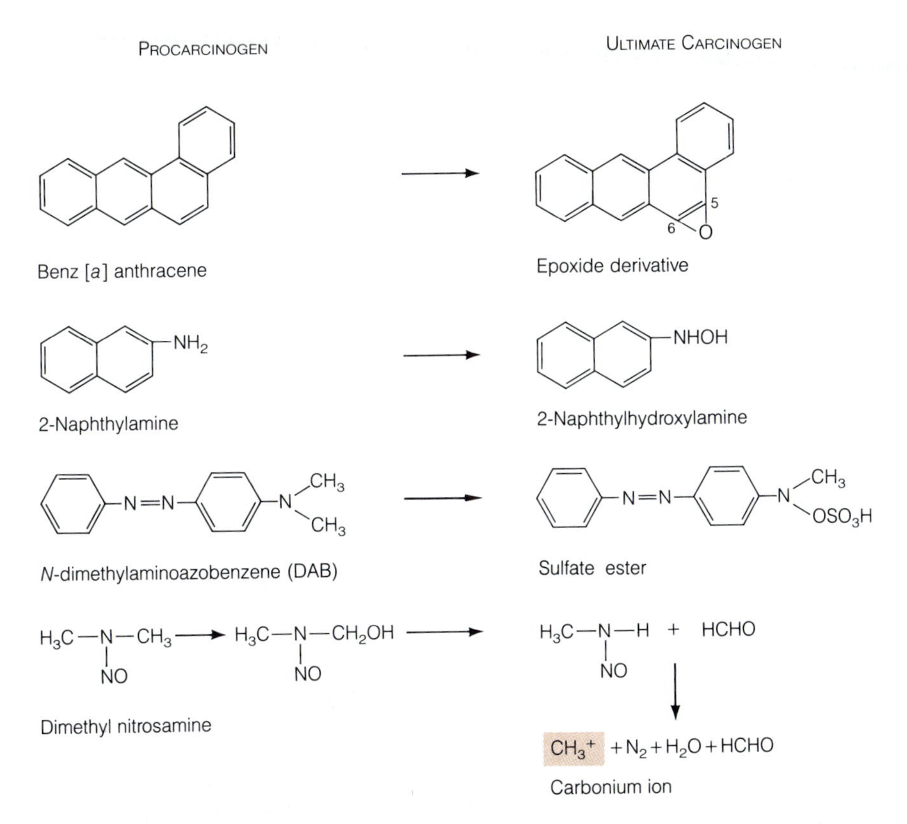

Figure 23-7 Metabolic Activation of Chemical Carcinogens. Most important chemical carcinogens are procarcinogens, chemicals that require metabolic activation into the form of the ultimate carcinogen to acquire carcinogenic potential. The enzymes that catalyze procarcinogen conversion are the mixed-function oxidases. The normal function of this enzymatic pathway is to add electrophilic groups to organic molecules to render the latter more soluble in the blood so that they can be cleared by the kidney. Acquisition of carcinogenic potential is an unfortunate by-product of this activity.

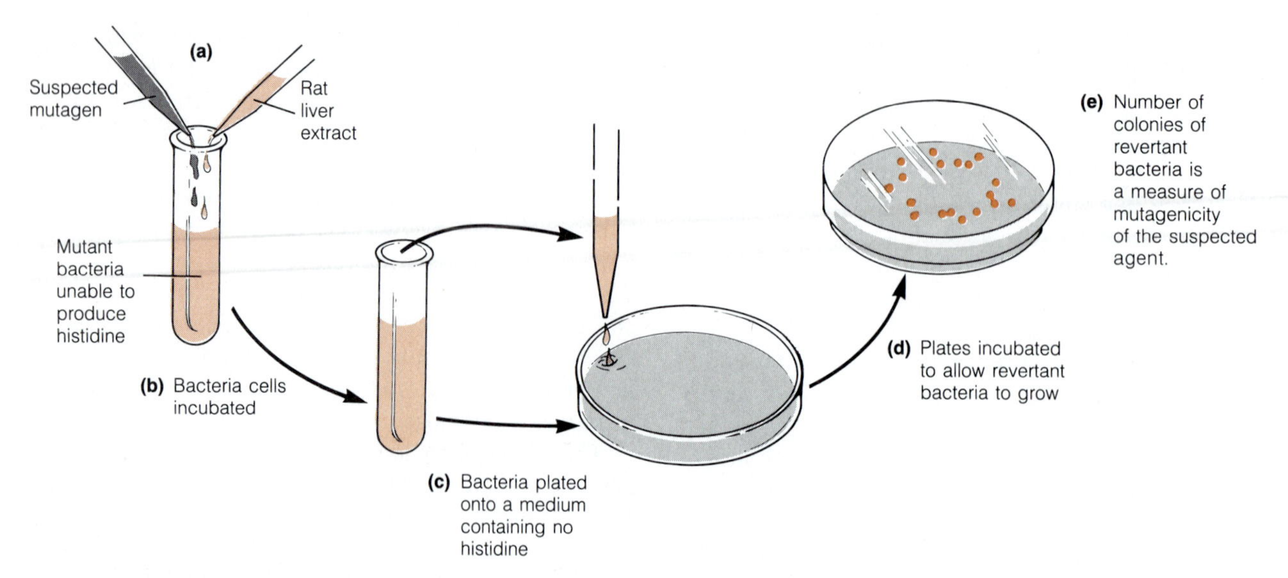

Figure 23-8 The Ames Test for Mutagens. The Ames test uses bacteria to determine the mutagenic activity of chemicals. (a) The suspected mutagenic chemical and a membrane-enriched extract of rat liver tissue are added to a mutant strain of *Salmonella* that is unable to produce histidine. (b) During the incubation period, mutations may occur in the bacterial cells. Some of the mutations may restore the faulty gene in the histidine biosynthetic pathway to a functional form. (c) The bacterial cells are then plated onto a medium lacking histidine. (d) During the incubation period, only revertant bacteria (i.e., bacteria that had experienced a back mutation that restores gene function) are able to grow on the histidine-free medium. (e) The number of colonies of bacteria is then determined and compared with the number of colonies on control plates inoculated with *Salmonella* cells that were not exposed to the mutagen. The number of colonies on the experimental plates is a measure of the mutagenicity of the chemical being tested. The rat liver membrane fraction added to the bacterial culture at step (a) ensures the presence of the mixed-function oxidase enzymes that are capable of converting procarcinogens into ultimate carcinogens.

these chemicals for carcinogenicity would require enormous expenditures and an army of pathologists to evaluate the histological slides. Development of the Ames test represents a major advance in our ability to prevent cancer by identifying cancer-causing chemicals.

The Genetic Basis of Neoplasia

With only a few exceptions, the neoplastic phenotype is heritable—that is, the daughter cells of a neoplastic cell also show abnormal growth regulation. As we have already seen, the principal initiating factors of the neoplastic transformation—gene mutations and chromosomal abnormalities, viruses, and environmental carcinogens–all act at the genetic level. This heritability of the neoplastic phenotype suggested to early investigators that genetic alterations play important roles in the establishment and progression of cancers. Currently, one of the most exciting areas of cancer biology is the study of the genetic basis of cancer. The sophisticated tools of virology, genetics, and molecular biology are enabling investigators to clarify the genetics and physiology of cellular growth regulation, resulting in a rapid expansion of our understanding of the details of both normal and neoplastic tissue growth. These studies have defined two categories of genes that are involved in the development of cancer: the *oncogene* and the *tumor-suppressor gene,* sometimes called the *anti-oncogene.*

Oncogenes

An **oncogene** is defined as a gene that contributes to the neoplastic transformation when it is introduced into a nonneoplastic cell. Two means of introduction have been widely used: infection with an oncogenic virus that carries an oncogene on the viral chromosome and the introduction of DNA isolated from a naturally occurring tumor into a nonneoplastic cell. The oncogene can be identified by determining the genetic element responsible for transformation by the oncogenic virus (Fig. 23-9) and by isolating and characterizing the tumor-derived DNA that transforms the transfected cells in the DNA transfection experiment

(Table 23-3). These two independent techniques have identified overlapping sets of genes that can operate as oncogenes (Table 23-4). One of the most unexpected findings of these studies, for which the Nobel Prize for Medicine was awarded to Michael Bishop and Harold Varmus in 1989, is that most oncogenes isolated from RNA tumor viruses and from spontaneous tumors are variants of genes normally present in the genome of the host animal. In the usual terminology, the overt oncogene is denoted by an italicized abbreviation, such as *src, myc,* or *ras,* and its normal cellular homologue is denoted by the same abbreviation with the prefix "c-"—c-*src,* c-*myc,* c-*ras.* Many of the normal cellular *homologues* of these oncogenes, also termed **proto-oncogenes,** are widely distributed in eukaryotes. Homologues of the *src* oncogene have been identified in many vertebrates and insects; homologues of the *ras* oncogene are present in humans and yeasts. The evolutionary conservation of the proto-oncogenes suggests that their gene products mediate important cellular processes.

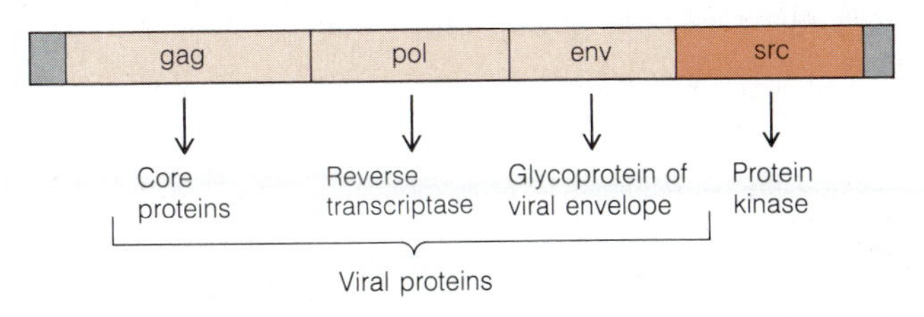

Figure 23-9 Diagram of the Chromosome of the Avian Sarcoma Virus, an Example of an Acutely Transforming Retrovirus. The wild-type progenitor of the avian sarcoma virus presumably contained in its genome only genes necessary for viral replication. These are *gag,* which codes for nuclear proteins of the virion; *pol,* which encodes reverse transcriptase; and *env,* which codes for proteins of the viral envelope. In addition, the virus contains a fourth gene, *src,* that is responsible for transformation of cells infected by the virus. All transformation-defective mutants of the avian sarcoma virus map to the *scr* locus. *Src* is the oncogene carried by avian sarcoma virus.

The presence of cellular homologues of overt oncogenes raises the question of the alterations that transform genes involved in the biology of normal, nonneoplastic cells into genes that cause loss of growth control and establishment of the neoplastic condition. Two basic categories of alteration have been postulated: *dosage effects* and *gene mutation.* A dosage effect is an overexpression of the gene such that the cell suffers excessive levels of the gene product. This effect does not require mutations that alter the basic functional characteristics of the gene product. The second type of alteration, gene mutation, is thought to result in the synthesis of a protein with altered function.

Certain tumors resulting from chromosomal translocation present examples of a dosage effect. In the human tumor *Burkitt's lymphoma,* a segment of chromosome 8 containing the cellular proto-oncogene c-*myc* is translocated to chromosome 2, 14, or 22. The sites of translocation are close to the locations of genes coding for subunits of the immunoglobulin proteins.* These immunoglobulin genes are transcriptionally active in the lymphocytic cell lineage: Approximately half of the total protein produced by the mature lymphocyte is immunoglobulin. It has been suggested that spatial proximity to a transcriptionally active immunoglobulin gene causes overexpression of the translocated c-*myc* gene, which, in turn, contributes to the establishment of Burkitt's lymphoma, a tumor of the lymphoblastic lineage.

A role for gene mutation has been demonstrated for the Ha-*ras* gene. Ha-*ras* was one of the first oncogenes identified by the cell transfection assay. The parent DNA

* The κ-light chain genes are on chromosome 2; the H-chain genes are on chromosome 14; and the λ-light chain genes are on chromosome 22.

Table 23-3 Identification of Oncogenes in Naturally Occurring Tumors by Gene Transfection

1. Transfection Step

   Nonneoplastic mouse fibroblasts of the NIH-3T3 cell line are presented with *Bam*1 endonuclease fragments of DNA isolated from cells obtained from a human bladder carcinoma under conditions in which the NIH-3T3 cells incorporate DNA fragments into their own chromosomes.

2. Transformation Step

   Transfected cells are grown in culture, and cell colonies with deviant growth characteristics are isolated and tested for neoplastic growth behavior by transplantation to mice. The conversion efficiency typically is low because DNA incorporation is random; most of the human DNA fragments contain genes that are irrelevant to the neoplastic transformation so the cells incorporating these DNA segments remain nonneoplastic.

3. Analysis Step

   The human DNA is isolated from clones of neoplastic NIH-3T3 cells and analyzed by molecular hybridization with probes generated from oncogenes isolated from retroviruses. The gene present in the human bladder carcinoma that was capable of transforming the NIH-3T3 cells was identified as Ha-*ras*.

was from a tissue culture cell line derived from a human bladder carcinoma. Under the experimental conditions used, normal c-Ha-*ras* was unable to produce the neoplastic transformation when transfected into the nonneoplastic host cells. The tumor cell-derived Ha-*ras* gene responsible for transformation had several point mutations along its length. To determine which of the point mutations of the oncogenic form of Ha-*ras* was responsible for its transforming activity, hybrid DNA molecules were prepared in vitro in which sections of the normal form of the Ha-*ras* proto-oncogene were replaced by corresponding sections from the tumor-derived homologue. It was found that transformation resulted from transfection with a hybrid molecule containing a short 5′ section of the tumor-derived Ha-*ras* that replaced the same 5′ segment of the normal homologue. The active 5′ segment of the tumor-derived gene contained only a single base-pair substitution differentiating it from the normal, nononcogenic form of the gene. The normal codon GGC, which codes for the amino acid glycine in position 12 of the Ha-*ras* protein, had been replaced by the codon GTC, which codes for valine. These observations offer dramatic proof that point mutations can, in selected cases, transform a normal cellular proto-oncogene into an overt oncogene. For further information on the *ras* oncogene, see the box on p. 732–733.

Tumor-Suppressor Genes

Oncogenes have the character of dominant cancer-causing agents. When oncogenes are present in an appropriately mutated or overexpressed form, transformation of the cell follows. In contrast to the dominant character of the oncogene, the neoplastic phenotype of spontaneous tumors is recessive. When neoplastic cells are fused in cell culture with nonneoplastic cells, the progeny, which contain the complete genotypes of both parental cells, are nonneoplastic. They show normal growth behavior in cell culture and, more important, fail to form tumors when introduced into appropriate host animals. In these characteristics, they resemble the nonneoplastic rather than the neoplastic par-

Table 23-4 Transforming Genes of Acutely Transforming Retroviruses

Gene/Gene Product	Prototype Virus	Isolation Source
Oncogenes That Code for Mitogens		
sis/p28^{sis}	Simian sarcoma virus	Wooley monkey
Oncogenes That Code for Protein Kinases		
src/p60^{src}*	Rous sarcoma virus	Chicken
fps†/p140*	Fujinami sarcoma virus	Chicken
fes†/p85*	Snyder-Theilen feline sarcoma virus	Cat
yes/p90*	Y73 sarcoma virus	Cat
erb-B/p65*	Avian erythroblastosis virus	Chicken
ros/p68*	UR2 sarcoma virus	Cat
fms/p140*	McDonough feline sarcoma virus	Cat
fgr/p70*	Gardner-Rasheed feline sarcoma virus	Cat
mos/p37‡	Moloney sarcoma virus	Mouse
raf/?‡	3611 murine sarcoma virus	Mouse
abl/p120*	Abelson leukemia virus	Mouse
Oncogenes That Code for *Ras* Proteins		
Ha-*ras*/p21-Ha-*ras*	Harvey sarcoma virus	Mouse
Ki-*ras*/p21-Ki-*ras*	Kirsten sarcoma virus	Mouse
Oncogenes That Code for Nuclear Proteins		
myc/p110	Myelocytomatosis virus	Mouse
myb/p48	Avian myeloblastosis virus	Mouse
fos/p55	FBJ osteosarcoma virus	Mouse

* Tyrosine protein kinase activity.
† Homologous genes.
‡ Serine, threonine protein kinase activity.

ent cell. After extended periods of growth in cell culture, variant clones of cells that have reverted to neoplasia frequently develop. These cells grow abnormally in cell culture and form growing tumors when transplanted into animals. In all cases, the reversion to neoplasia is accompanied by loss of particular chromosomes initially derived from the nonneoplastic parent cell. These observations imply that normal cells contain discrete genes that *suppress* unregulated cell growth. When present, these **tumor-suppressor genes** can slow the uncontrolled growth even of an overtly neoplastic cell. Once lost, however, the neoplastic cell reverts to uncontrolled growth behavior.

The best-characterized of the tumor-suppressor genes is the human gene *rbl*. Inactivation of this gene was first associated with the inherited tumor *bilateral retinoblastoma*. The tumor develops in young children from neuroblasts of the developing retina, and both retinas develop one or more metastatic nodules. The tumor is curable by early surgery, and many individuals who have had the disease have survived to maturity and have had children. The children of parents with retinoblastoma show a 50% incidence of the tumor, demonstrating that the tumor is heritable. Development of the tumor has been associated with chromosomal deletion at band 14 of the long arm of chromosome 13. It has been proposed that a gene, designated *rbl*, resides in this region of chromosome 13 and is necessary for the regulation of growth of the rapidly dividing neuroblasts in the retina. Half of the children with one parent who has experienced retinoblastoma would be expected to inherit one normal chromosome 13 from the disease-free parent and one chromosome 13 that lacks a functional allele of the *rbl* gene from the parent with retinoblastoma. During the many rounds of cell division in the tissue of the developing retina, occasional cells are produced in which the remaining normal chromosome 13 is lost or suffers a somatic deletion or mutation involving the *rbl* gene. This cell, which then lacks even one normal *rbl* gene, serves as the founder for a clone of neoplastic tissue that develops into the retinoblastoma tumor. In addition to the heritable, multifocal form of the tumor, instances of spontaneous retinoblastoma are also known. This spontaneous condition is monofocal and presumably results from the somatic mutation of both copies of the *rbl* gene within a single cell. This double mutation clearly would occur more rarely than the single somatic mutation required for inactivation of the second copy of the *rbl* locus in cells of individuals in whom all cells contained only one normal copy of the gene. This accounts for the rarity and the monofocal character of spontaneous retinoblastoma.

This hypothesis has recently received dramatic support with the identification, isolation, and characterization of the *rbl* gene. Cloned fragments of the normal chromosome 13 were tested for their ability to hybridize with fragments of chromosome 13 isolated from retinoblastoma cells. It was expected that the *rbl* gene would be present in DNA segments in the normal chromosome 13 but would be absent in the chromosome from the tumor. A candidate gene was identified that is expressed in normal retinoblasts but not in retinoblastoma cells. When the normal form of this gene was transfected into retinoblastoma cells, the cells lost the neoplastic phenotype and reverted to normal growth behavior. Thus, the *rbl* gene was identified and isolated. The gene has been sequenced and expressed to yield a *fusion protein* by transfecting the gene into the bacterium *E. coli* under conditions in which the bacterium synthesizes the protein product of the *rbl* gene. Immunocytochemical staining of normal retinoblasts with antibodies produced against the *rbl* fusion protein has shown that the *rbl* protein is localized in the cell's nucleus. The *rbl* protein contains stretches characteristic for DNA-binding proteins, suggesting that the *rbl* protein may function in the regulation of gene activity. The evidence all points to a necessary involvement of the *rbl* protein in cellular differentiation and/or growth regulation. When the protein is absent in the rapidly dividing retinoblasts of infants, the cells are unresponsive to normal regulatory influences and become neoplastic.

Tumor Dissemination

At the cellular level, the two distinguishing characteristics of a malignant neoplasm are *loss of positional regulation* and *loss of normal growth regulation*. We will begin by discussing positional regulation before going on to the regulation of tissue growth in the next section.

Tumor Invasion: Dissemination to Nearby Tissue

As mentioned previously, the anatomical relations of the tissues of the different organs of the body are highly stable—that is, cells remain within the confines of the parent tissues for the life of the organism. This organizational stability is disrupted by the onset of the invasive and metastatic behavior of cancerous tissue. *Invasion* is the intrusion of malignant tissue into neighboring normal tissues. Typically, the neoplastic tissue invades and replaces adjacent normal tissue, disrupting normal function of the afflicted organ (Fig. 23-10).

Several processes contribute to tumor invasion. For example, many tumors release a variety of degradative enzymes into the local environment. These may function to degrade the extracellular matrix of adjacent normal tissues, facilitating penetration by the neoplastic tissue. The regulation of motile behavior of invasive cells has also been shown to differ from that of noninvasive cells. Both normal and invasive tissue cells are capable of cellular migration in cell culture. When observed in monolayer cell culture, noninvasive cells typically demonstrate **contact paralysis of cellular locomotion,** in which the motile pseudopodia become immobilized when one cell comes into contact with another. Invasive tumor cells have been shown to lack contact paralysis following encounters with normal tissue cells in cell culture, allowing them to migrate freely over the surface of the contacted normal cell. It has been suggested that contact paralysis restrains the active migration of normal cells in coherent tissues and that its absence is a necessary condition for tumor invasion. Contact paralysis is apparently a consequence of a cell's ability to establish specialized adhesive contact junctions following cell-cell collision. Careful electron microscopic studies have determined that the adhesive junctions form within a few minutes after a collision between two noninvasive cells and fail to form when invasive cells collide with normal cells. It follows that the ability or inability to display contact paralysis is a function of the adhesive characteristics of the cells. The inability to display contact paralysis of locomotion may represent one way in which alterations in cell-cell adhesive interactions can initiate invasive behavior by tumor cells.

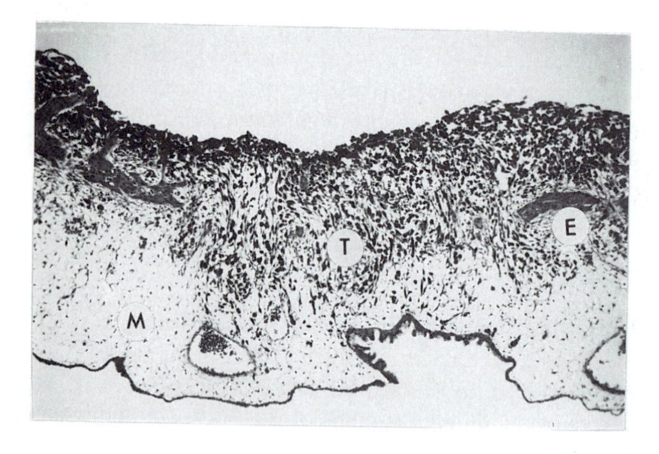

Figure 23-10 Tumor Invasion. To study tumor invasion, mouse tumor cells were transplanted to the surface of the chorioallantois, an extraembryonic organ of the chicken egg. After one day, the tumor cells (T) had begun to invade the tissues of the chorioallantois by migrating across the epithelium (E) at the surface and penetrating the underlying mesenchyme (M). Experimental systems such as this have allowed investigators to study the process of invasion in a controlled laboratory setting and to identify the underlying mechanisms of tumor invasion. (LM)

Metastasis: Dissemination to Distant Organs

As we have seen, tumor invasiveness accomplishes local dissemination of the malignant tissue from the parent tumor mass. *Metastasis,* on the other hand, serves to disseminate malignant tissue to distant sites in the body by the transport of clumps of tumor cells in the fluids of the body cavities and vasculature (Fig. 23-11). Metastatic transport can occur within the peritoneal cavity, the neural canal, the lymphatic system, or the vascular system. The last route is the most important for the dissemination of solid tumors in humans and will therefore be the basis for the discussion that follows.

Vascular metastasis begins with the establishment of a vascular supply for the solid tumor. The endothelial cells that line blood vessels actually invade the tumor from vascular capillaries of adjacent normal tissue. The tumor cells then gain access to the vascular lumen by invasion across the endothelium of these vessels. Small clumps of tumor cells that protrude into the vascular lumen dislodge from the main mass and are swept away in the blood. In advanced cancers, many thousands of such cell clumps may be shed every day. Although most of these cells die, the successful clumps must lodge in the capillaries of an appropriate organ and then invade across the capillary endothelium to begin growth in the parenchyma of the new host organ.

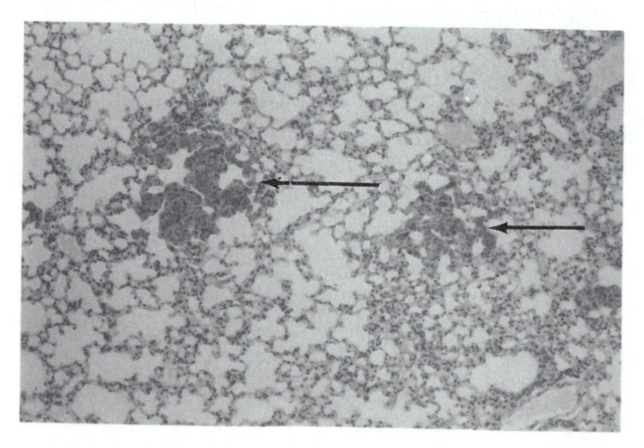

Figure 23-11 Metastasis. In this experimental investigation of metastasis, cells of a mouse melanoma were grafted beneath the skin of an isogeneic newborn mouse. The tumor grew at the site of transplantation and, within 1 month, had produced numerous metastases in organs distant from the parent tumor, including the lung. The melanoma metastases (arrows) are easily distinguished from healthy lung tissue by the pigment granules in the tumor cells. Invasion causes the local spread of cancer, and metastasis is responsible for the spread of tumors to distant sites. (LM)

As we have just described, metastasis is a complex process of vascularization, invasion, vascular transport, arrest, and invasion. The process of vascularization of solid tumors, termed **tumor angiogenesis,** is essential not only to metastasis but also to malignant progression and growth. In the absence of a vascular supply, solid tumors typically remain small and confined. The typical avascular tumor consists of an actively growing cortex at the surface and dead (necrotic) tissue at the core. The progression from the neoplastic but nonmalignant state of the tumor in situ to the progressively growing state of the malignant tumor requires a vascular bed to supply the interior of the tumor mass with nutrients and oxygen and to remove metabolic wastes. The initiation of tumor vascularization has been shown to depend on the release of diffusible factors from the tumor that stimulate the directed migration of the capillary endothelium from the surrounding normal tissues into the tumor. In this situation, the tumor experiences invasion by capillaries from adjacent tissues.

One of the experimental systems that has proved useful in the analysis of tumor angiogenesis involves implantation of small pieces of tumor tissue in the avascular connective tissue of the cornea of an experimental animal. Following implantation, the tumor establishes itself in the cornea as a tumor in situ—a nonvascularized nodule that remains small with an actively dividing surface layer and a necrotic interior. If placed within a few millimeters of the edge of the cornea, the tumor induces the migration of capillary sprouts from the vascularized tissue peripheral to the cornea. These capillary sprouts progressively invade the connective tissue of the cornea and finally contact and vascularize the tumor. Induction of corneal vascularization can also be produced by cell-free extracts of the tumor, suggesting that chemical factors are released by neoplastic tissue that attract migrating vascular endothelial cells. Several angiogenic proteins have recently been identified and characterized, including *angiogenin, basic fibroblast growth factor,* and *tumor necrosis factor-α.* The latter is interesting because it is produced not by the tumor cells themselves, but by host *macrophages* that have colonized the tumor. **Macrophages** are scavenging cells derived from a class of circulating blood cells known as **monocytes.** Most solid tumors contain significant populations of macrophages. Tumor necrosis factor-α is the principal secreted protein that enables the macrophages to kill tumor cells. The factor thus has multiple effects, including destruction of tumor cells by its direct tumoricidal activity and promotion of tumor growth by its ability to attract blood vessels into the tumor.

The consequence of the establishment of a vascular supply by a previously avascular tumor in situ is a spectacular increase in the rate of tumor enlargement. In the corneal implantation system, the tumor implant rapidly switches from the dormant state to one of rapid enlargement. This change in the tumor's rate of enlargement indicates that the ability to initiate angiogenesis can play an important role in tumor progression—the avascular solid tumor is generally benign, whereas the vascularized solid tumor is usually overtly malignant. This phenomenon suggests that inhibition of vascularization might produce a reversal of the progressive enlargement of solid tumors. Recent observations from the laboratory of Judah Folkman support this possibility. Folkman and colleagues have shown that combinations of steroidal anti-inflammatory agents and the polysaccharide heparin inhibit and reverse the vascularization of tumors in experimental animals and produce a dramatic reversal of tumor growth.

The Regulation of Tissue Growth

The stability of anatomical form is for many tissues a dynamic situation. This is best illustrated by the regulation of tissue volume. Constancy of tissue volume means that the rates at which cells are added by mitosis are equal to the rates at which cells are lost by cell death or emigration. In many organs, the rates of cell division can increase when injury occurs or when an appropriate physiological demand is made on the system. In rodents, for example, when a portion of the liver is surgically removed, the rate of cell division in the remaining liver tissue increases until the original volume is restored. During embryonic and larval development, and in normal cellular turnover and wound repair in the adult, the growth of the tissues is strictly regulated. The volumes of the various tissues are proportional to the size of the organism. Neoplasia results from the establishment of a tumor mass in which this strict regulation of tissue volume is no longer operative. The basic question posed by neoplastic growth is how tissue growth is controlled in the normal state and how this control is lost in the neoplastic state.

Mitogens: Chemical Regulators of Tissue Growth

The most useful system for the experimental study of the regulation of tissue growth has been that in which the mitotic behavior of both normal and cancerous cells is examined in *monolayer* tissue culture. Cells are placed in culture dishes and provided with nutrient culture medium.

On the Trail of the ras *Oncogene*

Although they appeared on the molecular scene only recently, oncogenes already have many investigators on their trail, and intensive molecular analysis is well underway. The research is spurred on by the many significant findings that have already been made and by the obvious implications of these results for our understanding of cancer and of normal growth and differentiation. A particularly intriguing trail is that of the *ras* gene family, so let's follow it to see where it leads.

Our trail begins with the RNA tumor viruses, because the *ras* gene was originally identified in two different retroviruses of the rat, the *Harvey sarcoma virus* and the *Kirsten sarcoma virus*. The *ras* genes of these two viruses are referred to as Ha-*ras* and Ki-*ras* (Table 23-4). *Ras* proto-oncogenes have since been discovered in a variety of species, including humans, rodents, fruit flies, and yeast. Like other proto-oncogenes, these *ras* genes are thought to be required in normal cells for the regulation of growth and development.

In mammalian cells, the *ras* gene codes for a protein with a molecular mass of approximately 21,000 that seems to be associated with the inner surface of the plasma membrane. The protein is known to bind guanosine triphosphate (GTP) and bears a close resemblance to *G* (for GTP-binding) *proteins* that regulate the activities of a number of important enzymes, including phospholipase C and adenylate cyclase, the enzyme responsible for the formation of cyclic AMP from ATP.

An important development in our understanding of oncogenes came when the *ras* oncogene trail led to human cancer. DNA isolated from two different human bladder carcinomas was shown to induce transformation in the NIH 3T3 line of mouse fibroblast cells. Transforming activity was traced to a single sequence that in both cases turned out to correspond to the *ras* genes of the rat sarcoma viruses. Since then, *ras* oncogenes have been detected in other human carcinomas, including lung and colon tumors.

Based on prior findings with the *src* gene and other viral oncogenes, it was initially thought that the *ras* gene probably acts by increasing the cellular level of its product in transformed cells. However, no such increase was found in NIH 3T3 cells following transformation by DNA from the human bladder carcinomas. In other words, transformation did not depend critically on the amount of *ras* gene product made by the cells.

How does the *ras* oncogene act, if not by elevating the level of its product? When the *ras* genes from the bladder carcinomas were analyzed further in the laboratories of Robert Weinberg and Mariano Barbacid, the trail led to a gene product with an altered amino acid sequence. DNA from both carcinoma and normal tissue of the same patient was digested with restriction enzymes, and the fragments containing the *ras* genes (along with flanking sequences) were sequenced. The sequences were identical except for a single position, where the normal DNA fragment (the proto-oncogene) had a G and the DNA fragments from the carcinomas (the oncogenes) had a T. This is, as Dr. Barbacid noted, "the first time that a mutated gene has been found in a primary tumor and not in a normal tissue from the same patient".

This substitution affects the twelfth amino acid from the N-terminus of the protein, which would be glycine in the normal cell and valine in the carcinoma. That difference is probably significant, because when the Ha-*ras* and Ki-*ras* oncogenes of the sarcoma viruses were sequenced, the same codon was affected. One of the viruses has an arginine in place of the glycine, and the other has a serine. It appears, therefore, that the glycine in position 12 is critical to the proper functioning of the normal protein and that any of several amino acid substitutions at this position is sufficiently disruptive of cellular function to induce malignancy.

Still more recently, the trail of the *ras* oncogene has broadened as a number of related *ras* genes have been sequenced, including those from lung and colon carcinomas and from a neuroblastoma, a tumor of embryonic nerve cells. In each case, the transforming activity could be attributed to small, specific changes in the nucleotide sequence. It appears that a *ras* gene can be activated by structural alterations in at least five sites. The sites are clustered in two regions of the gene, affecting the codons for amino acids 12 and 13 and for amino acids 59 to 63 of the protein.

The explanation of oncogene activation that emerges from these studies depends not on the overproduction of a gene product, but on small but apparently very important changes in the nucleotide sequence of the gene and hence in the amino acid sequence of the protein. These differences have been shown for tumor tissue and normal tissue from the same patient, strongly suggesting a somatic mutation as the cause of the tumor.

The implications of this mode of oncogene activation are clear, because most human cancers are thought to be caused by chemical carcinogens, all of which are known to be directly or indirectly mutagenic. In fact, the *ras* trail seems now to be leading to chemical carcinogenesis, if recent results from Barbacid's laboratory are any indication. A single dose of the carcinogen nitrosomethylurea administered to eight-week-old rats was enough to cause mammary tumors in almost all the animals within 6 to 12 months. In each case, the cancerous tissue contained a transformed *ras* gene, and the active oncogene differed from its normal counterpart by a single nucleotide, again in codon 12!

One of the most exciting advances in our understanding of *ras* gene function came in 1984 as a result of the discovery that *ras* genes are also present in yeast. Identification of oncogenes in yeast makes it possible to study these genes using biochemical and genetic methods that simply are not possible in higher organisms. An example of a manipulation that is possible with yeast cells is the replacement of specific genes with altered forms. By replacing one or both of the two *ras* genes in the yeast genome with nonfunctional genes, Michael Wigler and his colleagues were able to show that the yeast cell has an absolute requirement for at least one intact *ras* gene for spore germination. This finding strengthens the hypothesis that proto-oncogenes play an indispensable role in regulating cell division or differentiation.

Wigler and his colleagues also showed that gene function can be restored by substituting a human *ras* gene for an inactivated yeast gene, even though the proteins encoded by the *ras* genes are significantly larger in yeast cells than in human cells. They were then able to construct a mutant *ras* gene that mimicked one of the alterations known to activate the transforming potential of mammalian *ras* genes. When the mutant gene was inserted into yeast cells, the cells lost their ability to form spores, suggesting that the transformation of mammalian cells and the inability of yeast cells to form spores may have an explanation in common.

Clues to that explanation came when it was found that the mutant *ras* protein continuously activates adenylate cyclase, an enzyme important for cell regulation because of its role in the formation of cyclic AMP. Further support for an effect of the *ras* gene product on adenylate cyclase comes from similarities between the *ras* proteins and the two G proteins that are known to occur with adenylate cyclase on the inner surface of the plasma membrane. The G proteins are regulatory proteins that respond to signals generated by the binding of specific ligands to receptors on the cell surface. In many types of cells, the G proteins transduce such signals into changes in intracellular cAMP levels by stimulating or inhibiting the activity of adenylate cyclase.

Properties that *ras* proteins and G proteins share in common include their location on the inner surface of the plasma membrane, their affinity for GTP, and their ability to split GTP into GDP and inorganic phosphate. Moreover, the *ras* proteins and G proteins have amino acid sequences that are highly homologous, including the presence of glycine in position 12. This evidence suggests a connection between the *ras* proteins and adenylate cyclase regulation that is still somewhat tenuous, but already qualifies as another milestone of cancer research because it links an oncogene to a major regulatory mechanism of eukaryotic cells.

Clearly, much remains to be understood about the *ras* oncogene family. But the trail of discoveries that began with the oncogenes of two rat sarcoma viruses has already led to several human cancers, to some highly specific mutations in the corresponding human genes, and now to adenylate cyclase and membrane signal transduction as well. It seems a promising trail indeed, and it illustrates how the initially separate paths of oncogenes and cellular biology seem to be merging in ways that are almost certain to be as significant as they were unexpected.

Many different cells have been studied, including cells recently isolated from embryonic tissues, cultured cell lines (cells that have been maintained in culture for extended periods and that are specifically adapted to cell culture), and cells that have been transformed by infection with an oncogenic virus or by exposure to a chemical carcinogen.

One of the most important generalizations to emerge from the study of growth regulation in cell culture is that the progressive growth of nonneoplastic cells is dependent on a supply of specific growth-stimulatory proteins, known as **mitogens** (Table 23-5). In the usual culture situation, the relevant mitogens are supplied by the inclusion of **serum** in the culture medium. (Serum is the acellular fluid phase of the blood after the blood has been allowed to clot.) During clotting, the blood platelets release a variety of effector molecules into the serum, and certain of these molecules have mitogenic activity. One example is the protein **platelet-derived growth factor (PDGF)**, which is the most significant mitogen in serum for mesenchymal tissues. Another mitogen is the protein **epidermal growth factor (EGF)**. This protein, which is widely distributed in embryonic tissues, was initially isolated from the salivary glands of mice by Stanley Cohen, who received a Nobel Prize in 1987 for his pioneering investigations of mitogens.

A number of peptide mitogens have been purified from serum and other sources. The usual bioassay for the presence of a mitogen in a sample is to demonstrate the ability of the sample to stimulate cell division in a mitotically quiescent population of cultured cells. Certain mitogens, such as EGF, have broad spectra of activity, stimulating the growth of a wide variety of cell types. Other mitogens are more selective. Mitogens probably function in the stimulation of tissue growth during embryonic and larval development and the reparative growth that contributes to wound repair and regeneration in the adult. For example, release of the mitogen PDGF from blood platelets almost certainly plays a role in the reparative growth of connective tissue cells of injured blood vessels. Platelet aggregation and exocytotic release of PDGF and other factors take place in an early response to bleeding. The release of PDGF

Table 23-5 Mitogens

Growth Factor	Size in kDa (Structure)	Target Cells	Receptor
Epidermal growth factor (EGF)	6 (monomer)	Wide variety of epithelial and mesenchymal cells	c-*erbB* gene product; 170 kDa tyrosine kinase
Transforming growth factor-α (TGF-α)	5.6 (monomer)	Same as EGF	Same as EGF
Platelet-derived growth factor (PDGF)	32 (16 kDa B chain + 14–18 kDa A chain)	Mesenchyme, smooth muscle, trophoblast	185 kDa tyrosine kinase
Transforming growth factor-β (TGFβ)	25 (homodimer)	Fibroblastic cells	565–615 kDa complex: 2 × 280–290 kDa
Insulin-like growth factor-I (IGF-I)	7 (monomer)	Epithelium, mesenchyme	450 kDa complex
Insulin-like growth factor II (IGF-II)	7 (monomer)	Epithelium, mesenchyme	Single 260 kDa chain
Interleukin-2 (IL-2)	15 (monomer)	Cytotoxic T lymphocytes	55 kDa glycoprotein
Basic fibroblast growth factor (b-FGF)	14–16 (monomer)	Endothelium, mesenchyme	
β-Nerve growth factor (β-NGF)	26 (homodimer)	Neural crest derivatives	130 kDa
Colony stimulating factor-1 (CSF-1)	70 (homodimer)	Macrophage precursors	c-*fms* gene product; 170 kDa tyrosine kinase
Colony stimulating factor-2 (CSF-2)	15–28 (monomer)	Macrophage and granulocyte progenitors	
Interleukin-3 (IL-3)	28 (monomer)	Eosinophil, mast cell, granulocyte, macrophage progenitors	
Bombesin	1.6	Mesenchyme	

from platelets at sites of vascular wounding places it at the appropriate location to stimulate reparative growth of the connective tissue of the blood vessel wall.

The Mitogen Receptor

The interaction of peptide mitogens with target cells is initiated by the binding of the mitogen from solution to specific **mitogen receptors** at the cell surface. The existence of mitogen receptors can be demonstrated by the high affinity and saturable binding of radiolabeled mitogens to living cells (Fig. 23-12). Various mitogen receptors have been identified and characterized. Mitogen-receptor binding is selective: Each receptor binds one or only a few related mitogens. Specificity is determined by competition assays: The binding of the radiolabeled mitogen is carried out in the presence of a 10- to 100-fold excess of unlabeled mitogen. If the labeled and unlabeled mitogens are the same, the unlabeled mitogen molecules will occupy most of the receptor sites, thereby reducing the extent of binding of the radiolabeled molecules and the measured radioactivity bound to the cells. Thus, for example, excess unlabeled PDGF will reduce the extent of binding to the cell of radiolabeled PDGF. When one tests instead high concentrations of other unlabeled mitogens, such as EGF, the binding of radiolabeled PDGF is unaffected. From this we conclude that binding to the PDGF receptor is specific: The PDGF receptor binds PDGF but fails to interact with other mitogens. In this way, the mitogen's ability to compete for a given receptor allows the binding specificity of the different mitogen receptors to be characterized experimentally. The cellular specificity of the response to mitogens is mirrored by the patterns of expression of mitogen receptors: A cell type must express the receptor for a given mitogen for the cell to mount a growth response in the presence of that mitogen. Certain of the mitogen receptor proteins have been isolated and characterized. They are transmembrane proteins, with an extracellular portion containing the mitogen binding site, a transmembrane domain that traverses the plasma membrane, and a cytoplasmic region that may include a domain with enzymatic function catalyzing the transfer of the γ-phosphate of ATP to tyrosine groups of target proteins (**tyrosine protein kinase** enzymatic activity).

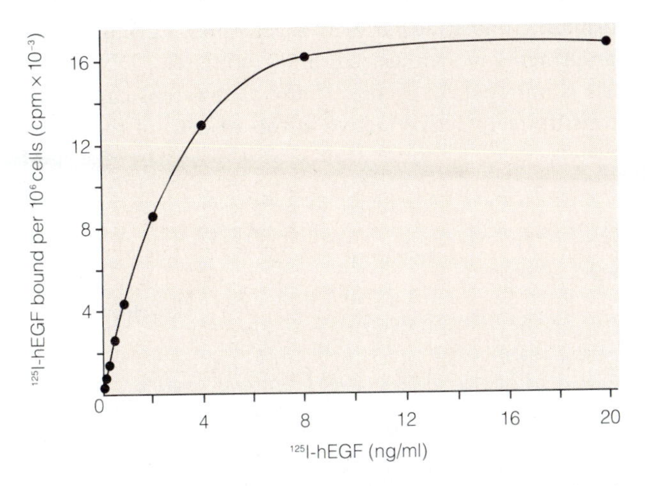

Figure 23-12 Demonstration of the Mitogen Receptor. The involvement of specific cell surface receptors in the interaction of soluble mitogens with target cells is shown by binding studies in which the purified mitogen molecule is tagged with a radioactive isotope of iodine, ^{125}I, and is then presented to cultures of cells. In this case, the binding of human epidermal growth factor (hEGF) to human fibroblasts has been detérmined. After a suitable incubation period, the unbound molecules are removed by careful washing, and the amount that is associated with the cells is estimated by determining the amount of cell-associated ^{125}I. A binding curve is produced by determining the amounts of cell-bound ^{125}I for different doses of ^{125}I-labeled mitogen. When the specific binding curve is known, the number of mitogen receptors per cell and the strength of binding, expressed as the dissociation constant or K_D for the binding, can be calculated. The K_D is the concentration of mitogen in the bathing medium at which half of the cell surface receptors have bound mitogen molecules. The lower the K_D, the stronger the binding. Typical values are 40,000–400,000 receptors per cell and $K_D = 1\text{–}10 \times 10^{-10} M$. The specificity of binding can be determined by comparing the abilities of different mitogens to displace the iodinated mitogen. If binding of iodinated molecules of one mitogen is significantly reduced by the presence of unlabeled molecules of the other mitogen, it can be concluded that both mitogens utilize the same receptors. If binding is unaffected, it can be concluded that the two mitogens utilize different and independent receptor systems.

Mitogen-Response Pathway

The binding of the mitogen to its receptor initiates a chain of events in the cytoplasm that culminates in the entry of the cell into the S phase of the cell cycle. Mitogen binding initiates expression of one or both of two distinct cytoplasmic signaling pathways. (1) Mitogen binding may activate the tyrosine-specific protein kinase enzymatic activity of the cytoplasmic domain of the mitogen receptor, leading to the phosphorylation of tyrosine residues of a variety of protein targets. (2) In addition, binding may activate the enzyme **phospholipase C,** which catalyzes the hydrolysis of phosphatidylinositol bisphosphate into inositol trisphosphate and diacylglycerol (Fig. 23-13). In the absence of bound mitogen, the tyrosine-specific protein kinase activity of the mitogen receptor is silent; binding of a mitogen molecule to the extracellular domain of the receptor activates the enzyme domain of the cytoplasmic end of the mitogen receptor. The consequence is phosphorylation of a number of target proteins in the cytoplasm. Activation of the phosphatidylinositol signaling pathway involves a class of proteins, the G proteins, that are associated with the mitogen receptors and with phospholipase C at the inner face of the plasma membrane (see Chapter 21). The binding of mitogen stimulates the G protein to bind one molecule of GTP and, as a result, to adopt its active configuration. This in turn allows the G protein to activate phospholipase C. Although some mitogens may activate both signaling pathways, others may activate only one or the other. The Chinese hamster fibroblast cell shows a mitogenic response to EGF, thrombin, and bombesin. Binding of EGF activates the tyrosine-specific protein kinase of the cytoplasmic domain of the EGF receptor, whereas thrombin and bombesin activate phospholipase C and the phosphatidylinositol bisphosphate metabolic pathway. Inhibitors of the latter pathway, such as pertussis toxin, have no effect on the mitogenic response of hamster fibroblasts to EGF, but block completely the response to thrombin and bombesin. Thus it appears that activation of the phosphatidylinositol bisphosphate metabolic pathway is essential for the response to bombesin and thrombin but is not essential when EGF is the stimulator of growth.

Although the individual steps further along on the response pathway have not been defined, an important event in the pathway appears to be the synthesis of a group of nuclear proteins. The mRNAs for these proteins appear between 0.5 and 3 hours after stimulation by mitogen.

Dysfunction of the Mitogen-Response Pathway in Neoplasia

One of the most important generalizations to emerge from recent investigations of the unregulated growth of neoplastic tissue is the hypothesis that this growth is a result of a dysfunctional expression of the mitogenic pathway. As indicated above, the mitogen-response pathway contains multiple steps: The mitogen must be present in the external environment; the mitogen must bind to specific high-affinity cell surface receptors; the tyrosine protein kinase of the cytoplasmic domain of the receptor and/or the enzyme phospholipase C become activated in response to ligand binding; target proteins are phosphorylated; the phosphatidylinositide pathway is activated; a group of nuclear proteins is synthesized; and, ultimately, DNA synthesis is initiated. In theory, if any of the steps cited here is constitutively active (active in the absence of the mito-

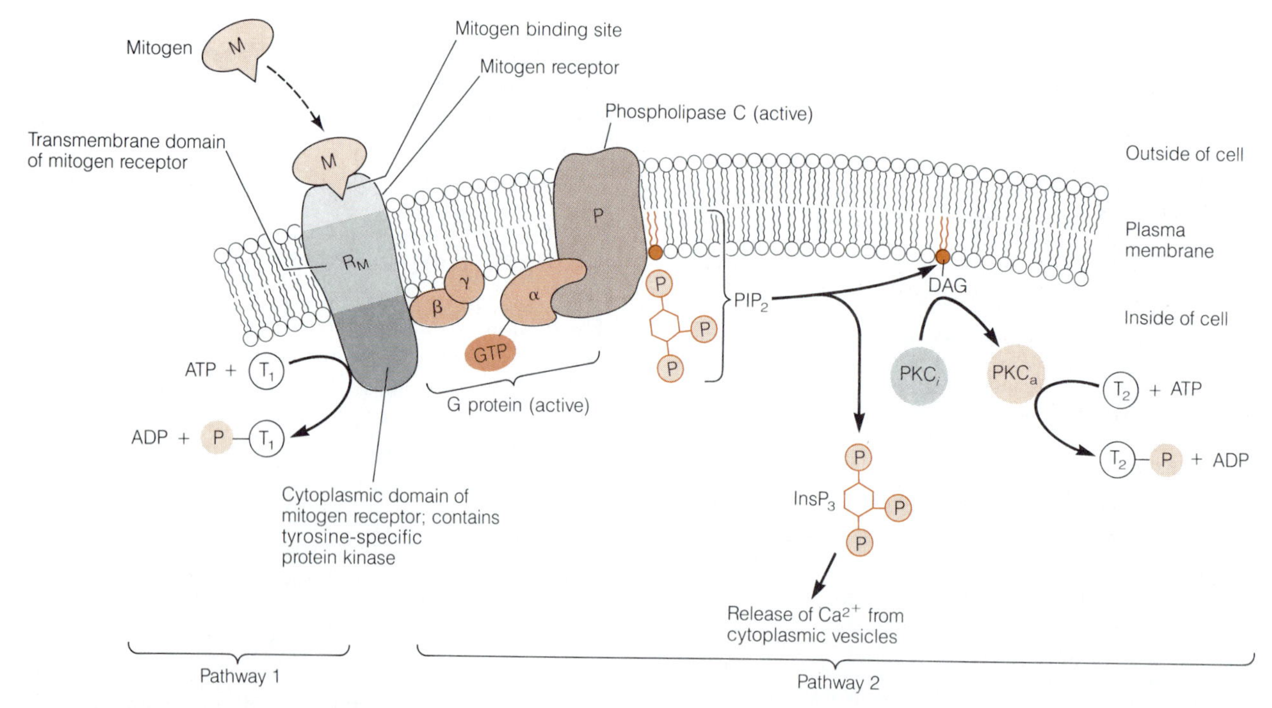

Figure 23-13 Binding of Mitogen Activates Two Cytoplasmic Signaling Pathways. A peptide mitogen (M) binds to its receptor (R), which is a protein embedded in the plasma membrane, and initiates one or both of two intracellular response pathways. In pathway 1, the binding of the mitogen activates the tyrosine-specific protein kinase located in the cytoplasmic domain of the mitogen receptor. This kinase can then catalyze the phosphorylation of target proteins (T_1). In pathway 2, the binding of mitogen to its receptor activates a G protein (G; see Figure 21-9), which in turn converts the inactive form of the membrane enzyme phospholipase C (PLC) to the activated form. The PLC enzyme then breaks down phosphatidylinositol bisphosphate (PIP_2) into inositol triphosphate ($InsP_3$) and diacylglycerol (DAG). These two products act as second messengers in the cytoplasm, stimulating the release of Ca^{2+} from cytoplasmic storage vesicles and activating protein kinase C (PKC). The active form of PKC (PKC_a), which is dependent on Ca^{2+}, then catalyzes the phosphorylation of its target proteins (T_2).

gen), the cell can be expected to divide continuously and independently of an external supply of the mitogen. In fact, examples have been described in which unregulated cell division is caused by constitutive expression of each of these steps of the mitogenic cascade. Furthermore, oncogenes are intimately involved—the gene products of a number of oncogenes are elements in the mitogen-response pathway. These findings illustrate a dramatic convergence of the molecular genetic and cell biological analyses of neoplasia that represents one of the outstanding successes of the last ten years of biological research.

One way in which a defective expression of the mitogen-response pathway could produce continual cycling of cells under conditions in which nonneoplastic cells are quiescent is for the neoplastic cells to produce their own supply of mitogen. This is the **autocrine secretion mode** of neoplastic growth. While the nonneoplastic cell requires an external supply of mitogen and will fail to divide in its absence, the cell that can produce and then respond to a growth stimulator will divide even in the absence of that external supply of the mitogen. An example of autocrine growth promotion is the NIH-3T3 fibroblast cell line transformed by infection with the simian sarcoma virus (the SSV-3T3 cell). The 3T3 cell is an example of a nonneoplastic cultured cell line. In culture, it shows regulated growth that is lost following transformation by SSV. Evidence that the SSV-3T3 cell produces autocrine growth factors includes the observation that virus-free culture medium that has been exposed to SSV-3T3 cells contains mitogenic activity and stimulates cell division when presented to quiescent populations of 3T3 cells. This activity is embodied in a molecule that is similar to the mitogen PDGF: SSV-3T3-conditioned medium eliminates binding of radiolabeled authentic PDGF to 3T3 cells, and antibodies to PDGF abolish the growth-stimulatory activity of SSV-3T3-conditioned medium. The precise nature of the relation of the growth-stimulatory activity of SSV-3T3-conditioned medium to PDGF became clear with the analysis of *sis*, the oncogene of the simian sarcoma virus. This gene was isolated and sequenced, and the predicted amino acid sequence was derived. This sequence turned out to be nearly identical to the peptide sequence of the B chain of PDGF. Thus, it has been concluded that introduction of the *sis* oncogene into a cell infected with the SSV retrovirus induces the synthesis of an active form of PDGF. The PDGF interacts with the infected cell to stimulate growth, in contrast to the normal situation, in which mesenchymal cells only are exposed to PDGF if blood platelets in their immediate vicinity release PDGF into the environment.

A second way in which the mitogen-response pathway could become constitutively active would be for a mitogen receptor to be altered such that the enzymatic activity of the C-terminal tyrosine protein kinase domain would be active even in the absence of mitogen binding. An apparent example of this situation is a cell transformed by the addition of the *erb-B* oncogene. This oncogene is present in the *erythroblastosis virus,* a retrovirus of avian origin that produces tumors of the erythroblastic lineage. When the *erb-B* oncogene was sequenced, it was found that the derived protein sequence was related to that of the EGF receptor. The *erb-B* protein included a modified version of the cytoplasmic and transmembrane domains of the EGF receptor, but lacked almost all of the extracellular domain. It has been suggested that the *erb-B* protein is a constitutively active form of the EGF receptor, which conducts the protein phosphorylation reaction independently of EGF. The normal form of the EGF receptor, of course, shows enzymatic activity only when EGF is bound to the appropriate site in the extracellular portion of the receptor protein.

Tyrosine-Specific Protein Kinases

The *erb-B* protein is one of a class of oncogene proteins that show tyrosine-specific protein kinase activity (Table 23-4). One of the important areas of investigation is the detailed elucidation of the mode of action of these proteins in promoting neoplastic growth. Of specific interest is the discovery and characterization of the target proteins for phosphorylation, as well as the elucidation of the way in which tyrosine phosphorylation affects function of the target proteins. Several targets of phosphorylation have been discovered, but the second goal has remained elusive. Detailed clarification of the mechanism of action of the protein kinase oncogenes has been complicated by the multiplicity of target proteins. Presumably, several of these may be merely adventitious targets, whose phosphorylation has no involvement in the regulation of growth. Discovering which targets matter and which do not has been difficult.

As we have discussed, there are two possible primary cytoplasmic responses to mitogen binding: stimulation of tyrosine-specific protein kinase activity and stimulation of phospholipase C, which then catalyzes the breakdown of phosphatidylinositol bisphosphate. The cellular response to the altered protein of the *erb-B* oncogene is an example of the consequences of constitutive expression of the protein kinase leg of the signaling pathway. An example of dysfunctional expression of the second signaling pathway is seen as a result of the introduction of one of the *ras* oncogenes into the cell. The gene products of the different *ras* oncogenes (Ki-*ras*, Ha-*ras*, and N-*ras*) all show sequence similarity to the G proteins normally involved in regulation of phospholipase C activity. Normal G proteins remain active only for seconds or minutes before they are inactivated by hydrolysis of the bound GTP to GDP. As we have already seen, the active forms of the *ras* oncogenes may differ by only a single amino acid replacement from the proto-oncogene. It now appears that these mutations produce a form of the *ras* protein in which the bound GTP is protected from hydrolysis. As a result, these forms of the *ras* proteins are continuously active. The expected consequence is continuous activity of phospholipase C and continuous breakdown of phosphatidylinositol bisphosphate.

Although full knowledge of the later steps in the mitogenic response pathway is not yet in hand, the induction of the synthesis of the inducible nuclear proteins appears to be important. These proteins normally appear soon after stimulation of cells by EGF and PDGF. The nuclear proteins are represented by the oncogenes *myc, fos,* and *myb,* suggesting that unregulated expression of one or another of these nuclear proteins produces neoplastic growth. Unregulated expression of the proto-oncogene c-*myc* can have a similar effect. When 3T3 cells are transfected with c-*myc* under the control of a strong promoter, they become independent of PDGF for growth. Investigation of the function of these nuclear proteins is currently an active area of research.

Growth-Inhibitory Proteins

The mitogen-response pathway is involved in the stimulation of cell proliferation. In addition to growth stimulators, a number of growth suppressors have been identified. One example is a protein called transforming growth factor-β (TGF-β). This is a 25-kDa homodimer with both growth-stimulating and growth-inhibiting properties. Its antiproliferative actions include antagonism of the mitogenic effects of peptide mitogens, such as EGF and PDGF. Interestingly, cell binding of TGF-β involves a family of three different cell surface receptors, all three of which are missing in retinoblastoma cells. Clearly, the gene product of the *rbl* gene is a nuclear protein, rather than a cell surface TGF-β receptor, but it is possible that the loss of the receptors may be important in the establishment of the neoplastic phenotype of this particular tumor.

Perspective

In summary, the study of the protein products of oncogenes has clarified our understanding of normal growth regulation, and the understanding of normal growth has informed the investigation of neoplastic growth. A basic lesson has been that any of a variety of possible perturbations that make the mitogen-response pathway constitutively active can produce the same basic phenotype, neoplastic growth. The constitutive activation of the mitogenic cascade can be at the level of the mitogen itself (autocrine stimulation), the mitogen receptor, or post-receptor steps (activation of the relevant G proteins, cytoplasmic protein kinases, or the relevant nuclear proteins).

It is important to realize that the abnormal biology of cancerous tissue is largely a consequence of an altered expression of normal cellular processes, rather than the expression of novel processes. Mitogens are important regulators of normal tissue growth during embryonic and larval development and in situations of wound repair. It is only when the mitogen response pathway is activated in tissues in which it is normally silent that unregulated cell proliferation becomes a problem. Certain normal tissues are invasive and even metastatic, especially during embryonic development. However, in the adult, these tissues remain in stable anatomical relations. The reinitiation of invasive and metastatic behavior in the adult is pathological. Angiogenesis is a normal event of embryonic development and of wound healing. Tumor angiogenesis is nothing more than the angiogenic response in an abnormal context. The abnormal behavior of the cancerous tissue has resulted from an exaggerated, unrestrained, or misplaced expression of a family of normal processes. It has been the purpose of this chapter to identify these processes and to show how cancer results from their dysfunctional expression.

Finally, a lesson of this chapter is the utility of the study of pathological processes to learn more about normal processes. Understanding the biology of cancerous tissue has provided valuable insights into the biology of normal tissues, since both classes of tissues operate with the same processes. The oncogenes and tumor-suppressor genes were first discovered in the context of neoplasia. Now we know that these genes play essential roles in normal tissues. Most of the early interest in angiogenesis derived from its importance to the growth of solid tumors, but recent interest has shifted to the role of angiogenesis in normal development. The recent history of the study of cancer has exemplified the power of parallel study of the biology of normal and pathological systems. The insights from study of the one inform our understanding of the other. Cancer has served as an experiment of nature whose investigation has significantly strengthened our understanding of the important problems of growth control and tissue organizational regulation.

Key Terms for Self-Testing

Cancer: A Loss of Normal Growth Regulation
tumor (p. 718)
neoplasm (p. 718)
benign tumor (p. 718)
malignant tumor (p. 718)
cancer (p. 718)
neoplastic transformation (p. 718)
invasion (p. 718)
metastasis (p. 718)
stem cells (p. 719)
neuroblasts (p. 719)
myoblasts (p. 719)
fibroblasts (p. 719)
neuroblastomas (p. 719)
myelomas (p. 719)
carcinoma (p. 719)
sarcoma (p. 719)
lymphomas (p. 719)
leukemias (p. 719)
myelomas (p. 719)
erythroblastomas (p. 719)

Causes of Neoplastic Transformation
chromosomal alterations (p. 719)
tumor progression (p. 719)
primary tumor (p. 719)
RNA tumor virus (p. 720)
retrovirus (p. 720)
provirus (p. 720)
reverse transcriptase (p. 720)
horizontal transmission (p. 720)
vertical transmission (p. 720)
environmental carcinogens (p. 722)
chemical carcinogens (p. 722)
oncogenic (p. 722)
epidemiology (p. 722)
procarcinogen (p. 723)
ultimate carcinogen (p. 723)
genetic mutations (p. 724)
mutagen (p. 724)
Ames test (p. 724)

The Genetic Basis of Neoplasia
oncogene (p. 726)
proto-oncogenes (p. 727)
tumor-suppressor genes (p. 729)

Tumor Dissemination
contact paralysis of cellular locomotion (p. 730)
tumor angiogenesis (p. 731)
macrophages (p. 731)
monocytes (p. 731)

The Regulation of Tissue Growth
mitogens (p. 734)
serum (p. 734)
platelet-derived growth factor (PDGF) (p. 734)
epidermal growth factor (EGF) (p. 734)
mitogen receptors (p. 735)
tyrosine protein kinase (p. 735)
phospholipase C (p. 735)
autocrine secretion mode (p. 737)

Suggested Reading

General Articles and Reviews

Albert B., D. Bray, J. Lewis, M. Raff, K. Roberts, and J. D. Watson. Cancer. In *Molecular Biology of the Cell,* 2d ed., chapter 21, p. 1187. London: Garland Publishing Co., 1989.

Becker, F. F. *Cancer, a Comprehensive Treatise.* New York: Plenum, 1975–1988.

Cairns, J. *Cancer: Science and Society.* San Francisco: Freeman, 1978.

Pitot, H. C. *Fundamentals of Oncology,* 3d ed. New York: Marcel Dekker, 1986.

Chromosomal Alterations

Croce, C. M., and G. Klein. Chromosome translocations and human cancer. *Sci. Amer.* 252 (March 1985): 54.

Kurzrock, R., J. U. Gutterman, and M. Talpaz. The molecular genetics of Philadelphia chromosome-positive leukemias. *N. Engl. J. Med.* 319 (1988): 990.

Leder, P., J. Battey, G. Lenoir, C. Moulding, W. Murphy, H. Potter, T. Stewart, and R. Taub. Translocations among antibody genes in human cancer. *Science* 222 (1983): 765.

Nowell, P. C. Chromosomal and molecular clues to tumor progression. *Semin. Oncol.* 16 (1989): 116.

Oncogenic Viruses

Gallo, R. C., and F. Wong-Staal. Current thoughts on the viral etiology of certain human cancers. *Cancer Res.* 44 (1984): 2743.

Mühlbock, O., and P. Bentvelzen. The transmission of mammary tumor viruses. *Perspect. Virol.* 6 (1968): 75.

Nazerian, K. Marek's disease: A neoplastic disease of chickens caused by a herpesvirus. *Adv. Cancer Res.* 17 (1973): 279.

Chemical Carcinogens

Ames, B. N. Identifying environmental chemicals causing mutations and cancer. *Science* 204 (1979): 547.

Miller, E. C. Some current perspectives on chemical carcinogenesis in humans and experimental animals. *Cancer Res.* 38 (1978): 1479.

Oncogenes

Bishop, J. M. The molecular genetics of cancer. *Science* 235 (1987): 305.

Jove, R., and H. Hanafusa. Cell transformation by the viral *src* oncogene. *Annu. Rev. Cell Biol.* 3 (1987): 31.

Reddy, E. P., R. K. Reynolds, E. Santos, and M. Barbacid. A point mutation is responsible for the acquisition of transforming properties by the T24 human bladder carcinoma oncogene. *Nature* 300 (1982): 149.

Tabin, C. J., S. M. Bradley, C. I. Bargmann, R. A. Weinberg, A. G. Papageorge, E. M. Scolnick, R. Dhar, D. R. Lowry, and E. H. Chang. Mechanism of activation of a human oncogene. *Nature* 300 (1982): 143.

Tumor-Suppressor Genes

Harris, H. The analysis of malignancy by cell fusion. *Cancer Res.* 48 (1988): 3302.

Huang, H.-J. S., J.-K. Yee, J.-Y. Shew, P.-L. Chen, R. Bookstein, T. Friedmann, E. Y.-H. P. Lee, and W.-H. Lee. Suppression of the neoplastic phenotype by replacement of the RB gene in human cancer cells. *Science* 242 (1988): 1563.

Lee, W.-H., J.-Y. Shew, F. D. Hong, T. W. Sery, L. A. Donoso, L.-J. Young, R. Bookstein, and E. Y.-H. P. Lee. The retinoblastoma susceptibility gene encodes a nuclear phosphoprotein associated with DNA binding activity. *Nature* 329 (1987): 642.

Weinberg, R. A. Finding the anti-oncogene. *Sci. Amer.* 259 (March 1988): 44.

Tumor Dissemination

Armstrong, P. B. Invasiveness of non-malignant cells. In *Invasion. Experimental and Clinical Implications* (M. Mareel and K. Calman, eds.). Oxford: Oxford University Press, 1984.

Bock, G., ed. *Metastasis* (Ciba Foundation Symposium No. 141), Chichester, U.K.: Wiley, 1988.

Folkman, J., and M. Klagsbrun. Angiogenic factors. *Science* 235 (1987): 235.

Leibovich, J., P. J. Polverini, H. M. Shepard, D. M. Wiseman, V. Shively, and N. Nuseir. Macrophage-induced angiogenesis is mediated by tumor necrosis factor-α. *Nature* 329 (1987): 630.

The Regulation of Tissue Growth

Chambard, J. C., S. Paris, G. L'Allemain, and J. Pouysségur. Two growth factor signalling pathways in fibroblasts distinguished by pertussis toxin. *Nature* 326 (1987): 800.

Cheifetz, S., J. A. Weatherbee, M. L.-S. Tsang, J. K. Anderson, J. E. Mole, R. Lucas, and J. Massagué. The transforming growth factor-β system, a complex pattern of cross-reactive ligands and receptors. *Cell* 48 (1987): 409.

Roberts, A. B., K. C. Flanders, P. Kondaiah, N. L. Thompson, E. Van Obberghen-Schilling, L. Wakefield, P. Rossi, B. De Crombrugghe, U. Heine, and M. B. Sporn. Transforming growth factor β: Biochemistry and roles in embryogenesis, tissue repair and remodeling, and carcinogenesis. *Recent Prog. Hormone Res.* 44 (1988): 157.

Sibley, D. R., J. L. Benovic, M. G. Caron, and R. J. Lefkowitz. Regulation of transmembrane signaling by receptor phosphorylation. *Cell* 48 (1987): 913.

Problem Set

1. **Telling the Difference.** Suggest an observation or an experimental test to distinguish between the two members of each of the following pairs.
   (a) Benign tumor; malignant tumor
   (b) Nontransformed 3T3 cells; 3T3 cells transformed with simian sarcoma virus
   (c) Viral oncogene; proto-oncogene
   (d) Mitogen that binds to the EGF receptor; mitogen that fails to bind to the EGF receptor
   (e) Normal granular leukocyte (granulocyte); chronic myelogenous leukocyte

2. **Oncogenic Viruses.** You suspect that a particular tumor found in a laboratory animal is caused by a particular virus. What experiments would you conduct to determine whether that tumor was indeed caused by the virus?

3. **Carcinogens and Mutagens.** Present the two basic categories of evidence that can be used to show that a particular substance, such as cigarette smoke, is carcinogenic. How would you show that the same substance is mutagenic?

4. **Vascular Metastasis.** List, in the proper temporal order, the steps of vascular metastasis of a solid tumor.

5. **Tumor-Suppressor Genes.** Outline the two basic categories of experimental evidence for the involvement of tumor suppressor genes in the development of the neoplastic phenotype.

6. **Carcinogenic Potential.** What is the evidence that leads us to suspect that the carcinogenic potential of most chemical carcinogens is dependent on their ability to act as mutagens? Cite two different kinds of evidence.

7. **Medical Possibilities.** Recent advances in understanding the basic biology of the cancerous cell have suggested attractive possibilities for the prevention and clinical management of cancer. Discuss the medical possibilities suggested by recent studies of:
   (a) Tumor angiogenesis
   (b) Techniques to identify chemical carcinogens

8. **Tumor Progression.** Tumor progression is the development over time of progressively more malignant characteristics in a tumor. What are some of the processes that contribute to tumor progression?

24

Cellular Aspects of the Immune Response

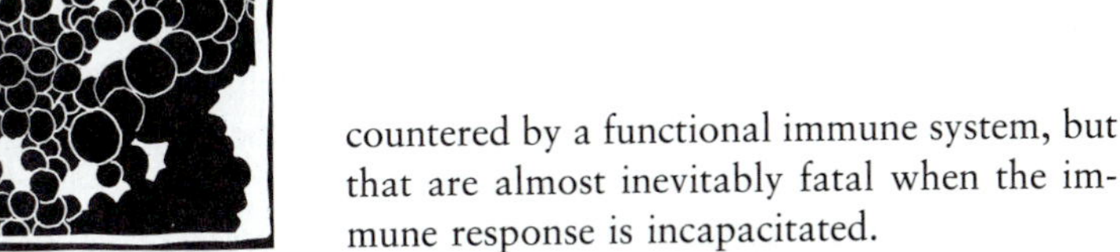

Our environment is filled with a host of infectious agents, including bacteria, viruses, parasites, and fungi. As a first line of defense against these pathogenic (disease-causing) invaders, vertebrates depend on **innate immunity,** which is nonspecific and includes physical barriers (skin, mucous membranes) as well as secretions with antimicrobial activity (mucus, tears containing the enzyme lysozyme). Pathogens that breach these physical barriers are often destroyed and eliminated by *phagocytes*—white blood cells that engulf foreign agents such as bacteria and fungi by the process of phagocytosis.

But physical barriers and phagocytes alone are not sufficient to ward off the onslaught of various pathogens. Fortunately, vertebrates also have an **adaptive immune system** as a specific, inducible defense mechanism. The immune system is essentially a means of surveillance, intended to protect the organism from infections and cancer by searching out and destroying the causative agents. As such, it is an indispensable defense mechanism.

To appreciate how vital the immune system is, we need only consider the fate of individuals with inherited or acquired deficiencies of the immune system. Infants born with a totally defective immune system can survive only if they are maintained in a sterile environment, completely free from microorganisms that might cause infection. Similarly, patients whose immune systems have been medically suppressed to prevent rejection of transplanted tissues or organs are abnormally susceptible to microbial infections against which immunosuppression has rendered them temporarily defenseless.

The importance of the immune system has been underscored in an especially devastating way by the recent worldwide outbreak of **acquired immunodeficiency syndrome** (AIDS). AIDS victims have an impaired immune system, due to infection with human immunodeficiency virus (HIV). Patients with AIDS are highly susceptible to infections and malignant diseases that can be successfully countered by a functional immune system, but that are almost inevitably fatal when the immune response is incapacitated.

Immunology, the study of the immune system, arose out of the observation that persons who have recovered from certain infections, such as smallpox or chicken pox, rarely, if ever, contract the same disease again. This protective **immunity** is highly specific and is generally long-lasting. That is, an individual who has survived an infection by the virus that causes smallpox is *immune* to that virus but not to other pathogens, such as the virus that causes mumps. Medical research has taken advantage of this specific immunity to develop vaccines against a variety of such pathogens. Smallpox, for example, once a major scourge of humankind, has now been completely eradicated.

In this chapter, we will first discuss the development of the immune system. Then we will describe what is known about *lymphocytes,* the cells of the immune system, and how they and their secreted products account for the memory and specificity of the immune response. Finally, we will describe how these lymphocytes interact with one another to regulate the immune response and to distinguish between self and foreign substances. As we become acquainted with the field of immunology, we will also see that it has contributed much to molecular and cell biology, including model systems of differentiation, cell-cell communication, and cell surface interactions that lead to proliferation and secretion.

The Immune Response

The immune system recognizes and eliminates foreign invaders or substances by a process known as the **immune response.** A substance capable of inducing an immune response is called an **immunogen** or, historically, an **antigen.**

The latter term is now frequently used for molecules capable of reacting with antibodies, but not necessarily capable of inducing antibody formation. However, the use of the term "antigen" when "immunogen" would be more appropriate is still prevalent in textbooks and even in the current literature. Therefore, in order to avoid confusion, the term *antigen* will be used uniformly to designate compounds capable of inducing an immune response as well as reacting with the products of the immune response. The most common antigens are foreign proteins and polysaccharides, although the immune system of an organism may sometimes also respond to substances produced by the organism itself, leading to disorders called **autoimmune diseases.**

The immune response is directed against small, discrete parts of an antigen perhaps five to ten amino acid or sugar residues in length, which are called **antigenic determinants** or **epitopes.** The number of epitopes on an antigen is a function of its molecular size and chemical complexity. Although most antigens have many epitopes, some are more effective than others in eliciting an immune response and may therefore dominate the response.

Characteristics of the Immune Response

The immune response to an antigen has four fundamental characteristics. First, the immune system responds *adaptively* to foreign intruders. When pathogens enter the body, they are carried by the circulatory system to specialized *lymphoid tissues*. There they encounter lymphocytes that respond to the foreign antigens, culminating in the demise of the invaders by several different mechanisms. Second, the immune response has exquisite *specificity*. It can often distinguish between antigens that are extremely similar to each other, such as proteins that differ by only a single amino acid.

Third, the immune response displays long-term *memory*. That is, the immune system can "remember" whether or not it has previously encountered a particular antigen. If it has, the immune system makes a *secondary response*, which is faster and more intense than the initial or primary response. And fourth, the immune system is able to *distinguish self from nonself*. It is, in other words, able to recognize which macromolecules are foreign and which are not. This capability is necessary to prevent the immune response from unleashing its destructive forces on the very organism of which it is a part. If it fails in this task, distinctive **autoimmune** (antiself) **reactions** may occur, as happen in some diseases.

These four characteristics—adaptivity, specificity, memory, and the ability to distinguish between self and nonself—set the immune system apart from all other systems of the body. Other systems may share some of these properties, but not all four. For example, the nervous system is also adaptive and displays specific memory, but it cannot recognize foreign from self.

Types of Immune Responses

There are two major types of immune responses: cell-mediated and humoral. **Cell-mediated immune responses** are produced by certain lymphocytes that react directly with foreign antigens such as viral proteins expressed on the surface of infected cells in the body. In a cell-mediated response, the infected host cells are killed. **Humoral immune responses** are mediated by special antigen-recognizing proteins, called **antibodies,** that are produced by other lymphocytes and circulate in the bloodstream. When antibody molecules come in contact with antigens that they recognize, they bind specifically to the antigens, thereby trapping the antigens and promoting their destruction and removal.

These two immune responses often work together. A cell-mediated response can kill a virally infected cell, thereby stopping further production of viruses, while any escaping viruses may be trapped by antibodies in the bloodstream.

The Cellular Basis of the Immune Response

Both the cellular and humoral types of immune response are mediated by white blood cells called **lymphocytes,** which circulate in the blood and the lymph. (*Lymph* is the colorless fluid in the lymphatic vessels, which connect the lymphoid organs with one another.) Lymphocytes are also found in specialized **lymphoid tissues** distributed throughout the body (Figure 24-1). The total cell mass of the lymphoid tissue is similar to that of the liver or brain and consists of about 2×10^{12} lymphocytes in humans. This network of lymphoid tissue facilitates capture of foreign intruders regardless of their portal of entry.

The two major classes of immune responses are mediated by two distinct classes of lymphocytes, B cells and T cells. **B cells,** or **B lymphocytes,** are the antibody-producing cells responsible for the humoral response, while **T cells,** or **T lymphocytes,** are responsible for cell-mediated responses. A cardinal feature of T cells, which will be discussed in detail later, is that they respond only to molecules expressed on the surface of other cells. Therefore, cellular immunity is effective against intracellular pathogens such as viruses and many parasites but ineffective against extracellular microorganisms such as most bacteria. On the other hand, antibodies produced in humoral immune responses can attack and eliminate extracellular antigens. Thus, the two types of immune response complement each other.

Lymphocyte Development

Both B cells and T cells develop from **hematopoietic stem cells** located in the **hematopoietic tissue**—mainly the liver in embryos and the bone marrow in adult organisms (Figure 24-2). Hematopoietic stem cells also give rise to all other types of blood cells. These stem cells are *pluripotent,* which means that they can give rise to distinctly different sets of progeny cells, depending on the particular tissue in which the differentiation occurs. Some stem cells undergo a commitment to a specific developmental pathway. These committed cells express certain surface receptors, and specific soluble or cell-bound factors then trigger the cells to differentiate further.

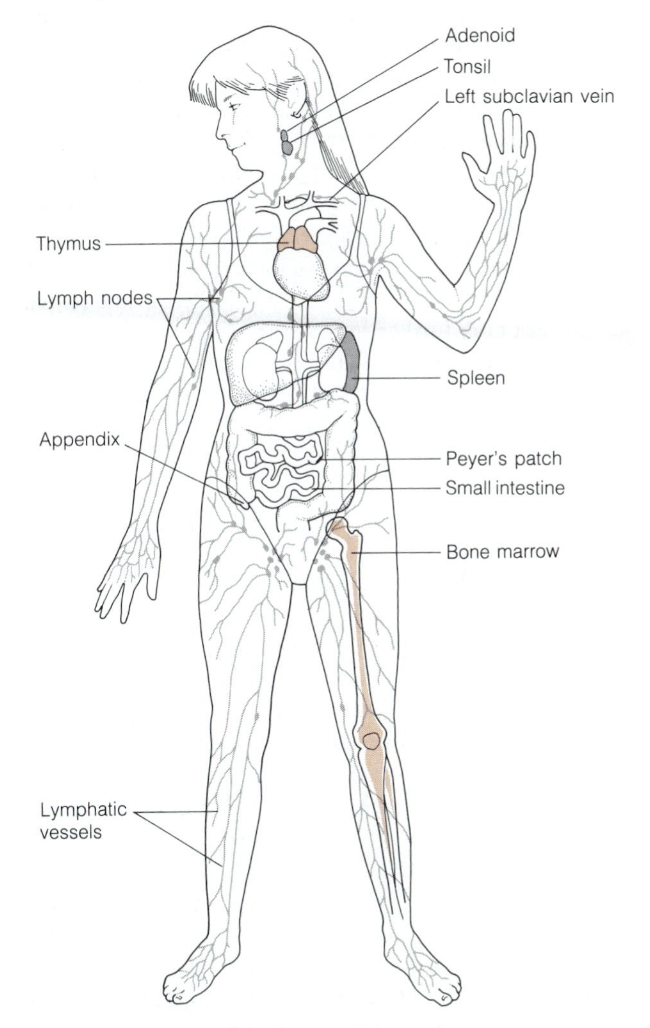

Figure 24-1 The Human Lymphoid System. The lymphoid system of the human body consists of the central lymphoid tissues (color) and the peripheral lymphoid tissues (gray). The central lymphoid tissues are the thymus and the bone marrow (in the long bones, only one of which is shown). The peripheral lymphoid tissues include the lymphatic vessels and lymph nodes, spleen, adenoids, tonsils, appendix, and the Peyer's patches of the small intestine. The lymphatic vessels collect the lymphocytes and antibody molecules from the tissues and lymph nodes and return them to the bloodstream at the left subclavian vein.

For lymphocytes to mature fully, progenitor cells must migrate from the hematopoietic tissue to the **central lymphoid tissue** (Figure 24-2a), where they encounter yet another set of cell-bound or soluble factors that induce further differentiation. The central lymphoid tissue consists of the *thymus,* the *bone marrow,* and, in birds, the *bursa of Fabricius,* an intestine-associated organ. Some lymphoid progenitor cells migrate by way of the blood to the thymus, where they proliferate and differentiate into T (for "thymus-derived") lymphocytes. In birds, other progenitor cells migrate to the bursa of Fabricius, where they develop into B (for "bursa-derived") lymphocytes. In mammals, which have no bursa of Fabricius, the bone marrow is probably the locus of B cell development.

B cell and T cell progenitors migrate via the bloodstream to **peripheral lymphoid tissues** such as *spleen, lymph nodes, Peyer's patches,* or *tonsils* (Figure 24-2b). The B cells and T cells seek out specific regions in these tissues. B lymphocytes reside principally in areas called *primary follicles,* while the T cells are found mainly in the *diffuse cortex.* At this point in development, the lymphocytes are mature and are ready to react with foreign antigens that enter the lymphoid tissues by way of the lymphatic system or the bloodstream. Antigen-activated B lymphocytes further differentiate into **plasma cells,** which secrete antibody.

From studies of animals with their central lymphoid tissues removed or absent, it is clear that thymus-derived T lymphocytes and bursa-derived B lymphocytes have different functions. Removal of the bursa of Fabricius from a newly hatched chick greatly impairs its ability to produce antibodies, but has little effect on its cell-mediated responses. On the other hand, removal of the thymus from a newborn animal profoundly impairs its cell-mediated immune responses, but has a less dramatic effect on its antibody responses. Moreover, the peripheral lymphoid tissues of an animal that has had its thymus removed at birth have no lymphocytes in the diffuse cortical regions where T cells normally reside.

Clonal Selection

The most remarkable feature of the immune system is its ability to respond specifically to an enormous variety of different antigens. What mechanism could possibly explain such remarkable versatility? A milestone in our understanding of the process was the exposition of the **clonal selection theory** in the 1950s by N. K. Jerne, F. M. Burnet, and others. Clonal selection is amply supported by compelling experimental evidence and is generally accepted by the scientific community. Before considering the theory, however, we should understand something about the antigen receptors of lymphocytes.

Antigen Receptors. Every lymphocyte carries membrane-bound molecules on its surface that serve as **antigen receptors.** The antigen receptors of B cells are antibodies. Those of T cells are similar, but not identical, to antibodies and have only recently been well characterized. One lymphocyte may have up to 10^5 receptor molecules on its surface, but all of the receptors on a given lymphocyte are identical in structure and, therefore, in ability to recognize and bind antigen. This commitment to express a particular antigen receptor is acquired by the lymphocyte during its development, without any exposure to antigen, by a molecular mechanism to be described later in the chapter.

Every lymphocyte, in other words, is predetermined to express a homogeneous set of membrane-bound receptors that have a particular specificity for antigen. The capacity to respond to an almost infinite variety of antigens depends on the presence in the lymphocyte population of a great diversity of cells with different receptor specificities. In the human body, for example, there are about 10^{12} lymphocytes, representing about 10^8 to 10^9 different receptor specificities. Moreover, most antigens have many epitopes, and a given epitope can be recognized by more than one kind of receptor, so the number of lymphocytes in the population that can respond to a given epitope is

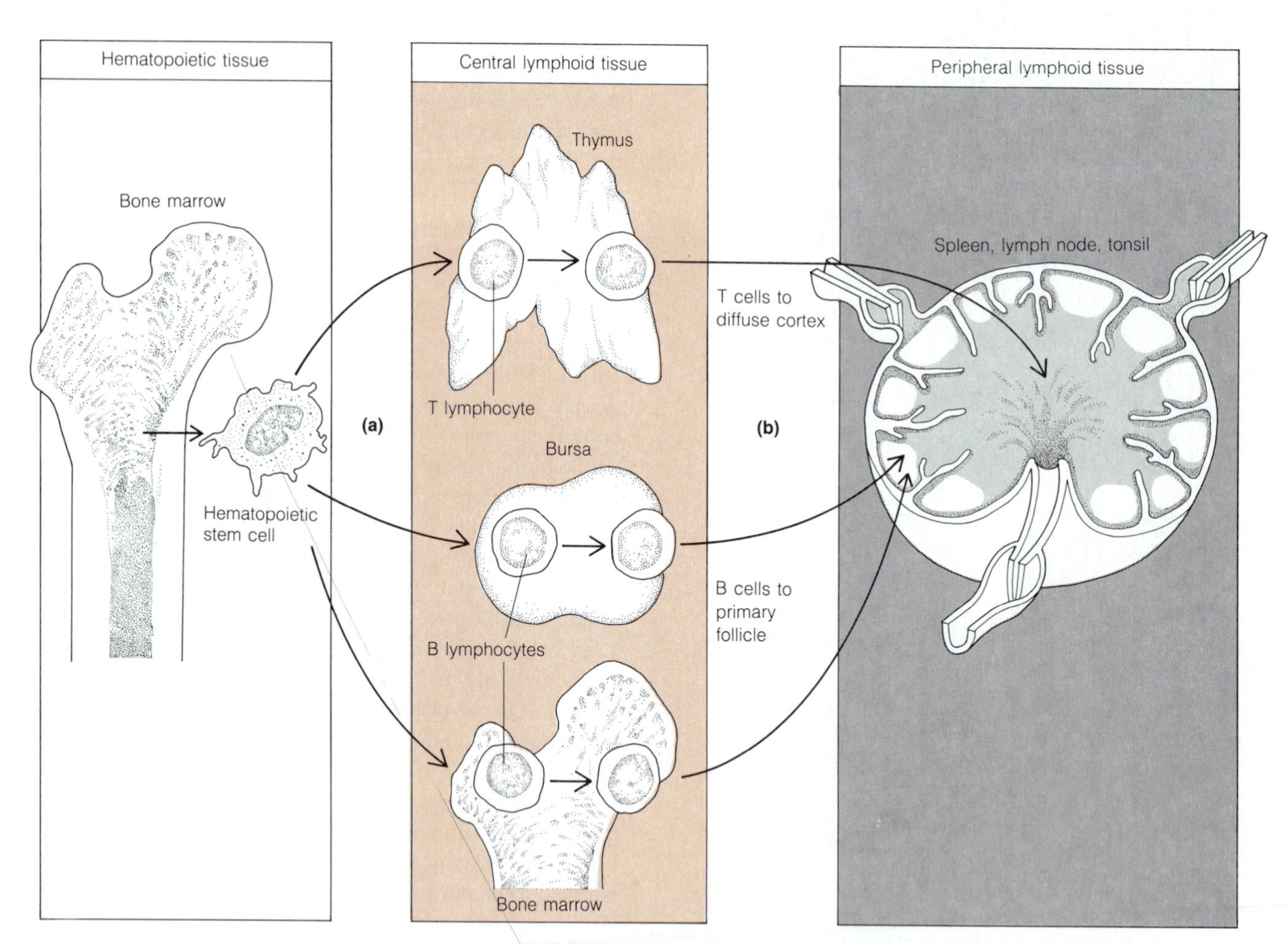

Figure 24-2 Lymphocyte Development. Lymphocytes develop from pluripotent stem cells in the hematopoietic tissue of the bone marrow. (a) Progenitor cells migrate from the bone marrow to the central lymphoid tissue (color), where they mature into either T cells (in the thymus) or B cells (in the bone marrow in mammals or the bursa of Fabricius in birds). (b) The B cells and T cells then migrate to the peripheral lymphoid tissue (gray), where they reside mainly in the primary follicles (B cells) or the diffuse cortex (T cells).

much larger than the number of cells possessing a specific kind of receptor. In other words, the response to even a single epitope is heterogeneous, involving the participation of a number of different clones of lymphocytes.

Clonal Selection. A key feature of the clonal selection theory is that the binding of an antigen to a lymphocyte with complementary receptors on its surface selectively activates that lymphocyte to proliferate and differentiate. The antigen, therefore, selects from a very large pool of lymphocytes the set capable of recognizing its epitopes. Activation results in **clones** of lymphocytes, the members of each clone being derived from the same ancestor and therefore all having the same epitope specificity. Thus, the immune system can be viewed as a collection of a vast number of clones of B and T lymphocytes, each descended from a single progenitor cell and therefore committed to the synthesis of the same surface receptor as the original cell.

According to clonal selection, lymphocytes display two phases of differentiation. During **antigen-independent differentiation** (Figure 24-3a), hematopoietic stem cells give rise to initially **uncommitted cells,** and each of these (and therefore the clone derived from each) becomes committed during differentiation to the synthesis of a particular surface receptor capable of reacting with a particular epitope or, more precisely, with a small set of structurally related epitopes. A few such clones of **committed cells** are numbered in Figure 24-3; in actuality, there are millions of them.

Antigen-dependent differentiation results when a particular antigen comes into contact with lymphocytes that carry surface receptors complementary to one of its epitopes. In Figure 24-3b, the antigen A is shown interacting with cells of three different clones, each presumably specific for a different epitope on the antigen. The cells to which the antigen binds are selectively stimulated to proliferate and differentiate into functionally active cells.

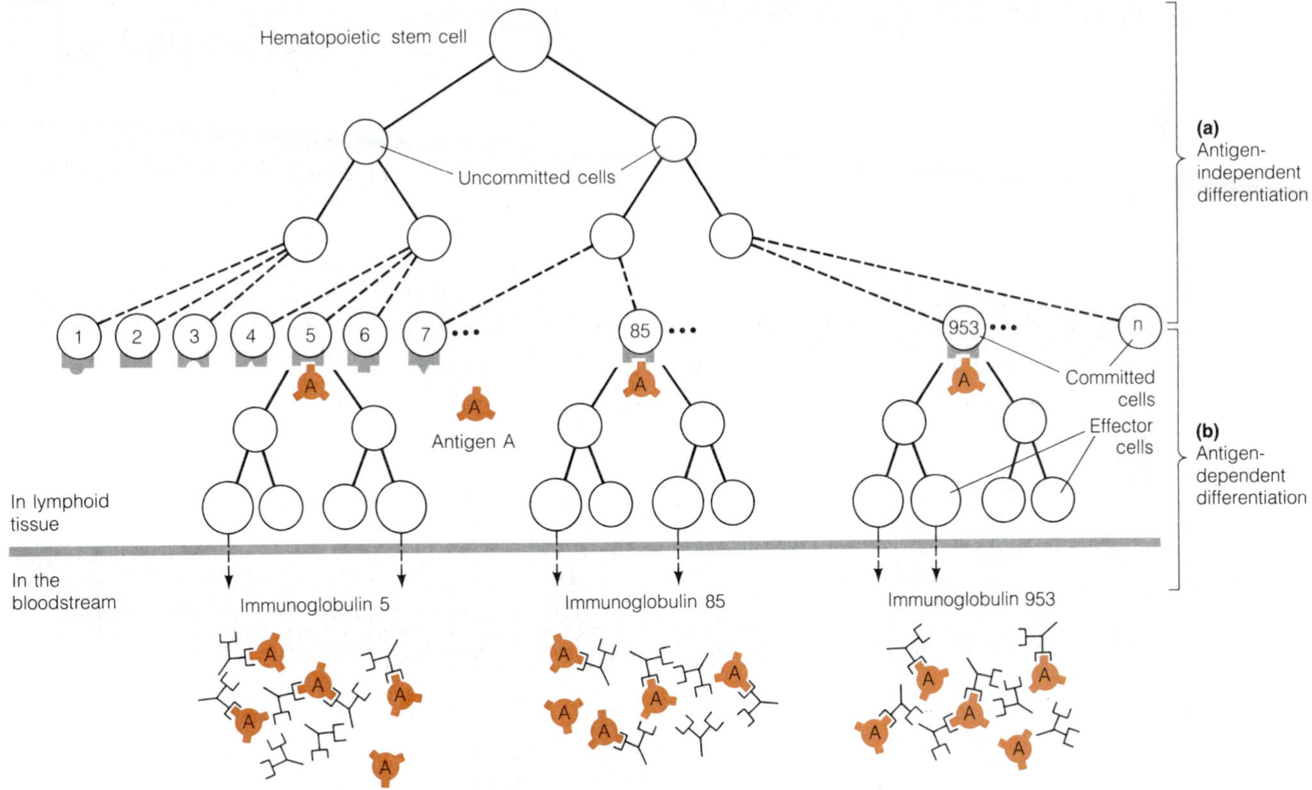

Figure 24-3 Formation of Clones and Their Selection by Antibodies. (a) Hematopoietic stem cells undergo an antigen-independent differentiation into uncommitted cells, each of which gives rise to a clone that is committed to the synthesis of one species of surface receptor (gray), capable of reacting with a specific antigen. Of the millions of cells with different specificities that can be generated, only a limited number are shown. (b) Binding of an antigen (color) to the surface of specific clones of cells initiates an antigen-dependent phase of differentiation that leads to the proliferation of clones of effector cells and clones of memory cells. In the case shown, three clones were committed to the production of a surface receptor that recognizes antigen A, so three clones of effector cells proliferate that produce antibodies, or immunoglobulins, capable of recognizing antigen A as they encounter the antigen in the bloodstream.

As we noted earlier, the terminal differentiation stage of the B lymphocyte is the *plasma cell,* which secretes antibodies with the same antigen-recognition specificity as its cell surface receptors. The plasma cell is larger and has a higher ratio of cytoplasm to nucleus than the B lymphocyte from which it originated. Its cytoplasm is loaded with rough endoplasmic reticulum and is, in essence, a protein-synthesizing machine that produces large quantities of antibody molecules.

Antigen-dependent differentiation of T lymphocytes gives rise to several functionally distinct types of cells. **Cytotoxic (killer) T cells** are responsible for lysing cells expressing foreign antigens. Other T cells are involved in regulating the immune response. One type of regulatory T cell suppresses immune responses, and such cells are therefore called **suppressor T cells.** Another subset of T cells has the opposite effect, and such cells are designated **helper T cells** because they induce other cells to proliferate and differentiate. We will discuss each of these classes of T cells in more detail later.

Immunological Memory

An important feature of the immune system is its ability to "remember." If you have ever had a viral disease such as chicken pox, you know that you are now immune to that particular disease and are unlikely ever to get it again. Somehow, your immune system "remembers" your initial encounter with that particular virus and responds accordingly. This phenomenon, called **immunological memory,** can be explained in terms of clonal selection.

Figure 24-4 illustrates the time course for the immune response of an animal that is injected for the first time with antigen A at day 0. The **primary immune response** appears after a lag of several days, rises rapidly, reaches a peak, and then falls again gradually. If the animal encounters the same antigen at some later time (day 28, in the case of Figure 24-4), the **secondary immune response** that this elicits is more rapid, of greater magnitude, and of longer duration. Clearly, the animal has "remembered" antigen A and responds differently in the second encounter. Moreover, the memory is specific for antigen A, because a different antigen (antigen B, for example) injected at day 28 elicits a primary and not a secondary immune response. The data of Figure 24-4 were obtained for a B cell response, but the same scenario applies to T cell responses.

The initial encounter causes specific B cell and T cell clones to proliferate and differentiate. The progeny lymphocytes include not only the **effector cells** that produce the primary immune response but also large clones of **memory cells** (Figure 24-5). Memory cells have the ability to produce both effector cells and more memory cells when they are stimulated by the same antigen later on. Effector cells have a short lifetime (a few days, usually), but memory

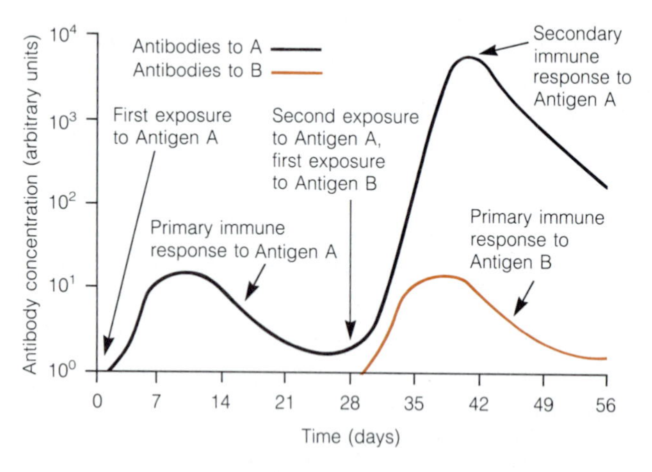

Figure 24-4 Primary and Secondary Immune Responses. Initial exposure to antigen A (at day 0) elicits a primary immune response with a lag phase, a rapid rise in antibody concentration, and then a gradual decrease. A second encounter with the same antigen (at day 28) results in a secondary immune response that is faster, stronger, and longer lasting than the primary response. Antigen B is included at day 28 to demonstrate the specificity of the immunological memory for antigen A.

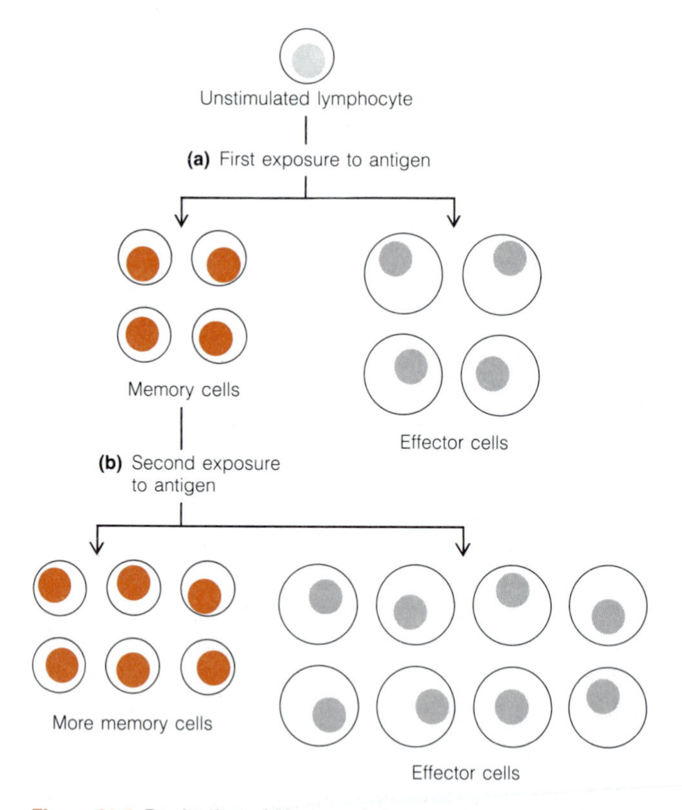

Figure 24-5 Production of Memory Cells. (a) Initial stimulation of a lymphocyte by its first exposure to antigen results in the proliferation of both effector cells (gray) and memory cells (color). The effector cells produce the primary immune response and are short-lived. The memory cells last much longer. (b) Upon encounter with the same antigen, the memory cells proliferate, and some differentiate into effector cells.

cells survive in the spleen and lymph nodes much longer—decades, in the case of humans. If enough memory cells are produced during the initial exposure to the antigen, they may persist for the lifetime of the organism, conferring permanent immunity.

Memory cells thus have three important properties:

1. They are produced along with effector cells as a result of lymphocyte proliferation in response to the initial encounter with the antigen.
2. They have a much longer life span than effector cells.
3. Because they are more numerous than the virgin lymphocytes from which they derived, the secondary response exhibits faster kinetics and greater magnitude than the primary response. Therefore, the difference between primary and secondary immune responses is a function of the numbers of cells poised to respond to a given antigen.

Telling the Players Apart: Differentiation Markers

All lymphocytes, whether B cells or T cells, look alike in the ordinary light microscope. Fortunately for immunologists, different lymphocyte types can be identified on the basis of differential expression of certain cell surface proteins.

Lymphocytes with a particular functional activity can be distinguished on the basis of certain *differentiation markers*. B lymphocytes express **immunoglobulins** (their antigen receptors) on their surface and can be distinguished from T lymphocytes on that basis. Although T cells also express antigen receptors, they are different from the immunoglobulins expressed by B cells. On the other hand, T cells express markers not found on B cells. All mature T cells express a set of polypeptide chains called the **CD3** complex. Helper T (T_h) cells also express another glycoprotein (**CD4**), whereas cytotoxic T (T_c) cells express a marker called **CD8.** Thus, T_h cells are phenotypically $CD3^+CD4^+CD8^-$, whereas T_c cells are $CD3^+CD4^-CD8^+$. Suppressor T (T_s) cells are phenotypically indistinguishable from T_c cells on the basis of markers identified thus far, but T_s cells do not appear to exert cytotoxic activity, so they seem to comprise a distinct subset of T cells.

It is noteworthy that the CD4 glycoprotein has been shown to be a receptor for the **human immunodeficiency virus (HIV)**, the causative agent of AIDS. The virus gains entry into the cell via the CD4 glycoprotein, replicates within the cell, and destroys it. This explains why the CD4 T cell population, which plays a very central role in the immune response (see Figure 24-9), becomes depleted in individuals suffering from AIDS, resulting in severe immunodeficiency.

The surface of a lymphocyte actually has many **differentiation markers,** specific gene products that are expressed only at characteristic stages during lymphocyte development. Some of these glycoproteins function as receptors for molecules on other cells with which they interact, or as receptors for hormone-like factors (*lymphokines* or *interleukins,* described later) necessary during a particular stage of differentiation. The importance of these receptors can be appreciated when we consider the differentiation of committed stem cells into mature functional cells. As lymphoid cells migrate from hematopoietic tissue into central lymphoid tissue and finally into peripheral lymphoid tissue, they encounter several different microenvironments in succession (Figure 24-2). How do these cells know where they are and what to do?

The answers to these questions may involve specialized sessile (nonmotile) cells present in each of these different microenvironments. Unlike the motile lymphoid cells, sessile cells are permanent residents in a specific lymphoid tissue. Examples of sessile cells include the *stromal cells* of the bone marrow, the *epithelial cells* of the thymus, and the *dendritic cells* of the lymph nodes. The sessile cells in a particular tissue may stimulate further differentiation or activation of lymphoid cells either by the release of soluble factors from the sessile cells or through direct, though poorly defined, cell-cell interactions between the sessile and lymphoid cells. In either case, the response of the lymphoid cell depends on the presence on its surface of differentiation markers that can serve as receptors for these various signals.

When developing lymphocytes leave the bone marrow or thymus and enter the bloodstream, they must be able to find their way to specific lymphoid tissues, such as the lymph nodes or the Peyer's patches of the intestine. Once again, receptor-mediated cell-cell interactions are involved. The lymphocytes have receptors on their surfaces that target them for preferential entry into particular lymphoid tissues. These receptors bind to complementary structures on specialized endothelial cells that line the tiny veins, or venules, of the lymphoid tissue. The lymphocytes then migrate across these endothelial walls and enter the lymphoid tissue.

Lymphocyte Activation Pathways

The immune response to an antigen consists of a complex sequence of events involving the interaction of different cells by direct contact or via soluble mediators (**interleu-**

kins) that promote proliferation or differentiation. In order to understand how the process works, we must first consider the gene products of the *major histocompatibility complex (MHC)*, which play a vital role in antigen recognition by T lymphocytes.

The Major Histocompatibility Complex

An important aspect of cell-mediated immune responses is the ability to distinguish between *self* and *nonself*. As mentioned earlier, T cells respond to antigens only when they are present on the surfaces of other cells. Thus, cell surface molecules play a critical role in the recognition of nonself antigens by T cells. Historically, the surface antigens involved in the distinction between self and nonself were discovered from studies of skin grafts. When skin is removed from one individual and grafted onto another individual of either the same species (*allograft*) or a different species (*xenograft*), the grafted skin is usually recognized as foreign and rejected by the immune system of the recipient. This phenomenon is called **graft rejection.** Transplantation of an organ such as a kidney or a heart usually leads to the same result unless the donor and recipient are genetically identical. It was, in fact, the immunological rejection of transplanted organs that provided a major driving force for the development of the field of immunology.

Graft rejection is mediated principally by T_c cells and T_c-like cells that respond to and react against the foreign antigens on the surface of the cells in the grafted tissue or organ. The cell surface antigens that elicit this response are called **transplantation, or histocompatibility, antigens.** (The prefix *histo* means "tissue.") The most important of these antigens are encoded by a complex of genes called the **major histocompatibility complex, or MHC.** In humans, the MHC antigens are called **human leukocyte-associated (HLA) antigens** and are encoded by the **HLA complex** on chromosome 6. In mice, they are called **histocompatibility-2 (H-2) antigens** and are encoded by genes in the **H-2 complex** on chromosome 17. Figure 24-6 illustrates the several loci of the HLA and H-2 complexes.

The MHC genes are expressed in a *codominant* manner, meaning that all the MHC genes inherited from both parents are expressed on the cell surfaces of the progeny. In outbred populations such as humans, there is great variation in the kinds of MHC genes. In other words, the MHC genetic loci are extremely **polymorphic.** That is, a great many alternative alleles at each locus exist in the gene pool of the species. Indeed, the gene loci of the MHC are the most polymorphic known. It is therefore quite unlikely that any two unrelated individuals will have the same set of MHC antigens. Since even a mismatch of a single antigen between donor and recipient can result in transplant rejection, the difficulties attending organ transplantation can be readily appreciated.

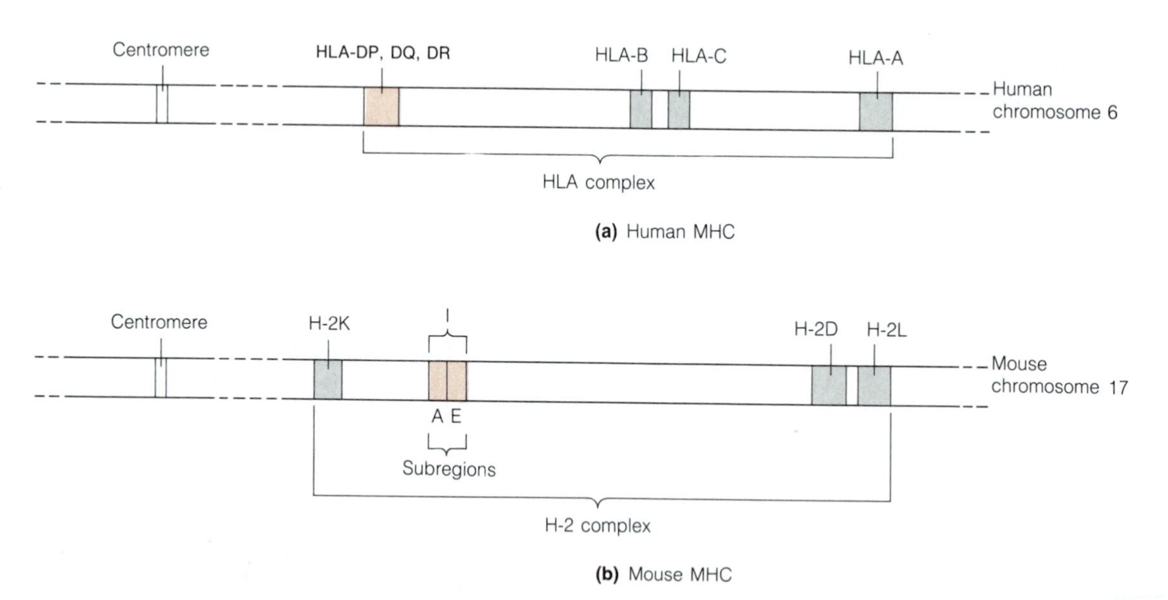

Figure 24-6 The Major Histocompatibility Complex (MHC) of the Human and Mouse Genomes. (a) The MHC of the human genome is the HLA complex on chromosome 6. (b) For the mouse genome, the MHC is the H-2 complex on chromosome 2. Each complex contains loci coding for class I MHC glycoproteins (gray) and for class II MHC glycoproteins (color). In the case of the mouse, the I region appears to contain several subregions that differ in their ability to code for the α and β chains of class II MHC glycoproteins.

Classes of MHC Antigens

The MHC genes encode two families of cell surface antigens called **class I antigens** and **class II antigens.** In humans, the loci HLA-A, HLA-B, and HLA-C code for class I antigens, while HLA-DP, HLA-DQ, and HLA-DR encode class II antigens (Figure 24-6a). In mice, the H-2D, H-2K, and H-2L loci code for class I antigens, and a complex locus called the I (immune response) region encodes class II antigens (Figure 24-6b). In histocompatibility reactions, cytotoxic T cells respond primarily against class I antigens, whereas helper T cells respond mainly against class II antigens. Both class I and class II MHC antigens are glycoproteins.

Class I Glycoproteins. The properties of class I and class II glycoproteins are quite different. Class I glycoproteins are encoded by three separate MHC loci, but the gene products are all integral membrane proteins ranging in molecular weight from 40,000 to 45,000. As shown in Figure 24-7, class I polypeptides are inserted through the plasma membrane. Characteristic of membrane proteins, they have a short hydrophilic tail inside the cell and a hydrophobic transmembrane segment. The extracellular portion of class I proteins is a large segment consisting of three domains with intrachain disulfide bridges. The carbohydrate group is attached to the exterior segment of the protein.

The domain closest to the membrane associates noncovalently with a small protein called **β2-microglobulin,** which also has an intrachain disulfide bridge. The outer two domains fold back toward each other to form a double-looped structure. The β2-microglobulin chain and one of the three outside domains in class I glycoproteins are homologous in amino acid sequence with regions of immunoglobulins containing disulfide bridges. This similarity may reflect the fact that both immunoglobulins and class I MHC glycoproteins are involved in cell-cell recognition and suggests that both are derived from a common ancestral gene. Class I MHC antigens are found on virtually all nucleated cells and are involved in presenting antigens to cytotoxic (CD8) T cells.

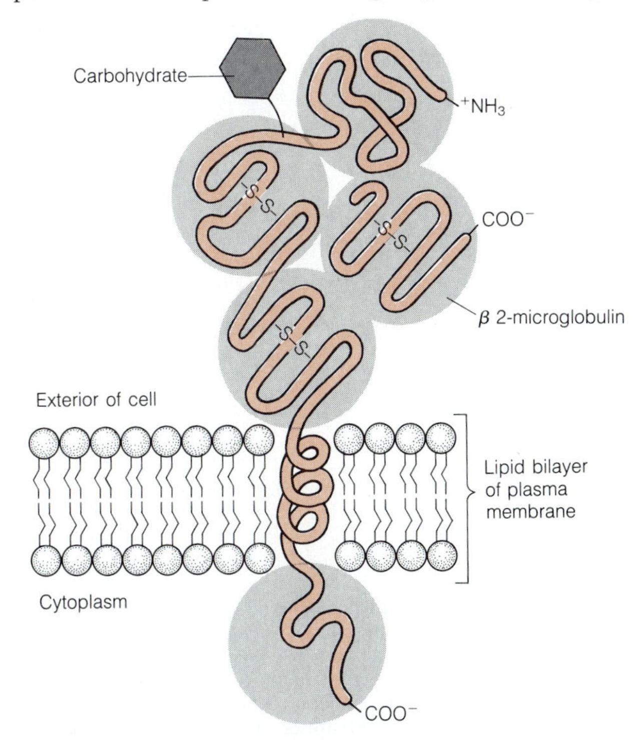

Figure 24-7 Structure and Orientation of a Class I MHC Glycoprotein. Class I MHC glycoprotein molecules are integral membrane proteins that associate noncovalently with β2-microglobulin on the external side of the membrane. Class I glycoproteins are found on the surface of virtually all nucleated cells, where they account for up to 1 percent of the total protein of the plasma membrane.

Class II Glycoproteins. Class II MHC antigens are also glycosylated integral membrane proteins (Figure 24-8). Each protein consists of two transmembrane polypeptides, an α chain with a molecular weight of about 33,000 and a β chain with a molecular weight of about 28,000. The two polypeptides are closely associated with each other, but they are not covalently joined. Each chain is glycosylated on the outside of the cell. Like class I antigens, these glycoproteins also contain domains that resemble those of immunoglobulin molecules.

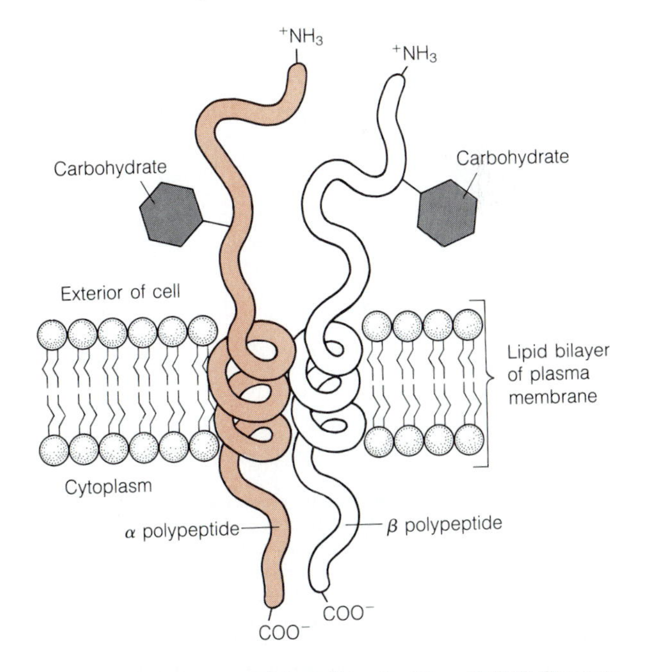

Figure 24-8 Structure and Orientation of a Class II MHC Glycoprotein. Class II MHC glycoprotein molecules are integral membrane proteins that consist of an α polypeptide noncovalently associated with a β polypeptide. Class II glycoproteins are found on the surface of all B lymphocytes and some macrophages.

Class II MHC antigens are less widely distributed than are class I antigens. They are expressed on most B lymphocytes, some macrophages, and certain other cells distributed throughout the body. Class II glycoproteins are involved in enabling helper (CD4) T cells to recognize and respond to antigen, which they, as well as CD8 T cells, can do only on the surface of a cell.

Antigen Processing and Presentation

Responses to most antigens require processing of the antigen by accessory cells called **antigen-presenting cells (APC)**. The reason for this is that T cells, which regulate the immune response and are essential for its initiation, recognize antigens only in the context of MHC glycoproteins on the surfaces of other cells. Consequently, the first steps in the immune response after entry of an antigen involve capture and processing of the antigen by APC and presentation of a "processed" form of the antigen to T_h cells (Figure 24-9a). Although the cell types that express class II MHC proteins are relatively limited, all such types are capable of acting as APC for T_h cells. The principal cell types serving this function are macrophages and B lymphocytes. Processing involves internalization of the an-

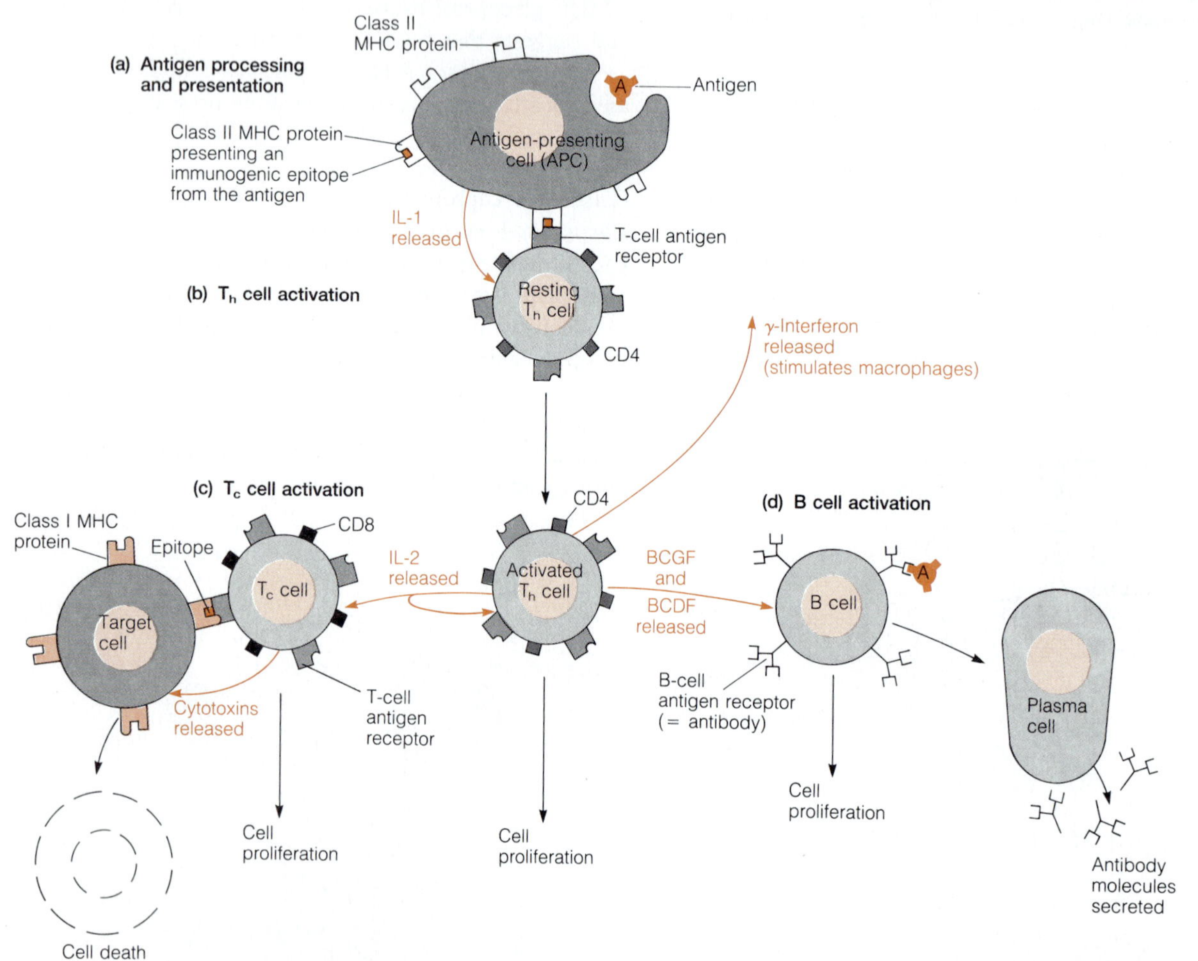

Figure 24-9 Pathways of the Immune Response. The response is initiated by the uptake of antigen by an antigen-presenting cell (APC), such as the macrophage shown here. (a) Antigen processing and presentation: The antigen is partially degraded within the APC and one or more fragments of it (epitopes) become associated with class II MHC proteins; the epitope-MHC complex is transported to the cell surface, where the epitope is "presented." (b) T_h cell activation: The epitope-MHC complex can bind to the antigen receptors of T_h cells. This reaction and the lymphokine interleukin-1 (IL-1) released by the APC activate the T_h cell. The activated T_h cell in turn releases interleukin-2 (IL-2), which stimulates the T_h cell itself to proliferate and to release a number of other lymphokines, including γ-interferon. (c) T_c cell activation: IL-2 also acts on T_c cells, activating them to kill target cells that present immunogenic epitopes complexed with class I MHC proteins. The killing is caused by cytotoxins released by the activated T_c cells. (d) B cell activation: Two other lymphokines released by activated T_h cells are B-cell growth factor (BCGF) and B-cell differentiation factor (BCDF). If the antigen receptors (antibody molecules) on the surface of a B cell are combined with antigen, BCGF and BCDF will activate the B cell to proliferate and to differentiate into antibody-secreting plasma cells. B cells, T_h cells, and T_c cells can be distinguished from one another by certain surface proteins; T_h cells have a surface protein called CD4, T_c cells have the CD8 protein, and B cells have neither. Note that the activation of all three types of lymphocyte requires (1) the action of one or more lymphokines and (2) the binding of an antigen (or antigen fragment) to an antigen receptor on the lymphocyte surface.

tigen by APC and degradation of the antigen in a compartment within the cell (*endosome*). Fragments of the original antigen become noncovalently associated with class II molecules. Such fragments are called **immunogenic epitopes** to distinguish them from epitopes that merely bind to antibodies. Only a limited number of the peptide fragments from a typical protein antigen possess the capacity to associate with class II molecules to form an immunogenic complex. It is believed that all immunogenic epitopes become bound to a single site on the class II protein. The peptide-class II complex is transported to the cell surface, where it is accessible to the T cell antigen receptor.

An important aspect of this process is that T cells do not respond to antigens in their native structural form. Instead, they recognize fragments of the original antigen if, and only if, the fragments are associated with self class II molecules. T cells that respond to antigen-class II complexes (CD4 T cells) are said to be **class II-restricted.**

It may seem curious that B lymphocytes, the precursors of antibody-secreting plasma cells, can present antigens to T cells. However, their capacity to do so is well documented. B cells express class II proteins and have a very efficient capturing mechanism for their specific antigens via their antigen receptors. Experimental evidence indicates that B cells are relatively poor activators of virgin T cells, possibly because such T cells require auxiliary activating mediators (interleukins) that B cells do not adequately supply. However, B cells require only about 1/1000 as much immunogen as do macrophages for activating memory T cells. Because B cells specific for a particular antigen are rare in an individual who has never encountered that antigen, and virgin T cells do not respond well to antigen presented by B cells, it is believed that macrophages play a greater role as APC in the initial or primary immune response, whereas B cells play a more dominant role in secondary responses.

Activation of Helper T Cells

The helper T cell is the conductor of the immunological orchestra, providing signals essential for the activation of macrophages, B cells, and T_c cells, the effector cells of the immune response responsible for the elimination of antigens. The activation of T_h cells (Figure 24-9b) occurs early in the immune pathway and requires at least two signals. One signal is furnished by the binding of T cell antigen receptors to class II-antigen complexes on APC. The second signal derives from a soluble protein mediator produced by the APC called **interleukin-1** (**IL-1**), for which the T_h cell has receptors. Macrophages produce much more IL-1 than do B cells, which may account for their heightened ability to activate virgin T cells. The two signals induce the T_h cell to express receptors for a lymphokine called **interleukin-2** (**IL-2**) and to produce IL-2 itself. The IL-2, by binding to IL-2 receptors on the T_h cell, further stimulates the cell to produce other lymphokines that are important for activating B cells and macrophages. The IL-2 produced by T_h cells is also stimulatory for T_c cells, and **γ-interferon,** another T_h cell product, activates macrophages (and exerts potent antiviral activity). The elaboration of this assortment of lymphokines, only some of which are described here, is how the T_h cell affects the other important cells of the immune network. The T_s cell has been omitted from the scheme in Figure 24-9 because the nature of its interactions with other cells in the network is still unclear.

Activation of Cytotoxic T Cells

We have already seen how the activated T_h cell is central to further progress along the immune pathway, notably to the triggering of T_c cells, whose major function is the killing of cells expressing foreign (nonself) antigens, and to the activation of B cells, which differentiate into antibody-producing plasma cells. The cytotoxic T cell is distinguished from the T_h cell by the expression of CD8 instead of CD4 and by its recognition of foreign antigen in the context of class I MHC proteins rather than class II proteins (Figure 24-9c). Thus, the T_c cell is class I-restricted, whereas the T_h cell is class II-restricted. CD4 and CD8 proteins have been shown to bind to MHC class II and class I proteins, respectively, and are believed to contribute to adhesion between T cells and APC.

T_c cells, like T_h cells, also require at least two activating signals. One signal is provided by interaction of the antigen receptor with a foreign epitope-class I complex on the **target cell,** which may be a virally infected cell, a tumor cell, or an allograft. The second signal is furnished by IL-2 produced by an activated T_h cell. The activated T_c cell then releases **cytotoxins,** which kill the target cell.

Activation of B Cells

The production of antibodies requires the activation of B lymphocytes and their differentiation into antibody-producing plasma cells. The sequence of events in this process is visualized as follows. While the T_h cell is being activated as described above, relevant B cells are also engaging epitopes on the antigen through their antigen receptors, which are membrane-bound forms of the antibodies they will later secrete (Figure 24-9d). Antigen binding is followed by endocytosis of the receptor complexed with antigen, which appears to furnish a suboptimal activating signal. Additional signals required for full activation of B cells are provided by activated T_h cells in the form of lymphokines.

These lymphokines have a short radius of activity, requiring proximity between the relevant antigen-specific T and B cells. Since B cells express class II proteins and can function as APC, they are believed to process endocytosed antigen and transport immunogenic epitopes associated with class II proteins to the cell surface. These can then serve as recognition structures for activated T cells (or they may directly activate memory T cells, as already discussed), which can then deliver the appropriate lymphokines at close range.

At least two lymphokines produced by T_h cells are needed for proliferation and differentiation of B cells. The first lymphokine is **B cell growth factor** (BCGF), which, in concert with antigen, stimulates B cells to proliferate. The second lymphokine, **B cell differentiation factor** (BCDF), induces activated B cells to differentiate into plasma cells. Thus, full activation of B cells requires at least three signals, one from the antigen and at least two from T_h cells. Additional T_h products may also be involved; the process has not yet been fully delineated.

A fraction of activated B cells proliferate but fail to differentiate into plasma cells, possibly because they receive insufficient BCDF. Such cells comprise the pool of memory B cells for that antigen and can respond to subsequent encounters with it. Some of the most important characteristics of the different cell types in the immune network are shown in Table 24-1.

The Structure and Function of Antibodies

The humoral immune system responds to a myriad of different antigens by making large numbers of highly specific antibodies. Antibodies constitute a class of proteins called **immunoglobulins,** abbreviated **Ig.** Immunoglobulins account for about 20% of the total plasma (noncellular) protein in the blood and are therefore one of the major classes of blood proteins.

The capacity to produce a specific antibody is acquired by every B lymphocyte during its development, without prior exposure to an antigen. One of the great mysteries of immunology has been how such a large number (10^8–10^9) of different antibody molecules can be made. The combination of molecular mechanisms that the immune system uses to meet this challenge is now well understood. But before we can appreciate the elegance of antibody diversity, we must first acquaint ourselves with the structure and function of antibodies.

The Antibody Molecule

Every antibody molecule has two functions: to recognize and bind to an antigen and to assist in the destruction and elimination of that antigen. This dichotomy of function is reflected in antibody structure, because every antibody molecule consists of discrete **domains** that participate in one of these two functions. Different antibodies must be able to recognize an almost unlimited variety of antigens. This task is alleviated by the fact that a given antibody can react with more than one epitope, provided the epitopes are closely related structurally. That is, antibody specificity for antigen is somewhat degenerate, but the binding affinity will vary depending on the particular epitope. The domains that bind antigen differ in amino acid sequence from one antibody to another and are called **variable** (V) **domains.** In contrast, only a few effector mechanisms are involved in antigen elimination, so only a few different kinds of

Table 24-1 Principal Cells of the Immune System

Cell Type	Surface Phenotype	MHC* Restriction	Antigen Specific	Function
Antigen-presenting cells				
Macrophages	Class II MHC	−	−	Process and present antigen to T lymphocytes
B lymphocytes	Class II MHC Ig**	−	+	Produce interleukin-1
T lymphocytes				
Helper	$CD3^+CD4^+CD8^-$	Class II	+	Positive regulation; produce interleukin-2, B cell growth factor, and B cell differentiation factor
Cytotoxic	$CD3^+CD4^-CD8^+$	Class I	+	Kill cells bearing foreign antigens; produce cytotoxins
Suppressor	$CD3^+CD4^-CD8^+$	?	+	Negative regulation; mode of action uncertain
B lymphocytes	Ig^+ MHC class II^+	−	+	Produce antibody; process and present antigen to T cells

* MHC = major histocompatibility complex
** IG = immunoglobulin

domains are involved in these functions. These are the **constant (C) domains.**

The simplest antibody molecule is a Y-shaped molecule with two identical **antigen binding sites** on its arms and an **effector site** on its leg (Figure 24-10). The antigen binding sites are the variable domains, whereas the effector site is the region of the constant domains. The two antigen-binding arms are connected to the leg of the molecule by flexible **hinge regions.** The structure of the hinge enables the arms of the molecule to be flexible and to bind to two antigen molecules (or to two antigenic determinants on the same molecule) simultaneously.

The basic antibody molecule consists of four polypeptide chains, two identical **light (L) chains** and two identical **heavy (H) chains,** organized into two **H-L pairs** (Figure 24-10). The subunit structure of such a molecule can therefore be represented as $(H\text{-}L)_2$. The molecular weight of each light chain is about 23,000, whereas a heavy chain has a molecular weight of about 55,000. Each light chain is linked to one heavy chain by a single disulfide bridge and by noncovalent associations. The heavy chains have oligosaccharide chains attached to them and are joined to each other by one or more disulfide bridges.

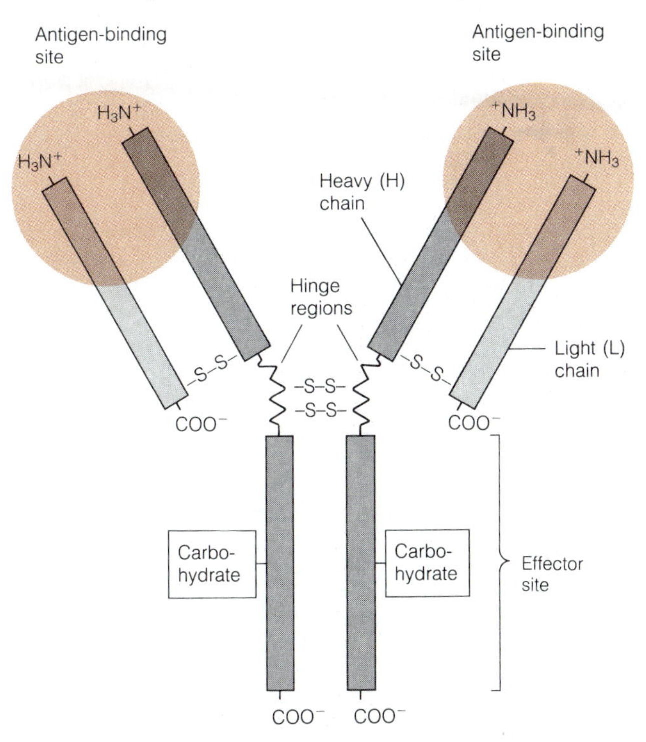

Figure 24-10 Structure of an Antibody. An immunoglobulin molecule consists of two identical heavy chains (dark gray) and two identical light chains (light gray) associated with one another in a Y-shaped configuration and cross-linked by disulfide bridges (—S—S—). The antigen binding site at the end of each arm (color) is formed from the amino-terminal regions of one light and one heavy chain, whereas the effector site in the leg of the molecule is formed from the carboxyl-terminal regions of the heavy chains only. Carbohydrates (oligosaccharides) are attached to the legs of both heavy chains.

Classes of Immunoglobulins

In mammals, there are five major classes (or **isotypes**) of immunoglobulins, each with its own specific heavy chain (Table 24-2). The five classes are *IgG*, *IgM*, *IgA*, *IgD*, and *IgE*, and the H chains found in each are γ, μ, α, δ, and ϵ, respectively. There are four distinct kinds, or subclasses, of IgG and two subclasses of IgA because of the existence of multiple forms of the γ and α chains. In addition to having their own characteristic H chains, the several classes and subclasses of mammalian immunoglobulins also differ in the number of disulfide bridges between H chains.

An antibody molecule of any class can have one of two kinds of light chains, either a κ or a λ chain. No functional differences between these two varieties of light chains are known. In general, a given B cell clone produces only one type of light chain and one of the five classes of heavy chains. At some stages in their differentiation, B cells may simultaneously produce two kinds of heavy chains and thus two classes of immunoglobulins, but no clone has ever been observed to produce more than a single type of light chain. The ratio of κ to λ immunoglobulins varies with animal species. For example, the $\kappa:\lambda$ ratio in humans is about 70:30, whereas in the mouse it is about 95:5.

Because of the differences in their constant domains, the various classes of immunoglobulins have different biological properties. A good example of a functional difference due to H chain variation is the differential transport of various antibodies across membranes. IgA molecules (but not other classes) may be transported across the epithelial lining of the intestine into the lumen, whereas only IgG molecules can pass across the placenta from the circulatory system of the mother to that of the developing fetus. Specific transport, like many other properties associated with a specific class or subclass of antibodies, is thought to involve binding of the effector site of the immunoglobulin molecule to specific cell surface receptors.

The major functional roles of the five classes of immunoglobulins are summarized in Table 24-2. We will discuss each briefly before going on to examine the fine structure of the immunoglobulin molecule.

IgG. IgG is the most abundant antibody produced in late primary and in secondary immune responses and is therefore the major class of immunoglobulins in the blood. It is effective in mediating the destruction of invading bacteria or viruses by phagocytosis because the effector site of the molecule binds to specific receptors on the surface of macrophages, thereby delivering to the macrophage the bacterial cell or virus to which the antigen binding sites have become attached (Figure 24-11). IgG, when it binds to antigen, can activate a family of proteins in the blood called the **complement system.** Activated complement has several important properties, including the enhancement of phagocytosis and an enzymatic activity that can punch holes in cell membranes, leading to destruction of target cells.

IgM. IgM is the first class of antibody to be produced during the development of a B cell, although most B cells will later switch to making other classes of antibody. IgM is also the major class of antibody secreted into the blood early in a primary immune response. In fact, low doses of antigen may stimulate IgM production only; higher doses are usually required to stimulate IgG production. Interestingly, IgM is the first class of immunoglobulin to appear in ontogeny (the developing individual) and phylogeny (evolution). Very primitive vertebrates, such as sharks, have only one class of immunoglobulins, and it is IgM. Thus, IgM appears to be the ancestral immunoglobulin from which the other classes descended. IgM activates the complement system very efficiently, which can facilitate the death and elimination of invading microorganisms.

The membrane-bound form of IgM is the basic $(H\text{-}L)_2$ structure. However, serum IgM molecules are pentamers, with five $(H\text{-}L)_2$ subunits held together by disulfide bridges. These pentameric molecules have ten antigen-binding sites each, allowing them to bind antigens very avidly. An accessory polypeptide called the **J (joining) chain** is associated with the pentameric form of IgM.

Because the monomeric form of IgM serves as a surface receptor, it must be sufficiently hydrophobic to remain anchored in the cell membrane. The pentameric form, on the other hand, is secreted into the bloodstream and must

Table 24-2 Properties of the Five Classes of Human Immunoglobulins

	Class of Immunoglobulin				
Properties	IgG	IgM	IgA	IgD	IgE
Heavy chain	γ	μ	α	δ	ϵ
Subclasses of heavy chain	$\gamma1, \gamma2, \gamma3, \gamma4$	None	$\alpha1, \alpha2$	None	None
Light chains	κ or λ	κ or λ	κ or λ	κ or λ	κ or λ
Formula of monomer	$\gamma_2\kappa_2$ or $\gamma_2\lambda_2$	$\mu_2\kappa_2$ or $\mu_2\lambda_2$	$\alpha_2\kappa_2$ or $\alpha_2\lambda_2$	$\delta_2\kappa_2$ or $\delta_2\lambda_2$	$\epsilon_2\kappa_2$ or $\epsilon_2\lambda_2$
Oligomeric form	None	Pentamer	Dimer	None	None
Accessory chains	None	J (pentamer only)	J (dimer only), secretory component	None	None
Serum concentration, mg/ml	12	1	3	0.1	0.001
Physiological function	Binds to phagoctyes; crosses placenta; activates complement	Activates complement	Crosses epithelial cells	?	Binds to mast cells and basophils

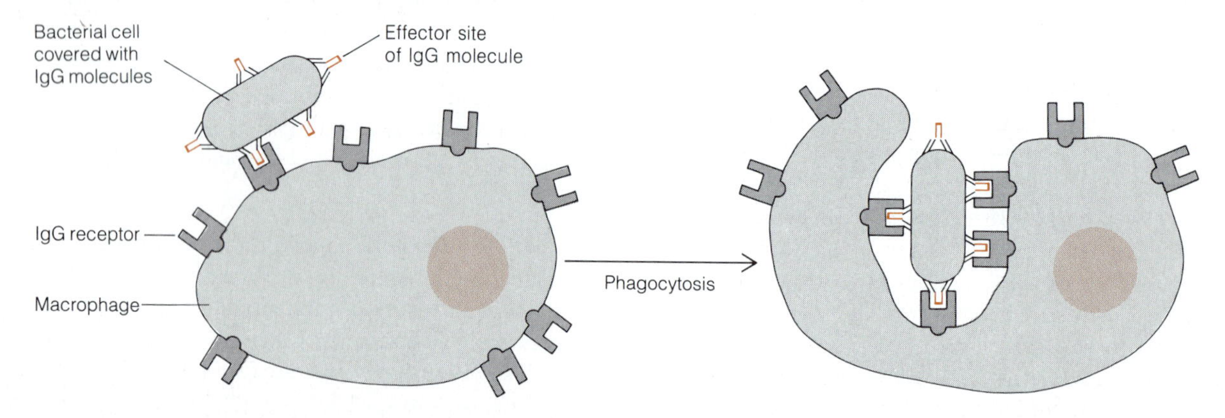

Figure 24-11 Role of IgG in Phagocytosis of a Bacterial Cell. IgG antibodies bind to antigenic determinants on an invading bacterial cell and deliver the bacterium to a macrophage that has surface receptors (gray) to which the effector site of the IgG molecules (color) can bind. The binding of the antibody-coated bacterium to the receptors on the macrophage activates the phagocytotic process.

therefore be sufficiently hydrophilic to circulate as a soluble molecule. This difference in solubility is due to a difference in the amino acid sequence at the carboxyl-terminal end of the μ heavy chain. The membrane-bound form of IgM has a hydrophobic sequence near the carboxyl terminus of its μ chain, which is absent from the secreted form of the protein.

IgA. **IgA** is the most prominent class of antibody found in external secretions such as tears, sweat, saliva, milk, and mucus of the intestinal and respiratory tracts. It is therefore the first line of immunological defense against bacterial and viral antigens. The soluble form of IgA is a dimer consisting of two $(H\text{-}L)_2$ units linked together by a J chain. In addition to the J chain, the dimer contains a polypeptide called **secretory component.** Secretory component becomes associated with the IgA dimer as a result of its passage across the epithelial cells that line secretory ducts. IgA molecules are moved across such cells by forming complexes with specific receptors on the surfaces of the cells. This transport process involves receptor-mediated endocytosis of the IgA-receptor complex on the external (nonluminal) side and exocytosis into the lumen on the other side of the cell (Figure 24-12). Once transport is complete, the receptor is cleaved, with a fragment of it, the secretory component, remaining associated with the secreted IgA dimer.

IgD. The function of **IgD,** a monomeric antibody, is not yet known. Along with IgM, IgD is a prominent class of antibody on the surface of mature resting B cells, but very few B cells secrete IgD, so it is usually present in the blood only in minute concentrations. Surface IgD molecules may play an important role in B cell triggering because B cells do not become responsive to antigens until they express surface IgD. Immature, antigen-unresponsive B cells express only surface IgM.

IgE. Like IgD, **IgE** is normally present in the blood in minute concentrations. IgE is involved in stimulating mast cells in connective tissues to release amines, such as histamine or serotonin, that cause dilation and increased permeability of blood vessels during allergic reactions such as hay fever or asthma. The effector site of the IgE molecule has a high affinity for receptor proteins on the surface of mast cells, but mast cell stimulation requires more than just the binding of the antibody to the cell surface. It is only when two adjacent surface-bound IgE molecules are cross-linked by an antigen molecule that the mast cells are triggered to release granules containing amines. The capacity of IgE to trigger inflammatory reactions is thought to be protective against parasitic infections, although it is not appreciated by hay fever sufferers.

Antibody Valence

The **valence** of an antibody is the number of identical antigen binding sites per molecule. For the basic monomeric structure shown in Figure 24-10, the valence is 2, and the molecule is said to be *bivalent.* IgG, IgD, and IgE are therefore always bivalent antibodies, whereas the dimeric form of IgA and the pentameric form of IgM have higher valences (4 and 10, respectively). The multiple valence of antibodies is very important, because it allows each antibody molecule to bind to at least two separate (though identical) epitopes. Antigens can thereby be cross-linked into chains and lattices that facilitate phagocytosis and complement activation.

The kind and size of antibody-antigen complex that forms depend on the number of different kinds of antibodies present, the valence of the antibodies, the valence of the antigen (number of antigenic determinants that can be recognized by the species of antibodies present), and the concentrations of both antibody and antigen. Figure

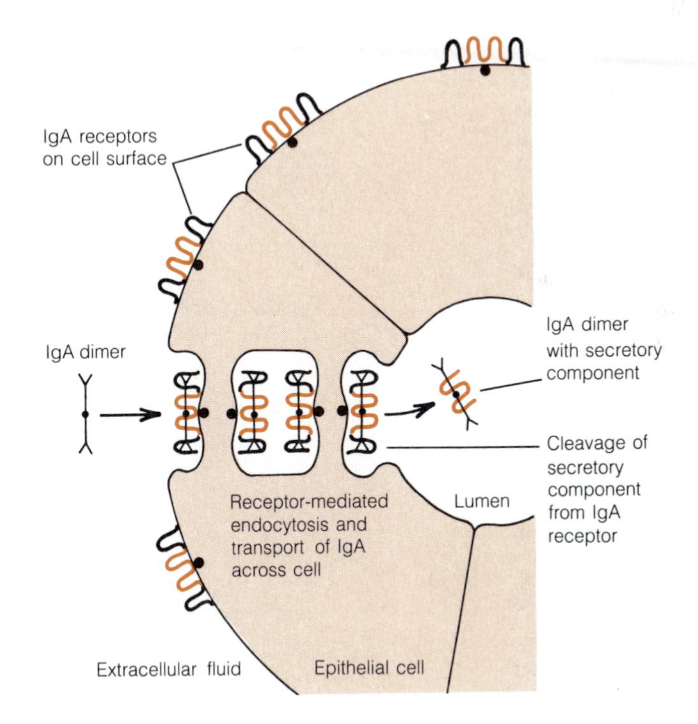

Figure 24-12 Transport of IgA Molecules Across Epithelial Cells of Secretory Tissue. IgA dimers from the blood are transported across epithelial cells from the extracellular fluid into the lumen of secretory tissue by a process that involves receptor-mediated endocytosis on the external surface and exocytosis on the luminal surface. IgA dimers bind to specific receptors on the external surface of the epithelial cells and are transported across the cell as IgA-receptor complexes. On the luminal side, the receptor is cleaved and a fragment of it, the secretory component, remains associated with the IgA dimer.

24-13 illustrates the effect of antigen valence on the kind of complex formed in the presence of a single species of bivalent antibody. Most antibody-antigen reactions involve not just one but many species of antibodies specific for different epitopes on the antigen and are therefore far more complex than the figure indicates.

The Fine Structure of Antibodies

What accounts for the specificity of antibodies, and how are so many diverse kinds of antibodies generated? A

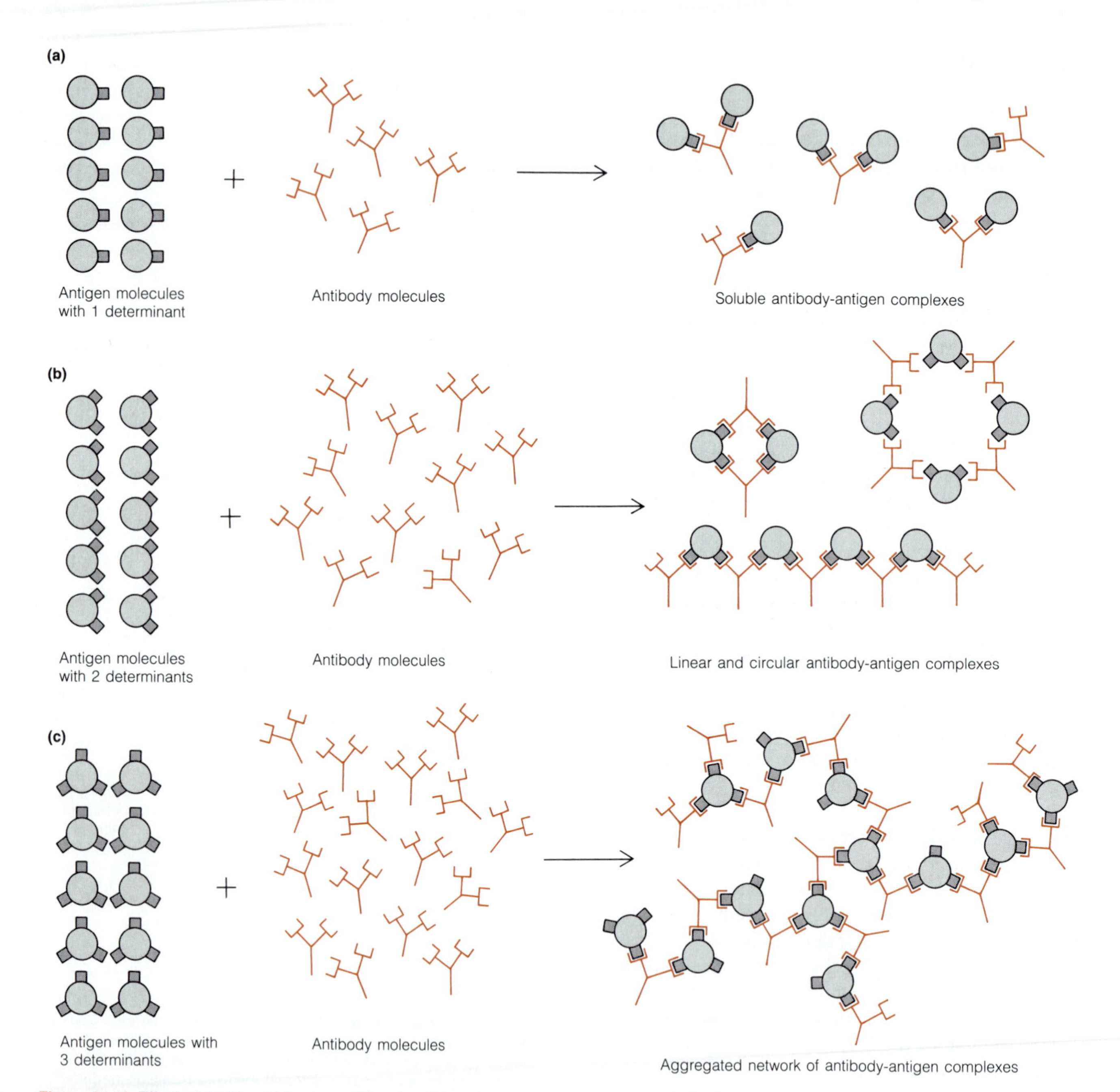

Figure 24-13 Effect of Antigen Valence on Complex Formation by a Single Species of Bivalent Antibody. (a) For antigens with a single antigenic determinant, the antibody-antigen complex is small and soluble, consisting of the antibody bound to one or two antigen molecules. (b) For antigens with two antigenic determinants, the antibody-antigen complex can form either linear or circular chains. (c) With three or more antigenic determinants on each antigen, a large complex can form that readily precipitates out of solution. If multiple species of antibody are present, the reaction is more complex, because different antibodies can react with different determinants on the antigen and thereby cooperate in cross-linking the antigen.

straightforward approach to answering these questions would be to compare the amino acid sequences of light and heavy chains of antibodies with different specificities. On the face of it, that may seem like an impossible feat, since the immunoglobulin fraction of blood contains millions of different antibodies in small quantities, making them hard to distinguish and separate from one another. How could a protein chemist ever hope to find a single antibody needle in so complex a haystack?

A breakthrough came when it was discovered that patients with *multiple myeloma* are a good source of homogeneous immunoglobulins. Multiple myeloma is a cancer of antibody-secreting plasma cells. This type of cancer usually results from the uncontrolled growth of a single antibody-secreting cell, so that the tumors formed consist of a clone of cells all secreting a single kind of homogeneous immunoglobulin, the **myeloma protein.** Large quantities of a single kind of myeloma protein are found in the blood of myeloma patients, making it possible to isolate the protein in pure form. Because myeloma proteins differ from one patient to another, it is possible to isolate a large number of different homogeneous myeloma proteins and compare their sequences.

A purified homogeneous immunoglobulin derived from a single clone of cells is called a **monoclonal antibody.** Apart from myeloma patients, there are now several other sources of monoclonal antibodies. Myeloma tumors can be experimentally induced in certain inbred strains of mice. Such tumors are very stable and produce large quantities of antibody. Moreover, these tumors can be transplanted from one mouse to others of the same inbred line to allow production of large quantities of a single immunoglobulin.

In addition, some mouse myelomas can be grown in vitro as cultured cell lines. Several myeloma cell lines have been selected for special properties and used to generate hybrid cell lines called **hybridomas** that produce specific monoclonal antibodies. Monoclonal antibody production by hybridomas is one of the most exciting and important technical advances in immunology in this century. In addition to facilitating the study of purified immunoglobulins of defined specificity, this technique has led to a number of important breakthroughs in medicine. Monoclonal antibodies are described in the box on pp. 758–760.

Variable and Constant Regions of L and H Chains

When the amino acid sequences of light chains from myeloma proteins or other sources are compared, a striking pattern is evident. All such polypeptides differ from one another in amino acid sequence, but virtually all of the differences are restricted to the N-terminal half of the polypeptides (Figure 24-14a). By contrast, the C-terminal halves of all the κ light chains are identical (or nearly so), as are the C-terminal halves of all the λ light chains. The N-terminal half of the L chain is therefore called the **variable (V_L) region** of the chain, and the C-terminal half is called **the constant (C_L) region.** The variable and constant regions of both the κ and λ light chains are each about 107 amino acids long.

A similar pattern was found when heavy chains were sequenced (Figure 24-14b). Again, the variable region (V_H) is located at the N-terminal end and the constant region (C_H) at the C-terminus. The variable region of the heavy chain is about 117 amino acids in length, approximately the same as the V_L region of the light chain. However, the C_H region of the heavy chain is about 330 or 440 amino acids long, depending on the class of immunoglobulin. Figure 24-15 indicates the variable and constant regions for both the light and heavy chains of a human IgG molecule.

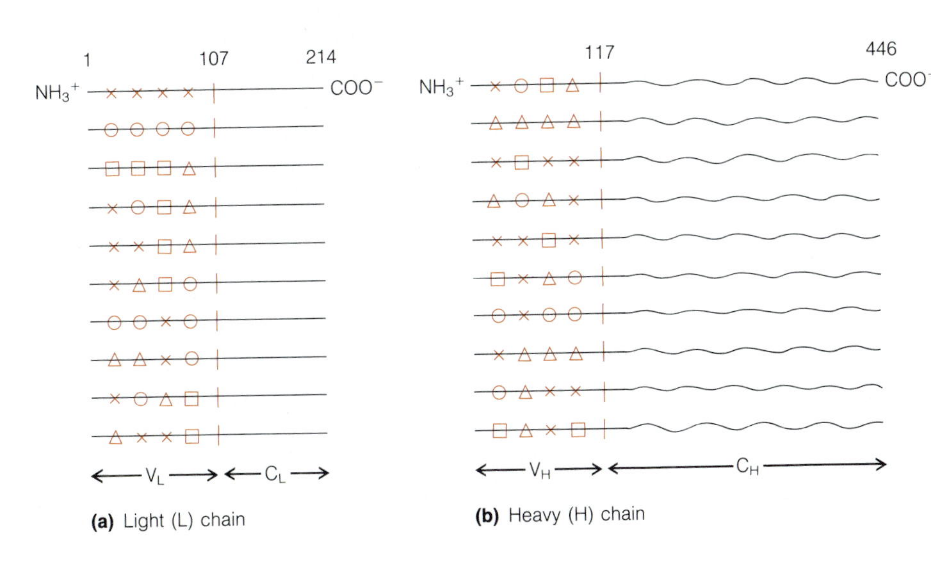

Figure 24-14 Variable and Constant Regions of the Heavy and Light Chains of Human IgG. Ten human IgG molecules are shown schematically, with differences in their amino acid sequences indicated by the symbols ×, ○, △, and □. The lengths of the variable (V) and constant (C) regions are indicated by amino acid number, beginning with amino acid 1 at the N-terminal end. (a) The light chains. (b) The heavy chains.

Monoclonal Antibodies: A Milestone in Immunology

Most antibody preparations are highly heterogeneous. Even when a very pure antigen is used, an animal will inevitably produce a great variety of immunoglobulin species because of the way in which antibodies are generated. Such highly heterogeneous antibody populations are of obvious advantage to the organism and are often useful to the researcher as well. Frequently, however, it is highly desirable to have a homogeneous population of identical antibodies, each with the same antigen binding site. For example, sequencing of immunoglobulins requires such a homogeneous preparation, as do various other research and clinical applications.

Myeloma patients are a valuable source of homogeneous antibody preparations, as are the myeloma tumors that can be induced in mice and the cell lines derived from them. Myeloma antibodies are of only limited use as analytical tools, however, because they are not generated in response to a specific antigen. Instead, a myeloma tumor develops from the random transformation of a normal lymphocyte, and the antibody that is produced by the tumorous clone is the species that happens to be synthesized by that particular lymphocyte. Myeloma tumors therefore provide large quantities of specific antibodies, but the researcher has no control over the species of immunoglobulin produced.

An exciting breakthrough came in 1975 when Cesar Milstein and Georges Köhler discovered a way to obtain homogeneous antibodies of almost any desired specificity, an approach that has already revolutionized the use of antibodies in both research and clinical medicine. For this pioneering work, Milstein and Köhler received the 1984 Nobel Prize in Medicine. (In fact, they shared the prize with Niels K. Jerne, a leading theoretician in immunology who was responsible in part for the clonal selection theory of antibody diversity.)

Milstein and Köhler's technique involves fusing an antibody-producing lymphocyte with a malignant myeloma cell. This approach takes advantage of specific properties of both cell types. A myeloma cell can multiply indefinitely in culture, but does not produce a specific antibody. By contrast, a lymphocyte from an immunized animal produces a specific antibody, but does not grow and divide in culture. When a lymphocyte is fused with a myeloma cell, the product is a hybrid cell called a *hybridoma* that can both proliferate indefinitely and produce large amounts of a single *monoclonal hybridoma antibody*.

The fusion technique is illustrated in Figure 24-A. The antigen of interest is injected into a mouse, causing the proliferation of specific antibody-producing cells. Several weeks later, the spleen of the mouse is removed and a suspension of B lymphocytes is prepared from it. Many of these lymphocytes will be making antibodies to the antigen, but they have only a limited life span in culture. However, they can be fused with cultured myeloma cells to generate hybrid cells that are in effect immortal. Cell fusion is generally induced by the addition of polyethylene glycol, and the resulting hybridomas are then selected by culturing the cells in a medium that does not allow unfused lymphocytes or myeloma cells to grow.

The ability to select hybridomas depends on a special property of the mutant myeloma cell line used in such fusions. The cell line is deficient in the enzyme hypoxanthine-guanine phosphoribosyltransferase (HGPRT), which catalyzes a reaction in one of two metabolic pathways for synthesis of guanine monophosphate (GMP), an indispensable substrate for nucleic acid biosynthesis. Cells deficient in HGPRT are nonetheless viable because they can still make GMP by the other pathway (which is in fact the normal pathway for GMP synthesis in most cells). However, these mutant myeloma cells will not survive if aminopterin is added to the medium, because aminopterin is a potent inhibitor of the normal pathway for GMP synthesis. By contrast, hybridoma cells will be able to grow in the presence of aminopterin, because they contain a functional HGPRT gene supplied by the lymphocyte genome.

This differential sensitivity to aminopterin is used to advantage in the selection procedure. The mixture of lymphocytes, myeloma cells, and hybridoma cells that results from the fusion step is cultured in the presence of a selective medium containing aminopterin. The myeloma cells cannot grow because of the aminopterin, and the lymphocytes grow only very slowly. However, the hybridoma cells grow rapidly, forming large, readily distinguishable colonies. Each such colony is then screened individually for production of antibody against the antigen of interest, and hybrid cells that produce the appropriate antibody can then be cloned either in culture or by growing as tumor cells in a recipient animal.

A tremendous advantage of the hybridoma technique is that monoclonal antibodies can be produced against unpurified molecules, even when they make up only a small fraction of a complex mixture of antigens. Regardless of how many different hybridomas are produced in the fusion step

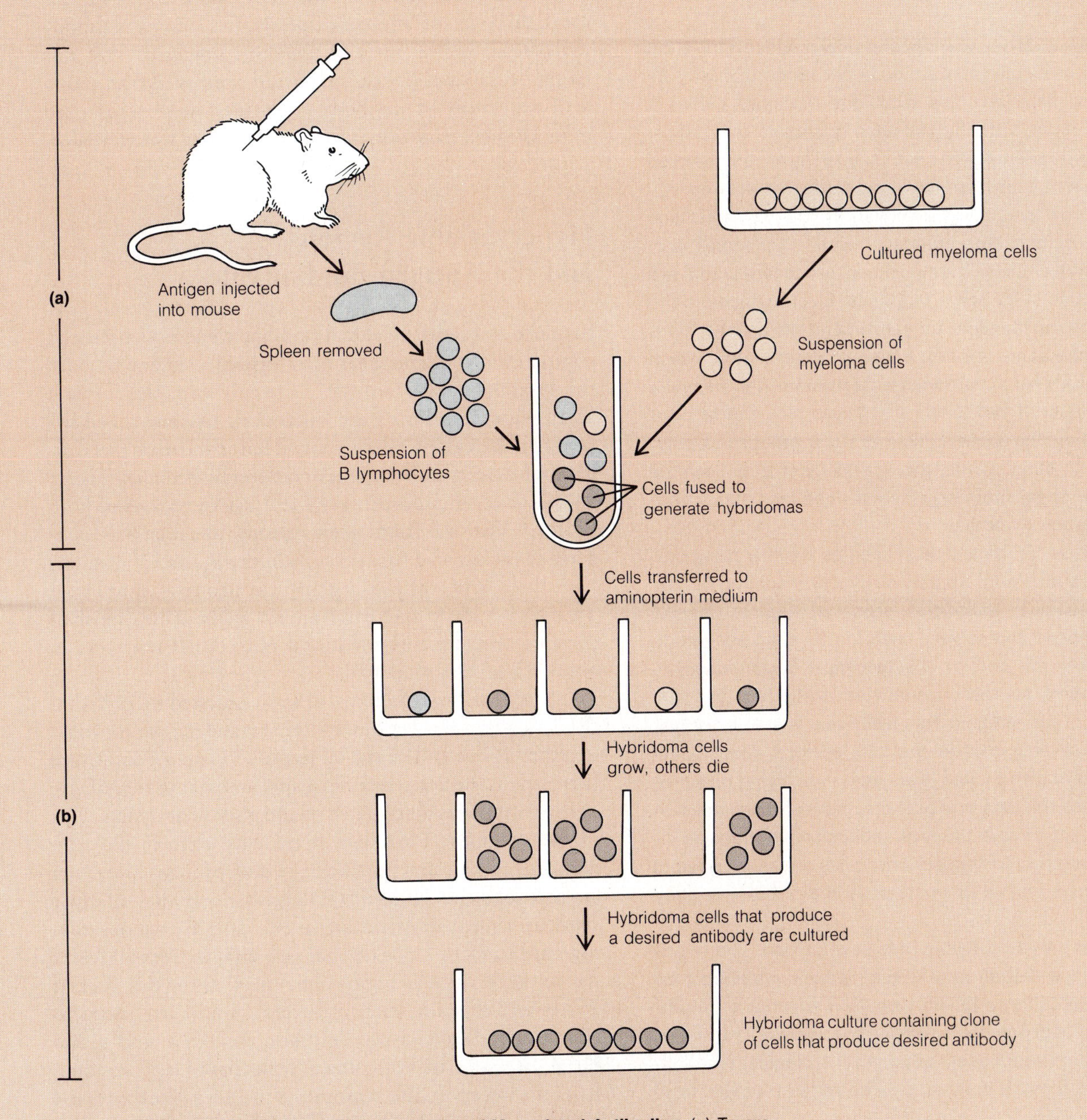

Figure 24-A Hybridoma Technique for Production of Monoclonal Antibodies. (a) To produce hybridoma cells, the spleen of a mouse injected with the antigen of interest is removed, and B lymphocytes (gray) are isolated. These are fused with cultured myeloma cells (light color) to generate hybridoma cells (dark color) that can be cultured indefinitely and produce a single species of antibody. (b) To select for fused hybridoma cells, aminopterin, an inhibitor of the guanine monophosphate (GMP) pathway, is added to the medium. The mutant myeloma cells used in the fusion process cannot survive in the presence of aminopterin because they are deficient in HGPRT, a key enzyme in the alternative pathway to GMP. Hybridoma cells can grow, however, because they have a functional HGPRT gene supplied by the lymphocyte genome.

and how great a variety of antibodies they produce, individual hybridomas producing the antibody of interest can still be detected and selected. This technique therefore makes it possible to obtain virtually unlimited quantities of homogeneous antibody preparations with specificity for any desired antigen, without ever even having to purify the antigen.

Monoclonal antibodies prepared in this way are much more specific than heterogeneous serum antibodies produced by conventional immunization procedures and are much more useful for some purposes. Each antibody in the population has an identical antigen binding site and therefore recognizes the same specific antigenic determinant, which is usually a very small portion of a macromolecule. This may involve a cluster of five to ten amino acid side chains on a protein or about the same number of sugar groups in a polysaccharide. In other words, the specificity of a monoclonal antibody can be far more precisely defined than that of conventional serum antibodies.

Many kinds of monoclonal antibodies are now being produced for use as highly specific analytical reagents. For example, monoclonal antibodies prepared against specific cellular components are very useful in locating such components within the cell by the technique of *immunocytochemistry*, using fluorescence-labeled antibodies to detect the position of the antigen. This technique is also useful for localization of antigens on the cell surface. Monoclonal antibodies have also proved invaluable in purifying rare proteins and other antigens to enable their structure and function to be studied. In clinical practice, monoclonal antibodies directed against specific drugs or hormones are used to detect and quantify very small amounts of such molecules in body fluids.

Although still a relatively new development, the hybridoma technique has already had a tremendous impact on many areas of biology. Given their twin advantages of immortality and antibody homogeneity, hybridomas are clearly here to stay and seem destined to affect biological research and medicine in ways that can only be called revolutionary.

The functional significance of these findings is clear and is related to the bifunctional nature of the antibody molecule: The main sequence differences between antibodies are in the domains responsible for antigen binding (Figure 24-15). Could the sequence heterogeneity in the variable regions of the H and L chains account for the great diversity of antigenic specificities among antibody molecules? Several lines of evidence suggested that this might be the case. But the real proof came when the amino acid sequences of antibody molecules were scrutinized more closely and regions of hypervariability were revealed.

Hypervariable Regions and the Antigen Binding Site

As more and more myeloma proteins were sequenced and compared, it became clear that variable regions of both the H and L chains are not uniformly variable. Instead, small segments of both the V_H and V_L regions, called **hypervariable regions,** are responsible for most of the variability between the V regions of different antibodies (Figure 24-16). The other portions of the variable regions of both chains, called the **framework regions,** are relatively constant in amino acid sequence. Both the V_H and V_L regions have three hypervariable regions, each consisting of about 5 to 10 (or 15, in one case) amino acids. Thus, most of the variability in both the L and H chains is restricted to about 15 to 30 amino acids.

Based on these findings, it was suggested in 1970 that the hypervariable regions of the H and L chains come together at one end of the molecule to form a pocket that functions as the antigen binding site, and that the specificity of the antibody is determined by the particular amino acids in these regions. Two lines of evidence confirm this suggestion. First, several purified myeloma proteins have been crystallized, and X-ray crystallographic studies of their structures have shown that the V_H and V_L regions fold together to form a pocket that can bind antigens. Five of the six hypervariable regions line the walls of this pocket. Moreover, crystallization of several antibodies bound to simple epitopes confirmed that the epitope is bound within the pocket and makes contact with amino acid residues from the hypervariable regions. Figure 24-17 illustrates the binding of an antibody to a simple antigen, a derivative of vitamin K_1. Notice how five of the six hypervariable regions of the antibody molecule (in gray) interact intimately with the antigen (in color) to hold it in the antigen-binding pocket.

A second line of evidence comes from studies in which small antigens were synthesized with a chemical group that could link covalently to the antibody molecule once the antigen was within the antigen binding site. This technique, called *affinity labeling*, showed that such synthetic antigens attached only to amino acids in the hypervariable regions of the L and H chains.

A significant feature of antibody structure is that the antigen binding site involves portions of two separate polypeptides, the L and H chains. This property contributes significantly to antibody diversity, because different H and L chains can presumably combine to generate different kinds of antigen binding sites. Theoretically, if any H chain can associate with any L chain to form a functional antibody, then 1000 different H chains and 1000 different L chains could combine to form 1000 × 1000, or 10^6, different antibody molecules. It is not yet known whether every possible combination in fact leads to a functional antibody, but even if only a fraction of all possible combinations are active, this ability to form **combinatorial associations** would be a great source of diversity.

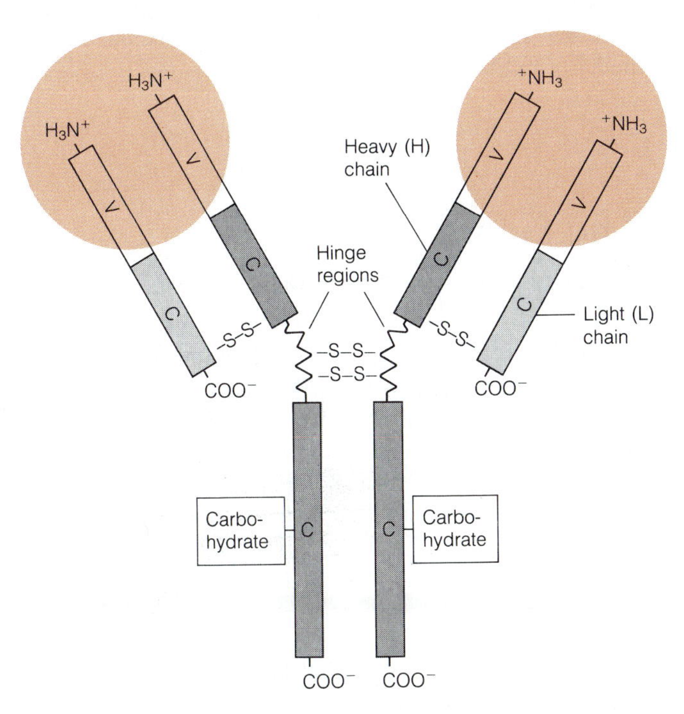

Figure 24-15 Variable and Constant Regions of the Immunoglobulin Molecule. A human IgG molecule, showing the variable (V) and constant (C) regions of both the light (L) and heavy (H) chains. Also shown are the interchain disulfide bridges, the carbohydrate groups, and the hinge regions of the heavy chains.

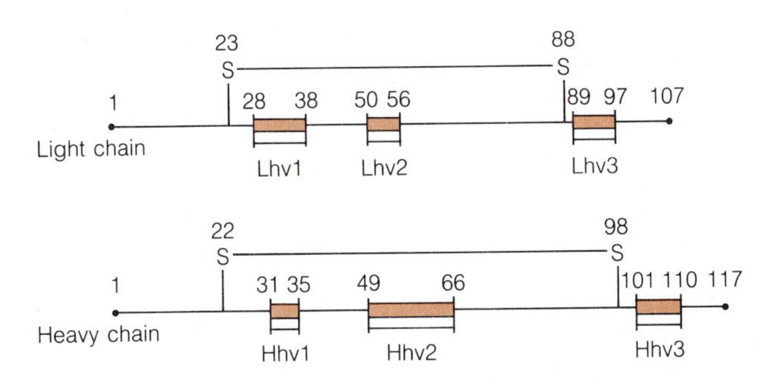

Figure 24-16 Hypervariable Regions of the Light and Heavy Immunoglobulin Chains. Linear map of the variable regions of both the light and heavy chains from human IgG, illustrating the localization of sequence variability to three small segments of each chain (color). These hypervariable (hv) regions are identified by an L (for light chain) or an H (for heavy chain) and are numbered 1, 2, and 3 from the amino-terminal end of each chain. Amino acids bracketing each hypervariable region are numbered, as are the cysteines that give rise to the disulfide bond in the variable region of each chain.

Immunoglobulin Fingerprints: Idiotypes

Each clone of B cells produces a homogeneous immunoglobulin with unique V_H and V_L sequences that distinguish it from all other monoclonal immunoglobulins. These

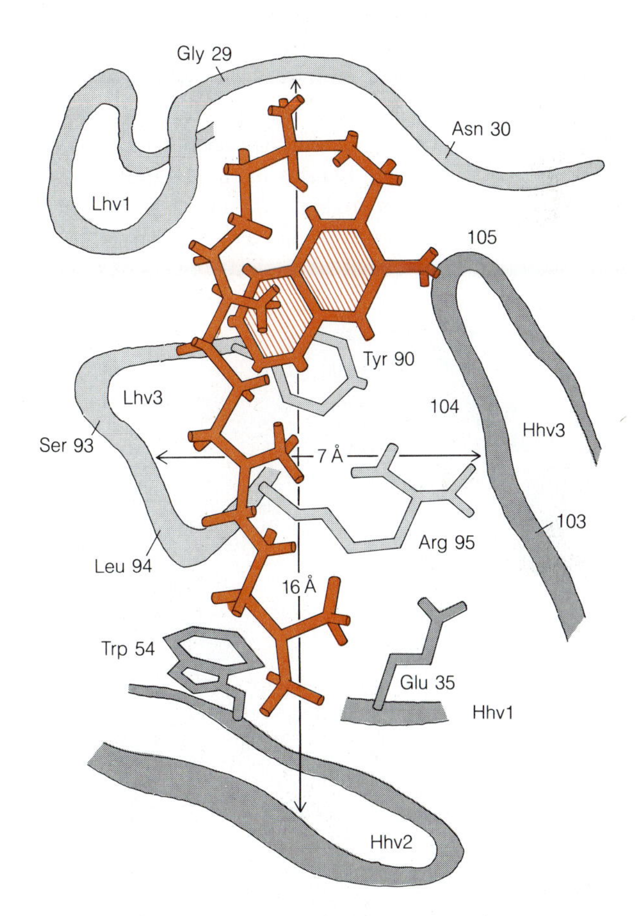

Figure 24-17 Role of Hypervariable Regions in Antigen Binding. The binding of a derivative of an antigen, a derivative of vitamin K_1, to a human IgG molecule. The approximate location of two of the hypervariable (hv) regions of the light chain (light gray) and the three hv regions of the heavy chain (dark gray) are indicated with respect to the antigen molecule (color). Several specific amino acids along each chain are numbered for cross-reference with Figure 24-16.

unique sequences comprise the **idiotype** of that immunoglobulin. An idiotype, then, is a fingerprint or signature of a single B cell clone. The pool of immunoglobulins in blood is composed of minute quantities of many millions of different clonal products or idiotypes. When an immunogen stimulates an antibody response, the concentration of those antibodies (and their idiotypes) becomes transiently elevated (Figure 24-4). Although idiotypes are really "self" rather than "foreign" epitopes, this radical change in concentration provides an antigenic stimulus, inducing the production of **anti-idiotype antibodies.** It has been shown that anti-idiotype antibodies appear during the course of normal immune responses, usually during the declining phase of the original antibody response (Figure 24-4), and they are believed to be involved in the regulation of the antibody response.

Folding of Immunoglobulin Molecules

Comparison of the amino acid sequences of various H and L chains has revealed many sequence similarities in different parts of the immunoglobulin molecule. For example, V_H and V_L, the variable regions in the H and L chains, respectively, are homologous to each other. Similarly, the constant region of the H chain consists of three homologous segments, each also homologous to the constant region of the L chain.

Based on these findings, it was suggested that the L and H chains of immunoglobulins are made up of repeating segments that fold into compact globular domains connected by short segments of more extended polypeptide chains. This prediction was confirmed by X-ray analysis of myeloma fragments and of an intact immunoglobulin molecule. Figure 24-18 illustrates the three-dimensional model of antibody structure that has emerged from such studies.

The Genetic Basis of Antibody Diversity

We already know that the immune system of vertebrates is capable of making an incredibly large number of antibodies—perhaps as many as 10^8 to 10^9 different molecules. This **antibody repertoire,** as it is called, can recognize virtually any foreign substance the animal will ever encounter. To understand how this large repertoire is generated has been a major focus of immunological research. From this research has come the remarkable finding that the DNA

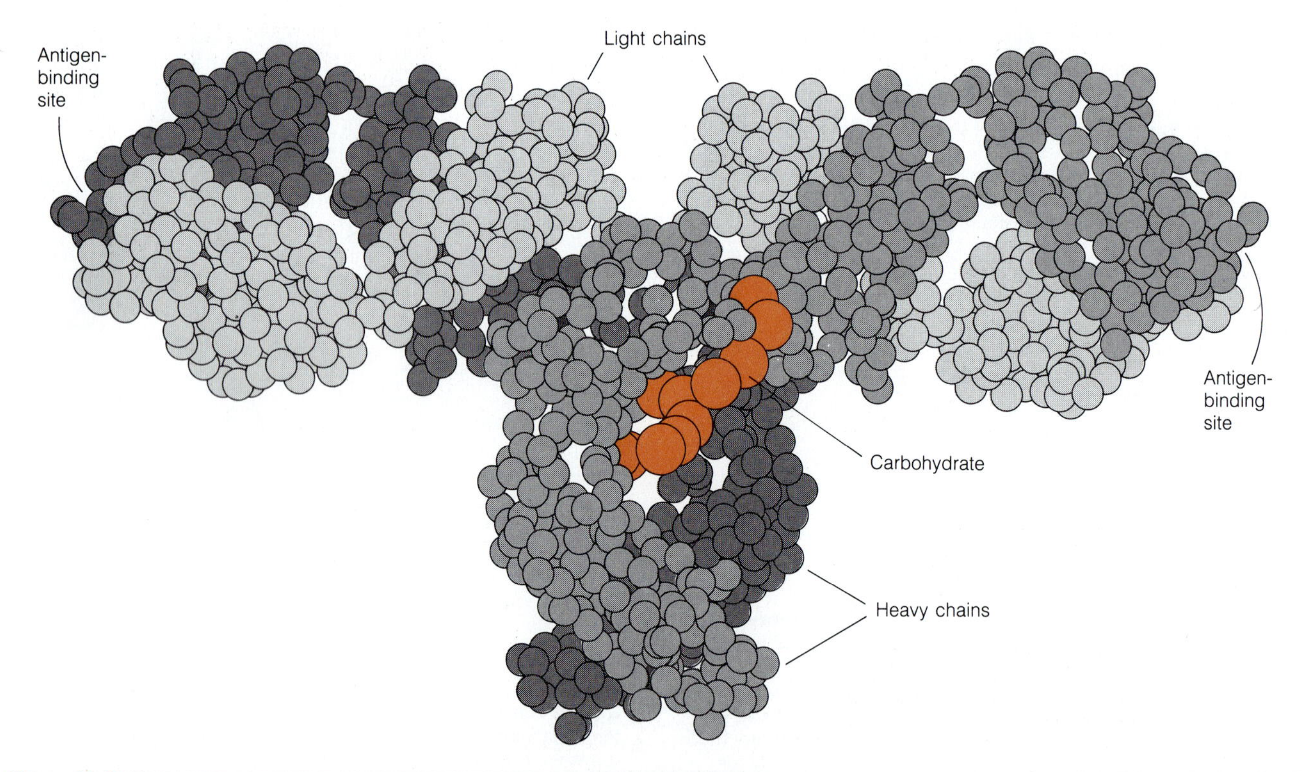

Figure 24-18 Three-Dimensional Structure of an Immunoglobulin Molecule. Each amino acid is drawn as a small sphere. The heavy chains are shown in dark gray, the light chains in light gray, and the carbohydrate in color.

sequences that encode an immunoglobulin molecule exist at several locations within the genome and are assembled during differentiation of B cells in a way that accounts for much of the diversity of antibodies.

Separate V Genes and C Genes Are Brought Together During B Cell Differentiation

We have already encountered an important clue to understanding the genetic basis of antibody diversity. Each L and H chain consists of distinct variable (V) and constant (C) regions. In 1965, William Dreyer and J. Claude Bennett proposed that the V and C regions are encoded by two separate genes that are somehow brought together during differentiation of B cells. The suggestion that eukaryotic genes could somehow be split in the genome and rearranged during development ("jumping genes") was revolutionary at the time, but has now been amply verified.

The first direct support came when it was found that the immunoglobulin genes present in mouse myeloma cells that produce antibodies are different from the genes present in embryonic cells that do not. To show this, DNA from the two cell types was first digested with a restriction nuclease and then hybridized to a radioactive mRNA probe corresponding to the V + C sequence of a specific L chain or to the C sequence only (Figure 24-19). The V_L and C_L coding sequences were found on different restriction fragments in embryonic DNA (Figure 24-19a), but on the same fragment when myeloma DNA was used (Figure 24-19b). The conclusion seemed inescapable: Immunoglobulin DNA is rearranged during B cell development. In other words, genes that start out far away from one another in the genome of an undifferentiated precursor cell are brought together at the right moment during B cell differentiation. Tsusumu Tonegawa was awarded the Nobel Prize in Medicine in 1986 for this landmark research.

The κ light chains, λ light chains, and all classes of heavy chains are each encoded in germ-line DNA by separate clusters of genes on different chromosomes. These three gene clusters are called the κ, λ, and H **gene families**, respectively (Figure 24-20). Each gene cluster has a set of different **V genes** (only two of V_λ, but about 250 each of V_κ and V_H in mice; the size of the gene families varies in different species) separated from one or more **C genes** by hundreds of kilobase pairs of DNA.

Multiple Gene Segments as a Source of Antibody Diversity

The three families of V genes in the germ line can account for a considerable amount of antibody diversity by combinatorial association of different L and H chains. If 250 L chains with different V regions can combine effectively

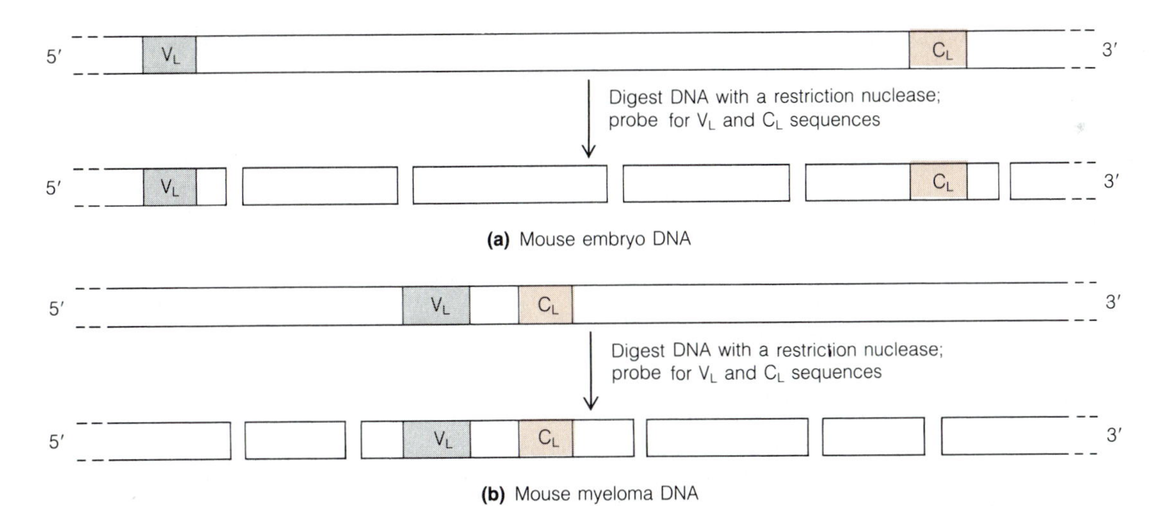

Figure 24-19 Proximity of V and C Genes in Myeloma and Embryo DNA. To determine the location of the genes coding for the V_L (gray) and C_L (color) regions of the mouse light chain, mouse DNA was digested with a restriction nuclease, and the digestion fragments were separated by electrophoresis on agarose. The restriction fragments were then analyzed for the presence of V_L and C_L coding sequences by hybridization with a radioactive mRNA probe corresponding either to the V + C sequences of a specific myeloma λ chain or to the C sequence only. (a) In DNA from a 13-day mouse embryo, the V_L and C_L genes are always found on separate fragments. (b) In DNA from a mouse myeloma synthesizing a specific λ light chain, the sequences are always found on the same DNA fragments.

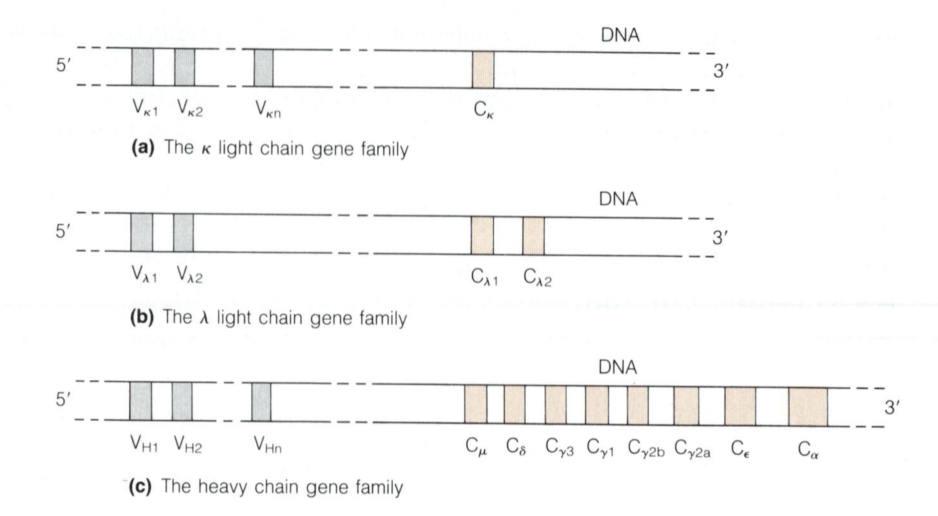

Figure 24-20 The κ, λ, and H Gene Families. In the mouse, the κ gene family contains several hundred V genes and one C gene (a), the λ family contains two V and two C genes (b), and the H gene family contains several hundred V genes and eight C genes (c). In each case, V genes are shown in gray and C genes in color.

with 250 different H chains, over 60,000 antibodies with different L-H combinations can be generated. This is already an impressive diversity, but it clearly falls short of accounting for the 10^8 to 10^9 antibodies in the total repertoire. Clearly, other mechanisms are necessary to generate the additional diversity.

An additional source of antibody diversity was discovered when the DNA base sequences of cloned L chain genes and H chain genes from embryonic cells and myeloma cells were compared as described above. It quickly became clear that the gene segments actually encoding the proteins are separated in embryonic cells. For instance, mouse embryonic DNA contains a V_λ gene segment that encodes for 97 of the 110 amino acids of the V region of the λ chain. The remaining 13 amino acids are encoded by a small DNA segment that is located not in the vicinity of the V gene, but near the C gene instead. Such a segment of DNA is called a **J (joining) gene segment** because it joins the V gene to the C gene in a differentiated B cell. J genes encode the last ten or so amino acids of both the L and H chains, including part of the third hypervariable region. (Be careful not to confuse the J gene with the J chain present in oligomeric IgA and IgM molecules; despite the common name, the two are not related in any way.)

Since both the V and J genes code for only part of the V region of an L chain, they are usually referred to not as genes, but as **gene segments.** The κ and λ gene families have sets of V and J gene segments and one or a few C genes (Table 24-3). The H chain is also encoded by several gene segments. In addition to a C gene and V and J gene segments, the H chain gene family has a set of very small sequences called **D (diversity) gene segments,** which code for the most variable portion of the heavy chain variable region (Table 24-3).

Table 24-3 Abundance of Genes and Gene Segments in the Three Immunoglobulin Gene Families of the Mouse and Human Genomes

Gene Family	Gene	Gene Segment	Copies per Haploid Genome: Mouse	Copies per Haploid Genome: Human
λ light chain	V_λ	V	2	2
		J	2	2
	C_λ		2	6
κ light chain	V_κ	V	200–300	200–300
		J	4	5
	C_κ		1	1
Heavy chain	V_H	V	200–300	200–300
		J	4	6
		D	10–20	10–20
	C_H		8	10

Gene Segments Are Assembled by Somatic Rearrangement

Once we understand that the L and H chains are both encoded by several separate gene segments, we can ask how these segments are brought together and expressed during differentiation. As Figure 24-21 shows, this is a multistep process, involving both the rearrangement of DNA sequences and the splicing of RNA transcripts. The initial event is a **somatic rearrangement** that assembles a complete V gene (Figure 24-21a and b, step 1). For the light-chain gene, this rearrangement joins a V segment to a J segment; for the heavy-chain gene, a D segment is first linked to a J segment, followed by the transposition of a V segment to the 5′ side of the DJ sequence, yielding a complete V_H gene.

These genetic rearrangements involve deletion of the intervening DNA between the gene segments, which proceeds by way of interactions between special **recognition sequences** adjacent to the V, D, and J segments. Thus, for assembly of a V_κ gene, the recognition sequence next to a V_κ segment interacts with a complementary recognition sequence next to a J_κ segment. The loop of DNA between the two segments is then excised as the V_κ and J_κ segments are spliced together. The variable segments of the heavy-chain genes are assembled in a similar manner, but by a two-stage process: A D_H segment is first spliced to a J_H segment, and the hybrid DJ_H segment is then joined to a V_H segment.

The next two steps in the expression of immunoglobulin genes involve the transcription of the C gene and the rearranged V gene as a single large transcript (Figure 24-21, step 2) and the splicing of that transcript to remove noncoding regions, thereby forming a functional mRNA molecule (step 3). The mRNA molecule is then translated to generate the immunoglobulin L chain and H chain (step 4), each of which is synthesized initially with a short N-terminal signal sequence that is cleaved as the polypeptide appears in the lumen of the rough endoplasmic reticulum.

Somatic rearrangement does not always occur at a precise location in the recognition sequences. In the case of the gene for the κ chain, for example, the same V_κ and J_κ segments can be joined in slightly different ways. Several different nucleotide sequences are generated in this way at the junction of the V and J segments, each capable of coding for a different amino acid. The amino acid variability that this introduces into the κ chain is particularly significant because the DNA sequence at the V-J juncture specifies part of the hypervariable region. The same flexibility is seen in heavy-chain V genes when D and J or DJ and V segments are joined. Thus, the flexibility of somatic recombination (**junctional diversity**) provides yet another contribution to the generation of antibody diversity.

How and when are these various immunoglobulin gene segments brought together and expressed? Recall that each B cell makes only one kind of antibody with a single antigen binding site. Thus, there must be a mechanism limiting B cells to the expression of a single V_H gene and a single V_L gene; otherwise, they would make a variety of H chains and L chains. The expression of only one H chain gene and one L chain gene is called **allelic exclusion,** which simply means that the alternative alleles fail to be expressed. Thus, although we are diploid organisms, a given B lymphocyte expresses only the paternal or maternal gene for an antibody polypeptide chain.

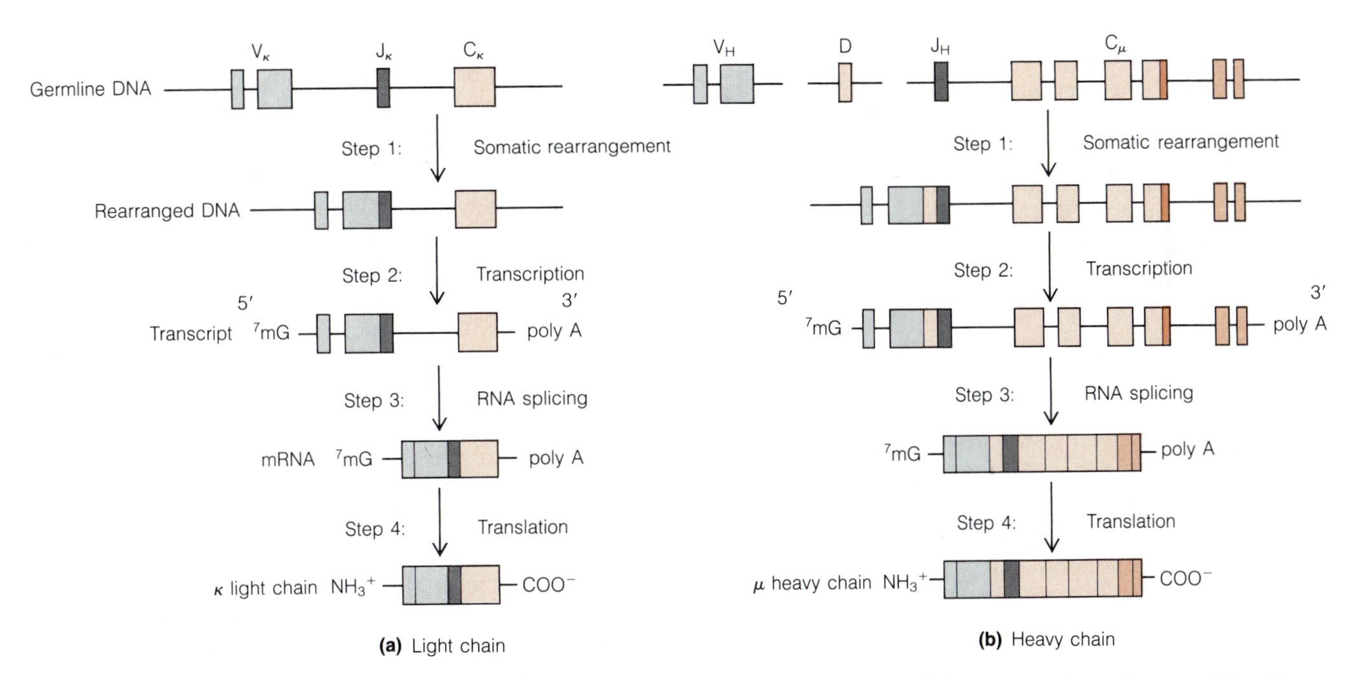

Figure 24-21 Rearrangement and Expression of Immunoglobulin Genes in B Cells. The genes for both (a) the light chain and (b) the heavy chain of an antibody molecule are assembled from several gene segments by genetic rearrangement of germ-line DNA during B cell differentiation (step 1). The particular light-chain and heavy-chain genes shown are those for the κ light chain and the μ heavy chain. The light-chain gene is assembled from V, J, and C segments; the heavy-chain gene requires a D segment in addition. For simplicity, only a single gene segment of each kind is represented. The rearranged DNA is then transcribed into an RNA transcript with a ^{7}mG cap at the 5′ terminus and a poly-A sequence at the 3′ terminus (step 2). RNA splicing eliminates introns and generates a functional mRNA molecule (step 3). The mRNA is then translated to synthesize the desired polypeptide (step 4).

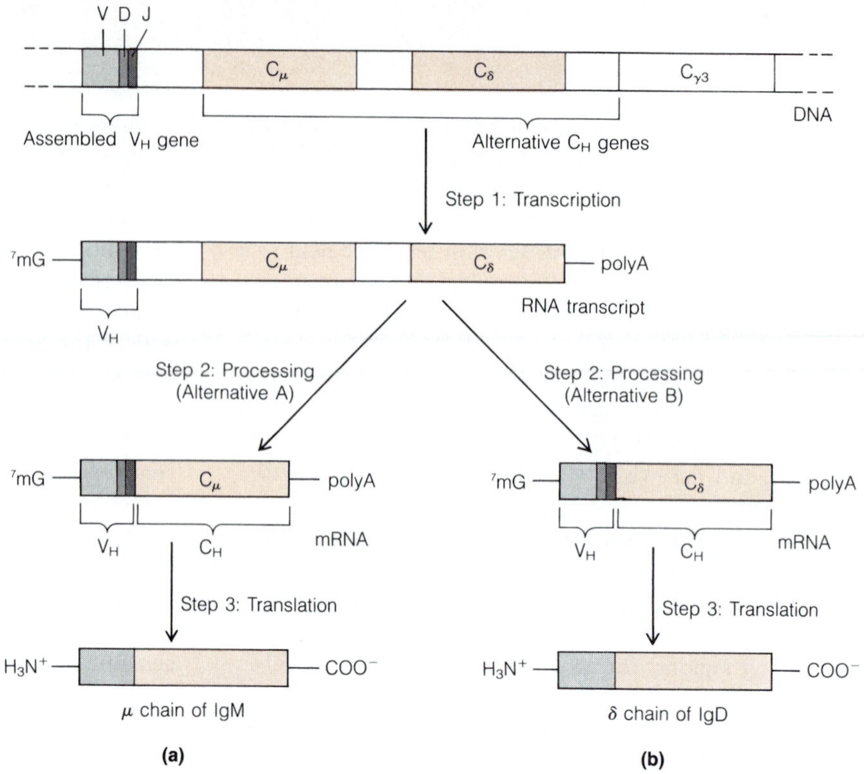

Figure 24-22 Expression of IgM and IgD by Alternative Splicing and Processing of RNA. The first step in the switch in antibody production from one class to another is thought to involve alternative splicing and processing of a single long RNA transcript that encodes both of the alternative C_H regions (and possibly others as well). In the case shown here, the switch from IgM production to IgD production involves a change in the processing of the same initial transcript. (a) Initially, processing is such that the transcript of the assembled V_H gene (consisting of V, D, and J segments) is linked to the C_μ sequence, generating the mRNA for the μ chain of IgM. (b) Later, processing is such that the V_H gene transcript becomes linked to the C_δ sequence, generating the mRNA for the δ chain of IgD.

Hypermutation as a Source of Antibody Diversity

Thus far, we have considered the contributions to antibody diversity made by multiple V gene segments, variation in joining sites of the segments (junctional diversity), and combinatorial association of H and L chains. Yet another mechanism that contributes significantly to antibody diversity is an unusually high mutation rate in the V gene segments, which has been called **hypermutation.** These somatic mutations, which occur during the expansion and differentiation of activated B lymphocytes, can increase the antibody repertoire by two to three orders of magnitude. The mechanism that accounts for hypermutation within V regions is not well understood as yet. The mutations accumulate with time and are therefore more numerous in B cells that have already switched from IgM production to the production of one of the later classes of immunoglobulins, such as IgG or IgA. Consequently, antibodies of classes other than IgM usually display many more mutations within the same V_H gene.

Class Switching

Having seen how the enormous diversity in antigen-binding sites arises genetically, we turn now to the question of how the switch from one class of immunoglobulins to another is made as a B lymphocyte matures. We will see that two distinct mechanisms are involved in the sequential expression of different classes of immunoglobulin.

All B lymphocytes begin by synthesizing an IgM molecule, which is initially expressed only as a membrane-bound antigen receptor. At this stage of its life cycle, the B cell is immature and is unresponsive to antigen. Responsiveness to antigen is acquired with the simultaneous surface expression of IgD as well as IgM. The two membrane immunoglobulins use the same L chain and V_H genes, changing only the C_H gene. This simultaneous expression of IgM and IgD occurs by **alternative splicing** of a large RNA transcript that contains the assembled V_H sequence and the C_μ and C_δ sequences, which are adjacent to each other on the chromosome (Figure 24-22). The transcript is spliced in two different ways (step 2), generating mRNAs that have the V_H sequence linked to one or the other of the two C_H sequences. Translation of these two different mRNAs (step 3) results in the synthesis of two different H chains, both having the same V region and therefore the same antigen binding site, but different C regions.

Following antigen-induced activation, B lymphocytes produce a secreted form of their IgM. As maturation proceeds, the B cell switches from IgM secretion to the secretion of another class of immunoglobulin, usually IgG but sometimes IgA or IgE. This class switch to H chains encoded by C_H genes downstream from C_δ has a mechanism that is different from alternative splicing but similar to the rearrangement of V gene segments. In this case, the **VDJ** segment is **transposed** to another C_H gene with elim-

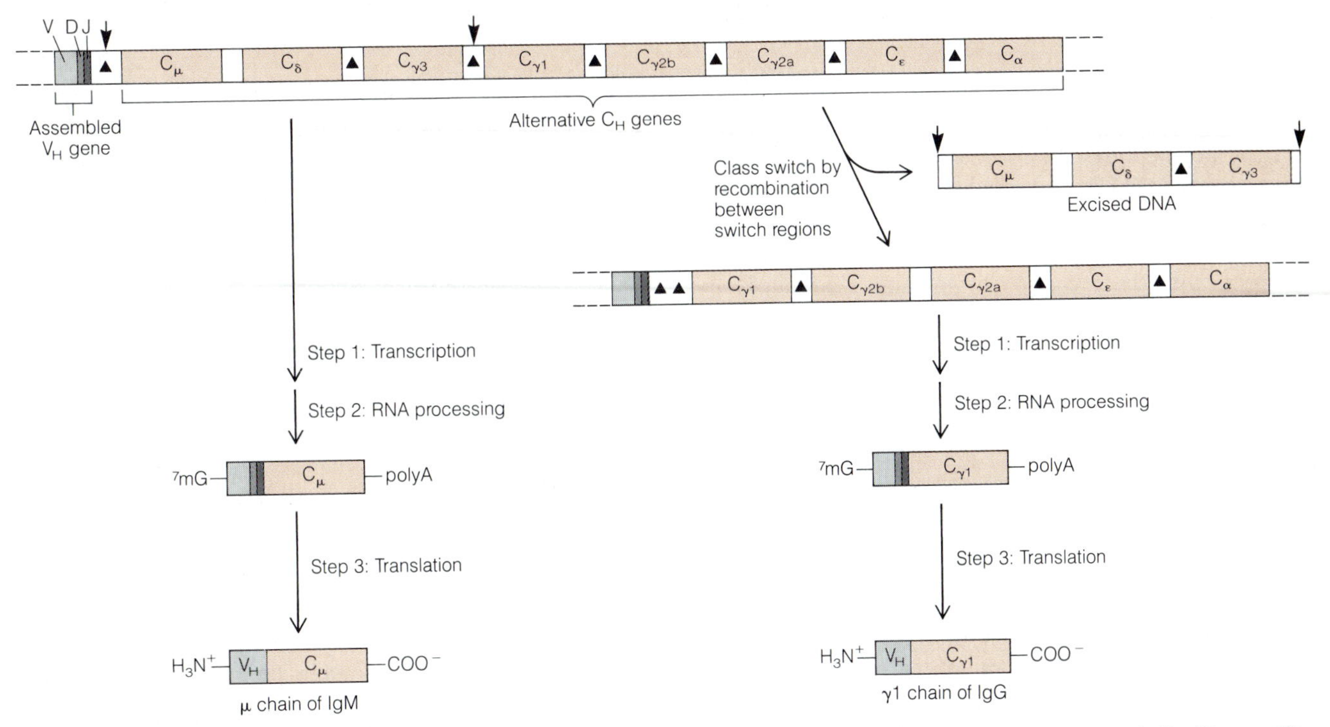

Figure 24-23 Class Switching at the DNA Level. A B cell initially makes IgM following the rearrangement of V, D, and J segments. When the cell switches to a new isotype (IgG_1 in this example), the DNA between VDJ and the C_H gene to be expressed is excised and discarded, bringing VDJ and $C_H\gamma 1$ into proximity. The specific switch regions located upstream of each C_H gene segment except C_δ are denoted by triangles and are the regions within which recombination takes place.

ination of the intervening DNA (Figure 24-23), which includes the C_H genes residing between VDJ and the new C_H gene to be expressed. Such deletions of DNA occur by genetic recombination between S (**switch**) **regions** that lie in front (upstream) of each C_H gene. These switch regions are composed of multiple copies of short sequences that apparently serve as recognition elements for recombination. Transcription and translation of the new H chain gene then occur as usual, yielding a new class or *isotype* of antibody with the original antigen binding site (Figure 24-23). An important ramification of this process is that in a given B cell class switching can occur sequentially in a downstream direction—for example, from μ to γ_3 to α (Figure 24-20c)—but the B cell cannot return to the synthesis of an upstream class because the gene encoding that C_H region has been excised and lost. Class switching provides the immune response with flexibility, inasmuch as antibodies with identical specificity for antigen but with different biological properties can be made against a particular pathogen.

The T Cell Antigen Receptor

T lymphocytes, like B lymphocytes, express clonally distributed specific receptors for antigen. The antigen receptors on the two types of lymphocytes are similar in many respects, but there are several important differences between them. For one thing, T cell antigen receptors are found only in a membrane-bound form, whereas immunoglobulins exist as membrane-bound and secreted proteins. For another, T cells recognize antigens only when they are presented in association with MHC molecules, whereas antibody specificity is directed exclusively at the foreign antigen. Consequently, the T cell receptor must recognize either a foreign epitope and self MHC with separate binding sites, or a complex epitope composed of parts of the foreign antigen and the self MHC protein with a single binding site. Experimental evidence favors the latter interpretation, so the T cell receptor can be visualized as possessing a single binding site specific for a foreign epitope plus self MHC.

The T cell receptor itself is composed of two disulfide-linked polypeptide chains of about equal size. Each of the chains, like immunoglobulin H and L chains, has an N-terminal variable domain and a C-terminal constant region. Two forms of the antigen receptor are found in the pool of T cells. The majority of T cells express a receptor composed of **α and β chains,** but a small fraction of cells found primarily in thymus, skin, and gut epithelium express a different receptor with **γ and δ chains.** The functional significance of the two receptor types is unknown.

At the DNA level, the four types of polypeptide chains are encoded by four separate gene families, much like the immunoglobulin H and L gene families. The analogy with immunoglobulins is carried further, inasmuch as the T cell receptor polypeptides are also encoded by V, D, J, and C gene segments, which are brought together by site-specific recombination during T cell differentiation in the thymus. T cells use the same mechanisms that B cells use to generate diversity, with the exception of somatic hypermutation, which has not been observed in the V regions of T cell receptors. Instead, junctional diversity seems to play a particularly important role in the generation of receptor diversity in T cells.

The α/β and γ/δ receptors are physically associated on the T cell surface with the CD3 marker found on all T cells (Table 24-1). This complex is believed to be essential for the transmission of the antigen-induced activation signal to the interior of the cell. All T cells, whether functionally cytotoxic (CD8) or helper (CD4), appear to use the same gene pools to construct their antigen receptors. There is recent evidence, although it is not yet conclusive, for biased selection of V genes from the pools depending on whether the T cell is MHC class I-restricted or class II-restricted. Thus, certain V genes may be preferentially expressed by CD4 T cells, while others are preferentially expressed by CD8 T cells.

Perspective

The immune system is a highly specific defense mechanism that enables vertebrates to recognize and eliminate foreign invaders. The immunogen to which the system responds is usually a foreign macromolecule with multiple antigenic determinants (epitopes). The response is mediated by lymphocytes that originate from hematopoietic stem cells, develop either in the thymus (for T cells) or the bone marrow (for B cells), and later migrate to the peripheral lymphoid tissues. Each lymphocyte carries identical antigen receptors on its surface that are specific for a structurally related set of epitopes.

Lymphocytes become committed to the synthesis of a particular antigen receptor before exposure to antigen. The total lymphocyte population is composed of millions of different clones, each expressing its particular receptor, thus endowing the system with enormous diversity. When a lymphocyte encounters an antigen that it recognizes, a complex sequence of events takes place that culminates in the differentiation of the lymphocyte into a mature functional cell. B lymphocytes become antibody-secreting plasma cells. T lymphocytes are subdivided along functional lines. Two types, helper T cells and suppressor T cells, regulate the immune response, while a third type, cytotoxic T cells, kill cells expressing foreign antigens on their surface. Whereas B cells recognize foreign antigens alone, T cells recognize foreign antigens only when they are associated with cell surface proteins encoded by genes of the major histocompatibility complex (MHC). Thus, T cells respond only to antigens on the surfaces of other cells. Activated T cells exert their functional activities by secreting soluble mediators called lymphokines or interleukins that act on other cells in the system.

The antibodies synthesized and released by plasma cells are Y-shaped immunoglobulin molecules consisting of two identical light (L) chains and two identical heavy (H) chains. Each arm of the molecule carries an antigen binding site and consists of one L chain and the N-terminal half of one H chain. The leg of the Y consists of the C-terminal halves of both H chains and is responsible for the effector functions that determine the physiological role of each class of antibody.

Both the H and L chains have a variable region V, consisting of about 110 amino acids at the N-terminal end, and a constant region C, comprising the rest of the molecule. The V and C regions of both chains are encoded by genes that are far apart in the germ-line DNA but are brought together by somatic recombination during lymphocyte differentiation. The genes for the variable regions of both chains are assembled randomly from gene segments (V and J for the light chain; V, D, and J for the heavy chain), giving rise to a great diversity of L and H chains. These chains can, in turn, combine in many possible ways to generate a still greater variety of antibodies. The sequence heterogeneity of the variable regions of both the L and H chains accounts for the great diversity of antigenic specificities among antibody molecules. For both chains, the greatest genetic variability within the V region is concentrated in three hypervariable regions that together form the antigen binding site of the immunoglobulin molecule.

A B lymphocyte expresses the same V-region genes throughout its life, as do all of its descendants, but the antigen specificity of antibodies can be altered by somatic mutations, which occur with high frequency in V regions. A B cell can switch to making a different class of antibody by changing from one type of C_H region to another without changing its V gene or its L chain. In this way, a lymphocyte can switch effector function without changing antibody specificity.

T cells also express antigen-specific receptors with variable and constant regions, but they are encoded by different gene families from those encoding antibodies. As with antibodies, the variable regions of T cell receptors are assembled from V, D, and J gene segments.

Key Terms for Self-Testing

innate immunity (p. 741)
adaptive immune system (p. 741)
acquired immunodeficiency syndrome (AIDS) (p. 741)
immunology (p. 741)
immunity (p. 741)

The Immune Response
immune response (p. 741)
immunogen (p. 741)
antigen (p. 741)
autoimmune diseases (p. 742)
antigenic determinants (p. 742)
epitopes (p. 742)
autoimmune reactions (p. 742)
cell-mediated immune responses (p. 742)
humoral immune responses (p. 742)
antibodies (p. 742)

The Cellular Basis of the Immune Response
lymphocytes (p. 742)
lymphoid tissues (p. 742)
B cells (B lymphocytes) (p. 742)
T cells (T lymphocytes) (p. 742)
hematopoietic stem cells (p. 743)
hematopoietic tissue (p. 743)
central lymphoid tissue (p. 743)
peripheral lymphoid tissues (p. 743)
plasma cells (p. 743)
clonal selection theory (p. 744)
antigen receptors (p. 744)
clone (p. 745)
antigen-independent differentiation (p. 745)
uncommitted cells (p. 745)
committed cells (p. 745)
antigen-dependent differentiation (p. 745)
cytotoxic (killer) T cells (p. 746)
suppressor T cells (p. 746)
helper T cells (p. 746)
immunological memory (p. 746)
primary immune response (p. 746)
secondary immune response (p. 746)
effector cells (p. 746)
memory cells (p. 746)
immunoglobulins (p. 747)
CD3 (p. 747)
CD4 (p. 747)
CD8 (p. 747)
human immunodeficiency virus (HIV) (p. 747)
differentiation markers (p. 747)

Lymphocyte Activation Pathways
interleukins (p. 747)
graft rejection (p. 748)
transplantation antigens (p. 748)
histocompatibility antigens (p. 748)
major histocompatibility complex (MHC) (p. 748)
human leukocyte-associated (HLA) antigens (p. 748)
HLA complex (p. 748)
histocompatibility-2 (H-2) antigens (p. 748)
H-2 complex (p. 748)
polymorphism (p. 748)
class I MHC antigens (p. 749)
class II MHC antigens (p. 749)
β2-microglobulin (p. 749)
antigen-presenting cells (APC) (p. 750)
immunogenic epitope (p. 751)
class II-restricted (p. 751)
interleukin-1 (IL-1) (p. 751)
interleukin-2 (IL-2) (p. 751)
γ-interferon (p. 751)
target cell (p. 751)
cytotoxins (p. 751)
B cell growth factor (BCGF) (p. 752)
B cell differentiation factor (BCDF) (p. 752)

The Structure and Function of Antibodies
immunoglobulins (Ig) (p. 752)
domains (p. 752)
variable (V) domains (p. 752)
constant (C) domains (p. 753)
antigen binding sites (p. 753)
effector site (p. 753)
hinge regions (p. 753)
light (L) chains (p. 753)
heavy (H) chains (p. 753)
H-L pairs (p. 753)
immunoglobulin classes (isotypes) (p. 753)
IgG (p. 753)
complement system (p. 753)
IgM (p. 754)
J (joining) chain (p. 754)
IgA (p. 755)
secretory component (p. 755)
IgD (p. 755)
IgE (p. 755)
valence (p. 755)

The Fine Structure of Antibodies
myeloma protein (p. 757)
monoclonal antibody (p. 757)
hybridomas (p. 757)
variable (V_L) region (p. 757)
constant (C_L) region (p. 757)
hypervariable regions (p. 760)
framework regions (p. 760)
combinatorial associations (p. 761)
idiotype (p. 762)
anti-idiotype antibodies (p. 762)

The Genetic Basis of Antibody Diversity
antibody repertoire (p. 762)
gene families (p. 763)
V genes (p. 763)
C genes (p. 763)
J (joining) gene segment (p. 764)
gene segments (p. 764)
D (diversity) gene segments (p. 764)
somatic rearrangement (p. 764)
recognition sequences (p. 765)
junctional diversity (p. 765)
allelic exclusion (p. 766)
hypermutation (p. 766)

Class Switching
alternative splicing of mRNA (p. 766)
VDJ transposition (p. 767)
S (switch) regions (p. 767)

The T Cell Antigen Receptor
α and β chains (p. 767)
γ and δ chains (p. 767)

Suggested Reading

General References and Review Articles

Golub, E. S. *The Cellular Basis of the Immune Response,* 2d ed. Sunderland, Mass.: Sinauer, 1981.

Hood, L. E., I. L. Weissman, W. B. Wood, and J. H. Wilson. *Immunology,* 2d ed. Menlo Park, Calif.: Benjamin/Cummings, 1984.

McConnell, I., A. Munro, and H. Waldmann. *The Immune System: A Course on the Molecular and Cellular Basis of Immunity,* 2d ed. Oxford, England: Blackwell, 1981.

Paul, W. E. *Fundamental Immunology.* New York: Raven, 1984.

The Immune System

Burnet, F. M. *The Clonal Selection Theory of Acquired Immunity.* Nashville, Tenn.: Vanderbilt University Press, 1959.

Cooper, M., and A. Lawton. The development of the immune system. *Sci. Amer.* 231 (November 1974): 59.

Jerne, N. K. The immune system. *Sci. Amer.* 229 (November 1973): 52.

Lymphocyte Development

Kincade, P. W. Experimental models for understanding B lymphocyte formation. *Adv. Immunol.* 41 (1987): 181.

Sprent, J., and S. R. Webb. Function and specificity of T cell subsets in the mouse. *Adv. Immunol.* 41 (1987): 39.

Swain, S. T cell subsets and the recognition of MHC class. *Immunol. Rev.* 74 (1983): 129.

The Major Histocompatibility Complex

Gotze, D. *The Major Histocompatibility System in Man and Animals.* New York: Springer-Verlag, 1977.

Hood, L., M. Steinmetz, and B. Malissen. Genes of the major histocompatibility complex of the mouse. *Annu. Rev. Immunol.* 1 (1983): 529.

Klein, J. The major histocompatibility complex of the mouse. *Science* 203 (1979): 516.

Matzinger, P., and R. Zamoyska. A beginner's guide to major histocompatibility function. *Nature* 297 (1982): 628.

Lymphocyte Activation

Bevan, M. J. Class discrimination in the world of immunology. *Nature* 325 (1987): 192.

Germain, R. N. The ins and outs of antigen processing and presentation. *Nature* 322 (1986): 687.

Kishimoto, T. Factors affecting B cell growth and differentiation. *Annu. Rev. Immunol.* 3 (1985): 133.

Weiss, A., and J. B. Imboden. Cell surface molecules and early events involved in human T lymphocyte activation. *Adv. Immunol.* 41 (1987): 1.

Antibody Structure and Function

Amzel, L. M., and R. J. Poljak. Three-dimensional structure of immunoglobulins. *Annu. Rev. Biochem.* 48 (1979): 961.

Capra, J. D., and A. B. Edmundson. The antibody combining site. *Sci. Amer.* 236 (January 1977): 50.

Richards, F., W. Konigsberg, R. Rosenstein, and J. Varga. On the specificity of antibodies. *Science* 189 (1975): 130.

The Genetic Basis of Antibody Diversity

Adams, J. M. The organization and expression of immunoglobulin genes. *Immunol. Today* 1 (1980): 10.

Baltimore, D. Molecular immunology: Growth into adolescence. *Trends Biochem. Sci.* 9 (1984): 137.

Dreyer, W. J., and J. C. Bennett. The molecular basis of antibody formation: A paradox. *Proc. Natl. Acad. Sci. USA* 54 (1965): 864.

Honjo, T. Immunoglobulin genes. *Annu. Rev. Immunol.* 1 (1983): 499.

Kurosawa, Y., and S. Tonegawa. Organization, structure and assembly of immunoglobulin heavy-chain diversity DNA segments. *J. Exp. Med.* 155 (1982): 201.

Leder, P. The genetics of antibody diversity. *Sci. Amer.* 246 (May 1982): 72.

Marcu, K. B., R. B. Lang, W. L. Stanton, and L. J. Harris. A model for the molecular requirements of immunoglobulin heavy-chain class switching. *Nature* 292 (1982): 87.

Tonegawa, S. Somatic generation of antibody diversity. *Nature* 302 (1983): 575.

The T Cell Antigen Receptor

Allison, J. P., and L. L. Lanier. Structure, function and serology of the T cell antigen receptor complex. *Annu. Rev. Immunol.* 5 (1987): 503.

Hedrick, S. M., et al. Isolation of cDNA clones encoding T cell-specific membrane-associated proteins. *Nature* 308 (1984): 149.

Hedrick, S. M., et al. Sequence relationships between putative T cell receptor polypeptides and immunoglobulins. *Nature* 308 (1984): 153.

Marrack, P., and J. Kappler. The T cell receptor. *Science* 238 (1987): 1073.

Monoclonal Antibodies

McMichael, A. J., and J. W. Fabre, eds. *Monoclonal Antibodies in Clinical Medicine.* New York: Academic Press, 1982.

Milstein, C. Monoclonal antibodies. *Sci. Amer.* 243 (October 1980): 66.

Yelton, D. E., and M. D. Scharff. Monoclonal antibodies: A powerful new tool in biology and medicine. *Annu. Rev. Biochem.* 50 (1981): 657.

Problem Set

1. **Immunological True or False.** Indicate whether each of the following statements is true (T) or false (F), and explain the error in each false statement.
   (a) A large antigen can have many different antibody molecules bound to it.
   (b) Immunological memory can last 20 years or more in humans.
   (c) The removal of the bursa of Fabricius from a chick at the time of hatching will impair both the humoral and the cell-mediated immune responses.
   (d) Antibody-producing cells capable of responding to a specific antigen are already present within the lymphoid tissues of an animal prior to exposure to the antigen.
   (e) A cell expressing class, I, but not class II, MHC proteins can present antigen to $CD4^+CD8^-$ T cells.

2. **The Role of Lymphocytes in the Immune Response.** Figure 24-24 depicts an experiment done in 1954 to establish the central role of lymphocytes in the immune response. A mouse that would otherwise show a normal immune response upon injection of an antigen (mouse 1) is unable to make a response if it is first heavily irradiated to destroy most of its white blood cells (mouse 2). The immune response is restored if an irradiated animal (mouse 3) is given lymphocytes from an unirradiated donor (mouse D) of the same inbred (genetically homogeneous) line; however, the immune response is not restored if other white blood cells from the same donor (mouse D) are given to an irradiated animal (mouse 4).
   (a) What conclusion can be drawn from the result observed with mouse 2?
   (b) Why were the experiments with mice 3 and 4 done? What conclusion can be drawn from them?
   (c) Which would you expect to find restored in mouse 3, the humoral immune response, the cellular immune response, or both? Explain.
   (d) How would the immune response of mouse 2 differ if, instead of being irradiated, the animal had had its thymus removed shortly after birth?
   (e) How important is it that the donor mouse for experiments 3 and 4 be of the same inbred line as the irradiated animal?
   (f) Mouse 3 probably would have died if lymphocytes from a genetically different animal had been injected into it. Why?

3. **The Instruction Hypothesis of Antibody Diversity.** Before the clonal selection theory was accepted by immunologists in the 1950s, antibody diversity was explained by the *instruction hypothesis*. According to this theory, antibodies are made as unfolded polypeptide chains, and their final conformation is determined by their association with a specific antigen.
   (a) Why do you suppose the instruction hypothesis was an attractive explanation of antibody diversity at one time?

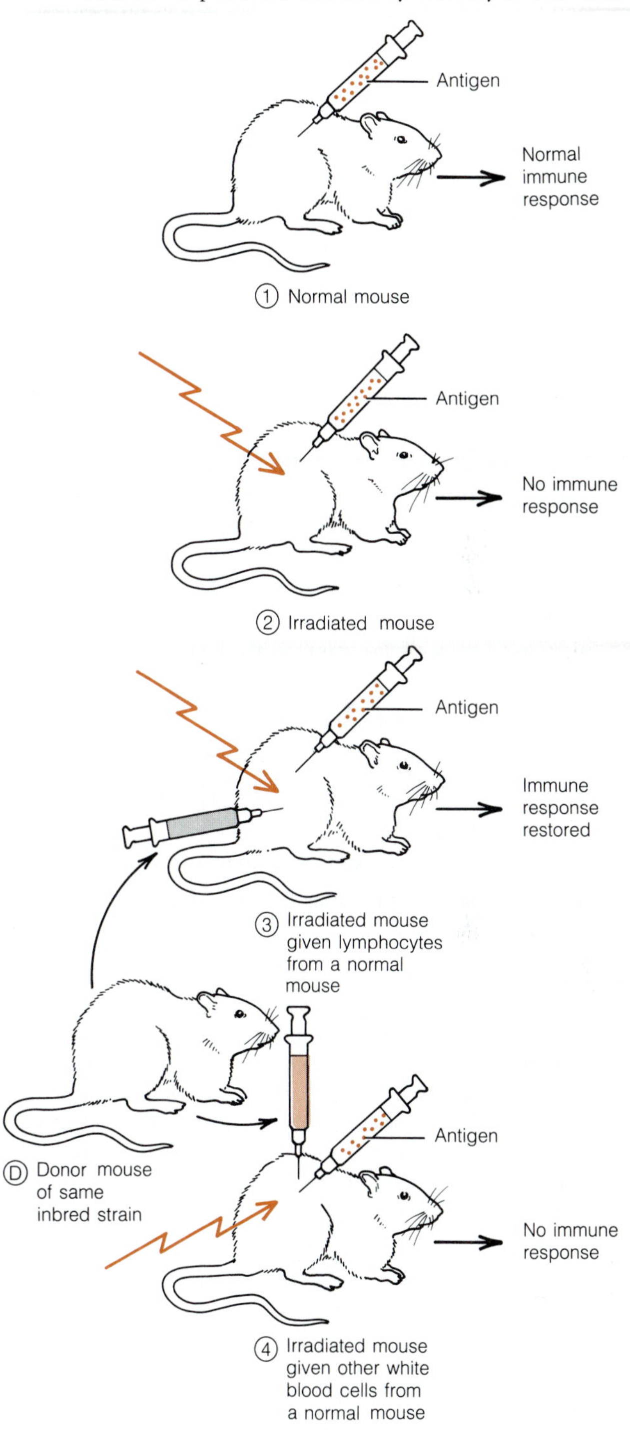

Figure 24-24 The Importance of Lymphocytes in the Immune Response.

(b) What crucial discovery about the three-dimensional structure of protein molecules do you suppose led to the demise of the instruction hypothesis?

(c) How could you disprove the instruction hypothesis directly?

4. **Antibody Structure.** Indicate whether each of the following is descriptive of V_H, V_L, C_H, C_L, or any combination of the four.

(a) The binding site for antigen

(b) The idiotype

(c) Transplacental passage

(d) Complement fixation

(e) Isotype or class

5. **Acquired Immune Deficiency Syndrome (AIDS).** The CD4 glycoprotein is a receptor for human immunodeficiency virus (HIV). Describe how this is related to the severe immunodeficiency seen in AIDS.

6. **Immunoglobulin Structure.** Indicate whether each of the following statements is true of the light chain only (L), of the heavy chain only (H), of both light and heavy chains (LH), or of neither chain (N).

(a) It is found in the arms but not in the leg of an immunoglobulin molecule.

(b) It contains both a variable and a constant region.

(c) It associates with secretory component when an IgA dimer crosses an epithelial cell.

(d) It has a length of about 110 amino acids.

(e) It has several short hypervariable segments scattered along about half of its length.

(f) It is involved in the antigen binding site.

(g) It exists in multiple forms, but not within the same antibody molecule.

7. **Ordering C_H Genes.** Unstimulated lymphocytes from the organism *Vaccinatia hypothetica* make only membrane-bound IgM, though their germ-line DNA is known to contain the genes for the constant regions of five different heavy chains, designated m, n, o, p, and q in this species. Myeloma cells secreting IgQ are found to lack the DNA sequences that code for C_m and C_o, while cells secreting IgN lack the sequences that encode all other classes of heavy-chain constant regions.

(a) How do you suppose it was determined that the DNA of specific myeloma cells was missing certain sequences?

(b) Suggest an order for the five C_H genes and explain why you ordered them as you did.

8. **Antibody Diversity.** Consider a genome that contains 500 V_κ and 5 J_κ light-chain gene segments and 400 V, 20 D, and 4 J heavy-chain gene segments. The genome also contains 2 C_κ genes and 6 C_H genes.

(a) Calculate the number of different V_L and V_H genes and antibody molecules that can be generated by all possible combinations of these gene segments.

(b) Why are the several different forms of C_κ and C_H genes not considered when calculating the number of different antigen binding sites that can be formed?

(c) Assume that the joining of V and J segments of the V_κ gene is imprecise, but always occurs at one of three adjacent bases near the codon for amino acid 95, and that the joining of D to J and of DJ to V is similarly imprecise in assembly of the V_H gene. What is the maximum amount of additional antibody diversity that can be attributed to the alternative reading frames introduced in this way?

(d) Suggest yet another means whereby further antibody diversity might be achieved in this system, and explain briefly.

(e) Compare mechanisms for the generation of diversity in antibodies and in T cell antigen receptors.

9. **The Role of MHC Glycoproteins.** A singularly important series of experiments performed in the mid-1970s led to our awareness of the role of class I MHC glycoproteins in cellular immune responses. Inbred mice of strain 1 were infected with virus X. One week later, activated T_c cells were recovered from the spleens of these mice and tested for their ability to kill virus-infected mouse fibroblasts in culture. Explain each of the following findings.

(a) Cultured fibroblasts from strain 1 that were infected with virus X were killed by the T_c cells within several hours.

(b) Cultured fibroblasts from strain 1 that were infected with virus Y were not killed by the T_c cells.

(c) Cultured fibroblasts from strain 2 that were infected with virus X were not killed by the T_c cells.

(d) If the experiment of part (a) was repeated with cultured fibroblasts that differed genetically from the infected mice at all loci except the class I MHC loci, the fibroblasts were killed, but if the fibroblasts and the infected mice were genetically identical at all loci except the class I MHC loci, the fibroblasts were not killed.

Appendix A

Principles and Techniques of Microscopy

Cell biologists need to examine the structure of cells and their components. The microscope is an indispensable tool for this purpose because most cellular structures are too small to be seen by the unaided eye. In fact, the beginnings of cell biology can be traced to the invention of the **light microscope,** which made it possible for scientists to see enlarged images of cells for the first time. The first generally useful light microscope was developed in 1590 by Z. Janssen and H. Janssen, an uncle and nephew team. Many important observations in biology were reported during the next century, notably those of Robert Hooke and Antonie van Leeuwenhoek in the last quarter of the seventeenth century. Since then, the light microscope has undergone numerous improvements and modifications, right up to the present time.

By contrast, the **electron microscope** is of much more recent vintage, dating from the early 1930s. Just as the invention of the light microscope heralded a wave of scientific achievement, the development of the electron microscope triggered a revolution in the exploration of cell structure and function, and ultimately in the way we think about cells.

Today, we are approaching the theoretical limits of resolution for light microscopy and, to a lesser extent, for electron microscopy. New ways to visualize biological specimens are being used, particularly x-ray and **electron diffraction** methods that provide information about molecular structure of crystallized macromolecules. The most recent arrivals include the scanning tunneling microscope and the atomic force microscope. These remarkable instruments are considerably less expensive than a typical electron microscope and yet are capable of providing information at the atomic scale of resolution. In this appendix, we will look at each of these methods in some detail. We will begin by exploring the fundamental principles of both light and electron microscopy. Then we will go on to examine a variety of techniques relevant to cell biology.

Image Formation in Light and Electron Microscopy

Regardless of the kind of microscope being used, three elements are always needed to form an image: a **source of illumination,** the **specimen** to be observed, and a system of **lenses** to focus the illumination on the specimen and enlarge the image. Figure A-1 illustrates these features for both a light microscope and an electron microscope. In a light microscope (Figure A-la), the source of illumination is visible light and the lens system consists of a series of glass lenses. The image can then be viewed directly through a lens or focused on a film for photography. In an electron microscope (Figure A-lb), the illumination source is a beam of electrons emitted by a heated tungsten filament and the lens system consists of a series of electromagnets. The electron beam is focused either on a fluorescent screen of zinc sulfide for direct visualization of the image or on photographic film.

Despite these differences in illumination source and instrument design, both types of microscope depend on the same principles of optics and form images in a similar manner. When a specimen is placed in the path of the light or electron beam, physical characteristics of the beam are changed in a way that can be interpreted by the human eye or recorded on a photographic plate. The image can often be enhanced by modifying specific properties of either the specimen or the illumination source, or both. We will discuss some of these modifications when we examine different types of specimen preparation and microscope design.

To understand the interaction between an illumination source and a specimen and the limits that the illumination source sets on the size of objects that can be detected, consider the simple analogy depicted in Figure A-2. If two individuals hold onto opposite ends of a slack rope and

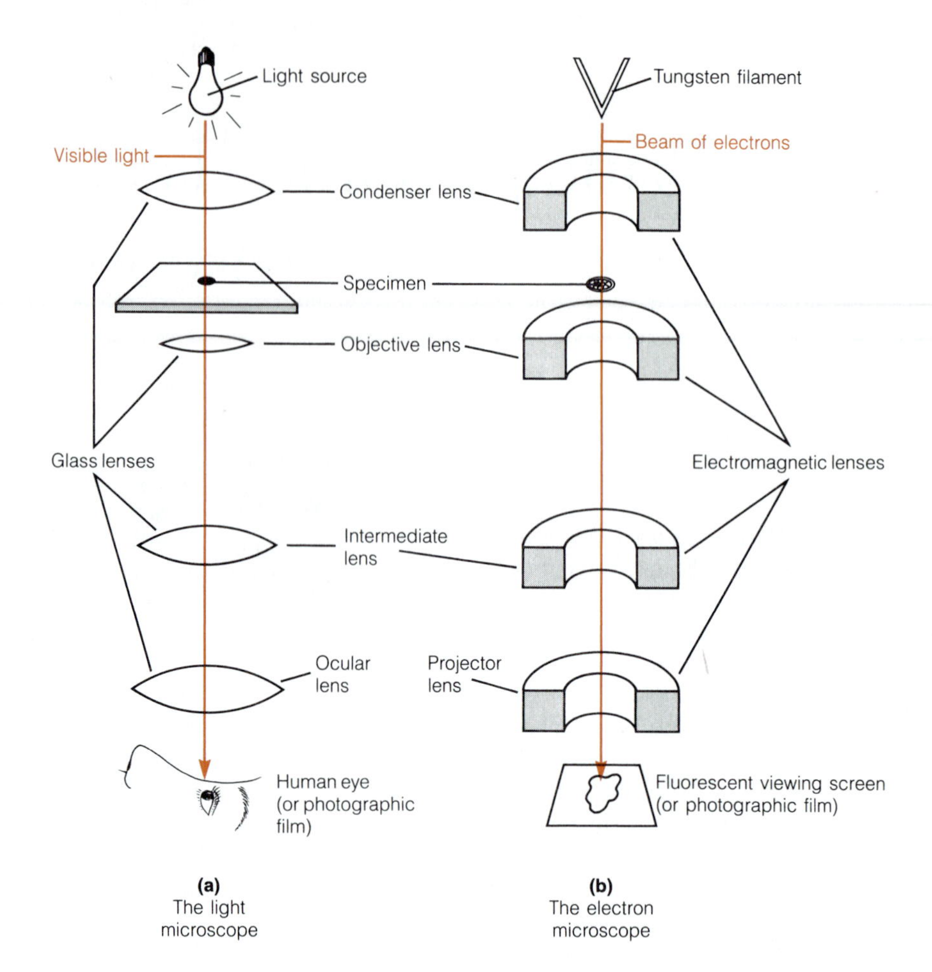

Figure A-1 The Optical Systems of the Light Microscope and the Electron Microscope. (a) The light microscope uses visible light and glass lenses to form an image of the specimen that can be either seen by eye or focused on photographic film. (b) The electron microscope uses a beam of electrons emitted by a tungsten filament and focused by electromagnetic lenses to form an image of the specimen on a fluorescent screen or on photographic film. (These diagrams have been drawn to emphasize the similarities in overall design between the two types of microscope. In reality, a light microscope is designed with the light source at the bottom and the ocular lens at the top, as shown in Figure A-6b.)

wave the rope with a rhythmic up-and-down motion, they will generate a long, regular pattern of movement in the rope called a **wave form** (Figure A-2a). The distance from the crest of one wave to the crest of the next is called the **wavelength.**

If someone standing to one side of the rope tosses a large object such as a beach ball against the rope, the ball may interfere with, or perturb, the wave form of the rope's motion (Figure A-2b). However, if a small object such as a marble is tossed against the rope, the movement of the rope probably will not be affected at all (Figure A-2c). If the rope holders take up the slack in the rope but continue their motion, the motion of the rope will still have a wave form, but the wavelength will be shorter (Figure A-2d). In this case, a marble tossed against the rope is very likely to perturb the rope's movement (Figure A-2e). This simple analogy illustrates an important principle: The ability of an object to perturb a wave motion depends crucially on the size of the object in relation to the wavelength of the motion.

This principle is a great importance in microscopy, because it means that the wavelength of the illumination source sets a limit on how small an object can be and still be seen. To understand this relationship, we need to recognize that the moving rope of Figure A-2 is analogous to the beam of photons or electrons that is used as an illumination source in a microscope. When photons or electrons encounter a specimen, the specimen alters the physical characteristics of the illuminating beam, just as the ball or marble alters the motion of the rope. And since an object can be detected only by its effect on the wave, the wavelength must be comparable in size to the object that is to be detected.

Once we understand this relationship between wavelength and object size, we can readily appreciate why very small objects can be seen only by electron microscopy: The wavelength of electrons is very much shorter than that of photons. Thus, objects such as viruses and ribosomes are too small to perturb a wave of photons, but they can readily interact with a wave of electrons. As we discuss different types of microscopes and specimen preparation techniques, you might find it helpful to ask yourself how the source and specimen are interacting and how the characteristics of both are modified to produce an image.

Optical Principles of Microscopy

Although light and electron microscopes differ in many ways, they use similar optical principles to form images. In a light microscope, glass lenses are used to direct the course of photons, whereas an electron microscope uses electromagnets as lenses to direct the course of electrons. Yet, both kinds of lenses have two basic properties in common: focal length and angular aperture. The **focal length** is the distance between the midline of the lens and the point at which rays passing through the lens converge to a focus (Figure A-3). The **angular aperture** is the half-angle α of the cone of light entering the objective lens of the microscope from the specimen (Figure A-4). Angular aperture is therefore a measure of how much of the illumination that leaves the specimen actually passes through the lens, which in turn determines the ability of the lens to convey information about the specimen. In the best microscopes, the angular aperture is about 70° (Figure A-4b).

Resolution

The most important characteristic of any lens is its **resolution.** Simply put, resolution is the ability to see two neighboring objects as distinct entities. In the light microscope, resolution is limited primarily by properties of visible light rather than by characteristics of the lens system. Specifically, resolution in light microscopy is governed by three factors: the wavelength of the light used to illuminate the specimen, the angular aperture, and the refractive index of the medium surrounding the specimen. (**Refractive index** is a measure of the change in the velocity of light as it passes from one medium to another.)

Resolution is described quantitatively by the following equation:

$$r = 0.61\lambda / n \sin \alpha \tag{A-1}$$

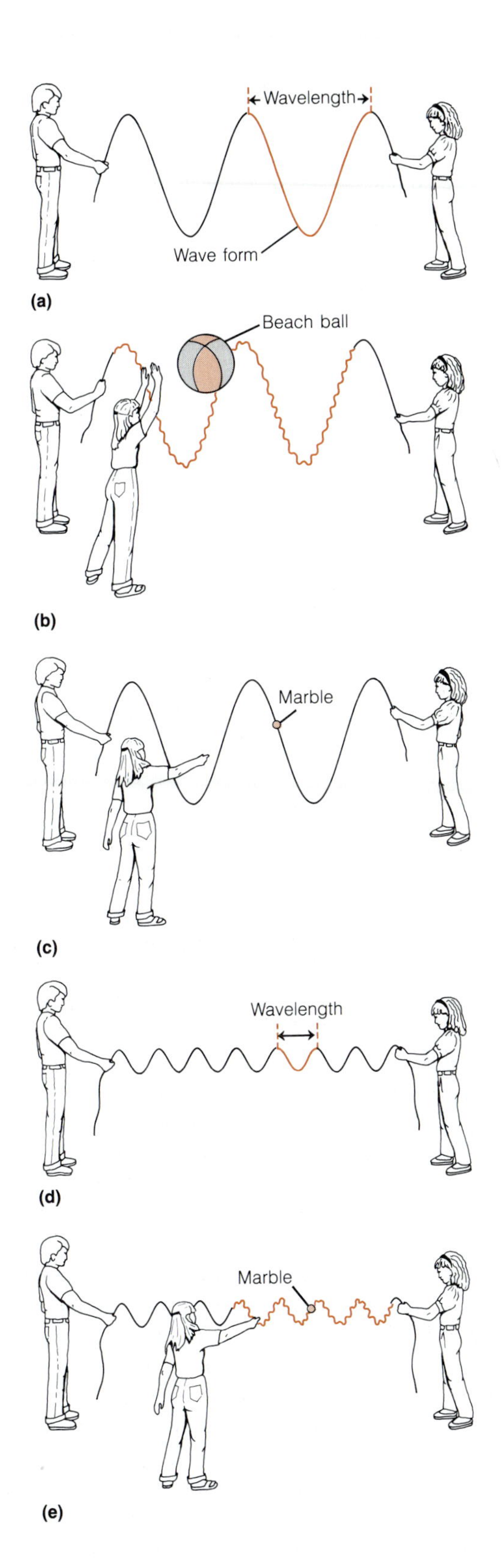

Figure A-2 Wave Motion, Wavelength, and Perturbations. The wave motion of a rope held between two individuals is analogous to the wave form of both visible light and electrons, and can be used to illustrate the effect of size on an object's ability to perturb wave motion. (a) Moving a slack rope up and down rhythmically will generate a wave form with a characteristic wavelength. (b) When thrown against a rope, a beach ball or other object with a diameter that is comparable to the wavelength of the rope will perturb the motion of the rope. (c) A marble or other object with a diameter significantly less than the wavelength of the rope will cause little or no perturbation of the rope. (d) If the rope is held more tautly, the wavelength will be reduced substantially. (e) A marble is now able to perturb the motion of the rope because its diameter is comparable to the wavelength of the rope.

where r is the resolution, λ is the wavelength of the light used for illumination, n is the refractive index of the medium between the specimen and the objective lens of the microscope, and α is the aperture angle as already defined. (The constant 0.61 describes the degree to which image points can overlap and still be recognized as separate points by an observer.)

The quantity $n \sin \alpha$ is called the **numerical aperture** of the objective lens, abbreviated *NA*. An alternative expression for resolution is therefore

$$r = 0.61\lambda/NA \qquad \text{(A-2)}$$

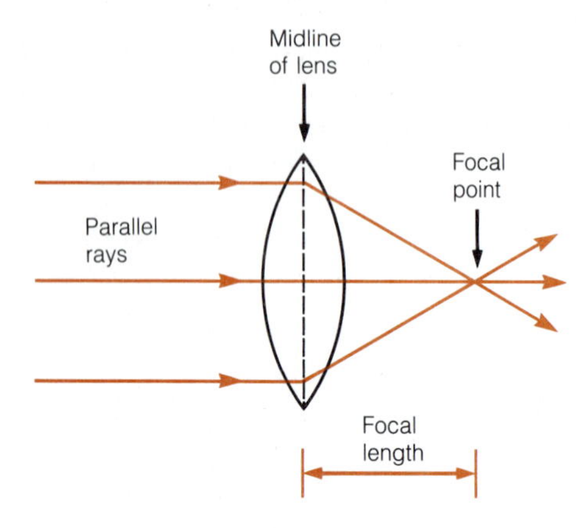

Figure A-3 Focal Length of a Lens. The focal length is the distance from the midline of the lens to the point at which rays passing through the lens converge to a focus.

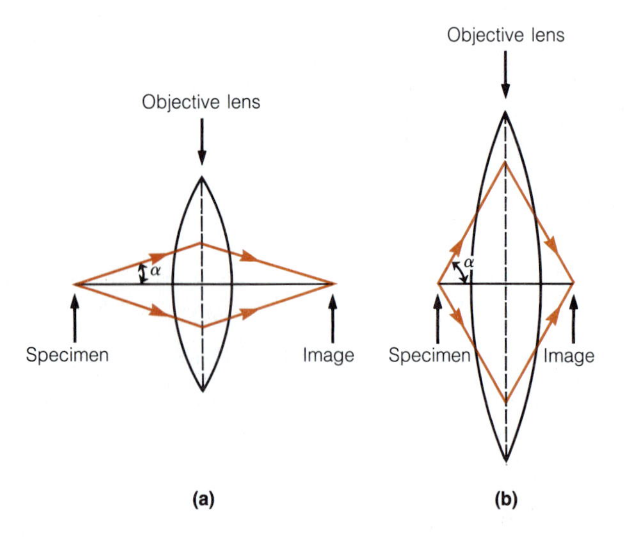

Figure A-4 Angular Aperture of a Lens. The angular aperture is the half-angle α of the cone of light entering the objective lens of the microscope from the specimen. (a) A low-aperture lens (α is small). (b) A high-aperture lens (α is large). The larger the angular aperture, the more information the lens can transmit. The best glass lenses have an angular aperture of about 70°.

Maximizing Resolution

Because r is a measure of how close two points can be and still be distinguished from each other, resolution improves as r becomes smaller. Thus, for best resolution, the numerator of equation A-2 should be as small as possible and the denominator should be as large as possible.

We will consider the numerator first. The wavelength range for visible light is about 400–700 nm, and the minimum value for λ is set by the shortest wavelength in this range that is usable for illumination, which is blue light of about 450 nm.

To maximize the denominator of equation A-3, recall that the numerical aperture is the product of the angular aperture and the refractive index. Both of these values must therefore be maximized for best possible resolution. Since the angular aperture for the best objective lenses is about 70°, the maximum value for $\sin \alpha$ is about 0.94. The refractive index of air is about 1, so for a lens designed for use in air, the maximum numerical aperture is about 0.94. Thus, for a lens with an aperture angle of 70°, the resolution in air for a sample that is illuminated with blue light of about 450 nm can be calculated as follows:

$$\begin{aligned} r &= 0.61\lambda/NA \\ &= (0.61)(450)/0.94 = 292 \text{ nm} \approx 0.3\ \mu\text{m} \end{aligned} \qquad \text{(A-3)}$$

As a rule of thumb, then, the limit of resolution for a glass lens used in air is about 300 nm, or 0.3 μm.

As a means of increasing the numerical aperture value, some microscope lenses are designed to be used with a layer of *immersion oil* between the lens and the specimen. Immersion oil has a higher refractive index than air and therefore allows the lens to receive more of the light transmitted through the specimen (Figure A-5). Since the refractive index of immersion oil is about 1.5, the maximum numerical aperture for an oil immersion lens is about $1.5 \times 0.94 = 1.4$. The resolution of an oil immersion lens is therefore about 0.2 μm:

$$\begin{aligned} r &= 0.61\lambda/NA \\ &= (0.61)(450)/1.4 = 196 \text{ nm} \approx 0.2\ \mu\text{m} \end{aligned} \qquad \text{(A-4)}$$

Thus, the **limit of resolution** (best possible resolution) for a microscope that uses visible light is about 0.3 μm in air and 0.2 μm with an oil immersion lens. By using ultraviolet light, the resolution can be pushed to about 0.1 μm because of the shorter wavelength(200–300 nm). However, the image must then be recorded on a photographic plate because ultraviolet light is invisible to the human eye. Moreover, expensive quartz lenses must be used because ordinary glass is opaque to ultraviolet light. In any case, these calculated values are theoretical limits to the resolution. In actual practice, such limits can rarely be reached because of technical flaws or *aberrations* in the lenses.

The limit of resolution for a lens sets the upper limit on the **useful magnification** that is possible with that lens. In general, the greatest useful magnification that can be achieved with a light microscope is about 1000 times the numerical aperture of the lens that is used. And since numerical aperture ranges from about 1.0 to about 1.4, the useful magnification of a light microscope is limited to about 1000 × in air and about 1400 × with immersion oil. Magnification greater than these limits is referred to as "empty magnification" because it provides no additional information about the object being studied.

Because the wavelength of an electron is so much shorter than that of a photon of visible light, the electron microscope has a theoretical limit of resolution much lower than that of the light microscope—about 0.1–0.2 nm instead of 200–300 nm. For biological samples, however, problems with specimen preparation and contrast are such that the practical limit of resolution is almost always much greater than the theoretical limit—generally about 2 nm. Practically speaking, therefore, resolution in an electron microscope is about 100 times greater than in the light microscope. As a result, the useful magnification of an electron microscope is about 100 times that of a light microscope, or about 100,000 × .

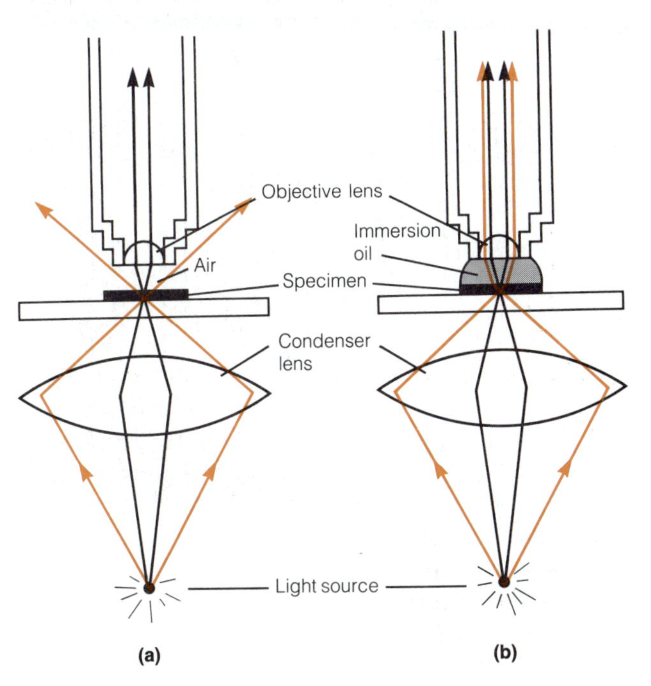

Figure A-5 Effect of Immersion Oil on the Transmission of Light to the Objective Lens. (a) With air between the specimen and the objective lens, some refracted light is transmitted to the objective lens (black lines), but much is lost (color lines). (b) With immersion oil between the specimen and the objective lens, more refracted light is transmitted to the objective lens (black and color lines) because of the high refractive index of immersion oil (about 1.5 versus 1.0 for air).

The Light Microscope

An important name in the history of light microscopy is that of Antonie van Leeuwenhoek, the Dutch shopkeeper who is generally regarded as the father of microscopy. Leeuwenhoek's lenses, which he manufactured himself, were of surprisingly high quality for his time and were capable of 300-fold magnification. His observations, made over a period of more than 25 years, were remarkable, especially in view of the limitations imposed by the single-lens microscopes used at the time.

Today, the instrument of choice for light microscopy uses several lenses in combination and is therefore called a **compound microscope** (Figure A-6). The optical path through a compound microscope is illustrated in Figure A-6b. The path begins with the source of illumination, usually a lamp located in the base of the instrument. The light rays from the source first pass through the **condenser lens,** a lens that directs the light toward the specimen that is mounted on a glass slide and positioned on the **stage** of the microscope. The **objective lens** is located immediately above the specimen and is responsible for forming the *primary image.* Most compound microscopes have several objective lenses of differing magnification mounted on a rotatable turret.

The primary image is further enlarged by the **ocular lens,** or *eyepiece.* (In some microscopes, an **intermediate lens** is juxtaposed between the objective and ocular lenses to accomplish still further enlargement.) We can calculate the overall magnification of the image by multiplying the enlarging powers of the objective lens, the ocular lens, and the intermediate lens (if present). Thus, a microscope with a 10 × objective lens, a 2.5 × intermediate lens, and a 10 × ocular lens will magnify a specimen 250-fold.

There are many kinds of light microscopy, but all of them are variations on the general theme of image formation that we have just described. Figure A-7 illustrates four different techniques by comparing the same blue-green algal cells as seen by (a) brightfield illumination, (b) phase-contrast microscopy, (c) darkfield illumination, and (d) differential interference contrast microscopy. We will look at these and several other important techniques.

Brightfield Microscopy

The elements of the microscope as they have been described so far represent the basic form of light microscopy, that of **brightfield microscopy** (Figure A-7a). Compared to other microscopes, the brightfield microscope is inexpensive and simple to align and use. However, the only specimens that can be seen directly by brightfield microscopy are those that possess color or have some other property that affects the amount of light that passes through. Many biological specimens lack these characteristics and must therefore be stained with dyes before they can be seen by brightfield optics.

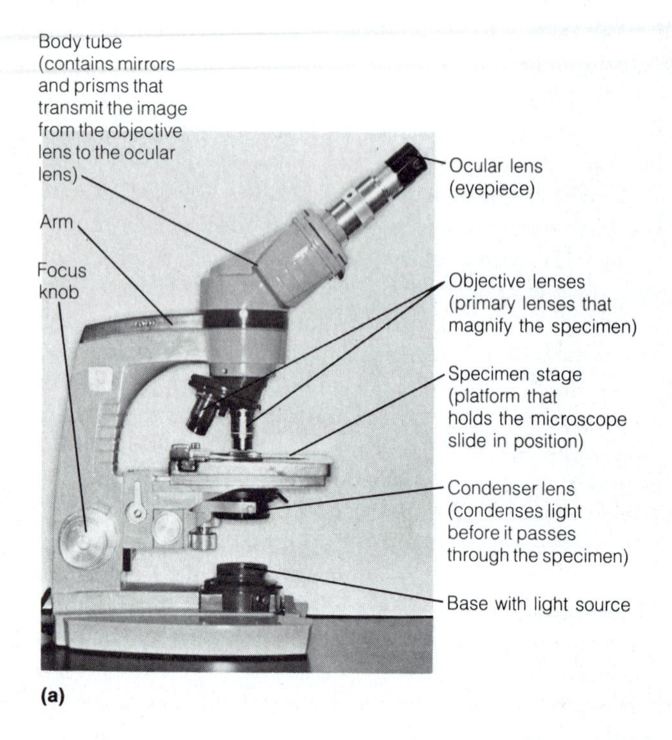

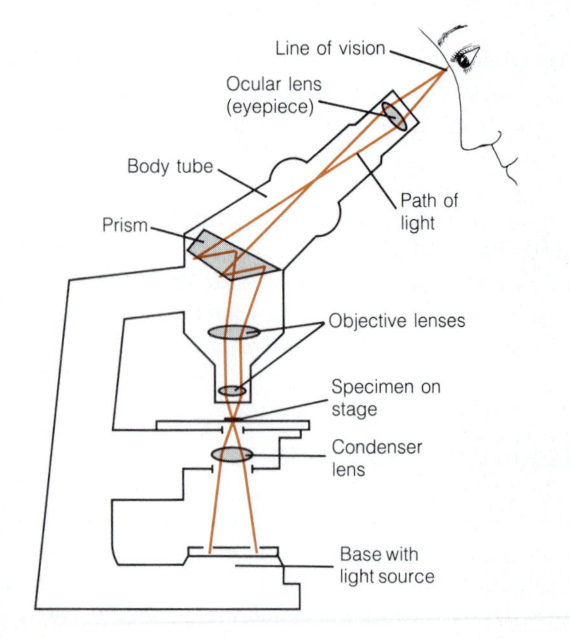

Figure A-6 The Compound Light Microscope. (a) A compound light microscope. (b) The path of light rays through the compound microscope.

Phase-Contrast Microscopy

Phase-contrast microscopy is illustrated in Figure A-7b. To understand the basis of this technique, we need to recognize that a beam of light is made up of many individual rays of light. As the rays pass from the light source through the specimen, their velocity may be affected by the physical properties of the specimen. Specifically, the rays are *diffracted* and their phase is changed to different extents by different regions of the specimen. As a result, the rays that pass through the specimen get out of phase with respect to rays of light that do not pass through the specimen.

We can exploit this difference by inserting into the light path an optical material that is capable of bringing the direct or undiffracted rays into phase with those that have been diffracted by the specimen. The resulting pattern of wavelengths intensifies the image. The phase-contrast microscope takes advantage of this effect by inserting a **phase plate** into the light path above the objective lens (Figure A-8). Phase contrast produces an image with highly contrasting bright and dark areas against a neutral gray background (Figure A-9). As a result, internal structures of cells are often better visualized by phase-contrast microscopy than with brightfield optics.

This approach to light microscopy is particularly useful for examining living, unstained specimens, since biological materials almost inevitably diffract light. Phase-contrast microscopy is widely used in microbiology and tissue culture research to detect bacteria, cellular organelles, and other small entities in living specimens. An artifact common to this type of microscopy is a white band called a *phase halo* around the edges of cells or other specimens (see Figures A-7b and A-9).

Darkfield Microscopy

Like phase-contrast microscopy, **darkfield microscopy** depends on the ability of a specimen to alter the characteristics of the light rays that pass through it. In the phase-contrast microscope, both the rays that pass through the specimen and those that do not are used to form the final image by bringing the latter into phase with the former. In darkfield optics, however, only the rays that are diffracted by the specimen are used to form the image. This is done by using an objective lens with a small aperture that allows only the rays diffracted by the specimen to enter (Figure A-10). The image therefore appears bright against a dark background (Figure A-7c).

Like phase-contrast microscopy, darkfield optics is useful for detecting very small entities such as bacteria and is therefore widely used in microbiology. As in phase-contrast microscopy, the resolving capabilities are more limited than in other forms of light microscopy.

Fluorescence Microscopy

To understand **fluorescence microscopy**, it is first necessary to understand the physical phenomenon of **fluorescence**. A chemical compound is said to be *fluorescent* if it is capable of absorbing ultraviolet light and reemitting the energy as visible light. Some fluorescent compounds occur naturally, whereas others are synthetic compounds, but all have the property of absorbing ultraviolet light while emitting visible light.

A fluorescent microscope has an *exciter filter* between the light source and the condenser lens that transmits only ultraviolet light (Figure A-11). The condenser lens then focuses the ultraviolet rays on the specimen, causing flu-

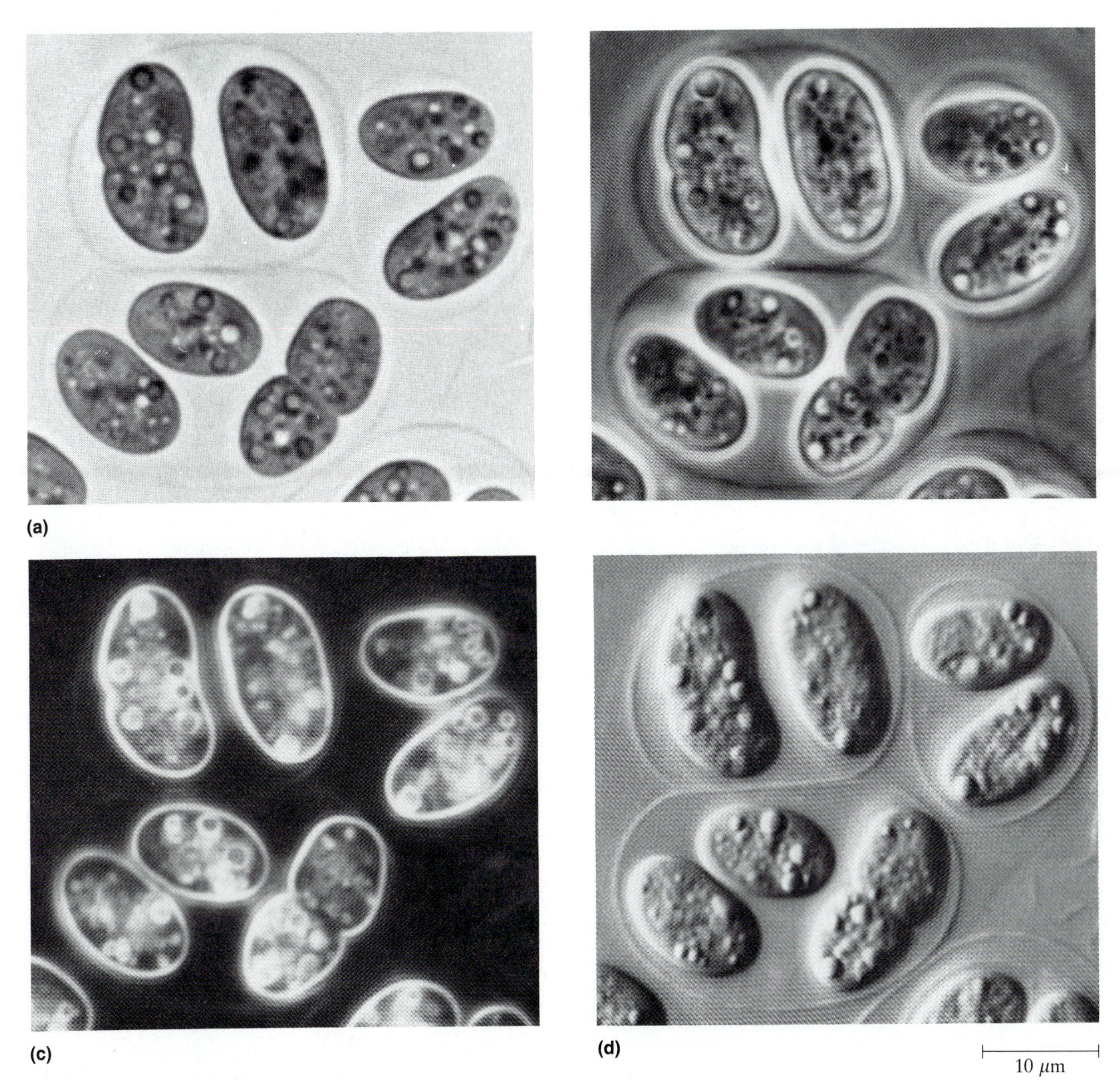

Figure A-7 Optical Techniques with the Light Microscope. These micrographs show the same cells of the cyanobacterium (blue-green alga) *Gloeocapsa* as visualized by several different light microscope techniques, including (a) brightfield illumination, (b) phase-contrast microscopy, (c) darkfield illumination, and (d) differential interference contrast microscopy. In each case, internal features of the cells are visible. Also visible with three of the four techniques is the transparent sheath that usually surrounds clusters of two or four recently divided *Gloeocapsa* cells. (all LMs)

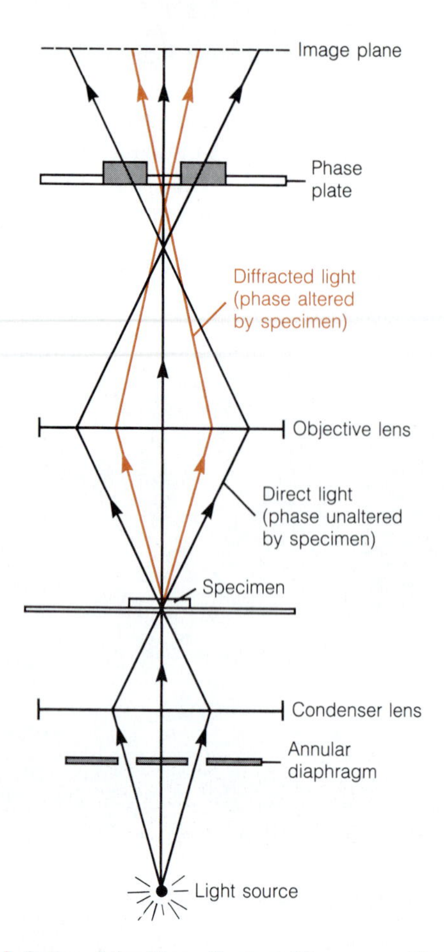

Figure A-8 Optics of the Phase-Contrast Microscope. The configuration of the optical elements and the path of light rays through the phase-contrast microscope. Color lines represent light diffracted by the specimen, whereas the black lines represent direct light.

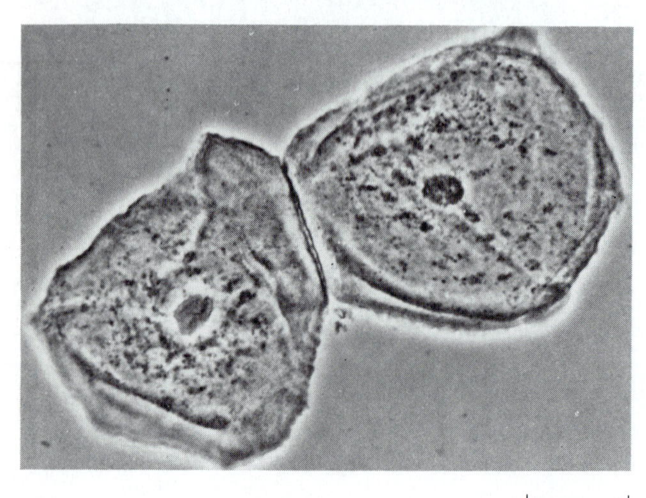

Figure A-9 Phase-Contrast Microscopy. A phase-contrast micrograph of epithelial cells from the mouth. The cells were observed unprocessed and unstained, a major advantage of phase-contrast microscopy. For a view of the same epithelial cells as visualized by differential interference contrast microscopy, see Figure A-14.

orescent compounds in the specimen to emit visible light. Both the ultraviolet light from the illuminator and the visible light generated by fluorescent compounds in the specimen then pass through the objective lens. As the light passes through the tube of the microscope above the objective lens, it encounters a *barrier filter* that specifically removes the ultraviolet wavelengths. This leaves only wavelengths of visible light to form the final image, which therefore appears bright against a dark background.

A variety of **fluorescent dyes** can be used to gain insight into the functions and metabolism of the specimen under study. The value of such dyes lies in their specificity for certain cellular components such as nucleic acids. Alternatively, nonspecific dyes (such as *rhodamine* or *fluorescein*) can be coupled to compounds that have a high affinity for specific biological molecules but are not themselves fluorescent. For example, antibody molecules can be labeled or "tagged" with fluorescent dyes, thereby making it possible to visualize the antibody molecules with the

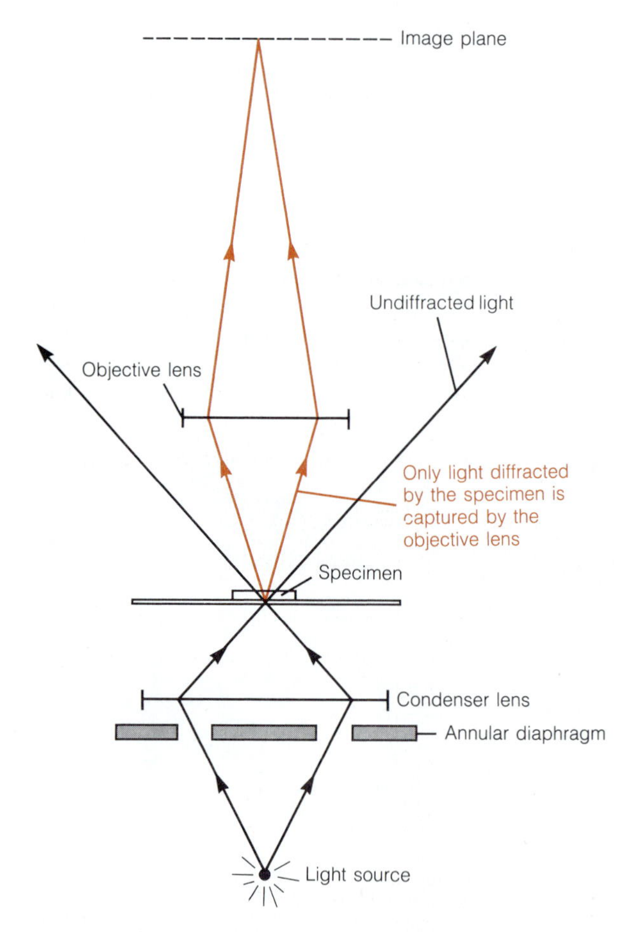

Figure A-10 Optics of the Darkfield Microscope. The configuration of the optical elements and the path of light rays through the darkfield microscope. Color lines represent light diffracted by the specimen, whereas the black lines represent direct light. Only light diffracted by the specimen is captured by the objective lens.

fluorescent microscope. When the antibody molecules recognize and bind to the appropriate antigen molecules within the cell, they in effect label the antigen molecules with a fluorescent "tag" that will cause that region of the cell to "light up" when illuminated with ultraviolet light.

Figure A-12 illustrates the use of fluorescence microscopy to detect structures in a cultured muscle cell that contain the proteins actin and tubulin. In Figure A-12a, the distribution of actin-containing structures in the cell is demonstrated by the use of rhodamine-labeled *phalloidin*, a toxic fungal alkaloid that binds very specifically to actin filaments. The same cell is shown in Figure A-12b, but treated with fluorescein-labeled antibody against tubulin, such that microtubule-containing structures are specifically visualized. Figure A-12c is a phase-contrast micrograph of the same cell to show the general morphology of the cell.

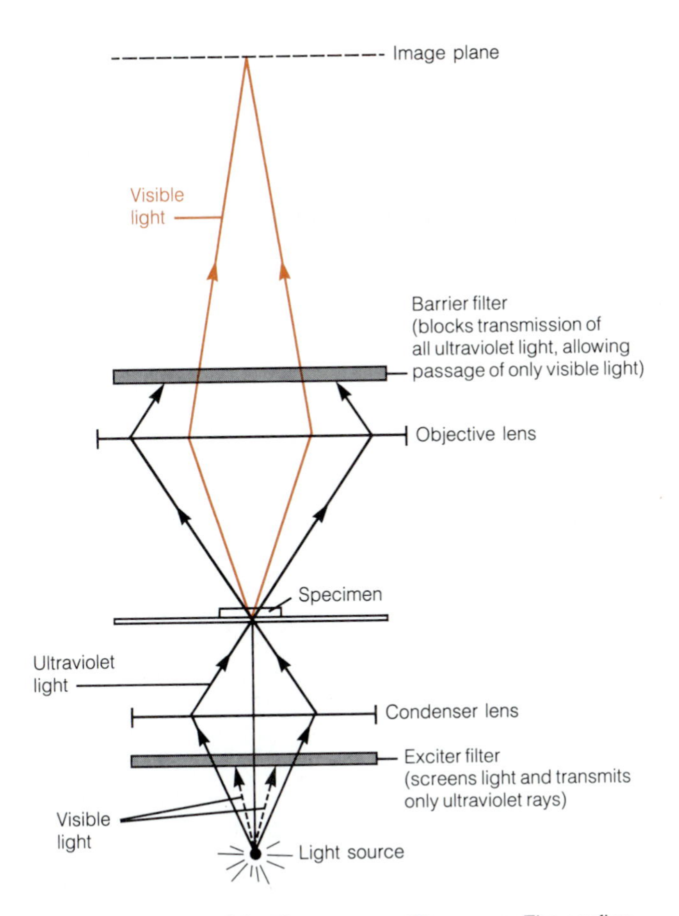

Figure A-11 Optics of the Fluorescence Microscope. The configuration of the optical elements and the path of light rays through the fluorescence microscope. Light from the source passes through an exciter filter that transmits only ultraviolet light (solid black lines). Illumination of the specimen with ultraviolet light induces fluorescent molecules in the specimen to emit visible light (color lines). The barrier filter subsequently removes the ultraviolet light, while allowing passage of the visible light. The image is therefore formed exclusively by visible light emitted by fluorescent molecules in the specimen.

Polarization Microscopy

Like fluorescence microscopy, **polarization microscopy** also depends on a modified form of illumination to gain specific information about the specimen. The design of a polarizing microscope is shown in Figure A-13. The optical

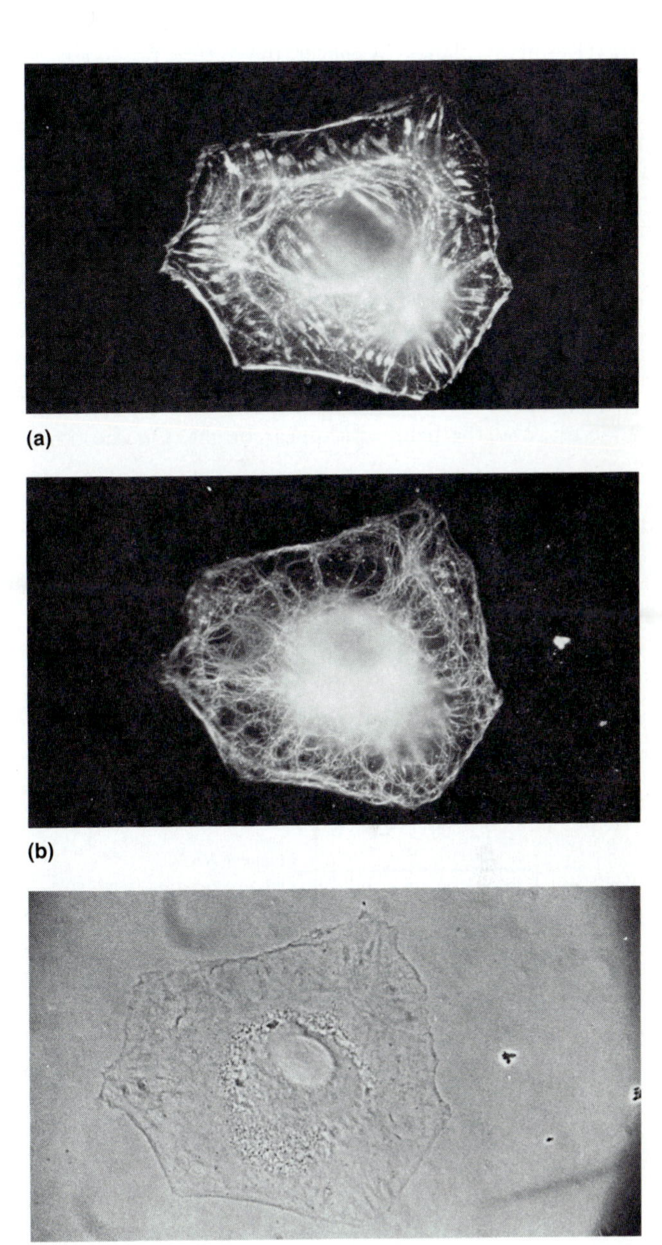

Figure A-12 Fluorescence Microscopy. (a) A cultured rat vascular smooth muscle cell that has been treated with rhodamine-labeled phalloidin. Phalloidin binds specifically to actin, causing actin-containing structures in the cell to "light up" in the fluorescence microscope. (b) The same cell, but treated with fluorescein-labeled antibody against tubulin, to cause microtubules to fluoresce. (c) A phase-contrast micrograph of the same cell. (all LMs)

path is similar to that of the brightfield microscope except for the presence of a *polarizer* just above the light source and an *analyzer* above the objective lens. Both the polarizer and the analyzer are rotatable pieces of optical material with molecular characteristics that lead to the *polarization* of the light as it passes through.

To understand polarization, we need to recognize that the waves of light radiating from a source of illumination travel in many planes. However, the optical properties of the polarizer are such that only waves that are traveling in one plane can pass through; waves traveling in all other planes are blocked. This **plane-polarized light** passes first through the specimen and the objective lens, then through the analyzer. Because the analyzer is made of the same material as the polarizer, only light waves in one plane can pass through the analyzer and reach the eyepiece to form an image.

If the polarizer and analyzer are oriented so that their transmission planes are parallel to each other, all light that passes through the polarizer will also pass through the analyzer, and the field will appear bright. On the other hand, if the transmission planes of the polarizer and analyzer are perpendicular to each other, the field will appear dark.

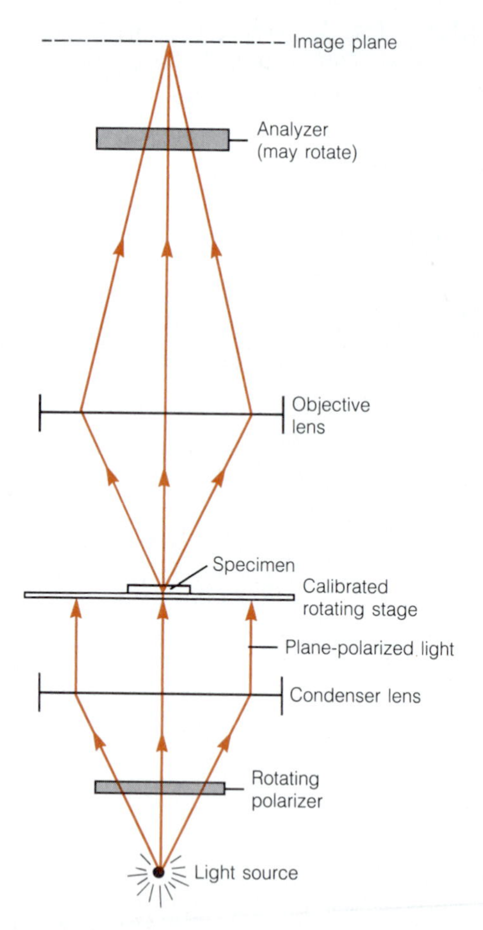

Figure A-13 Optics of the Polarizing Microscope. The configuration of the optical elements and the path of light rays through the polarizing microscope.

In polarization microscopy, the polarizer and analyzer are positioned perpendicular to each other, so we expect to see a dark field. However, some specimens are capable of rotating the plane of light as it passes through, thereby allowing the light to be transmitted by the analyzer and form an image. This property of rotating plane-polarized light is called **birefringence**. The polarizing microscope is therefore a device to detect and visualize birefringent structures in cells. Birefringence is usually a property of highly ordered structures. A chloroplast, for example, is a birefringent structure because the individual thylakoids in a granum are stacked together in an ordered manner.

An important feature of polarization microscopy is that it can be used to gain information about the structure of ordered cellular components even though the components themselves are well beyond the resolving power of the light microscope. A disadvantage of the technique is that the optics are very costly, and polarizing microscopes are therefore quite expensive.

Differential Interference Contrast Microscopy

Differential interference contrast (DIC) **microscopy** has recently come to the forefront as an optical technique in cell biology. This type of light microscopy, sometimes also called **Nomarski interference microscopy**, is a close relative

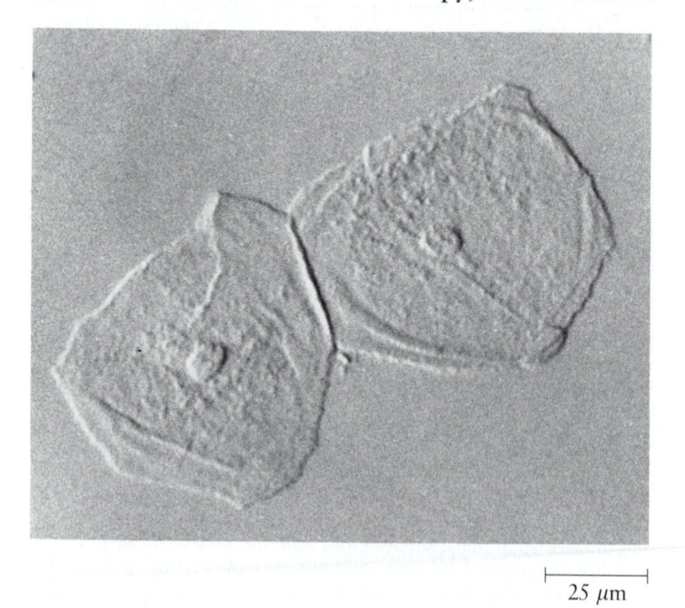

Figure A-14 Differential Interference Contrast (DIC) Microscopy. A DIC micrograph of the same epithelial cells seen in the phase-contrast micrograph of Figure A-9 (LM).

of polarization microscopy. Like the polarizing microscope, the DIC microscope has a polarizer and an analyzer. It also has *Wollaston prisms* positioned above the polarizer and below the analyzer to enhance contrast. As a result, the DIC microscope produces an illusion of three dimensions, accomplished by a shadow-casting effect in which the specimen appears to cast a shadow along one side. Figures A-7d and A-14 are both produced with DIC optics.

The shadow-casting effect can be moved around the object by rotating the stage, thus making interpretation of complex objects easier. Moreover, images with high resolution and contrast can be obtained from specimens that might otherwise be difficult to visualize. The DIC microscope is also used for **optical sectioning**, in which a specimen is viewed with high resolution in many different planes as the microscopist focuses down through the specimen.

The DIC system is useful for general examination of most biological specimens, but is especially valuable for living, unfixed, and unstained specimens. Because of the sophisticated optical elements required for DIC optics, such a microscope is often very expensive, which is probably its single greatest disadvantage.

Sample Preparation Techniques in Light Microscopy

One of the most attractive features of light microscopy is the ease with which most specimens can be prepared for examination. Preparation often involves nothing more than mounting a small piece of the specimen in a suitable liquid on a glass slide and covering it with a glass cover slip. The slide is then positioned on the specimen stage of the microscope and examined through the ocular lens. Magnification can be changed simply by rotating a turret that holds different objective lenses. Objective lenses capable of high magnification (40 × and greater) generally require the application of a small drop of immersion oil to the cover slip for optimal viewing. Specimens that are pathogenic or potentially pathogenic are often killed chemically prior to viewing by the application of a small amount of fixative (such as an aqueous buffered aldehyde solution) to the specimen suspension.

Specimen Processing: Fixation

Specimens to be viewed under a light microscope must be very thin and refractile for adequate resolution and contrast. In many cases, this means that **thin sections** of the specimen must be prepared and placed on glass slides. The first step in preparing the specimen is **primary fixation,** generally in a buffered aldehyde fixative. Fixation stabilizes the chemical components of the cells and hardens the specimen in anticipation of further processing and sectioning.

One way to fix a specimen is simply to immerse it in the fixative solution. An alternative approach for animal tissues is to pass the fixative through the bloodstream of the animal before removing the organs. This technique, called **perfusion**, often helps reduce *artifacts*, false or inaccurate representations of the specimen that result from chemical treatment or handling of the cells or tissues.

Embedding and Sectioning

After fixation, the specimen is embedded in a medium that will hold it rigidly in position while sections are cut. The specimen is usually embedded in paraffin wax. Since paraffin is insoluble in water, any water in the specimen must first be removed (by dehydration in alcohol, usually) and replaced by an organic solvent such as xylene, in which paraffin is soluble. The processed tissue is then placed in warm, liquefied paraffin and allowed to harden. Dehydration is less critical if the specimen is embedded in a water-soluble medium instead of in paraffin. Specimens may also be embedded in epoxy plastic resin.

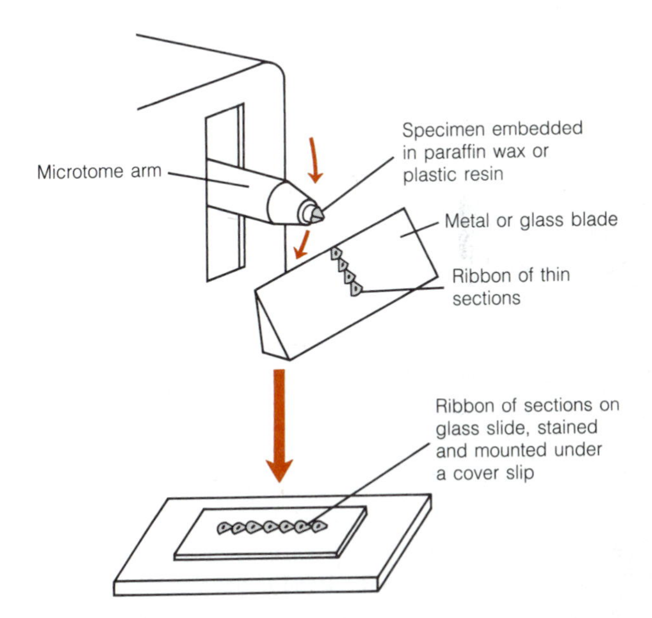

Figure A-15 Sectioning with a Microtome. The fixed specimen is embedded in paraffin wax or plastic resin and mounted on the arm of the microtome. As the arm moves up and down through a circular arc, successive sections are cut. These sections adhere to each other, forming a ribbon of thin sections that can be mounted on a glass slide, stained, and protected with a cover slip.

Next, the embedded specimen is sectioned into very thin slices, usually a few (1-10) micrometers thick. Sections are cut with a **microtome**, an instrument that operates rather like a meat slicer (Figure A-15, p. 783). The paraffin or plastic block containing the specimen is mounted on the arm of the microtome, which advances the block by small increments toward a metal or glass blade. As successive sections are cut, they usually adhere to one another, forming a ribbon of thin sections. These sections can be mounted on a glass slide, stained (if desired), and protected with a cover slip.

Staining

The purpose of **staining** is to give distinctive color characteristics to different kinds of cellular components. The general approach is to mount the fixed tissue on a microscope slide and then treat it with any of a great variety of dyes and stains that have been adapted for this purpose. Sometimes the tissue is treated with a single stain, but more often a series of stains is employed, each with an affinity for a different kind of cellular component. Figure A-16 illustrates the *Gram stain*, a technique widely used in microbiology as a means of assigning bacteria to one of two large groups, depending on whether they can be stained with a purple dye such as gentian violet (gram-positive species) or not (gram-negative species).

Autoradiography

Autoradiography is a technique that uses photographic film to determine where within a cell a radioactively labeled compound is at the time the cell is fixed and sectioned for microscopy. Autoradiography can therefore be used to localize cellular processes to specific structures and to provide information about cell function. The preparation of a sample for autoradiography is shown in Figure A-17. In essence, the process involves incubating the tissue with a radioactively labeled compound (step 1), fixing the tissue (step 2), sectioning it in the conventional way (step 3), and mounting the fixed tissue on a microscope slide (step 4).

The slide is then covered with a thin layer of photographic emulsion (step 5) and placed in a sealed box for the desired length of time, often for several weeks (step 6). During this time, the radioactivity in the cell will expose the emulsion directly above it, providing a photographic

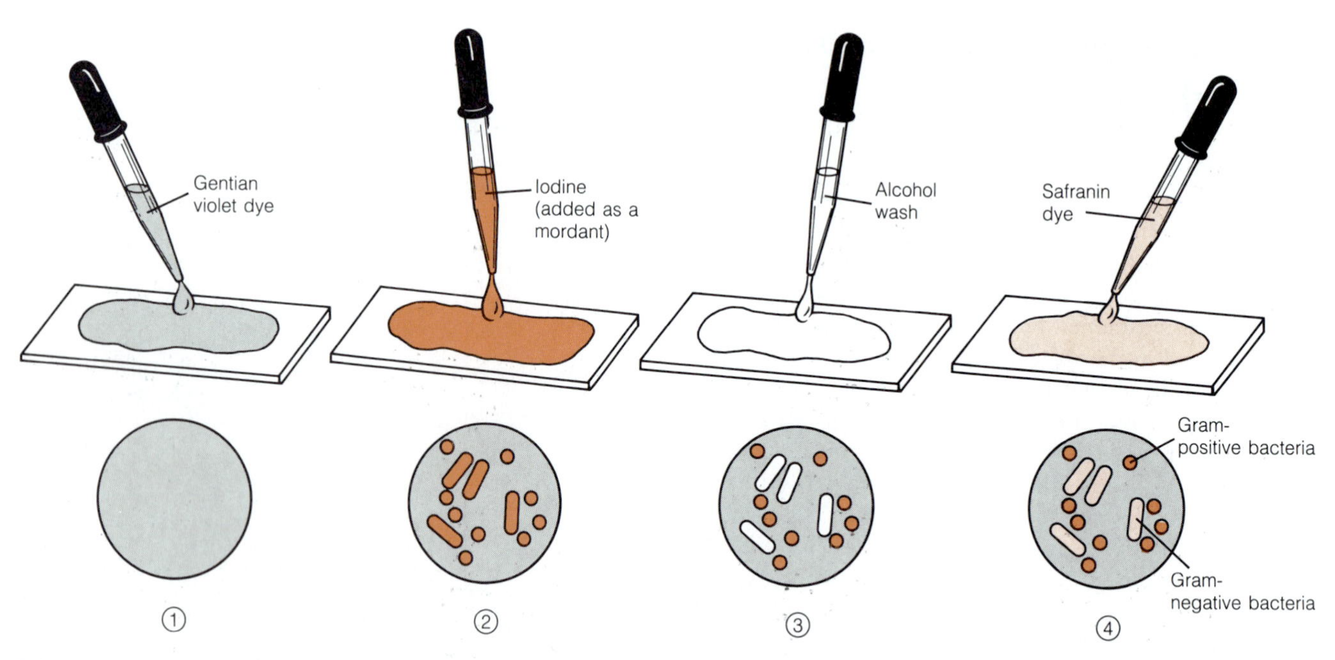

Figure A-16 The Gram Stain for Bacteria. The Gram stain is one of the most useful staining procedures because it divides bacteria into two large groups: gram-positive and gram-negative. (1) A purple dye such as gentian violet is applied to a heat-fixed preparation of bacterial cells, then washed off again. (2) Iodine is added as a mordant, then washed off. At this point, both gram-positive and gram-negative bacteria appear purple. (3) Next, the slide is washed with alcohol, then with water. After this step, gram-positive bacteria are still purple, but gram-negative cells are colorless. (4) Finally, the preparation is stained with safranin to stain the gram-negative bacteria light pink. The slide is then ready to be washed, dried, and examined with the microscope.

record of precisely where in the cell the radioactive compound is located. Thereafter, the slide containing the specimen and the emulsion is developed in much the same way as conventional black and white film (step 7) and is then ready for examination under the microscope (step 8).

Autoradiography can be applied to both light microscopy (Figure A-18a) and electron microscopy (Figure A-18b). In either case, it provides valuable information about the localization of specific molecules, structures, or processes within the cell.

The Electron Microscope: Design and Practice

The effect of electron microscopy on our understanding of cells can only be described as revolutionary. Yet, like light microscopy, electron microscopy has both strengths and weaknesses. In light microscopy, most specimens can be prepared easily and examined readily, but resolution of cellular ultrastructure is severely limited by the physical characteristics of light. In electron microscopy, resolution is much better, but specimen preparation and instrument operation are often more difficult.

Electron microscopes are of two basic designs: the *transmission electron microscope* and the *scanning electron microscope*. A third type of instrument, the *scanning transmission electron microscope*, is a hybrid of the two, as the name implies. The scanning and transmission electron microscopes are similar in that each uses a beam of electrons to produce an image. However, the instruments use quite different mechanisms to form the final image.

Transmission Electron Microscopy

A **transmission electron microscope** (TEM) is shown in Figure A-19. Most of the parts of the TEM are similar in name and function to their counterparts in the light microscope, although their physical orientation is reversed. We will look briefly at each of the major features.

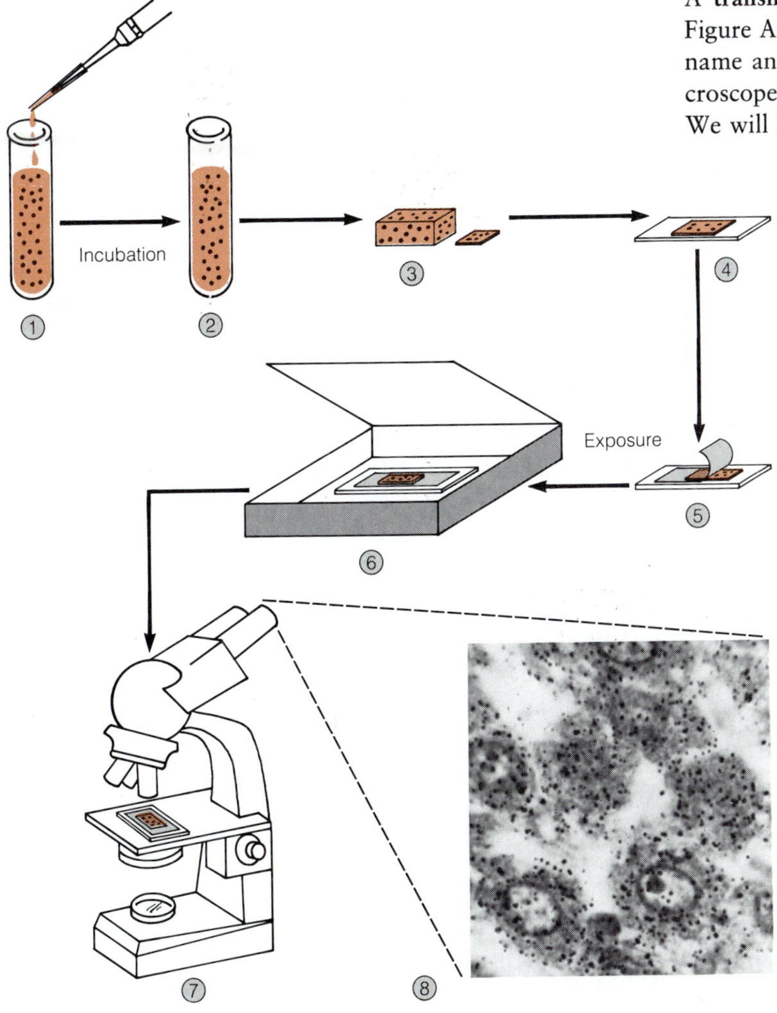

Figure A-17 The Technique of Autoradiography. Autoradiography is a means of localizing cellular processes to specific structures. (1) To a suspension of organisms or cells is added the desired radioactively labeled compound (color solution), followed by a period of incubation. (2) The incubation is then stopped, and the cells are rinsed, fixed, and recovered from the fluid. (3) The fixed cells are embedded and sectioned. (4) A section is washed and placed on a microscope slide. (5) The slide is covered with a thin layer of photographic emulsion (in a darkroom, of course). (6) The slide is then placed in a sealed box for the desired length of time, to allow the radioactivity in the cell to expose the film directly above it. (7) The slide is developed, rinsed, fixed, and washed. (8) The specimen is then ready for examination under the microscope. The inset is an autoradiograph of rat liver cells 12 hours after injection of the rat with ^{3}H-labeled cytidine, a precursor of RNA.

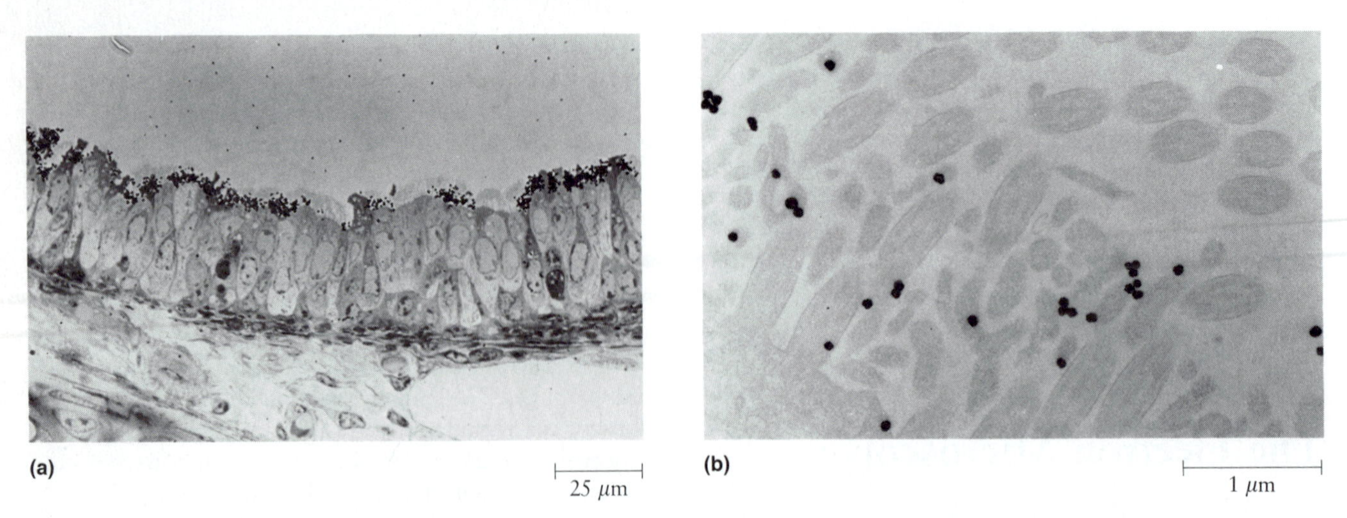

Figure A-18 Autoradiography. Autoradiography can be applied to both (a) light microscopy and (b) electron microscopy. A radioactively labeled microbial pathogen was allowed to infect epithelial cells of the respiratory tract of an animal. The microbes can be identified by the black dots, which represent areas where the radioactive label exposed silver grains in the overlying photographic emulsion. In (a), the limited resolving power of the light microscope precludes direct observation of the organisms, but their presence is indicated by the numerous black dots corresponding to exposed silver grains. In (b), the higher magnification and resolution of the electron microscope reveal both the microbial cells and their identifying label.

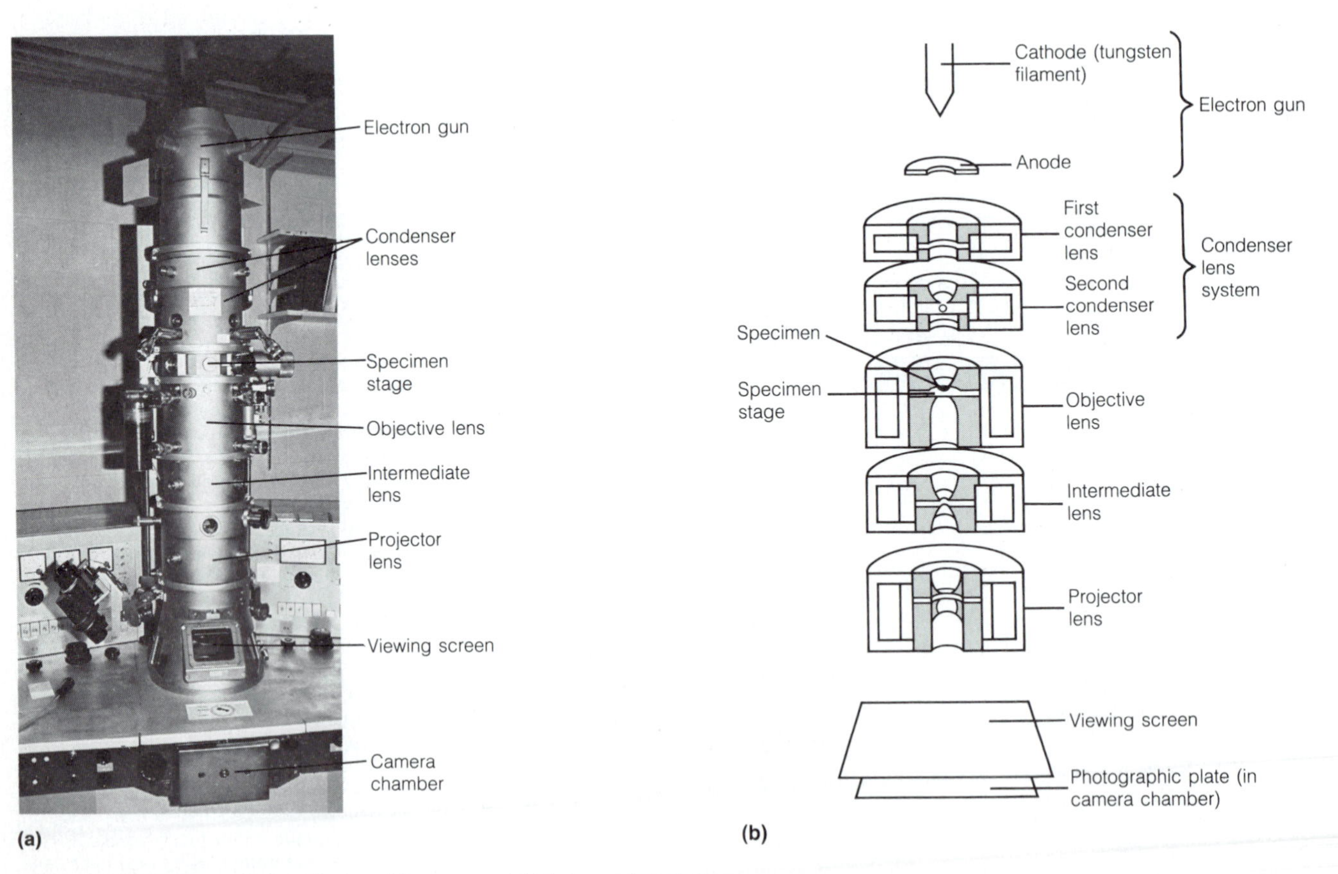

Figure A-19 The Transmission Electron Microscope. (a) Photograph and (b) schematic diagram of a transmission electron microscope.

The Vacuum System. Since electrons cannot travel very far in air, a strong vacuum must be maintained along the entire path of the electron beam. Two types of vacuum pumps work together to create the vacuum in the column of an electron microscope. A standard *rotary pump* is used to achieve the initial low vacuum when the instrument is first started up. The high vacuum required for operation is achieved by an *oil diffusion pump*. The diffusion pump is an oil-filled reservoir in which oil is vaporized by heating. As the oil vapor rises, it traps air molecules and is then condensed by condensing vanes, which are cooled by circulating cold water. The diffusion pump cannot function independently; it requires backup by the rotary pump to remove the trapped air molecules from the system.

On some TEMs, a device called a *coldfinger* is incorporated into the vacuum system to help establish a high vacuum. The coldfinger is a metal insert in the column of the microscope that is cooled by liquid nitrogen. The coldfinger attracts gases and random contaminating molecules, which then solidify on the cold metal surface. When functioning at their best, most modern electron microscopes maintain a vacuum of about 1×10^{-4} torr (1×10^{-4} mm Hg) or less.

The Electron Gun. The electron beam in a TEM is generated by the **electron gun,** an assembly of several components. The *cathode*, a tungsten filament similar to a light bulb filament, emits electrons from its surface when it is heated. The cathode tip is near a circular opening in a metal housing called the *Wehnelt cylinder*. A negative voltage on the cylinder helps control electron emission and shape the beam. At the other end of the cylinder is the *anode*. The anode is kept at 0 V, which is 50–100 kV less than the cathode. This difference in voltage is called the **accelerating voltage** because it causes the electrons to accelerate as they pass through the cylinder.

Electromagnetic Lenses and Image Formation. As the electron beam leaves the upper region of the condenser lens system, it enters a series of lenses made of electromagnets (Figure A-19b). The lens itself is simply a space influenced by an electromagnetic field. The focal length of each lens can be increased or decreased by varying the current applied to its energizing coils. Thus, when several lenses are arranged together, they can control illumination, focus, and magnification.

The **condenser lens** is the first lens to affect the electron beam. It functions in the same fashion as its counterpart in the light microscope to collimate the beam for illumination of the specimen. Most electron microscopes actually use a condenser lens system with two lenses to achieve better focus of the electron beam.

The next component, the **objective lens,** is the most important part of the electron microscope's sophisticated lens system. The specimen is positioned on the specimen stage within the objective lens (Figure A-19b). The objective lens, in concert with the **intermediate lens** and the **projector lens,** produces a final image on a *viewing screen* of zinc sulfide that fluoresces when struck by the electron beam.

How is an image formed from the action of these lenses on an electron beam? Recall that the electron beam generated by the cathode passes through the condenser lens system and impinges on the specimen. As the beam strikes the specimen, some electrons are scattered by the sample, whereas others continue in their paths relatively unimpeded. This scattering of electrons is a result of properties created in the specimen by the preparation procedure. Specimen preparation, in other words, imparts selective electron density to the specimen; that is, some areas become more opaque to electrons than others. Such electron-dense areas of the specimen will appear dark because few electrons pass through, whereas other areas will appear lighter because they permit the passage of more electrons.

The contrasting light, dark, and intermediate areas of the specimen create the image seen on the screen. The fact that the image is formed by differing extents of electron transmission through the specimen is reflected in the term *transmission electron microscope*.

The Photographic System. In addition to observing the image on a fluorescent screen, an electron microscopist can also record the image photographically as an **electron micrograph.** An electron micrograph is therefore a permanent photographic record of the specimen.

Most transmission electron microscopes have a *camera chamber* mounted directly beneath the viewing screen. The camera is little more than a box that allows *photographic plates* to be moved manually or automatically to the area immediately beneath the viewing screen. To photograph a specimen, the microscopist simply aligns the image on the screen, focuses the image with the objective lens control, adjusts the illumination to a predetermined intensity with the condenser adjustment, and makes the exposure. The exposure may be made automatically or by lifting the screen with a lever on the microscope console. Once the exposure is made, the plate is advanced out of the exposure position to a container where it is stored until retrieved from the instrument for later development and printing.

Clearly, an electron microscopist must have a good working knowledge of photography. For a full appreciation of photographic principles and techniques, consult appropriate references on photography.

Scanning Electron Microscopy

Scanning electron microscopy is a relatively recent development. It is an especially spectacular technique because of the sense of depth it gives to biological structures, thereby allowing surface topography to be studied. As the name implies, a **scanning electron microscope** (**SEM**) generates an image by scanning the specimen with a beam of electrons.

An SEM and its optical system are shown in Figure A-20. The vacuum system and electron source are similar to those found in the transmission electron microscope,

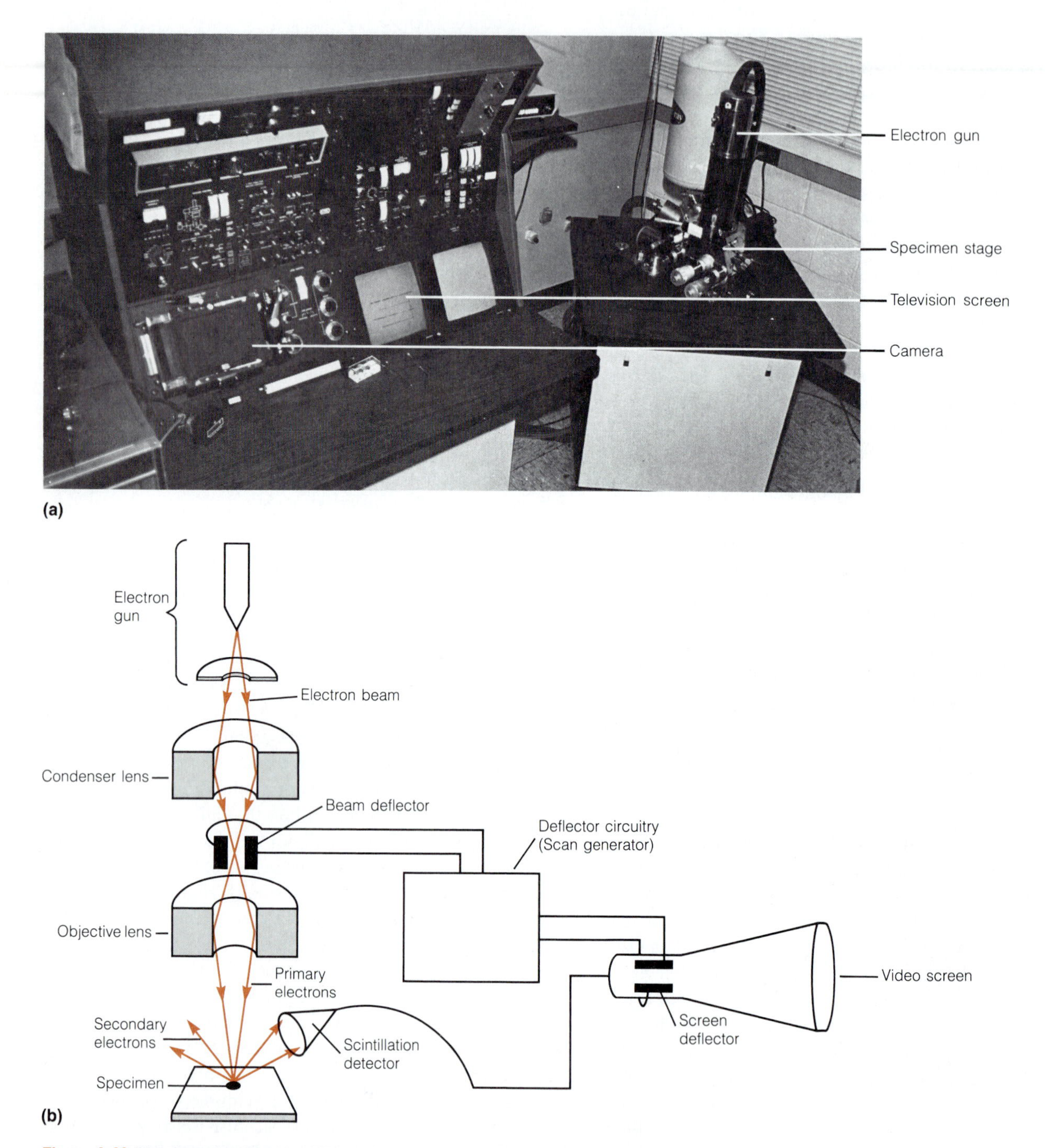

Figure A-20 The Scanning Electron Microscope. (a) Photograph and (b) schematic diagram of a scanning microscope. The image is generated by secondary electrons (color) that are emitted by the specimen as a focused beam of primary electrons sweeps rapidly over it. The signal to the video screen is synchronized to the movement of the primary electron beam over the specimen by the deflector circuitry of the scan generator.

although the accelerating voltage is much lower (about 5–30 kV). The significant difference between the two kinds of instruments lies in the way the image is formed. In the SEM, a magnetic lens system focuses the beam of electrons into an intense spot on the surface of the specimen. The spot is moved back and forth across the specimen by charged plates called *beam deflectors* located between the condenser lens and the specimen. The beam deflectors attract or repel the beam according to the signals sent to them by the deflector circuitry (Figure A-20b).

As the electron beam sweeps rapidly over the specimen, molecules in the specimen are excited to high energy levels and emit **secondary electrons.** These secondary electrons are used to form an image of the specimen surface. They are then captured by a detector that is located immediately above and to one side of the specimen. The essential component of the detector is a *scintillator*, which emits photons of light when excited by the electrons incident upon it. The photons are used to generate an electronic signal to a video screen. The image then develops point by point, line by line on the screen as the primary electron beam sweeps over the specimen. Photomicrographs such as the one shown in Figure A-21 are made by directly photographing the video screen, usually with a polaroid camera. Recall also the scanning electron micrographs of Figure 1-4 in Chapter 1.

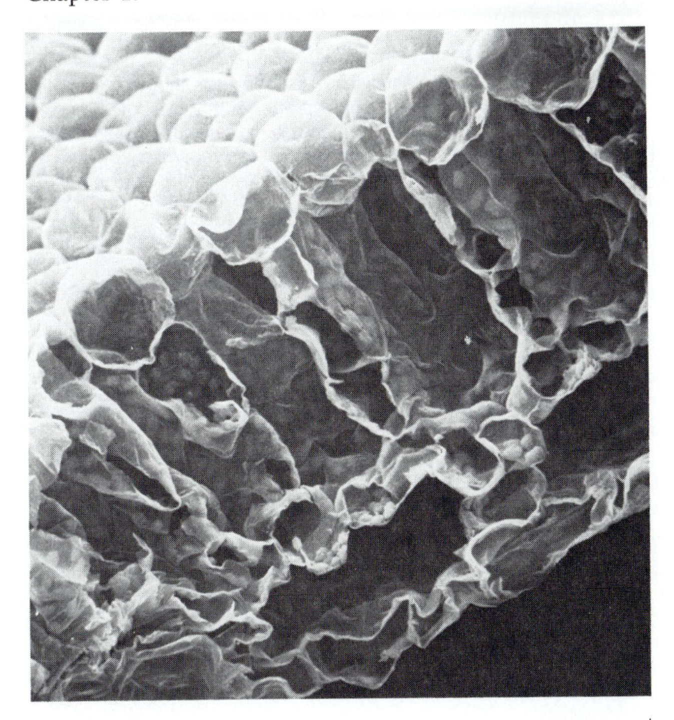

Figure A-21 Scanning Electron Microscopy. A transverse section through the leaf of the water fern *Salvinia*, as seen with a scanning electron microscope. Numerous chloroplasts can be seen inside the palisade cells in the center of the leaf. (SEM)

Scanning Transmission Electron Microscopy

A **scanning transmission electron microscope** (**STEM**) contains elements of both transmission and scanning electron microscopes (Figure A-22). Like the SEM, the STEM uses an electron beam that sweeps over the specimen. But the image is then formed by electrons transmitted through the specimen, as with the TEM. The STEM is capable of distinguishing specific characteristics of the electrons that are transmitted by the specimen, thus deriving information about the specimen not obtainable with a conventional TEM. However, an STEM is technically sophisticated, requires a very high vacuum, and is much more electronically complex than a TEM or an SEM.

High-Voltage Electron Microscopy

A **high-voltage electron microscope** (**HVEM**) is very similar to a transmission electron microscope except that its accelerating voltage is much higher. Whereas a TEM uses accelerating voltages of 50–100 kV, a high-voltage instrument uses voltages of about 200–1000 kV. Because of the high voltage and the greatly reduced chromatic aberration that it makes possible, relatively thick specimens can be examined with good resolution. As a result, cellular structure can be studied in sections as thick as 1 μm, about 10 times the thickness possible with an ordinary TEM. Figure A-23 shows a million-volt HVEM, and Figure A-24 shows a micrograph of a polytene chromosome from a fruit fly as visualized by high-voltage electron microscopy. For another example of a high-voltage electron micrograph, see Figure 18-3.

Sample Preparation Techniques in Transmission Electron Microscopy

Specimens for electron microscopy can be prepared in several different ways, depending on the type of microscope and the kind of information the microscopist wants to obtain. In each case, however, the method is complicated, time-consuming, and costly compared to methods used for light microscopy. Moreover, living specimens cannot be examined because of the vacuum to which specimens are exposed in the electron microscope.

Specimen Processing: Fixation

Specimens to be prepared for electron microscopy must first be chemically fixed and stabilized. This primary fixation kills the cells but keeps the cellular components much as they were in the living cell. Primary fixatives are usually buffered solutions of aldehydes. Glutaraldehyde is the most common fixative. Following primary fixation, the specimen is usually treated with a 1%–2% solution of buffered osmium tetroxide (OsO_4). The osmium tetroxide stains specific parts of the cell, making them more electron dense.

Embedding, Sectioning, and Poststaining

The next step in specimen preparation for transmission electron microscopy is to dehydrate the tissue by passing it through a series of alcohol solutions. The specimen is then placed into a fluid such as acetone or propylene oxide to prepare it for embedding in liquefied plastic epoxy resin. After the plastic has infiltrated the specimen, it is put into a mold and heated in an oven to harden the plastic. When the plastic is hardened, it is removed from the oven, and the area around the specimen is trimmed to make a face that is appropriate for sectioning.

The specimen is then sliced into the ultrathin sections required for examination with the transmission electron microscope. The instrument used for this purpose is an **ultramicrotome** (Figure A-25). The specimen is mounted firmly on the arm of the ultramicrotome, which then advances the specimen in small increments toward a glass or diamond knife (Figure A-25b). When the block reaches the knife blade, ultrathin sections (about 60–90 nm thick) are cut from the block face. The sections float from the blade onto a water surface, where they can be picked up on a circular copper specimen grid. The grid consists of a meshwork of very thin copper strips, which support the specimen while still allowing "windows" between adjacent strips through which the specimen can be observed.

Once in place on the grid, the sections are usually stained with solutions containing lead and uranium. This procedure, called **poststaining,** enhances the contrast of the specimen because the lead and uranium give still greater electron density to specific parts of the cell. After poststaining the specimen is ready for viewing or photography with the transmission electron microscope. Numerous examples of transmission electron micrographs are found throughout this text.

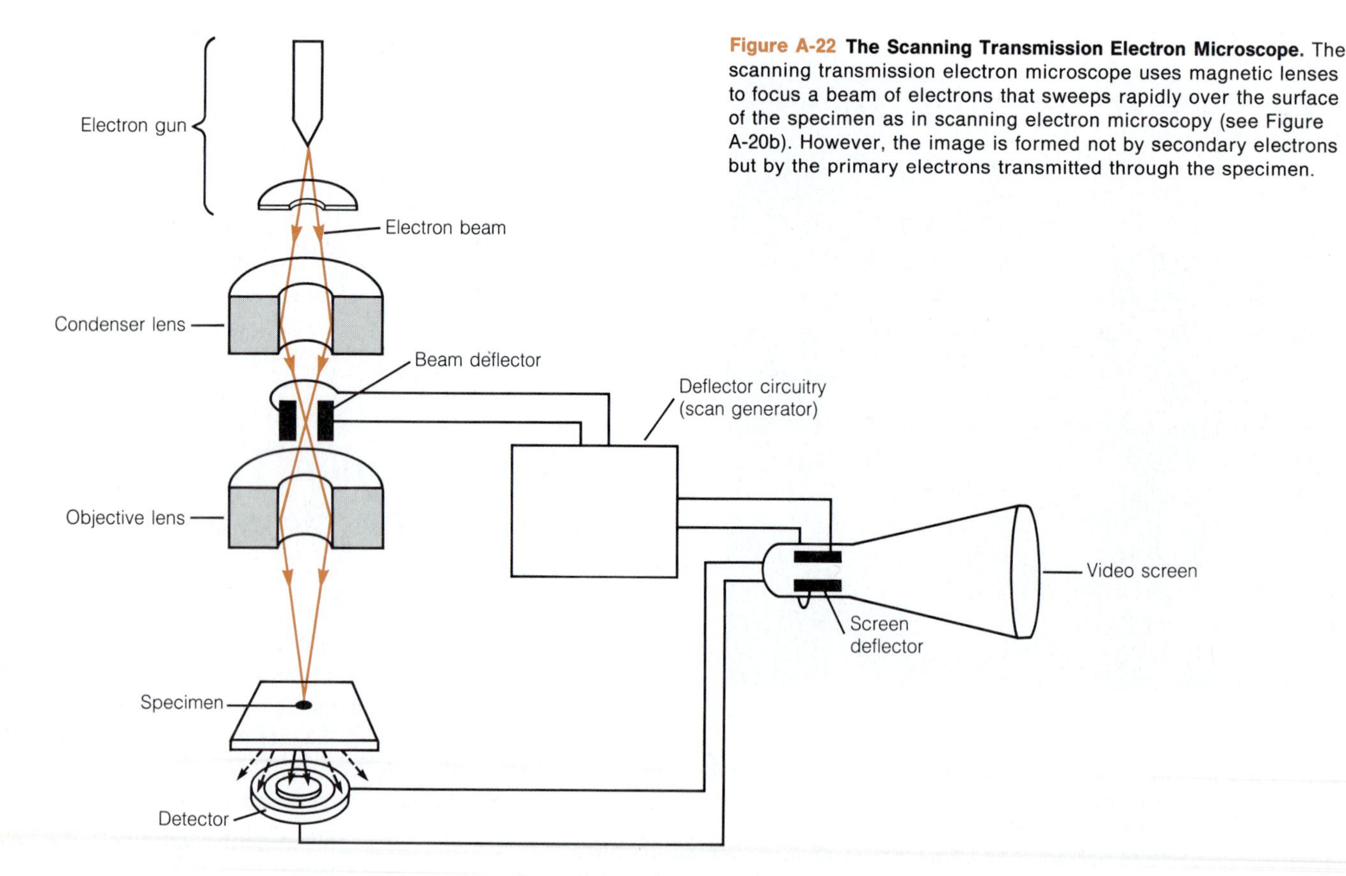

Figure A-22 The Scanning Transmission Electron Microscope. The scanning transmission electron microscope uses magnetic lenses to focus a beam of electrons that sweeps rapidly over the surface of the specimen as in scanning electron microscopy (see Figure A-20b). However, the image is formed not by secondary electrons but by the primary electrons transmitted through the specimen.

Electron Microscopic Autoradiography

The autoradiographic techniques described in our discussion of light microscopy can be used for transmission electron microscopy with only minor changes. For the TEM, the specimen containing the radioactively labeled compounds is examined in ultrathin sections on copper specimen grids instead of in thin sections on glass slides.

Figure A-23 The High-Voltage Electron Microscope. This million-volt electron microscope is located on the campus of the University of Wisconsin in Madison. The facility is supported by the Biotechnology Resources Program of the National Institutes of Health and is available free of charge to biomedical researchers.

Negative Staining

In contrast to the considerable effort necessary to prepare ultrathin sections, **negative staining** is one of the simplest techniques used in transmission electron microscopy. It is the preferred method for examining very small objects, such as viruses or isolated organelles.

For negative staining, the copper specimen grid must first be overlaid with an ultrathin plastic film. The specimen is then suspended in a small drop of liquid, applied to the overlay, and allowed to dry in air. After the specimen has dried on the grid, a drop of stain such as uranyl acetate or phosphotungstic acid is applied to the film surface. The edges of the grid are then blotted in several places with a piece of filter paper to absorb the excess stain. This draws the stain down and around the specimen and its ultrastructural features. When viewed in the TEM, the specimen is seen in **negative contrast** because the background is dark and heavily stained, whereas the specimen itself is lightly stained. Negative staining is illustrated by the electron micrograph shown in Figure A-26. Notice how well the swirled pattern of individual cellulose microfibrils can be visualized by this technique.

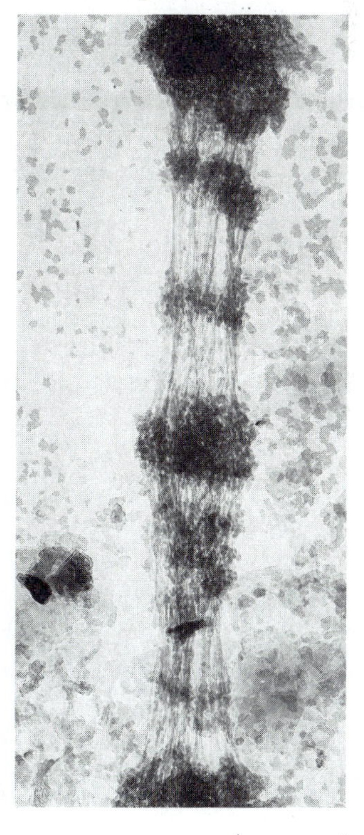

Figure A-24 High-Voltage Electron Microscopy. A polytene chromosome from the fruit fly *Drosophila melanogaster*, as seen with a high-voltage electron microscope operated at 1 million volts. For a three-dimensional view of this structure, see the stereo pair of Figure A-32. (TEM)

Shadowing

The same cell wall seen in Figure A-26 is shown in Figure A-27 as visualized by the technique of **shadowing.** Shadowing involves the deposition of a thin layer of an electron-dense metal such as gold or platinum on a biological

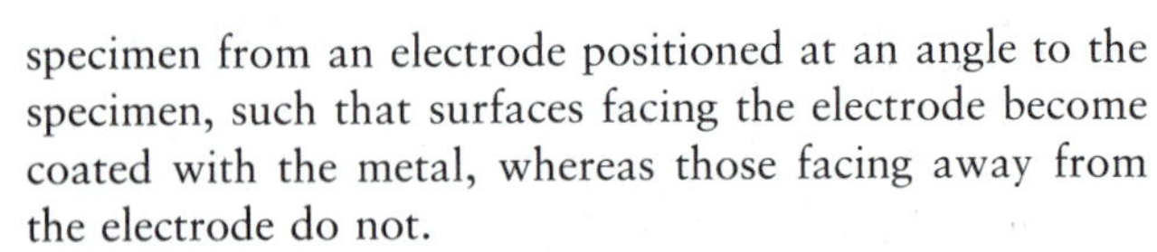

specimen from an electrode positioned at an angle to the specimen, such that surfaces facing the electrode become coated with the metal, whereas those facing away from the electrode do not.

Figure A-28a illustrates the shadowing technique. Specimens that can be suspended in water-based solutions

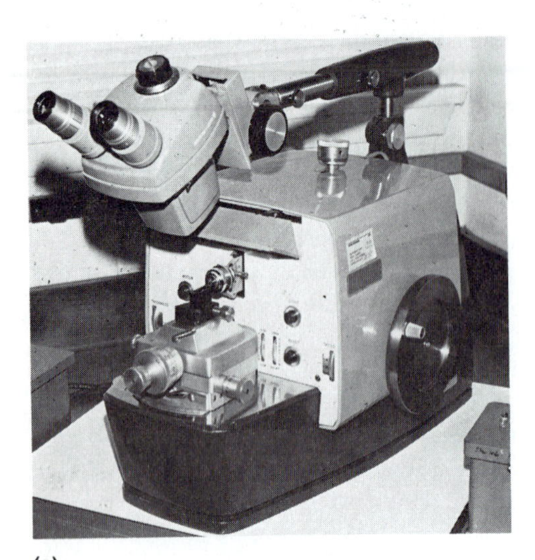

(a)

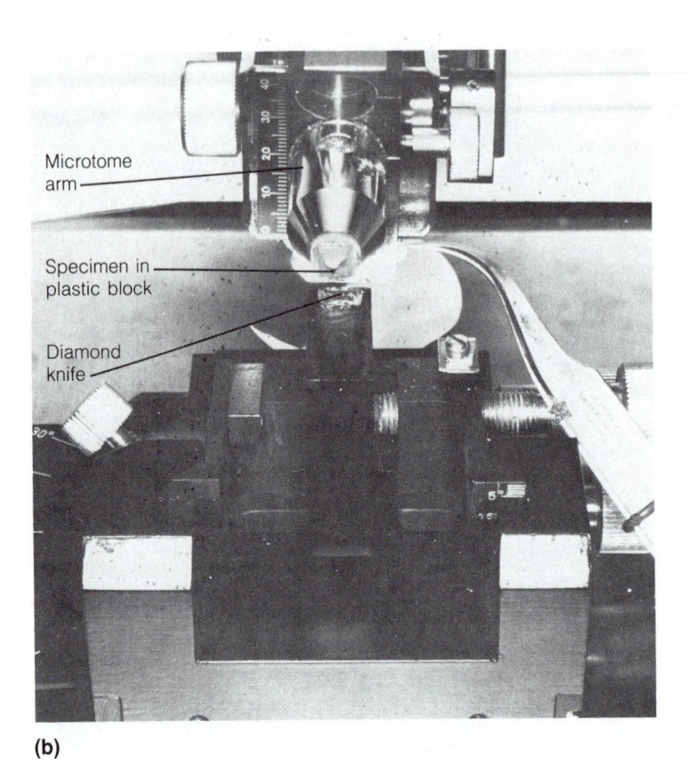

(b)

Figure A-25 An Ultramicrotome. (a) Photograph of an ultramicrotome. (b) Close-up view of the microtome arm, showing the specimen in a plastic block mounted on the end of the arm. As the microtome arm moves up and down, the block is advanced in small increments, and ultrathin sections are cut from the block face by the diamond knife.

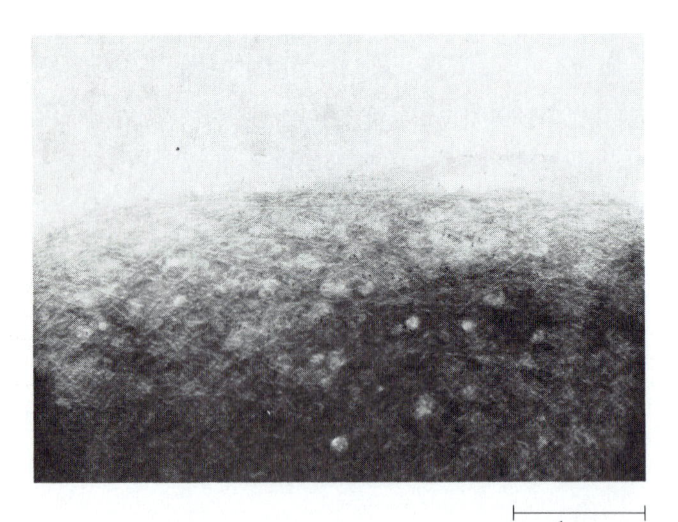

Figure A-26 Negative Staining. An electron micrograph of the cell wall of a flagellated freshwater alga as seen in a negatively stained preparation. For a shadowed preparation of the same specimen, see Figure A-27. (TEM)

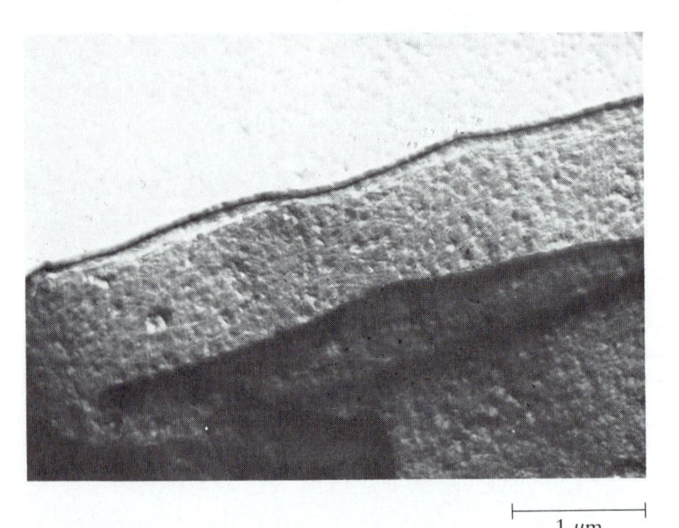

Figure A-27 Shadowing. An electron micrograph of the cell wall of a flagellated freshwater alga as seen in a shadowed preparation. For a negatively stained preparation of the same specimen, see Figure A-26. Notice that the same swirling pattern of cellulose microfibrils is seen in both preparations. (TEM)

are especially suitable for shadowing. The specimen is first spread on a clean mica surface and dried (step 1). It is then placed in a **vacuum evaporator,** a bell jar in which a vacuum is created by a system similar to that of an electron microscope (Figure A-28b). Also within the evaporator are two electrodes, one consisting of a carbon rod located directly over the specimen and the other consisting of a metal wire positioned at an angle of about 10°–45° relative to the specimen.

After a vacuum is created in the evaporator, current is applied to the metal electrode, causing the metal to evaporate from the electrode and spray over the surface of the specimen (Figure A-28a, step 2). Because of the angular positioning of the metal electrode, the metal will accumulate as a thin coating on the sides of any surface irregularities that face the electrode, generating a metal **replica** of the surface. These same irregularities prevent deposition of the metal on the side facing away from the electrode, thus producing contrast due to the shadow effect.

The carbon electrode is then fired, coating the specimen from directly overhead with evaporated carbon to give stability and support to the metal replica (step 3). The mica support containing the specimen is then removed from the vacuum evaporator and lowered gently onto a water surface, causing the replica to float away from the mica surface. The replica is transferred into an acid bath, which dissolves away remaining bits of specimen, leaving a clean metal replica of the specimen (step 4). The replica can then be returned to the water surface and retrieved on a standard copper grid (step 5). Replicas can be viewed in the TEM in the same way as ultrathin sections.

Freeze-Fracturing

Freeze-fracturing is a relatively recent technique that has proved very useful to cell biologists. It is especially valuable for studying the ultrastructure of biological membranes. Freeze-fracturing involves the cleavage of a frozen specimen under a vacuum, followed by platinum/carbon shadowing to create a replica of the fractured surface, which is often the interior of a membrane.

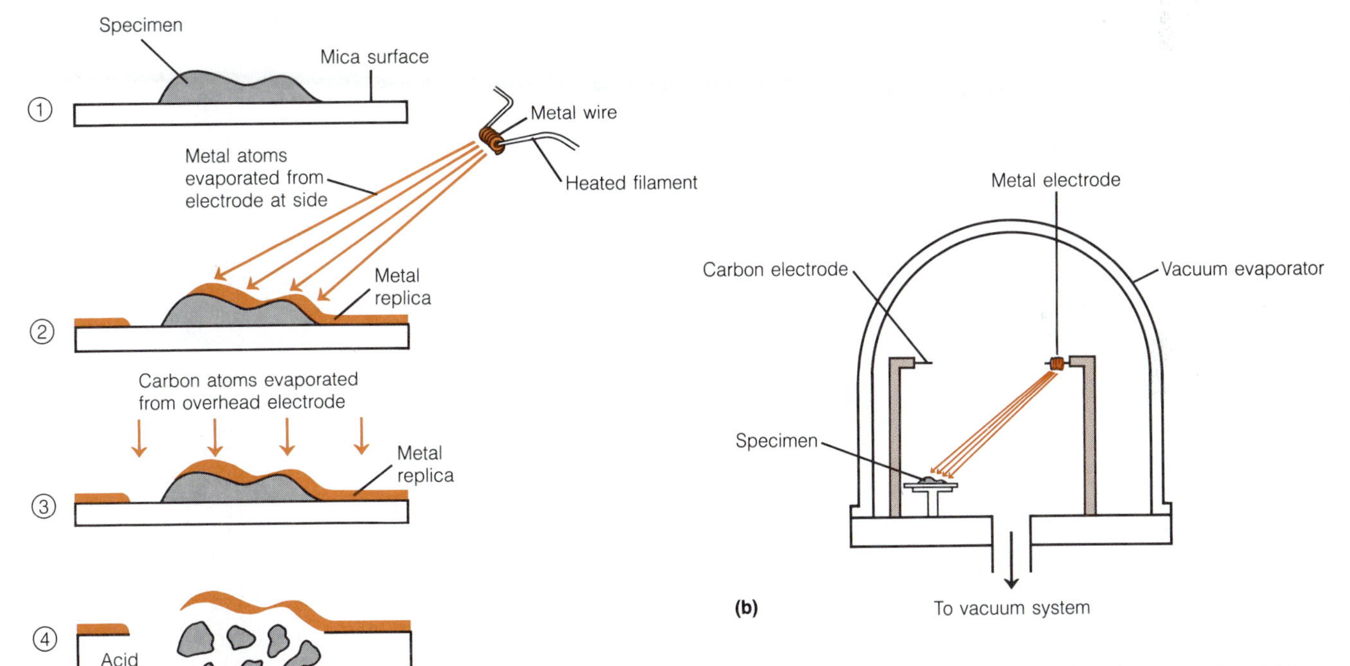

Figure A-28 The Technique of Shadowing. (a) Stepwise procedure for shadowing. (1) The specimen is spread on a mica surface and dried. (2) The specimen is shadowed by coating it with atoms of a heavy metal (platinum or gold, shown in color) that are evaporated from a heated filament located to the side of the specimen in a vacuum evaporator. This generates a metal replica (color), the thickness of which reflects the surface contours of the specimen. (3) Next, the specimen is coated with carbon atoms evaporated from an overhead electrode to strengthen and stabilize the metal replica. (4) The replica is then floated onto the surface of an acid bath to dissolve away the specimen, leaving a clean metal replica. (5) The replica is washed and picked up on a copper grid for examination in the transmission electron microscope. (b) The vacuum evaporator in which shadowing is done. The carbon electrode is located directly over the specimen, whereas the heavy metal electrode is off to the side.

Freeze-fracturing is illustrated in Figure A-29. It takes place in a modified vacuum evaporator with an internal microtome knife for fracturing the frozen specimen and with provision for precise control of the temperature of the specimen stage and the microtome arm and knife. Specimens are generally fixed prior to freeze-fracturing, although some living tissues can be frozen fast enough to keep them in almost lifelike condition. Because cells contain a lot of water, fixed specimens are usually treated with an antifreeze such as glycerol to provide **cryoprotection**—that is, to reduce the formation of ice crystals during freezing.

The cryoprotected specimen is mounted on a metal specimen support (Figure A-29, step 1) and immersed rapidly in freon cooled with liquid nitrogen (step 2). This procedure also reduces formation of ice crystals in the cells. With the frozen specimen positioned on the specimen stage in the vacuum evaporator (step 3), a high vacuum is established, the stage temperature is adjusted to around − 100°C, and the frozen specimen is fractured with a blow from the microtome knife (step 4). A replica of the fractured specimen is made by shadowing with platinum and carbon as described in the previous section (step 5), and the replica is then ready to be viewed in the transmission electron microscope (step 6).

Newcomers to the freeze-fracturing technique often misunderstand what a freeze-fracture replica represents. One might think that the fracture plane should pass through the specimen in a straight line as is clearly the case when a fixed and embedded sample is sectioned conventionally with an ultramicrotome (Figure A-30a). In actuality, however, the fracture line passes through the hydrophobic interior of membranes whenever possible, because this is the line of least resistance through the frozen specimen (Figure A-30b). As a result, a freeze-fracture replica is largely a view of the interiors of membranes, showing the inside of one or the other of the two monolayers of the membrane.

Freeze-fractured membranes appear as smooth surfaces studded with **intramembranous particles (IMPs)** that are either randomly distributed in the membrane or organized into ordered complexes. These are thought to be integral membrane proteins that have remained with one lipid monolayer or the other as the fracture plane passes through the interior of the membrane.

The electron micrograph shown in Figure A-31 illustrates the two faces of a plasma membrane as revealed by freeze-fracturing. The **P-face** is the interior face of the inner monolayer; it is called the P-face because this monolayer

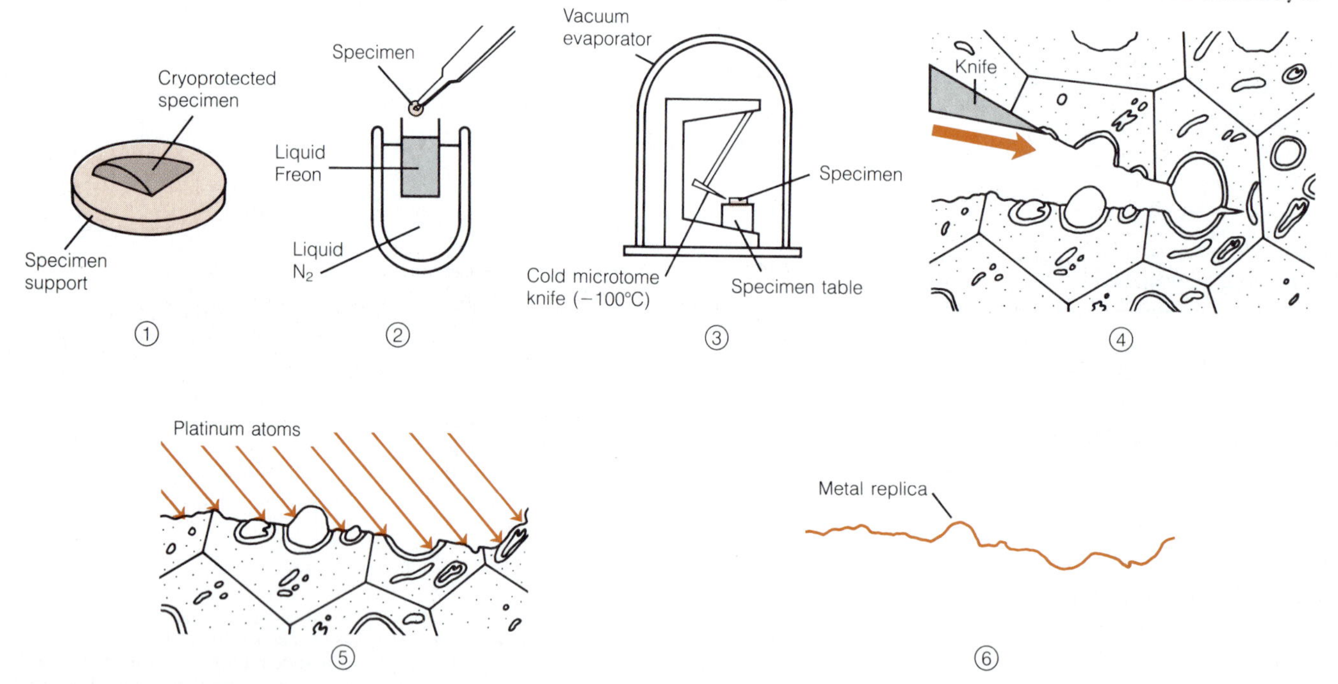

Figure A-29 The Technique of Freeze-Fracturing. (1) A cryoprotected specimen is mounted on a metal support. (2) The mounted specimen is immersed in liquid freon cooled in liquid nitrogen. (3) The frozen specimen is transferred to a vacuum evaporator and adjusted to a temperature of about − 100°C. (4) The specimen is fractured with a blow from the microtome knife. The fracture plane passes through the interior of lipid bilayers wherever possible, because this is the line of least resistance through the frozen specimen, as shown in Figure A-30. (5) The fractured specimen is shadowed with platinum and carbon as in Figure A-28 to make a metal replica of the specimen. (6) The metal replica is examined in the transmission electron microscope.

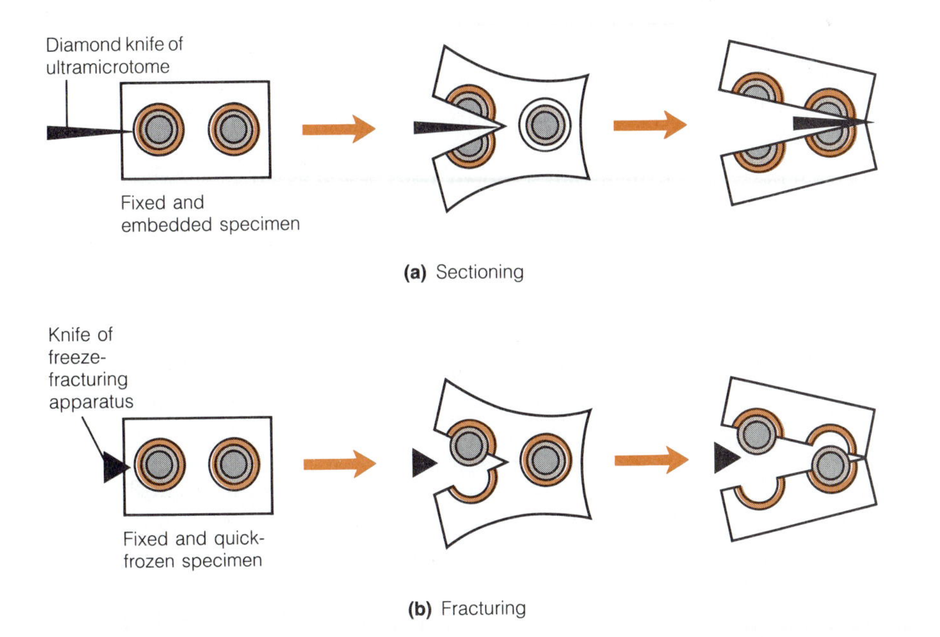

Figure A-30 Sectioning Versus Fracturing of Specimens. (a) When a fixed and embedded specimen is sectioned for conventional transmission electron microscopy, the edge of the diamond knife makes a clean, linear cut through the tissue. (b) When a fixed and quick-frozen specimen is fractured, the blow of the knife generates a fracture plane through the frozen sample that passes through the interiors of membranes whenever possible, because the hydrophobic interior of a phospholipid bilayer is more readily fractured than is the ice that surrounds it. The interiors of membranes are therefore exposed on the fracture surfaces, as Figure A-31 illustrates.

is on the *protoplasmic* side of the membrane. The **E-face** is the interior face of the outer monolayer; it is called the E-face because this monolayer is on the *exterior* side of the membrane. Notice that the P-face has far more intramembranous particles than does the E-face. In general, most of the particles in the membrane stay with the inner monolayer when the fracture plane passes down the middle of a membrane.

To have a P-face and an E-face appear side by side as in Figure A-31, the fracture plane must pass through two neighboring cells, such that one cell has its cytoplasm and the inner monolayer of its plasma membrane removed to reveal the E-face, while the other cell has the outer monolayer of its plasma membrane and the associated intercellular space removed to reveal the P-face. Accordingly, E-faces are always separated from P-faces of adjacent cells by a "step" (marked by the arrows in Figure A-31) that represents the thickness of the intercellular space.

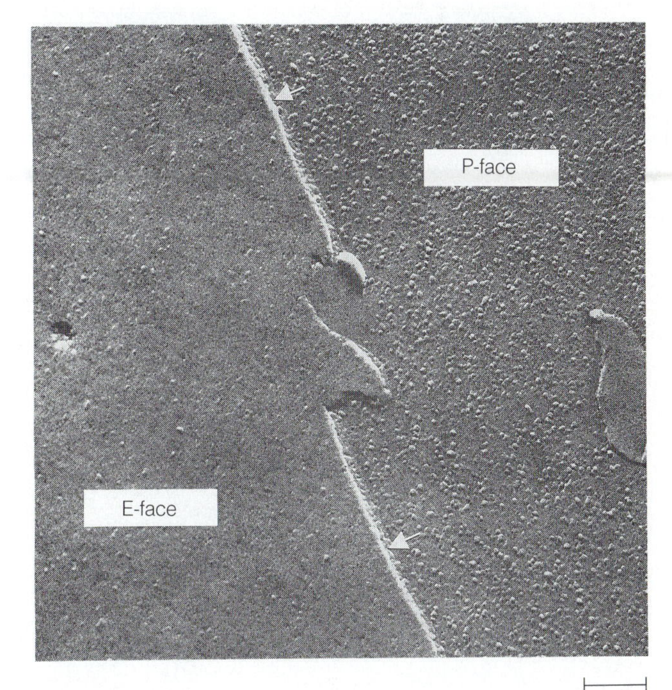

Figure A-31 Freeze-Fracturing of the Plasma Membrane. This electron micrograph shows the exposed faces of the plasma membranes of two adjacent endocrine cells from a rat pancreas as revealed by freeze-fracturing. The P-face is the inner surface of the lipid monolayer on the protoplasmic side of the plasma membrane. The E-face is the inner surface of the lipid monolayer on the exterior side of the plasma membrane. The P-face is much more richly studded with intramembranous particles than is the E-face. The arrows indicate the "step" along which the fracture plane passed from the interior of the plasma membrane of one cell to the interior of the plasma membrane of a neighboring cell. The "step" therefore represents the thickness of the intercellular space. (TEM)

Freeze-Etching

Although **freeze-etching** is related to freeze-fracturing, there is a considerable difference between the two techniques. Freeze-etching adds a further step to the conventional freeze-fracture procedure that makes the technique even more informative. Following the fracture of the specimen but prior to shadowing (that is, between steps 4 and 5 of Figure A-29), the microtome arm is placed directly over the specimen for a short time (a few seconds to several minutes). This maneuver causes a small amount of water to evaporate (sublime) from the surface of the specimen to the cold knife surface. Where the fracture has passed through a membrane, etching will cause small areas of the true cell surface around the periphery of the fracture face to stand out in relief against the background.

By using ultrarapid freezing techniques and a volatile cryoprotectant such as aqueous methanol, which sublimes very readily to a cold surface, the etching period can be extended and a much deeper layer of ice can be removed, exposing large areas of the specimen surface to view. This modification, called **deep-etching,** has provided a fascinating new look at cellular structure. For an example, see Figure 18-10 in Chapter 18 , an electron micrograph of the cytoskeleton as revealed by deep-etching.

Stereo Electron Microscopy

Electron microscopists frequently want to visualize their specimens in three dimensions. Techniques such as shadowing, freeze-fracturing, and scanning electron microscopy are useful for this purpose. However, they can be further enhanced by the use of stereophotography. Specifically, the same specimen is photographed from two different angles to generate a **stereo pair** of photos that are then fused optically. To do this, a special specimen stage is used that can be tilted relative to the electron beam. The specimen is first tilted in one direction and photographed, then tilted an equal amount in the opposite direction and photographed again.

The two micrographs are mounted side by side as a stereo pair. When you view a stereo pair through a stereoscopic viewer, your brain uses the two independent images to construct a three-dimensional view that gives a striking sense of depth to the structure under investigation. (Some people can achieve the same effect without a viewer by letting their eyes cross slightly, fusing the two micrographs into a single image.)

Figure A-32 is a stereo pair of the *Drosophila* polytene chromosome seen earlier in the high-voltage electron micrograph of Figure A-24. Here, however, it appears in three dimensions. The two micrographs shown in Figure A-32 were taken with the stage tilted first 5° to the right, then 5° to the left of the electron beam. A striking view of the chromosome can be achieved either with a stereo viewer or by allowing your eyes to fuse the two images visually.

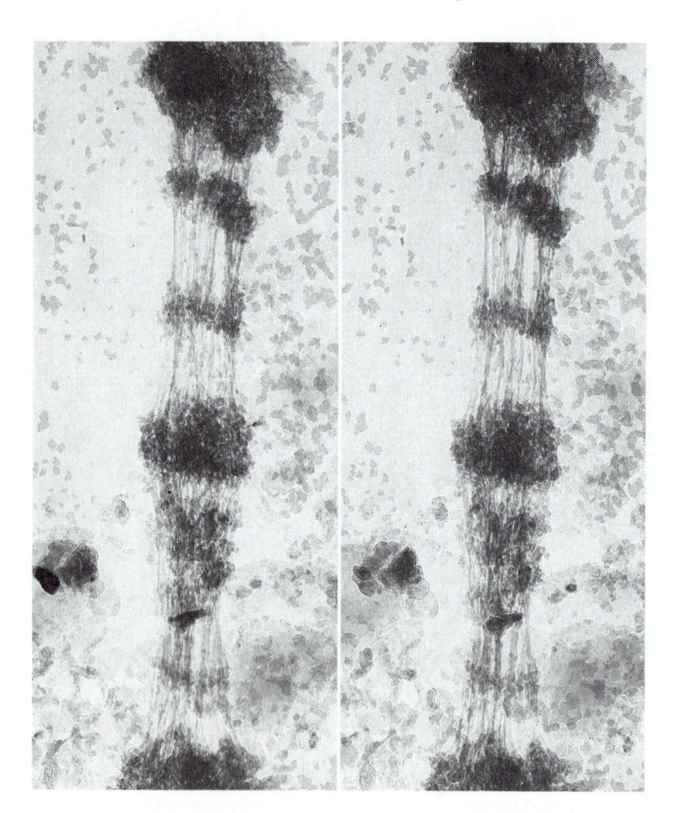

Figure A-32 Stereo Electron Microscopy. The polytene chromosome seen in the high-voltage electron micrograph of Figure A-24 is shown here as a stereo pair of photos that can be fused optically to generate a three-dimensional image. The two photographs were taken by tilting the specimen stage first 5° to the right, then 5° to the left of the electron beam. When viewed through a stereoscopic viewer, a stereo pair gives the viewer a three-dimensional view of the structure. To achieve the same effect without a viewer, simply let your eyes cross slightly, fusing the two micrographs into a single image. (TEM)

Sample Preparation Techniques in Scanning Electron Microscopy

When preparing a specimen for scanning electron microscopy, the goal is to preserve the structural features of the cell surface and to treat the tissue in a way that minimizes damage by the electron beam. The procedure is actually quite similar to the preparation of ultrathin sections for transmission electron microscopy. The tissue is fixed in aldehyde, postfixed in osmium tetroxide, and processed through a series of alcohol solutions for dehydration. The tissue is then placed in a fluid such as freon and transferred to a heavy metal canister called a **critical point bomb,** which is used to dry the specimen under conditions of controlled temperature and pressure. This helps keep structures on the surfaces of the tissue in much the same condition they were in before dehydration.

The dried specimen is then attached to a metal specimen mount with a metallic paste. The mounted specimen is coated with a layer of gold or a mixture of gold and palladium, using a modified form of vacuum evaporation called **sputter coating.** These procedures allow electrons to pass through the specimen more readily, thereby minimizing heating of the specimen when struck by the electron beam. Once the specimen has been mounted and coated, it is ready to be examined in the microscope.

As we have seen, preparing specimens for electron microscopy is often expensive and time-consuming. However, the high resolution and unique perspective on the structure and function of cells provide worthwhile insights into the biology of cells and tissues.

Other Imaging Methods

Light and electron microscopy are direct imaging methods, in that they use photons or electrons to produce actual images of a specimen. There are other methods of microscopy that are indirect imaging methods. To understand what is meant by indirect imaging, suppose you are given some object to handle with your eyes closed. You might feel six flat surfaces, twelve edges and eight corners, and if you then draw what you have felt, it would turn out to be a box. This is an example of an indirect imaging procedure.

The indirect imaging methods to be described include scanning tunneling microscopy, atomic force microscopy, and x-ray diffraction. Each method has the potential for showing molecular structures at near-atomic resolution, ten times better than the best electron microscope. Each method also has certain characteristics that limit its application to biological material, but when they have been successfully used, the resulting images provide exciting information that can be obtained in no other way.

Scanning Tunneling Microscopy and Atomic Force Microscopy

Although "scanning" is involved in both scanning electron microscopy and scanning tunneling microscopy, the two methods are in fact quite different. The **scanning tunneling microscope** (STM) does not use an electron beam, but instead depends on a tip made of a conducting material such as platinum-iridium. The tip is extremely sharp, ideally with its point composed of a single atom. It is under precise control of an electronic circuit that can move it in three dimensions over a surface. Two of the dimensions (x and y) scan the surface, while the z dimension governs the distance of the tip above the surface (Figure A-33).

The basic principle of the STM is electron tunneling. At the quantum-mechanical level, an electron has both wave-like and particle-like properties. These properties allow the electron to cross barriers that it cannot penetrate as a particle, but that it can penetrate in the form of a wave. This penetration is called tunneling. As the tip of the STM is moved across a surface, voltages from a few millivolts to several volts are applied. Under these conditions, if the tip is close enough to the surface and the surface is electrically conductive, electrons will begin to tunnel between tip and surface. The tunneling is highly dependent on the distance, so that even small irregularities in the size range of single atoms will affect the rate of electron conductance. The changes in conductance are used to regulate the distance between tip and surface so that tunneling current is kept constant. The resulting feedback current is electronically amplified and displayed on a video screen.

An important limitation of the STM is that the specimen must be electrically conductive. Therefore, the technique is better suited to producing images of physical surfaces rather than biological specimens, which often are good insulators. Nonetheless, some progress in biological imaging has been made. For instance, a short strand of DNA has been visualized with STM. (Recall the Box on p. 11 in Chapter 1.)

Atomic force microscopy (AFM) is related to STM, in that it uses a tip of atomic dimensions. It has the

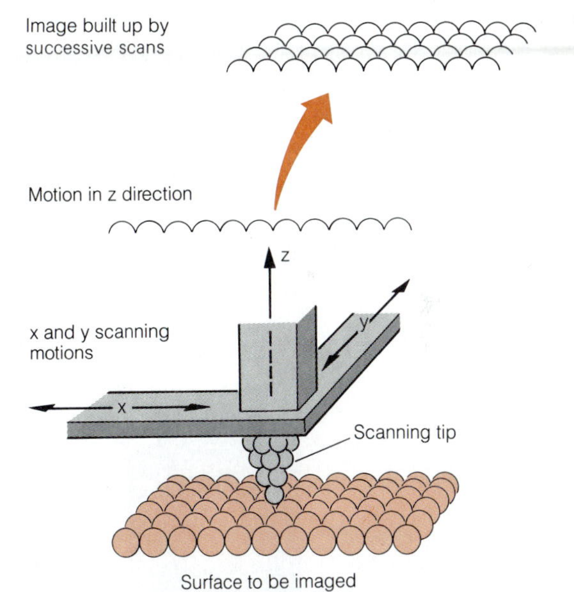

Figure A-33 Scanning Tunneling Electron Microscopy. The scanning tunneling microscope (STM) uses electronic methods to scan a metallic tip across the surface of a specimen. The tip is not drawn to scale in this illustration, but the point of the tip is ideally composed of one or a few atoms, shown here as balls. An electrical voltage is produced between the tip and the specimen surface. As the tip scans the specimen in the x and y directions, electron tunneling occurs at a rate dependent on the distance between the tip and the first layer of atoms in the surface. The instrument is designed to move the tip in the z direction to maintain a constant current flow. The movement is therefore a function of the tunneling current and is presented on a video screen. Successive scans then build up an image of the surface at atomic resolution.

important advantage that the specimen does not need to be an electrical conductor. This is because the tip is actually moved over the specimen surface, bumping along individual atoms in the specimen. The minute movements of the tip are followed with an optical system that senses deflection.

One of the most important potential applications of the STM and AFM instruments is measurement of dynamic changes in the conformation of a functioning biomolecule. Consider, for instance, how exciting it would be to "watch" a single enzyme molecule change its shape as it hydrolyzes ATP to provide the energy needed to transport ions across membranes. Such molecular eavesdropping is now entirely within the realm of possibility.

X-Ray Diffraction

X-ray diffraction does not involve microscopy, but instead reconstructs images from the diffraction patterns of x-rays

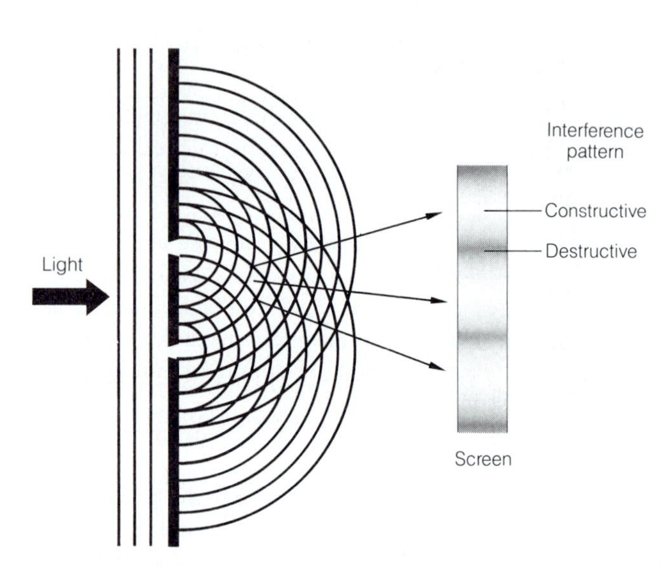

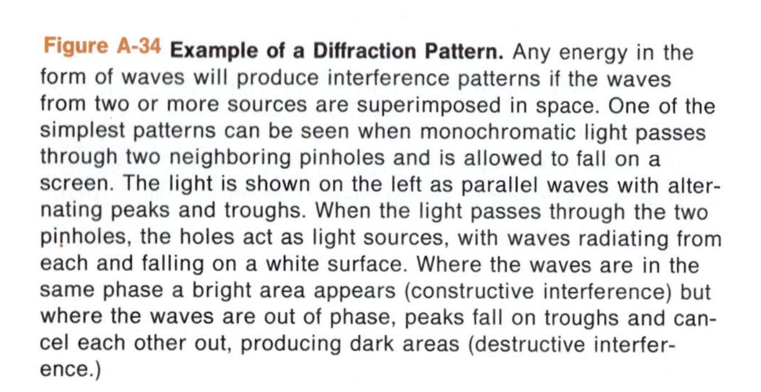

Figure A-34 Example of a Diffraction Pattern. Any energy in the form of waves will produce interference patterns if the waves from two or more sources are superimposed in space. One of the simplest patterns can be seen when monochromatic light passes through two neighboring pinholes and is allowed to fall on a screen. The light is shown on the left as parallel waves with alternating peaks and troughs. When the light passes through the two pinholes, the holes act as light sources, with waves radiating from each and falling on a white surface. Where the waves are in the same phase a bright area appears (constructive interference) but where the waves are out of phase, peaks fall on troughs and cancel each other out, producing dark areas (destructive interference.)

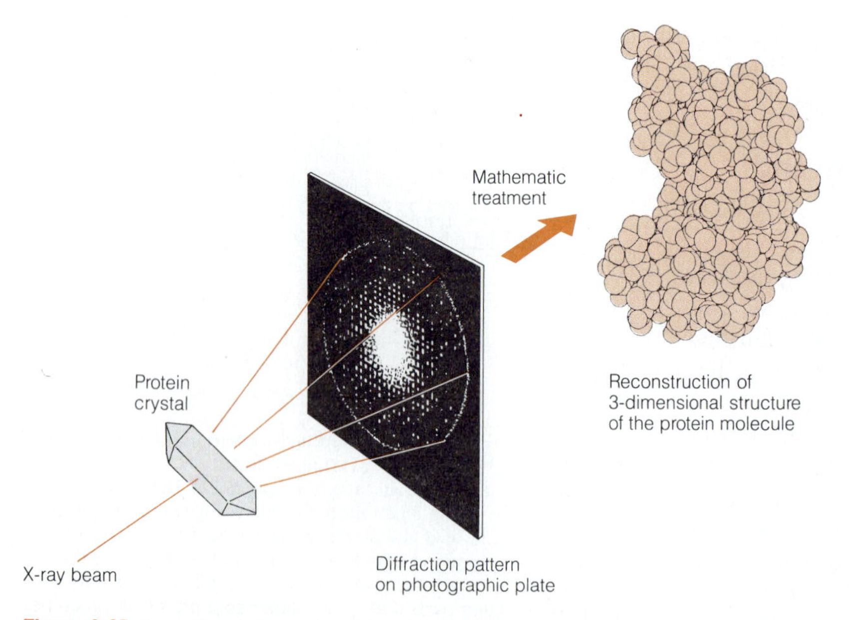

Figure A-35 X-ray Crystallography of Protein Structure. Just as light waves are diffracted by small holes, x-rays are diffracted by crystals. However, because x-rays have much smaller wavelengths, they can be diffracted by layers of atoms in the crystals, producing diffraction patterns that are recorded photographically. A highly complex mathematical treatment allows the investigator to reconstruct the molecular structure of the crystallized proteins at a resolution approaching 0.15 nm.

passing through a crystalline specimen. This method can be used to deduce molecular structure at the atomic level of resolution. A good way to understand x-ray diffraction is to draw an analogy with light. As discussed earlier, light has certain properties that are best described as wave-like. Whenever wave phenomena occur in nature, there exists the possibility for interaction between waves. If waves from two sources come into phase with one another, their total energy is additive *(constructive interference)* and if they are out of phase, their energy is reduced *(destructive interference)*. This effect can be seen when light passes through two pinholes in a piece of opaque material and then falls on a white surface. Interference patterns result, with dark regions where light waves are out of phase and bright regions where they are in phase (Figure A-34). If the wavelength of the light is known (red light would be in the range of 650 nm) one can measure the angle α between the original beam and the first diffraction peak and then calculate the distance between the two holes from the formula below:

$$d \text{ (distance)} = \text{wavelength}/\sin \alpha \qquad \text{(A-5)}$$

This is an important result, because the same approach allows us to calculate the distance between atoms in crystals such as proteins, and this is the only method to analyze protein structure at near-atomic resolution. Instead of two holes in a sheet of paper, imagine that we have multiple layers of atoms organized in a crystal. And instead of light, which has much too long a wavelength to interact with atoms, we will use a narrow beam of x-rays with wavelengths in the range of interatomic distances. As the x-rays pass through the crystal, they reflect off planes of atoms, and the reflected beams come into constructive and destructive interference. If the x-ray beams are allowed to fall onto photographic plates behind the crystal, distinctive diffraction patterns are produced that can be mathematically analyzed to determine the organization of atomic layers in the original crystal (Figure A-35).

The **x-ray crystallography** method was developed in 1912 by Sir William Bragg, who went on to establish the structure of relatively simple mineral crystals. Forty years later, Max Perutz and John Kendrew found ways to apply x-ray diffraction to crystals of hemoglobin and myoglobin, providing our first view of the intricacies of protein structure. Membrane proteins such as the photosynthetic reaction center are even more difficult to crystallize than water-soluble proteins like hemoglobin, but Henri Michel and Johannes Deisenhofer solved the problem in 1985, going on to describe the molecular organization of the photosynthetic reaction center at 0.3 nm resolution.

Key Terms for Self-Testing

light microscope (p. 773)
electron microscope (p. 773)
electron diffraction (p. 773)

Image Formation in Light and Electron Microscopy
source of illumination (p. 773)
specimen (p. 773)
lenses (p. 773)
wave form (p. 774)
wavelength (p. 774)

Optical Principles of Microscopy
focal length (p. 775)
angular aperture (p. 775)
resolution (p. 775)
refractive index (p. 775)
numerical aperture (p. 776)
limit of resolution (p. 776)
useful magnification (p. 777)

The Light Microscope
compound microscope (p. 777)
condenser lens (p. 777)
stage (p. 777)
objective lens (p. 777)
ocular lens (p. 777)
intermediate lens (p. 777)
brightfield microscopy (p. 777)
phase-contrast microscopy (p. 778)
phase plate (p. 778)
darkfield microscopy (p. 778)
fluorescence microscopy (p. 779)
fluorescence (p. 779)
fluorescent dye (p. 780)
polarization microscopy (p. 781)
plane-polarized light (p. 782)
birefringence (p. 782)
differential interference contrast (DIC) microscopy (p. 782)
Nomarski interference microscopy (p. 782)
optical sectioning (p. 783)

Sample Preparation Techniques in Light Microscopy
thin sections (p. 783)
primary fixation (p. 783)
perfusion (p. 783)
microtome (p. 784)
staining (p. 784)
autoradiography (p. 784)

The Electron Microscope: Design and Practice
transmission electron microscope (TEM) (p. 785)
electron gun (p. 787)
accelerating voltage (p. 787)
condenser lens (p. 787)
objective lens (p. 787)
intermediate lens (p. 787)
projector lens (p. 787)
electron micrograph (p. 787)
scanning electron microscope (SEM) (p. 788)

secondary electrons (p. 789)
scanning transmission electron microscope (STEM) (p. 789)
high-voltage electron microscope (HVEM) (p. 789)

Sample Preparation Techniques in Transmission Electron Microscopy
ultramicrotome (p. 790)
poststaining (p. 790)
negative staining (p. 791)
negative contrast (p. 791)
shadowing (p. 792)
vacuum evaporator (p. 793)
replica (p. 793)
freeze-fracturing (p. 793)
cryoprotection (p. 794)
intramembranous particles (IMPs) (p. 794)
P-face (p. 794)
E-face (p. 795)
freeze-etching (p. 795)
deep-etching (p. 796)
stereo pair (p. 796)

Sample Preparation Techniques in Scanning Electron Microscopy
critical point bomb (p. 796)
sputter coating (p. 796)

Other Imaging Methods
scanning tunneling microscope (STM) (p. 797)
atomic force microscopy (AFM) (p. 797)
x-ray diffraction (p. 798)
x-ray crystallography (p. 799)

Suggested Reading

General References

Barer, R. Microscopes, microscopy, and microbiology. *Ann. Rev. Microbiol.* 28 (1974): 371.

Gray, P., ed. *Encyclopedia of Microscopy and Microtechniques.* New York: Van Nostrand, 1973.

Porter, K., and M. A. Bonneville. *Fine Structure of Cells and Tissues.* Philadelphia: Lea and Febiger, 1973.

Rhodin, J. A. G. *Histology: A Text and Atlas.* New York: Oxford University Press, 1974.

Light Microscopy and Histology

Bancroft, J. D., and A. Stevens, eds. *Theory and Practice of Histological Techniques.* New York: Churchill Livingstone, 1977.

Bradbury, S. *The Evolution of the Microscope.* Elmsford, N.Y.: Pergamon, 1967.

Smith, A., and J. Bruton. *Color Atlas of Histological Techniques.* Chicago: Year Book, 1977.

Spencer, M. *Fundamentals of Light Microscopy.* Cambridge, England: Cambridge University Press, 1982.

Transmission and Scanning Electron Microscopy

Everhart, T. E., and T. L. Hayes. The scanning electron microscope. *Sci. Amer.* 226 (January 1972): 54.

Hayat, M. A. *Principles and Techniques of Electron Microscopy,* vols. 1–8. New York: Van Nostrand, 1970–1978.

Hayat, M. A. *Principles and Techniques of Scanning Microscopy,* vols. 1–6. New York: Van Nostrand, 1974–1978.

Kessel, R. G., and R. H. Kardon. *Tissues and Organs: A Text Atlas of Scanning Electron Microscopy.* San Francisco: W. H. Freeman, 1979.

Meek, G. A. *Practical Electron Microscopy for Biologists.* 2d ed. New York: Wiley, 1976.

Palade, G. E. Albert Claude and the beginning of biological electron microscopy. *J. Cell Biol.* 50 (1971): 5D

Pease, D. C., and K. R. Porter. Electron microscopy and ultramicrotomy. *J. Cell Biol.* 91 (1981): 287s.

Slayter, E. M. *Optical Methods in Biology.* New York: Wiley, 1970.

Weakley, B. S. *A Beginner's Handbook in Biological Transmission Electron Microscopy.* 2d ed. New York: Churchill Livingstone, 1981.

Wieschnitzer, S. *Introduction to Electron Microscopy.* 3d ed. Elmsford, N.Y.: Pergamon, 1981.

Freeze-Etching

Heuser, J. Quick-freeze, deep-etch preparation of samples for 3-D electron microscopy. *Trends Biochem. Sci.* 6 (1981): 64.

Orci, L., and A. Perrelet. *Freeze-Etch Histology: A Comparison Between Thin Sections and Freeze-Etch Replicas.* New York: Springer-Verlag, 1975.

Pinto da Silva, P., and D. Branton. Membrane splitting in freeze-etching. *J. Cell Biol.* 45 (1970): 598.

Scanning Tunneling Microscopy

Hansma, P. K., V. B. Elings, O. Marti, and C. E. Bracker. Scanning tunneling microscopy and atomic force microscopy: Application to biology and technology. *Science* 242 (1988): 209.

X-Ray Diffraction

Deisenhofer, J., O. Epp, K. Miki, R. Huber, and H. Michel. The structure of the protein subunits in the photosynthetic reaction centre of *Rhodopseudomonas viridis* at 3 Å resolution. *Nature* 318 (1985): 618.

Glusker, J. P., and K. N. Trueblood. *Crystal Structure Analysis: A Primer.* Oxford, England: Oxford University Press, 1985.

Kendrew, J. C. The three-dimensional structure of a protein molecule. *Sci. Amer.* 205 (December 1961): 96.

Perutz, M. F. The hemoglobin molecule. *Sci. Amer.* 211 (November 1964): 64.

Appendix B

Recombinant DNA Technology

Cell biology is in the midst of an era of extraordinary excitement and achievement. It is an era that has been made possible by the advent of **recombinant DNA technology,** a collection of techniques for isolating, replicating, and recombining genetic information with unprecedented ease. Arising from basic research on the molecular biology of bacteria, these techniques enable researchers to study and manipulate genes from both prokaryotes and eukaryotes with greater efficiency and precision than has ever been possible before. Any segment of DNA can now be excised from the genome, fused to an appropriate carrier DNA molecule, and introduced into bacterial cells, which then replicate the DNA as they proliferate.

This process, called **gene cloning,** makes it possible to generate large quantities of specific genes or other DNA segments—and possibly of their protein products as well, if the cloned genes can be read and expressed by the bacterial cells. Genes can also be introduced into yeast, plant, and animal cells, thereby generating *transgenic organisms.* These organisms can be useful not only for basic research, but also for practical applications.

To appreciate the importance of recombinant DNA technology and gene cloning, we need to grasp the magnitude of the problem that biologists faced as they tried to study the genomes of eukaryotic organisms. As Part Four of this text makes clear, most of our initial understanding of information flow in cells came from studies with bacteria and viruses. The genomes of bacteria, phages, and other viruses have been mapped and analyzed in great detail, and ingenious genetic techniques are available to manipulate bacterial and viral DNA.

Until recently, however, investigators despaired of ever being able to understand and manipulate eukaryotic genomes to the same extent. The size of the genome was especially intimidating. The typical eukaryote has at least 10,000 times as much DNA as the best-studied phages—truly an awesome haystack in which to find a gene-sized needle. The thought of ever being able to identify, isolate, sequence, or manipulate specific eukaryotic genes seemed unrealistic.

Yet in less than a decade, from the early to the late 1970s, all of that changed. Once considered one of the most intractable molecules to work with, DNA has now become one of the easiest. And what was once regarded as one of the most formidable objects of study—the eukaryotic genome—has become the source of an explosion of research discoveries.

What has made all this possible in such a short time? The answer to this question is to be found in recombinant DNA technology, the subject of this appendix. To explore this topic, we will look first at *restriction enzymes,* which play a crucial role in the isolation of specific DNA segments and the joining of these segments to other DNA. Then we will discuss the *mapping techniques* that these enzymes have made possible. Next we will consider the methods used to construct novel molecules of *recombinant DNA* and to *clone,* or replicate, specific DNA fragments. And finally, we will explore briefly the possibilities for *genetic engineering* of bacteria, plants, and animals, and the impact of this technology on agriculture and medicine.

Restriction Enzymes

Much of what we call recombinant DNA technology has been made possible by the discovery of **restriction enzymes.** Restriction enzymes are endonucleases that are present in most bacterial cells. These enzymes protect the bacterial cell from foreign DNA molecules—particularly those of bacteriophages—that may invade the cell. In fact, restriction enzymes get their name from their ability to *restrict* the ability of foreign DNA to take over the protein-synthetic machinery of the bacterial cell.

To protect its own DNA from being degraded, the bacterial cell has enzymes that methylate specific nucleotides that its own restriction enzymes would otherwise recognize. Once methylated, the nucleotides cannot be recognized by the restriction enzymes, and the bacterial DNA is therefore not digested by the restriction enzymes present in the same cell. In other words, restriction enzymes are part of a *restriction/methylation system* in the bacterial cell: foreign DNA is degraded by the restriction enzymes, and the bacterial genome is protected by methylation.

Restriction enzymes are named after the bacteria from which they are obtained. In each case, the name is derived by combining the first letter of the genus with the first two letters of the species. The particular strain of the bacterium is also indicated in the name, and if two or more enzymes have been isolated from the same species, the enzymes are numbered in order of discovery. Thus, the first restriction enzyme isolated from *E. coli* strain R is designated *Eco*RI, whereas the third enzyme isolated from *Hemophilus aegyptius* is called *Hae*III.

Recognition Sequences and Restriction Fragments

Restriction enzymes are specific for double-stranded DNA and always cleave both of the strands. Every restriction enzyme recognizes a specific DNA sequence, which usually consists of four or six (but may be as many as eight) nucleotide pairs. The **recognition sequence** is therefore a characteristic of the particular restriction enzyme, and dictates where the enzyme will cleave the DNA molecule. For example, the specific enzyme *Hae*III recognizes the tetranucleotide sequence GGCC and cleaves the DNA double helix as shown in Figure B-1a. Table B-1 indicates the recognition sequences for a number of common restriction enzymes.

The recognition sequences for most restriction enzymes are *palindromes,* which simply means that each such sequence reads the same in either direction. (The English word "radar" is a palindrome, for example.) The palindromic nature of a recognition sequence is due to the **twofold rotational symmetry** of the sequence: When rotated 180° in the plane of the paper, the sequence reads the same as it did before rotation. In other words, the recognition sequence has the same order of nucleotides on both strands but is read in opposite directions on the strands because of their antiparallel orientation.

The frequency with which a specific recognition sequence is likely to occur within a DNA molecule is statistically predictable. For a DNA molecule containing equal amounts of the four bases (A, T, C, and G), we can predict that, on the average, a recognition site with four nucleotide pairs will occur once every 256 (that is, 4^4) nucleotide pairs, whereas the likely frequency of a six-nucleotide sequence is once every 4096 (4^6) nucleotide pairs. Restriction enzymes therefore cleave most DNA into pieces that vary in length from a few hundred to a few thousand nucleotide pairs—gene-sized pieces, essentially. Such pieces are called **restriction fragments.** Because of the specificity of the recognition sequence, a particular restriction enzyme always cleaves DNA in a predictable manner and therefore generates a reproducible set of restriction fragments.

Some restriction enzymes cut both strands at the same point, generating restriction fragments with *blunt ends. Hae*III has such a cleavage pattern, as we have already seen (Figure B-1a). Other restriction enzymes cleave the two strands in an offset or staggered manner, generating short, single-stranded tails on both fragments. *Eco*RI is an example of such an enzyme; it recognizes the hexanucleotide sequence GAATTC and cuts the DNA molecule in an offset manner, leaving a —TTAA tail on both fragments (Figure B-1b). The restriction fragments generated by enzymes with this staggered cleavage pattern always have **cohesive** (or **"sticky") ends,** meaning that the single-stranded tail on the end of each such fragment can form

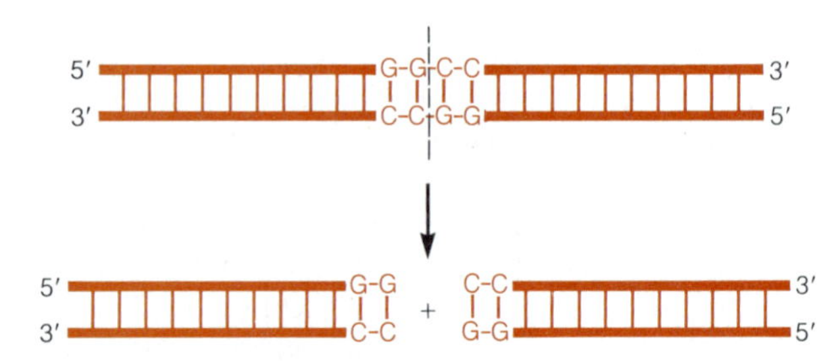

(a) Blunt-ended cleavage by the enzyme *Hae*III

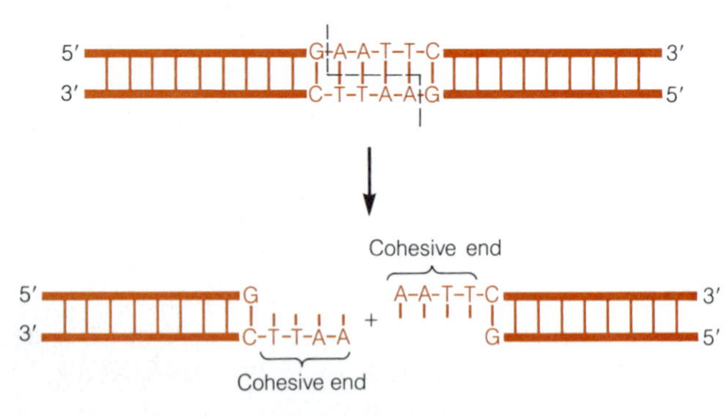

(b) Staggered cleavage by the enzyme *Eco*RI

Figure B-1 Cleavage of DNA by Restriction Enzymes. (a) Some enzymes, such as *Hae*III, cut both strands at the same point, generating blunt-ended fragments. (b) Other enzymes, such as *Eco*RI, make offset cuts, leading to fragments with complementary cohesive ends ("sticky" ends).

complementary base pairs with the tail at either end of any other fragment generated by the same enzyme, causing the fragments to stick to one another. Enzymes that generate such fragments are particularly useful because they are an important means of creating recombinant DNA molecules, as we will see shortly.

Separation of Restriction Fragments by Gel Electrophoresis

Digestion of a DNA sample by a specific restriction enzyme results in a collection of restriction fragments of different sizes. To determine the lengths of these fragments or to isolate them for further study, one must be able to separate the fragments from each other. The technique of choice for this purpose is **gel electrophoresis,** essentially the same method used for the separation of proteins and polypeptides (see Figure 7-12). In fact, the procedure for DNA is even simpler than for proteins, since DNA molecules have an inherent negative charge (due to their phosphate groups) and therefore have no need for treatment with a negatively charged detergent to make them move toward the anode.

For small pieces of DNA (less than about 500 nucleotides), gels of *polyacrylamide* are used, just as for electrophoresis of proteins and polypeptides. In fact, fragments that differ in length by as little as a single nucleotide can be separated from each other on such gels, a property that is very important for DNA sequencing (see Figure 13-14). For larger DNA fragments, however, the pores in polyacrylamide gels are too small to allow passage. Such fragments are therefore separated on more porous gels of *agarose,* a polysaccharide.

Figure B-2 illustrates the separation of restriction fragments of different sizes using gel electrophoresis. DNA samples are first digested with the desired restriction enzyme(s), and a small portion of each digestion mixture is placed in a separate well at the top of the gel. An electrical potential of several hundred volts is then applied across the gel, with the anode (the positive electrode) at the bottom of the gel. Because of the negative charge of their phosphate groups, DNA fragments migrate down the gel toward the anode. Smaller fragments are able to move through the gel with relative ease and therefore migrate rapidly, while larger fragments move more slowly. The technique therefore resolves DNA fragments based on their size.

When the electrophoretic process is complete, the DNA fragments on the gel can be visualized either by *staining* or, in the case of radioactively labeled DNA, by

Table B-1 Some Common Restriction Enzymes and Their Recognition Sequences

Enzyme	Organism	Recognition Sequence*
AvaI	*Anabena variabilis*	5′ C↓Py–C–G–Pu–G 3′ G–Pu–G–C–Py↑C
BamHI	*Bacillus amyloliquefaciens*	5′ G↓G– A–T–C^m–C 3′ C–C^m–T–A–G ↑G
EcoRI	*Escherichia coli*	5′ G↓A–A–T–T–C 3′ C–T–T–A–A↑G
HaeIII	*Hemophilus aegyptius*	5′ G–G↓C–C 3′ C–C↑G–G
HindIII	*Hemophilus influenzae*	5′ A^m↓A–G–C–T–T 3′ T –T–C–G–A↑A^m
PstI	*Providencia stuartii* 164	5′ C–T–G–C–A↓G 3′ G↑A–C–G–T–C
PvuI	*Proteus vulgaris*	5′ C–G–A–T↓C–G 3′ G–C↑T–A–G–C
PvuII	*Proteus vulgaris*	5′ C–A–G↓C–T–G 3′ G–T–C↑G–A–C
SalI	*Streptomyces albus* G	5′ G↓T–C–G–A–C 3′ C–A–G–C–T↑G

* The arrows within the recognition sequence indicate the points at which the restriction enzyme cuts the two strands of the DNA molecule. Abbreviations: Py = Pyrimidine (C or T), Pu = Purine (G or A), A^m = N^6-methyladenosine, C^m = 5-methylcytosine.

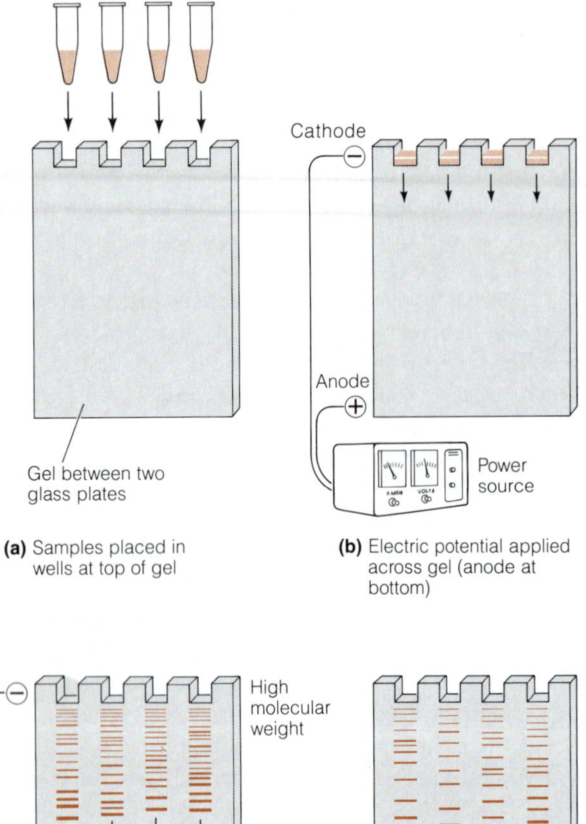

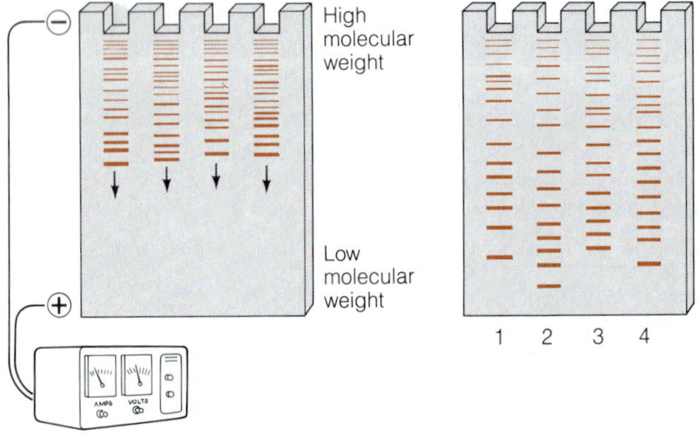

Figure B-2 Gel Electrophoresis of DNA. (a) To fractionate a DNA preparation containing fragments of various sizes, a small sample of the preparation is applied to the top of a gel of either polyacrylamide (for small DNA fragments) or agarose (for large DNA fragments). The gel is formed between two glass plates, and DNA solutions are placed in the wells at the top. (b) An electrical potential of several hundred volts is then applied across the gel, such that the anode (the positive electrode) is at the bottom of the gel and the cathode (the negative electrode) is at the top. (c) The DNA fragments in the applied sample migrate toward the anode at a rate that is inversely related to their size; the smaller fragments migrate more rapidly down the gel than do the larger ones. (d) After a predetermined amount of time, the gel is removed and stained with a dye such as ethidium bromide, which binds to the DNA fragments and causes them to fluoresce under ultraviolet light. (Alternatively, autoradiography can be used to locate the DNA bands on the gel, provided that the DNA is radioactively labeled.)

autoradiography. A common staining technique involves soaking the gel in the dye *ethidium bromide,* which binds to DNA fragments and fluoresces when exposed to ultraviolet light. If the DNA fragments contain a radioisotope, their locations on the gel can be determined by exposing the gel to photographic film. When the film is developed, the resulting autoradiogram will have a blackened area wherever the radioactivity in a DNA fragment has exposed the film.

Restriction Mapping Techniques

One of the first uses of restriction enzymes was in the construction of **restriction maps,** which indicate the location of restriction enzyme cleavage sites in relation to one another. To construct a restriction map for a specific genome or genomic region, the DNA is digested sequentially with two or more restriction enzymes, and the sizes of the restriction fragments produced by the various treatments are compared.

Restriction mapping is illustrated in Figure B-3, using two restriction enzymes designated as enzymes A and B. Because of its specificity, a given restriction enzyme will always cleave a specific DNA molecule into the same number of fragments, and each such fragment will have a characteristic length (step 1). The length of each fragment can be determined by electrophoresis on a polyacrylamide or agarose gel (step 2). In general, two restriction enzymes with different recognition sites will generate two quite different sets of fragments, although the sum of the fragment lengths will of course be the same in both cases.

All the fragments generated by one enzyme are then isolated—for example, by cutting the gel into slices and soaking each slice in a buffer solution. The fragments are then subjected to digestion by the second enzyme, producing two series of smaller fragments (step 3). Each such fragment is generated twice in the procedure, once when the fragments generated by enzyme A are further digested with enzyme B, and again when the fragments generated by enzyme B are further digested with enzyme A. It is therefore often possible to determine the order of both sets of fragments by looking for overlapping pieces common to two fragments, one in each set (step 4). The resulting restriction map depicts the location of each cleavage site with respect to the others. Ambiguous cases can usually be resolved by the use of a third enzyme, which generates yet another distinctive pattern of fragments.

Using this approach, restriction maps have been constructed for the genomes of phages, eukaryotic viruses, bacteria, and organelles, and for extensive segments of eukaryotic chromosomes as well. As an example, a simple restriction map for the circular genome of the maize chloroplast is shown in Figure B-4.

A restriction map is the physical equivalent of a genetic map constructed from recombination data. It is often possible to locate specific genetic markers on a restriction map, particularly if the RNA product of a gene is available as a *probe*. For example, scientists were able to localize the ribosomal genes to specific regions of the chloroplast genome of Figure B-4 by showing that chloroplast ribosomal RNA hybridizes (by complementary base pairing) with certain of the restriction fragments, which could then be located on the map.

Recombinant DNA and Cloning Techniques

Restriction enzymes that make staggered cuts in DNA are especially useful to molecular geneticists because the cohesive ends they generate provide a simple means for joining DNA fragments obtained from different sources. In es-

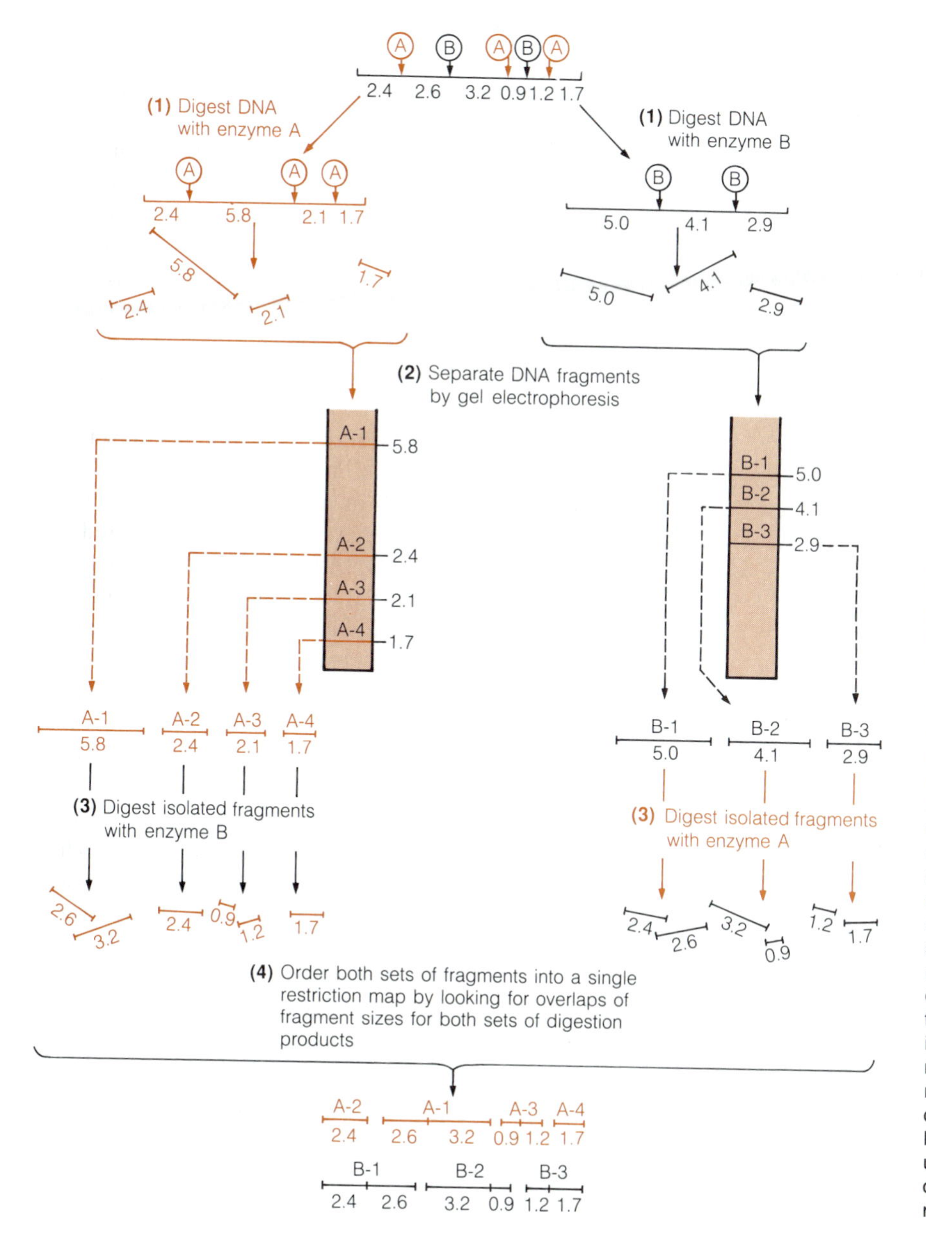

Figure B-3 Restriction Mapping. The DNA molecule shown at the top contains 12,000 nucleotide pairs. Based on probability, such a molecule should contain an average of $12{,}000/4^6 = 2.93$ recognition sites for restriction enzyme with a six-nucleotide recognition sequence. As indicated, the molecule contains three recognition sites for enzyme A and two sites for enzyme B. (The numbers between recognition sites represent thousands of nucleotide pairs.) Step 1: Upon digestion with these enzymes, the molecule is therefore cleaved at three points by enzyme A (left, color) or at two points by enzyme B (right, black). The fragments generated in this way are of characteristic length and are unique for the specific restriction enzyme. Step 2: The DNA fragments can be separated according to size by gel electrophoresis on polyacrylamide or agarose. The fragments are usually numbered in order from largest to smallest. Step 3: Each of these fragments can be recovered from the gel and digested with the other restriction enzyme. In some cases (such as fragments A-1, A-3, and all three B fragments), further cleavage occurs, because the fragment contains a recognition site for the second enzyme. In other instances (fragments A-2 and A-4), this is not the case. The number and sizes of such fragments should be the same for both digestions, but the fragments occur in different permutations because the order of digestion is different (A followed by B in one case, B followed by A in the other). Step 4: The two sets of fragments can then be ordered into a single restriction map by a trial-and-error process of looking for overlaps in fragment lengths for the two sets of digestion products. If ambiguities are found, they can frequently be resolved by use of a third restriction enzyme.

sence, any two DNA fragments generated by the same restriction enzyme can be joined together. Since the only property needed for the fusion of fragments is the complementarity of their cohesive ends, novel combinations of DNA are readily made.

Recombinant DNA Molecules

Figure B-5 illustrates the general approach for fusing DNA fragments from two different sources. DNA molecules from the two sources are first digested with a restriction enzyme known to generate fragments with cohesive ends (step 1). The fragments from the two sources are then mixed under conditions that favor base-pairing between the complementary cohesive ends of the fragments (step 2). Once joined in this way, DNA fragments can be covalently linked by *DNA ligase* (step 3), an enzyme that is normally involved in DNA replication and repair (see Chapter 14). The product is a **recombinant DNA molecule,** consisting of pieces that are not naturally a part of the same molecule.

The combined use of restriction enzymes and DNA ligase allows any two (or more) pieces of DNA to be fused, regardless of their origins. A piece of human DNA, for example, can be fused to bacterial or phage DNA just as easily as it can be linked to another piece of human DNA. It is possible, in other words, to form recombinant DNA molecules that have never existed in nature before, without any regard for the natural barriers that otherwise limit genetic recombination to genomes of the same or closely related species. Therein lies both the power of recombinant DNA technology—and the risk of the unknown, as well.

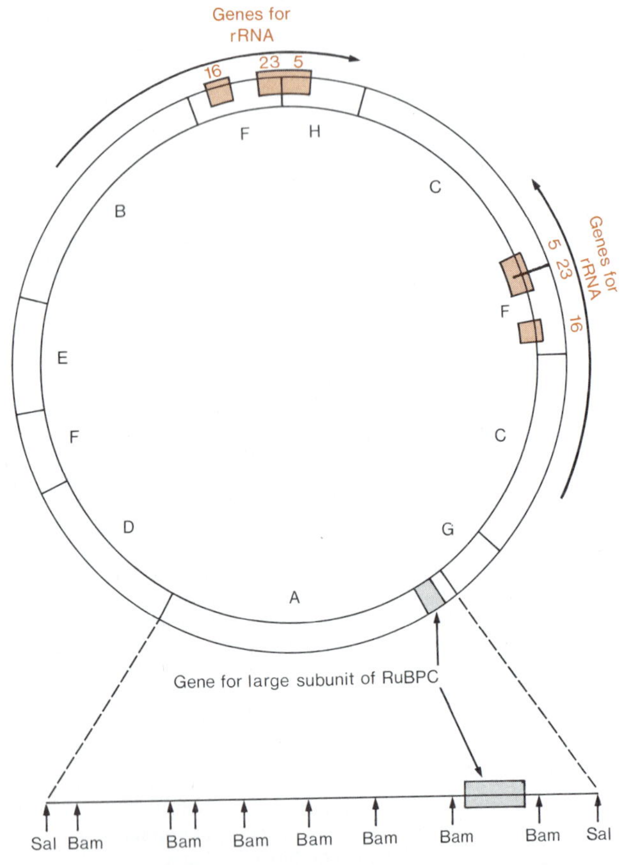

Figure B-4 Restriction Map of the Maize Chloroplast Genome. The chloroplast genome from maize (corn) has been mapped by the restriction mapping technique described in Figure B-3. Purified chloroplast DNA was digested with the restriction enzyme *SalI*, which cleaves the circular DNA molecule at 11 sites. The 11 digestion fragments were separated electrophoretically and labeled by letter (A, B, C, etc.) from the largest fragment to the smallest. The genome was found to contain two long repeated sequences, present in inverted order with respect to each other (arrows). The genes for 16S, 23S, and 5S ribosomal RNA (rRNA, in color) were localized to these inverted repeats by hybridization of ribosomal RNA to *Sal* digestion fragments. *Sal* fragment A was subjected to further digestion with the restriction enzyme *BamI*. The gene for the large subunit of ribulose bisphosphate carboxylase (RuBPC, gray) was localized to one of the *Bam* subfragments from *Sal* fragment A by hybridization of the large subunit mRNA to subfragments.

DNA Cloning

An essential feature of recombinant DNA technology is the ease with which a desired segment of DNA, often a restriction fragment containing a specific gene, can be introduced into a bacterial cell as part of a DNA molecule that will replicate as the bacteria proliferate. By identifying bacterial colonies that contain the desired DNA and then growing up such colonies in mass, one can obtain large quantities of the desired DNA. This process of generating many copies of specific DNA fragments is called **DNA cloning.** (In classical biology, a *clone* is a population of organisms derived from a single ancestor and hence homogeneous. By analogy, a *DNA clone* is a population of DNA molecules derived from a single molecule and hence identical to one another.)

The details vary, but DNA cloning always involves four stages: (1) insertion of the DNA into a cloning vector, (2) amplification of the resulting recombinant DNA molecule in bacterial cells, (3) selection of all bacterial cells that contain recombinant DNA, and (4) identification of the specific cells containing the DNA of interest. We will consider each of these stages in turn.

Stage 1: Insertion of DNA into a Cloning Vector. The first stage in cloning a desired fragment of DNA is its insertion into an appropriate **cloning vector**—a genetic element that can replicate autonomously in a bacterial cell. The cloning vector used for this purpose can be either a bacteriophage or a plasmid. A **bacteriophage** is a virus that infects bacterial cells; the bacteriophage *lambda* is used extensively as a cloning vector. A **plasmid** is a circular extrachromosomal genetic element present in many bacterial cells. Plasmids usually carry a small number of genes, which may include genes that confer resistance to specific antibiotics. The ability to confer antibiotic resistance upon bacterial cells is an important feature of plasmids used as cloning vectors, as we will see shortly.

Most vectors used for DNA cloning are themselves recombinant DNA molecules, designed specifically for this purpose to include antibiotic resistance genes and a variety of restriction sites. A popular vector for cloning in the bacterium *E. coli* is the plasmid *pBR322,* shown schematically in Figure B-6a. This plasmid carries genes that confer resistance to the antibiotics ampicillin (amp^R) and tetracycline (tet^R), with a number of restriction sites within each gene.

Introduction of foreign DNA at such an intragene site will create a recombinant plasmid with one of its antibiotic resistance genes disrupted and hence inactivated (Figure B-6b). This feature turns out to be important in subsequent selection steps, since bacterial cells containing recombinant plasmids will be resistant to one, but not both, antibiotics. Antibiotic resistance genes are therefore **selectable markers** that enable the researcher to discriminate between recombinant and nonrecombinant plasmids. In the example shown in Figure B-6, a DNA fragment generated by the enzyme *Hind*III is added at the single *Hind*III site on the plasmid. Because the *Hind*III site lies within the tet^R gene, that gene will be inactivated, but the amp^R gene will be unaffected.

Figure B-7 illustrates the general strategy for introducing a specific fragment of foreign DNA into a plasmid cloning vector, using a plasmid similar to pBR322 as the vector and a restriction enzyme that cleaves the plasmid at a single site. Incubation with the restriction enzyme opens the plasmid at that site (step 1). The same restriction enzyme is used to cleave the DNA molecule(s) containing the fragment(s) to be cloned (step 2). These sticky-ended fragments are incubated with the linearized vector molecules under conditions known to favor base pairing (step 3), followed by treatment with DNA ligase to link the molecules covalently by repairing the cuts that the restriction enzyme made (step 4). (To keep the diagram simple, Figure B-7 shows only the desired product of the cloning procedure—a recombinant plasmid containing the desired fragment of foreign DNA. In practice, however, a variety

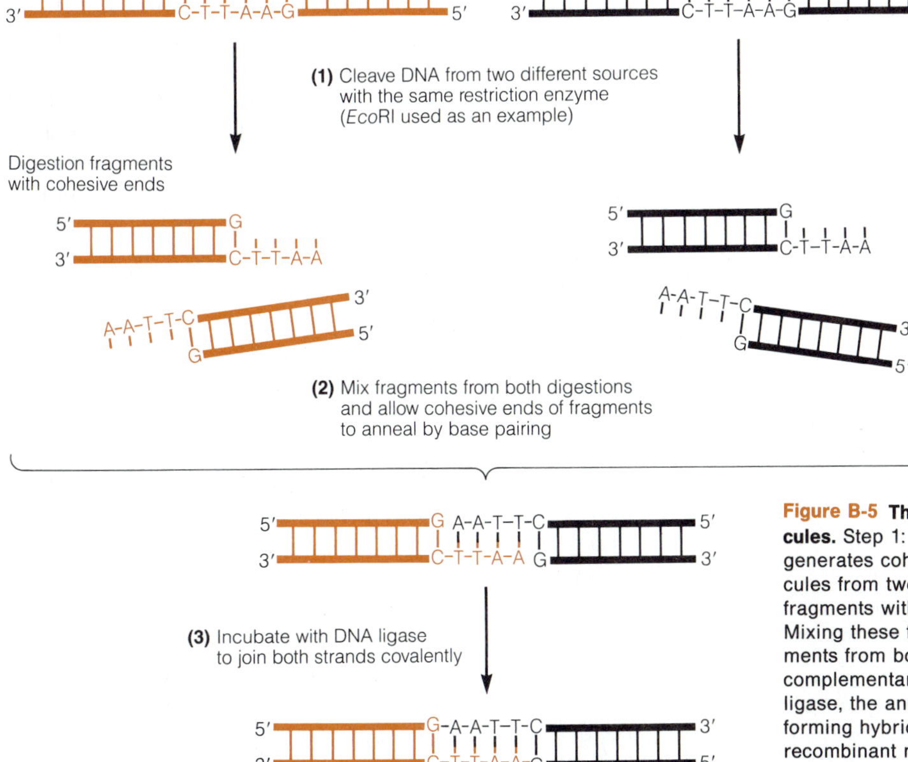

Figure B-5 The Generation of Recombinant DNA Molecules. Step 1: A restriction enzyme (such as *Eco*RI) that generates cohesive ends is used to cleave DNA molecules from two different sources, generating two sets of fragments with complementary cohesive ends. Step 2: Mixing these fragments allows cohesive ends of fragments from both sets to anneal by base pairing of the complementary ends. Step 3: Upon incubation with DNA ligase, the annealed strands are joined covalently, forming hybrid DNA molecules, some of which will be recombinant molecules because the fragments that have been joined came initially from two different sources.

of products will be present, including nonrecombinant plasmids and recombinant plasmids containing other fragments that were also generated by the initial digestion process.)

Stage 2: Amplification of Recombinant Vector Molecules in Bacterial Cells. Once foreign DNA has been inserted into a cloning vector, the recombinant DNA molecules must be *amplified* to obtain adequate quantities of the cloned DNA fragment. For amplification, the recombinant molecules are introduced into bacterial cells either by *transduction* (in the case of bacteriophage vectors) or by *transfection* (for plasmids). In either case, conditions are chosen to optimize uptake of the vector by the bacterial cells. The bacterial cells are then plated onto a nutrient medium, where they proliferate and form colonies, each of which is derived from a single cell. As the bacterial cells multiply, the recombinant vectors also replicate, producing an enormous number of circular DNA molecules containing the cloned DNA. This process is shown in Figure B-7 as step 5 (transfection of *E. coli* cells by recombinant plasmids) and step 6 (proliferation of the bacteria, with concomitant replication of the plasmids and amplification of the cloned DNA).

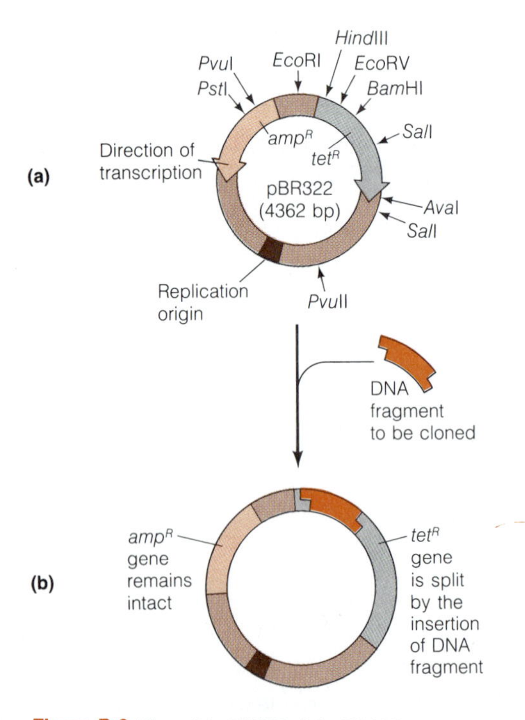

Figure B-6 Plasmid pBR322. (a) pBR322 is a plasmid with 4362 base pairs that is widely used as a cloning vector. The plasmid has been genetically engineered to include genes that confer resistance to the antibiotics ampicillin and tetracycline. There are a number of restriction enzyme recognition sequences within each antibiotic-resistance gene, as indicated by the small arrows. The large arrows show the direction of transcription of the ampicillin (*amp*^{R}) and tetracycline (*tet*^{R}) resistance genes. Part (b) shows the effect of inserting a foreign DNA fragment into the *Hin*dIII site located within the *tet*^{R} gene. Insertion disrupts and inactivates the gene, making the recombinant plasmid incapable of conferring tetracycline resistance on the host bacterial cell.

Stage 3: Selection of Bacterial Cells Containing Recombinant DNA. Regardless of the specific cloning vector used, an essential part of the cloning procedure is to select for bacterial cells that contain the recombinant DNA. In the case of plasmids, selection usually depends on the antibiotic resistance that the plasmid confers upon its host cell. In the example of Figure B-7, bacteria carrying recombinant plasmids will be resistant to an antibiotic, since all plasmids have an intact antibiotic-resistance gene. Such cells can therefore be selected by inclusion of the antibiotic in the culture medium. This stage is usually carried out simultaneously with stage 2 (step 6 in Figure B-7).

Stage 4: Identification of Bacterial Colonies Containing the DNA of Interest. When the desired level of amplification has been achieved, the cloned DNA fragments can be recovered by isolating vector DNA from the bacteria and digesting it with the same restriction enzyme used to prepare the recombinant plasmids initially. In most cases, however, many different kinds of DNA fragments will have been cloned, only one or a few of which are relevant to the desired research. In Figure B-7, for example, the bacterial colonies present on the petri plates after step 6 are likely to contain as many different kinds of fragments as there are enzyme recognition sites on the DNA molecules used in step 2. The final stage in any recombinant DNA procedure therefore involves the **screening** of the colonies to identify those that contain the specific DNA fragment of interest.

The techniques used to screen the colonies of bacteria depend on what the researcher knows about the gene being cloned. If the RNA product of the gene (messenger RNA, in most cases) has already been isolated, a radioactively labeled DNA copy can be made by reverse transcription of the RNA product and used as a nucleic acid probe to screen for the desired DNA using the *colony hybridization technique* shown in Figure B-8. A **nucleic acid probe** is a piece of labeled DNA or RNA that is used to identify a desired gene by base-pairing with it. Alternatively, if the protein that the gene encodes has been isolated, antibodies against that protein can be used as probes to check bacterial cultures for the presence of the protein. This second screening method obviously depends on the ability of the bacterial cells to produce a foreign protein encoded by a cloned gene, and will fail to detect cloned genes that are not expressed in the host cell. However, special *expression vectors* are available to increase the likelihood that foreign genes are transcribed and translated properly.

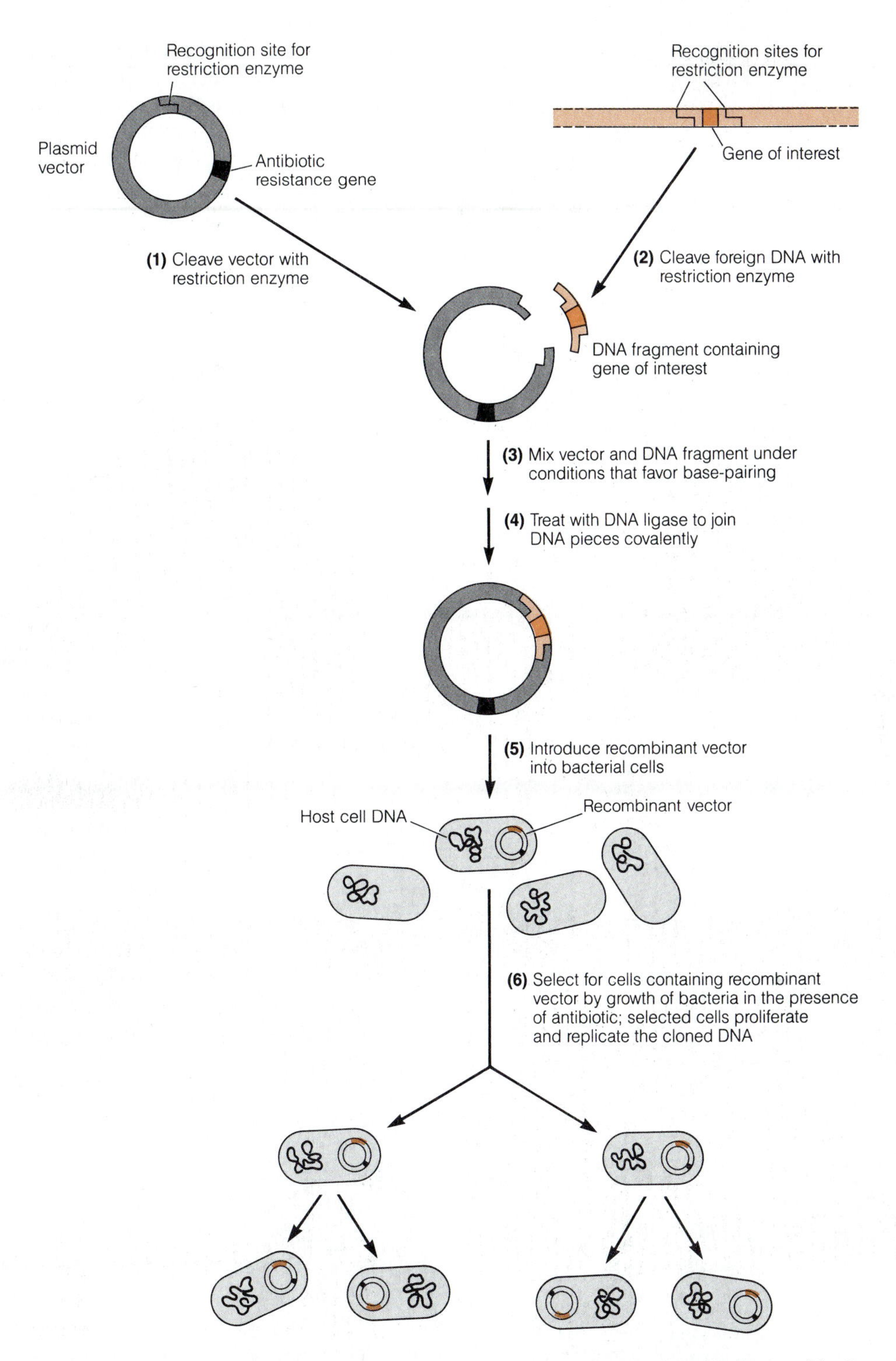

Figure B-7 DNA Cloning in Bacteria Using a Plasmid Vector. Step 1: A circular cloning vector is cleaved with a restriction enzyme known to recognize a single site on the vector. The example shown here is the plasmid pBR322, and the host is *E. coli*. Step 2: The same enzyme is used to cleave the DNA molecule(s) containing the fragment(s) of interest, thereby generating fragments that have the same cohesive ends as the linearized vector DNA. Step 3: These fragments are then mixed with the linearized vector DNA and incubated under conditions that favor base-pairing between the cohesive ends. Among the variety of expected products will be vector molecules linked by base-pairing to a single DNA fragment. Step 4: Upon incubation with the enzyme DNA ligase, some of these structures will circularize to regenerate the vector, but with a piece of foreign DNA integrated into it. Step 5: Recombinant vector molecules generated in this way are then introduced into bacterial cells, where they replicate as the bacterial cells proliferate. Step 6: If the bacterial cells are grown in the presence of antibiotic, only the cells containing recombinant vectors will survive to form colonies.

Genomic and cDNA Libraries

The cloning of foreign DNA in bacterial cells is now a routine procedure. In principle, two different approaches are used with respect to the DNA starting material. The procedure described above and depicted in Figure B-7 is the "shotgun" approach. In this approach, the whole genome of an organism (or some substantial portion of the genome) is cleaved into a large number of restriction fragments, and these are then introduced randomly into a large number of bacterial cells, such that the entire genome is represented in the collection, but cut randomly with respect to genes. Such a collection of cloned DNA is called a **genomic library,** since it contains cloned fragments representing most if not all of the genome.

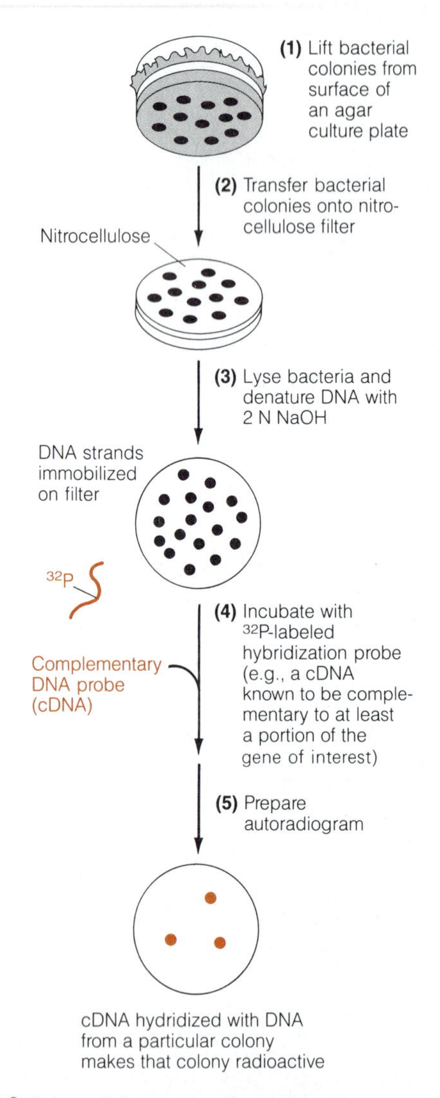

Figure B-8 Colony Hybridization Technique. This technique is used to screen bacterial colonies for the presence of DNA that is complementary in sequence to a complementary DNA (cDNA) probe. Step 1: Bacterial colonies are lifted from the surface of an agar culture plate with a velvet cloth and (step 2) transferred onto a nitrocellulose filter. Step 3: The filters are then treated with strong base (2 N NaOH) to lyse the bacteria and denature the DNA and (step 4) incubated with radioactively labeled cDNA, which will hybridize to any complementary DNA present on the filter. Step 5: The filter is then rinsed and subjected to autoradiography, which will visualize only those colonies containing DNA that is complementary in sequence to the cDNA probe.

Genomic libraries of eukaryotic DNA are valuable resources from which specific genes can be "fished out," provided only that a sufficiently sensitive selection technique is available. Once a rare bacterial colony that contains the desired DNA fragment has been identified, it can be grown on a nutrient medium to generate as many copies of the fragment as may be needed. Amplifications of a millionfold or more are easily achieved using the procedures already described. As you might guess, however, selection of the desired colonies for amplification is often the most difficult part of the cloning procedure.

Alternatively, **complementary DNA (cDNA)** can be synthesized by reverse transcription of mRNA, converted into double-stranded form by the enzyme DNA polymerase, inserted into a suitable vector, and cloned (Figure B-9). Each clone obtained in this way is called a **cDNA clone,** and the collection of clones derived from an mRNA preparation constitutes a **cDNA library.** Such a library can be screened by the same techniques used to probe a genomic library. Unlike a genomic library, however, a cDNA library will contain only those DNA sequences that are transcribed into RNA, presumably the active genes in the tissue from which the mRNA was prepared.

If mRNA corresponding to a specific protein has already been isolated, the corresponding cDNA can also be synthesized and cloned by the procedure shown in Figure B-9. In this case, no screening or selection is necessary, since all of the cDNA clones should correspond to the purified RNA.

DNA Amplification Using the Polymerase Chain Reaction

An even simpler technique for amplifying selected DNA sequences has been developed recently. The **polymerase chain reaction (PCR),** as it is called, is carried out in a test tube rather than a bacterial cell, and is therefore much quicker than standard cloning procedures. However, the PCR method is applicable only to DNA fragments for which at least a part of the nucleotide sequence is already known, because the technique requires the synthesis of short DNA oligonucleotides (containing 15 to 20 nucleotides each) corresponding to sequences that flank the segment to be amplified. These oligonucleotides serve as primers for in vitro synthesis of DNA. The DNA polymerase used for this purpose was first isolated from bacteria able to grow in thermal hot springs (70°–80°C) and is therefore a heat-stable enzyme.

The PCR technique is illustrated in Figure B-10. Each cycle of the reaction begins with a short heat treatment to denature the DNA double helix into its two strands (step 1). The DNA solution is cooled and incubated with the synthetic DNA oligonucleotides under conditions that allow them to hybridize with matching regions on the single-stranded DNA (step 2). DNA polymerase is then added, along with the four deoxyribonucleoside triphosphates (dATP, dTTP, dCTP, and dGTP). The DNA polymerase uses the short oligonucleotides as primers, thereby ensuring the selective synthesis of the regions of DNA downstream from the primers (step 3). Note that this basic cycle (Figure B-10, steps 1 to 3) results in a doubling of the desired DNA segment. Because the DNA polymerase is heat-stable, the mixture can be heated again to melt the new helices, and the cycle can be repeated to double the amount of the DNA again (Figure B-10, steps 1′ to 3′). This reaction cycle is then repeated as often as necessary, with each cycle doubling the amount of DNA from the previous cycle.

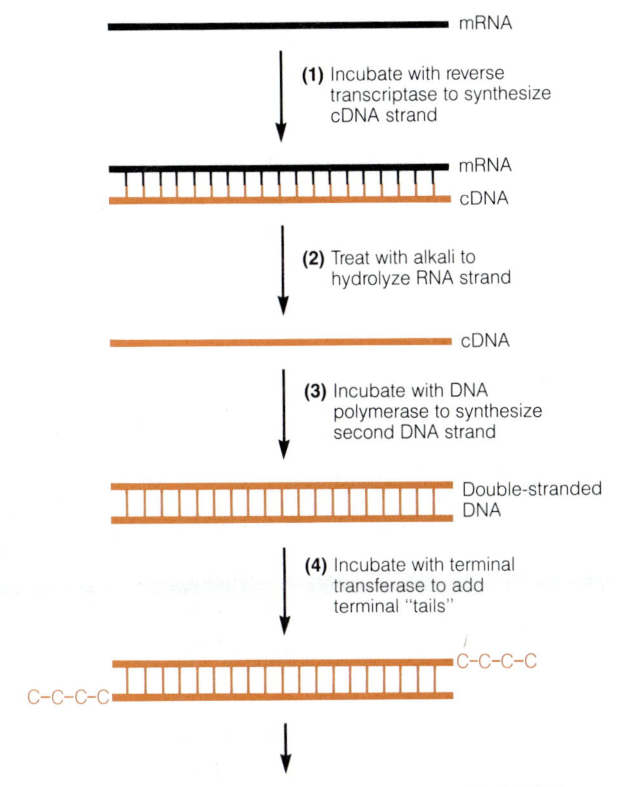

Figure B-9 Preparation of Complementary DNA for Cloning. Step 1: Messenger RNA (either total cellular mRNA or purified mRNA for a specific protein) is incubated with reverse transcriptase, which uses the mRNA as a template for synthesis of a complementary DNA (cDNA) strand. Step 2: The resulting mRNA–cDNA hybrid is then treated with alkali to hydrolyze the RNA, leaving the single-stranded cDNA. Step 3: DNA polymerase is then able to synthesize double-stranded DNA, using the single-stranded cDNA as a template. Step 4: The double-stranded DNA must have terminal tails that are complementary to those of the cloning vector. These can be added by incubation with terminal transferase, an enzyme that adds nucleotides one at a time to the ends of the molecule. Step 5: If short stretches of one nucleotide (cytosine, for example) are added to the double-stranded cDNA and short stretches of the complementary nucleotide (guanine, in this case) are added in the same way to a linearized cloning vector, recombinant molecules can be generated. These can then be sealed by DNA ligase and cloned in bacterial cells, as in Figure B-7. (As an alternative to steps 4 and 5, short synthetic "linker" molecules containing a variety of restriction sites can be ligated to the ends of both the double-stranded DNA and the blunt-ended cloning vector. The linkers are then cleaved with a restriction enzyme that generates cohesive ends, allowing recombinant molecules to be generated, as in Figure B-7.)

In most cases, 20 to 30 reaction cycles are needed, yielding an amplification of a millionfold or more ($2^{20} = 1.05 \times 10^6$; $2^{30} = 1.07 \times 10^9$). Each cycle requires only about 5 minutes, and the process can be automated to complete 20 or more cycles in a few hours, compared to the several days required for cloning DNA in bacteria using the conventional method described earlier. The PCR method therefore reduces substantially the time required for DNA amplification. Furthermore, the method can be used with as little as one molecule of DNA, and can be carried out with DNA from almost any source. For instance, DNA from the remains of a woolly mammoth that had been frozen for 40,000 years was recently amplified, enabling the mitochondrial genome from the mammoth to be compared with that of the present-day elephant!

Genetic Engineering

With the recognition that virtually any DNA fragment could be cloned in bacterial cells and that bacteria could express at least some foreign genes, attention quickly turned to the possibility of using this approach to produce desirable but rare gene products on a commercial scale. One of the first mammalian genes to be expressed in bacteria was that for rat insulin, an achievement reported in 1978. Since then, the gene for human insulin has also been cloned and expressed in bacteria, as has the gene for human growth hormone. Both of these hormones have been approved for use in treating human patients in the United States and are being produced commercially (Figure B-11).

These hormones were among the first success stories of **genetic engineering,** the application of recombinant DNA technology to practical problems, primarily in medicine and agriculture. The availability of human insulin produced by genetically engineered bacteria is good news to the two million diabetics in the United States, especially since the only insulin otherwise available is obtained from animal sources and causes adverse reactions in some people. Similarly, the availability of human growth hormone is a great benefit to children born with hypopituitarism, a syndrome in which the pituitary gland fails to provide an

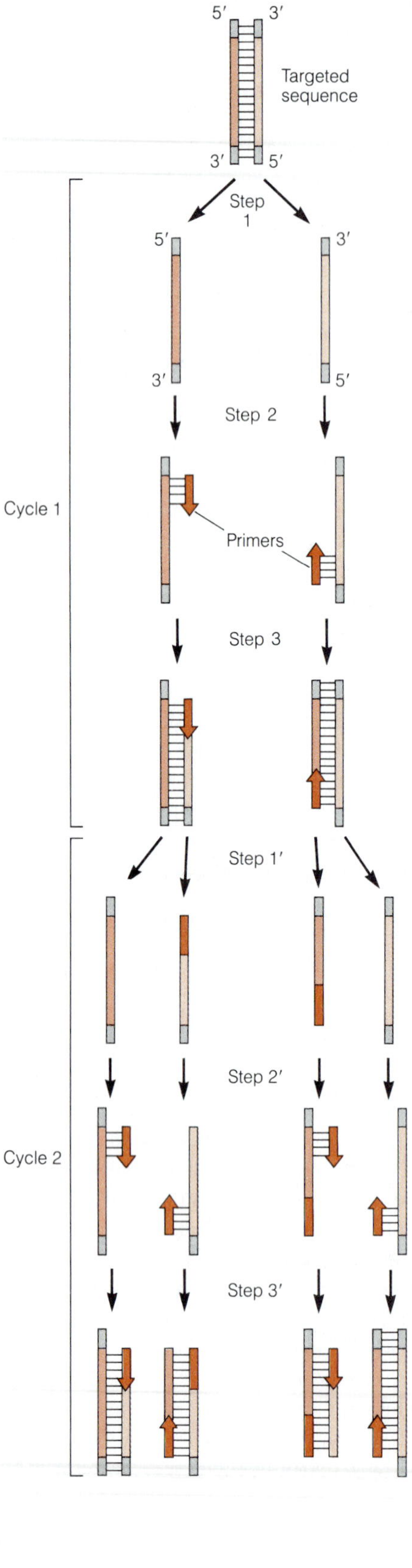

Figure B-10 DNA Cloning Using the Polymerase Chain Reaction. Specific DNA segments can be amplified in vitro by repeated cycles of replication, using as primers synthetic oligonucleotides that are complementary to sequences known to flank the DNA segment of interest. Each cycle consists of three steps: step 1, separation of the DNA double helix into its two strands by brief heat treatment; step 2, incubation of the single-stranded DNA with synthetic oligonucleotides that bind to the complementary sequences on either end of the desired segment; and step 3, incubation with deoxyribonucleotide triphosphates (dNTPs) and a heat-stable DNA polymerase, which uses the short oligonucleotides as primers for synthesis of the downstream regions on both strands. This three-step sequence is then repeated as often as necessary, with each cycle doubling the number of copies of the DNA segment bounded by the two oligonucleotide primers. (Steps 1′ to 3′ represent one such additional cycle.) The theoretical amount of amplification that is accomplished by n cycles is 2^n. The polymerase chain reaction works best when the DNA segment to be amplified (the region flanked by the two primers) has between 50 and 2000 nucleotides.

Figure B-11 Commercial Production of Human Growth Hormone. Seen in this photograph are bottles of human growth hormone produced through recombinant DNA techniques. The bottles represent the final stage in a manufacturing process that makes human growth hormone commercially available for treating children with growth hormone deficiencies.

adequate natural supply of the hormone. Other medical uses of recombinant DNA technology include the production of vaccines and the development of diagnostic tests for genetic disorders. Eventually, genetic engineering may also provide means for correcting rather than just treating genetic diseases. However, the prospects for such *human gene therapy* raise many questions, not only technical but also ethical and social.

Increasingly, recombinant DNA techniques are being extended to eukaryotes, not just as sources of genes to be cloned in bacteria, but also as targets of genetic engineering. This field is being pioneered with yeast cells, which can be made to take up DNA by transformation, and have plasmids as well. Plants and animals are also amenable to genetic engineering. The "supermouse" described in the box on pp. 814–815 is an especially dramatic example of a **transgenic animal,** an animal that carries genes from another species.

Exciting results of genetic engineering in higher organisms are being reported by plant molecular biologists, who enjoy the special advantage of being able to regenerate adult plants of many species from single cells grown in tissue culture. The vector of choice in most cases is a plasmid carried by the bacterium *Agrobacterium tumefaciens.* In nature, this bacterium is a pathogenic organism that causes tumors called *crown galls* in certain plant species. The tumors are induced by the plasmid, which is called the **Ti** (for **tumor-inducing**) **plasmid.** The significant feature of this plasmid for genetic engineering is that a segment of its DNA, called the *T DNA region,* becomes integrated into the genome of the host plant cell. This property has been exploited by plant molecular biologists, who have developed strategies for genetically engineering plants using the Ti plasmid.

The general approach for creating **transgenic plants** is shown in Figure B-12. The Ti plasmid is isolated from *Agrobacterium* cells (step 1), and the desired DNA (a specific gene, most commonly) is inserted into the T-DNA region of the plasmid using standard recombinant DNA techniques (step 2). The plasmid is put back into *Agrobacterium* (step 3), and the genetically engineered bacterium is then used to infect plant cells growing in culture (step 4). When the recombinant plasmid enters the plant cell, its T DNA becomes stably integrated into the plant genome and is passed on to both daughter cells at every cell division. The plants that are eventually regenerated from such cells (step 5) contain the T DNA—and therefore the desired DNA—in all of their cells. Although still in its infancy, plant genetic engineering holds great promise as a means of increasing agricultural production, since such diverse traits as herbicide resistance, stress tolerance, quality of seed proteins and lipids, and nitrogen fixation are potentially amenable to genetic modification.

We are, in short, seeing the emergence of a whole new technology that seems destined not only to enhance our understanding of the eukaryotic genome, but to have a profound impact on medicine and agriculture as well.

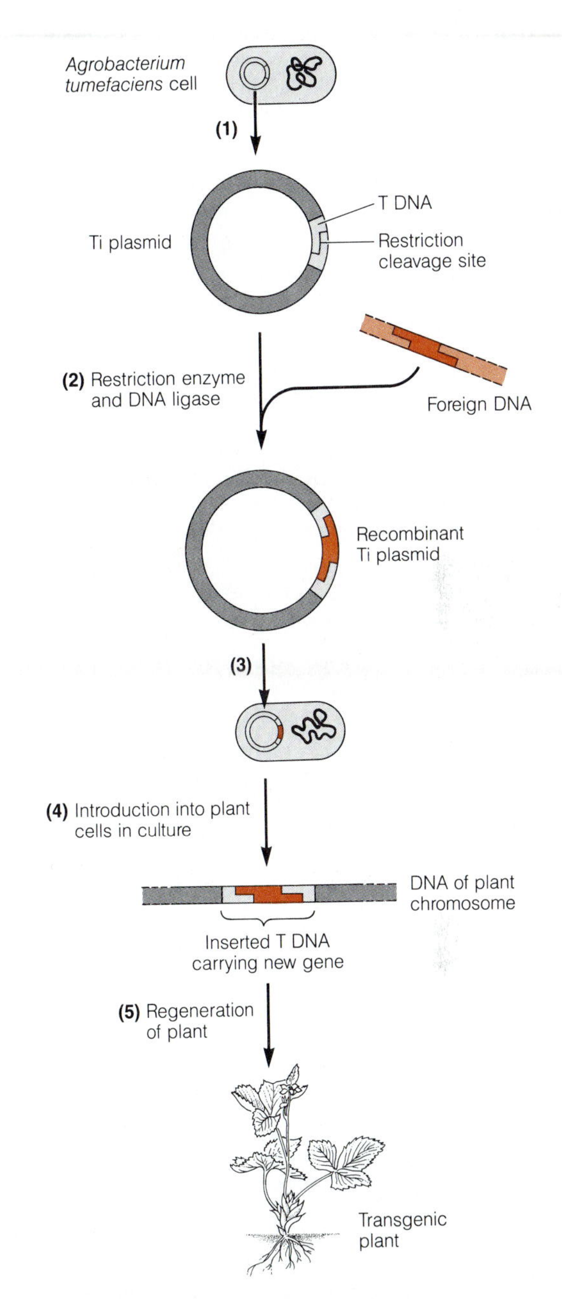

Figure B-12 Generation of Transgenic Plants. Most transgenic plants are generated using the Ti plasmid from *Agrobacterium tumefaciens.* The Ti plasmid is (step 1) isolated from *Agrobacterium* cells, (step 2) subjected to standard recombinant DNA techniques to insert the desired DNA into the T-DNA region of the plasmid, and then (step 3) put back into *Agrobacterium.* Cultured plant cells are (step 4) infected with bacterial cells containing the recombinant plasmid, and (step 5) used to regenerate whole plants. These are found to contain the recombinant T-DNA region stably integrated into the genome of every cell.

Supermouse, the Transgenic Triumph

> A DNA fragment containing the gene of rat growth hormone was microinjected into the pronuclei of fertilized mouse eggs. Of 21 mice that developed from these eggs, seven carried the gene and six of these grew significantly larger than their littermates (Palmiter et al. 1982, 611).

With these words, a team of investigators led by Richard Palmiter and Ralph Brinster reported how a controlled genetic variation can be introduced experimentally into mice without going through the usual procedure of "breeding," a process of sexual reproduction followed by selection of desired traits. The result was a "supermouse" that weighed almost twice as much as its littermates because of the *growth hormone* genes that had been injected into the fertilized egg from which the mouse developed (see photograph). The accomplishment was heralded as a significant breakthrough, since it proved the feasibility of applying genetic engineering to animals, with all the scientific and practical consequences such engineering is likely to have.

What Palmiter, Brinster, and their colleagues did to create supermouse is an intriguing story that begins with the isolation of the structural gene for growth hormone (GH) from a cloned library of rat DNA, using techniques similar to those described in this appendix. A fragment of the cloned GH gene from which the 5' regulatory region had been deleted was then fused to the regulatory portion of a mouse gene—the

Genetic Engineering in Mice. "Supermouse" (on the left) is significantly larger than its littermate because of the genes for growth hormone that were genetically engineered into the fertilized egg from which the mouse developed.

Key Terms for Self-Testing

recombinant DNA technology (p. 801)
gene cloning (p. 801)

Restriction Enzymes
restriction enzyme (p. 801)
recognition sequence (p. 802)
twofold rotational symmetry (p. 802)
restriction fragment (p. 802)
cohesive ("sticky") end (p. 802)
gel electrophoresis (p. 803)

Restriction Mapping Techniques
restriction map (p. 804)

Recombinant DNA and Cloning Techniques
recombinant DNA molecule (p. 806)
DNA cloning (p. 806)
cloning vector (p. 806)
bacteriophage (p. 806)
plasmid (p. 807)
selectable marker (p. 807)
screening (p. 808)
nucleic acid probe (p. 808)
genomic library (p. 810)
complementary DNA (cDNA) (p. 810)
cDNA clone (p. 810)
cDNA library (p. 810)
polymerase chain reaction (PCR) (p. 810)

Genetic Engineering
genetic engineering (p. 811)
transgenic animal (p. 813)
Ti (tumor-inducing) plasmid (p. 813)
transgenic plant (p. 813)

gene that codes for *metallothionein* (MT). MT is a small metal-binding protein that is normally present in most mouse tissues and appears to be involved in regulating the level of zinc in the animal. The advantage of MT as a "carrier" for the GH gene is that expression of the metallothionein can be specifically induced (turned on) by heavy metals such as zinc.

To make more copies of the MT-GH hybrid gene, it was cloned in *E. coli* cells as part of a circular plasmid called pMGH. After the plasmid DNA was recovered from the bacterial cells, the MT-GH region was excised by digestion of the plasmid DNA with two restriction enzymes, each of which has a unique digestion site on one end or the other of the desired DNA fragment. This fragment consisted of about 5000 nucleotide pairs and remained linear because it had two different ends and therefore could not form a circle.

About 600 copies of this fragment were microinjected into fertilized mouse eggs, in a volume of about 2 picoliters (0.000002 μL!). The DNA was injected into the male pronucleus, because the success rate for integration and retention of the MT-GH gene was found to be higher when the male pronucleus was used than when the DNA was injected into either the female pronucleus or the cytoplasm. Of the 170 eggs that were injected and implanted back into the reproductive tracts of foster mothers, 21 animals developed. Seven of them turned out to be "engineered" (or *transgenic*) mice, with MT-GH genes present in their cells. In at least one case, a transgenic mouse transmitted the MT-GH genes faithfully to about half of its offspring, suggesting that the genes were integrated stably into one of its chromosomes.

Because the growth hormone gene was linked to a metallothionein regulatory region, the mice were given a high level of zinc in their drinking water as a way of turning the hybrid genes on if they were present. Expression of the growth hormone genes was evidenced in three ways. When liver tissue was assayed for the presence of messenger RNA for GH, the results indicated about 800 to 3000 mRNA molecules per liver cell. Moreover, elevated levels of growth hormone were found in the blood: Four of the transgenic mice had blood GH levels that were 100 to 800 times higher than those of their nonengineered littermates! Of course, the most obvious evidence that the genes were expressed is that the transgenic mice grew significantly faster and got significantly bigger. They weighed about twice as much as normal mice, and during the period of maximum sensitivity to growth hormone (3 weeks to 3 months), they grew three to four times faster than normal littermates.

Clearly, it has become possible to introduce isolated, cloned genes into higher organisms and get those genes integrated into the genome, where they are passed on to offspring and expressed. As the authors pointed out when the report was published, "this approach has implications for studying the biological effects of growth hormone, as a way to accelerate animal growth, as a model for gigantism (a human growth abnormality caused by growth hormone), as a means of correcting genetic diseases, and as a method of farming valuable products." Whether (and how fast) these possibilities become realities remains, of course, to be seen. But certainly the potential benefits seem sufficiently attractive to trigger and sustain a great deal of research in the near future. Supermouse, it seems, is just the beginning.

Suggested Reading

General References

Drlica, K. *Understanding DNA and Gene Cloning*. New York: John Wiley, 1984.

Grobstein, C. The recombinant-DNA debate. *Sci. Amer.* 237 (July 1977): 22.

Sambrook, J., E. F. Fritsch, and T. Maniatis. *Molecular Cloning: A Laboratory Manual*, 2nd ed. Cold Spring Harbor, NY: Cold Spring Harbor Laboratory, 1989.

Singer, M., and P. Berg. *Genes and Genomes*. Mill Valley, Calif.: University Science Books, 1990.

Watson, J. D., J. Tooze, and D. T. Kurtz. *Recombinant DNA: A Short Course*. New York: Scientific American Books, 1983.

Watson, J., and J. Tooze. *The DNA Story: A Documentary History of Gene Cloning*. New York: W. H. Freeman, 1981.

Restriction Enzymes

Nathans, D., and H. O. Smith. Restriction endonucleases in the analysis and restructuring of DNA molecules. *Annu. Rev. Biochem.* 44 (1975): 273.

Roberts, R. Restriction and modification enzymes and their recognition sequences. *Nucleic Acids Res.* 10 (1982): r117.

Smith, H. O. Nucleotide sequence specificity of restriction endonucleases. *Science* 205 (1979): 455.

Recombinant DNA and Gene Cloning

Abelson, J., and E. Butz, eds. Recombinant DNA. *Science* 209 (1980): 1317 (An entire issue of *Science* devoted to this topic.)

Cohen, S. N. The manipulation of genes. *Sci. Amer.* 233 (July 1975): 24.

Gilbert, W., and L. Villa-Komoroff. Useful proteins from recombinant bacteria. *Sci. Amer.* 242 (April 1980): 74.

Howard, B. H. Vectors for introducing genes into cells of higher eukaryotes. *Trends Biochem. Sci.* 8 (1983): 209.

Sinsheimer, R. L. Recombinant DNA. *Annu. Rev. Biochem.* 46 (1977): 415.

Berger, S. L., and A. R. Kimmel. *Guide to Molecular Cloning Techniques* (*Meth. Enzymol.*, Vol. 152). San Diego, Calif.: Academic Press, 1987.

Genetic Engineering

Anderson, W. F., and E. G. Diakumkos. Genetic engineering in mammalian cells. *Sci. Amer.* 245 (July 1981): 106.

Brinster, R. L., and R. D. Palmiter. Induction of foreign genes in animals. *Trends Biochem. Sci.* 7 (1982): 438.

Capecchi, R. M. The new mouse genetics: Altering the genome by gene targeting. *Trends Genetics* 5 (1989): 70.

Chilton, M. D. A vector for introducing new genes into plants. *Sci. Amer.* (June 1983): 50.

Koshland, D. E., ed. The new harvest: Genetically engineered species. *Science* 24A (1989): 1275. (A special issue of *Science* featuring seven articles that summarize recent applications of gene cloning and genetic engineering.)

Marx, J. L. Gene therapy—so near and yet so far away. *Science* 232 (1986): 824.

Palmiter, R. D., R. L. Brinster, R. E. Hammer, M. E. Trumbauer, M. G. Rosenfeld, N. C. Birnberg, and R. M. Evans. Dramatic growth of mice that develop from eggs microinjected with metallothionein-growth hormone fusion genes. *Nature* 300 (1982): 611

Palmiter, R. D., and R. L. Brinster. Germ line transformation in mice. *Annu. Rev. Genet.* 20 (1986): 465.

Pestka, S. The purification and manufacture of human interferons. *Sci. Amer.* 249 (August 1983): 37.

Westphal, H. Transgenic mammals and biotechnology. *FASEB J.* 3 (1989): 117.

Glossary

Most Key Terms that occur in the text are defined here. Numbers in parentheses identify the chapter(s) in which a given term appears. The designations A and B refer to Appendices A and B, respectively.

A band (anisotropic band) Region of the sarcomere of a muscle fibril consisting of thick (myosin-containing) filaments and seen as a dark band in the electron microscope. (19)

Absolute refractory period Time interval in which sodium channels of a nerve cell are inactivated and therefore unable to respond to additional depolarization. (20)

Absorption spectrum Relative extent to which light of different wavelengths is absorbed by a pigment. (12)

Accelerating voltage Difference in voltage between the cathode and anode of an electron microscope, responsible for accelerating electrons prior to their emission from the electron gun. (A)

Accessory pigments Molecules such as carotenoids and phycobilins that confer enhanced light-gathering properties on photosynthetic tissue by absorbing light of wavelengths not absorbed by chlorophyll; accessory pigments give distinctive colors to plant tissue, depending on their specific absorption properties. (12)

Acetylcholine Common neurotransmitter at chemical synapses between nerve cells. (20)

Acetyl coenzyme A (acetyl CoA) High-energy thioester of acetic acid and coenzyme A; form in which carbon atoms from glycolysis and fatty acid oxidation enter the tricarboxylic acid cycle. (11)

Acquired immunodeficiency syndrome (AIDS) Nonhereditary loss of immune response that occurs among certain high-risk individuals, notably sexually active male homosexuals and intravenous drug users and their sexual partners, characterized by susceptibility to infections and malignancies and almost always fatal; thought to be the result of a deficiency in T_h lymphocytes. (24)

Acrosomal process Narrow channel through the egg surface coat formed by the polymerization of a pool of actin located behind the acrosomal vesicle of the sperm cell. (22)

Acrosomal reaction Exocytotic release of enzymes from the acrosomal vesicle when the sperm makes contact with the egg. (22)

Acrosomal vesicle Vesicle in the sperm head containing enzymes that catalyze the breakdown of the egg surface coat. (22)

Actin Principal protein of thin filaments of skeletal muscle and of microfilaments of many nonmuscle cells; synthesized as a globular monomer (G-actin) that polymerizes into long, linear filaments (F-actin). (18, 19)

Actin-binding proteins Proteins that bind to actin microfilaments, thereby regulating the length or assembly of microfilaments or mediating their association with each other or with other cellular structures, such as the plasma membrane. (18, 19)

Actin-bundling proteins Proteins such as fimbrin and villin that bind the parallel actin microfilaments at the core of a microvillus into tight bundles at regular intervals. (18)

α-Actinin Muscle protein that is thought to anchor the thin filaments to the Z line. (19)

Action potential Propagation, along the membrane of a nerve cell, of the electrical changes that serve as the means of transmission of a nerve impulse. (20)

Action spectrum Relative extent to which light of different wavelengths affects a particular light-dependent reaction or process. (12)

Activation (a) Increasing the potential of a molecule for biological reactivity, usually by chemical modification of one of its functional groups. (b) Turning on of a specific process or pathway, usually as a result of a conformational change or a covalent modification of an enzyme; also called up-regulation. (21, 22)

Activation energy Energy required to initiate a chemical reaction. (6)

Active site Region of an enzyme molecule at which the substrate binds and the catalytic event occurs; also called the catalytic site. (6)

Active transport Carrier-mediated movement of a substance through a membrane against a concentration or electrochemical gradient, an energy-requiring process. (7, 8)

Adaptive enzyme synthesis Regulation of the intracellular concentration of an enzyme by modulating synthesis of that enzyme in response to cellular needs. (17)

Adenine (A) Nitrogen-containing aromatic base, chemically designated as a purine, that serves as an informational monomeric unit when present in nucleic acids with other bases in a specific sequence. (3, 16)

Adenosine Compound formed by linking adenine and ribose. (3)

Adenosine diphosphate (ADP) Adenosine with two phosphates linked to each other by a phosphoanhydride bond and to carbon 5 of the ribose by a phosphoester bond. (3)

Adenosine monophosphate (AMP) Adenosine with a phosphate linked to carbon 5 of the ribose by a phosphoester bond. (3)

Adenosine triphosphate (ATP) Adenosine with three phosphates linked to each other by phosphoanhydride bonds and to carbon 5 of the ribose by a phosphoester bond; principal energy storage compound of most cells, with energy stored in the high-energy phosphoanhydride bonds. (3, 10)

Adenylate cyclase Enzyme that catalyzes the formation of cyclic AMP from ATP; located on the inner surface of the plasma membrane of many eukaryotic cells and is activated by specific ligand-receptor interactions on the outer surface of the membrane. *See also* Cyclic AMP. (21)

ADP *See* Adenosine diphosphate.

Adrenalin *See* Epinephrine.

Adrenergic synapse Chemical synapse that uses norepinephrine or epinephrine as the neurotransmitter. (20)

Adsorptive pinocytosis *See* Receptor-mediated endocytosis.

Aerobic Occurring in the presence of oxygen. (5)

A-face *See* P-face.

***a* gene** Structural gene in the *lac* operon of *Escherichia coli* that encodes thiogalactoside transacetylase, an enzyme involved in lactose metabolism. (3)

AIDS *See* Acquired immunodeficiency syndrome.

Alcoholic fermentation Anaerobic catabolism of carbohydrates with ethanol and carbon dioxide as the end products. (10)

Aldosugar Sugar with an aldehyde functional group. (3)

Allele One of two or more alternative forms of a gene. (15)

Allosteric effector Small molecule that causes a change in the configuration of an allosteric protein by binding to a site other than the active site. (6)

Allosteric protein (allosteric enzyme) Regulatory protein that has two alternative configurations, each with a different biological property; interconversion of the two configurations is mediated by the reversible binding of a specific small molecule to the effector site. (6, 7)

Allosteric regulation Control of a reaction pathway by the effector-mediated reversible interconversion of the two forms of an allosteric enzyme in the pathway. (6, 10, 17)

α helix Spiral-shaped secondary structure of protein molecules, consisting of a backbone of peptide bonds with R groups of amino acids jutting out. (3)

Alternative splicing Regulated utilization of different combinations of intron/exon splice junctions in a primary RNA transcript to produce messenger RNAs that differ in exon composition; results in production of variant protein isoforms from a single gene sequence. (17, 24)

Ames test Preliminary screening test for carcinogens, using a mutant strain of *Salmonella* that cannot grow on a histidine-deficient medium unless the original mutation is reversed by a second mutation caused by a carcinogen. (23)

Amide bond Covalent bond between an amino group of one molecule and a carboxyl group of another molecule. *See also* Peptide bond. (3)

Amino acid Monomeric unit of proteins, consisting of a carboxylic acid with an amino group and one of a variety of R groups attached to the α carbon; 20 different kinds of amino acids are normally found in proteins. (3)

Aminoacyl site (A site) Site on a ribosome at which the incoming tRNA-linked amino acid binds. (16)

Aminoacyl tRNA tRNA molecule with an amino acid covalently bound to it by an ester bond. (16)

Aminoacyl-tRNA synthetase Enzyme that joins an amino acid to its appropriate tRNA molecule by the formation of an ester bond, using energy provided by the hydrolysis of ATP. (16)

Amoeba Unicellular organism often characterized by pseudopodia that are used for motility and feeding. (19, 22)

Amoeboid movement Mode of locomotion that depends on pseudopodia, or extensions of cytoplasm. (19)

AMP *See* Adenosine monophosphate.

Amphibolic pathway Series of reactions that can function both in a catabolic mode and as a source of precursors for anabolic pathways. (11)

Amphipathic molecule Molecule having spatially separated hydrophilic and hydrophobic regions. (2)

Amylopectin Branched-chain form of starch consisting of $\alpha(1 \rightarrow 6)$ glycosidic bonds between successive glucose units, with $\alpha(1 \rightarrow 6)$ branches that occur every 12 to 25 units and consist of 20 to 25 glucose units. (3)

Amyloplast Plastid specialized for starch storage. (12)

Amylose Straight-chain form of starch consisting of $\alpha(1 \rightarrow 4)$ glycosidic bonds between successive glucose units. (3)

Anabolic pathway Series of reactions that results in the synthesis of one or more specific cellular components. (10)

Anabolism Synthesis of complex molecules from simpler units, almost always with the input of energy. (10)

Anaerobic Occurring in the absence of oxygen. (5)

Anaphase Third and shortest phase of mitosis, characterized by the separation of centromeres and movement of sister chromatids to opposite poles. (14)

Anaphase I Third stage of the first meiotic division, in which centromeres of each bivalent move toward opposite poles. (15)

Anastral mitosis Mitosis as it occurs in higher plants, characterized by the absence of an astral array of microtubules at the poles of the mitotic spindle. (14)

Angstrom (Å) Unit of length; 1 Å = 0.1 nm = 10^{-8} cm. (1)

Animal-vegetal axis Polarity in an animal egg that determines the overall orientation of the embryo. (22)

Anterior pituitary Front part of the pituitary gland; secretes a number of tropic hormones that regulate other endocrine glands, including adrenocorticotropin, thyroid-stimulating hormone (TSH), follicle-stimulating hormone (FSH), luteinizing hormone (LH), somatotropin (growth hormone), and prolactin. (21)

Antibody *See* Immunoglobulin.

Antibody repertoire Total collection of different kinds of antibodies available to an organism that accounts for the ability of the organism to recognize virtually any foreign substance it will ever encounter. (24)

Anticodon Trinucleotide sequence, located on one of the loops of a tRNA molecule, that recognizes the appropriate codon of the mRNA molecule by complementary base pairing. (16)

Antigen Substance capable of eliciting an immune response. (24)

Antigen binding site Variable domain located on the arms of the Y-shaped immunoglobulin molecule that recognizes and binds antigen. (24)

Antigen-dependent differentiation Phase of lymphocyte differentiation initiated when a specific antigen binds to the surface receptors of a lymphocyte and causes the lymphocyte to develop into an effector cell. (24)

Antigenic determinant Specific portion of an antigen molecule to which the immune system is capable of responding; also called an epitope. (24)

Antigen-independent differentiation Phase of lymphocyte differentiation in which hematopoietic cells give rise to uncommitted cells and each of these becomes committed to the synthesis of a specific antigen receptor capable of recognizing a specific antigen. (24)

Antigen receptor Membrane-bound molecule on the lymphocyte surface capable of responding to a single antigenic determinant. All receptors on a lymphocyte are identical. (24)

Antiport Coupled transport of two solutes across a membrane in opposite directions. (8)

Antisense RNA molecule RNA molecule with a nucleotide sequence complementary to that of a messenger RNA molecule; useful in the study of gene regulation because it hybridizes to mRNA, thereby inhibiting synthesis of the protein encoded by the mRNA. (16)

Arachidonic acid Twenty-carbon fatty acid with four double bonds that is released from membrane phospholipids by phospholipase A activity and serves as the precursor for the synthesis of prostaglandins, thromboxanes, and leukotrienes by the cyclooxygenase pathway. *See also* Phospholipase A; Prostaglandin. (21)

Archenteron Interior cavity of animal body that is formed by invagination of cells in gastrulation and eventually gives rise to the gastrointestinal tract; also called primitive gut. (22)

Asexual reproduction Form of reproduction in which a single parent organism is the only contributor to the genetic information of the new organism; reproduction in the absence of any sexual process. (15)

A site *See* Aminoacyl site. (16)

Aster Radial array of microtubules that is found at the spindle pole and originates from the centrioles. (14)

Astral mitosis Mitosis as it occurs in higher animals, characterized by the presence of an astral array of microtubules at each pole of the mitotic spindle. (14)

Asymmetric carbon atom Carbon atom that has four different substituents. Two different stereoisomers are possible for each asymmetric carbon atom in an organic molecule. (2)

ATP *See* Adenosine triphosphate.

ATP synthase complex Complex of an F_1 and an F_0 that results in the synthesis of ATP by linking the exergonic movement of protons through F_0 to ATP generation by F_1. (11)

Attenuation Means of regulating bacterial operons that are involved in amino acid synthesis. (17)

Attenuation site Nucleotide sequence located between the promotor and structural genes of certain prokaryotic operons, at which RNA polymerase molecules may cease transcribing the operon. (17)

Autoimmune reaction Antiself reaction in which the immune system attacks cells of the organism of which it is a part; results from an inability to distinguish self from nonself. (24)

Autolysis Self-destruction of a cell by the intracellular release of hydrolytic enzymes from its lysosomes. (9)

Autonomic nervous system Portion of the nervous system that controls involuntary activities. (20)

Autophagic lysosome Secondary lysosome that contains hydrolytic enzymes involved in the digestion of substances of intracellular origin. (9)

Autophagic vacuole Vacuole formed when an old or unwanted organelle or other cellular structure is wrapped in membranes derived from the endoplasmic reticulum prior to digestion by lysosomal enzymes. (9)

Autophagy Intracellular digestion of old or unwanted organelles or other cell structures as it occurs within autophagic lysosomes; "self-eating." (9)

Autoradiography Technique in which photographic film is used to localize a radioactive compound within a cell, often to determine the sequence of events in which the compound is involved. (A)

Axis of polarity Gradient of information (often proposed to be a morphogen gradient) that distinguishes one end of an organism or tissue from the other and thus establishes polarity for cells in between; a form of positional information. (22)

Axon Extension of nerve cell that conducts impulses away from the cell body. (20)

Axonal transport Movement of proteins and membrane-bounded vesicles along the axon of a nerve cell between the cell body and the synaptic knobs. (18)

Axonemal microtubules Microtubules present as a highly ordered bundle in the axoneme of a cilium or flagellum. *See also* Cytoplasmic microtubules. (18)

Axoneme Central shaft of a cilium or flagellum made up of a stable microtubular organization with nine outer doublets and a central pair of microtubules. (19)

Axoplasm The cytoplasm within the axon of a nerve cell. (20)

Backcross Genetic cross between a heterozygote and one of the original homozygous parental stocks. (15)

Bacterial flagellum Locomotory structure in bacteria consisting of a spiral filament attached by means of a hook to a rod that penetrates the membrane. (19)

Bacteriochlorophyll Type of chlorophyll found in bacteria that is able to extract electrons from donors other than water. (12)

Bacteriophage Virus that infects bacterial cells; also called phage. (4, 13, B)

Bacteriorhodopsin Transmembrane protein complexed with rhodopsin, capable of transporting protons across the bacterial cell membrane to create a light-dependent electrochemical proton gradient. (8, 11)

Bacterium Unicellular prokaryotic organism that has neither a true nucleus nor membrane-bounded organelles.

Basal body Microtubule-containing structure at the base of a eukaryotic flagellum or cilium that serves as an anchor for the appendage; identical in appearance to a centriole. (19)

Base pairing Complementary relationship between purines and pyrimidines involving hydrogen bonds and providing a mechanism for nucleic acids to recognize each other. (3)

B cell *See* B lymphocyte.

B-DNA Right-handed Watson-Crick helix of a DNA molecule. *See also* Z-DNA.

Belt desmosome Continuous region of tight adhesion between plasma membranes of adjacent cells. (7)

Benign tumor Tumor that does not proliferate indefinitely and does not usually adversely affect other tissues. (23)

β pleated sheet Extended sheetlike secondary structure of proteins in which adjacent polypeptides are linked by hydrogen bonds between imino and carbonyl groups. (3)

B-face *See* E-face.

Bidirectional replication Mode of DNA replication in a circular DNA molecule in which two replication forks are formed at a single origin, allowing replication to proceed simultaneously in both directions. (14)

Bindin Protein released by a sea urchin sperm during the acrosomal reaction; coats the surface of the acrosomal vesicle, enabling the sperm to bind to the vitelline membrane of the egg. (22)

Bioenergetics Area of science that deals with the application of thermodynamic principles to reactions and processes in the biological world. (5)

Bioluminescence Production of light by an organism as a result of the reaction of ATP with specific luminescent compounds. (5)

Biopolymer Biologically important macromolecule consisting of many monomeric subunits covalently bound in an identical manner as a result of the repetition of a short sequence of chemical condensation reactions. (2)

Biosynthesis Generation of new molecules through a series of chemical reactions within the cell. (5)

Birefringence Property of rotating the plane of plane-polarized light; characteristic of highly ordered structures. (A)

Bivalent Pair of homologous chromosomes that have synapsed in the first meiotic division. *See also* Tetrad. (15)

Blastocoel Fluid-filled cavity within the blastula of an animal embryo. (22)

Blastomere One of the many cells that make up the blastula of an animal embryo. (22)

Blastopore Area on the surface of the blastula where the invagination of gastrulation begins. Cells migrate from the surface of the blastula over the blastopore lip and into the interior of the embryo. (22)

Blastula Stage of animal embryogenesis in which cells formed by the cleavage of the zygote form a partly hollow ball of cells. (22)

Blood clotting Process triggered by injury and leading rapidly to the formation of a tangled mass of red blood cells, platelets, and fibers of the protein fibrin, thereby stopping bleeding and reducing blood loss. (21)

Blood platelets Round or oval nonnucleated disks, found in the blood of mammals, characterized by the presence of numerous dense granules containing platelet factors that are involved in the process of blood clotting; produced by fragmentation of bone-marrow cells called megakaryocytes. *See also* Blood clotting. (21)

Blue-green alga Unicellular phototropic organism with no membrane-bounded organelles; also called a cyanobacterium.

B lymphocyte (B cell) Antibody-producing white blood cell that is responsible for the humoral response of the immune system. (24)

Bond energy Amount of energy required to break one mole of a particular chemical bond. (2)

Brightfield microscopy Basic form of light microscopy in which the specimen possesses color, has been stained, or has some other property that affects the amount of light that passes through, thereby allowing the image to be formed. (A)

Bundle sheath cells Internal cells of the leaf of a C_4 plant located in close proximity to the vascular bundle of the leaf; site of carbon metabolism by the Calvin cycle. (12)

Buoyant density Point in a density gradient at which the density of an organelle or molecule is exactly equal to that of the solution surrounding it. (9)

Ca^{2+}-calmodulin complex Complex formed by the reversible binding to calmodulin of four calcium ions with concomitant changes in the conformation of calmodulin, thereby increasing its affinity for calmodulin-binding proteins that are sensitive to the intracellular calcium level. *See also* Calmodulin; Calmodulin-binding protein. (19, 21)

Calmodulin Calcium-binding protein involved in many calcium-regulated processes in eukaryotic cells. (19, 21)

Calmodulin-binding protein Any of a variety of proteins, mostly enzymes, that depend on the Ca^{2+}-calmodulin complex for activation and are thereby responsive to the intracellular calcium level. *See also* Ca^{2+}-calmodulin complex; Calmodulin. (19, 21)

calorie (cal) Unit of energy; amount of energy needed to raise the temperature of 1 gram of water 1°C. (2, 5)

Calvin cycle Cyclic series of reactions that results in the reductive fixation of carbon dioxide and the subsequent conversion of the fixed carbon to sugars; also called the light-independent reactions of photosynthesis. (12)

cAMP *See* Cyclic AMP.

Cancer Condition in which cellular control mechanisms no longer function normally, such that certain cells within the body proliferate uncontrollably, eventually destroying other cells and usually causing the death of the organism. (23)

Cap Methylated guanosine added posttranscriptionally to the 5′ end of a eukaryotic mRNA molecule; adds to the stability of the mRNA molecule and may be involved in alignment of the messenger in translation. (16)

CAP *See* Catabolite activator protein.

Carbohydrate General name given to molecules that contain carbon, hydrogen, and oxygen in a ratio $C_nH_{2n}O_n$; examples include starch, glycogen, and cellulose. (3)

Carcinogen Physical or chemical agent that can cause the transformation of a normal tissue into a tumor; cancer-causing agent. (9, 23)

Carcinoma Tumor derived from an epithelial cell. (23)

Cardiac muscle Striated muscle of the heart, highly dependent on aerobic respiration. (19)

Carotenoid Accessory pigment found in most plant species that absorbs in the violet-to-green region of the visible spectrum (400–500 nm) and is therefore yellow to red-orange in color. (12)

Carrier-mediated diffusion *See* Passive transport.

Carrier-mediated transport *See* Passive transport.

Carrier protein *See* Permease.

Catabolic pathway Series of reactions that results in the degradation of one or more specific cellular components. (10)

Catabolism Degradation of complex molecules into simpler substances. (10)

Catabolite activator protein (CAP) Regulatory protein that complexes with cAMP and then binds to a site adjacent to a promoter, thereby enhancing the binding of RNA polymerase and stimulating initiation of transcription of certain prokaryotic operons. (17)

Catabolite repression Decreased synthesis by bacterial cells of enzymes involved in specific catabolic pathways due to the presence of a preferred alternative substrate. (17)

Catalyst Agent that enhances the rate of a reaction by lowering the activation energy without itself being consumed; catalysts change the rate at which a reaction approaches equilibrium, but not the position of equilibrium. *See also* Enzyme. (6)

Catalytic site *See* Active site.

Catecholamine Compound that is derived from the amino acid tyrosine and that functions as a hormone and neurotransmitter. (20)

***cdc* genes** *See* Cell-division cycle genes. (22)

***cdc2*/MPF protein kinase** Enzyme that plays a critical role in regulating the cell cycle by phosphorylating unknown substrates, such that cells proceed through G-1 when the kinase is in the active form, but enter the quiescent G-0 phase when the kinase is inactive. The kinase is conserved in species as highly divergent as yeast and humans. (22)

cDNA *See* Complementary DNA.

cDNA probe Radioactively labeled complementary DNA molecule that is used to determine the presence and/or location of specific mRNA molecules. (16)

Cell body Portion of a nerve cell that contains the nucleus and other organelles and has branches, or processes, extending from it. (20)

Cell coat Layer of material consisting of glycoproteins and other substances that is exterior to the plasma membrane in animal cells. *See also* Glycocalyx. (4, 7)

Cell cortex Three-dimensional network of actin microfilaments and associated proteins located just beneath the plasma membrane of most animal cells; supports the plasma membrane, confers strength on the cell surface, and facilitates shape changes and cellular movement. (18)

Cell cycle Sequence of events by which all of the genetic information present in the nucleus of the parent cell is duplicated and parceled out to two daughter cells. (14)

Cell-division cycle (*cdc*) genes Any of approximately 50 genes in yeast that function at specific points in the cell cycle; shifting a yeast cell with a temperature-sensitive *cdc* to the restrictive temperature causes the cell to stop growing at a specific point in the cell cycle. (22)

Cell junction Specialized modification of the plasma membranes at the point of contact between two adjacent animal cells. (7)

Cell line Cells of a particular type that can be cultured indefinitely in the laboratory and are thus "immortal"; important tool for the study of differentiated cell types. (22)

Cell lineage The complete set of ancestral cells and cell divisions that give rise to a particular cell during development. (22)

Cell-mediated immune response Defense process in which certain cells react directly with foreign antigens, engulfing and destroying them or killing the cells to which they are bound. (24)

Cell membrane *See* Plasma membrane.

Cell plate Cell wall in the process of formation, leading to the separation of two daughter nuclei during the division of a plant cell. (14)

Cell theory Theory of cellular organization that states that all organisms consist of one or more cells, that the cell is the basic unit of structure for all organisms, and that all cells arise only from preexisting cells. (1)

Cell wall Rigid, nonliving structure exterior to the plasma membrane of bacterial, algal, fungal, and plant cells; plant cell walls consist of cellulose microfibrils embedded in a noncellulosic matrix. (4, 7)

Cellular oncogene (c-*onc*) *See* Proto-oncogene.

Cellulose Structural polysaccharide present in plant cell walls, consisting of β-D-glucose units linked by $\beta(1 \rightarrow 4)$ bonds. (3)

Central dogma of molecular biology The flow of genetic information from DNA to RNA (transcription) and then to protein (translation), as well as from DNA to DNA (replication). (16)

Central granule Structure found within the annulus of a nuclear pore; also called the central plug. (13)

Central lymphoid tissue Site of lymphocyte maturation to which progenitor cells migrate from the hematopoietic tissue, encountering cell-bound or soluble factors that induce further differentiation; consists of the thymus and bone marrow (or in birds, the bursa of Fabricius). (24)

Central nervous system (CNS) Sensory and motor cells of the brain and spinal cord. (20)

Central vacuole Large membrane-bounded organelle present in many plant cells; helps maintain turgor pressure of the plant cell, plays a limited storage role, and is also capable of a lysosome-like function in intracellular digestion. (4)

Centriole Microtubule-containing structure that lies near the nucleus of an animal cell and is involved in the formation of the mitotic spindle. *See also* Basal body. (14)

Centromere Point along a chromosome at which sister chromatids are held together prior to anaphase; also, the site of attachment of the kinetochore. (14)

Centrosome Microtubule-organizing center consisting of granular material surrounding the two centrioles of an animal cell; also called the cell center. (18)

CF_0 Component of the ATP synthase complex of the chloroplast that is embedded in the thylakoid membrane and serves as the proton translocator. (12)

CF_1 Component of the ATP synthase complex of the chloroplast; knoblike sphere that protrudes from the thylakoid membrane and serves as the site of ATP generation; also called the chloroplast coupling factor. (12)

C gene DNA sequence that encodes the constant region of the light or heavy chain of an immunoglobulin molecule. (24)

Channel former Ionophore that functions by forming a hydrophilic channel within the membrane, thereby allowing ions to cross the membrane. (8)

Charge repulsion Force separating two ions, molecules, or regions of molecules that are of the same electric charge. (10)

Chemical receptor Substance, usually an allosteric protein, that has a binding site for a specific ligand and initiates or mediates a particular event or series of events in response to ligand binding. *See also* Intracellular receptor; Ligand; Membrane receptor; Receptor protein. (21)

Chemical synapse Junction between two nerve cells at which specific chemicals called neurotransmitters are involved in the transmission of a nerve impulse across the junction. (20)

Chemiosmotic coupling model Model that postulates an electrochemical potential across the inner mitochondrial membrane as the link between electron transport and ATP generation. (11)

Chemotaxis Ability of an organism to detect and move toward a chemical attractant or away from a chemical repellent. (19)

Chemotroph Organism that is dependent on the bond energies of oxidizable food molecules to satisfy energy requirements; "chemical feeder." (5)

Chiasma X-shaped profile formed by contact between homologous chromatids during prophase I, possibly resulting in a crossover event between two nonsister chromatids; physical manifestation of recombination. (15)

Chitin Structural polysaccharide found in insect exoskeletons and crustacean cells; consists of *N*-acetylglucosamine units linked by $\beta(1 \rightarrow 4)$ bonds. (3)

Chlorophyll Light-absorbing molecule that donates photoenergized electrons to organic molecules, initiating photochemical events that lead to the generation of the NADPH and ATP required for the Calvin cycle; because of its absorption properties, chlorophyll gives plants their characteristic green color. (12)

Chloroplast Large plastid in a plant or algal cell that contains chlorophyll and the enzymes necessary to carry out photosynthesis. (12)

Chloroplast coupling factor *See* CF_1.

Cholesterol Lipid constituent of animal cell plasma membrane; serves as a precursor to the steroid hormones. (3)

Cholinergic synapse Chemical synapse that uses acetylcholine as the neurotransmitter. (20)

Chromatid One of the daughters of a replicated chromosome, still joined to its sister chromatid at the centromere.

Chromatin Nucleoprotein complex that makes up chromosomes; usually consists of nucleosomes spaced regularly along a DNA chain. (4)

Chromatin fiber Structural element of eukaryotic chromosomes, consisting of adjacent nucleosomes in close juxtaposition. (13)

Chromoplast Plastid that contains pigments and is responsible for the characteristic coloration of plant parts such as petals, fruits, and leaves. (12)

Chromosomal alteration Structural alteration of a chromosome, such as a deletion, translocation, duplication, or inversion. (23)

Chromosomal fiber Bundle of microtubules that extends outward from the kinetochore of a chromosome toward one of the two poles of the mitotic spindle; also called kinetochore fiber. (14)

Chromosome Structural unit of genetic material in the nucleus of a eukaryotic cell, consisting of a DNA molecule in association with histones. (1, 13, 14)

Chromosome mapping Determining the order and location of genes on a chromosome on the basis of crossover frequencies. (15)

Chromosome puff Uncoiled region of a polytene chromosome that is undergoing transcription. (17)

Chromosome theory of heredity Theory stating that hereditary factors are located on the chromosomes within the nucleus. (1)

Cilium Membrane-bounded appendage on the surface of a eukaryotic cell, composed of a specific arrangement of microtubules and responsible for motility of the cell or the environment around the cell. *See also* Flagellum. (19)

Cis-acting mutation Mutation that alters the nucleotide sequence of a DNA binding site for a transcriptional regulatory protein; the mutation affects only physically linked structural genes. *See also* Trans-acting mutation. (17)

Cis face *See* Forming face.

Cisterna Membrane-bounded flattened sac, such as in the endoplasmic reticulum; also called a saccule. (4, 9)

Cisternal space Internal region of the cisternae of the endoplasmic reticulum, the Golgi complex, the lysosomes, and the perinuclear space of the nuclear envelope. (9)

Cis-trans test Analysis used to determine whether a mutation in a bacterial operon affects a regulatory protein or its DNA-binding site. (17)

Cistron Nucleotide sequence in DNA that carries information for a single polypeptide, a single rRNA, or a single tRNA molecule. *See also* Gene. (16)

Citric acid cycle *See* Tricarboxylic acid cycle.

Class I MHC antigen Glycosylated integral membrane protein that elicits a response from cytotoxic T cells. (24)

Class II MHC antigen Glycosylated integral membrane protein that elicits a response from helper T cells. (24)

Clathrin Large protein that forms the bristlelike "cage" or "basket" around the coated vesicles and coated pits involved in endocytosis and other cellular and intracellular transport processes. (7, 9)

Cleavage Series of mitotic divisions that occur during early animal embryogenesis with little or no increase in the size of the embryo. (14, 22)

Cleavage furrow Groove formed during the division of an animal cell that deepens progressively and leads to separation of the daughter nuclei. (14, 19)

Clonal selection theory Explanation for immune system diversity, stating that every lymphocyte is predetermined to express a homogeneous set of membrane-bound receptors with a single antigenic specificity and that, upon binding an antigen, a lymphocyte may be triggered to proliferate to form large clones of progeny lymphocytes with surface receptors that have the same specificity as the parent cell. (24)

Clone Organism, cell, or molecule that is genetically identical to another organism, cell, or molecule from which it is derived. (17, 24)

Cloning Amplification of a specific DNA sequence by insertion of the sequence into a bacterial cell such that it is replicated along with the bacterial DNA. (B)

Cloning vector Genetic element to which a selected sequence of DNA can be added prior to replication of the vector in bacterial cells; phages and plasmids are the most common cloning vectors. (B)

CNS *See* Central nervous system.

Coated pit Specialized region of the plasma membrane with a bristlelike clathrin coat on the cytoplasmic side; upon endocytosis, coated pits invaginate to form coated vesicles. (9)

Coated vesicle Membrane-bounded structure with a bristlelike coat consisting of a polyhedral lattice of clathrin subunits; important in intracellular transport processes. (9)

Coat protein Protein that makes up the outer layer, or coat, of a virus. (2)

Coding strand Strand of DNA duplex that serves as the template for RNA synthesis at that point along the duplex. (16)

Codon Triplet of nucleotides that is present in an mRNA molecule and serves as the coding unit during protein synthesis. (16)

Coenzyme Small organic molecule that functions with an enzyme by serving as a carrier of electrons or functional groups. (10)

Coenzyme A (CoA) Organic molecule that serves as a carrier of acyl groups by forming a high-energy thioester bond with an organic acid. (11)

Coenzyme Q (CoQ) Nonprotein component of the mitochondrial electron transport chain that serves as the collection point for electrons from both NADH- and $FADH_2$-linked dehydrogenases of the cell; also called ubiquinone. (11)

Coenzyme Q–cytochrome *c* reductase Enzyme complex involved in the transfer of electrons from coenzyme Q to cytochrome *c* in the electron transport chain. (11)

Cohesive end Single-stranded end of a DNA fragment that is generated by cleavage with a restriction enzyme and that tends to reassociate with another fragment generated by the same restriction enzyme because of base complementarity; also called a sticky end. (B)

Colchicine Plant alkaloid that binds to tubulin and prevents its polymerization into microtubules. (18)

Colinear Having corresponding parts arranged in the same linear order. (16)

Collagen Fibrous protein with high tensile strength, prominent in connective tissue; component of the fuzzy layer around animal cells. (7)

Combinatorial model for gene regulation Model that suggests that complex regulatory patterns such as tissue-specific gene expression could be achieved by utilizing specific combinations of a relatively small number of cis-acting regulatory elements and their respective transcription factors, with transcription of a specific gene occurring when the set of transcription factors expressed in a cell match the array of regulatory elements for that gene. (17)

Commitment Condition whereby the developmental options of a cell are limited, usually to a single fate. (22)

Compartmentalization Localization of specific cellular functions within specific organelles. (4)

Competitive inhibition Inhibition of enzyme activity that results from the reversible binding of an inhibitor to the active site, thereby competing with the substrate for the active site and decreasing the generation of product. (6)

Complement Family of normal serum proteins that are activated by reaction with antigen-antibody complexes. Activated complement aids in the elimination of pathogenic microorganisms by direct killing (lysis) or by facilitating phagocytosis by macrophages. (24)

Complementary DNA (cDNA) Single-stranded DNA molecule synthesized from an mRNA template by reverse transcription. (22, B)

Compound microscope Light microscope that uses several lenses in combination; usually has a condenser lens, an objective lens, and an ocular lens. (A)

Concentration work Use of energy to transport ions or molecules across a membrane against an electrochemical or concentration gradient. (5)

Condensation reaction Chemical reaction that results in the joining of two molecules by the removal of a water molecule. (2)

Condenser lens Lens of a light microscope (or electron microscope) that directs the light rays (or electron beam) from the source toward the specimen. (A)

Condensing vacuole Vacuole that buds off the Golgi complex and gives rise to a mature secretory granule by the progressive loss of water. (16)

Conformation The three-dimensional structure of a polypeptide or other biological macromolecule. (3)

Conjugation Cellular mating process by which genetic information can be transferred from one bacterial cell to another. (15)

Connexon Assembly of six protein subunits with a hollow center that forms a channel through the plasma membrane at a gap junction. When connexons in the plasma membranes of two adjacent cells are aligned, a means of communication is established between the cells. (7)

Consensus binding site Most likely nucleotide sequence required for specific recognition and binding of a regulatory protein. (17)

Consensus sequence Order of nucleotides in a highly conserved segment of DNA that indicates the most likely nucleotide at each position. (16)

Constant region C-terminal region of the light or heavy chain of an immunoglobulin molecule, nearly identical in amino acid sequence to all other constant regions of the same type of immunoglobulin. (24)

Constitutive heterochromatin Chromosomal regions that are permanently condensed in all cells of an organism and therefore never genetically active in any cell. (13)

Contractile ring Beltlike bundle of actin microfilaments that forms beneath the plasma membrane and acts to constrict the cleavage furrow during the division of an animal cell. (14, 19)

Coordinate regulation of genes Simultaneous induction of a set of structural genes in response to a single signal; accomplished in prokaryotes by organization of several genes into a single physically linked regulatory unit and in eukaryotes by common regulatory elements that mediate coordinated transcriptional activation of physically unlinked genes. (17)

Core glycosylation Initial addition of carbohydrates to polypeptides that occurs in the endoplasmic reticulum. (16)

Core particle Histone octamer with 146 base pairs of DNA wound around it. *See also* Nucleosome. (13)

Corepressor Effector molecule required with the repressor to prevent transcription of a bacterial or viral operon. (17)

Cortical cytoplasm Region of cytoplasm just beneath the cell membrane. (22)

Cortical reaction Release of proteases and other enzymes from cortical vesicles, causing elevation and hardening of the vitelline membrane to prevent further sperm penetration following fertilization of an animal egg. (22)

Cortical vesicle Membrane-bounded structure, found in the cortex of an egg cell, that releases proteases and other enzymes exocytotically when stimulated by an elevation in the intracellular calcium concentration. (22)

Cotransductional mapping Method of determining the location of two markers in a bacterial genome by quantifying the frequency with which they accompany each other in a transducing phage. (15)

Cotranslational transport Transfer of a protein into or across a membrane as synthesis of the protein proceeds. (16)

Coupled transport Concomitant facilitated transport of two solutes across a membrane, either in the same or opposite directions. *See also* Antiport; Symport. (8)

Coupling factor *See* F_1.

Covalent bond Strong chemical bond in which two atoms share two or more electrons. (2)

C_3 plant Plant in which the immediate product of carbon dioxide fixation is the three-carbon compound 3-phosphoglycerate. (12)

C_4 plant Plant in which the immediate product of carbon dioxide fixation is the four-carbon compound oxaloacetate. (12)

Creatine phosphate High-energy compound in muscle cells that is used to regenerate the ATP needed for muscle contraction. (19)

Crista Infolding of inner mitochondrial membrane into the matrix of the mitochondrion, thereby increasing the total surface area of the inner membrane; contains the enzymes of electron transport and of oxidative phosphorylation. (4, 11)

Critical concentration Minimal concentration of subunits required for assembly into a polymer. (18)

Cross-bridge Structure formed by contact between the myosin heads of thick filaments and the thin filaments in muscle fibrils. (19)

Crossing over Physical exchange of genetic information between homologous chromosomes. (15)

Cryoprotection Reduction in the formation of ice crystals during freezing of a specimen for microscopy. (A)

C (carboxyl) terminus The end of a polypeptide characterized by a free carboxyl group; the first portion of a polypeptide to be synthesized during translation of a mRNA molecule. (3)

Cyanobacterium *See* Blue-green alga.

Cyclic AMP (cAMP) Molecule of adenosine monophosphate with the phosphate group linked to both the 3′ and 5′ carbons by phosphodiester bonds; a second messenger important in mediating the effects of hormones on certain cellular functions. (17, 21, 22)

Cyclic electron flow Transfer of photoexcited electrons from photosystem I along a transport chain that returns them to a chlorophyll molecule of the same photosystem, with the energy of transport used to drive ATP synthesis. (12)

Cyclosis Circular flow of cell contents around a central vacuole. (19)

Cytochalasins Family of drugs produced by certain fungi that inhibits a variety of cell movements by preventing actin polymerization. (19)

Cytochrome *c* oxidase Enzyme complex of the electron transport chain, involved in the transfer of electrons from cytochrome *c* to molecular oxygen. (11)

Cytochromes Heme-containing proteins of the electron transport chain, involved in the transfer of electrons from coenzyme Q to oxygen by the oxidation and reduction of the central iron atom of the heme group. (11)

Cytokinesis Division of the cytoplasm of a parent cell into two daughter cells; usually follows mitosis. (14, 19)

Cytology Study of cellular structure, based primarily on microscopic techniques. (11)

Cytoplasm Liquid colloidal material of a cell enclosed within the plasma membrane; in eukaryotic cells, membrane-bounded organelles and other structures are suspended in the cytoplasm. (4, 18)

Cytoplasmic determinant Substance or set of interacting substances that resides in a specific region of the cytoplasm of an egg cell and influences the developmental course of the cells that inherit that cytoplasm. (22)

Cytoplasmic dynein ATPase responsible for inward axonal transport of molecules or vesicles from the synaptic knob toward the cell body; originally called MAP-1C. (18)

Cytoplasmic microtubules Loosely organized, dynamic network of microtubules in the cytoplasm of eukaryotic cells. *See also* Axonemal microtubules. (18)

Cytoplasmic streaming Back-and-forth flow of cytoplasm, especially prominent in some plant and algal cells. (19)

Cytosine (C) Nitrogen-containing aromatic base, chemically designated as a pyrimidine, that serves as an informational monomeric unit when present in nucleic acids with other bases in a specific sequence. (3, 16)

Cytoskeleton Complex network of microtubules, microfilaments, and intermediate filaments that provides structure to the cytoplasm of a eukaryotic cell and plays an important role in cell movement; also called the cytoskeletal network. (4, 18)

Cytotoxic T (T_c) cell Activated T cell responsible for lysing foreign cells; also called killer T cell. (24)

DAG *See* Diacylglycerol.

Darkfield microscopy Light microscopic technique that uses only light rays diffracted by the specimen to produce a bright image against a dark background. (A)

Daughter cell One of the two cells that result from the division of a prior cell. (14)

Deamination Form of DNA damage caused by the hydrolysis of the amino group from a nucleotide base, with cytosine as the most susceptible base. (14)

Deep-etching Modification of freeze-etching in which an ultrarapid freezing technique and a volatile cryoprotectant are used to extend the etching period and remove a deeper layer of ice, thereby producing a larger exposed area of the specimen surface, which allows the interior structure of a cell to be examined in depth; also called quick-freeze deep-etch microscopy. (18, A)

Degenerate Describing the capability of the genetic code to specify a given amino acid by more than one triplet codon. (16)

Dehydrogenation Removal of hydrogen ions (protons) from an organic molecule; oxidation. (10)

Denaturation Loss of the natural three-dimensional structure of a macromolecule, usually resulting in a loss of its biological activity; caused by agents such as heat, extremes of pH, urea, salt, and other chemicals. (3, 13)

Dendrite Extension of a nerve cell that receives impulses and transmits them inward toward the cell body. (20)

Density gradient Solution in which the concentration of a solute increases from top to bottom, generating a range of densities. (9)

Deoxyribonucleic acid (DNA) Macromolecule that serves as the repository of genetic information in all cells; contains deoxyribose in nucleotides. (3)

Deoxyribose Five-carbon sugar present in DNA. (3)

Desmin Protein characteristic of the intermediate filaments found in muscle cells. (18)

Desmosome Region of tight adhesion between adjacent animal cells that gives the tissue structural integrity and allows the cells to function as a unit. (4, 7)

Desmotubule Tubular structure that lies in the central channel of the plasmodesma between two plant cells. (7)

Determination Commitment of a cell to a specific developmental pathway; once determined, the cell will develop in a specific fashion, not only in isolation but also if transplanted to another part of the embryo. (22)

Developmental potential Range of possible fates that a given cell of a developing organism may undertake. (22)

Diacylglycerol (DAG) Glycerol esterified to two fatty acids; formed, along with inositol trisphosphate, upon hydrolysis of phosphatidylinositol-4,5-bisphosphate by phospholipase C; remains membrane-bound after hydrolysis and functions as a second messenger by activating protein kinase C, which then phosphorylates specific serine and threonine groups on a variety of target proteins, depending on the cell type. *See also* Inositol trisphosphate; Phospholipase C; Protein kinase C. (21)

DIF *See* Differentiation inducing factor.

Differential centrifugation Technique used to isolate organelles or molecules that differ in size, shape, or density by subjecting cellular fractions to centrifugation at high speeds and separating particles based on their differential rates of sedimentation in response to the centrifugal force that is generated. (9)

Differential transcription Selective activation of transcription of unique subsets of genes in different cell types to produce distinct populations of cellular mRNAs. (17)

Differentiation Process in embryonic development by which cells become specialized in function and structure. (22)

Differentiation inducing factor (DIF) Lipid-like molecule involved in the differentiation of the stalk and spore cell types in the fruiting body of the slime mold *Dictyostelium.* (22)

Differentiation marker Specific gene product that is expressed only at a characteristic stage during lymphocyte development. (24)

Diffuse cortex Region in the peripheral lymphoid tissue where T lymphocytes reside. (24)

Diffusion Free, unmediated movement of a solute, with direction and rate dictated by the difference in solute concentration between two different regions. (8)

Diffusion constant (k_D) Measure of diffusibility of a specific solute dissolved in a solvent; related to v, the rate of diffusion of a solute across a membrane barrier, by the equation $v = k_D([X]_{outside} - [X]_{inside})$, where $[X]_{outside} - [X]_{inside}$ is the concentration gradient that provides the energy necessary to drive the diffusion process. (8)

Diglyceride Glycerol molecule linked to two fatty acids by ester bonds. (3)

Diploid Containing two sets of chromosomes and therefore two copies of each genetic locus; describes a cell or nucleus, or an organism consisting of such cells. (15)

Directionality Having two ends that are chemically different from each other; used to describe a polymer chain such as a protein, nucleic acid, or carbohydrate. (2, 8)

Disaccharide Carbohydrate consisting of two covalently linked monosaccharide units. (3)

Disulfide bond Covalent bond formed between two sulfur atoms by oxidation of sulfhydryl groups; important in stabilizing the tertiary structure of proteins. (3)

Division phase Period in cell cycle that consists of nuclear division (mitosis) and cytoplasmic division (cytokinesis). (14)

DNA *See* Deoxyribonucleic acid.

DNA cloning The process of generating multiple copies of a specific DNA sequence, either by replication of a recombinant plasmid or bacteriophage within bacterial cells or by use of the polymerase chain reaction. (13, A2)

DNA glycosidase Enzyme that recognizes a deaminated base and removes it from the DNA molecule by cleaving the glycosidic bond between the base and the deoxyribose sugar to which it is attached. (14)

DNA gyrase Enzyme required for unwinding of DNA duplex during DNA replication; a type II topoisomerase that can relax the positive supercoiling of a partially unwound DNA duplex and can also actively introduce negative supercoils that then promote strand separation. (13)

DNA ligase Enzyme that joins DNA fragments together by catalyzing the formation of phosphoester bonds between the 3′ and 5′ ends of adjacent fragments. (14)

DNA polymerase Enzyme responsible for adding successive nucleotides to a growing DNA strand. (14)

DNA replication Highly ordered process in which the DNA duplex unwinds and both strands serve as templates for the synthesis of two identical duplexes. (14)

DNA sequencing Determination of the order of nucleotides in an isolated, purified DNA molecule or fragment, based either on the use of chemical reagents that cleave DNA at specific bases or on the incorporation of chain-terminating nucleotides into a DNA strand. (13)

DNase I hypersensitivity site Specific location near an active gene that shows extreme sensitivity to digestion by the nuclease DNase I; thought to correlate with binding sites for transcriptional factors or other regulatory proteins. (17)

DNase I sensitivity Enhanced accessibility of active chromatin to the nuclease DNase I. (17)

Docking protein Membrane protein that is located in the rough endoplasmic reticulum and that serves as the receptor for the SRP-blocked ribosomal complex, thereby overcoming the translation block. *See also* Signal recognition particle (SRP). (16)

Dolichol phosphate Activated lipid carrier to which monosaccharides are added stepwise to form the core oligosaccharide of glycoproteins. (16)

Dominant Allele that is phenotypically expressed, whether present in the heterozygous or homozygous form. (15)

Dominant control region Region in the globin gene cluster that can influence the structure of surrounding chromatin and hence the level of gene transcription. (17)

Dosage effect Genetic effect whereby the phenotype of a cell is altered by an increased amount of a particular gene product. (23)

Double bond Chemical bond formed between two atoms as a result of the sharing of two pairs of electrons. (2)

Double helix Double-stranded helical structure of DNA. (3)

Double-reciprocal plot Graphic method for analyzing enzyme kinetic data by plotting $1/v$ versus 1/[S]. (6)

Down-regulation *See* Inactivation.

Drug detoxification Alteration in the chemical structure of a drug to enhance its removal from the body; usually involves hydroxylation in the smooth endoplasmic reticulum, thereby increasing the water solubility of the compound. (9)

Dynamic instability model Model for microtubule assembly that presumes two populations of microtubules, one growing in length by continued polymerization at the plus end and the other shrinking in length by depolymerization at the minus end. (18)

Dynein Protein that makes up the arms that reach out between adjacent microtubule doublets in the axoneme of a flagellum or cilium; the dynein arms have ATPase activity and are critical in motility. (19)

Ectoderm Primary germ layer that gives rise to the epidermis and the central nervous system of an animal; outer of the three germ layers of the embryo. (22)

Ectoplasm Thick, gel-like cytoplasm just beneath the plasma membrane of cells that undergo amoeboid movement. (19)

E-face Interior face of the outer monolayer of a membrane as revealed by the technique of freeze-fracturing; also called the B-face. (A)

Effector (a) Small organic molecule that functions as a regulator of a specific allosteric enzyme or other protein to which it binds in a reversible, concentration-dependent manner. (6) (b) Element in a feedback control loop that regulates the level of a specific product by changing the rate at which the product is generated in the system that is so regulated. *See also* Feedback control loop; Sensor. (17, 21)

Effector cell Cell type resulting from the differentiation of a lymphocyte, involved in the immune response; B lymphocyte effector cells are plasma cells that secrete antibodies, and T lymphocyte effector cells are activated T cells that lyse foreign cells, suppress the immune response, or induce an increased immune response. (24)

Effector site (a) Region of an allosteric protein at which the regulatory substance (allosteric effector) binds reversibly, bringing about a change in configuration that affects the functional properties of the protein; also called the modulator site. (6) (b) Constant domain located in the leg of the Y-shaped immunoglobulin molecule. (24)

EGF *See* Epidermal growth factor.

Egg Haploid gamete produced by the female. (15, 22)

Einstein One mole of photons. (2)

Electrical synapse Junction between two nerve cells that allows a nerve impulse to be transmitted by direct electrical connection, without the involvement of chemical neurotransmitters. (20)

Electrical work Use of energy to transport ions across a membrane against a potential gradient. (5)

Electrochemical gradient Transmembrane gradient of an ionic species, with both an electrical component due to charge separation and a concentration component. (11, 20)

Electrochemical proton gradient Transmembrane gradient of protons, with both an electrical component due to charge separation and a chemical component due to a difference in proton concentration (pH) across the membrane. *See also* Proton motive force. (11, 12)

Electronegative Describing an atom that tends to draw electrons toward it. (2)

Electron micrograph Permanent photographic record of a specimen produced by the exposure of a photographic plate to the image-forming electron beam of an electron microscope. (A)

Electron microscope Instrument that uses a beam of electrons to visualize cellular structures and thereby examine cellular architecture; the resolution is much greater than that of the light microscope, allowing detailed ultrastructural examination. (1, A)

Electron transport Process of coenzyme reoxidation under aerobic conditions, involving stepwise transfer of electrons to oxygen by means of a series of electron carriers. (11)

Electron transport chain Series of electron carriers arranged in order of increasing electron affinity to accomplish the transfer of electrons from donor to acceptor with an accompanying release of energy at each transfer step. (11)

Electrophilic substitution Reaction in which an electron-deficient (electropositive) molecule or region of a molecule attacks an electron-rich (electronegative) molecule or region of a molecule, resulting in the formation of a covalent bond. (6)

Embden-Meyerhof pathway *See* Glycolysis.

Embryogenesis Complex process of cell division and progressive cellular specialization and morphogenesis whereby a multicellular organism is formed from a zygote. (22)

Embryonic stem cells Totipotent cells isolated from very early stages of animal embryogenesis and cultured in vitro. When reintroduced into an embryo, they participate in normal development, giving rise to a mosaic organism. (22)

Emerson enhancement effect Achievement of greater photosynthetic activity with red light of two slightly different wavelengths than is possible by summing the activities obtained with the individual wavelengths separately. (12)

Endergonic Pertaining to an energy-requiring reaction characterized by a positive free energy change ($G > 0$). (5)

Endocrine hormone Substance that is synthesized and released by a specific set of cells (usually comprising a gland or tissue) and travels by means of the circulatory system to other sets of cells, where it regulates one or more specific functions. *See also* Paracrine hormone. (21)

Endocytosis Uptake of extracellular materials by infolding of the plasma membrane, followed by budding of a membrane-bound vesicle containing extracellular fluid and materials. (7, 9)

Endoderm Primary germ layer that gives rise to the digestive tract of an animal; inner of the germ layers of the embryo. (22)

Endoplasm Fluid, sol-like cytoplasm within the interior of a cell that undergoes amoeboid movement. (19)

Endoplasmic reticulum (ER) Network of interconnected membranes distributed throughout the cytoplasm and involved in the synthesis, processing, and transport of proteins in a eukaryotic cell. (4, 9)

Endoreplication Repeated replication of chromosomes without separation of the daughter chromatids. (17)

Endosome Vesicle within the cytoplasm of a eukaryotic cell formed by the fusion of small vesicles internalized by receptor-mediated endocytosis; also called a receptosome. (9, 24)

End-product repression Regulation of an anabolic pathway that depends on the ability of an end product to repress the further synthesis of enzymes involved in production of that end product. (17)

Energy Capacity to do work; ability to cause specific changes. (5)

Enhancer element Specific DNA region that contains a binding site(s) for transcription factors and confers transcriptional specificity (of time, place, or level of expression) on the associated structural gene. The element can be located some distance from the upstream promoter region, need not maintain a fixed orientation, and can be upstream or downstream from, or even within, the transcription unit. (17)

Entropy (S) Measure of the randomness or disorder in a system. (5)

Enucleation Removal of the nucleus from a eukaryotic cell. (22)

Enzyme Biological catalyst; protein that acts on one or more specific substrates, converting them to products with different molecular structures. (1, 6)

Enzyme kinetics Quantitative analysis of enzyme reaction rates and the manner in which they are influenced by a variety of factors. (6)

Epidermal growth factor (EGF) Peptide mitogen that stimulates epidermal cells and a variety of other cell types to divide and proliferate. (23)

Epinephrine Hormone secreted by the adrenal medulla in response to stress, capable of eliciting metabolic responses that enable the organism to deal with the stress; also called adrenalin. (21)

Epitope *See* Antigenic determinant.

EPSP *See* Excitatory postsynaptic potential.

Equilibrium constant (K_{eq}) Ratio of product activities to reactant activities for a given chemical reaction when the reaction has reached equilibrium. (5)

Equilibrium density centrifugation Technique used to separate cellular components by subjecting cellular fractions to centrifugation in a solution that increases in density from the top to the bottom of the centrifuge tube; during centrifugation, an organelle or molecule sediments to the density layer equal to its own density, at which point no further net force acts on the particle. (9, 14)

Equilibrium potential Membrane potential that will exactly offset the effect of the concentration gradient for a given ionic species. (20)

ER *See* Endoplasmic reticulum.

Erythroblastoma Tumor derived from erythroblast (red blood cell–forming) cells. (23)

Ester bond formation Condensation reaction in which the hydroxide of a carboxyl group and the hydrogen of an alcohol group of the same or different molecules are removed as water and a bond is formed. (3)

Estrogen Steroid hormone produced by the ovaries that generally results in development or maintenance of female characteristics in an animal. (21)

Euchromatin Diffuse, uncondensed chromatin prominent during interphase. *See also* Heterochromatin. (13)

Eukaryote Category of organisms that includes all plants, animals, fungi, and protozoa and most algae; characterized by the presence of a true membrane-bounded nucleus and other membrane-bounded organelles. (4)

Excitatory postsynaptic potential (EPSP) Change in the potential of the postsynaptic membrane upon binding of a neurotransmitter to its receptor; if the EPSP exceeds a threshold level, it may trigger an action potential. (20)

Exergonic Pertaining to an energy-releasing reaction characterized by a negative free energy change ($\Delta G < 0$). (5)

Exocrine tissue Tissue that secretes its product(s) directly into ducts that then transport the secretion to another part of the body. (21)

Exocytosis Fusion of vesicular membranes with the plasma membrane so that contents of the vesicle can be expelled or secreted to the extracellular environment. (9)

Exon Sequence of nucleotides in an initial RNA transcript that is preserved in the mature, functional RNA molecule. *See also* Intron. (16)

Extensin Protein found in the cell walls of plants and fungi. (4)

External fertilization Process whereby male and female gametes are brought in contact outside the female body, usually in an aqueous medium. (22)

Extracellular digestion Degradation of components outside a cell, usually by lysosomal enzymes that are released from the cell by exocytosis. (9)

F_0 Component of the ATP synthase complex of the mitochondrion that is embedded in the inner mitochondrial membrane and serves as the proton translocator. (11)

F_1 Component of the ATP synthase complex of the mitochondrion; a knoblike sphere that protrudes from the inner mitochondrial membrane and serves as the site of ATP generation; also called coupling factor. (11)

Facilitated transport Passage of substances across an otherwise impermeable membrane, mediated by permeases embedded in the membrane. (8)

F-actin Component of microfilaments consisting of G-actin monomers that have been polymerized into long, linear strands. (18, 19)

Facultative heterochromatin Chromosomal regions that have become permanently but specifically inactivated in a specific cell type. (13)

Facultative organism Organism that can function in either an anaerobic or aerobic mode. (10)

FAD *See* Flavin-adenine dinucleotide.

Fast block to polyspermy Rapid depolarization of the membrane of an animal egg; occurs as the tip of the acrosomal process makes contact with the egg membrane, thereby rendering the membrane insensitive to fusion with further sperm. (22)

Fat Triglyceride containing primarily saturated fatty acids, usually solid or semisolid at room temperature. *See also* Vegetable oil. (3)

Fate map Schematic diagram of an early stage in development (such as an oocyte or blastula) indicating the normal developmental fate of each region of cytoplasm or cells. (22)

Fatty acid Long, unbranched hydrocarbon chain that has a carboxyl group at one end and is therefore amphipathic; usually contains an even number of carbon atoms and may be of varying degrees of unsaturation. (3, 11)

Feedback control loop Regulatory system that controls the level of a specific product by regulating the generation of that product in response to changes in the level of the product; consists minimally of a sensor that detects the level of the product and an effector that regulates generation of the product. *See also* Effector; Sensor. (21)

Feedback inhibition Ability of the end product of a biosynthetic pathway to inhibit the activity of the first enzyme in the pathway, thereby ensuring that the functioning of the pathway is sensitive to the intracellular level of its product. (6, 17)

Fermentation Partial oxidation of carbohydrates by oxygen-independent pathways, resulting often (but not always) in the production of either ethanol and carbon dioxide or lactate. (5, 10)

Fertility factor (F factor) Plasmid present in some bacterial cells that confers maleness and is transferred from the male (donor) cell to the female (recipient) cell in the process of sexual conjugation. (13, 15)

Fertilization Union of two haploid gametes to form a diploid cell, the zygote, capable of developing into a new organism. (15)

F_1 generation First-generation offspring resulting from a cross between two parental strains. (15)

F_2 generation Second-generation offspring resulting from crosses between parental strains. (15)

Fibroblast Mesenchymal cell that is present in connective tissue and secretes large quantities of the major structural proteins of the extracellular matrix. (23)

Fibrous protein Protein with extensive α helix or β pleated sheet structure that confers a highly ordered, repetitive structure. (3)

First law of thermodynamics Law of conservation of energy; principle that energy can be converted from one form to another but cannot be created or destroyed. (5)

Fischer projection Model depicting the chemical structure of a molecule as a chain drawn vertically with the most oxidized atom on top and horizontal projections that are understood to be coming out of the plane of the paper. (3)

Flagellin Protein subunit that makes up the spiral filament of the bacterial flagellum. (19)

Flagellum Membrane-bounded appendage on the surface of a eukaryotic cell composed of a specific arrangement of microtubules and responsible for motility of the cell. *See also* Cilium. (19)

Flavin-adenine dinucleotide (FAD) Coenzyme that accepts two electrons and two protons from an oxidizable organic molecule to generate the reduced form, $FADH_2$; important electron carrier in energy metabolism. (10, 11)

Flavin mononucleotide (FMN) Prosthetic group that is important as a carrier of electrons in energy metabolism. (11)

Flavoprotein Protein that has a tightly bound flavin coenzyme and that serves as a biological electron donor or acceptor; electron carrier in the electron transport chain of the mitochondrion. (11)

Fluidity Property of membranes that allows lateral movement of proteins and lipids in the plane of the membrane; determined by the types of phospholipids present, the cholesterol content, and the temperature. (7)

Fluid mosaic model Model for membrane structure that postulates a lipid bilayer with proteins associated as discrete globular entities that penetrate the bilayer to various extents and are free to move laterally in the membrane. (7)

Fluorescence Property of molecules that absorb ultraviolet light and reemit the energy as visible light. (A)

Fluorescence microscopy Light microscopic technique that produces a bright image on a dark background by focusing ultraviolet rays on the specimen, thereby causing fluorescent compounds in the specimen to emit visible light. (A)

Fluorescent antibody Antibody with fluorescent dye molecules covalently linked to it such that the fluorescence of the antibody can be used to identify and localize antigen molecules. (7)

Fluorescent dye Chemical that emits visible light when ultraviolet light is absorbed. (A)

FMN *See* Flavin mononucleotide.,

Focal length Distance between the midline of a lens and the point at which rays passing through the lens converge to a focus. (A)

Follicle cell One of many cells arranged as an epithelial layer around a mammalian egg for the purpose of nourishing the developing oocyte by the passage of small molecules through gap junctions. (22)

Food cup Structure formed around a food particle by invagination and folding of the plasma membrane of a phagocytic protozoan. (9)

Forming face Convex face of the Golgi complex, usually oriented toward the rough endoplasmic reticulum; also called the cis face. (9)

***N*-Formylmethionine** Methionine that has been formylated on its nitrogen atom; the amino acid with which every polypeptide is initiated in prokaryotes. (16)

Frameshift mutation Insertion (or deletion) of one or more nucleotide base pairs into (or from) a DNA molecule, resulting in a change in the reading frame of the mRNA molecule and leading usually to a garbled message. (16)

Free energy (G) Thermodynamic function that measures the extractable energy content of a molecule; under conditions of constant temperature and pressure, the change in free energy is a measure of the ability of the system to do work; also called Gibbs free energy. (5)

Free energy change ($\Delta G°$) Thermodynamic parameter used to quantify the net free energy liberated or required by a reaction or process; measure of thermodynamic spontaneity. (5)

Freeze-etching Technique consisting of the cleavage of a quick-frozen specimen, as for freeze-fracturing, but with subsequent sublimation of ice from the specimen surface followed by etching to expose small areas of the true cell surface. (A)

Freeze-fracturing Technique consisting of the cleavage of a quick-frozen specimen under vacuum, followed by platinum/carbon shadowing to create a replica of the fractured surface, often the interior of a membrane. (7, A)

Functional group Group of elements covalently bonded to each other that confers characteristic chemical properties upon any molecule to which it is covalently linked. (2)

Fuzzy layer Layer of material exterior to the cell coat of animal cells, made up of collagen and glycosaminoglycans. *See also* Glycocalyx. (7)

G *See* Free energy.

GABA γ-Aminobutyric acid, the primary inhibitory neurotransmitter of the central nervous system. (20)

GABA receptor Binding site for GABA (γ-aminobutyric acid) on the surface of most nerve cell bodies; acts as an inhibitor of central nervous system (CNS) activity because GABA is the primary inhibitory neurotransmitter of the CNS. (20)

G-actin Globular monomeric form of actin, the protein of microfilaments; polymerizes to form F-actin filaments. (18, 19)

Gamete Haploid reproductive cell; for example, ovum or spermatozoon. (15, 22)

Gametogenesis Production of gametes. (15, 22)

Gametophyte Haploid generation in the life cycle of an organism that alternates between haploid and diploid forms; form that produces gametes. (15)

Gap junction Region where the plasma membranes of two adjacent animal cells are brought close together, allowing the exchange of chemicals and the transmission of electrical signals and thereby serving as a means of communication between the cells. *See also* Connexon. (4, 7)

Gastrulation Coordinated series of cell movements during animal embryogenesis, in which a portion of the outer layer invaginates into the interior of the embryo, generating a gastrula with three germ layers. (22)

Gating current Small electrical current arising from the movement of charged groups within the sodium channel protein prior to the flow of sodium ions in the transmission of a nerve impulse. (20)

Gene Nucleotide sequence in DNA to which a specific genetic function can be assigned. *See also* Cistron. (16)

Gene amplification Selective replication of specific DNA sequences within the genome. (17)

Generation time Length of the cell cycle. (14)

Genetic code Relationship between the sequence of bases in a DNA molecule and the order of amino acids in the polypeptide chain for which the DNA molecule codes. (16)

Genetic mutation Heritable change in the nucleotide sequence of a gene. *See also* Mutation. (23)

Genome Genetic content of a cell or virus consisting of the DNA (or RNA) that represents one complete set of its genetic material. (13)

Genomic control Regulatory mechanism in eukaryotes involving either the selective loss or the selective amplification of specific genetic information of the genome. (17)

Genophore Genetic material of viruses and prokaryotes, made up of DNA or RNA that is not integrally associated with proteins. (13)

Genotype Genetic makeup of an organism. (15)

Germ cell Cell specialized for the production of haploid gametes. (22)

Germ line Class of cells that will give rise to gametes. (17)

GFA protein *See* Glial fibrillary acidic protein. (18)

Gibbs free energy *See* Free energy.

Glial fibrillary acidic (GFA) protein Protein of the intermediate filaments present in the glial cells that surround and insulate nerve cells. (18)

Globular protein Protein with local regions of secondary structure but also with less ordered regions, such that the overall shape of the molecule is determined by tertiary interactions between distant parts of the polypeptide chain. (3)

Gluconeogenesis Synthesis of glucose from precursors such as amino acids, glycerol, or lactate; occurs in the liver via a pathway that is essentially the reverse of glycolysis. (10, 19)

Glucose An aldohexose with the formula $C_6H_{12}O_6$ that is widely used as the starting molecule in cellular energy metabolism. (3, 10)

D-Glucose Most stable of the 16 isomers of the aldohexose $C_6H_{12}O_6$; the D designates the specific optical isomer. (10)

α-D-Glucose Cyclic form of D-glucose with the C-1 hydroxyl pointing down in the Haworth model and to the right in the Fischer projection. (10)

β-D-Glucose Cyclic form of D-glucose with the C-1 hydroxyl pointing up in the Haworth model and to the left in the Fischer projection. (10)

Glycerol Three-carbon alcohol with a hydroxyl group on each carbon; serves as the backbone for triglycerides. (3, 11)

Glycerol-3-phosphate dihydroxyacetone phosphate shuttle The mechanism by which electrons and protons from cytoplasmic NADH are passed across the inner mitochondrial membrane to the electron transport chain without the movement of the coenzyme molecule across the membrane. (11)

Glycocalyx Complex of cell coat and fuzzy layer surrounding certain animal cells. (4, 7)

Glycogen Highly branched storage polysaccharide in animal cells consisting of α-D-glucose repeating subunits linked by $\alpha(1 \rightarrow 4)$ bonds and $\alpha(1 \rightarrow 6)$ bonds. (3)

Glycolate pathway Series of reactions occurring in photosynthetic cells by which phosphoglycolate generated in the chloroplast by the oxygenase activity of ribulose-1,5-bisphosphate carboxylase is converted oxidatively to 3-phosphoglycerate in the peroxisome and mitochondrion with the consumption of oxygen and the liberation of carbon dioxide. *See also* Photorespiration. (12)

Glycolipid Lipid molecule with a carbohydrate group linked to its hydrophilic head. (7)

Glycolysis Series of reactions by which glucose or some other monosaccharide is catabolized to pyruvate without the involvement of oxygen, generating two molecules of ATP per molecule of monosaccharide; also called the glycolytic or Embden-Meyerhof pathway. (5, 10)

Glycoprotein Polypeptide with one or more carbohydrate side chains attached. (7)

Glycosaminoglycan Carbohydrate with a repeating disaccharide unit; component of fuzzy layer of animal cells. (7)

Glycosidic bond Bond linking a sugar to another molecule, which may be another sugar molecule. (3)

Glycosylation Addition of sugar residues to a polypeptide chain, usually beginning in the lumen of the endoplasmic reticulum and completed in the Golgi complex. (16)

Glyoxysome Membrane-bounded organelle that contains enzymes responsible for the conversion of stored fat to carbohydrate in germinating seeds. (4)

Goldman equation Variation of the Nernst equation, used to calculate the resting potential of a membrane by summing the effects of all relevant ions, each weighted for its relative permeability. (20)

Golgi complex Stacks of flattened vesicles in eukaryotic cells that are important in the processing and packaging of secretory proteins and in the synthesis of complex polysaccharides; also called the Golgi apparatus. (9)

Gonad Reproductive organ in which gametes are generated; for example, ovary in the female and testis in the male. (15)

G-0 phase Resting phase in the cell cycle; normally an alternative to G-1. (14)

G-1 phase Phase in eukaryotic cell cycle between the end of the previous division and the onset of chromosomal DNA synthesis. (14)

G-2 phase Phase in eukaryotic cell cycle between the completion of chromosome replication and the onset of cell division. (14)

G protein Any of several GTP-binding regulatory proteins that are located in the plasma membrane and mediate a variety of signal transduction systems, usually by activating a specific membrane-bound enzyme. (21)

Graft rejection Response of the immune system to tissue that is recognized as foreign and incompatible. (24)

Granum Single stack of thylakoid disks within a chloroplast; part of the lamellar system. (4, 12)

Gray crescent Region on the surface of the zygote at which invagination occurs at the onset of gastrulation in amphibian embryogenesis. (22)

Growth factor Any of a number of highly specific serum proteins that stimulate cell division in particular types of mammalian cells; examples include platelet-derived growth fctor (PDGF), epidermal growth factor (EGF), nerve growth factor (NGF), and interleukin-2 (IL-2). *See also* Transmembrane protein kinase. (21)

Growth phase Phase in eukaryotic cell cycle during which the cell doubles its mass and duplicates all of its contents in preparation for the next round of division. (14)

Guanine (G) Nitrogen-containing aromatic base, chemically designated as a purine, which serves as an informational monomeric unit when present in nucleic acids with other bases in a specific sequence. (3, 16)

Gyrase Enzyme that promotes unwinding of the DNA helix during the process of replication by introducing negative supercoils ahead of the replication fork. (14)

Half-life Length of time required for the random degradation of 50% of the molecules on hand at a given time. (16)

Haploid Containing a single set of chromosomes and therefore only one copy of each gene; describes a cell or nucleus, or an organism consisting of such cells. (15)

Hatch-Slack pathway Series of reactions in C_4 plants in which carbon dioxide is fixed in the mesophyll cells and transported as a four-carbon compound to the bundle sheath cells, where subsequent decarboxylation results in a higher concentration of carbon dioxide and therefore a higher rate of carbon fixation by ribulose-1,5-bisphosphate carboxylase. (12)

Haworth projection Model that depicts the chemical structure of a molecule and suggests the spatial relationship of different parts of the molecule. (3)

H chain *See* Heavy chain, (b).

Heat Transfer of energy as a result of a temperature difference. (5)

Heavy chain (a) Polypeptide component of the myosin molecule. (19) (b) Polypeptide component of the immunoglobulin molecule, with a molecular weight of 55,000; commonly called the H chain. (24)

Heavy meromyosin (HMM) Fragment of myosin molecule that contains the ATPase heads of the molecule. *See also* Light meromyosin. (19)

Helicase Enzyme that utilizes the energy of ATP to unwind the parental helix as it moves down the DNA duplex in the $3' \rightarrow 5'$ direction, thereby exposing the single strands to replication enzymes; also called unwindase. (14)

Helix Natural spiral conformation of many regular biological polymers. (3)

Helix-destabilizing protein *See* Single-strand binding protein.

Helper T (T_h) cell T lymphocyte that promotes or augments the immune response of other lymphocytes. (24)

Hematopoietic stem cell Actively dividing cell that gives rise to all types of blood cells. (24)

Hematopoietic tissue Tissue containing the stem cells that give rise to blood cells; bone marrow in adult organisms, liver in some embryos. (24)

Heme Prosthetic group consisting of an iron-containing porphyrin ring capable of being reversibly oxidized and reduced. (11)

Hemicellulose Polymeric substances deposited with cellulose in the cell walls of plants and fungi to provide added strength. (11)

Hemoglobin Protein responsible for the transport of oxygen and carbon dioxide in the bloodstream; consists of two α chains and two β chains, each with an iron-containing heme group. (2)

Heptose A sugar that contains seven carbon atoms. (3)

Hermaphrodite An organism that has both male and female gonads and produces both sperm and oocytes. (22)

Herpesvirus Large DNA tumor virus responsible for a variety of diseases. (23)

Heterochromatin Chromosomal regions that remain condensed during interphase. *See also* Euchromatin. (13)

Heterochronic mutation Mutation that disrupts a particular cell lineage, causing particular cell divisions to be skipped or repeated, thereby accelerating or delaying terminal differentiation. (22)

Heterogeneous nuclear RNA (hnRNA) Product of transcription in eukaryotic nuclei, believed to be the precursor of messenger RNA. (16)

Heterophagic lysosome Secondary lysosome that contains hydrolytic enzymes involved in the digestion of substances of extracellular origin. (9)

Heterozygous Having two different alleles for a given trait. (15)

Hexose A sugar that contains six carbon atoms. (3)

Hfr (high frequency of recombination) bacterium F^+ bacterial cell in which the F factor has become integrated into the bacterial genophore so that the cell is now capable of transferring genomic DNA during conjugation. (15)

Hierarchical assembly Synthesis of biological structures from simple starting molecules to progressively more complex structures, usually by self-assembly. (2)

High-mobility group (HMG) nonhistones Acidic proteins that are associated with the chromatin of eukaryotic cells and show high mobility when subjected to electrophoresis; thought to include proteins that regulate gene expression. (17)

High-voltage electron microscope (HVEM) An electron microscope that uses accelerating voltages up to several thousand kilovolts, thereby allowing the examination of thicker samples than is possible with a conventional electron microscope. (1, 18, A)

Histocompatibility antigen *See* Transplantation antigen.

Histocompatibility-2 antigen (H-2 antigen) Cell-surface antigen in the mouse that is recognized by the immune system. (24)

Histocompatibility-2 complex (H-2 complex) Complex of genes that encode the mouse H-2 antigens. (24)

Histone Class of basic proteins found in eukaryotic chromosomes; an octamer of histones forms the core of nucleosomes. (13)

HLA *See* Human leukocyte–associated antigen.

H-L pair Dimeric component of an immunoglobulin molecule consisting of a heavy chain and a light chain linked by a single disulfide bond and by noncovalent associations. (24)

HMG *See* High-mobility group nonhistones.

HMM *See* Heavy meromyosin.

hnRNA *See* Heterogeneous nuclear RNA.

Holoenzyme Complete form of an enzyme. (14, 16)

Homeo box Highly conserved DNA sequence that encodes the site-specific domain of a DNA-binding protein. First found in the fruit fly Drosophila, proteins containing such domains are transcription factors and are frequently important regulators of gene expression during development. (22)

Homeotic gene *See* Segment identity gene.

Homeotic mutation Mutation that alters the developmental fate of a cell, group of cells, or entire body part, transforming it into a cell, group of cells, or body part normally found elsewhere in the organism and thereby causing a duplication of that structure. (22)

Homopolymer Polymer containing only one type of subunit. (16)

Homozygous Having two identical alleles for a given trait. (15)

Hormone Chemical that is synthesized in one organ, secreted into the blood, and able to cause a physiological change in cells or tissues of another organ. (21)

Hormone response element (HRE) Enhancer element that contains a binding site for a steroid hormone–receptor complex and functions as the cis-acting control region that confers hormone inducibility on a specific gene. (17)

***Hox* gene** Mouse gene that contains a homeo box. (22)

Human leukocyte–associated antigen (HLA) Human cell-surface antigen that is recognized by the immune system. (24)

Human leukocyte–associated antigen complex (HLA complex) Complex of genes that encode the HLA antigens. (24)

Humoral antibody Soluble antigen-recognizing protein that circulates in the bloodstream. (24)

Humoral immune response Defense process in which antigen-recognizing proteins circulating in the bloodstream bind to specific antigens, promoting their destruction and removal. (24)

HVEM *See* High-voltage electron microscope.

Hybrid Product of the cross of two genetically different parents. (15)

Hybridoma Hybrid cell line that produces specific monoclonal antibodies, generated by fusion of an antibody-producing B lymphocyte from the spleen with a myeloma cell. (24)

Hydrocarbon An organic molecule consisting only of carbon and hydrogen atoms; not generally compatible with living cells. (2)

Hydrogenation Addition of hydrogen ions (protons) to an organic molecule; reduction. (10)

Hydrogen bond Weak attractive interaction between an electronegative atom and a hydrogen atom that is covalently linked to a second electronegative atom. (2)

Hydrolase Enzyme capable of cleaving specific biological molecules by the process of hydrolysis. (4, 9)

Hydrolysis Splitting of one molecule into two or more smaller molecules by the addition of water. (3, 4)

Hydrophilic Describing molecules or regions of molecules that readily associate with water because of a preponderance of polar groups; "water-loving." (2)

Hydrophobic Describing molecules or regions of molecules that are poorly soluble in water because of a preponderance of nonpolar groups; "water-hating." (2)

Hydrophobic interactions Preferential association of hydrophobic molecules with one another in an aqueous environment, driven mainly by the tendency of water molecules to exclude molecules that disrupt hydrogen bonds. (2)

Hydroxylation reaction Chemical reaction in which molecular oxygen (O_2) is used to generate a hydroxyl group on an organic compound. (9)

Hypervariable region Small segment of the variable regions of the light and heavy chains of the immunoglobulin molecule that is responsible for most of the variability between the variable regions of different antibodies. (24)

Hypothalamus Part of the brain that is located just above the pituitary gland and regulates many basic body functions. *See also* Anterior pituitary. (21)

I band (isotropic band) Region of the sarcomere of a muscle fibril consisting of thin (actin-containing) filaments and seen as a light band in the electron microscope. (19)

I-cell disease Human genetic disorder in which cells synthesize all the normal lysosomal hydrolases but fail to incorporate them into their lysosomes. (9)

Idiotype Unique epitope present on a monoclonal immunoglobulin; formed by the variable regions of the immunoglobulin heavy and light chains and therefore distinctive for each monoclonal immunoglobulin. (24)

***i* gene** Regulator gene in the *lac* operon of *Escherichia coli*; nucleotide sequence that codes for the repressor of the *lac* operon. (17)

Immune Resistant to a pathogen as a result of prior exposure. (24)

Immune response Process by which the immune system recognizes and eliminates a specific virus, bacterium, fungus, or other foreign substance. (24)

Immune system Defense mechanism in vertebrates that allows foreign substances to be recognized and eliminated. (24)

Immunity Resistance to a particular pathogen as a result of prior exposure to that pathogen. (24)

Immunofluorescence microscopy Technique in which antibodies are labeled with a fluorescent dye to enable them to be identified and localized microscopically based on their fluorescence. (18, A)

Immunoglobulin (Ig) Protein molecule that binds to specific antigenic determinants and is a component of the humoral immune system; antibody. (24)

Immunology Study of the process by which resistance to foreign substances is developed in an organism. (24)

i^- mutant Mutant form of the *i* gene in the *lac* operon of *Escherichia coli;* gives rise to a mutant repressor that cannot recognize or bind the operator, such that the enzymes of the *lac* operon are synthesized constitutively. (17)

Inactivation Turning off of a specific process or pathway, usually as a result of a conformational change or a covalent modification of an enzyme; also called down-regulation. (21)

Indifferent gonad A primordial gonad with the capacity to differentiate into the mature gonad of either sex, depending on the influence of other tissues of the developing organism. (22)

Induced-fit model Model for enzyme function that postulates the active site of an enzyme to be relatively specific for the substrate before it binds, but even more so thereafter because of a conformational change in the enzyme induced by the substrate. (6)

Inducible enzyme Enzyme for which synthesis is regulated by the presence or absence of its substrate. (17)

Induction Developmental process in which one cell or tissue causes another cell or tissue to adopt a specific developmental fate that it would not have adopted in the absence of the inducer. (22)

Inductive interaction Regulatory mechanism in development in which the cells of a tissue can be influenced by surrounding cells or tissues to differentiate in a specific manner. (22)

Informational macromolecule Polymer of nonidentical subunits ordered in a nonrandom sequence that carries information important to the function or utilization of the macromolecule; nucleic acids and proteins are informational macromolecules. (2)

Inhibition Decreasing the potential of a molecule for biological function, either by a change in conformation or by chemical modification of one of its functional groups. (6)

Inhibitory postsynaptic potential (IPSP) Change in the potential of the postsynaptic membrane so that the membrane is hyperpolarized, thereby reducing the amplitude of an excitatory postsynaptic potential and possibly preventing the firing of an action potential. (20)

Inhibitory protein Protein that associates with the intracellular receptor for a steroid hormone, thereby masking the DNA binding site of the receptor, but is released from the receptor upon binding of the hormone to the receptor. *See also* Intracellular receptor; Steroid hormone. (21)

Initiation codon AUG codon recognized by the tRNA molecule that carries *N*-formylmethionine. (16)

Initiation factor (IF) Molecule involved in the formation of the initiation complex required for translation in prokaryotes; initiation factor 1 (IF1) is required for the binding of the 50S subunit to the 30S initiation complex, initiation factor 2 (IF2) mediates the binding of f-Met $tRNA^{f\text{-}Met}$, and initiation factor 3 (IF3) mediates the formation of a complex between the mRNA and a 30S ribosomal subunit. (16)

Inner membrane The inner of the two lipid bilayers that surround a mitochondrion, chloroplast, or nucleus. (4, 11, 12, 13)

Inositol trisphosphate ($InsP_3$) Triply phosphorylated inositol molecule formed from phosphatidylinositol by successive addition of two phosphate groups, followed by hydrolysis by phospholipase C; functions as a second messenger by triggering the release of calcium ions from intracellular storage sites. *See also* Diacylglycerol; Phospholipase C. (21)

In situ RNA hybridization analysis Use of labeled, in vitro synthesized antisense RNA to detect the distribution of a specific mRNA within an organism. (22)

$InsP_3$ *See* Inositol trisphosphate.

Insulin Peptide hormone synthesized by the β cells of the pancreas; stimulates carbohydrate storage and protein synthesis. (21)

Integral membrane protein Hydrophobic protein localized within the interior of the membrane, but usually having hydrophilic regions that protrude from the membrane on one or both sides. (7)

Interdoublet link Link between adjacent doublets in the axoneme of a cilium or flagellum, believed to limit the extent of doublet movement with respect to each other as the axoneme bends. (19)

Interleukin *See* Lymphokine.

Interleukin-1 (IL-1) Lymphokine secreted by macrophages that stimulates certain T_h cells to secrete interleukin-2. (24)

Interleukin-2 (IL-2) Lymphokine secreted by certain T_h lymphocytes that binds to cell-surface receptors of activated T lymphocytes and stimulates them to proliferate; also called T cell growth factor. (24)

Intermediate filaments Protein filaments that play a structural or tension-bearing role in the cytoskeleton of eukaryotic cells, with a diameter of 10–12 nm, which is intermediate between the diameters of actin microfilaments and microtubules. (4, 18)

Intermediate filament typing Technique of identifying cell type by determining the kind of intermediate filaments present in the cell; useful in research to trace cell lineages during development and in clinical pathology to classify tumors according to the tissue of origin. (18)

Intermembrane space Region of a mitochondrion or chloroplast between the inner and outer membranes. (12)

Internal energy (E) Total energy stored within a system; cannot be measured directly, but the change in internal energy, ΔE, is measurable. (5)

Internal fertilization Process whereby male and female gametes are brought in contact within the reproductive tract of the female. (22)

Interneuron Nerve cell that processes and integrates the information received from other nerve cells. (20)

Interphase Longest phase of mitosis, in which chromosomes are dispersed and not generally visible by microscopy. (14)

Intervening sequence *See* Intron.

Intracellular receptor Receptor protein located within the cytoplasm of a cell. *See also* Chemical receptor; Membrane receptor; Receptor protein; Steroid hormone. (21)

Intracellular transport Movement of substances across membranes of organelles inside the cell. (8)

Intrathylakoid space Space within the membranes of the thylakoids and the stroma lamellae. (12)

Intron Sequence of nucleotides in an RNA molecule that is part of the primary transcript but not the functional RNA molecule; also called intervening sequence. *See also* Exon. (16)

Invasion Intrusion of cells into a space within the body that is normally occupied by cells of another tissue. (22)

In vitro Term used to designate observations or experiments on isolated cells, tissues, or cell-free extracts; Latin: "in glass." (14)

In vivo Term used to designate observations or experiments in which the organism remains intact; Latin: "in life." (14)

Ion carrier Ionophore that functions by surrounding an ion with a hydrophobic coat such that the ion can diffuse through the hydrophobic interior of the membrane. (8)

Ion channel Integral protein in a membrane that permits the passage of specific ions through the membrane; generally regulated by either changes in membrane potential (voltage-gated channels) or binding of a specific ligand (ligand-gated channels). (8, 22)

Ionophore Antibiotic that increases the permeability of membranes for specific ions either by surrounding the ion with a hydrophobic coat or by providing a hydrophilic channel through the membrane. *See also* Channel former; Ion carrier. (8)

IPSP *See* Inhibitory postsynaptic potential.

Iron-protoporphyrin IX Form of heme found in hemoglobin and in cytochromes b, c, and c_1. (11)

Iron response element (IRE) A mRNA nucleotide sequence involved in mediating iron-dependent translation of ferritin mRNA and iron-dependent destabilization of transferrin receptor mRNA. (17)

Iron-sulfur protein Protein that contains iron and sulfur atoms complexed with four cysteine groups of the protein and serves as an electron carrier in the electron transport chain. (11)

Irreversible inhibitor Molecule that binds to an enzyme permanently, causing an irrevocable loss of catalytic activity. (6)

Isothermal Having a constant temperature. (5)

J chain *See* Joining chain.

Jelly coat Gelatinous matrix of polysaccharides and glycoproteins that surrounds the egg of amphibian and other animal species. (22)

J gene *See* Joining gene.

Joining chain (J chain) Accessory polypeptide chain present in the dimeric form of IgA molecules and the pentameric form of IgM molecules. (24)

Joining gene (J gene) DNA segment that joins the V gene to the C gene in a differentiated B lymphocyte. (24)

kcal *See* Kilocalorie.

Keratin filament Intermediate filament found in the epithelial cells that cover the body's surfaces and line its cavities. (18)

Keratins Proteins present in the intermediate filaments found in the epithelial cells that cover the body's surfaces and line its cavities. (18)

Ketosugar A sugar with a ketone functional group. (3)

Killer T cell *See* Cytotoxic T cell.

Kilocalorie (kcal) Unit of energy; amount of energy needed to raise the temperature of 1 kilogram of water 1°C. (2, 5)

Kinase Any of a variety of enzymes that phosphorylate their substrates. (19, 21)

Kinesin Enzyme that uses the energy of ATP hydrolysis to drive microtubule-based movements, such as axonal transport. *See also* Motor molecule. (18, 19)

Kinetochore Region of chromosome to which spindle microtubules attach during division. (13, 14)

Kinetochore fiber *See* Chromosomal fiber.

Krebs cycle *See* Tricarboxylic acid cycle.

Lactate fermentation Anaerobic catabolism of carbohydrates with lactate as the end product. (10)

Lactose Disaccharide consisting of glucose linked to galactose by an α glycosidic bond; milk sugar. (3)

Lactose (lac) operon Cluster of genes in *Escherichia coli*, consisting of structural genes, a regulatory gene, a promoter, and an operator, which functions coordinately in the induction of the enzymes involved in the metabolism of lactose. (17)

Lagging strand Strand of DNA that grows in the $3' \rightarrow 5'$ direction by discontinuous synthesis of short fragments in the $5' \rightarrow 3'$ direction, followed by ligation of adjacent fragments. *See also* Leading strand. (14)

Lampbrush chromosome Chromosome in the nucleus of an oocyte with extended loops of uncoiled DNA that are sites of active RNA synthesis. (17)

Large ribosomal subunit Component of a ribosome with a sedimentation coefficient of 60S in eukaryotes and 50S in prokaryotes; associates with a small ribosomal subunit to form a functional ribosome. (4)

Laser ablation Use of a highly focused laser beam to destroy specific cells in a developing organism; widely used in cell lineage studies with *Caenorhabditis*. (22)

Lateral diffusion Movement of lipid or protein molecules in the plane of a membrane. (7)

Laws of inheritance Conclusions reached by Gregor Mendel as a result of his genetic investigations with pea plants; Mendel's laws take into account the existence of discrete genetic factors (genes), the presence of pairs of determinants (alleles), and the phenomena of independent segregation and independent assortment of genetic factors. (15)

Laws of thermodynamics Statements of the principles governing energy flow. (15)

L chain *See* Light chain, (b).

Leader peptide Peptide chain produced by the leader sequence of a messenger RNA that codes for the enzymes of an amino acid biosynthetic pathway; the leader peptide contains multiple residues of the amino acid synthesized by the pathway to which the enzymes encoded by the operon belong. (17)

Leader sequence Nucleotide sequence at the 5′ end of polycistronic mRNA that is important in regulation of the transcription of the structural genes in the operon. *See* Leader peptide. (16, 17)

Leading strand Strand of DNA that grows by continuous addition of deoxyribonucleotides in the $5' \rightarrow 3'$ direction. *See also* Lagging strand. (14)

Leaf peroxisome Membrane-bounded organelle that contains enzymes involved in photorespiration. (4, 12)

Lecithin Phosphoglyceride with a choline on the phosphate group; most abundant phospholipid in animal tissue; also called phosphatidyl choline. (3)

Lectin Sugar-binding protein that is isolated from plant seeds and is used in studies of glycoproteins. (7)

Leukemia Tumor derived from leukocytes (white blood cells). (23)

Ligand Substance that binds to a specific receptor, thereby initiating the particular event or series of events for which that receptor is responsible. *See also* Chemical receptor; Membrane receptor; Receptor protein. (9, 21)

Light chain (a) Polypeptide component of the head of the myosin molecule. (19) (b) Polypeptide component of the immunoglobulin molecule, with a molecular weight of 23,000. (24)

Light-dependent reactions The photochemical reactions of photosynthesis responsible for generating ATP and NADPH; initiated whenever light of appropriate wavelengths is absorbed by chlorophyll molecules. (12)

Light-independent reactions *See* Calvin cycle. (12)

Light meromyosin (LMM) Fragment of the myosin molecule that does not contain the ATPase heads. *See also* Heavy meromyosin. (19)

Light microscope Instrument consisting of a source of visible light and a system of glass lenses that allows an enlarged image of a specimen to be viewed. (A)

Lignin Polymeric substance deposited with cellulose in the cell walls of plants and fungi to provide added strength. (7)

Limb bud Developing limb of an embryo; first appears as a small bulge of cells on the surface of the developing embryo, with growth of the limb bud caused by the progress zone of mesodermal cells underlying the apical ectodermal ridge at the tip of the bud. (22)

Lineweaver-Burk equation Linear equation obtained by the inversion of the Michaelis-Menten equation, useful in determining parameters V_{max} and K_m and in analysis of enzyme inhibition. (6)

Linkage group Group of genes that are transmitted, inherited, and assorted together. (15)

Linker DNA *See* Spacer DNA.

Lipid bilayer Unit of membrane structure, consisting of two layers of phospholipids arranged so that their hydrophobic tails face toward each other and the polar region of each faces the aqueous environment on one side or the other of the bilayer. (4, 7)

Lipid diffusion Translational and rotational motion of lipid molecules within the lipid bilayer of a membrane. (7)

Lipid monolayer Single layer of lipid molecules oriented so that the hydrophilic heads are in a polar environment and the hydrophobic tails are in a nonpolar environment. (7)

Lipids Large and chemically diverse class of organic compounds that are poorly soluble or insoluble in water but soluble in organic solvents. (3)

Lipophilicity The physical property of a molecule that makes the molecule readily soluble in a nonpolar environment, such as the lipid bilayer of a membrane. (7)

Liposome Closed vesicular membrane composed of one or more lipid bilayers; commonly used as a model for biological membranes. (8)

LMM *See* Light meromyosin.

Lumen Internal space or cavity formed by a membrane or cluster of cells. (4, 16, 18)

Lymphocyte Type of white blood cell that functions in the immune system; *see also* B lymphocyte, T lymphocyte. (24)

Lymphoid tissue Network of tissue distributed throughout the body that contains lymphocytes and that mediates the immune response by recognizing, binding, and eliminating antigens as they enter the body; includes bone marrow, thymus, spleen, tonsils, and lymph nodes. (24)

Lymphokine Substance secreted by T_h lymphocytes that activates white blood cells, including other lymphocytes; also called interleukin. (23)

Lymphoma Tumor derived from lymphoblast (lymphocyte-forming) cells. (23)

Lysosomal storage disease Disease resulting from a deficiency of one or more specific lysosomal enzymes and characterized by the undesirable accumulation and storage of excessive amounts of specific substances that would normally be degraded by the deficient enzymes. (9)

Lysosome Membrane-bounded organelle that segregates hydrolytic enzymes from the rest of the cellular contents. (4, 9)

Lytic growth Mode of bacteriophage growth that results in the production of many progeny phages and the death and lysis of the host cell; caused by infection with a virulent bacteriophage. (13)

Macromolecule Polymer consisting of small repeating monomer units, with molecular weights ranging from a few thousand to hundreds of millions. (2)

Macrophage Large phagocytic white blood cell that arises when a circulating monocyte arrives at a site of inflammation and undergoes differentiation. *See also* Monocyte. (9, 23)

Macropinocytosis Uptake mechanism in which the plasma membrane invaginates around the substance to be ingested, forming a pinocytotic vesicle that has a diameter in the range of 1–2 μm. (9)

Major groove Larger of the two grooves in the double-stranded helical structure of DNA, resulting from the way the two strands are twisted around each other. (13)

Major histocompatibility complex (MHC) Complex of genes that encode the cell-surface antigens that identify a cell to the immune system. (24)

Malignant tumor Tumor that proliferates indefinitely and destroys nearby tissue, eventually threatening the life of the organism; a cancer. (23)

Maltose Disaccharide consisting of two glucose molecules linked together by an α glycosidic bond. (3)

MAPs *See* Microtubule-associated proteins.

Map unit Unit used to measure the distance between loci on a chromosome based on crossover frequencies; 1% frequency of crossing over equals one map unit. (15)

Marker enzyme Enzyme known to be localized to a specific intracellular structure, such that the activity of the enzyme can be used to identify that structure. (7)

Maternal effect gene Gene encoding a product (RNA or protein) that is a normal and necessary constituent of the oocyte. If the gene product is needed only for oogenesis, the *mutant/mutant* female will develop normally, but her eggs will not develop normally upon fertilization. (22)

Mating bridge Structure necessary for the transfer of DNA from a male bacterial cell to a female cell. (15)

Mating type switch Change in cell mating type from *a* to α or vice versa in the progeny of a haploid yeast cell; involves DNA rearrangements at the *MAT* locus that determines mating type. (22)

Matrix Unstructured semifluid substance that fills the space inside a cell or organelle. (4, 11)

Maturation promoting factor (MPF) Cell cycle regulator in the amphibian *Xenopus*; critical for control of egg maturation and cleavage. *See also* *cdc2*/MPF protein kinase. (14, 22)

Maturing face Concave face of the Golgi complex, usually oriented toward the cell surface; also called the trans face. (9)

Maximum velocity (V_{max}) Upper limiting reaction rate approached by an enzyme-catalyzed reaction as the substrate concentration approaches infinity. (6)

Mechanical work Use of energy to bring about a physical change in the position or orientation of a cell or some part of it. (5)

Meiosis Process in which two successive nuclear divisions occur with only one duplication of chromosomes, producing four haploid daughter nuclei; also called reduction division. (15)

Melting (a) Increase in fluidity of a membrane with increased temperature. (7) (b) Thermal denaturation of double-stranded DNA into the two component strands. (13)

Melting temperature (T_m) Temperature at which one-half of the absorbance change has been achieved during thermal denaturation of double-stranded DNA. (13)

Membrane Permeability barrier surrounding and delineating cells and organelles; consists of a bilayer of phospholipids with associated proteins. (2, 7)

Membrane asymmetry Property of a membrane based on differences between the two monolayers and the proteins associated with each. (7)

Membrane permeability Relative ability of a specific solute molecule to traverse a membrane; membranes are highly permeable to lipophilic molecules (and, paradoxically, to water), but are relatively impermeable to ions and to polar molecules. (8)

Membrane potential The potential or charge gradient that exists across a membrane; usually, the inside of a cell is negatively charged with respect to the outside. (8)

Membrane receptor Integral membrane protein that has a binding site on the membrane surface for a specific ligand and initiates or mediates a particular intracellular event or series of events in response to ligand binding. *See also* Chemical receptor; Intracellular receptor; Ligand; Receptor protein. (21)

Membrane turnover Continual removal and replacement of the lipid and protein components of a membrane. (9)

Memory cell Long-living differentiated lymphocyte arising from B lymphocytes and T lymphocytes in response to stimulation by an antigen; memory cells are able to produce effector cells and more memory cells in response to further stimulation by an antigen that has been encountered previously, thereby conferring long-lasting immunity on the organism. (24)

Mendelian genetics Understanding of the genetic consequences of chromosome behavior during meiosis on the basis of the laws of inheritance determined by Gregor Mendel from his genetic experiments with pea plants. (15)

Mesenchyme Loose, space-filling, three-dimensional meshwork of cells that originates from the mesoderm of an embryo and gives rise to the muscles, skeleton, and connective tissue of the animal body by interactions with the epithelial ectoderm. (22)

Mesoderm Primary germ layer that gives rise to skeletal elements, muscle, connective tissue, gonads, and blood; middle of the three germ layers of the embryo. (22)

Mesophyll cell Outer cell in a leaf of a C_4 plant; site of carbon fixation by the Hatch-Slack pathway. (12)

Messenger RNA (mRNA) RNA molecule that provides the information required to specify the amino acid sequence of one or more polypeptides. (3, 16)

Metabolic pathway Series of cellular enzymatic reactions that convert one molecule to another via a series of intermediates. (10)

Metabolism All chemical processes occurring within a cell. (10)

Metacentric chromosome Chromosome with its centromere near the center such that the two arms are equal. (14)

Metamorphosis Series of dramatic morphological and physiological changes that transform an organism from one major pattern of body organization to another. (22)

Metaphase Second phase of mitosis, in which chromosomes are maximally condensed and become aligned at the metaphase plate in preparation for nuclear and cellular division. (14)

Metaphase plate Plane between the two poles of the cell, along which chromosomes align at metaphase. (14)

Metastable Describing potential reactants that are thermodynamically unstable but lack adequate kinetic energy to exceed the activation energy threshold required for reaction. (6)

Metastasis Spread of tumor cells to a part or parts of the body distant from the parent tumor; usually occurs via the circulatory system, but may also involve other body fluids. (23)

Michaelis constant (K_m) Specific substrate concentration at which an enzyme-catalyzed reaction is proceeding at one-half of its maximum velocity. (6)

Michaelis-Menten equation Relationship between velocity and substrate concentration that is widely used in enzyme kinetics: $V = V_{max}[S]/K_m + [S]$. (6)

Microbody Morphological term for a peroxisome. (9)

Microfibril Aggregate of long, linear cellulose rods that serves as a structural component of plant and fungal cell walls. (3)

Microfilament Polymer of actin that is an integral part of the cytoskeleton and that contributes to the support, shape, and mobility of a eukaryotic cell; microfilaments consist of double strands of polymerized G-actin. (4, 18)

Microfilament-based movement Type of motility based on microfilaments composed of actin and thicker filaments of myosin; includes muscle contraction, amoeboid movement, cytoplasmic streaming, and cell division. (19)

Microfilament cross-linking protein Protein that binds microfilaments together into a stable network with gel-like properties. (19)

Microfilament-severing protein Protein that breaks actin microfilaments, thereby disrupting microfilament networks and causing the gel-to-sol transition required for the process of cytoplasmic streaming. (18)

Micromere Small cell located at the far end of the vegetal hemisphere in a developing sea urchin embryo. (23)

Micrometer (μm) Unit of measure; 1 μm = 10^{-6} m; also called a micron (μ). (1)

Micropinocytosis Uptake mechanism in which the plasma membrane invaginates around the substance to be ingested, forming a pinocytotic vesicle that has a diameter of about 0.1 μm. (9)

Microsome Vesicle formed by fragments of endoplasmic reticulum when tissue is homogenized. (9)

Microtome Instrument used to slice an embedded biological specimen into thin sections for light microscopy. (1, A)

Microtubule Polymer of the protein tubulin that is an integral part of the cytoskeleton and that contributes to the support, shape, and motility of a eukaryotic cell; also found in the cilia and flagella of many eukaryotic cell types. (4, 18)

Microtubule-associated proteins (MAPs) Accessory proteins that bind along the entire length of a microtubule; MAPs appear to enhance the polymerization of tubulin subunits and may also be involved in establishing cross-links between microtubules and cellular structures. (18)

Microtubule-based movement Type of motility based on microtubules; includes motility due to cilia, flagella, and sperm tails, as well as chromosomal movements mediated by spindle fibers. (19)

Microtubule-organizing center (MTOC) Structure in the cytoplasm of eukaryotic cells from which microtubules radiate out toward the periphery of the cell; centrosomes and kinetochores are examples. (18)

Microvillus Fingerlike projection that increases effective membrane surface area; important in cells that have an absorption function, such as those that line the intestine. (18)

Minor groove Smaller of the two grooves in the double-stranded helical structures of DNA, resulting from the way the two strands are twisted around each other. (13)

Mitochondrion Large organelle that is the site of aerobic respiration and hence of ATP generation. (4, 11)

Mitogen Soluble substance, usually a protein, that induces mitosis in a resting population of cells, thereby causing the cells to resume proliferation. (23)

Mitogen receptor Transmembrane protein that binds a specific mitogen at the surface of the cell, thereby initiating mitosis and cell division. (23)

Mitosis Process in which two genetically identical daughter nuclei are produced from one nucleus as the chromosomes of the parent cell replicate and segregate into separate nuclei; usually followed by cell division. (14)

Mitotic index Fraction of cells in a culture that are in any stage of mitosis at a certain point in time; used to estimate the relative length of the mitosis phase of the cell cycle. (14)

Mitotic spindle Microtubular structure responsible for separating chromosomes during mitosis. (14)

Modulator site *See* Effector site.

Monocistronic mRNA Messenger RNA molecule that yields one polypeptide product upon translation. (16)

Monoclonal antibody Homogeneous immunoglobulin derived from a single clone of cells. (24)

Monoclonal tumor Tumor derived from a single progenitor cell that has undergone a permanent, heritable change. (23)

Monocyte Circulating white blood cell that can differentiate into a macrophage. *See also* Macrophage. (23)

Monoglyceride Glycerol molecule with a single fatty acid linked to it by an ester bond. (3)

Monomer Small, water-soluble organic molecule that serves as a subunit in the assembly of a macromolecule. (2)

Monomeric protein Protein that consists of a single polypeptide chain. (3)

Monosaccharide Simple sugar; the repeating unit of polysaccharides. (3)

Morphogen Substance secreted by one group of cells that causes specific changes in the cellular fate and morphogenesis of another group of cells; commonly hypothesized to form gradients in which cells adopt different developmental fates based on their location within the gradient. (22)

Morphogenesis Development of body pattern or form during embryogenesis. (22)

Motility Physical change in the location or orientation of an organism or cell that requires an expenditure of energy; movement. (19)

Motor molecule Any of three molecules (myosin, dynein, and kinesin) that interact with cytoskeletal elements (microtubules and microfilaments) to produce cell motility. (19)

Motor neuron Nerve cell that transmits impulses from the central nervous system to muscles or glands. (20)

MPF *See* Maturation promoting factor.

mRNA *See* Messenger RNA.

MTOC *See* Microtubule-organizing center.

Mucopolysaccharide Chemical component of the fuzzy layer of many animal cells. (7)

Mucoprotein Protein with a long and highly acidic carbohydrate chain; component of the fuzzy layer of many animal cells. (7)

Multimeric protein Protein that consists of more than one polypeptide chain. (3)

Multivesicular body Structure formed by inward budding of an uncoated vesicle membrane. (9)

Muscle fiber Long, thin, multinucleate cell that is specialized for contractile function. (19)

Mutagen Chemical or physical agent capable of inducing mutations. (16, 23)

Mutation Heritable change in the gene structure of a chromosome that can produce structural or functional variations in offspring. (14, 15)

Myelin sheath Concentric layers of membrane that surround an axon and serve as effective electrical insulation to allow rapid transmission of nerve impulses. (20)

Myeloma Tumor derived from stem cells that normally give rise to granular leukocytes. (23)

Myeloma protein Antibody secreted in large quantities by a cancerous plasma cell. (24)

Myoblast Stem cell that gives rise to muscle tissue. (22, 23)

Myofibril Functional unit of contraction, consisting of bundles of thick filaments and thin filaments. (19)

Myoglobin Protein used to bind and store oxygen in muscle cells. (19)

Myokinase Enzyme that catalyzes ATP formation by phosphorylating one ADP molecule at the expense of another. (19)

Myomesin A bundling protein that holds myosin molecules in place in the thick filament of a muscle cell. (19)

Myosin Muscle protein that makes up thick filaments and interacts with actin in muscle contraction. (19)

NAD^+ *See* Nicotinamide-adenine dinucleotide.

NADH dehydrogenase Enzyme complex of the electron transport chain involved in electron transfer from NADH to coenzyme Q. (11)

$NADP^+$ *See* Nicotinamide-adenine dinucleotide phosphate.

Nanometer (nm) Unit of measure: $1 \text{ nm} = 10^{-3} \mu\text{m} = 10^{-9} \text{ m}$. (1)

Nascent polypeptide Incomplete polypeptide still attached to a ribosome. (16)

Native Describing the original conformation of a macromolecule specified by its subunit structure and formed spontaneously because of its thermodynamic stability. (3, 13)

Negative contrast Optical system in which a lightly stained specimen is seen against a dark, heavily stained background. (A)

Negative control Genetic control in which the key regulatory element acts by turning off expression of the operon. (17)

Negative regulation Control mechanism in which a metabolic pathway is slowed or stopped by the presence of certain substances, usually end products, that allosterically inhibit one or more enzymes in the pathway. (6)

Negative staining Microscopic technique in which a stain is drawn around the specimen and its ultrastructural features, such that the specimen is seen lightly stained on a heavily stained background when viewed in the transmission electron microscope. (A)

Negative supercoil Coil in a circular DNA molecule formed by a left-handed twist of a relaxed molecule; form of most circular DNA molecules in nature. (13)

Neoplasm Tissue that continues to grow under conditions where normal tissue does not; a tumor. (23)

Neoplastic transformation Conversion of a tissue that shows normal growth regulation into a tumor that grows in a progressive, uncontrolled manner. *See also* Tumor. (23)

Nernst equation Equation useful for calculating the equilibrium potential for a particular species of ion; $E_x = (RT/2F) \ln [X_o/X_i]$. (20)

Nerve Anatomical structure leading from nervous tissue to another tissue such as muscle; composed of bundles of myelinated axons surrounded by connective tissue. (20)

Nerve impulse Signal transmitted along nerve cells by a wave of depolarization-repolarization events, propagated along the axonal membrane. (20)

Nerve terminal *See* Synaptic knob. (20)

Nervous system Group of cells, tissues, and organs that collect, process, and respond to information from the environment and from within the organism by the transmission of electrical impulses and exchange of chemical signals. (20)

Neural cleft *See* Neural groove.

Neural crest cells Cells that migrate away from the neural tube to specific locations along the periphery of the embryo, where they give rise to pigment cells, cartilage, and numerous other tissues. (22)

Neural groove Invagination of the ectoderm along the dorsal surface of the embryo, induced by the dorsal mesoderm during animal embryogenesis; also called the neural cleft. (22)

Neural tube Tube formed along the dorsal surface of the embryo by pinching off the invaginating tissue of the neural groove. (22)

Neuroblast Cell that gives rise to nerve cells. (23)

Neuroblastoma Tumor derived from neuroblast (nerve cell–forming) cells. (23)

Neurofilament Intermediate filament found in nerve cells. (18)

Neurofilament (NF) protein Protein of the intermediate filaments found in nerve cells. (18)

Neuron Specialized cell directly involved in the conduction and transmission of nerve impulses; nerve cell. (20)

Neurotoxin Toxic substance that disrupts the transmission of nerve impulses. (20)

Neurotransmitter Chemical responsible for the transmission of a nerve impulse across a synapse. (20)

Neurulation Process whereby the dorsal mesoderm of an animal embryo causes the overlying ectoderm to invaginate and form the neural cleft, which then pinches off to become the neural tube. (20)

Neutrophil *See* Polymorphonuclear leukocyte.

Nexin Protein that connects and maintains the spatial relationship of adjacent microtubule doublets in the axoneme of flagella. (19)

Nicotinamide-adenine dinucleotide (NAD^+) Coenzyme that accepts two electrons and one proton to generate the reduced form, NADH; important electron carrier in energy metabolism. (10)

Nicotinamide-adenine dinucleotide phosphate ($NADP^+$) Coenzyme that accepts two electrons and one proton to generate the reduced form, NADPH; important electron carrier in the Calvin cycle and other biosynthetic pathways. (12)

Node of Ranvier Small segment of bare axon between successive segments of the myelin sheath. (20)

Noncellulosic matrix Component of plant and fungal cell walls that consists of the polymers hemicellulose, pectin, and lignin and the protein extensin. (3)

Noncoding strand Strand of a DNA duplex that is not used as a template for RNA synthesis at that point along the duplex. (16)

Noncompetitive inhibition Inhibition of enzyme activity that results from the reversible binding of an inhibitor to a site other than the active site, thereby preventing the catalytic action of the enzyme on the substrate and decreasing the generation of product. (16)

Noncyclic electron flow Continuous, unidirectional flow of electrons from water to $NADP^+$ in phototrophs, with light providing the energy to drive the transfer. (12)

Nonhistone chromosomal protein Nuclear protein that forms a complex with DNA in chromatin, but not as part of the nucleosome structure; includes transcription factors, RNA polymerase, HMG proteins, etc. (13)

Nonsense mutation Change in nucleotide sequence converting a codon that previously coded for an amino acid into a stop codon. (16)

Nuclear cortex *See* Nuclear lamina.

Nuclear envelope Double membrane around the nucleus that is interrupted by numerous small pores. (4, 13)

Nuclear equivalence Principle that the nuclei of a multicellular organism are genetically identical and that regulatory mechanisms operating during development are therefore due to selective gene expression, not selective gene gain or loss. (22)

Nuclear lamina Electron-dense layer of fibrous material on the nucleoplasmic side of the inner nuclear membrane; also called the nuclear cortex. (13, 18)

Nuclear lamins Proteins found in the nuclear lamina. (18)

Nuclear matrix filament Structural filament within the nucleus. (13)

Nuclear pore Small opening in the nuclear envelope that allows water-soluble molecules to pass between the nucleus and cytoplasm. (13)

Nuclear run-on transcription Technique that permits identification of genes engaged in transcriptional activity in the nucleus by allowing nascent transcripts in isolated nuclei to be completed in vitro in the presence of radioactively labeled ribonucleotides. (17)

Nucleic acid An unbranched polymer composed of ribose or deoxyribose, phosphate groups, and the organic bases guanine, cytosine, adenosine, and thymine (for DNA) or uracil (for RNA). (3)

Nucleoid Region of cytoplasm in which the genetic material of a prokaryotic cell is located. (13)

Nucleolar organizer region (NOR) Site on certain chromosomes at which the genes for ribosomal RNA are located and nucleoli form; also called a secondary constriction. (13)

Nucleolus Large, discrete structure present in the nucleus of a eukaryotic cell; the site of ribosomal RNA synthesis and processing and of the assembly of ribosomal subunits. (13)

Nucleophilic substitution Reaction in which an electron-rich (electronegative) molecule or region of a molecule donates electrons to an electron-deficient (electropositive) molecule or region of a molecule, resulting in the formation of a covalent bond. (6)

Nucleoplasm Semifluid matrix that fills the interior of the nucleus. (4, 13)

Nucleoside Molecule containing a nitrogenous base (purine or pyrimidine) linked to a pentose sugar (ribose or deoxyribose); a nucleotide with the phosphate removed. (3)

Nucleoside monophosphate *See* Nucleotide.

Nucleosome Basic structural unit of chromosomes, consisting of about 200 base pairs of DNA associated with an octamer of histone proteins. (13)

Nucleotide Molecule containing a purine or pyrimidine, ribose or deoxyribose, and a phosphate group; also called a nucleoside monophosphate. (3)

Nucleus Membrane-bounded organelle that contains the genetic information of a eukaryotic cell. (4, 13)

Numerical aperture (NA) The quantity $n \sin \alpha$, where n is the refractive index of the medium between the specimen and the objective lens of a light microscope and α is the aperture angle. (A)

Obligate aerobe Organism that has an absolute requirement for oxygen as an electron acceptor and therefore cannot live under anaerobic conditions; also called a strict aerobe. (10)

Obligate anaerobe Organism that has an absolute requirement for an electron acceptor other than oxygen and therefore cannot live under aerobic conditions; also called a strict anaerobe. (10)

Okazaki fragment Short sequence of nucleotides synthesized on the lagging strand as a DNA replication intermediate. (14)

Oligodendrocyte Cell in the central nervous system that is responsible for forming the myelin sheath around a nerve axon. (20)

Oncogene Gene present in the genome of a retrovirus that contributes to tumor formation when introduced into a nontumor cell and is therefore responsible for the cancer-causing ability of the virus. (21, 23)

Oncogenic Cancer-causing. (23)

Oncogenic virus Cancer-causing virus. (23)

Oogenesis Process of oocyte (egg) development. (17, 22)

Oogonium Primordial germ cell that will eventually give rise to an ovum. (22)

Ootid Product of division of a secondary oocyte; gives rise to the ovum. (22)

Operator (*o* site) Nucleotide sequence in an operon that is recognized by a repressor molecule. (17)

Operator-constitutive mutant (o^c) Mutant form of an operator characterized by the production of gene products whether the inducer is present or not. (17)

Operon Cluster of structural and regulatory genes with coordinated function under the control of a single operator and a single promoter. (16, 17)

Organ Body part that consists of several tissue types grouped and organized into a structural and functional unit. (22)

Organelle Membrane-bounded structure that compartmentalizes functions within a eukaryotic cell; each organelle contains a specific enzyme complement and chemical composition related to its function. (1)

Organic molecule Molecule containing two or more covalently linked carbon atoms in addition to hydrogen and other atoms. (2)

Organogenesis Interactions between cells that lead to the formation of specific tissues and organs. (22)

Origin of replication Discrete point on a genophore or chromosome at which DNA replication is initiated. (14)

Origin of transfer Point on an F factor plasmid at which the transfer of the plasmid from an F^+ (male) bacterial cell to an F^- (female) recipient cell begins in conjugation. (15)

Ouabain Toxic plant steroid that inhibits the sodium-potassium pump of animal cells. (8)

Outer membrane The outer of the two lipid bilayers that surround a mitochondrion, chloroplast, or nucleus. (4, 11, 12, 13)

Ovary Female organ that produces egg cells, or ova; female gonad. (15, 22)

Ovulation Release of a mature ovum from the ovary. (22)

Ovum Haploid gamete produced by the female; egg cell. (15, 22)

Oxidation Chemical reaction involving the loss of electrons; oxidation of organic molecules frequently involves the removal of both electrons and hydrogen ions (protons) and is therefore also called a dehydrogenation reaction. (10)

β-Oxidation Series of reactions in which a fatty acid is catabolized to acetyl CoA by successive cycles of oxidation of the β carbon of an activated form of the fatty acid. (11)

Oxidation-reduction couple Pair of molecules between which electrons are transferred with a reduction potential measured with respect to a reference potential; also called a redox couple. (11)

Oxidative deamination Release of free ammonia from the carbon skeleton of an amino acid, with concomitant oxidation of the molecule to an α keto acid. (11)

Oxidative phosphorylation Formation of ATP from ADP and inorganic phosphate by coupling the exergonic reoxidation of reduced coenzyme molecules by oxygen to the phosphorylation of ADP, with an electrochemical proton gradient as the intermediate. (10, 11)

Oxygenic photoautotroph Organism that utilizes water as the electron donor in photosynthesis, with release of oxygen. (12)

Packing ratio Ratio of the length of a DNA molecule to the length of the chromosome or fiber into which it is packaged; used to quantify the extent of DNA coiling and folding in the chromosome. (13)

Palindromic Reading the same backward as forward; GCTTCG is a palindromic sequence. (B)

Paracrine hormone Substance that is synthesized and released by a specific set of cells but acts only on the cells in the immediate environment rather than on cells of target tissues in distant locations. *See also* Endocrine hormone. (21)

Passive transport Carrier-mediated movement of inorganic ions or small organic molecules across a membrane without expenditure of energy; thermodynamically feasible process that requires a carrier to overcome membrane impermeability; also called carrier-mediated diffusion or carrier-mediated transport. (7, 8)

Patch clamping Technique used to study individual membrane channels; a glass micropipette is placed on the cell surface and suction is applied gently to form a tight seal; current can enter and leave the pipette only through the small number of channels in the patch of membrane sealed to the pipette tip. (20)

Patching and capping Formation, within the plasma membrane of a cell that has been treated with a specific antibody to a surface protein, of patches of cross-linked membrane proteins, which then collect over one pole of the cell to form a cap. (7)

Pattern formation Generation of a specific plan of multicellular organization during embryogenesis. (22)

PDGF *See* Platelet-derived growth factor.

Pectin Polymeric substance deposited with cellulose in the cell walls of plants and fungi to provide added strength. (7)

Pentose A sugar that contains five carbon atoms. (3)

Peptidase Enzyme that degrades small peptide chains by hydrolyzing peptide bonds.

Peptide bond A covalent bond between the amino group of one amino acid and the carboxyl group of a second amino acid; a specific kind of amide bond. (3)

Peptidoglycan Component of bacterial cell walls that consists of polysaccharide chains cross-linked by small peptides. (7)

Peptidyl site (P site) Site on a ribosome at which the growing polypeptide chain is attached. (16)

Peptidyltransferase Enzyme that catalyzes formation of a peptide bond between the two amino acids attached to tRNAs in the A and P sites of the ribosome during the process of protein synthesis. (16)

Perinuclear space Fluid-filled compartment between the two nuclear membranes that is continuous with the internal spaces of the endoplasmic reticulum. (13)

Perinucleolar chromatin Fibrils of chromatin that surround the nucleolus and that contain the DNA coding for ribosomal RNA. (13)

Peripheral lymphoid tissue Location of mature lymphocytes in the organism; includes spleen, lymph nodes, and tonsils.

Peripheral membrane protein Hydrophilic protein located on the surface of a membrane. (7)

Peripheral nervous system (PNS) Sensory and motor neurons of the nervous system that control skeletal voluntary movement and involuntary activities of muscles and glands. *See also* Central nervous system (CNS). (20)

Permease Integral membrane protein that mediates the passage of specific substances across the membrane; also called a carrier protein or transport protein. (8)

Peroxisome Single membrane-bounded organelle that contains catalase and one or more hydrogen peroxide–generating oxidases and is therefore involved in the metabolism of hydrogen peroxide. (4, 9, 12)

P-face Interior face of the inner, or cytoplasmic, monolayer of a membrane as revealed by the technique of freeze-fracturing; also called the A-face. (A)

pH Measure of the hydrogen ion concentration of an aqueous solution on a logarithmic scale; $pH = -\log [H^+]$. (6)

Phage *See* Bacteriophage.

Phagocyte Specialized white blood cell that carries out phagocytosis as a defense mechanism. (9)

Phagocytic vacuole Membrane-bounded structure containing food particles that fuses with a primary lysosome to begin digestion. (9)

Phagocytosis Type of endocytosis in which particulate matter or even an entire cell is taken up from the environment and incorporated into vesicles for digestion; "cell eating." (7)

Phalloidin Drug that inhibits a variety of cell movements by preventing actin depolymerization. (18)

Phase-contrast microscopy Light microscopic technique that produces an image with highly contrasting bright and dark areas against a gray background by inserting into the light path an optical material that is capable of bringing undiffracted rays into phase with those that have been diffracted by the specimen. (A)

Phenotype Observable physical characteristics of an organism attributable to the genotype and to interactions between the organism and its environment during development. (15)

Pheromone Any chemical messenger that is produced and released by one organism and elicits a physiological response in another organism. (22)

Phosphatidic acid Basic component of phosphoglycerides consisting of two fatty acids and a phosphate group linked by ester bonds to glycerol; key intermediate in the synthesis of other phosphoglycerides. (3)

Phosphatidyl choline *See* Lecithin.

Phosphoanhydride bond High-energy bond between phosphate groups. (10)

Phosphodiesterase Enzyme that catalyzes the hydrolysis of cyclic AMP to adenosine-5′-monophosphate. *See also* Cyclic AMP. (21)

Phosphodiester bond A chemical linkage involving two alcohols joined covalently to a single phosphate group. (3)

Phosphoester bond Ester bond formed by removing a hydroxide ion from a phosphate group and a hydrogen ion from an alcohol group. (3, 10)

Phosphoglyceride Primary component of cell membranes consisting of a glycerol molecule esterified to two fatty acids and a phosphate group. (3)

Phospholipase A Enzyme that catalyzes the hydrolysis of ester bonds that link fatty acids to glycerol in phospholipids, thereby releasing free fatty acids, including arachidonic acid. *See also* Arachidonic acid. (21)

Phospholipase C Enzyme that catalyzes the hydrolysis of phosphatidylinositol bisphosphate into inositol trisphosphate and diacylglycerol; activation of this enzyme initiates the inositol phospholipid pathway. (21, 23)

Phospholipid Primary component of cell membranes consisting of a polar head region and a nonpolar tail of long hydrocarbon chains. (3)

Phospholipid bilayer *See* Lipid bilayer.

Phospholipid transfer protein Protein that recognizes a specific kind of phospholipid and mediates the transfer of that phospholipid from one membrane to another. (9)

Phosphorolytic cleavage Mobilization of storage polysaccharides involving cleavage by inorganic phosphate of $\alpha(1 \rightarrow 4)$ bonds between successive glucose units, which are liberated as glucose-1-phosphate. (10)

Phosphorylating transport Inward transfer of a sugar into a bacterial cell, with phosphorylation of the sugar as an inherent part of the uptake mechanism. (8)

Photoautotroph Organism capable of obtaining energy from the sun and able to use carbon dioxide as a source of carbon. (12)

Photochemical reduction Transfer of photoexcited electrons from one molecule to another. (12)

Photoexcitation Excitation of an electron to a higher energy level by the absorption of a photon of light. (12)

Photoheterotroph Organism capable of obtaining energy from the sun but dependent on organic compounds for carbon. (12)

Photolysis Light-dependent oxidative splitting of water. (12)

Photophosphorylation Light-dependent generation of ATP driven by an electrochemical proton gradient established and maintained as excited electrons of chlorophyll return to their ground state via an electron transport chain that is coupled to vectorial proton translocation across the thylakoid membrane. (12)

Photoreduction Light-dependent generation of NADPH by the transfer of energized electrons from photoexcited chlorophyll molecules to $NADP^+$ via a series of electron carriers. (12)

Photorespiration Light-dependent pathway responsible for the oxidative metabolism of glycolate and characterized by the uptake of oxygen and release of carbon dioxide, thereby detracting from the efficiency of photosynthesis. *See also* Glycolate pathway. (12)

Photosynthetic unit Cluster of 250 to 300 chlorophyll molecules, each capable of absorbing light but with only a few molecules at the reaction center able to participate in the actual photochemical reactions required to generate ATP and NADPH. (12)

Photosystem Assembly of chlorophyll and other pigment molecules that is embedded in the thylakoid membrane and that functions in the light-requiring reactions of photosynthesis. (12)

Photosystem I Photosystem containing a chlorophyll molecule (P700) that absorbs 700-nm red light maximally; light of this wavelength can excite electrons to an energy level that allows them to reduce $NADP^+$ to NADPH. *See also* Photosystem. (12)

Photosystem II Photosystem containing a chlorophyll molecule (P680) that absorbs 680-nm red light maximally; light of this wavelength can excite electrons donated by water (or some alternative source) to an energy level that allows them to reduce organic electron carriers. *See also* Photosystem. (12)

Phototroph Organism that is capable of utilizing the radiant energy of the sun to satisfy energy requirements; "light feeder." (5, 12)

Phragmoplast Cylindrical structure in the central region of a dividing plant cell, formed by a parallel array of microtubules and involved in cell plate formation. (14)

Phycobilin Accessory pigment found only in red algae and blue-green algae that absorbs visible light in the green-to-orange range of the spectrum (550–630 nm), thereby giving the algae their characteristic colors. (12)

Pigment Light-absorbing compound responsible for the color of a substance. (12)

Pilus Long, hairlike projection of an F^+ (male) bacterial cell that facilitates the transfer of DNA during conjugation between the F^+ cell and an F^- (female) cell. (15)

Pinocytosis Type of endocytosis in which soluble materials are taken up from the environment and incorporated into vesicles for digestion; "cell drinking." (9)

Plane-polarized light Light waves that travel in one plane only. (A)

Plaque Hole in a lawn of plated bacterial cells as a result of infection of the cells in that region of the plate with bacteriophage. (13)

Plasma cell Differentiated B lymphocyte that secretes the antibodies required to agglutinate and eliminate foreign substances. (24)

Plasma membrane Flexible bilayer of phospholipids and proteins that defines the boundary of the cell and regulates the flow of materials into and out of the cell; also called the cell membrane. (7)

Plasmid Relatively small extrachromosomal genetic element of bacteria that replicates autonomously. (B)

Plasmodesma Cytoplasmic channel through pores in the cell walls of two adjacent plant cells, allowing fusion of the plasma membranes and chemical communication between the cells. (4, 7)

Plastid A plant organelle derived from a proplastid that may serve any of several different functions; chloroplasts, amyloplasts, and chromoplasts are plastids. (4)

Plastocyanin Copper-containing protein that donates electrons to chlorophyll P700 of photosystem I in the light-requiring reactions of photosynthesis. (12)

Plastoquinone Molecule containing a quinone ring that is part of the electron transfer chain between photosystems I and II in the light-requiring reactions of photosynthesis. (12)

Platelet-derived growth factor (PDGF) A peptide mitogen released from blood platelets during blood clotting that stimulates cells of connective tissue to divide and proliferate. (23)

Pluripotent Describing a cell or nucleus that is able to give rise to distinctly different sets of progeny, depending on the environment in which differentiation occurs. (22)

PNS *See* Peripheral nervous system.

Polar body Product of cell division during oogenesis that receives a disproportionately small amount of cytoplasm and subsequently degenerates. (15, 22)

Polar center Region of cytoplasm that organizes an axis of polarity in a developing tissue or organism. (22)

Polar fiber Bundle of microtubules that extends inward from the spindle pole toward the equator of the mitotic spindle. (14)

Polarity (a) Directionality; orientation. (13, 18) (b) Property of a molecule that results from part of the molecule having a partial positive charge and another part having a partial negative charge, usually because one region of the molecule possesses one or more electronegative atoms that draw electrons toward that region. (2)

Polarity gene Gene encoding a protein product that is involved in the establishment of polar centers and axes of polarity in a developing tissue or organism. (22)

Polarization microscopy Light microscopic technique in which a polarizer located above the source of illumination and an analyzer located above the objective lens allow only waves traveling in one plane to pass through. (A)

Poly-A tail Repetitive sequence of adenylate groups added posttranscriptionally to the 3′ end of a eukaryotic mRNA molecule, apparently conferring stability on the molecule and possibly involved in its transport to the cytoplasm. (16)

Polycistronic mRNA A messenger RNA molecule that yields more than one polypeptide product upon translation. (16, 17)

Polymer Large molecule consisting of many monomeric subunits covalently bound to one another in an identical manner as a result of the repetition of a short sequence of chemical reactions. (3)

Polymerase chain reaction Reaction in which a specific segment of DNA is amplified by repeated cycles of (1) heat treatment to separate the two strands of the DNA double helix; (2) incubation with a large excess of two synthetic oligonucleotide primers, one complementary to each strand of the DNA; and (3) incubation with DNA polymerase for synthesis of the regions of DNA downstream from the primers; for effective DNA amplification, 20–30 cycles are required, with each cycle doubling the amount of DNA synthesized in the previous cycle. (B)

Polymorphonuclear leukocyte White blood cell that engulfs and digests invading microorganisms or foreign materials in the bloodstream; also called a neutrophil. (9)

Polymorphy Condition of having many different alternate alleles for each locus. (24)

Polynucleotide Linear chain of nucleotides linked by phosphodiester bonds. (3)

Polypeptide Linear chain of amino acids linked by peptide bonds. (2, 3)

Polyribosome Structure formed when two or more ribosomes attach to and simultaneously translate a single mRNA molecule. (16)

Polysaccharide Polymeric carbohydrate molecule consisting of monosaccharides linked by glycosidic bonds. (3)

Polyspermy Fertilization of an egg by more than one sperm. (22)

P/O ratio The number of molecules of ATP generated as a pair of electrons passes through the electron transport chain to reduce a single oxygen atom to water. (11)

Pore A small opening in a membrane or envelope. *See also* Nuclear pore. (4)

Positional information Any form of molecular information that directs a cell to adopt a particular developmental fate based on the cell's position within a developing embryo or tissue; includes cytoplasmic determinants, morphogen gradients, and axes of polarity. (22)

Positive regulation Control mechanism in which a pathway is activated by the presence of certain substances, usually substrates. (6)

Positive supercoil Coil in a circular DNA molecule formed by a right-handed twist of a relaxed molecule. (13)

Poststaining Procedure in which tissue that has been fixed, embedded, and sectioned for electron microscopy is stained, usually with lead or uranium, to enhance contrast. (A)

Postsynaptic membrane Membrane of an axon that has receptors for neurotransmitter molecules and that conducts a nerve impulse away from the synaptic cleft. (20)

Posttranscriptional control Mechanisms of genetic regulation involving selective expression of information already present in the cell as RNA transcripts; includes control of processing and transport, translation, and posttranslational events. (17)

Posttranscriptional processing Chemical modifications of an RNA molecule to generate the final RNA product from the primary transcript. (16)

Posttranslational control Mechanisms of gene regulation involving selective utilization of polypeptides that have already been synthesized; includes covalent modifications, removal of a portion of the polypeptide, association of subunits, and protein turnover. (17)

Presynaptic membrane Membrane of an axon that conducts an impulse to the synaptic cleft, leading to the release of neurotransmitter molecules into the synaptic cleft. (20)

Pribnow box Specific hexanucleotide sequence in the promoter of bacterial operons, responsible for determining the precise nucleotide with which transcription begins. (16)

Primary cell wall Cell wall of a growing plant cell. *See also* Secondary cell wall. (7)

Primary fixation Initial step in the preparation of a specimen for microscopy that stabilizes the chemical components of the cells and hardens the specimen for further processing. (A)

Primary germ layer Any of the three tissue layers of an animal embryo at the time of gastrulation; the primary germ layers are ectoderm, mesoderm, and endoderm. (22)

Primary immune response Initial immune response to a particular antigen, characterized by large amounts of IgM antibodies in the blood. *See also* Secondary immune response. (24)

Primary lysosome Lysosome possessing a full complement of hydrolytic enzymes, but not yet engaged in digestive activity. (9)

Primary oocyte Cell that is derived from an oogonium by mitotic division and that gives rise to an egg cell by meiosis. (22)

Primary spermatocyte Cell that is derived from a spermatogonium by mitotic division and that gives rise to sperm cells by meiosis. (22)

Primary structure Sequence of amino acids in a polypeptide chain or of nucleotides in a nucleic acid. (3)

Primary tumor Initial malignant mass of proliferating cells in an organism. (23)

Primase DNA-dependent RNA polymerase that synthesizes the RNA primers required for initiation of replication on both the lagging and leading strands of a DNA duplex. (14)

Primer Short sequence of RNA at the 5′ end of an Okazaki fragment onto which deoxyribonucleotides are added. (14)

Primer recognition factor Protein that causes priming, or the synthesis of a short piece of RNA at specific sites, on the lagging strand of a DNA duplex undergoing replication. (14)

Primitive gut *See* Archenteron. (22)

Primordial germ cell Cell that migrates to the genital ridge during animal embryogenesis and gives rise to either oogonia or spermatogonia, depending on the hormones released by somatic cells as development proceeds. (22)

Primosome Complex of proteins that is associated with primase and that contains recognition factors required for synthesis of the primers involved in the initiation of DNA replication. (14)

Procarcinogen Chemical that can act as a carcinogen after it has been activated by the mixed-function oxidase system. (23)

Process Extension or branch from a nerve cell body; an axon or a dendrite. (20)

Prokaryote Category of organisms that includes the bacteria and the cyanobacteria (blue-green algae) and that is characterized by the absence of a true nucleus and other membrane-bounded organelles. (4)

Promoter (a) Nucleotide sequence in DNA at which RNA polymerase binds to initiate transcription. (16, 17) (b) Chemical that causes a mutant cell to divide more rapidly than neighboring cells. (23)

Prophage Bacteriophage DNA that has been incorporated into a bacterial genophore and is replicated with the DNA of the host cell. (13)

Prophase First phase of mitosis, in which chromosomes have condensed into readily visible structures. (14, 15)

Prostaglandin Any of a series of paracrine hormones that are synthesized from arachidonic acid by the cyclooxygenase pathway and function by activating contraction of smooth muscles. *See also* Arachidonic acid. (21)

Prosthetic group Small organic molecule or metal ion component of an enzyme that plays an indispensable role in the catalytic activity of the enzyme. (6)

Protamines Small, positively charged proteins to which the DNA of sperm chromosomes is complexed in some species of animals. (13, 22)

Protease Enzyme that hydrolyzes peptide bonds of a polypeptide chain or protein, resulting in small peptides and some free amino acids. (11)

Protein Macromolecule that consists of one or more polypeptides folded into a conformation specified by the linear sequence of amino acids and that functions as an enzyme, a hormone, an antibody, or a structural component of the cell. (3)

Protein kinase C Enzyme that can phosphorylate specific serine and threonine groups on a variety of target proteins when activated by diacylglycerol. *See also* Diacylglycerol. (21)

Protein/lipid ratio Molar ratio of protein to lipid in a membrane; typically in the range of 1:100. (7)

Proteoliposome Artificial membrane formed by the incorporation of specific proteins into a phospholipid bilayer. (7)

Proteolysis Degradation of proteins by hydrolysis of the peptide bonds that link amino acids together in the polypeptide chain. (11)

Protofilament Linear polymer of spherical subunits of tubulin that serves as the structural component of microtubules. (18)

Proton motive force (pmf) Force exerted by an electrochemical gradient produced by a difference in proton concentration across the membrane and by the membrane potential. (11, 12)

Proton translocator Channel through which protons flow across a membrane, owing to an electrochemical gradient. *See also* CF_0; F_0. (11, 12)

Proto-oncogene Nononcogenic homologue of an oncogene that is present in normal (nonneoplastic) cells. *See also* Oncogene. (23)

Provirus Double-stranded DNA copy of a retroviral genophore that becomes integrated into a chromosome of a cell infected with the retrovirus. *See also* Retrovirus. (23)

Pseudoplasmodium Stage in the life cycle of a slime mold in which the amoebas have aggregated; also called a slug. (19)

Pseudopod Extension of a cell formed by cytoplasmic streaming; "false foot." (19)

P site *See* Peptidyl site.

Purine Two-ringed nitrogen-containing molecule; parent compound of the bases adenine and guanine. (3)

Pyrimidine Single-ringed nitrogen-containing molecule; parent compound of the bases cytosine, thymine, and uracil. (3)

Pyrimidine dimer formation Form of DNA damage in which a covalent bond is formed between adjacent pyrimidine bases, thereby blocking replication and transcription; caused by ultraviolet light. (14)

Quantum requirement Unit amount of radiant energy, in photons, necessary to bring about a specific change. (12)

Quaternary structure Level of protein structure involving interactions between two or more individual polypeptides to form a single multimeric protein. (3)

Quick-freeze deep-etch microscopy *See* Deep-etching.

Radial spokes Inward projections from each of the nine microtubule doublets to the center pair of microtubules in the axoneme of a cilium or flagellum, believed to be important in converting the sliding of the doublets into a bending of the axoneme. (19)

Random coil Region of unordered (irregular) secondary structure of a protein. (3)

Reaction center Portion of a photosynthetic unit containing the chlorophyll molecules that initiate electron transfer, utilizing the energy gathered by other chlorophyll molecules and accessory pigments in the photosynthetic unit. (12)

Receptor-mediated endocytosis (RME) Type of endocytosis initiated at coated pits and resulting in coated vesicles; believed to be a major mechanism for selective uptake of macromolecules and peptide hormones; also called adsorptive pinocytosis. (9)

Receptor protein Allosteric protein, located either on the plasma membrane or within the cytoplasm, that has a binding site for a specific ligand and initiates or mediates a particular event or series of events in response to ligand binding. *See also* Chemical receptor; Intracellular receptor; Ligand; Membrane receptor. (21)

Receptosome *See* Endosome.

Recessive Allele that is present in the genome but is phenotypically expressed only in the homozygous form; in the heterozygous form, a recessive allele is masked by a dominant allele. (15)

Recognition sequence DNA sequence adjacent to immunoglobulin gene segments, responsible for splicing together the V, J, and/or D segments to assemble a complete V gene. (24)

Recombinant DNA molecule Hybrid DNA molecule formed by the fusion of DNA fragments from different sources. (B)

Recombination Creation of new genotypes in offspring as a result of physical exchange of genetic information between homologous chromosomes when they are paired during prophase I. (15)

Red cell membrane Plasma membrane of the erythrocyte, or red blood cell; used for investigations of membrane structure because of its relative simplicity. (7)

Redox couple *See* Oxidation-reduction couple.

Reduction Chemical reaction involving the addition of electrons; reduction of organic molecules frequently involves the addition of both electrons and hydrogen ions (protons) and is therefore also called a hydrogenation reaction. (10)

Reduction division *See* Meiosis.

Refractive index Measure of the change in the velocity of light as it passes from one medium to another; calculated as the ratio of the velocity of light in a vacuum, c, to the velocity of light in a given medium, v: refractive index = c/v. (A)

Refractory period Time interval required between successive stimuli to allow the membrane of a nerve cell to return to its resting potential; during the refractory period, the membrane is unable to respond to a stimulus. (20)

Regulator gene Nucleotide sequence that codes for the repressor, which regulates transcription of a prokaryotic operon by binding to the operator site; usually located outside the operon. (17)

Regulatory gene Any of several kinds of nucleotide sequences involved in the control of the expression of structural genes. (17)

Releasing factor Any of several small neuropeptides that are produced by neurons in the hypothalamus and released into blood vessels leading to the anterior pituitary, where they control the release of other hormones. *See also* Anterior pituitary; Hypothalamus. (21)

Renaturation Return of a protein from a denatured state to the native conformation determined by its amino acid sequence, usually accompanied by restoration of physiological function. (3)

Repair endonuclease Enzyme that detects the absence of a base in a DNA molecule and nicks the phosphodiester bonds at that point, thereby allowing other enzymes to repair the strand. (14)

Repair synthesis Replacement of damaged DNA segments by the removal of defective nucleotides and insertion of new nucleotides. (14)

Replication bubble Structure seen in replicating DNA molecules that is formed as two replication forks move away from a common origin of replication. (14)

Replication fork Y-shaped structure that represents the site at which replication of a DNA duplex is occurring. (14)

Replicon Self-replicating unit of DNA possessing a site at which replication begins. (14)

Repression Ability of the end product of a pathway to prevent further synthesis of the enzymes involved in its formation; also called end-product repression. (17)

Repressor (R) Protein that binds to the operator site of an operon and prevents transcription of adjacent structural genes. (17)

Residual body Secondary lysosome in which digestion has ceased and only indigestible material remains. (9)

Resistance (R) factor Class of plasmids that carry the genes for drug resistance in *Escherichia coli* cells. (13)

Resistivity Characteristic electrical resistance of biological membranes. (7)

Resolution Measure of how close two objects can be and still be seen as distinct entities. (A)

Resonance stabilization Achievement of the most stable configuration of a molecule by maximal delocalization of π electrons over all possible bonds. (10)

Respiration Release of energy by the oxidation of organic molecules, with the electrons transferred eventually to oxygen; respiration occurs only under aerobic conditions and includes the tricarboxylic acid cycle, β-oxidation of fatty acids, electron transport, and ATP generation. (5, 10, 11)

Respiratory control Regulation of oxidative phosphorylation and electron transport by the availability of ADP. (11)

Respiratory metabolism *See* Respiration.

Resting potential Potential difference across the plasma membrane of an unstimulated nerve cell. (20)

Restriction enzyme Bacterial endonuclease that degrades foreign DNA molecules at or near a palindromic recognition sequence of four to six nucleotide pairs; used in recombinant DNA technology to cleave DNA molecules at specific, well-defined sites. (B)

Restriction map Diagrammatic ordering of the fragments produced by digestion of a DNA molecule, with overlapping fragments used to show the location of the restriction sites in relation to each other. (B)

Restriction point Point in the G-1 phase of the eukaryotic cell cycle beyond which a cell is committed to division. (B)

Retinal Derivative of vitamin A that is present in proteins such as rhodopsin and bacteriorhodopsin and undergoes a conformational change upon absorption of a photon of light. (8)

Retinoic acid Derivative of vitamin A that functions as a morphogen in the developing chick limb bud. (22)

Retrovirus Virus that contains RNA as its genetic information but uses double-stranded DNA as an intermediate in its propagation; certain retroviruses can cause tumors in infected cells. (23)

Reverse transcriptase Enzyme capable of using an RNA template to synthesize a complementary single-stranded DNA molecule. (16, 23)

Reverse transcription Process by which an RNA molecule is used as a template to make a single-stranded DNA copy. (16)

Reversible inhibitor Inhibitor molecule that causes a reversible loss of catalytic activity when bound to an enzyme; upon dissociation of the inhibitor, the enzyme regains biological function. (6)

Ribonuclease Enzyme that can cleave the phosphodiester bonds of an RNA molecule. (6)

Ribonucleic acid (RNA) Nucleic acid that contains ribose in each of its nucleotides and plays several different roles in the expression of genetic information. *See* Messenger RNA; Ribosomal RNA; Transfer RNA. (3)

Ribose Five-carbon sugar present in RNA. (3)

Ribosomal RNA (rRNA) RNA molecule that is an important constituent of ribosomes. (3, 16)

Ribosomal subunits Ribonucleoprotein particles that pair to form a functional ribosome; every ribosome consists of one large subunit and one small subunit, although the subunits of eukaryotic and prokaryotic cells differ in size and composition. (16)

Ribosome A small, complex structure that is composed of RNA and protein and that functions in protein synthesis; ribosomes consist of two subunits and are present in the cytoplasm of both eukaryotic and prokaryotic cells. (4, 16)

RME *See* Receptor-mediated endocytosis.

RNA *See* Ribonucleic acid.

RNA-DNA hybridization Association of the complementary sequences of a single-stranded DNA molecule and an RNA molecule. (16)

RNA polymerase Enzyme responsible for adding successive nucleotides to a growing RNA strand, with the order of nucleotides dictated by the DNA template. (16)

RNA splicing Excision of introns from an initial RNA transcript to generate the functional form of the RNA molecule. (16)

RNA tumor virus Cancer-causing virus that contains RNA as the genetic information but uses double-stranded DNA as an intermediate in propagation. *See also* Retrovirus. (23)

Rough endoplasmic reticulum (rough ER) Region of endoplasmic reticulum that is studded with ribosomes because of its involvement in protein synthesis. (4, 9)

rRNA *See* Ribosomal RNA.

Saccule *See* Cisterna.

Sarcoma Tumor derived from a connective tissue cell or other cell of mesodermal origin. (23)

Sarcomere Repeating unit of striated muscle that extends from one Z line to the next and that consists of two sets of thin (actin) and one set of thick (myosin) filaments. (19)

Sarcoplasm Cytoplasm of a muscle cell. (19)

Sarcoplasmic reticulum (SR) Endoplasmic reticulum of a muscle cell; intricate membranous network that stores calcium ions. (19)

Saturated fatty acid Fatty acid without double or triple bonds such that every carbon atom in the chain has the maximum number of hydrogen atoms bonded to it. (3)

Saturation Inability of higher substrate concentrations to increase the velocity of an enzyme-catalyzed reaction beyond a fixed upper limit determined by the finite number of enzyme molecules available. (6)

Scanning electron microscope (SEM) Microscope in which an electron beam sweeps over the specimen and an image is formed by secondary electrons emitted by molecules in the specimen in response to the primary electron beam. (A)

Schwann cell Specialized glial cell in the peripheral nervous system, responsible for forming the myelin sheath around a nerve axon. (20)

SDS-polyacrylamide gel electrophoresis (SDS-PAGE) Separation technique in which proteins are treated with the anionic detergent sodium dodecyl sulfate (SDS) and layered onto a gel of acrylamide polymers; a potential difference is applied across the gel, causing the SDS-coated polypeptides to migrate toward the anode at a rate inversely related to size. (7)

Secondary cell wall Wall deposited on the inner surface of the primary wall when a plant cell has achieved its final size and shape. *See also* Primary cell wall. (7)

Secondary constriction *See* Nucleolar organizer region.

Secondary electrons Electrons that are emitted by molecules in a biological specimen and that are used to form an image of the specimen surface in the scanning electron microscope. (A)

Secondary immune response Immune response to a previously encountered antigen that is more rapid, of greater magnitude, and of longer duration than the primary response and that is characterized by high levels of IgG antibody in the blood. *See also* Primary immune response. (24)

Secondary lysosome Structure formed by the fusion of a primary lysosome and a phagocytic vacuole; contains digestive enzymes and digestible foodstuffs. *See also* Autophagic lysosome; Heterophagic lysosome. (9)

Secondary oocyte Product of the first meiotic division of a primary oocyte, from which an egg cell will eventually arise. (22)

Secondary spermatocyte Product of the first meiotic division of a primary spermatocyte, from which sperm cells will eventually arise. (22)

Secondary structure Level of protein structure involving local interactions between continuous amino acids along the chain; may exist as an irregular chain or a repeating pattern. *See also* α helix; β pleated sheet. (3)

Secondary tumor Malignant mass of cells in an organism originating from a primary tumor located elsewhere in the body. (23)

Second law of thermodynamics The law of thermodynamic spontaneity; principle stating that all physical and chemical changes proceed in a manner such that the entropy of the universe increases. (5)

Second messenger Any of several substances, including cyclic AMP, calcium ion, inositol trisphosphate, and diacylglycerol, that act as intracellular mediators of specific extracellular signals. (21)

Secretory component Polypeptide chain that is derived from the IgA receptor on the membrane of epithelial cells of secretory ducts and that remains associated with the IgA molecule. (24)

Secretory granule Membrane-bounded compartment of a eukaryotic cell that carries secretory proteins from the Golgi complex to the plasma membrane for exocytosis and that may serve as a storage compartment for such proteins for some time before they are released; also called a secretory vesicle. (9)

Secretory protein Protein destined to be exported from the cell in which it was synthesized. (9)

Secretory vesicle *See* Secretory granule.

Sedimentation coefficient (S value) A measure of the rate at which a biological macromolecule or structure moves in a centrifugal force field. *See also* Svedberg unit. (4)

Segmentation gene Any of several genes encoding the proteins that establish the segmental body plan of the embryo of an insect such as the fruit fly *Drosophila.* (22)

Segment identity gene Any of several genes whose expression confers a particular developmental fate on the cells of an individual segment in the embryo of the fruit fly *Drosophila;* also called a homeotic gene. (22)

Selective permeability Characteristic of biological membranes that allows only specific molecules to cross. (2)

Self-assembly Principle that the information required to determine the structure of a macromolecule or of a supramolecular structure exists in the monomer sequence of the polymer itself. (2)

SEM *See* Scanning electron microscope.

Semiautonomous organelle Organelle, either a mitochondrion or a chloroplast, that contains DNA and is therefore able to encode some of its polypeptides, although it is dependent on the nuclear genome to encode most of them. (13)

Semiconservative replication Mode of replication in which one-half of a parent molecule is retained by each daughter molecule. (14)

Seminiferous tubule Tubule within the testis where spermatogenesis occurs. (22)

Sensor Element in a feedback control loop that detects the level of a specific product in the system that is so regulated. *See also* Effector; Feedback control loop. (21)

Sensory neuron Nerve cell that contains receptors or is connected to receptor cells and transmits impulses inward to the central nervous system. (20)

70S initiation complex Complex formed by the association of the 30S initiation complex, the 50S ribosomal subunit, and an initiation factor. (16)

Sexual reproduction Form of reproduction in which two parent organisms each contribute to the genetic information of the new organism; reproduction by the fusion of gametes. (15)

Shadowing Deposition of a thin layer of an electron-dense metal on a biological specimen from a heated electrode such that surfaces facing toward the electrode are coated, but surfaces facing away are not. (A)

Shuttle streaming Flow of cytoplasm that reverses direction with predictable periodicity. (19)

Sickle-cell anemia Inherited condition in which the β chain of hemoglobin has a single amino acid substitution (valine instead of glutamate at position 6) that changes the tertiary structure of the polypeptide and dramatically impairs the oxygen-carrying capability of the molecule. (3, 16)

Sigma factor (σ factor) Subunit of bacterial RNA polymerase that ensures the initiation of RNA synthesis at the correct site on the DNA strand. (16)

Signal hypothesis Model proposed to explain uptake of secretory and membrane proteins by the rough endoplasmic reticulum; according to the model, a signal sequence on a nascent polypeptide directs the ribosome-mRNA-polypeptide complex to the surface of the rough ER, allowing the polypeptide to be transported into or across the membrane of the ER as translation proceeds. (16)

Signal peptidase Enzyme that removes the signal sequence of a polypeptide as the polypeptide is transported into the lumen of the endoplasmic reticulum. (16)

Signal peptide *See* Signal sequence.

Signal recognition particle (SRP) Cytoplasmic factor that recognizes and binds to the signal sequence of a secretory or membrane protein. (16)

Signal sequence Sequence of 15 to 30 amino acids at the amino terminus of a secretory or membrane protein; directs the ribosome-mRNA-polypeptide complex to the surface of the rough endoplasmic reticulum; also called a signal peptide. (16)

Single bond Chemical bond formed between two atoms as a result of sharing a pair of electrons. (2)

Single-strand binding protein (SSB) Protein that interacts with the single strands of DNA at the replication fork to prevent the duplex from forming again; also called helix-destabilizing protein. (14)

Sister chromatid One of two replicated chromosomes that remain attached to each other until anaphase of mitosis. (14)

Sliding-microtubule model Model proposed to explain microtubule-based motility; according to the model, the length of microtubules remains unchanged, but adjacent outer doublets slide past each other, causing a localized bending due to lateral connections between adjacent microtubules and radial links to the center pairs that prevent free sliding of the microtubules past each other. (19)

Slow block to polyspermy Elevation and hardening of the vitelline membrane of an animal egg to prevent further penetration of the egg by other sperm. (22)

Slug *See* Pseudoplasmodium.

Small ribosomal subunit Component of a ribosome with a sedimentation coefficient of 40S in eukaryotes and 30S in prokaryotes; associates with a large ribosomal subunit to form a functional ribosome. (4)

Smooth endoplasmic reticulum (smooth ER) Region of endoplasmic reticulum that has no ribosomes and plays no direct role in protein synthesis; involved in packaging of secretory proteins and synthesis of lipids. (4, 9)

Sodium cotransport Use of the highly exergonic inward transport of sodium ions to drive active symport of organic solutes. (8)

Sodium-potassium pump Permease that couples ATP hydrolysis to the inward transport of potassium ions and the outward transport of sodium ions to maintain the Na^+ and K^+ gradients that exist across the plasma membrane of most animal cells. (8)

Solenoid Helical coil of chromatin fiber that serves as an intermediate structure in chromosome condensation. (13)

Solute Substance that is dissolved in a solvent, forming a solution; usually present at a lesser concentration than the solvent. (2)

Solvent Substance, usually liquid, in which other substances are dissolved, forming a solution; usually present at a greater concentration than the solute. (2)

Somatic cells Cells that make up most of the body of a multicellular organism and have no reproductive capabilities; "body cells." (17, 22)

Somatic nervous system Component of the peripheral nervous system that controls voluntary movements of skeletal muscles. (20)

Somatic rearrangement Process whereby the DNA of somatic cells is rearranged during the assembly of a complete V gene for the light or heavy chain of an immunoglobulin molecule. (24)

Somatostatin Small peptide hormone that functions mainly to inhibit growth hormone action. (21)

Somite Discrete cluster of mesodermal cells that are found in postneurula vertebrate embryos and give rise to such repeated structures as the vertebrae and ribs. (22)

Spacer DNA Lengths of DNA of eukaryotic chromosomes that are not associated with nucleosomes and are preferentially digested by deoxyribonuclease; also called linker DNA. (13)

Spacer sequence Transcribed sequence of nucleotides in an RNA molecule that is excised during RNA processing. (16)

S1 particle Head of myosin molecule released by the action of trypsin on heavy meromyosin; has ATPase activity and an actin binding site. (19)

Specific heat The amount of heat needed to raise the temperature of 1 gram of a substance 1°C. (2)

Sperm Flagellated haploid gamete produced by the male. (15, 22)

Spermatid Haploid product of second meiotic division of primary spermatocytes. (22)

Spermatogenesis Process of sperm development. (22)

Spermatogonium Primordial germ cell that will eventually give rise to spermatozoa. (22)

Spermatozoan Mature sperm cell. (15, 22)

Spermiogenesis Morphogenetic process in which spermatids are transformed into spermatozoa. (22)

S phase (synthetic phase) Phase in the eukaryotic cell cycle in which DNA is synthesized. (14)

Sphere of hydration Cluster of water molecules about an ion due to the interaction of the ion with the oppositely charged region of the polar water molecule; allows for the dissociation of ion pairs in solution. (2)

Sphingolipids Class of lipids based on the amine alcohol sphingosine, with a hydroxyl group available for binding to a wide variety of polar groups. (3)

Sphingosine Amine alcohol with a long hydrocarbon chain having a single site of unsaturation near its polar end; serves as backbone for sphingolipids. (3)

Spindle fiber An individual microtubular fiber associated with the mitotic spindle. *See also* Chromosomal fiber; Polar fiber. (14)

Spindle pole Focal point on either end of the mitotic spindle consisting of a pair of centrioles surrounded in animal cells by an aster of microtubules. (14)

Spontaneous Describing reactions and processes that are characterized by a negative change in free energy and are therefore capable of proceeding in the direction indicated without an input of energy. (5)

Spore Haploid product of meiosis in organisms that display an alternation of generations; gives rise to the haploid form of the organism upon spore germination. (15)

Sporophyte Diploid generation in the life cycle of an organism that alternates between haploid and diploid forms; form that produces spores by meiosis. (15)

Spot desmosome Localized point of tight adhesion between plasma membranes of adjacent cells. (7)

S region *See* Switch region.

SRP *See* Signal recognition particle.

SSB *See* Single-stranded binding protein.

Staining Incubation of fixed tissue specimens in a solution of dye, heavy metal, or other substance that binds specifically to selected cellular constituents, thereby giving those constituents a distinctive color or electron density. (A)

Standard free energy change ($\Delta G°$) Free energy change accompanying the conversion of 1 mole of reactants to 1 mole of products, with the temperature, pressure, pH, and concentration of all relevant species maintained at standard values. (5)

Standard reduction potential (E_0) Convention used to quantify the electron transport potential of oxidation-reduction couples relative to the H^+/H_2 redox couple, which is assigned an E_0 value of 0.0 V at pH 7.0. (11)

Standard state Most stable form of a substance when present at a temperature of 25°C (298 K) and a pressure of 1 atmosphere; if in solution, the substance is present at a concentration of 1 *M*. (5)

Starch Storage polysaccharide in plants, consisting of D-glucose repeating subunits linked by $\alpha(1 \rightarrow 4)$ bonds as well as $\alpha(1 \rightarrow 6)$ bonds. *See also* Amylopectin; Amylose. (3)

START The point in the G-1 phase of the cell cycle at which the decision is made either to progress through the cell cycle or to enter the G-0 phase. (22)

State Condition of a system defined by various properties, such as temperature, pressure, and volume. (5)

Steady state Nonequilibrium condition of an open system through which matter is flowing, such that all components of the system are present at constant, nonequilibrium concentrations. (5)

Stem cell Animal cell that is capable of generating specific kinds of cells and tissues continuously. *See also* Hematopoietic stem cell. (14, 23)

Stereoisomers Two molecules that have the same structural formula but are not superimposable; stereoisomers are mirror images of each other. (2)

Stereo pair Two photographs of the same object taken at two different angles, such that when viewed with a stereoscopic viewer, they create the impression of a three-dimensional image. (A)

Steroid Hydrophobic molecule derived from cholesterol; some steroids function as hormones. (3)

Steroid hormone Any of several steroids derived from cholesterol that function as regulatory signals, moving via the circulatory system to target tissues, where the hormone crosses the plasma membrane and interacts with an intracellular receptor to form a hormone-receptor complex capable of activating transcription of specific genes. *See also* Intracellular receptor. (17, 21)

Sticky end *See* Cohesive end.

Stop codon Sequence of nucleotides on an mRNA molecule that signals the end of the genetic message; UAG, UAA, and UGA are stop codons. (16)

Storage macromolecule Polymer that consists of one or a few kinds of subunits in no specific order and that serves as a storage form of monosaccharides; starch and glycogen are storage macromolecules. (2)

Stress fiber Temporary contractile bundle of actin microfilaments and myosin, resembling tiny myofibrils in structure and function; prominent component of the cytoskeleton of fibroblast cells in culture. (18)

Striated muscle Skeletal muscle responsible for voluntary movements of body parts. (19)

Strict aerobe *See* Obligate aerobe.

Strict anaerobe *See* Obligate anaerobe.

Stroma Unstructured semifluid matrix that fills the interior of the chloroplast. (4, 12)

Stroma lamellae Membranes that connect thylakoid disks within the chloroplast. (4, 12)

Structural gene Nucleotide sequence that codes for a specific polypeptide product. (17)

Structural macromolecule Polymer that consists of one or a few kinds of subunits in no specific order and that provides structure and mechanical strength to the cell; cellulose is a structural macromolecule. (2)

Substrate activation Role of the active site of an enzyme in making a substrate molecule maximally reactive by subjecting it to the appropriate chemical environment for catalysis. (6)

Substrate analogue Compound that resembles the true substrate of an enzyme enough to bind to the active site but is itself unable to undergo reaction. (6)

Substrate induction Regulatory mechanism for catabolic pathways in which the synthesis of enzymes involved in the pathway is activated in the presence of the substrate and inactivated in the absence of the substrate. (17)

Substrate-level phosphorylation Formation of ATP from ADP and inorganic phosphate as part of a specific reaction mechanism in a metabolic pathway. (10, 11)

Succinate-coenzyme Q reductase Enzyme complex of the electron transport chain involved in the transfer of electrons from $FADH_2$ to coenzyme Q. (11)

Sucrose Disaccharide consisting of glucose linked to fructose; common table sugar. (12)

Supercoil Extra twist in a circular DNA molecule that causes the helix to coil upon itself. (3)

Superrepressor mutant (i^s) Mutant form of the *i* gene in the *lac* operon of *Escherichia coli* that encodes a repressor that binds tightly to the operator under all conditions and therefore represses enzyme synthesis regardless of whether inducer is present. (17)

Suppressor strain Strain of bacteria with a mutant tRNA molecule that inserts an amino acid at what would otherwise be a stop codon. (16)

Suppressor T (T_s) cell T lymphocyte that suppresses or diminishes the immune response of other lymphocytes. (24)

Suppressor tRNA Mutant tRNA molecule that inserts an amino acid at what would otherwise be a stop codon. (16)

Supramolecular structure Component of a cell or an organelle consisting of macromolecules ordered into a variety of multimolecular assemblies. (2)

S value *See* Sedimentation coefficient.

Svedberg unit Unit for expressing the sedimentation coefficient of biological macromolecules; $1\ S = 10^{-13}$ second; in general, the greater the mass of a particle, the greater the observed sedimentation rate, though the relationship is not linear. (3)

Switch region (S region) DNA sequence composed of multiple copies of short, repeated elements that recombine and move the V_H gene from a position adjacent to one C_H gene to a position adjacent to another C_H gene; active site for class switching. (24)

Symport Coupled transport of two solutes across a membrane in the same direction. (8)

Synapse (a) To come together; pair. (b) Small gap between a neuron and another cell (neuron, muscle fiber, or gland cell), across which the nerve impulse is transferred by direct electrical connection or by chemicals called neurotransmitters. (15, 20)

Synaptic cleft Gap between the presynaptic and postsynaptic membranes at the junction between two nerve cells. (20)

Synaptic knob Structure present at the end of axons that is responsible for transmitting the impulse to the next cell; also called a nerve terminal or synaptic bouton. (20)

Synaptic vesicle Membrane-bounded structure that is found near the terminus of the presynaptic axon and that contains neurotransmitter molecules that are discharged exocytotically into the synaptic cleft upon arrival of a nerve impulse. (20)

Synaptonemal complex Structure present during prophase I of meiosis that allows the close pairing of homologous regions of synapsed chromosomes known to be essential for crossing over. (15)

Syncytial blastoderm Stage of *Drosophila* development in which the embryo is a single cell containing about 6000 nuclei. (22)

Synthetic work Use of energy to generate new molecules by forming new covalent bonds. (5)

Target tissue Tissue that is specifically affected by a particular hormone because it has receptors for that hormone located either on the plasma membrane or within the cytoplasm of the cell. (21)

TATA box Tetranucleotide sequence located about 30 nucleotides upstream from the initiation site for transcription by RNA polymerase II in eukaryotic cells; thought to be important in positioning RNA polymerase II molecules correctly on the DNA template. (16)

Tau protein Accessory microtubular protein that binds along the entire length of the microtubule and appears to enhance the polymerization of tubulin subunits. (18)

Tautomerization Change in the localization of a proton in a molecule accompanied by an alteration in the chemical properties of the molecule. (10)

Tay-Sachs disease Lysosomal storage disease that occurs as a recessive trait, mainly among Jews of eastern European origin. (9)

TCA cycle *See* Tricarboxylic acid cycle.

T cell *See* T lymphocyte.

T cell growth factor *See* Interleukin-2.

Tektin Protein component of axonemal microtubules that helps arrange tubulin molecules into the A and B tubules of the doublet. (19)

Telocentric chromosome Chromosome with its centromere at one end such that the chromosome appears to have only one arm. (14)

Telophase Final phase of mitosis, in which daughter chromosomes arrive at the poles of the spindle and begin to decondense, accompanied by the reappearance of the nuclear envelope and nucleoli. (14)

Telophase I Fourth stage in the first meiotic division, in which movement of homologous chromosomes to opposite poles is completed. (15)

TEM *See* Transmission electron microscope.

Temperate bacteriophage Bacteriophage that can either undergo lytic growth in a host cell or integrate its DNA into the bacterial genophore such that the bacteriophage DNA (prophage) is replicated with the bacterial DNA, often for many generations. (13)

Temperature-sensitive (Ts) mutation Mutation characterized by an altered protein that functions normally at one temperature, called the *permissive* temperature, but functions abnormally or not at all at another temperature, called the *restrictive* temperature; usually understood to mean that the restrictive temperature is higher than the permissive temperature, unless specifically identified as a *cold-sensitive* mutation. (22)

Temperature-stabilizing capacity Ability to maintain structure with the addition of heat. (2)

Template Preexisting molecule used as a pattern for the synthesis of a complementary molecule. (3)

Terminal glycosylation Modification of glycoproteins in the Golgi complex; removal and/or addition of sugars to the carbohydrate side chains formed by prior core glycosylation in the endoplasmic reticulum. (16)

Terminal oxidase Electron transfer intermediate that is capable of direct transfer of electrons to oxygen. (11)

Terminal web Specialized cortex at the apex of an intestinal epithelial cell to which the lower portion of the bundle of actin microfilaments of a microvillus is anchored; contains a dense meshwork of spectrin molecules that overlies a layer of intermediate filaments. (18)

Termination signal Point on the chromosome at which transcription is terminated for a specific transcriptional unit. (16, 17)

Tertiary structure Level of protein structure involving long-distance interactions between stretches of amino acids, often with three-dimensional folding. (3)

Testis Male organ that produces spermatozoa; male gonad. (15, 22)

Testosterone Steroid hormone produced by the adrenal cortex and testes that generally results in development and maintenance of male characteristics in an organism. (21, 22)

Tetrad Bundle of four chromatids formed by the synapsis of homologous chromosomes in the first meiotic division. *See also* Bivalent. (15)

Thermodynamics Area of science that deals with the laws governing the energy transactions that accompany all physical processes and chemical reactions. (5)

Thermodynamic spontaneity Probability of a reaction occurring as determined by the changes in entropy and free energy that accompany the reaction. (5)

Theta replication Bidirectional replication mode of circular DNA, so called because the intermediates look like the Greek letter theta. (14)

Thick filament Arrangement of myosin molecules in a staggered array with the heads of the myosin molecules projecting out in a repeating pattern; component of myofibrils of striated muscle cells. (19)

Thin filament Arrangement of two F-actin molecules in a helix associated with tropomyosin and troponin; component of myofibrils of striated muscle cells. (19)

Thin-layer chromatography Procedure for separating compounds by chromatography in a medium, such as silicic acid, that is bound as a thin layer to a glass or metal surface. (7)

30S initiation complex Complex formed by the association of mRNA, the 30S ribosomal subunit, an f-Met-tRNA$^{f\text{-Met}}$ molecule, and several initiation factors. (16)

Threshold stimulus Sufficiently strong stimulus that is able to depolarize the membrane by at least 20 mV and thereby initiate a nerve impulse. (20)

Thylakoid disk Flattened membranous sac in the granum of a chloroplast that contains the pigments, enzymes, and electron carriers involved in the light-requiring reactions of photosynthesis. (4)

Thymine (T) Nitrogen-containing aromatic base, chemically designated as a pyrimidine, which serves as an informational monomeric unit when present in DNA with other bases in a specific sequence. (3, 16)

Tight junction Contact between the plasma membranes of two adjacent animal cells that extends around the circumference of the cells and effectively seals off the two sides of the junction plane, such that materials must pass through the cells to get from one compartment to another. (4, 7)

Titin Muscle protein that links thick filaments to the Z band of the sarcomere. (19)

T lymphocyte (T cell) White blood cell involved in the cell-mediated responses of the immune system. (24)

Tonofilament Keratin-containing intermediate filament found in the epithelial cells that cover the body surfaces and line its cavities; thought to be involved in tensile strength. (7, 18)

Topoisomerase Enzyme that catalyzes the interconversion of the relaxed and supercoiled forms of DNA. (13)

Topological isomers Molecules that differ only in their state of supercoiling. (13)

Totipotent Describing a cell or nucleus that is not irreversibly committed to a single specific developmental fate and is therefore still able to direct the complete development of an embryo. (17, 22)

Trans-acting mutation Mutation that alters the function of a diffusible regulatory protein, thereby causing defects in regulation of structural genes that are not physically linked to the mutation. *See also* Cis-acting mutation. (17)

Transamination Transfer of an amino group from an amino acid to an α keto acid acceptor. (11)

Transcellular transport Transfer of substances across a cell, involving inward transport on one side and outward transport on the other side. (8)

Transcription Process whereby the genetic information of DNA is used to specify the nucleotide sequence of a complementary RNA strand. (16)

Transcriptional control Regulatory mechanisms operative at the level of transcription involving selective accessibility of chromosomal regions to transcription and the action of regulatory proteins at the transcription site. (17)

Transcriptional factor Sequence-specific DNA-binding protein that regulates the extent of gene transcription by binding to a DNA recognition site near the structural gene(s). (17)

Transcription unit Segment of DNA that is transcribed as a single continuous RNA molecule. (16)

Transducing bacteriophage Bacteriophage capable of carrying bacterial DNA from one host cell to another by reversibly incorporating host cell DNA into its own genome. (15)

Transduction Transport of a DNA segment from one cell to another by a virus. (15)

Trans face *See* Maturing face.

Transfer RNA (tRNA) RNA molecule that directs the correct amino acid to each successive codon of a messenger RNA in an activated form conducive to peptide bond formation. (3, 16)

Transformation (a) Genetic modification induced by the assimilation of exogenous DNA into a cell. (13, 15, B) (b) Incorporation of exogenous DNA into an existing genome. (13, 15) (c) Process whereby a normal cell is converted into an abnormal one. (23)

Transforming principle Factor present in pathogenic strains of *Pneumococcus* that confers pathogenicity upon previously nonpathogenic strains of the bacterium; shown by Avery, MacLeod, and McCarty to be DNA. (13)

Transgene DNA that is modified in vitro, then introduced into the genome of an organism, whether of the species from which the DNA was originally obtained or of a different species. (B)

Transgenic organism Any organism whose genome contains DNA, from the same species or a different species, that has been modified by the techniques of genetic engineering. (B)

Transition temperature Temperature at which a membrane will undergo a sharp increase in fluidity (melting), as determined by the kinds of fatty acid side chains present. (7)

Translation Process whereby the genetic information in a messenger RNA molecule is used to specify the sequence of amino acids in a polypeptide. (16)

Translational control Regulatory mechanisms operative at the level of translation, involving selective utilization of specific mRNAs, variations in rates of mRNA degradation, and differential availability of specific tRNAs. (17)

Translocation Shifting of a ribosome along an mRNA molecule to direct the binding of the next aminoacyl tRNA in the process of polypeptide chain elongation. (16)

Transmembrane α helix Portion of a polypeptide chain that contains relatively nonpolar amino acids and that spans a membrane in the form of an α helix. (7)

Transmembrane protein kinase Enzyme capable of phosphorylating tyrosine, threonine, or serine groups of specific proteins on the cytoplasmic side of the plasma membrane when activated by binding of a specific growth factor on the outer membrane surface. *See also* Growth factor. (21)

Transmission electron microscope (TEM) Type of electron microscope in which the image is formed by the varying extents to which electrons are transmitted by different parts of the specimen. (A)

Transplantation antigen Cell-surface antigen that elicits an immune response by T lymphocytes against grafted tissue or organ transplants; also called histocompatibility antigen. (24)

Transport protein *See* Permease.

Transport vesicle Coated vesicle formed by the pinching off of the rough endoplasmic reticulum; involved in the transfer of lipids and proteins synthesized in the rough ER to the Golgi complex or other subcellular location. (9)

Transverse diffusion Movement of a phospholipid or protein from one layer of the membrane to the other, a thermodynamically unfavorable and therefore infrequent event. (7)

Transverse tubule system (T-system) Invaginations of the plasma membrane that penetrate into a muscle cell and conduct electrical impulses rapidly into the interior of the cell, allowing calcium to be released onto all of the myofibrils at the same time. (19)

Treadmilling Process whereby a microfilament is assembled on the plus end by polymerization of G-actin molecules and simultaneously disassembled at the minus end by dissociation of actin monomers; individual actin molecules are therefore continually transferred from the plus end of the microfilament to the minus end, even though the net length of the microfilament does not change. (18)

Tricarboxylic acid cycle (TCA cycle) Cyclic series of reactions in which acetyl CoA is oxidized fully to carbon dioxide in the presence of oxygen; a component of aerobic respiration; also called the citric acid cycle or the Krebs cycle. (11)

Trigger protein Protein that by its rapid accumulation during the G-1 phase of the cell cycle causes the cell to pass into S phase and eventually to divide. (14)

Triglyceride Glycerol molecule linked by ester bonds to three fatty acids that can vary in chain length and degree of saturation. *See also* Fat; Vegetable oil. (3, 11)

Triple bond Chemical bond formed between two atoms as a result of sharing three pairs of electrons. (2)

Triplet code Code in which three units of information are read as a unit. (16)

Triskelion Structure of clathrin molecules consisting of three polypeptides radiating from a central vertex; the basic unit of assembly for clathrin coats. (9)

tRNA *See* Transfer RNA.

Tropomyosin Long, rodlike protein molecule that lies in the groove on either side of the actin helix; component of the calcium-sensitive switch that activates muscle contraction. (19)

Troponin Protein complex of three polypeptides that associates with a tropomyosin molecule; component of the calcium-sensitive switch that activates muscle contraction. (19)

Tryptophan (*trp*) operon Cluster of structural genes that code for enzymes involved in tryptophan synthesis as well as the control elements that regulate the synthesis of these enzymes in *Escherichia coli.* (17)

T-system *See* Transverse tubule system.

Tubulin Major protein component of microtubules; exists in microtubule as a dimer of α- and β-tubulin. (18)

Tumor Tissue that grows in a progressive, uncontrolled manner under conditions where normal tissue does not; a neoplasm. (23)

Tumor angiogenesis Establishment of a vascular system within a tumor. (23)

Tumorigenic Describing cells that have been transformed in culture and can cause malignant tumors when injected into animals.

Tumor progression Development over time of progressively more malignant phenotypic characteristics by a tumor. (23)

Tumor-suppressor gene Gene that limits the proliferative activity of a tissue and hence its ability to develop into a tumor; loss or inactivation of tumor-suppressor genes by mutation or chromosomal deletion is thought to play a major role in tumor development. (23)

Turnover Process whereby molecules are continuously degraded and replaced, thereby maintaining a relatively constant steady-state level. (5, 17)

Tyrosine protein kinase Enzyme that catalyzes transfer of the terminal phosphate group from ATP to a tyrosine in a target protein; several mitogen receptors have a tyrosine protein kinase activity that is located in the cytoplasmic portion of the receptor and is activated when the appropriate mitogen binds to the receptor. (21, 23)

Ubiquinone *See* Coenzyme Q. (11)

Ubiquitin Small protein that binds to damaged proteins, thereby marking them for proteolysis by endogenous proteases and lysosomes; found in virtually all eukaryotic cells. (10)

Ultimate carcinogen Reactive form of a procarcinogen that has been modified by the mixed-function oxidase system and can therefore react with cellular DNA to produce changes leading to tumor formation. *See also* Procarcinogen. (23)

Ultracentrifuge Instrument used to separate subcellular structures and macromolecules on the basis of size, shape, and density by centrifugation at very high speeds. (1)

Ultramicrotome Instrument used to slice an embedded biological specimen into ultrathin sections for electron microscopy. (1, A)

Uncoupler Substance capable of abolishing the link between electron transport and generation of ATP. (11)

Unit membrane Structure seen in electron micrographs of biological membranes, consisting of two dark electron-dense lines separated by an unstained space. (7)

Unsaturated fatty acid Fatty acid that contains one or more double or triple bonds. (3)

Unwindase *See* Helicase.

Up-regulation *See* Activation.

Upstream promoter element Region of DNA located immediately in front (i.e., at the 5′ end) of a eukaryotic gene and extending for about 100 base pairs; generally contains several regulatory elements that are necessary for basal levels of transcription but are common to many different genes and therefore do not account for specificity of expression in time or space. (17)

Uracil (U) Nitrogen-containing aromatic base, chemically designated as a pyrimidine, which serves as an informational monomeric unit when present in RNA with other bases in a specific sequence. (3, 16)

Vacuole Membrane-bounded organelle in the cytoplasm of a cell, used for temporary storage or transport. (4)

Vacuum evaporator Bell jar containing a metal electrode and a carbon electrode in which a vacuum can be created; used in the preparation of metal replicas of the surfaces of biological specimens. (A)

Valence (a) Charge on an ion. (2) (b) Number of identical antigenic determinants to which an immunoglobulin molecule can bind simultaneously. (24)

Variable domain Discrete region of an antibody molecule that functions to bind antigen; differences in variable domains among antibodies allow for recognition of a maximum number of antigens. (24)

Variable region N-terminal region of the light or heavy chain of an immunoglobulin molecule that differs greatly in amino acid sequence from the same region of other chains. (24)

Vectorial pumping Unidirectional movement of molecules across a membrane. (8)

Vegetable oil Triglyceride containing primarily unsaturated fatty acids, usually liquid at room temperature. *See also* Fat. (3)

Vesicle Small, spherical, membrane-bounded structure. (9)

V gene DNA sequence that codes for the variable regions of the light or heavy chain of the immunoglobulin molecule. (24)

Vimentin Major protein component of the intermediate filaments found in connective tissue and other cells of mesenchymal origin. (18)

Viral oncogene (v-*onc*) Cancer-causing gene present in the viral genome. *See also* Proto-oncogene. (23)

Virulent bacteriophage Bacteriophage that always undergoes lytic growth in a host cell, leading to the production of progeny bacteriophages and resulting in death and lysis of the host cell. (13)

Virus Subcellular parasite composed of a protein coat and DNA or RNA, incapable of independent existence. (2, 4)

Vitelline membrane Outer membrane of the egg of marine invertebrates; a specialized coat that only sperm can penetrate. (22)

Vitellogenesis Phase in animal oogenesis (egg development) in which yolk accumulates. (22)

Vitellogenin Major storage protein present in the yolk of animal eggs. (22)

Wavelength Distance between the crests of two successive waves. (A)

Wild type Normal, nonmutant form of an organism, usually the form found in nature. (15)

Wobble Flexibility in the binding of the base in the 5′ position of the anticodon to the base in the complementary 3′ position of the codon. (16)

Work Transfer of energy from one place or form to another place or form by any process other than heat flow. (5)

Yolk granule Concentrated storage form of protein and carbohydrate in an egg cell. (22)

Z-DNA Left-handed helical configuration of a DNA molecule; alternative to B-DNA. (13)

z gene Structural gene in the *lac* operon of *Escherichia coli* that codes for β-galactosidase, an enzyme involved in lactose metabolism. (17)

Zinc-finger protein A protein characterized by a conserved domain that forms a complex with a zinc ion and binds to specific sites on a DNA molecule. (17)

Z line Dark line in the middle of the I band of striated muscle fibrils that defines the length of a sarcomere. (19)

Zona pellucida Outer layer of the mammalian egg; a specialized coat that only sperm can penetrate. (22)

Zygote Diploid cell formed by the union of two haploid gametes. (15, 22)

Zymogen granule Secretory granule containing digestive enzyme precursors. (16)

Photograph and Illustration Credits

Chapter 1 Figure 1-3: From *Cell Ultrastructure*, by William A. Jensen and Roderic B. Park. Copyright © 1967 by Wadsworth Publishing Co., Inc. Used by permission of the authors and publisher. Figure 1-4a: Courtesy of G. G. Borisy. Figure 1-4b: Courtesy of C. Greaves and J. Croxdale; photo provided by J. Croxdale. Figures 1-4c and 1-4d: Courtesy of J. Croxdale. Figure 1-C: Courtesy of Lawrence Livermore National Laboratory. Thomas P. Beebe, Jr., Troy E. Wilson, D. Frank Ogletree, Joseph E. Katz, Rod Balhorn, Miquel B. Salmeron, Wigbert J. Siekhaus, *Science* 243(1989):370. Used with permission.

Chapter 2 Figure 2-13: Courtesy Don Fawcett, copyright © Photo Researchers, Inc. Figure 2-14: Adapted from S. J. Singer and G. L. Nicholson, *Science* 175(1972):720. Copyright © 1972 by the AAAS. Figure 2-15; upper left: Courtesy of G. F. Bahr. Figure 2-15; upper right: Courtesy of E. F Newcomb. Figure 2-15; middle right: Courtesy of E. Frei & R. D. Preston.

Chapter 3 Figure 3-4: Adapted from R. E. Dickerson and I. Geis. *The Structure and Action of Proteins*. Redwood City, Calif.: Benjamin/Cummings, 1969. Copyright © 1969 by I. Geis. Figure 3-19a: Courtesy of J. Neel and M. Costello. Figure 3-19b: Courtesy of G. E. Palade. Figure 3-20: Courtesy of E. Frei & R. D. Preston.

Chapter 4 Figure 4-2: Courtesy of S. Ito. Figure 4-3a: Courtesy of L. D. Simon, Waksman Institute of Microbiology, Rutgers University, New Brunswick, N.J. Figure 4-3b: Courtesy of E. H. Newcomb. Figure 4-4: Courtesy of R. Rodewald, University of Virginia/BPS. Figure 4-5: Courtesy of H. Ris. Figure 4-6: From O. L. Miller, Jr., B. A. Hamkalo, and C. A. Thomas, Jr. *Science*, 169(1970):392. Copyright © 1970 by the AAAS. Figure 4-7: Courtesy of R. Rodewald, University of Virginia/BPS. Figure 4-8b: Micrograph by W. P. Wergin; photograph provided by E. H. Newcomb. Figure 4-9c: H. C. Aldrich and J. H. Gregg, *Experiments in Cell Research* 81(1973):407. Copyright © 1973 by Academic Press. Figure 4-10b: Courtesy of R. Rodewald, University of Virginia/BPS. Figure 4-10c: From J. P. Strafstrom and L. A. Staehelin, *J. Cell Biol.* 98(1984):699. Reproduced by copyright permission of The Rockefeller University Press. Figure 4-11c: Courtesy of N. Simionescu. Figure 4-12b: From J. B. Rattner and B. R. Brinkley, *J. Ultrastructure. Res.* 32(1970):316. Copyright © 1970 by Academic Press. Figure 4-13b: Micrograph by W. P. Wergin; photograph provided by E. H. Newcomb. Figure 4-14b: Courtesy of H. S. Pankratz, Michigan State University/BPS. Figure 4-14c: Courtesy of J. F. King, University of California School of Medicine/BPS. Figure 4-15b: Courtesy of E. H. Newcomb. Figure 4-17b: R. Rodewald, University of Virginia/BPS. Figure 4-17c: Courtesy of N. Simionescu. Figure 4-18b: From S. E. Frederick and E. H. Newcomb, *J. Cell Biol.* 43(1969):343. Reproduced by copyright permission of The Rockefeller University Press; photo provided by E. H. Newcomb. Figure 4-19b: Micrograph by P. J. Gruber; photo provided by E. H. Newcomb. Figure 4-21: From Sigrid Regauer, Werner W. Franke, and Ismo Virtanen, *J. Cell Biol.* 100(1988):997–1009. Figure 4-23: From P. H. Raven, R. F. Evert, and H. A. Curtis, *Biology of Plants*, 2nd edition. New York: Worth, 1981. Photo provided by R. F. Evert. Figure 4-24: Courtesy of H. C. Aldrich, University of Florida, Gainesville. Figure 4-25: Courtesy of R. C. Williams and H. W. Fisher.

Chapter 5 Figure 5-3: From D. A. Cuppels and A. Kelman, *Phytopathology* 70(1980):1110. Photo provided by A. Kelman. Copyright © 1980 by the American Phytopathological Soc. Figure 5-6: Courtesy of H. H. Iltis.

Chapter 6 Figure 6-2: Courtesy of Richard J. Feldmann, National Institutes of Health. Figure 6-3: Reproduced with permission from D. M. Blow and T. A. Steitz in the *Annual Review of Biochemistry* Vol. 39, p. 63, copyright © 1970 by Annual Reviews, Inc.

Chapter 7 Figure 7-1: Courtesy of Steve Bretscher. Figure 7-3: Micrograph courtesy of J. David Robertson. Figure 7-4: Adapted from R. M. Dowben, *General Physiology*. 1969 by copyright permission of Rockefeller University Press. Figure 7-6: Courtesy of John Heuser. Figure 7-7: Adapted from S. J. Singer and G. L. Nicholson, *Science* 175(1972):720. Copyright © 1972 by the AAAS. Figure 7-12g: Courtesy of V. T. Marchesi. Figure 7-13: Micrographs of E and P faces courtesy of P. Claude. Figure 7-18b: From Douglas E. Kelly, *J. Cell Biol.* 28(1966):51. Reproduced by copyright permission of The Rockefeller University Press. Figure 7-19b: From L. Orci and A. Perrelet, *Freeze-Etch Histology*. Heidelberg: Springer-Verlag, 1975. Figures 7-19c and d: Courtesy of P. Claude. Figure 7-20b: From C. Peracchia and A. F. Dulhunty, *J. Cell Biol.* 70(1976):419. Reproduced by copyright permission of The Rockefeller University Press. Figure 7-20c: Courtesy of P. Claude. Figure 7-21: Courtesy of S. Ito. Figure 7-22: Courtesy of Jerome Gross. Figure 7-25a: Micrograph by B. A. Palevitz; photo provided by E. H. Newcomb. Figure 7-26a: Micrograph by W. P. Wergin; photo provided by E. H. Newcomb. Figure 7-26c: Courtesy of E. H. Newcomb.

Chapter 8 Figure 8-3: Adapted from R. Collander, *Trans. Faraday Soc.* 33(1937):986. Copyright © 1936 by The Royal Society of Chemistry. Figure 8-15: Adapted from E. Epstein, D. W. Rains, and O. E. Elzam, *Proc. Nat. Acad. Sci.* 49(1963):684. Figure 8-16: Adapted from E. Epstein, D. W. Rains, and O. E. Elzam, *Proc. Nat. Acad. Sci.* 49(1963):684.

Chapter 9 Figure 9-4: From L. Orci and M. Perrelet, *Freeze-Etch Histology*. Heidelberg: Springer-Verlag, 1975. Figure 9-5: Courtesy of D. S. Friend. Figure 9-6a: Courtesy of G. E. Palade. Figure 9-7: From R. Wetherbee, *Protoplasma* 95(1978):347. Copyright © 1978 by Springer-Verlag. Figure 9-9: From W. J. Brown and M. G. Farquhar, *Cell* 36(1984):295. Copyright © 1984 by MIT Press. Figure 9-10a: From R. A. Crowther and B. M. F. Pearse, *J. Cell Biol.* 91(1981):790. Reproduced by copyright permission of The Rockefeller University Press. Figure 9-14: From M. M. Perry and A. B. Gilbert, *J. Cell Sci.* 39(1979):257. Copyright © 1979 by The Company of Biologists Ltd. Figure 9-15b: Courtesy of John Heuser. Figure 9-18: Courtesy of Z. Hruban. Figures 9-22 and 9-23: Courtesy of H. Shio and P. B. Lazarow. Reproduced by copyright permission of The Rockefeller University Press; photos provided by P. B. Lazarow. Figure 9-A: Courtesy of E. Ungewickell and D. Branton; photo provided by D. Branton.

Chapter 11 Figure 11-1: Courtesy of Charles R. Hackenbrock. Figure 11-4b: Courtesy of K. R. Porter. Figure 11-5: From L. Packer, *Ann. N. Y. Acad. Sci.* 227(1974):166. Copyright © 1974 by the New York Academy of Sciences. Photo provided by H. T. Ngo. Figure 11-6: Courtesy of S. M. Wang. Figure 11-7: Courtesy of E. P. Gogol and R. A. Capaldi. Figure 11-A: Adapted from W. Stoeckenius, *Sci. Amer.* 234(1976):38. Copyright © 1976 by Scientific American, Inc. All rights reserved.

Chapter 12 Figure 12-1: Micrograph by M. W. Steer; photo provided by E. H. Newcomb. Figure 12-2: From U. G. Johnson and K. R. Porter *J. Cell Biol.* 38(1968):403. Reproduced by copyright permission of The Rockefeller University Press. Photo provided by Dr. Ursula Goodenough. Figures 12-3a and c: Micrographs by W. P. Wergin; photos provided by E. H. Newcomb. Figure 12-3d: Adapted from T. E. Weier, C. R. Stocking, and L. K. Shumway, *Brookhaven Symp. Biol.* 19(1961):371. Figure 12-5: Adapted from F. P. Zscheile and C. L. Comar, *Botan. Gaz.* 102(1941):463. Copyright © by The University of Chicago Press. Figure 12-6: Adapted from F. T. Haxo and L. R. Blinks, *J. Gen Physiol.* 33(1950):389. Reproduced by copyright permission of The Rockefeller University Press. Figures 12-10b and c: Courtesy of Dr. Kenneth Miller, Brown University. Figure 12-12: Courtesy of Dr. Kenneth Miller, Brown University. Figure 12-17: From E. Fritz, R. F. Evert, and W. Heyser, *Planta* 159(1983):193. Copyright © 1983 by Springer-Verlag. Photo provided by R. F. Evert. Figure 12-18: From S. E. Frederick and E. H. Newcomb, *J. Cell Biol.* 43(1969):343. Reproduced by copyright permission of The Rockefeller University Press. Photo provided by E. H. Newcomb. Figure 12-22: From R. N. Trelease, P. J. Gruber, W. M. Becker, and E. H. Newcomb, *Plant Physiology* 48(1971):461. Figure 12-A: Courtesy of James A. Bassham.

Chapter 13 Figure 13-6b: J. C. Wang, *Sci. Amer.* 247(1982):94. Figure 13-10: Adapted from J. Marmur and P. Doty, *Nature* 183(1959):1428. Figure 13-18: From Donald E. Olins and Ada L. Olins, *Amer. Sci.* 66(1978):704. Reprinted by permission of American Scientist, the journal of Sigma Xi. Photo courtesy of Donald and Ada Olins, University of Tennessee-Oak Ridge. Figure 13-21: From D. R. Wolstenholme, K. Koike, and P. Cochran-Fouts, *Cold Spring Harbor Symp. Quant Biol.* 38(1973):267. Copyright © 1973 by Cold Spring Harbor Press. Figure 13-22: Adapted from L. A. Grivell, *Sci. Amer.* 248(1983):78. Copyright © 1983 by Scientific American, Inc. All rights reserved. Figure 13-23a: From L. Orci and A. Perrelet, *Freeze-Etch Histology*. Heidelberg: Springer-Verlag, 1975. Figure 13-23b: Micrograph by S. R. Tandon; photo provided by E. H. Newcomb. Figure 13-24a: L. Orci and A. Perrelet, *Freeze-Etch Histology*. Heidelberg: Springer-Verlag, 1975. Figure 13-25: From L. Orci and A. Perrelet, *Freeze-Etch Histology*. Heidelberg: Springer-Verlag, 1975. Figure 13-27: From A. C. Faberge, *Cell Tiss. Res.* 15(1974):403. Heidelberg: Springer-Verlag, 1974. Figure 13-29: Courtesy of G. L. Scott. Figure 13-30: From Sasha Koulish and Ruth G. Kleinfeld, *J. Cell Biol.* 23(1964)39. Reproduced by copyright permission of The Rockefeller University Press. Figure 13-B: Courtesy of Robley C. Williams, Virus Laboratory, University of California, and Harold W. Fisher, University of Rhode Island. Figure 13-C: Courtesy of Dr. L. D. Simon, Waksman Institute of Microbiology, Rutgers University, New Brunswick, N.J. Figure 13-D: Courtesy of Dr. R. Humbert, Stanford University/BPS.

Chapter 14 Figure 14-8: From J. H. Taylor, ed., *Molecular Genetics*, Part I. New York: Academic Press, 1963. Copyright © 1963 by Academic Press. Figure 14-9: Courtesy of J. Cairns. Figures 14-12a and b: From D. J. Burks and P. J. Stambrook, *J. Cell Biol.* 77(1978):762. Reproduced by permission of The Rockefeller University Press. Photos provided by P. J. Stambrook. Figure 14-21: Courtesy of W. T. Jackson. Figure 14-22a: Courtesy of H. L. Wedberg. Figure 14-22b: Courtesy of G. F. Bahr. Figure 14-23: From E. Karsenti, J. Newport, R. Hubbl, and M. Kirschner, *J. Cell Biol.* 98(1984):1730. Reproduced by copyright permission of The Rockefeller University Press. Photo provided by M. Kirschner. Figure 14-24: Courtesy of P. Kronebusch and G. G. Borisy. Figure 14-25: From H. W. Beams and R. G. Kessel, *Amer. Sci.* 64(1976):279. Reprinted by permission of American Scientist, the journal of Sigma Xi. Figure 14-26: Micrograph by B. A. Palevitz; photo provided by E. H. Newcomb.

Chapter 15 Figure 15-9: From P. B. Moens, *Chromosoma* 23(1968): 418. Copyright © 1968 by Springer-Verlag. Figure 15-10: Courtesy of James L. Walters. Figure 15-22: Courtesy of Charles C. Brinton, Jr., and Judith Carnehan.

Chapter 16 Figure 16-16: From S. H. Kim et al., *Science* 185(1974):435. Copyright © 1974 by the AAAS. Figure 16-17: From O. L. Miller, Jr., B. A. Hamkalo, and C. A. Thomas, Jr., *Science* 169(1970):392. Copyright © 1970 by the AAAS. Figure 16-33: From L. Orci and A. Perrelet, *Freeze-Etch Histology*. Heidelberg: Springer-Verlag, 1975.

Chapter 17 Figure 17-14: From D. D. Brown and I. B. Dawid, *Science* 160(1968):272. Copyright © 1968 by the AAAS. Figures 17-24a and b: Reprinted with permission from S. R. C. Elgin, *CRC Crit. Rev. Biochem.* 3(1982):1. Copyright © by CRC Press, Inc., Boca Raton, Fla. Figure 17-24c: From J. J. Bonner and M. L. Pardue, *Cell* 12(1977):227. Copyright © 1977 by MIT Press. Figure 17-25: From Joseph G. Gall, *Methods in Cell Physiology* (D. M. Prescott, ed.) vol. 2, p. 37. New York: Academic Press, 1966. Copyright © 1966 by Academic Press.

Chapter 18 Figure 18-1: From J. E. Heuser and M. W. Kirscher, *J. Cell Biol.* 86(1980):212–234. Figure 18-2: Courtesy of Dr. John Heuser. Figure 18-3: From H. Ris, *J. Cell Biol.* 100(1984):1474. Reproduced by copyright permission of The Rockefeller University Press. Figure 18-4b: Courtesy of L. E. Roth, Y. Shigenaka, and D. J. Pihlaja/BPS. Figures 18-6a and b: Courtesy of Helen Kim, Lester Binder, and J. L. Rosenbaum. Figure 18-9: From Nobutaka Hirokawa, *J. Cell Biol.* 94(1982):129. Reproduced by copyright permission of The Rockefeller University Press. Figure 18-10: Courtesy of John Heuser. Figure 18-11b: Courtesy of R. Niederman and J. Hartwig. Figure 18-12: Courtesy of J. Roger Craig. Figure 18-16: Courtesy of S. Ito. Figure 18-17: From Nobutaka Hirokawa, *J. Cell Biol.* 94(1982):425. Reproduced by copyright permission of The Rockefeller University Press. Figure 18-18: From M. W. Aynardi, P. M. Steinert and R. D. Goldman, *J. Cell Biol.* 98(1984):1407. Reproduced by copyright permission of The Rockefeller University Press. Photo provided by R. D. Goldman.

Chapter 19 Figure 19-3: Courtesy of J. Roger Craig. Figure 19-6b: Courtesy of H. Ris. Figures 19-7 and 19-8a: Courtesy of C. Franzini-Armstrong. Figure 19-13: Courtesy of H. Ris. Figure 19-19: Courtesy of Don Fawcett, copyright © Photo Researchers, Inc. Figures 19-22a and b: From W. Jeon and M. S. Jeon, *J. Protozool.* 23(1976):83. Figure 19-24: From U. W. Goodenough, *J. Cell Biol.* 96(1983):1610. Reproduced by copyright permission of The Rockefeller University Press. Figure 19-25a: Courtesy of C. J. Brokaw, California Institute of Technology. Figure 19-26: Courtesy of William L. Dentler. Figure 19-27: Courtesy of G. G. Borisy. Figure 19-30: From D. A. Cuppels and A. Kelman, *Phytopathology* 70(1980):1110. Copyright © 1980 by the American Phytopathological Society. Photo provided by A. Kelman.

Chapter 20 Figure 20-2b: Courtesy of Ed Reschke. Figure 20-3d: Adapted from B. B. Boycott, L. Peichl, and H. Wässle, *Proc. Roy. Soc. London B.* 203(1978):229. Figure 20-4: Courtesy of G. L. Scott, J. A. Feilbach, and T. A. Duff. Figure 20-20: From J. Cartaud, E. L. Bendetti, A. Sobel, and J. P. Changeux, *J. Cell Sci.* 29(1978):313. Copyright © 1978 by The Company of Biologists, Ltd.

Chapter 21 Figure 21-A: From Richard G. Kessel and Randy H. Kardon, *Tissues and Organs: A Text-Atlas of Scanning Electron Microscopy*. New York: W. H. Freeman & Co., Copyright © 1979.

Chapter 22 Figure 22-5: Courtesy of Peter Devreotes. Figure 22-9b: Courtesy of D. M. Phillips, The Population Council. Figure 22-10: A. Tsuiki, K. Hoshi, A. Saito, K. Kyono, H. Hoshiai, and M. Suzuki, *Acta Obstet. Gynecol. Jpn.* 35(1983):1705. Photos courtesy of Akira Tsuiki and Kazuhiko Hoshi. Figure 22-13: Courtesy Lionel Jaffe, *J. Cell Biol.* 76(1978):448. Figure 22-25: Jean Paul Thiery and Michael Sémériva, *Nature* 325(1987):349. Figure 22-29: D. A. Melton, Cell 51(1987). Figure 22-30: Courtesy of Robert Rugh and Landrum B. Shettles. Figure 22-31a: Adapted from B. T. Wakimoto and T. C. Kaufman, *Dev. Biol.* 81(1981):51. Copyright © 1981 by Academic Press. Figure 22-33: From P. W. Ingham, *Nature* 335(1988):29. Figure 22-34: Courtesy of E. B. Lewis. Figure 22-36: From John Sulston, *Developmental Biology* 56(1977):111. Figure 22-A: Courtesy of B. Kay Simandl and John F. Fallon. Figure 22-D: From Michael Hoffman, *Genes and Development*, Hoffman and Goodman, 1987, vol. 2, pp. 615–625.

Chapter 23 Figure 23-1: Courtesy of Peter Armstrong, University of California, Davis. Figure 23-4: From E. Norrby, H. Marysk, and C. Orvell, *J. Virol.* 6(1970):237. Copyright © 1970 by the American Society for Microbiology. Figures 23-10 and 23-11: Courtesy of Peter Armstrong, University of California, Davis.

Chapter 24 Figure 24-3: Adapted from G. M. Edelman, *Sci. Amer.* 223(1970):34. Copyright © 1970 by Scientific American, Inc. All rights reserved. Figure 24-17: Adapted from L. M. Amzel, R. J. Poljak, F. Saul, J. M. Varga, and F. F. Richards, *Proc. Nat. Acad. Sci.* 71(1974):1427. Figure 24-18: Adapted from E. W. Silverton, M. A. Navia, and D. R. Davies, *Proc. Nat. Acad. Sci.* 74(1977):5140.

Appendix A Figure A-6: Courtesy of Unitron Instruments, Inc., Plainview, N.Y. Figure A-7: Courtesy of J. R. Waaland. Figure A-9: Courtesy of J. L. Carson, Biological Education Consultants. Figure A-12: Courtesy of K. Fujiwara. Figure A-14: Courtesy of J. L. Carson, Biological Education Consultants. Figure A-17(inset): From Sasha Koulish and Ruth G. Kleinfeld, *J. Cell Biol.* 23(1964):39. Reproduced by copyright permission of The Rockefeller University Press. Figures A-18, A-19A, and A-20a: Courtesy of J. L. Carson, Biological Education Consultants. Figure A-21: Courtesy of J. Croxdale. Figures A-23 and A-24: Courtesy of H. Ris. Figures A-25, A-26, and A-27: Courtesy of J. L. Carson, Biological Education Consultants. Figure A-31: L. Orci and M. Perrelet, *Freeze-Etch Histology*, Heidelberg: Springer-Verlag, 1975. Figure A-32: Courtesy of H. Ris.

Appendix B Copyright © Genentech. p. 814 "Supermouse" photo: Reprinted from the cover of *Nature* 300(1982). Photo provided by R. L. Brinster.

Index

NOTE: A *t* following a page number indicates tabular material; an *f* following a page number indicates a figure; and a *g* following a page number indicates a glossary entry.

B

F

G

J

K

L

N

O

P

Q

R